PIPELINE RULES OF THUMB HANDBOOK

NINTH EDITION

PIPELINE RULES OF THUMB HANDBOOK

A Manual of Quick, Accurate Solutions to Everyday Pipeline Engineering Problems

Mark J. Kaiser
Louisiana State University, Baton Rouge, LA, United States
E.W. McAllister
Houston, TX, United States

ELSEVIER
Gulf Professional Publishing
An imprint of Elsevier

Gulf Professional Publishing is an imprint of Elsevier
50 Hampshire Street, 5th Floor, Cambridge, MA 02139, United States
The Boulevard, Langford Lane, Kidlington, Oxford, OX5 1GB, United Kingdom

Notices

Knowledge and best practice in this field are constantly changing. As new research and experience broaden our understanding, changes in research methods, professional practices, or medical treatment may become necessary.

Practitioners and researchers must always rely on their own experience and knowledge in evaluating and using any information, methods, compounds, or experiments described herein. In using such information or methods they should be mindful of their own safety and the safety of others, including parties for whom they have a professional responsibility.

To the fullest extent of the law, neither the Publisher nor the authors, contributors, or editors, assume any liability for any injury and/or damage to persons or property as a matter of products liability, negligence or otherwise, or from any use or operation of any methods, products, instructions, or ideas contained in the material herein.

ISBN: 978-0-12-822788-6

For information on all Gulf Professional Publishing publications visit our website at https://www.elsevier.com/books-and-journals

Publisher: Charlotte Cockle
Senior Acquisitions Editor: Katie Hammon
Editorial Project Manager: Sara Valentino
Production Project Manager: Prasanna Kalyanaraman
Cover Designer: Matthew Limbert

Typeset by TNQ Technologies

Printed in India

Last digit is the print number 9 8 7 6 5 4 3 2 1

Contents

List of Contributors xv

About the Authors xvii

Preface xix

1: General Information 1

System Schematics 2

Schematic of hazardous liquid pipeline system networks 2
Schematic of natural gas pipeline system networks 2
Hydrocarbon classification of gases and products 3
Typical pipeline construction sequence 3

U.S. Pipeline Maps 4

Gas transmission and hazardous liquid pipelines circa 2020 4
Hazardous liquid pipeline accident heat map, 2010-2020 5
Gas transmission pipeline incident heat map, 2010-2020 6
LNG plants connected to natural gas pipeline systems, 2020 7
CO_2-EOR operations and infrastructure in the U.S. circa 2020 8
Permian Basin CO_2 pipeline infrastructure circa 2020 9
Gulf Coast CO_2 pipeline infrastructure circa 2020 10
Rocky Mountain CO_2 pipeline infrastructure circa 2020 11
Mid-Continent CO_2 pipeline infrastructure circa 2020 11

U.S. Statistics 12

Annual mileage for hazardous liquid or carbon dioxide systems in the United States, 2004-2019 12
Annual mileage for natural gas transmission and gathering systems in the United States, 1984-2019 13
Annual mileage for gas distribution systems in the United States, 1984-2019 14

Billion barrels delivered by liquids pipeline in the United States, 2014-2018 ... 14
Liquefied natural gas facilities in the United States circa 2020 ... 15

Cables and Ropes ... 16

Estimating strength of cable ... 16
Find the working strength of Manila rope ... 16
How large should drums and sheaves be for wire rope? ... 16
Find advantages of block and tackle with pull out friction ... 17
Safe loads for wire rope ... 17
Stress in guy wires ... 18
Strength and weight of popular wire rope ... 20
Measuring the diameter of wire rope ... 20
Wire rope: field troubles and their causes ... 21
Capacity of drums ... 22

Belts and Shafts ... 22

Determine length of a V-belt ... 22
Calculate stress in shaft key ... 23
Estimate the horsepower that can be transmitted by a shaft ... 23

Miscellaneous ... 24

Area formula ... 24
Surface and volume formula ... 25
Recommended drill sizes for self-tapping screws ... 26
Determine pulley speed ... 26
ASTM standard reinforcing bars ... 27
Pressure rating for carbon steel flanges ... 27
How to estimate length of material contained in roll ... 27
Convenient antifreeze chart for winterizing cooling systems ... 28
How to determine glycol requirements for desired protection ... 28
Weight in pounds of round steel shafting ... 29
Properties of shafting ... 30
Tap drills and clearance drills for machine screws ... 31
Common nails ... 32
Drill sizes for pipe taps ... 32
Carbon steel—color and approximate temperature ... 32
Bolting dimensions for flanges ... 33
Flange bolt tightening sequence ... 37
Steel fitting dimensions ... 38
ANSI forged steel flanges ... 39
Trenching and excavation safety ... 40
Reuniting separated mercury in thermometers ... 43
Typical wire resistance ... 43
How to cut odd-angle long radius elbows ... 44
How to read land descriptions ... 45
Size an air receiver for engine starting ... 47
Dimensions of hex nuts and hex jam nuts ... 48
Color codes for locating underground utilities ... 49
Approximate angle of repose for sloping sides of excavations ... 49
Wind chill chart ... 50
Fluid power formulas ... 51
How is pipeline manufactured? ... 51
How to designate and classify PE piping materials ... 55
Pipeline books published after 2000 ... 58

2: Construction ... 61

Right of Way ... 62

Project scoping data worksheet for major facilities ... 62
Clearing and grading right-of-way: labor/equipment considerations ... 64
Estimating man hours for removing trees ... 64
Estimating man hours for removing tree stumps ... 65
Clearing and grading right-of-way ... 65

Ditching and Trenching ... 66

How many cubic yards of excavation in a mile of ditch? ... 66
Shrinkage and expansion of excavated and compacted soil ... 66
Ditching and trenching: labor/equipment considerations ... 66
Trench excavations ... 67
Determination of soil type ... 68
Test equipment and methods for evaluating soil type ... 68
OSHA trench box requirements ... 76

Laying and Lowering ... 78

How to determine the degrees of bend in a pipe that must fit a ditch calling for a bend in both horizontal and vertical planes ... 78
How to bend a pipe to fit a ditch ... 78
Pipe bending computations made with a calculator ... 79
Calculate maximum bend on cold pipe ... 80
Determine length of a pipe bend ... 82
Length of pipe in arc subtended by any angle ... 82
Linear feet of pipelay per 10 h day in trench in level, rock-free terrain ... 83
Linear feet of pipelay per 10-hour day on waist-high supports on level ground ... 84
Allowable pipe span between supports ... 84

How engineers make pipe fit the ditch 85
How to lower an existing pipeline that is still in service 88

Welding 91

When should steel be preheated before welding? 91
Welding and brazing temperatures 92
Mechanical properties of pipe welding rods 92
Lens shade selector 93
Reasons for rejecting a pipeweld and ways to reduce reject rates 93
How many welds will the average welder make per hour? 102
How much welding rod is required for a mile of schedule 40 pipeline? 102
How many pounds of electrodes are required per weld on line pipe? 102
Welding criteria permit safe and effective pipeline repair 103
Cross-country pipeline—vertical down electrode consumption, pounds of electrode per joint 109

Directional Crossing 110

Guidelines for a successful directional crossing bid package 110

Polyethylene Piping 118

Simplified installation guidelines for PE pipe 118
How to construct a trench for PE piping 120

3: Pipe Design 125

Steel pipe design 126
Properties of pipe 131
Length of pipe in bends 134
Calculation of pipe bends 135
Spacing of pipe supports 137
Chart of deflection of horizontal pipelines 138
American standard taper pipe threads (NPT) 139
British standard taper pipe threads 140
Normal engagement between male and female threads to make tight joints 141
Calculate pipe weight, contents, velocity 141
Formulas and constants for tubular goods problems 142
How to calculate the contraction or expansion of a pipeline 143
Estimate weight of pipe in metric tons per kilometer 143
How to find pipe weight from outside diameter and wall thickness 144
What is the maximum allowable length of unsupported line pipe? 144
Identify the schedule number of pipe by direct measurement 144
Determine buoyancy of bare steel pipe 145
Determine buoyancy of bare and concrete-coated steel pipe in water and mud 145
Weights of piping materials 146
Allowable working pressure for carbon steel pipe 147
Find the stress in pipe wall due to internal pressure 148
How to calculate stress in above ground/below ground transitions 148
How to identify the series number of flanged fittings 152
Dimensions of three-diameter ells with tangents 152
Spectacle blind thicknesses 152
Carbon dioxide pipeline design 153

4: Electrical Design 159

Electrical design 160
Hazardous locations 161
NEMA enclosure types 162
Size portable electric generators 163
Typical wattages for tools and appliances 164
Knockout dimensions 164
National Electrical Code tables 165
Electrical formulas 169
Full-load currents—single-phase transformers 169
Conduit size for combinations of cables with different outside diameters 170
Minimum bending radius for insulated cables for permanent training during installation 170
Full-load currents—three-phase transformers 172
Motor controller sizes 172
Voltage drop on circuits using 600 V copper conductors in steel conduit 173
Determine the most economical size for electric power conductors 173
How to find the resistance and weight of copper wires 174
What you should remember about electrical formulas 174
How to calculate microwave hops on level ground 174
Quick determination of hp per ampere for induction motors (3 phase) at different voltages 175

Chart of electric motor horsepower for pumping units 175
Pumping stations 176
Floodlighting calculations 177
Point-by-point method 177
Beam-lumen design method 178
Conductor size conversion chart—metric to AWG 179
Commonly used switchgear device numbers 180
Bonding the grounding system to building and structure foundations 181

5: Hydrostatic Testing........ 183

History of gas pipeline pressure test requirements 184
The benefits and limitations of hydrostatic testing 189
Hydrostatic testing for pipelines 199
Volume of water required to fill test section 204
Volume of water required at test pressure 205
How to use charts for estimating the amount of pressure change for a change in test water temperature 206
Compressibility factor for water 209
Hydrostatic test records 209
Pressure testing and recordkeeping requirements – reconciling historical practices with new requirements 209

6: Pigging, Cleaning & Drying........................... 217

A brief history of pipeline pigging 218
Why pig a pipeline? 219
Cleaning and construction utility pigs 226
Pipeline dewatering, cleaning, and drying 232
Moisture content of air 236
Commissioning ethylene pipelines 238
Vacuum drying 241
Pig trap design 244

7: Control Valves............... 247

Control valve sizing formulas 248
Sizing control valves for throughput 252
Control valve selection 257
Relief valve sizing, selection, installation, and testing 259
Rupture disk sizing methodologies 263
Rupture disk sizing using the resistance to flow method (K_R) 264
Variable orifice rotary control valves 266
Basic valve flow-capacity coefficient (C_v) 268
Visualize pump and control valve interaction 272
Avoid cavitation in butterfly valves 278
How to read a regulator flow curve 281

8: Corrosion and Coatings 285

Types of corrosion and control methods 286
Fundamentals of pipeline corrosion 288
NAPCA specification designations for coal tar enamel coatings 302
NAPCA standard applied pipe coating weights 302
NAPCA specifications for pipeline wrappers 304
How much primer for a mile of pipe? 304
How much coal-tar enamel for a mile of pipe? 305
How much wrapping for a mile of pipe? 305
Estimating coating and wrapping materials per mile of pipe 305
Coefficient of friction for pipe coating materials 306
Troubleshooting magnesium anode cathodic protection systems 308
Cathodic protection for pipelines 309
Estimate the pounds of sacrificial anode material required for offshore pipelines 317
Comparison of other reference electrode potentials with that of copper-copper sulfate reference electrode at 25 °C 319
Chart aids in calculating ground bed resistance and rectifier power cost 320
How can output of magnesium anodes be predicted? 321
How to determine the efficiency of a cathodic protection rectifier? 321
How to calculate the voltage drop in ground bed cable quickly 322
What spacing for test leads to measure current on a pipeline? 322
How to estimate the number of magnesium anodes required and their spacing for a bare line or for a corrosion "hot spot"? 323
How can resistivity of freshwater be determined from chemical analysis? 323
What will be the resistance to earth of a single graphite anode? 324

What will be the resistance to earth of a group of graphite anodes, in terms of the resistance of a single anode? 324
How can the current output of magnesium rod used for the cathodic protection of heat exchanger shells be predicted? 324
How many magnesium anodes are needed for supplementary protection to a short-circuited bare casing? 324
Group installation of sacrificial anodes 325
How to find the voltage rating of a rectifier if it is to deliver a given amount of current through a given ground bed? 325
Determining current requirements for coated lines 326
Determine current requirements for coated lines when pipe-to-soil potential values are estimated 326
HVDC effects on pipelines 327
Troubleshooting bed cathodic protection systems 331
How to control corrosion at compressor stations 332
Project leak growth 333
Advances in pipeline protection 334
Estimate the number of squares of tape for pipe coating 339
Estimate the amount of primer required for tape 340
Tape requirements for fittings 340
Induced AC voltages on pipelines may present a serious hazard 341
Minimizing shock hazards on pipelines near HVAC lines 343
Corrosion of low-velocity, high water cut oil emulsion pipelines 344
Pipeline external corrosion direct assessment process 346

9: Gas—General 353

Know the gas laws 354
Calculate gas properties from a gas analysis 356
Physical properties of selected hydrocarbons and other chemicals and gases 360
How to calculate the specific gravity of natural gas 368
What are the four different types of heating value? 368
How to calculate the heating value of natural gas 369
Nomograph for calculating density and specific volume of gases and vapors 370
Considerations for selecting energy measurement equipment 371
Facts about methane and its behavior 377
Conversion table for pure methane 381
Introduction to LNG 382
Glossary of common gas industry terms 396

10: Gas—Compression 399

Compressor definitions 400
Performance calculations for reciprocating compressors 401
Estimate suction and discharge volume bottle sizes for pulsation control for reciprocating compressors 403
Compression horsepower determination 405
Generalized compressibility factor 407
Nomograph aids in diagnosing compressor cylinder ills 408
Centrifugal compressor performance calculations 409
Nomographs for estimating compressor performance 413
Estimate horsepower required to compress natural gas 418
Estimate compressor horsepower where discharge pressure is 1000 psi 418
Calculate brake horsepower required to compress gas 419
How can we find the size of a fuel gas line for a compressor station? 419
Estimate engine cooling water requirements 420
Performance testing guidelines for centrifugal compressors 421

11: Gas—Hydraulics 433

Gas pipeline hydraulics calculations 434
Equivalent lengths for multiple lines based on Panhandle A 435
Determine pressure loss for a low-pressure gas system 436
Nomograph for determining pipe-equivalent factors 437
How much gas is contained in a given line section? 438
How to estimate equivalent length factors for gas lines 438
Estimating comparative capacities of gas pipelines 439
Determination of leakage from gas line using pressure drop method 439
A quick way to determine the size of gas gathering lines 440
Energy conversion data 440
How to estimate time required to get a shut-in test on gas transmission lines 441
How to approximate a maximum acceptable pressure loss for new lines 441
Estimate pipe size requirements for increasing throughput volumes of natural gas 442
Calculate line loss using cross-sectional areas table when testing mains with air or gas 443
Flow of fuel gases in pipelines 444

Calculate the velocity of gas in a pipeline 445
Determining throat pressure in a blow-down system 445
Estimate the amount of gas blown off through a line puncture 446
A practical way to calculate gas flow for pipelines 446
How to calculate the weight of gas in a pipeline 447
Estimate average pressure in gas pipeline using upstream and downstream pressures 447
Chart for determining viscosity of natural gas 448
Fanning equation for gas flow 448
Multiphase flow 452
Nomograph for calculating Reynolds number for compressible flow friction factor for clean steel and wrought iron pipe 457

12: Liquids—General......... 461

Determining the viscosity of crude 462
Liquid gravity and density conversion chart 463
Nomograph for calculating viscosities of liquid hydrocarbons at high pressure 464
Calculate viscosity of a blend 465
Viscosity of a mixture of liquids using blending index 465
Calculate specific gravity of a blend 466
Convert viscosity units 466
Convert specific gravity to API gravity 466
Viscosities of hydrocarbon liquids 467
Calculate bulk modulus 469
Nomograph for calculating viscosity of slurries 469
Nomograph for calculating velocity of liquids in pipes 471
Nomograph for calculating velocity of compressible fluids in pipes 471
Nomograph for calculating velocity of liquids in pipes 472
Table gives velocity heads for various pipe diameters and different rates of discharge 474
Derivation of basic ultrasonic flow equations 475
How fast does oil move in a pipeline? 477
Estimate the volume of a pipeline per linear foot using the inside diameter 477
What is the linefill of a given pipe in barrels per mile? 477
Estimate leakage amount through small holes in a pipeline 478
Carbon capture and sequestration processes 478
CO_2 pipeline infrastructure in the United States 479
Effect of variability on CO_2 pipeline operation 486

13: Liquids—Hydraulics..... 491

Marine hose pressure loss 492
Pressure loss in hose string 495
Pressure drop formula for rubber hose 497
Example of pressure drop calculations for rubber hose 497
Formulas for calculating pressure drop and flow rates for pipelines 497
Using and computing hydraulic gradients 499
What are equivalent lengths? 502
Pressure drop in series systems 503
Pressure drop in looped systems 504
Calculate pressure loss in annular sections 505
Calculate pressure and temperature loss for viscous crudes $\geq$ 1,000 cP 505
Determine batch injection rate as per enclosure 508
Pressure loss through valves and fittings 509
Nomograph for calculating Reynolds number for clean steel and wrought iron pipe 515
Nomograph for calculating pressure drop of liquids in lines for turbulent flow 517
Drag-reducing agents 521
How to estimate the rate of liquid discharge from a pipe 524
Predict subsurface temperature ranges 524
Sizing pipelines for water flow 525
How approximate throughput of a line can be estimated from pipe size 525
Estimate crude gathering line throughput for a given pipe diameter 526
How to determine head loss due to friction in ordinary iron pipeline carrying clear water 526
How to size lines, estimate pressure drop, and estimate optimum station spacing for crude systems 527
How to size crude oil and products lines for capacity increases 527
What is the hydrostatic pressure due to a column of liquid H feet in height? 527
How to determine the maximum surge pressure in liquid-filled pipeline when a valve is suddenly closed 528
Transient pressure analysis 528

Hydraulics calculations for multiphase systems, including networks 538

14: Pumps 545

Centrifugal pumps 546
Speed torque calculation 558
Variable speed or impeller trim? 559
Pulsation control for reciprocating pumps 560
Rotary pumps on pipeline services 569
Centrifugal pump parameters and applications 574
How to examine centrifugal pump systems 579
Estimate the discharge of a centrifugal pump at various speeds 584
How to estimate the head for an average centrifugal pump 585
Find the reciprocating pump capacity 585
How to estimate the hp required to pump at a given rate at a desired discharge pressure 585
Nomograph for determining reciprocating pump capacity 586
Nomograph for determining specific speed of pumps 587
Nomograph for determining horsepower requirement of pumps 588
How to select motors for field-gathering pumps 588
How a reciprocating pump works 589
Understanding the basics of rotary screw pumps 598
How to evaluate VFD speed on hydraulics 604
Progressive cavity pumps 606
Pump curve accuracy 607
The basics of NPSH and pump operating regions 614
Reading a centrifugal pump curve 617
Navigating pump curves: a roadmap to reliability 620

15: Measurement 625

Multiphase flow meter 626
Pipeline flow measurement factors 627
Liquid measurement orifice plate flange taps 630
Mass measurement light hydrocarbons 632
Pipeline measurement of supercritical carbon dioxide 633
Master meter proving orifice meters in dense phase ethylene 639
Gas or vapor flow measurement—orifice plate flange taps 646
Properties of gas and vapors 648
Determine required orifice diameter for any required differential when the orifice and differential are known 650
Estimate the temperature drop across a regulator 651
Estimate natural gas flow rates 651
How to estimate the average pressure differential on the remaining meter runs of a parallel system 652
Sizing a gas metering run 652
Determine the number of purges for sample cylinders 653
Find the British thermal units when the specific gravity of a pipeline gas is known 653
Make quick calculation of the gas flow through orifice meters 654
How to measure high pressure gas 654
Four ways to calculate orifice flow in field 658
Practical maintenance tips for positive displacement meters 661
Sizing headers for meter stations 665
Measuring flow of high-viscosity liquids 668
Matching the flow meter to the application 673
Use liquid ultrasonic meters for custody transfer 680
Handling entrained gas 685
How do Coriolis flow meters work? 687
Proving liquid ultrasonic flow meters for custody transfer measurement 689
V-Cone® flow meter 700
Sampling and testing program for unrefined petroleum-based products 706

16: Instrumentation 709

Types of control systems 710
Developments in pipeline instrumentation 712
Automatic tank gauging systems and proof testing 726
Instrumentation and sensors for automatic tank gauging 737

17: Integrity Management 745

Overview of leak detection technologies 746
Inspection tools and case studies 750
Factors to consider when evaluating damage on pipelines 760
Understanding mechanical damage in pipelines 766
Understanding stress corrosion cracking in pipelines 774

When does a pipeline need revalidation? 783
Modeling pipeline risk assessment 792
A risk management tool for establishing budget priorities 797
Floating roof failure modes and risk 802
Floating roof tanks rainfall risk 812

18: Tanks 819

Classification of storage tanks 820
What is the difference between fixed and floating tank types? 826
Tank styles and appurtenances 830
Floating roof applications and components 834
Summary of federal requirements for aboveground storage tanks 844
Design and containment for atmospheric storage tanks 845
Overview of petroleum storage tank standards 851
Tank weight estimation tool using diameter as input variable 870
How to estimate roof and framing weight for fixed cone tank 871
Budgetary pricing cost curve of cone roof tanks 872
How to estimate flow rate from a leaking tank hole 873
Charts give vapor loss from internal floating-roof tanks 875
Estimating the contents of horizontal cylindrical tanks 878
How to gauge a horizontal cylindrical tank 879
Correct the volume of light fuels from actual temperature to a base of 60°F 880
Volume of liquid in vertical cylindrical tanks 880
Chart gives tank's vapor formation rate 881
Tank farm line sizing 882
Tank farm management 884
Managing oil movement and storage in tank farms 892
Standard construction practices for LNG storage tanks 902
Choosing the best containment for LNG storage 908
A review of API 2350 overfill protection, 4th and 5th edition 911
Manual tank gauging procedures 916
How to apply vapor cloud analysis methodology to tank overfills 923
Simplified procedure to calculate emissions for liquid organic storage tanks 932
Calculating tank emissions with BREEZE TankESP 936

19: Economics 939

Time Value of Money 940

Simple interest versus compound interest 940
Nominal interest rate versus effective annual interest rate 941
Present value of a single cash flow to be received in the future 941
Future value of a single investment 942
The importance of cash flow diagrams 942
Analyzing and valuing investments/projects with multiple or irregular cash flows 942
Perpetuities 943
Future value of a periodic series of investments 944
Annuities, loans, and leases 944
Gradients (payouts/payments with constant growth rates) 945
Analyzing complex investments and cash flow problems 946

Decision and Evaluation Criteria 947

Payback method 947
Accounting rate of return (ROR) method 948
Internal rate of return (IRR) method 949
Net present value (NPV) method 950

Sensitivity Analysis 951
Decision Tree Analysis 952
Accounting Fundamentals 956

20: Cost 961

Cost Estimation 962

Cost estimation methods, normalization and data 962
How to use time and location cost indices 966
How to adjust pipeline construction cost across time 967

Onshore Construction 968

Onshore pipeline construction process 968
Onshore pipeline construction cost factors 971
Onshore FERC gas pipeline cost statistics in the United States, 2014–2019 974
How to estimate onshore pipeline construction cost using unit costs? 977

How to estimate cost bounds in onshore pipeline construction in the United States? 978
Onshore gas pipeline construction cost in the United States, 2017–2019 978
How to develop construction cost models using regression 980

Offshore Construction 984

Offshore pipeline construction process 984
Offshore pipeline construction cost factors 986
U.S. Gulf of Mexico pipeline cost statistics, 1995–2019 991
How to estimate shallow water pipeline construction cost bounds in the U.S. Gulf of Mexico using empirical data? 995
Offshore pipeline construction cost estimation using work decomposition 996

Offshore Decommissioning 1002

Offshore pipeline decommissioning process in the U.S. Gulf of Mexico 1002
Offshore pipeline decommissioning cost factors 1004
Offshore pipeline decommissioning cost statistics in the U.S. Gulf of Mexico, 1980–2019 1005
How to estimate the cost to decommission shallow water pipeline in the U.S. Gulf of Mexico? 1007
How to estimate the cost to decommission hurricane-damaged pipeline in the U.S. Gulf of Mexico? 1007
How to estimate deepwater pipeline decommissioning cost in the U.S. Gulf of Mexico? 1008
How to estimate deepwater umbilical, flowline, and riser decommissioning cost in the U.S. Gulf of Mexico? 1008
Typical workflows for offshore pipeline decommissioning 1009

Gas Compression Stations 1010

Onshore FERC gas transmission compressor station cost statistics in the United States, 2014–2019 1010
How to estimate gas transmission compressor station cost in the United States? 1013
How to compare gas transmission compressor station cost in the United States? 1013

Scaling Rules 1013

Equipment and process scaling rules 1013
How to estimate equipment cost curves 1014
How to estimate equipment cost using two-factor models 1015
How to estimate equipment cost curves using two data points 1017

Tariff Rates 1017

What are reservation and commodity charges? 1017
How to calculate firm service with fixed maximum daily quantity and reservation charge? 1019
How is the cost of service formula applied? 1019
How to compute price adjustments in pipeline system quality banks 1023

21: Health, Safety and Environment 1025

Health 1026

Grading the nation: state disclosure policies for lead pipes 1026

Safety 1031

Trenching and excavation safety 1031
Crane and hoist safety 1033
Electrical welding safety 1035
Tank fire types and causes 1038
Firefighting protection systems for atmospheric storage tanks 1040
Carbon dioxide pipeline safety and integrity 1042

Environment 1044

U.S. natural gas standards and regulations development 1044
What are the key performance indicators in the U.S. pipeline industry? 1045
Oil spills in the United States, 2002–2016 1047
Estimating oil spill thickness on water visually 1049
Fast discharge volume calculator for offshore leaking pipeline 1051
How to estimate discharge volume for offshore leaking pipeline (fast method) 1053
Advanced discharge volume calculator for offshore leaking pipeline 1053
How to estimate discharge volume for offshore leaking pipeline (advanced method) 1056
Guidelines on oil characterization to inform spill planning and decision making 1058
Oil spills inland response good practice guidelines 1066
Calculating volatile organic compounds flash emissions from crude oil and condensate tanks at oil and gas production sites 1083

22: Specifications 1089

Pipeline Steel Specifications 1090

API 5L pipeline specifications ... 1090
Main differences between API 5L PSL 1 and PSL 2 pipeline ... 1091
Chemical composition for API 5L PSL 1 welded pipe ... 1092
Chemical composition for API 5L PSL 2 welded pipe ... 1093
Tensile property requirements for pipe body of seamless and welded pipe for PSL 1 ... 1094
Tensile property requirements for pipe body of seamless and welded pipe for PSL 2 ... 1094

Pipeline Material Coating Specifications ... 1095

External application procedures for plant applied fusion bonded epoxy coatings and abrasion resistant overlay coatings to steel pipe ... 1095
External application procedures for coal tar epoxy protective coatings to steel pipe ... 1099
External application procedures for polyolefin pipe coating applied by the cross head extrusion method or the side extrusion method to steel pipe ... 1102
Exterior application procedures for plant applied tape coatings to steel pipe ... 1107
Suggested procedures for coating field joints, fittings, connections, and prefabricated sections using tape coatings ... 1110
Suggested procedures for coating field joints using heat-shrinkable materials ... 1111
External application procedures for hot applied coal tar coatings to steel pipe ... 1113

Fuel Specifications ... 1117

Types of fuel specifications ... 1117
Transportation fuel specifications ... 1119
Fuel quality ... 1119
Gasoline fuel specifications ... 1121
Aviation gasoline fuel specifications ... 1124
Jet fuels specifications ... 1126
Diesel fuel specifications ... 1131

Pipeline Product Specifications ... 1136

NGL products, propane ... 1136
NGL products, butane, normal butane ... 1137
NGL products, butane, refinery grade butane ... 1138
NGL products, iso-butane ... 1139
Gasoline, conventional, 85.5 octane (no ethanol) ... 1140
Gasoline, conventional, 87 octane (no ethanol) ... 1141
Gasoline, subgrade, 87 octane after 10% ETOH addition ... 1142
Distillate, #1 diesel fuel, ultra-low sulfur (15 ppm max) ... 1143
Distillate, #1 diesel fuel, ultra-low sulfur (15 ppm max) ... 1144
Distillate, #2 diesel fuel, ultra-low sulfur (15 ppm max) ... 1145

23: Conversion Factors.... 1147

Metric conversions—metric to English, English to metric ... 1148
Density conversions ... 1149
General liquid density nomograph ... 1149
Chart gives specific gravity/temperature relationship for petroleum oils ... 1151
Weight density and specific gravity of various liquids ... 1151
True vapor pressure of crude oil stocks with a Reid vapor pressure of 2–15 psi ... 1152
Low temperature vapor pressures for light hydrocarbons ... 1153
High temperature vapor pressures for light hydrocarbons ... 1154
Hydrocarbon gas viscosity ... 1155
Viscosity—equivalents of kinematic and Saybolt Universal Viscosity ... 1156
Viscosity—equivalents of kinematic and Saybolt Furol Viscosity at 122 °F and 210 °F ... 1156
Viscosity—general conversions ... 1157
ASTM standard viscosity temperature chart for liquid petroleum products ... 1159
Pressure conversion chart ... 1160
Energy conversion chart ... 1161
Flow conversion chart ... 1162
Petroleum fuels volume, weight and calorific value factors ... 1162
Crude oil approximate conversion factors ... 1163
Refined products approximate conversion factors ... 1164
Natural gas and LNG approximate conversion factors ... 1164
Units and calorific equivalents ... 1165
Weight of water per cubic foot at various temperatures ... 1165

Index ... 1167

For additional information on the topics covered in the book, visit the companion site:
https://www.elsevier.com/books-and-journals/book-companion/9780128227886

List of Contributors

Michael S. Choi
Process Engineering Fellow in ConocoPhillips, Houston, TX, United States

Philip E. Myers
Founder and Director, PEMY Consulting, Pleasant Hill, CA, United States

Brock Trotter
E.I.T., PEMY Consulting, Oklahoma City, OK, United States

George G. Woodworth
Data Analyst, PEMY Consulting, Pleasant Hill, CA, United States

About the Authors

Mark J. Kaiser is a Professor at the Center for Energy Studies at Louisiana State University in Baton Rouge, Louisiana, where he has focused mainly on the oil, gas, and refining industry. His areas of specialization include cost evaluation, economic analysis, field development, production, and infrastructure. He has worked in the United States and internationally on numerous infrastructure and regulatory projects and started at the Center in 2001.

Edward W. McAllister has more than 50 years of experience in the pipeline industry as an Engineer, Administrator, and Project Manager. He has worked for Chevron Pipe Line Company and Gulf Pipe Line Company and has directed his own engineering services consulting firm in Houston, Texas, since 1983.

Preface

Pipeline Rules of Thumb columns began in the 1960s in *Pipeline Industry Magazine*, where it was conceived and written by Don Taylor, the first editor of the magazine. Don had been a pipeline engineer before transitioning to the publishing business, and he thought it would be helpful to provide veteran and new engineers some basic tips on various aspects of pipelining.

This was years before the pipeline technology that is available today. SCADA systems weren't in wide use at that time. Desktop computers were in someone's dreams, and engineers were still using slide rules. The Office of Pipeline Safety had not been established, and the pipeline infrastructure, especially the natural gas systems, was just starting to expand across North America.

Initially, the *Rules of Thumb* columns were published each month as a brief item, covering not only oil, gas, and product pipelines but also the power, water, utility, and chemical process industries. Early subjects dealt with the basics, such as how to determine the depth of cover for a new pipeline, the pressure required for a particular length and diameter line, the degrees of a pipe bend, etc.

By the late 1970s, there were sufficient tips to produce a book, and the first printing of the *Pipeline Rules of Thumb Handbook (RoT)* was by Gulf Publishing Company in 1978. In 1986, an advertisement for someone to update RoT was placed in *Pipeline Industry Magazine*. After retiring from Gulf Pipeline Company after 33 years in the pipeline business, Ed McAllister was looking for a new project and answered the ad. After interviewing with B.J. Lowe, who later formed Clarion Technical Conferences, Ed agreed to take on the task of updating the Handbook.

The second edition of the Handbook was published in 1988 and doubled in size to about 400 pages. Subsequent editions appeared every five years or so. The third edition in 1993 was released with a software version, and the 602 page fourth edition was printed in 1998. The seventh edition was printed in 2009 by Gulf Professional Publishing, an imprint of Elsevier, and the eighth edition in 2014 came with an online companion website hosting interactive data tables.

In this ninth edition, Mark J. Kaiser, Professor at the Center for Energy Studies at Louisiana State University, joins as co-editor. After 31 years in academia and not quite ready for retirement, MJK's goal was to update RoT with the latest technical information while maintaining its basic organizational structure and layout.

Today, the pipeline industry is a multibillion-dollar global enterprise that keeps growing with new products, new methods, new technologies, and new standards to improve economics, safety, and reliability. In the future, pipelines will continue to play an important role in the efficient use and distribution of gas and liquid products, and an integral component in the transition to cleaner burning fuels, the sequestration of carbon dioxide, the transport of hydrogen, and new uses we haven't yet imagined.

The Handbook's enduring popularity and success is undoubtedly due to its focus on: (1) short practical quantitative rules of thumb, (2) clear and abundant illustrations, and (3) broad coverage of all topics related to pipelines, from A to Z. In this edition, we build upon each of these features, enhancing and expanding coverage on most major subjects, and expunging antiquated material where needed.

Four new chapters have been added on Cost, Integrity Management, Safety and Environment, and Specifications, along with a significantly expanded Tanks chapter. There is also new material on CO_2 pipelines, corrosion fundamentals, hydrostatic testing, lead pipes, LNG, mechanical damage, PE piping, pigging, reading pump curves, risk management, sampling standards, proof testing sensors, stress corrosion cracking, trench shoring, and more. RoT 9th edition checks in at a healthy and hunky 6.2 pounds 2.5 inches.

Long-time readers already know that RoT contains many different types of information in many different forms in addition to rules of thumb—worked examples, design guidelines and operational tips, specification sheets, nomographs, physical properties, numerical simulations, ... Methods described range from back of envelope calculations to the results of mathematical modeling, and the wide variety of technical papers and excerpts provide detailed information to complement engineering basics.

Written by engineers for engineers, managers, and regulators, RoT aims to serve both working professionals and those new to the industry. We hope you agree that one of the pleasures of thumbing through the handbook is in the discovery of hidden gems written by the experts themselves. One cannot help feel sublime awe in the presence of such vast accumulated knowledge and experience.

Readers' comments are welcome.

Mark J. Kaiser
Baton Rouge, LA

E.W. McAllister
Houston, TX

August 2022

1: General Information

System Schematics 2
Schematic of hazardous liquid pipeline system networks 2
Schematic of natural gas pipeline system networks 2
Hydrocarbon classification of gases and products 3
Typical pipeline construction sequence 3

U.S. Pipeline Maps 4
Gas transmission and hazardous liquid pipelines circa 2020 4
Hazardous liquid pipeline accident heat map, 2010-2020 5
Gas transmission pipeline incident heat map, 2010-2020 6
LNG plants connected to natural gas pipeline systems, 2020 7
CO_2-EOR operations and infrastructure in the U.S. circa 2020 8
Permian Basin CO_2 pipeline infrastructure circa 2020 9
Gulf Coast CO_2 pipeline infrastructure circa 2020 10
Rocky Mountain CO_2 pipeline infrastructure circa 2020 11
Mid-Continent CO_2 pipeline infrastructure circa 2020 11

U.S. Statistics 12
Annual mileage for hazardous liquid or carbon dioxide systems in the United States, 2004-2019 12
Annual mileage for natural gas transmission and gathering systems in the United States, 1984-2019 13
Annual mileage for gas distribution systems in the United States, 1984-2019 14
Billion barrels delivered by liquids pipeline in the United States, 2014-2018 14
Liquefied natural gas facilities in the United States circa 2020 15

Cables and Ropes 16
Estimating strength of cable 16
Find the working strength of Manila rope 16
How large should drums and sheaves be for wire rope? 16
Find advantages of block and tackle with pull out friction 17
Safe loads for wire rope 17
Stress in guy wires 18
Strength and weight of popular wire rope 20
Measuring the diameter of wire rope 20
Wire rope: field troubles and their causes 21
Capacity of drums 22

Belts and Shafts 22
Determine length of a V-belt 22
Calculate stress in shaft key 23
Estimate the horsepower that can be transmitted by a shaft 23

Miscellaneous 24
Area formula 24
Surface and volume formula 25
Recommended drill sizes for self-tapping screws 26
Determine pulley speed 26
ASTM standard reinforcing bars 27
Pressure rating for carbon steel flanges 27
How to estimate length of material contained in roll 27
Convenient antifreeze chart for winterizing cooling systems 28
How to determine glycol requirements for desired protection 28
Weight in pounds of round steel shafting 29
Properties of shafting 30
Tap drills and clearance drills for machine screws 31
Common nails 32
Drill sizes for pipe taps 32
Carbon steel—color and approximate temperature 32
Bolting dimensions for flanges 33
Flange bolt tightening sequence 37
Steel fitting dimensions 38
ANSI forged steel flanges 39
Trenching and excavation safety 40
Reuniting separated mercury in thermometers 43
Typical wire resistance 43
How to cut odd-angle long radius elbows 44
How to read land descriptions 45
Size an air receiver for engine starting 47
Dimensions of hex nuts and hex jam nuts 48
Color codes for locating underground utilities 49
Approximate angle of repose for sloping sides of excavations 49
Wind chill chart 50
Fluid power formulas 51
How is pipeline manufactured? 51
How to designate and classify PE piping materials 55
Pipeline books published after 2000 58

SYSTEM SCHEMATICS

Schematic of hazardous liquid pipeline system networks

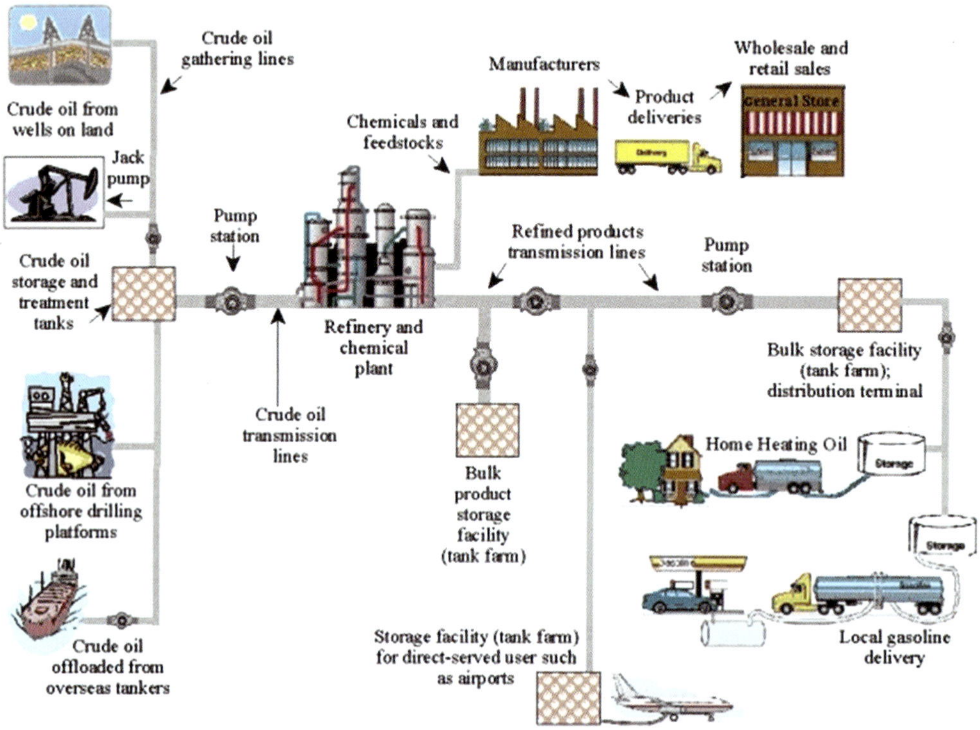

Figure 1. Schematic of hazardous liquid pipeline systems from the wellhead to the consumer. Source: PHMSA

Schematic of natural gas pipeline system networks

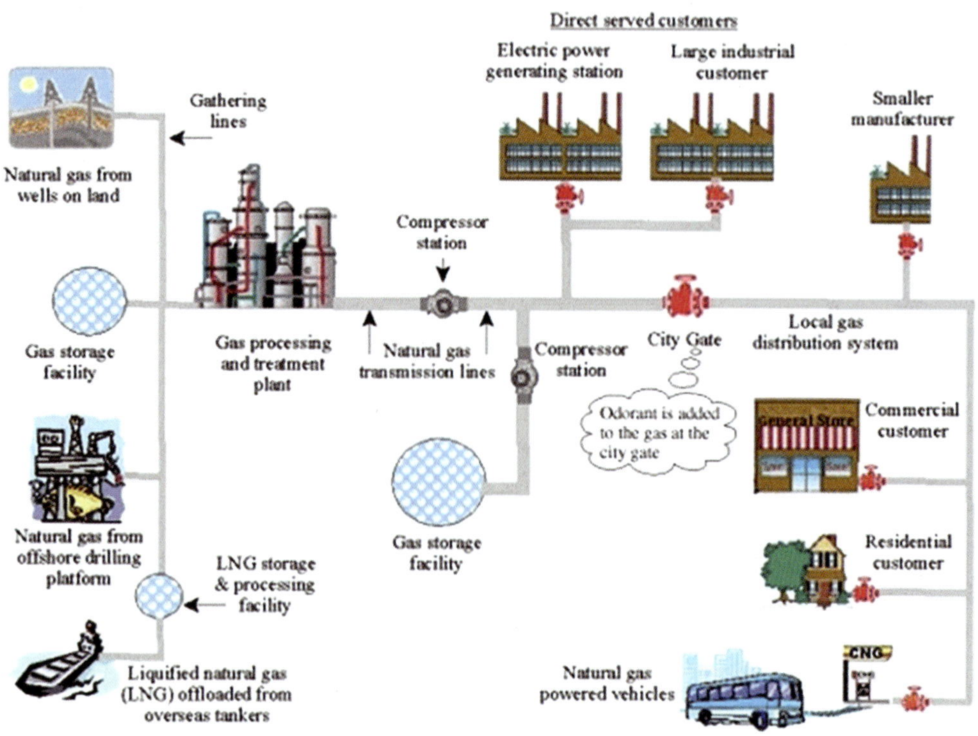

Figure 2. Schematic of natural gas pipeline systems from the wellhead to the consumer. Source: PHMSA

Hydrocarbon classification of gases and products

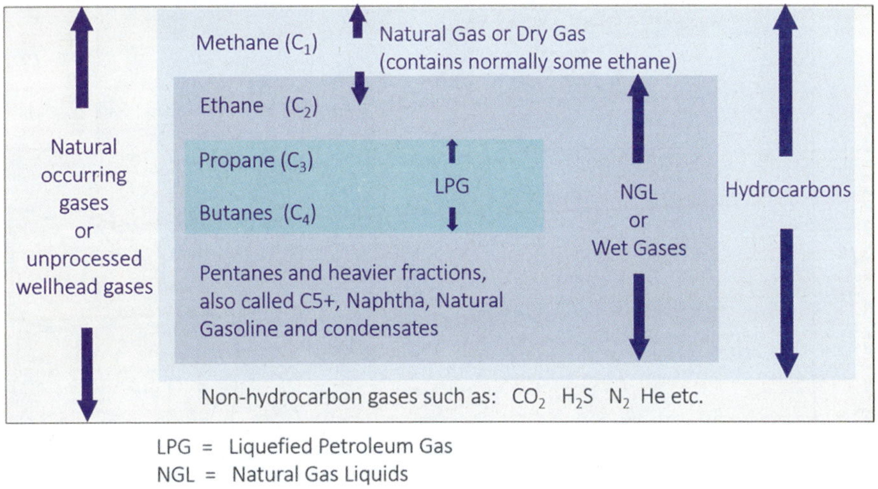

Figure 3. Hydrocarbon classification of gases and products. Source: Norwegianpetroleum.no IEA/MPE

Typical pipeline construction sequence

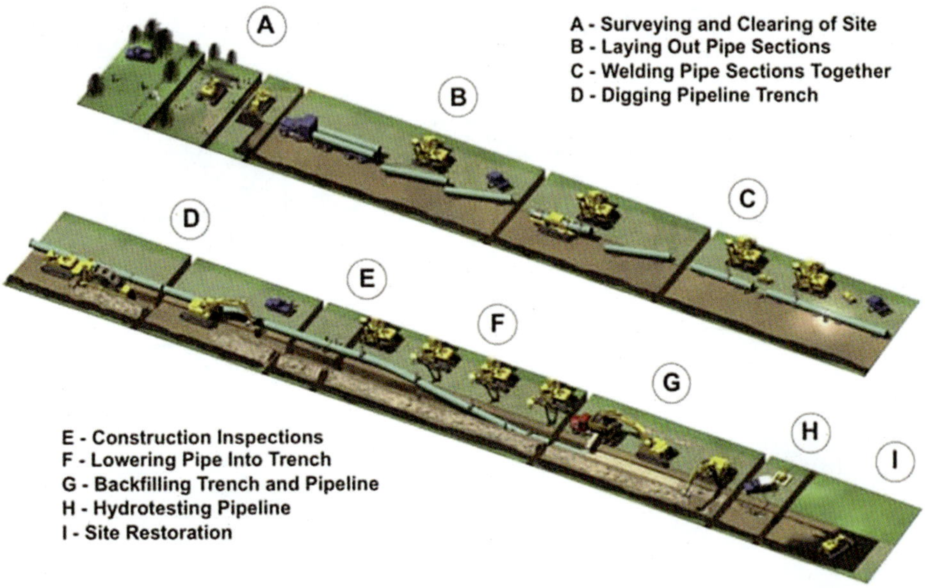

Figure 4. Typical pipeline construction sequence. Source: AOPL

U.S. PIPELINE MAPS

Gas transmission and hazardous liquid pipelines circa 2020

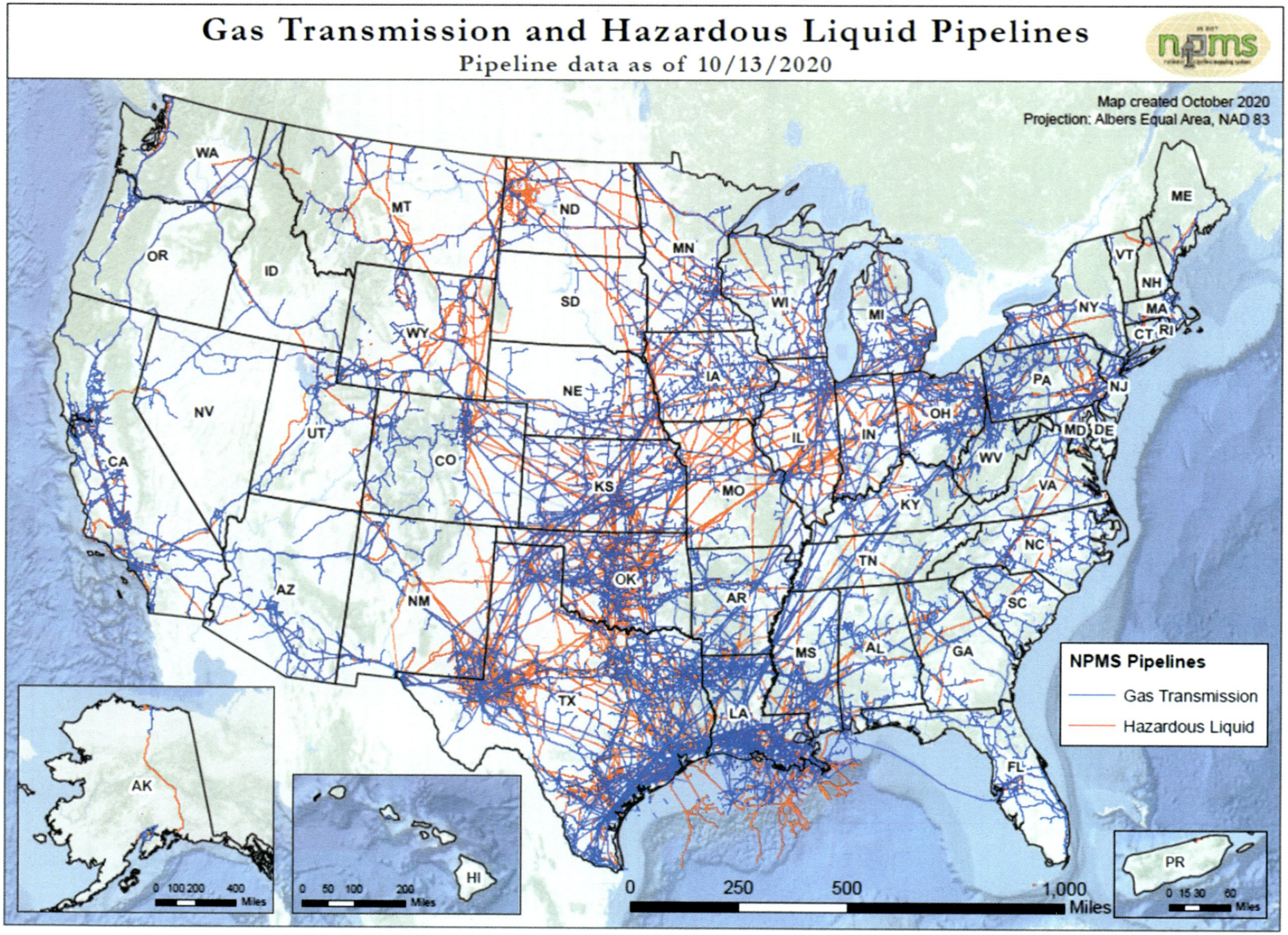

Figure 5. Gas transmission and hazardous liquid pipelines systems circa 2020. Source: NPMS

Hazardous liquid pipeline accident heat map, 2010-2020

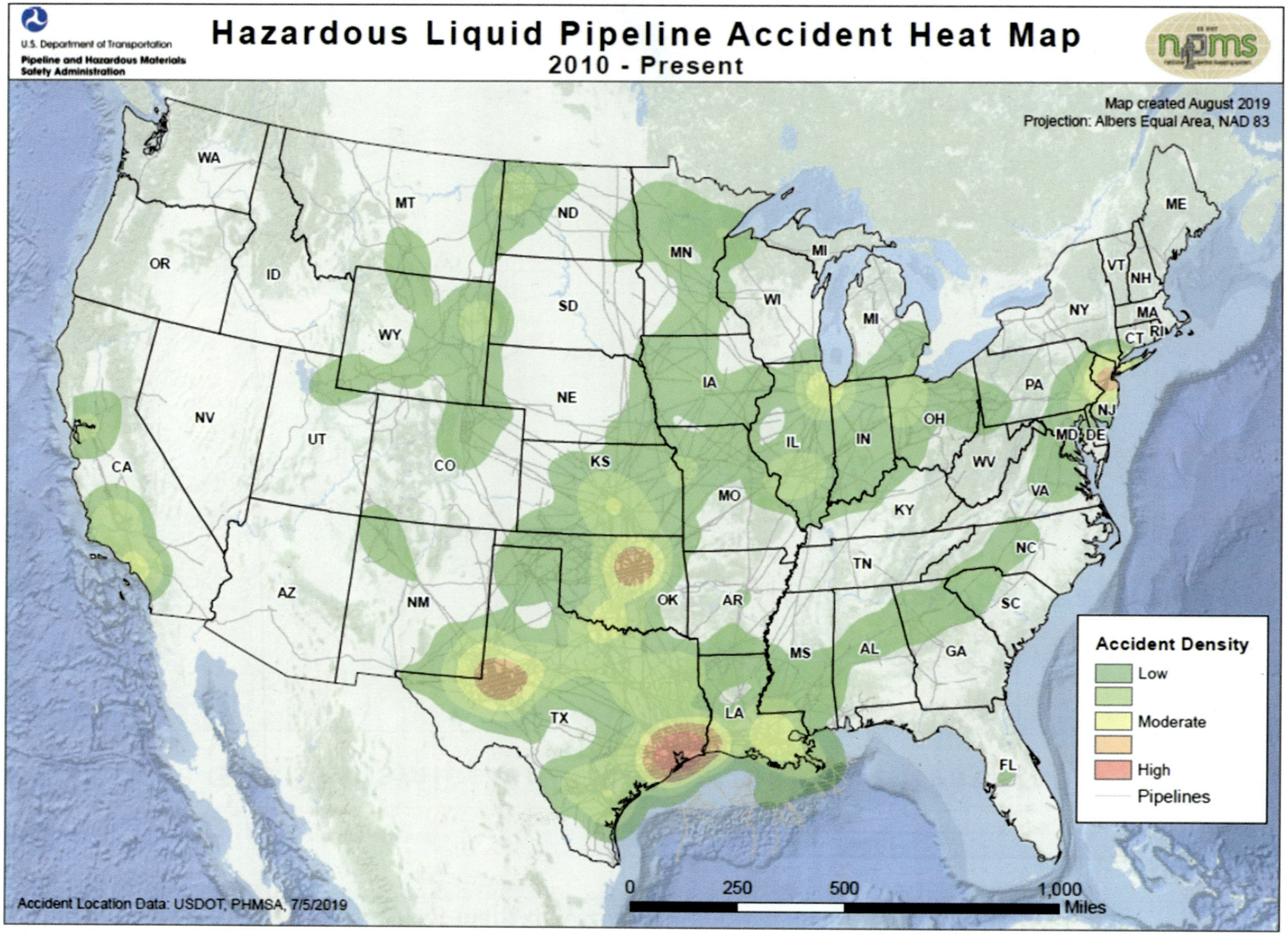

Figure 6. Hazardous liquid pipeline accident heat map, 2010-2020. Source: NPMS

Gas transmission pipeline incident heat map, 2010-2020

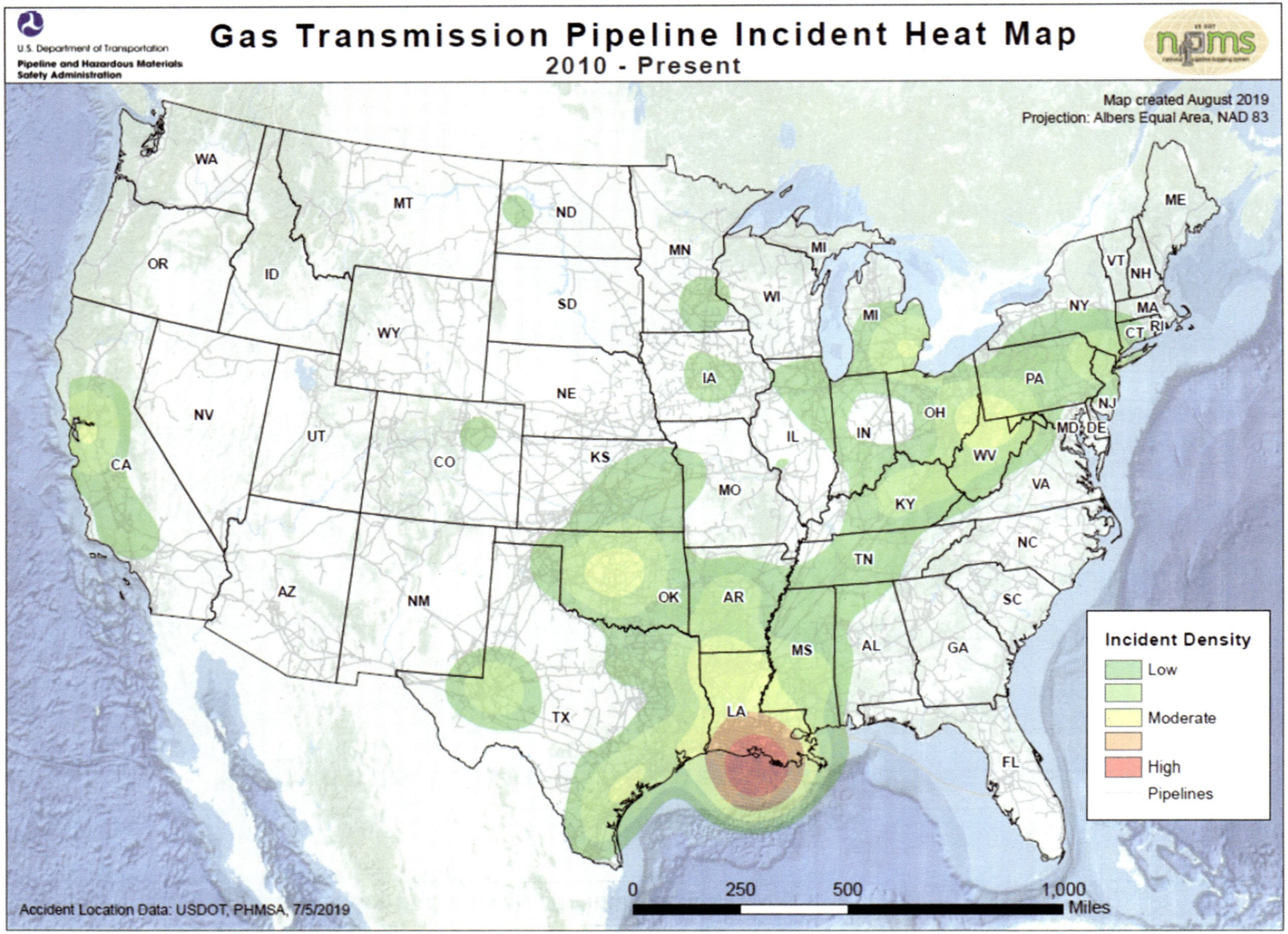

Figure 7. Gas transmission pipeline incident heat map, 2010-2020. Source: NPMS

LNG plants connected to natural gas pipeline systems, 2020

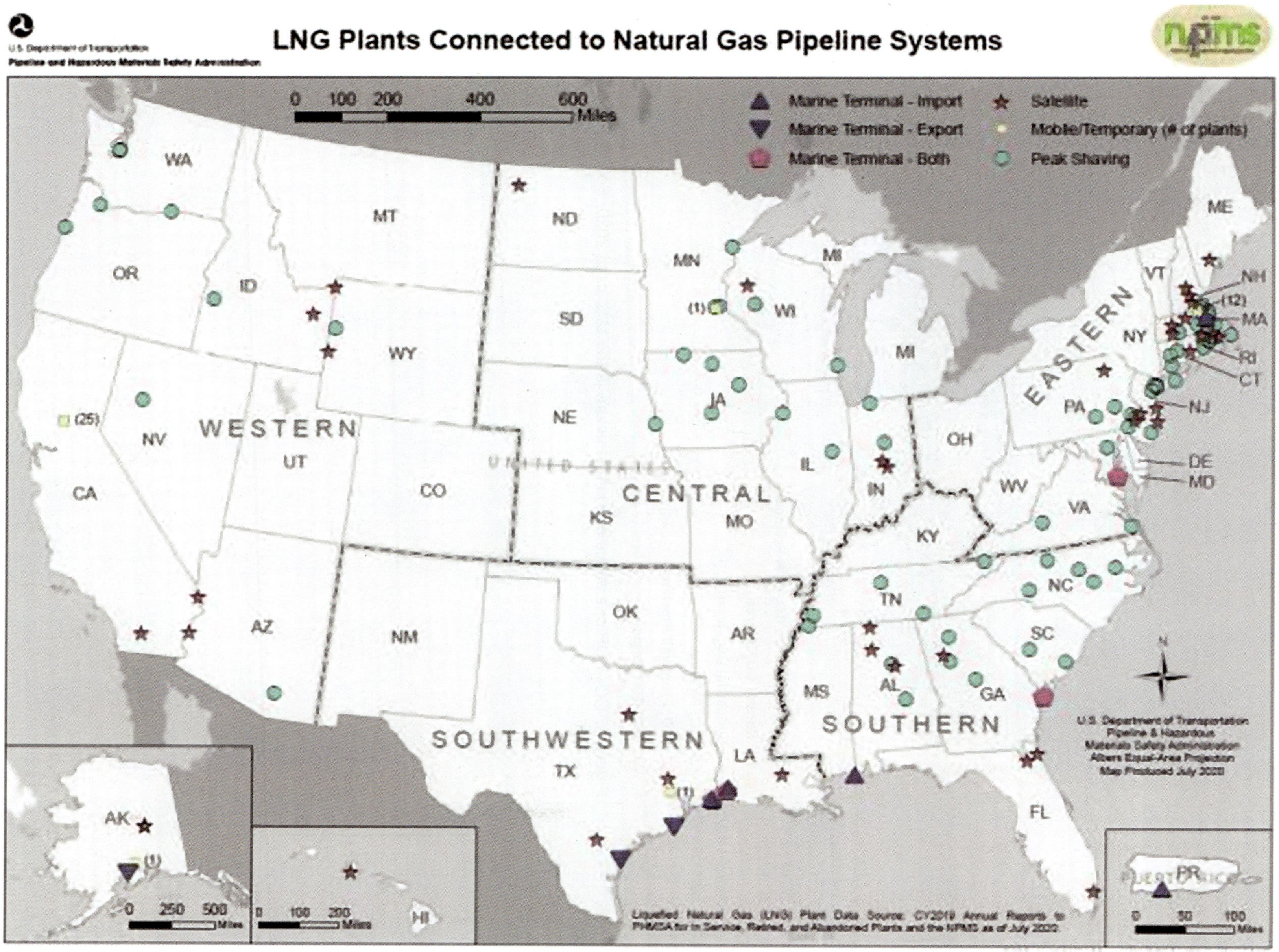

Figure 8. LNG plants connected to natural gas pipeline systems, 2020. Source: NPMS

CO_2-EOR operations and infrastructure in the U.S. circa 2020

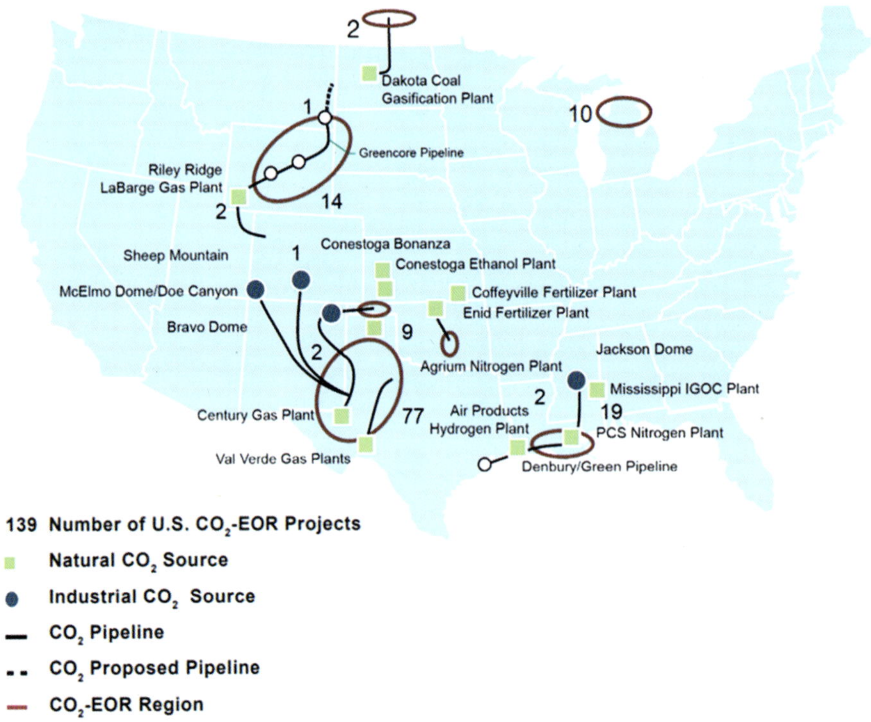

Figure 9. CO_2-EOR operations and infrastructure in the U.S. circa 2020. Source: DOE/NETL

Permian Basin CO_2 pipeline infrastructure circa 2020

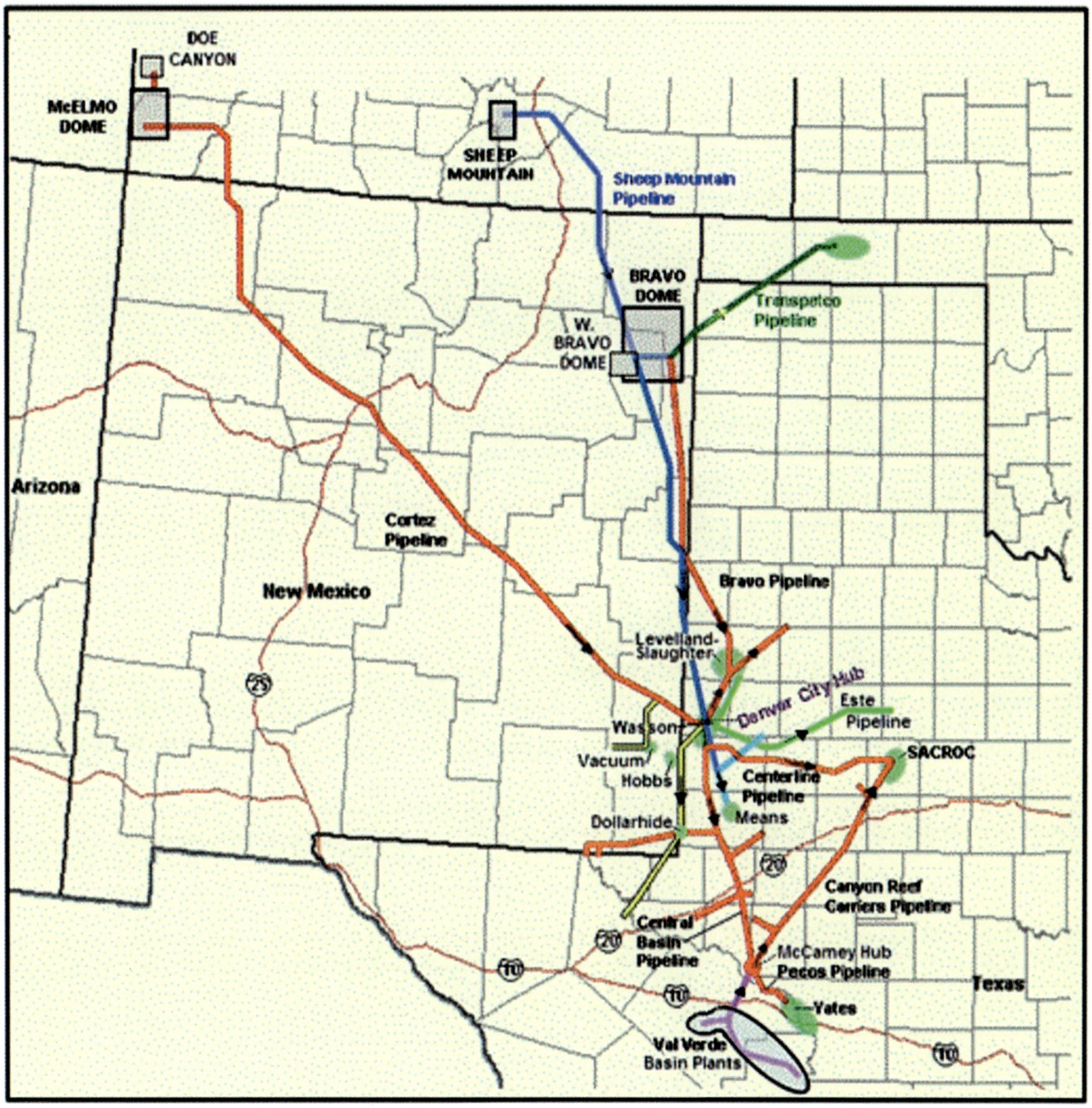

Figure 10. Permian Basin CO_2 pipeline infrastructure circa 2020. Source: DOE/NETL

Gulf Coast CO_2 pipeline infrastructure circa 2020

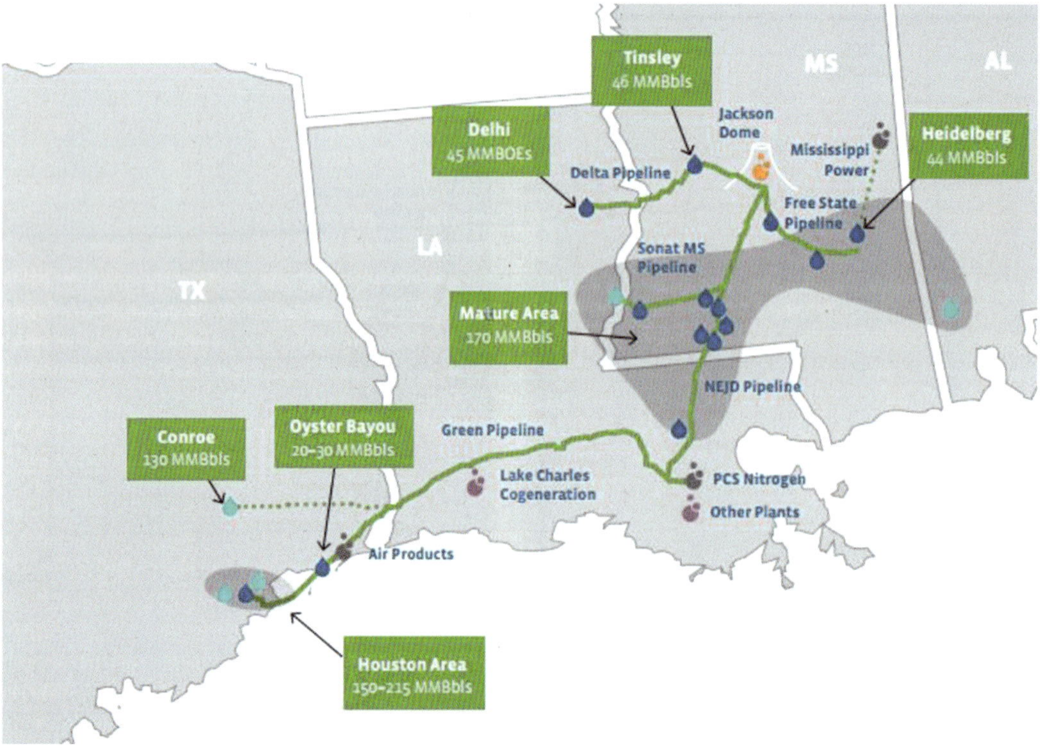

Figure 11. Gulf Coast CO_2 pipeline infrastructure circa 2015 and potential proved and produced-to-date tertiary reserves estimates circa 2013

Rocky Mountain CO_2 pipeline infrastructure circa 2020

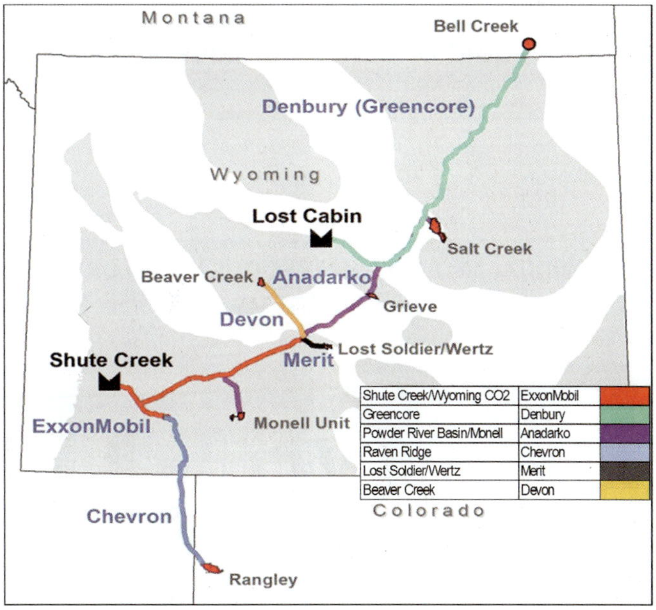

Figure 12. Rocky Mountain CO_2 pipeline infrastructure circa 2020. Source: DOE/NETL

Mid-Continent CO_2 pipeline infrastructure circa 2020

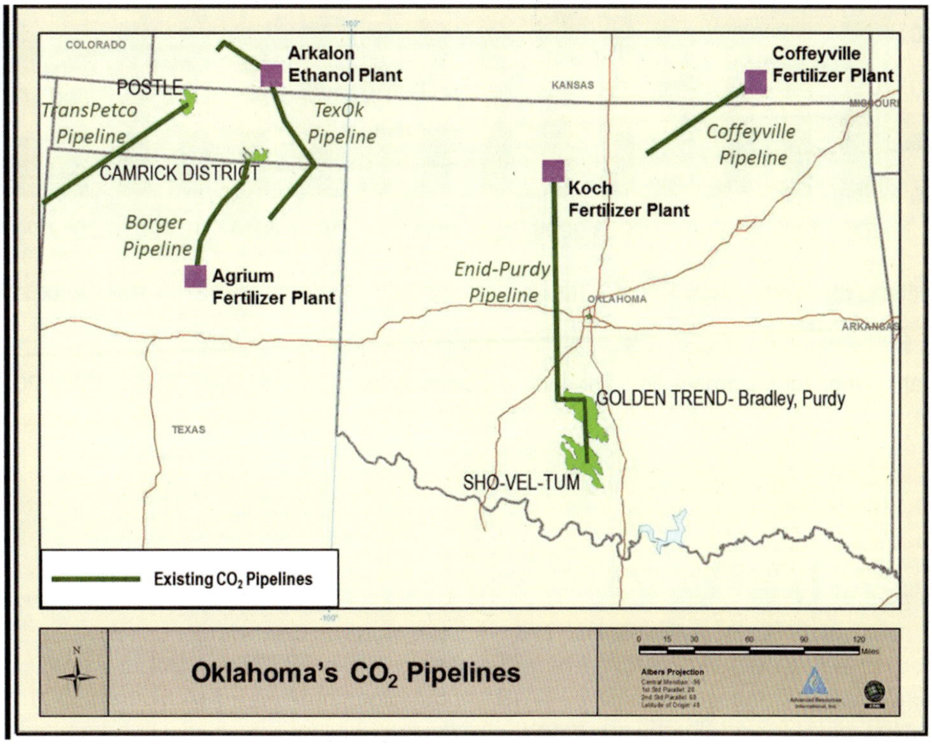

Figure 13. Mid-Continent CO_2 pipeline infrastructure circa 2020. Source: DOE/NETL

U.S. STATISTICS

Annual mileage for hazardous liquid or carbon dioxide systems in the United States, 2004-2019

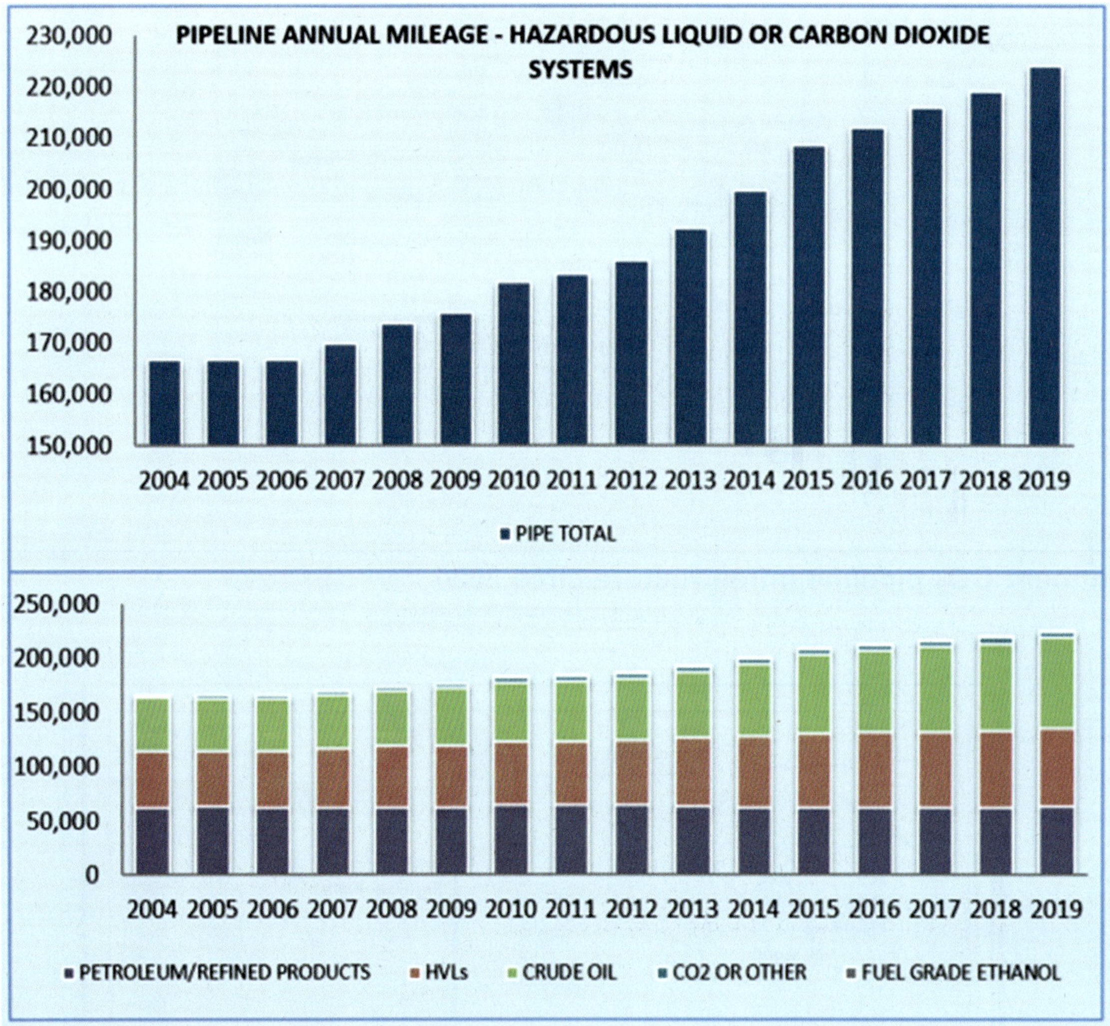

Figure 14. Annual mileage for hazardous liquid or carbon dioxide systems in the United States, 2004-2019. Source: PHMSA

Annual mileage for natural gas transmission and gathering systems in the United States, 1984-2019

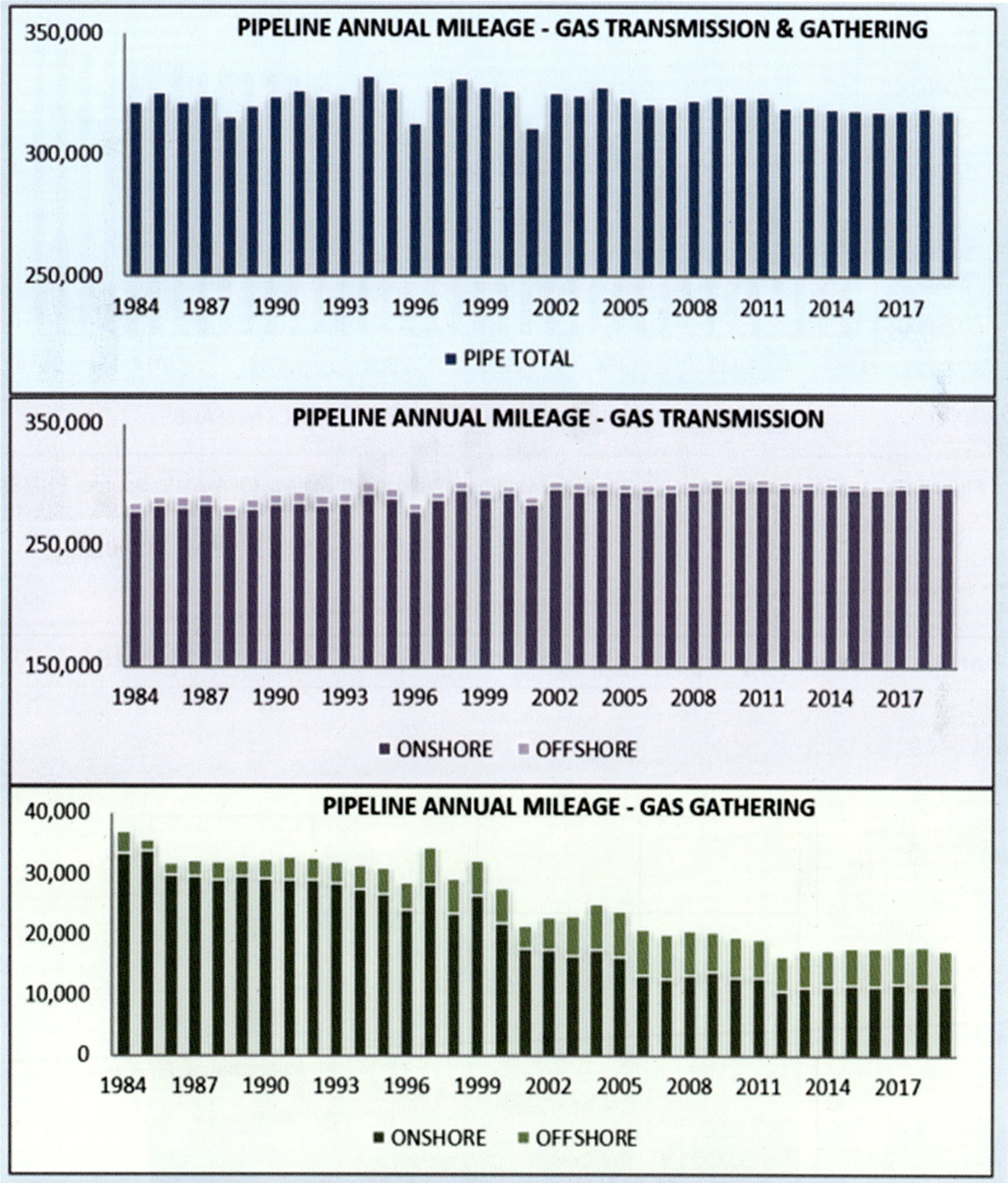

Figure 15. Annual mileage for natural gas transmission and gathering in the United States, 1984-2019. Source: PHMSA

Annual mileage for gas distribution systems in the United States, 1984-2019

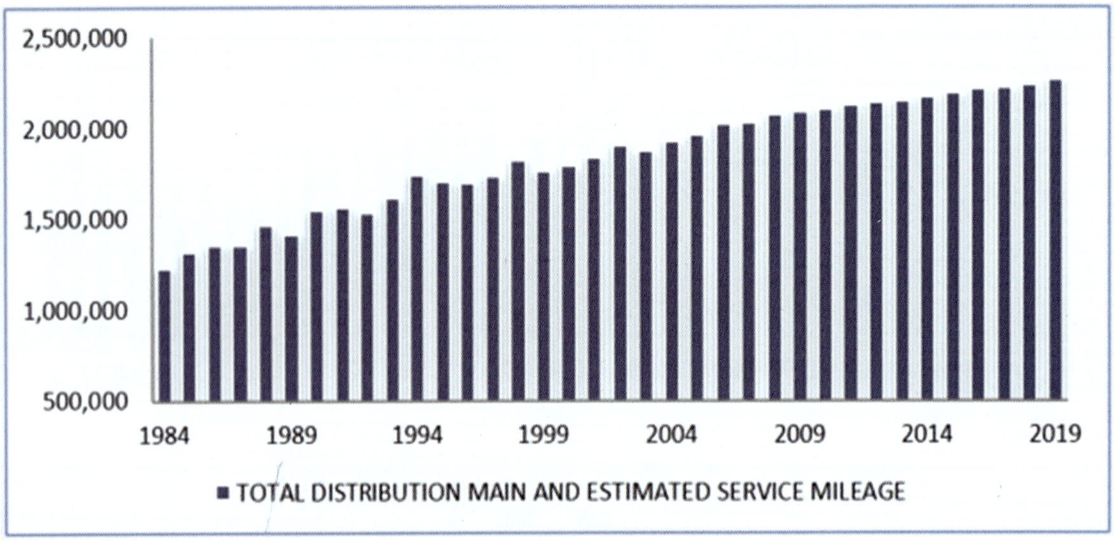

Figure 16. Annual mileage for gas distribution systems in the United States, 1984-2019. Source: PHMSA

Billion barrels delivered by liquids pipeline in the United States, 2014-2018

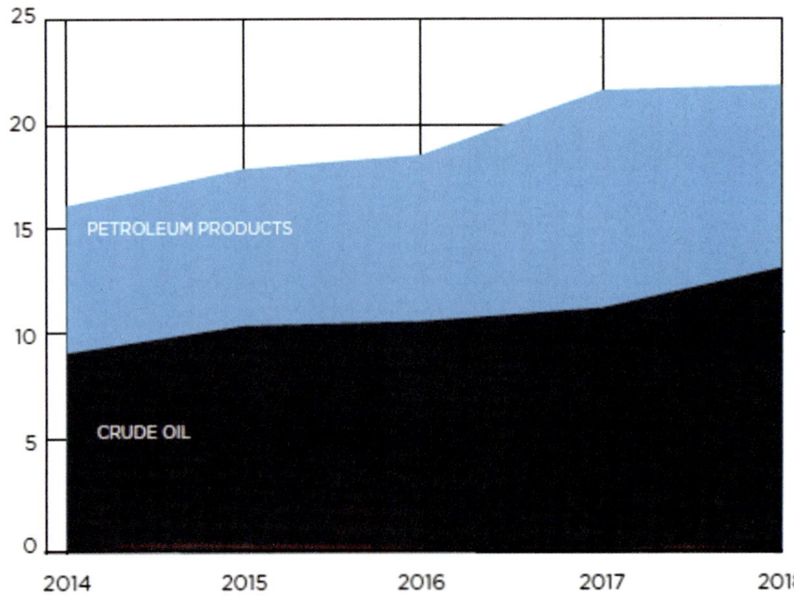

Figure 17. Billion barrels delivered by liquids pipeline in the United States, 2014-2018. Source: AOPI, API

Liquefied natural gas facilities in the United States circa 2020

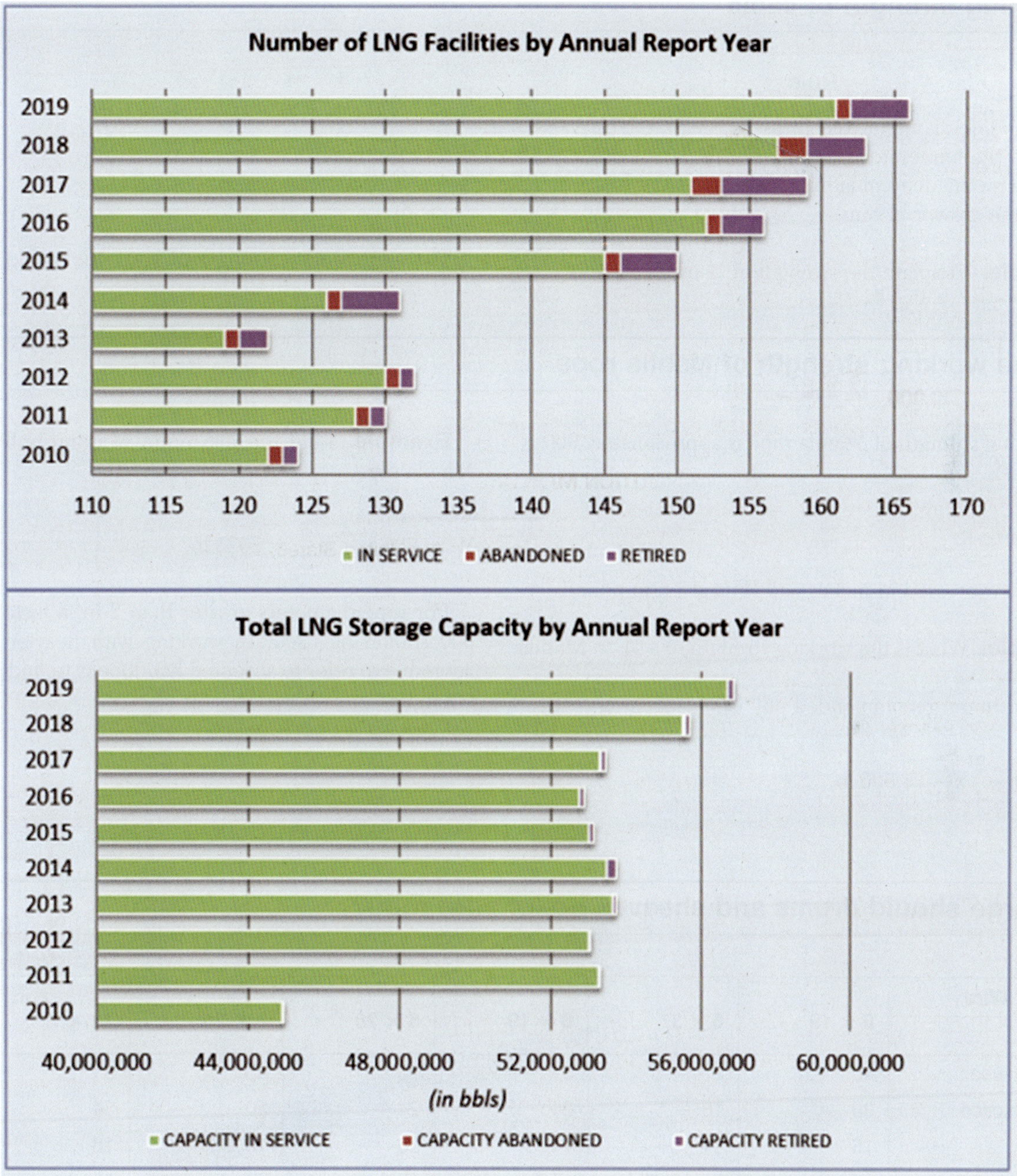

Figure 18. Liquefied natural gas facilities and total storage capacities in the United States circa 2020. Source: PHMSA

CABLES AND ROPES

Estimating strength of cable

Rule

1. Change line diameter to eighths
2. Square the numerator
3. Divide by the denominator
4. Read the answer in tons

Example. Estimate the strength of ½-in. steel cable:

$$\text{Diameter} = \frac{1}{2} = \frac{4}{8}$$

$$\frac{4^2}{8} = \frac{16}{8} = 2$$

The approximate strength of ½-in steel cable is 2 tons.

Find the working strength of Manila rope

The working strength of Manila rope is approximately 900 × (diameter)^2:

$$W = 900d^2,$$

where d is expressed in inches and W is given in pounds.

Example. What is the working strength of a ¾ in Manila rope?

The maximum recommended pull is

$$W = 900 \times \frac{3}{4} \times \frac{3}{4} = 506 \text{ lb.}$$

Example. Find the maximum working pull for a 1½ in Manila rope.

$$W = 900 \times \frac{3}{2} \times \frac{3}{2} = 2025 \text{ lb.}$$

For rope diameters greater than 2 in, a factor lower than 900 should be used. In working with heavier rigging it is advisable to refer to accepted handbooks to find safe working strengths.

How large should drums and sheaves be for wire rope?

Type of Wire Rope	6 × 19	6 × 37	8 × 19	5 × 28	6 × 25	18 × 7	6 × 7
For best wear	45	27	31	36	45	51	72
Good practice	30	18	21	24	30	34	42
Critical	16	14	14	16	16	18	28

Construction Methods and Machinery, by F. H. Kellogg, Prentice-Hall, Inc., 1954.

The diameter of sheaves or drums should preferably fall within the table for most efficient utilization of the wire rope.

Example. What size should the hoisting drum on a dragline be, if the wire rope is 6 × 19 construction, ¾ in in diameter?

From the table, good practice calls for 30 diameters, which in this instance would be 22½ in. Loads, speeds, bends, and service conditions will also affect the life of wire rope, so it is better to stay somewhere between the "good practice" and "best wear" factors in the table.

Find advantages of block and tackle with pull out friction

The efficiency of various sheaves differs. For one with roller bearings the efficiency has been estimated at 96.2%. For plain bearing sheaves a commonly used figure is 91.7%. The following formula will give close results:

$$\mathrm{MA} = \frac{W}{\omega} = E\frac{1-E^n}{1-E},$$

where

MA = Mechanical advantage,
W = Total weight to be lifted by the assembly,
ω = Maximum line pull at the hoist,
n = Number of working parts in the tackle,
E = Efficiency of individual sheaves.

It is assumed that the line leaving the upper block goes directly to the hoist without additional direction change (requiring a snatch block).

Example. Find the mechanical advantage of a four-part block and tackle using upper and lower blocks having journal bearings, which have an efficiency of 91.7%.

$$\mathrm{MA} = 0.917\frac{1-0.917^4}{1-0.917} = 0.917,$$

$$\left(\frac{1-0.707^4}{1-0.917}\right) = 0.917\left(\frac{0.293}{0.083}\right),$$

$$\mathrm{MA} = 3.25.$$

If the load weighed 3250 lb, what pull would be required on the lead line?

$$\frac{W}{\omega} = \mathrm{MA},$$

$$\frac{3250}{\omega} = 3.25,$$

$$\omega = 1000 \text{ lb}$$

Safe loads for wire rope

Safe Loading Factors For Wire Rope and Chains

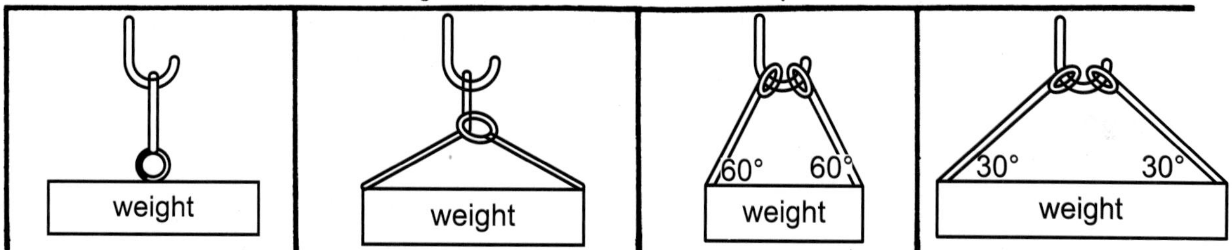

Dia. In.	SINGLE PART LINE Wire Rope (Tons)	SINGLE PART LINE Alloy Chain (Tons)	SINGLE PART LINE Wrought Chain (Tons)	SINGLE PART-CHOKER HITCH Wire Rope (Tons)	SINGLE PART-CHOKER HITCH Alloy Chain (Tons)	SINGLE PART-CHOKER HITCH Wrought Chain (Tons)	TWO PARTS (60°) Wire Rope (Tons)	TWO PARTS (60°) Alloy Chain (Tons)	TWO PARTS (60°) Wrought Chain (Tons)	TWO PARTS (30°) Wire Rope (Tons)	TWO PARTS (30°) Alloy Chain (Tons)	TWO PARTS (30°) Wrought Chain (Tons)
1/4	.5	....	....	.35	....	....	.6	....	....	.35	....	....
9/32	.5	1.3	....	....	1	....	....	1.8	....	....	1	....
3/8	1	2.3	.8	.75	1.8	.63	1.4	3.1	1.1	.8	1.8	.64
7/16	1	3	1.1	....	2.4	.9	....	4.2	1.5	....	2.4	.9
1/2	1.8	4	1.4	1.4	3.2	1.1	2.4	5.6	1.9	1.4	3.2	1.1
9/16	1.8	5	....	....	4	....	....	7	....	....	4	....
5/8	2.8	6	2.2	2.1	4.8	1.8	3.7	8.2	3	2.2	4.8	1.8
3/4	3.9	7.8	3.2	2.8	6.2	2.6	5.2	10.7	4.4	3	6.2	2.6
7/8	5.2	10.4	4.3	3.7	8.3	3.4	7	14.4	6	4	8.3	3.4
1	6.6	13.8	5.7	4.6	10.8	4.6	9	18.7	7.8	5.2	10.8	4.6
1 1/8	6.6	17.8	6.7	....	14.2	5.4	....	24.7	9.3	....	14.2	5.4
1 1/4	9.7	23.	8.3	6.6	18.2	6.6	13	31.9	11.5	7.5	18.4	6.6
1 3/8	9.7	....	10	....	....	8	....	....	13.9	....	....	8
1 1/2	13.2	....	12	9.2	....	9.6	17.3	....	16.6	10	....	9.6
1 5/8	13.2	....	14	....	....	11.2	....	....	19.4	....	....	11.2
1 3/4	17	....	16	12.1	....	12.8	21.5	....	22.2	12.4	....	12.8
2	20.6	....	21	15.4	....	16.8	26.7	....	29.1	15.4	....	16.8

There are various kinds of wire rope and chain available, and where it is possible strength factors should be reviewed before using any particular kind. Tables here may be used on construction jobs. They allow for a safety factor of four as based on mild plow steels, etc. Where the wire rope is made of improved plow steel, etc., it will have a greater safety factor.

Stress in guy wires

Guys are wire ropes or strands used to hold a vertical structure in position against an overturning force. The most common types of guyed structures are stacks, derricks, masts for draglines, reversible tramways, and radio transmission towers.

As a general rule, stresses in guys from temperature changes are neglected, but in structures such as radio masts this is an important feature and must be subject to special analysis. The number of guys used for any particular installation is contingent on several variable factors such as type of structure, space available, contour of the ground, etc., and is not a part of this discussion.

It is desirable to space guys uniformly whenever possible. This equalizes the pull, P, on each guy insofar as possible, particularly against forces that change in direction, as when a derrick boom swings in its circle.

It is also desirable to equalize the erection tensions on the guys. When no external force is acting on the structure, the tension in each guy should be the same. A tension indicator is sometimes used to determine the tension in guys. If this instrument is not available, the tension can be very closely approximated by measuring the deflection at the center of the span from the chord drawn from the guy anchorage to the point of support on the structure. A good average figure to use for erection tension of guys is 20% of the maximum working tension of the guy.

This discussion outlines the method for determining the stresses in guys. One of the first considerations is the location of the guy anchorages. The anchorages should be so located that the angle α, between the horizontal plane and the guy line, is the same for all guys (to equalize erection tensions). Angle α, in good practice, seldom exceeds 45°, with 30° being commonly used. The tension in the guys decreases as angle α becomes smaller. The direct load on the structure is also less with a smaller value of α.

To find the maximum extra tension, T, that will be applied to any single guy by the force, F, first determine the pull, P, which is the amount required along the guys, in the same vertical plane as the force to resist the horizontal component

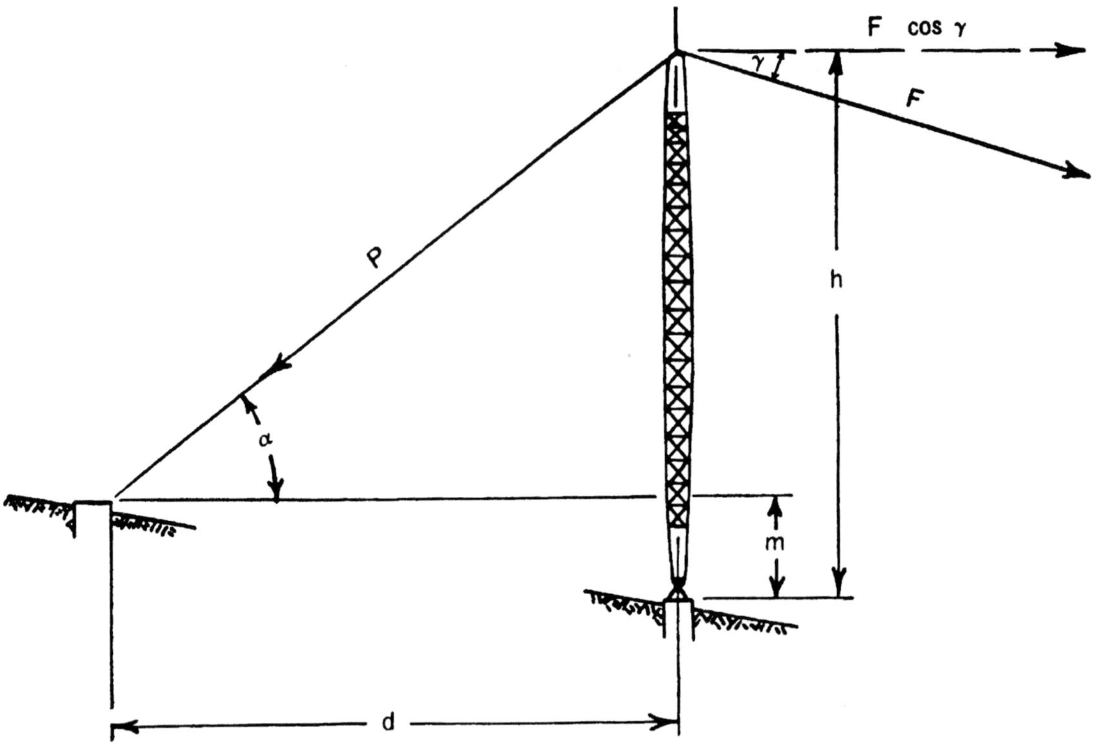

Figure 1.

of the force. This pull is entirely independent of the number of guys. Assume that the following are known:

F = The total resultant external force acting on the structure,
G = The angle between the horizontal plane and the force F,
h = The height of the structure,
d = The horizontal distance from structure to guy anchorage,
m = The vertical height of anchorage above or below the base of the structure,
α = The angle whose tangent is $(h \pm m) \div d$. m is added if the anchorage is below the base of the structure and subtracted if it is above.

P represents the total pull acting along the guys at an angle, α, with the horizontal, and in the same vertical plane as the force, F:

$$P = F \cos \gamma \div \cos \alpha$$

As $\cos \alpha$ is always less than 1, P is always greater than $F \cos \gamma$, the horizontal component of force F.

If only one guy were used, P would represent the extra tension, T. In practice, however, a number of guys are always used, and, therefore, the pull on any one guy will not be equal to P. The following table gives factors for any number of guys from 3 to 15, equally spaced about a central structure. To find the maximum extra tension, T, that will be applied to any single guy by the force, F, capable of rotating 360° around a vertical axis, it is only necessary to multiply the value of P, as determined above, by the factor for the number of guys used. It must be clearly understood in using this table that the guys are uniformly spaced and under equal tension when no load is acting on the structure.

Table 1

No. of Guys	Factors*
3	1.15
4	1.00
5	0.90
6	0.75
7	0.65
8	0.55
9	0.50
10	0.45
11	0.40
12	0.37
13	0.35
14	0.32
15	0.30

*These factors are for average conditions. If the guys are erected under accurately measured tensions of not less than 20% of the working load, the factors for five or more guys may be reduced by 10%. If the erecting tensions are low or not accurately equalized, the factors for five or more guys should be increased 10%.

Example. A derrick mast 90 ft high is supported by nine equally spaced guys anchored at a horizontal distance of 170 ft from the mast, and the elevations of the guy anchorages are 10 ft below the base of the mast. The load on the structure is equivalent to a force of 10,000 lb, acting on an angle of 10° below the horizontal. What is the maximum pull on any single cable?

From Figure 1:

$h = 90$ ft,
$d = 170$ ft,
m = 10 ft,
g = 10°00′,
F = 10,000 lb,

$$\tan \alpha = \frac{90 + 10}{170} = 100 = 0.588,$$

$$\alpha = 30^\circ 28',$$

$$P = \frac{F \cos \gamma}{\cos \alpha} = \frac{10,000 \times 0.985}{0.862} = 11,427 \text{ lb}.$$

From Table 1, $T = 11{,}427 \times 0.50 = 5714$ lb.

If erection tension is 10% of total working tension, 5714 is 90% of total working tension. Therefore, working tension = (5714 × 100)/90 = 6349 lb.

Strength and weight of popular wire rope

The following tables give the breaking strength for wire rope of popular construction made of improved plow steel.

Size	6 × 19	
	Breaking strength	Weight
1/4	5480	0.10
5/16	8520	0.16
3/8	12,200	0.23
7/16	16,540	0.31
1/2	21,400	0.40
9/16	27,000	0.51
5/8	33,400	0.63
1 3/4	47,600	0.90
7/8	64,400	1.23
1	83,600	1.60
1 1/8	105,200	2.03
1 1/4	129,200	2.50
1 3/8	155,400	3.03
1 1/2	184,000	3.60
1 5/8	214,000	4.23
1 3/4	248,000	4.90
1 7/8	282,000	5.63
2	320,000	6.40

To apply the above table to wire rope of other construction, multiply by the following factors:

Wire Rope Construction	6 × 19	6 × 29	6 × 37	18 × 7
Strength factors	1.00	0.96	0.95	0.92
Weight factors	1.00	0.97	0.97	1.08

Example. Find the breaking strength of 6 × 29 improved plow steel wire rope 2 in in diameter.

$$\text{Strength} = 320,000 \times 0.96 = 307,000 \text{ lb.}$$

The weight can be found the same way.

Measuring the diameter of wire rope

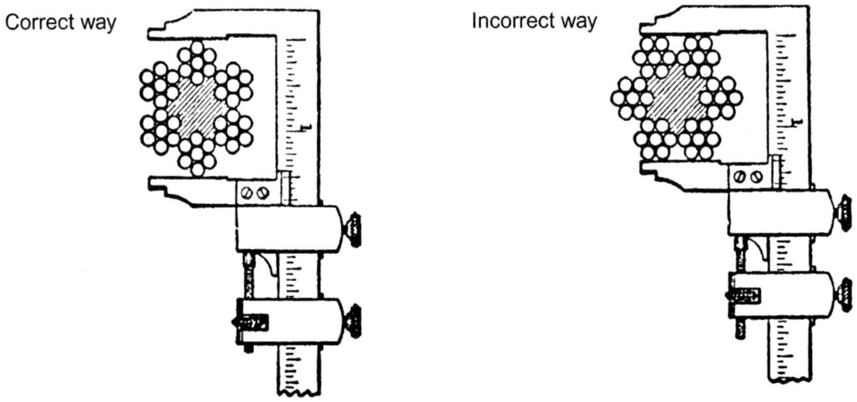

Wire rope: field troubles and their causes

All wire rope will eventually deteriorate in operation or have to be removed simply by virtue of the loads and reversals of load applied in normal service. There are, however, many conditions of service or inadvertent abuse that will materially shorten the normal life of a wire rope of proper construction although it is properly applied. The following field troubles and their causes give some of the field conditions and practices that result in the premature replacement of wire rope (it should be borne in mind that in all cases the contributory cause of removal may be one or more of these practices or conditions):

	Wire-Rope Trouble	Cause
a.	Rope broken (all strands)	Overload resulting from severe impact, kinking, damage, localized wear, weakening of one or more strands, or rust-bound condition and loss of elasticity.
b.	One or more whole strands parted	Overloading, kinking, divider interference, localized wear, or rust-bound condition. Fatigue, excessive speed, slipping, or running too loosely. Concentration of vibration at dead sheave or dead-end anchor.
c.	Excessive corrosion	Lack of lubrication. Exposure to salt spray, corrosive gases, alkaline water, acid water, mud, or dirt. Period of inactivity without adequate protection.
d.	Rope damage in hauling to the well or location	Rolling reel over obstructions or dropping from car, truck, or platform. The use of chains for lashing, or the use of lever against rope instead of flange. Nailing through rope to flange.
e.	Damage by improper socketing	Improper seizing, which allows slack from one or more strands to work back into rope; improper method of socketing or poor workmanship in socketing, frequently shown by rope being untwisted at socket, loose or drawn.
f.	Kinks, dog legs, and other distorted places	Kinking the rope and pulling out the loops such as in improper coiling or unreeling. Improper winding on the drum. Improper tie-down. Open-drum reels having longitudinal spokes too widely spaced. Divider interference. The addition of improperly spaced cleats to increase the drum diameter. Stressing while rope is over small sheave or obstacle.
g.	Damage by hooking back slack too tightly to girt	Operation of walking beam causing a bending action on wires at clamp and resulting in fatigue and cracking of wires, frequently before rope goes down into hole.
h.	Damage or failure on a fishing job	Rope improperly used on a fishing job, resulting in damage or failure as a result of the nature of the work.
i.	Lengthening of lay and reduction of diameter	Frequently produced by some type of overloading, such as an overload resulting in a collapse of the fiber core in swabbing lines. This may also occur in cable-tool lines as a result of concentrated pulsating or surging forces, which may contribute to fiber-core collapse.
j.	Premature breakage of wires	Caused by frictional heat developed by pressure and slippage, regardless of drilling depth.
k.	Excessive wear in spots	Kinks or bends in rope due to improper handling during installation or service. Divider interference; also, wear against casing or hard shells or abrasive formations in a crooked hole. Too infrequent cut-offs on working end.
l.	Spliced rope	A splice is never as good as a continuous piece of rope, and slack is liable to work back and cause irregular wear.
m.	Abrasion and broken wires in a straight line	Injury due to slipping rope through clamps.
n.	Drawn or loosened strands	Rapid fatigue breaks.
o.	Reduction in tensile strength or damage	Excessive heat due to careless exposure to fire or torch.
p.	Distortion of wire rope	Damage due to improperly attached clamps or wire-rope clips.
q.	High strands	Slipping through clamps, improper seizing, improper socketing or splicing, kinks, dog legs, and core popping.

	Wire-Rope Trouble	Cause
r.	Wear by abrasion	Lack of lubrication. Slipping clamp unduly. Sandy or gritty working conditions. Rubbing against stationary object or abrasive surface. Faulty alignment. Undersized grooves and sheaves.
s.	Fatigue breaks in wire	Excessive vibration due to poor drilling conditions, i.e. high speed, rope slipping, concentration of vibration at dead sheave or dead-end anchor, undersized grooves and sheaves, and improper selection of rope construction. Prolonged bending action over spudder sheaves, such as that due to hard drilling.
t.	Spiraling or curling	Allowing rope to drag or rub over pipe, sill, or any object during installation or operation. It is recommended that a block with sheave diameter 16 times the nominal wire-rope diameter, or larger, be used during installation of the line.
u.	Excessive flattening or crushing	Heavy overload, loose winding on drum, or cross winding. Too infrequent cut-offs on working end of cable-tool lines. Improper cutoff and moving program for cable-tool lines.
v.	Bird-caging or core-popping	Sudden unloading of line such as hitting fluid with excessive speed. Improper drilling motion or jar action. Use of sheaves of too small diameter or passing line around sharp bend.
w.	Whipping off of rope	Running too loose.
x.	Cutting in on drum	Loose winding on drum. Improper cutoff and moving program for rotary drilling lines. Improper or worn drum grooving or line turn-back late.

Capacity of drums

The capacity of wire line drums may be figured from the following formula:

$$M = (A+B) \times A \times C \times K.$$

Rope Size	*K*	Rope Size	*K*
$^3/_8$	1.86	1	0.262
7/16	1.37	$1^1/_8$	0.207
½	1.05	$1^1/_4$	0.167
9/16	0.828	$1^5/_8$	0.138
$^5/_8$	0.672	$1^1/_2$	0.116
$^3/_4$	0.465	$1^5/_8$	0.099
$^7/_8$	0.342	$1^3/_4$	0.085

where

M = Rope capacity of drum, inches,
A = Depth of flange, inches,
B = Diameter of drum, inches,
C = Width of the drum between flanges, inches,
K = Constant depending on rope size

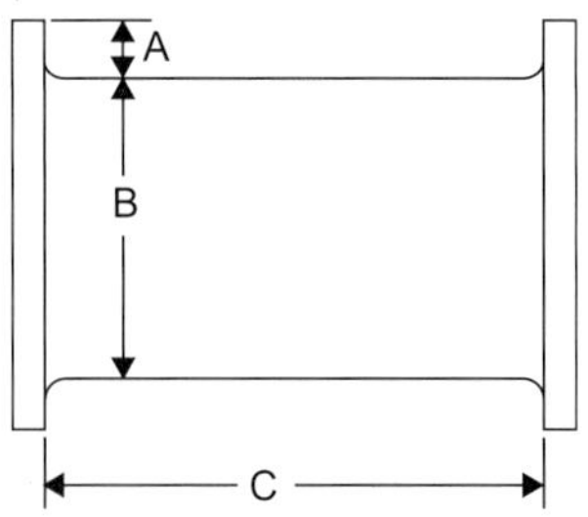

Figure 1. Capacity of wire line drums.

BELTS AND SHAFTS

Determine length of a V-belt

Rule: To find the nominal length of a V-belt, lay the belt on a table and place within it a pair of circular objects of the same diameter (flanges, tin cans, or whatever may be at hand). Pull them apart until the belt is fully extended without stretching. Then measure the shortest distance between the two circles, in inches. The belt size is then twice this figure, plus 5.14 times the diameter of the circles. This value, for any standard belt, should be a whole number of inches, which is the belt size.

V-belts are made in four standard sections, classified as A, B, C, and D; the widths (at the widest part) are ½, ⅝, ⅞, and 1⅛ in, respectively. The complete designation of the belt is the letter showing the width, followed by the length in inches; thus, an A26 belt is ½-in. wide and 26 in. long on the inside edge. The pitch length of the belt is measured along a median section and corresponds to the length that runs on the pulley diameter, which determines the actual speed ratio—about half of the depth of the groove. Pitch lengths for A, B, C, and D belts are greater than their nominal lengths by 1.3, 1.8, 2.9, and 3.3 in, respectively.

Calculate stress in shaft key

The shear and compressive stresses in a key are calculated using the following equations:

$$Ss = \frac{2T}{d \times W \times L}, \quad Sc = \frac{2T}{d \times h_1 \times L}$$

$Ss =$ Shear stress in psi,
$Sc =$ Compressive stress in psi,
$T =$ Shaft torque inch-pounds,
$d =$ Shaft diameter, inches (for taper shafts, use average diameter),
$w =$ Width key, inches,
$L =$ Effective length of key, inches,
$h_1 =$ Height of key in the shaft or hub that bears against the keyway, inches,
$h_1 = h_2$ for square keys. For designs where unequal portions of the key are in the hub or shaft, h_1 is the minimum portion.

Key material is usually AISI 1018 or AISI 1045 with the following allowable stresses:

Material	Heat Treatment	Allowable Stresses—psi	
		Shear	Compressive
AISI 1018	None	7500	15,000
AISI 1045	225–300 Bhn	15,000	30,000

Reprinted with permission: The Falk Corporation.

Example. Determine key stresses for these conditions: 300 HP @ 600 RPM; 3″ diameter shaft, $^3/_4 \times {}^3/_4$ key, 4″ key engagement length.

$$T = \text{Torque} = \frac{\text{HP} \times 63,000}{\text{RPM}} = 31.500 \text{ in.-lbs}$$
$$Ss = \frac{2T}{d \times W \times L} = \frac{2 \times 31,500}{3 \times 3/4 \times 4} = 7000 \text{ psi},$$
$$Sc = \frac{2T}{d \times h_1 \times L} = \frac{2 \times 31,500}{3 \times 3/8 \times 4} = 14,000 \text{ psi}.$$

The AISI 1018 key should be used since it is within allowable stresses. *Note*: If shaft had been $2^3/_4$″ diameter (4″ long hub), the key would be ⅝ × ⅝ and Ss = 9200 psi, Sc = 18,400 psi, and a heat-treated key, AISI 1045, would have been required.

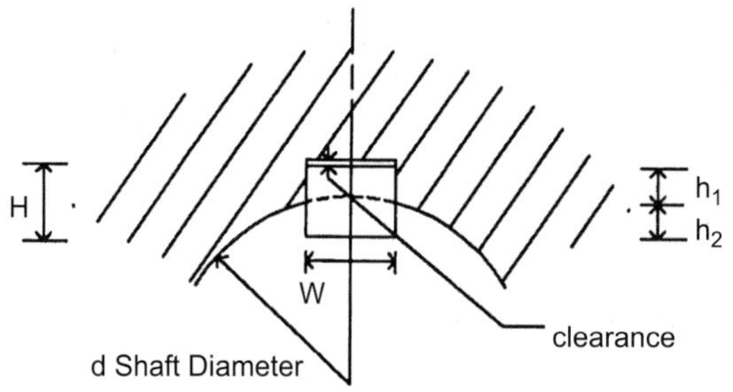

Estimate the horsepower that can be transmitted by a shaft

1. Where there are no stresses due to bending, weight of the shaft, pulleys, gears, or sprockets, use

$$\text{HP} = \frac{D^3 N}{50},$$

where

$D =$ diameter of shaft, inches,
$N =$ revolutions per minute.

2. For heavy duty service use

$$\text{HP} = \frac{D^3 N}{125}.$$

Example. What horsepower can be transmitted to an atmospheric cooling coil by a 2 in shaft turning at 1800 revolutions per minute?

$$\text{HP} = \frac{(2)^3(1800)}{125} = 115.2 \text{ horsepower}.$$

MISCELLANEOUS

Area formula

Form		Method of Finding Area
Triangle	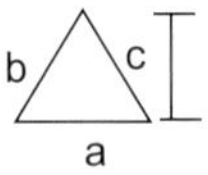	Base x 1/2 perpendicular height $\sqrt{s(s-a)(s-b)(s-c)}$ s = 1/2 sum of the three sides a,b,c
Trapezium	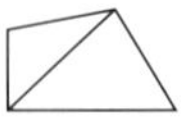	Sum of area of the two triangles
Trapezoid		1/2 sum of parallel sides x perpendicular height
Parallelogram		Base x perpendicular height
Reg. polygon		1/2 sum of sides x inside radius
Circle	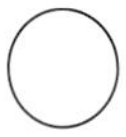	πr^2 = 0.78540 x diam^2. = 0.07958 x circumference^2
Sector of a circle		$\frac{\pi r^2 A^\circ}{360}$ = 0.0087266r^2A° = arc x 1/2 radius
Segment of a circle		$r^2\left(\frac{\pi A^\circ}{180} - \sin A^\circ\right)$
Circle of same area as a square	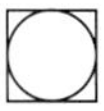	Diameter = side x 1.12838
Square of same area as a circle	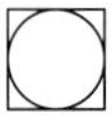	Side = diameter x 0.88623
Ellipse	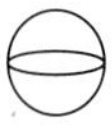	Long diameter x short diameter x 0.78540
Parabola	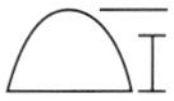	Base x 2/3 perpendicular height
Irregular plane surface	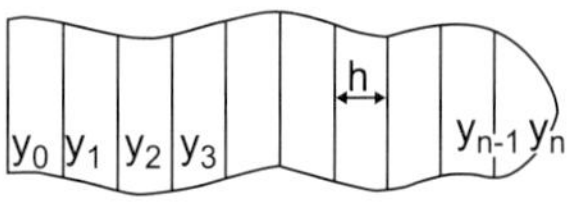	The larger the value of n, the greater the accuracy of approximation (Simpson's Rule). $A = \frac{h}{3}\left[(y_o + y_n) + 4(y_1 + y_3 + \ldots + y_{n-1}) + 2(y_2 + y_4 + \ldots + y_{n-2})\right]$

Surface and volume formula

Method of Finding Surfaces and Volumes of Solids
S = lateral or convex surface V = Volume

Parallelopiped

S = perimeter P perp. to sides x lat. length, l
V = area of base x perpendicular height, h
V = area of section A perp. to sides x lat. length, l

Prism-right, oblique, regular or irregular

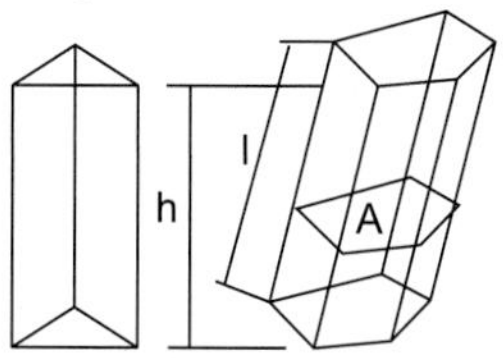

S = perimeter P perp. to sides x lat length l
V = area of base x perp. height, h
V = area of section A perp. to sides x lat. length, l

Cylinder-right, oblique, circular or elliptic, etc.

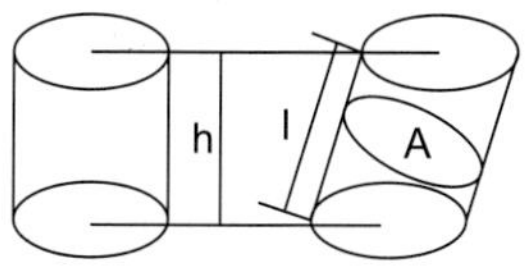

S = perimeter of base x perp. height
V = area of base x perp. height
V = area of section A perp. to sides x lat. length, l

Frustum of any prism or cylinder

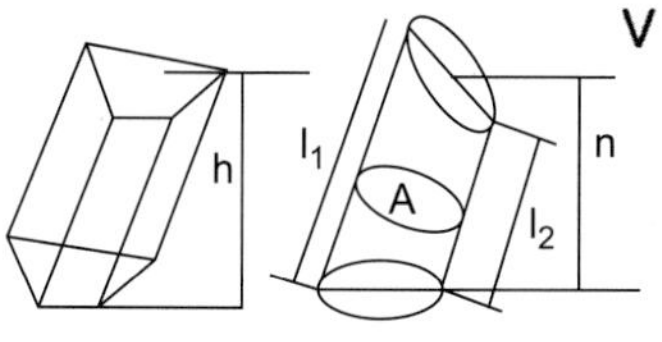

V = area of base x perpendicular distance h, from base to center of gravity of opposite face. For cylinder, $1/2A(l_1+l_2)$

Pyramid or Cone, right or regular

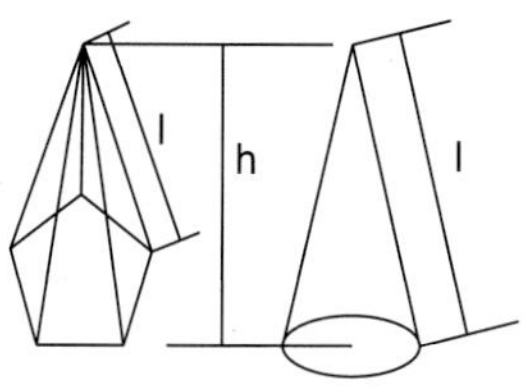

S = perimeter of base x 1/2 slant height, l

V = area of base x 1/3 perp. height, h

Pyramid or Cone-right, oblique, regular or irregular

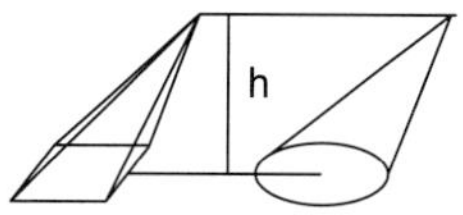

V = area of base x 1/3 perp. height, h
V = 1/3 vol. of prism or cylinder of same base and perp height
V = 1/2 vol. of hemisphere of same base and perp. height

Frustum of pyramid or cone, right and regular, parallel ends

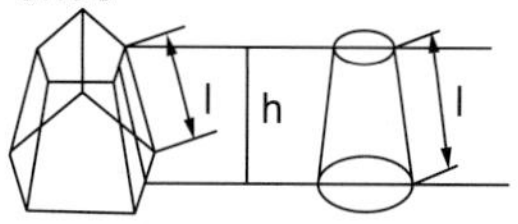

S = (sum of perimeter of base and top) x 1/2 slant height, l

$$V = \frac{h}{3}\left(A_1 + A_2 + \sqrt{A_1 x A_2}\right)$$

Where A_1 and A_2 are the areas of the bases

Frustum of any pyramid or cone, parallel ends

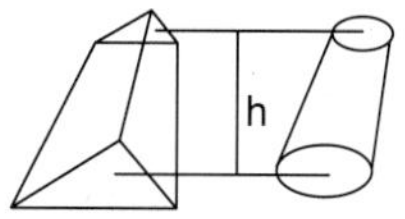

$$V = \frac{h}{3}\left(A_1 + A_2 + \sqrt{A_1 x A_2}\right)$$

Where A_1 and A_2 are the areas of the bases

Sphere

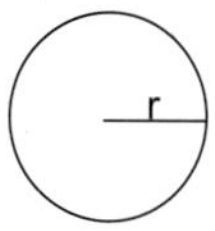

$S = 4\pi r^2$
$V = 4/3\pi r^3$

Spherical Sector

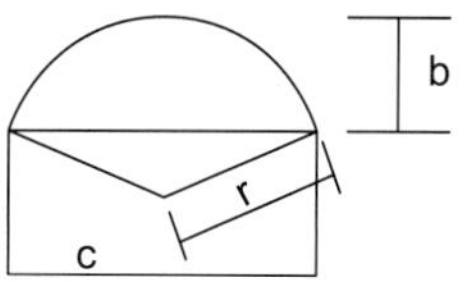

$V = 2/3\pi r^2 b$
$S = 1/2\pi r(4b+c)$

Spherical Zone

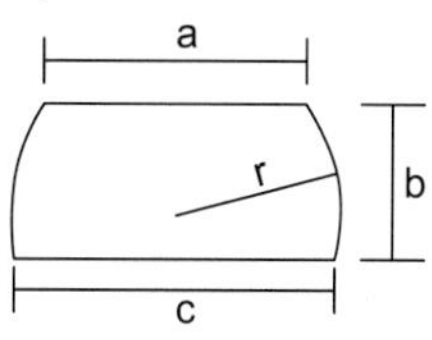

$S = 2\pi rb$
$V = 1/24\pi b(3a^2+3c^2+4b^2)$

Recommended drill sizes for self-tapping screws

Self-Tapping Screw Size			Major Thread Diameter			Minor Thread Diameter			Heavy Metals		Light Metals	
No.	Threads/In	OD	Max	Mean	Min	Max	Mean	Min	Drill Size		Drill Size	
2	32	0.086	0.088	0.0850	0.082	0.064	0.0620	0.0060	49	0.0730	49	0.0730
4	24	0.112	0.114	0.1110	0.108	0.086	0.0840	0.082	41	0.0960	41	0.0960
5	20	0.125	0.130	0.1265	0.123	0.094	0.0920	0.090	36	0.1065	36	0.1065
6	20	0.138	0.139	0.1355	0.132	0.104	0.1015	0.099	32	0.1160	32	0.1160
7	19	0.151	0.154	0.1505	0.147	0.115	0.1120	0.109	30	0.1285	30	0.1285
8	18	0.164	0.166	0.1625	0.159	0.122	0.1190	0.116	28	0.1405	29	0.1360
10	16	0.190	0.189	0.1855	0.182	0.141	0.1380	0.135	20	0.1610	21	0.1590
12	14	0.216	0.215	0.2115	0.208	0.164	0.1605	0.157	13	0.1850	14	0.1820
1/4	14	0.250	0.246	0.2415	0.237	0.192	0.1885	0.185	3	0.2130	4	0.2090
5/16	12	0.313	0.315	0.3105	0.306	0.244	0.2400	0.236	I	0.2720	H	0.2660
3/8	12	0.375	0.380	0.3755	0.371	0.309	0.3040	0.299	R	0.3390	Q	0.3320

Determine pulley speed

Speed of Driven Pulley Required:

Diameter and speed of driving pulley and diameter of driven pulley are known.

D_1 = Diameter of driving pulley 15 in,
RPM_1 = 180 (driving pulley speed),
d_2 = Diameter of driven pulley 9 in,
RPM_2 = Speed of driven pulley,

$$RPM_2 = \frac{15 \times 180}{9} = 300 \text{ RPM}.$$

Diameter of Driven Pulley Required:
Diameter and speed of driving pulley and speed of driven pulley are known.

D_1 = Diameter of driving pulley 24 in,
RPM_1 = 100 (driving pulley speed),

$$d_2 = \frac{24 \times 100}{600} = 4 \text{ in},$$

RPM_2 = Speed of driven pulley = 600.

Diameter of Driving Pulley Required:

D_1 = in,
d_2 = 36 in,
RPM_2 = 150,
RPM_1 = 600,

$$D_1 = \frac{36 \times 150}{600} = 9 \text{ in}.$$

Speed of Driving Pulley Required:

D_2 = 4 in,
RPM_2 = 800,
D_1 = 26 in,

$$RPM_1 = \frac{4 \times 800}{26} = 123.$$

Speed of Driven Pulley in Compound Drive Required:

$RPM_A = 260$,
$RPM_D = 720$ (required),

$$\frac{260}{720} = 13/36 \text{ required speed ratio.}$$

Resolve 13/36 into two factors $\frac{1 \times 13}{2 \times 18}$.
Multiply by trial numbers 12 and 1:

$$\frac{(1 \times 12) \times (13 \times 1)}{(2 \times 12) \times (18 \times 1)} = \frac{12 \times 13}{24 \times 18}.$$

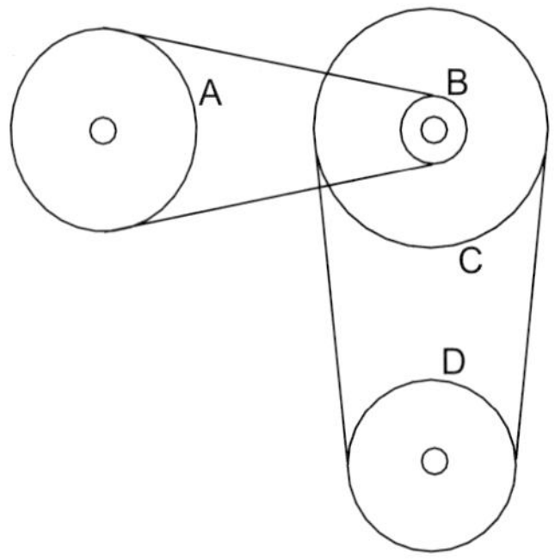

The values 12 and 13 in the numerator represent the diameter of driven pulleys B and D, and the values 24 and 18 in the denominator represent the diameter of the driving pulleys A and B.

ASTM standard reinforcing bars

Soft Metric Size	Nom Diam (mm)	Area (mm^2)	Weight Factors		Imper Size	Nom Diam (in)	Area (in^2)	Weight Factors	
			kg/m	kg/ft				kg/m	kg/ft
10	9.5	71	0.560	0.171	3	0.375	0.11	0.376	1.234
13	12.7	129	0.994	0.303	4	0.500	0.20	0.668	2.192
16	15.9	199	1.552	0.473	5	0.625	0.31	1.043	3.422
19	19.1	284	2.235	0.681	6	0.750	0.44	1.502	4.928
22	22.2	387	3.042	0.927	7	0.875	0.60	2.044	6.706
25	25.4	510	3.973	1.211	8	1.000	0.79	2.670	8.760
29	28.7	645	5.060	1.542	9	1.128	1.00	3.400	11.155
32	32.3	819	6.404	1.952	10	1.270	1.27	4.303	14.117
36	35.8	1006	7.907	2.410	11	1.410	1.56	5.313	17.431
43	43.0	1452	11.384	3.470	14	1.693	2.25	7.650	25.098
57	57.3	2581	20.239	6.169	18	2.257	4.0	13.600	44.619

Pressure rating for carbon steel flanges

Temperature (°F)	Flange Class						
	150	300	400	600	900	1500	2500
	150	300	400	600	900	1500	2500
<100	285	740	990	1480	2220	3705	6170
200	260	675	900	1350	2025	3375	5625
300	230	655	875	1315	1970	3280	5470

How to estimate length of material contained in roll

Where material of uniform thickness, like belting, is in a roll, the total length may be obtained by the following rule:

Measure the diameter of the hole in the center, and of the outside of the roll, both measurements in inches; count the number of turns; multiply the sum of the two measured diameters by the number of turns, and multiply this product by 0.13; the result is the total length of the material in feet.

Example. A roll of belting contains 24 turns. The diameter of the hole is 2 in, and that of the outside of the roll is 13 in.

$$(2 + 13) \times 24 \times 0.13 = 46.8$$

The roll contains 46.8 ft of belting.

Note: The rule can even be applied to materials as thin as pipeline felt; counting the turns is not as difficult as it might appear without a trial.

Convenient antifreeze chart for winterizing cooling systems

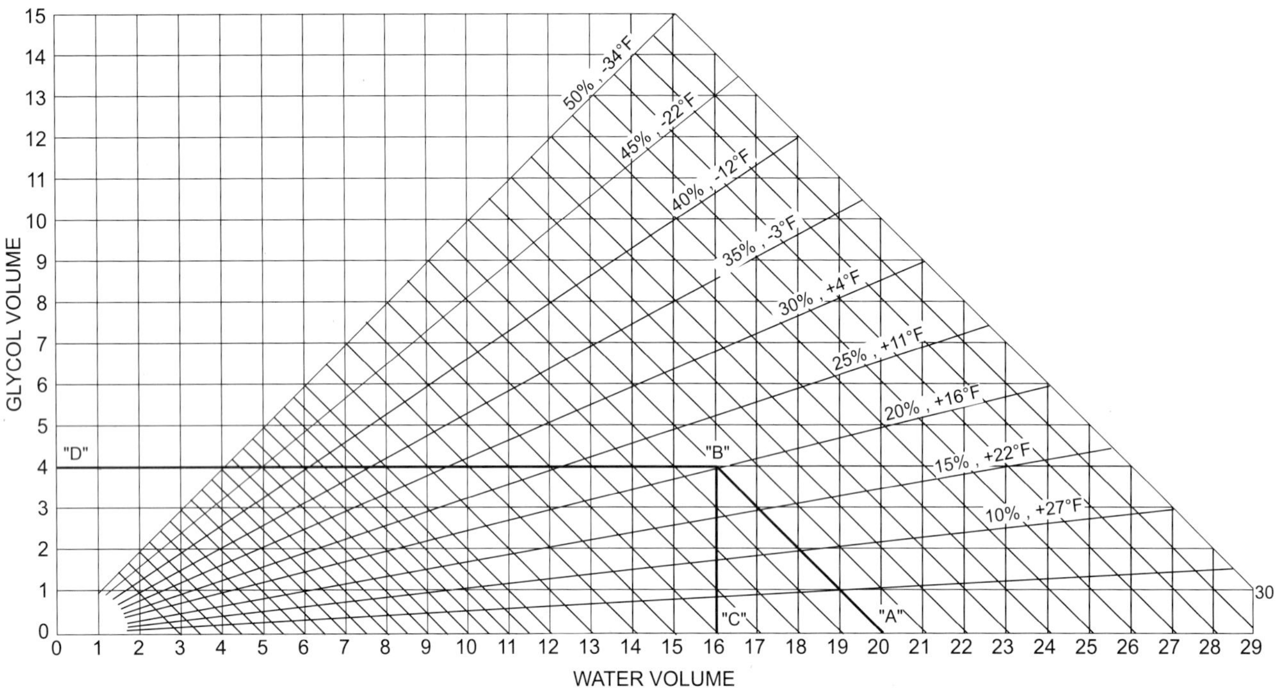

This plot of water volume versus glycol volume at various conditions of temperature and percent of glycol in the system makes winterizing field engines relatively easy.

Example. Determine the amount of glycol to be used in a 200 gallon system for protection to 16 °F.

To use the chart, first find the total system capacity, 200 gallons, at point A on the water volume axis. Point A also can represent 2, 20, 2000, etc. Proceed along the 45° line from point A to point B at the intersection with the 16 °F line. Then, move from point B to point C and read 160 on the water volume axis. This is the amount of water the system should contain. Move from point B to point D on the glycol axis and read 40. This is the amount of glycol that must be added to the system.

How to determine glycol requirements for desired protection

Solve the equation below to find the number of gallons of aqueous solution that must be removed and replaced with glycol.

$$d = \frac{D}{B},$$

where

d = Gallons of aqueous solution to be removed from the system and replaced with an equal number of gallons of glycol,
D = Total gallons of glycol required for protection at desired temperature less gallons of glycol in original system,
B = 1.0 − fraction of glycol in system.

Example. Find the number of gallons of existing aqueous solution that must be removed and replaced with glycol in a 1200 gallon system to give protection to −10 °F. Glycol in the system determined by a hydrometer test is 28% (0.28 fraction). The published figure for −10 °F protection is 519 gallons of glycol in the 1200 gallon system.

Then:

Gallons of glycol in system = 1200 × 0.28 = 336 gallons,
$D = 519 - 336 = 183$ gallons difference,
$B = 1.0 - 0.28 = 0.72.$

$$\text{Thus } d = \frac{183}{0.72} = 254 \text{ gallons.}$$

Therefore, 254 gallons of existing aqueous solution need to be removed and replaced with 254 gallons of glycol for the desired protection to −10 °F.

Weight in pounds of round steel shafting

Diameter of Shaft (in)	Length in Feet																	Weight per Inch
	1	2	3	4	5	6	7	8	9	10	12	14	16	18	20	22	24	
$^{15}/_{16}$	2	5	7	9	12	14	16	19	21	23	28	33	37	42	47	51	56	0.195
$1^{3}/_{16}$	4	8	11	15	19	23	26	30	34	38	45	53	60	68	76	83	91	0.314
$1^{7}/_{16}$	6	11	17	22	28	33	39	44	50	55	66	77	88	99	110	121	133	0.460
$1^{11}/_{16}$	8	15	23	30	38	46	53	61	68	76	91	107	122	137	152	167	183	0.634
$1^{15}/_{16}$	10	20	30	40	50	60	70	80	90	100	120	140	161	181	201	221	241	0.835
$2^{3}/_{6}$	13	26	38	51	64	77	90	102	115	128	153	179	205	230	256	281	307	1.06
$2^{7}/_{16}$	16	32	48	63	79	95	111	127	143	159	190	222	254	286	317	349	381	1.32
$2^{11}/_{16}$	19	39	58	77	97	116	135	154	174	193	232	270	309	348	386	425	463	1.61
$2^{15}/_{16}$	23	46	69	92	115	138	161	184	208	231	277	323	369	415	461	507	553	1.92
$3^{3}/_{16}$	27	54	81	109	136	163	190	217	244	272	326	380	435	489	543	598	652	2.26
$3^{7}/_{16}$	32	63	95	126	158	189	221	253	284	316	379	442	505	568	631	695	758	2.63
$3^{11}/_{16}$	36	73	109	146	182	218	255	291	328	364	437	510	582	655	728	801	874	3.03
$3^{15}/_{16}$	41	83	124	166	207	248	290	331	373	414	497	580	662	745	828	911	994	3.45
$4^{7}/_{16}$	53	105	158	210	263	315	368	421	473	526	631	736	841	946	1052	1157	1262	4.38
$4^{15}/_{16}$	65	130	195	260	326	391	456	521	586	651	781	911	1041	1172	1302	1432	1562	5.43
5	67	134	200	267	334	401	467	534	601	668	801	935	1068	1202	1335	1469	1602	5.56
$5^{7}/_{16}$	79	158	237	316	395	474	553	632	711	790	947	1105	1263	1421	1579	1737	1894	6.58
$5^{1}/_{2}$	81	162	242	323	404	485	565	646	727	808	969	1131	1292	1454	1615	1777	1938	6.73
$5^{15}/_{16}$	94	188	282	377	471	565	659	753	847	941	1130	1318	1506	1695	1883	2071	2259	7.85
6	96	192	288	385	481	577	673	769	865	961	1154	1346	1538	1731	1923	2115	2307	8.01
$6^{1}/_{2}$	113	226	338	451	564	677	790	903	1015	1128	1354	1579	1805	2031	2256	2482	2708	9.40
7	131	262	393	524	654	785	916	1047	1178	1309	1507	1832	2094	2355	2677	2878	3140	10.90
$7^{1}/_{2}$	150	300	451	601	751	901	1051	1202	1352	1502	1803	2103	2103	2704	3004	3305	3605	12.52
8	171	342	513	684	855	1024	1196	1367	1538	1709	2051	2051	2303	3076	3418	3760	4102	14.24

Properties of shafting

Diameter of Shaft (in)	Area of Shaft	Weight per Inch	Weight per Foot	Section Modulus		Moment of Inertia		Radius of Gyration
				Bending	Torsion	Bending	Torsion	
15/16	0.6903	0.195	2.34	0.0809	0.1618	0.0379	0.0758	0.2343
1 3/16	1.1075	0.314	3.76	0.1644	0.3288	0.0976	0.1952	0.2969
1 7/16	1.6230	0.460	5.52	0.2916	0.5832	0.2096	0.4192	0.3594
1 11/16	2.2365	0.634	7.61	0.4718	0.9435	0.3981	0.7961	0.4219
1 15/16	2.9483	0.835	10.03	0.7140	1.4281	0.6971	1.3835	0.4844
2 3/16	3.7583	1.06	12.78	1.0276	2.0553	1.1240	2.2480	0.5469
2 7/16	4.6664	1.32	15.86	1.4218	2.8436	1.7328	3.4656	0.6094
2 11/16	5.6727	1.61	19.29	1.9057	3.8113	2.5607	5.1215	0.6719
2 15/16	6.7771	1.92	23.04	2.4885	4.9770	3.6549	7.3099	0.7344
3 3/16	7.9798	2.26	27.12	3.1794	6.3589	5.0672	10.1345	0.7969
3 7/16	9.2806	2.63	31.56	3.9878	7.9755	6.8539	13.7079	0.8594
3 11/16	10.680	3.03	36.31	4.9226	9.8452	9.0761	18.1521	0.9219
3 15/16	12.177	3.45	41.40	5.9932	11.9865	11.7992	23.5984	0.9844
4 7/16	15.466	4.38	52.58	8.5786	17.1571	19.0337	38.0674	1.1094
4 15/16	19.147	5.43	65.10	11.8174	23.6348	29.1742	58.3483	1.2344
5	19.635	5.56	66.76	12.2718	24.5437	30.6796	61.3592	1.2500
5 7/16	23.221	6.58	78.96	15.7833	31.5666	42.9108	85.8217	1.3594
5½	23.758	6.73	80.78	16.3338	32.6677	44.9180	89.8361	1.3750
5 15/16	27.688	7.85	94.14	20.5499	41.0999	61.0077	122.0153	1.4844
6	28.274	8.01	96.12	21.2058	42.4115	63.6172	127.2345	1.5000
6½	33.183	9.40	112.82	26.9612	53.9225	87.6240	175.2481	1.6250
7	38.485	10.90	130.84	33.6739	67.3479	117.8588	235.7176	1.7500
7 1/2	44.179	12.52	150.21	41.4175	82.8350	155.3155	310.6311	1.8750
8	50.265	14.24	170.90	50.2655	100.5310	201.0619	402.1239	2.0000

Steel shaft sizes calculated for strength

Base formula : $d = \sqrt{\frac{C(HP)}{N}}$

Base Fiber Stress = 60,000 lb sq in.

Diameter of Shaft(in)		$\frac{HP}{N}$		
Fractional	Decimal	$C = 50$ Saf. Fac. 9.4	$C = 75$ Saf. Fac. 14.1	$C = 100$ Saf. Fac. 18.8
1/16	0.0625	0.000004883	0.000003256	0.000002492
1/8	0.125	0.00003905	0.00002604	0.00001953
3/16	0.1875	0.0001318	0.00008788	0.00006592
1/4	0.25	0.0003124	0.0002083	0.0001563
5/16	0.3125	0.0006105	0.0004070	0.0003053
3/8	0.375	0.001054	0.0007030	0.0005272
7/16	0.4375	0.001675	0.001116	0.0008375
1/2	0.5	0.002500	0.001666	0.001250
9/16	0.5625	0.003558	0.002372	0.001779
5/8	0.625	0.004883	0.003256	0.002442
11/16	0.6875	0.006500	0.004333	0.003250
3/4	0.75	0.008439	0.005626	0.004220
13/16	0.8125	0.01074	0.007155	0.005367
7/8	0.875	0.01340	0.008932	0.006699
15/16	0.9375	0.01647	0.01098	0.008235

Diameter of Shaft(in)		$\frac{HP}{N}$		
Fractional	Decimal	$C = 50$ Saf. Fac. 9.4	$C = 75$ Saf. Fac. 14.1	$C = 100$ Saf. Fac. 18.8
1	1.0	0.02000	0.01333	0.01000
1 1/16	1.0625	0.02396	0.01598	0.01199
1 1/8	1.125	0.02846	0.01898	0.01423
1 3/16	1.1875	0.03353	0.02236	0.01677
1 1/4	1.25	0.03905	0.02604	0.01953
15/16	1.3125	0.04519	0.03019	0.02259
1 3/8	1.375	0.5199	0.03465	0.02599
1 7/16	1.4375	0.05940	0.03960	0.02970
1 1/2	1.5	0.06750	0.04500	0.03375
1 9/16	1.5625	0.07628	0.05086	0.03815
1 5/8	1.625	0.08578	0.05719	0.04289
1 11/16	1.6875	0.09524	0.06409	0.04807
1 3/4	1.75	0.1072	0.07144	0.05358
1 13/16	1.8125	0.1191	0.07939	0.05955
1 7/8	1.875	0.1318	0.08788	0.06592

Diameter of Shaft(in) $\frac{HP}{N}$				
Fractional	**Decimal**	**$C = 50$ Saf. Fac. 9.4**	**$C = 75$ Saf. Fac. 14.1**	**$C = 100$ Saf. Fac. 18.8**
$1^{15}/_{16}$	1.9375	0.1456	0.09707	0.07281
2	2.0	0.1600	0.1066	0.07998
$2^{1}/_{16}$	2.0625	0.1755	0.1169	0.08774
$2^{1}/_{18}$	2.125	0.1918	0.1278	0.09592
$2^{3}/_{16}$	2.1875	0.2095	0.1397	0.1048
$2^{1}/_{4}$	2.25	0.2278	0.1519	0.1140
$2^{5}/_{16}$	2.3125	0.2474	0.1649	0.1237
$2^{3}/_{8}$	2.375	0.2678	0.1785	0.1339
$2^{7}/_{16}$	2.4375	0.2897	0.1932	0.1449
2½	2.5	0.3124	0.2083	0.1563
$2^{9}/_{16}$	2.5625	0.3363	0.2243	0.1682
$2^{7}/_{8}$	2.625	0.3616	0.2411	0.1808
$2^{11}/_{16}$	2.6875	0.3884	0.2589	0.1942
$2^{3}/_{4}$	2.75	0.4159	0.2772	0.2079
$2^{13}/_{16}$	2.8125	0.4449	0.2966	0.2225
$2^{7}/_{8}$	2.875	0.4754	0.3170	0.2378
$2^{15}/_{16}$	2.9375	0.5074	0.3383	0.2537
3	3.0	0.5399	0.3683	0.2700
$3^{1}/_{16}$	3.0625	0.5742	0.3836	0.2871
$3^{1}/_{8}$	3.125	0.6105	0.4070	0.3053
$3^{3}/_{16}$	3.1875	0.6479	0.4319	0.3240
$3^{1}/_{4}$	3.25	0.6866	0.4577	0.3434
$3^{5}/_{16}$	3.3125	0.7266	0.4844	0.3634
$3^{3}/_{8}$	3.375	0.7685	0.5123	0.3842
$3^{7}/_{16}$	3.4375	0.8127	0.5418	0.4061
$3^{1}/_{2}$	3.5	0.8567	0.5718	0.4288
$3^{9}/_{16}$	3.5625	0.9038	0.6026	0.4520

Diameter of Shaft(in) $\frac{HP}{N}$				
Fractional	**Decimal**	**$C = 50$ Saf. Fac. 9.4**	**$C = 75$ Saf. Fac. 14.1**	**$C = 100$ Saf. Fac. 18.8**
$3^{5}/_{8}$	3.625	0.9526	0.6351	0.4763
$3^{11}/_{16}$	3.6875	1.003	0.6688	0.5017
$3^{3}/_{4}$	3.75	1.054	0.7030	0.5272
$3^{13}/_{16}$	3.8125	1.108	0.7382	0.5538
$3^{7}/_{8}$	3.875	1.163	0.7759	0.5820
$3^{15}/_{16}$	3.9375	1.222	0.8143	0.6108
4	4.0	1.280	0.8535	0.6401
$4^{1}/_{16}$	4.0625	1.340	0.8933	0.6701
$4^{1}/_{8}$	4.125	1.404	0.9356	0.7018
$4^{3}/_{16}$	4.1875	1.469	0.9792	0.7345
$4^{1}/_{4}$	4.25	1.536	1.023	0.7678
$4^{5}/_{6}$	4.3125	1.603	1.069	0.8019
$4^{3}/_{8}$	4.375	1.675	1.116	0.8375
$4^{7}/_{16}$	4.4375	1.748	1.165	0.8742
$4^{1}/_{2}$	4.5	1.823	1.214	0.9111
$4^{9}/_{16}$	4.5625	1.900	1.266	0.9497
$4^{5}/_{8}$	4.625	1.978	1.319	0.9893
$4^{11}/_{16}$	4.6875	2.059	1.373	1.030
$4^{3}/_{4}$	4.75	2.143	1.429	1.072
$4^{13}/_{16}$	4.8125	2.228	1.486	1.114
$4^{7}/_{8}$	4.875	2.316	1.544	1.158
$4^{15}/_{16}$	4.9375	2.407	1.604	1.203
5	5.0	2.500	1.666	1.250

C = 50 corresponds to very little bending; C = 75, average conditions of bending; C = 100, severe conditions of bending Haven-Swett "Treatise on Leather Belting," American Leather Belting Association.

Tap drills and clearance drills for machine screws

Screw Size	Coarse Thread		Fine Thread		Clearance Drill
	TPI	Drill	TPI	Drill	
4	40	43	48	42	32
5	40	38	44	37	30
6	32	36	40	33	27
8	32	29	36	29	18
10	24	25	32	21	9
12	24	16	28	14	2
14	20	10	24	7	D
¼	20	7	28	3	F
$^{5}/_{16}$	18	F	24	1	P
$^{3}/_{8}$	16	$^{5}/_{16}$	24	Q	W
$^{7}/_{16}$	14	U	20	$^{25}/_{64}$	$^{29}/_{64}$
½	13	$^{27}/_{64}$	20	$^{29}/_{64}$	$^{33}/_{64}$
$^{9}/_{16}$	12	$^{31}/_{64}$	18	$^{33}/_{64}$	$^{37}/_{64}$
$^{5}/_{8}$	11	$^{17}/_{32}$	18	$^{37}/_{64}$	$^{41}/_{64}$
$^{3}/_{4}$	10	$^{21}/_{32}$	16	$^{11}/_{16}$	$^{49}/_{64}$

Common nails

Size	Length	Gauge No.	Head Diameter	Approx. No./lb
2d	1.00	15	11/64	845
3d	1.25	14	13/64	540
4d	1.50	12.5	¼	290
5d	1.75	12.5	¼	250
6d	2.00	11.5	17/64	265
7d	2.25	11.5	17/64	250
8d	2.50	10.25	9/32	200
9d	2.75	10.25	9/32	90
10d	3.00	9	5/16	65

Size	Length	Gauge No.	Head Diameter	Approx. No./lb
12d	3.25	9	5/16	60
16d	3.50	8	11/32	45
20d	4.00	6	11/32	30
30d	4.50	5	7/16	20
40d	5.00	4	15/32	17
50d	5.50	3	½	13
60d	6.00	2	17/32	10

Drill sizes for pipe taps

Tap Size (in)	TPI	Drill Diameter (in)
1/8	27	11/32
¼	18	7/16
3/8	18	37/64
½	14	23/32
3/4	14	59/64
1	11½	1 5/32
1¼	11½	1½
1½	11½	1 49/64
2	11 ½	2 3/16

Tap Size (in)	TPI	Drill Diameter (in)
2½	8	2 9/16
3	8	3 3/16
3½	8	3 11/16
4	8	4 3/16
4½	8	4 3/4
5	8	5 5/16
6	8	6 5/16

Carbon steel—color and approximate temperature

Color	Temperature (°F)
Black red	990
Dark blood red	1050
Dark cherry red	1175
Medium cherry red	1250
Light cherry, scaling	1550

Color	Temperature (°F)
Salmon, free scaling	1650
Light salmon	1725
Yellow	1825
Light yellow	1975
White	2220

Bolting dimensions for flanges

ASME/ANSI B16.5 provides weld neck, slip-on, socket weld, threaded, lap joint and blind flange dimensions for pipes ranging 1/2" to 24" in classes ranging 150 to 2500.

Nominal Pipe Size NPS (inches)	Class 150				
	Diameter of Flange (inches)	No. of Bolts	Diameter of Bolts (inches)	Diameter of Bolt Holes (inches)	Bolt Circle (inches)
1/4	3-3/8	4	½	0.62	2-1/4
1/2	3-1/2	4	½	0.62	2-3/8
3/4	3-7/8	4	½	0.62	2-3/4
1	4-1/4	4	½	0.62	3-1/8
1-1/4	4-5/8	4	½	0.62	3-1/2
1-1/2	5	4	½	0.62	3-7/8
2	6	4	5/8	0.75	4-3/4
2-1/2	7	4	5/8	0.75	5-1/2
3	7-1/2	4	5/8	0.75	6
3-1/2	8-1/2	8	5/8	0.75	7
4	9	8	5/8	0.75	7-1/2
5	10	8	¾	0.88	8-1/2
6	11	8	¾	0.88	9-1/2
8	13-1/2	8	¾	0.88	11-3/4
10	16	12	7/8	1	14-1/4
12	19	12	7/8	1	17
14	21	12	1	1.12	18-3/4
16	23-1/2	16	1	1.12	21-1/4
18	25	16	1-1/8	1.25	22-3/4
20	27-1/2	20	1-1/8	1.25	25
24	32	20	1-1/4	1.38	29-1/2

Nominal Pipe Size NPS (inches)	Class 300				
	Diameter of Flange (inches)	No. of Bolts	Diameter of Bolts (inches)	Diameter of Bolt Holes (inches)	Bolt Circle (inches)
¼	3-3/8	4	½	0.62	2-1/4
½	3-3/4	4	½	0.62	2-5/8
¾	4-5/8	4	5/8	0.75	3-1/4
1	4-7/8	4	5/8	0.75	3-1/2
1-1/4	5-1/4	4	5/8	0.75	3-7/8
1-1/2	6-1/8	4	¾	0.88	4-1/2
2	6-1/2	8	5/8	0.75	5
2-1/2	7-1/2	8	¾	0.88	5-7/8

Nominal Pipe Size NPS (inches)	Class 300				
	Diameter of Flange (inches)	No. of Bolts	Diameter of Bolts (inches)	Diameter of Bolt Holes (inches)	Bolt Circle (inches)
3	8-1/4	8	¾	0.88	6-5/8
3-1/2	9	8	¾	0.88	7-1/4
4	10	8	¾	0.88	7-7/8
5	11	8	¾	0.88	9-1/4
6	12-1/2	12	¾	0.88	10-5/8
8	15	12	7/8	1	13
10	17-1/2	16	1	1.12	15-1/4
12	20-1/2	16	1-1/8	1.25	17-3/4
14	23	20	1-1/8	1.25	20-1/4
16	25-1/2	20	1-1/4	1.38	22-1/2
18	28	24	1-1/4	1.38	24-3/4
20	30-1/2	24	1-1/4	1.38	27
24	36	24	1-1/2	1.62	32
Nominal Pipe Size NPS (inches)	**Class 400**				
	Diameter of Flange (inches)	**No. of Bolts**	**Diameter of Bolts (inches)**	**Diameter of Bolt Holes (inches)**	**Bolt Circle (inches)**
¼	3-3/8	4	½	0.62	2-1/4
½	3-3/4	4	½	0.62	2-5/8
¾	4-5/8	4	5/8	0.75	3-1/4
1	4-7/8	4	5/8	0.75	3-1/2
1-1/4	5-1/4	4	5/8	0.75	3-7/8
1-1/2	6-1/8	4	¾	0.88	4-1/2
2	6-1/2	8	5/8	0.75	5
2-1/2	7-1/2	8	¾	0.88	5-7/8
3	8-1/4	8	¾	0.88	6-5/8
3-1/2	9	8	7/8	1	7-1/4
4	10	8	7/8	1	7-7/8
5	11	8	7/8	1	9-1/4
6	12-1/2	12	7/8	1	10-5/8
8	15	12	1	1.12	13
10	17-1/2	16	1-1/8	1.25	15-1/4
12	20-1/2	16	1-1/4	1.38	17-3/4
14	23	20	1-1/4	1.38	20-1/4
16	25-1/2	20	1-3/8	1.5	22-1/2
18	28	24	1-3/8	1.5	24-3/4
20	30-1/2	24	1-1/2	1.62	27
24	36	24	1-3/4	1.88	32
Nominal Pipe Size NPS (inches)	**Class 600**				
	Diameter of Flange (inches)	**No. of Bolts**	**Diameter of Bolts (inches)**	**Diameter of Bolt Holes (inches)**	**Bolt Circle (inches)**
¼	3-3/8	4	½	0.62	2-1/4
½	3-3/4	4	½	0.62	2-5/8
¾	4-5/8	4	5/8	0.75	3-1/4

Nominal Pipe Size NPS (inches)	Class 600				
	Diameter of Flange (inches)	No. of Bolts	Diameter of Bolts (inches)	Diameter of Bolt Holes (inches)	Bolt Circle (inches)
1	4-7/8	4	5/8	0.75	3-1/2
1-1/4	5-1/4	4	5/8	0.75	3-7/8
1-1/2	6-1/8	4	$^3/_4$	0.88	4-1/2
2	6-1/2	8	5/8	0.75	5
2-1/2	7-1/2	8	$^3/_4$	0.88	5-7/8
3	8-1/4	8	$^3/_4$	0.88	6-5/8
3-1/2	9	8	7/8	1	7-1/4
4	10-3/4	8	7/8	1	8-1/2
5	13	8	1	1.12	10-1/2
6	14	12	1	1.12	11-1/2
8	16-1/2	12	1-1/8	1.25	13-3/4
10	20	16	1-1/4	1.38	17
12	22	20	1-1/4	1.38	19-1/4
14	23-3/4	20	1-3/8	1.5	20-3/4
16	27	20	1-1/2	1.62	23-3/4
18	29-1/4	20	1-5/8	1.75	25-3/4
20	32	24	1-5/8	1.75	28-1/2
24	37	24	1-7/8	2	33

Nominal Pipe Size NPS (inches)	Class 900				
	Diameter of Flange (inches)	No. of Bolts	Diameter of Bolts (inches)	Diameter of Bolt Holes (inches)	Bolt Circle (inches)
½	4-3/4	4	$^3/_4$	0.88	3-1/4
$^3/_4$	5-1/8	4	$^3/_4$	0.88	3-1/2
1	5-7/8	4	7/8	1	4
1-1/4	6-1/4	4	7/8	1	4-3/8
1-1/2	7	4	1	1.12	4-7/8
2	8-1/2	8	7/8	1	6-1/2
2-1/2	9-5/8	8	1	1.12	7-1/2
3	9-1/2	8	7/8	1	7-1/2
4	11-1/2	8	1-1/8	1.25	9-1/4
5	13-3/4	8	1-1/4	1.38	11
6	15	12	1-1/8	1.25	12-1/2
8	18-1/2	12	1-3/8	1.5	15-1/2
10	21-1/2	16	1-3/8	1.5	18-1/2
12	24	20	1-3/8	1.5	21
14	25-1/4	20	1-1/2	1.62	22
16	27-3/4	20	1-5/8	1.75	24-1/2
18	31	20	1-7/8	2	27

Nominal Pipe Size NPS (inches)	Class 900				
	Diameter of Flange (inches)	No. of Bolts	Diameter of Bolts (inches)	Diameter of Bolt Holes (inches)	Bolt Circle (inches)
20	33-3/4	20	2	2.12	29-1/2
24	41	20	2-1/2	2.62	35-1/2

Nominal Pipe Size NPS (inches)	Class 1500				
	Diameter of Flange (inches)	No. of Bolts	Diameter of Bolts (inches)	Diameter of Bolt Holes (inches)	Bolt Circle (inches)
1/2	4-3/4	4	3/4	0.88	3-1/4
3/4	5-1/8	4	3/4	0.88	3-1/2
1	5-7/8	4	7/8	1	4
1-1/4	6-1/4	4	7/8	1	4-3/8
1-1/2	7	4	1	1.12	4-7/8
2	8-1/2	8	7/8	1	6-1/2
2-1/2	9-5/8	8	1	1.12	7-1/2
3	10-1/2	8	1-1/8	1.25	8
4	12-1/4	8	1-1/4	1.38	9-1/2
5	14-3/4	8	1-1/2	1.62	11-1/2
6	15-1/2	12	1-3/8	1.5	12-1/2
8	19	12	1-5/8	1.75	15-1/2
10	23	12	1-7/8	2	19
12	26-1/2	16	2	2.12	22-1/2
14	29-1/2	16	2-1/4	2.38	25
16	32-1/2	16	2-1/2	2.62	27-3/4
18	36	16	2-3/4	2.88	30-1/2
20	38-3/4	16	3	3.12	32-3/4
24	46	16	3-1/2	3.62	39

Nominal Pipe Size NPS (inches)	Class 2500				
	Diameter of Flange (inches)	No. of Bolts	Diameter of Bolts (inches)	Diameter of Bolt Holes (inches)	Bolt Circle (inches)
1/2	5-1/4	4	3/4	0.88	3-1/2
3/4	5-1/2	4	3/4	0.88	3-3/4
1	6-1/4	4	7/8	1	4-1/4
1-1/4	7-1/4	4	1	1.12	5-1/8
1-1/2	8	4	1-1/8	1.25	5-3/4
2	9-1/4	8	1	1.12	6-3/4
2-1/2	10-1/2	8	1-1/8	1.25	7-3/4
3	12	8	1-1/4	1.38	9
4	14	8	1-1/2	1.62	10-3/4
5	16-1/2	8	1-3/4	1.88	12-3/4
6	19	8	2	2.12	14-1/2
8	21-3/4	12	2	2.12	17-1/4
10	26-1/2	12	2-1/2	2.62	21-1/4
12	30	12	2-3/4	2.88	24-3/8

Flange bolt tightening sequence

Flange Bolt Torque Sequence

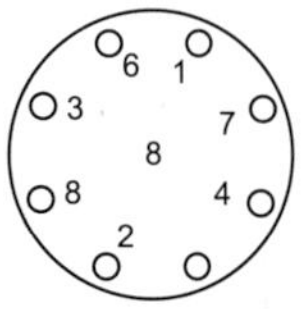

Sequential	Rotational Order
1-2	1
3-4	2
5-6	3
7-8	4
	5
	6
	7
	8

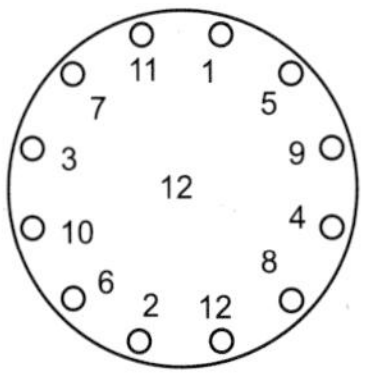

Sequential Order	Rotational Order
1-2	1
3-4	5
5-6	9
7-8	3
9-10	7
11-12	11
	2
	6
	10
	4
	8

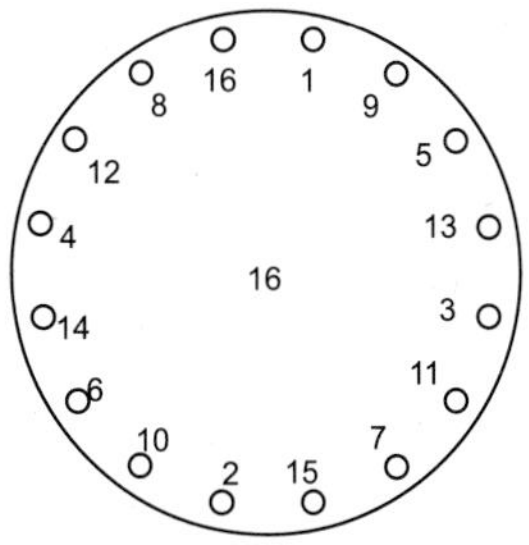

Sequential Order	Rotational Order	
1-2	1	2
3-4	9	10
5-6	5	6
7-8	13	14
9-10	3	4
11-12	11	12
13-14	7	8
	15	16
	2	
	10	
	6	

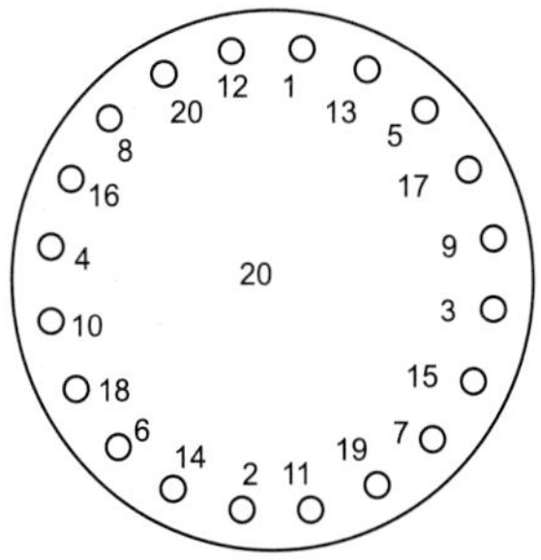

Sequential Order	Rotational Order	
1-2	1	2
3-4	13	14
5-6	5	6
7-8	17	18
9-10	9	10
11-12	3	4
13-14	15	16
15-16	7	8
17-18	19	20
19-20	11	12

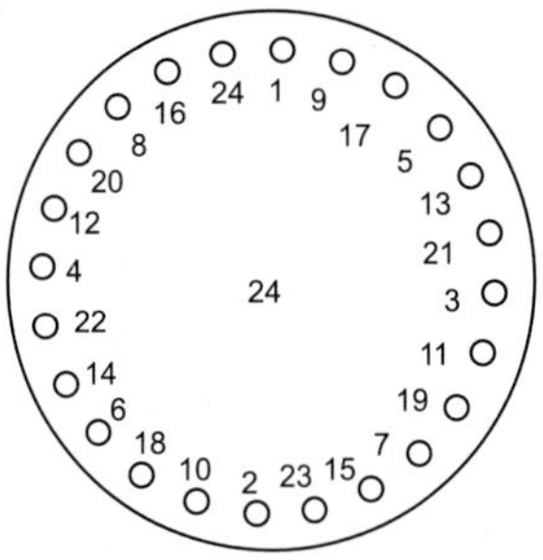

Sequential Order	Rotational Order	
1-2	1	2
3-4	9	10
5-6	17	18
7-8	5	6
9-10	13	14
11-12	21	22
13-14	3	4
15-16	11	12
17-18	19	20
19-20	7	8
21-22	15	16
23-24	23	24

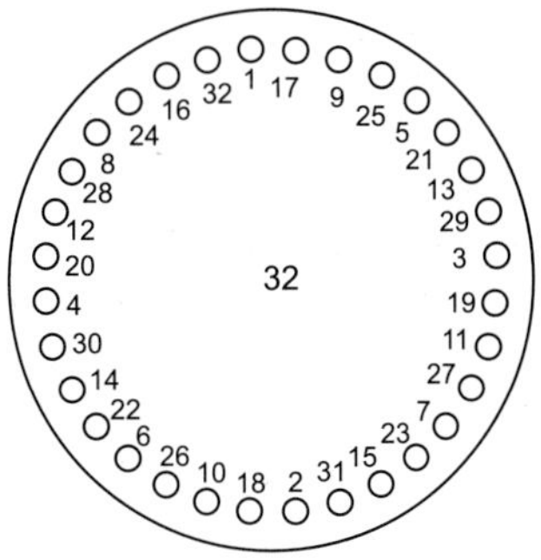

Sequential Order	Rotational Order	
1-2	1	2
3-4	17	18
5-6	9	10
7-8	25	26
9-10	5	6
11-12	11	12
13-14	13	14
15-16	29	30
17-18	3	4
19-20	19	20
21-22	11	12
23-24	27	28
25-26	7	8
27-28	23	24
29-30	15	16
31-32	31	32

used with permissino of Sypris Technologies, Inc, Tube Turns Division, Louisville, KY

Steel fitting dimensions

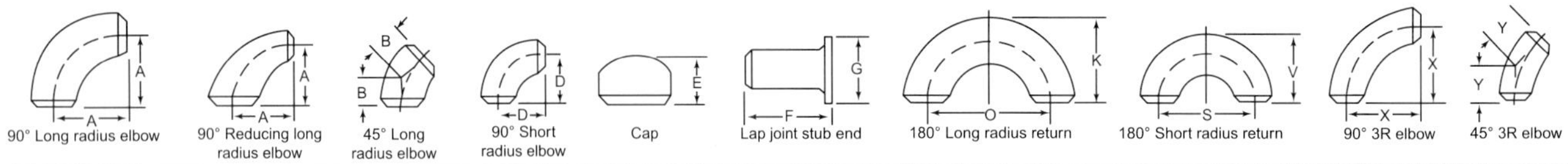

Nom. Pipe Size	Pipe O.D. (In.)	WALL THICKNESS Light Wall	SCH 20	SCH 30	STD	SCH 40	SCH 60	X STG	SCH 80	SCH 100	SCH 120	SCH 140	SCH 160	XX STG	A	B	D	E	F ANSI	F MSS	G	O	K	S	V	X	Y	Nom. Pipe Size
1/2	.84				.109	.109		.147	.147				.188	.294	1.50	.62		1.00	3.00	2.00	1.38	3.00	1.88					1/2
3/4	1.05				.113	.113		.154	.154				.219	.308	1.12	.44		1.00	3.00	2.00	1.69	2.25	1.69					3/4
1	1.32				.133	.133		.179	.179				.250	.358	1.50	.88	1.00	1.50	4.00	2.00	2.00	3.00	2.19	2.00	1.62			1
1 1/4	1.66				.140	.140		.191	.191				.250	.382	1.88	1.00	1.25	1.50	4.00	2.00	2.50	3.75	2.75	2.50	2.06			1 1/4
1 1/2	1.90				.145	.145		.200	.200				.281	.400	2.25	1.12	1.50	1.50	4.00	2.00	2.88	4.50	3.25	3.00	2.44			1 1/2
2	2.38				.154	.154		.218	.218				.344	.436	3.00	1.38	2.00	1.50	6.00	2.50	3.62	6.00	4.19	4.00	3.19	6.00	2.50	2
2 1/2	2.88				.203	.203		.276	.276				.375	.552	3.75	1.75	2.50	1.50	6.00	2.50	4.12	7.50	5.19	5.00	3.94			2 1/2
3	3.50	.188			.216	.216		.300	.300				.438	.600	4.50	2.00	3.00	2.00	6.00	2.50	5.00	9.00	6.25	6.00	4.75	9.00	3.75	3
3 1/2	4.00				.266	.266		.318	.318						5.25	2.25	3.50	2.50	6.00	3.00	5.50	10.50	7.25	7.00	5.50			3 1/2
4	4.50	.188			.237	.237		.337	.337		.438		.531	.674	6.00	2.50	4.00	2.50	6.00	3.00	6.19	12.00	8.25	8.00	6.25	12.00	5.00	4
5	5.56				.258	.258		.375	.375		.500		.625	.750	7.50	3.12	5.00	3.00	8.00	3.00	7.31	15.00	10.31	10.00	7.75			5
6	6.62	.219			.280	.280		.432	.432		.562		.719	.864	9.00	3.75	6.00	3.50	8.00	3.50	8.50	18.00	12.31	12.00	9.31	18.00	7.50	6
8	8.62	.219	.250	.277	.322	.322	.406	.500	.500	.594	.719	.812	.906	.875	12.00	5.00	8.00	4.00	8.00	4.00	10.62	24.00	16.31	16.00	12.31	24.00	10.00	8
10	10.75	.219	.250	.307	.365	.365	.500	.500	.594	.719	.844	1.000	1.125	1.000	15.00	6.25	10.00	5.00	10.00	5.00	12.75	30.00	20.38	20.00	15.38	30.00	12.50	10
12	12.75	.250	.250	.330	.375	.406	.562	.500	.688	.844	1.000	1.125	1.312	1.000	18.00	7.50	12.00	6.00	10.00	6.00	15.00	36.00	24.38	24.00	18.38	36.00	15.00	12
14	14.00	.250	.312	.375	.375	.438	.594	.500	.750	.938	1.094	1.250	1.406		21.00	8.75	14.00	6.50	12.00	6.00	16.25	42.00	28.00	28.00	21.00	42.00	17.50	14
16	16.00	.250	.312	.375	.375	.500	.656	.500	.844	1.031	1.219	1.438	1.594		24.00	10.00	16.00	7.00	12.00	6.00	18.50	48.00	32.00	32.00	24.00	48.00	19.88	16
18	18.00	.250	.312	.438	.375	.562	.750	.500	.938	1.156	1.375	1.562	1.781		27.00	11.25	18.00	8.00	12.00	6.00	21.00	54.00	36.00	36.00	27.00	54.00	22.38	18
20	20.00	.250	.375	.500	.375	.594	.812	.500	1.031	1.281	1.500	1.750	1.969		30.00	12.50	20.00	9.00	12.00	6.00	23.00	60.00	40.00	40.00	30.00	60.00	24.88	20
24	24.00	.250	.375	.562	.375	.688	.969	.500	1.219	1.531	1.812	2.062	2.344		36.00	15.00	24.00	10.50	12.00	6.00	27.25	72.00	48.00	48.00	36.00	72.00	29.81	24
30	30.00	.312	.500	.625	.375			.500							45.00	18.50	30.00	10.50				90.00	60.00	60.00	45.00	90.00	37.25	30
36	36.00	.312	.500	.625	.375	.750		.500							54.00	22.25	36.00	10.50						72.00	54.00	108.00	44.75	36
42	42.00				.375			.500							63.00	26.00	42.00	12.00								126.00	52.19	42
48	48.00				.375			.500							72.00	29.88	48.00	13.50								144.00	59.69	48

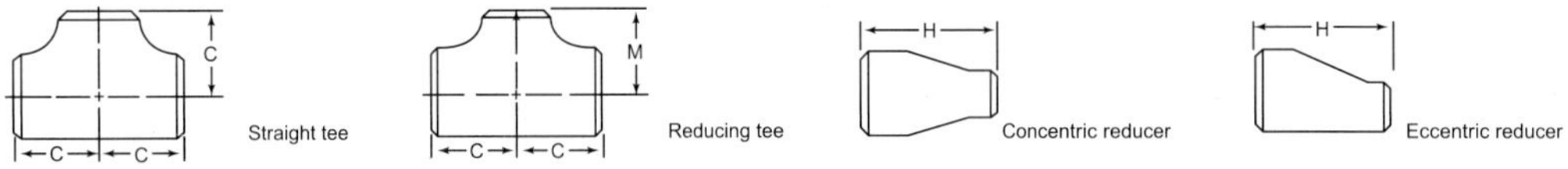

Nom. Pipe Size Run	Outlet	C	M	H
3/4	3/4	1.12		
	1/2	1.12	1.12	1.50
1	1	1.50		
	3/4	1.50	1.50	2.00
	1/2	1.50	1.50	2.00
1 1/4	1 1/4	1.88		
	1	1.88	1.88	2.00
	3/4	1.88	1.88	2.00
	1/2	1.88	1.88	2.00
1 1/2	1 1/2	2.25		
	1 1/4	2.25	2.25	2.50
	1	2.25	2.25	2.50
	3/4	2.25	2.25	2.50
	1/2	2.25	2.25	2.50
2	2	2.50		
	1 1/2	2.50	2.38	3.00
	1 1/4	2.50	2.25	3.00
	1	2.50	2.00	3.00
	3/4	2.50	1.75	3.00
2 1/2	2 1/2	3.00		
	2	3.00	2.75	3.50
	1 1/2	3.00	2.62	3.50
	1 1/4	3.00	2.50	3.50
	1	3.00	2.25	3.50
3	3	3.38		
	2 1/2	3.38	3.25	3.50
	2	3.38	3.00	3.50
	1 1/2	3.38	2.88	3.50
	1 1/4	3.38	2.75	3.50
3 1/2	3 1/2	3.75		
	3	3.75	3.62	4.00
	2 1/2	3.75	3.50	4.00
	2	3.75	3.25	4.00
	1 1/2	3.75	3.12	4.00
4	4	4.12		
	3 1/2	4.12	4.00	4.00
	3	4.12	3.88	4.00
	2 1/2	4.12	3.75	4.00
	2	4.12	3.50	4.00
	1 1/2	4.12	3.38	4.00
5	5	4.88		
	4	4.88	4.62	5.00
	3 1/2	4.88	4.50	5.00
	3	4.88	4.38	5.00
	2 1/2	4.88	4.25	5.00
	2	4.88	4.12	5.00
6	6	5.62		
	5	5.62	5.38	5.50
	4	5.62	5.12	5.50
	3 1/2	5.62	5.00	5.50
	3	5.62	4.88	5.50
	2 1/2	5.62	4.75	5.50
8	8	7.00		
	6	7.00	6.62	6.00
	5	7.00	6.38	6.00
	4	7.00	6.12	6.00
	3 1/2	7.00	6.00	6.00
10	10	8.50		
	8	8.50	8.00	7.00
	6	8.50	7.62	7.00
	5	8.50	7.50	7.00
	4	8.50	7.25	7.00
12	12	10.00		
	10	10.00	9.50	8.00
	8	10.00	9.00	8.00
	6	10.00	8.62	8.00
	5	10.00	8.50	8.00
14	14	11.00	11.00	
	12	11.00	10.62	13.00
	10	11.00	10.12	13.00
	8	11.00	9.75	13.00
	6	11.00	9.38	13.00
16	16	12.00		
	14	12.00	12.00	14.00
	12	12.00	11.62	14.00
	10	12.00	11.12	14.00
	8	12.00	10.75	14.00
	6	12.00	10.38	
18	18	13.50		
	16	13.50	13.00	15.00
	14	13.50	13.00	15.00
	12	13.50	12.62	15.00
	10	13.50	12.12	15.00
	8	13.50	11.75	
20	20	15.00		
	18	15.00	14.50	20.00
	16	15.00	14.00	20.00
	14	15.00	14.00	20.00
	12	15.00	13.62	20.00
	10	15.00	13.12	
	8	15.00	12.75	
24	24	17.00		
	22	17.00	17.00	20.00
	20	17.00	17.00	20.00
	18	17.00	16.50	20.00
	16	17.00	16.00	20.00
	14	17.00	16.00	
	12	17.00	15.62	
	10	17.00	15.12	
30	30	22.00		
	24	22.00	21.00	24.00
	20	22.00	20.00	24.00
	18	22.00	19.50	
	16	22.00	19.00	
	14	22.00	19.00	
36	36	26.50		
	30	26.50	25.00	24.00
	24	26.50	24.00	24.00
	20	26.50	23.00	
	18	26.50	22.50	
	16	26.50	22.00	
42	42	30.00		
	36	30.00	28.00	24.00
	30	30.00	28.00	24.00
	24	30.00	26.00	
	20	30.00	26.00	
48	48	35.00		
	42	35.00	32.00	28.00
	36	35.00	31.00	
	30	35.00	30.00	
	24	35.00	29.00	

NOTES:
Fittings are to ASTM A234 WPB with dimensions to ANSI B16.9 except that short radius elbows and returns are to ANSI B16.28 and 3R elbows are to MSS-SP75.

Other types, sizes and thicknesses of fittings on application.

ANSI forged steel flanges

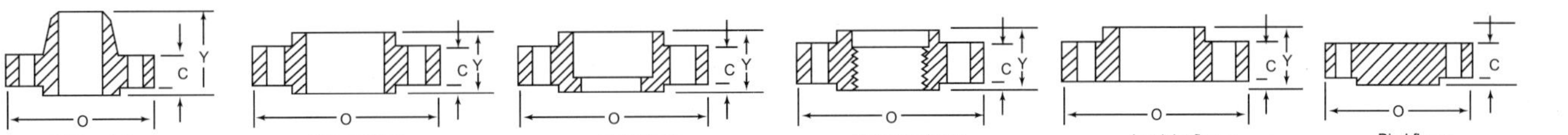

Weld neck flange — Slip-on flange — Socket type — Threaded flange — Lap joint flange — Bind flange

CLASS 150 FLANGES

Nom. Pipe Size	O	C†	Y† Weld Neck	Y† Slip-on Thrd. Socket	Y† Lap Joint	Bolt Circle	No. and Size of Holes
1/2	3.50	.44	1.88	.62	.62	2.38	4-0.62
3/4	3.88	.50	2.06	.62	.62	2.75	4-0.62
1	4.25	.56	2.19	.69	.69	3.12	4-0.62
1 1/4	4.62	.62	2.25	.81	.81	3.50	4-0.62
1 1/2	5.00	.69	2.44	.88	.88	3.88	4-0.62
2	6.00	.75	2.50	1.00	1.00	4.75	4-0.75
2 1/2	7.00	.88	2.75	1.12	1.12	5.50	4-0.75
3	7.50	.94	2.75	1.19	1.19	6.00	4-0.75
3 1/2	8.50	.94	2.81	1.25	1.25	7.00	8-0.75
4	9.00	.94	3.00	1.31	1.31	7.50	8-0.75
5	10.00	.94	3.50	1.44	1.44	8.50	8-0.88
6	11.00	1.00	3.50	1.56	1.56	9.50	8-0.88
8	13.50	1.12	4.00	1.75	1.75	11.75	8-0.88
10	16.00	1.19	4.00	1.94	1.94	14.25	12-1.00
12	19.00	1.25	4.50	2.19	2.19	17.00	12-1.00
14	21.00	1.38	5.00	2.25	USE SLIP-ON DIMS.	18.75	12-1.12
16	23.50	1.44	5.00	2.50		21.25	16-1.12
18	25.00	1.56	5.50	2.69		22.75	16-1.25
20	27.50	1.69	5.69	2.88		25.00	20-1.25
24	32.00	1.88	6.00	3.25		29.50	20-1.38
30	38.75	2.94	5.38			36.00	28-1.38
36	46.00	3.56	6.19			42.75	32-1.62
42	53.00	3.81	6.75			49.50	36-1.62
48	59.50	4.25	7.56			56.00	44-1.62
54	66.25	4.75	8.50			62.75	44-1.88
60	73.00	5.19	9.44			69.25	52-1.88

CLASS 300 FLANGES

Nom. Pipe Size	O	C†	Y† Weld Neck	Y† Slip-on Thrd. Socket	Y† Lap Joint	Bolt Circle	No. and Size of Holes
1/2	3.75	.56	2.06	.88	.88	2.62	4-0.62
3/4	4.62	.62	2.25	1.00	1.00	3.25	4-0.75
1	4.88	.69	2.44	1.06	1.06	3.50	4-0.75
1 1/4	5.25	.75	2.56	1.06	1.06	3.88	4-0.75
1 1/2	6.12	.81	2.69	1.19	1.19	4.50	4-0.88
2	6.50	.88	2.75	1.31	1.31	5.00	8-0.75
2 1/2	7.50	1.00	3.00	1.50	1.50	5.88	8-0.88
3	8.25	1.12	3.12	1.69	1.69	6.62	8-0.88
3 1/2	9.00	1.19	3.19	1.75	1.75	7.25	8-0.88
4	10.00	1.25	3.38	1.88	1.88	7.88	8-0.88
5	11.00	1.38	3.88	2.00	2.00	9.25	8-0.88
6	12.50	1.44	3.88	2.06	2.06	10.62	12-0.88
8	15.00	1.62	4.38	2.44	2.44	13.00	12-1.00
10	17.50	1.88	4.62	2.62		15.25	16-1.12
12	20.50	2.00	5.12	2.88		17.75	16-1.25
14	23.00	2.12	5.62	3.00	USE SLIP-ON DIMS.	20.25	20-1.25
16	25.50	2.25	5.75	3.25		22.50	20-1.38
18	28.00	2.38	6.25	3.50		24.75	24-1.38
20	30.50	2.50	6.38	3.75		27.00	24-1.38
24	36.00	2.75	6.62	4:19		32.00	24-1.62
30	43.00	3.62	8.25			39.25	28-1.88
36	50.00	4.12	9.50			46.00	32-2.12
42	50.75	4.69	7.88			47.50	32-1.75
48	57.75	5.25	8.81			54.00	32-2.00
54	65.25	6.00	9.94			61.00	28-2.38
60	71.25	6.44	10.75			67.00	32-2.38

CLASS 400 FLANGES

Nom. Pipe Size	O	C†	Y† Weld Neck	Y† Slip-on Thrd. Socket	Y† Lap Joint	Bolt Circle	No. and Size of Holes
1/2	3.75	.56	2.06	.88	.88	2.62	4-0.62
3/4	4.62	.62	2.25	1.00	1.00	3.25	4-0.75
1	4.88	.69	2.44	1.06	1.06	3.50	4-0.75
1 1/4	5.25	.81	2.62	1.12	1.12	3.88	4-0.75
1 1/2	6.12	.88	2.75	1.25	1.25	4.50	4-0.88
2	6.50	1.00	2.88	1.44	1.44	5.00	8-0.75
2 1/2	7.50	1.12	3.12	1.62	1.62	5.88	8-0.88
3	8.25	1.25	3.25	1.81	1.81	6.62	8-0.88
3 1/2	9.00	1.38	3.38	1.94	1.94	7.25	8-1.00
4	10.00	1.38	3.50	2.00	2.00	7.88	8-1.00
5	11.00	1.50	4.00	2.12	2.12	9.25	8-1.00
6	12.50	1.62	4.06	2.25	2.25	10.62	12-1.00
8	15.00	1.88	4.62	2.69	2.69	13.00	12-1.12
10	17.50	2.12	4.88	2.88		15.25	16-1.25
12	20.50	2.25	5.38	3.12		17.75	16-1.38
14	23.00	2.38	5.88	3.31	USE SLIP-ON DIMS.	20.25	20-1.38
16	25.50	2.50	6.00	3.69		22.50	20-1.50
18	28.00	2.62	6.50	3.88		24.75	24-1.50
20	30.50	2.75	6.62	4.00		27.00	24-1.62
24	36.00	3.00	6.88	4.50		32.00	24-1.88
30	43.00	4.00	8.62			39.25	28-2.12
36	50.00	4.50	9.88			46.00	32-2.12
42	52.00	5.25	8.81			48.25	32-2.00
48	59.50	6.00	10.12			55.25	28-2.38
54	67.00	6.69	11.38			62.25	28-2.62
60	74.25	7.31	12.56			69.00	32-2.88

CLASS 600 FLANGES

Nom. Pipe Size	O	C†	Y† Weld Neck	Y† Slip-on Thrd. Socket	Y† Lap Joint	Bolt Circle	No. and Size of Holes
1/2	3.75	.56	2.06	.88	.88	2.62	4-0.62
3/4	4.62	.62	2.25	1.00	1.00	3.25	4-0.75
1	4.88	.69	2.44	1.06	1.06	3.50	4-0.75
1 1/4	5.25	.81	2.62	1.12	1.12	3.88	4-0.75
1 1/2	6.12	.88	2.75	1.25	1.25	4.50	4-0.88
2	6.50	1.00	2.88	1.44	1.44	5.00	8-0.75
2 1/2	7.50	1.12	3.12	1.62	1.62	5.88	8-0.88
3	8.25	1.25	3.25	1.81	1.81	6.62	8-0.88
3 1/2	9.00	1.38	3.38	1.94	1.94	7.25	8-1.00
4	10.75	1.50	4.00	2.12	2.12	8.50	8-1.00
5	13.00	1.75	4.50	2.38	2.38	10.50	8-1.12
6	14.00	1.88	4.62	2.62	2.62	11.50	12-1.12
8	16.50	2.19	5.25	3.00	3.00	13.75	12-1.25
10	20.00	2.50	6.00	3.38	USE SLIP-ON DIMS.	17.00	16-1.38
12	22.00	2.62	6.12	3.62		19.25	20-1.38
14	23.75	2.75	6.50	3.69		20.75	20-1.50
16	27.00	3.00	7.00	4.19		23.75	20-1.62
18	29.25	3.25	7.25	4.62		25.75	20-1.75
20	32.00	3.50	7.50	5.00		28.50	24-1.75
24	37.00	4.00	8.00	5.50		33.00	24-2.00
30	44.50	4.50	9.75			40.25	28-2.12
36	51.75	4.88	11.12			47.00	28-2.62
42	55.25	6.62	11.00			50.50	28-2.62
48	62.75	7.44	12.44			57.50	32-2.88
54	70.00	8.25	13.75			64.25	32-3.12
60	78.50	9.19	15.31			71.75	28-3.62

CLASS 900 FLANGES

Nom. Pipe Size	O	C†	Y† Weld Neck	Y† Slip-on Thrd. Socket	Y† Lap Joint	Bolt Circle	No. and Size of Holes
1/2	4.75	.88	2.38	1.25	1.25	3.25	4-0.88
3/4	5.12	1.00	2.75	1.38	1.38	3.50	4-0.88
1	5.88	1.12	2.88	1.62	1.62	4.00	4-1.00
1 1/4	6.25	1.12	2.88	1.62	1.62	4.38	4-1.00
1 1/2	7.00	1.25	3.25	1.75	1.75	4.88	4-1.12
2	8.50	1.50	4.00	2.25	2.25	6.50	8-1.00
2 1/2	9.62	1.62	4.12	2.50	2.50	7.50	8-1.12
3	9.50	1.50	4.00	2.12	2.12	7.50	8-1.00
3 1/2							
4	11.50	1.75	4.50	2.75	2.75	9.25	8-1.25
5	13.75	2.00	5.00	3.12	3.12	11.00	8-1.38
6	15.00	2.19	5.50	3.38	3.38	12.50	12-1.25
8	18.50	2.50	6.38	4.00		15.50	12-1.50
10	21.50	2.75	7.25	4.25	USE SLIP-ON DIMS.	18.50	16-1.50
12	24.00	3.12	7.88	4.62		21.00	20-1.50
14	25.25	3.38	8.38	5.12		22.00	20-1.62
16	27.75	3.50	8.50	5.25		24.25	20-1.75
18	31.00	4.00	9.00	6.00		27.00	20-2.00
20	33.75	4.25	9.75	6.25		29.50	20-2.12
24	41.00	5.50	11.50	8.00		35.50	20-2.62
30	48.50	5.88	12.25			42.75	20-3.12
36	57.50	6.75	14.25			50.75	20-3.62
42	61.50	8.12	14.62			54.75	24-3.62
48	70.25	9.19	16.50			62.50	24-4.12
54							
60							

CLASS 1500 FLANGES

Nom. Pipe Size	O	C†	Y† Weld Neck	Y† Slip-on Thrd. Socket	Y† Lap Joint	Bolt Circle	No. and Size of Holes
1/2	4.75	.88	2.38	1.25	1.25	3.25	4-0.88
3/4	5.12	1.00	2.75	1.38	1.38	3.50	4-0.88
1	5.88	1.12	2.88	1.62	1.62	4.00	4-1.00
1 1/4	6.25	1.12	2.88	1.62	1.62	4.38	4-1.00
1 1/2	7.00	1.25	3.25	1.75	1.75	4.88	4-1.12
2	8.50	1.50	4.00	2.25	2.25	6.50	8-1.00
2 1/2	9.62	1.62	4.12	2.50	2.50	7.50	8-1.12
3	10.50	1.88	4.62	2.88	2.88	8.00	8-1.25
3 1/2							
4	12.25	2.12	4.88	3.56	3.56	9.50	8-1.38
5	14.75	2.88	6.12	4.12	4.12	11.50	8-1.62
6	15.50	3.25	6.75	4.69	4.69	12.50	12-1.50
8	19.00	3.62	8.38	5.62	5.62	15.50	12-1.75
10	23.00	4.25	10.00	6.25		19.00	12-2.00
12	26.50	4.88	11.12	7.12		22.50	16-2.12
14	29.50	5.25	11.75			25.00	16-2.38
16	32.50	5.75	12.25			27.75	16-2.62
18	36.00	6.38	12.88			30.50	16-2.88
20	38.75	7.00	14.00			32.75	16-3.12
24	46.00	8.00	16.00			39.00	16-3.62

CLASS 2500 FLANGES

Nom. Pipe Size	O	C†	Y† Weld Neck	Y† Slip-on Thrd. Socket	Y† Lap Joint	Bolt Circle	No. and Size of Holes
1/2	5.25	1.19	2.88	1.56	1.56	3.50	4-0.88
3/4	5.50	1.25	3.12	1.69	1.69	3.75	4-0.88
1	6.25	1.38	3.50	1.88	1.88	4.25	4-1.00
1 1/4	7.25	1.50	3.75	2.06	2.06	5.12	4-1.12
1 1/2	8.00	1.75	4.38	2.38	2.38	5.75	4-1.25
2	9.25	2.00	5.00	2.75	2.75	6.75	8-1.12
2 1/2	10.50	2.25	5.62	3.12	3.12	7.75	8-1.25
3	12.00	2.62	6.62	3.62	3.62	9.00	8-1.38
4	14.00	3.00	7.50	4.25	4.25	10.75	8-1.62
5	16.50	3.62	9.00	5.12	5.12	12.75	8-1.88
6	19.00	4.25	10.75	6.00	6.00	14.50	8-2.12
8	21.75	5.00	12.50	7.00	7.00	17.25	12-2.12
10	26.50	6.50	16.50	9.00	9.00	21.25	12-2.62
12	30.00	7.25	18.25	10.00	10.00	24.38	12-2.88

NOTES:

Flange material is ASTM A105 unless specified otherwise.

Flange dimensions are to ANSI B16.5 for sizes 24" and smaller. Sizes over 24" are to MSS-SP44.

Always specify bore or matching pipe wall when ordering.

Other types, sizes and facings on application.

† Includes .06 raised face in Class 150 and Class 300 standards. Does **NOT** include .25 raised face in Class 400 and heavier standards.

Trenching and excavation safety

Excerpted from OSHA 2226-10R 2015, Trenching and Excavation Safety, U.S. Department of Labor Occupational Safety and Health Administration.

Excavation and trenching are among the most hazardous construction operations. The Occupational Safety and Health Administration's (OSHA) Excavation standards, 29 Code of Federal Regulations (CFR) Part 1926, Subpart P, contain requirements for excavation and trenching operations.

What is the difference between an excavation and a trench?

OSHA defines an excavation as any man-made cut, cavity, trench, or depression in the Earth's surface formed by earth removal. A trench is defined as a narrow excavation (in relation to its length) made below the surface of the ground. In general, the depth of a trench is greater than its width, but the width of a trench (measured at the bottom) is not greater than 15 feet (4.6 m).

What are the dangers of trenching and excavation operations?

Trenching and excavation work presents serious hazards to all workers involved. Cave-ins pose the greatest risk and are more likely than some other excavation-related incidents to result in worker fatalities. One cubic yard of soil can weigh as much as a car. An unprotected trench can be an early grave. Employers must ensure that workers enter trenches only after adequate protections are in place to address cave-in hazards. Other potential hazards associated with trenching work include falling loads, hazardous atmospheres, and hazards from mobile equipment.

OSHA Excavation Standards

OSHA's Excavations standards at 29 CFR Part 1926, Subpart P. The *Occupational Safety and Health Act* requires employers to comply with safety and health standards promulgated by OSHA or by a state with an OSHA-approved state plan. The standards apply to all open excavations made in the Earth's surface, including trenches. Following the requirements of the standards will prevent or greatly reduce the risk of cave-ins and other excavation-related incidents.

What are the soil classification categories?

Some of the compliance methods permitted under the Excavation standards require a competent person to classify soil and rock deposits as:

- Stable rock;
- Type A soil;
- Type B soil; or
- Type C soil.

See Appendix A to Subpart P of Part 1926 paragraph (b) for a detailed definition of soil type.

Stable Rock — Natural solid mineral matter that can be excavated with vertical sides and remain intact while exposed.

Type A — Cohesive soils with an unconfined compressive strength of 1.5 tons per square foot (tsf) (144 kPa) or greater.

Examples include: clay, silty clay, sandy clay, and clay loam. Certain conditions preclude soil from being classified as Type A. For example, no soil is Type A if it is fissured or has been previously disturbed.

Type B — Includes cohesive soil with an unconfined compressive strength greater than 0.5 tsf (48 kPa) but less than 1.5 tsf (144 kPa) and granular cohesionless soils (such as angular gravel, similar to crushed rock, silt, silt loam, sandy loam, and, in some cases, silty clay loam and sandy clay loam).

Type C — Cohesive soil with an unconfined compressive strength of 0.5 tsf (48 kPa) or less, granular soils (including gravel, sand, and loamy sand), submerged soil or soil from which water is freely seeping, submerged rock that is not stable, or material in a sloped, layered system where the layers dip into the excavation or with a slope of four horizontal to one vertical (4H:1V) or steeper.

Protective Systems

OSHA generally requires that employers protect workers from cave-ins by:

- Sloping and benching the sides of the excavation;
- Supporting the sides of the excavation; or
- Placing a shield between the side of the excavation and the work area.

What types of protective systems can employers use to protect workers from cave-ins?

In many cases the type of protective system needed is well known and simple to use. At other times employers will undertake the more complex process of designing a protective system. Designing a protective system requires consideration of many factors, including: soil classification, depth of cut, water content of soil, weather and climate, and other operations in the vicinity. Employers are free to choose the most practical design that will provide the necessary protections. Any system used must meet the required performance criteria. The Excavation standards describe various methods and approaches that can be used to design protective systems.

Examples of protective systems that can be used to comply with the Excavation standards include:

- Sloping the sides of the excavation to an angle not steeper than 1½:1 (for every foot of depth, the trench must be excavated back 1½ feet). A slope of this gradation is safe for any type of soil.
- Designing a sloping and benching system in accord with tabulated data, such as tables and charts, approved by a registered professional engineer. This data must be in writing and must include enough explanatory information (including the criteria for selecting a system and the limits on the use of the data) for the user to be able to select an appropriate protective system. At least one copy of the data, which identifies the registered professional engineer who approved it, must be kept at the worksite during construction of the protective system. After the system is completed, the data may be stored away from the jobsite, but a copy must be provided upon request to an OSHA compliance officer.
- Using a trench box or shield approved by a registered professional engineer or designed in accord with tabulated data approved by a registered professional engineer.

Trenching and Excavation Safety

The Excavation standards do not require a protective system when an excavation is made entirely in stable rock or when an excavation is less than 5 feet (1.52 meters) deep and a competent person has examined the ground and found no indication of a potential cave-in.

What other precautions do employers need to take to protect workers from cave-ins?

The Excavation standards require employers to provide support systems, such as shoring, bracing, or underpinning, when necessary to ensure that adjacent structures (including adjoining buildings, walls, sidewalks and pavements) remain stable for the protection of workers. The standards also prohibit excavation below the base or footing of any foundation or retaining wall that could be reasonably expected to pose a hazard to workers unless:

- The employer provides a support system, such as underpinning;
- The excavation is in stable rock; or
- A registered professional engineer determines that the structure is far enough away from the excavation that it would not be affected by the excavation activity or that the excavation work will not pose a hazard to workers.

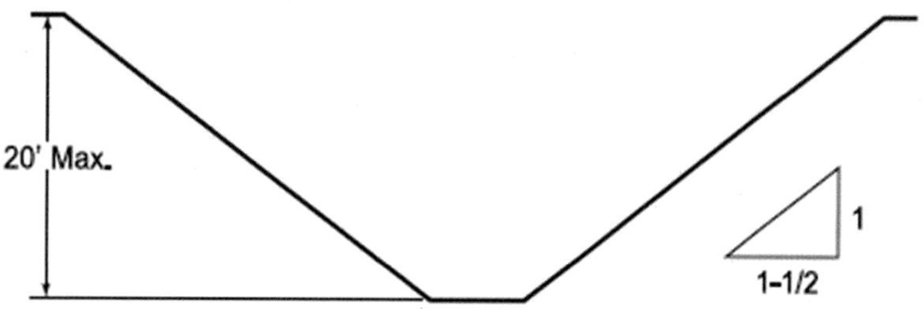

Figure 1. Slope of 1½ : 1

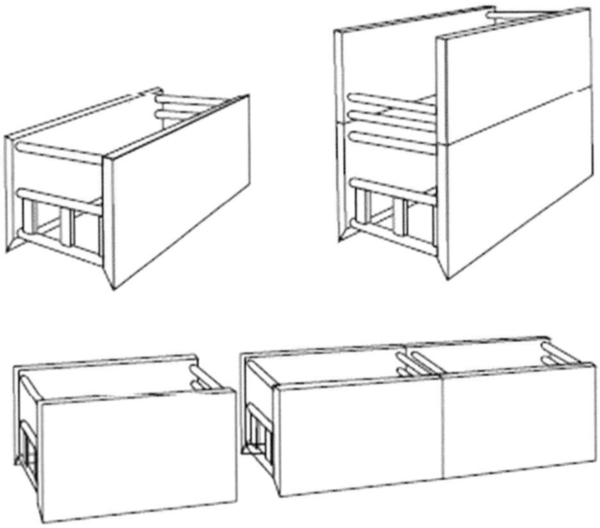

Figure 2. Trench shields

Excavations that would undermine sidewalks, pavement, and appurtenant structures are prohibited unless the employer provides an appropriately designed support system or another effective method of protecting workers from the possible collapse of those structures.

What are the requirements for safely installing and removing protective systems?

The Excavation standards require employers to take certain steps to protect workers when installing and removing support systems. For example:

- Members of support systems must be securely connected to prevent sliding, falling, kickouts or predictable failure.
- Support systems must be installed and removed in a manner that protects workers from cave-ins and structural collapses and from being struck by members of the support system.
- Members of support systems must not be overloaded.
- Before temporary removal of individual members, additional precautions are required, such as installing other structural members to carry loads imposed on the support system.
- Removal must begin at, and progress from, the bottom of the excavation.
- Backfilling must progress together with the removal of support systems from excavations.

In addition, the standards permit excavation of 2 feet (0.61 meters) or less below the bottom of the members of a support system, but only if the system is designed to resist the forces calculated for the full depth of the trench and there are no indications, while the trench is open, of a possible loss of soil from behind or below the bottom of the support system. Employers must coordinate the installation of support systems with the excavation work.

Employer Responsibilities

Employers are responsible for maintaining materials and equipment used for protective systems. Defective and damaged materials and equipment can cause protective systems to fail and lead to other excavation hazards.

Employers must ensure that:

- Materials and equipment are free from damage or defects that might impair their proper function;
- Manufactured materials and equipment are used and maintained consistent with the manufacturer's recommendations and are used so as to prevent worker exposure to hazards;
- A competent person examines any damaged materials or equipment to evaluate its suitability for continued use; and
- If a competent person cannot assure that damaged material or equipment can support the intended loads or is otherwise suitable for use, the materials and equipment are removed from service until evaluated and approved by a registered professional engineer.

Reuniting separated mercury in thermometers

The largest single cause for failure of precision thermometers is separated mercury columns. This can occur in transit or in use. The mercury may be reunited by cooling the thermometer in a solution of solid CO_2 (dry ice) and alcohol so that the mercury column retreats slowly into the bulb. Do not cool the stem or mercury column. Keep the bulb in the solution until the main column, as well as the separated portion, retreats into the bulb. Remove and swing the thermometer in a short arc, forcing all of the mercury into the bulb.

Most mercury thermometers can be reunited using this method regardless of range (with the exception of deep immersion thermometers) provided only the bulb is immersed in the CO_2.

Caution: Do not touch the bulb until it has warmed sufficiently for the mercury to emerge from the bulb into the capillary.

Never subject the stem or mercury column to the CO_2 solution, as it will freeze the mercury column in the capillary and may cause the bulb to fracture.

Typical wire resistance

(Standed Copper Conductors at 59 °F).

Wire Size AWG	Resistance (ohm/ft)
0000	0.00005
000	0.00006
00	0.00008
0	0.00010
1	0.00012
2	0.00016
4	0.00025
6	0.00039
8	0.00063
10	0.00098
12	0.00160

How to cut odd-angle long radius elbows

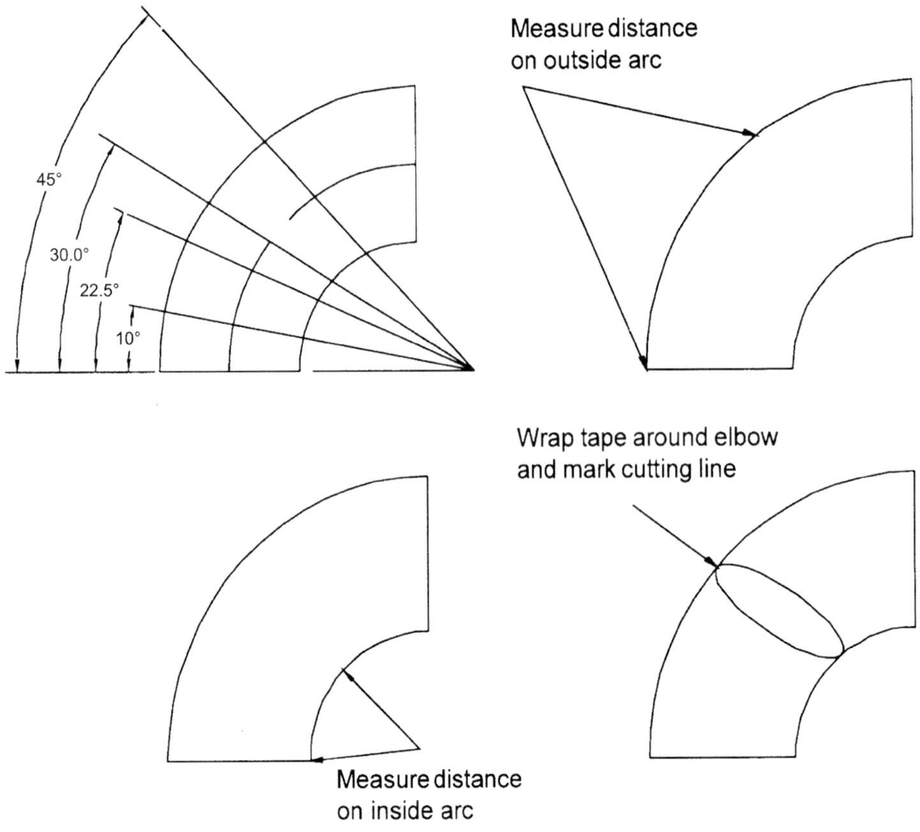

Nominal Pipe Size	Outside Arc Length*							Inside Arc Length*						
	1°	5°	10°	15°	22.5°	30°	45°	1°	5°	10°	15°	22.5°	30°	45°
2"	5/64	3/8	23/32	1-3/32	1-21/32	2-3/16	3-9/32	1/32	5/32	5/16	15/32	23/32	15/16	1-7/16
2 1/2"	3/32	7/16	29/32	1-11/32	2-1/32	2-23/32	4-1/16	3/64	3/16	13/32	19/32	29/32	1-7/32	1-13/16
3"	7/64	9/16	1-1/8	1-5/8	2-15/32	3-9/32	4-29/32	3/64	1/4	1/2	23/32	1-3/32	1-7/16	2-5/32
3 1/2"	1/8	5/8	1-9/32	1-29/32	2-27/32	3-13/16	5-11/16	1/16	9/32	9/16	27/32	1-9/32	1-11/16	2-9/16
4"	9/64	23/32	1-7/16	2-5/32	3-1/4	4-5/16	6-15/32	1/16	5/16	21/32	3-1/32	1-15/32	1 31/32	2-15/16
5"	3/16	29/32	1-25/32	2-11/16	4-1/32	5-3/8	8-1/16	5/64	13/32	13/16	1-1/4	1-27/32	2-15/32	3-23/32
6"	7/32	1-1/16	2-5/32	3-7/32	4-27/32	6-7/16	9-21/32	3/32	1/2	1	1-1/2	2-7/32	2-31/32	4-15/32
8"	9/32	1-7/16	2-27/32	4-9/32	6-13/32	8-17/32	12-13/16	1/8	11/16	1-11/32	2	3-1/32	4-1/32	6-1/32
10"	11/32	1-25/32	3-9/16	5-11/32	8	10-21/32	16	5/32	27/32	1-11/16	2-17/32	3-25/32	5-1/32	7-9/16
12"	7/16	2-1/8	4-1/4	6-3/8	9-9/16	12-3/4	19-5/32	7/32	1	2-1/32	3-1/16	4-9/16	6-3/32	9-1/8
14"	1/2	2-7/16	4-7/8	7-5/16	11	14-21/32	22	1/4	1-7/32	2-7/16	3-21/32	5-1/2	7-11/32	11
16"	9/16	2-13/16	5-19/32	8-3/8	12-9/16	16-3/4	25-1/8	9/32	1-13/32	2-13/16	4-3/16	6-9/32	8-3/8	12-5/8
18"	5/8	3-1/8	6-9/32	9-7/16	14-1/8	18-27/32	28-9/32	5/16	1-9/16	3-1/8	4-23/32	7-1/16	9-7/16	14-1/8
20"	11/16	3-1/2	7	10-15/32	15-23/32	20-15/16	31-13/32	11/32	1-3/4	3-1/2	5-1/4	7-27/32	10-15/32	15-11/16
22"	3/4	3-27/32	7-11/16	11-17/32	17-9/32	23-1/32	34-9/16	3/8	1-29/32	3-27/32	5-3/4	8-5/8	11-17/32	17-9/32
24"	27/32	4-3/16	8-3/8	12-9/16	18-27/32	25-1/8	37-11/16	13/32	2-3/32	4-3/16	6-9/32	9-7/16	12-9/16	18-27/32
26"	29/32	4-17/32	9-3/32	13-5/8	20-13/32	27-7/32	40-27/32	15/32	2-9/32	4-17/32	6-13/16	10-7/32	13-5/8	20-13/32
30"	1-1/32	5-1/4	10-15/32	15-3/4	23-9/16	31-13/32	47-1/8	17/32	2-5/8	5-1/4	7-27/32	11-25/32	15-23/32	23-9/16
36"	1-7/32	6-9/32	12-17/32	18-7/8	28-7/32	37-11/16	56-17/32	5/8	2-13/16	6-1/4	9-7/16	14-1/8	18-27/32	28-1/4

*Rounded to nearest practical fraction.

How to read land descriptions

A land description is a description of a tract of land, in legally acceptable terms, that defines exactly where the tract of land is located and how many acres it contains.

Table 1
Land Measurements

Linear Measure			
1 in	0.833 ft	16½ ft	1 rod
7.92 in	1 link	5½ yards	1 rod
12 in	1 ft	4 rods	1 link
1 vara	33 in	66 ft	1 chain
$2^3/_4$ ft	1 vara	80 chains	1 mile
3 ft	1 yard	320 rods	1 mile
25 links	16½ ft	8000 links	1 mile
25 links	1 rod	5280 ft	1 mile
100 links	1 chain	1760 yards	1 mile
Square Measure			
144 sq in	1 sq ft	43,560 sq ft	1 acre
9 sq ft	1 sq yd	640 acres	1 sq mile
30¼ sq yd	1 sq rod	1 sq mile	1 section
10 sq rods	1 sq chain	36 sq miles	1 township
1 sq rod	172¼ sq ft	6 miles sq	1 township
1 sq chain	4356 sq ft	208 ft 8 in sq	1 acre
10 sq chains	1 acre	80 rods sq	40 acres
160 sq rods	1 acre	160 rods sq	160 acres
4840 sq yd	1 acre		

In nonrectangular land descriptions, distance is usually described in terms of either feet or rods (this is especially true in surveying today), while square measure is in terms of acres. Such descriptions are called "metes and bounds" descriptions and will be explained in detail later. In rectangular land descriptions, square measure is again in terms of acres, and the location of the land is in such terms as N ½ (north one half), SE ¼ (south east one fourth or quarter), etc., as shown in Figures 2–5.

Meandered water and government lots

A meandered lake or stream is water next to which the adjoining land owner pays taxes on the land only. Such land is divided into divisions of land called government lots. The location, acreage, and lot number of each such tract of land was determined, surveyed, and plotted by the original government surveyors.

The original survey of your county (complete maps of each township, meandered lakes, government lots, etc.) is in your courthouse and is the basis for all land descriptions in your county. See Figure 1.

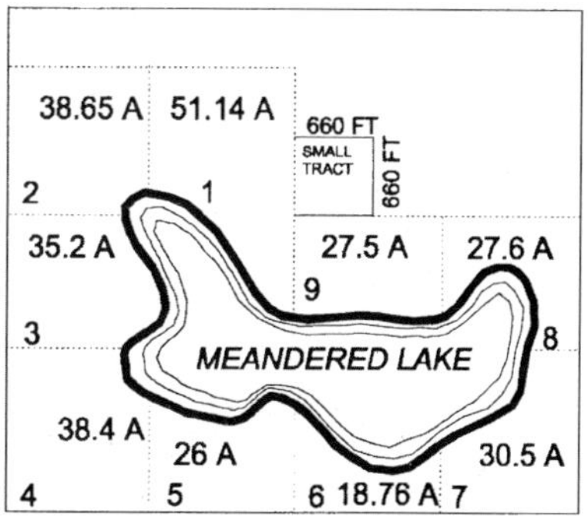

Figure 1.

The government lot number given to a piece of land is the legal description of that tract of land.

How can you tell whether water is meandered or privately owned? If you find government lots adjoining a body of water or stream, those waters are meandered. If there are no government lots surrounding water, that water is privately owned; the owner is paying taxes on the land under the water and controls the hunting, fishing, trapping rights, etc., on that water within the regulations of state and federal laws. Note that where such water is deemed navigable, other rulings may sometimes pertain.

As a generality, meandered water is public water that the public may use for recreational purposes, fishing, hunting, trapping, etc., provided that the public can reach the waters without trespassing. There still is much litigation concerning this that will have to be settled in court.

Reading land descriptions

Descriptions of land always read first from either the north or the south. In Figures 2–5, notice that they all start with N (north) or S (south), such as NW, SE, etc. They are never WN (west north), ES (east south), etc.

It is simple for anyone to understand a description. The secret is to read or analyze the description from the rear, or backwards.

Example. Under Figure 4, the first description reads E ½, SE ¼, SW ¼, SW ¼. The last part of the description reads SW ¼, which means that the tract of land we are looking for is somewhere in that quarter (as shown in Figure 2). Next back, we find SW ¼, which means the tract we are after is somewhere in the SW ¼ SW ¼ (as shown in Figure 3). Next back, we find the SE ¼, which means that the tract is in the SE ¼ SW ¼ SW ¼ (as shown in Figure 5). Next back and the last part to look up is the E ½ of the above, which is the location of the tract described by the whole description (as shown in Figure 4).

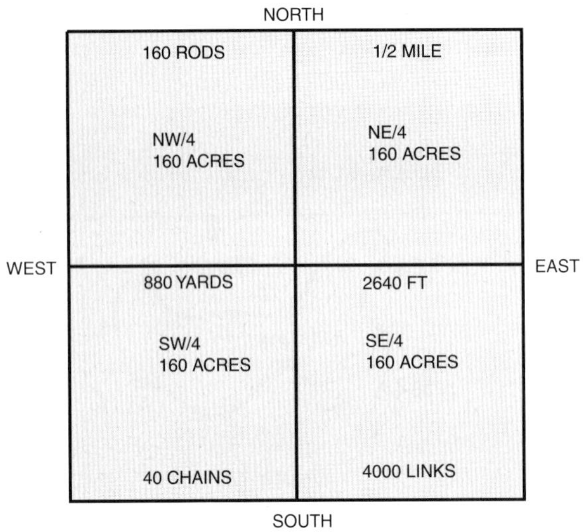

Figure 2.

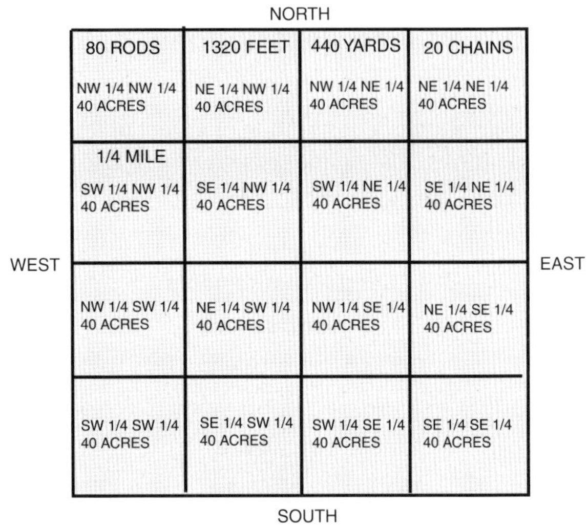

Figure 3.

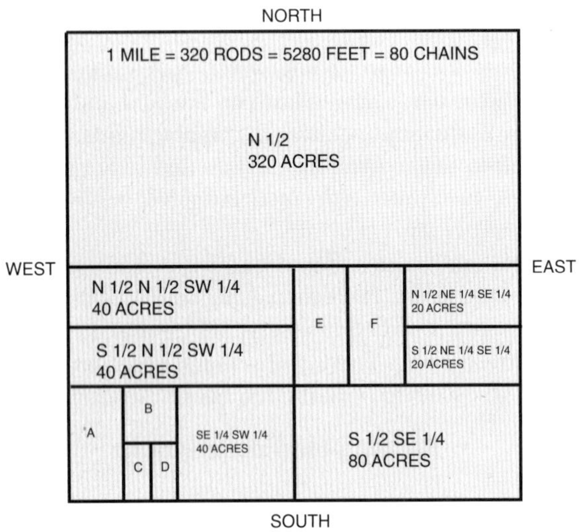

A = W 1/2 SW 1/4 SW 1/4 20 ACRES
B = NE 1/4 SW 1/4 SW 1/4 10 ACRES
C = W 1/2 SE 1/4 SW 1/4 SW 1/4 5 ACRES
D = E 1/2 SE 1/4 SW 1/4 SW 1/4 5 ACRES
E = W 1/2 NW 1/4 SE 1/4 20 ACRES
F = E 1/2 NW 1/4 SE 1/4 20 ACRES

Figure 4.

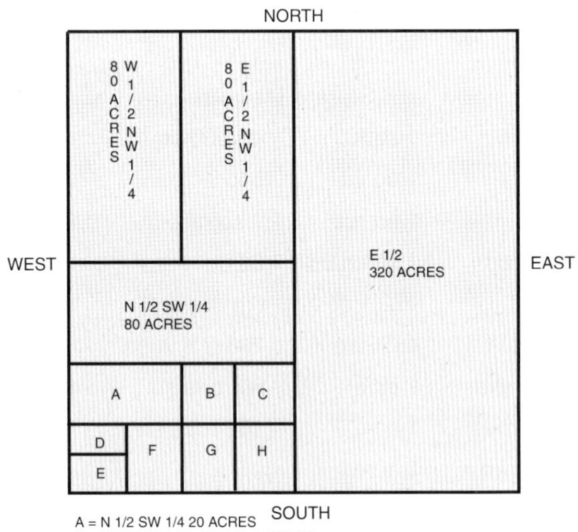

A = N 1/2 SW 1/4 20 ACRES
B = NW 1/4 SE 1/4 SW 1/4 10 ACRES
C = NE 1/4 SE 1/4 SW 1/4 10 ACRES
D = N 1/2 SW 1/4 SW 1/4 SW 1/4 5 ACRES
E = S 1/2 SW 1/4 SW 1/4 SW 1/4 5 ACRES
F = SE 1/4 SW 1/4 SW 1/4 10 ACRES
G = SW 1/4 SE 1/4 SW 1/4 10 ACRES
H = SE 1/4 SE 1/4 SW 1/4 10 ACRES

Figure 5.

Metes and bounds descriptions

A metes and bounds description is a description of land obtained by starting at a given point, running so many feet in a certain direction, so many feet in another direction, etc., back to the point of beginning.

Example. In Figure 1, notice the small tract of land outlined. A typical metes and bounds description of this tract of land would be as follows: "Begin at the center of the section, thence north 660 ft, thence east 660 ft, thence south 660 ft, and thence west 660 ft, back to the point of beginning and containing 10 acres, being a part of Section No. 2."

Important: To locate a tract of land from a metes and bounds description, start from the point of beginning and follow it out (do not read backwards as in the case of a rectangular description).

The small tract of land just located by the above metes and bounds description could also be described as the SW ¼, SW ¼, NE ¼ of the section. In most cases, the same tract of land may be described in different ways. The rectangular system of describing and locating land as shown in Figures 2–5 is the simplest and is almost always used when possible.

In land descriptions, degree readings are not a measure of distance. They are combined with either north or south to show the direction a line runs from a given point.

How to read descriptions that show directions in terms of degrees

In Figure 6, the north–south line and the east–west line divide the circle into four equal parts with each part containing 90°. Several different directions are shown in this diagram, with the number of degrees each varies east or west from the north and south starting points (remember that all descriptions read from the north or south). North-west is a direction that is halfway between north and west. In terms of degrees, the direction north-west would read north 45° west.

Notice the small tract of land in Figure 6. The following metes and bounds description will locate this tract: "Begin at the beginning point, thence N 20° west—200 ft, thence N 75° east—1320 ft, thence S 30° east—240 ft, thence S 45° west—420 ft, thence west—900 ft back to the point of beginning, containing so many acres, etc."

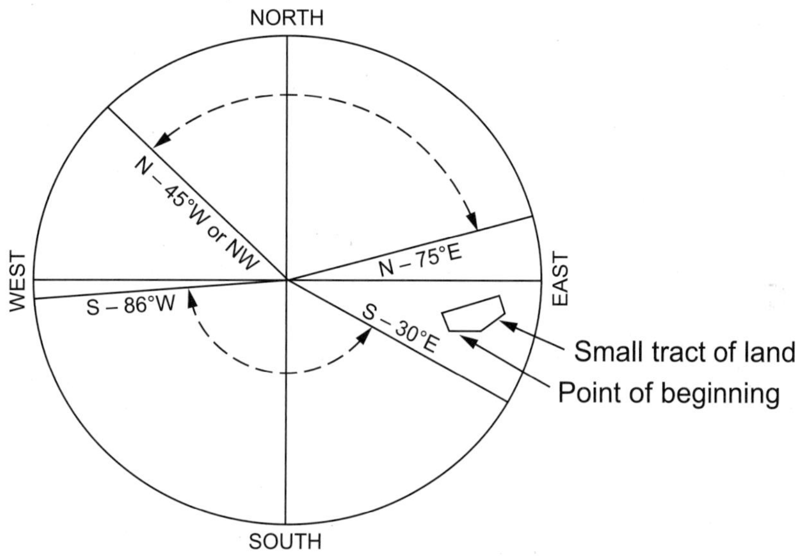

Figure 6.

Size an air receiver for engine starting

Assume that the engine starter requires 16 standard cubic feet per second, that 10 s is required for starting, and that you want to store enough air in the receiver for three starts. Assume the initial pressure in the receiver is 215 psia and the final pressure is 115 psia. Define:

V = Receiver volume, cu ft,
R = 53.5 for air,
T = Temperature (assumed constant at 520 °R),
M = Weight of air required to start the engine based on density of air = 0.0763 pounds per cu ft,
P = Pressure, lb/sq ft.

Example.

$$M = 16 \text{ cu ft/s} \times 10 \text{ s/start} \times 3 \text{ starts} \times 0.0763 \text{ lb per cu ft}$$
$$= 36.624 \text{ lb},$$

$$V = \frac{53.5 \times 520 \times 36.264}{144 \times (215 - 115)},$$
$$V = 71.76 \text{ cu ft.}$$

Assume receiver is 30″ or 2.5 ft inside diameter. Find the height of the receiver.

$$V = \frac{\pi \times \text{ID}^2}{4} \times H,$$

where

V = Volume to be contained in receiver, cu ft,
ID = Inside diameter of the receiver, ft,
H = Height of receiver, ft.

Solving for H yields 14.6 ft. Use two receivers, each 7.3 ft high.

Dimensions of hex nuts and hex jam nuts

Nominal size	F			G		H			J		
	Width Across Flats			Width Across Corners		Thickness Hex Nuts			Thickness Hex Jam Nuts		
	Basic	Max	Min	Max	Min	Basic	Max	Min	Basic	Max	Min
¼	7/16	0.438	0.428	0.505	0.488	7/32	0.263	0.212	5/32	0.163	0.150
5/16	½	0.500	0.489	0.557	0.577	17/64	0.273	0.258	3/16	0.195	0.180
3/8	9/16	0.562	0.551	0.650	0.628	21/64	0.337	0.320	7/23	0.227	0.210
7/16	11/16	0.688	0.675	0.794	0.768	3/8	0.385	0.365	¼	0.260	0.240
½	3/4	0.750	0.736	0.855	0.840	7/16	0.448	0.427	5/16	0.323	0.302
9/16	7/8	0.875	0.861	1.010	0.982	31/64	0.496	0.473	5/16	0.324	0.301
5/8	15/16	0.938	0.922	1.083	1.051	35/64	0.559	0.535	3/8	0.387	0.363
3/4	1 1/18	1.125	1.088	1.299	1.240	41/64	0.665	0.617	27/64	0.446	0.398
7/8	1 5/16	1.312	1.269	1.516	1.447	3/14	0.776	0.724	31/64	0.510	0.458
1	1½	1.500	1.450	1.732	1.653	55/64	0.887	0.831	35/64	0.575	0.519
11/8	1 11/16	1.688	1.631	1.949	1.859	31/32	0.999	0.939	39/64	0.639	0.579
1 1/4	1 7/8	1.875	1.812	2.165	2.066	1 1/16	1.094	1.030	32/32	0.751	0.687
13/18	2 1/16	2.062	1.994	2.382	2.273	1 11/64	1.206	1.138	25/32	0.815	0.747
1½	2¼	2.250	2.175	2.598	2.480	1 9/52	1.317	1.245	27/32	0.880	0.808

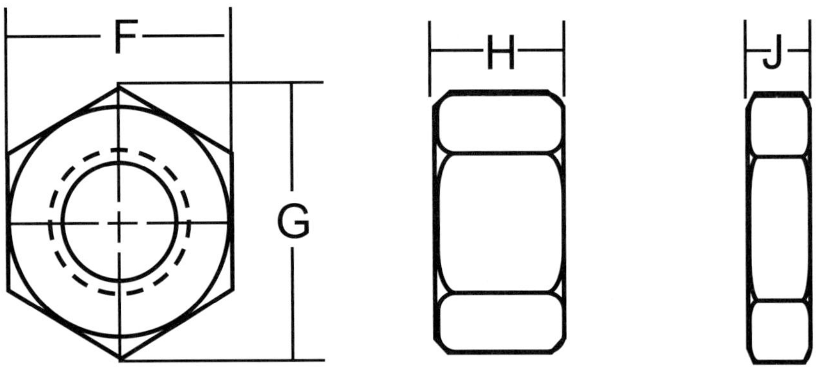

Reprinted with permission—Barnhill Bolt Co., Inc. Albuquerque, NM.

Color codes for locating underground utilities

Color	Use
WHITE	**White:** Pre-marking of the outer limits of the proposed excavation or marking the centerline and width of proposed lineal installations of buried facilities.
PINK	**Pink:** Temporary Survey Markings.
RED	**Red:** Electric power lines, cables or conduit, and lighting cables.
YELLOW	**Yellow:** Gas, oil, steam, petroleum, or gaseous materials.
ORANGE	**Orange:** Communication, alarm or signal lines, cables or conduits, and fiber.
BLUE	**Blue:** Potable water.
PURPLE	**Purple:** Slurry, irrigation and reclaimed water.
GREEN	**Green:** Sewers, drainage facilities or other drain lines.

Approximate angle of repose for sloping sides of excavations

Where excavation activities are regulated by a regulatory body, the user should consult and use the regulations promulgated by the regulatory body.

Note: Clays, SIMS, loams, or non-hermorporus soils require shoring and bracing. The presence of groundwater requires special treatment.

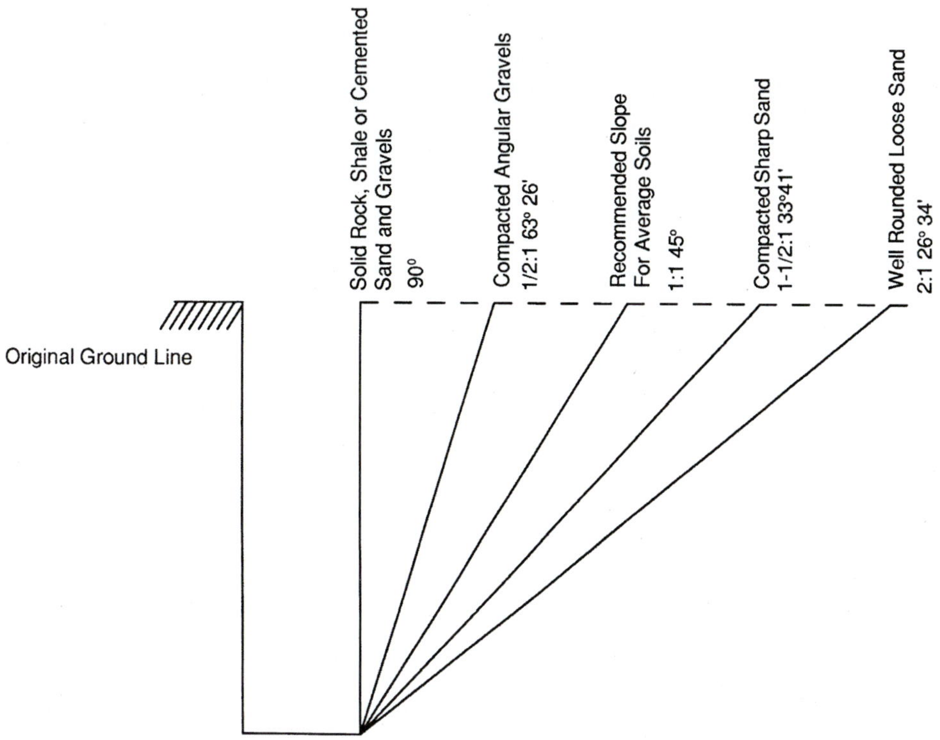

Wind chill chart

Wind Speed		Cooling Power of Wind Expressed as "Equivalent Chill Temperature," °F																				
Knots	MPH																					
Calm	Calm	40	35	30	25	20	15	10	5	0	−5	−10	−15	−20	−25	−30	−35	−40	−45	−50	−55	−60
Equivalent Chill Temperature																						
3–6	5	35	30	25	20	15	10	5	0	−5	−10	−15	−20	−25	−30	−35	−40	−45	−50	−55	−65	−70
7–10	10	30	20	15	10	5	0	−10	−15	−20	−25	−35	−40	−45	−50	−60	−60	−70	−75	−80	−90	−95
11–15	15	30	15	10	0	−5	−10	−20	−25	−30	−40	−45	−50	−60	−65	−70	−80	−85	−90	−100	−105	−110
16–19	20	25	10	5	0	−10	−15	−25	−30	−35	−45	−50	−60	−65	−75	−80	−85	−95	−100	−110	−115	−120
20–32	25	20	10	0	−5	−20	−20	−30	−35	−45	−50	−60	−65	−75	−80	−90	−95	−105	−110	−120	−125	−135
24–28	30	15	5	0	−10	−25	−25	−30	−40	−50	−55	−65	−70	−80	−85	−95	−100	−110	−115	−125	−130	−140
29–32	35	10	5	−5	−10	−30	−30	−35	−40	−50	−60	−65	−75	−80	−90	−100	−105	−115	−120	−130	−135	−145
33–36	40	10	0	−5	−10	−20	−30	−35	−45	−55	−60	−70	−75	−85	−95	−100	−110	−115	−125	−130	−140	−150
Winds above 40 °F have little additional effect		Little danger					Increasing danger (Flesh may freeze within 1 min)						Great danger (Flesh may freeze within 30 s)									

Fluid power formulas

Heat equivalent of fluid power:
BTU/*H* = PSI × GPM × 1½

GPM = gallons per minute,
PSI = pounds per square inch.

Hydraulic cylinder piston travel speed:
S = CIM ÷ A

S = piston travel speed, inches per minute,
CIM = oil flow into cylinder, cubic inches per minute.

Force for piercing or shearing sheet metal:
***F* = *P* × *T* × PSI**

F = force required, pounds,
P = perimeter around area to be sheared, inches,
T = sheet thickness, inches,
PSI = shear strength rating of the material, pounds per square inch.

Side load on pump or motor shaft:
***F* = (HP × 63,024) ÷ (RPM × *R*)**

F = side load, pounds, against shaft,
R = pitch radius of sheave on pump shaft, inches,
HP = driving power applied to shaft.

Effective force of a cylinder working at an angle to direction of the load travel: *F* = *T* × sin A

T = total cylinder force, pounds,
F = effective force, pounds,
A = least angle between cylinder axis and load direction, degrees.

Relationship between displacement and torque of a hydraulic motor: *T* = *D* ÷ PSI ÷ 24π

T = torque in foot pounds,
D = displacement in cubic inches per revolution,
PSI = pressure difference across the motor.

Thrust or force of a cylinder: *T* = A ÷ PSI

T = thrust, pounds,
A = effective area of cylinder, square inches,
PSI = pounds per square inch gauge pressure.

Recommended maximum oil velocity in hydraulic lines

Pump suction lines: 2–4 feet per second (fps),
Pressure lines to 500 psi: 10 to 15 fps,
Pressure lines to 3000 psi: 15 to 20 fps,
Pressure lines over 3000 psi: 25 fps.

How is pipeline manufactured?

Steel production

Steel is produced via the blast furnace-basic oxygen furnace (BF-BOF) and electric arc furnace (EAF), and modern secondary metallurgy practices including ladle refining and vacuum degassing (Figure 1).

The BF-BOF route produces steel using raw materials such as iron ore, coal, limestone and scrap steel. First, iron ores are reduced to iron (hot metal, or pig iron), and then the iron is converted to steel in the BOF. After casting and rolling and coating, the steel is delivered as strip, plate, sections or bars. About 75% of the steel in the world circa 2020 is produced using the BF-BOF route (Worldsteel 2020).

The blast furnace dates to the Bessemer converter which pumped air through molten iron to drive out the impurities in pig iron. Rather than cooling the iron, the air reacted with impurities such as carbon, manganese and silicon, causing them to oxidize. This raised the temperature further, igniting even more impurities, and producing the violent display with sparks and flames (Figure 2). A basic oxygen furnace uses oxygen instead of air to drive off the excess carbon from pig iron to produce steel. Modern BOFs can convert an iron charge of up to 350 tonnes into steel in less than 40 min.

The EAF process uses electricity to melt scrap steel and other sources of metallic iron such as direct-reduced iron or hot metal, and additives are used to adjust the steel composition to the desired specifications. Casting, reheating and rolling processes are like those found in the BF-BOF route. About 25% of steel produced in the world circa 2020 is via the EAF route.

Up to the 1950s, steel was poured into stationary molds forming ingots that were then rolled into sheets or smaller shapes and sizes. In continuous casting, liquid steel is fed continuously in a conveyor belt process and then cut into slabs or blooms. The blocks of steel are then rolled into finished and semi-finished pieces.

For a detailed description on all aspects of steel manufacturing, see one of the later editions of US Steel's remarkable "Making Steel" 1500-page treatise (e.g., Lankford, et al., 1985). For concise high-level treatments, Worldsteel (2012), Billington (1996), and Kiefner and Trench (2001) are recommended.

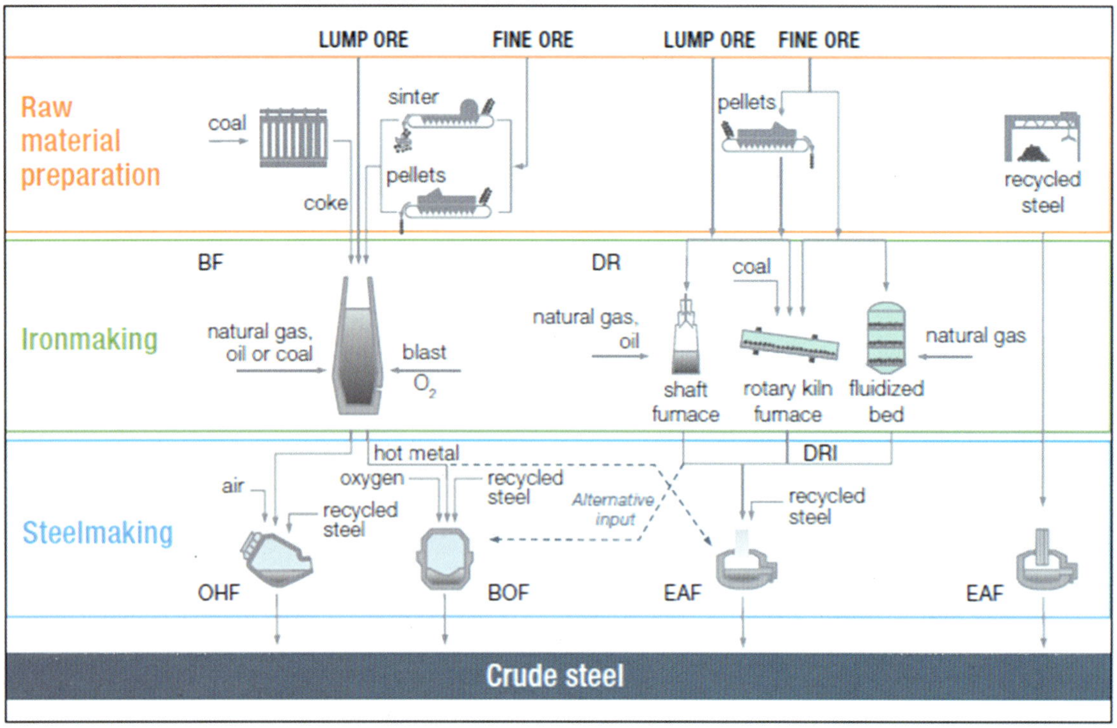

Figure 1. Steelmaking routes. Source: worldsteel.org

Figure 2. Charging ladle moving hot metal into blast furnace-basic oxygen furnace. Source: worldsteel.org/Seong Joon Cho

Pipeline manufacturing

Mills produce two types of line pipe, seamless and welded (Figure 3). Welded pipe is classified based on how it is formed and the type of welding technique used (ICF 2017).

Seamless pipe

Seamless pipe is formed from a cylindrical bar of hot steel (Figure 4). The bar is heated to a high temperature and then a probe is inserted to create a hole through the cylinder (Figure 5). The cylinder is then transferred to rollers which size the cylinder to the specified diameter and wall thickness (Figure 6). A few mills can produce seamless pipe up to 24-in in diameter. Seamless manufacturing methods are used for small diameter pipe but have higher cost and limited availability, and as diameter increases welded pipes are more economical.

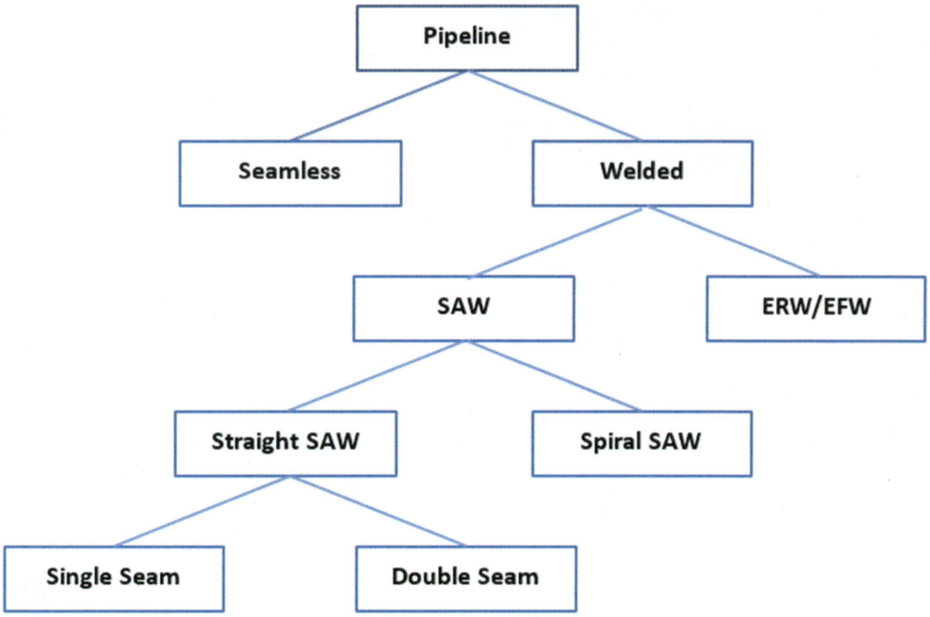

Figure 3. Pipe classification by manufacturing process.

Figure 4. Seamless pipe before insertion.
Source: worldsteel.org/Seong Joon Cho

Figure 6. Seamless pipe on cooling rack.
Source: worldsteel.org/Seong Joon Cho

Figure 5. Seamless pipe after formation.
Source: worldsteel.org/Seong Joon Cho

Figure 7. Stacks of finished seamless pipe.
Source: worldsteel.org/Hristo Shondov

Figure 8. Steel plate on cooling bed. Source: worldsteel.org/Seong Joon Cho

Figure 9. API Spec 5L 355 mm x 9.53 mm grade X52 PSL 2 welded pipeline. Source: AOPL

Welded pipe

In welded pipe, welding is used to close the seam after forming a steel plate or coil into a cylindrical shape (Figures 8-9). The mill uses ultrasonic and/or radiological inspection methods to ensure the quality of the weld seam and initiates pressure tests on each joint of pipe to levels that exceed the proposed operating pressure.

Shaping techniques

There are five commonly used shaping techniques that accept steel plate and coil as inputs (Figure 10).

U-O-E forming: A flat plate is first bent into a U-shape using hydraulic presses by pushing down on the center of the plate along the length. The U-shaped plate is then formed into an O-shape in a die that forces the two edges together. The two edges are welded and the pipe is expanded using a mandrel or hydraulic pressure. Heat treatment is the final step to relieve pipe stresses.

J-C-O forming: In this shaping technique a flat plate is bent into a J-shape, then C-shape and finally O-shape before welding the two edges together to form a pipe. For final shaping and sizing the pipe is expanded using a mandrel or hydraulic pressure and then heat treated to relieve pipe stresses.

Pyramid rolling: Three rollers are placed in a pyramidal arrangement where one roller on the top center pushes down on the center of the plate along the length as it is rolled by the two bottom rollers. Successive rolling motions along the width result in a hollow pipe structure that is welded along the open edges. The pipe is then expanded using a mechanical mandrel or hydraulic pressure and heat treated to relieve stresses.

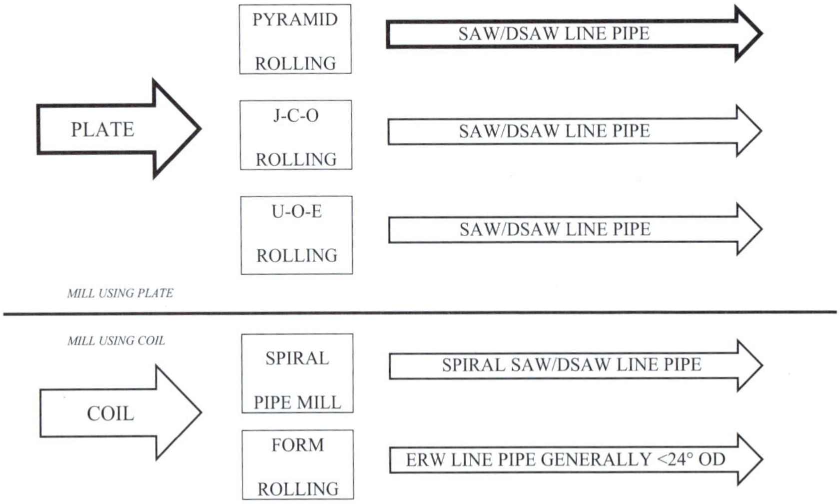

Figure 10. Inputs and outputs in pipeline manufacturing

Spiral rolling: In this rolling technique a continuous coil known as skelp is bent so that the edge along the length of the coil forms a helical pattern along the length of the pipe. The helical edges are welded to produce spiral pipe. This pipe is expanded for final shaping and sizing and heat treated to relieve stresses.

Form rollers: Smaller diameter line pipe 24 in or less is usually formed by passing a continuous steel plate coil sequentially through forming rollers that turn it into a cylindrical shape. The edges are welded to produce pipe and heat treated to relieve stresses.

Welding techniques

Electric resistance welding (ERW) and submerged arc welding (SAW) are commonly applied welding techniques.

Electric resistance welding: In ERW an electrode may be used to generate an electric arc or high frequency electromagnetic energy can be used to induce electricity flow to locally heat the two metal edges that need to be joined such that they melt and fuse together. An inert gas is often used to shield the arc as it impinges on the metal to avoid contamination of the weld joint. Today, ERW is mostly done in the form of high frequency welding (HFW) without electrodes and is used primarily to make longitudinal welds in form roller manufacturing of line pipe of 24-inch diameters and below.

Submerged arc welding: In SAW a consumable wire electrode is used to generate an arc between the joint and the electrode. The wire melts and fuses with the metal joint to provide a welded joint. Flux deposited ahead of the electrode is used to shield the arc and weld from contamination. Some of the flux melts as the weld progresses and is removed as slag. A SAW performed on both sides of the joint is referred to as double SAW or DSAW. The SAW process can be used on both longitudinal welds (L-SAW) as in the case of U-O-E, J-C-O or pyramid processes of spiral/helical welds (S-SAW, H-SAW).

References

D.P. Billington, The Innovators. John Wiley & Sons, Inc., New York, 1996.

ICF, Feasibility and impacts of domestic content requirements for U.S. oil and gas pipelines. Report prepared for the American Petroleum Institute, Fairfax, VA, 2017, May.

J.F. Kiefner, C.J. Trench, Oil pipeline characteristics and risk factors: illustrations for the decade of construction. A Report Prepared for API, Washington, D.C, 2001, Dec.

W.T. Lankford, N.L. Samways, R.E. Craven, H.E. McGannon, The Making, Shaping and Treating of Steel, United States Steel, 10th ed., Published by Association of Iron and Steel Engineers, Pittsburgh, PA.

World Steel Association, The white book of steel, Brussels, Belgium. worldsteel.org, 2012, [online]. http://worldsteel.org/steelstory Accessed March 28, 2020.

How to designate and classify PE piping materials

Excerpted from Chapter 5, Standard Specifications, Standard Test Methods, and Codes for PE (Polyethylene) Piping System, from The Plastics Pipe Institute Handbook of Polyethylene Pipe, 2nd Ed, 2008.

Standard requirements for PE piping materials

Polyethylene (PE) is a complex polymer with properties that can be optimized based on the desired end use. Such modifications are effected by choice of catalyst system, polymerization conditions and, the use of a small quantity of comonomer (a monomer or monomers other than ethylene). All these changes allow PE to be tailor made to a wide range of processing and performance requirements.

For classifying this wide array of property variations that find use in piping applications, ASTM issued standard D 3350, "Standard Specification for Polyethylene Plastic Pipe and Fittings Materials". This standard recognizes six properties that are considered important in the manufacture of PE piping, in the heat fusion joining of this material and, in defining its long-term performance capabilities. Each property is assigned into a "Cell" and, each cell consists of a number of "Classes". A cell number covers a narrow range of the larger overall range that is covered by a property "cell". These D 3350 property cells and classes are identified in Table 1.

In addition, by means of a Code letter, ASTM D3350 designates whether the material includes a colorant and also, the nature of the stabilizer that is included for protecting the material against the potential damaging effects of the ultraviolet (UV) rays in sunlight. Table 2 lists the Code letters that are used in D 3350 and what they represent.

For designating a PE material in accordance with ASTM D 3350 the cell number for each cell property is identified, and this is done in the same order as shown in Table 1. This is then followed by an appropriate Code letter to indicate color and stabilization as shown in Table 2. An example of this

Table 1
Cell Classification System from ASTM D 3350-06[1,2]

Property	Total Method	0	1	2	3	4	5	6	7	8
Density, g/cm^3	D1505	un-specified	≤0.925	>0.925–0.940	>0.940–0.947	>0.947–0.955	>0.955		specify value	
Melt Index	D1238	un-specified	>1.0	1.0–0.4	<0.4–0.15	<0.15	A		specify value	
Flexural Modulus, MPa (psi), 2% secant	D790	un-specified	<138 (<20,000)	138–<276 (20,000–<40,000)	276–<552 (40,000–<80,000)	552–<758 (80,000–<110,000)	758–<1103 (110,000–<160,000	>1103 (>160,000)	specify value	
Tensile Strength at Yield, MPa (psi)	D638	un-specified	<15 (<2000)	15–<18 (2200–<2600)	18–<21 (2600–<3000)	21–<24 (3000–<3500)	24–<28 (3500–<4000)	>28 (>4000)	specify value	
Slow Crack Growth Resistance I. ESCR	D1693	un-specified								
			A	B	C	C				
a. Test condition										specify
b. Test duration, hours			48	24	192	600				value
c. Failure, max, %			50	50	20	20				
Slow Crack Growth Resistance II. PENT (hours) Molded Plaque, 80°C, 2.4MPa, notch depth Table 1	F 1473	un-specified				10	30	100	500	specify value
Hydrostatic Strength Classification I. Hydrostatic design basis, MPa (psi), (23°C)	D2837	NPR[a]	5.52 (800)	6.89 (1000)	8.62 (1250)	11.03 (1600)				
Hydrostatic Strength Classification II. Minimum required strength, MPa (psi), (23°C)	ISO 12162						8 (1160)	10 (1450)		

Notes: A: Refer to 10.1.4.1 (ASTM D 3350) B: NPR = Not Pressure Rated, 1.) D 3350 is subject to periodic revisions, contact ASTM to obtain the latest version, 2.) The property and density are measured on the PE base resin; all the other property values are measured on the final compound.

Table 2
Code Letter Representation

Code Letter	Color and UV Stabilizer
A	Natural
B	Colored
C	Black with 2% minimum carbon black
D	Natural with UV stabilizer
E	Colored with UV stabilizer

Table 3
Properties of a Cell Number PE445574 Material

Digit Designating the Applicable Property Cell[(1)]	Class Number or Code Letter	Corresponding Value of Property (from Table 1)
1st Digit - Density of PE base resin, gm/cm^3	4	>0.947 – 0.955
2nd Digit - Melt Index of compound, gm/10 minutes	4	<0.15
3rd Digit - Flexural Modulus of compound, psi (MPa)	5	110,000 – <160,000 (758 – <1103)
4th Digit - Tensile Strength at Yield of compound, psi (MPa)	5	3500 – <4000 (24 – <28)
5th Digit - Resistance to Slow Crack Growth of compound (SCG), hrs	7	500 minimum based on PENT test
6th Digit - Hydrostatic Design Basis for water at 73°F (2323), psi of compound (MPa)	4	1600 (11.03)
Code Letter	C	Black with 2% minimum carbon black

[(1)]The density is that of the PE resin. All the other properties are determined on the final compounded material.

material designation system is presented in Table 3 for the case of a PE material having designation code PE445574C.

A PE material that complies with the Table 3 cell designation i.e. PE445574C would be a higher density (higher crystallinity), lower melt index (higher molecular weight) material that exhibits exceptionally high resistance to slow crack growth. In addition, it offers a hydrostatic design basis (HDB) for water at 73°F (23°C) of 1600 psi (11.03 MPa). Finally, it would be black and contain a minimum of 2% carbon black.

The cell classification system provides the design engineer with a very useful tool in specifying the requirements of PE materials for piping projects.

Standard PE piping material designation code

While all PE piping standards specify minimum material requirements based the on the cell requirements of ASTM D3350, a simpler, short-hand, ASTM recognized material designation code is commonly used for quickly identifying the most significant engineering properties of a PE pipe material. An important feature of this designation code is that it identifies the maximum recommended hydrostatic design stress (HDS) for water, at 73°F (23°C). Originally, this designation code was devised to only apply to materials intended for pressure piping. However, there is a recognition that even in non-pressure applications stresses are generated which makes it prudent to use a stress rated material. This has led to the common practice of using this material designation code for quickly identifying all PE piping materials intended for pipes of solid wall or, of profile wall construction.

This code is defined in ASTM F412, "Standard Terminology Relating to Plastic Piping Systems", under the definition for the term code, thermoplastic pipe materials designation. It consists of the ASTM approved abbreviation for the pipe material followed by four digits (e.g., PE4710). The information delivered by this code is as follows:

- The ASTM recognized abbreviation for the piping material. PE, in the case of polyethylene materials.
- The first digit identifies the density range of the base PE resin, in accordance with ASTM D3350, that is used in the material. As discussed in Chapter 3, the density of a PE polymer reflects the polymer's crystallinity which, in turn, is the principal determinant of the final material's strength and stiffness properties.
- The second digit identifies the compound's resistance to slow crack growth (SCG), also in accordance with ASTM D3350. A material's resistance to SCG relates very strongly to its long-term ductility, a property that defines the material's capacity for safely resisting the effects of localized stress intensifications.
- The last two numbers identify the compound's maximum recommended hydrostatic design stress (HDS) category (1) for water, at 73°F(23°C). This recommendation is established in consideration of various factors but, primarily the following: The capacity for safely resisting the relatively well distributed stresses that are generated only by internal pressure, and, the capacity for safely resisting add-on effects caused by localized stress intensifications.

The Standard Designation Codes for materials which are recognized as of this writing by current ASTM, AWWA, CSA and other standards are listed in Table 4. This table gives

Table 4
Standard Designation Codes for Current Commercially Available PE Piping Compositions

<table>
<tr><th rowspan="2"></th><th colspan="3">What the Digits in the Code Denote</th></tr>
<tr><th>The 1st Digit</th><th>The 2nd Digit</th><th>The last two Digits[(1)]</th></tr>
<tr><th>Standard Designation Code</th><th>Cell number based on the density cell in accordance with ASTM D3350 (See Table 1)</th><th>Cell number based on the resistance to SCG cell in accordance with ASTM D3350[(2)] (See Table 1)</th><th>Recommended standard hydrostatic design stress (HDS) category, for water, at 73°F (23°C) (psi)</th></tr>
<tr><td>PE2406</td><td rowspan="2">Cell number 2</td><td>Cell number 4</td><td>630</td></tr>
<tr><td>PE2708</td><td>Cell number 7</td><td>800</td></tr>
<tr><td>PE3408</td><td rowspan="4">Cell number 3</td><td>Cell number 4</td><td rowspan="2">800</td></tr>
<tr><td>PE3608</td><td>Cell number 6</td></tr>
<tr><td>PE3708</td><td rowspan="2">Cell number 7</td><td>800</td></tr>
<tr><td>PE3710</td><td>1000</td></tr>
<tr><td>PE4708</td><td rowspan="2">Cell number 4</td><td rowspan="2">Cell number 7</td><td>800</td></tr>
<tr><td>PE4719</td><td>1000</td></tr>
</table>

[(1)]The last two digits code the Standard HDS Category in units of 100 psi. For example, 06 is the code for 630psi and 10 is the code for 1000 psi.
[(2)]It should be noted that the lowest Cell number for SCG resistance for pipe is 4. Based on research and experience a rating of at least 4 has been determined as sufficient for the safe absorption of localized stresses for properly installed PE pipe.

a brief explanation of the significance of the code digits. It should be recognized that a new material may be commercialized which qualifies for a code designation that has not been recognized as of this writing.

For a listing of the most current recognized code designations the reader is invited to consult the periodically updated PPI publication TR-4. Contact PPI via their website, www.plasticpipe.org.

Reprinted with permission - PPI. For additional details and the complete Handbook, see The Plastics Pipe Institute Handbook of Polyethylene Pipe, 2nd Ed., 2008 at the website https://plasticpipe.org/publications/pe-handbook.html.

Pipeline books published after 2000

Here is a list of notable pipeline and pipeline-related books published after 2000.

B.E. Abulnaga, Slurry Systems Handbook, 2nd Ed. McGraw-Hill, New York, 2021, p. 960.

Z. Ahmad, Principles of Corrosion Engineering and Corrosion Control. Butterworth-Heinemann, Waltham, MA, 2006, p. 672.

D. Alkazraji, A Quick Guide to Pipeline Engineering. Woodhead Publishing, Waltham, MA, 2008, p. 176.

A. Aloqaily, Cross Country Pipeline Risk Assessments and Mitigation Strategies. Gulf Professional Publishing, Boston, 2018, p. 198.

C. An, M. Duan, S.F. Estefen, S. Jian, Structural and Thermal Analyses of Deepwater Pipes. Springer International Publishing, New York, 2021, p. 234.

G.A. Antaki, Piping and Pipeline Engineering: Design, Construction, Maintenance, Integrity, and Repair. Marcel Dekker, New York, 2003, p. 564.

M.A. Ayoade, Disused Offshore Installations and Pipelines: Towards Sustainable Decommissioning. Kluwer Law, New York, 2002, p. 244.

Y. Bai, Pipelines and Risers, Volume 3, Elsevier, New York, 2001, p. 520.

Y. Bai, Q. Bai, Subsea Engineering Handbook. Gulf Professional Publishing, Boston, 2018, p. 968.

Y. Bai, Q. Bai, Subsea Pipelines and Risers. Elsevier, New York, 2005, p. 840.

Y. Bai, Q. Bai, Subsea Pipeline Integrity and Risk Management. Gulf Professional Publishing, Boston, 2014, p. 428.

Q. Bai, Y. Bai, Subsea Pipeline Design, Analysis, and Installation. Gulf Professional Publishing, Boston, 2014, p. 824.

Q. Bai, Y. Bai, W. Ruan, Flexible Pipes: Advances in Pipes and Pipelines. John Wiley & Sons Inc., Hoboken, NJ, 2017, p. 640.

H. Bannwarth, Liquid Ring Vacuum Pumps, Compressors and Systems: Conventional and Hermetic Design. John Wiley & Sons Inc., Hoboken, NJ, 2006, p. 512.

E.N. Barkanov, A. Dumitrescu, I.A. Parinov, Non-destructive Testing and Repair of Pipelines. Springer International Publishing, New York, 2018, p. 451.

G. Barker, The Engineer's Guide to Plant Layout and Piping Design for the Oil and Gas Industries. Gulf Professional Publishing, Boston, 2017, p. 532.

M. Borremans, Pumps and Compressors. John Wiley & Sons Inc., Hoboken, NJ, 2019, p. 512.

M.W. Braestrup, et al. (Eds.), Design and Installation of Marine Pipelines, Blackwell Science: ASME Press, Fairfield, NJ, 2005, p. 342.

G. Chilingar, M. Ryan, G. Al-Qahtani, The Fundamentals of Corrosion and Scaling for Petroleum and Environmental Engineers. Gulf Professional Publishing, Boston, 2008, p. 276.

Y.F. Cheng, Stress Corrosion Cracking of Pipelines. John Wiley & Sons Inc., Hoboken, NJ, 2013, p. 288.

W.C. Chin, Computational Rheology for Pipeline and Annular Flow. Gulf Professional Pub., Boston, 2001, p. 257.

M. Chowdhury, A. Apon, K. Dey, Data Analytics for Intelligent Transportation Systems. Elsevier, New York, 2017, p. 344.

M. Ciszewski, M. Giergiel, T. Buratowski, P. Małka, Modeling and Control of a Tracked Mobile Robot for Pipeline Inspection. Springer International Publishing, New York, 2020, p. 132.

J. Cordell, H. Vanzant, Pipeline Pigging Handbook. Clarion Technical Publishers, Houston, TX, 2003, p. 282.

M.G. Davies, A.B. Lumsden, W.E. Kline, I. Kakadiaris, Pumps and Pipes: Proceedings of the Annual Conference. Springer International Publishing, New York, 2011, p. 257.

R.M. Denys, Pipeline Technology. Elsevier, New York, 2000.

G. Desjardins, R. Sahney, Encyclopedic Dictionary of Pipeline Integrity. Clarion Technical Publishers, Houston, TX, 2012, p. 314.

J.-P. Duroudier, Fluid Transport. ISTE Press – Elsevier, London, 2016, p. 366.

P. Ellenberger, Piping and Pipeline Calculations Manual. Butterworth-Heinemann, Waltham, MA, 2014, p. 412.

K. Escoe, Piping and Pipelines Assessment Guide. Gulf Professional Publishing, Boston, 2006, p. 560.

M. Farshad, Plastic Pipe Systems: Failure Investigation and Diagnosis. Elsevier, New York, 2006, p. 224.

A.A. Fernandes, A.M.P. de Jesus, R.N. Jorge, Monotonic and Ultra-Low-Cycle Fatigue Behaviour of Pipeline Steels: Experimental and Numerical Approaches. Springer International Publishing, New York, 2018, p. 524.

J. Fink, Guide to the Practical Use of Chemicals in Refineries and Pipelines. Gulf Professional Publishing, Boston, 2016, p. 266.

W.E. Forsthoffer, Compressors. Elsevier, Oxford, 2006, p. 415.

P. Girdhar, Performance Evaluation of Pumps and Compressors. Deutsche Babcock-Borsig AG, Berlin, 2008, p. 228.

A. Groysman, Corrosion in Systems for Storage and Transportation of Petroleum Products and Biofuels: Identification, Monitoring and Solutions. Springer International Publishing, New York, 2014, p. 297.

J.F. Gülich, Centrifugal Pumps. Springer International Publishing, New York, 2020, p. 1264.

B. Guo, S. Song, A. Ghalambor, T. Lin, Offshore Pipelines. Gulf Professional Publishing, Boston, 2013, p. 400.

R.T. Hardee, J.L. Sines, Piping System Fundamentals, 2nd Ed. ESI Press, Lacey, WA, 2012, p. 254.

M. Henrie, P. Carpenter, R.E. Nicholas, Pipeline Leak Detection Handbook. Gulf Professional Publishing, Boston, 2016, p. 340.

G.W. Hiltscher, Industrial Pigging Technology: Fundamentals, Components, Applications. Wiley-VCH, 2003, p. 334.

G. Hiltscher, W. Mühlthaler, J. Smits, Industrial Pigging Technology: Fundamentals, Components, Applications. John Wiley & Sons Inc., Hoboken, NJ, 2006, p. 352.

M.J. Kaiser, The Offshore Pipeline Construction Industry. Gulf Professional Publishing, Boston, 2020, p. 502.

S.A. Karamanos, A.M. Gresnigt, G.J. Dijkstra, Geohazards and Pipelines: State-of-the-Art Design Using Experimental, Numerical and Analytical Methodologies. Springer International Publishing, New York, 2021, p. 204.

V.M. Karbhari, Rehabilitation of Pipelines Using Fiber-reinforced Polymer (FRP) Composites. Woodhead Publishing, Waltham, MA, 2014, p. 314.

S. Kyriakides, E. Corona, Mechanics of Offshore Pipelines. Elsevier, Amsterdam, 2007, p. 448.

S. Kyriakides, L.-H. Lee, Mechanics of Offshore Pipelines, Volume 2, Gulf Professional Publishing, Boston, 2020, p. 452.

Z. Li, Pipeline Spatial Data Modeling and Pipeline WebGIS: Digital Oil and Gas Pipeline: Research and Practice. Springer International Publishing, New York, 2020, p. 154.

H. Liu, Pipeline engineering. Boca Raton, Fla: Lewis Publishers, c2003. p. 420.

M.V. Lurie, E. Sinaiski (Translator), Modeling of Oil Product and Gas Pipeline Transportation. John Wiley & Sons Inc., Hoboken, NJ, 2008, p. 234.

M. Mahmoodian, Reliability and Maintainability of In-Service Pipelines. Gulf Professional Publishing, Boston, 2018, p. 186.

E.W. McAllister, Pipeline Rules of Thumb Handbook. Gulf Professional Publishing, Boston, 2014, p. 791.

E.S. Menon, Liquid Pipeline Hydraulics. Marcel Dekker, New York, 2004, p. 274.

E.S. Menon, P.S. Menon, Gas Pipeline Hydraulics. Trafford, Boca Raton, 2013, p. 534.

E.S. Menon, Transmission Pipeline Calculations and Simulations Manual. Gulf Professional Publishing, Boston, 2014, p. 612.

T.O. Miesner, Local Distribution Pipelines in Nontechnical Language. PennWell, Tulsa, OK, 2016, p. 390.

T.O. Miesner, W.L. Leffler, Oil & Gas Pipelines in Nontechnical Language, 2nd Edition. PennWell, Tulsa, OK, 2020, p. 395.

M. Mohitpour, H. Golshan, A. Murray, Pipeline Design & Construction: A Practical Approach. ASME Press, New York, 2003, p. 656.

M. Mohitpour, J. Szabo, T. Van Hardeveld, Pipeline Operation & Maintenance: A Practical Approach. ASME Press, New York, 2005, p. 653.

M. Mohitpour, K.K. Botros, T. Van Hardeveld, Pipeline Pumping and Compression Systems: A Practical Approach. ASME Press, New York, 2008, p. 525.

A.P. Moser, S. Folkman, Buried Pipe Design. McGraw-Hill, New York, 2008, p. 601.

W.K. Muhlbauer, Pipeline Risk Management Manual. Gulf Professional Publishing, Boston, 2003, p. 395.

NACE International, In-line Inspection of Pipelines. NACE International, Houston, TX, 2012, p. 42.

M. Najafi, Trenchless Technology: Pipeline and Utility Design, Construction, and Renewal. New York: McGraw-Hill, c2005. p. xix, 489.

M. Najafi, Trenchless Technology: Planning, Equipment, and Methods. McGraw Hill, New York, 2013, p. 608.

N.L. Nayyar, Piping Handbook. McGraw Hill, New York, 2000, p. 780.

M. Orazem, Underground Pipeline Corrosion. Woodhead Publishing, Waltham, MA, 2014, p. 340.

A.C. Palmer, R.A. King, Subsea Pipeline Engineering, 2nd Edition. PennWell, Tulsa, OK, 2008, p. 650.

S. Papavinasam, Corrosion Control in the Oil and Gas Industry. Gulf Professional Publishing, Boston, 2013, p. 1020.

A.N. Papusha, Beam Theory for Subsea Pipelines: Analysis and Practical Applications. John Wiley & Sons Inc., Hoboken, NJ, 2015, p. 320.

Pigging Products, Services Association, An Introduction to Pipeline Pigging. Pipes & Pipelines International, Beaconsfield, UK, 2020, p. 92.

S. Razvarz, R. Jafari, A. Gegov, Flow Modelling and Control in Pipeline Systems: A Formal Systematic Approach. Springer International Publishing, New York, 2021, p. 198.

M.B. Reagan, Pipeline Safety: Incident Response and Automated Valve Use. Nova Publishers, New York, 2013, p. 104.

C. Reed, A.J. Robinson, D. Smart, Techniques for monitoring structural behaviour of pipeline systems. Awwa Research Foundation and American Water Works Association, Denver, CO, 2004, p. 246.

R.W. Revie (Ed.), Oil and Gas Pipelines: Integrity and Safety Handbook, John Wiley & Sons Inc., Hoboken, NJ, 2015, p. 856.

P. Roberge, Corrosion Engineering. McGraw Hill, New York, 2008, p. 790.

W.T. Shaw, Cybersecurity for SCADA Systems, 2nd Edition. PennWell, Tulsa, OK, 2020, p. 499.

O. Shoham, Mechanistic Modeling of Gas-Liquid Two-Phase Flow in Pipes. Society of Petroleum Engineers, Richardson, TX, 2006, p. 396.

B. Silowash, Piping Systems Manual. McGraw-Hill, New York, 2009, p. 416.

R. Singh, Arctic Pipeline Planning. Gulf Professional Publishing, Boston, 2013, p. 136.

R. Singh, Pipeline Integrity. Gulf Professional Publishing, Boston, 2017, p. 334.

R. Singh, Pipeline Integrity Handbook. Gulf Professional Publishing, Boston, 2013, p. 320.

K. Sotoodeh, A Practical Guide to Piping and Valves for the Oil and Gas Industry. Gulf Professional Publishing, Boston, 2021, p. 992.

M. Stewart, Surface Production Operations: Volume III: Facility Piping and Pipeline Systems. Gulf Professional Publishing, Boston, 2015, p. 1090.

M. Stewart, Surface Production Operations: Volume IV: Pumps and Compressors. Gulf Professional Publishing, Boston, 2018, p. 934.

P. Sulzer, Centrifugal Pump Handbook. Butterworth-Heinemann, Waltham, MA, 2010, p. 302.

W. Tillner, H. Fritsch, R. Kruft, W. Lehmann, H. Louis, G. Masendorf, Avoidance of Cavitation Damage: Principles, Methods of Test, Applications, Experience. John Wiley & Sons Inc., Hoboken, NJ, 2005, p. 267.

S. Timashev, A. Bushinskaya, Diagnostics and Reliability of Pipeline Systems. Springer International Publishing, New York, 2016, p. 407.

J. Tiratsoo (Ed.), Pipeline Pigging and Integrity Technology, Clarion Technical Publishers, Houston, TX, 2013, p. 514.

M. Toghraei, Piping and Instrumentation Diagram Development. John Wiley & Sons Inc., Hoboken, NJ, 2019, p. 496.

C. Verde, L. Torres, Modeling and Monitoring of Pipelines and Networks: Advanced Tools for Automatic Monitoring and Supervision of Pipelines. Springer International Publishing, New York, 2017, p. 264.

D.A. Willoughby, R.D. Woodson, R. Sutherland, Plastic Piping Handbook. McGraw-Hill, New York, 2002.

2: Construction

Right of Way 62
Project scoping data worksheet for major facilities 62
Clearing and grading right-of-way: labor/equipment considerations 64
Estimating man hours for removing trees 64
Estimating man hours for removing tree stumps 65
Clearing and grading right-of-way 65

Ditching and Trenching 66
How many cubic yards of excavation in a mile of ditch? 66
Shrinkage and expansion of excavated and compacted soil 66
Ditching and trenching: labor/equipment considerations 66
Trench excavations 67
Determination of soil type 68
Test equipment and methods for evaluating soil type 68
OSHA trench box requirements 76

Laying and Lowering 78
How to determine the degrees of bend in a pipe that must fit a ditch calling for a bend in both horizontal and vertical planes 78
How to bend a pipe to fit a ditch 78
Pipe bending computations made with a calculator 79
Calculate maximum bend on cold pipe 80
Determine length of a pipe bend 82
Length of pipe in arc subtended by any angle 82
Linear feet of pipelay per 10 h day in trench in level, rock-free terrain 83
Linear feet of pipelay per 10-hour day on waist-high supports on level ground 84
Allowable pipe span between supports 84
How engineers make pipe fit the ditch 85
How to lower an existing pipeline that is still in service 88

Welding 91
When should steel be preheated before welding? 91
Welding and brazing temperatures 92
Mechanical properties of pipe welding rods 92
Lens shade selector 93
Reasons for rejecting a pipeweld and ways to reduce reject rates 93
How many welds will the average welder make per hour? 102
How much welding rod is required for a mile of schedule 40 pipeline? 102
How many pounds of electrodes are required per weld on line pipe? 102
Welding criteria permit safe and effective pipeline repair 103
Cross-country pipeline—vertical down electrode consumption, pounds of electrode per joint 109

Directional Crossing 110
Guidelines for a successful directional crossing bid package 110

Polyethylene Piping 118
Simplified installation guidelines for PE pipe 118
How to construct a trench for PE piping 120

RIGHT OF WAY

Project scoping data worksheet for major facilities

Project Name

____

Location______________________________

Latitude______________________________

Longitude______________________________

☐ Offshore ☐ Onshore ☐ Arctic

☐ Marsh Site elevation, ft ______________________

Project description ______________________

____

____

____

1.0 Type of contract required

1.1 Describe the type of contract that the project will be based on.
1.2 What is the completion date?
1.3 What type of permits will be required and who will obtain them?
1.4 Will an environmental impact statement be required?
1.5 Describe currency requirements if in a foreign country.
1.6 If the project is located in a foreign country, does the country have regulations/codes that will need to be complied with?

2.0 General engineering

2.1 Does the client have engineering standards and specifications, or will it be necessary to develop these, or can the Engineering Contractor's specifications and standards be used?
2.2 Will in-country engineering concerns, such as design institutes, be required to provide engineering services?
2.3 If in-country engineering facilities are required, how much assistance will be required?
2.4 Describe any existing above-ground or underground obstructions at the job site.
2.5 Obtain copies of any applicable local codes dealing with air and/or water pollution, sanitary systems, electrical systems, and structures. Obtain copies of any regulations dealing with waste disposal.
2.6 Are soil bearing data available?
2.7 If the project involves a pipeline that transports heat-sensitive fluid, are any soil thermal conductivity data available?
2.8 Describe type of soil—sandy, rocky, caliche, etc.
2.9 Does the site drain naturally? If fill material is required, what is the availability? Is sand-asphalt mix available?

3.0 Drafting

3.1 Will the drawings need to be sealed by a professional engineer?
3.2 Determine drawing forms to be used and type of drawings required—single line or double line.

4.0 Structural and architectural

4.1 Will all exposed steel need to be galvanized or have other protective coating?
4.2 Is sand, rock, cement, and ready-mixed concrete available locally? Price?
4.3 Requirements for building design:
Steel
Concrete/brick
Cooling and heating requirements
Other requirements

5.0 Piping

5.1 How will storm water, waste water, and sewage be disposed of? Will an oil-water separator be required?
5.2 Will connections to any existing pipelines be required?
5.3 Are there any facilities outside the job site that must be given consideration in the development of a plot plan?
5.4 What is the pressure rating for any existing manifolds, suction lines, or outgoing pipelines?
5.5 Are there any special requirements for isolating the facility?
5.6 Will scraper traps be required? Incoming, outgoing, or both?
5.7 How many relief headers are required?
5.8 If system relief valves are required, where will they relieve to?
5.9 Will a fire water system be required? Is a source of water available? If a source is available, what is the capacity of the source?
5.10 Is the water fresh, brackish, or seawater?

6.0 Electrical

6.1 Is power available from the local power grid? If yes, specify voltage, frequency, phase, and system capacity.
6.2 If power generation will be required in connection with the project, will it be a requirement to connect to a local grid?
6.3 If local power is available, define the approximate location where the power will enter the job site.
6.4 Will the local power entity furnish and install any substation equipment that will be required?
6.5 If power is to be supplied from a local power grid, attempt to determine the short-circuit capacity of the system.
6.6 Will across-the-line starting of large electric motors be permitted?
6.7 If across-the-line starting is not permitted, what is the maximum horsepower that will be permitted for across-the-line starting?

7.0 Instrumentation equipment

7.1 What type of control panel will be required?
7.2 Define electrical classification that will be required for a control house if a control house is required.
7.3 Are local communications facilities available? Describe.

8.0 Equipment

8.1 Have any long-delivery items been placed on order? List items and delivery schedule.
8.2 Who will supply general equipment specifications?
8.3 Will in-country purchasing be required?
8.4 List design criteria for equipment sizing.
8.5 Define spare equipment requirements. If equipment is purchased outside the country where the project is located, consideration should be given to transit time from the country of origin to the job site.
8.6 Define the design life for the facility.
8.7 Will oil storage tanks be required? Will secondary seals be required? Will tank mixers be required?

9.0 Material purchasing

9.1 If a preferred vendor list is available, obtain a copy of the list.
9.2 If such a list is not available, will it be a requirement to develop a list for the client's approval?
9.3 Determine the shipping address.
9.4 If material is shipped to a foreign location, who will handle receipt of the material and clearance with local customs?
9.5 Is the location near an established seaport? If not, determine the nearest seaport and logistics for moving material from the seaport to the job site. This should include any weight limitations, load width restrictions, and requirements for moving heavy equipment at night during periods of light road traffic.
9.6 Where will material be stored pending installation?
9.7 What procedure will be used for turning material over to the contractor?

10.0 Special considerations

10.1 Is heavy equipment available locally and does it have the capacity to lift the heaviest piece of material?
10.2 Is an adequate supply of sufficiently skilled local labor available?
10.3 If the project is in a foreign location, is there a limit on the use of expatriates?
10.4 What is the procedure for obtaining a work permit?
10.5 What is the availability of contractors that may be working in the area?
10.6 Will a construction camp be required?
10.7 If a construction camp is required, what is the availability of a local catering service?

11.0 Environmental information

11.1 Ambient Temp. (°F) Max.________Min.________
11.2 Relative humidity
Wet bulb (°F) Max.______Min.______
Dry bulb (°F) Max.______Min.______
11.3 Water temp. surface (°F) Max.______Min.______
11.4 Water temp. bottom (°F) Max.______Min.______
11.5 Potential for icing Y or N
11.6 Rainfall (in/h) Max.______Min.___
11.7 Prevailing wind direction
11.8 Velocity Max.______Min.___
11.9 Weather window (time of year)______

12.0 Fluid characteristics

12.1 Crude oil
12.2 Sediment and water, %________________________
12.3 Free water________________________
12.4 Salt content (#/1000 bbls)________________________
12.5 API gravity ____________ @ ____________ °F and ____________ @ ____________ °F (2 required)
12.6 Viscosity ____________ cP @ ____________ °F and ____________ cP @ ____________ °F (2 required)
12.7 Oil pour point ____________ °F
12.8 Reid vapor pressure ________________________
12.9 Produced water: oil & grease (ppm)____
12.10 Total suspended solids (mg/kg oil)______
12.11 Natural Gas
12.12 MOL % inerts________________
12.13 MOL % CO_2________________
12.14 ppm H_2S or grains/100 SCF __________
12.15 Water content__________#/MMSCF or dew point ____________°F
12.16 Max. heating value (BTU/SCF)______
12.17 Min heating value (BTU/SCF)________
12.18 NGL content _____ gal/MCF or dew point________°F

Clearing and grading right-of-way: labor/equipment considerations

To estimate labor crew and equipment spread for clearing and grading operations, the following items, as they may apply to a given project, should be given consideration:

1. Removal of trees, brush, and stumps.
2. Grubbing and removal of stumps that are in the way of the ditch.
3. Disposal of all debris, including method of disposal and length of haul.
4. Clearing area spoil a sufficient distance from the ditch line so that the spoil-bank from the ditching operations will not fall in any foreign material that might become mixed with the excavated spoil.
5. Cutting of merchantable timber into standard lengths and stacked along the right-of-way for disposition by others if specifically required by the right-of-way agreement.
6. Providing temporary walks, passageways, fences, or other structures so as not to interfere with traffic.
7. Providing sufficient and proper lighting where required.
8. Providing guards where required.
9. Preserving all trees, shrubs, hedges, and lawns where required.
10. Grading irregularities where required.
11. Preserving topsoil for replacement, through all cultivated or improved fields and pastures, to its original position.
12. Proper grading of the terrain so as to allow passage of loaded trucks and equipment hauling materials and so ditching operations can be properly performed.
13. Protecting and preserving existing drainage facilities.
14. Protecting any existing structures or pipelines.
15. Protecting any telephone or utility lines and keeping them in service.
16. Cutting through fences and hedges where required and replacing these when necessary.
17. Installing gates and fences where required.

Estimating man hours for removing trees

Net Man Hours—Cross-Cut Saws

Average Tree Diameter in Inches	Softwood Trees		Hardwood Trees	
	Open Area	Congested Area	Open Area	Congested Area
4	1.49	1.86	1.88	2.35
6	2.26	2.83	2.82	3.53
8	3.24	4.02	4.00	4.96
10	4.10	5.08	5.00	6.20
12	4.98	6.18	6.00	7.44
14	6.39	7.86	7.70	9.47
16	7.39	9.09	8.80	10.82
18	8.32	10.23	9.90	12.18
20	10.58	12.91	12.60	15.37
24	12.71	15.51	15.12	18.45
30	17.09	20.85	20.10	24.52
36	20.50	25.01	24.12	29.43

Net Man Hours—Chain Saws

Average Tree Diameter in Inches	Softwood Trees		Hardwood Trees	
	Open Area	Congested Area	Open Area	Congested Area
4	0.37	0.46	0.47	0.59
6	0.57	0.71	0.71	0.89
8	0.81	1.00	1.00	1.24
10	1.03	1.28	1.25	1.55
12	1.25	1.55	1.50	1.86
14	1.60	1.97	1.93	2.37
16	1.85	2.28	2.20	2.71
18	2.08	2.56	2.48	3.05
20	2.65	3.23	3.15	3.84
24	3.18	3.88	3.78	4.61
30	4.28	5.22	5.03	6.14
36	5.13	6.26	6.03	7.36

Man hours include ax trimming, cutting down with cross-cut saws or chain saws, and cutting into 4-ft lengths for the tree diameter sizes. Man hours are average for various heights of trees.

Man hours do not include hauling, piling, and burning of trees or branches, or the removal of stumps.

Estimating man hours for removing tree stumps

Net Man Hours—Grub & Removal by Hand

Item	Laborer	Oper. Engr.	Powder Man	Total
8″–12″ diameter	6.00	-	-	6.00
14″–18″ diameter	7.50	-	-	7.50
20′–24″ diameter	9.00	-	-	9.00
26″–36″ diameter	11.20	-	-	11.20

Net Man Hours—Blast & Pull by Tractor

Item	Laborer	Oper. Engr.	Powder Man	Total
8″–12″ diameter	0.83	0.11	1.50	2.44
14″–18″ diameter	1.05	0.23	2.33	3.61
20″–24″ diameter	1.50	0.30	3.38	5.18
26″–36″ diameter	2.11	0.42	4.76	7.29

Man hours include excavating and removing by hand or blasting and removing with cables and tractors. Man hours do not include burning or removal from premises.

Clearing and grading right-of-way

Equipment Spread

Equipment Description	Number of Units for											
	50 Linear ft Width				80 Linear ft Width				100 Linear ft Width			
	L	M	MH	H	L	M	MH	H	L	M	MH	H
D8 Tractor W/Dozer	1	1	1	1	1	1	2	2	1	2	3	3
D7 Tractor W/Dozer	0	0	1	1	1	1	1	1	1	1	2	2
Truck—2½ Ton Dump	2	2	2	2	2	2	2	2	2	2	3	3
Truck—Pick-up	1	1	2	2	1	2	3	3	2	3	4	4
Ripper or Brushrake	1	1	1	1	1	1	1	1	1	2	2	2

Labor Crew

Personnel Description	Number of Men for											
	50 Linear ft Width				80 Linear ft Width				100 Linear ft Width			
	L	M	MH	H	L	M	MH	H	L	M	MH	H
Foreman	1	1	2	2	1	2	3	3	2	3	4	4
Operator	1	1	2	2	1	2	3	3	2	4	5	5
Mechanic	1	1	1	1	1	1	1	1	1	1	1	1
Swamper	1	1	2	2	1	2	3	3	2	4	5	5
Truck Driver	2	2	2	2	2	2	2	2	2	2	3	3
Laborer	10	15	20	30	15	25	30	40	20	35	40	50
Total Crew	16	21	29	39	21	34	42	52	29	49	58	68

L = Light—light brush and grass, no trees.
M = Medium—considerable brush of larger size.
MH = Medium Heavy—large brush and small trees.
H = Heavy—much small brush, many small trees, and occasional large trees.

Above equipment spread should be ample for clearing and grading 1 mile of right-of-way per 10-hour day for the width and conditions outlined. Haul trucks are based on a round-trip haul of 2 miles. If brush and trees are to be burned on site, omit the above dump trucks. Small tools such as saws, axes, etc., must be added as required for the individual job.

Above total crew should be ample for clearing and grading 1 mile of right-of-way per 10-hour day for the width and conditions outlined. Crew spread includes cutting, stacking or piling, loading, and hauling a round-trip distance of 2 miles. If burning is necessary or permitted, substitute fire tenders for dump truck drivers.

Source

Page, J.S., *Cost Estimating Man-Hour Manual for Pipelines and Marine Structures*, Gulf Publishing Co., Houston, Texas, 1977.

DITCHING AND TRENCHING

How many cubic yards of excavation in a mile of ditch?

Multiply the width in inches by the depth in inches by 1.36; the answer is cubic yards per mile.

Example. How many cubic yards of excavation in a mile of 12-in ditch 30-in deep?

$12 \times 30 \times 1.36 = 490$ cubic yards per mile.

The rule is correct within about 1/5 of 1%. Actually, the errors in depth and width are much greater than this. To get the cubic yards per 1000 ft, as in computing rock ditching, use 0.257 instead of 1.36.

Shrinkage and expansion of excavated and compacted soil

Ever notice how the spoil from a ditch occupies a greater volume than the ditch itself? There's a reason. Excavate sand and it expands about 10%. Ordinary soil expands about 25% and clay expands about 40%.

Here's a summary of the bulk you can expect from excavated soil:

Type soil	Undisturbed	Excavated	Compacted
Sand	1.00	1.11	0.95
Ordinary earth	1.00	1.25	0.90
Clay	1.00	1.43	0.90

Example. Find the volume of loose spoil from a pipeline ditch 42 in wide, 60 in deep. The excavation is through clay.

Volume of undisturbed clay soil:

$3.5 \times 5.0 = 17.5$ cu ft per lineal foot of ditch.

Volume of the spoil will be 143% of the undisturbed volume.

Volume of spoil $= 1.43 \times 17.5 = 25$ cu ft per lineal ft.

Through mechanical compaction, this volume can be reduced to 15.75 cu ft.

Ditching and trenching: labor/equipment considerations

In determining the labor crew and equipment spread for ditching and trenching operations, the following should be given consideration should they apply to the particular project:

1. Ditching or trenching for buried pipelines should be in accordance with the following table of minimum width and coverage for all soil formations.
2. In rock, cut ditches at least 6 in wider.
3. If dirt-filled benches are used, ditch should be excavated deeper to obtain proper coverage.
4. Trench should be excavated to greater depth when required for proper installation of the pipe where the topography of the country warrants same.
5. Repair any damage to and maintain existing natural or other drainage facilities.
6. Do not open ditch too far in advance of pipelay crew.
7. Obtain permits for blasting.
8. When blasting, use extreme caution and protection.
9. Clean up blasted rock to prevent damage to coated pipe.

Nominal Pipe Size Inches	Minimum Width Inches	Normal Minimum Coverage Inches
4	22	30
6	26	30
8	26	30
10	26	30
12	30	30
14	32	30
16	36	30
18	38	30
20	40	30
24	44	30
30	50	30
36	52	36
42	58	36

Trench excavations

Excerpted and edited from OSHA Technical Manual (OTM). OSHA Instruction TED 01-00-015 [TED 1-0.15A]. Section V: Chapter 2. Excavations: Hazard Recognition in Trenching and Shoring.

Excavating is recognized as one of the most hazardous construction operations. OSHA recently revised Subpart P, *Excavations*, of 29 CFR 1926.650, 29 CFR 1926.651, and 29 CFR 1926.652 to make the standard easier to understand, permit the use of performance criteria where possible, and provide construction employers with options when classifying soil and selecting employee protection methods.

Figure 1. Tension crack.

Soil Mechanics

A number of stresses and deformations can occur in an open cut or trench. For example, increases or decreases in moisture content can adversely affect the stability of a trench or excavation. The following diagrams show some of the more frequently identified causes of trench failure.

Tension Cracks. Tension cracks usually form at a horizontal distance of 0.5 to 0.75 times the depth of the trench, measured from the top of the vertical face of the trench. See Fig. 1.

Sliding or sluffing may occur as a result of tension cracks, as illustrated in Fig. 2.

Toppling. In addition to sliding, tension cracks can cause toppling. Toppling occurs when the trench's vertical face shears along the tension crack line and topples into the excavation as illustrated in Fig. 3.

Subsidence and Bulging. An unsupported excavation can create an unbalanced stress in the soil, which, in turn, causes subsidence at the surface and bulging of the vertical face of the trench. If uncorrected, this condition can cause face failure and entrapment of workers in the trench (Fig. 4).

Heaving or Squeezing. Bottom heaving or squeezing is caused by the downward pressure created by the weight of adjoining soil. This pressure causes a bulge in the bottom of the cut, as illustrated in Fig. 5. Heaving and squeezing can occur even when shoring or shielding has been properly installed.

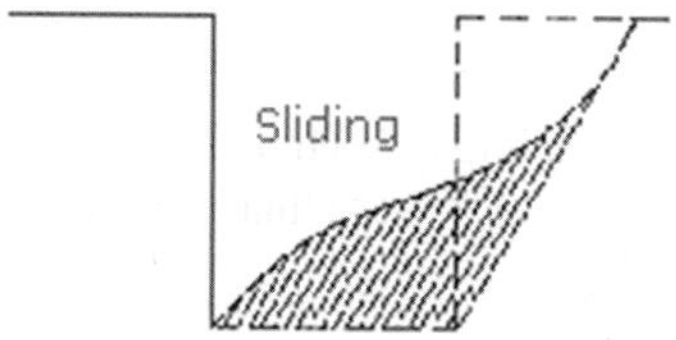

Figure 2. Sliding.

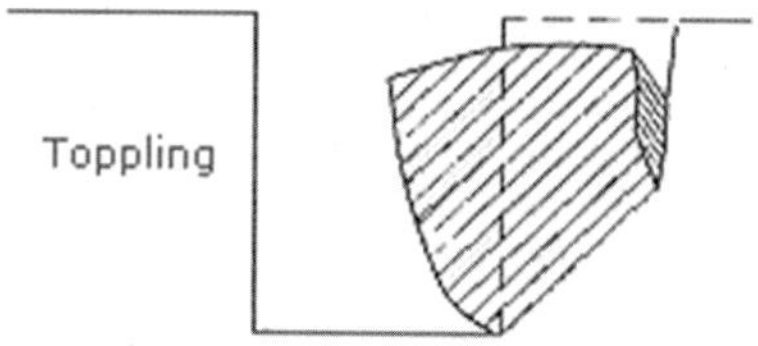

Figure 3. Toppling.

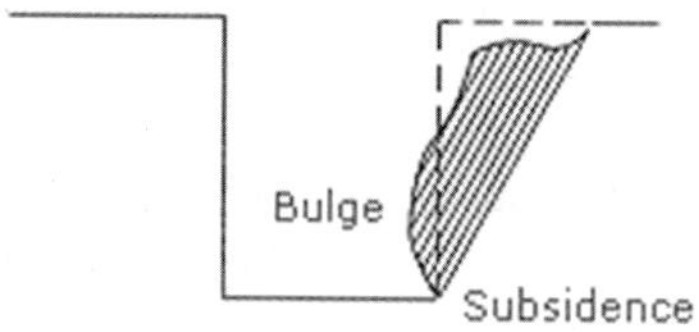

Figure 4. Subsidence and bulging.

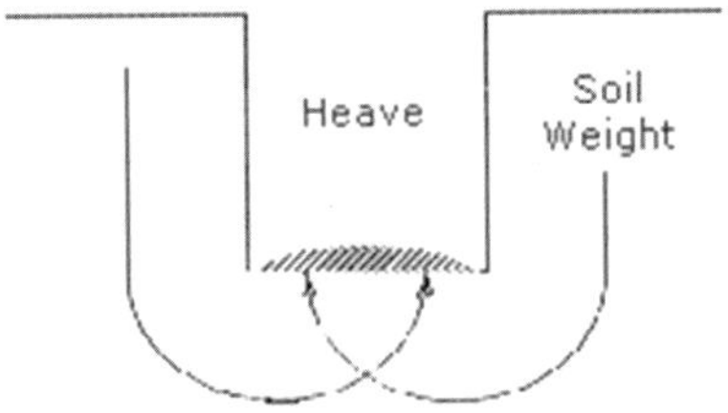

Figure 5. Heaving or squeezing.

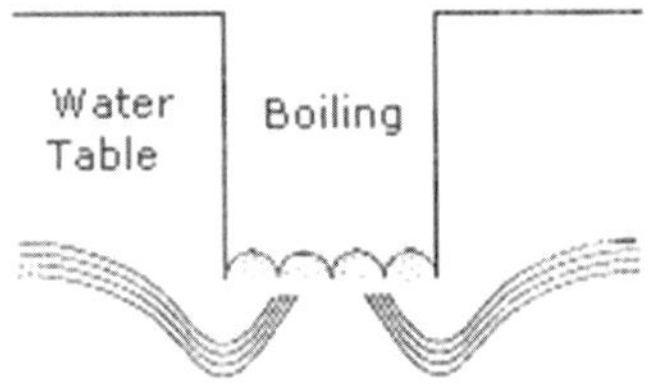

Figure 6. Boiling.

Boiling is evidenced by an upward water flow into the bottom of the cut (Fig. 6). A high water table is one of the causes of boiling. Boiling produces a "quick" condition in the bottom of the cut, and can occur even when shoring or trench boxes are used.

Unit Weight of Soils refers to the weight of one unit of a particular soil. The weight of soil varies with type and moisture content. One cubic foot of soil can weigh from 110 pounds to 140 pounds or more, and one cubic meter (35.3 cubic feet) of soil can weigh more than 3000 pounds.

Determination of soil type

OSHA categorizes soil and rock deposits into four types:

Stable Rock is natural solid mineral matter that can be excavated with vertical sides and remain intact while exposed. It is usually identified by a rock name such as granite or sandstone. Determining whether a deposit is of this type may be difficult unless it is known whether cracks exist and whether or not the cracks run into or away from the excavation.

Type A Soils are cohesive soils with an unconfined compressive strength of 1.5 tons per square foot (tsf) (144 kPa) or greater. Examples of Type A cohesive soils are often: clay, silty clay, sandy clay, clay loam and, in some cases, silty clay loam and sandy clay loam. (No soil is Type A if it is fissured, is subject to vibration of any type, has previously been disturbed, is part of a sloped, layered system where the layers dip into the excavation on a slope of 4 horizontal to 1 vertical (4H:1V) or greater, or has seeping water.

Type B Soils are cohesive soils with an unconfined compressive strength greater than 0.5 tsf (48 kPa) but less than 1.5 tsf (144 kPa). Examples of other Type B soils are: angular gravel; silt; silt loam; previously disturbed soils unless otherwise classified as Type C; soils that meet the unconfined compressive strength or cementation requirements of Type A soils but are fissured or subject to vibration; dry unstable rock; and layered systems sloping into the trench at a slope less than 4H:1V (only if the material would be classified as a Type B soil).

Type C Soils are cohesive soils with an unconfined compressive strength of 0.5 tsf (48 kPa) or less. Other Type C soils include granular soils such as gravel, sand and loamy sand, submerged soil, soil from which water is freely seeping, and submerged rock that is not stable. Also included in this classification is material in a sloped, layered system where the layers dip into the excavation or have a slope of four horizontal to one vertical (4H:1V) or greater.

Layered Geological Strata. Where soils are configured in layers, i.e., where a layered geologic structure exists, the soil must be classified on the basis of the soil classification of the weakest soil layer. Each layer may be classified individually if a more stable layer lies below a less stable layer, i.e., where a Type C soil rests on top of stable rock.

Test equipment and methods for evaluating soil type

Many kinds of equipment and methods are used to determine the type of soil prevailing in an area.

Pocket Penetrometer

Penetrometers are direct-reading, spring-operated instruments used to determine the unconfined compressive strength of saturated cohesive soils. Once pushed into the soil, an indicator sleeve displays the reading. The instrument is calibrated in either tons per square foot (tsf) or kilograms per square centimeter (kPa). Penetrometers have error rates in the range of ± 20-40%.

1. **Shearvane (Torvane).** To determine the unconfined compressive strength of the soil with a shearvane, the blades of the vane are pressed into a level section of undisturbed soil, and the torsional knob is slowly turned until soil failure occurs. The direct instrument reading must be multiplied by 2 to provide results in tons per square foot (tsf) or kilograms per square centimeter (kPa).
2. **Thumb Penetration Test.** The thumb penetration procedure involves an attempt to press the thumb firmly into the soil in question. If the thumb makes an indentation in the soil only with great difficulty, the soil is probably Type A. If the thumb penetrates no further than the length of the thumb nail, it is probably Type B soil, and if the thumb penetrates the full length of the thumb, it is Type C soil. The thumb test is subjective and is therefore the least accurate of the three methods.
3. **Dry Strength Test.** Dry soil that crumbles freely or with moderate pressure into individual grains is granular. Dry soil that falls into clumps that subsequently break into smaller clumps (and the smaller clumps can be broken only with difficulty) is probably clay in combination with gravel, sand, or silt. If the soil breaks into clumps that do not break into smaller clumps (and the soil can be broken only with difficulty), the soil is considered unfissured unless there is visual indication of fissuring.

Plasticity or Wet Thread Test

This test is conducted by molding a moist sample of the soil into a ball and attempting to roll it into a thin thread approximately 1/8 inch (3 mm) in diameter (thick) by 2 inches (50 mm) in length. The soil sample is held by one end. If the sample does not break or tear, the soil is considered cohesive.

Visual Test

A visual test is a qualitative evaluation of conditions around the site. In a visual test, the entire excavation site is observed, including the soil adjacent to the site and the soil being excavated. If the soil remains in clumps, it is cohesive; if it appears to be coarse-grained sand or gravel, it is considered granular. The evaluator also checks for any signs of vibration.

During a visual test, the evaluator should check for crack-line openings along the failure zone that would indicate tension cracks, look for existing utilities that indicate that the soil has previously been disturbed, and observe the open side of the excavation for indications of layered geologic structuring.

The evaluator should also look for signs of bulging, boiling, or sluffing, as well as for signs of surface water seeping from the sides of the excavation or from the water table. If there is standing water in the cut, the evaluator should check for "quick" conditions. In addition, the area adjacent to the excavation should be checked for signs of foundations or other intrusions into the failure zone, and the evaluator should check for surcharging and the spoil distance from the edge of the excavation.

Shoring Types

Shoring is the provision of a support system for trench faces used to prevent movement of soil, underground utilities, roadways, and foundations. Shoring or shielding is used when the location or depth of the cut makes sloping back to the maximum allowable slope impractical. Shoring systems consist of posts, wales, struts, and sheeting. There are two basic types of shoring, timber (Fig. 7) and aluminum hydraulic (Fig. 8).

Hydraulic Shoring

The trend today is toward the use of hydraulic shoring, a prefabricated strut and/or wale system manufactured of aluminum or steel. Hydraulic shoring provides a critical safety advantage over timber shoring because workers do not have to enter the trench to install or remove hydraulic shoring. Other advantages of most hydraulic systems are that they:

- Are light enough to be installed by one worker;
- Are gauge-regulated to ensure even distribution of pressure along the trench line;
- Can have their trench faces "preloaded" to use the soil's natural cohesion to prevent movement; and
- Can be adapted easily to various trench depths and widths.

All shoring should be installed from the top down and removed from the bottom up. Hydraulic shoring should be checked at least once per shift for leaking hoses and/or cylinders, broken connections, cracked nipples, bent bases, and any other damaged or defective parts.

Pneumatic Shoring

Pneumatic shoring works in a manner similar to hydraulic shoring (Fig. 9). The primary difference is that pneumatic shoring uses air pressure in place of hydraulic pressure. A disadvantage to the use of pneumatic shoring is that an air compressor must be on site.

1. **Screw Jacks.** Screw jack systems differ from hydraulic and pneumatic systems in that the struts of a screw jack system must be adjusted manually. This creates a hazard because the worker is required to be in the trench in order to adjust the strut. In addition, uniform "preloading" cannot be achieved with screw jacks, and their weight creates handling difficulties.
2. **Single-Cylinder Hydraulic Shores.** Shores of this type are generally used in a water system, as an assist to timber shoring systems, and in shallow trenches where face stability is required.

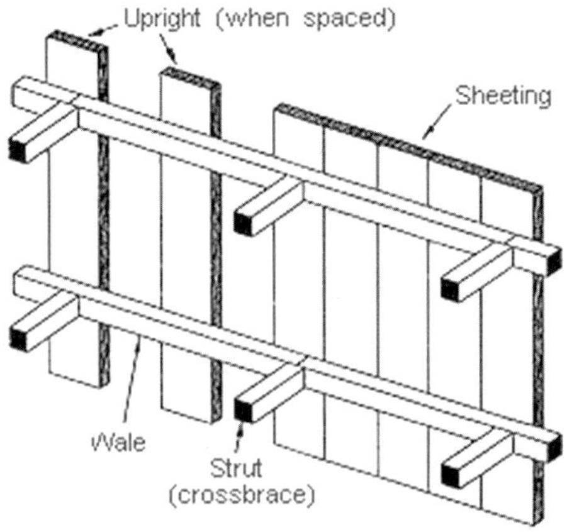

Figure 7. Timber shoring.

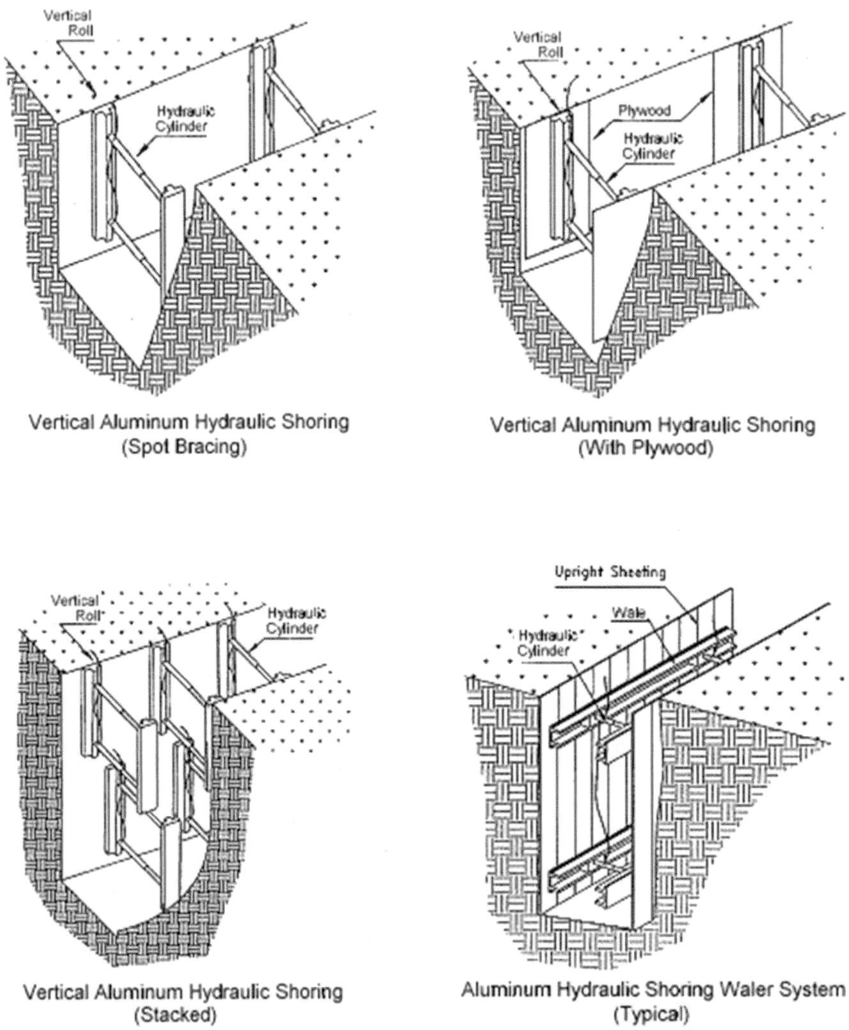

Figure 8. Shoring variations: typical aluminum hydraulic shoring installations.

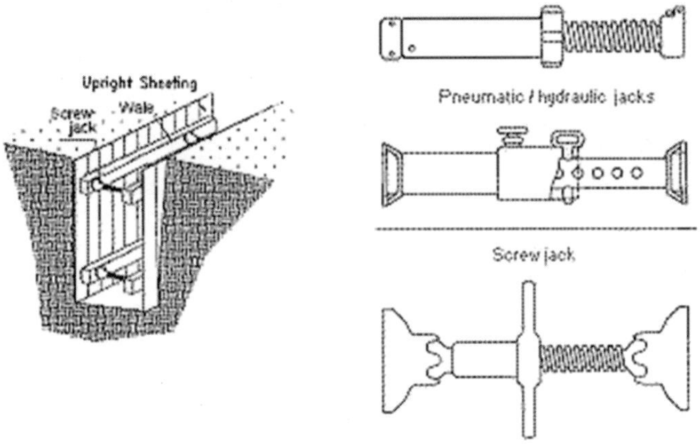

Figure 9. Shoring variations.

3. **Underpinning.** This process involves stabilizing adjacent structures, foundations, and other intrusions that may have an impact on the excavation. As the term indicates, underpinning is a procedure in which the foundation is physically reinforced. Underpinning should be conducted only under the direction and with the approval of a registered professional engineer.

Shielding Types

Trench Boxes are different from shoring because, instead of shoring up or otherwise supporting the trench face, they are intended primarily to protect workers from cave-ins and similar incidents (Figs. 10, 11). The excavated area between the outside of the trench box and the face of the trench should be as small as possible. The space between the trench boxes and the excavation side are backfilled to prevent lateral movement of the box. Shields may not be subjected to loads exceeding those which the system was designed to withstand.

Combined Use. Trench boxes are generally used in open areas, but they also may be used in combination with sloping and benching (Fig. 12). The box should extend at least 18 in (0.45 m) above the surrounding area if there is sloping toward excavation. This can be accomplished by providing a benched area adjacent to the box.

Earth excavation to a depth of 2 ft (0.61 m) below the shield is permitted, but only if the shield is designed to resist the forces calculated for the full depth of the trench and there are no indications while the trench is open of possible loss of soil from behind or below the bottom of the support system). Conditions of this type require observation on the effects of bulging, heaving, and boiling as well as surcharging, vibration, adjacent structures, etc., on excavating below the bottom of a shield. Careful visual inspection of the conditions mentioned above is the primary and most prudent approach to hazard identification and control.

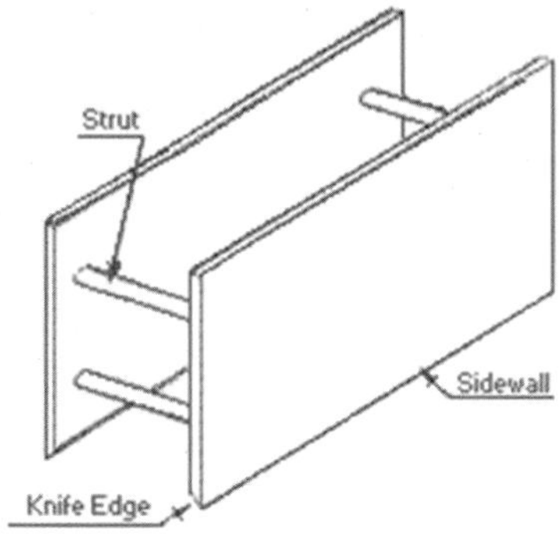

Figure 10. Trench shield.

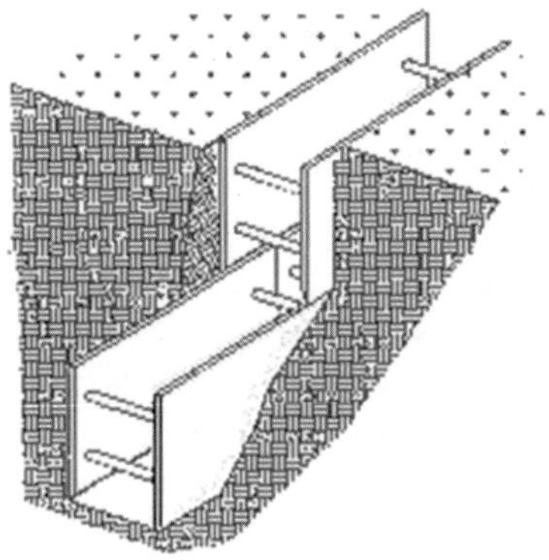

Figure 11. Trench shield, stacked.

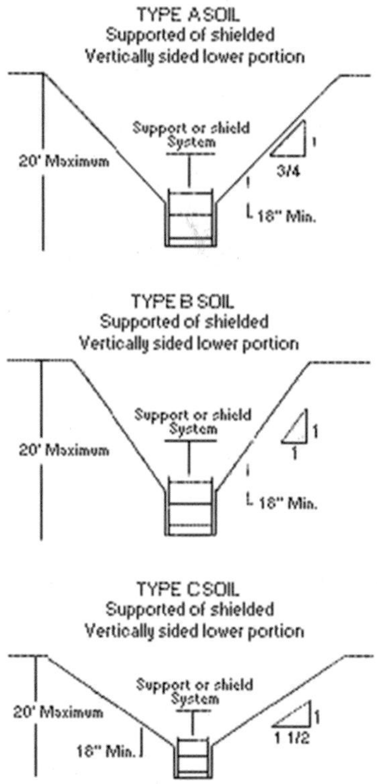

Figure 12. Slope and shield configurations.

Sloping and Benching

Sloping

Maximum allowable slopes for excavations less than 20 ft (6.09 m) based on soil type and angle to the horizontal are as described in Table 1 and illustrated in Figs. 13-15.

Table 1
Allowable Slopes

Soil type	Height:Depth ratio	Slope angle
Stable Rock	Vertical	90°
Type A	¾ :1	53°
Type B	1:1	45°
Type C	1½:1	34°
Type A(short-term)	½:1	63°

For a maximum excavation depth of 12 ft

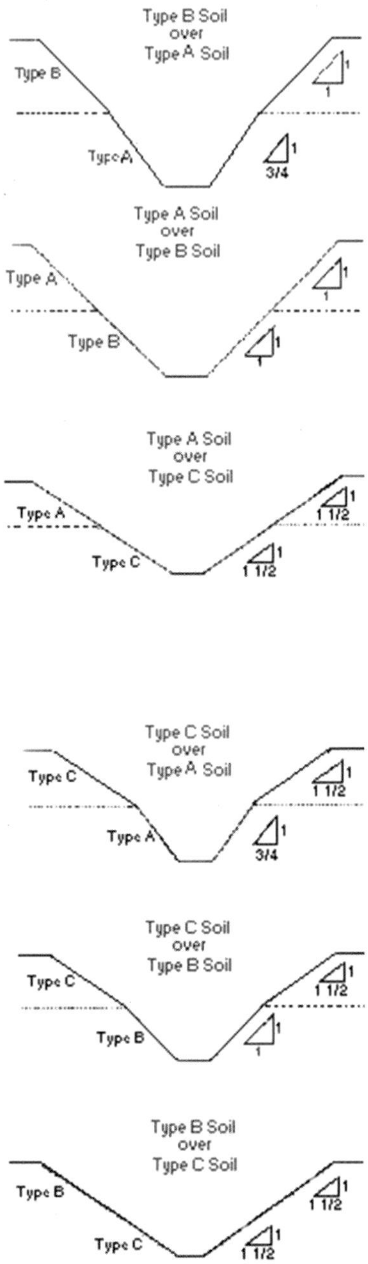

Figure 13. Slope configurations: excavations in layered soils.

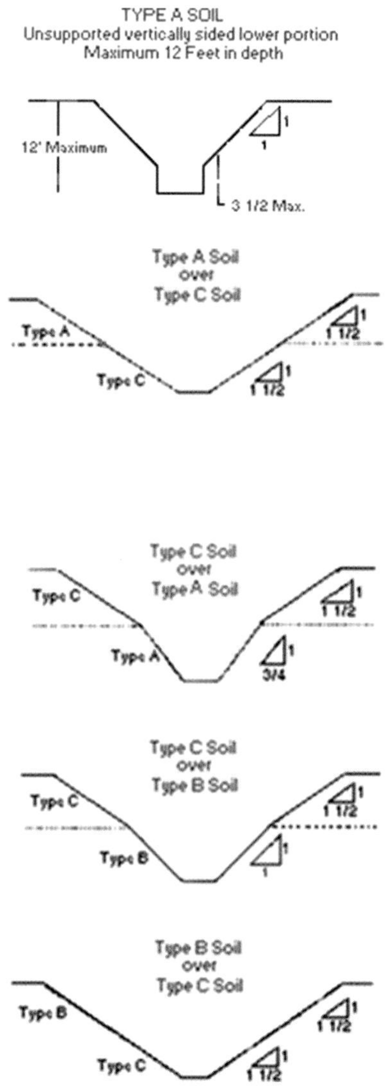

Figure 14. Excavations made in type A soil.

Benching

There are two basic types of benching, simple and multiple. The type of soil determines the horizontal to vertical ratio of the benched side. As a general rule, the bottom vertical height of the trench must not exceed 4 ft (1.2 m) for the first bench. Subsequent benches may be up to a maximum of 5 ft (1.5 m) vertical in Type A soil and 4 ft (1.2 m) in Type B soil to a total trench depth of 20 ft (6.0 m). All subsequent benches must be below the maximum allowable slope for that soil type. For Type B soil the trench excavation is permitted in cohesive soil only.

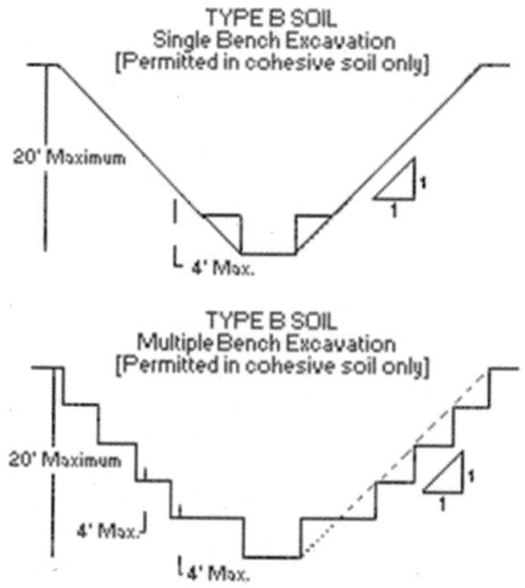

Figure 15. Excavations made in type B soil.

Spoil

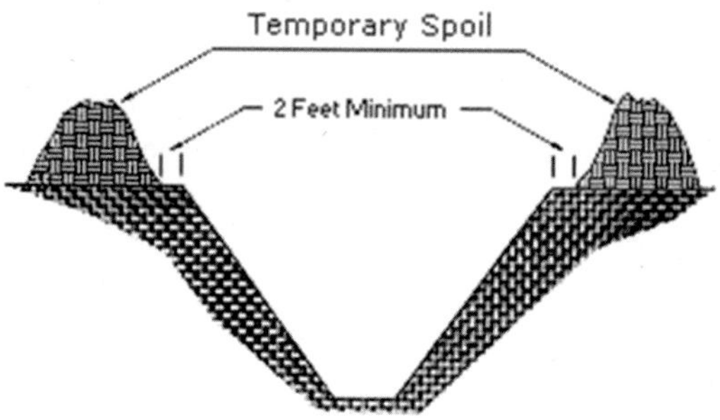

Figure 16. Temporary spoil.

Temporary Spoil

Temporary spoil must be placed no closer than 2 ft (0.61 m) from the surface edge of the excavation, measured from the nearest base of the spoil to the cut (Fig. 16). This distance should not be measured from the crown of the spoil deposit. This distance requirement ensures that loose rock or soil from the temporary spoil will not fall on employees in the trench.

Spoil should be placed so that it channels rainwater and other run-off water away from the excavation. Spoil should be placed so that it cannot accidentally run, slide, or fall back into the excavation.

Permanent Spoil

Permanent spoil should be placed at some distance from the excavation. Permanent spoil is often created where underpasses are built or utilities are buried. The improper placement of permanent spoil, i.e. insufficient distance from the working excavation, can cause an excavation to be out of compliance with the horizontal-to-vertical ratio requirement for a particular excavation. This can usually be determined through visual observation. Permanent spoil can change undisturbed soil to disturbed soil and dramatically alter slope requirements.

Special Health and Safety Considerations

Competent Person

The designated competent person should have and be able to demonstrate the following:

- Training, experience, and knowledge of:
  - soil analysis;
  - use of protective systems; and
  - requirements of 29 CFR Part 1926 Subpart P
- Ability to detect:
  - conditions that could result in cave-ins;
  - failures in protective systems;
  - hazardous atmospheres; and
  - other hazards including those associated with confined spaces.
- Authority to take prompt corrective measures to eliminate existing and predictable hazards and to stop work when required.

Surface Crossing of Trenches

Surface crossing of trenches should be discouraged; however, if trenches must be crossed, such crossings are permitted only under the following conditions:

- Vehicle crossings must be designed by and installed under the supervision of a registered professional engineer.
- Walkways or bridges must be provided for foot traffic. These structures shall:
  - have a safety factor of 4;
  - have a minimum clear width of 20 in (0.51 m);
  - be fitted with standard rails; and
  - extend a minimum of 24 in (0.61 m) past the surface edge of the trench.

Ingress and Egress

Access to and exit from the trench require the following conditions:

- Trenches 4 ft or more in depth should be provided with a fixed means of egress.
- Spacing between ladders or other means of egress must be such that a worker will not have to travel more than 25 ft laterally to the nearest means of egress.
- Ladders must be secured and extend a minimum of 36 in (0.9 m) above the landing.
- Metal ladders should be used with caution, particularly when electric utilities are present.

Exposure to Vehicles

Procedures to protect employees from being injured or killed by vehicle traffic include:

- Providing employees with and requiring them to wear warning vests or other suitable garments marked with or made of reflectorized or high-visibility materials.
- Requiring a designated, trained flagperson along with signs, signals, and barricades when necessary.

Exposure to Falling Loads

Employees must be protected from loads or objects falling from lifting or digging equipment. Procedures designed to ensure their protection include:

- Employees are not permitted to work under raised loads.
- Employees are required to stand away from equipment that is being loaded or unloaded.
- Equipment operators or truck drivers may stay in their equipment during loading and unloading if the equipment is properly equipped with a cab shield or adequate canopy.

Warning Systems for Mobile Equipment

The following steps should be taken to prevent vehicles from accidentally falling into the trench:

- Barricades must be installed where necessary.
- Hand or mechanical signals must be used as required.
- Stop logs must be installed if there is a danger of vehicles falling into the trench.
- Soil should be graded away from the excavation; this will assist in vehicle control and channeling of run-off water.

Hazardous Atmospheres and Confined Spaces

Employees shall not be permitted to work in hazardous and/or toxic atmospheres. Such atmospheres include those with:

- Less than 19.5% or more than 23.5% oxygen;
- A combustible gas concentration greater than 20% of the lower flammable limit; and
- Concentrations of hazardous substances that exceed those specified in the *Threshold Limit Values for Airborne Contaminants* established by the ACGIH (American Conference of Governmental Industrial Hygienists).

All operations involving such atmospheres must be conducted in accordance with OSHA requirements for occupational health and environmental controls (see 29 CFR Part 1926 Subpart D) for personal protective equipment and for lifesaving equipment (see 29 CFR Part 1926 Subpart E). Engineering controls (e.g., ventilation) and respiratory protection may be required.

When testing for atmospheric contaminants, the following should be considered:

- Testing should be conducted before employees enter the trench and should be done regularly to ensure that the trench remains safe.
- The frequency of testing should be increased if equipment is operating in the trench.
- Testing frequency should also be increased if welding, cutting, or burning is done in the trench.

Employees required to wear respiratory protection must be trained, fit-tested, and enrolled in a respiratory protection program. Some trenches qualify as confined spaces. When this occurs, compliance with the Confined Space Standard is also required.

Emergency Rescue Equipment

Emergency rescue equipment is required when a hazardous atmosphere exists or can reasonably be expected to exist. Requirements are as follows:

- Respirators must be of the type suitable for the exposure. Employees must be trained in their use and a respirator program must be instituted.
- Attended (at all times) lifelines must be provided when employees enter bell-bottom pier holes, deep confined spaces, or other similar hazards.
- Employees who enter confined spaces must be trained.

Standing Water and Water Accumulation

Methods for controlling standing water and water accumulation must be provided and should consist of the following if employees are permitted to work in the excavation:

- Use of special support or shield systems approved by a registered professional engineer.
- Water removal equipment, i.e. well pointing, used and monitored by a competent person.
- Safety harnesses and lifelines used in conformance with 29 CFR 1926.104.
- Surface water diverted away from the trench.

- Employees removed from the trench during rainstorms.
- Trenches carefully inspected by a competent person after each rain and before employees are permitted to re-enter the trench.

Inspections

Inspections shall be made by a competent person and should be documented. The following guide specifies the frequency and conditions requiring inspections:

- Daily and before the start of each shift;
- As dictated by the work being done in the trench;
- After every rainstorm;
- After other events that could increase hazards, e.g. snowstorm, windstorm, thaw, earthquake, etc.;
- When fissures, tension cracks, sloughing, undercutting, water seepage, bulging at the bottom, or other similar conditions occur;
- When there is a change in the size, location, or placement of the spoil pile; and
- When there is any indication of change or movement in adjacent structures.

Bibliography

29 CFR 1926, Subpart P. Excavations.

Construction Safety Association of Ontario. *Trenching Safety*. 74 Victoria St., Toronto, Ontario, Canada M5C2A5.

International Labour Office (ILO). *Building Work: A Compendium of Occupational Safety and Health Practice*. International Occupational Safety and Health Information Centre (CIS): ILO, Geneva, Switzerland.

National Safety Council. *Accident Prevention Manual for Industrial Operations, Engineering and Technology*, 9th ed., Chicago, IL: National Safety Council.

National Safety Council. Protecting Worker's Lives: A Safety and Health Guide for Unions. Chicago, IL: National Safety Council.

National Safety Council. Industrial Data Sheets: I-482, *General Excavation*, and I-254, Trench Excavation, Chicago, IL: National Safety Council.

National Utility Contractors Association, *Competent Person Manual*-1991.

NBS/NIOSH, *Development of Draft Construction Safety Standards for Excavations*, Volume I (April 1983) NIOSH 83-103, Pub. No. 84-100-569. Volume II, April 1983. NIOSH 83-2693, Pub. No. 83-233-353.

A.J. Scardino Jr., *Hazard Identification and Control–Trench Excavation*. Carlton Press, Lagrange, TX, 1993.

Site Assessment Questions

During first and subsequent visits to a construction or facility maintenance location, the compliance officer (or the site's safety officer or other competent person) may find the following questions useful.

1. Is the cut, cavity, or depression a *trench* or an *excavation*?
2. Is the cut, cavity, or depression more than 4 ft (1.2 m) in *depth*?
3. Is there *water* in the cut, cavity, or depression?
4. Are there adequate means of *access* and *egress*?
5. Are there any *surface encumbrances*?
6. Is there exposure to *vehicular traffic*?
7. Are *adjacent structures stabilized*?
8. Does *mobile equipment* have a *warning system*?
9. Is a *competent person in charge* of the operation?
10. Is *equipment operating* in or around the cut, cavity, or depression?
11. Are procedures required to monitor, test, and *control hazardous atmospheres*?
12. Does a competent person *determine soil type*?
13. Was a *soil testing device* used to determine soil type?
14. Is the *spoil* placed 2 ft (0.6 m) or *more from the edge* of the cut, cavity, or depression?
15. Is the *depth* 20 ft (6.1 m) *or more* for the cut, cavity, or depression?
16. Has a *registered professional engineer* approved the procedure if the depth is more than 20 ft (6.1 m)?
17. Does the procedure require *benching* or *multiple benching*? *Shoring*? *Shielding*?
18. If provided, *do shields extend at least 18 in (0.5 m)* above the surrounding area if it is sloped toward the excavation?
19. If shields are used, is the depth of the cut *more than 2 ft (0.6 m) below the bottom of the shield*?
20. Are any required *surface crossings* of the cut, cavity, or depression the *proper width and fitted with hand rails*?
21. Are means of *egress* from the cut, cavity, or depression *no more than 25 ft (7.6m) from the work*?
22. Is *emergency rescue equipment* required?
23. Is there *documentation of the minimum daily excavation inspection*?

Definitions

Accepted Engineering Practices are procedures compatible with the standards of practice required of a registered professional engineer.

Adjacent Structures Stability refers to the stability of the foundation(s) of adjacent structures whose location may create surcharges, changes in soil conditions, or other disruptions that have the potential to extend into the failure zone of the excavation or trench.

Competent Person is an individual who is capable of identifying existing and predictable hazards or working conditions that are hazardous, unsanitary, or dangerous to employees, and who *has authorization to take prompt corrective measures to eliminate or control these hazards and conditions.*

Confined Space is a space that, by design and/or configuration, has limited openings for entry and exit, unfavorable natural ventilation, may contain or produce hazardous substances, and is not intended for continuous employee occupancy.

Excavation. An **Excavation** is any man-made cut, cavity, trench, or depression in an earth surface that is formed by earth removal. A **Trench** is a narrow excavation (in relation to its length) made below the surface of the ground. In general, the depth of a trench is greater than its width, and the width (measured at the bottom) is not greater than 15 ft (4.6 m). If a form or other structure installed or constructed in an excavation reduces the distance between the form and the side of the excavation to 15 ft (4.6 m) or less (measured at the bottom of the excavation), the excavation is also considered to be a trench.

Hazardous Atmosphere is an atmosphere that by reason of being explosive, flammable, poisonous, corrosive, oxidizing, irritating, oxygen-deficient, toxic, or otherwise harmful may cause death, illness, or injury to persons exposed to it.

Ingress and Egress mean "entry" and "exit," respectively. In trenching and excavation operations, they refer to the provision of safe means for employees to enter or exit an excavation or trench.

Protective System refers to a method of protecting employees from cave-ins, from material that could fall or roll from an excavation face or into an excavation, and from the collapse of adjacent structures. Protective systems include support systems, sloping and benching systems, shield systems, and other systems that provide the necessary protection.

Registered Professional Engineer is a person who is registered as a professional engineer in the state where the work is to be performed. However, a professional engineer who is registered in any state is deemed to be a "registered professional engineer" within the meaning of Subpart P when approving designs for "manufactured protective systems" or "tabulated data" to be used in interstate commerce.

Support System refers to structures such as underpinning, bracing, and shoring that provide support to an adjacent structure or underground installation or to the sides of an excavation or trench.

Subsurface Encumbrances include underground utilities, foundations, streams, water tables, transformer vaults, and geological anomalies.

Surcharge means an excessive vertical load or weight caused by spoil, overburden, vehicles, equipment, or activities that may affect trench stability.

Tabulated Data are tables and charts approved by a registered professional engineer and used to design and construct a protective system.

Underground Installations include, but are not limited to, utilities (sewer, telephone, fuel, electric, water, and other product lines), tunnels, shafts, vaults, foundations, and other underground fixtures or equipment that may be encountered during excavation or trenching work.

Unconfined Compressive Strength is the load per unit area at which soil will fail in compression. This measure can be determined by laboratory testing, or it can be estimated in the field using a pocket penetrometer, by thumb penetration tests, or by other methods.

Definitions That Are No Longer Applicable. For a variety of reasons, several terms commonly used in the past are no longer used in revised Subpart P. These include the following:

1. **Angle of Repose.** Conflicting and inconsistent definitions have led to confusion as to the meaning of this phrase. This term has been replaced by **Maximum Allowable Slope.**
2. **Bank, Sheet Pile, and Walls.** Previous definitions were unclear or were used inconsistently in the former standard.
3. **Hard Compact Soil** and **Unstable Soil.** The new soil classification system in revised Subpart P uses different terms for these soil types.

OSHA trench box requirements

OSHA 29 CFR 1926.650,29 CFR 1926.651, and 29 CFR 1926.652, Subpart P, Excavations, states that the greatest construction site risk is excavating. To reduce the probability of injury and potential for death, contractors are required by the Occupational Safety and Health Administration, to follow OSHA trench box requirements, to prevent soil from collapsing during trenching and excavation operations.

Trench boxes offer a safe, OSHA compliant work environment to expedite cast iron pipe, conduit, fiberglass tank, and reinforced concrete pipe installation in trenches, restricted right-of-ways, and unstable soil areas. The boxes don't prevent cave-ins but decrease the probability of a soil collapse trapping a worker. Trench boxes fortify a trench against unstable soil resulting from tension cracks, sliding, bulging and heaving.

The Occupational Safety and Health Administration requires that a trench box must be used for all trenching or excavations five feet or greater in-depth. Trench boxes must extend eighteen inches above the surface and be no more than two feet from the bottom of the trench. Should the trench depth exceed twenty feet, an OSHA trench box requirements stipulate a certified, professional engineer must design the trench box.

Prior to a contractor excavating a site, the contractor must inspect and classify the soil using visual and manual test procedures. Once the tested soil content is verified, the contractor can use contemporary shoring equipment or approved sloping methods to provide additional support for the trench box.

There are four classifications of soil:

Type A Soils: standard, silty, sandy clay and clay loam with a 1.5 tons per square foot (tsf) (144 kPa) or greater of compressive strength.

Type B Soils: angular gravel, silt loam with 0.5 tsf (48 kPa) but less than 1.5 tsf (144 kPa) of compressive strength.

Type C Soils: gravel, sand and loamy sand, submerged soil, water seeping soil, with 0.5 tsf (48 kPa) or less of compressive strength.

Layered Geological Strata: soils are configured in different layers

Another one of OSHA trench box requirements is that the contractor must provide safe access and egress with the trench box. Ladders, steps, and ramps are to be situated within twenty-five feet of workers to allow a quick exit should the soil collapse. Workers may not remain in the trench box when it's being installed, taken out, nor moved vertically.

The Excavation standards do not require a protective system when an excavation is made entirely in stable rock or when an excavation is less than 5 feet (1.52 meters) deep and a competent person has examined the ground and found no indication of a potential cave-in.

The following safe work practices will help keep trenches safe for workers:

- PREPARE a safe trench.
  - Provide safe entry and exit before starting work.
  - Keep materials at least 2 feet away from the edge.
- PROTECT workers from a cave-in by using protective systems.
  - Sloping or benching trench walls, or
  - Shoring trench wall with supports, or
  - Shielding trench walls with trench boxes.
- INSPECT the trench for hazards.
  - Look for standing water and other environmental hazards.
  - Never enter a trench unless it has been inspected and approved by the competent person.

Employers should also ensure there is a safe way to enter and exit the trench. Keep materials away from the edge of the trench. Look for standing water or atmospheric hazards. Never enter a trench unless it has been properly inspected.

Figure 1.

Figure 2.

LAYING AND LOWERING

How to determine the degrees of bend in a pipe that must fit a ditch calling for a bend in both horizontal and vertical planes

Rule. To find the number of degrees in the combination bend, square the side bend and the sag or overbend; add them together and extract the square root. The answer will be the number of degrees necessary to make the pipe fit the ditch.

Example. Determine the bend to make in a pipe whose ditch has a 3° overbend and a 4° side bend. Let X = the unknown angle:

$$X^2 = 3^2 + 4^2$$
$$X^2 = 9 + 16$$
$$X^2 = 25$$
$$X = 5^\circ$$

Example. Determine the bend to make in a pipe whose ditch has a 9° sag and a 12° side bend. Let X = the unknown angle:

$$X^2 = 9^2 + 12^2$$
$$X^2 = 225$$
$$X = 15^\circ$$

How to bend a pipe to fit a ditch

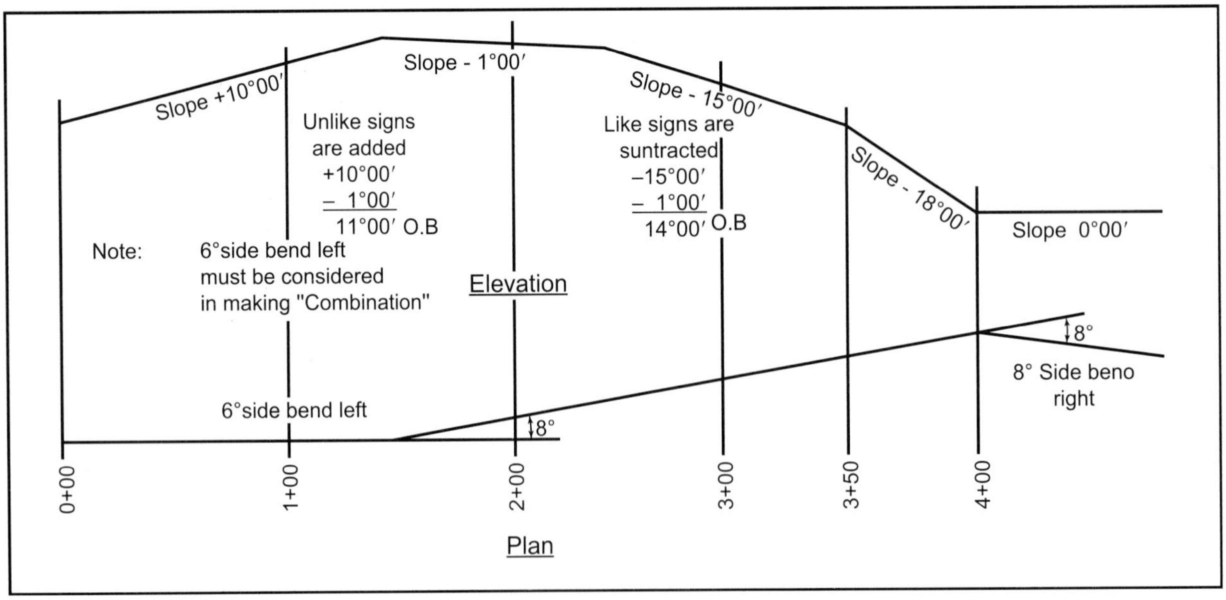

To make straight sags or overbends fit the ditch (see drawing), add angles of unlike signs and subtract those of like signs.

Example. (Sta. 2 + 40)

Slope − 1500′
Slope − 100′
(sub.) 1400′ overbend

In cases involving either a sag or overbend, in addition to a side bend, the rule becomes: *Make the combination bend equal to the largest angle plus $^1/_3$ of the smallest.*

Example. (Sta. 1 + 40)

Slope + 1000′
Slope − 100′
(add) 1100′ overbend(1)
600′ overbend(2)

Combination bend $= 11^\circ + (^1/_3 \times 6^\circ) = 13^\circ 00'$ overbend left.
Note: This rule gives an error of approximately 1° for a maximum bend of 18.5°.

Pipe bending computations made with a calculator

Cold bending pipe is subject to provisions of Liquid Petroleum Transportation Piping Systems, B31.4, and Gas Transmission and Distribution Piping Systems, B31.8. These codes stipulate a minimum bending radius for each size pipe in addition to requirements for thinning, flattening, etc.

The minimum bending radius for cold field bends is specified in ASME B31.4 (Liquid Lines) and ASME B31.8 (Gas Pipelines) ASME B31.4.

Nominal Pipe Size	Minimum Radius in Pipe Diameters
≤12	18 D
14	21
16	24
18	27
≥20	30

Nominal Pipe Size	Deflection of Longitudinal Axis in Degrees	Minimum Radius in Pipe Diameters
≤12	Refer to 841.2.3 (a)(4) in ASME B31.8	18 D
12	3.2	18
14	2.7	21
16	2.4	24
18	2.1	27
≥20	1.9	30

A 6 ft tangent is preferred; however, some bending machines may accommodate a shorter tangent.

The following equations address the geometry of fabricating offset bends, sloping scraper traps, connections, and direction changes common to all pipelines. Minimum code radii are calculated for each size of oil or gas line. The vertical distance between below- and above-ground center lines is matched to a design distance with given tangents (5 ft is a convenient tangent for pipe bending machines). The bend angle, overall horizontal distance, and total length of pipe can be calculated for each configuration.

(The code states "longitudinal axis of the pipe shall not be deflected more than 1½ degrees in any length along the pipe axis equal to the diameter of the pipe.")

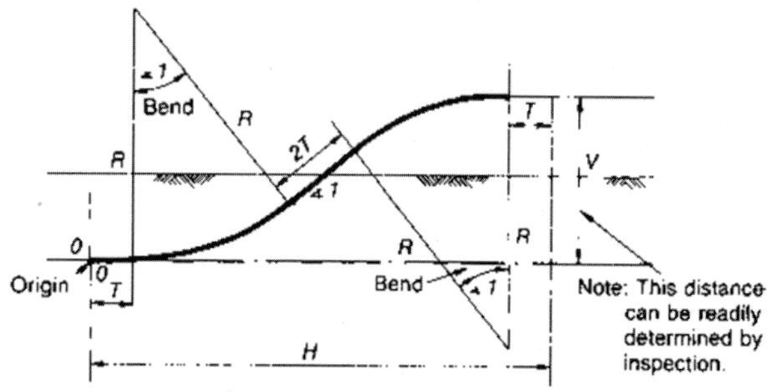

Figure 1. Typical offset bend.

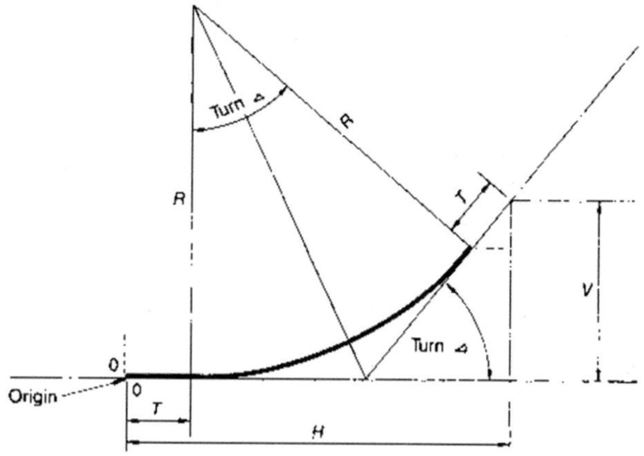

Figure 2. A conventional direction change.

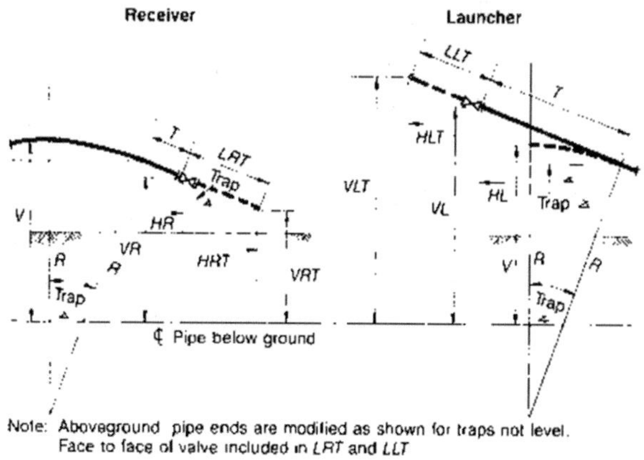

Figure 3. Receiver and launcher scraper trap connections, where the traps are not level.

Formulas for reverse bends

$$V = 2R(1 - \cos\measuredangle 1) + 2T\sin\measuredangle 1$$

$$H = 2R\sin\measuredangle 1 + 2T\cos\measuredangle 1 + 2T$$

$$L = 2R\frac{\measuredangle 1}{57.2958} + 4T$$

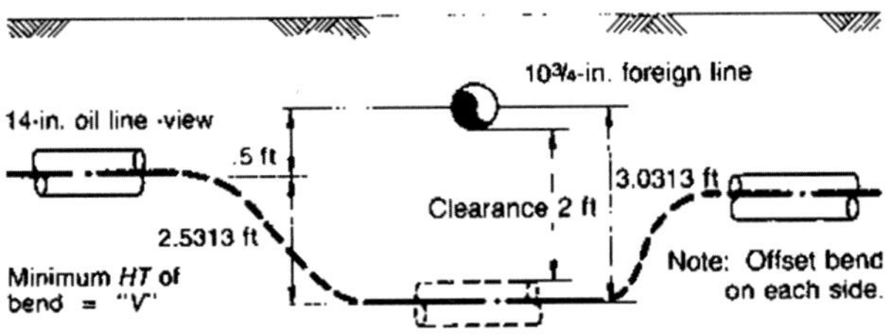

Figure 4. A crossing under a foreign line, where offset bends are used.

Formulas for scraper trap connections

Receiver.

$$VR = V - R(1 - \cos\text{Trap}\measuredangle)T\sin\text{Trap}\measuredangle$$
$$HR = H - T + R\sin\text{Trap}\measuredangle + T\cos\text{Trap}\measuredangle$$
$$LL = L - R\frac{\text{Trap}\measuredangle}{57.2958}$$

Launcher.

$$VL = V - R(1 - \cos\text{Trap}\measuredangle)T\sin\text{Trap}\measuredangle$$
$$HL = H - T + R\sin\text{Trap}\measuredangle + T\cos\text{Trap}\measuredangle$$
$$LL = L - R\frac{\text{Trap}\measuredangle}{57.2958}$$

Receiver trap.

$$VRT = VR - LRT\sin\text{Trap}\measuredangle$$
$$HRT = HR + LRT\cos\text{Trap}\measuredangle$$

Launcher trap.

$$VLT = VR - LLT\sin\text{Trap}\measuredangle$$
$$HLT = HL + LLT\cos\text{Trap}\measuredangle$$

Formulas for direction change

$$V = R(1 - \cos\text{Turn}\measuredangle) + T\sin\text{Turn}\measuredangle$$
$$H = R(\sin\text{Turn}\measuredangle) + T(1 + \cos\text{Turn}\measuredangle)$$
$$L = 2T + R\frac{\text{Turn}\measuredangle}{57.2958}$$

Assume a trial angle, say 20° for a small pipe. Let $T = 0.5$ ft.

For noncode work: user may employ a radius that is suitable in his judgment. Caution: Be aware that short radii for large pipes, say 8 in and larger, soon get into a hot bend category.

Example 1. Offset bend

$12^3/_4$-in oil line. Let $V = 8$ ft, $T = 5$ ft. Smaller diameters have larger bend $\measuredangle$. Try 25.4°. Minimum radius for 12″ is 18 × 12.750″ = 19.125 ft. sin 25.4 = 0.428935 cos 25.4 = 0.903335 $V = 2R(1 - \cos\measuredangle 1) + 2T\sin\measuredangle 1$ $V = 2 \times 19.125\,(1 - 0.903335) + 2 \times 5 \times 0.428935$ $V = 7.9868$ ft.

Example 2. Direction change

24-in gas line, say 30° turn. Let $R = 80$ ft, $T = 5$ ft. $V = R(1 - \cos\text{Turn}\measuredangle) + T\sin\text{Turn}\measuredangle$ cos 30 = 0.866025, sin 30 = 0.5000 $V = 60(1 - 0.866025) + 5 \times 0.5000$ $V = 10.53850$ ft.

Example 3. Crossing

A 14-in oil line crosses under $10^3/_4$ in foreign line. Take 2 ft, 0 in min. clearance. Center distance = 3.0313 ft. Min. height of bend = 2.513 ft = "V". Take 10° as trial $\measuredangle$, $T = 5$ ft. $V = 2R(1 - \cos\measuredangle 1) + 2T\sin\measuredangle 1$ sin 10° = 0.173648, cos 10° = 0.984808 $V = 2 \times 24.5\,(1 - 0.984808) + 2 \times 5 \times 0.173648$ $V = 2.4809$ ft. Increment angle to 10.1° sin 10.1° = 0.175367, cos 10.1° = 0.984503 $V = 2 \times 24.5(1 - 0.984503) + 2 \times 5 \times 0.175367$ $V = 2.513$ ft.

Note: This is a possible solution. However, this may be an alternative solution. Builders will take advantage of sag bends where possible, depending on site conditions, space available, pipe size, etc.

Calculate maximum bend on cold pipe

How much of a bend can a cold straight pipe take without overstressing? In other words, in laying straight pipe cross-country, what is the smallest radius of curvature possible within the allowable stress?

The following is an analytical solution to these questions.

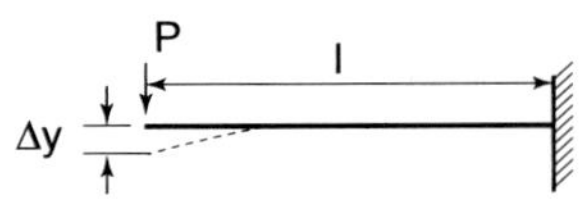

The deflection equation for a length of straight pipe used as a cantilever beam with a concentrated end load is

$$\Delta y \frac{pl^3}{3EI} \tag{1}$$

The maximum stress in the beam is

$$s = \frac{M}{Z} \tag{2}$$

The maximum bending moment is

$$M = Pl \tag{3}$$

If the allowable stress is S_A, rewriting Equation (2) gives

$$S_A = \frac{M}{Z} \tag{4}$$

$$S_A = \frac{Pl}{Z} \tag{5}$$

$$Pl = ZS_A \tag{6}$$

Rewriting Equation (1):

$$\Delta y = \frac{(Pl)l^2}{3EI} = \frac{ZS_A l^2}{3EI} \tag{7}$$

Taking the first and second derivatives of Equation (7):

$$\frac{d\Delta y}{dl} = \frac{2lZS_A}{3EI} \tag{8}$$

$$\frac{d^2\Delta y}{dl^2} = \frac{2ZS_A}{3EI} \tag{9}$$

The radius of curvature is given by the equation

$$R = \frac{\left[1 - (d\Delta y/dl)^2\right]^{3/2}}{d^2\Delta y/dl^2} \tag{10}$$

Substituting in Equation (10):

$$R = \frac{\left[1 + (2lZS_A/3EI)^2\right]^{3/2}}{2ZS_A/3EI} \tag{11}$$

$$\lim_{1\to 0} R = \frac{1^{8/2}}{2ZS_A/3EI} = \frac{3EI}{2ZS_A} \tag{12}$$

Notation

Δy	Deflection, inches
l	Length, inches
P	Load, pounds
E	Modulus of elasticity, pounds per square inch
I	Cross-sectional moment of inertia, inch^4
S	Stress, pounds per square inch
S_A	Allowable stress, pounds per square inch from Code for Pressure Piping
M	Bending moment, inch-pounds
Z	Section modulus, inch^3
R	Radius of curvature, inches

Example. What is the smallest radius of curvature that can be used without overstressing straight 24-in Schedule 20 seamless steel pipe ASTM 53? The pipe is at 70 °F and is to be used for oil outside of refinery limits.

$$R = \frac{3EI}{2ZS_A}$$

$E = 29 \times 10^6$ lb/in^2
$I = 1943$ in^4
$Z = 162$ in^3
$S_A = 25{,}500$ psi (from Code for Pressure Piping)

$$R = \frac{3 \times 20 \times 10^6 \times 1.943}{2 \times 162 \times 25,500} = 20,460 \text{ in or } 1705 \text{ ft}$$

The section modulus $= Z = 2I/D$, and may be simplified to $Z = I/r$ and substituted for Z in Equation (12).

I = Moment of inertia
r = Distance from center pipe
D = Outside diameter of pipe

Equation (12) then becomes

$$R = 3Er/2S_A \tag{13}$$

$$R = \left(21.75 \times 10^6 / S_A\right) \times D \tag{14}$$

S_A, the allowable stress, is determined by the design factor permitted by the design code used. Equation (14) may be further simplified by using the following table.

Example. What is the minimum radius of curvature that can be used without overstressing 42″ API 5LX X-65 line pipe at 72°F used in type C construction (gas line)? The design factor from ANSI B31.8 is 0.5; therefore, use the column headed $F = 0.5$.

$R = 669 \times 42'' = 28.098''$ or 2.342 ft

For liquid lines designed in accordance with ANSI B31.4, use the column headed $F = 0.72$.

Minimum Radius of Curvature, *R*

Yield Stress (psi)	*F* = 0.72	*F* = 0.6	*F* = 0.5	*F* = 0.4
35,000	863 D	1036 D	1243 D	1553 D
42,000	719 D	863 D	1036 D	1295 D
46,000	657 D	788 D	946 D	1182 D
52,000	581 D	697 D	836 D	1046 D
56,000	539 D	647 D	777 D	971 D
60,000	503 D	604 D	725 D	906 D
65,000	465 D	558 D	669 D	836 D
70,000	431 D	518 D	621 D	776 D
80,000	377 D	453 D	544 D	680 D

Determine length of a pipe bend

For smooth bends (other than wrinkle bends) use the following formula:

$$L = R \times D \times 0.017 + T$$

where

L = Length of bend
R = Radius of bend (center line of pipe)
D = Number of degrees in the bend
T = Total length of the two tangents at either end

Example. Find the length of a piece of pipe required to make a 90° bend with a radius of 60 in. There should be a 4-in tangent at either end.

$L = ?$
$R = 60$
$D = 90$
$T = 8$

$$L = 60 \times 90 \times 0.01745 + 8$$
$$= 94.23 + 8$$
$$= 102.23 \text{ in}$$

Length of pipe in arc subtended by any angle

Find the length of pipe subtended by an angle of 42°35′ with a radius of 24 in.

Assuming $R = 1$ in
Angle of 42° subtends an arc = 0.7330 in
Angle of 35′ subtends an arc = 0.01018 in
Angle of 42°35′ subtends an arc = 0.7432 in

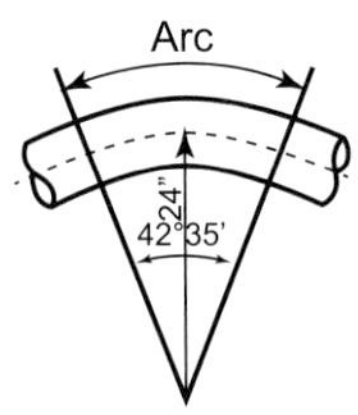

The arc subtended for any angle is directly proportional to the radius:

$$\frac{1 \text{ in}}{24 \text{ in}} = \frac{0.7432 \text{ in}}{\text{Arc}}$$

$$\text{Arc} = 0.7432 \times 24 = 17^{13}/_{36} \text{ in.}$$

Degrees						Minutes			
Degrees	**Length Of Arc**	**Degrees**	**Length Of Arc**	**Degrees**	**Length Of Arc**	**Minutes**	**Length Of Arc**	**Minutes**	**Length Of Arc**
1	.0175	31	.5411	61	1.0647	1	.000291	31	.009018
2	.0349	32	.5585	62	1.0821	2	.000582	32	.009308
3	.0524	33	.5760	63	1.0996	3	.000873	33	.009599
4	.0698	34	.5934	64	1.1170	4	.001164	34	.009890
5	.0873	35	.6109	65	1.1345	5	.001454	35	.010181
6	.1047	36	.6283	66	1.1519	6	.001745	36	.010472
7	.1222	37	.6458	67	1.1694	7	.002036	37	.010763
8	.1396	38	.6632	68	1.1868	8	.002327	38	.011054
9	.1571	39	.6807	69	1.2043	9	.002618	39	.011345
10	.1745	40	.6981	70	1.2217	10	.002909	40	.011636
11	.1920	41	.7156	71	1.2392	11	.003200	41	.011926
12	.2094	42	.7330	72	1.2566	12	.003491	42	.012217
13	.2269	43	.7505	73	1.2741	13	.003782	43	.012508
14	.2443	44	.7679	74	1.2915	14	.004072	44	.012799
15	.2618	45	.7854	75	1.3090	15	.004363	45	.013090
16	.2793	46	.8029	76	1.3265	16	.004654	46	.013381
17	.2967	47	.8203	77	1.3439	17	.004945	47	.013672
18	.3142	48	.8378	78	1.3614	18	.005236	48	.013963
19	.3316	49	.8552	79	1.3788	19	.005527	49	.014254
20	.3491	50	.8727	80	1.3963	20	.005818	50	.014544
21	.3665	51	.8901	81	1.4137	21	.006109	51	.014835
22	.3840	52	.9076	82	1.4312	22	.006400	52	.015126
23	.4014	53	.9250	83	1.4486	23	.006690	53	.015417
24	.4189	54	.9425	84	1.4661	24	.006981	54	.015708
25	.4363	55	.9599	85	1.4835	25	.007272	55	.015999
26	.4538	56	.9774	86	1.5010	26	.007563	56	.016290
27	.4712	57	.9948	87	1.5184	27	.007854	57	.016581
28	.4887	58	1.0123	88	1.5359	28	.008145	58	.016872
29	.5061	59	1.0297	89	1.5533	29	.008436	59	.017162
30	.5236	60	1.0472	90	1.5708	30	.008727	60	.017453

Linear feet of pipelay per 10 h day in trench in level, rock-free terrain

Linear Feet of Pipelay Per 10 h Day In Trench in Level, Rock-Free Terrain						
Nominal Pipe Size	**Pipe Wall Thickness in Inches**					
	0.000–0.250	**0.251–0.375**	**0.376–0.500**	**0.501–0.625**	**0.626–0.750**	**0.756–1.000**
4	10,800	10,500	10,300	-	-	-
6	9600	9400	9210	-	-	-
8	9120	8900	8720	8550	-	-
10	8900	8760	8580	8410	-	-
12	8600	8400	8230	8070	7910	-
14	8400	8230	8160	8000	7840	-
16	8000	7840	7680	7530	7380	7230

Linear Feet of Pipelay Per 10 h Day In Trench in Level, Rock-Free Terrain						
Nominal Pipe Size	Pipe Wall Thickness in Inches					
	0.000–0.250	0.251–0.375	0.376–0.500	0.501–0.625	0.626–0.750	0.756–1.000
18	7800	7640	7500	7350	7200	7050
20	7400	7250	7100	6960	6820	6680
24	-	6600	6470	6340	6210	6080
30	-	6000	5880	5760	5640	5530
36	-	-	5400	5300	5200	5090
42	-	-	4800	4700	4600	4500

Productivity will vary where different types of terrain or rock are encountered. The above footage is based on installing double random joints of pipe.

Linear feet of pipelay per 10-hour day on waist-high supports on level ground

Linear Feet of Pipelay Per 10-hour Day on Waist-High Supports on Level Ground						
Nominal Pipe Size	Pipe Wall Thickness in Inches					
	0.000-0.250	0.251-0.375	0.376-0.500	0.501-0.625	0.626-0.750	0.756-1.000
4	11,320	11,100	10,880	-	-	-
6	10,080	9880	9680	-	-	-
8	9560	9370	9190	9010	-	-
10	9390	9200	9020	8840	-	-
12	8800	8620	8500	8330	8160	-
14	8700	8570	8400	8250	8090	-
16	8040	7880	7730	7580	7430	7280
18	7800	7650	7500	7350	7200	7060
20	7550	7400	7250	7100	6960	6820
24	-	6720	6590	6460	6330	6200
30	-	6120	6000	5880	5770	5660
36	-	-	5440	5350	5240	5140
42	-	-	4840	4740	4650	4560

Productivity will vary with contour of terrain and accessibility. Above footage is based on installing double random joints of pipe.

Allowable pipe span between supports

When it is desired to run pipelines on supports above the ground, the engineer must often calculate the allowable pipe span between supports from a structural viewpoint. The formula for the span length can be obtained by equating the maximum bending moment caused by the loading to the resisting moment of the pipe. This formula is

$$L\left(\frac{SI}{WC}\right)^{0.5}$$

where

L = Span length, ft
C = Radius to outside of pipe, inches
I = Moment of inertia of pipe, inches^4
S = Maximum allowable stress of material, pounds per inch^2
W = Uniform loading per unit length of pipe, pounds per foot

The above formula applies to pipes greater than 2 in in diameter. Reference 1 gives calculated pipe span data, as well as estimated pipe loading for different pipe diameters, taking into account the fluid load, dead load, and the horizontal component of the wind loading.

The design of the pipe span can be quickened up with Figure 1.

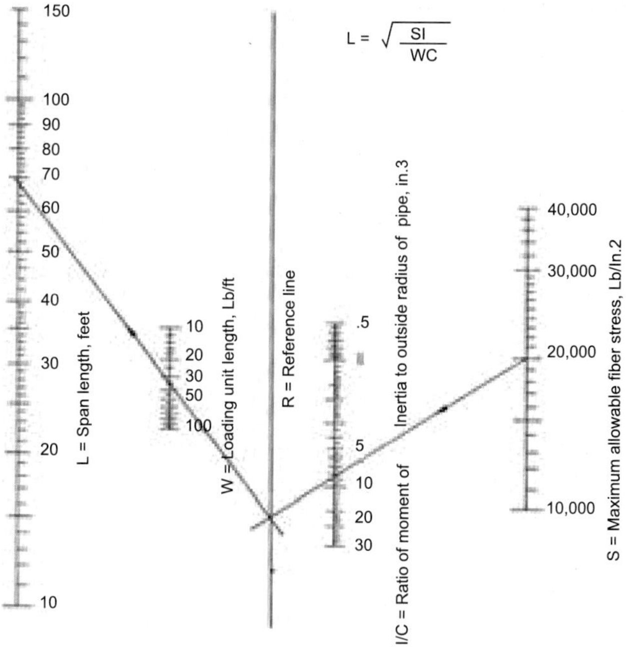

Figure 1. Allowable pipe span between supports.

Example. Find the allowable span length for a 6-in continuous steel pipe strung between supports. The estimated loading (sum of pipe, fluid, dead, and wind loads) is 36 lb/ft. The allowable fiber stress for steel can be taken at 20,000 lb/in^2.

From any suitable handbook, C for a 6-in pipe is 6.625/2 = 3.3175 in, I is 28.14 in.^{4} Now calculate $I/C = 28.14/3.3175 = 8.5$ in.^{3} Connect $S = 20{,}000$ lb/in^2 on the S scale with 8.5 on the I/C scale to the intersection with the reference line. With the reference line as a pivot, connect the pivot point with $W = 36$. At the intersection with the L scale, obtain a pipe span of 68 ft.

Check the amount of deflection to make sure that it does not exceed your piping practices or applicable design code.

Source

J.F. Kuong, *Petroleum Refiner* 39, No. 10, 196 (1960).

Reference

R.J. Sweeney, *Chem. Eng.*, 63, No. 3 199 (1956).

How engineers make pipe fit the ditch

The five-men crews of engineers that keep track of the pipe and calculate side bends, overbends, sags, and combination bends can make money—or lose money—for any contractor crossing rough terrain. If they function smoothly, the bending crew and pipe gang have an easy time and the lowering-in can be a breeze. But when they make mistakes, the welded-up pipe may not fit the ditch, and this spells trouble for the contractor.

Bending crew

As a rule, there are two rod men, two chairmen, and one instrument man in the measurement crew. Their duties are to measure each joint of pipe, number it, and stake its position in the line. In addition, they calculate degrees of bends, etc. Here is how they perform these duties:

Measuring and numbering joints

First, the crew measures all joints of pipe and numbers them from 0 to 100. When 100 is reached, the numbering starts over. Next, they stake the position of each joint of pipe in the line. They use stakes that are numbered according to the pipe, with the number of the joint behind on front of the stake and the joint ahead on back of the stake.

Stakes are set back

Because the setup gang will move the pipe back half a joint's length and the pipe gang back another half, the engineers set the stakes from one to two joints' length behind the corresponding numbered pipe.

In Figure 1, note that joint 25 was measured 39.9 ft, and the gap between stakes 24 and 25 and 25 and 26 was set at 29.9 ft to allow a 10-ft lap for tie-in. This leaves a pup long enough to be carried ahead and worked into the line. If the pups are distributed over sections of bends, there is a possibility of saving a cut in a full joint. If the stakes catch up with the pipe, one joint is carried ahead of the stringing.

Calculating bends

The method used to determine the size of overbend and sags is illustrated in Figure 2. Note that at 0 + 00 the instrument is set up on the ditch bottom at a height of 5 ft. The rod is set on the ditch bottom ahead of the radius of overbend. (Care should be taken to set neither the transit nor the rod on the arc of the change in slope.) The transit is moved behind the radius of overbend at 1 + 00, and the height of the instrument is measured to be 4.90 ft.

The rod is set just ahead of the change in radius of overbend and is read at 4.90 ft, establishing a slope of −1°. The magnitude of bend can now be determined.

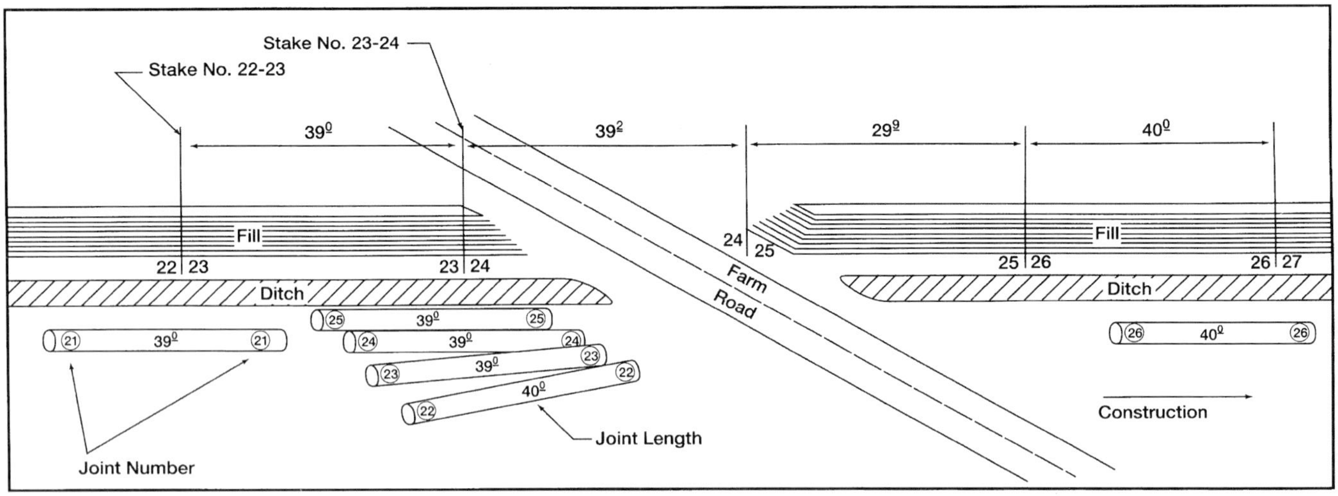

Figure 1.

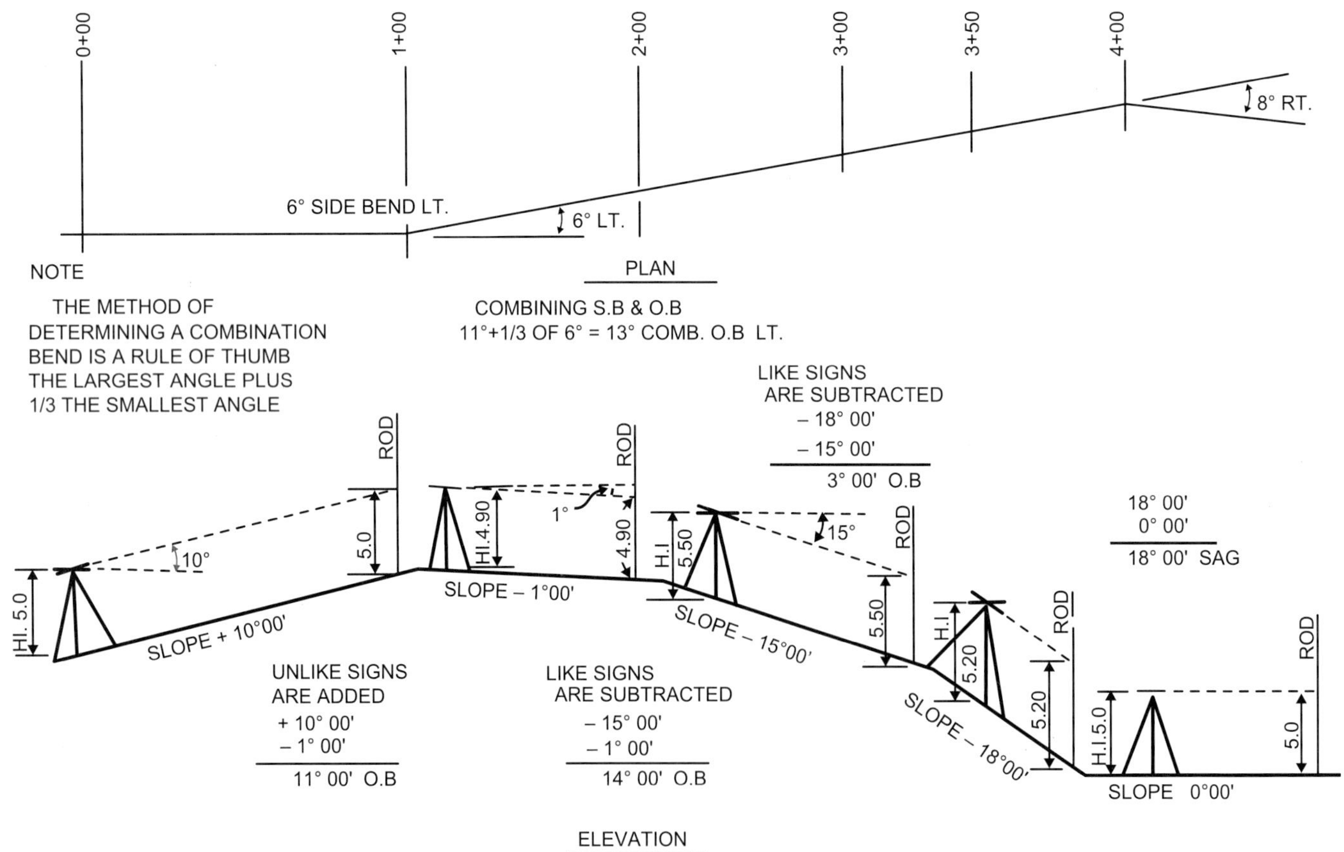

Figure 2.

Add unlike angles

If the slope is positive and the tangent of the slope ahead is decreasing, an overbend is established. If the slope is positive and the tangent ahead is greater than the positive slope, a sag is noted. The reverse holds true for a change in negative slope: if the tangent ahead is increasing, the transition is an overbend; if it is decreasing, the transition is a sag. The unlike algebraic signs are added and like signs are subtracted. Figure 2, station 2 + 00, shows a positive 10° slope approaching the overbend and a negative 1° slope leaving it. Since the signs are unlike, the sum of the two angles is 11°, and since the slope is positive and decreasing, the notation overbend is used.

Side bend angles

Side bends are determined by "eyeballing" a P.I. in the ditch and setting the instrument up over it. A back sight is taken on the ditch or preceding P.I., whichever is more easily read. The scope is flipped and a foresight is taken on the ditch ahead. Referring to Figure 2 at station 1 + 00, the deflection angle is 6° left; station 4 + 00 shows a deflection of 8° right.

Combination bends

At points where side bends and vertical bends occur, the angles are combined to make one angle rolled. The method of calculating is a rule of thumb: *the larger angle plus one-third the smaller angle*.

This rule is accurate enough to give satisfactory results in actual practice. The actual derivation of this rule comes from an approximation of the following:

$$\sin(\text{combination angle}) = \sqrt{\sin^2 A + \sin^2 B}.$$

Angles A and B are the horizontal and vertical angles. Figure 3 shows a plot of the ⅓ rule and of the equation. From the plot it can be seen that for the maximum bend of 18.5, an error of slightly over 1° is introduced. The maximum error evidenced on the graph is due to the side and vertical bends being equal.

Marking pipe

The final step of the measurement crew is marking the joints to be bent. After the bend is established, its position between the pipe stakes is then transferred to the pipe joint number matching the pipe stakes number. The bend size, the centerline of the bend, and arrows indicating the direction of construction are marked on the joint (see Figure 5). If a bend falls too close to the end of the pipe, the joint would have to be moved so as to leave enough room for the bend. A pup could be inserted immediately behind the joint in question to give better bend positioning.

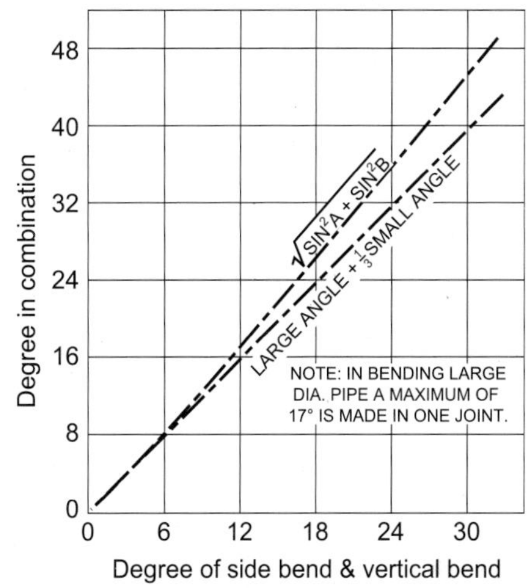

Figure 3.

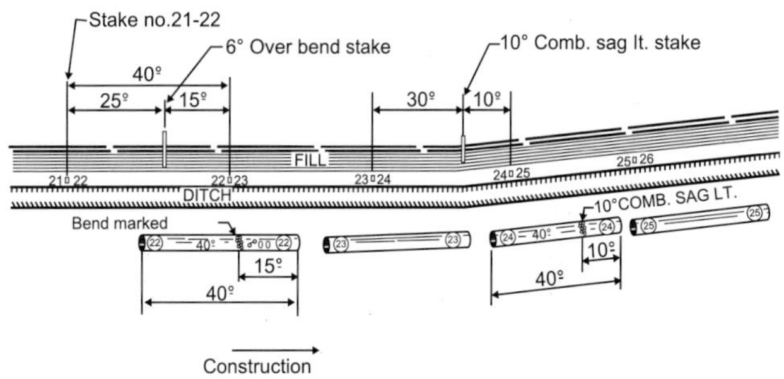

Figure 4.

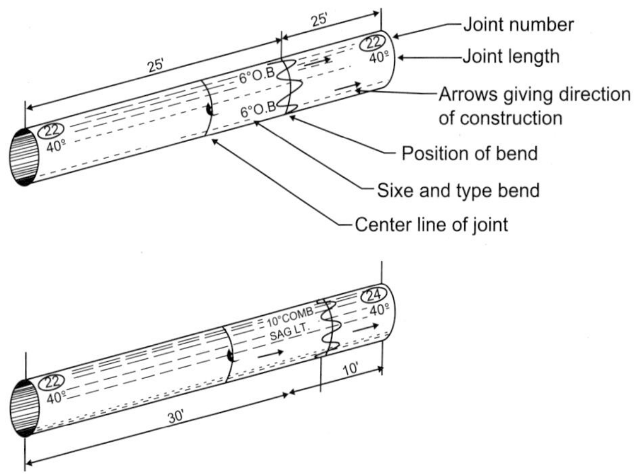

Figure 5.

How to lower an existing pipeline that is still in service

Marshall D. Cromwell, Senior Project Engineer, PT Caltex Pacific Indonesia, Jakarta, Indonesia

Lowering an existing line is a dirty job, but it can have big cost benefits. The line can be lowered while remaining in service with no lost production, and the cost of lowering an existing pipeline section is relatively cheap.

No expenses are incurred for new pipe, valves, stopples, and fittings. Construction is much faster than cutting and relocating a pipeline section.

Construction of new highways, buildings, airport runways, and other facilities is often planned at locations where above-ground pipelines are present. Relocating such lines can be extremely expensive in terms of downtime and new pipeline materials.

Long-delivery items, such as hot tap equipment, valves, and fittings, can add to the pipeline construction time, delaying an already tight schedule for the new facility. The alternative is to lower the existing lines, with adequate protection, so as to eliminate the obstruction.

The following information describes how to calculate an optimum profile for smooth transitions, while keeping the pipe stresses within allowable limits, and how to physically lower the pipeline in the field to match this profile.

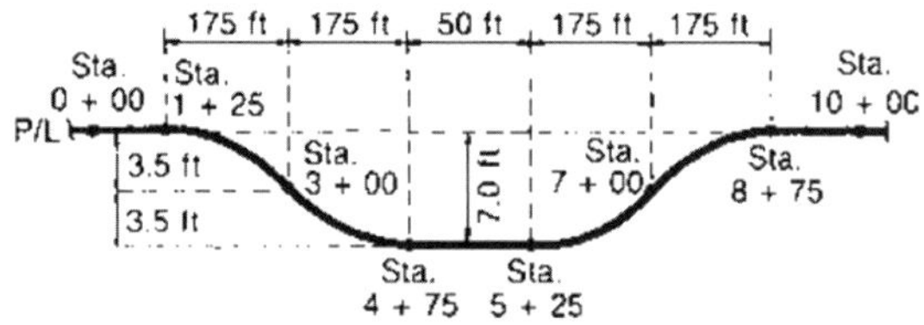

Figure 1. Road crossing profiles.

Table 1
Profile Elevations

Station	X, ft	Deflection, ft	Top of Pipe Existing Elevation	Top of Pipe New Elevation
1 + 25	0	0.00	From field survey	Existing elevation minus deflection
1 + 50	25	0.07		
1 + 75	50	0.29		
2 + 00	75	0.65		
2 + 25	100	1.15		
2 + 50	125	1.80		
2 + 75	150	2.59	(Another column can be added for trench-bottom elevations if desired.)	
3 + 00	175	3.50		
3 + 25	200	4.41		
3 + 50	225	5.20		
3 + 75	250	5.85		
4 + 00	275	6.35		
4 + 25	300	6.71		
4 + 50	325	6.93		
4 + 75	350	7.00		
5 + 00	375	7.00		
5 + 25	400	7.00		
5 + 50	425	6.93		
5 + 75	450	6.71		
etc.	etc.	etc.		

After elevations are received from the field, end-points (Sta 1 + 25 and 8 + 75) should be checked by use of equations (7) and (4) if the pipeline is at the uphill slope.

Precise engineering

There is more to lowering an existing line than just digging a hole and letting gravity take over. Engineering and design of a new profile must be precise so that the line is not dropped too suddenly, causing a buckle or rupture (the larger the pipeline, the more critical the profile).

For instance, in Saudi Arabia, an extensive highway modernization program resulted in scores of large-diameter line pipelines being lowered. A few line sections ruptured during the pioneer stages due to the fact that each situation

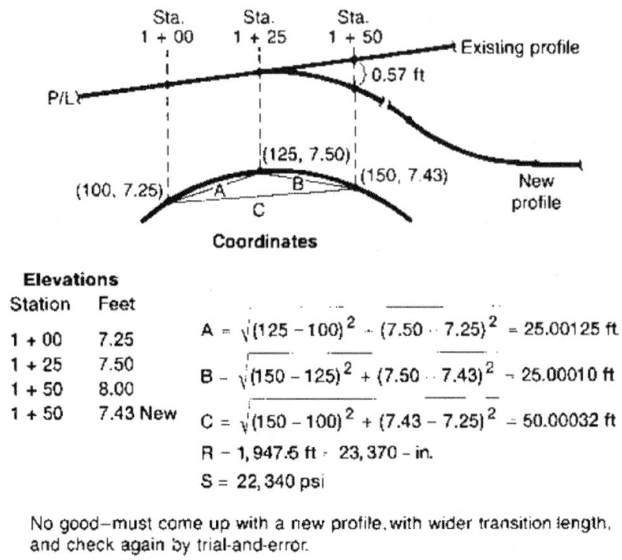

Figure 2. Elevation coordinates.

$$1.\ R = \frac{x^2 + y^2}{2y}$$

$$2.\ x = \sqrt{2Ry - y^2}$$

$$3.\ y = R - \sqrt{R^2 - x^2}$$

$$4.\ R = \frac{EC'}{S} \text{ or } S = \frac{EC'}{R}$$

$$5.\ y = \frac{EC'}{S} - \sqrt{\left(\frac{EC'}{S}\right)^2 - x^2}$$

$$6.\ x = \sqrt{2y\frac{EC'}{S} - y^2}$$

$$7.\ R = \frac{A\ \mathrm{SIN}\left[90 - \mathrm{COS}^{-1}\frac{B^2 + C^2 - A^2}{2BC}\right]}{\mathrm{SIN}\left[2\ \mathrm{COS}^{-1}\frac{B^2 + C^2 - A^2}{2BC}\right]}$$

(R = Feet here)

Note that formulas are independent of w.t., I, and cross-sectional area (cancelled out).

R = Radius of bend (in.)
x = Horizontal distance (in.)
y = Deflection or drop from the horizontal (in.)
S = Bending stress, psi
C' = One half diameter of pipeline (in.)
E = Young's modulus of elasticity (29×10^6 psi for steel)
A, B & C = Dimensions of triangle (ft.)

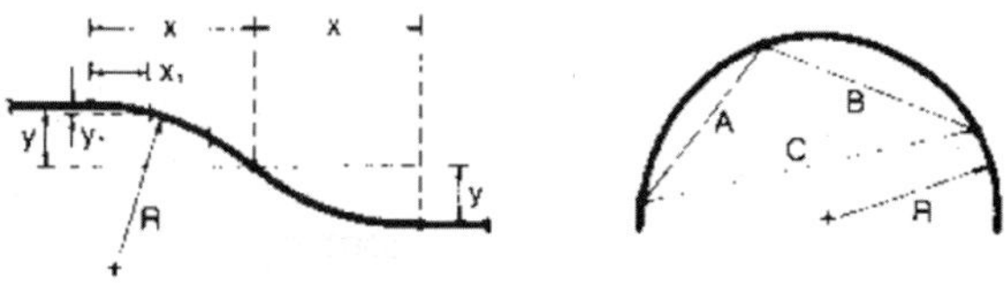

P/L profile

3 points on a circle

• = Inflection points
x_1 = 25', x_2 = 50', etc.

Figure 3. Formulas for changes in slope.

was not properly evaluated in terms of a critical engineering problem.

Excavation begins. Ten-foot soil plugs remained at 50-ft centers to act as temporary pipe supports for the 36-in.

The key to the problem's solution is to provide a smooth profile with constant radius bends and to include a detailed list of instructions with the design package. Careful monitoring by survey equipment (levels) will assure smooth transitions in the field.

The following approach has been used effectively on numerous occasions in Saudi Arabia and Indonesia and can be recommended. Design steps to be included are as follows:

- Determine how much soil cover there should be over the top of the pipeline to assure protection from construction equipment and wheel loads (see API-RP-1102). Four feet of cover is adequate for pipelines 12 in and smaller. Larger pipelines may require additional protection by way of more cover, pipe sleeves, or concrete slabs.
- Determine allowable bending stress on pipeline, based on grade of pipe. Normally, 75% of allowable stress is acceptable for cross-country pipelines (ANSI B31.4, Section 419.6). However, 10,000-psi stress is very conservative and often preferred. If the pipeline is completely restrained (anchors on each end and no expansion loops), thermal stresses must also be considered.
- Once the total required drop of pipeline is established, calculate the minimum bending radius based on allowable bending stress. Then calculate the overall transition length from one end to the other.
- Obtain elevations on top of the pipeline every 25 ft (Table 1), marking station numbers and elevations with paint. If the line is on supports, also include grade elevations at each station.
- Starting at each end of transition, calculate the drop at 25-ft stations. Draw up a profile and fit it on the existing profile. Add stations and new elevations (top of pipe and trench bottom) to the profile drawing.
- Check the transition points at the beginning and end of the new profiles. If the pipeline was going uphill, and suddenly it needs to go downhill, the end points may be critical in terms of stress.

Since any three points not on the same line form an arc of constant radius, the stress may be approximated by Equations (7) and (4) (Figure 3). The end point should become the center of three points for calculation. If overstressed, widen the transition length and check again.

Example 1. A new village was being built in the jungle, but access was obstructed by an above-ground, 36-in pipeline. The village head requested that the line be buried or relocated so that bicycles, becaks, and water buffalo would have easier access.

It was determined that 4 ft of cover was required to protect the line from excessive wheel loads (people-packed buses and trucks). This means the line must be lowered 7 ft (Figure 1). The pipe is Grade B, allowable bending stress is 10,000 psi, and crossing width is 50 ft. To calculate the minimum bending radius, total transition length, and deflections at 25-ft intervals, use:

$$R = EC'/S = (29,000,000)(18)/10,000 = 52,200 \text{ in}$$
$$= 4350 \text{ ft(min)}$$

Total transition length = 4(X) + 50 ft.

$Y = 7.0 \text{ ft}/2 = 3.5 \text{ ft}.$

$$X = \sqrt{2RY - Y^2}.$$

$$X = \sqrt{(2)(4,350)(3.5) - (3.5)^2} = 175 \text{ ft}.$$

Total transition length = 750 ft.

It is suggested that field personnel should start the survey (elevations) 500 ft left of the crossing centerline and stop at station 10 + 00.

Deflections at 25-ft intervals are calculated from Equation (3) (Figure 3), and a table is prepared accordingly (Table 1). Note that calculations for Y are required for 175 ft only. Since the curve is then inverted (Sta 3 + 00 to Sta 4 + 75), results are subtracted from 7 ft, working backwards, to arrive at true deflection or drop. For example, deflection at Sta 4 + 50 = 7.0 − 0.07 = 6.93, etc.

Example 2. Given the following elevations, is pipeline being dropped too suddenly, and will it become overstressed between stations 1 + 00 and 1 + 50 (Figure 2)?

Construction

The design package to the field should include a drawing of new profiles, stations, tables of new pipeline elevations, trench elevations, and a very detailed construction sequence. The following may be used as a guide for construction.

- Starting at one end of transition, dig 40 ft of trench under the line. Leave 10 ft of soil for pipe support. Dig another 40 ft, repeat until the opposite end is reached. When complete, 10-ft soil plugs are supporting the line at 50 ft centers.
- Clean the line between soil supports, take ultrasonic test (UT) readings, and install a sleeve where required. Add primer and tape wrap between supports, maintaining station numbers and elevation points on the line.
- Add 12 in of sand or soil/sand premix to the bottom of the trench and compact to proper trench bottom elevation (profile elevation minus pipeline diameter).
- Install 6-in × 6-in wooden blocks every 50 ft (between soil plugs) to support the line.
- Remove soil plugs and complete line cleaning, UT readings, sleeve installations, tape wrapping, and trench preparation.
- Check elevations of new trench bottom and adjust as required. Check tape wrap with a holiday detector if desired.
- Starting at one end, use six sidebooms to raise approximately 250 ft of line off its supports. Remove 6 in (or less) of blocks on outside supports and 12 in of blocks on all inner supports. Lower line onto supports.

Hand cleaning. The 30-in. and 36-in. crude lines were cleaned by hand to expose corroded areas and prepare for tape wrapping.

Lowering lines. The two lines, full of crude oil, are gradually lowered with the use of six sidebooms.

Gradual lowering. A full sweep of the total transition length is made by lowering each line 6 in. or less at a time. The procedure is repeated until the lines finally reach desired profile elevations.

- Move convoy of sidebooms 250 ft, lift the line, and remove 6 in block at beginning and end supports and 12 in of blocks at inner supports.
- Repeat until total transition length is lowered 12 in, except for several end supports, which are lowered less than 6 in to match required deflections. Then come back and do it again, repeating several times until the pipeline is completely lowered. Note that each repeat becomes shorter. Check final pipeline elevations if desired.
- Check tape wrap for possible damage, add 12 in of sand or premix around pipe, backfill, and compact.

Surprisingly, this operation goes very quickly but must be well coordinated. One person should signal the sideboom operators to lower the pipeline concurrently, so as to eliminate excessive bends and sags. The pipeline is heavy when full of liquid, and if not handled carefully, it could easily rupture. Pressure should be reduced, and extreme caution exercised during the actual lowering process.

Strain calculations

A new profile means the pipeline has been stretched a certain amount, assuming the line was horizontal originally. (If the pipeline was convex, lowering it will put the steel into compression.) Fortunately, steel is very forgiving either way, and the added strain is generally not a problem. It can be checked, however, by use of Young's Modulus of Elasticity (E = stress over strain).

The new pipe length can be calculated from the following formula:

Arc Length $= \pi R\theta/180$, where $\theta = \sin^{-1} X/R$.

A comparison of the new pipe length with the original pipe length will result in a figure for strain as a result of the new profile. Knowing the strain, additional longitudinal stress can be easily calculated.

In Example 1, $\theta = \sin^{-1}$ 175 ft/4350 ft = 2.3056°.

Arc Length (AL) = π(4350 ft)(2.3056°)/180 = 175.047 ft.

Length of new line = 4(AL) + 50 = 750.19 ft.

So the pipeline was actually stretched 0.19 ft or 2.28 in.

Strain = $\Delta L/L$ = (750.19 − 750.00)750.00 = 0.000253.

E = stress/strain, so longitudinal stress = $E \times$ Strain or LS = $29 \times 10^6 \times 0.000253 = 7347$ psi.

This is well within the 35,000-psi yield strength of Grade B pipe.

Reprinted with permission - *Pipe Line Industry*, July 1986.

WELDING

When should steel be preheated before welding?

From the chemistry of the steel, determine the carbon equivalent:

$$\text{Carbon equivalent} = C + \frac{\text{Mn}}{4}.$$

If it exceeds 0.58, the steel may be crack sensitive and should be preheated before welding in ambient temperatures below 40 °F.

Example. If steel pipe having a carbon content of 0.25 and manganese content of 0.70 is to be welded in springtime temperatures, ranging from 40 to 80 °F, is preheat necessary?

$$\text{Carbon equivalent} = 0.25 + \frac{0.70}{4} = 0.25 + 0.175 = 0.45.$$

It is not necessary to preheat this particular steel before welding it.

But for another example:

Carbon = 0.20

Manganese = 1.60

$$\text{Carbon equivalent} = 0.20 + \frac{1.60}{4} = 0.60.$$

This steel should be preheated, particularly for early morning welding.

Why does preheating prevent cracking? It slows the cooling rate and reduces the amount of austenite retained as the weld cools. This prevents microcracking. Other alloying elements and pipe wall thickness may also influence when joints of high strength pipe should be preheated.

Welding and brazing temperatures

Carbon Steel Welding	2700–2790 °F
Stainless Steel Welding	2490–2730 °F
Cast Iron Welding	1920–2500 °F
Copper Welding and Brazing	1980 °F
Brazing Copper-Silicon with Phosphor-Bronze	1850–1900 °F
Brazing Naval Bronze with Manganese Bronze	1600–1700 °F
Silver Solder	1175–1600 °F
Low Temperature Brazing	1175–1530 °F
Soft Solder	200–730 °F
Wrought Iron	2700–2750 °F

Reprinted with permission-Tube Turns, Inc.

Mechanical properties of pipe welding rods

Mechanical properties of electrodes depend upon the chemistry of the pipe, the welding procedures, and the rate of cooling, and because these factors vary between actual applications and testing conditions, the mechanical properties may also vary.

Electrode	AWS Class	Tensile psi	Yield psi	% Elong. in 2″	Tensile psi	Yield psi	% Elong. in 2″
Fleetweld 5P+	E6010	62–86,000[1]	50–75,000[1]	22-28[1]	67–78,000	51-67,000	30-34
Fleetweld 5P	E6010	62–75,000[1]	50–64,000[1]	22-30[1]	60–69,000	46–56,000	28–36
Shield-Arc 85	E7010-A1	70–78,000[1]	57–71.000[1]	22–26[1]	70–83,000	57–72,000	22–28
Shield-Arc 85P	E7010-A1	70–78,000[1]	57–63,000[1]	22–27[1]	70–77,000	57–68,000	22–25
Shield-Arc HYP	E7010-G	70–82,000[1]	60–71.000[1]	22–28[1]	72–81,000	60–72,000	25–29
Shield-Arc 70+	E8010-G	80–86,000[1]	67–74,000[1]	19–29[1]	76–80,000	68–72,000	23–29
Jetweld LH-70	E7018[4]	72–86,000	60–76,000	22–30	65–74,000	55–60,000	24–34
Jetweld LH-75MR	E8018-B2	72–87,000	60–74,000	22–30	70–82,000	70–82,000	29–32
Jetweld LH-90MR	E8018-C1	97–107,000	84–97,000	17–24	80–105,000[2]	67–93,000[2]	19–23[2]
Jet-LH8018-C1MR	E8018-C3	80–95,000	67–81,000	19–25	80–86,000	67–75,000	19–31
JEt-LH8018-C3MR	H8018-C3	80–90,000	68–80,000	24–30	75–84,000	66–73,000	24–32
Jetweld LH-100M1MR	MIL10018-M1	95–101,000	82–91,000	20–27	93–96,000[5]	80–90,000[5]	20–28[5]
Jetweld LH-110MMR	E11018-M	110–123,000	98–109,000	20–24	110–120,000[3]	95–107,000[3]	20–25[3]

[1]Aged 48 h @ 220 °F.
[2]Stress relieved @ 1275 °F.
[3]Stress relieved @ 1025 °F.
[4]Imprinted identification 7018-1 to indicate meeting minimum CVN impact requirement of 20 ft-lbs at 50 °F.
[5]Stress relieved @ 1125 °F.

Reprinted with permission—Lincoln Electric Co.

Lens shade selector

Operation	Shade No.	Operation	Shade No.
Soldering	2		
Torch Brazing	3 or 4		
Oxygen Cutting		Nonferrous Gas Tungsten-Arc Welding	
Up to 1 in	3 or 4	$^1/_{16}$, $^3/_{32}$, $^1/_8$, $^5/_{32}$ in electrodes	11
1–6 in	4 or 5		
6 in and over	5 or 6		
Ferrous Gas Welding		Ferrous Gas Metal-Arc Welding	
Up to $^1/_8$ in	4 or 5	$^1/_{16}$, $^3/_{32}$, $^1/_8$, $^5/_{32}$ in electrodes	12
$^1/_8$% –$^1/_2$ in	5 or 6	Shielded Metal-Arc Welding	
½ in and over	6 or 8	$^3/_{16}$, $^7/_{32}$, ¼ in electrodes	12
Shielded Metal-Arc Welding		$^5/_{16}$, $^3/_8$ in electrodes	14
$^1/_{16}$, $^3/_{32}$, $^1/_8$, $^5/_{32}$ in electrodes	10	Atomic Hydrogen Welding	10–14
		Carbon-Arc Welding	14

Reprinted with permission—Tube Turns, Inc

Reasons for rejecting a pipeweld and ways to reduce reject rates

Pipelines are inspected more critically than ever before, and today's radiographic equipment and techniques produce clearer radiographs with greater sensitivity than in the past. Although codes have not changed drastically, interpretation standards have been upgraded. The combination of more rigorous inspection, better testing methods, and high acceptability standards often approaches an attitude requiring zero defects.

This poses some serious problems because the job of welding cross-country pipelines under typical conditions has always been an extreme challenge requiring specialized and highly developed skills. Now that the demands are greater, even the best welding operators are having trouble. Rejectable defects usually require cutting out the entire weld. This is expensive and can cost competent pipeline welders their jobs.

The purpose of this bulletin is to discuss some of the more common reasons given for rejecting a pipeweld and to suggest what may be done to correct some of these conditions.

Change in attitude

Any discussion on this subject should begin by accepting the fact that the standards are higher than they were a few years ago. Therefore, it will take a change in the attitude of the welders and everyone else involved if this new level of quality and workmanship is to be met. In turn, a commitment to new methods, equipment, and theory will be required by all concerned.

Importance of joint preparation

It must be recognized that all joint preparation details (Figure 6) are critical and any variation could directly contribute to rejected welds.

Without good cooperation from those who prepare the pipe edges and the line-up crew, the welder has very little chance of meeting today's rigid inspection requirements. Too often the attitude exists that variations in fit-up and joint preparation are permissible and that the welder can compensate for them. This attitude cannot be tolerated. It puts too much responsibility on the welder and inevitably leads to rejects.

Recommended procedures for properly cleaning pipe

For lowest cost and highest quality, careful attention to the cleaning of pipe joint surfaces is critically important.

Courtesy of Pipeline & Gas Industry.

Today's pipe welders face pipe that may be covered with a variety of coatings; these include primers, epoxy, tar, paper, varnish, rust, scale, or moisture.

While joint cleanliness is important in all welding, it is especially so in the root pass of pipe. Even a thin film of contaminants, which may be difficult to see with the naked eye, may be present. Failure to recognize and properly clean joints can result in a hollow bead or other welding defects. Follow these instructions to minimize costly defects:

1. Remove all moisture and condensations of any type prior to welding. The joint must be completely dry.
2. Clean BOTH ends of the pipe INSIDE AND OUT to remove traces of paint, rust, scale, oil, epoxy, or organic materials. The area to be cleaned should extend at least 1 in (25 mm) from the end of the bevel (per ANSI/AWS D10.11-80 Page 1) on both the INSIDE and OUTSIDE surfaces.
3. A recommended method for cleaning in the field as described above is the use of a heavy-duty straight shaft grinder with a rubber expanding wheel and a carbide coated sleeve. The small shaft and reduced overall weight allows easy access to the inside and outside surfaces of the pipe.

Internal undercut

See Figures 1–5 for X-rays of the various defects that can cause rejects. Of these, one of the most common and troublesome is internal undercut (undercut on the inside of the pipe). This is understandable because it occurs on the "blind side" and the operator cannot see it happening. Consequently, he cannot immediately do anything to correct it. To make matters worse, he seldom gets a chance to see the X-ray negative or the actual weld itself. All of this makes it difficult for him to correct the situation.

Undercut may be the direct result of poor joint preparation or fit-up. If the joint does not conform to the details specified in the established procedures (see Figure 6), every reasonable effort should be made to correct it before starting to weld.

Internal undercut will tend to occur if

1. the root face (land) is too small
2. the root opening is too large
3. a severe high-low condition exists
4. the current is too high.

When any undesirable variation occurs in the joint preparation, the normal reaction is to compensate for it by

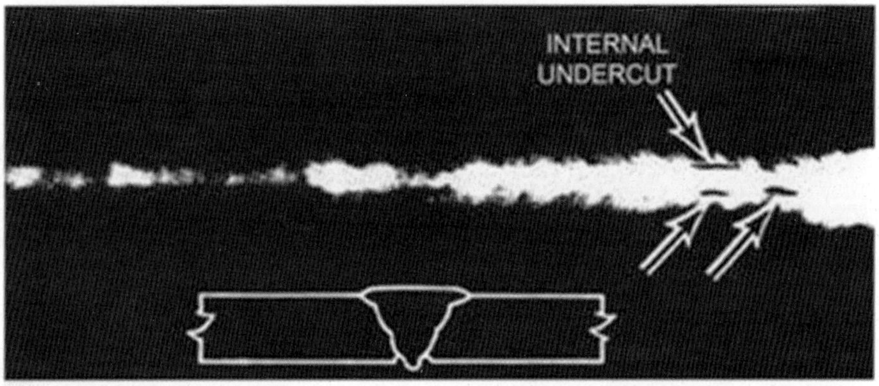

Figure 1. Radiograph of internal undercut. This defect may be intermittent or continuous (Figure 2) and either on one side or both sides of the weld centerline.

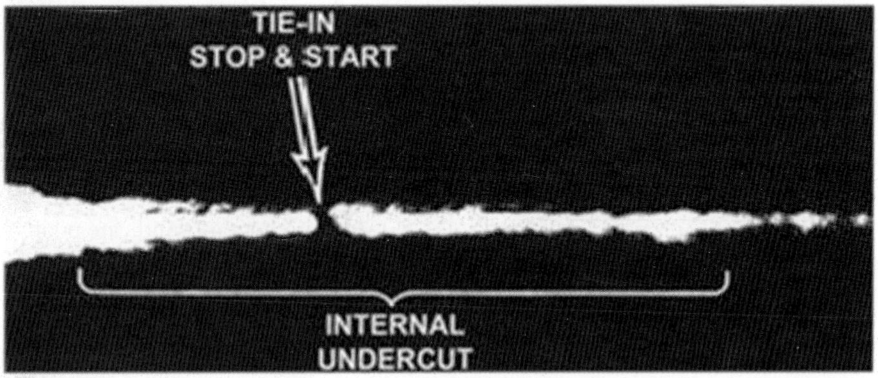

Figure 2. Radiograph of internal undercut and the lack of penetration that tends to occur at a stop and start.

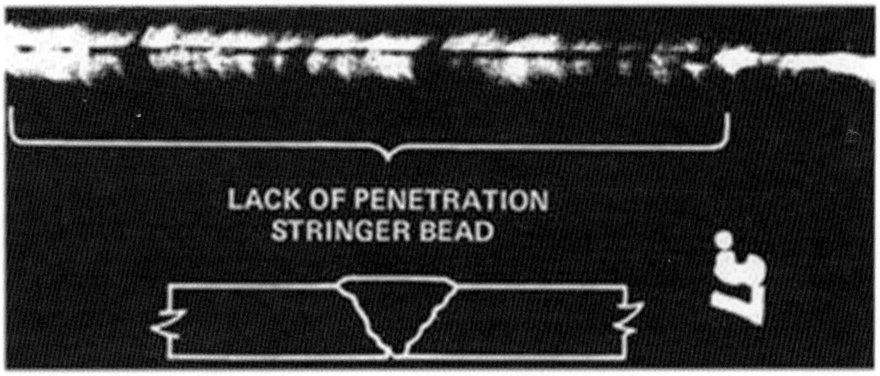

Figure 3. Radiograph of lack of penetration on the stringer bead appears as a single, straight, dark line.

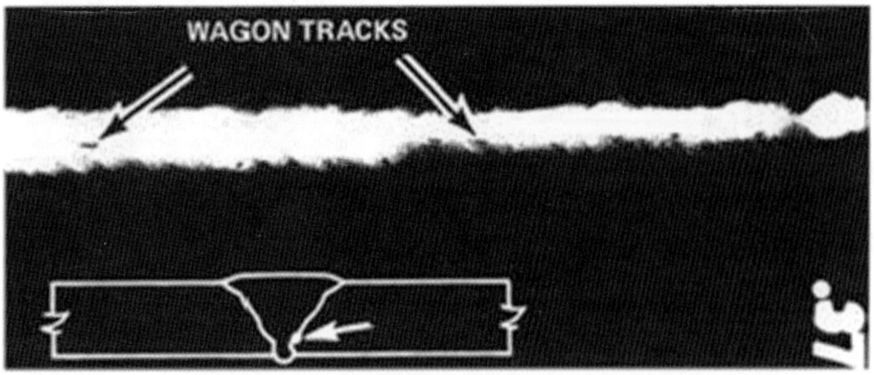

Figure 4. Radiograph of unfilled wagon tracks.

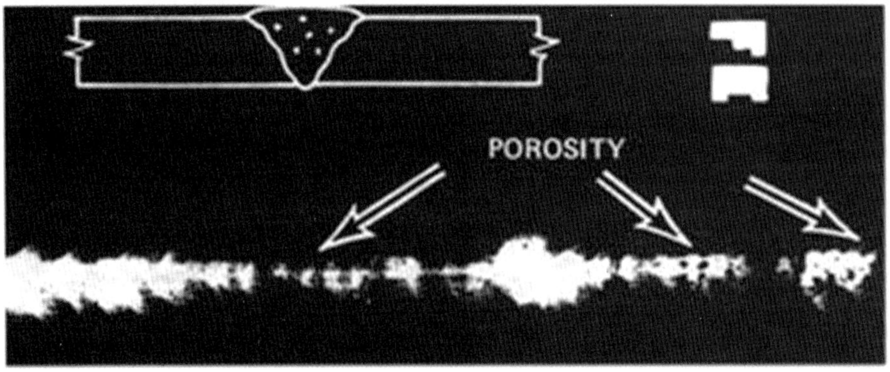

Figure 5. Radiograph of porosity.

juggling the root spacing. Within limits, this can be fairly effective. For example:

Condition	Change in Root Spacing
Land too small	Decrease root spacing
Land too large	Increase root spacing
High-low condition	Decrease root spacing
Bevel too small	Increase root spacing

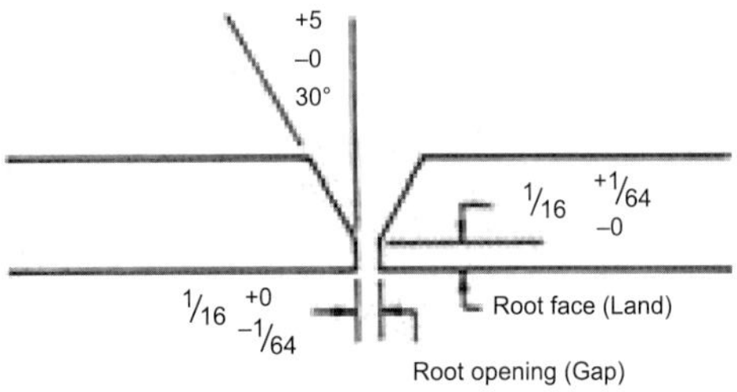

Figure 6. Recommended joint preparation and typical procedures—operate within these tolerances to help ensure good-quality welds.

A skillful line-up crew can be helpful in juggling the root spacing to bring about the most favorable results, but this has limitations. And it is at best a poor substitute for a uniform and consistent joint preparation.

When radiographs indicate poor workmanship, best results can be obtained by showing the radiographs to the welder so that they can understand what is being done wrong and correct it.

Pass	Electrode[(1)] in. (mm)	Current[(2)] Amps DC(+)[(s)]	Altemate stringer Bead Procedure Amps DC(–)[(s)]
Stringer Bead	1/8 (3.2)[(3)] or 5/32 (4.0)[(3)]	90 – 120 140 – 165[(4)]	85 – 115 115 – 140
Hot Pass	5/32 (4.0)	160 – 185	–
Fills and Stripper	3/16 (4.8)	160 – 190	–
Cap Pass	3/16 (4.8)	150 – 180	–

(1) Fleetweld 5P+, Shield-Arc[c] HYP or 70+ suggested depending upon the pipe steel requirements and contractor specifications.

(2) The ideal current should be selected from within the range shown. In general, the best quality will be obtained by operating at the lower end of the range. Hot pass currents up to 200 amps are often used but cause early coating breakdown and large stub losses.

(3) Use 1/8" (3.2 mm) diameter electrode for stringer beads when the wall thickness is .250" (6.35 mm) or less and when the gap is too small to permit use of the 5/32" (4.0 mm) size [1/8"(3.2 mm) diameter electrode should also be used on other passes when the wall thickness is .250" (6.35 mm) or less; however, 5/32" (4.0 mm) diameter can also be used on fill and cap passes on .250" (6.35 mm) wall].

(4) Weld stringer bead at 23-25 arc volts and 10-16 in/min arc speed. Measure arc voltage between the pipe and electrode holder.

(5) Polarity—DC (–) (negative polarity) should be used for stringer bead welding when burn-through, internal undercut and hollow bead defects are a problem. These problems generally occur on thin wall pipe steels containing over .1% silicon. Lower currents are used with DC(–) which helps to reduce these problems. Travel speeds with DC(–) will be equal to travel speeds with DC(+).

Hot pass and all other passes should be run DC(+) (positive polarity).

Welding DC(–) (negative polarity) on the stringer bead will not be harmful to either mechanical or metallurgical properties.

General Guidelines for Power Source Adjustment (SA- and SAE Machines)

In general, the "Current Range" or "Current Control" switch (whether a tap or continuous control), should always be set as low as possible to get the current wanted, and the "Fine Current Adjustment" or "Job Selector" control (which controls open circut voltage) should be set as high as possible. It is usually better to set this control (for OCV) at the mid-point or higher for best arc stability and fewest popouts.

When using a Lincoln SA-250 engine driven power source, it will normally be necessary to set the "Current Range Selector" tap switch one position lower than used on a Lincoln SA-200 power source to get the same operating characteristics.

Internal chamfer

It is important to remove any burr or overhang on the inside edge of the pipe after the root face has been machined (or ground). However, this clean-up operation should never produce an internal chamfer. A condition such as this is almost certain to result in "internal undercut" (Figure 7). See Recommended Procedures for Cleaning Pipe on the previous page.

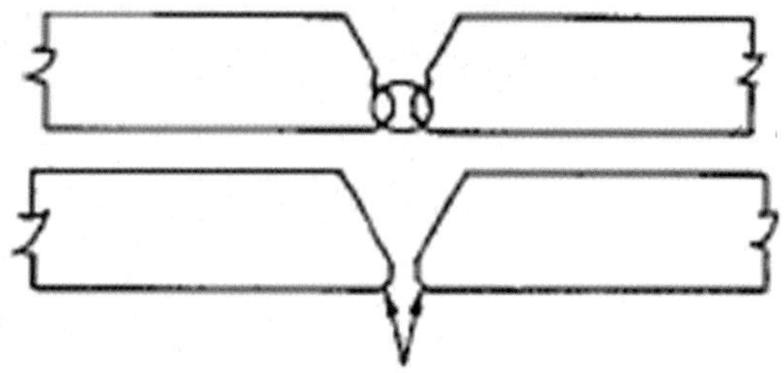

Figure 7. An internal chamfer will tend to leave an unfilled area on one or both sides of the stringer bead. On the X-ray this will be interpreted as internal undercut.

Undercut vs. welding current

Excessive current produces internal undercut. Further, the incidence of internal undercut is more prevalent in the 4—6—8 o'clock portions of the pipe. In this bottom area it is possible to use a considerably higher current than is necessary without getting burn-through, and this results in overheating and undercutting. Knowing this, it would appear to be a simple matter to specify an ideal current that would not give undercut. However, this is extremely difficult because of the many variables of the many interrelated factors.

A recommended current range is given in Figure 6, but it is necessarily rather broad to cover all reasonable variations. From a practical standpoint, the correct current should be determined by starting within the recommended range and then "fine tuning" to get precisely what "feels good" and produces the required results. This final current value will vary slightly from one operator to another and with different pipe materials, diameters, wall thickness, etc.

Because of inaccuracies in ammeters, the effect of arc voltage on current, the inability to accurately read a bobbing ammeter, etc., it is impractical to hold to an arbitrary current value. For the accomplished stringer bead welder, the selection of the ideal current is not too difficult—but he will be doing it primarily by "feel and touch."

Keyhole size vs. current and undercut

To get a good inside bead it is highly desirable to maintain a small, visible keyhole at all times. If the keyhole is allowed to get too big, it will result in internal undercut and/or windows (Figure 8).

Assuming that the joint preparation is correct, the keyhole size is a function of current, electrode angle, and pressure. The current should be "fine-tuned" to produce a small keyhole when the electrode is dragged lightly using the normal electrode angle. If minor variations occur in the keyhole size (for any reason), the electrode angle and pressure can be manipulated to maintain the proper keyhole size (Figure 9).

In general, a keyhole that is about $\frac{1}{8}''$ (3.2 mm) in length is ideal; if it becomes 5/32″ (4.0 mm) or longer, undercut and windows are imminent.

Frequently, a good inside bead is obtained without having a visible keyhole. For example, at a tight spot, the keyhole may disappear and the arc goes "inside the pipe." With the proper manipulative skill, this condition is tolerable and a good inside bead will be obtained. However, when it is done in this manner, the welder is largely dependent on the sound of the arc inside the pipe. A small, visible keyhole is easier to work with and is a much more controllable condition.

Significance of arc voltage

It is recommended that meters be used since a fairly good correlation can be established between the arc voltage and the keyhole size. For example, under controlled conditions,

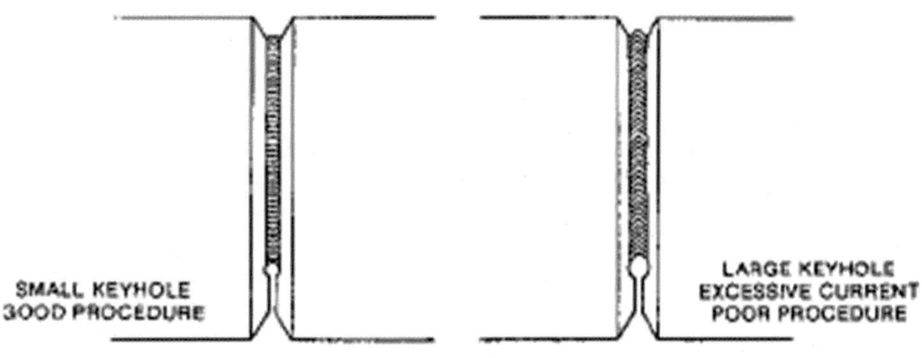

Figure 8. Maintaining a small keyhole will help assure a good inside bead. The larger keyhole at the right is almost certain to give internal undercut and/or windows.

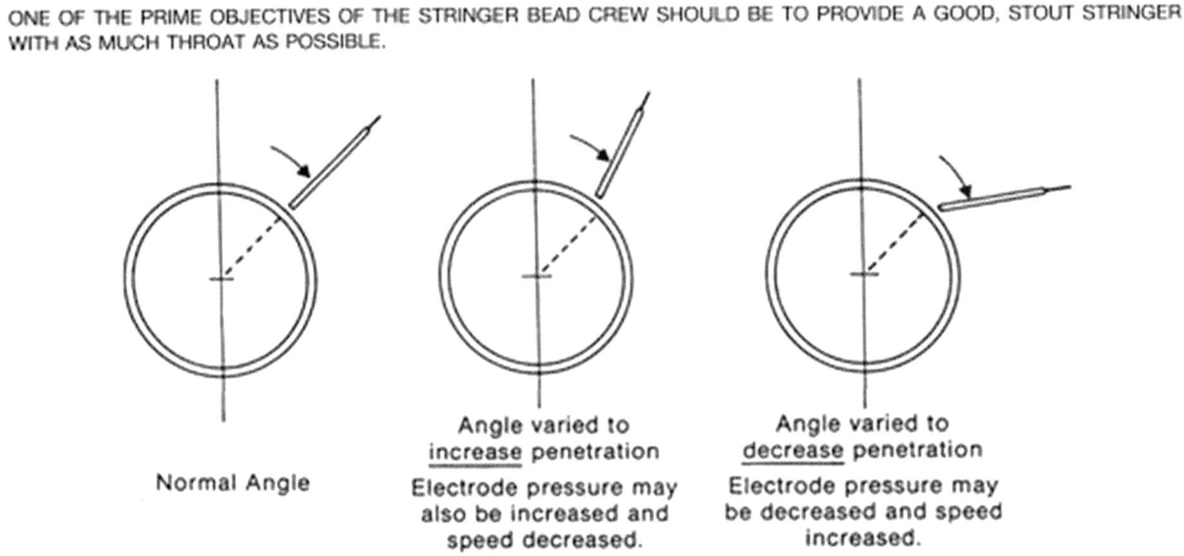

Figure 9. The penetration and keyhole size can be effectively controlled by varying the electrode angle as shown—if the joint preparation is uniform.

an ideal keyhole size is consistently accompanied by an arc voltage of no more than 25 volts. When keyhole size is increased by increasing current, increasing the gap, changing the electrode angle, etc., the arc voltage increases to 26–28 volts. This is reasonable because the arc voltage reflects arc length. At the higher arc voltages, internal undercut and wagon tracks may occur.

The same arc voltage numbers quoted above may not occur in all instances. However, it is readily apparent that there is a dependable correlation between keyhole size and arc voltage. A maximum tolerable arc voltage can be determined by experimentation.

It should be further noted that this arc voltage is determined while welding and it should not be confused with the open circuit voltage. Also, it can be affected by the welding technique; changes in electrode angles and drag techniques should not be allowed to cloud the issue.

Varying current/varying gap

Greater root penetration naturally occurs at the top and bottom portions of the pipe, and the least penetration tends to occur at the sides. This being the case, it would be desirable to alter the root spacing accordingly, but this obviously is not practical. It should also be noted that a condition which permits maximum spacing on top and bottom and minimum spacing on the sides is not tolerable.

Assuming a uniform gap all the way around, the ideal condition would be to vary the current as the weld progresses from 12 to 6 o'clock to compensate for the changes in penetration due to position. Instead, penetration changes are usually controlled by the manipulative skills of the welder (see Figure 9). In an extreme case, a second pair of hands may be required to adjust the current up or down on a signal from the welder. This requires a good communication system and a well-coordinated effort to avoid overshooting or undershooting the current. In some cases, this becomes a survival technique to make up for other undesirable variations.

Undercut—real or imaginary?

If visual inspection were practical for determining internal undercut, it would be easy to accurately determine its presence and severity. However, this is not the case. Radiography is the judge of internal conditions, and this has led to many questionable interpretations. If internal undercut is present, it will show up on the film as a very narrow, dark line immediately adjacent to the stringer bead—on one or both sides. The darkness of the line will vary with the depth of the undercut (Figures 1 and 2).

Proper identification of undercut is sometimes difficult because its location is directly in line with the "wagon track" area (Figure 4). Distinguishing one from the other may be difficult.

Correct interpretation of internal undercut may be further complicated by the fact that a "stout" inside bead will stand out clearly (lighter on the film) against the adjacent area, which is thinner and therefore darker. The thinner, darker area will be further darkened by the presence of any fairly deep, widely spaced surface ripples in the cap pass. The net effect is to produce a dark shading in the area where internal undercut might occur.

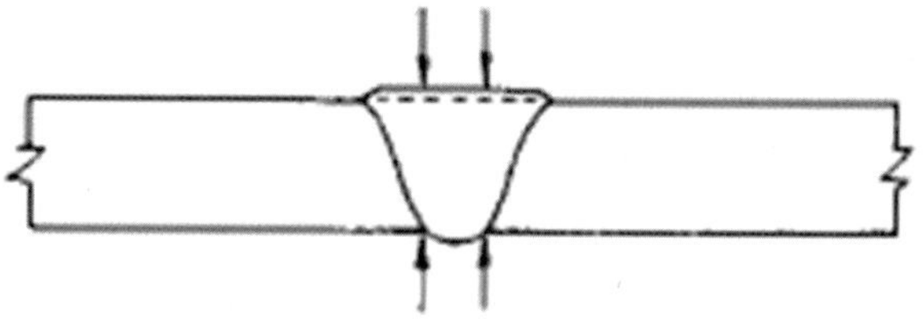

Figure 10. Any deep surface ripples located directly above a point adjacent to the inside bead will contribute to a dark shading of the film at this area. This has been misinterpreted by some as internal undercut.

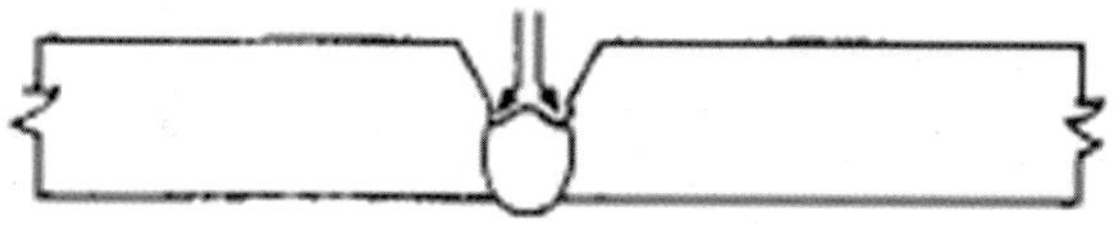

Figure 11. "Wagon tracks" are located on either side of the stringer bead.

Although this darker area is considerably wider than any undercut, it has been mistaken for undercut and resulted in cutting out good welds. Normally, a "stout" inside bead is considered good, but in this instance, the stouter the bead, the darker the adjacent area would appear (Figure 10).

Wagon tracks

It is not possible to completely eliminate the sidewall undercut produced by the stringer bead. This condition is generally referred to as "wagon tracks" (Figures 4 and 11).

It is the responsibility of the stringer bead crew to minimize wagon tracks and the responsibility of the hot pass crew to burn them out. This is normally done with 5/32″ (4.0 mm) electrode at about 175–180 amps. The secret in melting out wagon tracks lies primarily in the skill of the hot pass crew and the penetration factor of the electrode. The majority of the hot pass men use a rapid stitching technique that exposes the wagon track momentarily and permits them to direct the arc into the undercut area and melt them out.

Condition	Results
Bevel too small (Figure 12)	Increases depth of W.T.
Root spacing (gap) too small	Increases depth of W.T.
Current and/or speed too high	Increases depth of W.T.

The depth of the wagon tracks may be affected by the following:

In extremes, it may be necessary to use a grinder to open up the sidewalls to minimize deep "wagon tracks" or, if they do occur, to grind the stringer to eliminate the high peaked center (Figure 13). In all cases, a $\frac{5}{32}$″ (4.0 mm) thickness disc grinder should be used to grind root beads on all pipe from 12″ (304.8 mm) up irrespective of wall thickness (a 1/8″ (3.2 mm) disc grinder will grind the center and roll the sides over on the wagon track unless side pressure is applied). This will make it easier to melt them out.

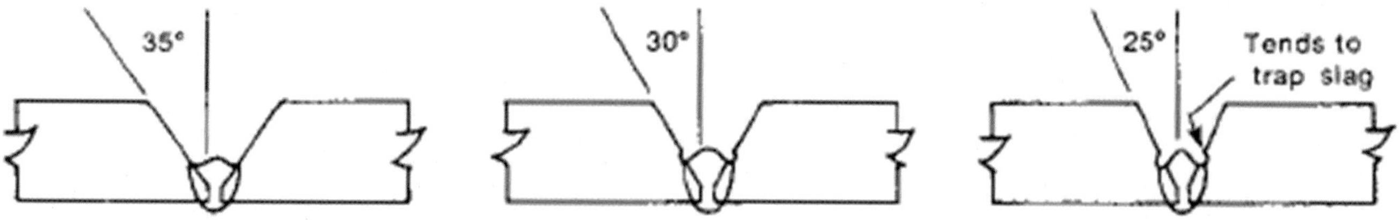

Figure 12. The depth of the wagon tracks will vary inversely with the bevel angle.

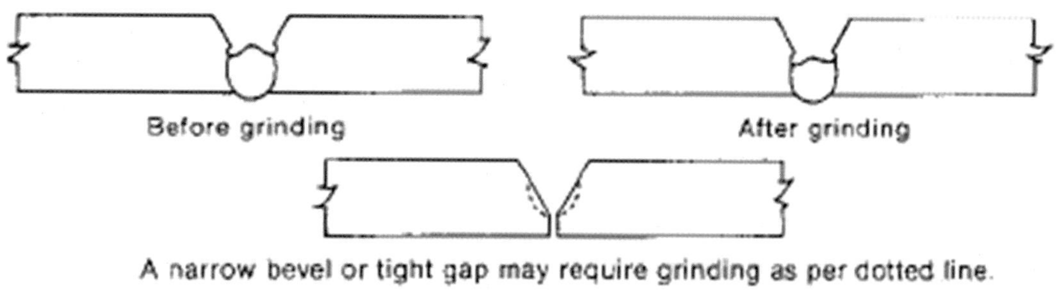

Figure 13. A disc grinder, if used with restraint, can be helpful in correcting the condition shown above.

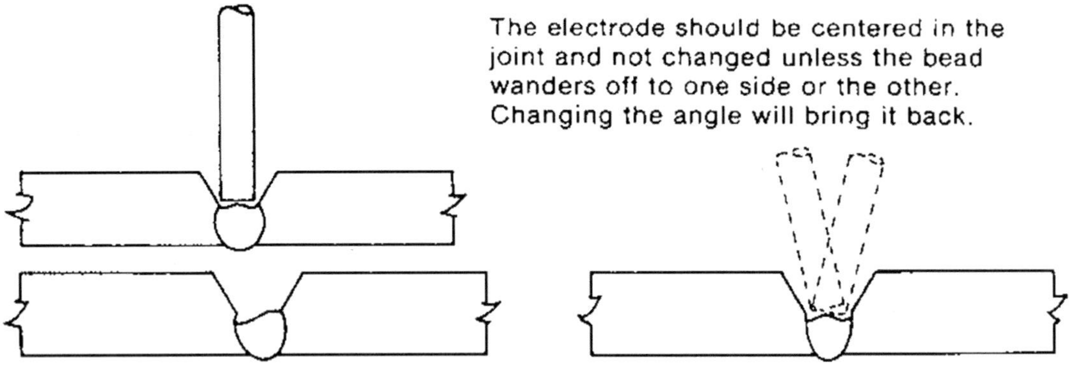

Figure 14. The correct angle can only be determined by observing the shape and location of the bead. If the bead drifts to one side for any reason, it will leave a deep undercut on the opposite sidewall. Tilting the electrode just a few degrees toward the undercut side will straighten the bead up. This change must take place rapidly if the correct bead placement is to be maintained.

If the stringer bead tends to wander to one side or the other, it will leave a deep undercut on the shallow side. This condition should be corrected immediately by changing the electrode angle (Figure 14).

The hollow bead

The stringer bead defect shown in this sketch is known by several names including the hollow bead, vermicular porosity, and wormhole porosity. Its length varies from a fraction of an inch to several inches. Radiography exposes the presence of the problem clearly.

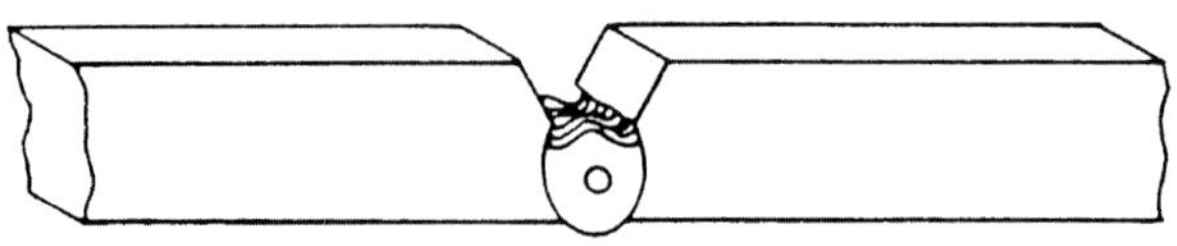

Vermicular porosity occurs most readily when welding high silicon pipe—generally above 0.10% silicon. It is aggravated by excessive travel speeds and high currents. Welding the stringer bead with a $\frac{5}{32}''$ (4.0 mm) electrode at 130–165 amps DC(+) and 12 to 14 in/min (0.3–0.4 m/min) travel speed minimizes its occurrence. DC(−) (negative polarity) should be used for stringer bead welding when burn-through, internal undercut, and hollow bead defects are a problem. These problems generally occur on thin wall pipe and on pipe steels containing over 0.1% silicon. Lower currents can be used with DC(−), which helps to reduce these problems. Travel speed with DC(−) will be equal to travel speeds with DC(+).

Hot pass and all other passes should be run DC(+) (positive polarity).

Welding DC(−) (negative polarity) on the stringer bead will not be harmful to either mechanical or metallurgical properties.

Hollow bead or wormhole porosity may also be caused by poor joint preparation and/or cleanliness. See Recommended Procedures for Properly Cleaning Pipe.

Filler, stripper, and cap

In general, there is little fault found with the fillers, strippers, and caps. Most of this welding today is being done with $\frac{3}{16}''$ (4.8 mm) Shield-Arc® HYP or 70+ at 160–180 amps, and the results have been excellent.

A reasonably competent crew of firing line welders armed with FW5P+, HYP, or 70+ can do a very fine job of finishing the weld if the stringer and hot pass crew have been doing their work properly.

The size and consistency of the final weld will have its influence on the X-ray radiograph. Thus, the firing line welders should be encouraged to produce a cap pass that is as uniform as possible with neatly stitched close ripples and as much reinforcement as required.

Welding cracks

Since the advent of the higher tensile pipe steels (5LX 52, 60, 65, 70 etc.), it has been necessary to exercise better procedural control to eliminate the possibility of weld and heat-affected zone (HAZ) cracks. To do this effectively, all of the following factors must be controlled.

1. Joint preparation and root spacing must be as specified in the approved procedure. (See Figure 6.)
2. Hi-Low conditions must be held to a minimum.
3. Weld 5LX70 with Shield-Arc® 70+. Weld 5LX65 and lower with Shield-Arc HYP or Fleetweld 5P+. Fleetweld 5P+ is recommended for all stringer beads when lower hardness is a consideration. Lower stringer bead hardness will result in improved resistance to HAZ cracking.
4. The need for preheat varies considerably between applications. Cracking tendencies increase with higher carbon and alloy content, thicker pipe wall and diameter (faster quench rate), and lower ambient temperature. Preheat cold pipe to at least 70 °F (21 °C). Preheat to as much as 300 °F (149 °C) may be required to slow the quench rate and prevent cracking when welding high-strength, large-diameter, or heavy-wall pipe. Specific preheat requirements must be determined for each situation based on these considerations.
5. Ideally the line-up clamp should not be removed nor should any movement of the pipe take place until the stringer bead is 100% completed. Any movement of the pipe could cause a partially completed stringer bead to be overstressed and to crack. The tendency for such cracking varies with the chemistry of the pipe, its diameter and wall thickness, and its temperature. Under favorable conditions, it may be possible to remove the line-up clamp with as little as 60% of the stringer bead completed, but this should be done only when it has been clearly demonstrated that this practice does not cause cracks to occur.
6. After removal of the line-up clamp, the pipe must be gently and carefully lowered on skids.
7. Use only enough current on the stringer to get a good inside bead and travel slowly to get the maximum weld cross-section.
8. Restrict lack of penetration on the inside bead at tie-ins to ¼″ (6.4 mm) or less. Use a disc grinder to improve this situation on starts and stops only.
9. Remove slag from each bead by power wire brushing. Grinding should not be necessary except possibly to clean up a lumpy start or humped up center or perhaps to improve a crater condition. Note: Grinding of the stringer bead should be done with a $\frac{5}{32}''$ (4.0 mm) disc. Excessive grinding can be detrimental.
10. Weld stringer beads with two or more persons welding on opposite sides to equalize stress. Use three welders on 20–30″ (508–762 mm) pipe and four welders on larger pipe.

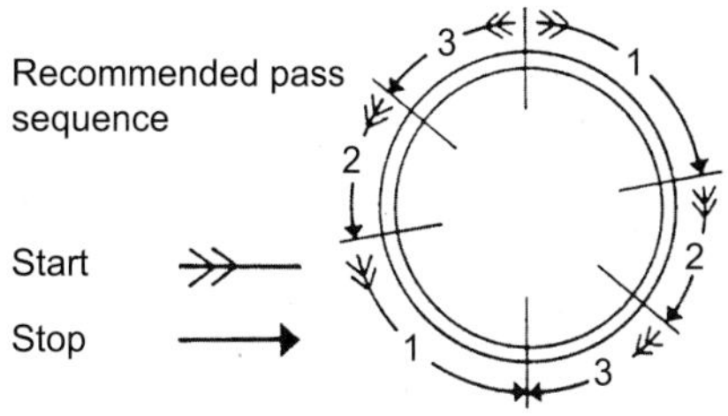

11. Start the hot pass immediately after completion of the stringer—always within 5 min. At least two hot pass welders should be used on each joint, and to put this pass in as soon as possible after the stringer bead it may require a total of four hot pass welders leap-frogging each other to keep up.
12. Minimize the wagon tracks—this area is highly vulnerable to cracking until the hot pass is completed (Figure 15).

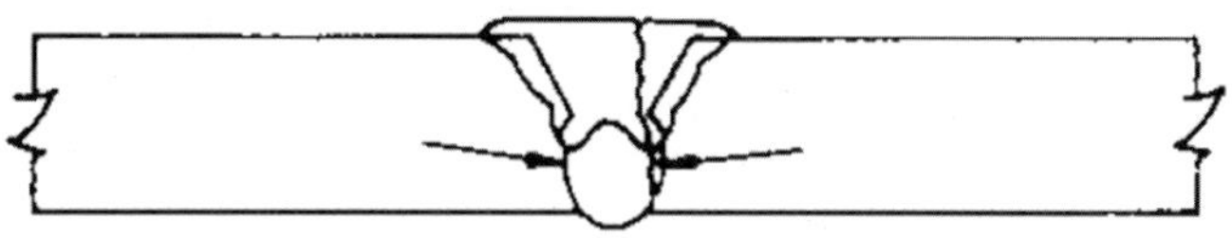

Figure 15. Cracks tend to occur at the area indicated on one side or the other of the stringer bead. This could eventually propagate up through the weld. A properly controlled procedure and good workmanship can eliminate this condition.

Reprinted with permission—Lincoln Electric Co.

How many welds will the average welder make per hour?

Where stringer beads have been run, the average welder can complete about 140 in of weld per hour on ordinary ¼-in wall line pipe. To find the average number of welds per hour, divide the circumference of the pipe into 140.

$N = \frac{140}{C}$, where circumference (C) = pipe diameter in inches times 3.14

Example. How many welds per hour will a welder complete on ¼-in wall $10^3/_4$ -in line pipe, if the stringer beads have been run?

$$N = \frac{140}{(10.75)(3.14)}$$

$N = 4.15$ welds per hour or about 42 welds per 10-h day.

On large diameter pipe where wall thickness has been increased, welders average about 100 in of completed weld per hour.

Example. How many welds per hour will a welder complete on 30-in line pipe if the stringer beads have been run?

$$N = \frac{100}{C}$$

$N = \frac{100}{(30)(3.14)} = \frac{100}{94.2} = 1.1$ welds per hour or 11 welds per 10-h day

How much welding rod is required for a mile of schedule 40 pipeline?

Rule. For 40-ft joints of schedule 40 pipe, multiply the nominal pipe diameter by 22½; the answer is pounds of electrode per mile of pipeline.

Example. 6-in pipe:
6 × 22½ = 135 lb per mile.

Example. 10-in pipe, 30-ft joints:

10 × 22½ = 225 lb per mile.
For 40-ft joints:

$225 \times \frac{40}{30} = 300$ lbper mile for 30-ft joints.

The rule is based on standard weight (schedule 40) pipe and the usual number of passes, which varies from two for small pipe up to four for 16 in; larger pipe is usually thinner wall. The rule includes an allowance for wastage.

How many pounds of electrodes are required per weld on line pipe?

Divide the nominal pipe size by 2 and multiply the result by one-fourth pound.

Pounds of electrode $= \frac{N}{2} \times 0.25$.

Example. How many pounds of electrode will be used per weld on 10-in line pipe?

Pounds of electrode $= \frac{10}{2} \times 0.25 = 1.25$ lb per weld.

Example. How many pounds of electrode will be used per weld on 24-in line pipe?

Pounds of electrode = 12 × 0.25 or 3 lb per weld.

Welding criteria permit safe and effective pipeline repair

J. F. Kiefner, Senior Research Engineer, Battelle. Houston, Texas

Research conducted by the American Gas Association (AGA)[1–5] and others[6–8] showed that pipeline defects can be repaired without removing the lines from service. Two repair methods evaluated were full-encirclement sleeves and direct deposition of weld metal on defects.* To ensure minimum risk when using these methods, criteria for repairing pipelines in service have been formulated (Table 1).

The criteria in the table are intended as a repair guide for lines that operate at 40% of their specified minimum yield strength (SMYS) or more. The criteria may by unnecessarily restrictive for lines that operate at or below 40% of SMYS.

Note that defects in pressurized pipelines, especially pneumatically pressurized lines, can cause sudden catastrophic ruptures. When the procedures are carefully planned, risks are minimized. Nevertheless, repair errors could reduce the safety margin of these criteria. Consequently, users of Table 1 are urged to exercise caution when performing repairs.

Choice of repair method

The repair method options included in Table 1 are as follows:

- Type A sleeve
- Type B sleeve
- Deposited weld metal
- Removal
- Removal by hot-tapping.

Type A sleeve repair consists of placing a reinforcing band concentric with the pipe while leaving the band ends unsealed. The sleeve strengthens a defective area by restraining bulging that would otherwise occur when the weakened pipe shell is pressurized. Such a sleeve may also carry a small portion of the hoop stress, but in no case can it become the sole pressure-containing element since its ends are not sealed. As a result, it cannot be used to repair a leaking defect. If a gap exists over the defect, filling of the gap is required. Two classes of Type A sleeve exist: no filler and with filler.

Type B sleeve repair requires a pressure-tight reinforcing concentric band. Its ends are sealed to the carrier pipe, and it can contain pressure. It can be used to repair leaking defects. In addition, a Type B sleeve can also be used as a Type A sleeve. Three classes of Type B sleeves exist: no filler, with filler, and pressurized.

Deposited weld metal completely fills and eliminates clearly visible defects such as corrosion. Access to an entire defect is required so that weld metal can penetrate and bond to sound metal. It may be possible to enlarge pits, laps, or undercut by grinding a groove wide enough to permit access. Deposited weld metal can be selected when requirements to prevent burn-through are met. These requirements are discussed later.

Removal of a damaged pipe section is another way of eliminating a defect. The line must be taken out of service, purged, removed, and replaced with a sound tie-in piece. Removal is often a poor economic choice in terms of wasted product and interrupted service. Thus, removal is recommended in Table 1 with the understanding that it may not be practical in certain instances.

Removal by hot-tapping takes advantage of a widely accepted method for making branch connections on live pipelines. The coupon removed from the line must contain the entire defect. Limitations of this technique will be discussed later.

Types of defects

Classes of defects listed in Table 1 are as follows:

- Manufacturing defects—cracks, undercut, lack of fusion, and lack of penetration in seam welds; laps, pits, cracks, and rolled-in slugs in the pipe body; hard spots.
- Environmentally caused defects—selective corrosion and hydrogen stress cracking in electric resistance welded or flash welded seams; general pipe corrosion; pitting corrosion; stress corrosion cracks; hydrogen stress cracking in hard spots.
- Defects caused by outside forces including dents and gouges.
- Construction defects.

Submerged arc-welded (SAW) seam defects, both straight seam and spiral, include undercut, incomplete fusion, incomplete penetration, and cracks. Repair methods recommended for these defects are shown in Table 1. No filler is required with sleeves, because after the weld reinforcement is ground flush, no gap between the pipe and sleeve should exist. Removal by hot-tapping is not recommended because the tap would involve cutting through the seam, a practice considered unacceptable by many companies. Repair by deposited weld metal is applicable only for undercuts and is subject to special requirements given in footnotes of Table 1.

Type B pressurized sleeves used on nonleaking defects can be pressurized only by using a tapping nipple. Since other means of sleeve repair are entirely adequate when used as directed, the intentional pressurization of a sleeve when no leak or near leak is present is not necessary.

Inside or interior defects require special consideration since they are not readily visible. Removal may be the best alternative unless one can be reasonably certain of the extent of the defects. Hot-tapping is not recommended because of the uncertainty of the extent of an inside or interior defect.

Electric resistance welded (ERW) or flash welded (FW) seam defects include upturned fiber imperfections, incomplete fusion, penetrators, cold welds, and cracks. These can only be repaired with pressurized sleeves and removal since the welds are susceptible to brittle fracture or low-resistance ductile fracture initiation. The value of restraint of bulging from either type of nonpressurized sleeve is uncertain. Hence they are not recommended for these defects. Hot-tapping also is not recommended because of the involvement of the seam weld. It is not recommended that weld metal be deposited because of possible low ductility. Required grinding before such welding would involve creating or enlarging a defect in a potentially low-toughness material. Only removal or use of a Type B pressurized sleeve that stress-relieves the defect is recommended.

Other seam defects include lap welds and furnace butt welds. These must be dealt with on a case-by-case basis. Such welds usually appear in older or smaller and lower pressure pipelines where fracture resistance requirements are less stringent. On the other hand, some of these older materials can be quite susceptible to low-toughness fracture behavior. It is probably best to treat these materials with the same caution as ERW and FW seams as they are being operated at stress levels exceeding 40% of SMYS.

Laps, pits, seams, cracks, and rolled-in plugs should be repaired subject to these limitations: special requirements of I.D. and interior defects and leaks must be observed; hot-tapping is acceptable on nonleaking, O.D. defects as long as the entire defect is removed with the coupon; deposited weld metal may be used to repair pits or laps if they can be entirely exposed by grinding.

Hard spots created in the plate by accidental quenching on the run-out table may become flat spots in the pipe since they do not yield when the plate is formed to pipe. Such hard spots do not fail spontaneously unless they contain quench cracks or unless they undergo hydrogen stress cracking. Such spots should be repaired if they are not cracked. Type A or B sleeves without filler would not be acceptable since they would not restrain flat spots from bulging.

Hard spots can be remedied by several methods if they occur as outside nonleaking defects. Sleeve repair methods with filler not only provide strengthening but shield the hard spot from the environment, which may cause it to become cracked.[a] Hard spots at the I.D. surface should not be repaired with Type A sleeves since they may be susceptible to cracking in certain types of product service. Such a sleeve would not prevent a leak if the resulting crack grew through the wall.

Selective corrosion and hydrogen stress cracking in ERW and FW seams are subject to low toughness fracture behavior. It is recommended that they be removed or repaired with pressurized Type B sleeves to relieve stress.

General or widespread corrosion is that which covers too wide an area to be repairable by means of hot-tapping or deposited weld metal. Any other repair means are acceptable if they cover the critically affected area. When internal corrosion is present, its extent must be reasonably well known and further corrosion must be prevented. Removal may sometimes be the only suitable choice.

Pitting corrosion may involve isolated pits or groups of pits. Generally, isolated single pits will not require repair if further corrosion can be halted. ASME guidelines can be used to determine whether or not a corroded region requires repair. Groups of interacting pits can substantially lower remaining strength. If strength has been lowered, the pipe should be repaired. Any of the methods of Table 1 are suitable if the noted precautions are observed.

Stress-corrosion cracks occur in clusters and often cover large areas. It is difficult to repair these cracks by hot-tapping or deposited weld metal. Repair methods for stress-corrosion cracks are confined to those which can strengthen the entire affected area.

Hydrogen stress cracking appears in hard spots attacked by hydrogen emitted from bacteria external to the pipe and from cathodic protection. Internally the hard spots can be attacked by certain types of products—especially those containing hydrogen sulfide. Because flatness often occurs near hard spots, nonfilled sleeves are not recommended. Deposited weld metal is not recommended because grinding of a hydrogen stress crack in a hard—and usually brittle—spot while the pipe is under pressure is not safe. Interior hydrogen stress cracking should not be repaired by a Type A sleeve since such a sleeve cannot prevent leaks if the crack grows through the wall.

Dents and gouges and a combination gouge-in-dent result from external encroachment by mechanical excavating equipment, other kinds of equipment, or rocks. Plain dents are usually innocuous unless they are quite large[b]—2% or more of the diameter—or unless they involve a seam or girth weld. When the dent is large or welds are included in the

[a] One method of protecting hard spots from hydrogen stress cracking that so far, has proven adequate, is using a concentric band of sheet metal spaced away from the pipe by rubber seals. The annular space is filled with coal tar to exclude ground water, and the metal band (which is coated) shields the hard spot from hydrogen generating cathodic protection current.

[b] Present requirements of federal regulations Part 192 dictate removal of dents extending over 2% of the pipe diameter.

Table 1

Criteria for Selecting an Appropriate Repair Method (X Indicates an Acceptable Repair)

Repair Method	Type of Defect								
	Manufacturing Defect				Environmentally Caused Defect in ERW or FW, Seam Weld in Body, SAW Seam or Girth Weld				
	In Seam Weld		In the Body of the Pipe						
	SAW Undercut, Incomplete Fusion, Incomplete Penetration, Crack	ERW and FW Upturned Fiber Imperfection, Incomplete Fusion, Penetrator, Cold Weld Crack	Lap, Pit, Seam, Crack Rolled-In Slug	Hard Spot Exceeding $R_c > 35$ and 2 in in Extent or More	Selective Corrosion, Hydrogen Stress Cracking in Weld Zone	General Corrosion	Pitting Corrosion	Stress-Corrosion Cracking	Hydrogen Stress Cracking in Hard Spot
Nonleaking O.D. defect[a]									
Type A Sleeve, No Filler	X	…	X	…	…	X	X	X	…
Type A Sleeve, with Filler	…	…	…	X[g]	…	X	X	…	X
Type B Sleeve, No Filler	X	…	X	…	…	X	X	X	…
Type B Sleeve, with Filler	…	…	…	X	…	X	X	…	X
Type B Sleeve, Pressurized	X[b]	X[b]	X[b]	X[b]	X[b]	X[b]	X[b]	X[b]	X[b]
Deposited Weld Metal (h)	X[c]	…	X[c]	…	…	…	X[c]	…	…
Removal	X	X	X	X	X	X	X	X	X
Removal by Hot Tap	…	…	X[d]	X[d]	…	…	X[d]	…	X[d]
Nonleaking interior or I.D. defect[a]									
Type A Sleeve, No Filler	X	…	X	…	…	X	X	…	…
Type B Sleeve, No Filler	X	…	X	X	…	X	X	…	X
Type B Sleeve, Pressurized	X[b]	X[b]	X[b]	X[b]	X[b]	X[b]	X[b]	…	X[b]
Removal	X	X	X	X	X	X	X	…	X
Leaking defects									
Type B Sleeve Pressurized	X	X	X	X	X	X	X	X	X
Removal	X	X	X	X	X	X	X	X	X

Corrosion or other defect that is known to exceed 80% of the wall thickness in depth should be treated as though it were a leak. Pressurization to be accomplished upon completion of repair by drilling hole through carrier pipe through a small tapping nipple. Only for undercut, laps, and pits where the defect can be safely enlarged to a weldable groove and subject to remaining wall thickness rules. Only if the branch coupon entirely contains and removes defect. Only with a humped sleeve or if a flat sleeve is used. Girth weld must be ground flush. Do not use if extreme high-low condition exists.

Defect Caused by Outside Force					Construction Defect
In Seam or Other Weld		In Body of Pipe			In Girth Weld
Plain Dent or Gouge-in-Dent, SAW Seam or Girth Weld	Plain Dent, or Gouge-in-Dent ERW or FW Seam	Plain Dent Greater than 2% or Pipe Diameter	Gouge (Also Including Gouges in SAW Seams or Girth Welds)	Gouge-in-Dent	Undercut, Incomplete, Fusion, Incomplete Penetration, Crack
...	...	...	...	...	...
X	...	X	...	X	...
...	...	...	X	...	X(e)
X	...	X	...	X	...
X(b)	X(b)	X(b)	X(b)	X(b)	X(b,e)
...	...	...	X(b)	...	X(c)
X	X	X	X	X	X
...	...	...	X(d)	...	...
...	...	...	...	...	...
...	...	...	...	...	X(e)
...	...	...	...	...	X(b,e)
...	...	...	...	...	X
X	X	X	X	X	X(e)
X	X	X	X	X	X

One method of protecting hard spots from hydrogen stress cracking, which so far has proven adequate, is that of using a concentric band of sheet metal spaced away from the pipe by rubber seals. The annular space is filled with coal tar to exclude ground water, and the metal band (which is itself coated) shields the hard spot from hydrogen-generating cathodic protection current. The role of this band is merely to shield the pipe from the cathodic protection and not to strengthen the pipe. Use of low hydrogen electrodes is recommended.

dent or gouge-in-dent, the repair should prevent its outward movement. Sleeves with fillers or pressurized sleeves are required. If the included weld is an ERW or FW seam, the defect should be removed or the stress relieved with a pressurized sleeve.

Gouges without dents in the body of the pipe or in SAW seams or girth welds may be repaired by any means except deposited weld metal. Gouges-in-dents may be repaired only by means that prevent outward movement of the dent. Sleeves without fillers are not acceptable.

When dents are involved, hot-tapping is not recommended, since it may not remove the dent entirely. For any gouge or gouge-in-dent, repair by deposited weld metal is not recommended, since concealed cracks may exist. When a dent is present, weld metal may not have sufficient ductility to withstand the severe strains that accompany outward dent movement.

Construction defects in girth weld include undercut, incomplete fusion, incomplete penetration, and cracks. Deposited weld metal may be one of the best ways to repair undercut or other externally connected defects in girth welds that can be ground for access. Sleeves, if used, should have a special shape such as a central hump to avoid interference with the girth weld reinforcement. Only sleeves with welded ends are recommended. Such sleeves tend to strengthen the defective girth joint, whereas those with nonweld ends do not.

Burn-through

Burn-through into a live pipeline would defeat the purpose of a repair, would probably cause the line to be shut down, and might create a serious safety hazard to the repair crew.

An AGA study revealed that the following parameters control burn-through: remaining wall-thickness, welding heat

Table 2
Limitations on Remaining Wall-Thickness for Repairing Without a Burn-Through

Values are the minimum recommended thicknesses in inches with natural gas as the pressurizing medium at the pressures and flows shown

Maximum welding voltage, 20 volts

Maximum welding current, 100 amps

Pressure, psia	Gas Flow Rate, ft/s			
	0	5	10	20
15	0.320	-	-	-
500	0.300	0.270	0.240	0.205
900	0.280	0.235	0.190	0.150

input, and the nature and state of the pressurization media in the pipe. Consequently, it is possible to prescribe safe limits for avoiding burn-through while depositing weld metal on a defect (Table 2).

Table 2 shows a relationship between I.D. surface temperature during welding and remaining wall thickness for various pressure and flow rates of natural gas. These values were taken from AGA curves. The table is based on the following: heat input characteristics of the $\frac{3}{32}$-in or $\frac{1}{8}$ -in low hydrogen electrode, 80—100 amps welding current at 20 volts DC, and an average electrode travel speed of 4—5 in/min.

Energy input should not exceed these limits. Low hydrogen electrodes are recommended for reasons explained later. Under these conditions, repairs can safely be made in the flat, horizontal, or overhead positions provided flow or pressure is controlled. Analytic results predict that the I.D. wall temperature will not exceed 2000 °F and a burn-through is highly unlikely when repairs are made in this manner.

Values shown in Table 2 are believed to be quite conservative. Repairs were made without burn-throughs on 0.180-in remaining wall with air at ambient pressure and no flow inside the pipe. This conservatism provides an extra safety margin and allows linear extrapolations for values between those shown in the table. The minimum wall-thickness of 0.150 in is established on the basis of the experimental results. This thickness is also recommended by the British Gas Corp. Within these limits, repairs may be made by deposited weld metal (Table 1).

Underbead cracking

Underbead cracking can be minimized by making repairs with low hydrogen electrodes in a manner that avoids hard weld HAZs.

One way to help assure that extremely hard weld zones are not formed during repair welding is to limit repairs to well-known carbon equivalent ranges. Unfortunately, chemistries of specific samples needing repairs will seldom be known. Low hydrogen electrodes prevent hydrogen from being present in the welding atmosphere, but they do not prevent formation of hard weld zones when adverse chemistries are present. Making crack-free weldments requires careful control of heat input to avoid rapid quenching or post-repair treatments to assure that extreme hardness does not remain.

In the case of sleeves, the critical area in which creaking most often occurs is the fillet-weld at the ends. At least two procedures for making these welds, with a minimum risk of cracking, are available.[1,4,5,8] In the case of deposited weld metal repairs, heat input must be kept low to avoid burn-through during the first or second passes. During later passes, however, higher heat inputs can be used to soften the resulting repair metal microstructure and the HAZ of base metal. A high heat input final pass can be made with a nonfusing tungsten electrode as suggested by the British Gas Corp.8 An alternative procedure is to make an extra pass with high heat input using a conventional electrode. This pass can be ground off, since its purpose is merely to soften the heat-affected microstructure of previous passes.

Reprinted with permisison - *Pipe Line Industry*, January 1980.

References

1. J.F. Kiefner, A.R. Duffy, A Study of Two Methods for Repairing Defects in Line Pipe. Pipeline Research Committee of the American Gas Association, 1974, Catalog No. L22275.
2. J.F. Kiefner, A.R. Duffy, J.S. Bunn, L.E. Hanna, Feasibility and Methods of Repairing Corroded Line Pipe, Materials Protection and Performance III (10) (Oct., 1972).
3. J.F. Kiefner, "Corroded Pipe: Strength and Repair Methods," Fifth Symposium on Line Pipe Research. American Gas Association, Houston, Texas, 1974, Catalog No. L30174.
4. J.F. Kiefner, Repair of Line Pipe Defects by Full- Encirclement Sleeves, Welding Journal 56 (No. 6) (June, 1977).
5. J.F. Kiefner, G.R. Whitacre, R.J. Eiber, "Further Studies of Two Methods for Repairing Defects in Line Pipe," to Pipeline Research Committee of the American Gas Association, NG-18 Report No. 112 (March 2, 1978).
6. B.A. Cassie, The Welding of Hot-Tap Connections to High Pressure Gas Pipelines, in: J. W. Jones Memorial Lecture, Pipeline Industries Guild. British Gas Corp., October, 1974.

Terminology[9–11]

Defect parameters

Defect: A crack, pit, crevice, gouge, dent, metallurgical anomaly, or combination of two or more of the above that is known or suspected to reduce the effective pipe strength level to less than 100% of its specified minimum yield strength (SMYS).

O.D. defect: A defect emanating at and extending radially inward from the outside surface but not entirely through the wall of the pipe.

I.D. defect: A defect emanating at and extending radially outward from the inside surface but not entirely through the wall of the pipe.

Interior defect: A defect emanating in the interior of the pipe wall but not of sufficient radial extent to be connected with either the inside or the outside surface.

Leaking O.D. defect: A defect that was initially an O.D. defect but which has grown through the wall to become a leak.

Leaking I.D. defect: A defect that was initially an I.D. defect but which has grown through the wall to become a leak.

Superficial defect: A lap, crevice, pit, group of pits, metallurgical anomaly, or plain dent (i.e., without scratches, gouges, or cracks) that is of insufficient extent to reduce the effective strength level of the pipe below 100% of SMYS.

Kinds of defects

*Definitions marked by an asterisk are from API Bulletin 5T1 Nondestructive Testing Terminology, Third Edition, April, 1972.

Not every conceivable kind of defect in pipe is covered by this list. The list is limited to those that are likely to be encountered in an in-service pipeline.

Defects originating from pipe manufacture

a. Defects not necessarily in the seam weld (primarily in the body of the pipe)
   - * Lap: Fold of metal that has been rolled or otherwise worked against the surface of rolled metal but has not fused into sound metal.
   - * Pit: A depression resulting from the removal of foreign material rolled into the surface during manufacture.
   - * Rolled-in slugs: A foreign metallic body rolled into the metal surface, usually not fused.
   - * Seam: Crevice in rolled metal that has been more or less closed by rolling or other work but has not been fused into sound metal.
   - * Hard spot: An area in the pipe with a hardness level considerably higher than that of surrounding metal; usually caused by localized quenching.
   - * Crack: A stress-induced separation of the metal that, without any other influence, is insufficient in extent to cause complete rupture of the material.

b. Defects in the seam weld
   - * Incomplete fusion: Lack of complete coalescence of some portion of the metal in a weld joint.
   - * Incomplete penetration: A condition where weld metal does not continue through the full thickness of the joint.
   - * Undercut: Undercutting on submerged-arc-welded pipe is the reduction in thickness of the pipe wall adjacent to the weld where it is fused to the surface of the pipe.
   - * Weld area crack: A crack that occurs in the weld deposit, the fusion line, or the HAZ. (Crack: A stress-induced separation of the metal that, without any other influence, is sufficient in extent to cause complete rupture of the material.)
   - * Upturned fiber imperfection: Metal separation, resulting from imperfections at the edge of the plate or skelp, parallel to the surface, which turn toward the I.D. or O.D. pipe surface when the edges are upset during welding.
   - * Penetrator: A localized spot of incomplete fusion.
   - * Cold weld: A metallurgically inexact term generally indicating a lack of adequate weld bonding strength of the abutting edges, due to insufficient heat and/or pressure. A cold weld may or may not have separation in the weld line. Other more definitive terms should be used whenever possible.

Defects originating from external or internal environmental degeneration of the pipe

a. Seam weld defects

Selective corrosion: Preferential corrosion in the fusion line of an electric resistance-welded or flashwelded longitudinal seam.

Hydrogen stress cracking: Environmentally stimulated cracking of the weld metal or HAZ of the longitudinal seam.

b. Defects not in the body of the pipe (possibly in the seam weldbut not specifically because of the seam weld)

7. R.D. Morel, Welded Repairs on API 5LX-52 Pipe. 13th Annual Petroleum Mechanical Engineering Conference, Denver, Colorado, September 24, 1958.
8. J.R. Christian, B.A. Cassie, Analysis of a Live Welded Repair on An Artificial Defect,. ERS.C.96, October, 1975.
9. R.B. Smith, R.J. Eiber, "Field Failure Survey and Investigation," Fourth Symposium on Line Pipe Research. American Gas Association, 1969, Catalog No. L30075.
10. R.J. Eiber, "Field Failure Investigations," Fifth Symposium on Line Pipe Research. American Gas Association, 1974, Catalog No. L30174.
11. R.R. Wright, Proper Inspection Methods Minimize Pipeline Failures, Oil and Gas Journal (May 23, 1977) 51–56.

Cross-country pipeline—vertical down electrode consumption, pounds of electrode per joint

Wall Thickness	5/16″			3/8″			1/2″			5/8″			3/4″		
Elec. Diameter	5/32″	3/16″	Total	5/32″	3/16″	Total	5/32″	3/16″	Total	5/32″	3/16″	Total	5/32″	3/16″	Total
Nom. Pipe Diameter, in	lb			lb			lb			lb			lb		
6	0.34	0.54	0.88	0.34	0.86	1.2									
8	0.45	0.71	1.2	0.45	1.1	1.6	0.45	2.2	2.7						
10	0.56	0.88	1.4	0.56	1.4	2.0	0.56	2.7	3.3						
12	0.67	1.0	1.7	0.67	1.7	2.4	0.67	3.2	3.9						
14	0.73	1.1	1.8	0.73	1.8	2.5	0.73	3.5	4.2	0.73	5.7	0.73	8.3	9.0	
16	0.84	1.3	2.1	0.84	2.1	2.9	0.84	4.0	4.8	0.84	6.5	7.3	0.84	9.4	10.2
18	0.94	1.5	2.4	0.94	2.3	3.2	0.94	4.5	5.4	0.94	7.3	8.2	0.94	10.6	11.5
20	1.10	1.6	2.7	1.10	2.6	3.7	1.1	5.0	62.71	1.1	8.1	9.2	1.1	11.8	12.9
22	1.20	1.8	3.0	1.20	2.9	4.1	1.2	5.5	6.7	1.2	8.9	10.1	1.2	13.0	14.2
24	1.30	2.0	31.3	1.30	3.1	4.4	1.3	6.0	7.3	1.3	9.7	11.0	1.3	14.2	15.5
26	1.40	2.1	3.5	1.40	3.4	4.8	1.4	6.5	7.9	1.4	10.5	11.9	1.4	15.3	16.7
28	1.50	2.3	3.8	1.50	3.7	5.2	1.5	7.0	8.5	1.5	11.3	12.8	1.5	16.5	18.0
30	1.60	2.5	4.1	1.60	3.9	5.5	1.6	7.5	95.51	1.6	13.71	13.7	1.6	17.7	19.3
32	1.70	2.6	4.3	1.70	4.2	5.9	1.7	8.0	9.7	1.7	13.0	14.7	1.7	18.9	20.6
34	1.80	2.8	4.6	1.80	4.4	6.2	1.8	8.6	10.4	1.8	13.8	15.6	1.8	20.1	21.9
36	1.90	2.9	4.8	1.90	4.7	6.6	1.9	4.61	11.0	1.9	14.6	16.5	1.9	21.3	23.2
38	2.00	3.1	5.1	2.00	2.0	7.0	2.0	9.6	11.6	2.0	22.4	22.4	2.0	22.4	24.4
40	2.10	3.3	5.4	2.10	5.42	7.3	2.1	42.1	123.62	2.1	16.2	18.3	2.1	23.6	25.7
42				2.20	5.5	7.7	2.2	10.6	12.8	2.2	17.0	192	2.2	24.8	27.0
44				2.30	5.7	8.0	2.3	11.1	13.4	2.3	17.8	20.1	2.3	26.0	28.3
46				2.40	6.0	8.4	2.4	11.6	14.0	2.4	18.6	21.0	2.4	27.2	29.6
48				2.50	6.3	8.8	2.5	12.1	14.6	2.5	19.4	21.9	2.5	28.3	30.8

Figures in the table include 4-in stub lengths. These figures will vary with different stub loss practices. Quantities required for the 5/32" size will vary based on travel speeds of the stringer bead and hot pass. Slow travel speeds may increase these quantities by up to 50%.

Reprinted courtesy of the Lincoln Electric Company.

DIRECTIONAL CROSSING

Guidelines for a successful directional crossing bid package

The Directional Crossing Contractors Association (DCCA) has been addressing the issue of what information should be made available to contractors and engineers so that future projects proceed as planned. Crossings of rivers and other obstacles using directional drilling techniques are increasingly being utilized around the world. As in any construction project, it is necessary for the contractor to have as much information as possible to prepare a competitive and comprehensive proposal and to be able to successfully install the crossing. Better preconstruction information also allows the work to be undertaken more safely and with less environmental disturbance.

Development and uses

Originally used in the 1970s, directional crossings are a marriage of conventional road boring and directional drilling of oil wells. The method is now the preferred method of construction. Crossings have been installed for pipelines carrying oil, natural gas, petrochemicals, water, sewerage and other products. Ducts have been installed to carry electric and fiber-optic cables. Besides crossing under rivers and waterways, installations have been made crossing under highways, railroads, airport runways, shore approaches, islands, areas congested with buildings, pipeline corridors, and future water channels.

Technology limits

The longest crossing to date has been about 11,000 ft. Pipe diameters of up to 48 in have been installed. Although directional drilling was originally used primarily in the U.S. Gulf Coast through alluvial soils, more and more crossings are being undertaken through gravel, cobble, glacial till, and hard rock.

Advantages

Directional crossings have the least environmental impact of any alternate method. The technology also offers maximum depth of cover under the obstacle, thereby affording maximum protection and minimizing maintenance costs. River traffic is not interrupted, because most of the work is confined to either bank. Directional crossings have a predictable and short construction schedule. Perhaps most significantly, directional crossings are in many cases less expensive than other methods. Bundles of multiple pipes can be pulled into a common bore.

Technique

Pilot Hole—A pilot hole is drilled beginning at a prescribed angle from horizontal and continues under and across the obstacle along a design profile made up of straight tangents and long radius arcs. A schematic of the technique is shown in Figure 1. Concurrent to drilling a pilot hole, the contractor may elect to run a larger diameter "wash pipe" that will encase the pilot drill string. The wash pipe acts as a conductor casing providing rigidity to the smaller diameter pilot drill string and will also save the drilled hole should it be necessary to retract the pilot string for bit changes. The directional control is brought about by a small bend in the drill string just behind the cutting head. The pilot drill string is not rotated except to orient the bend. If the bend is oriented to the right, the drill path then proceeds in a smooth radius bend to the right. The drill path is monitored by an electronic package housed in the pilot drill string near the cutting head. The electronic package detects the relation of the drill string to the earth's magnetic field and its inclination. These data are transmitted back to the surface where calculations are made as to the location of the cutting head. Surface location of the drill head also can be used where there is reasonable access.

Pre-ream—Once the pilot hole is complete, the hole must be enlarged to a suitable diameter for the product pipeline. For instance, if the pipeline to be installed is 36-in diameter, the hole may be enlarged to 48-in diameter or larger. This is accomplished by "pre-reaming" the hole to successively larger diameters. Generally, the reamer is attached to the drill string on the bank opposite the drilling rig and pulled back into the pilot hole. Joints of drill pipe are added as the reamer makes its way back to the drilling rig. Large quantities of slurry are pumped into the hole to maintain the integrity of the hole and to flush out cuttings.

Pullback—Once the drilled hole is enlarged, the product pipeline can be pulled through it. The pipeline is prefabricated on the bank opposite the drilling rig. A reamer is attached to the drill string, and then connected to the pipeline pullhead via a swivel. The swivel prevents any translation of the reamer's rotation into the pipeline string, allowing for a smooth pull into the drilled hole. The drilling rig then begins the pullback operation, rotating and pulling on the drill string and once again circulating high volumes of drilling slurry. The pullback continues until the reamer and pipeline break ground at the drilling rig.

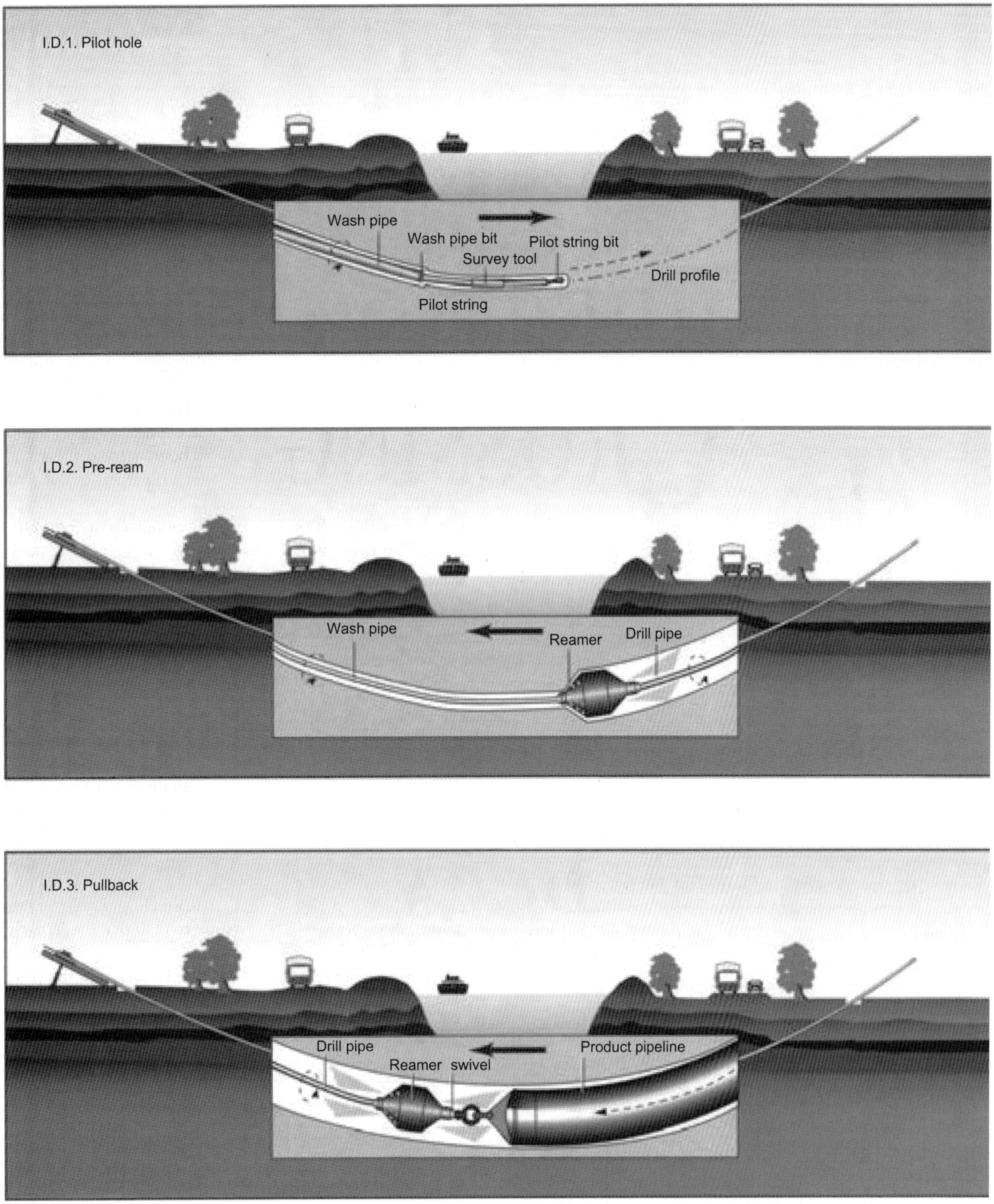

Figure 1. Technique of directional crossing - pilot hole followed by pre-reaming and pullback.

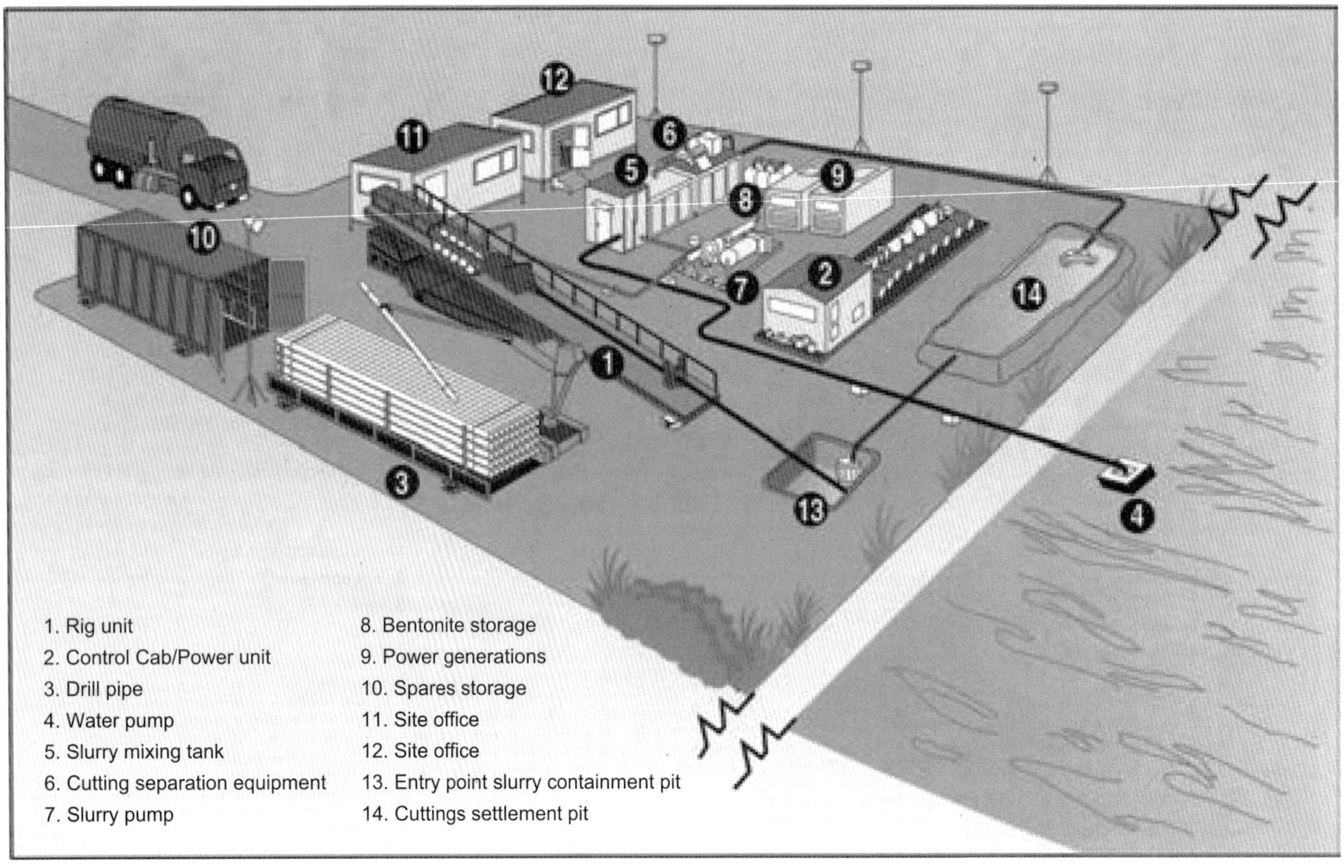

Figure 2. Rig side work space.

Layout and design

Access—Heavy equipment is required on both sides of the crossing. To minimize cost, access to either side of the crossing should be provided with the least distance from an improved road. Often the pipeline right-of-way is used for access. All access agreements should be provided by the owner. It is not practical to negotiate such agreements during the bid process.

Work space

Rig Side—The rig spread requires a minimum 100-ft-wide by 150-ft-long area. This area should extend from the entry point away from the crossing, although the entry point should be at least 10 ft inside the prescribed area. Since many components of the rig spread have no predetermined position, the rig site can be made up of smaller irregular areas. Operations are facilitated if the area is level, hardstanding, and clear of overhead obstructions. The drilling operation requires large volumes of water for the mixing of the drilling slurry. A nearby source of water is necessary (Figure 2).

Pipe Side—Strong consideration should be given to provide a sufficient length of work space to fabricate the product pipeline into one string. The width will be as necessary for normal pipeline construction, although a work space of 100-ft-wide by 150-ft-long should be provided at the exit point itself. The length will assure that during the pullback the pipe can be installed in one uninterrupted operation. Tie-ins of successive strings during the pullback operation increase the risk considerably, because the pullback should be continuous (Figure 3).

Profile survey

Once the work locations have been chosen, the area should be surveyed and detailed drawings prepared. The eventual accuracy of the drill profile and alignment is dependent on the accuracy of the survey information.

Profile design parameters

Depth of Cover—Once the crossing profile has been taken and the geotechnical investigation complete, a determination

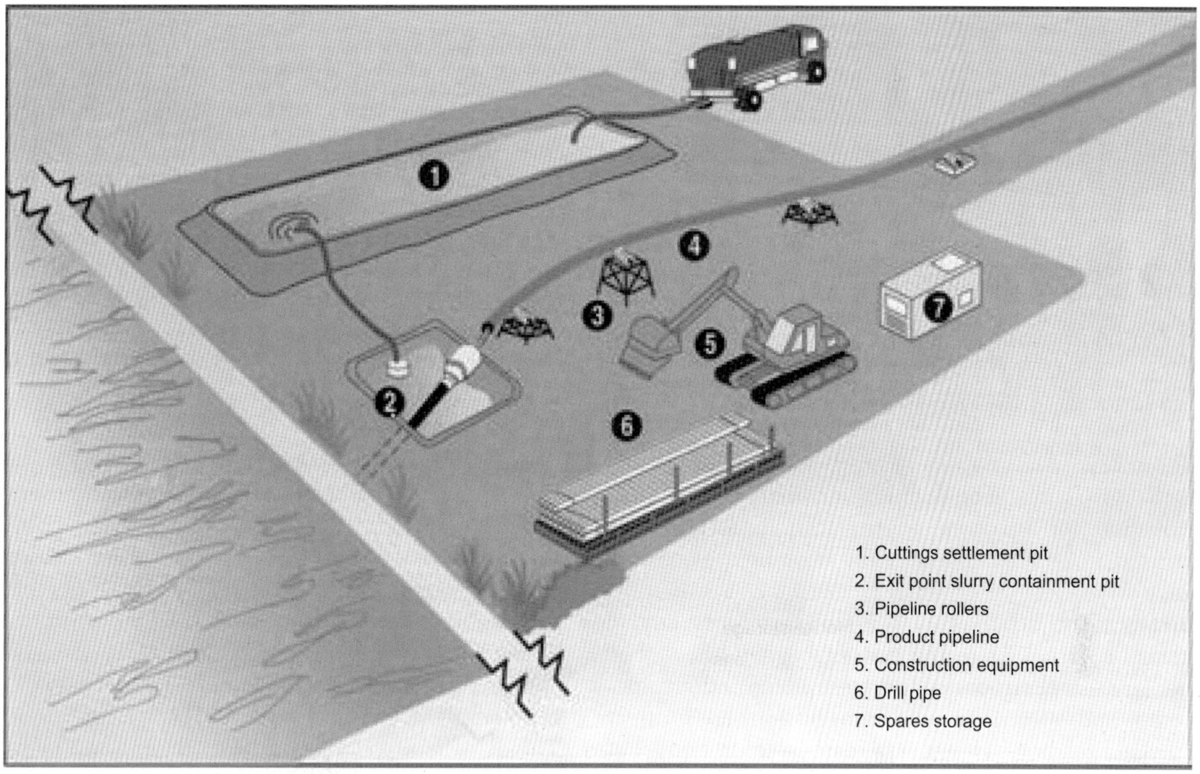

Figure 3. Pipe side work space.

of the depth of cover under the crossing is made. Factors considered may include flow characteristics of the river, the depth of scour from periodic flooding, future channel widening/deepening, and whether pipeline or cable crossings already exist at the location. It is normally recommended that the minimum depth of cover be 20 ft under the lowest section of the crossing. While 20 ft is a recommended depth of cover on a river crossing, crossings of other obstacles may have differing requirements.

Penetration Angles and Radius of Curvature—An entry angle between 8° and 20° can be used for most crossings. It is preferable that straight tangent sections are drilled before the introduction of a long radius curve. The radius of the curve is determined by the bending characteristic of the product pipeline, increasing with the diameter. A general "rule-of-thumb" for the radius of curvature is 100 ft/1-in diameter for steel line pipe. The curve usually brings the profile to the elevation providing the design cover of the pipeline under the river. Long horizontal runs can be made at this elevation before curving up toward the exit point. Exit angle should be kept between 5° and 12° to facilitate handling of the product pipeline during pullback.

Calculate minimum radius of curvature:

$$R = \frac{EC}{FS - \frac{PD}{4t}}$$

$$C = \frac{D}{12} \times 0.5$$

$$R = \frac{30,000,000 \times 0.58}{0.9 \times 35,000 - \frac{1440 \times 14}{4 \times 0.5}} = 816.99$$

R = Minimum radius, ft
E = Modulus of elasticity, 30,000,000
C = One half of pipe OD in ft
F = Design factor
S = Pipe SMYS, psi
P = Design pressure , psig
D = Pipe outside diameter, inches
t = Pipe wall thickness, inches

Example.

$D = 14$ in
$t = 0.500$ in
$F = 0.9$
$S = 35{,}000$ psi
$E = 30{,}000{,}000$

$$C = \frac{D}{12} \times 0.5 = 0.58$$

$$R = \frac{30{,}000{,}000 \times 0.58}{0.9 \times 35{,}000 - \dfrac{1440 \times 14}{4 \times 0.5}} = 817 \text{ ft}$$

Drill Survey—Most downhole survey tools are electronic devices that give a magnetic azimuth (for "right/left" control) and inclination (for "up/down" control). Surface locators can also be used in conjunction with the downhole electronic package.

Accuracy—The accuracy of the drill profile is largely dependent on variations in the earth's magnetic field. For instance, large steel structures (bridges, pilings, other pipelines, etc.) and electric power transmission lines affect magnetic field readings. However, a reasonable drill target at the pilot hole exit location is 10 ft left or right, and −10 ft to +30 ft in length.

As-Built Drawings—Normally, survey calculations are conducted every 30 ft during pilot hole operations. As-built drawings that are based on these calculations should be provided by the contractor. Alternate methods such as gyroscoping, ground penetrating radar, or "intelligent" pigs may also be used to determine the as-built position.

Geotechnical investigation

Number of Borings—The number of exploration holes is a function of the proposed crossing length and the complexity of the strata. If the crossing is about 1000 ft, a borehole made on each side of the crossing may suffice. If an examination of these borings indicates that conditions are likely to be homogeneous on both sides, it may not be necessary to conduct further sampling. If the report indicates anomalies discontinuity in the strata or the presence of rock or large concentrations of gravel, it is advisable to make additional borings to better define the strata. Longer crossings (especially large-diameter pipelines) that indicate gravel, cobble, boulders, or rock should have samples taken about 600—800 ft apart unless significant anomalies are identified that might necessitate more borings. All borings should be located on the crossing profile along with their surface elevations being properly identified. If possible, the borings should be conducted at least 25 ft off of the proposed centerline. The bore holes should be grouted upon completion. This will help prevent the loss of drilling slurry during the crossing installation.

Depth of Borings—All borings should be made to a minimum depth of 40 ft below the lowest point in the crossing or 20 ft below the proposed depth of the crossing, whichever is greater. In some instances, it may be beneficial to the owner and the contractor to install the crossing at a greater depth than the owner requires for his permit. It is suggested that all borings be through the same elevation to better determine the consistency of the underlying material and note any patterns that may be present.

Standard Classification of Soils—A qualified technician or geologist should classify the material in accordance with the Unified Soil Classification System and ASTM Designations D-2487 and D-2488. It is beneficial to have a copy of the field-drilling log completed by the field technician or driller. These logs include visual classifications of materials, as well as the driller's interpretation of the subsurface conditions between samples.

Standard Penetration Test (SPT)—In order to better define the density of granular materials, the geotechnical engineer generally uses the Standard Penetration Test (SPT) in general accordance with ASTM Specifications D-1586. This is a field test that involves driving a 2-in split spoon sampler into the soil by dropping a hammer of a specific weight (usually 140 lb) a specified distance (usually 30 in) to determine the number of blows necessary to drive the sampler 12 in. In very dense soils, the field technician may note the number of blows required to drive the sampler less than the required 12 in (i.e., 50 blows for 3 in). The number obtained is the standard penetration resistance value (N) and is used to estimate the in situ relative density of cohesionless soils. Some geotechnical firms will conduct these penetration tests in cohesive materials and rock, and to a lesser extent, the consistency of cohesive soils and the hardness of rock can be determined.

Thinwalled "Shelby" Tube Sampling—Most geotechnical firms prefer to use a Thin-walled Tube Sampling method for obtaining samples of cohesive materials. These tests are conducted in general accordance with ASTM Specification D-1587. This test is similar to the Standard Penetration test except the sample is collected by hydraulically pushing a thin-walled seamless steel tube with a sharp cutting edge into the ground. The hydraulic pressure required to collect the sample is noted on the field log. This produces a relatively undisturbed sample that can be further analyzed in the laboratory. These samples can be field tested with hand-held penetrometers, but more accurate readings of density and consistency can be obtained by performing unconfined compressive strength tests where the results are noted in tons per square foot. Generally, for directional drilling contractors a standard penetration test using the split spoon sampler described earlier will suffice in both materials.

Sieve Analysis of Granular Materials—A sieve analysis is a mechanical test of granular materials performed on samples collected in the field during the standard penetration test with the split spoon sampler. The split spoon samples are taken to the laboratory and processed through a series of screens. The sample provides a percentage analysis of the granular material by size and weight. It is one of the most important tests undertaken.

Rock Information—If rock is encountered during the soil investigation borings, it is important to determine the type, the relative hardness, and the unconfined compressive strength. This information is typically collected by the geotechnical drilling firm by core drilling with a diamond bit core barrel. The typical core sample recovered with this process has a 2-in diameter. The type of rock is classified by a geologist. The geologist should provide the Rock Quality Designation (RQD), which rates the quality of the rock based on the length of core retrieved in relation to the total length of the core. The hardness of the rock (Mohs' Scale of Hardness) is determined by comparing the rock to ten materials of known hardness. The compressive strength is determined by accurately measuring the core and then compressing the core to failure. This information pertaining to the underlying rock formation is imperative to determine the type of downhole equipment required and the penetration rates that can be expected.

Pipe material selection

Wall Thickness—D/t "Rule-of-Thumb"—The following table provides generalized recommendations for the selection of steel pipe wall thicknesses relative to pipe diameter. These recommendations are meant to be used only as a starting point in the design. It is recommended that in the final design, specific stresses be calculated and compared with allowable limits.

(For high-density polyethylene (HDPE) pipe, a standard dimension ratio of *D/t*, SDR, of 11 or less is recommended and the pipe manufacturer should be consulted.)

Diameter (D)	Wall Thickness (t)
6″ and smaller	0.250″
6–12″	0.375″
12–30″	0.500″
>30″	$D/t < 50$

Stress Analysis—In finalizing the design, the stresses imposed during construction and in-service must be calculated and checked to be within allowable limits for the grade of material. The stresses at each stage must be considered acting individually and in combination. Stresses result due to spanning between rollers prior to pullback, the hydrostatic testing pressures, pulling forces during installation, radius of curvature as the pipe enters the ground, the drilling profile curvature, external pressures in the drilled hole, and the working pressure.

1. Pre-installation
   a. Hoop and longitudinal stresses resulting from hydrostatic testing are calculated.
   b. Using the known distance between rollers as the free spanning distance, the maximum hogging and sagging moments can be calculated. Considering the greater of these two moments, the maximum spanning stress is calculated. Note: During hydrostatic testing the pipeline will be full of water; therefore, the additional weight of water must be included in these calculations.
2. Installation
   a. The spanning stresses calculated in stage 1.b. also apply in this installation phase.
   b. The theoretical pulling force must be determined in order to provide the stresses that will result. An assumed downhole friction factor of 1.0 is recommended to provide conservative results and to include the effect of the pipeline being pulled around a curve. The maximum predicted pulling force should then be used in calculating the resulting longitudinal stress.
   c. Allowing for a 10% drilling tolerance leads to the use of a radius of curvature 90% of the design radius when calculating the longitudinal curvature stresses.
   d. External pressure from static head in the drilled hole and/or overburden pressures must be considered. It is recommended that the static head resulting from the maximum envisaged drilling fluid density should be used with a factor of safety of 1.5 to provide conservative estimations of resulting hoop and longitudinal stresses.
3. Post-installation
   a. The longitudinal curvature stresses calculated for stage 2.c. above are used again here.
   b. External pressure stresses from 2.d. apply.
   c. Hoop and longitudinal stresses resulting from the final hydrostatic test are calculated.
4. In-service
   a. Curvature—see 2.c.
   b. External pressure—see 2.d.

The maximum working pressure of the pipeline is used in calculating longitudinal and hoop stresses that will be imposed during service.

Allowable Stresses—Having determined the individual and combined stresses at each stage of construction and those for the in-service condition, they must be compared with allowable limits.

1. ASME B31.8—1992, Table A842.22 provides the following limits
   - Maximum allowable longitudinal stress: 80% SMYS.
   - Maximum allowable hoop stress: 72% SMYS.
   - Maximum allowable combined stress: 90% SMYS.

(Where SMYS is the Specified minimum Yield Strength of the pipe material.)

2. Regulatory bodies may impose additional limits to those specified above—owner companies should identify any such further constraints and ensure the adequacy of the design.

Pipeline coating

Coatings are applied to provide a corrosion barrier and an abrasion barrier. Directional crossings generally encounter varying materials and often can be exposed to extra abrasion during the pullback. An outer abrasion-resistant overcoat is often warranted. To facilitate the pullback of the pipeline, the coating should bond well to the pipe to resist soil stresses and have a smooth, hard surface to reduce friction and maintain the corrosion barrier. As in any pipeline construction, the recommended external coating system should be compatible with any specifications for the field joint coating or any internal coating.

The recommended pipe coating is mill-applied fusion bonded epoxy (FBE). The recommended minimum thickness is 20 mils.

Joint Coating—The coating application of the weld area is the most critical field operation to maintain a smooth abrasion-resistant pipe string. It is recommended that the girth weld be coated with FBE powder utilizing the induction heating coil and powder application machine to a minimum dry film thickness of 25 mils. As an alternate, two component catalyzed liquid epoxy may be applied to the girth weld area to a minimum dry film thickness of 25 mils using a paintbrush or roller. Tape should never be used for joint coating on the pullback portion of a directional crossing.

Coating Repair—It is recommended that small coating damaged areas be repaired with a polymeric melt stick patching material. Coating Repair—It is recommended that small coating damaged areas be repaired with a polymeric melt stick patching material. Holidays larger than 1 in in diameter should be repaired utilizing the two-component catalyzed liquid epoxy applied with a paintbrush or roller. Tapes should never be used for repair of coating damaged areas on the pullback portion of a directional crossing. Shrink sleeves designed for directional drilling may be used to repair holidays in the coating.

Abrasion—Resistant Overcoat—As an extra abrasion-resistant barrier for crossings that may encounter stones. Abrasion-Resistant Overcoat—As an extra abrasion-resistant barrier for crossings that may encounter stones, boulders, or solid rock, it is recommended the FBE-coated pipeline be overcoated with an epoxy-based polymer concrete. The material should be applied at a mill or with a portable yard coating machine to a minimum thickness of 40 mils. Girth weld and coating damaged areas should be field coated using an epoxy-based polymer concrete compatible with the overcoat material. The patch material should be applied so that the material tapers uniformly and feathers into the original coating. Stability of the pipeline in drilled crossings is not normally a concern, so a Portland cement-type concrete coating is not recommended.

Drilling slurry—containment, recycling, and disposal

The directional crossing process requires the use of large volumes of slurry that provide the following functions. The directional crossing process requires the use of large volumes of slurry that provide the following functions:

1. Hydraulic cutting with a jet.
2. Provide energy to the drill motor.
3. Lubricate the cutting head.
4. Transport drill cuttings to the surface.
5. Stabilize the hole against collapse.
6. Guard against loss of slurry into surrounding formations.

Slurry Composition—The slurries most commonly used are bentonite-based. Bentonite is a naturally occurring Wyoming clay known for its hydrophilic characteristics. Often polymer extenders are also added to enhance certain characteristics. Material Safety Data Sheets (MSDS) are readily available from suppliers and can be presented to regulatory/disposal authorities.

Containment—The slurry is pumped downhole and circulates back to the surface and collected in "return pits." These pits typically have a volume of at least 500 cu. ft. Depending on the nature of the project, the slurry is pumped from the return pits to a "settling and containment pit." These pits vary in size depending on pumping rates and contain the slurry for recycling or disposal.

Recycling Slurry—Slurry that has been circulated downhole and collected in the containment pit is then passed through machinery that separates the cuttings from the slurry. This process involves a series of shaking sieves and various size hydroclones.

Slurry and Cuttings Disposal—Significant amounts of slurry are normally disposed of at the end of a project.

Economics for disposal is extremely site specific. This slurry can be disposed of by:

1. Use at another drilling location.
2. Spread onto raw land for water retention improvement.
3. Evacuate to a dumpsite.

If working in an area of contaminated ground. If working in an area of contaminated ground, the slurry should be tested for contamination and disposed of in a manner that meets governmental requirements.

Cost mitigation for the owner

With prebid planning and research. With prebid planning and research, the owner can realize significant savings in slurry disposal. It is in the owner's interest to define and specify all disposal issues. In particular:

1. Define an approved disposal site as part of the project specifications.
2. Because it is difficult to estimate disposal quantities, disposal should be a separate bid item as either "cost plus" or on "unit rates."
3. Inadvertent returns are not uncommon and difficult to predict. The issue should be fairly represented to permitting bodies prior to construction. Contingency plans for containment and disposal of inadvertent returns should be priced as a separate bid item and agreed prior to construction.

Conditions of the contract

Always utilize a written contract to maximize communication and minimize controversy. Always utilize a written contract to maximize communication and minimize controversy. A contract should be used to anticipate what the parties intend to do if a problem occurs on the job. The contract should be readable and understandable.

Bid package

A proposal presented by a contractor to the owner is an offer by the contractor that becomes a binding contract if accepted by the owner. A proposal presented by a contractor to the owner is an offer by the contractor that becomes a binding contract if accepted by the owner. The parties, price, and performance must be specified. Define the project to be undertaken by detailing the scope of work and incorporate all plans and specifications from the bid package.

Differing ground conditions and walkaway provision

Owners should accept the responsibility of performing an adequate geotechnical investigation. Owners should accept the responsibility of performing an adequate geotechnical investigation. Despite adequate testing of ground conditions, unknown, unusual, and unexpected ground conditions may be encountered. The contract should provide solutions when the project encounters differing ground conditions. The walkaway provision in the contract should entitle the contractor to stop work and walk away from the job without the owner having the right to take over the contractor's equipment. The contractor should be entitled to receive compensation for demobilization, lost profits, and work performed prior to walking away. If the project is completed, the contractor should be paid on a cost-plus basis. Assumption of risk of unforeseen ground conditions by the contractor affects the bid price.

Environmental concerns

Before the project begins. Before the project begins, address environmental concerns because owners and contractors are included as potentially responsible parties when environmental damages and clean-up costs are assessed. Federal, state, and local laws must be evaluated, and licensing, permitting, and other regulations must be followed. Directional crossings that damage soil or water may cause liability.

1. Turbidity of Water and Inadvertent Returns—As these events are difficult to predict and work stoppage may occur, the contract should offer a mechanism to mutually address and mitigate the problem. Liabilities are generally shared by both the contractor and owner and many times can be insured.
2. Slurry Disposal—Comply with the regulations of the area regarding slurry disposal. Slurry disposal should be referred to in the contract and bid as a separate line item on a cost plus or unit price basis.

Allocation of risk of loss

Evaluate and allocate risks of loss that may occur during the project. Evaluate and allocate risks of loss that may occur during the project. Owners should share the risk of loss rather than shifting all the losses through the indemnification to the contractor, because the bid price is directly affected by contingent losses. Insurance may provide coverage by third parties for losses from differing ground conditions or environmental losses.

Dispute resolution

Provide for dispute resolution in the event of controversy by including mediation or arbitration provisions in the contract, 1940 Provide for dispute resolution in the event of controversy by including mediation or arbitration provisions in the contract. Disputes should be resolved in the following order: (1) negotiation, (2) nonbinding mediation through a third party, (3) landing arbitration, and lastly, (4) litigation. Determine whom should be parties to the resolution, what law will be used, and where the dispute will be resolved.

POLYETHYLENE PIPING

Simplified installation guidelines for PE pipe

Excerpted from Chapter 7, Underground Installation of PE Piping, The Plastics Pipe Institute Handbook of Polyethylene Pipe, 2nd Ed, 2008.

Flexible Pipe Installation Theory

PE piping is considered "flexible" pipe. Flexible pipes can deflect up to their allowable deflection limit without damage. Most PE pipes can withstand large amounts of deflection without damage but for practical purposes PE pipes are limited to 7.5% deflection or less depending on the DR and application. For PE pipes, flexibility is directly proportional to the Dimension Ratio (DR). Low DR pipes such as DR 7.3 have high resistance to deflection because their flexibility is very low, or conversely their stiffness is high. DR 7.3 has a Pipe Stiffness (PS) per ASTM D2412 of about 1600 psi. On the other hand DR 32.5 pipe has a PS of about 12.5 psi. Such a wide range in flexibility across a product line means different installation requirements for different DR's may be necessary to achieve successful and economical installations. The depth of cover and anticipated surface loads also affect the particular installation requirements. Therefore the engineer has to make an assessment of the application and site conditions to determine the best and most economical installation design.

In general there are two objectives to achieve in an installation. The first is to provide an envelope of embedment to protect the pipe from mechanical damage from impact or hard objects (cobbles, boulders) in the soil. The second is to provide support against earth and live load pressures, where this is required. The envelope surrounding the pipe is referred to as the "embedment". The earth and live loads are supported by the combination of the pipe's stiffness and the embedment's stiffness. Lower DR pipes will carry more of the load and require less support from the soil. When support from the embedment is needed by the pipe to resist earth and live loads, the embedment material is often compacted. The trench backfill placed on top of the embedment material may also be compacted. Compaction of trench backfill immediately above the pipe facilitates the redistribution of some of the load away from the pipe and into the side-fill soil.

Terminology of Pipe Embedment Materials

The materials enveloping a buried pipe are generally identified, as shown by their function or location (Figure 1).

Foundation - A foundation is required only when the native trench bottom does not provide a firm working platform for placement of the pipe bedding material.

Initial Backfill - This is the critical zone of embedment soil surrounding the pipe from the foundation to at least 6 inches over the pipe. The pipe's ability to support loads and resist deflection is determined by the quality of the embedment material and the quality of its placement. Within the initial backfill zone are bedding, haunching, primary, and secondary zones.

Bedding - In addition to bringing the trench bottom to required grade, the bedding levels out any irregularities and ensures uniform support along the length of the pipe.

Haunching - The backfill under the lower half of the pipe (haunches) distributes the superimposed loadings. The nature of the haunching material and the quality of its placement are one of the most important factors in limiting the deformation of PE pipe.

Primary Initial Backfill - This zone of backfill provides the primary support against lateral pipe deformation. To ensure such support is available, this zone should extend from trench grade up to at least 75 percent of the pipe diameter. Under some conditions, such as when the pipe will be permanently below the ground water table, the primary initial backfill should extend to at least 6 inches over the pipe.

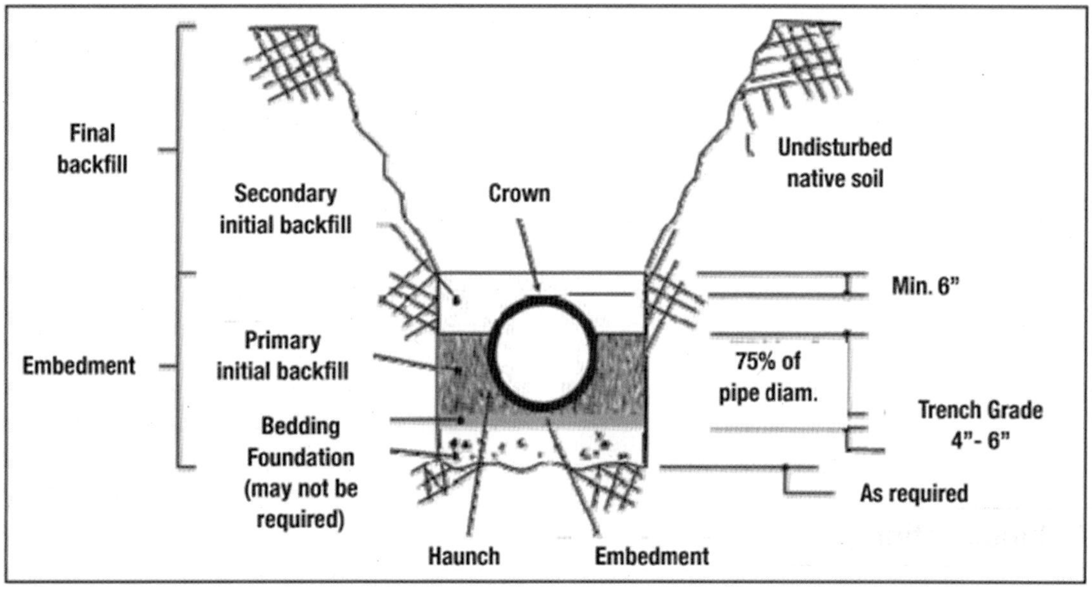

Figure 1. Pipe trench. Note: When groundwater levels are expected to reach above the pipe, the secondary initial backfill should be a continuation of the primary initial backfill in order to provide optimum pipe support. Minimum trench width will depend on site conditions and embedment materials.

Secondary Initial Backfill - The basic function of the material in this zone is to distribute overhead loads and to isolate the pipe from any adverse effects of the placement of the final backfill.

Final Backfill - As the final backfill is not an embedment material, its nature and quality of compaction has a lesser effect on the flexible pipe. However, arching and thus a load reduction on the pipe is promoted by a stiff backfill. To preclude the possibility of impact or concentrated loadings on the pipe, both during and after backfilling, the final backfill should be free of large rocks, organic material, and debris. The material and compaction requirements for the final backfill should reflect sound construction practices and satisfy local ordinances and sidewalk, road building, or other applicable regulations.

Simplified Installation Guidelines

The engineer must evaluate the site conditions, the subsurface conditions, and the application objectives to determine the extent of support the pipe may need from the surrounding soil. Where the pipe burial depth is relatively deep, where subsurface soil conditions are not supportive of pipe, where surface loads or live loads are present, or where the pipe DR is high, the engineer will generally want to prepare a specific installation specification. Guidelines for what to include in that specification are given in later sections of this chapter. On the other hand there are many applications that meet the criterion below for using Simplified Installation Guidelines. These applications would include many rural transmission and distribution water lines, many force main sewer lines, and many process water lines. Typically these lines contain pressure pipes installed at shallow depths which are sufficiently stiff to resist the minimal earth load. In some cases a pipeline may contain sections that require specific engineering such as a section that crosses a road.

Simplified Installation Guidelines for Pressure Pipe

(Small diameter pressure pipes usually have adequate stiffness and are usually installed in such shallow depths that it is unnecessary to make an internal inspection of the pipe for deflection.)

A quality job can be achieved for most installations following the simple steps that are listed below. These guidelines apply where the following conditions are met:

1. Pipe diameter of 24-inch or less
2. SDR equal to or less than 26
3. Depth of cover between 2.5 feet and 16 feet
4. Groundwater elevation never higher than 2 feet below the surface
5. The route of the pipeline is through stable soil

Stable soil is an arbitrary definition referring to soil that can be cut vertically or nearly vertically without significant sloughing, or soil that is granular but dry (or de-watered) that can stand vertical to at least the height of the pipe. These soils must also possess good bearing strength. (Quantitatively, good bearing capacity is defined as a minimum unconfined compressive strength of 1000 psf for cohesive soils or a minimum standard penetration resistance of 10 blows per ft for coarse grained soils.) Examples of soils that normally do not possess adequate stability for this method are mucky, organic, or loose and wet soils.

Where the above conditions are met, the specifier can write installation specifications from the following steps. The specifier should insure that all OSHA, state and local safety regulations are met.

The following are general guidelines for the installation of PE pipe. Other satisfactory methods or specifications may be available. This information should not be substituted for the judgment of a professional engineer in achieving specific requirements.

Simplified Step-by-Step Installation

Trenching. Trench collapses can occur in any soil and account for a large number of worker deaths each year. In unbraced or unsupported excavations, proper attention should be paid to sloping the trench wall to a safe angle. Consult the local codes. All trench shoring and bracing must be kept above the pipe. (If this is not possible, consult the more detailed installation recommendations.) The length of open trench required for fused pipe sections should be such that bending and lowering the pipe into the ditch does not exceed the manufacturer's minimum recommended bend radius and result in kinking. The trench width at pipe grade should be equal to the pipe outer diameter (O. D.) plus 12 inches.

De-watering. For safe and proper construction the groundwater level in the trench should be kept below the pipe invert. This can be accomplished by deep wells, well points or sump pumps placed in the trench.

Bedding. Where the trench bottom soil can be cut and graded without difficulty, pressure pipe may be installed directly on the prepared trench bottom. For pressure pipe, the trench bottom may undulate, but must support the pipe smoothly and be free of ridges, hollows, and lumps. In other situations, bedding may be prepared from the excavated material if it is rock free and well broken up during excavation. The trench bottom should be relatively smooth and free of rock. When rocks, boulders, or large stones are encountered which may cause point loading on the pipe, they should be removed and the trench bottom padded with 4 to 6 inches of tamped bedding material. Bedding should consist of free-flowing material such as gravel, sand, silty sand, or clayey sand that is free of stones or hard particles larger than one-half inch.

Placing Pipe in Trench. PE pressure pipe up to about 8" in diameter and weighing roughly 6 lbs per ft or less can usually be placed in the trench by hand. Heavier, larger diameter pipe will require handling equipment to lift, move, and lower the pipe into the trench. Pipe must not be dumped, dropped, pushed, or rolled into the trench. Appropriate safety precautions must be observed whenever persons are in or near the trench

Pipe Embedment. The embedment material should be a coarse grained soil, such as gravel or sand, or a coarse grained soil containing fines, such as a silty sand or clayey sand. The particle size should not exceed one-half inch for 2 to 4-inch pipe, three-quarter inch for 6 to 8-inch pipe and one inch for all other sizes. Where the embedment is angular, crushed stone may be placed around the pipe by dumping and slicing with a shovel. Where the embedment is naturally occurring gravels, sands and mixtures with fines, the embedment should be placed in lifts, not exceeding 6 inches in thickness, and then tamped. Tamping should be accomplished by using a mechanical tamper. Compact to at least 85 percent Standard Proctor density as defined in ASTM D698, Standard Test Methods for Laboratory Compaction Characteristics of Soil Using Standard Effort, (12 400 ft-lbf/ft3 (600 kN-m/m3))." Under streets and roads, increase compaction to 95 percent Standard Proctor density.

Leak Testing. If a leak test is required, it should be conducted in accordance with the procedure in Chapter 2 after the embedment material is placed.

Trench Backfill. The final backfill may consist of the excavated material, provided it is free from unsuitable matter such as large lumps of clay, organic material, boulders or stones larger than 8 inches, or construction debris. Where the pipe is located beneath a road, place the final backfill in lifts as mentioned earlier and compact to 95 percent Standard Proctor Density.

How to construct a trench for PE piping

Excerpted from Chapter 7, Underground Installation of PE Piping, The Plastics Pipe Institute Handbook of Polyethylene Pipe, 2nd Ed, 2008.

Trench Construction

Trenches should be excavated to line and grade as indicated by contract documents and in accordance with applicable safety standards. Excavation should proceed upgrade. Excessive runs of open trench should be avoided to minimize such problems as trench flooding, caving of trench walls and the freezing of trench bottom and backfill material, and to minimize hazards to workmen and traffic. This can be accomplished by closely coordinating excavation with pipe installation and backfilling. native soil supporting and containing the pipe and its embedment soil, stability of trench walls, and water accumulation in the trench. When encountering unstable soils or wet conditions, they should be controlled by providing an alternate foundation, sloping or bracing the trench walls, de-watering the trench bottom, or some other such measure.

Trench Width

Since flexible pipe has to support, at most, only the weight of the "prism" or vertical column of soil directly over the pipe, the precaution of keeping the trench as narrow as possible is not the concern that it is for a rigid pipe, which can be subjected to the weight of the soil beside the prism as well as the prism itself. With PE pipe, widening the trench will generally not cause a loading greater than the prism load on the pipe. Trench width in firm, stable ground is determined by the practical consideration of allowing sufficient room for the proper preparation of the trench bottom and placement and compaction of the pipe embedment materials, and the economic consideration of the costs of excavation and of imported embedment materials. Trench width in firm, stable ground will generally be determined by the pipe size and the compacting equipment used. Table 1 gives minimum trench width values.

The trench width may need to be increased over the values in Table 1 to allow for sufficient clearance between the trench sidewalls and the pipe for compaction equipment. Typically for large diameter pipe (18" and larger), this required clearance will vary from 12 to 18 inches. If two or more pipes are laid in the same trench, sufficient space must be provided between the pipes so that embedment material can be compacted.

Table 1
Minimum Trench Width in Stable Ground vs. Pipe Size

Nominal Pipe Size (in)	Minimum Trench Width (in)
< 3	12
3 to 24	Pipe O.D. + 12
$>$ 24 to 63	Pipe O.D. + 24

Note: Minimum trench widths do not apply to trenching techniques that use chain or wheel trenchers or plows to lay PE pipe. Chain and wheel trenching techniques feed PE pipe over the earth-cutting machine and lay the pipe immediately into the earth-cut. These techniques use round-bottom chain or wheel trenching machines that match pipe radius and do not require extra trench width to place embedment in the pipe haunches below the pipe springline. Plowing techniques feed smaller diameter PE pipe or tubing through a chute that is integrated into an earth plow. Plowing may not require backfilling.

Trench Length

Table 2 lists the recommended lengths of trench openings for each placement of continuous lengths of fused pipe, assembled above the trench. When the trench sidewalls are significantly sloped, somewhat shorter trench openings may be used. When space or ground conditions do not permit these suggested trench openings, the pipe lengths may be joined within the trench, using a joining machine or flanged couplings. When bell-and-spigot jointed pipe or flange-end pipe is used, the trench opening needs to be only long enough to accommodate placement and assembly of a single pipe length.

Stability of the Trench

Although the native soil in which PE pipe is installed need not be as strong and stiff as the pipe embedment materials, it should provide adequate support and stable containment of the embedment material so that the density of the embedment material does not diminish. If the trenching conditions present construction problems such as trench sidewalls that readily slough off or a soft trench floor that will not support workers or compaction, it is termed unstable. The instability is usually a condition of the trench and not the soil. Most often the primary cause of the instability is high groundwater, not the soil. Even soft or loose soils can provide good support for the pipe if they are confined. The problem with unstable conditions generally occurs during the installation. When the trench is opened where groundwater is present, most soils, except firm, cohesive soils (firm clays) or cemented soils, tend to slough off the trench wall. This results in a trench that keeps widening, with loose material falling into the trench floor.

Table 2
Suggested Length of Minimum Trench Opening (Feet) for Installation of Joined Lengths of PE Pipe

Nominal Pipe Size (in)	Depth of Trench (Feet)					
	3	5	7	9	11	13
½ to 3	15	20	25	30	35	40
4 to 8	25	30	35	40	45	50
10 to 14	35	40	45	50	55	60
16 to 22	45	50	55	60	65	70
24 to 42		60	65	70	75	80
48			80	90	100	110

Soil formations that commonly lead to unstable trenching conditions include materials with fine grain soils (silts or clays) saturated with water and uncemented sands saturated with water. In some cases, where the soil has an extremely high water content, such as with peat or with clay (or silt) having a water content beyond the liquid limit, the soil behaves "hydraulically", that is, the water in the soil controls the soil's behavior. Here, the backfill must be designed to sustain all the pressure from the pipe without support from the in-situ soil. These conditions may occur in saturated fine-grained soils where the unconfined compressive strength of the soil is less than 500 psf, or in saturated, sandy soils where the standard penetration value, N, is less than 6 blows per ft. In this case, an engineering evaluation should be made to determine the necessity for special procedures such as a "wide" trench or permanent trench sheeting of the trench width.

As mentioned above, most trench stability problems occur in trenches that are excavated below the groundwater level. (However, the designer and the contractor should keep in mind that all trenches pose the risk of collapse and therefore workers should not be in trenches that are not adequately braced or sloped.) Stability can be improved by lowering the water table through deep wells, well-points, or other such means. In some ground the permeability is such that the only option is to remove the water after it has seeped out of the trench walls. Here the contractor will use underdrains or sumps on the trench floor. De-watering should continue throughout the pipe laying operation until sufficient cover is placed over the pipe so that it will not float.

Stability of Trench Floor

Trench floor stability is influenced by the soils beneath the trench. The floor must be stable in order to support the bedding material. A stable bedding minimizes bending of the pipe along its horizontal axis and supports the embedment enveloping the pipe. Generally, if the trench floor can be walked on without showing foot prints it is considered stable.

In many cases the floor can be stabilized by simply dewatering. Where dewatering is not possible or where it is not effective, stabilization of the trench floor may be accomplished by various cost-effective methods which can be suited to overcome all but the most difficult soil conditions. Included among these are the use of alternate trench foundations such as wood pile or sheathing capped by a concrete mat, or wood sheathing with keyed-in plank foundation; stabilization of the soil by the use of special grout or chemicals; geofabric migration barriers; or ballasting (undercutting). A cushion of bedding material must be provided between any special foundation and the pipe. Permanently buried timber should be suitably treated.

Stabilization by ballasting (undercutting) is the removal of a sufficient quantity of undesirable material. This technique is frequently employed to stabilize randomly encountered short sections of unstable soil. The extent of required over-excavation and details of accompanying construction requirements will be determined by the engineer in consideration of the qualities of the unstable soil and the specific design requirements. The following are general guidelines:

The trench bottom should be over-excavated over the full trench width from 18 to 36 inches below the pipe grade (depending on the soil strength and pipe diameter) and then brought back to grade with a foundation of ballast material topped with Class I material. An appropriate bedding should then be placed on the foundation. The grading of the foundation material should be selected so that it acts as an impervious mat into which neither the bedding, other embedment material, nor the surrounding native soil will migrate.

These guidelines are suitable for most situations except for extremely weak soils (such as quicksands, organic silts, and peats) which may call for further over excavation, or other special treatment.

Stability of Trench Walls

In order to control deflection, the embedment material must be placed from undisturbed trench sidewall to undisturbed trench sidewall. Where trench walls are unstable, it may be necessary to use trench shields, bracing, or permanent sheeting to achieve a stable sidewall while installing the pipe. Where material sloughs into the trench it should be removed. This technique often leads to widening the trench.

Walls of trenches below the elevation of the crown of the pipe should be maintained as vertical as possible. The shape of the trench above the pipe will be determined by the stability of the trench walls, excavation depth, surface loadings near the trench, proximity of existing underground structures, presence of groundwater or runoff water, safety and practical considerations. These will determine if the trench walls may be vertical, excavated with slope or benched sides, or shored. When trench walls are shored or otherwise stabilized, the construction scheme must allow for the proper placement and compaction of pipe embedment materials. Some suggested trench construction schemes follow. The final procedure must be in compliance with all applicable safety regulations.

Sloping of trench walls in granular and cohesionless soils should be provided whenever the walls are more than about four feet in depth or otherwise required by state, local or federal regulations. For safety, if the walls are not sloped, they should be stabilized by alternate means such as shoring or bracing. The slope should be no greater than the angle of repose of the materials being excavated and should be approved by the engineer.

Shoring or bracing will frequently be required in wet fine grained cohesive type soils and clays. Bracing or sheathing that is constructed of treated timber, steel or other acceptable material may be used to stabilize trench walls either permanently or temporarily. Wherever possible, sheathing and bracing should be installed so that its bottom extends no lower than about one-quarter of the pipe diameter below the pipe crown. When so installed, pulling the sheathing will minimally disturb the embedment material and the side support it provides. Sheathing that is installed to project below the pipe springline should be left in place unless, as with some thinner sheathing, it is designed to be pulled and removed without disturbing the embedment next to the pipe. In this case, the trench width should be increased by 12 to 24 inches depending on the pipe diameter to allow for minor disturbance to the embedment near the sheathing. Vibratory placement or extraction of sheeting is not advised. This method can cause severe disturbance to the bedding and liquefaction of the surrounding soils. Where steel sheet piling is used as sheathing and is to be removed or pulled, to minimize disturbance to the pipe embedment, it should be installed so that it is not closer than one pipe diameter or 18 inches, whichever is larger, from either side of the pipe. The void left by removal of the sheathing should be filled with embedment material.

3: Pipe Design

Steel pipe design ... 126
Properties of pipe ... 131
Length of pipe in bends ... 134
Calculation of pipe bends ... 135
Spacing of pipe supports ... 137
Chart of deflection of horizontal pipelines ... 138
American standard taper pipe threads (NPT) ... 139
British standard taper pipe threads ... 140
Normal engagement between male and female threads to make tight joints ... 141
Calculate pipe weight, contents, velocity ... 141
Formulas and constants for tubular goods problems ... 142
How to calculate the contraction or expansion of a pipeline ... 143
Estimate weight of pipe in metric tons per kilometer ... 143
How to find pipe weight from outside diameter and wall thickness ... 144
What is the maximum allowable length of unsupported line pipe? ... 144
Identify the schedule number of pipe by direct measurement ... 144
Determine buoyancy of bare steel pipe ... 145
Determine buoyancy of bare and concrete-coated steel pipe in water and mud ... 145
Weights of piping materials ... 146
Allowable working pressure for carbon steel pipe ... 147
Find the stress in pipe wall due to internal pressure ... 148
How to calculate stress in above ground/below ground transitions ... 148
How to identify the series number of flanged fittings ... 152
Dimensions of three-diameter ells with tangents ... 152
Spectacle blind thicknesses ... 152
Carbon dioxide pipeline design ... 153

Steel pipe design

The maximum allowable design pressure stress will depend upon the intended service for the pipeline.

Pipelines to be used for transporting liquid petroleum are covered by ANSI/ASME B31.4—"Liquid Petroleum Transportation Piping Systems." Pipelines used for transporting gas are covered by ANSI/ASME B31.8—"Gas Transmission and Distribution Systems." Other design codes will apply to pipelines transporting other materials such as carbon dioxide and hydrogen.

Pipelines that must be operated in compliance with the Federal Pipeline Safety Regulations will also need to comply with the applicable parts of these regulations: Part 192 for gas transportation systems and Part 195 for liquid transportation systems.

Gas pipelines—ANSI/ASME B31.8

The maximum allowable pressure is calculated by the following equation:

$$P = (2St/D) \times F \times E \times T$$

where

P = Design pressure, lb/in^2,
S = Specified minimum yield strength, lb/in^2 (see Table 1),
t = Nominal wall thickness, in,
D = Nominal outside diameter, in,
F = Design factor (see Table 2),
E = Longitudinal joint factor (see Table 3),
T = Temperature derating factor (see Table 4).

Class location definitions may be obtained from p. 192.111 of Part 192 of the Federal Pipeline Safety Regulations. A typical calculation is as follows:

Pipe: 16″ OD × 0.250″ wt API 5LX X52 ERW
Location: Class 1, therefore $F = 0.72$ (see Table 2)
Temperature: 90 °F, Temp. factor $T = 1$ (see Table 4)
Joint Factor: $E = 1.0$ (see Table 3)

$$P = (2 \times 52,000 \times 0.250/16.0) \times 0.72 \times 1 \times 1$$
$$P = 1170\ \text{lb/in}^2\ \text{gauge}$$

Table 1
Specified Minimum Yield Strength for Steel and Iron Pipe Commonly Used in Piping System

Specification	Grade	Type[1]	SMYS, psi
API 5L	A25	BW, ERW, S	25,000
API 5L	A	ERW, FW, S, DSA	30,000
API 5L	B	ERW, FW, S, DSA	35,000
API 5LS [Note (2)]	A	ERW, DSA	30,000
API 5LS	B	ERW, DSA	35,000
API 5LS	X42	ERW, DSA	42,000
API 5LS	X46	ERW, DSA	46,000
API 5LS	X52	ERW, DSA	52,000
API 5LS	X56	ERW, DSA	56,000
API 5LS	X60	ERW, DSA	60,000
API 5LS	X65	ERW, DSA	65,000
API 5LS	X70	ERW, DSA	70,000
API 5LX [Note (2)]	X42	ERW, FW, S, DSA	42,000
API 5LX	X46	ERW, FW, S, DSA	46,000
API 5LX	X52	ERW, FW, S, DSA	52,000
API 5LX	X56	ERW, FW, S, DSA	56,000
API 5LX	X60	ERW, FW, S, DSA	60,000
API 5LX	X65	ERW, FW, S, DSA	65,000
API 5LX	X70	ERW, FW, S, DSA	70,000
ASTM A53	Open Hrth. Bas. Oxy., Elec. Furn.	BW	25,000
ASTM A53	Bessemer	BW	30,000

(table continued on next page)

Table 1
Specified Minimum Yield Strength for Steel and Iron Pipe Commonly Used in Piping System (continued)

Specification	Grade	Type[1]	SMYS, psi
ASTM A53	A	ERW, S	30,000
ASTM A53	B	ERW, S	35,000
ASTM A106	A	S	30,000
ASTM A106	B	S	35,000
ASTM A106	C	S	40,000
ASTM A134	—	EFW	[Note (3)]
ASTM A135	A	ERW	30,000
ASTM A135	B	ERW	35,000
ASTM A139	A	ERW	30,000
ASTM A139	B	ERW	35,000
ASTM A333	1	S, ERW	30,000
ASTM A333	3	S, ERW	35,000
ASTM A333	4	S	35,000
ASTM A333	6	S, ERW	35,000
ASTM A333	7	S, ERW	35,000
ASTM A333	8	S, ERW	75,000

Reproduced from ANSI/ASME Code B31-8-1982, Appendix D. Reprinted courtesy of The American Society of Mechanical Engineers.

Table 2
Values of Design Factor F

Construction Type (See 841.151)	Design Factor *F*
Type A	0.72
Type B	0.60
Type C	0.50
Type D	0.40

Reproduced from ANSI/ASME Code B31-8-1982, Table 841.1A. Reprinted courtesy of The American Society of Mechanical Engineers.

Table 3
Longitudinal Joint Factor E

Spec. Number	Pipe Class	*E* Factor
ASTM A53	Seamless	1.00
	Electric resistance welded	1.00
	Furnace welded	0.60
ASTM A106	Seamless	1.00
ASTM A134	Electric fusion arc welded	0.80
ASTM A135	Electric resistance welded	1.00
ASTM A139	Electric fusion welded	0.80
ASTM A211	Spiral welded steel pipe	0.80
ASTM A381	Double submerged-arc-welded	1.00
ASTM A671	Electric fusion welded	1.00*
ASTM A672	Electric fusion welded	1.00*
API 5L	Seamless	1.00
	Electric resistance welded	1.00
	Electric flash welded	1.00
	Submerged arc welded	1.00
	Furnace butt welded	0.60
API 5LX	Seamless	1.00
	Electric resistance welded	1.00
	Electric flash welded	1.00
	Submerged arc welded	1.00
API 5LS	Electric resistance welded	1.00
	Submerged welded	1.00

Note: Definitions for the various classes of welded pipe are given in 804.243.

*Includes Classes 12, 22, 32, 42, and 52 only.

Reproduced from ANSI/ASME Code B31-8-1982, Table 841.1B. Reprinted courtesy of The American Society of Mechanical Engineers.

Table 4
Temperature Derating Factor T for Steel Pipe

Temperature, °F	Temperature Derating Factor *T*
250 or less	1.000
300	0.967
350	0.933
400	0.900
450	0.867

Note: For Intermediate temperatures, interpolate for derating factor.

Reproduced from ANSI/ASME Code B31-8-1982, Table 841.1C. Reprinted courtesy of The American Society of Mechanical Engineers.

Liquid pipelines—ANSI/ASME B31.4

The internal design pressure is determined by using the following formula:

$$P = (2St/D) \times E \times F$$

where

P = Internal design pressure, lb/in^2 gauge,
S = Specified minimum yield strength, lb/in^2 (see Table 5),
t = Nominal wall thickness, in,
D = Nominal outside diameter of the pipe, in,
E = Weld joint factor (see Table 6),
F = Design factor of 0.72.

Note: Refer to p. 195.106 of Part 195 Federal Pipeline Safety Regulations for design factors to be used on offshore risers and platform piping and cold worked pipe.

A typical calculation of the internal design pressure is as follows:

Pipe: 26″ OD × 0.3125″ wt API 5LX X52 ERW
Weld joint factor $E = 1.0$ (see Table 6)
Design factor $F = 0.72$

$$P = (2 \times 52,000 \times 0.3125/26) \times 1 \times 0.72$$
$$P = 900 \text{ lb/in}^2 \text{ gauge}$$

Table 5
Allowable Stresses for Reference Use in Piping Systems

Allowable stress values (*S*) shown in this table are equal to 0.72 × E (weld joint factor) × specified minimum yield strength of the pipe. Allowable stress values shown are for new pipe of known specification. Allowable stress values for new pipe of unknown specification, ASTM A 120 specification or used (reclaimed) pipe shall be determined in accordance with 402.3.1.
For some code computations, particularly with regard to branch connections [see 404.3.1 (d) (3)] and expansion, flexibility, structural attachments, supports, and restraints (Chapter II, Part 5), the weld joint factor E need not be considered.
For specified minimum yield strength of other grades in approved specifications, refer to that particular specification.
Allowable stress value for cold worked pipe subsequently heated to 600 °F (300 °C) or higher (welding excepted) shall be 75% of value listed in table.
Definitions for the various types of pipe are given in 400.2.
(Metrics stress levels are given in MPa [1 Megapascal = 1 million pascals])

Specification	Grade	SMYS psi (MPa)	Notes	(*E*) Weld Joint Factor	(*S*) Allowable Stress psi (MPa)
Seamless					
API 5L	A25	25,000 (172)	(1)	1.00	18,000 (124)
API 5L, ASTM A53, A106	A	30,000 (207)	(1)(2)	1.00	21,600 (149)
API 5L, ASTM A53, A106	B	35,000 (241)	(1)(2)	1.00	25,200 (174)
ASTM A106	C	40,000 (278)	(1)(2)	1.00	28,800 (199)
ASTM A524	I	35,000 (241)	(1)	1.00	25,200 (174)
ASTM A524	II	30,000 (207)	(1)	1.00	21,600 (149)
API 5LU	U80	80,000 (551)	(1)(4)	1.00	57,600 (397)
API 5LU	U100	100,000 (689)	(1)(4)	1.00	72,000 (496)
API 5LX	X42	42,000 (289)	(1)(2)(4)	1.00	30,250 (208)
API 5LX	X46	46,000 (317)	(1)(2)(4)	1.00	33,100 (228)
API 5LX	X52	52,000 (358)	(1)(2)(4)	1.00	37,450 (258)
API 5LX	X56	56,000 (386)	(1)(4)	1.00	40,300 (278)
API 5LX	X60	60,000 (413)	(1)(4)	1.00	43,200 (298)
API 5LX	X65	65,000 (448)	(1)(4)	1.00	46,800 (323)
API 5LX	X70	70,000 (482)	(1)(4)	1.00	50,400 (347)
Furnace Welded-Butt Welded					
ASTM A53		25,000 (172)	(1)(2)	0.60	10,800 (74)
API 5L Class I & Class II	A25	25,000 (172)	(1)(2)(3)	0.60	10,800 (74)
API 5L (Bess) A53 (Bess)		30,000 (207)	(1)(2)(5)	0.60	12,950 (89)

Table 5
Allowable Stresses for Reference Use in Piping Systems (continued)

Specification	Grade	SMYS psi (MPa)	Notes	(*E*) Weld Joint Factor	(*S*) Allowable Stress psi (MPa)
Furnace Welded-Lap Welded					
API 5L Class I		25,000 (172)	(1)(2)(6)	0.80	14,400 (99)
API 5L Class II		28,000 (193)	(1)(2)(6)	0.80	16,150 (111)
API 5L (Bessemer)		30,000 (207)	(1)(2)(6)	0.80	17,300(119)
API 5L Electric furnace		25,000 (172)	(1)(2)(6)	0.80	14,400 (99)
Electric Resistance Welded and Electric Flash Welded					
API 5L	A25	25,000 (172)	(1)(7)(2)	1.00	18,000 (124)
API 5L, ASTM A53, A135	A	30,000 (207)		0.85	18,360 (127)
API 5L, 5LS, ASTM A53, A135	A	30,000 (207)	(1)	1.00	21.600 (149)
API 5L, ASTM A53, A135	B	35,000 (241)	(2)	0.85	21,420 (148)
API 5L, 5LS, ASTM A53, A135	B	35,000 (241)	(1)	1.00	25,200 (174)
API 5LS, API 5LX	X42	42,000 (289)	(1)(2)(4)	1.00	30,250 (208)
API 5LS, API 5LX	X46	46,000 (317)	(1)(2)(4)	1.00	33,100 (228)
API 5LS, API 5LX	X52	52,000 (358)	(1)(2)(4)	1.00	37,450 (258)
API 5LS, API 5LX	X56	56,000 (386)	(1)(4)	1.00	40,300 (279)
API 5LS, API 5LX	X60	60,000 (413)	(1)(4)	1.00	43,200 (297)
API 5LS, API 5LX	X65	65,000 (448)	(1)(4)	1.00	46,800 (323)
API 5LS, API 5LX	X70	70,000 (482)	(1)(4)	1.00	50,400 (347)
API 5LU	U80	80,000 (551)	(1)(4)	1.00	57,600 (397)
API 5LU	U100	100,000 (689)	(1)(4)	1.00	72,000 (496)
Electric Fusion Welded					
ASTM A 134	—	—		0.80	—
ASTM A 139	A	30,000 (207)	(1)(2)	0.80	17,300 (119)
ASTM A 139	B	35,000 (241)	(1)(2)	0.80	20,150 (139)
ASTM A 155	—	—	(2)(8)	0.90	—
ASTM A 155	—	—	(1)(8)	1.00	—
Submerged Arc Welded					
API 5L, API 5LS	A	30,000 (207)	(1)	1.00	21,600 (149)
API 5L, API 5LS	B	35,000 (241)	(1)	1.00	25,200 (174)
API 5LS, API 5LX	X42	42,000 (289)	(1)(2)	1.00	30,250 (208)
API 5L, API 5LS	B	35,000 (241)	(1)	1.00	25,200 (174)
API 5LS, API 5LX	X42	42,000 (289)	(1)(2)(4)	1.00	30,250 (208)
API 5LS, API 5LX	X46	46,000 (317)	(1)(2)(4)	1.00	33,100 (228)
API 5LS, API 5LX	X52	52,000 (358)	(1)(2)	1.00	37,450 (258)
API 5LS, API 5LX	X56	56,000 (386)	(1)(4)	1.00	40,300 (278)
API 5LS, API 5LX	X60	60,000 (413)	(1)(4)	1.00	43,200 (298)
API 5LS, API 5LX	X65	65,000 (448)	(1)(4)	1.00	46,800 (323)
API 5LS, API 5LX	X70	70,000 (482)	(1)(4)	1.00	50,400 (347)
API 5LU	U80	80,000 (551)	(1)(4)	1.00	57,600 (397)
API 5LU	U100	100,000 (689)	(1)(4)	1.00	72,000 (496)
ASTM A 381	Y35	35,000 (241)	(1)(2)	1.00	25,200 (174)
ASTM A 381	Y42	42,000 (290)	(1)(2)	1.00	30,250 (209)
ASTM A 381	Y46	46,000 (317)	(1)(2)	1.00	33,100 (228)
ASTM A 381	Y48	48,000 (331)	(1)(2)	1.00	34,550 (238)
ASTM A 381	Y50	50,000 (345)	(1)	1.00	36,000 (248)
ASTM A 381	Y52	52,000 (358)	(1)	1.00	37,450 (258)
ASTM A 381	Y60	60,000 (413)	(1)	1.00	43,200 (298)
ASTM A 381	Y65	65,000 (448)	(1)	1.00	46,800 (323)

Notes: (1) Weld joint factor E (see Table 402.4.3) and allowable stress value are applicable to pipe manufactured after 1958; (2) Weld joint factor E (see Table 402.4.3) and allowable stress value are applicable to pipe manufactured before 1959; (3) Class II produced under API 5L 23rd Edition, 1968, or earlier has a specified minimum yield strength of 28,000 psi (193 MPa); (4) Other grades provided for in API 5LS, API 5LU, and API 5LX not precluded; (5) Manufacture was discontinued and process deleted from API 5L in 1969; (6) Manufacture was discontinued and process deleted from API 5L in 1962; (7) A25 is not produced in electric flash weld; (8) See applicable plate specification for yield point and refer to 402.3.1 for calculation of (S).

Table 6
Weld Joint Factor

Specification Number	Pipe Type (1)	Weld Joint Factor *E*	
		Pipe Mfrd. Before 1959	Pipe Mfrd. After 1958
ASTM A53	Seamless	1.00	1.00
	Electric-resistance-welded	0.85 (2)	1.00
	Furnace lap-welded	0.80	0.80
	Furnace butt-welded	0.60	0.60
ASTM A106	Seamless	1.00	1.00
ASTM A134	Electric-fusion (arc)-welded single or double pass	0.80	0.80
ASTM A135	Electric-resistance-welded	0.85 (2)	1.00
ASTM A139	Electric-fusion-welded single or double pass	0.80	0.80
ASTM A155	Electric-fusion-welded	0.90	1.00
ASTM A381	Electric-fusion-welded, double sub arc-welded	—	1.00
API 5L	Seamless	1.00	1.00
	Electric-resistance-welded	0.85 (2)	1.00
	Electric-flash-welded	0.85 (2)	1.00
	Electric-induction-welded	—	1.00
	Submerged arc-welded	—	1.00
	Furnace lap-welded	0.80	0.80 (3)
	Furnace butt-welded	0.60	0.60
API 5LS	Electric-resistance-welded	—	1.00
	Submerged arc-welded	—	1.00
API 5LX	Seamless	1.00	1.00
	Electric-resistance-welded	1.00	1.00
	Electric-flash-welded	1.00	1.00
	Electric-induction-welded	—	1.00
	Submerged arc-welded	1.00	1.00
API 5LU	Seamless	—	1.00
	Electric-resistance-welded	—	1.00
	Electric-flash-welded	—	1.00
	Electric-induction-welded	—	1.00
	Submerged arc-welded	—	1.00
Known	Known	(4)	(5)
Unknown	Seamless	1.00 (6)	1.00 (6)
Unknown	Electric-resistance or flash-welded	0.85 (6)	1.00 (6)
Unknown	Electric-fusion-welded	0.80 (6)	0.80 (6)
Unknown	Furnace lap-welded or over NPS 4	0.80 (7)	0.80 (7)
Unknown	Furnace butt-welded or NPS 4 and smaller	0.60 (8)	0.60 (8)

Notes: (1) Definitions for the various pipe types (weld joints) are given in 400.2; (2) A weld joint factor of 1.0 may be used for electric-resistance-welded or electric-flash-welded pipe manufactured prior to 1959 where (a) pipe furnished under this classification has been subjected to supplemental tests and/or heat treatments as agreed to by the supplier and the purchaser, and such supplemental tests and/or heat treatment demonstrate the strength characteristics of the weld to be equal to the minimum tensile strength specified for the pipe, or (b) pipe has been tested as required for a new pipeline in accordance with 437.4.1; (3) Manufacture was discontinued and process deleted from API 5L in 1962; (4) Factors shown above for pipe manufactured before 1959 apply for new or used (reclaimed) pipe, if pipe specification and pipe type are known and it is known that the pipe was manufactured before 1959, or not known whether it was manufactured after 1958; (5) Factors shown above for pipe manufactured after 1958 apply for new or used (reclaimed) pipe if pipe specification and pipe type are known and it is known that the pipe was manufactured after 1958; (6) Factor applies for new or used pipe of unknown specification and ASTM A120 if type of weld joint is known; (7) Factor applies for new or used pipe of unknown specification and ASTM A120 if type of weld joint is known to be furnace lap-welded, or for pipe over NPS 4 if type of joint is unknown; (8) Factor applies for new or used pipe of unknown specification and ASTM A120 if type of weld joint is known to be furnace butt-welded, or for pipe NPS 4 and smaller if type of joint is unknown.

Reproduced from ANSI/ASME Code B31-4-1979, Table 402.4.3. Reprinted courtesy of The American Society of Mechanical Engineers.

Properties of pipe

D	T	d	W	A_o	V_m	A_f	I	Z
4.500	0.125	4.250	5.840	1.178097	92.644911	0.098516	4.12	1.83
	0.156	4.188	7.240		89.961575	0.095662	5.03	2.24
	0.172	4.156	7.950		88.592057	0.094206	5.49	2.44
	0.188	4.124	8.660		87.233042	0.092761	5.93	2.64
	0.219	4.062	10.010		84.629845	0.089993	6.77	3.01
STD	0.237	4.026	10.790		83.136406	0.088405	7.56	3.36
6.625	0.188	6.249	12.93	1.734421	200.292532	0.212985	19.7	6.0
	0.219	6.187	14.99		196.337808	0.208779	22.6	6.8
	0.250	6.125	17.02		192.422518	0.204616	25.5	7.7
STD	0.280	6.065	18.98		188.671072	0.200627	28.1	8.5
	0.312	6.001	21.04		184.710235	0.196415	30.9	9.3
	0.375	5.875	25.03		177.035128	0.188254	36.1	10.9
8.625	0.188	8.249	16.94	2.258020	349.016785	0.371133	44.4	10.3
	0.219	8.187	19.66		343.790038	0.365575	51.1	11.9
	0.250	8.125	22.36		338.602723	0.360059	57.7	13.4
	0.277	8.071	24.70		334.116868	0.355289	63.4	14.7
	0.312	8.001	27.70		328.346391	0.349153	70.5	16.3
STD	0.322	7.981	28.56		326.706916	0.347410	72.5	16.8
	0.344	7.937	30.43		323.114514	0.343590	76.9	17.8
	0.375	7.875	33.05		318.086203	0.338243	82.9	19.2
10.750	0.219	10.312	24.63	2.814343	545.418061	0.579980	87.0	16.2
	0.250	10.250	28.04		538.879221	0.573027	113.7	21.2
	0.279	10.192	31.20		532.797939	0.566560	125.9	23.4
	0.307	10.136	34.24		526.959102	0.560352	137.5	25.6
STD	0.365	10.020	40.49		514.966704	0.547599	160.8	29.9
	0.500	9.750	54.74		487.587921	0.518486	212.0	39.4
12.750	0.219	12.312	29.31	3.337942	777.500935	0.826770	169.3	26.6
	0.250	12.250	33.38		769.690071	0.818464	191.9	30.1
	0.281	12.188	37.43		761.918639	0.810200	214.1	33.6
	0.312	12.126	41.45		754.186639	0.801978	236.0	37.0
STD	0.375	12.000	49.57		738.594720	0.785398	279.0	43.8
	0.438	11.874	57.60		723.165661	0.768991	321.0	50.4
	0.500	11.750	65.42		708.140511	0.753014	363.0	56.7
14.000	0.219	13.562	32.24	3.665191	943.389822	1.003171	226.0	32.3
	0.250	13.500	36.72		934.783943	0.994020	255.0	36.5
	0.281	13.438	41.18		926.217495	0.984910	285.1	40.7
	0.312	13.376	45.62		917.690481	0.975843	315.0	45.0
	0.344	13.312	50.18		908.929763	0.966527	344.3	49.2
STD	0.375	13.250	54.58		900.482886	0.957545	373.0	53.3
	0.438	13.124	63.45		883.438151	0.939420	429.0	61.4
	0.500	13.000	72.10		866.822970	0.921752	484.0	69.1

D	*T*	*d*	*W*	A_o	V_m	A_f	*I*	*Z*
16.000	0.219	15.562	36.92	4.188790	1242.151387	1.320864	338.0	42.3
	0.250	15.500	42.06		1232.273483	1.310360	384.0	48.0
	0.281	15.438	47.18		1222.435011	1.299899	429.0	53.6
	0.312	15.376	52.28		1212.635972	1.289479	474.0	59.3
	0.344	15.312	57.53		1202.562197	1.278766	519.0	64.8
STD	0.375	15.250	62.59		1192.843296	1.268432	562.0	70.3
	0.438	15.124	72.81		1173.213479	1.247558	649.0	81.1
	0.500	15.000	82.78		1154.054250	1.227185	732.0	91.5
18.000	0.250	17.500	47.40	4.712389	1570.796063	1.670335	549.0	61.0
	0.281	17.438	53.18		1559.685567	1.658520	614.2	68.2
	0.312	17.376	58.95		1548.614504	1.646747	679.0	75.5
	0.344	17.312	64.88		1537.227671	1.634639	744.0	82.7
STD	0.375	17.250	70.60		1526.236746	1.622952	807.0	89.6
	0.438	17.124	82.16		1504.021848	1.599329	932.0	103.6
	0.500	17.000	93.46		1482.318570	1.576250	1053.0	117.0
20.000	0.250	19.500	52.74	5.235988	1950.351683	2.073942	756.6	75.7
	0.281	19.438	59.19		1937.969163	2.060775	846.5	84.6
	0.312	19.376	65.61		1925.626075	2.047650	935.5	93.5
	0.344	19.312	72.22		1912.926185	2.034145	1026.5	102.6
STD	0.375	19.250	78.61		1900.663236	2.021105	1113.8	111.4
	0.438	19.124	91.52		1875.863256	1.994733	1288.6	128.9
	0.500	19.000	104.14		1851.615930	1.968950	1457.2	145.7
24.000	0.250	23.500	63.42	6.283185	2832.562043	3.012056	1315.7	109.6
	0.281	23.438	71.19		2817.635474	2.996184	1473.1	122.8
	0.312	23.376	78.94		2802.748338	2.980354	1629.2	135.8
	0.344	23.312	86.92		2787.422334	2.964056	1789.1	149.1
STD	0.375	23.250	94.63		2772.615336	2.948311	1942.8	161.9
	0.438	23.124	110.23		2742.645193	2.916442	2251.3	187.6
	0.500	23.000	125.51		2713.309770	2.885247	2550.0	212.5
26.000	0.250	25.500	676	6.806784	3335.216783	3.546564	1676.8	129.0
	0.281	25.438	77.20		3319.018190	3.529339	1878.0	144.5
	0.312	25.376	85.61		3302.859030	3.512155	2077.7	159.8
	0.344	25.312	94.27		3286.219968	3.494462	2282.3	175.6
STD	0.375	25.250	102.64		3270.140946	3.477364	2479.1	190.7
	0.438	25.124	119.59		3237.585722	3.442746	2874.5	221.1
	0.500	25.000	136.19		3205.706250	3.408846	3257.8	250.6
30.000	0.250	29.500	79.44	7.853982	4463.625383	4.746477	2585.8	172.4
	0.281	29.438	89.20		4444.882741	4.726547	2897.5	193.2
	0.312	29.376	98.94		4426.179533	4.706659	3207.1	213.8
	0.344	29.312	108.97		4406.914357	4.686173	3524.7	235.0
STD	0.375	29.250	118.66		4388.291286	4.666370	3830.4	255.4
	0.438	29.124	138.30		4350.565898	4.626254	4445.7	296.4
	0.500	29.000	157.55		4313.598330	4.586943	5043.5	336.2

(table continued on next page)

D	*T*	*d*	*W*	A_o	V_m	A_f	*I*	*Z*
32.000	0.250	31.500	84.78	8.377580	5089.379243	5.411884	3143.2	196.4
	0.281	31.438	95.20		5069.364577	5.390601	3522.7	220.2
	0.312	31.376	105.60		5049.389344	5.369360	3899.9	243.7
	0.344	31.312	116.32		5028.811111	5.347478	4287.0	267.9
STD	0.375	31.250	126.67		5008.916016	5.326322	4659.7	291.2
	0.438	31.124	147.66		4968.605547	5.283457	5410.3	338.1
	0.500	31.000	168.23		4929.093930	5.241442	6140.2	383.8
34.000	0.250	33.500	90.11	8.901179	5756.166143	6.120924	3775.3	236.0
	0.281	33.438	101.19		5734.879453	6.098289	4231.9	264.5
	0.312	33.376	112.25		5713.632196	6.075695	4685.9	292.9
	0.344	33.312	123.65		5691.740905	6.052416	5151.8	322.0
STD	0.375	33.250	134.67		5670.573786	6.029908	5600.7	350.0
	0.406	33.188	145.67		5649.446099	6.007442	6047.1	377.9
	0.438	33.124	157.00		5627.678235	5.984294	6505.2	406.6
	0.469	33.062	167.95		5606.630686	5.961913	6946.5	434.2
	0.500	33.000	178.89		5585.622570	5.939574	7385.4	461.6
36.000	0.250	35.500	95.45	9.424778	6463.986083	6.873597	4487.0	280.4
	0.281	35.438	107.20		6441.427369	6.849609	5030.4	314.4
	0.312	35.376	118.92		6418.908087	6.825663	5570.9	348.2
	0.344	35.312	131.00		6395.703739	6.800988	6125.9	382.9
STD	0.375	35.250	142.68		6373.264596	6.777127	6660.6	416.3
	0.406	35.188	154.34		6350.864885	6.753308	7192.6	449.5
	0.438	35.124	166.35		6327.783964	6.728764	7738.7	483.7
	0.469	35.062	117.97		6305.464391	6.705031	8265.0	516.6
	0.500	35.000	189.57		6283.184250	6.681339	8788.5	549.3
42.000	0.344	41.312	153.04	10.995574	8753.790482	9.308503	9767.7	610.5
STD	0.375	41.250	166.71		8727.535266	9.280584	10624.3	664.0
	0.406	41.188	180.35		8701.319482	9.252707	11477.1	717.3
	0.438	41.124	194.42		8674.299389	9.223974	12353.3	772.1
	0.469	41.062	208.03		8648.163743	9.196183	13198.3	824.9
	0.500	41.000	221.61		8622.067530	9.168433	14039.4	877.5
48.000	0.344	47.312	175.08	12.566371	11481.174585	12.208716	14625.3	914.1
STD	0.375	47.250	190.74		11451.103296	12.176739	15912.4	994.5
	0.406	47.188	206.37		11421.071439	12.144805	17194.3	1074.6
	0.438	47.124	222.49		11390.112175	12.111883	18512.4	1157.0
	0.469	47.062	238.08		11360.160456	12.080034	19784.2	1236.5
	0.500	47.000	253.60		11330.248170	12.048226	21050.9	1315.7

D = Outside diameter, ins.; *t* = Wall thickness, ins.; *d* = Inside diameter, ins.; *W* = Pipe weight #/ft.; 1 = 0.0491 ($D^4 - d^4$); Z = Section modulus, in^3; *Ao* = Outside surface area sq ft/ft; *Vm* = Displacement, bbls/mile; A_f = Flow area, sq ft; *I* = Moment of Inertia, in^4; $Z = 0.0982\ (D^4 - d^4)/D$; Gals/ft = $d^2 \times 0.0408$; Gals/mile = $d^2 \times 215.4240$; Bbls/mile = $d^2 \times 5.1291$.

Length of pipe in bends

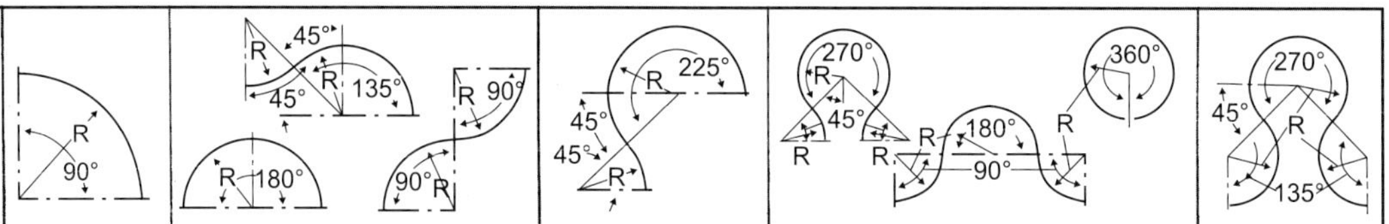

Radius of Pipe Bends: Inches	Feet	90° Bends	180° Bends	270° Bends	360° Bends	540° Bends
1		1½″	3″	4¾″	6¼″	9½″
2		3	6¼	9½	12½	18¾
3	¼	4¾	9½	14¼	18¾	28¼
4		6¼	12½	18¾	25¼	37¾
5		7¾	15¾	23½	31½	47¼
6	½	9½	18¾	28¼	37¾	56½
7		11	22	33	44	66
8		12½	25¼	37¾	50¼	75½
9	¾	14¼	28¼	42½	56½	84¾
10		15¾	31½	47¼	62¾	94¼
11		17¼	34½	51¾	69	103¾
12	1	18¾	37¾	56½	75½	113
12.5		19¾	39¼	59	78½	117¾
14		22	44	66	88	132
15	1¼	23½	47	70¾	94¼	141¼
16		25¼	50¼	75½	100½	150¾
17.5		27½	55	82½	110	165
18	1½	28¼	56½	84¾	113	169¾
20		31½	62¾	94¼	125¾	188½
21	1¾	33	66	99	132	198
24	2	37¾	75½	113	150¾	226¼
25		39¼	78½	117¾	157	235½
30	2½	47¼	94¼	141¼	188½	282¾
32		50¼	100½	150¾	201	301½
36	3	56½	113	169½	226¼	339¼
40		62¾	125¾	188½	251¼	377
48	4	75½	150¾	226¼	301½	452½
50		78½	157	235½	314¼	471¼
56		88	176	264	351¾	527¾
60	5	94¼	188½	282¾	377	565½
64		100½	201	301½	402	603¼
70		110	220	329¾	439¾	659¾
72	6	113	226¼	339¼	452½	678½
80		125¾	251¼	377	502¾	754
84	7	132	263¾	395¾	527¾	791½
90	7½	141¼	282¾	424	565½	848¼
96	8	150¾	301½	452½	603	904¾
100		157	314¼	471¼	628¼	942½
108	9	169½	339¼	509	678½	1017¾
120	10	188½	377	565½	754	1131
132	11	207¼	414¾	622	829½	1244
144	12	226¼	452½	678½	904¾	1357¼
156	13	245	490	735¼	980¼	1470¼
168	14	263¾	527¾	791½	1055½	1583½
180	15	282¾	565½	848¼	1131	1696½
192	16	301½	603	904¾	1206¼	1809½
204	17	320½	640¾	961¼	1281¾	1922½
216	18	339¼	678½	1017¾	1357¼	2035¾
228	19	358	716¼	1074½	1432½	2148¾
240	20	377	754	1131	1508	2262

To find the length of pipe in a bend having a radius not given above, add together the length of pipe in bends whose combined radii equal the required radius.

Example: **Find length of pipe in 90° bend of 5′ 9″ radius.**
Length of pipe in 90° bend of 5′ radius = 94¼″
Length of pipe in 90° bend of 9″ radius = 14¼″

Then, length of pipe in 90° bend of 5′ 9″ radius = 108½″

Reprinted with permission—Crane Company.

Calculation of pipe bends

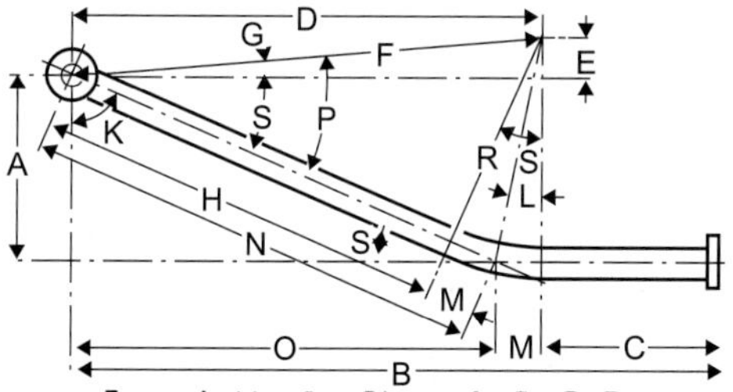

Example No. 1—**Given A, B, C, R**

$$D = B - C$$
$$E = R - A$$
$$F = \sqrt{D^2 + E^2}$$
$$\frac{E}{F} = \sin \angle G$$
$$H = \sqrt{F^2 - R^2}$$
$$\frac{R}{F} = \sin \angle P$$
$$\angle S = \angle P - \angle G$$
$$\angle K = 90^\circ - \angle S$$
$$\angle L = \tfrac{1}{2} \angle S$$
$$M = \tan \angle L \times R$$
$$N = H + M$$
$$O = B - C - M$$

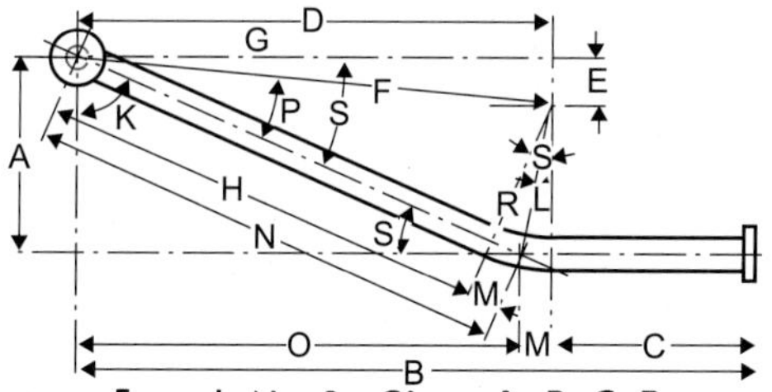

Example No. 2—**Given A, B, C, R**

$$D = B - C$$
$$E = A - R$$
$$F = \sqrt{D^2 + E^2}$$
$$\frac{E}{F} = \sin \angle G$$
$$H = \sqrt{F^2 - R^2}$$
$$\frac{R}{F} = \sin \angle P$$
$$\angle S = \angle P + \angle G$$
$$\angle K = 90^\circ - \angle S$$
$$\angle L = \tfrac{1}{2} \angle S$$
$$M = \tan \angle L \times R$$
$$N = H + M$$
$$O = B - C - M$$

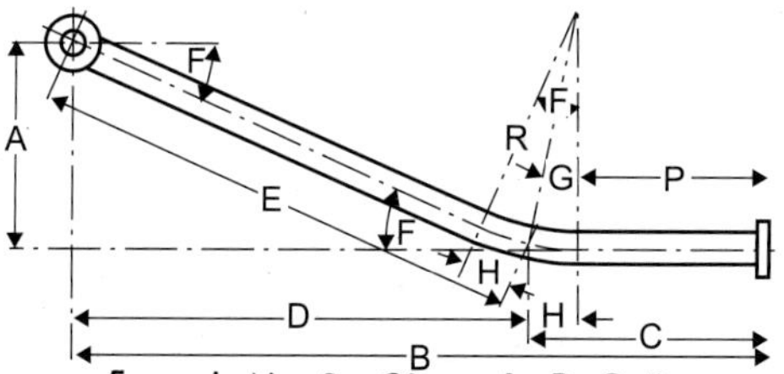

Example No. 3—**Given A, B, C, R**

$$D = B - C$$
$$E = \sqrt{A^2 + D^2}$$
$$\frac{A}{E} = \sin \angle F$$
$$\angle G = \tfrac{1}{2} \angle F$$
$$H = \tan \angle G \times R$$
$$P = C - H$$

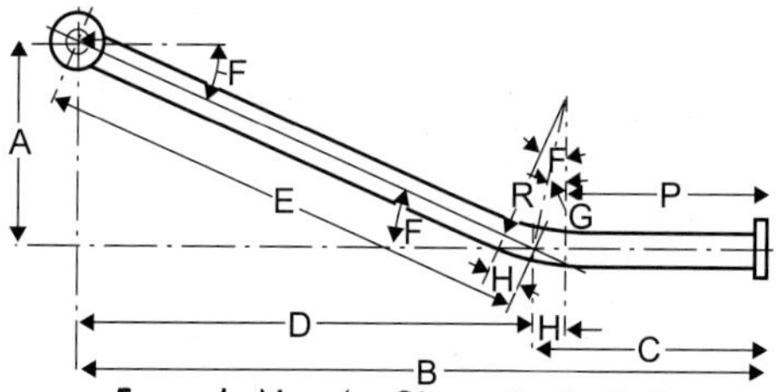

Example No. 4—**Given A, B, C, R**

$$D = B - C$$
$$E = \sqrt{A^2 + D^2}$$
$$\frac{A}{E} = \sin \angle F$$
$$\angle G = \tfrac{1}{2} \angle F$$
$$H = \tan \angle G \times R$$
$$P = C - H$$

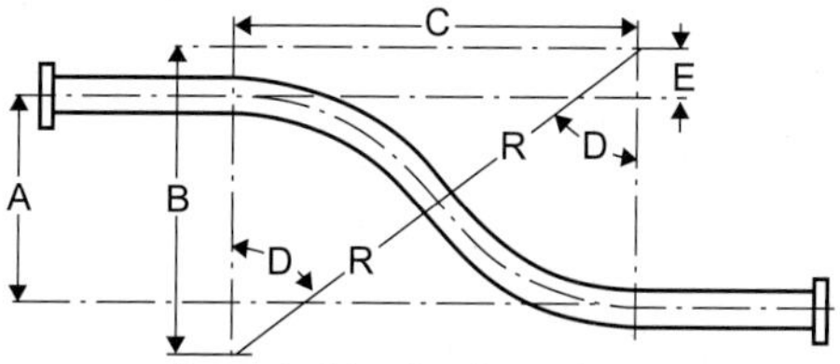

Example No. 5—**Given A, R**

$$B = 2R - A$$
$$C = \sqrt{(2R)^2 - B^2}$$
$$\frac{C}{2R} = \sin \angle D$$

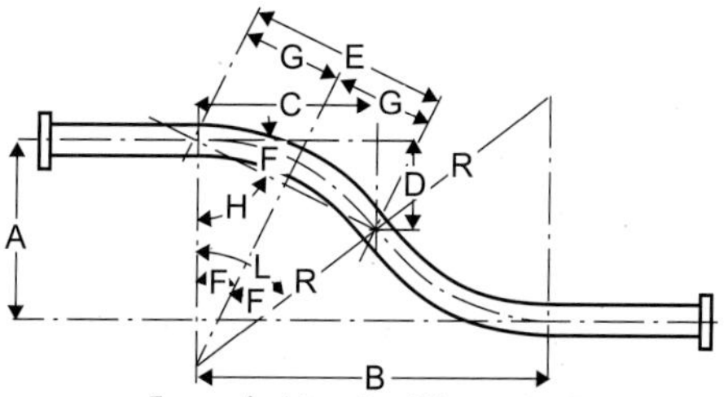

Example No. 6—**Given A, B**

$$C = \tfrac{1}{2} B$$
$$D = \tfrac{1}{2} A$$
$$E = \sqrt{C^2 + D^2}$$
$$\frac{D}{E} = \sin \angle F$$
$$G = \tfrac{1}{2} E$$
$$\angle H = 90^\circ - \angle F$$
$$R = \frac{A^2 + B^2}{4A}$$
$$L = 2F$$

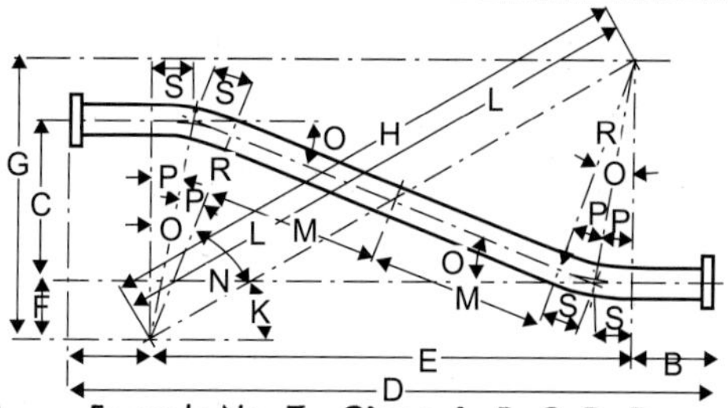

Example No. 7—**Given A, B, C, D, R**

$$E = D - A - B$$
$$F = R - C$$
$$G = R + F$$
$$H = \sqrt{E^2 + G^2}$$
$$G/H = \sin \angle K$$
$$L = \tfrac{1}{2} H$$
$$M = \sqrt{L^2 - R^2}$$
$$M/L = \sin \angle N$$
$$\angle O = 90^\circ - \angle K - \angle N$$
$$\angle P = \tfrac{1}{2} \angle O$$
$$S = \tan \angle P \times R$$

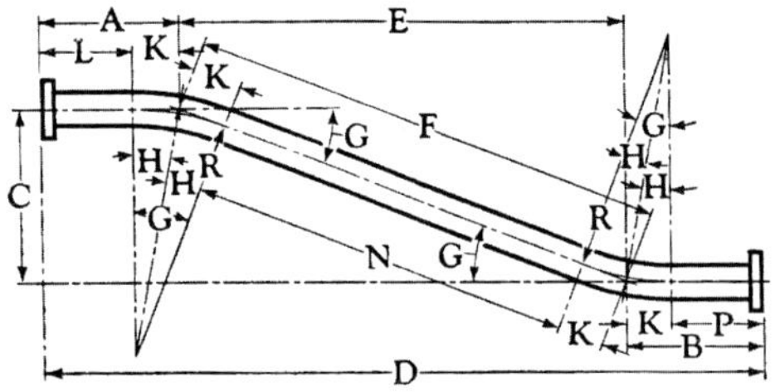

Example No. 8—**Given A, B, C, D, R**

$E = D - A - B$

$F = \sqrt{E^2 + C^2}$

$\frac{C}{F} = \sin \angle G$

$\angle H = \frac{1}{2} \angle G$

$K = \tan \angle H \times R$

$L = A - K$

$P = B - K$

$N = F - 2K$

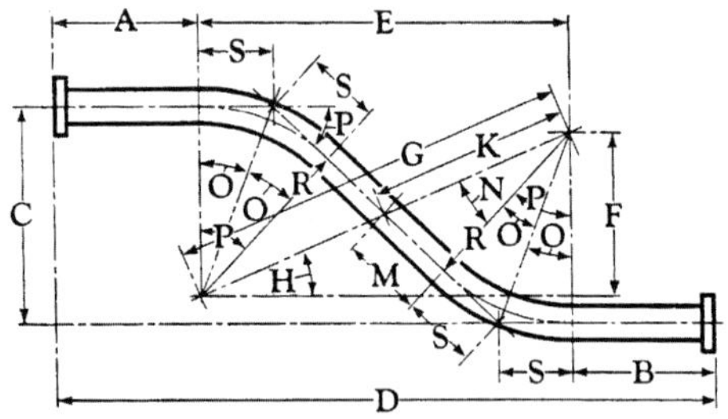

Example No. 9—**Given A, B, C, D, R**

$E = D - A - B$

$F = 2R - C$

$G = \sqrt{E^2 + F^2}$

$\frac{F}{E} = \tan \angle H$

$K = \frac{1}{2} G$

$M = \sqrt{K^2 - R^2}$

$\frac{M}{K} = \sin \angle N$

$\angle P = 90° - \angle H - \angle N$

$\angle O = \frac{1}{2} \angle P$

$S = \tan \angle O \times R$

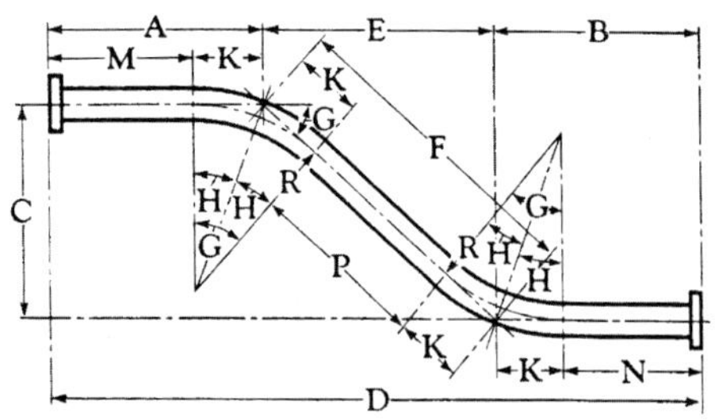

Example No. 10—**Given A, B, C, D, R**

$E = D - A - B$

$F = \sqrt{C^2 + E^2}$

$\frac{C}{F} = \sin \angle G$

$\angle H = \frac{1}{2} \angle G$

$K = \tan \angle H \times R$

$M = A - K$

$N = B - K$

$P = F - 2K$

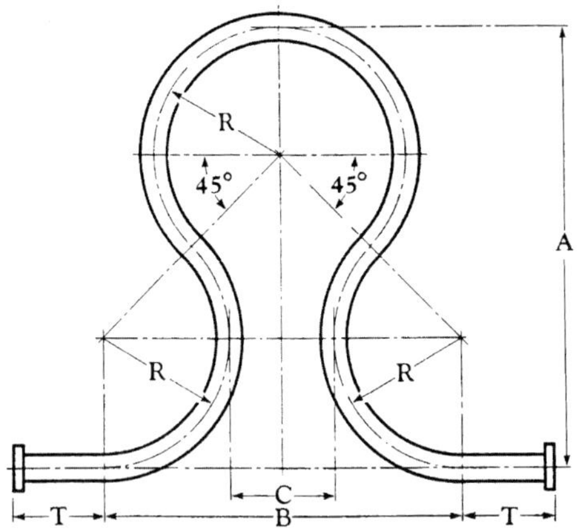

Example No. 11—**Given R and 45° Angles**

$A = 3.414 \times R$

$B = 2.828 \times R$

$C = 0.828 \times R$

T = Tangent

Length of pipe in bend = $9.425 \times R + 2T$

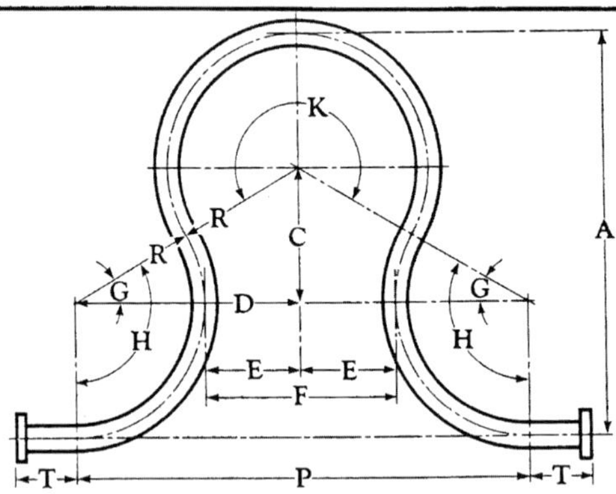

Example No. 12—**Given A, R**

$C = A - 2R$

$D = \sqrt{(2R)^2 - C^2}$

$P = 2D$

$E = D - R$

$F = 2E$

$C/2R = \sin \angle G$

$\angle H = 90° + \angle G$

$\angle K = 180° + 2 \angle G$

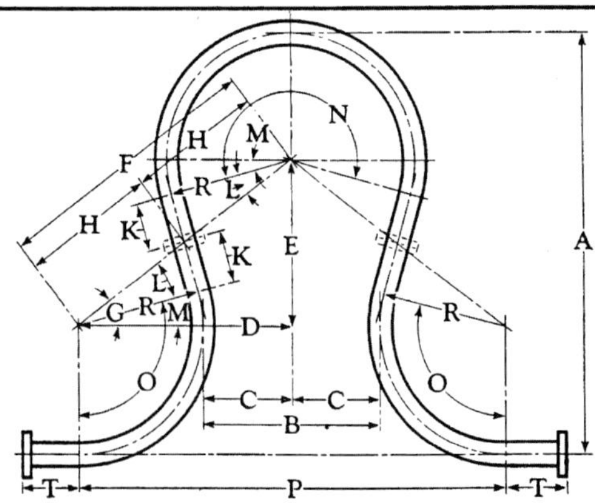

Example No. 13—**Given A, B, R**

$C = \frac{1}{2} B$

$D = R + C$

$E = A - 2R$

$F = \sqrt{D^2 + E^2}$

$E/F = \sin \angle G$

$H = \frac{1}{2} F$

$K = \sqrt{H^2 - R^2}$

$K/H = \sin \angle L$

$\angle M = \angle G - \angle L$

$\angle N = 180° + 2 \angle M$

$\angle O = 90° + \angle M$

$P = 2D$

Spacing of pipe supports

When a horizontal pipeline is supported at intermediate points, sagging of the pipe occurs between these supports, the amount of sag being dependent upon the weight of the pipe, fluid, insulation, and valves or fittings that may be included in the line. If the pipeline is installed with no downward pitch, pockets will be formed in each span, in which case condensation may collect if the line is transporting steam. In order to eliminate these pockets, the line must be pitched downward so that the outlet of each span is lower than the maximum sag.

Crane has conducted tests to determine the deflection of horizontal standard pipelines filled with water, in pipe sizes 3/4 to 4″ inclusive, the results of which have indicated that for pipes larger than 2″ and with supports having center to center dimensions greater than 10 ft, the resultant deflection is less than that determined by the use of the formula for a uniformly loaded pipe fixed at both ends. For pipe sizes 2″ and smaller, the test deflection was in excess of that determined by the formula for pipe having fixed ends and approached, for the shorter spans, the deflection as determined by the use of the formula for pipelines having unrestrained ends.

The deflection of horizontal standard pipelines filled with water, for varying spans, based upon the results obtained from tests for sizes 2″ and smaller and upon the formula for fixed ends for the larger sizes of pipe is described previously. The deflection values given on the chart are twice those obtained from test or calculation, to compensate for any variables including weight of insulation, etc.

The formula given below indicates the vertical distance that the span must be pitched so that the outlet is lower than the maximum sag of the pipe.

$$h = \frac{144^2 y}{36S^2 - y^2}$$

where

h = Difference in elevation of span ends, inches,
S = Length of one span, feet,
y = Deflection of one span, inches.

By eliminating the inconsequential term "$-y^2$" from the denominator, the formula reduces to:

$$h = 4y$$

The pitch of pipe spans, called the Average Gradient, is a ratio between the drop in elevation and the length of the span. This is expressed as so many inches in a certain number of feet.

$$\text{Average gradient} = \frac{4y}{S}$$

Reprinted with permission—Crane Company.

The dotted lines as shown on the chart on the opposite page are plotted from the above formula and indicate average gradients of 1″ in 10′, 1″ in 15′, 1″ in 20′, 1″ in 30′, and 1″ in 40′.

Example. What is the maximum distance between supports for a 4″ standard pipeline assuming a pitch or average gradient of 1″ in 30 ft?

Using the chart on the opposite page, find the point where the diagonal dotted line for an average gradient of 1″ in 30 ft intersects the diagonal solid line for 4″ pipe. From this point, proceed downward to the bottom line where the maximum span is noted to be approximately 22 ft.

Code for Pressure Piping: The Code for Pressure Piping, ASA B 31.1, makes the following statements relative to installations within the scope of the code:

"**605 (g)** Supports shall be spaced so as to prevent excessive sag, bending and shear stresses in the piping, with special consideration given to those piping sections where flanges, valves, etc., impose concentrated loads. Where calculations are not made, suggested maximum spacing of hangers or supports for carbon steel piping operating at 750 °F and lower are given in Table 21a (*see the table below*)."

"Where greater distance between supports, concentrated loads, higher temperatures, or vibration considerations are involved, special consideration should be given to effects of bending and shear stresses."

"**623 (b)** The design and spacing of supports shall be checked to assure that the sum of the longitudinal stresses due to weight, pressure, and other sustained external loading does not exceed the allowable stress (S value) in the hot condition."

Suggested Maximum Spacing Between Pipe Supports (at Maximum Operating Temperature of 750 °F)

Nominal Pipe Size Inches	Maximum Span Feet	Nominal Pipe Size Inches	Maximum Span Feet
1	7	8	19
1½	9	10	22
2	10	12	23
2½	11	14	25
3	12	16	27
3½	13	18	28
4	14	20	30
5	16	24	32
6	17	—	—

Note: The values in the table apply for straight runs of standard wall and heavier pipe but do not apply where there are concentrated loads between supports such as flanges, valves, etc.

Spacing is based on a combined bending and shear stress of 1500 psi when pipe is filled with water and the pitch of the line is such that a sag of 0.1 in between supports is permissible.

Chart of deflection of horizontal pipelines

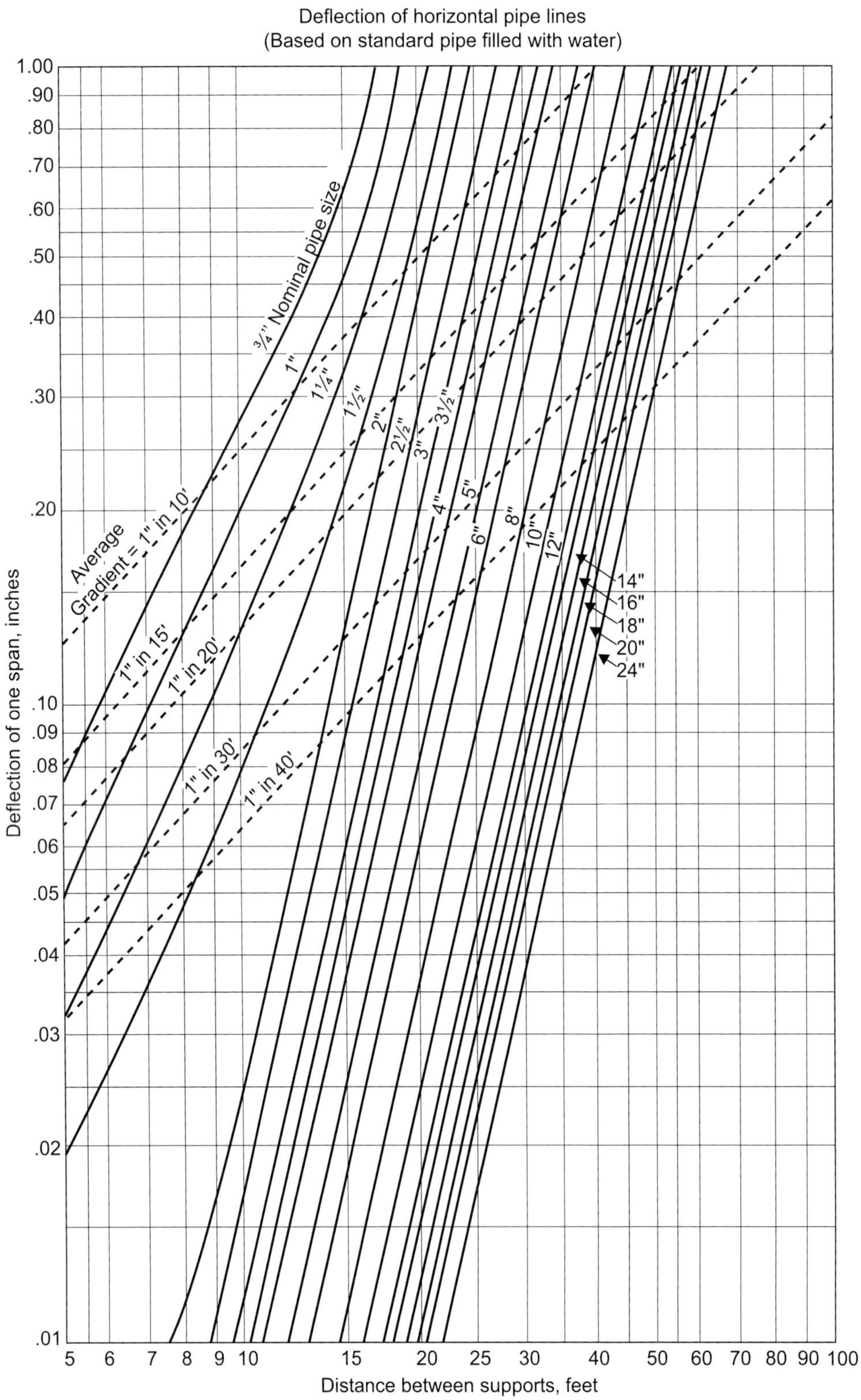

Reprinted with permission—Crane Company.

American standard taper pipe threads (NPT)

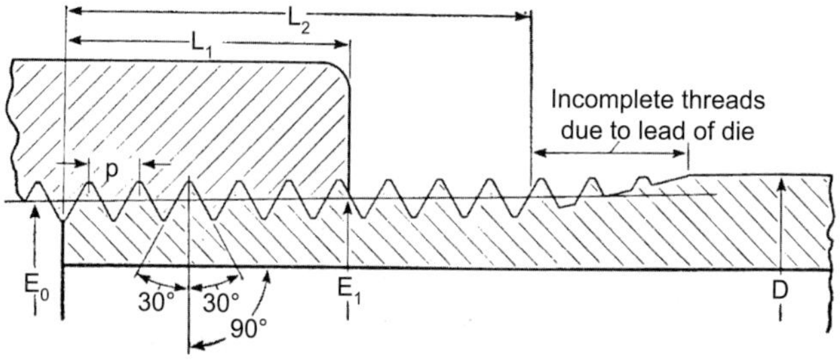

$E_0 = D - (0.050D + 1.1)\, p$
$E_1 = E_0 + 0.0625\, L_1$
$L_2 = (0.80D + 6.8)\, p$

p = Pitch
Depth of thread = 0.80p
Total Taper ¾ - in.per foot

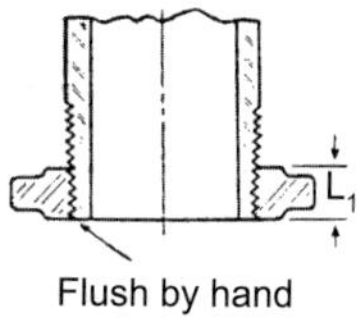

Tolerance on product

One turn large or small from notch on plug gauge or face of ring gauge.

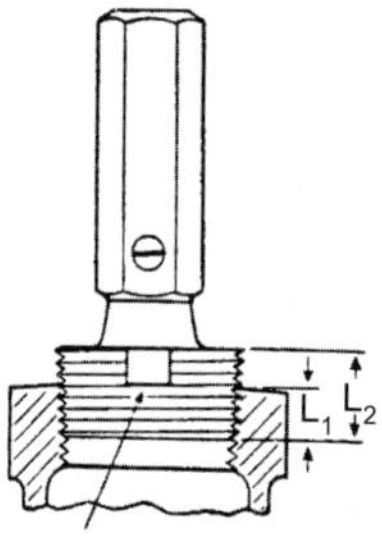

Notch flush with face of fitting. If chamfered, notch flush with bottom of chamfer.

Dimensions, in Inches

Nominal pipe size	D Outside diameter of pipe	E_0 Pitch diameter at end of external thread	E_1† Pitch diameter at end of internal thread	L_1¶ Normal engagement by hand between external and internal threads	L_2§ Length of effective thread	p Pitch of thread	Depth of thread	Number of threads Per Inch
⅛	0.405	0.36351	0.37476	0.180	0.2639	0.03704	0.02963	27
¼	0.540	0.47739	0.48989	0.200	0.4018	0.05556	0.04444	18
⅜	0.675	0.61201	0.62701	0.240	0.4078	0.05556	0.04444	18
½	0.840	0.75843	0.77843	0.320	0.5337	0.07143	0.05714	14
¾	1.050	0.96768	0.98887	0.339	0.5457	0.07143	0.05714	14
1	1.315	1.21363	1.23863	0.400	0.6828	0.08696	0.06957	11½
1¼	1.660	1.55713	1.58338	0.420	0.7068	0.08696	0.06957	11½
1½	1.900	1.79609	1.82234	0.420	0.7235	0.08696	0.06957	11½
2	2.375	2.26902	2.29627	0.436	0.7565	0.08696	0.06957	11½
2½	2.875	2.71953	2.76216	0.682	1.1375	0.12500	0.10000	8
3	3.500	3.34062	3.38850	0.766	1.2000	0.12500	0.10000	8
3½	4.000	3.83750	3.88881	0.821	1.2500	0.12500	0.10000	8
4	4.500	4.33438	4.38712	0.844	1.3000	0.12500	0.10000	8
5	5.563	5.39073	5.44929	0.937	1.4063	0.12500	0.10000	8
6	6.625	6.44609	6.50597	0.958	1.5125	0.12500	0.10000	8
8	8.625	8.43359	8.50003	1.063	1.7125	0.12500	0.10000	8
10	10.750	10.54531	10.62094	1.210	1.9250	0.12500	0.10000	8
12	12.750	12.53281	12.61781	1.360	2.1250	0.12500	0.10000	8
14 O.D.	14.000	13.77500	13.87262	1.562	2.2500	0.12500	0.10000	8
16 O.D.	16.000	15.76250	15.87575	1.812	2.4500	0.12500	0.10000	8
18 O.D.	18.000	17.75000	17.87500	2.000	2.6500	0.12500	0.10000	8
20 O.D.	20.000	19.73750	19.87031	2.125	2.8500	0.12500	0.10000	8
24 O.D.	24.000	23.71250	23.86094	2.375	3.2500	0.12500	0.10000	8

†Also pitch diameter at gauging notch.
§Also length of plug gauge.
¶Also length of ring gauge, and length from gauging notch to small end of plug gauge.

British standard taper pipe threads

Whitworth 55° Form of Thread

Total taper ¾-inch per ft

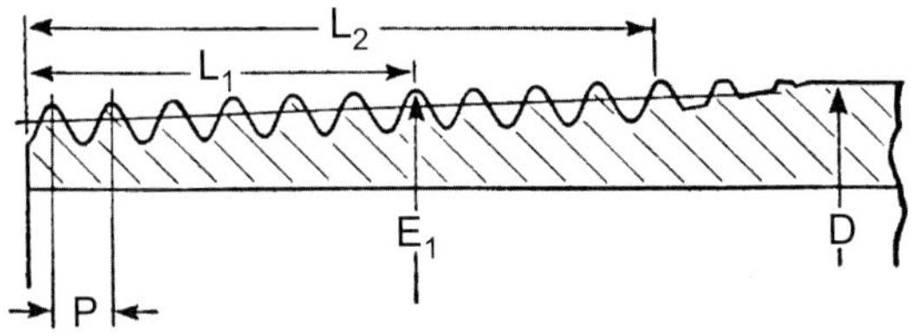

This information is taken from the British Engineering Standards Association report no. 21-1938.

Dimensions, in Inches

Pipe size Nominal Bore	D Approximate outside diameter of pipe	E_1 Pitch diameter at end of internal thread	L_1 Normal engagement by hand between external and internal threads	L_2 Length of effective thread	p Pitch of threads	Depth of threads	Number of threads Per Inch
1/8	13/32	0.3601	0.1563	0.2545	0.03571	0.0229	28
1/4	17/32	0.4843	0.2367	0.3814	0.05263	0.0337	19
3/8	11/16	0.6223	0.2500	0.3947	0.05263	0.0337	19
1/2	27/32	0.7793	0.3214	0.5178	0.07143	0.0457	14
3/4	1 1/16	0.9953	0.3750	0.5714	0.07143	0.0457	14
1	1 11/32	1.2508	0.4091	0.6591	0.09091	0.0582	11
1 1/4	1 11/16	1.5918	0.5000	0.7500	0.09091	0.0582	11
1 1/2	1 29/32	1.8238	0.5000	0.7500	0.09091	0.0582	11
2	2 3/8	2.2888	0.6250	0.9204	0.09091	0.0582	11
2 1/2	3	2.9018	0.6875	1.0511	0.09091	0.0582	11
3	3 1/2	3.4018	0.8125	1.1761	0.09091	0.0582	11
3 1/2	4	3.8918	0.8750	1.2386	0.09091	0.0582	11
4	4 1/2	4.3918	1.0000	1.4091	0.09091	0.0582	11
5	5 1/2	5.3918	1.1250	1.5795	0.09091	0.0582	11
6	6 1/2	6.3918	1.1250	1.5795	0.09091	0.0582	11
7	7 1/2	7.3860	1.3750	1.9250	0.10000	0.0640	10
8	8 1/2	8.3860	1.5000	2.0500	0.10000	0.0640	10
9	9 1/2	9.3860	1.5000	2.0500	0.10000	0.0640	10
10	10 1/2	10.3860	1.6250	2.1750	0.10000	0.0640	10
11	11 1/2	11.3700	1.6250	2.3125	0.1250	0.0800	8
12	12 1/2	12.3700	1.6250	2.3125	0.1250	0.0800	8

Normal engagement between male and female threads to make tight joints

The normal amount of engagement to make a tight joint for various types of screwed material is given in the table. These dimensions have been established from tests made under practical working conditions.

The normal engagement specified for American Standard Pipe Thread and API Line Pipe Thread joints is based on parts being threaded to the American Standard for Pipe Threads or the API Standard for Line Pipe Threads.

In order to obtain the thread engagements listed in the table, it is necessary to vary the torque or power applied according to the size, metal, and weight of material used. For example, it requires considerably less power to make up a screwed joint using a light bronze valve than a high pressure steel valve.

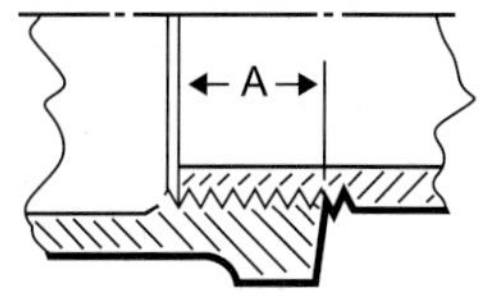

American standard and API line pipe threads

Dimensions, in Inches			
Dimensions Given Do Not Allow for Variations in Tapping or Threading			
Size	**A**	**Size**	**A**
1/8	1/4	2½	15/16
1/4	3/8	3	1
3/8	3/8	3½	$1^1/_{16}$
1/2	1/2	4	$1\ ^1/_8$
3/4	9/16	5	1¼
1	11/16	6	$1^5/_{16}$
1¼	11/16	8	$1^7/_{16}$
1 ½	11/16	10	$1^5/_8$
2	3/4	12	$1^3/_4$

Reprinted with permission—Crane Company.

Calculate pipe weight, contents, velocity

Pipeline engineers, accountants, and system operators are deeply involved with questions concerning line pipe such as:

How much does it weigh (lb/ft, tons/mi)?
How much will it contain (bbl/mi, gal/100 ft, cu m/km)?
What are velocities at various feed rates (ft/sec, mi/hr, m/hr)?

Results for the weight category are calculated by an API formula and given as: pounds (mass) per ft; kg per m; short tons (2000 lb) per mile; metric tons per km.
Content results are described by: bbl/mi, gal (U.S.)/100 ft, cu ft/mi, L/m, cu m/km.
Flowrates are described as: b/h or b/d; gal/hr or gal/d; cu ft/min, /h, /d; cu m/min, /h, /d.
Velocity results: mph; fps; km/h; m/s.

Basic data calculated are used for:

- Weight: Cost of construction, hauling and stringing, investments.
- Content: Amount of water for hydrostatic test, line fill, oil stocks. Amount of one grade to be displaced for another to reach a station, shut-off valve, or terminal. When a segment must be drained, barrels or gallons to be removed.
- Velocity: Time for interface or scraper to traverse distance between two points. Time to accumulate or withdraw a given quantity of fluid. Locating head of column of a grade of oil or a scraper from given station after a certain time. Determining expected time of arrival (ETA) of interface or scraper at a given point.

Formulas

Pipe weight LBM/FT (plain end) is calculated by formula used by API:

$$\text{LBM/FT} = 10.68\left(Dt - t^2\right)$$

D = Pipe OD, in,

t = pipe wall thickness, in.

Weights in other units are found by multiplying by conversion factors:

$$\text{BBL/MI} = 5.129131(D - 2t)^2$$

Basic velocity relation:

$$\frac{BBL}{HR} \div \frac{BBL}{MI} = \frac{MI}{HR}$$

This is converted to other units.

Example 1. Calculate pounds mass per foot.

$D = 12.750$ in

$t = 0.375$ in

$$\text{Pounds/foot} = 10.68(12.750 \times 0.375 - 0.375^2)$$
$$= 49.56 \text{ pounds per foot}$$

Example 2. Calculate barrels per mile (use pipe data from example 1).

$$\text{BBL/MILE} = 5.129131(12.750 - 2 \times 0.375)^2$$
$$= 738.59 \text{ bbls/mile}$$

Formulas and constants for tubular goods problems

Area of Metal in Tubular Section:

Area in square inches $= 3.1416(D - t)t$

where

D = Outside diameter, in,
t = Wall thickness, in.

Capacity:

$$B_{100} = 0.1273A = 0.0972D^2$$

where

B_{100} = Barrels per 100 ft,
A = Internal area of pipe, in^2
D = Internal diameter, in

$$F_B = \frac{808.5}{A} = \frac{1029.4}{D^2}$$

where

F_B = Number of feet filled by one barrel,
A = Internal area of pipe, in^2,
D = Internal diameter, in.

Steel constants (applicable to tubular goods):

One cubic inch = 0.2833 lb

One cubic foot = 489.542 lb

Specific gravity = 7.851

Approximate coefficient of expansion for commercial casing, drill pipe, and tubing is

Coefficient $= 6.9 \times 10^{-6}$ per°F, over range from 0 to 400°F.

Formula for expansion of pipe due to change of temperature is as follows:

$$L_t = L_o(1 + 0.0000069t)$$

where

L_o = Length at atmospheric temperature,
L_t = Length after change of temperature t.

Modulus:

$$\text{Section Modulus} = \frac{\pi}{64}(D^4 - D_1^4)$$

$$\text{Polar Modulus} = \frac{\pi}{32}(D^4 - D_1^4)$$

where

D = Outside diameter,
D_1 = Inside diameter.

Relation between twisting effort, torsional strain, and stress of cylindrical shaft or tube

$$\frac{q}{r} = \frac{T}{J} = \frac{C\theta}{L}$$

How to calculate the contraction or expansion of a pipeline

Steel pipe contracts or expands approximately 0.8 in per 100 °F temperature change for each 100 ft of pipe. For construction in the United States, slack is not normally needed in welded lines for contraction or expansion unless abnormal variations are encountered. However, slack is often provided near road crossings where the pipe may have to be lowered at some future time.

$$\text{Contraction} = 0.8 \times \frac{\text{length of pipe (feet)}}{100} \times \frac{\text{temp change (°F)}}{100}$$

Example. Calculate the amount a 1000-ft section of pipeline would contract if laid at a temperature of 100 °F and cooled to a low of 0 °F during winter operation (assuming the line were free of soil or other resisting loads).

$$\text{Contraction} = 0.8 \times \frac{1000}{100} \times \frac{100}{100} = 0.8 \times 10 \times 1 = 8 \text{ in.}$$

Large temperature changes may arise on exposed line or in lines downstream from compressor stations.

When slack is desired in a line, the amount of sag or rise needed may be estimated quickly. To provide for 0.8 in of movement longitudinally along the pipe, a sag of 21 in would be needed in a 100-ft loop. For a 150-ft loop, 31 in are necessary. Such slack put into the line at time of construction might allow for lowering the pipe under roadbeds without putting excessive stress upon the pipe.

Location of overbends, sags, and side bends should be considered when laying slack loops. When loops are placed, parts of the line should be backfilled to serve as a tie-down. Loops should be lowered during the coolest part of the day when the line is the shortest. After lowering in, sags should rest on the bottom of the ditch; overbends should "ride high." Side bends should rest on the bottom of the ditch and against the outside wall. The line should be lowered so that all sections of the pipe are in compression.

Estimate weight of pipe in metric tons per kilometer

To estimate the weight of pipe in metric tons per kilometer, multiply the nominal diameter by the number of sixteenths of an inch in wall thickness.

Example. Find the weight of pipe in metric tons per kilometer for 20-in diameter pipe, wall thickness 1/4-in.

$4 \times 20 = 80$ tons (metric) per kilometer

Actual answer from pipe tables is 79 metric tons per kilometer.

This rule of thumb is based on a density of 490 lb. per cubic foot for steel. For larger diameter thin wall pipe, this approximation gives an answer usually about 1% low. The accompanying table provides a comparison between actual weights in metric tons per kilometer, as compared to that calculated by this rule of thumb.

Pipe Dia., in	Wall Thickness, in	Actual Tons, metric	R.O.T. Tons, metric
10	3/16	31.5	30
12	5/16	62	60
16	7/16	109	112
20	1/4	79	80
20	5/16	99	100
20	7/16	137	140
24	5/16	118	120
24	1/2	189	192
28	5/16	139	140
30	5/16	148	150
32	7/16	221	224
36	1/4	143	144
36	1/2	285	288

How to find pipe weight from outside diameter and wall thickness

$$\text{Weight (pounds per foot)} = (Dt - t^2) \times 10.68$$

where

D = outside diameter, in,
t = wall thickness, in.

Example. What is the weight in pounds per foot for a 4.5 in O.D., 0.25 in wall thickness pipe?

$$W = (4.5 \times 2.5 - 2.5^2) \times 10.68$$
$$= (1.125 - 0.625) \times 10.68$$
$$= 1.0625 \times 10.68 = 11.35 \text{ lb per ft}$$

The above equation is based on a density of 490 lb. per cubic foot for the steel. High-yield-point, thin-wall pipe may run slightly heavier than indicated.

What is the maximum allowable length of unsupported line pipe?

For Schedule 40 pipe, the maximum span between supports is given by

$$S = 66\sqrt{D}$$

where S is the span length in feet and D is the outside diameter of the pipe in inches. For pipe smaller than 12 in, the nominal size may be used, and

$$S = 7\sqrt{N}$$

where N is the nominal outside diameter in inches.

Example. How wide can a ditch be spanned with 4-in line pipe without the use of intermediate supports?

$$S = 66\sqrt{4.5} = 6.6 \times 2.12 = 13.99(14 \text{ ft})$$
$$S = 7\sqrt{4} = 7 \times 2 = 14 \text{ ft.}$$

The above rule is approximate, but conservative. It does not assume that the pipe is held fixed at either end; this condition can make longer spans safe. Also, it does not guarantee that the pipe will not be damaged by floating logs or debris. This hazard must be evaluated from the conditions encountered.

Identify the schedule number of pipe by direct measurement

Add 3 in to the actual inside diameter; divide this by the wall thickness; the schedule number can then be identified by the following table:

Schedule	Range
30	40 to 50
40	29 to 39
60	25 to 29
80	20 to 23
100	16 to 18
120	13 to 15
140	11 to 13
160	9 to 11

Example. Pipe measures 12.5 in inside diameter, wall thickness 0.75 in; 15.5/0.75 = 20.2; pipe is Schedule 80, 14 in outside diameter.

The rule fails on small-diameter pipe (under 6 in), and also on schedules 10 and 20 because of the use of "standard" thicknesses in this range.

Determine buoyancy of bare steel pipe

Multiply the square of the outside diameter of the pipe, in inches, by 0.341, and subtract the pipe weight in lb/ft; the answer is the buoyancy per foot in fresh water.

Example. 12.750-in OD × 0.250-in w.t., 33.4#/ft

$$\begin{aligned}\text{Buoyancy} &= 12.750^2 \times 0.341 - 33.4\\ &= 162.56 \times 0.341 - 33.4\\ &= 22\ \text{lb/ft}\end{aligned}$$

Because all of the coating materials in common use are heavier than water (but not as much as steel), coated pipe will have less buoyancy than indicated by the rule. For sea water, use 0.350 for the constant.

Determine buoyancy of bare and concrete-coated steel pipe in water and mud

To find the buoyancy of bare steel and concrete coated pipe in water in lb/ft:

For bare pipe:

$$\text{Buoyancy}(B) = \frac{D}{3}(D - 32t) + 11t^2$$

For coated pipe:

$$\text{Buoyancy}(B) = \frac{D}{3}(D - 32t) + t_1 D\left(\frac{63 - W_c}{48}\right)$$

where

D = outside diameter of pipe, in,
t = wall thickness of pipe, in,
t_1 = thickness of concrete coating, in,
W_c = weight of concrete, lb/ft^3.

To find the buoyancy of bare steel and coated pipe in mud in lb/ft:

For bare pipe:

$$\text{Buoyancy}(B) = 10.7D\left[\left[\frac{DW_m}{2000} - t\right] + 11t^2\right]$$

For coated pipe:

$$B = 10.7D\left[\left[\frac{DW_m}{2000} - t\right] + t_1 D\left[\frac{W_m - W_c}{48}\right]\right]$$

where W_m is weight of mud, lb/ft^3.

Example 1. Find the buoyancy of steel pipe with 20″ OD × 0.500″ w.t. steel pipe in water. What thickness of concrete coating will be required to give this pipe a negative buoyancy of 100 lb/ft ($W_c = 143$)?

$$\text{Buoyancy}(B) = \frac{D}{3}(D - 32t) + 11t_1^2$$

$$\text{Buoyancy}(B) = \frac{20}{3}(20 - 32 \times 0.5) + 11 \times 0.5^2$$

$$= 29.5 \text{ ft positive buoyancy.}$$

To give this pipe a negative buoyancy of 100 lb/ft:

$$-100 = \frac{20}{3}(20 - 32 \times 0.5) + t_1 20\left[\frac{63 - 143}{48}\right]$$

$$-100 = 26.7 - 33.3t_1$$

$t_1 = 3.8$ in of concrete coating.

The error introduced by this method is about 15%, but would be less as the thickness of coating decreases.

Example 2. Bottom conditions at a certain crossing require the pipe to have a negative buoyancy of at least 50 lb/ft Determine if 10.750″ pipe with 2″ of concrete ($W_c = 143$) has sufficient negative buoyancy

$$B = \frac{10}{3}(10 - 32 \times 0.5 + 2 \times 10)\left[\frac{63 - 143}{48}\right]$$

$$B = -20 - 33.3$$

$= -53.3$ lb/ft; so coated pipe meets design requirements.

Weights of piping materials

The weight per foot of steel pipe is subject to the following tolerances:

Specification	Tolerance
ASTM A – 53 ASTM A – 120	STD WT +5%, –5% XS WT +5%, –5% XXS WT +10%, –10%
ASTM A –106	SCH 10-120 + 6.5%, –3.5% SCH 140-160 + 10%, –3.5%
ASTM A–158 ASTM A–206 ASTMA A–280	12″ and under +6.5%, –3.5% Over 12″ +10%, –5%
API 5L	All sizes +65%, –3.5%

Weight of tube $= F \times 10.6802 \times T \times (D - T)$ lb/ft

where

T = wall thickness, in,
D = outside diameter, in,
F = relative weight factor.

The weight of tube furnished in these piping data is based on low carbon steel weighing 0.2833 lb/in^3.

Relative Weight Factor F of Various Metals

Metal	F Factor
Aluminum	0.35
Brass	1.12
CastIron	0.91
Copper	1.14
Lead	1.44
Ferritic Stainless Steel	0.95
Austenitic Stainless Steel	1.02
Steel	1.00
Tin	0.93
Wrought Iron	0.98

Weight of contents of tube $= G \times 0.3405 \times (D - 2T)^2$ lb/ft

where

G = specific gravity of contents,
T = tube wall thickness, in,
D = tube outside diameter, in.

The weight of welding tees and laterals is for full size fittings. The weight of reducing fittings is approximately the same as for full size fittings.

The weight of welding reducers is for one size reduction and is approximately correct for other reductions.

Pipe covering temperature ranges are only intended as a guide and do not constitute a recommendation for specific thickness of material.

Pipe covering thicknesses and weights indicate average conditions and include all allowances for wire, cement, canvas, bands, and paint. The listed thicknesses of combination covering is the sum of the inner and the outer layer thickness. When specific inner and outer layer thicknesses are known, add them, and use the weight for the nearest tabulated thickness.

To find the weight of covering on fittings, valves, or flanges, multiply the weight factor (light faced subscript) by the weight per foot of covering used on straight pipe. All flange weights include the proportional weight of bolts or studs required to make up all joints.

Lap joint flange weights include the weight of the lap.

Welding neck flange weights are compensated to allow for the weight of pipe displaced by the flange. Pipe should be measured from the face of the flange.

All flanged fitting weights include the proportional weight of bolts or studs required to make up all joints.

To find the approximate weight of reducing-flanged fittings, subtract the weight of a full size slip-on flange and add the weight of reduced size slip-on flange.

Weights of valves of the same type may vary because of the individual manufacturer's design. Listed valve weights are approximate only. When it is possible to obtain specific weights from the manufacturer, such weights should be used.

To obtain the approximate weight of flanged end steel valves, add the weight of two slip-on flanges of the same size and series to the weight of the corresponding welding-end valves.

Allowable working pressure for carbon steel pipe

Based on formula No. 7 of the ASA Code for Pressure Piping, Section 3B, this table was computed for 6⅝-in OD to 36-in OD carbon steel seamless pipe, API Specifications 5L Grade A, 5L Grade B, and API 5LX Grade X-42.

The formula is:

$$t = \frac{PD}{2(S + YP)} + u + c$$

where

t = wall thickness, in,
P = allowable working pressure, psi,
D = outside diameter, in,
S = allowable stress, psi,
Y = a coefficient,
u = under-thickness tolerance, in,
c = allowance for threading, mechanical strength, and corrosion, in.

Note: In the calculation of the table, the values of $Y = 0.4$, $u = 0.125t$, and $c = 0.050$ in were used.

Substituting in formula and solving for P:

$$P = \frac{1.75S(t - 0.057)}{D + 0.04 - 0.7t}.$$

The S-values are:

Spec. API 5L Grade A—25,500 psi
Spec. API 5L Grade B—29,750 psi
Spec. API 5LX Grade X-42—35,700 psi

Example. Find pipe specification for a 12-in line operating at 1078 psi: In the 12¾ -in column these conditions will be met using Grade X-42 pipe with 0.281-in wall thickness, Grade B pipe with 0.330-in wall thickness, or Grade A pipe with 0.375-in wall thickness.

O.D. / Wall Thickness	6⅝	8⅝	10¾	12¾	14	16	18	20	22	24	26	28	30	32	34	36	O.D. / Wall Thickness
0.188	895 1044 1253	685 799 959	548 640 768	....	....	....	....	....	....	....	....	....	....	....	....	....	0.188
0.203	.....	1070	857	721	.....	.....	.....	.....	.....	.....	.....	.....	.....	.....	.....	.....	0.203
0.219	1110 1295 1554	849 991 1189	680 793 952	572 667 801	 729	 637	....	....	....	....	....	....	....	....	....	....	0.219
0.250	1327 1548 1858	1014 1184 1420	811 947 1136	683 797 956	621 725 870	543 633 760	482 562 675	434 506 607	 551	 505	 466	 433	 404	 378	 356	 336	0.250
0.281	....	....	....	794 926 1111	722 842 1010	631 736 883	560 654 784	504 588 705	 641	 587	 542	 503	 469	 439	 414	 390	0.281
0.312	1765 2059 2471	1347 1572 1886	....	905 1056 1267	823 961 1153	719 839 1007	638 745 894	574 670 804	521 608 730	478 557 669	441 514 617	 573	 534	 501	 471	 445	0.312
0.344	1994 2326 2791	1520 1774 2128	1214 1416 1700	1021 1190 1429	928 1083 1299	810 946 1135	720 839 1007	647 755 906	587 685 823	538 628 753	 695	 645	 602	 564	 530	 501	0.344
0.375	2216 2568 3103	1689 1970 2364	....	1133 1322 1586	1030 1202 1442	900 1049 1259	798 931 1118	718 837 1005	652 760 912	597 696 836	551 642 771	 715	 667	 625	 588	 555	0.375
0.406	.....	.....	.....	1744	.....	.....	1228	1104	1002	918	847	786	733	687	646	610	0.406
0.438	....	2034 2373 2848	1622 1892 2271	1362 1589 1907	1238 1444 1733	1081 1261 1513	959 1119 1342	862 1005 1206	782 913 1095	716 836 1003	661 771 925	 858	 801	 750	 706	 666	0.438
0.500	....	 3328	 2651	 2225	1444 1685 2022	1260 1470 1764	1118 1304 1565	1004 1171 1406	911 1063 1276	834 974 1168	770 898 1077	 1000	 932	 873	 822	 775	0.500
Wall Thickness / O.D.	6⅝	8⅝	10¾	12¾	14	16	18	20	22	24	26	28	30	32	34	36	Wall Thickness / O.D.

Find the stress in pipe wall due to internal pressure

Multiply ½ by diameter by pressure and divide by wall thickness to get stress in psi.

Example. Find the wall stress due to internal pressure in a 24-in pipeline having a 1/4-in wall thickness if the pressure gauge reads 800 psi:

$1/2 \times 24 \times 800 \div 1/4 = 38,400$ psi

What wall thickness should be specified for a 30-in discharge line where pressure runs 1000 psi and wall stress is to be limited to 30,000 psi?

$1/2 \times 30 \times 1000 \div \text{thickness} = 30,000$ psi

Wall thickness = 1/2 in:

This method of calculating transverse tensile stress due to internal pressure on the pipe is called the Barlow formula. Actually, there are two accepted variations, and some designers use inside diameter of the pipe in calculating this stress. This gives a slightly lower figure for the transverse tensile stress and consequently gives less safety factor. The method used here is based on the assumption that maximum stress occurs at the outside diameter of the pipe. The results give a conservative value for safety factor calculations.

How to calculate stress in above ground/below ground transitions

P. J. Schnackenberg, Design Engineer, Iran-Texas Engineering Co., Tehran, Iran

Stresses and deflections occur in pipelines at the transition from the belowground (fully restrained) to the aboveground (unrestrained) condition.

Analysis of the stresses and deflections in transition areas, resulting from internal pressure/temperature change, is necessary in determining anchor block requirements and design. Longitudinal deflections are used to determine whether an anchor block is required. Anchor block forces required to maintain the pipe in a fully constrained condition are then determined.

Iran-Texas Engineering Co. recently completed an analysis for a gas gathering/compression/reinjection project in southern Iran under contract to Oil Service Co. of Iran. Each of the two fields, Bibi-Hakimeh and Rag-E-Safid have two gas/oil separation plants, and at present, the gas is being flared. The broad scope of the project is as follows:

- Provision of gas gathering lines from existing and new wellhead separation to a production unit.
- Design of four production units (each adjacent to the GOSP) comprising knockout facilities, compressor stations, and dehydration plant.
- Provision of transfer pipelines to send gas plus condensate as a two-phase flow to a natural gas liquid separation plant (under design by Foster Wheeler).
- Provision of reinjection system to inject lean gas back into a number of wells.
- Some 9.4 MMm^3/d (330 MMscfd) of gas plus 15,000 bpd condensate will be going to the NGL plant. LPG will be used as feedstock and for export.

The following is a brief review of the analysis that resulted in more accurate solutions for deflection and anchor block forces. Sample calculations are for line sizes up to 41-cm (16-in) CD, pressure to 193 bars (2800 psig), and temperatures to 72 °C (162 °F).

Case I—no anchor. Considering the pipeline as shown in Figure 1, the section up to point A is fully restrained, the section A—B is a transition from fully restrained to unrestrained, and beyond point B is unrestrained. (Fully restrained is taken to be a condition of zero longitudinal strain.)

Analyzing a short section of fully restrained pipe, subject to internal pressure and a temperature change, there is a tensile stress due to a Poisson effect of

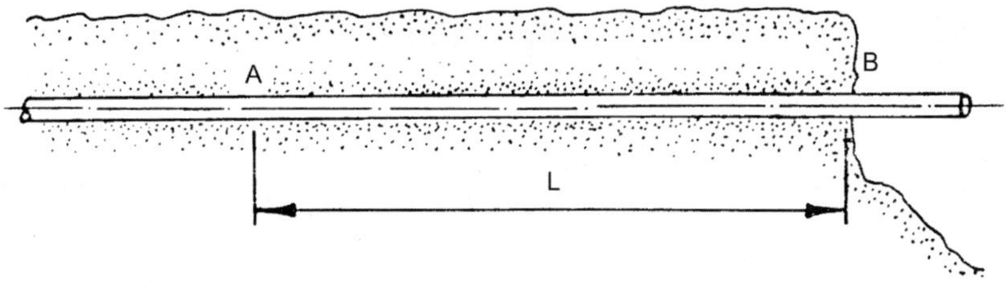

Figure 1. Area A—B depicts transition from fully restrained (left) to fully unrestrained (right).

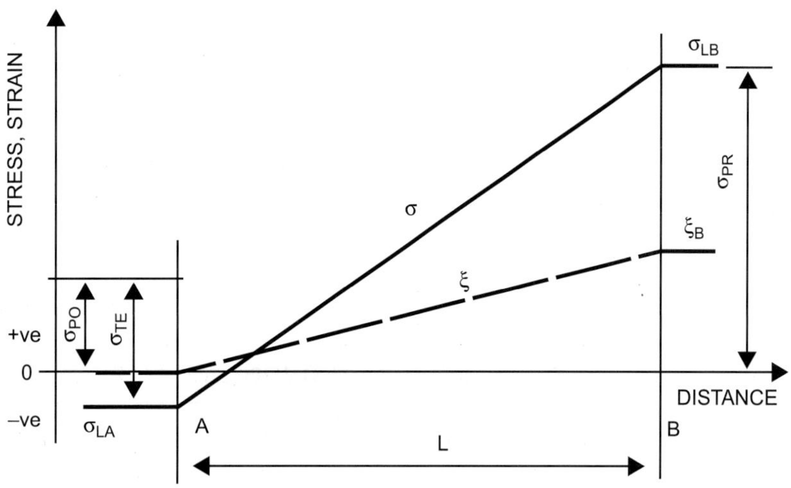

Figure 2. Transition of stress and strain between points *A* and *B* varies as linear function of length.

$$\sigma_{PO} = \upsilon\sigma$$

where,

$$\sigma h = \frac{PD_i}{2t}$$

$$\sigma_{TE} = E\alpha\Delta T.$$

The net longitudinal stress, therefore, at point A will be

$$\sigma_{LA} = \upsilon\sigma_H - E\alpha\Delta T \text{ psi} \tag{1}$$

and the strain at point A, being fully restrained, will be zero

$$\epsilon_A = 0. \tag{2}$$

In a section of unrestrained pipe at B, the longitudinal stress caused by only internal pressure will be half the hoop stress.

$$\sigma_{LB} = \frac{\sigma_H}{2} \text{psi} \tag{3}$$

The associated strain will be made up of a

(i) Temperature effect

$$\epsilon_{TE} = \alpha\Delta T$$

(ii) Pressure effect

$$\epsilon_{PR} = \frac{\sigma_{LB}}{E} = \frac{\sigma_H}{2E}$$

(iii) Poisson effect

$$\epsilon_{PO} = \frac{\upsilon\sigma_H}{E}.$$

The net longitudinal strain at point *B* will therefore be

$$\epsilon_B = \sigma\Delta T = \frac{\sigma_H}{E}(1/2 - \upsilon) \text{ in/in} \tag{4}$$

The transition of stress and strain between points *A* and *B* is assumed to vary as a linear function of length as shown in Figure 2.

In order to establish the length *L* over which the transition occurs, the longitudinal resistance of the soil needs to be known. For this part, it is assumed that any tendency to move will be counteracted by constant and opposite soil force. Wilbur[1] has recommended a design value for average soils of

$$F_s = 80\left(\frac{D_o}{12}\right) \text{ lbf/ft.} \tag{5}$$

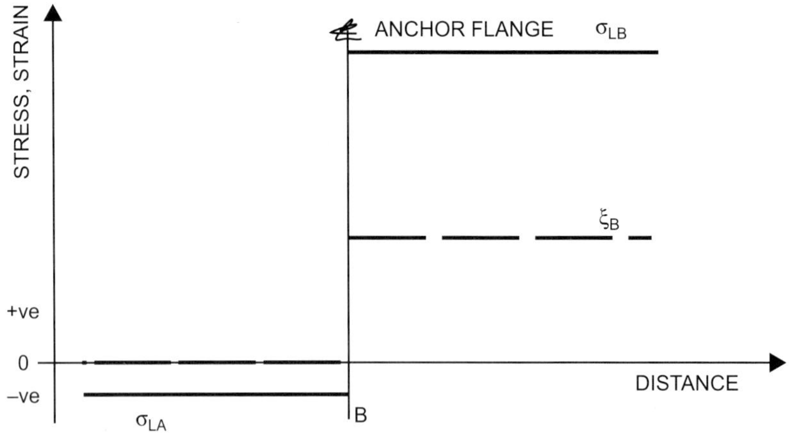

Figure 3. Distribution of stress where anchor is required to contain longitudinal deflections.

Between A and B, equilibrium of forces exists and therefore,

$$F_sL = A_m(\sigma_{LB} - \sigma_{LA})$$

from which

$$L = \frac{A_m(\sigma_{LB} - \sigma_{LA})}{F_s} \text{ ft.} \tag{6}$$

Total movement at B will be the average strain between A and B over length L or

$$\delta = \frac{\epsilon_B}{2}(12L) \text{ in} \tag{7}$$

Cell II—with anchorWhere an anchor is required to contain longitudinal deflections, the stress distribution will be as shown in Figure 3. The transition from being fully restrained to unrestrained will occur at the anchor. Resultant force on the anchor will simply be the difference in stress on each side times the pipe metal area (equilibrium of forces) or

$$F = (\sigma_{LB} - \sigma_{LA})A_m \text{ lb} \tag{8}$$

Note that for the case of an increase in wall thickness at and beyond the anchor block, the result will be essentially the same—the decrease in stress being compensated by an increase in pipe metal area.

It can be shown that this force is equal to that to produce a deflection of 2δ in Equation 7 over length L.

Table 1 is a summary of results for a number of high pressure gas transmission lines currently under design in Iran. Lines 1 through 4 are transfer lines from a number of production units to an NGL separation plant. Lines 5 through 7 are injection lines from the NGL separation plant to a number of wellheads. The following values were used for the various parameters:

Youngs modulus E 29×10^6, psi
Poisson's ratio υ, 0.3
Thermal expansion coefficient 6.5×10^{-6}, in/in °F
Installation temperature, T_i, 80 °F (minimum summer temperature when construction is in progress).

Negative stress values denote a compressive stress.

Discussion

As shown in Table 1, anchor block forces can be quite large; therefore, careful attention must be paid to the design. In order to minimize the size of the block, upper design limit

Table 1
Summary of Results for High Pressure Gas Transmission Lines Under Design in Iran

No.	Pipe (All API-5LX-60)	P psig	T_O °F	F_S lb./ft.	σ_H psi	σ_{LA} psi	σ_{LB} psi	L ft.	δ in.	F lb.
1	16″ OD × 0.312″ w.t.	1150	162	142.2	28,334	−6957	14,167	2285	9.81	324,890
2	20″ OD × 0.344″ w.t.	1000	162	222.2	28,070	−7036	14,035	2014	8.62	447,550
3	16″ OD × 0.281″ w.t.	1000	162	142.2	27,470	−7216	13,735	2045	8.71	290,800
4	16″ OD × 0.250″ w.t.	880	162	142.2	27,280	−7273	13,640	1819	7.73	258,690
5	16″ OD × 0.688″ w.t.	2800	150	142.2	29.758	−4268	14,879	4457	17.66	633,770
6	10″ NB × 0.500″ w.t.	2800	150	64.2	28,700	−4585	14,350	4728	18.60	304,850
7	6″ NB × 0.432″ w.t.	2800	150	24.4	18,670	−7594	9,335	5828	20.41	142,200

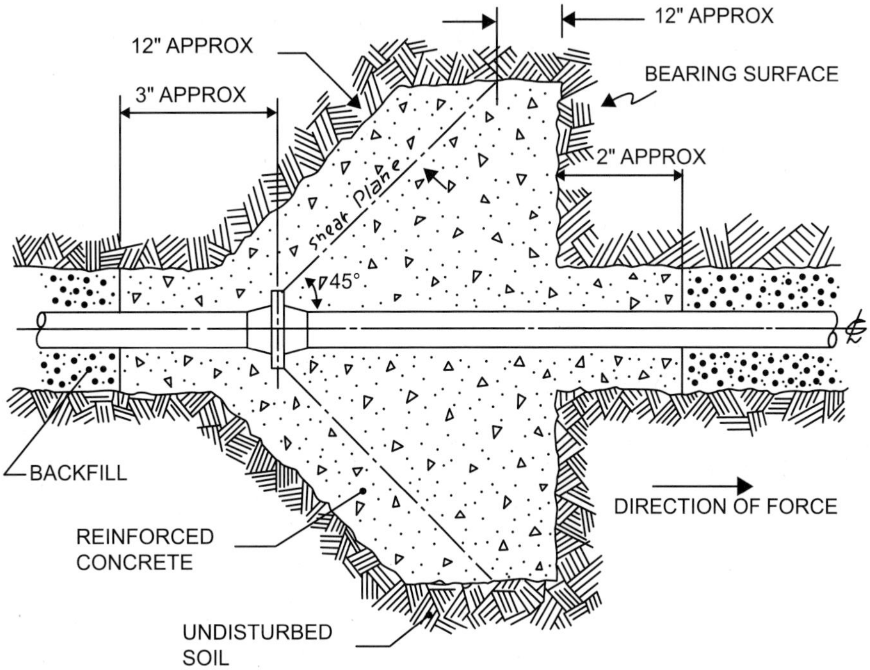

Figure 4. Sectional plan for anchor block design.

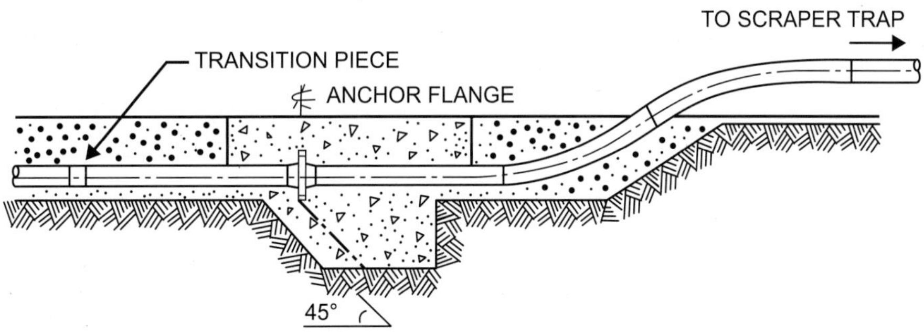

Figure 5. Sectional elevation for anchor block design.

values of lateral soil bearing pressure with due consideration to possible slip have been used.

For large blocks, friction between the block and the soil may also be figured in obtaining its resistance to imposed forces. The block should be reinforced concrete cast against undisturbed soil. Figures 4 and 5 indicate in broad detail the anchor block design.

Care should also be taken to ensure that connected surface piping has sufficient flexibility to absorb a degree of lateral anchor movement. Scraper traps should be installed such that they can move with the piping rather than being rigidly attached to blocks. Numerous examples of traps together with their support blocks being displaced a few inches are not uncommon.

References

1. Wilbur, W. E., *Pipe Line Industry,* February 1963.
2. Timoshenko, S., Theory of Elasticity.

Nomenclature

A_m	Pipe metal area $= \pi(D_o = t)t$, sq in
D_o	Pipe outside diameter, in
D_i	Pipe inside diameter, in
t	Pipe wall thickness, in
E	Young's modulus of elasticity, pi
F	Anchor force, lb
F_s	Soil resistance, lb/ft
L	Transition length, ft
P	Design pressure, psig
T_i	Installation temperature, °F
T_o	Operating temperature, °F
ΔT	Temperature difference $= (T_o - T_i)$, °F
α	Coefficient of thermal expansion, in/in °F
σ	Stress, psi
ϵ	Strain, in/in
υ	Poisson's ratio
δ	Deflection, in

Subscripts (for Greek letters only)

L	Longitudinal
H	Hoop
A	Point A, fully restrained
B	Point B, unrestrained
PR	Pressure
TE	Temperature
PO	Poisson

How to identify the series number of flanged fittings

Divide the flange thickness, expressed in sixteenths of an inch, by the nominal pipe size + 12. If the answer is less than 1.00, the flange is Series 15; if between 1.15 and 1.33, it is Series 30; if between 1.37 and 1.54, it is Series 40; and if it is between 1.60 and 1.85, it is Series 60.

Example. A 6-in flange has a thickness of $^{17}/_{16}$ inch = $^{23}/_{16}$ in.

$^{23}/_{18}$ = 1.28; the flange is series 30.

The rule is not applicable to fittings below 4 in in size; it also fails for a 4-in Series 60, which comes out 1.50.

Dimensions of three-diameter ells with tangents

Here are face-to-face and center-to-face dimensions of the three-diameter radius weld ells being used in natural gas compressor station piping.

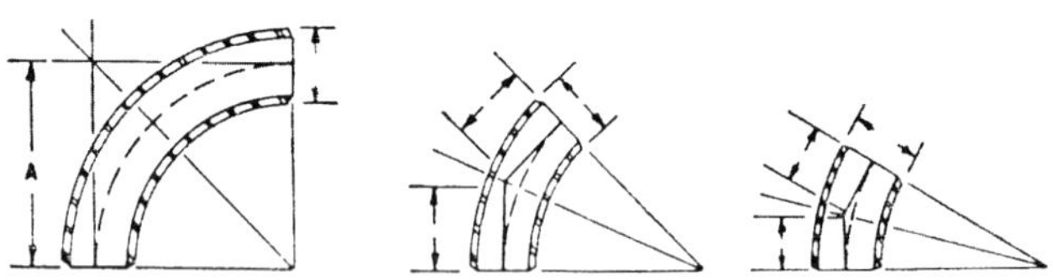

Degree Ell	Outside Diameter	A (Center To Face)	A With 4″ Tan	A With 6″ Tan
30°	16″	1′-0 7/8″	1′- 4 7/8″	1′- 6 7/8″
30°	20″	1′-4 1/16″	1′- 8 1/16″	1′-10 1/16″
30°	24″	1′-7 5/16″	1′-11 5/16″	2′- 1 5/16″
30°	26″	1′-8 7/8″	2′- 0 7/8″	2′- 2 7/8″
30°	30″	2′-0 1/8″	2′- 4 1/8″	2′- 6 1/8″
45°	16″	1′-7 7/8″	1′-11 7/8″	2′- 1 7/8″
45°	20″	2′-0 7/8″	2′- 4 7/8″	2′- 6 7/8″
45°	24″	2′-5 13/16″	2′- 9 13/16″	2′-11 13/16″
45°	26″	2′-8 5/16″	3′- 0 5/16″	3′- 2 5/16″
45°	30″	3′-1 1/4″	3′- 5 1/4″	3′- 7 1/4″
90°	16″	4′-0″	4′- 4″	4′- 6″
90°	20″	5′-0″	5′- 4″	5′- 6″
90°	24″	6′-0″	6′- 4″	6′- 6″
90°	26″	6′-6″	6′-10″	7′- 0″
90°	30″	7′-6″	7′-10″	8′- 0″

Spectacle blind thicknesses

The thickness of blinds is calculated using Formula 16, page 23, section 304.5.3 from Petroleum Refinery Piping ANSI B31.3-1973, sponsored by the American Society of Mechanical Engineers:

$$t = d\sqrt{\frac{3P}{16SE}} + ca$$

where:

t = Thickness, inches,
D = Pipe inside diameter, inches,
P = Maximum pressure blind is to contain, psig,
SE = SMYS of blind material,
ca = Corrosion allowance.

Example:

Pipe OD	8.625 in
d	7.981 in
P	1440 psig
SE	42,000 psi
ca	0.125 in

$$t = 7.981\sqrt{\frac{3 \times 1440}{16 \times 42,000}} + 0.125$$

$$t = 0.7649 \text{ in.}$$

Carbon dioxide pipeline design

Excerpts from World Resources Institute (WRI). CCS Guidelines: Guidelines for Carbon Dioxide Capture, Transport, and Storage, Washington, DC, 2008.

CO_2 pipeline operating experience

In the United States, significant CO_2 pipeline operating experience exists in the EOR industry. Since the early 1970s, pipeline companies have been successfully operating a substantial CO_2 pipeline infrastructure, transporting an estimated 0.78 trillion cubic feet of CO_2 per year through an estimated 3,900 miles of infrastructure, through pipelines of varying diameters, mainly for use in EOR. The Permian Basin region of West Texas and New Mexico remains the center of CO_2-based EOR activity. The oldest long distance CO_2 pipeline in the United States is the 140-mile Canyon Reef Carriers pipeline, which began service in 1972 for EOR in regional Texas oil fields. The longest CO_2 pipeline, the 502-mile Cortez pipeline, has been delivering about 20 million metric tons of CO_2 per year to the CO_2 hub in Denver City, Texas, since 1984.

CO_2 composition

CO_2 used for carbon dioxide capture and storage is typically in the supercritical stage, where the density resembles a liquid but it expands to fill space like a gas. Supercritical CO_2 is purchased, as a commodity, for use in many industrial processes. In the climate change context CO_2 is most often classified as an important greenhouse gas, an emission, or—in some countries—a waste. There is concern that the classification of CO_2 under various U.S. regulatory programs (e.g., air, waste, drinking water protection) may trigger unintended requirements that impose increased cost without increasing project performance or safety.

In response to this concern, the Interstate Oil and Gas Compact Commission recommends that CO_2 not be classified as a waste and, in fact, suggests that states adopt legislation recognizing CO_2 of certain purity as a commodity. At least one state (Oklahoma) has followed this lead. International frameworks have taken varied approaches. For example, Australia regulations use "greenhouse gas substances" and the London Protocol specifies "carbon dioxide streams from carbon dioxide capture processes for sequestration." Some stakeholders have advocated for setting a CO_2 purity standard of >90 percent, but many feel that there is enough uncertainty regarding the precise composition of the CO_2 stream that it is best to simply design projects with materials and procedures that account for any co-constituents in the gas stream.

Pipeline CO_2 composition

Prior to transport, captured CO_2 is conditioned to remove impurities and compressed into supercritical form. The U.S. Department of Transportation's (DOT's) Office of Pipeline Safety (OPS) defines pipeline CO_2 as a fluid consisting of more than 90 percent CO_2 molecules compressed to a supercritical state. There are currently no composition requirements (e.g., moisture or co-constituents) for the transport and geologic storage of CO_2. While there is no established standard for permitted levels of impurities in CO_2 for CCS, the pipeline-quality CO_2 compositions adhered to by the major EOR pipeline operators constitute best practice. Currently, these requirements are built into contracts between the supplier and the transporter and between the transporter and the end user.

Captured CO_2 may contain impurities like water vapor, H_2S, N_2, methane (CH_4), O_2, mercury, and hydrocarbons that may require specific handling or treatment. Before transport, the CO_2 is dehydrated to levels below 50 ppm of water. Presence of water above this level is not desirable from an operational standpoint (Aspelund and Jordal 2007). CO_2 reacts with water to form carbonic acid, which is corrosive. Additionally, under the appropriate thermodynamic conditions, hydrates (solid ice-like crystals) can form and plug the pipeline (Barrie et al. 2004). H_2S is toxic, even at low concentrations of 200 ppm. Pipelines containing H_2S will require extra due diligence, particularly near populations. However, it is important to note that it is possible to safely store H_2S with CO_2; facilities in Canada have been disposing of H_2S through injection in geologic formations since 1989 (Heinrich et al. 2004). Injection of acid gas currently occurs at 39 active operations in Alberta and northeastern British Columbia. Since surface desulfurization through the Claus process is generally uneconomical, and the surface storage of the produced sulfur constitutes a liability, more operators are turning to acid gas disposal by injection into deep geologic formations. Dehydration is particularly important in these cases, because H_2S reacts with water to form sulfuric acid, which is highly corrosive and may also result in pipeline cracking, increasing the potential for leaks.

The presence of CH_4 affects the exhibited vapor pressure of CO_2 and complicates the accurate prediction of flow (Svensson 2004). In EOR applications, in particular where organic materials are present for bacteria, oxygen is tolerable only in minute quantities (10 ppm). Even in deep saline formations organics may be present, and significant quantities of oxygen in the gas stream could allow for formation of bacterial colonies, affecting the injection operations. Additionally and significantly, mercury is present in coal and is a natural byproduct of the combustion process; it could condense in the pipeline system and create operational issues as well as implications for storage.

While not strictly a transport issue, the impact of injection of co-constituents with CO_2 is unknown on a large scale and will most likely affect the requirements for CO_2 purity. For the sole purpose of storage, the threshold for impurities could be different; hence, the processing requirements and pipeline standards could be uniquely prescribed for a particular project (e.g., Type I or Type II). On the other hand, interest in developing a network of interconnectable pipelines, for maximum utilization of geologic storage sites with or without oil recovery opportunities, may indicate the need for of a set of CO_2 specifications for pipelines similar to the ones in use today for EOR.

Pipeline operating temperature and pressure

The most efficient way to transport CO_2 is in a supercritical phase. In the supercritical state, CO_2 has the characteristics of both a liquid and a gas, maintaining the compressibility of a gas and some of the properties of a liquid, such as density. Low viscosity is important for pipeline transport. In the supercritical phase, the viscosity of CO_2 is the same as in the gas phase, which is 100 times lower than in the liquid phase. Important from a cost standpoint, supercritical transport allows for substantially higher throughput through a given pipe cross-section than transport as a lower-pressure gas.

The critical point at which CO_2 exists in a supercritical phase is 1,070 psi (73 atmospheres (atm)) and 88°F (31°C) (Figure 1). CO_2 is generally transported at temperature and pressure ranges between 55°F and 110°F and 1,250 psi (85 atm) and 2,200 psi (149.6 atm), respectively (Mohitpour et al. 2007). The upper pressure limit is mostly due to economic concerns, and is set to the ASME-ANSI 900# flange rating (the maximum pressures for ANSI 900# flange is material dependent). ASME/ANSI (American Society of Mechanical Engineers/American National Standards Institute) pipe flanges that are made to standards called out by

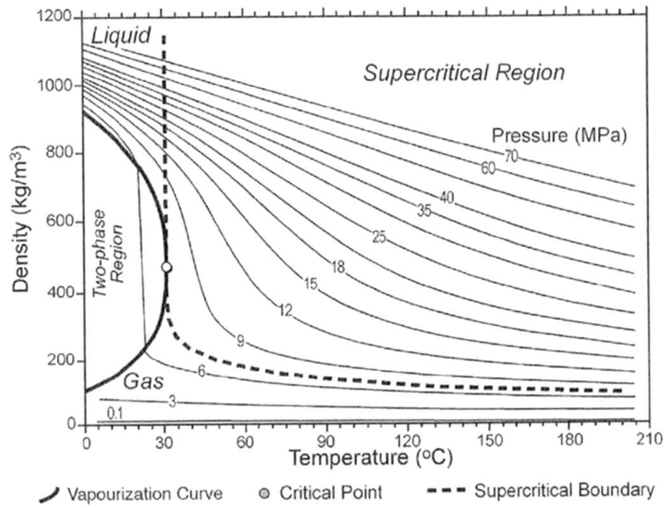

Figure 1. Variation of CO_2 density as a function of temperature and pressure.

ASME/ANSI B16.5 or ASME/ANSI B16.47 are typically made from forged materials and have machined surfaces. They are typically in "Pressure Classes," such as 150#, 300#, 600#, and 900#, and 1500#. These Pressure Classes have both pressure and temperature ratings for specific materials.

The lower pressure limit is set by the phase behavior of CO_2, and should be sufficient to maintain supercritical condition. The upper temperature limit is determined by the compressor-station discharge temperature and the temperature limits of the external pipeline coating material. The lower temperature limit is set by winter ground temperature (Farris 1983).

It is important for operators to maintain single-phase flow in CO_2 pipelines by avoiding abrupt pressure drops. In a two-phase flow, two physical phases are present in the pipeline simultaneously (e.g., liquid and gas, or supercritical fluid and gas), which creates problems for compressors and other transport equipment, increasing the chances of pipeline failure. At pressures very close to the critical point, a small change in temperature or pressure yields a very large change in the density of CO_2, which could result in a change of phase and fluid velocity, resulting in slug flow. Slug flow is encountered when gas and liquid flow simultaneously in a pipe and is characterized by long "Taylor" bubbles, also called gas slugs, rising and nearly filling a pipe cross-section. In a slugging column, with flowing gas and liquid, the flow field is extremely complex. Transmission pipelines may experience changing temperatures because of both weather and pipeline conditions. Operators should include a wide margin of safety above the rated critical pressure of CO_2 to avoid complications.

EOR Industry CO_2 Purity Specifications

For enhanced oil recovery, the CO_2 concentration in the gas transportable via pipeline typically ranges from 95 to 99 percent. At pressure in a reservoir, CO_2 can combine with components in the oil to create miscibility, wherein the fluid combination moves through the reservoir with a viscosity like that of a liquid rather than a gas. For this to happen in the reservoir, the CO_2 should be quite pure. Depending on the depth of the reservoir and properties of the oil, this may be 90 percent purity or higher. Other constituents can also be important. Nitrogen and methane raise the pressure at which the dense phase is reached, as well as the minimum miscibility pressure.

A subtle feature of using highly purified CO_2 (>95%) that is not readily apparent at first glance is its ability, when compressed and cooled, to form a supercritical fluid. Should significant amounts of noncondensable gases such as oxygen, nitrogen, or methane be present in the CO_2 stream, it may not be possible to practically produce a supercritical fluid. Thus, for any proposed gas composition, the pipeline designer should conduct appropriate compositional simulations to guarantee that supercritical phase behavior can be achieved at proposed pipeline operating conditions. Additionally, oxygen may also lead to overheating at the injection point due to reaction with oil and formation of bacterial colonies.

Sources: Zhang et al. 2004; Aspelund and Jordal 2007.

CO_2 pipeline design

There are existing design and safety criteria to ensure safe and reliable transport of CO_2. Pipeline designers consider the pressure, temperature, and properties of the fluid; the elevation or slope of the terrain; dynamic effects, such as earthquakes, waves, currents, live and dead loads, and thermal expansion and contraction; and the relative movement of connected components. The compressibility and density of CO_2 undergo significant nonlinear variation in normal pipeline operating conditions (within normal pipeline pressure and temperature ranges). Therefore, the design of CO_2 pipelines requires point-by-point estimation of fluid properties using computational models.

For pipeline construction, selection of pipe diameter, wall thickness, material strength, and toughness depends on the transmissible fluid's temperature, pressure, composition, and flow rate. For example, fluid flow rates are lower in larger-diameter pipes. Lower fluid flow rates result in fewer pressure drops, allowing a pipeline designer to consider reducing the pressure requirements for CO_2 entering the pipeline, or reducing the number of compressors along the pipeline. However, the installation costs of pipelines rise with increases in diameter. A designer will consider the economic tradeoff of increasing pipeline diameter with the cost of CO_2 compression.

Compression is the largest operating cost for the transmission system. Compressors convert the transmissible gas from atmospheric pressure to supercritical state, the desired transmissible phase. Moreover, depending on the length and terrain of pipeline, recompression or decompression of CO_2 may be required to maintain supercritical phase CO_2. The CO_2 pipeline industry currently uses centrifugal, single-stage, radial-split pumps for recompression, rather than compressors (Mohitpour et al. 2007). These booster pumping stations are installed as required to maintain sufficient pressure at high elevation points, in order to ensure a single-phase CO_2 flow. For reference, the compression unit at the Great Plains Synfuel Plant consists of two 8-stage compressors. Feedgas is taken at 3 pound per square inch gauge (psig) and compressed to 2,700 psig, which is in the supercritical range for CO_2.

Avoiding initiation and propagation of longitudinal-running fractures is also essential. Fracture arresters are typically installed every 500 meters (545 yards), and lower-strength steel and thicker-wall pipe are employed (Mohitpour et al. 2007). The pipelines for CO_2 transportation are usually constructed of steel (60,000–80,000 psi yield strength), such as American Petroleum Institute (API) X60- or X80-grade material. The optimum strength and wall thickness are determined based on the aforementioned factors, as well as fabrication and handling considerations. To reduce the chances of corrosion, CO_2 pipelines typically have an external coating of fusion-bonded epoxy or polyurethane with full cathodic protection; internal pipeline coatings are also available and can be applied where appropriate.

The main components of a pipeline include valves, compressors, booster pumps, pig launchers and receivers, batching stations and instrumentation, metering stations, and Supervisory Control and Data Acquisition (SCADA) systems. Valves are typically used for control functions around compressor and metering stations and at the injection sites.

One important consideration in pipeline design is the distance between block valves. Block valves are used to isolate sections of pipe in the event of a leak or for maintenance. Block valves are spaced every 16–32 kilometers (10–20 miles), depending on the location of the pipe, and are installed more frequently near critical locations, such as road and river crossings and urban areas. Installing block valves more frequently increases both the cost of the pipeline and the risk of leakage from the valves themselves. The farther apart the valves are installed, the greater the volume contained between the valves, which increases the distance from the pipeline required for the gas to dissipate to a safe level in the event of a pipeline rupture (Gale and Davidson 2004). Current pipeline design safety standards already take into consideration valve spacing as a function of pipeline diameter and surrounding land use.

Instrumentation along the pipeline is typically used to measure the flow rate, pressure, and temperature of the CO_2 and provides sufficient information for the pipeline's normal operation. The instrumentation is located at compressor and metering stations and sometimes at the block valves. SCADA systems are used for remote monitoring and operation of the compressor stations and the pipeline. These systems are designed to provide operators at a central control center with sufficient data on the status of the pipeline to enable them to control the flows through the compressors and the pipeline as necessary. Metering is used for computational pipeline monitoring (CPM) leak-detection systems for single-phase lines (without gas in the liquid). Currently CO_2 pipelines are not required to have CPM, mainly because it is technically difficult. Other leak-detection methods, such as pressure point analysis and aerial and visual surveys, may be used to ensure safe CO_2 transport.

The majority of onshore CO_2 pipelines are buried over most of their length, to a depth of 1-1.2 meters (3-4 feet), except at metering or pumping stations, and most offshore lines are also usually buried below the shallow water seabed. In deeper water, only pipelines with a diameter of less than 400 millimeters (16 inches) are trenched and sometimes buried to protect them against damage by fishing gear. The exact depth varies based on project-specific needs, and variances can be granted where appropriate.

Experience from decades of pipeline operations suggests that designing and operating CO_2 pipelines do not pose any new challenges. The optimum solution is site specific and will depend on different factors, including volumes of gas to be transmitted, gas composition, local population density, topography, and meteorological conditions.

Primary types of CO_2 pipelines

Existing CO_2 pipelines operate at pressures ranging from 1,250 to 2,200 pounds per square inch (psi.) Since most natural gas pipelines operate at pressures at or below 1,200 psi, CO_2 pipelines are constructed specifically for transporting CO_2 and are normally listed as either Type II or Type III pipelines. Acceptable CO_2 and co-constituent concentrations for both pipeline types are shown in Table 1.

The majority of the CO_2 pipelines in North America can be listed as Type II pipelines, which serve multiple sources and user lines and have a strictly limited composition. Less common Type III pipelines have relaxed composition standards when compared to Type II pipelines. The best example of a Type III pipeline is the pipeline that connects the Dakota Gasification plant near Beulah, North Dakota with the Weyburn enhanced oil recovery (EOR) project in southern Saskatchewan. This pipeline carries a CO_2 mix that has a relatively higher hydrogen sulfide (H_2S) concentration. It should be noted that extra operational precautions are required at both the source and the sinks when there are high H_2S concentrations. As a result of the different composition standards, it will not be easy for operators to connect Type III pipelines to Type II pipelines.

Type I pipelines do not exist in today's CO_2 EOR industry, but can be developed for a specific single use (i.e., a single CCS project). These pipelines can be applied to situations wherein case-by-case specifications for CO2 composition are appropriate based upon the capture system, but they cannot be connected with the existing CO_2 EOR pipeline network of Type II pipelines.

Table 1
Acceptable CO_2 and Co-Constituent Concentrations

Parameter	Type I	Type II	Type III
CO_2 (% by volume)	>95%	>95%	>96%
H_2S (ppmbw)	<10	<20	<10,000
Sulphur (ppmbw)	<35	<30	
Total hydrocarbons (% by volume)	<5	<5	
CH_4 (% by volume)			<0.7
C_2+ hydrocarbons (% by volume)			<23,000
CO (% by volume)			<1,000
N_2 (% by volume)	<4	<4	<300
O_2 (ppm by weight)	<10	<10	<50
H_2O (#/mmcf* or ppm by volume**)	<25*	<30*	<20**

ppmbw = ppm by weight
Source: Steve Melzer.

References

A. Aspelund, K. Jordal, Gas Conditioning—The Interface Between CO_2 Capture and Transport, International Journal of Greenhouse Gas Control 1 (3) (July 2007) 343–354.

J. Barrie, K. Brown, P.R. Hatcher, H.U. Schellhase, Carbon Dioxide Pipelines: A Preliminary Review of Design and Risks, in: Proceedings of the 7th International Conference on Greenhouse Gas Control Technologies. September 5–9. Canada, Vancouver, 2004.

C.B. Farris, Unusual Design Factors for Supercritical CO_2 Pipelines, Energy Progress 3 (3) (September 1983) 153–158.

Reprinted with permission. For complete citations and full text, see the report World Resources Institute (WRI). CCS Guidelines: Guidelines for Carbon Dioxide Capture, Transport, and Storage. 2008. Washington, DC: WRI. Available at www.wri.org

J. Gale, J. Davidson, "Transmission of CO_2—Safety and Economic Considerations, Energy 29 (9-10) (2004) 1319–1328.

J.J. Heinrich, H.J. Herzog, D.M. Reiner. March, Environmental Assessment of Geologic Storage of CO_2. MITLFEE 2003-002. Prepared for the Clean Air Task Force (2004).

M. Mohitpour, H. Golshan, A. Murray, Pipeline Design & Construction: A Practical Approach, 3rd ed. ASME Press, New York, NY, 2007.

R. Svensson, M. Odenberger, F. Johnsson, L. Strömberg, Transportation Systems for CO_2—Application to Carbon Capture and Storage, Energy Conversion and Management 45 (2004) 2343–2353.

P.Y. Zhang, S. Huang, S. Sayegh, X.L. Zhou, April, Effect of CO2 Impurities on Gas Injection EOR Process, in: Prepared for the 2004 SPE/DOE Fourteenth Symposium on Improved Oil Recovery. Society of Petroleum Engineers, Tulsa, OK, 2004.

4: Electrical Design

Electrical design 160
Hazardous locations 161
NEMA enclosure types 162
Size portable electric generators 163
Typical wattages for tools and appliances 164
Knockout dimensions 164
National Electrical Code tables 165
Electrical formulas 169
Full-load currents—single-phase transformers 169
Conduit size for combinations of cables with different outside diameters 170
Minimum bending radius for insulated cables for permanent training during installation 170
Full-load currents—three-phase transformers 172
Motor controller sizes 172
Voltage drop on circuits using 600 V copper conductors in steel conduit 173
Determine the most economical size for electric power conductors 173
How to find the resistance and weight of copper wires 174
What you should remember about electrical formulas 174
How to calculate microwave hops on level ground 174
Quick determination of hp per ampere for induction motors (3 phase) at different voltages 175
Chart of electric motor horsepower for pumping units 175
Pumping stations 176
Floodlighting calculations 177
Point-by-point method 177
Beam-lumen design method 178
Conductor size conversion chart—metric to AWG 179
Commonly used switchgear device numbers 180
Bonding the grounding system to building and structure foundations 181

Electrical design

Most pipeline facilities use electrical equipment of some sort or another, ranging from simple power circuits of a few amperes capacity to sophisticated supervisory control and data acquisition systems. It is beyond the scope of this manual to provide a complete section on electrical information for pipeline facilities. This section is intended to provide some basic data that will prove useful to field personnel responsible for electrical installations.

Electric motor selection

Based on known facts and calculations, the best selection is made after a close study of the installation, operation, and servicing of the motor. Basic steps in proper selection are numbered 1–8 and are briefly described.

1. **Power supply**
   a. Voltage—National Electrical Manufacturers Association (NEMA) has recommended the following standards.

Nominal Power System (V)	Motor Nameplate (V)
240	230
480	460
600	575

   b. Frequency—Motors rated 200 horsepower or less can vary not to exceed 5% above or below its rated frequency.
   c. Phases—Three-phase power supplies are found in most industrial locations; for most residential and rural areas, only single-phase power is available.
2. **HP and duty requirements**
   a. Continuous duty—means constant load for an indefinite period (about 90% of all motor applications).
   b. Intermittent duty—means alternate periods of load and no-load, or load and rest.
   c. Varying duty—means both the load and time operation vary to a wide degree.
3. **Speeds**—Single speed motors are the most common, but when a range of speeds is required multispeed motors will give two, three, or four fixed speeds.
4. **Service factors**—Open, general purpose motors have a service factor depending upon the particular rating of the motor (usually between 1.15 and 1.25).
5. **Selection of motor type**

   There are three main types:

   a. DC motors are designed for industrial drives requiring a controlled speed range and constant torque output on adjustable voltage control systems. Because the speed of rotation controls the flow of current in the armature, special devices must be used for starting DC motors.
   b. Single-phase alternating-current motors also require some auxiliary arrangement to start rotation.
   c. Polyphase motors are alternating-current types (squirrel cage or wound rotor).

Applications—Some driven machines require a low-starting torque that gradually increases to full-load speed; others require higher-starting torque. NEMA code letters A, B, C, D, etc., on the motor nameplate designate the locked rotor kVA per horsepower of that particular motor design. The diagram shows representative speed-torque curves for polyphase and single-phase NEMA design motors Types A–D.

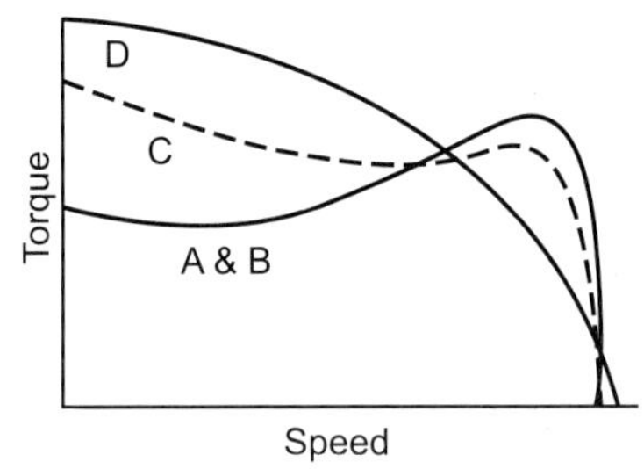

6. **Torque definitions**
   a. Locked-rotor torque, or "starting torque," is the minimum-torque that the motor will develop at rest for all angular positions of the rotor.

Code Letter	Locked Rotor (kVA/hp)
A	0–3.14
B	3.15–3.54
C	3.55–3.99
D	4.0–4.49
E	4.5–4.99
F	5–5.59

   b. Breakdown torque is the maximum-torque at rated voltage and frequency without abrupt speed drop.
   c. Full-load-torque is the torque necessary to produce rated horsepower at full-load speed.

d. Locked-rotor current is the steady-state current of the motor with the rotor locked at rated voltage and frequency.

7. **Selection of enclosure**—The two general classifications are open, which permits passage of air over and around the windings, and totally enclosed, which prevents exchange of air between inside and outside of frame (but is not strictly airtight).
   a. Open Drip-Proof—means that liquid or solid particles falling on the motor at an angle not greater than 15° from vertical cannot enter the motor.
   b. NEMA Type 1—a weather-protected machine with its ventilating passages constructed to minimize the entrance of rain, snow, and airborne particles to the electric parts and having its ventilating openings so constructed to prevent the passage of a cylindrical rod 3/4 in in diameter.
   c. NEMA Type II—a weather-protected machine that has in addition to the enclosure defined for a Type I weather-protected machine, its ventilating passages at both intake and discharge so arranged that high velocity air and airborne particles blown into the machine by storms or high winds can be discharged without entering the internal ventilating passages directly leading to the electric parts of the machine itself. The normal path of the ventilating air that enters the electric parts of the machine shall be so arranged by baffling or separate housings as to provide at least three abrupt changes in direction, none of which shall be less than 90°. In addition, an area of low velocity, not exceeding 600 ft/min, shall be provided in the intake air path to minimize the possibility of moisture or dirt being carried into the electric parts of the machine.
   d. Totally Enclosed Nonvent—means that a motor is not equipped for cooling by external means.
   e. Totally Enclosed Fan-Cooled—means the motor has a fan integral with the motor but not external to the enclosed parts.
   f. Explosion-Proof—means the enclosure is designed to withstand an explosion of a specified gas that may occur within the motor and to prevent the ignition of gas around the motor.
8. End shield mountings

Three types of end shields with rabbets and bolt holes for mounting are standard in the industry:

a. Type C Face provides a male rabbet and tapped holes.
b. Type D Flange has a male rabbet with holes for through bolts in the flange.
c. Type P base has a female rabbet and through bolts for mounting in the flange (used for mounting vertical motors).

Hazardous locations

Definition of class locations

500—5. Class I Locations. Class I locations are those in which flammable gases or vapors are or may be present in the air in quantities sufficient to produce explosive or ignitible mixtures. Class I locations shall include those specified in (a) and (b) below.

a. **Class I, Division 1.** A Class I, Division 1 location is a location: (1) in which ignitible concentrations of flammable gases or vapors can exist under normal operating conditions; or (2) in which ignitible concentrations of such gases or vapors may frequently exist because of repair or maintenance operations or because of leakage; or (3) in which breakdown or faulty operation of equipment or processes might release ignitible concentrations of flammable gases or vapors and might also cause simultaneous failure of electric equipment.

(FPN): This classification usually includes locations where volatile flammable liquids or liquefied flammable gases are transferred from one container to another; interiors of spray booths and areas in the vicinity of spraying and painting operations where volatile flammable solvents are used; locations containing open tanks or vats of volatile flammable liquids; drying rooms or compartments for the evaporation of flammable solvents; locations containing fat and oil extraction equipment using volatile flammable solvents; portions of cleaning and dyeing plants where flammable liquids are used; gas generator rooms and other portions of gas manufacturing plants where flammable gas may escape; inadequately ventilated pump rooms for flammable gas or for volatile flammable liquids; the interiors of refrigerators and freezers in which volatile flammable materials are stored in open, lightly stoppered, or easily ruptured containers; and all other locations where ignitible concentrations of flammable vapors or gases are likely to occur in the course of normal operations.

b. **Class I, Division 2.** A Class I, Division 2 location is a location: (1) in which volatile flammable liquids or flammable gases are handled, processed, or used, but in which the liquids, vapors, or gases will normally be confined within closed containers or closed systems from which they can escape only in case of accidental rupture or breakdown of such containers or systems, or in the case of abnormal operation of equipment; or (2) in which ignitible concentrations of gases or vapors are normally prevented by positive mechanical ventilation and which might become hazardous through failure or abnormal operation of the ventilating equipment; or (3) that is adjacent to a Class I, Division 1 location, and to which ignitible concentrations of gases or vapors might occasionally be communicated unless such communication is prevented by adequate positive-pressure ventilation from a source of clean air, and effective safeguards against ventilation failure are provided.

(FPN No. 1): This classification usually includes locations where volatile flammable liquids or flammable gases or vapors are used, but which, in the judgment of the authority having jurisdiction, would become hazardous only in case of an accident or of some unusual operating condition. The quantity of flammable material that might escape in case of accident, the adequacy of ventilating equipment, the total area involved, and the record of the industry or business with respect to explosions or fires are all factors that merit consideration in determining the classification and extent of each location.

(FPN No. 2): Piping without valves, checks, meters, and similar devices would not ordinarily introduce a hazardous condition even if used for flammable liquids or gases. Locations used for the storage of flammable liquid or of liquefied or compressed gases in sealed containers would not normally be considered hazardous unless also subjected to other hazardous conditions.

Group A: Atmospheres containing acetylene.

Group B: Atmospheres containing hydrogen, fuel, and combustible process gases containing more than 30% hydrogen by volume or gases or vapors of equivalent hazard such as butadiene,[1] ethylene oxide,[2] propylene oxide,[2] and acrolein.[2]

Group C: Atmospheres such as cyclopropane, ethyl ether, ethylene, or gases or vapors of equivalent hazard.

Group D: Atmospheres such as acetone, ammonia,[3] benzene, butane, ethanol, gasoline, hexane, methanol, methane, natural gas, naptha, propane, or gases of vapors of equivalent hazard.

[1] Group D equipment may be used for this atmosphere if such equipment is isolated in accordance with Section 501-5(a) by sealing all conduits 1/2-in or larger.
[2] Group C equipment may be used for this atmosphere if such equipment is isolated in accordance with Section 501-5(a) by sealing all conduits 1/2-in or larger.
[3] For classification of areas involving ammonia atmosphere, see Safety Code for Mechanical Refrigeration, ANSI/ASHRAE 15-1978, and Safety Requirements for the Storage and Handling of Anhydrous Ammonia, ANSI/CGA G2.1-1981.

NEMA enclosure types

Nonhazardous locations

Type	Intended Use and Description
1	Enclosures are intended for indoor use primarily to provide a degree of protection against limited amounts of falling dirt.
2	Enclosures are intended for indoor use primarily to provide a degree of protection against limited amounts of falling water and dirt.
3	Enclosures are intended for outdoor use primarily to provide a degree of protection against rain, sleet, windblown dust, and damage from external ice formation.
3R	Enclosures are intended for outdoor use primarily to provide a degree of protection against rain, sleet, and damage from external ice formation.
3S	Enclosures are intended for outdoor use primarily to provide a degree of protection against rain, sleet, windblown dust, and to provide for operation for external mechanisms when ice laden.
4	Enclosures are intended for indoor or outdoor use primarily to provide a degree of protection against windblown dust and rain, splashing water, hose-directed water, and damage from external ice formation.

4X Enclosures are intended for indoor or outdoor use primarily to provide a degree of protection against corrosion, windblown dust and rain, splashing water, hose-directed water, and damage from external ice formation.

5 Enclosures are intended for indoor use primarily to provide a degree of protection against settling airborne dust, falling dirt, and dripping noncorrosive liquids.

6 Enclosures are intended for indoor or outdoor use primarily to provide a degree of protection against hose-directed water, the entry of water during occasional temporary submersion at a limited depth, and damage from external ice formation.

6P Enclosures are intended for indoor or outdoor use primarily to provide a degree of protection against hose-directed water, the entry of water during prolonged submersion at a limited depth, and damage from external ice formation.

12 Enclosures are intended for indoor use primarily to provide a degree of protection against circulating dust, falling dirt, and dripping noncorrosive liquids.

12K Enclosures with knockouts are intended for indoor use primarily to provide a degree of protection against circulating dust, falling dirt, and dripping noncorrosive liquids.

13 Enclosures are intended for indoor use primarily to provide a degree of protection against dust, spraying of water, oil, and noncorrosive coolant.

Hazardous (classified locations)

7 Enclosures are intended for indoor use in locations classified as Class I, Groups A, B, C, or D, as defined in the National Electrical Code.

8 Enclosures are intended for indoor use in locations classified as Class I, groups A, B, C, or D, as defined in the National Electrical Code.

9 Enclosures are intended for indoor use in locations classified as Class II, groups E, F, or G, as defined in the National Electrical Code.

10 Enclosures are constructed to meet the applicable requirements of the Mine Safety and Health Administration.

Reference NEMA Standards Publications/No. 250-1991.
Used with permission—National Electrical Manufacturer's Association.

Size portable electric generators

A power plant must be properly sized for the required load. Electric motors are particularly difficult for portable plants, because they require two to three times motor nameplate amps or wattage.

A portable power plant's surge capacity is limited by engine horsepower and inertia of its rotating parts. Thus, a current surge of short duration can be supplied by a power plant, but a current demand of longer duration, such as a heavily loaded motor starting a high inertia load, can overload the power plant and possibly damage the power plant and the motor. A 3450 rpm air compressor is a prime example of this type of load.

Example:
Load requirements:

Electric heater	1000 W
11–100 W lamps	100 W
Motor	600 W
Total motor load	600 × 3 = 1800 W
Total other load	2100 W
Total load	3100 W
Add 25% for future	975 W
Total load for sizing power plant	4875 W

A generator capable of continuously supplying 5000 W will be adequate.

When more than one motor is connected to a power plant, always start the largest motor first. If the total load is connected to one receptacle on the power plant, be sure the ampere rating of the receptacle is not exceeded.

Typical wattages for tools and appliances

Equipment	Running Watts
1/2″ Drill	1000
1″ Drill	1100
6″ Circular saw	800
10″ Circular saw	2000
14″ Chain saw	1100
Radio	50–200
Television	200–500
Toaster, coffeemaker	1200
Water heater (storage type)	1100–5500
Hot plate or range (per burner)	1000
Electric range (small/large)	1200/2000
Skillet	1200
Fan	50–200
Floodlight	1000
Water pump	500
Vacuum cleaner	200–300
Refrigerator (conventional)	200–300
Refrigerator (freezer-combination)	250–600
Furnace fan blower	500–700

Knockout dimensions

Conduit size, in	Knockout Diameter, in		
	Minimum	Nominal	Maximum
1/2	0.859	0.875	0.906
3/4	1.094	1.109	1.141
1	1.359	1.375	1.406
1¼	1.719	1.734	1.766
1½	1.958	1.984	2.016
2	2.433	2.469	2.500
2½	2.938	2.969	3.000
3	3.563	3.594	3.625
3½	4.063	4.125	4.156
4	4.563	4.641	4.672
5	5.625	5.719	5.750
6	6.700	6.813	6.844

Reference NEMA Standards Publications/No. 250-1991.
Used with permission—National Electrical Manufacturer's Association.

National Electrical Code tables

Table 430–150
Full-Load Current* Three-Phase Alternating-Current Motors

	Induction Type Squirrel-Cage and Wound-Rotor Amperes							Synchronous Type †Unity Power Factor Amperes			
HP	115V	200V	208V	230V	460V	575V	2300V	230V	460V	575V	2300V
½	4	2.3	2.2	2	1	.8					
¾	5.6	3.2	3.1	2.8	1.4	1.1					
1	7.2	4.1	4.0	3.6	1.8	1.4					
1½	10.4	6.0	5.7	5.2	2.6	2.1					
2	13.6	7.8	7.5	6.8	3.4	2.7					
3		11.0	10.6	9.6	4.8	3.9					
5		17.5	16.7	15.2	7.6	6.1					
7½		25.3	24.2	22	11	9					
10		32.2	30.8	28	14	11					
15		48.3	46.2	42	21	17					
20		62.1	59.4	54	27	22					
25		78.2	74.8	68	34	27		53	26	21	
30		92	88	80	40	32		63	32	26	
40		119.6	114.4	104	52	41		83	41	33	
50		149.5	143.0	130	65	52		104	52	42	
60		177.1	169.4	154	77	62	16	123	61	49	12
75		220.8	211.2	192	96	77	20	155	78	62	15
100		285.2	272.8	248	124	99	26	202	101	81	20
125		358.8	343.2	312	156	125	31	253	126	101	25
150		414	396.0	360	180	144	37	302	151	121	30
200		552	528.0	480	240	192	49	400	201	161	40

*These values of full-load current are for motors running at speeds usual for belted motors and motors with normal torque characteristics. Motors built for especially low speeds or high torques may require more running current, and multispeed motors will have full-load current varying with speed, in which case the nameplate current rating shall be used.

†For 90 and 80% power factor the above figures shall be multiplied by 1.1 and 1.25%, respectively. The voltages listed are rated motor voltages. The currents listed shall be permitted for system voltage ranges of 110–120, 220–240, 440–480, and 550–600 V.

Table 310.16
Allowable Ampacities of Insulated Conductors Rated 0–2000 V, 60–90 °C (140–194 °F), Not More Than Three Current-Carrying Conductors in Raceway, Cable, or Earth (Direct Buried), Based on Ambient Temperature of 30 °C (86 °F)

Size AWG or kcmil	Temperature Rating of Conductor (See Table 310.13.)						Size AWG or kcmil
	60°C (140°F)	75°C (167°F)	90°C (194°F)	60°C (140°F)	75°C (167°F)	90°C (194°F)	
	Types TW, UF	Types RHW, THHW, THW, THWN, XHHW, USE, ZW	Types TBS, SA, SIS, FEP, FEPB, MI, RHH, RHW-2, THHN, THHW, THW-2, THWN-2, USE-2, XHH, XHHW, XHHW-2, ZW-2	Types TW, UF	Types RHW, THHW, THW, THWN, XHHW, USE	Types TBS, SA, SIS, THHN, THHW, THW-2, THWN-2, RHH, RHW-2, USE-2, XHH, XHHW, XHHW-2, ZW-2	
	Copper			**Aluminium or Copper-Clad Aluminium**			
18	—	—	14	—	—	—	—
16	—	—	18	—	—	—	—
14*	20	20	25	—	—	—	—
12*	25	25	30	20	20	25	12*
10*	30	35	40	25	30	35	10*
8	40	50	55	30	40	45	8
6	55	65	75	40	50	60	6
4	70	85	95	55	65	75	4
3	85	100	110	65	75	85	3
2	95	115	130	75	90	100	2
1	110	130	150	85	100	115	1
1/0	125	150	170	100	120	135	1/0
2/0	145	175	195	115	135	150	2/0
3/0	165	200	225	130	155	175	3/0
4/0	195	230	260	150	180	205	4/0
250	215	255	290	170	205	230	250
300	240	285	320	190	230	255	300
350	260	310	350	210	250	280	350
400	280	335	380	225	270	305	400
500	320	380	430	260	310	350	500
600	355	420	475	285	340	385	600
700	385	460	520	310	375	420	700
750	400	475	535	320	385	435	750
800	410	490	555	330	395	450	800
900	435	520	585	355	425	480	900
1000	455	545	615	375	445	500	1000
1250	495	590	665	405	485	545	1250
1500	520	625	705	435	520	585	1500
1750	545	650	735	455	545	615	1750
2000	560	665	750	470	560	630	2000
Correction Factors							
Ambient Temp. (°C)	For ambient temperatures other than 30°C (86°F), multiply the allowable ampacities shown above by the appropriate factor shown below.						Ambient Temp. (°F)
21–25	1.08	1.05	1.04	1.08	1.05	1.04	70–77
26–30	1.00	1.00	1.00	1.00	1.00	1.00	78–86
31–35	0.91	0.94	0.96	0.91	0.94	0.96	87–95
36–40	0.82	0.88	0.91	0.82	0.88	0.91	96–104
41–45	0.71	0.82	0.87	0.71	0.82	0.87	105–113
46–50	0.58	0.75	0.82	0.58	0.75	0.82	114–122
51–55	0.41	0.67	0.76	0.41	0.67	0.76	123–131
56–60	—	0.58	0.71	—	0.58	0.71	132–140
61–70	—	0.33	0.58	—	0.33	0.58	141–158
71–80	—	—	0.41	—	—	0.41	159–176

*See 240.4(D).

Table C8(A)
Maximum Number of Compact Conductors in Rigid Metal Conduit

Compact Conductors													
		Metric Designator (Trade Size)											
Type	Conductor Size (AWG/kcmil)	16 (½)	21 (¾)	27 (1)	35 (1¼)	41 (1½)	53 (2)	63 (2½)	78 (3)	91 (3½)	103 (4)	129 (5)	155 (6)
THW,	8	2	4	7	12	16	26	38	59	78	101	158	228
THW-2,	6	1	3	5	9	12	20	29	45	60	78	122	176
THHW	4	1	2	4	7	9	15	22	34	45	58	91	132
	2	1	1	3	5	7	11	16	25	33	43	67	97
	1	1	1	1	3	5	8	11	17	23	30	47	68
	1/0	1	1	1	3	4	7	10	15	20	26	41	59
	2/0	0	1	1	2	3	6	8	13	17	22	34	50
	3/0	0	1	1	1	3	5	7	11	14	19	29	42
	4/0	0	1	1	1	2	4	6	9	12	15	24	35
	250	0	0	1	1	1	3	4	7	9	12	19	28
	300	0	0	1	1	1	3	4	6	8	11	17	24
	350	0	0	1	1	1	2	3	5	7	9	15	22
	400	0	0	1	1	1	1	3	5	7	8	13	20
	500	0	0	0	1	1	1	3	4	5	7	11	17
	600	0	0	0	1	1	1	1	3	4	6	9	13
	700	0	0	0	1	1	1	1	3	4	5	8	12
	750	0	0	0	0	1	1	1	3	4	5	7	11
	1000	0	0	0	0	1	1	1	1	3	4	6	9
THHN,	8	—	—	—	—	—	—	—	—	—	—	—	—
THWN,	6	2	5	8	13	18	30	43	66	88	114	179	258
THWN-2	4	1	3	5	8	11	18	26	41	55	70	110	159
	2	1	1	3	6	8	13	19	29	39	50	79	114
	1	1	1	2	4	6	10	14	22	29	38	60	86
	1/0	1	1	1	4	5	8	12	19	25	32	51	73
	2/0	1	1	1	3	4	7	10	15	21	26	42	60
	3/0	0	1	1	2	3	6	8	13	17	22	35	51
	4/0	0	1	1	1	3	5	7	10	14	18	29	42
	250	0	1	1	1	2	4	5	8	11	14	23	33
	300	0	0	1	1	1	3	4	7	10	12	20	28
	350	0	0	1	1	1	3	4	6	8	11	17	25
	400	0	0	1	1	1	2	3	5	7	10	15	22
	500	0	0	0	1	1	1	3	5	6	8	13	19
	600	0	0	0	1	1	1	2	4	5	6	10	15
	700	0	0	0	1	1	1	1	3	4	6	9	13
	750	0	0	0	1	1	1	1	3	4	5	9	13
	1000	0	0	0	0	1	1	1	2	3	4	6	9
XHHW,	8	3	5	9	15	21	34	49	76	101	130	205	296
XHHW-2	6	2	4	6	11	15	25	36	56	75	97	152	220
	4	1	3	5	8	11	18	26	41	55	70	110	159
	2	1	1	3	6	8	13	19	29	39	50	79	114
	1	1	1	2	4	6	10	14	22	29	38	60	86
	1/0	1	1	1	4	5	8	12	19	25	32	51	73
	2/0	1	1	1	3	4	7	10	16	21	27	43	62
	3/0	0	1	1	2	3	6	8	13	17	22	35	51
	4/0	0	1	1	1	3	5	7	11	14	19	29	42
	250	0	1	1	1	2	4	5	8	11	15	23	34
	300	0	0	1	1	1	3	5	7	10	13	20	29
	350	0	0	1	1	1	3	4	6	9	11	18	25
	400	0	0	1	1	1	2	4	6	8	10	16	23
	500	0	0	0	1	1	1	3	5	6	8	13	19
	600	0	0	0	1	1	1	2	4	5	7	10	15
	700	0	0	0	1	1	1	1	3	4	6	9	13
	750	0	0	0	1	1	1	1	3	4	5	8	12
	1000	0	0	0	0	1	1	1	2	3	4	7	10

Definition: *Compact stranding* is the result of a manufacturing process where the standard conductor is compressed to the extent that the interstices (voids between strand wires) are virtually eliminated.

Table C9
Maximum Number of Conductors or Fixture Wires in Rigid PVC Conduit, Schedule 80

Conductors													
		Metric Designator (Trade Size)											
Type	Conductor Size (AWG/kcmil)	16 (½)	21 (¾)	27 (1)	35 (1¼)	41 (1½)	53 (2)	63 (2½)	78 (3)	91 (3½)	103 (4)	129 (5)	155 (6)
RHH,	14	3	5	9	17	23	39	56	88	118	153	243	349
RHW,	12	2	4	7	14	19	32	46	73	98	127	202	290
RHW-2	10	1	3	6	11	15	26	37	59	79	103	163	234
	8	1	1	3	6	8	13	19	31	41	54	85	122
	6	1	1	2	4	6	11	16	24	33	43	68	98
	4	1	1	1	3	5	8	12	19	26	33	53	77
	3	0	1	1	3	4	7	11	17	23	29	47	67
	2	0	1	1	3	4	6	9	14	20	25	41	58
	1	0	1	1	1	2	4	6	9	13	17	27	38
	1/0	0	0	1	1	1	3	5	8	11	15	23	33
	2/0	0	0	1	1	1	3	4	7	10	13	20	29
	3/0	0	0	1	1	1	3	4	6	8	11	17	25
	4/0	0	0	0	1	1	2	3	5	7	9	15	21
	250	0	0	0	1	1	1	2	4	5	7	11	16
	300	0	0	0	1	1	1	2	3	5	6	10	14
	350	0	0	0	1	1	1	1	3	4	5	9	13
	400	0	0	0	0	1	1	1	3	4	5	8	12
	500	0	0	0	0	1	1	1	2	3	4	7	10
	600	0	0	0	0	0	1	1	1	3	3	6	8
	700	0	0	0	0	0	1	1	1	2	3	5	7
	750	0	0	0	0	0	1	1	1	2	3	5	7
	800	0	0	0	0	0	1	1	1	2	3	4	7
	1000	0	0	0	0	0	1	1	1	1	2	4	5
	1250	0	0	0	0	0	0	1	1	1	1	3	4
	1500	0	0	0	0	0	0	1	1	1	1	2	4
	1750	0	0	0	0	0	0	0	1	1	1	2	3
	2000	0	0	0	0	0	0	0	1	1	1	1	3
TW,	14	6	11	20	35	49	82	118	185	250	324	514	736
THHW,	12	5	9	15	27	38	63	91	142	192	248	394	565
THW,	10	3	6	11	20	28	47	67	106	143	185	294	421
THW-2	8	1	3	6	11	15	26	37	59	79	103	163	234
RHH*,	14	4	8	13	23	32	55	79	123	166	215	341	490
RHW*,	12	3	6	10	19	26	44	63	99	133	173	274	394
RHW-2*	10	2	5	8	15	20	34	49	77	104	135	214	307
	8	1	3	5	9	12	20	29	46	62	81	128	184
RHH*,	6	1	1	3	7	9	16	22	35	48	62	98	141
RHW*,	4	1	1	3	5	7	12	17	26	35	46	73	105
RHW-2*,	3	1	1	2	4	6	10	14	22	30	39	63	90
TW,	2	1	1	1	3	5	8	12	19	26	33	53	77
THW,	1	0	1	1	2	3	6	8	13	18	23	37	54
THHW,	1/0	0	1	1	1	3	5	7	11	15	20	32	46
THW-2	2/0	0	1	1	1	2	4	6	10	13	17	27	39
	3/0	0	0	1	1	1	3	5	8	11	14	23	33
	4/0	0	0	1	1	1	3	4	7	9	12	19	27
	250	0	0	0	1	1	2	3	5	7	9	15	22
	300	0	0	0	1	1	1	3	5	6	8	13	19
	350	0	0	0	1	1	1	2	4	6	7	12	17
	400	0	0	0	1	1	1	2	4	5	7	10	15
	500	0	0	0	1	1	1	1	3	4	5	9	13
	600	0	0	0	0	1	1	1	2	3	4	7	10
	700	0	0	0	0	1	1	1	2	3	4	6	9
	750	0	0	0	0	0	1	1	1	3	4	6	8
	800	0	0	0	0	0	1	1	1	3	3	6	8
	900	0	0	0	0	0	1	1	1	2	3	5	7
	1000	0	0	0	0	0	1	1	1	2	3	5	7
	1250	0	0	0	0	0	1	1	1	1	2	4	5
	1500	0	0	0	0	0	0	1	1	1	1	3	4
	1750	0	0	0	0	0	0	1	1	1	1	3	4
	2000	0	0	0	0	0	0	0	1	1	1	2	3

(table continued on next page)

Table C9
Maximum Number of Conductors or Fixture Wires in Rigid PVC Conduit, Schedule 80 (continued)

Conductors

Type	Conductor Size (AWG/kcmil)	16 (½)	21 (¾)	27 (1)	35 (1¼)	41 (1½)	53 (2)	63 (2½)	78 (3)	91 (3½)	103 (4)	129 (5)	155 (6)
THHN, THWN, THWN-2	14	9	17	28	51	70	118	170	265	358	464	736	1055
	12	6	12	20	37	51	86	124	193	261	338	537	770
	10	4	7	13	23	32	54	78	122	164	213	338	485
	8	2	4	7	13	18	31	45	70	95	123	195	279
	6	1	3	5	9	13	22	32	51	68	89	141	202
	4	1	1	3	6	8	14	20	31	42	54	86	124
	3	1	1	3	5	7	12	17	26	35	46	73	105
	2	1	1	2	4	6	10	14	22	30	39	61	88
	1	0	1	1	3	4	7	10	16	22	29	45	65
	1/0	0	1	1	2	3	6	9	14	18	24	38	55
	2/0	0	1	1	1	3	5	7	11	15	20	32	46
	3/0	0	1	1	1	2	4	6	9	13	17	26	38
	4/0	0	0	1	1	1	3	5	8	10	14	22	31
	250	0	0	1	1	1	3	4	6	8	11	18	25
	300	0	0	0	1	1	2	3	5	7	9	15	22
	350	0	0	0	1	1	1	3	5	6	8	13	19
	400	0	0	0	1	1	1	3	4	6	7	12	17
	500	0	0	0	1	1	1	2	3	5	6	10	14
	600	0	0	0	0	1	1	1	3	4	5	8	12
	700	0	0	0	0	1	1	1	2	3	4	7	10
	750	0	0	0	0	1	1	1	2	3	4	7	9
	800	0	0	0	0	1	1	1	2	3	4	6	9
	900	0	0	0	0	0	1	1	1	3	3	6	8
	1000	0	0	0	0	0	1	1	1	2	3	5	7
FEP, FEPB, PFA, PFAH, TFE	14	8	16	27	49	68	115	164	257	347	450	714	1024
	12	6	12	20	36	50	84	120	188	253	328	521	747
	10	4	8	14	26	36	60	86	135	182	235	374	536
	8	2	5	8	15	20	34	49	77	104	135	214	307
	6	1	3	6	10	14	24	35	55	74	96	152	218
	4	1	2	4	7	10	17	24	38	52	67	106	153
	3	1	1	3	6	8	14	20	32	43	56	89	127
	2	1	1	3	5	7	12	17	26	35	46	73	105
PFA, PFAH, TFE	1	1	1	1	3	5	8	11	18	25	32	51	73
PFA, PFAH, TFE, Z	1/0	0	1	1	3	4	7	10	15	20	27	42	61
	2/0	0	1	1	2	3	5	8	12	17	22	35	50
	3/0	0	1	1	1	2	4	6	10	14	18	29	41
	4/0	0	0	1	1	1	4	5	8	11	15	24	34
Z	14	10	19	33	59	82	138	198	310	418	542	860	1233
	12	7	14	23	42	58	98	141	220	297	385	610	875
	10	4	8	14	26	36	60	86	135	182	235	374	536
	8	3	5	9	16	22	38	54	85	115	149	236	339
	6	2	4	6	11	16	26	38	60	81	104	166	238
	4	1	2	4	8	11	18	26	41	55	72	114	164
	3	1	2	3	5	8	13	19	30	40	52	83	119
	2	1	1	2	5	6	11	16	25	33	43	69	99
	1	0	1	2	4	5	9	13	20	27	35	56	80
XHH, XHHW, XHHW-2, ZW	14	6	11	20	35	49	82	118	185	250	324	514	736
	12	5	9	15	27	38	63	91	142	192	248	394	565
	10	3	6	11	20	28	47	67	106	143	185	294	421
	8	1	3	6	11	15	26	37	59	79	103	163	234
	6	1	2	4	8	11	19	28	43	59	76	121	173
	4	1	1	3	6	8	14	20	31	42	55	87	125
	3	1	1	3	5	7	12	17	26	36	47	74	106
	2	1	1	2	4	6	10	14	22	30	39	62	89

Table C9
Maximum Number of Conductors or Fixture Wires in Rigid PVC Conduit, Schedule 80 (continued)

Conductors

Type	Conductor Size (AWG/kcmil)	16 (½)	21 (¾)	27 (1)	35 (1¼)	41 (1½)	53 (2)	63 (2½)	78 (3)	91 (3½)	103 (4)	129 (5)	155 (6)
XHH, XHHW, XHHW-2	1	0	1	1	3	4	7	10	16	22	29	46	66
	1/0	0	1	1	2	3	6	9	14	19	24	39	56
	2/0	0	1	1	1	3	5	7	11	16	20	32	46
	3/0	0	1	1	1	2	4	6	9	13	17	27	38
	4/0	0	0	1	1	1	3	5	8	11	14	22	32
	250	0	0	1	1	1	3	4	6	9	11	18	26
	300	0	0	1	1	1	2	3	5	7	10	15	22
	350	0	0	0	1	1	1	3	5	6	8	14	20
	400	0	0	0	1	1	1	3	4	6	7	12	17
	500	0	0	0	1	1	1	2	3	5	6	10	14
	600	0	0	0	0	1	1	1	3	4	5	8	11
	700	0	0	0	0	1	1	1	2	3	4	7	10
	750	0	0	0	0	1	1	1	2	3	4	6	9
	800	0	0	0	0	1	1	1	1	3	4	6	9
	900	0	0	0	0	0	1	1	—	3	3	5	8
	1000	0	0	0	0	0	1	1	1	2	3	5	7
	1250	0	0	0	0	0	1	1	1	1	2	4	6
	1500	0	0	0	0	0	0	1	1	1	1	3	5
	1750	0	0	0	0	0	0	1	1	1	1	3	4
	2000	0	0	0	0	0	0	1	1	1	1	2	4

Fixture Wires

Type	Conductor Size (AWG/kcmil)	16 (½)	21 (¾)	27 (1)	35 (1¼)	41 (1½)	53 (2)
FFH-2, RFH-2, RFHH-3	18	6	11	19	34	47	79
	16	5	9	16	28	39	67
SF-2, SFF-2	18	7	14	24	43	59	100
	16	6	11	20	35	49	82
	14	5	9	16	28	39	67
SF-1, SFF-1	18	13	25	42	76	105	177
RFH-1, RFHH-2, TF, TFF, XF, XFF	18	10	18	31	56	77	130
RFHH-2, TF, TFF, XF, XFF	16	8	15	25	45	62	105
XF, XFF	14	6	11	20	35	49	82
TFN, TFFN	18	16	29	50	90	124	209
	16	12	22	38	68	95	159
PF, PFF, PGF, PGFF, PAF, PTF, PTFF, PAFF	18	15	28	47	85	118	198
	16	11	22	36	66	91	153
	14	8	16	27	49	68	115
HF, HFF, ZF, ZFF, ZHF	18	19	36	61	110	152	255
	16	14	27	45	81	112	188
	14	10	19	33	59	82	138
KF-2, KFF-2	18	28	53	88	159	220	371
	16	19	37	62	112	155	261
	14	13	25	43	77	107	179
	12	9	17	29	53	73	123
	10	6	11	20	35	49	82
KF-1, KFF-1	18	33	63	106	190	263	442
	16	23	44	74	133	185	310
	14	16	29	50	90	124	209
	12	10	19	33	59	82	138
	10	7	13	21	39	54	90
XF, XFF	12	3	6	10	19	26	44
	10	2	5	8	15	20	34

Note: This table is for concentric stranded conductors only. For compact stranded conductors, Table C9(A) should be used.

*Types RHH, RHW, and RHW-2 without outer covering.

Article 344—Rigid Metal Conduit

Metric Designator	Trade Size	Nominal Internal Diameter		Total Area 100%		2 Wires 31%		Over 2 Wires 40%		1 Wire 53%		60%	
		mm	in	mm^2	in^2	mm^2	in^2	mm^2	in^2	mm^2	in^2	mm^2	in^2
12	³⁄₈	-	-	-	-	-	-	-	-	-	-	-	-
16	½	16.1	0.632	204	0.314	63	0.097	81	0.125	108	0.166	122	0.188
21	¾	21.2	0.836	353	0.549	109	0.170	141	0.220	187	0.291	212	0.329
27	1	27.0	1.063	573	0.887	177	0.275	229	0.355	303	0.470	344	0.532
35	1¼	35.4	1.394	984	1.526	305	0.473	394	0.610	522	0.809	591	0.916
41	1½	41.2	1.624	1333	2.071	413	0.642	533	0.829	707	1.098	800	1.243
53	2	52.9	2.083	2198	3.408	681	1.056	879	1.363	1165	1.806	1319	2.045
63	2½	63.2	2.489	3137	4.866	972	1.508	1255	1.946	1663	2.579	1882	2.919
78	3	78.5	3.090	4840	7.499	1500	2.325	1936	3.000	2565	3.974	2904	4.499
91	3½	90.7	3.570	6461	10.010	2003	3.103	2584	4.004	3424	5.305	3877	6.006
103	4	102.9	4.050	8316	12.882	2578	3.994	3326	5.153	4408	6.828	4990	7.729
129	5	128.9	5.073	13050	20.212	4045	6.266	5220	8.085	6916	10.713	7830	12.127
155	6	154.8	6.093	18821	29.158	5834	9.039	7528	11.663	9975	15.454	11292	17.495

Electrical formulas

Single phase

kVA = VA/1000
hp = $(VA \times eff \times PF)/746$
kw = $(VA \times PF)/1000$
PF = kW/kVA
Torque(ft-lb) = $hp \times 5250/rpm$
Motor kVA = $(hp \times 0.746)/eff \times PF$
Motor kW = $(hp \times 0.746)/eff$
V = line-to-line volts
A = amperes
eff = efficiency (decimal)

PF = power factor (decimal)
kVA = kilovolt amperes
kW = kilowatts
hp = horsepower
rpm = revolutions per minute

Three phase

$(3^{1/2}V \times A)/1000$
$(3^{1/2}VA \times PF \times eff)/746$
$(3^{1/2}VA \times PF)/1000$
$(kW \times 1000)/3^{1/2} \times VA$

Full-load currents—single-phase transformers

kVA Rating	Voltage-Volts					
	120	240	480	2400	4160	7200
1.0	8.3	4.2	2.1	0.4	0.2	0.1
3.0	25.0	12.5	6.3	1.3	0.7	0.4
5.0	41.7	20.8	10.4	2.1	1.2	0.7
7.5	62.5	31.3	15.6	3.1	1.8	1.0
10.0	83.3	41.7	20.8	4.2	2.4	1.4
15.0	125.0	62.5	31.3	6.3	3.6	2.1
25.0	208.3	104.2	52.1	10.4	6.0	3.5
37.5	312.5	156.3	78.1	15.6	9.0	5.2
50.0	416.7	208.3	104.2	20.8	12.0	6.9
75.0	625.0	312.5	156.3	31.3	18.0	10.4
100.0	833.3	416.7	208.3	41.7	24.0	13.9
125.0	1041.7	520.8	260.4	52.1	30.0	17.4
167.5	1395.8	697.9	349.0	69.8	40.3	23.3
200.0	1666.7	833.3	416.7	83.3	48.1	27.8
250.0	2083.3	1041.7	520.8	104.2	60.1	34.7
333.0	2775.0	1387.5	693.8	138.8	80.0	46.3
500.0	4166.7	2083.3	1041.7	208.3	120.2	69.4

Conduit size for combinations of cables with different outside diameters

The conduit size required for a group of cables with different outside diameters may be determined by calculating the equivalent outside diameter d'' and then finding the size required for this diameter in Table 1 below. For one to four cables:

$$d'' = \sqrt{\frac{n_1 d_1^2 + n_2 d_2^2 + n_m d_m^2 + \ldots}{n_1 + n_2 n_m + \ldots}}$$

where:

d'' = Equivalent diameter of same number of cables all of same outside diameter having total area equal to total area of group of cables of different sizes (a fictitious diameter appearing in the column corresponding to that number of cables ($n_1 + n_2 + n_m + \ldots$)
n_1 = number of cables of diameter d_1
n_2 = number of cables of diameter d_2
n_m = number of cables of diameter d_m, etc.

Example. Find the size conduit for two neoprene-sheathed cables having diameters of 1.20″ and one cable having a diameter of 0.63″.

$$d'' = \sqrt{\frac{2 \times (1.20)^2 + 1 \times (0.63)^2 \ldots}{2 + 1}}$$

$$d'' = 1.045$$

Refer to column 3 in Table 1 and note that the diameter of 1.045″ is between 0.901 and 1.120″. Therefore, a 3″ conduit will be required.

To determine the size conduit required for any number (n) of equal diameter cables in excess of four, multiply the diameter of one cable by $\sqrt{n/4}$. This will give the equivalent diameter of four such cables, and the conduit size required for (n) cables may then be found by using the column for four cables.

Table 1
Maximum Allowable Diameter (in Inches) of Individual Cables in Given Size of Conduit

Nonmetallic Jacketed Cable (All Cables of Same Outside Diameter)				
Nominal Size Conduit	Number of Cables Having Same OD			
	1	2*	3	4
½	0.453	0.244	0.227	0.197
¾	0.600	0.324	0.301	0.260
1	0.763	0.412	0.383	0.332
1¼	1.010	0.542	0.504	0.436
1½	1.173	0.633	0.588	0.509
2	1.505	0.812	0.754	0.653
2½	1.797	0.970	0.901	0.780
3	2.234	1.206	1.120	0.970
3½	2.583	1.395	1.296	1.121
4	2.930	1.583	1.470	1.273
5	3.675	1.985	1.844	1.595
6	4.416	2.385	2.215	1.916

*Diameters are based on percent fill only. The jam ratio, conduit OD to cable ID should be checked to avoid possible jamming.

Minimum bending radius for insulated cables for permanent training during installation

These limits do not apply to conduit bends, sheaves, or other curved surfaces around which the cable may be pulled under tension while being installed. Larger radii bends may be required for such conditions to limit sidewall pressure. In all cases the minimum radii specified refer to the inner surface of the cable and not to the axis of the cable.

The minimum bending radii for both single and multiple conductor cable with or without lead sheath and without metallic shielding or armor are given in Table 2.

Table 2
Minimum Bending Radius for Conductor Cable

Thickness of Conductor Insulation, Inches	Overall Diameter of Cable, Inches		
	Minimum Bending Radius as a Multiple of Cable Diameter		
	1.000 and Less	1.001–2.000	2.001 and Over
0.156 and less	4	5	6
0.157 to 0.315	5	6	7
0.316 and over		7	8

Table 3
Twisted Pair Instrumentation Cable

Type of Cable	Minimum Bending Radius as a Multiple of Cable Diameter
Armored, wire type, or corrugated sheath or interlocked type	7*
Nonarmored without shielded conductors	6
Nonarmored, metallized-polyester, or braid shielded	8
Nonarmored, flat, or corrugated tape shielded	12†

*With shielded conductors.
†For longitudinally applied corrugated shield with PVC jacket.

Table 4
Power and Control Cables with Metallic Shielding or Armor

Type of Cable	Minimum Bending Radius as a Multiple of Cable Diameter	
	Power	Control
Armored, flat tape, or wire type	12	12
Armored smooth aluminum sheath, up to		
0.75 in cable diameter	10*	10*
0.76 in cable diameter	12	12
>1.5 in cable diameter	15	15
Armored corrugated sheath or interlocked type	7	7
With shielded single conductor	12	12
With shielded multiconductor	†	†
NFC tape shielded single conductor	12	12
Tape shielded multiconductor	†	†
Multiconductor overall tape shield	12	12
LCS with PVC jacket	15	15
Nonarmored, flat strap shielded	8	–
Nonarmored, wire shielded	Refer to twisted pair instrumentation cable (Table 3)	

NFC = nonarmored, flat, or corrugated.
LCS = longitudinally applied corrugated shield.
*With shielded conductors, 12.
†12 times single conductor diameter or 7 times overall cable diameter, whichever is greater.

Full-load currents—three-phase transformers

kVA Rating	Voltage (Line-to-Line)				
	240	480	2400	4160	12,470
3	7.2	3.6	0.7	0.4	0.1
6	14.4	7.2	1.4	0.8	0.3
9	21.7	10.8	2.2	1.2	0.4
15	36.1	18.0	3.6	2.1	0.7
30	72.2	36.1	7.2	4.2	1.4
37.5	90.2	45.1	9.0	5.2	1.7
45	108.3	54.1	10.8	6.2	2.1
50	120.3	60.1	12.0	6.9	2.3
75	180.4	90.2	18.0	10.4	3.5
100	240.6	120.3	24.1	13.9	4.6
112.5	270.6	135.3	27.1	15.6	5.2
150	360.9	180.4	36.1	20.8	6.9
200	481.1	240.6	48.1	27.8	9.3
225	541.3	270.6	54.1	31.2	10.4
300	721.7	360.9	72.2	41.6	13.9
450	1082.6	541.3	108.3	62.5	20.8
500	1202.8	601.4	120.3	69.4	23.2
600	1443.4	721.7	144.3	83.3	27.8
750	1804.3	902.1	180.4	104.1	34.7
1000	2405.7	1202.8	240.6	138.8	46.3
1500	3608.5	1804.3	360.9	208.2	69.5
2000	4811.4	2405.7	481.1	277.6	92.6
2500	6014.2	3007.1	601.4	347.0	115.8
5000	12,028.5	6014.2	1202.8	694.0	231.5
10,000	24,057.0	12,028.5	2405.7	1387.9	463.0

Motor controller sizes

Polyphase motors

	Maximum Horsepower Full Voltage Starting	
NEMA Size	230 V	460–575 V
00	1.5	2
0	3	5
1	7.5	10
2	15	25
3	30	50
4	50	100
5	100	200
6	200	400
7	300	600

Single-phase motors

NEMA Size	Maximum Horsepower Full Voltage Starting (Two Pole Contactor)	
	115 V	230 V
00	1.3	1
0	1	2
1	2	3
2	3	7.5
3	7.5	15

Voltage drop on circuits using 600 V copper conductors in steel conduit

To determine the voltage drop, multiply the length in feet of one conductor by the current in amperes and by the number listed in the table for the type of system and power factor and divide the result by 1,000,000 to obtain the voltage loss. This table takes into consideration reactance on AC circuits and resistance of the conductor. Unless otherwise noted, the table is based on 60 Hz.

Wire Size	Direct Current	Three-Phase Lagging Power Factor					Single-Phase Lagging Power Factor				
		100%	90%	80%	70%	60%	100%	90%	80%	70%	60%
14	6100	5280	4800	4300	3780	3260	6100	5551	4964	4370	3772
12	3828	3320	3030	2720	2400	2080	3828	3502	3138	2773	2404
10	2404	2080	1921	1733	1540	1340	2404	2221	2003	1779	1547
8	1520	1316	1234	1120	1000	880	1520	1426	1295	1159	1017
6	970	840	802	735	665	590	970	926	850	769	682
4	614	531	530	487	445	400	614	613	562	514	462
3	484	420	425	398	368	334	484	491	460	425	385
2	382	331	339	322	300	274	382	392	372	346	317
1	306	265	280	270	254	236	306	323	312	294	273
0	241	208	229	224	214	202	241	265	259	247	233
00	192	166	190	188	181	173	192	219	217	209	199
000	152	132	157	158	155	150	152	181	183	179	173
0000	121	105	131	135	134	132	121	151	156	155	152
250M	102	89	118	123	125	123	103	136	142	144	142
300M	85	74	104	111	112	113	86	120	128	130	131
350M	73	63	94	101	105	106	73	108	117	121	122
400M	64	55	87	95	98	100	64	100	110	113	116
500M	51	45	76	85	90	92	52	88	98	104	106
600M	43	38	69	79	85	87	44	80	91	98	101
700M	36	33	64	74	80	84	38	74	86	92	97
750M	34	31	62	72	79	82	36	72	83	91	95
800M	32	29	61	71	76	81	33	70	82	88	93
900M	28	26	57	68	74	78	30	66	78	85	90
1000M	26	23	55	66	72	76	27	63	76	83	88

Determine the most economical size for electric power conductors

To quickly calculate the most economical size copper wire for carrying a specified current load, use the formula:

$$A = (59)(I)\sqrt{\frac{C_e \times t}{C_c \times F}}$$

where:

A = Conductor cross-sectional area (circular mils)
I = Current load (amperes)
C_e = Cost of electrical power (cents per kWh)
C_c = Cost of copper (cents per lb)
t = Hours of service per year
F = Factor for fixed charges (amortization, insurance, taxes, etc.)

Selection of the proper size electric conductor will depend on the load to be carried and the mechanical strength required, as well as economics. Once these are considered, the most economical size conductor is that for which the annual energy cost equals the copper cost. This method can be used for other metallic conductors.

Example. Determine the most economical size copper conductor for an installation operating 365 days a year, 8 h per day, with a 100 A load. Energy costs are \$0.00875 per kWh. Fixed charges are 23%, considering a 5-year amortization with 3% for insurance and taxes. Copper costs \$0.46 per lb.

$$A = (59)(100)\sqrt{\frac{0.00875 \times 365 \times 8}{0.46 \times 0.23}}$$

$$= 91,686 \text{ circular mils}$$

How to find the resistance and weight of copper wires

It is easy to mentally calculate the approximate resistance and weight of standard sizes of copper wires, by remembering the following rules:

Rule 1: Number 10 wire has a resistance of 1 Ω per 1000 ft. Number 2 wire weighs 200 lb per 1000 ft.
Rule 2: An increase of 10 numbers in the size is an increase of 10 times in the resistance, and a decrease to one-tenth of the weight.
Rule 3: An increase of three numbers in the size doubles the resistance and halves the weight.
Rule 4: An increase of one number in the size multiplies the resistance by 5/4 and the weight by 4/5.

Always proceed from the known value (No. 10 for resistance, No. 2 for the weight) in the smallest number of jumps; that is, go by tens, then by threes, and finally by ones; the jumps do not all have to be in the same direction.

Example. What is the weight and resistance of 200 ft of No. 18? Apply the rules in order, as follows; resistance first:

Number 10 wire has a resistance of 1 Ω per 1000 ft (Rule 1)
Number 20 wire has a resistance of 10 Ω per 1000 ft (Rule 2)
Number 17 wire has a resistance of 5 Ω per 1000 ft (Rule 3)
Number 18 wire has a resistance of 6.25 Ω per 1000 ft (Rule 4)
Number 18 wire has a resistance of 1.25 Ω per 200 ft (Answer)

Number 2 wire weighs 200 lb per 1000 ft (Rule 1)
Number 12 wire weighs 20 lb per 1000 ft (Rule 2)
Number 22 wire weighs 2 lb per 1000 ft (Rule 2)
Number 19 wire weighs 4 lb per 1000 ft (Rule 3)
Number 18 wire weighs 5 lb per 1000 ft (Rule 4)
Number 18 wire weighs 1 lb per 200 ft (Answer)

The weight determined by the above rules is that of bare copper wire. Weight of insulated wire varies widely according to the type of insulation. Resistance values are correct for bare or insulated wire and for solid or stranded. Errors resulting from the use of the rules will rarely exceed 2%.

What you should remember about electrical formulas

Ohm's law

Easily **R**emember **I**t, or E = RI.

Power

DC **P**ower **I**s **E**asy, or P = IE.
It **R**eally **I**s, or P = IRI = I^2R

Other facts

The number of *watts in one hp* is equal to the year Columbus discovered America divided by two, or 1492/2 = 746 W/hp.

The horsepower of a reciprocating engine necessary to drive a three-phase, 60-cycle electric generator can be determined by multiplying kW by 1.5.

How to calculate microwave hops on level ground

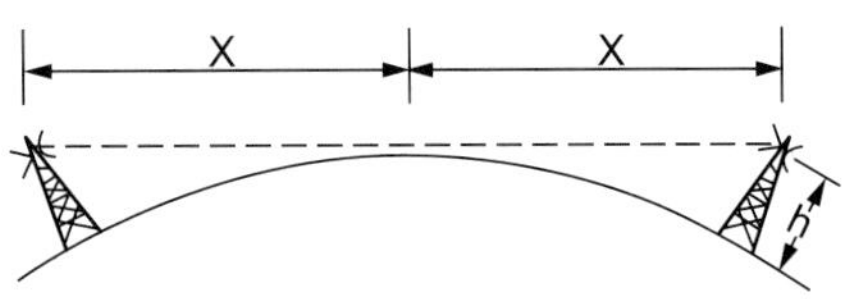

Excluding obstructions such as hills, the microwave (line-of-sight) distances between two towers will be twice the distance from the transmitter to the horizon. The line-of-sight distance in miles from the top of the tower to the horizon can be found by adding the square root of the tower height to the square root of the first square root (the latter, of course, is the fourth root).

$$x = \sqrt{h} + \sqrt{\sqrt{h}} \text{ or } \sqrt{h} + \sqrt[4]{h}$$

Example. How far apart can 100-ft-high towers be spaced to provide line-of-sight transmission if there are no obstructions between them?

$X = \sqrt{100} + \sqrt{10}$

$X = 10 + 3$ or 13 miles

Distance between towers can be 2X or 26 miles.

This same formula can be used to estimate distances in an airplane.

Example. How far is the line-of-sight distance to the horizon in an airplane flying at an altitude of 4900 ft?

$(\sqrt{4900} = 70)$

$X = \sqrt{4900}\sqrt{70}$

$X = 70 + 8$

Distance to horizon is 78 miles.

Quick determination of hp per ampere for induction motors (3 phase) at different voltages

Voltage	hp/A
480	1
2400	5
4160	8

Chart of electric motor horsepower for pumping units

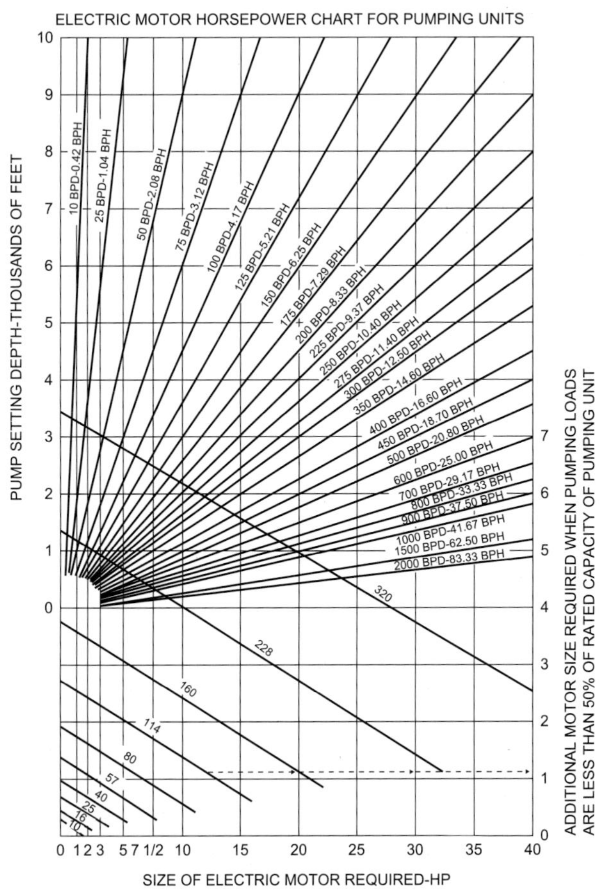

This chart provides a means of determining the power requirements for a beam pumping installation powered by an electric motor based on a fluid with a specific gravity of 1, with fluid at the pump. The power requirement determined by the chart includes a mechanical efficiency factor of 0.45 and a cyclic factor of 0.75, which are factors frequently applied to motors used in sucker rod pumping service.

An arrangement is available for correcting the power requirement in the case of an underloaded pumping unit.

After the power requirements are determined from the chart, the next higher size of commercially available motor is used.

Pumping stations

Capacitor multipliers for kilowatt loads for different desired power factor improvements

Original Power Factor, Percent	DESIRED POWER FACTOR—Percent				
	100	95	90	85	80
50	1.732	1.403	1.248	1.112	0.982
51	1.687	1.358	1.203	1.067	0.937
52	1.643	1.314	1.159	1.023	0.893
53	1.600	1.271	1.116	0.980	0.850
54	1.559	1.230	1.075	0.939	0.809
55	1.518	1.190	1.035	0.898	0.769
56	1.480	1.151	0.996	0.860	0.730
57	1.442	1.113	0.958	0.822	0.692
58	1.405	1.076	0.921	0.785	0.655
59	1.369	1.040	0.885	0.749	0.619
60	1.333	1.004	0.849	0.713	0.583
61	1.299	0.970	0.815	0.679	0.549
62	1.266	0.937	0.782	0.646	0.516
63	1.233	0.904	0.749	0.613	0.483
64	1.201	0.872	0.717	0.581	0.451
65	1.169	0.840	0.685	0.549	0.419
66	1.138	0.809	0.654	0.518	0 388
67	1.108	0.779	0.624	0.488	0.358
68	1.078	0.749	0.594	0.458	0.328
69	1.049	0.720	0.565	0.429	0.299
70	1.020	0.691	0.536	0.400	0.270
71	0.992	0.663	0.508	0.372	0.242
72	0.964	0.635	0.480	0.344	0.214
73	0.936	0.607	0.452	0.316	0.186
74	0.909	0.580	0.425	0.289	0.159
75	0.882	0.553	0.398	0.262	0.132
76	0.855	0.526	0.371	0.235	0.105
77	0.829	0.500	0.345	0.209	0.079
78	0.802	0.473	0.318	0.182	0.052
79	0.776	0.447	0.292	0.156	0.026
80	0.750	0.421	0.266	0.130	....
81	0.724	0.395	0.240	0.104	....
82	0.698	0.369	0.214	0.078	....
83	0.672	0.343	0.188	0.052	....
84	0.646	0.317	0.162	0.026	....
85	0.620	0.291	0.136	....	....
86	0.593	0.264	0.109	....	....
87	0.567	0.238	0.083	....	....
88	0.540	0.211	0.056	....	....
89	0.512	0.183	0.028	....	....
90	0.484	0.155			
91	0.456	0.127			
92	0.426	0.097			
93	0.395	0.066			
94	0.363	0.034			
95	0.329	....			
96	0.292	....			
97	0.251	....			
98	0.203	....			
99	0.143	....			

Example

Assume total plant load is 100 kw at 60 percent power factor. Capacitor kvar rating necessary to improve power factor to 80 percent is found by multiplying kw (100) by multiplier in table (0.583), which gives kvar (58.3). Nearest standard rating (60 kvar) should be recommended.

Courtesy General Electric Company.

Floodlighting calculations

Floodlighting encompasses many variations. Since the location of the floodlight relative to the object to be lighted can be in any plane and at any distance from the source, a floodlighting application is often considered the most complex and difficult of all lighting techniques.

Candela or Candlepower—Light sources do not project the same amount of light in every direction. The directional characteristic of a lamp is described by the candlepower in specific directions. This directional strength of light or luminous intensity is measured in candelas.

Lumen—Light quantity, irrespective of direction, is measured in lumens. Lamp lumens are the quantity of light produced by a lamp. Average light level calculations use total lamp lumens as a basis and then adjust for all factors that lower this quantity. The amount of useful light in a floodlight beam is measured in beam lumens.

Footcandle (fc)—Specifications are usually based on density of light or level of illumination, which is measured in footcandles. Footcandles are the ratio of quantity of light in lumens divided by the surface area in square feet on which the lumens are falling. A density of one lumen per square foot is one footcandle. One footcandle is equal to 10.76 lux.

Lux (lx)—The SI unit of illuminance. This metric measurement is based upon the density of lumens per unit surface area similar to footcandle, except one lux is one lumen per square meter. One lux is equal to 0.09 footcandle.

Light Loss Factor (LLF)—These factors are used to adjust lighting calculations from laboratory test conditions to a closer approximation of expected field results. The I.E.S. Lighting Handbook, 1984 Reference Volume, defines LLF as follows: "a factor used in calculating illuminance after a given period of time and under given conditions. It takes into account temperature and voltage variations, dirt accumulations on luminaire and room surfaces, lamp depreciation, maintenance procedures and atmosphere conditions."

The most commonly used systems for floodlight calculations are the point-by-point method and the beam-lumen method.

Point-by-point method

The point-by-point method permits the determination of footcandles at any point and orientation on a surface and the degree of lighting uniformity realized for any given set of conditions.

In such situations, the illumination is proportional to the candlepower of the source in a given direction and inversely proportional to the square of the distance from the source. See Fig. 1. Footcandles on plane normal to the light ray is described by the following formula:

$$\text{Footcandles on plane} = \frac{\text{Candlepower of light ray}}{\text{Distance squared}}$$

$$E = 1/D^2 \tag{1}$$

When the surface on which the illumination to be determined is tilted, the light will be spread over a greater area, reducing the illumination in the ratio of the area of plane A to the area of plane B as shown in Fig. 2. This ratio is equal to the cosine of the angle of incidence; thus:

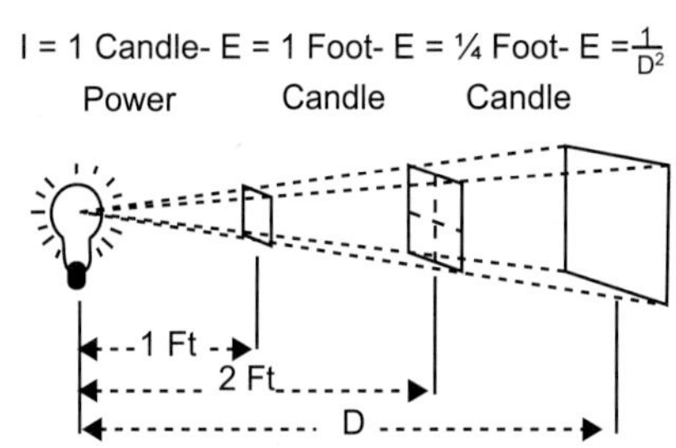

Figure 1. Inverse square law.

$$\text{Footcandles on plane B} = \frac{\text{Candlepower of light ray}}{\text{Distance squared}} \times \text{Cosine of angle } \beta$$

$$E = I/D^2 \times \cos\beta \quad (2)$$

Then, β equals the angle between the light ray and a perpendicular to the plane at that point.

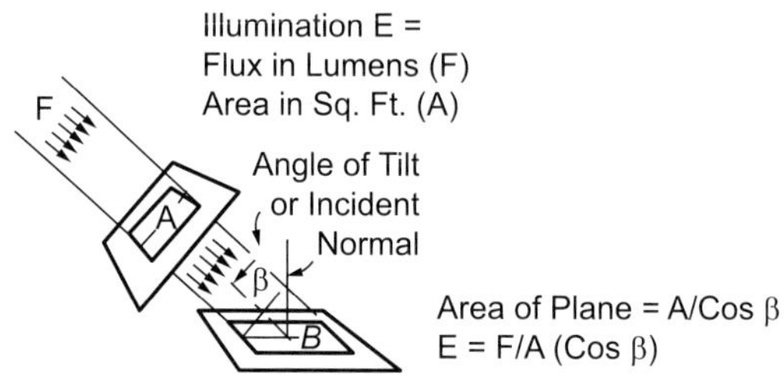

Figure 2. Cosine of angle law.

Beam-lumen design method

The beam-lumen method is quite similar to the method used for interior lighting except that the utilization factors must take into consideration the fact that floodlights are not usually perpendicular to the surface and therefore not all of the useful light strikes the surface.

Beam lumens are defined as the quantity of light that is contained within the beam limits as described as "beam spread." Beam lumens equal the lamp lumens multiplied by the beam efficiency of the floodlight.

Coverage—It is recommended that sufficient point-by-point calculations be made for each job to check uniformity and coverage.

Light Loss Factor—The maintenance or light loss factor is an allowance for depreciation of lamp output with age and floodlight efficiency due to the collection of dirt on the lamp, reflector, and cover glass. The total factor may vary from 0.65 to 0.85 depending on the type of lamp and luminaire used and may include losses due to lamp orientation, or "tilt."

Example. Step 1—Determine the level of illumination. See Table 1 for some typical levels of illumination (fc). The basic formula is:

$$\text{fc} = \frac{N \times \text{BL} \times \text{CBU} \times \text{LLF}}{A} \quad (3)$$

where:

N = quantity of luminaries
A = area in square feet
BL = beam lumens
CBU = coefficient of beam utilization
LLF = light loss factor

Step 2—Determine type and location of floodlights. Regardless of light source, there are industry standards on beam spreads. See Table 2.

Step 3—Determine the coefficient of beam utilization. This factor, CBU, written as a decimal fraction, is expressed in the following ratio:

$$\text{CBU} = \frac{\text{Utilized lumens}}{\text{BL}} \quad (4)$$

The exact CBU can be graphically determined by projecting the outline of the area to be lighted upon the photometric data and totaling the utilized lumens. This procedure is detailed in the I.E.S. Lighting Handbook. Typical CBU values are given in Table 3.

Table 1
Typical Illumination Levels

Area	fc
Building exteriors entrances	
Active	5
Inactive	1
Vital locations	5
Catwalks	3
Stairs	5
Barge or truck unloading	5
Storage tanks	1
Pump areas	2
Parking	1
Substation	2
Storage yards	
Active	20
Inactive	1

Table 2
Outdoor Floodlight Luminaire Designations

Beam Spread, Degrees	NEMA Type
10–18	1
>18–29	2
>29–46	3
>46–70	4
>70–100	5
>100–130	6
>130 up	7

The following general principles apply in the choice of beam spread.
1. The greater the distance from the floodlight to the area to be lighted, the narrower the beam spread desired.
2. Since by definition the candlepower at the edge of a floodlight beam is 10% of the candlepower near the center of the beam, the illumination level at the edge of the beam is one-tenth or less of that at the center. To obtain reasonable uniformity of illumination, the beams of individual floodlights must overlap each other, as well as the edge of the surface to be lighted.
3. The percentage of beam lumens falling outside the area to be lighted is usually lower with narrow-beam units than with wide-beam units. Thus, narrow-beam floodlights are preferable where they will provide the necessary degree of uniformity of illumination and the proper footcandle level.

As an approximation, the average CBU of all the floodlights in an installation should fall within the range of 0.6–0.9. If less than 60% of the beam lumens are utilized, a more economical lighting plan should be possible by using different locations or narrower beam floodlights. If the CBU is greater than 0.9, it is probable that the beam spread selected is too narrow and the resultant illumination will be spotty. An estimated CBU can be determined by experience or by making calculations for several potential aiming points and using the average figure thus obtained.

Table 3
Typical CBUs

Area	CBU	Area	fc
Baseball		Tennis	0.75
Infield	0.65	Large parking lots	0.80
Outfield	0.85	Building facades	0.65
Football	0.60		

Step 4—Determine the quantity of floodlights (N) required. Rearrange the basic formula in Step 1 as follows:

$$N = \frac{A \times \text{fc}}{\text{BL} \times \text{CBU} \times \text{LLF}} \tag{5}$$

Conductor size conversion chart—metric to AWG

Metric Size	AWG Size
2.5 mm^2	#14
4 mm^2	#12
6 mm^2	#10
10 mm^2	#8
16 mm^2	#6
25 mm^2	#31
35 mm^2	#2
35 mm^2	#1*
50 mm^2	#1/0
70 mm^2	#2/0
95 mm^2	#3/0
95 mm^2	#4/0*
120 mm^2	250 kcmil
150 mm^2	300 kcmil
185 mm^2	350 kcmil
185 mm^2	400 kcmil*

Metric Size	AWG Size
240 mm^2	500 kcmil
300 mm^2	600 kcmil
400 mm^2	700 kcmil
400 mm^2	750 kcmil
400 mm^2	800 kcmil*

*This AWG size is the closest equivalent to the metric size.

Commonly used switchgear device numbers

Number	Function
12	Overspeed—Usually a direct-connected speed switch that functions on machine overspeed.
15	Speed or frequency matching that functions to match and hold the speed or the frequency of a machine or system equal to or approximately equal to that of another machine, source, or system.
21	Distance relay that functions when the circuit admittance, impedance, or reactance changes beyond determined limits.
25	Synchronizing or synchronism check device that operates when two AC circuits are within the desired limits of frequency, phase angle, or voltage, to permit paralleling two circuits.
32	Directional power relay that functions on a desired value of power flow in a given direction or upon reverse power.
38	Bearing protective device that functions on excessive bearing temperature or other abnormal conditions associated with the bearing such as undue wear that may ultimately result in excessive bearing temperature or failure.
39	Mechanical condition monitor that functions when an abnormal mechanical condition (except conditions associated with bearings covered under device 38) such as excessive vibration, eccentricity, expansion, shock, tilting, or seal failure.
40	Field relay that functions on a given or abnormally low value or failure of machine field current, or due to excessive value of the reactive component of the armature current in an AC machine indicating abnormally low field excitation.
41	Field circuit breaker functions to apply or remove the field excitation of a machine.
43	Manual Transfer or Selector device is a manually operated device that transfers the control circuits in order to modify the plan of operation of switching equipment or of some of the devices.
46	Reverse phase or phase balance current relay operates when the polyphase currents are of reverse phase sequence or when the currents are unbalanced or contain negative phase sequence components greater than a set value.
47	Phase sequence voltage relay operates upon a predetermined value of polyphase voltage in the desired sequence.
50	Instantaneous overcurrent or rate-of-rise relay instantaneously operates on an excessive value of current or on an excessive rate of current in an AC circuit exceeds a predetermined value.
51	AC time overcurrent relay that has a definite or inverse time characteristic that operates when the current in an AC circuit exceeds a predetermined value.
52	AC circuit breaker that is used to close and interrupt an AC power circuit under normal conditions or under fault or emergency conditions.
59	Overvoltage relay that operates on a given value of overvoltage.
60	Voltage or current balance relay that operates on a given difference in voltage or current input or output of two circuits.
62	Time delay stopping or opening relay that serves in conjunction with the device that initiates the shutdown, stopping, or opening operation in an automatic sequence or protective relay system.
64	Ground protective relay operates on failure of the insulation of a machine, transformer, or other apparatus to ground, or on flashover of a DC machine to ground. This device is assigned only to a relay that detects the flow of current from the frame of a machine or enclosing case or structure of a piece of apparatus to ground, or detects a ground on a normally ungrounded winding or circuit. It is not applied to a device connected in the secondary circuit of a current transformer, or in the secondary neutral of current transformers, connected in the power circuit of a normally grounded system.
65	Governor is the assembly of fluid, electrical, or mechanical control equipment used for regulating the flow of water, steam, or other medium to the prime mover for such purposes as starting holding speed, load, or stopping.
67	AC directional overcurrent relay that operates on a desired value of AC overcurrent flowing in a predetermined direction.
70	AC reclosing relay that controls the automatic reclosing and locking out of an AC circuit interrupter.
81	Frequency relay operates on a predetermined value of frequency (either under or over or on normal system frequency or rate of change of frequency.
86	Lock-out relay is an electrically operated hand or reset relay or device that operates to shutdown or hold an equipment out of service, or both, when an abnormal condition occurs.
87	Differential protective relay that operates on a percentage or phase angle or other quantitative difference of two currents or of some other electrical quantities.

Suffix letters permit a manifold multiplication of available function designations for the large number and variety of devices that may be used. They may also serve to denote individual or specific parts or auxiliary contacts of these devices or certain distinguishing features, characteristics, or conditions that describe the use of the device or its contacts in the equipment.

Suffix letters should be used only when they accomplish a specific purpose. For example, when all of the devices in an equipment are associated with only one kind of apparatus, such as a feeder or motor or generator, it is common practice in order to retain maximum simplicity in device function identification, not to add the respective suffix letter F or M or G to any of the device function numbers.

Some commonly used suffix designations are as follows:

A	Automatic
M	Manual
27R	Voltage restraint
27C	Voltage controlled
R	Reverse
RV	Reverse VARS
G	Generator or ground
T	Transformer

Bonding the grounding system to building and structure foundations

Whitt Trimble and Edie Guidy, NFPA

During the 2005 National Electrical Code (NEC) cycle, a subtle change was made to Section 250.50. The words "if available" were removed from the beginning of the paragraph to clarify the objective to include all concrete-encased electrodes in the foundations of new buildings and structures.

Looking at it from a strictly electrical standpoint, this seems like a good idea. Without a doubt, concrete-encased electrodes make excellent grounding electrodes. If the concrete stays moist and the reinforcing steel bars are conductive (that is, not coated with epoxy or anything else) the results, for grounding purposes, will be exceptional. Even without moist conditions, the concrete-encased system will provide a plane of equal potential when bonded to the building steel and other grounding electrodes indicated in Section 250.52. However, what appeared to be a step in the right direction as far as safety is concerned may have some potential drawbacks in some installations.

In large industrial refineries and petrochemical plants, the typical grounding electrode system usually consists of 2/0 AWG, 4/0 AWG, up to 500 kcmil bare cooper cables laid out in a grid around structural steel, pumps, compressors, and other equipment. Multiple ground rods installed throughout the grid system are typically copper-clad steel 3/4 of an inch (1.9 cm in diameter in lengths of 10–20 ft (25–51 m). Nearly every piece of equipment is bonded to the grounding electrode system through taps of smaller-gauge cable, such as 2 AWG or 2/0 AWG. This allows the formation of an equipotential plane that prevents a lightning strike or an electrical fault from creating an unsafe condition resulting from the subsequent potential difference between neighboring vessels, motors, steel, and pipes.

In the past, the concrete-encased electrodes in the pipe rack piers and vessel footings were not usually bonded to the grid, even though the above-ground structural steel was. Blindly doing so now without considering the potential for corrosion of the structural rebar in the concrete can be costly.

Corrosion is an electrochemical process involving the flow of electrons and ions that occurs when a corrosion cell is formed. A corrosion cell consists of an anode, a cathode, a common electrolyte containing the anode and cathode, and a metallic path between the anode and cathode. Corrosion occurs at the anode. When the anode and cathode are of dissimilar metals, current is generated between the two due to the inherent difference in their refined energy levels. Electrons flow in the metallic path and ions flow in the electrolyte. This is the basis for the operation of the common dry cell or battery.

In a copper-based grounding system of a facility constructed of steel, the copper will behave as a cathode and the steel will behave as an anode. The resulting corrosion of the steel reduces the structural strength of the rebar and destroys the rebar/concrete bond. In addition, the corrosion products are physically larger than the original reactants and will thus exert tremendous pressure on the surrounding concrete, causing the concrete to fracture and fail. This is referred to as spalling.

For facilities constructed in compliance with the 2005 edition of the NEC, there are several corrosion-mitigation options available.

Option one

The first option is to eliminate the corrosion cell circuit by breaking the metallic path. This can be done by isolating the copper grounding system from underground steel objects such as piping and rebar. To eliminate the dissimilar-metal corrosion problem, the isolation must be guaranteed by proper construction practices and periodic testing. Isolation is normally maintained using insulating flange kits, dielectric unions, nonmetallic conduits and cable trays, dielectric pads and supports, and separate grounding systems. Care has to be taken in the instrumentation and electrical designs so as not to short the intended isolation.

This isolation option is also practiced when cathodically protected steel structures come into physical contact with nonprotected steel structures. However, isolating the copper grounding system in this manner does not comply with NEC Section 250.50, and the method should not be used where electrical systems are present and installation is within the scope of NEC Section 90.2(A). This option may be used in facilities, such as a utility power generation facility, that are not subject to NEC requirements.

As a practical matter, complete electrical isolation is very difficult to achieve and maintain.

Option two

The second option is to eliminate the voltage difference between the anode and cathode using components other than copper in the soil and concrete electrolyte. Copper grounding electrode cables can be insulated from the soil as they travel from rod to rod and all taps in between, or a material other than copper, such as zinc-clad or stainless steel, can be used in the rods. Of course, stainless steel or nonferrous ground rods less than 5/8 of an inch (1.6 cm) in diameter should be listed per NEC 250.52 (A)(5)(b). Listed zinc-galvanized ground rods are available. Stainless steel rods should only be used after a thorough soils investigation, as high-chloride content in the soil can badly corrode stainless steel.

One disadvantage of this option is the loss of contact between the copper cables and soil throughout the grid. Since insulated cables do not lower the overall resistance to ground, more rods will be needed to achieve the same ohmic levels. The cables in this case become nothing more than jumpers between the rods, allowing a convenient place to tap to the equipment.

This option may be the most economical and practical choice for many industrial facilities, provided a regimented maintenance plan is put into effect. Periodic testing of the grounding system should be conducted as the grounding components are no longer cathodically protected. This option also complies with NEC Section 250.50.

Option three

The third option is to provide cathodic protection for both the grounding system and structural steel (see Fig. 4). Cathodic protection reduces the corrosion of a metal surface by making the surface the cathode of an electrochemical cell. Before applying cathodic protection, corroding structures will have both cathodic and anodic areas where corrosion occurs. This is one reason grounding practitioners prefer to use copper: it appears to be very stable when, in fact, it is being cathodically protected by the steel anode in the corrosion cell circuit.

Because grounding system components are generally bare to maximize surface area and minimize the resistance to remote earth and because cathodic protection depends on an adequate current flow to all bare metal surfaces, protecting the facility in its entirety often requires a tenfold increase in the amount of cathodic protection current required.

In the case of a utility power generation facility, where the cost of such an installation can be justified and both grounding performance and integrity are of great importance, option three may be the best alternative.

In many industrial plants, however, combining options two and three may work best.

By using noncopper grounding rods interconnected with insulated conductors and including the grounding system in the cathodic protection system's zone of protection, the grounding system, buried steel piping, and concrete reinforcing steel are all protected. Eliminating exposed copper from the grounding system will reduce the required capacity of the cathodic protection system.

Will rebar corrosion be a problem in residential and commercial foundations? Probably not. These types of structures typically have limited amounts of copper grounding electrodes in the soil, and circuit resistance of the electrolytic path will be relatively high. Minor corrosion of rebar and limited foundation cracking and spalling will not seriously affect their design lives.

There is certainly more to be learned about corrosion and cathodic protection than can be covered here. The purpose of this article is to make the electrical design engineer aware of the possible adverse consequences and to suggest that a qualified corrosion expert be consulted when designing copper grounding systems that are bonded to structural steel rebar.

5: Hydrostatic Testing

History of gas pipeline pressure test requirements............................ 184
The benefits and limitations of hydrostatic testing............................ 189
Hydrostatic testing for pipelines.. 199
Volume of water required to fill test section 204
Volume of water required at test pressure 205
How to use charts for estimating the amount of pressure change for a change in test water temperature 206
Compressibility factor for water 209
Hydrostatic test records 209
Pressure testing and recordkeeping requirements – reconciling historical practices with new requirements................................... 209

History of gas pipeline pressure test requirements

Michael J Rosenfeld, Kiefner & Associates Inc./Applus-RTD, Worthington, Ohio and **Rick W Gailing**, Southern California Gas Co, Los Angeles, CA

Advent of hydrostatic testing

Hydrostatic pressure testing[1] is now a standard practice for commissioning a pipeline today but this was not always the case. The concept of pressure testing as a means of establishing the ability of pipe to safely contain pressure in operation was adopted from the vessel industry, which had begun to implement that practice prior to 1900. However, pressure testing a natural gas pipeline that is many miles long with water is much more difficult than filling a vessel with water; these differences posed serious challenges to early pipeline operators, for a couple of reasons. One is that the large quantity of clean water necessary to fill a cross-country pipeline was difficult to obtain and manage in any location and particularly so in dry-climate regions where many early large pipelines were constructed. The second problem was dewatering, since the methods and tools to accomplish that had yet to be developed. (This was also the case with other complications such as bleeding trapped air, rupture isolation/containment, and refilling or transferring test water from section to section.) Similar limitations affect gas distribution systems: the quantities of water required are still large, the networked nature of the systems complicates dewatering, and residual water in distribution piping is a problem for customers. Consequently, through the 1940s, if a pressure test was performed at all, it was usually accomplished using the transported commodity, natural gas in the case of gas pipelines, or crude oil or petroleum products in the case of liquid transmission pipelines. Owing to concerns for the consequences of a failure when testing with product (loss of product in the case of liquids, and loss of extensive quantities of pipe due to fracture propagation in the case of testing with natural gas), operators typically limited test pressures to between 5 psig and 50 psig above, or at most 10% above, their intended operating pressure[2,3,4,5].

The first large-scale use of proof testing long-distance gas pipelines with water was carried out by the Texas Eastern Transmission Corporation (TETCO) in 1950[6,7]. In 1947, TETCO acquired the two War Emergency Pipelines built to transport crude oil and fuel from Texas to New Jersey during World War II, and converted them to transport natural gas. TETCO experienced many service failures due to original pipe manufacturing defects that may have enlarged while in petroleum transportation service, and also due to corrosion because parts of the line were installed uncoated to save time. In 1950, TETCO completed an ambitious program to revalidate the integrity of the pipelines by pressure testing them with water to levels well above the MAOP and in some cases up to yielding. TETCO was able to do this because they had already developed cleaning pigs that were inserted into traps and propelled by gas pressure to sweep accumulated liquids out of the line as part of the process of converting the lines from liquid to gas[17]. Although they experienced hundreds of pipeline breaks during testing[8], the tested pipelines were reliable in subsequent years and portions of them are still in service today[9]. As a result of TETCO's experience, the industry performed scientific studies between 1953 and 1968 to better understand the benefits, limitations, and mechanics of hydrotesting[10,11,12]. Over time, other operators began to adopt the practice of pressure testing with water to higher stress levels than had previously been customary.

The evolution of test requirements for commissioning a new pipeline system as they pertain to transmission pipelines constructed from steel pipe is summarized briefly below. Testing requirements are not discussed herein for: low- and high-stress distribution piping, mains, and service lines; piping fabricated from plastic or cast iron pipe; testing for purposes of uprating; and testing to accommodate changes in location classes. The reason for omitting these requirements is they introduce significant complexity in details that are not central to the issue at hand. Figure 1 gives a timeline for major changes in pressure testing requirements.

B31.8 Standard, predecessors, and sequels

Two important eras can be defined with respect to pressure testing requirements in B31.8: pre-1955 and 1955-and-later, because the 1955 edition of the Code marks the first time that pressure testing a pipeline after construction was made a requirement to complying with the standard. The basis for this conclusion is discussed below.

ASA B31.1 Prior to 1955

Section 2 of the 1935 B31.1 defined two categories of pipe based on location: Division 1 piping was air or gas piping constructed within power plants, gas plants, or

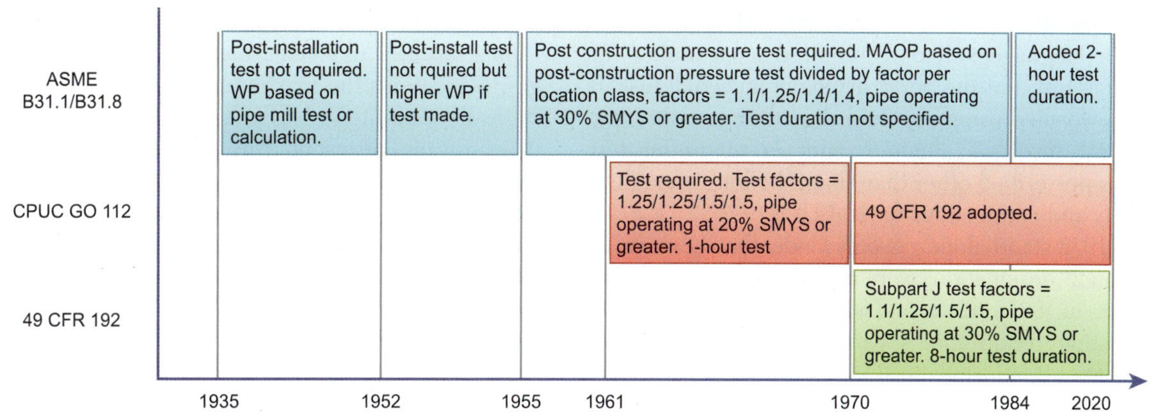

Figure 1. Timeline of pipeline hydrostatic pressure test requirements.

manufacturing plants, or within the boundaries of cities or villages; Division 2 piping was constructed in compressor stations, installed cross-country, or outside boundaries of cities or villages. Within both divisions, before installation, valves and fittings were to be "capable of withstanding a hydrostatic shell test" to designated pressures based on pressure rating classes similar to present-day pressure ratings for valves and flanged fittings. Pipe used in Division 1 service was also to be "capable of meeting the hydrostatic test requirements" contained in listed pipe product specifications, but pipe used in Division 2 was to be "subjected to and safely withstand a mill pressure test" in accordance with the pipe product specification (but not in excess of 90% of the yield point or yield strength of the material).

Some parties have suggested that the words "capable of withstanding" a pressure test to some level is means that a post-construction test was required. However, the Code used clearly different language to indicate an actual test requirement (e.g., pipe in the mill) as opposed to a capability for withstanding a test, which is a design requirement that is met through specifying an adequate combination of wall and metal strength. This is consistent with language used in contemporaneous standards for wrought fittings that also required that items be "capable of withstanding a pressure test to 1.5 times the working pressure", followed immediately by language stating that an actual test of each item was not required[13,14]. This is further supported by recent interpretations of similar pressure capability language in PHMSA regulations[15].

After installation, Division 1 piping systems containing welded joints were to be "capable of withstanding a hydrostatic test" to 1.5 times the service pressure, with the test to be applied "where practical". It was further stated that "if a test is performed" welds were to be struck by hammer blows to jar them during the hydrostatic pressure test, something that can only be done with exposed piping, not a buried pipeline. With Division 2 piping, there were no pressure test requirements post- installation because such a test was deemed unnecessary: fittings were designed to be as strong as matching pipe, and pipe was required to have been tested at the mill. The working pressure was 80% of the pipe mill test pressure, or a percentage of the yield strength calculated as the seam joint efficiency factor divided by 1.4.

In no case was the working pressure established in the 1935 Code on the basis of a post-installation pressure test. The 1935 Code was understood to mean that testing of the pipe after installation was discretionary for Division 1 piping and not required for Division 2 piping. Most pipeline operators made this same interpretation until such time as testing became a clearly stated requirement in the 1955 edition[16].

The 1942 edition slightly revised the post-installation testing to requirements to be "capable of withstanding a test pressure" of 150% of the service pressure for Division 1 piping or 50 psig greater than the maximum service pressure for Division 2 piping. A test after installation "may be made with air or gas" which "need not exceed 120% of the maximum allowable working pressure" for Division 1 piping or "shall not exceed 120% of the maximum allowable working pressure" for Division 2 piping. Clearly, these test limits conflict with an interpretation that a requirement to be "capable of withstanding a pressure test" is synonymous with an actual requirement to carry out such a test, to 150% of the service pressure. The duration of a pressure test, if performed, was not specified. The Code stated that "where an actual internal pressure test is made" (implying the existence of places where an "actual internal pressure test" was not made), the test pressure should be maintained for long enough to inspect the joints and connections. This requirement implies that the test's primary purpose was a leak test of flanged, threaded, or welded connections, not a proof of the strength of the pipe. Nowhere were working pressures established on the basis of a post-installation test. They were based on the mill test or an engineering calculation.

The 1947 Addendum to the 1942 B31.1 standard did not change the testing requirements for gas and air piping. The 1951 B31.1 standard slightly revised the post-installation testing provision to read "Where an internal fluid pressure test is made, it shall not exceed" 150% of the maximum allowable working pressure for Division 1 piping, and for Division 2 piping, 120% of or 50 psig greater than the maximum allowable working pressure, whichever was greater. The language still only required a capability for withstanding a test, not the performance of an actual test. If a test was performed using any fluid (liquid or gaseous), the maximum test level was limited, and no minimum test duration was prescribed other than that it be long enough to inspect joints and connections for leaks.

With the 1952 stand-alone gas-only Code, the "Division 1" and "Division 2" designations were replaced with description of the systems in §807(c)(2)(a) and §807(c)(1)(a), respectively. Pressure testing requirements were found in Chapter 5, "Requirements after installation" and were identical to the 1951 Code. However, for cross-country pipelines working pressures did depend on whether a post-installation test was performed. For pipelines installed prior to 1952, the allowed working pressure was either 80% of the pipe mill test pressure or a maximum of 72% of the yield strength ("Y") multiplied by a joint efficiency factor ("E"). For steel pipe outside of compressor stations, the allowed working pressure was either 80% of the pipe mill test pressure or 60% of Y times E for pipe not tested after installation, or 85% of Y times E for pipe tested after installation. Working pressure limits dependent upon whether a post-installation test was performed conclusively indicates that a post-installation test was not a requirement.

B31.8 Post-1955

The 1955 Code introduced the concept of 4 location class factors based on density of land development adjacent to the pipeline, each with different maximum allowable operating stress levels, and different pressure test requirements following installation. The precise definitions of the classes in terms of building or dwelling counts and the dimensions of the reference area were somewhat different than today, but the intended meanings of the classes were the same as today (e.g. Class 1 being rural, and so on) and the allowed operating stresses were also the same.

Testing requirements were stated in §841.3 "Testing after construction". All mains and services were to be tested, except tie-ins where individual test sections were eventually joined after testing. This was the first time in the gas piping standard that testing after installation became a firm requirement, but no minimum test duration was specified. The design requirement for a capability to withstand a pressure test was moved to Chapter 3 "Piping System Components and Fabrication Requirements", §831 "Piping System Components", where components were to be designed to withstand the system pressure test without failure, leakage, or impairment of their serviceability. Moving the "capability to withstand" language to Chapter 3 further substantiates the fact that a post-construction test was not intended or required by that wording.

Pressure test requirements were given in §841.4 "Test requirements". All pipelines and mains to be operated at a hoop stress of 30% or more of the specified minimum yield strength (SMYS) "shall be given a field test to prove strength after construction and before being placed in operation". Piping installed in Class 1 areas was to be tested with air or gas to 1.1 times the maximum operating pressure or hydrostatically tested to at least 1.1 times the maximum operating pressure; piping installed in Class 2 areas was to be tested with air to 1.25 times the maximum operating pressure or hydrostatically tested to at least 1.25 times the maximum operating pressure; and piping installed in Class 3 and 4 areas was to be hydrostatically tested to at least 1.4 times the maximum operating pressure.

The hydrotest requirement for Class 3 and 4 piping was waived if the ground temperature at the time of the test was or might fall below a temperature of 32 F, or water of satisfactory quality was not available in sufficient quantity. In that case, an air test to 1.1 times the maximum operating pressure could be performed and the test pressure ratio of 1.4 did not apply. Air testing of Class 3 and 4 pipe was allowed in any case, provided strict hoop stress limits were observed, the pipe was not operated at more than 80% of the test pressure, and the pipe had a seam joint efficiency factor of 1.00.

Paragraph §841.5 "Safety during tests" advised the user to give due regard to the safety of employees and the public during pressure tests. When air or gas is the test medium, steps were required to remove persons not involved in conducting the test when the test hoop stress level exceeds 50% SMYS.

Pressure test requirements in the 1958, 1963, 1967, 1968, 1975, and 1982 standards and their addenda were the same as in the 1955 standard.

The 1984 Addenda to the 1982 edition specified that the pressure test of all piping intended to operate at hoop stress levels of 30% SMYS or greater be held for a minimum duration of 2 hours. This was the first occurrence of a specified minimum test duration in B31.8. Test levels were the same as previously. The pressure test requirements in the 1986 edition were the same as the 1984 Addenda.

The 1989 standard introduced a new operating stress level in excess of the traditional maximum operating stress level of 72% SMYS in Class 1, up to a maximum operating stress of 80% SMYS. Pipe in this category was referred to as Class 1, Division 1, and was to be pressure tested to a minimum stress level of 100% SMYS, with water as the only permitted test fluid. The traditional maximum operating stress of 72% SMYS was referred to as Class 1, Division 2. The same test requirements applied for Class 1, Division 2, and for Classes 2, 3, and 4 as in previous editions. The requirements from the 1989 edition remained unchanged in the 1992, 1995, 1999, 2003, and 2007 editions.

Important revisions were made to the pressure testing requirements with the 2010 edition. The minimum test ratio for Class 1, Division 2 pipe (with a maximum operating stress level up to 72% SMYS) was raised to 1.25, regardless of test medium, and the minimum test ratio for Class 3 and 4 piping was raised to 1.50. Also, significant additional guidance on test planning, execution, and risk mitigation is provided.

49 CFR 192

The first full set of federal pipeline regulations were issued in 1970. Subpart J – Test Requirements, §192.501 through §192.517 set forth requirements for pressure testing of pipelines after construction. An important new requirement relative to those contained in preceding or contemporaneous editions of B31.8 or GO 112 was §192.505(c) stipulating maintaining the strength test pressure for at least 8 hours. As originally proposed, the specified minimum test duration was 24 consecutive hours, a practice that was observed by some but not all pipeline operators. This was reduced to 8 hours on the recommendation of the TPSSC because there was no evidence that a longer test was a superior test[17].

Aside from limitations based on maximum hoop stress levels, maximum allowable operating pressure was based on dividing the pressure test by a minimum specified factor, given in Subpart L – Operations, Clause 192.619(a)(2)(ii). For pipe installed after November 11, 1970, test pressure ratios were 1.1, 1.25, and 1.5 in Classes 1, 2, and 3 or 4, respectively. For pipe installed and tested prior to November 12, 1970, the test ratio for Classes 3 and 4 was 1.4, based on the requirements in the interim Federal standard between 1968 and 1970, which were the same as B31.8, and based on B31.8 being the de facto national standard prior to 1968 (except in California and perhaps a few other states).

These requirements for testing after construction have remained static in subsequent years.

Grandfathered pipelines

The term "grandfathered pipelines" refers to those pipelines for which the operating pressure was established on the basis of operating history rather than pressure testing in accordance with Subpart L. The origin and basis are described in the Preamble to the first full issuance of Title 49 – Transportation published in the Federal Register[18].

In the original proposal for Part 192, no recognition was given for piping installed prior to 1955[19] on the basis of very loose testing requirements, and for piping already operating at hoop stress levels greater than 72% SMYS. The Federal Power Commission (FPC) wrote to OPS pointing out that there were thousands of miles of pipeline already in service, installed in accordance with prevailing standards and practices, that could not continue operating at their then-current levels and comply with the proposed regulations. The FPC also stated that based on a review of the operating records of interstate pipelines, no improvement in safety would be gained by reducing the operating pressures of existing pipelines "which have been proven to be capable of withstanding present operating pressures through actual operation." In response, OPS included a "grandfather" clause to permit continued operation of pipelines at the highest operating pressure the pipeline had experienced in service during the 5 years preceding July 1, 1970 (even if the pipe had previously been subjected to a hydrostatic pressure test to qualify a higher MAOP but the pipe had not operated at that level during the specified 5-year interval).

The concept that new or evolving requirements concerning materials, design, construction, and the establishment of the MAOP are not retroactive to existing facilities that are already in operation was recognized in the federal pipeline regulations from the outset. This concept is embodied in §192.13 and is fully expressed in the discussion of the retroactive effect on existing pipelines in the Preamble to Part 192. Fed. Reg., pg. 13250, on the subject of the retroactive effect on existing pipelines, quotes the Natural Gas Pipeline Safety Act, Section 3(b): "Not later than 24 months after the enactment of this Act, and from time to time thereafter, the Secretary shall, by order, establish minimum Federal safety standards for the transportation of gas and pipeline facilities. Such standards may apply to the design, installation, inspection, testing, construction, extension, operation, replacement, and maintenance of pipeline facilities. Standards affecting the design, installation, construction, initial inspection and initial testing shall not be applicable to pipeline facilities in existence on the date such standards are adopted."

Compliance matrix of pressure testing

Figure 2 presents a matrix of compliance in terms of pressure test activity versus time. The time scale is divided between Pre-1955, 1955 to 1961, 1961 to 1970, and Post-1970. The pressure test activities are grouped as "No test", "Probably tested but no records", "Partial test records available", and "Full test records available". Situations where the indicated combination of testing history and documentary completeness are likely to comply with applicable requirements of the era, or not, are suggested in the matrix. Other situations fall in a gray area that should be evaluated making a reasonable accounting of circumstances. The burden of revalidating confidence in the existing pipeline system is likely to be tied in part to the perceived degree of compliance to requirements of the era. Figure 2 can serve as a guide to that process.

Summary

This paper reviewed the history of industry standards and regulatory requirements in the areas of hydrostatic pressure testing. The review shows that hydrostatic pressure testing after construction was not required by applicable industry standards (ASME B31.8) until 1955. The technology for pressure testing cross-country gas pipelines was not developed until 1950. Prior to 1955, operating pressure was established by the pipe mill test or an engineering calculation. Post-construction tests were discretionary, and generally for detecting leaks at flanged or welded joints above ground. Industry standards were referenced by some state regulations (e.g., GO 112 in 1961) until the issuance of federal safety regulations in 1970.

	Before 1955	1955-1961	1961-1970	1970-2020
Full test records currently available	*Likely in compliance.*	*Likely in compliance.*	*Likely in compliance.*	*Likely in compliance.*
Partial test records currently available	*Likely in compliance.*	*Likely in compliance.*	*May or may not be in compliance.*	*May or may not be in compliance.*
Probably tested but no records currently available	*Likely in compliance.*	*May or may not be in compliance.*	*Likely not in compliance.*	*Likely not in compliance.*
No pressure test	*Likely in compliance.*	*May or may not be in compliance.*	*Likely not in compliance.*	*Likely not in compliance.*

Figure 2. Pressure test compliance matrix.

References

1. "Hydrostatic testing" means conducting a pressure test of a pipe or vessel using water as the pressurizing medium. However, the term has often been used historically and today to refer to pressure testing using any fluid including gaseous media such as air, nitrogen, or natural gas. Here, "hydrotesting" is used in the incorrect but colloquial form to indicate a pressure test using any fluid except where a distinction is made with respect to the test medium.
2. J.F. Kiefner, "Section 4, Hydrostatic Testing", GRI Guide for Locating and Using Pipeline Industry Research. Prepared by Kiefner & Associates, Inc. for Gas Research Institute, March 2001, GRI-00/0192.04.
3. F.A. Hough, "The New Gas Transmission and Distribution Piping Code (ASA B31 Section 8)", Gas Magazine, Series in 8 Parts, January through September 1955.
4. W.B. McGehee, Maximum Allowable Operating Pressure (MAOP) Background & History. Report for Gas Research Institute, March 5, 1998.
5. T.M. Shires, M.R. Harrison, Development of the B31.8 Code and Federal Pipeline Safety Regulations: Implications for Today's Natural Gas Pipeline System (December 1998) GRI-98/0367.1.
6. C.J. Castaneda, J.A. Pratt, From Texas to the East: A Strategic History of Texas Eastern Corporation. Texas A&M University Press, 1993.
7. Castaneda and Pratt.
8. S.A. Bergman, Why Not Higher Operating Pressure for Lines Tested to 90% SMYS? Pipeline and Gas Journal (December 1974).
9. Kiefner.
10. McGehee.
11. Kiefner.
12. Bergman.
13. American Standards Association, Steel Butt Welding Fittings (1940) ASA B16.9.
14. American Society for Testing and Materials, Standard Specifications (Tentative) for Factory-Made Wrought Carbon-Steel and Carbon-Molybdenum-Steel Welding Fittings. ASTM A234, 1940.
15. F.A. Gale, Office of Hazardous Materials Standards, PHMSA, letter to Fox, M., Chemical Accidents Reconstruction Services, Inc., response to inquiry concerning 49 CFR 173.306, Ref. No. 04-0203, March 18, 2005. "Section 173.306(a)(3)(ii) requires a metal aerosol container to be capable of withstanding without bursting a pressure of one-and-one-half times the pressure of the content at 130 degrees F. The [Hazardous Materials Regulations] do not specify a method for demonstrating that the container is capable of withstanding the specified pressure. You may demonstrate that the container meets the standard by testing or design specifications."
16. Hough (1955).
17. Hough (1955).
18. Fed. Reg., pg. 13248-13276.
19. In its comments to the original docket, the TPSSC referred to 1952 as the first year that the ASME B31.1.8 gave minimum test pressures. However, that new test requirement occurred in 1955, not 1952. The TPSSC comments are interpreted accordingly herein.

Excerpted with permission - Applus+, Energy & Industry Division, 3 Sugar Creek Center Blvd. Suite 600, Sugar Land, TX. For full paper and citations, go online at www.applus.com. Paper "Pressure testing and recordkeeping: reconciling historic pipeline practices with new requirements" presented at Pipeline Pigging and Integrity, Management Conference, Houston, Texas, 13-14 February, 2013.

The benefits and limitations of hydrostatic testing

John F. Kiefner and Willard A. Maxey, Kiefner and Associates, Inc. Worthington, Ohio

The purpose of this discussion is to clarify the issues regarding the use of hydrostatic testing to verify pipeline integrity. There are those who say it damages a pipeline, especially if carried out to levels of 100% or more of the specified minimum yield strength (SMYS) of the pipe material. These people assert that if it is done at all, it should be limited to levels of around 90% of SMYS. There are those who insist that pipelines should be retested periodically to reassure their serviceability. The reality is that if and when it is appropriate to test a pipeline, the test should be carried out at the highest possible level that can feasibly be done without creating numerous test failures. The challenge is to determine if and when it should be done, the appropriate test level, and the test-section logistics that will maximize the effectiveness of the test.

The technology to meet these challenges has been known for 30 years. Nothing has arisen in the meantime to refute this technology. The problem is that people both within and outside the pipe industry either are not aware of the technology or have forgotten it, or for political reasons are choosing to ignore it.

In this discussion we show the following:

- It makes sense to test a new pipeline to a minimum of 100% of SMYS at the highest elevation in the test section.
- Pipe that meets the specified minimum yield strength is not likely to be appreciably expanded even if the maximum test pressure is 110% of SMYS.
- If hydrostatic retesting is to be conducted to revalidate the serviceability of a pipeline suspected to contain defects that are becoming larger with time in service, the highest feasible test pressure level should be used.
- If the time-dependent defects can be located reliably by means of an in-line inspection tool, using the tool is usually preferable to hydrostatic testing.

We also note the following as reminders:

- When a pipeline is tested to a level in excess of 100% of SMYS, a pressure–volume plot should be made to limit yielding.
- A test may be terminated short of the initial pressure target, if necessary, to limit the number of test breaks as long as the MOP guaranteed by the test is acceptable to the pipeline's operator.

And, we suggest that:

- Test-section length should be limited to prevent elevation differences within a test section from exceeding 300 ft.

The pressure level for verifying integrity can be higher than the level needed to validate the MOP of the pipeline, and the integrity test to a level above 1.25 times MOP, if used, needs to be no longer than $\frac{1}{2}$ h.

Background

The concept and value of high-pressure hydrostatic testing of cross-country pipelines were first demonstrated by Texas Eastern Transmission Corporation. Texas Eastern sought the advice of Battelle in the early 1950s as they began to rehabilitate the War Emergency Pipelines and to convert them to natural gas service. Prior to testing, these pipelines exhibited numerous failures in service due to original manufacturing defects in the pipe. The Battelle staff recommended hydrostatic testing to eliminate as many of these types of defects as possible. After being tested to levels of 100–109% of SMYS, during which time "hundreds" of test breaks occurred, not one in-service failure caused by a manufacturing defect was observed. The news of this successful use of hydrostatic testing spread quickly to other pipeline operators, and by the late 1960s the ASA B31.8 Committee (forerunner of ASME B31.8) had established an enormous database of thousands of miles of pipelines that had exhibited no in-service ruptures from original manufacturing or construction defects after having been hydrostatically tested to levels at or above 90% of SMYS.[1] These data were used to establish the standard practice and ASA B31.8 Code requirement that, prior to service, each gas pipeline should be hydrostatically tested to 1.25 times its maximum allowable operating pressure. Later, a similar requirement for liquid pipelines was inserted into the ASME B31.4 Code. When federal regulations for pipelines came along, the precedent set by the industry of testing to 1.25 times the MOP was adopted as a legal requirement.

Both field experience and full-scale laboratory tests have revealed much about the benefits and limitations of hydrostatic testing. Among the things learned were the following:

- Longitudinally oriented defects in pipe materials have unique failure pressure levels that are predictable on the basis of the axial lengths and maximum depths of the defects and the geometry of the pipe and its material properties.[2]
- The higher the test pressure, the smaller will be the defects, if any, that survive the test.

- With increasing pressure, defects in a typical line-pipe material begin to grow by ductile tearing prior to failure. If the defect is close enough to failure, the ductile tearing that occurs prior to failure will continue even if pressurization is stopped and the pressure is held constant. The damage created by this tearing when the defect is about ready to fail can be severe enough that if pressurization is stopped and the pressure is released, the defect may fail upon a second or subsequent pressurization at a pressure level below the level reached on the first pressurization. This phenomenon is referred to as a pressure reversal.[3,4]
- Testing a pipeline to its actual yield strength can cause some pipe to expand plastically, but the number of pipes affected and the amount of expansion will be small if a pressure—volume plot is made during testing and the test is terminated with an acceptably small offset volume or reduction in the pressure-volume slope.[5]

Test-pressure-to-operating-pressure ratio

The hypothesis that "the higher the test-pressure-to-operating-pressure ratio, the more effective the test," is validated by Figure 1. Figure 1 presents a set of failure-pressure-versus-defect-size relationships for a specific diameter, wall thickness, and grade of pipe. A great deal of testing of line-pipe materials over the years has validated these curves.[2] Each curve represents a flaw with a uniform depth-to-wall-thickness ratio. Nine such curves are given (d/t ranging from 0.1 to 0.9).

Consider the maximum operating pressure (MOP) for the pipeline (the pressure level corresponding to 72% of SMYS). That pressure level is represented in Figure 1 by the horizontal line labeled MOP. At the MOP, no defect longer than 10 in. and deeper than 50% of the wall thickness can exist. Any such defect would have failed in service. Similarly, no defect longer than 4 in. and deeper than 70% of the wall thickness can exist, nor can one that is longer than 16 in. and deeper than 40% of the wall thickness.

By raising the pressure level above the MOP in a hydrostatic test, the pipeline's operator can assure the absence of defects smaller than those that would fail at the MOP. For example, at a test pressure level equivalent to 90% of SMYS, the largest surviving defects are determined in Figure 1 by the horizontal line labeled 90% of SMYS. At that level, the longest surviving defect that is 50% through the wall can be only about 4.5 in. Compare that length to the length of the longest possible 50%-through flaw at the MOP; it was 10 in. Alternatively, consider the minimum survivable depth at 90% SMYS for a 10-in.-long defect (the size that fails at the MOP if it is 50% through the wall). The survivable depth is only about 32% through the wall. By a similar process of reasoning, one can show that even smaller flaws are assured by tests to 100% or 110% of SMYS (the horizontal lines drawn at those pressure levels on Figure 1).

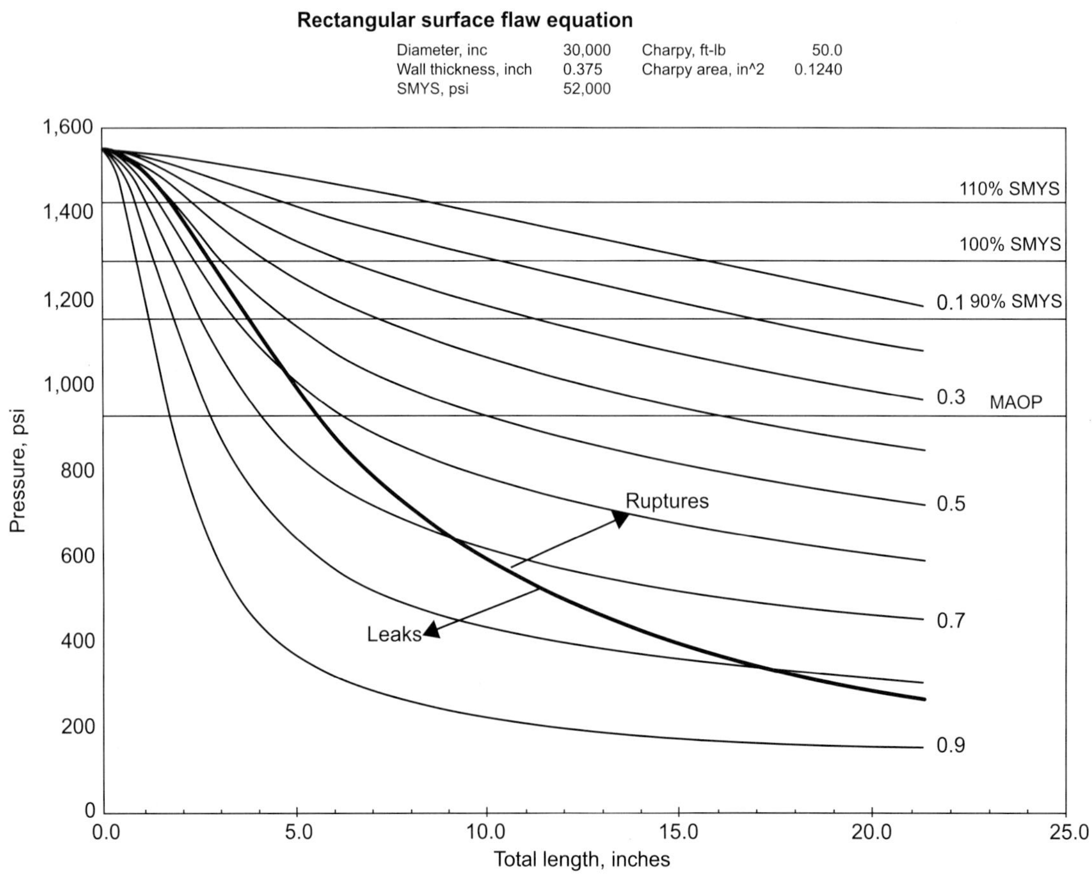

Figure 1. Impact of test pressure levels on margin of safety.

The point is that the higher the test pressure (above MOP), the smaller will be the possible surviving flaws. This fact means a larger size margin between flaw sizes left after the test and the sizes of flaws that would cause a failure at the MOP. If surviving flaws can be extended by operating-pressure cycles, the higher test pressure will assure that it takes a longer time for these smaller flaws to grow to a size that will fail at the MOP. Thus, Figure 1 provides proof of the validity of the hypothesis (i.e. the higher the test-pressure-to-operating-pressure ratio, the more effective the test).

Testing to levels above 100% of SMYS

Given the previous argument for testing to the highest feasible level, one needs to consider practical upper limits. In the case of a new pipeline constructed of modern high-quality, high-toughness line pipe, the maximum test level can generally be in excess of 100% of SMYS. Reasons why this will not cause significant yielding of the pipe are as follows.

First, as shown in Figure 2, the average yield strength of an order of pipe is usually well above the minimum specified value. Very few pieces will have yield strengths low enough to cause yielding at 100% of SMYS.

Secondly, when a buried pipeline is pressurized, it is restrained by the soil from shortening in the axial direction. This causes an axial tensile stress equal to Poisson's ratio times the hoop stress (Poisson's ratio is 0.3 for steel). What this means to testing in excess of 100% of SMYS is shown in Figure 3.

The tensile test commonly used to assess the yield strength of line pipe is a transverse, flattened uniaxial specimen designed to test the circumferential (hoop) direction tensile properties. A test of such a specimen reveals a unique value of yield strength at a certain value of applied stress. In Figure 3, that value is represented as 1.0 on the circumferential tensile stress to uniaxial yield stress ratio (vertical) axis. If one were to test the same type of specimen using a longitudinal specimen, the unique yield strength measured in the test could be plotted on Figure 3 at 1.0 on the longitudinal tensile stress uniaxial yield strength (horizontal axis). Negative numbers on the horizontal axis represent axial compressive stress. The typical line-pipe material exhibits an elliptical yield-strength relationship for various combinations of biaxial stress.[4] As shown in Figure 3, this results in yielding at a higher value of circumferential tensile stress to uniaxial yield strength ratios than 1.0. In tests of pressurized pipes, the ratio for a buried pipeline (longitudinal tensile stress to circumferential tensile stress ratio of 0.3) was found to be about 1.09.[6] So, this effect also suppresses yielding in a hydrostatic test of a pipeline to a pressure level in excess of 100% of SMYS.

To resolve how much yielding actually takes place, Texas Eastern designed a gauging pig in the mid 1960s to measure diametric expansion.[5] In 300 miles of 30-in OD X52 pipe, tested to a maximum of 113% of SMYS, they found only 100 joints of pipe (out of 40,000) that had expanded as much as 1.0%. In 66 miles of 36-inch OD X60 pipe tested to a maximum of 113% of SMYS, they found 100 joints of pipe (out of 6600) that had expanded as much as 1.0%, still not a lot of expanded pipe.

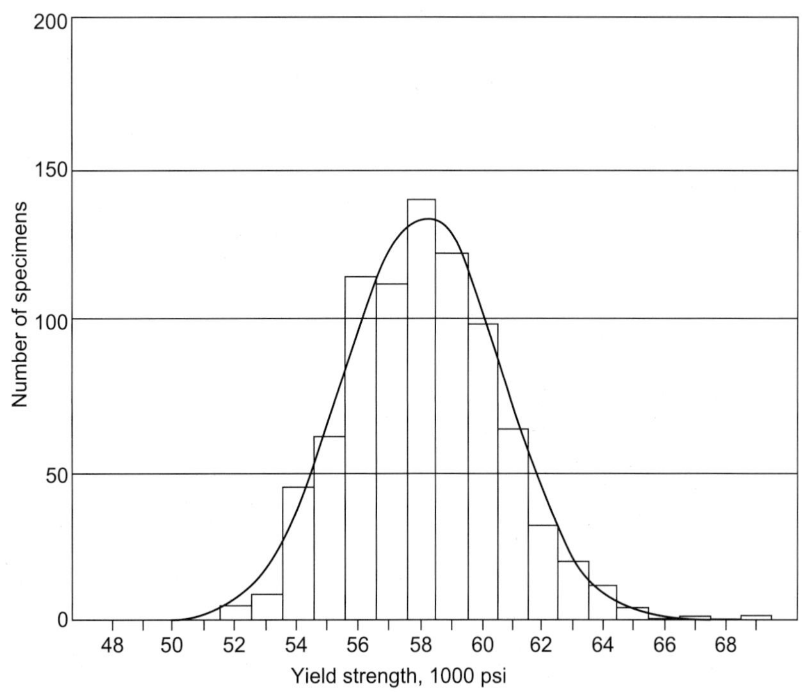

Figure 2. Normal curve fitted to frequency distribution of yield strength for 844 pipe specimens from 30-Inh by 0.375-inch X52 pipe

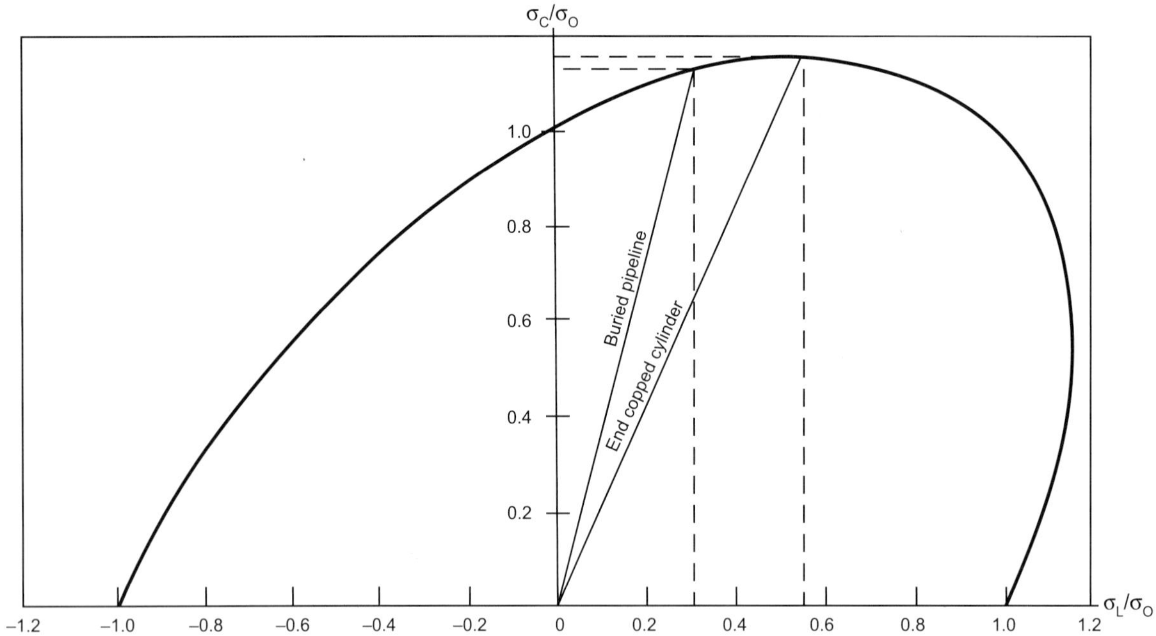

Figure 3. Explanation for higher yield strength with biaxial tensile stress.

Pressure–volume plots

Figure 4 shows a pressure-versus-pump-stroke plot of an actual hydrostatic test. The plot is created by recording the number of pump strokes of a positive displacement pump as each 10 psig increase of pressure is attained. Prior to beginning the plot, it is prudent to hold the test section at a constant pressure to assure that there are no leaks. After it is established that there is no leak, the plot should be started at a pressure level no higher than 90% of SMYS for the low elevation point in the test section in order to establish the "elastic" slope of the plot. By projecting the elastic slope lines across the plot as shown, one can then record pump strokes and compare the evolving plot to those slopes. If and when the actual plot begins to deviate from the elastic slope, either some pipe is beginning to yield or a leak has developed. The pressurization can be continued in any event until the "double-the-strokes" point is reached. This is the point at which it takes twice as many strokes to increase the pressure 10 psi as it did in the elastic range. Also, we suggest stopping at 110% of SMYS if that level of pressure is reached before the double-the-stroke point. Once the desired level has been reached, a hold period of 30 min should establish whether or not a leak has developed. Some yielding can be taking place while holding at the maximum pressure. Yielding will cease upon repeated repressurization to the maximum pressure, whereas a leak likely will not.

Testing an existing pipeline to a level at which yielding can occur may or may not be a good idea. It depends on the number and severity of defects in the pipe, the purpose of the test, and the level of maximum operating pressure that is desired. More will be said about this in the next section.

Finally, on the subject of testing to actual yield, the following statements apply:

- Yielding does not hurt or damage sound pipe. If it did, no one would be able to make cold-expanded pipe or to cold bend pipe.
- Yielding does not damage the coating. If it did, one could not field-bend coated pipe or lay coated pipe from a reel barge.
- Very little pipe actually undergoes yielding in a test to 110% of SMYS.
- Those joints that do yield do not affect pipeline integrity, and the amount of yielding is small.
- The only thing testing to a level in excess of 100% of SMYS may do is to void a manufacturer's warranty to replace test breaks if such a warranty exists.

Testing existing pipelines

Testing of an existing pipeline is a possible way to demonstrate or revalidate its serviceability. For a variety of reasons, retesting of an existing pipeline is not necessarily the best means to achieve confidence in its serviceability, however. First, a pipeline operator who elects to retest a pipeline must take it out of service and purge it of product. The downtime represents a loss of revenue and a disruption to shippers.

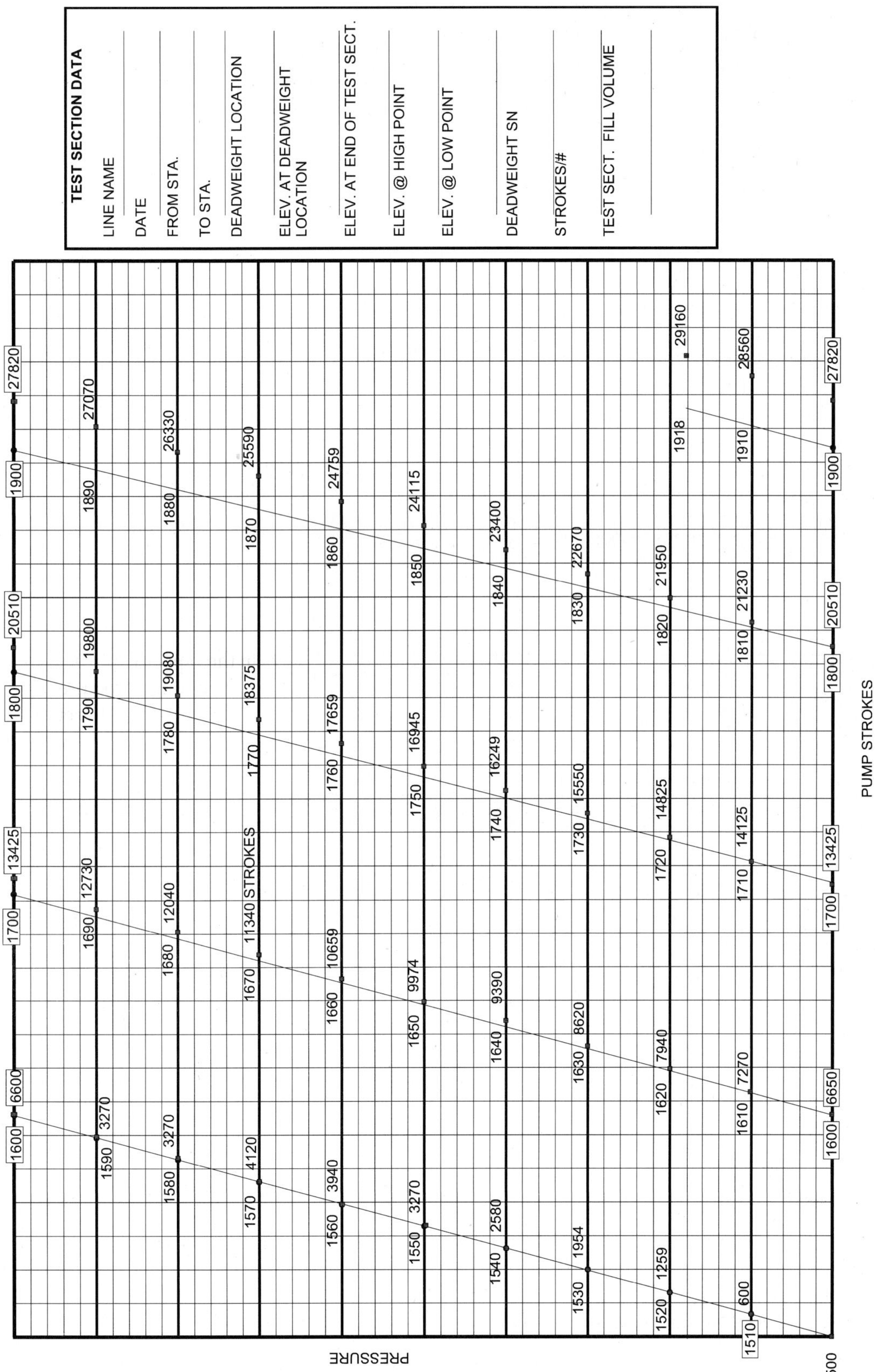

Figure 4. Pressure volume plot.

Second, the operator must obtain test water. To fill 30 miles of a 16-in. pipeline, an operator would need nearly 40,000 barrels of water. This is equivalent to a 100 × 100-ft pond, 22 ft in depth. For 30 miles of 36-in. pipe, the volume required would be five times as large. After the test, the water is considered a hazardous material because of being contaminated with product remaining in the pipeline. And, a test break, if one occurs, releases contaminated water into the environment. Aside from these issues, some problematic technical considerations exist.

The most important reason why a hydrostatic test may not be the best way to validate the integrity of an existing pipeline is that in-line inspection is often a better alternative. From the standpoint of corrosion-caused metal loss, this is almost certainly the case. Even with the standard resolution tools that first emerged in the late 1960s and 1970s this was true. Consider Figure 5. This figure shows the relative comparison between using a standard resolution in-line tool and testing a pipeline to a level of 90% of SMYS. The assumption is made in this case that the operator excavates and examines all "severe" and "moderate" anomalies identified by the tool, leaving only the "lights" unexcavated. In terms of the 1970s technology, the terms "light," "moderate," and "severe" meant the following:

- Light indications: metal loss having a depth less than or equal to 30% of the wall thickness.
- Moderate indication: metal loss having a depth more than 30% of the wall thickness but less than 50% of the wall thickness.
- Severe indication: metal loss having a depth more than 50% of the wall thickness.

In Figure 5, the boundary between "light" and "moderate" is nearly at the same level of failure pressure as the 90% of SMYS test for long defects, and it is well above that level for short defects. Because even the standard resolution tools have some defect—length-indicating capability, an in-line inspection on the basis represented in Figure 5 gives a better assurance of pipeline integrity than a hydrostatic test to 90% of SMYS. With the advent of high-resolution tools, the advantage shifts dramatically in favor of using in-line inspection instead of hydrostatic testing to validate the serviceability of a pipeline affected by corrosion-caused metal loss.

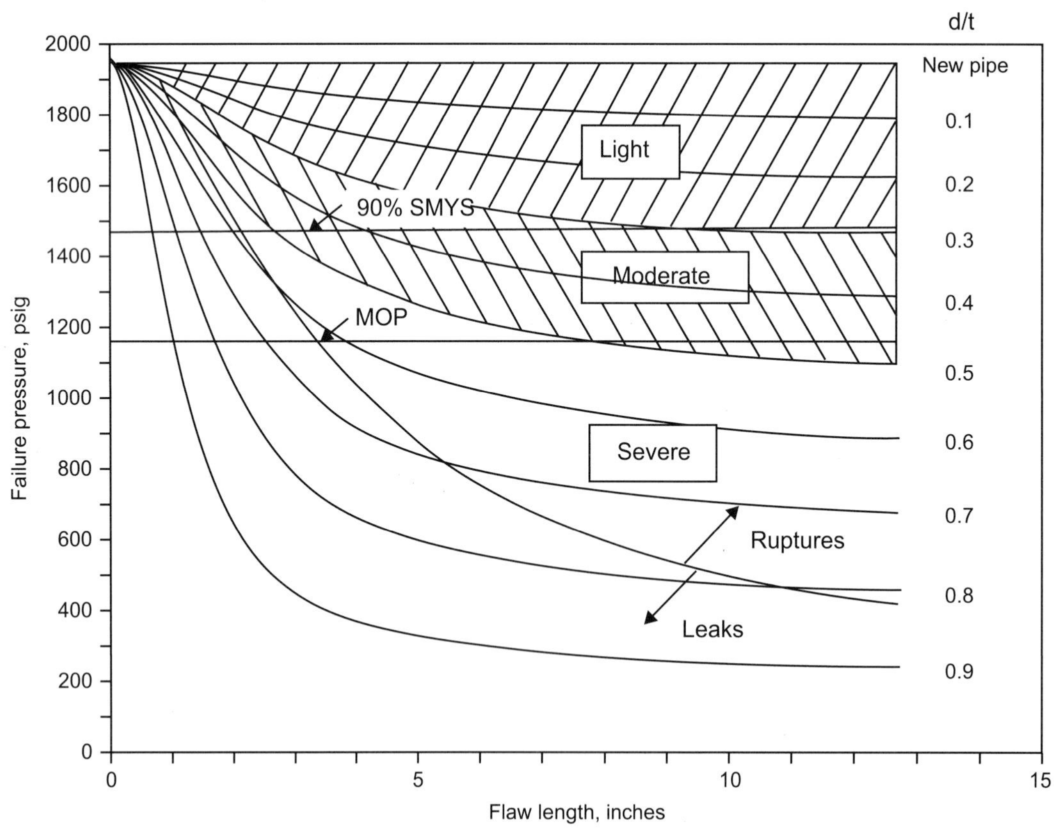

Figure 5. Sizes of flaws located by in-line inspection (correction) 16 inch by 0.250 inch X52, blunt defect.

From the standpoint of other types of defects, the appropriate in-line inspection technology is evolving rapidly and, in some cases, it has proven to be more effective than hydrostatic testing. One example is the use of the elastic-wave tool for detecting seam-weld defects in submerged arc-welded pipe.[7] Another is the use of transverse-field magnetic-flux-leakage inspection to find seam anomalies alongside or in the seams of electric-resistance-welded (ERW) pipe.[8] In these cases, the particular tools revealed defects that were too small to have been found by a hydrostatic test to any reasonable level up to and including 110% of SMYS. When a tool has established this kind of track record, a pipeline operator can justify using the tool instead of hydrostatic testing.

The concept of using in-line tools to detect flaws invariably raises the question about defects possibly not being detected. The reasonable answer is that the probability of nondetection is small (acceptably small in the authors' opinion) but not zero. In the same context, one must also recognize that hydrostatic testing is not foolproof either. One issue with hydrostatic testing is the possibility of a pressure reversal (discussed later). The other issue is that because hydrostatic testing can leave behind defects that could be detected by in-line inspection, the use of hydrostatic testing often demonstrates serviceability for only a short period of time if a defect-growth mechanism exists. This possibility is discussed in our companion paper to this article.[9]

Pressure reversals

A pressure reversal is defined as the occurrence of a failure of a defect at a pressure level that is below the pressure level that the defect has previously survived due to defect growth produced by the previous higher pressurization and possible subsequent damage upon depressurization. Pressure reversals were observed long before their probable cause was identified.[3] The pipeline industry supported a considerable amount of research to determine the causes of pressure reversals. The most complete body of industry research on this subject is Kiefner, Maxey, and Eiber.[4] Figure 6, taken from that source, reveals the nature of experiments used to create and demonstrate pressure reversals. It shows photographs of highly magnified cross sections of the tips of six longitudinally oriented flaws that had been machined into a single piece of 36-in. OD by 0.390 in. w.t. X60 pipe. Each flaw had the same length, but each was of a different depth, giving a graduation in severities. When the single specimen containing all six flaws was pressurized to failure, the deepest flaw (Flaw 1) failed. By calculations based on their lengths and depths, the surviving flaws were believed to have been pressurized to the percentages of their failure pressures shown in Table 1.

Table 1
Test Pressure at Failure of Flaw as a Percentage of the Calculated Failure Pressure

Flaw Number	% Failure Pressure
2	97
3	94
4	91
5	89
6	87

As one can see, the tips of Flaws 2, 3, and 4 exhibit some crack extension as a result of the pressurization to failure. The nearer the defect to failure, the more crack extension it exhibited. In fact, due to its extension during the test, Flaw 2 is now deeper than Flaw 1 was at the outset. Logic suggests that if we could have pressurized the specimen again, Flaw 2 would have failed at a level below that which it experienced during the testing of Flaw 1 to failure. Indeed, in similar specimens designed in a manner to allow subsequent pressurizations, that is exactly what often occurred. This type of testing led to an understanding of pressure reversals in terms of ductile crack extension occurring at near-failure pressure levels where the amount of crack extension is so great that crack closure upon depressurization does further damage, leading to the inability of the flaw to endure a second pressurization to the previous level. The pressure reversal is expressed as a percentage:

$$\text{Pressure reversal} = \frac{(\text{original pressure} - \text{failure pressure})}{(\text{original pressure} \times 100)}.$$

Once the cause was known, the next key question was: What is the implication of the potential for pressure reversals on confidence in the safety margin demonstrated by a hydrostatic test? This question has been answered in particular circumstances, and the answer comes from numerous examples of actual hydrostatic tests.[a] Figure 7 is an example of an analysis of pressure reversals in a specific test. This figure is a plot of sizes of pressure increases or decreases (reversals) on subsequent pressurizations versus the frequency of occurrence of that size of increase or decrease. Among other things, these data show that upon pressurization to the target test-pressure level, a 1% pressure reversal (34 psi) can be expected about once in every 15 pressurizations, a 2% pressure reversal (68 psi) can be expected about once in every 100 pressurizations, and a 3% pressure reversal (102 psi) can be expected about once in every 1000 pressurizations. For a target test pressure level of

[a] In actual hydrostatic tests, direct evidence that pressure reversals are the result of the type of flaw growth shown in Figure 6 has seldom been obtained. However, a few such cases have been documented and it is assumed that defect growth is responsible for all such cases.

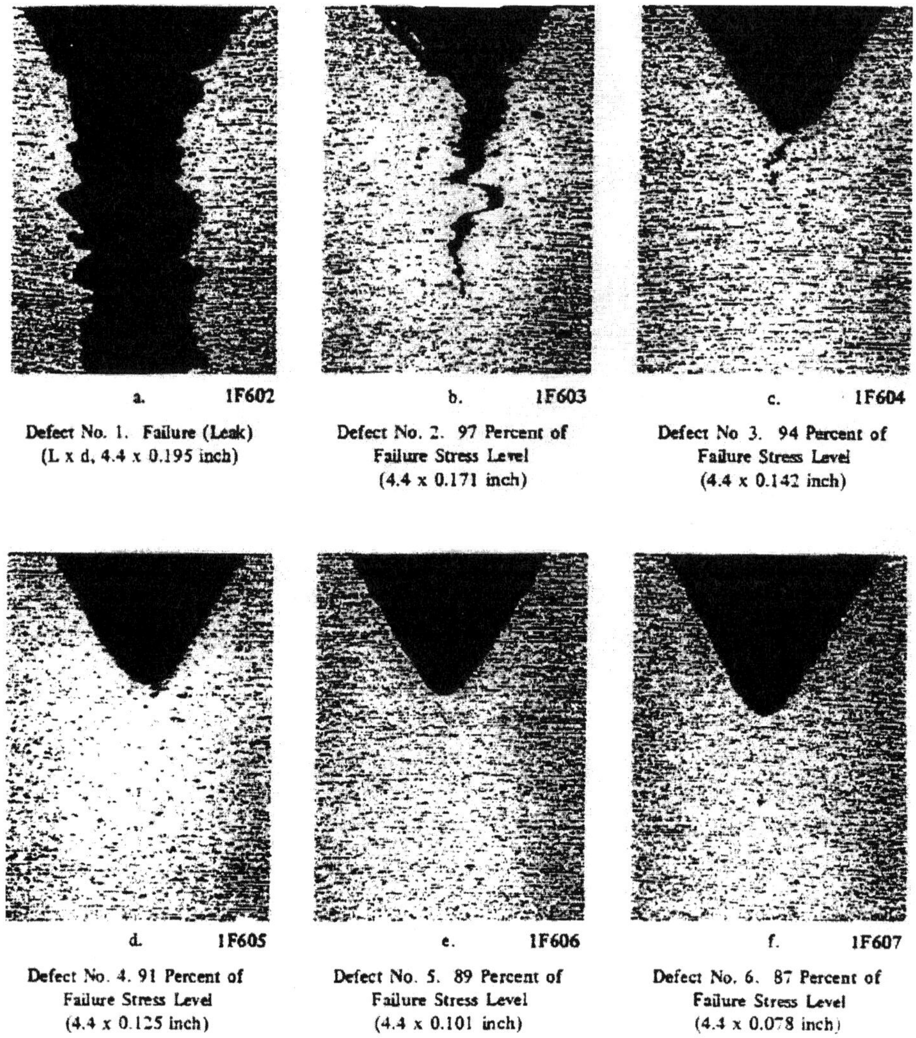

Note: Loading consisted of –
1st cycle – 0 → 1330 psig with 30 sec hold
2nd cycle – 0 → 1300 psig with 30 sec hold
3rd cycle – 0 → 1230 psig with 30 sec hold

Figure 6. Flaw growth in 4.4-inch-long part-through flaws in 36 $\times$ 0.390-inch X60 pipe.

1.25 times the MOP, the expectation of a 20% pressure reversal (enough to cause failure at the MOP) is off the chart, that is, it is an extremely low-probability event (but not an impossible event).

There have been a handful of pipeline service failures in which a pressure reversal is the suspected but unproven cause. There is also one case of a large pressure reversal (62%) that was unequivocally demonstrated because it occurred on the fifth cycle of a five-cycle hydrostatic test.[4] It should be noted that most of the experiences of numerous and large pressure reversals in actual hydrostatic tests have involved flaws associated with manufacturing defects in or near ERW seams, particularly in materials with low-frequency welded (generally pre-1970) ERW pipe. But in most cases where numerous reversals occurred, the sizes of the actual pressure reversals observed are small (less than 5%). One thing seems clear: if a hydrostatic test can be successfully accomplished without the failure of any defect, the likelihood of a pressure reversal will be extremely small. It is the tests in which numerous failures occur that have the highest probabilities of reversals. And, when the number of reversals becomes large, the probability of a reversal of a given size can be estimated as was done on the basis of Figure 7.

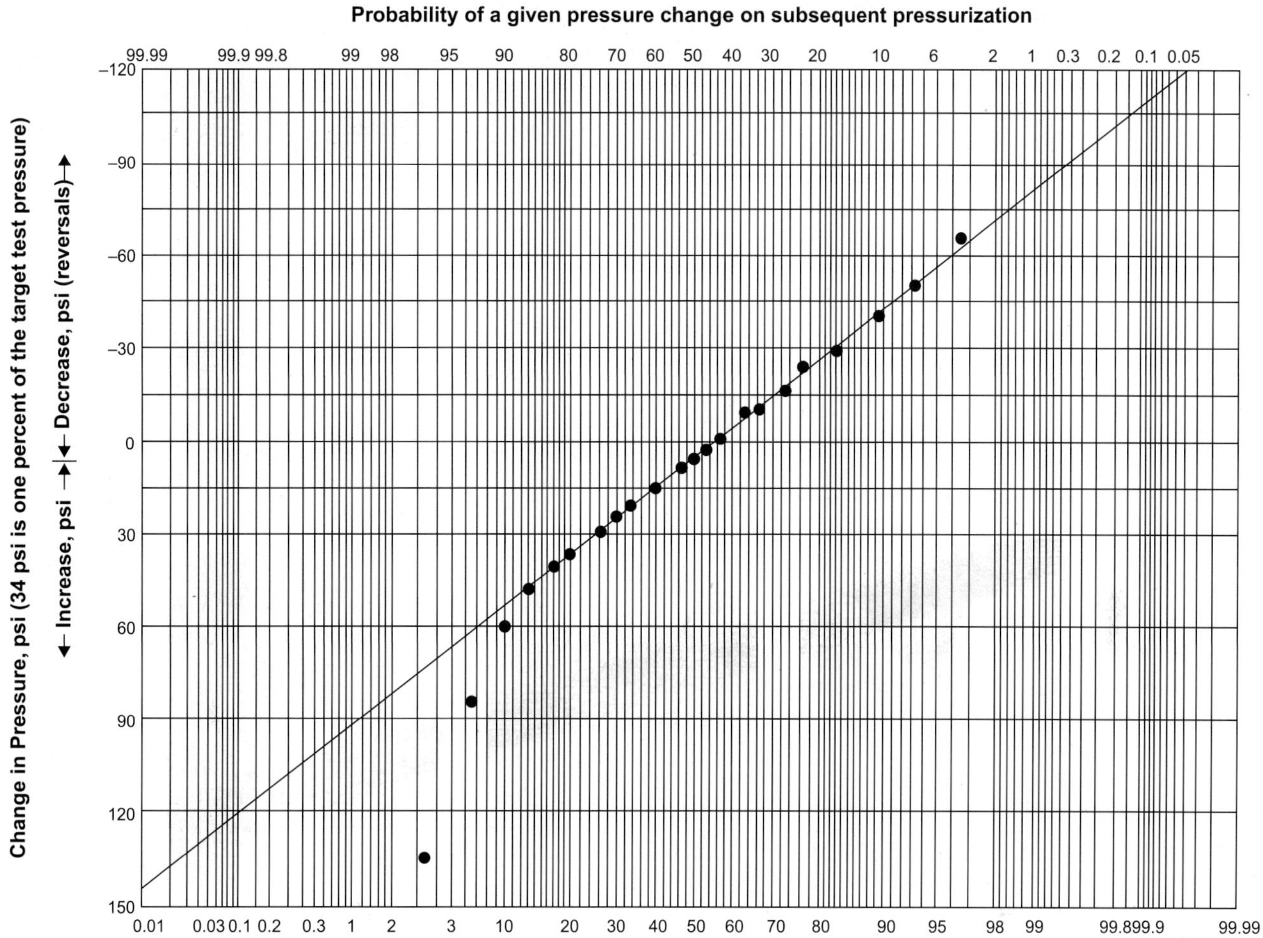

Figure 7. Evaluation of probability of having a pressure reversal.

Practical considerations

For new pipeline materials made to adequate specifications with adequate inspection and pipe-mill testing, one does not expect test failures even at pressure levels corresponding to 100% or more of SMYS. Therefore, there is no reason not to test a pipeline constructed of such materials to levels in excess of 100% of SMYS. As has been shown, the higher the ratio of test pressure to operating pressure, the more confidence one can have in the serviceability of a pipeline. In the case of existing pipelines, especially the older ones, such test levels may be impossible to achieve, and if numerous test failures occur, the margin of confidence may become eroded by the potential for pressure reversals. Weighing against low test-pressure-to-operating-pressure ratios, on the other hand, is the fact that such tests, by definition, generate lower levels of confidence and buy less time between retests if the issue of concern is time-dependent defect growth. A discussion of some possible responses to this dilemma follows.

First and foremost, as has already been mentioned, the use of an appropriate in-line inspection tool is always to be preferred to hydrostatic testing if there is sufficient confidence in the ability of the tool to find the defects of significance. Most of the pipe in a pipeline is usually sound. Therefore, it makes sense to use a technique that will find the critical defects and allow their repair as opposed to testing the whole pipeline when it is not necessary. The industry now has access to highly reliable tools for dealing with corrosion-caused metal loss, and tools are evolving rapidly to detect and characterize cracks. As has been noted, some uses of these tools have already proven their value and, in those cases, their use in lieu of hydrostatic testing makes good sense.

There are still certain existing pipelines for which hydrostatic testing remains the best (in some cases the only) means to revalidate their serviceability. In those cases, the following advice may be useful. Determine the mill hydrostatic test level for the pipe. The mill-test certificates will show the level applied if such certificates can be found. Also, search the records for prior hydrostatic tests at or after the time of construction. Review the pressure levels and causes of mill-test or in-place test failures if they exist. If none of these records is available, look up the API 5L specification applicable to the time the pipe was manufactured. This will reveal the standard mill-test pressure for the pipe. Do not assume, if you do not know, that the pipe was tested in the mill to 90% of SMYS. This was not always the case especially for non-X grades and smaller diameter pipe materials. If you decide to test the pipe to a level in excess of its mill-test pressure for the first time ever, anticipate test failures. If you cannot tolerate test failures, consider testing to a level just below the mill-test pressure. This may mean, of course, that the MOP you validate is less than 72% of SMYS (the minimum test pressure must be at least 1.25 times the MOP for 4 h plus 1.10 times MOP for 4 h for a buried pipeline; see Federal Regulation Part 195). Alternatively, if you can tolerate at least one test failure, pressurize to a level as high as you wish or until the first failure, whichever comes first. If you then conduct your 1.25-times MOP test at a level at least 5% below the level of the first failure, a second failure will be highly improbable.

It is always a good idea to conduct an integrity test as a "spike" test. This concept has been known for many years,[4] but more recently it has been advocated for dealing with stress-corrosion cracking.[10] The idea is to test to as high a pressure level as possible, but to hold it for only a short time (5 min is good enough). Then, if you can live with the resulting MOP, conduct your 8-h test at a level of at least 5% below the spike-test level. The spike test establishes the effective test-pressure-to-operating-pressure ratio; the rest of the test is only for the purpose of checking for leaks and for meeting the requirements of Part 195.

Summary

To summarize, it is worth reporting the following:

- Test-pressure-to-operating-pressure ratio measures the effectiveness of the test.
- In-line inspection is usually preferable to hydrostatic testing.
- Testing to actual yield is acceptable for modern materials.
- Pressure reversals, if they occur, tend to erode confidence in the effectiveness of a test but usually not to a significant degree.
- Minimizing test-pressure cycles minimizes the chance for pressure reversals.

References

1. S.A. Bergman, "Why Not Higher Operating Pressures for Lines Tested to 90% SMYS?" *Pipeline and Gas Journal*, December 1974.
2. J.F. Kiefner, W.A. Maxey, R.J. Eiber, and A.R. Duffy, "Failure Stress Levels of Flaws in Pressurized Cylinders," *Progress in Flaw Growth and Toughness Testing*, ASTM STP 536, American Society for Testing and Materials, pp. 461–481, 1973.
3. L.E. Brooks, "High-Pressure Testing—Pipeline Defect Behavior and Pressure Reversals," ASME, 68-PET-24, 1968.
4. J.F. Kiefner, W.A. Maxey, and R.J. Eiber, *A Study of the Causes of Failure of Defects That Have Survived a Prior Hydrostatic Test*, Pipeline Research Committee, American Gas Association, NG-18 Report No. 111, November 3, 1980.
5. A.R. Duffy, G.M. McClure, W.A. Maxey, and T.J. Atterbury, *Study of Feasibility of Basing Natural Gas Pipeline Operating Pressure on Hydrostatic Test Pressure*, American Gas Association, Inc., Catalogue No. L30050, February 1968.
6. W.A. Maxey, R.J. Podlasak, J.F. Kiefner, and A.R. Duffy, "Measuring the Yield Strength of Pipe in the Mill Expander," Research report to A.G.A./PRC, A.G.A. Catalog No L22273, 61 pages text plus 2 appendices, 23 pages, December 1973.
7. W.A. Maxey, R.E. Mesloh, and J.F. Kiefner, "Use of the Elastic Wave Tool to Locate Cracks Along the DSAW Seam Welds in a 32-Inch (812.8 mm) OD Products Pipeline," *International Pipeline Conference*, Volume 1, ASME, 1998.
8. K. Grimes, "A Breakthrough in the Detection of Long Seam Weld Defects in Steel Pipelines," Pipeline Integrity International.
9. J.F. Kiefner, and W.A. Maxey, "Periodic Hydrostatic Testing or In-line Inspection to Prevent Failures from Pressure-Cycle-Induced Fatigue," Paper Presented at API's 51st Annual Pipeline Conference and Cybernetics Symposium, New Orleans, Louisiana, April 18–20, 2000.
10. B.N. Leis, and F.W. Brust, *Hydrotest Strategies for Gas Transmission Pipelines Based on Ductile-Flaw-Growth Considerations*, Pipeline Research Committee, American Gas Association, NG-18 Report No. 194, July 27, 1992.

Hydrostatic testing for pipelines

After construction of a pipeline is completed, or if pipe has been replaced or relocated, it is necessary to hydrostatically test the pipeline to demonstrate that the pipeline has the strength required to meet the design conditions, and to verify that the pipeline is leak free.

The U.S. Federal Safety Regulations for Pipelines require that pipelines used to transport hazardous or highly volatile liquids be tested at a pressure equal to 125% of the maximum allowable operating pressure (MAOP) for at least 4 continuous hours and for an additional 4 continuous hours at a pressure equal to 110% or more of the MAOP if the line is not visually inspected for leakage during the test. A design factor of 72% of the specified minimum yield strength (SMYS) of the pipe is used to determine the maximum allowable operating pressure. The requirement to test to 125% of the MAOP will therefore cause the pipe to be tested to a pressure equal to 90% of the SMYS of the pipe. See Section 3—Pipe Design for additional information on calculating the MAOP.

The regulations for gas lines specify design factors based on the class location of the pipeline. The class location is determined by the number of buildings in a specified area on either side of the pipeline. Refer to Part 192.111 of the Minimum Federal Safety Standards for Gas Lines for the details on how to determine the class location. See Section 3, Pipe Design for a listing of the design factors for the different class locations. The regulations specify that the test pressure must be maintained for at least 8 h and must be equal to at least 125% of the maximum allowable operating pressure. The regulations should be consulted for the specific testing requirements, as the regulations are subject to change.

Usually, operators will specify a test pressure range from 90% to 95% of the SMYS of the pipe. Some will allow test pressures as high as 100% of the SMYS of the pipe and some will test to slightly beyond the SMYS of the pipe. Specifying a test pressure at least equal to 90% of the SMYS of the pipe will qualify it for the maximum allowable operating pressure. In some cases, the test pressure will be based on the minimum yield strength determined from the mill test reports.

Hydrostatic testing a pipeline is certainly a major operation and should be carefully planned. Most companies have hydrostatic test manuals that detail the procedures to be followed to complete the test. Usually, this work is performed by a hydrostatic testing contractor hired by the pipeline owner, or hydrostatic testing may be included as a part of the main construction contract.

One of the first steps in planning the hydrostatic test operation is to examine the elevation gradient. The gradient, along with the location of the water source, and the pipe design data, will be used to determine the length and number of test sections. Figure 1 shows a typical pipeline elevation gradient.

Where the pipeline traverses hilly terrain, the elevation gradient must be carefully considered in selecting the pipeline test segments. Different companies have differing philosophies on how to do this. Some limit the amount of elevation difference while others may specify a range of allowable percentages of the SMYS of the pipe—i.e. 90–95% or 90–100% of the SMYS of the pipe. In any case, the test gradient should be plotted to be sure the test pressure falls within the specified pressure limits.

The test gradient must be based on water head in feet, if water is used as the test medium. Water pressure may be converted to head by dividing the pressure by 0.433. If sea water is being used as the test medium, a factor of 0.445 should be used. This assumes a specific gravity of 1.02 for sea water. In any case, the factor of 0.433 should be modified according to the specific gravity of the test medium.

Let's assume that we have a line that is to be tested and the elevation at the test site is 1000 feet and the elevation at the end of the line is 1200 ft. Fresh water will be used as the test medium, and it is desired to test the pipe to a minimum of 90% and a maximum of 96% of the SMYS of the pipe. The pipe is 30-in OD × 0.390″ wt API 5LX X60. There is a difference in elevation of 200 ft and this is equal to 86.6 psig (200 ft × 0.433 = 86.6 psig). A test pressure equal to 90% of SMYS is 1404 psig. Since the test site is lower than the high end of the line, the 86.6 psig is added to 1404 psig to obtain a test site pressure of 1491 psig. The pressure at the end of the line will be 1404 psig, which equates to 90% of SMYS. The pressure at the low point equates to 96% of SMYS.

Now, let's assume a line has a high point elevation of 1100 ft, a low elevation point of 1000 ft, and the elevation at the test site is 1050 ft. The test pressure at the high point will need to

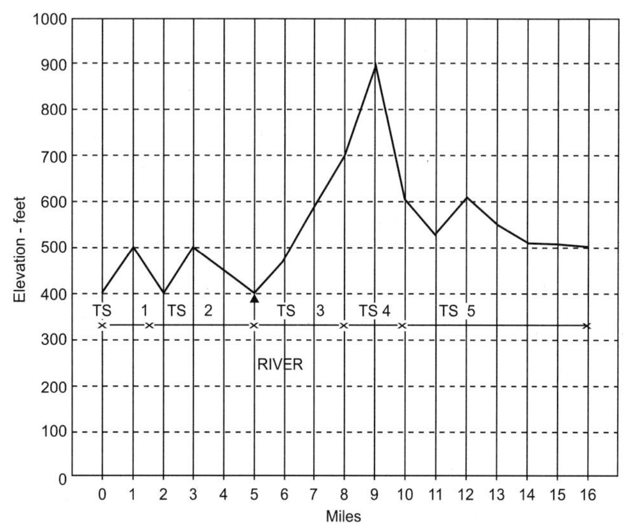

Figure 1. Pipeline elevation profile.

be 1404 psig in order to meet the 90% of SMYS requirement. The pressure at the low point will be 1447 psig, and the pressure at the test site will be 1426 psig.

When testing offshore lines, the pressure at or below the water surface will be the same as at the low elevation point due to the offsetting external subsea pressure head. If the line previously described was laid offshore, the test pressure at the water surface would be 1404 psig. The elevations are as follows:

Top of riser	+11 ft
Test site	+7 ft
Water surface	0 ft
Pipe depth	−168 ft

The test pressure at the test site would be 1404 − (7 × 0.445) or 1401 psig. At the top of the riser, the test pressure would be 1404 − (11 × 0.445) or 1399 psig.

The typical profile shown in Figure 1 represents a pipeline that requires testing. The pipeline crosses a river at approximately MP 5, and river water will be used to test the line. Test sections 1–5 have been chosen as indicated. The lengths of sections 3 and 4 are limited by elevation difference. This line was designed to operate at 936 psig or 72% of the SMYS of 30 in × 0.375 in wt API 5LX X52 pipe for Class I locations, and 60% of the SMYS of 30 in × 0.390 in wt API 5LX X60 pipe for Class II locations, and 50% of the SMYS of 30 in × 0.438 in wt API 5LX X65 for Class III locations.

The pipeline crosses one railroad, one highway, and the river. It also includes one main line block valve assembly. The valve assembly, river crossing, and road crossing were pretested before installation and are tested again after installation.

Four test manifolds were installed to facilitate filling the line and for isolating the test sections during the test operation. See Figure 2 for a typical pressure sectionalizing manifold. Two-way pigs were loaded at each of the intermediate manifolds. The pigs are moved by the fill water and are necessary to remove the air from the line.

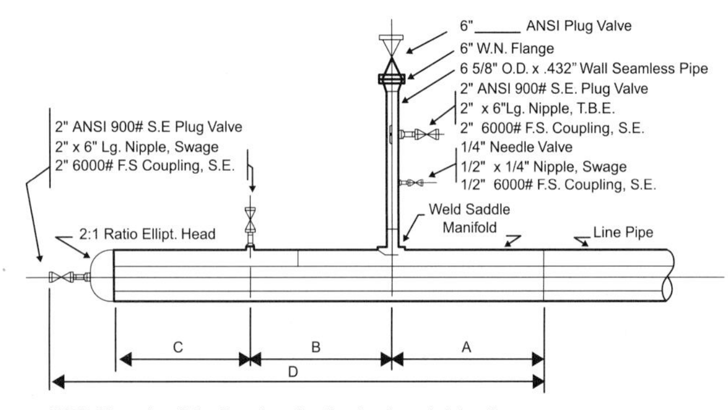

Figure 2. Typical pressure manifold. Courtesy Milbar Hydrotest.

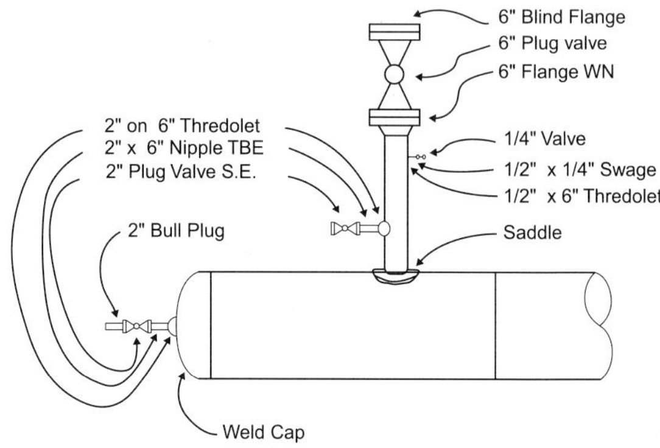

Figure 3. Typical test manifold. Courtesy Milbar Hydrotest.

It is generally a good idea to test the section most distant from the water source first. If it should rupture, then testing on intermediate sections can continue while repairs on the failed section are completed.

The pumps used to fill the line should have sufficient capacity to fill the line at a rate of about 1/2 mile per hour. Water filters should be used and are normally equipped with 100 mesh screens. See Figure 4 for a typical fill site arrangement. Some companies may specify a finer mesh screen. The filling unit should be equipped with a flow meter to measure the amount of water pumped into the pipeline. Temperature recorders will be used to record ambient temperature, the temperature of the pipe and water, and ground temperature. See Figure 5 for typical test thermometer placement. A pressure recorder will be used to provide a recording of the pipeline test pressure. A deadweight tester will be used to calibrate the pressure recorder. A high pressure pump capable of delivering 70–150 gpm at a pressure exceeding the required test pressure will be used in the final

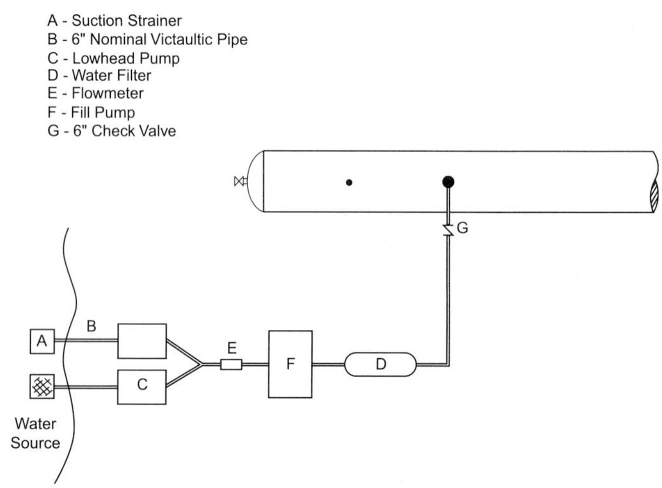

Figure 4. Typical fill site. Courtesy Milbar Hydrotest.

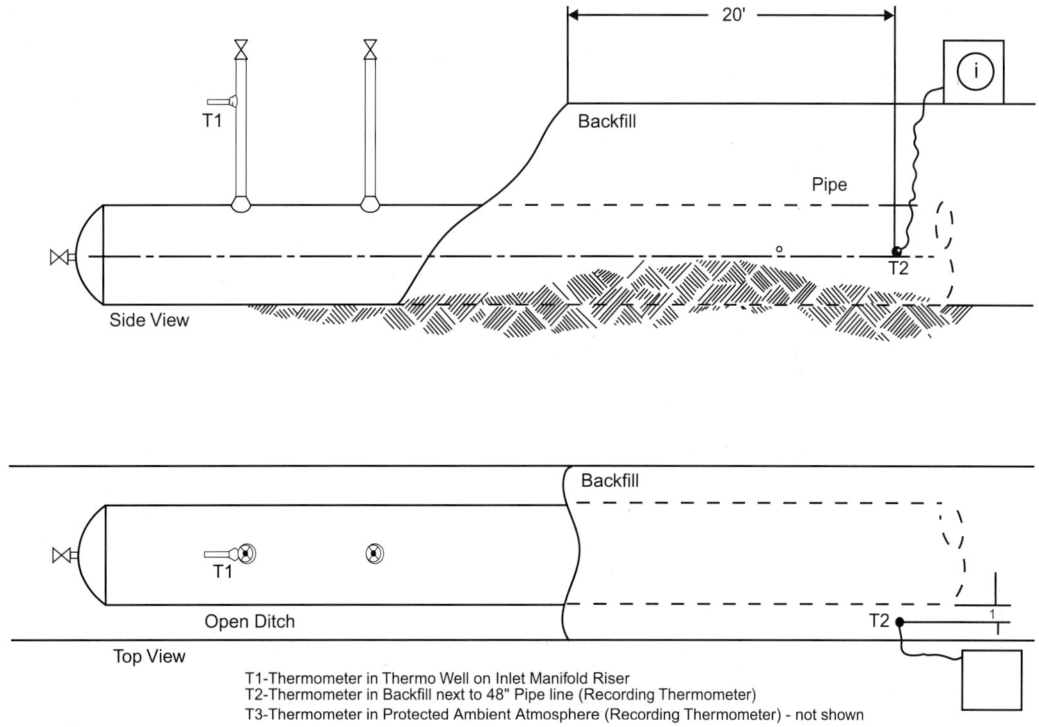

Figure 5. Test thermometer placement. Courtesy Milbar Hydrotest.

pressuring operation. Small high pressure units with a capacity of 6–30 gpm may be used for short sections of large diameter lines and smaller diameter lines.

Deadweight pressure and temperature readings are recorded after the prescribed test pressure has been reached and the pressures and temperatures have stabilized. A pressure vs. time plot may also be made. Readings are usually made at 15-min intervals for the first hour and at 30-min intervals thereafter. The procedure for accepting a leakage test will vary from company to company. Some will accept the leakage test if there is no pressure drop in a 3-h period.

A pressure—volume plot may be required, especially if the test pressures approach or exceed the SMYS of the pipe. The plot is made manually during the pressuring operation by recording pump strokes on the *X*-axis and pressure on the *Y*-axis. A straight line will be produced until the plastic range of the pipe is reached or until a leak occurs. This type of plot is also useful should a joint of lighter wall thickness or lower yield strength pipe be inadvertently placed in the pipeline.

Small leaks during the testing operation can be difficult to locate. A change in the water/pipe temperature may give the appearance of a leak. If the temperature of the pipe/water decreases, the test pressure will decrease. An increase in water/pipe temperature will cause the test pressure to increase. The effect of a temperature change may be estimated using the equations and data contained in Appendices A and B. To achieve any degree of accuracy in these calculations, it will be necessary to have accurate temperature and pressure readings.

The pipeline test temperature may be affected by river/creek crossings along the pipeline route. Also, if cold river water is used, there will be a changing temperature gradient. Air, either trapped or entrained, will also affect the pressure-temperature calculations since the coefficient of expansion for air is not the same as for water. If the calculations are made and the decrease in pressure cannot be attributed to a temperature change, then it may be necessary to sectionalize the test section by manifolding and repressuring the shorter segments to facilitate locating the leak. Once the leak has been located and repaired, testing operations can resume.

Air compressors will be required to remove the water once the testing is complete. The maximum head for this particular project was 450 ft (195 psig). Allowance will need to be included for friction loss in the cross-over manifolds. The compressors will need to have sufficient capacity to remove the water at about the same rate as the fill rate. Most portable compressors are limited to approximately 125 psig. In this case, booster compressors will be needed for the 195 + psig. See Figure 7 for a typical dewatering connection.

Upon completion of testing, the test pressure is bled off to leave approximately 10–20 psig at the high point in the test section. Dewatering may be accomplished using the air compressors discussed earlier. Some companies may elect to displace the test water with the material that will normally be pumped through the line. This is usually the case with crude oil and refined product pipelines. Pipelines used to transport natural gas and certain chemicals are usually dewatered using compressed air. Care must be

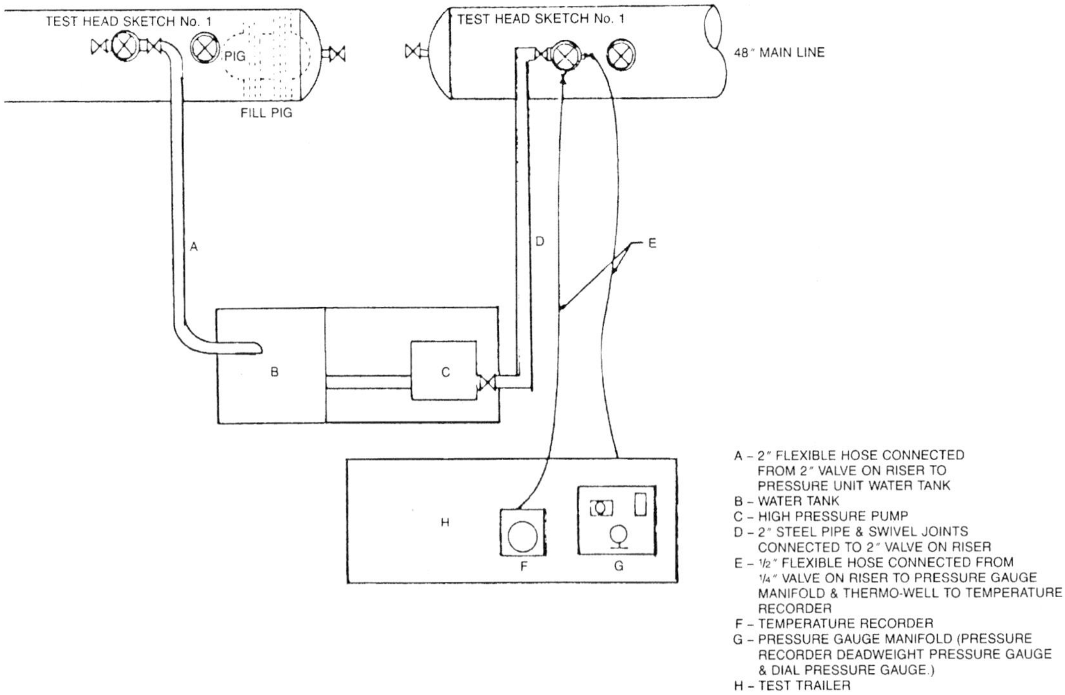

Figure 6. Typical pressure unit setup. Courtesy Milbar Hydrotest.

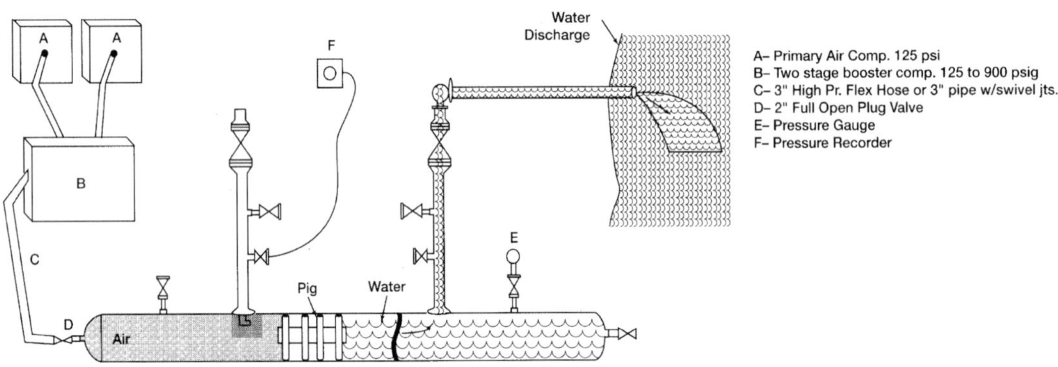

Figure 7. Dewatering arrangement. Courtesy Milbar Hydrotest.

exercised in the dewatering operation to make sure that no air is introduced into the test section, thus minimizing the possibility of air locks. An air lock is probably the most severe problem involved in dewatering. Air locks are caused by air accumulating in the downhill leg and water accumulating in the uphill leg, which creates a manometer in the pipeline. In some instances, extremely high pressures may be required to overcome the manometer. Care should be exercised to be sure the maximum allowable pipeline operating pressure is not exceeded. It may also be necessary to tap the line and vent air at the high points.

If it is necessary to pipe the water away from the right-of-way, welded pipe should be used. The pipe must be securely anchored. The use of tractors and skids will probably provide sufficient anchoring.

Always use a valve to control the amount of water being bled from the pipeline.

If there will be a delay in placing the pipeline in operation, and it is decided to leave the test water in the line until operations begin, consideration should be given to the use of a corrosion inhibitor. Long-term storage of the line may necessitate the use of oxygen scavengers in addition to the inhibitor.

If after dewatering, residual moisture in the pipeline poses a problem, it will be necessary to clean and dry the pipeline. See Section 6 for drying details.

A permit to discharge the water may be required, especially if there is a possibility the water may enter a stream of water.

This should be determined and permits obtained, if required, in the planning phase of the testing operation. A "splash" plate or bales of hay may be used when dewatering to prevent erosion.

The most important aspect of any test operation is complete and accurate documentation, since it will become a permanent record that must be retained for as long as the facility tested remains in operation. Federal Pipeline Safety Regulations for pipelines transporting liquids require that these records include the following information:

- the pressure recording charts;
- test instrument calibration data;
- the names of the operator, of the person responsible for making the test, and of the test company used, if any;
- the date and time of the test;
- the minimum test pressure;
- the test medium;
- a description of the facility tested and the test apparatus;
- an explanation of any pressure discontinuities, including test failures, that appear on the pressure recording charts; and,
- where the elevation differences in the section under test exceed 100 ft, a profile of the pipeline that shows the elevation and test sites over the entire length of the test section.

The records required by the Federal Pipeline Safety Regulations for pipelines transporting gas are similar to those required for pipelines used to transport liquids.

A typical test report is shown in Figure 8. If failures occur during the test, a report documenting the failure and the suspected reasons for the failure should be completed. A typical form is shown in Figure 9.

A daily operating log should be used to record activities associated with the test, operating status of the test equipment in use, engine rpm, and any unusual circumstances that occur during the test. A typical form is shown in Figure 10.

FIELD PRESSURE & TEST REPORT

Page ______ of ______

Contractor or Company: ______

Pipe Line Description: ______

Section Tested: From: ______ To: ______

Test Section No.: ______ Length: ______

Type and Size of Pipe: ______ O. D. x ______ W.T. x Grade ______ Manufacturer: ______

Pressure Unit No.: ______ Gallons per Stroke: ______ Fill Unit No.: ______

Pressure Unit Location: ______ Water Source: ______

Time and Date Test Started: ______ AM PM ______ Test Pressure (Maximum): ______ PSIG

Time and Date Test Ended: ______ AM PM ______ Pressure Volume Plot: Yes ______ No ______

Section Accepted ______ Section Leaking ______ Section Ruptured ______

DEADWEIGHT RECORDED READING (PSIG)

TIME	PRESSURE	TEMP DEG.F	TIME	PRESSURE	TEMP DEG. F	TIME	PRESSURE	TEMP DEG. F
Date:								
AM PM			AM PM			AM PM		

Strokes per 10 psi: ______ Remarks: ______

Milbar Supt. ______ Inspector ______

Witness: 1. ______ 2. ______

Gauges and Recorders Last Tested ______ Deadweight Serial No. ______

Figure 8. Typical test report. Courtesy Milbar Hydrotest.

PIPE LINE FAILURE REPORT

1. Company ____________________
2. Section tested ____________________
3. Time of failure __________ AM, PM. Date ____________________
4. Location of failure: ____________________
5. Pressure, psig, at point of failure ____________________
6. Description of failure: Leak __________, Break __________, Length of failure __________
7. If leak, fill in blanks: __________ gallons lost per hour, __________ pounds lost per hour.
8. Describe any peculiarities or defects on failed part, such as corrosion or evidence of prior damage, etc. ____________________
9. Possible cause of failure ____________________
10. Pipe size: __________ OD x __________ Wt., Grade __________ Mfg. by __________
11. Repairs: Pipe installed __________ OD x __________ Wt., Grade __________
    Mfg. by __________, Length of joint or pup __________
12. Date repaired __________ 19 ____ By __________
13. Remarks (Damage to property, persons injured, etc.) ____________________

Figure 9. Pipeline failure report. Courtesy Milbar Hydrotest.

DAILY OPERATING REPORT

COMPANY & CONTRACTOR ____________________ DATE __________

LOCATION (City) ____________________ CONTRACT NO. __________

PIPE LINE DESCRIPTION: ____________________

PERSONNEL: ____________________

EQUIPMENT: ____________________

ACTIVITIES: ____________________

COMPANY REP. __________ CONTRACTOR REP. __________ MILBAR SUPT. __________

Figure 10. Daily operation report. Courtesy Milbar Hydrotest.

Volume of water required to fill test section

$$V = L \times \left(3.1416 \times d^2/4\right) \times 12/231 \quad (1)$$

$$V = 0.0408 \times d^2 \times L$$

where

L = Length of test section, feet,
d = Inside diameter of pipe, inches,
V = Gallons required to fill @ 0 psig.

Volume of water required at test pressure

$$V_{tp} = V \times F_{wp} \times F_{pp} \times F_{pwt} \quad (2)$$

V_{tp} = Gallons contained in test section at test pressure P, and temperature T, °F,

F_{wp} = Factor to correct for the compressibility of water due to increasing pressure from 0 psig to test pressure P, psig,

F_{pp} = Factor to correct for volume change in pipeline due to pressure increase from 0 psig to test pressure P, psig,

F_{pwt} = Factor to correct for the change in water volume and pipe volume due to change in pipe and water temperature from base of 60 °F to pipe and water test temperature, T, °F.

$$F_{wp} = 1/\left[1 - \left(4.5 \times 10^{-5}\right) \times (P / 14.73)\right] \quad (3)$$

P = Test pressure, psig.

$$F_{pp} = 1 + \left[(D/t) \times \left(0.91P / 30 \times 10^{6}\right)\right] + \left[3.6 \times 10^{-6}(T - 60)\right] \quad (4)$$

D = Outside diameter of pipe, inches,
t = Pipe wall thickness, inches,
T = Pipe temperature, °F.

$$F_{pwt} = F_{pt}/F_{wt} \quad (5)$$

F_{pt} = Factor to correct for change in pipe volume due to thermal expansion of the pipe from base temperature of 60°F.

$$F_{pt} = 1 + \left[(T - 60) \times 18.2 \times 10^{-6}\right] \quad (6)$$

F_{wt} = Factor to correct for thermal change in specific water volume from 60°F to test water temperature. Refer to Table 1.

Table 1
Factor to Correct for the Thermal Change in the Specific Volume of Water from 60°F to the Test Water Temperature

Temp. (°F)	F_{wt}	Temp. (°F)	F_{wt}
35	0.9990777	70	1.0010364
36	0.9990590	71	1.0011696
37	0.9990458	72	1.0012832
38	0.9990375	73	1.0014229
39	0.9990340	74	1.0015420
40	0.9990357	75	1.0016883
41	0.9990421	76	1.0018130
42	0.9990536	77	1.0019657
43	0.9990694	78	1.0021222
44	0.9990903	79	1.0022552
45	0.9991150	80	1.0024178
46	0.9991451	81	1.0025561
47	0.9991791	82	1.0027251
48	0.9992168	83	1.0028684
49	0.9992599	84	1.0030435
50	0.9993061	85	1.0031919
51	0.9993615	86	1.0033730
52	0.9994112	87	1.0035573
53	0.9994715	88	1.0037133
54	0.9995322	89	1.0039034
55	0.9996046	90	1.0040642
56	0.9996683	91	1.0042601
57	0.9997488	92	1.0044357
58	0.9998191	93	1.0046271
59	0.9999074	94	1.0047972
60	1.0000000	95	1.0050043
61	1.0000803	96	1.0052142
62	1.0001805	97	1.0053915
63	1.0002671	98	1.0056067
64	1.0003746	99	1.0057884
65	1.0004674	100	1.0060090
66	1.0005823	101	1.0061949
67	1.0006811	102	1.0064207
68	1.0008031	103	1.0066108
69	1.0009290	104	1.0068417

Sample calculation

Pipe Size: 10.750 in. OD × 0.279 in. w.t. X-52
Length: 5 miles
Test pressure: 2430 psig
Temperature: 50°F

Use equation 1 to determine the initial line fill volume, V.

$$V = 0.0408 \times (10.75 - (2 \times 0.279))^{2} \times 5280 \times 5 = 111,888 \text{ gallons}$$

Volume required to achieve test pressure $= V_{tp} = V \times F_{wp} \times F_{pp} \times F_{pwt}$.

$$F_{wp} = 1/1 - \left[\left(4.5 \times 10^{-5}\right) \times (2430 / 14.73)\right] = 1.007479.$$

$$F_{pp} = 1 + \left[(10.75 / 0.279) \times \left(0.91 \times 2430 / \left(30 \times 10^{6}\right)\right)\right] + \left[3.6 \times 10^{-6} \times (50 - 60)\right] = 1.002804$$

$$F_{pwt} = F_{pt}/F_{wt}$$

$$F_{pt} = 1 + \left[(50 - 60) \times 18.2 \times 10^{-6}\right] = 0.999818$$

$$F_{wt} = 0.9993061 \text{ (from Table 1)}$$

$$F_{pwt} = 0.999818/0.9993061 = 1.000512$$

$$V_{tp} = 111,888 \times 1.007457 \times 1.002804 \times 1.000512$$
$$= 113,099 \text{ gals.}$$

Incremental volume required to reach test pressure, P.

$$= 113,099 - 111,888 = 1211 \text{ gals.}$$

After a period of time, the test pressure, P has decreased to 2418 psig and the temperature of the pipe and test water has decreased to 48°F. The calculation procedure previously described may be repeated to determine if the pressure decrease is attributable to the decrease in temperature.

$V = 111,888$ gals,
$P_1 = 2422$ psig,
$T_1 = 48°F$,
$F_{wp1} = 1.007454$,
$F_{pp1} = 1.002796$,
$F_{pt1} = 0.999781$,
$F_{wt1} = 0.999217$ (from Table1),
$F_{pwt1} = 0.999781/0.999217 = 1.000565$,

$$V_{pt1} = 111,888 \times 1.007454 \times 1.002796 \times 1.00565 \text{ (@2422 psig and 48°F)},$$

$V_{tpl} = 113,101$ (Volume at reduced temperature, represents volume that would be required to raise test procedure to original value at 48°F),

$V_{tp} = 113,099$ (Volume at initial test conditions).

To restore the pressure to the original test pressure, 2 gals of water could be added to the test section. If more than two gallons are required to restore the test pressure, there is the possibility that a leak may exist.

When making these calculations, care must be exercised to be sure that accurate temperature and pressure data is available. Where test sections are long, it is very likely that the temperature will not be uniform throughout the test section and thus affect the accuracy of the calculated results.

How to use charts for estimating the amount of pressure change for a change in test water temperature

Charts to estimate the amount of pressure change for a change in test water temperature are based on the following formula and are presented in four figures.

$$\Delta P = \frac{\beta - 2\alpha}{\frac{D \times (1 - \nu^2)}{E \times t + C}}$$

where

ΔP = psig change per °C,
β = Coefficient of expansion for water,
α = Coefficient of linear expansion for steel 1.116×10^6 per °C,
D = Pipe OD inches,
ν = Poisson's ratio 0.3,
E = Young's modulus for steel 30×10^6,
t = Pipe wall thickness, inches,
C = Compressibility factor for water, cu in/cu in/psig.

$$\beta \times 10^6 = -64.268 + (17.0105 \times T) - \left(0.20369 \times T^2\right) + \left(0.0016048 \times T^2\right)$$

T = water temperature, °C.

Example

For pipe 18 in OD × 0.375 in wt, 1800 psig test pressure, and water temperature 70°F at beginning of test, estimate the pressure change for water temperature at a later time of 66°F.

Calculate D/t:

$$D/t = 18/0.375 = 48$$

where

D = Pipe OD, in,
t = Pipe wall thickness, in.

Calculate average temperature:

$$\text{Average temperature} = (70 + 66)/2 = 68°F$$

Use the chart in Figure 3. Enter at 68 °F and at the intersection with D/t line representing 48, read 23 psig/°F.

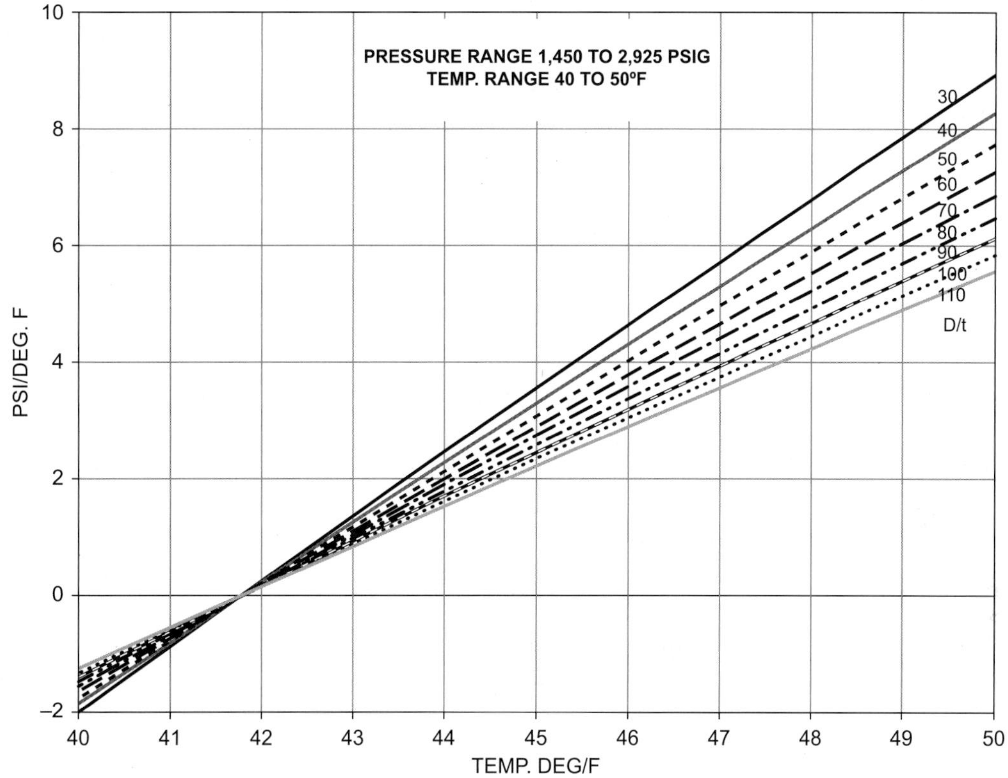

Figure 1. Pressure change for a change in temperature between 40 to 50 °F.

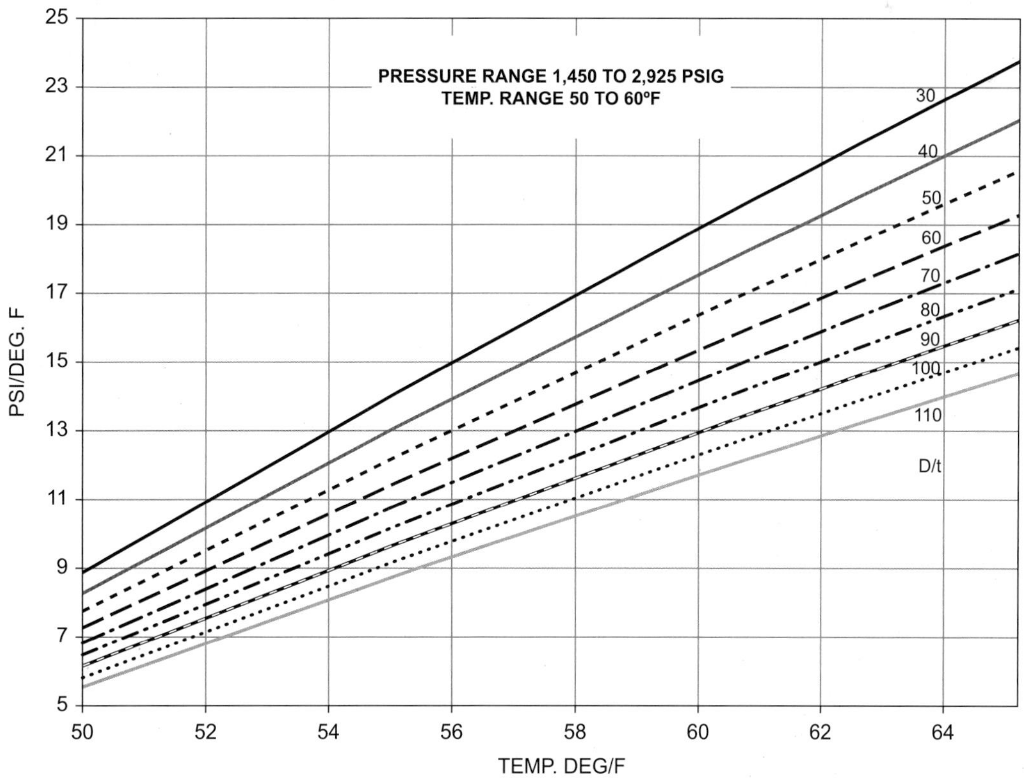

Figure 2. Pressure change for a change in temperature between 50 to 60 °F.

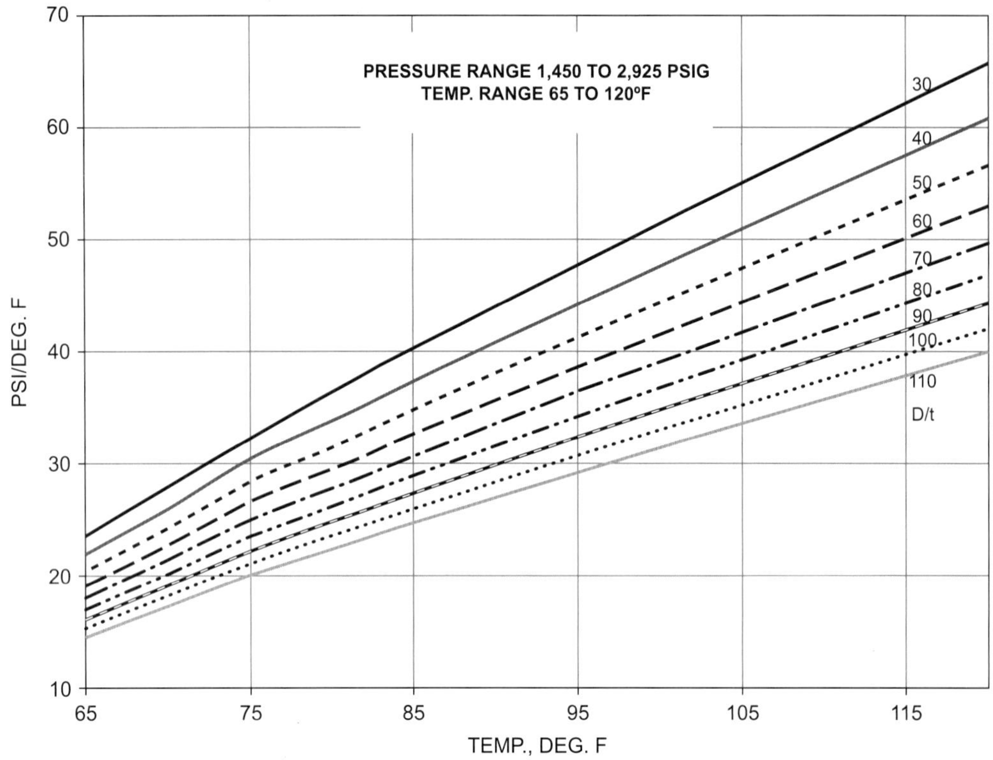

Figure 3. Pressure change for a change in temperature between 65 to 120 °F.

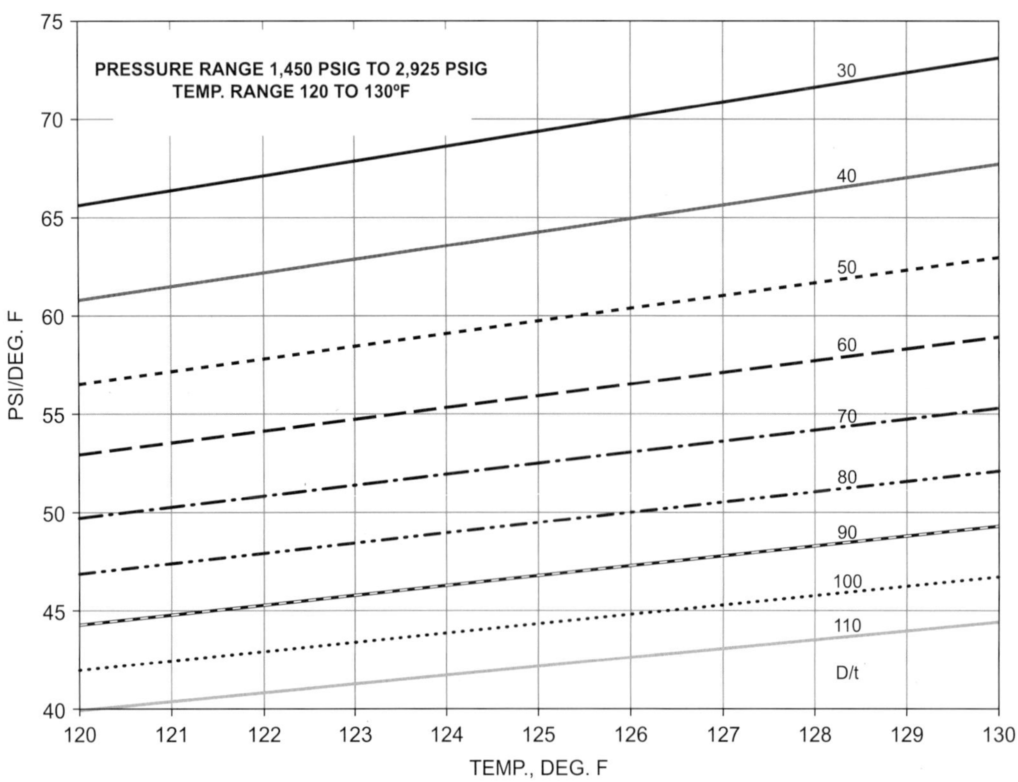

Figure 4. Pressure change for a change in temperature between 120 to 130 °F.

Compressibility factor for water

Pressure range 1450 psig to 2925 psig

Temperature (°C)	Compressibility factor cu in/cu in/psig
0	3.35×10^{-6}
10	3.14×10^{-6}
20	3.01×10^{-6}
50	2.89×10^{-6}

From *PIPECALC 2.0: PRACTICAL Pipeline Hydraulics*, Gulf Publishing Co., Houston, Texas.

Hydrostatic test records

At a minimum, the following records should be made for a hydrostatic test.

- Calculations supporting basis for hydrostatic test. This should take into account the high and low point elevations in the test section.
- Pressure recording charts.
- Test instrument calibration. This should include the pressure recorder and temperature recorder. Record the serial numbers of the instruments used for the test.
- Date and time of the test.
- Temperature of the test medium.
- Test medium used.
- Explanation of any pressure or temperature discontinuities.
- Explanation of any failures that occurred during the test.
- Weather conditions during test.
- Description of pipe in the test section.
- Any permits required for the test.
- Permits for discharge of test water.
- Any other records required by agencies having jurisdiction over the testing/operation of pipelines.
- All records should be signed and dated by the engineer in charge of the test.

Pressure testing and recordkeeping requirements – reconciling historical practices with new requirements

Michael J Rosenfeld, Kiefner & Associates Inc./Applus-RTD, Worthington, Ohio and **Rick W Gailing,** Southern California Gas Co, Los Angeles, CA

Recordkeeping requirements prior to 1955

Recordkeeping requirements specified in engineering standards for gas pipeline prior to 1955 were few and focused on welding. The 1935 B31.1 standard required employers of welders to maintain records of their welding operators showing dates of employment, results of welding tests, and their assigned identifying mark. Welders were required to stamp their identifying mark adjacent to welds they made on pipe. The 1942 B31.1 standard, Appendix I, Part I required that records of welding procedure qualification testing and copies of the record for each qualified welder were to be kept by the manufacturer or contractor. No record retention period was specified, and no other recordkeeping requirements were expressed.

No provisions or requirements for recordkeeping of any kind dealing with welding or installation were specified in the

1951 B31.1. Similarly, none were given in the 1952 B31.1, Section 8 in its entirety.

It would be reasonable to expect that a variety of documents related to the design and construction of a pipeline facility be retained long-term. However, retention of technical documents was not addressed by the engineering standards of the day. It was generally thought that a copy of the specifications under which the pipeline was built (and supplemented by commercial documents, e.g. contracts and purchase orders) would generally be adequate to provide evidence of the work that was done[1].

Recordkeeping requirements 1955 to 1961

The 1955 B31.1.8 Chapter II "Welding" required that records of welding procedure qualification tests be retained for as long as the welding procedure is in use. Further, the pipeline operator or contractor (presumably whoever employed the welders) was required, during construction, to maintain a record of the welders qualified, their dates of employment, and test results.

Chapter IV "Design, Installation, and Testing" required maintaining records showing the type of fluid used for pressure testing and the test pressure of pipelines that operate at a hoop stress of 30% or more of SMYS. The specified retention period was the useful life of the facility. This recordkeeping requirement was not stated under general testing provisions applicable to all pipelines, nor under subsequent paragraphs that presented separate pressure test requirements for pipe operating at less than 30% of SMYS but more than 100 psig, leak test requirements for pipe operating at 100 psig or more, or leak test requirements for pipe operating at less than 100 psig, respectively. Thus, an operator might reasonably not have retained records for tests performed in accordance with those paragraphs.

The 1955 edition was the first B31 piping standard to extend its scope beyond design, construction, and commissioning of the piping system to include operation and maintenance. Accordingly, additional recordkeeping language was introduced in Chapter V, "Operating and Maintenance Procedures". "Basic requirements" therein stated that "each operating company having gas transmission or distribution facilities ... shall: (a) Have a plan covering operating and maintenance procedures...(c) Keep records necessary to administer the plan properly." Further, records "should" be made of pipeline inspections for external or internal corrosion, listing several items of potential interest, and records "should" be made covering leaks and repairs. In addition, leakage survey records, line patrol records and other records relating to routine or unusual inspections "should" be kept on file as long as the section of line remains in service. The operator was required to have plans for inspecting pipe-type and bottle-type gas holders, and to keep records detailing the inspection and test work done and the results.

The terms "shall" and "should" were used throughout B31.1.8 and its sequels. "Shall" is understood to mean an action is required, while "should" is understood to mean an action is recommended but not required. Records adequate to effectively execute the pipeline operation and maintenance were required, but specific records were merely recommended and what was actually required was left to the operator. The possibility was not precluded that data different than or in addition to what the standard said "should" be recorded might be necessary in order to fulfill the requirement to "keep records necessary to administer" the operation and maintenance plan. Note also that the Code has historically given leave to not follow specific requirements where the operator can show by experience, testing, or analysis that an alternative is safe and reliable[2]. An operator could conceivably set forth a position that maintaining some kinds of records is unnecessary based on experience.

Recordkeeping requirements 1961 to 1970

The 1958, 1963, and 1968 editions of ASME B31.8 did not differ from the 1955 edition with respect to recordkeeping. The 1968 edition included certain enhancements such as the weld inspection requirements but without the accompanying weld inspection recordkeeping requirement. On the other hand, the corrosion inspection and leak investigation recordkeeping provisions became required, not recommended.

Recordkeeping requirements post-1970

Complete federal safety standards for gas pipelines were introduced in 1970. Although some technical content was based on the 1968 edition of B31.8, the regulatory provisions went well beyond B31.8 in terms of inspections and recordkeeping. All provisions were required, not merely recommended ("shall", not "should"), such as:

- Subpart E – Welding: §192.243(f), where nondestructive testing (i.e., radiography) of welds is performed, a record is required showing the number of girth welds made, the number tested, the number rejected, and their disposition by location (e.g., milepost), for the life of the pipeline. Also §192.225(c), requires a record of the details of each qualification of a welding procedure, to be retained for as long as the procedure is used.
- Subpart J – Test Requirements: §192.517, a record is required of each test performed on pipelines operating at a hoop stress of 30% or more of SMYS or above 100 psig but below 30% of SMYS. The record must indicate the following 7 items: (1) the names of the operator, the responsible employee, and the test company (if any); (2) the test medium used; (3) the test pressure; (4) the test duration; (5) pressure readings; (6) elevation variations if they are significant; and (7) leaks or failures. Such records must be retained for the useful life of the facility.
- Subpart K – Uprating: §192.553(b), a record is required of each investigation (e.g., review of the design, and operating and maintenance history), work done, and each pressure test in connection with the uprate. The record must be retained for the life of the uprated segment.

- Subpart L – Operations: §192.619(a) sets forth criteria for establishing the MAOP, as the lowest of the design pressure of the weakest components or pipe based on specified attributes, the pressure obtained by dividing the post-construction test pressure by a specified factor, the highest actual operating pressure during 5 years preceding July 1, 1970, for furnace butt-welded pipe a pressure equal to 60% of the mill test pressure, for other pipe a pressure equal to 85% of the highest test pressure the pipe experienced in the field or pipe mill, or the maximum safe pressure determined in consideration of the condition and operating history of the pipeline.
- Subpart M – Maintenance: §192.709, a record is required of each leak discovered, repair made, line break, leak survey, line patrol, and inspection of transmission pipelines for as long as the line remains in service. Records have to be retained at least until the next round of inspections (e.g., 5 years).
- Numerous other activities (sampling of odorant, valve maintenance, vault maintenance, distribution leakage surveys, and others) must occur at specified periodic intervals. No recordkeeping was specified in connection with those activities.

The 1970 issuance of Part 192 added Subpart I on corrosion control, which required installation and criteria for the cathodic protection (CP) of buried steel pipelines, periodic monitoring of the effectiveness of the CP system, monitoring of internal corrosion, and monitoring of atmospheric corrosion. Recordkeeping requirements as of July 31, 1972 are discussed below.

- Subpart I – Corrosion Control: §192.491(a), each operator was required to maintain records or maps showing the location of cathodically protected pipe, CP facilities (e.g., rectifiers or anodes), and other structures bonded to the pipe. Also §192.491(b), each record or map from (a) plus records of each test or inspection of the CP system in sufficient detail to show adequacy of corrosion control were required to be retained as long as the facility is in service.

Important and extensive new recordkeeping requirements were put in place to support operator qualification (OQ) in 1999, integrity management planning (IMP) for transmission pipelines in high consequence areas (HCAs) in 2004, and distribution system IMP in 2009, as discussed below.

- Subpart N – Qualification of Pipeline Personnel: §192.807, requires the operator to maintain qualifications of personnel performing covered tasks. The qualification records must include identification of the individuals, the covered tasks each individual is qualified for, the dates of qualification, and the qualification method. The records must be maintained while the person is performing the covered task and for 5 years after.
- Subpart O – Gas Transmission Pipeline Integrity Management: §192.947, requires the operator to maintain records demonstrating compliance to Subpart O. The required items listed are (a) a written integrity management plan, (b) documents supporting the threat identification and risk assessment, (c) a written baseline assessment plan (BAP), (d) documents supporting each decision, analysis or process of each element of the BAP and IMP, (e) personnel training program and records, (f) prioritized assessment mitigation schedule, (g) documents supporting the Direct Assessment (DA) plan, (h) documents supporting the Confirmatory Direct Assessment (CDA) plan, and (i) verification of notifications made to OPS or any state regulator as required by Subpart O.
- Subpart P – Distribution Pipeline Integrity Management: §192.1011, requires the operator to maintain records that demonstrate compliance to the requirements of Subpart P, for at least 10 years. The records must include any superseded copies of the IMP.

Clearly, the regulations are punctuated by several major new additions to recordkeeping requirements, and today's recordkeeping requirements cannot be presumed to have applied at all times prior to the introduction of new requirements. A timeline of introduction of major recordkeeping requirements is shown in Figure 1.

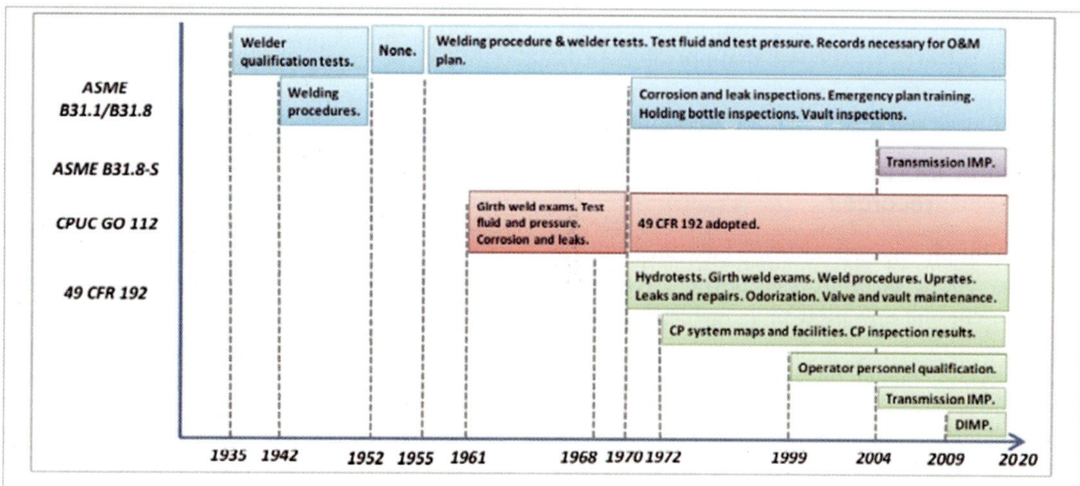

Figure 1. Timeline of significant recordkeeping requirements

Unbroken chain of documentation not the rule

The practical significance of the "grandfather" rule was that it was not necessary for an existing pipeline already in service to have been pressure tested to the minimum specified ratio of the MAOP. In fact, clause §192.619 offered four possible alternatives for establishing the MAOP:

- §192.619(a)(1) recognized the design pressure of the weakest component in accordance with Subparts C and D. In this case the MAOP would be based on manufacturer's component pressure ratings or engineering calculations using specified material strength and wall thickness.
- §192.619(a)(3) recognized the highest pressure to which the pipeline had been subjected during the 5 years preceding July 1, 1970.
- §192.619(a)(4) recognized 85% of the highest test pressure to which the pipe had been subjected, either in the pipe mill or in the field. If no field test was documented, the mill test would govern. The operator could determine the pipe mill test pressure from the pipe product specification.
- §192.619(a)(5) allowed the operator to determine the maximum safe pressure considering the history of the segment, known corrosion, and actual operating pressure. This might be used, for example, with an uncoated pipeline that had experienced general wall thinning due to corrosion. (It is notable that this language existed prior to the use of in-line inspection for conducting integrity assessment, so an operator might not have had complete information about the extent of corrosion.)

None of the above methods for establishing the MAOP necessarily required documentation of a prior post-installation pressure test. In fact, the method given in clause (a)(3) requires knowing no information about the specified grade or wall thickness of the pipe. That these alternative methods of establishing MAOP were allowed indicates that OPS accepted that records of testing or of pipe physical attributes were not always available. In particular, an operator was not prohibited from using clause (a)(3) even if a test had been performed and test records had been lost for some reason. These alternatives have been in Part 192 from 1970 to today, so OPS/PHMSA has since 1970 accepted that not all records need necessarily be present, or if present, need necessarily be complete or represent an unbroken chain of traceability.

Can an operator be "prudent" while missing records?

It is not uncommon for pipeline operators to have incomplete or inaccurate data about attributes of portions of their pipeline systems, including specified pipe material grades, specified nominal wall thicknesses, seam types, pipe manufacturers, coating types, flange or valve pressure classes, installation dates, construction specifications, welding procedures, pressure tests, corrosion control, and historic operating pressures.

The likelihood of records being incomplete increases with the age of the system, particularly with systems built prior to 1970 when the more-extensive records requirements of Part 192 came into effect. Nationwide, 37% of natural gas transmission pipelines now in service was installed before 1960, and 61% was installed before 1970[3], thus a sizable proportion of existing pipelines was installed at a time when only minimal provisions for recordkeeping were found in standards and regulations. While the likelihood of gaps in the data increases with age, compromised data exist in systems built in many eras, including those built after 1970. Whether a lack of certain documents constitutes violation of regulations or indicates operator imprudence has become central to whether shareholders or rate payers pay for costly retesting or replacement of pipe.

There are many innocuous causes for loss of records including: an individual not recognizing the importance of a document or collection of documents, change of facility ownership, loss event (fire, flood), clerical mishandling, or misplacement in offsite storage, to list a few. Certainly back-up copies in one form or another can offset the loss of originals, but consider that photocopy technology was not widely available until the mid-1960s, perhaps after some original documents were already lost, and the back-up process is not without risk either[4].

Loss of useful records for any reason is not desirable, but past failure to preserve records does not necessarily imply operator imprudence or irresponsibility, neither does operating a pipeline while gaps exist in some records. Not all records are important to safely operating a pipeline day-to-day once the primary purpose of the record has been satisfied; prudence is exercised in making good choices with the information available. Consider that a pressure test of a pipeline following construction had been performed and that all stakeholders (owner, state or federal regulator, lender, insurer) were satisfied that the pipeline had been properly designed, constructed, and commissioned. The MAOP is entered into a ledger, a memo stating the MAOP is issued to control room operator, pressure control set points are confirmed, and operating procedures are updated. Consider next that the actual pressure test records become lost some years later. How does the loss of that record affect any of the numerous activities a prudent operator is obliged to carry out day after day, such as: controlling pressure within established set points, marking the line for excavators, conducting damage prevention and public education programs, periodically testing valves, performing leakage surveys, repairing leaks, conducting line surveillance, maintaining cathodic protection, or training operations personnel, to name a few? The answer is that it does not. Once the MAOP has been correctly established using any one of the allowed methods, those records have little bearing on day to-day operation of the line.

The foregoing discussion is not meant to suggest that all records losses or data gaps are inconsequential. In fact, accurate and readily available data of some kinds are essential for safe and efficient operation. The authors support the industry's efforts to respond to the San Bruno incident by evaluating the accuracy of its records. There is value in good records, however, the authors also believe that more has been made of the role of recordkeeping as a cause of that incident than may have been warranted. The conclusion that the incorrect identification of the pup that originated the failure as "30-inch seamless" in the pipe inventory database would have led any operator to enact a series of decisions culminating in the removal of the pipe decades after its installation stretches credibility. The industry, including regulators and other stakeholders, should contemplate whether any amount of retrospective records analysis can offer complete protection against "unknown unknowns", particularly where they originated many decades ago[5]. That argues in favor of the CPUC Decision to require replacement or retesting where adequate test records are lacking. It also leans toward ratepayers carrying the financial burden of achieving the added assurance provided by pipe replacement or retesting unless it can be shown that an operator's behavior went well beyond the lack of historic documentation.

Gaps in data that validate the MAOP severely limit an operator's options for addressing a change in location class, pressure uprate, or request for regulatory waiver or special permit, which is as it should be. Data quality also has implications for integrity management. Certain elements of an IMP, notably the integrity threat identification and risk assessment tasks, are facilitated by having reasonably complete and accurate historical and technical data. ASME B31.8S recognizes that data important or useful to these tasks may be missing: §4.2.1 "Data Requirements: Prescriptive Integrity Management Programs" states that if listed data elements relevant to an integrity threat are not available, the integrity threat must be assumed to apply; §4.4 "Data Collection, Review, and Analysis" states that unavailability of data cannot be used to justify excluding an integrity threat; §5.9 "Data Collection for Risk Assessment" advises that if significant data are not available, the risk model may need to be modified based on an analysis of the impact of the data being unavailable; Appendix A, the paragraph "Gathering, reviewing, and Integrating Data" states that where the operator is missing data, conservative assumptions shall be used with the risk assessment or the segment shall be prioritized higher for each integrity threat listed. Part 192, Subpart O, §192.917 requires the operator to perform integrity threat identification and risk assessment in accordance with B31.8S, Sections 4 and 5, respectively, which incorporate the above provisions concerning how to compensate for unavailable data. By referencing these sections, the regulations clearly contemplate that data important to an IMP may be unavailable.

"Traceable, verifiable, and complete" represents new requirements on recordkeeping

It has been suggested in the course of public debate that the criteria for document reliability, being "traceable, verifiable, and complete" do not represent new standards for the quality of natural gas pipeline records. While the attributes of "traceable, verifiable, and complete" are certainly desirable, and reasonably expected in modern times, they are not standardized thresholds for data quality for pipelines of all eras, and have no basis in regulation. They represent new documentary criteria.

The words "traceable, verifiable, and complete" appear nowhere in any issuances of B31.8 or 49 CFR Part 192 prior to SR P-10-2. The words, as applied to documents related to the design, construction, or operation of a natural gas pipeline, did not originate with the federal pipeline regulatory agency or any other state pipeline regulatory agency. The terminology "traceable, verifiable, and complete", as used in connection with gas pipelines, originated with the NTSB's Safety Recommendation (SR) P-10-2 issued to Pacific Gas and Electric (PG&E)[6]. The NTSB recommendation that records "should" (not "shall" or "must") be traceable, verifiable, and complete applied to "all as-built drawings, alignment sheets, and specifications, and all design, construction, inspection, testing, maintenance, and other related records…relating to pipeline system components, such as pipe segments, valves, fittings, and weld seams for Pacific Gas and Electric Company natural gas transmission lines in class 3 and class 4 locations and class 1 and class 2 high consequence areas that have not had a maximum allowable operating pressure established through prior hydrostatic testing". The NTSB did not extend the recommendation to all pipeline facilities in all locations, nor to any facilities anywhere that have in fact been pressure tested.

In the gas pipeline context, the terms originated with NTSB as stated above, though NTSB may have assimilated the terms from applications outside the pipeline industry. The NTSB is not a regulatory agency, but rather an independent agency of the United States government that has no responsibility for writing regulation and no powers of regulatory enforcement. Based on accident investigations that it is authorized to perform, the Board offers opinions and recommendations that may or may not be observed[7]. The NTSB recommendations almost certainly were not made in consultation with PHMSA[8].

Shortly after issuance of SR P-10-2, PHMSA issued ADB-2011-01 advising operators that records they rely on for establishing the MAOP "must be reliable" and that the records "shall be traceable, verifiable, and complete"[9]. It took PHMSA another 16 months to develop guidance to the industry as to how to interpret the terminology in order to satisfy the new requirements[10]. In the meantime, the industry attempted to articulate what was necessary to meet these requirements by issuing white papers[11] and developing individual company processes in the hope of meeting the regulator's unspoken criteria. No guidance was found in the "GPTC Guide"[12], a widely used reference guide to the interpretation of and compliance to Part 192. The fact that no guidance could be found in any common external publication is consistent with the position that "traceable, verifiable, and complete" represented new criteria.

From ADB-2012-06, "traceable" records are tied to original documents. It should not be surprising that some documents that predated 1961 or even 1970 might not have been retained if there was no regulatory requirement to retain them. It is also possible, with the passage of time, for original documents to become lost through any of the possible loss mechanisms described above. In these or similar circumstances, it becomes impossible to meet the "traceable" test. While not optimal, losses of traceability are not uncommon.

- "Verifiable" records are those in which data is confirmed by other separate documentation. ADB-2012-06 appears to require that any record used to establish the MAOP must be confirmed by another record. Nowhere in the historical or current regulatory language reviewed above does agreement between multiple data sources appear as a requirement.
- "Complete" records are finalized by a signature or date. ADB-2012-06 gives, as an example: "a complete pressure testing record should identify a specific segment of pipe, who conducted the test, the duration of the test, the test medium, temperatures, accurate pressure readings, and elevation information as applicable." This example lists two items that are not specified in §192.517, namely the specific segment of pipe, and temperatures. Thus meeting PHMSA's requirements for recordkeeping since 1970 actually does not meet the test for completeness in ADB-2012-06, so clearly a new requirement has been imposed.

The language of SR P-10-2 is clearly made in reference to "grandfathered pipelines" that are now in Class 3 or 4 areas. As explained in Part D above, gaps in documentation could well occur in connection with many "grandfathered pipelines". Therefore, the notion that the criteria of SR P-10-2 represent thresholds of data reliability that have always existed in regulations is inconsistent with established fact.

What does this mean for policy?

The authors do not dispute the need to modernize the nation's pipeline infrastructure, verify the integrity of "grandfathered" pipelines as well as any other category of pipeline, and enhance data as well as is practical in order to manage risk as well as is practical. We believe that pipeline operators should be accountable for failures to meet regulatory obligations. However, we further believe that regulators should not give in to the temptation (perhaps driven by public pressure) to penalize operators retrospectively for practices that previously were not considered deficient as judged by standards of the era. Past practices must be gauged against an accurate and reasonable interpretation of historic standards, regulations, and accepted practices. Moreover, proceedings to establish financial penalties for past failure to comply with regulations should be separated from ratemaking proceedings designed to regulate future behavior[13]. When weighing financial penalties, consideration should also be given to the intensity of the regulator's past focus on regulatory provisions deemed in retrospect to have been contravened. (In other words, if the regulatory violation was a failure to comply with recordkeeping requirements, did the regulator recognize an infraction in the past? If so, what was done about it? If not, has a reasonable "statute of limitations" rendered it moot?) Finally, customers should pay rates that fully reflect the cost of providing the goods or services[14]. Rates should therefore reflect the cost of complying with new regulatory requirements. This includes the cost of retesting or replacing pipe not thought to meet modern standards of integrity, where those activities are necessitated by new regulations.

Summary

This paper reviewed the history of industry standards and regulatory requirements in the areas of recordkeeping. Recordkeeping requirements began with welding quality control. In 1955, hydrostatic test records and those necessary for executing the operator's O&M plan were required. Of test records, only test fluid and test pressure were required, and then that requirement only pertained to pipelines operating at a hoop stress of 30% or more of SMYS. It was thought at that time that engineering specifications and commercial documents were adequate to demonstrate practices. Additional recordkeeping requirements were imposed by federal safety regulations in 1970, including for hydrostatic pressure testing.

When federal regulations were issued in 1970, several options were available for establishing the MAOP of existing ("grandfathered") pipelines that did not rely on a documented post-construction pressure test. This establishes that for the past 50 years, the regulators have accepted that documents supporting the MAOP could be incomplete. In that context, the tests set forth in ADB-2012-06 represent new standards that should not be used to judge whether an operator has complied with prior standards for that purpose as part of the ratemaking process.

References

1. F.A. Hough, "The New Gas Transmission and Distribution Piping Code (ASA B31 Section 8)", Gas Magazine, Series in 8 Parts, January through September 1955.
2. This includes but is not limited to: materials and equipment selection, fittings and components design, above

ground piping design, longitudinal stresses in buried pipelines, valve spacing, and cathodic protection criteria.
3. http://www.phmsa.dot.gov/pipeline/library/data-stats.
4. An anecdote reported to the authors was an occasion where a clerical worker, instructed to photocopy hydrostatic test records, first separated the pressure charts from the test report forms which had been stapled together into separate piles, irreversibly breaking the link between pressure records and test segments. In another case, documents sent by an operator to a third-party long-term storage facility were misplaced by the storage contractor.
5. Certainly intensive QA auditing efforts soon after construction can potentially reveal incorrect actions or wrongdoing.
6. NTSB, Safety Recommendation (SR) P-10-2 (January 3, 2011).
7. The NTSB's reports are not admissible in court, 49 USC 1154(b): "No part of a report of the Board, related to an accident or an investigation of an accident, may be admitted into evidence or used in a civil action for damages resulting from a matter mentioned in the report"; although its investigators' factual reports are, 49 USC 835.
8. 49 USC 1131: "The Board shall provide for appropriate participation by other departments, agencies, or instrumentalities in the investigation. However, those departments, agencies, or instrumentalities may not participate in the decision of the Board about the probable cause of the accident."
9. PHMSA Advisory Bulletin, "Establishing Maximum Allowable Operating Pressure or Maximum Operating Pressure Using Record Evidence, and Integrity Management Risk Identification, Assessment, Prevention, and Mitigation", ADB-2011-01.
10. PHMSA Advisory Bulletin, "Pipeline Safety: Verification of Records", ADB-2012-06.
11. American Gas Association, Industry Guidance on Records Review for Re-affirming Transmission Pipeline MAOPs (October 2011).
12. GPTC Guide for Gas Transmission and Distribution Piping Systems", ANSI Z380.1, various years.
13. CPUC Rulemaking 11-02-019, Exhibit 21, Pacific Gas and Electric, Pipeline Safety Enhancement Plan (Implementation Plan), prepared testimony of Susan F. Tierney, "Principles to Align Safety and Regulatory Ratemaking Policy" (February 28, 2012).
14. Tierney.

6: Pigging, Cleaning & Drying

A brief history of pipeline pigging 218
Why pig a pipeline? 219
Cleaning and construction utility pigs 226
Pipeline dewatering, cleaning, and drying 232
Moisture content of air 236
Commissioning ethylene pipelines 238
Vacuum drying 241
Pig trap design 244

A brief history of pipeline pigging

J. Frank Gray, Pipeline Dehydrators, Inc.

Pigs are devices which are inserted into, and travel throughout the length of a pipeline, driven by the product flow. They fall into two categories: 'utility' pigs - which perform a function such as cleaning, separating or dewatering the pipeline, and in-line inspection, or more simply, 'ILI tools' (sometimes referred to as 'intelligent pigs' or 'smart pigs') - which provide information on the condition of the line as well as the extent and location of any problems.

The history of pipeline pigging is interesting and probably lacking in full truth; however, the following is usually accepted as an answer to many questions about the history of pipeline pigs. The first pigging operation took place around the year of 1870, a few years after Colonel Drake discovered oil in Titusville, Pennsylvania.

Before pipelines were used for transporting it, the oil was trucked to the refinery by horse-drawn tank wagons. This proved to be very difficult during winter months because of heavy snows and frozen wagon tracks, and in wet weather when wagons would sink in the mud. To improve upon this method of transportation, a pipeline was constructed, the material of which is not recorded, but each length of pipe was almost certainly joined by the bell-and-spigot method that we see today in plastic pipe. After transporting crude oil for a year or two through this pipeline, the flows began to decrease, and the pumping pressure increased, indicating that there were deposits building up on the inside walls of the pipe. Many things were tried to remove the paraffin deposits, but nothing worked effectively for any period of time. Eventually the idea of pumping something through the pipeline was considered. It has been suggested that a bundle of rags tied in a ball was used, and with positive results. Later, bundles of leather were used in place of the rags. Leather will swell when wet, so it created a tight seal going through the pipeline.

Another story concerning the history of pipeline pigging takes place around 1904, when a 4-in steel gas line was installed in Montana. During the construction, there was a rock slide which covered the line, and it was not known whether there was a buckle or a flat spot, so a rubber ball was made to pump through the pipeline to find out. Upon exiting, it was determined that if there was any damage to the pipe, it was not sufficient to restrict the flow. It was also discovered that a lot of debris was left in the pipeline such as rocks, dirt, sand, tools, etc. Rubber balls (spheres) have been used ever since as displacement pigs and for meter proving.

The question most people ask is "How did a pig get its name?" One story goes that two pipeliners were told to go out to the pipeline and listen for the pipe cleaner to go by. This pipe cleaner consisted of a steel pipe or mandrel body with flanges welded on both ends. Discs made of leather sheets were stacked together to provide thickness, and attached to the flanges. As the pipe cleaner travelled down the line pushing out debris, it made a squealing, scraping, noise. As the cleaner went by, one pipeliner made a comment to the other about "hearing that pig squeal". These stories are generally accepted as possibly being true.

Whether they are true or not, they at least indicate that pigs have been around for a long time. For more than half of a century pigs consisted of steel bodies and rubber, leather, or urethane cups or discs. They were equipped with wire brushes, scrapers, knife blades, and other devices for ploughing. Until 1960, most pipe cleaning was limited to the oil and gas industry. Then the foam bullet-shaped pig was developed; it was referred to as the Polly Pig because it was made of polyurethane foam. Although the oil and gas industry remains the largest user of the foam pig, many new industries such as municipal water and sewer, processing industries, petrochemical, mining, and other industries are now using pigs in their pipelines, realizing gains such as energy savings, increased flows, decreased pumping pressures, cleaner product, and salvaged product.

The lessons learned from standard pigging operations to clean, dewater, fill, and displace product from pipelines, and the pressures, speeds, and problems incurred, have contributed greatly to the development of instrumented pigs. These were introduced in the late 1960s, and development is continuing to the present day.

The development and improvement of all types of pipeline pigs has been a continuing process, and this trend is set to continue. There are now literally hundreds of different types, some with specific or limited use and others which are standard products. Apart from the main functions of sweeping, drying, wiping, cleaning, scraping, inspection, and integrity monitoring, 'semi-intelligent' pigs now perform additional functions such as alerting and initiating actions involving pumps and valves, and making an input in computerized operations, sometimes by through pipe-wall communications.

Excerpted from An Introduction to Pipeline Pigging, 13th edition, 2020. Pigging Products & Services Association and Jim Cordell/Hershel Vanzant. Suffolk United Kingdom, http://www.ppsa-online.com. Reprinted with permission.

Why pig a pipeline?

Excerpted from An Introduction to Pipeline Pigging, 13th edition, 2020. Pigging Products & Services Association and Jim Cordell/Hershel Vanzant.

It is generally agreed that a pipeline should be pigged, but the reasons for doing so are not always fully appreciated. This may result in improper pigging programs and/or the use of the wrong type of pig, and this could have an adverse effect on the pipeline's operating and maintenance costs.

Safeguarding assets and optimizing efficiency

Many operators only resort to pigging when a problem becomes apparent and they have no other choice. Often, this is too late. The damage has been done and a multi-million dollar investment is put at risk.

Use of the correct type of pig and a proper pigging program will help to maintain the integrity and optimum efficiency of the pipeline, safeguarding both the environment and the assets of the pipeline owner.

A pipeline is arguably the most efficient method yet devised for transporting fluids - whether gases or liquids. Its relative efficiency, though, depends upon two fundamental requirements:

1. it must operate continuously; and
2. the required throughput must be obtained with the least capital investment and the lowest operating costs.

Most pipelines are designed and constructed on these principles. Even so, the capital and operating costs are often very high, particularly for large-diameter, long-distance, transport pipelines.

While many pipelines are constructed for purely financial reasons (they are simply the cheapest means of transporting a certain volume of fluid between certain points), others are built for strategic (often military), political, or international trading purposes.

Whatever the reason, it is certain that the decision to build a major pipeline has been taken at a very high (often the highest) level, and that the decision is based on the two fundamental considerations mentioned above.

What is not always appreciated is that such decisions presuppose that these two fundamental requirements can be maintained during the life of the pipeline. It is not sufficient to simply design and construct the pipeline on the basis of continuous operation and optimum efficiency: it must be kept that way.

Pigs play a major role in both obtaining and maintaining the 'two fundamentals':

1. they help to ensure continuous operation by:
   - removing any substance which might damage the pipeline process systems;
   - helping to prevent the formation of corrosion cells;
   - providing timely information of any developing problems;
   - providing data on any perceived problems to enable informed decisions to be made;
   - providing an alternative to shutting down for statutory periodic testing.
2. they help to ensure maximum efficiency by:
   - removing any debris or foreign matter in the line;
   - removing any deposits, either liquid or solid, which might otherwise build up and restrict the flow;
   - by monitoring the operating and/or physical conditions of the line.

In short, pipeline pigs help to ensure that the pipeline is constructed properly and that it stays that way.

However, it must be remembered that the level of the contribution made to both the efficiency of the pipeline and to its protection will depend upon the pigging program and the effectiveness of the pigs which are used. Under any given set of circumstances there will be significant variations in the performance of different types of pig and even between different makes of the same type. However, the use of virtually any pig will be better than none at all.

Some elementary calculations will prove the point. Consider a simple, short, straight section of a 12-in (300-mm) water line. If the inside diameter is reduced by only 5% (15mm) by a smooth deposit, the loss of throughput at a given pressure would be over 10%. To bring the throughput back to its original level would require the pressure to be increased by almost 30%. However if, as is more likely, the deposit was uneven, the resulting turbulent flow may cause an effective reduction of 15%. In this case, throughput could be reduced by some 35%, while the increase in the pressure to overcome these losses may need to be increased by more than 140%. Admittedly, this example is based on theoretical calculations, but similarly-dramatic effects have been noted in practice.

Perhaps one of the most-common and yet costly mistakes is to rely totally on corrosion monitoring and automatic corrosion control systems to decide when to run a pig. The reliability of any monitoring system depends upon where the probes, coupons, and/or sampling points are located - and this is difficult, if not impossible, to achieve on subsea lines or to determine accurately on land lines, particularly on those which have any significant gradients.

Once corrosion, particularly pitting or channel corrosion, has occurred, it is unlikely that a pig will be able to remove the water which will accumulate in these recesses. It is therefore advisable to include batch inhibition (running a slug of inhibitor between two pigs) as part of any corrosion-control program.

Government regulations

The US Department of Transportation, Pipeline and Hazardous Materials Administration requires gas transmission, hazardous liquid and carbon dioxide pipeline operators, under their pipeline integrity management plans, to identify those regulated pipelines located in high consequence areas and periodically perform internal inspections by pigging or equivalent methods. The DOT regulations also require new pipelines, replacement pipe and components to accommodate pigs. There are few international regulations requiring pigging, however some countries licenses to construct and operate pipelines are dependent upon them being piggable.

Practical requirements

The earliest pigs were used simply to remove any large deposits of wax or dirt in order to maintain a flow through the pipeline. Today, pigging is required during each phase in the life of a pipeline - for many different reasons, typically:

During construction

i) removing construction debris from the line
ii) acceptance testing (waterfill, dewatering, etc.)
iii) commissioning

During operation

i) pipe-wall cleaning
ii) condensate removal
iii) product separation (batching)
iv) applying inhibitors

For inspection

i) to check for physical damage (geometry)
ii) to detect corrosion, laminations, or cracking
iii) leak detection
iv) sampling
v) line cover and spanning (subsea)

For general maintenance and repair

i) corrosion inhibition
ii) pre-inspection cleaning
iii) decommissioning
iv) isolation
v) recommissioning

During renovation/rehabilitation

i) gel pigging
ii) applying *in situ* coatings
iii) chemical cleaning
iv) scale removal
v) cleaning for product conversion

Decommissioning

i) product removal
ii) pipe-wall cleaning
iii) inspecting/testing
iv) inerting

When considering any of the above applications, it should be remembered that each pipeline is different. They have different: diameters, lengths, contents, geometry, operating pressures and temperatures, materials, wall thickness, locations, etc.... They are also built by different contractors and have different owners - both of whom will have different philosophies and operating procedures. They will also have to be constructed and operated in accordance with different Codes and comply with requirements of different Authorities. Pig selection and the pigging program must be tailored to suit the particular need as well as the characteristics of the individual pipeline and its operation.

Pigging during construction

Debris removal

During the construction of any new pipeline, debris, sand, stones, welding rods, rags, etc., will invariably be left inside the pipe and, in spite of the use of 'night caps', animals like mice and rabbits will frequently find their way inside! All of these foreign materials must be removed prior to performing cleaning or gauging runs, so the first pigs through the line will usually be the contractor's cleaning pigs. The type of pig used will depend on various factors associated with the pipeline such as its diameter and length, the topography, whether it is onshore or offshore, sandy, rocky, etc. Several debris-removal pig runs may be necessary and, because of the very onerous duty, it is good practice to consider the use of wear discs on the front of the pig, particularly if the pipeline is relatively long.

During the debris-removal stage it is normal to use compressed air to push the pig through the pipeline. Each end of the pipeline section is usually fitted with 'test ends'. These are, in effect, temporary pig traps for launching and receiving the pig. The receiving test end should be sufficiently long to accommodate the debris that is likely to be removed. 'Pig catchers' may occasionally also be used, which are very basic 'traps' designed to allow the debris to exit, but to catch the pig as it emerges. Strict safety measures are necessary when pigging with compressed air, and personnel must be kept well clear.

Offshore, things are somewhat different. Because offshore operations are much more costly, several steps may be combined. In spite of more-strict specifications, some debris will still accumulate, but the line is usually capped and laid down on the sea bed, then tied in at some later stage.

The 'lay-down head' used at each end of the section on offshore lines is similar to the test end on a land line, and is basically a very long pig trap. It may have to hold a large number of pigs, sometimes of different types, which need to be launched and received separately. To do this requires a manifold comprising numerous connections and very complex pigging procedures. These operations are therefore normally sub-contracted to a competent pigging service company.

Cleaning

Following the debris-removal pig runs, a wire-brush pig is run through the pipeline to remove more stubborn debris including mill scale, weld bead slag, etc., from the pipe wall. More than one run of the cleaning pig is expected. Compressed air is usually used to push the cleaning pig through the pipeline. For a series of cleaning runs, the launcher and receiver test ends should be long enough to accommodate the number of pigs used, and a manifolding system similar to that used offshore may be required.

Initial gauging

After debris removal and cleaning have been satisfactorily completed, gauging pigs are run. Gauging runs are necessary to determine that there is nothing protruding inside the pipeline which might cause an obstruction, and to ensure that the ovality of the pipeline is within accepted tolerances.

For this, the pigs are fitted with a circular mild steel or, more commonly, an aluminum plate which is machined to some specified diameter, typically 95% of the smallest diameter of the installed pipeline. This is known as the 'gauging plate'. These 'gauging pigs' are then pushed through the pipeline, usually with compressed air. If the gauging plate sustains any damage, an investigation should be carried out to establish the possible cause(s) of the damage and remedial action taken.

Geometry or caliper pigging

There is an inherent problem in relying on gauging-pig runs to detect any anomalies in the bore of a pipeline. All too often, when the gauging plate is damaged, there is no way of finding out what caused the damage. It is usually also impossible to determine the location of the fault and, in severe cases, the pig may not emerge at all and be stuck in the pipeline. This has been overcome by the use of instrumented pigs which can pass severe obstructions while measuring and recording both the extent and location of the reduction. These are known as 'geometry' or 'caliper' pigs.

Running this type of pig through the pipeline will provide a catalogue of information on the entire pipeline. It will record the number of joints used between the launcher and the receiver, any changes in inside diameter, (which may be due simply to changes in the wall thickness), the exact location of any joint which is outside the allowable tolerances for ovality, the location of any protrusions into the pipeline, and whether the valves are fully open. Under careful control, these pigs can be driven through the line with compressed air, but their accuracy may be affected by irregular speeds, so they are best run in a liquid. Unless there is an immediate problem to resolve, they are therefore often run during dewatering of the pipeline following hydrostatic testing, and then periodically during the operational life of the pipeline. This provides an 'as-laid' record and a means of monitoring the condition of the line.

Hydrostatic testing

Every pipeline is subjected to a hydrostatic pressure test - always prior to commissioning, and often at periodic intervals throughout its life - to reconfirm its integrity or to upgrade its MAOP.

For pressure testing it is important to ensure that the pipe is filled with a solid column of water. This is achieved by inserting a pig, or pigs, in the launcher before pumping starts. The pig(s) in front of the hydrotest-fill water will ensure all the air is removed. A regulating air vent valve is installed at the receiver to enable the movement of the pig(s) to be controlled by the operator at the receiving end. Air vents should also be installed at all high points of the pipeline system.

There are a number of reasons for having to ensure all the air is removed. One is to avoid pressure variations due to vapor locks, and another is for safety in the event of a failure. In mountainous regions particularly, where there are significant variations in the elevation of the line, any air trapped at the high points will eliminate the syphon effect and cause the static heads of the individual liquid columns which are thus created to become additive. The pressure required to drive a pig may then exceed the safe working pressure of the pipeline and/or the capacity of the pumps. Tappings must then be made at the high points to bleed off the trapped air and create a single column of liquid.

Dewatering

Once the hydrostatic test is completed, the water inside the pipe is removed by a series of pigs normally pushed through the pipeline by compressed air. For relatively-long pipelines, particularly in arid regions, it is not uncommon for the pipeline to be hydrostatic tested in sections, with the hydrotest water transferred from one section of pipe to the next section during the dewatering operation.

The first step of the dewatering operation is to study the features of the pipeline that may affect running a pig through the pipeline with compressed air while displacing the hydrostatic test water. These include the length, diameter, wall thickness, internal configuration and location of main line valves, the piping configuration at the ends of the line, changes in pipeline elevation, and the internal condition of the pipeline. The type of dewatering pig used will be determined based on the pipeline features discussed above, and the air compressor(s) should be capable of running the dewatering pig(s) at a speed of at least 1mph under these conditions. The discharge outlet and connecting discharge pipe or hose should be as large as possible and have a water discharge valve connected directly to the pipeline in order to be able to completely shut off the discharge, or throttle the rate of discharge to safe levels. It is also extremely important to prepare the receiving end for any violent surges of discharge water which may occur as the dewatering pig approaches the receiving end if some compressed air has by-passed the pig. It must be emphasized that pigging with compressed air, or any other gaseous medium, can be highly dangerous, and should only be carried out by fully-qualified personnel.

Where practicable, it is usual to run a series of foam pigs to swab up any water remaining in the low points of the pipeline as well as some of the residual water adhering to the pipe wall. However, if the product to be transported in the pipeline requires a dry line (e.g. natural gas) then, after dewatering, the line can be dried using various techniques.

Drying

A typical method of drying a pipeline is to blow dry air through the line, usually in conjunction with foam pigs. Other methods include the use of nitrogen, or to use 'vacuum drying'.

The dry air and foam pig method is primarily for onshore pipelines, and is capable of achieving a degree of dryness and cleanliness that will prevent contamination of most pipeline-transported products. First, lightweight polyurethane foam swab pigs are run with extremely dry compressed air, typically with a -90°F atmospheric dewpoint temperature. These swab pigs wipe out and soak up the residual water, while the dry air absorbs as much water as it can under the existing conditions and carries it out of the pipeline.

The swab pigs are run continuously until they are received dry. In a cross-country pipeline, this may require 200 or even 300 runs. Once dry swabs are received, the next step is to scratch loose the loosely adhered rust and mill scale and break up any packed dirt that may have settled out of the hydro-test water. Polyurethane pigs equipped with fine steel wire brushes are run continuously until they are received without any material packed into the bristles. To remove the loose material that remains in the pipeline in the form of very fine dust, more foam swab pigs are run. The swabs are run continuously until they are received clean enough to meet a specified criterion, or to ensure that a dewpoint specification has been met. This phase may also require 200 or 300 swabs. By using super-dry air to propel all pigs, the pipeline will be extremely dry when it is clean, as evidenced by the condition of the final swabs run through it.

Vacuum drying is useful for offshore pipelines. It works on the principle that the boiling point of water is determined by the pressure. Reducing the line to below atmospheric pressure causes the remaining water film on the pipe wall, and elsewhere, to boil. This water is then removed from the line in the form of steam. The drying process continues until it reaches some predetermined dewpoint value. Once this dewpoint value is reached and accepted, the pipeline is ready to commence commissioning procedures.

Commissioning

'Commissioning' is the stage when the completed pipeline is filled with the product and brought to the point where it is ready for operation. It will almost always involve the use of pigs, and may often require 'slugs' of other substances, such as methanol, glycol, diesel, etc., to be pumped through ahead of the product. Slugs, if used, are contained between two or more pigs and are usually designed to avoid product contamination as, particularly with hydrocarbons, there may otherwise be a risk of explosion or perhaps serious blockages due to hydrate formation. Because of the complexity, commissioning is another specialist pigging activity which is almost always entrusted to a qualified pigging service company.

Pigging during operation

Debris removal

In previous sections, it has been demonstrated that pipe-wall cleaning is essential to protect both the investment and the operational efficiency of the pipeline. This applies regardless of whether the debris is wax, scale, or some other form of solid material.

There is probably no one reason for regularly pigging a line which is more important than any other, but there is one which is perhaps less well appreciated than it should be - and that is to prevent corrosion.

Free water, even in crude oil lines, collects in the bottom and must be both swept out and inhibited. An inhibition program on its own is generally inadequate. If pigs are not used to remove the surface debris such as dirt, sand, wax, corrosion products, etc., then water will collect under it and it will prevent the inhibitors from properly treating the active corrosion area. The same situation will arise if there are bacterial corrosion problems caused, for example, by SRBs (sulphate-reducing bacteria).

Once the pipe wall is corroded locally, especially if it is pitting corrosion, it will be extremely difficult, if not impossible, to remove any accumulated water or debris from the pits. This creates a progressively-worsening situation which may require quite drastic and costly inhibition programs to contain it - and it can never be reversed.

Condensate removal

Condensate drop-out is a constant problem on many gas pipelines. The rate at which it drops out depends to some extent upon the flow, and most gas pipeline operators use the flowrate to determine when they need to pig.

Condensate rarely threatens to block a line because as it settles out in the low spots and reduces the effective bore, this restriction causes the gas flow to increase locally and pick up the condensate once more. This situation is not acceptable however, because the turbulence will cause an increase in the pressure drop, thus reducing throughput, and the constant pick-up and drop-out at the same point could conceivably give rise to problems of erosion.

When a particular section of line becomes saturated, the condensate will be carried over to the next, and so on, until it arrives at the terminal. This is always unexpected, and is usually in very large slugs. The slug catchers are sized to be able to handle the maximum likely volume and many of them are very large indeed. Even so, the slug which arrives sometimes exceeds their capacity, and when it does, the condensate gets into the processing plant and invariably causes both damage and a shutdown.

Condensate drops out of untreated gas continuously, so pigging is not designed to remove all the condensate. Rather, it is intended to remove the condensate in a controlled and regulated manner and that is why, in spite of their relatively-poor sealing qualities, spheres are often used. Spheres lend themselves readily to the type of automated, unmanned, equipment required for condensate control, especially in gas-gathering systems, which usually comprise pipelines of progressively-increasing diameter.

Product separation ('Batching')

Batching is primarily used on multi-product pipelines which transport different products in the same line at the same time, typically refined products such as petroleum, fuel oil, kerosene, and jet fuel.

The problem is that the product at the end of one batch mixes with the product at the head of the next. This is called 'interface mixing' and results in contamination, the volume of which will depend on such things as the pipeline configuration, pumping rates, and the overall length. To reduce this interface mixing, a 'batching pig' is inserted in between each batch of product. However, the problem is to get the pig in between the batches and out again at precisely the right moment. Spheres lend themselves to predictable and automatic handling, and it is for this reason that they are commonly used for batching in multi-product pipelines.

Batch inhibition

Running a batch of inhibitor between two pigs is one way to be reasonably sure that the whole of the internal surface of a pipeline is wetted, but even then there is no way of guaranteeing this - especially at the top center or 12 o'clock position. As a result, although the technique is well known, it is not very often applied.

Inspection pigging

Utility pigs, which are used during the construction and operational phases, are intended to perform a function such as cleaning, dewatering, etc. On the other hand, intelligent pigs, normally referred to as 'In-line Inspection', or simply ILI tools, provide information on the condition of the pipe and/or its contents. They are not sold but are, in effect, the tools which are used by specialized inspection pigging companies to gather the data on the pipeline. This data is then analyzed by their engineers and technicians to determine and report on the condition of the line.

The information provided by these ILI services covers a wide range of inspection and trouble-shooting needs, including: diameter/geometry measurement, curvature monitoring, temperature/pressure recording, bend detection and measurement, metal loss/corrosion detection, photographic inspection, crack detection, wax-deposition measurement, leak detection, and product sampling.

Many inspection surveys, particularly those for metal loss and geometry, are usually run during or soon after commissioning in order to provide a 'base-line survey'. This enables subsequent inspections to be compared with the original results, and makes it possible to detect any trends, as well as to establish the average rate at which any changes are taking place. Such information is invaluable for both maintenance and production planning.

Pigging for general maintenance and repair

Pigs are frequently used as part of a maintenance and repair program. Batch inhibition has already been mentioned, but there are a number of other very important uses for pigs during maintenance and repair programs. Possibly the two most important are for pre-inspection cleaning and for isolating a section of line for repair.

Pre-inspection cleaning

To obtain optimum performance from ILI tools, it is essential that the pipeline is clean - really clean! Experience has shown that, in some cases, standard pigs are not capable of cleaning the pipeline sufficiently well and if it is not, the ILI survey may have to be aborted due to dirt, and more especially, wax. The problem is that magnetic flux-leakage tools with their powerful brush/magnets, remove the residual film of wax which has been left behind by the cleaning pig. This residue fouls and sometimes even destroys the sensors. Ultrasonic pigs also experience severe problems due to the attenuation of the signals by residual wax.

This does not mean that a standard cleaning pig does not do an effective job. For routine maintenance, a standard pig does not need to remove all the wax. In fact, a thin film left behind is beneficial in further reducing the friction factor at the same time as helping to protect the pipe wall from corrosive products. Pre-inspection cleaning therefore requires a completely separate pigging program, often using specially-designed pigs, and this is now normally carried out by the inspection company themselves.

Isolation

To travel through a pipeline, a pig should ideally have an interference fit within the pipe bore to create a seal. If it is stopped within the line and fixed in some way, it is obviously a means of isolating sections of the line for modification, maintenance, or repairs.

Although simple in principle, there are a number of fundamental problems in using pigs for this purpose. A longitudinal weld bead, for example, may create a leak path past the seal. Possibly the most-difficult aspect is to restrain the pig against the loads imposed upon it by the differential pressure, particularly in large-diameter lines.

Over-inflated spheres have been used for many years to provide a seal within the bore of a pipe which is being repaired, thus preventing any flammable products from being exposed to welding and cutting torches. Special types known as hyperbaric spheres are used almost as standard during subsea tie-ins which involve the use of hyperbaric chambers. However, because of their shape, spheres have a very limited area in contact with the pipe wall and so can only be used where there is no differential pressure.

In subsea lines especially, due to temperature and tidal changes, it is often difficult to maintain a balanced pressure either side of the pig, so work has gone into the development of 'high-differential' pigs. These are pigs with a multiplicity of oversize polyurethane driving discs which are generally stacked closely together and often of varying hardnesses. Even so, these pigs can only withstand a very limited pressure differential.

The last few years have seen the development of highly sophisticated plugging pigs. These are designed to stop at some predetermined point, to be remotely actuated in order to create a seal, then subsequently to be deactivated so they may either continue their journey or be reverse pumped back to their starting point. Until quite recently, because of the difficulties in developing fool-proof remote operation, most plugging pigs were attached to an umbilical or 'tether', and were thus limited to operations close to an access or 'launch' point.

Pigging during renovation/rehabilitation

Gel pigging

'Gel pigs' are not 'designed' in the generally-accepted manner: they are 'formulated'. Based upon a thick jelly-like substance, similar to wallpaper paste, they can be made to almost any density and range from pre-formed pigs which are usually shipped to site within their metal mold (which can be as simple as an empty oil drum), to a light foam which is sprayed into the line. Most applications, though, require it in a jelly-like state. This material can be combined with a wide range of other chemicals such as foaming and degreasing agents or corrosion inhibitors to perform various tasks.

Gel pigging is invariably expensive as the gel is usually required in very large quantities and so it must be manufactured 'on site'. It is also usually bio-degradable, and so has a very limited life. Even so, there are a number of situations where it has proved invaluable. It was first used as a major component in cleaning the FLAGS pipeline in the North Sea where the volume of debris to be removed after construction was thought to be too great for conventional methods. The gel was trapped between pigs to form 'slugs' which picked up the debris, held it in suspension, and literally carried it out of the line. It has also been very useful in removing 'stuck' pigs, creating high-sealing pigs, and even for temporary isolation of a pipeline for repairs.

Other than when using the pre-formed type of gel pig (which is, in effect, a 'soluble pig'), the formulation and work of gel pigging is always carried out by qualified contractors.

Applying in situ coatings

Three critical factors influence the success of any coating project: surface preparation, coating selection, and coating application. The wrong choice in any area may cause premature failure of the coating.

Surface preparation is the first step of any coating job. It is essential to thoroughly clean the inside of the pipe to properly prepare its surface. This is to remove all deposits from the line, including rust, scale, and salts that could interfere with the coating bond. After cleaning, the line should be completely dry and under a purge of an inert gas to prevent flash rusting. A typical cleaning process involves scouring the inside of the pipeline with an abrasive material, such as flint, which is propelled in a low-pressure, high-velocity, stream of nitrogen. The cleaning particles impact the wall of the pipe at a low angle of incidence, gouging and/or chipping away the deposit. Removed material is then carried through the line with the nitrogen, and collected at the outlet. After abrasive cleaning, pigs and/or solvents are used to remove any remaining dust.

Occasionally, the cleaning process may uncover very thin, hard deposits, such as magnetite, which are more economically cleaned with chemicals. Also, by removing rust or scale, cleaning may expose leaks that must be repaired before coating.

Coating selection is the second critical aspect. A wide variety of coatings have been used to internally coat in-place pipelines, but they must generally have specific thixotropic properties to enable the coating to be spread onto the pipe wall and then immediately 'gel' to prevent it from running or sagging. The most commonly-used coating is a two part polyamide-cured epoxy. This polyamide coating is recommended for lines carrying potable, fresh, and salt water, as well as crude oils, transportation fuels, natural gas, and some solvents. However, the final coating selection must first be discussed with a professional *in situ* coating service company, and in most cases, it is good practice to install a test spool cleaned and coated under field conditions to assure compatibility with the intended service.

Coating application to in-place pipelines is achieved by placing the coating material between two pigs and propelling the 'pig train' through the line. Various types of pigs, including multiple-cup-and disc, bi-directional disc, and spheres, are commonly used.

The coating thickness is affected primarily by the size of the cup pigs and the coating train speed. By controlling the differential pressure across the coating train and hence the train's speed, the required coating thickness can be achieved. Nitrogen is used as both the driving force and back pressure.

Chemical cleaning

If conventional pigging methods are unable to clean the line sufficiently, then chemical treatment may be considered. The type of chemical(s), their strength, contact time, and overall procedures, will be dictated by the circumstances, and each individual situation will be different. It is therefore vital to seek professional advice and assistance before adopting this approach, as mistakes may be irreversible and errors can be very costly indeed.

Scale removal

Scale can usually be removed by the proper use of suitably-aggressive cleaning pigs, such as the various grades of wire-brush foam pigs, and utility pigs equipped with brushes or steel blades. Some special pigs have been developed for this purpose, typically the 'studded foam pigs' and the 'pin-wheel pig'. Typical deposits may include paraffin, asphaltene, sediments, silica, coke, iron sulphides, calcium, or other corrosive products which are often extremely difficult to remove.

The 'studded foam pig' is a relatively-simple pipeline pig which can remove hard deposits from the interior of a pipe. The pig, being symmetrical, permits travel in both directions which is important when cleaning industrial heaters and pipelines. The body is formed of high-strength polyethylene, polypropylene, rubber, or polyurethane. The flexibility and hardness of the body are adapted to the intended use. This body has circumferential and longitudinal ribs, forming square recesses which retain an internally-threaded anchor. This allows different types of studs to be screwed in. Studs are available in various sizes, shapes, and hardness, and are selected on the basis of the hardness of the deposit, the extent of the build-up, the location of the deposit, and the tube or pipe material.

In use, the pig is inserted into a pipe, and is propelled by a fluid, usually water. The pressure acting on one end of the pig causes it to expand, forcing the studs into contact with the pipe wall and dislodging the scale. If they become worn, the studs can be removed and replaced. Furthermore, the stud height can be adjusted using shims. This controls the rate of deposit removal and also sets the amount of liquid by-pass around the pig body which flushes out the loosened deposits, reducing the possibility of blocking the pipe.

The 'pin-wheel pig' is generally used on the longer and larger diameter pipelines, and is used as part of a service rather than being sold outright. Further details are included in later sections of this publication, but basically the cleaning assemblies consist of a number of heavy-duty polyurethane discs which have an outside diameter significantly less than the inside diameter of the pipeline.

Protruding radially from the circumferential edge of each disc are a number of steel pins with hardened tips which are radiused to prevent damage to the pipe wall. The diameter across any two opposite pins is greater than the inside diameter of the pipeline, so that when the pig is travelling through the line, the pins are bent back at a slight angle. This both assists in the cleaning action and also compensates for any wear.

None of the wax or scale removed from the pipe wall will actually be pushed forward by the pig itself. This will be left behind in the line for removal by another type of specialist pig as part of the overall program.

Cleaning for product conversion

Some pipelines are designed from the outset to transport different products. 'Multi-products pipelines' carrying such things as petroleum spirit, kerosene, jet fuel, and other refined hydro-carbons are typical. Similar systems on a much smaller scale are used in a wide range of industries manufacturing things such as pharmaceuticals, foodstuffs, paint, confectionery, etc. Many of these industries change products daily if not hourly, and generally they do not require cleaning as such, but they may require a batch of some other product to prevent contamination or to sterilize the system.

Large-diameter, long-distance pipelines present a different situation. They often have a life which far exceeds the reserves of the product they were designed to transport. Similarly, as demand changes, they may outlive their usefulness in terms of transporting the original product but may prove to be extremely cost-effective if converted to transport some other product, maybe even in the opposite direction.

Typical of these situations is the conversion of an oil pipeline to gas service, or vice versa. This has now been done on a number of occasions, and the techniques used will depend upon the specific circumstances. It is always difficult to generalize, but it is usually easier to convert a gas pipeline to carry crude oil than to convert a crude oil line to gas transmission.

Decommissioning

Decommissioning is still very much in its infancy, and a great deal of discussion is going on about the methods and requirements for abandonment of pipelines. Typical procedures for abandoning a hydrocarbon pipeline in environmentally-sensitive areas might include product removal, cleaning, inspecting/testing and finally, inerting. It is possible that some evidence of the condition of the pipeline might be needed to obtain the approval to abandon it, and this may well involve the use of one or other of the ILI tools.

If it is a hydrocarbon-carrying pipeline, then clearly a basic requirement would be to remove the remaining product. The product used to propel the pig during this phase of the operation will depend upon the overall procedure, but nitrogen would probably be high on the list of options.

If the line is to be abandoned without inerting, it may be necessary to ensure that it is clean, because eventually it would corrode away and any remaining contents would pollute the surrounding area. But however well cleaned, a hydrocarbon pipeline would inevitably constitute some hazard due to the fumes which would emanate. It is therefore more likely that it would be filled with an inert substance, and this too would require pigs.

Excerpted from An Introduction to Pipeline Pigging, 13th edition, 2020. Pigging Products & Services Association and Jim Cordell/Hershel Vanzant. Suffolk United Kingdom. For the complete report, see: www.ppsa-online.com. Reprinted with permission.

Cleaning and construction utility pigs

Excerpted from An Introduction to Pipeline Pigging, 13th edition, 2020. Pigging Products & Services Association and Jim Cordell/Hershel Vanzant.

Utility pigs can be divided into two types:

a) Cleaning pigs - for removing accumulated solids or debris;
b) Sealing pigs - for removing accumulated liquids, separating dissimilar fluids, filling, dewatering, etc.

Cleaning pigs are often referred to as 'scrapers', while sealing pigs may be called 'batching' or sometimes 'swabbing' pigs. Utility pigs are provided in four different forms. Typical examples of each are shown in Figure 1:

Figure 1. An assortment of pipeline pigs. *Source: Contractresources.*

1) *Mandrel pigs* which are assembled from a number of component parts and for which spare parts are available. Mandrel pigs are available in most sizes.
2) *Foam pigs* which are molded from polyurethane foam with various configurations of solid polyurethane strips and/or wire bristles, studs, etc., permanently bonded to them. Foam pigs are available in most sizes.
3) *Solid-cast pigs* which are molded in one piece – usually from polyurethane. These pigs are normally only available in the smaller sizes.
4) *Spheres* normally filled with water/glycol, these pigs can be inflated to the optimum diameter.

Standard pigs can only traverse the line in one direction, while 'bi-di' pigs are capable of travelling in either direction (i.e. bi-directionally). Multi-diameter (or 'dubl-diameter') pigs can traverse a pipeline which has more than one nominal diameter (normally only one or two line sizes different).

Cleaning pigs

The type of pig used for cleaning a pipeline will depend largely upon the nature of the material which is to be removed. As there is an almost infinite number of different sediments, deposits, and other substances which can cause reduced throughput, there is similarly a very large number of different types or configurations of pig which might be used to combat these problems. Often a single configuration of pig may not be the optimum for even one pipeline, let alone all pipelines, so discussion will be restricted to those most commonly used.

Mandrel pigs

These are pigs which are made up of a number of component parts which are mounted on a body tube so that they may be replaced or re-configured as the need arises. With careful design, a relatively small number of standard components can be assembled in different ways in order to build pigs suitable for different types of pipeline.

The pig bodies are usually of steel, although some manufacturers produce the whole pig in polyurethane or some of the other tougher plastic materials. The driving cups (or seals) are invariably in polyurethane, whilst the cleaning elements are in a material and configuration which is determined by the nature of the deposit. For hard deposits, wire brushes or sometimes steel scraper blades may be needed, while for soft materials such as some waxes, sludge, etc., elastomer 'plough' blades may be used. Heavy-duty discs machined to the line ID may also be used to good effect, those fitted to 'bi-di' pigs (Figure 2) being typical examples.

'By-pass' is often essential for an efficient cleaning operation. By utilizing the differential pressure across the pig, a fluid flow is deliberately created from the back to the front. This results in severe turbulence ahead of the pig which keeps the solids which the pig removes in suspension, preventing the line from being plugged. Also, if the by-pass is directed through or directly at the cleaning elements, it will flush them free from any accumulated debris, so maintaining their cleaning efficiency. Care should be exercised when deciding on the amount of by-pass to be used, particularly in low-flow conditions. If the line is very dirty, then high differentials could occur. This could increase the flow through the by-pass ports to a point where the pig remains stationary and in effect becomes 'stuck'.

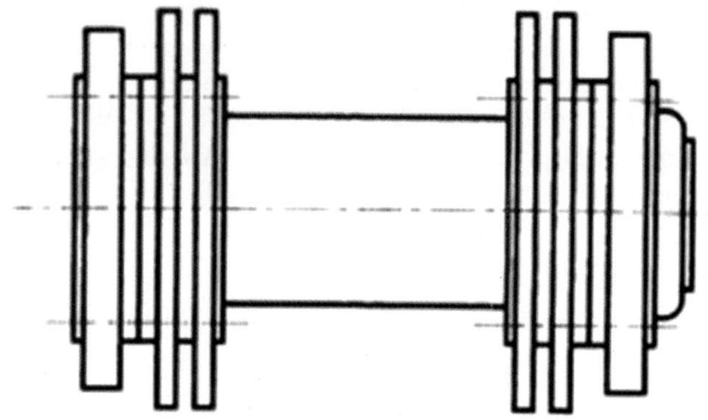

Figure 2. Example of a typical 'bi-di' pig, showing the heavy duty discs front and rear.

Foam pigs

As with so many other products, the principle advantages of using plastic materials for pipeline pigs were apparent before the materials themselves were really sufficiently developed. Low cost price, zero maintenance, and reduced risk of 'sticking', encouraged many operators to use them for purposes for which they were not suitable. As a result, some of the early foam pigs gained a questionable reputation, which in most cases was undeserved.

Today, the technology has developed to a point where foam materials have a very high tear and wear resistance; most manufacturers now include some foam pigs in their range, but those who specialize in these products may carry over 30 different standard types. Foam pigs are widely used in the pipeline industry, especially for rehabilitation work. They are also particularly useful when developing a pigging program for a line which has not been regularly pigged and which may contain some unexpected problems.

Solid-cast pigs

Although solid-cast pigs have been used for a long time as 'sealing pigs', they have only recently been developed for cleaning. Solid-cast pigs resulted from the fact that the labour cost of assembling and replacing parts on small pigs is often prohibitive. A 2-in diameter pig may take as long to assemble as a 6-in or 8-in size; in fact, the point at which it is cheaper to replace the whole pig rather than a single component is usually around the 12-in size.

The most important factor, though, is the effectiveness of a pig - not its cost. A very slight improvement in pipeline efficiency can easily offset the cost of two or three man-hours, and operators would be well advised to take this into account before deciding on the type of pig to use. A saving will only be achieved if the pig selected is at least as effective as any of the alternatives - and probably the best way to determine this is to compare the results.

Spheres

These are almost exclusively used as 'sealing pigs', but there have been a limited number of spherical pigs designed for removing solids. Spheres can negotiate 'zero radius bends' and they roll freely. This means they can travel along a lateral, drop through a tee into a larger-diameter line, and then be pushed into the trap by another pig sized to suit the main line. This versatility may help to solve the problems of pigging pipelines with complex configurations such as gas-gathering systems.

Sealing pigs

As with the cleaning pigs, the final configuration of a sealing pig will depend upon its purpose. Typical applications of sealing pigs include:

- removal of condensates/liquids from gas lines
- separation of dissimilar fluids in multi-product lines
- line filling prior to pressure testing
- dewatering after pressure testing
- commissioning (separating slugs of methanol, nitrogen, etc.)
- *in situ* lining/coating
- product removal prior to shutdown

These pigs also fall into much the same categories as cleaning pigs, so similar groupings will be used.

Mandrel pigs

Mandrel-type sealing pigs are made up of a number of standard component parts; often the same parts as used for cleaning pigs - but without the cleaning elements. But cleaning pigs do not need a very effective seal - in fact as already mentioned, a by-pass is deliberately created. For sealing pigs, the most important aspect is the design of the driving cups (which will be referred to as 'seals').

Seal configurations fall roughly into the three types shown in Figure 3:

1) the 'standard' seal, which was the earliest type and was derived from the hydraulic piston seals.
2) the 'conical' seal, which had its origins in the early ILI tools.
3) the 'disc' seal which originated with the bi-directional pig.

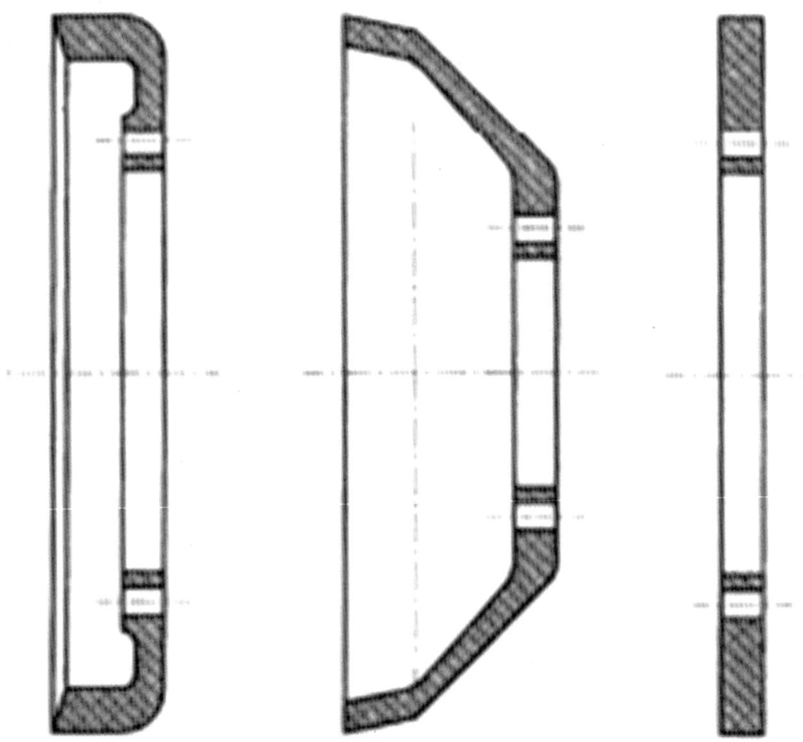

Figure 3. Seal (or 'driving cup') configurations (left to right): standard seal, conical seal, disc seal.

The basic requirement for any seal is that it must maintain close contact with the pipe wall. This presupposes that it is designed and fitted properly so as to prevent any leakage between it and the pig body (a factor frequently overlooked), and that there is no leakage through the pig body itself.

Close contact with the pipe wall creates wear which eventually reduces the sealing efficiency. Materials play a major part in overcoming this problem, but whatever material or design is used, wear will always occur, so some form of wear compensation must be incorporated in the design. This is usually achieved by using the differential pressure to expand the seal into contact with the pipe wall. Skillful design is necessary, because if the radial force created by the differential pressure is greater than necessary, then excessive wear will result.

As has already been mentioned, the seal materials are almost exclusively polyurethane, but care must be taken to ensure the correct grade is used. Polyurethane is made up of various combinations of polyols and isocyanates from which a vast range of hardnesses and properties can be obtained. The two main types of polyol used for pig seals are polyether and polyester. Polyethers have a good resistance to water (which at elevated temperatures will break down a polyurethane material), while polyesters have a better resistance to hydrocarbons. Most pig seals fall within a hardness range of 60 to 85 Shore 'A'.

The early bi-directional pigs utilized four (or more) solid discs which were made oversize. Often, these discs were made of natural or synthetic (neoprene or nitrile) rubber. Wear rates were extremely high, in many cases giving less than 50 miles. But as they were generally used for filling and dewatering short sections of newly-constructed line, they were adequate.

The modern 'bi-di' pig has overcome these problems. It will be noted in Figure 2 that the pig is made up of two different types of disc. One type is thin and soft and made 5-10% larger than the inside diameter of the line to provide a good seal with plenty of wear compensation. The other is thicker and much harder and is machined to the actual inside diameter. These discs carry the weight of the pig and provide a simple yet effective cleaning element. This type of pig can therefore be used for cleaning and for sealing.

Care needs to be taken in selecting the type of seal best suited to a particular pipeline. Quality of the polyurethane material is a fundamental requirement. Of the three types mentioned, the standard seal is usually the least expensive. Conical seals are very effective in thin-wall pipe as they compensate very well for any out-of-roundness. The new bi-di pigs are very effective for both sealing and cleaning provided there are no severe out-of-roundness problems. But there are many other factors to consider and the suppliers will provide guidance in this respect.

Foam pigs

To seal well, foam pigs must also maintain good contact with the pipe wall, and they too must have some compensation for wear if they are to perform well over any distance.

Being of cylindrical shape, they contact the wall over their entire length and this reduces the unit loading. However, it is impossible to make a foam which will have the same tear and abrasion resistance as a solid-cast urethane, so the manufacturers often apply various coatings of (usually) solid-cast urethane. By applying these in criss cross or spiral patterns, a greater degree of both wear resistance and wear compensation is achieved. Many types can be used for both cleaning and sealing.

Solid-cast pigs

Solid-cast sealing pigs are very common in the smaller sizes. Most small-diameter lines are not very long, and the pigs are not very heavy, so wear is not usually a problem. By making the pigs a little oversize, the wear compensation is usually adequate and in any event, replacement costs are minimal. These types of pig are particularly useful in process-plant systems. Some companies now produce this type of pig in much larger sizes, and some types are available in standard sizes up to 36 in.

Spheres

These are probably the oldest, best-known, and most widely used sealing pigs. Some of their characteristics have already been discussed. Their shape makes them ideal for use in automatic systems, and this is the main reason why they are widely used for removing condensates. In many gas-gathering lines, for example, condensates are dropping out continuously, so it is necessary to have very frequent pig runs to control the condensate levels and it is in this situation that spheres are particularly useful.

Because they roll easily, they are very easy to automate and large numbers can be installed in a trap and automatically launched at predetermined intervals. They have the added advantage that they can be discharged into other, larger, lines and be collected at some main receiving point to be sorted and reloaded into their original traps.

Most spheres are capable of being inflated to the required diameter. This may be from the nominal pipe inside diameter to 1% or 2% oversize, depending on the requirements and the experience with a particular pipeline. Because they are spherical and have only one sealing surface, particular attention must be paid to the design of the pipeline to ensure that the sphere is guided past any voids or obstructions and can maintain its seal at all times. Specially-designed 'flow tees' must be installed in all laterals of a line designed to run spheres.

Gauging pigs

Although too small a group to justify a separate category, gauging pigs cannot be classified as either cleaning pigs or sealing pigs. Gauging pig runs are necessary to determine that there is nothing protruding inside the pipeline which might cause an obstruction and to ensure that the ovality of the pipeline is within accepted tolerances. For this, utility pigs are fitted with a circular mild steel, or more commonly, an aluminum plate which is machined to some specified diameter, typically 95% of the smallest diameter of the installed pipeline. This is known as the 'gauging plate'. These 'gauging pigs' are then run through the pipeline. If the gauging plate sustains any damage, an investigation is carried out to establish the possible cause(s) of the damage and remedial action taken.

Dual-diameter pigs

Dual diameter, or multi-diameter, pigs are another small but important group which fall outside the normal classifications. They are designed to traverse pipeline systems which have been built with more than one nominal diameter in between pig traps. This situation is common in water lines where the valves are often undersized.

Many older hydrocarbon pipelines were designed with two diameters so that the smaller pipe could be transported inside the larger one to reduce shipping costs. Very few new hydrocarbon pipelines are now laid with more than one diameter between traps. However, the increasing need to pig subsea production systems by looping from one line into another, and to tie-in marginal fields to existing export pipelines, is reversing this trend again, and there is growing interest in pigs which are capable of more effective multi diameter performance.

Foam pigs are used widely in dual-diameter pipelines, due primarily to their ability to conform to difficult pipe configurations. They have the ability to enter a pipeline through a tee, traverse zero-radius bends, and can pass through valves as much as 50% less than the pipeline diameter.

For hydrocarbon lines, a commonly-used configuration is shown in Figure 4 comprising a solid disc for the smaller diameter, with a pair of flaps to provide drive in the larger diameter. Unless they are of dual-module design (the two modules being linked by a universal joint), multi-diameter pigs of this type may require very large bend radii. Generally, dual-diameter pigging should be avoided if possible. Where there is no option but to install a dual diameter system, then a gradual transition from one line to the other is essential, particularly if there are low pressures or low flows involved. Sudden changes could cause the pig to hold up and either

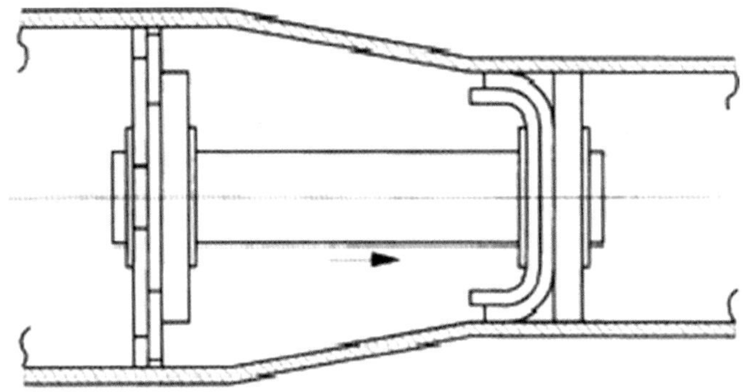

Figure 4. Typical dual-diameter pig entering a section of smaller-diameter pipe.

cause an uncontrolled 'speed excursion' as the pig breaks free, or if the pressure builds up sufficiently, it could cause the flaps to reverse and the pig would then become stuck.

Industrial pigging systems

In the pipeline industry, after ensuring that the pipeline concerned is suitable for pigging, attention then focuses mainly on the traps and facilities at each end. But for industrial pigging systems, things are generally far more complicated - even if they are much smaller in diameter.

Many process-plant pigging systems are extremely complex. They have evolved from the lessons learned in pipeline pigging and, like a number of other technologies, they have advanced to the point where some of the know-how is now proving useful to the pipeliners.

Apart from the quick-disconnect couplings and the special hygienic fittings needed for foodstuffs which are features of industrial pigging systems, there are a number of other things which are worth noting. Often, ingenious manifold systems are used which permit pigging from any one of a number of lines into any one of a number of others. Other interesting features are the integral launch/receive stations and special diverter valves.

Specialist utility pigs

While the majority of operational pipelines can be successfully pigged using standard products, there are many occasions where a specialist pig is required and there are some companies that specialize in this type of 'one-off' product rather than having a range of standard pigs. The following is typical of the many specialist pigs which have been developed.

Magnetic-cleaning pig

The presence of ferrous debris, such as welding rods, metallic dust, and steel shot, does not generally present a major problem in an operational pipeline, but it is essential that such debris is removed if on-line inspection is to take place. Many pig manufacturers can provide simple magnetic-cleaning pigs, but if there is a particularly high volume of ferrous debris to be removed, multiple runs may be necessary. In these situations, a specially-designed pig would have reduced the number of runs considerably.

These specialist magnetic-cleaning pigs are the result of a long development program which has determined not only the best type of magnet, but also the method of mounting them and the optimum orientation to provide maximum efficiency and performance. It is also usual for these pigs to consist of two, or even three, modules if necessary. It should be remembered that because the volume of debris cannot be predicted until cleaning actually begins, magnetic-cleaning programs should take place well in advance of any form of on-line inspection operation.

Pressure by-pass pig

The front of the 'pressure by-pass pig' is fitted with what is effectively a pressure-relief valve, having a diameter of around 40% of the internal bore of the pipeline, and set to open at a pre-chosen differential pressure. If, during a proving or cleaning run, the pig builds up a large accumulation or slug of debris ahead of it, the differential pressure across the pig will rise as the pig begins to work harder in order to push the debris. Once the pre-set differential pressure is reached, the by-pass valve opens, resulting in the debris being jetted or blown away from the front of the pig. This causes the differential pressure required to run the pig to drop, which then closes the by-pass valve again. In a particularly-dirty pipeline this sequence may take place many hundreds of times during a run before the pig reaches the receiver.

This feature also makes it highly unlikely that the pressure by-pass pig can ever totally block the pipeline, since the by-pass allows a continuous flow, giving time for the problem to be investigated and rectified.

Pin-wheel pig

The 'pin-wheel' pig (Figure 5) cleaning assemblies consist of a number of heavy-duty polyurethane discs which have an outside diameter significantly less than the inside diameter of the pipeline. Protruding radially from the circumferential edge of each disc are a number of steel pins with hardened tips which are radiused to prevent damage to the pipe wall. The diameter across any two opposite pins is greater than the inside diameter of the pipeline, so that when the pig is travelling through the line, the pins are bent back at a slight angle. This both assists in the cleaning action and also compensates for any wear.

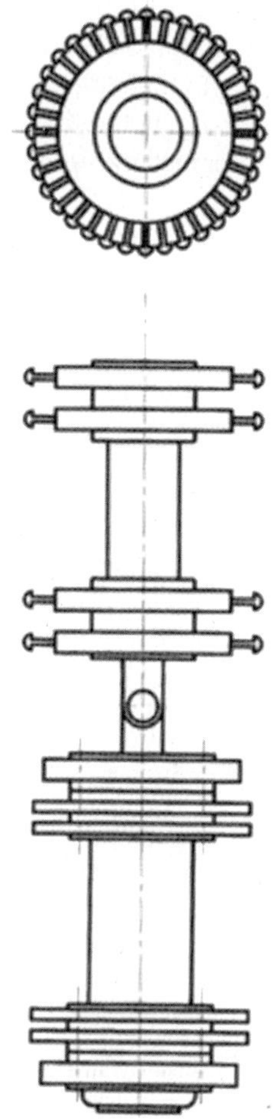

Figure 5. The pin-wheel pig cleaning assemblies consist of a number of heavy-duty polyurethane discs.

None of the wax or scale removed from the pipe wall will actually be pushed forward by the pig itself, being left behind in the line for removal by another type of specialist pig as part of the overall program. The number of pin-wheel discs is determined by the size of pipeline, and the fact that the pins are removable and the hardness of the discs can be varied, allows the 'aggressivity' to be tailored to suit the circumstances.

Shunting pig

The shunting pig is basically a three-module articulated pig which has been specifically developed for the removal of stuck or lost pigs from pipelines. Experience has shown that if a pig does become stuck or lost in the pipeline, there is little point in running a second pig of similar or identical design to push out the first pig, since this pig is also likely to fail for the same reasons and simply add to the problem.

What often happens to a pig which is required to push a stuck or lost pig is that the additional effort to remove it causes the second pig to become damaged itself. Using a three-module articulated pig, the first module will probably become damaged to a considerable extent as it pushes the debris ahead of it, but drive will be maintained because of the second and third modules which never come into contact with the debris being pushed.

In principle it is simple, but much attention is paid to the design of a shunting pig. The weight can be critical and metal-to-metal contact between the shunting pig and the debris is avoided by fitting a hard polyurethane bumper to the front of the first module. The shunting pig is also equipped with permanent magnets for tracking purposes, together with a battery-operated electromagnetic transmitter for positive location when stationary.

Excerpted from An Introduction to Pipeline Pigging, 13th edition, 2020. Pigging Products & Services Association and Jim Cordell/Hershel Vanzant. Suffolk United Kingdom. For the complete report see: www.ppsa-online.com. Reprinted with permission.

Pipeline dewatering, cleaning, and drying

Marvin D. Powers, Pipeline Dehydration, Inc., Houston, Texas

Newly constructed pipelines are typically hydrostatically tested, using water as the test medium, upon completion of construction. Older lines already in service may be re-tested either to confirm an earlier test or to qualify the line for a higher operating pressure. Once the hydrostatic testing has been completed, it is necessary to remove the water from the line and place the line in service. Dewatering can be a simple process or, if the procedure is not properly planned, a difficult one. Pipelines used to transport crude oil and/or refined products will probably only require removal of the test water before the line is placed in service. If the pipeline will be used to transport materials that must meet a specified dryness requirement, the pipeline will need to be dewatered, cleaned, and dried. Pipelines used to transport natural gas will need some drying, depending on the operating pressure and the location of the line, to prevent the formation of hydrates. Other pipelines may require drying to protect the pipe from internal corrosion caused by the formation of corrosive acids, such as carbonic acid in the case of carbon dioxide pipelines.

Dewatering

Dewatering is considered to commence with the running of the first pig after hydrostatic testing is completed and begins with the insertion of a displacer, commonly referred to as a pig, in the pipeline. The dewatering pig may be pushed through the pipeline with crude oil or other petroleum product if no drying is required. If the pipeline is to be cleaned and/or dried, the pig will be pushed by either compressed air or gas. In either case, proper precautions must be taken to be sure the test water is properly disposed of and that any required water discharge permits are obtained ahead of the dewatering operation.

Several types of pigs may be used for the dewatering phase, and pig selection should depend upon the design characteristics of the pipeline and the degree of cleaning, if any, that is desired. Ideally, the pig will form a perfect seal with the inner periphery of the pipe and will not allow any of the material behind the pig to leak past the pig and co-mingle with the test water ahead of the pig.

Normally, the pig will move through the pipeline without difficulty. However, there are many opportunities for problems to develop, causing the pig to stick or even disintegrate.

Sticking may be caused by any or a combination of the following things:

- Pig is incorrect length and cannot negotiate check valves, tees, and bends;
- Pig is too large for heavy wall sections of the line;
- Pipeline may be equipped with reduced opening valves;
- Excess debris or construction material may be left in the pipeline;
- Air/gas may bypass the pig and create an air lock condition; and,
- Other unknown reasons.

Air locks are more likely to occur in hilly country than in flat land. Air locks occur when the accumulated static heads are greater than the available displacing pressure. In some cases, a pressure greater than the maximum allowable pipeline pressure would be required to overcome the air lock. Usually, the air lock occurs where air/gas has bypassed the pig and the downhill legs of the pipeline are filled with air and the uphill legs of the pipeline are filled with water. The air/gas can get in front of the dewatering pig in several ways, such as:

- Poor filling techniques;
- Poor dewatering procedure such as draining water from the line at low points;
- Air/gas bypassing the pig because the pig is too small for the pipeline;
- Air/gas bypassing the pig before it is launched;
- Air/gas bypassing the pig in a fitting such as in a tee, or a steel shaft pig in a short radius bend; and,
- Other unknown reasons.

When an air lock condition occurs, it is necessary to either increase the displacing pressure or remove air/gas through existing vents or other connections at high points in front of the pig.

For these reasons, it is important that the dewatering process be carefully planned, especially for pipelines located in hilly country.

Cleaning pipelines

It is believed there are no written specifications to define the degree of cleaning for a pipeline since there is a question of the definition of a clean pipeline, and furthermore, a method for measuring the cleanliness has not been developed. It is known, however, that cleaning a pipeline does at least four good things. Cleaning the pipeline will:

- Improve flow efficiency because of a smoother pipe wall
- Reduce product contamination and formation of hydrates
- Reduce abrasive damage to pipeline appurtenances such as valves and instruments
- Facilitate pipeline drying

Internal cleaning of the pipeline may be accomplished by any of a combination of the following methods:

- Running a brush pig with air, gas, or liquid
- Internal sand blasting
- Chemical cleaning
- Purging with air or gas followed by a liquid flush

Brush pig run with gas

One brush pig pushed by gas or air will not usually remove a significant amount of small debris from the pipeline, particularly if the debris is wet. However, the scratching action of a good brush pig will more evenly distribute the loose debris and leave a smoother pipe wall, which will improve pipeline flow efficiency.

Brush pig run with liquid

A brush pig displaced by water is a very efficient means for removing debris from the pipeline provided certain precautions are taken. Experience has shown that a pig velocity greater than 3 ft per second is desired. A pig, such as the Power Brush Pig, that allows some of the liquid to bypass the pig to keep the bristles clean and move the loose debris into suspension in the liquid in front of the pig is also desirable. The debris must be free of any sticky material such as oil, grease, or paraffin.

It is not necessary that the pipeline be completely filled with liquid. Short slugs of liquid batched between two pigs are usually preferable as long as there is enough water to keep the pigs apart and maintain a velocity greater than 3 ft per second. Short slugs will minimize the water/debris disposal problems and will allow for higher velocities with lower pressures.

Disposal of the dirty liquid may present a problem, depending upon the location. As indicated earlier, environmental aspects should be investigated thoroughly during the pre-planning phase. Usually, rust will settle out of still water in about 1 hour, leaving an almost clear liquid. If the dumping of semi-clear water is acceptable, then the problem may be solved by the use of holding tanks.

The debris-holding ability of the propelling fluid is proportional to the viscosity/density of the propelling fluid. The greater the viscosity/density, the higher the holding ability. It has been reported that gels have been developed and used that are supposedly more effective than water.

Internal sand blasting

Internal sand blasting is an extremely effective method for cleaning pipelines. This method was extensively used when natural gas was inexpensive and there were no environmental controls. Some short, small-diameter pipelines are still being internally sand blasted by using air or nitrogen to propel the sand. This procedure, if properly executed, will leave the pipeline dry upon completion.

After the pipeline has been dewatered, clay is blown into short sections of the pipeline with a dry gas to absorb the remaining water. The correct type of sand is then blown through the dust dry section at a high velocity to remove the clay, rust, and mill scale. Caution must be taken to ensure that short sections are cleaned each time so that high velocities may be maintained; that large quantities of dry air or gas are available to sustain the high velocities; that the pipeline is dust dry; that all sand has been removed from the pipeline after the sand blasting is complete; and, that the pipeline facilities will not be damaged by the high-velocity sand.

Chemical cleaning

Cleaning with chemicals is typically used for smaller diameter lines. A properly planned chemical cleaning operation can produce a very clean pipeline. Chemical cleaning will remove rust from any internal pits. Chemical cleaning is accomplished by pushing a batch of hydrochloric acid and water for rust removal, a neutralizer, and a passivator, all separated by pigs, through the pipeline. This method will leave a bright metal finish on the interior of the pipeline. If a pipeline is being converted for a different service, such as a crude oil line being converted to chemical service, a batch of detergent can be added behind the cleaning process to remove any traces of the crude oil or other product.

Typically, chemical cleaning is more involved than other cleaning processes in that some special equipment will be required to safely handle the chemicals. Disposing of the spent material will also present a problem. Care must be exercised in the selection of pigs to be sure they will not be consumed by the cleaning solution. Personnel safety must be carefully considered when choosing this method of pipeline cleaning.

Pipeline drying

Pipelines used to transport petrochemicals such as propylene and ethylene must be dried in order for the delivered product to meet moisture specifications. Natural gas pipelines are usually dried to a lesser extent to prevent the formation of hydrates. It is not unusual for a petrochemical line to be dried to a dew point of −80°F. A typical dew point for a propylene pipeline will be −70°F. A carbon dioxide pipeline might be typically dried to a −40°F dew point. The natural gas industry specifies dryness in pounds of water per million standard cubic feet of gas. The table at the end of this section will enable you to convert from one method to the other.

Dew point, by definition, is the temperature at which water vapor begins to condense out of a gas at atmospheric pressure. For example, at atmospheric pressure, water vapor begins to condense out of a gas that has a moisture content of 7 lb. of water per million standard cubic feet at −39°F. Therefore, a gas that has a moisture content of 7 lb. of water per million standard cubic feet has a dew point of −39°F.

Air at +70°F with a humidity of 100% has a dew point of +70°F and holds approximately 1,253 pounds of water (150 gallons) per million standard cubic feet.

The most common methods for drying pipelines are as follows:

- Drying with super dry air
- Drying with methanol
- Drying with inert gas such as nitrogen
- Internal sand blasting
- Drying with the medium to be transported
- Vacuum drying

All of these methods may be applied to pipeline drying depending on the particular line and amount of dryness required. No single method can be considered ideal for all situations. Many times, a combination of two or more methods will be used to achieve a dry pipeline at the least cost. The first three methods are probably the most economical and technically feasible for most pipeline drying applications.

Before any type of drying operation commences, it will be necessary to clean the line using one of the previously described processes. If the rust and mill scale are not removed from the pipe wall, moisture will remain trapped and will bleed out over a long period of time. A pipeline can be dried without cleaning; however, the cost and time required will be great.

In *drying with super dry air*, soft foam pigs pushed by dry air are used to absorb any free water remaining in the pipeline after dewatering. After the line is dust dry, wire brush pigs are run to remove any water bearing debris from the pipe wall. The wire brush pigs are then followed by soft foam pigs to absorb the loosened debris. Near the end of this phase, the pigs may be weighed prior to insertion and after removal to monitor the amount of debris that is being removed. The first pigs run through the line will naturally weigh much more than their clean weight. As the cleaning and drying progress, the pig weights will approach their original weight. Pig color will also give some indication of how the debris removal is proceeding.

Dew-point readings will need to be made to determine when the line has been dried to the specified dew point.

It will be necessary to give special consideration to laterals, by-passes, and valve body cavities as any free water trapped here could affect the final dew point readings.

Drying with super dry air provides internal corrosion protection if the line is to remain out of service for some period of time before it is placed in service.

Methanol drying relies on the hygroscopic effect of the methanol. Any remaining moisture in the line will be absorbed by batches of methanol pushed through the line with either gas or dry air. Pigs are used to separate the methanol batches from the displacing medium.[1] Methanol drying usually requires fewer pig runs and, consequently, less line cleaning is accomplished. Pure methanol is expensive, and sometimes a 96% methanol/water mix is used. Since the methanol mix contains water, some water will be left in the pipeline. Some of the methanol will vaporize in the pipeline and will be absorbed by the displacing medium. Toward the end of the line, the moisture content of the methanol will increase, which in turn reduces the amount of water that it can absorb. If the pressure used in the drying operation is too high, hydrate formation can occur, usually at the far end of the line. If natural gas is being used to push the methanol batches, it will probably be necessary to flare some of the line fill volume to be sure that no methanol impurities are contained in the gas.[1]

Methanol run with a dry gas will absorb most of the water and facilitate the vaporization of the remaining water. Soft swabs run through a line with a dry purge gas will accelerate the evaporation of remaining methanol/water solution.

Also of great concern with the methanol method of drying is the fact that explosive mixtures can easily be formed, whether gas or air is used to displace the methanol batches. It may be desirable to use an inert gas such as nitrogen to buffer the methanol batch from the air or gas used to displace the methanol batch. However, many pipelines have been dried without the use of nitrogen buffers with no adverse results. The air/methanol mixture is also highly poisonous and corrosive.[1]

Plans will need to be made for proper disposal of the spent methanol.

If internal corrosion protection is desired, then another drying method should be considered. If the pipeline is to be used to transport sour gas, the methanol drying method should be carefully evaluated before it is used.

Drying with nitrogen is accomplished in much the same manner as when using super dry air. Nitrogen drying will cost more than super dry air by a factor of approximately 50%. Also, super dry air is plentiful and non-polluting.

Internal sand blasting will leave the pipeline clean and dry. The procedure for internal sand blasting is described earlier in this section.

Drying with natural gas requires large volumes of gas. This method is slow and not very effective unless the line is thoroughly cleaned by one of the cleaning processes previously described. If, however, the gas being used to dry with can be blended with another dry gas stream and sold or used, then this is an economical method for pipeline drying. The cost of the gas that will be used to purge the line during the drying process should be weighed against the cost of using super dry air. The Btu value of a volume of natural gas is approximately 15 times greater than that of the fuel required to produce an equal volume of super dry air.

Vacuum drying is a slow process. All free water should be removed from the pipeline before drying begins. This method appears to be used infrequently, and perhaps only offshore.

If the pipeline has been properly cleaned by the water slug method using brush pigs run with liquid, drying can be accomplished by running soft foam pigs with dry air or gas to remove any free water left in the pipeline. This will usually produce a pipeline dry enough for natural gas operations. If additional drying is desired, it can be accomplished by using methanol or super dry air.

Reference

1. Kopp. *Pipe Line Industry*, October 1981, January 1986.

Based on a paper entitled "The State of the Art of Drying Large Diameter Gas Pipelines after Hydrotest," by Marvin D. Powers, Pipeline Dehydrators, Inc., Houston, Texas.

Moisture content of air

Dew Point °F @ 14.7 PSIA	Dew Point °C @ 14.7 PSIA	PPM by Volume	PPM by Weight	LBS. Water MMSCF Dry Air
-110	-78.9	0.63	0.39	.032
-109	-78.3	0.69	0.43	.034
-108	-77.8	0.75	0.47	.038
-107	-77.2	0.82	0.51	.041
-106	-76.7	0.88	0.55	.044
-105	-76.1	1.00	0.62	.050
-104	-75.5	1.08	0.67	.054
-103	-75.0	1.18	0.73	.059
-102	-74.4	1.29	0.80	.065
-101	-73.9	1.40	0.87	.070
-100	-73.3	1.53	0.95	.077
-99	-72.8	1.66	1.03	.083
-98	-72.2	1.81	1.12	.091
-97	-71.7	1.96	1.22	.098
-96	-71.1	2.15	1.34	.108
-95	-70.6	2.35	1.46	.118
-94	-70	2.54	1.58	.127
-93	-69.4	2.76	1.72	.138
-92	-68.9	3.00	1.86	.150
-91	-68.3	3.28	2.04	.164
-90	-67.8	3.53	2.19	.177
-89	-67.2	3.84	2.39	.192
-88	-66.7	4.15	2.58	.221
-87	-66.1	4.50	2.80	.225
-86	-65.6	4.78	2.97	.239
-85	-65.0	5.30	3.30	.265
-84	-64.4	5.70	3.50	.285
-83	-63.9	6.2	3.9	.31
-82	-63.3	6.6	4.1	.33
-81	-62.8	7.2	4.5	.36
-80	-62.2	7.8	4.8	.39
-79	-61.7	8.4	5.2	.42
-78	-61.1	9.1	5.7	.46
-77	-60.6	9.8	6.1	.49
-76	-60.0	10.5	6.5	.53
-75	-59.4	11.4	7.1	.57
-74	-58.9	12.3	7.6	.62
-73	-58.3	13.3	8.3	.67
-72	-57.9	14.3	8.9	.72
-71	-57.2	15.4	9.6	.77

Dew Point °F @ 14.7 PSIA	Dew Point °C @ 14.7 PSIA	PPM by Volume	PPM by Weight	LBS. Water MMSCF Dry Air
-70	-56.7	16.6	10.3	.83
-69	-56.1	17.9	11.1	.96
-68	-55.6	19.2	11.9	1.0
-67	-55.0	20.6	12.8	1.03
-66	-54.4	22.1	13.7	1.1
-65	-53.9	23.6	14.7	1.2
-64	-53.3	25.6	15.9	1.3
-63	-52.8	27.5	17.1	1.4
-62	-52.2	29.4	18.3	1.5
-61	-51.7	31.7	19.7	1.6
-60	-51.1	34.0	21.1	1.7
-59	-50.6	36.5	22.7	1.8
-58	-50.0	39.0	24.2	2.0
-57	-49.4	41.8	26.0	2.1
-56	-48.9	44.6	27.7	2.2
-55	-48.3	48.0	30	2.4
-54	-47.8	51	32	2.6
-53	-47.2	55	34	2.8
-52	-46.7	59	37	2.9
-51	-46.1	62	39	3.1
-50	-45.6	67	42	3.4
-49	-45.0	72	45	3.6
-48	-44.4	76	47	3.8
-47	-43.9	82	51	4.1
-46	-43.3	87	54	4.4
-45	-42.8	92	57	4.6
-44	-42.2	98	61	4.9
-43	-41.7	105	65	5.3
-42	-41.1	113	70	5.7
-41	-40.6	119	74	6.0
-40	-40.0	125	80	6.4
-39	-39.4	136	85	6.8
-38	-38.9	144	89	7.2
-37	-38.3	153	95	7.7
-36	-37.9	164	102	8.2
-35	-37.2	174	108	8.7
-34	-36.7	185	115	9.3
-33	-36.1	196	122	9.8
-32	-35.6	210	131	10.5
-31	-35.0	222	140	11.1

Dew Point °F @ 14.7 PSIA	Dew Point °C @ 14.7 PSIA	PPM by Volume	PPM by Weight	LBS. Water MMSCF Dry Air
-30	-34.4	235	146	11.8
-29	-33.9	250	155	12.5
-28	-33.3	265	165	13.3
-27	-32.8	283	176	14.2
-26	-32.2	300	186	15.0
-25	-31.7	317	197	15.9
-24	-31.1	338	210	16.9
-23	-30.6	358	223	17.9
-22	-30.0	378	235	18.9
-21	-28.4	400	249	20.0
-20	-28.9	422	212	21.1
-19	-28.3	448	278	22.4
-18	-27.8	475	295	23.8
-17	-27.2	500	311	25.0
-16	-26.7	530	329	26.5
-15	-26.1	560	348	28.0
-14	-25.6	590	367	29.5
-13	-25.0	630	391	31.5
-12	-24.4	660	410	33.0
-11	-23.9	700	435	35.0
-10	-22.3	740	460	37.0
-9	-22.8	780	485	39.0
-8	-22.2	820	509	41.0
-7	-21.7	870	541	43.5
-6	-21.1	920	572	46.0
-5	-20.6	970	603	48.5
-4	-20.0	1020	634	51.0
-3	-19.4	1087	675	54.4
-2	-18.9	1153	716	57.7
-1	-18.3	1220	758	61.0

(table continued on next page)

DEW POINT °F @ 14.7 PSIA	DEW POINT °C @ 14.7 PSIA	PPM BY VOLUME	PPM BY WEIGHT	LBS. WATER/MMSCF DRY AIR	DEW POINT °F @ 14.7 PSIA	DEW POINT °C @ 14.7 PSIA	PPM BY VOLUME	PPM BY WEIGHT	LBS. WATER/MMSCF DRY AIR	DEW POINT °F @ 14.7 PSIA	DEW POINT °C @ 14.7 PSIA	PPM BY VOLUME	PPM BY WEIGHT	LBS. WATER/MMSCF DRY AIR
0	-17.8	1299	807	64.9	40	4.4	8874	5514	444	80	26.7	34844	21650	1742
1	-17.2	1378	856	68.9	41	5.0	9272	5761	464	81	27.2	36138	22454	1807
2	-16.7	1458	906	74.3	42	5.6	9670	6008	484	82	27.8	37432	23258	1872
3	-16.1	1537	955	76.9	43	6.1	10068	6256	503	83	28.3	38727	24062	1936
4	-15.6	1616	1004	80.8	44	6.7	10466	6503	522	84	28.9	40021	24866	2001
5	-15.0	1695	1053	84.8	45	7.2	10824	6725	541	85	28.4	41315	25670	2066
6	-14.4	1775	1103	88.8	46	7.8	11182	6948	559	86	30.0	42609	26474	2130
7	-13.9	1854	1152	92.7	47	8.3	11540	7170	577	87	30.6	43903	27278	2195
8	-13.3	1933	1201	96.7	48	8.9	11898	7393	595	88	31.1	45198	28083	2260
9	-12.8	2053	1276	103	49	9.4	12255	7614	613	89	31.7	46492	28887	2325
10	-12.2	2174	1351	109	50	10.0	12613	7837	631	90	32.2	48146	29914	2407
11	-11.7	2294	1425	115	51	10.6	12971	8059	649	91	32.8	49801	30943	2490
12	-11.1	2415	1501	121	52	11.1	13329	8282	666	92	33.3	51455	31971	2573
13	-10.6	2535	1575	127	53	11.7	13687	8504	684	93	33.9	53109	32998	2655
14	-10.0	2656	1650	133	54	12.2	14268	8865	713	94	34.4	54764	34027	2738
15	-9.4	2776	1725	139	55	12.8	14849	9226	742	95	35.0	56419	35055	2821
16	-8.9	2897	1800	145	56	13.3	15430	9587	772	96	35.6	58023	36083	2904
17	-8.3	3017	1875	151	57	13.9	16011	9948	801	97	36.1	59728	37111	2986
18	-7.8	3197	1986	160	58	14.4	16593	10310	830	98	36.7	61382	38139	3069
19	-7.2	3376	2098	169	59	15.0	17174	10671	859	99	37.2	63477	39440	3174
20	-6.7	3556	2209	178	60	15.6	17754	11031	889	100	37.9	65571	40741	3279
21	-6.1	3735	2321	187	61	16.1	18336	11393	917	101	38.3	67666	42043	3383
22	-5.6	3915	2432	196	62	16.7	18917	11754	946	102	38.9	69760	43344	3488
23	-5.0	4094	2544	205	63	17.2	19685	12231	984	103	39.4	71855	44646	3593
24	-4.4	4274	2656	214	64	17.8	20452	12707	1023	104	40.0	73950	45948	3698
25	-3.9	4453	2767	223	65	18.3	21220	13185	1061	105	40.6	76045	47249	3802
26	-3.3	4633	2879	232	66	18.9	21987	13661	1099	106	41.1	78139	48550	3907
27	-2.8	4883	3034	244	67	19.4	22755	14138	1138	107	41.7	80234	49852	4012
28	-2.2	5133	3189	357	68	20.0	23522	14615	1176	108	42.2	83861	51484	4193
29	-1.7	5383	3345	269	69	20.6	24290	15092	1215	109	42.8	85489	53117	4274
30	-1.1	5633	3400	282	70	21.1	25057	15569	1253	110	43.3	88116	54749	4406
31	-0.56	5883	3655	294	71	21.7	25825	16046	1291					
32	0	6133	3811	307	72	22.2	26827	16668	1341					
33	0.56	6383	3966	319	73	22.8	27829	17291	1391					
34	1.1	6633	4121	332	74	23.3	28831	17914	1442					
35	1.7	6883	4277	344	75	23.9	29833	18536	1492					
36	2.2	7281	4524	364	76	24.4	30836	19159	1542					
37	2.8	7679	4771	384	77	25.0	31838	19782	1592					
38	3.3	8077	5019	404	78	25.6	32840	20405	1642					
39	3.9	8475	5266	424	79	26.1	33842	21027	1692					

Commissioning ethylene pipelines

Marvin D. Powers, Pipeline Dehydration, Inc., Houston, Texas

Pipelines carrying a variety of gaseous and volatile liquid products are extensively employed in today's industry as a safe and efficient means of transportation. New pipelines must be placed into service initially (i.e., commissioned), and older lines occasionally are taken out of service in order to perform some desired maintenance and are then recommissioned. Among the reasons for taking a pipeline out of service (i.e., decommissioning) are: hydrostatic testing to recertify or upgrade the pipeline's ability to be used at higher operating pressures; performance of construction work on the pipeline; and a change in the product transported by the pipeline. Pipelines, or sections of a pipeline, may need to be relocated because of highway work, the necessity to deepen a canal, or because of increases in the population surrounding the pipeline. It may also be necessary to replace valves, fittings or a damaged section of the pipeline or add a new connection to service a customer or supplier.

Typically, decommissioning and recommissioning a petrochemical pipeline will include the steps of decommissioning by removing the product from the pipeline, and flaring any remaining residual product. Any necessary construction, upgrading, or cleaning of the pipeline can then be performed. Usually, the pipeline is then filled with water for hydrostatic testing. After pressure testing, the water is removed, the pipeline is cleaned and dried to a specific low dewpoint (to avoid the problems of water contamination of the product), and the pipeline is inerted with nitrogen for recommissioning. Refer to the section on Pipeline Drying for drying techniques.

In order to recommission the pipeline, the nitrogen must be displaced by the desired petrochemical. Before returning to service, product purity must be established and the line safely filled to operating pressure. The terms "commissioning" and "recommissioning" are used interchangeably to refer to a process whereby a first inerting gas, normally nitrogen, in a pipeline is replaced with the desired product at the desired purity and pressure.

Safety and economics are two primary concerns for any proposed pipeline operation. Thus, a commissioning process that brings the purity and pressure of potentially explosive products such as ethylene or propylene up to specification quickly, but at the risk of damage to the pipeline or reduced safety to operating personnel, is not acceptable. Similarly, a process that uses large quantities of product to push nitrogen from a line and results in wasted product/nitrogen mixtures that must be flared or otherwise disposed of would also be unacceptable. Improvement over known methods requires consideration of its safety and economic benefits (considering both the cost of wasted product and the cost attributed to the time the line must remain out of service) as well as its ability to bring the line back into service with product at desired pressures and purities.[1]

The main problems posed by recommissioning are (1) the fast and economical purging of nitrogen so as to obtain uncontaminated products in the line, and (2) possible damage to the pipeline due to cold temperatures to which it may be subjected during the process. The latter can be a problem when the pipeline is to be recommissioned with a product at a pressure significantly higher than that of the nitrogen inerted line. The pressure drop of a petrochemical product entering a pipeline can cause large drops in temperature and, consequently, potential damage to the carbon steel pipeline.

Due to the demands of the expanding petrochemical industry, and the aging of the present pipeline system, the need for decommissioning, repairing, cleaning, drying, and recommissioning pipelines is increasing. Given the value of today's petrochemical products, a recommissioning procedure, that brings the product purity to acceptable levels as quickly as possible, is desirable. Furthermore, processes and equipment which reduce the possibility for damage to the pipeline due to cold temperature conditions which can occur during the recommissioning process are also highly desirable.[2]

Due to the unusual properties of ethylene, decommissioning and recommissioning ethylene pipelines require special consideration. Ethylene is a colorless, flammable, gaseous, unsaturated hydrocarbon obtained by the pyrolysis of petroleum hydrocarbons. The relative density of ethylene is 0.9686 (air = 1), and its molecular weight is 28.054. The molecular weight of air is 28.964, and the molecular weight of nitrogen is 28.013.

Ethylene is subject to thermal decomposition (self propagating reaction zone, explosion, or flame) under certain circumstances. Thermal decomposition occurs when the temperature of a substance is raised above the value needed to cause it to begin to self-heat because of a change in molecular structure at a rate high enough to result in combustion. An external source of ignition is not required to precipitate this process, nor is a source of oxygen needed. Thermal decomposition is usually initiated by a source of heat from events such as sudden compression, external source of heat, or the thermal runaway of a heater.[4] Heaters are sometimes used in ethylene measurement stations to maintain the ethylene at an optimum temperature for measurement.

Ethylene is routinely handled under conditions where decomposition may be initiated. The point at which decomposition begins is dependent upon system parameters such as line or vessel size, operating pressure, and operating temperatures. Generally, ethylene decomposition is initiated in pipeline facilities by a rapid rise in temperature associated with sudden compression in the presence of a diatomic gas such as nitrogen or oxygen, exposure of the pipe to flame, possibly hot tapping operations.

The autoignition temperature (AIT) for ethylene is 914°F at atmospheric pressure.[3] Experience of others suggests that at 1,000 to 1,500 psig, the AIT could drop to as low as 400° to 700°F.[3,4] As the concentration of oxygen increases, the decomposition initiation temperature decreases.[4] From this, it may be concluded that a mixture of low oxygen concentration and ethylene could be very hazardous. For this reason, ethylene pipelines, which have been evacuated for maintenance or repairs, are usually filled with nitrogen to displace oxygen from the pipe. Compressing nitrogen with ethylene may result in high nitrogen temperatures—especially if the compression is rapid. Also, when the pressure of the ethylene source is high, refrigeration across the throttling valve used to control the ethylene may result in extremely low temperatures and could cause the valve to fail due to excessively low temperature.

Commissioning an ethylene pipeline with warm ethylene (±70°F) at a low pressure (±50 psig) is quick, easy, inexpensive, and very safe when the ethylene is heated before it is admitted to the pipeline being commissioned.

A patented process used by Pipeline Dehydrators can be used to safely commission ethylene pipelines. The process uses a shell and tube heat exchanger to heat the high pressure ethylene before the pressure is reduced to the low pressure desired for line commissioning.

The methods currently used to commission ethylene pipelines are potentially hazardous, expensive, and time consuming. The two most severe problems are potential damage to the carbon steel in the pipeline due to the cold temperature of the expanded ethylene and contamination of the ethylene from the nitrogen used to inert the pipeline during commissioning.

An ethylene pipeline that is being commissioned has usually been dried to a −70°F dew point or more and inerted with nitrogen. The nitrogen pressure left on the line for commissioning varies from 10 to 1,000 psig depending on the pressure of the ethylene source and the commissioning procedure used and the pipeline owner's preference.

The source of ethylene used for commissioning an ethylene pipeline is usually more than 900 psig and may be up to 2,200 psig.

If the pressure of the ethylene source used for commissioning is high and the nitrogen pressure in the pipeline being commissioned is low, the resulting temperature of the ethylene, due to the pressure drop, may be well below the −20°F design temperature of most carbon steel pipelines.

The carbon steel in some pipelines is subject to becoming brittle at temperatures below −20°F and may fail at a very low internal pressure because of the low temperature.

At low pressures, the very cold ethylene will be much more dense than nitrogen, thereby producing an extensive product interface during commissioning or trapping pockets of the less dense nitrogen at each high place in the pipeline.

In this case, the amount of ethylene burned to produce product purity may be expensive and time consuming and may create environmental problems in addition to subjecting the steel in the pipeline to sub-design temperatures.

If the pressure of the ethylene source used for commissioning is high, and the nitrogen pressure in the pipeline being commissioned is also high, cold temperatures will not be encountered. However, even at warm temperatures, high-pressure ethylene is still much more dense than nitrogen at the same pressure. Therefore, the extensive product interface and the tapped nitrogen at the high points will still occur, resulting in an expensive and time consuming purification process plus the cost of the large volume of nitrogen. Also, there is a large inventory of ethylene in a pipeline at high pressure. If a pipeline blow-down is required because of a leak, lack of purity, or for whatever reason, it will be a major product loss and expense.

The Pipeline Dehydrators system utilizes a tube and shell heat exchanger and a water–glycol mix that is used as the heat transfer fluid. A pump forces the heat transfer fluid through the fired heater and then through the shell side of the exchanger. The ethylene passes through the tubes of the heat exchanger and is safely and properly warmed by the heat transfer fluid without entering the direct fired heater, thereby eliminating the possibility of an ethylene decomposition due to heater thermal runaway.

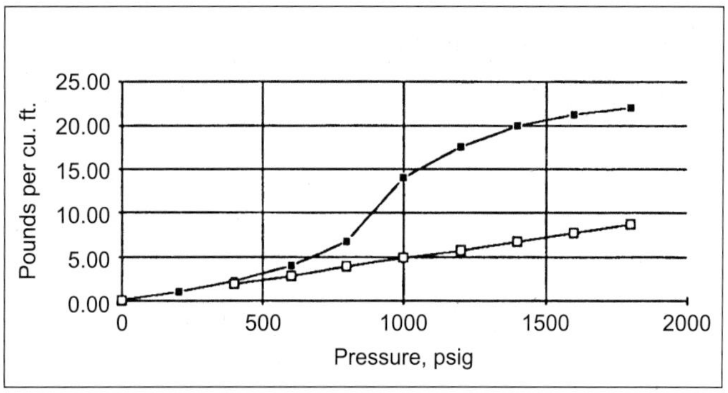

Figure 1. Ethylene-Nitrogen density comparison @ 80 °F.

The heater will produce ethylene at ground temperature (±70°F) at ±50 psig corresponding to the conditions of the nitrogen in the pipeline. Under these conditions, ethylene and nitrogen have basically the same density; therefore, the product interface during commissioning will be insignificant; no nitrogen will be trapped at the high points in the pipeline; the steel in the pipeline will not be subjected to cold temperatures; purification will be simple, short, and inexpensive; and the ethylene inventory in the pipeline will be

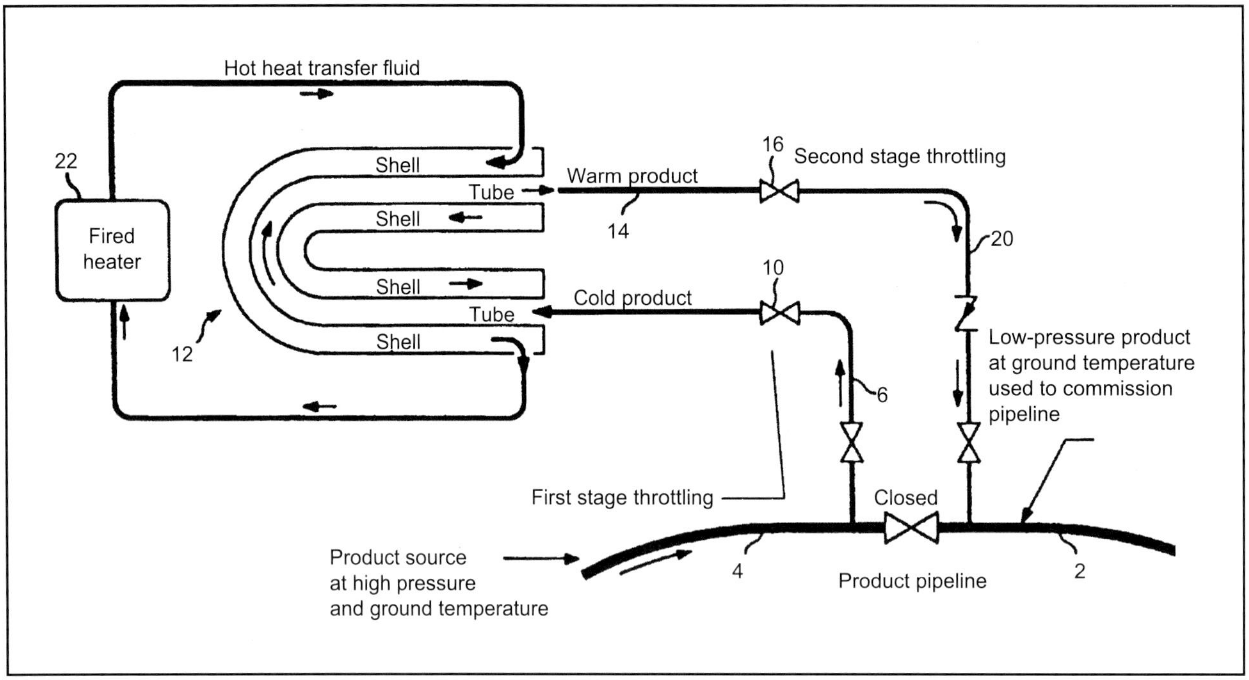

Figure 2. Pipeline dehydrator system utilizing a tube and shell heat exchanger and a water-glycol mix.

minimal in case a pipeline blow-down is required. The savings on nitrogen will also be significant.

Two pressure drops will be taken on the ethylene, one on each side of the tube-shell heat exchanger. The pressure will be dropped to approximately 600 psig prior to the ethylene entering the heat exchanger. The 600 psig ethylene will be warmed to approximately 140°F. Then a second pressure drop will occur, allowing the pressure and temperature of the ethylene to equalize with the pressure and temperature of the nitrogen in the pipeline (±50 psig @ ±70°F). Under these conditions, PLD's 6 million BTU per hour heater can handle up to 45,000 lb. of ethylene per hour. This high flow rate will normally be used only during the pressuring up phase of the process and not used during the purification process. The warm ethylene at low pressure, traveling about 10 mph in the small pipelines that are typically used to transport ethylene, will produce excellent results during the purification phase of the process.

The minimum ethylene temperature encountered in this commissioning process is well above 0°F, and the maximum temperature is 160°F. This is a very safe temperature range to handle ethylene in carbon steel piping systems.

Purity can be established at 50 psig or less, with very little loss of product, and the system can be checked for leaks with ethylene in the pipeline while the pressure is low and the dollar value of ethylene in the pipeline is still small.

After purity has been established at ±50 psig and potential leak points have been checked, the major remaining effort will be to warm enough ethylene to bring the pipeline up to approximately 600 psig. Beyond this point, the ethylene can be throttled into the pipeline without heating, because the cooling due to pressure drop will no longer be great enough to cause damage to the pipeline and equipment. During this pressure-up phase of the commissioning operation, the ethylene does not have to be heated to +70°F temperature and the pipeline pressure will be increasing, both resulting in a reduction in the amount of heat required per pound of ethylene. Therefore, for this phase of the commissioning operation, PLD's new ethylene heater will handle considerably more than 45,000 lb. of ethylene per hour.

This system provides exscellent conditions for commissioning an ethylene pipeline quickly, inexpensively, and safely.

This heater can also be utilized to commission other product pipelines where cold temperatures are encountered due to pressure drops such as with propylene, ethane, propane, and carbon dioxide.[2]

References

1. US Patent No. 5,027,842 July. 2, 1991 Powers.
2. M.D. Powers. "*Ethylene Heater For Commissioning Pipelines*" 5—89 Pipeline Dehydrators.
3. C.J. Hilado and S.W. Clark. "*Autoignition Temperatures of Organic Chemicals*" Chemical Engineering September 4, 1972, p 76.
4. F.F. McKay, G.R. Worrell, B.C. Thornton, and H.L. Lewis. "*How To Avoid Decomposition In Ethylene Pipelines*," Pipe Line Industry August 1978, p 80.

Excerpts from the paper on Ethylene Heater For Commissioning Pipelines by Marvin D. Powers. Reprinted with permission.

Vacuum drying

The vacuum drying process is shown graphically in Figure 1 and consists of three separate phases. Corrosion is generally inhibited at relative humidity (R. H.) levels below 30%, but in the presence of hygroscopic dirt (present in millscale) corrosion can occur at R.H. levels of 20%. Therefore, systems should be thoroughly drained and then vacuum dried to lower than 20% R.H.

Target Dewpoint (D.P.)

$$\text{R.H} = \frac{\text{Water vapor pressure}}{\text{Saturated Vapor Pressure (SVP)}}$$

for a constant temperature. Assume the lowest average temperature of the system is 0°C (during the winter months), which gives

$$\text{SVP} = 0.6 \text{ KPaA}$$

The vacuum level required for water vapor at 20% R.H. is:

$$0.2 \times \text{SVP} = 0.2 \times 0.6 = 0.12 \text{ KPaA}$$

This corresponds to a dewpoint of −18°C. A safety allowance should be provided for some desorption from the pipe walls; and therefore a dewpoint of −20°C would be used.

Phase 1–Evacuation

During this phase, the pressure in the pipeline is reduced to a level where the ambient temperature of the pipeline will cause the free water to boil and change to water vapor. This pressure level corresponds to the saturated vapor pressure of the free water in the pipeline, which is dependent upon the ambient temperature of the pipeline.

The approximate pressure value is calculated in advance but is easily recognized on site by a fall in the rate of pressure reduction, which is noted from the plot of pressure against time.

At some convenient point in time a "leak test" is carried out by stopping the vacuum equipment and observing the pressure, usually for a period of 4 hours. Any "air-in" leaks on flanges, fittings, or hoses are rectified at this time, although leaks are not a common occurrence.

Phase 2–Evaporation

Once the saturated vapor pressure has been reached, then evaporation of the free water into water vapor will commence. During this phase, the vacuum equipment is carefully controlled to maintain the pressure at a constant level until all the free water has been converted into water vapor. This phase may take several days to complete depending on: (1) the amount of water to be evaporated; (2) the size of the vacuum equipment; and (3) the ambient temperature of the pipeline. The end of the evaporation phase will be observed on site by a noticeable decrease in pressure.

At this time it is prudent to carry out a "soak test" to ensure that all the free water has in fact evaporated. The vacuum equipment is temporarily isolated from the pipeline, usually for a period of 12 hours, and a careful note made of the pressure. If all free water has evaporated then the pressure will remain constant and the final drying phase can be commenced.

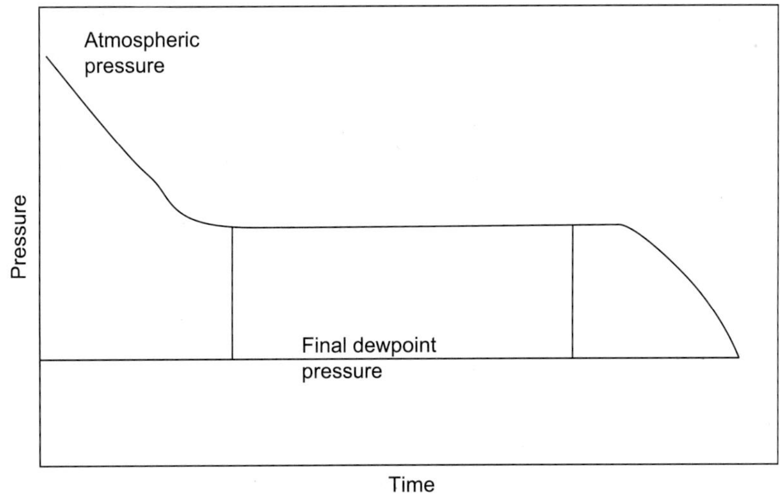

Figure 1. Vacuum drying pressure vs time curve.

Phase 3–Final drying

Once the free water has been converted into water vapor, the majority of it must be removed from the pipeline in order to reach the required dryness level. This is achieved by reducing the pressure in the pipeline still further which has the effect of drawing the water vapor out of the pipeline through the vacuum equipment. Obviously, the more water vapor removed, the drier the pipeline will become.

During this phase a careful watch is kept on the slope of the final drying line to ensure that it follows the calculated value, since a shallower slope would indicate the continuing presence of some free water still remaining in the pipeline.

The necessary calculations to determine the time required for each phase of vacuum drying are detailed in Figure 2.

Evacuation:

$$t = \frac{V}{Se} \log_e \frac{Pi - Pv}{Pf - Pv}$$

$$Se = \frac{1}{C} + \frac{1}{Sp}$$

Evaporation:

$$t = \frac{M}{SeDv}$$

Final drying:

$$t = \frac{V}{Se} \log_e \frac{Pv}{Preq}$$

Pi = Initial pressure
Pf = Final pressure
Pv = Vapor pressure
Se = Effective pump speed
V = System volume
Sp = Pump speed
C = Conductance
M = Mass of water
Dv = Vapor density
Preq = Required dewpoint pressure

Figure 2. Vacuum drying calculations.

What is dryness?

The dryness of a pipeline is measured in terms of dewpoint, which is the temperature at which mist or dew will begin to form. A convenient method of measuring dewpoint is to use an instrument called a mirror hygrometer where the water vapor is passed across a polished surface that is slowly cooled until dew forms. The temperature at which the dew forms is the dewpoint of the water vapor and is normally expressed in degrees centigrade. The drier the air, the lower the temperature at which dew will form.

In terms of a pipeline being vacuum dried, the lower the pressure in the pipeline, the lower the dewpoint will be. For example, at a pressure level of 0.26 kPaA, the equivalent dewpoint of the pipeline would be −10°C. If the pressure were further reduced to 0.104 kPaA, then the dewpoint would be −20°C.

For gas pipelines a dewpoint level of −20°C is generally considered to be adequate and the 0.10 kPaA bar pressure level required to achieve this dewpoint is readily attainable using the portable vacuum equipment described previously.

For example, consider a 100-mile-long 36-inch-diameter pipeline that, prior to drying, contained 10,000 gallons of water as a film on the inside surface of the pipe. At a dewpoint of −10°C, the quantity remaining would be reduced to only 46.5 gallons and at −20°C to 19.7 gallons. Also, this water would not be free water but rather water vapor, and it would only revert to free water if the ambient temperature of the pipeline were further reduced. This water vapor can subsequently be removed from the pipeline during the purging operation.

The relations between pressure and dewpoint are shown in Figure 3 and Table 1.

Proving the dryness

Immediately following the final drying phase, a dry gas purge using atmospheric air or nitrogen is carried out to prove the dryness of the pipeline. It is possible, under certain circumstances, for a small amount of free water to still remain in the pipeline. Usually this water will have turned to ice due to the chilling effect of the vacuum drying process and may not be apparent during the final drying phase or soak test.

Nitrogen or atmospheric air is allowed to enter the pipeline through a valve at the end remote from the vacuum equipment until the pressure has risen to the SVP equivalent of the target dewpoint.

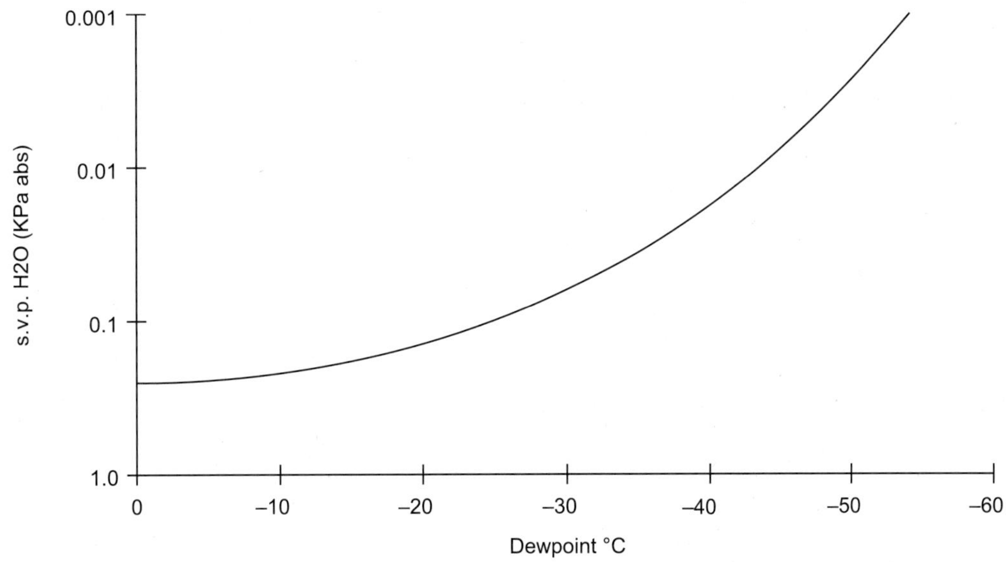

Figure 3. Dewpoint temperature versus saturated vapor pressure.

Table 1
Water Vapor Pressure Table

T°C	SVP (kPaA) (mb × K 10^{-1})	Vapor Density (g/m^3)
−50	0.0039 (0.039 mb)	0.038
−45	0.0128	0.119
−35	0.0223	0.203
−30	0.0308	0.339
−25	0.0632	0.552
−20	0.1043	0.884
−15	0.1652	1.387
−10	0.2597	2.139
−5	0.4015	3.246
0	0.6108	4.847
5	0.8719	6.797
10	1.2270	9.399
15	1.7040	12.830
20	2.3370	17.300
25	3.1670	23.050
30	4.2430	30.380

Once this pressure level has been reached, the vacuum equipment is started and that pressure level maintained. This has the effect of drawing gas through the pipeline under vacuum at a relatively constant dewpoint equal to the final dewpoint required.

At some point in time the purge gas, now under vacuum at a dewpoint of say, −20°C, will reach the vacuum equipment and be pulled through it. The dewpoint at both ends of the pipeline is carefully monitored and compared. If there is no free water remaining in the pipeline then the dewpoint at the vacuum equipment end will be the same as the dewpoint at the remote end. However, if there is any free water present then the dry air passing through the pipeline under vacuum will absorb the water hygroscopically. The dry gas purge operation must then continue to remove the remaining free water until the dewpoints at both ends are equal, at which time purging is discounted. The pipeline has now been vacuum dried to the required dewpoint level, and the dryness proved.

Purging and commissioning

Once the dryness has been attained and proved, the pipeline is ready for commissioning. It is possible to introduce the gas directly into the vacuum or to relieve the vacuum using dry nitrogen.

Pig trap design

The primary purpose of a pig trap is to provide a means for safely and without flow interruption inserting and launching or receiving and retrieving a pig from a pipeline.

For sphere launchers utilizing an automatic retractable flap system, a larger oversize may be required. Consult the manufacturer. The length of the barrel will need to be sized for the number of spheres to be handled.

The reducer between the major barrel and the minor barrel should be eccentric for horizontal traps (with the bottom of the major and minor barrel at the same level) and concentric for vertical traps.

The launcher should be designed for the same working pressure as the pipeline and to the same design code. This will usually be ASME B 31.4 for pipelines transporting liquids and ASME B 31.8 for pipelines transporting natural and other gas. It may be necessary to design to ASME B 31.3 for pipelines located within the boundaries of a processing plant. If the pipeline is subject to regulation by a governmental agency, it will be necessary to conform to the requirements of the regulation. In the US, the regulations for pipelines transporting hazardous liquids can be found in 49 CFR Part 195 and the regulations for transporting natural and other gases can be found in 49 CFR Part 192. Other design codes may apply depending on the location. If any electrical apparatus is attached to the trap, e.g. a pig sig, the device must be compatible with the product being handled. For example, devices rated for Group D would be suitable for crude oil. Systems handling ethylene will require devices suitable for Group C materials. Consult NFPA 70 for complete information on the group classification.

Two types of pigs (maintenance pigs and intelligent pigs) are typically used in pipeline operations. The trap shown here is designed for handling intelligent pigs that are used for performing in-line inspection.

The trap should be designed to accommodate the length of the longest pig (plus 10%) anticipated to be used. This is usually an intelligent pig. The major barrel is normally two pipe sizes larger than the size of the pipeline for pipelines with a diameter less than 20 in and for pipelines larger than 20 in in diameter, the oversize should be 4 in. A quick opening closure is welded to the end of the major barrel to allow for the quick insertion and removal of pigs. The closure must be fitted with a device to prevent opening the closure until the pressure in the barrel has been relieved. The closure must have a pressure and temperature rating equal to the pipe to which it is attached. The closure should be designed in accordance with the ASME Boiler and Pressure Vessel Code, section VIII, Pressure Vessels, Division 1. If intelligent pigs will be used infrequently, e.g. every 5–10 years, the major barrel may be shortened to facilitate handling maintenance type pigs. If the barrel is to be shortened, a flange should be installed between the hinged closure and the major barrel. The barrel can be extended to accommodate intelligent pigs by inserting a flanged extension.

A balance line is used to enable filling and pressurizing the barrel on both sides of the pig at the same time. This will prevent the pig from possibly moving forward and damaging the pig trap valve or from moving backward and losing the seal in the reducer. Refer to Table 1 for balance line sizing.

A vent valve should be provided near the trap valve for a launcher and near the closure for a receiver. For liquid systems, a 2″ vent is normally adequate. For high pressure gas systems, consideration should be given to a blow down line

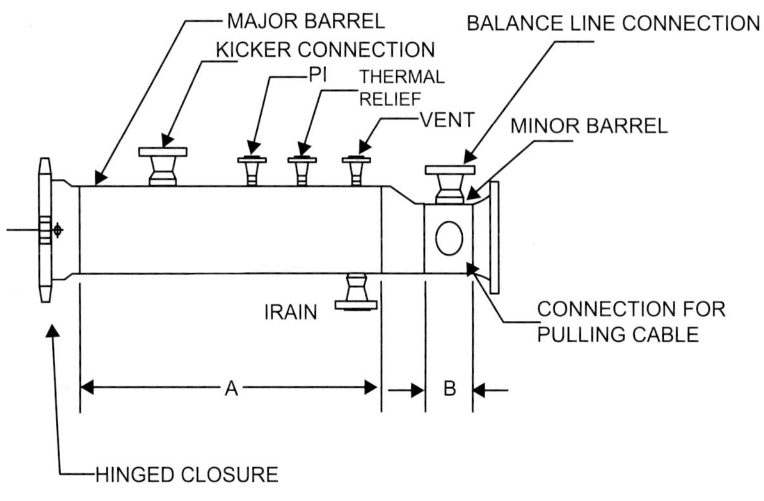

Figure 1. Typical pig launcher.

Pipeline Diameter	Bypass Line	Kicker Line	Balance Line	Drain Line	Major Barrel Diameter	Major Barrel "A"	Minor Barrel "B"	Major Barrel "A"	Minor Barrel "B"
						Launcher		Receiver	
4	3	2	2	2	8	110	20	110	110
6	4	2	2	2	10	110	59	110	110
8	4–6	4	2	2	12	161	59	154	154
10	6	4	2	2	14	169	59	169	169
12	6–8	4	2	2	16	169	59	169	169
14	6–10	4	4	2	16	189	59	189	189
16	8–12	6	4	4	20	200	59	200	200
18	10–12	8	4	4	22	200	59	200	200
20	10–16	8	4	4	24	200	59	200	200
24	12–18	8	4	4	28	224	59	224	224
28	16–20	10	4	4	32	228	59	228	228
30	16–24	10	4	4	36	236	59	236	236
32	16–24	10	4	4	36	260	59	260	260
36	18–28	10	4	4	40	260	59	201	260
38	20–28	12	4	4	42	260	59	216	260
40	20–32	12	4	4	44	260	59	216	260
42	20–36	16	4	4	46	260	59	252	260
48	24–36	18	4	4	52	260	59	260	260
56	32–40	20	4	4	60	260	59	260	260

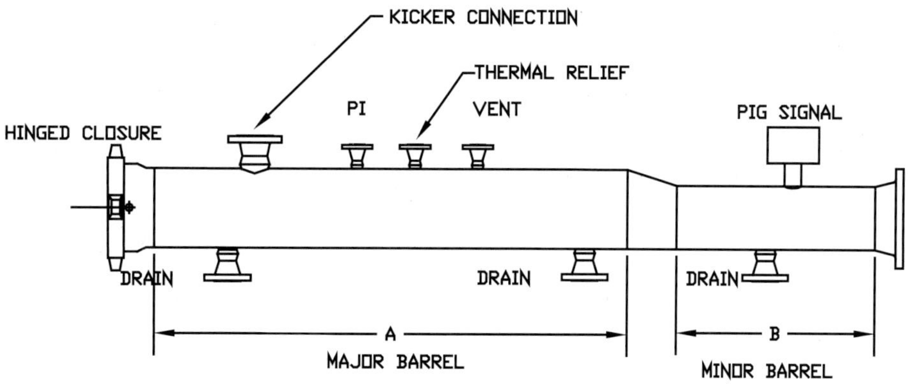

Figure 2. Typical pig receiver.

equipped with a valve that can be used for depressurization of the trap. For systems handling other materials such as ethylene, consideration will need to be given to the reduced temperature as a result of throttling a high pressure to a low pressure. For high pressure gas systems, consideration needs to be given to the thrust developed during the blow down operation.

The pig trap and bypass valves should be motor operated for 10 in and larger valves. The pig trap and kicker valves should be tight shutoff valves to allow for complete isolation of the trap from the main line. Refer to Figure 3 for valve identification.

The closure should be hinged or held by a built-in davit to prevent dropping and to facilitate handling during operation. The "O" ring seal should be carefully examined every time the closure is opened. If the pipeline is in HVL service, the O ring should be replaced every time the closure is opened.

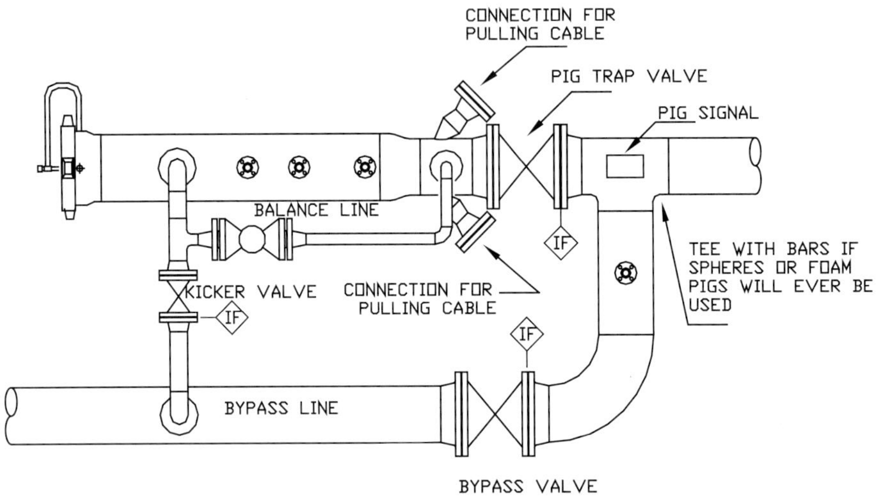

Figure 3. Typical launcher configuration.

For horizontal traps, the reducer between the major and minor barrels should be an eccentric type with the flat side down so that the bottom of both barrels will be at the same level.

Flange insulating kits should be used to isolate the pipeline system from station/terminal system for cathodic protection purposes.

7: Control Valves

Control valve sizing formulas 248
Sizing control valves for throughput 252
Control valve selection 257
Relief valve sizing, selection, installation, and testing 259
Rupture disk sizing methodologies 263
Rupture disk sizing using the resistance to flow method (K_R) 264
Variable orifice rotary control valves 266
Basic valve flow-capacity coefficient (C_v) 268
Visualize pump and control valve interaction 272
Avoid cavitation in butterfly valves 278
How to read a regulator flow curve 281

Control valve sizing formulas

Liquid formulas (noncompressible fluids).

Volume basis:

$$C_v = q\sqrt{G/\Delta P} \tag{1}$$

Weight basis:

$$C_v = W/(500\sqrt{\Delta P \times G}) \tag{2}$$

Equations 1 and 2 exclude control valve calculations where flashing, high-viscosity, two-phase flow, or high pressure drops are involved.

Gas and vapors (other than steam).

Volume basis:

$$C_v = (Q\,/\,1360)\sqrt{2GT/[\Delta P(P_1 + P_2)]} \tag{3}$$

Equation 3 does not take into consideration gas compressibility, mixed phase flow or high pressure drop. Compressibility is considered in Equations 12 and 13. Usable pressure drop for valve sizing calculations for gas, vapor, and steam flows is always limited to half of the absolute inlet pressure of the control valve. At that approximate value, critical (sonic) velocity is reached and any further reduction of the downstream pressure will not increase the velocity through a valve, thus, there will be no increase in flow.

Steam formula.

$$C_v = \frac{W(1.0 + 0.0007\ \mathrm{Tsh})}{2.1\sqrt{\Delta P(P_1 + P_2)}} \tag{4}$$

Special conditions

Equations are given below to correct for flashing, cavitation, high-viscosity fluids, and mixed phase flow, where the simplified equations may lead to inaccurate sizing. These additional expressions, together with the FCI equations, provide satisfactory solutions for most control valve sizing applications.

Liquids

Flashing service. When flashing occurs, the simplified liquid sizing equation gives erroneous results. Flashing occurs when liquids enter a valve at or near their boiling points, and they begin vaporizing as their pressures decrease, due to conversions of static pressure to velocity heads. With the advent of high recovery valves, this condition has been aggravated. High recovery valves take a much higher pressure drop from inlet to valve orifice than standard control valves (see Figure 1). When the pressure at the orifice drops to the vapor pressure of the fluid at flowing temperature, the maximum valve capacity is reached. This is true, because flashing occurs across the valve orifice and a flow condition similar to the critical flow of vapors exists.

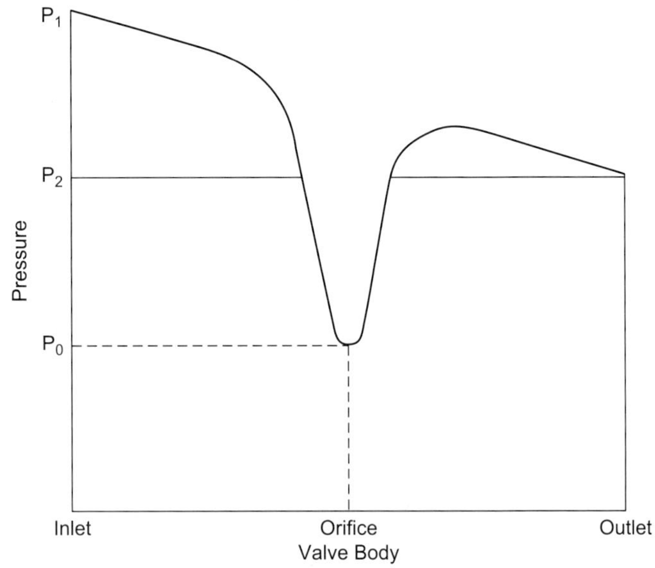

Figure 1. Typical control valve pressure gradient. The ΔP between P_1 and P_2 is much less for high recovery valves.

Two approaches take into account the flashing condition:

1. Hans D. Baumann, in a paper presented in *ISA Transaction*, Volume 2, Number 2 (April, 1963), introduced an equation using a C_f factor to help determine the valve capacity when flashing occurs. The equation, based also on the use of liquid vapor pressure (P_v), is

$$C_v = \left(Q\,/\,C_f\right)\sqrt{G/(P_1 - P_v)} \tag{5}$$

2. C. B. Schuder, Fisher Controls, in a paper, "How to Size High Recovery Valves Correctly," proposed an equation to solve for the maximum pressure drop to be used for liquid sizing:

$$\Delta P_m = K_m(P_1 - r_c P_v) \tag{6}$$

(See Figure 2 for values for r_c).

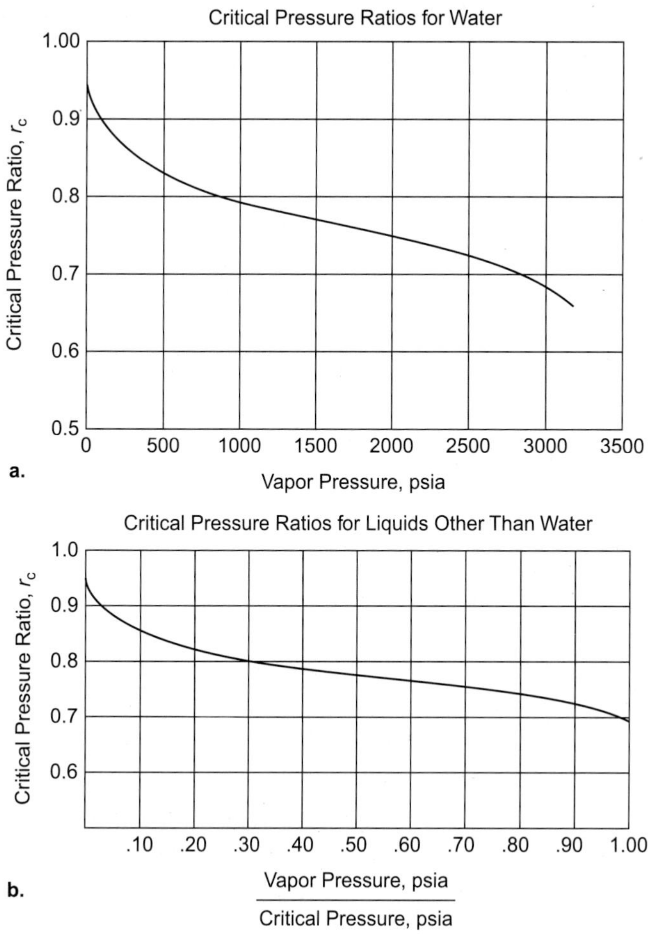

Figure 2. Critical pressure ratios (r_c) for use with Equation 6.

When this limiting pressure drop from Equation 6 is used in the liquid flow formula, Equation 1 gives a valve size in which flashing will not occur unless the pressure drop increases above the calculated ΔP_m. When the ΔP exceeds ΔP_m, flashing and possible cavitation can occur. Cavitation is the effect caused by the sudden condensing of vapors when the static pressure increases to a point above the boiling point of the liquid.

When the possibility of cavitation exists, the process should be reviewed and the condition eliminated if at all possible. A high recovery valve should not be used in this application. This second approach is suggested as a good solution to the flashing problem.

Viscosity Correction. The liquid sizing formulas in general use have no factor applied to correct for viscosity. This correction is a complex function involving the Reynolds number, which varies with the valve size, port area, velocity, and viscosity. No correction is required for a viscosity value below 100 SSU. Above viscosity values of 100 SSU, two equations are given to obtain correction factors:

$$R = (10,000 \times q)/\sqrt{C_v} \times cs \tag{7}$$

This equation is used for viscosity values between 100 and 200 SSU; the viscosity must be converted to centistokes. When viscosities exceed 200 SSU, Equation 8 applies:

$$R = (46,500 \times q)/\sqrt{C_v} \times \text{SSU} \tag{8}$$

In either equation, the following steps should be followed:

1. Solve for C_v, assuming no viscosity effects.
2. Solve for factor R using Equation 7 or 8.
3. From the viscosity correction curve, Figure 3, read the correction factor at the intercept of factor R on the curve.
4. Multiply the C_v obtained in Step 1 by the correction factor obtained in Step 3 for the corrected C_v.

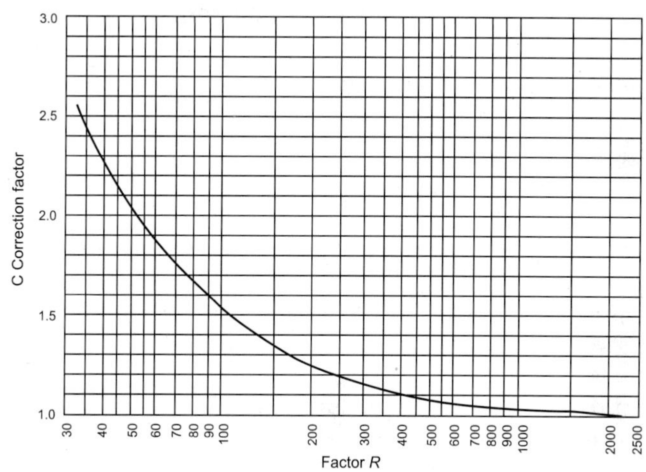

Figure 3. Viscosity correction curve for use with Equation 8.

Gases and vapors

The simplified equation for gases and vapors (Equation 3) favorably agrees with experimental data when used for conventional globe valves, but agreement with experimental data is poor when used with high recovery valves.

A universal equation proposed by C. B. Schuder, Fisher Controls, in *Instrumentation Technology*, February, 1968, reportedly showed excellent agreement with experimental results covering a wide range of body plug configurations and pressure drop ratios.

Volume basis:

$$Q = \sqrt{\frac{520}{GT}} C_1 C_2 C_v P_1 \sin\left[\frac{3417}{C_1 C_2}\sqrt{\frac{\Delta P}{P_1}}\right] \text{Deg.} \tag{9}$$

Weight basis:

$$W = 1.06\sqrt{dP_1ZC_1C_2C_v}\,\sin\left[\frac{3417}{C_1C_2}\sqrt{\frac{\Delta P}{P_1}}\right]\text{Deg.} \qquad (10)$$

The C_1 factor in Equations 9 and 10 is currently available from Fisher Controls for their valves. Other companies presumably do not publish this factor. (The C_2 factor is shown in Figure 4.)

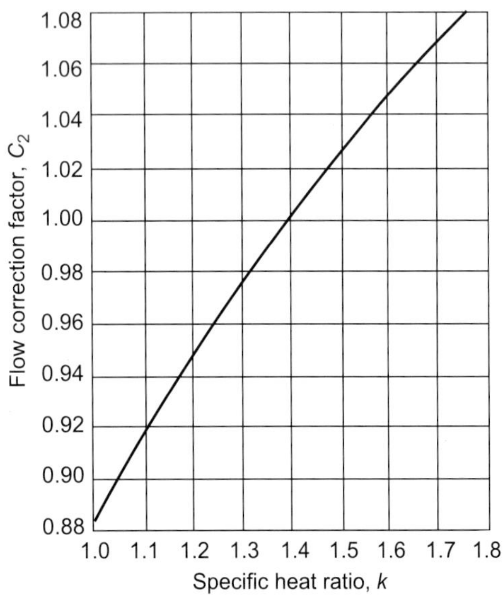

Figure 4. C_2 factor applicable only to Equations 9 and 10, the Fisher universal equations for gas and vapor flows.

An equation for gases and vapors proposed by the Foxboro Company (Foxboro TI Bulletin 31-4K) is identical with the FCI equation, except that P_2 is used instead of $(P_1 + P_2)/2$:

$$C_v(Q\,/\,1360)\sqrt{GT/\Delta P(P_2)} \qquad (11)$$

A comparison of the Fisher Universal, FCI, and Foxboro equations is made in the TI Bulletin 31-4K. A plot is given showing the results of these three equations using the same data (Figure 5).

The Foxboro equation gives C_v values essentially the same as the universal equation at pressure drops up to 25% of P_1 and gives 7% larger values at a pressure drop of 50% of P_1. At the higher end of $\Delta P/P_1$ ratios, a slightly larger valve should be helpful to allow for the expanding vapors. The same agreement is claimed for steam flows, although no plot is given.

To correct for deviation from the universal gas law, the compressibility factor is inserted in Equation 11, becoming:

$$C_v = (Q\,/\,1360)\sqrt{G_bTZ/(\Delta P) \times P_2} \qquad (12)$$

On a weight basis, this equation becomes:

$$C_v = (W\,/\,104)\sqrt{TZ/(\Delta P \times P_2 \times G_b)} \qquad (13)$$

Mixed phase flow. Mixed phase flow is generally experienced when:

1. Liquid entrainment is carried by a gaseous flow.
2. Liquid at or near its boiling point vaporizes as it flows through a line and its restrictions.

In the discussion under "Flashing Service," it is explained that flashing can be reduced or eliminated by limiting the ΔP across the valve. When two-phase flow already exists or when it occurs in the control valve, two methods are given to size the valve.

1. A rule of thumb that has been used for years is to calculate the C_v based on liquid only and increase the valve body to the next size. This is satisfactory in many cases, but it obviously does not take into account varying ratios of liquid to vapor.
2. A preferred method is to determine the amount of vapor and liquid, then calculate and add the C_v of each phase to get the overall C_v requirement. The percent vaporization can be calculated by the responsible engineer if it is not furnished by the process people.

Example.

Fluid—Propane

100 gpm @ 300 psia and 130°F
s.g. @ 130° F $= 0.444$
s.g. @ 60°F $= 0.5077$
s.g. @ P_2(200 psia) $= 0.466$ @ 104°F
$\Delta P = 100$ psi
% vaporization = 15.05%
M.W. = 44
$Z = 0.8$
$G_b = 44/29 = 1.518$
Flow = 100 gpm × 0.444 = 22,200 lb/h
Vapor = 22,200 × 0.1505 = 3340 lb/h
Liquid = 22,200 − 3340 = 18,860 lb/h

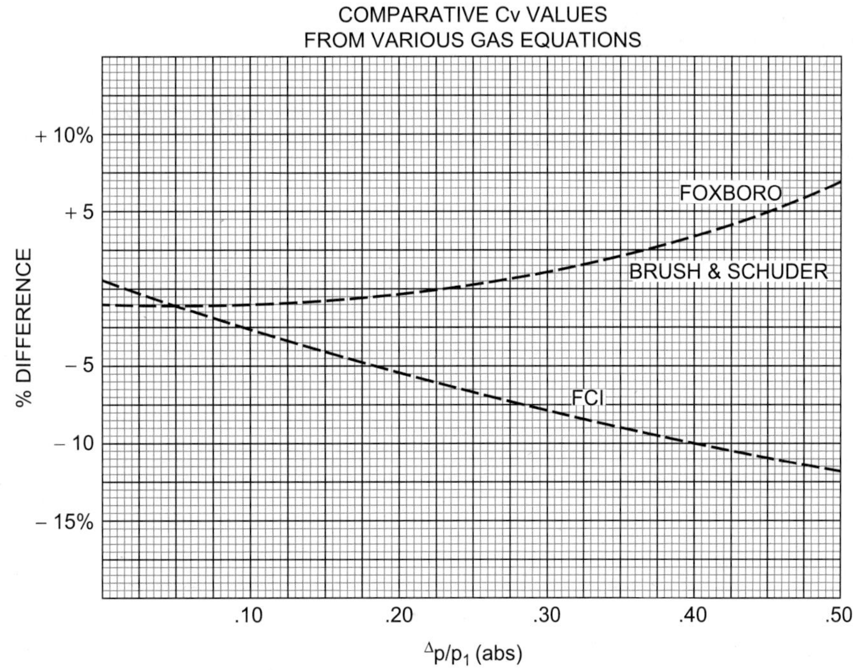

Figure 5. Curves showing the close agreement (particularly for low pressure drops) of the Foxboro, FCI, and Fisher universal equations for gas flows. Courtesy of the Foxboro Co.

Using Equation 13:

$$
\begin{aligned}
C_v(\text{vapor}) &= (W/104)\sqrt{(T \times Z)/(\Delta P \times G_b \times P_2)} \\
&= (3340/104)\sqrt{(564)(8)/100(1.518)200} \\
&= 3.92
\end{aligned}
$$

Using Equation 2:

$$
\begin{aligned}
C_v(\text{liquid}) &= (W/500)/(18,860/\sqrt{\Delta P \times G}) \\
&= 18,860/(500\sqrt{100 \times 0.466}) \\
&= 5.53
\end{aligned}
$$

$$C_v(\text{total}) = 3.92 + 5.53 = 9.45$$

The value of P must never exceed $P_2/2$ for sizing calculation.

Nomenclature

C_v = Valve flow coefficient
G = Specific gravity of fluid at flowing temperature
P = Pressure drop (P_1–P_2), psi
W = Flow of fluid, lb/hr
q = Flow rate, gpm
T = Flowing temperature, °R
T_{sh} = Degrees of superheat, °F
C_f = Critical flow factor, from valve manufacturer
P_m = Maximum useable P (choked flow occurs), psi
K_m = Valve recovery coefficient
r_c = Theoretical critical pressure ratio at P_v
P_v = Fluid vapor pressure at flowing temperature, psia
R = Factor for use with Figure 3
cs = Viscosity at flowing temperature, centistokes
SSU = Viscosity at flowing temperature, Saybolt Seconds Universal
C_1 = Valve sizing factor
C_2 = Specific heat correction factor (see Figure 4)
P_1 = Valve upstream pressure, psia
P_2 = Valve downstream pressure, psia
d = Density of vapors at valve inlet lb/ft^3
d_1 = Density of vapors at valve outlet lb/ft^3
Z = Compressibility factor
G_b = Specific gravity of gas or vapor at base conditions

Source

Applied Instrumentation in the Process Industries, Gulf Publishing Company, Houston, Texas.

Sizing control valves for throughput

W. G. Birkhead, Gas Measurement Engineer, United Gas Pipeline Co., Shreveport, Louisiana

Pressure losses through piping fittings, once given little consideration in the sizing of control valves, have become important as more efficient control valves have been developed. With the development of high capacity valves and the desire for a more exacting flow formula and efficiency, designers and users look to consideration of minor losses through elbows, tees, swages, or reduced ported valves. The capacity of control valves and regulators was once calculated using the basic hydraulic head equation with an assumed efficiency. This is no longer the case.

The head loss in any closed filled conduit has been accepted as proportional to its length, diameter, and fluid velocity head, and can be calculated as

$$H_L = f\frac{LV^2}{D2g} \tag{1}$$

where,

H_L = Head loss of flowing fluid (ft of fluid)
f = Friction factor
L = Length of pipe, ft
D = Diameter of pipe, ft
V = Average fluid velocity, ft/s
g = Acceleration of gravity, ft/s/s.

Sometimes *L/D* is referred to as the "equivalent length in diameters." Valves, fittings, etc., have often been expressed in equivalent diameters. If the friction factor *f* is combined with *L/D*, a constant *K* is derived for any pipe, valve, or fitting. If $K = f(L/D)$, then

$$H_L = K\left(\frac{V^2}{2g}\right) \tag{2}$$

The term *K* is called the "head loss factor" or "pressure drop factor."

Most control valves are rated with a capacity term called C_v. C_v is the number of gallons of water per minute that will flow through the valve with a 1 lb/in^2 pressure drop across the valve.

$$C_v = Q/\sqrt{\Delta P} \tag{3}$$

Q = Quantity of water in gpm
ΔP = Pressure drop in psi

Rearranging Equation 3:

$$\Delta P = Q^2/(C_v)^2 \tag{3a}$$

This equation may be rearranged so that C_v in terms of our *K* factor may be expressed as:

$$C_v = \frac{29.9(d)^2}{\sqrt{K}} \tag{4}$$

where d = inside diameter of the pipe or fitting, inches.

The head loss or pressure drop of a fluid flowing through various pipes, valves, or fittings in series can be stated:

$$(\Delta P)_{\text{Total}} = (\Delta P)_1 + (\Delta P)_2 + (\Delta P)_3 + \ldots \tag{5}$$

Substituting Equation 3a in Equation 5:

$$\frac{1}{(C_v)^2_{\text{Total}}} = \frac{1}{(C_v)^2_1} + \frac{1}{(C_v)^2_2} + \frac{1}{(C_v)^2_3} + \ldots \tag{6}$$

This equation is unwieldy if several fittings are used. By equating Equation 4, substituting in Equation 6 and dividing through by $[29.9(d)^2]^2$:

$$K_{\text{Total}} = K_1 + K_2 + K_3 + \ldots = \sum K \tag{7}$$

Substituting in Equation 4 for K_T an overall C_v can be derived:

$$C_v = \frac{29.9\left(d^2\right)}{\sqrt{\sum K}}$$

K factors for various valves and fittings

For full open valves, pipes or fittings, the *K* factor is approximately equal to *f*(*L/D*) or Equation 1, where

L = Equivalent length of valve or pipe or fitting, inches (or feet)
D = Diameter of the valve, inches (or feet)
f = Friction factor.

Some of the recognized K factors for various welding fittings are shown in Table 1.

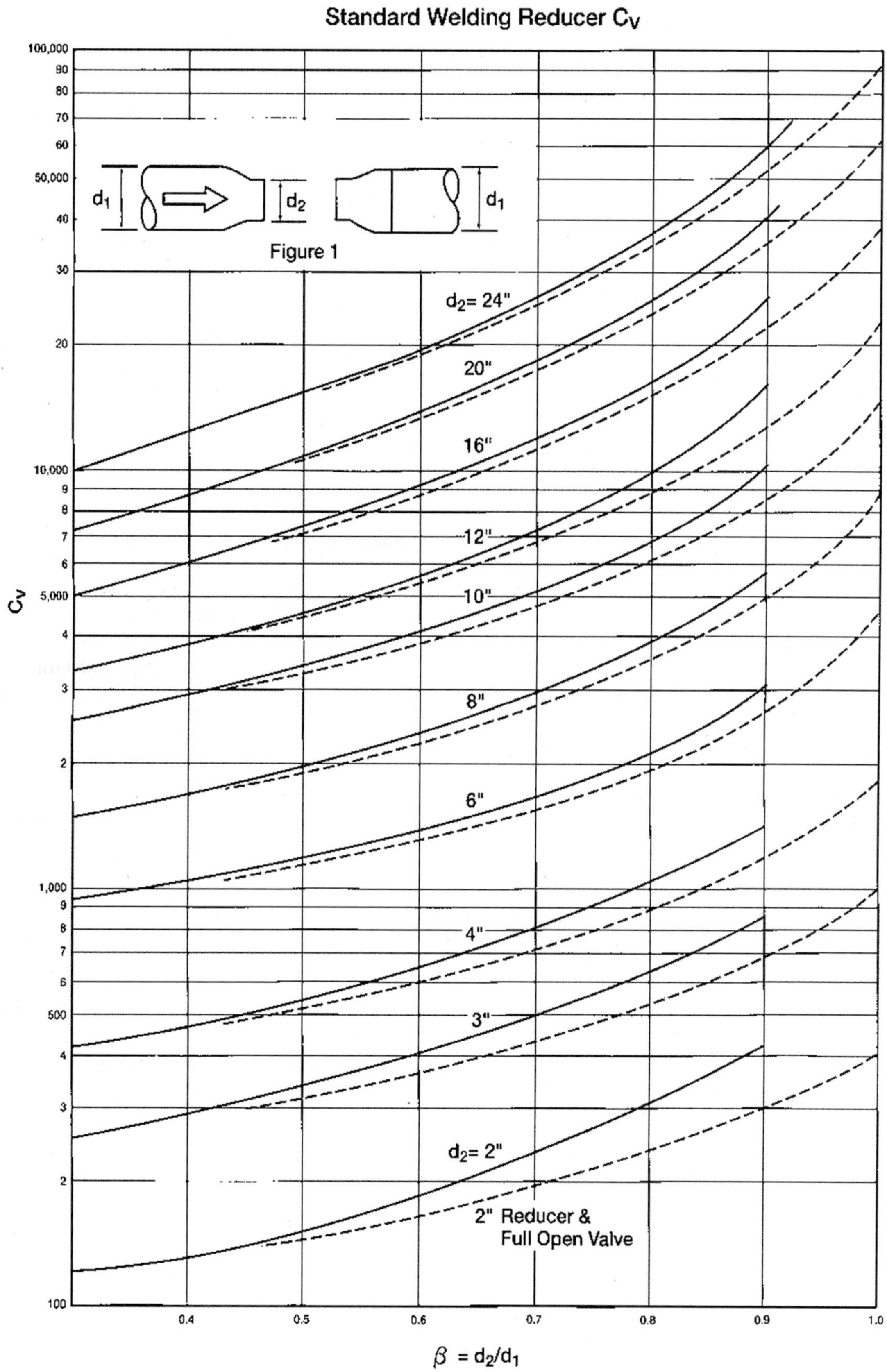

Figure 1. C_v for standard welding reducers.

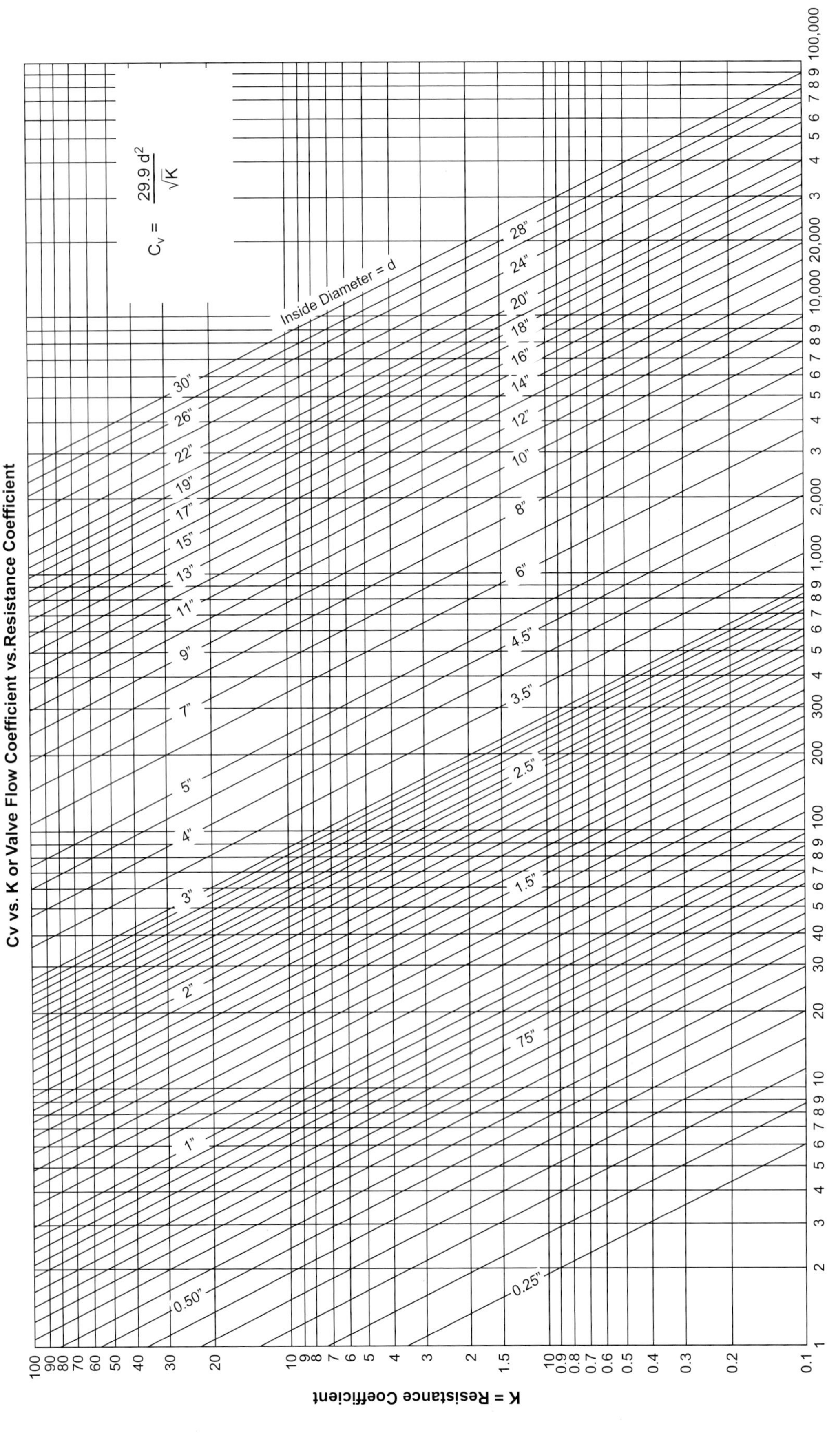

Figure 2. C_v versus K values for valves.

Installation procedures

Safety relief valve installation requires that careful consideration be given to inlet piping, discharge piping, pressure sense lines (where used), and bracing if required. Any marginal installation practice can render the safety relief valve inoperable, at worst, or severely restrict its design capacity at best. Either condition compromises the safety of the installation.

Inlet piping

The proper design of inlet piping is extremely important.

Very often, safety relief valves are added to an installation at the most physically convenient location, with little regard to flow considerations. Pressure loss during flow in a pipe always occurs. Depending upon the size, geometry, and inside surface condition of the pipe, the pressure loss may be large (20, 30, or 40%) or small (less than 5%).

API RP 520, Part 2, recommends a maximum inlet pipe pressure loss to a safety relief valve of 3%. This pressure loss shall be the sum total of the inlet loss, line loss, and when a block valve is used, the loss through it. The loss should be calculated using the maximum rated flow through the pressure relief valve.

The 3% maximum inlet loss is a commendable recommendation but often very difficult to achieve. If it cannot be achieved, then the effects of excessive inlet pressure should be known. These effects are rapid or short cycling with direct spring operated valves or resonant chatter with pilot operated relief valves. In addition, on pilot operated relief valves, rapid or short cycling may occur when the pilot pressure sensing line is connected to the main valve inlet. Each of these conditions results in a loss of capacity.

Pilot operated valves can tolerate higher inlet losses when the pilot senses the system pressure at a point not affected by inlet pipe pressure drop. However, even though the valve satisfactorily operates, reduced capacity will still occur because of inlet pipe pressure losses. The sizing procedure should consider the reduced flowing inlet pressure when the required orifice area, "A," is calculated.

A conservative guideline to follow is to keep the equivalent *L/D* ratio (length/diameter) of the inlet piping to the relief valve inlet to 5 or less.

Rapid cycling

Rapid valve cycling with direct spring operated valves occurs when the pressure at the valve inlet excessively decreases at the start of relief valve flow. Figure 3 is a schematic of system pressures that is shown before and during flow. Before flow begins, the pressure is the same in the tank and at the valve inlet. During flow, the pressure at the valve inlet is reduced due to the pressure loss in the inlet pipe to the valve. Under these conditions, the valve will cycle at a rapid rate.

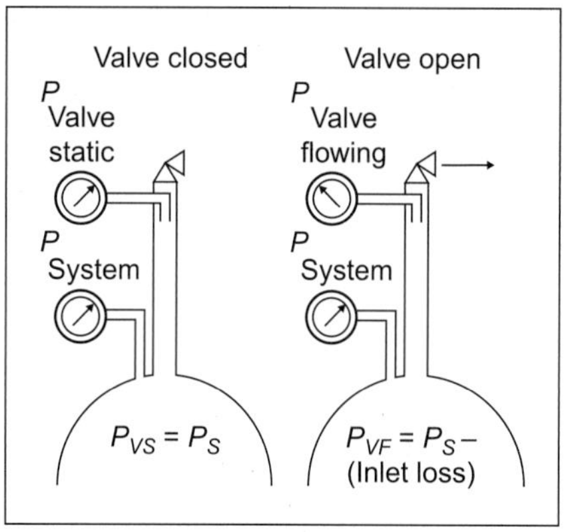

Figure 3. System pressures before and after flow.

The valve responds to the pressure at its inlet. When that pressure decreases during flow to below the valve's reseat point, the valve will close. However, as soon as the flow stops, the inlet pipe pressure loss becomes zero and the pressure at the valve inlet once again rises to tank pressure. If the tank pressure is still equal to or greater than the relief valve set pressure, the valve will open and close again. Rapid cycling reduces capacity and is sometimes destructive to the valve seat in addition to subjecting all of the moving parts in the valve to excessive wear.

On pilot operated relief valves with the pilot valves' sense line connected at the main valve inlet, rapid cycling can also occur for the same reasons described earlier; namely the pressure at the relief valve inlet is substantially reduced during flow (Figure 4). With the pilot sense line located at the main valve inlet, the pilot valve responds to the pressure at

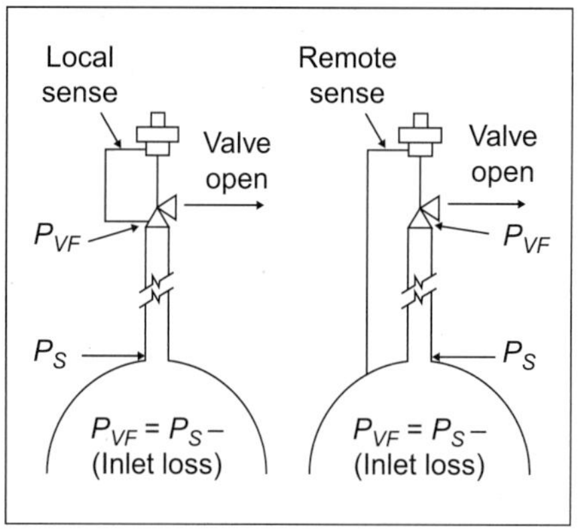

Figure 4. Schematic of rapid cycling.

the location and therefore closes the main valve. If the pressure loss during flow is small (less than the reseat pressure setting of the pilot), the main valve may or may not partially close depending on the type of pilot design used.

In actual application, because of the long response time of a pilot operated valve compared with a direct spring operated valve, the main valve may only go into partial life since the inlet pipe pressure loss occurs before the pilot has sufficiently vented the dome area for full piston travel.

As with direct spring operated valves, capacity is reduced but under severe inlet pressure loss conditions, resonant chatter may be induced in pilot operated relief valves. This chatter is extremely destructive and may result in complete valve failure.

On existing installations, the corrective action is limited. For direct spring operated valves, increasing the blowdown will minimize or eliminate rapid cycling if the blowdown can be adjusted to below the initial flowing pressure.

Resonant chatter

Resonant chatter occurs with pilot operated relief valves when the inlet piping produces excessive pressure losses at the valve inlet. The higher the set pressure or the greater the inlet pipe pressure loss, the more likely resonant chatter will occur.

Unlike the rapid cycling noted earlier, resonant chatter is uncontrolled; that is, once started, it cannot be stopped unless the pressure is removed from the valve inlet. In actual application, however, the valve will self-destruct before a shutdown can take place because of the magnitude of the forces involved in a resonant mode.

Discharge piping

Discharge piping for direct spring operated valves is more critical than for pilot operated valves. As with inlet piping, pressure losses occur in discharge headers with large equivalent L/D ratios. Excessive backpressure will reduce the lift of a direct spring operated valve, and enough backpressure (15–20% of set + overpressure) will cause the valve to reclose. As soon as the valve closes, the backpressure in the discharge header decreases and the valve opens again with the result that rapid cycling can occur.

Pilot operated relief valves with the pilot vented to the atmosphere or with a pilot balanced for backpressure are not affected by backpressure. However, if the discharge pressure can even exceed the inlet pressure, a back flow preventer must be used.

The valve relieving capacity for either direct or pilot operated relief valves can be affected by backpressure if the flowing pressure with respect to the discharge pressure is below critical (subsonic flow).

Balanced bellows valves (direct spring operated) have limitations on maximum permissible backpressure due to the collapse pressure rating of the bellows element. This limitation will, in some cases, be less than the backpressure limit of a conventional valve. Manufacturer's literature should be consulted in every case. If the bellows valve is used for systems with superimposed backpressure, the additional built up backpressure under relieving conditions must be added to arrive at maximum backpressure.

The performance of pressure relief valves, both as to operation and flow capacity, can be achieved through following good discharge piping practices:

- Discharge piping must be at least the same size as the valve outlet connection and may be increased when necessary to larger sizes.
- Flow direction changes should be minimized, but when necessary use long radius elbows and gradual transitions.
- If a valve has a drain port on the outlet side, it should be vented to a safe area. Avoid low spots in discharge piping, preferably pitch piping away from the valve outlet to avoid liquid trap at the valve outlet.
- Proper pipe supports to overcome thermal effects, static loads due to pipe weight, and stresses that may be imposed due to reactive thrusts forces must be considered.

Reactive force

On large orifice, high-pressure valves, the reactive forces during valve relief are substantial and external bracing may be required (Figure 5).

API RP 520, Part 2 gives the following formula for calculating this force.

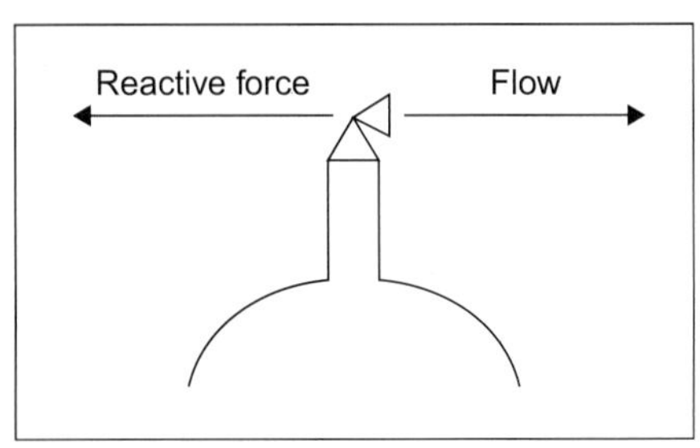

Figure 5. Reactive forces during valve relief.

$$F = \frac{Q_h}{366}\sqrt{\frac{kT}{(k+1)M}} \quad (1)$$

where

F = Reactive force at valve outlet centerline (lb)
Q_h = Flow capacity (lb/h)
k = Ratio of specific heats (C_p/C_v)
T = Inlet temperature, absolute (°F + 460)
M = Molecular weight of flowing media.

If bracing is not feasible, a dual outlet valve (available in some pilot operated safety relief valves) can be used. The reactive forces are equal but opposite, resulting in zero force on the valve outlet, but if redirected can still impose loads that must be reacted in some manner.

Testing

Using DOT Title 49, Part 192.739 as a guideline, each valve should be inspected at least once a year to determine that it:

- Is in good mechanical condition
- Will operate at the correct set pressure
- Has adequate capacity and is operationally reliable
- Is properly installed and protected from dirt, liquids, or other conditions that might prevent proper operation

The effectiveness of a safety relief valve installation greatly depends on proper sizing and selection of the valve type, suitable installation conditions, and proper and timely testing of the valve.

Rupture disk sizing methodologies

Changes to Section VIII, Division 1 of the ASME Code have impacted the methods used to size pressure relief systems that include rupture disks.

Terminology

- K_D (Coefficient of Discharge)—This is a unitless factor used to derate the theoretical flowing capacity of a device or system. For rupture disks, this value is defined by the Code as 0.62.
- K_R (Resistance to Flow Factor)—This is a unitless factor used to characterize the resistance to flow of a rupture disk device. The larger the K_R value, the more restrictive the device. The K_R value has no direct correlation to the K_D, coefficient of discharge.
- Combination Capacity Factor—This is a unitless factor used to derate the capacity of a pressure relief valve, when a rupture disk is installed upstream. The combination capacity factor may be the default value of 0.9, or a higher value, if the disk/valve combination has been tested.
- MNFA (Minimum Net Flow Area)—This is a derived area based on the area of the device, and/or the piping area, and is validated during ASME certification testing. The MNFA is used only in the coefficient of the discharge method.

Sizing methodologies

The code describes three methods to provide adequate relieving capacity for a pressure vessel. These methodologies are applied based on the characteristics of the relieving system, not individual components.

When a rupture disk is installed upstream of a pressure relief valve, the code requires the capacity of the valve to be derated by a factor of 0.9, or by a *combination capacity factor* determined by test for that disk/valve combination. This method assumes that the disk has a capacity equal to or greater than the valve and does not interfere with the normal functioning of the valve. From the rupture disk standpoint, this means the rupture disk should be the same nominal size as the relief valve inlet (or larger) and that the disk be of a full opening design. This method is unchanged from the previous versions of Section VIII.

The *coefficient of discharge method* (K_D) is not a rupture disk sizing method but a method for sizing a relatively simple relief system. The calculations for this method remain unchanged from previous versions of the code, but guidelines for use have been added. This method calculates the theoretical capacity of system and derates it by the coefficient of discharge, $K_D = 0.62$. This method is used only when each of the following conditions is true:

(1) the rupture disk discharges to atmosphere
(2) the rupture disk is installed within eight pipe diameters of the vessel
(3) there are no more than five pipe diameters of discharge piping on the outlet of the device.

The reason this is considered a system analysis method is because it takes into account the entrance conditions from the vessel to the discharge piping, eight pipe diameters of piping to the rupture disk, and five pipe diameters of piping, from the rupture disk to the atmosphere.

The *resistance to flow method* (K_R) is used if neither of the first two methods apply, such as systems with long runs of piping, discharging into manifolds, or flare systems. The resistance to flow method is not a rupture disk sizing method, but a system sizing method. This analysis takes into account the resistance to flow of all of the elements within the relieving system, including the resistance of the rupture disk. The code then requires that the system capacity then be derated by a factor of 0.9 to account for uncertainties in the method.

The code does not describe the detailed analysis method, other than to say "accepted engineering practices." Commonly used references for this type of analysis are: API RP521 Guide for Pressure Relieving and Depressuring Systems of Crane Paper 410.

As a point of reference, the following table is an excerpt from API RP521 listing typical K values for various piping components. The rupture disk K_R values available from the manufacturer are used in the same way as the K values for these common piping components. In many cases the rupture disk is not a limiting or even significant factor in the capacity of a relieving system. Note the API generic rupture disk K value listed here is not necessarily conservative and actual tested values may be significantly higher or lower than 1.5.

Fitting	K
Globe valve, open	9.7
Angle valve, open	4.6
Swing check valve, open	2.3
90° mitered elbow	1.72
Rupture disk	1.5
Welded tee flowing through branch	1.37
15 feet of 3 in sch 40 pipe	1.04
90° std threaded elbow	0.93
90° long sweep elbow	0.59

Reprinted with permission—The Fike Corp. Blue Springs, Missouri.

Rupture disk sizing using the resistance to flow method (K_R)

An excerpt from Fike Technical Bulletin TB8102:

This is a sizing method for relief systems that have a rupture disk as the relief device. Rupture disk devices can be characterized as to their respective resistance to fluid flow. The K_R value represents the velocity head loss due to the rupture disk device. This head loss is included in the overall system loss calculations to determine the capacity of the relief system.

ASME PTC25 provides standardized test methods to measure the K_R of rupture disk devices. By quantifying this performance characteristic, rupture disk devices may be accounted for in piping system sizing calculations in the same way as piping and piping components (exit nozzles on vessels, elbows, tees, reducers, valves).

Sample calculation using K_R method

The example shown is based on Crane TP-410 methods. It assumes a steady-state relieving condition where the vessel volume is large relative to the relieving capacity.

Piping Component or Feature	Flow Resistance Value (K)	Reference
Entrance—sharp edged	$K_1 = 0.50$	Crane 410 Pg A-29
1 ft of 3 in sch 40 pipe	$K_2 = 0.07$	$K= fL/D$; $f= 0.018$ (Crane 410 Pg A-26) $L= 1$ ft, ID $= 3.068$ in/12 ft
Fike 3 in SRX-GI Rupture Disk	$K_R = 0.99$	National Board Cert. No. FIK-M80042
20 ft of 3 in sch 40 Pipe	$K_3 = 1.41$	$K= fL/D$; $f= 0.018$ (Crane 410 Pg A-26) $L= 20$ ft, ID $= 3.068$ in/12 ft
3 in sch 40 standard 90° elbow	$K_4 = 0.54$	Crane 410 Pg A-29
40 ft of 3 in sch 40 Pipe	$K_5 = 2.82$	$K= fL/D$; $f=0.018$ (Crane 410 Pg A-26) $L= 40$ ft, ID $= 3.068$ in/12 ft
Pipe exit—sharp edged	$K_6 = 1.00$	Crane 410 Pg A-29
Total system flow resistance	$K_T = 7.33$	$K_T = K_1 + K_2 + K_R + K_3 + K_4 + K_5 + K_6$

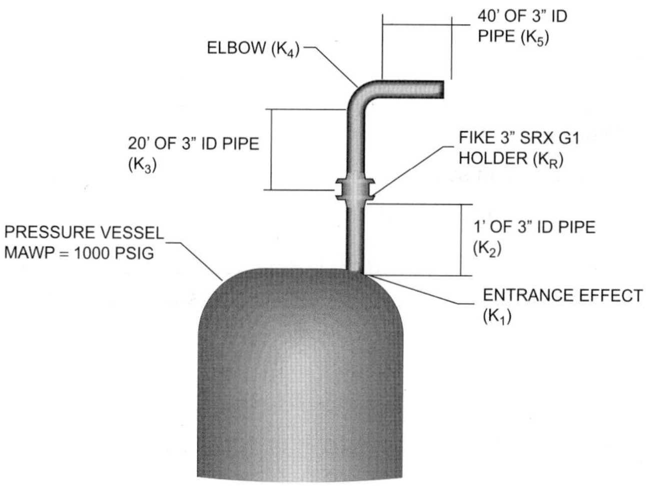

Given information:

1. Pressure vessel MAWP = 1000 psig.
2. Relieving pressure as allowed by ASME Section VII Div. 1 + 110% × MAWP = 1114.7 psia = $P'1$
3. Back pressure (outlet pressure) = 14.7 psia = $P2$
4. Working fluid—air ($k = Cp/Cv = 1.4$)
5. Air temperature at disk rupture = 500 °F = 960 °R = $T1$
6. Maximum flow rate into the vessel = 40,000 SCFM-Air
7. Rupture disk—Fike 3 in SRX-GI $K_R = 0.99$

The Darcy equation defines the discharge of compressible fluids through valves, fittings, and pipes. Since the flow rate into the example vessel is defined in SCFM, the following form of the Darcy equation is used:

$$q'_m = 678 \times Y \times d^2 \sqrt{\frac{\Delta P \times P'_1}{KT_1 S_g}}$$

q'_m = Rate of flow in cubic feet per minute at standard conditions (14.7 psia and 60 °F)
Y = Net expansion factor for compressible flow through orifices, nozzles, and pipes (Crane 410 Pg A-22)
Δd = Internal diameter of pipe (inches)
ΔP = Change in pressure from state 1 to 2 (entrance to exit).
P'_1 = Pressure at state 1 (entrance) in psia
K = Loss coefficient
T_1 = Absolute temperature at state 1 (entrance) in degrees Rankine (°R)
S_g = Specific gravity relative to air = 1.0 for this example

To determine Y, first it must be determined if the flow will be sonic or subsonic. This is determined by comparing the *actual* Δ/P'_1 to the *limiting* Δ/P'_1 for sonic flow. Crane Table A-22 shows limiting factors for $k = 1.4$ for sonic flow at a known value of K_T. If $(\Delta P / P'_1)_{\text{sonic}}$ is greater than $(\Delta / P'_1)_{\text{actual}}$, then the flow will be sonic.

Limiting Factors for Sonic Velocity ($k = 1.4$).

K	$\Delta P / P'_1$	Y
1.2	0.552	0.588
1.5	0.576	0.606
2.0	0.612	0.622
3	0.622	0.639
4	0.697	0.649
6	0.737	0.671
8	0.762	0.685
10	0.784	0.695
15	0.818	0.702
20	0.839	0.710
40	0.883	0.710
100	0.926	0.710

Excerpted from Crane 410 P4 A-22.

For this example: $(\Delta P/P'_1)_{\text{actual}} = \dfrac{114.7 - 14.7}{1114.7} = 0.9868$

From Table A-22 at $K_T = 7.33$

$(\Delta P/P'_1)_{\text{sonic}} = 0.754$

Since $(\Delta P/P'_1)_{\text{sonic}}$ is greater than $(\Delta P/P'_1)_{\text{actual}}$ the flow is sonic in the piping. Thus, we will use $Y_{\text{sonic}} = 0.680$ in the flow calculations (Crane 410 Eq. 3–20).

Since $(\Delta P/P'_1)_{\text{sonic}} = 0.754$, then $\Delta P = 0.754{*}P'_1 = 0.754{*}1114.7 = 840.5$ psi.

Calculating the system capacity is completed by substituting the known values into Crane 410 Equation 3–20.

$$q'_m = 678 \times Y \times d^2 \sqrt{\frac{\Delta P \times P'_1}{KT_1 S_g}}$$

$$q'_m = 678 \times 0.680 \times (3.068)^2 \sqrt{\frac{840.5 \times 1114.7}{7.33 \times 960 \times 1}}$$

$$q'_m = 50,074 \text{ SCFM}_{\text{Air}}$$

The ASME Pressure Vessel Code, Section VIII, Division 1, paragraph UG-127(a)(2)(b), also requires that the calculated system capacity using the resistance to flow method must also be multiplied by a factor of 0.90 to account for uncertainties inherent with this method.

$$q'_{m-\text{ASME}} = 50,074 \times 0.90 = 45,066 \text{ SCFM}_{\text{Air}}$$

Thus, the system capacity is greater than the required process capacity (40,000 SCFM_{Air}).

Reprinted with permission—The Fike Corp. Blue Springs, Missouri and *Flow Control Magazine*.

Variable orifice rotary control valves

Flow coefficient (C_v). The flow coefficient, which describes the flow capacity of a valve, is an empirically determined constant equal to the valve's flow capacity in U.S. gallons per minute of water at 60 °F when a pressure drop of 1 psi exists across the valve in its most wide-open position. Table 1 lists C_v for all sizes of DANTROL-270 valves, with and without trim.

The flow capacity of a valve in a partially open position may be calculated by selecting the appropriate C_v from Table 1 and multiplying it times the percent of maximum C_v at the desired position on the selected flow curve in Figure 2.

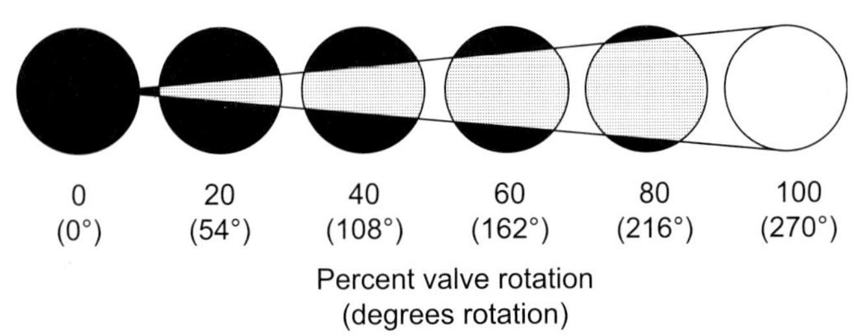

Figure 1. Orifice opening.

Table 1
Maximum Flow Coefficients (C_v)

Valve Size	Ball Dia.	Slot Length	Plates Per Slot	C_v w/Trim	C_v w/o Trim
1	4.25	7.40	8	45	53
1 ½	5.50	9.58	16	98	120
2	6.50	11.32	16	195	220
3	8.00	13.93	18	410	480
4	9.75	16.58	20	765	860
6	13.25	23.08	25	1900	1990

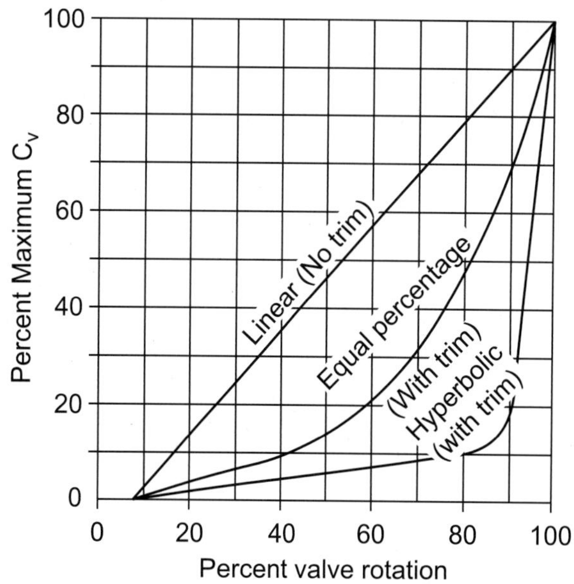

Figure 2. Percent of maximum C_v versus valve rotation.

Sizing valves for liquid flow

These equations for single-phase Newtonian liquid flow assume that the liquid is nonviscous and neither cavitating nor flashing. (Factors and constants are subsequently outlined.)

$$C_v = \frac{Q}{N_1 F_p E_R}\sqrt{\frac{G_F}{(P_1 - P_2)}}$$

or

$$C_v = \frac{W}{N_6 F_p E_R \sqrt{(P_1 - P_2)\gamma}}$$

The piping geometry factor (F_P)

The piping geometry factor (F_P) is equal to 1.0 for line-size valves. With reducers, use the value of F_P from the "Valve Reference Data"—Table 2 or 4.

The Reynold's number factor (F_R)

The Reynold's number factor (F_R) has a value of 1.0 for most flows in the process industry. However, sizing errors due to low Reynold's numbers (RE_v) can occur when the fluid viscosity is high or flow velocities low. In either case, RE_v can be calculated from these equations:

$$\mathrm{RE}_v = \frac{N_3 W}{v\sqrt{F_{LP} C_v}}\left[\frac{(F_{LP} C_v)^2}{N_2 D^4}\right]^{0.25}$$

or

$$\mathrm{RE}_v = \frac{N_3 Q}{v\sqrt{F_{LP} C_v}}\left[\frac{(F_{LP} C_v)^2}{N_2 D^4}\right]^{0.25}$$

Using the calculated RE_v, the value of F_R can be found from the graph in Figure 3. Substitute F_L for F_{LP} with line-size valves. (Values for F_L and F_{LP} are in the "Valve Reference Data"—Table 2 or 4.)

Choked flow

Choked flow is a limiting flow that occurs as a result of liquid vaporization when the internal valve pressure falls below the liquid's vapor pressure. Cavitation has then fully developed within the valve, and an increase in pressure drop across the valve will not produce an increase through the valve if

$$\Delta P \geq F_{LP}^2 (P_1 - F_F P_v)$$

Substitute F_L for F_{LP} with line-size valves. If the valve must be operated under choked conditions, use the following choked-flow equations in place of the basic equation for valve flow capacity:

$$C_v = \frac{Q}{N_1 F_{LP}}\sqrt{\frac{G_F}{(P_1 - F_F P_v)}}$$

or

$$C_v = \frac{Q}{N_6 F_{LP}}\sqrt{\frac{W}{(P_1 - F_F P_v)\gamma}}$$

where,

$$F_F = 0.96 - 0.28\frac{P_v}{P_c}$$

(use F_L for F_{LP} for line-size valves).

The cavitation index (K_c)

The cavitation index (K_c) relates to a condition just prior to flow choking where cavitation has only partially developed and flow rate is no longer proportional to the square root of pressure drop. This condition occurs when:

$$\Delta P \geq K_c(P_1 - P_v)$$

Values for the cavitation index (K_c) can be found in the Valve Reference Data—Table 2 or 4.

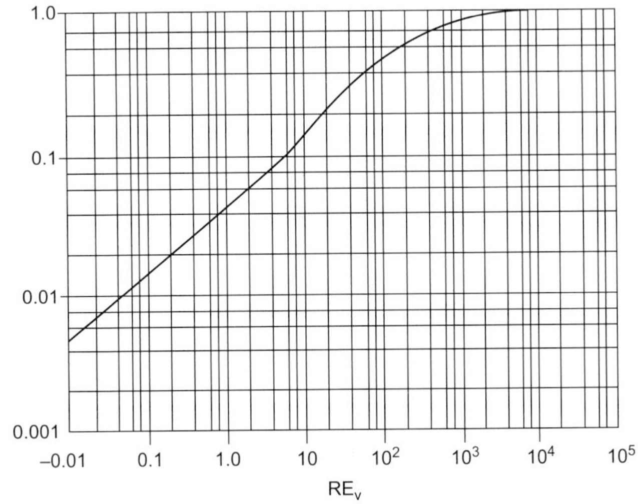

Figure 3. Reynolds number factor.

Anticavitation trim

Generally, operating a valve at conditions corresponding to K_c or choked flow will result in damaging cavitation, noise, and vibration.

If cavitation exists, a valve with anticavitation trim will prevent or lessen the effects of cavitation. When anticavitation trim is used, recalculate the flow-capacity coefficient (C_v) using factors for "With Trim" in the Valve Reference Data—Table. If noise is a concern, see the procedure under "Hydrodynamic Noise Prediction" in the manufacturer's literature.

Basic valve flow-capacity coefficient (C_v)

The basic valve flow coefficient (C_v) for gas or vapor flow through a variable orifice valve may be calculated using any of the following equivalent formulas for nonchoked flow:

$$C_v = \frac{Q}{N_7F_PY}\sqrt{\frac{G_GT_1Z}{X}}$$

or

$$C_v = \frac{Q}{N_6F_PY\sqrt{XP_1\gamma_1}}$$

or

$$C_v = \frac{Q}{N_9F_PP_1Y}\sqrt{\frac{MT_1Z}{X}}$$

$$C_v = \frac{W}{N_8F_PP_1Y}\sqrt{\frac{T_1Z}{XM}}$$

where

$$F_k = \frac{K}{140} \text{ and } X = \frac{P_1 - P_2}{P_1}$$

and

$$Y = \frac{1-X}{3F_KX_{TP}} \leq 0.667$$

Substitute X_T for X_{TP} with line-size valves. Values for F_P, X_T, and X_{TP} are found in the "Valve Reference Data"—Table 2 or 4.

The compressibility factor (Z)

The compressibility factor (Z) can be found from Figure 4 after the reduced pressure (P_R) and reduced temperature (T_R) have been calculated from these equations:

$$P_R = \frac{\text{Inlet Pressure(Absolute)}}{\text{Critical Pressure(Abolsute)}}$$

$$T_R = \frac{\text{Inlet Temperature(Absolute)}}{\text{Critical Temperature(Abolsute)}}$$

$$(^\circ\text{R} = \ ^\circ\text{F} + 460,\ ^\circ\text{K} = \ ^\circ\text{C} + 273)$$

Choked flow

Gas flow through valves can choke when the gas velocity reaches the velocity of sound. The pressure drop ratio at this point has the value:

$$X = F_K X_{TP}$$

Substitute X_T for X_{TP} with line-size valves. If this valve is to be used under choked conditions, Y is fixed at 0.667 regardless of the pressure drop. Substitute 0.667 for Y and $F_K X_T$ or $F_K X_{TP}$ for X in a basic value flow-capacity coefficient equation.

Valve outlet velocity

Sonic velocity at the valve's outlet will cause shock waves that produce vibration and noise problems. Therefore, you should make sure the outlet velocity does not reach sonic velocity for gas and steam service. The actual valve outlet diameter must be larger than the calculated diameter from the following equations:

$$d = N_{10}\sqrt{\frac{Q\sqrt{G_G T}}{P_2}} \text{ for gas}$$

$$d = N_{11}\sqrt{\frac{W(1 + N_{12}T_{SH})}{P_2}} \text{ for steam}$$

Low-noise trim

When the pressure-drop ratio (X) is near or exceeds the minimum value for choked flow, check the valve for excessive noise using the procedure under "Aerodynamic Noise Prediction." If noise is excessive, use antinoise trim; recalculate the flow-capacity coefficient (C_V) using factors from "With Trim" in the "Valve Reference Data"—Table 4. For procedures on noise-level prediction, refer to the manufacturer's literature.

Reprinted with permission - Daniel Valve Company.

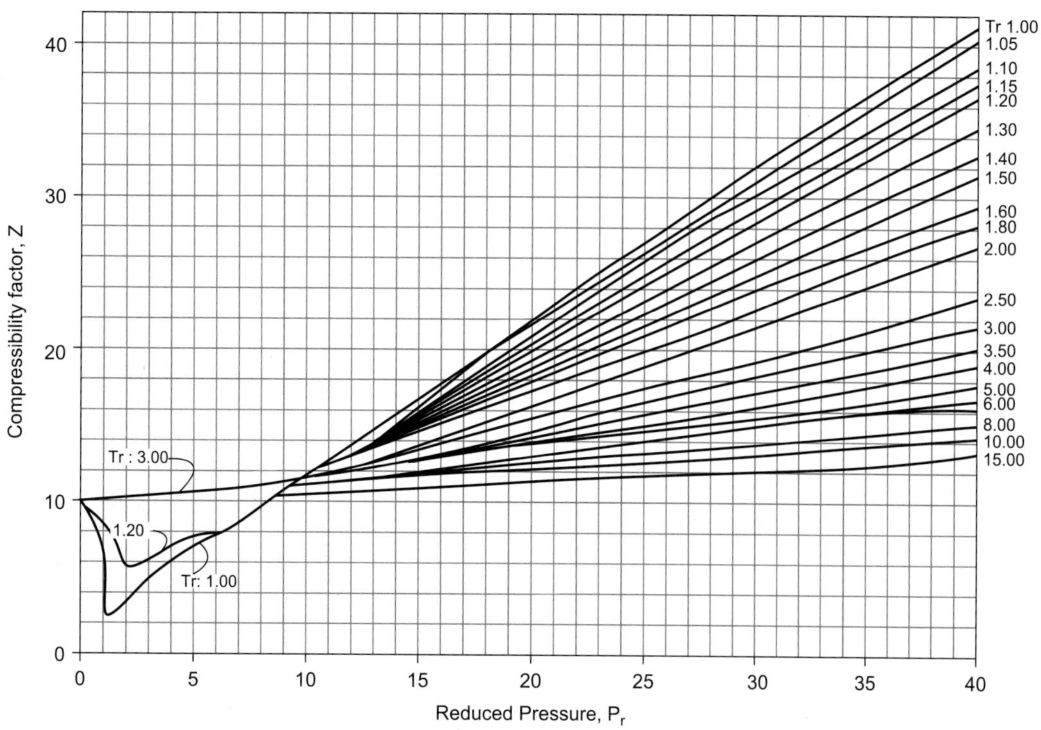

Figure 4. Compressibility versus reduced pressure.

Table 2
Valve Reference Data (No Trim)

Value	Percent Valve Rotation											D/d
	10	20	30	40	50	60	70	80	85	90	95	100	
F_L	0.95	0.94	0.93	0.91	0.89	0.85	0.78	0.69	0.65	0.60	0.56	0.51	
X_T	0.86	0.85	0.83	0.80	0.75	0.65	0.49	0.32	0.24	0.17	0.10	0.08	1.00
K_C	0.78	0.77	0.73	0.69	0.61	0.55	0.48	0.41	0.36	0.31	0.25	0.18	
F_P	1.0	1.0	1.0	1.0	1.0	0.99	0.98	0.95	0.91	0.86	0.78	0.62	
F_{IP}	0.95	0.94	0.93	0.91	0.88	0.83	0.76	0.64	0.59	0.52	0.45	0.35	1.25
X_{TP}	0.86	0.85	0.83	0.80	0.75	0.65	0.50	0.33	0.26	0.20	0.14	0.10	
F_P	1.0	1.0	1.0	1.0	1.0	0.99	0.97	0.92	0.89	0.83	0.74	0.56	
F_{IP}	0.95	0.94	0.93	0.91	0.88	0.82	0.75	0.64	0.57	0.50	0.42	0.34	1.33
X_{TP}	0.86	0.85	0.83	0.80	0.75	0.65	0.50	0.35	0.27	0.21	0.16	0.12	
F_P	1.0	1.0	1.0	1.0	1.0	0.99	0.96	0.91	0.86	0.76	0.64	0.50	
F_{IP}	0.95	0.94	0.93	0.91	0.88	0.82	0.74	0.63	0.56	0.49	0.40	0.32	1.50
X_{TP}	0.86	0.85	0.83	0.80	0.75	0.65	0.51	0.36	0.29	0.25	0.19	0.16	
F_P	1.0	1.0	1.0	1.0	1.0	0.99	0.94	0.85	0.78	0.68	0.56	0.44	
F_{IP}	0.95	0.94	0.93	0.91	0.88	0.81	0.73	0.62	0.55	0.46	0.38	0.29	1.67
X_{TP}	0.86	0.85	0.83	0.80	0.75	0.65	0.49	0.32	0.24	0.17	0.10	0.08	
F_P	1.0	1.0	1.0	1.0	1.0	0.97	0.93	0.83	0.76	0.65	0.50	0.40	
F_{LP}	0.95	0.94	0.93	0.91	0.88	0.80	0.72	0.60	0.53	0.45	0.36	0.27	2.00
X_{TP}	0.86	0.85	0.83	0.80	0.75	0.66	0.53	0.40	0.35	0.32	0.29	0.28	
F_P	1.0	1.0	1.0	1.0	1.0	0.95	0.89	0.75	0.64	0.46	0.38	0.33	
F_{LP}	0.95	0.94	0.93	0.91	0.88	0.80	0.72	0.60	0.53	0.45	0.36	0.24	2.67
X_{TP}	0.86	0.85	0.83	0.80	0.75	0.66	0.54	0.46	0.41	0.38	0.36	0.34	

Table 3
Constants in Liquid Sizing Equations

N		W	Q	ΔP, P	D, d	γ	V
N_1	0.865	—	m³/h	bar	—	—	—
	1.000	—	gpm	psia	—	—	—
N_2	0.00214	—	—	—	mm	—	—
	890	—	—	—	in	—	—
N_3	76.0	kg/h	—	—	mm	—	cSt
	34.5	lb/h	—	—	in	—	cSt
N_4	76,000	—	m³/h		mm		cSt
	17,300	—	gpm	—	in	—	cSt
N_6	27.3	kg/h	—	bar	—	kg/m³	—
	63.3	lb/h	—	psia	—	lb/ft³	—

Table 4
Valve Reference Data (With Trim)

Value	Percent Valve Rotation								D/d
	10–50	60	70	80	85	90	95	100	
F_L	1.0	1.0	1.0	0.92	0.85	0.77	0.67	0.54	
X_T	1.0	1.0	1.0	0.87	0.76	0.61	0.37	0.20	1.00
K_C	1.0	1.0	1.0	0.90	0.83	0.69	0.46	0.27	
F_P	1.0	1.0	1.0	0.98	0.97	0.95	0.92	0.87	
F_{IP}	1.0	1.0	1.0	0.90	0.84	0.75	0.63	0.48	1.25
X_{TP}	1.0	1.0	1.0	0.89	0.80	0.65	0.41	0.22	
F_P	1.0	1.0	1.0	0.98	0.97	0.95	0.92	0.87	
F_{IP}	1.0	1.0	1.0	0.90	0.84	0.75	0.62	0.47	1.33
X_{TP}	1.0	1.0	1.0	0.89	0.80	0.66	0.44	0.23	
F_P	1.0	0.99	0.98	0.95	0.93	0.89	0.85	0.78	
F_{IP}	1.0	1.0	1.0	0.89	0.83	0.74	0.61	0.46	1.50
X_{TP}	1.0	1.0	1.0	0.89	0.81	0.68	0.47	0.27	
F_P	1.0	0.99	0.98	0.94	0.91	0.86	0.80	0.71	
F_{IP}	1.0	1.0	1.0	0.89	0.83	0.74	0.61	0.45	1.67
X_{TP}	1.0	1.0	1.0	0.89	0.81	0.69	0.50	0.30	
F_P	1.0	0.99	0.98	0.94	0.90	0.84	0.77	0.66	
F_{IP}	1.0	1.0	1.0	0.89	0.83	0.73	0.60	0.44	2.00
X_{TP}	1.0	1.0	1.0	0.90	0.82	0.71	0.55	0.36	
F_P	1.0	0.99	0.98	0.94	0.89	0.83	0.75	0.64	
F_{IP}	1.0	1.0	1.0	0.89	0.83	0.73	0.59	0.43	2.33
X_{TP}	1.0	1.0	1.0	0.90	0.82	0.72	0.57	0.39	
F_P	1.0	0.99	0.98	0.93	0.88	0.81	0.73	0.61	
F_{IP}	1.0	1.0	1.0	0.88	0.81	0.71	0.57	0.42	2.50
X_{TP}	1.0	1.0	1.0	0.92	0.87	0.80	0.71	0.60	
F_P	1.0	0.99	0.98	0.92	0.86	0.79	0.70	0.59	
F_{IP}	1.0	1.0	1.0	0.88	0.81	0.70	0.55	0.41	2.67
F_{TP}	1.0	1.0	1.0	0.93	0.88	0.82	0.75	0.65	

Table 5
Constants in Gas/Vapor Sizing Equations

N		*W*	*Q*	ΔP, *P*	*D*, *d*	γ	*T*	T_{SH}
N_6	27.3	kg/h	—	bar	—	kg/m^3	—	—
	63.3	lb/h	—	psia	—	lb/ft^3	—	—
N_7	417	—	m^3/h	bar	—	—	°K	—
	1360	—	scfh	psia	—	—	°R	—
N_8	94.8	kg/h	—	bar	—	—	°K	—
	19.3	lb/h	—	psia	—	—	°R	
N_9	2250	—	m^3/h	bar	—	—	°K	—
	7320	—	scfh		—	—	°R	—
N_{10}	0.25	—	m^3/h	bar	mm	—	°K	—
	63.3	—	lb/h	psia	in	—	°R	—
N_{11}	37.1	kg/h	—	bar	mm	—	—	°C
	0.119	lb/h	— —	psia	in	—	—	°F
N_{12}	0.00126	kg/h	—	bar	mm	—	—	°C
	0.0007	lb/h	—	psia	in	—	—	°F

Visualize pump and control valve interaction

G. Unnikrishnan, Kuwait Oil Company, Kuwait

Control valves are the final control elements of a process control system and play the key role of controlling the process variables within the safe operational envelope. It is the dynamic element in the system that translates the control system logic to mechanical action. The fact is that little attention is usually paid to this aspect while designing the process plant. During the typical project design stage for a pump and control system, the process design engineers usually assign a certain pressure drop for the control valve and develop the pump discharge pressures in a pumping circuit. Instrument design engineers select a suitable control valve based on the process data furnished by the process design engineers, the most important being the parameters at minimum, normal, and maximum flows. The required C_v value is calculated from the process parameters and the control valve is selected such that the ratio of calculated C_v to selected control valve C_v is in the 0.6 to 0.75 range. Final selection and confirmation is usually from the manufacturer.

One of the most important factors that govern dynamic behavior of the control valve is the control valve gain. Gain is defined as:

Gain = (Change in the flow rate)/(Change in valve position).

Ideally this should not change much; it should be constant at least in the normal control operating region. In other words, the valve travel when the flow changes from, say, 30–40% should be same as when the flow changes from 50% to 60%. The dynamic nature of the valve gain can be seen from the valve equations and formula for valve position shown in Table 1. Further certain key points to understand control valve dynamics are given in Table 2. Once the valve gain is computed and presented in a suitable form, as a graph, control valve behavior can be easily visualized. This will save the time required for full-scale control system modeling.

Table 1
Valve Types and Equations

Valve type	Valve equation	Valve position, %
Linear $f(L) = L$	$Q = Cvsel\, L\sqrt{\frac{\Delta P}{sg}}$	$L = \frac{Cvinst}{Cvsel}$
Equal-percentage $f(L) = Rn^{(L-1)}$	$Q = Cv\, Rn^{(L-1)}\sqrt{\frac{\Delta P}{sg}}$	$L = 100\left[1 + \frac{\ln(B)}{\ln(Rn)}\right]^2$

This article shows a way to solve the problem using well-established graphical methods. The capabilities of popular spreadsheets make the effort easy. Following are the main steps in easily visualizing and understanding the behavior of typical pump and control valve interaction.

The graphs and spreadsheet are set up in the following sequence:

Table 2
Key Points to Understand Control Valve Dynamics

1. The Cv_1 values found in the control valve manufacturer's catalog refer to the coefficients when there is a constant pressure drop of 1 psi across the control valve and nothing else. This is the inherent Cv of the control valve.
2. The inherent Cv will be modified when the control valve interacts with the system in which it operates in a dynamic way depending on the type of control valve selected—linear, equal-percentage, quick-opening, etc. The installed characteristics will be different from the inherent characteristics. For example, a valve with an equal- percentage inherent characteristic will not show the same behavior when installed. In fact, equal-percentage valves exhibit more linear characteristics when interacting with the system.
3. Control valve gain ideally should be the same, that is, it should be constant for the flows around the operating region. Literature says that the gain at any flow should not be less than 0.50.
4. Try to match the valve type, its characteristic and gain with the process system gain under consideration. There is extensive literature available on the subject. The results are worth the effort.
5. Control valves are mechanical devices and need time to react to the commands from the controllers that act quite fast. The controllers do not wait to see the results of the control valve action and issue further commands to the valve. Effectively, the control valve struggles to catch up with the controller commands resulting in continuing cycles of opposite movements by the control valve, resulting in the so called "induced limit cycle." An oversized control valve has high process gain and small changes to inputs tend to make large changes in the process parameter. Interested readers can see Ref. [3].
6. The whole control loop including the control valve behaves nonlinearly. For critical control loops, review of the nonlinear behavior will definitely be beneficial.

A The pump and system curves

A.1 Pump curve. The characteristic for a preliminary selected pump is plotted in a spreadsheet graph. The plot will show pump head on the y axis and flow rate on the x axis.

Basic pump parameters are calculated while designing the system. The rated flow rate and pump differential pressure are known (this includes the pressure loss assigned to the control valve at rated flow). If the pump curve is not available, a preliminary curve can be plotted by fixing two more points. The shutoff head at zero flow is usually taken as 1.25 times the head at the rated flow. One more point for pump head can be fixed at the other end of the curve as a suitable fraction of the head at maximum flow expected over and above that of the rated flow. We now have three points for the head versus flow relationship for the pump to plot the graph.

A.2 System curve. Fix the static head. This will have to be added as a constant to the variable pressure loss through the system comprising piping, pipe fittings, and other equipment. The variation in pressure drop with flow can be ascertained by a detailed calculation or can be simplified by using the relationship

$$Q_1/Q_2 = K(\Delta P_1/\Delta P_2)^2$$

A reasonable approximation is enough at this stage. The system curve does not include the pressure drop for the control valve. Get at least three points for the pressure loss through the system.

Plot these points on the same graph (Figure 1).

A.3 Fitting equations to the curves. Use the "Add Trend Line" feature available in the spreadsheet program to fit polynomial equations to both these curves (two coefficient equations will be enough to give a good fit). Use the Options tab to display the equations and the R^2 values. The R^2 values give a measure of the goodness of fit. That is, the closer the values of R^2 to 1, the better the fit. The fitted equations are displayed on the graph (Figure 1). Now set up a spreadsheet table as shown in Table 3. The spreadsheet layout shows the type of entry in each column. Note these fitted equations and enter the same in columns 2 and 3 of Table 3.

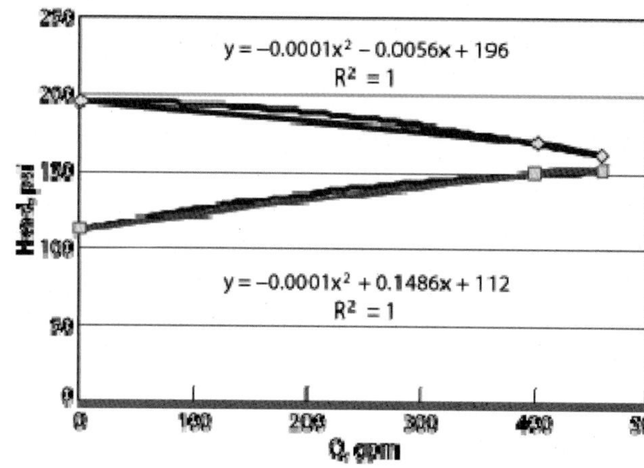

Figure 1. Pump and system curves.

Table 3
Spreadsheet Layout

1	2	3	4	5	6	7	8	9	10	11
Q	P_p	P_s	ΔP	C_{vinst}	$L\%$	$Q\%$	Gain	L%	Q%	Gain
Manual entry	Formula from equation for trend line for pump curve or manual entry	Formula from equation for trend line for system curve or manual entry	Difference of columns 2 and 3 (Step B.1)	Formula for installed *Cv*, *Cvinst* (Equation 2)	Valve opening for equal-percentage control valve, % (Equation 6)	Flow rate as percentage of maximum flow (Step B.3)	Formula for gain for equal-percentage control valve obtained by differentiating equation for trend line (Figure 2)	Valve opening for linear control valve, % maximum flow	Flow rate as percentage of	Formula for gain for linear control valve obtained by differentiating equation for trend line

B Control valve positions for various flows

This is where the graphing capabilities of the spreadsheet can be used. It consists of two parts.

B.1 Computing the absorbed pressure drop. The control valve on the pump discharge will absorb the difference in pressures between that developed by the pump and that required by the system. This difference can be calculated with the equations obtained from Step A.3 by entering the difference in column 4. Fill the blank cells for different flow rates by copying the formulas down. (Note: If sufficient data points are available, Step A.3 can be avoided. Just plot the graphs and write the formula for the absorbed pressure directly in the table.)

While computing the differences, omit the intersection point of the pump and system curves. At this point, the valve is fully open theoretically with zero pressure drop but in actuality it will still drop some pressure across it depending on the internals. Therefore, it is better to understand this point as a system with no valve.

B.2 Preliminary selection of control valve. Preliminary selection of a suitable control valve has to be done at this stage. Using the process parameters available from process design, calculate the required *Cv* (denoted as *Cvr*) of the valve at the *rated* flow—at this point a reasonable pressure drop has to be assigned to the control valve based on the pressure differences available from the table. Select a suitable size control valve based on calculated *Cvr* from the manufacturer's catalogs such that the ratio of the *Cvr* to the *Cvsel* of the selected valve is in the 0.60 to 0.75 range. Remember that this *Cvsel* value is based on a 1 psi pressure drop across the valve and represents the inherent characteristic of the valve. Add one more column (column 5) in Table 3 for *Cv* at installed conditions (*Cvinstal*) at different flow rates. The values can be calculated by manipulating the formula for flow through the control valve to get *Cvinstal*:

$$Q = Cinstal\sqrt{\Delta P/sg} \tag{1}$$

$$Cvinstal = \frac{Q}{\sqrt{\Delta P/sg}} \tag{2}$$

The formula can be copied down the column for various flow rates. The calculations are implemented in the spreadsheet table to show the valve opening for various flows.

B.2.1 Calculating the control valve opening. Add one more column (column 6) for the percentage valve opening based on the installed characteristic equation. The characteristic generic equation for a control valve is typically:

$$Q = Cvsel\, f(L)\sqrt{\Delta P/sg} \tag{3}$$

f(*L*) is a function of the valve opening, *L* (from 0 fully closed to $L = 1$ full open). The valve behavior is altered as a result of varying flow and pressure drop across the valve.

As an example, let us highlight the steps for a control valve with equal-percentage trim. Equation 3 will have to be modified for the equal-percentage valve. For an equal-percentage valve, $f(L) = Rn^{(L-1)}$. Rn = rangeability (ratio of maximum to minimum flow—usually in the 20 to 50 range for globe valves).

The equations satisfying the flow, pressure drop, and valve opening for a particular flow *Q* will be:

$$Q = Cvsel\, Rn^{(L-1)}\sqrt{\Delta P/sg} = Cvinst\sqrt{\Delta P/sg} \tag{4}$$

Thus,

$$R^{(L-1)} = Cvinst/Cvsel \tag{5}$$

Manipulate the equation to get valve opening in percentage as a function of the pressure drop through the valve as:

$$L = 100[1 + \ln(B)\,/\,\ln(Rn)] \tag{6}$$

$$B = Cvinst/Cvsel \tag{7}$$

Enter this formula in column 6 of Table 3 and copy it down to get the opening for various flow rates.

This is the equation that will be implemented in the graph from which the valve gain will be obtained.

Similar equations and steps for a linear type valve are shown in Table 3.

B.3 Flow rates as percentage of maximum. Convert the flow rates as a percentage of maximum ($100Q/Q_o$) and add a column (column 7) for the same. Now we have the percentage valve opening values for different percentage flow rates.

B.4 Plotting percentage valve opening. Plot the percentage flow rates and percentage valve opening values in a graph with flow rate on the y axis and valve opening on the x axis using the values in columns 6 and 7 of Table 3 (Figure 2). Again, use the "Add Trend Line" feature to fit the equation to the graph. Use a three-coefficient equation for a good fit. Note the equation.

C Calculating the control valve gain

C.1 Remember that control valve gain is the ratio of flow rate change to valve position change. We can derive this by *differentiating the equation obtained* in Step B.4 with respect to L. Implement the differentiated equation in Table 3 by adding another column (column 8) and copying down the formula. We now have a set of values for valve gain with respect to percentage flow rates.

C.2 Plot these values from C.1 in a spreadsheet graph (Figure 3). [For comparison, Step B.4 can be repeated for a linear valve by using the equations in Table 1. Readers can compare the valve behavior for the two types (Figures 4 and 5)].

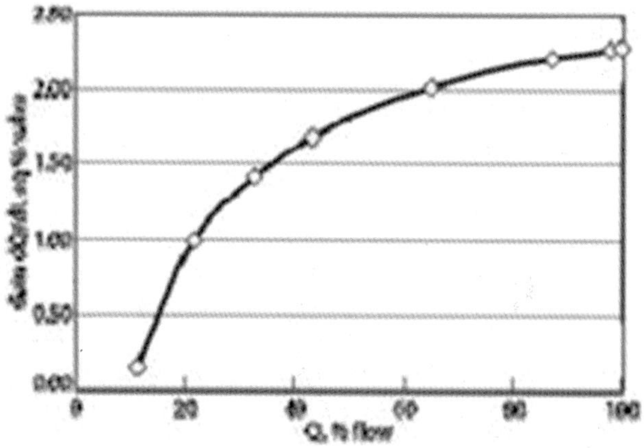

Figure 3. Gain versus percent flow for an equal-percentage valve.

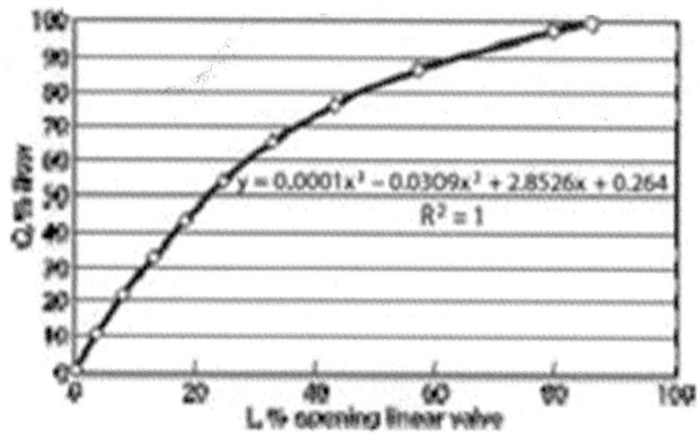

Figure 4. Percent flow versus percent valve opening for a linear valve.

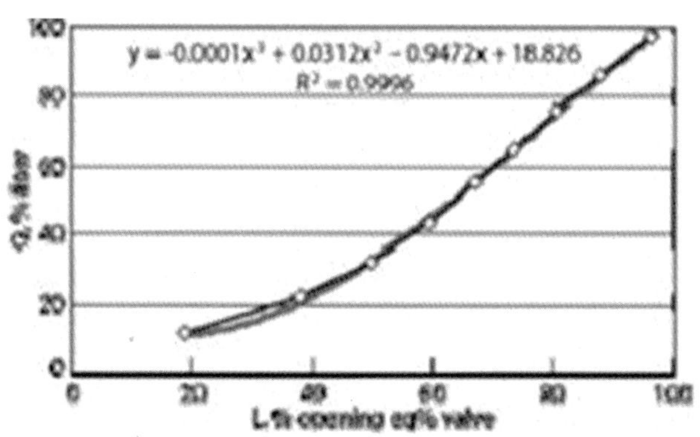

Figure 2. Percent flow vx percent valve opening for an equal percentage valve.

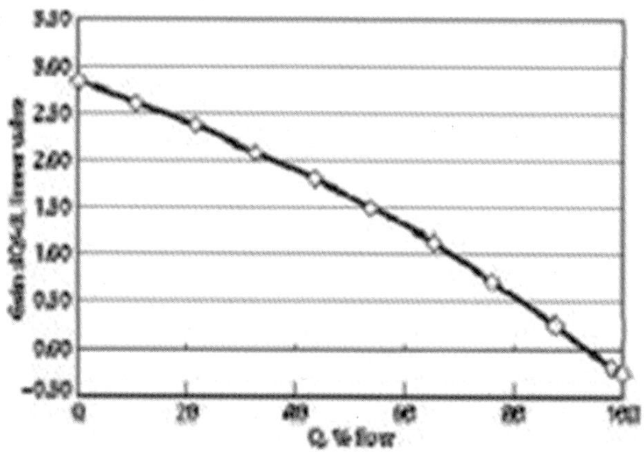

Figure 5. Gain versus percent flow for a linear valve.

Table 4
Spreadsheet

Fluid sg				1	Selected valve Cvsel @ rated flow equal percentage			Selected valve Cvsel @ rated flow linear		
Values used for plotting				Cvr						
Q	Pp	Ps	ΔP	85			136			148
0.00	196.00	112.00	84.00							
400.00	170.00	150.00	20.00							
460.00	162.00	152.00	10.00							
Spreadsheet (Refer to Table 3 for layout)					Selected valve equal %R=	50			Selected valve linear	
1	2	3	4	5	6	7	(8) gain	9	10	(11) gain
Q	Pp	Ps	ΔP	Cvinst	L%	Q%	dQ/dL	L%	Q%	dQ/dL
	Values from graph									
0.00	196.00	112.00	84.00	0.00	0.00	0.00	−0.95	0.00	0.00	2.85
50.00	195.47	119.18	76.29	5.72	19.02	10.87	0.13	3.87	10.87	2.62
100.00	194.44	125.84	68.58	12.08	38.10	21.74	0.99	8.16	21.74	2.37
150.00	192.91	132.04	60.87	19.23	49.99	32.61	1.42	12.99	32.61	2.10
200.00	190.88	137.72	53.16	27.43	59.08	43.48	1.69	18.53	43.48	1.81
250.00	188.35	142.90	45.45	37.08	66.78	54.35	1.88	25.06	54.35	1.49
300.00	185.32	147.58	37.74	48.83	73.82	65.22	2.02	33.00	65.22	1.14
350.00	181.79	151.76	30.03	63.87	80.68	76.09	2.13	43.15	76.09	0.74
400.00	177.76	155.44	22.32	84.67	87.89	86.96	2.22	57.21	86.96	0.30
450.00	173.23	158.62	14.61	117.73	96.31	97.83	2.28	79.55	97.83	−0.17
460.00	172.26	159.20	13.07	127.25	98.30	100.00	2.29	85.98	100.00	−0.24

These graphs will show valve behavior for different flow rates. Ideally, the valve gain should be constant near the operating flow rates. Also, one has to particularly look for the behavior at minimum and maximum flows. Check if this behavior is reasonable with respect to the process requirements. Otherwise, the valve selection, pumping pressure, and line size need to be reviewed at this stage.

With some effort, the steps can be implemented for a gas compressor and its control valve. Table 4 shows a spreadsheet with a numerical example and Table 5 describes steps to be followed that are described in the article and the corresponding equations.

Nomenclature

C_{vr} = Control valve capacity coefficient required at rated flow
C_{vsel} = Capacity coefficient of selected control valve
C_{vinst} = Control valve capacity coefficient at installed condition
ΔP_1 *and* ΔP_2 = System pressure drop at flow rates Q_1 and Q_2
K = Constant multiplier
L = Valve opening, fraction
DP = Pressure drop across control valve, psi
P_p = Pump head, psi
P_s = System head, psi
Q = Flow rate, gpm
Q_o = Flow rate when control valve is full open, gpm
Q_1 and Q_2 = Flow rate at particular operating conditions 1 and 2, gpm
$Q\%$ = Flow rate as percentage of maximum
Rn = Rangeability
sg = Specific gravity

References

1. J.R. Connell, "Realistic Control Valve Pressure Drops," *Chemical Engineering, September*, 1987 123–127.
2. L. Driskell, *Control Valve Selection and Sizing*. Instrument Society of America, 1983.
3. EnTech Control Engineering Inc., *Control Valve Dynamic Specification* (November 1998).
4. M. Fehr, "Improve Valve and Pump Sizing," *Hydrocarbon Processing* (July 1998) 53–57.

Reprinted with permission - *Hydrocarbon Processing*, August 2007 issue.

Table 5
Numerical Example (Step Numbers are the Same as in the Text)

A Pump and system curve

A.1 Pump curve

Assuming that the actual pump curve is not available, rated flow and differential head are calculated as $Q = 400$ gpm at $H = 170.0$ psi

Head at shut-off head at $Q = 0 = 1.25 \times 170.00 = 196.00$ psi

Head at maximum flow at $Q = 460 = 0.95 \times 170.00 = 162.00$ psi

The above factors can be varied based on the actual context

Plot the points in a spreadsheet (Figure 1).

A.2 System curve

Calculate values of flow versus pressure drop through the system (excluding the control valve) for:

Pressure at $Q = 0$ static head only $= 112$ psi

Pressure required at $Q = 400$ gpm $= 150$ psi

Pressure required at $Q = 460$ gpm $= 152$ psi

Plot the points in Figure 1.

A.3 Fitting equations to the curves

Right click on the curve to get the “Add Trend Line” feature. Select a two-polynomial equation. Also, check the boxes for show equation and R^2 values. OK all of the above to get the equation in terms of $y = f(x)$ displayed on the graph. Repeat the same for the system curve. The equations are:

$Y_p = -0.0001\ X^2 - 0.0056X + 196$ pump curve

$Y_s = -0.0001\ X^2 + 0.1486X + 112$ system curve

Divide the flow rate in suitable increments to get sufficient number of points and enter them manually at Column 1 of Table 1.

For column enter the system head values manually. Implement the difference of equations obtained in A.2 ($Y_p - Y_s$) in Column 3 to get the pressure drop.

B Control valve positions for various flows

B.1 Pressure drop absorbed by the control valve is obtained as noted above

B.2 Select control valve:

Calculate *Cvr* of the control valve at $Q = 400$ gpm and ΔP of 22 psi as 85 using formula:

$Cvr = \frac{Q}{\sqrt{\Delta P/sg}}$, which will be $= \frac{400}{\sqrt{22/2}} = 85$ (taking sg as 1.00)

B.2.1 From a manufacturer’s catalog obtain the *Cvsel* for a 4 in equal-percentage valve for about 72% valve opening as 136. Check on the ratio:

Cvr/Cvsel = 85/136 = 0.63

Implement Equation 6 for opening of equal-percentage control valve *L* in %.

Copy the formula down to fill all rows (Column 6).

B.3 Convert the flow rates as percentages of maximum flow and Implement in Column 7.

B.4 Plot another graph (Figure 2) with flow Q% on the y axis and L% opening on the x axis.

Repeat the “Add Trend Line” procedure and get a three-coefficient equation as:

$Y = -0.0001\ X^3 + 0.0312\ X^2 - 0.9472\ X + 18.826$ which is

$Q = -0.0001\ L^3 + 0.0312\ L^2 - 0.9472\ L + 18.826$ in terms of Q and L.

This equation gives flow rate, *Q*% (*Y*) as a function of valve opening, L% (*X*).

C. Calculating the control valve gain

C.1 Differentiate the above equation with respect to *L* to get d*Q*/d*L* (rate of change of flow with respect to rate of change of valve opening) to get gain as:

$dQ/dL = -0.0003\ L^2 + 0.0624\ L - 0.9472$

The right-hand side of the above equation is written on top of Figure 2.

Implement the above equation in Column 8 and fill all rows with gain values at different flow rates.

C.2 Plot Figure 3 using the values of flow percentage, *Q*%, on the *x* axis and gain on the *y* axis to visualize the behavior of the equal-percentage-type control valve.

The above steps are repeated for a linear-type valve with the equations shown in Table 1 in Columns 9, 10, and 11 to get the gain versus flow rates (Figures 4 and 5).

Avoid cavitation in butterfly valves

E. Casado Flores, EEAA, Madrid, Spain

Butterfly valves may be used to control flow in pipes if they are throttled in angles where the valve works stable without noise or vibration, i.e. cavitation free. The initial cavitation coefficients, C_i, presented in this article are compared with those predicted by Refs [1] and [2], suggesting the C_i values for different valve sizes and the maximum allowable pressure drop and flow rates through the valves to run without cavitation.

They are based on tests performed in a Spanish nuclear power plant on 24 in butterfly valves installed in the service water system. The goal was to know the throttling positions free of cavitation, to have the system running safely and to obtain the head-flow curves of the new pumps.

Tests performed on 24 in valves. To determine that the throttling positions were free of cavitation and the butterfly valves were completely quiet, we performed different tests for valve opening angles from 25° to 90° (valve 100% open). Figure 1 shows one of the tested valves. In some throttling positions the valves were quiet. In other positions, significant noise was generated. Finally, in the rest of the throttling positions, the typical cavitation noise (like stones dragged inside the pipe) began to be heard. In these positions, a slight opening of the valve (approximately 1°) eliminated the noise. More than 50 tests were performed for different valve openings. The cases where the cavitation noise begins correspond to the initial cavitation.

Figure 1. More than 50 tests were performed for different valve openings on 24 in butterfly values. The data collected during the tests were: Pressures upstream and downstream of the valves with 0.5% accuracy manometers and 0–10 kg/cm^2 range; flow through the valves with a 1.12% accuracy Annubar and 0–5000 m^3/h range.

Results obtained for C_i. The coefficient, C_i when cavitation noise begins is defined as

$$C_i = \frac{P_1 - P_v}{P_1 - P_2}$$

where

P_2 = downstream valve pressure, kg/cm^2 absolute
P_1 = upstream valve pressure, kg/cm^2 absolute
P_v = fluid vapor pressure, kg/cm^2 absolute.

It is possible to obtain the different C_i values from the pressure measurements. The C_i values are represented in Figure 1 as a function of C_d with a 10% conservatism.

C_d is the discharge valve coefficient defined as

$$C_d = \frac{1}{\left[(17183.71/C_p)^2 + 1\right]^{\frac{1}{2}}}$$

C_v is the valve flow coefficient defined as

$$C_v = \frac{1.17Q}{(P_1 - P_2)^{\frac{1}{2}}}$$

where Q = water flow, m^3/h.

Water temperature during the tests was 20 °C, so P_v is approximately 0.

NOTE: There is no cavitation in the valve if the values of P_1, P_2, and P_v during its running fulfill the following expression: $(P_1 - P_v)/(P_1 - P_2)\ C_i$.

Comparison with the data. In Ref. [1], a 6 in butterfly valve was tested at $P_1 = 5.8$ kg/cm^2 and presented the corresponding C_i values as a function of C_d similar to Figure 2. For other pressures and valve sizes, Ref. [1] proposes a group of correction equations to account for the scale effects.

Applying this to the cavitation level curve of Ref. [1], named as critical cavitation, it is possible to obtain the critical cavitation values for 24 in valves and the test pressure predicted by Ref. [1] as shown in Figure 3. The critical cavitation of Ref. [1] is very similar to the C_i values in Figure 2.

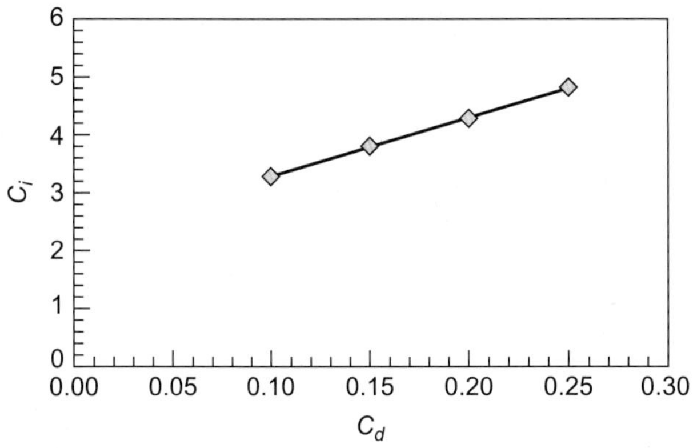

Figure 2. C_i as a function of C_d.

In Ref. [2], the author predicts the cavitation coefficient for 24 in butterfly valves starting from the data measured in 6 in valves for the incipient damage level and calculated sound level using 85 dBA as a limiting condition. The proposed cavitation values for 24 in butterfly valves according to Ref. [2] are also shown in Figure 3.

So, Figure 3 shows the comparison of the C_i tested values with those theoretically predicted by Refs [1] and [2] for 24 in butterfly valves. The C_i values for C_d greater than 0.3 are assumed to be in the curve parallel to that of Ref. [1]. The result is that the tested C_i values are practically the same as that of Ref. [1] for the critical level and higher than that predicted by Ref. [2], but taking into account, in this case, that they correspond to an incipient damage level.

Another conclusion is that the incipient cavitation level proposed in Ref. [1] (higher values than critical cavitation level) may be very conservative to determine the industrial butterfly valves' throttling limits to perform without cavitation. For this reason, based on my experience with these types of valves, I proposed the C_i curves shown in Figure 4 as adequate to establish the initial cavitation level.

C_i proposed values. Based on the C_i values obtained in the tests for 24 in butterfly valves—since they are slightly higher than the critical cavitation values of Ref. [1]—I assume that, for 6 in valves, the corresponding C_i values will also be very similar to the critical cavitation values of Ref. [1] for this size. So, it is possible by interpolation for other sizes to obtain the C_i curves shown in Figures 4 and 5.

The curves of Figures 4 and 5 are base reference curves for water and liquids with P_1 (pressure before the valve) equal to or below 10 kg/cm^2 abs and vapor pressures, P_v, equal to or greater than 0.

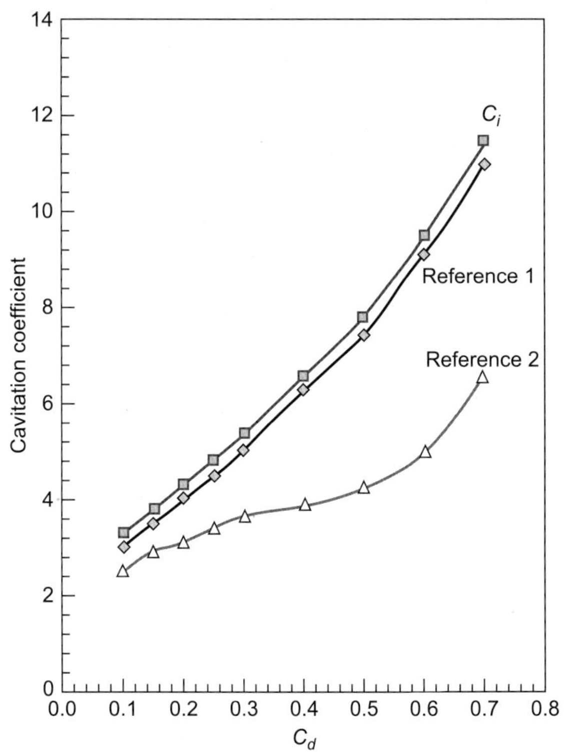

Figure 3. C_i comparison values for 24 in butterfly valves.

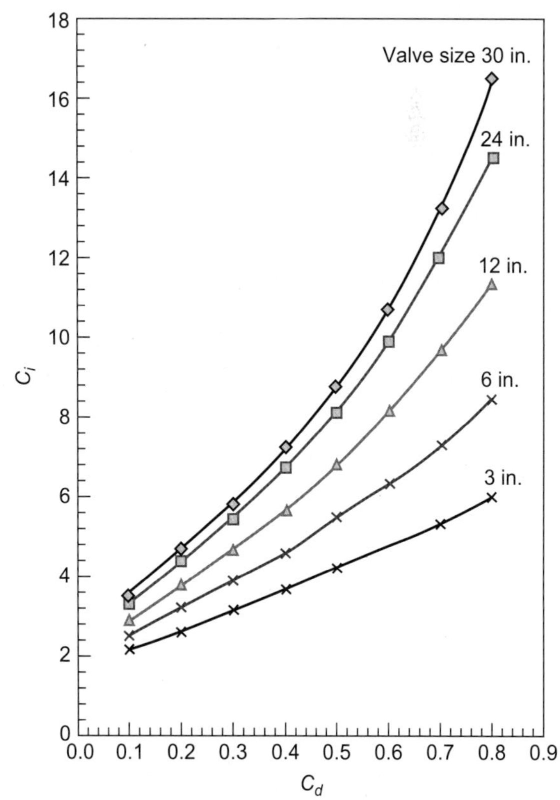

Figure 4. Proposed C_i values to establish the initial cavitation.

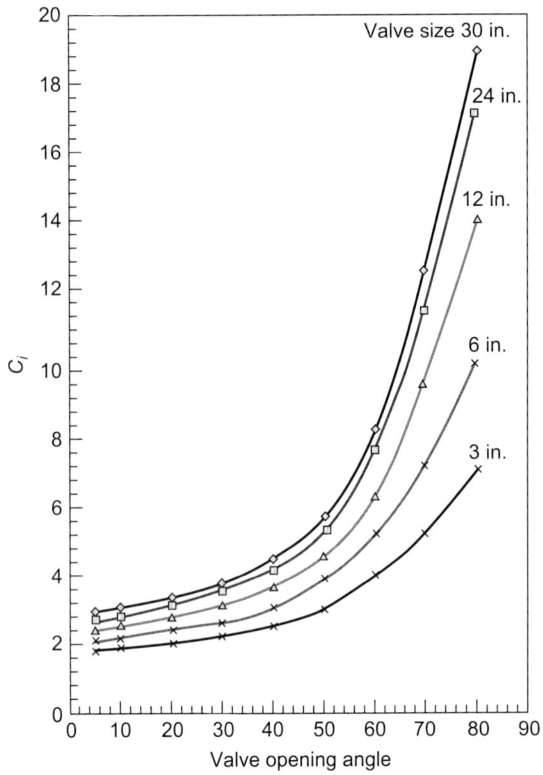

Figure 5. C_i values as a function of valve opening angle.

From the C_i curves in Figure 5, it is possible to obtain the maximum admissible pressure drop and flow for the different opening angles of the valve as follows:

Pressure drop, $DP = P_1 - P_2 = (P_1 - P_v)/C_i$
For the base reference values, $P_1 = 10$ kg/cm^2 abs and $P_v = 0$, $DP = 10/C_i$
Maximum admissible flow, $Q = 0.85\ C_v\ DP^{1/2} = 0.85\ C_v\ (10/C_v)^{1/2}$
The curves for DP and Q correspond to Figures 6 and 7.

Figure 6 shows the maximum allowable pressure drop through the valves without cavitation as a function of the opening angle.

Figure 7 shows the maximum allowable flow through the valves corresponding to the pressure drops of Figure 6.

Corrections for pressures. The corrections factors deduced from Ref. [1] are:

PV for $P_v > 0$ is $1 - P_v/P_1$
PSE for $P_1 > 10$ kg/cm^2 abs is$(P_1/10)^{0.3} = 0.5\ P_1^{0.3}$

These factors are applied to the C_i reference values, maximum allowable pressure drop, DP, and maximum

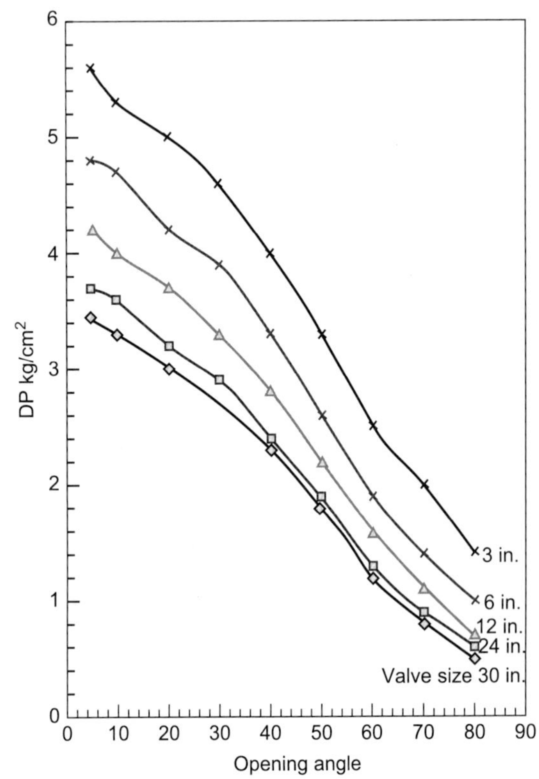

Figure 6. Maximum allowable pressure valves through the valves without cavitation.

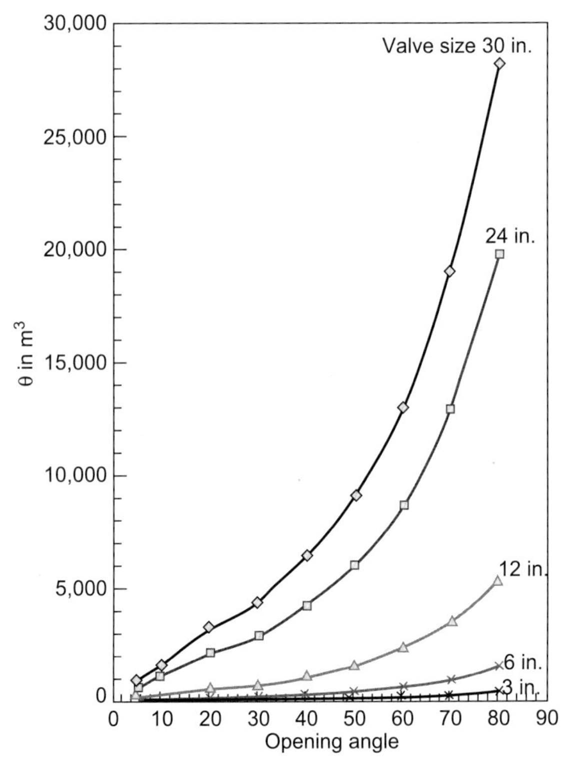

Figure 7. Maximum allowable flow through the valves.

allowable flow, Q, respectively obtained from Figures 5–7 to get the corrected ones: C_{ip}, DP', and Q':

$$C_{ip} = 1 + PSE(C_i - 1) = 1 + 0.5P_1^{0.3}(C_i - 1)$$

$$DP' = \frac{PVP_1}{1 + PSE(10/DP - 1)}$$

$$Q' = Q(PV)^{0.5}\left[\frac{P_1}{DP(1 - PSE) + 10PSE}\right]^{0.5}$$

where DP' is kg/cm^2 and Q' is m^3/h.

The C_i proposed values for butterfly valves have sufficient conservatism to avoid cavitation in their throttling positions.

Figures 4 and 5 directly allow obtaining the C_i values for different valve sizes, avoiding the use of the size corrections proposed in Ref. [1].

Figures 6 and 7 help plant operators know how much they may throttle the butterfly valves to work without cavitation.

References

1. J.P. Tullis, Cavitation Guide for Control Valves. *NUREG/CR-6301*, April 1993.
2. H.D. Baumann, Preventing Cavitation in Butterfly Valves. *Chemical Engineering*, March 1985.

How to read a regulator flow curve

Bill Menz, Swagelok Company

The best way to select a regulator for your application is to examine its flow curve, which is often provided by the manufacturer. "Flow curve" is a misleading name. You could easily call it a "pressure curve" instead, since a regulator controls pressure, not flow. The curve represents the range of pressures that a regulator will maintain given certain flow rates in a system.

When selecting a regulator, you are not just looking for the right size. You are looking for a set of capabilities, which is a function of the regulator's design. A flow curve illustrates the regulator's range of capabilities at one glance. Once you understand a flow curve—and we will explain how—it is easy and very quick to read.

Unfortunately, a more common way to select a regulator is to consult its flow coefficient (C_v). If the system flow is within range of the C_v, some people may believe that the regulator is the right "size." But that is not necessarily true. The C_v represents the regulator's maximum flow capacity. At maximum flow, a regulator can no longer control pressure. If you are expecting flow rates to reach the regulator's C_v, it is probably not the right regulator for your system.

Let us discuss how to read a flow curve. We will cover the basics first and then some of the complexities, including droop, choked flow, seat load drop or lock-up, hysteresis, and accumulation.

The basics

A regulator's main purpose is to maintain a constant pressure on one side of the regulator even though there is a different pressure or fluctuating pressure on the other side. In the case of a pressure-reducing regulator, you control the downstream pressure. In the case of a backpressure regulator, you control the upstream pressure.

For now, let us talk in terms of pressure-reducing regulators since they are more common. Later, we will provide some direction for selecting backpressure regulators.

A flow curve illustrates a regulator's performance in terms of outlet pressure (Y axis) and flow rate (X axis). Flow is not controlled by the regulator. It is controlled downstream by a valve or flow meter. The curve shows you how a regulator will respond as flow in the system changes.

Let us look at the top curve in Figure 1. The curve starts at 400 pounds per square inch gauge (psig) (27.5 bar). This is the original set pressure for the regulator. No adjustments are made to the regulator, yet the curve shows a change in pressure.

The regulator is making adjustments, trying to maintain the original set pressure as flow changes. But, it is not perfect. No regulator is. As system flow increases, pressure downstream of the regulator drops. The important question is: How much does it drop?

When reading a flow curve, first identify the range of flows that you can expect to see in your system. Mark these on the graph. Then, look to see what the corresponding changes in outlet pressure will be. Is that range of pressures acceptable to you? If not, you need to look for a different regulator.

Ideally, you want to operate the regulator on the flattest part of the curve. There, the regulator will maintain relatively constant pressures even with significant changes in flow. At the far ends of the curve, there are steep drops, where pressure will dramatically change with even the slightest

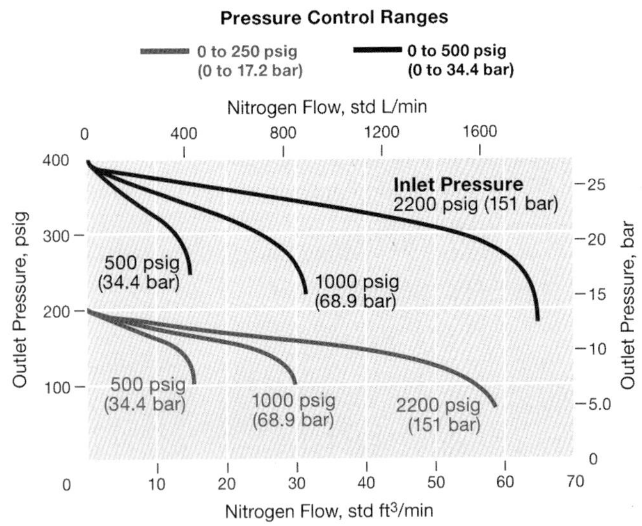

Figure 1. Manufacturers often provide multiple flow curves for the same regulator at different inlet pressures to provide a range of the regulator's operating capabilities.

change in flow. You do not want to operate the regulator at these locations.

Any given regulator can produce a nearly infinite number of curves, so you need to make sure you are looking at the right one. For every set pressure, there will be a different curve. In Figure 1, there are two main sets of curves: one based on a set pressure of 400 psig (27.54 bar) and one for a set pressure of 200 psig (13.7 bar). It is helpful when a manufacturer provides more than one set of curves representing the range of set pressures possible with a particular regulator. If your set pressure lies between the curves, you can interpolate. Note that the two curves are close to the same shape but in different locations on the graph.

There is one additional variable that affects the shape of a curve—inlet pressure (i.e. pressure going into a pressure-reducing regulator on the upstream side). Note that for each of the two sets of curves in Figure 1, there are three curves representing a range of inlet pressures.

In sum, to locate the right curve for your system, look for (1) the right set pressure, (2) the right inlet pressure, and (3) the right range of flows.

Finally, make sure you are looking at the right units. Pressure readings are provided most commonly as psig or bar. Flow rate units vary depending on the system media, so be sure to note whether the regulator is rated for liquid or gas service. Liquid flow is typically expressed as gallons per minute (gal/min) or liters per minute (L/min), while gas flow is conveyed as standard cubic feet per minute (std ft^3/min) or standard liters per minute (std L/min).

Curves are usually created using air or nitrogen (for gas service) or water (for liquid service). If your system media is a gas, you may need to make an adjustment in the manufacturer's curve. Gases compress at different rates, so you may need to multiply the flow curve's volume units by a gas correction factor. For example, hydrogen's correction factor is 3.8; meaning 3.8 molecules of hydrogen have the same volume as one molecule of air. Therefore, the point on a flow curve showing an air flow volume of 100 std ft^3/min (2831 std L/min) indicates a comparable hydrogen flow of 380 std ft^3/min (10 760 std L/min). The curve stays the same, but the flow scale will change.

For liquids, the difference in flow between water and a different medium is not as dramatic due to the incompressibility of liquids.

Droop, choked flow, and other complexities

As mentioned above, it is best to operate along the flattest—or most horizontal—part of a flow curve. Indeed, the ideal flow curve would be a flat line. However, no regulator can produce a perfectly flat line over the full range of pressures because the internal components of the regulator have limitations. In a springloaded regulator, longer springs produce flow curves with broader horizontal sections. Domeloaded regulators, which use a trapped volume of gas instead of a spring, produce even broader horizontal sections. Electrically controlled pneumatic-loaded regulators and domeloaded regulators with external feedback produce the broadest horizontal sections.

Typically, a flow curve consists of three parts: (1) a relatively flat part in the middle, (2) a steep drop on the far left, and (3) a steep drop on the far right. The flat part in the middle is not perfectly flat. Usually, it slopes downward, and this is called droop. As flow increases, pressure will drop some—or a lot depending on the regulator design. While droop is relatively modest along the flat part of the curve, it is quite steep at the far ends of the curve.

Choked flow occurs on the far right of a curve. See the choked-flow area in Figure 2, where pressure begins to droop sharply at 140 std ft^3/min (3964 std L/min). Eventually, near 150 std ft^3/min (4247 std L/min), pressure drops to zero. At this point, the flow demand has exceeded the pressure-controlling capabilities of the regulator. Here, the regulator is wide open and is no longer regulating pressure. It has essentially changed from a pressure-controlling device to an open orifice. Increasing downstream flow to this point or beyond renders the regulator ineffective. It is not advisable to operate a regulator in the choked-flow area due to the sharp pressure drop.

Note that C_v is measured at the regulator's fully choked point, and that is why it is not an especially useful measurement. It is a poor indicator of a regulator's overall performance.

Seat load drop occurs on the far left of the regulator curve (Figure 2), where there is initially a steep drop in pressure. If you are reading the curve from left to right, imagine that the

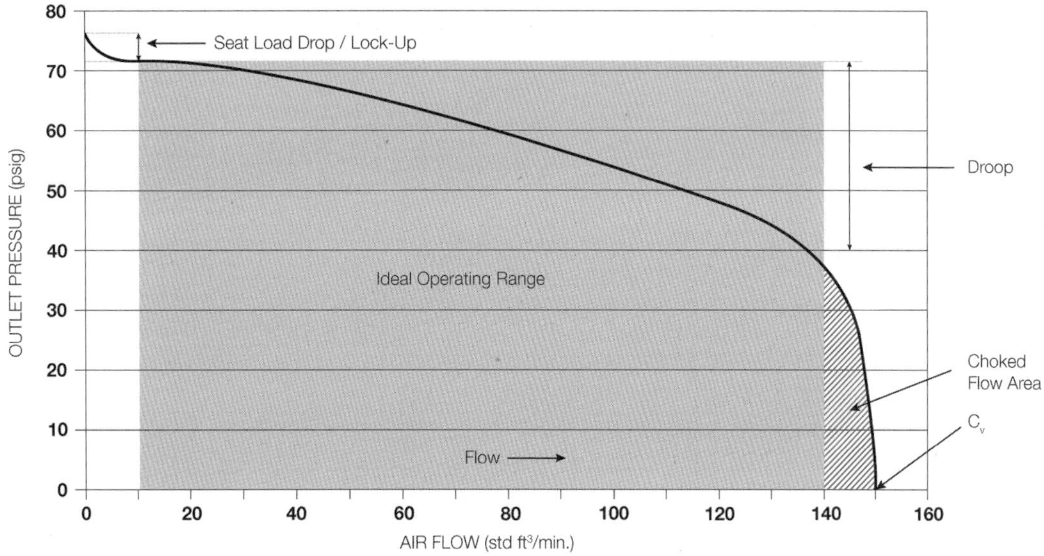

Figure 2. Typical flow curve for a pressure reducing regulator demonstrates several phenomena, including the ideal operating are, droop, choked flow, seat load drop or lock-up, and flow coefficient (*Cv*).

system is in a no-flow state. The regulator is set to a certain pressure, but there is no flow. Then, imagine that an operator slowly opens a downstream valve to initiate flow. Right away, there is a sharp drop in pressure because it is difficult for a regulator to maintain pressure at this location. If your regulator is operating along this steep drop in the curve, you may hear chattering or pulsating as the regulator fluctuates between flow and no-flow conditions.

Now let us read the curve from right to left. Imagine that the system is operating along the flat part of the curve. Then, imagine that an operator is slowly closing a downstream valve, reducing flow to near zero. We are moving up the curve. As we near the no-flow state, the regulator has difficulty maintaining the set pressure. Again, you may hear chattering. Eventually, the regulator snaps shut, stopping flow. This is called **lock-up**.

The terms seat load drop and lock-up are essentially interchangeable. Sometimes, people use lock-up to describe both conditions. It is not advisable to operate a regulator in this area.

Hysteresis

It matters whether you are reading a flow curve from left to right or right to left. When reading left to right, flow is increasing. And the reverse is true when reading right to left. Depending on whether flow is increasing or decreasing, the curve will be slightly different. Outlet pressure will not follow the same "droop line," nor end at the original set pressure. This phenomenon, which is illustrated in Figure 3, is called **hysteresis**. The upper line is for decreasing flow, while the bottom line is for increasing flow.

Hysteresis, which results from dynamic friction forces within the regulator, is usually not an issue when evaluating the performance of a regulator. However, it can be a point of confusion during system operation. Suppose an operator sets up a system to deliver an outlet pressure of 50 psig (3.4 bar) at 110 ft^3/min (3115 L/min). The next day, he notices that the pressure is now 50.5 psig (3.48 bar), but the flow is still 110 ft^3/min (3115 L/min). Hysteresis is to blame. It is likely that something in the system temporarily created more flow demand downstream. Moving from left to right on the curve, the temporary flow increase slightly reduced the outlet pressure. Then, as the flow demand returned to 110 ft^3/min (3115 L/min), hysteresis caused the outlet pressure to return to a point slightly higher than the initial set point.

This scenario illustrates the need for operators to periodically check systems to ensure regulators are set to the proper outlet pressures. It is recommended to approach your set pressure from a lower pressure. Another best practice is to employ pressure gauges in a system to help fine tune regulator settings to achieve desired operating pressures.

Backpressure regulators and accumulation

Flow curves for backpressure regulators reveal a phenomenon known as accumulation (see Figure 4), which is the opposite of droop. Because a backpressure regulator controls inlet pressure, pressure will go up as flow increases instead of down as in a pressure-reducing regulator. This phenomenon occurs because the backpressure regulator acts as a restriction and essentially limits flow at the inlet. Your objective is the same as it is with pressure-reducing regulators: to operate on the flat part of the curve.

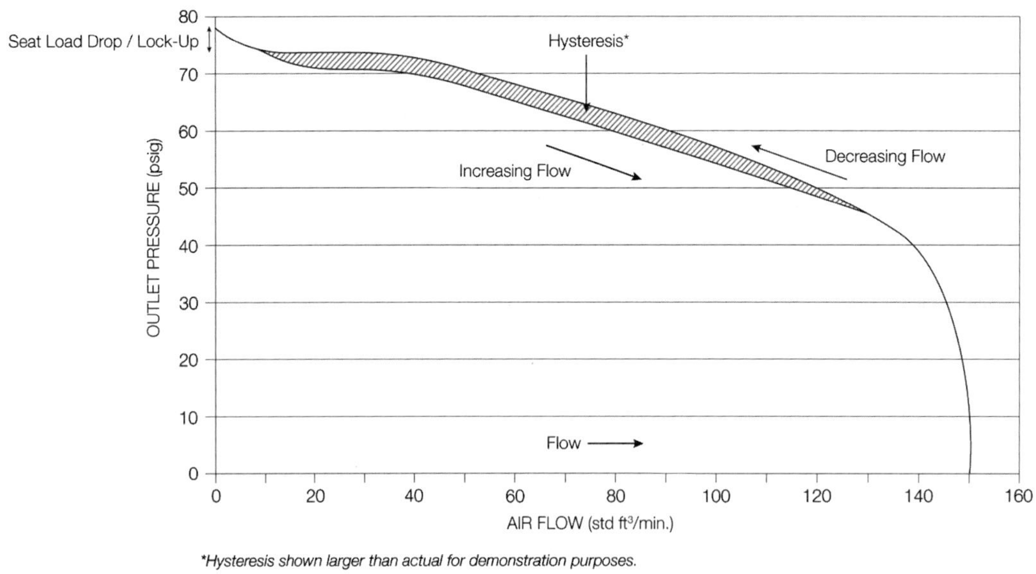

Figure 3. Phenomenon of hysteresis reveals which will be higher at the same flow volume when decreasing flow compared to when increasing flow. (NOTE: Hysteresis shown larger than actual for demonstration purposes.)

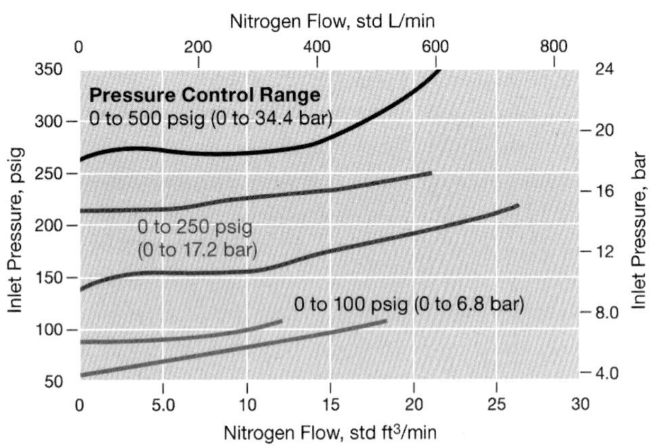

Figure 4. Flow curve for a backpressure regulator indicates accumulation, which is the opposite of droop.

Conclusion

When selecting a regulator for your application, consult the flow curve, not the C_v. Make sure you are looking at the right curve. Does it reflect the pressure you will set your regulator to (your "set pressure")? Does it reflect the right inlet pressure range? Are you working with the right units? Do you need to calculate any adjustments if your system media is a gas?

Once these issues have been addressed, you can start reading your curve. On the curve, identify the range of flows you can expect in your system. Given that range, the curve will tell you what pressures you can expect the regulator to maintain. Ideally, you want the regulator to operate along the relatively flat part of the curve. That is where the regulator will perform most consistently. You want to avoid operating the regulator on the far ends of the curve where undesirable conditions like lock-up and choked flow occur.

If a flow curve is not available or if you need help selecting a regulator, consult with a knowledgeable sales and service representative who can provide guidance on properly sizing a regulator for your application.

Terminology

- **Accumulation**: The increase in inlet pressure required to obtain a specified flow rate for a backpressure regulator.
- **Choked-Flow Area**: The area of a flow curve in which flow demand is greater than the pressure-controlling capabilities of the regulator.
- **Droop**: The outlet pressure change from the set pressure that occurs as flow rate increases for a pressure-reducing regulator.
- **Hysteresis**: The difference between flow curves for increasing and decreasing flow.
- **Lock-Up**: The outlet pressure increase that occurs as flow is decreased to zero.
- **Seat Load Drop**: The initial pressure drop experienced when a regulator starts from a no-flow state.

Reprinted with permission - Swagelok Company and *Flow Control Magazine*, August 2011 issue.

8: Corrosion and Coatings

Types of corrosion and control methods ... 286
Fundamentals of pipeline corrosion ... 288
NAPCA specification designations for coal tar enamel coatings ... 302
NAPCA standard applied pipe coating weights ... 302
NAPCA specifications for pipeline wrappers ... 304
How much primer for a mile of pipe? ... 304
How much coal-tar enamel for a mile of pipe? ... 305
How much wrapping for a mile of pipe? ... 305
Estimating coating and wrapping materials per mile of pipe ... 305
Coefficient of friction for pipe coating materials ... 306
Troubleshooting magnesium anode cathodic protection systems ... 308
Cathodic protection for pipelines ... 309
Estimate the pounds of sacrificial anode material required for offshore pipelines ... 317
Comparison of other reference electrode potentials with that of copper-copper sulfate reference electrode at 25 °C ... 319
Chart aids in calculating ground bed resistance and rectifier power cost ... 320
How can output of magnesium anodes be predicted? ... 293
How to determine the efficiency of a cathodic protection rectifier? ... 293
How to calculate the voltage drop in ground bed cable quickly ... 322
What spacing for test leads to measure current on a pipeline? ... 322
How to estimate the number of magnesium anodes required and their spacing for a bare line or for a corrosion "hot spot"? ... 323
How can resistivity of freshwater be determined from chemical analysis? ... 323
What will be the resistance to earth of a single graphite anode? ... 324
What will be the resistance to earth of a group of graphite anodes, in terms of the resistance of a single anode? ... 324
How can the current output of magnesium rod used for the cathodic protection of heat exchanger shells be predicted? ... 324
How many magnesium anodes are needed for supplementary protection to a short-circuited bare casing? ... 324
Group installation of sacrificial anodes ... 325
How to find the voltage rating of a rectifier if it is to deliver a given amount of current through a given ground bed? ... 325
Determining current requirements for coated lines ... 326
Determine current requirements for coated lines when pipe-to-soil potential values are estimated ... 326
HVDC effects on pipelines ... 327
Troubleshooting bed cathodic protection systems ... 331
How to control corrosion at compressor stations ... 332
Project leak growth ... 333
Advances in pipeline protection ... 334
Estimate the number of squares of tape for pipe coating ... 339
Estimate the amount of primer required for tape ... 340
Tape requirements for fittings ... 340
Induced AC voltages on pipelines may present a serious hazard ... 341
Minimizing shock hazards on pipelines near HVAC lines ... 343
Corrosion of low-velocity, high water cut oil emulsion pipelines ... 344
Pipeline external corrosion direct assessment process ... 346

Types of corrosion and control methods

Types of corrosion

There are several types of pipeline corrosion. Most corrosion failures occur due to a combination of more than one of these types.

General Attack Corrosion

This type of corrosion occurs more or less uniformly over an exposed surface compared to localized corrosion, resulting in fairly uniform thinning of pipe material. The mechanism is an electrochemical process that occurs at the surface of the material.

Localized Corrosion

Localized corrosion occurs at a specific area of the pipe surface. This includes pitting, crevice, and filiform corrosion. Pitting results in deep, narrow attack causing rapid penetration of the pipe wall thickness. Crevice corrosion occurs at localized sites such as crevices where materials meet. Filiform corrosion is a form of oxygen cell corrosion that occurs underneath organic or metallic coatings on pipe materials.

Galvanic Corrosion

Galvanic corrosion occurs due to the potential differences between metals when they are in electrical contact and exposed to an electrolyte. Dissimilar metals accelerate the process. This is due to electrochemical action that causes one metal to be the anode and the other the cathode. The anode corrodes and the cathode is protected.

Environmental Cracking

Environmental cracking occurs very rapidly and can be catastrophic. It is the brittle failure of a ductile material due to the combined effect of corrosion and tensile stress. This type includes stress corrosion cracking, hydrogen-induced cracking, liquid metal, and corrosion fatigue.

Flow-Assisted Corrosion

Flow-assisted corrosion occurs due to the combined action of fluid flow and corrosion. It includes erosion-corrosion, impingement, and cavitation.

Intergranular Corrosion

Intergranular corrosion occurs at the grain boundaries of a metal when the grain boundaries or areas directly adjacent to them are anodic to the surrounding grain materials, caused by differences in impurity levels or the strain energy of the misalignment of atoms in the grain boundaries.

Dealloying

Dealloying occurs when one constituent of an alloy is removed preferentially, thus creating an altered residual structure. Since alloys consist of mixtures of elements, one element acts as an anode compared to another. Thus, by galvanic action, selective corrosion occurs.

Fretting Corrosion

Fretting corrosion occurs due to metal deterioration resulting from repetitive slip at the interface between two surfaces in contact. The movement between surfaces can remove protective films or, combined with the abrasive action of corrosive products, mechanically remove material from surfaces.

High-Temperature Corrosion

This type of corrosion occurs due to chemical reactions resulting in the deterioration of metals at high temperatures.

Corrosion control methods

Four common methods used to control corrosion on pipelines are protective coatings and linings, cathodic protection, materials selection, and inhibitors.

Protective Coatings

External pipe coatings are used for corrosion protection of the exterior of the pipeline. Internally coated pipe (lining) is used to protect the interior of the pipe from caustic and abrasive fluids transported through the pipeline. Internally coated pipe also reduces friction in the pipeline and hence improves the possible flow rate. Protective coatings are used in conjunction with cathodic protection systems to provide the most cost-effective protection for pipelines.

External coatings, applied on the outside of the pipe to form a barrier between the pipe material and the environment consist of the following:

- Bitumastic materials such as coal tar, asphalt, or bitumen used with steel, cast iron, and concrete pipes
- Epoxies such as fusion based epoxy and other polyester materials
- Wrappings consisting of tape applied around the pipe used on coated and uncoated pipes
- Shrink sleeves used to protect welded joints of pipe

Common coating application methods include brush or roller, spray, and dipping. In addition to proper coating selection and application methods, substrate preparation is critical to the success of the coating. The majority of coating failures are caused either completely or partially by faulty surface preparation, such as leaving contaminants on the surface or having an inadequate anchor (sand blast) pattern.

Cathodic Protection

The concept of cathodic protection goes back to 1824, in London, when the copper sheeting on British ships was being corroded. Experiments were conducted that resulted in protecting copper against corrosion from seawater utilizing iron anodes. Since then, cathodic protection has grown to have many uses in marine and underground structures, water storage tanks, gas pipelines, oil platform supports, and many other facilities exposed to corrosive environments.

Cathodic protection is a technique by which a buried pipeline is made as the cathode and another metal is used as the anode that gets corroded by electrochemical process instead of the pipe material, as shown in Figure 1.

The electric circuit requires a source of protective current. This is provided by either an active (impressed current) or a passive (sacrificial) system of galvanic anodes (usually magnesium, aluminum, or zinc).

A direct current system is used to counteract the normal external corrosion of a metal pipeline resulting from the potential difference between the pipe and the soil.

Cathodic protection basically reduces the corrosion rate of a metallic structure by reducing its corrosion potential, bringing the metal closer to an immune state. The two main methods of achieving this goal are by either sacrificial anodes or the impressed current method:

- **Sacrificial anodes:** This involves connecting the pipe to a zinc or magnesium electrode and the surrounding soil in a buried pipeline. This creates a galvanic cell shown in Figure 2 in which the pipe becomes the cathode and the zinc or magnesium electrode becomes the anode that gets corroded. Hence, the reason for the term sacrificial anodes.
- **Impressed current method:** This method requires the use of an external DC source (rectifier). The negative terminal of the rectifier is connected to the steel pipe and the positive terminal is connected to the ground through an electrode. The pipe thus becomes the cathode as shown in Figure 3. When a copper sulfate electrode is used, the impressed current rectifier is adjusted so that the pipe to soil potential is maintained at −0.85 V. This potential should be maintained at various points along the pipeline.

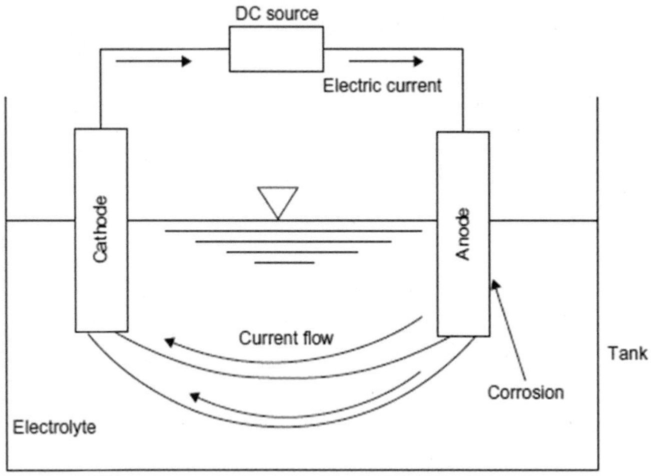

Figure 1. Electrolytic cell.

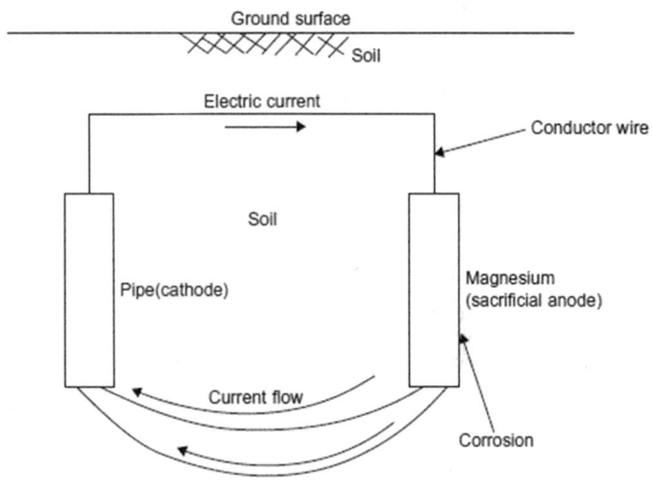

Figure 2. Galvanic cell.

Materials Selection and Design

Corrosion can be mitigated by proper selection of materials for the specific application. Corrosion-resistant materials

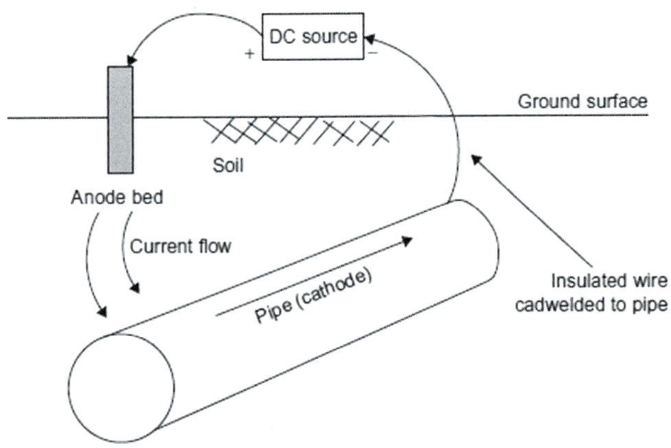

Figure 3. Impressed current system.

include stainless steels, plastics, and special alloys used in conjunction with pipelines. Materials selection must take into account the design life of the pipeline and the environment surrounding it.

Corrosion Inhibitors

Corrosion inhibitors such as acids, cooling waters, and steam are used to decrease the rate of corrosion by adding them to the environment. These inhibitors control corrosion of the metal by forming thin films at the metal surface.

Source

E.S. Menon, Pipeline Planning and Construction Field Manual. Gulf Professional Publishing, Waltham, MA, 2011.

Fundamentals of pipeline corrosion

Excerpts from the report "Pipeline Corrosion" performed by Michael Baker Jr., Inc. for the U.S Department of Transportation, Office of Pipeline Safety, November 2008.

Understanding the process

Although there are various definitions of corrosion, the definition used by NACE International (NACE), the primary support organization in the corrosion industry, is "The deterioration of a material, usually a metal, which results from a reaction with its environment." With respect to pipeline corrosion, the metal is line-pipe steel, primarily comprised of iron with one to two percent alloy for strength and toughness (alloys have been determined essentially irrelevant to the corrosion process). In regards to external corrosion, the environment would be groundwater or moist soil for onshore pipelines and seawater for offshore pipelines. For internal corrosion, the environment would be water containing sodium chloride (salt), hydrogen sulfide, and/or carbon dioxide. The deterioration would be dissolution of the iron into the environment, which reduces the strength of the pipeline.

When iron dissolves, it does so as a positively charged ion. The process, represented as follows, is referred to as an anodic reaction (Figure 1):

$$Fe \rightarrow Fe^{++} + 2e^{-}$$

The electrons produced from the reaction move through the metal pipe to another location where they are in turn consumed in a reaction that produces hydroxyl ions. The specific reaction depends on the nature of the electrolyte, but typically is one of the following:

$$O_2 + 2H_2O + 4e^{-} \rightarrow 4OH^{-}$$

$$2H_2O + 2e^{-} \rightarrow H_2 + 2OH^{-}$$

The reactions represented are referred to as cathodic reactions. Movement of the ions through the electrolyte completes the electrical circuit. The iron ions typically react with the water and/or oxygen to form a corrosion deposit of

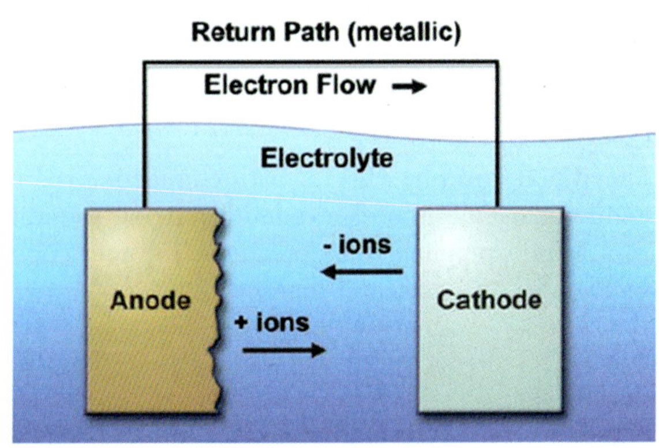

Figure 1. Basic corrosion cell.

rust or some other iron oxide, but, in some cases, they may react with carbon dioxide or hydrogen sulfide to form iron carbonate or iron sulfide.

The anode and cathode components of a corrosion cell can be next to each other or separated by many feet.

Uniform vs. localized corrosion

Pitting

Typically, corrosion in pipelines manifests as pits rather than as a uniform reduction of the wall thickness. This is because the environment at an anodic area tends to become more acidic, since the iron ions in solution react with the hydroxyl ions of the water to leave an excess of hydrogen ions.

$$Fe^{++} + H_2O \rightarrow FeOH + H^+$$

Conversely, as indicated in Figure 1, hydroxyl ions will be produced at the cathodic areas, making the environment more alkaline and less corrosive. Therefore, once a pit starts to form, subsequent corrosive attack tends to be concentrated at that location.

Sometimes the pits will be isolated from each other, and, other times, they will be so close together that they overlap and produce a general but irregular thinning of the pipe wall. Both features are illustrated in Figure 2. Isolated pits may be seen in the top half of the picture. The overlapping pits shown in the bottom portion of the picture have caused enough thinning of the pipe wall to result in a rupture.

Figure 2. Pipeline damage from pitting.

Selective Seam Corrosion

Although seamless pipes have been used in some systems, most line pipe contains a longitudinal weld, or seam. The long seam, as it is called, most frequently is made by submerged-arc welding or upset-butt welding. A submerged-arc weld contains a filler metal that has a composition slightly different from that of the body of the pipe, and the heat-affected zone next to the weld metal has a microstructure different from that of the rest of the pipe. Although upset-butt welds, which can be either electric-resistance welds or flash welds, do not contain filler metal, they also have a heat affected zone that has a different microstructure. Because these different microstructures can be more susceptible to corrosion than the surrounding metal, selective corrosion at the seam can sometimes occur with little adjacent corrosion-related damage. If the seam also contains cracks or discontinuities that, by themselves, are not large enough to cause a failure, corrosion in the same area might enlarge such flaws to critical size and precipitate a leak or rupture. Certain vintages of pipe, including pre-1971 manufactured low frequency electric weld resistance (ERW) pipe, have exhibited seam-related problems that might be particularly susceptible to selective seam corrosion.

Microbial Corrosion

MIC (Microbiologically influenced corrosion) is caused by microbes whose actions initiate the corrosion cycle. There are several types of microbes that, while producing different effects, have been found to promote either external or internal corrosion. The main types are sulfate-reducing bacteria (SRB) and acid-producing bacteria (APB). Bacteria can promote external corrosion by depolarizing the pipe through the consumption of hydrogen gas formed at the pipe surface by the cathodic protection currents. Once the pipe is depolarized, corrosion can take place. One way that internal corrosion can occur is by bacteria forming an acidic biofilm that traps electrolytes and acids. Internal MIC is more common in liquid pipelines than in gas pipelines. The corrosion mechanism is illustrated schematically in Figure 3. Part of a pipeline that has experienced MIC is shown in Figure 4.

External corrosion

Factors that Affect External Corrosion

Onshore Buried Pipelines. A number of correlations between the characteristics of a soil and its corrosivity have been found through research and practical experience.

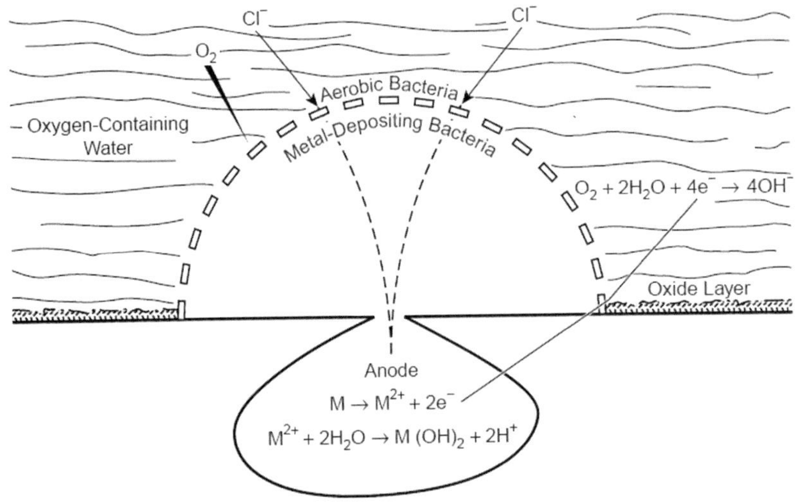

Figure 3. Possible reactions under tubercles created by metal depositing bacteria.

Figure 4. Disbonded pipeline coating associated with external localized microbiologically influenced corrosion.

Since corrosion requires movement of ions through the electrolyte (soil), factors that increase the electrical conductivity of the soil tend to increase its corrosivity. Thus, higher moisture contents, poor drainage, and high salt contents tend to increase corrosivity. Higher oxygen contents also tend to increase corrosivity. However, corrosivity is only one of many factors, and not even a primary factor, that determine the rate of external corrosion on a pipeline.

One of the main factors that influence the rate of external corrosion is the differences in the characteristics of the soil from place to place along a pipeline, as well as from top to bottom. Differences in aeration, moisture content, and soil composition in these areas can produce strong driving forces for corrosion.

Although low-conductivity soils tend to exhibit relatively low corrosivity, they also retard the flow of cathodic-protection currents to the pipe and thus may precipitate pipeline corrosion to a greater extent than anticipated.

The soil type also can affect the rate of deterioration of the coating. For example, heavy clay soils can pull the coating away from the pipe as they expand and contract with changes in moisture content. Rocky soils, while expected to be well drained and therefore of low corrosivity, might puncture the coating, and their high resistivity might block the cathodic protection currents from the pipe at those locations.

If there is a second pipeline in the area, cathodic protection currents in the soil going to or from the second pipeline might jump onto and off of the first pipeline. Places where the current discharges from the first pipeline would be highly anodic.

As detailed below, the two most important factors in reducing or preventing the development of external corrosion on a pipeline are the level of cathodic protection and the quality of the pipeline coating.

Offshore Pipelines. While not a focus of this study, it is important to contrast the issue of external corrosion in onshore buried pipelines with external corrosion of offshore pipelines. Although salt water is much more corrosive than most soil environments, cases of significant external corrosion on offshore pipelines are extremely rare. The ability to control external corrosion has been mastered to a high degree, as compared to onshore performance. This is particularly due to the homogeneity of the offshore environment, and the predictability of coating and cathodic protection. The offshore environment is very uniform in composition and of high conductivity, thus enabling the uniform and consequently effective application cathodic protection. Furthermore, the alkaline environment produced by the cathodic protection causes calcareous deposits (primarily magnesium carbonate) to precipitate at the coating holidays (holes), essentially plugging the holidays and separating the steel from the water.

Failures that do occur on offshore pipelines occur predominately on the riser as a result of corrosion. The consistent wetting and drying in the splash zone combined with defects in the coatings are the usual contributors to the problem. Risers will fail often, but the failure is rarely catastrophic and downtime is usually minimal as compared with onshore pipeline failures due to corrosion.

Methods to Prevent or Mitigate External Corrosion on Buried Pipelines

The primary method of preventing or mitigating external corrosion on buried pipelines involves a combination of cathodic protection and coatings. Cathodic protection involves applying a current to the pipeline through the soil from an external source and thus overriding the local anodes, rendering the entire exposed pipeline surface cathodic. Coatings function to separate the steel from the electrolyte, and thus prevent corrosion.

The application of cathodic protection alone to protect against corrosion would not be practical, because the amount of current required is proportional to the exposed area, and it would be too expensive to cathodically protect a long, bare pipeline. Therefore, coatings are needed to reduce the amount of exposed area as much as possible, and are therefore, the primary method of corrosion control and prevention. Coatings by themselves also would not be totally effective, because it is impossible to produce a perfect coating over an entire pipeline. As well, some damage during construction and degradation over time are inevitable. Therefore, cathodic protection is needed to prevent corrosion at the breaks (holidays) in the coating.

Coatings. Pipeline coatings have undergone dramatic technological changes over the past two decades. Coatings now must perform at higher in-service operating temperatures, must not be damaged in handling during construction or in operation by soil stress or soil movement, and must provide exceptional corrosion protection. Coatings also must be user friendly and must be able to be applied in a mill or in the field.

Since the 1950s, corrosion coatings have continuously improved. During the period of 1950 to 1960, common coatings included coal tar-based coatings, asphalt coatings, early grease coatings, early cold-and hot-applied tapes, and the first coal tar enamel coatings. All of these coatings were designed to isolate the pipe from its environment and to prevent water from reaching the pipe surface. In addition, they offered some dielectric strength (i.e., resistance to the passage of electrical current from the environment to the pipe surface). Unfortunately, many of these coatings were difficult to apply and almost always did not uniformly bond to the pipe surface, resulting in voids, pinholes, and other imperfections. Most of these coatings also degraded in the ground over time. Degradation reduced their ability to resist soil strain and moisture. After a number of years, many would become porous, or worse, disbond from the surface of the pipe.

In the 1960s, newer and more long-lasting coatings were produced and applied, and the use of certain kinds of grease wraps was discontinued during this time. Also, the adhesive capability of coal tar products was improved.

In the 1970s and 1980s, the early epoxy products and new cold-applied tapes were introduced. New wax or grease products were introduced to accommodate irregularly shaped features such as valves and fittings. Improvements continued in the development of the other types of coatings, including durability when subjected to soil strain, moisture, and other environmental stresses.

In the 1990s, better and more effective dry powder and wet-applied epoxies appeared and were widely utilized. These coatings possessed good dielectric properties and degraded relatively slowly over time from environmental exposure. However, they lacked mechanical strength. Rough handling would cause coating voids; moisture would subsequently penetrate the voids, forming blisters. Polyolefin coatings were also developed which adhered very well to the pipe surface. In addition, polyolefin provided good soil and mechanical damage resistance. However, polyolefin coatings were also expensive, and, if they did disbond, shielding could potentially occur. Two- and three-layer polyolefin coatings have become particularly popular outside of the United States.

Fusion-bonded epoxy (FBE) coating, sometimes called thin-film epoxy, is an epoxy-based powder coating that is currently widely used to protect pipelines, including valves. FBE coatings are thermoset polymer coatings. The name is derived from resin cross-linking and method of application, which is different from that of a conventional liquid paint. FBE coatings are made from dry powder. The resin and hardener parts of the dry powder remain unreacted at normal storage conditions. At typical coating application temperatures, usually in the range of 180°C to 250°C (350°F to 480°F), the contents of the powder melt and transform to a liquid. The liquid FBE film flows onto and wets the steel surface and, assisted by heating, soon becomes a solid coating by chemical cross-linking. This process is known as "fusion bonding." The chemical cross-linking reaction that takes place in this case is irreversible, which means that, once the curing takes place, the coating cannot be converted back to its original form by any means. The coating process is illustrated in Figure 5.

In the 1950s and 1960s, many of the coatings - especially coal tar, asphalt, and tape - were applied continuously by machine in the field after the pipe joints were welded together and cleaned by scraping or wire brushing. Those cleaning methods were not particularly effective, and the remaining dust and mill scale interfered with a good bond between the coating and the pipe. More recently, many of the coatings, including FBE, have been applied in the coating mill after the pipe surface was cleaned by grit blasting. This resulted in a far superior surface finish and better bonding. However, the ends of each joint had to be coated in the field after the girth welds were made. These field applied coatings for joints and other fittings have evolved from very simple tar to multipart epoxy and heat-shrink sleeves. Each has its strengths and weakness.

The earliest field-applied joint coating was a "granny ragged" hot tar coating applied using a saturated cloth. Its application involved little or no surface preparation of the joint. Because the hot tar would run down the pipe, cloth reinforcement was required in some cases. Nevertheless, the coating at the top of the pipe was considerably thinner than at the bottom, and it could also contain pin holes. Moisture could enter the pin holes but could not drain because the bottom was waterproofed. Therefore, pockets of water resulted, with the potential to create a corrosive environment. Besides not being very effective, this method of coating field joints was very labor intensive and subject to contamination by dirt.

Cold-applied tapes, another type of field-applied coating, had limited success. The tapes were wrapped over the field joint, and, depending on the adequacy of surface preparation, could either adhere or disbond. If the tape adhered well to the joint, no problems would result. However, if tapes disbonded, they could adhere to one another and also establish a shielded environment, which would promote localized corrosion. Soil strain and the manner in which the pipe was lowered into the trench were other concerns associated with the application of tape coatings. In some cases, especially during warm weather, the slings used to lower a pipeline with fittings into a trench could cause cold-applied tape to bunch, creating voids and areas that exposed pipe surfaces.

In the 1990s, a field-applied epoxy system was used to complement the FBE coatings on piping systems (Figure 6). Similar to the factory-applied FBE, when the field-applied coatings fail, they do not cause shielding. Instead, they allow the cathodic protection currents to reach the bare pipe.

Also at this time, shrink sleeves were widely used (Figure 7). Shrink-sleeve coating systems consist of

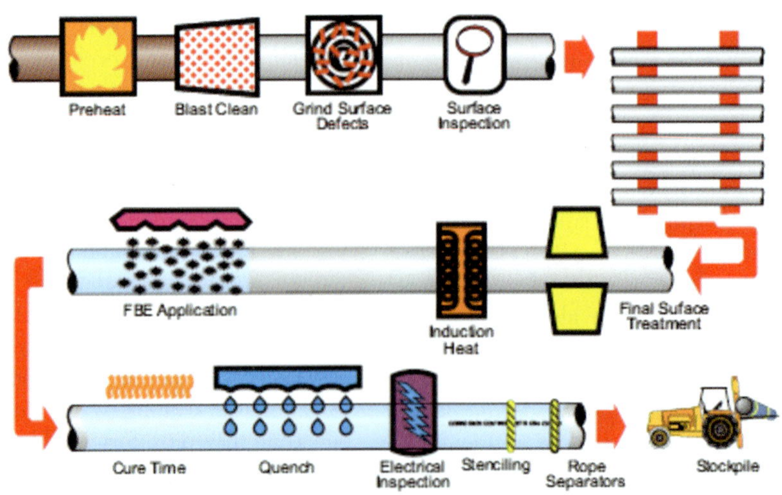

Figure 5. Fusion Bonded Epoxy powder application schematic.

Figure 6. Field applied joint coating.

thermoset plastic that, when applied to the joint and heated, will shrink around the joint. Some shrink sleeves, particularly those utilized in directionally drilled pipe installations, may require epoxy primer application or surface preparation. If the surface is not properly prepared, a shrink sleeve will not adhere, the coating will fail, and water will collect in the failed coating voids. Moreover, the shrink wrap will shield the voids from cathodic protection and the pipeline could corrode. If the installation is done incorrectly or if the pipe is not sufficiently heated, voids can form beneath the sleeve. The sleeve will subsequently disbond from the pipe, causing shielding. If a pin hole is present, then moisture can enter, and because of the shielding, a corrosion cell can form.

Oftentimes, offshore pipelines also incorporate concrete coatings to offset buoyancy as well as to prevent fatigue damage and flow-induced vibrations due to wave or current forces by providing stability across irregular seafloor topography or where there is scouring or local differences in load bearing capacities of the soil.

Figure 7. Shrink sleeve application.

The use of thermal sprayed metallic coatings (Aluminum and Zinc) has been gaining popularity in submerged offshore applications, including some pipelines. This coating offers the benefits of low cost with excellent corrosion protection over a wide range of temperature and conditions. While thermal sprayed coatings do provide a measure of cathodic protection, they still require supplemental anodes but the quantity can be significantly reduced. These coatings are still under development for broad-based pipeline applications but expect to see continued progress.

Cathodic Protection. As stated earlier, anodic areas, where corrosion occurs, discharge current to the cathodic areas, where corrosion does not occur. By driving current to the pipe through the environment, as illustrated in Figure 8, the anodic currents can be overridden, and the entire pipe can be made to behave like a cathode. At the same time, a benign alkaline environment is produced at the pipe surface. In sea water and many soils, a calcareous deposit will form in that alkaline environment and further protect the steel.

There are two types of cathodic protection systems: sacrificial anode and impressed current anode. Sacrificial anode systems utilize an externally connected sacrificial metal with a relative activity value greater than steel (iron) and thereby protect steel from corrosion. Alloys of zinc and magnesium are the sacrificial metals most commonly employed. The sacrificial anode is connected to the pipeline via a wire and placed some distance from the pipeline. The current flows from the anode into the surrounding soil (electrolyte) and is picked up by the pipeline at coating holidays. The circuit is completed by a wire that connects the anode to the pipe. The number and placement of anodes is based on the site-specific requirements of the particular pipeline that is to be protected. A well-coated pipeline with a few small holidays does not require many anodes.

Impressed-current anode systems, shown in Figure 9, involve the application of direct-current voltage between an anode and the pipeline. Impressed-current anodes can be made from graphite, high-silicon cast iron, lead-silver alloys, precious metals, mixed-metal oxides, or steel. As with sacrificial anodes, the shapes, locations, and number depend on the geology of the area and the nature of the pipeline system.

Other Preventive/Mitigative Measures. Placing a metallic pipe within another pipe that contains nonconductive filler in its annular space is yet another method to

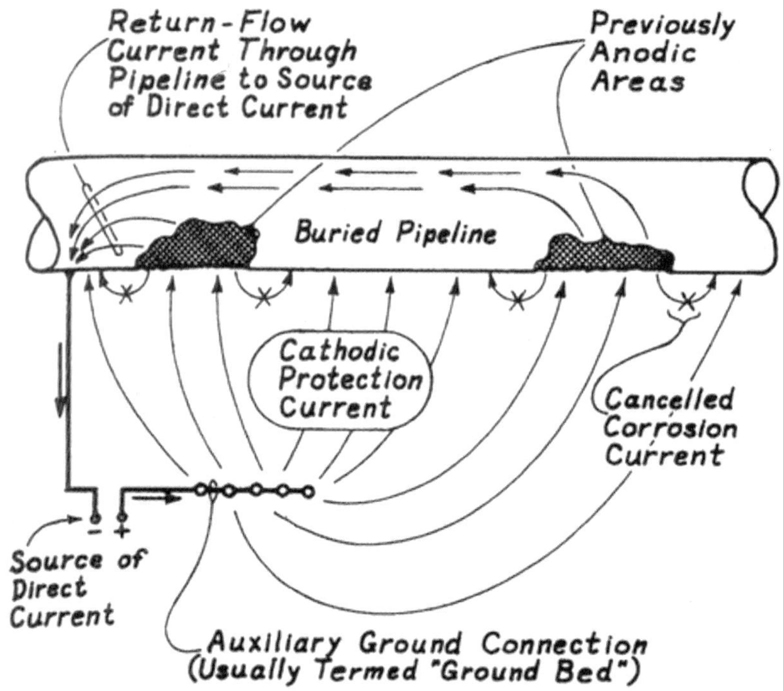

Figure 8. Basic concept of cathodic protection.

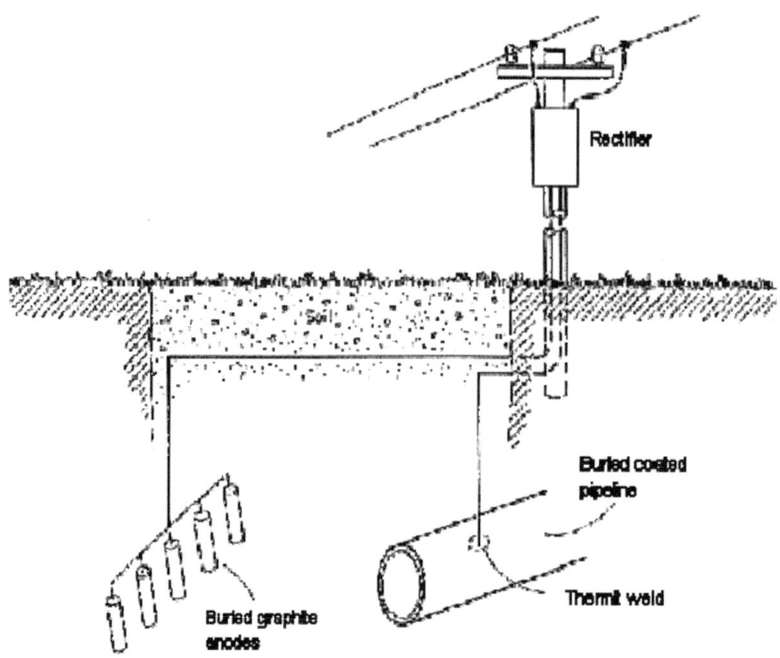

Figure 9. Impressed-current cathodic protection of a buried pipeline.

Figure 10. Installing casing insulators on product pipe prior to installation in casing pipe, 4300-foot HDD Colville River Crossing.

protect the structure from environmental impact and thus reduce or eliminate its potential for corrosion. This "pipe-in-pipe" method may be suitable for short sections of pipeline, such as those that pass under streams or rivers. This technique was used in the horizontally directionally drilled (HDD) 4,300-foot crude oil pipeline crossing of the Colville River on the North Slope of Alaska, as shown in Figure 10.

Monitoring Techniques for External Corrosion

Just as a battery, which consists of two different metals in contact with an electrolyte, produces a voltage, a steel pipeline in contact with the soil will produce a voltage between itself and another piece of metal that is in the soil and in electrical contact with the pipe. The voltage will depend on the relative differences in corrosion tendency of the two metals in the environments that surround each. This concept is the basis for the method that is used to monitor external corrosion on buried pipeline without excavating it. The pipeline in contact with the soil or seawater is one half of the "battery," and a piece of copper in contact with a saturated solution of copper sulfate is commonly used as a standard reference for the other half of the "battery." The potential (voltage) generated between a pipeline and the standard reference can be correlated with the tendency of the pipeline to corrode. The numerical value of the voltage is referred to as the pipe-to-soil potential. The primary way to make sure that the pipeline is not undergoing significant corrosion is to monitor the pipe-to-soil potential. If the pipe-to-soil potential is found to be insufficient, the cathodic-protection system should be adjusted by increasing the output of the rectifiers for an impressed-current system or by increasing the number of sacrificial anodes.

Although the exact value of pipe-to-soil potential necessary to prevent significant corrosion will be different for different environments, experience has shown that a value more negative than −850 mV versus the copper/copper sulfate electrode (CSE) or at least 100 mV more negative than the native (freely corroding) potential would be adequate in almost all soils. However, if MIC is involved, a potential of −950 mV versus CSE or a shift of at least 300 mV is recommended.

49 CFR part 192 requires that pipe-to-soil readings for cathodically protected pipeline systems must be taken annually at all test stations, at intervals not to exceed 15 months from the previous read. Typically, these test stations are installed at convenient locations, such as road crossings. Occasionally, more detailed surveys, as illustrated in Figure 11, may be conducted every few feet along the pipeline. Some operators also will momentarily interrupt the cathodic protection system once or twice a minute to obtain readings free from errors caused by the current flowing in the soil between the position of the reference electrode and the pipe. The reading with the current flowing is called the "on potential," and the reading with the cathodic protection interrupted is referred to as the "off potential" or the "instant off potential." The difference between the two values is termed the "IR drop," because it is equal to the current, I, in the soil times the resistance, R.

Internal Corrosion

Internal corrosion generally cannot occur in a pipeline unless there is an electrolyte to complete the corrosion cell. Water or other aqueous materials (such as glycols from dehydration processes) are needed to form the electrolyte. Also, other chemicals usually must be present: for example, carbon dioxide (CO_2) for the formation of dilute organic and inorganic acids or sulfur for the formation of acid or growth of bacteria. Once introduced, the corrosive materials may continue to damage the pipeline until they are removed, or until they are consumed in corrosion reactions.

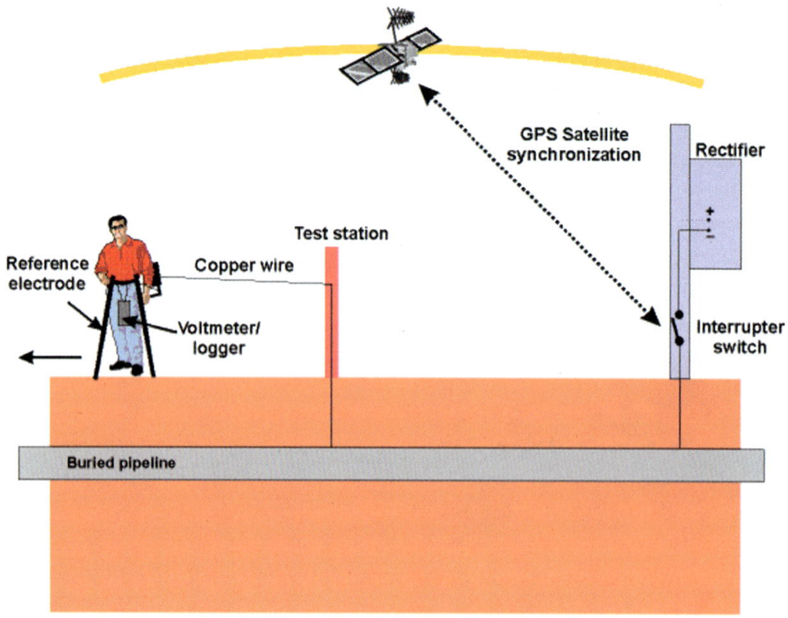

Figure 11. Close-interval pipe-to-soil potential survey.

Gas Pipelines

Typically, sales-quality dry gas will not corrode pipeline interior surfaces. However, natural gas, as it comes from the well, may contain small amounts of contaminants such as water, carbon dioxide, and hydrogen sulfide. If the water condenses, it can react with the carbon dioxide or hydrogen sulfide to form an acid that might collect in a low spot and cause internal corrosion.

Liquid Pipelines

Similarly, internal corrosion can occur in hazardous liquid pipelines carrying corrosive liquids or liquids containing corrosive contaminants. Liquid pipelines can experience internal corrosion anywhere along their length where electrolytes or solids drop out and wet the surface or provide a place for electrolytes to collect. An example of internal corrosion in a liquid pipeline is shown in Figure 12.

Figure 12. Internal corrosion of a crude oil pipeline.

Preventive/Mitigative Measures for Internal Corrosion

Dehydration. Dehydration is the most commonly applied measure to protect against internal corrosion in gas pipelines (and also in liquid pipelines that contain oil with free water or other electrolytes). Dehydration removes condensation and free water that, if permitted to remain, would allow internal corrosion to occur at points where water droplets precipitate from the gas stream to either form liquid puddles at the bottom of the pipe, or adhere to the top of the pipe. Where the gas stream is usually dry, topside corrosion rarely takes place. Complete dehydration is very effective, but, because the systems are neither 100 percent effective nor 100 percent dependable, there always is the potential to introduce water and other electrolytes into a gas pipeline.

Several methods may be used to dry gas. The most common method is dehydration, which involves the use of devices to chemically "scrub" the gas stream to remove moisture and reduce the gas dew point. Glycol is the scrubbing agent most frequently used. Provided that the glycol is removed from the gas stream, the environmental conditions that promote corrosion are eliminated.

There also are physical methods of eliminating entrained moisture – the most commonly utilized are scrubbers that employ cyclone separators to remove the water or glycol droplets from the gas stream. However, physical methods cannot significantly reduce the dew point, so periodic upsets can occur. Most gas gathering system and storage operators utilize dehydration devices to scrub gas of humidity by removing the moisture present in the produced gas. The process reduces the gas stream dew point so that the gas becomes "tariff gas" that contains less than seven pounds of moisture per million cubic feet.

Liquid pipelines that contain free water also can experience internal corrosion. The fluid can sometimes be treated to remove both free water and dissolved water. When water is present, pigging is required to move the water from areas where it accumulates due to local flow conditions. Water is removed from crude oil by gravity separation, but there always is a percentage of water remaining in the product crude oil. Hydrocarbon products may use a salt dryer or other means of removing water; however, because the systems are not totally reliable or efficient, there is always the possibility of introducing water or an electrolyte into the pipeline. These upset conditions, and the corrosion that they can cause, must be monitored and mitigated.

Inhibitors. Inhibitors are chemicals that can be added to a pipeline to reduce the rate of corrosion. They can adsorb onto the metal surface or react with it to form a protective film, or they may react with the corrodent to make it less corrosive. Many different chemicals are available commercially. The choice will depend on the type of product in the pipeline and the type of corrodent. Other considerations include cost, availability, toxicity, and environmental friendliness.

Coatings. Internal coatings have been used on some gas transmission pipelines to improve product flow by reducing drag and eliminating dust. Such coatings can be somewhat effective in controlling internal corrosion, but they are very difficult to apply uniformly, which impacts their effectiveness. In lieu of coatings, some operators have attempted to install plastic or high-density polyethylene liners or inserts in their pipelines. Plastic liners are an effective barrier against corrosion but are not fail-safe. Problems occur if pinholes are present and allow corrosive materials to migrate behind a pipe's liner. Relying solely on liners or coatings may not be prudent since it is very unlikely that the problems associated with each can be remedied. Many operators who do use liners or coatings also apply additional preventive measures.

Buffering. In principle, buffering agents that change the chemical composition of fluids that remain in the pipeline can be utilized to prevent internal corrosion. The introduction of a buffering agent, such as a mild or dilute alkaline mixture, can significantly reduce the corrosivity of any standing liquid, predominantly by raising its pH value above seven (neutral), so that it turns from acidic to alkaline. Alkaline liquids cause virtually no harm to steel. In general, buffering is not very effective because it is difficult to cover the entire pipe surface.

Cleaning Pigs. The frequent use of cleaning pigs to scour the internal surfaces of a pipeline is another viable preventive measure. There are many types of cleaning pigs. The choice of which type of pig to use depends on the product carried by the pipeline and the contaminant to be removed. Although their application may preclude the use of internal corrosion direct assessment (ICDA) models, cleaning pigs can effectively direct both liquids and corrosive solids to pig traps for removal from the pipeline. It is noted that the buildup of solids also can create internal corrosion, since the solids can entrap corrosive or low-pH liquids in a corrosive matrix.

Routine pigging will channel any liquid pools away from low points and, if performed properly, out of the entire pipeline. Cleaning pigs also will displace the solids and remove them from the pipeline via the pig trap at the end of the pipeline, provided that the pipeline is properly configured (i.e. no dead legs or other features that would trap liquids or solids and prevent cleaning pig access). Pigging effectiveness is a function of the pigging velocity, pigging distance, and characteristics of the materials targeted by the pig run.

Biocides. Biocides can be used in the pipeline to inhibit the corrosive actions of the microbes that cause MIC and thereby reduce or eliminate MIC. Biocide is injected into the pipeline in the stream of a non electrolytic carrier. In many cases, the biocide is added to the buffering agent. There are also other active agents that can be added, such as film-formers that aid in forming a passive barrier at the pipe surface and agents that promote the evaporation of electrolytes. Many of these agents are expensive, and, depending on the gas flow at the time of injection or use, may or may not reach the location where the electrolytes and microbes are trapped.

Additional Preventive Measures. System designs that include features to trap contaminants can prevent corrosion within a complete piping system. The simplest design alternative consists of the installation of separators at the system's entry point or at a location just downstream of the point where liquids could be entering the gas stream (such as at the end of gathering systems or outlets from storage fields). Separators, which are typically designed to handle a range of flow rates, can be configured in parallel to facilitate their addition or removal from the equipment train, according to flow velocity requirements.

Drip legs or logs (also known as traps/drips) are another design mechanism to prevent internal corrosion. These devices are designed to trap liquid contaminants and prevent them from traveling downstream. The drips can be pumped so the liquids can be removed before significant internal corrosion can take place. However, if the liquids are not removed in a timely manner, the drips themselves can corrode. Some drips are filled with heavy oils that permit an aqueous layer to form overtop and, in this manner, enable removal of the contaminants. The use of drips began at the time when manufactured gas was the only gas available and was stored in tanks with water seals. The use of the water seals, in combination with the manufacturing process, yielded gas saturated with water vapor and tars. The water vapor and tars were actually beneficial in preserving the integrity of the early pipelines that were constructed of wood, but proved detrimental to the operation of cast iron, wrought iron, and steel pipelines. As the gas progressed through a metallic pipeline, the water vapor and tars would condense and coat the inside of the pipe. Their accumulation at low points would result in the development of a corrosion cell.

Preventive/Mitigative Measures for Selective Seam Corrosion. Most pipelines in the United States are constructed with the longitudinal seam weld in the top half of the pipe to keep water from contacting the seam on the inside of the pipe. Other methods focus on either alleviating the integrity problems of the seam by replacing the pipe or inhibiting corrosion by applying one or several of the measures previously discussed.

Preventive/Mitigative Measures for MIC on Gas Pipelines. To prevent or correct internal MIC, the internal environment must be changed by eliminating the electrolyte at the pipe wall. This can be accomplished by drying the gas and eliminating any design features in which liquids or wet solids can accumulate. The use of buffers or other pH-altering chemicals also can modify the environment and possibly eliminate the growth of the bacteria that cause MIC. he gas stream is needed.

Monitoring Internal Corrosion

There are several monitoring methods that are effective in evaluating internal corrosion as well as the conditions conducive to its development. One method involves the installation and periodic examination of removable corrosion coupons in areas of the pipeline known to be susceptible to internal corrosion. This method can provide data on corrosion rates and the conditions that are causing the corrosion. Correct placement of the coupons is critical to the accuracy of the evaluation. Coupons must be placed as close as possible to the entry point of the electrolyte and in direct contact with it. There are several designs to accomplish this, but, at present, there is no industry guidance or standard to direct operators in proper coupon placement. To highlight the disparity, some operators have suggested that the removable corrosion coupons be installed at low points in the pipeline, raised just above the bottom surface of the pipe, so that the coupons would be exposed to any liquids that might have become trapped; however, a contrary view is that this does not work well, because the water present may only be a wetting film on the surface of the pipe, and the coupon would never touch the water. The operator then would have falsely non-conservative information, thinking no corrosion is occurring and not investigating farther.

Probes also can be effective in monitoring for internal corrosion if placed in the proper location and oriented where they will contact any electrolytes present. Probes that measure liquid conductivity are used to determine if electrolytes are present, while pH probes can be used to determine the pH of the liquid, which may or may not indicate the corrosion potential. Ultrasonic probes installed around the pipe can be used to determine if metal loss is occurring.

Permanently placed ultrasonic measuring probes can be utilized to measure the pipe wall in areas believed to be at risk for internal corrosion. The Pipeline and Hazardous Materials Safety Administration (PHMSA) notes that this is a very "iffy" type of monitoring, unless the corrosion is general corrosion, as in acid corrosion. Such probes probably would not reliably detect pitting corrosion, but instead would yield a false non-conservative reading, leading to the misconception that corrosion is occurring. The probes measure the remaining pipe wall and can provide real-time data on the amount of internal corrosion occurring in a particular section of pipeline but only if it is general overall corrosion or the probe happens to be placed on an active pit that is still corroding.

The use of an indirect monitoring technique involving continuous moisture monitors enables operators to determine whether electrolytes are being introduced into the pipeline from entry or supply points. The moisture monitors only detect water vapor in the gas; they do not monitor for liquids. Moisture monitoring must be conducted in conjunction with a means of detecting liquid water/electrolyte in the pipeline.

Environmentally assisted cracking

Another form of degradation that would fall within the NACE definition of corrosion is environmentally assisted cracking (EAC), in which the combined action of a tensile stress and a corrosive environment causes cracks to form in the metal. There are a number of different cracking mechanisms within the category of EAC, including stress corrosion cracking (SCC), corrosion fatigue, hydrogen-stress cracking, hydrogen-induced cracking (HIC), hydrogen-induced loss of ductility, and sulfide-stress cracking (SSC). Compared to pitting corrosion, EAC results in relatively few significant incidents.

Stress-Corrosion Cracking (SCC)

External SCC. Since its discovery in 1965 as a possible cause of failures in pipelines, SCC has caused, on average, one to two failures per year in the U. S. Two types of environments have been associated with external SCC. High-pH SCC is caused by a concentrated solution of sodium carbonate and sodium bicarbonate with a pH that typically is between nine and 10.5. Near-neutral-pH SCC is caused by a relatively dilute solution of carbon dioxide and sodium bicarbonate with a pH that typically is between six and seven.

Both forms of SCC have similar appearances. As is illustrated in Figure 13, the cracks occur in clusters, and the fracture surface contains a black corrosion deposit that corresponds to the size of the cracks just prior to failure. However, the two forms can be distinguished by metallographic examination: high-pH stress-corrosion cracks are intergranular (the cracks follow the grain boundaries of the metal), while near-neutral stress-corrosion cracks are transgranular (the cracks run through the grains of the metal) and also exhibit more corrosion along their sides.

Typically, the cracks are oriented in the axial direction, because they grow perpendicular to the direction of largest tensile stress, which usually is the hoop stress due to the internal pressure. However, there have been some instances of circumferential cracks in areas, particularly on slopes, where soil movement caused high longitudinal stresses.

Although high-pH SCC can occur at normal ground temperatures, the crack growth rate increases with increasing temperature. Therefore most high-pH SCC has been found within 20 miles (32 km) downstream of gas compressor stations and rarely on liquid pipelines. Near neutral-pH SCC is not strongly affected by temperature, and such cracks have been observed on both gas and liquid pipelines and at all distances from compressor and pump stations. However, the majority of in-service failures have occurred within the first valve section (typically 15 to 20 miles [24 to 32 km]) downstream from compressor or pump stations. This is thought to be due to the more rapid coating deterioration downstream of compressor stations and larger pressure fluctuations near pump stations. Neither form of SCC has been detected on offshore pipelines.

External SCC has been observed only under field-applied coatings. Despite the fact that some mill applied coatings, particularly FBE, have been in service for more than 40 years, no SCC has been found beneath those coatings. This is thought to be due to several factors, including better surface preparation, compressive residual stresses from grit blasting, and improved coating properties. It is widely accepted that the use of such coatings is an effective way to prevent SCC.

Internal SCC. To date, there have been no reported cases of internal SCC in North America. However, since ethanol is increasingly being used as an additive to gasoline, the pipeline industry is considering shipping denatured ethanol in its pipelines. Recently, concern has been raised

Figure 13. Stress-corrosion cracking colony on a large-diameter, high-pressure transmission gas pipeline.

about the possible development of internal SCC in pipelines that would transport ethanol, because SCC has been observed inside storage tanks and user terminals that contain fuel-grade ethanol. The determination of safe conditions for the transport of ethanol is currently the subject of intense debate.

Corrosion Fatigue

When a metal within a corrosive environment is subjected to cyclic stresses, fatigue cracks will grow more rapidly and at lower stress levels than they would in the presence of dry air. Corrosion fatigue rarely is reported as a cause of failure of a pipeline. However, because fatigue cracks are transgranular and thus difficult to distinguish from near-neutral-pH SCC, it has been speculated that corrosion fatigue may have contributed to some failures that were attributed to near-neutral-pH SCC.

Hydrogen Embrittlement

The term "hydrogen embrittlement" is sometimes used loosely to refer to one of several different ways that hydrogen can degrade the properties of a metal. With respect to pipelines, three forms of hydrogen damage are of interest: hydrogen-stress cracking, hydrogen-induced cracking, and loss of ductility.

Hydrogen-Stress Cracking. Hydrogen-stress cracking is a delayed-failure mechanism that sometimes occurs in high-strength steels that have absorbed hydrogen that was produced at the surface through an electrochemical reaction (corrosion or cathodic protection). Line-pipe steels with normal properties for grades up to and including at least X80 are not considered to be susceptible to hydrogen-stress cracking. However, some hydrogen-stress cracking failures have occurred in unusually hard regions of X52 pipe. Those failures usually are called hard-spot failures. The hardness pattern associated with one such failure is shown in Figure 14. A Brinell hardness of approximately 180 would be typical for X52 pipe. The hardness values near the center of the hard spots were equivalent to an ultimate tensile strength in excess of 200,000 psi, which is more than twice as high as would be expected. The origin of such hard spots has been attributed to upsets during the hot rolling of the steel plate in the steel mill.

Hard spots can be detected with magnetic-flux leakage in-line inspection pigs, and the preferred method of preventing hard-spot failures is to locate and remove the hard spots rather than try to eliminate the source of hydrogen.

Hydrogen-Induced Cracking. Some cathodic reactions, either from corrosion or cathodic protection, can deposit atomic hydrogen on the external or internal surface of a pipeline. Some of the hydrogen atoms combine with other hydrogen atoms to form molecular hydrogen gas (H2), which leaves the surface and enters the soil around the pipe or the

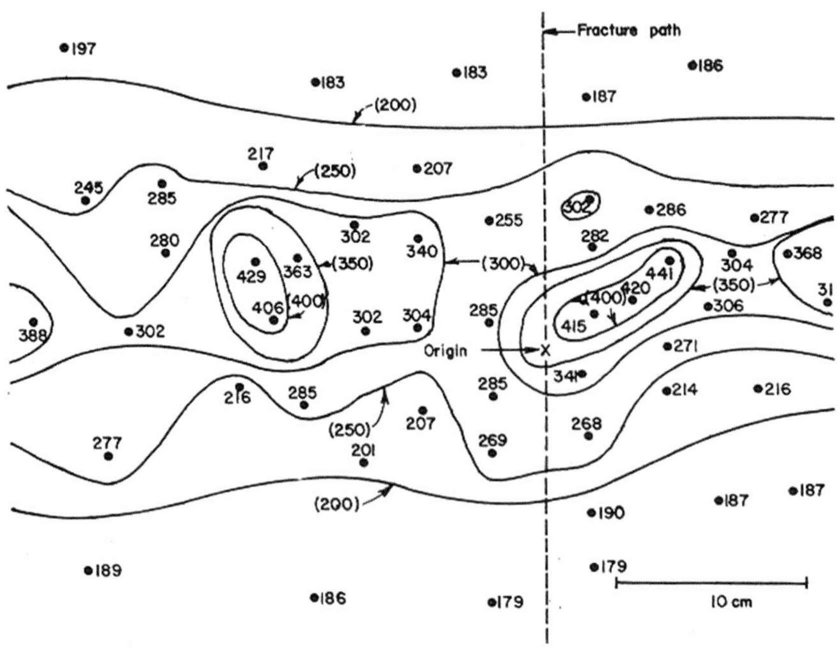

Figure 14. Brinell hardness pattern associated with a hard-spot failure in X52 pipe.

fluid inside the pipe. Some of the atoms will be absorbed by the steel and diffuse to the opposite surface where they will then combine with other hydrogen atoms and leave as a gas molecule. However, if the hydrogen atoms encounter an internal discontinuity such as a lamination or large inclusion in the steel, they may combine to form hydrogen gas at that location. By that mechanism, it is possible to create pressures up to 2,700 psi (18.7 MPa), which is sufficient to cause internal cracks in the pipe wall, even in the absence of an externally applied stress. Such cracks usually are oriented parallel to the surfaces of the pipe, because most discontinuities are parallel to the surfaces. At times, the pressures are large enough to create a blister in the pipe wall. Occasionally, the stresses at the ends of the blister can increase enough to cause a through-wall defect. If there are multiple cracks at various positions through the wall thickness and the pipe experiences a high hoop stress, the cracks sometimes join to form a through-wall crack; this phenomenon is termed stress-oriented hydrogen-induced cracking.

Loss of Ductility. When pulled in tension, line-pipe steels typically can be stretched 10 to 20 percent before they will break. However, if the steel contains a large amount of hydrogen, the steel will not be able to exhibit that level of plastic deformation. This form of hydrogen embrittlement is not of concern for operating pipelines because they never undergo any significant amount of plastic deformation.

Implications for Pipelines in a National Hydrogen Economy. Hydrogen is foreseen by many as an important energy carrier in the future sustainable energy society. Currently, hydrogen is transported primarily by way of railroad cars, tanker trucks, tanker ships and a limited amount of pipeline built specifically for hydrogen transport. Over the road transportation is very expensive due to the energy density of hydrogen. For long distance transportation, pipelines will be the most economical transportation mode. As discussed above, hydrogen's ability to embrittle steel has been well documented. However, the long term effect of high pressure hydrogen transmission through commercial grade steel pipelines of various qualities and properties has not been thoroughly explored.

Sulfide-Stress Cracking

SSC is a type of spontaneous brittle failure experienced by steels and other high-strength alloys when they are in contact with moist hydrogen sulfide and other sulfidic environments. SSC is also referred to as hydrogen sulfide cracking, sulfide cracking, sulfide corrosion cracking, and sulfide stress corrosion cracking. The variation of the name is due to the lack of agreement regarding the cracking mechanism. Some researchers consider SSC a type of SCC, while others consider it a type of hydrogen-stress cracking.

SSC in pipelines can occur from two sources: internally, from transporting wet, sour products, or from water containing sulfate-reducing bacteria (SRB); and externally, from SRB in soil or water in contact with the pipe. Reported failures due to SSC are relatively few. Internal SSC is far more common than external, which is rare.

Susceptibility to SSC is a function of a number of variables: two of the more important are strength or hardness of the steel and the level of tensile stresses. For any steel, there is a minimum level of applied stress, called the threshold stress, below which failure due to SSC will not occur. The threshold stress decreases as the steel's strength level increases. For example, a steel with a yield strength of 200,000 psi might fail at a stress of only 30,000 psi, while a steel with a yield strength of 80,000 psi must be stressed above 80,000 psi before it will fail. Therefore, the common way to prevent SSC failures in steel pipelines that could be exposed to wet hydrogen sulfide environments is to ensure that the pipe design incorporates steel with a maximum strength level of 80,000 psi. It also is important to control the welding processes to make sure that they do not induce regions of high hardness and high residual stress.

Extracted from the U.S. Department of Transportation, Pipeline and Hazardous Materials Safety Administration, Office of Pipeline Safety report "Pipeline Corrosion" by Michael Baker Jr., Inc. and Raymond R. Fessler, BIZTEK Consulting, Inc., November 2008. For the full report and citations, go online at www.phmsa.dot.gov.

NAPCA specification designations for coal tar enamel coatings

NAPCA Bulletin 1-65-94 recommended specification designations for coal tar enamel coatings:

Specification Symbol	*These are recommended enamel coating system designations except for specialty requirements.*								
* Part #1 **Part #2	Abrasive Blast	Primer	93.75 Mils Min Enamel	Fiberglass Innerwrap	Felt Wrapper	Seal Coat	Outerwrap	Kraft Paper	Electrical Inspection
TF –	X	X	X		X			X	X
TG –	X	X	X	X				X	X
TGF –	X	X	X	X	X			X	X
TGO –	X	X	X	X			X	X	X
TGSF–	X	X	X	X	X	X		X	X

*Part #1 Symbol — Enamel & Wrappers
**Part #2 Symbols Grades of Enamel
T — Coal Tar
G — Fiberglass Innerwrap
F — Coated Glass Felt or Mineral Felt
O — Fiberglass Outerwrap
S — Seal Coat
3 — Fully-Plasticized Coat Tar
(Ordered by penetration which is dictated by ambient temperature)
4 — High temperature service coal tar

Example: Full-plasticized coal tar enamel wrapped with fiberglass, felt, and kraft - the symbol would be TGF - 3.

The intended use of the coatings identified is to provide corrosion protection for buried pipelines. Above ground storage of coated pipe in excess of six months without additional Ultraviolet protections is not recommended.

Reprinted courtesy of National Association of Pipe Coating Applicators.

NAPCA standard applied pipe coating weights

NAPCA Bulletin 2-66-94 standard applied pipe coating weights for coating specifications:

Symbol		TF-3,4	TG-3,4	TGF-3,4	TGSF-3,4	0(1)	PE-Std	PPStd	FBE/mil
Pipe Sze	Weight in Pounds per 100 Linear Feet								
1/2"	ID	19	17	20	25	0.38	3.74	3.61	0.16
3/4"	"	24	21	24	31	0.48	4.67	4.51	0.21
1"	"	30	26	31	39	0.60	5.85	5.65	0.26
1-1/4"	"	38	33	29	49	0.76	7.39	7.13	0.33
1-1/2"	"	44	38	44	56	0.87	8.46	8.16	0.37
2-3/8"	OD	55	48	55	71	1.09	12.12	11.63	0.47
2-7/8"	"	66	58	67	85	1.32	14.68	14.08	0.56
3-1/2"	"	81	70	82	104	1.60	20.16	17.13	0.69
4"	"	92	80	93	119	1.83	23.04	19.58	0.79
4-1/2"	"	104	90	105	134	2.06	25.92	22.03	0.88
5-9/16"	"	128	112	130	165	2.55	35.54	34.08	1.09

Symbol		TF-3,4	TG-3,4	TGF-3,4	TGSF-3,4	0(1)	PE-Std	PPStd	FBE/mil
6-5/8"	"	153	133	154	197	3.04	42.32	40.59	1.30
8-5/8"	"	199	173	201	256	3.95	55.10	52.84	1.69
10-3/4"	"	248	215	251	310	4.93	68.17	65.86	2.11
12-3/4"	"	294	256	297	380	5.84	81.45	78.11	2.50
14"	"	322	281	326	416	6.41	89.43	85.77	2.75
16"	"	368	321	373	475	7.33	102.21	98.02	3.14
18"	"	415	361	420	534	8.25	114.98	110.27	3.53
20"	"	461	401	466	594	9.16	127.76	122.52	3.93
22"	"	407	441	513	653	10.08	169.33	161.84	4.32
24"	"	453	481	560	713	11.00	184.73	176.56	4.71
26"	"	599	521	606	772	11.91	200.12	191.27	5.11
28"	"	645	561	653	831	12.83	215.51	205.98	5.50
30"	"	691	601	699	891	13.74	230.91	220.70	5.89
32"	"	737	641	746	950	14.66	246.30	235.41	6.28
34"	"	783	681	793	1009	15.58	261.70	250.12	6.68
36"	"	829	722	839	1069	16.49	277.09	264.84	7.07
42"	"	967	842	979	1247	19.24	377.15	360.66	8.25
48"	"	1105	962	1119	1425	21.99	306.62	294.05	9.42
54"	"	1244	1082	1259	1603	24.74	344.95	330.81	10.60
60"	"	1382	1203	1399	1781	27.49	383.28	367.57	11.78

(1)For this outerwrap, add the indicated amount per 100 LF to the other basic symbol weight per 100 LF.

Weight per 100 LF is for calculating estimated shipping weights only and is based on the following nominal weights per square foot.

ENAMEL AND PRIMER (T)	1/32	0.2513	FIBERGLASS(G) INNERWRAP	0.011
	2/32	0.4948		
	3/32	0.7326		
	4/32	0.9761	GLASS FELT (F)	0.125
	5/32	1.2196		
FIBERGLASS OUTERWRAP (O)		0.0175	MASTIC & PE (10/40 MIL)	0.244
		0.022	MASTIC & PP (10/40 MIL)	0.234
			FBE-1 MIL	0.0075

Reprinted courtesy of National Association of Pipe Coating Applicators.

NAPCA specifications for pipeline wrappers

NAPCA Bulletin 5-69-94 describes pipeline wrapper specifications.

1. **General** These specifications cover the properties of pipeline wrappers to be used as part of a coating system employed above or below ground for the prevention of corrosion.
2. **Scope** This specification covers pipeline wrappers used as a reinforcement or protective shield for coal tar coating systems.
3. **Requirements** The specifications and tests shown in Table 1, 2 and 3 describe a finished product which can be handled and applied satisfactorily, using good operating practice.

Consolidated Tables	Table 1 Fiberglass Innerwraps (G)		Table 2 Felt (F)		Table 3 Other Outerwraps (O)
Property	Reinforced	Non-reinforced	Glass Felt(9)	Mineral Felt(10)	Fiberglass(11) Outerwrap
Weight (lb/100 sqft, min)	0.84(1)	0.84(1)	12(1)	12(1)	1.75(2)
Saturation (%)				22-32(1)	
Ash Content (%)			70(1)	60(1)	
Thickness/Caliper (in,min)	0.013(4)	0.013(4)	0.030(3)	0.018(3)	0.025(2,12)
Tensil Strength (lb/in) (md, md/cd)	13(1)/4(5)	4(1)1/4(5)	27(1) /35(5)	25(1)	25(1)/15(5) min
Loss of Heating (%, max)			2(6)	10(6)	
Pliability	no cracking(1)	no cracking(1)	no cracking	no cracking	
Elmendorf Tear Strength (md/cd)		0.22/0.22(7)			
Porosity (in. water loss)	0.022(8)	0.022(8)			

[1]ASTM D-146-78a
[2]Nominal
[3]TAPPIT-411
[4]ASTM D-146, modified
[5]ASTM-D882 (cd)
[6]AWWA C-203
[7]TAPPIT-414 (md/cd)
[8]ASTM D-737-75 modified
[9]Coal Tar Coated
[10]Coal Tar Saturated (Heavier available)
[11]Non-Coated
[12]40 mil available

Reprinted courtesy of National Association of Pipe Coating Applicators.

How much primer for a mile of pipe?

Multiply pipe O.D. by $2\frac{2}{3}$ to get gallons of primer per mile.

Examples. Pipe 24 in in diameter will require $24 \times 2\frac{3}{4}$ or 66 gal of primer per mile; Pipe 16 in in diameter will require $16 \times 2\frac{3}{4}$ or 44 gal of primer per mile; and Pipe $4\frac{1}{2}$ in diameter will take $4\frac{1}{2} \times 2\frac{3}{4}$ or 123/8 gal of primer per mile (12 gal in round figures).

The given multiplier is applicable only where the pipe is new and in good condition. For rough pipe the multiplying factor should be increased from $2\frac{3}{4}$ to 3.45. In other words, rough 20-in pipe will require 20×3.45 or 69 gal of primer per mile.

It is hard to estimate the amount of spillage and waste. Temperature and relative humidity affect the way the primer goes on the pipe.

How much coal-tar enamel for a mile of pipe?

For an average cover of 3/32 in, multiply outside diameter of pipe by 0.6. The answer is in tons of enamel per mile. Add 15% to take care of gate valves, drips, bends, etc. Where reconditioned pipe is being coated, it is recommended that the amount of enamel be increased by about 20%.

Examples. How many tons of coal-tar enamel will be required to apply a 3/32-in coat to a 26-in pipeline?

$26 \times 0.6 = 15.6$ tons per mile

Add 15% to take care of gate valves, bends etc.

$15.6 + 2.3 = 17.9$ or 18 tons per mile

How many tons of coal-tar enamel will be required to apply a 3/32-in coat to an 8.625-in pipeline?

$t8.625 \times 0.6 = 5.2$ tons per mile

$t5.2 \times 15\% = 0.8$

$5.2 + 0.8 = 6$ tons per mile

The following factors strongly affect the amount of coating used per mile on pipelines: 1. Temperature of the pipe and ambient temperatures. 2. The type of dope machine. 3. Humidity. 4. The number of bends per mile. 5. The number of valves, crossovers, etc. 6. The condition of the pipe.

How much wrapping for a mile of pipe?

For a 14-in to 48-in pipe, multiply the diameter in inches by 15 to get the number of squares $(100\ ft^2)$. For pipe sizes of 12¾ in and smaller, add 1 to the nominal size of pipe and multiply by 15.

Examples. How many squares of asbestos felt would be required to wrap 1 mile of 30-in pipeline?

$30 \times 15 = 450$ squares of felt

How many squares of felt would be required to wrap 1 mile of 4-in pipe?

$(4 + 1) \times 15 = 75$ squares of felt

These figures are conservative and do not allow for wrapping drips, patching holidays, and for short ends that are usually discarded. They allow for machine wrapping using ¾-in lap on diameters up to 10¾ in and using 1-in lap on larger diameters.

For smaller diameters of pipe, this figure should be increased about 5% to take care of additional wastage and patching.

Estimating coating and wrapping materials per mile of pipe

Primer and coal-tar enamel

Gallons of primer and tons of coal-tar enamel are based on covering new pipe in good condition. For rough pipe, increase primer quantities by 25% and the coal-tar quantities by 20%. No allowance is included for spillage or waste.

Wrappings

Squares are based on machine wrapping, allowing 3/4-in laps on pipe sizes through 10 in and 1-in laps on pipe sizes greater than 10 in. No allowance is included for wrapping drips, patching holidays, or short ends, which are usually discarded.

NPS	Primer Gallons	3/32" Coal-tar Enamel Tons	Glass Wrap Squares	Kraft Paper Squares	15-Pound Felt Squares
4	11	3	75	75	75
6	17	4	105	105	105
8	22	6	135	135	135
10	28	7	165	165	165
12	33	8	195	195	195
14	39	10	210	210	210
16	44	11	240	240	240
18	50	13	270	270	270
20	55	14	300	300	300
24	66	17	360	360	360
30	83	21	450	450	450
36	99	25	540	540	540
42	116	29	630	630	630

Coefficient of friction for pipe coating materials

J. B. Ligon, Assistant Professor, Mechanical Engineering-Engineering Mechanics Dept., Michigan Technological University, Houghton, Mich., and **G. R. Mayer**, Project Engineering Manager, Bechtel, Inc., San Francisco

A major factor in the stress analysis of buried pipelines is the movement that pipe undergoes in the presence of temperature and pressure differentials during its life. This movement is highly dependent upon friction resistance of the soil.

Although ample information is available on the static coefficient of friction for many materials, there is a lack of data on friction between soils and various coatings used in the pipeline industry. In the past, friction coefficient information was extrapolated from data in the literature that were believed to have a similarity to the external pipe coating to soil interface.

However, with the development of thin-film epoxy resin coating systems and the increasing use of these systems in the pipeline industry, a change from a conventional coal tar felt coating to a thin-film epoxy coating would indicate a significant change in the friction coefficient design criteria due to the extreme contrast in the surface texture of the two materials.

To evaluate the effect of the difference in surface texture on a pipeline system, test procedures were developed to determine the coefficient of friction for both coal tar felt and thin-film epoxy to various soils and to obtain more reliable information for future pipeline designs using these coating materials.

Static friction tests were conducted to find the coefficient of friction between coal tar felt and thin-film epoxy pipe coating and eight representative backfill soil samples from typical locations along a pipeline right of way. The results indicate that the friction coefficients are significantly larger than those previously extrapolated from literature and that coal tar has a higher friction resistance in respect to anchorage of a pipeline.

For the coal tar felt coating, the coefficient varies from 0.59 to 0.91 depending on the soil and moisture content. The thin-film epoxy coating varies from 0.51 to 0.71 under the same conditions.

Definitions

Static coefficient of friction. The theoretical longitudinal soil force acting on the pipe surface can be calculated from the relation:

$$F = \mu \int_A p \, dA$$

where:

F = Longitudinal soil friction force (lb)
μ = Coefficient of friction (dimensionless)
p = Normal soil pressure acting on the pipe surface (psi)
dA = Soil to pipe differential contact area (in^2)
$\int_A p\,dA$ = Total normal soil force on pipe surface (lb).

The above relation is independent of the pressure per unit area of the mating surfaces as long as very high contact pressures are not encountered, which is the case for a buried pipeline.

In terms of a buried pipeline supporting a soil burden up to three pipe diameters of depth, the soil force relation becomes:

$$F = \mu \left[2D\gamma \left(H - \frac{D}{2} \right) + Wp \right]$$

where:

Wp = Weight of pipe and contents (lb/in)
tD = Pipe diameter (in)
H = Depth of the pipe centerline (in)
γ = Specific weight of the soil (Ib/in^3)

The sensitivity of the soil friction forces to the coefficient of friction is readily apparent from the above relation. Since this force is inversely proportional to the active length that a pipeline moves as a result of temperature and pressure expansion, the coefficient of friction becomes a major factor in pipeline stress design.

Test system. The theoretical soil friction force acting on the surface of a coated plate can be calculated from the relation:

$$F = \mu N$$

where:

N = Normal force acting on the plate surfaces (lb)
W = Total weight of plate and external load (lb).

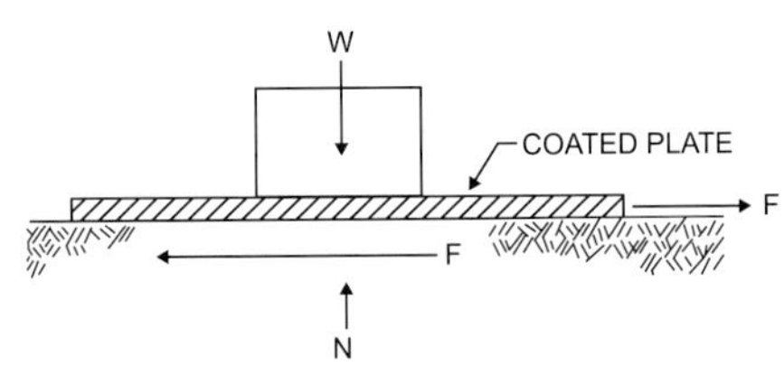

Since the coefficient of the friction parameter is the same in all of the previous expressions, the plate test system can be used to simulate the soil-to-pipeline interface.

Coal tar felt. The coal tar coating, enamel, and felt combination, was prepared to simulate an actual coated pipeline.

The surface of a steel plate was coal tar coated and then wrapped in a 15-lb felt. Therefore, the soil surface contact was predominantly with the felt material.

Thinfilm epoxy. The epoxy coating simulated a thin-film epoxy resin pipeline coating applied by the fusion method.

Plates used in the tests were coated by the fluidized bed method. However, the surface texture was the same as that expected by coating plant production equipment.

Results

Coal tar to soils. Figure 1 illustrates that the friction coefficient varies from 0.59 to 0.91 for the various soils tested at field moisture content.

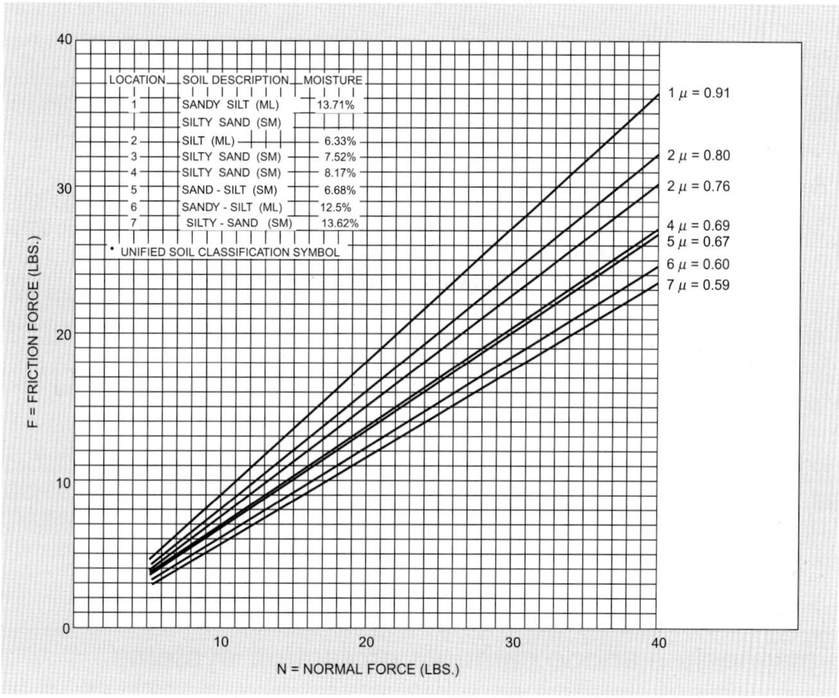

Figure 1. Coefficient of friction for coal tar coating to soil.

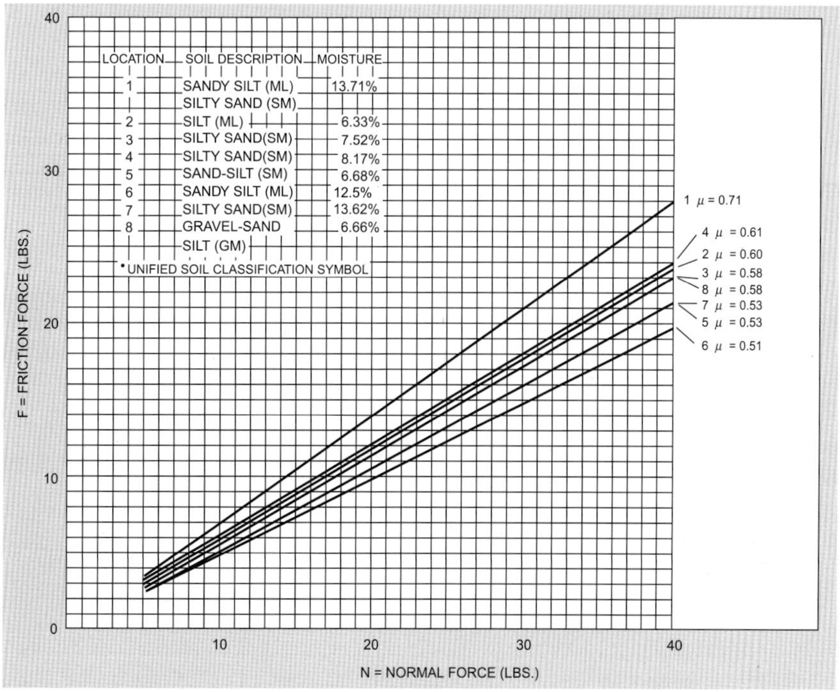

Figure 2. Coefficient of friction for thin film epoxy to soil.

The mechanism by which the friction coefficient changes among the various soils is not apparent from the results of these tests, since the soil samples used did not vary considerably in description. However, the range over which the factor changes is considerably greater than that previously extrapolated from the literature, as follows:

Soil Description	Coefficient of Friction Commonly Used
Silt	0.3
Sand	0.4
Gravel	0.5

The tests also indicate that the moisture content alters the friction factor to some extent, as would be expected.

In order to investigate the influence of temperature, tests were conducted with coal tar felt wrapping heated to 120 °F. Only a slight softening beneath the coating surface was observed, and it is believed that temperatures up to this range will not significantly affect the magnitude of the friction coefficient.

Thinfilm epoxy to soils. As expected, results shown in Figure 2 indicate that the friction factor range of 0.51 to 0.71 for the epoxy is somewhat lower than that of the coal tar.

Again, the mechanism is not clear, but the results are fairly consistent with the friction coefficient increasing in most instances in the same order of magnitude as for coal tar. Since such information on this coating was not available in the literature, no comparison to similar tests could be made.

Temperatures in the 120 ° F range should have little or no effect on the coefficient of friction of epoxy to soil.

Conclusions

Although it is virtually impossible to precisely simulate the surface contact situation of a pipeline in a back-filled trench, the test procedure and apparatus reported here are a means of approximating it. The results indicate that coal tar coatings have a higher friction resistance than epoxy coatings as far as anchoring capabilities of the soil are concerned.

The selection of a coating, based on its soil friction resistance, could be of economic value in reducing those situations where extreme expansions call for reinforced fittings or elaborate culverts to overcome excessive stress levels in a pipeline system.

The test results show that previous values commonly used for the coefficient of friction were conservative for similar soils, and it is suggested that some conservatism still be incorporated in future analyses.

Since the tests indicated that the friction coefficient of the epoxies was similar to those previously used for coal tar coatings, their continued use for epoxy-coated pipelines should be valid. A more conservative approach to these values is recommended in soils where the presence of excessive moisture could substantially change the friction factor when interfaced with a smooth epoxy coating.

Troubleshooting magnesium anode cathodic protection systems

First, the current output of stations nearest the point of low potential should be checked; if these are satisfactory, a similar check should be extended in both directions until it is clear that the trouble must be on the line. When a given anode group shows a marked drop in current output, the cause may be drying out, shrinkage of backfill, or severed or broken lead wires. If the current is zero, the pipe-to-soil potential of the lead wire will show whether it is still connected to the pipe or the anode, and thus indicate the direction to the failure. If the current is low, there may be a loss of one or more anodes by a severed wire; a pipe-to-soil potential survey over the anodes will show which are active, just as in the case of rectifier anodes.

Trouble indicated on the line, rather than at the anode stations, is tracked down in the same manner as that used for line protected by rectifiers–first, investigating locations where something that has been done might be responsible; next, checking potentials; and finally, making a line current survey.

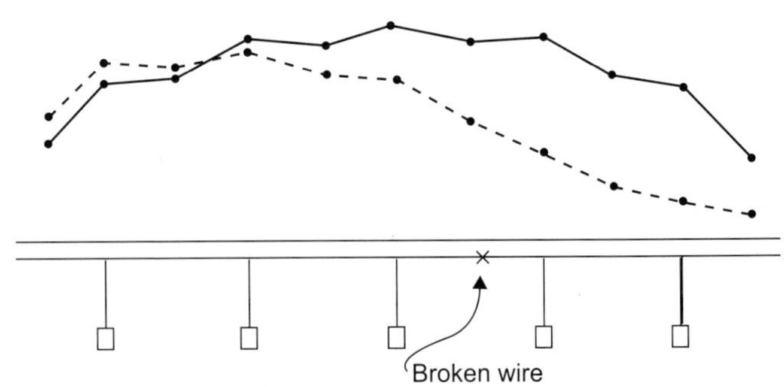

Figure 1. Locating idle anodes by surface potentials. The solid line shows the potentials found along a line of anodes when all are delivering current; the dotted line exhibits the change when there is a break in the anode lead at the point indicated. Single disconnected anodes may also be located by this method. A driven ground rod, a pipe lead, or even a rectifier terminal may be used for the reference ground; all readings should be referred to the same reference.

Cathodic protection for pipelines

Estimate the rectifier size required for an infinite line. Refer to Figure 1. If coating conductance tests have not been performed, pick a value from Table 1.

Table 1
Typical Values of Coating Conductance

Micromhos/ft^2	Coating Condition
1–10	Excellent coating—high resistivity soil
10–50	Good coating—high resistivity soil
50–100	Excellent coating—low resistivity soil
100–250	Good coating—low resistivity soil
250–500	Average coating—low resistivity soil
500–1000	Poor coating—low resistivity soil

Use a value of 0.3 volts for ΔE_x. This is usually enough to raise the potential of coated steel to about 0.85 volts.

Use a value of 1.5 volts for AE at the drain point. Higher values may be used in some circumstances; however, there may be a risk of some coating disbondment at higher voltages.

Calculate I at the drain point.

$$\Delta I_A = \text{amp/in} \times \Delta E_x/0.3 \times D,$$

where D = pipe OD, inches.

Example. 30-in OD line with coating conductivity = 100 micromhos/ft^2. What is the current change at the drain point,

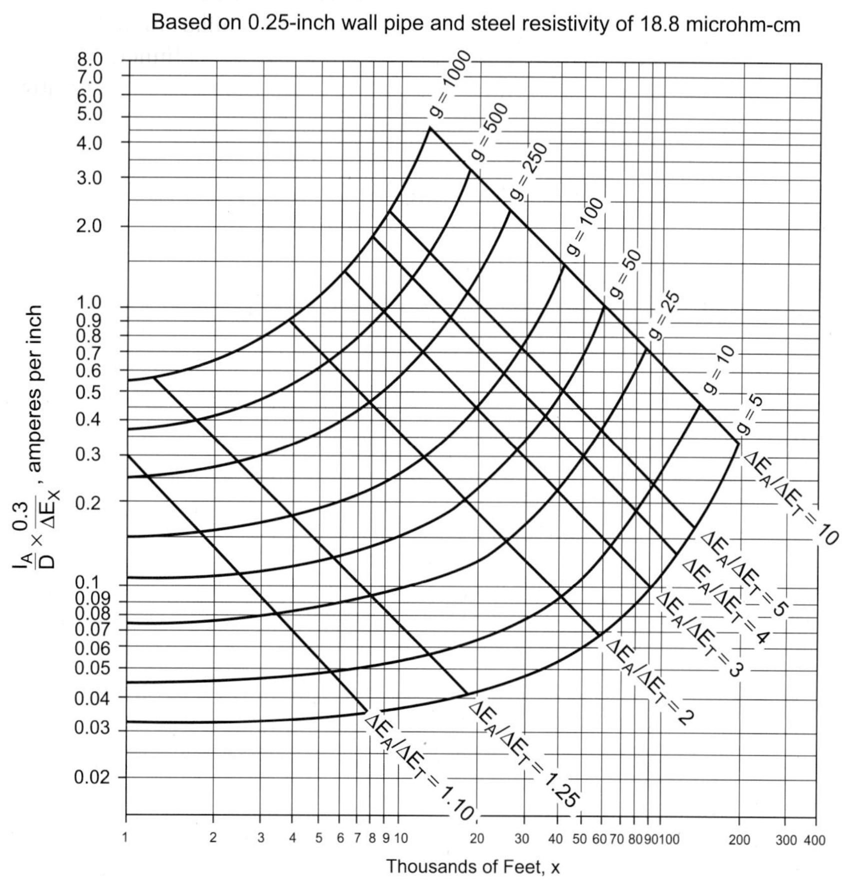

Figure 1. Drainage current versus distance for a coated infinite line.

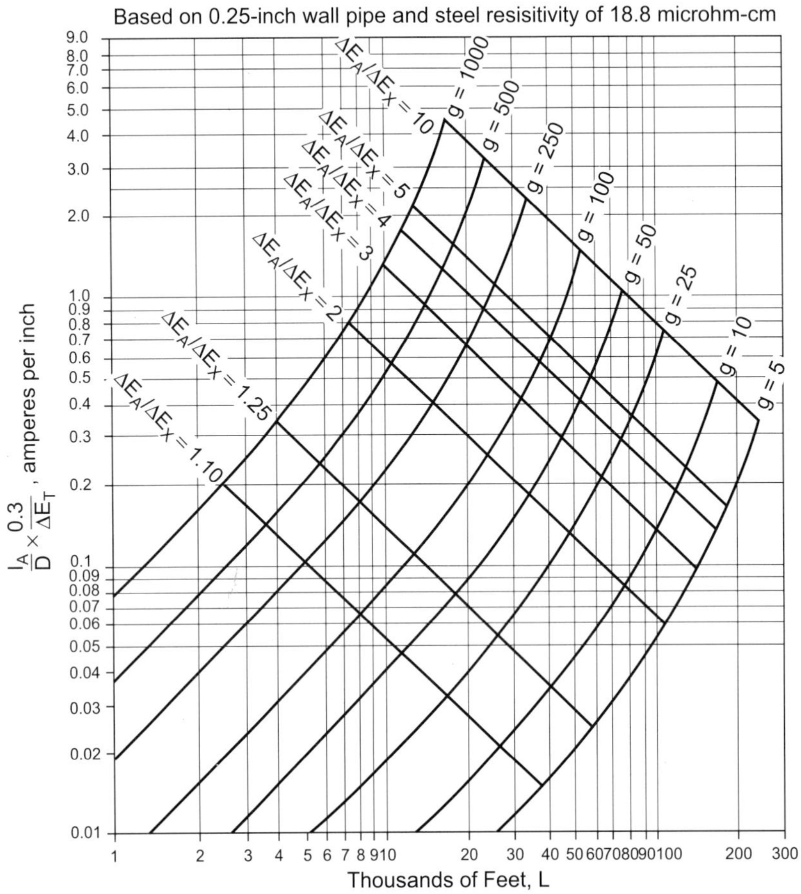

Figure 2. Drainage current versus distance for a coated finite line.

and how far will this current protect an infinitely long pipeline?

$\Delta E_A = 1.5$

$\Delta E_x = 0.3$

$\Delta E_A/\Delta E_x = 5.0$

Refer to Figure 1 and read

$L = 30{,}000$ ft

$\Delta I_A/D \times 0.3/\Delta E_x = 0.7$ (from Figure 1)

$\Delta I_A = 0.7 \times 30 = 21$ amp.

21 amps will protect the line for 30,000 ft in either direction from the drain point.

Estimate rectifier size for a finite line. Refer to Figure 2. A 3½-in OD line 20,000 ft long is protected by insulating flanges at both ends and has a coating conductivity of 500 micromhos/ft^2. What is $\Delta E_A/E_T$ and what ΔI_A is required? Assume that

$E_T = 0.3$

From Figure 2, read

$\Delta E_A/E_T = 5.8$

$\Delta E_A = 1.74$

$\Delta I_A/D \times 0.3/E_T = 1.8$ amp/in

$\Delta I_A A = 1.8 \times 3.5 = 6.3$ amp.

Estimate ground bed resistance for a rectifier installation.

Example. A ground bed for a rectifier is to be installed in 1000 ohm-cm soil. Seven 3-in x 60-in vertical graphite anodes with backfill and a spacing of loft will be used. What is the resistance of the ground bed?

Refer to Figure 3. The resistance is 0.56 ohms.

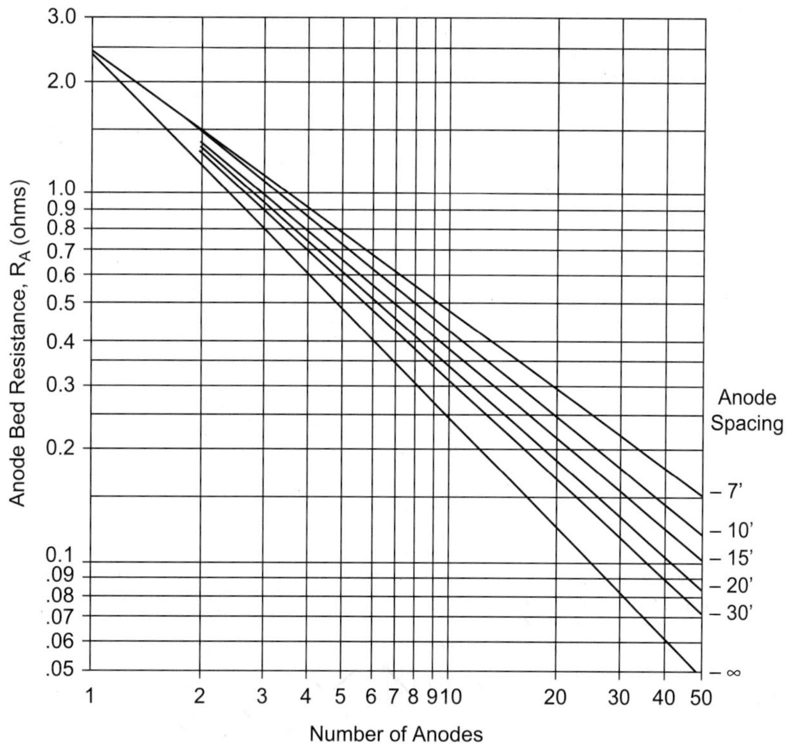

Figure 3. Anode bed resistance versus number of anodes. 3-in × 60-in vertical graphite anodes in backfill.

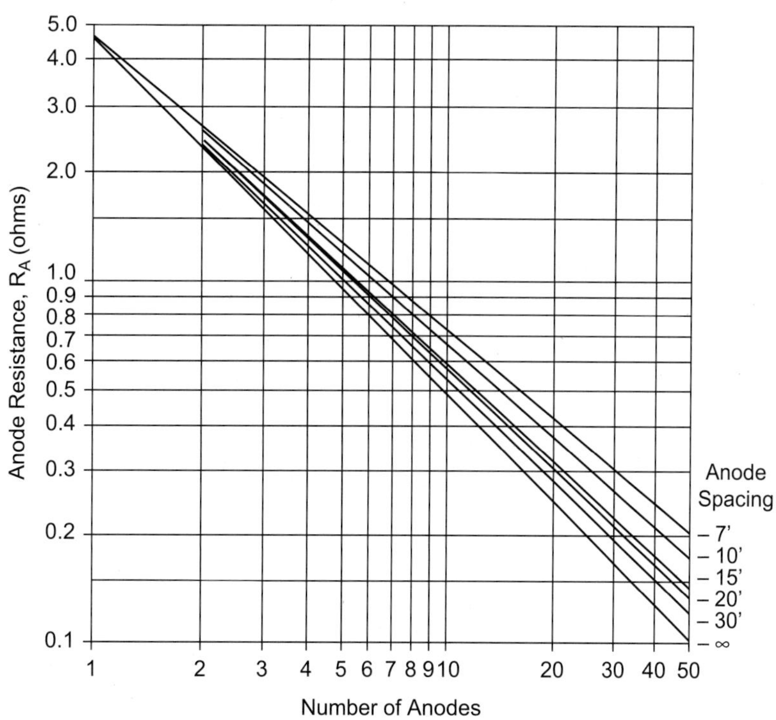

Figure 4. Anode bed resistance versus number of anodes. 2-in × 60-in vertical bare anodes.

Figure 3 is based on a soil resistivity of 1000 ohms-cm. If the soil resistivity is different, use a ratio of the actual soil resistivity divided by 1000 and multiply this by the reading obtained from Figure 3.

A rectifier ground bed is to be composed of 10 – 2 × 60-in bare "Duriron" silicon anodes spaced 20 ft apart. The soil resistivity is 3000 ohms-cm. What is the ground bed resistance? Refer to Figure 4. The resistivity from the chart is 0.55 ohms. Since the chart is based on soil resistivity of 1000 ohms-cm, the ground bed resistivity is 3000/1000 × 0.55 or 1.65 ohms.

The resistance of multiple anodes installed vertically and connected in parallel may be calculated with the following equation:

$$tR = 0.00521\ P/NL \times (2.3 \log 8L/d - 1 + 2L/S \log 0.656N) \quad (1)$$

where:

R = Ground bed resistance, ohms
P = Soil resistivity, ohm-cm
N = Number of anodes
d = Diameter of anode, feet
L = Length of anode, feet
S = Anode spacing, feet.

If the anode is installed with backfill such as coke breeze, use the diameter and length of the hole in which the anode is installed. If the anode is installed bare, use the actual dimensions of the anode.

Figure 5 is based on Eq. 1 and does not include the internal resistivity of the anode. The resistivity of a single vertical anode may be calculated with Eq. 2.

$$R = 0.00521\ P/L \times (2.3 \log 8L/d - 1) \quad (2)$$

If the anode is installed with backfill, calculate the resistivity using the length and diameter of the hole in which the anode is installed. Calculate the resistivity using the actual anode dimensions. The difference between these two values is the internal resistance of the anode. Use the value of *P*, typically about 50 ohm-cm, for the backfill medium.

Figure 5 is based on 1000 ohm-cm soil and a 7-ft × 8-in hole with a 2-in × 60-in anode.

Example. Determine the resistivity of 20 anodes installed vertically in 1500 ohm-cm soil with a spacing of 20 ft.

Read the ground bed resistivity from Figure 5.

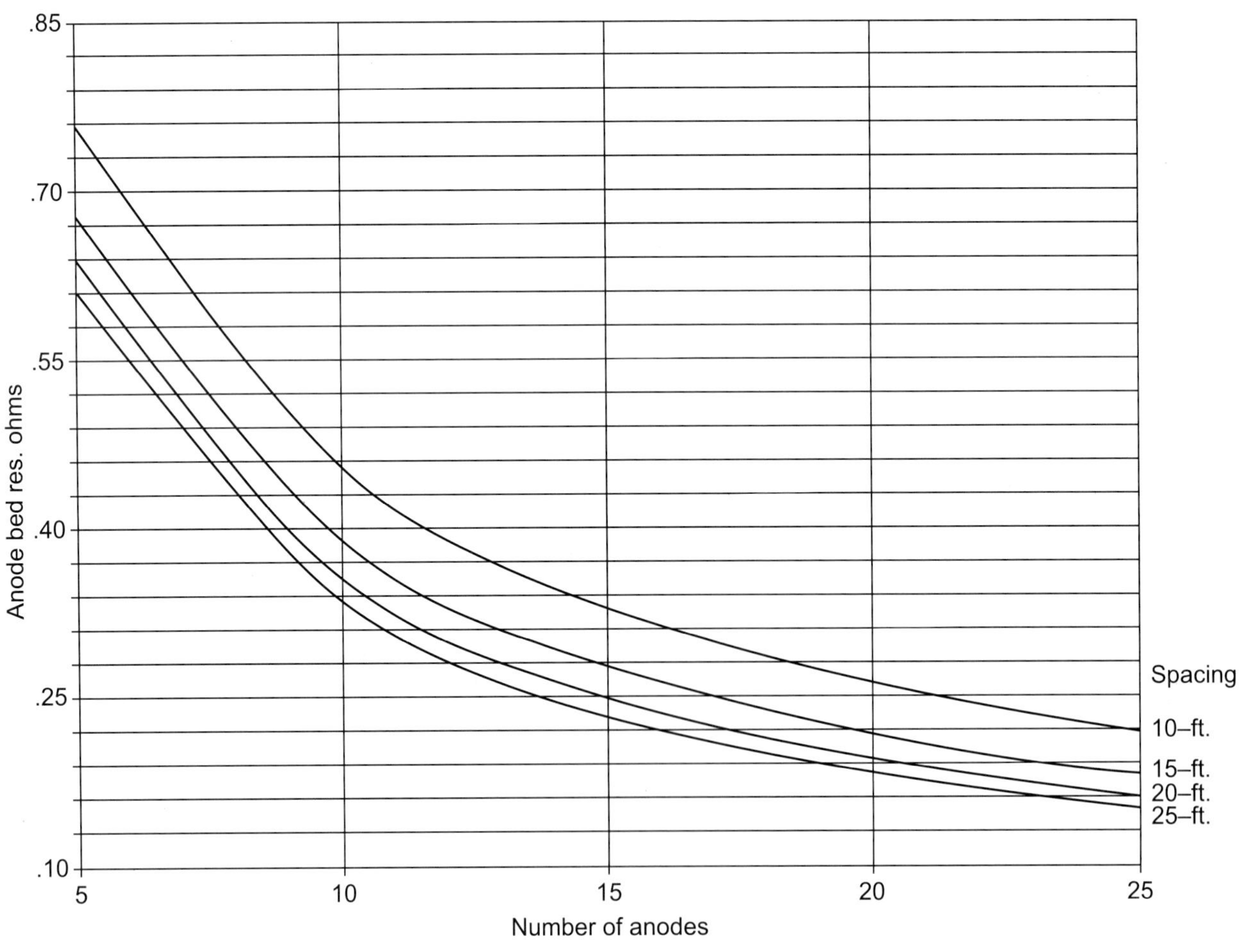

Figure 5. Anode bed resistance.

R = 0.202 ohm.

Since the anodes are to be installed in 1500 ohm-cm soil and Figure 5 is based on 1000 ohm-cm soil, multiply *R* by the ratio of the actual soil resistivity to 1000 ohm-cm.

$$R = 0.202 \times 1500/1000$$

$$R = 0.303 \text{ ohm.}$$

The internal resistivity for a single 2-in × 60-in vertical anode installed in 50 ohm-cm backfill (7 ft × 8-in hole) is 0.106 ohm.

Since 20 anodes will be installed in parallel, divide the resistivity for one anode by the number of anodes to obtain the internal resistivity of the anode bank.

$$0.106 / 20 = 0.005 \text{ ohm}$$

The total resistivity of the 20 anodes installed vertically will therefore be 0.308 ohm (0.303 + 0.005).

Galvanic anodes

Zinc and magnesium are the most commonly used materials for galvanic anodes. Magnesium is available either in standard alloy or high purity alloy. Galvanic anodes are usually prepackaged with backfill to facilitate their installation. They may also be ordered bare if desired. Galvanic anodes offer the advantage of more uniformly distributing the cathodic protection current along the pipeline, and it may be possible to protect the pipeline with a smaller amount of current than would be required with an impressed current system, but not necessarily at a lower cost. Another advantage is that interference with other structures is minimized when galvanic anodes are used.

Galvanic anodes are not an economical source of cathodic protection current in areas of high soil resistivity. Their use is generally limited to soils of 3000 ohm-cm except where small amounts of current are needed.

Magnesium is the most-used material for galvanic anodes for pipeline protection. Magnesium offers a higher solution potential than zinc and may therefore be used in areas of higher soil resistivity. A smaller amount of magnesium will generally be required for a comparable amount of current. Refer to Figure 6 for typical magnesium anode performance data. These curves are based on driving potentials of —0.70 volts for H-1 alloy and —0.90 volts for Galvomag working against a structure potential of —0.85 volts referenced to copper sulfate.

The driving potential with respect to steel for zinc is less than for magnesium. The efficiency of zinc at low current levels does not decrease as rapidly as the efficiency for magnesium. The solution potential for zinc referenced to

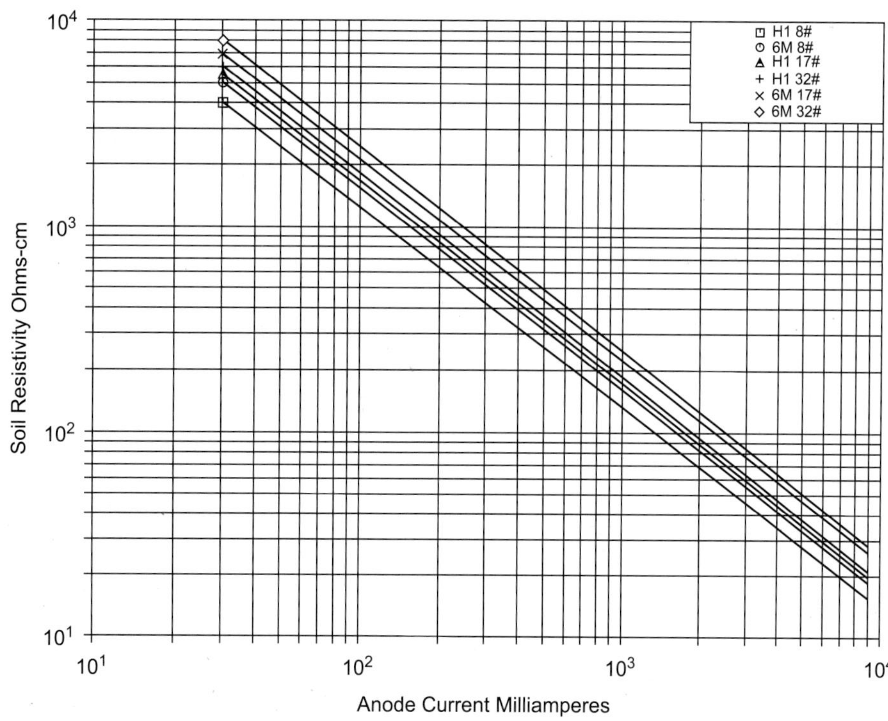

Figure 6. Magnesium anode current.

a copper sulfate cell is −1.1 volts; standard magnesium has a solution potential of −1.55 volts; and high purity magnesium has a solution potential of −1.8 volts.

If, for example, a pipeline is protected with zinc anodes at a polarization potential of −0.9 volts, the driving potential will be −1.1 − (−0.9) or −0.2 volts. If standard magnesium is used, the driving potential will be −1.55 − (−0.9) or −0.65 volts. The circuit resistance for magnesium will be approximately three times as great as for zinc. This would be handled by using fewer magnesium anodes, smaller anodes, or using series resistors.

If the current demands for the system are increased due to coating deterioration, contact with foreign structures, or by oxygen reaching the pipe and causing depolarization, the potential drop will be less for zinc than for magnesium anodes. With zinc anodes, the current needs could increase by as much as 50% and the pipe polarization potential would still be about 0.8 volts. The polarization potential would drop to about 0.8 volts with only a 15% increase in current needs if magnesium were used.

The current efficiency for zinc is 90%, and this value holds over a wide range of current densities. Magnesium anodes have an efficiency of 50% at normal current densities.

Magnesium anodes may be consumed by self-corrosion if operated at very low current densities. Refer to Figures 7(a) and 7(b) and Figures 8(a) and 8(b) for zinc anode performance data. The data in Figures 7(a) and 7(b) are based on the anodes being installed in a gypsum-clay backfill and having a driving potential of −0.2 volts. Figures 8(a) and 8(b) are based on the anodes being installed in water and having a driving potential of −0.2 volts.[1] In Figures 7(a), 7(b) and 8(a), spacing is 15 ft for 1.4 in × 1.4 in by 5 ft long anodes.

Example. Estimate the number of packaged anodes required to protect a pipeline.

What is the anode resistance of a packaged magnesium anode installation consisting of nine 32-lb anodes spaced 7 ft apart in 2000 ohm-cm soil?

Refer to Figure 9. This chart is based on 17-lb packaged anodes in 1000 ohm-cm soil. For nine 32-lb anodes, the resistivity will be:

$$1 \times 2000/1000 \times 0.9 = 1.8 \text{ ohm.}$$

See Figure 10 for a table of multiplying factors for other size anodes.

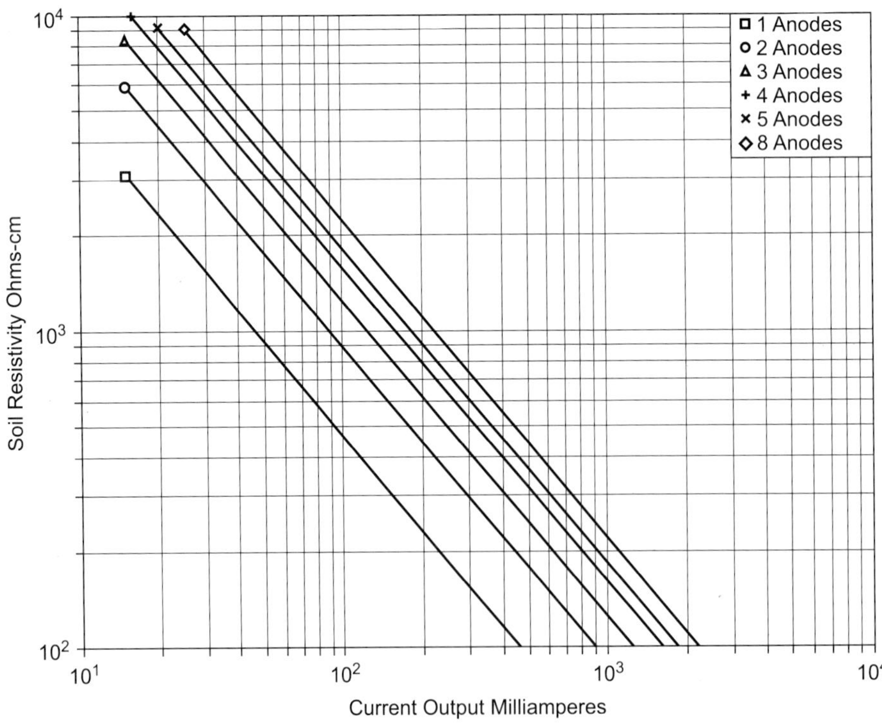

Figure 7a. Current output zinc anodes.

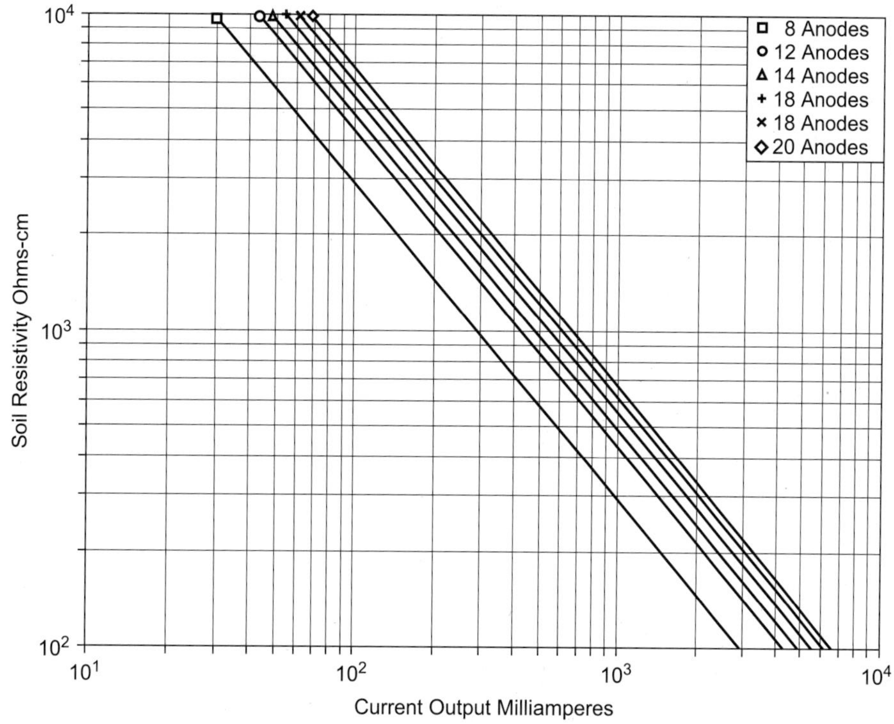

Figure 7b. Current output zinc anodes.

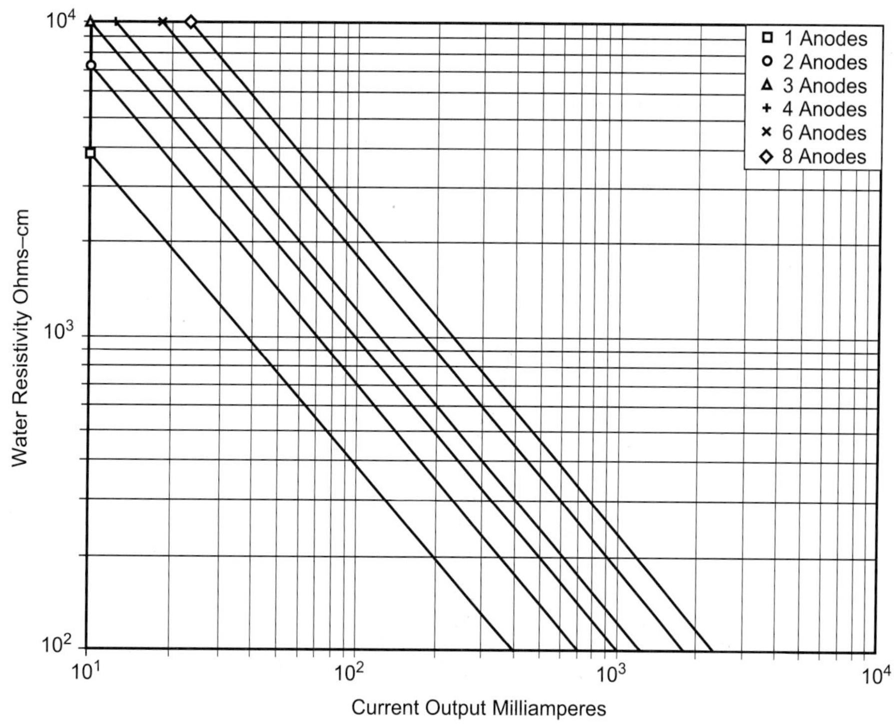

Figure 8a. Current output zinc anodes.

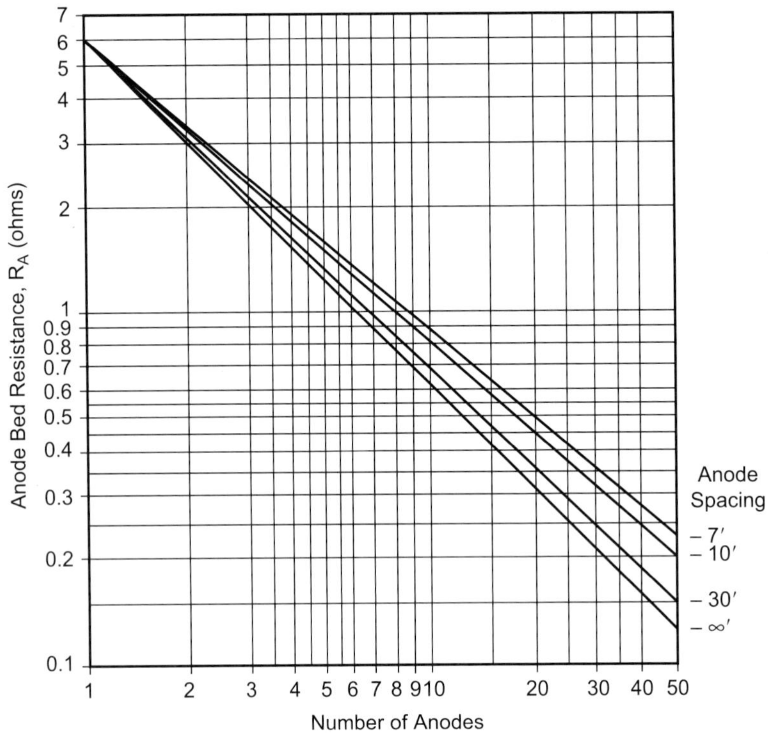

Figure 8b. Current output zinc anodes.

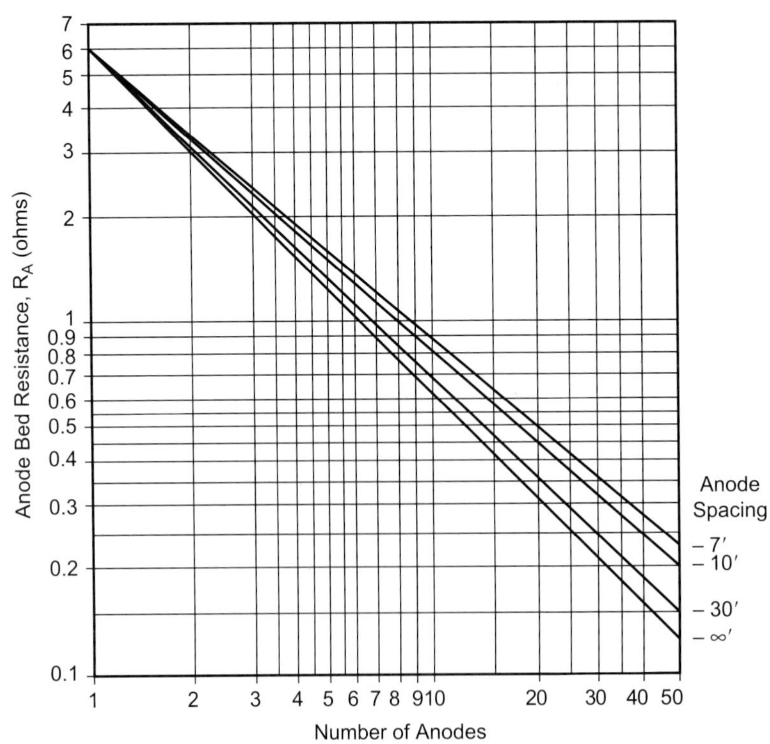

Figure 9. Anode bed resistance versus number of anodes; 17-lb packaged magnesium anodes.

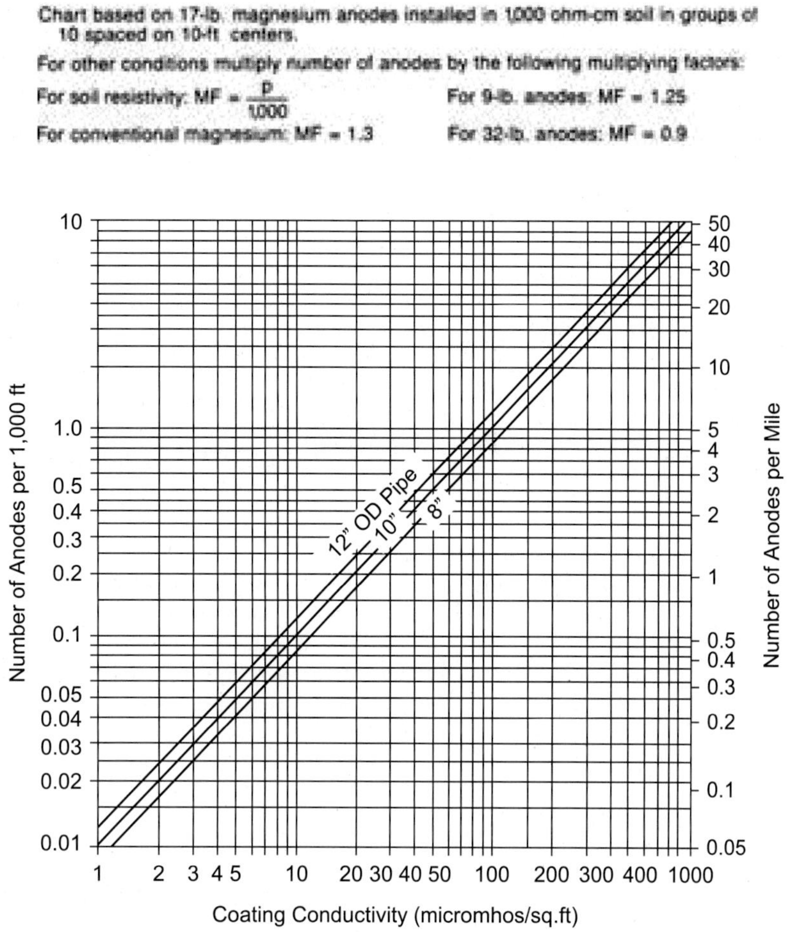

Figure 10. Number of anodes required for coated line protection.

Example. A coated pipeline has a coating conductivity of 100 micromhos/sq ft and is 10,000 ft long, and the diameter is 10 ¾ in. How many 17-lb magnesium anodes will be required to protect 1000 ft?

Refer to Figure 7 and read 2 anodes per 1000 ft a total of twenty 17 lb anodes will be required for the entire line.

Source

Pipeline Corrosion and Cathodic Protection, Gulf Publishing Company, Houston, Texas.

Reference

1. From data prepared for the American Zinc Institute.

Estimate the pounds of sacrificial anode material required for offshore pipelines

This rule of thumb is based on the following assumptions:

- 5% of pipeline considered bare
- 3 mA/ft^2 required for protection
- 30-year design life

Basis for rule: sq ft of surface area of pipeline/linear ft × % of bare pipe × 3.0 mA/ft^2 × lb/amp-year × design life in years = the lb of galvanic anode material that will be required.

$$\text{Multiplier} = \text{OD} \times \text{lb/amp-yr} \times (12\ \text{PI}/144) \times 30 \times 0.05 \times 0.003 \times 1000$$

or:

Pipe OD × lb/amp-yr × 1.1781

= lb of anode material/1000 linear ft of pipeline.

Example. How many pounds of zinc will be required for a 30-in OD line installed in seawater and operating at 80 °F? Pounds of zinc per ampere-year = 24.8 (seawater).

30 × 24.8 × 1.1781 = 877 lb/1000 linear ft of pipeline.

How many pounds of Galvalum® III will be required for a 30-in OD line installed in seawater and operating at 80 °F? Pounds of Galvalum® III per ampere-year in seawater = 7.62 (32—95 °F).

30 × 7.62 × 1.1781 = 269 lb/1000 linear ft of pipeline.

How many pounds of Galvalum® III will be required for a 30-in OD line installed in saline mud and operating at 80 °F? Pounds of Galvalum® III per ampere-year in saline mud = 9.22 (32—95 °F).

30 × 9.22 × 1.1781 = 326 lb/1000 linear ft of pipeline.

How many pounds of Galvalum® III will be required for a 30-in OD line installed in saline mud and operated at a temperature of 160 °F? Pounds of Galvalum® III per ampere-year in saline mud = 21.9 (140—170 °F).

30 × 21.9 × 1.1781 = 774 lb/1000 linear ft of pipeline.

It is always a good idea to check with the bracelet vendor to determine the nearest size mold available to avoid the need for a special size mold. Also, the mold size will be a factor in determining anode spacing.

These calculations assume a utilization factor of 100%, and the designer should select the appropriate utilization factor for the specific application. For estimating, you may want to use 85%. Temperatures used in the examples are assumed to be the anode operating temperature.

Generally, zinc anodes should not be used if the pipeline will be operated at temperatures above 90 °F. See ASTM B418-80 and NACE Paper No. 311 by Elwood G. Haney presented at CORROSION/85.

See Table 3 for typical anode performance data. All performance data shown here are suitable for estimating. For final design purposes, the anode manufacturer or distributor should be consulted for actual anode performance data required for a specific application.

Galvalum® III is a registered trademark of Oronzio De Nora S. A., Lugano (Switzerland). Galvomag® is a registered trademark of Dow Chemical Company.

Typical anode specifications

Tables 1—4 provide guideline information suitable for general use. The characteristics of a specific application must be considered, and it may be necessary to contact the anode manufacturer or distributor to obtain detailed anode performance data before finalizing a design.

Table 1
Zinc Anodes

Material	MIL-A-18001H	ASTM Type 1	ASTM Type II
Lead	0.006 max	—	—
Iron	0.005 max	0.005 max	0.0014 max
Cadmium	0.025—0.15	0.03—0.1	0.003 max
Copper	0.005 max	—	—
Aluminum	0.10—0.50	0.1—0.4	0.005 max
Silicon	0.125 max	—	—
Zinc	Remainder	Remainder	Remainder

Table 2
Cast Magnesium Anodes

Material	H-1 Grade A Type III	H-1 Grade B Type II	H-1 Grade C Type I	Galvomag®
Al	5.3–6.7	6.3–6.7	5.0–7.0	0.010 max
Zn	2.5–3.5	2.5–3.5	2.0–4.0	0.05 max
Mn	0.15 min	0.15 min	0.10 min	0.5–1.3
Fe	0.003 max	0.003 max	0.003 max	0.03 max
Ni	0.002 max	0.003 max	0.003 max	0.001 max
Cu	0.02 max	0.05 max	0.10 max	0.02 max
Si	0.10 max	0.30 max	0.30 max	0.05 max
Others	0.30 max	0.30 max	0.30 max	0.05 max, each 0.30 max, total

Reprinted courtesy of Dow Chemical U.S.A.

Table 3
Typical Anode Performance Data

Anode Type	Amp-h/lb-in Seawater	Potential Negative Volts Ag/AgCl	Approx. Current Density mA/ft^2
Zinc	354	1.03	36–600
Galvalum® 111	1150–1204 (32–95 °F) in saline mud	1.10	100–1000
	950 (32-95 °F)	1.10	100
	930 (104 °F)	1.08	100
	600 (140 °F)	1.05	100
	400 (176 °F)	1.04	100
	400 (221 °F) in earth	1.02	100
Magnesium	400–500	1.53	36–600
Galvomag®	500–540	1.75	36–600

Table 4
Typical Environments for Sacrificial Anodes

Material	Driving Potential	Environment
Magnesium	0.9 volt	Underground Freshwater 200–10,000 ohm-cm
Aluminum	0.25	Seawater Brine Saline mud <200 ohm-cm
Zinc	0.25	Seawater Brine Underground <1500 ohm-cm

Comparison of other reference electrode potentials with that of copper-copper sulfate reference electrode at 25 °C

Type of Comparative Reference Electrode	Structure-to-Comparative Reference Electrode Reading Equivalent to −0.85 volt with Respect to Copper Sulfate Reference Electrode	To Correct Readings between Structure And Comparative Reference Electrode to Equivalent Readings with Respect to Copper Sulfate Reference Electrode
Calomel (saturated)	−0.776 volt	Add −0.074 volt
Silver-silver chloride (0.1 N KCl solution)	−0.822	Add −0.028
Silver-silver chloride (silver screen with deposited silver chloride)	−0.78	Add −0.07
Pure zinc (special high grade)	+0.25*	Add −1.10

Based on zinc having an open circuit potential of −1.10 volt with respect to copper sulfate reference electrode. (From A. W. Peabody, Chapter 5, p. 59, *NACE Basic Corrosion Course,* Houston, TX, 1973.)

Chart aids in calculating ground bed resistance and rectifier power cost

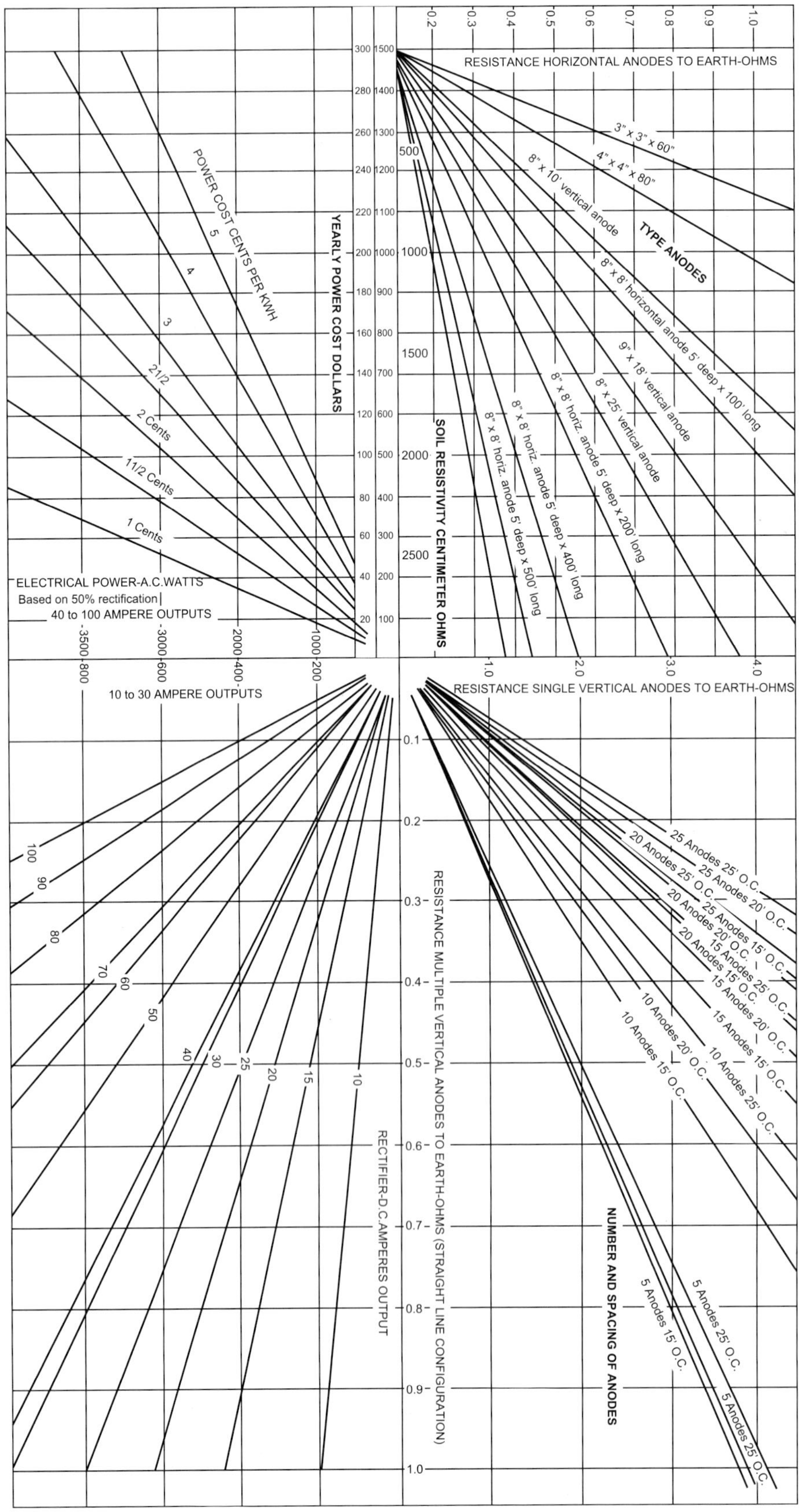

For conventional-type ground beds, this chart permits a rapid determination of the resistance to earth of various ground beds and the yearly power costs based on 50% efficient rectification. These curves are specifically designed for standard 3-in by 60-in and 4-in by 80-in graphite anodes centered in a vertical or horizontal excavation and backfilled to the dimensions indicated with well-tamped coal coke breeze.

Example.

Soil resistivity = 1000 ohm-cm

Power rate = 3 cents per kWh

Rectifier rating = 20-amp

To determine the optimum size ground bed, enter the chart at the "Soil Resistivity-ohm-cm" line and extend a vertical line from the 1000-ohm-cm point to intersect all the anode types. Proceed horizontally to the left for the resistance to earth of horizontal type anodes and horizontally to the right for vertical type anodes. Various groups of vertical anodes with 15-, 20-, and 25-ft spacings are represented by curves.

Resistance values for horizontal anodes obtained from the ordinate of the upper left-hand section should be transferred to the abscissa of the upper right-hand section to determine power consumption and annual costs from the lower right-hand and lower left-hand section. In obtaining power costs for the two lower sections, the ordinate shows power consumption in AC watts based on 50% efficient rectification, and the abscissa of points on the lower left-hand section indicates annual power costs in dollars.

How can output of magnesium anodes be predicted?

For a single 17-lb. magnesium anode, working against a fully protected pipe, divide the soil resistivity in ohm-cm units into 180,000; the answer is the current output in mA.

Example. Soil resistivity 3200 ohm-cm

$$t180,000/3200 = 56.25$$

Current output will be about 56 mA.

The rule works for *single* anodes against fully protected pipe; anode groups will have lower current output per anode, and if the pipe is not fully protected, the current output will be greater. For 32-lb anodes use 195,000; for 50-lb anodes use 204,000.

How to determine the efficiency of a cathodic protection rectifier?

Multiply DC voltage output times DC amperes times 100 and divide by the AC watts input. The answer is % efficiency.

Example. A rectifier has an output of 50 amps DC at 15 volts. The wattmeter gives a direct reading of an AC input of 1155 watts.

$$\text{Efficiency} = \frac{(\text{DC volts})(\text{DC amperes})(100)}{(\text{AC watts input})}$$

$$t = \frac{(15)(50)(100)}{(1,155)} = 65\%$$

On those rectifier units that do not have a wattmeter, the AC input may be obtained from the watt hour meter mounted on the pole. To do this, count the number of revolutions made by the watt hour meter in so many seconds, multiply by the meter constant stamped on the meter times 3600, and divide by the number of seconds timed. This gives AC input in watts. Then, go ahead and use the formula as above.

Example. A rectifier unit does not have a wattmeter. The wheel of the watt hour meter revolves 11 times in one minute. Meter constant is shown as 1.8. Output of the rectifier is the same as in the first case, 50 amps at 15 volts DC.

$$\text{AC watts} = \frac{\text{K} \times \text{N} \times 3600}{\text{T}}$$

where *K* is the meter constant and *N* is the number of meter revolutions in *T* seconds.

$$\text{AC watts} = \frac{(1.8)(11)(3600)}{60} = 1188 \text{ watts AC}$$

$$\text{Efficiency} = \frac{(15)(50)(100)}{1188} = 63\%$$

How to calculate the voltage drop in ground bed cable quickly

Use the following formula for calculating the voltage drop in a copper ground bed cable:

$$tV_T = \frac{R_{1000}}{1000}\left[I_T D_{GR} + D_A(N-1)(I_T) - I_A(1+2+3+4+\ldots+N-1)\right]$$

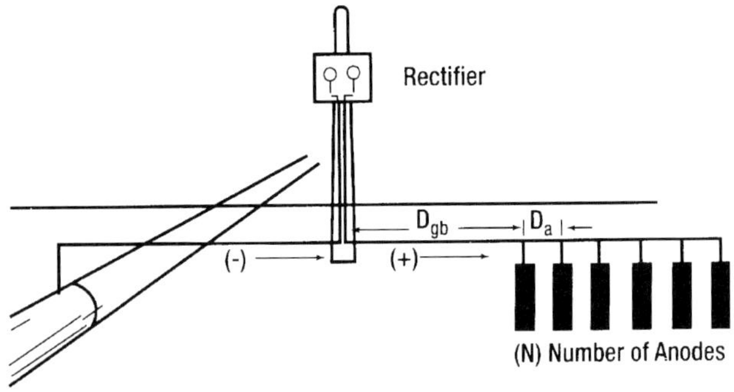

where:

V_T = Total voltage drop in volts
R_{1000} = Cable resistance per 1000 ft
I_T = Total current in amps
I_A = I_T/N current per anode
D_{GB} = Distance from rectifier to first anode in ft
D_A = Distance between anodes in ft
N = Number of anodes in ground bed.

Example. What is the voltage drop in a copper wire ground bed cable in a cathodic protection system having a rectifier output of 225 amps? The distance from rectifier to ground bed is 100 ft, and 16 anodes are spaced on 20-ft centers. The cable used is No. 4 copper having a resistance of 0.2487 ohm per 1000 ft.

Then:

R_{1000} = 0.2487 ohm
I_T = 225 amps
I_A = 225/16 = 14 amps
D_{GB} = 100 ft
D_A = 20 ft
N = 16

$$V_T = \frac{0.2487}{1000}[(225)(100) + 20[(15)(225) - (14)(120)]]$$

$$V_T = \frac{0.2487}{1000}[22500 + 20(1695)]$$

$$V_T = \left(\frac{0.2487}{1000}\right)(56,400)$$

$$V_T = 14.00 \text{ volts drop.}$$

What spacing for test leads to measure current on a pipeline?

Multiply the pipe weight, in lb/ft, by 4; the result is a spacing that will give a resistance of 0.001 ohm, or 1 amp per millivolt IR drop.

Example. Pipe weight is 40.48 lb/ft
40.5 × 4 = 162
Install test leads at 162 ft spacing.

The rule works for welded lines of grade B seamless; variations in steel may introduce errors as great as 15%. It is always better to calibrate a line section if accurate results are to be had, but if the above rule is used, the spacing will not be too small for satisfactory measurement, or too great for convenience in installation.

How to estimate the number of magnesium anodes required and their spacing for a bare line or for a corrosion "hot spot"?

Soil and pipeline conditions will vary widely, but where these given assumptions hold, an estimate can be made for test installations using this method. First, find the area of pipe to be protected, multiply by the milliamperes per square feet necessary to protect the line, and divide by design output of an anode in this type soil. This will give the number of anodes required; to find spacing, divide number of anodes into the length of line.

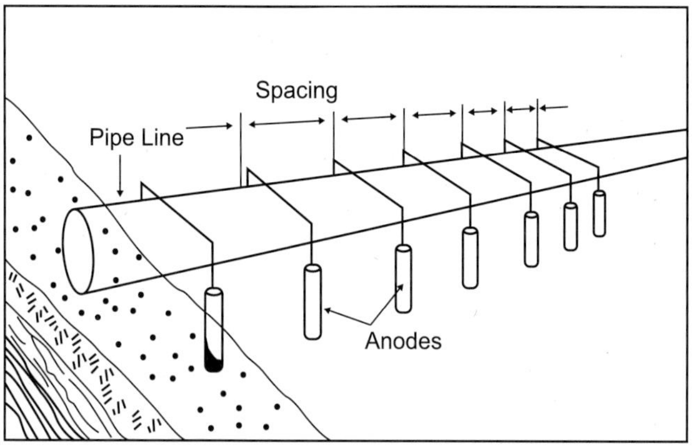

Example. Find the number of 17-lb. magnesium anodes needed and spacing used to protect 10,000 ft of bare 4-in pipeline in a corrosive soil having a resistance of 1000 ohm per cubic centimeter. Assume a current requirement of 1 mA/ft^2, and that the anode output curve shows an output of 100 mA per anode in this type soil.

$$\text{Area of pipe} = \frac{4 \times 3.14}{12} \times 10,000 = 10,500 \text{ ft}^2$$

$$\text{Total current requirement} = 10,500 \times 1.0 = 10,500 \text{ mA}$$

$$\text{Anodes required} = \frac{10,500 \text{ mA}}{100 \text{ mA/anode}}$$

$$t = 105 \text{ anodes distributed}$$

$$\text{Spacing} = \frac{10,000}{105} = 95 \text{ ft}$$

Similarly, find the number of anodes required to protect ten 100-ft long "hot-spots" on a 4-in line assuming a 100-mA output from a 17-lb magnesium anode. Estimate as follows:

$$\text{Area of each hot spot} = \frac{4 \times 3.14}{12} = 105 \text{ ft}^2$$

$$\text{Anodes required} = \frac{105 \text{ ft}^2 \times 1.0 \text{ mA/ft}^2}{100 \text{ mA per anode}} = 1.05$$

One 17-lb magnesium anode will produce approximately enough current in 1000-ohm soil to protect one of these "hot-spots."

Ten anodes are required for "hot-spots" on this section of line.

Note: For these rules, the protective criterion is −0.85 volts pipe-to-soil potential using a copper-copper sulfate reference electrode; or a negative change of potential of as much as −0.25 volts. Soil and pipeline conditions vary widely, so consult an experienced corrosion engineer in the design of a complete cathodic protection system.

How can resistivity of freshwater be determined from chemical analysis?

Divide the total solids expressed in parts per million into 500,000; the result is the resistivity in ohm-centimeter units.

Example. Total solid content is 400 ppm.
500,000 ÷ 400 = 1250 ohm centimeters resistivity.

This rule of thumb will give very accurate results for any naturally occurring freshwaters; severely contaminated waters may differ markedly, however, depending on the kind of contamination.

What will be the resistance to earth of a single graphite anode?

For the most common size, 3-in by 60-in, installed in 10 ft of coke breeze backfill in an 8-in hole, multiply the soil resistivity in ohm-cm by 0.002; this will give the resistance to earth in ohm.

Example. 1800 ohm-cm soil

$$1800 \times 0.002 = 3.6 \text{ ohm}$$

This assumes that the resistivity is uniform to considerable depth; if it increases with depth, the resistance will be slightly higher.

What will be the resistance to earth of a group of graphite anodes, in terms of the resistance of a single anode?

When the spacing is 20 ft (the most common value), the resistance of the group will be equal to the resistance of a single anode, divided by the square root of the number of anodes.

Example. Resistance of 4 anodes = ½ resistance of single; resistance of 16 anodes = ¼ resistance of single, etc.

This rule is reasonably accurate for soils that are uniform to a considerable depth; if the resistivity increases with depth, the "crowding" effect will be greater, whereas if it decreases, the individual anodes will not affect each other so much.

How can the current output of magnesium rod used for the cathodic protection of heat exchanger shells be predicted?

Divide the total dissolved solids content of the water by 13; the answer is the current output per ft of 1.315 in (1-in pipe size) magnesium rod to be expected.

This is an approximation to the current flow that will be found after stabilized conditions have been reached, if protection is obtained. If insufficient magnesium is installed, the current flow will probably be greater, and if too much is installed, the current will be less.

How many magnesium anodes are needed for supplementary protection to a short-circuited bare casing?

For 17-lb anodes, use $A\rho/50{,}000$ to give the number needed for the casing alone; A is the area of the casing, in square feet, and ρ is the soil resistivity in ohm-centimeters. The life of the anodes may be estimated from $\rho/180$. If the life in years, as given by this expression, is too short, refigure for 32-lb anodes, using $A\rho/55{,}000$ for the number and $\rho/100$ for the life. If this is still too short, figure for 50-lb anodes; number is $A\rho/60{,}000$ and life is $\rho/65$.

Example. Suppose the pipeline has shorted out to 45 ft of 24-in bare casing. Soil resistivity is 1100 ohm-cm.

How many 17-lb anodes would be required for supplementary protection?

$$\text{Number of anodes} = \frac{A\rho}{50,000}$$

$$A = 2 \times 3.14 \times 45 = 283$$

$$\text{Number of anodes} = \frac{(283)(1100)}{50,000} = 6.21$$

Place three anodes at each end of casing. Their life expectancy will be 1100/180 = 6 years.

Group installation of sacrificial anodes

When anodes are installed in groups, a spacing correction factor must be taken into account to determine the current output. The chart below shows the correction factor for multiple anode installations.

Example. Single anode 8-in by 60-in in 1,500 ohm soil.
Resistance = 4.849 ohm
Potential = 0.85 volts

$$\text{Current output} = \frac{E}{R} = \frac{17.5 - 0.85}{4.849} = 0.185 \text{ amps}$$

Current output of 5 anodes on 15 ft spacing under above conditions.

$$\text{Combined resistance of group} = \frac{4.849}{5} \times 1.22$$

Correction factor = 1.17 ohm

$$\text{Total current output} = \frac{1.75 - 0.85}{1.17} = 0.769 \text{ amps}$$

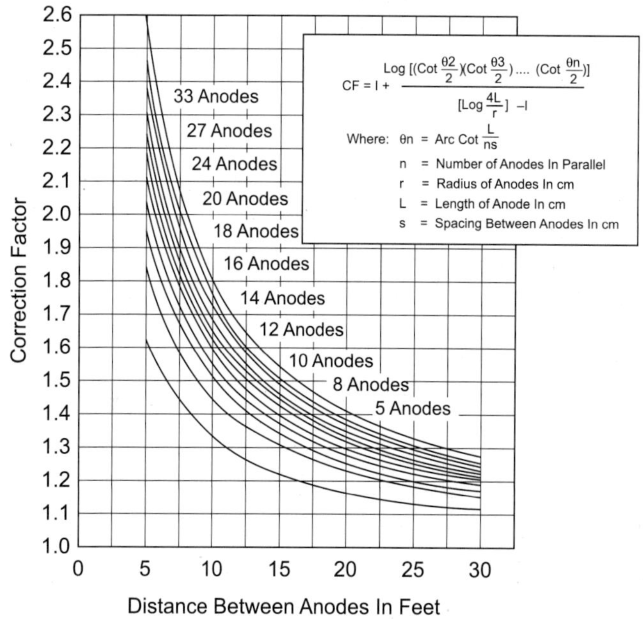

How to find the voltage rating of a rectifier if it is to deliver a given amount of current through a given ground bed?

Multiply the desired current, in amperes, by the total resistance of the ground bed, in ohm (either calculated or measured); add 2 volts to overcome the galvanic difference between graphite and protected steel.

Example. A current of 20 amps is desired from a ground bed whose resistance is 0.9 ohm.

$$(20 \times 0.9) + 2 = 20 \text{ volts}$$

Specify the next larger size, which will probably be (depending upon the manufacturer) 24 volts.

Note: Where the resistance is measured rather than calculated, and is the total loop resistance between pipe and ground bed, this method is quite precise; for calculated values, caution should be observed.

Determining current requirements for coated lines

Survey along the line after a polarization run of three or four h, and locate the two points whose pipe-to-soil potentials are 1.0 volt and 0.8 volt. The section of line lying between these two points will then be at an average potential of about 0.9 volt, which is approximately the average of a line protected with distributed magnesium anodes. Then, if, the line current is measured at each of these two points, the difference between these two line currents will be the total amount of current picked up on the line in the section. This quantity, divided by the length of the section, will then give the current requirements of the line, in amperes per mile. Naturally this figure can only be used for an entire line if the section chosen is truly representative; judgment must be used at this point.

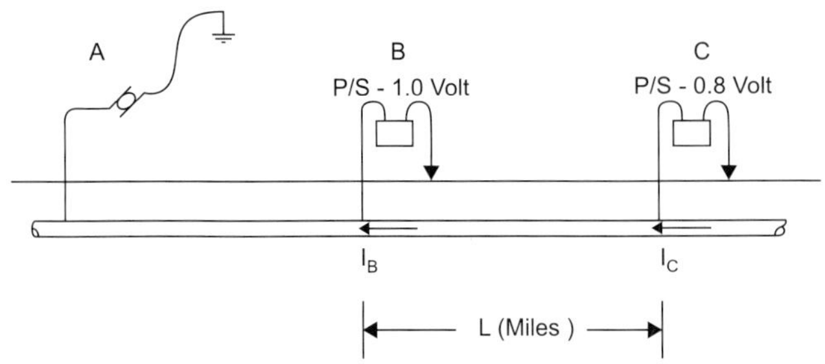

Determine current requirements for coated lines when pipe-to-soil potential values are estimated

If it is not possible or convenient to find the points that have the exact values of 1.0 and 0.8, then readings can be taken that approximate these values, and both potential and line current can then be adjusted to find the desired quantity; this is illustrated in Figure 1; readings were taken at points "*A*" and "*B*," and, by assuming a static potential of 0.6 volt and linear attenuation, it is determined that points "*C*" and "*D*" have the values of 1.0 and 0.8; since line current varies according to the same laws as potential, it is possible to determine the current values at the two latter points, and so to arrive at the desired quantity.

It is not to be expected that the estimate thus made will be highly accurate; it is intended to be merely preliminary.

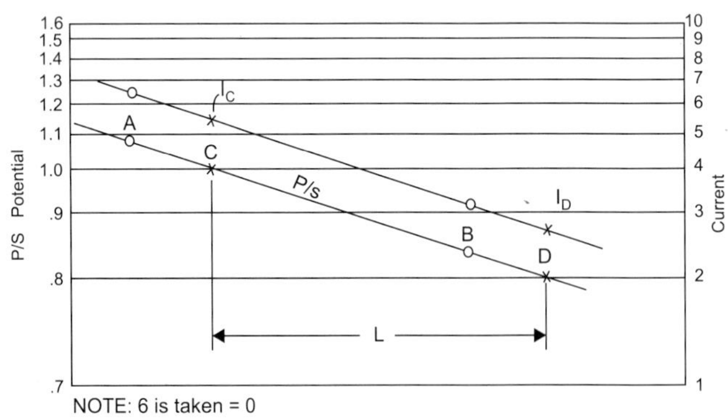

HVDC effects on pipelines

Thomas B. Hays, George W. Silkworth, Robert D. Schilling, Cornell, Howell, Hayes & Merryfield, Inc.

A pipeline collects, conducts, and discharges electric current of High Voltage Direct Current (HVDC) origin as a result of the electric field established by the DC current transmitted through the earth and by the electrical properties of the pipeline itself.

The electric field

It is assumed that the electrical properties of the earth and pipeline are uniform and that the HVDC effects are isolated from other stray or induced current effects. This assumption makes possible the analysis of this portion of the total protection problem and assists in visualizing the unique effects of HVDC earth currents.

Ultimately, the total electrical current conducted by a pipeline will be the sum of all the various components. This includes protective current, telluric current, and other stray or interference currents.

Conduction of a direct current through the earth is an electrolytic phenomena, characterized by an electric field that surrounds each of the two earth electrodes, and is directly proportional to the HVDC earth current (Figure 1). When the electrodes are spaced several hundred miles apart, the field appears to be concentric around each ground electrode.

The fields surrounding the opposite electrodes are of the opposite polarity. If the polarity of the ground current reverses, the surrounding electric fields will also reverse. The potential at any point in the electric field with respect to remote earth (V), due to an electrode current (I) at a distance (r) from an electrode in a homogeneous earth of the resistivity (p) is found by the relationship:

$$V = \frac{\rho I}{2\pi r}$$

This relationship shows a simplified representation of what actually occurs. The earth's resistivity is not uniform at all depths or horizontal distances, which causes irregularities to the ideal voltage distribution existing around an electrode in a homogeneous earth.

To refine the earth current analysis and to account for different resistivities, computer techniques have been developed to estimate the earth's gradients existing for conditions of multilayers of earth of different resistivities. Pipeline effects have also been predicted by this method. Reasonable agreement between these models and field data has been reported. Techniques are available to calculate the quantitative characteristics of both the earth's gradient and HVDC effects on pipeline.

HVDC effects on a pipeline

Whether the HVDC component of the total current existing on the line causes a corrosion problem depends on the level of protection along the line and the effect of the added component

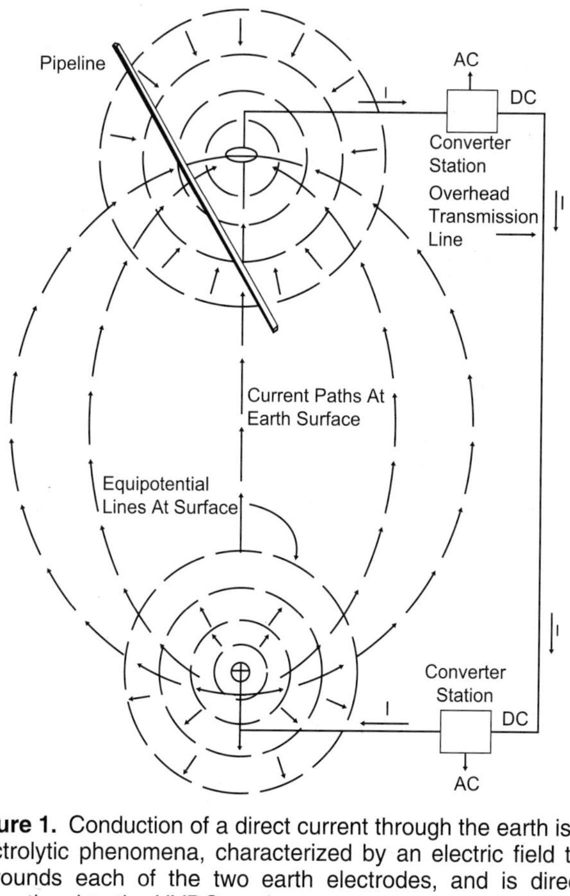

Figure 1. Conduction of a direct current through the earth is an electrolytic phenomena, characterized by an electric field that surrounds each of the two earth electrodes, and is directly proportional to the HVDC earth current.

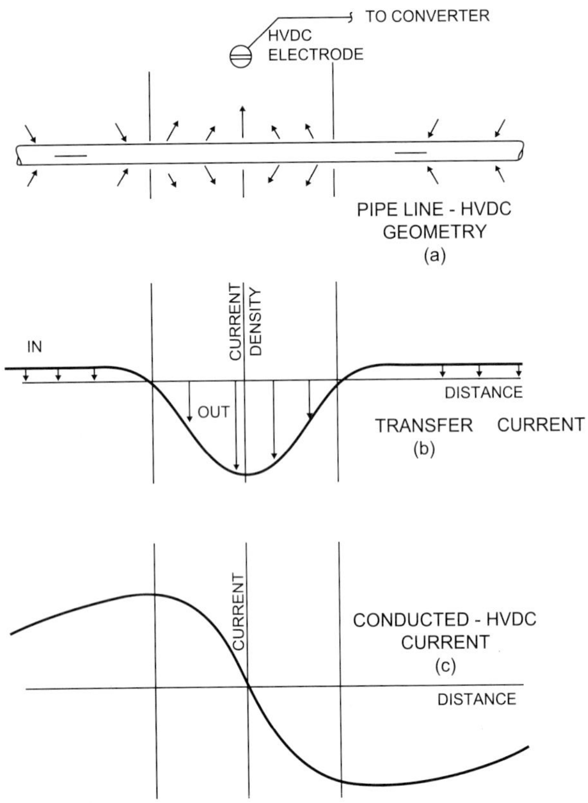

Figure 2. (a) Geometric plan between the HVDC electrode and the pipeline. Arrows indicate the direction of the HVDC component current. These would be reversed if the polarity of the electrode were reversed. (b) Curve shows characteristics of the transfer of HVDC into and out of the pipeline along its length. The transfer current is equivalent to current density. (c) Conducted current in the pipeline, which results in the summation of the transfer current along the pipe.

to the level of protection. Figure 2(a) illustrates the HVDC component of the total conducted current along the pipeline for a given separation distance under the condition where the HVDC electrode is collecting current from the earth.

In determining protection requirements for the pipeline, the characteristic of the current transfer relationship is important (Figure 2(b)). In those regions of the pipeline where the current transfer is from the pipeline to the soil—along the center in this case—corrosion could occur if the HVDC component of the total transfer current to the soil is great enough to exceed any local level of protection against corrosion.

The pipeline HVDC arrangement shown in Figure 2(a) will serve as a basis to develop and illustrate the performance requirements for protecting a pipeline from the detrimental effects of HVDC earth currents. The hypothetical problem, as shown, is one that may be encountered in the field. In this example, the following assumptions are made:

- The pipeline is long, straight, electrically continuous, and uniform in wall thickness, diameter, material, and coating.
- It is approximately 30 miles away from the nearest HVDC earth electrode.
- The earth's resistivity is uniform over the general area.
- The maximum value of the HVDC earth current in either polarity causes the pipeline effects.

While these conditions may not exist in any actual pipeline HVDC system situation, they will serve to identify and evaluate the requirements for protection against HVDC effects in an actual situation.

Figure 3 shows the range or envelope of effects as the HVDC varies from its maximum value in one polarity to the maximum value of the other polarity. The coordinates on this graph are DISTANCE along the *X* axis and VOLTAGE on the *Y* axis.

The voltage corresponds to the effect that would usually be measured on an actual pipeline: pipe-to-soil voltage measured with a copper-copper sulfate half-cell. The value of the voltage effect at any point on the line is the product of the current transfer, and the electrical resistance between the pipe material and earth over the same area.

Figure 3 also shows reasonable hypothetical scale values illustrating the various combinations of protection requirements that may exist in an actual installation. The voltage values give the change in pipe-to-soil reading that would be caused by the HVDC transfer current. If this HVDC is considered as the only source of corrosion, then a pipe-to-soil voltage reading with a copper—copper sulfate half-cell would be approximately 0.85 volts at no transfer current. This scale is shown on the left side of the figure. Corrosion occurs for any reading below this axis, since the current is leaving the pipeline and entering the soil to cause this change in the pipe-to-soil reading.

At locations on the pipeline where HVDC causes conditions that increase the volt-meter reading, the current is collecting on the pipeline. The horizontal axis, therefore, represents a corrosion limit. Above this axis, a coating disbonding limit can exist where the density of the current transferred from the earth to the pipeline causes gas evolution, which may blister the coating.

In this example, a pipe-to-soil reading with a copper—copper sulfate half-cell of 1.5 volts is shown as the coating disbonding limit. The difference between these limits gives an assumed safe operating band width of 0.65

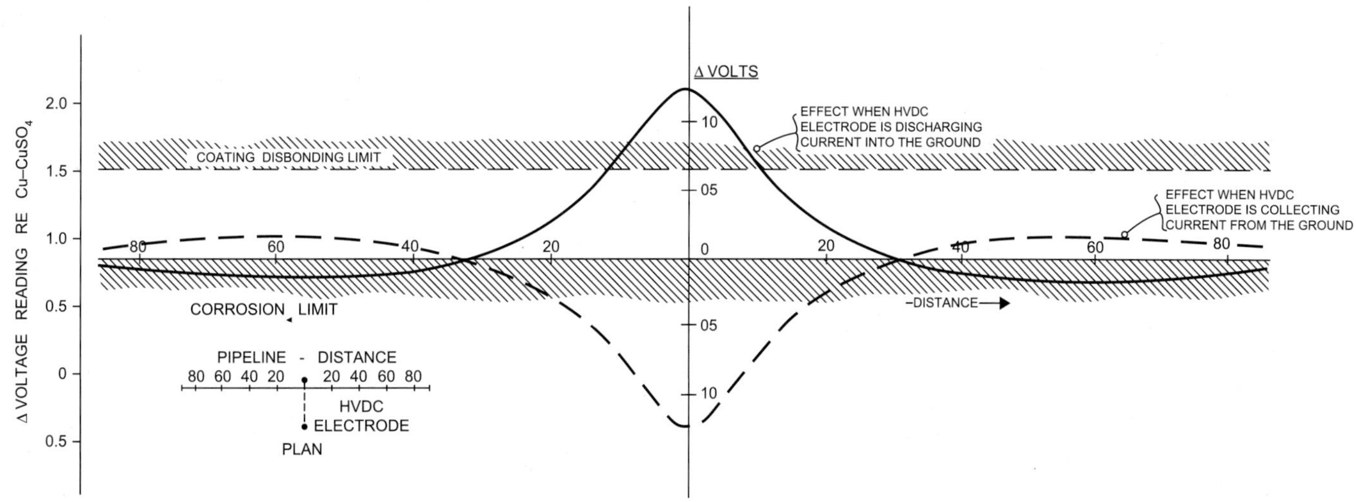

Figure 3. Range or envelope of effects which an HVDC component can have along the length of a pipeline as the HVDC varies from its maximum value in one polarity to the maximum value of the other polarity.

volts. Values other than these may be used for specific applications. These limits, however, will serve to illustrate the principles.

Protective system limits

The requirements for protection along the line from either corrosion or disbonding are developed from examination of Figure 3, and the results are plotted in the subsequent figures. The upper limit of protection is the maximum amount of current that can be collected by the pipeline without exceeding the disbonding limit. The lower limit is the minimum amount of current that can be collected by the pipeline to keep it above the corrosion limit.

At this value of HVDC (Figure 4), the uncorrected pipeline effect is shown by Curve 1. No protection is needed along either end of the pipeline in Regions *A-B* or *E-F*. Corrosion protection is needed between *B* and *E*. The minimum protection effect along the line is shown by Curve 2. The resultant of the HVDC and this protection effect lies on the

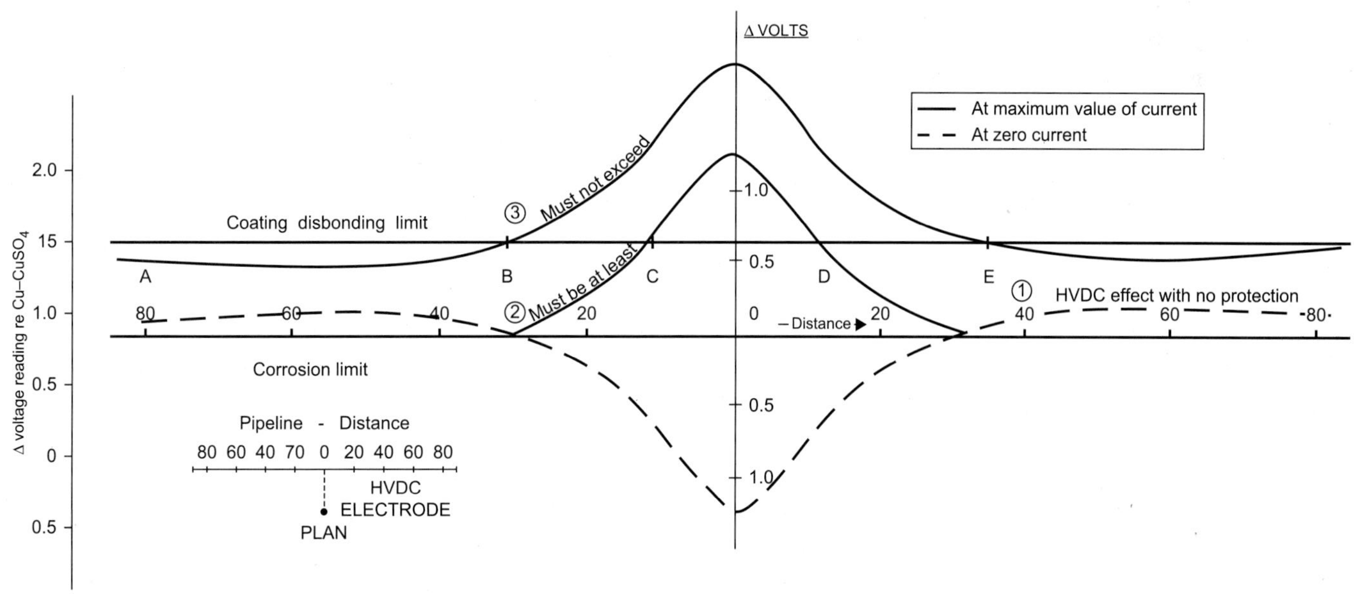

Figure 4. Limits for the assumed condition of the HVDC system when it is collecting current from the earth at the highest value possible.

corrosion limit. The maximum protective effect that can be applied along the line, without exceeding the coating disbonding limit, is shown by Curve 3. Under this maximum value and polarity of HVDC current, the region between these two protective limits, therefore, defines the range of acceptable protective system performance at any location on the pipeline.

When the HVDC effect decreases, due to a reduction in the ground current, the required level of protection on the line also decreases. Along the pipeline length from *C* to *D*, the protection *must* be reduced to avoid exceeding the disbonding limit. At zero HVDC effect, the protective system effect must fall between the corrosion and disbonding limits all along the line.

Examination of the limits for these two values of the HVDC system operation shows that the protective effect must be varied in response to the HVDC effect over the length of the pipeline from *C* to *D*: automatic regulation of the protective current is needed over this length of the pipeline. With only this polarity of HVDC current, a fixed level ofprotection can be applied along the pipeline from *A* to *C* and from *D* to *F*.

When the earth's current can reverse polarity, as with the Pacific Northwest (PNW)-Pacific Southwest (PSW) intertie, then the same process that was used to develop the limits, shown in Figure 4, can be repeated for the opposite HVDC polarity. The limits on protection along the pipeline when the HVDC electrode is discharging current, at its maximum value, into the earth are shown in Figure 5.

The combination of these two sets of limiting characteristics is shown in Figure 6. The graph shows the length along the pipeline where a fixed level of cathodic protection can be used and those regions where controlled protection is required for any value of HVDC earth current. In this particular example, variable protection is required along the pipeline for approximately 20—30 miles on either side of the point closest to the HVDC electrode. Cathodic protection is required when the HVDC electrode is collecting current from the earth. That is, the HVDC component of current, conducted by the pipeline, is discharging from the line near the earth electrode. The required protective current along the pipeline to counteract the HVDC effect under this condition is show as Region 1 in Figure 6.

When the HVDC polarity is reversed and the magnitude of the ground current is increased to the maximum value, a different problem is encountered. The electric field which is established around the electrode and pipeline causes current to enter the pipeline through the coating. In this example, the disbonding limit is exceeded over the length of the pipeline from *C* to *D* at the maximum value of HVDC earth current. The required correction for this detrimental effect is similar to the effect for the previous HVDC polarity: namely, the

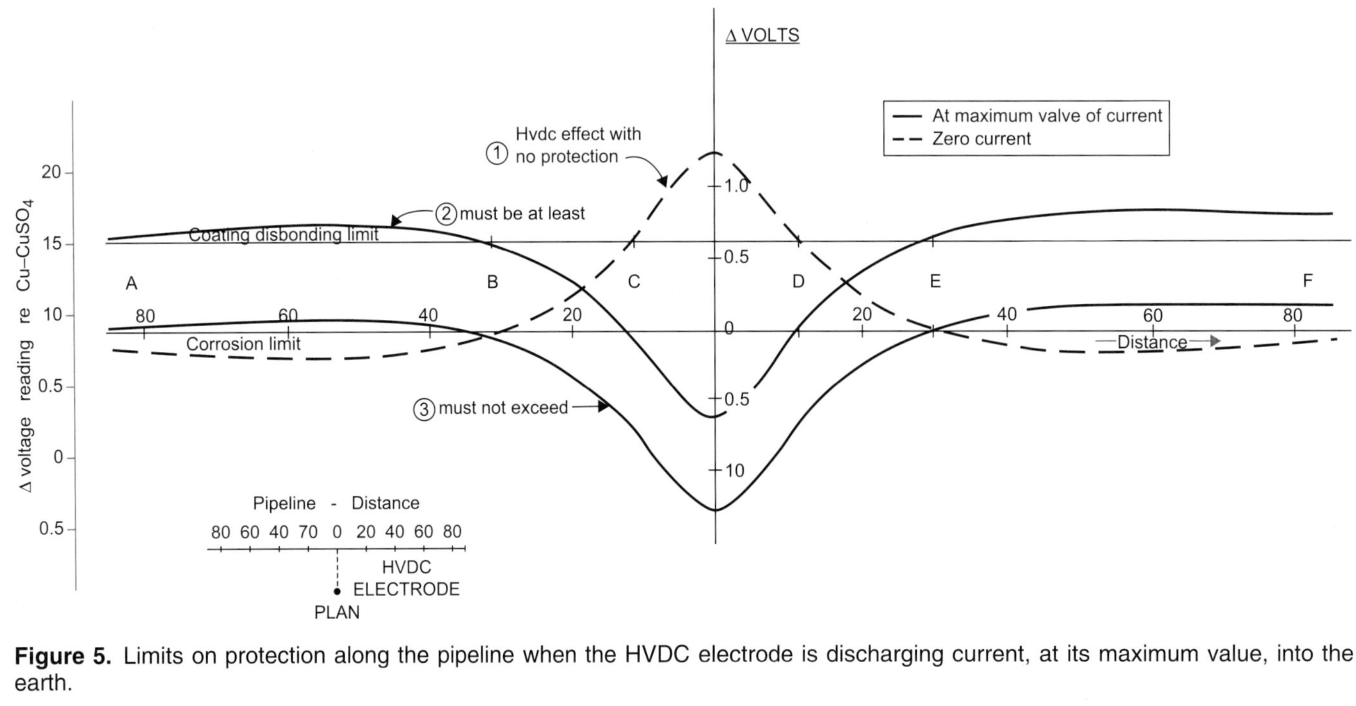

Figure 5. Limits on protection along the pipeline when the HVDC electrode is discharging current, at its maximum value, into the earth.

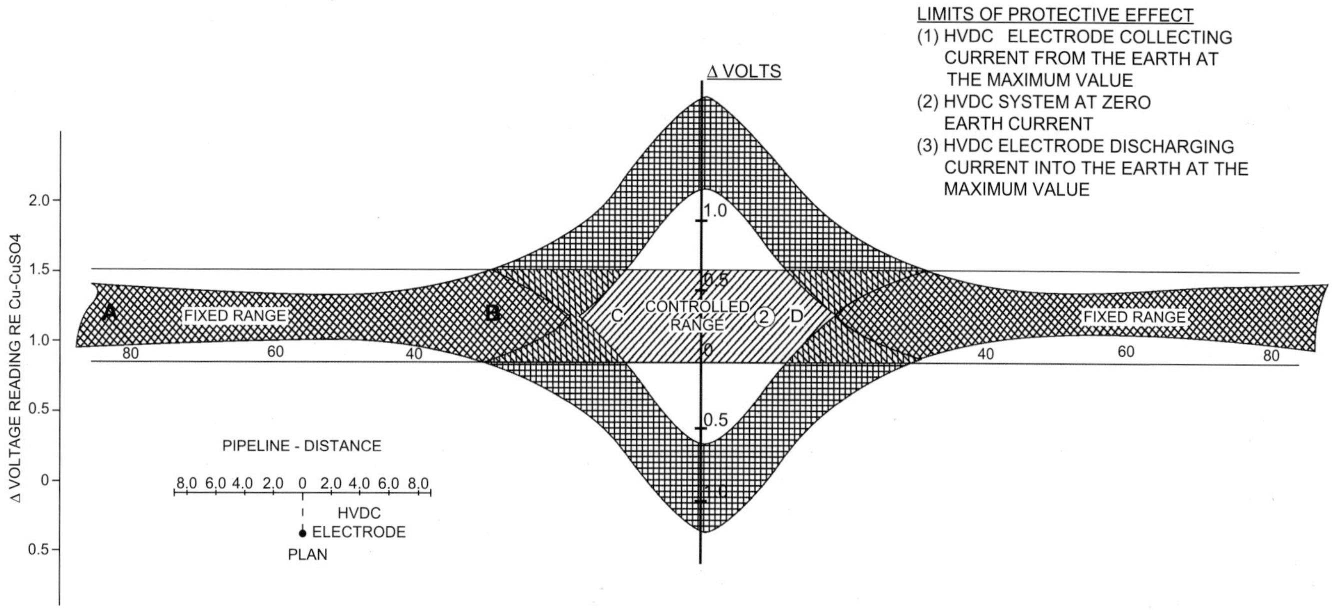

Figure 6. Combination of Figures 4 and 5. The length along the pipeline where a fixed level of cathodic protection can be used and those regions where controlled protection is required for any value of HVDC earth current. In this example, variable protection is required along the pipeline for approximately 20—30 miles on either side of the point closest to the HVDC electrodes.

uncorrected electric field must be modified by an added field to give an acceptable total. Theoretically, in this situation, a portion of the HVDC earth current can be collected in a protective ground bed and conducted to the pipeline to satisfactorily modify the original field into an acceptable one.

Troubleshooting bed cathodic protection systems

When the monitor, or a pipe-to-soil potential survey, indicates inadequacy of protection, the first place to look is at the protective unit. The current output of the rectifier should be checked; if it is normal, the trouble is on the line itself; if it is high, and accompanied by low voltage, the trouble is certainly on the line, and is caused either by increased current demand or by a short-circuit to parasitic metal. If the current output is low, with normal or high voltage, the trouble is in the ground bed or connecting cables. A pipe-to-soil potential over the ground bed will show a peak over every anode that is working; a disconnected anode will not show at all. This test is particularly useful if comparison can be made to a similar test made at the time of the original installation. When inactive anodes are found, only digging will uncover the cause.

If the rectifier and its anode bed appear to be performing satisfactorily, the source of the low potentials must be sought on the line itself. Any area in which work has been recently done should be investigated; for example, if a new lateral has been connected, the insulation should certainly be checked. If investigation of such suspected sources discloses nothing, then a more detailed search must be made. First, the pipe-to-soil potentials should be studied, to see if the failure seems to be localized. A more thorough, but slower, approach is that of making a detailed line current survey; find out where the drained current is coming onto the line. This will be much more easily interpreted if a similar survey, made when the line was in satisfactory condition, is on record; a comparison will often quickly locate the offending parasite. On the other hand, if the current collected in each section is greater than on the earlier survey, with no pronounced differences, then the trouble is simply that of increased overall current requirement, probably due to coating deterioration.

How to control corrosion at compressor stations

R. E. Hodge, Superintendent of Corrosion Control, Natural Gas Pipeline Co. of America

Corrosion failures at compressor stations result from carelessness on the part of the user, or poor choice of material/configuration by the designer.

Design engineers should consider the following:

- Coat everything underground, except ground rods, with a coating properly selected for the conditions.
- Use proper coating application and inspection.
- Use ground rods anodic to steel and insulated (coated) ground wires.
- Avoid shielding as much as possible and allow space for corrosion control additions.
- Use a minimum of insulating fittings.
- Protect against water system corrosion.
- Select materials carefully, with corrosion in mind, as well as the more obvious considerations.
- Do not retire facilities in place—take them out.

Coatings

Coatings are one of the most important considerations for control of underground corrosion. Generally, all underground metallic structures, except ground rods, should be coated. This includes gas piping, control lines, tubing, water lines, conduit, air lines, braces, etc.

There is enough buried conduit around a compressor station to soak up large amounts of cathodic protection if not coated. Nonmetallic carriers can be used in some cases instead of coating.

Proper selection and use of coatings are important. The best of coatings are no good if used in the wrong place or improperly applied. The emphasis here is on the selection of coating for the temperature and environment expected.

For gas discharge lines, temperature becomes a dominant consideration. Be sure to get a coating that will withstand gas discharge temperature. Be careful, because sag temperatures listed in coating literature are not maximum operating temperatures. Very few coatings on the market can withstand operating temperatures above 120° without damage. Any bare metal underground becomes a drain on the cathodic protection system and a potential shielding problem.

There are other areas for coating consideration besides underground (e.g., tank interiors and atmospheric coatings). These must also be matched to the situation.

Grounding

Grounds should be of materials anodic to steel. These might include magnesium or zinc anodes, or galvanized ground rods.

Copper ground rods should not be used since it is cathodic to steel. Some compressor stations have been installed with massive bare copper grounding systems. These are large galvanic cells with the steel gas piping as the anode (corroding element). Bare ground wires and rods also take cathodic protection current. This bad situation can be avoided by using galvanized rods, for example, and using copper ground wire with insulation. The junction of copper ground wire and galvanized ground rod must be coated. Grounding systems should be checked periodically for proper functioning.

Shielding

Shielding poses a major problem at compressor stations. It could be described as the blocking of cathodic protection due to geometric configuration.

Congested areas, where buried metal concentration is high, are difficult to cathodically protect. Piping, or other metal structures, will tend to shield cathodic protection currents from reaching all of the buried structures to be protected.

Cathodic protection ground beds of various styles have been used at compressor stations. Remote conventional beds and deep beds are popular for many applications. However, we have found that distributed anodes seem to be the best way to get adequate cathodic protection to all the shielded areas in a station. This type of bed can be described as a ground bed with the anodes distributed in a nongeometrical pattern, interspersed through with piping.

Designers should allow space between lines where possible; avoid congestion. This would cause less shielding and allow space for distributed anode installations where required. It would also allow space for access to work equipment for anode installations.

Insulation

Use insulating fittings (flanges, couplings, unions, etc.) only where absolutely necessary.

Hold insulating fittings to a minimum. These can often cause more problems than they solve. You might want to insulate water wells from the adjacent piping for control of cathodic protection current to the wells. This may not even be necessary if distributed ground beds are used.

Some companies insulate compressor stations from the lines entering and leaving the station and other companies do not. This is a matter of individual company preference.

Water systems

Each water system (jacket water, piston water, boiler water, etc.) should have corrosion monitoring provisions designed into the system, such as coupons or corrosion rate probes. If there is a gas treating plant in conjunction with the compressor station, corrosion monitoring provisions should be designed into that system also.

Water treatment for corrosion control would have to be considered, depending on individual circumstances. Some treatment facilities may be designed into the systems.

Provisions should be considered for cathodic protection of the internal surfaces of storage tanks and water softeners. One problem often encountered on smaller tanks is adequate openings. Enough openings of sufficient size should be allowed on any tank for possible future work.

Water should be kept out of some systems. For instance, hair dryers may be used on control lines.

Materials selection

Proper selection of materials is important throughout design.

Two key considerations from a corrosion standpoint are: (1) avoid the use of dissimilar metals; (2) match the material to its environment. Examples of these would include the use of a brass valve in a buried steel water line or use of mechanically strong bolts in the wrong electrolyte, resulting in subsequent failure from stress corrosion cracking.

Select materials carefully, with corrosion in mind, as well as the more obvious considerations. If design engineers do this, many subsequent failures can be eliminated.

Retirements

Many compressor station designs involve retiring something that is already there.

If something must be retired, especially underground, take it out. Many times a company unnecessarily pays to continue protection of facilities retired in place.

Even if the retired facilities can be completely disconnected from the protective systems, shielding is still a problem. Shielding is a problem by foundations also, so if they are retired, they should be removed.

Thus, attacking the corrosion problems requires a truly interdisciplinary effort. In fact, corrosion offers one of the richest fields for interdisciplinary activity.

Based on a paper, *"Construction Dsign to Facilitate Corrosion Control—Compressor Stations,"* presented at AGA Distribution Conference, May 8—10, 1972, Atlanta, Ga.

Project leak growth

The numbers of leaks in older pipeline systems tend to increase exponentially. Exponential growth of leaks will fit the following equation:

$$N = N_1 e^{nA}$$

where:

N = Number of leaks for the study area in any year considered
N_1 = Number of leaks in the first year for which leak data are available
e = 2.718
n = nth year for which the number of leaks is to be estimated
A = Leak growth-rate coefficient

The coefficient A can be calculated from the graph drawn of the leak data for the study period. Preferably, the study period should be in the range of 5—10 years. Any data less than this will not give enough points to accurately determine the coefficient A. Before a graph is drawn, leak report quantities must be adjusted for any changes to the system during the study period. In other words, if a township was the study area, then pipe replaced, added, abandoned, or installed should be tallied. If this is not done, errors will result. For example, if leak data were collected on a township basis, and during the study period considered, 10% of the township's pipe was renewed, leak frequency in the old area will continue to rise while in the renewed area there should be none, possibly giving an overall number of leaks less than that of the year prior to the replacement. For an accurate picture and a determination of the leak growth-rate coefficient, the number of leaks in the years following the replacement should be accounted for as follows:

Let x = Percentage of replacement completed.
Let y = Actual number of leaks for the remainder of the township for the year considered

Adjusted number $N = y/(1 - x)$.

The leak growth rate for some pipelines may not follow the exponential curve.

For example, in a particular case, a linear increase with time may be predicted. In that case, use a linear equation instead of the exponential form.

Example.

N_1 = Number of leaks in first year (i.e., 2007) = 140
N = Number of leaks in year considered (2014) = 310
n = 14−7 = 7
$310 = 140e^{7A}$

Hence, coefficient A = 0.113. The estimated number of leaks in 2018 will be:

$$N = 140e^{11(0.113)} = 485 \text{ leaks}$$

Source

Hayat Ahmad, *Gas Pipeline Renewal Insertion Technology*, 3 and 4 (1990).

Advances in pipeline protection

M. D. Allen, CEng, MIM, MICorrST; **N. R. Barnes,** BTech, CEng, MIM, MiCorrST, Spencer and Partners, U.K.

The external corrosion protection of a pipeline is commonly achieved by a system comprising an insulating coating and cathodic protection. The performance of this system is normally assessed by regular monitoring of pipe to soil potentials at selected intervals along the pipeline. Three more detailed pipeline survey procedures are available to provide additional monitoring information and the case histories described indicate the experience and value gained from their selective use.

Introduction

The external corrosion protection of a buried steel pipeline is generally achieved by a system comprising an insulating coating and cathodic protection. On modern pipelines, the coating is regarded as the primary means of protection with the cathodic protection systems providing control of corrosion where the coating has failed or has been damaged.

The performance of the cathodic protection system is usually assessed by regular monitoring of pipe-to-soil potentials at selected intervals along the pipeline.

From these results, conclusions may be drawn concerning the level of cathodic protection being achieved and by inference the performance of the insulating coating. Under most circumstances, such measurements at selected locations will provide an acceptable indication of the overall level of corrosion protection being achieved on the pipeline.

These measurement locations, however, commonly at least 1 km apart, are generally selected for ease of access and thus only provide a valid assessment where pipeline coating is consistent, ensuring a constant attenuation of cathodic protection levels between each measurement point. Where the coating is variable in quality, it cannot be assumed that satisfactory levels of protection are maintained between measurement locations and a deficiency in the protection system may remain undetected allowing corrosion to occur.

Coating defect surveys have also been used to identify areas where pipeline coating quality is questionable. Historically, these surveys have almost universally taken the form of Pearson-type surveys. These surveys have entailed a two-man survey team walking the length of a pipeline identifying coating defects and subsequently marking the points with pegs for further investigation.

Surveys of this type have two inherent disadvantages, namely that the survey procedures are time consuming and no indication of cathodic protection levels at the defect location is obtained.

With these shortcomings in mind, two other survey procedures have been developed, the signal attenuation coating survey and the close interval potential survey. The former provides a rapid assessment and record of pipeline coating condition and more readily identifies areas where a defect or defects are present. Identified areas may then be surveyed in detail by close interval potential techniques that cannot only confirm coating defect locations, but also establish the status of the cathodic protection system at close intervals throughout the chosen areas.

Pearson survey

The Pearson survey is an aboveground technique used to locate coating defects on buried pipelines. An AC signal is injected onto the pipeline from a power source connected between a cathodic protection test post or valve and an earth pin. The AC frequency selected is dependent upon the type of pipeline coating in order to minimize current loss from the pipeline through sound coating. At coating defects, the AC signal leaks to earth and flows via the soil to the earth spike to complete the circuit. The localized current flow from the pipeline at the defect creates a potential gradient through the soil, and this is detected by measurement at a ground surface.

In practice, this is achieved by two surveyors walking in tandem over the entire pipeline, or section of pipeline. Earth contacts, either ski poles or foot cleats, are connected to a receiver via a cable harness. As the first surveyor approaches a coating fault, an increased voltage gradient is observed, either audibly through headphones, or visually on the receiver signal level meter. As the lead surveyor passes the defect, the signal fades, and then increases again as the rear surveyor's contacts pass over the defect.

The defect may be recorded on a record sheet, prepared beforehand, complete with a measured distance from a fixed reference point or indicated by a marker peg or nontoxic paint. A recent development in Pearson equipment has been the introduction of a system that continuously records, through data logging equipment, the potential difference between each earth contact. In this way, data may be graphically presented.

Signal attenuation coating survey

This method is similar in many respects to the Pearson survey, the essential difference being the method of defect detection. The AC current injected onto the pipeline creates a magnetic field around the pipe, the strength of this field being proportional to the amplitude of the AC signal. Measurements of the field strength are made by recording the voltage generated in a sensing coil and indicated on a dB scale, when the coil is held over the pipeline. Loss of current from a coating defect creates a change in the field strength due to a change in the pipe current and this is detected as a change in output from the sensing coil.

In practice, this is achieved by one surveyor measuring field strengths at discrete, preset intervals. At each location the field strength and pipeline chainage are recorded. A graph of distance versus field strength may then be produced. Sections of pipeline having poor coating quality are indicated by the steepest gradient on the line plot. The results may also be expressed as histograms to aid interpretation. Typical results are indicated in Figure 1. Any area indicating high signal losses may then be resurveyed using the same procedures to more closely locate the coating defect(s).

Close interval potential survey

The close interval potential survey is an aboveground technique used to provide a detailed profile of the pipe-to-soil potential with distance. This profile can be used to determine detailed information concerning the performance of the cathodic protection system, the coating system, and to a lesser degree, interaction effects. The advent of portable microprocessor-based measuring equipment and computer-assisted data handling, together with accurate synchronized high-speed current interruption devices, have allowed these surveys to provide accurate, practical data and collection methods.

To eliminate errors in the pipe-to-soil potentials recorded, arising from the inclusion of volt drops caused by the flow of cathodic protection current in the soil, synchronized current interrupters are employed in the cathodic protection units. Pipe-to-soil potentials are then instantly measured after the current is interrupted, before significant chemical depolarization has occurred, thus providing an "error free" polarized potential.

In practice, the precision current interrupters are best fitted in the DC output circuit of the cathodic protection power supplies, prior to commencing the survey to interrupt the cathodic protection current for short cyclic periods. The cycle time is chosen with regard to pipeline characteristics, although a range of "ON" and "OFF" periods and "ON/OFF" ratios is available.

For this survey, a variety of purpose built-voltmeter/microprocessors are available that record and store the pipe to soil potential data with an additional facility to store coded entries of topographical or pipeline features. This information is transferred to floppy disc storage from which graphs of ON and OFF pipe-to-soil potentials against pipeline chainage are produced. A typical close interval potential graph is illustrated in Figure 2.

It will be apparent from the description of the pipeline surveys that the form of results is different for each survey.

Pearson

Results are not generally presented in graphical form. Coating defects are marked in the field; in addition, the detection of defects is dependent upon operator skill. Data can only be presented in graphical form if the recording Pearson technique is used. At present, experience of this technique is limited.

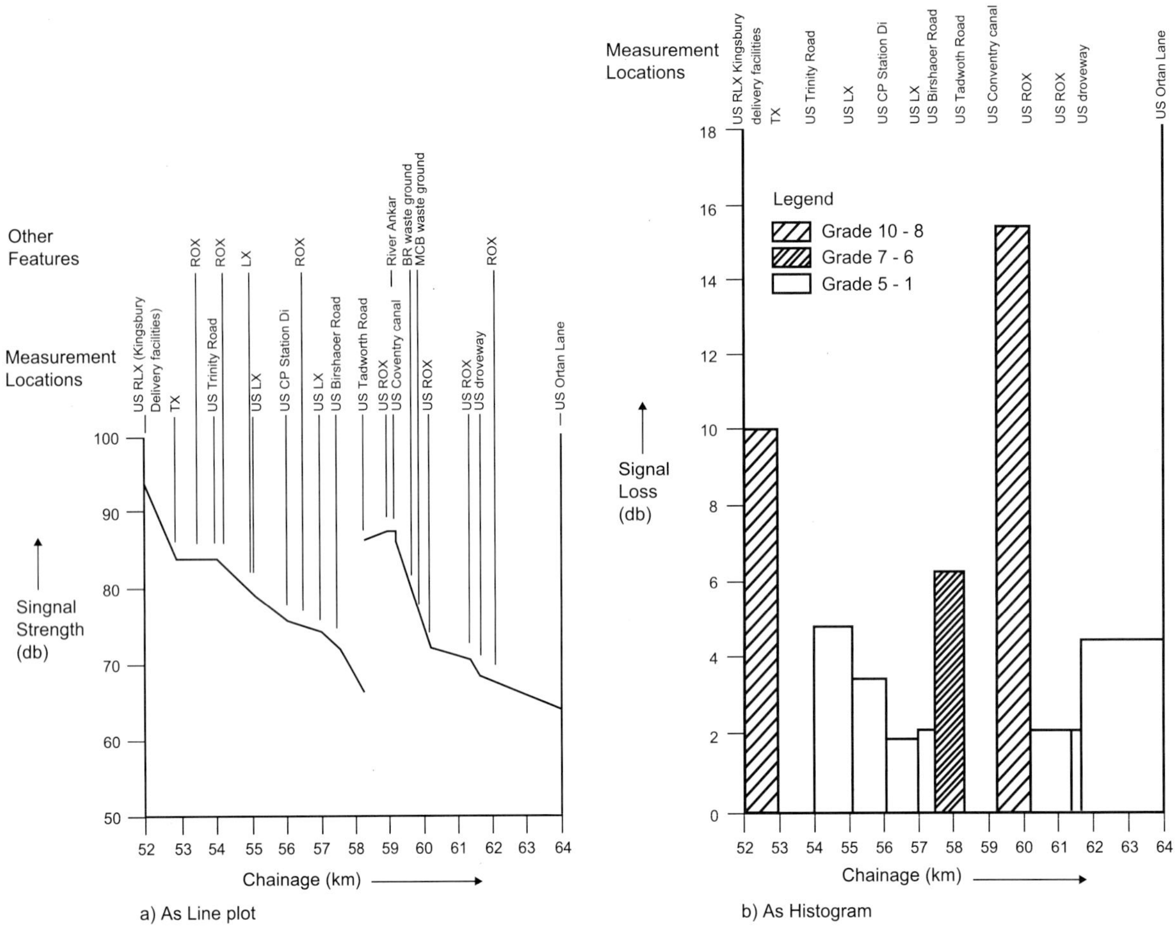

Figure 1. Typical signal attenuation coating graphs.

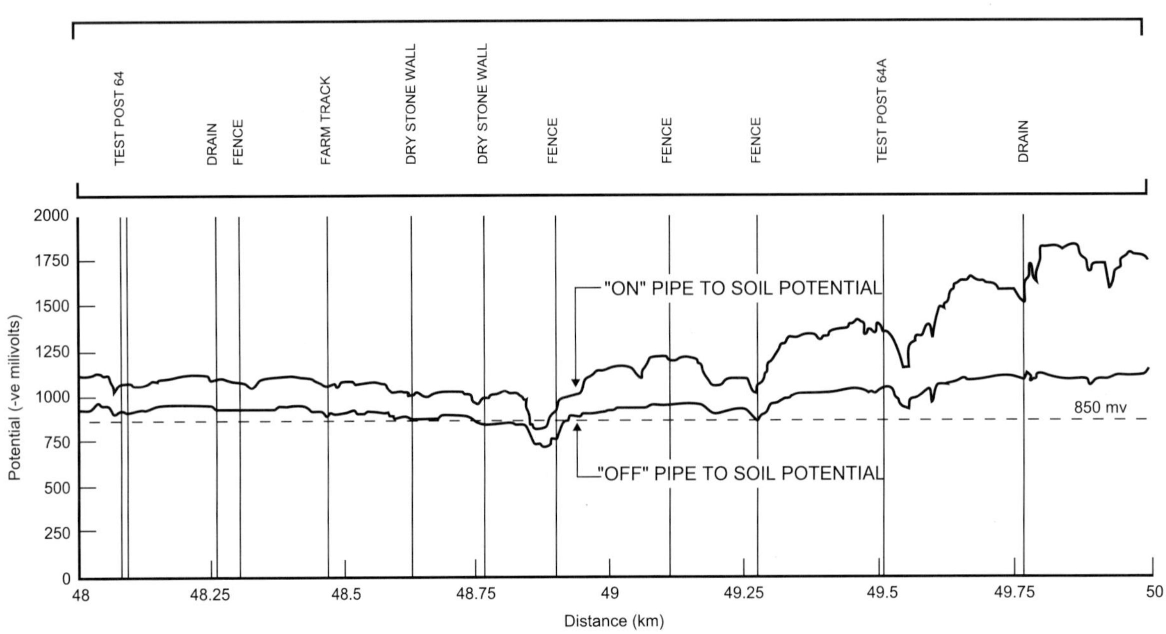

Figure 2. Typical close interval potential graph.

Signal attenuation coating survey

Results are presented in graphical form, although quick analysis may be performed in the field from raw data. Areas of poor coating quality may be readily identified from final graphs.

Following the production of graphs and histograms, an evaluation system employed in pipeline coating quality assessment may assign values to the level of signal loss and the measurement distance interval. The sum of the two values referred to as the "dB factor" and the "km factor" may be used to indicate the future action required and the level of priority for each section surveyed.

The evaluation of the coating quality is illustrated in Figure 1. It should be noted, however, that gradings assigned to one particular pipeline cannot be compared to those assigned to another.

Close interval potential survey

Results are shown in graphical form. Defect location is undertaken after the survey and requires experienced personnel to interpret the results.

The achievement of cathodic protection on the pipeline is indicated by the "ON" and "OFF" potential, both being more negative than the accepted criteria.

Pipeline locations with inadequate cathodic protection will be indicated by the graph "ON" and "OFF" potentials shifting to a less negative level and become cause for concern if the "OFF" potentials are more positive than, say, −0.85 V.

It is considered that a localized coating defect would normally be indicated by a sharp positive shift of the "ON" potential plot coupled with a plot shift of smaller amplitude of the "OFF" potential, i.e., that the potential difference (IR drop) between the "ON" and "OFF" reading decreases, indicating a reduced coating resistance at that point.

The "OFF" potential plot change would not be as large as that of the "ON" potential, as the "OFF" potential would be the polarized (or partially polarized) potential of the bare steel. This differs from the case where insufficient current for the achievement of polarization is available.

In all these cases, the IR drop measured between the "ON" potentials and the "OFF" potentials includes, along with the resistance of the coating, the resistance of the soil. The resistance of the coating is assumed to be the largest component of this resistance value.

Current is also a factor of the IR value. The current density impressed on the pipeline will increase as the cathodic protection stations are approached, giving the largest "*I*" value of the IR drop near the location of the negative connection.

In order to interpret the data fully obtained, the anomalies observed in the plotted results may be classified into a number of categories relating to "point" anomalies or "general area" anomalies. For example, the following categories may be used.

Classification	Definition
Type I	Both the "ON" and "OFF" potentials are more positive than preset criteria. There is little or no "IR" voltage. The potentials are unsatisfactory, and the locations are not cathodically protected.
Type II	The "OFF" potential is more positive than a preset limit, the "ON" potential more negative. There is a reduced level of protection indicated at "point" locations, which may be improved by increased output from the nearest cathodic protection station, although where this occurs over long pipeline lengths, it indicates high-resistance coating or soils.
Type III	Positive change in the "ON" and "OFF" potential plot where both remain more negative than the selected criteria. This indicates increased current demand at these locations, but protection is maintained at a satisfactory level.
Type IV	Unusual potential variations due to outside influences, e.g., stray currents, station earthing, AC induced voltages.
Type V	A change in level of the "ON" potential at a test post due to the change in the length of that part of the measuring circuit passing current, i.e., increasing "IR" volt drop due to increasing nine length in circuit.

Of the five categories listed above, the first three are indicative of coating defects. Types IV and V are commonly observed on close interval potential survey results and require careful interpretation.

It is important that these considerations are born in mind when selecting the survey or combination of surveys to be employed and, in addition, the choice of which survey to employ and when to employ it is dependent on a number of factors, namely:

- the purpose of the survey
- the condition of the pipeline
- the access over the pipeline
- the time allowed.

As the protection system relies on both the external coating and a cathodic protection system, these may be considered as first and second lines of defense against corrosion. It is possible to take the position that concern over damaged coating is not warranted as backup from the cathodic protection system is available, and therefore the integrity of the pipeline is not at risk. If this attitude is taken, then obviously close interval potential surveys would seem the most appropriate and perhaps the only survey system to consider.

Experience has shown, however, that the repair of significant coating defects will benefit the overall level of cathodic protection achieved, possibly permitting reduction in output levels and/or increasing the distance protected from anyone protection station. In addition, the maintenance of a "two-level" protection system also ensures the highest integrity is maintained for the pipeline. With this purpose in mind, the advantages of the coating surveys become apparent.

If we consider the requirements of a newly laid pipeline, it has been considered important to ensure as near perfect a coating as is possible at the start of the pipeline's life. It has therefore been common practice to "Pearson" survey the pipeline after installation. As described earlier, this survey requires a two-man team and commonly progresses at some 4—6 km per day. This procedure may be accelerated by the use of the signal attenuation coating survey. Bearing in mind the high-quality coating of a new pipeline, and the comparison basis employed by the signal attenuation coating survey, even small coating defects will become readily noticeable.

At the other extreme, with poorly coated old pipeline, a Pearson survey may have an almost continuous response, making it impossible to discern individual defects or the most significant coating breakdown. The signal attenuation coating survey would, however, be able to give a rapid assessment of coating condition and indicate where limited rewrapping would be of greatest benefit.

In general, therefore, experience has shown that it is cost effective for the more detailed surveys of coating condition and cathodic protection to be undertaken in conjunction with a signal attenuation coating survey. A number of case histories are described below.

Case histories

1. Pipeline	Multiproducts UK
Length	115 kilometers
Built	1982
Coating	Coal tar reinforced enamel with self-adhesive tape at field joints
Survey	Signal attenuation coating survey
Survey period	Nine days (four days for disconnection of all "temp" magnesium anodes, five days for survey)

Survey results:

Survey indicated that over some 80 km of the pipeline the coating was satisfactory and further survey work was not recommended for those areas. Signal losses over these sections ranged between 1.5—3 dB/km. Some nine sections, amounting to a total of some 10 km were given first priority for further investigation. Signal losses were as high as 17.6 dB over distances of 1.5 km. Two further sections, amounting to some 21 km, were classified as second-priority items.

It was decided that this information gathered be correlated with the results obtained from a proposed intelligent pig survey before further action.

2. Pipeline	White Oil products, UK
Length	115 kilometers
Built	Approx 1963
Coating	Coal tar enamel, tape, or bitumen flood field joints
Survey	Signal attenuation coating survey Pearson survey
	Close interval potential survey
Survey period	10 days—Signal attenuation coating survey
	1 day—Pearson survey
	4 days—Close interval potential survey

Survey results:

Initial signal attenuation coating survey identified 30 locations of significantly high signal losses; of these, 12 were attributed to interpipeline bonds and sacrificial anodes, and the remaining 18 were subject to a more detailed investigation.

Close interval potential surveys showed that a satisfactory level of cathodic protection was being achieved at each of the locations where defects had been pinpointed using Pearson equipment.

Three particular locations were excavated; in the first, a coating defect of some 800 mm associated with a welding arc burn was found within an area previously showing a signal loss of 6 dB over less than 100 m section. In the second area, showing similar signal attenuation, a number of small coating defects were found together with a box of welding rods touching the pipe. At the third location, a badly poured bitumen field joint had been indicated by a signal attenuation loss of 10 dB.

3. Pipeline	Oil, UK
Length	210 kilometers
Built	Approx. 1970
Coating	Coal tar enamel with tape-wrapped field joints.
Survey	Close interval potential survey Recording both "ON" and "OFF" potentials at 5 meter intervals
Survey period	40 days for three man team

Survey results:

Review of all the close interval potential results highlighted some 28 locations where both the "ON" and "OFF" potentials were less negative than the protection criteria. Of these 28 locations, 80% would not have been apparent from potential measurements recorded at adjacent test posts, thereby allowing the integrity of the pipeline to be put at risk where these unsatisfactory levels of cathodic protection coincided with coating defects.

4. Pipeline	Aviation Fuel 6", UK
Length	5 kilometers
Built	1960
Coating	Asphalt
Survey	Signal attenuation survey
Survey period	Two days

Survey results:

The survey showed consistently high signal losses, indicating a generally deteriorated coating. Signal loss values were of the order of 40 dB per km.

More than 50% of the total signal loss was concentrated over four areas, amounting to some 1720 meters of pipe. Correlating close interval potential survey results over the same areas allowed 315 meters to be selected for rewrapping.

5. Pipeline	Crude Oil, Middle East
Length	132 kilometers
Built	1986
Coating	Extruded polyethylene
Survey	Signal attenuation coating survey
Survey period	Four days

Survey results:

Generally, signal losses were found to average less than 1 dB per kilometer. One location was found to indicate a signal loss of 14 dB over 40 m. More detailed signal attenuation survey work located two possible coating defects with signal losses of 10 dB over 20 m and 48 dB over 20 m. A subsequent Pearson survey and two excavations revealed five small defects in a shrink sleeve weld joint coating for the 10 dB loss and one defect less than 100 mm^2 in the factory-applied coating for the 4.8 dB.

Conclusion

It is considered that the discussion and case histories illustrated have shown that each of the aboveground survey procedures offers certain advantages and disadvantages, and that many of the technical and economic penalties of choosing any one specific survey may be overcome if used in conjunction with another.

References

Argent, C.J. et al. "Coating defect location by above ground surveys"-proceedings of UK corrosion, 1986.

CCEJV "Cathodic Protection Monitoring of Buried Pipelines"—State of the Art Report by Task Group E4-2.

Frost N., Monitoring Pipeline Corrosion. *Pipes and Pipelines International*, May 1986.

Glyn J, and Thron, J., *Advances in Pipeline Protection*, Gulf Publishing Co., 1990.

Pearson J., *Petroleum Engineer* 12 (10) (1941).

Reprinted with permission - Paper presented at the 7th International Conference on the Internal and External Protection of Pipes, Cranfield, Bedford, MK43 OAJ, England 1987.

Estimate the number of squares of tape for pipe coating

$$\text{Squares of coating} = \frac{W \times A}{(W - O) \times 100\ ft^2/square}$$

where:

W = Width of coating, inches
A = Pipe surface area, square feet
O = Overlap, inches

Example.
Pipe OD = 6⅝″
Length = 1 mile
Tape width = 6″
Overlap = 1″

Pipe surface area = 1.73442 ft^2/ft

Total surface area = 1.73442 × 5280 = 9157 ft^2

$$\text{Square of coating} = \frac{6 \times 9157}{(6 - 1) \times 100} = 110\ \text{squares}$$

Estimate the amount of primer required for tape

Multiply pipe surface area by 0.001047 to obtain the number of gallons of primer required per mile of pipe.

Example.
Pipe OD = 6 ⅝″

Length = 1 mile
Pipe surface area = 1.73442 ft^2/ft
Total surface area = 1.73442 × 5280 = 9157 ft^2
Gals of primer = 9157 × 0.001074 = 9.8 gals

Tape requirements for fittings

Reducer Size	Tape Width	Linear Feet	Square Feet	Tee Size	Tape Width	Linear Feet	Square Feet	90° LR Ell Size	Tape Width	Linear Feet	Square Feet
3 x 2	2	9.2	1.5	2	2	13	2.2	2	2	9	1.5
4 x 3	3	8.4	2.1	3	2	21	3.5	3	2	11	1.8
6 x 4	4	12	4	4	2	28	4.6	4	3	14.6	3.6
8 x 4	6	14	17	6	3	38	9.5	6	4	26	8.6
10 x 6	6	18	9	8	4	45	15	8	4	40	16.5
12 x 8	6	25	12	10	6	34	17	10	6	48	24
14 x 10	6	26	13	12	6	50	25	12	6	64	32
16 x 12	6	38	19	14	6	60	30	14	6	81	45
18 x 14	6	50	25	16	6	71	35.5	16	6	101	50.5
20 x 16	6	66	33	18	6	85	42.5	18	6	123	61.5
				20	6	100	50	20	6	146	73
				22	6	126	63	22	6	173	86.5
				24	6	132	66	24	6	201	100.5

90° SR Ell Size	Tape Width	Linear Feet	Square Feet	45° Ell Size	Tape Width	Linear Feet	Square Feet
2	2	7	1.5	2	2	5.6	0.93
3	2	10	1.8	3	2	9	1.5
4	3	13	3.25	4	3	13	3.2
6	4	21	7	6	4	17.3	5.75
8	4	38	13	8	4	36	12
10	6	46	23	10	6	34	17
12	6	60	30	12	6	48	22
14	6	76	38	14	6	52	26
16	6	96	48	16	6	64	32
18	6	118	59	18	6	76	38
20	6	136	68	20	6	96	48
22	6	166	83	22	6	110	55
24	6	190	95	24	6	126	63

Induced AC voltages on pipelines may present a serious hazard

Earl L. Kirkpatrick, ELK Engineering Associates, Inc., Fort Worth, TX, and Kirk Engineering Co., Oneonta, Alabama

The problem of induced AC voltages on pipelines has always been with us. Early pipeline construction consisted of bare steel or cast iron pipe, which was very well grounded. Bell and spigot, mechanical, or dresser-style joint couplings were often used, creating electrically discontinuous pipelines that are less susceptible to AC induction. Although induced, AC affects any pipeline parallel to a high-voltage alternating current (HVAC) power line, the effects were not noticeable on bare pipelines. With the advent of welded steel pipelines, modern Cathodic Protection (CP) methods and materials, and the vastly improved quality of protective coatings, induced AC effects on pipelines have become a significant consideration on many pipeline rights-of-way.

In the past 2—3 decades, we have been seeing much more joint occupancy of the same right-of-way by one or more pipelines and power lines. As the cost of right-of-way and the difficulty in acquisition, particularly in urban areas, have risen, the concept of joint occupancy rights-of-way has become more attractive to many utility companies. Federal and state regulations usually insist on joint-use right-of-way when a utility proposes crossing regulated or publicly owned lands, wherever there is an existing easement. Such joint use allows the induced AC phenomena to occur and may create electrical hazards and interference to pipeline facilities. Underground pipelines are especially susceptible if they are well-coated and electrically isolated for CP.

Induced AC mechanisms

There are three coupling modes ofprimary concern by which voltages are induced in pipelines.

Electrostatic coupling, commonly known as "capacitive" coupling, is caused by the electrostatic field surrounding the energized conductor. This electrostatic field is proportionally produced by the voltage in the overhead conductor. The closer the pipeline is to the energized overhead conductor, the stronger the electrostatic voltage induced in the pipeline.

Electrostatic coupling is of primary concern when a pipeline is under construction near an overhead HVAC system. The length of the exposed conductor is also a factor in the electrostatic charge induced in the structure. A relatively short conductor isolated from ground, such as a single joint of line pipe supported by a nonmetallic sling or on a rubber-tired vehicle on the right-of-way, will become charged like a capacitor. (A rubber-tired vehicle on the right-of-way should have drag chains to ground the vehicle).

Once the pipeline is backfilled, electrostatic coupling is no longer of any real concern because the pipeline coating will allow sufficient leakage to earth through pinholes or holidays, which will effectively ground out the electrostatic effects.

Electromagnetic coupling

Electromagnetic coupling is also known as "transformer action." When current flows in an energized conductor, it produces an electromagnetic field at right angles to the conductor. In AC power systems, electric current flowing in the conductor changes direction 120 times per second for 60-Hz and 100 times per second to 50-Hz systems. Thus, the electromagnetic field surrounding the energized conductor is constantly expanding and contracting.

Whenever electromagnetic lines of force cut through another conductor, a voltage is induced in that conductor. This is the principle upon which power transformers, alternators, and generators function. A well-insulated pipeline parallel to an overhead HVAC line then becomes the secondary of an air core transformer. With sufficient parallel length, significant voltages and currents can be inducted into the pipeline. These voltages can be hazardous to anyone who comes in contact with the exposed pipeline or appurtenances. The parallel power lines may induce sufficient power into the pipeline to damage the pipeline or related facilities.

Pipeline hazards

A products pipeline failed because of lightning and fault current actually penetrating the pipeline wall. This was a clear case of arc burn penetration of the pipe wall. There are other case histories where actual pipe wall penetration has occurred with or without a fire being started. People have been killed by contacting an energized pipeline under construction. We must be aware of these electromagnetic and electrostatic effects on influenced pipelines and know how to avoid injury and damage. An excellent source of information on this subject is NACE Standard RP0177-95. Additional information is contained in IEEE Standard 80.

Resistive coupling

Resistive coupling is another mode of electrical coupling between a pipeline and a parallel overhead power system. This is a concern during "fault conditions" on a power system. If lightning strikes one of the energized conductors on the overhead power line, the resulting voltage rise on the wire will exceed the Breakdown Insulation Level (BIL) of the insulator at the nearest tower. When the BIL is exceeded, a flashover will occur from the energized conductor to the tower and then fault current will flow through the tower to the tower ground. Current will flow from the energized conductor to the tower structure via the ionized gases (plasma) generated by the lightning. Fault current will flow through the tower and the tower ground into the earth for a fraction of a second, until the circuit protection device has a chance to operate.

As a result of these fault currents, the current will radiate from the tower foundation and grounds in all directions from the faulted tower. A severe potential gradient will occur across the earth, radial to the faulted tower. If there is a nearby pipeline in the earth, the gradient field will be distorted and accentuated. This effect is greatest with bare pipelines and pipelines that are near the tower foundation. Consequently, a significant portion of the fault current may flow in the pipeline as a result of resistive coupling.

Physical separation between the pipeline and the power line tower is an important factor in determining the extent of damage that will occur on the pipeline as the result of a fault.

Predicting AC voltages on a pipeline

Induced AC voltage on a pipeline can be measured by methods similar to those used to conduct a DC pipe-to-soil cathodic protection potential survey. A multimeter is placed on the appropriate AC voltage range and a steel pin is used in place of the copper/copper-sulfate reference electrode. A detailed AC pipe-to-soil potential survey should be conducted over any area where induced AC voltages are suspected. It is important to note the time of each reading because many power companies maintain a record of currents in the circuit. The voltage induced in the pipeline is directly proportional to the line currents in the overhead conductors. This can provide valuable information to assist in calculating the peak voltages that can be anticipated at each location under maximum line loading conditions.

Whenever a coated pipeline and an HVAC transmission circuit are near each other, the magnetic field associated with the currents in the power transmission line will induce a voltage in the pipeline. The actual magnitude of the induced AC voltage depends on many factors, including the overall configuration of all the structures involved, soil resistivity, pipe coating effectiveness or resistance to remote earth, pipeline propagation constant, magnitude of the line currents in the power circuit(s), and any current imbalance between the phases.

The magnitude of steady-state AC potentials induced on an underground pipeline by parallel high-voltage transmission lines can be accurately estimated using the appropriate mathematical formulas. The formulas characterize the circuit in terms of "steady-state" line currents, phase relationships, pipeline-to-conductor distances, pipeline propagation constants, characteristic impedances, soil resistivity, and other factors. The technique can predict, with reasonable accuracy, the areas where the maximum AC potentials will occur and approximate the actual induced voltage at that point. These calculations require the use of advanced mathematics, which can best be done on a computerized workstation by an experienced professional.

Mitigation techniques

Mitigation techniques that can control induced voltages on an influenced pipeline include the following:

- Supplemental grounding of the pipeline with sacrificial anodes or other grounding means.
- Bonding the pipeline to individual power line pole grounds or towers through the use of polarization cells.
- Installation of parallel mitigation wires bonded to the pipeline at regular intervals.
- Bonding the pipeline to purposefully designed "made" grounds.
- Changing phase relationships between multiple power line circuit conductors.
- Use of Faraday cages with sacrificial anodes.
- Relocation of the pipeline or power line to provide greater separation from the influencing power system.
- Installation of a nonmetallic pipeline such as high-density polyethylene pipe, if design pressures permit.
- Installation of gradient control electrodes or mats at all aboveground appurtenances.
- Security fencing around aboveground appurtenances does not mitigate induced voltages, but they do limit access to the structure.

Whenever there is a pipeline closely parallel to a power line for any significant distance (e.g., closer than 500 ft or more than 2000 ft), further investigation is warranted. If five volts AC or more is measured at any point on the parallel section, this voltage should be related to line loading to estimate the pipeline-induced AC voltage at peak line-loading conditions. The induced voltage should be mitigated if the calculated values approach or exceed 15 volts AC. Fault conditions and step-and-touch voltages also must be considered.

Personnel safety

On any construction or maintenance project, safety is an attitude. This attitude is developed by proper training. A valuable source for such information is Section 4 of NACE Standard RPO177. NACE also offers an audiovisual presentation that deals with the effects of electric shock in such situations. RP0177 should be included in the construction specifications whenever a pipeline is built or exposed for maintenance on an energized HVAC right- of-way. One of the inspectors should be designated in charge of electrical safety. This individual must be familiar with and properly equipped to test for safe levels of induced voltage in the pipeline. A safety meeting should be held prior to construction with all construction employees to discuss electrical safety requirements.

The right-of-way must be vacated if thunderstorms come within 10 miles of the pipeline right-of-way or the power line(s) that are influencing the pipeline. Never work on an influenced pipeline during a thunderstorm because of the potential for a direct lightning strike. Power system fault currents will flow in the earth if lightning causes flashover of a power line insulator. A person on the right-of-way is in danger, even if they are not touching the pipeline.

Corrosion and maintenance personnel should be very cautious about stringing test lead wire on the right-of-way where the lead wire parallels the influencing HVAC line. Extremely hazardous voltages can be induced on a significant length of test lead wire laid on top of the ground.

The practice of using insulating rubber gloves should also be discouraged unless the work is being performed by a trained power company employee. "Rubber Goods" must be specially cared for and tested to ensure their reliability.

The only safe alternative in the pipeline environment is to test the level of induced AC on the pipeline before contacting anything that may be a conductor. If voltages are safe, normal measurement or repair and maintenance techniques may be employed. If unacceptably high voltages are encountered, one must work on a ground mat. A ground mat can be as simple as a piece of 6-ft road mesh laid on top of the ground and bonded to the pipeline or appurtenance with an automotive jumper cable. Since the individual, the ground mat, and the pipeline appurtenance are all at equal potential, it matters little as to how high the actual measured voltage is on the pipeline.

Reprinted with permission—*Pipeline & Gas Journal.*

Minimizing shock hazards on pipelines near HVAC lines

As right-of-way for pipelines becomes more difficult to obtain, especially in congested areas, rights-of-way are being shared by pipelines and high voltage transmission lines. Sharing a common right-of-way can pose problems during operation, and especially during pipeline construction unless special precautions are taken.

Considerable voltage can be induced in pipelines aboveground on skids unless the section of pipeline is grounded. Without grounding, a lethal shock hazard to construction personnel can occur. To minimize the voltage on the pipe, drive ground rods at each end of the section of aboveground pipe. The ground rods should be driven at least 4 ft in the ground and connected to the aboveground section of pipe with insulated copper wire and a C-clamp. If the aboveground section is longer than 1/2 mile, additional ground rods should be driven and connected to the pipe.

Work should be discontinued if thunderstorms are occurring in the vicinity (approximately 10 miles) of the pipeline construction. If lightning should strike the power line, even at a distant location, while personnel are in contact with the pipe, they could suffer a severe shock or even electrocution. To eliminate this danger, personnel should always stand on a 3-ft × 3-ft (minimum) steel mat made of 1″ square (maximum) hardware cloth that is electrically well grounded to the nearest electrical tower ground via a copper cable. The mat should be fabricated so that personnel can easily move it from tower to tower. This can be done by bolting a suitable length of flexible welding cable firmly to the metallic mat. The other end of the cable should be equipped with a suitable copper or brass clamp that can easily be connected to the tower ground. Personnel should always place the mat on the ground and stand on it while connecting the cable clamp to the tower ground. When work is complete, personnel should remain standing on the mat while disconnecting the cable clamp from the tower ground wire. This will make it safe to carry the mat to the next tower ground connection site.

When tie-ins are made, additional personnel protection is required. Use a metallic mat or driven ground rod connected to the pipe at each end of the tie-in, and connect an insulated copper cable (bond) across the tie-in point. The pipe and ground mat will be essentially at the same potential if a high-voltage surge occurs during the tie-in operation. The bond will prevent buildup of voltage between the unconnected ends at the tie-in location.

After the pipeline has been buried and placed in service, personnel should always connect a bonding cable across any location where the pipe is to be separated, i.e., a valve to be removed or a section of pipe is to be removed and replaced. The bonding cable should remain in place until all connections have been completed.

Equipment with steel tracks will not require any special grounding procedures for traveling along the power line right-of-way, but machinery with rubber tires should have a heavy-duty chain attached to the vehicle frame with a good electrical connection so that approximately 3 ft of chain will drag along the ground when moving or standing along the power line right-of-way. Any time there is a possibility of a lightning strike, personnel should not grasp any rubber-tired machinery while standing on the ground. Personnel should hop on or off when mounting or dismounting. Care should be exercised to prevent injury.

Corrosion of low-velocity, high water cut oil emulsion pipelines

Michael F. McDonald, CORR_X Engineering Ltd., Calgary, AB, Canada

In addition to water cut, the effect of fluid velocity has long been recognized as a key contributing factor to corrosion in oil and gas pipelines. Low velocities encourage water hold-up and the accumulation of solid debris. This condition can cause accelerated attack from concentration cells such as under-deposit corrosion. Likewise, high velocities can lead to Flow-Induced Localized Corrosion (FILC).[1] In this case, corrosion is caused by the formation of corrosion products, in combination with critical sheer stresses when the "erosional velocity"[2] is reached and exceeded. This mechanism is another problem and it is the author's experience that for many years, corrosion personnel have pondered how to determine a criterion for selecting pipelines in low-velocity, high water cut oil emulsion systems in which to apply corrosion mitigation measures. Ideally, utilizing a well-designed, field-wide corrosion monitoring program will greatly assist in determining the corrosion potential of the lines and assist in prioritizing lines in terms of failure risk. It is often difficult to justify the cost of such a monitoring program, however, because of monitoring costs, the number of lines in a field, and the average oil production for each line.

For fields in which pipeline corrosion is a serious issue and requires remedial action, most operators select a simple solution, usually based on failure history and water cut; they apply a threshold water cut value in which all pipelines at or over this value receive corrosion mitigation measures. Investigation of internal corrosion failures in several oil systems indicates that applying a water cut criterion to prioritize corrosion-prone pipelines is an overly simplistic and conservative approach to cost-effective corrosion control.

Significance of fluid velocity

Low velocities encourage water hold-up and the accumulation of solid debris. This condition can cause accelerated attack from concentration cells such as under-deposit corrosion. Likewise, high velocities can lead to FILC.[1] In this case, corrosion is caused by the formation of corrosion products, in combination with critical sheer stresses when the "erosional velocity"[2] is reached and exceeded. This mechanism is another problem and it is the author's experience that corrosion caused by very low velocities in oil pipelines is a more common problem in onshore production facilities than corrosion in high-velocity pipelines. The challenge, therefore, is to combine both parameters—water cut and velocity—into a practical corrosion mitigation model for prioritizing pipelines in a field with similar fluid properties.

Development of the strategy

Several years ago, the author was involved in developing a corrosion mitigation strategy for a large oil-gathering system in southeast Alberta. The field had experienced more than 10 internal corrosion pipeline failures during a three to four year period. The field consists of 10 batteries and ~150 producing wells, with the attendant flow lines and group lines, the majority of which are internally bare steel pipelines. The fluid properties are fairly common throughout the field and are characterized by fairly high levels of the HCO_3^- anion (~5000 mg/L) and low levels of the Cl^- anion (~2500 mg/L). This observation supports the premise that the corrosion being experienced in the oil flow lines and group lines in the field was caused by HCO_3^- anions and carbon dioxide (CO_2) partial pressure (~30 psig).

Although the Cl^- anion concentration level is relatively low, thereby reducing the overall electrolyte strength, the HCO_3^- anion concentration is roughly five to 10 times the concentration typically found in oilfield produced waters. Work by Schmidt, et al.[3] reported that bicarbonate ions will actually initiate and accelerate the CO_2 corrosion mechanism believed to be the primary cause of corrosion in the field. It is believed that the high concentration of bicarbonate anions cause localized precipitation of scales at the bottom of the pipe and facilitates an under deposit mode of attack.

The previous corrosion mitigation strategy employed by the operator involved installing High-Density Polyethylene (HDPE) liners in pipelines that experienced internal corrosion failures, but were still considered economically viable. The operator wished to upgrade corrosion control in this field by utilizing a proactive, yet cost-effective approach and requested the pipeline integrity consultant to create a "liner priority matrix" for the ~150 internally bare steel pipelines in the field. This "liner priority matrix" consisted of a schedule in which 10—12 pipelines would be budgeted for lining every quarter. This staged lining approach permitted acceptable budgetary planning within operating expense limits; if engineered and implemented correctly, it was intended to provide adequate corrosion mitigation combined with tolerable cost control in a large, mature gathering system with a serious internal corrosion control problem.

The challenge in developing a sound, reliable method for prioritizing the lines was to establish a relationship between the parameters that are variable from line to line. These variable parameters were water cut, fluid velocity, and whether or not a particular line had failed from internal corrosion. All other corrosion risk factors such as fluid properties and the conditions and age of the lines were fairly similar and as such would be considered irrelevant. A series of graphs were prepared showing the water cuts and fluid velocities of the lines that were known to have failed from internal corrosion over the last four to five years. The graphs were plotted using various scales and normal, semi-log, and log formats.

Next, the water cut and velocities of the other internally bare steel lines in the field were plotted on the same graphs. The lines that fell outside the "failure risk curve" were judged to be at high risk of internal corrosion failure. These lines were designated "high risk priority 1". The lines that fell directly on the curve were designated "high risk priority 2". The lines that fell just inside the curve were designated "moderate risk priority 3".

The long-term plan was to review the production of all wells in the field every one to two years to determine if the updated water cuts (which generally increase) and fluid velocities (which usually decrease provided no changes occur, such as installing new pumps) of the lines alter their position on the "failure risk curve."

Effect of hydrogen sulfide and water chemistry

The operator had hydrogen sulfide (H_2S) measurements taken on the solution gas from most wellheads in the field, including the fluid from all failed lines. The range of H_2S concentrations varied from 20 to 200 mg/L and no correlation between failures and H_2S levels was in evidence. It is strongly suspected that the higher than normal bicarbonate levels (typically 5000 mg/L) in the produced water is the dominant species in contributing to the corrosion mechanism. The effect of the bicarbonate ion in the corrosion process is well documented.[3]

Validation of the corrosion mitigation strategy

The development of the "failure risk curve" and "liner schedule matrix" were accepted by the operator. When the internal corrosion potential for this field was revisited by the author several years later, it was discovered that of the 13 lines originally deemed "high risk priority 1," seven lines were later shut-in because of declining production, three lines were lined with HDPE, and three lines failed. These results validated the strategy (virtually 100% correlation) and the relationship among water cut, fluid velocity, and failures. This approach has since been applied to several smaller fields, and the results have indicated that, to the author's knowledge, one of these fields has since experienced a single failure; this occurred on a pipeline that fell into the "high risk priority 1" category.

Limitations of the strategy

It is not proposed to utilize this strategy for all oil fields with a documented internal corrosion failure history. This strategy can be applied only to fields that have similar fluid properties and conditions throughout the field, and the ages of the pipelines should ideally be about the same. If the ages of the lines vary considerably, a certain amount of judgment is required in prioritizing the lines. Furthermore, if a pigging program is in place, then all lines in question should have the same pigging frequency. In addition, this strategy should be employed with great care if any type of corrosion inhibition program is being applied. If, however, a failure risk curve can be generated that looks somewhat predictable (such as an exponential curve) regardless of any previous or current corrosion mitigation measures in place, then the operator can predict with some degree of confidence the corrosion behavior of all pipelines in the field that have similar fluid properties.

The above strategy has been applied only to systems under primary production. It is not known if the corrosion behavior of lines under secondary or water flood production would be as predictable as described in this article. If, however, a predictable "failure risk curve" can be generated for the lines (including group lines), this strategy may be employed, with some caution initially, to prioritize the lines as to failure risk.

Other potential applications of this strategy

The application of a similar strategy on other types of systems, such as low-velocity sweet gas lines, where it is not economical to employ a comprehensive, field-wide monitoring program, should be investigated. Although fluid velocity will undoubtedly be one of the parameters, the parameter representing water production would need to be thought out and a certain amount of trial and error experimentation would be required.

References

1. W. H. Swain, "Measuring Unwanted Alternating Current in Pipe," W. H. Swain Co. Web site: www.SwainMeter.com.
2. R.A. Gummow, et al., "AC Corrosion—A Challenge to Pipeline Integrity, MP 38 (1999) 2.
3. Private communication from B, Sheehan at San Diego County Water Authority (August 28, 2000).

Pipeline external corrosion direct assessment process

David H. Kroon, Olagoke Olasisi, Larry G. Rankin, James T. Carroll, Dale D. Lindemuth, and Marlane L. Miller,
Corrpro Companies, Inc., Houston, Texas

For cased pipe, improvements to the External Corrosion Direct Assessment (ECDA) process include identification of applicable new technologies, development of techniques for the application of existing technologies, and creation of a threat matrix for cased pipes. The research led to a recommended ECDA methodology for cased pipes.

The purpose of the cased pipe study was to determine the applicability of existing and emerging technologies to assess cased pipes for external corrosion using ECDA. This included examination of existing ECDA processes, best practices of pipeline operators, and emerging technologies. The project findings are significant for gas transmission pipeline operators in the United States because the integrity of all pipe in High-Consequence Areas (HCA) must be assessed by December 17, 2012, including those segments of pipe in casings. There is an industry need for a methodology to assess cased pipe where in-line inspection and pressure testing are neither possible nor practical. There are currently a number of emerging technologies that could be used on cased pipe such as:

- Electromagnetic wave
- Pulsed eddy current
- Conformable array
- Magnetometer electrochemical impedance spectroscopy
- Robotic cameras and sensors

Although these technologies are promising, they will likely take years to develop. Both past and current research and testing have demonstrated that a CP current can reach cased pipes when an electrolyte is present in the annulus between the pipe and casing, even if the casing is electrically shorted to the pipe.[1] Consequently, standard CP and protective coating survey techniques should provide valuable information regarding the condition of the protective coating and the level of CP on pipe within a casing, recognizing the limitations that may be created by shorted conditions, coated casings, and no filling, intermittent filling, or partial filling of the annulus with groundwater.

A matrix was developed to identify external corrosion threats for pipes in casings. The threat matrix can be used to prioritize cased pipe for assessment. It incorporates construction, operations, and maintenance parameters such as:

- Electrical isolation
- CP
- Groundwater level and fluctuations
- Annulus atmosphere
- Age
- Pipe coating
- Casing coating
- High temperature

The developed cased-pipe ECDA methodology follows the general steps in the process delineated in NACE SP0502-2008.[2]

Preassessment

The preassessment step requires compiling ad analyzing construction, operation, and maintenance information for the cased pipe under study. Annual CP survey, including pipe-to-soil (P/S) and casing-to-soil potential measurements and Close Interval Survey (CIS) should be reviewed. Possible external corrosion mechanisms should be identified and ECDA regions defined. A feasibility analysis needs to be performed to include a critical determination of conditions that could prevent or alter electrical measurements on the pipe within the casing, such as casings that are electrically shorted to the pipe, coated casings, and lack of electrolyte within the casing. The ECDA Indirect Inspection (IDI) tools must then be selected to accommodate pipe and casing conditions.

Indirect inspection

The IDI tools for cased pipe ECDA are the same as those used for direct buried pipe. A minimum of two tools are required. At least one tool should address the protective coating condition and defects, and the other should evaluate CP levels and detect CP indications with CIS. IDI tools used for cased pipe are primarily CIS, direct current voltage gradient (DCVG), alternating current (AC) attenuation (ACCA), and AC voltage gradient (ACVG). The methodology requires that IDI be performed over the entire length of cased pipe plus 300 ft (91 m) beyond both ends.

For this project, data were collected and analyzed for more than 200 cased pipes and adjacent uncased pipe for six transmission and distribution system operators.

The data from the 200 cased pipes that were tested clearly indicate that IDI surveys can be used for ECDA of pipelines within casings, subject to certain limitations. Further, the research data were used to develop the IDI severity ranking.

Direct examination

Referencing Table 2, all cased pipes with an action priority of "immediate" are to be examined within the period to time commensurate with the indicated conditions. Cased-pipe locations prioritized as "schedule" are to be examined before the next integrity assessment. Cased-pipe locations with action priority "monitor" are to be monitored at appropriate time intervals until the next integrity assessment is performed. No action is necessary for those locations indicated as such. Direct examination techniques include visual inspection, ILI, pressure test, Guided Wave Ultrasonic Testing (GWUT), electromagnetic wave, and other acceptable techniques either individually or in combinations.

Post assessment

Evaluations of the direct examination results of cased pipes with action priorities of "immediate" and "schedule" determine the need to perform additional direct examinations on those cased-pipe segments with a priority of "monitor." Reassessment intervals are then set independently of adjacent, uncased, buried pipe. A very rigorous evaluation of the effectiveness of the ECDA methodology for the cased-pipe segments studied should then be performed.

Table 1
IDI Severity Ranking

IDI Survey Tools	Severity Classifications			
	None	Minor	Moderate	Severe
AC attenuation survey	Uniform attenuation profile with no significant changes	Small change in attenuation profile over short length of pipe inside or near casing	Moderate change in attenuation profile over short length of pipe inside or near casing	Large change in attenuation profile over short length of pipe inside or near casing
DC or AC voltage gradient survey	No indications on adjacent buried pipe and no indications on cased pipe	Few indications on adjacent buried pipe but no indications on cased pipe	Several indications on adjacent buried pipe but no indications on cased pipe	Numerous indications on adjacent buried pipe or indications on cased pipe
Close interval potential survey	Uniform potential profile with no significant depressions and all potentials more negative than -850 mV	Minor potential depressions but all potentials more negative than -850 mV	Moderate potential depressions but all potentials more negative than -850 mV	Large potential depressions or any potentials less negative than -850 mV

Table 2
Action Priorities for IDI Indications

Action Prioritization for Cased Pipe Segments Based on IDI Survey Severity Classifications	**CP Severity Classifications Based on Close Interval Potential Survey Results**					
	No Indications	**Minor Indications**	**Moderate Indications**	**Severe Indications**		
Coating condition severity classifications	Based on AC current attenuation survey results	No indications	No Action	Monitor	Schedule	Immediate
		Minor indications	Monitor	Monitor	Schedule	Immediate
		Moderate indications	Monitor	Schedule	Schedule	Immediate
		Severe Indications	Schedule	Schedule	Immediate	Immediate
	Based on DC or AC voltage gradient survey results	No indications	No Action	Monitor	Schedule	Immediate
		Minor indications	Monitor	Monitor	Schedule	Immediate
		Moderate indications	Monitor	Schedule	Schedule	Immediate
		Severe Indications	Schedule	Schedule	Immediate	Immediate

Results

The cased-pipe portion of the project concluded:

- ECDA can be used to evaluate the integrity of cased pipe in accordance with a recommended methodology.
- The methodology makes use of ECDA IDI survey techniques currently being used for uncased, buried pipe, as part of the process for identifying and ranking direct examination priorities and selecting the most effective assessment tools.
- Direct examination techniques include visual inspection ILI, pressure test, GWUT, and electromagnetic wave either individually or in combinations.
- It is possible for cased pipe with a water-filled annulus to be categorized as "no action," required, or "monitor."
- There are limitations to the use of ECDA as a tool for integrity management of cased pipes that need to be recognized and understood.

Severity ranking

The purpose of the severity ranking portion of the study is to enhance the existing Tables 3 and 4 in NACE SP0502-2010.[1] The existing tables in the standard are very general, which result in varying interpretations and inconsistencies in application under the current practice. The project goals were to identify improvements that could be made and develop an enhanced severity ranking methodology.

Data from five transmission and distribution system operators were compiled, sorted, and analyzed. This included the results of 400 direct examinations with complete, applicable data sets, including soil analysis. Fifty percent (200) of the data sets used in the study demonstrated measurable external corrosion. ILI data were also analyzed, which covered 14,000 joints of pipe where CIPS and ACCA surveys had been performed. These data included 4000 joints of pipe with measureable corrosion and 100 excavations. To capture best current industry practices, this phase of the project was discussed with 10 qualified and experienced operator and service provider professionals with a total of over 300 years of experience.

In developing improvements to Tables 3 and 4 in NACE SP0502-2010, specific numerical criteria were developed, which covered a wide range of definable conditions. The work included analysis of Rupture Pressure Ratio (RPR) and percent wall loss relative to above grade measurements and individual IDI indications. During the course of evaluating soil data at IDI indications, it was noticed that soil conditions appeared to correlate with the severity ranking as defined by the numerical ranges. This, of course, is consistent with what we have all observed, but have only loosely been able to tie to specific soil properties such as soil resistivity, pH, active ion concentration, and moisture content. A broader

Table 3
Enhanced Severity Classification Criteria of Indirect Inspections (IDI) (Modification of Table 3 in NACE SP0502-2010)

IDI Classification			
Measure	**Minor**	**Moderate**	**Severe**
	IDI Tool = Close Interval Potential Survey		
A = off (polarized) potential (mV)	−950 mV < A < −850 mV OR	−950 mV < A < −650 mV OR	−650 mV < A AND
B = on potential (mV)	−1.000 mV < B < −950 mV	−950 mV < B < −850 mV	−850 mV < B
C = on/off convergence (mV)	50 mV < C < 70 mV	30 mV < C < 10 mV	10 mV < C
D = on and/or off profile depression within 100 ft (30.5 m) (mV/span)	50 < C < 100 mV/span	100 < D < 200 mV/span	200 mV/span < D
	IDI Tool = AC Current Attenuation		
E = current 98 Hz frequency signal loss (-) (mdB(mA)/ft)	7 < E < 3 mdb(mA)/ft AND/OR	12 < E < 7 mdb(mA)/ft AND/OR	12 mdb(mA)/ft < E AND/OR
F = current 4 Hz frequency signal loss (-) (mdb(mA)/ft	20 < F < 40 mdb(mA)/ft	40 < F < 60 mdb(mA)/ft	60 mdb(mA)/ft < F
CP level modifier	Adequate CP level	Adequate to marginal CP level	All indications with inadequate CP level
	IDI Tool = AC Voltage Gradient		
G = voltage signal loss (-) 9 dB(mV)	44 dB(mV) < G < 60 dB(mV) AND	60 dB(mV) < G < 78 dB(mV) AND	78 dB(mV) < G OR
CP level modifier	Adequate CP level	Adequate to marginal CP level	All indications with inadequate CP
	IDI Tool = DC Voltage Gradient		
H = coating defect size (% IR)	5% IR < H < 20% IR AND	20% IR < H < 50% IR OR	50% IR < H OR
I = corrosion state assessment (normal conditions)	I = Cathodic/cathodic or cathodic/neutral AND	All indications 5% IR < H < 50% IR where I = cathodic/anodic AND	All indications where I = anodic/anodic OR
CP level modifier	Adequate CP level	Adequate to marginal CP level	All indications with inadequate CP level

Table 4
Enhanced Prioritization Criteria for Indirect Inspection Indications (Modification of Table 4 in NACE SP0502-2010)

USDA Soil Texture Modifier	IDI Tool 2 Classification	IDI Tool 1 Classification		
		Severe	Moderate	Minor
Severe	Severe	Immediate	Immediate	Scheduled
Severe	Moderate	Immediate	Scheduled	Scheduled
Severe	Minor	Scheduled	Scheduled	Monitored
Moderate	Severe	Immediate	Scheduled	Scheduled
Moderate	Moderate	Scheduled	Scheduled	Monitored
Moderate	Minor	Monitored	Monitored	Monitored
Minor	Severe	Immediate	Scheduled	Monitored
Minor	Moderate	Scheduled	Monitored	Monitored
Minor	Minor	Monitored	Monitored	Monitored

characterization of soil "texture" was postulated to provide a better indicator of corrosive conditions. This correlation was investigated using the U.S. Department of Agriculture (USDA) Web Soil Survey,[2] which provides significant detail regarding soil texture and physical/chemical properties across the United States. It has the additional advantage of being easily accessible on the Internet, free for anyone to use. The investigation concluded that percent clay content (expressed as a percentage of total composition) was a parameter that correlated with the presence of external pipeline corrosion.

The data were then further analyzed by ranking the leak and rupture hazards as a percent of clay for the data set. The threat of leaks was indicated by wall loss while the threat of rupture was expressed as rupture pressure ratio. The data clearly illustrated that as the percent clay in the soil increases, so does the threat of both rupture and leak. A soil modifier was then applied to the Severity Classification Criteria of Indirect Inspections and the Prioritization Criteria for Indirect Indications.

Results

Improved tables for Severity Classification and Prioritization Criteria for Indirect Inspection Indications were developed, which provide a more consistent assessment of the external corrosion threat.

Methodologies developed represent an enhancement to NACE SP 0502-2010 for quantification and qualification of IDI indications, effective use of available soils data, and introduction of a soil texture modifier.

The new methodologies quantified and verified both the project research data and industry knowledge and experience.

Potentials in paved areas

Current industry practices for collecting potential measurements in paved areas are to drill through the pavement, collect potentials offset from the location of the pipeline, surface wetting, or simply skipping data collection in paved areas. The purpose of the potential in the paved areas portion of the project was to develop a methodology to collect more reliable data in a more user friendly, efficient, and safe manner. The goal was for the methodology to be applicable to both transmission and distribution systems, and ultimately to provide for more data collection in paved areas, thereby enhancing pipeline integrity assessment.

We considered gravel, asphalt, and concrete surfaces. Variability in thickness, aggregate, sub-base, and construction yields a nearly infinite number of conditions. If basic electrical measurements could characterize a pavement, then decisions and guidelines could be developed regarding the validity of potential measurements with reference electrode placement on the pavement. As the research progressed, this postulation was tested and refined. The result was a simple test procedure that can be used at the onset of a potential survey to determine if on-paving measurements can be made accurately.

The research approach consisted of reviewing prior work, running large scale laboratory tests, and collecting field data on operating distribution and transmission pipelines.

Resistance measurements to characterize the pavement were made using a digital meg-ohmmeter having a maximum 1000-V DC source specifically manufactured for high-resistance circuits. Most measurements were made with one terminal of the meg-ohmmeter connected to an 8-in × 8-in (203–mm × 203–mm) metal plate electrode on the paved surface and the other terminal collected to a nearby electrical ground used as an earth electrode. No surface wetting was done for these measurements, other than to use a damp towel directly under the metal plate electrode.

Many CIS surveys were performed in the field that compared current on and instant-off pipe-to-soil (P/S) potential measurements with the cell placed on dry pavement, wet pavement, or drilled holes through the pavement.

For gravel and asphalt, a procedure has been developed for measurement of the resistance through the pavement using a metallic electrode on the paved surface and a MU resistance meter. Based on analysis of the data collected, a threshold normalized resistance of 2×10^5 Ω-ft^2 has been established. That is, when gravel or asphalt paving exhibits a resistance of $200,000$ Ω-ft^{2}or less, a reliable potential measurement can be made with the reference electrode on the pavement.

For concrete pavement, the research concludes there is no clear, consistent method for making reliable P/S potential measurements without placing the reference electrode in direct contact with the underlying soil (e.g., by drilling holes through the pavement), P/S potentials with the reference electrode on a concrete surface are either more negative or more positive than when in contact with the underlying soil. While P/S potential measurements are not valid with a reference electrode on the concrete pavement, DCVG measurements may be.

Results

For gravel and asphalt pavement:

A simple, straightforward, pre-survey surface resistance measurement can be used to determine ifpotentials recorded with the reference electrode placed on the pavement will provide accurate data.

A threshold of $200,000$ Ω-ft^2 has been identified, below which potentials on pavement demonstrated accuracy.

A standard 3-in (76-mm) diameter reference electrode with a wetted towel or sponge is adequate to minimize contact resistance.

For concrete pavement:

No clear, consistent method for recording accurate potential measurements on concrete pavement was identified except by drilling through the pavement to facilitate reference electrode contact with the underlying soil.

References

1. Rankin L. R., and H.M. Al Manrous, External Corrosion Probability Assessment for Carrier Pipes Inside Casings. Gas Technology Institute, March 2005, GRI 05/0020.
2. NACE Standard SP0200-208, Steel Cased Pipeline Practices. NACE International, Houston, Texas, 2008.
3. NACE Standard SP0169-2007, Control of External Corrosion on Underground or Submerged Metallic Piping Systems. Texas NACE, Houston, 2007.
4. NACE SP0502-2010, Pipeline External Corrosion Direct Assessment Methodology. Texas NACE International, Houston, 2010.
5. U.S. Department of Agriculture Web Soil Survey. http://websoilsurvey.nrcs.usda.gov/app/HomePage.htm.

Bibliography

J. Carroll, D. Lindemuth, M. Miller, O. Olabisi, L. Rankin, "PHMSA-Sponsored Research; Improvements to ECDA Process Severity Ranking." Corrosion 2010, paper no. 10054. Texas NACE, Houston, 2010.

J. Carroll, D. Lindemuth, M. Miller, O. Olabisi, L. Rankin, "PHMSA-Sponsored Research; Improvements to ECDA Process Severity Ranking." Corrosion 2010, paper no. 10055. NACE, Houston, Texas, 2010.

J. Carroll, D. Lindemuth, M. Miller, O. Olabisi, L. Rankin, "PHMSA-Sponsored Research; Improvements to ECDA Process Severity Ranking." Corrosion 2010, paper no. 10056. NACE, Houston, Texas, 2010.

"External Corrosion Probability Assessment for Carrier Pipes Inside Casings (Casing Corrosion Direct Assessment—CCDA)" gri-05/0020, Des Plaines, Illinois, Gas Technology Institute.

NACE SP0200-2008, Steel-Cased Pipeline Practices. NACE, Houston, Texas, 2008.

9: Gas—General

Know the gas laws 354
Calculate gas properties from a gas analysis 356
Physical properties of selected hydrocarbons and other chemicals and gases 360
How to calculate the specific gravity of natural gas 368
What are the four different types of heating value? 368
How to calculate the heating value of natural gas 369
Nomograph for calculating density and specific volume of gases and vapors 370
Considerations for selecting energy measurement equipment 371
Facts about methane and its behavior 377
Conversion table for pure methane 381
Introduction to LNG 382
Glossary of common gas industry terms 396

Know the gas laws

The symbols that will be used in the mathematical expression of the various gas laws are:

V_1 = Volume of the gas in cubic ft with original conditions
V_2 = Volume of gas in cubic ft under the second set of conditions
T_1 = Initial temperature of the gas in °R (°F + 460)
T_2 = Temperature of gas under second conditions in °R
P_1 = Pressure of gas in psia under original conditions
P_2 = Pressure of gas in psia under second conditions

Any other symbols used in mathematically expressing the gas laws will be explained at the time of their introduction.

Boyle's law. If the temperature remains the same, the volume of a given quantity of gas will vary inversely as the absolute pressure. This may be expressed mathematically as:

$$\frac{V_1}{V_2} = \frac{P_2}{P_1}$$

Inasmuch as in the application of Boyle's Law we are generally interested in the volume at a second set of pressure conditions, a rearrangement of the formula more readily used is:

$$V_2 = V_1 = \frac{P_1}{P_2} \tag{1}$$

Example. A quantity of gas at a pressure of 42 psig has a volume of 1,000 cubic ft. If the gas is compressed to 100 psig, what volume would it occupy? Assume the barometric pressure to be 14.2 psia and the temperature to remain the same. Substituting in the second arrangement of the formula for Boyle's Law above would give:

$$V_2 = 1,000 \times \frac{42 + 14.2}{100 + 14.2}$$
$$= 492.1 \text{ cubic ft}$$

Charles' law. (sometimes called Gay-Lussac's Law). If the pressure remains the same, the volume of a given quantity of gas will vary directly as the absolute temperature.

This may be expressed mathematically as:

$$\frac{V_1}{V_2} = \frac{T_1}{T_2}$$

Again, since we are usually more interested in the volume at a second set of temperature conditions than any other information, an arrangement of the formula that would be handy is:

$$V_2 = V_1 \times \frac{T_2}{T_1} \tag{2}$$

A second part of Charles' Law is: If the volume of a quantity of gas does not change, the absolute pressure will vary directly as the absolute temperature.

Expressed mathematically this is:

$$\frac{P_1}{P_2} = \frac{T_1}{T_2}$$

In this instance, we would probably be more interested in the pressure at a second temperature condition, which could be expressed as:

$$P_2 = P_1 \times \frac{T_2}{T_1} \tag{3}$$

Example. A given weight of gas has a volume of 450 cubic ft when the temperature is 45°F and the pressure is 10 psig. If the pressure remains the same, but the temperature is changed to 90°F, what will be the volume of the gas?

Table 1
Molecular Weights of Hydrocarbons and Other Compounds Associated with Natural Gas

Compound	Atomic Formula	Molecular Weight
Methane	CH_4	16.043
Ethane	C_2H_6	30.070
Propane	C_3H_8	44.097
Butane	C_4H_{10}	58.124
Pentane	C_5H_{12}	72.151
Hexane	C_6H_{14}	86.178
Heptane	C_7H_{16}	100.205
Carbon Dioxide	CO_2	44.011
Nitrogen	N_2	28.016
Oxygen	O_2	32.00
Water	H_2O	18.016
Air		28.967

Note: Molecular weights based on following values of atomic weights: hydrogen 1.008, carbon 12.011, nitrogen 14.008, oxygen 16.000, and argon 39.944. Air was assumed to consist of 78.09% nitrogen, 20.95% oxygen, 0.93% argon, and 0.03% carbon dioxide.

Substitution in the formula for the first part of the Charles' Law gives:

$$V_2 = 450 \times \frac{(90+460)}{(45+460)}$$
$$= 490 \text{ cubic ft}$$

It is desired to determine what the new pressure would be for the gas in the above example if the volume remains the same and the temperature changes from 45°F to 90°F as indicated. (Atmospheric pressure is 14.4 psia.)

Substitution in the formula gives:

$$P_2 = (10+14.4) \times \frac{(90+460)}{(45+460)}$$
$$= 26.6 \text{ psia or } 12.2 \text{ psig}$$

A convenient arrangement of a combination of Boyle's and Charles' laws that is easy to remember and use can be expressed mathematically as:

$$\frac{P_1V_1}{T_1} = \frac{P_2V_2}{T_2} \qquad (4)$$

One can substitute known values in the combination formula and solve for any one unknown value. In cases where one of the parameters, such as temperature, is not to be considered, it may be treated as having the same value on both sides of the formula, and consequently it can be cancelled out.

Avogadro's law. This law states that equal volumes of all gases at the same pressure and temperature conditions contain the same number of molecules.

From this it may be seen that the weight of a given volume of gas is a function of the weights of the molecules and that there is some volume at which the gas would weigh, in pounds, the numerical value of its molecular weight.

The volume at which the weight of the gas in pounds is equal to the numerical value of its molecular weight (known as the "mol-volume") is 378.9 cubic ft for gases at a temperature of 60°F and a pressure of 14.73 psia. Table 1 gives the atomic formula and molecular weights for hydrocarbons and other compounds frequently associated with natural gas. Reference to the table reveals that the molecular weight for methane is 16.043. Going back to the mol-volume explanation shows that 378.9 cubic ft of methane at 60°F and a pressure of 14.73 psia would weigh 16.043 lb.

Avogadro's law ties in closely with what is usually known as the ideal gas law.

The ideal gas law. Although expressed in many slightly different arrangements, this law is most frequently expressed as:

$$PV = nRT$$

where: P = Pressure of the gas
V = Volume of the gas
n = Number of lb-mols of gas
R = The universal gas constant, which varies depending upon the units of pressure, volume, and temperature employed

Since the number of lb-mols of a gas would be equal to the weight of the gas divided by the molecular weight of the gas, we can express the ideal gas law as:

$$PV = 10.722 \times \frac{W}{M} \times T \qquad (5)$$

where: P = Pressure of the gas, psia
V = Volume of the gas, cubic ft
W = Weight of the gas, lb
M = Molecular weight of the gas
T = Temperature of the gas, °R

The constant 10.722 is based upon the generally used value for the universal gas constant of 1,544 when the pressure is expressed in lb/sq. ft absolute.

This formula can be used in many arrangements. An arrangement that may be used to determine the weight of a quantity of gas is:

$$W = 0.0933 \times \frac{MVP}{T} \qquad (6)$$

when the symbols and units are as above.

Example. It is desired to find the weight of a gas in a 1,000-cubic-ft container if the gas is at a pressure of 150 psig and a temperature of 90°F. The molecular weight of the gas is known to be 16.816, and the barometric pressure is 14.3 psia.

Substitution in the formula gives:

$$W = \frac{0.0933 \times 16.816 \times 1,000 \times (150+14.3)}{(90+460)}$$
$$= 468.7 \text{ lb}$$

The formula above may be used when the molecular weight of a gas is known; however, at times it is desirable to determine the weight of a given volume of gas when the molecular weight is not known or cannot be readily determined. This may be accomplished if the specific gravity of the gas is known by using the formula:

$$W = 2.698 \times \frac{GVP}{T}$$

where G = the specific gravity of the gas (air equals 1.000) and the other symbols and units are as previously given.

Example. If the molecular weight of the gas in the preceding example was unknown, but the specific gravity was known to be 0.581, substitution in the formula would give:

$$W = 2.698 \times \frac{0.581 \times 1,000 \times (150 + 14.3)}{(90 + 460)}$$

$$= 468.3 \text{ lb}$$

Calculate gas properties from a gas analysis

In most practical applications involving gas calculations, the gas will consist of a mixture of components. Properties of the components are known; however, the properties of the mix must be determined for use in other calculations, such as compressor performance calculations. The gas molecular weight, K-value (isentropic exponent), and compressibility of the gas mix will be determined in the calculation to follow. Other data that may be needed for compressor performance determination will likely come from the process design and from the manufacturer's data.

Sample calculation 1

A gas mixture at 30 psia and 60°F consisting of ethane, propane and n-butane is given in a mole ratio of 0.05, 0.80 and 0.15 with properties shown in Table 1. In Table 2, the individual component contributions are computed by multiplying the mole percentage by each property.

Table 1

Gas	Mol %	Mol wt	P_c	T_c	c_p
Ethane	5	30.070	706.5	550	12.27
Propane	80	44.097	616.0	666	17.14
n-Butane	15	58.123	550.6	766	22.96
Total	100				

Table 2

Gas	Mol %	Mol wt	P_c	T_c	c_p
Ethane	5	1.50	35	28	0.61
Propane	80	35.28	493	533	13.71
n-Butane	15	8.72	83	115	3.44
Total	100	45.5	611	676	17.76

Therefore, the properties of the mix will be:

Molecular weight = MW = 45.5
Critical pressure = P_c = 611
Critical temperature = T_c = 676
Specific heat at constant pressure = c_p = 17.76

The ratio of specific heats for the mixture are calculated from Equation 1.

$$k_{mix} = c_{p_{mix}} / \left(c_{p_{mix}} - 1.986\right) \quad (1)$$

$$k_{mix} = 17.76/(17.76 - 1.986)$$
$$= 1.126$$

Determine the compressibility of the mix by calculating the reduced pressure and reduced temperature by using Equations 2 and 3.

$$P_r = P/P_c \quad (2)$$

$$T_r = T/T_c \quad (3)$$

$$P_r = 30/611$$
$$= 0.049$$

$$T_r = (60 + 460)/676$$
$$= 0.769$$

Refer to Figure 1 to determine the compressibility using the values of P_r and T_r. Z = 0.955.

If the mixture of components is given in mols of component or mass flow of components, these must be converted to mol % before calculating the properties of the mix.

Sample calculation 2

Mols of components are given:

Ethane	400 mols/hr
Propane	4,800 mols/hr
n-Butane	900 mols/hr
Total Flow	6,100 mols/hr

Mol % ethane = 400/6,100 = 6.56%

Table 3

Component	Mol/hr	Mol %
Ethane	400	6.56
Propane	4,800	78.69
n-Butane	900	14.75
Totals	6,100	100.00

The mol % values listed in Table 4 can be used to calculate the properties of the mixture following the procedure detailed in Sample Calculation 1.

Sample calculation 3

Mass flow of components is given.

Table 4

Component	Mass flow (kg/h)	Mol wt	Mol flow (kmol/h)	Mol %
Ethane	15,000	30.070	499	11.08
Propane	150,000	44.097	3,402	75.55
n-Butane	35,000	58.123	602	13.37
Totals			4,503	100.00

$$\text{Mol flow of ethane} = 15,000/30.07 = 499$$

$$\text{Mol \% of ethane} = 499/4,503 = 11.08\%$$

When a dry gas is saturated with water, it will be necessary to consider the effects of the water on the molecular weight of the mixture. The water will also affect the total mass flow of the mixture.

Sample calculation 4

Given: Compress 1,000 lbm/min dry CO_2, which is initially water saturated. MW = 44.01, P_{inlet} = 16 psia, T_{inlet} = 100°F.

Calculate the molecular weight of the mixture and the total mass flow for the required 1,000 lbm/min of dry CO_2.

$$\text{Mol fraction of water vapor} = y = P_w/P \tag{4}$$

P_w = partial pressure of water vapor. It is exactly equal to the saturation pressure if the gas is saturated.

$$MW_{mix} = yMW_{mix} + (1-y) \times MW_{dg} \tag{5}$$

where w = water vapor
dg = dry gas

$$m_{total} = m_{dg}\left[1 + (MW_w - P_w)/\left(MW_{dg} \times P_{dg}\right)\right] \tag{6}$$

Saturated pressure of water vapor at 100°F is equal to 0.95 psia, and since the mixture is saturated, the partial pressure is also equal to 0.95 psia.

Calculate mol % of water vapor using Equation 4.

y = 0.96/16 = 5.94 mol % water vapor

Calculate MW_{mix} using Equation 5.

$$MW_{mix} = 0.0594(18.02) + (1 - 0.059) \times 44.01 = 42.48$$

Calculate the mass flow of the mix using Equation 6.

$$M_{total} = 1,000[1 + (18.02 \times 0.95)/(0.44 \times (16 - 0.95))] = 1,026 \text{ lbm/min}$$

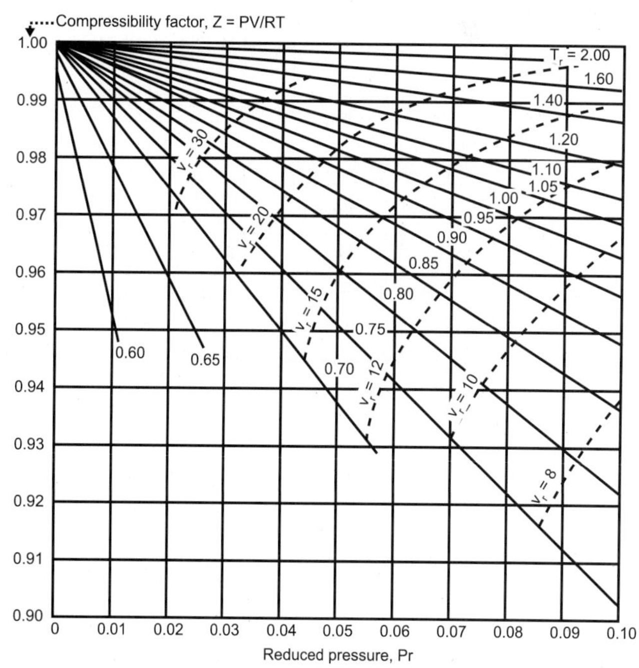

Figure 1. Compressibility chart for very low values of reduced pressure. Reproduced by permission of *Chemical Engineering*, McGraw Hill Publications Company, July 1954.

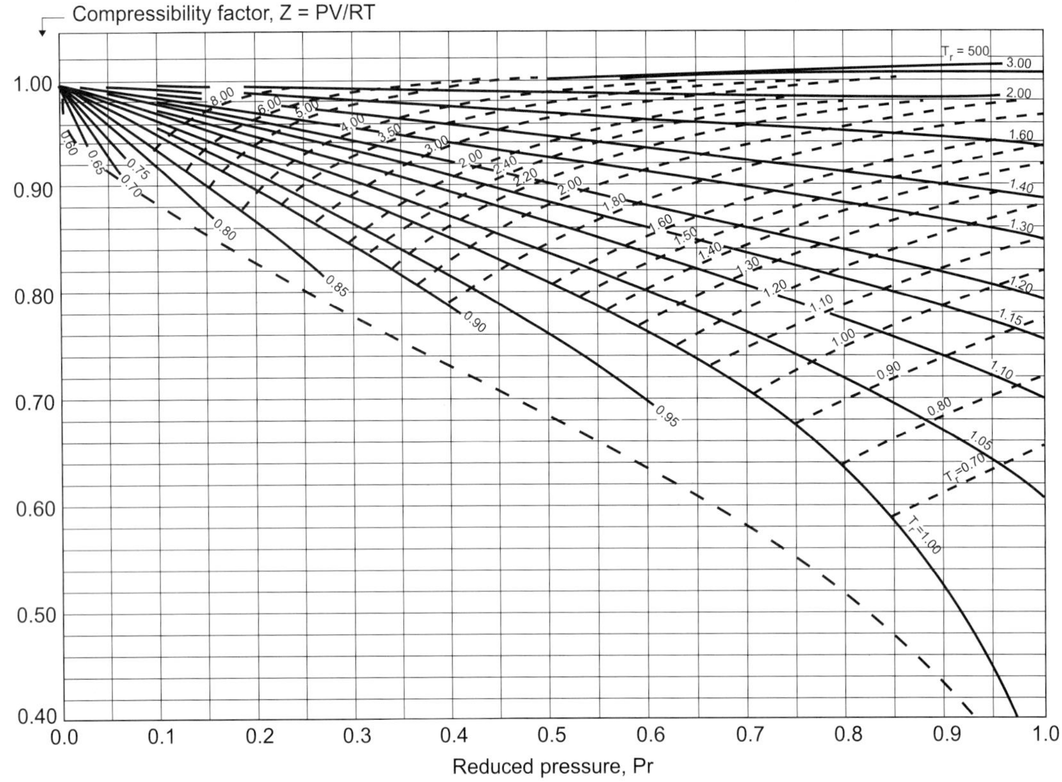

Figure 2. Compressibility chart for low values of reduced pressure. Reproduced by permission of *Chemical Engineering*, McGraw Hill Publications Company, July 1954.

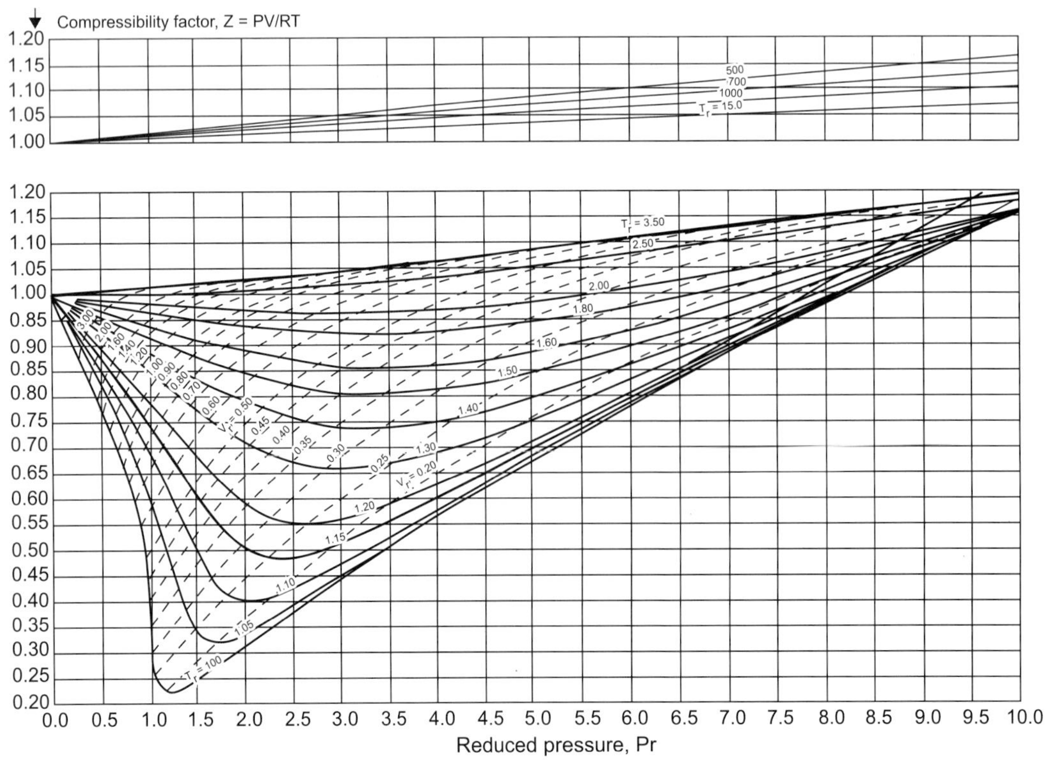

Figure 3. Compressibility chart for low to high values of reduced pressure. Reproduced by permission of *Chemical Engineering*, McGraw Hill Publications Company, July 1954.

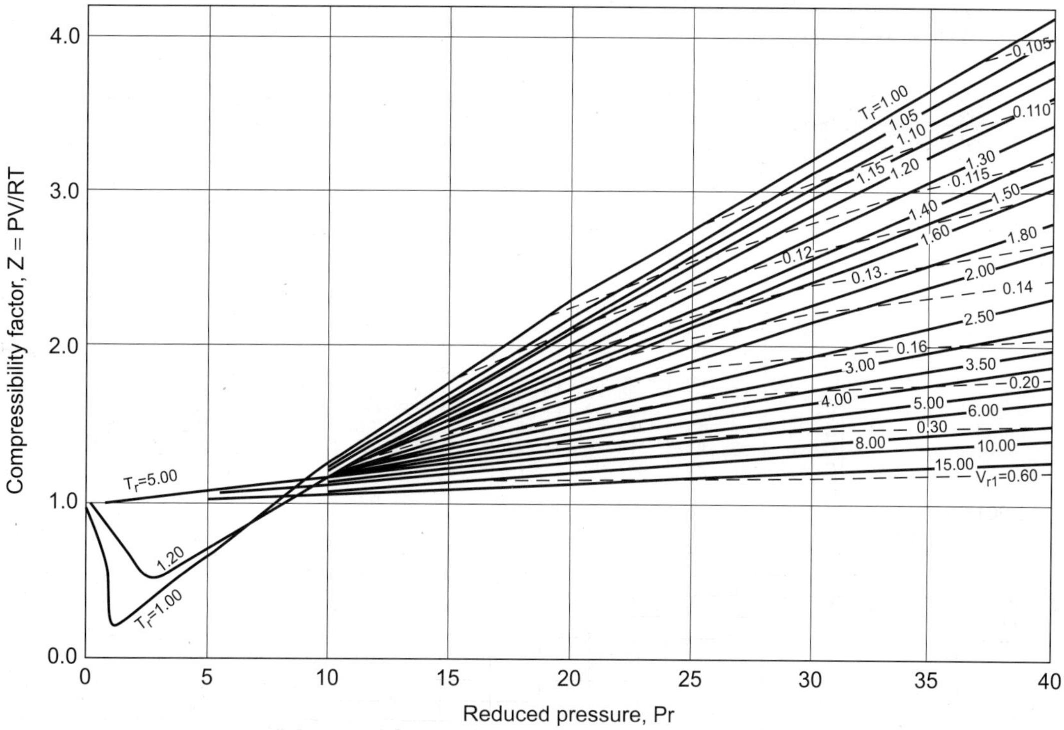

Figure 4. Compressibility chart for low to very high values of reduced pressure. Reproduced by permission of *Chemical Engineering*, McGraw Hill Publications Company, July 1954.

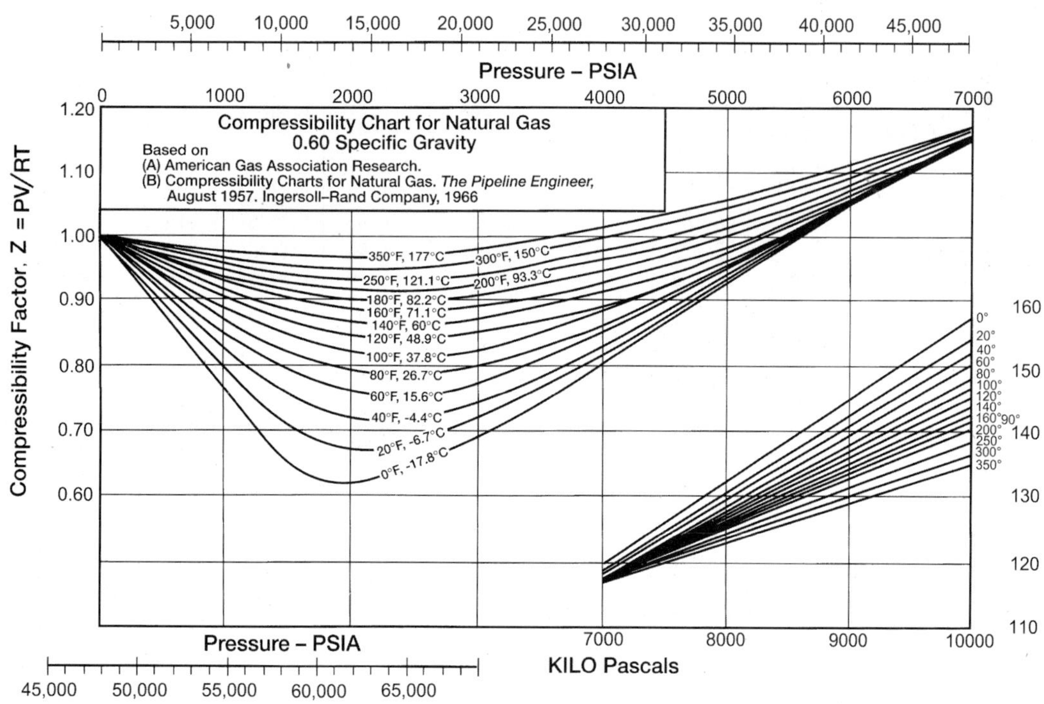

Figure 5. Compressibility chart for natural gas. Reprinted by permission and courtesy of Ingersol Rand.

Physical properties of selected hydrocarbons and other chemicals and gases

Physical Constants

*See the Table of Notes and References.

Number	Compound	Formula	Molar mass (molecular weight)	Boiling point, °F 14.696 psia	Vapor pressure, psia 100 °F	Freezing point, °F 14.696 psia	Refractive index, n_D 60 °F	Critical constants: Pressure, psia	Critical constants: Temperature, °F	Critical constants: Volume, ft³/lbm	Number
	See Note No. ->		A.	B.		C.	D.				
1	Methane	CH_4	16.043	−258.73	(5000)*	−296.44*	1.00042*	666.4	−116.67	0.0988	1
2	Ethane	C_2H_6	30.070	−127.49	(800)*	−297.04*	1.20971*	706.5	89.92	0.0783	2
3	Propane	C_3H_8	44.097	−43.75	188.64	−305.73*	1.29480*	616.0	206.06	0.0727	3
4	Isobutane	C_4H_{10}	58.123	10.78	72.581	−255.28	1.3245*	527.9	274.46	0.0714	4
5	n-Butane	C_4H_{10}	58.123	31.08	51.706	−217.05	1.33588*	550.6	305.62	0.0703	5
6	Isopentane	C_5H_{12}	72.150	82.12	20.445	−255.82	1.35631	490.4	369.10	0.0679	6
7	n-Pentane	C_5H_{12}	72.150	96.92	15.574	−201.51	1.35992	488.6	385.8	0.0675	7
8	Neopentane	C_5H_{12}	72.150	49.10	36.69	2.17	1.342*	464.0	321.13	0.0673	8
9	n-Hexane	C_6H_{14}	86.177	155.72	4.9597	−139.58	1.37708	436.9	453.6	0.0688	9
10	2-Methylpentane	C_6H_{14}	86.177	140.47	6.769	−244.62	1.37387	436.6	435.83	0.0682	10
11	3-Methylpentane	C_6H_{14}	86.177	145.89	6.103	———	1.37888	453.1	448.4	0.0682	11
12	Neohexane	C_6H_{14}	86.177	121.52	9.859	−147.72	1.37126	446.8	420.13	0.0667	12
13	2,3-Dimethylbutane	C_6H_{14}	86.177	136.36	7.406	−199.38	1.37730	453.5	440.29	0.0665	13
14	n-Heptane	C_7H_{16}	100.204	209.16	1.620	−131.05	1.38989	396.8	512.7	0.0691	14
15	2-Methylhexane	C_7H_{16}	100.204	194.09	2.272	−180.89	1.38714	396.5	495.00	0.0673	15
16	3-Methylhexane	C_7H_{16}	100.204	197.33	2.131	———	1.39091	408.1	503.80	0.0646	16
17	3-Ethylpentane	C_7H_{16}	100.204	200.25	2.013	−181.48	1.39566	419.3	513.39	0.0665	17
18	2,2-Dimethylpentane	C_7H_{16}	100.204	174.54	3.494	−190.86	1.38446	402.2	477.23	0.0665	18
19	2,4-Dimethylpentane	C_7H_{16}	100.204	176.89	3.293	−182.63	1.38379	396.9	475.95	0.0668	19
20	3,3-Dimethylpentane	C_7H_{16}	100.204	186.91	2.774	−210.01	1.38564	427.2	505.87	0.0662	20
21	Triptane	C_7H_{16}	100.204	177.58	3.375	−12.81	1.39168	428.4	496.44	0.0636	21
22	n-Octane	C_8H_{18}	114.231	258.21	0.53694	−70.18	1.39956	360.7	564.22	0.0690	22
23	Diisobutyl	C_8H_{18}	114.231	228.39	1.102	−132.11	1.39461	360.6	530.44	0.0676	23
24	Isooctane	C_8H_{18}	114.231	210.63	1.709	−161.27	1.38624	372.4	519.46	0.0656	24
25	n-Nonane	C_9H_{20}	128.258	303.47	0.17953	−64.28	1.40746	331.8	610.68	0.0684	25
26	n-Decane	$C_{10}H_{22}$	142.285	345.48	0.06088	−21.36	1.41385	305.2	652.0	0.0679	26
27	Cyclopentane	C_5H_{10}	70.134	120.65	9.915	−136.91	1.40896	653.8	461.2	0.0594	27
28	Methylcyclopentane	C_6H_{12}	84.161	161.25	4.503	−224.40	1.41210	548.9	499.35	0.0607	28
29	Cyclohexane	C_6H_{12}	84.161	177.29	3.266	43.77	1.42862	590.8	536.6	0.0586	29
30	Methylcyclohexane	C_7H_{14}	98.188	213.68	1.609	−195.87	1.42538	503.5	570.27	0.0600	30
31	Ethene(Ethylene)	C_2H_4	28.054	−154.73	(1400)*	−272.47*	(1.228)*	731.0	48.54	0.0746	31
32	Propene(Propylene)	C_3H_6	42.081	−53.84	227.7	−301.45*	1.3130*	668.6	197.17	0.0689	32
33	1-Butene(Butylene)	C_4H_8	56.108	20.79	62.10	−301.63*	1.3494*	583.5	295.48	0.0685	33
34	cis-2-Butene	C_4H_8	56.108	38.69	45.95	−218.06	1.3665*	612.1	324.37	0.0668	34
35	trans-2-Butene	C_4H_8	56.108	33.58	49.87	−157.96	1.3563*	587.4	311.86	0.0679	35
36	Isobutene	C_4H_8	56.108	19.59	63.02	−220.65	1.3512*	580.2	292.55	0.0682	36
37	1-Pentene	C_5H_{10}	70.134	85.93	19.12	−265.39	1.37426	511.8	376.93	0.0676	37
38	1,2-Butadiene	C_4H_6	54.092	51.53	36.53	−213.16	———	(653.)*	(340.)*	(0.065)*	38
39	1,3-Butadiene	C_4H_6	54.092	24.06	59.46	−164.02	1.3975*	627.5	305.	0.0654	39
40	Isoprene	C_5H_8	68.119	93.31	16.68	−230.73	1.42498	(558.)*	(412.)*	(0.065)*	40
41	Acetylene	C_2H_2	26.038	−120.49*	———	−114.5*	———	890.4	95.34	0.0695	41
42	Benzene	C_6H_6	78.114	176.18	3.225	41.95	1.50396	710.4	552.22	0.0531	42
43	Toluene	C_7H_8	92.141	231.13	1.033	−139.00	1.49942	595.5	605.57	0.0550	43
44	Ethylbenzene	C_8H_{10}	106.167	277.16	0.3716	−138.966	1.49826	523.0	651.29	0.0565	44
45	o-Xylene	C_8H_{10}	106.167	291.97	0.2643	−13.59	1.50767	541.6	674.92	0.0557	45
46	m-Xylene	C_8H_{10}	106.167	282.41	0.3265	−54.18	1.49951	512.9	651.02	0.0567	46
47	p-Xylene	C_8H_{10}	106.167	281.07	0.3424	55.83	1.49810	509.2	649.54	0.0570	47
48	Styrene	C_8H_8	104.152	293.25	0.2582	−23.10	1.54937	587.8	(703.)*	0.0534	48
49	Isopropylbenzene	C_9H_{12}	120.194	306.34	0.1884	−140.814	1.49372	465.4	676.3	0.0572	49
50	Methyl alcohol	CH_4O	32.042	148.44	4.629	−143.79	1.33034	1174.	463.08	0.0590	50
51	Ethyl alcohol	C_2H_6O	46.069	172.90	2.312	−173.4	1.36346	890.1	465.39	0.0581	51
52	Carbon monoxide	CO	28.010	−312.68	———	−337.00*	1.00036*	507.5	−220.43	0.0532	52
53	Carbon dioxide	CO_2	44.010	−109.257*	———	−69.83*	1.00048*	1071.	87.91	0.0344	53
54	Hydrogen sulfide	H_2S	34.08	−76.497	394.59	−121.88*	1.00060*	1300.	212.45	0.0461	54
55	Sulfur dioxide	SO_2	64.06	14.11	85.46	−103.86*	1.00062*	1143.	315.8	0.0305	55
56	Ammonia	NH_3	17.0305	−27.99	211.9	−107.88*	1.00036*	1646.	270.2	0.0681	56
57	Air	N_2+O_2	28.9625	−317.8	———	———	1.00028*	546.9	−221.31	0.0517	57
58	Hydrogen	H_2	2.0159	−422.955*	———	−435.26*	1.00013*	188.1	−399.9	0.5165	58
59	Oxygen	O_2	31.9988	−297.332*	———	−361.820*	1.00027*	731.4	−181.43	0.0367	59
60	Nitrogen	N_2	28.0134	−320.451	———	−346.00*	1.00028*	493.1	−232.51	0.0510	60
61	Chlorine	Cl_2	70.906	−29.13	157.3	−149.73*	1.3878*	1157.	290.75	0.0280	61
62	Water	H_2O	18.0153	212.000*	0.9501	32.00	1.33335	3198.8	705.16	0.0497$_5$	62
63	Helium	He	4.0026	−452.09	———	———	1.00003*	32.99	−450.31	0.2300	63
64	Hydrogen chloride	HCl	36.461	−121.27	906.71	−173.52*	1.00042*	1205.	124.77	0.0356	64

12/12/86

Physical Constants

*See the Table of Notes and References.

Number	E. Density of liquid 14.696 psia, 60°F: Relative density (specific gravity) 60°F/60°F	E. lbm/gal.	E. gal./lb mole	F. Temperature coefficient of density, 1/°F	G. Acentric factor, ω	H. Compressibility factor of real gas, Z 14.696 psia, 60°F	I. Ideal gas 14.696 psia, 60°F: Relative density (specific gravity) Air = 1	I. ft³ gas/lbm	I. ft³ gas/gal. liquid	J. Specific Heat 60°F 14.696 psia Btu/(lbm·°F): C_p, Ideal gas	J. C_p, Liquid	Number
1	(0.3)*	(2.5)*	(6.4172)*	——	0.0104	0.9980	0.5539	23.654	(59.135)*	0.52669	——	1
2	0.35619*	2.9696*	10.126*	——	0.0979	0.9919	1.0382	12.620	37.476*	0.40782	0.97225	2
3	0.50699*	4.2268*	10.433*	−0.00162*	0.1522	0.9825	1.5226	8.6059	36.375*	0.38852	0.61996	3
4	0.56287*	4.6927*	12.386*	−0.00119*	0.1852	0.9711	2.0068	6.5291	30.639*	0.38669	0.57066	4
5	0.58401*	4.8690*	11.937*	−0.00106*	0.1995	0.9667	2.0068	6.5291	31.790*	0.39499	0.57272	5
6	0.62470	5.2082	13.853	−0.00090	0.2280	——	2.4912	5.2596	27.393	0.38440	0.53331	6
7	0.63112	5.2617	13.712	−0.00086	0.2514	——	2.4912	5.2596	27.674	0.38825	0.54363	7
8	0.59666*	4.9744*	14.504*	−0.00106*	0.1963	0.9582	2.4912	5.2596	26.163*	0.39038	0.55021	8
9	0.66383	5.5344	15.571	−0.00075	0.2994	——	2.9755	4.4035	24.371	0.38628	0.53327	9
10	0.65785	5.4846	15.713	−0.00076	0.2780	——	2.9755	4.4035	24.152	0.38526	0.52732	10
11	0.66901	5.5776	15.451	−0.00076	0.2732	——	2.9755	4.4035	24.561	0.37902	0.51876	11
12	0.65385	5.4512	15.809	−0.00076	0.2326	——	2.9755	4.4035	24.005	0.38231	0.51367	12
13	0.66631	5.5551	15.513	−0.00076	0.2469	——	2.9755	4.4035	24.462	0.37762	0.51308	13
14	0.68820	5.7376	17.464	−0.00068	0.3494	——	3.4598	3.7872	21.729	0.38447	0.52802	14
15	0.68310	5.6951	17.595	−0.00070	0.3298	——	3.4598	3.7872	21.568	0.38041	0.52199	15
16	0.69165	5.7664	17.377	−0.00070	0.3232	——	3.4598	3.7872	21.838	0.37882	0.51019	16
17	0.70276	5.8590	17.103	−0.00069	0.3105	——	3.4598	3.7872	22.189	0.38646	0.51410	17
18	0.67829	5.6550	17.720	−0.00070	0.2871	——	3.4598	3.7872	21.416	0.38594	0.51678	18
19	0.67733	5.6470	17.745	−0.00073	0.3026	——	3.4598	3.7872	21.386	0.39414	0.52440	19
20	0.69772	5.8170	17.226	−0.00067	0.2674	——	3.4598	3.7872	22.030	0.38306	0.50138	20
21	0.69457	5.7907	17.304	−0.00068	0.2503	——	3.4598	3.7872	21.930	0.37724	0.49920	21
22	0.70696	5.8940	19.381	−0.00064	0.3977	——	3.9441	3.3220	19.580	0.38331	0.52406	22
23	0.69793	5.8187	19.632	−0.00067	0.3564	——	3.9441	3.3220	19.330	0.37571	0.51130	23
24	0.69624	5.8046	19.679	−0.00065	0.3035	——	3.9441	3.3220	19.283	0.38222	0.48951	24
25	0.72187	6.0183	21.311	−0.00061	0.4445	——	4.4284	2.9588	17.807	0.38246	0.52244	25
26	0.73421	6.1212	23.245	−0.00057	0.4898	——	4.9127	2.6671	16.326	0.38179	0.52103	26
27	0.75050	6.2570	11.209	−0.00073	0.1950	——	2.4215	5.4110	33.856	0.27199	0.42182	27
28	0.75349	6.2819	13.397	−0.00069	0.2302	——	2.9059	4.5090	28.325	0.30100	0.44126	28
29	0.78347	6.5319	12.885	−0.00065	0.2096	——	2.9059	4.5090	29.452	0.28817	0.43584	29
30	0.77400	6.4529	15.216	−0.00062	0.2358	——	3.3902	3.8649	24.940	0.31700	0.44012	30
31	——	——	——	——	0.0865	0.9936	0.9686	13.527	——	0.35697	——	31
32	0.52095*	4.3432*	9.6889*	−0.00173*	0.1356	0.9844	1.4529	9.0179	39.167*	0.35714	0.57116	32
33	0.60107*	5.0112*	11.197*	−0.00112*	0.1941	0.9699	1.9373	6.7636	33.894*	0.35446	0.54533	33
34	0.62717*	5.2288*	10.731*	−0.00105*	0.2029	0.9665	1.9373	6.7636	35.366*	0.33754	0.52980	34
35	0.60996*	5.0853*	11.033*	−0.00106*	0.2128	0.9667	1.9373	6.7636	34.395*	0.35574	0.54215	35
36	0.60040*	5.0056*	11.209*	−0.00117*	0.1999	0.9700	1.9373	6.7636	33.856*	0.37690	0.54839	36
37	0.64571	5.3834	13.028	−0.00089	0.2333	——	2.4215	5.4110	29.129	0.36351	0.51782	37
38	0.65799*	5.4857*	9.8605*	−0.00101*	0.2540	(0.969)	1.8677	7.0156	38.485*	0.34347	0.54029	38
39	0.62723*	5.2293*	10.344*	−0.00110*	0.2007	(0.965)	1.8677	7.0156	36.687*	0.34120	0.53447	39
40	0.68615	5.7205	11.908	−0.00082	0.1568	——	2.3520	5.5710	31.869	0.35072	0.51933	40
41	(0.41796)	(3.4842)	(7.473)	——	0.1949	0.9930	0.8990	14.574	——	0.39754	——	41
42	0.88448	7.3740	10.593	−0.00067	0.2093	——	2.6971	4.8581	35.824	0.24296	0.40989	42
43	0.87190	7.2691	12.676	−0.00059	0.2633	——	3.1814	4.1184	29.937	0.26370	0.40095	43
44	0.87168	7.2673	14.609	−0.00056	0.3027	——	3.6657	3.5744	25.976	0.27792	0.41139	44
45	0.88467	7.3756	14.394	−0.00052	0.3942	——	3.6657	3.5744	26.363	0.28964	0.41620	45
46	0.86875	7.2429	14.658	−0.00053	0.3257	——	3.6657	3.5744	25.889	0.27427	0.40545	46
47	0.86578	7.2181	14.708	−0.00056	0.3216	——	3.6657	3.5744	25.800	0.27471	0.40255	47
48	0.91108	7.5958	13.712	−0.00053	(0.2412)	——	3.5961	3.6435	27.675	0.27110	0.41220	48
49	0.86634	7.2228	16.641	−0.00055	0.3260	——	4.1500	3.1573	22.804	0.29170	0.42053	49
50	0.79626	6.6385	4.8267	−0.00066	0.5649	——	1.1063	11.843	78.622	0.32316	0.59187	50
51	0.79399	6.6196	6.9595	−0.00058	0.6438	——	1.5906	8.2372	54.527	0.33222	0.56610	51
52	0.78939*	6.5812*	4.2561*	——	0.0484	0.9959	0.9671	13.548	89.163*	0.24847	——	52
53	0.81802*	6.8199*	6.4532*	−0.00583*	0.2667	0.9943	1.5196	8.6229	58.807*	0.19911	——	53
54	0.80144*	6.6817*	5.1005*	−0.00157*	0.0948	0.9846	1.1767	11.135	74.401*	0.23827	0.50418	54
55	1.3974*	11.650*	5.4987*	——	0.2548	0.9802	2.2118	5.9238	69.012*	0.14804	0.32460	55
56	0.61832*	5.1550*	3.3037*	——	0.2557	0.9877	0.5880	22.283	114.87*	0.49677	1.1209	56
57	0.87476*	7.2930*	3.9713*	——	——	1.0000	1.0000	13.103	95.557*	0.23988	——	57
58	0.071070*	0.59252*	3.4022*	——	−0.2202	1.0006	0.06960	188.25	111.54*	3.4038	——	58
59	1.1421*	9.5221*	3.3605*	——	0.0216	0.9992	1.1048	11.859	112.93*	0.21892	——	59
60	0.80940*	6.7481*	4.1513*	——	0.0372	0.9997	0.9672	13.546	91.413*	0.24828	——	60
61	1.4244*	11.875*	5.9710*	——	0.0878	(0.9875)	2.4482	5.3519	63.554*	0.11377	——	61
62	1.00000	8.33712	2.1609	−0.00009	0.3443	——	0.62202	21.065	175.62	0.44457	0.99974	62
63	0.12510*	1.0430*	3.8376*	——	0.	1.0006	0.1382	94.814	98.891*	1.2404	——	63
64	0.85129*	7.0973*	5.1373*	−0.00300*	0.1259	0.9923	1.2589	10.408	73.869*	0.19086	——	64

12/12/86

Physical Constants

*See the Table of Notes and References.

Number	Compound	K. Heating value, 60°F — Net: Btu/ft³ ideal gas, 14.696 psia	K. Net: Btu/lbm Liquid	K. Gross: Btu/ft³ ideal gas, 14.696 psia	K. Gross: Btu/lbm Liquid	K. Gross: Btu/gal Liquid	L. Heat of vaporization 14.696 psia at boiling point, Btu/lbm	M. Air required for combustion ideal gas ft³(air)/ft³(gas)	Flammability limits, vol % in air mixture: Lower	Flammability limits: Higher	ASTM octane number: Motor method D-357	ASTM octane number: Research method D-908	Number
1	Methane	909.4	—	1010.0	—	—	219.45	9.548	5.0	15.0	—	—	1
2	Ethane	1618.7	20277.*	1769.6	22181.*	65869.*	211.14	16.710	2.9	13.0	+0.05	+1.6*	2
3	Propane	2314.9	19757.*	2516.1	21489.*	90830.*	183.01	23.871	2.0	9.5	97.1	+1.8*	3
4	Isobutane	3000.4	19437.*	3251.9	21079.*	98917.*	157.23	31.032	1.8	8.5	97.6	+0.1*	4
5	n-Butane	3010.8	19494.*	3262.3	21136.*	102911.*	165.93	31.032	1.5	9.0	89.6*	93.8*	5
6	Isopentane	3699.0	19303.	4000.9	20891.	108805.	147.12	38.193	1.3	8.0	90.3	92.3	6
7	n-Pentane	3706.9	19335.	4008.9	20923.	110091.	153.57	38.193	1.4	8.3	62.6*	61.7*	7
8	Neopentane	3682.9	19235.*	3984.7	20822.*	103577.*	135.58	38.193	1.3	7.5	80.2	85.5	8
9	n-Hexane	4403.8	19232.	4755.9	20783.	115021.	143.94	45.355	1.1	7.7	26.0	24.8	9
10	2-Methylpentane	4395.2	19202.	4747.3	20753.	113822.	138.45	45.355	1.18	7.0	73.5	73.4	10
11	3-Methylpentane	4398.2	19213.	4750.3	20764.	115813.	140.05	45.355	1.2	7.7	74.3	74.5	11
12	Neohexane	4384.0	19163.	4736.2	20714.	112916.	131.23	45.355	1.2	7.0	93.4	91.8	12
13	2,3-Dimethylbutane	4392.9	19195.	4745.0	20746.	115246.	136.07	45.355	1.2	7.0	94.3	+0.3	13
14	n-Heptane	5100.0	19155.	5502.5	20679.	118648.	136.00	52.516	1.0	7.0	0.0	0.0	14
15	2-Methylhexane	5092.2	19133.	5494.6	20657.	117644.	131.58	52.516	1.0	7.0	46.4	42.4	15
16	3-Methylhexane	5096.0	19146.	5498.6	20671.	119197.	132.10	52.516	(1.01)	(6.6)	55.8	52.0	16
17	3-Ethylpentane	5098.3	19154.	5500.7	20679.	121158.	132.82	52.516	(1.00)	(6.5)	69.3	65.0	17
18	2,2-Dimethylpentane	5079.6	19095.	5481.9	20620.	116606.	125.12	52.516	(1.09)	(6.8)	95.6	92.8	18
19	2,4-Dimethylpentane	5084.2	19111.	5486.7	20635.	116526.	126.57	52.516	(1.08)	(6.8)	83.8	83.1	19
20	3,3-Dimethylpentane	5086.4	19119.	5488.8	20643.	120080.	127.20	52.516	(1.04)	(7.0)	86.6	80.8	20
21	Triptane	5081.2	19103.	5483.5	20628.	119451.	124.21	52.516	(1.08)	(6.8)	+0.1	+1.8	21
22	n-Octane	5796.1	19096.	6248.9	20601.	121422.	129.52	59.677	0.8	6.5	—	—	22
23	Diisobutyl	5780.5	19047.	6233.5	20552.	119586.	122.83	59.677	(0.92)	(6.3)	55.7	55.2	23
24	Isooctane	5778.8	19063.	6231.7	20568.	119389.	112.94	59.677	0.95	6.0	100.0	100.0	24
25	n-Nonane	6493.2	19054.	6996.5	20543.	123634.	124.36	66.839	0.7	5.6	—	—	25
26	n-Decane	7189.6	19018.	7742.9	20494.	125448.	119.65	74.000	0.7	5.4	—	—	26
27	Cyclopentane	3512.1	18825.	3763.7	20186.	126304.	167.33	35.806	(1.48)	(8.3)	84.9*	+0.1	27
28	Methylcyclopentane	4199.4	18771.	4501.2	20132.	126467.	148.54	42.968	1.0	8.35	80.0	91.3	28
29	Cyclohexane	4179.7	18675.	4481.7	20036.	130873.	153.03	42.968	1.2	8.35	77.2	83.0	29
30	Methylcyclohexane	4863.6	18640.	5215.9	20002.	129071.	136.30	50.129	1.1	6.7	71.1	74.8	30
31	Ethene(Ethylene)	1499.1	—	1599.8	—	—	207.41	14.323	2.7	36.0	75.6	+0.03	31
32	Propene(Propylene)	2181.8	(19858.)	2332.7	(21208.)	(92113.)	188.19	21.484	2.0	11.7	84.9	+0.2	32
33	1-Butene(Butylene)	2878.7	19309.*	3079.9	20670.*	103582.*	167.96	28.645	1.6	10.	80.8*	97.4	33
34	cis-2-Butene	2871.0	19241.*	3072.2	20602.*	107724.*	178.89	28.645	1.6	10.	83.5	100.0	34
35	trans-2-Butene	2866.8	19221.*	3068.0	20582.*	104666.*	174.37	28.645	1.6	10.	—	—	35
36	Isobutene	2859.9	19182.*	3061.1	20543.*	102830.*	169.47	28.645	1.6	10.	—	—	36
37	1-Pentene	3575.0	19184.	3826.5	20545.	110602.	154.48	35.806	1.3	10.	77.1	90.9	37
38	1,2-Butadiene	2789.0	19378.*	2939.9	20437.*	112111.*	191.88	26.258	(1.62)	(10.3)	—	—	38
39	1,3-Butadiene	2729.0	18967.*	2879.9	20025.*	104717.*	185.29	26.258	2.0	12.5	—	—	39
40	Isoprene	3410.8	18832.	3612.1	19953.	114141.	163.48	33.419	(1.12)	(8.5)	81.0	99.1	40
41	Acetylene	1423.2	(20887.)	1473.5	(21613.)	(75204.)	151.90	11.935	1.5	100.	—	—	41
42	Benzene	3590.9	17256.	3741.8	17989.	132651.	169.24	35.806	1.2	8.0	+2.8	—	42
43	Toluene	4273.6	17421.	4475.0	18250.	132661.	154.83	42.968	1.2	7.1	+0.3	+5.8	43
44	Ethylbenzene	4970.5	17593.	5222.2	18492.	134387.	144.02	50.129	1.0	8.0	97.9	+0.8	44
45	o-Xylene	4958.2	17544.	5209.9	18444.	136036.	149.10	50.129	1.0	7.6	100.0	—	45
46	m-Xylene	4956.3	17541.	5207.9	18440.	133559.	147.24	50.129	1.0	7.0	+2.8	+4.0	46
47	p-Xylene	4957.1	17545.	5208.8	18444.	133131.	145.71	50.129	1.0	7.0	+1.2	+3.4	47
48	Styrene	4829.8	17414.	5031.1	18147.	137841.	152.85	47.742	1.1	8.0	+0.2	>+3.*	48
49	Isopropylbenzene	5660.9	17709.	5962.8	18662.	134792.	134.24	57.290	0.8	6.5	99.3	+2.1	49
50	Methyl alcohol	766.1	8559.	866.7	9751.	64731.	462.58	7.161	5.5	44.0	—	—	50
51	Ethyl alcohol	1448.1	11530.	1599.1	12770.	84539.	359.07	14.323	3.28	19.0	—	—	51
52	Carbon monoxide	320.5	—	320.5	—	—	92.77	2.387	12.50	74.20	—	—	52
53	Carbon dioxide	0.0	—	0.0	—	—	246.47*	—	—	—	—	—	53
54	Hydrogen sulfide	586.8	6337.*	637.1	6897.*	46086.*	235.63	7.161	4.30	45.50	—	—	54
55	Sulfur dioxide	0.0	—	0.0	—	—	167.22	—	—	—	—	—	55
56	Ammonia	359.0	—	434.4	—	—	589.48	3.581	15.50	27.00	—	—	56
57	Air	0.0	—	0.0	—	—	88.20	—	—	—	—	—	57
58	Hydrogen	273.8	—	324.2	—	—	192.74	2.387	4.00	74.20	—	—	58
59	Oxygen	0.0	—	0.0	—	—	91.59	—	—	—	—	—	59
60	Nitrogen	0.0	—	0.0	—	—	85.59	—	—	—	—	—	60
61	Chlorine	—	—	—	—	—	123.75	—	—	—	—	—	61
62	Water	0.0	*	*	0.0	0.0	970.18	—	—	—	—	—	62
63	Helium	0.0	—	0.0	—	—	—	—	—	—	—	—	63
64	Hydrogen chloride	—	—	—	—	—	190.43	—	—	—	—	—	64

12/12/86

(Cont'd)
NOTES AND REFERENCES FOR THE TABLE OF PHYSICAL CONSTANTS

	See Note No. ->		A.	B.		C.	D.				
								Critical constants			
Number	Compound	Formula	Molar mass (molecular weight)	Boiling point, °F 14.696 psia	Vapor pressure, psia 100 °F	Freezing point, °F 14.696 psia	Refractive index, n_D 60 °F	Pressure, psia	Temperature, °F	Volume, ft³/lbm	Number
1	Methane	CH_4	19	4	a,b	c,57	i,57	4	4	4	1
2	Ethane	C_2H_6	19	35	a,b	c,57	h,55	35	35	35	2
3	Propane	C_3H_8	19	34	34	c,57	h,55	34	34	34	3
4	Isobutane	C_4H_{10}	19	30	32	57	h,55	32	32	32	4
5	n-Butane	C_4H_{10}	19	33	33	57	h,55	33	33	33	5
6	Isopentane	C_5H_{12}	19	57	57	57	57	43	43	43	6
7	n-Pentane	C_5H_{12}	19	57	57	57	57	43	43	43	7
8	Neopentane	C_5H_{12}	19	57	57	57	h,57	43	43	43	8
9	n-Hexane	C_6H_{14}	19	57	57	57	57	43	43	43	9
10	2-Methylpentane	C_6H_{14}	19	57	57	57	57	43	43	43	10
11	3-Methylpentane	C_6H_{14}	19	57	57	f,22,29	57	43	43	43	11
12	Neohexane	C_6H_{14}	19	57	57	57	57	43	43	43	12
13	2,3-Dimethylbutane	C_6H_{14}	19	57	57	57	57	43	43	43	13
14	n-Heptane	C_7H_{16}	19	57	57	57	57	43	43	43	14
15	2-Methylhexane	C_7H_{16}	19	57	57	57	57	43	43	43	15
16	3-Methylhexane	C_7H_{16}	19	57	57	f,38	57	43	43	43	16
17	3-Ethylpentane	C_7H_{16}	19	57	57	57	57	43	43	43	17
18	2,2-Dimethylpentane	C_7H_{16}	19	57	57	57	57	43	43	43	18
19	2,4-Dimethylpentane	C_7H_{16}	19	57	57	57	57	43	43	43	19
20	3,3-Dimethylpentane	C_7H_{16}	19	57	57	57	57	43	43	43	20
21	Triptane	C_7H_{16}	19	57	57	57	57	43	43	43	21
22	n-Octane	C_8H_{18}	19	57	57	57	57	43	43	43	22
23	Diisobutyl	C_8H_{18}	19	57	57	57	57	43	43	43	23
24	Isooctane	C_8H_{18}	19	57	57	57	57	43	43	43	24
25	n-Nonane	C_9H_{20}	19	57	57	57	57	43	43	43	25
26	n-Decane	$C_{10}H_{22}$	19	57	57	57	57	43	43	43	26
27	Cyclopentane	C_5H_{10}	19	57	57	57	57	57	57	57	27
28	Methylcyclopentane	C_6H_{12}	19	57	57	57	57	57	57	57	28
29	Cyclohexane	C_6H_{12}	19	57	57	57	57	57	57	57	29
30	Methylcyclohexane	C_7H_{14}	19	57	57	57	57	57	57	57	30
31	Ethene(Ethylene)	C_2H_4	19	63	b	c,63	h,26	63	63	63	31
32	Propene(Propylene)	C_3H_6	19	5	5	5	h,26	g,5	44	57	32
33	1-Butene(Butylene)	C_4H_8	19	57	57	57	h,26	57	57	57	33
34	cis-2-Butene	C_4H_8	19	57	25	57	h,26	57	57	57	34
35	trans-2-Butene	C_4H_8	19	57	25	57	h,26	57	57	57	35
36	Isobutene	C_4H_8	19	57	57	57	h,26	57	57	57	36
37	1-Pentene	C_5H_{10}	19	57	57	57	26	57	57	57	37
38	1,2-Butadiene	C_4H_6	19	57	57	56		a	a	a	38
39	1,3-Butadiene	C_4H_6	19	57	25	56	h,26	57	57	57	39
40	Isoprene	C_5H_8	19	57	57	56		a	a	a	40
41	Acetylene	C_2H_2	19	d,1	b	c,56		57	57	57	41
42	Benzene	C_6H_6	19	14	57	14	57	14	14	14	42
43	Toluene	C_7H_8	19	57	57	15	57	57	57	57	43
44	Ethylbenzene	C_8H_{10}	19	57	57	15	57	57	57	57	44
45	o-Xylene	C_8H_{10}	19	57	57	15	57	57	57	57	45
46	m-Xylene	C_8H_{10}	19	57	57	15	57	57	57	57	46
47	p-Xylene	C_8H_{10}	19	57	57	15	57	57	57	57	47
48	Styrene	C_8H_8	19	56	13	56	57	56	a,11	56	48
49	Isopropylbenzene	C_9H_{12}	19	57	r,57	57	57	57	57	57	49
50	Methyl alcohol	CH_4O	19	58	58	58	58	58	58	58	50
51	Ethyl alcohol	C_2H_6O	19	58	58	58	58	1	58	58	51
52	Carbon monoxide	CO	19	58	b	c,58	i,58	1	1	1	52
53	Carbon dioxide	CO_2	19	d,3	b	c,3	i,58	3	3	3	53
54	Hydrogen sulfide	H_2S	19	31	31	c,31	i,58	31	31	31	54
55	Sulfur dioxide	SO_2	19	1	23	c,58	i,58	58	58	58	55
56	Ammonia	NH_3	19	1	23	c,58	i,58	1	1	1	56
57	Air	N_2+O_2	19,40	45	b		i,60	45	45	45	57
58	Hydrogen	H_2	19	p,58	b	p,c,58	i,58	p,58	p,58	p,58	58
59	Oxygen	O_2	19	u,10	b	c,u,10	i,58	46	46	46	59
60	Nitrogen	N_2	19	6	b	c,6	i,58	6	6	6	60
61	Chlorine	Cl_2	19	1	2	c,58	h,58	2	2	2	61
62	Water	H_2O	19	u,10	37	58	58	37	37	37	62
63	Helium	He	19	7	b		i,58	7	s,7	7	63
64	Hydrogen chloride	HCl	19	1	23	c,58	i,58	58	58	58	64

12/12/86

(Cont'd)
NOTES AND REFERENCES FOR THE TABLE OF PHYSICAL CONSTANTS

Number	E. Density of liquid 14.696 psia, 60°F: Relative density (specific gravity) 60°F/60°F	E. lbm/gal.	E. gal./lb mole	F. Temperature coefficient of density, 1/°F	G. Acentric factor, ω	H. Compressibility factor of real gas, Z 14.696 psia, 60°F	I. Ideal gas 14.696 psia, 60°F: Relative density (specific gravity) Air = 1	I. ft³ gas/lbm	I. ft³ gas/gal. liquid	J. Specific Heat 60°F 14.696 psia Btu/(lbm·°F): Cp, Ideal gas	J. Cp, Liquid	Number
1	b,a	b,a,v	b,a	b,a	4	57			b,a	8	b	1
2	h,35	h,v,35	h,35	h,35	57	57			h	8	53	2
3	h,34	h,34	h,34	h,34	34	57			h	8	34	3
4	h,n,32	h,n,32	h,n,32	h,n,32	32	57			h	30	32	4
5	h,33	h,33	h,33	h,33	33	57			h	8	33	5
6	57	57	57	57	57					8	36	6
7	57	57	57	57	57					8	47	7
8	h,57	h,57	h,57	h,57	57	57			h	8	24	8
9	57	57	57	57	57					8	47	9
10	57	57	57	57	57					8	57	10
11	57	57	57	57	57					8	57	11
12	57	57	57	57	57					8	57	12
13	57	57	57	57	57					8	57	13
14	57	57	57	57	57					57	47	14
15	57	57	57	57	57					57	57	15
16	57	57	57	57	57					57	38	16
17	57	57	57	57	57					57	57	17
18	57	57	57	57	57					57	57	18
19	57	57	57	57	57					57	57	19
20	57	57	57	57	57					57	57	20
21	57	57	57	57	57					57	57	21
22	57	57	57	57	57					57	47	22
23	57	57	57	57	57					57	57	23
24	57	57	57	57	57					57	57	24
25	57	57	57	57	57					57	47	25
26	57	57	57	57	57					57	47	26
27	57	57	57	57	57					8	22,59	27
28	57	57	57	57	r,57					57	22,59	28
29	57	57	57	57	57					8	9	29
30	57	57	57	57	57					57	22,59	30
31	b	b	b	b	57	57			b	8	b	31
32	h,5	h,5	h,5	h,5	5	57			h	8	17	32
33	h	h	h	h	57	57			h	57	24	33
34	h,r	h,r	h,r	h,r	25	57			h	57	17	34
35	h,r	h,r	h,r	h,r	25	57			h	57	r,17	35
36	h,r	h,r	h,r	h,r	57	57			h	57		36
37	57	57	57	57	57					57	57	37
38	h,r	h,r	h,r	h,r	r,57	a			h	57	57	38
39	h,r	h,r	h,r	h,r	25	a			h	57	54,59	39
40	r	r	r	r	57					57	48	40
41	b	b	b	b	r,57	57			b	8		41
42	57	57	57	57	57					57	15	42
43	57	57	57	57	57					16	15	43
44	57	57	57	57	57					16	15	44
45	57	57	57	57	57					16	15	45
46	57	57	57	57	57					16	15	46
47	57	57	57	57	57					16	15	47
48	j,50	j,50	j,50	j,52	a,51					57		48
49	57	57	57	57	57					61	42	49
50	58	58	58	58	58					62	62	50
51	62	62	62	62	58					62	62	51
52	k,39	k,39	k,39		39				t	8	b	52
53	h,3	h,3	h,3	h,3					h	8		53
54	h,31	h,v,31	h,31	h,31	31	58			h	8	31	54
55	h	h	h		23	58			h	8		55
56	h	h	h		23	58			h	58		56
57	k,45	k,45	k,45						t	45		57
58	k,p,21	k,p,21	k,p,21		p,23	58			t	8	b	58
59	k,46	k,46	k,46		q,58	58			t	8	b	59
60	k,6	k,6	k,6		6	58			t	8	b	60
61	h	h	h		2	a			h	58		61
62	37	37	37	41	r,58					27	28	62
63	k	k	k		58	58			t	8	b	63
64	h	h	h	h	58	58			h	58		64

12/12/86

(Cont'd)
NOTES AND REFERENCES FOR THE TABLE OF PHYSICAL CONSTANTS

Number	See Note No. ->	K.					L.	M.	Flammability limits, vol % in air mixture		ASTM octane number		Number
		Heating value, 60°F											
		Net		Gross									
	Compound	Btu/ft^3 Ideal gas, 14.696 psia	Btu/lbm Liquid	Btu/ft^3 ideal gas, 14.696 psia	Btu/lbm Liquid	Btu/gal. Liquid	Heat of vaporization 14.696 psia at boiling point, Btu/lbm	Air required for combustion ideal gas $ft^3(air)/ft^3(gas)$	Lower	Higher	Motor method D-357	Research method D-908	
1	Methane	57	b	57	b	b	4		x	x			1
2	Ethane	57	h,57	57	h,57	h,57	35		x	x	e	e,g	2
3	Propane	57	h,57	57	h,57	h,57	34		x	x		e,g	3
4	Isobutane	57	h,57	57	h,57	h,57	32		x	x		e,g	4
5	n-Butane	57	h,57	57	h,57	h,57	33		x	x	g	g	5
6	Isopentane	57	57	57	57	57	57		x	x			6
7	n-Pentane	57	57	57	57	57	57		x	x	g	g	7
8	Neopentane	57	h,57	57	h,57	h,57	57		x	x			8
9	n-Hexane	57	57	57	57	57	57		x	x			9
10	2-Methylpentane	57	57	57	57	57	57		x	x			10
11	3-Methylpentane	57	57	57	57	57	57		x	x			11
12	Neohexane	57	57	57	57	57	57		x	x			12
13	2,3-Dimethylbutane	57	57	57	57	57	57		x	x		e	13
14	n-Heptane	57	57	57	57	57	57		x	x			14
15	2-Methylhexane	57	57	57	57	57	57		x	x			15
16	3-Methylhexane	57	57	57	57	57	57		a,x	a,x			16
17	3-Ethylpentane	57	57	57	57	57	57		a,x	a,x			17
18	2,2-Dimethylpentane	57	57	57	57	57	57		a,x	a,x			18
19	2,4-Dimethylpentane	57	57	57	57	57	57		a,x	a,x			19
20	3,3-Dimethylpentane	57	57	57	57	57	57		a,x	a,x			20
21	Triptane	57	57	57	57	57	57		a,x	a,x	e	e	21
22	n-Octane	57	57	57	57	57	57		x	x			22
23	Diisobutyl	57	57	57	57	57	57		a,x	a,x			23
24	Isooctane	57	57	57	57	57	57		x	x			24
25	n-Nonane	57	57	57	57	57	57		x	x			25
26	n-Decane	57	57	57	57	57	57		a,x	a,x			26
27	Cyclopentane	57	57	57	57	57	57		a,x	a,x	g	e	27
28	Methylcyclopentane	20	20	20	20	20	57		x	x			28
29	Cyclohexane	20	57	57	57	57	57		x	x			29
30	Methylcyclohexane	20	20	20	20	20	57		x	x			30
31	Ethene(Ethylene)	57	b	57	b	b	63		x	x		e	31
32	Propene(Propylene)	57	a	57	a	a	57		x	x		e	32
33	1-Butene(Butylene)	57	h,57	57	h,57	h,57	57		x	x	g		33
34	cis-2-Butene	57	h,57	57	h,57	h,57	57		a,x	a,x			34
35	trans-2-Butene	57	h,57	57	h,57	h,57	57		a,x	a,x			35
36	Isobutene	57	h,57	57	h,57	h,57	57		a,x	a,x			36
37	1-Pentene	57	57	57	57	57	57		x	x			37
38	1,2-Butadiene	57	h,57	57	h,57	h,57	56		a,x	a,x			38
39	1,3-Butadiene	57	h,57	57	h,57	h,57	56		x	x			39
40	Isoprene	57	57	57	57	57	56		a,x	a,x			40
41	Acetylene	57	a	57	a	a	56		x	x			41
42	Benzene	57	57	57	57	57	14		x	x	e		42
43	Toluene	57	57	57	57	57	57		x	x	e	e	43
44	Ethylbenzene	57	57	57	57	57	57		x	x		e	44
45	o-Xylene	57	57	57	57	57	57		x	x			45
46	m-Xylene	57	57	57	57	57	57		x	x	e	e	46
47	p-Xylene	57	57	57	57	57	57		x	x	e	e	47
48	Styrene	20	20	20	20	20	56		x	x	e	e	48
49	Isopropylbenzene	57	57	57	57	57	57		x	x		e	49
50	Methyl alcohol	62	62	62	62	62	58		x	x			50
51	Ethyl alcohol	62	62	62	62	62	58		x	x			51
52	Carbon monoxide	8	b	8	b	b	58		x	x			52
53	Carbon dioxide						58						53
54	Hydrogen sulfide	18	h,58	18	h,58	h,58	31		x	x			54
55	Sulfur dioxide						56						55
56	Ammonia	58		58			58		x	x			56
57	Air						45						57
58	Hydrogen	58		58			p,58		x	x			58
59	Oxygen						46						59
60	Nitrogen						58						60
61	Chlorine						56						61
62	Water	8	w,8	w,8			37						62
63	Helium												63
64	Hydrogen chloride						56						64

12/12/86

(Cont'd)

NOTES FOR THE TABLE OF PHYSICAL CONSTANTS

a. Values in parentheses are estimated values.
b. The temperature is above the critical point.
c. At saturation pressure (triple point).
d. Sublimation point.
e. The + sign and number following specify the number of cm^3 of TEL added per gallon to achieve the ASTM octane number of 100, which corresponds to that of Isooctane (2,2,4-Trimethylpentane.)
f. These compounds form a glass. An earlier investigator [12] reported a value of -244°F for the freezing point of 3-Methylpentane.
g. Average value from octane numbers of more than one sample.
h. Saturation pressure and 60°F.
i. Index of refraction of the gas.
j. The temperature difference correction from [51].
k. Densities of the liquid at the normal boiling point.
m. Heat of sublimation.
n. Equation 2 of the reference was refitted to give: $a = 0.7872957$; $b = 0.1294083$; $c = 0.03439519$.
p. Normal hydrogen (25% para, 75% ortho).
q. The vapor pressure [5] at the critical temperature [43].
r. An extrapolated value.
s. The value for the critical point of helium depends upon the temperature scale chosen. See reference 7.
t. Gas at 60°F and the liquid at the normal boiling point.
u. Fixed points on the 1968 International Practical Temperature Scale (IPTS-68).
v. Densities at the normal boiling point are: Ethane, 4.540 [34]; Propane, 4.848 [33]; Propene, 5.083 [5]; Hydrogen Chloride, 9.948 [56]; Hydrogen Sulfide, 7.919 [30]; Ammonia, 5.688 [56]; Sulfur Dioxide, 12.20 [56].
w. Technically, water has a heating value in two cases: net (-1060. Btu/lbm) when water is liquid in the reactants and gross (+50.313 Btu/ft^3) when water is gas in the reactants. The value is the ideal heat of vaporization (enthalpy of the ideal gas less the enthalpy of the saturated liquid at the vapor pressure). This is a matter of definition; water does not burn.
x. Extreme values of those reported by ref. 24.

A. Molar mass (molecular weight) is based upon the following atomic weights: C = 12.011; H = 1.00794; O = 15.9994; N = 14.0067; S = 32.06, Cl = 35.453. The values were rounded off after calculating the molar mass using all significant figures in the atomic weights.
B. Boiling point: the temperature at equilibrium between the liquid and vapor phases at 14.696 psia.
C. Freezing point: the temperature at equilibrium between the crystalline phase and the air saturated liquid at 14.696 psia.
D. The refactive index reported refers to the liquid or gas and is measured for light of wavelength corresponding to the sodium D-line (589.26 nm).
E. The relative density (specific gravity): ρ(liquid, 60°F)/ρ(water, 60°F). The density of water at 60°F is 8.33718 lbm/gal.
F. The temperature coefficient of density is related to the expansion coefficient by:

$$\frac{1}{\rho}\left(\frac{\partial \rho}{\partial T}\right)_P = -\frac{1}{V}\left(\frac{\partial V}{\partial T}\right)_P, \quad \text{in units of } 1/^\circ F$$

G. Pitzer acentric factor:

$$\omega = -\log_{10}(P/P_c) - 1, \quad P \text{ at } T = 0.7\,T_c$$

H. Compressibility factor of the real gas, $Z = (PV)/(RT)$, is calculated using the second virial coefficient equation.
I. The density of an ideal gas relative to air is calculated by dividing the molar mass of the of the gas by 28.9625, the calculated average molar mass of air. See ref. 39 for the average composition of dry air. The specific volume of an ideal gas is calulated from the ideal gas equation. The volume ratio is: V(ideal gas)/V(liquid in vacuum).
J. The liquid value is not rigorously C_P, but rather it is C_S, the heat capacity at saturated conditions, defined by:

$$C_S = C_P - T\left(\frac{\partial V}{\partial T}\right)_P \frac{dP_S}{dT}$$

When dealing with liquids far from the critical point, $C_S \approx C_P$.
K. The heating value is the negative of the enthalpy of combustion at 60°F and 14.696 psia in an ideal reaction (one where all gasses are ideal gasses). For an arbitrary organic compound, the combustion reaction is:

$$C_nH_mO_hS_jN_k(s,l,or\ g) + (n + \frac{m}{4} - \frac{h}{2} + j)O_2(g) \longrightarrow nCO_2(g) + \frac{m}{2}H_2O(g\ or\ l) + \frac{k}{2}N_2(g) + jSO_2(g)$$

where s, l,and g denote respectively solid, liquid and ideal gas. For gross heating values, the water formed is liquid; for net heating values, the water formed is ideal gas. Values reported are on a dry basis. To account for water in the heating value, see GPA 2172. The Btu/lbm or gal. liquid column assumes a reaction with the fuel in the liquid state, while the Btu/ft^3 ideal gas column assumes the gas in the ideal gas state. Therefore, the values are not consistent if used in the same calculation, e.g., a gas plant balance.
L. The heat of vaporization is the enthalpy of the saturated vapor at the boiling point at 14.696 psia minus the enthalpy of the saturated liquid at the same pressure.
M. Air required for the combustion of ideal gas for compounds of formula $C_nH_mO_hS_jN_k$ is:

$$V(air)/V(gas) = (n + \frac{m}{4} - \frac{h}{2} + j)/0.20946$$

COMMENTS

Units - all dimensional values are reported in terms of the following three units, which are defined in terms of the corresponding SI units:

mass - pound (avdp), lbm = 0.45359337 kg
length - foot, ft = 0.3048 m
temperature - degree fahrenheit, (t/°F) = 32 + [1.8(t/°C)]
Temperature of the Celsius scale is defined by the International Practical Temperature of 1968 (IPTS-68).

Other derived units are:

volume - cubic foot, ft^3 = 0.02831685 m^3
gallon = 231 in^3 = 0.003785412 m^3
pressure - pound per square inch absolute = 6894.757 Pa
energy - British thermal unit (I.T.), Btu = 251.9958 cal (I.T.) = 1055.056 J

Physical constants:

gas constant, R = 1.98598 Btu(I.T.)/(°R·lb mol)
10.7316 ft^3 psia/(°R·lb mol)
8.31448 J/(K · mol)

Conversion factors:

1 ft^3 = 7.480520 gal.
1 lbm/ft^3 = 0.1336806 lbm/gal = 16.018462 kg/m^3
1 psia = 0.06804596 atm
1 atm = 14.69595 psia = 760 Torr
1 Btu (I.T.) = 252.1644 cal_{th}

REFERENCES FOR THE TABLE OF PHYSICAL CONSTANTS

1. Ambrose, D., National Physical Laboratory, Teddington, Middlesex, England: Feb. 1980, NPL Report Chem 107.
2. Ambrose, D.; Hall, D. J.; Lee, D. A.; Lewis, G. B.; Mash, C. J., J. Chem. Thermo., 11, 1089 (1979).
3. Angus, S.; Armstrong, B.; de Reuck, K. M., Eds. "Carbon Dioxide. International Thermodynamic Tables of the Fluid State-3"; Pergamon Press: Oxford, 1976.
4. Angus, S.; Armstrong, B.; de Reuck, K. M., Eds. "Methane. International Thermodynamic Tables of the Fluid State-5"; Pergamon Press: Oxford, 1978.
5. Angus, S.; Armstrong, B.; de Reuck, K. M., "Propylene (Propene). International Thermodynamic Tables of the Fluid State-7"; Pergamon Press: Oxford, 1980.
6. Angus, S.; de Reuck, K. M.; Armstrong, B., Eds. "Nitrogen. International Thermodynamic Tables of the Fluid State-6"; Pergamon Press: Oxford, 1979.
7. Angus, S.; de Reuck, K. M.; McCarthy, R. D., Eds. "Helium. International Thermodynamic Tables of the Fluid State-4"; Pergamon Press: Oxford, 1977.
8. Armstrong, G. T.; Jobe, T. L., "Heating Values of Natural Gas and its Components," NBSIR 82-2401, May 1982.
9. Aston, J. G.; Szasz, G. J.; Fink, H. L., J. Am. Chem. Soc., 65, 1135 (1943).
10. Barber, C. R., Metrologia 5, 35 (1969).
11. Boundy, R. H.; Boyer, R. F., (Eds.), "Styrene, Its Polymers, Copolymers and Derivatives," A.C.S. monograph No. 115, Reinholt, N.Y., 1952.
12. Bruun, J. H.; Hicks-Bruun, M. M., J. Res. NBS, 5, 933 (1930).
13. Chaiyavech, P.; Van Winkle, M., J. Chem. Eng. Data, 4, 53 (1959).
14. Chao, J., "Benzene," Key chemical data books, Thermodynamics Research Center, Texas A & M University, College Station, Texas (1978).
15. Chao, J., Hydrocarbon Processing, 59(11), 295 (1979).
16. Chao, J.; Hall, K. R., Thermochimica acta, 72, 323 (1984).
17. Chao, J.; Hall, K. R.; Yao, J., Thermochimica acta, 64, 285 (1983).
18. CODATA Task Group on Key Values for Thermodynamics, CODATA Special Report No. 7, 1978.
19. Commission on Atomic Weights and Isotopic Abundances, Pure and Appl. Chem. 49, 1102 (1983).
20. Cox, J. D.; Pilcher, G., "Thermochemistry of Organic and Organometallic Compounds," Academic Press, London, 1970.
21. Dean, J. W., "A Tabulation of the Properties of Normal Hydrogen from Low Temperature to 300 K and from 1 to 100 Atmospheres," NBS Tech. Note 120, November 1961.
22. Douslin, D. R.; Huffman, H. M., J. Am. Chem. Soc., 68, 1704 (1946).
23. Edwards, D. G., "The Vapor Pressure of 30 Inorganic Liquids Between One Atmosphere and the Critical Point," Univ. of Calif., Lawrence Radiation Laboratory, UCRL-7167. June 13, 1963.
24. Engineering Sciences Data Unit, "EDSU, Engineering Sciences Data," EDSU International Ltd., London.
25. Flebbe, J. L.; Barclay, D. A.; Manley, D. B., J. Chem. Eng. Data, 27, 405 (1982).
26. Francis, A. W., J. Chem. Eng. Data, 5, 534 (1960).
27. Friedman, A. S.; Haar, L., J. Chem. Phys. 22, 2051 (1954).
28. Ginnings, D. C.; Furukawa, G. T., J. Am. Chem. Soc. 75, 522 (1953).
29. Glasgow, A. R.; Murphy, E. T.; Willingham, C. B.; Rossini, F. D., J. Res. NBS, 37, 141 (1946).
30. Goodwin, R. D., "Isobutane: Provisional Thermodynamic Functions from 114 to 700K at Pressures to 700 Bar," NBSIR 79-1612, July 1979.
31. Goodwin, R. D., "Hydrogen Sulfide Provisional Thermochemical Properties from 188 to 700 K at Pressures to 75 MPa," NBSIR 83-1694, October 1983.
32. Goodwin, R. D.; Haynes, W. M., "Thermophysical Properties of Isobutane from 114 to 700 K at Pressures to 70 MPa," NBS Tech. Note 1051, January 1982.
33. Goodwin, R. D.; Haynes, W. M., "Thermophysical Properties of Normal Butane from 135 to 700 K at Pressures to 70 MPa," NBS Monograph 169, April 1982.
34. Goodwin, R. D.; Haynes, W. M., "Thermophysical Properties of Propane from 85 to 700 K at Pressures to 70 MPa," NBS Monograph 170, April 1982.
35. Goodwin, R. D.; Roder, H. M.; Straty, G. C.; "Thermophysical Properties of Ethane, 90 to 600 K at Pressures to 700 bar," NBS Tech. Note 684, August 1976.
36. Guthrie, G. B.; Huffman, H. M., J. Am. Chem. Soc., 65, 1139 (1943).
37. Haar, L.; Gallagher, J. S.; Kell, G. S., "NBS/NRC Steam Tables," Hemisphere Publishing Corporation, Washington, 1984.
38. Huffman, H. M.; Park, G. S.; Thomas, S. B., J. Am. Chem. Soc., 52, 3241 (1930).
39. Hust, J. G.; Stewart, R. B., "Thermodynamic Property Values for Gaseous and Liquid Carbon Monoxide from 70 to 300 Atmospheres," NBS Technical Note 202, Nov. 1963.
40. Jones, F. E., J. Res. NBS, 83, 419 (1978).
41. Kell, G. S., J. Chem. Eng. Data, 20, 97 (1975).
42. Kishimoto, K.; Suga, H.; Syuzo, S., Bull. Chem. Soc. Japan, 46, 3020 (1973).
43. Kudchadker, A. P.; Alani, G. H.; Zwolinski, B. J., Chem. Rev. 68, 659 (1968).
44. Marchman, H.; Prengle, H. W.; Motard, R. L., Ind. Eng. Chem., 41, 2658 (1949).
45. "The Matheson Unabridged Gas Data Book"; Matheson Gas Products; New York, 1974.
46. McCarty, R. D.; Weber, L. A., "Thermophysical Properties of Oxygen from the Freezing Liquid Line to 600 R for Pressures to 5000 Psia", NBS Technical Note 384, July 1971.
47. Messerly, J. F.; Guthrie, G. B.; Todd, S. S.; Finke, H. L., J. Chem. Eng. Data, 12, 338 (1967).
48. Messerly, J. F.; Todd, S. S.; Guthrie, G. B., J. Chem. Eng. Data, 15, 227 (1970).
49. Miller, A.; Scott, D. W., J. Chem. Phys., 68(3), 1317 (1978).
50. Miller, L.P.; Wachter, N. W.; Fried, V., J. Chem. Eng. Data, 20, 417 (1975).
51. Ohe, S., "Computer Aided Data Book of Vapor Pressure," Data Book Publishing Co., Tokyo, Japan, 1976.
52. Riddick, J. A.; Bunger, W. B., "Organic Solvents, Physical Properties and Methods of Purification," Wiley-Interscience: New York, 1970.
53. Roder, H. M., "Measurements of the Specific Heats, C_s, and C_v, of Dense Gaseous and Liquid Ethane," J. Res. Nat. Bur. Stand. (U.S.) 80A, 739 (1976).
54. Scott, R. B.; Meyers, C. H.; Rands, R. D.; Brickwedde, F. G.; Bekkedahl, N., J. Res. NBS, 35, 39 (1945).
55. Sliwinski, P., Z. Physik. Chem. (Frankfurt) 63, 263 (1969).
56. Stull, D. R.; Westrum, E. F.; Sinke, G. C., "The Chemical Thermodynamics of Organic Compounds," John Wiley & Sons, Inc., New York, 1969.
57. "TRC Thermodynamic Tables – Hydrocarbons," Thermodynamics Research Center, Texas A&M University System: College Station, Texas.
58. "TRC Thermodynamic Tables–Non-Hydrocarbons," Thermodynamics Research Center, Texas A&M University System: College Station, Texas.
59. Vargaftik, N. B.; "Tables on the Thermophysical Properties of Liquids and Gases," Hemisphere Publishing Corporation, Washington D.C., 1975.
60. Weast, R. C., "Handbook of Chemistry and Physics," The Chemical Rubber Co: Cleveland, Ohio, 1969.
61. Wilhoit, R. C.; Aljoe, K. L.; Adler, S. B.; Kudchadker, S. A., "Isopropylbenzene and 1-Methyl-2-, -3-, and -4-isopropylbenzene," API Monograph Series No. 722, American Petroleum Institute, 1984.
62. Wilhoit, R. C.; Zwolinski, B. J., J. Phys. Chem. Ref. Data, 2, Suppl. no. 1 (1973).
63. Younglove, B. A., J. Phys. Chem. Ref. Data, 11, Suppl. no. 1 (1982).

How to calculate the specific gravity of natural gas

A gas mixture is composed of molecules of various sizes and its molecular weight is calculated by summing the molar mass M_i of each component in the gas in proportion to its mole/volume composition y_i

$$M_c = \sum y_i M_i$$

For example, dry air consists essentially of nitrogen, oxygen, and small amounts of other gases in the approximate mole fraction 78% nitrogen, 21% oxygen, and 1% argon. Its molecular weight is estimated at 29.0 lb/lb mole:

$$0.78(28.02) + 0.21(32.00) + 0.01(39.94) = 28.97 \text{ lb/lb mole.}$$

The specific gravity of a gas is defined as the ratio of the density of the gas to the density of dry air with both measured at the same temperature and pressure.

Example. Calculate the specific gravity of a gas of the following composition:

Component	Mole fraction
Carbon Dioxide	0.002
Nitrogen	0.005
Methane	0.922
Ethane	0.028
Propane	0.022
n-Butane	0.008

Solution. The molecular weight of the gas is computed as 17.3 lb/lb mole, and the specific gravity γ is obtained by dividing by the molecular weight of air:

$$\gamma = \frac{17.29}{29.0} = 0.60$$

Component	Mole fraction, y_i	Molar mass, M_i	y_i M_i
CO_2	0.002	44.01	0.09
N_2	0.005	28.01	0.14
C_1	0.922	16.04	14.79
C_2	0.028	30.07	0.84
C_3	0.022	44.10	0.97
nC_4	0.008	58.12	0.46
			17.29

What are the four different types of heating value?

The heating value of a gas is the quantity of heat produced when the gas is burned completely to carbon dioxide and water. Four adjectives are used to describe heating values - wet, dry, gross, and net – depending on the nature of the gas prior to combustion and the condition of the water of combustion after burning.

Wet and dry refer to the condition of the gas prior to combustion, whether the gas is saturated with water vapor (wet) or contains no water vapor (dry or bone dry).

Gross and net refer to the condition of the water of combustion after burning. Gross heating value is the heat produced with the combustion products cooled to standard conditions and the water condensed to a liquid state, while in net heating value the water remains vapor at standard conditions. Gross heating value is the most frequently used in the petroleum industry.

How to calculate the heating value of natural gas

Ideal gas assumptions are usually adequate in calculations, since the change to real gas is usually very small (less than one percent) under normal conditions and is often ignored. The heating value of an ideal gas is calculated by summing the heating value of each component in the gas L_i in proportion to its mole/volume composition:

$$L_c = \sum y_i L_i$$

Heating values in the U.S. are usually expressed as Btu/scf. To convert ideal gas values to real gas values at standard conditions, the ideal value is divided by the compressibility factor z of the gas mixture,

$$L_c = \frac{L_{c,ideal}}{z}$$

where,

$$z = 1 - \left(\sum y_i (1 - z_i)\right)^2$$

Heating values L_i and z_i factors for molecules are obtained from reference texts.

Example. Calculate the heating value of the gas in Table 1 at ideal conditions and as a real gas. How much do they differ? Gross heating values and compressibility factors at standard conditions are given.

Solution. The heating value of the gas is computed as 1062.2 Btu/scf. The value of z for the mixture is computed as

$$z = 1 - (0.0484)^2 = 0.9977.$$

Thus, the gross heating value of the gas as a real gas is

$$L_{real} = \frac{L_{c,ideal}}{z} = \frac{1062.2}{0.9977} = 1064.6 \text{ Btu/scf}.$$

Table 1

Component	Mole fraction, y_i	Gross Heating Value, L_i	$y_i L_i$	Standard Condition	
				z_i	$y_i\sqrt{1-z_i}$
CO_2	0.002	0.0	0.0	0.9943	0.0002
N_2	0.005	0.0	0.0	0.9997	0.0001
C_1	0.922	1010.0	931.2	0.9998	0.0412
C_2	0.028	1769.6	49.5	0.9919	0.0025
C_3	0.022	2516.1	55.4	0.9825	0.0029
nC_4	0.008	3263.0	26.1	0.9667	0.0015
	0.987		1062.2		0.0484

Nomograph for calculating density and specific volume of gases and vapors

The density of a gas or vapor can be calculated from the equation for an ideal gas:

$$\rho = \frac{144P}{RT} = \frac{MP}{10.72T} = \frac{2.70PS_g}{T}$$

where:

ρ = Density of gas, lb/cu ft
P = Absolute pressure, psi (psig + 14.7)
R = Individual gas constant (10.72/M)
M = Molecular weight of gas
T = Temperature, °R
S_g = Specific gravity of gas relative to air

The above equation and corresponding nomograph (Figure 1) can be used to calculate the density (or specific volume) of gases at low pressures and high temperatures, where the ideal gas law holds. Note that the pressure scale is calibrated in psi gauge so that the correction of 14.7 psi is not needed when using the nomograph.

Example. What is the density of dry methane if its temperature is 100°F and its pressure is 15 psig?

Connect	With	Mark or Read
M = 16	t = 100°F	Index
Index Mark	P = 15 psig	ρ = 0.08 lb/cu ft

Source

"Fluid Flow Through Pipe, Fittings and Valves," *Technical Paper No. 410*, A-11, Crane Company, Chicago, IL (1957).

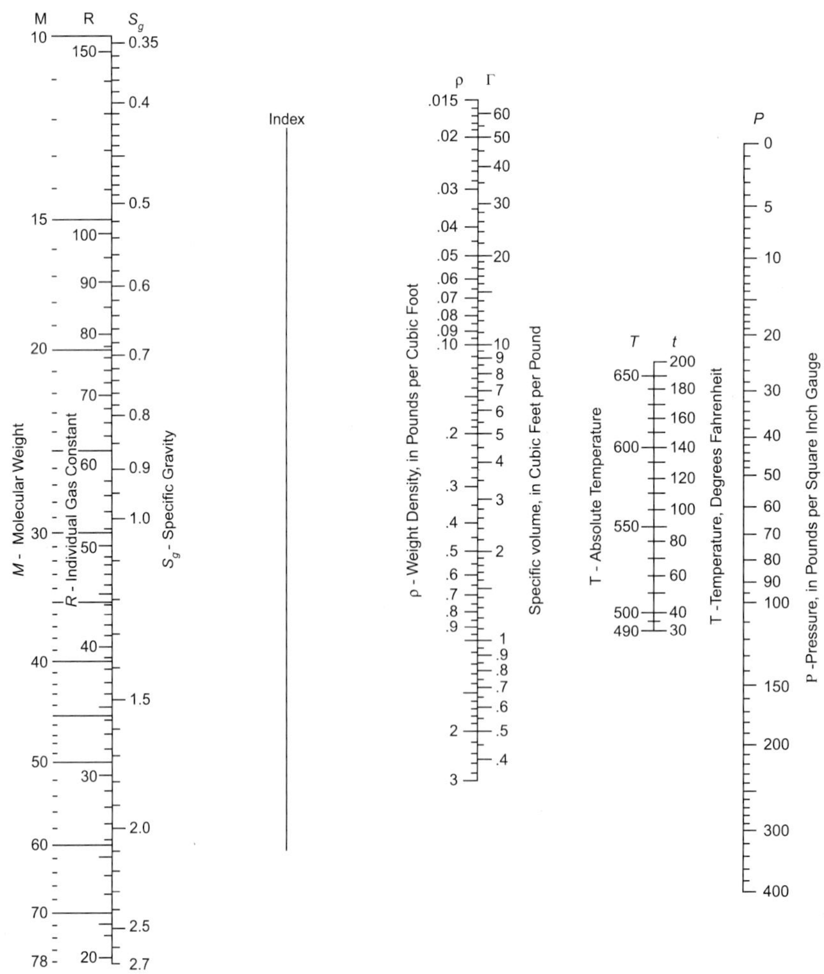

Figure 1. Weight density and specific volume of gases and vapors.

Considerations for selecting energy measurement equipment

Paul E. Kizer, Applied Automation, Inc., Houston, Texas

With natural gas energy determination becoming an integral part of the transmission and distribution business in today's market, companies are placing increased emphasis on choosing the correct measuring device. Most current contracts have a Btu specification, and many others use MMBtu instead of gas volume.

Here is what happens at the burner tip. The following is the exothermic reaction of methane and oxygen (80% to 95% of natural gas)[1]:

$$CH_4 + 2O_2 \rightarrow CO_2 + 2H_2O + \text{Heat (1010 Btu/cf)}$$

for the case of ethane (up to 3% of natural gas):

$$2C_2H_6 + 7O_2 \rightarrow 4CO_2 + 6H_2O + \text{Heat (1769 Btu/cf)}$$

and for the case of propane (usually up to 0.5%, except where propane/air peak shaving is used):

$$C_3H_8 + 5O_2 \rightarrow 3CO_2 + 4H_2O + \text{Heat (2516 Btu/cf)}$$

Since natural gas is a mixture of these three hydrocarbons plus small amounts of other compounds, these three chemical reactions account for 90% to 99% of the heat generated when the gas is burned. However, one mole (a mole is a standard number of molecules [6.023×10^{23}, Avogadro's number])[17] of propane will generate more than twice the heat of one mole of methane ($CH_4 = 1010$ Btu/cf vrs $C_3H_8 = 2516$ Btu/cf at 60°F and 414.696 psia). A small error in the heavier or longer chain hydrocarbon components of natural gas leads to a larger error in the Btu calculation. Consequently, a small shift in the composition of any component can affect the heating value of a cubic foot of carefully measured gas.

The term MMBtu first came into wide use after the 1978 Natural Gas Policy Act changed the method of custody transfer from Mcf to MMBtu. This method was upheld in a court case involving FERC Order 356 (Jan. 1984).[2]

How much is it worth to keep track of the Btu? Postulate a ±5% error between doing a lab determination of the heating value on a spot sample of gas and on-line (nearly continuous) monitor of heating value, which results in a ±50 Btu difference at a 1,000 Btu/cf level. A thousand Btu/cf is fairly average for natural gas.

On a station that has 50 MMcf/d at $2.50 per Mcf or MMBtu, this is $125,000 worth of gas per day. Five percent of this is $6,250 per day. If a process chromatograph, $50,000 installed cost, is used to determine the energy content, a payout of less than 10 days is obtained on a 50 MMcf/d station. This explains why most major interconnects have Btu measurement on-line today.

General methods

First, consider the direct methods of energy determination, since this is the oldest method in use today. All direct methods involve the complete oxidation (burning) of gas to determine its energy content. There are two basic classes of direct measurement devices, calorimeter and thermtitrator.

Early on, the term representing 100,000 Btus was one therm of energy. One dekatherm (10 therms) is equal to 1,000,000 Btu or 1 MMBtu.[21] Perhaps this is where the name thermtitrator came from. However, as discussed earlier, MMBtu[2] has supplanted the therm unit of measurement for now.

The calorimeter was adapted from a laboratory physical chemistry method called "bomb calorimetry" These were laboratory instruments that yield precise results when used under controlled conditions. However, each determination required 4 to 8 hours to complete. This is fine for a research project, but in a production setting it is easy to see how gas sample cylinders would start piling up. The American calorimeter[3] provided laboratory type results in less time, but the lab conditions had to be tightly controlled with continually manned equipment. For on-line energy content measurement, the Cutler-Hammer Recording Calorimeter stood alone for decades.[3,4]

While this device still requires rigid environmental conditions, it runs unattended for long periods of time (weeks). It also provides a recorder output, which allows the correlation of the heating value with the volume of gas. The recording calorimeter's stated precision is ±5 Btu out of 1,000 Btus, although many operators can do better.

With the introduction of solid-state electronics, micro calorimetry devices have been tried, though not with much commercial success. Hart Scientific, Provo, Utah, has used a catalyst bed to perform a controlled isothermal oxidation on a very precise volume of gas. Initial results seem to show ±0.5 Btu at 1,000 Btu precision.[5]

The thermtitrator devices utilize a more elegant chemical method, stoichiometry—that is, air to gas ratios. One instrument, the PMI thermtitrator, uses the temperature difference between two flames to maintain the gas flow ratios at stoichiometric conditions.[6,7]

Accuracy is around ±1 Btu at 1,000 Btus and compares favorably with the recording calorimeter.[8] The other stoichiometric device is the Honeywell HVT. It uses a zirconium oxide oxygen sensor to determine stoichiometric conditions and automatically adjusts air/gas ratios to minimize excess O. Both devices yield rapid reaction to step changes in Btu content and are useful in process control applications such as boiler feed forward controls.

There are two classes of indirect Btu or energy measurement, both chromatographic devices, relying on the separation, identification, and quantification of all the components that make up the natural gas mixture.[9] The two classes are: laboratory chromatographs and on-line or process chromatographs. Through the years a number of different people have compared calorimetry to chromatography with varied results.[10,11]

The laboratory chromatograph (Figure 1) is just that. It is located in a laboratory. Somehow a gas sample must be obtained and transported in a cylinder to the lab to be analyzed.

How the sample is obtained is the main difference between lab and on-line gas chromatographs (GCs). The natural gas samples analyzed in a lab today usually are called spot or grab samples. A cleaned vessel (cylinder), sometimes N_2 filled, is purged and filled at least three times for reliable sampling. The pressured-up bottle is carried to a lab to be analyzed. More and more samples today are being collected using a continuous sampler. This device takes a small aliquot of gas at a predetermined time interval so that the bottle is filled at the end of the sampling cycle required by the gas contract.

With recent developments, these devices have been tied to electronic flow computers to give flow-weighted composite samples rather than time-weighted ones. An example of such an installation is in Figure 2.

Flow weighting generally gives even better results than time-weighted samples if there are variations in the flowrate at the sampled station. Great care should be exercised when analyzing a sample cylinder of natural gas obtained by any of the previous methods. For instance, the sample bottle should be heated to 120°F for at least 2 hours before the analysis is made to ensure that all the heavier components are analyzed. A complete discussion of the problems associated with the recommendations for bottle sample analysis have been published previously.[9,12]

The process or on-line gas chromatograph (GC) (Figure 3) is designed to continually purge a small amount of the gas flowing through the station or pipeline to provide a representative sample at the time the GC is ready to make a new analysis. Typically, this is once every 6 minutes.

The main advantage of continuous sampling and analysis is the documentation of the duration and the amount of deviation of the energy content that has occurred during the measurement period. It is possible to assign a heating value to a specific volume of gas if the on-line GC is tied directly to an electronic flow computer. This, of course, allows actual billing on an MMBtu custody transfer basis.

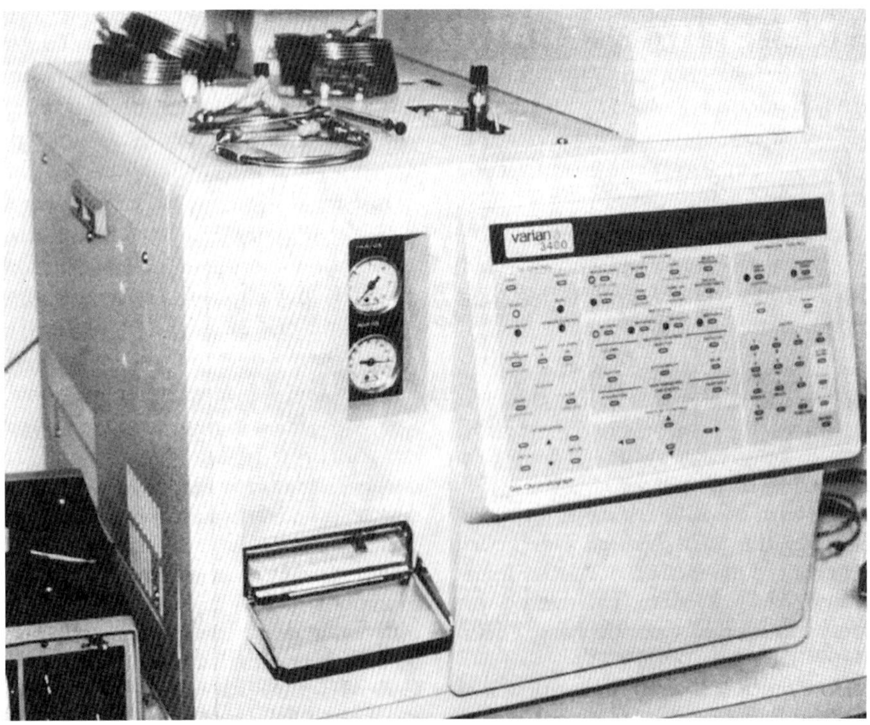

Figure 1. Laboratory chromatograph used to measure gas samples.

Figure 2. Example of on-line sampler connected to electronic flow computers.

Figure 3. Process sampler continually purges gas moving through a station or pipeline.

GCs—lab or on-line—using the computing power of today's modern microprocessor are capable of repeatability values of ±0.5 Btu at 1,000 Btu/cf. While the precision of a GC makes it well suited for custody transfer applications, the 6- to 10-minute analysis time makes it difficult to use for burner feed forward control. Generally, the on-line analyzers do not require much environmental protection at all. Most require protection from rain.

How chromatographs work

The way a modern microprocessor controlled on-line chromatograph works has been thoroughly discussed in previous papers.[13–15] However, it would be helpful to review briefly just how a GC works. Again, except for the sample conditioning system, this discussion generally applies to lab GCs as well as on-line GCs. For purposes of this discussion, the on-line GC has been divided into the following four sections:

- Sample conditioning system
- Gas chromatograph oven
- Chromatograph controller
- Input/output chassis

The sequence also is the route the sample takes from the pipeline to the printer. This makes it a logical flow for discussion as well.

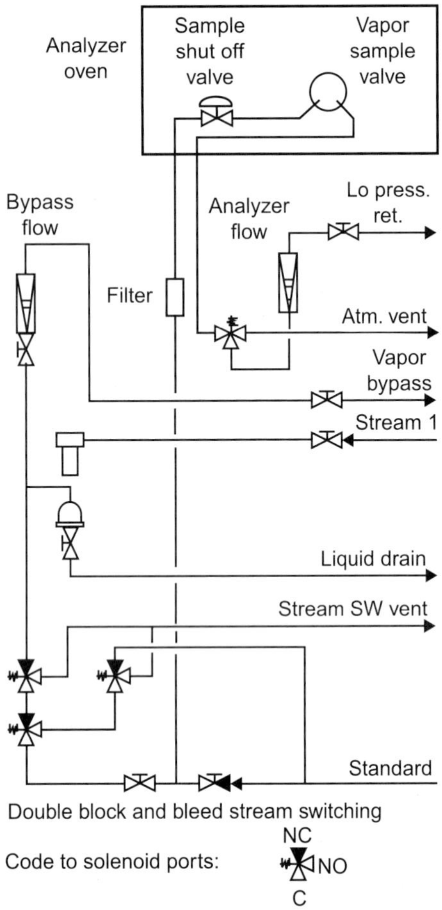

Figure 4. Bypass loop reduces sampling lag time.

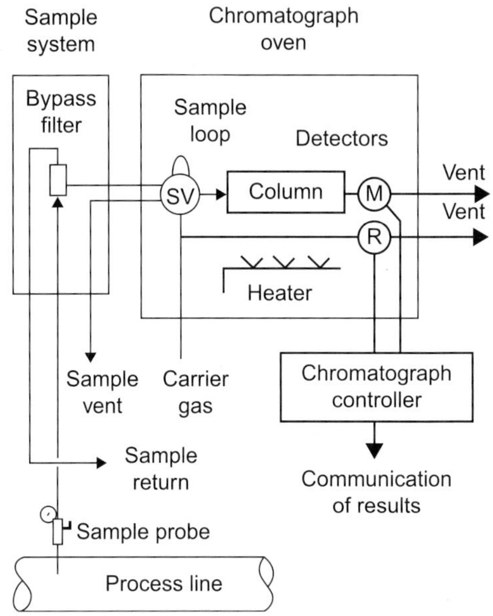

Figure 5. Block diagram illustrates gas chromatograph oven.

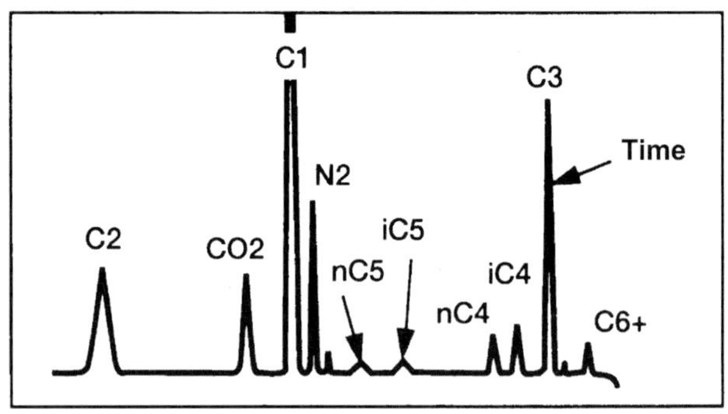

Figure 6. Separated components move through cell detectors for quantifying sample.

Sample conditioning system

To obtain a representative sample of gas, a probe should be inserted into the middle third of the pipeline. This procedure is as essential for good sampling for on-line GCs as it is for spot or composite samples. A major difference, however, is that the gas pressure is reduced to 15 psig at the probe for transport to the analyzer to ensure no heavy component (C^6+) is lost. With spot or composite samples, care is taken to maintain pipeline conditions. Sometimes it is useful to use a sample bypass loop to reduce sampling lag times, as shown in Figure 4.

Note that filtering and liquid coalescing is accomplished, if needed, to protect the integrity of the analyzer columns. Sometimes double block and bleed sample switching is used to prevent cross-contamination by the standard or other sample streams.

Gas chromatograph oven

Once the gas enters the oven, a repeatable gas sample is measured by a sample inject valve. The chromatograph controller then blocks the sample flow and allows it to equilibrate to atmospheric pressure. The sample is then swept onto a set of carefully selected chromatographic columns by a stream of helium gas for sequential separation of each component in the natural gas mixture. A block diagram is shown in Figure 5.

As each column does its specified separation job, it is switched out of the helium stream. The separated components make their way through the thermal conductivity cell detector where they can be quantified. An example of a detector, output is shown in Figure 6.

These peaks must be gated and the area under each peak integrated to be quantified, which is the job of the chromatograph controller.

Chromatograph controller

All the timing functions such as sample inject, column reverse, auto-calibration, and log results must be performed by the microprocessor controller board. In addition, this device must detect each peak, open and then close integration gates, identify the peak, and assign the correct response factor. The controller must then calculate the mole fraction of each component. Since these peaks are quite arbitrary in size due to equipment variability setup and environment, high-quality standards should be used to calibrate these instruments. This calibration assures accurate determination of mole fraction possible. All Btu determination methods require high quality standard samples.

On a Btu GC, the peaks also must be normalized to compensate for barometric pressure changes (which cause sample inject size variations). This allows the calculation of Btu to repeatability of ±0.5 Btu at 1,000 Btu/cf. This processor must also calculate the specific gravity, compressibility, condensable liquids, wobbie index, as well as the Btu. Calculation results must be made available to the outside world. This is the job of the input/output chassis.

Input/output chassis

The device houses communications option cards such as printer interface cards, analog output cards, and computer interface cards. These usually can be configured based on the customer's needs. However, caution is advised as specific computer protocols may or may not be supported by one vendor or another. One function for this input/output chassis, for example, is to hold an RS232 output card. This card would log results to an analyzer interface unit (AIU), which would then be connected to up to 30 flow computers (Fig. 7).

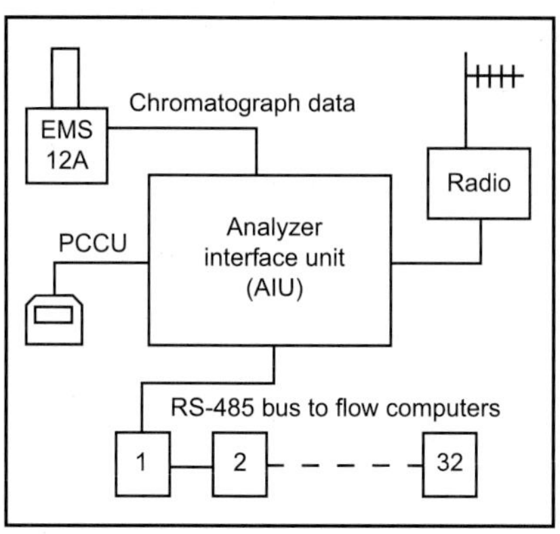

Figure 7. Output card logs results in an analyzer interface unit connected to at least 30 computers.

How Btu is calculated

The methods for calculation of the natural gas Btu from analysis data are well documented. Calculations have been published previously.[2,3,9,16,18] Nevertheless, it is good to review these methods. Although they look complicated, they really are straightforward. Below is the general formula for the Btu calculation:

$$\text{Btu/cf} = X_1 * C_1 + X_2 * C_2 + X_3 * C_3 \ldots + X_{11} * C_{11}$$

where:

X_1 = Mole fraction of C_1 or methane
X_2 = Mole fraction of C_2 or ethane
X_3 = Mole fraction of C_3 or propane

" "
" "
" "

X_{11} = Mole fraction of C_6+ hexanes and heavier

and:

C_1 = Molar heating value of C_1, or methane (1010 Btu/cf)
C_2 = Molar heating value of C_2, or ethane (1769 Btu/cf)

" "
" "
" "

C_{11} = Molar heating value of C_6+ or hexanes and heavier (4943 Btu/cf)

The calculation of specific gravity and condensable liquids is handled in the same manner. For example:

$$\text{Specific gravity} = X_1 * G_1 + X_2 * G_2 + \ldots + X_{11} * G_{11}$$

where: X_1, X_2, is the mole fraction of each natural gas component and: G_1, G_2, is the molar gravity contribution of each respective component.

This calculation yields gross or total heating value and is not compensated for compressibility Z. A short method for calculation of Z is:

$$Z = 1.00369 - (0.010 * SG) + (0.007 * X_{inert})$$

Total heating value would then be expressed by the formula:

$$\frac{\text{Btu/cf}}{\text{Z}} = \text{Btu/cf Dry}$$

in order to convert Dry Btu to saturated:

$$\text{Btu saturated} = \frac{\text{Btu Dry}}{1120 \text{ saturation factor } (1.0177)}$$

Some process gas chromatographs calculate all the above values at a pressure base of 14.696 psia only. Others allow user selection of a different contract pressure base, and all the constants are then factored by this new pressure base.

Care should be exercised when comparing Btu, SG, and others that all these values are calculated at the same temperature base and pressure base.

Conclusions

There are valid reasons for choosing one energy measurement device over the other. Thermtitrators are fast if control capability is needed. They are not suited for hazardous locations and require more rigorous attention to their operational environment. The modern microprocessors used on most chromatographs make them extremely capable in terms of computer interfaces.

Chromatographs are suitable for hazardous area installations. They are somewhat slower reacting, having 6-minute cycle times. However, they generally require less maintenance than calorimeters even though they seem more technically complicated. With the AGA-8 supercompressibility calculations, there will most likely be a requirement for a total analysis of the gas to do this part of the gas volume calculations. Thermtitrators cannot do this. GCs that only present the flow computer with four analog signals, proportional to Btu, SG, CO_2, and N_2, may not be much more help.

One company has an interface unit that will pick up and store hourly averages of all 12 components from a gas chromatograph. These hourly averages can be stored for up to three streams for 35 days each so that each hour's flow in the flow computer can be correlated with each hour's gas composition. This allows composition data for not only MMBtu computation but also AGA-8 volume calculations. From an audit standpoint, this allows downstream corrections and verification that has heretofore not been available.

References

1. GPA Standard 2145–84, p. 2.
2. GPA Reference Bulletin, Heating Value As a Basis for Custody Transfer of Natural Gas (1984 Draft).
3. AGA Gas Measurement Manual, Section 11A.2 Determination of Heating Value of Gas, p. IIA2.1 ff.
4. Installation and Operation of Recording Calorinators, A. E. Kersey ISHM #64, 1989, p. 93 ff.
5. Field Deployable Natural Gas Energy Meter, Lewis and Greefield, Sales Information 1989.
6. "Determination of Calorific Values of Natural Gas by Combustion Instruments," R. L. Howard, ISHM #64, 1989, p. 509.
7. "Fundamentals of Energy Management," R. L. Howard, GCMSC #24, 1989, p. 43.
8. "The Therm-titrator, A Comparison with You-Know-What," John Light, GPA #60, 1981.
9. Analysis of Natural Gas by Gas Chromatography," ASTM D1945, p. 133 ff.
10. Comparison of Calculated and Measured Heating Values of Natural Gas, Donald C. Melrose, Appalachian Gas Measurement Short Course (1975).
11. Measuring Gas; Chromatography Bests Calorimetry, Leisey, Potter, andMcKoy, Oil and Gas Journal (July 1977).
12. Sampling and Analysis of Natural Gas," Schepens, Kilmer and Bernos, GPA proceedings #61.
13. "Energy Measurement Utilizing On-Line Chromatographs," L. N. Cox, Prod. of Inter. School of Hydrocarbon Meas. #64, 1989, p. 469 ff.
14. "Btu Measurement in Natural Gas Using Process Gas Chromatography," Kizer and Sund, Proc. of Gulf Coast Meas. Soc. #24, 1989, p. 72 ff.
15. "How to Measure Btu Effectively", Staff, *Gas Industries*, August 1989.
16. "Calculations of Gross Heating Value, Relative Density and Compressibility Factor for Natural Gas Mixtures from Compositional Analysis," GPA Std. 2171–86.
17. *College Chemistry*, Bruce H. Mahan, Addison Wesley., 1966, p. 23.
18. "Gas Heating Value: What It Is and How To Measure, Calculate," D. N. McClanahan, *Oil and Gas Journal*, Feb. 20, 1967, p. 84 ff.
19. Campbell, "436 Pushes Measurement into the Electronics Age," *American Oil and Gas Report*, May 1987, p. 15 ff.
20. Hagar, "Pipelines, Producers Cope with Problems under FERC Order 436," *Oil and Gas Journal*, May 1987, p. 17 ff.
21. Paul, "A Multi Btu Approach to Fuels Procurement," Nymex Energy in the News, Spring, 1990, p. 4.

Facts about methane and its behavior

Dr. Robert O. Parker and **Dr. William H. Kapfer**, New York University, New York

The physical, transport and thermodynamic properties of liquefied natural gas (LNG) requires that the physical, transport, and thermodynamic properties of methane, the principal constituent of natural gas, be readily available. This article will present values most frequently used in process calculations and in other studies associated with the storage and processing of natural gas. In addition, sources of more detailed information are listed.

Methane, CH_4, is the hydrocarbon of lowest molecular weight (16.042). The molecule is best visualized as a regular tetrahedron with the carbon atom at the geometric center and the hydrogen atoms at each of the four apices. Methane is an end product of the anaerobic (without air) decay of plants, that is, of the breakdown of certain very complicated molecules. As such, it is the major constituent (up to 97%) of natural gas. It is the firedamp of the coal mines and can be seen as marsh gas bubbling to the surface of swamps.

Properties

The physical, transport, and thermodynamic properties required are:

Physical

1. Critical constants
2. Equation of state (PVT relationships)
3. Vapor pressure, including bp and fp
4. Vapor and liquid density

Transport

1. Viscosity of liquid and gas
2. Thermal conductivity of liquid and gas
3. Diffusivity

Thermodynamic

1. Heat capacity
2. Latent heat
3. Enthalpy
4. Entropy

They are required for process calculations: material and energy balances; heat transfer, mass-transfer and pressure loss calculations.

Critical constants

$T_c = -115.8°F\ (-82.1°C)$
$P_c = 673$ psia (45.8 atm)
$d_c = 10.17\ lb/ft^3$ (0.1630 g/cc)

The main utility of critical constants[1,2] to the engineer is related to the fact that correlations of many of the physical, transport, and thermodynamic properties of hydrocarbons are best expressed in terms of reduced temperature, T_R, and pressure, P_R:

$$T_R = \frac{T}{T_c},\ P_R = \frac{P}{P_c}$$

This is simply an application of the "principle of corresponding states" and is recommended only in the absence of sufficient reliable experimental data.

Equations of state

For ordinary conditions, low to moderate pressures and temperatures not too close to the normal boiling point, i.e., when inspection of a compressibility factor chart shows it to be nearly unity, the ideal gas law is:

$$PV = \frac{W}{M}RT \tag{1}$$

where: P = Pressure, psia
V = Volume, ft^3
W = Weight of gas, lb
M = Molecular weight (here 16.042)
R = Gas constant = 10.73 psia (ft^3)/lb mole (°R)
T = Temperature, °R

is sufficient. For most other conditions, Equation 1 is modified by inclusion of the compressibility factor, Z, in the right-hand member, viz:

$$PV = \frac{W}{M} ZRT \tag{1a}$$

The writers recommend compressibility factors of Nelson and Obert.[3] If an analytical equation of state is preferred to Equation 1a the writers recommend the 8 constant equation of Benedict, Webb, and Rubin.[4]

$$P = RTd + \left(B_oRT - A_o - C_o / T^2\right)d^2 + (bRT - a)d^3 + a\alpha d^6 + \frac{cd^3\left(1 + \gamma d^2\right)e^{-\gamma d^2}}{T^2}$$

where d = density, g-moles/l.
For methane the constants[4] are:

$B_o = 0.042600$
$A_o = 1.85500$
$C_o = 22{,}570.0$
$b = 0.00338004$
$a = 0.0494000$
$c = 2{,}545.00$
$\gamma = 0.0060000$
$\alpha = 0.000124359$

With these constants, R = 0.0820 liter-atm/(°K) (gram mole), P is in atmospheres, T is in °K, and e is the base of natural logarithms. Constants in this equation are available for ethane, ethylene, propane, propylene, n-butane, i-butane, i-butene, n-pentane, i-pentane, n-hexane, and n-heptane.[4] Using Equation 2 to predict the pressure corresponding to 89 data points,[5,6] it was found that the average absolute and maximum deviations for methane were 0.376% and +2.25% respectively. Reference 6 discusses a generalization of the Benedict—Webb—Rubin equation.

Vapor pressure

Freezing point in air at 1 atm = −296.46°F (182.48°C)
Boiling point at 29.921 in Hg = −258.68°F (−161.49°C)
Vapor pressures are tabulated on p. 121 of Reference 2 from 10 mm to 1,500 mm Hg and are represented by the Antoine equation.

$$\log_{10}P = A - B/(C + t) \tag{3}$$

or:

$$t = B/\left(A - \log_{10}P\right) - C \tag{3a}$$

with P in mm Hg and t in °C.
A = 6.61184
B = 389.93
C = 266.00

Transport properties

Viscosity (liquid)[2]

T°C	μ, centipoises
−185	0.226 (subcooled)
−180	0.188
−175	0.161
−170	0.142
−165	0.127
−165	0.115 (at sat. pressure)

Viscosity (vapor)[7] at 1 atm pressure

T°C	μ, centipoises
−181.6	0.00348
−78.5	0.00760
0	0.01026
20	0.01087
100	0.01331
200.5	0.01605
284	0.01813
380	0.02026
499	0.02264

Thermal conductivity

Thermal conductivity (liquid)

T°F	K, Btu/hr ft°F
−280	0.0841
−258.68	0.0790
−200	0.0568
−115.8	0.029

Values for liquid thermal conductivity represent estimates by the authors using the method of Viswanath,[9] which seems to be quite reliable for organic liquids (halogenated compounds excepted).

Thermal conductivity (vapor)[7]

T°F	K, Btu/hr ft°F
−300	0.00541
−200	0.00895
−100	0.01265
−40	0.0149
−20	0.0157

T°F	K, Btu/hr ft°F
0	0.01645
20	0.01725
40	0.01805
60	0.0190
80	0.01985
100	0.0208
120	0.0216
200	0.0258

Diffusivity

The transport property, mass diffusivity (D), is of importance in calculations usually entering computations in combination with viscosity and density as the Schmidt No., μ/ρ D. Reliable estimates of D_{12}, the diffusivity of species 1 through species 2, are best made by the method of Hirschfelder, Bird, and Spotz.[11]

$$D_{12} = \frac{0.001858T^{3/2}[M_1 + M_2]/[M_1M_2]^{1/2}}{P\sigma_{12}^2\Omega_D} \quad (4)$$

where:

D = Diffusivity, cm^2/sec
T = Temperature, °K
M = Molecular weight, M_1 of species 1, M_2 of species 2
P = Pressure, atm
$\sigma_{12} = (\sigma_1 + \sigma_2)/2$ = collision diameter, A°
Ω_D = Collision integral = function of KT/ε_{12}
K = Boltzmann's constant
ε_{12} = Force constant, $\sqrt{\varepsilon_1\varepsilon_2}$

Force constant and collision diameters for air and methane are:

	ε/K, °K	σ, A°
Air	97	3.617
Methane	136.5	3.822

These are known as the Lennard—Jones force constants, and an extensive table is given.[11] Ω_D, which depends upon the temperature through KT/ε, is tabulated in the reference.[12] A few values are:

KT/ε	Ω_D
0.30	2.662
0.50	2.066
1.00	1.439
1.50	1.198
2.00	1.075
4.00	0.8836
10	0.7424
50	0.5756
100	0.5130
400	0.4170

If the calculation requiring diffusivities is a crucial one, we recommend that the engineer consult the principal literature reference.[11]

Thermodynamic properties

The most extensive, reliable compilation of the thermodynamic properties is the one by Matthews and Hurd.[13] In addition to a pressure—enthalpy diagram, tabulations are provided for saturated methane from −280°F to −115.8°F, the critical temperature, and for superheated methane to 500°F and 1,500 psia. This tabulated information is reproduced in the Chemical Engineers' Handbook 3rd Ed.[14]

Corresponding to the pressure—temperature pair, values are given for:

a. Ratio fugacity (where fugacity[13] refers to "escaping tendency") to pressure
b. Specific volume of liquid and vapor
c. Enthalpy of liquid, vaporization, and vapor
d. Entropy of liquid and vapor

A few numbers of immediate interest to the process engineer are:

a. Latent heat of vaporization at normal boiling point = 219.22 Btu/lb
b. Liquid density at normal boiling point = 26.6 lb/ft^3
c. Vapor density at normal boiling point = 0.113 lb/ft^3

The liquid density is required to compute the volume of a diked area to contain a spill. Latent heat is useful in estimating the time required for a spill to be vaporized. The vapor density points up the fact that, although methane has a specific gravity referred to air of about 0.55, vapors generated at a spil l are about 1.5 times as heavy as ambient air; a matter of importance in various studies.

Heat capacities

If heat capacities are required for process calculations, they can be extracted from the thermodynamic tables mentioned above by the following procedure:

$$c_p = \frac{H_1 - H_2}{T_1 - T_2}, \text{Btu/lb°F} \quad (5)$$

Example. Find the heat capacity for superheated methane at 15 psia and 100°F.

Figure 1. Cold box where liquefaction of natural gas is accomplished.

From the Matthews and Hurd superheat table at 15 psia:[13,14]

$$T_1 = 110 \qquad H_1 = 426.6$$
$$T_2 = 90 \qquad H_2 = 415.8$$

Then

$$c_p = \frac{426.6 - 415.8}{110 - 90} = 0.54 \ \text{Btu/lb}^\circ\text{F}$$

The procedure given above is adequate for liquid heat capacities when enthalpy values are taken from the table for saturated methane. This works because the effect of pressure on liquid heat capacity is essentially negligible.

Alternatively, instantaneous values of heat capacity may be determined.[15]

$$c_p^0 = 0.213 + 1.11 \times 10^{-3}T - 2.59 \times 10^{-7}T^2 \ \text{Btu/lb}^\circ\text{F} \tag{6}$$

Note that Equation 6 yields the heat capacity at low pressure, 1 atm; at elevated pressures a correction may be required. If Equation 5 above is used, no pressure correction is necessary.

Heat capacity ratio

The ratio $c_p/c_v = \gamma$, useful in process calculations (adapted from reference 14), is for atmospheric pressure:

T°F	$\gamma = c_p/c_v$
−150	1.39
−100	1.36
−50	1.33
0	1.31
50	1.28
100	1.27
150	1.25

Flammability characteristics

Since methane is a combustible gas, consideration of any process involving its use or storage must include studies of the associated fire hazard. Basic data are flammable limits and the autoignition temperature. For methane in air these are[16,17] at atmospheric pressure:

a. Lower Flammable, Limit @ 25°C about 5% by vol
b. Upper Flammable, Limit @ 25°C about 15% by vol
c. Autoignition Temperature 1,004°F

These figures mean that mixtures of methane and air having less than 5% methane by volume are too lean to burn, those with more than 15% are too rich. The autoignition temperature is the minimum temperature at which ignition can occur. Figures on flammable limits refer to "homogeneous" mixtures. Industrially, few mixtures are completely homogeneous. In safety computations, therefore, the engineer must establish that the "highest local concentration" is below the lower flammable limit by a suitable margin before saying there is no fire hazard.

References

1. Kobe K.A., Lynn R.E., *Chem. Revs.* 52 (1953) 117.
2. American Petroleum Institute, "Selected Values of Physical and Thermodynamic Properties of Hydrocarbons and Related Compounds," Project 44, Carnegie Press, Pittsburgh, Pa. 1953 (NBS–C–461).
3. Nelson L.C., Obert E.F., *Trans.* ASME 76 (1954) 1057.
4. Benedict M., Webb G.B., Rubin L.C., Friend L., *Chem. Eng. Prog.* 47 (1951) 419, 449, 571, 609.
5. Kvalnes H.M., Gaddy V.L., *J. Ann. Chem. Soc.* 53 (1931) 394.
6. Edmister W.C., Vairogs J., Klekers A.J., *AIChE Journal* 14 (No. 3) (1968) 479.
7. *Handbook of Chemistry and Physics, 47th Edition*, 1966–67.
8. Huang E.T.S., Swift G.W., Kurata F., *AIChE Journal* 12 (1966) 932.
9. Viswanath D.S., *AIChE Journal* 13 (1967) 850.
10. Gamson B.W., *Chem. Eng. Prog.* 45 (1949) 154.
11. Hirschfelder J.O., Bird R.B., Spotz E.L., *Trans. ASME* 71 (1949) 921.
12. Hirschfelder J.O., Curtiss C.F., Bird R.B., *Molecular Theory of Gases and Liquids*. John Wiley and Sons, Inc., New York, 1954.
13. Matthews C.S., Hurd C.O., *Trans. AIChE* 42 (1946) 55 ibid., pp. 78 (1009).
14. J.H. Perry (Ed.), *Chemical Engineers Handbook, 3rd ed.* McGraw-Hill, New York, 1950.
15. Reid R.C., Sherwood T.K., *The Properties of Gases and Liquids*. McGraw-Hill, New York, 1958.
16. Coward H.F., Jones G.W., U.S. Bureau of Mines, Bulletin 503 (1952).
17. Zabetakis M.G., U.S. Bureau of Mines, Bulletin 627 (1965).

Conversion table for pure methane

		Long tons liq.	ft^3 liquid	ft^3 gas	m^3 liquid	m^3 gas	U.S. bbl liq.	Therms	Thermies
long ton liquid	=	1	84.56	52,886	2.394	1,419.7	15.06	5352	13,491
ft^3 liquid	=	0.011826	1	625.43	0.028311	16.789	0.1781	6.329	159.54
ft^3 gas $\times 10^6$	=	18.91	1,599	1×10^6	45.27	26,847	284.785	10,121	255,115
m^3 liquid	=	0.4177	35.315	22,090	1	593	629	223.55	5,635.2
m^3 gas $\times 10^9$	=	704,374	59.562×10^6	37252×10^9	1.686×10^6	1×10^9	10.608×10^6	0.377×10^9	9.503×10^9
U.S. bbl liquid	=	0.0664	5.615	3,512	0.15896	94.268	1	35.54	895.8
Therm $\times 10^6$	=	1,868	157,958	98.791×10^6	2.652×10^6	1×10^6	25.201×10^6		
Thermie $\times 10^6$	=	74.12	6,268	3.92×10^6	177.44	105,228	1,1162	39,669	1×10^6

Notes

1. ft^3 measured at 60°F, 14.696 lbf/in^2, dry
2. m^3 measured at 0°C, 760 mmHg, dry
3. Gross heat content of methane is 1012 Btu/ft^3 in gaseous forms: Specific gravity 0.554 (Air = 1) Boiling point −258.7°F (14.696 lbf/in^2)
4. 1×10^6 ft^3 gas = 24.42 long tons (0.4417 b/d) fuel oil equivalent
5. The above factors are based on the properties of pure methane; LNG as shipped in practice will vary in composition according to the presence of heavier hydrocarbons (C_2, C_3, C_4,) in the gas stream.

Introduction to LNG

Michael Choi, Process Engineering Fellow in ConocoPhillips, Houston, Texas

Among fossil fuels currently powering the world, natural gas is the most environmentally friendly option. While remote parts of the world and North America are awash with excess gas, the industrialized West and developing economies of the East cannot get enough of what some consider as the "transitional fuel" to renewable sources. To transport this compressible fluid long distances to markets greater than 1,500 miles across bodies of water, liquefied natural gas (LNG) has proved to be the most economic option. By refrigerating natural gas (primarily methane) to −260°F (−162°C), thereby shrinking its volume by 600:1, LNG can be transported in large insulated atmospheric-pressure tankers at reasonable cost.

Natural gas liquefaction is a series of refrigeration systems like the air conditioning system in a typical home consisting of a compressor, condenser, and evaporator to chill and condense the gas. The difference is in the scale and magnitude of the refrigeration. A typical single-train LNG plant may cost billions of dollars and consume 6-8% of the inlet gas as fuel. Since many of the impurities (water vapor, carbon dioxide, hydrogen sulfide, etc.) and heavier hydrocarbon compounds in natural gas would freeze at LNG temperatures, they must first be removed, and disposed or marketed as separate products.

This article is intended to provide an overview of an entire LNG liquefaction facility, from inlet gas receiving to LNG storage and loading, with emphasis on the liquefaction process and equipment. Differences among the commercially available liquefaction processes will be discussed. Although majority of LNG plants are located onshore, one floating LNG (FLNG) liquefaction facility has been installed through 2020. No doubt there will be others as reliability of the technology and expected improved project and field development economics are demonstrated. The innovations unique to FLNG will be introduced. The aim is to provide readers with a clear understanding of the technologies, equipment and process choices required for a successful LNG project, regardless of gas source or producing field circumstances.

Introduction

The liquefied natural gas industry underwent explosive growth during the first two decades of the twenty-first century due to a convergence of technology and market. The reduction in cost of liquefaction and shipping due to technological advances have made LNG economically attractive to both producer and consumer. Although LNG growth, and that of fossil fuels industry in general have slowed recently due to mounting evidence that burning of fossil fuels is the primary cause of Global Climate Change and the Covid-19 induced worldwide economic recession, many LNG projects continue to progress unhindered. The reason is that natural gas is "clean burning", emitting far less particulates, US EPA listed Hazardous Air Pollutants (HAPs), and heat trapping carbon dioxide than other fossil fuels. It is viewed by some as the "transitional fuel" to renewable sources.

Historically, LNG plants are in remote petroleum producing regions of the world with the LNG product exported to the industrialized West. Prior to the onset of large-scale shale gas production in North America in the 2000's, the US natural gas demand was expected to exceed supply and a number of LNG receiving/regasification terminals were built to accommodate LNG imports to meet the shortfall. However, with the enormous success of shale gas production in North America, there is currently an excess of natural gas in the US and the economics have reversed. A few "greenfield" natural gas liquefaction plants are being built to export some of the excess. All the previous receiving terminals have been converted to liquefaction service, thereby repurposing the LNG tanks, and loading facilities for export.

Air Products & Chemicals Inc.'s (APCI) Propane-pre-cooled Mixed Refrigerant and the Phillips Optimized Cascade processes are the dominant base load LNG technologies on the market with majority of the existing plant capacity. However, development of the Linde/Statoil Mixed Fluid Cascade, Shell Double Mixed Refrigerant, APCI APX and other processes to meet special field circumstances and economic requirements, has given producers options. The competition has fostered innovations in plant design, train size and equipment utilization. In this article, the gas pretreatment required, thermodynamics of natural gas liquefaction, features of the different liquefaction processes and the major equipment used are discussed.

Many large commercially attractive gas reservoirs are located offshore in remote parts of the world where the gas is destined for the LNG export market. For decades, operators have recognized the potential cost savings of having a floating LNG (FLNG) facility to: produce the gas via subsea tie-back well systems, process and convert the gas to LNG, and store and offload the LNG to tankers for transport to remote markets. That vision was finally realized in 2017 with the start-up of Shell's Prelude FLNG, the world first floating LNG vessel moored offshore Western Australia. A brief discussion of this technology is included for completeness.

Background

Natural gas, unlike oil, is compressible and occupies huge volumes at atmospheric pressure. The most economic means of transporting the fuel to market is through high pressure pipelines. As a result, the gas industry was developed primarily around regional or intra-continental pipeline networks. Until the 1940's, there was no intercontinental or trans-ocean trade of natural gas. In the 1960's, Japan, an island nation, began importing LNG in limited quantities. By refrigerating to about −260°F (−162°C) through various means, natural gas (which is primarily methane) is transformed into a liquid, thereby shrinking its volume by 600:1. The liquefied natural gas which has a specific gravity of 0.45 is transported by large ocean-going tankers. Majority of the current world fleet of these highly insulated cryogenic tankers is about 140,000 cubic meters. They carry the equivalent of 3.1 Bscf (87 million standard cubic meter) of gas, or about 512,000 BOE. Figure 1 is a photo of such an LNG tanker being loaded at the Oman LNG plant. The larger QMax and QFlex LNG tankers with capacity of 266,000 cubic meters, developed for the longer trade routes from Qatar to Europe and North America allow greater economy of scale should project economics dictate.

Worldwide LNG trade underwent explosive growth during the first two decades of the twenty-first century due to a convergence of technology and market. The reduction in cost of liquefaction and shipping due to technological advances have made LNG economically attractive to both producer and consumer. New plants in Australia, Egypt, Nigeria, Norway, Qatar, Trinidad, and other petroleum producing provinces came onstream. To achieve greater economy of scale, mega-size LNG trains with capacity of upward of 9 Mtpa (1.2 Bscf per day of gas equivalent) have been built. In 2020, most gas exporting countries have an LNG plant in service, under construction or in the planning stage. This includes the entry of North America which found itself with surplus natural gas due to the success of shale production.

To many petroleum engineers whose focus is finding and getting hydrocarbons out of the ground, LNG is a bit of a mystery. The plants are huge with tall columns resembling refineries serviced by our downstream colleagues. In fact, many of these columns serve the same purpose as those in the refinery. Some of them are absorbers, while some are distillation columns. However, they have nothing to do with the function of liquefying natural gas. They are there primarily to condition the produced fluid for liquefaction. An LNG plant would simply be a big refrigeration system if the reservoir was able to produce pure methane (no water vapor, heavy hydrocarbons, or other impurities). The refrigeration would consist of the same major components as the air conditioning (AC) system in our home. It would have a compressor/driver, refrigerant condenser, and evaporator (process heat exchanger). The liquefaction equipment in an LNG plant would only differ from a home AC system in size and operating temperature.

However, Mother Nature is never that simple. All produced gas is saturated with water vapor and would freeze and plug heat exchangers at cryogenic temperatures. In

Figure 1. Oman LNG tanker being loaded.

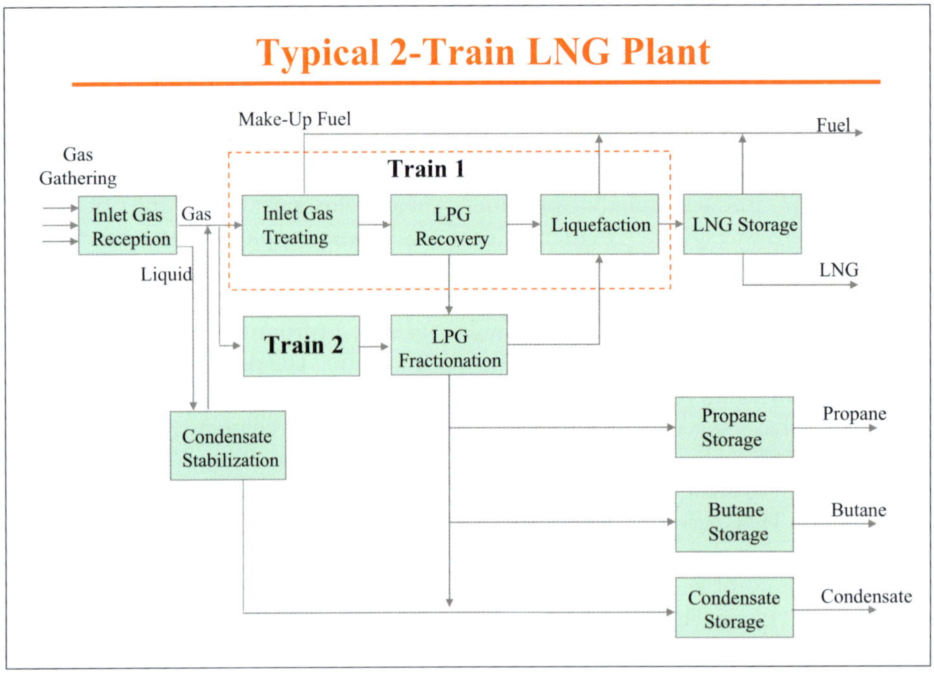

Figure 2. LNG plant block flow diagram.

addition, reservoir gas typically contains varying amounts of H_2S, CO_2, mercury, and heavy hydrocarbons. To meet customers' specifications and prevent damage to liquefaction equipment, plant inlet gas goes through a series of pre-treatment steps as required by the nature of the reservoir fluids. Figure 2 is a block flow diagram of a typical 2-train LNG plant showing these pre-treatment steps. For LNG plants, particularly those in North America which process "pipeline" quality gas, some of the pre-treatment steps such as Inlet Gas Treating and LPG Recovery, may not be necessary if the pipeline gas satisfy the inlet requirements for processing.

LNG Plant Components

Inlet Gas Reception at a minimum consists of pipeline manifold and an inlet separator. To accommodate offshore production or long production flowlines, pig receivers and a slug catcher may also be required. Water separated in the slug catcher and/or inlet separator would be treated and disposed. Hydrocarbon liquids would be sent to Condensate Stablization.

Condensate Stablization consists of a multi-stage fractionation column and a vapor compressor. Typical specifications for the stablized condensate is 10 psia Reid Vapor Pressure (RVP). RVP is the vapor pressure of the condensate at 100°F determined by the Reid method. Overhead vapor from the stabilizer is recompressed and sent back to comingle with the separator gas for further processing. For sour produced fluids, the concentration of mercaptans in the stabilized condensate may be too high and require a de-odorization system. The same may apply to total sulfur content. However, the price differential between high and low sulfur condensate usually does not justify the cost of desulfurization of the condensate. Desulfurization may be achieved more cost-effectively at the refinery where the product is sold.

Inlet Gas Treating is a prerequisite for natural gas liquefaction. H_2S concentration must be reduced to less than 4 ppm to meet typical gas sales specification. CO_2 must be less than 50 ppm to avoid freezing and deposition in the cryogenic heat exchangers. An amine system is typically employed to remove both H_2S and CO_2 from the inlet gas. If more than 50 tons per year of H_2S is removed, the H_2S will have to be converted to elemental sulfur in a sulfur recovery unit (SRU). Since most host countries have regulations that specify recovery of 99.5%+ of all sulfur compounds, a Claus off-gas clean-up system will also be required along with the SRU. Most countries allow unlimited CO_2 venting and atmospheric discharge of SO_2 up to 100 tons per year. Therefore if total amount of sulfur (H_2S) in the inlet gas is adequately low, acid gas (H_2S and CO_2) from amine regeneration may be incinerated and discharged into the atmosphere.

Most cryogenic heat exchangers are made of aluminum. Mercury is very corrosive to aluminum even in minute quantities and must be removed prior to exposure. Because the consequences are so severe, most LNG plant operators will install a mercury removal system even if no mercury is detected in the gas. Mercury adsorption guard beds are non-regenerative and are sized for the amount of mercury expected between change-outs. Another must-have pre-treatment is molecular sieve dehydration even if the plant is processing pipeline spec gas. Whereas pipelines generally limit water concentration to 5-7 lbs-of-water per million standard cubic feet of gas, water vapor concentration must be reduced to less than 100 ppb to avoid freezing at cryogenic temperatures. To reduce load on molecular sieve beds, sweet inlet gas is usually first cooled by the first stage of refrigeration with ethylene gylcol injection for hydrate control. Bulk of the water vapor will condense and be removed by separation, thus reducing the size of the more costly molecular sieve dehydration system. Pretreatment with Tri-ethylene gylcol dehydration system may also be used for this purpose.

LPG Recovery can be as simple as a scrubber or knock-out drum after one or two stages of refrigeration, depending on gas composition and vapor-liquid equilibrium. The purpose is to condense and remove butane and heavier hydrocarbons in the gas to prevent freezing in the low temperature parts of the cryogenic process. For gas with high ethane content, a turbo-expander LPG recovery system may be necessary to achieve the high propane recovery, sometime as high as 99%+, that is required to meet US and European heating value limits. Most US buyers limit gross heating value at 1070 btu/scf, while European GHV spec may be as high as 1100 btu/scf. Typically, there are also limits on Wobbe index and nitrogen content that prevent reducing heating value through dilution with nitrogen. Ethane is often left in the LNG with the methane since it is difficult to transport as a separate product due to its high vapor pressure and unfavorable economics to do otherwise. Therefore very high propane recovery is required to allow significant amount of ethane to remain in the LNG and still meet heating value limits.

LPG Fractionation is often required to handle the recovered LPG. After LPG is extracted from the gas, methane and ethane in solution with the liquid product will have to be removed. A de-ethanizer column is needed to fractionate (separate) out these light ends. De-ethanizer overhead vapor, containing the methane and ethane, is recompressed and sent to liquefaction along with the lean gas. Depending on marketing strategy, the raw LPG stream (de-ethanizer bottoms) may have to be fractionated into separate products. If separate propane, butanes and condensate (containing the pentanes and heavier compounds) products are desired, two or more columns in the form of de-propanizer and de-butanizer are required. Depending on the relative amount of iso-pentane, normal-pentane and heavier hydrocarbons, a de-isopentanizer may also be needed to reduce RVP of the condensate product to 10 psia or below. Commercial grade propane limits ethane content to 2% by volume and a vapor pressure of 200 psig at 100°F, and contains a minimum of 95% propane. Butanes product must be at least 95% and have a vapor pressure not to exceed 70 psig. Both propane and butanes are limited to less than 0.5 ppm H_2S and 15 ppm of total sulfur compounds. Since the small amount of residue of H_2S and sulfur that may have escaped through the amine system (Inlet Gas Treating) tends to concentrate in the propane and butanes, these products may require further treatment by molecular sieve adsorption to meet sufur specifications. Molecular sieve regeneration gas, containing the removed sulfur compounds, is usually sent to the amine system in the Inlet Gas Treating section for handling.

LNG Storage is by far the single largest item in a plant. Figure 3 shows a typical LNG tank farm. LNG is stored in near atmospheric pressure tanks. LNG from liquefaction at elevated pressure is normally pressure reduced to near atmospheric pressure and then pumped to storage. This reduces the need for gas handling and recompression at the storage tanks to normal boil-off gas.

Figure 3. Typical LNG tank farm.

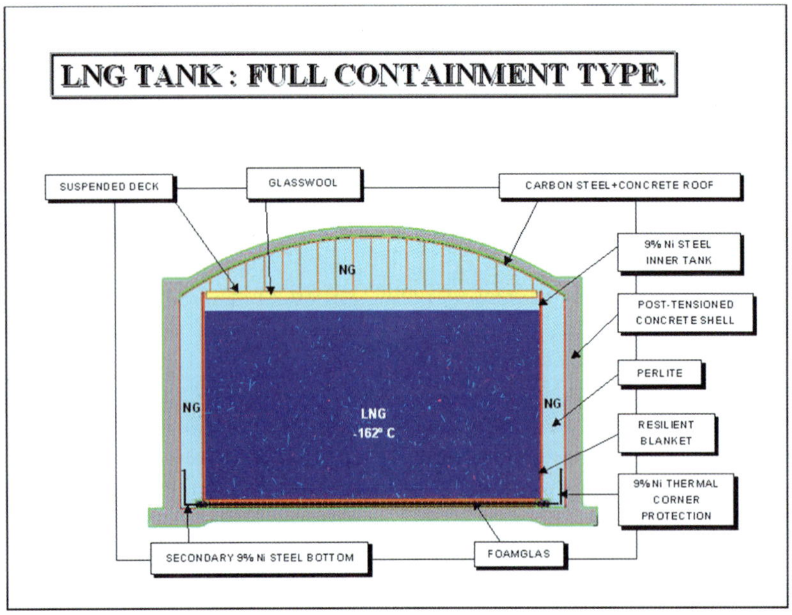

Figure 4. Schematic of full containment LNG tank.

Storage capacity is based on the volume of a tanker plus three to four days of production depending on the size of the farm. Majority of the LNG tankers built in the last few decades have capacity in the 140,000 cubic meter (m^3) range. To handle the new class of "super" tankers QMax and QFlex, designed for long haul trade of 10,000+ miles (i.e. Qatar to US or Europe around the Cape of Good Hope), storage tanks can be as large as 265,000 m^3. At the current time, the largest LNG tank available is in the range of 180,000 m^3. Its dimensions are approximately 75 m (246 feet) diameter × 40 m (131 feet) high. Most LNG plants will have two or more storage tanks in order to meet tanker load requirement and swing volumes needed between loadings.

LNG tank design has gone through an evolution toward greater safety and space efficiency. Early tanks were "single containment" and then "double containment" types. Most operators have opted for the "full containment" design. The full containment type tank has an inner tank made of 9% nickel steel. It is the primary cryogenic liquid barrier. Then there is an outer shell of post-tensioned concrete. Concrete being a cryogenic temperature resistance material constitutes the secondary liquid barrier. This design avoids the need for a large liquid spill containment area around the tank(s). In between the inner and outer shells is an insulation system consisting of perlite and other material. A typical LNG tank is insulated to reduce liquid boil-off to less than 0.05% of the tank content per day. Suspended over the inner tank is a deck with glass wool insulation on the top side. This deck is seal against the walls of the inner tank to prevent gas bypass. Full containment tank has an outer dome made of carbon steel and concrete as illustrated in Figure 4. The supended roof and insulation will maintain temperature in the upper dome to no less than −50°F, the working limit of carbon steel. Boil-off gas is recovered and compressed for use as fuel.

Fill and off-take lines, as well as boil-off vapor lines for the tank are all top entry. All these pipes penetrate the steel/concrete dome and suspended deck, and the penetrations are sealed against vapor communication. Electric submersible pumps are used to transfer LNG out of the tank. The pumps are mounted in caissons within the tank as illustrated in Figure 5. There are no penetrations on the sides or bottom of the tank to reduce the potential for leaks.

Cryogenic in-tank pumps transfer LNG from tanks to the loading system as in Figure 6. Central to the loading system are the Chiksan™ type loading arms. There are usually five arms in the system: three LNG arms plus a spare, and a vapor return arm. At the current time, the cryogenic arms are limited to 16" in diameter. Each arm has a capacity of about 3500 m^3/hr. With three arms in operation, total loading rate is about 10,500 m^3/hr. Some plants are installing four operating LNG arms for a total capacity of 14,000 m^3/hr to load the larger QMax and QFlex tankers within a reasonable period. This higher capacity necessitates a corresponding increase in LNG loading and vapor return lines size.

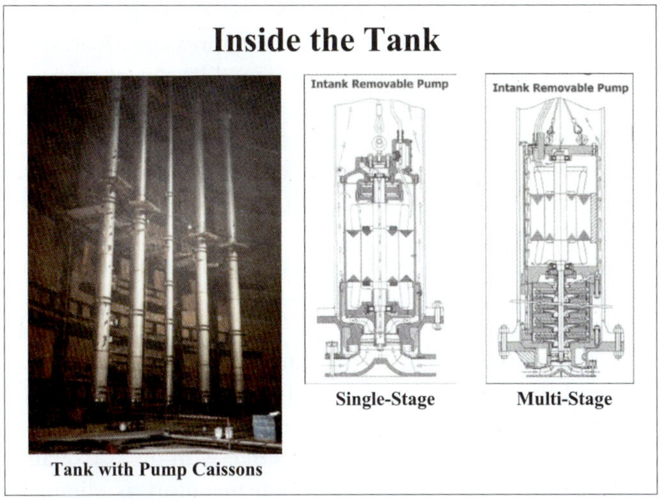

Figure 5. Pump caissons and pump schematics.

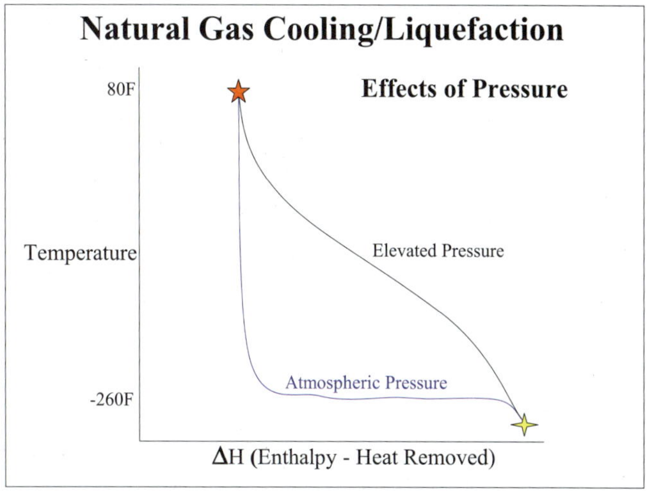

Figure 7. Cooling / liquefaction curves.

Figure 6. LNG loading system.

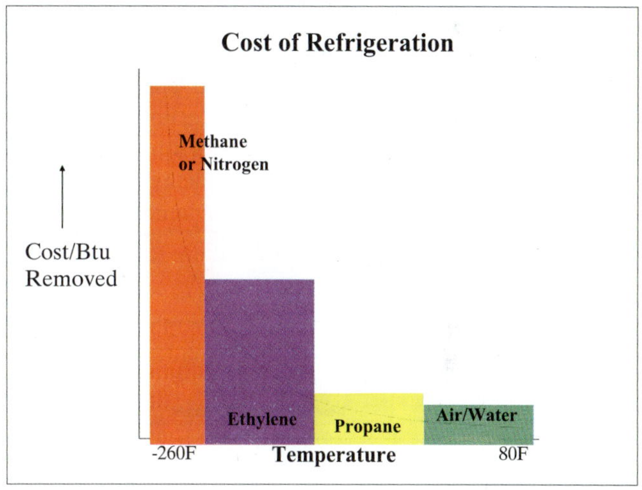

Figure 8. Relative cost of refrigeration.

Liquefaction Technology and Cost

Natural Gas Liquefaction can be achieved most cost efficiently at elevated pressures. Although gas can be liquefied at atmospheric pressure as illustrated in Figure 7, majority of the heat must be removed at an extremely low temperature (the condensing temperature of methane at atmospheric pressure is −260°F). Heat removal can also be achieved at higher temperatures as depicted in the "Elevated Pressure" curve by increasing gas pressure. Why is this important?

Cost of refrigeration is inversely proportional to temperature; lower the temperature, the higher the cost of removing a unit of heat. It stands to reason that the cost of cooling to say 80°F with air or water would be significantly lower than cooling to −30°F with propane, or −260°F with nitrogen. This principle is qualitatively illustrated in Figure 8. Therefore, the goal of the process designer is to utilize the highest level of refrigeration (lowest cost) to achieve the necessary cooling and liquefaction. Elevating inlet gas pressure is the primary method to lower condensing temperature, but there are practical mechanical and thermodynamic limits. For most processes, the practical range is about 600 to 1200 psig. Another method is to formulate the refrigerant(s) to duplicate the condensing curve of the gas.

In theory, the most thermodynamically efficient liquefaction process is one with a refrigerant system that can duplicate the shape of the natural gas condensing curve at operating pressure as shown in Figure 9. The refrigerant would have to be colder by a minimum temperature difference to provide driving force for heat transfer. Of course, the

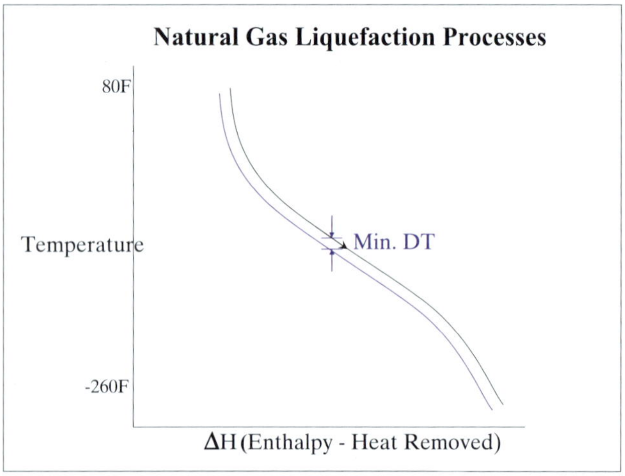

Figure 9. Minimum delta T.

smaller the temperature difference between the process and refrigerant, the more heat exchange area is required for the duty. Therefore, LNG process design is an excise of optimization among refrigerant selection and composition, heat exchanger design and transfer area, and matching refrigerant characteristics to available compressor/driver capacity.

Some licensors aim for process simplicity in this optimization as Black & Veatch's PRICO® single mixed refrigerant process. Other such as ConocoPhillips with its Phillips Optimized Cascade process utilize commonly available plate-fin exchangers and well proven compressor/driver sets to achieve cost efficiency. APCI designed its mixed refrigerant processes around the capabilities of its spiral wound cryogenic heat exchanger. Linde, another spiral wound heat exchanger supplier, combined the features of mixed refrigerant in a cascade process in its MFC® (Mixed Fluid Cascade) technology.

Some companies have proposed a nitrogen refrigeration process for offshore floating LNG (FLNG) applications. Nitrogen being non-flammable is perceived to be a safer alternative refrigerant to propane and other hydrocarbons. One of the main concerns with FLNG is the large inventory of hydrocarbon refrigerants in the relatively small confines of the floating vessel. The thinking is that with the use of an all-nitrogen process, that hydrocarbon inventory is eliminated. Nitrogen which has a normal boiling temperature of -320°F is certainly capable of liquefying methane at −260°F. However, it has been ruled out by most due to the thermodynamic inefficiency of a single nitrogen refrigerant, even if it is staged at various pressure levels.

The PRICO® process, shown in Figure 10, strives for simplicity with a single mixed refrigerant. In theory, the composition of the mixed refrigerant would be formulated to have a vaporization curve that duplicates the natural gas condensing curve, but at a lower temperature to provide driving force for heat transfer (Figure 11). However, this is almost never achieved in practice. With one refrigerant loop,

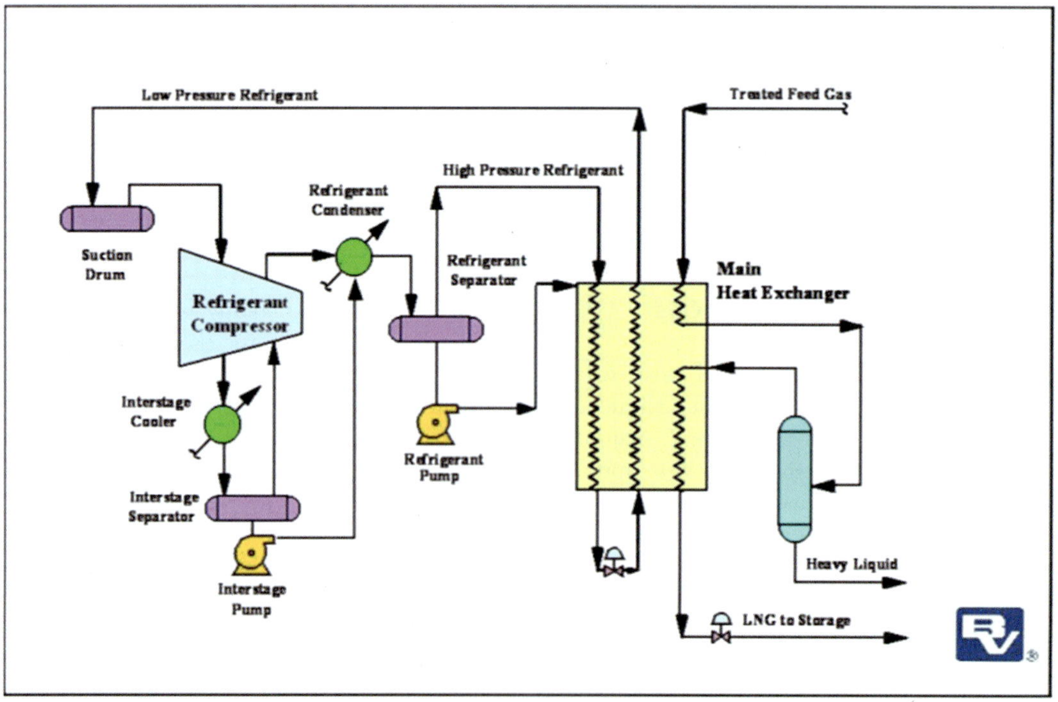

Figure 10. PRICO® schematic.

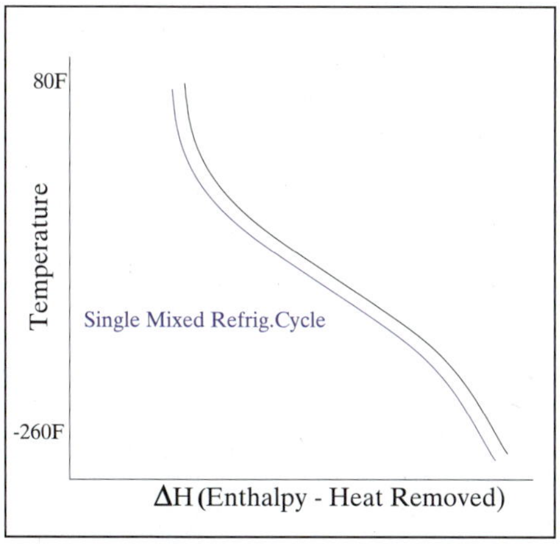

Figure 11. Theoretical condensing curve.

the number of equipment in the system is much reduced. But because it utilizes only one refrigerant compressor, capacity of the process is limited by the throughput of the compressor. Consequently, PRICO® is used primarily in peak shaving LNG plants. Peak shavers are plants that have small liquefaction capacity, typically in the 1-2 MMscfd range. During periods of high gas demand or when pipeline delivery is constrained, stored LNG is vaporized to satisfy the peaks.

Phillips Optimized Cascade process took the stair step approach to liquefaction by busing three successively colder, pure refrigerant, i.e., propane, ethylene and methane as shown in Figure 12 and 13. To improve thermodynamic efficiency and to better match compressor performance, each refrigerant circuit is further divided into two or three pressure steps. This, along with super-heating the ethylene and methane refrigerants with the use of plate-fin exchangers, the process can better match a given natural gas condensing curve.

Propane refrigerant duty is provided by heat of vaporization of the boiling propane at three pressure levels. Operations of the Optimized Cascade process is inherently stable due to the predictability of the pure component refrigerants. ConocoPhillips' Kenai plant has been in continuous operation since 1969 and has not missed a cargo delivery in that period. With its "2-in-1" design utilizing two compressors/drivers on each refrigerant service, the process is cost effective for large base-load plants with single-train capacity upward of 5 Mtpa. Prior to the late 1990's, use of the Phillips Cascade Process was limited to the Kenai plant. Since then, it has captured enough market share to nearly match those of APCI's Propane Pre-cooled Mixed Refrigerant process.

APCI's Propane Pre-cooled Mixed Refrigerant is the most widely used liquefaction process on the market. Figures 14 and 15 describe the process. Since its first plant in Algeria in 1972, the process has come to dominate base load LNG technology with majority of the existing plant capacity. The process utilizes a three or four stage propane refrigerant cycle (depending on ambient temperature and gas composition), like a cascade system, to precool the gas and condense some

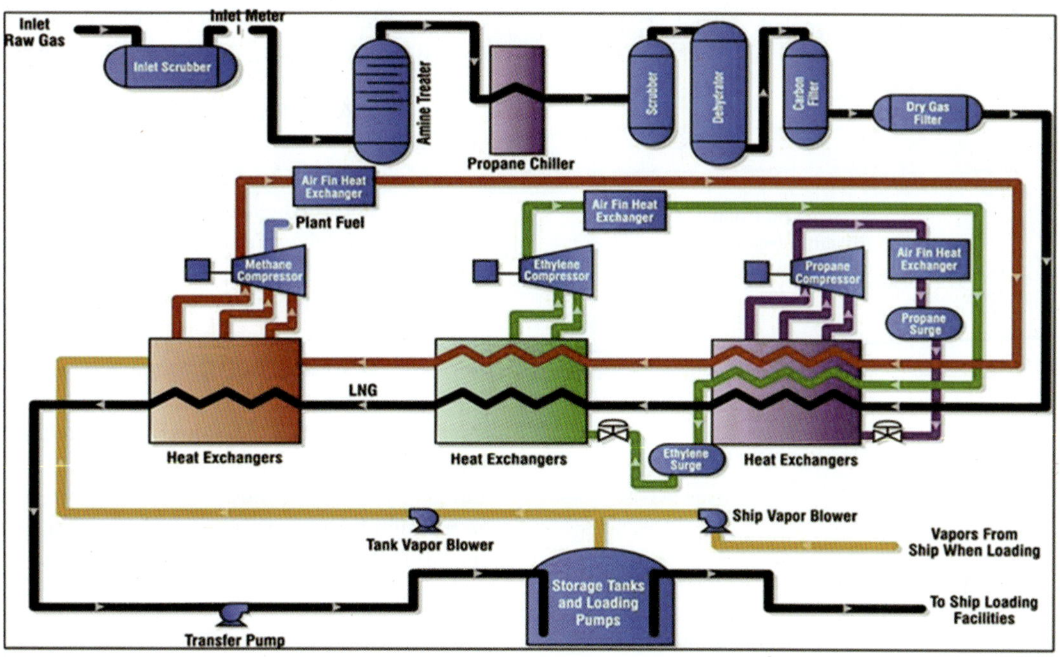

Figure 12. Phillips Optimized Cascade.

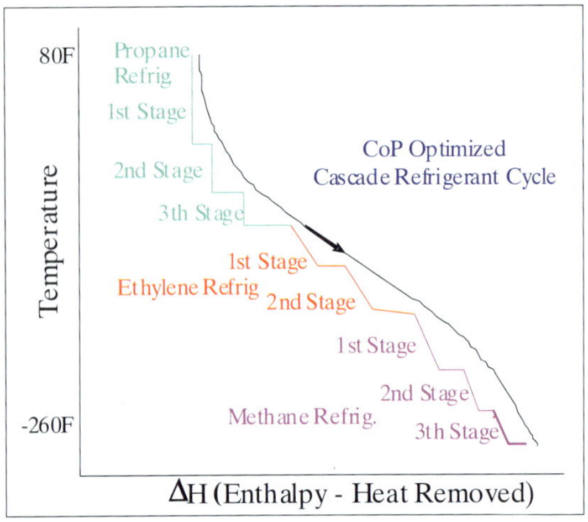

Figure 13. Optimized Cascade condensing curve.

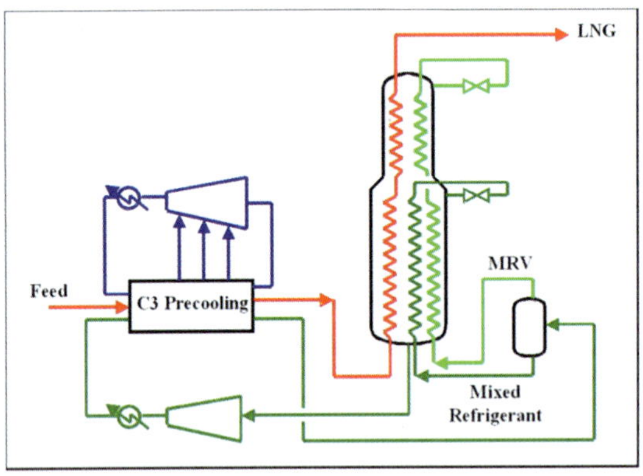

Figure 14. APCI Propane Precooled Mixed Refrigerant process.

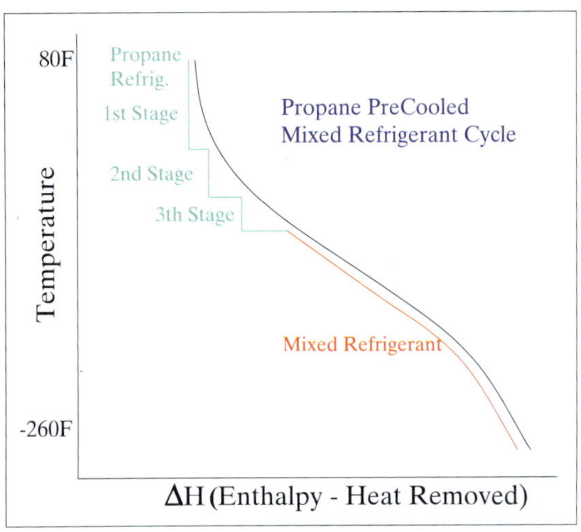

Figure 15. APCI C3-MR condensing curve.

of the heavier hydrocarbons. After precooling, a mixed refrigerant circuit is used to condense and sub-cool the LNG. The degree of LNG subcooling is dictated by fuel gas demand. Majority of plant fuel is derived from flash gas resulting from pressure let down of the high pressure, sub-cooled LNG. APCI licenses the process to all EPC contractors. KBR, Chiyoda, JGC Foster-Wheeler, and others have built plants based on APCI's design.

The heart of the APCI process is the spiral wound heat exchanger. Through the years, the industry has progressively increased train size in the drive to gain greater economy of scale. APCI reached the LNG capacity limit of the largest spiral wound exchanger at around 5 Mtpa. To bypass this problem, APCI developed the AP-X process, a hybrid of the C3-MR.

In the AP-X process, APCI split the LNG condensing and subcooling duties into two separate spiral wound exchangers as shown in Figure 16 and 17. By doing so, it was able to achieve single train capacity in the range of 8 to 10 Mtpa. The process still utilizes a propane circuit for precooling. To provide additional power required by the higher capacity refrigerant compressors, the system depends on three General Electric Fame 9 gas turbines. Each turbine rated at nominally 120 MW of power services each of the three refrigerant loops: propane, mixed refrigerant, and nitrogen. Qatargas II Train 1 is the first to utilize the AP-X process. It started up in 2009. All 6 AP-X trains built and committed to date are in Qatar.

Shell, among other companies, tried to closely duplicate the condensing thereby achieve higher thermodynamic efficiency with the Double Mixed Refrigerant (Double MR) Process. Figures 18 and 19 provide a brief description of its basic principles.

Another one-of-a-kind use is the Linde Mixed Fluid Cascade (MFC®) Process. Figure 20 is a schematic of the process.

The Linde MFC® process is like the cascade system with three distinct refrigerant loops. However instead of pure component refrigerants, the process utilizes mixtures to better duplicate the LNG condensing curve. It was designed for large base load plants. To date, only one plant has been built based on this process. The plant is in a remote part of Norway, Melkoya Island, Hammerfest. It processes gas from Statoil's offshore Snohvit field. The plant was built by Aker in a joint venture with Linde and has a nameplate capacity of 4.3 Mtpa. Due to novelty of the process, remoteness of the site and many first-of-a-kind innovations, the plant has experienced significant start-up problems and continued to face operational challenges years after first gas.

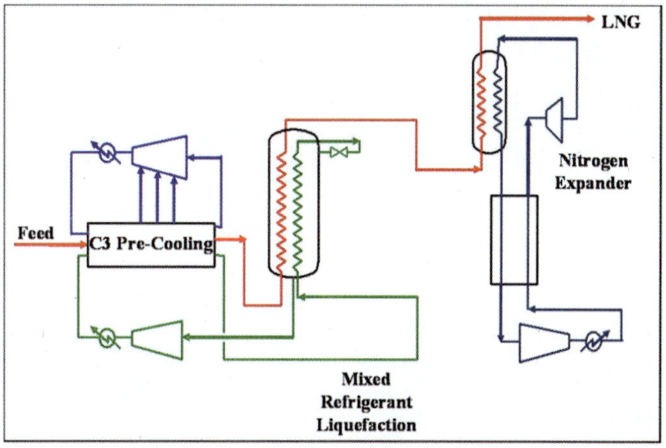

Figure 16. APCI AP-X process.

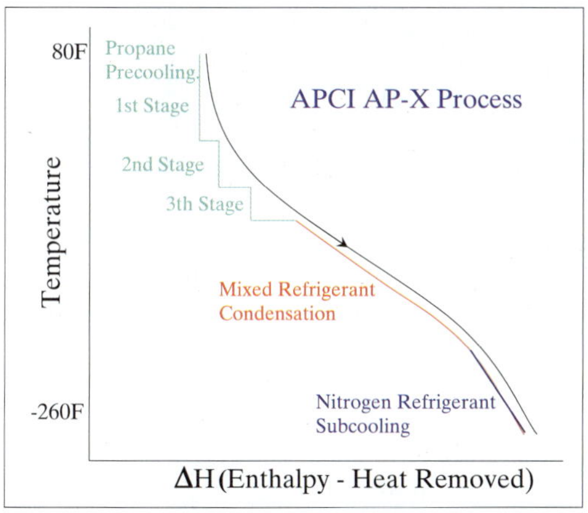

Figure 17. AP-X condensing curve.

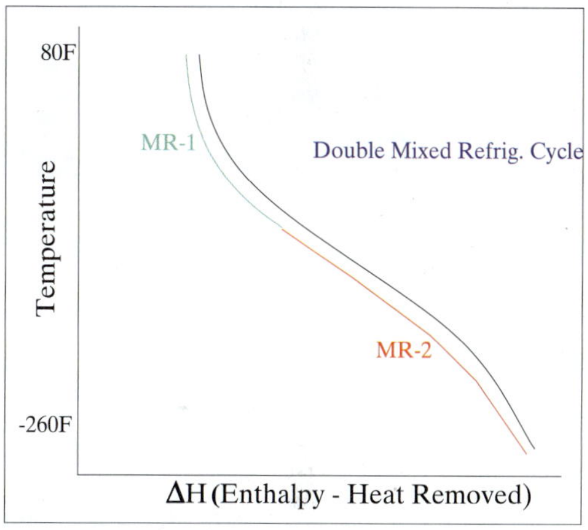

Figure 19. Double MR condensing curve.

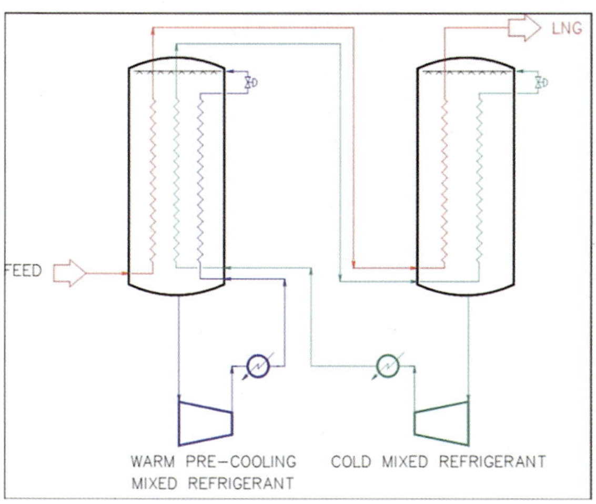

Figure 18. Double Mixed Refrigerant process.

Process Similarities

All processes use similar equipment for liquefaction. The three main components of the refrigeration system used for natural gas liquefaction: compressor/driver, refrigerant condenser, and evaporator (process heat exchanger) are functionally the same as the AC system in our home. Most modern plants use large gas turbines, sometime in combined cycle to drive the refrigerant compressors. Most applications depend on industrial type gas turbines like the GE Frame 7 shown in Figure 21. However, more compact and fuel efficient, but lower horsepower, areo-derivative gas turbines such as GE LM 2500 have been used as well.

Some older and smaller plants have steam turbine drivers. But due to boiler feed water issues, poor fuel efficiency and system complexity, there have not been any recent plants using steam turbine drivers. Many peak shaver LNG plants

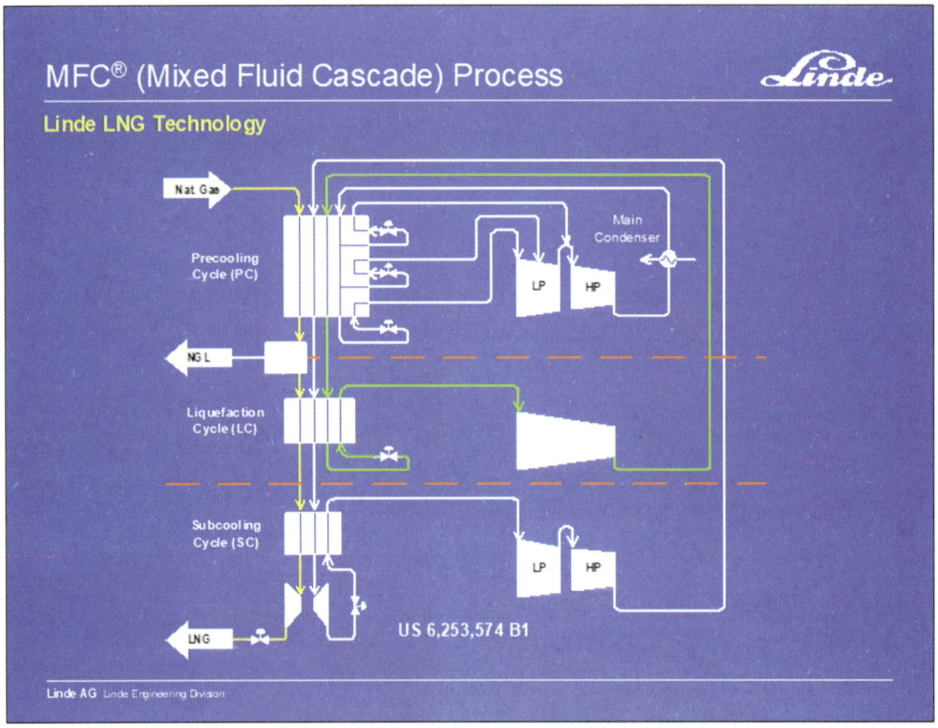

Figure 20. Linde Mixed Fluid Cascade process.

Figure 21. GE MS7001 FB industrial gas turbine.

use electric motor to drive their refrigerant compressors. With recent advances in large motors (upward of 100 MW) and variable speed control, many operators have considered all-electric motor drivers for base load plants. So, only Statoil has taken the plunge with its Melkoya Island facility.

Refrigerant compressors found in LNG plants are standard centrifugal type selected to meet the flow and head requirement of the process. They are the same basic compressors used in refineries, chemical plants and even pipeline applications. Air and water are both used as coolant for refrigerant condensing. Since nearly all LNG plants are located near an ocean, direct and in-direct seawater cooling is always a consideration. In most cases, seawater can offer a lower condensing temperature, which can significantly improve the overall fuel efficiency of the plant. Since all the heat extracted in the liquefaction process is eventually rejected to the ambient coolant, the final heat rejection temperature can have significant impact on the process. However, air cooled condenser system is less expensive than seawater, simpler to implement and less prone to operational fouling. The decision to use air or water cooling for condenser duty is typically based on life-cycle economics, ambient temperature variance of air and water, and environmental considerations.

The main equipment difference among liquefaction processes is the design of the cryogenic heat exchanger. APCI and Linde supply spiral wound type exchangers as shown in Figure 22, while cascade and other processes utilize the standard plate fin aluminum brazed exchanger (Figure 23) commonly found in air separation and gas processing plants.

Spiral wound exchangers are made of miles and miles of aluminum tubing wound around a central stainless-steel structural core. The complete bundle is then slipped into an outer pressure vessel in a vertical configuration. Warm natural gas enters the bottom of the exchanger and is directed

Figure 22. Spiral wound cryogenic heat exchanger.

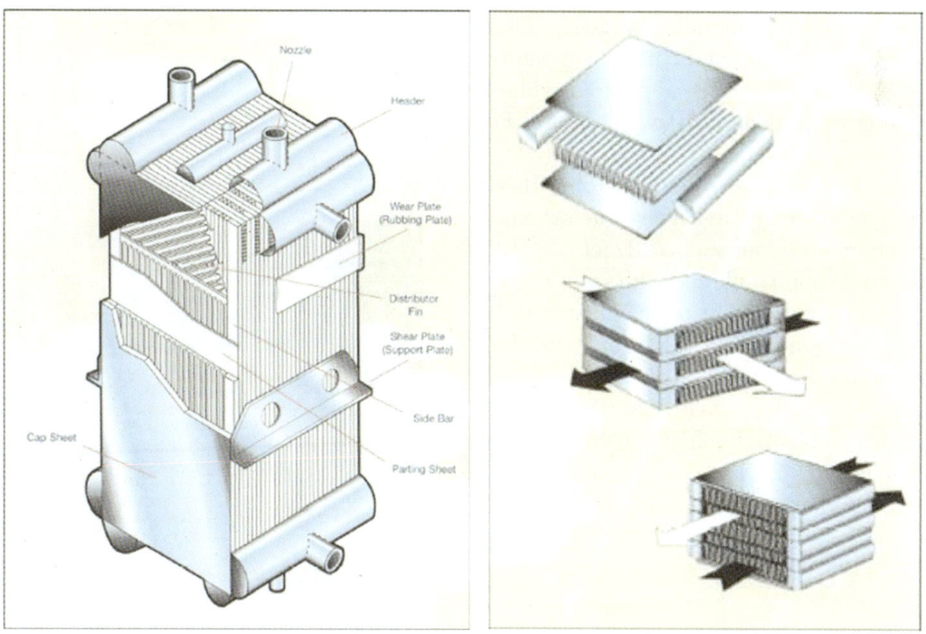

Figure 23. Plate fin brazed aluminum heat exchanger.

into one of thousands of aluminum tubes in the bundle through a header arrangement. As warm gas progresses up the tube, it is cooled, condensed and sub-cooled by falling vaporizing refrigerant on the shell side. The spiral wound design is in-effect a vertically mounted, counter-current flow shell and tube exchanger with the refrigerant and process gas in the shell and tube sides, respectively.

Floating LNG facility

For decades, numerous Floating Production, Storage and Offloading (FPSO) vessels have been servicing offshore oil fields all around the world. The technology has proven to be reliable and economically attractive for remote, offshore production where long pipeline(s) are required. Associated produced gas is either flared onsite, reinjected into the reservoir or pipelined to shore for processing. Although operators have recognized the potential cost savings of FLNG, implementation has lagged far behind due to the complexity and relatively large footprint of gas processing and LNG liquefaction facilities.

The vision of exchanging wellhead/processing platform(s) with a floating storage and offtake (FSO) vessel for natural gas developments has been dreamed for decades. The floating vessel treats the raw gas streams from the wellheads, liquifies the methane and recovers and stabilizes the condensate and NGL products, avoiding a pipeline to an onshore LNG plant. That vision was finally realized with the launching of Shell's Prelude FLNG hull in 2013 and installation offshore Western Australia in 2017. Other projects have also been announced and currently in various stages of development. Figure 24 is a photo of Shell's Prelude vessel and Figure 25 is an artist concept of a Petronas FLNG.

Onshore gas production can also benefit from FLNG. Among onshore discoveries, many are in river deltas where LNG export by tankers require long and costly jetty to reach acceptable water depth. When combined with extensive site preparation and continuous dredging to maintain a viable shipping channel, the economics and safety of an onshore plant can be severely impaired. These challenges can be eliminated by transporting the gas via a short pipeline to a FLNG moored at near-shore location adequate for convenient tankers access. Figures 26 and 27 illustrate the size difference of a land-based plant and a near-shore FLNG concept.

Figure 24. Shell's Prelude FLNG hull float out of dry dock in Dec. 2013.

Figure 25. Artist image of Petronas FLNG.

Figure 26. NLNG river delta site, Bonney Island, River State, Nigeria.

Figure 27. Artist concept of near-shore FLNG with tank loading.

With large LNG projects in remote areas, labor and logistical infrastructure can be severely challenged. An example is Western Australia around the 2000's period when multiple projects created shortage of resources of every kind. The results are massive cost over-run and significant schedule delays. FLNG can ease the burden by offering established organization and discipline work force, existing supply chain infrastructure and efficiency of construction in a shipyard. Other intangible cost savings and shorter project duration can be further incentives for FLNG.

Conclusions

Despite the recent slowdown in the petroleum industry, the demand for natural gas continued to be strong. To transport this compressible fluid long distances to markets greater than 1500 miles across bodies of water, LNG has proven to be the most economic option. Although there are many mystiques around liquefaction processes, the basic system is no different from the AC unit in our home, consisting of a compressor/driver, refrigerant condenser, and evaporator.

Liquefaction processes design is an exercise of optimization among:

- Refrigerant selection and composition
- Heat exchanger design and transfer area
- Matching refrigerant characteristics to available compressor/driver capacity

All the major liquefaction processes available under license are well proven, robust, and can be optimized for any application. The most cost-effective process depends on LNG plant size, gas composition, ambient conditions, sale specifications and other commercial considerations.

Although liquefaction is the heart of a LNG plant, complexity and size of the gas pre-treatment facilities often dwarfs the liquefaction system. All LNG plants must have molecular sieve dehydration and mercury removal systems. Depending on production field configuration and liquids content of the produced gas, extensive facilities for inlet gas reception, gas treating, and LPG recovery may be required. To produce marketable hydrocarbon liquid products, which can contribute significantly to the economics of the project, condensate stabilization and LPG fractionation may also be required. For remote offshore gas field developments, recent advances in FLNG technologies and design have introduced a potential economically attractive option to the traditional onshore LNG plant.

Acknowledgments

The author thanks all the licensors of the LNG liquefaction processes discussed in the paper: Air Products & Chemicals, Black & Veatch, ConocoPhillips, Linde, Shell, Statoil for information and material provided through the years in the form of vendor literature and presentations.

Until retirement in 2015, Michael Choi was a Process Engineering Fellow in ConocoPhillips' (COP) Global Production Department located in Houston, Texas. His specialties are production facilities, sour gas treating and LNG. He was the lead process design engineer for CoP's Qatargas3 LNG project. Prior to joining Conoco in 1985, Michael worked in various engineering capacities with El Paso Natural Gas, Aminoil and Getty Oil/Texaco. His work has led to several Society of Petroleum Engineers (SPE) publications and six US patents for separator design, emissions control system for glycol dehydrators, and subsea processing and storage systems. Michael has been active in SPE as member of the PF&C Committee, program chairman and member of the ATCE and other international conferences and forums. He graduated from the University of Southern California in 1974 with a BS degree in chemical engineering.

Glossary of common gas industry terms

(italics denote cross reference)

absolute pressure *gauge pressure plus atmospheric pressure.*

atmospheric pressure the pressure of the weight of air and water vapor on the surface of the earth. Approximately 14.7 lbf/in.^{2} at sea level.

blue water gas also called "water gas." Made in a cyclic process in which an incandescent bed of coke or coal is alternately subjected to blasts of air and steam. The gas consists mainly of equal proportions of carbon monoxide and hydrogen and has a gross heat content of about 300 Btu/ft^3 (see also *carbureted water gas*).

British thermal unit the heat required to raise the temperature of 1 lb of water 1° F.

burner capacity the maximum Btu/hr that can be released by a burner while burning with a stable flame.

calorie the quantity of heat required to raise the temperature of 1 gram of water 1°C (centigrade).

calorific value the quantity of heat released by the combustion, in a calora meter at constant pressure of 1 atmosphere, of a unit quantity of fuel measured under given conditions. In the case of gas, the conditions of temperature, pressure, and water vapor content under which the unit volume was measured affect the calorific value.

calorimeter an instrument for determining the heating value of a fuel.

carbureted water gas (carbureted blue gas) the gas resulting from the enrichment of *blue water* gas during its manufacture by a simultaneous process of light distillate, gas oil, or fuel oil gasification. The gas has a gross heat content of about 500 to 550 Btu/ft^3.

city gate a measuring station, which may also include pressure regulation, at which a distributing gas utility receives gas from a natural gas pipeline company or the transmission system.

coal gas or coke oven gas a *manufactured gas* made by destructive distillation ("carbonization") of bituminous coal in a gas retort or by-product coke oven. Its chief components are methane (20 to 30%) and hydrogen (about 50%). This gas generally has a gross heating value of 500 to 550 Btu/ft^3. When the process takes place in a closed oven (with gas as a by-product of coke production) it is generally designated as "Coke Oven Gas," and when produced in retorts it is called "Coal Gas."

commodity charge that portion of the charge or rate based upon the total volume of gas consumed or billed.

condensate a liquid/hydrocarbon mixture of 45 to 65° API gravity, which may be recovered at the surface from some *non-associated* gas reservoirs.

connection charge an amount to be paid by the customer in a lump sum, or in installments, for connecting the customer's facilities to the supplier's facilities.

contract demand the volume that the supplier of gas service agrees to deliver and, in general, the amount that the customer agrees to take or pay for.

cubic foot and cubic meter (gas) common units of measurement of gas volume. They are the amounts of gas required to fill a volume of 1 cubic ft or 1 cubic meter under stated conditions of temperature, pressure, and water vapor content.

customer charge a fixed amount to be paid periodically by the customer without regard to *demand or commodity* consumption.

cycling or recycling the process by which a *non-associated gas* reservoir is produced only for the recovery of condensate. The gas, after removal of the condensate, is reinjected into the formation.

daily average send-out the total volume of gas delivered during a period of time divided by number of days in the period.

daily peak the maximum volume of gas delivered in any one day during a given period, usually a calendar year.

degree day a measure of the extent to which the mean daily temperature falls below an assumed base, say 65°F. Thus, each degree by which the mean temperature for any day is less than 65°F would represent one degree day. (In Continental Europe, °C are used instead of °F, and the assumed base temperature is generally taken as 16° C equivalent to 60.8°F.)

demand (load) rate of flow of gas in a specified time interval, usually expressed in cubic ft (or cubic meters) per hour, per day, or per year.

demand charge that portion of the charge or rate for gas service based upon the customer's *demand* characteristics.

distribution company or gas utility a company that obtains the major portion of its gas operating revenues from the operation of a retail gas *distribution system* and which operates no transportation system other than incidental connections to a transportation system of another company. For purposes of American Gas Association (AGA) statistics, a distribution company obtains at least 95% of its gas operating revenues from the operation of its retail gas distribution system (see also *transmission company*).

distribution system feeders, mains, services, and equipment which carry or control the supply of gas from the point or points of local supply (usually the *city gate* station) to and including the consumer *meters*.

diversity factor the ratio of the sum of the non-coincident maximum demands of two or more loads to their coincident maximum demand for the same period.

dyne the standard centimeter-gram-second unit of force, equal to the force that produces an acceleration on one centimeter per second on a mass of one gram.

erg the centimeter-gram-second unit of work or energy, equal to the work done by a force of 1 dyne when its point of application moves through a distance of 1 centimeter in the direction of the force.

flat rate schedule a rate schedule that provides for a specified charge irrespective of the quantity of gas used or the *demand*.

gauge pressure the pressure generally shown by measuring devices. This is the pressure in excess of that exerted by the atmosphere (see *absolute pressure*).

gross calorific (heating) value the gross calorific value at constant pressure of a gaseous fuel is the number of heat units produced when unit volume of the fuel, measured under standard conditions, is burned in excess air in such a way that the materials after combustion consist of the gases carbon dioxide, sulfur dioxide, oxygen, and nitrogen, water vapor equal in quantity to that in the gaseous fuel and the air before combustion, and liquid water equal in quantity to that produced during combustion, and that the pressure and temperature of the gaseous fuel, the air, and the materials after combustion are 1 standard atmosphere and 25°C.

inch of mercury a unit of pressure (equivalent to 0.491154 lbf/in^2) by which 1 lb/in.^{2} equals 2.036009 in. of mercury column at 0°C.

inch of water a unit of pressure (equivalent to 0.03606 lbf/in^2) by which 1 lb/in^2 equals 27.68049 in of water column at 4°C.

interruptible gas gas made available under agreements which permit curtailment or cessation of delivery by the supplier.

joule the work done when the point of application of a force of 1 newton is displaced through a distance of one meter in the direction of the force. It is equal to 107 *ergs*, and in electrical units is the energy dissipated by one watt in a second.

kilocalorie 1,000 *calories*.

kilowatt-hour a unit of energy equivalent to the energy transferred or expanded in 1 hour by 1 kilowatt of power.

line pack a method of *peak-shaving* by withdrawing gas from a section of a pipeline system in excess of the input into that section, i.e., normally the difference between the actual volume of gas in the pipeline at low flow (increased pressure) and that at normal flow.

liquefied natural gas (LNG) natural gas that has been liquefied by cooling to minus 258°F (—161°C) at atmospheric pressure.

liquefied petroleum gas (LPG) any hydrocarbon mixture, in either the liquid or gaseous state, the chief components of which consist of propane, propylene, butane, iso-butane, butylene, or mixtures thereof in any ratio.

load curve a graph in which the *send-out* of a gas system, or segment of a system, is plotted against intervals of time.

load density the concentration of gas load for a given area, expressed as gas volume per unit of time and per unit of area.

load duration curve a curve of loads, plotted in descending order of magnitude, against time intervals for a specified period. The coordinates may be absolute quantities or percentages.

load factor the ratio of the average load over a designated period to the peak load occurring in that period. Usually expressed as a percentage.

low pressure gas distribution system a gas *distribution system*, or the mains of a segment of a distribution system, operated at low pressures of less than 15 *inches water column*.

LPG-air mixtures mixtures of *liquefied petroleum gas* and air to obtain a desired heating value and capable of being distributed through a *distribution system* also used for stand-by and *peak-shaving* purposes by gas utilities.

manufactured gas combustible gases derived from primary energy sources by processes involving chemical reaction. For instance, gas produced from coal, coke or liquid hydrocarbons (see also *town gas*).

meter (gas) a mechanical device for automatically measuring quantities of gas. Some of the more important types are:

Diaphragm Meter a gas meter in which the passage of gas through two or more chambers moves diaphragms that are geared to a volume-indicating dial. Sizes up to 20,000 cubic ft per hour and 1,000 lb/in^2 working pressure are available.

Mass Flowmeter a type of meter in which the gas flow is measured in terms of mass by utilizing the velocity and density of the gas stream to calculate the mass of a gas flowing in unit time.

Orifice Meter meter for measuring the flow of fluid through a pipe or duct by measurement of the pressure differential across a plate having a precision machined hole in its center. Ratings up to very high throughputs and pressures are available.

Rotary Displacement Meter the positive-pressure blower principle in reverse is used in this meter in which gas pressure turns twin matching impellers and the quantity of gas passing through is proportional to the number of revolutions. These meters are made in the range of 3,000 to 1,000,000 cubic ft per hour and for pressures up to 1,200 lbf/in^2

Turbine Flowmeter a type of meter in which there is a turbine wheel or rotor made to rotate by the flowing gas. The rate or speed at which the turbine wheel revolves is a measure of the velocity of the gas.

Venturi Meter a meter in which the flow is determined by measuring the pressure drop caused by the flow through a Venturi throat. The flow rate is proportional to the square root of the pressure drop across the throat.

minimum charge a clause providing that the charge for a prescribed period shall not be less than a specified amount.

natural gasoline those liquid hydrocarbon mixtures containing essentially pentanes and heavier hydrocarbons that have been extracted from natural gas.

natural gas liquids a term used for those mixtures of hydrocarbon fractions that can be extracted in liquid form from natural gas. In the widest sense, they can comprise any combination of the following: ethane, propylene, propane, butylene, isobutane, n-butane, pentanes, hexanes, heptanes, and octanes.

net calorific (heating) value the net calorific value at constant pressure of a gaseous fuel is the number of heat units produced when unit volume of the fuel, measured under standard conditions, is burned in excess air in such a way that the materials after combustion consist of the gases carbon dioxide, sulfur dioxide, oxygen, nitrogen, and water vapor, and that the pressure and temperature of the gaseous fuel, the air and the materials after combustion are 1 standard atmosphere and 25°C. This value is derived from the measured gross figure.

off-peak the period during a day, week, month, or year when the load being delivered by a gas system is not at or near the maximum volume delivered by that system.

peak or peak load the maximum load consumed or produced by a unit or group of units in a stated period of time.

peak shaving the practice of augmenting the normal supply of gas during peak or emergency periods from another source where gas may have either been stored during periods of low demand or manufactured specifically to meet the peak demand.

pressure maintenance (repressing) a process in which natural gas is injected into a formation capable of producing crude petroleum to aid in maintaining pressure in an underground reservoir for the purpose of assisting in the recovery of crude.

producer gas a gas manufactured by burning coal or coke with a regulated deficiency of air, normally saturated with steam. The principal combustible component is carbon monoxide (about 30%), and the gross heat content is between 120 and 160 Btu/ft^3.

recycling see *cycling*.

reforming the process of thermal or catalytic cracking of natural gas, *liquefied petroleum gas*, refinery gas or gas from oil, resulting in the production of a gas having a different chemical composition.

refinery gas a gas resulting from oil refinery operations consisting mainly of hydrogen, methane, ethylene, propylene, and the butylenes. Other gases such as nitrogen and carbon dioxide may also be present. The composition can be highly variable, and the heat content can range from 1,000 to 2,000 Btu/ft^3.

send-out the quantity of gas delivered by a plant or system during a specified period of time.

specific gravity of gas the ratio of the density of gas to the density of dry air at the same temperature and pressure.

standard metering base standard conditions, plus agreed corrections, to which all gas volumes are corrected for purposes of comparison and payment.

summer valley the decrease that occurs in the summer months in the volume of the daily load of a gas *distribution system*.

therm 100,000 Btus.

thermie the heat required to raise the temperature of 1 metric ton of water 1°C (centigrade).

transmission company a company that operates a natural gas transmission system and which either operates no retail distribution system, or, as defined by the American Gas Association (AGA), receives less than 5% of its gas operating revenues from such retail distribution system (see also *distribution company or gas utility*).

unaccounted for gas the difference between the total gas available from all sources and the total gas accounted for as sales, net interchange and company use. The difference includes leakage or other actual losses, discrepancies due to *meter* inaccuracies, variations of temperature, and pressure and other variants.

utilization factor the ratio of the maximum *demand* on a system or part of a system to the rated capacity of the system or part of the system under consideration.

water gas see *blue water gas*.

watt the meter-kilogram-second unit of power, equivalent to one joule per second and equal to the power in a circuit in which a current on one ampere flows across a potential difference on one volt.

wobbe index the gross calorific value of the gas divided by the square root of the density of the gas as compared with air.

10: Gas—Compression

Compressor definitions 400
Performance calculations for reciprocating compressors 401
Estimate suction and discharge volume bottle sizes for pulsation control for reciprocating compressors 403
Compression horsepower determination 405
Generalized compressibility factor 407
Nomograph aids in diagnosing compressor cylinder ills 408
Centrifugal compressor performance calculations 409
Nomographs for estimating compressor performance 413
Estimate horsepower required to compress natural gas 418
Estimate compressor horsepower where discharge pressure is 1000 psi 418
Calculate brake horsepower required to compress gas 419
How can we find the size of a fuel gas line for a compressor station? 419
Estimate engine cooling water requirements 420
Performance testing guidelines for centrifugal compressors 421

Compressor definitions

The following definitions are applied in the approximation of horsepower needed for the compression of gas.

Definitions

The **"N value"** of gas is the ratio of the specific heat at constant pressure (C_p) to the specific heat at constant volume (C_v).

If the composition of gas is known, the value of N for a gas mixture may be determined from the molal heat capacities of the components. If only the specific gravity of the gas is known, an approximate N value may be obtained by using the chart in Figure 1.

Piston displacement of a compressor cylinder is the volume swept by the piston with the proper deduction for the piston rod. The displacement is usually expressed in cubic feet per minute.

Clearance is the volume remaining in one end of a cylinder with the piston positioned at the end of the delivery stroke for this end. The clearance volume is expressed as a percentage of the volume swept by the piston in making its full delivery stroke for the end of the cylinder being considered.

Ratio of compression is the ratio of the absolute discharge pressure to the absolute inlet pressure.

Actual capacity is the quantity of gas compressed and delivered, expressed in cubic feet per minute, at the intake pressure and temperature.

Volumetric efficiency is the ratio of actual capacity, in cubic feet per minute, to the piston displacement, in cubic feet per minute, expressed in percent.

Adiabatic horsepower is the theoretical horsepower required to compress gas in a cycle in which there is no transfer of sensible heat to or from the gas during compression or expansion.

Isothermal horsepower is the theoretical horsepower required to compress gas in a cycle in which there is no change in gas temperature during compression or expansion.

Indicated horsepower is the actual horsepower required to compress gas, taking into account losses within the cylinder, but not taking into account any loss in frame, gear, or power transmission equipment.

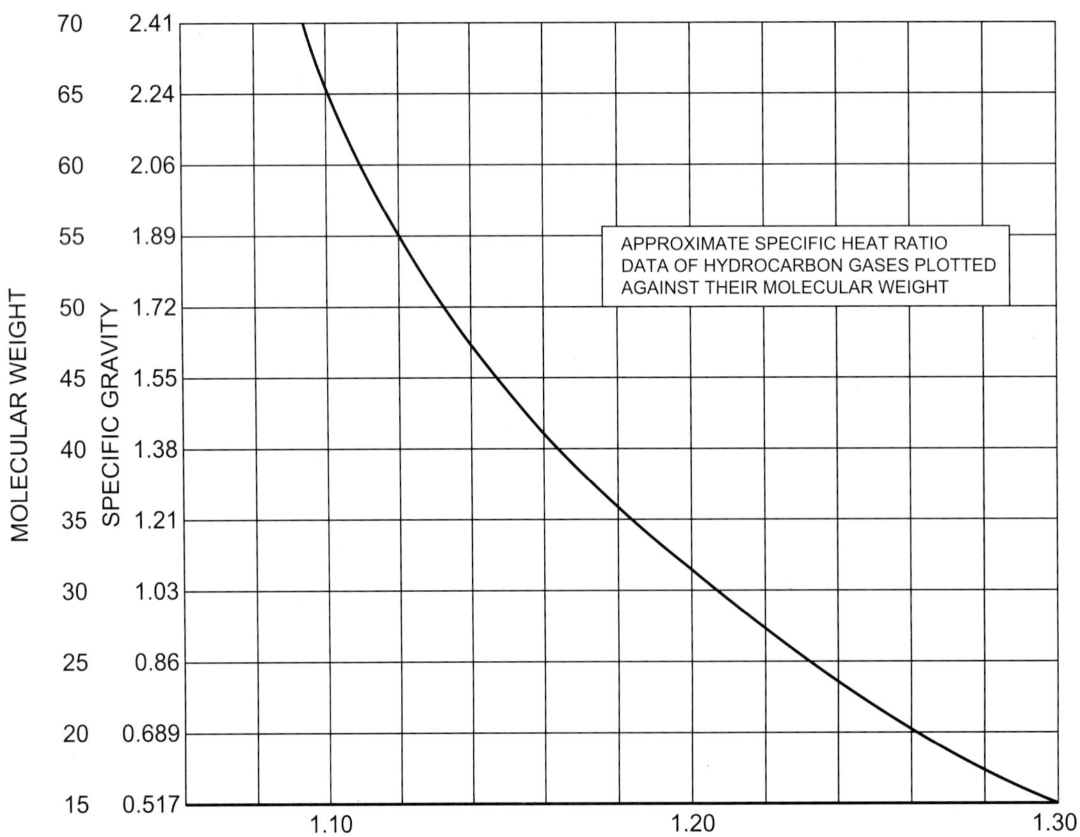

Figure 1. Ratio of specific heat (N-value).

Compression efficiency is the ratio of the theoretical horsepower to the actual indicated horsepower required to compress a definite amount of gas. The efficiency, expressed in percent, should be defined in regard to the base at which the theoretical power was calculated, whether adiabatic or isothermal.

Mechanical efficiency is the ratio of the indicated horsepower of the compressor cylinder to the brake horsepower delivered to the shaft in the case of a power-driven machine. It is expressed in percent.

Overall efficiency is the product, expressed in percent, of the compression efficiency and the mechanical efficiency. It must be defined according to the base, adiabatic isothermal, which was used in establishing the compression efficiency.

Piston rod gas load is the varying, and usually reversing, load imposed on the piston rod and crosshead during the operation, by different gas pressures existing on the faces of the compressor piston.

The maximum piston rod gas load is determined for each compressor by the manufacturer, to limit the stresses in the frame members and the bearing loads in accordance with mechanical design. The maximum allowed piston rod gas load is affected by the ratio of compression and also by the cylinder design, i.e. whether it is single- or double-acting.

Performance calculations for reciprocating compressors

Piston displacement

Single-acting compressor:

$$P_d = \left[S_t \times N \times 3.1416 \times D^2\right]/[4 \times 1728] \tag{1}$$

Double-acting compressor without a tail rod:

$$P_d = \left[S_t \times N \times 3.1416 \times \left(2D^2 - d^2\right)\right]/[4 \times 1728] \tag{2}$$

Double-acting compressor with a tail rod:

$$P_d = \left[S_t \times N \times 3.1416 \times 2 \times \left(D^2 - d^2\right)\right]/[4 \times 1728] \tag{3}$$

Single-acting compressor compressing on the frame end only:

$$P_d = \left[S_t \times N \times 3.1416 \times \left(D^2 - d^2\right)\right]/[4 \times 1728] \tag{4}$$

where: P_d = Cylinder displacement, cu ft/min,
S_t = Stroke length, in,
N = Compressor speed, number of compression strokes/min,
D = Cylinder diameter, in,
d = Piston rod diameter, in.

Volumetric efficiency

$$E_v = 0.97 - \left[(1/f)r_p^{1/k-1}\right]C - L \tag{5}$$

where: E_v = Volumetric efficiency,
f = Ratio of discharge compressibility to suction compressibility Z_2/Z_1,
r_p = Pressure ratio,
k = Isentropic exponent,
C = Percent clearance,
L = Gas slippage factor.

Let $L = 0.3$ for lubricated compressors
Let $L = 0.07$ for nonlubricated compressors (6)

These values are approximations, and the exact value may vary by as much as an additional 0.02—0.03.

Note: A value of 0.97 is used in the volumetric efficiency equation rather than 1.0 since even with 0 clearance, the cylinder will not fill perfectly.

Cylinder inlet capacity

$$Q_1 = E_v \times P_d \tag{7}$$

Piston speed

$$P_S = [2 \times S_t \times N]/12 \tag{8}$$

Discharge temperature

$$T_2 = T_1\left(r_p^{(k-1)/k}\right) \tag{9}$$

where: T_2 = Absolute discharge temperature °R,
T_1 = Absolute suction temperature °R.

Note: Even though this is an adiabatic relationship, cylinder cooling will generally offset the effect of efficiency.

Power

$$W_{cyl} = \left[144P_1Q_1/33{,}000n_{cyl}\right] \times [k/(k-1)] \times \left[r_p^{k-1/k} - 1\right] \tag{10}$$

where: n_{cyl} = Efficiency,
W_{cyl} = Cylinder horsepower.

See Figure 1 for the curve of the efficiency versus pressure ratio. This curve includes a 95% mechanical efficiency and a valve velocity of 3000 ft per minute. Tables 1 and 2 permit a correction to be made to the compressor horsepower for specific gravity and low inlet pressure. While it is recognized that the efficiency is not necessarily the element affected, the desire is to modify the power required per the criteria in these figures. The efficiency correction accomplishes this. These corrections become more significant at the lower pressure ratios.

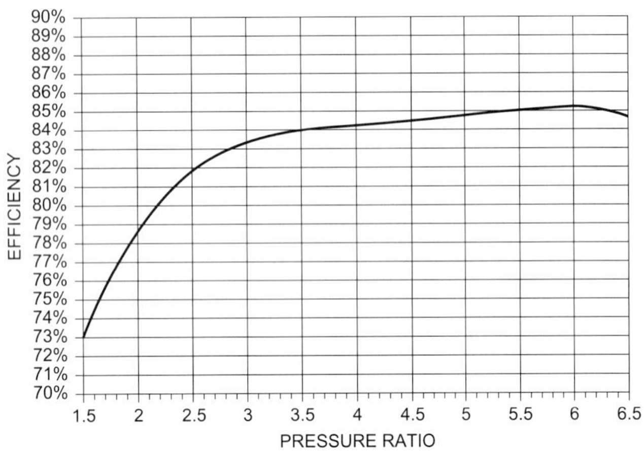

Reciprocating compressor efficiencies plotted against pressure ratio with a valve velocity of 3,000 fpm and a mechanical efficency of 95 percent.

Figure 1. Reciprocating compressor efficiencies.

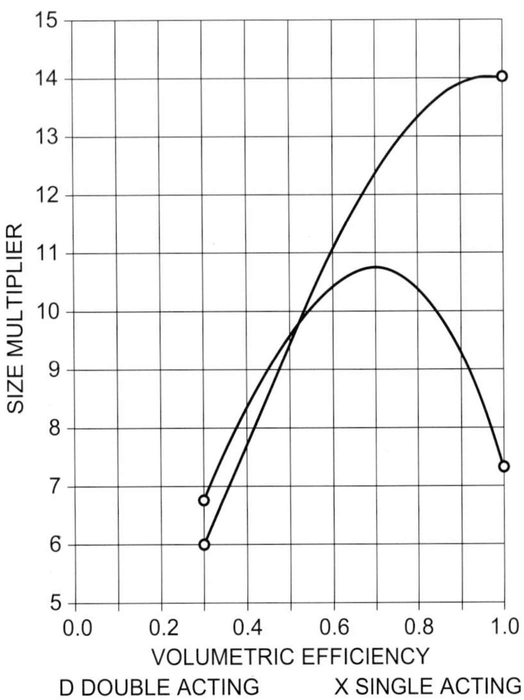

Figure 2. Volume bottle sizing.

Inlet valve velocity

$$V = 288 \times P_d / A \tag{11}$$

where

V = Inlet valve velocity,
A = Product of actual lift and the valve opening periphery; it is the total for inlet valves in a cylinder expressed in square in. (This is a compressor vendor furnished number.)

Example. Calculate the following:

Suction capacity Horsepower
Discharge temperature Piston speed

Given:

Bore	= 6 in,
Stroke	= 12 in,
Speed	= 300 rpm,
Rod diameter	= 2.5 in,
Clearance	= 12%,
Gas	= CO_2,
Inlet pressure	= 1720 psia,
Discharge pressure	= 3440 psia,
Inlet temperature	= 115°.

Calculate piston displacement using Equation 2.

$$P_d = 12 \times 300 \times 3.1416 \times \left[2(6)^2 - (2.5)^2\right]/1728 \times 4 = 107.6 \text{ cfm}$$

Calculate volumetric efficiency using Equation 5. It will first be necessary to calculate *f*, which is the ratio of discharge compressibility to suction compressibility.

$$T_1 = 115 + 460 = 575^\circ\text{R}$$

$$T_r = T/T_c = 575/548 = 1.05$$

where

T_r = Reduced temperature,
T_c = Critical temperature = 548° for CO_2,
T = Inlet temperature.

$$P_r = P/P_c = 1720/1071 = 1.61$$

where

P_r = Reduced pressure,
P_c = Critical pressure = 1071 psia for CO_2,
P = Suction pressure.

From the generalized compressibility chart

$$Z_1 = 0.312$$

Determine discharge compressibility. Calculate discharge temperature by using Equation 9.

$$r_p = 3440/1720 = 2$$

$$k = cp/cv = 1.3 \text{ for } CO_2$$

$$T_2 = 575 \times \left[2.0^{13-1/1.3}\right]$$
$$= 674.7^\circ R$$
$$= 214^\circ F$$

$$T_r = 674.7/548 = 1.23$$

$$P_r = 3440/1071 = 3.21$$

From the generalized compressibility chart

$$Z_2 = 0.575$$

$$f = 0.575/1.61 = 1.843$$

$$E_v = 0.97 - \left[(1/1.843) \times 2.0^{1/1.3} - 1\right] \times 0.12 - 0.05$$
$$= 0.929$$
$$= 93\%$$

Note: A value of 0.05 was used for L because of the high differential pressure.

Calculate suction capacity.

$$Q_1 = E_v \times P_d$$
$$= 0.93 \times 107.6$$
$$= 100.1 \text{ cfm}$$

Calculate piston speed.

$$P_S = [2 \times S_t \times N]/12$$
$$= [2 \times 12 \times 300]/12$$
$$= 600 \text{ ft/min}$$

Table 1
Efficiency Multiplier for Low Pressure

r_p	Pressure psia							
3.5	10	14.7	20	40	60	80	100	150
3.0	0.990	1.00	1.00	1.00	1.00	1.00	1.00	1.00
2.5	0.980	0.985	0.990	0.995	1.00	1.00	1.00	1.00
2.0	0.960	0.965	0.970	0.980	0.990	1.00	1.00	1.00
1.5	0.890	0.900	0.920	0.940	0.960	0.980	0.990	1.00

Source: Modified courtesy of the Gas Processors Suppliers Association and Ingersoll-Rand.

Table 2
Efficiency Multiplier for Specific Gravity

r_p	SG				
2.25	1.5	1.3	1.0	0.8	0.6
2.0	0.99	1.0	1.0	1.0	1.01
1.75	0.97	0.99	1.0	1.01	1.02
1.5	0.94	0.97	1.0	1.02	1.04

Source: Modified courtesy of the Gas Processors Suppliers Association.

Estimate suction and discharge volume bottle sizes for pulsation control for reciprocating compressors

Pressure surges are created as a result of the cessation of flow at the end of the compressor's discharge and suction stroke. As long as the compressor speed is constant, the pressure pulses will also be constant. A low pressure compressor will likely require little, if any, treatment for pulsation control; however, the same machine with increased gas density, pressure, or other operational changes may develop a problem with pressure pulses. Dealing with pulsation becomes more complex when multiple cylinders are connected to one header or when multiple stages are used.

API Standard 618 should be reviewed in detail when planning a compressor installation. The pulsation level at the outlet side of any pulsation control device, regardless of type, should be no more than 2% peak-to-peak of the line pressure of the value given by the following equation, whichever is less.

$$P\% = 10/P_{line}^{1/3} \quad (12)$$

When a detailed pulsation analysis is required, several approaches may be followed. An analog analysis may be performed on the Southern Gas Association dynamic compressor simulator, or the analysis may be part of the compressor purchase contract. Regardless of who performs the analysis, a detailed drawing of the piping in the compressor area will be needed.

The following equations are intended as an aid in estimating bottle sizes or for checking sizes proposed by a vendor for simple installations—i.e., a single cylinder connected to a header without the interaction of multiple cylinders. The bottle type is the simple unbaffled type.

Example. Determine the approximate size of suction and discharge volume bottles for a single-stage, single-acting, lubricated compressor in natural gas service.

Cylinder bore	= 9 in,
Cylinder stroke	= 5 in,
Rod diameter	= 2.25 in,
Suction temp.	= 80 °F,
Discharge temp.	= 141°F,
Suction pressure	= 514 psia,
Discharge pressure	= 831 psia,
Isentropic exponent, k	= 1.28,
Specific gravity	= 0.6,
Percent clearance	= 25.7%.

Step 1. Determine suction and discharge volumetric efficiencies using Equations 5 and 13.

$$E_{vd} = [E_v \times f]/r_p^{1/k} \quad (13)$$

$$r_p = 831/514 = 1.617$$

$$Z_1 = 0.93$$

$$Z_2 = 0.93$$

$$f = 0.93/0.93 = 1.0$$

Calculate suction volumetric efficiency using Equation 5:

$$E_v = 0.97 - \left[(1/1) \times (1.617)^{1/1.28} - 1\right] \times 0.257 - 0.03$$
$$= 0.823$$

Calculate discharge volumetric efficiency using Equation 13:

$$E_{vd} = 1 \times [0.823]/\left[1.617^{1/1.28}\right]$$
$$= 0.565$$

Calculate volume displaced per revolution using Equation 1:

$$P_d/N = \left[S_t \times 3.1416 \times D^2\right]/[1728 \times 4]$$
$$= \left[5 \times 3.1416 \times 9^2\right]/[1728 \times 4]$$
$$= 0.184 \text{ cu ft or } 3.18 \text{ cu in.}$$

Refer to Figure 2, volume bottle sizing, using volumetric efficiencies previously calculated, and determine the multipliers.

Suction multiplier = 13.5,
Discharge multiplier = 10.4,
Discharge volume = 318 × 13.5 = 3308 cu in,
Suction volume = 318 × 10.4 = 4294 cu in.

Calculate bottle dimensions. For elliptical heads, use Equation 14.

$$\text{Bottle diameter } d_b = 0.86 \times \text{volume}^{1/3} \quad (14)$$

Volume = suction or discharge volume

$$\text{Suction bottle diameter} = 0.86 \times 4294^{1/3}$$
$$= 13.98 \text{ in}$$

$$\text{Discharge bottle diameter} = 0.86 \times 3308^{1/3}$$
$$= 12.81 \text{ in}$$

$$\text{Bottle length} = L_b = 2 \times d_b \quad (15)$$

$$\text{Suction bottle length} = 2 \times 13.98$$
$$= 27.96 \text{ in}$$

$$\text{Discharge bottle length} = 2 \times 12.81$$
$$= 25.62 \text{ in}$$

Source

Compressors—Selection & Sizing, Gulf Publishing Company, Houston, Texas.

Compression horsepower determination

The method outlined below permits the determination of approximate horsepower requirements for the compression of gas.

1. From Figure 1, determine the atmospheric pressure in psia for the altitude above sea level at which the compressor is to operate.
2. Determine intake pressure (P_s) and discharge pressure (P_d) by adding the atmospheric pressure to the corresponding gauge pressure for the conditions of compression.
3. Determine the total compression ratio $R = P_d/P_s$. If the ratio R is more than 5 to 1, two or more compressor stages will be required. Allow for a pressure loss of approximately 5 psi between stages. Use the same ratio for each stage. The ratio per stage, so that each stage has the same ratio, can be approximated by finding the nth root of the total ratio, when n = number of stages. The exact ratio can be found by trial and error, accounting for the 5 psi interstage pressure losses.
4. Determine the N value of gas.
5. Figure 3 gives horsepower requirements for compression of 1 million cu ft per day for the compression ratios and N values commonly encountered in oil-producing operations.
6. If the suction temperature is not 60°F, correct the curve horsepower figure in proportion to the absolute temperature. This is done as follows:

$$\text{HP} \times \frac{460^\circ + T_s}{460^\circ + 60^\circ\text{F}} = \text{hp (corrected for suction temp.)}$$

where T_s is suction temperature in °F.

7. Add together the horsepower loads determined for each stage to secure the total compression horsepower load. For altitudes greater than 1500 ft above sea level, apply a multiplier derived from the following table to determine the nominal sea level horsepower rating of the internal combustion engine driver.

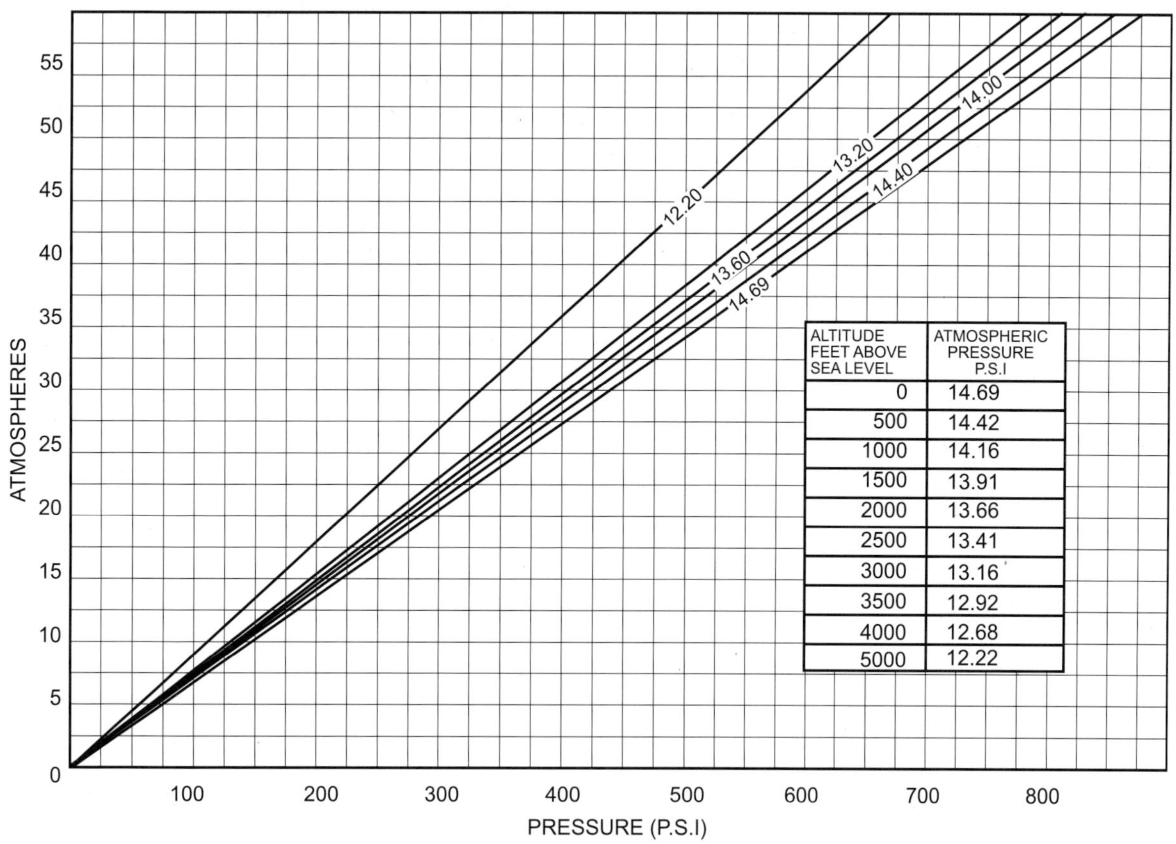

Figure 1. Atmospheres at various atmospheric pressures. *From Modern Gas Lift Practices and Principles*, Merla Tool Corp.

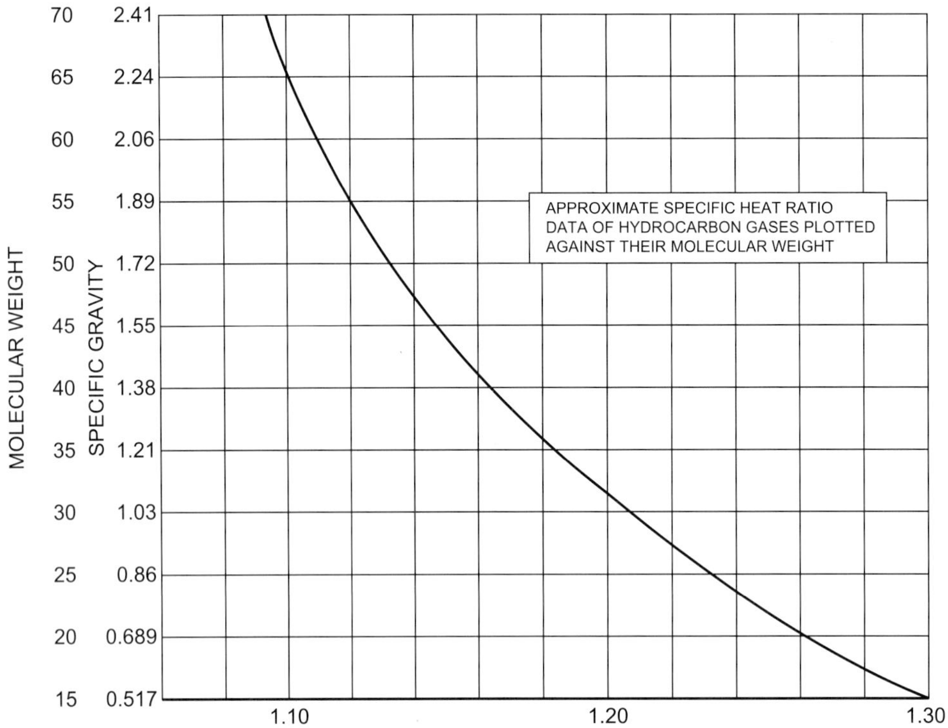

Figure 2. Ratio of specific heat (*N*-value).

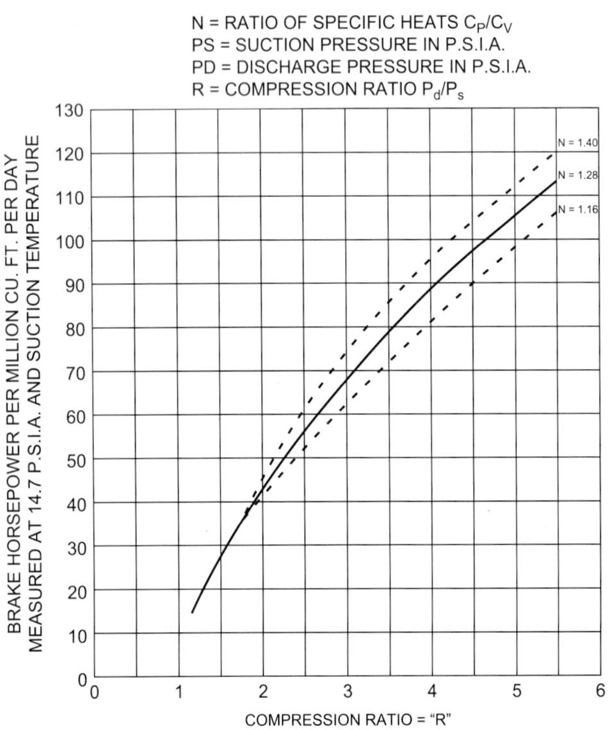

Figure 3. Brake horsepower required for compressing natural gas.

Altitude	Multiplier	Altitude	Multiplier
1500 ft	1.000	4000 ft	1.12
2000 ft	1.03	4500 ft	1.14
2500 ft	1.05	5000 ft	1.17
3000 ft	1.07	5500 ft	1.20
3500 ft	1.10	6000 ft	1.22

8. For a portable unit with a fan cooler and pump driven from the compressor unit, increase the horsepower figure by 7½%.

The resulting figure is sufficiently accurate for all purposes. The nearest commercially available size of compressor is then selected.

The method does not take into consideration the supercompressibility of gas, and it is applicable for pressures up to 1000 psi. In the region of high pressure, neglecting the deviation of gas behavior from that of the perfect gas may lead to substantial errors in calculating the compression horsepower requirements. The enthalpy—entropy charts may be used conveniently in such cases. The procedures are given in References 1 and 2.

Example. What is the nominal size of a portable compressor unit required for compressing 1,600,000 standard cu ft of gas per 24 h at a temperature of 85°F from 40 psig pressure to 600 psig pressure? The altitude above sea level is 2500 ft. The *N* value of gas is 1.28. The suction temperature of stages, other than the first stage, is 130°F.

Solution

$$P_s = 53.41 \text{ psia}$$
$$P_d = 613.41 \text{ psia}$$
$$R_d = P_d/P_s = 11.5 = (3.4)^2 = (2.26)^3$$

Try a solution using a 3.44 ratio and two stages.

$$\text{1st stage: } 53.41 \text{ psia} \times 3.44 = 183.5 \text{ psia discharge}$$

$$\text{2nd stage: } 178.5 \text{ psia} \times 3.44 = 614 \text{ psia discharge}$$

Horsepower from curve, Figure 3 = 77 hp for 3.44 ratio

$$\frac{77 \text{ hp} \times 1,600,000}{1,000,000} = 123.1 \text{ (for } 60^\circ\text{F suction temp.)}$$

$$\text{1st stage: } 123.1 \text{ hp} \times \frac{460 + 85^\circ}{460 + 60^\circ} = 129.1 \text{ hp}$$

$$\text{2nd stage: } 123.1 \text{ hp} \times \frac{460 + 130^\circ}{460 + 60^\circ} = 139.7 \text{ hp}$$

$$1.05(129.1 \text{ hp} + 139.7 \text{ hp}) = 282 \text{ hp}$$

$$1.075 \times 282 \text{ hp}$$

Nearest nominal size compressor is 300 hp.

Centrifugal compressors

The centrifugal compressors are inherently high-volume machines. They have extensive application in gas transmission systems. Their use in producing operations is very limited.

References

1. *Engineering Data Book*, Natural Gasoline Supply Men's Association, 1957.
2. Dr. George Granger Brown: "A Series of Enthalpy—Entropy Charts for Natural Gas," *Petroleum Development and Technology*. Petroleum Division AIME, 1945.

Generalized compressibility factor

The nomogram (Figure 1) is based on a generalized compressibility chart.[1] It is based on data for 26 gases, excluding helium, hydrogen, water, and ammonia. The accuracy is about 1% for gases other than those mentioned.

To use the nomogram, the values of the reduced temperature (T/T_c) and reduced pressure (P/P_c) must be calculated first, where:

T = Temperature in consistent units
T_c = Critical temperature in consistent units
P = Pressure in consistent units
P_c = Critical pressure in consistent units

Example. $P_r = 0.078$, $T_r = 0.84$. What is the compressibility factor, z? Connect P_r with T_r and read $z = 0.948$.

Source

Davis, D. S., *Petroleum Refiner*, 37, No. 11, (1961).

Reference

1. Nelson, L. C., and Obert, E. F., *Chem. Engr.*, 203 (1954).

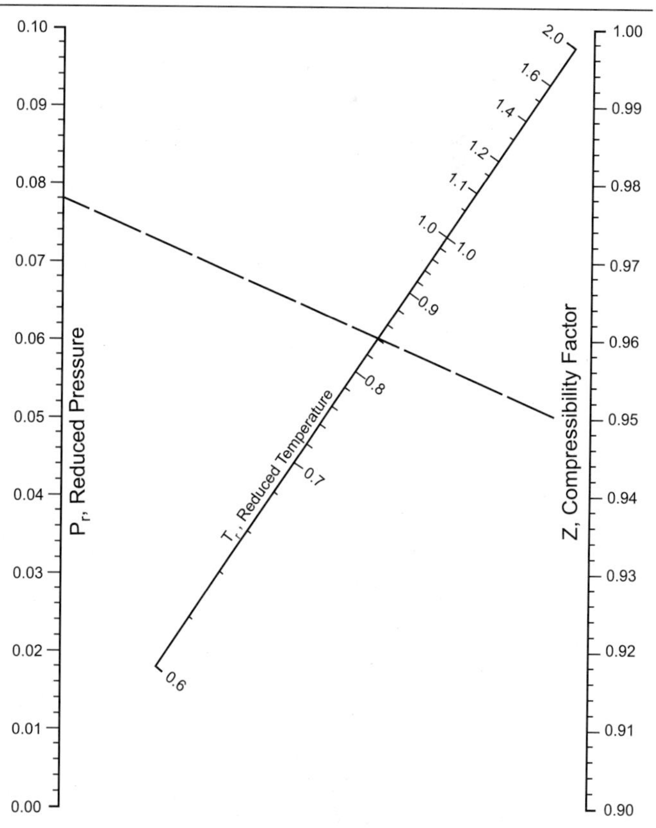

Figure 1. Generalized compressibility factor. (Reproduced by permission, *Petroleum Refiner*, Vol. 37, No. 11, copyright 1961, Gulf Publishing Co., Houston.)

Nomograph aids in diagnosing compressor cylinder ills

B. W. Robertson, Tennessee Gas Pipeline Co., Houston, Texas

Compressor cylinders, like people, develop a fever when they are sick. Unlike humans, however, the normal temperature (suction) is not constant. The trick of using temperatures to determine malfunctions lies in determining the correct starting position.

To determine the starting position, it is necessary to know the true k value of the product being pumped. The k value may be obtained from measured pressures and temperatures according to the formula:

$$T_2/T_1 = (P_2/P_1)^{\frac{k-1}{k}}$$

where the subscript 2 refers to discharge conditions and the 1 refers to suction, and all values of P and T are absolute.[1]

The preferred time to obtain this k would be with new equipment or from data obtained when the equipment was new. The next best time would be with newly overhauled equipment. Lacking either of these data, any values of k must be considered to contain some malfunctions, and any unnecessarily high numbers should be discarded.

Nomograph 1 has been successfully used for a number of years, although has been largely supplanted by computational algorithms. Nomograph 2 may be finished by completing the suction and discharge temperature scales according to the following formulas:

$$\text{Suction: } t_i = 8.557\rho \ \text{Log}(t_s + 460) - 23.135\rho + 2.51 \text{ in above base line}$$

$$\text{Discharge: } t = 23.067\rho \ \text{Log}(t_d + 460) - 62.65\rho \text{ in above base line}$$

where

$\rho = \dfrac{k}{k-1}$ and logs are to the base 10.

Reference

1. Virgil Moring Faires, *Applied Thermodynamics,* 11th Printing, 1946, p. 43.

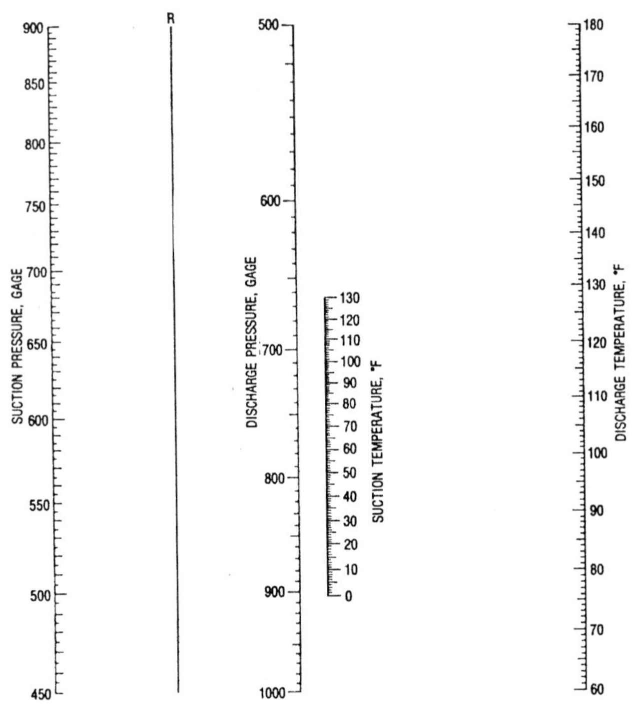

Nomograph 1. To determine starting position, it is necessary to know the true "k" value of the product being pumped. The "k" value may be obtained from measured pressure and temperatures according to the formula given.

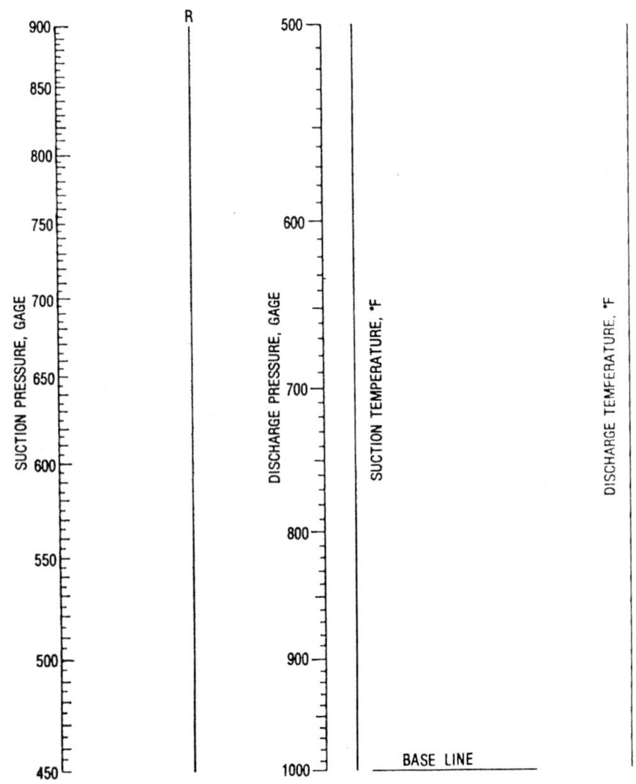

Nomograph 2. This nomograph may be finished by completing the suction and discharge temperature scales according to the formulas given.

Centrifugal compressor performance calculations

Centrifugal compressors are versatile, compact, and generally used in the range of 1000—100,000 inlet cu ft per minute (ICFM) for process and pipeline compression applications.

Centrifugal compressors can use either a horizontal or a vertical split case. The type of case used will depend on the pressure rating with vertical split casings generally being used for the higher pressure applications. Flow arrangements include straight through, double flow, and side flow configurations.

Centrifugal compressors may be evaluated using either the adiabatic or polytropic process method. An adiabatic process is one in which no heat transfer occurs. This does not imply a constant temperature, only that no heat is transferred into or out of the process system. Adiabatic is normally intended to mean adiabatic isentropic. A polytropic process is a variable-entropy process in which heat transfer can take place.

When the compressor is installed in the field, the power required from the driver will be the same whether the process is called adiabatic or polytropic during design. Therefore, the work input will be the same value for either process. It will be necessary to use corresponding values when making the calculations. When using adiabatic head, use adiabatic efficiency, and when using polytropic head, use polytropic efficiency. Polytropic calculations are easier to make even though the adiabatic approach appears to be simpler and quicker.

The polytropic approach offers two advantages over the adiabatic approach. The polytropic approach is independent of the thermodynamic state of the gas being compressed, whereas the adiabatic efficiency is a function of the pressure ratio and therefore is dependent upon the thermodynamic state of the gas.

If the design considers all processes to be polytropic, an impeller may be designed, its efficiency curve determined, and it can be applied without correction regardless of pressure, temperature, or molecular weight of the gas being compressed. Another advantage of the polytropic approach is that the sum of the polytropic heads for each stage of compression equals the total polytropic head required to get from state point 1 to state point 2. This is not true for adiabatic heads.

Sample performance calculations

Determine the compressor frame size, number of stages, rotational speed, power requirement, and discharge temperature required to compress 5000 lb-in/min of gas from 30 psia at 60 °F to 100 psia. The gas mixture molar composition is as follows:

Ethane	5%
Propane	80%
n-Butane	15%

This is the same gas mixture that was considered in Sample Calculation 1, where the gas properties were calculated. The properties of this mixture were determined to be as follows:

$$\begin{aligned} \text{MW} &= 45.5 \\ P_c &= 611 \text{ psia} \\ T_c &= 676^\circ R \\ c_p &= 17.76 \\ k_1 &= 1.126 \\ Z_1 &= 0.955 \end{aligned}$$

Before proceeding with the compressor calculations, let us review the merits of using average values of Z and k in calculating the polytropic head.

The inlet compressibility must be used to determine the actual volume entering the compressor to approximate the size of the compressor and to communicate with the vendor via the data sheets. The maximum value of θ is of interest and will be at its maximum at the inlet to the compressor where the inlet compressibility occurs (although using the average compressibility will result in a conservative estimate of θ).

Compressibility will decrease as the gas is compressed. This would imply that using the inlet compressibility would be conservative since as the compressibility decreases, the head requirement also decreases. If the variation in compressibility is drastic, the polytropic head requirement calculated by using the inlet compressibility would be practically useless. Compressor manufacturers calculate the performance for each stage and use the inlet compressibility for each stage. An accurate approximation may be substituted for the stage-by-stage calculation by calculating the polytropic head for the overall section using the average compressibility. This technique results in overestimating the first half of the impellers and underestimating the last half of the impellers, thereby calculating a polytropic head very near that calculated by the stage-by-stage technique.

Determine the inlet flow volume, Q_1:

$$Q_1 = m_1[(Z_1RT_1)/(144P_1)]$$

where

m = Mass flow,
Z_1 = Inlet compressibility factor,
R = Gas constant = 1545/MW,
T_1 = Inlet temperature °R,
P_1 = Inlet pressure.

$$Q_1 = 5000[(0.955)(1545)(60+460)/(45.5)(144)(30)] = 19{,}517 \text{ ICFM}$$

Refer to Table 1 and select a compressor frame that will handle a flow rate of 19,517 ICFM. A Frame C compressor will handle a range of 13,000–31,000 ICFM and would have the following nominal data:

$$H_{pnom} = 10{,}000 \text{ ft-lb/lbm (nominal polytropic head)}$$

$$n_p = 77\% \text{ (polytropic effectively)}$$

$$N_{nom} = 5900 \text{ rpm}$$

Determine the pressure ratio, r_p.

$$r_p = P_2/P_1 = 100/30 = 3.33$$

Determine the approximate discharge temperature, T_2.

$$n/n-1 = [k/k-1]n_p = [(1.126/1.000)](0.77) = 6.88$$

$$T_2 = T_1(r_p)^{(n-1)/n} = (60+460)(3.33)^{1/6.88} = 619^\circ\text{R} = 159^\circ\text{F}$$

where

T_1 = Inlet temp.

Determine the average compressibility, Z_a.

$$Z_1 = 0.955 \text{ (from gas properties calculation)}$$

where

Z_1 = Inlet compressibility.

$$(P_r)_2 = P_2/P_c = 100/611 = 0.614$$

$$(T_r)_2 = T_2/P_c = 619/676 = 0.916$$

Table 1
Typical Centrifugal Compressor Frame Data*

Frame	Nominal Inlet Volume Flow		Nominal Polytropic Head		Nominal Polytropic Efficiency (%)	Nominal Rotational Speed (RPM)	Nominal Impeller Diameter	
	English (ICFM)	Metric (m^3/h)	English (ft-lbf/lbm)	Metric (k-Nm/kg)			English (in)	Metric (mm)
A	1000–7000	1700–12,000	10,000	30	76	11,000	16	406
B	6000–18,000	10,000–31,000	10,000	30	76	7700	23	584
C	13,000–31,000	22,000–53,000	10,000	30	77	5900	30	762
D	23,000–44,000	39,000–75,000	10,000	30	77	4900	36	914
E	33,000–65,000	56,000–110,000	10,000	30	78	4000	44	1120
F	48,000–100,000	82,000–170,000	10,000	30	78	3300	54	1370

*Based on a survey of available equipment.

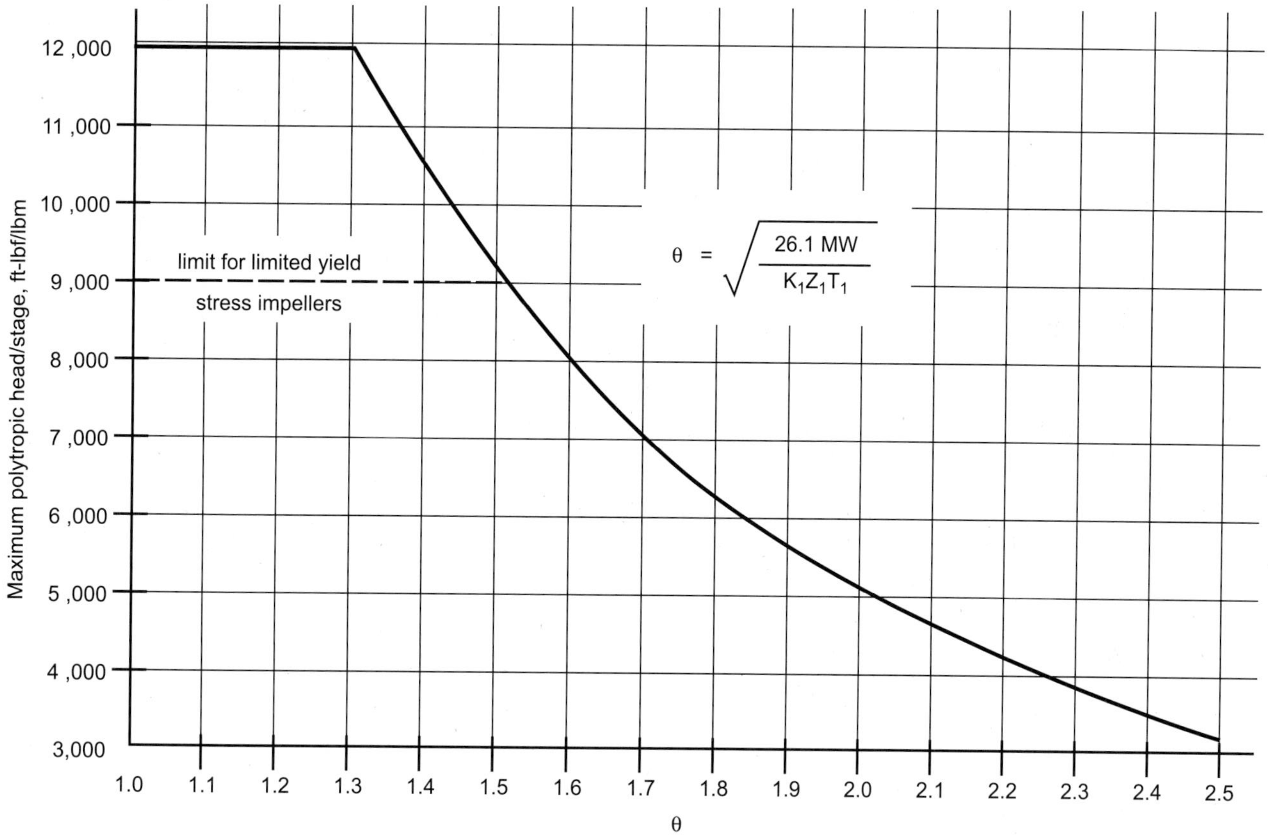

Figure 1. Maximum polytropic head per stage—English system.

As described earlier, find discharge compressibility Z_2.

$$Z_2 = 0.925$$

$$Z_a = (Z_1 + Z_2)/2 = 0.94$$

Determine average k-value. For simplicity, the inlet value of k will be used for this calculation. The poly- tropic head equation is insensitive to the k-value (and therefore the n-value) within the limits that k normally varies during compression. This is because any errors in the $n/(n-1)$ multiplier in the polytropic head equation tend to balance corresponding errors in the $(n-1)/n$ exponent. Discharge temperature is very sensitive to k-value. Since the k-value normally decreases during compression, a discharge temperature calculated by using the inlet k-value will be conservative, and the actual temperature may be several degrees higher—possibly as much as 25–50 °F Calculating the average k-value can be time-consuming, especially for mixtures containing several gases, since not only must the mol-weighted c_p of the mixture be determined at the inlet temperature but also at the estimated discharge temperature.

The suggested approach is as follows:

1. If the k-value is felt to be highly variable, one pass should be made at estimating discharge temperature based on the inlet k-value; the average k-value should then be calculated using the estimated discharge temperature.
2. If the k-value is felt to be fairly constant, the inlet k-value can be used in the calculations.
3. If the k-value is felt to be highly variable, but sufficient time to calculate the average value is not available, the inlet k-value can be used (but be aware of the potential discrepancy in the calculated discharge temperature).

$$k_1 = k_a = 1126$$

Determine average $n/(n-1)$ value from the average k-value. For the same reasons discussed earlier, use $n/(n-1) = 6.88$.

Table 2
Approximate Mechanical Losses as a Percentage of Gas Power Requirement*

Gas Power Requirement		Mechanical Losses, L_m (%)
English (hp)	Metric (kW)	
0–3000	0–2500	3
3000–6000	2500–5000	2.5
6000–10,000	5000–7500	2
10,000+	7500+	1.5

*There is no way to estimate mechanical losses from gas power requirements. This table will, however, ensure that mechanical losses are considered and yield useful values for estimating purposes.

Determine polytropic head, H_p

$$H_p = \left[Z_a RT_1(n/n-1)r_p^{(n-1)/n} - 1\right]$$
$$= (0.94)(1545/45.5)(520)(6.88)\left[(3.33)^{1/6.88} - 1\right]$$
$$= 21,800 \text{ ft-lbf/lbm}$$

Determine the required number of compressor stages, θ:

$$\theta = [(26.1 \text{ Mw})/k_1 Z_1 T_1]^{0.5}$$
$$= [(26.1)(45.5)/(1.126)(0.955)(520)]^{0.5} = 1.46$$

max H_p/stage from Figure 1 using $\theta = 1.46$.

$$\text{Number of stages} = H_p/\text{max}.H_p/\text{stage} = 21,800/9700$$
$$= 2.25 = 3 \text{ stages.}$$

Determine the required rotational speed:

$$N = N_{nom}\left[H_p/H_{p\ nom} \times \text{no. stages}\right]^{0.5}$$
$$= 5900[21,800/(10,000)(3)]^{0.5} = 5030 \text{ rpm.}$$

Determine the required shaft power:

$$PWR_g = mH_p/33,000\ n_p$$
$$= (5000)(21,800)/(33,000)(0.77) = 4290 \text{ hp}$$

$$\text{Mechanical losses } (L_m) = 2.5\% \text{ (from Table 2)}$$
$$L_m = (0.025)(4290) = 107 \text{ hp.}$$

$$PWR_s = PWR_g + L_m = 4290 + 107 = 4397 \text{ hp.}$$

Determine the actual discharge temperature:

$$T_2 = T_1\left(r_p\right)^{(n-1)/n} = 520(3.33)^{1/6.88} = 619°R$$
$$= 159°F.$$

The discharge temperature calculated in the last step is the same as that calculated earlier only because of the decision to use the inlet k-value instead of the average k-value. Had the average k-value been used, the actual discharge temperature would have been lower.

Source

Estimating Centrifugal Compressor Performance, Gulf Publishing Company, Houston.

Nomographs for estimating compressor performance

Convert SCFM to inlet CFM.

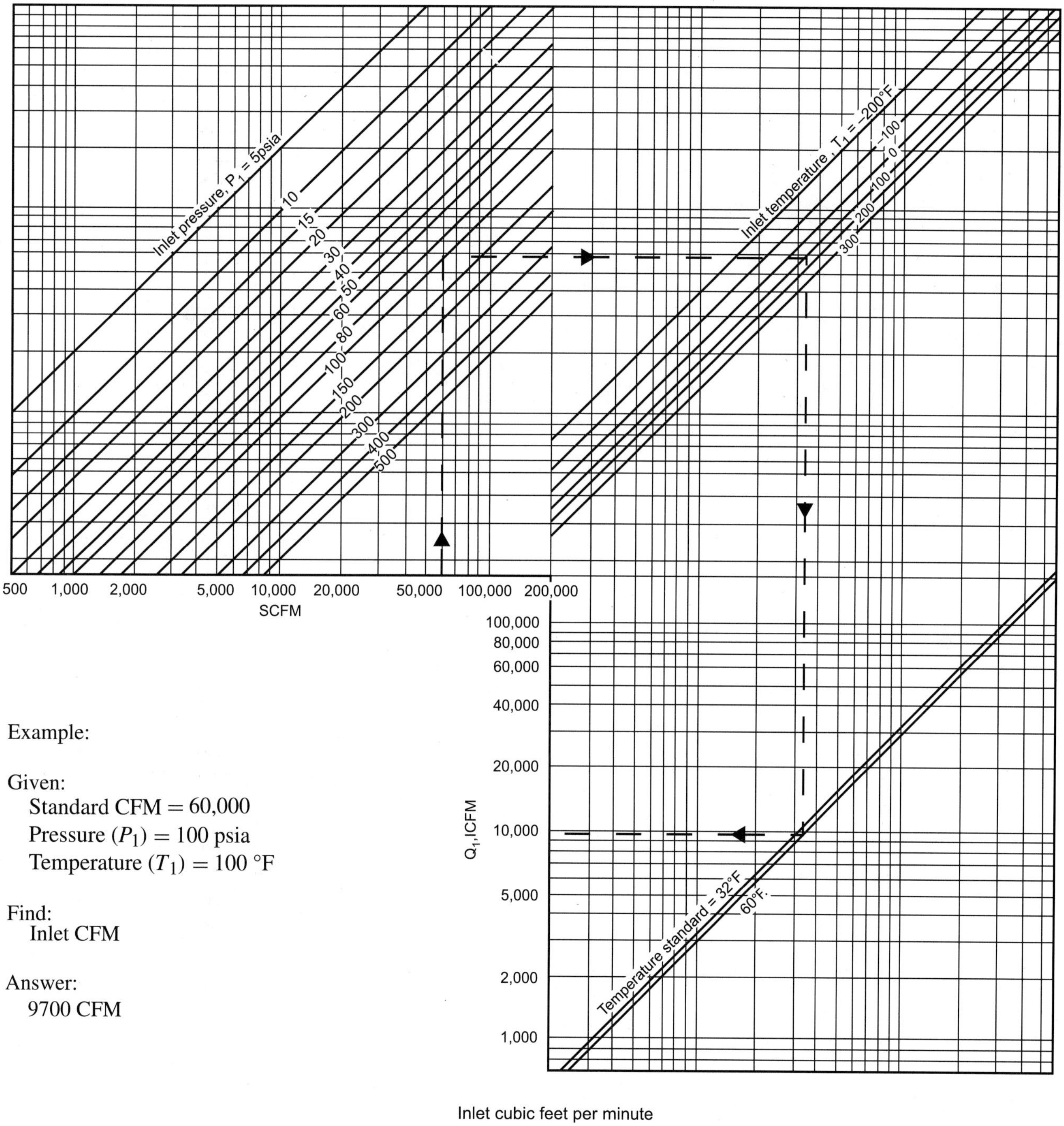

Example:

Given:
Standard CFM = 60,000
Pressure (P_1) = 100 psia
Temperature (T_1) = 100 °F

Find:
Inlet CFM

Answer:
9700 CFM

Reprinted by permission of Gas Processors Suppliers Association.

Convert weight flow to inlet CFM.

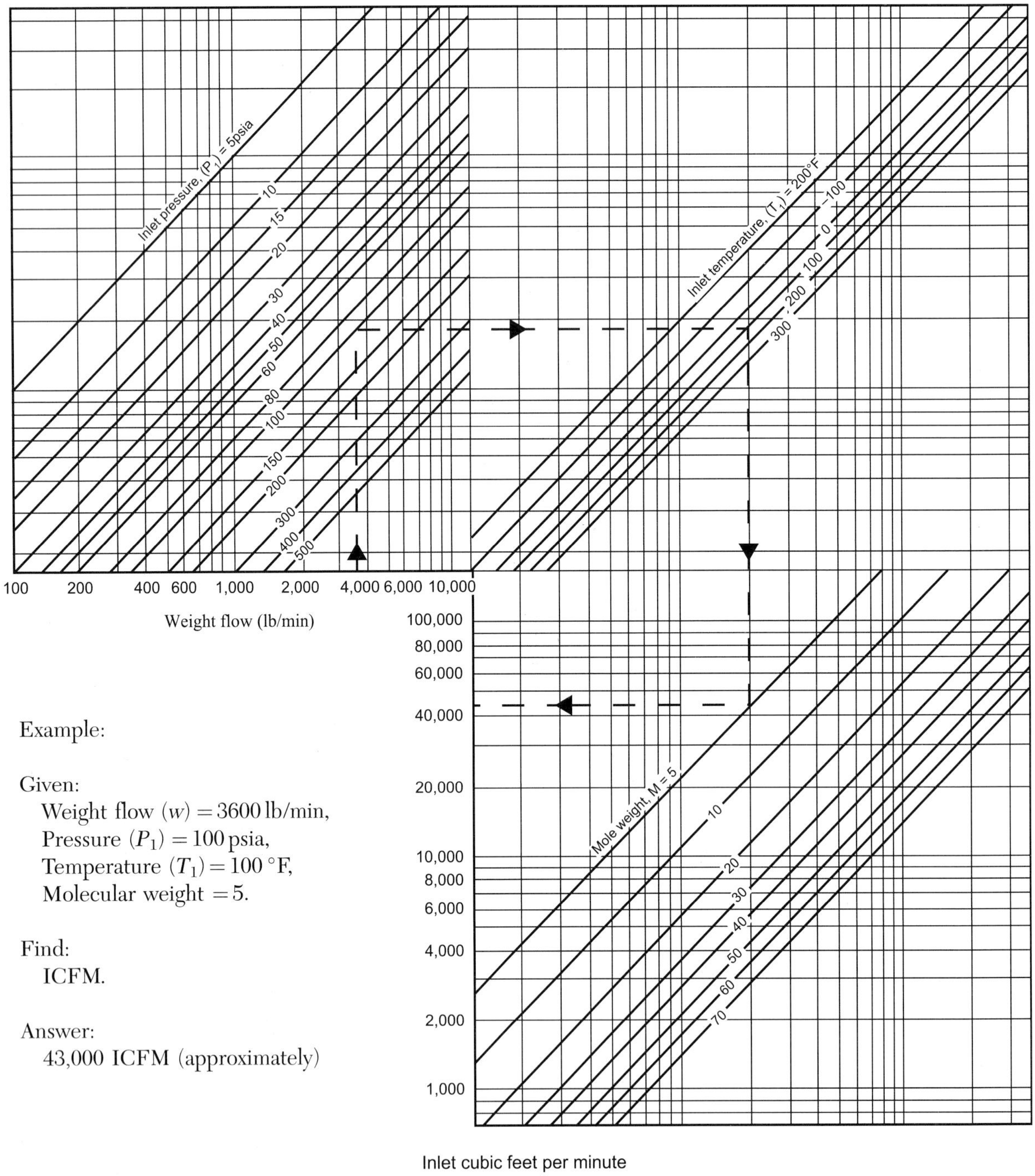

Example:

Given:
Weight flow $(w) = 3600$ lb/min,
Pressure $(P_1) = 100$ psia,
Temperature $(T_1) = 100\,°F$,
Molecular weight $= 5$.

Find:
ICFM.

Answer:
43,000 ICFM (approximately)

Convert weight compression ratio to head.

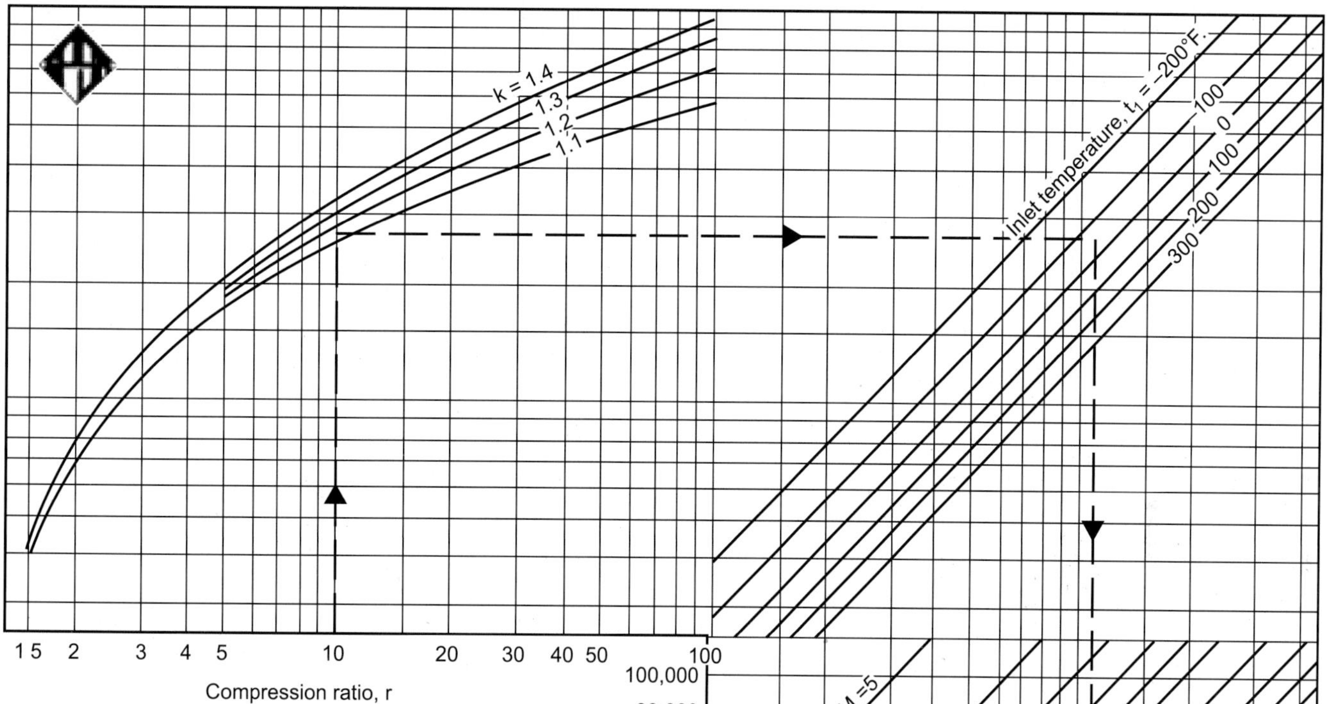

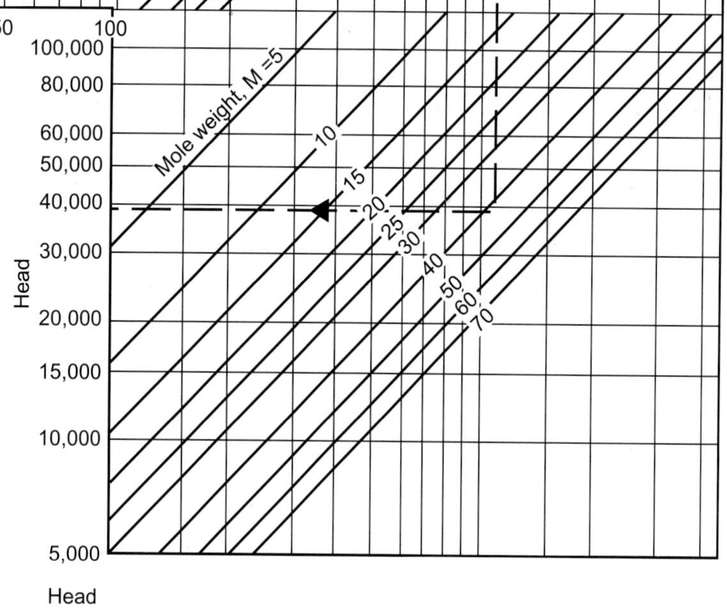

Example:

Given:
Compression ratio = 10,
$k = 1.15$,
Inlet temperature = −50 °F,
Molecular weight = 42.

Find:
Head.

Answer:
39,000 ft-lb/lb (approximately).

Convert compression ratio to discharge temperature.

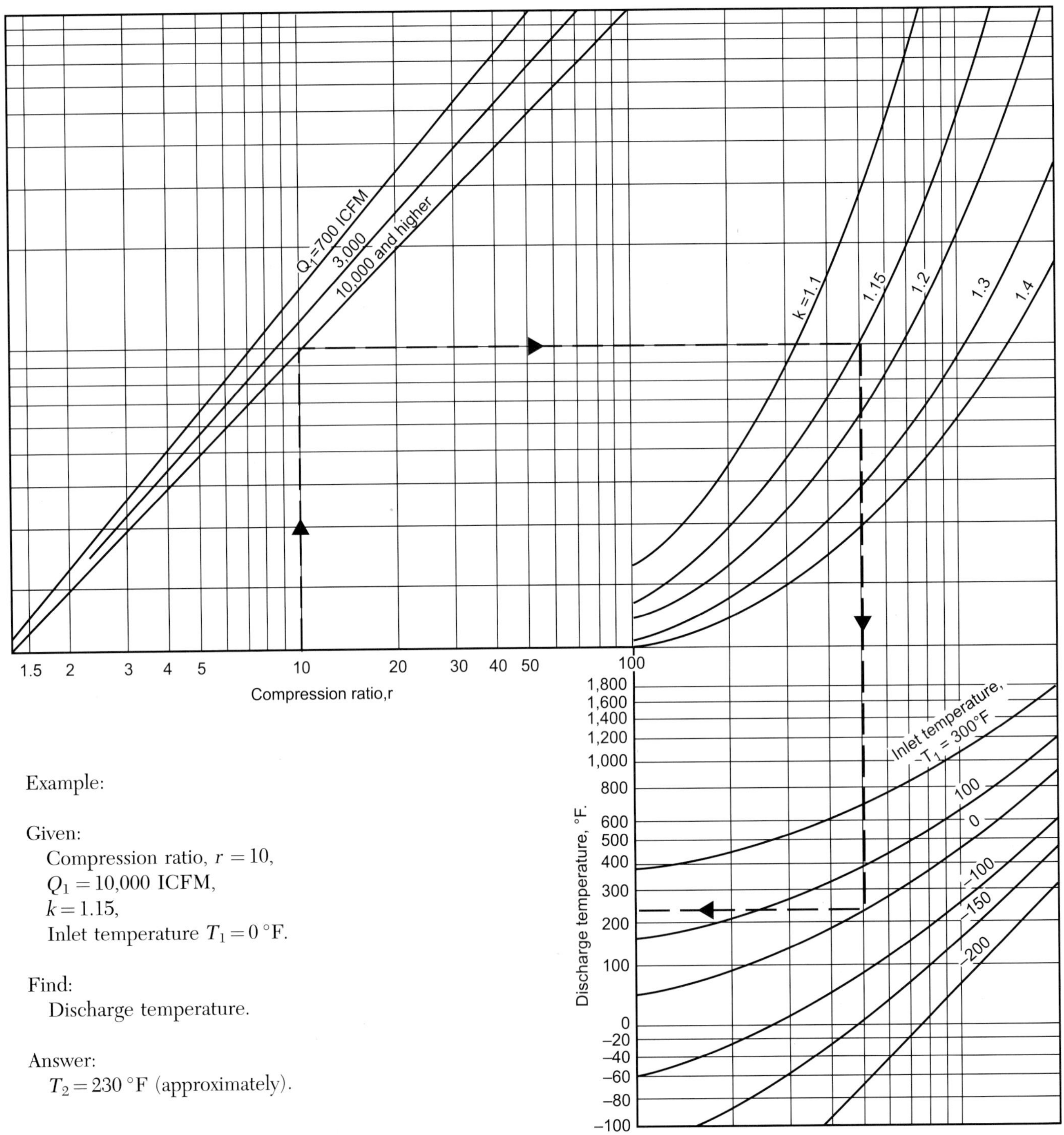

Example:

Given:

Compression ratio, $r = 10$,

$Q_1 = 10{,}000$ ICFM,

$k = 1.15$,

Inlet temperature $T_1 = 0\ °F$.

Find:

Discharge temperature.

Answer:

$T_2 = 230\ °F$ (approximately).

Discharge temperature

Reprinted by permission of Gas Processors Suppliers Association.

Horsepower determination.

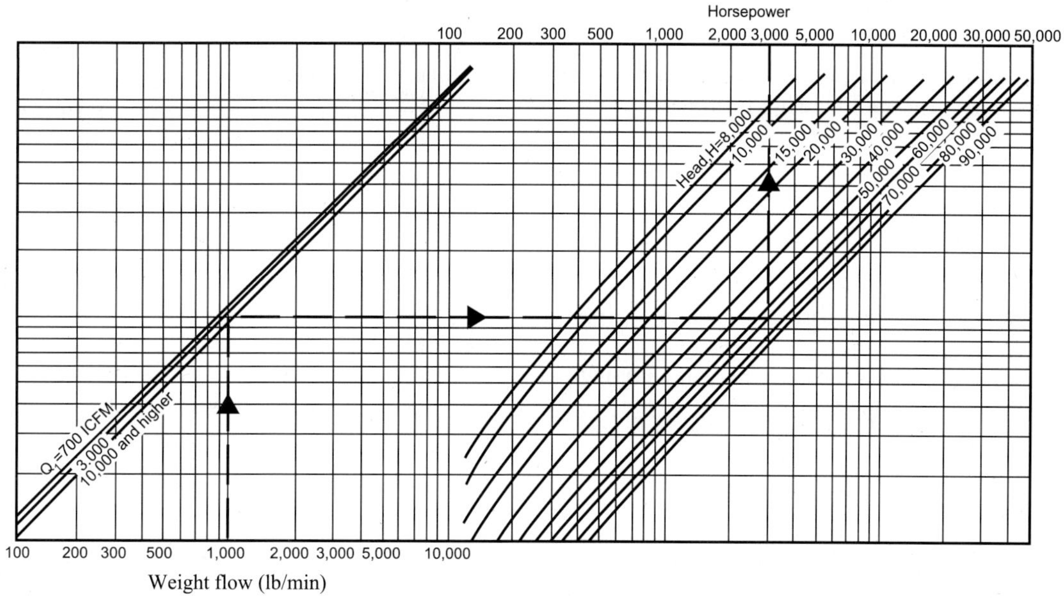

Example:

Given:

Weight flow, w = 1000 lb/min,
Head = 70,000 ft-lb/lb.
Q_1 = 10,000 ICFM

Find:

Horsepower.

Answer:

$H_p = 3000$.

Efficiency conversion.

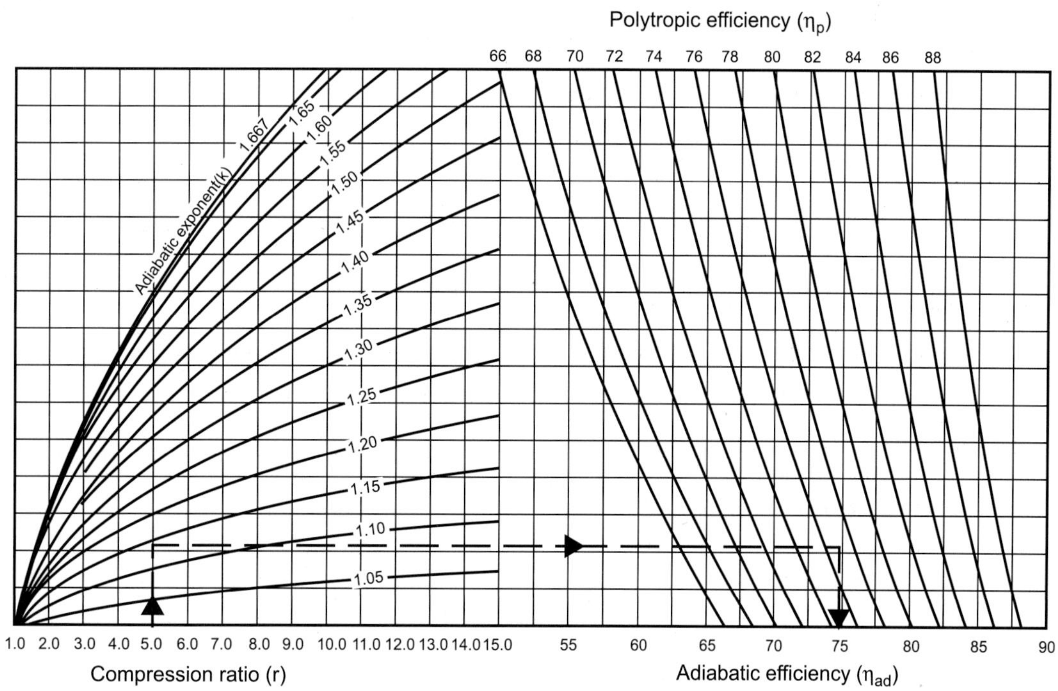

Reprinted by permission of Gas Processors Suppliers Association.

Estimate horsepower required to compress natural gas

To estimate the horsepower to compress a million cubic ft of gas per day (MMcfd), use the following formula:

$$\text{BHP/MMcfd} = \frac{R}{R + RJ}\left(\frac{5.16 + 124\text{Log}\ R}{0.97 - 0.03R}\right)$$

where

R = Compression ratio. Absolute discharge pressure divided by absolute suction pressure,

J = Supercompressibility factor—assumed 0.022 per 100 psia suction pressure.

Example. How much horsepower should be installed to raise the pressure of 10 million cu ft of gas per day from 185.3 psi to 985.3 psi?

Solution. This gives absolute pressures of 200 and 1000.

$$R = \frac{1000}{200} = 5.0$$

Substituting in the formula:

$$\begin{aligned}\text{BHP/MMcfd} &= \frac{5.0}{5.0 + 5 \times 0.044} \times \frac{5.15 + 124 \times .699}{0.97 - 0.03 \times 5}\\ &= 106.5\ \text{hp} = \text{BHP for 10 MMcfd}\\ &= 1065\ \text{hp}\end{aligned}$$

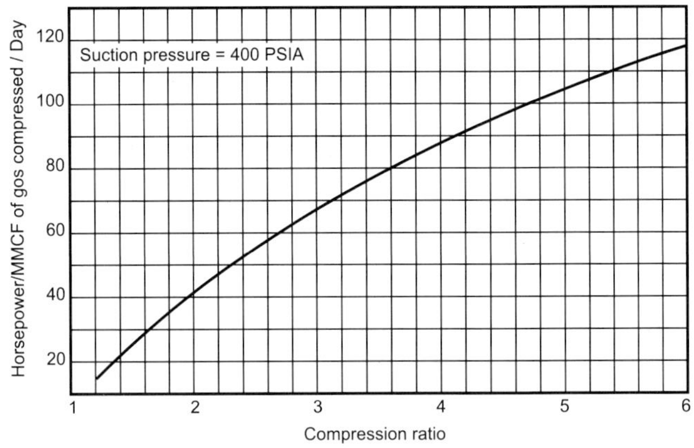

where the suction pressure is about 400 psia, and the brake horsepower per MMcfd can be read from the chart.

The previous formula may be used to calculate horsepower requirements for various suction pressures and gas physical properties to plot a family of curves.

Estimate compressor horsepower where discharge pressure is 1000 psi

Rule. Use the following formula for reciprocating compressors where the compressor ratio is between 1.20 and 1.40.

Brake horsepower per MMcfd = $11 + (42.5)(R - 1.20)$

where R is compression ratio.

For a centrifugal compressor the formula would be:

Brake horsepower per MMcfd = $10 + (49)\ (R - 1.20)$

Example. How much horsepower would have to be installed in a field compressor station where the discharge pressure would be 1055 psia and suction would be 800 psia? The station must handle 10,000,000 cu ft per day.

Solution. The compression ratio is $\frac{1055}{800} = 1.32$.

Substituting in the formula:

Brake horsepower per MMcfd = $11 + (42.5)(1.32 - 1.20)$
$= 11 + 5.1 = 16.1$ hp.

The station would require 161 hp.

Example. How much horsepower should be installed in a centrifugal compressor station handling 100,000,000 cu ft per day if the suction and discharge are the same as above?

Solution.

Brake horsepower per MMcfd = $10 + (49)(0.12) = 15.9$

For 100 MMcfd the horsepower would be 1590.

Calculate brake horsepower required to compress gas

The approximate brake horsepower per million cubic feet of gas to be compressed may be obtained from this table. The total brake horsepower required is calculated by multiplying the total million cubic feet of gas compressed. In this estimation, gas volumes are expressed at 14.7 psia and 60 °F, specific gravity as 0.6, and K (the ratio of specific heats) as 1.27. The suction temperature is assumed as 80 °F and interstage temperature as 130 °F. The relative humidity of the air equals 100%.

Example. What is the brake horsepower required to compress 5.0 million cu ft of gas from 200 psig suction pressure to 1000 psig discharge pressure?

On the chart, read opposite a suction pressure of 200 and under a discharge pressure of 1000 psig, that 100 brake horsepower is required per million cubic feet of gas compressed. Multiply 5.0 times 100 = 500 bhp.

Horsepower Selection Chart Natural Gas Compression											
		Discharge Pressure									
		250 psig	**375 psig**	**500 psig**	**750 psig**	**1000 psig**	**1250 psig**	**1500 psig**	**2000 psig**	**3500 psig**	**6000 psig**
Suction	0 psig	205	235	260	280	300	320	330	360	390	—
Pressure	10 psig	165	195	215	250	260	275	290	310	345	390
	25 psig	130	155	175	205	230	250	255	275	320	370
	50 psig	95	125	140	170	190	205	215	245	280	325
	75 psig	70	95	120	145	165	180	195	215	250	300
	100 psig	50	75	95	120	145	160	175	195	230	275
	200 psig	—	—	55	80	100	120	130	150	190	240
	300 psig	—	—	30	55	75	90	105	120	160	205
	500 psig	—	—	—	25	40	55	65	85	125	165
	750 psig	—	—	—	—	15	30	40	60	100	140
	1000 psig	—	—	—	—	—	—	25	40	75	125
	1500 psig	—	—	—	—	—	—	—	—	50	90
	2000 psig	—	—	—	—	—	—	—	—	—	70
	2500 psig	—	—	—	—	—	—	—	—	—	55

How can we find the size of a fuel gas line for a compressor station?

Suppose that an 8000 hp compressor building is added to an existing compressor plant. What size fuel line should be run from the meter building 1600 ft distant? Take-off pressure at the meter building will be 50 lb/in^2, and it is desirable to maintain 25 psi pressure on the large engine fuel header in the building. Assume that the gas is heated prior to expansion so that outlet temperature is 60 °F.

The first step is to estimate the ultimate compressor-building horsepower. Assume in the foreseeable future the building will be expanded to 12,000 hp.

Next, estimate the fuel needs.

$$\begin{aligned} Q &= 12,000 \times 240 \text{ cu ft per day} \\ &= 2,880,000 \text{ cu ft per day} \end{aligned}$$

To estimate the size of the line that must be run for fuel, use this formula:

$$Q = \frac{(36,000)d^3\sqrt{P_1^2 - P_2^2}}{\sqrt{l}}$$

where

Q = Cubic feet of gas per 24 h
d = Pipe ID in inches
l = Length of pipe in feet
P_1 = Psi (abs) at starting point
P_2 = Psi (abs) at ending point

Using this formula, try several pipe diameters to determine the approximate size. For instance, if 6-in ID pipe were used, the approximate flow could be found as follows:

$$Q = \frac{36,300 \times 6^3 \sqrt{50 + 15^2 - (25 + 15)^2}}{\sqrt{1600}}$$

$$Q = \frac{36,300 \times 216\sqrt{65^2 - 40^2}}{40}$$

Q = approximately 10, 036, 000 cu ft per 24 h.

From this it is obvious that 6-in pipe would be considerably oversized, especially if the fuel gas header in the compressor building is of large diameter.

Try substituting 4-in diameter pipe in this formula to find an approximate size pipe.

$$Q = \frac{36,300 \times 4^3 \sqrt{50 + 15^2 - (25 + 15)^2}}{\sqrt{1600}}$$

$$Q = \frac{36,300 \times 64\sqrt{65^2 - 40^2}}{40}$$

$$Q = 2,973,700 \text{ cu ft per 24 h}$$

Apparently 4-in pipe will be the correct size. To check it, use this more accurate formula, which is Weymouth formula:

$$Q = \frac{63,320 \times d^{8/3}\sqrt{P_1^2 - P_2^2}}{\sqrt{l}}$$

$$Q = \frac{63,320 \times 41\sqrt{65^2 - 40^2}}{\sqrt{1600}}$$

Four-in pipe is adequate.

$$Q = 3,322,000 \text{ cu ft per day}$$

Note. This latter formula is based on the Weymouth formula, for gas with a specific gravity of 0.60 and a temperature of 60 °F. For every 10° variation in temperature, the answer will be approximately 1% in error. For every 0.01 variation in specific gravity, the answer will be three-fourths percent in error.

Estimate engine cooling water requirements

This equation can be used for calculating engine jacket water requirements, as well as lube oil cooling water requirements:

$$\text{GPM} = \frac{\text{H} \times \text{BHP}}{500\Delta t}$$

where

H = Heat dissipation in Btu per BHP/h.
BHP = Brake horsepower
Δt = Temperature differential across engine. Usually manufacturers recommend this not exceed 15 °F; 10 °F is preferable.

This will vary for different engines; where they are available, the manufacturers' values should be used. Otherwise, you will be safe in substituting the following values in the formula: For engines with water-cooled exhaust manifolds—engine jacket water = 2200 Btu per BHP/h. Lube oil cooling water = 600 Btu per BHP/h.

For engines with dry-type manifolds (so far as cooling water is concerned), use 1500 Btu/BHP/h for the engine jackets and 650 Btu/BHP/h for lube oil cooling water requirements.

Example. Find the jacket water requirements for a 2000-hp gas engine that has no water jacket around the exhaust manifold.

Solution.

$$\text{GPM} = \frac{1500 \times 2000}{500 \times 10}$$

$$\text{GPM} = \frac{3,000,000}{5000} = 600 \text{ gallons per min}$$

The lube oil cooling water requirements could be calculated in a like manner.

Performance testing guidelines for centrifugal compressors

E. Wilcox, Chevron Energy Technology Company, Houston, Texas

Detailed performance analysis ofcentrifugal compressors in the field is essential to evaluate their existing condition. Current performance of a compressor can also be a valuable tool in evaluating its reliability. A decrease in compressor performance can be an excellent indication of internal wear or fouling, which, if allowed to continue, may result in unscheduled outages or reduced throughput. In contrast, perceived performance problems may be a result of a compressor operating far from its original design. However, obtaining accurate performance data in the field can be very challenging.

This article will explain the relative importance that different process variables have in performance calculations, as well as specify the necessary instrumentation to obtain process data with an acceptable uncertainty. Since original design conditions almost never match actual operating conditions, how to compare actual field data to design data using nondimensional head and efficiency will be demonstrated. Likewise, the limits on these comparisons will be outlined for users. Several example field performance evaluations are discussed with some common pitfalls that can be avoided. Examples of the effects of inaccurate process data are also included in the discussion.

Obtaining accurate field data. Accurate centrifugal compressor performance measurement depends on the quality of the field data. In general, testing should follow the conditions set forth in ASME PTC 10–1997. However, most petrochemical plants do not have the instrumentation specified in this test code. Compressor piping should be designed to accommodate flowmeter runs and to meet location requirements for pressure and temperature as well. Small inaccuracies (in certain areas) can make a large difference between the measured versus actual conditions.

The field data required for an accurate performance evaluation are:

- Suction and discharge pressures,
- Suction and discharge temperatures,
- Mass flowrate,
- Gas composition,
- Rotational speed,
- Driver load.

The most important factors in obtaining accurate field data are:

- Pressure and temperature transmitters should be located within a few feet of the compressor case at least 10 pipe diameters from any valve, tee, elbow, or other obstruction.
- Transmitters (both pressure and temperature) should be calibrated prior to the test.
- Data should only be taken during steady-state conditions. This is usually specified by requiring the discharge temperature to remain constant over a 15-min interval.
- RTDs should be used instead of thermocouples, if possible, due to improved accuracy.
- Test points at different flowrates or speeds are incredibly valuable. Inaccuracies in flow, pressure, or temperature may be difficult to determine with only one test point. But a map of points will usually reveal the problem.

Flowrate. The flowrate reported by the plant DCS is usually not absolutely correct. The meter factor, *K*, which is used to convert the measured differential pressure into a flowrate, is always a function of the gas pressure, temperature, and molecular weight (Equation 1). The flowrate reported by the meter must be corrected for the actual process conditions. However, the correction depends on the type of flow units. The correction equations for standard volume, mass, and actual volume flows are given in Equations 2–4, respectively. It is common for flowmeters to be "compensated" in the DCS for the actual pressure and temperature. However, it is rare for the compensation to include the gas molecular weight, since this requires an online gas analyzer. The flowmeter design and location should meet the requirements of the guidelines established by ASME PTC-10.

$$Q = K\sqrt{\Delta P}$$

where

$$K = f(P, T, Mw) \tag{1}$$

$$Q_A = Q\sqrt{\frac{P_A T_D Z_D Mw_D}{P_D T_A Z_A Mw_A}} \text{ for std. volume flow} \tag{2}$$

$$\dot{m}_A = \dot{m}\sqrt{\frac{P_A T_D Z_D Mw_A}{P_D T_A Z_A Mw_D}} \text{ for mass flow} \tag{3}$$

$$Q_A = Q\sqrt{\frac{P_A T_A Z_A Mw_D}{P_A T_D Z_D Mw_A}} \text{ for actual voulme flow} \tag{4}$$

Example—incorrect flowmeter. An incorrect flow measurement will cause the compressor to appear either low or high in head because the operating point is marked incorrectly on the performance map. The best method to determine if a flow measurement is incorrect is to obtain several data points to compare against the entire curve. As can be seen in Figure 1, the maximum head should remain the same. The curve is just shifted to the right or left.

Gas composition. The gas composition is the most important data required to evaluate a compressor's performance. Likewise, in a wet gas application, an accurate gas composition is also the most difficult piece of test data to obtain. At least two gas samples should be taken during the performance test. Two samples are required to validate the gas analysis (for comparison) as well as when one of the samples is lost or invalid (i.e. without the gas composition, the other process data are worthless). The sample should be obtained using a free-flowing arrangement (Figure 2 shows two examples). An insertion probe should be used, if possible, to obtain the gas sample instead of a wall tap. Samples obtained from wall taps will generally be leaner in the higher molecular weight components due to the boundary layer effect. The sample bomb should be heated to the process temperature before it is analyzed. This is to prevent any condensation of liquids that could alter the gas composition. Obviously, this is much more important for higher molecular weight services such as FCC wet gas compared to reformer or hydrocracker recycle hydrogen.

As an example, a sample taken from the discharge of a coker wet gas compressor was analyzed at the lab ambient temperature (approximately 75 °F) and at 275 °F

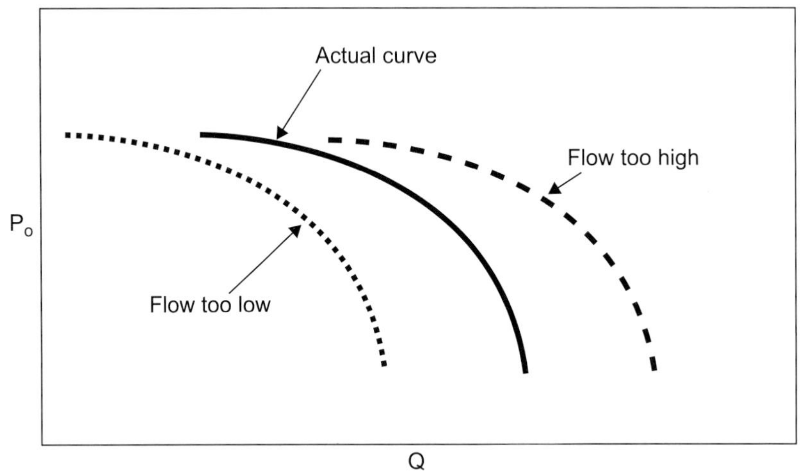

Figure 1. Effects of incorrect flow measurement.

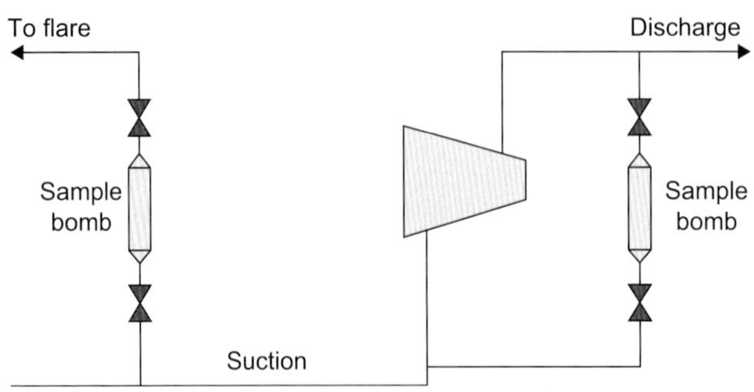

Figure 2. The sample should be obtained using a free-flowing arrangement.

(sample temperature, Table 1). As can be seen, the incorrect gas composition has a pronounced effect on the calculated gas horsepower. This effect is magnified because the molecular weight is used to calculate the head as well as correct the flow.

The gas composition should be reviewed to determine if it is feasible. One of the most common methods used to determine the gas sample composition is a gas chromatograph. A gas chromatograph determines the components by burning them in the presence of a carrier gas. For this reason, a gas chromatograph will not show any water vapor (i.e. water will not burn). However, it is very common for process gases to be saturated with water. For this reason, the measured gas composition must be adjusted if it is indeed saturated with water. The mole fraction of any constituent in a mixed gas stream is defined by Equation 5. Since the gas stream temperature is known, the water vapor partial pressure is equal to its saturation pressure at the given temperature. After the water vapor mole fraction is found, the normalized gas composition must be adjusted so that the total mole fraction equals 1.0.

$$\text{mole fraction} = \frac{\text{constituent partial pressure}}{\text{mixture total pressure}} \tag{5}$$

Likewise, for a multisection compressor, the gas composition will typically become leaner as liquids are condensed out in the intersection knockouts. If the gas samples do not reflect this, there is usually a problem. Also, certain components (i.e. H_2 or H_2S) are normally noncondensable. The mass fractions of these components should actually increase as the heavier liquids are removed from the gas stream.

Example—incorrect gas composition. If the measured gas composition is lower than the actual gas composition in the compressor, it will have the most pronounced effect on both the corrected flowrate and calculated polytropic head. An incorrect low molecular weight will cause the corrected flowrate (if it is measured in standard cubic feet) to be higher (Equation 2). Likewise, the low molecular weight will cause the calculated polytropic head to be higher than it actually is for the given compression ratio. A molecular weight that is too low will also cause the calculated polytropic efficiency to be higher, though not as pronounced. Additionally, the lower molecular weight will lower the calculated mass flow, which will in turn lower the calculated horsepower. Note, these are the effects produced by an incorrect gas composition being used as the input for a field test. These are not the results of a compressor that is operating in a gas that is actually lower in molecular weight than its design. If the measured gas composition is too high, the results are just the inverse of the above (i.e. the flow, head and efficiency are all lower). A summary is shown in Table 2.

Table 1
Effects of Incorrect Gas Composition on Corrected Flow and Calculated Horsepower

Sample	Correct Gas Sample, 275 °F	Incorrect Gas Sample, 75 °F
Molecular weight	48	34
Flow, MMscfd	27.4	32.5
Shaft horsepower, hp	5775	7036

Calculating and evaluating performance parameters. Once the field performance data are obtained, the correct performance parameters must be accurately calculated and evaluated to ensure that the field data are realistic. The most critical step in calculating performance parameters is determining the inlet and outlet density, enthalpy, and entropy. For hydrocarbon gas mixtures, performance programs that use equations of state, such as Lee-Kesler, Redlich-Kwong, or Peng-Robinson, will provide much better results than approximations from Mollier diagrams or ideal gas relationships. Process simulators can be used as well. Once the gas properties are calculated, the correct parameters must be selected to adequately evaluate the compressor performance. Evaluating the results of these calculations should be made to determine the validity of both the calculations and the field data (see Schultz for a listing of the equations for these parameters).

- Shaft horsepower—If a torque meter is available, the shaft horsepower may be calculated directly. However, since this is not commonly the case, the horsepower can be calculated by the heat balance method, as given in ASME PTC-10 and compared to the calculated driver horsepower. This comparison between driver and driven power is the most important indicator of accuracy of a performance test (i.e. if the two values are not close, something is wrong with the test data).

Applying the first law of thermodynamics to the control volume around the compressor is shown in Figure 3.

$$\text{Shp} = \frac{(\dot{m}_1 - \dot{m}_{S1})(h_2 - h_1) + Q_R}{2545} + \text{hp}_{\text{MECH}} \tag{6}$$

Table 2
Effects of Incorrect Gas Composition on Calculated Performance Parameters

Mw	H_P	Effy.	hp	mf
↑	↓	↓	↑	↑
↓	↑	↑	↓	↓

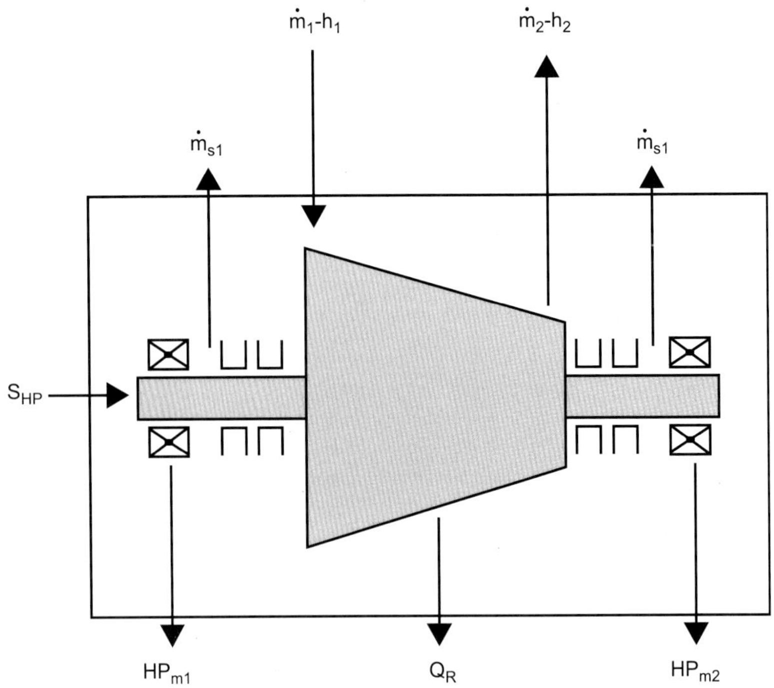

Figure 3. The first law of thermodynamics is applied to the control volume around the compressor.

$$\text{Shp} = \text{hp}_{\text{GAS}} + \text{hp}_{\text{MECH}};\text{hp}_{\text{MECH}} = \text{hp}_{M1} + \text{hp}_{M2} \tag{7}$$

The radiant heat loss, Q_R, is normally negligible, but it can be approximated by dividing the compressor case into axial sections and approximating the heat transfer from each section. Likewise, the seal leakage on the inlet is normally less than 1%, but it can be easily calculated at the orifice in the vent off the seal pots. Note, the internal seal losses (i.e. balance piston and impeller labyrinth seals) do not affect the calculated shaft horsepower. However, they do affect the calculated head and efficiency, which in turn affects the discharge pressure and temperature. The effects of seal losses will be discussed later.

- Polytropic head—This value is limited to approximately 10,000—12,000 ft-lbf/lbm per impeller for most closed impellers. The maximum is set by the impeller tip speed which is limited by the gas sonic velocity as well as the impeller yield stress. Impellers designed to operate in highly corrosive processes (H_2S, CO_2) often require a maximum yield stress of 90 kpsi and Rockwell C of less than 22, which limits them to approximately 9000 ft-lbf/ lbm. For example, if a performance test is done on a compressor and the calculated head per impeller is 20,000 ft-lbf/lbm, then either the measured compression ratio is too high or the measured molecular weight is too low.
- Polytropic efficiency—The maximum value depends on the impeller specific speed, but is limited to approximately 75—78% for older 2D impellers with vaneless diffusers and 80—85% for newer 3D impellers with vaned diffusers.

Losses. Losses are generally grouped into three distinct areas: mechanical, seal, and aerodynamic. Mechanical losses include power dissipated through bearings, oil or gas seals, shaft-driven lube oil pumps, and gearboxes. Seal losses are the decrease in the amount of energy available to convert into pressure head due to internal recirculation inside the compressor. Aerodynamic losses include effects such as friction and pressure losses in the impellers and diffusers. For the purpose of this article, aerodynamic losses are included in the impeller efficiencies and not discussed.

Mechanical losses. Mechanical losses are mostly a function of size and speed. Larger bearings and seals at higher speeds dissipate more power. Mechanical losses are simply added to the calculated gas horsepower. These losses can be approximated from:

- Measuring the flowrate and lube oil temperature increase and using Equation 8. This may be difficult due to many different lube oil return lines and pressure controllers that spill back to the reservoir.
- OEM-supplied curves for different bearings and seals.
- Bearing rotordynamic computer models that calculate horsepower losses as well.

Table 3
Mechanical Losses

Gas Power Requirement	Mechanical Losses, %
0–3000	3.0
3000–6000	2.5
6000–10,000	2.0
10,000+	1.5

Table 4
Gearbox Efficiencies

Gear Type	Efficiency, %
Helical	97–99
Herringbone	6–99
Straight bevel	95–98
Spiral bevel	96–98

- Tables based on compressor gas horsepower (Table 3).

Approximate gearbox efficiencies are given in Table 4.

$$hp_{MECH} = \frac{\dot{m} c_p \Delta T_{oil}}{33,000} \tag{8}$$

Seal losses. Balance piston or division wall leakage is the only seal loss evaluated in this discussion since they are usually much larger than impeller labyrinth seal leakage. Seal losses are much more difficult to estimate than mechanical losses because they are not just added to the calculated gas horsepower. Leakage through the balance piston seal to the compressor suction increases the volume flow through the impellers as well as increases the inlet temperature, both of which decrease the compressor discharge pressure (Figure 4). Balance piston leakage causes the measured head and efficiency to decrease, but does not increase the calculated gas horsepower. This does not mean that increased balance piston seal leakage does not affect compressor power. In the operating world, compressors have to put up a required discharge pressure to pump forward. As the balance piston seal leakage increases, either the compressor speed or suction pressure must be increased to maintain the desired discharge pressure and flow. This causes the power consumed by the compressor to increase.

A feel for the amount of balance piston leakage can be established by measuring the differential pressure across the balance piston line. Most OEMs design for this to be less than 2 or 3 psid. Anything above this usually means that a balance piston seal is leaking excessively. A differential pressure gauge is almost always required to make this measurement due to the small differential.

Seal leakage rate can be estimated by using a leakage equation. Equation 9 applies for adjacent teeth in the labyrinth. This is used to iteratively solve for the labyrinth cavity pressures and seal leakage rate. If the flow is choked, Equation 10 is used for the last labyrinth.

$$\dot{m} = \mu_0 \mu_1 H \sqrt{\frac{P_{i-1}^2 - p_i^2}{ZRT}} \tag{9}$$

$$\dot{m}_{NT} = \frac{0.51 \mu_0 P_{NC} H}{\sqrt{ZRT}} \tag{10}$$

Once the leakage rate is determined, the measured field data must be adjusted to find the actual conditions. In the case of a straight-through compressor with a balance piston seal, the hot balance piston seal leakage is mixed with the gas at the inlet using a mass and enthalpy balance to determine the actual inlet temperature to the first impeller (Figure 5). This will change the polytropic head and efficiency calculations. Adding in the model for the balance piston seal leakage allows us to see why the compressor efficiency is not as designed.

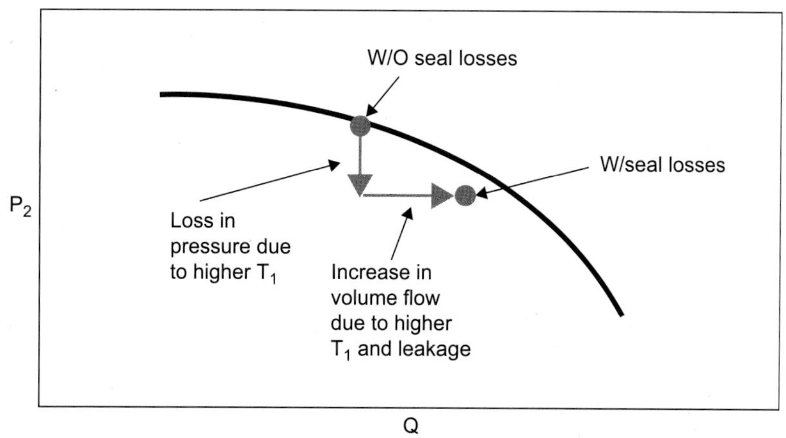

Figure 4. Effect of balance piston leakage on discharge pressure.

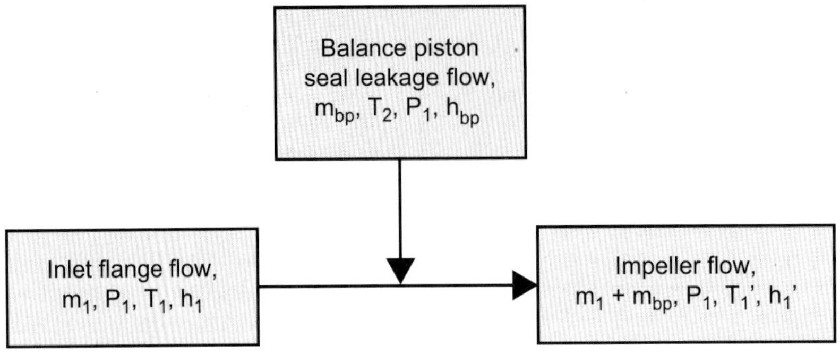

Figure 5. Mass and heat balances for balance piston seal leakage.

Comparing measured field performance to shop test or predicted performance. Since field conditions never exactly match the original design, certain nondimensional parameters must be calculated so that the field performance can be compared to the OEM shop test or predicted performance data. While these nondimensional parameters will enable "apple-to-apple" comparisons for different conditions, they have very real limitations based on the aerodynamic characteristics of the impellers.

These nondimensional parameters include: Polytropic head coefficient, μ_p

$$\mu_P = \frac{H_P}{(\pi DN)^2} \tag{12}$$

Polytropic efficiency, η_P

$$\eta_P = \frac{H_P}{h_s - h_1} \tag{13}$$

Flow coefficient, Φ

$$\Phi = \frac{Q}{ND^3} \text{ or } \frac{Q}{N} \tag{14}$$

Impeller Mach number, M

$$M = \frac{\pi DN}{\sqrt{k_1 Z_1 R T_1}} \tag{15}$$

The first step in any comparison is to obtain a set of nondimensional curves for the shop test or predicted data. If they were not provided by the OEM, they can be obtained by iteratively calculating the μ_P and η_P values from the given values of discharge pressure and shaft horsepower. This can be accomplished by guessing a discharge temperature for the given discharge pressure until the correct shaft horsepower has been reached. Once the correct discharge temperature is known, the polytropic head coefficient and efficiency can be calculated to give a set of nondimensional curves (Figure 6). These curves will predict the compressor performance for the given Mach number. They can be used to compare against the existing operating conditions if the field Mach number is close enough to the Mach number for the curves. The ASME

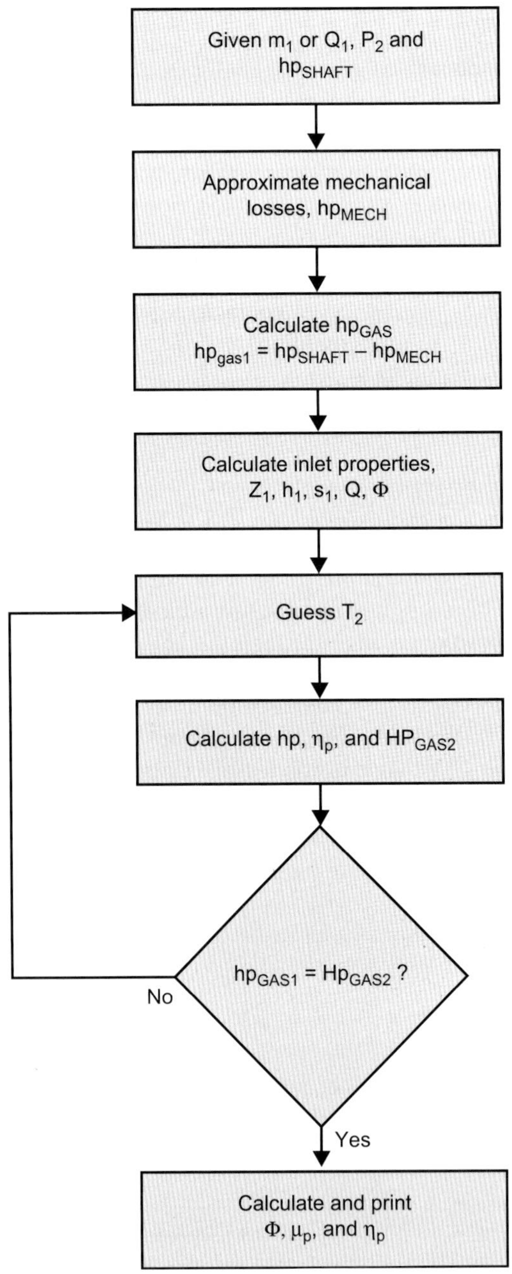

Figure 6. The polytropic head coefficient and efficiency can be calculated to give a set of nondimensional curves.

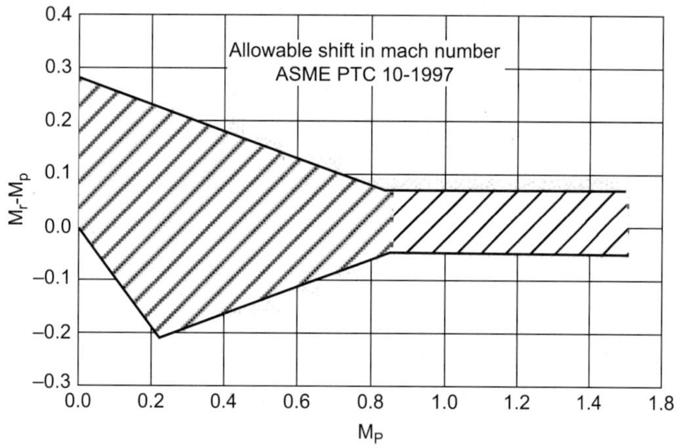

Figure 7. Allowable shift in test Mach number from design Mach number.

PTC-10 test code defines the maximum shift in Mach number ratio for a certified shop performance test (Figure 7). These limitations are good to apply in the field as well. If the Mach number shift is too large, the comparison may be inaccurate. If this is the case, a new set of performance curves, that match the actual inlet conditions, should be obtained from the OEM.

The polytropic head coefficient and efficiency at the existing operating conditions can be plotted on the nondimensional graphs to determine if the compressor is operating on its curve. Also, the nondimensional curves can be used to calculate the field discharge conditions (pressure, temperature, horsepower, etc.) based on the field inlet conditions and the performance curves (Figure 8). Even though seal losses do not affect the calculated horsepower (i.e. they are included in the lower measured efficiency), they do greatly affect the predicted horsepower. The seal losses increase the calculated value of *F*, which moves the predicted operating point farther to the right on the performance curves, which always increases the horsepower.

Example performance evaluation. Reformer hydrogen recycle compressor configuration:

- Six impellers, straight-through, barrel-type,
- Vaneless diffusers,
- Interlocking aluminum labyrinth balance piston seal,
- Motor-driven through a speed-increasing gearbox,
- Contact-type oil face seals.

Original design:

- 6370 hp,
- 7940 rpm,
- Inlet flow = 14,600 cfm,
- Molecular weight = 5.18,
- $T_1 = 100\ °F$,
- $P_1 = 187$ psia,
- $P_2 = 275$ psia.

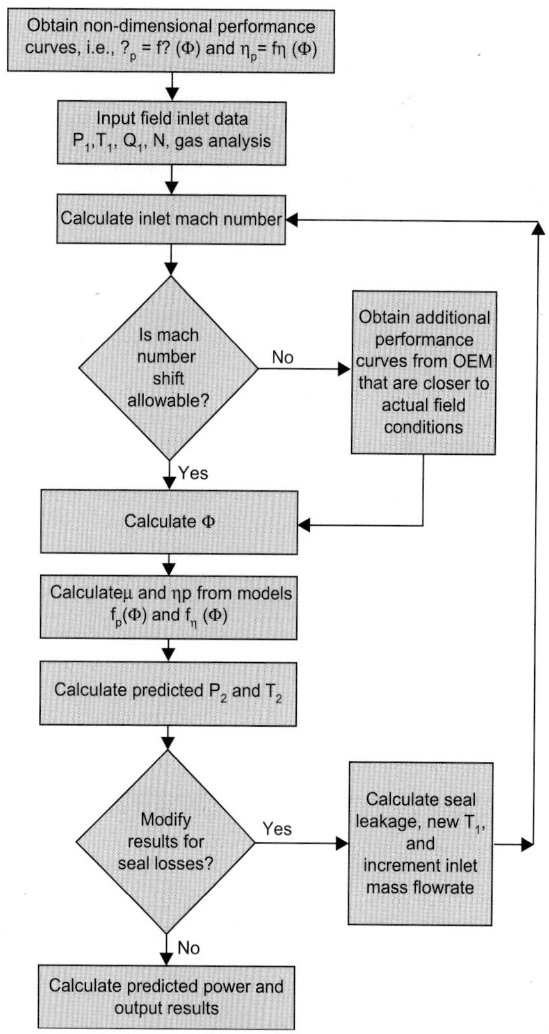

Figure 8. The nondimensional curves can be used to calculate the field discharge conditions.

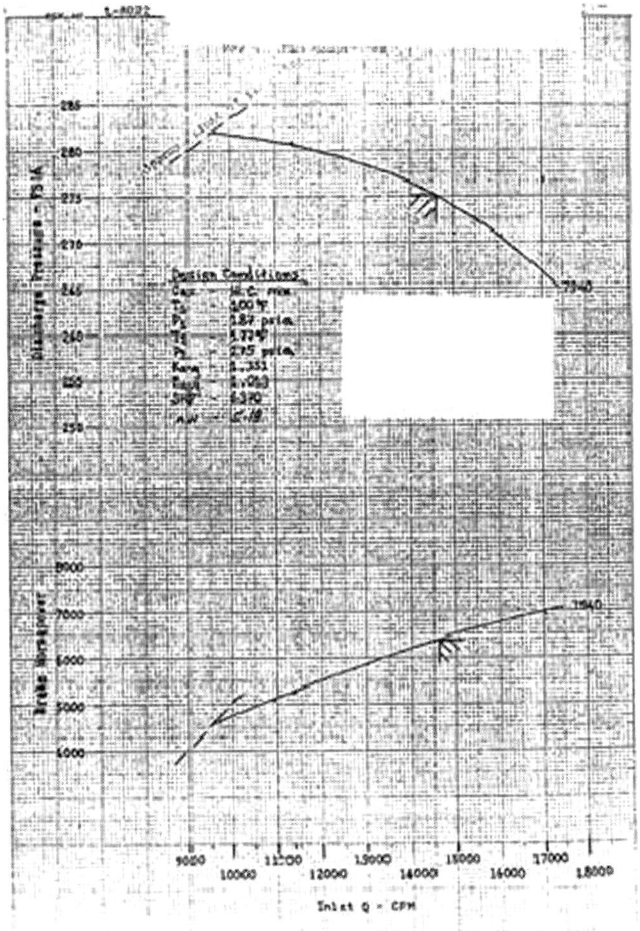

Figure 9. The OEM supplied this performance curve showing predicted discharge pressure and shaft horsepower.

Balance piston labyrinth dimensional data:

Stationary labyrinth: nine teeth, 0.375-in equal spacing, 0.1875 in tall, 14.595 in internal diameter.
Rotating labyrinth: eight teeth, 0.375 in equal spacing, 0.1875 in tall, 14.937 in outside diameter.
Calculated seal leakage rate = 120 lbm/min.

Measured performance data:

$P_1 = 150$ psia,
$T_1 = 80\ ^\circ\mathrm{F}$,
$P_2 = 248$ psia,
$T_2 = 191\ ^\circ\mathrm{F}$,
Balance chamber pressure = 156 psia,
Flow = 190 MMscfd,
Speed = 7949 rpm,
Molecular weight = 7.1.

Driver data:

Amperage = 978,
Voltage = 4000,
Speed = 1784,
Pf = 0.92,
$E = 0.957$.

Calculate the horsepower supplied to the compressor

$$\mathrm{hp}_{\mathrm{motor}} = \frac{VI_eP_f}{431} = \frac{4000 \times 978 \times 0.957 \times 0.92}{431} = 7988\ \mathrm{hp}$$

If the gearbox efficiency is assumed to be 97%, the calculated shaft horsepower delivered to the compressor is:

$$\mathrm{hp}_{\mathrm{comp}} = 7988 \times 0.97 = 7750\ \mathrm{hp}$$

The OEM supplied a performance curve showing predicted discharge pressure and shaft horsepower (Figure 9). These curves are valid for only the design conditions listed previously (i.e. you cannot plot the measured discharge pressure and shaft horsepower on these curves). Likewise, they are only predicted curves (i.e. the compressor was not shop performance tested). These curves must be converted into a nondimensional form so that they can be used to predict the existing field performance (Figure 10).

These curves must be converted into a nondimensional form so that they can be used to predict the existing field performance (Figure 10).

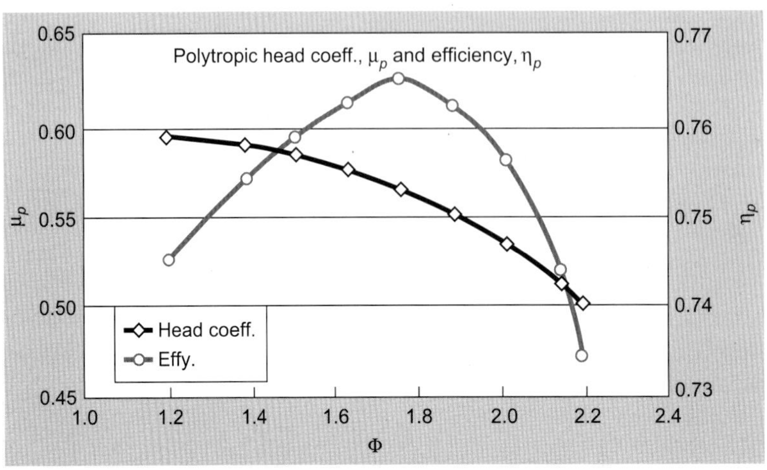

Figure 10. Nondimensional polytropic head coefficient and efficiency.

Table 5
Reformer Hydrogen Recycle Compressor Performance before Cleaning

Parameter	Measured Field Performance	Predicted Performance (w/o seal losses)	Predicted Performance (w/seal losses)
P_2	248	258	252
T_2	191	176	182
Head, lbf-ft/lbm	64,504	68,898	66,417
Efficiency, %	65	79	76
Shp	7647	6729	7030

Table 6
Example Test and Repredicted Field Performance Data

Title Hydrogen Recycle Compressor						
Test Data			**Corrected Test Data**		**Predicted Data**	
Suction Conditions:		**Discharge Conditions:**	**Inlet**	**Outlet**	**Inlet**	**Discharge**
Press, psia	150	248.0	150.0	248.0	150.0	252.1
Temp, °F	80	191.0	84.9	191.0	84.6	182.1
Flow	190 0 0	MMscfd (only one flow is required) Icfm lbm/min				
Flowmeter design data	1 Location (1 - suction, 2—discharge) 0 Mole weight 0 Temp., °F 0 Press., psia					
Speed	7940 rpm				7940	
Impeller	Number 6	Diameter, 23 in.				
		Calculated data				
Mole weight	7.2					
Properties	**Inlet**	**Outlet**	**Inlet**	**Outlet**	**Predicted Inlet**	**Predicted Outlet**
Compressibility	1.0026	1.0059	1.0026	1.0059	1.0026	1.0059
Enthalpy, Btu/lbm	556.1	683.4	561.6	683.4	561.3	673.0
Specific volume, ft^3/lbm	5.39	3.94	5.44	3.94	5.44	3.83
Specific volume, ft^3/lbm	5.39	3.94	5.44	3.94	5.44	3.83
C_p/C_v	1.33	1.32	1.33	1.32	1.33	1.32
Entropy, Btu/lbm, °F	3.76	3.84	3.77	3.84	3.77	3.82
Flow (corrected)				**Flow (corrected)**		**Flow (predicted)**
MMscfd	190.0					
Icfm	13,456.9	Seal losses, lbm/min		114.0		120.1
lbm/min	2496.0	Impeller flow, lbm/min		2610.0		2616.1
Q/N	1.695	Impeller *q/n*		1.789		1.792
	Performance			**Performance (corrected)**		**Predicted Performance**
Polytropic exponent	1.61			1.56		1.48
Polytropic head	64,509			64,795		66,417
Polytropic head coeff.	0.545			0.547		0.561
Polytropic wf.	1.000			1.000		1.000
Polytropic effy.	0.65			0.68		0.76
Gas horsepower	7497			7497		6892
Shaft horsepower	7647			7647		7030
Mach no.	0.357			0.359		0.355
Volume ratio	1.367			1.379		1.421

The discharge conditions are repredicted with and without seal losses to compare against the measured field results (Tables 5 and 6). The horsepower for the existing field performance data must be close to the horsepower supplied by the driver for the results to be considered valid. As you can see, the compressor efficiency is considerably lower than what is predicted by the OEM performance curve. Adding the seal losses (approximately 4–5% of the total flow) to the predicted curves brings the predicted and actual conditions closer together. However, the measured efficiency is just too low to be a balance piston problem alone. Likewise, the thrust bearing temperature and axial position were relatively low. Based on the past fouling history in this compressor as well as the high balance piston differential pressure, the loss in efficiency was thought to be a result of fouling. The compressor was still meeting the desired discharge pressure, but the low efficiency was causing excessive horsepower consumption, which was limiting unit charge rate. Because the motor was oversized and the loss in efficiency had been gradual, operations was unaware that the problem was in the compressor and not the motor (i.e. they thought the motor was dirty).

The compressor was pulled because it could not be washed in place due to a lack of adequate case drains. A large amount of ammonia chloride buildup was found in the compressor stationary components (the diffuser channels had approximately 40% blockage, Figures 11 and 12). Note: The synchronous vibration amplitudes were relatively low (<1 mil) because the fouling was mostly on the stationary components; the only fouling on the rotor was on the inside diameter of the impeller eyes.

After the compressor was reinstalled, the measured field performance was within 3% of the predicted.

Figure 11. Fouled inlet guide vanes on hydrogen recycle compressor.

Figure 12. Diaphragm from hydrogen recycle compressor.

Nomenclature.

C_p	specific heat, constant pressure, Btu/lbm, °F
C_v	specific heat, constant volume, Btu/lbm, °F
D	impeller diameter, in
h	enthalpy, Btu/lbm
H	head, ibf-ft/lbm
hp	horsepower
k	specific heat ratio, c_p/c_v
K	meter factor
m	mass flow, lbm/s
M	Mach number, non-dim.
MMscfd	million standard cubic feet per day
MW	mole weight, lbm/lb-mole
N	speed, rpm
NS	specific speed, non-dim.
P	pressure, psia
Q	volume flow, cfm
Q_r	radiative heat transfer, Btu
R	gas constant
T	temperature, °F
v	specific volume, ft3/lbm
Z	compressibility

Subscripts

1	inlet conditions
2	outlet conditions
bp	balance piston
D	design conditions
MECH	mechanical
M	mechanical
P	polytropic state
s	isentropic state
S	corresponds to seal conditions

Symbols

D	differential
h	efficiency, non-dim.
F	sonic velocity ratio, non-dim.
m	head coefficient, non-dim.
$m_{0,1}$	labyrinth seal entrance coefficient, non-dim.

References

1. W. Aicher, "Test of Process Turbocompressors Without CFC Gases," *Proceedings of the Twenty-Second Turbomachinery Symposium, Turbomachinery Laboratory*, Texas A&M University, College Station, Texas, 1993, pp. 179–184.
2. J.C. Bartos, "Predicting Centrifugal Compressor Performance from Impeller Dimensional Data," Power Machinery and Compression Conference, 1990, Houston, Texas.
3. M.P. Boyce, "Principals of Operation and Performance Estimation of Centrifugal Compressors, " *Proceedings of the Twenty-Second Turbomachinery Symposium*, Turbomachinery Laboratory, Texas A&M University, College Station, Texas, 1993, pp. 161–175.
4. W.A. Clark, "Measure Power by the Simplest Method and Use for Onstream Computer Monitoring, " *Proceedings of the Tenth Turbomachinery Symposium*, Turbomachinery Laboratory, Texas A&M University, College Station, Texas, 1981, pp. 151–153.
5. H.M. Davis, "Compressor Performance, Field Data Acquisition and Evaluation," *Proceedings of the Second Turbomachinery Symposium*, Turbomachinery Laboratory, Texas A&M University, College Station, Texas, 1973, pp. 15–22.
6. M.T. Gresh, Compressor Performance—Selection, Operation, and Testing of Axial and Centrifugal Compressors, Butterworth-Heinemann, Boston, 1991.
7. E. Heftmann, ed., Chromatography—A Laboratory Handbook of Chromatographic and Electrophoretic Methods, Van Nostrand Reinhold Co., New York, 1975.
8. D.J. Hill, "Centrifugal Compressor Rerates and Redesigns," Power Machinery and Compression Conference, Houston, Texas, 1995.
9. D.J. Hunt, "Field Testing of Centrifugal Compressors," Proceedings of the Tenth Turbomachinery Symposium, Turbomachinery Laboratory, Texas A&M University, College Station, Texas, 1981.
10. D. Japikse, "Efficiency: The Silent Partner of Machine Performance," Proceedings of the Eleventh Turbomachinery Symposium, Turbomachinery Laboratory, Texas A&M University, College Station, Texas, 1982, pp. 105–111.
11. T.W. Kerlin, and E.M. Katz, "Temperature Measurement in the 1990s," INTECH, April 1990, pp. 40–43.
12. R.P. Lapina, Estimating Centrifugal Compressor Performance, Gulf Publishing, Houston, Texas, 1982.
13. B.I. Lee, and M. G. Kelser, "A Generalized Thermodynamic Correlation Based on Three-Parameter Corresponding States, " AIChE Journal, Vol. 21, No. 3, May 1975, pp. 510–527.
14. T. Matthews, "Field Performance Testing to Improve Compressor Reliability," *Proceedings of the Tenth Turbomachinery Symposium*, Turbomachinery Laboratory, Texas A&M University, College Station, Texas, 1981, pp. 165–167.
15. N.S. Nathoo, and W. G. Gottenberg, "Measuring the Thermodynamic Performance of Multistage Compressors Operating on Mixed Hydrocarbon Gases," *Proceedings of the Tenth Turbomachinery Symposium*, Turbomachinery Laboratory, Texas A&M University, College Station, Texas, 1981, pp. 15–22.
16. R.B. Pais, and G.J. Jacobik, "Field Performance Testing of Centrifugal Compressors," *Proceedings of the Third Turbomachinery Symposium*, Turbomachinery Laboratory, Texas A&M University, College Station, Texas, 1974, pp. 84–90.

17. J.M. Schultz, "The Polytropic Analysis of Centrifugal Compressors," Journal of Engineering for Power, Vol. 1, pp. 69–83.
18. ASME PTC 10–1997 Performance Test Code on Compressors and Exhausters, American Society of Mechanical Engineers, New York.
19. F.L. Van Lanigham, "Guidelines for Performance Testing Centrifugal Compressors," *Proceedings of the Tenth Turbomachinery Symposium*, College Station, Texas, 1981, pp. 169–171.
20. "Bulletin TS-102, " Minco Product Catalog.

Reprinted with permission-*Hydrocarbon Processing*, August 2007.

11: Gas—Hydraulics

Gas pipeline hydraulics calculations 434
Equivalent lengths for multiple lines based on Panhandle A 435
Determine pressure loss for a low-pressure gas system 436
Nomograph for determining pipe-equivalent factors 437
How much gas is contained in a given line section? 438
How to estimate equivalent length factors for gas lines 438
Estimating comparative capacities of gas pipelines 439
Determination of leakage from gas line using pressure drop method 439
A quick way to determine the size of gas gathering lines 440
Energy conversion data 440
How to estimate time required to get a shut-in test on gas transmission lines 441
How to approximate a maximum acceptable pressure loss for new lines 441
Estimate pipe size requirements for increasing throughput volumes of natural gas 442
Calculate line loss using cross-sectional areas table when testing mains with air or gas 443
Flow of fuel gases in pipelines 444
Calculate the velocity of gas in a pipeline 445
Determining throat pressure in a blow-down system 445
Estimate the amount of gas blown off through a line puncture 446
A practical way to calculate gas flow for pipelines 446
How to calculate the weight of gas in a pipeline 447
Estimate average pressure in gas pipeline using upstream and downstream pressures 447
Chart for determining viscosity of natural gas 448
Fanning equation for gas flow 448
Multiphase flow 452
Nomograph for calculating Reynolds number for compressible flow friction factor for clean steel and wrought iron pipe 457

Gas pipeline hydraulics calculations

Equations commonly used for calculating hydraulic data for gas pipelines are as follows:

Panhandle A.

$$Q_b = 435.87 \times (T_b/P_b)^{1.0788} \times D^{2.6182} \times E \times \left[\frac{P_1^2 - P_2^2 - \dfrac{\left(0.0375 \times G \times (h_2 - h_1) \times P_{avg}^2\right)}{T_{avg} \times Z_{avg}}}{G^{0.8539} \times L \times T_{avg} \times Z_{avg}}\right]^{0.5394}$$

Panhandle B.

$$Q_b = 737 \times (T_b/P_b)^{1.020} \times D^{2.53} \times E \times \left[\frac{P_1^2 - P_2^2 - \dfrac{\left(0.0375 \times G \times (h_2 - h_1) \times P_{avg}^2\right)}{T_{avg} \times Z_{avg}}}{G^{0.961} \times L \times T_{avg} \times Z_{avg}}\right]^{0.51}$$

Weymouth.

$$Q_b = 433.5 \times (T_b/P_b) \times \left[\frac{P_1^2 - P_2^2}{GLTZ}\right]^{0.5} \times D^{2.667} \times E$$

$$P_{avg} = \frac{2}{3}\left[P_1 + P_2 - \frac{P_1 \times P_2}{P_1 + P_2}\right]$$

Nomenclature for Panhandle equations

Q_b = Flow rate, SCFD
P_b = −Base pressure, psia
T_b = Base temperature, °R
T_{avg} = Average gas termperature, °R
P_1 = Inlet pressure, psia
P_2 = Outlet pressure, psia
G = Gas specific gravity (air = 1.0)
L = Line length, miles
Z = Average gas compressibility
D = Pipe inside diameter, in
h_2 = Elevation at terminus of line, ft
h_1 = Elevation at origin of line, ft
P_{avg} = Average line pressure, psia
E = Efficiency factor

$E = 1$ for new pipe with no bends, fittings, or pipe diameter changes
$E = 0.95$ for very good operating conditions, typically through first 12–28 months
$E = 0.92$ for average operating conditions
$E = 0.85$ for unfavorable operating conditions

Nomenclature for Weymouth equation

Q = Flow rate, SCFD
T_b = Base temperature, °R
P_b = Base pressure, psia
G = Gas specific gravity (air = 1)
L = Line length, miles
T = Gas temperature, °R
Z = Gas compressibility factor
D = Pipe inside diameter, in
E = Efficiency factor. (See Panhandle nomenclature for suggested efficiency factors)

Sample calculations

$Q = ?$
$G = 0.6$
$T = 100\ °F$
$L = 20$ miles
$P_1 = 2000$ psia
$P_2 = 1500$ psia
Elev diff = 100 ft
$D = 4.026$ in
$T_b = 60\ °F$
$P_b = 14.7$ psia
$E = 1.0$

$$P_{avg} = 2/3\left[2000 + 1500 - \frac{2000 \times 1500}{2000 + 1500}\right]$$

$$P_{avg} = 1762 \text{ psia}$$

Panhandle A.

$$Q_b = 435.87 \times (520/14.7)^{1.0788} \times 4.026^{2.6182} \times 1 \times \left[\frac{(2000)^2 - (1500)^2 - \dfrac{\left(0.0375 \times 0.6 \times 100 \times (1762)^2\right)}{560 \times 0.835}}{(0.6)^{0.8539} \times 20 \times 560 \times 0.835}\right]^{0.5394}$$

$$Q_b = 16{,}577 \text{ MCFD}$$

Panhandle B.

$$Q_b = 737 \times (520/14.7)^{1.020} \times 4.026^{2.53} \times 1$$

$$\times \left[\frac{(2000)^2 - (1500)^2 - \dfrac{(0.0375 \times 0.6 \times 100 \times (1762)^2)}{560 \times 0.835}}{(0.6)^{0.961} \times 20 \times 560 \times 0.835} \right]^{0.51}$$

$$Q_b = 17,498 \text{ MCFD}$$

Weymouth.

$$Q = 0.433 \times (520/14.7) \times \left[(2000)^2 - (1500)^2/(0.6 \times 20 \times 560 \times 0.835)\right]^{1/2} \times (4.026)^{2.667}$$

$$Q = 11,101 \text{ MCFD}$$

Source

Pipecalc 4.02, Gulf Publishing Company, Houston, Texas. *Note:* Pipecalc 4.02 will calculate the compressibility factor, the minimum pipe inner diameter (ID), the upstream pressure, the downstream pressure, and the flow rate for Panhandle A, Panhandle B, Weymouth, AGA, and Colebrook—White equations.

Equivalent lengths for multiple lines based on Panhandle A

Condition I

Condition I is a single pipeline that consists of two or more different diameter lines.

Let

L_E = Equivalent length
$L_1, L_2, \ldots, L_n$ = Length of each diameter
$D_1, D_2, \ldots, D_n$ = Internal diameter of each separate line corresponding to $L_1, L_2, \ldots, L_n$
D_E = Equivalent internal diameter

$$L_E = L_1 \left[\frac{D_E}{D_1}\right]^{4.8539} + L_2 \left[\frac{D_E}{D_1}\right]^{4.8539} + \ldots + L_n \left[\frac{D_E}{D_n}\right]^{4.8539}$$

Example. A single pipeline 100 miles in length consists of 10 miles $10\frac{3}{4}$-in outer diameter (OD); 40 miles $12\frac{3}{4}$-in OD, and 50 miles of 22-in OD lines.

Find equivalent length (L_E) in terms of 22-in OD pipe.

$$L_E = 50 + 40\left[\frac{21.5}{12.25}\right]^{4.8539} + 10\left[\frac{21.5}{10.25}\right]^{4.8539}$$

$$= 50 + 614 + 364$$

$$= 1028 \text{ miles equivalent length of 22-in OD}$$

Condition II

Condition II is a multiple pipeline system consisting of two or more parallel lines of different diameters and different length.

Let

L_E = Equivalent length
$L_1, L_2, L_3, \ldots, L_n$ = Length of various looped sections
$d_1, d_2, d_3, \ldots, d_n$ = ID of the individual line corresponding to length L_1, L_2, L_3, and L_n

$$L_E = L_1 \left[\frac{d_E^{2.6182}}{d_1^{2.6182} + d_2^{2.6182} + d_3^{2.6182} + \ldots + d_n^{2.6182}}\right]^{1.8539} + \ldots$$

$$+ L_n \left[\frac{d_E^{2.6182}}{d_1^{2.6182} + d_2^{2.6182} + d_3^{2.6182} + \ldots + d_n^{2.6182}}\right]^{1.8539}$$

Let

L_E = Equivalent length
L_1, L_2, L_3, and L_n = Length of various looped sections
d_1, d_2, d_3, and d_n = Internal diameter of individual line corresponding to lengths L_1, L_2, L_3, and L_n

$$L_E = L_1 \left[\frac{d_E^{2.6182}}{d_1^{2.6182} + d_2^{2.6182} + d_3^{2.6182} + \ldots + d_n^{2.6182}}\right]^{1.8539}$$

$$+ \ldots$$

$$+ L_n \left[\frac{d_E^{2.6182}}{d_1^{2.6182} + d_2^{2.6182} + d_3^{2.6182} + \ldots + d_n^{2.6182}}\right]^{1.8539}$$

when

L_1 = Length of unlooped section
L_2 = Length of single looped section
L_3 = Length of double looped section, and
$d_E = d_1 = d_2$.

Then:

$$L_E = L_1 + 0.27664L_2 + L_3\left[\frac{d_E^{2.6182}}{2d_1^{2.1682} + d_3^{2.6182}}\right]^{1.8539}$$

when $d_E = d_1 = d_2 = d_3$
then $L_E = L_1 + 0.27{,}664L_2 + 0.1305L_3$

Example. A multiple system consisting of a 15 miles section of 3—8⅝-in OD lines and 1—10¾ -in OD line, and a 30 mile section of 2—8⅝-in lines and 1—10¾ -in OD line.
Find the equivalent length in terms of single 12-in ID line.

$$L_E = 15\left[\frac{12^{2.6182}}{3(7.981)^{2.6182} + 10.02^{2.6182}}\right]^{1.8539}$$

$$+30\left[\frac{12^{2.6182}}{2(7.981)^{2.6182} + 10.02^{2.6182}}\right]^{1.8539}$$

$$= 5.9 + 18.1 = 24.0 \text{ miles equivalent of 12-in ID pipe}$$

Example. A multiple system consisting of a single 12-in ID line 5 miles in length and a 30 mile section of 3×12-in ID lines.
Find equivalent length in terms of a single 12-in ID line.

$$L_E = 5 + 0.1305 \times 30$$
$$= 8.92 \text{ miles equivalent of single 12-in ID line}$$

Determine pressure loss for a low-pressure gas system

Use the Spitzglass equation for systems operating at less than 1 psig.

$$Q_h = 3550\sqrt{\frac{\Delta h_w d^5}{S_g L(1 + (3.6/d) + 0.03d)}}$$

where:

Q_h = Rate of flow, in cubic feet per hour at standard conditions (14.7 psia and 60 °F), scfh
h_w = Static pressure head, in inches of water
S_g = Specific gravity of gas relative to air, the ratio of the molecular weight of the gas to that of air
d = Internal diameter of the pipe, inches, and
L = Length of the pipe, feet.

Example. Given the following conditions, find the flow in the system:

h_w = 50 in of water
S_g = 0.65
L = 500 ft, and
d = 6.187 in.

$$Q_h = 3550\sqrt{\frac{50 \times 6.187^5}{0.65 \times 500(1 + (3.6/6.187) + 0.03 \times 6.187)}}$$

$$Q_h = 99.723 \text{ scfh}$$

Nomograph for determining pipe-equivalent factors

For turbulent flow, this handy nomograph will save a great deal of time in flow calculations where different sizes of pipe are connected. The advantage over a table of values is in having all sizes of pipe expressed in relation to any other size.

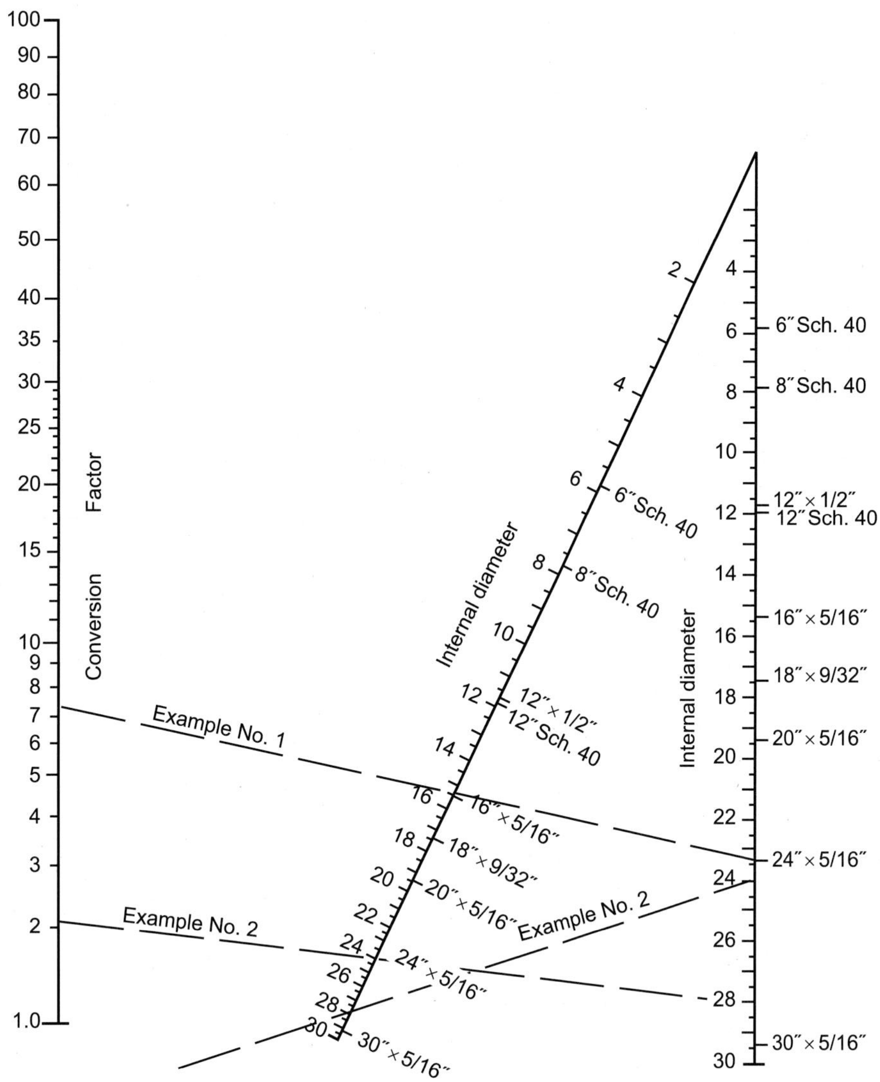

Example 1. 16 in × 5/16 in terms of 24 in x 5/16 in.

$$\left(\frac{23.375}{15.375}\right)^{4.735} = 1.5203^{4.735} = 7.26$$

Example 2. Reciprocal values can be found when chart limits are exceeded. 28-in ID in terms of 24-in ID falls beyond the lower limits of the chart; however, 24-in ID in terms of 28-in ID gives a conversion factor of 2.1.

How much gas is contained in a given line section?

Multiply the square of the inside diameter, in inches, by the gauge pressure, in lb/in^2; multiply this by 0.372; the answer is the approximate number of cubic feet of gas (standard conditions) in 1000 ft of line.

Example. How much gas in 1000 ft of 16-in schedule 30 pipe at 350 lb pressure?

$15.25 \times 15.25 \times 350 \times 0.372 = 30{,}280$ cu. ft of gas.

Example. Approximately how much gas in 8 miles of 0.250 wall 24-in pipe if the pressure is 400 psi?

$$\begin{aligned}\text{Cu. ft} &= (23.5)(23.5)(400)(8)(5.28)(0.372)\\ &= 3,740,000 \text{ cu. ft}\end{aligned}$$

When a section of line is blown down from one pressure to another, the total gas lost may be computed by the difference in the contents at the two pressures, using the above rule.

How to estimate equivalent length factors for gas lines

Internal Diameter (ID)	Nominal Diameter	36″ 35.250	34″ 33.250	30″ 29.375	26″ 25.375	24″ 23.375	24″ 23.000	22″ 21.500	20″ 19.500	18″ 17.500	16″ 15.500	$12^3/_4$″ 12.250	$10^3/_4$″ 10.250	$8^5/_8$″ 8.125
35.250″	36″	1	0.758	0.416	0.204	0.137	0.126	0.092	0.056	0.034	0.019	0.006	0.002	0.0008
33.250	34	1.32	1	0.55	0.270	0.181	0.167	0.121	0.075	0.044	0.025	0.008	0.003	0.001
29.375	30	2.41	1.82	1	0.492	0.330	0.305	0.22	0.136	0.081	0.045	0.014	0.006	0.002
25.375	26	4.90	3.71	2.03	1	0.671	0.622	0.45	0.278	0.164	0.092	0.029	0.012	0.004
23.375	24	7.25	5.54	3.04	1.49	1	0.922	0.67	0.415	0.246	0.137	0.044	0.018	0.006
23.000	24	7.90	6.00	3.29	1.62	1.08	1	0.72	0.450	0.267	0.148	0.048	0.020	0.0064
21.500	22	10.93	8.31	4.55	2.24	1.50	1.37	1	0.624	0.369	0.205	0.065	0.027	0.009
19.500	20	17.60	13.33	7.31	3.59	2.41	2.22	1.60	1	0.590	0.328	0.105	0.044	0.014
17.500	18	29.8	22.5	12.35	6.08	4.075	3.78	2.72	1.69	1	0.555	0.177	0.074	0.024
15.500	16	52.8	40.5	22.2	10.9	7.345	6.75	4.88	3.04	1.80	1	0.319	0.134	0.043
12.250	$12^3/_4$	168	127	69.8	34.3	23.01	21.2	15.35	9.55	5.65	3.13	1	0.420	0.136
10.250	$10^3/_4$	400	303	167	81.7	54.67	50.6	36.5	22.7	13.4	7.47	2.38	1	0.324
8.125	$8^5/_8$	1235	935	513	252	168.8	156.2	112.5	70.5	41.5	23.0	7.35	3.08	1
6.125	$6^5/_8$	4860	3680	2020	992	666	615	443	276	163	90.7	29.0	12.2	3.94

This table shows the equivalent length factors for pipelines of different diameters. It is based on the Panhandle formula, and the factors were obtained by varying the lengths of the pipe and keeping the other variables constant.

Example. Find the equivalent length of a 24-in pipeline as compared with a $12^3/_4$ -in line.

Enter the chart at the top at 24 in and proceed downward to the 12.250 parameter. The equivalent length is 21.2.

This means that under the same conditions of temperatures, pressures, specific gravities, etc., 1 mile of $12^3/_4$ -in pipe will flow the same amount of gas that 21.2 miles of 24-in pipe will flow. Another way of putting it, the pressure drop in 1 mile of 24-in pipe will be the same as that in 0.048 miles of $12^3/_4$ -in pipe if the pressures, volumes, etc., are the same.

Where a line is composed of several sizes of pipe diameters, or where the lines are looped, this chart should prove useful. In solving problems of this sort, the line is reduced to an equivalent length of some arbitrarily selected size pipe and treated as a single line of uniform size.

Estimating comparative capacities of gas pipelines

To estimate capacity of one diameter line in terms of another, read downward in the appropriate column to the desired diameter line. The values depicted give capacity of one diameter pipe as a percent of another. This chart is based on diameter factors using the Panhandle equation. Equivalent inlet and outlet pressures, lengths of line section, and uniform flow are assumed for both lines.

Example. A 24-in (23.375-in ID) pipeline has a capacity 160% that of a 20-in (19.500-in ID) line.

ID	Nominal Diameter	36″ 35.250″	34″ 33.250″	30″ 29.3125″	26″ 25.375″	24″ 2.375″	24″ 23.000″	20″ 19.500″	18″ 17.500″	16″ 15.500″	12¾ 12.250″
35.250″	36″	100	85.7	62.7	43.6	35.2	33.7	22.0	16.5	12.0	6.5
33.250″	34″	116.8	100	73.1	50.8	41.0	39.3	25.7	19.2	14.0	7.6
29.3125″	30″	159.7	137	100	69.6	56.1	53.8	35.7	26.3	19.2	10.4
25.375″	26″	229	197	149	100	80.7	77.5	50.2	37.8	27.6	14.8
23.375″	24″	285	244	178	124	100	96.0	62.3	46.9	34.2	18.5
23.000″	24″	298	255	186	129	104	100	65.1	49.0	35.7	19.2
19.500″	20″	458	392	287	199	160	154	100	75.3	35.0	29.6
17.500″	18″	608	519	379	264	213	204	132	100	73.2	39.3
15.500″	16″	835	715	522	363	293	281	182	127	100	53.8
12.250″	12¾″	1545	1322	968	673	543	521	238	254	185	100

Determination of leakage from gas line using pressure drop method

To determine the leakage in MMcf per year at a base pressure of 14.4 psia and a temperature of 60 °F, take as a basis a 1 mile section of line under 1 h test. Use the formula:

$$L_y = 9 \times D^2 \left[\frac{P_1}{460 + t_1} - \frac{P_2}{460 + t_2}\right]$$

where:

D = Inside diameter of the pipe, inches
P_1 = Pressure at the beginning of the test in psia
t_1 = Temperature at the beginning of the test in °F
P_2 = Pressure at end of the test, and
t_2 = Temperature at end of the test.

The error in using this formula can be about 4%.

Example. A section of a 10-in pipeline 1 mile in length is tested using the pressure drop method. ID is equal to 10.136 in. Pressure and temperature at the start of the test were 190.5 psig and 52 °F. After 1 h, the readings were 190.0 psig and 52 °F. Calculate the leakage in MMcf/year.

The atmospheric pressure for the area is 14.4 psia. Thus:

$$P_1 = 190.5 + 14.4 = 204.9 \text{ psia}$$

$$P_2 = 190.0 + 14.4 = 204.4 \text{ psia}$$

$$L_y = 9(10.136)^2\left[\frac{204.9}{460 + 52} - \frac{204.4}{460 + 52}\right]$$

$$= 0.903 \text{ MMcf per mile per year.}$$

A quick way to determine the size of gas gathering lines

Here is a shortcut to estimate gas flow in gathering lines. For small gathering lines, the answer will be within 10% of that obtained by more accurate formulas.

$$Q = \frac{(500)\left(d^3\right)\sqrt{P_1^2 - P_2^2}}{\sqrt{L}}$$

where:

Q = Cubic feet of gas per 24 h
d = Pipe ID in inches
P_1 = Psi (abs) at starting point
P_2 = Psi (abs) at ending point, and
L = Length of line in miles.

Example. How much gas would flow through four miles of 6-in ID pipe if the pressure at the starting point is 485 psi and if the pressure at the downstream terminus is 285 psi?

Solution.

$$Q = \frac{(500)\left(6^3\right)\sqrt{(485+15)^2 - (285-15)^2}}{\sqrt{4}}$$

$$Q = \frac{(500)(216)\sqrt{500^2 - 300^2}}{2}$$

$$Q = 21,600,000 \text{ cu. ft per 24 h}$$

Using a more accurate formula (a version of Weymouth's):

$$Q = \frac{(871)\left(d^{8/3}\right)\sqrt{P_1^2 - P_2^2}}{\sqrt{L}}$$

$$Q = \frac{(871)(118.94)(400)}{2}$$

$$Q = 20,719,417 \text{ cu. ft per 24 h}$$

The error in this case is 880,582.8 cu. ft per 24 h, or about 4.25%.

Energy conversion data

The accompanying table gives the high heating values of commonly used fuels along with their relation one to the other.

Example. If light fuel oil will cost a potential industrial customer 3 dollars per gallon and natural gas will cost him 4 dollars per Mcf, which fuel will be more economical?

Solution. Assume the operator would use 10,000 gallons of light fuel oil per year at 3 dollars per gallon. The cost would be $30,000. From the table, 10,000 gallons of fuel oil are equivalent to 10,000 × 134 = 1,340,000 cu. ft of natural gas. At 4 dollars per Mcf this would cost 1340 × 4 = $5360.

The burning efficiency of both fuels is the same. The savings by using natural gas would come to $24,640 per year.

Fuel	Heating Value	PERCENT BURNING EFFICIENCY			Unit	NUMBER OF CUBIC FEET OF NATURAL GAS EQUIVALENT TO 1 UNIT OF OTHER FUEL		
		Domestic	Commercial	Industrial		Domestic	Commercial	Industrial
Natural Gas	1000 B.T.U. per Cubic Foot	75	75	78	Cubic Foot	1	1	1
Propane	91,600 B.T.U. per U.S. gallon	75	75	78	Gallon	91.5	91.5	91.5
Butane	102,000 B.T.U. per U.S. gallon	.....	75	78	Gallon	.....	102	102
Light Fuel Oil	134,000 B.T.U. per U.S. gallon	65–75	75	78	Gallon	117–134	134	134
Heavy Fuel Oil	146,000 B.T.U. per U.S. gallon	.....	75	78	Gallon	.....	146	146
Hydro-electricity	3,412 B.T.U. per Kilowatt Hour	100	100	100	Kilowatt Hour	4.549	4.549	4.549
*Coal	8,700–11,000 B.T.U. per pound	55–60	65	73	Pound	6.37–8.07	7.55–9.57	8.45–10.7

*Coal qualities vary considerably. The figures given are average for Western Coals; slack, run-of-the-mine; potentially available at lowest cost.

How to estimate time required to get a shut-in test on gas transmission lines

This rule applies to air or gas testing of gas transmission lines. It is not applicable for hydrostatic tests. Values of the constants used by different transmission companies may vary with economic considerations, line conditions, and throughputs.

To determine the minimum time required to achieve a good shut-in test after the line has been charged and becomes stabilized, use the following formula:

$$H_m = \frac{3 \times D^2 \times L}{P_1}$$

where:

H_m = Minimum time necessary to achieve an accurate test in hours
D = Internal diameter of pipe in inches
L = Length of section under test in miles, and
P_1 = Initial test pressure in lb/in^2 gauge.

Example. How long should a 14 mile section of 26-in (25.375-in ID) pipeline be shut in under a 1050 psi test pressure to get a good test?

$$H_m = \frac{3 \times (25.375)^2 \times 14}{1.050}$$

$$= 25.75 \text{ h}$$

How to approximate a maximum acceptable pressure loss for new lines

When a new line is being shut in for a test, the following formula may be used to evaluate whether or not the line is "tight":

$$P_{ad} = \frac{H_t \times P_1}{D \times 949}$$

where:

P_{ad} = Acceptable pressure drop in lb/in^2 gauge
H_t = Shut-in test time in hours
D = Internal diameter of pipe in inches, and
P_1 = Initial test pressure in lb/in^2 gauge.

Example. What would be the maximum acceptable pressure drop for a new gas transmission line under air or gas test when a 44 mile section of 30-in (29.250-in ID) pipeline has been shut in for 100 h at an initial pressure of 1100 psig?

$$P_{ad} = \frac{100 \times 1100}{29.250 \times 949} = 4 \text{ lb/in}^2$$

If the observed pressure drop after stabilization is less than 4 psig, the section of line would be considered "tight."

Note: Corrections must he made for the effect of temperature variations upon pressure during the test.

Estimate pipe size requirements for increasing throughput volumes of natural gas

The following nomograph reflects the capacity-to-pipe size relationship when L (length), P_1 (inlet pressure), and P_2 (outlet pressure) are constant.

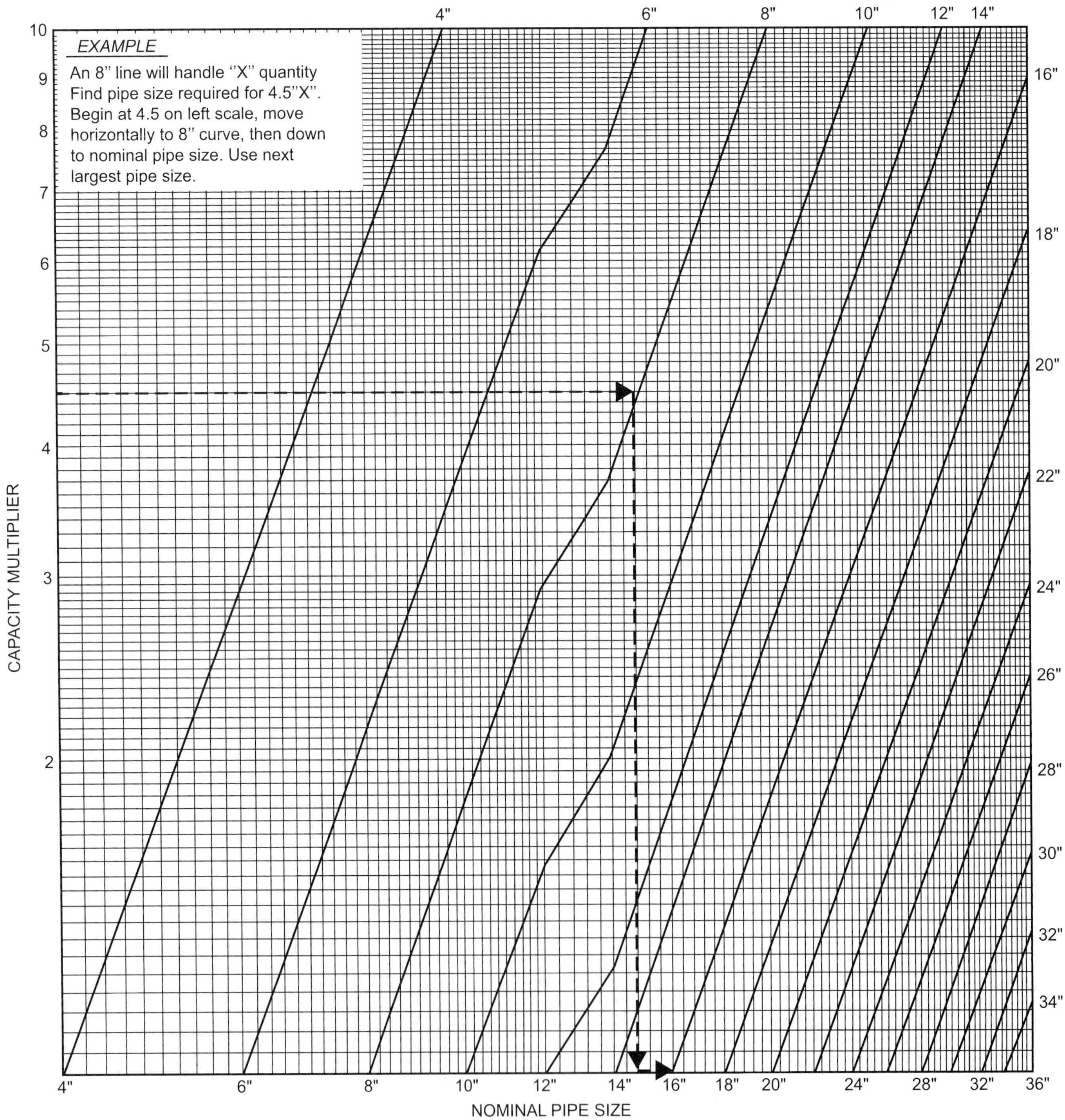

Calculate line loss using cross-sectional areas table when testing mains with air or gas

$$\text{Initial volume} = \frac{P_i}{P_B}(AL)$$

$$\text{Final volume} = \frac{P_f}{P_B}(AL)$$

where:

P_i = Initial pressure, or pressure at start of test, psia
P_f = Final pressure, or pressure at end of test, psia
P_B = Base pressure, psia
A = Cross-sectional area of inside of pipe in ft^2, and
L = Length of line in ft.

The loss is the difference of the volumes, found by the above formula, written as:

$$\text{Loss during test} = AL\frac{\left(P_i - P_f\right)}{P_B}$$

To facilitate calculations, Table 1 contains the cross-sectional areas in ft^2 for some of the more popular wall thickness pipe from 1–24 in nominal size.

Example. Assume 3562 ft of 8.125 in inside diameter pipe is tested with air for a period of 26 h. The initial pressure was 116 psig, the final pressure 112 psig, and the barometer was 14.4 psia in both instances. It is desired to calculate the volume of loss during the test using a pressure base of 14.65 psia.

If the area "A" is obtained from Table 1, and information as given above is substituted in the formula, we would have:

$$\begin{aligned}\text{Loss during test} &= 0.360 \times 3562 \\ &\times \frac{[(116 + 14.4) - (112 + 14.4)]}{14.65} \\ &= 350.1 \text{ cu. ft}\end{aligned}$$

Table 1
Characteristics of Pipe

Nominal Pipe Size D_n	Outside Diameter	Wall Thickness	Inside Diameter	"A" Inside Area in Sq. Feet
1	1.315	.133	1.049	.006
		.179	.957	.005
1¼	1.660	.140	1.380	.010
		.191	1.278	.009
1½	1.900	.145	1.610	.014
		.200	1.500	.012
2	2.375	.154	2.067	.023
		.218	1.939	.021
2½	2.875	.203	2.469	.033
		.276	2.323	.029
3	3.500	.125	3.250	.058
		.156	3.188	.055
		.188	3.124	.053
		.216	3.068	.051
		.250	3.000	.049
3½	4.000	.125	3.750	.077
		.156	3.688	.074
		.188	3.624	.072
		.226	3.548	.069
		.250	3.500	.067
4	4.500	.125	4.250	.099
		.156	4.188	.096
		.188	4.124	.093
		.219	4.062	.090
		.237	4.026	.088
		.250	4.000	.087
5	5.563	.156	5.251	.150
		.188	5.187	.147
		.219	5.125	.143
		.258	5.047	.139
6	6.625	.188	6.249	.213
		.219	6.187	.209
		.250	6.125	.205
		.280	6.065	.201
8	8.625	.188	8.249	.371
		.219	8.187	.366
		.250	8.125	.360
		.277	8.071	.355

Nominal Pipe Size D_n	Outside Diameter	Wall Thickness	Inside Diameter	"A" Inside Area in Sq. Feet
10	10.750	.188	10.374	.587
		.219	10.312	.580
		.250	10.250	.573
		.279	10.192	.567
		.307	10.136	.560
12	12.750	.219	12.312	.827
		.250	12.250	.818
		.281	12.188	.810
		.312	12.126	.802
		.330	12.090	.797
14	14.000	.250	13.500	.994
		.281	13.438	.985
		.312	13.376	.976
		.344	13.312	.966
		.375	13.250	.958
16	16.000	.250	15.500	1.310
		.281	15.438	1.300
		.312	15.376	1.289
		.344	15.312	1.279
		.375	15.250	1.268
18	18.000	.250	17.500	1.670
		.281	17.438	1.658
		.312	17.376	1.647
		.344	17.312	1.635
		.375	17.250	1.623
20	20.000	.250	19.500	2.074
		.281	19.438	2.061
		.312	19.376	2.048
		.344	19.312	2.034
		.375	19.250	2.021
22	22.000	.312	21.376	2.492
		.344	21.312	2.477
		.375	21.250	2.463
24	24.000	.312	23.376	2.980
		.344	23.312	2.964
		.375	23.250	2.948

Flow of fuel gases in pipelines

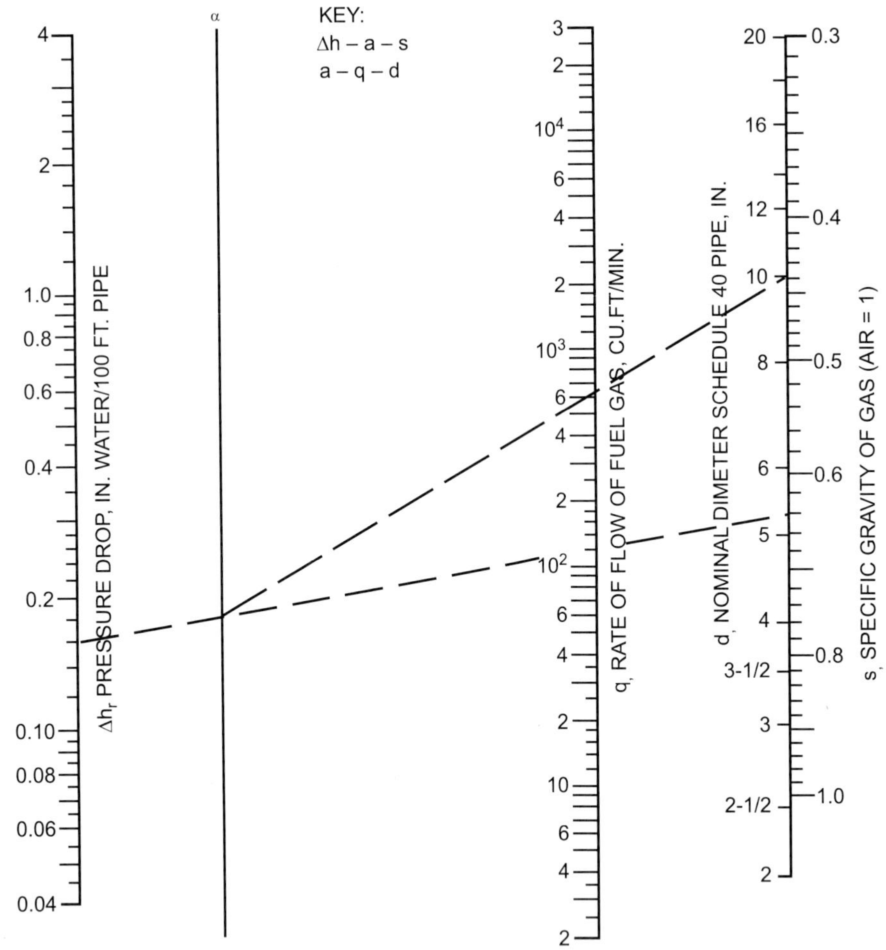

The rate of flow of fuel gases at low pressures, ordinary temperatures, and under turbulent conditions in pipelines is reflected by an equation based on sound theoretical considerations:

$$q = \frac{3.185 \Delta h^{0.543} d^{2.63}}{s^{0.468}}$$

where:

q = Rate of flow of gas, cf/min
Δh = Pressure drop, inches of water per 100 ft of pipe
d = Actual ID of Schedule 40 steel pipe, inches, and
s = Specific gravity of gas at the prevailing temperature and pressure relative to air at 68 °F and 30 in of mercury.

The equation can be solved by means of the accompanying monograph.

The use of the chart is illustrated as follows. At what rate will fuel gas with a specific gravity of 0.640 flow in a 10-in Schedule 40 pipeline if the pressure is 0.160 in of water per 100 ft of pipe? Following the key, connect 0.160 on the Δh scale and 0.640 on the s scale with a straight line and mark the intersection with the α axis. Connect this point with 10 on the d scale for nominal diameters and read the rate of flow at 624 cf/min on the q scale.

Some authorities give the exponent of s as 0.408 when, for the data given, the rate of flow as read from the chart would be decreased by 2.7% (17 cf/min) to attain 607 cf/min.

Source

D.S. Davis, Nomography and Empirical Equations, 2nd ed. Reinhold Publishing Corp., New York, 1962, Chapter 6.

Calculate the velocity of gas in a pipeline

$$V = \frac{1440ZQT}{d^2P}$$

where:

V = Velocity of gas, ft/s
Z = Compressibility
Q = Volume, million cf/h at standard conditions
T = Operating temperature, °R
d = ID of pipe, in
P = Operating pressure, psia

Example.

Q = 5 MMcfh
T = 60 °F
d = 15.5 in
P = 500 psig (514.7 psia)
Z = 0.94 (from compressibility chart)
T = 520 °R

$$V = \frac{1440 \times 0.94 \times 5 \times 520}{240.25 \times 514.7}$$

$$V = 28 \text{ ft/s}$$

Consideration should be given to acceptable noise levels (60–80 ft/s) and erosion, especially where the gas may contain solid particles. Refer to API RP 14E for erosion and corrosion considerations. If the gas contains liquids, the minimum velocity should be limited to about 10 ft/s to help sweep liquids from the pipe and prevent slugging.

Determining throat pressure in a blow-down system

In designing a gas blow-down system, it is necessary to find the throat pressure in a blow-off stack of a given size with a given flow at critical velocity. The accompanying nomograph is handy to use for preliminary calculations.

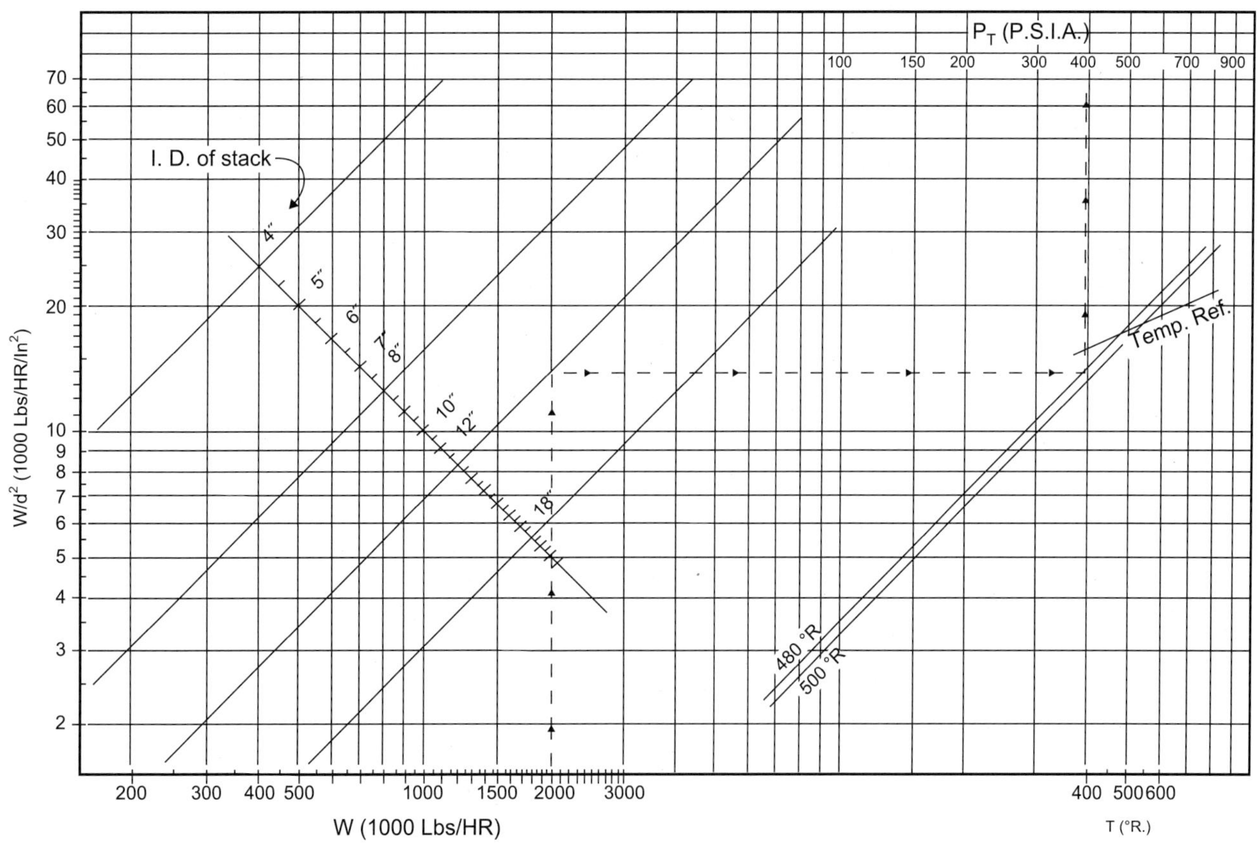

The nomograph was derived from the equation:

$$P_t = \frac{W}{d^2} \times \frac{\sqrt{\frac{RT_1}{K(K-1)}}}{11,400}$$

where:

P_t = Pressure in the throat, psia
W = Weight of flowing gas, lb/h
d = Actual ID of stack, in
R = Gas constant, ft lb/°F
$K = C_p/C_v$
C_p = Specific heat at constant pressure, Btu/lb/°F
C_v = Specific heat at constant volume, Btu/lb/°F
T_1 = Absolute temperature of gas flowing, °R = °F + 460

Using the following values for natural gas, $R = 88.888$, and $K = 1.31$, the above equation reduces to:

$$P_t = \frac{W}{d^2} \times 0.0012978\sqrt{T_1}$$

Example. To use the accompanying nomograph, assume that the rate of flow is known to be 2.00×10^6 lb/h, the ID of the stack is 12 in, and the flowing temperature is 480 °R. Start from the abscissa at the rate of flow, W, and proceed upward to the ID of the pipe. From this intersection proceed to the right to the flowing temperature of 480 °R. Turn upward from this intersection to the P_t scale and read the answer 395 psia.

Estimate the amount of gas blown off through a line puncture

To calculate the volume of gas lost from a puncture or blowoff, use the equation:

$Q = D^2P_1$

where:

Q = Volume of gas in Mcf/h at a pressure of 14.7 psi, 60 °F, and with a specific gravity of 0.60
D = Diameter of the nipple or orifice in inches
P = Absolute pressure in lb/in^2 at some nearby point upstream from the opening

Example. How much gas will be lost during a 5 min blowoff through a 2-in nipple if the upstream pressure is 1000 psi absolute?

$Q = D^2P_1$
$Q = 2^2 \times 1000$
$Q = 4000$ Mcf/h, and
$Q_{5\ min} = 4000 \times 5/60 = 333$ Mcf

A practical way to calculate gas flow for pipelines

Here is a shortcut to calculate gas flow in pipelines. It is based on the Weymouth formula. At 60 °F and specific gravity of 0.60, the answer will be accurate. For every 10° variation in temperature, the answer will have 1% error. For every 0.01 variation in specific gravity, the answer will be three-fourths percent in error.

Formula:

$$Q = \frac{(871)\left(d^{8/3}\right)\sqrt{P_1^2 - P_2^2}}{\sqrt{L}}$$

where:

Q = Cubic feet of gas per 24 h
d = Pipe ID in
P_1 = Psi (abs) at starting point
P_2 = Psi (abs) at ending point, and
L = Length of line in miles.

Example. How much gas would flow through 1 mile of 6-in ID pipe if the pressure at the starting point is 485 psi and if the pressure at the downstream terminus is 285 psi?

Solution.

$$Q = \frac{(871)\left(6^{8/3}\right)\sqrt{(485+15)^2 - (285+15)^2}}{\sqrt{L}}$$

$$= \frac{(871)(120)\sqrt{500^2 - 300^2}}{\sqrt{L}}$$

$$= 41,800,000 \text{ cu. ft per 24 h}$$

Example. What would be the flow through 4 miles of 10¾-in quarter-inch wall pipe using the same pressures?

Solution.

$$Q = \frac{(871)\left(10.25^{8/3}\right)\sqrt{500^2 - 300^2}}{\sqrt{4}}$$

$$Q = 86,350,000 \text{ cf/day}$$

How to calculate the weight of gas in a pipeline

Rule. Find the volume of the pipe in cubic feet and multiply by the weight of the gas per cubic feet. To find the latter, multiply the absolute pressure of the gas times 3 and divide by 1000.

The basis for the latter is that gas with a specific gravity of 0.60 at 70 °F weighs 3.06 lb/cf at 1000 psia. And everything else remaining equal, weight of the gas is proportional to absolute pressure. Thus, to find the weight of gas at, say, 500 psia, multiply 3 × 500/1000 = 1.5 lb.

Example. Find the weight of gas in a 1250 ft aerial river crossing where the average pressure reads 625.3 on the gauges; the temperature of the gas is 70 °F, and specific gravity of gas is 0.60.

Solution. Change psig to psia:
625.3 + 14.7 = 640 psia.
Weight of gas per cubic feet:

$$3.06 \times \frac{640}{1000} = 1.959\ \text{lb/cf}$$

Volume of 1250 ft of 23¼-in ID pipe = 3685 cf.
Weight of gas = 7217 lb.

This method is fairly accurate; here is the same problem calculated with the formula:

$$W = \frac{(V)(144)(P_{abs})}{RT}$$

where:

W = Weight of gas in lbs
V = Volume of pipe
P_{abs} = Absolute pressure of gas
R = 1544/molecular weight of gas, and
T = Temperature of gas in °R

To find the molecular weight of gas, multiply specific gravity × molecular weight of air, or in this case, 0.6 × 28.95 = 17.37.

$$W = \frac{(3685)(144)(640)}{(1544 \div 17.37)(70+460)}$$

$$W = 7209\ \text{lb}$$

Estimate average pressure in gas pipeline using upstream and downstream pressures

To find the average gas pressure in a line, first divide downstream pressure by upstream pressure and look up the value of the factor X in the table shown. Then multiply X times the upstream pressure to get the average pressure.

Example. Find the average pressure in a pipeline if the upstream pressure in 775 psia and the downstream pressure is 420 psia.

$$\frac{P_d}{P_u} = \frac{420}{775} = 0.54$$

Interpolating from the table,

$X = 0.78 + 4/5(0.02) = 0.796.$
$P_{avg} = 0.796 \times 775 = 617$ psia

Using the formula:

$$P_{avg} = \frac{2}{3}\left(P_u + P_d - \frac{P_uP_d}{P_u + P_d}\right) = \frac{2}{3}\left(1195 - \frac{325,500}{1195}\right)$$

$$P_{avg} = 615\ \text{psia}$$

P_d/P_u	X
.10	.67
.15	.68
.20	.69
.25	.70
.30	.71
.35	.725
.40	.74
.45	.76
.50	.78
.55	.80
.60	.82
.65	.84
.70	.86
.75	.88
.80	.90
.85	.925
.90	.95
.95	.975
1.00	1.00

Chart for determining viscosity of natural gas

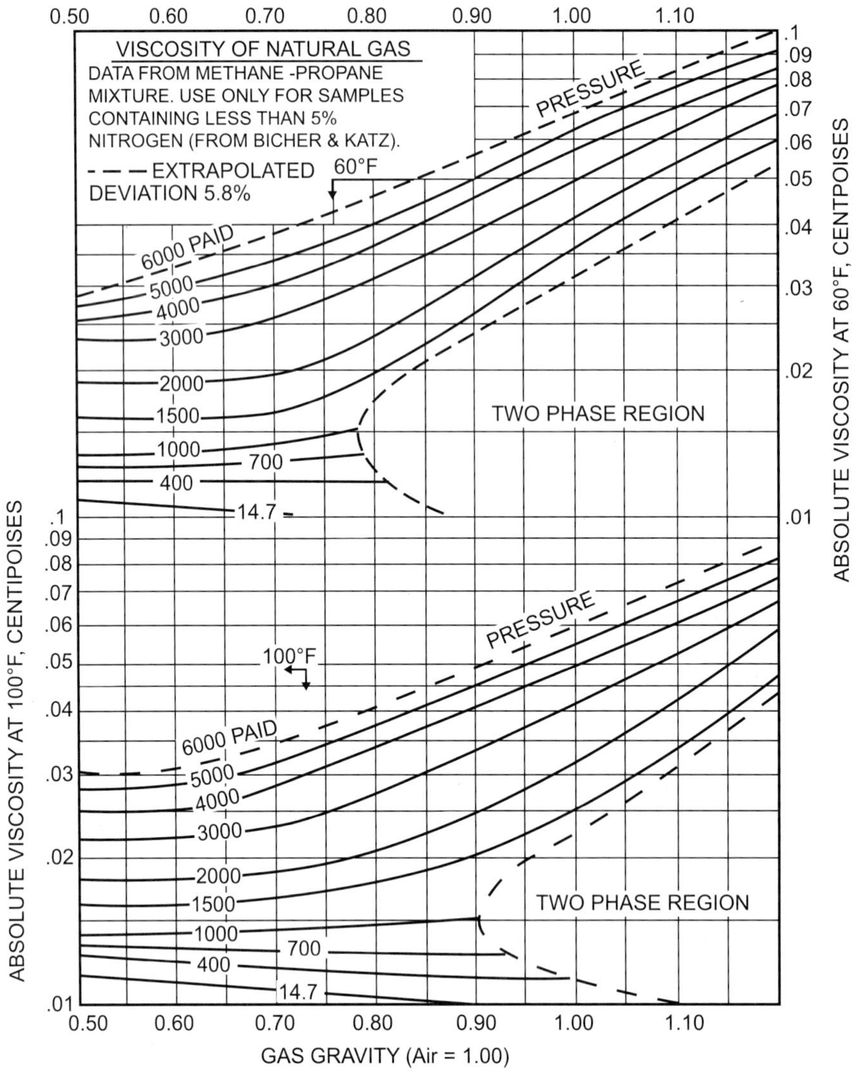

From Carlton Beal, "The Viscosity of Air, Water, Natural Gas, Crude Oil and Its Associated Gases at Oil Field Temperatures and Pressures," *Petroleum Development and Technology*, AIME, 1946.

Fanning equation for gas flow

For gas flow problems as encountered in oil field production operations, the Fanning equation for pressure drop may be used. A modified form of this equation employing units commonly used in oil field practice is[1-6]:

$$W = \frac{FL(MCFD)^2 STZ}{20,000\, d^5 P_{avg}}$$

and for determining the friction factor F:

$$R_W = \frac{20.14(\mathrm{MCFD})S}{d\mu}$$

where:

P = Pressure drop, psia
F = Friction factor, dimensionless
L = Length of pipe, ft
$MCFD$ = Gas flow at standard condition
S = Specific gravity of gas
T = Absolute temperature (°F + 460)
P_{avg} = Average flowing pressure, psia
d = Internal pipe diameter, inches
Re = Reynolds number, dimensionless, and
μ = Viscosity, centipoises.

In practice, empirical flow formulas are used to solve the gas flow problems of field production operations. The Weymouth formula is the one most frequently used since results obtained by its use agree quite closely with actual values. Recent modification of the formula, by including the compressibility factor, Z, made the formula applicable for calculation of high-pressure flow problems.

The modified formula is as follows:

$$Q = 433.45\frac{T_s}{P_s} \times d^{2.667} \times \left(\frac{P_1^2 - P_2^2}{LSTZ}\right)^{1/2}$$

where:

Q_S = Rate of flow of gas in cubic feet per 24 h measured at standard conditions

d = Internal diameter of pipe in inches.

P_1 = Initial pressure, psi absolute

P_2 = Terminal pressure, psi absolute

L = Length of line in miles

S = Specific gravity of flowing gas (air = 1.0)

T = Absolute temperature of flowing gas.

Standard conditions for measurements:

T_s = Standard absolute temperature

P = Standard pressure, psi absolute, and

Z = Compressibility factor.

Values of expression $d^{2.667}$ for different inside diameter of pipe are given in Table 1.

Table 1

Values of $d^{2.667}$ for Use in Weymouth's Gas-Flow Formula

(Increments of tenths of an inch with certain standard pipe sizes inserted)

d	$d^{2.667}$	d	$d^{2.667}$	d	$d^{2.667}$	d	$d^{2.667}$
1.0	1.0	6.2	129.9	11.5	673.8	16.9	1881
1.1	1.3	6.3	135.4	11.6	689.5	17.0	1911
1.2	1.6	6.4	141.2	11.7	705.5	17.1	1941
1.3	2.0	6.5	147.1	11.8	721.7	17.2	1971
1.4	2.4	6.6	153.3	11.9	738.1	17.250	1988
1.5	2.9	6.7	159.5	12.000	754.8	17.3	2002
1.6	3.5	6.8	166.0	12.1	771.7	17.4	2033
1.7	4.1	6.9	172.6	12.2	788.8	17.5	2064
1.8	4.8	7.0	179.3	12.3	806.1	17.6	2096
1.9	5.5	7.1	186.2	12.4	823.7	17.7	2128
2.0	6.3	7.2	193.3	12.5	841.6	17.8	2160
2.067	6.9	7.3	200.5	12.6	859.7	17.9	2193
2.1	7.2	7.4	207.9	12.7	878.0	18.0	2225
2.2	8.2	7.5	215.5	12.8	896.5	18.1	2259
2.3	9.2	7.6	223.3	12.9	915.3	18.2	2292
2.4	10.3	7.7	231.2	13.0	934.4	18.3	2326
2.5	11.5	7.8	239.3	13.1	953.7	18.4	2360
2.6	12.8	7.9	247.6	13.2	973.2	18.5	2394
2.7	14.1	7.981	254.5	13.250	983.9	18.6	2429
2.8	15.6	8.0	256.0	13.3	993.0	18.7	2464
2.9	17.1	8.1	264.6	13.4	1013	18.8	2499
3.0	18.7	8.2	273.4	13.5	1033	18.9	2535
3 068	19.9	8.3	282.4	13.6	1054	19.0	2570
3.1	20.4	8.4	291.6	13.7	1075	19.1	2607
3.2	22.2	8.5	300.9	13.8	1096	19.182	2639
3.3	24.1	8.6	310.5	13.9	1117	19.2	2643
3.4	26.1	8.7	320.2	14.0	1138	19.250	2664
3.5	28.2	8.8	330.1	14.1	1160	19.3	2680
3.6	30.4	8.9	340.2	14.2	1182	19.4	2717
3.7	32.7	8.941	344.6	14.3	1205	19.5	2755
3.8	35.2	9.0	350.5	14.4	1227	19.6	2793
3.9	37.7	9.1	361.0	14.5	1250	19.7	2831
4.0	40.3	9.2	371.6	14.6	1273	19.8	2869
4.026	41.0	9.3	382.5	14.7	1297	19.9	2908
4.1	43.1	9.4	393.6	14.8	1321	20.0	2947
4.2	45.9	9.5	404.8	14.9	1344	20.1	2987
4.3	48.9	9.6	416.3	15.0	1368	20.2	3026
4.4	52.0	9.7	428.0	15.1	1393	20.3	3067
4.5	55.2	9.8	439.8	15.2	1418	20.4	3107
4.6	58.5	9.9	451.9	15.250	1431	20.5	3148
4.7	62.0	10.0	464.2	15.3	1443	20.6	3196
4.8	65.6	10.020	467.0	15.4	1468	20.7	3230
4.9	69.3	10.1	476.6	15.5	1494	20.8	3272
5.0	73.1	10.2	489.3	15.6	1519	20.9	3314
5.1	77.1	10.3	502.2	15.7	1545	21.0	3357
5.2	81.2	10.4	515.3	15.8	1572	21.1	3400
5.3	85.4	10.5	528.7	15.9	1599	21.1	3443
5.4	89.8	10.6	542.2	16.0	1625	21.250	3468
5.5	94.3	10.7	555.9	16.1	1653	21.3	3486
5.6	98.9	10.8	569.9	16.2	1680	21.4	3530
5.7	103.7	10.9	584.1	16.3	1708	21.5	3574
5.8	108.6	11.000	598.5	16.4	1736	21.6	3619
5.9	113.7	11.1	613.1	16.5	1765	21.7	3663
6.0	118.9	11.2	627.9	16.6	1793	21.8	3709
6.065	122.4	11.3	643.0	16.7	1822	21.9	3754
6.7	124.2	11.4	658.3	16.8	1851	22 0	3800

—The Oil & Gas Journal

Another formula quite commonly used is the one determined by F. H. Oliphant:

$$Q_s = 42a\sqrt{\frac{P_1^2 - P_2^2}{L}}$$

where:

Q = Discharge in cf/h
P_1 = Initial pressure in lb/in^2, absolute
P_2 = Final pressure in lb/in^2, absolute
L = Length of line in miles, and
a = Diameter coefficient.

This formula assumes a specific gravity of gas of 0.6. For any other specific gravity S, multiply the final result by

$$\sqrt{\frac{0.6}{S}}$$

The values of diameter coefficients for different sizes of pipes are given as:

¼ in	0.0317
½ in	0.1810
¾ in	0.5012
1 in	1.0000
1½ in	2.9300
2 in	5.9200
2½ in	10.37
3 in	16.50
4 in	34.10
5 in	60.00
5⅝ in	81.00
6 in	95.00
8 in	198
10 in	350
12 in	556
16 in	1160
18 in	1570

For pipes greater than 12 in in diameter, the measure is taken from the outside and for pipes of ordinary thickness, the corresponding inside diameters and multipliers are as follows:

Outside	Inside	Multiplier
15 in	14¼ in	863
16 in	15¼ in	1025
18 in	17¼ in	1410
20 in	19¼ in	1860

To determine gas flow rates for specific pressure drops for pipe sizes used in production operations, Tables 3–5 may be used. The tables are calculated by the Weymouth formula for listed inside diameters of standard weight threaded line pipe for the sizes shown.

Tables 3–5 give the hourly rates of flow of 0.70 specific gravity gas, flowing at 60 °F and measured at a standard temperature of 60 °F and at a standard pressure base of 14.4 psi + 4 oz. gauge pressure of 14.65 psi. For specific gravity of gas other than 0.70 and for flowing temperature other than 60 °F, correction can be made by use of the factors shown in Table 2.

To find from Tables 3–5 the volume of gas delivered through any length L, the volume found from the table should be multiplied by $1/\sqrt{L}$. L must be expressed as a multiple of the length heading the table in use. For instance, if the delivery is to be determined from a 4500 ft line, the volume found for a 1000 ft line in Table 3 should be multiplied by $1/\sqrt{4.5}$.

Example. What is the delivery of a 2-in gas line, 3600 ft long, with an upstream pressure of 200 psi and a downstream pressure of 50 psi? The specific gravity of gas is 0.07; the flowing temperature, 70 °F.

From Table 3 the volume for 1000 ft of line, for 0.70 specific gravity gas and 60 °F flowing temperature for 200 and 50 psi up and downstream pressure, respectively, is 109.5 MCF/h. Correction factor for the pipe length, from the same table is 0.53. Correction factor, from Table 2 for specific gravity and flowing temperature is 0.991. Therefore:

Delivery of gas = 109.5 × 0.53 × 0.991 = 57,512 cf/h.

Table 2
Temperature—Specific Gravity Multipliers

Specific Gravity	FLOWING TEMPERATURE—DEGREES F.												
	0	10	20	30	40	50	60	70	80	90	100	110	129
0.60	1.15	1.14	1.12	1.11	1.10	1.09	1.08	1.07	1.06	1.05	1.04	1.03	1.02
0.70	1.06	1.05	1.04	1.03	1.02	1.01	1.00	.991	.981	.972	.964	.955	.948
0.80	.995	.984	.974	.964	.954	.945	.935	.927	.918	.910	.901	.894	.886
0.90	.938	.928	.918	.909	.899	.891	.882	.874	.965	.858	.850	.842	.835
1.00	.890	.880	.871	.862	.853	.845	.837	.829	.821	.814	.806	.799	.792

Table 3
Volume of Gas Delivered—Thousands of Cubic Feet Per Hour Through 1000- and 2000-ft Lines

0.7 SP. GR. 60° F.

GAGE PRESSURE Up Stream P.S.I.	GAGE PRESSURE Down Stream P.S.I.	1000 FOOT LINE, Nominal Pipe Size* 2	2½	3	4	6	2000 FOOT LINE, Nominal Pipe Size* 2	2½	3	4	6
		Thousands of Cu. Ft. per Hour					Thousands of Cu. Ft. per Hour				
400	10	221.5	355.8	635.0	1311.	3910.	156.6	251.6	449.0	926.8	2765.
	50	219.2	352.1	628.4	1297.	3869.	155.0	248.9	444.3	917.2	2736.
	100	213.2	342.5	611.4	1262.	3764.	150.8	242.2	432.3	892.4	2662.
	350	105.6	169.7	302.9	625.2	1865.	74.70	120.0	214.2	442.1	1319.
350	10	194.7	312.7	558.1	1152.	3436.	137.6	221.1	394.6	814.6	2430.
	50	192.0	308.5	550.6	1136.	3390.	135.8	218.1	389.3	803.6	2397.
	100	185.2	297.5	531.0	1096.	3270.	131.0	210.4	375.5	775.1	2312.
	300	98.63	158.4	282.8	583.7	1741.	69.74	112.0	200.0	412.8	1231.
300	10	167.8	269.6	481.2	993.1	2962.	118.7	190.6	340.2	702.3	2095.
	50	164.8	264.6	472.4	975.0	2908.	116.5	187.1	334.0	689.4	2056.
	100	156.8	251.9	449.5	927.9	2768.	110.9	178.1	317.9	656.1	1957.
	250	91.08	146.3	261.1	539.0	1608.	64.40	103.5	184.7	381.1	1137.
250	10	141.0	226.4	404.1	834.2	2488.	99.67	160.1	285.8	589.8	1759.
	50	137.3	220.5	393.6	812.5	2424.	97.08	155.9	278.3	574.5	1714.
	100	127.6	205.0	365.9	755.3	2253.	90.24	145.0	258.7	534.0	1583.
	200	82.84	133.1	237.5	490.3	1462.	58.58	94.09	167.9	346.7	1034.
200	10	114.0	183.2	327.0	674.9	2013.	80.64	129.5	231.2	477.2	1424.
	50	109.5	175.9	313.9	647.9	1933.	77.42	124.4	222.0	458.2	1367.
	100	97.08	155.9	278.3	574.5	1714.	68.65	110.3	196.8	406.3	1212.
	150	73.68	118.4	211.3	436.0	1301.	52.10	83.69	149.4	308.3	919.7
150	10	87.04	139.8	249.6	515.1	1537.	61.55	98.87	176.5	364.2	1086.
	25	85.45	137.3	245.0	505.7	1508.	60.42	97.06	173.2	357.6	1067.
	50	80.98	130.1	232.2	479.3	1430.	57.26	91.98	154.2	338.9	1011.
	100	63.21	101.5	181.2	374.1	1116.	44.70	71.80	128.2	264.5	789.0

GAGE PRESSURE Up Stream P.S.I.	GAGE PRESSURE Down Stream P.S.I.	1000 FOOT LINE, Nominal Pipe Size* 2	2½	3	4	6	2000 FOOT LINE, Nominal Pipe Size* 2	2½	3	4	6
		Thousands of Cu. Ft. Per Hour					Thousands of Cu. Ft. per Hour				
100	10	59.84	96.12	171.6	354.1	1056.	42.31	67.97	121.3	250.4	746.9
	25	57.50	92.36	164.9	340.3	1015.	40.66	65.31	116.6	240.6	717.7
	50	50.62	81.32	145.1	299.6	893.6	35.79	57.50	102.6	211.8	631.9
	75	38.22	61.39	109.6	226.2	674.6	27.02	43.41	77.5	159.9	477.0
80	5	49.46	79.45	141.8	292.7	873.1	34.97	56.18	100.3	207.0	617.4
	10	48.82	78.43	140.0	288.9	861.8	34.52	55.46	98.98	204.3	609.4
	30	44.50	71.64	127.9	264.0	787.3	31.54	50.66	90.42	186.6	556.7
	55	34.26	55.03	98.23	202.8	604.8	24.23	38.91	69.46	143.4	427.6
60	5	38.45	61.77	110.3	227.6	678.8	27.19	43.68	77.96	160.9	480.0
	10	37.63	60.45	107.9	222.7	664.3	26.61	42.74	76.29	157.5	469.7
	20	35.32	56.73	101.3	209.0	623.5	24.97	40.12	71.61	147.8	440.9
	40	27.17	43.65	77.91	160.8	479.7	19.21	30.86	55.09	113.7	339.2
40	5	27.21	43.71	78.01	161.0	480.3	19.24	30.91	55.16	113.9	339.6
	10	26.03	41.81	74.63	154.1	549.5	18.41	29.57	52.77	108.9	324.9
	15	24.50	39.36	70.26	145.0	432.6	17.33	27.83	49.68	102.5	305.9
	25	20.08	32.26	57.58	118.8	354.5	14.20	22.81	40.71	84.04	250.7
30	5	21.38	34.35	61.30	126.5	377.4	15.12	24.29	43.35	89.48	266.9
	10	19.86	31.90	56.94	117.5	350.6	14.04	22.56	40.26	83.11	247.9
	15	17.81	28.61	51.07	105.4	314.5	12.60	20.23	36.11	74.54	222.4
	20	15.03	24.14	43.09	88.94	265.3	10.63	17.07	30.47	62.89	187.6
20	5	15.21	24.43	43.61	90.00	268.5	10.75	17.28	30.83	63.65	189.8
	10	12.98	20.85	37.22	76.83	229.2	9.180	14.75	26.32	54.33	162.1
10	5	7.923	12.73	22.72	46.89	138.9	5.603	9.000	16.06	33.16	98.90

MULTIPLIERS TO CONVERT 1000 FOOT LINE VALUES TO LENGTHS LISTED:

Length of Line—Feet	200	400	600	800	1000	1200	1400	1600	1800	2000	2200	2400	2600	2800	3000	3200	3400	3600	3800	4000	4200	4400	4600	4800	5000	5200	Feet—Length of Line
Multipliers	2.24	1.58	1.29	1.12	1.00	.91	.85	.79	.75	.71	.67	.65	.62	.60	.58	.56	.54	.53	.51	.50	.49	.48	.47	.46	.45	.44	Multipliers

For remarks regarding use of the Table refer to page 78 of source.

Courtesy Armco Steel Corporation.

Table 4
1- and 2½-mile Lines

0.7 SP. GR. 60° F.

GAGE PRESSURE Up Stream P.S.I.	GAGE PRESSURE Down Stream P.S.I.	1 MILE LINE, Nominal Pipe Size* 2	2½	3	4	6	2½ MILE LINE, Nominal Pipe Size* 2	2½	3	4	6
		Thousands of Cu. Ft. per Hour					Thousands of Cu. Ft. per Hour				
400	10	96.39	154.8	276.4	570.4	1701.	60.96	97.92	174.8	360.8	1076.
	50	95.38	153.2	273.5	564.5	1684.	60.32	96.90	173.0	357.0	1065.
	100	92.80	149.1	266.1	549.2	1638.	58.69	94.28	168.3	347.3	1036.
	350	45.98	73.85	131.8	272.1	811.6	29.08	46.71	83.37	172.1	513.3
350	10	84.71	136.1	242.9	501.3	1495.	53.58	86.06	153.6	317.1	945.8
	50	83.57	134.2	239.6	494.6	1475.	52.85	84.90	151.5	312.8	933.0
	100	80.61	129.5	231.1	477.1	1423.	50.98	81.90	146.2	301.7	900.0
	300	42.92	68.95	123.1	254.0	757.7	27.15	43.61	77.84	160.7	479.2
300	10	73.03	117.3	209.4	432.2	1289.	46.19	74.20	132.4	273.4	815.4
	50	71.70	115.2	205.6	424.3	1266.	45.35	72.84	130.0	268.4	800.5
	100	68.23	109.6	195.6	403.8	1204.	43.15	69.32	123.7	255.4	761.8
	250	38.64	63.67	113.6	234.6	699.7	25.07	40.27	71.88	148.4	442.5
250	10	61.34	98.53	175.9	363.0	1083.	38.80	62.32	111.2	229.6	684.8
	50	59.75	95.98	171.3	353.6	1055.	37.79	60.70	108.3	223.6	667.1
	100	55.54	89.21	159.2	328.7	980.4	35.13	56.42	100.7	207.9	620.1
	200	36.05	57.91	103.4	213.4	636.4	22.80	36.63	65.37	134.9	402.5
200	10	49.63	79.72	142.3	293.7	876.1	31.39	50.42	90.00	185.8	554.1
	50	47.65	76.54	136.6	282.0	841.1	30.13	49.41	86.40	178.3	532.0
	100	42.25	67.87	121.1	250.0	745.8	26.72	42.92	76.61	158.1	471.7
	150	32.06	51.51	91.94	189.8	566.0	20.28	32.58	58.14	120.0	358.0
150	10	37.88	60.85	108.6	224.2	668.7	23.96	38.48	68.69	141.8	422.9
	25	37.19	59.74	106.6	220.1	656.5	23.52	37.78	67.44	139.2	415.2
	50	35.24	56.61	101.0	208.6	622.1	22.29	35.80	63.91	131.9	393.5
	100	27.51	44.19	78.87	162.8	485.6	17.40	27.95	49.88	103.0	307.1

GAGE PRESSURE Up Stream P.S.I.	GAGE PRESSURE Down Stream P.S.I.	1 MILE LINE, Nominal Pipe Size* 2	2½	3	4	6	2½ MILE LINE, Nominal Pipe Size* 2	2½	3	4	6
		Thousands of Cu. Ft. per Hour					Thousands of Cu. Ft. per Hour				
100	10	26.04	41.83	74.67	154.1	459.7	16.47	26.46	47.22	97.47	290.7
	25	25.02	40.20	71.75	148.1	441.7	15.83	25.42	45.38	93.66	279.4
	50	22.03	35.39	63.16	130.4	388.9	13.93	22.38	39.95	82.46	246.0
	75	16.63	26.72	47.68	98.4	293.6	10.52	16.90	30.16	62.25	185.7
80	5	21.53	34.58	61.72	127.4	380.0	13.61	21.87	39.03	80.57	240.3
	10	21.25	34.13	60.92	125.7	375.1	13.44	21.59	38.53	79.53	237.2
	30	19.41	31.18	55.65	114.9	342.6	12.28	19.72	35.20	72.65	216.7
	55	14.91	23.95	42.75	88.24	263.2	9.430	15.15	27.04	55.81	166.5
60	5	16.74	26.88	47.98	99.04	285.4	10.58	17.00	30.35	62.64	186.8
	10	16.38	26.31	46.95	96.91	289.1	10.36	16.64	29.70	61.29	182.8
	20	15.37	24.69	44.07	90.96	271.3	9.721	15.62	27.87	57.53	171.6
	40	11.83	19.00	33.91	69.98	208.8	7.479	12.01	21.44	44.26	132.0
40	5	11.84	19.02	33.95	70.08	209.0	7.489	12.03	21.47	44.32	132.2
	10	11.33	18.20	32.48	67.04	200.0	7.165	11.51	20.54	42.40	126.5
	15	10.66	17.13	30.58	63.11	188.3	6.745	10.83	19.34	39.92	119.1
	25	8.740	14.04	25.06	51.72	154.3	5.527	8.879	15.85	32.71	97.57
30	5	9.315	14.95	26.68	55.07	164.3	5.885	9.453	16.87	34.83	103.9
	10	8.643	13.88	24.78	51.15	152.6	5.466	8.781	15.67	32.35	96.49
	15	7.752	12.45	22.23	45.88	136.8	4.903	7.875	14.06	29.02	86.55
	20	6.540	10.51	18.75	38.71	115.5	4.137	6.645	11.86	24.48	73.02
20	5	6.619	10.63	18.98	39.17	116.8	4.186	6.724	12.00	24.77	73.90
	10	5.650	9.076	16.20	33.44	99.74	3.573	8.740	10.25	21.15	63.08
10	5	3.448	5.539	9.886	20.41	60.87	2.181	3.503	6.253	12.91	38.50

MULTIPLIERS TO CONVERT 1 MILE LINE VALUES TO LENGTHS LISTED:

Length of Line—Miles	1.2	1.4	1.6	1.8	2.0	2.2	2.4	2.6	2.8	3.0	3.2	3.4	3.6	3.8	4.0	4.2	4.4	4.6	4.8	Miles—Length of Line
Multipliers	.91	.85	.79	.75	.71	.67	.65	.62	.60	.58	.56	.54	.53	.51	.50	.49	.48	.47	.46	Multipliers

For remarks regarding use of the Table refer to page 78 of source.

Courtesy of Armco Steel Corporation.

Table 5
5- and 10-mile Lines

0.7 SP. GR. 60° F.

Gage Pressure Up Stream P.S.I.	Gage Pressure Down Stream P.S.I.	5 Mile Line 2	5 Mile Line 2½	5 Mile Line 3	5 Mile Line 4	5 Mile Line 6	10 Mile Line 2	10 Mile Line 2½	10 Mile Line 3	10 Mile Line 4	10 Mile Line 6
		Thousands of Cu. Ft. Per Hour					Thousands of Cu. Ft. Per Hour				
400	10	43.10	69.24	123.6	255.1	760.9	30.48	48.96	87.39	180.4	538.0
	50	42.66	68.52	122.3	252.4	753.0	30.16	48.45	86.48	178.5	532.4
	100	41.50	66.67	119.0	245.6	732.6	29.35	47.14	84.14	173.7	518.0
	350	20.56	33.03	58.95	121.7	363,0	14.54	23.35	41.69	86.04	256.7
350	10	37.88	60.86	108.6	224.2	668.8	26.79	43.03	76.81	158.5	472.9
	50	37.37	60.03	107.2	221.2	659.7	26.43	42.45	75.77	156.4	466.5
	100	36.05	57.91	103.4	213.3	636.4	25.49	40.95	73.09	150.9	450.0
	300	19.20	30.84	55.04	113.6	338.9	13.57	21.80	38.92	80.33	239.6
300	10	32.66	52.46	93.64	193.3	576.6	23.09	37.10	66.22	136.7	407.7
	50	32.07	51.51	91.94	189.8	566.0	22.67	36.42	65.01	134.2	400.2
	100	30.51	49.02	87.49	180.6	538.7	21.58	34.66	61.86	127.7	380.9
	250	17.73	28.47	50.82	104.9	312.9	12.53	20.13	35.94	74.18	221.3
250	10	27.43	44.07	78.65	162.3	484.3	19.40	31.16	55.62	114.8	342.4
	50	26.72	42.92	76.61	158.1	471.7	18.89	30.35	54.17	111.8	333.5
	100	24.84	39.90	71.21	147.0	438.5	17.56	28.21	50.36	103.9	310.0
	200	16.12	25.90	46.23	95.41	284.6	11.40	18.31	32.69	67.47	201.2
200	10	22.20	35.65	63.64	131.4	391.8	15.69	25.21	45.00	92.88	277.0
	50	21.31	34.23	61.10	126.1	376.2	15.07	24.20	43.20	89.17	266.0
	100	18.89	30.35	54.17	111.8	333.5	13.36	21.46	38.31	79.07	235.8
	150	14.34	23.03	41.11	84.86	253.1	10.14	16.29	29.07	60.01	179.0
150	10	16.94	27.21	48.57	100.3	299.0	11.98	19.24	34.35	70.89	211.5
	25	16.63	26.72	47.68	98.42	293.6	11.76	18.89	33.72	69.60	207.6
	50	15.76	25.32	45.19	93.27	278.2	11.14	17.90	31.95	65.96	196.7
	100	12.30	19.76	35.27	72.81	217.2	8.699	13.97	24.94	51.48	153.6

0.7 SP. GR. 60° F.

Gage Pressure Up Stream P.S.I.	Gage Pressure Down Stream P.S.I.	5 Mile Line 2	5 Mile Line 2½	5 Mile Line 3	5 Mile Line 4	5 Mile Line 6	10 Mile Line 2	10 Mile Line 2½	10 Mile Line 3	10 Mile Line 4	10 Mile Line 6
		Thousands of Cu. Ft. Per Hour					Thousands of Cu. Ft. Per Hour				
100	10	11.65	18.71	33.39	68.92	205.6	8.235	13.23	23.61	48.74	145.4
	25	11.19	17.98	32.09	66.23	197.5	7.913	12.71	22.69	46.83	139.7
	50	9.852	15.83	28.25	55.31	173.9	6.966	11.19	19.97	41.23	123.0
	75	7.438	11.95	21.32	44.02	131.3	5.259	8.448	15.08	31.12	92.84
80	5	9.626	15.46	27.60	56.97	169.9	6.807	10.93	19.52	40.28	120.2
	10	9.502	15.26	27.24	56.23	167.7	6.719	10.79	19.26	39.76	118.6
	30	8.680	13.94	24.89	51.37	153.2	6.138	9.860	17.60	36.32	108.4
	55	6.668	10.71	19.12	39.46	117.7	4.715	7.574	13.52	27.90	83.23
60	5	7.484	12.02	21.46	44.29	132.1	5.292	8.501	15.17	31.32	93.42
	10	7.324	11.76	21.00	43.34	129.3	5.179	8.319	14.85	30.65	91.42
	20	6.873	11.04	19.71	40.68	121.3	4.861	7.808	13.94	28.77	85.80
	40	5.288	8.495	15.16	31.30	93.36	3.740	6.007	10.72	22.13	66.01
40	5	5.296	8.507	15.18	31.34	93.48	3.745	6.015	10.74	22.16	66.10
	10	5.066	8.138	14.53	29.98	89.43	3.582	5.755	10.27	21.20	63.24
	15	4.769	7.661	13.67	28.22	84.19	3.372	5.417	9.669	19.96	59.53
	25	3.908	6.278	11.21	23.13	69.00	2.764	4.439	7.924	16.36	48.79
30	5	4.161	6.685	11.93	24.63	73.46	2.943	4.727	8.437	17.41	51.94
	10	3.865	6.209	11.08	22.87	68.23	2.733	4.390	7.836	16.17	48.25
	15	3.467	5.569	9.940	20.52	61.20	2.451	3.938	7.029	14.51	43.27
	20	2.925	4.698	8.386	17.31	51.63	2.068	3.322	5.930	12.24	36.51
20	5	2.960	4.755	8.487	17.52	52.25	2.093	3.362	6.001	12.39	36.95
	10	2.527	4.059	7.244	14.95	44.60	1.787	2.870	5.123	10.57	31.54
10	5	1.542	2.477	4.421	9.126	27.22	1.090	1.752	3.126	6.453	19.25

MULTIPLIERS TO CONVERT 10 MILE LINE VALUES TO LENGTHS LISTED.

Length of Line—Miles	5.2	5.4	5.6	5.8	6.0	6.2	6.4	6.6	6.8	7.0	7.2	7.4	7.6	7.8	8.0	8.2	8.4	8.6	8.8	9.0	9.2	9.4	9.6	9.8	Miles—Length of Line
Multipliers	1.39	1.36	1.34	1.31	1.29	1.27	1.25	1.23	1.21	1.20	1.18	1.16	1.15	1.13	1.12	1.10	1.09	1.08	1.07	1.05	1.04	1.03	1.02	1.01	Multipliers

For remarks regarding use of the Table refer to page 78 of source.

Courtesy of Armco Steel Corporation.

Multiphase flow

As stated at the beginning of this chapter, the problems of the simultaneous flow of oil and gas or of oil, gas, and water through one pipe are complex because the properties of two or more fluids must be considered and because of the different patterns of fluid flows. These patterns, in any given line, may change as the flow conditions change, and they may be coexistent at different points of the line.[3,5,7]

Different investigators recognize different flow patterns and use different nomenclature. Those generally recognized are:

1. Bubble flow—bubbles of gas flow along the upper part of the pipe with about the same velocity as the liquid.
2. Plug flow—the bubbles of gas coalesce into large bubbles and fill out the large part of the cross-sectional area of the pipe.
3. Laminar flow—the gas–liquid interface is relatively smooth with gas flowing in the upper part of the pipe.
4. Wavy flow—same as above except that waves are formed on the surface of the liquid.
5. Slug flow—the tops of some waves reach the top of the pipe. These slugs move with high velocity.
6. Annular flow—the liquid flows along the walls of the pipe, and the gas moves through the center with high velocity.
7. Spray flow—the liquid is dispersed in the gas.

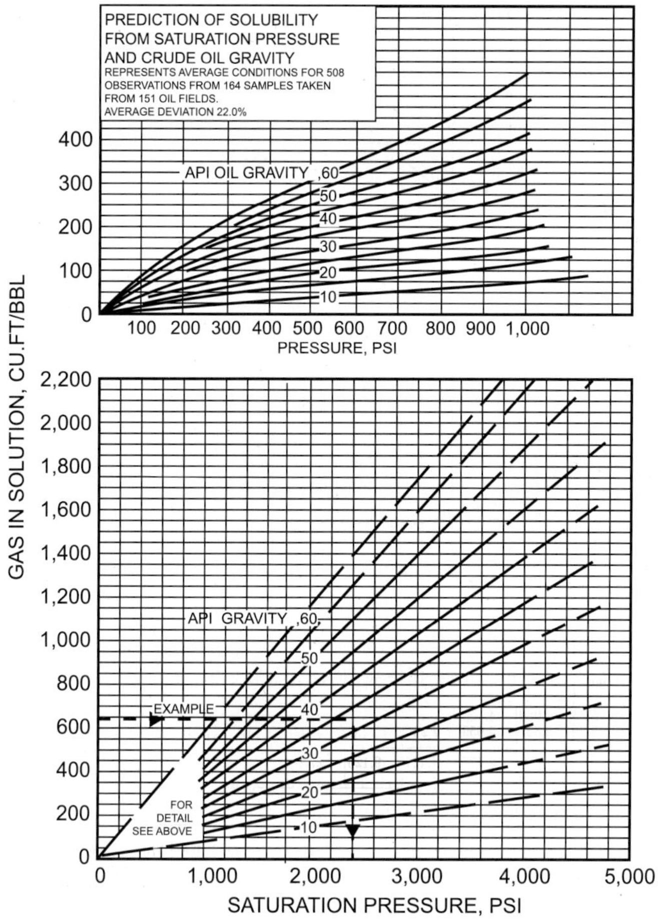

Figure 1. Prediction of solubility from saturation pressure and gravity of crude oil. (Ref. 12.)

The above description of flow patterns has been given to emphasize the first reason why the pressure drop must be higher for the multiphase than the single-phase flow. In the latter the pressure drop is primarily the result of friction. In the multiphase flow, in addition to friction, the energy losses are due to gas accelerating the fluid, due to compression of gas and formation of waves and slugs. These losses are different for different patterns, with patterns changing as a result of changes in flow conditions. The second reason is the fact that with two phases present in the pipe, less cross-sectional area is available for each of the phases. As previously discussed, the pressure drop is inversely proportional to the fifth power of pipe diameter.

Two-phase horizontal flow

More than 20 correlations have been developed by different investigators for predicting the pressure drop in the two-phase flow. Under research, sponsored jointly by the American Petroleum Institute and American Gas Association at the University of Houston, five of these correlations were tested by comparing them with 2620 experimental measurements culled from more than 15,000 available.

The statistical part of the report on the project concludes that of the five correlations tested, the Lockhart–Martinelli correlation shows the best agreement with the experimental data, particularly for pipe sizes commonly used in oil field production operations. The method may be summarized as follows[3,4]:

1. Determine the single-phase pressure drops for the liquid and the gas phase as if each one of these were flowing alone through the pipe. This is done by use of previously given equations.
2. Determine the dimensionless parameter:

$$X = \sqrt{\frac{\Delta P_L}{\Delta P_G}}$$

where ΔP_L and ΔP_G are single-phase pressure drops for liquid and gas, respectively.

3. The method recognizes four regimens of the two-phase flow as follows:

Flow regimes

Gas	Liquid
Turbulent	Laminar
Turbulent	Turbulent
Laminar	Laminar
Laminar	Turbulent

The type of flow of each of the phases is determined by its Reynolds number (Figure 2).

4. Determine factor Φ from a chart as a function of X for the appropriate flow regimen.
5. The two-phase pressure drop is then:

$$\Delta P_{TP} = \Phi^2 P_L \text{ of } G$$

The single-phase pressure drop of either gas or liquid phase can be used. The results are the same.

For the two-phase flow problems in oil field production operations, certain simplifications can be used if only approximate estimates are desired:

1. Information that would permit determining the volume of oil and gas under flow conditions from equilibrium flash calculation is usually unavailable. The following procedure may be used:

   Gas volume is determined by use of the chart in Figure 1, which shows gas in solution for oil of different

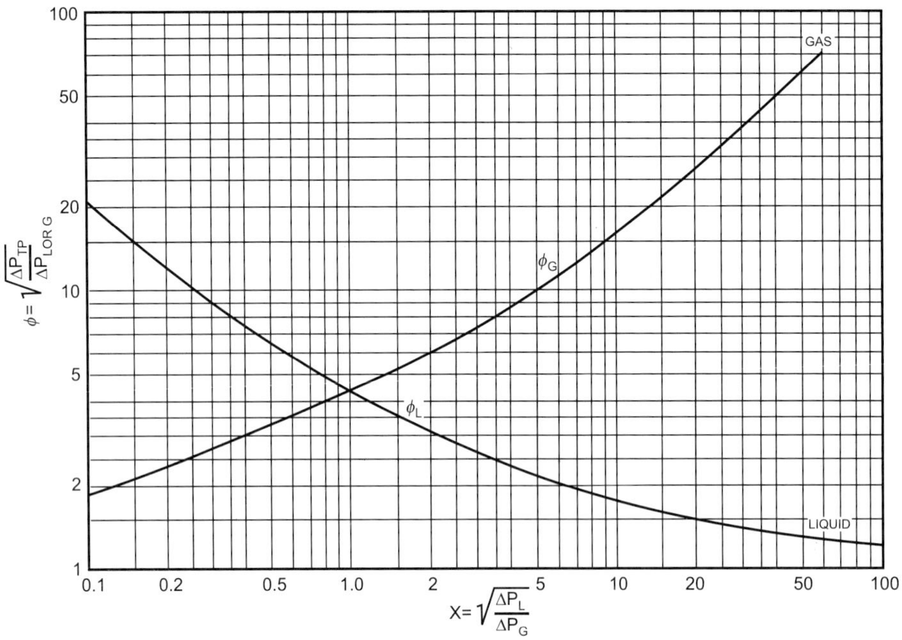

Figure 2. Parameter Φ for Lockhart–Martinelli correlation for two-phase flow. (Ref. 9.)

gravities and different saturation pressures (in this case the flowline pressures). If the gas–oil ratio is known for one pressure, the ratio for another pressure can be determined from the chart as follows:

Assume a gas–oil ratio of 600 cf/bbl at 900 psi for a 30°API oil. Determine the gas–oil ratio at 1300 psi. From the chart, at 900 psi, the gas in solution is 205 cu. ft/bbl and at 1300 psi is 300 cf/bbl, an increase of 95 cf/bbl. Therefore, at 1300 psi the gas–oil ratio will be 600–95 = 505 cf/bbl.

Increase in volume of oil because of the increase of the dissolved gas can be obtained from charts for calculation of formation volume of bubble point liquids. However, for approximate estimates in problems involving field flowlines, this step frequently can be omitted without seriously affecting the validity of results.

2. In a majority of cases here under consideration, the flow regimen is turbulent–turbulent. Therefore, the chart in Figure 2 shows the factor for only this regimen.

The procedure is illustrated by the following examples.

Example 1. Determine two-phase pressure drops for different flow ratios and different flow pressures under the following conditions:

Oil	31°API
Viscosity	5 cp at 100 °F
Separator pressure	900 psig
Gas–oil ratio	670 cf/bbl
Flow line	2 in

Wellhead pressure of 1300 psig was assumed, and calculations were made for flow rates of 100 and 50 bbl per day. The single-phase pressure drops were calculated for the gas and liquid phases, each of them flowing alone in the line, according to the Fanning equations.

Calculation of pressure drop for liquid phase:

$$H_f = \frac{f\,L(\text{bbl/day})^2(\text{lb/gal})}{181,916\,D^5}$$

where:

f = Friction factor
L = Length of pipe, ft, and
D = Inside diameter of pipe, inches

Calculation of pressure drop for gas phase.

For the two flow rates and the two pressures, these single-phase pressure drops were calculated to be as follows:

Flow Rate bbl/day	100		50	
Pressure, psia	**1300**	**900**	**1300**	**900**
ΔP_G psi/100 ft	7.9×10^{-4}	2.0×10^{-3}	2.4×10^{-4}	7.7×10^{-4}
ΔP_1 psi/100 ft	$2. \times 10^{-4}$	2.7×10^{-2}	1.3×10^{-2}	1.3×10^{-2}
$\Delta X = \sqrt{\frac{\Delta P_L}{\Delta P_G}}$	5.9	3.8	7.4	4.2
From Figure 2	2.0	2.4	1.9	2.3
Pressure drop $\Delta P_{TP} = \phi_L \times \Delta P_L$	0.12	0.16	0.048	0.069

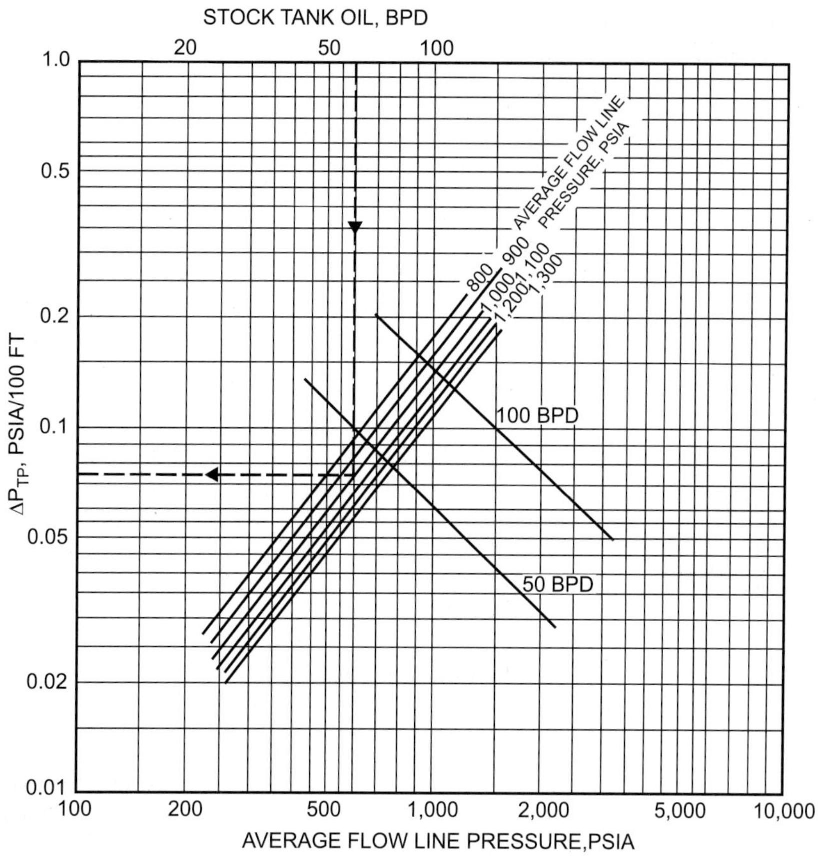

Figure 3. Determining two-phase flow pressure drops.

With two pressure drops determined for two flow line pressures for each of the flow rates, lines can be drawn showing pressure drop for different flow line pressures for these two rates (Figure 3). From these two lines relationships can be established for pressure drops for other rates and flow line pressures. The procedure is as follows:

Start with, for instance, 800 psia line. For the rate of 50 bpd at 800 psia flow line pressure, the pressure drop from the chart is 0.074 psia/100 ft. From the 50 bpd rate on the scale in the upper part of the chart, go downward to the line representing 0.074 psia pressure drop. This gives one point of the 800 psia line.

For the rate of 100 bpd the chart indicates that at 800 psia flowline pressure, the pressure drop is 0.175 psia/100 ft. From the 100 bpd rate in the upper part of the chart, go downward to the 0.175 psia/100 ft line. This gives the second point of the 800 psia line. The line can now be drawn through these two points. Lines for other pressures can be constructed in a similar way.

The chart is used as follows: For conditions represented by this particular chart, what is the two-phase pressure drop for the rate of 60 bpd and average flow line pressure of 1000 psia?

Start at the rate of 60 bpd on the upper scale of the chart. Go downward to the intersection with the 1000 psia pressure, then to the left, and read the two-phase pressure drop as 0.075 psia/100 ft.

Example 2. If pipe diameter is to be determined for the desired pressure drop and the known length of line, flow rate, and the properties of fluids, then the problem cannot be solved directly. The procedure is as follows:

An estimate is made of the diameter of the two-phase line. The best approach is to determine, for the conditions, the diameter of the line for each of the phases flowing separately. In a majority of cases, the sum of these two diameters will be a good estimate of the diameter for the two-phase line. Calculations are then made for the two-phase pressure drop using the assumed diameter. If this drop is reasonably close to the desired one, the estimate is correct. If not, a new estimate must be made and the calculations repeated.

Inclined two-phase flow

In hilly terrain additional pressure drop can be expected in case of the two-phase flow. Little published information is available on the subject. Ovid Baker suggested the following empirical formula[10]:

$$\Delta P_{TPH} = \Delta P_{TP} + \frac{1.61NH\rho_L}{144V_g^{0.7}}$$

where:

ΔP_{TPH} = Pressure drop, corrected for hilly terrain in psi
ΔP_{TP} = Calculated two-phase pressure drop for horizontal pipe, psi
N = Number of hills
H = Average height of hills, ft
ρ_L = Density of liquid phase, lb/ft^3, and
V_g = Velocity of gas, ft/s (based on 100% pipe area).

No test data are available to determine the accuracy of this approach. Until more information is developed on the subject, the method may be considered for approximate estimates.

Horizontal three-phase flow

Very little is known about the effect on the pressure drop of addition to the gas–oil system of a third phase, an immiscible liquid, and water. Formation of emulsion results in increased viscosities. Formulas are available for approximate determination of viscosity of emulsion, if viscosity of oil and percent of water content of oil is known. The point is, however, that it is not known what portion of the water is flowing in emulsified form.

From one published reference[11] and from some unpublished data, the following conclusions appear to be well founded:

1. Within the range of water content of less than 10% or more than 90%, the flow mechanism appears to approach that of the two-phase flow.
2. Within the range of water content of from 70% to 90%, the three-phase pressure drop is considerably higher than for the two-phase flow.

For solution of three-phase flow problems of oil field flowlines, the following approach has been used; the oil and water are considered as one phase and the gas as the other phase. Previously given calculation of a two-phase drop is used. The viscosity of the oil-water mixture is determined as follows:

$$\mu_L = \frac{V \times \mu_O + V_W \times \mu_W}{V_O + V_W}$$

where: μ_L, μ_O and μ_W are viscosities of mixture of oil and of water, respectively, and V_O and V_W represent the corresponding volumes of oil and water.

References

1. *Cameron Hydraulic Data*, 13th ed., Ingersoll-Rand Company, New York, 1962.
2. *Flow of Fluids Through Valves, Fittings and Pipe*, Technical Paper No. 410, Engineering Division, Crane Company, Chicago, 1957.
3. Streeter, Victor L., *Handbook of Fluid Dynamics*, McGraw-Hill Book Company, Inc., New York, 1961.
4. Prof. W. E. Durand's Discussions in *Handbook of the Petroleum Industry*, David T. Day; John Wiley & Sons, Inc., New York.
5. Baker, Ovid, "Simultaneous Flow of Oil and Gas," *The Oil and Gas Journal*, July 26, 1954.
6. Campbell, Dr. John M., "Elements of Field Processing," *The Oil and Gas Journal*, December 9, 1957.
7. Campbell, Dr. John M., "Problems of Multi-phase Pipe Line Flow," in "Flow Calculations in Pipelining," *The Oil and Gas Journal*, November 4, 1959.
8. Duckler, A. E., Wicks, III, Moye, and Cleveland, R. G., "Frictional Pressure Drop in Two-Phase Flow," *AIChE Journal*, January, 1964.
9. Lockhart, R. W., and Martinelli, R. C., "Proposed Correlation of Data for Isothermal Two-Phase, Two-Component Flow in Pipes," *Chemical Engineering Progress* 45: 39–48, (1949).
10. Ovid Baker's discussion of article "How Uphill and Downhill Flow Affect Pressure Drop in Two-Phase Pipe Lines," by Brigham, W. E., Holstein, E. D., and Huntington, R. L.; *The Oil and Gas Journal*, November 11, 1957.
11. Sobocinski, D. P., and Huntington, R. D., "Concurrent Flow of Air, Gas, Oil and Water in a Horizontal Line," *Trans. ASME* 80: 1,252, 1958.
12. Beal, Carlton, "The Viscosity of Air, Water, Natural Gas, Crude Oil and Its Associated Gases at Oil Field Temperatures and Pressures," *Petroleum Development and Technology*, AIME, 1946.
13. Switzer, F. G., *Pump Fax*, Gould Pumps, Inc.
14. *Pump Talk*, The Engineering Department, Dean Brothers Pumps, Inc., Indianapolis.
15. Switzer, F. G., *The Centrifugal Pump*, Goulds Pumps, Inc.
16. Zalis, A. A., "Don't Be Confused by Rotary Pump Performance Curves," *Hydrocarbon Processing and Petroleum Refiner*, Gulf Publishing Company, September, 1961.

Nomograph for calculating Reynolds number for compressible flow friction factor for clean steel and wrought iron pipe

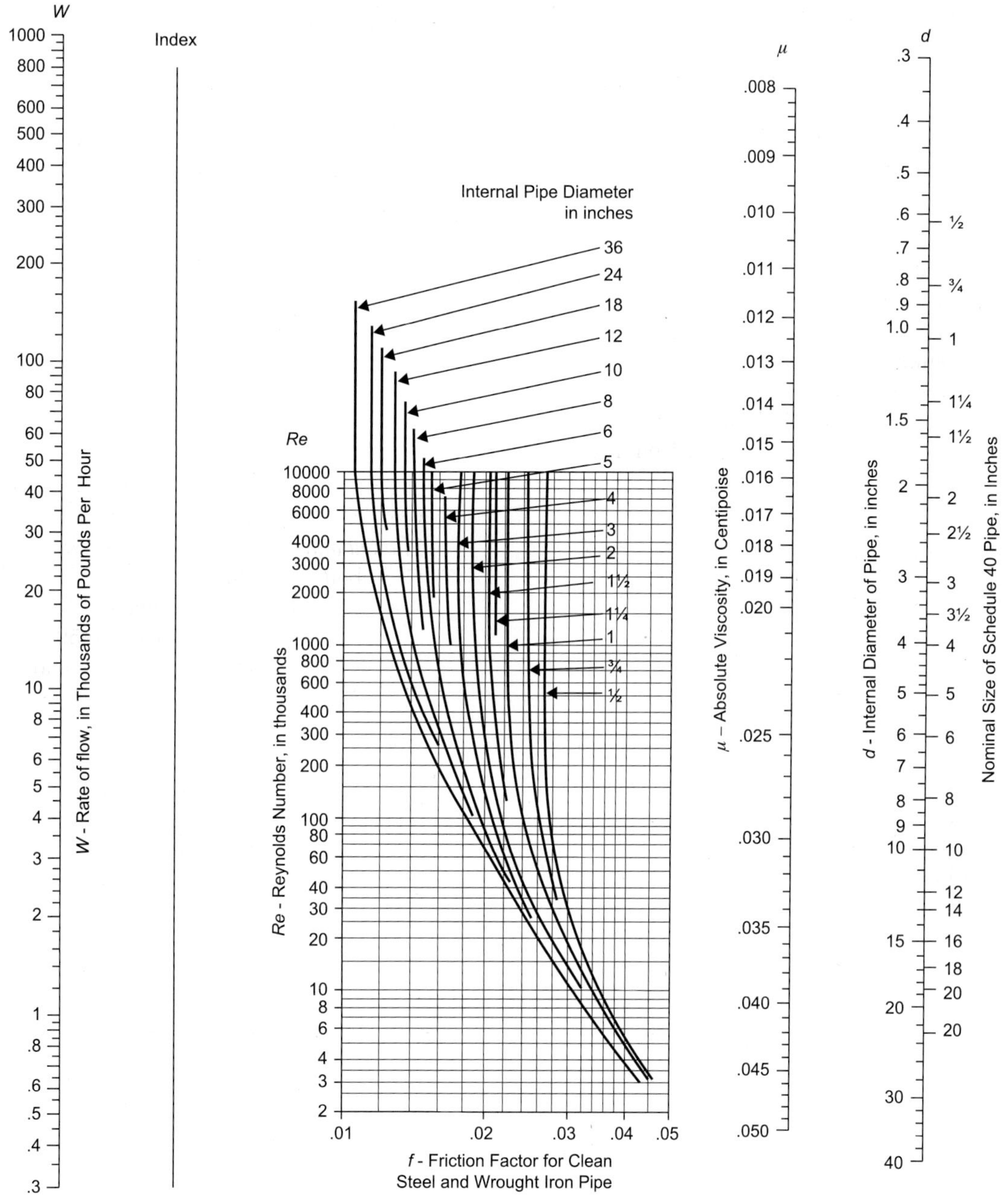

Figure 1. Reynolds number for compressible flow friction factor for clean steel and wrought iron pipe.

The nomograph in Figure 1 permits calculation of the Reynolds number for compressible flow and the corresponding friction factor. It is based on the equation:

$$\mathrm{Re} = 6.31\frac{W}{d\mu} = 0.482\frac{q_s \times S_g}{d\mu}$$

where:

W = Flow rate, lb/h

d = Internal diameter of pipes, inches

S_g = Specific gravity of gas relative to air

q_s = Volumetric flow rate, cf/s (at 14.7 psia and 60 °F), and

μ = Fluid viscosity, centipoise.

If the flow rate is given in lb/h, the nomograph can be used directly without resorting to Figure 2 to obtain q_s. Figure 2

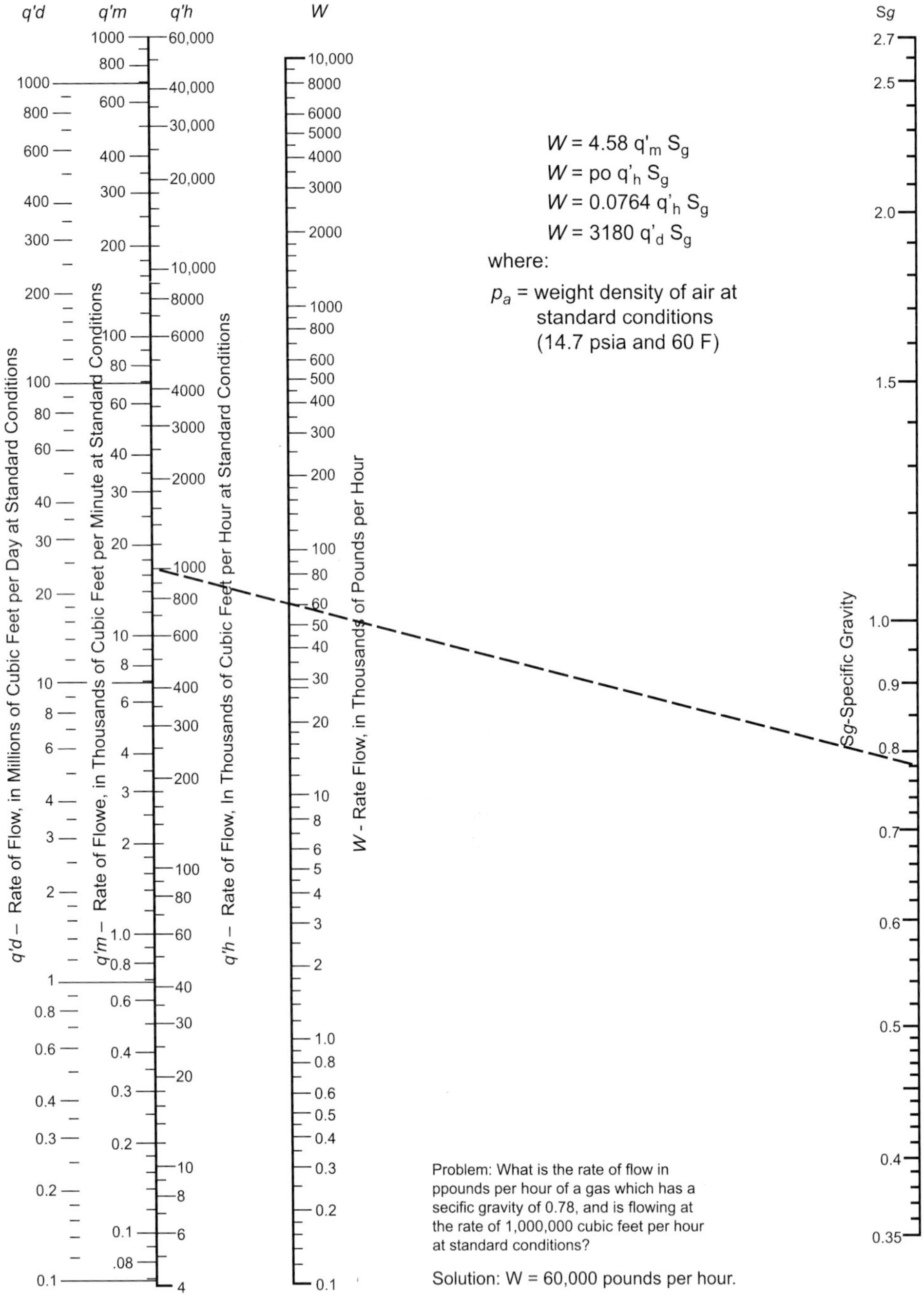

Figure 2. Equivalent volume and weight flow rates of compressible fluids.

converts volume flow to weight flow rates if the specific gravity of the fluid is known.

Example. Natural gas at 250 psig and 60 °F with a specific gravity of 0.75 flows through an 8-in schedule 40 clean steel pipe at a rate of 1,200,000 cf/h at standard conditions.

Find the Reynolds number and the friction factor.

At $S_g = 0.75$ obtain $q = 69{,}000$ from Figure 1, $\mu = 0.011$.

Connect	With	Mark or Read
W = 69,000	$\mu = 0.011$	Index
Index	$d = 8$ in	Re = 5,000,000
Re = 5,000,000	Horizontally to 8 in	$f = 0.014$

Source

Flow of Fluids Through Valves, Fittings, and Pipes, Technical Paper No. 410, 3–19, Crane Company, Chicago, (1957).

12: Liquids—General

Determining the viscosity of crude 462
Liquid gravity and density conversion chart 463
Nomograph for calculating viscosities of liquid hydrocarbons at high pressure 464
Calculate viscosity of a blend 465
Viscosity of a mixture of liquids using blending index 465
Calculate specific gravity of a blend 466
Convert viscosity units 466
Convert specific gravity to API gravity 466
Viscosities of hydrocarbon liquids 467
Calculate bulk modulus 469
Nomograph for calculating viscosity of slurries 469
Nomograph for calculating velocity of liquids in pipes 471
Nomograph for calculating velocity of compressible fluids in pipes. 471
Nomograph for calculating velocity of liquids in pipes 472
Table gives velocity heads for various pipe diameters and different rates of discharge 474
Derivation of basic ultrasonic flow equations 475
How fast does oil move in a pipeline? 477
Estimate the volume of a pipeline per linear foot using the inside diameter 477
What is the linefill of a given pipe in barrels per mile? 477
Estimate leakage amount through small holes in a pipeline 478
Carbon capture and sequestration processes 478
CO_2 pipeline infrastructure in the United States 479
Effect of variability on CO_2 pipeline operation 486

Determining the viscosity of crude

If the viscosity of a gas-saturated crude oil at the saturation (bubble-point) pressure is known, using this nomograph you can quickly estimate the viscosities at higher pressures.

Example. Find the viscosity at 4200 psia for a crude oil when its viscosity is 30 cp at the saturation pressure of 1200 psia. Notice that 4200 psia is 3000 psi above the saturation pressure. Connecting 3000 on the pressure difference scale (left) with 30 on the curved scale for viscosity at the bubble-point pressure, the intersection with the scale on the right at 48 cp is the desired value.

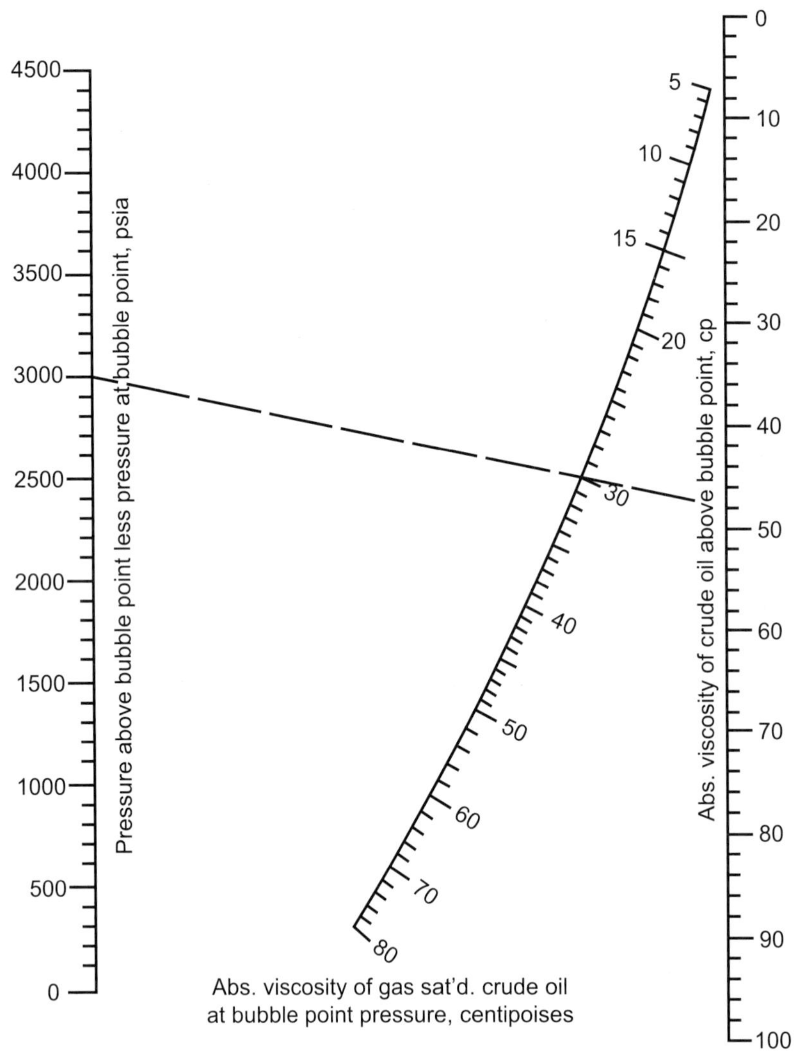

Liquid gravity and density conversion chart

This line chart provides an easy method for converting units of liquid gravity and density. Draw a horizontal line perpendicular to the scale line through a known value and read the equivalent value on any other scale.

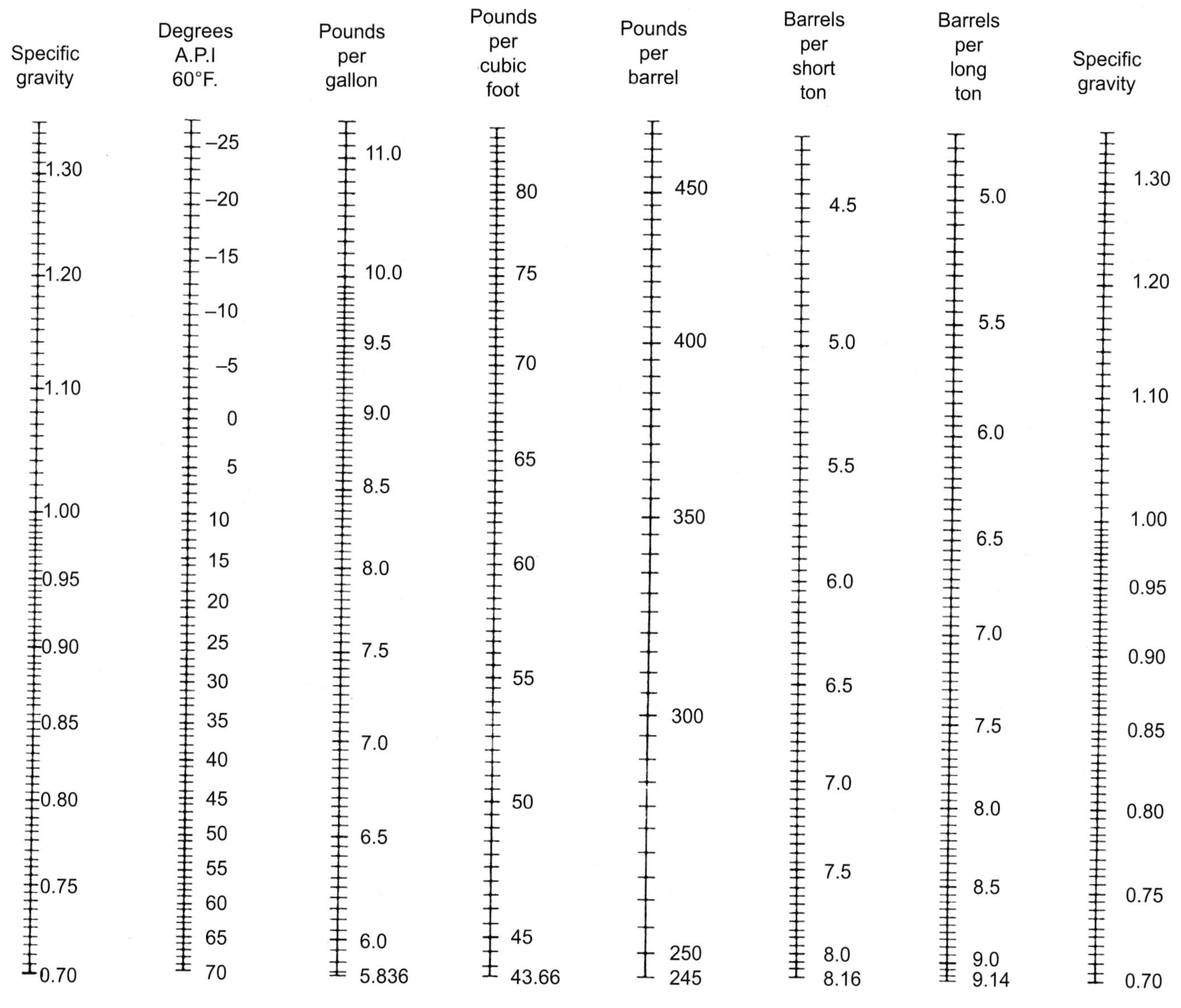

Liquid gravity and density

Nomograph for calculating viscosities of liquid hydrocarbons at high pressure

Lockhart and Lenoir developed a graphical correlation showing the effect of pressure on viscosity of liquid hydrocarbons. This correlation is shown in Figure 1, which is based primarily on data of Hersey and Hopkins, including pure hydrocarbons, lubricating oils, bright stocks, and distillates.

To use the nomograph, the characterization factor of Watson, K_W, and the viscosity of the liquid at atmospheric pressure are required.

The accuracy of the correlation decreases as pressure increases.

Example. What is the viscosity of an oil at 5400 psia, if its characterization factor is 11.8 and its viscosity at atmospheric conditions is 90 cP?

Enter 5400 psia in the pressure scale to the viscosity line of 90 and proceed horizontally to the middle reference scale. Follow the curve lines to intersect the vertical line drawn at $K_W = 11.8$ and read the ratio of viscosity on the extreme left scale at 2.6. The viscosity of the oil is (2.6) (90) = 234 cP.

Source

Lockhart F.J., Lenoir J.M., *Petroleum Refiner* 40 (No. 3) (1961) 209.

Hersey M.D., Hopkins R.F., *Viscosity of Lubricants under Pressure*. ASME, 1954.

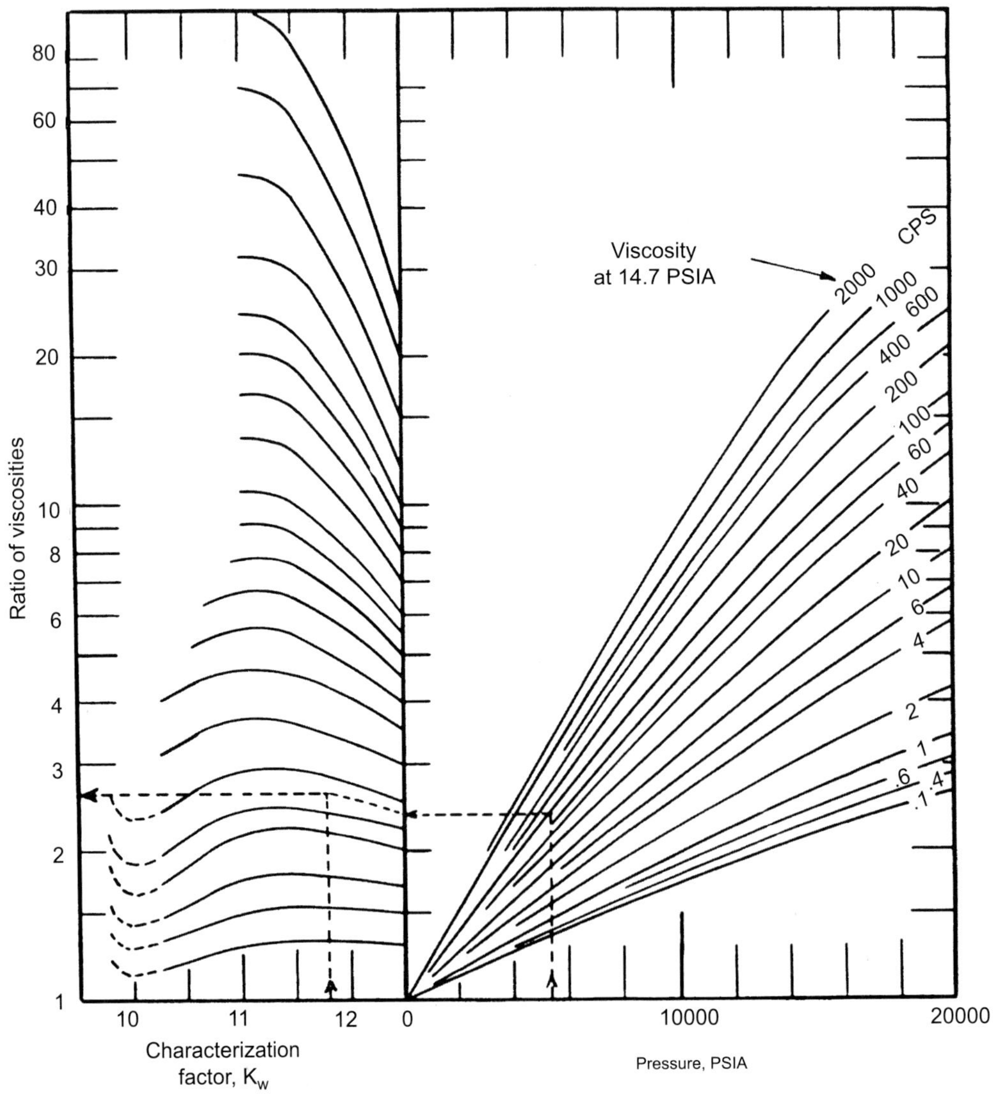

Figure 1. Isothermal effect of pressure on viscosities of liquid hydrocarbons at low reduced temperatures. Reproduced by permission *Petroleum Refiner,* Vol. 40, No. 3 page 209, copyright 1961, Gulf Publishing Co., Houston.

Calculate viscosity of a blend

Viscosity is not an additive function of percentage composition. The viscosity of a blend can be calculated with the following equation:

$$V_b = \left[\frac{G_1 + G_2 + G_3}{\frac{G_1}{\sqrt{V_1}} + \frac{G_2}{\sqrt{V_2}} + \frac{G_3}{\sqrt{V_3}}}\right]^2$$

where

V_b = Viscosity of blend,
G = Volume (gals, barrels, etc.),
V = Viscosity in SUS.

Example.

$G_1 = 10{,}000$ bbls $\quad V_1 = 45$ SUS
$G_2 = 20{,}000$ bbls $\quad V_2 = 50$ SUS
$G_3 = 5000$ bbls $\quad V_3 = 37$ SUS

$$V_b = \left[\frac{10000 + 20000 + 5000}{\frac{10000}{\sqrt{45}} + \frac{20000}{\sqrt{50}} + \frac{5000}{\sqrt{37}}}\right]^2,$$

$V_b = 46.4$ SUS.

Viscosity of a mixture of liquids using blending index

A blending index is frequently used to estimate the viscosity of a mixture. In this method, an index is assigned to each liquid based on its viscosity. Next, the blending index of the mixture is calculated from the individual blending indices by using the weighted average of the composition of the mixture. Finally, the viscosity of the blended mixture is calculated using the blending index of the mixture. The equations used are as follows:

$$H = 40.73 - 46.414 \log_{10} \log_{10}(V + B) \quad (1)$$

$$B = 0.931(1.72)^V \text{ for } 0.2 < V < 1.5 \quad (2)$$

$$B = 0.6 \text{ for } V \geq 1.5 \quad (3)$$

$$H_m = [H_1(\text{pct}_1) + H_2(\text{pct}_2) + \ldots]/100 \quad (4)$$

where:

$H, H_1, H_2, \ldots$ – Blending index of liquids
H_m – Blending index of mixture
B – Constant in blending index equation
V – Viscosity in cSt
$\text{pct}_1, \text{pct}_2, \ldots$ – Percentage of liquids 1, 2, … in blended mixture.

Example. Three hydrocarbons of viscosity 0.4 cSt, 1.4 cSt and 1.8 cSt are blended in a 20:30:50 ratio. Compute the viscosity of the blend using blending indices.

Solution. First, compute the blending index of each liquid using equations (2) and (3) to obtain $B_1 = 1.16$, $B_2 = 1.99$, $B_3 = 0.6$. Then, compute the H-values for each liquid using equation (1): $H_1 = 73.94$, $H_2 = 53.51$, $H_3 = 60.22$. Finally, compute H of the mixture using equation (4): 0.61 cSt.

V (cSt)	B	H	%	%H
0.4	1.158	73.94	0.2	14.78834
1.4	1.991	53.51	0.3	16.05393
1.8	0.6	60.22	0.5	30.11136
				0.61 cSt

Source

Menon E.S., Pipeline Planning and Construction Field Manual. Gulf Professional Publishing, Waltham, MA, 2011.

Calculate specific gravity of a blend

The specific gravity of a blend may be directly calculated by a ratio of the specific gravity of each component in the blend.

Example. Assume a blend consisting of the following parameters:

Partial specific gravity $= (10{,}000/45{,}000) \times 0.8551$
$= 0.1900.$

Specific gravity = sum of partial blends.

Volume, bbls	Specific Gravity	Partial Specific Gravity
10,000	0.8551	0.1900
15,000	0.8337	0.2779
20,000	0.8481	0.3769
Total 45,000		**0.8448**

Convert viscosity units

Convert SUS to centistokes:

$V = 0.2260\ V_{SUS} - 135/V_{SUS}$ where $V_{SUS} > 100$

$V = 0.2260\ V_{SUS} - 195/V_{SUS}$ where $V_{SUS} \leq 100$

Convert centistokes to centipoises (cP)
$cP = V_{CS} \times$ density, gm/cm^3

$SUS = cS \times 4.6673$, if $cS > 500$, $SUS = cS \times 4.6347$

$cSt = 2.24(SSF - 184/SSF)$ for $25 < SSF < 40$

$cSt = 2.16(SSF - 60/SSF)$ for $SSF > 40$.

Convert specific gravity to API gravity

$$API_{60} = \frac{141.5}{S_{60}} - 131.5$$

where S_{60} = specific gravity @ 60 °F

The reverse calculation, from API gravity to specific gravity, follows from:

$$S_{60} = \frac{141.5}{API_{60} + 131.5}$$

Viscosities of hydrocarbon liquids

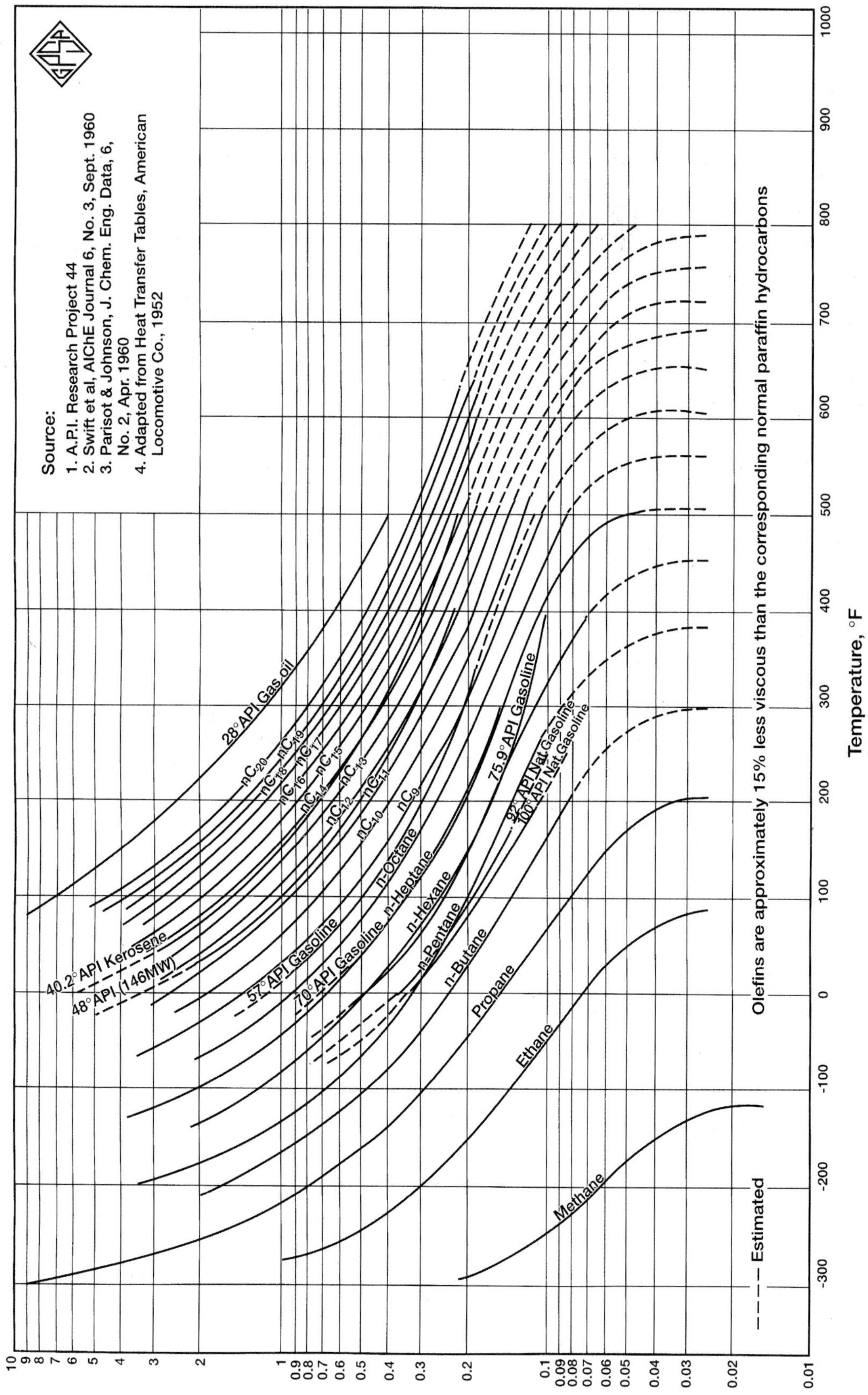

Reprinted with Permission—Gas Prrocessors Suppliers Association, *Tenth Edition Engineering Data Book.*

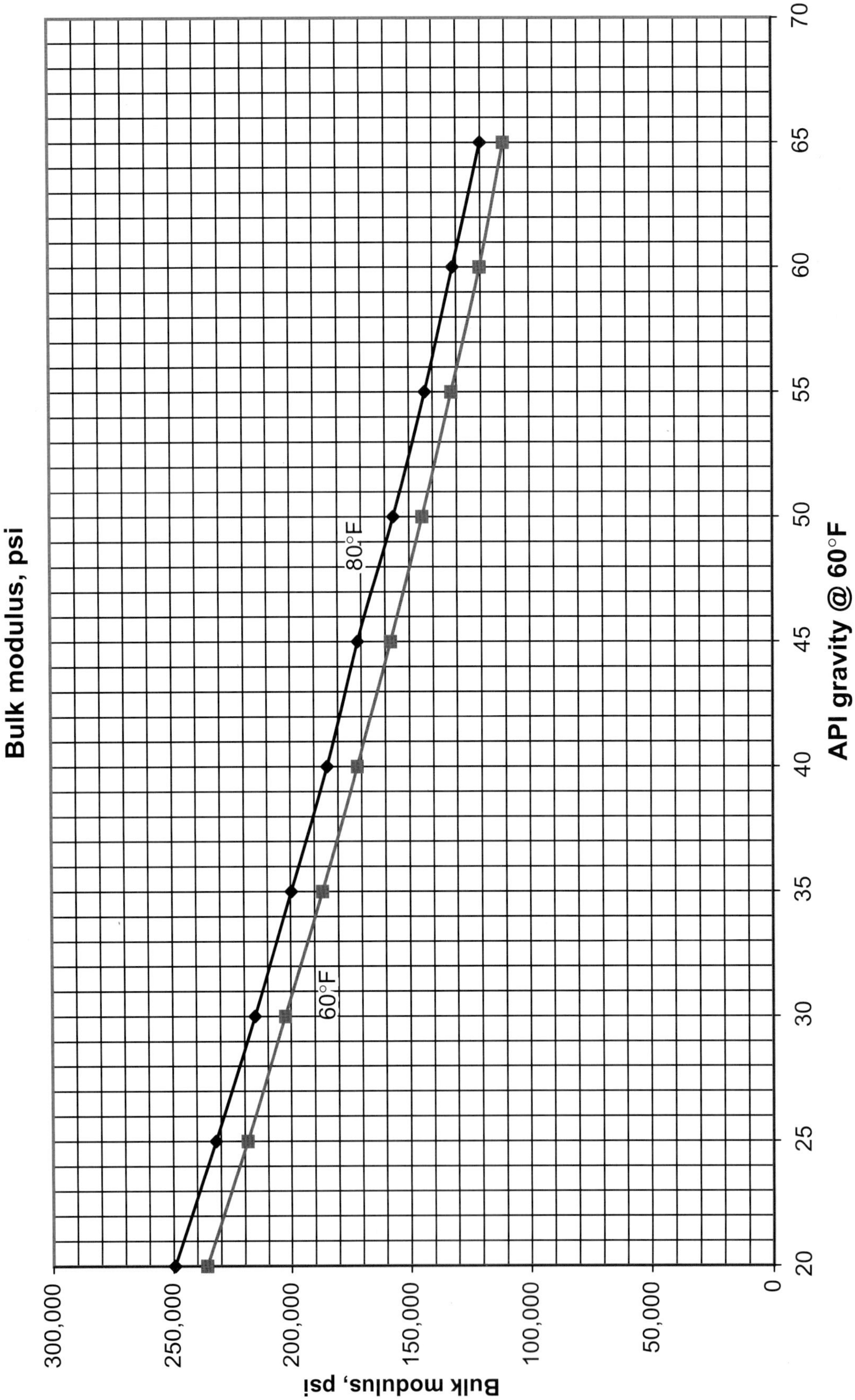

Based on Chapter 11.2.1, API MPMS, Compressibility Factors for Hydrocarbons, May 21, 1992.

Calculate bulk modulus

The bulk modulus, K, of a liquid is the reciprocal of its compressibility. The bulk modulus for water is $K = 300{,}000$ psi. The following relation, known as the ARCO formula, can be used to calculate the bulk modulus for crude oil.

$$K = 1.286(10^6) + 13.55P - 4.122(10^4)T^{1/2} - 4.53(10^3)\mathrm{API} - 10.59\mathrm{API}^2 + 3.228T(\mathrm{API})$$

where

P = Average pressure.
T = Temperature (°F)
API = API gravity at 60 °F

Source

Fluid Flow Consultants, Inc., Grove, Oklahoma

Nomograph for calculating viscosity of slurries

This nomogram is based on the Hatschek equation for estimating the viscosity of slurries in an aqueous suspension.

$$\mu = \frac{\mu_w}{1 - x^{0.333}},$$

where

μ = Viscosity of slurry, centipoises,
μ_w = Viscosity of water at the temperature of the slurry, centipoises,
x = Volume fraction of dry solids in the slurry.

The nomogram (Figure 1) shows a scale calibrated in terms of the temperature of the water instead of its viscosity. Thus, the actual water viscosity is not needed.

Example. What is the viscosity of a slurry having a volume fraction of solids of 0.06 at a temperature of 39 °C?

Connect $T = 39$ °C with $x = 0.06$ and read $\mu_w = 1.1$ cP on the middle scale.

Source

Davis, D. S., *Brit. Chem. Eng.*, 4, 9, 478 (1959).

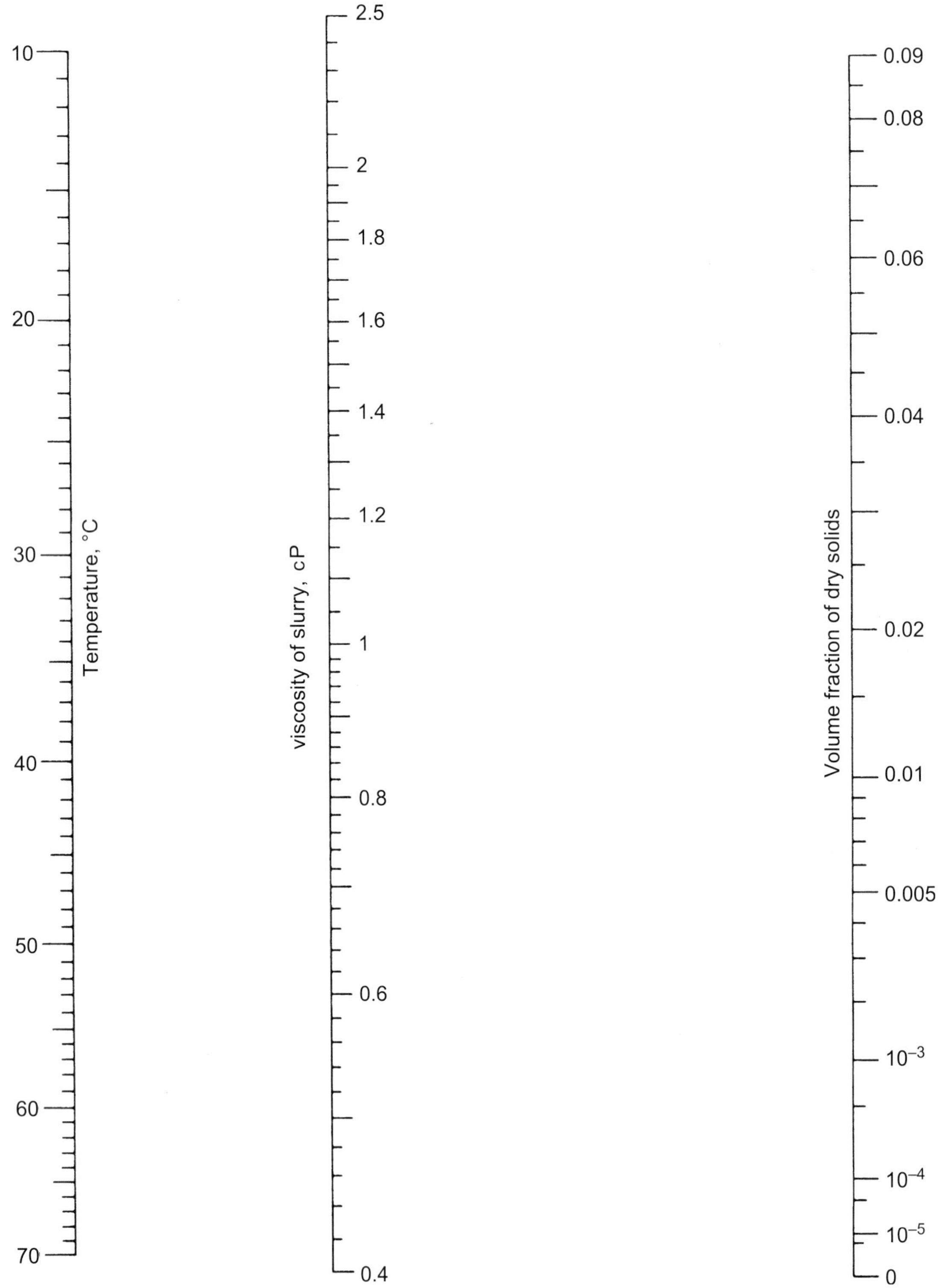

Figure 1. Viscosity of slurries. (Reproduced by permission, *British Chemical Engineering,* vol. 4, page 478, copyright 1959).

Nomograph for calculating velocity of liquids in pipes

The mean velocity of any liquid flowing in a pipe of circular cross section can be calculated from the following formula:

$$v = 183.3\frac{q}{d^2} = 0.408\frac{Q}{d^2} = 0.0509\frac{W}{d^2p}$$

where

v = Average fluid velocity, feet per second,
d = Inside diameter of pipe, inches,
q = Rate of flow, cubic feet per second,
Q = Rate of flow, gallons per minute,
W = Rate of flow, pounds per hour,
p = Fluid density, pounds per cubic foot.

The nomograph in Fig. 1 for calculating velocity of liquids in pipes can be used to calculate the liquid velocity when the rate of flow is in cubic feet, gallons, or thousands of lb. Conversely, knowing the flow rate and velocity, the pipe diameter may be calculated.

Example. What is the velocity of fuel oil at 60 °F flowing through a 2-in schedule 40 pipe at a rate of 45,000 lb/h. The oil density is 56.02 lb/gal.

Use Figure 1

Connect	With	Mark or Read
$W = 45$	$p = 56.02$	q = 100 cu ft/s
q = 100	d = 2-in schedule 40	v = 10 ft/s

Source

Flow of Fluids through Valves, Fittings, and Pipe, Technical Paper No. 410, 3–7, Crane Company, Chicago, (1957).

Nomograph for calculating velocity of compressible fluids in pipes

The mean velocity of a compressible fluid in a pipe can be computed by the following formula, which is obtained after dividing the rate of flow in appropriate units by the cross-sectional area of the pipe:

$$V = \frac{3.06W\overline{V}}{d^2} = \frac{3.06W}{d^2p}$$

where

V = Mean velocity of flow, feet per minute,
W = Rate of flow, in pounds per hour,
$\overline{V}$ = Specific volume of fluid, cubic feet per pound,
d = Internal diameter of pipe, inches,
p = Density of fluid, pounds per cubic foot.

Example. Steam at 600 psig and 850 °F is to flow through a schedule 80 pipe at a rate of 30,000 lb/h. Find the pipe size if the velocity is to be limited to 8000 ft/min.

Use Figure 2

Connect	With	Mark or Read
850 °F	600 psig line (vertically)	600 psig
600 psig	Horizontally to	$\overline{V}$ = 1.21
$\overline{V}$=1.21	W = 30,000	Index
Index	V = 8000	d = 3.7, use 4-in schedule 80 pipe

If a 4-in schedule 80 pipe is used, the actual velocity is found by connecting the Index with 4-in schedule 80 to get V = 7600 ft/min.

Source

Flow of Fluids through Valves, Fittings, and Pipes, Technical Paper No. 410, 3–16, Crane Company, Chicago, (1957).

Nomograph for calculating velocity of liquids in pipes

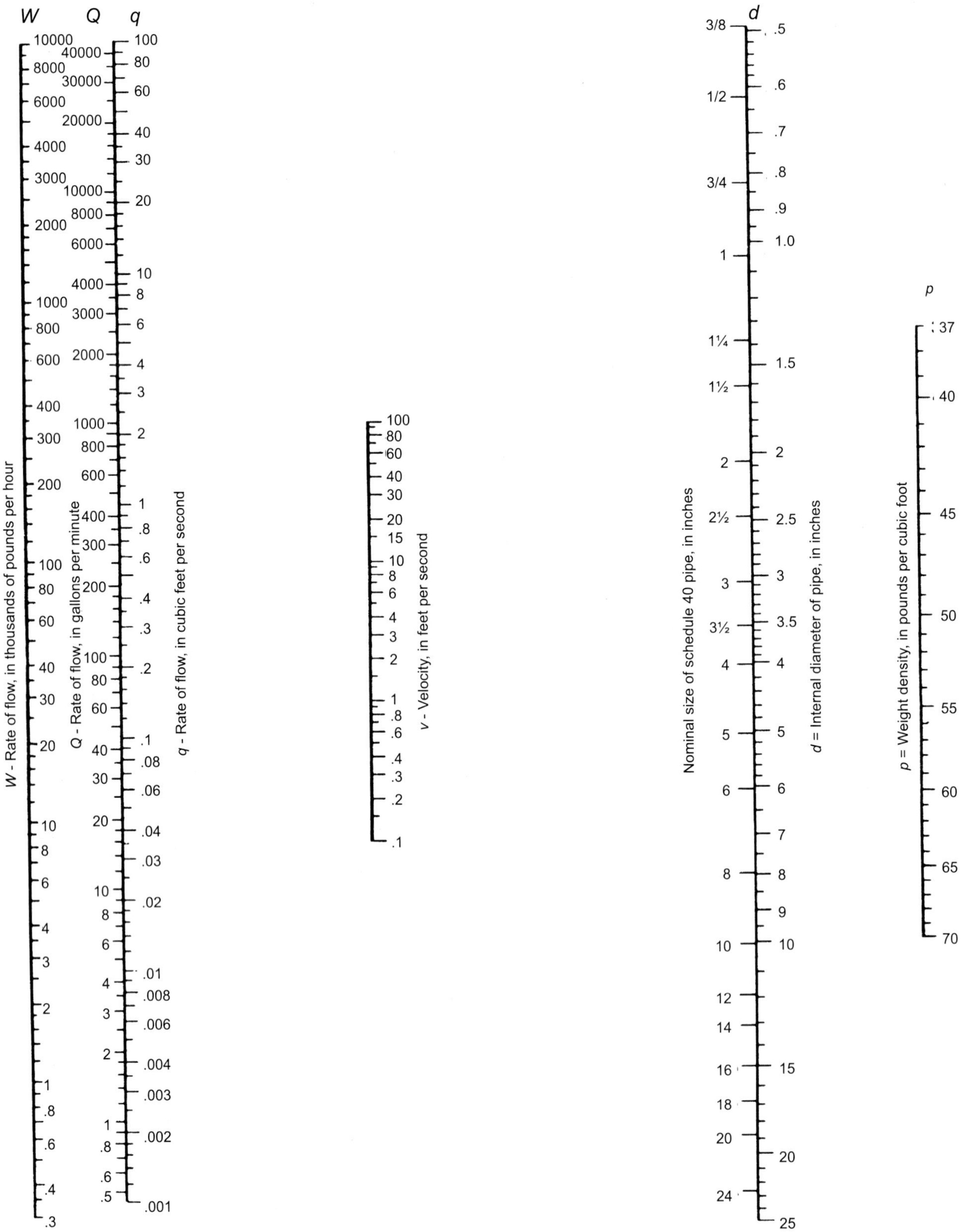

Figure 1. Velocity of liquids in pipe.

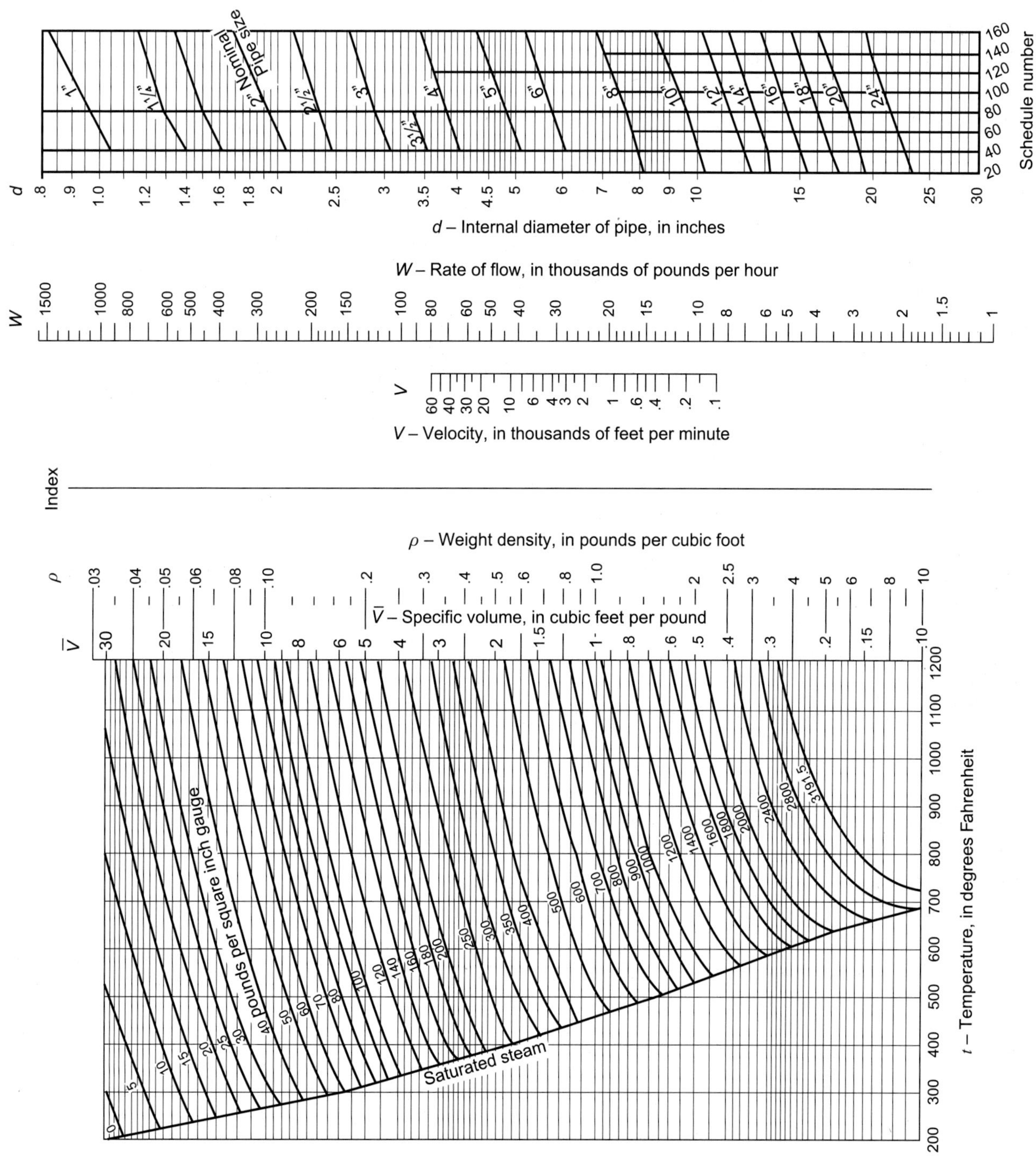

Figure 2. Velocity of compressible fluids in pipes.

Table gives velocity heads for various pipe diameters and different rates of discharge

Rate of Flow in Gallons per Minute	NOMINAL DIAMETER OF PIPE (Velocity Heads in Feet)														Rate of Flow in Gallons per Minute
	⅛"	¼"	⅜"	½"	¾"	1"	1¼"	1½"	2"	2½"	3"	4"	5"	6"	
1	.488	.148	.043	.017											1
2		.590	.175	.069	.022										2
3			.392	.155	.050	.020									3
4			.698	.276	.090	.035	.012	.006							4
5			1.089	.430	.141	.054	.018	.010							5
10				1.720	.564	.215	.071	.038	.016	.007	.003				10
15					1.265	.488	.160	.087	.036	.015	.007				15
20					2.250	.861	.286	.154	.065	.027	.013				20
25						1.345	.447	.241	.101	.041	.020				25
30						1.933	.643	.346	.146	.060	.029				30
35						2.635	.877	.472	.198	.082	.039				35
40						3.343	1.145	.618	.259	.106	.052	.016			40
45							1.548	.779	.329	.134	.065	.021			45
50							1.787	.963	.406	.166	.080	.026			50
70							3.503	1.888	.795	.326	.157	.050	.020		70
75								2.165	.912	.373	.180	.057	.023		75
100								3.851	1.621	.665	.320	.101	.041	.020	100
120								5.548	2.333	.956	.462	.146	.060	.029	120
125									2.528	1.035	.498	.158	.065	.031	125
150									3.649	1.493	.736	.229	.093	.045	150
175									4.954	2.021	.980	.309	.127	.061	175
200	8"								6.483	2.656	1.281	.406	.166	.080	200
225										3.359	1.618	.514	.209	.101	225
250	.040									4.131	1.992	.633	.259	.125	250
270	.047										2.333	.740	.302	.146	270
275	.048										2.429	.769	.313	.151	275
300	.057										2.883	.912	.373	.180	300
350	.078	10"										1.242	.509	.246	350
400	.102											1.621	.665	.320	400
450	.129	.053										2.053	.839	.406	450
470	.141	.057	12"									2.239	.917	.443	470
475	.144	.059										2.276	.934	.453	475
500	.159	.065	.031									2.535	1.038	.502	500
550	.193	.079	.038										1.256	.606	550
600	.229	.093	.045										1.493	.721	600
				14"											
650	.269	.109	.053	.026									1.752	.846	650
700	.312	.127	.062	.031									2.035	.980	700
750	.358	.146	.071	.035									2.337	1.123	750
800	.407	.166	.080	.040										1.281	800
850	.462	.187	.090	.045										1.548	850
					15"										
900	.516	.209	.102	.050										1.618	900
950	.574	.234	.113	.056	.043									1.813	950
1000	.637	.259	.125	.062	.048									1.999	1000
1050	.703	.296	.138	.069	.053									2.202	1050
1100	.771	.313	.152	.075	.058									2.421	1100
1150	.841	.343	.166	.082	.064	16"								2.648	1150
1200	.917	.373	.181	.090	.069									2.880	1200
1250	.995	.406	.196	.097	.075	.062								3.135	1250
1500	1.433	.582	.282	.145	.108	.088									1500
2000	2.547	1.035	.502	.249	.193	.158									2000
							20"	24"	30"						
2500		1.618	.784	.391	.301	.246									2500
3000		2.329	1.128	.562	.434	.355	.147								3000
3500		3.224	1.540	.766	.590	.502	.199	.096							3500
4000			2.007	1.000	.771	.633	.261	.125	.051						4000
4200			2.209	1.102	.851	.698	.288	.137	.056						4200
4500			2.539	1.265	.975	.799	.330	.158	.065						4500
5000			3.135	1.561	1.204	.985	.407	.195	.079	36"					5000
5500				1.884	1.460	1.196	.493	.227	096						5500
6000				2.239	1.733	1.421	.586	.281	.115	.056					6000
6500				2.627	2.035	1.669	.690	.329	.134	.065					6500
7000				3.056	2.360	1.936	.797	.384	.156	.076					7000
7200					2.709	2.056	.846	.404	.165	.080					7200
7500					3.047	2.224	.917	.440	.180	.087					7500
8000							1.045	.498	.204	.099					8000
8500							1.177	.564	.230	.112					8500
9000							1.319	.633	.259	.125	42"				9000
9500							1.469	.702	.289	.140					9500
10000							1.630	.779	.319	.154	.083				10000
10500								.858	.352	.169	.092				10500
11000								.943	.387	.186	.101				11000
11500								1.033	.423	.204	.110				11500
12000								1.123	.460	.222	.120				12000
12500								1.218	.500	.240	.130				12500
13000								1.316	.541	.260	.141				13000
13500								1.421	.582	.281	.152				13500

F. G. Switzer "Pump-Fax," Goulds Pumps, Inc.

Derivation of basic ultrasonic flow equations

The actual measurements performed by ultrasonic flowmeters are of flow velocity, with all other variable effects due to temperature, pressure, density, viscosity, etc., being cancelled via a differential sensing technique.

Multiplication by the cross-sectional pipe area readily enables the output to be conditioned to read in volumetric flow rate scaled to engineering units. Refinement of the technique enables density and other physical properties to be measured as well so that mass flow rates and other parameters pertaining to various process variables can be determined.

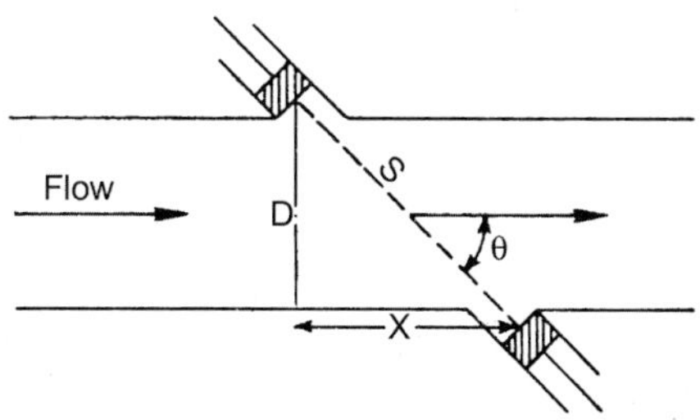

Let

$D =$ Pipe diameter,
$X =$ Axial displacement of transducers,
$S =$ Path length between transducers,
$\theta =$ Angle of transducers to flow stream,
$c =$ Velocity of sound in the flowing medium,
$v =$ Velocity of flowing medium,
$t_1 =$ Flight time of upstream pulse,
$t_2 =$ Flight time of downstream pulse,

then

$$t_2 = \frac{S}{c + v \cos \theta} \tag{1}$$

$$t_1 = \frac{S}{c - v \cos \theta} \tag{2}$$

To cancel the effects of variations in the velocity of sound c in the medium itself, the circuitry is arranged to compute the reciprocal of each flight time and to subtract one from the other. This difference of the flight time reciprocals leads to the relationship:

$$\frac{1}{t_2} - \frac{1}{t_1} = \frac{2v \cos \theta}{S} \tag{3}$$

Solving for v

$$v = \frac{S}{2 \cos \theta}\left(\frac{1}{t_2} - \frac{1}{t_1}\right) \tag{4}$$

$$\text{but } \cos \theta = \frac{X}{S} \tag{5}$$

$$\text{Therefore, } v = \frac{S^2}{2X}\left(\frac{1}{t_2} - \frac{1}{t_1}\right) \tag{6}$$

The circuitry that performs the reciprocal time computation produces an output frequency, since:

$$\frac{1}{t} = f \tag{7}$$

$$\text{Therefore, } \frac{1}{t_2} = f_2 \tag{8}$$

$$\text{and } \frac{1}{t_1} = f_1 \tag{9}$$

$$\text{and } \frac{1}{t_2} - \frac{1}{t_1} = f_2 - f_1 = \Delta f \tag{10}$$

Substituting in Equation 6 we have the basic velocity equation:

$$v = \frac{S^2}{2X}\Delta f \tag{11}$$

or

$$v = k_1 \Delta f \tag{11a}$$

$$\text{The relationship } k_1 = \frac{S^2}{2X}\Delta f \tag{12}$$

is the flow velocity scaling factor and is usually expressed in ft per second or ft per minute. Note that it is a function of the square of the distance between the transducers. This is an important consideration when providing for re-zeroing of the meter calibration each time the transducers are removed and re-inserted.

For volumetric flow, the velocity must be multiplied by the cross-sectional area of the pipe, and the following considerations are involved:

$$A = \pi\left(\frac{D}{2}\right)^2 \tag{13}$$

Volumetric flow

$$V - vA \tag{14}$$

Therefore, volumetric flow

$$V\pi = \frac{S^2 D^2 \Delta f}{8X} \tag{15}$$

The relationship

$$k_2 = \frac{S^2D^2}{8X}\pi \tag{16}$$

is the volumetric flow scaling factor and is usually expressed in cubic ft per minute.

Further information on the physical properties of the flowing medium is also acquired in the process of the basic measurement. Specifically, the velocity of sound in the medium itself can be measured independent of the velocity of its motion.

Equation 3 computed the difference of the reciprocals of the upstream and downstream flight times. In the instrument, this is performed in the delta computing channel. Supposing an addition of the reciprocals of the respective upstream and downstream flight times is performed in a separate computing channel, called the sigma channel. The following relationships ensue:

$$\frac{1}{t_2}+\frac{1}{t_1} = \frac{c+v\cos\theta}{S}+\frac{c-v\cos\theta}{S} = \frac{2c}{S} \tag{17}$$

$$\text{then } c = \frac{S}{2}\sum f \tag{18}$$

$$\text{Let } k_3\frac{S}{2} \tag{19}$$

be the sound velocity scaling factor.

$$\text{Then } c = k_3\sum f \tag{20}$$

The speed of sound c in a medium is a function of its bulk modulus and its density and is expressed by the following equation:

$$c = \sqrt{\frac{B}{\rho}} \tag{21}$$

where

B = Bulk modulus of elasticity,
ρ = Density.

From this, the equation for density may be derived thus

$$c^2 = \frac{B}{\rho} \tag{22}$$

$$\text{and } \rho = \frac{B}{c^2} \tag{23}$$

Substituting the value of c from Equation 20 into 23 gives

$$\rho = \frac{B}{(k_3\sum f)^2} \tag{24}$$

For a given gas, B is essentially constant.

$$\text{Therefore, } \frac{B}{k_3 2} = k_4 \tag{24a}$$

is a fair representation.

$$\text{Then } \rho(\text{density}) = k_4\sum f^{-2} \tag{25}$$

An interesting relationship may be derived at this point, namely, an equation for mass flow. Since

$$\text{massflow} = \rho V \tag{26}$$

$$\text{and since } V = k_1\Delta f \tag{11a}$$

$$\text{and } \rho = \frac{k_4}{\sum f^2}$$

$$\text{then mass flow} = k_1k_4\frac{\Delta f}{\sum f^2} \tag{27}$$

or, letting $k_1k_4 = k_5$

$$\text{mass flow} = k_5\frac{\Delta f}{\sum f^2} \tag{28}$$

Further, the density of a gas is related by the equation

$$\rho = \frac{P}{T} \tag{29}$$

where

P = Pressure,
T = Temperature (°K).

Therefore, if the pressure is known, the temperature may be extracted from the sigma channel in this way:

$$\text{From Equation 25, } \frac{P}{T} = k_4\sum f^{-2} \tag{30}$$

$$\text{and } T = k_4P^{-1}\sum f^{-2} \tag{31}$$

Let us go back to Equation 24. If the chemical properties of the gas or the ratio of the mixture between two or more gases change, then B is not a constant. If separate transducers are used to measure density, say via the pressure/temperature relationship, then the variation of B can also be measured.

$$\text{Thus, } \rho = B\left(k_3\sum f\right)^{-2} \tag{32}$$

$$\text{and } B = \rho\left(k_3\sum f\right)^2 \tag{33}$$

The above relationship holds for liquids or gases. In the case of a gas, we may further say:

$$B = \frac{P}{T}\left(k_3\sum f\right)^2 \tag{34}$$

How fast does oil move in a pipeline?

Divide the throughput, in barrels per day, by the square of the nominal diameter, in inches; divide this result by 100 and subtract 1 from the answer. The answer is the speed of the oil column in miles per hour.

Example. A 12-in line has a throughput of 70,000 bpd,

$70{,}000 \div 144 = 486,$
$486 \div 100 = 4.86,$
$4.86 - 1.00 = 3.86.$

The oil moves about 4 miles per hour.

This rule is rough, but seldom is an exact answer needed for this problem. Greater accuracy can be had by using the true inside diameter of the line in the following formula:

$$V = 0.0081 \times Q/d^2$$

where

V = Speed, mph,
Q = Throughput, bpd,
d = Inside diameter of pipe, in.

Take the above example: If the wall thickness of the pipe is 3/8 in, then

$$V = (0.0081)(70,\ 000 \div 144)$$

$$V = 3.94$$

or about 4 miles per hour.

Example. A 20-in pipeline has a throughput of 200,000 barrels per day. How fast does the oil move inside the pipeline?

By the first method,

$200{,}000 \div 400 = 500,$
$500 \div 100 \div 1 = 4$ miles per hour.

By the second method,

$$V = (0081)\left(200{,}000 \div 19.25^2\right)$$

$V = 4.37$, or about 4 miles per hour.

Estimate the volume of a pipeline per linear foot using the inside diameter

To find the volume of a pipeline in gallons per ft, square the inside diameter in inches and multiply by 4%. A more accurate answer may be obtained by adding 2% of this result to the first answer.

Example. Find the volume of a pipeline having an ID of 7.981 in expressed in gallons per ft.

$$\text{Volume} = (0.04) \text{ x } (7.981)^2 = 2.548 \text{ gallons per ft.}$$

For greater accuracy, add 2% of 2.548.

$$2.548 + 0.051 = 2.599 \text{ gallon per ft.}$$

Actual volume = 2.599 gallons per ft.

To find the volume of the pipeline in barrels per ft, square the inside diameter in inches and divide by 1000. For greater accuracy, subtract 3% of the first answer.

Example. Find the volume of a pipeline with an ID of 7.981 in in barrels per ft.

$$\text{Volume} = \frac{(7.981)^2}{1000} = 0.0637 \text{ barrels per ft.}$$

For greater accuracy, subtract 3% of 0.0637,

$$0.0637 - 0.0019 = 0.0618 \text{ barrels per ft.}$$

Actual volume = 0.0619 barrels per ft.

What is the linefill of a given pipe in barrels per mile?

Multiply the inside diameter of the pipe, in inches, by itself, and multiply the result by 5.13; the answer is the linefill in barrels per mile.

Example. Six-in schedule 40 pipe; inside diameter is 6.065 in.

$$6.065 \times 6.065 \times 5.13 = 188.7 \text{ barrels per mile.}$$

If the correct inside diameter is used, the rule gives the correct linefill in standard 42-gallon barrels per mile of pipe.

Estimate leakage amount through small holes in a pipeline

The amount of fluid lost through a small hole in a pipeline can be estimated using the following equation:

$$Q = 0.61A\sqrt{2gh}$$

where

Q = Flow in cu ft/s,
A = Cross sectional area, sq ft,
g = Gravitational constant, ft/s/s,
h = Head, feet.

Example. Assume the following conditions:

Hole diameter = 0.125 in,
Pressure = 30 psig,
Sp. gr. of fluid = 0.85,
$A = 0.0000852$ sq ft,
$h = 81.5$ ft,
$g = 32.2$,

$$Q = 0.61 \times 0.0000852\sqrt{2 \times 32.2 \times 81.5},$$

$$Q = 0.003766 \text{ cu ft/s}.$$

Carbon capture and sequestration processes

Carbon capture and sequestration is essentially a three-part process involving a CO_2 source facility, a long-term CO_2 storage site, and an intermediate mode of CO_2 transportation.

Capture

The first step in direct sequestration is to produce a concentrated stream of CO_2 for transport and storage. Currently, three main approaches are available to capture CO_2 from large-scale industrial facilities or power plants:

Pre-combustion, which separates CO_2 from fuels by combining them with air and/or steam to produce hydrogen for combustion and CO_2 for storage,

Post-combustion, which extracts CO_2 from flue gases following combustion of fossil fuels or biomass, and

Oxyfuel combustion, which uses oxygen instead of air for combustion, producing flue gases that consist mostly of CO_2 and water from which the CO_2 is separated.

These approaches vary in terms of process technology and maturity, but all yield a stream of extracted CO_2 which may then be compressed to increase its density and make it easier (and cheaper) to transport. Although technologies to separate and compress CO_2 are commercially available, they have not been applied to large-scale CO_2 capture from power plants for the purpose of long-term storage.

Transportation

Pipelines are the most common method for transporting large quantities of CO_2 over long distances. CO_2 pipelines are operated at ambient temperature and high pressure, with primary compressor stations located where the CO_2 is injected and booster compressors located as needed further along the pipeline. In overall construction, CO_2 pipelines are similar to natural gas pipelines, requiring the same attention to design, monitoring for leaks, and protection against overpressure, especially in populated areas. Many analysts consider CO_2 pipeline technology to be mature, stemming from its use since the 1970s for enhanced oil recovery (EOR) and in other industries. Marine transportation may also be feasible when CO_2 needs to be transported over long distances or overseas; however, many manmade CO_2 sources are located far from navigable waterways, so such as scheme would still likely require pipeline construction between CO_2 sources and port terminals. Rail cars and trucks can also transport CO_2, but these modes would be logistically impractical for large-scale CCS operations.

Sequestration in Geological Formations

In most CCS approaches, CO_2 would be transported by pipeline to a porous rock formation that holds (or previously held) fluids where the CO_2 would be injected underground. When CO_2 is injected over 800 meters deep in a typical storage formation, atmospheric pressure induces the CO_2 to become relatively dense and less likely to migrate out of the formation. Injecting CO_2 into such formations uses existing technologies developed primarily for oil and natural gas production which potentially could be adapted for long-term storage and monitoring of CO_2.

Other underground injection applications in practice today, such as natural gas storage, deep injection of liquid wastes, and subsurface disposal of oil-field brines, also provide potential technologies and experience for sequestering CO_2. Three main types of geological formations are being considered for carbon sequestration: (1) oil and gas reservoirs, (2) deep saline reservoirs, and (3) unmineable coal seams. The overall capacity for CO_2 storage in such formations is potentially huge if all the sedimentary basins in the world are considered. The suitability of any particular site, however, depends on many factors, including proximity to CO_2 sources and other reservoir-specific qualities like porosity, permeability, and potential for leakage.

Source

Parfomak P.W., Folger P., Carbon Dioxide (CO2) Pipelines for Carbon Sequestration: Emerging Policy Issues. CRS Report for Congress, 2007.

CO_2 pipeline infrastructure in the United States

Excerpts from the report "A Review of CO_2 Pipeline Infrastructure in the U.S." performed by Energy Sector Planning and Analysis for the U.S Department of Energy, Office of Energy Policy and System Analysis and the National Energy Technology Laboratory, April 2015.

The initial large-scale CO_2 pipeline in the U.S., the Canyon Reef pipeline, was built in the 1970s. Much of the remainder of the current CO_2 pipeline infrastructure was built between the 1980s and 1990s. In 2015, there were nearly 50 CO_2 transportation pipelines in the U.S. with a combined length of over 4,500 miles, operated by over a dozen different companies.

About 80 percent of CO_2 used for EOR in the U.S. is from natural sources. However, CO_2 supplies from industrial sources (natural gas processing plants, other chemical processing plants, and electric power facilities) are expected to increase their application in the future. Table 1 illustrates the major CO_2 transport pipelines that exist in the U.S. circa 2015. Figure 1 shows the CO_2-EOR operations and infrastructure in the U.S.

A number of industrial CO_2-capture facilities have been proposed and partially developed for delivering CO_2 to EOR fields over the past several decades. However, the significant amount of capital required by many of these projects has inhibited a number of them from meeting their announced CO_2-capture goals on time, or coming online entirely. But, as new industrial CO_2-capture projects begin to provide greater volumes of CO_2 to the EOR industry, it is anticipated that development costs will begin to decrease. Proven industrial CO_2-capture technology should lower the perceived risk of providing CO_2 supplies to the EOR industry.

Table 1
Geographic Areas with Large-Scale CO_2 Pipeline Systems in the United States circa 2015

Large-Scale CO2 Pipeline Systems	Miles of Pipeline
Permian Basin (West TX, NM, and South CO)	2600
Gulf Coast (MS, LA, and East TX)	740
Rocky Mountains (North CO, WY, and MT)	730
Mid-Continent (OK and KS)	480
Other (ND, MI, Canada)	215

Permian Basin

The Permian Basin contains the largest network of CO_2 pipelines in the U.S. Over 2,600 miles of CO_2 pipelines in this region carry both natural and industrial CO_2 supplies to CO_2-EOR projects throughout the region.

Three main pipelines deliver CO_2 from four natural sources of CO_2 to the Permian Basin (Figure 2). The Cortez pipeline delivers CO_2 from McElmo Dome and Doe Canyon in southwestern Colorado. The Sheep Mountain pipeline delivers CO_2 from the Sheep Mountain CO_2 field in central Colorado, and the Bravo pipeline delivers CO_2 from Bravo Dome in northeast New Mexico to the Permian Basin. All three of these major pipelines meet at the Denver City CO_2 hub, where CO_2 is dispersed through a network of smaller CO_2 pipelines to various oil fields and their CO_2-EOR projects. A smaller pipeline, the TransPetco/Bravo pipeline, transports a modest amount of CO_2 to the Postle CO_2-EOR operation in western Oklahoma.

Three other important CO_2 pipelines round out the large-scale pipeline system of the Permian Basin:

- The Canyon Reef Carrier CO_2 pipeline, the initial large-scale CO_2 pipeline, links the CO_2 captured from the gas processing plants in the Val Verde Basin (West Texas) with the pioneering Scurry Area Canyon Reef Operators Committee (SACROC) CO_2-EOR project, 170 miles to the northeast.
- The Centerline and Central Basin CO_2 pipelines deliver natural CO_2 from the Denver City CO_2 hub to the oil fields in West Texas and New Mexico.

Table 2 lists the CO_2 transportation pipelines installed in the Permian Basin region.

Gulf Coast

The 740 mile Gulf Coast CO_2 pipeline network is owned and operated by Denbury Onshore LLC (Figure 3). Two main pipelines service the region, the North East Jackson Dome (NEJD) Pipeline and the Green Pipeline. These two pipelines connect the natural CO_2 source in Jackson Dome, Central Mississippi, to Denbury's CO_2-EOR projects in Mississippi, Louisiana, and East Texas. Several industrial sources of CO_2 are connected to the Green Pipeline for delivery to CO_2-EOR. Table 3 lists CO_2 transportation pipelines installed in the Gulf Coast region circa 2015.

Rocky Mountains

The CO_2-EOR operations in the Rocky Mountain region are serviced by two major sources of CO_2: the Shute Creek natural gas processing plant and the Lost Cabin Gas Plant (Figure 4). The Shute Creek pipeline, operated by ExxonMobil, is the central trunkline (i.e., a pipeline that originates at a transshipment node) for several smaller pipelines, which deliver CO_2 to CO_2-EOR projects in central Wyoming, as well as the Rangely CO_2-EOR project in northwest Colorado.

Denbury completed construction of the Greencore pipeline in 2012, which delivers CO_2 supplies from the Lost Cabin Gas Plant to the Salt Creek, Bell Creek, and other CO_2-EOR projects in the Rocky Mountain region.

Table 4 lists the CO_2 transportation pipelines installed in the Rocky Mountain region, including a short, 40-mile delivery pipeline from McElmo Dome to the Aneth CO_2-EOR project in Utah.

Mid-Continent

The Mid-Continent CO_2 pipeline system (Figure 5) is mainly a set of fragmented source-to-field pipelines supplying

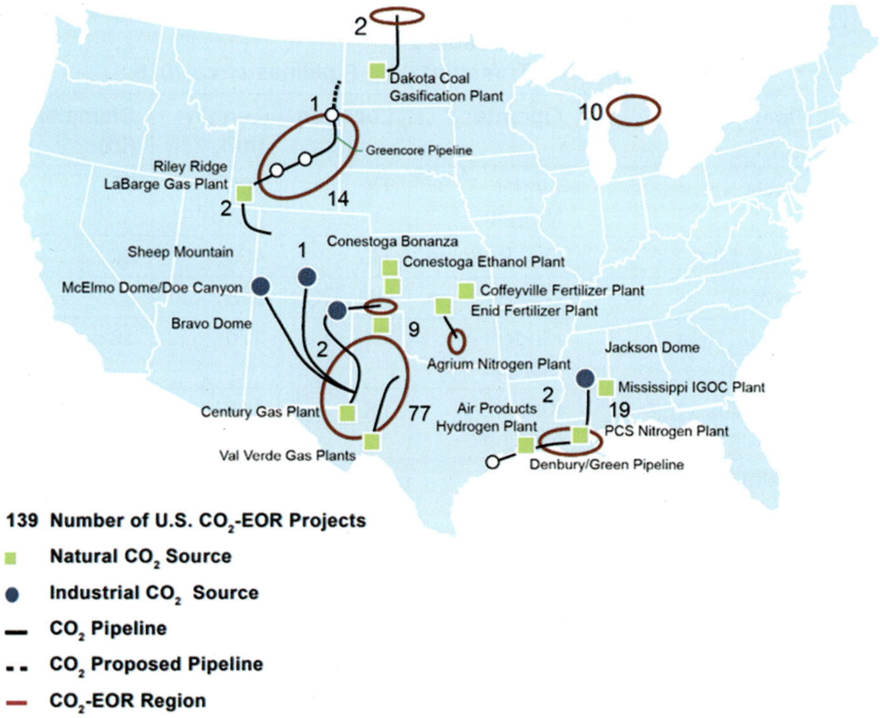

Figure 1. CO_2-EOR operations and infrastructure in the U.S. circa 2015.

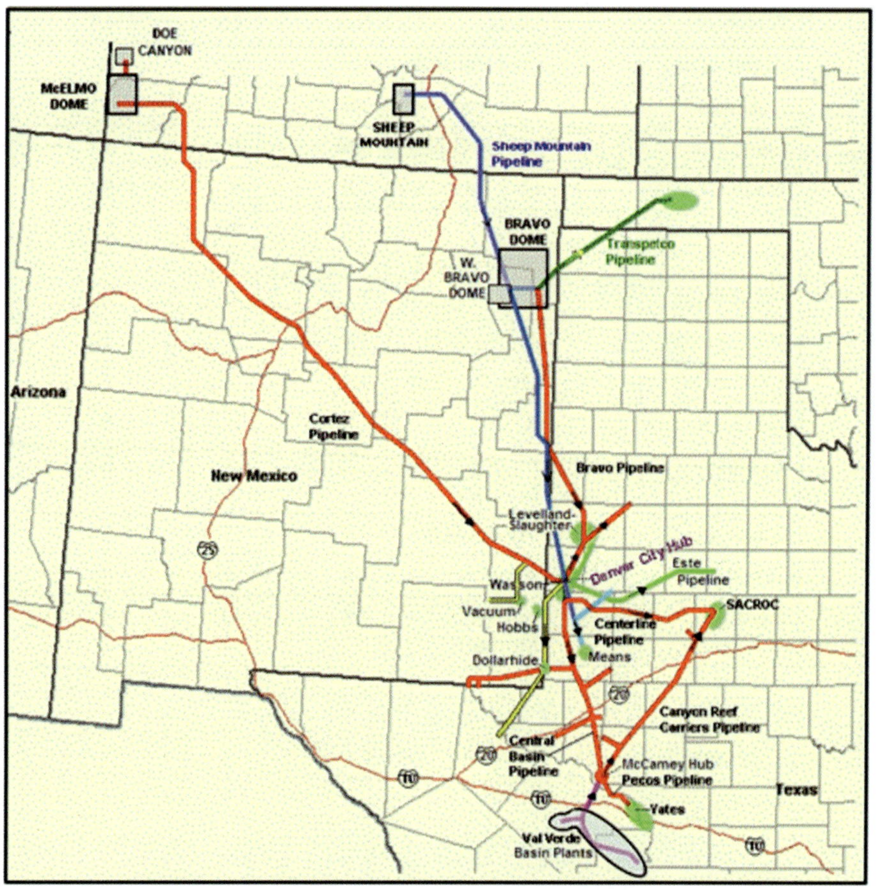

Figure 2. Permian Basin CO_2 pipeline infrastructure circa 2015.

Table 2
Permian Basin CO_2 Transportation Pipelines circa 2015

Scale	Pipeline	Operator	Location	Length (mi)	Diameter (in)	Flow Capacity (MMcfd)
Large-scale trunklines	Cortez	Kinder Morgan	TX	502	30	1,300
	Sheep Mtn	Oxy Permian	TX	408	24	590
	Bravo	Oxy Permian	NM, TX	218	20	380
	Canyon Reef Carriers	Kinder Morgan	TX	170	16	220
	Centerline	Kinder Morgan	TX	113	16	220
	Central Basin	Kinder Morgan	TX	143	16	220
Small-scale distribution systems	Este I - to Welch, TX	ExxonMobil, et al	TX	40	14	180
	Este II - to Salt Creek Field	Oxy Permian	TX	45	12	130
	Means	ExxonMobil	TX	35	12	130
	North Ward Estes	Whiting	TX	26	12	130
	Slaughter	Oxy Permian	TX	35	12	130
	Mabee Lateral	Chevron	TX	18	10	110
	Val Verde	Oxy Permian	TX	83	10	110
	Rosebud	Hess	NM	50*	12	100
	Anton Irish	Oxy Permian	TX	40	8	80
	Dollarhide	Chevron	TX	23	8	80
	Llano	Trinity CO2	NM	53	12	80
	North Cowden	Oxy Permian	TX	8	8	80
	Pecos County	Kinder Morgan	TX	26	8	80
	Pikes Peak	Oxy Permian	TX	40	8	80
	West Texas	Trinity CO2	TX, NM	60	12	80
	Comanche Creek	Oxy Permian	TX	120	6	70
	Cordona Lake	XTO	TX	7	6	70
	El Mar	Kinder Morgan	TX	35	6	70
	Wellman	Trinity CO2	TX	25	6	70
	Adair	Apache	TX	15	4	50
	Ford	Kinder Morgan	TX	12	4	50

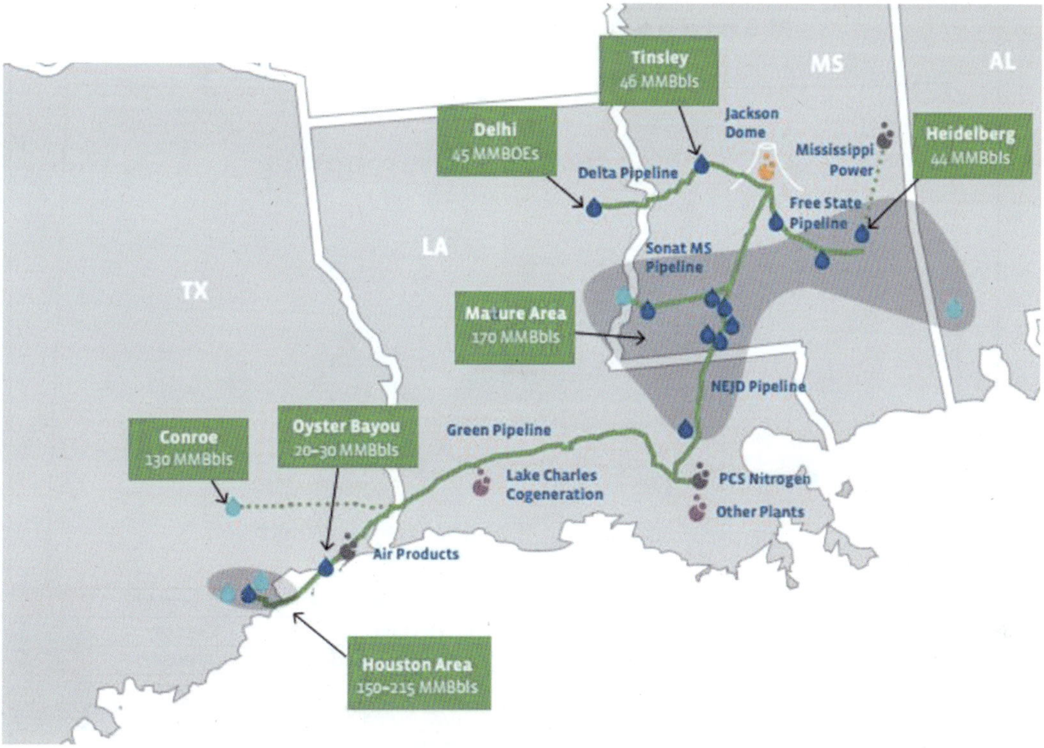

Figure 3. Gulf Coast CO_2 pipeline infrastructure circa 2015 and potential proved and produced-to-date tertiary reserves estimates circa 2013.

Table 3
Gulf Coast CO_2 Transportation Pipelines circa 2015

Scale	Pipeline	Operator	Location	Length (mi)	Diameter (in)	Flow Capacity (MMcfd)
Large-scale distribution systems	Green Line	Denbury	LA, TX	314	24	930
	Delta	Denbury	MS, LA	108	24	590
	NE Jackson Dome	Denbury	MS, LA	183	20	360
Distribution line	Free State	Denbury	MS	85	20	360
	Sonat	Denbury	MS	50	18	170

captured CO_2 from industrial sources to individual CO_2-EOR operations. Chaparral owns and operates most of the smaller pipelines while Anadarko controls the Enid-Purdy pipeline in Central Oklahoma. A small amount of natural CO_2 from Bravo Dome is delivered to the Postle CO_2-EOR operation via the TransPetco Pipeline. These CO_2 pipelines are listed in Table 5.

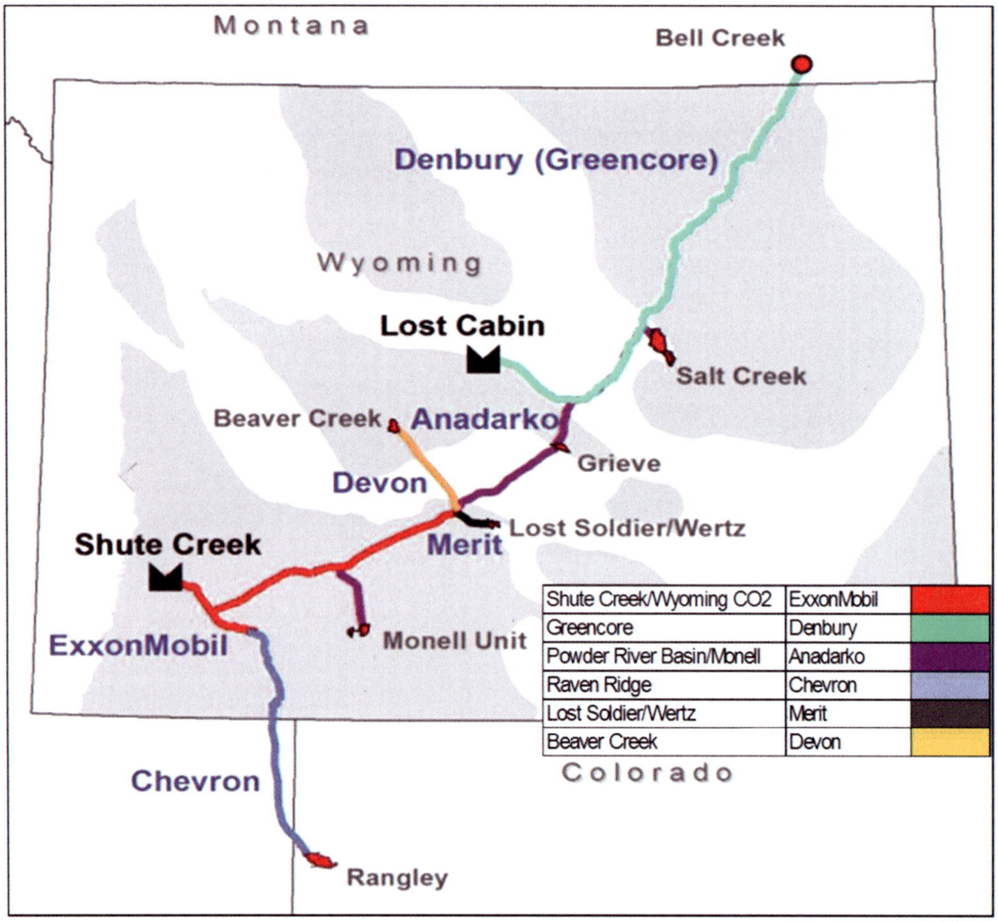

Figure 4. Rocky Mountain CO_2 pipeline infrastructure circa 2015.

Table 4
Rocky Mountain CO_2 Transportation Pipelines circa 2015

Scale	Pipeline	Operator	Location	Length (mi)	Diameter (in)	Flow Capacity (MMcfd)
Large-scale trunklines	Shutr Creek/Wyoming CO_2	ExxonMobil	WY	142	30–20	1220
	Greencore	Denbury	WY, MT	230	22	720
Small-scale distribution systems	Powder River Basin CO_2	Anadarko	WY	125	16	220
	Raven Ridge	Chevron	WY, CO	160	16	220
	McElmo Creek	Kinder Morgan	CO, UT	40	8	80
	Monell	Anadarko	WY	33	8	80
	Loist Soldier/Wertz	Merit	WY	30	16	43
	Beaver Creek	Devon	WY	53	8	30

Other U.S. CO_2 pipeline networks

Two other CO_2 pipeline networks exist, one in North Dakota and one in Michigan. The Dakota Gasification pipeline delivers captured CO_2 from the Great Plains Synfuels plant to the Weyburn CO_2-EOR project in Saskatchewan, Canada. The White Frost pipeline delivers captured CO_2 from the Antrim Gas Processing plant to several small-scale CO2-EOR projects in Otsego County, Michigan. These CO2 pipelines are listed in Table 6.

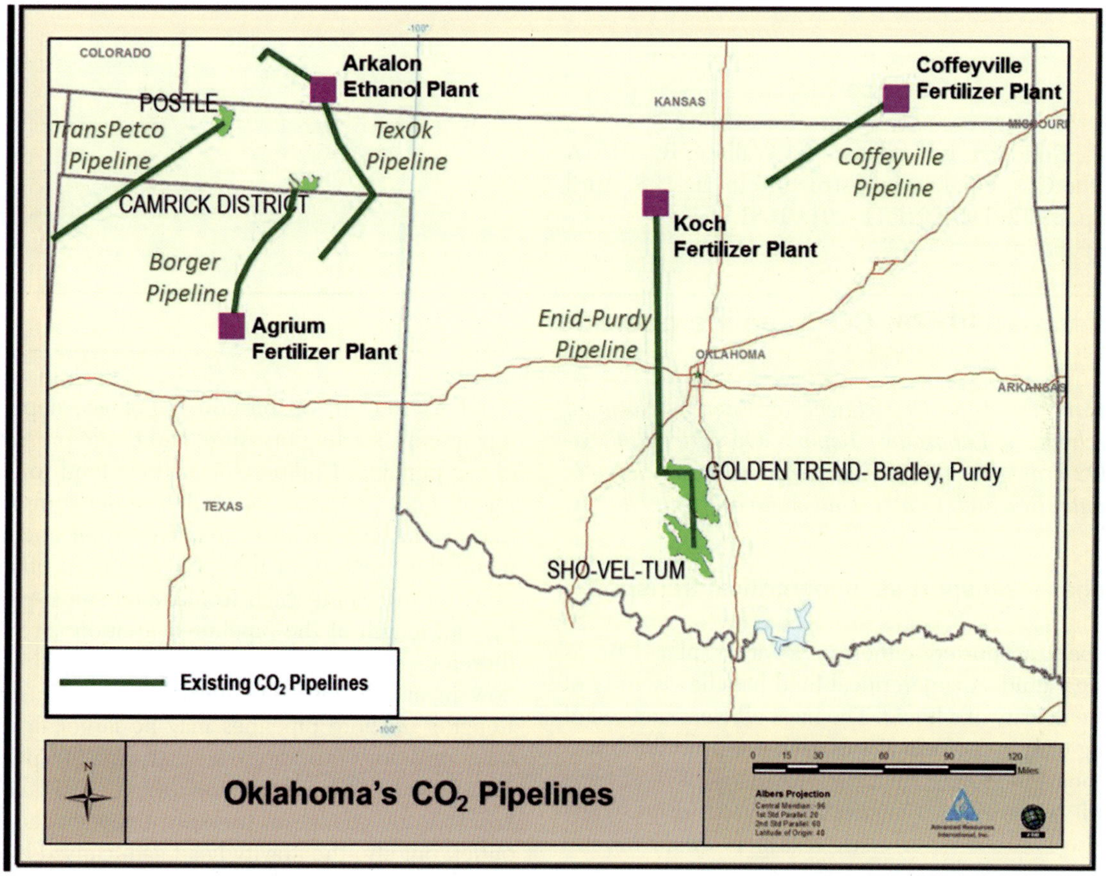

Figure 5. Mid-Continent CO_2 pipeline infrastructure circa 2015.

Table 5
Mid-Continent CO_2 Transportation Pipelines circa 2015

Scale	Pipeline	Operator	Location	Length (mi)	Diameter (in)	Flow Capacity (MMcfd)
Small-scale distribution systems	Coffeyville-Burbank	Chaparral	KS, OK	68	8	80
	Enid-Purdy	Anadarko	OK	117	8	80
	TransPetco	TransPetco	TX, OK	110	8	80
	TexOk	Chaparral	OK	95	6	70
	Bprger	Chaparral	TX, OK	86	4	50

Table 6
Other CO_2 Transportation Pipelines in the U.S. circa 2015

Pipeline	Operator	Location	Length (mi)	Diameter (in)	Flow Capacity (MMcfd)
Dakota Gasification	Dakota Gasification	ND, SK	204	14	130
White Frost	Core Energy, LLC	MI	11	6	70

Source

Wallace, M., Goudarzi, L., Callahan, K., Wallace, R. 2015. A Review of the CO_2 Pipeline Infrastructure in the U.S. April 21, 2015. pp. 3–12. DOE/NETL-2014/1681.

Effect of variability on CO_2 pipeline operation

Excerpts from the U.S. Department of Energy, National Energy Technology Laboratory Report "Operational Flexibility of CO2 Transport and Storage" performed by Energy & Environmental Research Center, University of North Dakota, 2014.

Vapor-phase transport vs. supercritical transport

CO_2 can be transported either in a vapor phase or as a supercritical fluid. A supercritical fluid has the viscosity of a gas but the density of a liquid. The most efficient method of transporting CO_2 in a pipeline is as a supercritical fluid near its critical point. It is likely that the economics of most CCS projects will require that CO_2 be transported in its supercritical phase because vapor-phase transport would require considerably larger diameter pipelines for the same mass flow rate and would experience high pressure drops. Vapor-phase transport is not used for pipelines that carry significant quantities of CO_2 for long distances.

Offshore vs. onshore pipelines

Compressor stations are used to boost the CO_2 stream pressure so as to maintain single-phase flow and ensure that the CO_2 stream arrives at the geologic sink at the expected pressure. Compressor stations are typically found every 161 km (100 mi) on onshore pipelines, depending on the topography, starting pressure, and target pressure at the end of the pipeline. Compressor stations tend to be impractical for offshore pipelines. Instead of using compressor stations that would be difficult to maintain in an undersea environment, the pressure of the CO_2 within an offshore pipeline must be sufficiently high to maintain single-phase flow over the full length of the pipeline. Variations in pressure could therefore cause more issues with maintaining single-phase flow in an offshore pipeline than in an onshore pipeline. Lengthy offshore pipelines may be larger in diameter than their onshore equivalents so that pipeline pressure can be maintained. Offshore pipeline design must take into account not only the change in pressure between the inlet and the outlet, but also the gravity head gain caused by the decrease in elevation. Therefore, the allowable pressure drop is greater for offshore than onshore pipelines, and this may actually reduce the diameter required for some offshore pipelines.

Seasonal variations in temperature are likely to affect onshore pipelines more than offshore pipelines where the temperature of the water changes little relative to the air or ground temperature. The temperature of the ground surrounding a buried onshore pipeline varies seasonally and therefore impacts the amount of CO_2 that is transported through the pipeline.

Effect of flow rate variability

Temperature affects pipeline capacity indirectly as well as directly because the operating temperature affects not only the amount of CO_2 that can be compressed to fit the pipeline, but also other factors, such as viscosity. Temperature changes to which the pipeline is exposed and the subsequent changes in pressure can affect the ability to maintain pipeline pressure within an acceptably narrow operating margin. Supercritical-phase conditions should be retained at the designed operating conditions. Therefore, pipeline pressure control systems must be designed and operated in a manner that ensures maintenance of the operating conditions as well as at pipeline shut-in conditions. When the topography over which the pipeline is built changes significantly (such as up and down mountains and valleys), it is possible for the CO_2 pressure to change, potentially forming multiphase flow. If this happens, it can cause transport difficulties such as when liquid CO_2 pools in the low spots simultaneously when vapor-phase CO_2 forms. Two-phase flow is more likely to occur when the pipeline is oversized relative to the amount of CO_2 that is transported.

On the other hand, two-phase flow need not necessarily be avoided in pipelines and injection wells. As knowledge about the Sleipner reservoir has increased and there is a better understanding of the flow conditions, it is apparent that the flow at the wellhead is in two-phase flow and that there is a gradual phase change inside the well itself. It should be remembered that Sleipner has a short transport distance and no rotating equipment is installed that could be damaged by the two-phase flow.

Pressure oscillations may cause gas bubbles to form in the dense phase, altering its behavior and potentially damaging the pipeline. Cavitation may occur when the pressure suddenly drops lower than the bubble point of the CO_2.

Capture facilities–power plant load variations

Traditional power plants have been designed for base-load operation, but there is discussion that power plants (at least those in Europe) may soon be required to operate more flexibly, i.e., at more variable loads in response to the changing market. Capture of CO_2 from a facility operating at variable loads will mean that the quantity of the CO_2 stream that is fed to the pipeline could vary significantly with time. This can be dealt with partly by turning down the compressors but there is a point at which the compressors cannot be turned down further. In this case, it would be useful to have multiple [smaller] CO_2 compressors or variable frequency drives to enable additional turndown.

Simulations have been used to study transient variations in terms of pressure, temperature, and phase. During injection with full load from the power plant, researchers have found that CO_2 would be in the supercritical state in both the pipeline and the well. The pressure and temperature drop below the critical point (7.1 MPaA and 31°C) during shutdown and the phase changes from dense phase to gas and liquid in both the upper part of the well and the pipeline. Low loads of less than 20% of full load will have difficulty maintaining supercritical conditions in the pipeline and well because of the combination of lower pressure and temperature. Their simulations also indicated that the pressure in the reservoir near the well area reacts almost instantaneously to a sudden discontinuation in the CO_2 injection.

Effects of impurities

The composition of an anthropogenic CO_2 stream depends on both the process producing the stream and the method of capturing that stream. The presence of impurities:

- Changes the physical and transport properties of CO_2.
- Affects the CO_2 hydraulics, which in turn affects the number of compressors, compressor power, temperature transients, etc.
- Changes other aspects such as fracture propagation, corrosion, nonmetallic component deterioration, and the formation of hydrates and clathrates.
- Changes the capacity of the pipeline itself.

Impurities change the critical temperature and pressure of the CO_2 stream, typically by increasing them. Hydrogen, for example, increases the critical temperature and pressure, which reduces the optimum pipeline operating region. Because operation in the two-phase region should be minimized so as to optimize cost and throughput, the pipeline pressure must be maintained above the CO_2 stream's critical point.

The number and interaction of the impurities must be taken into account. The critical temperature and pressure increase with the presence of additional impurities as well as with increasing impurity concentration. The impurities can have a significant effect not only on the critical temperature and pressure, but also on the density and viscosity of the fluid. Some combinations (e.g., if H_2 or N_2 are present) cause higher pressure and temperature drops for a given length of pipeline than other combinations. Sudden temperature drops can cause embrittlement and/or hydrate formation, either of which can damage a pipeline. Variability in CO_2 composition can also affect the injection infrastructure and/or the geologic storage formation. In some cases, this may require that a well be shut in or vented.

The U.S. Department of Energy National Energy Technology Laboratory has issued CO_2 Impurity Design Parameters as part of their Quality Guidelines for Energy System Studies. In the guidelines, they provide recommended limits for CO_2 stream impurities required by the pipeline, EOR applications, and by a saline reservoir. Table 1 presents these limits. Several contaminants have common specific issues relative to CO_2 stream composition and its transport and/or end use. For example:

- N_2, CH_4, and H_2 all have a lower critical temperature than CO_2, which would require increased pipe strength to minimize ductile fracture potential.

Table 1
Recommended Limits for CO_2 Stream Compositions

Component	Unit[a]	Carbon Steel Pipeline		EOR		Saline Reservoir Sequestration		Saline CO_2 and H_2S Cosequestration		Venting Concerns
		Concept	Literature Range	Concept	Literature Range	Concept	Literature Range	Concept	Literature Range	
CO_2	vol%	95	90–99.8	95	90–99.8	95	90–99.8	95	20–99.0	Yes; IDLH[b] 40,000 ppmw
H_2O	ppmw[c]	300	20–650	300	20–650	300	20–650	300	20–650	
N_2	vol%	4	0.01–7	1	0.02–2	4	0.01–7	4	0.01–7	
O_2	vol%	4	0.01–4	0.01	0.001–1.3	4	0.01–4	4	0.01–4	
Ar	vol%	4	0.01–4	1	0.01–1	4	0.01–4	4	0.01–4	
CH_4	vol%	4	0.01–4	1	0.01–2	4	0.01–4	4	0.01–4	Yes; asphyxiate, explosive
H_2	vol%	4	0.01–4	1	0.01–1	4	0.01–4	4	0.01–4	Yes; asphyxiate, explosive
CO	ppmv[d]	35	10–5000	35	10–5000	35	10–5000	35	10–5000	Yes; IDLH 1200 ppmv
H_2S	vol%	0.01	0.002–1.3	0.01	0.002–1.3	0.01	0.002–1.3	75	10–77	Yes; IDLH 100 ppmv
SO_2	ppmv	100	10–50,000	100	10–50,000	100	10–50,000	100	10–50,000	Yes; IDLH 100 ppmv
NO_x	ppmv	100	20–2500	100	20–2500	100	20–2500	100	20–2500	Yes; IDLH NO=100 ppmv, NO_2=200 ppmv
NH_3	ppmv	50	0–50	50	0–50	50	0–50	50	0–50	Yes; IDLH 300 ppmv
COS	ppmv	Trace	Trace	5	0–5	Trace	Trace	Trace	Trace	Lethal at high concentrations (>1000ppmv)
C_2H_6	vol%	1	0–1	1	0–1	1	0–1	1	0–1	Yes; asphyxiant, explosive
C_3+	vol%	<1	0–1	<1	0–1	<1	0–1	<1	0–1	
Particulate	ppmv	1	0–1	1	0–1	1	0–1	1	0–1	
HCl	ppmv	N.I.[e]	N.I	N.I.	N.I.	N.I.	N.I.	N.I.	N.I.	Yes; IDLH 50 ppmv
HF	ppmv	N.I.	N.I.	N.I.	N.I.	N.I.	N.I.	N.I.	N.I.	Yes; IDLH 30 ppmv
HCN	ppmv	Trace	Trace	Trace	Trace	Trace	Trace	Trace	Trace	Yes; IDLH 50 ppmv
Hg	ppmv	N.I.	N.I.	N.I.	N.I.	N.I.	N.I.	N.I.	N.I.	Yes; IDLH 2 mg/m^3 (organo)
Glycol	ppbv	46	0–174	46	0–174	46	0–174	46	0–174	
MEA	ppmv	N.I.	N.I.	N.I.	N.I.	N.I.	N.I.	N.I.	N.I.	MSDS Exp. Limits 3 ppmv, 6 mg/m^3
Selexol	ppmv	N.I.	N.I.	N.I.	N.I.	N.I.	N.I.	N.I.	N.I.	

[a] Maximum unless otherwise noted.
[b] Immediately dangerous to life and health.
[c] Parts per million by weight.
[d] Parts per million by volume.
[e] N.I. = Not enough information.

- Noncondensables (e.g., N_2, O_2, Ar, CH_4, and H_2) should be limited to less than 4 vol% because their presence increases the amount of compression work.
- The toxicity of CO and H_2S limits their allowable concentrations because of the potential for inadvertent releases.
- The concentration of O_2 in CO_2 that will be used for EOR should be limited so as to eliminate exothermic reactions with the hydrocarbons; H_2S and SO_2 should also be limited as they can be reproduced at the pumping well when the CO_2 front breaks through.
- Moisture content requirements vary and depend upon the formation of acids with CO_2 or SO_2 that can corrode standard piping.
- In the presence of free water, O_2 can increase cathodic reactions, causing corrosion induced thinning of the CO_2 pipeline. According to some researchers, this may occur at very low oxygen concentrations.
- The presence of N_2, H_2, and CH4 increase the miscibility pressure during EOR activities and should thus be limited.
- NO_x and COS can also be reproduced at the pumping well when the CO_2 front breaks through.

Effects of corrosion

CO_2 readily dissolves in water to form carbonic acid, which can corrode a pipeline at a rate of 1-2 mm within 2 weeks. Depending on the temperature, supercritical CO_2 can store several hundred ppm of water. Both experimental work and experience in the field have shown that pure CO_2 with a dissolved water concentration that is below the saturation limit is not corrosive to carbon steel at pipeline operating conditions. The water may form hydrates that can cause plugs that clog the pipeline and is dependent on the pressure, temperature, and water content of the CO_2. The risk of hydrate formation is higher when the pressure is high. However, if the pressure falls, water may precipitate out and create carbonic acid. Clearly, variations in the temperature and pressure within a CO_2 pipeline caused by variability in flow rate can have a substantial effect on corrosion if any water is present in the CO_2 stream. These effects could be amplified if the CO_2 stream contains impurities that lead to changes in phase behavior at pipeline pressure.

Pipeline first fill, start-up, and shutdown procedures

Initial Start-Up

Following pipeline commissioning, the pipeline and compressor section are filled with dry nitrogen or dry air to a minimum pressure of 0.5 MPa (this is enough pressure to prevent infiltration of the pipeline by wet air or water). As soon as possible, dry CO_2 enters the pipeline at the minimum pressure and flow rate possible by the compressor train. It is difficult to know when the pipeline has been purged of inert gas, so it is recommended that a pig be mounted in the pipeline to provide physical separation between the O_2/N_2 and the CO_2 entering the pipeline from the compressor. Upon receipt of the pig at the platform, the pipeline is allowed to pressurize to low reservoir pressure for the start-up of injection.

Normal Start-Up

Following a planned shutdown, there are two modes of start-up that can be followed. The first mode involves pressurizing the pipeline using the full compressor flow to a pressure where flow within the pipeline is a single phase. The valves at the platform/injection site are opened to allow the gas to enter the well piping. The disadvantage of this method is that the flow will reach very high velocities in the tail end of the well piping, potentially causing vibrations and damaging the pipe. In addition, the temperature of the wellhead material must be monitored to avoid low temperatures.

The second mode is preferable. In this approach, the well pressure control valve is opened as soon as the pressure in the pipeline has reached the wellhead pressure. The pipeline may still contain a considerable amount of liquid CO_2, but because of the low flow rate, the liquid slugs should not do any harm. Because of the low pressure drop near the valve at the wellhead, a minimal quantity will flow into the well piping. The low flow should not cause vibration or erosion issues in the downhole well piping. As the pipeline pressure increases, the flow rate into the well will increase.

Planned Shutdown

Start-up will be faster when the pipeline is pressurized and heated with less content. Therefore, it is advantageous to heat the pipeline content to the highest temperature possible prior to a planned shutdown. Once the "best" conditions have been reached, the compressor discharge valve is closed. The pipeline is emptied into the well as long as positive flow can be maintained. When the pressure in the well nears that of the pipeline, the valve should be shut so as to ensure that there is no backflow from the well into the pipeline.

During stoppage of a CO_2 capture facility, the pressure of the CO_2 stream will drop and the pipeline will cool gradually until it is the same temperature as its surroundings. This may result in the formation of gas-phase CO_2 (and therefore two-phase flow) in the pipeline.

A planned shutdown of the CO_2 source requires coordination with the geologic sink. Interviews with industry indicate that sufficient advance notice allows for greater flexibility in operations. If the sink is an oil field under EOR, the oil field can simply recycle the CO_2 from the recovered oil to maintain continuity of operations. If the outage lasts for an extended period of time, CO_2 injection and/or production may need to stop, which can result in undesirable consequences.

The procedure for shutting down the pipeline would be to shut down the compressor(s) and then close the valves on the pipeline. The pipeline would then be allowed to sit at full pressure. Over time, the CO_2 will cool down to ground temperature, with the pressure dropping gradually as this happens. During extended outages, the pressure of the CO_2 within the pipeline may naturally drop to close to 6.9 MPa. At 5.5–6.2 MPa, the supercritical CO_2 will begin to solidify. Therefore, prior to the pipe pressure dropping that low, the pipeline is emptied and refilled when the CO_2 source outage is over. In the case of the Sleipner pipeline, the hydrate control strategy is based on insulation of the pipeline so as to maintain a fluid temperature that is above the hydrate equilibrium temperature during normal operations. For lengthy planned shutdowns, the pipeline is inhibited with MEG. If the shutdown is unplanned and lengthy, the pipeline will be depressurized. Other inhibitors (wax, scale, and asphaltene) are required as well.

Unplanned Shutdown

If an emergency at the capture plant dictates an unplanned shutdown of the pipeline, the compressor discharge valve is closed to protect the contents of the pipeline. The pipeline is emptied as described under planned shutdown, except that because the pipeline may contain more CO_2 than during normal shutdown activity. In this case, start-up after the cooldown may take longer and care should be taken to avoid too much CO_2 flow into the well at too high a velocity.

Transient Conditions

Complete Depressurization. Pipeline depressurization, such as can occur during an accident or during planned maintenance, will cause phase transition and a strong Joule-Thomson cooling of the pipe. Depressurization of a pipeline carrying supercritical CO_2 can result in the deposit of a significant proportion of the CO_2 as a solid at low points in the pipeline. The temperature of these solids at atmospheric pressure is −78°C, which could cause metallurgical damage, such as embrittlement of pipe material or valves, rupture, injury, or material and infrastructure damage. If the solid is warmed rapidly, the pipeline could over-pressurize because of the rapid increase in volume as the solid sublimes to the vapor phase. This could happen by the reintroduction of supercritical CO_2. Unless the sink could adapt and take up the additional CO_2, it might be necessary to vent it. The presence of impurities only makes it more difficult to model the needed conditions for safe depressurization.

Onshore pipelines utilize regular vent stations but long offshore pipelines can only be depressurized at the compression site, and it can be a lengthy process to safely depressurize while keeping the temperature above design limits. Heat-transfer coefficients are important in determining the total depressurization time. Following depressurization for maintenance purposes, the pipeline will probably be started up by repressurizing the pipeline with warm, high-pressure (supercritical) CO_2 from the compressor train. The CO_2 in the pipe will change phase from gas to liquid and then to supercritical phase as the pressure and temperature are increased. Simulation predicts that liquid slugs can be expected that may travel through the pipeline at relatively high speed. The details of the dynamics of this process are still uncertain and more experience is required to better understand the nature of the flow, the range of slug speeds, and the induced stresses on the pipeline.

Design Considerations. Solids and cooling issues must be considered when designing a pipeline to carry supercritical CO_2. Generally, solid formation and excessive material cooling can be avoided during normal pipeline operations but uncontrolled depressurization (as from a leak) must be considered when designing the pipeline. Hydrates may cause plugs that could clog the pipeline. There is some uncertainty whether free water contained in supercritical CO_2 would form hydrates before it forms carbonic acid, but there is a dependency on the CO_2 pressure, temperature, and water content.

The primary differences between onshore and offshore pipelines are in their design and construction. Offshore pipelines are made of pipe having a thicker wall than onshore pipelines to resist external pressure, tend to have a concrete outer coating for protection, and employ sacrificial anodes to minimize corrosion.

Source

Melanie D. Jensen, Steven M. Schlasner, James A. Sorensen, John A. Hamling, Operational Flexibility of CO2 Transport and Storage. Energy & Environmental Research Center, University of North Dakota. 2014-EERC-12-17, December 2014, For the complete report and citations, go online at: https://undeerc.org.

13: Liquids—Hydraulics

Marine hose pressure loss 492
Pressure loss in hose string 495
Pressure drop formula for rubber hose 497
Example of pressure drop calculations for rubber hose 497
Formulas for calculating pressure drop and flow rates for pipelines 497
Using and computing hydraulic gradients 499
What are equivalent lengths? 502
Pressure drop in series systems 503
Pressure drop in looped systems 504
Calculate pressure loss in annular sections 505
Calculate pressure and temperature loss for viscous crudes $\geq$ 1,000 cP 505
Determine batch injection rate as per enclosure 508
Pressure loss through valves and fittings 509
Nomograph for calculating Reynolds number for clean steel and wrought iron pipe 515
Nomograph for calculating pressure drop of liquids in lines for turbulent flow 517
Drag-reducing agents 521
How to estimate the rate of liquid discharge from a pipe 524
Predict subsurface temperature ranges 524
Sizing pipelines for water flow 525
How approximate throughput of a line can be estimated from pipe size 525
Estimate crude gathering line throughput for a given pipe diameter 526
How to determine head loss due to friction in ordinary iron pipeline carrying clear water 526
How to size lines, estimate pressure drop, and estimate optimum station spacing for crude systems 527
How to size crude oil and products lines for capacity increases 527
What is the hydrostatic pressure due to a column of liquid H feet in height? 527
How to determine the maximum surge pressure in liquid-filled pipeline when a valve is suddenly closed 528
Transient pressure analysis 528
Hydraulics calculations for multiphase systems, including networks 538

Marine hose pressure loss

How to calculate the length of hose, and pressure loss, through marine hose is described.

CALM system

The overall length of the floating hose string is designed as follows:

$$TL = C + RX(1 + E) + B/2 + L + H + M + G \qquad (1)$$

where:

TL = Required overall length
C = Diameter of buoy
R = Length of mooring hawser
E = Max. elongation of mooring hawser
B = Beam of tanker
L = Distance between bow and manifold
H = Maximum (light ballast) freeboard (approx. 0.8 × tanker depth)
M = Distance between rail and manifold
G = Height of manifold from deck

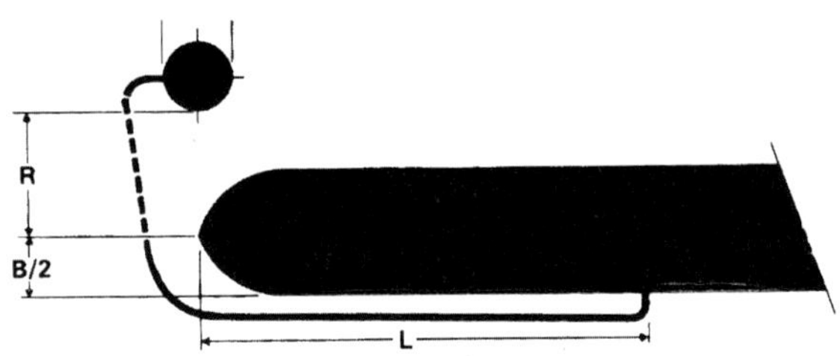

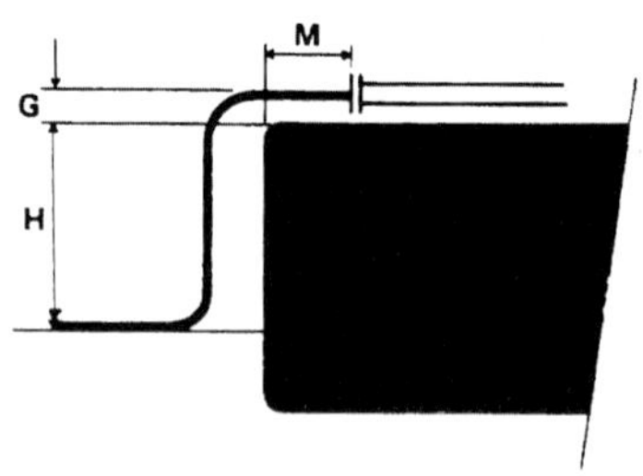

SALM system

The required overall length of hose string is designed as follows:

$\ell_1 < R$

where:

ℓ_1 = Distance between hose arm* and floating hose
R = Length of mooring hawser

$$TW = \sqrt{\ell_1 + W^2}$$

W = Maximum water depth at hose arm

$$TL = HM + RX(1 + E) + TW + (B/2) + L + H + M + G + \ell_1 \qquad (2)$$

where:

TL = Required overall length of hose string
HM = Max. horizontal movement of buoy
R = Length of mooring hawser
E = Max. elongation of mooring hawser
B = Beam of tanker
L = Distance between bow and manifold
H = Maximum light (ballast) freeboard (approx. 0.8 × tanker's depth)
M = Distance between rail and manifold
G = Height of manifold from deck

The hose arm is defined as the horizontal (I_1) or the vertical (W) distance between the underwater hose to swivel connection and the commencement of the main floating line.

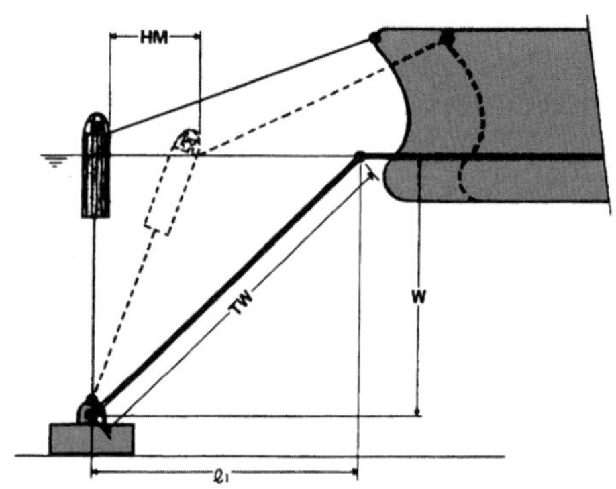

Tandem system

The required overall length of hose string is designed as follows:

$$TL = LF + HF + R + (1 + E) + LT + HT + MT + GT \quad (3)$$

where:

TL = Maximum water depth at hose arm
LF = Distance between manifold and stern of floating storage vessel (FSV)
HF = Max. height of manifold of FSV from sea-surface
R = Length of mooring hawser
E = Max. elongation of mooring hawser
LT = Distance between bow and manifold of off-take-tanker (OTT)
HT = Maximum light (ballast) freeboard of OTT (approx. 0.8 × tanker's depth)
MT = Distance between rail and manifold of OTT
GT = Height of manifold from deck of OTT

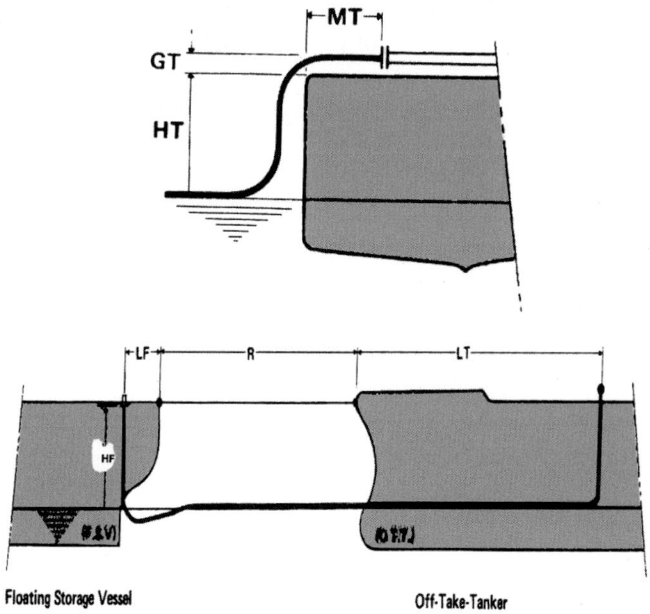

Multi-point mooring system

The required overall length of hose string is designed as follows:

$$TL = 1.5 \times L + D + H + G + M + \sqrt{(FA/2)^2 + (SW/2)^2} \quad (4)$$

where:

TL = Required overall length
L = Distance between PLEM and tanker hull
D = Maximum water depth
H = Maximum light (ballast) freeboard (approx. 0.8 × tanker's depth)
M = Distance between rail and manifold
G = Height of manifold from deck
FA = Max. fore and aft excursion of tanker
SW = Max. sideways excursion of tanker

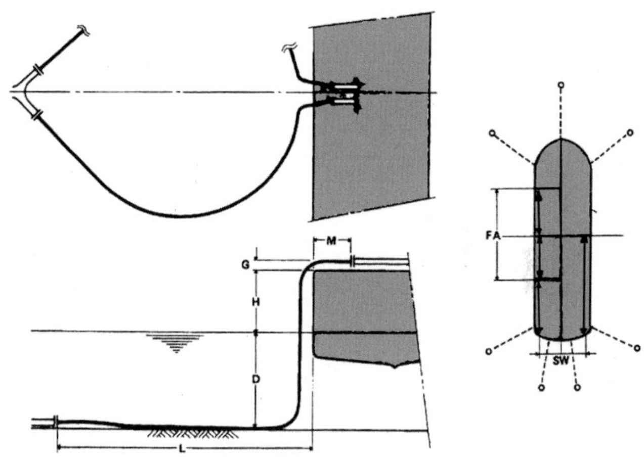

Reprinted with permission Bridgestone Engineered Products Co.

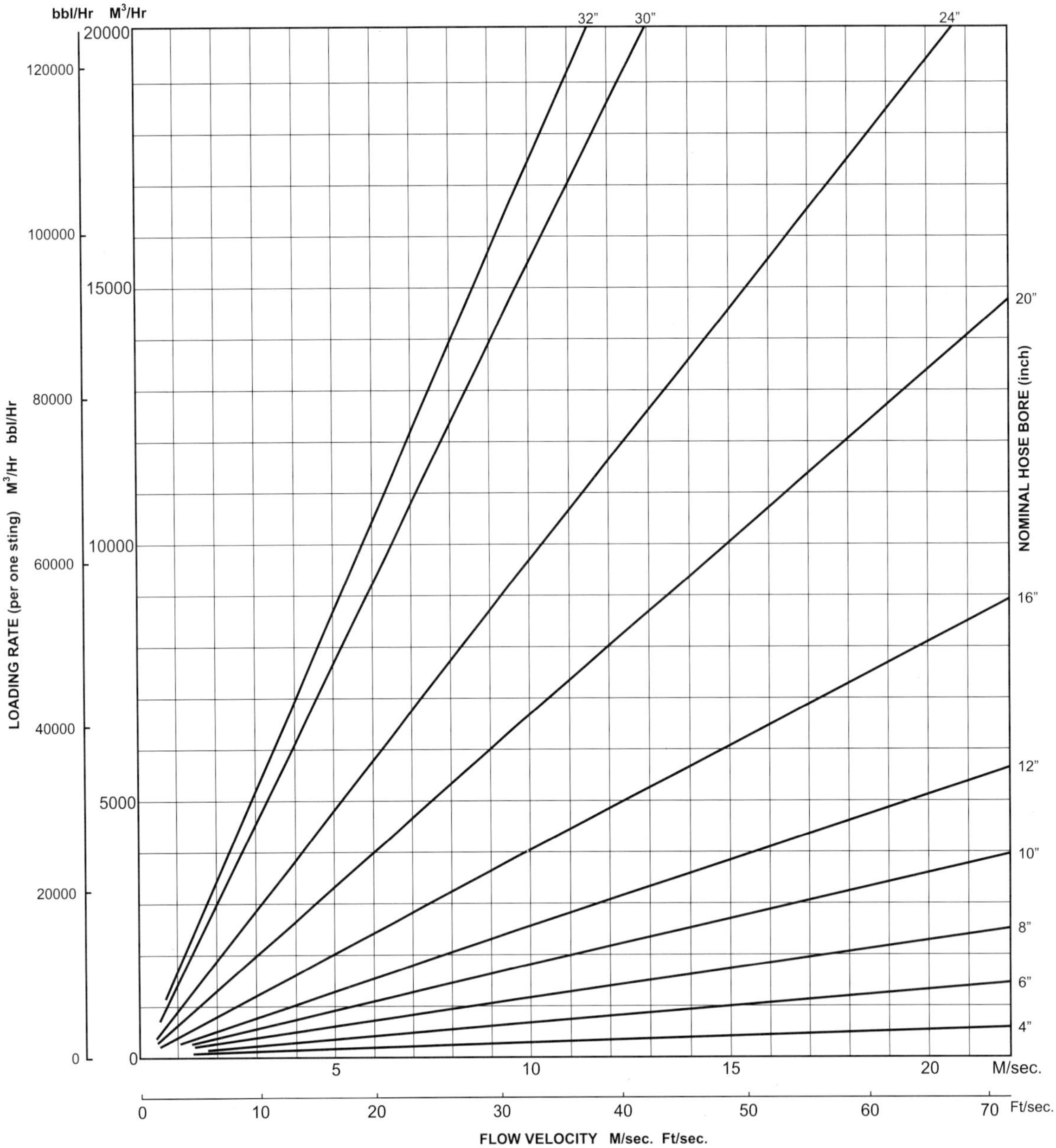

Figure 1. Loading rate and flow velocity.

Pressure loss in hose string

The pressure loss in a hose string is calculated as follows:

$$\Delta P = \frac{\lambda \times L \times V^2 \times \gamma \times 10^{-4}}{2 \times g \times D} \qquad (1)$$

where:

ΔP = Pressure loss (kg/cm^2)
λ = Friction factor
L = Length of hose string (m)
V = Flow velocity (m/sec)
γ = Specific weight of oil (kg/m^3)
g = Acceleration of gravity (m/sec^2)
D = Inside diameter of hose (m)

λ is calculated by using the following Mises' experimental formula:

$$\lambda = 0.0096 + 5.7 \times \sqrt{K/D} + 1.7 \times \sqrt{1/Re} \qquad (2)$$

where:

K = Coefficient of wall roughness by Mises' experiment (m)
K = 0.3×10^{-6} for rubber hose
Re = Reynolds number

Re is calculated as follows:

$$Re = \frac{D \times V}{\nu} \qquad (3)$$

where:

ν = Kinematic viscosity (m^2/sec)

Input Data:

Viscosity and specific weight of oil depend on temperature, and the values at the operating temperature sometimes need to be converted from values at other given temperatures.

Specific weight can be converted from other units as follows:
Specific weight (kg/m^3) = 1000 × specific gravity

$$\text{Specific gravity} = \frac{141.5}{131.5 + {}^\circ API} \qquad (4)$$

Specific gravity at operating temperature can be calculated using the following formula:
S.G. operating = S.G. measured − A × (Temp. operating − Temp. measured).
Note: Coefficient of "A" depends on the specific gravity of the oil and is given in Table 1.
Kinematic viscosity is converted from other units as follows:

1 cSt = 1 cP/S.G. (cP is Absolute viscosity)
1 cSt = 0.245 × R (Redwood Sec.)
1 cSt = 7.60 × E (Engler Deg.)
1 cSt = 0.220 × S (Saybolt Universal Sec.)

Conversion of kinematic viscosity at a different temperature depends on the characteristics of the cargo and can be calculated from data at two different temperatures.

$$\log_{10} \log_{10}(n + 0.6) = \nu - m \times \log T \qquad (5)$$

where:

T = Temperature (°K)
ν = Kinematic viscosity (cSt)

Flow velocity is determined by the internal diameter of the hose and the loading rate as follows:

$$V = \frac{4 \times Q}{3600 \times \pi \times D^2} \qquad (6)$$

where:

V = Flow velocity (m/sec)
Q = Loading rate (m^3/hr)
D = Inside diameter of hose (m)

Table 1
Change of Specific Gravity of Oil

S.G. at meas. temp.	Coeff. "A"	S.G. at meas. temp.	Coeff. "A"
0.640–0.643	0.00095	0.754–0.759	0.00078
0.644–0.648	0.00094	0.760–0.765	0.00077
0.649–0.654	0.00093	0.766–0.771	0.00076
0.655–0.661	0.00092	0.772–0.777	0.00075
0.662–0.667	0.00091	0.778–0.783	0.00074
0.668–0.674	0.00090	0.784–0.790	0.00073
0.675–0.681	0.00089	0.791–0.799	0.00072
0.682–0.688	0.00088	0.800–0.808	0.00071
0.689–0.696	0.00087	0.809–0.818	0.00070
0.697–0.703	0.00086	0.819–0.828	0.00069
0.704–0.711	0.00085	0.829–0.838	0.00068
0.712–0.719	0.00084	0.839–0.852	0.00067
0.720–0.726	0.00083	0.853–0.870	0.00066
0.727–0.734	0.00082	0.871–0.890	0.00065
0.735–0.741	0.00081	0.891–0.970	0.00064
0.742–0.747	0.00080	0.971–1.000	0.00063
0.748–0.753	0.00079		

Example

Conditions:

Loading rate:	40,000 bbl/hr
Hose size:	20″ & 16″
Hose length:	100 m for 20″ hose 50 m for 16″ hose
Operating temp.:	30°C
Specific gravity of oil	34.4° API at 60°F
Kinematic viscosity:	6.90 cSt at 30°C

Calculations:
Loading rate: Q = 40,000 × 0.15899 = 6,360 m^3/hr
Flow velocity for 20″ hose:

$$V_{20''} = \frac{4 \times 6,360}{3600 \times \pi \times 0.486^2} = 4.83\ \mathrm{m/sec}$$

$$V_{16''} = \frac{4 \times 6,360}{3600 \times \pi \times 0.380^2} = 5.92\ \mathrm{m/sec}$$

Specific weight:

$$\mathrm{S.G.}_{60^\circ F} = \frac{141.5}{131.5 + 34.4} = 0.853\ (\text{at } 15.56^\circ C)$$

$$\mathrm{S.G.}_{30^\circ C} = 0.853 - 0.00066 \times (30 - 15.56) = 0.843$$

Specific weight = 843 kg/m^3

Kinematic viscosity:

$$\nu = 6.9\ \mathrm{cSt} = 6.9 \times 10^{-6}\ \mathrm{m^2/sec}$$

Reynolds number:

$$\mathrm{Re} = \frac{D \times V}{\nu}$$

$$\mathrm{Re}_{20''} = \frac{0.486 \times 9.52}{6.9 \times 10^{-6}} = 670,539$$

$$\mathrm{Re}_{16''} = \frac{0.380 \times 15.6}{6.9 \times 10^{-6}} = 859,130$$

Friction factor:

$$\lambda = 0.0096 + 5.7 \times \sqrt{\frac{K}{D}} + 1.7 \times \sqrt{\frac{1}{\mathrm{Re}}}$$

$$\lambda_{20''} = 0.0096 + 5.7 \times \sqrt{\frac{0.3 \times 10^{-6}}{0.486}} + 1.7 \times \sqrt{\frac{1}{670539}} = 0.01615$$

$$\lambda_{16''} = 0.0096 + 5.7 \times \sqrt{\frac{0.3 \times 10^{-6}}{0.380}} + 1.7 \times \sqrt{\frac{1}{859130}} = 0.1650$$

Pressure loss:

$$\Delta P = \frac{\lambda \times L \times V^2 \times \gamma \times 10^{-4}}{2 \times g \times D}$$

$$\Delta P_{20''} = \frac{0.01615 \times 200 \times 9.52^2 \times 843 \times 10^{-4}}{2 \times 9.81 \times 0.486} = 2.59\ \mathrm{kg/cm^2}$$

$$\Delta P_{16''} = \frac{0.01650 \times 50 \times 15.6^2 \times 843 \times 10^{-4}}{2 \times 9.81 \times 0.380} = 2.27\ \mathrm{kg/cm^2}$$

Total pressure loss = 2.59 + 2.27 = 4.86 kg/cm^2

Pressure drop formula for rubber hose

The general flow formula for rubber hose is as follows:

$$\Delta P = \frac{0.0134 \times f \times S \times L \times Q^2}{d^5}$$

where: ΔP = Pressure drop, psi
f = Friction factor
S = Specific gravity
L = Length of line, ft
Q = Flow, gpm
d = Inside diameter of line, in

$$N_R = \frac{3,160 \times Q}{V_C \times d}$$

where: N_R = Reynolds number
Q = Flow, gpm
V_C = Fluid viscosity, centistokes
d = Inside diameter of line, in

For laminar flow, $f = \frac{64}{N_R}$.

Source

Gates Rubber Company.

Example of pressure drop calculations for rubber hose

Flow rate = Q = 1,000 gpm
Hose ID = d = 6.0 in
Length = L = 100 ft
Fluid = Gasoline specific gravity = S = 0.71

$$\text{Viscosity} = V_C = 0.476 \text{ centipoise}/0.71$$
$$= 0.6704 \text{ centistokes}$$

$$\text{Reynolds number} = N_R = [3,160 \times Q]/[V_C \times d]$$
$$= [3,160 \times 1,000]\,[0.6704 \times 6]$$
$$= 785,600$$

From Figure 1, f = 0.012 for N_R = 785,600

$$P = \left[0.0134 \times f \times s \times L \times Q^2\right]/d^5$$
$$P = [0.0134 \times 0.012 \times 0.71 \times 100 \times 1,000,000]/7,776$$
$$P = 1.468 \text{ psi per } 100 \text{ ft}$$

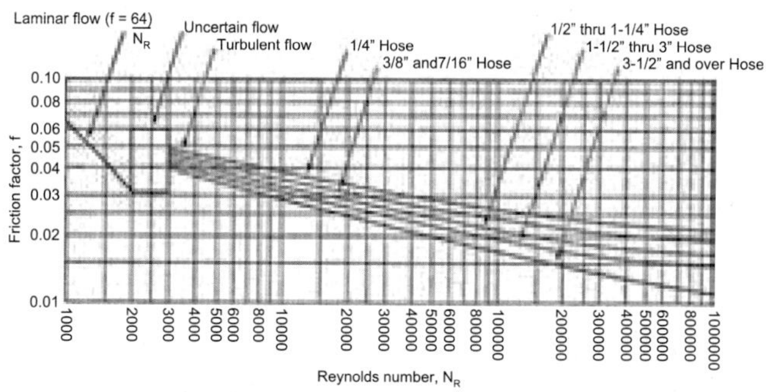

Figure 1. Friction factors for rubber hose. Reproduced with permission of Gates Rubber Company.

As a rule of thumb, loss for each set of couplings equals 5% of the loss per 100 ft of hose. For two sets of couplings,

$$\text{Pressure loss} = 2 \times 0.05 \times 1.468 = 0.15 \text{ psig}$$

Formulas for calculating pressure drop and flow rates for pipelines

Shell/MIT

$$P = \left(0.241\ fsQ^2\right)/d^5$$

For viscous flow,

$$f = 0.00207 \times (1/r)$$

For turbulent flow,

$$f = 0.0018 + 0.00662 \times (1/r)^{0.355}$$
$$r = 0.0119Q/dv$$

For viscous flow,

$$r = 0.1 \text{ to } 0.135$$

Turbulent flow.

R > 0.4

Indeterminate region, r = 0.135 to 0.4

Benjamin Miller

$$B = (0.1692)\left(d^5 P/s\right)^{0.5}\left(\log\left(d^3 sP / z^2\right) + 4.35\right)$$

T. R. Aude

$$B = 0.871 \times K\left(\frac{1}{z^{0.104}}\right)\left(\frac{\left[\left(P^{0.552} \times d^{2.656}\right)\right]}{s^{0.448}}\right)$$

$$P = \left[\frac{Bz^{0.104}s^{0.448}}{0.871\ Kd^{2.656}}\right]^{1.812}$$

Hazen-Williams

$$Q = \frac{0.148 \times d^{2.63} \times d^{0.54}}{s^{0.54}}$$

where:

Q = Barrels per day
B = Barrels per hour
P = Pressure loss, psi per mile
d = Pipe inside diameter, inches
z = Absolute viscosity, centipoises
s = Specific gravity
K = Efficiency factor (Aude's formula)
f = Friction factor
r = Reynolds number modified Re/7,742
C = Hazen-Williams coefficient (use 120 for new steel pipe; 10 for ordinary 15-year-old steel pipe; and 150 for fiberglass pipe)
SUS = Saybolt Universal Seconds
v = Kinetic viscosity, centistokes

Examples

Shell/MIT. Find the pressure loss for a crude oil pipeline where the flow rate is 30,000 barrels per day with the following conditions:

Pipe: 10.750″ × 0.307″ w.t.
Specific gravity = 0.85
Viscosity = 80 SUS

Convert viscosity (80 SUS) to centistokes

$$V_{cs} = 0.226\ V_{SUS} - 135/V_{SUS}$$
$$V = 0.226(80) - 135/80$$
$$V = 16.39\ cS$$

$$r = 0.0119\ Q/dv$$
$$r = (0.0119 \times 30,000)/(10.136 \times 16.39)$$
$$r = 2.148$$

$$f = 0.0018 + 0.00662 \times (1/r)^{0.355}$$
$$f = 0.0018 + 0.00662 \times (1/2.148)^{0.355}$$
$$f = 0.006846$$

$$P = \left(0.241\ fsQ^2\right)/d^5$$
$$P = \left(0.241 \times p.006846 \times 0.85 \times 30000^2\right)/10.136^5$$
$$P = 11.798\ psi/mile$$

T. R. Aude. Find the pressure loss, psi/mile, when pumping gasoline with the following conditions

B = 2,000 barrels per hour
Pipe = 8.625″ × 0.322″ w.t.
z = 0.59 centipoise
s = 0.733
K = 0.94 (pipe roughness/efficiency factor)

$$P = ((Bz^{0.104}\ s^{0.448})/(0.871\ Kd^{2.656}))^{1.812}$$
$$P = ((2000 \times 0.59^{0.104} \times 0.733^{0.448})/(0.871 \times 0.94 \times 7.981^{2.656}))^{1.812}$$
$$P = 44.1\ psi/mile$$

Note: Aude's formula is based on data taken from 6″ and 8″ lines and no extension data are available.

Benjamin Miller. Find the pressure loss, psi/mile, when pumping gasoline with the following conditions:

B = 4,167 barrels per hour
Pipe = 16″ × 0.3125″ w.t.
z = 0.59 centipoise
s = 0.733
d = 15.375″
$B = (0.169176)(d^5P/s)^{0.5}(\log(d^3sP/z^2)+4.35)$

Let P = 6.4 psi/mile for first try:

$$4167 = (0.1691676)(15.375^5 6.4/0.733)^{0.5}(\log(15.375^3 \times 0.733 \times 6.4)/(0.59^2) + 4.35)$$
$$4167 = 0.1691676 \times 2738 \times (4.69+4.35)$$
$$4167 \neq 4188$$

Let P = 6.34 psi/mile for second try:

$$4,167 = 0.169167 \times 2,726 \times (4.68 + 4.35)$$
$$4,167 = 4,167$$

Therefore, P = 6.34 psi/mile.

Hazen-Williams. Typically used for water system hydraulic calculations. Find the pressure drop, psi/mile, in a water system where the flow rate is 20,000 barrels per day.

The pipe is 4.5-in × 0.226-in w.t. new steel. Since the flowing medium is water, s = 1.0.

$20{,}000 = (0.148 \times 4{,}048^{2.63} \times 120 \times P^{0.54})$

$$P^{0.54} = 20{,}000/702.23 = 28.48$$

$$P = 493 \text{ psi/mile}$$

All of the preceding examples have assumed no difference in elevation. Where a difference in elevation exists, this must be considered in determining the total pressure required for a given flow rate.

Using and computing hydraulic gradients

To use pressure drop formulas to calculate pressure drop and flow rates for pipelines, it is first necessary to determine the characteristics of the fluid to be transported—primarily the specific gravity, viscosity, and either the flow rate or allowable pressure drop.

Before the system design is completed, it is necessary to know the elevation of the terrain to be traversed by the pipeline. Once the elevation of the system is known and a pipe size has been selected, the pressure loss data may be used to calculate or plot a hydraulic gradient. A hydraulic gradient is the height that a column of fluid will rise to if small tubes were installed along the pipeline as shown in Figure 1. Feet of head may be converted to pressure by dividing by 2.31 and multiplying by the specific gravity. Conversely, pressure is converted to feet of head by multiplying pressure by 2.31 and dividing by the specific gravity of the fluid.

The slope of the hydraulic gradient will represent the amount of pressure drop per unit of linear distance. The pressure drop per unit of length is uniform as long as there is no change in pipe size or specific gravity, and therefore the hydraulic gradient will be a straight line with a slope proportional to the system flow rate.

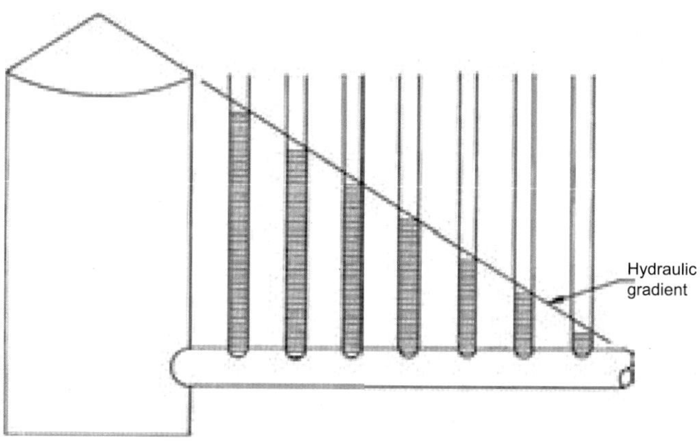

Figure 1. Schematic for hydraulic gradient.

Since the pipeline elevation profile is usually plotted in feet, it is helpful to plot the hydraulic gradient in feet rather than pressure. The hydraulic gradient may be plotted by converting pressure at the origin and end points to feet using the specific gravity of the fluid in the pipeline. The end point in this case will be the end of the pipeline if there are no intermediate booster stations. If the system is equipped with intermediate booster stations, then the suction of the booster station in the first segment will be the end point for that segment of the line. Don't forget to consider the suction pressure requirement for the booster station. These two points are then connected with a straight line, which will represent the pressure (in feet) at any intermediate point along the pipeline. The pressure at any point may be determined by taking the difference between the gradient and pipeline elevation and converting feet to psi.

Figure 2 shows a typical elevation profile plot with two hydraulic gradients—one for no booster station and one for an intermediate booster station and no change in elevation. In Figure 3 there is an elevation change, and for a hydraulically balanced system, it was necessary to locate the booster station at a point other than the geographic midpoint of the line. The location of the booster station in this example corresponds to the hydraulic mid-point of the system. A hydraulically balanced system is one in which all stations develop the same differential head. In Figure 2 where there was no elevation change, the booster station location is at the geographic mid-point of the system, which also corresponds to the hydraulic mid-point of the system. This is not meant to imply that a system must absolutely be hydraulically balanced. A booster station may be located off the hydraulic center; however, its differential head requirement will differ from that of the origin station and system throughput capacity will be affected. The amount of difference in head required, for a booster station which is located off hydraulic center, will

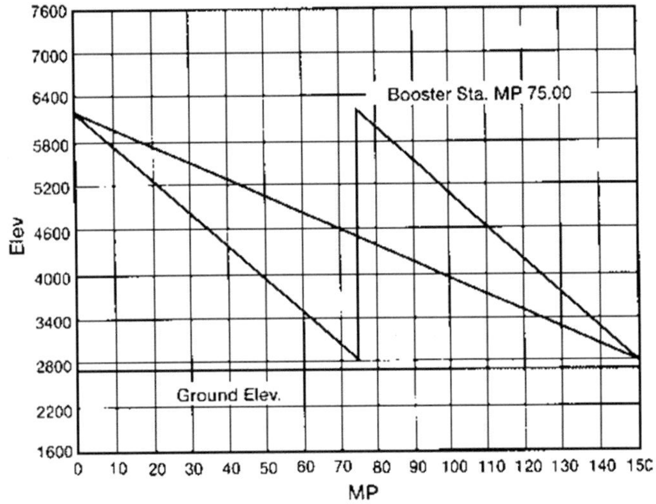

Figure 2. Profile with no elevation change.

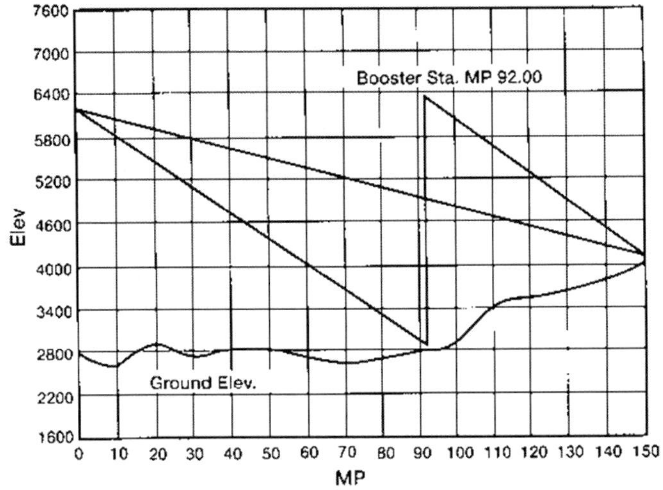

Figure 3. Profile with elevation change.

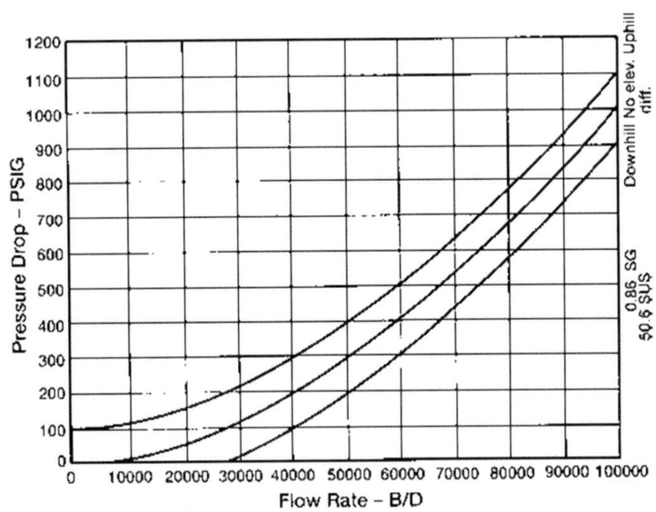

Figure 4. System curve. 30 mi., 12.750″ × 0.250″ pipe.

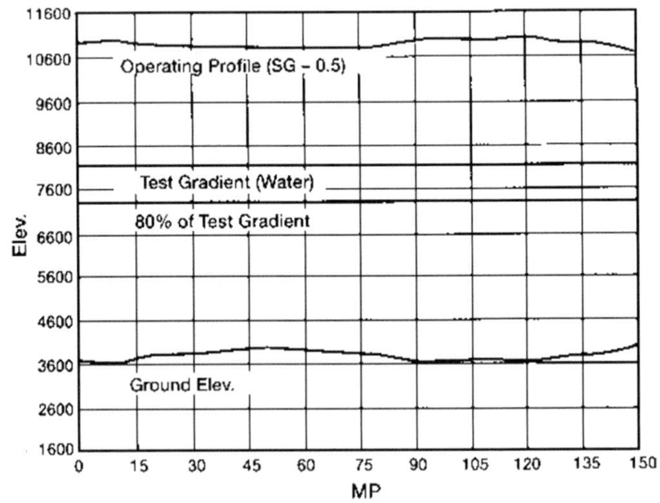

Figure 5. Pipeline elevation and test gradients and operating profile.

depend upon the amount of system elevation change and back pressure maintained at the end of the line.

The total system friction loss may be divided equally among an optimum number of booster stations. The discharge pressure at each station may be calculated with the following equation:

$$Pdp = (Pt + S + (N - 1) \times Ps + Pr)/N$$

where:

Pdp = Station discharge pressure
Pt = Total system friction loss
S = System static pressure
N = Number of booster stations
Ps = Station suction pressure
Pr = Residual or back pressure at end of line

In Figure 4, three system curves are shown to illustrate how elevation will affect the capacity of a system. With a constant input pressure of 750 psig, the capacity will be 77,500 barrels per day pumping uphill (272 ft); 84,000 barrels per day with no elevation change; and 90,000 barrels per day pumping downhill (272 ft). Note that where there is a difference in elevation, the amount of displacement of the curves, 100 psi in this case, is uniform.

The hydraulic gradient may also be used to analyze a system to assure that it is operated within the pressure limits determined by the hydrostatic test and the pipe design parameters. Figure 5 shows the test and maximum allowable operating gradients for a system. The hydrostatic test point was located at MP 150, and the test pressure was 1,800 psig, which in this case corresponds to 90% of the specified minimum yield strength (SMYS) of the pipe. Water was used as the test medium, and therefore 1,800 psig equates to 4,157 ft of head. The test gradient is drawn as a straight line across the profile from 8,157 ft (4,157 ft of head + 4,000 ft of pipeline elevation). This system may be operated at 80% of the test pressure, since this represents 72% of the SMYS of the pipe.

The 80% gradient (still based on water) is drawn as a straight line across the profile from 7,325 ft. The maximum allowable operating profile is now developed by taking the difference between the 80% of test pressure gradient, in feet, and the pipeline elevation and converting this difference to feet of head for the fluid to be transported. For this example, a specific gravity of 0.5 was used. At MP 150, the difference between pipeline elevation and the 80% of test gradient is 3,325 ft of head (water), and for a specific gravity of 0.5, this will convert to 6,650 ft of head. Therefore, the maximum operating pressure in ft of head for a fluid with a specific gravity of 0.5 will be 10,650 ft (6,650 ft of head + 4,000 ft pipeline elevation) at MP 150.

The maximum allowable operating profile will not be a straight line but will be a displaced mirror image of the pipeline elevation profile. The remainder of the maximum allowable operating profile may be developed by following this procedure at other points along the pipeline. When fluids with different specific gravities are transported, an operating profile will need to be developed for each specific gravity. If the line is hydrostatically tested in several segments rather than in one continuous length, an operating profile will need to be developed for each of the test segments.

Once the maximum allowable operating profile has been established, the hydraulic gradient may be superimposed. If the hydraulic gradient crosses the maximum allowable operating profile, the pipe will be overpressured in the segments where the gradient crosses the profile.

Figure 6 illustrates the effect hills have on the operation of a system. In this case, a hill is located near the end of the line and, due to its elevation, it will control the system capacity, since adequate back pressure must be maintained (if the vapor pressure of the fluid is above atmospheric pressure) at the end of the line to assure the gradient will "clear" the hill. If no back pressure is maintained at the end of the line, the gradient will intersect the side of the hill and, theoretically, a siphon action will occur if the elevation difference is not too great. In reality, the system flow rate will change of its own

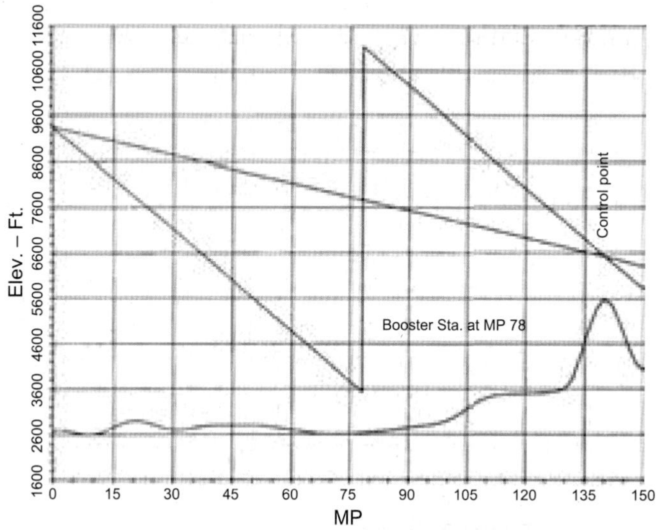

Figure 6. Pipeline elevation profile and hydraulic gradients.

accord to cause the gradient to intersect the top of the hill. The flow from that point to the end of the line will cascade or run away from the flow on the other side of the hill. If the vapor pressure of the fluid is very low, this condition can probably be tolerated. If segregated batches of material are handled or if the fluid vapor pressure is above atmospheric pressure, adequate back pressure must be maintained to keep the batches from mixing or forming vapor. Inadequate back pressure will cause difficulties in getting the flow in versus flow out to balance on a short-term basis as would be required for a leak detection system.

Two conditions are illustrated in Figure 6 to show the effect the flow rate will have on the amount of back pressure required to clear the hill. Some safety factor should be included when determining the amount of pressure to maintain in the system—especially when handling high vapor pressure materials. The flowing temperature will also need to be considered in making this determination.

For this example, the fluid has a specific gravity of 0.5 and its vapor pressure requires a minimum of 200 psig in the system. Therefore, it will be necessary to maintain 200 psig (924 ft) at the top of the hill located at MP 140. All gradients must be drawn through this point. If the specific gravity changes, the control point will move up or down depending upon the direction of the specific gravity change. The gradient connecting MP 0 to the end of the line passes through the control point and intersects the end of the line at 6,300 ft or 498 psig. The second condition, which assumes a booster station at MP 78, illustrates what happens to the back pressure requirement when the flow rate changes. The gradient from the discharge of the booster station must also pass through the control point, and since the flow rate is greater, the slope of the gradient is steeper and intersects the end of the line at 5,800 ft or 390 psig. It follows that as the system flow rate changes, it will be necessary to adjust the amount of back pressure. If the back pressure is permanently set to the higher value, system capacity will be penalized.

What are equivalent lengths?

Some pipeline systems are not made up of one single uniform pipe size. As the pressure in a system decreases, it may be desirable to use pipe with a thinner wall thickness to reduce the investment cost, or where deliveries are made along the pipeline, it is sometimes practical to use a smaller diameter pipe for the lower flow rate. Conversely, where injections are made into a system, it may be necessary to use a larger diameter pipe for the increased flow rate.

The capacity of a system may be increased by installing loop lines. Loop lines are two or more pipelines that are operated in parallel. Loop lines are not practical where segregated batches are pumped, since there will be a problem in getting the batches to arrive simultaneously. Also, if there are injections or deliveries along such a system, it will be almost impossible to balance the amount injected or delivered between the two lines. The most practical method for operating loops where segregated batches are handled is to operate each line separately. A product system might, for example, use one line for distillates and the other line for gasoline. If the gasoline volume is too large for one line, the excess could be batched through the other system. One other caution on loops is that the lengths of the legs may not be exactly the same.

Hydraulic systems are analogous to electrical circuits as shown in Figure 1, and this sometimes helps in analyzing a pipeline system. In a series electrical circuit, the same current flows through all of the circuit elements, and the voltage drops across the elements are additive. This same logic may be applied to a pipeline system with segments of different pipe sizes connected in a series. The pressure drops through the various segments are additive, and the sum of the pressure drops will equal the total system pressure drop.

A loop system is analogous to a parallel electrical circuit as shown in Figure 1. Since a common pressure is applied to all the legs of the loop system, the pressure drop across each leg of a loop system must be the same and the flow will divide in proportion to the friction loss in each leg.

An examination of the Shell/MIT flow formula reveals that the friction factor is dependent upon flow rate, pipe ID, and viscosity. Since the flow through each leg of a loop system will change as the total system flow changes, this will have an impact on the calculated equivalent length factor. For best accuracy, the equivalent length factor should be determined as near as possible to the expected operating parameters.

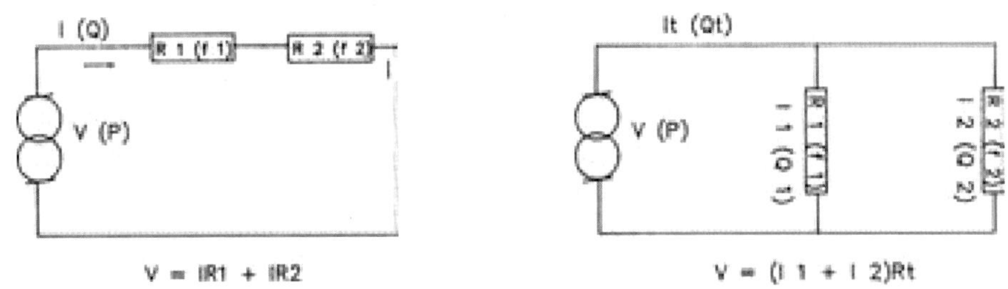

Series System

Loop System

Note: Hydraulic equivalents shown in ( )
P = pressure, Q = flow, f = friction
V = voltage, I = current, R = resistance

Figure 1. Series and loop system analogy.

Pressure drop in series systems

Figure 1 shows a system curve for a pipeline system 30 miles in length and consisting of 15 miles of 12.750-in OD × 0.250-in pipe in series with 15 miles of 10.750-in OD × 0.250-in pipe. System friction curves were developed for each pipe size and for the combination of the two sizes. Inspection shows that the curve for the combination is the sum of the 12.750-in OD curve and the 10.750-in OD curve. Pressure drop is additive on lines of constant flow for different pipe sizes connected in series.

The step of developing the individual curves for 12.750-in OD and 10.750-in OD pipe may be eliminated by calculating the equivalent length of 12.750-in OD or 10.750-in OD pipe. A system friction loss curve may then be calculated using the equivalent length of one pipe size or the other. For this example, the fluid is crude oil with a specific gravity of 0.86 and a viscosity of 50.6 SUS.

At a flow rate of 40,000 barrels per day, the pressure drop through 12.750-in OD × 0.250-in pipe is 6.618 psi/mile, and the pressure drop through 10.750-in OD × 0.250-in pipe is 15.469 psi/mile. The ratio of these two pressure drops is 6.618/15.469 or 0.428. This ratio times the length of the 12.750-in OD pipe equals the equivalent amount of 10.750-in OD pipe that will be required to yield the same pressure drop as 15 miles of 12.750-in OD pipe. Therefore, the equivalent length of 10.750-in OD pipe is 15 + (0.428 × 15) or 21.41 equivalent miles of 10.750-in OD pipe. If it is desired to use 12.750-in OD pipe as the base size, the same procedure may be followed except that the pressure ratio will be reversed and the reciprocal of 0.428 or 2.3364 multiplied by 15 miles of 10.750-in OD pipe to obtain 35.046 equivalent miles of 12.750-in OD pipe. The total equivalent length will be 15 miles + 35.046 miles, or 50.046 equivalent miles of 12.750-in OD pipe.

This may be checked as follows:

At a flow rate of 40,000 b/d, the pressure drop through 15 miles of 10.750-in OD is

$$15.469 \times 15 = 232 \text{ psig}$$

The pressure drop through 15 miles of 12.750-in OD is

$$6.618 \times 15 = 99 \text{ psig}$$

The total pressure drop will be

$$232 + 99 = 331 \text{ psig}$$

The pressure drop through 21.42 equivalent miles of 10.750-in OD is

$$21.42 \times 15.469 = 331 \text{ psig}$$

The pressure drop through 50.046 equivalent miles of 12.750-in OD is

$$50.046 \times 6.618 = 331 \text{ psig}$$

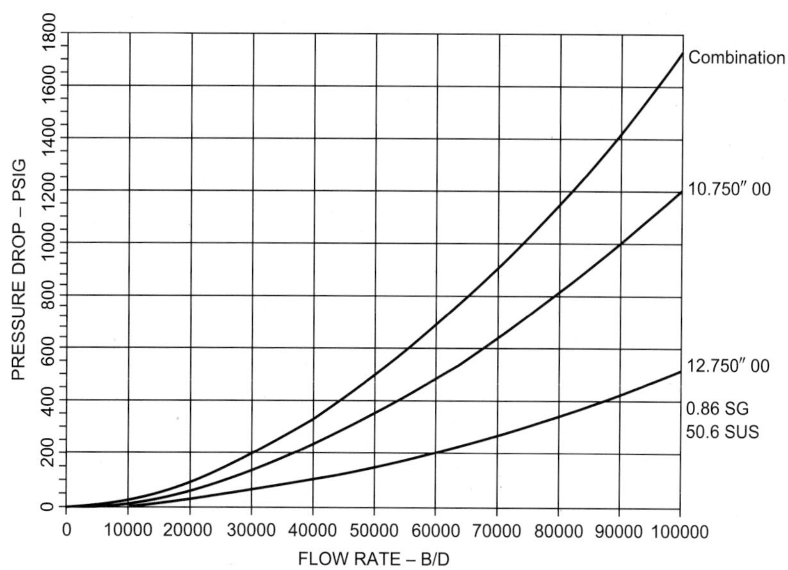

Figure 1. System curve. 15 mi., 12.750″ × 0.250″ pipe in series with 15 mi., 10.750″ × 0.250″ pipe.

Pressure drop in looped systems

The ratio of pressure drops may also be used for calculating the equivalent lengths of pipeline loops; however, the procedure is slightly different from that used for series systems.

Figure 1 shows the friction curves for a looped system, which is shown schematically in Figure 2. For reference, individual curves are shown for each pipe size and the combination of the two lines in parallel. The total friction loss for the system may be determined graphically using a system of curves such as shown in Figure 1, or the ratio of pressure drops may be used to determine an equivalent length for the looped system.

To use the ratio of pressure drops to convert the loop to a single line of equivalent size, determine the pressure drop through the larger size pipe (if this is to be the base size) using a chosen flow rate. In this example, a flow rate of 60,000 b/d of crude oil with a specific gravity of 0.86 and a viscosity of 50.6 SUS is used. The pressure drop through the 12.750-in OD pipe at 60,000 b/d is 13.54 psi/mile. Next, find a flow rate through the 10.750-in OD pipe that will yield the same pressure drop of 13.54 psi/mile. The required flow rate through the 10.750-in OD pipe is 37,100 b/d for a drop of 13.54 psi/mile. The sum of these two rates is 97,100 b/d. Use this rate, 97,100 b/d, and determine the pressure drop/mile through the 12.750-in OD pipe. The pressure drop/mile is 31.91 psi/mile. The factor for converting to an equivalent length of 12.750-in OD × 0.250-in pipe is 13.54/31.91 or 0.4243. The equivalent length of the loop combination will be 15 miles × 0.4243 or 6.3645 equivalent miles of 12.750-in OD × 0.250-in pipe.

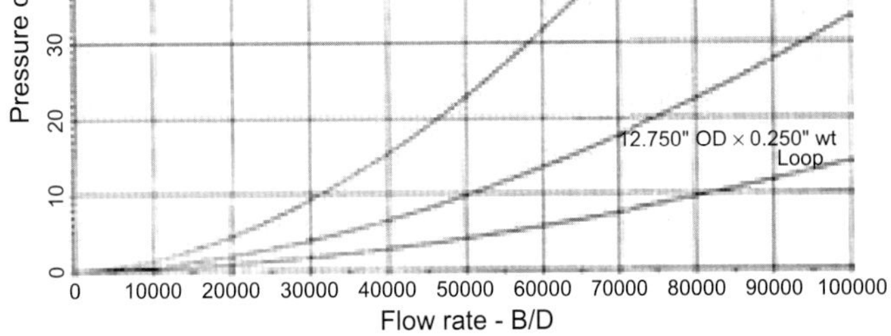

Figure 1. Friction curves. 10.750″ × 0.250″ and 12.750″ × 0.250″ and looped combination.

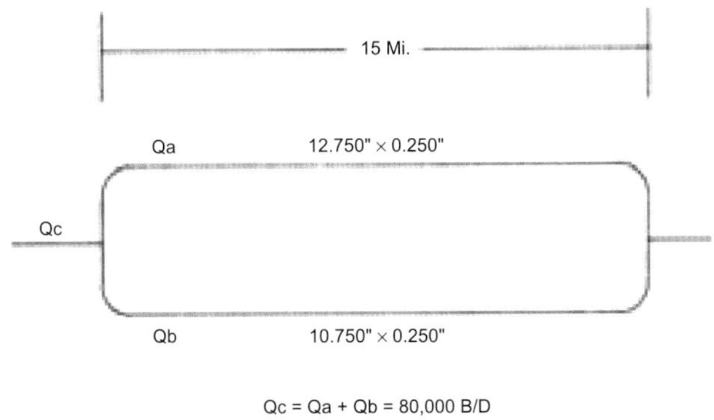

Figure 2. Looped pipelines.

If it is desired to convert the loop combination to a base diameter of 10.750-in OD, select a flow rate and determine the pressure drop in the 10.750-in OD leg. At 30,000 b/d, the pressure drop in the 10.750-in OD leg will be 9.32 psi/mile. Determine the flow rate through the 12.750-in OD leg to produce a pressure drop of 9.32 psi/mile. The rate will be 48,560 b/d. The sum of these two rates is 78,560 b/d. This rate through the 10.750-in OD leg will produce a pressure drop of 51.29 psi/mile. The equivalent length factor for 10.750-in OD pipe is 9.32/51.29 = 0.1817. The equivalent length will be 0.1817 × 15 or 2.7257 equivalent miles of 10.750-in OD pipe.

This may be checked as follows:

The pressure drop through the loop combination at a flow rate of 80,000 b/d is

$$9.62\ \text{psi/mile} \times 15\ \text{miles} = 144\ \text{pisg}$$

The flow through the 10.750-in OD leg will be 30,550 b/d and 49,450 b/d through the 12.750-in OD leg.

The pressure drop for 6.3645 equivalent miles of 12.750-in OD pipe at a flow rate of 80,000 b/d is

$$6.3645\ \text{equivalent miles} \times 22.584\ \text{psi/mile} = 144\ \text{pisg}$$

The pressure drop for 2.7257 equivalent miles of 10.750-in OD pipe at a flow rate of 80,000 b/d is

$$2.7257\ \text{eqivalent mass} \times 52.982\ \text{psi/miles} = 144\ \text{pisg}$$

The ratio-of-pressures method will work in conjunction with any of the flow formulas listed in this section.

Calculate pressure loss in annular sections

$P = 0.000027\ flpv^2/M$

where:

f = Friction factor
l = Length of section, ft
p = Density of fluid, lb/ft^3
v = Average velocity, fps
M = Mean hydraulic radius, ft
P = Pressure loss of length, psi
u = Absolute viscosity, ft/lb-sec (0.000672 centipose)
R_n = Reynolds number
Q = Volume, cu. ft per second
e/d = Relative roughness
$v = Q/0.7854{d^2}_A$
M = Area/wetted perimeter $= (d_1 - d_2)/4$

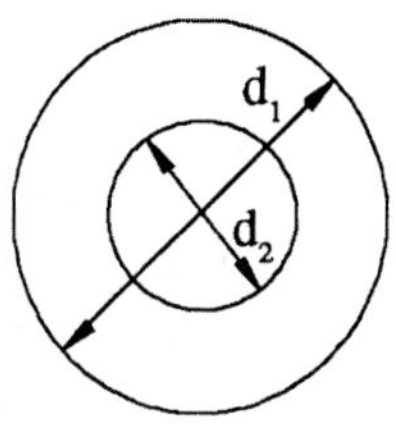

Example:

Throughput 60,000 barrels per day = 6.44 fps
Specific gravity 0.68
Viscosity 0.50 centipoise
$d_1 = 12.615''$
$d_2 = 7.000''$
$l = 2{,}000$ ft
$d_A = (d_2^1 - d_2^2)^{0.5} = (12.615^2 - 7.0^2)^{0.5}$
$M = (d_1 - d_2)/4 = (1.051 - 0.583)/4 = 0.117$ ft
$v = 6.44$ fps
$p = 62.4 \times 0.68 = 42.432$ lb/ft^3
$u = 0.000672 \times 0.50 = 0.000336$ ftlb-sec
$R_n = (4 \times 0.1178 \times 6.44 \times 42.432) \div 0.0000336 = 383{,}200$
$e/d = 0.00017$
$f = 0.015$
$P = (0.000027 \times 0.015 \times 2000 \times 42.432 \times 6.44^2) \div 0.1178$
$P = 12.1$ psi

Calculate pressure and temperature loss for viscous crudes ≥ 1,000 cP

When calculating pressure and temperature losses for highly viscous Newtonian crude oils, accurate flowing temperature data are required. Usually, crudes of this nature are flowing in the laminar region, and a change of a few degrees in temperature can change the viscosity by as much as 50% or more and thus cause a significant change in the friction factor.

Figure 1 illustates a typical pressure and temperature profile for an uninsulated pipeline.

The data typically required for determining a pressure/temperature profile for a viscous crude oil pipeline is shown in Table 1.

Once the data have been tabulated, a computer program such as GENNET-M can be used to calculate the temperature and pressure profile of the pipeline. The program performs a rigorous energy balance, i.e., static head, friction, and kinetic mechanical energy, plus friction, compression, and stored and transferred heat are included. Thermal resistances of the inside film, pipe, any insulation, and the outside environment (soil, air, or water) are calculated.

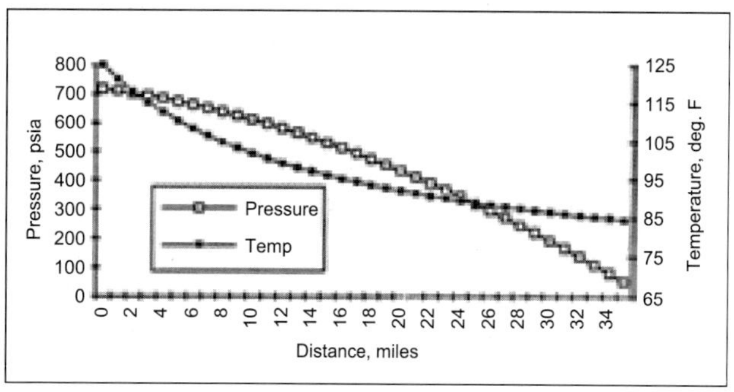

Figure 1. Uninsulated pipeline.

$$P_2 = P_1 - s\Delta x\left[\frac{C_Q Q^2 f}{d^5} + \frac{C_H(e_2 - e_1)}{\Delta x}\right]$$

where:

P_1 = Pressure at the upstream end of the calculation segment

Table 1
Data Required for Pressure/Temperature Loss Calculation

Length	Heat transfer coefficient (fluid)	Inlet temperature
Size, OD	Insulation thickness	Inlet pressure
Wall thickness	Thermal conductivity of insul.	Throughput
Pipe wall roughness	Thermal conductivity of soil	Sp. gr. at two temp.
Thermal conductivity of pipe	Pipe burial depth	Viscosity at two temp.
Upstream elevation	Soil temperature	Specific heat of liquid at two temp.
Downstream elevation	Vapor pressure at two temp.	Thermal conductivity of liquid at two temp.
Number of calculation segments	Soil K factor	

P_2 = Pressure at the downstream end of the calculation segment
Δx = Length of the calculation segment
f = Darcy–Weisbach friction factor
d = Inside diameter of the pipe
s = Liquid specific gravity
e_1 = Elevation at the upstream end of the calculation segment
e_2 = Elevation at the downstream end of the calculation segment
C_Q = Flow conversion constant
C_H = Head conversion constant

An energy balance is performed on each calculation segment and its surroundings, resulting in the next equation:

$$t_2 = \left[wC_{p1}t_1 - UA(LMTD) + w(h_1 - h_2)/J\right] \div \left(wC_{p2}\right)$$

where:

t_1 = Flowing temp. at the upstream end of the calculation segment
t_2 = Flowing temp. at the downstream end of the calculation segment
h_1 = Head at the upstream end of the calculation segment
h_2 = Head at the downstream end of the calculation segment
C_{p1} = Heat capacity at constant pressure evaluated at t_1
C_{p2} = Heat capacity at constant pressure evaluated at t_2
w = Total mass flow rate
U = Overall heat transfer coefficient
A = Surface area of the pipe in calculation segment
LMTD = Log mean temp. difference between the average flowing temp. and the average ambient temp.
J = Mechanical equivalent of heat

Example:

Pipeline length	35 miles
Size	14.0000″
Wall thickness	0.250″
Pipe wall roughness	0.0018″
Thermal conductivity (pipe)	29 Btu/hr-ft-F
Upstream elevation	1,500 feet
Downstream elevation	1,650 feet
Length	35 miles
Number of calculation segments	35
Heat transfer coefficient (calculated by program)	
Insulation thickness	0″ (Case 1)
Insulation thermal conductivity	0 Btu/hr-ft-F
Burial depth	36″
Soil temperature	50°F
Soil K factor	1.2
Inlet temperature	125°F
Inlet pressure	calculate
Outlet pressure	50 psia
Throughput	8,000 bpd

Liquid properties:

	Value	Temp.
Specific gravity	0.95	77
	0.90	210
Viscosity, cP	3260	80
	90	180
Specific heat, Btu/lbm-F	0.5	60
	0.5	80
Thermal conductivity, Btu/hr-ft-F	0.2	60
	0.22	100
Vapor pressure, psia	0.4	100

The pressure/temperature profile for an uninsulated line, based on the above conditions, is shown in Figure 1 and in Table 3.

If the pipeline crosses areas where the parameters such as soil thermal conductivity, pipe size, etc. change, GENNET-M will allow the system to be segmented into additional pipes with each pipe being configured with its own input data.

Assume that 2″ of urethane foam insulation is added to the pipeline previously described. The new pressure/temperature profile is shown in Figure 2 and in Table 4.

Typical soil K factors are shown in Table 2.

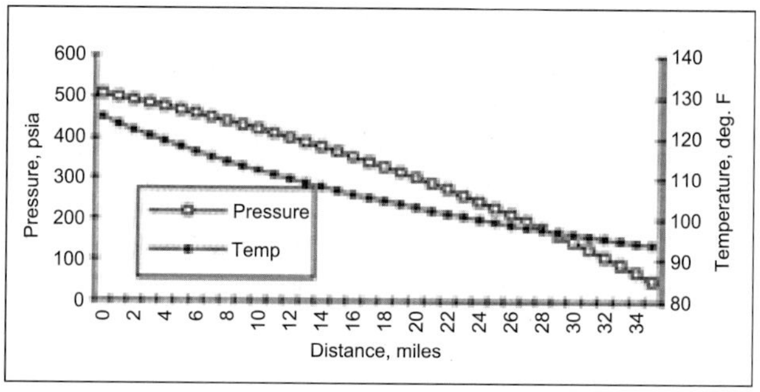

Figure 2. Insulated pipeline–2″ polyurethane.

Table 2
Soil K Factors Btu/hr/ft²/°F

Soil	Condition	Cover, In	K
Sandy	Dry	24	0.25 to 0.40
Sandy	Moist	24	0.50 to 0.60
Sandy	Soaked	24	1.10 to 1.30
Clay	Dry	24	0.20 to 0.30
Clay	Moist	24	0.40 to 0.50
Clay	Moist to wet	24	0.60 to 0.90
Sandy	Dry	8	0.60 to 0.70
Sandy	Moist to wet	8	1.20 to 2.40
River	Water	60	2.00 to 2.50
Air	Exposed	–	2.0 (approx.)

Table 3
Uninsulated Pipe

Distance miles	Pressure psia	Temp. deg. F
0	718	125
1	710.59	121.3
2	702.57	118.1
3	693.79	115.3
4	684.25	112.8
5	673.93	110.5
6	662.83	108.4
7	650.95	106.5
8	638.3	104.8
9	624.89	103.3
10	610.73	101.9
11	595.82	100.6
12	580.18	99.3
13	563.81	98.2
14	546.74	97.2
15	528.97	96.2
16	510.51	95.2
17	491.38	94.4
18	471.6	93.6
19	451.18	92.8
20	430.13	92.1
21	408.46	91.4
22	386.21	90.7
23	363.37	90.1
24	339.97	89.5
25	316.02	88.9
26	291.55	88.4
27	266.55	87.9
28	241.06	87.4
29	215.09	86.9
30	188.65	86.5
31	161.76	86
32	134.43	85.6
33	106.68	85.2
34	78.54	84.9
35	50	84.5

Table 4
Insulated Pipe

Distance miles	Pressure psia	Temp. deg. F
0	506.76	125
1	499.69	123.4
2	492.32	121.9
3	484.64	120.5
4	476.65	119.1
5	468.33	117.8
6	459.68	116.5
7	450.7	115.3
8	441.38	114.1
9	431.71	112.9
10	421.68	111.8
11	411.3	110.7
12	400.56	109.7
13	389.45	108.7
14	377.98	107.8
15	366.13	106.8
16	353.9	105.9
17	341.3	105.1
18	328.32	104.3
19	314.96	103.5
20	301.22	102.7
21	287.1	101.9
22	272.59	101.2
23	257.71	100.5
24	242.44	99.8
25	226.8	99.2
26	210.77	98.5
27	194.38	97.9
28	177.6	97.3
29	160.46	96.7
30	142.95	96.2
31	125.08	95.6
32	106.84	95.1
33	88.25	94.6
34	69.3	94.1
35	50	93.6

Reprinted with permission: Fluid Flow Consultants, Grove, Oklahoma

Determine batch injection rate as per enclosure

In the operation of a pipeline system, it may be necessary to inject a specific volume into a passing block of liquid. When it is desirable to maintain the same flow rate downstream of the injection and inject at the minimum rate or into most of the passing block, the following equations can be used to determine the upstream injection rates.

$$Q_U = \frac{B + H_O Q_D + Q - \left[(B + H_O Q_D + Q)^2 - 4 \times H_O \times B \times Q\right]^{0.5}}{2 \times H_O}$$

$$Q_I = Q_D - Q_U$$

where:

Q_D = Downstream rate that is to be held constant, bph
Q_I = Injection rate to be determined, bph
Q_U = Upstream pumping rate to be determined, bph
Q = Volume to be injected, barrels
B = Volume of passing block to be injected into, barrels
H_O = Block outage in hours. This is the time that the head and tail end of the block interface is passing the injection location, during which time injection is not allowed.

Example:

$$Q = 15{,}000 \text{ bbls for injection}$$

$$Q_I = \text{Injection rate bph}$$

$$B = 73{,}500 \text{ bbls}$$

$$Q_D = 15{,}600 \text{ bph (to be held constant)}$$

$$H_O = 0.5 \text{ hours}$$

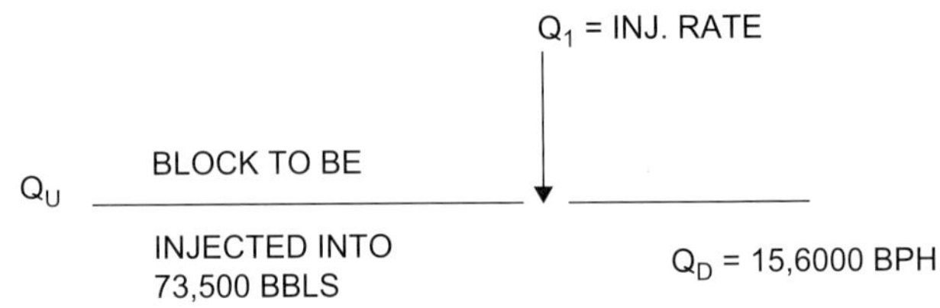

$$Q_U = \frac{73{,}500 + 0.5 \times 15{,}600 + 15{,}000 - \left[(73{,}500 + 0.5 \times 15{,}600 - 15{,}000)^2 - 4 \times 0.5 \times 73{,}500 \times 15{,}600\right]^{0.5}}{2 \times 0.5}$$

$$Q_U = 12{,}751 \text{ bph}$$

$$Q_I = 15{,}600 - 12{,}751 = 2{,}849 \text{ bph}$$

Pressure loss through valves and fittings

"K" FACTOR TABLE—SHEET 1 of 4

Representative Resistance Coefficients (K) for Valves and Fittings

("K" is based on use of schedule pipe as listed on page 2-10)

PIPE FRICTION DATA FOR CLEAN COMMERCIAL STEEL PIPE WITH FLOW IN ZONE OF COMPLETE TURBULENCE

Nominal Size	½″	¾″	1″	1¼″	1½″	2″	2½, 3″	4″	5″	6″	8-10″	12-16″	18-24″
Friction Factor (f_T)	.027	.025	.023	.022	.021	.019	.018	.017	.016	.015	.014	.013	.012

FORMULAS FOR CALCULATING "K" FACTORS*
FOR VALVES AND FITTINGS WITH REDUCED PORT
(Ref: Pages 2-11 and 3-4)

- **Formula 1**

$$K_2 = \frac{0.8\left(\sin\frac{\theta}{2}\right)(1-\beta^2)}{\beta^4} = \frac{K_1}{\beta^4}$$

- **Formula 2**

$$K_2 = \frac{0.5\,(1-\beta^2)\sqrt{\sin\frac{\theta}{2}}}{\beta^4} = \frac{K_1}{\beta^4}$$

- **Formula 3**

$$K_2 = \frac{2.6\left(\sin\frac{\theta}{2}\right)(1-\beta^2)^2}{\beta^4} = \frac{K_1}{\beta^4}$$

- **Formula 4**

$$K_2 = \frac{(1-\beta^2)^2}{\beta^4} = \frac{K_1}{\beta^4}$$

- **Formula 5**

$$K_2 = \frac{K_1}{\beta^4} + \text{Formula 1} + \text{Formula 3}$$

$$K_2 = \frac{K_1 + \sin\frac{\theta}{2}\left[0.8\,(1-\beta^2) + 2.6\,(1-\beta^2)^2\right]}{\beta^4}$$

- **Formula 6**

$$K_2 = \frac{K_1}{\beta^4} + \text{Formula 2} + \text{Formula 4}$$

$$K_2 = \frac{K_1 + 0.5\sqrt{\sin\frac{\theta}{2}}\,(1-\beta^2) + (1-\beta^2)^2}{\beta^4}$$

- **Formula 7**

$$K_2 = \frac{K_1}{\beta^4} + \beta\,(\text{Formula 2} + \text{Formula 4}) \text{ when } \theta = 180^\circ$$

$$K_2 = \frac{K_1 + \beta\left[0.5\,(1-\beta^2) + (1-\beta^2)^2\right]}{\beta^4}$$

$$\beta = \frac{d_1}{d_2}$$

$$\beta^2 = \left(\frac{d_1}{d_2}\right)^2 = \frac{a_1}{a_2}$$

Subscript 1 defines dimensions and coefficients with reference to the smaller diameter.

Subscript 2 refers to the larger diameter.

*** Use "K" furnished by valve or fitting supplier when available.**

SUDDEN AND GRADUAL CONTRACTION	SUDDEN AND GRADUAL ENLARGEMENT
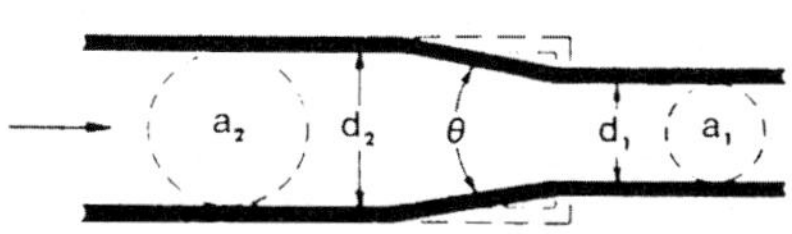	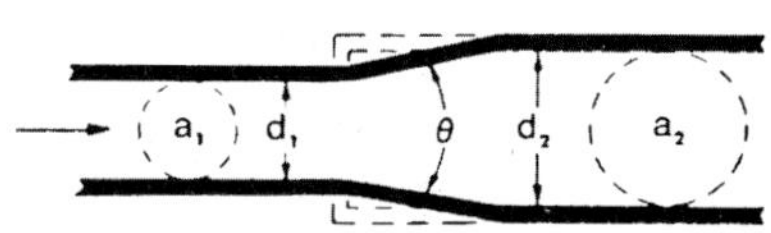
If: $\theta \leqq 45^\circ$ .... K_2 = Formula 1	If: $\theta \leqq 45^\circ$ .... K_2 = Formula 3
$45^\circ < \theta \leqq 180^\circ$ ... K_2 = Formula 2	$45^\circ < \theta \leqq 180^\circ$ ... K_2 = Formula 4

Reprinted by permission of Crane Co. from Technical Report 410

"K" FACTOR TABLE—SHEET 2 of 4

Representative Resistance Coefficients (K) for Valves and Fittings

(for formulas and friction data, see page A-26)

("K" is based on use of schedule pipe as listed on page 2-10)

GATE VALVES

Wedge Disc, Double Disc, or Plug Type

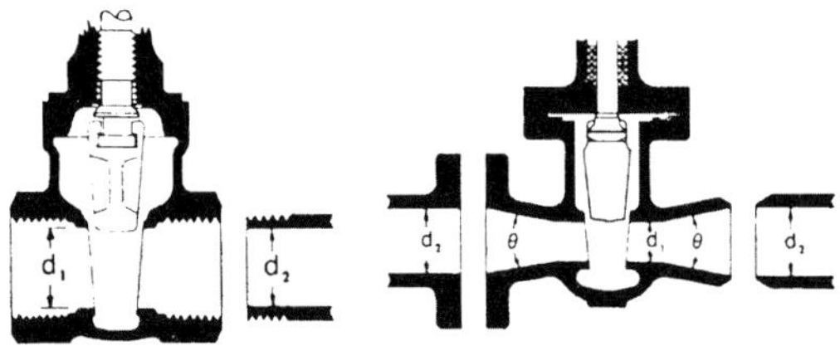

If: $\beta = 1,\ \theta = 0 \ldots\ldots K_1 = 8\,f_T$

$\beta < 1$ and $\theta \leqq 45° \ldots\ldots K_2 =$ Formula 5

$\beta < 1$ and $45° < \theta \leqq 180° \ldots K_2 =$ Formula 6

SWING CHECK VALVES

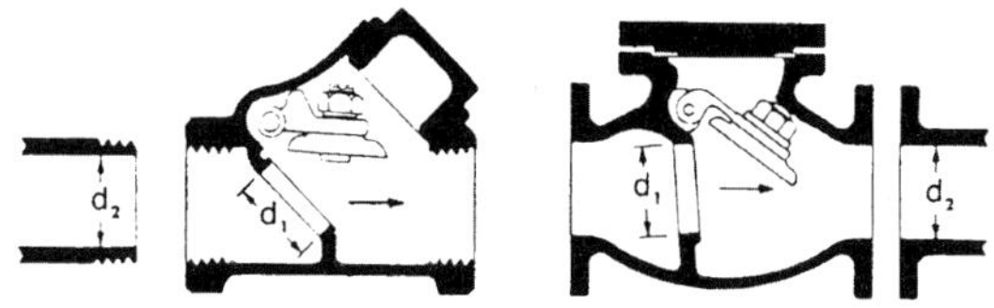

$K = 100\,f_T$	$K = 50\,f_T$
Minimum pipe velocity (fps) for full disc lift $= 35\sqrt{\bar{V}}$	Minimum pipe velocity (fps) for full disc lift $= 60\sqrt{\bar{V}}$ except U/L listed $= 100\sqrt{\bar{V}}$

GLOBE AND ANGLE VALVES

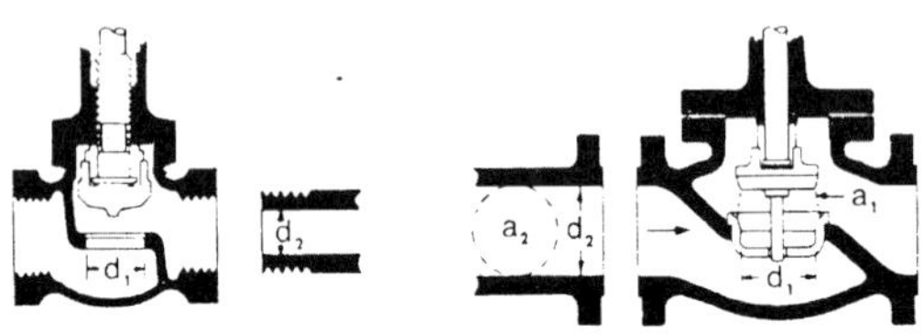

If: $\beta = 1 \ldots K_1 = 340\,f_T$

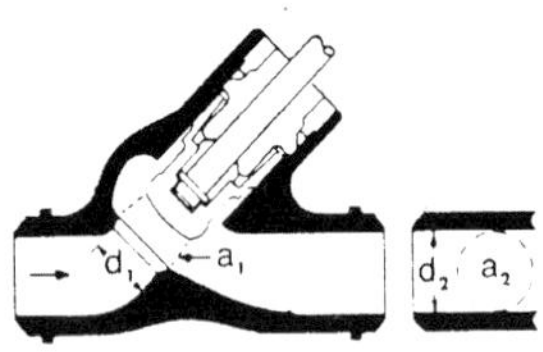

If: $\beta = 1 \ldots K_1 = 55\,f_T$

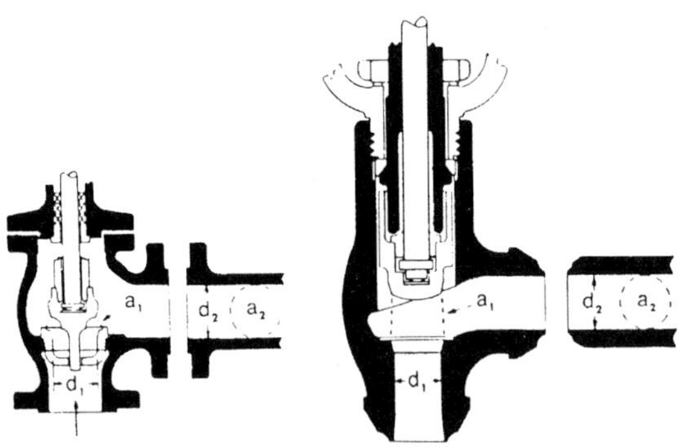

If: $\beta = 1 \ldots K_1 = 150\,f_T$ If: $\beta = 1 \ldots K_1 = 55\,f_T$

All globe and angle valves, whether reduced seat or throttled,

If: $\beta < 1 \ldots K_2 =$ Formula 7

LIFT CHECK VALVES

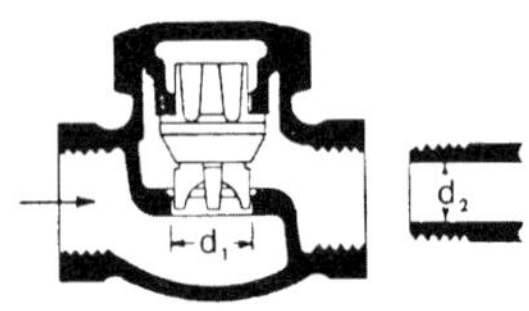

If: $\beta = 1 \ldots K_1 = 600\,f_T$

$\beta < 1 \ldots K_2 =$ Formula 7

Minimum pipe velocity (fps) for full disc lift $= 40\,\beta^2\sqrt{\bar{V}}$

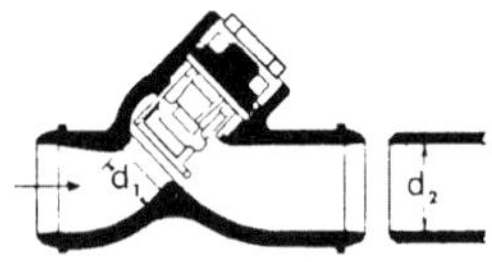

If: $\beta = 1 \ldots K_1 = 55\,f_T$

$\beta < 1 \ldots K_2 =$ Formula 7

Minimum pipe velocity (fps) for full disc lift $= 140\,\beta^2\sqrt{\bar{V}}$

TILTING DISC CHECK VALVES

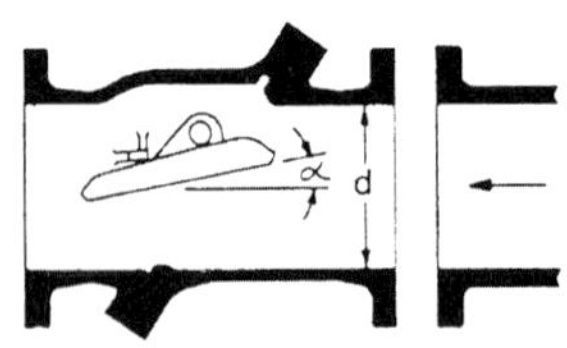

	$\alpha = 5°$	$\alpha = 15°$
Sizes 2 to 8"... K =	$40\,f_T$	$120\,f_T$
Sizes 10 to 14"... K =	$30\,f_T$	$90\,f_T$
Sizes 16 to 48"... K =	$20\,f_T$	$60\,f_T$
Minimum pipe velocity (fps) for full disc lift =	$80\sqrt{\bar{V}}$	$30\sqrt{\bar{V}}$

"K" FACTOR TABLE—SHEET 3 of 4

Representative Resistance Coefficients (K) for Valves and Fittings

(for formulas and friction data, see page A-26)

("K" is based on use of scheduled pipe as listed on page 2-10)

STOP-CHECK VALVES (Globe and Angle Types)

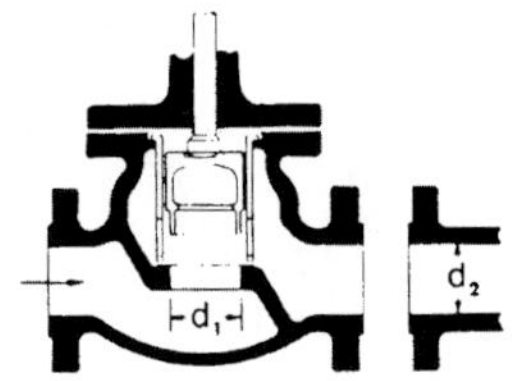

If:
$\beta = 1 \ldots K_1 = 400\, f_T$
$\beta < 1 \ldots K_2$ = Formula 7

Minimum pipe velocity for full disc lift $= 55\,\beta^2 \sqrt{\bar{V}}$

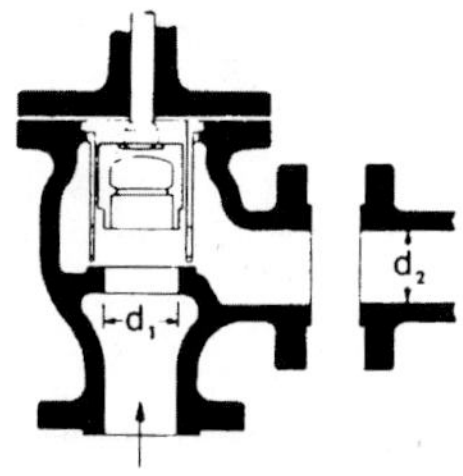

If:
$\beta = 1 \ldots K_1 = 200\, f_T$
$\beta < 1 \ldots K_2$ = Formula 7

Minimum pipe velocity for full disc lift $= 75\,\beta^2 \sqrt{\bar{V}}$

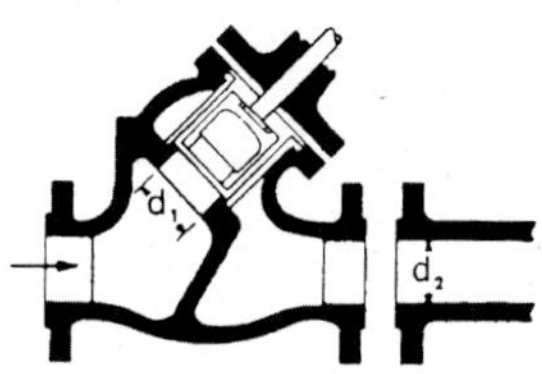

If:
$\beta = 1 \ldots K_1 = 300\, f_T$
$\beta < 1 \ldots K_2$ = Formula 7

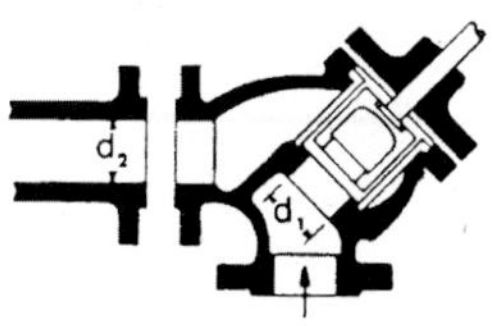

If:
$\beta = 1 \ldots K_1 = 350\, f_T$
$\beta < 1 \ldots K_2$ = Formula 7

Minimum pipe velocity (fps) for full disc lift $= 60\,\beta^2 \sqrt{\bar{V}}$

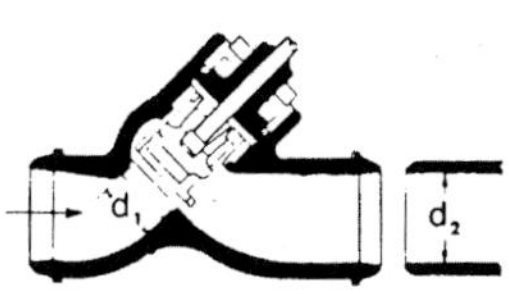

If:
$\beta = 1 \ldots K_1 = 55\, f_T$
$\beta < 1 \ldots K_2$ = Formula 7

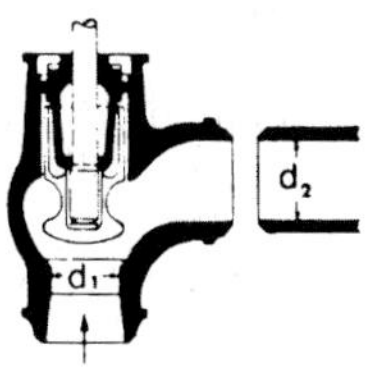

If:
$\beta = 1 \ldots K_1 = 55\, f_T$
$\beta < 1 \ldots K_2$ = Formula 7

Minimum pipe velocity (fps) for full disc lift $= 140\,\beta^2 \sqrt{\bar{V}}$

FOOT VALVES WITH STRAINER

Poppet Disc

$K = 420 f_T$

Minimum pipe velocity (fps) for full disc lift $= 15 \sqrt{\bar{V}}$

Hinged Disc

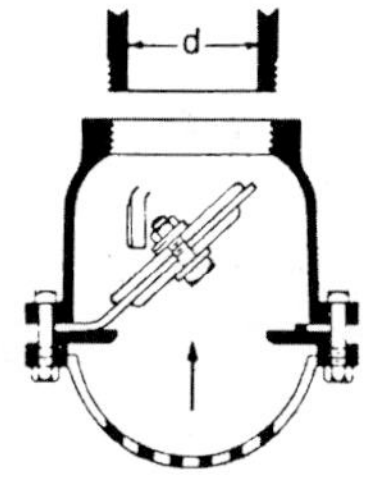

$K = 75 f_T$

Minimum pipe velocity (fps) for full disc lift $= 35 \sqrt{\bar{V}}$

BALL VALVES

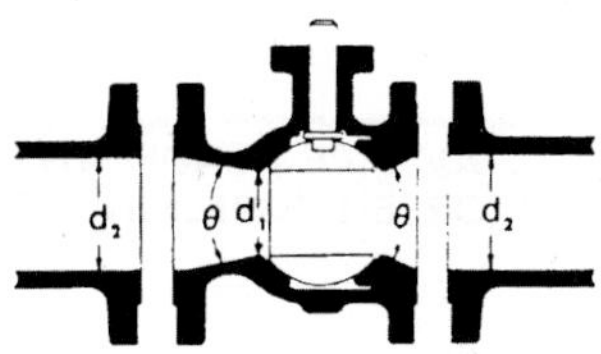

If: $\beta = 1,\ \theta = 0 \ldots\ldots\ldots\ldots K_1 = 3\, f_T$
$\beta < 1$ and $\theta \leqq 45° \ldots\ldots K_2$ = Formula 5
$\beta < 1$ and $45° < \theta \leqq 180° \ldots K_2$ = Formula 6

BUTTERFLY VALVES

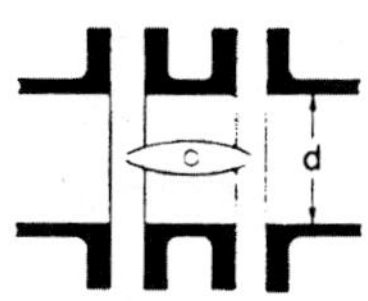

Sizes 2 to 8″... $K = 45\, f_T$
Sizes 10 to 14″... $K = 35\, f_T$
Sizes 16 to 24″... $K = 25\, f_T$

"K" FACTOR TABLE—SHEET 4 of 4

Representative Resistance Coefficients (K) for Valves and Fittings

(for formulas and friction data, see page A-26)
("K" is based on use of schedule pipe as listed on page 2-10)

PLUG VALVES AND COCKS

Straight-Way

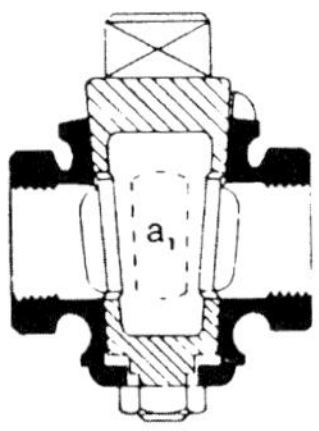

If: $\beta = 1$, $K_1 = 18 f_T$

3-Way

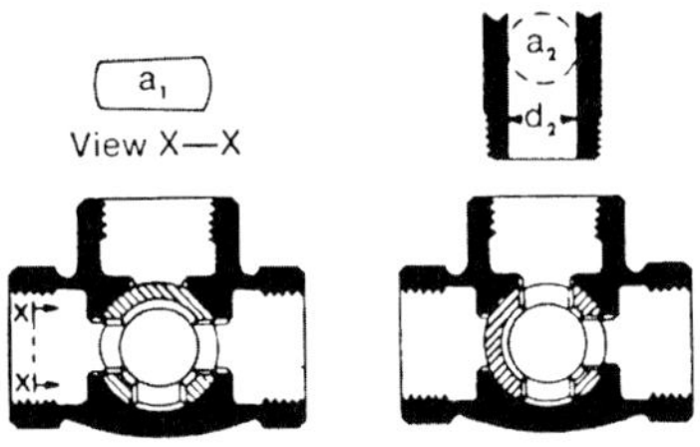

If: $\beta = 1$, $K_1 = 30 f_T$

If: $\beta = 1$, $K_1 = 90 f_T$

If: $\beta < 1 \ldots K_2 =$ Formula 6

MITRE BENDS

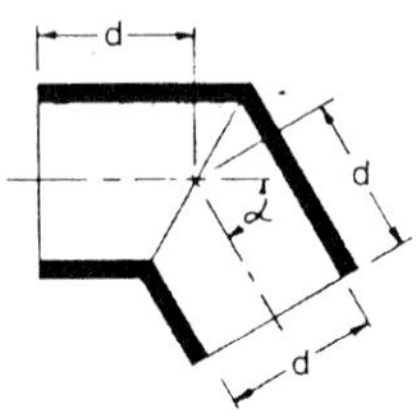

α	K
0°	2 f_T
15°	4 f_T
30°	8 f_T
45°	15 f_T
60°	25 f_T
75°	40 f_T
90°	60 f_T

90° PIPE BENDS AND FLANGED OR BUTT-WELDING 90° ELBOWS

r/d	K	r/d	K
1	20 f_T	8	24 f_T
1.5	14 f_T	10	30 f_T
2	12 f_T	12	34 f_T
3	12 f_T	14	38 f_T
4	14 f_T	16	42 f_T
6	17 f_T	20	50 f_T

The resistance coefficient, K_B, for pipe bends other than 90° may be determined as follows:

$$K_B = (n - 1)\left(0.25\ \pi\ f_T \frac{r}{d} + 0.5\ K\right) + K$$

n = number of 90° bends
K = resistance coefficient for one 90° bend (per table)

CLOSE PATTERN RETURN BENDS

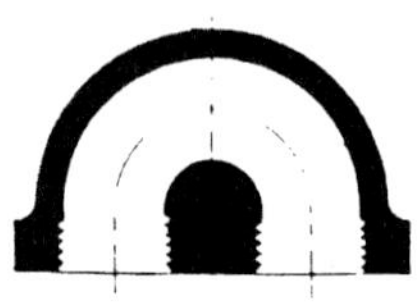

$K = 50 f_T$

STANDARD ELBOWS

90°

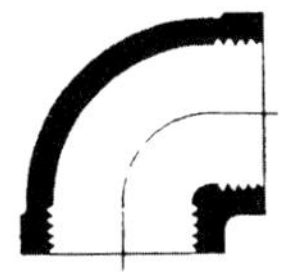

$K = 30 f_T$

45°

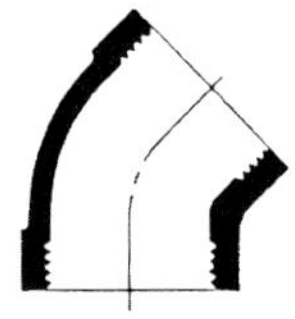

$K = 16 f_T$

STANDARD TEES

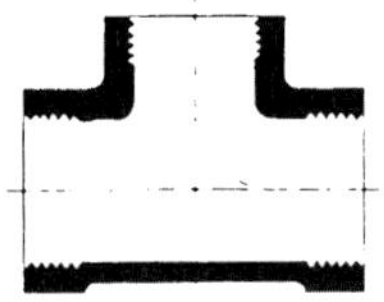

Flow thru run. . . . . . . $K = 20 f_T$
Flow thru branch. . . . $K = 60 f_T$

PIPE ENTRANCE

Inward Projecting

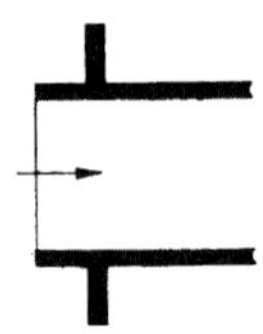

$K = 0.78$

r/d	K
0.00*	0.5
0.02	0.28
0.04	0.24
0.06	0.15
0.10	0.09
0.15 & up	0.04

*Sharp-edged

Flush

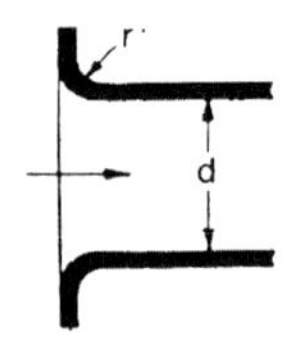

For K, see table

PIPE EXIT

Projecting

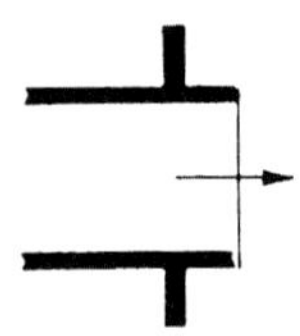

$K = 1.0$

Sharp-Edged

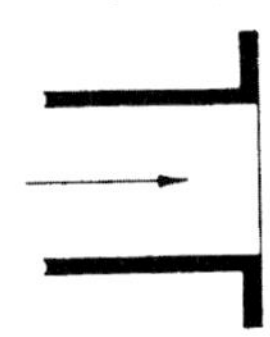

$K = 1.0$

Rounded

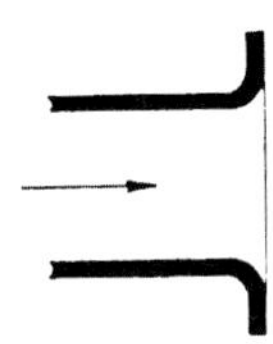

$K = 1.0$

Equivalent Lengths L and L/D and Resistance Coefficient *K*

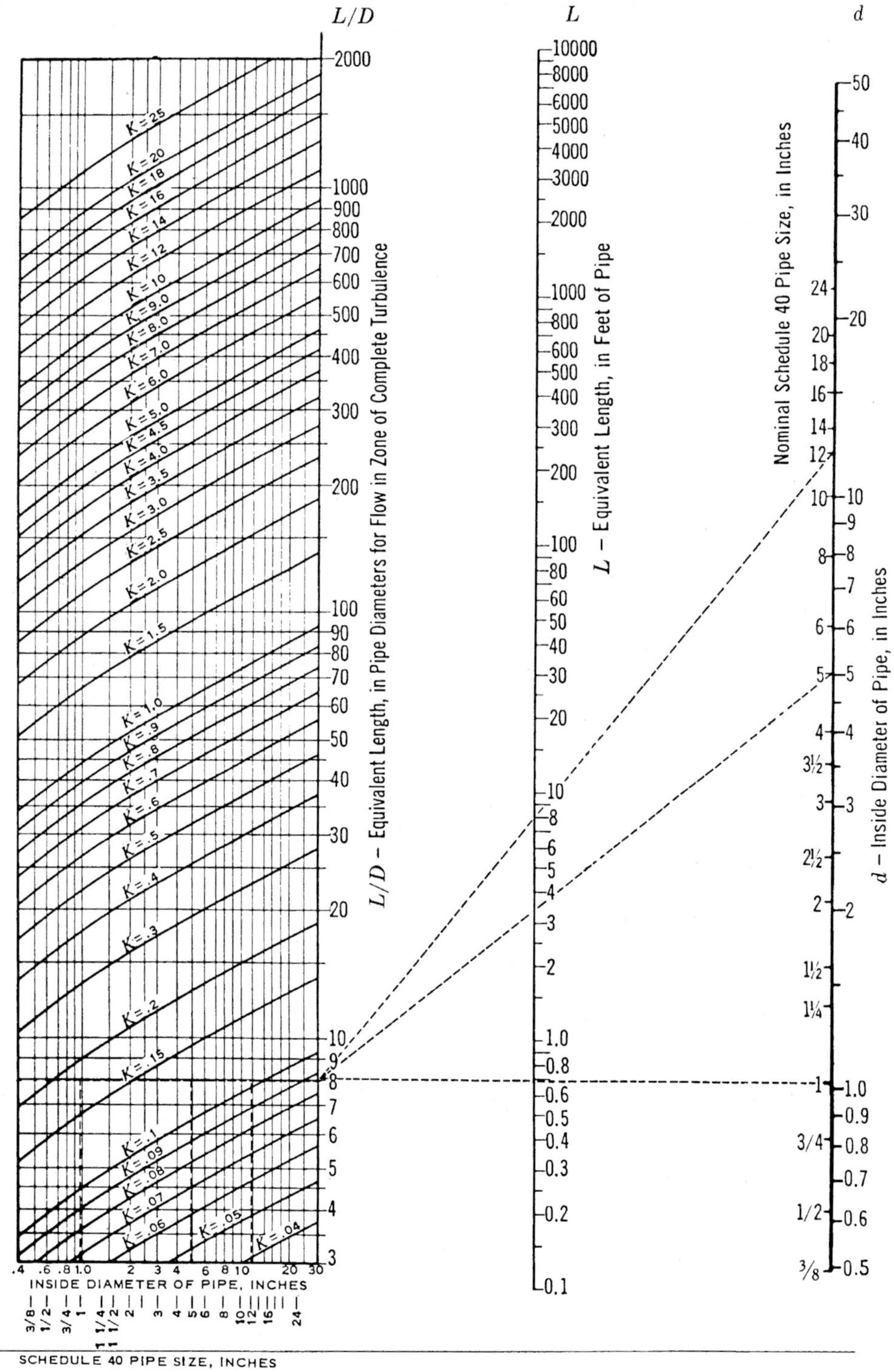

Problem: Find the equivalent length in pipe diameters and feet of Schedule 40 clean commercial steel pipe, and the resistance factor *K* for 1, 5, and 12-inch fully-opened gate valves with flow in zone of complete turbulence.

Solution

Valve Size	1″	5″	12″	*Refer to*
Equivalent length, pipe diameters	8	8	8	Page A-27
Equivalent length, feet of Sched. 40 pipe	0.7	3.4	7.9	Dotted lines on chart.
Resist. factor *K*, based on Sched. 40 pipe	0.18	0.13	0.10	

Equivalents of Resistance Coefficient *K* and Flow Coefficient C_v

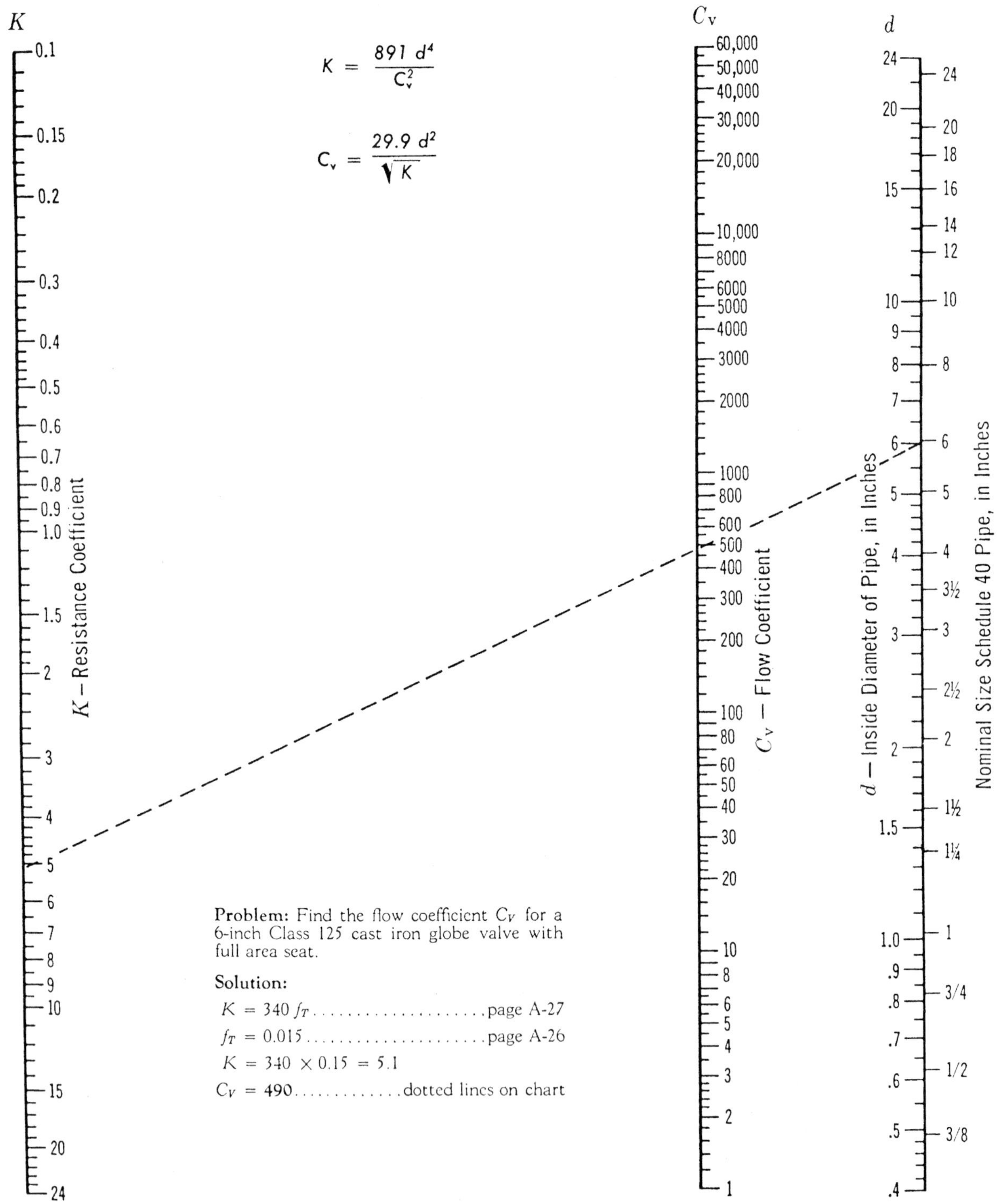

Problem: Find the flow coefficient C_V for a 6-inch Class 125 cast iron globe valve with full area seat.

Solution:

$K = 340\ f_T$ .... page A-27

$f_T = 0.015$ .... page A-26

$K = 340 \times 0.15 = 5.1$

$C_V = 490$ .... dotted lines on chart

Reprinted by permission of Crane Co. from Technical Report 410.

Nomograph for calculating Reynolds number for clean steel and wrought iron pipe

The nomograph (Figure 1) permits calculation of the Reynolds number for liquids and the corresponding friction factor for clean steel and wrought iron pipe.

The Reynolds number is defined as:

$$\mathrm{Re} = 22,700\frac{q\rho}{d\mu} = 50.6\frac{Q\rho}{d\mu} = 6.31\frac{W}{d\mu} = \frac{dv\rho}{\mu a}$$

where: Re = Reynolds number
d = Inside pipe diameter, in
q = Volumetric flow rate, cu. ft/sec
Q = Volumetric flow rate, gal/min
ρ = Fluid density, lb/cubic ft
μ = Fluid viscosity, centipoise
μa = Fluid viscosity, lb/ft-sec
v = Average fluid velocity, ft/sec
w = Mass rate of flow, lb/hr

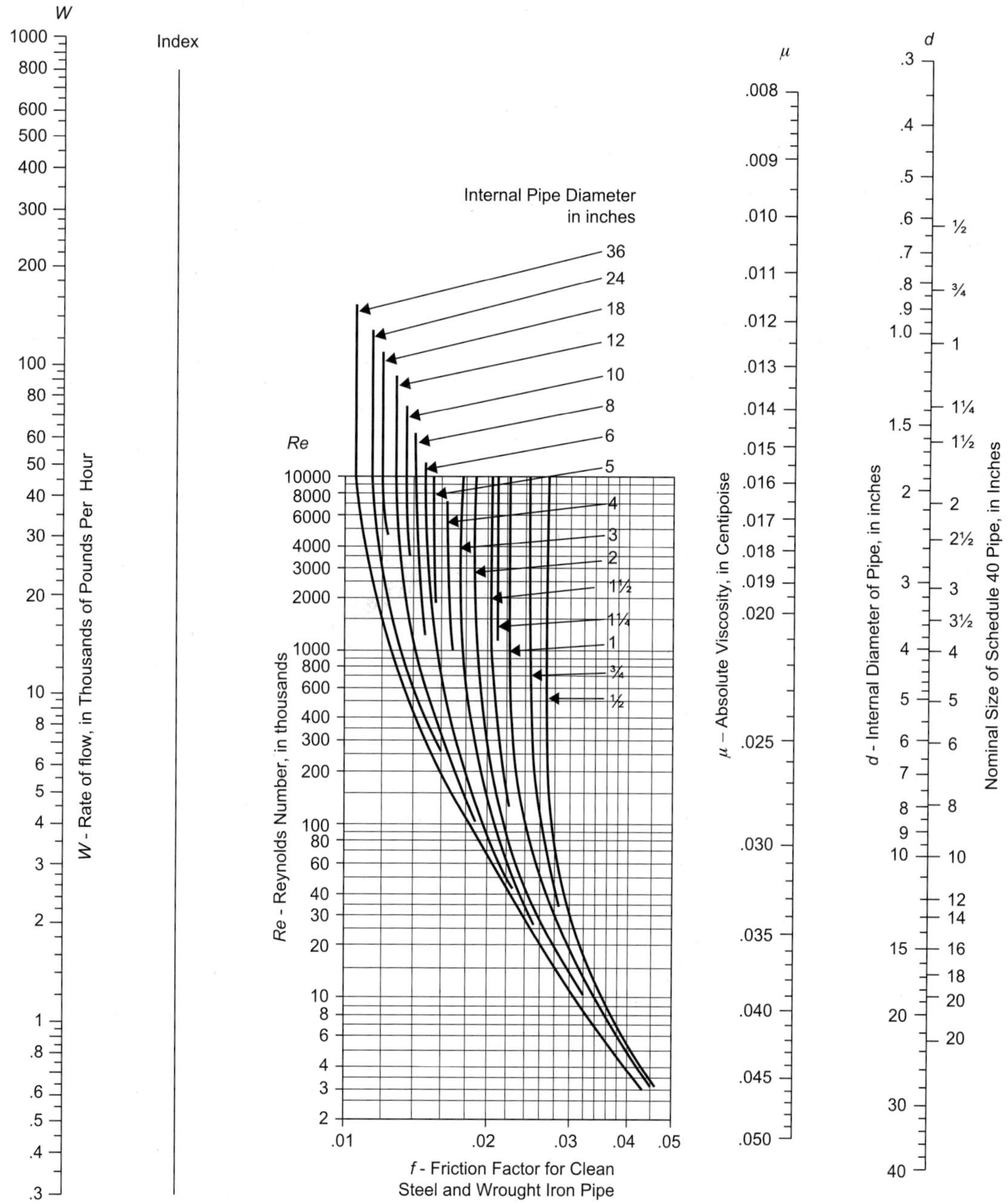

Figure 1. Reynolds number for liquid flow friction factor for clean steel and wrought iron pipe. (Reproduced by permission, *Tech. Paper, 410*, Crane Co., copyright 1957).

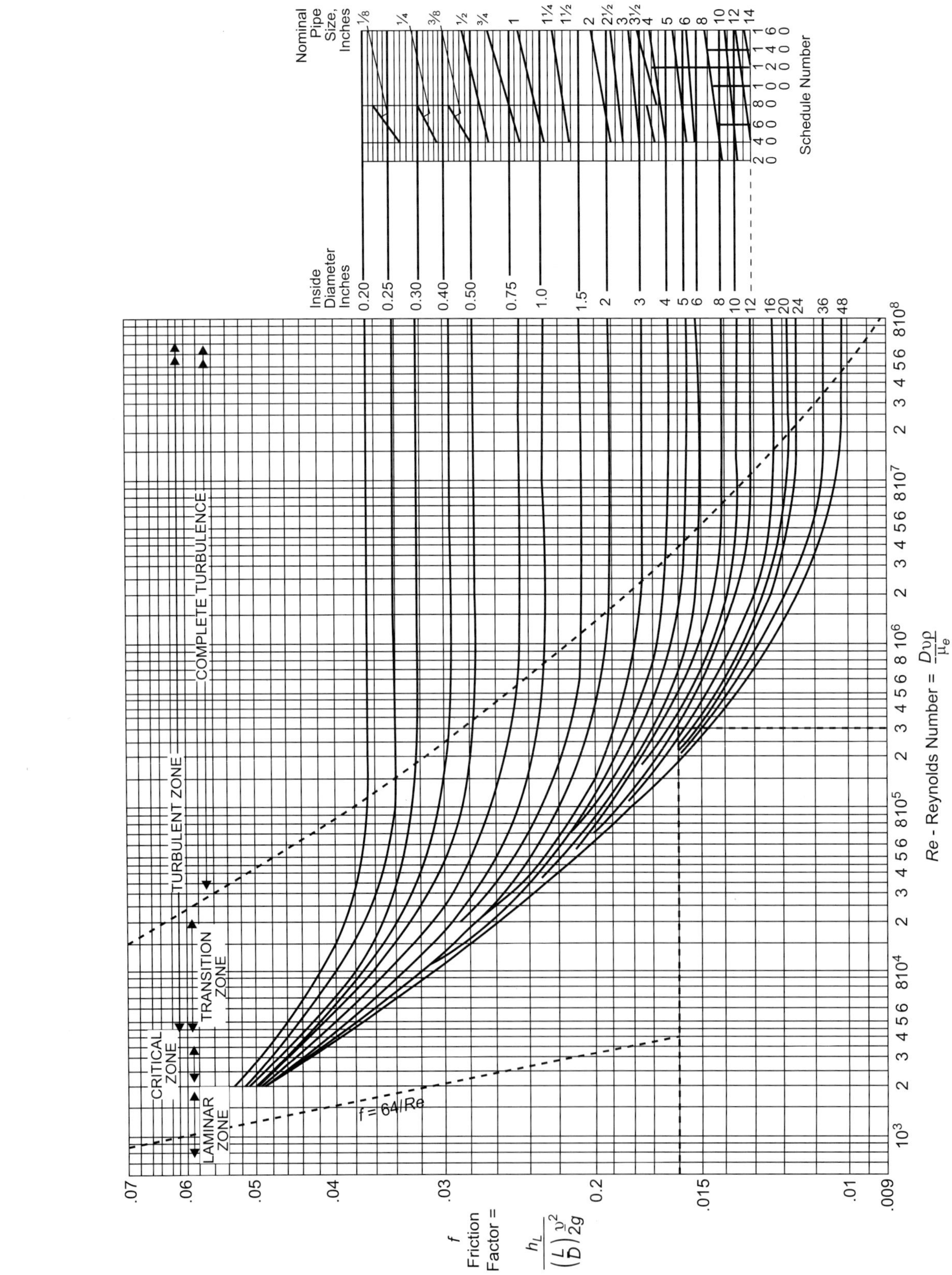

Figure 2. Friction factors for clean commercial steel and wrought iron pipe. (Reproduced by permission, *Tech. Paper, 410*, Crane Co., 1957.)

Example. Given water at 200°F and:

$d = $ 4-in schedule 40 steel pipe
$Q = 415$ gal/min
$\rho = 60.13$ lb/cu. ft
$\mu = 0.3$ centipoises

Obtain the Reynolds number and the friction factor.

Connect	With	Mark or Read
Q = 415	$\rho = 60.13$	W = 200,000
W = 200,000	d = 4-in schedule 40	Index
Index mark	$\mu = 0.3$	Re = 1,000,000

At Re = 1,000,000 read f = 0.017 from the graph on the center of the nomograph.

Figure 2 is an enlarged version of the friction factor chart, based on Moody's data for commercial steel and wrought iron. This chart can be used instead of the smaller version shown in Figure 1.

Reference

Moody, L. F., *Transactions, American Society of Mechanical Engineers*, 66, 671 (1944).

Nomograph for calculating pressure drop of liquids in lines for turbulent flow

The nomograph in Figure 1 is based on a modified version of the Fanning equation. It gives the pressure drop per 100 ft of pipe for liquids flowing through a circular pipe. The basic equation is:

$$P_{100} = \frac{0.0216\, f\rho Q^2}{d^2} = 4,359\frac{f\rho q^2}{d^2} = 0.000336\frac{fW^2}{d^2\rho}$$
$$= 0.1294\frac{fv^2\rho}{d}$$

where:

P_{100} = Pressure loss per 100 ft of pipe, psi
f = Friction factor ($4 \times f'$ in Fanning equation)
ρ = Fluid density, lb/cu. ft
d = Tube internal diameter, in
W = Flow rate, lb/hr
Q = Flow rate, gal/min
q = Flow rate, cu. ft/sec

The Moody friction factor and relative roughness generalized charts (Figures 2 and 3) are included for use with the nomogram.

Example. The estimated Reynolds number is 100,000 when water is flowing at a rate of 100 gal/min and a temperature of 70°F. Find the pressure drop per 100 ft of 2-in diameter pipe, the relative roughness from Figure 3 is 0.0009.

At e/d = 0.0009 and Re = 100,000, the friction factor, f, is 0.022 from Figure 2.

Using Figure 1:

Connect	With	Mark or Read
f = 0.022	$\rho = 62.3$ lb/cu. ft	Intersection at Index 1
Index 1	Q = 100 gal/min	Intersection at Index 2
Index 2	d = 2-in	P = 8 psi per 100 ft of pipe

Source

Flow of Fluids Through Valves, Fittings and Pipe, Technical Paper No. 410, 3-11, Crane Company, Chicago (1957).
Anon., *British Chemical Engineering*, 2, 3, 152 (1957).

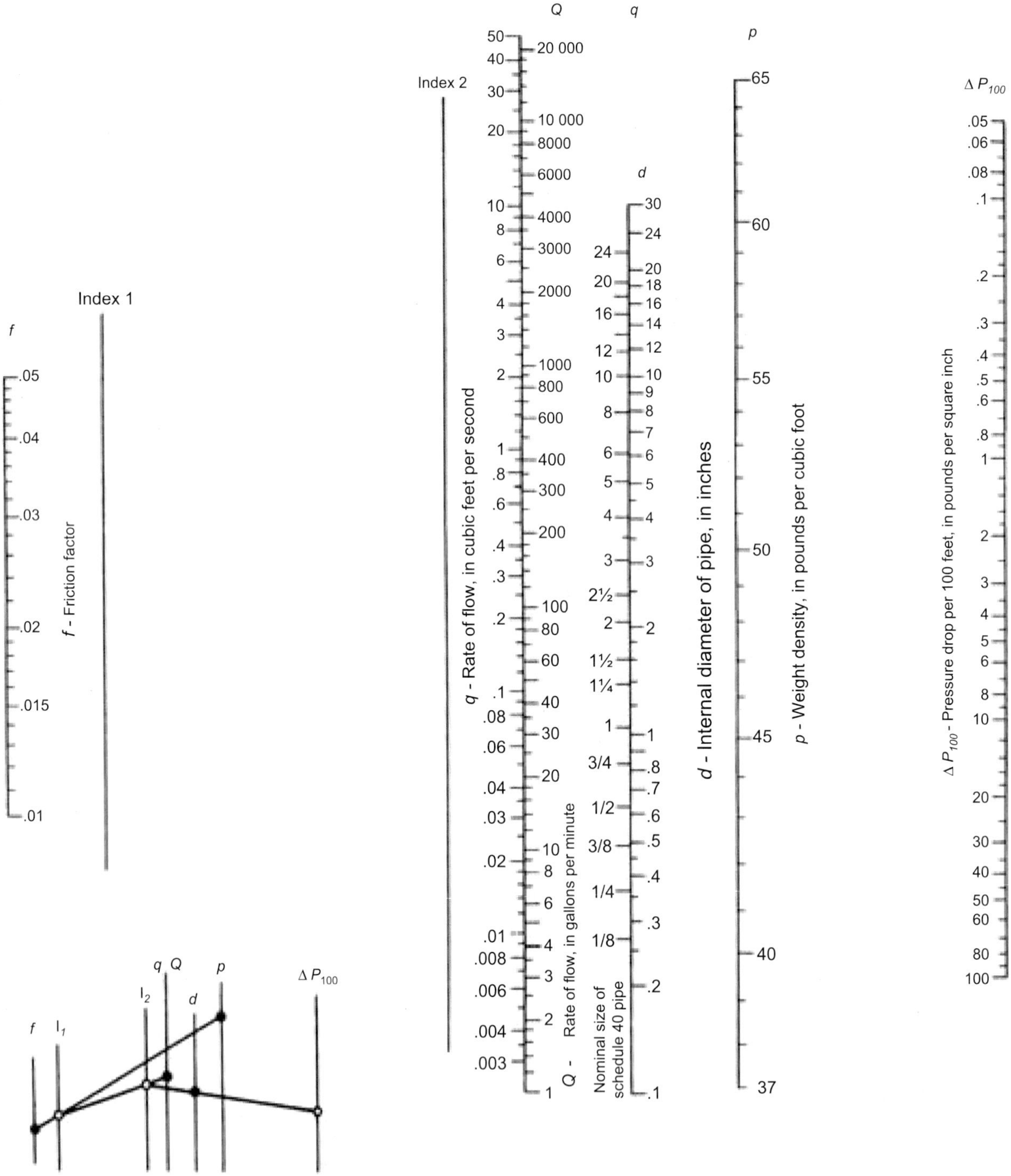

Figure 1. Pressure drop in liquid lines for turbulent flow.

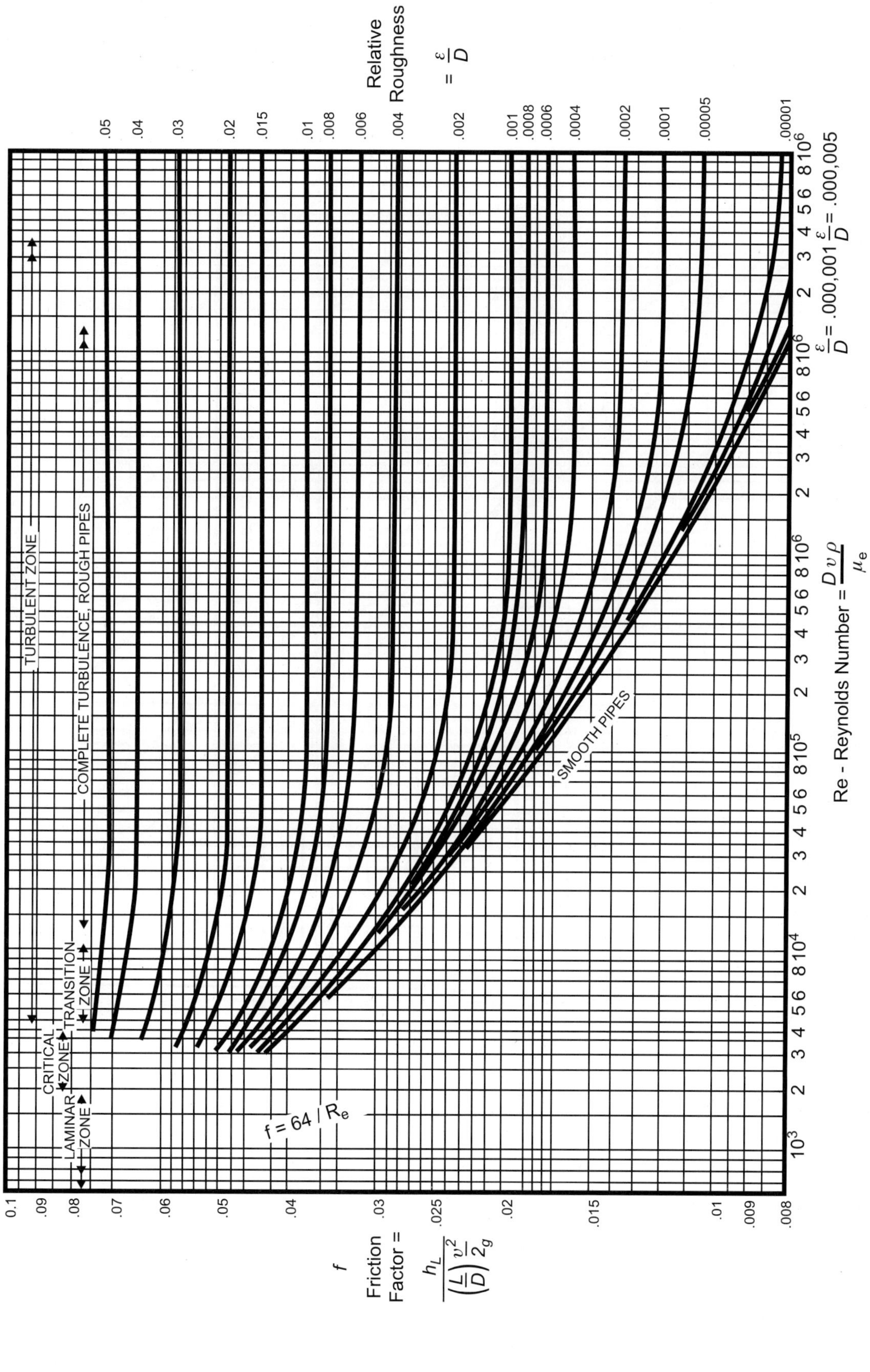

Figure 2. Friction factors for any type of commercial pipe.

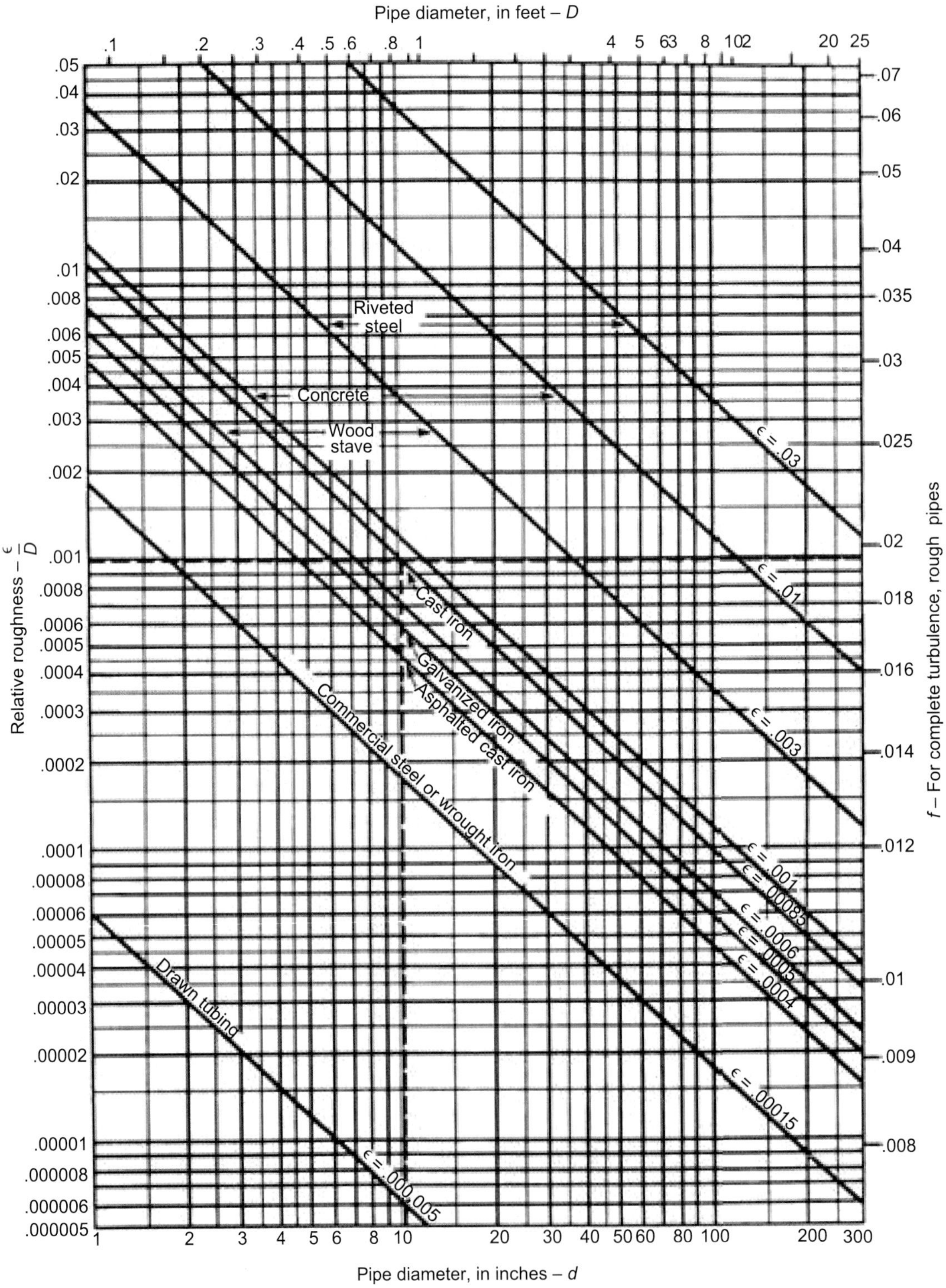

Figure 3. Relative roughness of pipe materials and friction factors for complete turbulence.

Drag-reducing agents

Drag-reducing agents (DRAs) can substantially reduce friction loss in most pipelines when flowing most hydrocarbon liquids.

When a DRA is dissolved in a solvent, it produces a solution that in laminar flow has the same pressure drop, and therefore the same Newtonian viscosity, as the solvent. But when the solution is in turbulent flow, a DRA produces a pressure drop smaller than that which would occur with untreated solvent moving at the same flow rate. The turbulent flow friction of the solution is less than that of the solvent.

DRAs are highly viscoelastic and thixotropic. They may have a viscosity at low shear rates in excess of 3,000,000 cP, with a viscosity 100 to 1,000 times less at higher rates. The viscosity is somewhat affected by temperature and increases with decreasing temperature. DRA-solvent solutions are viscoelastic, time-independent, shear degradable, non-Newtonian fluids. A DRA-solvent solution handles as an ordinary hydrocarbon except in turbulent flow, when the reduced friction becomes evident.

DRAs in present use in oil and products pipelines are themselves hydrocarbons and thus should have no effect on refining processes or refined products. Millions of barrels of crude treated with DRAs on the Alyeska system have been refined without incident. Drag-reduction installations need not involve large-diameter pipelines. The 200,000 b/d increase in the Alyeska system resulting from the injection of some thousands of gallons per day of DRA in this 48-in line is dramatic.

However, the offshore operator who obtains a 20% increase in capacity in his 8-in flowline by injecting a few dozen gallons of DRA per day is happy: Without the DRA, he would have to shut in production or loop the flowline. Either alternative would be orders of magnitude more expensive than the injection of a DRA.

As production decays over time, the amount of DRA injected will be reduced, with the eventual result that production will match the untreated flowline capability and operations can proceed as before but with no shut in wells in past history and no sterilized investment in a loop to be carried on the books in future years.

Early applications

The drag-reduction effect as applied to solutions was accidentally discovered by B. A. Toms in the United Kingdom in 1947.[1] It had been known for years that concentrated water–coal slurries and wood pulp–water slurries could produce a friction factor lower than that of water alone. But Dr. Toms's 1948 publication was the first indication of drag reduction in otherwise Newtonian solutions.

The first use of DRAs in the oil industry was to reduce pressure loss during pumping of fluids downhole to fracture tight formations. The quantities of DRA used are proportionally large, but the savings are substantial. Data published on one DRA-kerosene solution with 600 weight ppm of DRA in kerosene showed an 80% drag reduction over untreated kerosene.

The ongoing use of DRAs to debottleneck the Alyeska 48-in crude line has been described in several published articles.[2] There are many other uses on commercial pipelines—one considerably larger than the Alyeska example—and the list of operators using DRA is growing. Use of DRAs in multistation pipelines is a special application, because the drag-reduction effect does not survive shear stresses raised by centrifugal pumps and appears to be badly degraded going through most positive displacement pumps.

The general rule is that DRAs are always injected on the discharge side of a pump station. If drag reduction is required along a pipeline with N pump stations, the DRA must be injected at N points.

The effectiveness of a DRA depends to a certain extent on the viscosity of the untreated oil and decreases as viscosity increases. It varies slightly with diameter and decreases as diameter increases. The effect varies greatly with flow velocity, which, for a given oil and line diameter, means it varies greatly with Reynolds number. Drag reduction increases as Reynolds number increases. However, dilute solutions are believed to disintegrate at shear rates found in commercial pipelines.

There are some as yet little understood practical aspects of DRA application. For instance, there seems to be some relationship between the physical characteristics of the oil and of the DRA. A certain DRA will produce good results in a given crude yet produce poor results in a seemingly similar oil.

Untreated non-Newtonian oils may produce non-Newtonian DRA–oil solutions, and a particular DRA may increase, rather than decrease, flowing friction. Sometimes an additional, specially compounded additive is required to obtain the desired drag reduction. In addition, there appears to be something in certain DRA–solvent combinations that has to do with the solubility of the DRA in the oil. Test loop studies verify that for these combinations drag reduction appears to increase asymptotically as the distance from the point of injection increases. Specially designed DRA injection nozzles help in these cases.

DRA activity

The precise mechanism describing how a DRA works to reduce friction is not established. It is believed these agents work by directly reducing turbulence or by absorbing and later returning to the flowing stream energy that otherwise would have been wasted in producing the crossflows that comprise turbulence. The absorb-and-return theory has a certain intuitive attractiveness, because one of the characteristics of a viscoelastic fluid is its ability to do this on a physically large scale. Whether the effect exists at the molecular level is unknown.

Pipe liners generally work with the rational flow formula. The friction factor therein, which is a scaling factor applied to velocity head to yield friction loss, is a macro factor in that it lumps all of the friction-producing effects acting in a flowing system into a single numerical value. The friction factor concept does not provide an insight into how a DRA might work. It is necessary to look at the velocity distribution pattern in a pipeline in turbulent flow to understand the bases that support the concept of the f-factor and understand better how drag reduction might take place.

The concept of turbulent flow in pipelines provides that the point flow may be directed in any direction while maintaining net pipe flow in the direction of decreasing pressure. All flow not in the direction of net pipe flow absorbs energy; the more turbulent the flow becomes, the more energy absorbed in these crossflows. In flow in hydraulically rough pipes, the f-factor does not vary with increasing flow, but the unit friction loss increases as the square of the velocity of net pipe flow.

The model of turbulent pipe flow we will use is based on the Universal Law of the Wall, which applies to both smooth and rough pipe flow. It presumes two general modes of flow in a pipe: the wall layer and the turbulent core.

The wall layer is presumed to contain all of the flow where the point velocity varies with the distance of the point from the pipe wall. The turbulent core carries all of the flow where variations in point velocity are random and independent of this distance. The wall layer itself is presumed to be composed of three sublayers: the laminar sublayer, the buffer zone, and the turbulent sublayer.

The laminar sublayer has as its outer boundary the wall of the pipe, and flow in the entire sublayer is laminar. This definition requires that immediately at the wall lies a film of liquid that is not moving.

The increase in point velocity as a point moves away from the wall is a linear function of the distance from the wall and directed parallel to the wall in the direction of pipe flow. There are no crossflows in this sublayer. The contribution of the laminar sublayer to total friction arises in viscous shear generated in the sublayer.

Pipe roughness acts to inhibit laminar flow; therefore, the definition is only statistically true for rough or partially rough pipe.

In the turbulent sublayer, which has as its inner boundary the turbulent core, it is presumed that the velocity varies logarithmically with distance from the wall, but that crossflows in the sublayer may be statistically large and nearly as great as in the core. The contribution of the turbulent sublayer to total friction lies in the turbulent shear generated in this sublayer.

In the buffer zone, which lies between the laminar and turbulent sublayers, variation of point velocity with point position is not established. Ideally, it should be the same as the laminar sublayer at the outer boundary and the same as the turbulent sublayer at the inner boundary. Any number of mathematical functions can be written to approximate these two requirements. Both viscosity and density contribute to shear losses in the buffer zone.

The contribution of the turbulent core to total friction loss lies entirely in turbulent shear; DRAs are presumed not to act on this component of total friction. Therefore, the action of a DRA must take place in the laminar sublayer, the buffer zone, or the turbulent sublayer.

That the action of a DRA increases with increasing flowing velocity would indicate that DRA action could not be primarily in the laminar sublayer. It sometimes would appear that this is so: Liquids treated with a DRA flowing under conditions in which the pipe is hydraulically rough and viscosity presumed to make no contribution to total friction can yield excellent drag reduction. This kind of result would lead to the presumption that a DRA works in the buffer zone.

On the other hand, large doses of a DRA have been observed to keep a liquid in laminar flow at Reynolds numbers much higher than those expected of the same liquid without DRA, which leads to the conclusion that a DRA acts in the laminar sublayer.

This conclusion suggests that a DRA may inhibit the action of crossflows, whether generated in the laminar sublayer or the buffer zone, which move inward into the turbulent sublayer and core and absorb energy that otherwise would go toward producing pipe flow. It appears this is the case.

Reduction in the energy absorbed by crossflows originating at the pipe wall can be accomplished by reducing their number, size, or velocity. The energy in the crossflows may also be absorbed by a DRA and later returned to the flowing stream for reuse, which, in effect, means reducing their effect if not their size, number, or velocity. It is here that theory breaks down; no one really knows which effect predominates, or if either does.

The following is known:

- That DRAs act to reduce turbulence.
- That to do so, they seem to act in the laminar sublayer or the buffer zone by effectively enlarging them or their effect.
- That the net effect is to reduce total friction.
- That the f-factor is less than would be calculated from the pipe liner's usual sources.

Experimental data taken on liquids flowing without a DRA do not translate directly to liquids that have been dosed with a DRA. The rational flow formula still applies, but the usual f-factor relationship does not.[3]

At a given flow rate, the percentage of drag reduction is defined as follows:

$$\%DR = \frac{\Delta Po - \Delta Pi}{Po} \times 100$$

where:

%DR = Percentage drag reduction
Po = Pressure drop of untreated fluid
Pi = Pressure drop of fluid containing drag reducer.

Percentage throughput increase can be estimated as follows:

$$= \left[\left(\frac{1}{1 - \frac{\%DR}{100}}\right)^{0.55} - 1\right] \times 100$$

where %DR is the same as previously defined.

Given a constant Q (flow rate), elevation 1 = elevation 2, and baseline pressures P_1 and P_2, DPf = $P_2 - P_1$ (without DRA).

The volumetric capacity and/or horsepower of the existing pumps may limit the amount of throughput increase that can be achieved. The installation of additional pump capacity may be required in order to obtain the full benefits of DRA use.

Example

Pipe OD = 10.750 in
Pipe w.t. = 0.250 in
Length = 50 miles
Upstream elev. = 500 ft
Downstream elev. = 500 ft
Fluid crude oil
Oil temperature = 85°F
Discharge (source) pressure = 1,440 psig
Residual pressure (sink) = 50 psig
Base throughput (untreated) = 54,000 b/d

Oil properties

Temp.	60°F	85°F	90°F
Visc.	60 SUS	49 SUS	48 SUS
SG	0.8528	0.8423	0.8398
Visc.	8.70 cP	5.99 cP	5.72 cP
Visc.	10.20 cSt	7.11 cSt	6.81 cSt

Desired throughput increase = 20% or 64,800 b/d.

$$20\% = \left[\left(\frac{1}{1 - \frac{\%DR}{100}}\right)^{0.55} - 1\right] \times 100$$

Solve for %DR

%DR = 28.25%

Using a program such as GENNET-M, the calibration factor on the pipe data input can be adjusted as follows:

1 − 0.2825 = 0.7175. Once this is done, run the program to find the new flow rate of 64,900 b/d.

References

1. Toms, B. A., "Some observations on the flow of linear polymer solutions through straight pipe at large Reynolds numbers," *Proceedings of the International Congress on Rheology*, Scheveningen, The Netherlands, vol. 2, pp. 135–141 (1948).
2. Burger, Edward D., Monk, Werner R., and Wahl, Harry A., "Flow increase in the Trans Alaska pipeline using a polymeric drag reducing additive," *Society of Petroleum Engineers*, 9419 (1980).
3. *Oil and Gas Journal*, February 18, 1985.

How to estimate the rate of liquid discharge from a pipe

A quick and surprisingly accurate method of determining liquid discharge from an open-ended horizontal pipe makes use of the equation:

$$\text{gpm} = \frac{2.56 \times d^2}{\sqrt{Y}}$$

where:

gpm = Gallons per minute
d = Diameter of pipe, in
X = Inches
Y = Inches

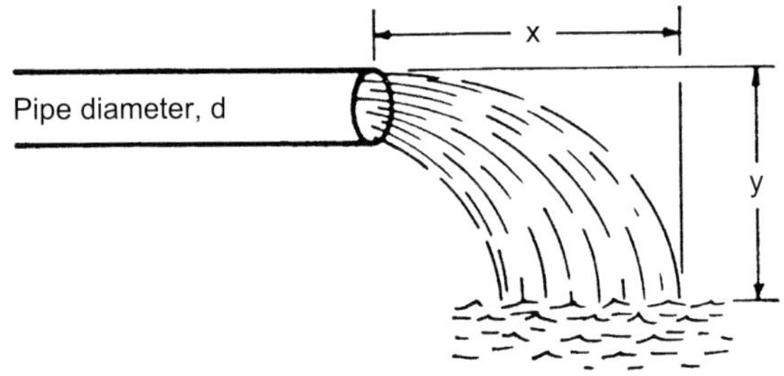

Example. For a 2 in diameter pipe that discharges approximately 24 inch away from the pipe end at a height of 24 in yields a discharge rate:

$$\text{gpm} = \frac{(2.56)(24)2^2}{\sqrt{25}} = \frac{245.76}{5} = 49.15$$

Predict subsurface temperature ranges

Ground temperatures are vital to pipeline design for two reasons. First, the heat exchange between the contents of the line and the ground are greatly affected by the differential temperatures of the pipe and the soil. Second, the temperature of the pipe at times of installation will have a considerable effect on stresses, particularly where the subsequent ground temperatures will vary considerably from the installation temperatures.

For example, in the construction of a 34-in line that was to operate at temperatures well over 100°F, care was exercised to limit bends when it was installed at temperatures ranging from 50°F to 70°F and the underground temperature will range from about 50°F to 80°F over a year's time.

Generally speaking, most areas will have somewhat similar curves from 3 ft on down. Of course, with depth, soil temperature approaches the temperature of underground water, and the summer–winter variation ceases.

To adapt for U.S. work. As a comparison, the following temperature ranges are encountered in winter and summer at depths of 3 ft in various areas in the United States.

California	51°F to 81°F
Colorado plains	35°F to 65°F
Texas (Panhandle)	42°F to 75°F
Texas (Coast)	60°F to 80°F
Mississippi	57°F to 80°F

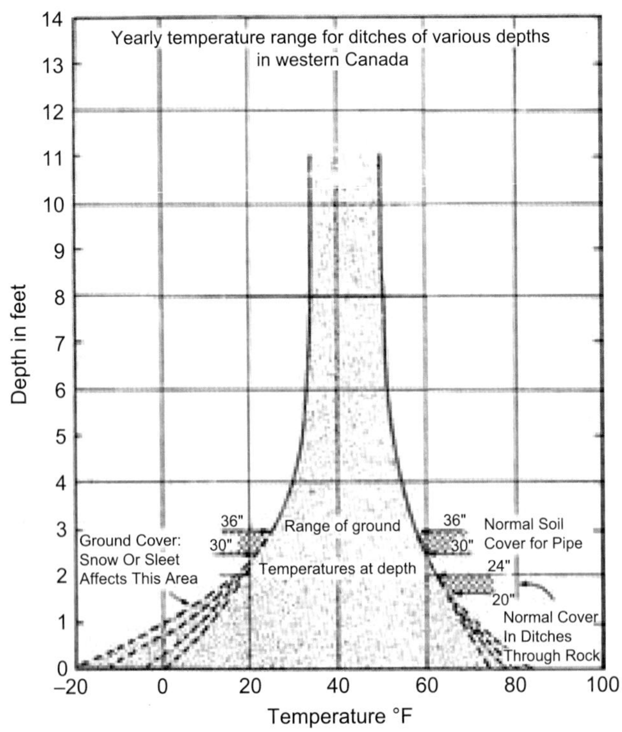

Figure 1. This curve shows the range of underground temperatures in the prairie regions of Canada. At shallow depths, there is some variation and uncertainty in temperature because of snow or sleet covering, but from about 3 ft down, the temperature stabilizes.

Idaho ....37°F to 56°F
Illinois ....41°F to 56°F
Montana ....33°F to 54°F
Michigan ....38°F to 60°F
Wyoming ....26°F to 65°F

Generally, a curve plotted for any one of the above will closely parallel the curve for Canada.

Sizing pipelines for water flow

This formula gives a reasonable degree of accuracy for pipe size under the assumptions described.

$$d^5 = \frac{(gpm)^2(D)}{(10)(H)(100)}$$

where:

d = Inside diameter of pipe in inches
gpm = Gallons per minute
D = Equivalent distance in feet from standpipe (or pump) to outlet
H = Total head lost due to friction in feet

Example. What size line should be run from a water tower to a cooling tower for a minimum of 160 gpm if the equivalent distance (actual distance in ft plus allowances for valves and fittings) is 1,000 ft and the minimum head supplied by the tower will always be at least 40 ft?

Solution.

$$d^5 = \frac{(160)^2(1,000)}{(10)(40)(100)} = 640 \text{ in}$$

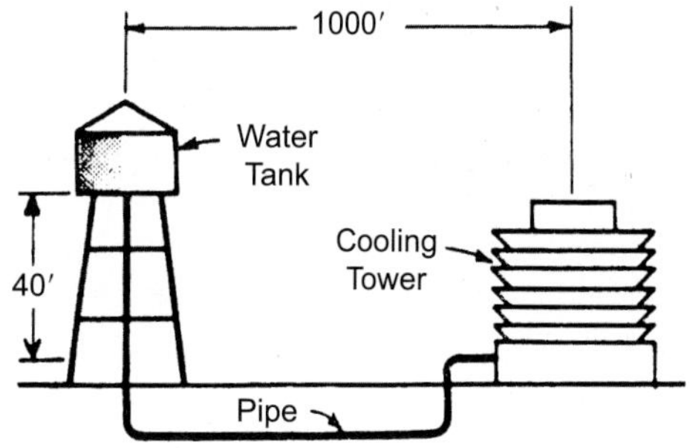

Rather than extract the fifth root of 640, simply raise inside diameters of logical pipe sizes to the fifth power and compare them with 640.

If 3-in ID pipe were used

$d^5 = 3^5 = 243$

This is obviously too small. Trying 4-in pipe:

$d^5 = 4^5 = 1,024$

How approximate throughput of a line can be estimated from pipe size

Multiply the square of the nominal pipe size by 500.

Example. The throughput of a 12-in line is approximately 144 by 500 = 72,000 barrels per day.

Naturally, the results of this rule are at best only a very rough approximation, since no account is taken of gravity, viscosity, pressures, or even of spacing between pump stations. The constant (given as 500) usually ranges between 300 and 600. The rule does *not* apply to short lines, such as loading lines, or runs within a refinery yard, but only to cross-country pipelines.

Estimate crude gathering line throughput for a given pipe diameter

To select the proper size line or estimate throughput for crude oil gathering lines, an engineer with a leading oil company has devised this handy rule: Throughput in barrels per hour is roughly equal to a constant times the square root of pressure drop per mile (psi/mile). Constants were derived from field data for various diameter lines between 2 and 10 in

Pipe diameter (in)	Constant (K)
2	7.8
3	19.2
4	37.8
6	105
8	209
10	385

Example. What size gathering line would be required to move 240 barrels per hour of crude through a 10-mile line using a pump discharge pressure of 400 psi? Assuming no significant changes in head due to elevation and a pressure drop of 40 psi per mile, for the first trial calculation a 4-in line appears to be the best choice, so try the constant 37.8.

$$\text{Throughput (4-in)} = 37.8\sqrt{40}$$
$$= 241 \text{ barrels per hour}$$

So a 4-in gathering line will be adequate.

How to determine head loss due to friction in ordinary iron pipeline carrying clear water

The form below gives a reasonable degree of accuracy:

$$S = \frac{G^2}{10\,d^5}$$

where:

S = Head loss due to friction in ft per 100 ft length of pipeline
G = Rate of flow in gallons per minute
d = Nominal diameter of pipe in inches

The exact head loss due to friction in pipelines carrying clear water is given by the Williams and Hazen formula:

$$S = \left[\frac{V}{1.318\,C\,R^{0.63}}\right]1.85$$

where:

S = Loss of head due to friction in ft per 100 ft
V = Average velocity in ft per second
C = Coefficient depending on the roughness inside the pipe
R = The mean hydraulic radius = D/4, where D is the diameter in ft

The average velocity V is given by Q/a where Q is the rate of flow in ft/sec and a is the cross-sectional area of the pipe.

C is a function of the material and age of the pipe and the type of water. For smooth new iron pipe C = 120. For 15-year-old ordinary pipe C = 100. For 25-year-old ordinary pipe C = 90.

Example. What will be the head loss per 100 ft if a 30-in line flows 10,000 gpm?

Solution.

$$S = \frac{(10,000)^2}{10(30)^5} = \frac{100}{243} = 0.41 \text{ ft}$$

Exact answer using the Williams and Hazen formula:

$$S = 0.34 \text{ ft}$$

Remarks. This rule of thumb could be satisfactorily used for pipe diameters 3 in and larger and a discharge greater than 400 gpm.

This rule applies only to wrought and cast iron pipes (20 years service). If new smooth brass or steel pipe is used, a factor should be used = 0.6.

For 30-year-old ordinary pipe factor = 1.2.

For pipe diameters 2 in and less and discharges less than 200 gpm, this rule of thumb does not give satisfactory results.

How to size lines, estimate pressure drop, and estimate optimum station spacing for crude systems

This table will give "horseback" estimates as to optimum line size, pressure drop, and pump station spacing for crude oil of approximately 40°API gravity and 60 SSU viscosity.

The depreciation was assumed to be 6% straight line, return on the investment 8%, and capital and operating costs based on Mid-Continent averages.

The pump station spacing assumes 800 psi discharge pressure.

For	Use		
Throughputs (bbl per day)	Pipe Diameter (in OD)	Pressure Drop (psi/mile)	Pump Station Spacing (miles)
0 to 2,000	3 ½	—	—
2,000 to 3,000	4 ½	32	25
3,000 to 7,500	6 ⅝	16	50
7,500 to 16,500	8 ⅝	10.5	76
16,500 to 23,500	10 ¾	8.5	94
23,500 to 40,000	12 ¾	7	112

How to size crude oil and products lines for capacity increases

A new pipeline for crude oil or products service should be sized to carry ⅔ of its ultimate capacity.

Example. Design a pipeline for an initial capacity of 100,000 barrels per day. This could be done by laying a 16-in line and spacing 14 pump stations along its length. To increase the capacity by 50% or to 150,000 barrels per day would require adding 14 intermediate booster stations, resulting in excessive operating costs. If the line had been designed so that the 100,000 barrels per day was ⅔ of its ultimate capacity, a 20-in line with five stations would be installed. To increase the capacity to 150,000 barrels per day would require adding five stations. This is based on the theory that when expanding the capacity of a line, install a loop line if the capacity is to be greater than 50% of the original capacity. If the expansion is to be less, add horsepower. Both of these approaches are based on statistical results rather than economic studies.

What is the hydrostatic pressure due to a column of liquid H feet in height?

$$P = \frac{H \times 61.41}{API + 131.5}$$

Example. A column of 37° oil is 20 ft high; what is the pressure at the base of the column?

$$P = \frac{20 \times 61.41}{37 + 131.5}$$

$$P = 7.28 \text{ psi}$$

How to determine the maximum surge pressure in liquid-filled pipeline when a valve is suddenly closed

The maximum surge pressure that occurs in a liquid pipeline when a valve is suddenly closed may be quickly determined by using the simple rule of thumb: Line pressure equals 0.8 times the weight per cubic foot of liquid times its velocity in feet per second. This compares favorably to the results when using the following equation.

$$P = \frac{12wV\sqrt{\frac{32.2}{w}\left(\frac{1}{\frac{1}{K}+\frac{D}{tE}}\right)}}{144 \times 32.2}$$

where:

P = Surge pressure, psi
w = Weight of liquid, lb/ft^3
V = Velocity change, fps
K = Bulk modulus of the liquid, psi
D = Outside diameter of the pipe, inches
t = Pipe wall thickness, inches
E = Young's modulus of the pipe material, psi

Example. An 8.625-in × 0.250-in w.t. products pipeline transports kerosene at 4.5 fps. The weight of the kerosene is 51.12 lb/ft^3, and the API gravity is 41.1 (rel. density = 0.8198). K = 171,000 psi.

Using the simplified equation, $P = 0.8\,wV = 0.8 \times 51.12 \times 4.5 = 184$ psi

Using the full equation, P = 178 psi; the difference is approximately 3%.

Transient pressure analysis

Transient pressure analysis is the study of a pipeline system's response to a time-dependent upset condition such as a valve closing or a pump shutting down. When a valve is closed downstream of the source pressure, the pressure will increase on the upstream side of the valve while the downstream pressure decreases. The valve closure time will determine how fast the upstream pressure increases and the downstream pressure decreases. The high upstream pressure "surge" will propagate upstream while the low pressure surge moves downstream. If the downstream side is open to a tank, the downstream surge will quickly dissipate due to the surge being absorbed in the tank. Similarly, when a pump shuts down, the pressure downstream of the pump decreases and the pressure upstream of the pump will increase. Other causes of pressure surges include tank changes, manifold changes, or leaks.

Steady-state versus transient conditions

Steady-state means that the pressure gradients and flow rates in the pipeline are not changing. Transient means the gradients and flows are a function of time. The event that causes conditions to be time dependent is the upset condition such as valve closure or pump failure. Transient conditions must be considered when determining if the pipeline will be subjected to abnormally high operating pressures.

Fluid properties

All pressures and flows, whether steady-state or transient, depend on the fluid properties. This example deals with a pipeline filled with batches of motor gasoline and diesel fuel. The program used for the analysis is designed for almost incompressible fluids such as crude oil, products, and water.

The properties required for surge analysis include: specific gravity, viscosity, bulk modulus, and vapor pressure. All properties should be at the same base temperature. Multiple batches can be handled and PVT data will be needed for each batch.

Pipe

The pipe data, such as outside diameter, wall thickness, specified minimum yield strength (SMYS), Poisson's ratio, and Young's modulus of elasticity are required. The length and elevation profile of the pipeline are also required.

Pipe environment

Since an isothermal surge analysis is being performed, the only environmental information required is how the pipeline is anchored. Is the pipe above ground or under water? Does it have expansion loops throughout, or is it just constrained at the source? Multiple pipes can be handled, and each pipe may be constrained differently.

Equipment

The data required for a centrifugal pump includes rated flow, rated head differential, shutoff head differential (PSOH), maximum allowed discharge pressure, minimum allowed suction pressure, discharge pressure trip setting, initial speed, final speed, time it takes to go from initial to final speed (ramp-down time), time when the pump starts to ramp down, and whether the pump is protected by a check valve in the pump discharge line.

The data required for valves includes valve closure time, time or pressure when the valve starts to close, closure curve shape exponent (the exponent is different for gate valves, ball valves, etc.), and the amount of valve closure.

Data required for pressure relief valves include the set pressure and the back-pressure on the relief line.

Initial conditions

Since surge is a time-dependent phenomenon, we need the conditions that exist before any upset occurs. The required initial conditions are: the pressure at the delivery end of the pipeline. The program will establish the initial steady-state flow rate automatically.

Perturbations that cause surges

Any time-dependent change, such as a leak, valve closure, or pump shutdown will cause surges in the system. Cascading upset events, such as internal pumps tripping, may cause other pumps to trip and can be modeled.

The fluid–pipe–equipment interaction

The liquid and the pipe, according to its environment, expand and contract as the surges travel. The pumps, valves, and other equipment change the pressures and/or flows as the surges reach them. The surges travel at the speed of sound in the fluid–pipe system. The fluid itself is what relays the surge energy to the various pipe sections and the equipment. The fluid, pipe, and equipment are inextricably linked, and their behavior must be determined by a model that employs simultaneous fluid–pipe and equipment solution procedures. The Method of Characteristics, which NATASHA PLUS© uses, is a simultaneous technique and therefore models "surge" phenomenon rigorously.

To illustrate the use of the program, the pipeline system shown in Figure 1 is equipped with a valve at the end of the pipeline that is caused to close at time T = 300 seconds. The valve travels from full open to full closed in 5 seconds. A pressure relief valve is connected to the system and is set to relieve at 275 psig. The system consists of six pipes configured as shown in Table 1.

Simulation

The operation of the system was simulated for 1,200 seconds, and the results are shown graphically in Figure 2. The program outputs a detailed tabular listing showing pressures and flow rates at various locations along the pipeline at selected time steps. A graphics file containing selected node pressures and flow rates is also produced by the program. Figure 2 was prepared from the graphics file produced by the program. A typical time page report is shown in Table 2. The program also prints a maximum pressure summary which is shown in Table 3. A summary of the relief valve operation is shown in Table 4, indicating that the total volume released was 855 bbls during the simulation.

The pump is not shut down in this simulation because of the action of the relief valve. A review of the maximum pressure summary indicates that the pipeline MOP is exceeded at two locations. In order to prevent overpressure, it will be necessary to make changes to the system such as changing the relief valve setting or slowing the closing time for the valve. NAASHA-PLUS© can then be used to simulate the changes to confirm that pipeline MOP is not exceeded.

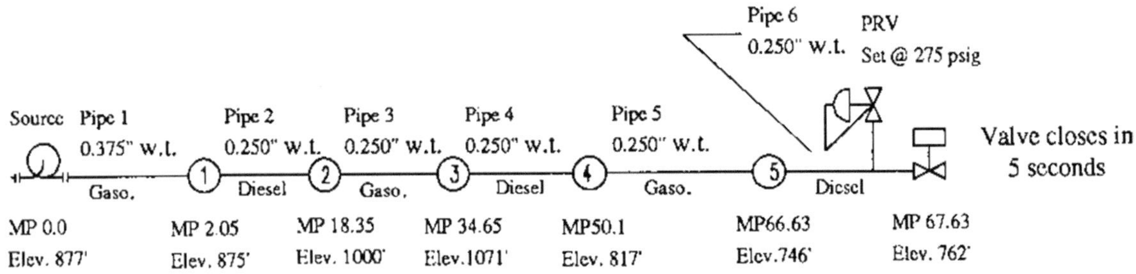

Product parameters:

Product	sp. gr.	visc. cSt	bulk modulus psi
Diesel	0.79	2.15	156,495
Gasoline	0.741	0.622	121,505

Initial flow rate is 92,500 BPD
Downstream pressure is 40 psig.

Figure 1. Sample pipeline for transient analysis study.

Table 1
Data Required by the Program

Pipe Design Data

Pipe No.	Length Mi.	OD in	W. T. in	Roughness in	SMYS psig	Design Fact.
1	2.05	12.75	0.375	0.0018	35000	72%
2	16.3	12.75	0.25	0.0018	52000	72%
3	16.3	12.75	0.25	0.0018	52000	72%
4	15.45	12.75	0.25	0.0018	52000	72%
5	16.53	12.75	0.25	0.0018	52000	72%
6	1.0	12.75	0.25	0.0018	52000	72%

Liquid Properties

Pipe No.	Sp. Gr.	Vis. cSt	Bulk Mod. psi	Vapor Press psia
1	0.741	0.622	0.122	5.7
2	0.79	2.15	0.156	10
3	0.741	0.622	0.122	5.7
4	0.79	2.15	0.156	10
5	0.741	0.622	0.122	5.7
6	0.79	2.15	0.156	10

Initial Conditions

Pipe No.	Flow BPD	Upstream Press psig
1	92500	1398.22
2	92500	1444.19
3	92500	958.58
4	92500	659.75

Initial Conditions				
Pipe No.		**Flow BPD**		**Upstream Press psig**
5		92500		363.1
6		92500		68.76
Pipe No.	**Friction Factor**	**Seg. Length Length, mi.**	**Press. Drop psi/mi.**	**Head Drop ft/mi**
1	0.014	1.03	21.5907	67.2098
2	0.0157	1.02	23.2816	67.978
3	0.014	0.91	19.4468	60.5359
4	0.0157	1.03	23.2816	67.978
5	0.014	0.92	19.4468	60.5359
6	0.0157	1.00	23.2816	67.978

PUMP PARAMETERS		
PUMP AT NODE 7 THAT DISCHARGES INTO PIPE 1		
RATED FLOW	=	92,500 bbl/day
ZERO FLOW HEAD DIFFERENTIAL	=	4520.00 ft-head
RATED HEAD DIFFERENTIAL	=	4228.02 ft-head
SUCTION PRESSURE	=	40.00 psig
MAXIMUM DISCHARGE CONTROLLED PRESSURE	=	1440.00 psig
DISCHARGE TRIP PRESSURE	=	1490.00 psig
INITIAL SPEED	=	3565.00 rpm
FINAL SPEED	=	3565.00 rpm
CURVE EXPONENT	=	0.5000

VALVE PARAMETERS		
PRESSURE RELIEF VALVE AT NODE 5		
MAX PRESS. BEFORE RELIEF	=	275.00 psig
BACK PRESSURE ON RELIEF LINE	=	0.00 psig
VALVE AT NODE 6		
VALVE STARTS TO CLOSE AT T	=	300.00 sec
VALVE CLOSES IN	=	5.00 sec
VALVE STARTS TO CLOSE ABOVE P	=	99999.00 psig
FINAL FRACTIONAL AREA OPEN TO FLOW	=	0.00
CLOSURE SHAPE EXPONENT	=	1.00

Elevation Data				
Pipe No.	**Point**	**MP mi.**	**Elev. ft.**	**MOP psig**
1	1	0	877	1440
	2	1.59	885	1440
	3	2.05	875	1440
	1	0	875	1440
	2	1.13	890	1440
	3	1.27	860	1440
	4	4.37	940	1440
2	5	4.96	890	1440
	6	10.96	990	1440
	7	11.98	1060	1440
	8	13.03	980	1440
	9	16.3	1000	1440
	1	0	1000	1440
3	2	1.71	1065	1440
	3	4.66	1010	1440
	4	6.42	1085	1440
	5	9.9	930	1440
	6	10.21	1040	1440
3	7	11.46	970	1440
	8	14.21	1040	1440
	9	16.3	1071	1440

(table continued on next page)

Table 1
Data Required by the Program (continued)

Elevation Data

Pipe No.	Point	MP mi.	Elev. ft.	MOP psig
4	1	0	1071	1440
	2	3.76	1050	1440
	3	5.53	1005	1440
	4	5.92	1020	1440
	5	6.32	1020	1440
	6	15.45	817	1440
5	1	0	817	1440
	2	4.05	770	1440
	3	5.45	875	1440
	4	5.68	810	1440
	5	6.26	915	1440
	6	7.13	805	1440
	7	8.31	930	1440
	8	12.36	755	1440
	9	16.53	746	1440
6	1	0	746	275
	2	0.28	760	275
	3	0.62	762	275
	4	1	762	275

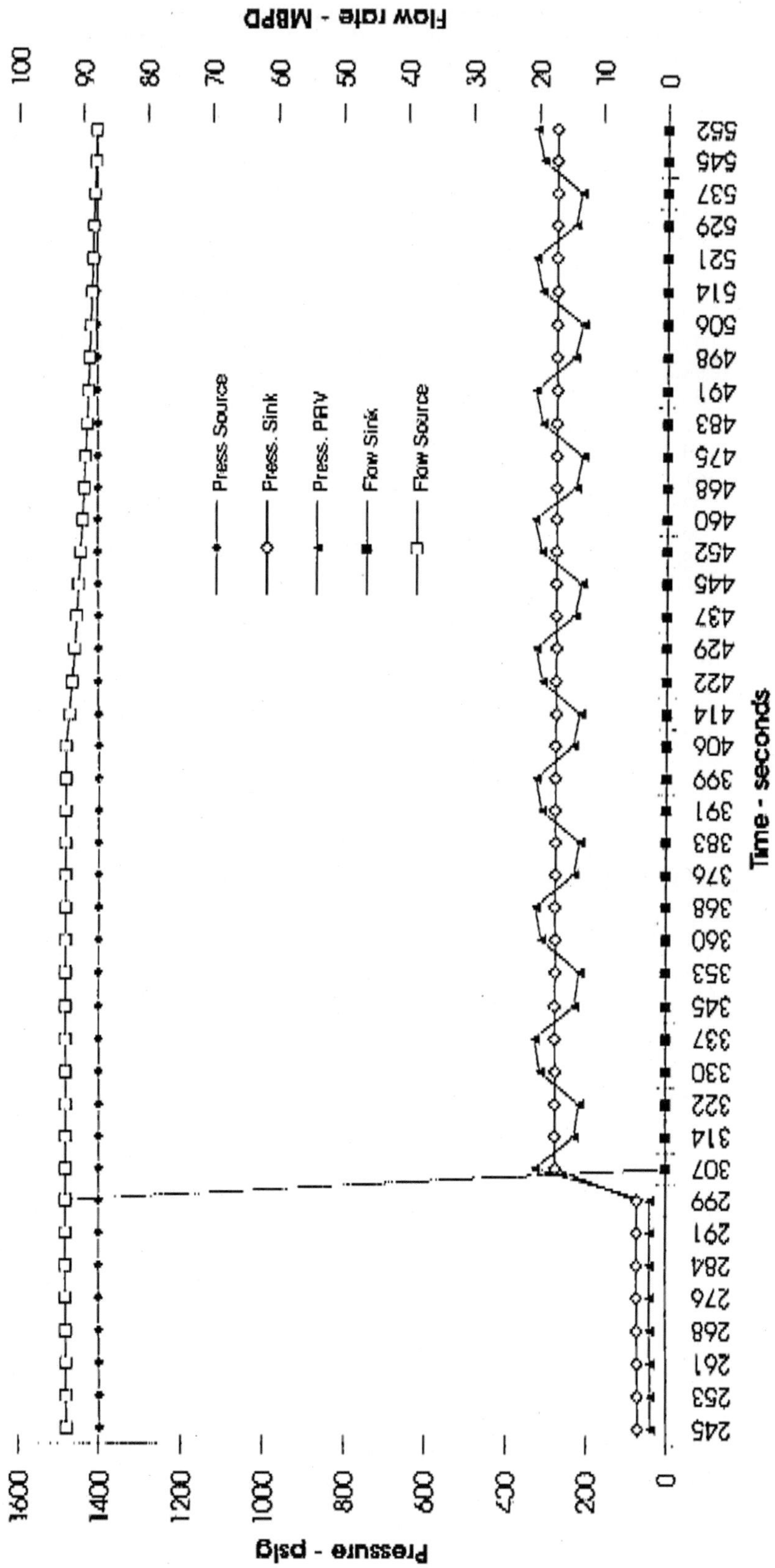

Figure 2. Transient analysis.

Table 2
Typical Time Page Report for T = 0 seconds

Pipe No.	Loc. mi.	Elev Ft.	Up Flow M bbl/day	Down Flow M bbl/day	Head Ft.	Press, psig
1	0	877	93	93	5230	1398
1	1	882	93	93	5161	1374
1	2	875	93	93	5092	1355
2	0	875	93	93	5092	1444
2	1	889	93	93	5023	1416
2	2	880	93	93	4953	1395
2	3	906	93	93	4884	1362
2	4	932	93	93	4815	1330
2	5	892	93	93	4746	1320
2	6	909	93	93	4676	1290
2	7	926	93	93	4607	1261
2	8	943	93	93	4538	1231
2	9	960	93	93	4468	1202
2	10	977	93	93	4399	1172
2	11	1007	93	93	4330	1138
2	12	1041	93	93	4261	1103
2	13	981	93	93	4191	1099
2	14	988	93	93	4122	1074
2	15	994	93	93	4053	1048
2	16	1000	93	93	3984	1022
3	0	1000	93	93	3984	959
3	1	1034	93	93	3929	930
3	2	1063	93	93	3874	903
3	3	1046	93	93	3820	891
3	4	1029	93	93	3765	879
3	5	1012	93	93	3710	867
3	5	1043	93	93	3655	839
3	6	1082	93	93	3600	809
3	7	1048	93	93	3545	802
3	8	1008	93	93	3491	798
3	9	968	93	93	3436	793
3	10	952	93	93	3381	780
3	11	1003	93	93	3326	746
3	12	978	93	93	3271	737
3	13	1001	93	93	3217	712
3	14	1024	93	93	3162	687
3	14	1044	93	93	3107	663
3	15	1058	93	93	3052	641
3	16	1071	93	93	2997	619
4	0	1071	93	93	2997	660
4	1	1065	93	93	2927	638
4	2	1059	93	93	2857	616
4	3	1054	93	93	2787	594
4	4	1041	93	93	2717	574
4	5	1015	93	93	2647	559
4	6	1020	93	93	2577	533
4	7	1000	93	93	2507	516
4	8	977	93	93	2437	500
4	9	954	93	93	2367	484
4	10	932	93	93	2297	468
4	11	909	93	93	2227	452
4	12	886	93	93	2157	435
4	13	863	93	93	2087	419
4	14	840	93	93	2017	403
4	15	817	93	93	1947	387

(table continued on next page)

Table 2
Typical Time Page Report for T = 0 seconds (continued)

Pipe No.	Loc. mi.	Elev Ft.	Up Flow M bbl/day	Down Flow M bbl/day	Head Ft.	Press, psig
5	0	817	93	93	1947	363
5	1	806	93	93	1892	349
5	2	796	93	93	1836	334
5	3	785	93	93	1781	320
5	4	774	93	93	1725	305
5	5	811	93	93	1669	276
5	6	858	93	93	1614	243
5	6	894	93	93	1558	213
5	7	828	93	93	1503	217
5	8	925	93	93	1447	168
5	9	892	93	93	1391	160
5	10	853	93	93	1336	155
5	11	813	93	93	1280	150
5	12	773	93	93	1225	145
5	13	754	93	93	1169	133
5	14	752	93	93	1113	116
5	15	750	93	93	1058	99
5	16	748	93	93	1002	82
5	17	746	93	93	947	64
6	0	746	93	93	947	69
6	1	762	93	93	879	40

Table 3
Maximum Pressure Summary

Pipe No.	MP mi.	Elev. ft.	Hmax ft.	Pmax psig	MOP psig	Time sec.	
1	0	877	5269.51	1411.07	1440	1199.44	
1	1.03	882.16	5209.78	1390.22	1440	1200.98	
1	2.05	875	5150.04	1373.33	1440	1199.44	
2	0	875	5150.04	1464.14	1440	1199.44	× (Exceeds MOP)
2	1.02	888.52	5089.72	1438.85	1440	1200.98	
2	2.04	879.81	5029.4	1421.18	1440	1199.44	
2	3.06	906.1	4969.08	1391.52	1440	1200.98	
2	4.07	932.39	4908.75	1361.85	1440	1199.44	
2	5.09	892.23	4848.44	1354.95	1440	1200.98	
2	6.11	909.21	4788.11	1328.47	1440	1199.44	
2	7.13	926.19	4727.8	1302	1440	1200.98	
2	8.15	943.17	4667.47	1275.53	1440	1199.44	
2	9.17	960.15	4607.16	1249.05	1440	1200.98	
2	10.19	977.13	4546.84	1222.58	1440	1199.44	
2	11.21	1006.9	4486.53	1191.73	1440	1200.98	
2	12.22	1041.33	4426.21	1159.27	1440	1199.44	
2	13.24	981.31	4365.9	1159.18	1440	1200.98	
2	14.26	987.54	4305.58	1136.38	1440	1199.44	
2	15.28	*993.77*	4245.27	1113.6	1440	1200.98	
2	16.3	1000	4184.95	1090.8	1440	1199.44	
3	0	1000	4184.95	1023.15	1440	1199.44	
3	0.91	1034.42	4137.43	996.82	1440	1200.98	
3	1.81	1063.11	4089.9	972.34	1440	1199.44	
3	2.72	1046.23	4042.38	962.49	1440	1200.98	
3	3.62	1029.35	3994.85	952.65	1440	1199.44	
3	4.53	1012.47	3947.33	942.81	1440	1200.98	
3	5.43	1042.95	3899.8	917.74	1440	1199.44	
3	6.34	1081.54	3852.29	890.08	1440	1200.98	
3	7.24	1048.28	3804.76	885.5	1440	1199.44	
3	8.15	1007.95	3757.25	883.2	1440	1200.98	
3	9.06	967.61	3709.72	880.89	1440	1199.44	
3	9.96	951.68	3662.21	870.74	1440	1200.98	
3	10.87	1003.23	3614.69	838.92	1440	1199.44	
3	11.77	977.95	3567.18	831.77	1440	1200.98	
3	12.68	1001	3519.66	809.1	1440	1199.44	
3	13.58	1024.05	3472.15	786.44	1440	1200.98	
3	14.49	1044.14	3424.63	764.72	1440	1199.44	
3	15.39	1057.57	3377.13	745.15	1440	1200.98	
3	16.3	1071	3329.61	725.57	1440	1199.44	
4	0	1071	3329.61	773.55	1440	1199.44	
4	1.03	1065.25	3268.67	754.64	1440	1200.98	
4	2.06	1059.49	3207.72	735.74	1440	1199.44	
4	3.09	1053.74	3146.78	716.84	1440	1200.98	
4	4.12	1040.85	3085.82	700.38	1440	1199.44	
4	5.15	1014.66	3024.89	688.48	1440	1200.98	
4	6.18	1020	2963.94	665.77	1440	1199.44	
4	7.21	1000.21	2903	651.68	1440	1200.98	
4	8.24	977.31	2842.06	638.65	1440	1199.44	
4	9.27	954.41	2781.13	625.63	1440	1200.98	
4	10.3	931.51	2720.18	612.6	1440	1199.44	
4	11.33	908.61	2659.25	599.57	1440	1200.98	
4	12.36	885.7	2598.31	586.55	1440	1199.44	
4	13.39	862.8	2537.39	573.52	1440	1200.98	
4	14.42	839.9	2476.45	560.5	1440	1199.44	
4	15.45	817	2415.53	547.47	1440	1200.98	

(table continued on next page)

Table 3
Maximum Pressure Summary (continued)

Pipe No.	MP mi.	Elev. ft.	Hmax ft.	Pmax psig	MOP psig	Time sec.	
5	0	817	2415.53	513.52	1440	1200.98	
5	0.92	806.34	2367.36	501.47	1440	1199.44	
5	1.84	795.69	2319.21	489.42	1440	1200.98	
5	2.76	785.03	2271.05	477.37	1440	1199.44	
5	3.67	774.37	2222.9	465.33	1440	1200.98	
5	4.59	810.63	2174.74	438.21	1440	1199.44	
5	5.51	858.04	2126.59	407.51	1440	1200.98	
5	6.43	893.72	2078.43	380.58	1440	1199.44	
5	7.35	827.95	2030.29	386.24	1440	1200.98	
5	8.27	925.23	1982.13	339.52	1440	1199.44	
5	9.18	892.26	1933.99	334.65	1440	1200.98	
5	10.1	852.58	1885.84	331.93	1440	1199.44	
5	11.02	812.9	1837.7	329.21	1440	1200.98	
5	11.94	773.22	1789.55	326.49	1440	1199.44	
5	12.86	753.93	1741.41	317.22	1440	1200.98	
5	13.78	751.95	1693.26	302.39	1440	1199.44	
5	14.69	749.96	1645.13	287.57	1440	1200.98	
5	15.61	747.98	1596.98	272.73	1440	1199.44	
5	16.53	746	1548.9	257.93	1440	742.37	
6	0	746	1548.9	274.98	275	742.37	
6	1	762	1721.75	328.7	275	460.14	× (Exceeds MOP)

Table 4
Pressure Relief Valve Summary

TOTAL VOLUME RELEASED AT NODE 5 = 855.4117bbl
MAX VOLUME RATE RELEASED AT NODE 5 = 102.958 1000 bbl/day

Hydraulics calculations for multiphase systems, including networks

The calculation of pressure, temperature, and volumes in multiphase systems (pipelines, wells, and pipe networks), is a complicated function of fluid composition, pressure, temperature, flow rate, fluid properties, pipe properties, angle of inclination (elevation differences), properties of the medium surrounding the pipes, and pipe connectivity. Therefore, for all but the simplest of systems, a computer program is required.

GENNET-M is a Multiphase Flow program from Fluid Flow Consultants, Inc., of Grove, Oklahoma. GENNET-M handles pipeline, well, and pipe NETWORK models (one or more sources, one or more sinks, and zero or more loops) of up to 500 pipes, has a liquids removal ("pigging") option, a pressure-drop calibration option, a viscosity curve input option, and "DP+DT" equipment (pressure and/or temperature change device, i.e., a pump, compressor, heater, cooler, minor loss across a valve or fitting, pressure drop from an IPR curve, etc.). GENNET seeks to solve for the flow rates in each pipe of an arbitrarily connected network subject to constant boundary (source/sink) pressures and/or flows and the resistance (friction and elevation differences) in the pipes.

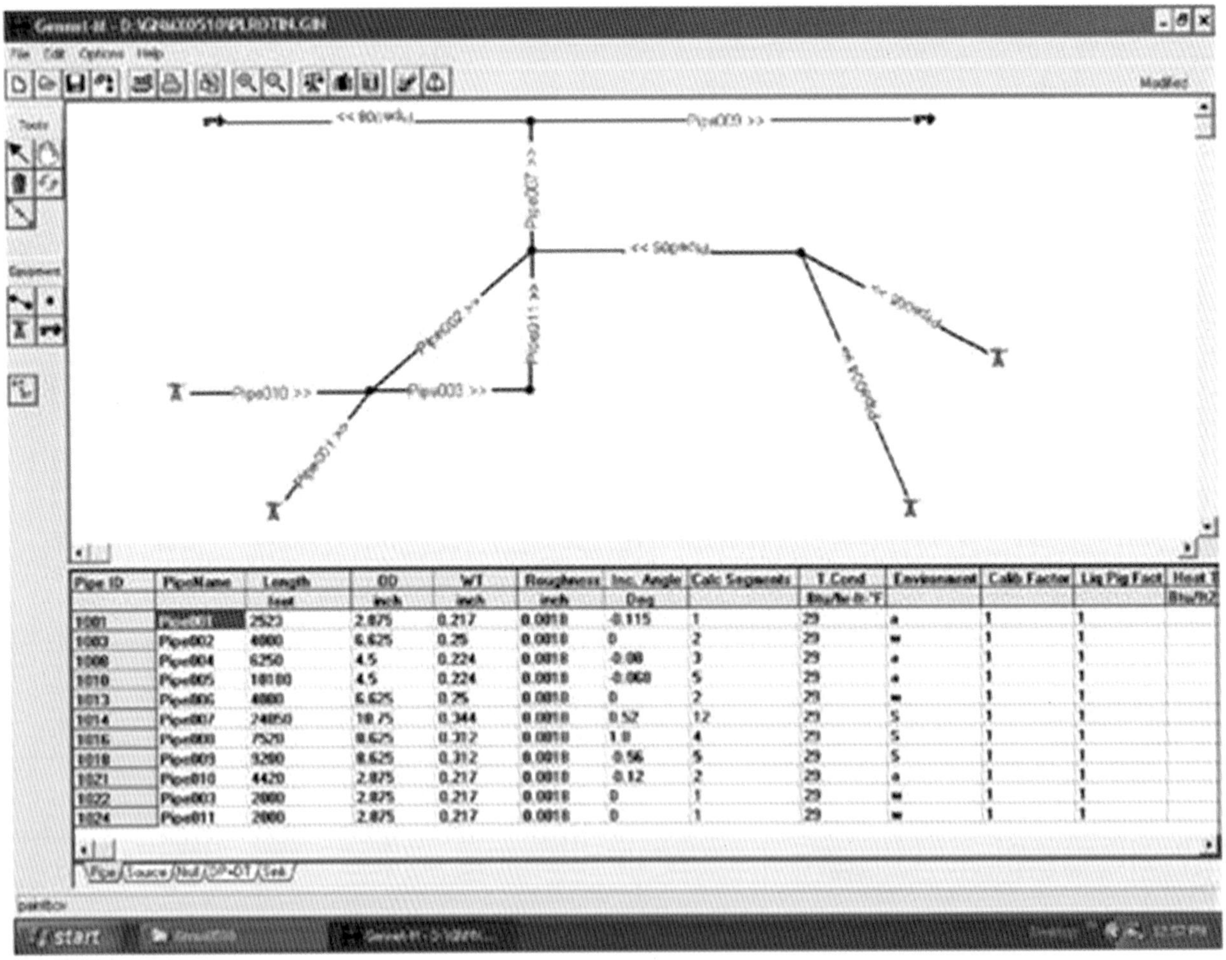

Figure 1. Diagram of network and pipe data input screen for the example.

Example. The GENNET-M input file name is PLROT.GIN. The text input file is PLROTIN.TXT. The text output file name is PLROTOUT.TXT.

This example (PLROT) is a multiphase (oil, gas, and water) problem with four sources, two sinks, and one loop. The source flows are known (GENNET-M calculates the source pressures), and the sink pressures are known (GENNET-M calculates the sink flows). Each source has a different flow rate and composition (oil API; oil viscosity curve; gas gravity; and CO_2, N_2, and H_2S content). GENNET-M mixes the fluids according to the flow rate in each pipe and then calculates the various fluid properties it needs as it calculates the flowing pressures and temperatures. All source pipes (Pipe001, Pipe010, Pipe004, and Pipe005) are in air exposed to a 10 mph wind. The loop pipes (Pipe002, Pipe003, and Pipe011) and Pipe006 are submerged in water exposed to a 10 mph cross-current. The sink pipes (Pipe008 and Pipe009) and Pipe007 are buried in soil at a depth of 36 in. The complete input and output data are shown.

Input

GENNET-M 5.4.23

GENeral NETwork Multiphase flow analyzer

DATE: 01/30/2004 TIME: 12:13

General Data
Title:
Company: Fluid Flow Consultants
Number of Pipes: 11
Fluid Type: Oil
Global Units: English.
Pipe Length Units: Feet

Flow pattern

Some correlations define flow patterns, and others do not. The modified Beggs and Brill-Moody correlations, which GENNET-M uses, define the following patterns:

1. Segregated (usually Stratified or Stratified-Wavy for near horizontal segments; usually Annular for near vertical segments)
2. Intermittent (Slug or Plug)
3. Distributed (Bubble for high liquid loadings; Mist for low liquid loadings)
4. Transition (somewhere between Segregated and Intermittent)

Because any optimization method requires continuity across flow pattern boundaries, three more "transition" zones were added to the Beggs and Brill flow pattern map:

5. D-S (transition zone between Distributed and Segregated)
6. D-T (transition zone between Distributed and Transition)
7. D-I (transition zone between Distributed and Intermittent)

It is important to note that the Beggs and Brill correlations do not necessarily predict the correct flow pattern if the angle of inclination is more than a few degrees from horizontal. If the segment is not close to being horizontal, the flow pattern is used only as a correlating parameter for the holdup.

PIPE PROPERTIES 1

Pipe Number	Pipe Name	Outside Diameter (in)	Wall Thickness (in)	Incl Angle (Deg)	Rough (in)
1	Pipe010	2.875	0.217	-0.1	0.0018
2	Pipe002	6.625	0.250	0.0	0.0018
3	Pipe004	4.500	0.224	-0.1	0.0018
4	Pipe005	4.500	0.224	-0.1	0.0018
5	Pipe006	6.625	0.250	0.0	0.0018
6	Pipe007	10.750	0.344	0.5	0.0018
7	Pipe008	8.625	0.312	1.8	0.0018
8	Pipe009	8.625	0.312	-0.6	0.0018
9	Pipe001	2.875	0.217	-0.1	0.0018
10	Pipe003	2.875	0.217	0.0	0.0018
11	Pipe011	2.875	0.217	0.0	0.0018

Adapted from Fluid Flow Consultants, Inc. Grove, Oklahoma. Reprinted with permission.

PIPE PROPERTIES 2

Pipe Number	Pipe Name	Thermal Conductivity (Btu/'-h-F)	Length (ft)	Calc Segs	Env	Calibration Factor	Liq Holdup Factor
1	Pipe010	29.00000	4420.0	2	Air	1.00000	1.00000
2	Pipe002	29.00000	4000.0	2	Water	1.00000	1.00000
3	Pipe004	29.00000	6250.0	3	Air	1.00000	1.00000
4	Pipe005	29.00000	10180.0	5	Air	1.00000	1.00000
5	Pipe006	29.00000	4000.0	2	Water	1.00000	1.00000
6	Pipe007	29.00000	24050.0	12	Soil	1.00000	1.00000
7	Pipe008	29.00000	7520.0	4	Soil	1.00000	1.00000
8	Pipe009	29.00000	9200.0	5	Soil	1.00000	1.00000
9	Pipe001	29.00000	2523.0	1	Air	1.00000	1.00000
10	Pipe003	29.00000	2000.0	1	Water	1.00000	1.00000
11	Pipe011	29.00000	2000.0	1	Water	1.00000	1.00000

PIPE CONNECTIVITY

Pipe Number	Upstream Node	Pipe Name	Downstream Node
1	8	Pipe010	1
2	1	Pipe002	2
3	6	Pipe004	3
4	7	Pipe005	3
5	3	Pipe006	2
6	2	Pipe007	4
7	4	Pipe008	9
8	4	Pipe009	10
9	5	Pipe001	1
10	1	Pipe003	11
11	11	Pipe011	2

PIPE ENVIRONMENT

Pipe Number	Pipe Name	Heat Transfer (Btu/ft2-hr-F)	Insul Thick (in)	Insul T.Cond (Btu/'-h-F)	Amb Temp (F)	A/W Vel (mph)	Burial Depth (in)	Soil T.COND (Btu/'-h-F)
1	Pipe010	0.00000	0.00	0.30	60.0	10.0	36.00	0.80
2	Pipe002	0.00000	0.00	0.30	60.0	10.0	36.00	0.80
3	Pipe004	0.00000	0.00	0.30	60.0	10.0	36.00	0.80
4	Pipe005	0.00000	0.00	0.30	60.0	10.0	36.00	0.80
5	Pipe006	0.00000	0.00	0.30	60.0	10.0	36.00	0.80
6	Pipe007	0.00000	0.09	0.30	60.0	10.0	36.00	0.80
7	Pipe008	0.00000	0.09	0.30	60.0	10.0	36.00	0.80
8	Pipe009	0.00000	0.09	0.30	60.0	10.0	36.00	0.80
9	Pipe001	0.00000	0.00	0.30	60.0	10.0	36.00	0.80
10	Pipe003	0.00000	0.00	0.30	60.0	10.0	36.00	0.80
11	Pipe011	0.00000	0.00	0.30	60.0	10.0	36.00	0.80

SOURCE CONDITIONS

```
Source Name: Source004,    Pipe Number:  1,    Pipe Name: Pipe010
  Flowing Temperature     (F):                      124.0
  Flowing Pressure        (psig):                  1200.49
  Oil Rate                (    bbl/day     ):      4350.000
  Gas/Oil Ratio           (  std ft3/bbl     ):     448.000
  Water Cut               (     Volume %     ):       8.200
  Gas Gravity:                                        0.695
  Separator Pressure      (psig):                     0.000
  Separator Temperature   (F):                       60.000
  Nitrogen, N2            (mole%):                    2.120
  Carbon Dioxide, CO2     (mole%):                    3.210
  Hydrogen Sulfide, H2S   (mole%):                    0.082
  Oil Gravity             (API):                     24.550
  Bub Pt Press            (psig):                     0.000
  Water Spec Grav         (FRAC):                     1.060

Source Name: Source002,    Pipe Number:  3,    Pipe Name: Pipe004
  Flowing Temperature     (F):                      136.0
  Flowing Pressure        (psig):                   721.52
  Oil Rate                (    bbl/day     ):      4400.000
  Gas/Oil Ratio           (  std ft3/bbl     ):    1000.000
  Water Cut               (     Volume %     ):      22.300
  Gas Gravity:                                        0.685
  Separator Pressure      (psig):                     0.000
  Separator Temperature   (F):                       60.000
  Nitrogen, N2            (mole%):                    0.125
  Carbon Dioxide, CO2     (mole%):                    1.080
  Hydrogen Sulfide, H2S   (mole%):                    0.015
  Oil Gravity             (API):                     32.800
  Bub Pt Press            (psig):                     0.000
  Water Spec Grav         (FRAC):                     1.100

Source Name: Source003,    Pipe Number:  4,    Pipe Name: Pipe005
  Flowing Temperature     (F):                      142.0
  Flowing Pressure        (psig):                   851.61
  Oil Rate                (    bbl/day     ):      4500.000
  Gas/Oil Ratio           (  std ft3/bbl     ):    1026.000
  Water Cut               (     Volume %     ):      23.800
  Gas Gravity:                                        0.669
  Separator Pressure      (psig):                     0.000
  Separator Temperature   (F):                       60.000
  Nitrogen, N2            (mole%):                    0.095
  Carbon Dioxide, CO2     (mole%):                    0.098
  Hydrogen Sulfide, H2S   (mole%):                    0.006
  Oil Gravity             (API):                     34.600
  Bub Pt Press            (psig):                     0.000
  Water Spec Grav         (FRAC):                     1.090

Source Name: Source001,    Pipe Number:  9,    Pipe Name: Pipe001
  Flowing Temperature     (F):                      128.0
  Flowing Pressure        (psig):                  1238.40
  Oil Rate                (    bbl/day     ):      5100.000
  Gas/Oil Ratio           (  std ft3/bbl     ):     750.000
  Water Cut               (     Volume %     ):       6.300
  Gas Gravity:                                        0.723
  Separator Pressure      (psig):                     0.000
  Separator Temperature   (F):                       60.000
  Nitrogen, N2            (mole%):                    1.430
  Carbon Dioxide, CO2     (mole%):                    4.230
  Hydrogen Sulfide, H2S   (mole%):                    0.065
  Oil Gravity             (API):                     28.600
  Bub Pt Press            (psig):                     0.000
  Water Spec Grav         (FRAC):                     1.060
```

OIL VISCOSITY

Source Name: Source004, Pipe Name: Pipe010:

Pipe	Temperature (F)	Viscosity (cp)
1	100.000	1.62000
1	210.000	0.88000

Power: Visc(cp) = 0.7154E+02 * Temp(F) ^ -.8225E+00

Source Name: Source002, Pipe Name: Pipe004:

Pipe	Temperature (F)	Viscosity (cp)
3	100.000	1.20000
3	210.000	0.44500

Power: Visc(cp) = 0.5666E+03 * Temp(F) ^ -.1337E+01

Source Name: Source003, Pipe Name: Pipe005:

Pipe	Temperature (F)	Viscosity (cp)
4	100.000	1.05000
4	210.000	0.22600

Power: Visc(cp) = 0.1451E+05 * Temp(F) ^ -.2070E+01

Source Name: Source001, Pipe Name: Pipe001:

Pipe	Temperature (F)	Viscosity (cp)
9	100.000	1.45000
9	210.000	0.95000

Power: Visc(cp) = 0.2001E+02 * Temp(F) ^ -.5699E+00

SINK CONDITIONS

Sink Name	Pipe Number	Pipe Name	Flowing Pressure (psig)	Flow Rate (bbl/day)
Sink001	7	Pipe008	100.00	11274.102
Sink002	8	Pipe009	200.00	7075.899

Output

The calculated output can be found in PLROTOUT.TXT.

GENNET - M 5.4.23

GENeral NETwork Multiphase flow analyzer
DATE: 01/30/2004 TIME: 12:12

Pipe No	Pipe Name	Dist (ft)	Flow (bbl/day)	Pressure (psig)	Temp (øF)	Liq Holdup (Bbl)	Vel mix (ft/sec)	Flow Patt
9	Pipe001	0.0	0.510E+04	1238.40	128.00			
9	Pipe001	2523.0	0.510E+04	465.89	109.24	6.9	30.3	Intr
						6.9		
3	Pipe004	0.0	0.440E+04	721.52	136.00			
3	Pipe004	2083.3	0.440E+04	654.40	115.02	13.6	15.3	Intr
3	Pipe004	4166.7	0.440E+04	583.04	99.37	13.1	16.2	Intr
3	Pipe004	6250.0	0.440E+04	504.91	87.90	12.5	17.6	Intr
						39.2		
4	Pipe005	0.0	0.450E+04	851.61	142.00			
4	Pipe005	2036.0	0.450E+04	790.08	120.64	14.4	14.0	Intr
4	Pipe005	4072.0	0.450E+04	726.25	104.37	14.1	14.4	Intr
4	Pipe005	6108.0	0.450E+04	658.77	92.18	13.7	15.1	Intr
4	Pipe005	8144.0	0.450E+04	585.88	83.18	13.1	16.2	Intr
4	Pipe005	10180.0	0.450E+04	504.91	76.64	12.3	17.8	Intr
						67.7		

Pipe No	Pipe Name	Dist (ft)	Flow (bbl/day)	Pressure (psig)	Temp (øF)	Liq Holdup (Bbl)	Vel mix (ft/sec)	Flow Patt
1	Pipe010	0.0	0.435E+04	1200.49	124.00			
1	Pipe010	2210.0	0.435E+04	880.72	106.14	7.9	16.6	Liq
1	Pipe010	4420.0	0.435E+04	465.89	92.76	6.7	21.5	Intr
						14.6		
2	Pipe002	0.0	0.869E+04	465.89	101.70			
2	Pipe002	2000.0	0.869E+04	444.63	60.00	29.9	11.2	Intr
2	Pipe002	4000.0	0.869E+04	423.41	60.00	30.1	11.0	Intr
						60.0		
5	Pipe006	0.0	0.890E+04	504.91	82.17			
5	Pipe006	2000.0	0.890E+04	465.45	60.00	26.5	16.8	Intr
5	Pipe006	4000.0	0.890E+04	423.41	60.00	25.7	17.5	Intr
						52.2		
10	Pipe003	0.0	0.760E+03	465.89	101.70			
10	Pipe003	2000.0	0.760E+03	444.74	60.00	4.8	6.2	Intr
						4.8		
11	Pipe011	0.0	0.760E+03	444.74	60.00			
11	Pipe011	2000.0	0.760E+03	423.41	60.00	4.8	6.0	Intr
						4.8		
6	Pipe007	0.0	0.184E+05	423.41	60.00			
6	Pipe007	2004.2	0.184E+05	409.86	60.00	73.2	11.6	Intr
6	Pipe007	4008.3	0.184E+05	396.04	60.00	72.0	11.9	Intr
6	Pipe007	6012.5	0.184E+05	381.94	60.00	70.7	12.3	Intr
6	Pipe007	8016.7	0.184E+05	367.52	60.00	69.3	12.7	Intr
6	Pipe007	10020.8	0.184E+05	352.75	60.00	67.9	13.1	Intr
6	Pipe007	12025.0	0.184E+05	337.60	60.00	66.4	13.7	Intr
6	Pipe007	14029.2	0.184E+05	321.99	60.00	64.9	14.2	Intr
6	Pipe007	16033.3	0.184E+05	305.89	60.00	63.2	14.9	Intr
6	Pipe007	18037.5	0.184E+05	289.21	60.00	61.5	15.6	Intr
6	Pipe007	20041.7	0.184E+05	271.86	60.00	59.6	16.4	Intr
6	Pipe007	22045.8	0.184E+05	253.71	60.00	57.6	17.4	Intr
6	Pipe007	24050.0	0.184E+05	234.60	60.00	55.5	18.6	Intr
						781.7		
7	Pipe008	0.0	0.113E+05	234.60	60.00			
7	Pipe008	1880.0	0.113E+05	206.43	60.00	31.1	19.8	Intr
7	Pipe008	3760.0	0.113E+05	175.85	60.00	28.9	22.5	Intr
7	Pipe008	5640.0	0.113E+05	141.48	60.00	26.3	26.6	Intr
7	Pipe008	7520.0	0.113E+05	99.99	60.00	22.9	33.7	Intr
						109.3		
8	Pipe009	0.0	0.708E+04	234.60	60.00			
8	Pipe009	1840.0	0.708E+04	228.19	60.00	31.0	11.9	Intr
8	Pipe009	3680.0	0.708E+04	221.55	60.00	30.6	12.2	Intr
8	Pipe009	5520.0	0.708E+04	214.65	60.00	30.1	12.6	Intr
8	Pipe009	7360.0	0.708E+04	207.47	60.00	29.6	12.9	Intr
8	Pipe009	9200.0	0.708E+04	199.99	60.00	29.1	13.3	Intr
						150.3		

END

14: Pumps

Centrifugal pumps ... 546
Speed torque calculation ... 558
Variable speed or impeller trim? ... 559
Pulsation control for reciprocating pumps ... 560
Rotary pumps on pipeline services ... 569
Centrifugal pump parameters and applications ... 574
How to examine centrifugal pump systems ... 579
Estimate the discharge of a centrifugal pump at various speeds ... 584
How to estimate the head for an average centrifugal pump ... 585
Find the reciprocating pump capacity ... 585
How to estimate the hp required to pump at a given rate at a desired discharge pressure ... 585
Nomograph for determining reciprocating pump capacity ... 586
Nomograph for determining specific speed of pumps ... 587
Nomograph for determining horsepower requirement of pumps ... 588
How to select motors for field-gathering pumps ... 588
How a reciprocating pump works ... 589
Understanding the basics of rotary screw pumps ... 598
How to evaluate VFD speed on hydraulics ... 604
Progressive cavity pumps ... 606
Pump curve accuracy ... 607
The basics of NPSH and pump operating regions ... 614
Reading a centrifugal pump curve ... 617
Navigating pump curves: a roadmap to reliability ... 620

Centrifugal pumps

Centrifugal pumps are best suited for large volume applications or for smaller volumes when the ratio of volume to pressure is high. The selection of the proper pump will depend on the system throughput, viscosity, specific gravity, and head requirements for a particular application. Where a variety of products with different characteristics are handled, it maybe necessary to utilize multiple pumps either in series or parallel, or a series—parallel combination, or pumps with variable speed capabilities to achieve the desired throughput at least cost. Economics will ultimately determine the type and number of pumps to be used.

The first step in selecting centrifugal pumps is to analyze the pipeline system and determine its characteristics such as the pressure required to move the desired flow rate. Initial and future conditions should be evaluated so that pumps with sufficient flexibility to handle both conditions with minor changes may be selected.

Pump manufacturers publish pump performance maps similar to the one shown in Figure 1 to aid in pump selection. The performance map shows the range of differential head and capacity that will be developed by a family of pumps. Once the system data have been established, a pump may be

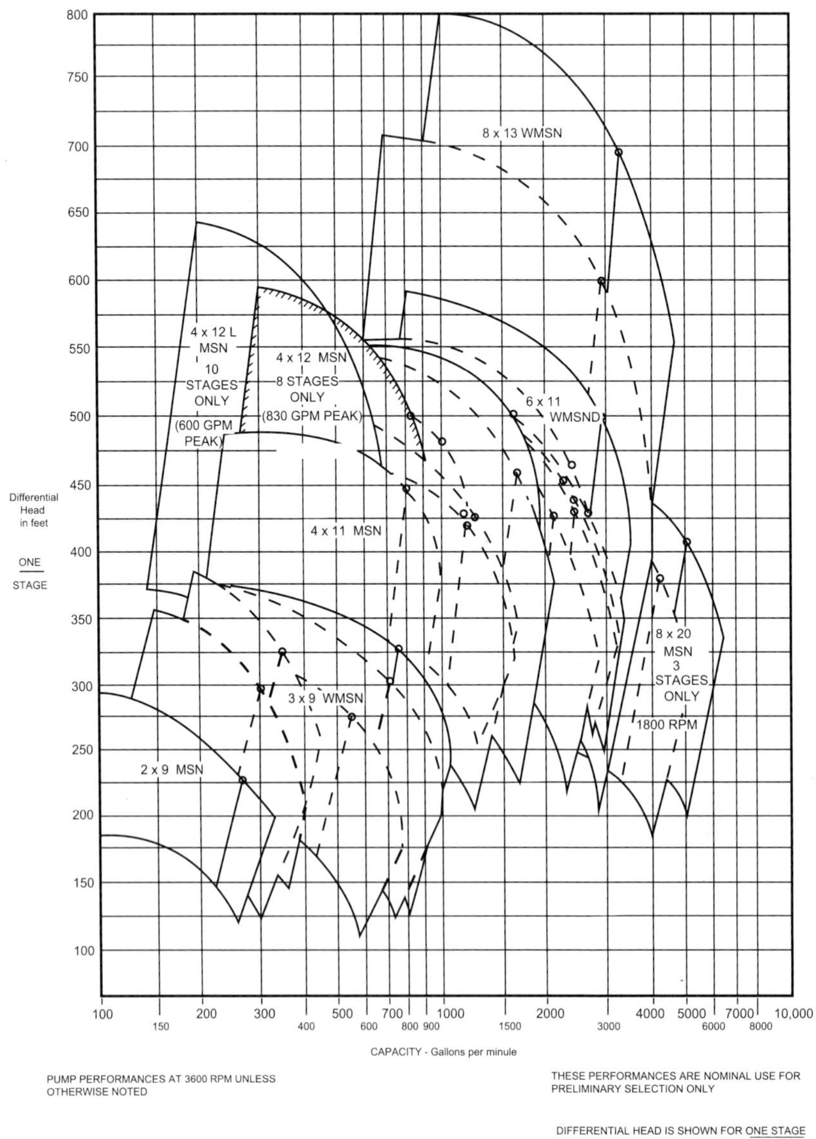

Figure 1. Typical performance map for centrifugal pumps. Courtesy United Centrifugal Pumps.

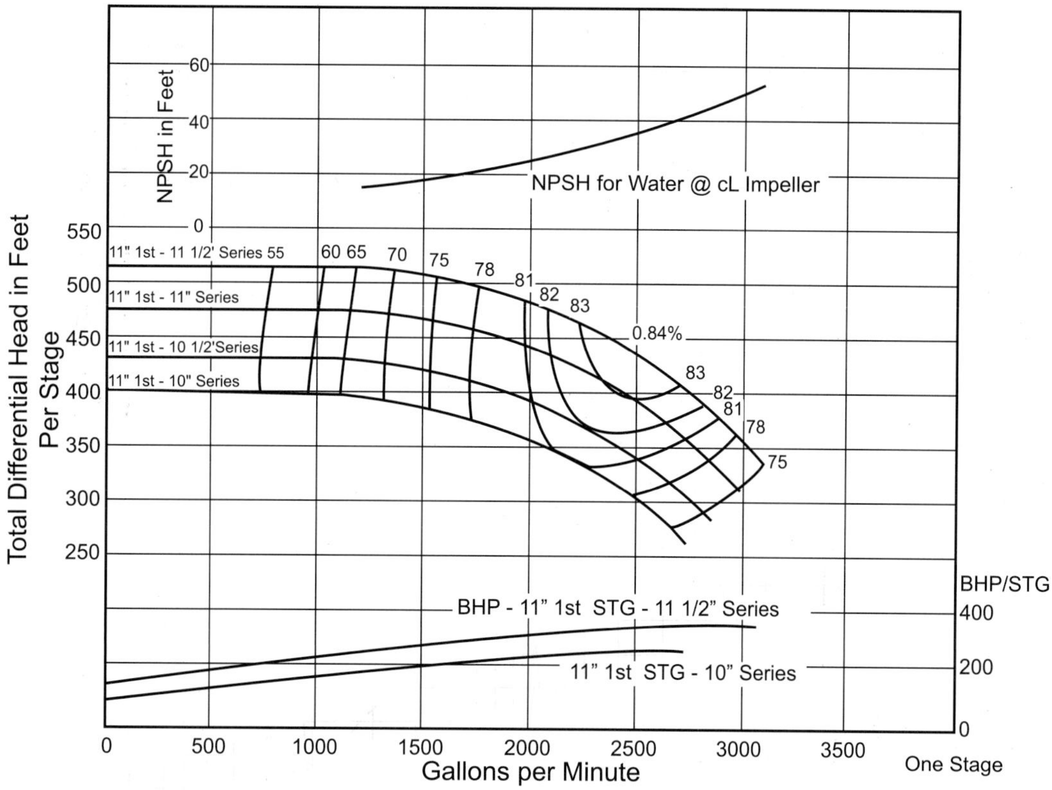

Figure 2. Typical performance curve for centrifugal pumps. Courtesy United Centrifugal Pumps.

selected from the map. The selection is further refined by referring to a curve such as that shown in Figure 2, which shows a typical pump manufacturer's catalog curve for a specific pump showing the performance for various impeller size, as well as the brake horsepower, NPSH requirement, and pump efficiency. Pump performance curves are plotted in feet and for a specific gravity of 1.0. When pumping a nonviscous liquid, the pump will develop the same head shown on the performance curve; however, the *pressure* will change with specific gravity, as will the horsepower required to drive the pump. Pressure may be calculated by multiplying the pump differential head by the specific gravity and by 0.4329. The horsepower required to drive the pump will also vary directly with the specific gravity of the fluid being pumped.

The viscosity of the fluid will influence the performance of the centrifugal pump. This is discussed in more detail later.

Centrifugal pumps are available in a number of different configurations such as horizontal, vertical, radial split, and axial split casings. The final choice will depend on the system hydraulic conditions (pressure and flow rate), the ease of maintenance desired, and amount of space available for installing the pump. The pressure rating of the case and material selection will be determined by the system requirements and the fluid to be pumped.

It is extremely important that sufficient net positive suction head (NPSH) above the vapor pressure of the liquid being pumped is available for safe and reliable pump operation. NPSH is always specified in feet of head above vapor pressure required at the center line of the impeller. The NPSH required will increase as the capacity of the pump increases, which makes it necessary for the user to make a best effort to furnish the pump manufacturer with an accurate range of flow rate conditions. The pump suction nozzle geometry, pump speed, and the impeller inlet will determine the NPSH that will be required. If too much margin of safety is included in the pump specifications, the pump will probably be oversized to obtain a low NPSH and will not operate at its best efficiency point. Figures 8-10 show three examples of how to calculate available NPSH. As a rule of thumb, the NPSH available should be at least 10% greater than the NPSH required by the pump. It is difficult and costly to correct NPSH problems after the pump is placed in service. NPSH is discussed in detail in *Centrifugal Pumps: Design and Application* by Ross and Lobanoff, Gulf Publishing Co., and it's recommended reading for those who are involved in the selection and application of centrifugal pumps.

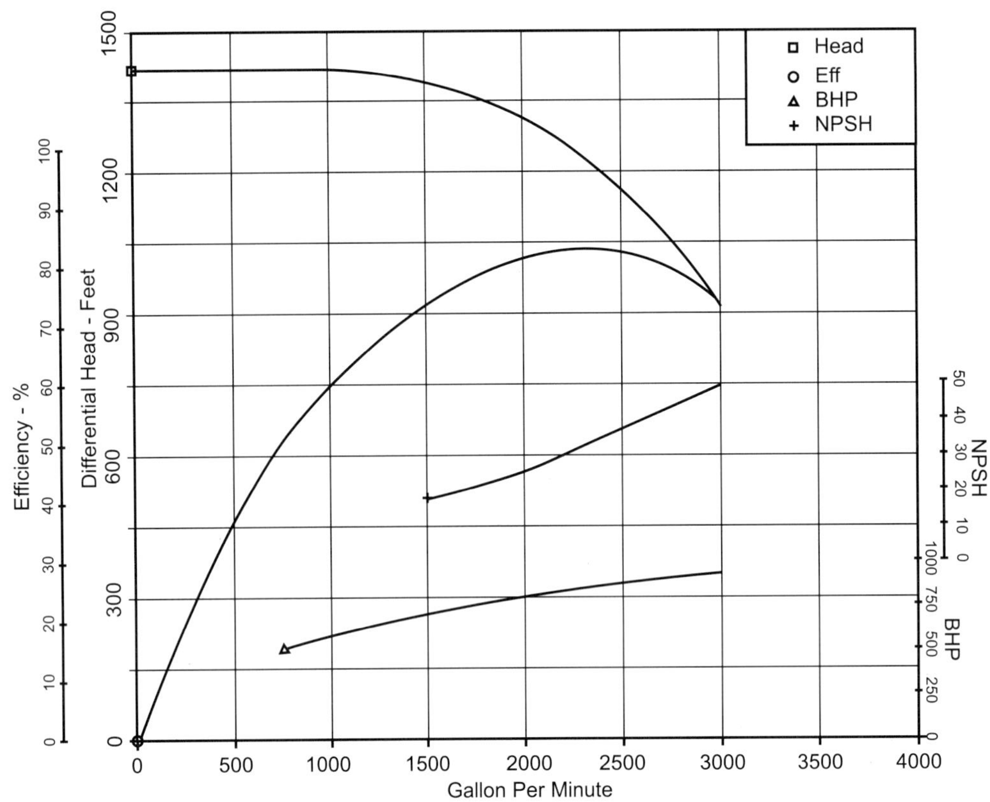

Figure 3. Centrifugal pump performance data.

Table 1 gives some conversion factors and formulas that will be useful in problems dealing with pumps.

All centrifugal pumps, except those pumps that are specifically designed for low suction pressure applications such as tank farm booster pumps, should be protected by a low suction pressure shut down switch. The value of pressure to be used to set the switch will be determined by the amount of NPSH required at the maximum expected operating flow rate, the vapor pressure of the fluid being pumped, and the atmospheric pressure at the pump location. Once the maximum expected operating flow rate is known, the amount of NPSH required may be obtained from the pump performance curve. The NPSH curve is plotted in feet of head, and this may be converted to psi for switch setting purposes as follows:

$$\text{PSIG} = \text{NPSH} \times 0.433 \times \text{specific gravity} + \text{vapor pressure} - \text{atmospheric pressure}$$

Figure 3 shows the performance data for a centrifugal pump that was chosen from the performance map. The impeller diameter was chosen from the curve shown in Figure 2. In this case, an impeller diameter of 11 in. was chosen.

Centrifugal pumps may be operated singly, in series, in parallel, or in a combination of series and parallel. When pumps are operated in series, the heads are additive at a given rate of flow. When pumps are operated in parallel, flow is additive at a given value of head. This is illustrated in Figure 3. When pumps are operated in a series—parallel combination, these same rules will apply to determine the total head developed by the combination of pumps. Figure 4 shows performance data for one pump and for series and parallel operation of two pumps.

Pumps connected in series have the discharge of one pump connected to the suction of the downstream pump. Pumps connected in parallel use a common suction line and common discharge line (See Figure 4a).

Pumps that are to be operated in parallel must be matched so that both develop the same head at the same flow rate. They should also have constantly rising head curves. Pumps with a head curve that has a hump (one that develops less head at shutoff than at some other point on the head curve) should be avoided since this could possibly preclude the second pump from pumping.

Table 1
Conversion Factors and Formulas

$$gpm = 0.07 \times \text{boiler hp}$$

$$gpm = 449 \times \text{cfs}$$

$$gpm = 0.0292 \times \text{bbl/day}$$

$$gpm = 0.7 \times \text{bbl/hr}$$

$$gpm = 4.4 \times \text{cubic meters/hr}$$

$$gpm = \frac{\text{lb per hr}}{500 \times \text{specific gravity}}$$

$$H = \frac{P \times 2.31}{\text{specific gravity}}$$

$$V = \frac{Q \times 0.321}{A}$$

$$U = \frac{\text{Diameter(in.)} \times N}{229}$$

$$h_v = \frac{v^2}{2g}$$

$$\text{whp} = \frac{Q \times H \times \text{specific gravity}}{3,960}$$

$$\text{bhp} = \frac{Q \times H \times \text{specific gravity}}{3,960 \times e}$$

$$\text{bhp} = \frac{Q \times P}{1,715 \times e}$$

$$T = \frac{\text{bhp} \times 5,250}{N}$$

$$N_s = \frac{N\sqrt{Q}}{H_1^{3/4}} = \frac{N\sqrt{\sqrt{H_1} \times \sqrt{Q}}}{H_1}$$

$$S = \frac{N\sqrt{Q}}{h_{sv}^{3/4}} = \frac{N\sqrt{\sqrt{h_{sv}} \times \sqrt{Q}}}{h_{sv}}$$

$$t_r = \frac{H\left(\frac{1}{e} - 1\right)}{780 \times C}$$

where
- N = speed in rpm
- N_s = specific speed in rpm
- S = suction specific speed in rpm
- Q = capacity in gpm
- P = pressure in psi
- H = total head in ft
- H_1 = head per stage in ft
- h_{sv} = net positive suction head in ft
- h_v = velocity head in ft
- whp = water horsepower
- bhp = brake horsepower
- U = peripheral velocity in ft per second
- g = 32.16 ft per second squared
- cfs = cubic ft per second
- bbl = barrel (42 gallons)
- C = specific heat
- psi = lb/in^2
- gpm = gallons per minute
- e = pump efficiency in decimal
- D = impeller diameter in inch
- V = velocity in ft per second
- t = torque in ft-lb
- t = temperature in °F
- t_r = temperature rise in °F
- A = area in in^2

Courtesy United Centrifugal Pumps.

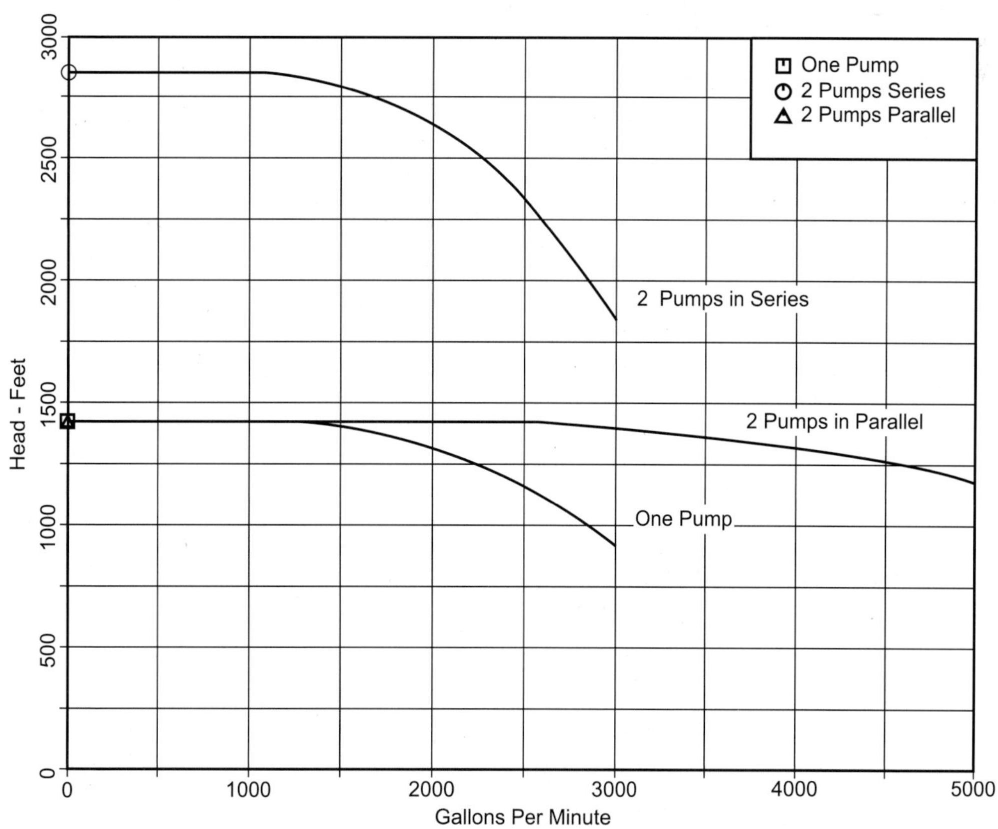

Figure 4. Series/parallel combinations—centrifugal pumps.

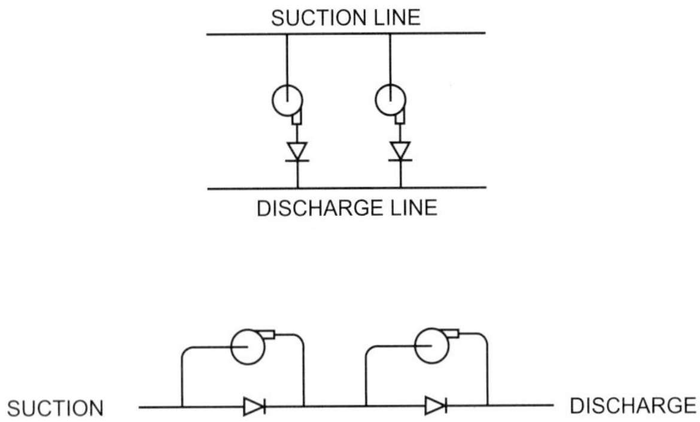

Figure 4a. Pumps connected in parallel/in series.

The high cost of energy will in almost all cases justify the additional cost of multiple pumping units or variable speed drivers. The use of multiple units will allow most pumping conditions to be met without throttling and therefore wasting power. If as many as three series units can be justified, the units should be sized as follows:

One unit—full head
One unit—3/4 of the head of the full head unit
One unit—1/2 of the head of the full head unit

This will allow the following pump combinations to be used:

Units	Total head	Units	Total head
½	½	½ ,1	½
¾	¾	¾,1	1 ¾
1	1	½,¾,1	2 ¼
½,¾	1 ¼	¾	

Increments of ¼ head can be achieved up to the last ½-head increment. This results in maximum flexibility when constant rpm drivers are used.

Figure 5 shows the performance of series versus parallel operation of pumps. In this particular example, series

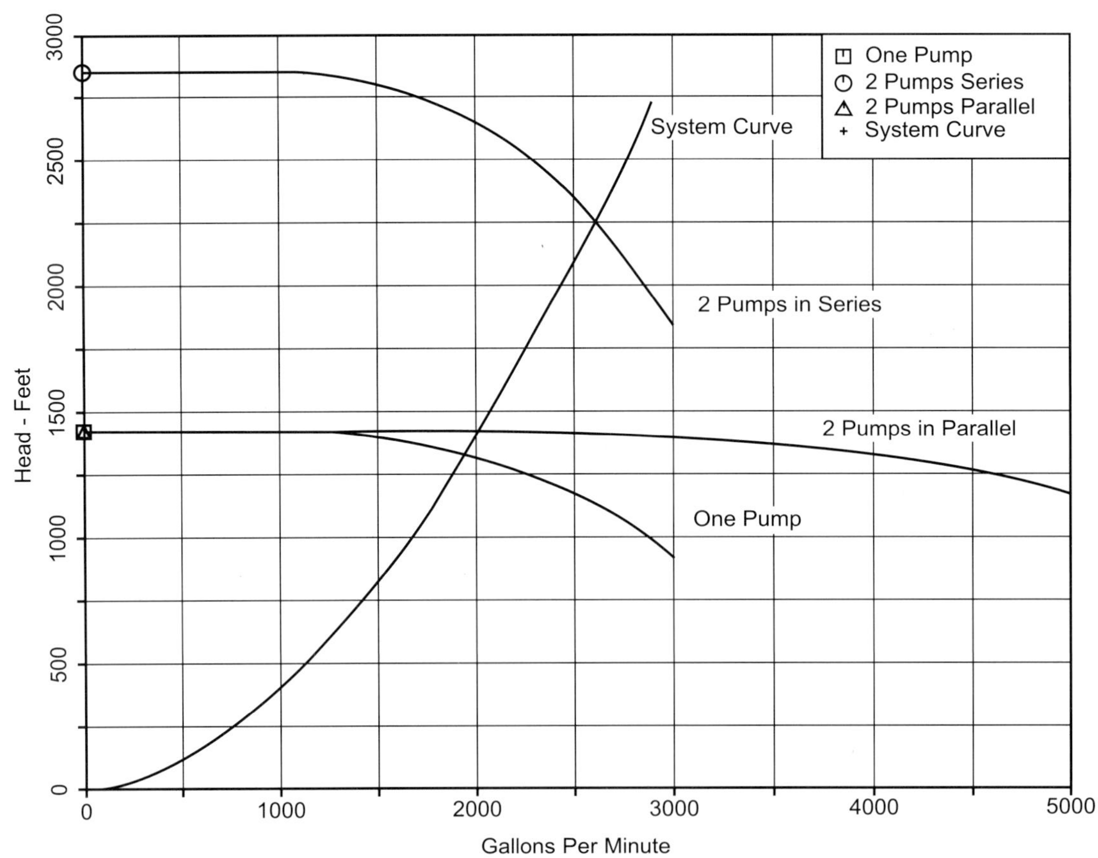

Figure 5. Series/parallel combinations with system curve—centrifugal pumps.

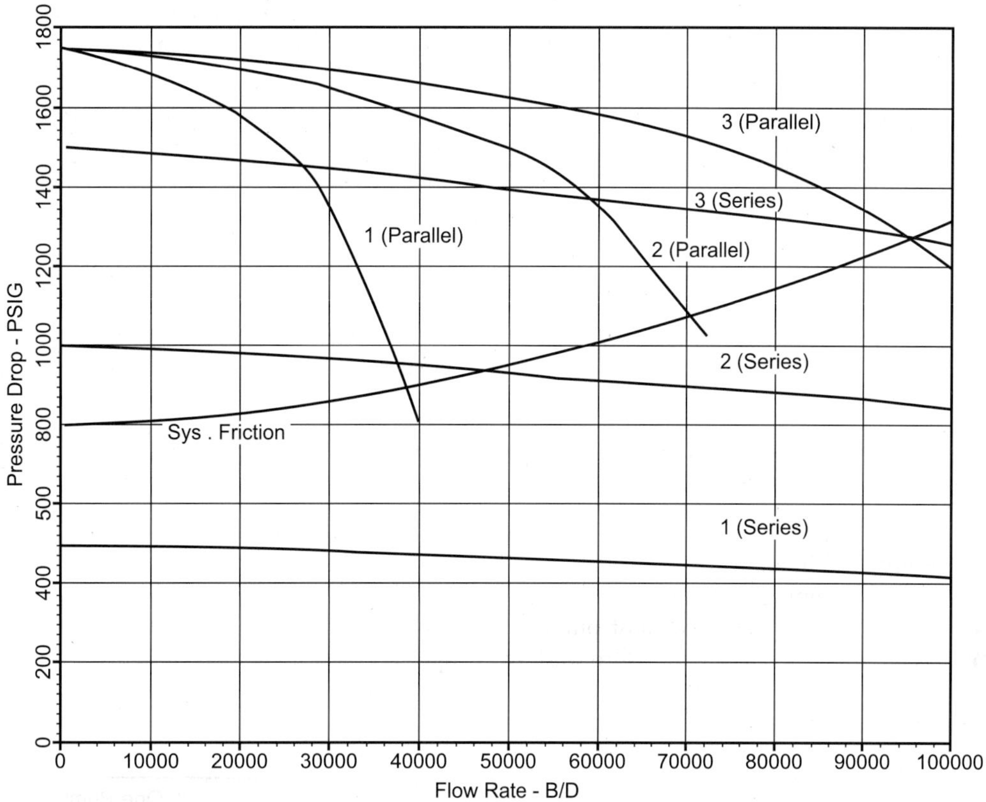

Figure 5a. Series and parallel operation—centrifugal pumps.

operation will provide a higher pumping rate than provided with a parallel operation, and two pumps in parallel will pump only slightly more than one single pump. The maximum flow rate with two pumps in series will be 2594 gpm or 89,439 b/d. The maximum flow rate for one pump will be 1917 gpm or 66,092 b/d.

For system rates between 66,092 b/d and 89,439 b/d, pressure throttling will be necessary to match the pump performance to the system curves, unless the pumps are equipped with variable speed drives. If throttling is necessary, it may be minimized by operating at a high flow rate for part of the day and at a lower rate for the balance of the day. If a system rate of 70,000 b/d is desired with the system shown in Figure 5, one pump operating 20 h per day and two pumps operating 4 h per day would achieve this rate; however, this may not be the most economical method of operation. It will be necessary to calculate and compare the cost of throttling pressure versus the cost to operate the second unit for 4 h. System rates less than 66,092 b/d will require pressure throttling unless the pump speed may be varied. It is not uncommon for pipeline pump stations to contain several pumps, each of which may be sized differently to increase the amount of flow rate flexibility.

The use of multiple pumps affords some protection against throughput reductions due to one or more pumps being out of service.

Parallel pump operation is usually best suited for systems where the residual static pressure is high compared to the line friction loss. Such systems usually have a flat system curve. An example of this is shown in Figure 5a.

For this system, three parallel pumps would offer the most flexibility, since one unit will always be able to overcome the system friction and static and would pump almost as much as two series pumps. One could make the argument that one large pump could be selected to meet the 2 (series) condition shown in Figure 5a and a ½-head pump could be used for the maximum volume condition. This argument is valid as long as the large pump is available. When it is not available, the system will be down, since the ½-head unit will not develop enough head to overcome the static and friction losses. Three pumps operating in parallel will offer the most flexibility for the system shown in Figure 5a.

Changes in system throughputs may require resizing the pump impellers to meet the new conditions. Some types of multistage pumps may be destaged or upstaged by adding or taking out impellers, depending upon the number of stages in

Table 2
Formulas for Recalculating Pump Performance with Impeller Diameter or Speed Change

Q_1, H_1, bhp_1, D_1 and N_1 = Initial Capacity, Head, Brake Horsepower, Diameter, and Speed. Q_2, H_2, bhp_2, D_2 and N_2 = New Capacity, Head, Brake Horsepower, Diameter, and Speed.		
Diameter Change Only	Speed Change Only	Diameter and Speed Change
$Q_2 = Q_1 = \left(\frac{D_2}{D_1}\right)$	$Q_2 = Q_1 = \left(\frac{N_2}{N_1}\right)$	$Q_2 = Q_1 = \left(\frac{D_2}{D_1} \times \frac{N_2}{N_1}\right)$
$H_2 = H_1 = \left(\frac{D_2}{D_1}\right)^2$	$H_2 = H_1 = \left(\frac{N_2}{N_1}\right)^2$	$H_2 = H_1 = \left(\frac{D_2}{D_1} \times \frac{N_2}{N_1}\right)^2$
$bhp_2 = bhp_1 = \left(\frac{D_2}{D_1}\right)^3$	$bhp_2 = bhp_1 = \left(\frac{N_2}{N_1}\right)^3$	$bhp_2 = bhp_1 = \left(\frac{D_2}{D_1} \times \frac{N_2}{N_1}\right)^3$

Courtesy United Centrifugal Pumps.

operation and the amount of head change required. If it is contemplated that it may be necessary in the future to destage or upstage, this should be specified when the pump is purchased. It is also possible to optimize the efficiency point by chipping the pump volutes and under-filing the pump impeller(s) or adding volute inserts. The pump manufacturer should be contacted for details on how to accomplish this.

The pump affinity laws shown in Table 1 may be used to predict the performance of the pump at different operating conditions. These laws are commonly used to determine new impeller diameters required for nominal changes in flow rates or performance at different speeds. The affinity laws will allow the translation of a point on the performance curves but will not define the operation of the pump in the system. A new pump performance curve may be constructed using these laws and then applied to the system curve to determine the actual operation of the pump.

The system shown in Figure 5, with both pumps configured with 11-in diameter impellers, will pump a maximum of 89,439 b/d. The system throughput requirements have now changed such that the maximum desired throughput is 87,931 b/d, and it is desired to trim the impellers in one of the pumps to meet this new condition. The new $H-Q$ condition for one pump will be $H_2 = 955$ ft and $Q_2 = 2550$ gpm. In order to establish the new impeller diameter, it will be necessary to translate this point to the existing pump curve to determine H_1 and Q_1. Select flow rates on both sides of the new rate and calculate the corresponding head. These data are plotted, and the resulting line will cross the existing pump head curve as

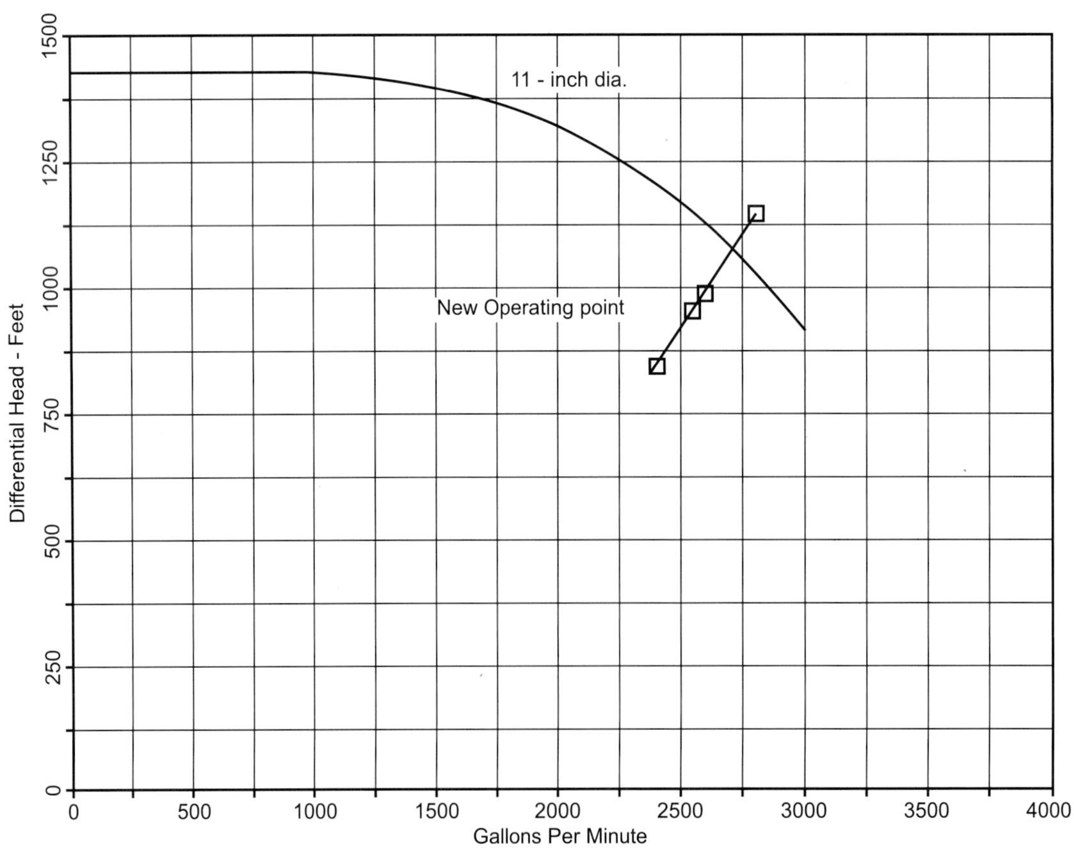

Figure 6. Impeller sizing—centrifugal pumps.

shown in Figure 6. Determine the head and flow rate at the point where this line crosses the existing head curve, and these values will be H_1 and Q_1. For this calculation, flow rates were selected as follows:

Selected Q	Calculated H
2800	1151
2600	993
2400	846

The head was calculated as follows:

$$H = H_2 \times (Q/Q_2)^2$$

This line crosses the existing head curve at 2700 gpm and 1080 ft, and these values will be the values to use for Q_1 and H_1, respectively.

Use the pump affinity law to calculate the new diameter as follows:

$$H_2/H_1 = (D_2/D_1)^2$$
$$D_2 = D_1 \times (H_1/H_2)^{1/2}$$
$$D_2 = 11.0 \times (955/1080)^{1/2}$$
$$D_2 = 10.34 \text{ in}$$

Since the pump affinity laws are not quite exact for a diameter change, it will be necessary to apply a correction factor to the calculated value of D_2 to determine the final diameter. The correction factor is determined by calculating the ratio of the calculated diameter to the original diameter. This value, expressed as a percentage, is then used to enter Figure 7 to determine the actual required diameter as a percentage of the original diameter.

$$D\text{ calc}/D\text{ orig} = 10.34/11 = 94\%$$

Enter the chart with 94% and read 94.5% as the actual required diameter as a percentage of the original diameter.

The new diameter will therefore be 0.945 x 11.0 = 10.395 in. The impellers would probably be trimmed to 10 7/16 in.

Once the new diameter is established, the new pump head curve may be calculated by using the affinity laws to calculate new $H-Q$ values.

The pump affinity laws are exact for a speed change. Figure 7a shows the effect of a speed change on the pump performance. The pump manufacturer should be consulted for data on the range of speeds over which the pump may be efficiently operated.

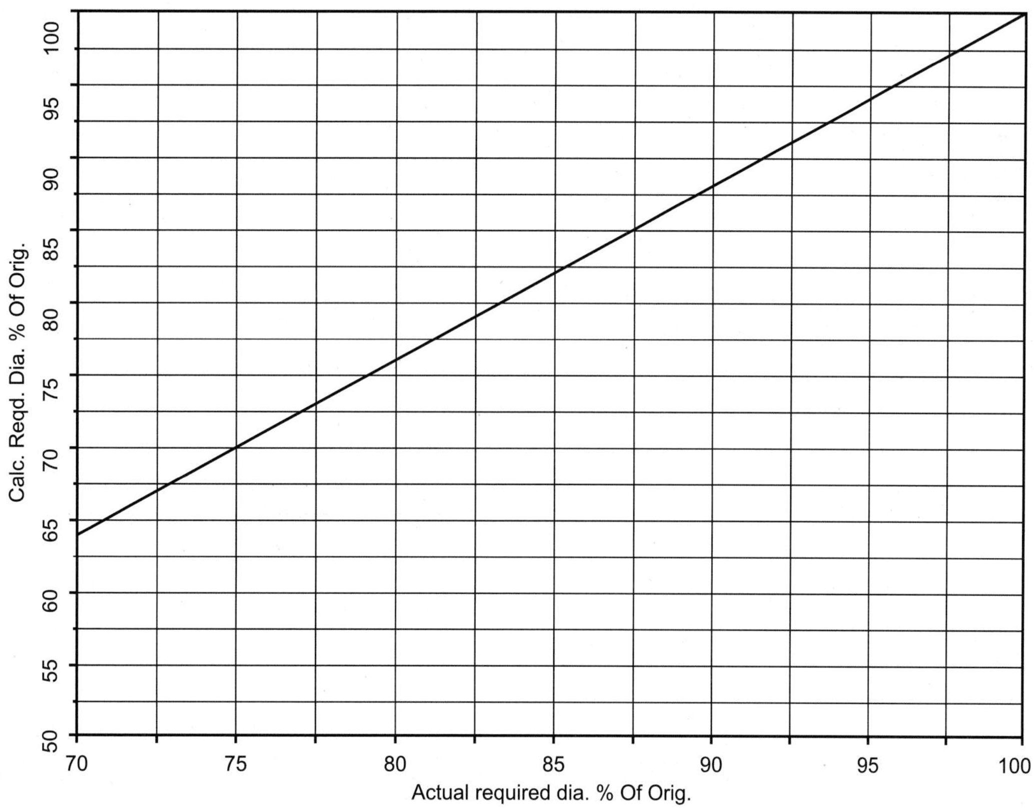

Figure 7. Impeller trim correction.

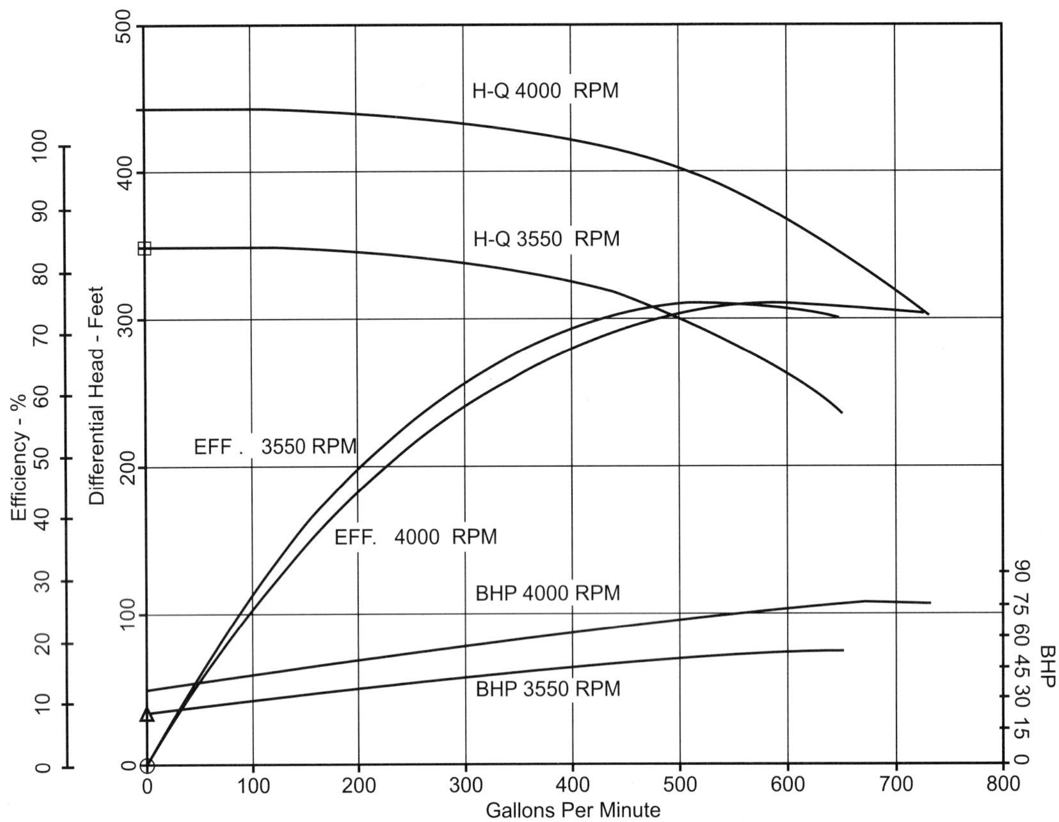

Figure 7a. Centrifugal pump performance data for speed.

Very viscous liquids will have an impact on the performance of centrifugal pumps. Figure 11 compares pump performance when pumping water to pumping a liquid with a viscosity of 6000 SUS. The effects of viscosity on pump performance may be predicted by the use of Figures 12 and 13. These charts are used as follows. Refer to Figure 11 and select the water capacity at the BEP (best efficiency point). In this example, the water capacity at the BEP is 23,000 gpm. Refer to Figure 12 and enter the chart with 23,000 gpm and move vertically to intersect the desired impeller diameter, 27 in, and then horizontally to intersect the 6000 SUS line. Move vertically to intersect the line representing the water efficiency (85%), then horizontally to read the viscous efficiency 62% (not a correction factor). The previous vertical line is extended to intersect the head correction and capacity correction curves. The capacity correction factor is 93%, and the head correction factor is 95%. The viscous head at shutoff will be the same as the shutoff head for water. The viscous head will be 617 ft at 21,400 gpm. Plot this point and then draw the head-capacity curve through this point. The viscous

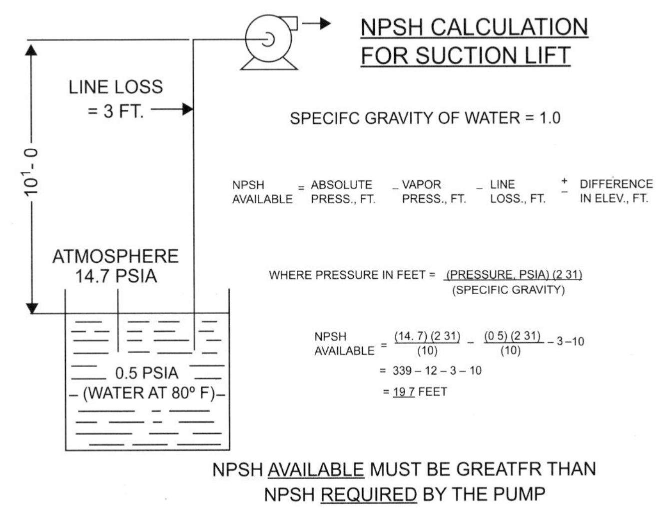

Figure 8. NPSH calculation for suction lift.

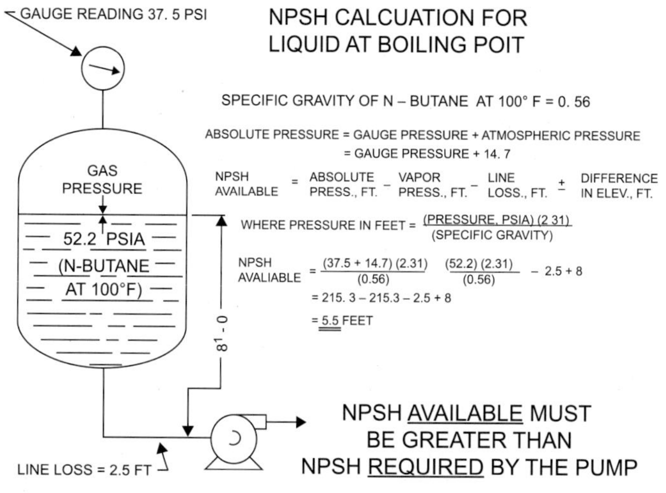

Figure 9. NPSH calculation for liquid at boiling point.

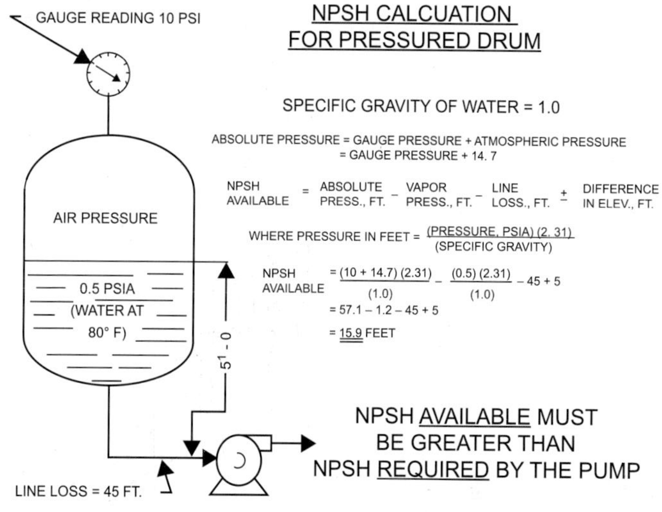

Figure 10. NPSH calculation for pressured drum.

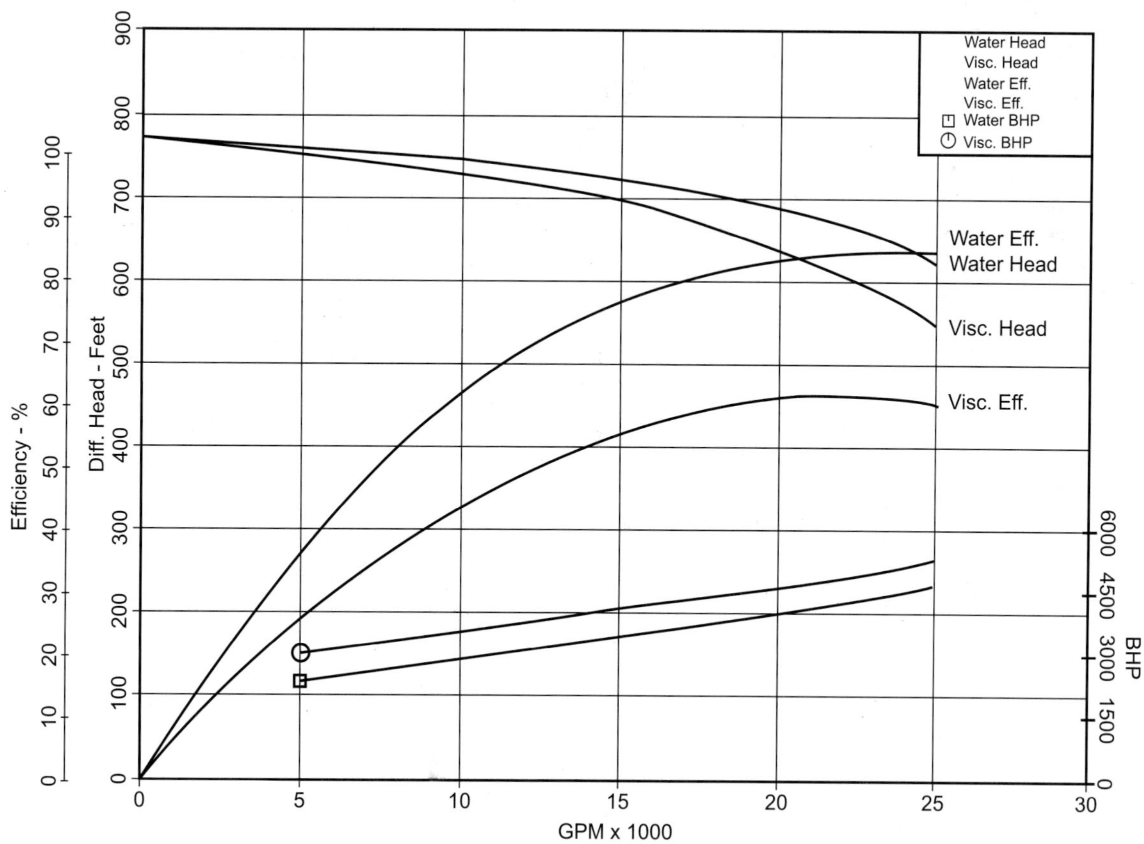

Figure 11. Effects of viscosity on centrifugal pump performance.

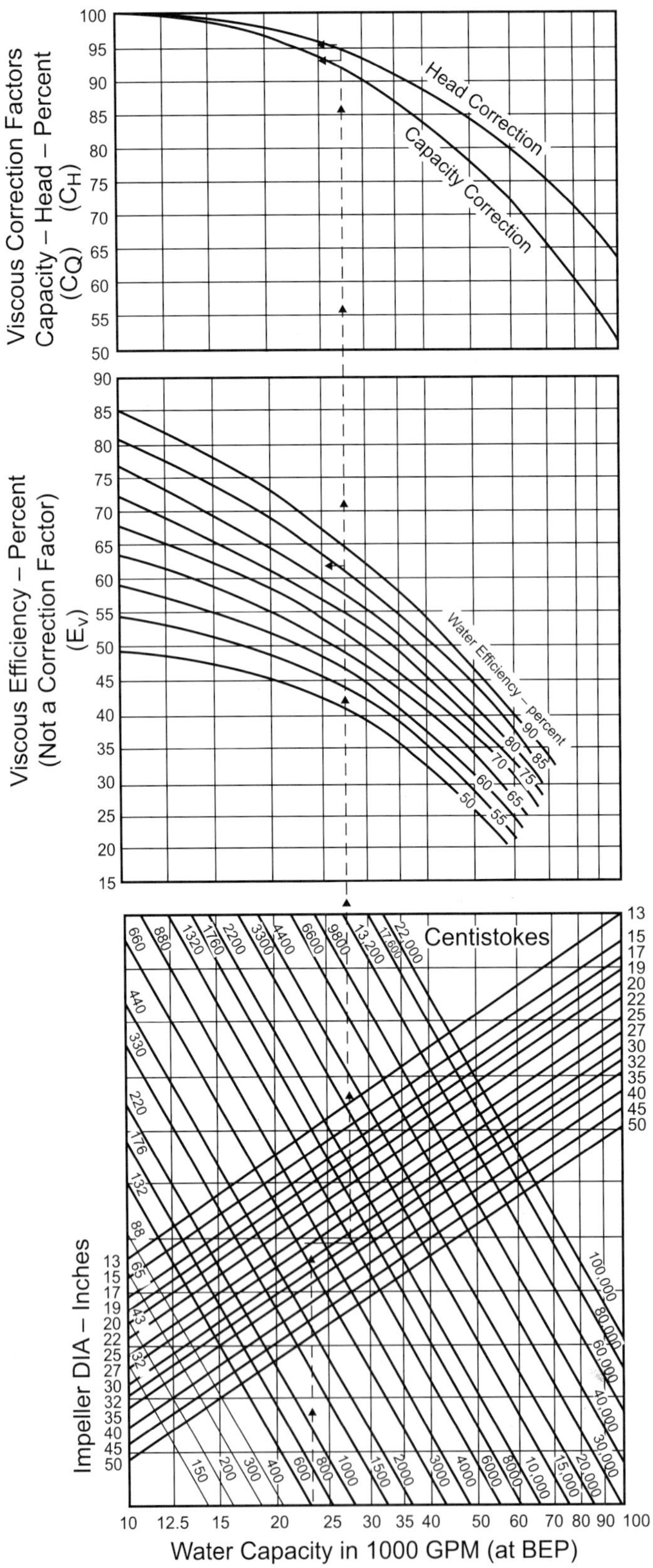

Figure 12. Viscosity correction chart—capacities above 10,000 gpm. Courtesy United Centrifugal Pumps.

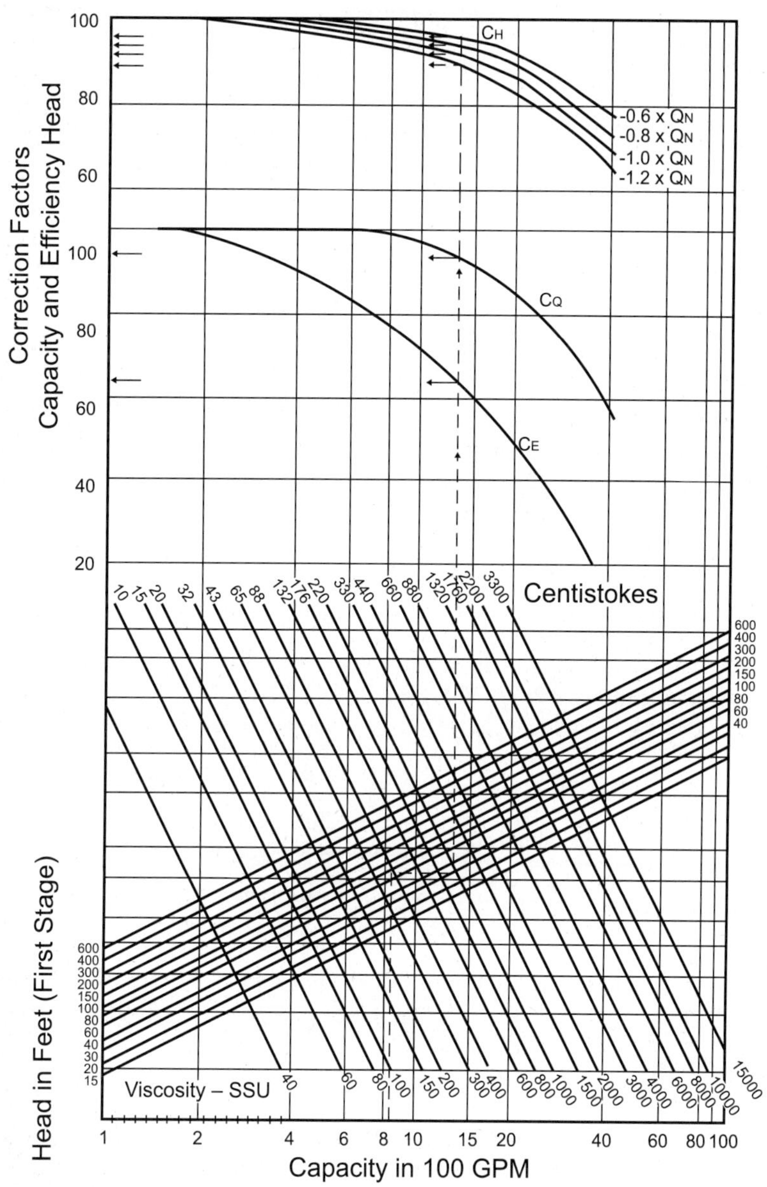

CORRECTION APPLIES TO PEAK EFFICIENCY ONLY. FOR OTHER CAP DRAW CORRECTED H.P. CURVE PARALLEL TO WATER H.P. CURVE AND CALCULATE EFFICIENCIES BASED ON ASSUMED HEAD - CAP CURVE. ADJUST AS REQUIRED TO OBTAIN A SMOOTH CURVE. USE CORRECTIONS OBTAINED FOR MAXIMUM DIAMETER FOR CUT DIAMETERS.

Figure 13. Viscosity correction chart—capacities up to 10,000 gpm. Courtesy United Centrifugal Pumps.

brake horsepower is calculated at the BEP (62% at 21,400 gpm). Draw the viscous brake horsepower curve through this point. The slope of the line will follow the slope of the water brake horsepower curve. The viscous efficiency will be 62% at 21,400 gpm, and the viscous efficiency curve may now be calculated using the viscous brake horsepower, viscous head and capacity, and the viscous specific gravity, which in this example is 0.9.

Figure 13 is used for pumps with capacities up to 10,000 gpm, and Figure 12 is used for pumps with capacities greater than 10,000 gpm.

There may be instances where it is necessary to calculate the pump speed torque requirements to determine if the pump driver will be able to accelerate the pump to operating speed. An example illustrates this calculation procedure, and the results are shown in Figure 14.

Speed torque calculation

The formula for calculating the speed-torque relation for a pump with a given specific gravity and viscosity is:

$$\text{Torque(lb-ft)} = \frac{5250 \times \text{HP}}{\text{RPM}}$$

The following data are required to plot the speed-torque curve:

1. HP @ Shutoff.
2. HP @ Open Valve (use one of the following):
   2a HP @ 120% of peak capacity.
   2b HP rating of driver if greater than HP @ 120% of peak capacity.
   2c Depending on pump application maximum horsepower on performance curve which is greater than driver rated horsepower.

NOTE: Using "2a," or "2b" will cover most of the applications. However on some applications "2c" must be considered.

3. Pump RPM.
4. Torque @ Shutoff.
5. Torque @ "2a," "2b" or "2c". From the above data the torque with the valve closed (using HP @ Shutoff), and the torque with the valve open (using HP @ "2a," "2b" or "2c") at full speed can be computed.

The torque varies as the square of the speed; therefore to obtain the torque at:

3/4 Speed—multiply full speed torque by 0.563
1/2 Speed—multiply full speed torque by 0.250
1/4 Speed—multiply full speed torque by 0.063
1/8 Speed—multiply full speed torque by 0.016

From the figures a speed-torque curve for both closed and open valve, covering the complete speed range, can be plotted (Figure 14).

	Example
HP @ shutoff=80	Torque @ Shutoff= $\frac{5250 \times 80}{3550}$ =118 lb ft @ 3550 RPM
HP @ 120% of peak capacity=150	
Pump RPM=3550	Torque @ 120% of peak cap.= $\frac{5250 \times 150}{3550}$ =222 lb ft @ 3550 RPM

Speed RPM	**Torque-Valve Open (150 HP)**	**Torque-Valve Closed (80 HP)**
Full–3550	Full–222	Full–118
3/4–2660	0.563–125	0.563–67
1/2–1775	1/2–1775	0.250–30
1/4–885	0.063–14	0.063–8
1/8–445	0.016–4	0.016–2

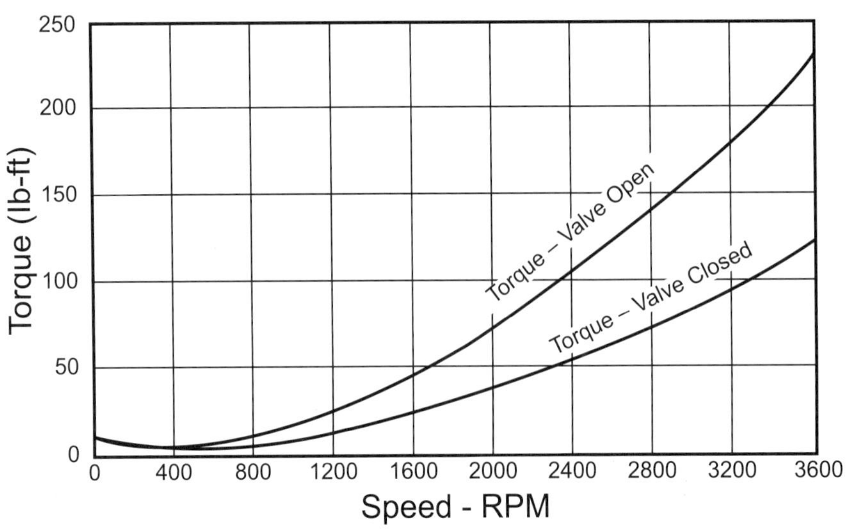

Figure 14. Calculated pump speed torque requirements.

Variable speed or impeller trim?

Dr, Lev Nelik, P.E., APICS Pumping Machinery LLC, Atlanta, Georgia

The simplest way to change pump flow is to throttle the discharge by the valve. Even in this situation, the impeller diameter cannot be arbitrarily chosen. The diameter should be selected to ensure that the pressure (head) is not excessive and that the power consumed provides the required coverage for the anticipated flow range.

Impeller size selection is straightforward and can be easily found in manufacturer manuals. Figure 1 shows pump performance curves at several impeller trims. Such representation is called a combined curve (versus a single line curve) and is intended to help the pump user select the appropriate impeller diameter for the rated point (and around it) and evaluate efficiency, power, and NPSHR.

Unfortunately, if the conditions of service change to require more flow and head than originally specified, the trimmed impeller size would limit such flexibility. Moving to 100 gpm and 120 ft of head for a rated point of 70 gpm at 100 ft (as shown in Figure 1) is impossible; the 5.12-in impeller would not allow it.

Changing the operating conditions by a variable speed drive can be a flexible alternative, as shown in Figure 2. Inefficiently throttling the valve to change the flow is unnecessary, and the entire range of flow and head expands dramatically.

The affinity laws calculate the required pump's speed if a flow range equivalent to that in Figure 1 is desired. Each operating flow-head point scales up or down by simply multiplying the flow by the ratio of the diameter (or speed) and the head by the square of that ratio. This is how Figure 2 is constructed. In our example, several points on the H-Q curve shown in Figure 1 are picked, for example 0, 40, 80, and 120 gpm with heads at these points being 15, 150, 138, and 112 feet.

If we wanted to make the new speeds correspond to exactly the same impeller diameters given in Figure 1 (6, 0.6, 5.0, 4.5, 4.0 and 3.5 in), we simply calculate the speed ratios corresponding to the ratios of the impeller trim at a given curve to the impeller diameter at the maximum OD.

The main advantage of the VFD with regard to efficiency and energy savings is that efficiency essentially remains the same for a given "scale-down" of speed. Trimming the impeller diameter, while allowing the same modification to the head-flow curve, can be inefficient (compare Figure 1 to Figure 2).

For example, BEP (best efficiency point) efficiency at maximum diameter is 58% and drops to about 43% when the impeller is trimmed from 6 in to 3.5 in. With equivalent reduction of speed from 3500 rpm to 2014 rpm, efficiency essentially remains the same. You can easily calculate the energy saved by this method.

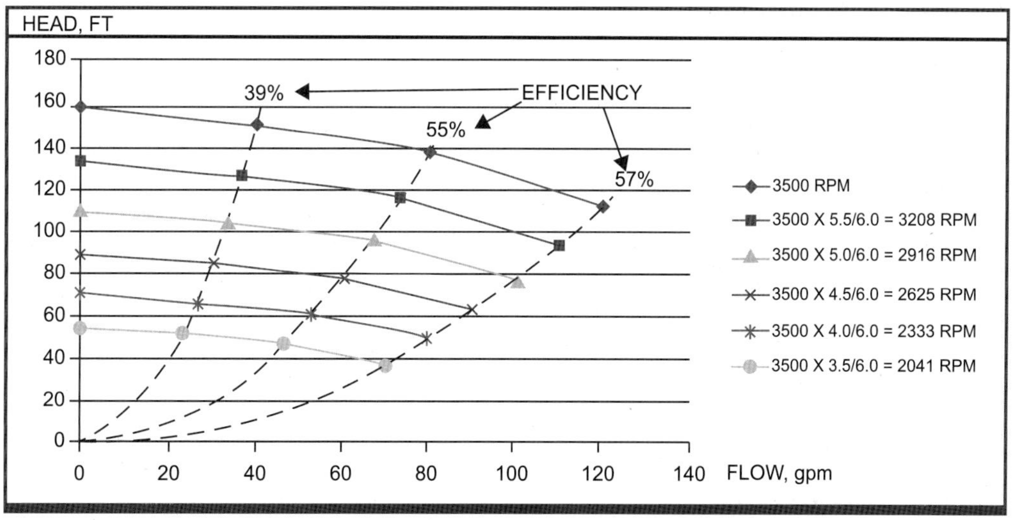

Figure 1.

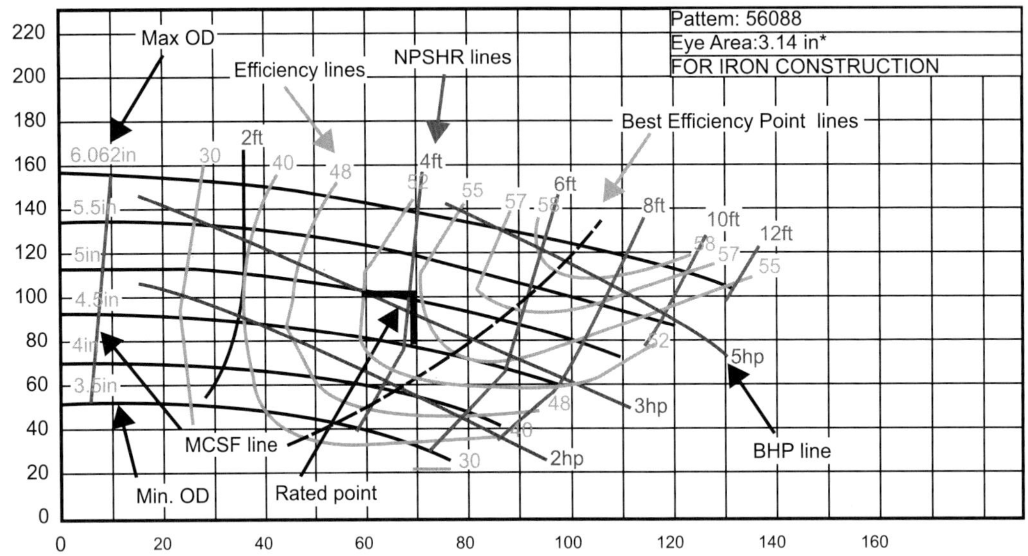

Figure 2.

Scaling down the H-Q curve by trimming the impeller follows the affinity laws only approximately, and certain correction factors must be applied. Efficiency decreases with a reduction in speed. However, this reduction in efficiency with speed is small.

In general, the higher the power, the more money saved—as long as you properly account for all variables, including initial purchase cost of the VFD, installation and maintenance, and efficiency difference.

Pulsation control for reciprocating pumps

V. Larry Beynart, Fluid Energy Controls, Inc., Los Angeles, California

This article reviews the specific requirements for the selection and sizing of pulsation dampeners to reduce the fluid pressure pulsations generated by positive displacement reciprocating pumps. It includes analyses of the nature of pressure pulsation and explains the advantages of using pulsation dampeners and suction stabilizers. Sizing procedures for dampeners illustrate how their application improves suction and discharge control in reciprocating pumps, reduces pressure pulsations, increases Net Positive Suction Head (NPSH), and makes the service life of pumps longer and more efficient.

Combating pulsation

The primary purpose of pulsation control for reciprocating pumps is to attenuate or filter out pump-generated pressures that create destructive forces, vibration, and noise in the piping system.

Every reciprocating pump design has inherent, built-in pressure surges. These surges are directly related to the crank-piston arrangement (Figure 1). Fluid passing through a reciprocating pump is subjected to the continuous change in piston velocity as the piston accelerates, decelerates, and stops with each crankshaft revolution. During the suction stroke, the piston moves away from the pump's head, thereby reducing the pressure in the cylinder. Atmospheric pressure, which exists on the surface of a liquid in a tank, forces liquid into the suction pipe and pump chamber. During the discharge stroke, the piston moves toward the pump's head, which increases the pressure in the cylinder and forces the liquid into the discharge pipe. This cycle is

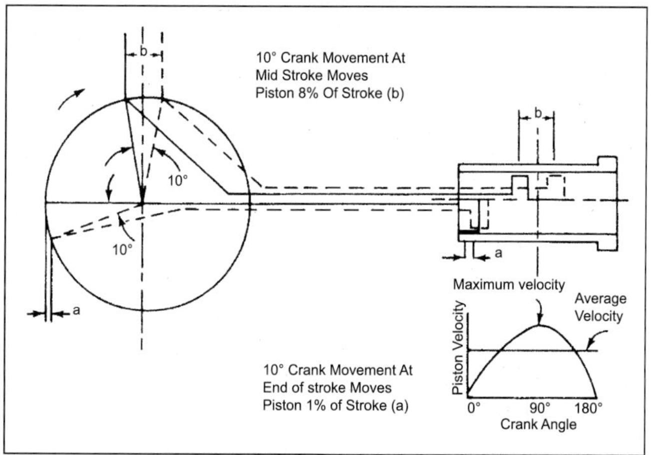

Figure 1. The rotary motion of the crank movement is transformed into the reciprocating motion of the piston.

		Frequency, Hz		
RPM	Pump	Duplex Pulse[a]	Triplex Pulse[a]	Quint Pulse[a]
50	0.8	3.2	2.4	4.0
100	1.7	6.8	5.1	8.5
150	2.5	10.0	7.5	12.5
200	3.3	13.2	9.9	16.5

[a]Pulse = pump RPM x number of cyliders/60

(a)

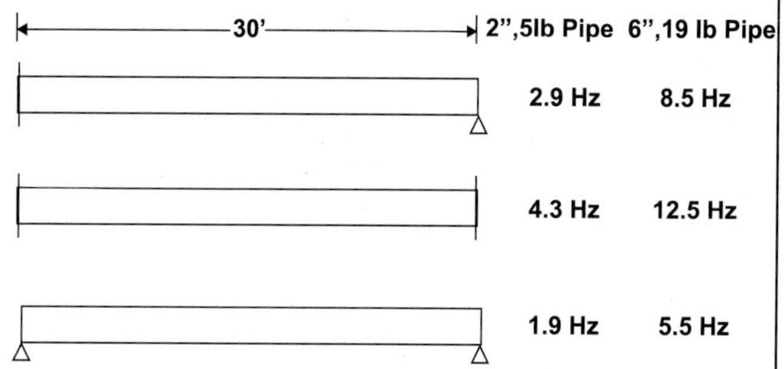

(b)

Figure 2. (a) Typical pump and pulse frequencies. (b) Typical piping natural frequencies.

repeated at a frequency related to the rotational speed of the crank.

At every stroke of the pump, the inertia of the column of liquid in the discharge and suction lines must be overcome to accelerate the liquid column to maximum velocity. At the end of each stroke, inertia must be overcome and the column decelerated and brought to rest. Figure 2 compares pump pulsation frequencies to pipe span natural frequencies of vibration.

A key benefit of pulsation control is the reduction of fatigue in the pump's liquid end and expendable parts. Reducing the pressure peaks experienced by the piston will in turn reduce the power-end peak loading.

The source of fluid pressure pulsations

It is almost mandatory that multiple pump installations have well-designed individual suction and discharge pulsation control equipment. While multiple pumps may be arranged to run at slightly different speeds, it is impossible to prevent them from frequently reaching an "in-phase" condition where all the pump flow or acceleration disturbances occur simultaneously. Having more than one pump can multiply the extent of the pipe vibration caused by such disturbances because the energy involved is similarly increased.

The reduction of the hydraulic pressure pulses achieved by using pulsation control equipment is usually reported in total pulsation pressure "swing" as a percentage of average pressure. This method is used widely within the industry and is often recommended as a standard. Referring to Figure 3, the percent of residual pulsation pressure can be calculated as follows:

Case I: $\Delta P/\text{Ave} \times 100 = 460/1000 \times 100 = 46\%$

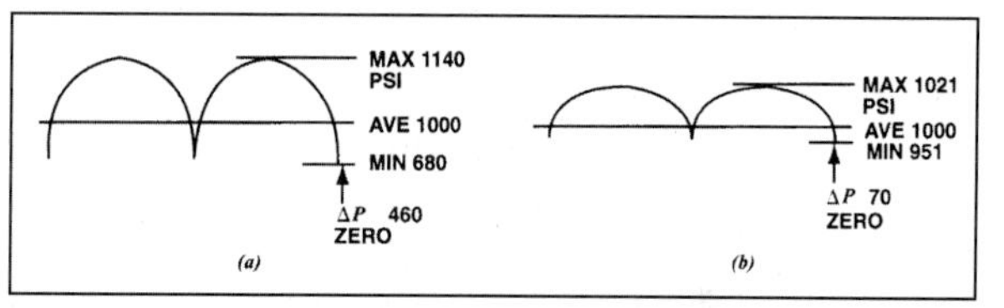

Figure 3. Reporting the degree of pulsation and control. (a) Non-dampened waveform from flow variation (Case I). (b) Dampened waveform from flow variation (Case II).

Case II: $\Delta P/\text{Ave} \times 100 = 70/1000 \times 100 = 7\%$

It is very important that any reference to degree of pulsation should apply to the total excursion in terms of pressure and percentage. In Case I, the total pulsation of 460 psi (46%) implies that the excursion is from 680 psi (32% below the average) to 1140 psi (14% above the average).

Reciprocating pumps introduce into the suction and discharge systems three apparently unrelated pressure disturbances, which are illustrated in Figure 4. These include:

- A low-frequency disturbance at A, based on the rate at maximum flow velocity pressures
- A higher frequency due to maximum acceleration pressure at the beginning of each piston stroke at B
- A pressure disturbance at the point of flow velocity change (valley) at C

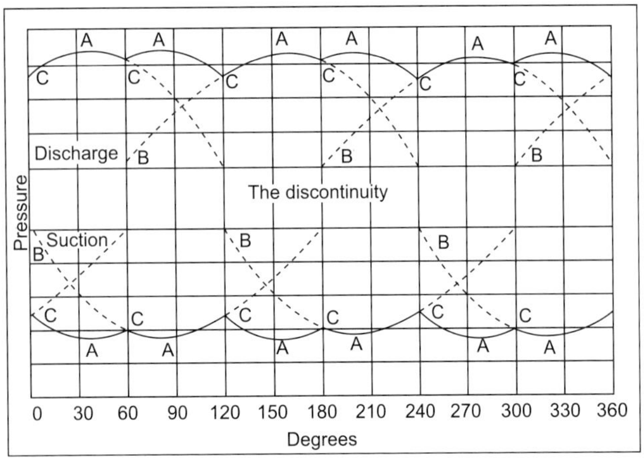

Figure 4. Triplex single-acting pump-points of induced pressure disturbances.

Major problems caused by pressure pulsations

Unsteady flow. The most obvious problem caused by pulsation is that flow is not constant, which can lead to problems in processes where a steady flow rate is required, such as applications involving spraying, mixing, or metering. Flow rates can be difficult to measure with some conventional types of flowmeters.

Noise and vibration. Many reciprocating pump installations suffer from problems that can lead to excessive maintenance costs and unreliable operation. Typical examples are noise and vibration in the piping and the pump. Vibration can lead to loss of performance and failure of valves, crossheads, crankshafts, piping, and even pump barrels. High levels of pulsation can occur when the pulsation energy from the pump interacts with the natural acoustic frequencies of the piping. A reciprocating pump produces pulsation at multiples of the pump speed, and the magnified pulsation in the system is generally worse at multiples of the plunger frequency. Most systems have more than one natural frequency, so problems can occur at differing pump speeds.

Shaking. Shaking forces result from the pulsation and cause mechanical vibration of the piping system. These forces are a function of the amplitude of the pulsation and the cross-sectional area of the piping. When the exciting frequency of the pulsation coincides with a natural frequency of the system, amplification and excessive vibration occur. Amplification factors can be as high as 40 for pulsation resonance and 20 for mechanical resonance. If the mechanical resonance coincides with the acoustic resonance, a combined amplification factor of 800 is possible.

Wear and fatigue. Pulsation can lead to wearing of pump valves and bearings and, if coupled with mechanical vibration, will often lead to loosening of fittings and bolts and thus to leakage. In severe cases, mechanical fatigue can lead to total failure of components and welded joints, particularly when the liquid being pumped is corrosive.

Cavitation. Under certain pumping conditions, areas of low system pressure can occur. If this pressure falls below the liquid's vapor pressure, local boiling of the liquid will occur and bubbles of vapor will form. If the liquid has dissolved gases, they will come out of solution before the liquid boils.

Pressure waves summary

The earlier analysis of the source and nature of fluid pressure pulsations enables us to build the idealized pressure wave pattern (Figure 5). Although identical forces are acting on the liquid in the suction and discharge sides of a pump, the effects of these forces vary greatly. The major difference is that acceleration effects in the suction at B tend to separate the liquid into vapor or "cavitation" pockets that collapse with a high-magnitude pressure pulse. Acceleration on the discharge side of the piston at B, however, tends to compress the liquid and create an impulse pressure pulse. This requires separate consideration of the suction and discharge systems when choosing and installing pulsation control and suction stabilizing equipment.

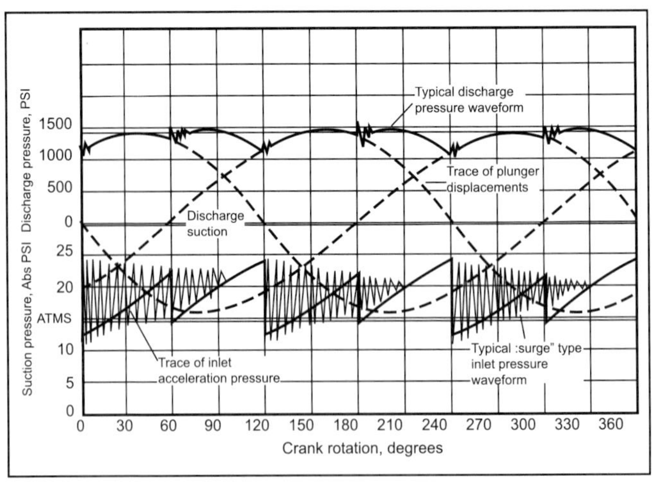

Figure 5. Idealized triplex single-acting pump pressure waveform showing the effect of surge-like acceleration pressures.

Pulsations in suction systems

The friction losses in a suction system are usually low because of the relatively short length and large diameter of the piping involved. Accordingly, the flow-induced A-type pressures are of low magnitude compared to the accepted percentage of change (Figure 4). For example, if the suction pressure is a static 20 psi, the 23% variation of a triplex single-acting pump would generate a theoretical A-type pressure variation of only 9.2 psi. This is hardly enough energy to set pipes in motion. The same flow variation at 1400 psi working pressure, however, creates more than enough energy—a 644 psi A-type pressure variation in the discharge line.

Even so, the forces of acceleration become overwhelming disturbances in the suction. Pressure pulses of more than 25 psi are often encountered in pumps—even in systems with short suction pipes. A small amount of dampening of the flow-induced pressure can reduce pulsation to a negligible amount, leaving the 25-psi C-type acceleration pulsations. These remaining pulsations can create damage by enabling cavitation (Figure 4).

Cavitation. In practice, cavitation occurs at a pressure slightly higher than the vapor pressure of the liquid. Because vapor pressure is a function of temperature, a system may run without cavitation during the winter months but be subject to cavitation problems during the summer when temperatures are higher. Cavitation is connected with the presence of microscopic gas nuclei, which are in the solid materials bonded in the liquid. It is because of these nuclei that liquids cannot withstand tension. Without them, it is estimated that water could transmit a tension of approximately 150,000 psi.

During the beginning stages of cavitation, the nuclei give rise to the formation of gas bubbles, which are swept along to an area where the pressure is higher. The bubbles then collapse, producing high-intensity pressure waves. The formation of successive bubbles quickly follows. This repeating cycle lasts for only a few milliseconds, but during this time the localized pressures can be as high as 60,000 psi and temperatures can rise as much as 1500 °F. Cavitation can occur at a pump's inlets, inside the barrels, at the draft tubes of turbines and on propellers, mixing blades, Venturi meters, and other places where changes in fluid velocity occur. The cavitation effects in rotary positive displacement pumps are similar to those in rotodynamic pumps. The usual effects are wear of surfaces in contact due to erosion, slow increase in clearances, reduction in performance, noise, and vibration. Cavitation damage in reciprocating pumps can happen very quickly and be catastrophic. A cavitating 60 hp pump can break a major component every 350 h.

Suction control. Suction control is an essential step in the design of any pump system and is equally important for reciprocating and centrifugal pumps. This process requires the following major steps:

1. Compare Net Positive Suction Head available (NPSHa) to Net Positive Suction Head required (NPSHr). (NPSH is a feature of the system layout and liquid properties. It is often referred to simply as NPSHa or NPSH available. Pump manufacturers usually quote NPSHr. This is a function of the pump's performance and is related to the pump's susceptibility to cavitation. To calculate the NPSHa to the pump, it is necessary to take into account the friction losses of the suction piping. NPSHa is then the difference between the total pressure on the inlet side of the pump and the sum of the vapor pressure of the liquid and the friction losses of the suction pipes.)
2. NPSHa is greater than NPSHr, the general suction conditions are under control, and their improvement (with respect to NPSHa) by installation of a suction pulsation dampener is not required. In this situation, the installation of a pulsation dampener would be required only for pressure fluctuation and noise reduction purposes.
3. If NPSHa is less than NPSHr and the piping rearrangement of a suction system is impossible or insufficient, installation of a pulsation dampener before the pump's inlet is essential.
4. Choose the appropriate dampener design based on steps 1—3. Using this process will help you achieve maximum reduction of pressure fluctuation.

Pulsations in discharge systems

The same forces at work on the suction side also affect discharge systems, but the pressure at A due to flow-induced pulsations becomes overwhelming at 460 psi (Figure 4). The 25 psi contribution from acceleration at C is a small percentage (2.5%) of the total discharge pressure. One exception is when the pump is delivering into a low-friction, high-pressure system such as a short vertical discharge system (Figure 6a). An example is a mine dewatering application, where the pump is delivering to an already pressurized system—a pressurized pipeline—through a short connecting pipe. Another exception is when the pump is delivering to already pressurized systems such as hydraulic press accumulators (Figure 6c). In those two cases, the acceleration pressures can become the overwhelming disturbance, particularly if the piping system is relatively long compared to a suction system (but considerably shorter than a "pipeline").

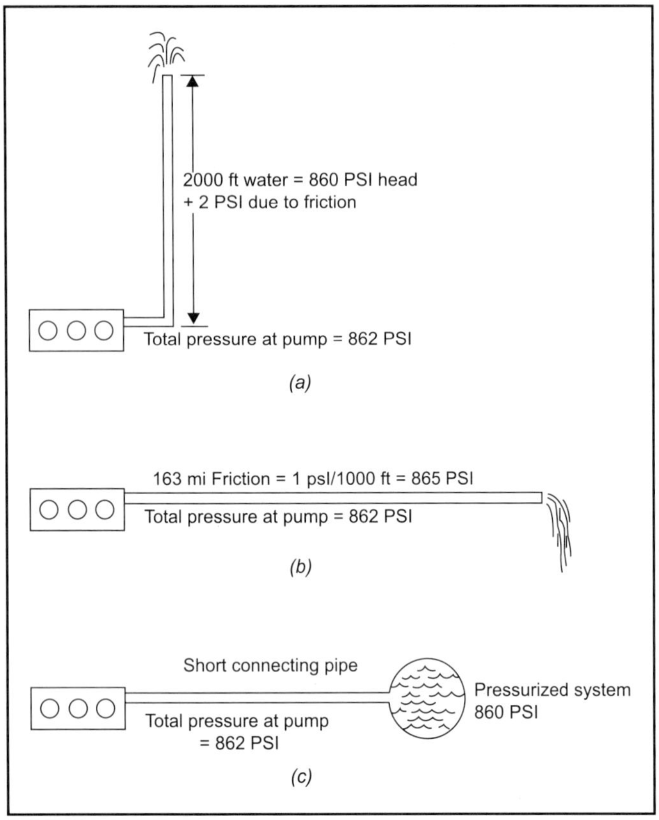

Figure 6. Dynamic differences in discharge systems. (a) Vertical discharge. (b) Horizontal discharge. (c) Pressurized systems.

Control of pulsation with dampeners

Successful control of damaging pulsation requires careful selection of pulsation dampeners and proper location of those dampeners in a piping system. The earlier pressure pulsation analysis showed that dampeners will reduce the level of pulsation and consequently pressure fluctuations. With lower pipe vibration and noise, wear at the pump, risk of cavitation, and metal fatigue will be less as well. (There are several OSHA standards dedicated to the control of industrial noise in the workplace. One of the most practical and inexpensive ways to meet OSHA noise level requirements and achieve the overall performance improvement of reciprocating pumps is installation of pulsation dampeners on suction and discharge systems.)

Dampener types. After evaluating the causes and types of pressure pulsation effects in reciprocating pumps, we can review which available fluid pressure pulsation control equipment is best for specific conditions.

Hydropneumatic pulsation dampeners are gas-type energy-absorbing design that uses a gas-filled bladder to absorb the "A" flow on peaks and to give back that flow on the "lows," thereby reducing flow variations beyond the pump and consequently reducing the flow-induced pressure pulsations. This is the most efficient, low-cost, and widely used type of pulsation control equipment.

This discussion is limited to the two major designs of bladder-type hydropneumatic pulsation dampeners: appendage and flow-through.

Appendage dampeners. Figure 7 shows a typical appendage design suction pulsation dampener. It consists of a homogenous shell and a nitrogen-filled flexible rubber bladder. The fluid port can be threaded (NPT or SAE) or flanged. This design is recommended for installation on pumping systems where suction is flooded. The dampener should be installed as close as possible to the pump inlet, and the nitrogen in the bladder should be precharged to 60% of the pump's inlet pressure. Figure 8 shows a large capacity version of this design.

The same type of suction pulsation dampener (150 psi) was installed on a 120 gpm and 30 psi Wheatley triplex pump with 3″ piston bore, 3-1/2″ stroke. Another two discharge 1500 psi pulsation dampeners were used with a 263 gpm and 642 psi Wheatley quintuplex pump with 3-3/8″ bore, 4-1/2″ stroke.

Flow-through dampeners. This design should be installed on a pump's suction system when inlet pressure (NPSH) is very low and the fluid can contain entrained gases such as air, carbon dioxide, sulfur dioxide, etc. Flow-through units have significant advantages over the appendage design for a few reasons. First, the inlet and outlet diameters of the dampener are equal to the pipe diameter. Second, the inline installation enables reduction of the inlet fluid velocity into the pump. In addition, the

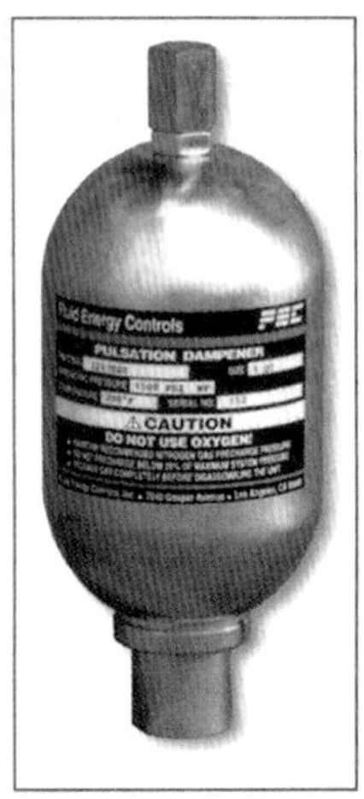

Figure 7. Appendage type 60 in^3 1500 psi stainless steel pulsation dampener with threaded fluid port.

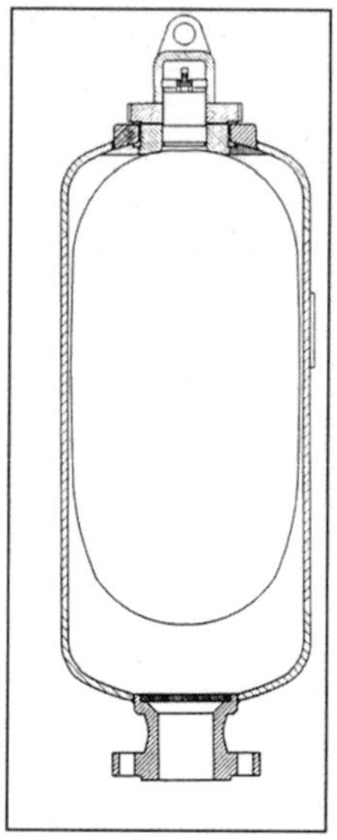

Figure 8. Typical large capacity appendage type flanged pulsation dampener.

bladder response time is faster than an appendage dampener, due to the minimum distance between the shell inlet and pipe centerline. The flow-through design can also be used to improve the NPSHa of systems with low NPSH because it acts as a suction stabilizer.

Figure 9 shows an actual installation of a discharge flow- through 2.5 gallon capacity dampener. This carbon steel, 3000 psi, 3″ flanged unit was installed on the discharge system of an existing triplex single-acting reciprocating pump with 2.88″ piston bore diameter and 5″ stroke—one of seven such units. The owner, Ormet Aluminum Mill Products Corporation, is a Hannibal Rolling Mill Division, Hannibal, Ohio. Currently, the company is scheduled to install another two pulsation dampeners of the same type. The installation of pulsation dampeners for the remaining pumps is scheduled for sometime in 2000.

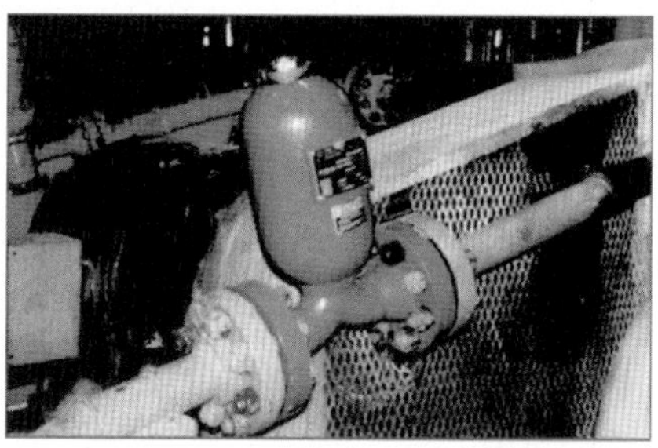

Figure 9. Flow-through pulsation dampener. Courtesy of Ormet Aluminum Mills Products Corporation, a Hannibal Rolling Mill Division.

The abilities of pulsation dampeners

Hydropneumatic bladder pulsation dampeners are able to prevent the generation of the most destructive low-frequency pulses such as those generated by the pump's rotary motion and the combination of flow from each of the pump's cylinders. (This is basically rpm multiplied by the number of cylinders.) Accordingly, based on a maximum of 500 rpm for most small pumps, the maximum frequency involved should not be more than about 50 Hz.

Performance comparisons. Figure 10 compares the performance of the two types of pulsation dampeners on a typical pump. In this particular case, installation of hydro- pneumatic dampeners offers overall improved performance.

An analysis of reciprocating pump pulsations is based on the premise that pressure or pressure variation (pulsation) due to flow velocity (flow-velocity variation) is a function of the square of such velocity. And the system must present enough resistance to flow—due to pipe friction, bends, fittings, chokes—to enable the generation of a pressure or pulsation in the first place. The magnitude of that pressure is directly related to the resistance. The analysis is also based on the premise that the acceleration of the liquid at the beginning of an increase in velocity (start of stroke) generates a pressure pulse, and the magnitude is directly related to the length of pipe connected to the pump suction or discharge. It is also inversely related to the diameter of the pipe and directly related to the rate of change of velocity.

Because the inherent flow variations of a reciprocating pump are of the same intensity (but out of phase) in both the suction and discharge, for many years it was believed that a pulsation dampener of the same size as the discharge should be used on the suction, and many such installations can be found. However, from the standpoint of flow-induced pulsations alone, it is logical that a smaller dampener can be used on the suction side. For example, a typical triplex single- acting pump has a flow variation of about 23% in both the suction and the discharge. If all of this flow variation were converted to pressure pulsations, the pulsations would be a maximum of 46% of the average pump pressure. In our example, the maximum change in pressure at the discharge would be about 460 psi at 1000 psi average pressure. If there is a 50 psi suction pressure, the change in pressure there would be only 23 psi; that is,

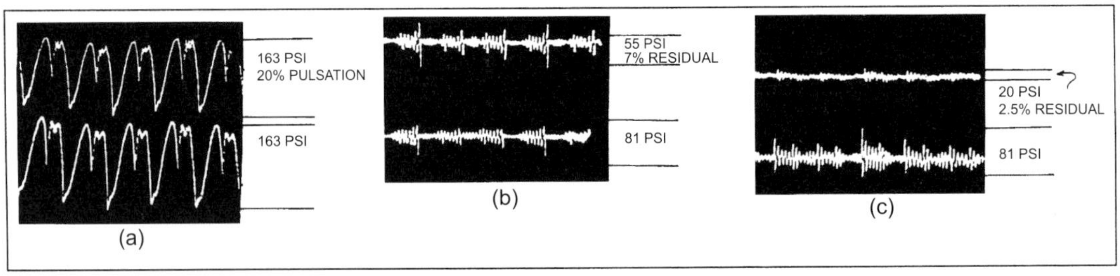

Figure 10. (a) No dampener installed. (b) Appendage type hydropneumatic dampener. (c) flow-through type dampener. Top chart: downstream of dampener. Bottom charts: upstream of dampener.

there would be about 20 times less energy available to set the piping in motion.

In addition, the acceleration due to velocity variations is the same on the suction and discharge sides. In the suction system, however, such disturbances usually overwhelm the disturbances due to flow variation, and conversely, in the discharge system the flow-induced pressure pulses usually overwhelm the acceleration pressure pulses.

There is evidence that only a flow disturbance or an acceleration disturbance (not both) can exist during any suction or discharge stroke of a pump. One will overwhelm and reduce the effect of the other. A plausible explanation is the theory that in pumps of greater than about 300 fpm piston speed, a disturbance of either type will momentarily reduce the volumetric efficiency to such an extent that it cannot recover fast enough to enable the flow rate to reach an adequate value to affect another disturbance in the short period of time (about 3 ms) between the disturbances. There is no such worry with slow speed pumps because few if any disturbances are generated.

It is possible for a pump with high inherent flow variations to actually deliver an almost pressure-pulseless liquid into an open-ended (in effect), short, relatively large-diameter horizontal discharge pipe. Some typical cases are pumps with short connecting pipes feeding a large-diameter already-pressurized system such as an existing pipeline or hydraulic press supply. If the discharge pipe is vertical, as in a mine dewatering system, the acceleration of the liquid column due to gravity becomes the predominant generator of pressure pulses. This great dynamic and hydraulic difference is shown in Figure 6, where the pump "sees" the same discharge pressure under three different physical arrangements. Energy-absorbing or hydropneumatic pulsation dampeners may not solve pulsation problems in cases (a) and (c) of Figure 6.

Application of a large-capacity, flange type of surge suppressors should be carefully considered (Figure 8).

Advantages of bladder type pulsation dampeners

- Positive sealing between the gas and fluid chambers, such as the bladder, serves as a permanent wall between them
- Zero leakage between both chambers due to bladder barrier
- Quick pressure response times for pressure control pump pulsation and shock-dampening applications are facilitated by the light weight of the bladder
- Simplicity of design reduces overall maintenance and operating cost of pumping system

Sizing considerations

Suction pulsation dampeners and suction stabilizers. The following sizing procedure will serve two purposes:

- Evaluation of the pump suction system and subsequent sizing of the pulsation dampener and suction stabilizer with respect to the reduction of pressure fluctuation
- Improvement of pump suction conditions represented by NPSHa

System data

Piping

- The maximum flow of 200 gpm originates from the storage tank.
- The tank is open to atmosphere.
- The length of the 8″ cast iron suction pipe is 106 ft.
- The difference of elevation between the storage tank and centerline of pump suction inlet is 23 ft.

- The fittings are: (1) 8″-90° elbows, (1) 8″-45° elbows, (4) 8″ Tees.

Pump (4 each)

- Single action triplex
- Max speed = 45 rpm
- Flow rate = 200 gpm (800 gpm for all 4 pumps)
- Min NPSH = 23 ft. of column of water
- Bore = 9.0 in
- Stroke = 6.0 in

Fluid

- Anaerobically digested municipal sludge conditioned with 10% lime and 5% ferric chloride.
- Fluid temperature = 100 °F

Analysis. In order to determinate the net positive suction head available (NPSHa), the acceleration head and the friction losses along the suction piping must be calculated. Acceleration head can be calculated as follows:

$$h_a = \frac{LVnC}{Kg}$$

where

h_a = Acceleration head, ft of fluid,
L = Length of suction line, ft,
V = Velocity in the suction line, fps,
n = Pump speed, rpm,
C = Pump type constant, for triplex single-acting pump 0.066,
K = 1.4 for hot water,
g = 32.2 ft/s^2.

Pipe friction loss factor

$$k_1 = f\frac{L}{D} = \frac{0.020 \times 106}{0.336} = 6.31,$$

Valve and fitting friction loss factor

$$\begin{aligned} K_2 &= (4 \text{ valves}) \times 1.8 + (1 - 90° \text{ elbow}) \times 1.5 \\ &\quad +(1 - 45° \text{ elbow}) \times 0.8 \\ &= 9.50 \end{aligned}$$

Total valve and fitting frictional loss factor

$$K_t = 15.81$$

Total pipe frictional losses

$$h_t = \frac{K_t V^2}{2g} = \frac{(15.81) \times (5.12)^2}{2 \times 32.2} = 6.44 \text{ ft of w.c.}$$

Acceleration head

$$h_a = \frac{(106) \times (5.12)^2 \times 45 \times 0.066}{1.4 \times 32.2} = 34.96 \text{ ft of w.c.}$$

The maximum static elevation head hs = 23.0 ft of water column.

Net Positive Suction Head Required NPSHr = 9.87 psig @ 45 rpm = 22.8 ft of w.c., say 23 ft of w.c.

$$\text{P}_{\text{task}} = 0 \text{ psig} = 0 \text{ ft of w.c.}$$

Net Positive Suction Head Available

$$\begin{aligned} \text{NPSHa} &= h_s + P_{\text{tank}} - h_a + h_f \\ &= (23 + 0) - (34.96 + 6.44) = -18.40 \text{ ft of w.c.} \end{aligned}$$

Since NPSHr = 23.0 ft of w.c., the NPSHa = 18.40 ft of w.c. and is less than NPSHr. The actual suction head is not adequate to operate the pump at the 23 ft of w.c. NPSH requirement.

This analysis shows the necessity of installing the suction stabilizer to improve suction conditions. Such stabilizers can reduce the acceleration head, thereby increasing the NPSHa. The unit will also eliminate the acceleration head by minimizing the mass of fluid in the suction pipeline that must be accelerated. The suction stabilizer installed at the pump inlet will reduce the fluid column that must be accelerated to the fluid between the bladder and the pump suction manifold. This fluid mass is normally very small and requires a low-pressure fluctuation. The losses correspond to the acceleration effects and are about 67.5% of the operating pressure. Acceleration losses

$$h_a = (0.075) \times (23 \text{ ft of w.c.}) = 1.73 \text{ ft of w.c.}$$

NPSHa for the system with the suction stabilizer is

$$\text{NPSHa} = (23.0 + 0) - (1.73 + 6.44) = 14.84 \text{ ft of w.c.}$$

As can be seen from the earlier analysis, the installation of a suction stabilizer will effectively reduce the acceleration head and provide a much greater NPSHa to the pump. This will prevent local cavitation and increase pump efficiency. To provide the required Net Positive Suction Head in our example, the sludge conditioning tank should be raised only 8.0 ft (versus to 18.4 feet) to compensate for the friction losses between the tank and pump suction manifold.

The actual volume of pulsation dampener/suction stabilizer can be calculated using the following sizing procedure:

$$\text{Volume} = 10 \times \text{displaced flow/revolution}$$

$$\text{Displaced flow} = \left(0.7854 \times B^2 \times \text{stroke} \times \text{no.pistons} \times \text{action}\right)/231 \text{ gallons}$$

where

B = Pump cylinder bore, in
stroke = Pump piston stroke, in
no. piston = 1 for simplex, 2 for duplex, 3 for triplex
suction = 1 for single acting pump, 2 for double acting pump.

By installing the pump data in the formula, we get the required actual volume of pulsation dampener/suction stabilizer.

$$\begin{aligned}\text{Volume} &= 10 \times \left(0.7854 \times 9^2 \times 6 \times 3 \times 1\right)/1 \times 231 \\ &= 49.6 \text{ gallons}\end{aligned}$$

The 50-gallon suction stabilizer installed at the pump inlet will reduce the pressure fluctuation on the inlet pipeline to 1.73 ft of w.c., thereby providing an NPSHa of 33.24 ft, greater than without a suction stabilizer.

Sizing of discharge pulsation dampeners

In order to achieve optimum results from the dampener, its size (capacity) must be properly calculated. The following simple formula can be applied for sizing a pulsation dampener for either a simplex, duplex, or triplex single-acting pump.

$$\begin{aligned}\text{Size of dampener} = {} & \text{plunger area(sq in)} \\ & \times \text{plunger stroke(in)} \times 50 \\ & \times \text{pump type factor}\end{aligned}$$

Notes:

1. 50 = pressure factor for about 65% pulsation control
2. pump type factor = 0.25 for duplex single-acting pump

Example. Determine the size of a pulsation dampener to control pulsations caused by a duplex single-acting plunger chemical metering pump handling 50% sodium chloride.

Data

- Plunger diameter (D) = $1^3/_4$ in
- Plunger stroke (L) = 3 in
- Max. pressure = 150 psig
- Pump type factor = 0.25

Size of pulsation dampener = plunger area (sq in) × plunger stroke (in) × 50 × pump type factor = $\pi D^2/4 \times 3 \times 50 \times 0.25 = 90.2$ cu in. Therefore, a 0.5-gallon size dampener should be selected.

Material and design specification guidelines for pulsation dampeners and suction stabilizers

The next step is proper specification of the dampener's material of construction with respect to working conditions. The diaphragm or bladder and all other seals in the dampener shall be EPR (ethylene propylene) compound compatible with 50% sodium hydroxide (NaOH). The metal parts of the dampener shall be of 316 stainless steel. The selected dampener in this example shall dampen pulsations within 5%—10% of the mean line pressure.

Conclusion

The following factors must be thoroughly considered during the selection process of pulsation dampeners.

1. Maximum design pressure rating of the dampener must be within the operating pressure limit of the system.
2. Liquid port size of the dampener should be equal to the discharge port of the pump or equal to the discharge line size.
3. Dampener should be charged with a neutral gas (usually nitrogen) to approximately 60% of the mean system pressure.
4. Appendage type dampeners: pipe or nipple connecting the dampener to the line should be as short as possible. The diameter of connecting pipe should not be smaller than the size of the dampener liquid port.
5. The dampener should be installed as close as possible to the pump discharge port.
6. A periodic check of nitrogen gas pressure in the dampener should be made to make sure the dampener is in operable condition.
7. The dampener's rubber and metal parts must be compatible with the pumped liquid.

Rotary pumps on pipeline services

James R. Brennan, Imo Pump, Monroe, North Carolina

Pipelines handling liquids in excess of about 500 SSU viscosity will normally use rotary, positive displacement pumps both for optimum efficiency and, frequently, lower initial cost. Here we provide insight into the application of rotary pumps to some of the unique requirements facing users of pipelines in crude oil transport, electric utility power generation, and general industrial needs. Coverage includes pumping system designs, flow control on variable flow systems, serial pump operation on multistation pipelines, and key components that help ensure reliable pump operation and maintainability.

Pipelines are used to transport a broad range of liquids, including crude oils, fuel oils, and refined products such as gasoline, as well as water, specialty chemicals, and others. Lines can be as short as several dozen feet to hundreds or even thousands of miles long. They can cross state or country boundaries. The modern world relies upon safe, efficient operation of all manner of pipelines, both liquid and gas. Oil producers, pipeline transport companies, refineries, and power plants are probably the largest operators of pipelines.

Types of pipeline services

Pipeline services can be broadly categorized as loading/unloading, local transport, and long-distance transport. Examples of load/unload include pumping a relatively short distance from or to rail cars, tank trucks, barges, or tanker ships. These services are typically intermittent, perhaps 8 or fewer hours per day, perhaps not every day of the week. Differential pump pressures are usually low, less than 125 psid. The combination of low power and intermittent use makes pump operating efficiencies less important than in other services.

Local transport would be defined as a single pumping station installation with line lengths less than 50 miles. Design operating pressures can reach 1450 psig or more. Local or national regulation may dictate maximum pressures. These pipelines can be intermittent or continuous duty services and can range from modest power levels to many hundreds of horsepower or more.

Long-distance transport is usually more than 50 miles and includes multiple pumping stations along the length of the pipeline. These pipelines normally operate around the clock at partial or full capacity. Controls and instrumentation tend to be more sophisticated and automated. Long-distance pipelines will frequently operate at variable station flow rates, depending on product demand at the end terminal. Pumping stations will normally have three or more pumps with one in standby mode (three half-capacity pumps). If pipeline capacity is to be increased over time, more pumps may be added to each station.

Pumping stations

Pumping stations for liquid pipelines can be as simple as a single pump located adjacent to a supply tank pumping liquid down the pipeline to a receiving tank (Figure 1). The supply and receiving tank level switches can control the pumping cycle by starting or stopping the pump at appropriate minimum and maximum tank liquid levels. Longer pipelines needing intermediate pumping stations can be handled in one of two basic arrangements. The first is to use storage tank(s) at each station (Figure 2). The upstream station pumps to the downstream station's tanks. The downstream station pump(s) take suction from these tanks and send the liquid farther downstream. This is a simple,

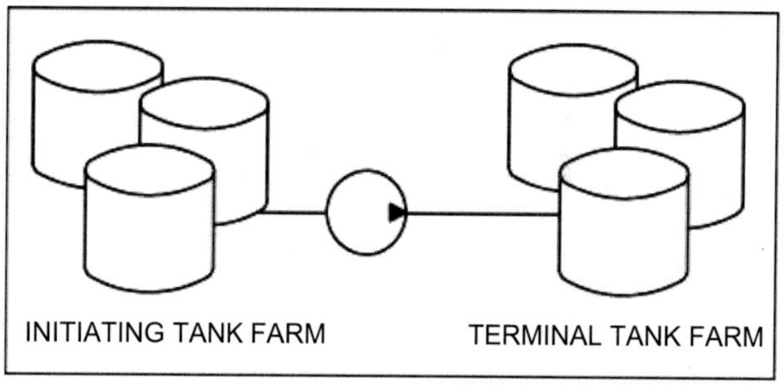

Figure 1. Simple pipeline system.

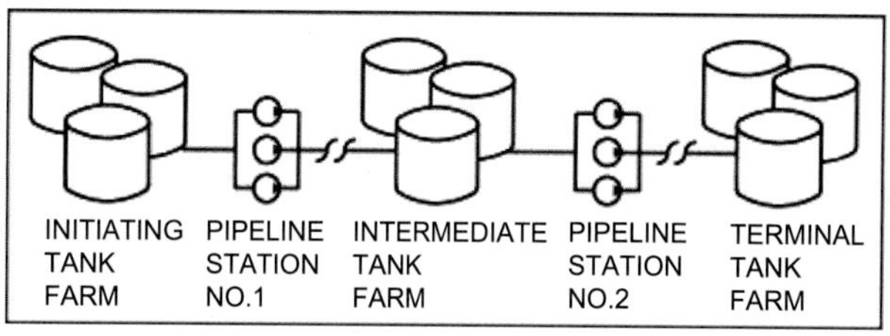

Figure 2. Multistation, tank system.

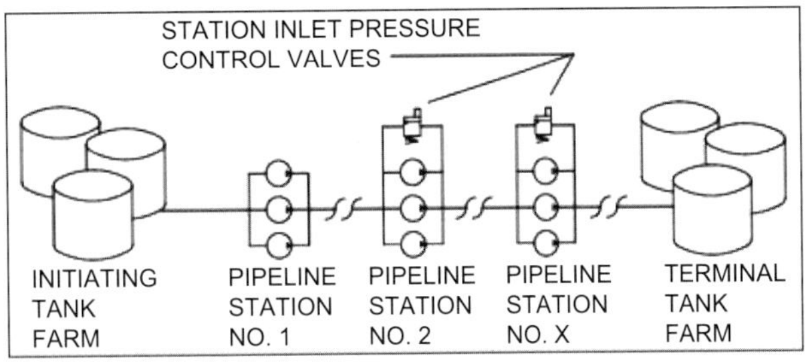

Figure 3. Multistation, tight system.

reliable method oflong-distance transport. Pumps are started or stopped depending on liquid level in tanks. This has several disadvantages. The cost of storage tanks, dykes, and tank condition monitoring are not small. There is also product in "inventory" in the tanks that does not help abusiness's cash flow. The alternate arrangement is a "tight" system, leaving the pumping stations in series with each station's pump(s) taking suction from the pipeline (Figure 3). This has the advantage of eliminating the need for storage tanks but requires control over each pumping station in the pipeline such that exactly the same flow rate is going through each station—more on that later.

Types of pumps applied

Most pipeline pumps are centrifugal units. When the product being transported is always a low viscosity liquid like water, gasoline, diesel oil, or very light crude oil, centrifugal pumps are cost-effective, reliable, and efficient. However, as the liquid viscosity increases, the frictional losses within a centrifugal pump quickly reduce pumping efficiency dramatically (Figure 4). For this reason, rotary, positive displacement pumps are often used when products such as heavy crude oil, bunker fuels (no. 6 fuel oil), low sulfur fuels, asphalt, Orimulsion®* (a manufactured boiler fuel emulsion of 30% water and 70% bitumen), and similar need to be transported. The energy costs of operating a pipeline can represent close to 50% of total operating costs, so pump operating efficiencies play a critical role in profitability. Table 1 illustrates differential annual operating costs for two real examples of centrifugal pumps versus rotary pumps.

Although there are a great many types of rotary pumps, the most common on pipeline services are two- and three-screw pumps, gear pumps (internal and external), and vane pumps for low pressure services. Two- and three-screw pumps are used for higher pressure services, especially at medium to high flow rates. Table 2 shows the approximate maximum flow and pressure capability for each type, not necessarily achievable concurrently.

* Orimulsion is a registered trademark of Petroleos de Venezuela S. A.

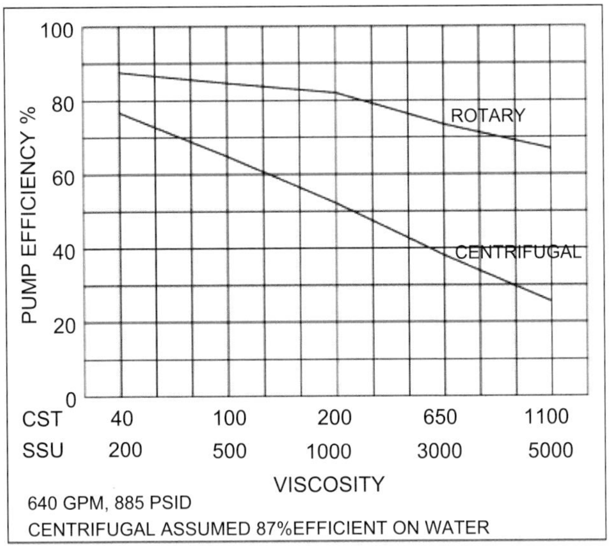

Figure 4. Rotary versus centrifugal efficiency.

Table 1
Operating Costs Comparison

Parameter	Centrifugal	Rotary
Type	6 stage	3 screw
rpm	3500	1750
Cst	200	200
psid	765	765
gpm	410	410
bhp	395	229
Efficiency %	45	81
Annual energy cost at $0.07/kWh	$180,449	$104,615

Table 2
Rotary Pump Flow and Pressure Ranges

Type	Maximum gpm	Maximum psid
Gear	1500	250
Vane	1000	250
2 Screw	5500	1500
3 Screw	3400	2000

Gear and vane pumps (Figure 5) tend to have the lower first cost and may or may not have the best lifecycle cost. They will normally operate at relatively low speeds, requiring speed reducers on all but the very smallest sizes. Multiple-screw pumps (Figure 6) move liquid axially rather than in a radial direction, so fluid velocities are relatively low even when

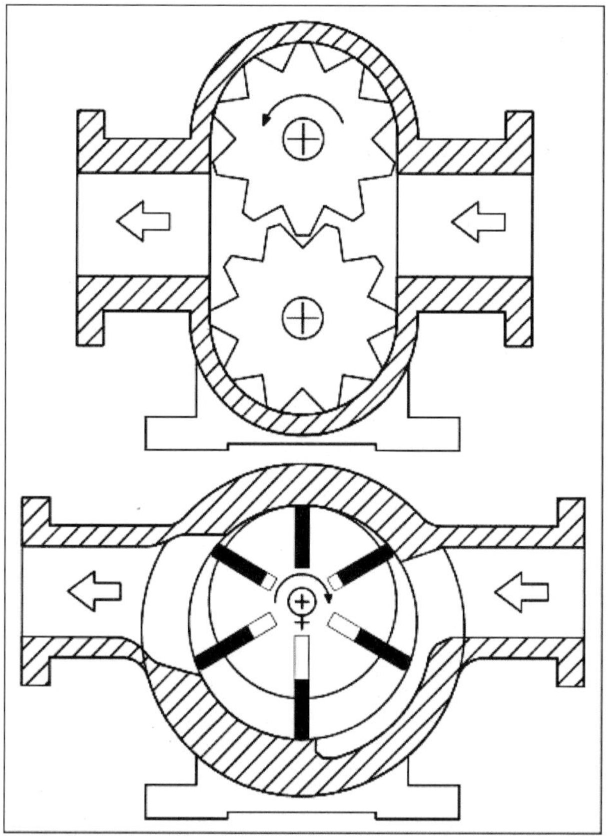

Figure 5. Gear (top) and vane pump.

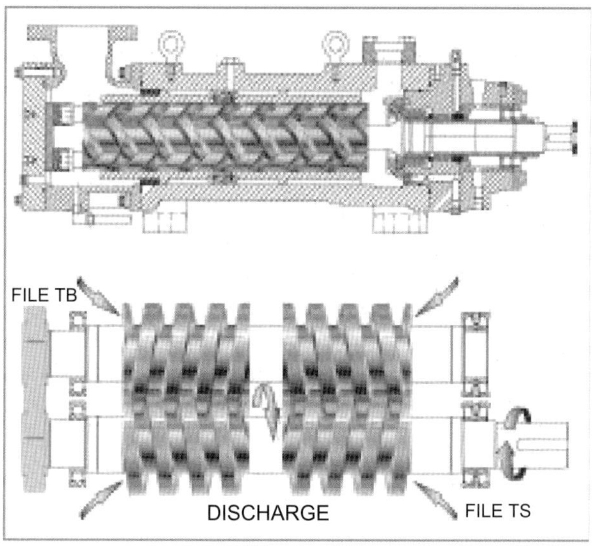

Figure 6. Multiple-screw pipeline pumps.

directly driven at four- and six-pole motor speeds (1800 and 1200 rpm at 60 Hz, 1500 and 1000 rpm at 50 Hz). Efficiencies remain high over a much broader range of differential pressure than centrifugal pumps. Figure 7 shows 12,000 hp of

Figure 7. Fifteen twin-screw pumps for subsea pipeline.

Figure 8. Diesel-driven three-screw pump at 1400 psi.

installed two-screw pumps (15 pumps in all) that are used to load *Orimulsion* tankers 8 miles offshore. Figure 8 illustrates a diesel engine—driven three-screw crude oil pipeline pump operating in Guatemala at 1400 psig.

Specified operating requirements

Obviously, pump and driver specifications will be needed to acquire equipment that will perform to requirements. Unless your organization has these talents in house, an engineering firm will be needed to explore the pipeline requirements. As it relates to rotary pumps, at a minimum the following will need to be defined:

1. Number of pumping stations
2. Number of pumps per station

3. Possible range of
   - Pump inlet pressures
   - Pump discharge pressures
   - Liquid temperature range
   - Liquid viscosity range
   - Per pump flow rate range
   - Particulate content
   - Upset conditions
   - Static conditions
4. Driver requirements
5. Pump instrumentation
6. Pump leak detection

Particulate content can range from unknown (take your chances) to specific weight percentage of sediment, sand, etc. For liquids containing abrasives, pump life is inversely proportional to something between the square and the cube of pump speed. Operating larger pumps at lower speeds will increase initial cost but can save on maintenance costs over many years. Another proven technique to extend pump life in abrasive environments is to select pumps with effectively more stages than needed just for pressure capability. For example, every wrap in a screw pump acts as a pressure stage, similar to a multistage centrifugal pump. More stages reduce the pressure rise per stage, reduce the slip flow (volumetric inefficiency), and spread out the wear. The result is extended pump service life and, frequently, higher average efficiency over time.

Product leak detection could be a critical element in both pumping station safety and environmental contamination. Unless there are observant attendants on site at all times, automatic alarm or shutdown should be considered in the event of product leakage from any source. Rotary pumps most commonly use mechanical seals to contain pumped product within the pump casing. A mechanical seal will wear and can fail catastrophically, and product under pressure can leak large volumes in short time periods. Many pumps used for pipeline services can be equipped with various seal-leak detection schemes to alert/alarm or shut down in the event of excessive seal leakage.

Driver selection depends on available power sources. In remote areas, diesel engines are usually chosen. They provide some range of speed and, therefore, flow control of the pumps. By far, the most common drive is a fixed speed electric motor, although variable frequency drives are becoming more popular. They can improve station efficiency by avoiding the necessity of drawing power to pump flow that is to be bypassed around the station (wasted energy). Spacer couplings are highly recommended between the pump and its drive. They allow access to the pump shaft-end seal and bearing, as well as the motor shaft-end bearing.

Upset conditions and static conditions need to be assessed during pipeline design. Most rotary, high-pressure pumps cannot tolerate anything close to discharge pressure at their inlet (suction) port. In a tight system, if an upstream pumping station is started and the downstream station is not started in very close synchronization, the downstream station may see excessive pressure on its inlet side. A low pressure set inlet side relief valve may be necessary, together with an overspill tank and appropriate instrumentation.

Instrumentation tends to be a very owner-specific issue. Some systems have only a few pressure gauges. Others include bearing temperature detectors, vibration detection, inlet and outlet pressure transducers, and temperature transducers, all sufficient to get a good picture of what is happening at a pumping station, sometimes from many hundreds of miles away.

Pump heating and/or cooling

If the product being pumped is so viscous that heating it before pumping is necessary, then it will probably be prudent or essential to provide a means of preheating the pumps so they are not faced with sudden introduction of hot liquid into cold pumps. Depending on available utilities, pumps can be jacketed for use of heating mediums such as steam or hot oil. Heat tracing with electric thermal wire is also an option. The outside of the pump will need to be insulated to contain the heat. Heating and insulation systems must take into account pump maintainability if the heating or insulating system must be removed to service the pump. Be sure to avoid applying heat to timing gear cases or pump bearings. In some instances, these pump components may actually need cooling to operate for extended periods.

Pump protection

Like most other pumps, rotary pumps are vulnerable to premature wear or failure from too little or too much inlet pressure, too high a discharge pressure, excessive inlet temperature, dry running, and ingestion of foreign material (weld rod, wire pieces from cleaning brushes, etc.). In addition, if the liquid pumped is inherently “dirty,” like many crude oils and bunker fuels, then pump wear will be a normal part of operating pipeline pumps and one should prepare for that eventuality.

Systems need to be instrumented to at least alarm if any of these parameters approach a limiting value. If pumps are unattended most of the time, then automatic shutdown is the only protection from catastrophic pump failure. Use of inlet strainers is highly desirable, provided they are instrumented to detect excessive pressure drop and cleaned when needed. Without regular monitoring and cleaning, strainers eventually will accumulate debris to the point that the pressure loss of liquid going through the strainer will allow pump inlet pressure to fall below the minimum allowable. Worse still, that pressure drop can cause the strainer element to collapse. On collapse, all the accumulated debris, as well as the strainer element, will arrive at the pump inlet. Nothing good will come of this.

As with any positive displacement pump, a pressure relief device is essential to safe operation. The "shutoff head" of a positive displacement pump can be enormous, far above the pressure rating of any system component. A relief valve is normally installed at each pump discharge with the valve outlet connected back to the supply tank (preferred) or to the pump inlet piping as far from the pump as practical.

System considerations

When starting a rotary pipeline pump with the pipeline already full, the system must provide for the time required to accelerate the liquid from zero velocity to final velocity. There may be 50 miles of product that needs to get to 10 ft/sec. This is normally accomplished with an electric-operated valve around each pump that bypasses discharge flow either back to the supply tank (preferred) or back to pump suction. The valve is open on pump startup and closed gradually as the flow velocity in the pipeline builds. If bypassing back to the pump inlet piping, go as far away from the pumps as practical. This will maximize the mass of liquid available to absorb the driver power draw that is not yet going down the pipeline while the bypass is not fully closed.

Flow control can be achieved in several ways. Variable-frequency drives are becoming more affordable in larger sizes. Because rotary pump flow is directly proportional to speed, control is fairly straightforward. In tight systems with fixed-speed drivers, pumping station variable flow is achieved first by how many of the station's pumps are in operation and second by bypassing flow around the station to control how much total flow leaves the station. Figure 9 is a simplified schematic of such a system. The bypass valve senses station inlet pressure. It bypasses whatever flow is necessary to control station inlet pressure at set point. If the station inlet pressure rises, less flow will be bypassed and more flow will therefore leave the station.

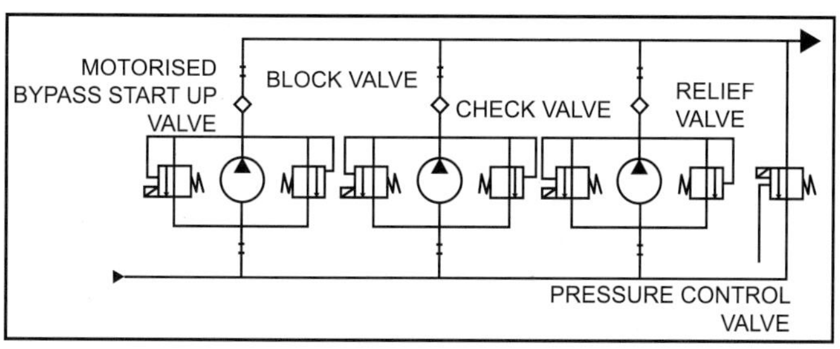

Figure 9. Tight series pump operation.

Maintenance

Maintenance for rotary pumps is not very different from any other pump. Hopefully, the installation provided both space to access the pump for service as well as some shelter from the elements. Because rotary pumps have relatively close internal running clearances, be especially careful to achieve and maintain good pump-to-motor shaft alignment. Avoid pipe strain at all costs for the same reason. If pumping station downtime is a serious matter (usually), then have an installed standby pump ready to take over, as well as critical spare parts. If a pump "goes down," try to identify the cause before starting another one. Sadly, too many times a series of failures is allowed to occur before enough time and effort is expended on root cause analysis.

Many pipeline pumps have service parts that need to be removed from both ends of the pump. Spacer couplings go a long way in easing the drive end service. Keeping walls or objects away from the pump end opposite the driver is usually necessary to avoid the need to disconnect piping and remove pumps to a bench area.

Some pumps are, in fact, better maintained on a bench in a maintenance area, where space to perform the work is more available, components can be more easily assessed for their condition, and new parts can be installed more correctly.

Conclusions

Rotary, positive displacement pumps can play key roles in reliable, efficient, effective transport of liquids over short or very long distances. They are an optimum choice where efficiency is a strong driver of pump selection and operation. There is an extensive history of successful applications of rotary pumps to pipeline transport.

Centrifugal pump parameters and applications

Doug Kriebel, P.E., Kriebel Engineered Equipment, King of Prussia, Pennsylvania

The purpose of this article is to familiarize the reader with centrifugal pump manufacturers' performance curves, show how a pump system head curve is constructed, and demonstrate how to analyze the effect of changes in the system.

A pumping system operates where the pump curve and the system resistance curve intersect. This important fact is one of the most overlooked concepts when designing or troubleshooting a pumping system.

The pump curve has a head/capacity portion, which indicates the change in flow with respect to head at the pump. It is generated by tests performed by the pump manufacturer. We will review the key facts that this curve offers.

The system resistance curve is the change in flow with respect to head of the *system*. It must be plotted by the user based upon the conditions of service. These include physical layout, process conditions, and fluid characteristics.

Anyone operating a pumping system must know where the pump is operating. Otherwise, it is like driving a car and not knowing how fast it's going.

The pump curve

The pump curve consists of a head/capacity curve (curve A of Figure 1) that shows the total differential head (H is expressed in feet of pumped liquid). This is the difference between the total discharge head (Hd) at the pump discharge and the total suction head (Hs) at the pump inlet at any given flow rate (Q), expressed in gpm (gallons per minute).

$$H = \mathrm{Hd} - \mathrm{Hs} \quad (1)$$

All pressure is in feet of liquid.

At this point, it's important to revisit some pressure units:

psi = pound/in^2
psia = pound/in^2 absolute
psig = pound/in^2 gauge
psia = psig + barometric pressure

The conversion from psi to feet of liquid is to divide the pressure, in psi, by its density:

$$(\mathrm{lb/in^2})/(\mathrm{lb/ft^3}) \times (144\ \mathrm{in^2/ft^2}) = \text{feet of pumped fluid}$$

For water with a specific gravity (SG) of 1, you substitute its density (62.4 lb/ft^3) into the above equation and the result is:

psi × 2.31/SG = feet of liquid See Figure 2 for illustration of feet, psia/psig/SG.

The pump curve also includes the pump efficiency (in percent—curve B), BHP (brake horsepower—curve C), and NPSHr (net positive suction head required—curve D) at any

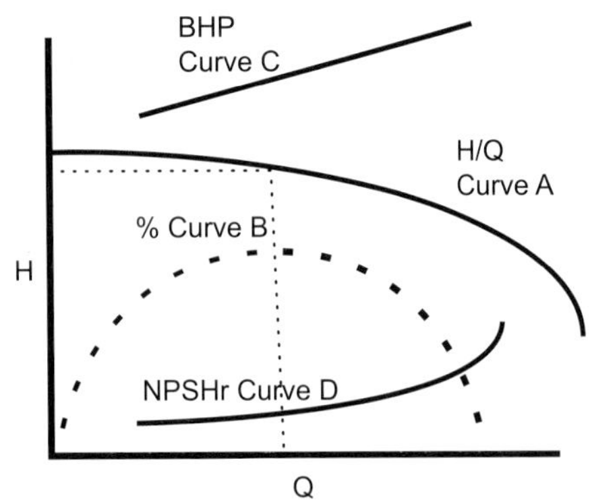

Figure 1. Performance curve data.

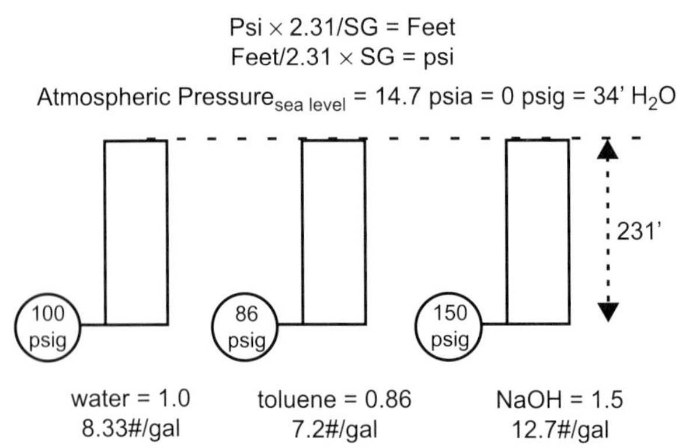

Figure 2. Pressure equivalent.

flow rate. This curve is plotted for a constant speed (rpm) and a given impeller diameter (or series of diameters).

Figure 1 is a single line curve; Figure 3 is an actual manufacturer's performance curve showing the same information but expressed as a series of curves for different impeller diameters (from 12-5/8″ to 15-15/16″). The pump impeller can be trimmed to any dimension between the maximum and minimum diameters.

The H, NPSHr, efficiency, and BHP all vary with flow rate, Q. The best efficiency point (BEP) is the point on the curve where the efficiency is at a maximum. For Figure 3, it's 82% at 1300 gpm and 100′ TDH.

BHP can be read from the curve at any flow rate; however, it is usually calculated from the efficiency, head, and flow using Equation 2. Pump curves are based on a specific gravity of 1.0. Other liquids' specific gravity must be considered.

$$\mathrm{BHP} = Q \times H \times \mathrm{SG}/(3960 \times \%\ \mathrm{eff}) \tag{2}$$

The NPSHr is read from the bottom curve from flows, and the NPSHr in feet is read on the lower right. The following examples apply to Figure 3.

Example 1. For an impeller diameter of 15-15/16″ (maximum), what are the head, efficiency, NPSHr, and BHP at 900 gpm? (Answer: 110′; 76.5%, 5′ NPSHr, and 32 BHP).

Example 2. What are the head, efficiency, NPSHr, and BHP at 1600 gpm? (Answer: 85′, 78%, 7.5′ NPSHr, and 44 BHP).

In a pumping system, the capacity of the pump moves along the pump curve based upon the differential head across the pump. For instance, for Figure 3, with the maximum impeller diameter, if the flow is 900 gpm and you want to increase it, you reduce the head to 85′, and the flow will increase to 1600 gpm. It's by controlling the pump differential head that you control the flow of the pump. If you wanted 1300 gpm, you would have to throttle the head back to 100′.

The slopes of pump curves vary depending on the pump-specific speed (Ns). How they vary depends onNs andis shown in Figure 4. The Ns is a dimensionless number calculated as:

$$\mathrm{Ns} = \mathrm{rpm} \times \left(Q^{1/2}\right)/H^{3/4} \text{ all at BEP} \tag{3}$$

The Ns for Figure 3 is 1311.

NPSHr is net positive suction head required by the pump—based on test.

NPSHa is net positive suction head available by the system—must be calculated.

$$\mathrm{NPSHa} = P1 + \mathrm{hs} - \mathrm{hvp} - \mathrm{hfs} \tag{4}$$

$P1$ = Pressure of suction vessel (atmospheric in open tanks)
hvp = Vapor pressure of liquid at its pumping temperature, t

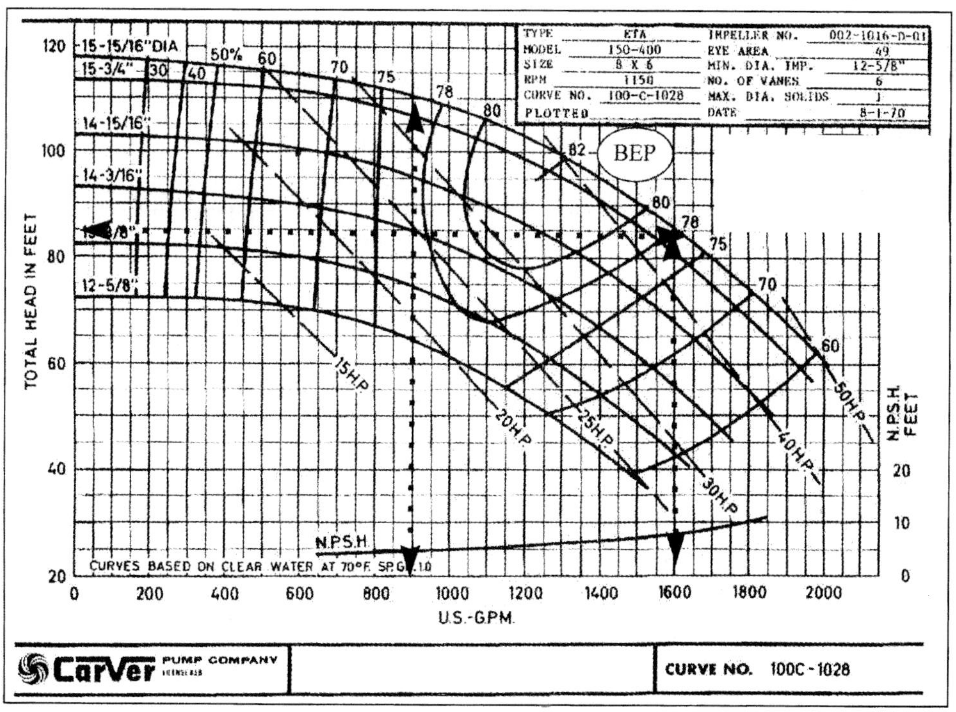

Figure 3. Manufacturer's performance curve.

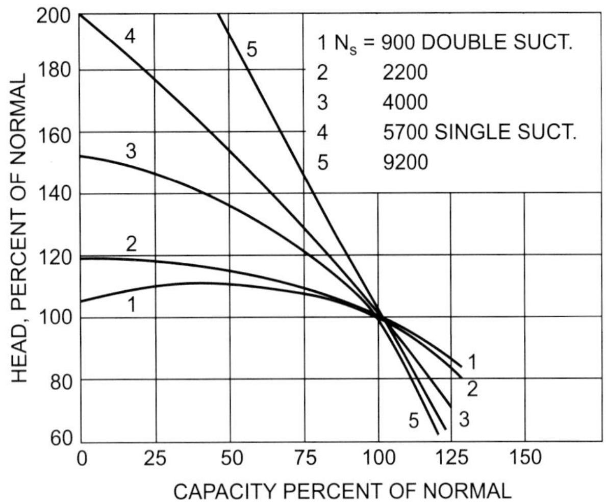

Figure 4. Head versus capacity for various speeds Ns.

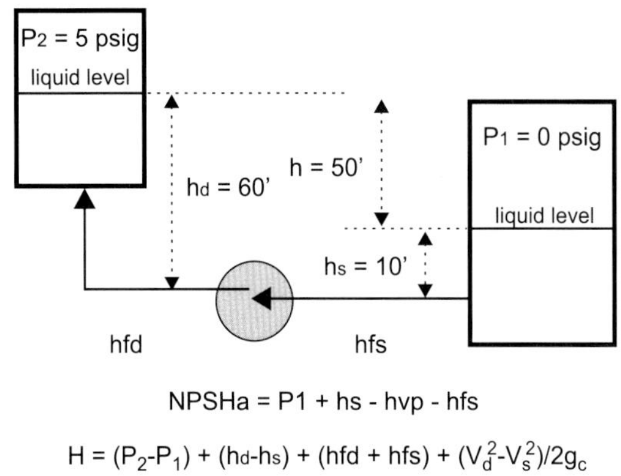

Figure 5. A typical pump system.

hs = Static suction head, the vertical distance between the eye of the first stage impeller centerline and the suction liquid level. (This may be a positive or negative number, depending on whether the suction liquid is above the pump suction eye or below it, as on a "lift.")

hsf = Friction and entrance losses on the suction side

Pump curves are generated by manufacturers and show a range of impeller diameters from minimum to maximum size for the casing size selected. These are based on a single speed (rpm). Head/capacity relationship changes with speed or impeller diameter per the affinity laws. Affinity laws are discussed in all the references. Other information often found is suction specific speed (Nss), which is a dimensionless number calculated at the BEP as

$$\text{Nss} = \text{rmp} \times \left(\text{gpm}_{\text{eye}}\right)^{1/2} / \text{NPSHr}^{3/4} \tag{5}$$

(*Note:* The gpm is per impeller eye, so for a double suction impeller, the flow is half the flow of single suction impeller.) The Nss in Figure 3 is 10,800.

Nss is a function of NPSHr. To reduce NPSHr, the impeller eye is modified. If the eye is modified too much, it causes a reduction in the low flow capabilities of the pump and limits the low end point of operation.

System head curves (Figure 5)

Step one in constructing a system/head curve is to consider the fluid energy requirement.

At any point in a pumping system, there are four components making up its total energy (head in feet):

- Pressure
- Velocity
- Elevation
- Heat (ignored, since pumping is usually at a constant temperature across the pump)

The difference in energy levels (differential head, *H*) between the suction (Hs) head and discharge (Hd) head of a pump can be expressed using Bernoulli's equation and adding the friction loss.

The friction loss is the loss in energy due to resistance to flow as it moves through a system. This resistance is due to viscous shear stresses within the liquid and turbulence at the walls of the pipe and fittings. It varies with liquid properties (viscosity, solids entrained); size of the pipe (velocity); roughness of the pipe's internals; length of travel; and losses at fittings, control valves, and the entrance and exits.

$$H\ \text{system} = \text{Hd} - \text{Hs} = (P_2 - P_1) + (\text{hd} - \text{hs}) + \left(V_d^2 - V_s^2\right)/2g_c + (\text{hfs} + \text{hfd}) \tag{6}$$

where

P_1 and P_2 = The entrance and exit pressures,

hd and hs = The elevations from the pump impeller eye to the entrance and exit liquid levels,

V_s and V_d = The entrance and exit velocities in ft/sec,

hfs and hfd = The frictional losses in the suction and discharge pipe system, expressed in feet.

To simplify:

$$\text{Hf} = \text{hfs} + \text{hfd is the total friction loss (at each flow rate) through the system} \quad (7)$$

$$h = \text{hd} - \text{hs is the difference in elevation from discharge level to suction level} \quad (8)$$

$$\text{hv} = V_d^2 - V_s^2/2g_c \text{ is the velocity head component} \quad (9)$$

Velocity head is often ignored, since in most cases the velocity is low (in tanks maintained at constant level, it's 0). However, if the discharge is coming out through a nozzle or other orifice, its velocity could be high and, therefore, significant if the other components are low. This should not be confused with entrance and exit head losses, which are accounted for in the Hf calculations.

Velocity head (hv = $V^2/2g_c$) is a confusing concept to many—it's the head (distance) required for a liquid to fall in order to reach a velocity, v. It costs energy to move (accelerate) from some low velocity to a higher velocity.

All components of Equation 6 must be consistent and usually in feet of liquid. Refer to Figure 5 for the following example.

For the pressure component

$$P_2 - P_1 = \frac{(5 \text{ psig} - 0 \text{ psig}) \times \dfrac{2.31 \text{ ft}}{\text{psi}}}{1.0} = 11.5'$$

It could have been

$$29.7 \text{ psia} - 14.7 \text{ psia} = 5 \text{ psia} \times \frac{2.31}{1.0} = 11.5$$

Use either; just be consistent. For open tanks there is 0 psig, 14.7 psia (at sea level).

For the static head component

$$h = \text{hd} - \text{hs} = 60 - 10 = 50'$$

Note that this is from liquid level to liquid level. This layout was for illustration; most industrial applications do not have the discharge pipe located beneath the liquid level because it would back siphon when the pump stops. If the pipe goes overhead and empties above the liquid line (or more often uses a vacuum breaker), use the centerline elevation of the discharge pipe at the exit for the hd. The pressure and static head components are constant with flow and plot as a straight line.

For the velocity head components

Assume the tanks are either being filled/emptied at the pumping rate or they are large enough that the downward and upward velocities are very low, or they are the same diameter and the velocity heads cancel. Therefore, the velocity head component is 0.

The friction component

Hf includes pipe losses and minor losses, as follows.

Pipe losses

There are several ways to calculate these losses:

1. Published data such as the Crane Handbook, the Cameron Hydraulic Data Book, or KSB's Centrifugal Pump Design.
2. Using the Hazen Williams formula:

$$\text{Hf} = \left[3.022(v)^{1.85}L\right] / \left[C^{1.85} \times D^{1.165}\right] \quad (10)$$

where

L = Length of pipe in feet
v = Velocity ft/sec
D = Pipe diameter in feet
C = Constant for various pipe materials

This is a simple way to go. The disadvantage is that it assumes constant NRe (Reynolds Number) in the turbulent range and a viscosity of 1.13 cs (31.5 SSU), which is water at 60 °F. This can yield errors of 20% or more for the range of 32-212 °F.

3. Using the Darcy formula:

$$\text{Hf} = fLv^2/2Dg_c \quad (11)$$

where
f = the friction factor based on the Reynolds number NRe:

$$\text{NRe} = D \times v/\text{vis} \quad (12)$$

where
vis = Viscosity in ft^2/s

$$\text{In laminar flow } (\text{NRe} \leq 2300)\ f = 64/\text{NRe} \quad (13)$$

In turbulent flow (NRe > 4000), f varies with NRe and the roughness of the pipe. This is expressed as "relative roughness" = e/D; where e is absolute roughness (in feet) found in tables and f is found in a Moody Friction Factor Chart using NRe and e/D.

If the range of NRe is between 2300 and 4000 (the transition zone), use turbulent flow design.

4. Using computerized flow simulators. (You need to know limits and assumptions upon which the program is based.)

Minor losses

These include obstructions in the line, changes in direction, and changes in size (flow area). The two most common methods of calculating these losses are

1. *Equivalent lengths:* uses tables to convert each valve or fitting into an "equivalent length," which is then added to the pipe length in *L,* and Hf is calculated in Darcy (Equation 11 above).
2. *Loss coefficients:* uses published resistance coefficients *(K)* in the formula: $hf = Kv^2/2g_c$. *K* values for many types of fittings are published. Loss coefficients are the way to determine entrance and exit losses and losses due to sudden reductions or enlargements.

Control valve losses

These losses are calculated using the Cv for the valve:

Cv = gpm pressure drop(psi)

The Hf increases as flow increases and is a parabolic curve. The Hf for the system is calculated using one of the above methods. Assume the suction pipe is 10″ diameter, 20′ long, with two elbows and one gate valve. The discharge pipe is 8″, 1000′ long, with eight elbows, one gate valve, and one check valve. The calculations yield Hf values of

1. Hf = 40′ @ 400 gpm
2. Hf = 12′ @ 800 gpm
3. Hf = 27′ @ 1200 gpm
4. Hf = 37′ @ 1400 gpm

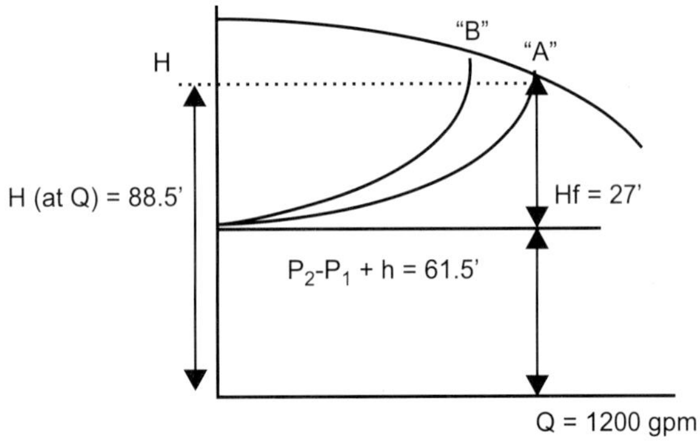

Figure 6. System head curve.

With the system head calculated at several flows (the pressure and height components are not affected by flow), we can plot the results (Figure 6).

The total differential head at 1200 gpm is $H = 11.5' + 50' + 27' = 88.5'$.

The pump's head capacity curve is plotted on the same graph. Where they intersect is the operating point ("A"). Note the pressure and static head components (66.50) are constant with flow, while the Hf component increases as flow increases.

If a control valve in the discharge piping were used, the system head curve could be throttled back to a higher head and therefore lower flow ("B").

What can go wrong?

A centrifugal pump will run where the system head curve tells it to run. The point where the centrifugal pump performance curve and system head curve intersect is the operating point. It's important to understand where the operating point is at all times and to avoid the dangerous areas of pump operation. What are some of the dangerous areas to consider?

Horsepower. A pump must always have a driver with more hp than it consumes. With most centrifugal pumps, as the pump flow increases, the BHP increases to a maximum, and there must always be enough hp available from the driver to supply the needs of the pump. This means as TDH falls, the flow goes up and so does the BHP. Undersizing overloads the driver.

NPSH. As the pump flow increases, the NPSHr increases. As flow increases, NPSHa decreases (hfs increases with flow). There must always be more NPSHa than NPSHr or the pump cavitates. Cavitation is where the pressure at a given point in the pump, prior to entering the impeller eye, falls below the vapor pressure of the liquid pumped, turning it to vapor. As the pressure increases through the impeller, the bubble of vapor collapses. This causes capacity reduction, noise, vibration, pitting of the impeller and, eventually, failure of a component.

No flow. It is possible for the TDH to rise to a point where there is no *flow*. This is called *shutoff* and means no flow is going through the pump. At this point, the temperature will rise in the pump because the inefficiencies of the pump are causing heat to go into the pumpage. This raises the temperature and will eventually cause the pumpage to vaporize, and the close clearances such as ring fits and mechanical seal faces will not receive lubrication. Eventually the seals fail or the pump seizes. High temperatures also cause the pump to distort and change clearances. Failure of some component will occur.

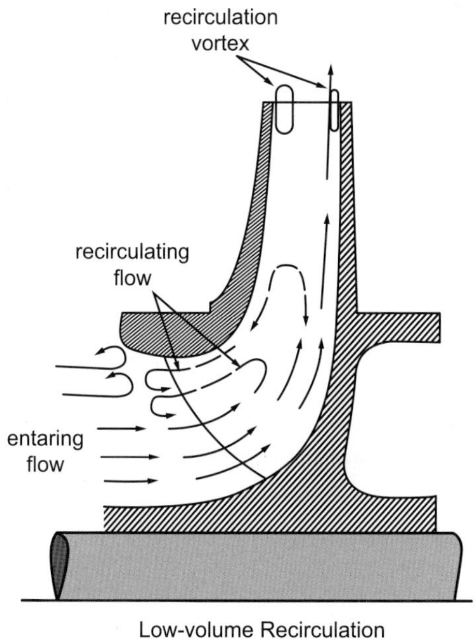

Figure 7. Low flow recirculation.

Low flow. It is possible for the TDH to rise to a point of low flow where damage other than no flow occurs. Such circumstances include:

- Low flow temperature rise—As the flow decreases, the efficiency falls and there is more BHP needed to drive the pump than is needed to move the liquid through the TDH. The excess BHP goes into the pump as heat. The flow is lower (less to carry away the heat), and therefore the temperature rise is greater as flow drops. It is possible to reach a point where the temperature rises to the point that it reduces the NPSHa below the NPSHr and flashing occurs as above. This point is the "thermal minimum flow" and is a function of not only the pump, but also the system characteristics.
- Low flow recirculation/separation damage—As flow decreases, the liquid begins to recirculate in both the impeller eye and impeller discharge (at the volute cutwater or diffuser). This recirculation will cause separation, eddy currents, and turbulence that will cause the same noise, vibration, and damage as cavitation (Figure 7). The minimum flow required to prevent low flow recirculation separation damage is called the "continuous minimum flow." It is specified by the pump manufacturer based on the impeller eye geometry and Nss.
- Radial or axial thrust damage—All pumps use bearings and balancing devices to prevent the rotating parts from contacting the stationary parts. These are pressure and flow related, and they may limit the flow of the pump to a minimum or maximum. The pump manufacturer must make known the limits based on the pump design.

References

1. Carver Pump Company. *Performance Curves*, Muscatine, IA.
2. Lindeburg P.E, Michael, *Mechanical Engineering Reference Manual*, 8th *Ed.* Professional Publications, Belmont, CA, 1990.
3. Karassik Igor, *Pump Handbook*. McGraw-Hill, New York, 1976.
4. Westaway C.R., *Cameron Hydraulic Data Book*, 16th *Ed.* Ingersoll-Rand, Woodcliff Lake, NJ, 1984.

Reprinted with permission - Kriebel Engineered Equipment, King of Prussia, Pennsylvania, and *Pumps and Systems Magazine*.

How to examine centrifugal pump systems

Doug Kriebel, P.E., Kriebel Engineered Equipment, King of Prussia, Pennsylvania

Each system must be analyzed to determine what happens to the pump at not only design, but also every other possible point of operation—start-up, off peak, unusual modes, "best case," and "worst case." The following examples illustrate how a system change can create either low flow, high flow, or NPSH problems. Refer to Figure 1, which shows a pump system. We will first look at the suction side to determine the NPSHa. The suction side is 20′ of 5″ pipe with two elbows and a gate valve. At 300 gpm the hfs is 1′. The pump selected is shown in Figure 2.

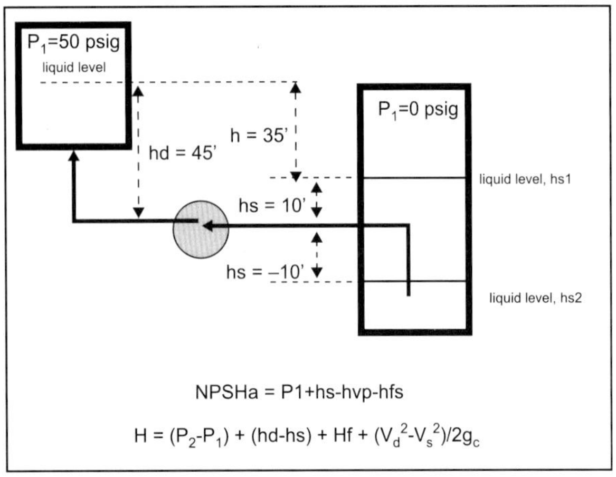

Figure 1. Pump system for Examples 1–6.

Example 1. The tank is open to atmosphere, 0 psig, 34′ water = P1; pumping water at 70 °F, hvp = 0.36 psi = 0.8′; from hs1 = 10′ then:

$$\begin{aligned}\text{NPSHa} &= P1 + \text{hs} - \text{hvp} - \text{hfs} = 34 + 10 - 0.8 - 1\\ &= 42.2'\end{aligned}$$

Referring to the pump curve Figure 2, at 300 gpm, the NPSHr = 23′ so we have adequate NPSHa.

Example 2. All conditions the same except the tank is drawn down to level hs2 = −10′

Then the NPSHa = 34 − 10 − 0.8 − 1 = 22.2′ and the NPSHa is less than the NPSHr and the pump cavitates.

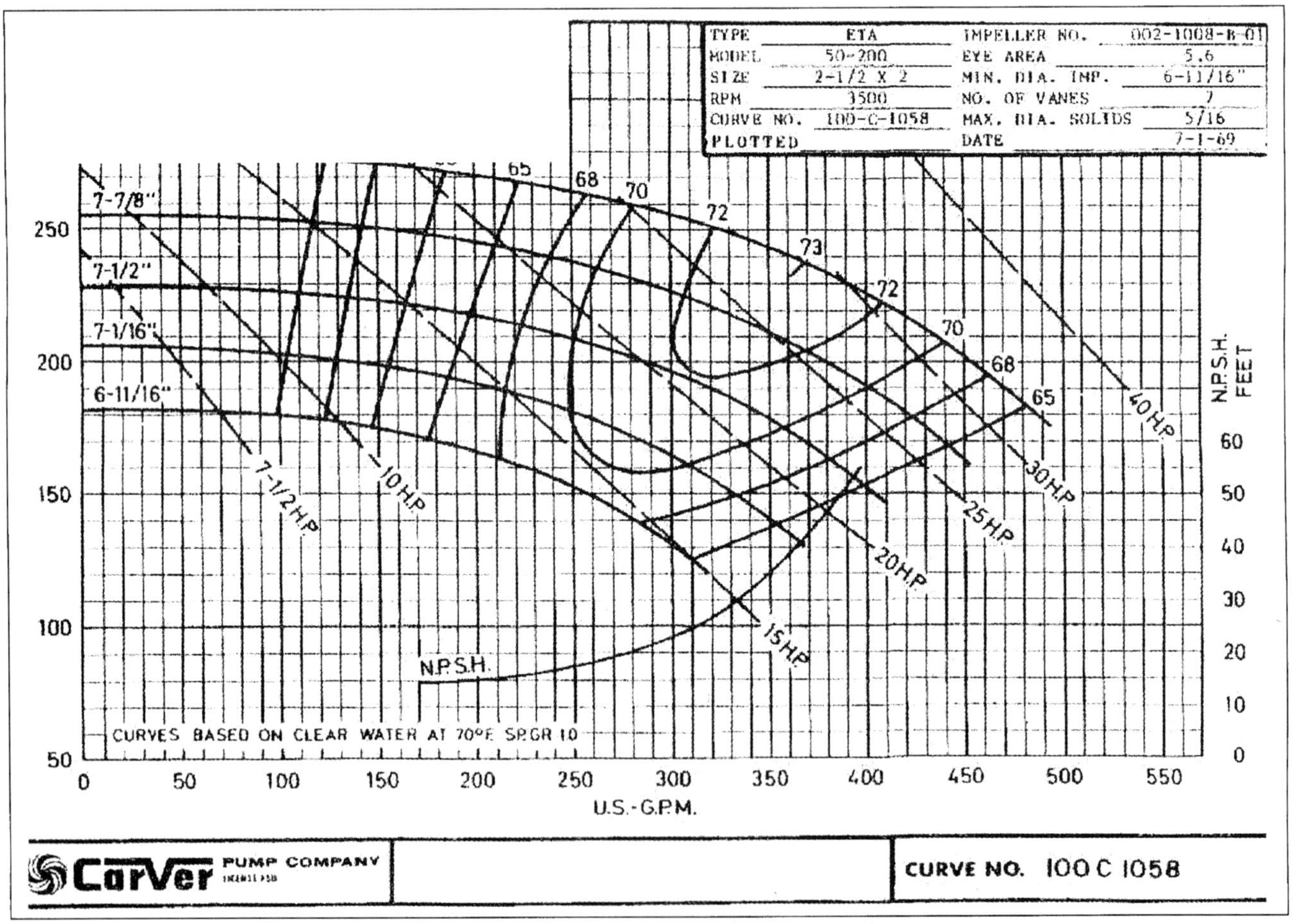

Figure 2. The selected pump, Example 1.

Example 3. All conditions the same as Example 1 except the water is now 190 °F; hvp = 9.34 psi = 9.34 × 2.31/ 0.966 = 22.3′.

Then NPSHa = 34 + 10 – 22.3 – 1 = 22.7′ and the pump cavitates.

Example 4. Same as Example 3 except from hs2 = –10′ Then NPSHa = 34 – 10 – 22.3 – 1 = 0.7′ and the pump cavitates.

Example 5. Conditions same as any above except the suction pipe is installed as 3″ instead of 5″: hfs = 10′ (instead of 1′). All the above NPSHa would be reduced by 90′!

The lesson to be learned on the *suction side:* Have as much static suction head as possible (hs), locate the pump as close as possible to the suction vessel (with a minimum suction side frictional pressure drop (hfs)), and be careful of liquids near their vapor pressure (hvp).

Let's go further and use the Example 1 conditions and look at the discharge side: Assume 150′ of 4″ pipe with six elbows, one gate valve, and one check valve: hfd = 14′ @ 300 gpm.

Example 6. Assume the levels in both suction tank and discharge tank remain constant with suction at hs1 then:

$$
\begin{aligned}
H &= (P_1 - P_2) + (\text{hd} - \text{hs}) + \left(V_2^2 - V_1^2\right)/2g_c + (\text{hfd} + \text{hfs}) \\
&= (50\ \text{psig} - 0\ \text{psig})2.31/1.0 + (45' - 10') + 0 + (14' + 1') \\
&= 115.5' + 35' + 15' = 165.5'
\end{aligned}
$$

The pump selected is shown in Figure 3. The operating point is "A".

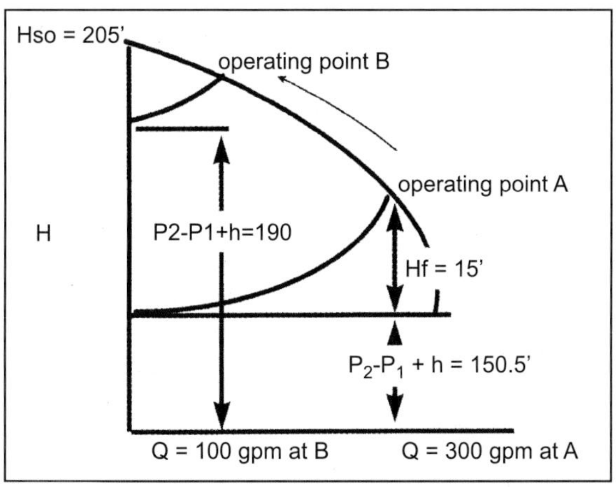

Figure 3. System head curved–Example 6.

What happens if the levels don't remain constant and the suction tank pulls down from level hs1 = +10′ to level hs_2 = –10′ and the discharge tank fills up an equal 20′ ?

The "*h*" component (hd – hs) would go up from 35′ to 75′ and the "*P*" + "*h*" components would rise from 150.5 to (115.5 + 75) 190.5′. This would cause the pump to "walk" back on the curve to some lower flow, point B.

All this means is that the pump will pump a lower flow rate. This is not necessarily a bad thing as long as it meets the process needs and is above the minimum flow for the pump. If it does go below minimum flow, damage will occur.

If the pressure (*P*2) is increased to 75 psig, the head goes up (75 – 0)2.31/1.0 = 173 plus *h* = 35′ to *H* = 208, which is above the shutoff of the pump (205′). The pump would be running with no flow. This could be an upset condition that causes overpressure to the discharge tank. It is also a situation where an inline filter is installed into an existing installation and the pressure drop is underestimated or the filter is not maintained (cleaned). This example cost an East Coast utility company more than $45,000 in damage to a pump when maintenance installed a filter to protect some heat exchangers without taking into account the pressure drop when fouled!

Example 7. More often the pump is selected for a "worst case" condition, (sometimes called the "design condition"). This example uses the same system as above except use 75 psig for the design *P*2, assuming "worst case" situation (Figure 4, Point A). This accounts for an upset, overpressure situation in the discharge tank or other circumstance.

However, when the system actually operates at 50 psig, the pump runs out to point B. This could overload the motor,

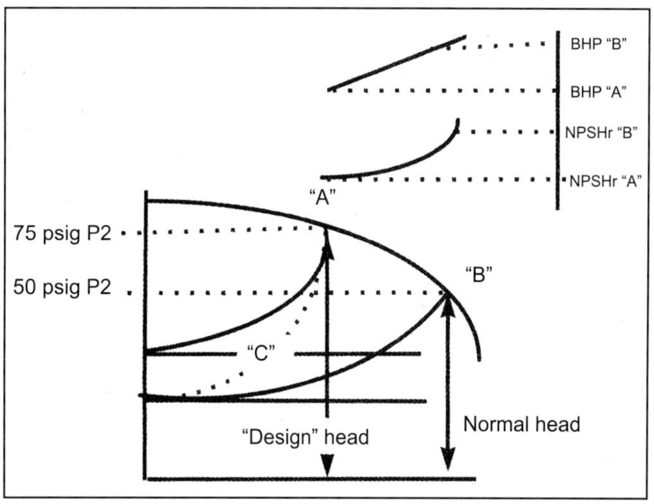

Figure 4. Example 7–changes in head and throttling.

since more BHP is required. In addition, the increased flow may cause the NPSHr to exceed the NPSHa (which decreases due to increased hfs). This may cause the pump to cavitate.

Most systems that are designed for higher *P*2 than normal have a control valve that will increase the hfd and "walk" the pump back on the curve to the correct operating point (back to "A" via line C).

A major chemical company using sealless pumps and no control valves (they were handling toxic material and did not want to risk leaks) designed a system for a specific H based on a *P*2 set by the process. After startup, which went without a hitch, the reactor pressure (*P*2) was decreased to improve yield. Without control valves, the pump ran out on the curve and destroyed itself by cavitation. It took $70,000 in repairs until someone ran the calculations to determine how the system change caused the pump failures!

A similar situation occurs if excess safety factor is utilized (Figure 5). If the actual design point is calculated as 300 gpm at 165′ (point A) and the pump is selected for 345 gpm and 200′ head (point B) and the calculations are correct, the pump will run out to point C. This would pump more than required. So the pump would be throttled back to point D to pump the 300 gpm. This means 215 — 165 = 50′ head is throttled out, which costs 5.2 BHP. Since you pay for BHP, this is wasted energy. In this case it would amount to $3100 per year at a rate of $0.085/kWh for service 24 h × 360 days. Higher heads or flows would magnify the costs.

There is nothing wrong with safety factors, and it is always better safe than sorry, but after installation, check the hydraulics. If gross over-sizing was the case, trimming the impeller will save considerable money in operations.

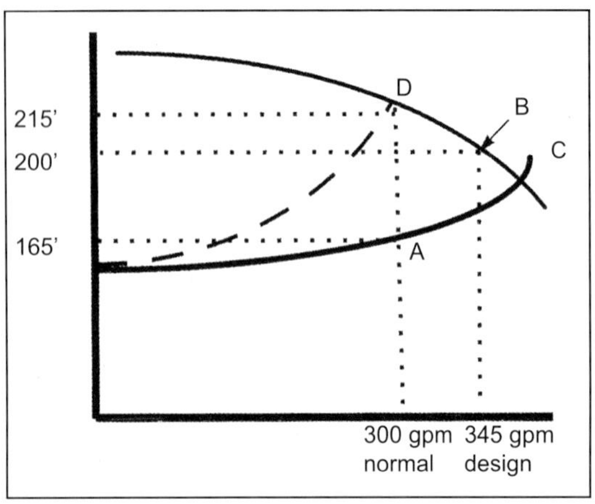

Figure 5. System head curve: design/normal/runout/throttle.

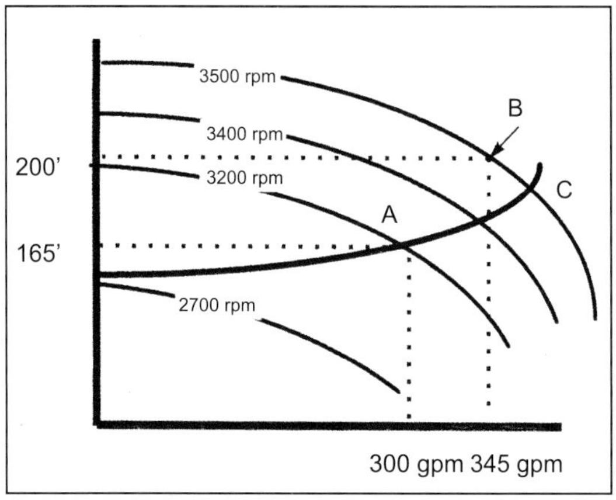

Figure 6. Example 8–variable speed operation.

Example 8 (variable speed operation). An economical alternative to throttling with a valve is to use variable speed to change the pump curve to meet variable demands of a system head curve. This is illustrated in Figure 6. The system head curve is plotted, a pump is picked to meet the worst case condition, and reduced speed is used to ride down the system curve to meet varying flows at reduced pressure. In some cases this can be very economical when compared to the wasted BHP of throttling control.

If the normal point is the same as above (300 gpm at 165′ point A on Figure 6) and with safety factors (345 gpm at 200′ point B) the pump operates on the uppermost curve for 3500 rpm, where it intersects the system head curve at point C. By slowing down to 3200 rpm, it goes through point A without having to waste energy by throttling the head. Any point can be reached on the system head curve by reducing speed. This is a valuable method of operating when flows or heads have large changes in magnitude.

The danger comes if you reduce the pump speed to a point where the system head curve causes the pump to operate at a point of too low flow or no flow (i.e. beneath the system head curve, such as 2700 rpm). It is possible to operate too high flow where BHP or NPSH becomes a factor, or at a speed where other damage could occur. Examples of other speed-related damages are if the pump reaches a critical speed, a speed where auxiliary equipment no longer functions (a shaft driven lube oil pump may not provide proper lubrication at too low a speed, or a motor fan motor may turn too slowly to provide adequate cooling).

Example 9 (parallel pumping applications). It's not uncommon to use two or more pumps in parallel. The flows are additive at a constant head (Figure 7). (The flow, Q, at head H for one pump at A is added to a second pump at the same head to get twice the flow, $2Q$, at the intersection of the two pump curves, the system head curve, B. It's important to draw the system head curve to make sure that if one pump operates alone it does not run out to point which will overload the motor or exceed the NPSH requirements. Point B is the operating point of two pumps in parallel, and point C is for one pump at runout (estimated here as $1.2Q$). The one pump runout must be checked to provide sufficient BHP, NPSHa, submergence against vortexing, and mechanical design for the higher flow.

This is more common in vertical cooling water service where several pumps are used in parallel and a control valve is not used on each pump. If one pump trips, the system allows the remaining pump(s) to run out.

Another potential for parallel pump problems is where two dissimilar pumps (dissimilar curves) are used (Figure 8). The pump curve for pump A is added to pump B at constant head. The system head is drawn. $Q1$ is the flow with only pump A running, $Q2$ is the flow with only B running, $Q3$ is the flow with both A + B running. *Note:* This is not $Q1 + Q2$. If the system is throttled back (by any means), the system head could cause the pump with the lower head to run below a minimum flow. Pump A is shut off at point C.

Example 10 (a pump for a given service is used for an alternate service). Figure 9 is the performance curve for a horizontal split case pump put in as a fire protection pump that is to run at 3000 gpm. The plant engineer lost his air compressor cooling water pump and decided to use the fire pump to supply cooling water for the compressor inner and after coolers. (If you need 300 gpm, a 3000 gpm pump will certainly do!)

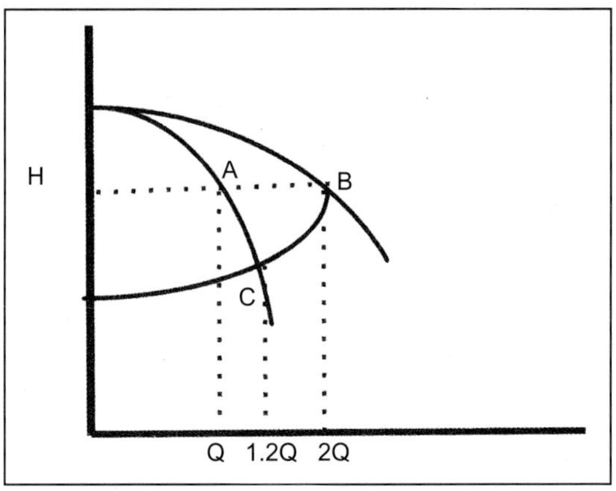

Figure 7. Example 9—parallel pumps.

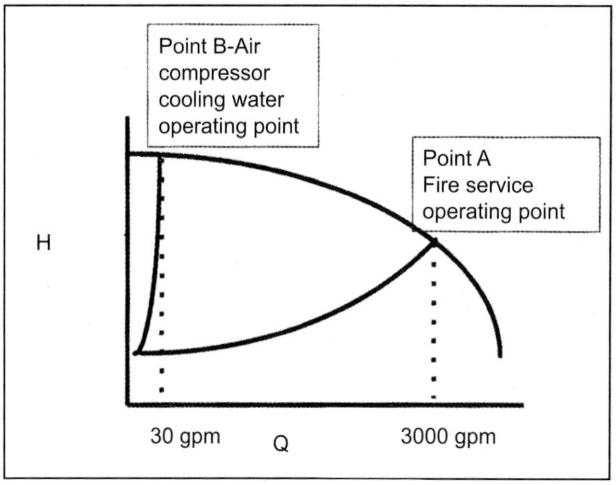

Figure 8. Parallel pumps.

The pump had six failures of the radial bearings. The maintenance group replaced them, then redesigned the pump and added angular contact back-to-back bearings in place of the double row bearings. They then added oil lubrication to the existing grease bearings and were about to change to oil mist when someone finally asked why the fire pump was running when there was no fire!

Point A is the design point, but point B is where it operated to supply the air compressor's cooling water. This was too far back on the curve and caused high radial loads that were the cause of failure. There was no way any bearing or lubrication upgrade would enable this pump to operate at point B.

In conclusion, the purpose of this article is to provide operations and maintenance people who do not normally do process calculations with the tools to analyze the pump systems they have to operate. This will enable them to operate the systems more economically and to determine the

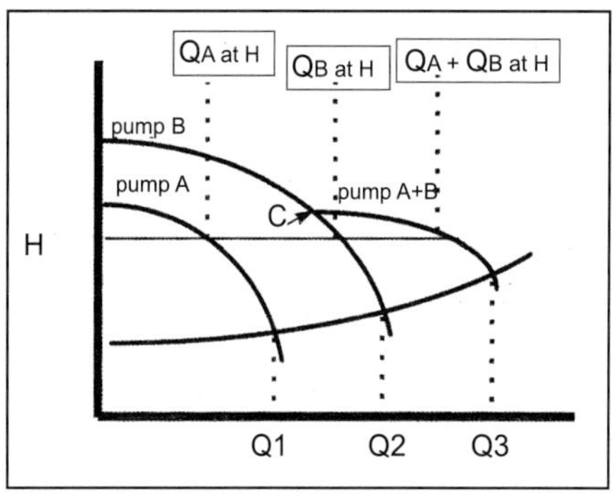

Figure 9. Example 10.

"real" root cause of failures. Failures are mechanical in nature, but the cause of the mechanical failure is a "system" incompatibility failure. In today's business world, systems are often designed by process engineers who are not available when the system becomes operational, and the system may not have been installed, operated, or maintained the way it was intended to be, or the correct method of operation is no longer the way it was originally designed.

References

1. Carver Pump Company. *Performance Curves,* Muscatine, IA.
2. P.E. Lindeburg, Michael, *Mechanical Engineering Reference Manual*, 8th *Ed.* Professional Publications, Belmont, CA, 1990.
3. Igor Karassik, *Pump Handbook*. McGraw-Hill, New York, 1976.
4. C.R. Westaway, *Cameron Hydraulic Data Book*, 16th *Ed.* Ingersoll-Rand, Woodcliff Lake, NJ, 1984.

Nomenclature

H = Total differential head
Hd = Total discharge head
Hs = Total suction head
SG = Specific gravity
Q = Flowrate
NPSH = Net positive suction head (r required, a available)
BHP = Brake horsepower
Ns = Specific speed
Nss = Suction specific speed
P = Pressure (1 suction, 2 discharge)
hvp = Vapor pressure of liquid
hs = Static suction head
hd = Static discharge head
h = hd — hs
hfs = Friction losses, suction side
hfd = Friction losses, discharge side
Hf = hfs + hfd
vd = Discharge velocity
vs = Suction velocity
v = Velocity (ft/sec)
hv = Celocity head
g_c = Gravitational constant (32.2 ft/s^2)
L = Pipe length in feet
D = Pipe diameter in feet
C = Hazen Williams constant for pipe materials
NRe = Reynolds number
f = Friction factor
vis = Viscosity
e = Absolute roughness
K = Resistance coefficient for pipe fittings

Estimate the discharge of a centrifugal pump at various speeds

Within the design limits of a pump, the discharge will be approximately proportionate to the speed.

$$Q_2 = Q_1\frac{N_2}{N_1}$$

where

Q_1 is the gallons per minute at a particular speed N_1, and Q_2 is the gallons per minute at N_2.

Example. If a pump delivers 350 gpm at 1200 rpm, what will it deliver at 1800 rpm?

$$Q_2 = 350 \times \frac{1,800}{1,200} = 525 \text{ gpm}$$

Similarly, for a given pump, design head varies as the square of the impeller speed and power as the cube of the speed.

How to estimate the head for an average centrifugal pump

By slightly modifying the basic pump formula:

$$H = \left(\frac{DN}{1,840\ K_u}\right)^2$$

it is possible to find a means of estimating the head a pump will develop for each impeller at various speeds. The terms are defined as follows:

H = Head, ft
D = Diameter of the impeller, in
N = rpm of the impeller
K_u = Pump design coefficient

The values of this expression may vary from 0.95 to 1.15. For estimating purposes, substitute 1900 for 1840 K_u. The formula becomes:

$$H = \left(\frac{DN}{1900}\right)^2 \times (\text{the number of impellers})$$

Example. What head will be developed by a well-designed centrifugal pump with four 11.75-in impellers turning at 1200 rpm?

$$H = 4 \times \left(\frac{11.75 \times 1,200}{1900}\right)^2 = 4 \times 7.4^2 = 220 \text{ ft of head}$$

Find the reciprocating pump capacity

Multiply the diameter of the plunger by itself, multiply this by the stroke, then by the number of strokes per minute, and finally by 0.11; the result is the capacity in barrels per day.

$$C = B^2 \times S \times N \times 0.11$$

Example. Four-in. bore, 6-in stroke, 80 strokes per minute.

$$4 \times 4 \times 6 \times 80 \times 0.11 = 845 \text{ barrels per day}$$

Note. The rule is approximate only; it allows about 5% for "slip," which is actually somewhat variable. Also note that it is applicable only to single-acting pumps; for a double-acting piston type pump, N must be twice the rpm, and a reduction must be made for the area of the piston rod on the crank end of the pump.

How to estimate the hp required to pump at a given rate at a desired discharge pressure

Ignoring effects of temperature, viscosity, etc.

$$\text{hp} = \text{psi} \times \frac{\text{barrels per day}}{1,000} \times 0.022$$

The factor 0.022 is arrived at by using a pump efficiency of 80% and a gear efficiency of 97%.

Example. What horsepower would be required to pump hp 15,000 barrels per day with an 800 psi discharge?

$$\begin{aligned}\text{hp} &= 800 \times 15 \times 0.022 \\ &= 264 \text{ horsepower}\end{aligned}$$

Nomograph for determining reciprocating pump capacity

The theoretical capacity of reciprocating pumps can be calculated by the following equation:

$$C = 6/55D^2L\,N$$

where

C = Theoretical pump capacity (delivery), gal/min
D = Diameter of cylinder, ft
L = Length of stroke, ft
N = Number of strokes/min

The nomograph (Figure 1) is based on this equation. The diameter and length-of-stroke scales are calibrated in inch units instead of feet. The nomograph also gives the delivery in British imperial gallons.

Example. What is the theoretical capacity of a reciprocating pump having the following characteristics?

$D = 3$ in
$L = 3$ in
$N = 60$ strokes/min

Connect	With	Mark or Read
$D = 3$ in	$L = 3$ in	Reference scale
Ref. Point	$N = 60$	$C = 5$ gal/min

Source

Macdonald, J. O. S., *British Chemical Engineering*, 1, 6, 301 1956).

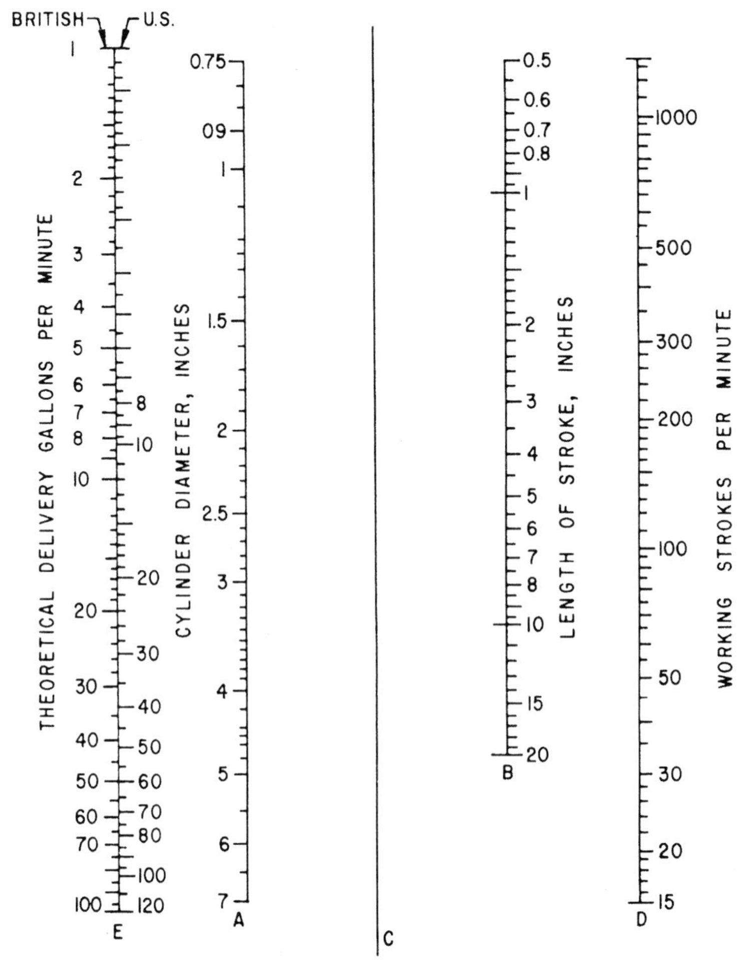

Figure 1. Theoretical capacity of reciprocating pumps.

Nomograph for determining specific speed of pumps

The specific speed is a criterion used in the selection of pumps for a given service. It can be used to (a) avoid cavitation troubles, (b) select the most economic pump for a given process layout, and (c) assist in redesigning process conditions.

For single-stage, side-entering pumps, the NPSH is defined by the following equation:

$$S = N\frac{Q^{1/2}}{H^{3/4}} \qquad (1)$$

where

S = Specific speed, no units
Q = Pump capacity
N = Pump speed, rev/min
H = Net positive suction head above the liquid vapor pressure (NPSH) measured at the pump suction flange, ft of fluid

H (in ft of liquid) is given by

$$H = h_a \pm h_h - h_p - h_v \qquad (2)$$

where

h_a = Ambient pressure at the liquid surface
h_h = Static head between liquid surfaces
h_p = Combined entrance, piping and fitting friction loss
h_v = Proper pressure at pumping temperature

The specific speed of centrifugal pumps may vary between 1100 and 15,000. Values above 12,000 are not considered practical and, where possible, the use of a specific speed below 8500 is recommended. Specific speed limits for each type of pump are usually available from manufacturers of pumps.

The nomograph (Figure 1) solves Equation 1.

The specific speed is also defined in terms of a dimensionless number, using consistent units, by the following equation:

$$S = \frac{Q^{0.5}N}{(\Delta H \times g_c)^{0.75}} \qquad (3)$$

where

g_c = gravitational conversion constant, 32.2 ft/s^2

The range for this equation is 0.03 to 0.87. Results of Equation 1 can be converted to these units by dividing by 17,200.

Example. The following data are given for a single-suction reboiler pump:

Q = 400 gal/min
N = 3500 rev/min

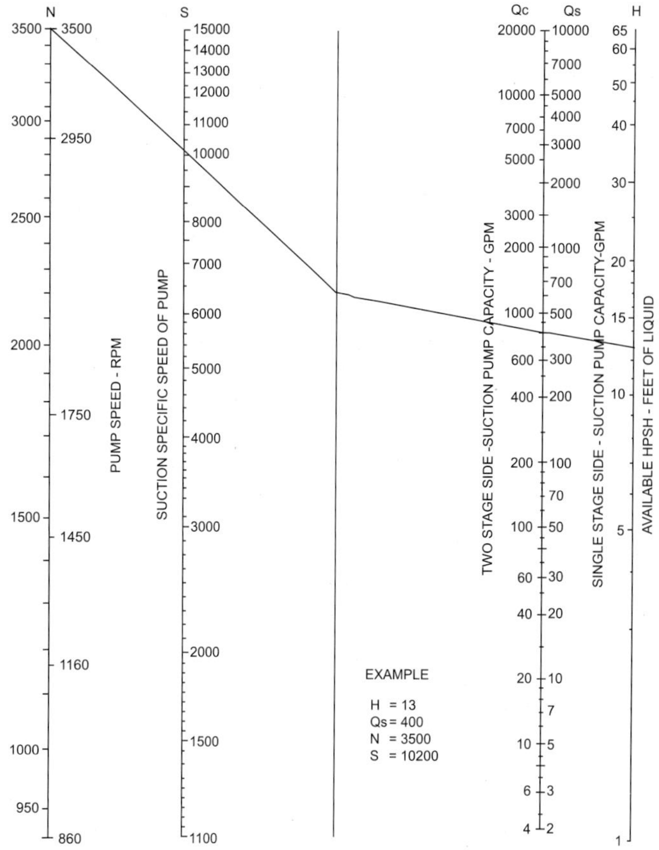

Figure 1. Pump-specific speed from net positive suction head. Reproduced by permission, *Petroleum Refiner*, Vol 36, No. 10, copyright 1957.

The reboiler liquid level is 18 ft above the pump, and friction from entrance, pipe, and fittings loss add up to 5 ft of liquid.

For a reboiler, the liquid is at its boiling point, so $h_a = h_v$, using Equation 2, H = 13 ft.

Connect	With	Mark or Read
NPSH = 13	Q = 400 gal/min	Reference line
Ref. line pivot	N = 3500	S = 10,200

Source

Jacobs, J. K., *Petroleum Refiner*, 36, 10, 145 (1957).

Reference

1. *Hydraulic Institute Standards*, 10th Edition, New York.
2. Brown, et al., *Unit Operations*, 190, John Wiley and Sons, New York, 1950.

Nomograph for determining horsepower requirement of pumps

In the selection of pumps it is necessary to calculate the horsepower requirement to handle a specified delivery rate of fluid against the specific head. The formula for the horsepower required by a pump is as follows:

$$hp = \frac{G \times H \times \text{specific gravity}}{3960 \times e}$$

where:

hp = Pump horsepower
G = Flow rate, gal/min
H = Total dynamic head (change in head), ft
e = Pump efficiency

Figure 1 solves this equation.

Example. Given:

G = 200 U.S. gal/min
H = 60 ft
Specific gravity = 1.1
e = 0.80

What is the required horse power?

Connect	With	Mark or Read
e = 0.80	H = 60 ft	C
C	G = 200 gal/min	hp = 0.78 horsepower

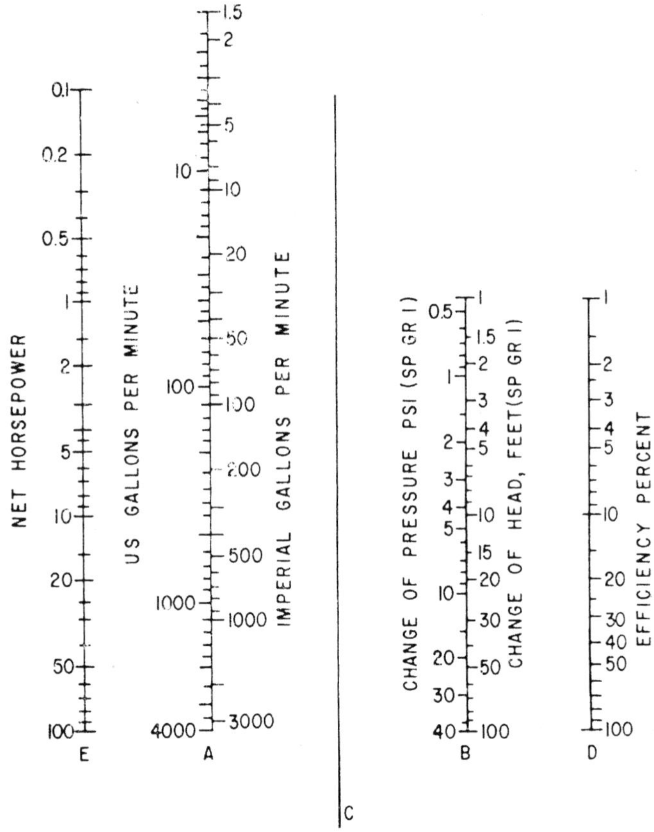

Figure 1. Pump horsepower.

Source

Macdonald, J. O. S., *British Chemical Engineering*, 1, 6, 290 (1956).

How to select motors for field-gathering pumps

The following simple formulas will provide a reasonably close approximation of the motor hp necessary to drive duplex double-acting pumps normally used in the field.

Rule.

1. $hp = \dfrac{bpd \times psi}{48,000}$
2. $hp = \dfrac{bpd \times psi}{2,000}$
3. $hp = \dfrac{gpm \times psi}{1,400}$

Motors selected with these formulas will fit delivery rates used in the formula if they are within capacity of the pump. If anticipated rate is below pump capacity, a savings may be made in first cost and power bills by selecting a motor corresponding to the actual rate rather than to full pump capacity.

It should be noted that net bpd values are to be used in Formula 1. That is, if 600 barrels are to be pumped each day, but it is desired to accomplish this in only 8 h, the rate to be used in the formula is 1800 bpd.

How a reciprocating pump works

A reciprocating pump is a positive displacement mechanism with liquid discharge pressure being limited only by the strength of the structural parts. Liquid volume or capacity delivered is constant regardless of pressure and is varied only by speed changes.

Characteristics of a GASO reciprocating pump are (1) positive displacement of liquid, (2) high pulsations caused by the sinusoidal motion of the piston, (3) high volumetric efficiency, and (4) low pump maintenance cost.

Plunger or piston rod load

Plunger or piston "rod load" is an important power and design consideration for reciprocating pumps. Rod load is the force caused by the liquid pressure acting on the face of the piston or plunger. This load is transmitted directly to the power frame assembly and is normally the limiting factor in determining maximum discharge pressure ratings. This load is directly proportional to the pump gauge discharge pressure and proportional to the square of the plunger or 'piston diameter.

Occasionally, allowable liquid end pressures limit the allowable rod load to a value below the design rod load. It is important that liquid end pressures do not exceed published limits.

Calculations of volumetric efficiency

Volumetric efficiency (E_v) is defined as the ratio of plunger or piston displacement to liquid displacement. The volumetric efficiency calculation depends upon the internal configuration of each individual liquid body, the piston size, and the compressibility of the liquid being pumped.

Tools for liquid pulsation control, inlet and discharge

Pulsation control tools (PCT, often referred to as "dampeners" or "stabilizers") are used on the inlet and discharge piping to protect the pumping mechanism and associated piping by reducing the high pulsations within the liquid caused by the motions of the slider-crank mechanism. A properly located and charged pulsation control tool may reduce the length of pipe used in the acceleration head equation to a value of 5–15 nominal pipe diameters. Figure 1 is a suggested piping system for power pumps. The pulsation control tools are specially required to compensate for inadequately designed or old/ adapted supply and discharge systems.

Acceleration head

Whenever a column of liquid is accelerated or decelerated, pressure surges exist. This condition is found on the inlet side

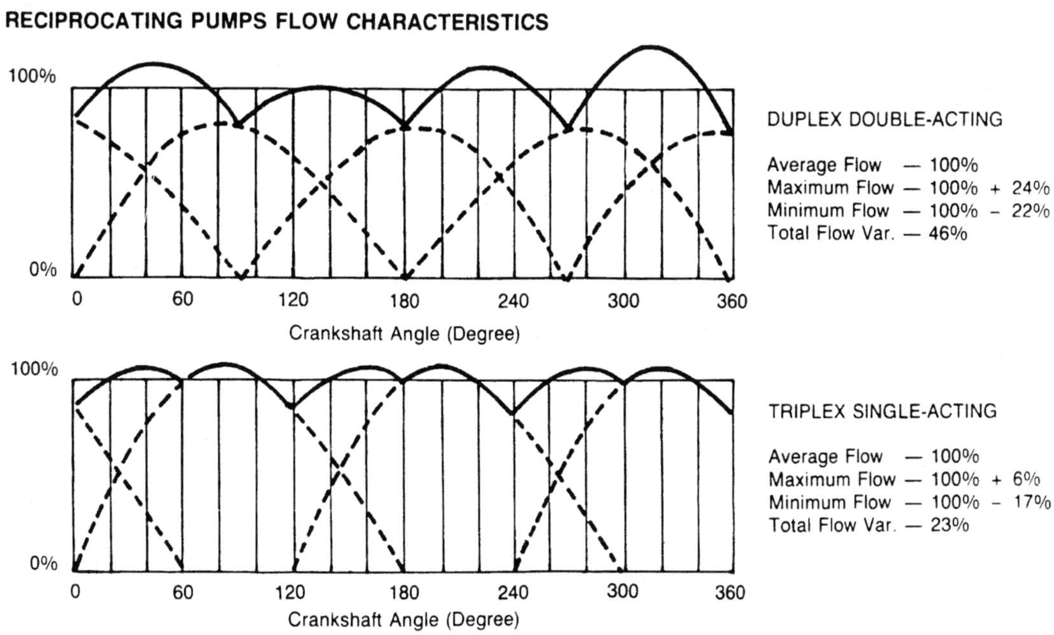

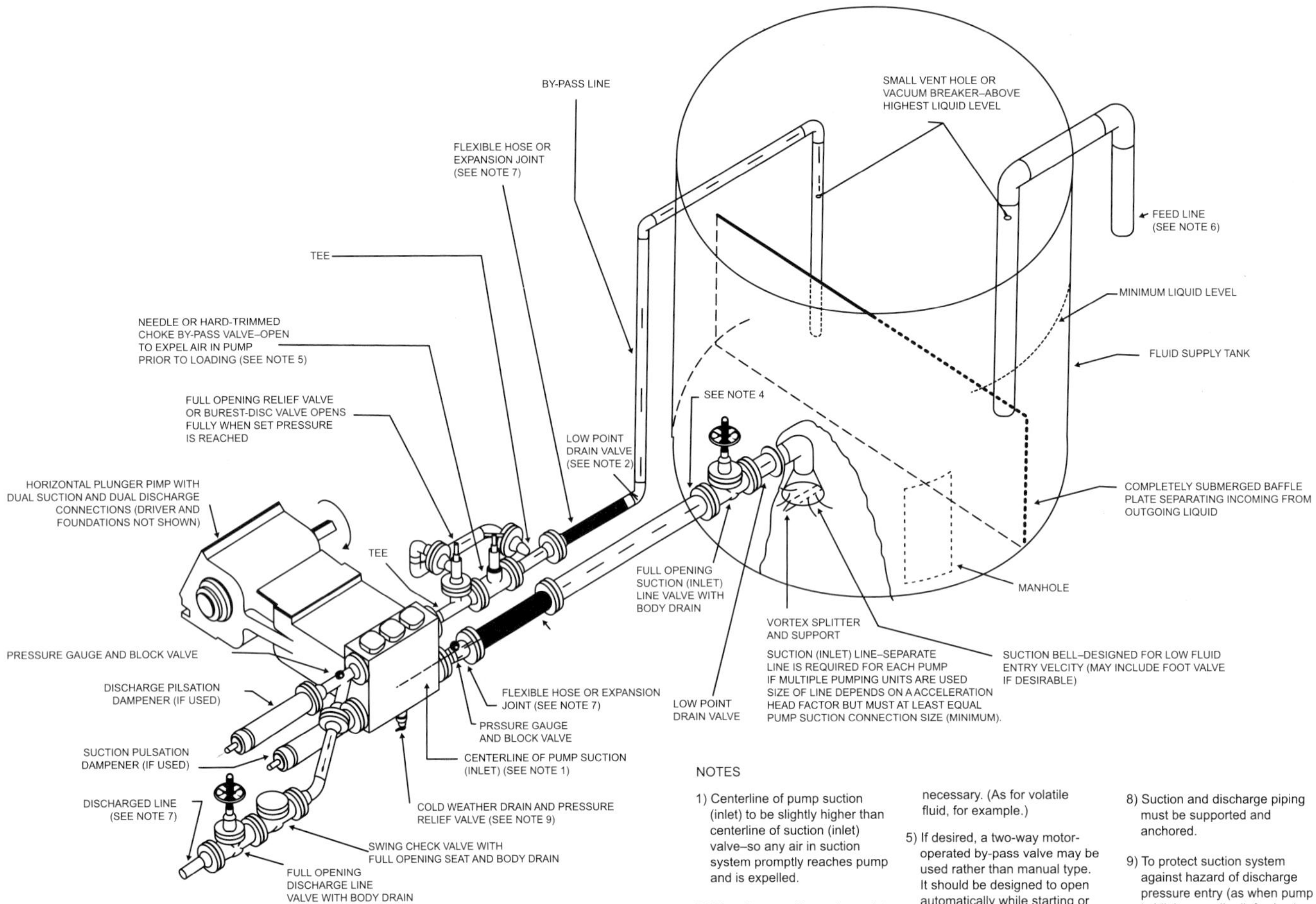

NOTES

1) Centerline of pump suction (inlet) to be slightly higher than centerline of suction (inlet) valve–so any air in suction system promptly reaches pump and is expelled.

2) Slope by-pass line so low point drain will fully empty relief and choke valves and all liquid in by–pass circuit.

3) Do not locate piping or other equiment in front of or above pump liquid end–preventing servicing.Refer to pump manufacturer for minimum

4) Locate charging pump at point shown–if a charging pump is necessary. (As for volatile fluid, for example.)

5) If desired, a two-way motor-operated by-pass valve may be used rather than manual type. It should be designed to open automatically while starting or stopping.

6) Feed line and by-pass line should feed liquid into tank belo minium liquid level.

7) To remove piping strain and vibration, a flexible hose, expansion joint, or swivel joint pair should be positioned to minimize effects of piping thermal expansion, contraction, and piping weight.

8) Suction and discharge piping must be supported and anchored.

9) To protect suction system against hazard of discharge pressure entry (as when pump is idle), a small relief valve is often connected here.

10) All system components must have adequate pressure ratings for operating, starting, and upset conditions. In order to reduce potential hazards, particular attention is recommended for the surge condition that will result downstream of the relief valve when normal discharge is blocked.

Figure 1. Suggested piping system for power pumps.

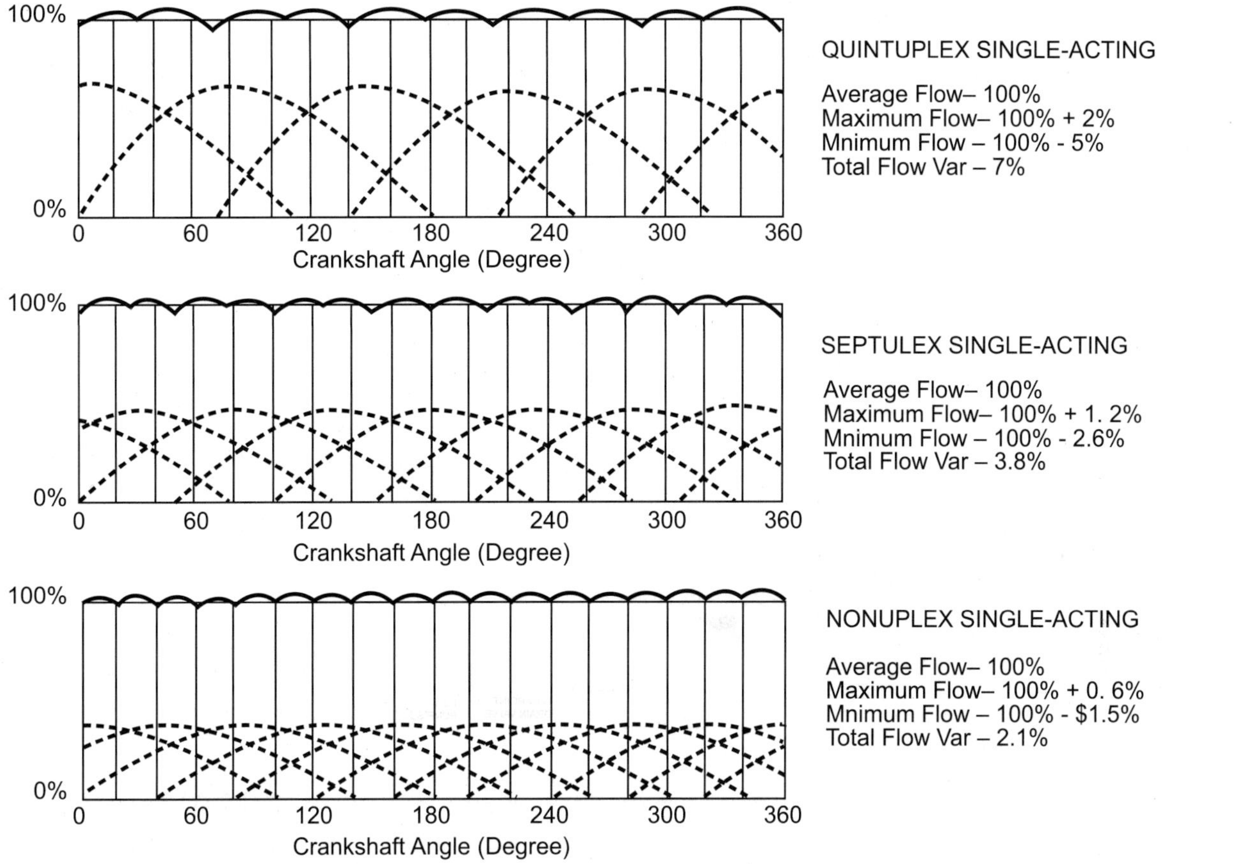

of the pump as well as the discharge side. Not only can the surges cause vibration in the inlet line, but they can restrict and impede the flow of liquid and cause incomplete filling of the inlet valve chamber. The magnitude of the surges and how they will react in the system is impossible to predict without an extremely complex and costly analysis of the system. Since the behavior of the natural frequencies in the system is not easily predictable, as much of the surge as possible must be eliminated at the source. Proper installation of an inlet pulsation control PCT will absorb a large percentage of the surge before it can travel into the system. The function of the PCT is to absorb the "peak" of the surge and feed it back at the low part of the cycle. The best position for the PCT is in the liquid supply line as close to the pump as possible or attached to the blind flange side of the pump inlet. In either location, the surges will be dampened and harmful vibrations reduced.

Acceleration head

$$h_a = \frac{LVNC}{Kg} \quad V = \frac{\text{GPM}}{(2.45)(\text{ID})^2}$$

where

h_a = Acceleration head (in feet)
L = Length of liquid supply line (in line)
V = Average velocity in liquid supply line (in fps)
N = Pump speed (revolutions per minute)
C = Constant depending on the type of pump
$C = 0.200$ for simplex double-acting
$= 0.200$ for duplex single-acting
$= 0.115$ for duplex double-acting
$= 0.066$ for triplex single- or double-acting
$= 0.040$ for quintuplex single- or double-acting
$= 0.028$ for septuplex, single- or double-acting
$= 0.022$ for nonuplex, single- or double-acting
K = Liquid compressibility factor
$K = 2.5$ for relatively compressible liquids (ethane)
$= 2.0$ for most other hydrocarbons
$= 1.5$ for amine, glycol, and water
$= 1.4$ for liquids with no compressibility (hot water)
g = Gravitational constant = 32.2 ft/s^2
d = Inside diameter of pipe (inches)

Stroke

One complete unidirectional motion of piston or plunger. Stroke length is expressed in inches.

Pump capacity (*Q*)

The capacity of a reciprocating pump is the total volume throughput per unit of time at suction conditions. It includes both liquid and any dissolved or entrained gases at the stated operating conditions. The standard unit of pump capacity is the US gallon per minute.

Pump displacement (*D*)

The displacement of a reciprocating pump is the volume swept by all pistons or plungers per unit time. Deduction for piston rod volume is made on double-acting piston type pumps when calculating displacement. The standard unit of pump displacement is the U.S. gallon per minute. For single-acting pumps:

$$D = \frac{Asnm}{231}$$

For double-acting piston pumps:

$$D = \frac{(2A - a)\, snm}{231}$$

where

A = Plunger or piston area, square inch
a = Piston rod cross-sectional area, square inch (double-acting pumps)
s = Stroke length, inch
n = RPM of crankshaft
m = Number of pistons or plungers

Plunger or piston speed (*v*)

The plunger or piston speed is the average speed of the plunger or piston. It is expressed in feet per minute.

$$v = \frac{ns}{6}$$

Pressures

The standard unit of pressure is the pound force per square inch.

Discharge pressure (Pd)—The liquid pressure at the centerline of the pump discharge port.

Suction pressure (Ps)—The liquid pressure at the centerline of the suction port.

Differential pressure (Ptd)—The difference between the liquid discharge pressure and suction pressure.

Net positive suction head required (NPSHR)—The amount of suction pressure, over vapor pressure, required by the pump to obtain satisfactory volumetric efficiency and prevent excessive cavitation.

The pump manufacturer determines (by test) the net positive suction head required by the pump at the specified operating conditions.

NPSHR is related to losses in the suction valves of the pump and frictional losses in the pump suction manifold and pumping chambers. Required NPSHR does not include system acceleration head, which is a system-related factor.

Slip (*S*)

Slip of a reciprocating pump is the loss of capacity, expressed as a fraction or percent of displacement, due to leaks past the valves (including the back-flow through the valves caused by delayed closing) and past double-acting pistons. Slip does not include fluid compressibility or leaks from the liquid end.

Power (*P*)

Pump power input (Pi)—The mechanical power delivered to a pump input shaft, at the specified operating conditions. Input horsepower may be calculated as follows:

$$\mathrm{Pi} = \frac{Q \times \mathrm{Ptd}}{1714 \times \eta p}$$

Pump power output (Po)—The hydraulic power imparted to the liquid by the pump, at the specified operating conditions. Output horsepower may be calculated as follows:

$$\mathrm{Pi} = \frac{Q \times \mathrm{Ptd}}{1714}$$

The standard unit for power is the horsepower.

Efficiencies (n)

Pump efficiency (ηp) (also called pump mechanical efficiency)—The ratio of the pump power output to the pump power input.

$$\eta p = \frac{\text{Po}}{\text{Pi}}$$

Volumetric efficiency (ηv)—The ratio of the pump capacity to displacement.

$$\eta \text{v} = \frac{Q}{D}$$

Plunger load (single-acting pump)

The computed axial hydraulic load, acting upon one plunger during the discharge portion of the stroke, is the plunger load. It is the product ofplunger area and the gauge discharge pressure. It is expressed in pounds force.

Piston rod load (double-acting pump)

The computed axial hydraulic load, acting upon one piston rod during the forward stroke (toward head end), is the piston rod load.

It is the product of piston area and discharge pressure, less the product of net piston area (rod area deducted) and suction pressure. It is expressed in pounds force.

$$\text{Liquid pressure}\left(\frac{\text{pounds}}{\text{square inch}}\right)\text{or}(\text{psi})$$

$$= \text{Cylinder area (square inches)}$$

$$= (3.1416) \times (\text{Radius(inches)})^2$$

$$= \frac{(3.1416) \times (\text{diameter(inches)})^2}{4}$$

$$\text{Cylinder force (pounds)} = \text{pressure (psi)} \times \text{area(square inches)}$$

Cylinder speed or average liquid velocity through

$$\text{piping (feet/second)} = \frac{\text{flow rate(gpm)}}{2.448 \times (\text{inside diameter (inch)})^2}$$

Reciprocating pump displacement (gpm)

$$= \frac{\text{rpm} \times \text{displacement (cubic in/revolution)}}{231}$$

Pump input horsepower – see horsepower calculations

$$\text{Shalt torque (foot-pounds)} = \frac{\text{horsepower} \times 5252}{\text{shaft speed (rpm)}}$$

$$\text{Electric motor speed (rpm)} = \frac{120 \times \text{frequency (Hz)}}{\text{number of poles}}$$

Three phase motor horsepower (output)

$$= \frac{1.73 \times \text{amperes} \times \text{volts} \times \text{efficiency} \times \text{power factor}}{746}$$

$$\text{Static head of liquid (feet)} = \frac{2.31 \times \text{static pressure (psig)}}{\text{specific gravity}}$$

$$\text{Velocity head of liquid (feet)} = \frac{\text{liquid velocity}^2}{\text{g}}$$

Absolute viscosity (centipoise)

$$= \text{specific gravity} \times \text{Kinematic viscosity (centistrokes)}$$

Kinematic viscosity (centistrokes)

$$= 0.22 \times \text{Saybolt viscosity (ssu)}$$

$$= \frac{180}{\text{Saybolt viscosity (ssu)}}$$

Absolute pressure (psia)

$$= \text{Local atmospheric pressure} + \text{gauge pressure (psig)}$$

Gallon per revolution

$$= \frac{\text{Area of plunger (sq in)} \times \text{length of stroke (in)} \times \text{number of plungers}}{231}$$

$$\text{Barrels per day} = \text{gal/rev} \times \text{pump speed (rpm)} \times 34.3$$

Example: Find the volumetric efficiency of a reciprocating pump with the following conditions:

Type of pump	3 in diam plunger × 5 in stroke Triplex
Liquid pumped	Water
Suction pressure	Zero psig
Discharge pressure	1785 psig
Pumping temperature	140 F
c	127.42 cu in
d	35.343 cu in
S	0.02

Find βt from Table of Water Compressibility (Table 1).

Table 1
Compressibility Factor $\beta t \times 10^{-6}$ = Contraction in Unit Volume Per Psi Pressure
Compressibility from 14.7 psia, 32 °F to 212 °F and Saturation Pressure Above 212 °F

Temperature																			
Pressure	0 °C	20 °C	40 °C	60 °C	80 °C	100 °C	120 °C	140 °C	160 °C	180 °C	200 °C	220 °C	240 °C	260 °C	280 °C	300 °C	320 °C	340 ° C	360 °C
psia	32 °F	63 °F	104 °F	140 °F	176 °F	212 °F	248 °F	284 °F	320 °F	356 °F	392 °F	428 °F	464 °F	500 °F	536 °F	572 °F	608 °F	644 ° F	680 °F
200	3.12	3.06	3.06	3.12	3.23	3.40	3.66	4.00	4.47	5.11	6.00	7.27							
400	3.11	3.05	3.05	3.11	3.22	3.39	3.64	3.99	4.45	5.09	5.97	7.21							
600	3.10	3.05	3.05	3.10	3.21	3.39	3.63	3.97	4.44	5.07	5.93	7.15	8.95						
800	3.10	3.04	3.04	3.09	3.21	3.38	3.62	3.96	4.42	5.04	5.90	7.10	8.85	11.6					
1000	3.09	3.03	3.03	3.09	3.20	3.37	3.61	3.95	4.40	5.02	5.87	7.05	8.76	11.4	16.0				
1200	3.08	3.02	3.02	3.08	3.19	3.36	3.60	3.94	4.39	5.00	5.84	7.00	8.68	11.2	15.4				
1400	3.07	3.01	3.01	3.07	3.18	3.35	3.59	3.92	4.37	4.98	5.81	6.95	8.61	11.1	15.1	23.0			
1600	3.06	3.00	3.00	3.06	3.17	3.34	3.58	3.91	4.35	4.96	5.78	6.91	8.53	10.9	14.8	21.9			
1800	3.05	2.99	3.00	3.05	3.16	3.33	3.57	3.90	4.34	4.94	5.75	6.87	8.47	10.8	14.6	21.2	36.9		
2000	3.04	2.99	2.99	3.04	3.15	3.32	3.56	3.88	4.32	4.91	5.72	6.83	8.40	10.7	14.3	20.7	34.7		
2200	3.03	2.98	2.98	3.04	3.14	3.31	3.55	3.87	4.31	4.89	5.69	6.78	8.33	10.6	14.1	20.2	32.9	86.4	
2400	3.02	2.97	2.97	3.03	3.14	3.30	3.54	3.85	4.29	4.87	5.66	6.74	8.26	10.5	13.9	19.8	31.6	69.1	
2600	3.01	2.96	2.96	3.02	3.13	3.29	3.53	3.85	4.28	4.85	5.63	6.70	8.20	10.4	13.7	19.4	30.5	61.7	
2800	3.00	2.95	2.96	3.01	3.12	3.28	3.52	3.83	4.26	4.83	5.61	6.66	8.14	10.3	13.5	19.0	29.6	57.2	238.2
3000	3.00	2.94	2.95	3.00	3.11	3.28	3.51	3.82	4.25	4.81	5.58	6.62	8.08	10.2	13.4	18.6	28.7	53.8	193.4
3200	2.99	2.94	2.94	3.00	3.10	3.27	3.50	3.81	4.23	4.79	5.55	6.58	8.02	10.1	13.2	18.3	27.9	51.0	161.0
3400	2.98	2.93	2.93	2.99	3.09	3.26	3.49	3.80	4.22	4.78	5.53	6.54	7.96	9.98	13.0	17.9	27.1	48.6	138.1
3600	2.97	2.92	2.93	2.98	3.09	3.25	3.48	3.79	4.20	4.76	5.50	6.51	7.90	9.89	12.9	17.6	26.4	45.4	122.4
3800	2.96	2.91	2.92	2.97	3.08	3.24	3.47	178	4.19	4.74	5.47	6.47	7.84	9.79	12.7	17.3	25.8	44.5	110.8
4000	2.95	2.90	2.91	2.97	3.07	3.23	3.46	3.76	4.17	4.72	5.45	6.43	7.78	9.70	12.5	17.1	25.2	42.8	101.5
4200	2.95	2.90	2.90	2.96	3.06	3.22	3.45	3.75	4.16	4.70	5.42	6.40	7.73	9.62	12.4	16.8	24.6	41.3	93.9
4400	2.94	2.89	2.90	2.95	3.05	3.21	3.44	3.74	4.14	4.68	5.40	6.36	7.68	9.53	12.2	16.5	24.1	40.0	87.6
4600	2.93	2.83	2.89	2.94	3.05	3.20	3.43	3.73	4.13	4.66	5.37	6.32	7.62	9.44	12.1	16.3	23.6	38.8	82.3
4800	2.92	2.87	2.88	2.94	3.04	3.20	3.42	3.72	4.12	4.64	5.35	6.29	7.57	9.36	12.0	16.0	23.2	37.6	77.7
5000	2.91	2.87	2.87	2.93	3.03	3.10	3.41	3.71	4.10	4.63	5.32	6.25	7.52	9.28	11.8	15.8	22.7	36.6	73.9
5200	2.90	2.85	2.87	2.92	3.02	3.18	3.40	3.69	4.09	4.61	5.30	6.22	7.47	9.19	11.7	15.6	22.3	35.6	70.3
5400	2.90	2.85	2.86	2.91	3.01	3.17	3.39	3.68	4.07	4.59	5.27	6.19	7.41	9.12	11.6	15.3	21.9	34.6	66.9

$\beta t = 0.00000305$ at 140 °F and 1800 psia. Calculate volumetric efficiency:

$$\text{Vol. Eff.} = \frac{1 - \left[P_{td}\beta t\left(1 + \frac{c}{d}\right)\right]}{1 - P_{td}\beta t} - S$$

$$= \frac{1 - (1785 - 0)(0.00000305)\left[1 + \frac{127.42}{35.343}\right]}{1 - (1785 - 0)(0.00000305)}$$

$$= 0.96026 = 96 \text{ percent}$$

$$\text{Specific gravity (at 60 °F)} = \frac{141.5}{131.5 + \text{API gravity (degree)}}$$

$$\text{Bolt clamp load (lb)} = 0.75 \times \text{proof strength (psi)} \times \text{tensile stress area (in}^2)$$

$$\text{Bolt torque (ft-lb)} = \frac{0.2 (\text{or } 0.15)}{12} \times \text{nominal diameter in inches} \times \text{bolt clamp load (lb)}$$

0.2 for dry washers
0.15 for lubricated, plating, and hardened washers

Calculating volumetric efficiency for water

The volumetric efficiency of a reciprocating pump, based on capacity at suction conditions, using table of water compressibility, shall be calculated as follows:

$$\text{Vol. Eff.} = \frac{1 - P_{td}\beta t\left(1 + \frac{c}{d}\right)}{1 - P_{td}\beta t} - S$$

where

βt = Compressibility factor at temperature t (degrees Fahrenheit or centigrade). (See Tables 1 and 2).
c = Liquid chamber volume in the passages of chamber between valves when plunger is at the end of discharge stroke in cubic inches
d = Volume displacement per plunger in cubic inches
P_{td} = Discharge pressure minus suction pressure in psi
S = Slip, expressed in decimal value

Calculating volumetric efficiency for hydrocarbons

The volumetric efficiency of a reciprocating pump based on capacity at suction conditions, using compressibility factors for hydrocarbons, shall be calculated as follows:

$$\text{Vol. Eff.} = 1 - \left[S - \frac{c}{d}\left(1 - \frac{p_d}{p_s}\right)\right]$$

where

c = Fluid chamber volume in the passages of chamber between valves, when plunger is at the end of discharge strike, in cubic inches
d = Volume displacement per plunger, in cubic inches
P = Pressure in psia (P_s = suction pressure in psia; P_d = discharge pressure in psia)
P_c = Critical pressure of liquid in psia
P_r = Reduced pressure

$$= \frac{\text{Actual pressure in psia}}{\text{Critical pressure in psia}} = \frac{P}{P_c}$$

P_{rs} = Reduced suction pressure = P_s/P_c
P_{rd} = Reduced discharge pressure = P_d/P_c
S = Slip expressed in decimal value
t = Temperature, in degrees Rankine (F + 460)
T_c = Critical temperature of liquid, in degrees Rankine (See Table 3)
T_r = Reduced temperature

$$= \frac{\text{Actual temp. in degrees Rankine}}{\text{Critical temp. in degrees Rankine}}$$

$$= \frac{t}{T_c} \text{ (See Figure 2)}$$

T_{rs} = Reduced suction temperature

$$= \frac{t_s}{T_c}$$

T_{rd} = Reduced discharge temperature

$$= \frac{t_d}{T_c}$$

Vol. Eff. = Volumetric efficiency expressed in decimal value.

$$= \frac{1}{1} \times \ldots \times 62.4 \;=\; \text{density of liquid in lb per cu ft}$$

ρ_s = Density in lb per cu ft at suction pressure

Table 2
Compressibility Factor $\beta t \times 10^{-6}$ = Contraction in Unit Volume Per Psi Pressure
Compressibility from 14.7 psia at 68 °F and 212 °F and Saturation Pressure at 392 °F

Temperature				Temperature			
Pressure	**20 °C**	**100 °C**	**200 °C**	**Pressure**	**20 °C**	**100 ° C**	**200 °C**
psia	**38 °F**	**212 °F**	**392 °F**	**psia**	**38 °F**	**212 °F**	**392 °F**
6000	2.84	3.14	5.20	22,000	2.61	2.42	3.75
7000	2.82	3.10	5.09	23,000	2.59	2.38	3.68
8000	2.80	3.05	4.97	24,000	2.58	2.33	3.61
9000	2.78	3.01	4.87	25,000	2.57	2.29	3.55
10,000	2.76	2.96	4.76	26,000	2.56	2.24	3.49
11,000	2.75	2.92	4.66	27,000	2.55	2.20	3.43
12,000	2.73	2.87	4.57	28,000	2.55	2.15	3.37
13,000	2.71	2.83	4.47	29,000	2.54	2.11	3.31
14,000	2.70	2.78	4.38	30,000	2.53	2.06	3.26
15,000	2.69	2.74	4.29	31,000	2.52	2.02	3.21
16,000	2.67	2.69	4.21	32,000	2.51	1.97	3.16
17,000	2.66	2.65	4.13	33,000	2.50	1.93	3.11
18,000	2.65	2.60	4.05	34,000	2.49	1.88	3.07
19,000	2.64	2.56	3.97	35,000	2.49	1.84	3.03
20,000	2.63	2.51	3.89	36,000	2.48	1.79	2.99
21,000	2.62	2.47	3.82				

ρ_d = Density in lb per cu ft at discharge pressure = Expansion factor of liquid

$\frac{1}{1}$ = Characteristic constant in grams per cubic centimeter for any one liquid that is established by density measurements and the corresponding values of (See Table 3).

Table 3
Critical Parameters of Liquids and Characteristic Constants

Carbon Atoms	Name	T_c Degrees Rankine	P_c lb per sq in	pl/wl grams per cc
1	Methane	343	673	3.679
2	Ethane	550	717	4.429
3	Propane	666	642	4.803
4	Butane	766	544	5.002
5	Pentane	847	482	5.128
6	Hexane	915	433	5.216
7	Heptane	972	394	5.285
8	Octane	1025	362	5.340
9	Nonane	1073	332	5.382
10	Decane	1114	308	5.414
12	Dodecane	1185	272	5.459
14	Tetradecane	1248	244	5.483

Example: Find volumetric efficiency of the previous reciprocating pump example under the following new conditions:

Type of pump	3 inch dia plunger × 5 inch stroke triplex
Liquid pumped	Propane
Suction temperature	70 °F
Discharge temperature	80 °F
Suction pressure	242 psig
Discharge pressure	1911 psig

Find density at suction pressure:

$$T_{rs} = \frac{t_s}{T_c} = \frac{460+70}{666} = 0.795$$

$$P_{rs} = \frac{P_s}{P_c} = \frac{257}{642} = 0.4$$

$$\frac{\rho 1}{\omega 1} = 4.803 \text{ (From Table 3, propane)}$$

$$\omega = 0.1048 \text{ (From Figure 2)}$$

$$\rho_s = \frac{\rho 1}{\omega 1} \times \omega \times 62.4 = 4.803 \times 0.1048 \times 62.4$$

$$= 31.4 \text{ lb per cu ft}$$

Find density at discharge pressure:

$$T_{rd} = \frac{t_d}{T_c} = \frac{460+80}{666} = 0.81$$

$$P_{rd} = \frac{P_d}{P_c} = \frac{1926}{642} = 3.0$$

$$\omega = 0.1089 \text{ (From Figure 2)}$$

$$\rho_d = \frac{\rho 1}{\omega 1} \times \omega \times 62.4 = 4.803 \times 0.1089 \times 62.4$$

$$= 32.4 \text{ lb per cu ft}$$

Therefore

$$\text{Vol. Eff.} = 1 - \left[S - \frac{c}{d}\left(1 - \frac{\rho_d}{\rho_s}\right)\right]$$

$$= 1 - \left[0.02 - \frac{127.42}{35.343}\left(1 - \frac{32.64}{31.4}\right)\right] = 0.8376$$

$$= 83.76\%$$

Reprinted with permission—Goulds Pumps.

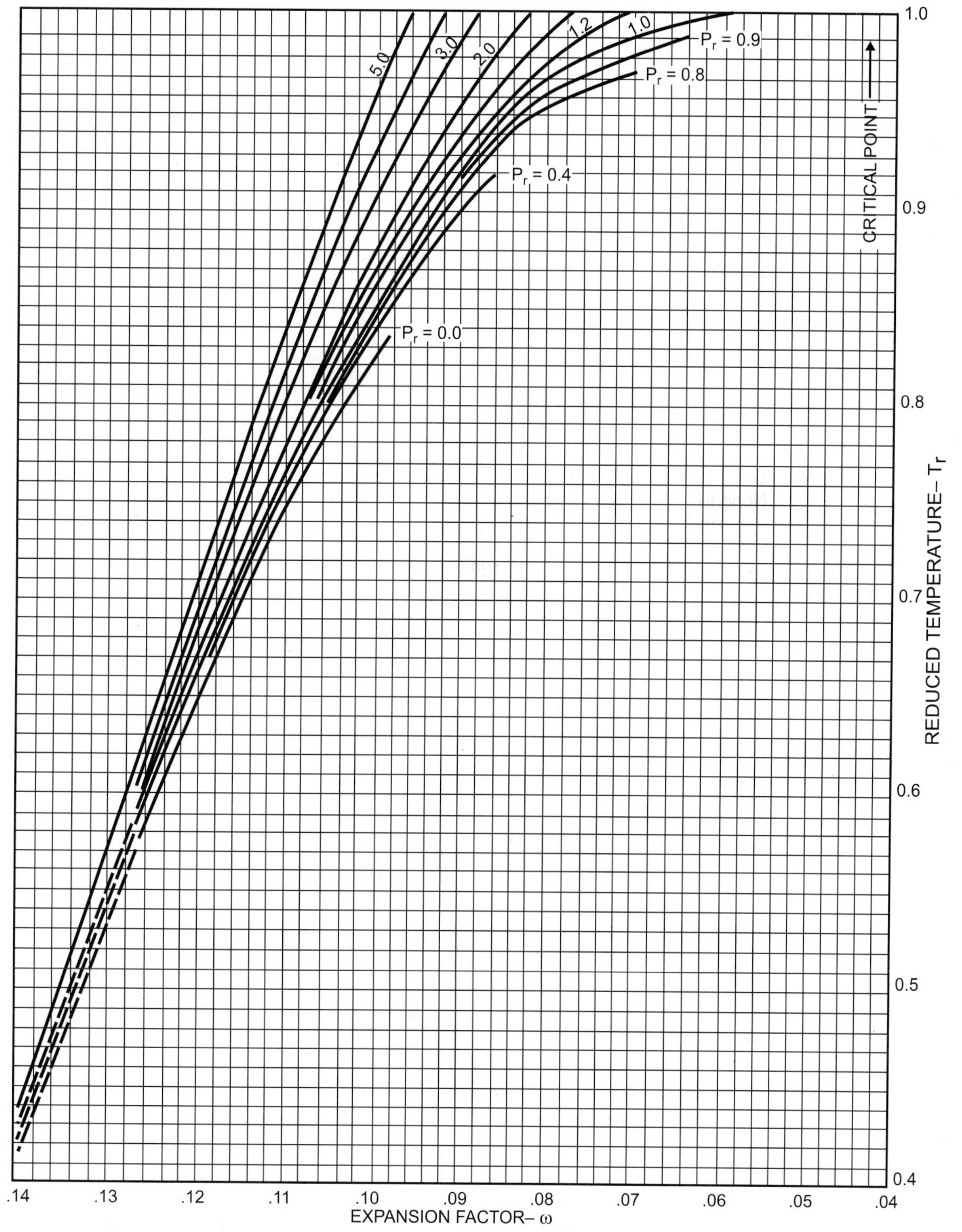

Figure 2. Thermal expansion and compressibility of liquids.

Understanding the basics of rotary screw pumps

Albert A. Zalis, Vice President Engineering, Warren Pumps, Inc., IMO Industries, Inc.

Basically only two categories of pumps need be considered when a pumping application is evaluated: centrifugal pumps and positive-displacement pumps. Centrifugal pumps depend on rotational speed to establish head and capacity characteristics and are capable of moderately high discharge heads and very high pumping capacities. Centrifugal pump design configurations include single and multistage, enclosed and open impeller, single and double suction, and horizontal and vertical.

When the application of a centrifugal pump is considered, in addition to the normal requirements of head and capacity, the pumped medium fluid characteristics must be known in detail:

- Abrasiveness
- Consistency and viscosity
- System head/capacity variations
- Net positive suction head available (NPSHA)
- Corrosiveness
- Required construction materials

Although centrifugal pumps successfully handle all types of slurries and viscous fluids, a suitable relationship between head, capacity, and fluid characteristics must exist. A simple end suction centrifugal pump with an open impeller can be used for handling both clear water and slurries; see Figure 1. Typical centrifugal pump performance curves are shown in Figure 2. Centrifugal pump performance can be evaluated using the affinity laws: With the impeller diameter held constant, capacity Q varies as the ratio of the change in speed N:

$$\frac{Q_2}{Q_1} = \frac{N_2}{N_1}$$

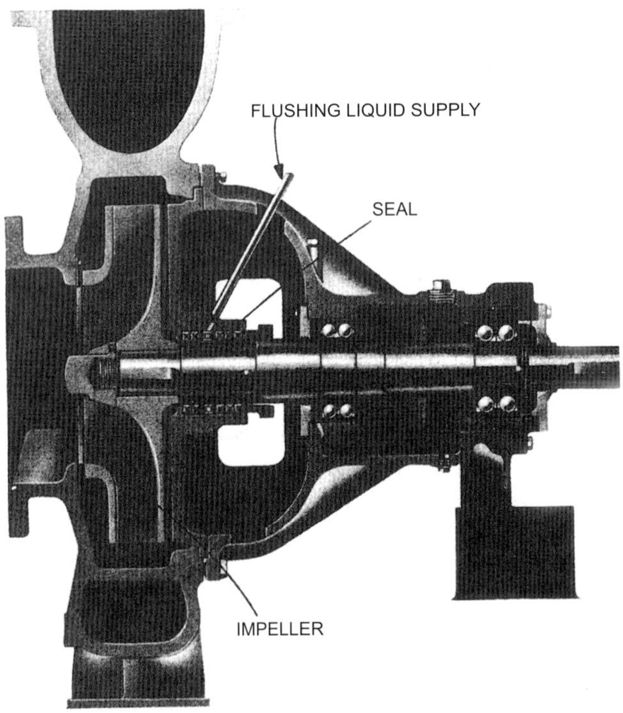

Figure 1. The end suction centrifugal pump is a general-purpose pump that is widely used for pumping a broad range of liquids. Seal may be flushed with clean liquid when corrosive or abrasive slurries are pumped.

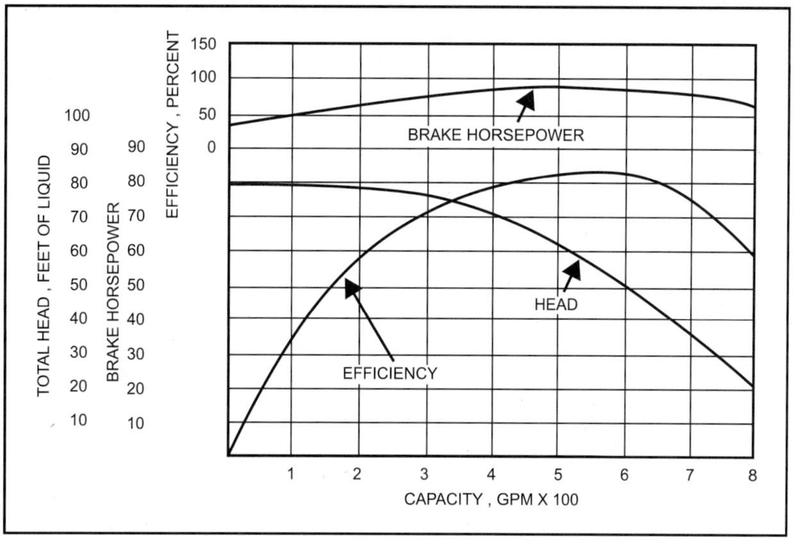

Figure 2. Typical performance curves for a centrifugal pump plot the brake horsepower, efficiency, and head (discharge pressure) as functions of capacity. The plots of horsepower and head are not straight lines, a fact that increases the difficulty of accurately predicting performance.

Head H varies as the square of the speed change ratio:

$$\frac{H_2}{H_1} = \left(\frac{N_2}{N_1}\right)^2$$

Brake horsepower (bhp) varies as the cube of the speed change ratio:

$$\frac{\text{bhp}_2}{\text{bhp}_1} = \left(\frac{N_2}{N_1}\right)^3$$

These laws make it simple to correct for varying speeds. However, these relationships assume that efficiency remains constant. Tests have shown that the correction for head, with regard to speed, is not precisely correct when head is reduced more than the amount calculated to result from a speed increase. Also, a speed increase yields a reduction in efficiency exceeding that calculated using the brake horsepower and head formula. However, the deviation of values is not great, and the affinity laws are a valuable tool when used with good engineering judgment.

Positive-displacement pumps include several pump designs. The major classifications are reciprocating and rotary pumps. Reciprocating pumps include power and steam driven, single and double acting, and plunger and piston.

Reciprocating pumps have high head capability, easily handle all types of slurries, and have high volumetric efficiency, permitting low-capacity/high-head relationships.

When the application of a reciprocating pump is considered in addition to the usual head and capacity data, the following conditions related to the fluid must be known:

- NPSHA
- Allowable pulsation levels
- Corrosiveness or required construction materials

Rotary pumps cover a broad range, including screw, gear, vane, and radial piston types.

Screw pumps can be further classified as single (progressive cavity), two screw (timed), and three screw (untimed). Gear pumps include external and internal gear designs.

Rotary pumps operate in generally the same pressure ranges as centrifugal pumps, except at lower capacities. Rotary pumps are not particularly suitable for handling severely abrasive fluids, except for the single-screw progressive cavity pump, which performs well in pumping abrasive slurries. Rotary pumps are generally capable of handling a wide range of fluid viscosities, at medium pressures, over a broad range of flow rates.

The remainder of this article is confined to a discussion of screw pumps, primarily the two-screw timed pump that utilizes timing gears to drive the screws and prevent driving contact between the intermeshing screws. This pump uses bearings to support the shafts and absorb some or all of the radial load, giving it a broad field of application suitable for pumping corrosive and noncorrosive fluids, lubricating and nonlubricating fluids, clean fluids and slurries, and Newtonian and non-Newtonian fluids.

Relatively high capacities, for positive-displacement pumps, can be handled by screw pumps. The primary limitation is the production of screws having the displacement required to produce a high discharge rate.

Performance characteristics of a screw pump are functions of screw design, provided it is used with properly designed inlet and outlet ports. In a particular pump size, performance

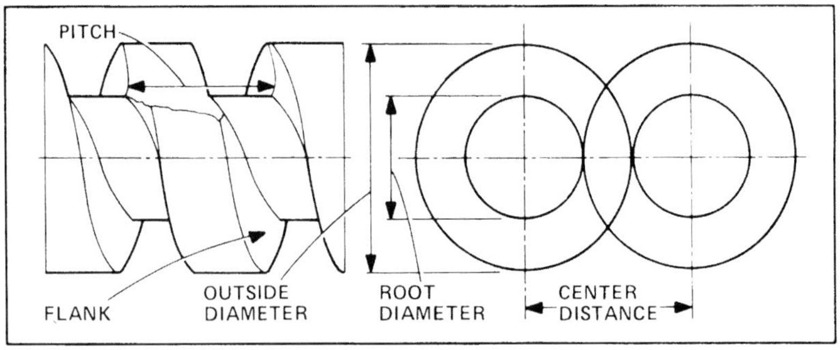

Figure 3. Four basic dimensions establish the pump displacement characteristics. One of the flanks must have an epicycloidal profile to establish a seal with the mating screw. The mating flank may have an epicycloidal curve or any form of convex curve that will achieve a sealed running mesh.

characteristics can be radically changed by altering the screw design. When a screw pump performs poorly in a specific application, the application parameters should be rechecked and the screw design examined.

The capacity of a screw pump is a function of four critical screw dimensions that control pump displacement (Figure 3). For the pump to deliver maximum displacement efficiently, the profiles of the intermeshing threads must be given careful attention. The outside diameter of the screw and the root diameter are cylindrical. Clearances of these dimensions average 0.0015 in. per in. of diameter. The flanks (thread sides) are profiled and must be carefully controlled to obtain good pumping efficiency. One of the two flanks must have a profile defined as an epicycloid. The other flank may have an epicycloidal curve or any form of convex curve that will form a continuous seal when it is mated with the meshing screw. The total axial or flank clearance normally equals the total diametrical clearance between the screw outside diameter and the casing bore.

There must be small clearances between the screws and between the screws and casing. These clearances allow a small amount of leakage from the pump outlet cavity to the pump inlet cavity. This leakage, identified as slip, is a function of pump outlet pressure, fluid viscosity, and fluid characteristics. If the screws are designed with epicycloidal curves on both flanks, the slip will be 70% greater than if the design provides for one epicycloidal curve and one convex curve. However, for certain applications the double epicycloid has distinct advantages.

In a pump design utilizing four pumping screws, the arrangement usually allows the fluid to enter the screws on each end and discharge in the center (Figure 4). When the

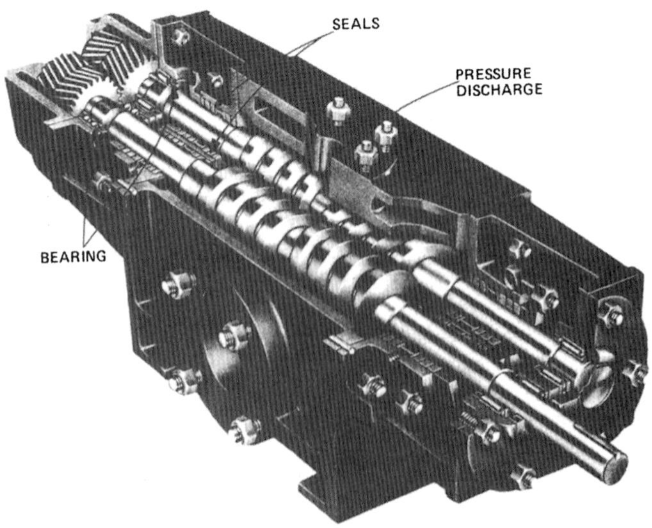

Figure 4. In a two-shaft, four-screw pump, the material enters at each end and discharges in the center. End thrust forces are balanced. Radial forces are functions of outlet pressure and screw dimensions and must be carried by the bearings at each end.

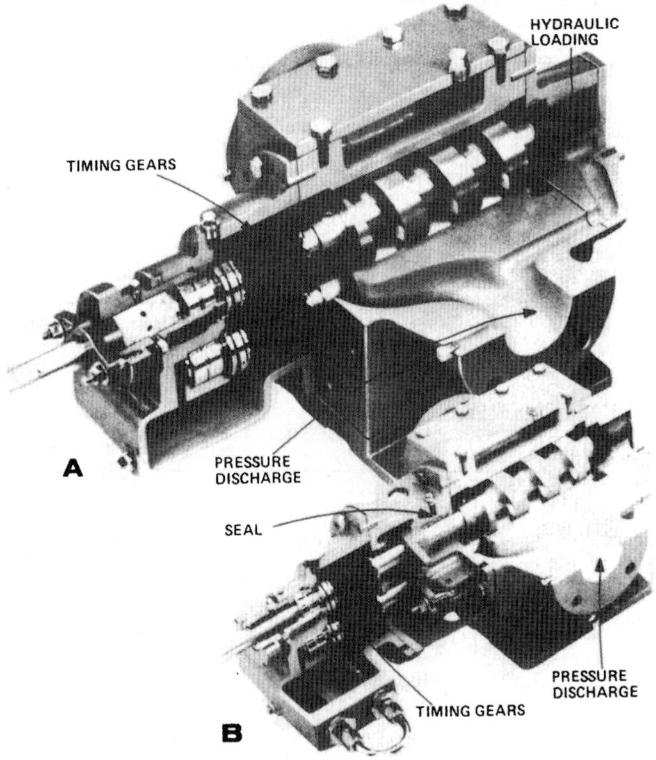

Figure 5. Pumps with two screws require timing gears to maintain constant relationship between screws. Axial forces are normally counterbalanced by hydraulic loading. The gear set may be product lubricated (a) or provided with an external lubrication system (b).

screws are all the same diameter, axial thrust forces are equal and opposite and therefore cancel each other. Although other arrangements for balancing radial forces are available, they are, at best, only partly effective and sometimes introduce other problems. Radial forces in screw pumps are functions of outlet pressure and screw dimensions. If the epicycloid/convex screw flank design is used, proper location of the profiles for the two screws on each shaft will partially balance radial loads.

Timed screw pumps are available in which only two pumping screws are used (Figure 5). In these pumps, strategic location of the flank profiles has no effect on radial forces. Axial forces are of appreciable magnitude and are normally counterbalanced hydraulically.

Screw pumps are capable of discharge pressures from atmospheric to more than 300 psig. However, the same screw design cannot be used throughout this range of pressures. In a centrifugal pump, high discharge heads are achieved either by high speeds or, more often, by staging a number of impellers to operate in series; each impeller carries a proportionate share of the discharge head.

Discharge head in a screw pump is regulated by controlling the number of screw turns. The minimum number of screw turns that can be used is two. Fewer than two turns results in erratic performance and excessive slip. Unlike the centrifugal pump impeller, each screw turn does not generate a proportionate share of the outlet pressure. The additional screw turns assist in reducing slip, thereby increasing volumetric efficiency.

A higher volumetric efficiency can mean higher mechanical efficiency, and with today's power costs it would appear desirable to use screws with many turns to achieve a high volumetric efficiency. This is not necessarily true. The greater the number of screw turns, the longer the screw (screw length equals the number of turns multiplied by the screw pitch). Screw length has a proportional bearing on the horsepower required to drive the pump at zero outlet pressure.

If a low-viscosity fluid is being pumped, this horsepower will be of small magnitude. However, the horsepower requirement increases rapidly as viscosity increases. Therefore, a slightly lower volumetric efficiency could result in

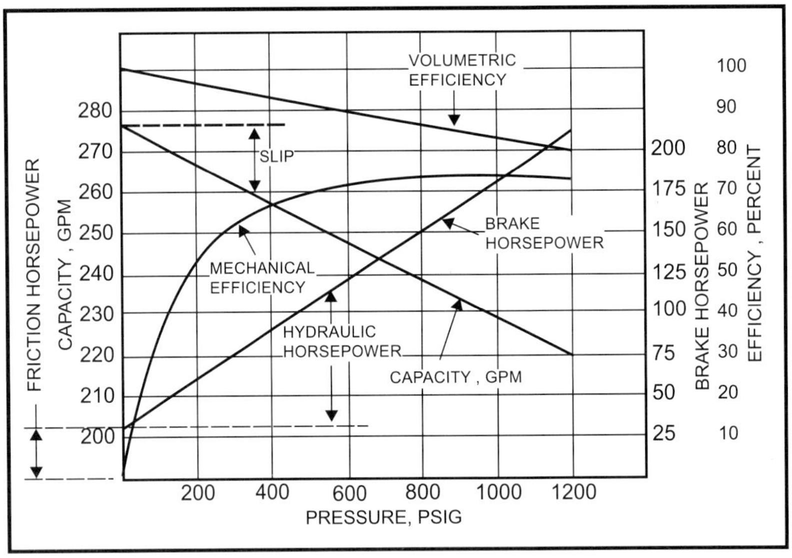

Figure 6. In contrast to performance curves for a centrifugal pump, the screw pump performance curves for capacity and brake horsepower are normally straight lines. Capacity, brake horsepower, mechanical efficiency, and volumetric efficiency are plotted as functions of discharge pressure.

a higher mechanical efficiency. Mechanical efficiency controls power costs; consequently, the application engineer, when selecting a screw pump, must consider both slip and friction horsepower to determine the best arrangement.

The remaining parts of a screw pump are subject to conventional mechanical design: Shafts must be designed for bending and torsional loads, and bearings must match the center distance between the two parallel shafts. Timing gears should be designed in accordance with American Gear Manufacturers Association standards. If spur gears are used, both shafts should be rigidly located axially. If herringbone gears are used to reduce noise level, the driven shaft should be allowed to float axially and be positioned by the apex of the gear. Velocities through the fluid inlet passages should be kept low, normally not more than 5 fps for average viscosities of 150 to 1000 SSU and decreased as the viscosity increases.

The performance characteristics of a screw pump are similar to those of other positive-displacement pumps. Capacity, horsepower, and efficiency are plotted as functions of discharge pressure. For centrifugal pumps, horsepower, efficiency, and discharge pressure are plotted as functions of capacity. With centrifugal pumps, plots of horsepower and head against capacity are not straight lines. Therefore, accurately predicting performance is difficult, and more testing and modeling are required. For positive-displacement pumps, plots of capacity and brake horsepower against pressure normally result in straight lines when coordinate paper is used. However, there are instances in which the capacity versus pressure line is slightly curved. This characteristic usually results from excessive clearance in the pump permitting deflection and eccentric operation of the rotating screw.

Performance conditions can be manipulated to reflect operation at speeds or pressures other than those shown in Figure 6, using these fundamental relationships:

Friction horsepower is the power required to overcome losses including packing seal friction, timing gear inefficiencies, bearing friction, and pumped fluid viscous shear. All of the losses are so small when compared to the viscous drag of the pumped fluid that it can be safely assumed that friction horsepower is entirely dependent on fluid viscosity and pump speed. Even pump rotor clearances, within the limits of efficient pump operation, have such a minor influence on the power consumed that extremely sensitive instrumentation is required for taking measurements.

Figure 7 shows the relationship between friction horsepower and viscosity at 1200 rpm. If, for example, it is necessary to determine friction horsepower fhp at 1800 rpm, the following relationship exists:

$$\text{fhp}_2 = \text{fhp}_1\left(\frac{1800}{1200}\right)^{1.5}$$

With today's programmable calculators and computers, the curves shown in Figure 7 may be easier to work with if they are reduced to mathematical formulas. Because the equation will take the form $x = ay^n$, by means of simultaneous equations the friction horsepower equation becomes

$$\text{fhp}_2 = 2.9541 \times (\text{viscosity in SSU})^{0.42065}$$

Hydraulic horsepower is the measure of the total hydraulic work done by the pump. This term should not be confused with water horsepower or work horsepower, the measure of the useful work done by the pump.

These terms can be defined by the following formulas:

$$\text{hydraulic horsepower} = \frac{(\text{displaced gpm})(\text{psi})}{1714}$$

$$\text{water horsepower} = \frac{(\text{displaced gpm})(\text{psi})}{1714}$$

Displaced gpm is the capacity at zero psi discharge pressure. Delivered gpm is the capacity at the rated or required discharge pressure, or the displaced gpm minus the slip gpm.

Referring to Figure 6, the brake horsepower is the sum of the friction horsepower and the hydraulic horsepower. This approach yields a fairly accurate brake horsepower through a wide range of operating conditions.

Slip is the loss of capacity from the higher pressure area through the internal clearances. Slip is a function of viscosity and differential pressure. Figure 7 is a slip/viscosity curve for 500 psi operating pressure. Slip varies in direct proportion to the change in pressure. Similarly, with the friction horsepower curve, slip can be expressed by the following formula:

$$\text{slip, gpm} = 294.95 \times (\text{viscosity in SSU})^{-0.4660}$$

Mechanical efficiency is a calculated quantity that until recently received very little attention. The energy crisis has made it more important. Mechanical efficiency is calculated by

$$\text{mechanical efficiency} = \frac{(\text{displaced gpm})(\text{psi})}{1714(\text{bhp})} \times 100$$

Volumetric efficiency is an infrequently used measure of internal slip and can be calculated by

$$\text{volumetric efficiency} = \frac{\text{delivered gpm}}{\text{displaced gpm}} \times 100$$

It may be well to review a suggested approach to the pumping of non-Newtonian fluids. One fairly common application for screw pumps is pumping pseudoplastic polymers of various types. A pseudoplastic fluid is one whose viscosity decreases as the shear rate in the pump increases. Data on pseudoplastic fluids are usually presented as a curve plotting apparent viscosity versus shear rate in reciprocal seconds. The determination of shear rate in a positive-displacement pump is a widely discussed topic. It is difficult, if not impossible, to arrive at one formula that is applicable to

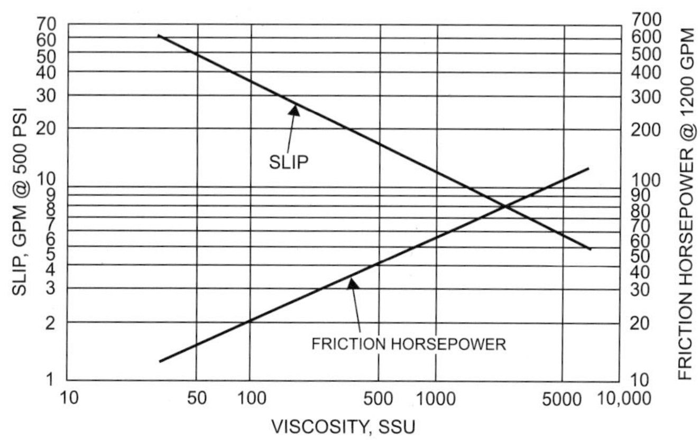

Figure 7. Slip is a loss of capacity from the higher pressure area to the lower pressure area and is a function of viscosity and pressure differential. Slip varies in direct proportion to the change in pressure. At a specific pressure as the viscosity increases, slip decreases, while the friction horsepower increases.

all pumps. For the timed screw pump, successful application has been made using the following formula:

$$\text{shear rate, s}^{-1} = \frac{\pi(\text{rotor diameter})(\text{rpm})}{(\text{radial clearance})(60)}$$

where

rotor diameter = Pumping screw diameter, inches
radial clearance = Clearance between screw diameter and casing bore, inches

This formula gives consideration to only the area of maximum shear in the pump. Experience has shown that refining this formula to include all shear surfaces yields values that are grossly incorrect. The shear rate calculated from the formula will permit a satisfactory pump selection; however, care should be taken to consider the shear rate in the inlet piping. This relatively small quantity results in a much higher viscosity and thereby influences the inlet pressure requirements.

Sometimes it becomes necessary to take a surplus pump and apply it to a new application. This often happens in chemical plants where processes are frequently subject to change. Earlier discussions dealt in some detail with screw profile variations, importance of screw turns, slip, and friction horsepower. All these factors must be considered for the new application. Performance characteristics can be changed radically by installation of a new rotating element in an existing casing.

This article has not covered all there is to learn about screw pumps and their applications but has discussed the more important areas on which there is little, if any, literature published.

How to evaluate VFD speed on hydraulics

Dr, Lev Nelik, P.E., APICS Pumping Machinery LLC, Atlanta, Georgia

Few would dispute that variable frequency drives (VFDs) save energy, but the exact amount depends on the system. Hydraulically speaking, the main difference between a variable frequency (speed) drive and a discharge valve is that a VFD only changes a *pump* curve, while a valve only changes a *system* curve. A pump operates at the intersection between its H-Q curve and a system curve, and a change in either moves the operating point to a new intersection.

Consider a centrifugal pump operating at 4000 gpm and producing 300 ft of head at 1800 rpm. According to its curve (Figure 1), pump efficiency is 82 percent. This happens to be at the best efficiency point (BEP) point, but, in general, the same logic would apply to any operating point.

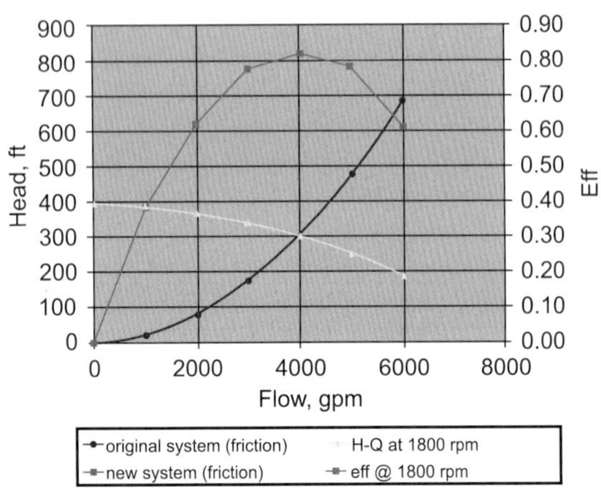

Figure 1. Flow control by throttling discharge valve.

Suppose we want to reduce the flow by 50 percent, to 2000 gpm. By closing the discharge valve, we would change a system curve, which would intersect the *same pump curve* at 370 ft of head, 61 percent efficiency. Such control process is simple, quick and easy. However, it has other issues, which we will review. Hydraulically, the system in Figure 1 is referred to as *predominantly friction* (no static head). Such system is ideal for VFD application for flow control. The primary reason is at the heart of the affinity laws, which state

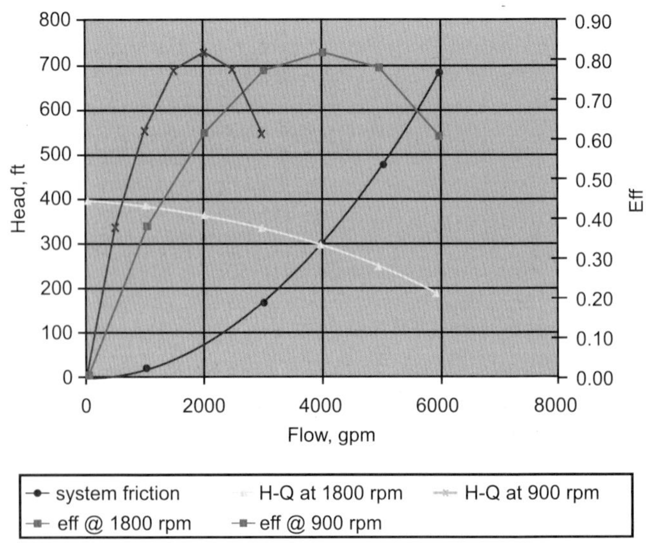

Figure 2. Flow control by VFD speed reduction, friction system head example.

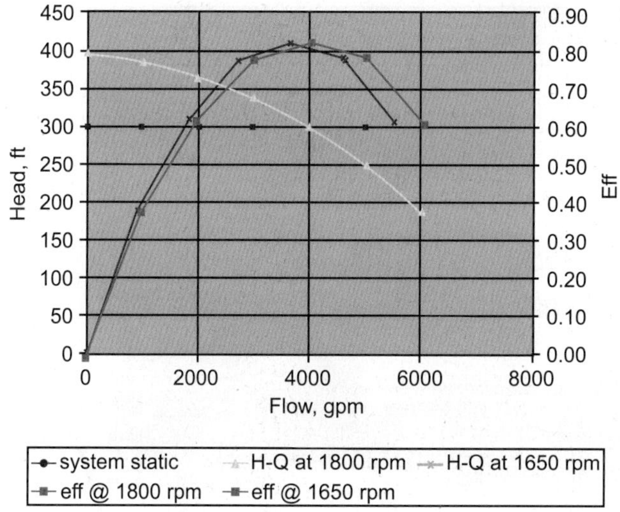

Figure 3. Flow control by VFD, static system head example.

that pump flow changes directly with the speed ratio, head as a square and power as cube. Therefore, the relationship between head and power as cube. Therefore, the relationship between head reduced as a square of a speed ratio (900/1800)^2, i.e. 0.25 x 300 = 75 ft. Notice that efficiency does not change as the entire efficiency curve "slides" to the left and remains at a peak of 82 percent.

What would happen if a pump operates against constant (static, or mostly static) head, as in a lift station application? Consider Figure 3.

In Figure 3, 300 ft is a constant static head, with friction losses assumed to be negligible in comparison. A range of RPM variation is significantly reduced, so the entire pump *H-Q* curve does not drop below the contact $H = 300$ ft. By applying affinity laws, we will find RPM = 1650, a new pump curve *(H-Q)* intersects a constant 300-ft head line at desired 2000 gpm and efficiency at that point is 62 percent.

Table 1 compares these three cases.

Table 1
Water Compressibility Comparison of Valve Throttling, VFD (Friction) and VFD (Static Head) Examples

	Throtting	VFD (Friction)	VFD (Static)
rpm	1800	900	1650
Flow, gpm	2000	2000	2000
Head, feet	370	75	300
Eff. %	*61%*	*82%*	*62%*
Power, hp	306	46	244
Energy cost, $/kW-h	0.10	0.10	0.10
Energy used per year, $	$200,193	$30,187	$159,701
BEP flow, gpm	4000	2000	3667
Operation %BEP	50%	100%	55%

Observations

- For *friction-dominated* systems (long pipes, flow transfer cases), a VFD saves a substantial amount of energy, and operates pumps reliably due to close proximity to BEP flow (100 percent in example shown)
- For *static-dominated* systems (injection against constant pressure, lifting against constant head), the energy savings are substantially less, and the pump surprisingly operates substantially *off-BEP* position, not significantly different from a valved flow control case
- For cases where both systems are present, an in-between scenario would result. VFD drives are expensive devices and require certain knowledge of controls, proper electrical and electronic maintenance and care. Flow control by VFD requires programming and training and is more complex than simple control by throttling a valve. These are some downsides to VFDs, but they offer the convenience of remote control (as do some valves), energy savings when operated in friction-dominated systems and more reliable pump operation. Operating closer to the BEP means less loads, better seal life, lower vibrations and improved suction recirculation conditions.

The user must have a detailed understanding of a system's hydraulics, unique *H-Q* curves and flow control requirements to make sure the VFD investment is justified. Organizations implementing VFDs must be prepared to deal with the added complexity but also appreciate their usefulness. The anticipated benefits of a VFD hinge on an organization's efforts to train users to operate and maintain such systems. For organizations leaving these issues to chance, the anticipated benefits can quickly become headaches.

Reprinted with permission—Dr. Lev Nelik, P.E., APICS Pumping Machinery, LLC, Atlanta, Georgia, www.PumpingMachinery.com and *Pumps & Systems Magazine*.

Progressive cavity pumps

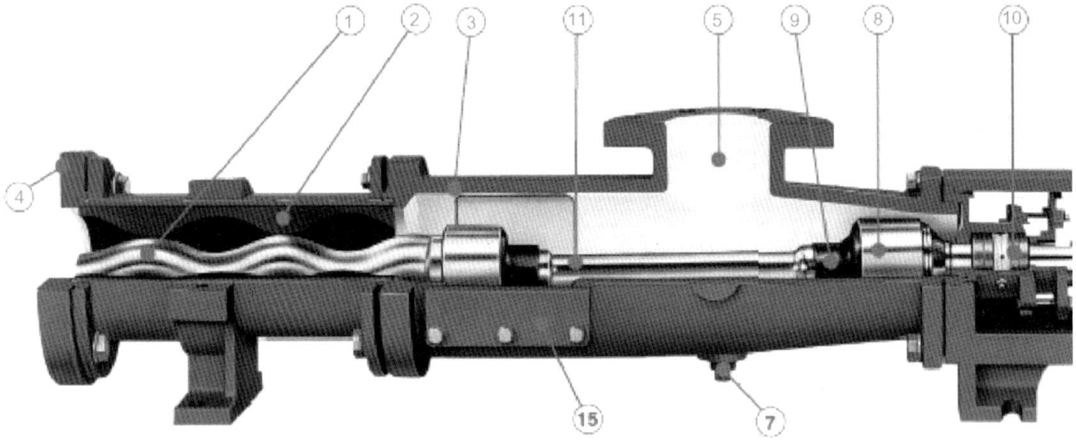

Progressive cavity pump parts

A progressive cavity (PC) pump is a positive displacement pump but does not rely on valves for operation. The parts identified in the above figure are as follows:

1—Rotor
2—Stator (usually elastomer material)
3—Wettable body parts
4—Pump discharge flange (outlet)
5—Pump suction flange (inlet)
7—Drain plug to drain suction housing
8—Universal joint
9—Universal joint seal
10—Pump packing
11—Connecting rod
15—Inspection port

They are especially suited for tough, nasty, multiphase fluids with gas and solids in suspension, at relatively moderate temperatures. PC pumps are ideal for shear sensitive liquids. They are capable of producing reasonably accurate repeatable flow and the output is fairly independent of head. Operating without valves, the PC delivers its output free of pulsations. The PC pump will not air lock and because it does not rely on valves as a conventional positive displacement pump, the NPSHr is low.

The stator is made of an elastomer and the rotor can be made of carbon or stainless steel. Carbon steel rotors are usually coated for added wear resistance. PC pumps have a single shaft that requires sealing.

This type of pump transfers fluid by means of the progress, through the pump, of a sequence of small, fixed shape, discrete cavities, as its rotor is turned. This leads to the volumetric flow rate being proportional to the rotation rate (bidirectionally) and to low levels of shearing being applied to the pumped fluid. Hence these pumps have application in fluid metering and pumping of viscous or shear sensitive materials. It should be noted that the cavities taper down toward their ends and overlap with their neighbors, so that, in general, no flow pulsing is caused by the arrival of cavities at the outlet, other than that caused by compression of the fluid or pump components.

PC pump difficulties

Care must be exercised in selecting the stator material to be sure that it is compatible with the fluid being pumped. Erosion is another difficulty. When erosion is a problem rotor material and/or rotor coating that is harder than the base metal must be selected to withstand erosion. The pump should not be allowed to run dry as this may heat the stator above its maximum operating temperature. In certain instances, a temperature probe may be inserted

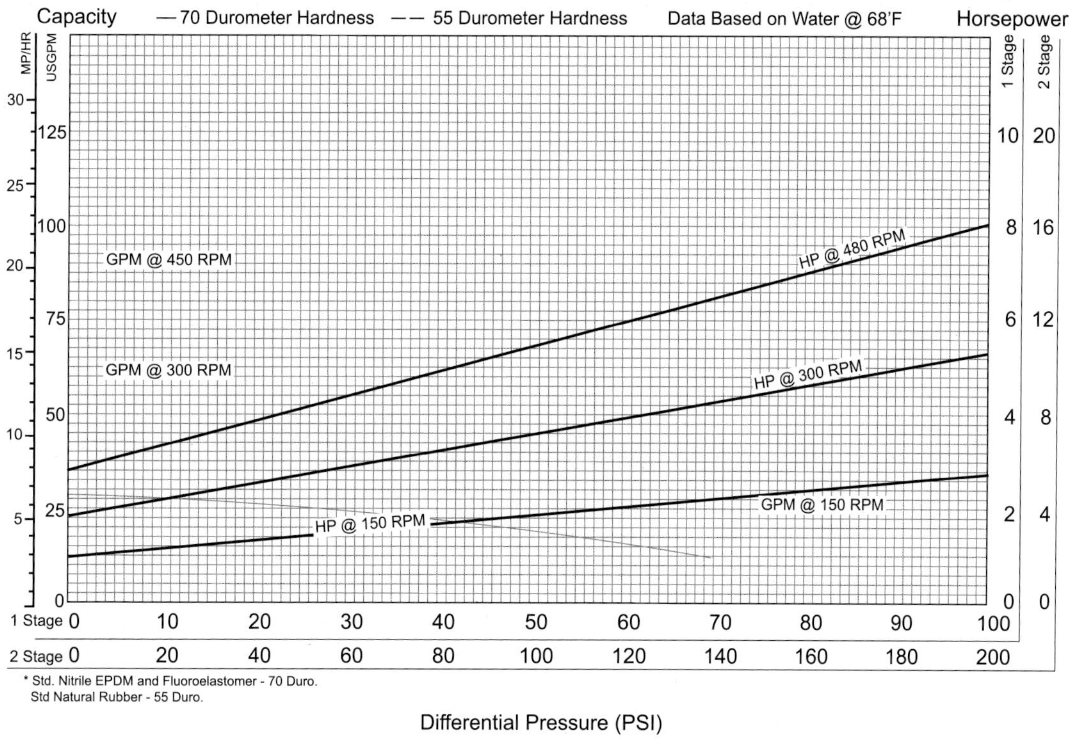

into the stator and used to shut the pump down before a temperature that can damage the stator is reached. Flow switches may also be used to detect "no flow" and shut the pump down.

The pumping rate can be varied over a range by controlling the speed of rotation of the rotor. A variable frequency drive is well suited for use with at PC pump. When selecting the VFD, the application should be carefully specified so that the driver will supply the required torque at start up as the start up torque is greater than the running torque.

Pump curve accuracy

George Stephens, Engineered Software, Inc., Lacey, Washington

Pump manufacturers present their pump performance data electronically. These digital pump catalogs can be found online, within stand-alone pump selection software, and through select hydraulic analysis software packages. Thousands of engineers worldwide use these catalogs to evaluate and select the pumps that best meet their needs.

Based on experience in digitizing more than 135,000 pump curves for manufacturers throughout the industry, this article outlines how digital pump curves are created, as well as some of the accuracy issues that may arise from the methods of curve digitization. It covers pump performance curves and the accuracy ramifications from different digitization methods. As the industry moves toward electronic pump catalogs, end users should know how accurate they are and how they are generated.

Pump selection programs

The main function of a pump selection program is the ability to digitally reproduce the manufacturer's pump performance data accurately and to create a selection list of pumps that meet the customer's criteria. All pump manufacturers have "paper curves" representing the performance of their pumps, and most also have these curves available in pdf format. These curves reproduce the manufacturer's bench test data and are the most accurate depictions of the pump's performance.

However, for a software program to understand and use this pump performance data, the data must first be converted into a form that can be read by the program. This is referred to as digitization of the pump curves. There are various methods of digitizing pump performance data, but occasionally, depending on the method used, some vital information can get lost in the translation. The accuracy of this translation can be critical to the pump selection and evaluation process.

To better understand the performance of the digitized curves, consider an example of a typical published pump curve from a manufacturer. Figure 1 shows a detailed curve that includes the flow versus head performance curves as well as isometric curves for efficiency, power, and net positive suction head required (NPSHR). These curves accurately represent the test performance of the indicated pump in accordance with the *ANSI/HI 1.6-2000 Centrifugal Pump Tests* standard.

If a pump were required for system design conditions of 800 U.S. gallons per minute and 35 feet of head and this particular pump were being evaluated, the system designer could feel confident that this pump with a 10-inch impeller would successfully fit the bill. The pump selection and evaluation software must accurately reproduce the manufacturer's published pump curve to aid the customer's evaluation of the pump for the application. The key to generating accurate digital curves is the method with which the pump performance data are saved and displayed.

All forms of pump curve digitization involve obtaining data points from the pump manufacturers. These points can either be extracted from the published paper curves or gathered directly from the manufacturers.

One simple method is to take several points along the performance curve and perform a polynomial expression describing the performance curve, which can be saved and used in a pump selection program.

The original data pints are then discarded, and the pump performance data are now represented solely by a curve drawn from that equation. Depending on the shape of the original published curve, this polynomial expression may or may not be an accurate representation.

A second, more accurate method is to collect as many data points as necessary to accurately fit the shape of the published performance curve and use these saved data points to represent the original published curve in the pump selection program. Storing the original data points is an integral step in maintaining the pump curve's accuracy, and interpolation between those data points is the most accurate method of properly sizing a pump.

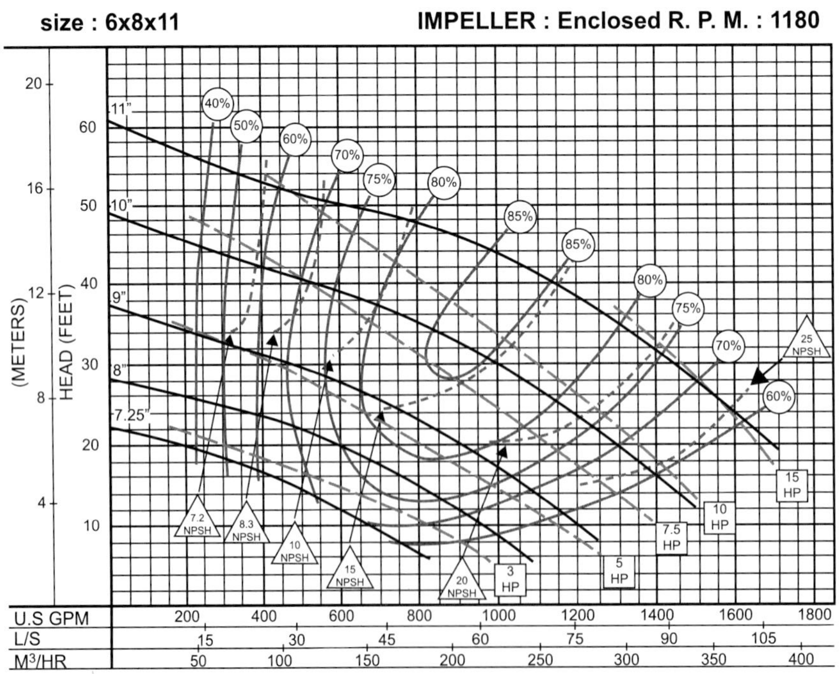

Figure 1.

When it comes to displaying the pump curve in the software, a simple polynomial regression curve can have a tendency to miss the original pump curve at critical points. On the other hand, if a more robust numerical analysis is performed, such as a cubic spline regression, then a curve can be drawn that precisely represents the original published curve.

Polynomial regression

A closer look at some actual pump curve data illustrates the problems associated with the simple polynomial regression method of curve digitization. Figure 2 depicts flow versus head performance curves, which are the primary search parameters for sizing a pump. The black curves are the actual tested performance curves provided by a pump manufacturer. Overlaid in pink are the performance curves generated by a pump selection tool using a simple polynomial regression. The inset is a close-up view of the anomalies in this case study.

Notice that the pump performance curves with anomalies, such as inflection points, are not well represented by the polynomial regression. The polynomial regression has a tendency to smooth out the anomalies such as those shown in Figure 2. The consequences of this curve characterization can be minor, or they can be significant. In this case, if an end user were sizing for a design point of 300 U.S. gallons per minute, at a total head of 126 feet, then this pump might not even be considered for evaluation.

A more detrimental scenario could occur if the polynomial regression curve rose significantly above the actual curve. In this case, the program may suggest a viable pump that might not meet the design point criteria. In addition, if the pump performs in areas where the polynomial regression curve deviates from the manufacturer's published performance curve, an end user will get erroneous energy usage figures and operating costs.

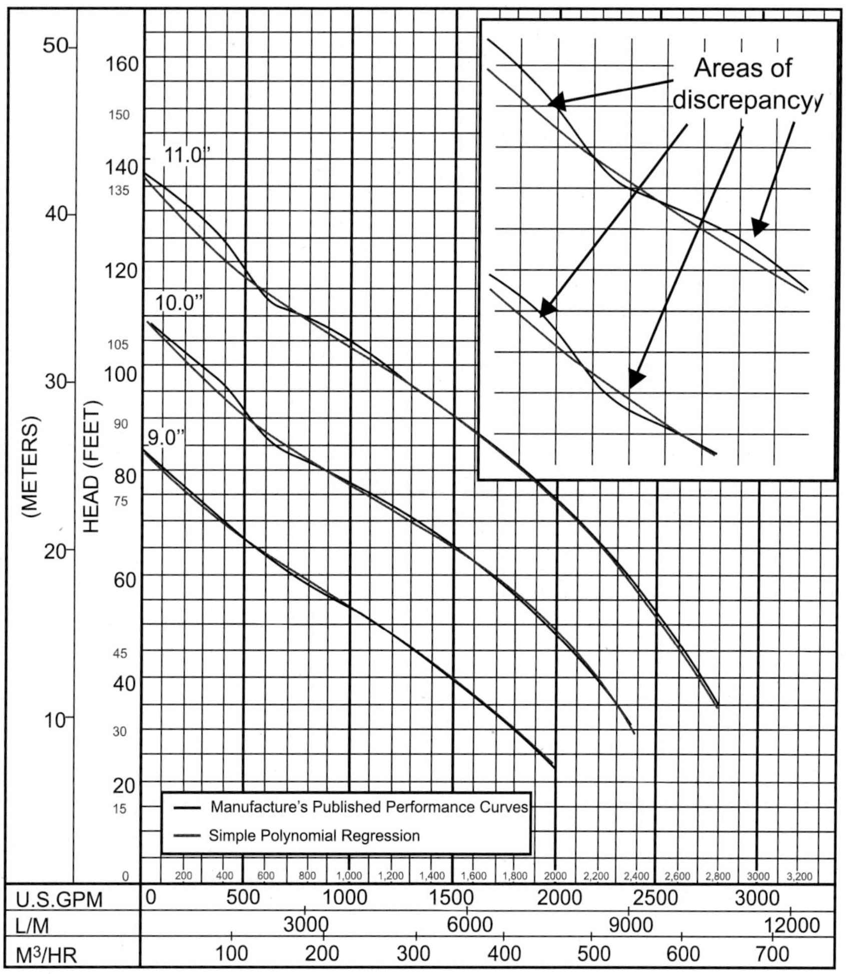

Figure 2.

Cubic spline regression

Figure 3 depicts the same manufacturer's performance curves in black. The green overlaid curves are generated by a pump selection tool that uses a cubic spline regression analysis. This method forces the curves to pass through every collected data point. It then smoothes the curves from point to point to provide clean, accurate pump performance curve displays. Flow and head are never significantly overstated or understated at any point in the curve range. End users can use these displays with confidence when performing pump evaluations.

Even the best numerical analysis techniques cannot predict anomalies between data points. If the performance curve to be digitized has these types of dips or abnormalities, it is imperative that enough data points are collected in and around that range. At a minimum, data points should be collected at the edges and at the peak (or valley) of the anomaly. Ideally, a couple more points should be collected between them. This allows the selection tool to display the most accurate representation of the original performance data for the valuation and selection process to proceed without any false selections.

Power and efficiency curves

In addition to the standard flow versus head performance curves, many manufacturers also include curves for pump power and pump efficiency. It must be stressed that pump efficiency and pump power are function of each other. The formulas relating power and efficiency are

$$\text{bhp} = \frac{Q \times H \times \rho}{247{,}000 \times \eta_p} \quad \text{and} \quad \eta_p = \frac{Q \times H \times \rho}{247{,}000 \times \text{bhp}},$$

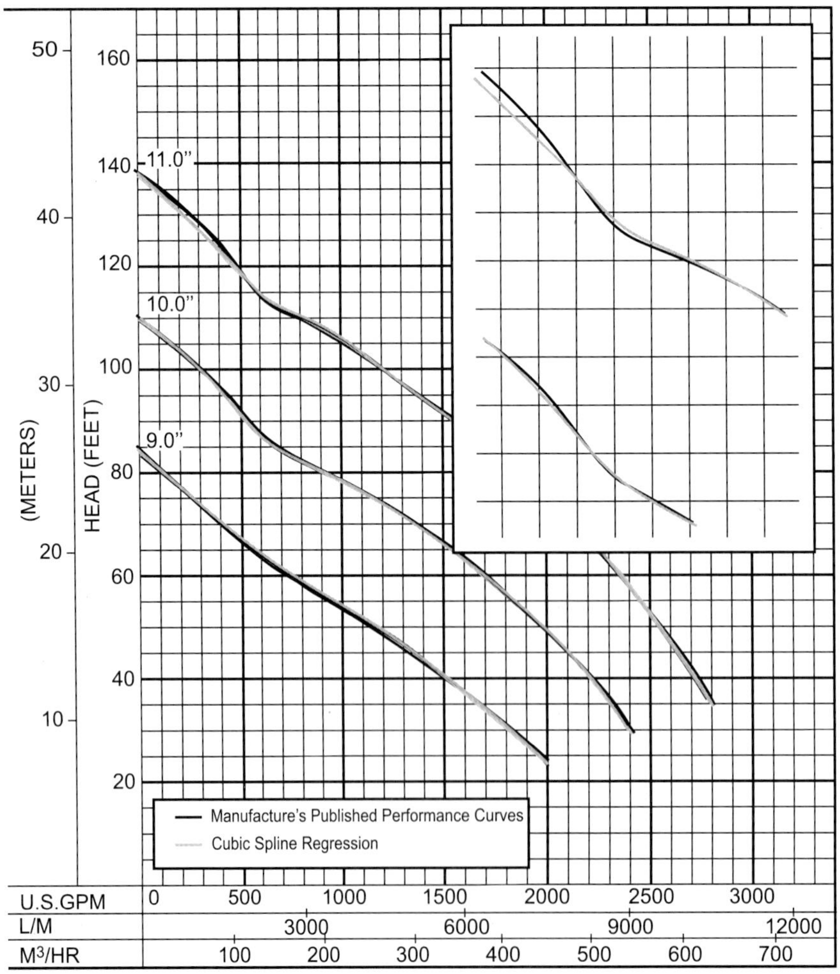

Figure 3.

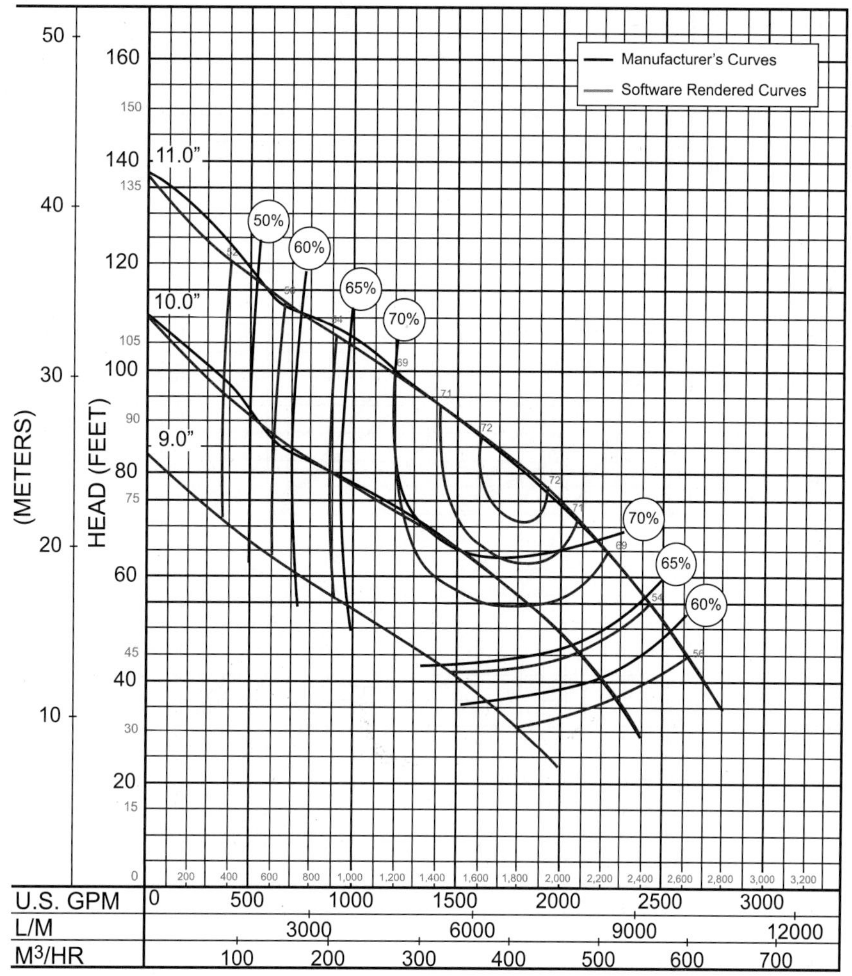

Figure 4a.

where

bhp = Brake horsepower,
η_p = Pump efficiency (expressed as a decimal),
Q = Flow rate (gpm),
H = Total head (ft), and
ρ = Fluid density (lb/ft^2).

Power is calculated with efficiency as a parameter and vice versa. To reproduce accurate and trustworthy results, one set of data must be gathered from the pump manufacturer, and the other set of data must be calculated. When generating curves, it is *not* accurate for a software program to try to collect and reproduce both sets of data from the manufacturer. This would leave open the possibility of errors. An end user might select a flow, head, and efficiency from the graph and then perform a hand calculation of the corresponding power to find that the calculated power does not match up with the power shown on the graph.

For this reason, only one set of empirical data, efficiency *or* power, should be taken from the manufacturer. The other should be calculated. Also, the efficiency data reported by most pump manufacturers typically has much more granularity than the power data. That is to say, there is typically a wider range of efficiency values over the range of pump operation than the range of power values. For this reason, it is usually more accurate to digitize the manufacturer's efficiency data, and then calculate the power data, than the other way around.

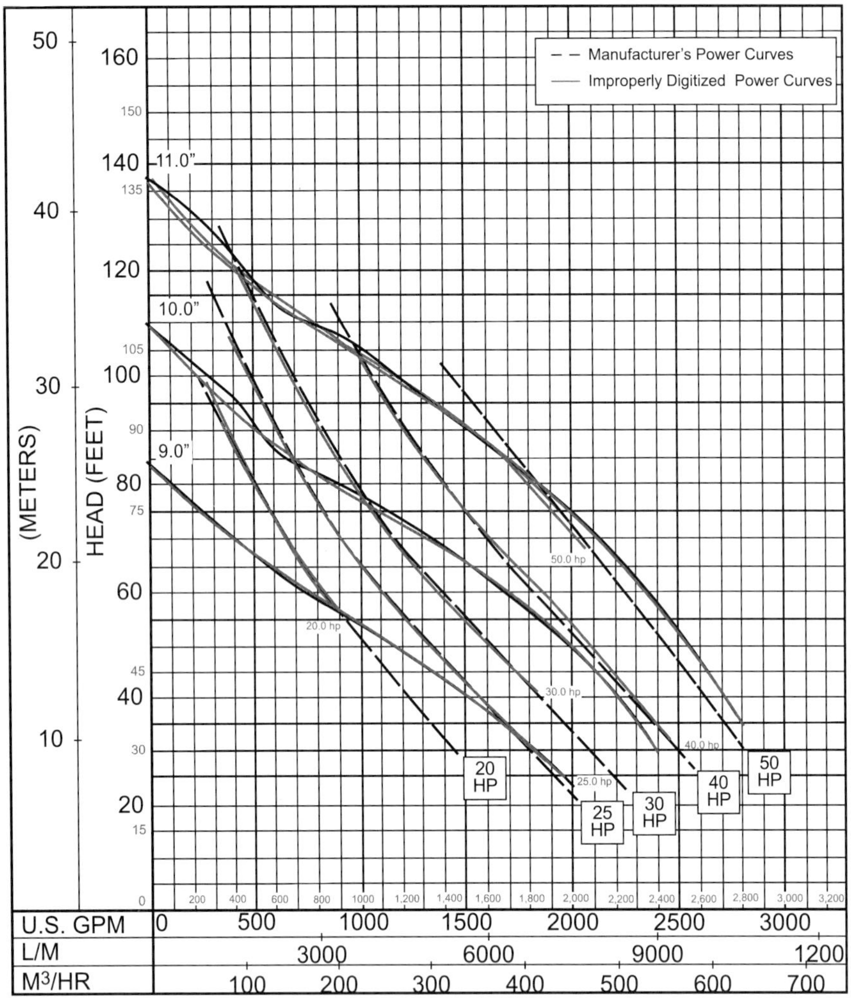

Figure 4b.

Example. To see the effect that digitizing can have on power and efficiency curve generation, consider the curves in Figures 4a and 4b. In this example, the pump selection program attempted to digitize the isometric power curves and then calculate isometric efficiency curves from the power. Notice that the software rendered efficiency curves in Figure 4a are well off the mark, particularly in and around the 70%-and-up operating range. Not only are the isometric efficiency curves misaligned, but the percentages clearly do not match up with the published curve. The pump's overall efficiency is overstated by up to 3%.

If a customer were to download these performance curves (Figure 4a in black) from the pump manufacturer's technical literature and then pull up the performance curves (Figure 4a in pink) from the pump selection software, it would be virtually impossible to tell that the curves are from the same pump.

Example. Now consider the curves in Figures 5a and 5b. In this case, the pump selection properly digitized the efficiency curves from the manufacturer using a cubic spline regression analysis. The isometric efficiency curves are dead accurate at every point in the graph, and the

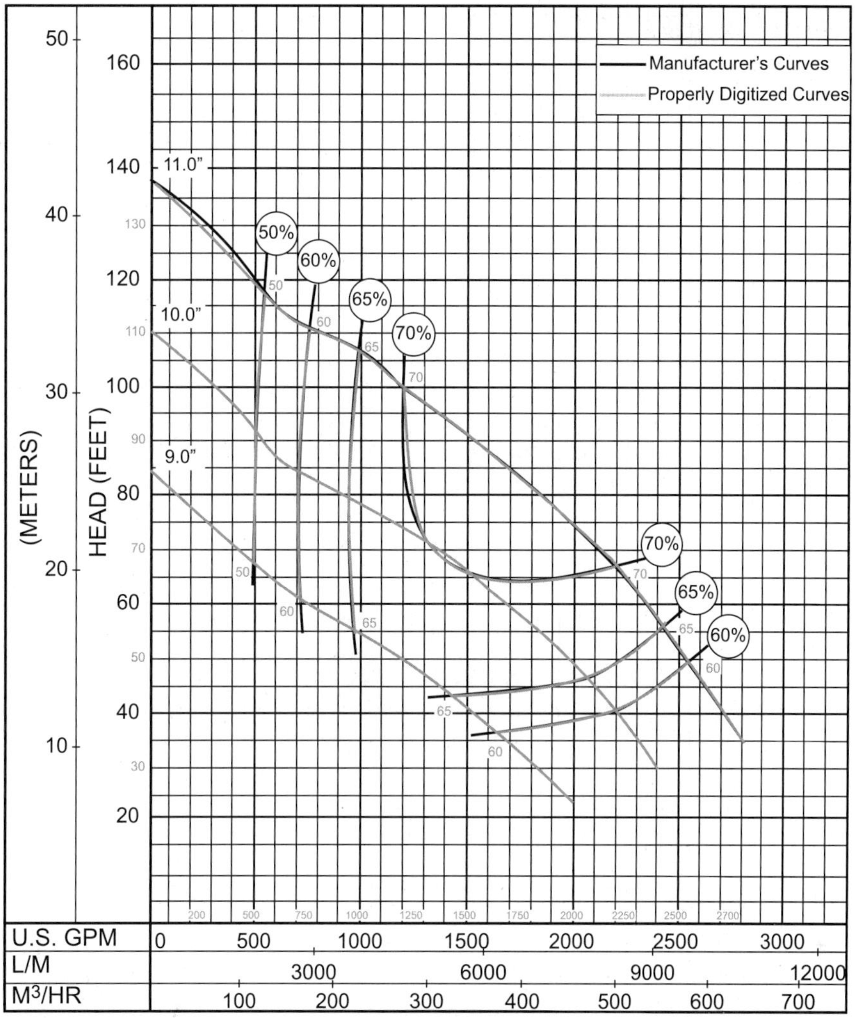

Figure 5a.

efficiency is never overstated. In Figure 5b, the power curves also track extremely well with the manufacturer's, given that they are calculated power curves, and more importantly, the endpoints at the run out flow rates are accurate.

Conclusion

Using a pump selection program that accurately represents the manufacturer's pump performance is a necessity.

Inaccuracies in curve digitization and rendering can lead to improper pump and motor selections as well as faulty operating cost calculations. Users should be aware of the digitization techniques employed by their pump selection software and the information in these articles should aid them in making an informed decision when choosing a software package.

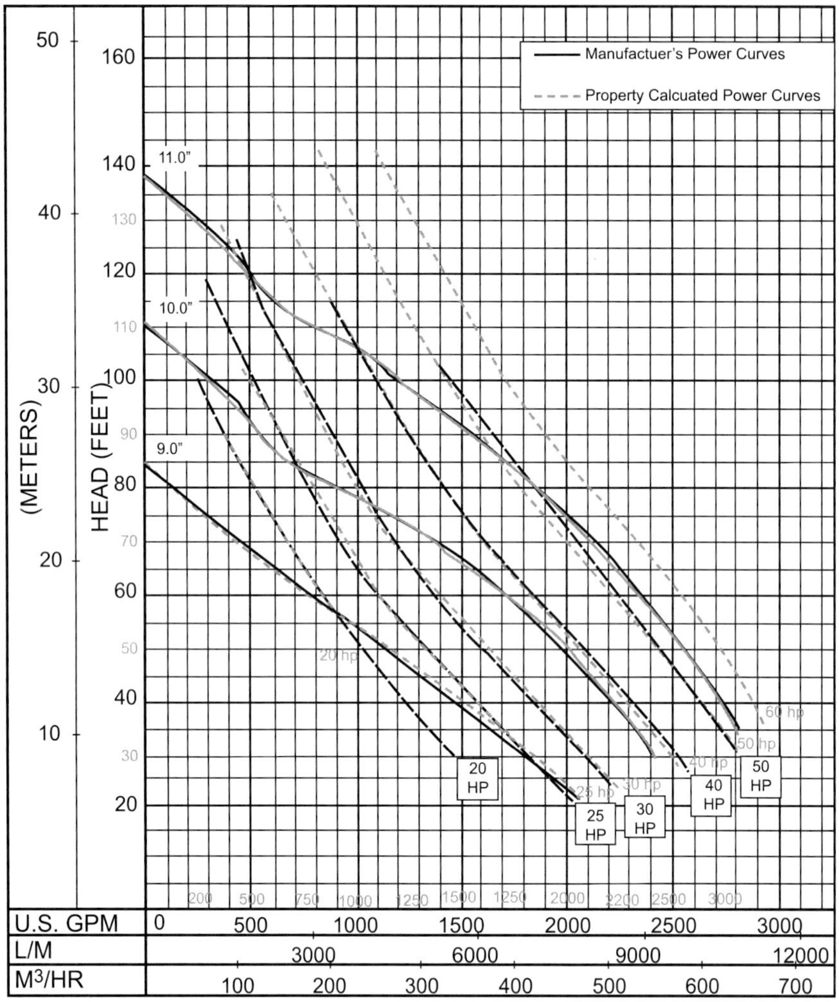

Figure 5b.

The basics of NPSH and pump operating regions

Peter Gaydon, Hydraulic Institute

The terms net positive suction head (NPSH), preferred operating region (POR) and allowable operating region (AOR) are common in the pumping industry, but many pump users do not fully understand the concepts or their impact on pump reliability. This article is aimed at defining the terms so that pump users understand their meanings. Users can also learn about resources available to better understand the requirements and why it is important to operate a pump within its POR and with adequate NPSH margin.

Industry standards

To address these two important issues of required NPSH margin and operating regions, the Hydraulic Institute published and recently revised two American National Standards covering the topics, providing recommendations on the NPSH margin required for pumps by market used design and the POR for pumps of different designs.

- ANSI/HI 9.6.1-2017 Rotodynamic Pumps – Guideline for NPSH Margin

- ANSI/HI 9.6.3-2017 Rotodynamic Pumps – Guideline for Operating Regions

Pressure and NPSH terminology

To understand these points, the terminology used to describe energy in a system is critical. In the rotodynamic pump industry, energy is described by total head in meters (m) or feet (ft) added to a system. There is a direct relationship between head and pressure. If we focus on U.S. customary units, 1 pound per square inch (psi) is equivalent to 2.31 feet of water at 68 F (specific gravity = 1.0). To adjust for other liquids or water at a different temperature, specific gravity is used and 1 psi is equal to 2.31/(specific gravity) (feet).

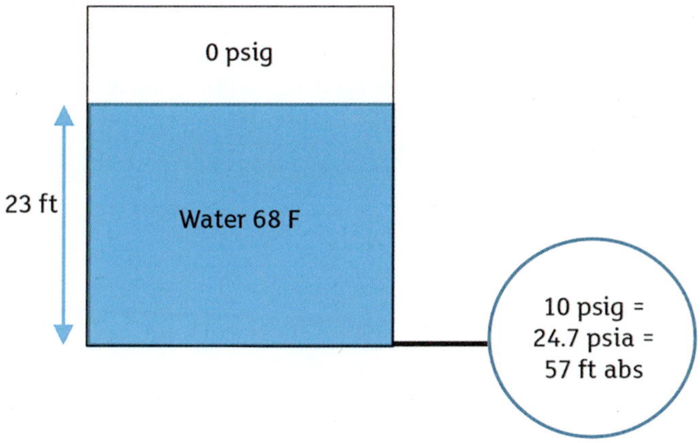

Figure 1. Pressure & head for static system (Courtesy of Hydraulic Institute).

This is described in Figure 1, showing that a tank with a 23-ft level of water will have a pressure of 10 psi gauge (psig) at the bottom of the tank when there is 0 psig at the water surface. The pressure units are psig, indicating that they are "gauge pressure," meaning they are in addition to the atmospheric pressure around us. The atmospheric pressure varies slightly day to day and changes predictably with elevation, but at sea level the standard atmospheric pressure is 14.7 psi absolute (psia). The 10 psig at the bottom of the tank is equivalent to 24.7 psia (10 psig +14.7 psia) or 57 ft absolute (24.7 psia * 2.31). This is an important distinction to consider as we start to discuss NPSH.

The NPSH available (NPSHA) describes the total head at a point in a system—measured in meters or feet absolute—greater than the vapor pressure of a liquid at its operating temperature, essentially indicating the relative closeness of a liquid to vaporizing. In Figure 1, the NPSHA at the bottom of the tank is 57 ft minus the 0.34 psia (0.78 ft absolute) vapor pressure of the water resulting in 56.2 ft of NPSHA. To cause the water to vaporize at the measurement point, the NPSHA would need to be reduced by 56.2 ft. In a static illustration such as this, it could occur by any combination of reducing the pressure at the surface of the water, reducing the level of the water or increasing the temperature of the water, which increases vapor pressure. In a pumping application, the piping friction head loss between the pressure measurement and the pump will also contribute to a reduced NPSH at the pump location.

When applied to a pumping system, it is important to evaluate the NPSHA at the pump suction. When the liquid enters the impeller, there is a point of reduced head in the impeller eye before the energy imparted by the impeller increases the head. The reduced head in the impeller eye can result in the fluid fallng below its vapor pressure, causing it to vaporize and suddenly change back to liquid in a higher pressure region of the impeller. The formation of vapor bubbles blocks flow area in the impeller, which results in reduced pump performance, and the repeated collapsing of the vapor bubbles results in erosion wear of the impeller. This phenomenonon is referred to as cavitation. To avoid cavitation and maximize the reliability of a pump, it is imperative that proper NPSH margin exists.

Operating regions

Pump impellers are designed to operate at a specific flow rate where there is a zero incidence angle between the inlet impeller vanes and the approaching liquid at a particular rate of flow, called shockless entry flow. The flow of shockless entry is typically near the pumps best efficiency point (BEP), and BEP is used as a proxy because it is more easily determined by people who did not design the impeller.

Higher or lower rates of flow relative to the BEP cause a mismatch between the angle of the approaching liquid and the impeller vane inlet tips. The greater the incidence angle, the greater the potential for flow separation and cavitation to occur. Design characteristics for both performance and service life are optimized near the BEP, and the pump operates with maximum hydraulic efficiency. Therefore, operating as close to BEP as possible will result in maximum pump reliability.

The Hydraulic Institute defines two important pump regions of operation. The preferred operating region (POR) is the range of rates of flow to on either side of BEP within which the hydraulic efficiency and operational reliability of the pump are not substantially degraded. The allowable operating region (AOR) is a wider range of flow, outside the POR, over which the service life of a pump is acceptable. When pumps are operated outside their POR, the flow through the pump is no longer uniform. This results in areas of flow recirculation and separation that can cause additional loading, flow induced vibration and local areas of cavitation, all of which result in reduced reliability. The AOR is impacted by factors such as hydraulic loads, temperature rise, vibration, noise, power limits, liquid velocity, pump and potential for clogging, NPSH margin, head flow curve shape, suction recirculation, and pump size. The user should consult with the pump manufacturer to determine this value.

NPSH margin and operating region

To understand the NPSH margin required, it is important to understand the NPSH requirements (NPSHR) of the pump. NPSHR is defined as a minimum NPSHA, provided by the manufacturer, which is required of a pump to achieve a specified performance at a specified rate of flow, speed and pumped liquid. This definition requires engineering judgement and is not easily documented. Therefore, as early as 1932, the Hydraulic Institute implemented a 3 percent reduction in head at a constant flow rate caused by a reduced suction head as the NPSHR of the pump. This is because this value was the smallest head drop that could be consistently and practically measured (see Figure 2). Historically, and still today, manufacturers' pump curves show plots of NPSH3 and pump flow rate.

If a pump is applied with an NPSHA equivalent to the NPSH3, the pump is operating with reduced head due to cavitation. The general consensus in 1932 was that pumps operating under conditions of 3 percent head drop would achieve generally acceptable service life. This was probably true at the time, when pumps for a given application were typically larger and slower than pumps for the same application today. Today's higher speed, higher energy density pumps may not achieve acceptable service life under suction conditions without an adequate NPSH margin above the NPSH3. For this reason the guidance provided in ANSI/HI 9.6.1 should be followed to ensure reliable and efficient operation.

The topics of operating regions and NPSH margin go hand in hand because when a pump is operated away from BEP and outside the POR, the required NPSH margin is increased. See Figure 3, which illustrates the amount of NPSHR required needed to limit the head drop to specific levels. Additionally, Figure 4 illustrates the POR and AOR for a pump along with the NPSH3 curve and the NPSH margin. To limit the potential cavitation damage that can occur, the NPSH margin must be increased at flows beyond the POR. Reference ANSI/HI 9.6.3 to determine the POR.

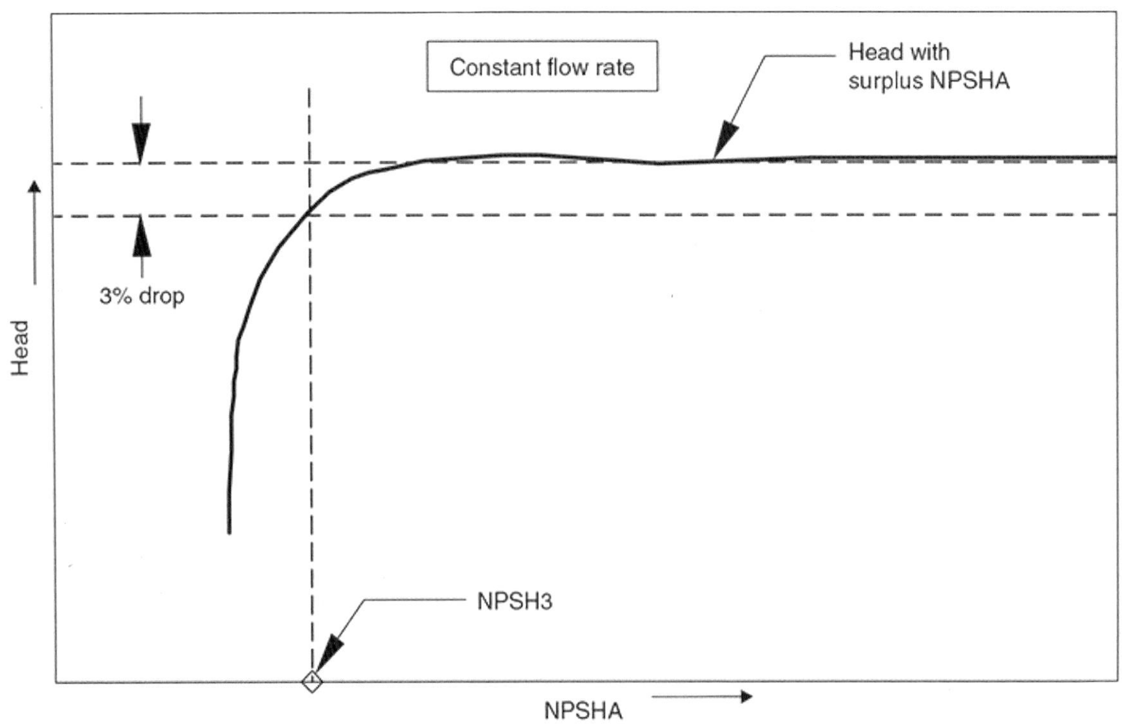

Figure 2. Determination of NPSH3.

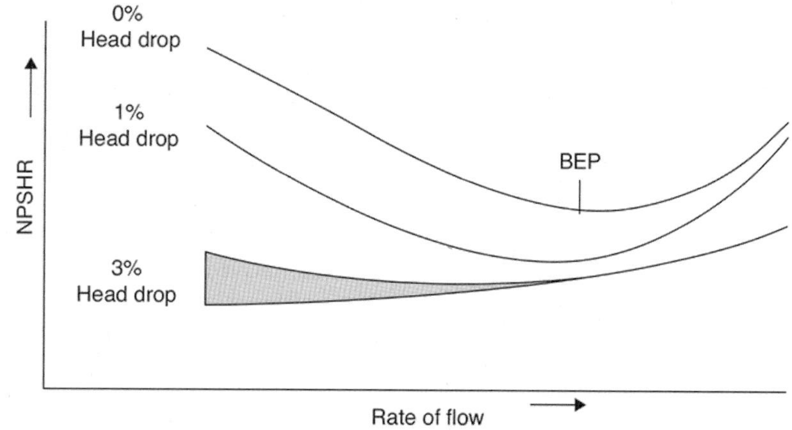

Figure 3. Head drop as a function of BEP flow rate due to NPSH.

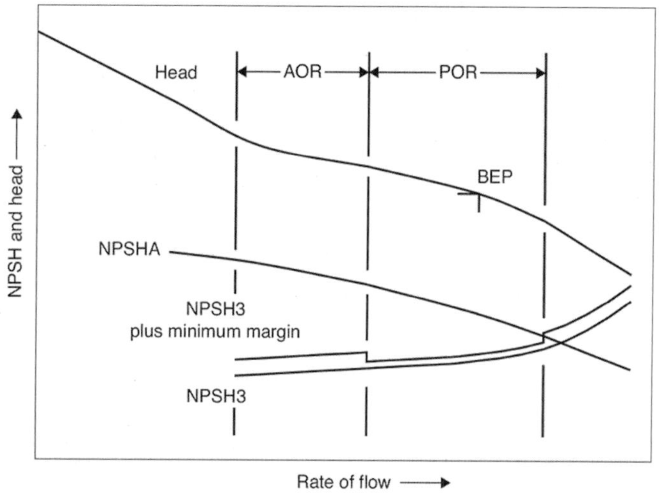

Figure 4. Illustration of AOR, POR and NPSH margin.

Reprinted by permission – *Pumps and Equipment*, July 2017.

Reading a centrifugal pump curve

Jim Swetye, Grundfos

Choosing the correct centrifugal pump for your application is critical for maximizing long-term performance. The wrong pump will not only operate inefficiently, but can fail prematurely because it is not ideally suited to the application's conditions.

Identifying the best pump for your application starts with examining the pump curve, which indicates how a pump will perform against specific rates of pressure head and flow. The proper interpretation of this data is the only way to make informed decisions on the choice of pump, motor sizing, power consumption strategies and other factors. Before reading a pump curve, the following information must be gathered for a given system:

- the required rate of flow in gallons per minute (gpm)
- the appropriate pipe sizes and system components
- the system head in feet

After making these calculations, it is time to find a pump that will operate efficiently within the system parameters.

Pump curve, BEP, and system curves

A *pump curve* denotes flow on the x-axis (horizontal) and head pressure on the y-axis (vertical). The curve begins at the point of zero flow, or shutoff head, and gradually descends until it reaches the pump runout point or maximum flow rate.

The pump's operating "sweet spot," or best efficiency point (BEP), is generally located near the middle of the curve. Pumps are the most efficient and have their highest life expectancy when they can run near their BEP, as determined by the manufacturer. Typically, the area on the curve between 70 and 120 percent of the BEP is known as the preferred operating region (POR) for the pump.

A second curve, called the system curve, is used in conjunction with the pump curve and can be overlaid on the same graph. The system curve represents the system head in your specific application at various flow rates and is calculated by determining the system's static head and friction loss.

On a system curve, as the flow rate increases there is a corresponding rise in system head, or pressure required to make the liquid move. The energy used to overcome flow resistance is referred to as the head (or pressure) loss due to friction.

Overlaying the system curve on the pump curve indicates how the pump will perform given a specific flow rate and head pressure, depending on the position of the pump control valve and the impeller diameter. The point at which the pump curve and system curve intersect on the graph indicates the pump's actual operating point in that particular system.

To illustrate, consider the following example and refer to the single speed curve displayed in Figure 1:

- a flow of 9,000 gpm
- a head of 180 feet

Locate 9,000 gpm on the x-axis, and follow it up until it intersects with 180 feet of head on the y-axis. The intersection point would fall toward the middle of the curve and likely be within the POR, making the pump a good choice for this example application.

It would be important to confirm that it does in fact fall within the POR by checking the manufacturer's guidelines.

The same graph can represent how the pump's performance will change if the impeller diameter is reduced or enlarged. The diameter is expressed in inches neighboring its respective curve. A change in impeller diameter does not affect the system curve, which only shifts if there is a change in system head, such as a closed valve. The intersection points are where the pump will perform at each diameter.

Note that it is permissible to change the impeller diameter, as well as system conditions, so long as the pump performance still falls within the POR. There is a broader range of the pump curve identified as the allowed operating region (AOR), in which it may be permitted and beneficial to operate the pump. It normally falls between the minimum continuous stable flow (MCSF) line and the runout line. If pump performance falls outside that zone, look for another pump.

Other pump curve elements

In addition to plotting the pump and system curves, a pump curve graph provides other elements important for choosing the correct product for your application.

Efficiency curve: The pump efficiency curve represents a pump's efficiency across its entire operating range. Efficiency is expressed in percentages on the right of the curve graph. The BEP is represented by the efficiency curve's peak, with efficiency declining as the curve arcs away, either right or left, from the BEP. Knowing the efficiency percentage will also help calculate horsepower required for an application.

ISO efficiency lines: International Organization for Standardization (ISO) lines are concentrically elliptical curves indicating equal efficiency on a pump curve graph. They are used as another means of representing how efficiency levels change along a pump curve as it moves away from the BEP or if the impeller diameter is reduced.

Power curve: The power curve represents the load the pump imposes on the driver at a given point on the pump curve and helps with proper motor sizing. It is represented as a separate curve graph and gradually rises toward its peak load, which is typically close to the BEP with most rotodynamic pump types. Afterward, it declines as it approaches the runout point.

Net positive suction head curve: The net positive suction head required (NPSHr) indicates how much force is needed to push liquid into the eye of the pump impeller. It is displayed in feet beneath the main pump curve graph. Knowing the correct amount of NPSHr will prevent the pump from cavitating, vibrating and failing prematurely.

Variable speed pumps

Thus far, only fixed, single-speed pumps have been considered. Now let us take a brief look at the variable speed curve, shown in Figure 2.

When noted on the graph, the various speeds are represented in rpm by separate curves. As the speed is reduced, the variable speed pump curves help predict what the corresponding reductions will be to flow and head levels, based on their intersection points with the system curve. Along the system curve arc, the reductions continue until the flow and head eventually reach zero and the pump rpm stops. Additional system curves can also be added to illustrate, for

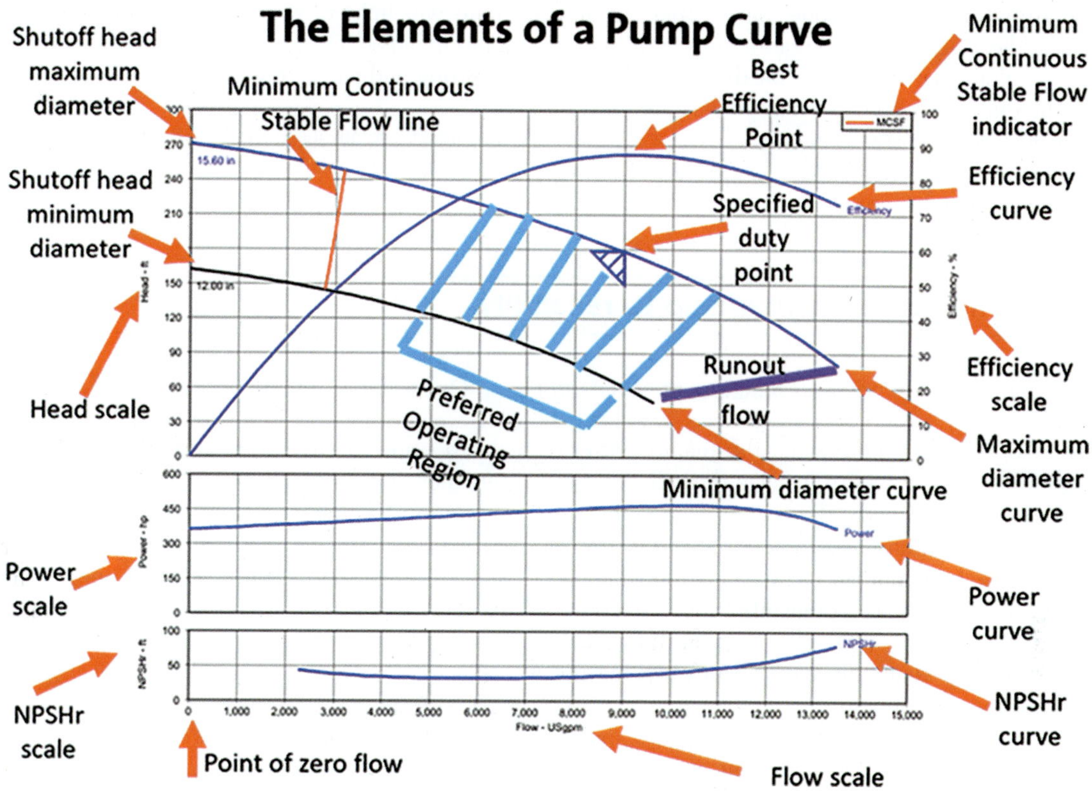

Figure 1. Single speed curve (Courtesy of Grundfos).

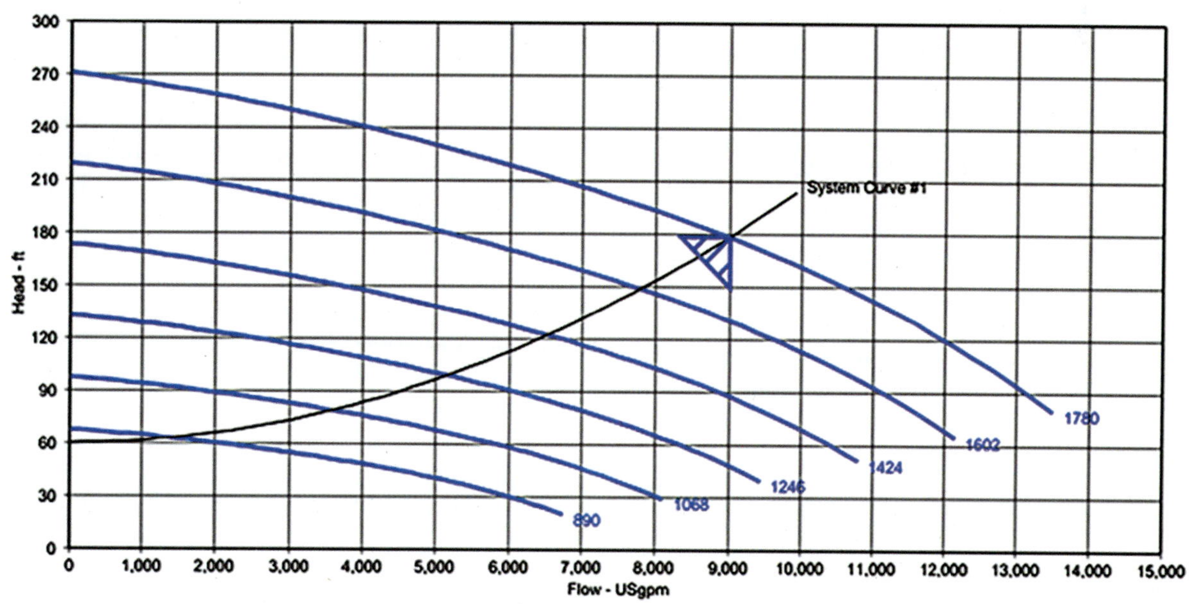

Figure 2. Variable speed curve (Courtesy of Grundfos).

example, the impact of various zone valves opening and closing as in a heating, ventilation and air conditioning (HVAC) system. Note that depending on the type of speed control used, the pump may operate on a control curve that is different than the system curve.

The above terms and considerations cannot cover every scenario a specifier may encounter. However, understanding them will help provide a solid foundation for interpreting a pump curve. This, in turn, will foster sound decisions when selecting and specifying pumps.

Reprinted by permission – *Pumps and Equipment*, February 2018.

Navigating pump curves: a roadmap to reliability

Jim Elsey, Summit Pump, Inc., and principal of MaDDog Pump Consultants, LLC

Frequently overlooked by end users, the manufacturer's pump curve can be a useful tool to manage the reliability of the equipment and, subsequently, the system. Depending on your technical knowledge and experience, the pump curve may be perceived as both simple and obvious information, and/or it may be viewed as confusing "gobbledygook," consequently to be discarded or ignored. The intention of this article is to illustrate how to use this handy dandy tool in ways that you have not thought of before.

As an added perspective, the curves are also the manufacturer's promise to you-a sort of covenant, if you please-based on actual performance tests, empirical data and calculations conducted at their facility.

See Figure 1 for a simple example of a typical published pump curve. Ask yourself, how many things does the curve really tell you? Let's examine a list of some obvious and potentially not-so-obvious information that is communicated by this simple plot.

This is not an article on how to read a pump curve, though it may assist you in that endeavor but more about looking at the pump curve in a new and different way.

The head vs. capacity curve

A simple pump curve can provide a plethora of useful information and there are many data points to be directly gleaned-there are things the curve tells you and there are things it does not. The curve will tell you the potential range of head and flow rates where the pump is capable of operating, but it doesn't tell you specifically where it will operate on that curve. You will need help from the system curve for that information, because the pump will operate exactly where the system curve tells it to operate, if it is capable of operating at that point.

Basic points defined

The head-capacity curve illustrates and defines the head and corresponding flow rate that the pump is capable of for the stated impeller diameters and speeds (in this example the impeller is trimmed to 14.625 inches, just short of the maximum 15-inch impeller). You may think of how the set of curves illustrates an "operating envelope" per se (see yellow shaded area). To take the concept one step further, it also partially defines the acceptable, reliable and safe operating envelope.

Shut-off head

From the set of head-capacity curves you can easily determine the shut-off (SO) head for a given impeller diameter and speed. SO head is that point of maximum developed head at zero flow and at the far left of the curve, where the curve intersects the y-axis (vertical). In our example, the 50 head is 226 feet. This is a parameter usually garnered in the field by operating the pump with a closed discharge valve. which is also not a good place to operate the pump for more than a few seconds. But, in that brief and interim operating position at SO, it does tell you a lot of information about the pump. including the health of the unit. While not 100 percent conclusive, if a pump meets the published SO head, it confirms that the impeller diameter and speed are likely what you expected and as stated on the curve. Indirectly, it also tells you the direction of rotation is correct (or not. because a pump operating in the wrong direction will not come close). While it will not tell you the temperature of the pumped fluid, the information confirms the specific gravity (SG) if you know the fluid temperature. (PSI × 2.31, divided by the head will yield the SG).

A satisfactory result from the SO head test also indicates the clearances in the pump are healthy, meaning they are at or near the manufacturer's specification. What it will not confirm for you is if there is an issue in the system with air entrainment, submergence, net positive suction head available (NPSHa), or the fluid's viscosity. You will need to operate farther out on the curve to investigate and collect the evidence for those issues.

If you are not confident in operating the pump at or near SO, consult with the manufacturer, their representative or an expert. Not all pump types, systems or energy levels lend themselves to this specific test without negative consequence.

Minimum flow

The point of allowable minimum flow will normally be defined on the curve. Do not operate the pump below this minimum flow rate for more than a few minutes. Some manufacturers refer to pump operation in the flow range just above the minimum flow point as MCSF, or minimum continuous stable flow. At flow rates below this stated point, the fluid will typically recirculate in the impeller eye and will be unstable. In Figure 1, minimum flow is defined as near 1,450 gallons per minute (gpm).

The minimum flow rate for centrifugal pumps is based on two criteria: the first being mechanical limits of the machine due to the resultant and exponentially increasing radial forces that occur when operating at the low flows, mostly dependent on casing design and, to a smaller extent, on impeller vane overlap and specific speed (NS). The second reason is the thermal limit, where the pumped liquid will overheat due to recirculation and potentially flash. The manufacturer calculates both scenarios and then sets the boundaries. The higher of the two will become the pump's minimum flow rate.

American Petroleum Institute (AP!) 610 states that MCSF will be no less than 60 percent as allowable value (but 70 percent is the preferred). It is up to the manufacturer to design the pump.to meet these criteria. American National Standards Institute (ANSI) B73.1-2011 states in paragraph 5.1.6 that, "Pumps shall be designed to operate continuously between 110% percent of BEP [best efficiency point] and the minimum flows."

Best efficiency point

The best efficiency point (BEP) is simply the design point for the pump. It is the hydraulic condition point of both head and flow rate where the pump is most efficient. In our example, the BEP for the pump at maximum impeller is approximately 3,450 gpm at 190 feet.

All commercial centrifugal pumps are essentially designed for one single point (BEP); any other operating points in the operating envelope are a commercial compromise. BEP is the point of highest efficiency. It is also the point of lowest radial

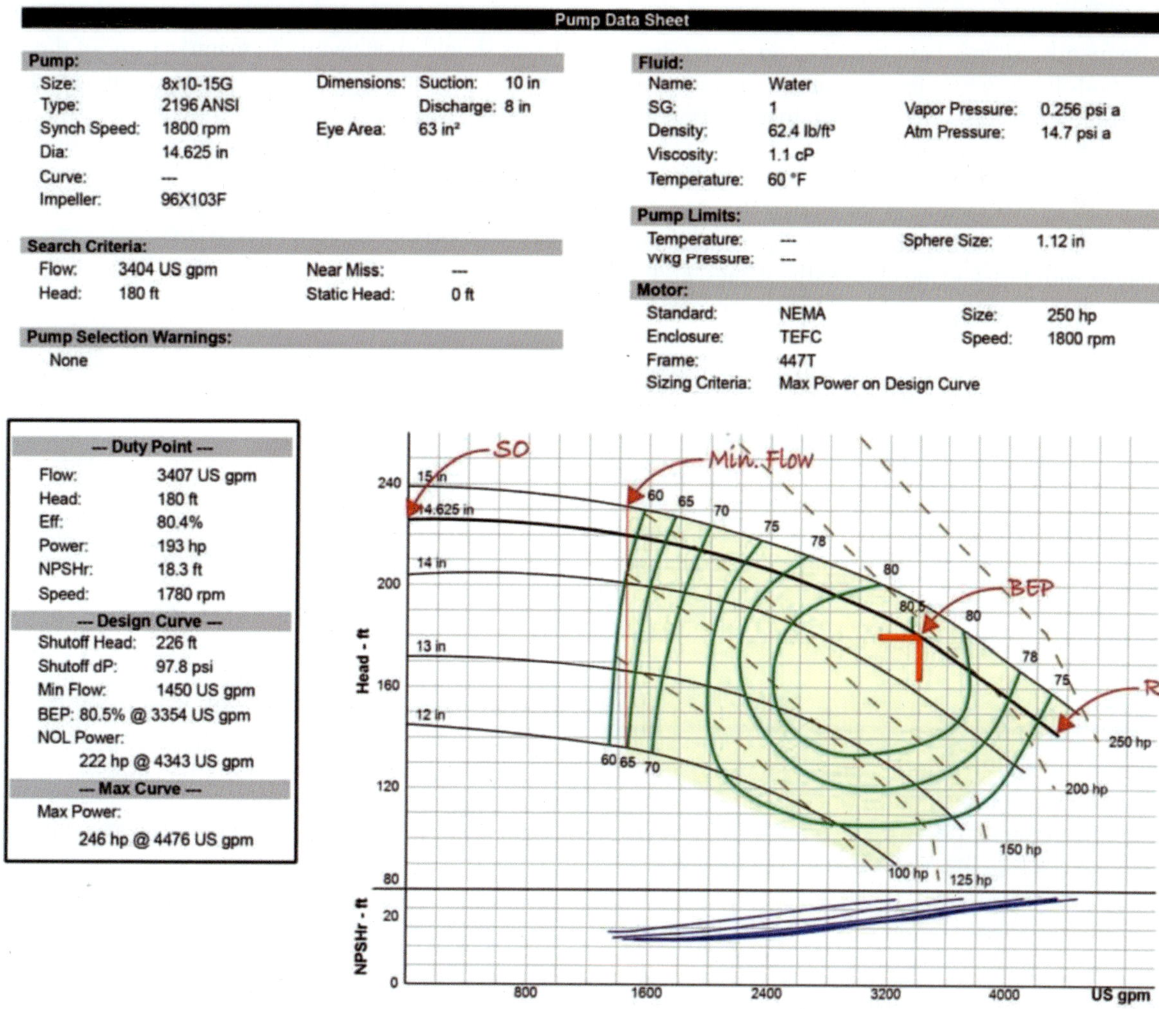

Figure 1.

forces, shaft deflection, noise and vibration. For some of us. it is also the operating point at or near shockless entry for the fluid entering the impeller eye.

Preferred vs. allowable

The preferred operating region (POR) is normally defined as flow rates between 70 and 120 percent of the flow at BEP, and the wider allowable operating region (AOR) is usually defined by the manufacturer and will vary with pump type, casing design and other factors including the shaft. Note that for pumps with high specific speed impellers that the POR is normally reduced to 80 to 110 percent of BEP.

Runout or end of the curve

The point of runout (RO) is also commonly referred to as the end of the curve (EOC), which is all the way at the right of the curve. In our example, flows approaching 4,500 gpm would be the RO point. This is the point of maximum flow for the curve and also the lowest head for that impeller diameter and speed. To attempt operating at higher flow rates would yield severe cavitation in the pump regardless of the net positive suction head (NPSH) margin. If the pump was capable of operating at higher flow rates, the manufacturer would surely publish that information.

People frequently ask me about the EOC operating region and the questions fall into two distinct camps. The first group is comprised of folks stating: Why not operate here? Followed by: Why won't the pump operate past the EOC? The other group are people that for various reasons know or perceive the potential or imagined issues at the end of the curve and will not operate their pump anywhere near this area. In a way it is like dealing with fire. Get too close to the flames, and you will be burned or damaged. But if you understand and carefully manage the fire as a useful tool-and if you maintain the correct distance-it will provide warmth and light.

If you don't know the details of your system curve and how it changes with time and dynamic conditions, then you should not operate in this region at the EOC. In my pump classes, I refer to this region as one filled with dragons and wizards whose sole purpose is to make your life miserable. A simpler alternative explanation may be, it is a world without bacon.

Implied or title block information

If you laboratory test 25 different impellers of the same design in 25 different casings where all impellers and casings are from the same respective patterns, you will get 25 different curves that are all close, but they will be different. Also, if you test a given sample impeller pattern today, the results will be different than they were a year ago (earlier in the pattern life), and they will be different next year (later in the pattern life). Note that using investment castings will reduce this variance substantially, but not all sizes and models lend themselves economically to the process. For this reason and other nuances, manufacturers publish their hydraulic performance with stated tolerances and restrictions for specific pump applications. Refer to ANSI/Hydraulic Institute specification 14.6 -2011 for details.

Information that is inferred, unless otherwise stated/published, is that the curve is based on hydraulic performance while pumping water at 60° F or 68° F. If the curve is based on a fluid different from water, and therefore a potentially different viscosity, the manufacturer must make that statement.

Viscosity-corrected curves will normally state that they are corrected. Additionally, the standard curve is based on a liquid with a specific gravity (SG) of 1.0. SGs other than 1.0 will result in different brake horsepower (BHP) requirements.

There is potential for confusion on exactly which temperature is the standard water temperature and SG that pump manufacturers should use for the basis of published curves. Users should check with the manufacturer of the pump and/or see Chapter 4 of the "Cameron Hydraulic Data book, 19th edition.

What may not be patently obvious is that the information required to calculate several additional parameters is also present, such as BHP specific speed (NS), suction specific speed (NSS) and suction energy (SE).

$$BHP = \frac{hqSG}{3960\ Eff} \tag{1}$$

BHP = Brake horsepower
h = Head, ft
q = Flow rate, gpm
SG = Specific gravity
Eff = Pump efficiency

Brake horsepower

Most published curves will offer the BHP information and my recommendation is to do the actual calculation anyway before anything is purchased to be safe. Again, note that the BHP published on the curve will be based on water at 68° F unless noted.

Knowing the efficiency and, subsequently, the BHP, you can calculate your operating expenses and total cost of ownership. From that cost data you can determine if you need to operate differently, change the system design, or

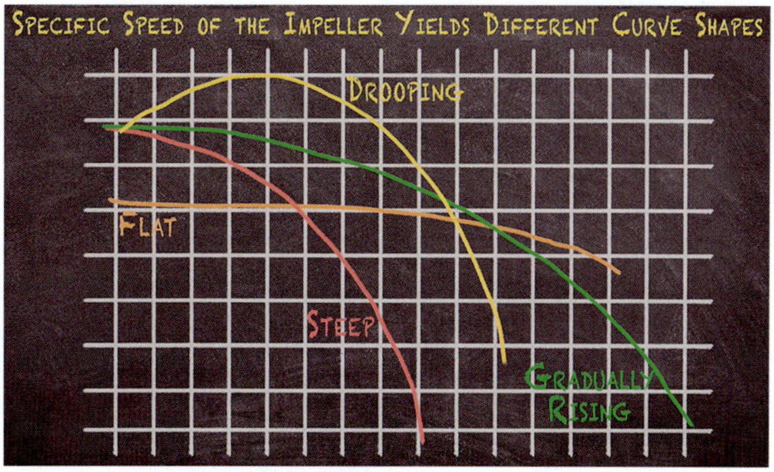

Figure 2.

even perhaps replace the pump with another model or convert to variable speed. You can also use the BHP formula to solve for other parameters.

Mathematically, you can easily solve the equation for any one unknown. In Equation 1, I already know the BHP, but not the flow rate, so I set up the equation to solve for flow (where percent efficiency is expressed as a decimal). Following this logic, you could also use this equation to solve for SG, head or efficiency.

Understanding specific speed and suction specific speed

The manufacturer may or may not offer this information somewhere in the title block, but to the experienced pump person the slope, size and shape of the curve will lend clues as to the NS range. See Figure 2 for a simple schematic on how curve shapes change as a result of their geometry and NS. Suffice to say that the number and shape/geometry of the impeller vanes will paint an illustrated story as to the NS.

If you know the pump's head and low at BEP (and also the pump speed), then you can calculate the NS. Further, if you know the NS and the NPSH required (NPSHr), you can calculate the NSS. For dual suction impellers (two eyes), use half of the flow at BEP.

Why is any of this information important to you? Well, depending on your system curve and where it intersects, the pump curve can make a big difference in the proper pump selection. You would not want to use a pump with a drooping curve if it was going to operate at lower flows, but if it was markedly higher in efficiency at high flows you would. If you had a system that used variable speed pumps, would you want a flat curve or a steep curve? For a boiler feed system in a base-loaded power plant, what curve shape would you want?

Pump curve from the manufacturer's perspective

As mentioned earlier, the published pump curve is a promise, a sort of manufacturer guarantee that the pump will operate on that curve as indicated.

It does not mean the pump will operate where you wish and hope it to run, unless that point is exactly where the system curve intersects the pump curve. For emphasis, I again point out that if the fluid has different properties not accounted for, or the system has issues with NPSH available (NPSHa), air entrainment, a miscalculated system curve or if the pump clearances are not set properly, the pump will operate somewhere else off the curve.

The pump curve is a statistical representation (albeit a conservative one) of empirical data collected and composed from the manufacturer's test facility. The curve is a published proof of the manufacturer's ability and it holds them accountable to the pump purchaser and end user.

Since most industrial pumps are constructed from castings, there are variables; and because no two castings are exactly alike when a curve is published, there are stated and acceptable industry tolerances.

Conclusion

I strongly recommend that you know where your pump is operating on its curve. If you don't know where it is operating, then please try your best to find out.

If you have any management by objectives (MBO), quality assurance/quality control (QA/QC), Kaizen or Six Sigma training, you already know (and I am paraphrasing here) that if you don't measure the pump performance then you can't manage it. Left unmanaged, the pump will prematurely fail, and prior to that day it will needlessly cost you time and money.

Reprinted with permission – *Pumps and Systems Magazine*, January and February, 2020.

15: Measurement

Multiphase flow meter 626
Pipeline flow measurement factors 627
Liquid measurement orifice plate flange taps 630
Mass measurement light hydrocarbons 632
Pipeline measurement of supercritical carbon dioxide 633
Master meter proving orifice meters in dense phase ethylene 639
Gas or vapor flow measurement—orifice plate flange taps 646
Properties of gas and vapors 648
Determine required orifice diameter for any required differential when the orifice and differential are known 650
Estimate the temperature drop across a regulator 651
Estimate natural gas flow rates 651
How to estimate the average pressure differential on the remaining meter runs of a parallel system 652
Sizing a gas metering run 652
Determine the number of purges for sample cylinders 653
Find the British thermal units when the specific gravity of a pipeline gas is known 653
Make quick calculation of the gas flow through orifice meters 654
How to measure high pressure gas 654
Four ways to calculate orifice flow in field 658
Practical maintenance tips for positive displacement meters 661
Sizing headers for meter stations 665
Measuring flow of high-viscosity liquids 668
Matching the flow meter to the application 673
Use liquid ultrasonic meters for custody transfer 680
Handling entrained gas 685
How do Coriolis flow meters work? 687
Proving liquid ultrasonic flow meters for custody transfer measurement 689
V-Cone® flow meter 700
Sampling and testing program for unrefined petroleum-based products 706

Multiphase flow meter

Principle of operation

The major parts of the TopFlow meter are the Venturi insert and the electrodes inside the Venturi throat. The flow rates of oil, water, and gas are calculated based on the measurements performed by the electrodes and the measurement of the differential pressure across the Venturi inlet.

The TopFlow meter utilizes a capacitance sensor for oil-continuous liquid mixtures, where the capacitance caused by the dielectric constant of the fluid mixture is measured in the throat of the Venturi. Similarly, it utilizes a four-electrode conductance sensor for water-continuous liquid mixtures, where the conductance is caused by the conductivity of the fluid mixture. All the necessary electrodes are incorporated within a Venturi that is somewhat modified. Advanced data processing is utilized to have continuous readings of the flow rates of oil, water, and gas.

The Venturi has a minimum of pressure drop. Nearly all multiphase meter vendors use a Venturi in one way or the other, and the reason is that this is probably the most reliable device for multiphase measurement because it measures the mass flow rate independent of what happens. One objection is the relatively limited turndown ratio that is specified, especially for single-phase measurements within fiscal uncertainties. However, because most wells produce in the high GVF range, where the total mass flow rate does not vary much, there is no general problem connected to the use of this device. Especially when installing one meter each single well and not having to deal with varying flow rates, the Venturi is very suitable for multiphase measurement. Erosion will not give any major difficulties, because the TopFlow meter uses a Venturi in terms of an insert into the pipe. This means that the insert itself can be made of a wear-resistant material like high-quality steel or even ceramics in cases when erosion is considered to be a problem. In extreme cases, due to the simplicity of the TopFlow meter, the insert itself can easily be replaced. However, in the latter case, the other parts of the production pipeline, such as bends and valves, would also be eroded to destruction and would have to be replaced. A Venturi is able to give a good picture of the dynamics of the flow (i.e. it is not sensitive to the flow regime effects for measuring the instant mass flow rate accurately). Modern differential pressure transmitters are also able to follow the fast variations through high-frequency response.

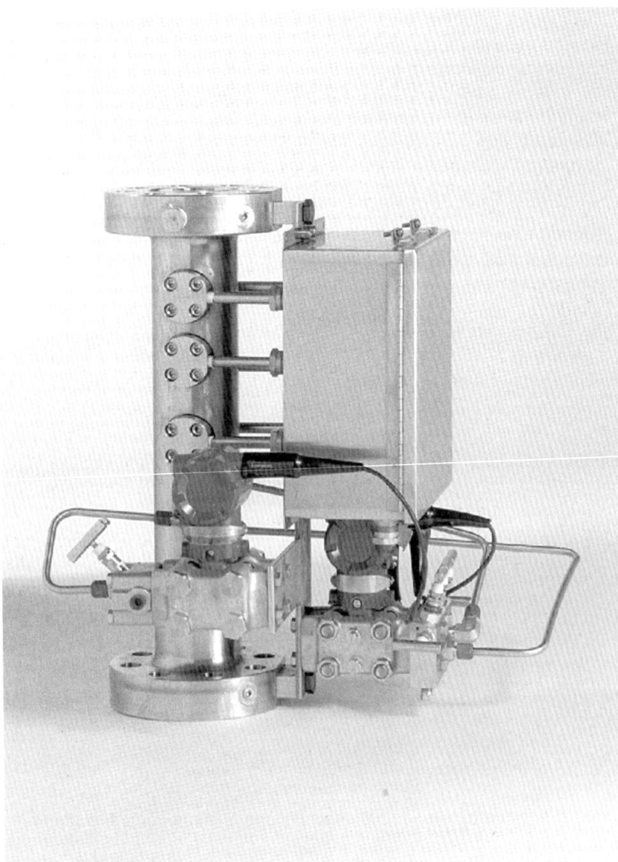

Figure 1. TopFlow multiphase flow meter.

The most important flow regime effects are mainly due to the ratio between gas and liquid. One is the longitudinal effect along the axis of the flow conduit, represented by the slug regime, where longer or shorter liquid plugs are separated by longer or shorter gas pockets. One challenge with this is that the liquid plug is not entirely liquid, but contains varying amounts of small gas bubbles. In addition, the gas pockets are not entirely gas but contain larger or smaller amounts of liquid. In general, the liquid and gas are also flowing with different velocities. It is of very high importance to employ sensors with fast time response to follow the rapid changes in slug flow. It is likewise important to measure the velocity often, because there is neither a constant, distinct gas velocity, nor a corresponding constant liquid velocity. Thereby, the velocity "components" are measured in real time and the slip between liquid and gas is handled automatically, without the need for theoretical or empirical slip models.

Another flow regime aspect is the radial effect, concerning the distribution of the liquid and gas over the cross-section of the flow conduit. In general, this plays a part in all multiphase flow conditions, but especially at high gas volume fractions, where the slip can be significant, the gas tends to gather in the center of the cross section, along the axis of the flow conduit. This is because of the lower flow resistance in the turbulent zone in this center. In plain words, the gas races through this channel, whereas the liquid is pushed aside and flows in contact with the pipe wall. The velocity difference is quite substantial and increases at increasing gas volume fraction. However, a part of the liquid flows as droplets following the gas phase. Taking these matters into consideration, it is obvious that the main source of uncertainty in multiphase flow rate measurement can be found in the complexity of the flow itself and not necessarily in the type of sensor used for the purpose. The crucial matter is therefore to choose appropriate technology and develop good physical models and reliable interpretation tools to solve the difficult tasks.

Adapted from FlowSys AS, Norway. Reprinted with permission.

Pipeline flow measurement factors

E. L. Upp, Flow Measurement Consultant

There are many considerations that influence our choice of methods and equipment in the measurement of liquids and gases in pipelines. A review of these problems is timely, since we have changing influences that are affecting our choices to accomplish the job of flow measurement. These are accuracy, electronic equipment, economics, fluid characteristics, standards and governmental influences, and testing or proving practices. All of these factors have had importance to some degree in the past; however, their relative importance to flow measurement and more importantly to a pipeline's overall operation has changed considerably as we have seen the prices of the fluids handled make dramatic changes along with supply and demand fluctuations of considerable magnitude versus a steady growth pattern of just a few years back.

Accuracy

Today's demand—better flow measurement accuracy. In analyzing the economics of the flow measurement industry, we find the customers are no longer in an expansion mode. With the turndown in their economies, they are more concerned with improving efficiencies in their present installations. Better measurement accuracy is one way to achieve this. The users are asking the manufacturer to improve his present devices as well as develop new meters and new technology to do the job "better." No one can fault these desires. However, measurement is a very conservative, "tradition-following" market. Industry metering standards are slow to change, and in the past it has been difficult to get the users to adopt new meters and products.

In the gas industry, contracts are written by lawyers who are not measurement experts and who tend to follow a contract form that was successful in the past. Manufacturers tried to switch users to electronic readout devices some 20 years ago. Electronic computers and flow recorders have been available to industry all this time, though admittedly not of today's quality. Yet today in the United States—for pipeline large-volume natural gas measurement—less than 2% of the stations installed utilize electronic computer controls and readouts. The industry has been more successful in applying electronic digital readout devices to liquid metering in conjunction with the turbine meter and is beginning to see production fields being automated for both liquid and gas measurement.

If the industry had a new meter today that offered absolute accuracy; infinite rangeability; met all the environmental requirements, codes, and governmental regulations; could be installed and operated by present personnel with minimal maintenance; was priced low compared to present devices on the market—it would still be difficult to sell it in sufficient quantities in the marketplace to make it profitable to manufacturers. The companies would at best be willing to try the meter for a year and compare it with present meters. Then and only then would they add it to their accepted list. The manufacturers are asking the user to change his cash register. The user has to be convinced that he is on solid legal, engineering, and procedural grounds.

Costs of accuracy. Many companies are taking a hard look at all capital expenditures in this time of high interest rates. Management people feel that the high price of the fluids they are handling allows a larger investment in order to get accurate measurement. Simple expenditure of capital does not eliminate the fact that normally the more expensive the device, the "better answer" we will get. However, careful consideration must be given to what the "better answer" will mean to our operations.

Quite often the flow measurement industry is called upon to solve flow measurement problems. Sometimes the end result is not exactly what the users had in mind. A new seller of natural gas or fluids that has a meter reading low by 1% would be delighted to use the new "more accurate meter." However, a purchaser who was just as interested in his balance sheet and who had his flow figure raised by 1% would not be too happy with a decision to use this "more accurate meter."

Many times, users have evaluated electronic flow computers, and when compared with the existing meter, will read differently. There are good, solid technical arguments why the computer is more accurate, but there are also political and economic reasons why customers cannot use more accuracy "if it lowers my billing." When two parties and a contract are involved, you must keep in mind that you cannot have a more accurate metering device that will raise measurement for both sides of the equation. As manufacturers, the only path that can be followed is to sell the most accurate metering device they know how to make and to let the customer work with the political and economic consequences. These influences certainly affect the industry's ability to bring out new products and have them accepted based on accuracy.

Their place in flow measurement. There is no denying that the high technology age is upon us. This basically refers to electronic calculating and data handling systems as far as the pipeline flow measurement problems are concerned. These tools have been available for several decades; however, we are finally seeing the beginning of the significant use of electronics to replace our previous mechanical methods of converting meter readings to custody transfer volumes. Basically, this is due to the availability of low-cost, reliable electronic devices that will meet the field conditions where flow measurement takes place and a growing awareness of the user on the requirements to successfully apply these devices. We have moved through stages of applications from "flow computers can do anything," to "flow computers do not operate in the field," to the stage we are in now where properly applied flow computers are a viable consideration for accomplishing the job we have done in the past, plus additional timely uses of the flow data to improve operations. However, this decision is not made without a serious look at the economic consequences.

Economics of flow measurement

Accuracy versus cost. As the cost of the handled fluids goes up, there is a demand by all parties involved in its purchase and sale to "reduce the tolerances," "get more accurate measurement," or "get more sophisticated equipment to get a better answer." Capital investment required to upgrade measurement is getting a hard look by pipeline management, as are all of their expenditures. So, our desire for better measurement is tempered by available capital and the resulting savings to be expected. The definition of "more accurate answers" of various measuring devices, both in terms of theoretical and practical use, must be reviewed to improve our ability to get the best measurement for the dollars invested.

If you are producing a well at 0.25 MMcfd, one level of equipment and procedure investment is justified, whereas a station delivering 25 or 250 MMcfd may require another. Similar thinking applies to liquid measurement. Physical limitations such as available power, cost of space to install equipment, and effect on the environment must also be considered. On offshore platforms, where space costs a premium, meters that can be installed with a minimum of piping have been used because of the realized savings. A large portion of the new metering devices' electronic readouts require good quality and dependable power. In some remote locations where commercial power is not available, the cost of getting the proper power may exceed the cost of the metering. In this case, the cost of power may dictate the use of mechanical or pneumatic readout devices or some local generated source such as solar or thermoelectric. These power sources are normally low-wattage devices, which may restrict the capabilities of the electronics or limit the choices of devices.

Personnel requirements. A very definite consideration affecting the economics of flow measurement is the personnel required to accomplish a given job. This may mean at least a retraining, if not a replacing of present measurement people. Generally, they will be in an upgraded position that requires both electronic and mechanical skills along with a solid understanding of flow measurement. Otherwise, you may end up with an electronics-oriented person chasing the details of proper programs and round-off procedures of the measurement readout system while not recognizing that the basic meter may be improperly applied or functioning and the total system is in error. The reverse can be true if a mechanically-oriented person is lacking in electronic knowledge. Fortunately, in the past few years most new workers have had good exposure to electronic devices and are more comfortable with electronic flow approaches than some of our older personnel. This problem exists at all levels of our company's operations and to a large degree will be the deciding factor of the final success of our system. It has a definite economic impact on our choice of changing our flow measurement systems.

Improving present flow measurement. Since most pipelines are in a nonexpansion mode, if not a reduction in throughput, operating people are spending a large portion of their efforts in improving efficiencies of their systems. This includes much closer attention to the efficiencies of compressor or pump stations. We are not talking about large percentage reductions in efficiencies being significant cost savings.

As an example, some gas pipelines have as much as 80% of their total cost of operation coming from fuel costs, so even a 5% reduction relates to a reduction in operation cost of 4%. To study these efficiencies for real results requires very accurate flow measurement including individual measurement on each compressor or pump and engine. This is a case where the use of flow measurement to reduce costs is the driving factor rather than the custody transfer service that we generally worry about.

Fluid characteristics

A changing problem. For years we have had the luxury of handling fairly simple fluids with well-defined and/or accepted correction factors to reduce fluids at flowing conditions to fluids at base conditions. We now are being faced with pipelines handling "dense fluids" such as carbon dioxide and ethylene and liquids such as nonstabilized crudes, light hydrocarbons operating at their vapor pressures, and refrigerated LPG. In these cases, the choice of the basic meter is less of a concern than the proper preparation of the fluid prior to passing through the meter, and the proper factors for making corrections to base conditions.

Sometimes we are trapped by our previous practices established with well-behaving fluids and can find large errors when these practices are applied to ill-defined fluids. In these cases the problems are not in the basic flow meter, but in the flow meter system application of the correcting factors based on inadequate knowledge. The proper answer to these problems has not been worked out to give the accuracy of flow measurement we desire, but with the recognition of the problem, we have been able to minimize some of the errors while continuing to seek new flow meters or methods of correcting for the fluid characteristics to solve the problem.

Standards and governmental effects on flow measurement

Outside influences on flow measurement. Industry practices, standards, and contracts, along with the ever increasing governmental requirements, must be considered to complete the picture of present day pipeline flow measurement. It is the belief of some of the bodies that measurement can be accomplished by legislation.

All orifice plates in some countries, for example, must be stamped by a government inspector, which makes them acceptable and correct. The manufacturer or user then can adopt the attitude that if the standards body or government finds it correct, it is not his worry and he really does not need to know much about flow measurement. Fundamentally, this has a serious flaw, since it removes the one item effective in obtaining accuracy—personal commitment and responsibility to accomplish the job. The manufacturers and users in countries where this is acceptable practice have passed the responsibility back to the government. They have reduced the caring as to whether or not their metering is accurate or properly installed and whether or not it maintains this accuracy in long-term service.

As we get further into the governmental agencies controlling our business, we have seen that the cost of doing the job is of secondary importance to meeting DOEE, EPA, or OSHA standards, as an example. Unfortunately, the governmental bodies are looking at getting further in the measurement of fluids business with every passing day. Experience tells us that present economics of fluid measurement will be less of a controlling factor as this continues to occur.

Testing or proving flow meters

Present practices. The term "proving" in flow measurement means a throughput test and comes to us from the liquid measurement people who developed the mechanical displacement prover (now termed the "pipe prover") for proving turbine and positive displacement meters on crude oil and petroleum products. Prior to this development, we had critical flow and low flow provers used on PD meters on gas, but these had relatively limited use. On the other hand, "proving head meters" was never done. We substituted mechanical inspection of the primary devices and calibration of the secondary devices for the throughput test.

Proving is the checking of the throughput of a meter versus a standard determination of volume from another device. The purpose of all inspection and calibration or proving is to establish a meter's performance accuracy to some tolerance. This tolerance is dependent on the basic accuracy of the standard used plus the conditions of the test. The allowable tolerance is dependent on the value and quantity of the product handled. Because of this, some metering installations cannot justify a few extra dollars in cost. In those cases, we continue to use inspection and calibration to determine the meter's accuracy.

This proves that meters can be quite expensive in terms of permanently installed provers. The decision is based on "cost considerations" and is balanced by the degree of "accuracy needed" and operations and maintenance money available to be spent in obtaining and maintaining the "accuracy." Subsequently, here, are several influences that affect our measurement testing and proving: present practice, standards, accuracy desired, and economics of the measurement job.

Improvement of present practices. Present practices will be continued in concept; however, equipment development has improved our tools such as air deadweight devices for differential pressure tests and precision test gauges for static pressure for gas meters. Liquid meters have improved equipment available for sampling and fluid characteristic testing.

The standards for flow measurement have had a great deal of study done recently to improve the accuracy and application range with such programs as the American Petroleum Institute work on the fluid characteristics, the development of the small provers for liquids, and the basic flow coefficient research on the orifice by the American Petroleum Institute and the Gas Processors Association. This work of upgrading standards to improve accuracy is continuing at a rapid pace. The accuracy requirements for all flow measurement is improved because of the high cost of the measured fluids, and we anticipate that this demand will increase.

Summary

The business of pipeline flow measurement has assumed a more important role in the overall pipeline operation picture because the economics of the job to be done requires: more accuracy; better knowledge of the fluids being handled; wider application of electronics to accomplish the basic flow measurement plus other uses of these data to improve operations; the adaptation to changing standards and governmental rules; and the use of new equipment and techniques to determine the accuracy of a flow meter. The people who are responsible for flow measurement have a challenge to respond to these changing influences to improve their company's flow measurement.

Liquid measurement orifice plate flange taps

Sequence of calculations

1. Calculate the S value.

$$S = \frac{W_m}{D^2\left(2831 + 0.0567T_f\right)\sqrt{G_f h_m}}$$

2. Calculate Reynolds number.

$$Re_D = \frac{4.42\ W_m}{D_\mu}$$

3. Iterate to find uncorrected beta ratio.

$$S = 0.589\beta^2 + 0.01\beta^3 + 0.518/\beta^{6.425}$$

$$fx = 0.598\beta^2 + 0.01\beta^3 + 0.518\beta^{6.425} - S$$

$$f^1x = 1.196\beta + 0.03\beta^2 + 3.3282\beta^{5.425}$$

Start with $\beta = 0.5$ and calculate fx/f^1x. If greater than 0.0001, subtract fx/f^1x from current value of β and repeat until the error is less than 0.00001.

4. Calculate Reynolds number correction factor. If $R_D < 350\beta^{1.3}$:

$$F_c = \left[A - \left(A\left(0.5^{(\beta^{25} - 0.12)}\right)\sqrt{1 - \frac{700\beta^{1.3}}{1200}}\right)\right] \times \left[\frac{\sin\left(90\sqrt{\dfrac{R_D}{350\beta^{1.3}}}\right)}{\beta^{0.15}}\right] + 0.997$$

where $A = (\beta - 0.2)^2 + \left(8.7e^{3.9\beta}\right)\left(700\beta^{1.3}\right)^{-0.89}$

When $\beta - 2 > 0$

$$(\beta - 0.2)^2 = 0$$

If $350^{1.3} < R_D < 700\beta^{1.3}$:

Table 1
Table of Equal Thermal Expansion Temperatures, °F, for Vario[us] Orifice Plate Materials

Type 304/316	Aluminum	Copper	Type 430	2% CRMO	5% CRMO	Bronze	Steel	Mone[l]
	−298					−360		
	−230	−378				−273		
−325	−178	−266				−216		
−226	−131	−194				−163		−28
−152	−85	−132				−111		−19
−86	−41	−72				−60		−10
−24	+3	−15				−8		−3
+38	+47	+42	+24	+28	+27	443	+31	+3
+101	+91	+99	+112	+108	+110	+95	+106	+10
+158			+202	+184	+191	+149	+181	+16
213			290	260	272	200	254	23
266			374	331	352	249	324	29
318			456	400	426	297	392	34
367			531	468	493	345	457	40
414			604	534	560	393	520	46
463			677	597	628	440	542	51
512			749	660	696		644	57
560			817	720	763		705	62
607			884	780	827		765	68
655			952	842	888		823	73
703			1020	899	948		879	78
750			1087	956	1009		934	83
796			1155	1011	1071		988	88
840			1223	1065	1133		1041	93

Use the appropriate factor for the orifice plate material.

$$F_c = \left[A - \left(A\left(0.5^{(\beta^{.25}-0.12)}\right)\sqrt{\frac{1-700\beta^{1.3}}{1200}}\right)\right] \times \left[\sqrt{\frac{\sin\left(90\sqrt{\frac{R_D}{350\beta^{1.3}}}\right)}{\beta^{0.15}}}\right] + 0.997$$

If $700\beta^{1.3} < R_D < 1200$:

$$F_c = \left[A - \left(A\frac{(350\beta^{1.3})^{(\beta^{.25}-0.12)}}{R_D}\sqrt{1-\frac{R_D}{1200}}\right)\right] + 0.997$$

When $R_D > 1200$:

$$F_c = (A + 0.997)$$

5. Divide original S by F_c for viscosity correction.
6. Repeat step 3 for new β for corrected S.
7. Repeat step 4 for final F_c for corrected β.
8. Repeat step 5 for final S.
9. Repeat step 3 for final β.
10. Calculate orifice bore.

$$d = \beta D \text{ in}$$

11. Calculate flow rate for an even-sized orifice plate.

$$\beta = \frac{d}{D}$$

Calculate $K_o\beta^2$ using the equation in step 2.

$$S = 0.598\beta^2 + 0.01\beta^3 + 0.581\beta^{6.425}.$$

$$\text{new flow} = \frac{\text{new } S}{\text{original } S} \times \text{original flow, lb/h}$$

Nomenclature

A = Partial calculation of F_c dimensionless
D = Inside diameter of pipe, in
d = Diameter of orifice bore, in
F_c = Reynolds number correction factor, dimensionless
G_f = Specific gravity of liquid at flowing temperature, dimensionless
h_m = Maximum differential range, in of water
R_D = Reynolds number, dimensionless
S = Beta ratio index, dimensionless
T_f = Flowing temperature, °F
W_m = Maximum rate of flow, lb/hr
μ = Absolute viscosity, centipoises
β = Beta ratio d/D, dimensionless

Reference

Spink, L. K., *The Principles and Practice of Flowmeter Engineering*, The Foxboro Company.

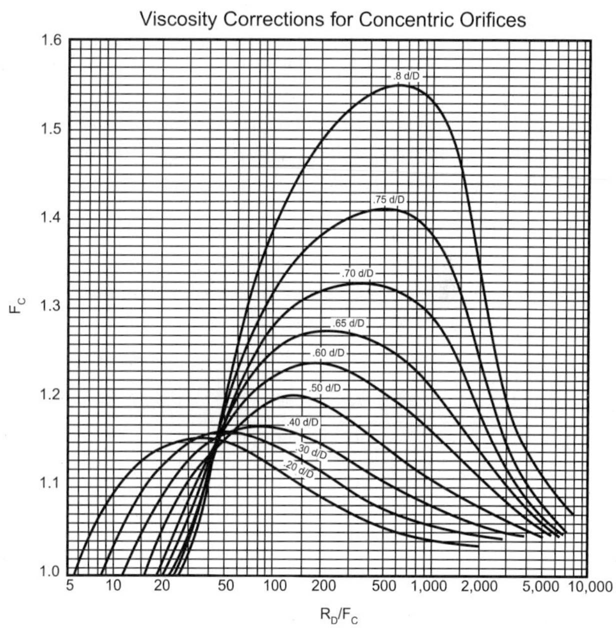

Mass measurement light hydrocarbons

Streams that transport mixed natural gas liquids require the use of mass measurement to accurately account for the volume of the components which make up the mix. When natural gas liquid components are mixed, nonideal mixing occurs. Compressibility factors are difficult to determine for mixes. Mass measurement eliminates the effects of nonideal mixing and the need for compressibility factors.

Certain precautions must be noted in using the mass measurement method. The density and volume must be measured at the same flowing conditions. An accurate composite sample of the flowing stream must be collected and analyzed with care. On systems where there are a number of input points that will be compared to an output point, it may be desirable to have all composite samples analyzed at the same laboratory.

The following data are needed for the calculation of quantities of natural gas liquid components:

- Metered volume at operating conditions
- Meter factor, if applicable
- Density at flowing conditions
- Composite sample and sample analysis for metered volume
- Component lb mass for each component in the stream
- Density—lb mass per U.S. barrel for each component in the stream

Sample calculation

Metered volume = 25,000 barrels
Meter factor = 1.0015
Density = 0.480

Composite sample analysis

Component	mol %
Carbon dioxide	0.08
Methane	1.90
Ethane	41.30
Propane	38.00
Isobutane	5.00
Normal butane	9.30
Isopentane	1.50
Normal pentane	1.10
Hexanes+	1.82
Total	100.00

See Table 1 for the details of the calculation.

Table 1
Mass Measurement–Light Hydrocarbons

Component	Mol %	Mol Wt	Mol % × Mol Wt	Wt Frac.	# Mass	Density (#/bbl)	U.S. bbl
CO_2	0.08	44.010	3.5208	0.000852	3588	286.4358	13
C_1	1.90	16.043	30.4817	0.007381	31,063	105.0000	296
C_2	41.30	30.070	1241.891	0.300739	1,265,556	124.7232	10,147
C_3	38.00	44.097	1675.686	0.405787	1,707,617	177.5256	9619
IC_4	5.00	58.123	290.615	0.070375	296,153	197.0934	1503
NC_4	9.30	58.123	540.5439	0.130899	550,844	204.4980	2694
IC_5	1.50	72.150	108.225	0.026208	110,287	218.7444	504
NC_5	1.10	72.150	79.365	0.019219	80,877	220.9914	366
C_6^+	1.82	87.436	159.1335	0.038536	162,166	244.2300	664
	100.00		4129.461		4,208,151		25,805

Total lb mass = 25,000 × 1.0015 × 0.480 × 8.337 × 42 = 4,208,151.
Column 1–From Sample Analysis.
Column 2–From Table of Hydrocarbon Properties (Figure 1).
Column 3–Column 1 × Column 2.
Column 4–Mol % × Mol Wt divided by sum of Column 3.
Column 5–Column 4 × total lb mass.
Column 6–From Table of Hydrocarbon Properties (Figure 1).
Column 7–Column 5 divided by Column 6.

Reference: GPA Standard 8173-83, Method for Converting Natural Gas Liquids and Vapors to Equivalent Liquid Volumes. GPA Standard 8182-82, Tentative Standard for the Mass Measurement of Natural Gas Liquids.

Pipeline measurement of supercritical carbon dioxide

G. W. Marsden, Shell Pipeline Corporation and **D. G. Wolter,** Shell Oil Company

The Shell Companies have been involved in carbon dioxide (CO_2) systems used for enhanced oil recovery for a number of years. Pilot programs existed in the late 1970s using recovered combustion generated carbon dioxide. In 1981, a plan was in place for a large-scale commercial pipeline transportation system connecting carbon-dioxide producing fields in southwestern Colorado and the oil fields of the Permian Basin in West Texas (see Figure 1). The Cortez Pipeline Company was set up as the ownership company and Shell Pipeline Corporation was established as the operator.

The 810-km (502 mi) long, 760-mm (30-in) diameter Cortez pipeline was designed to operate under supercritical conditions of 9000 to 19,000 kPa (1300–2700 psig), 286 to 322 K (55–120 °F), and with the nominal stream composition shown in Table 1. Design flow rates were 4.6–13.83 kmol/s (330–1000 MMscfd). Line fill was completed in May 1984 and since then, the system has been in continuous operation. There are 10 receipt meters and nine delivery meters currently in operation, each with a custody transfer measurement facility.

Table 1
Stream Composition

Major Components	m^3/m^3	% Vol
CO_2 minimum	0.95	95
Nitrogen, maximum	0.04	4
Methane	0.01–0.05	1–5
Contaminants		
Hydrogen sulfide	0.00002	0.002
Water	0.48 g per m^3 at standard conditions	30 lb/MMscf

Definitions

"Carbon dioxide," as referred to here and as used in the pipeline industry, is a CO_2-rich mixture normally having nitrogen and methane as principal volumetric contaminants. A minimum quality of 95% carbon dioxide purity is used by many pipeline carriers, while Cortez experience thus far is circa 98% CO_2 purity. The measured quantity is the total of all mixture constituents.

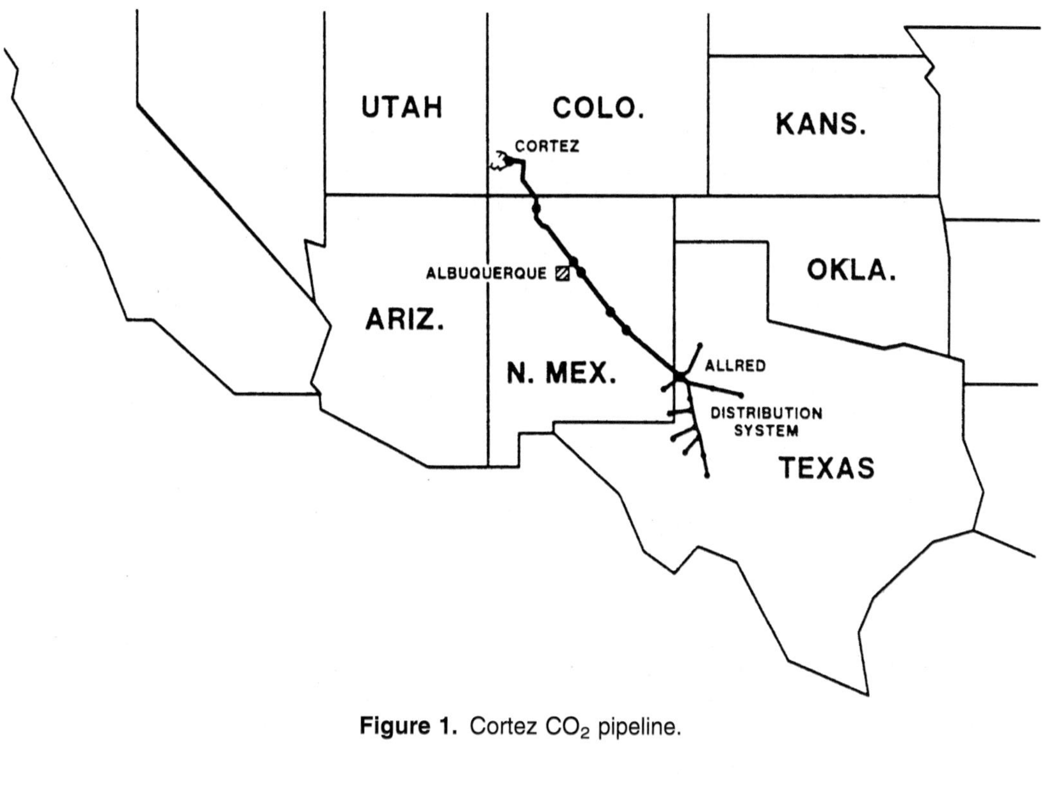

Figure 1. Cortez CO_2 pipeline.

Another term used by liquid pipeline operators is "custody transfer." Liquid pipeline operators typically do not own the fluid being transported, but instead accept custody at pipeline origin (receipt) and transfer that custody back to the owner or other party at the pipeline destination (delivery).

Thus, pipelines have custody transfer measurement facilities, which are composed of generally accepted, commercially available, and accurate measurement equipment. Additionally, the operation of this equipment is accomplished in a uniform manner acceptable to all parties to the transaction: buyers, transporters, and sellers. The *API Manual of Petroleum Measurement Standards* is widely used in the design and operation of these facilities.

These brief definitions provide insight into the selection of custody transfer measurement methods and equipment. Now, let us return to the CO_2-rich fluid we want to address.

Physical properties

CO_2, in its pure form, has a critical point of 304 K (88 °F) and 7415 kPa (1071 psia). Within practical limits, the characteristics of pipeline-quality CO_2-rich mixtures will approximate those of pure CO_2. Refer to the pressure enthalpy diagram for CO_2 (Figure 2), which illustrates the area of normal measurement for Cortez Pipeline.

The most economical pipeline operation is in the supercritical single-phase area. Fortunately, this operating area also coincides with our preferred measurement envelope. There is no pressure letdown for the delivery; in practice, the receiving party prefers the fluid with a pressure level similar to the pipeline operating pressure. This leads to measurement systems where pressure conditions are quite similar at both the inlet to and the outlet from the pipeline.

Product temperature does vary from the inlet measurement facility, which is downstream of compression equipment, to the delivery stations where the CO_2 stream approaches the ambient ground temperature. Future increases in throughput will require additional pumping/compression equipment, which will tend to create measurement operating temperatures that are similar at both pipeline inlet and outlet.

The upshot of these temperature and pressure conditions is the measurement envelope shown in Figure 2. This envelope is considerably more confined during normal measurement operations than is the pipeline operating envelope. The operating envelope includes the changes in hydraulic conditions that occur along the route of the pipeline.

Looking again at the P–E diagram (Figure 2), observe that the density of the CO_2-rich stream is in the range of 640–960 kg/m^3 (40–60 lb/ft^3). Additionally, note that the prime determinant of a change in density will be a change in temperature. This sensitivity of change in density with temperature is of concern in the design of the measurement facilities, and it has led to special attention being given to temperature loss/gain between elements of the measuring system. There is an approximate 0.9% volume change for each °C (0.5%/°F).

The change of density with pressure change in the measurement envelope is considerably less prominent, being on the order of one tenth as severe as temperature.

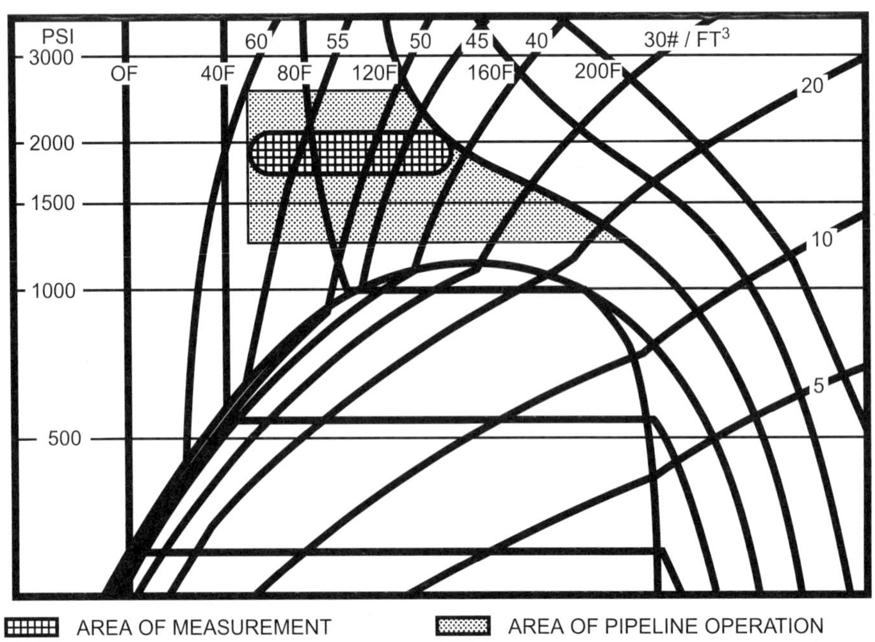

Figure 2. Carbon dioxide pressure enthalpy diagram.

Measurement methods in use today

The commercial planning and development of the Cortez carbon dioxide project preceded the engineering activity, and the supply and sales contract negotiations were well advanced before custody transfer flow measurement methods were investigated. The resulting contracts, based on existing natural gas pipeline systems, dictated the use of MMscfd volumetric units for gas phase equivalent volume of carbon dioxide.

Custody transfer measurement of carbon dioxide in large volumes at high pressures presented significant technological hurdles in the methods, performance, and materials areas. First, the probable stream compositions and the variations of the source field carbon dioxide were uncertain. This uncertainty of composition, and thus the physical properties, led to the selection of mass measurement techniques that would provide flexibility and operability when product conditions changed. Shell's decade-long experience with mass measurement of variable mix hydrocarbon streams and with supercritical streams suggested mass measurement as one of the strongest contenders. Mass measurement, however, would not directly provide the volumetric quantities that had been established in the preceding business negotiations.

The selection of meter systems to meet the above requirements resulted in two systems that are used today in the carbon dioxide transportation industry. The orifice meter, common in the natural gas industry, and the turbine meter, similar to those used in liquid measurement, were the dominant choices. As with most other choices, there are advantages and disadvantages (see Table 2) to each method. Both are operated as mass meter systems with subsequent conversion to volumetric terms for contract compliance and accounting requirements.

Fortunately, Shell Pipeline had several years of operating experience with mass measurement of petrochemical products, i.e. ethane and ethylene, as background in the orifice measurement system. The orifice meter was selected based on analysis of the factors shown in Table 2. The units conversion process, however, required additional facilities and methods for routinely and rapidly determining the composition of the streams, the calculation of molecular weight for the equivalent gas, and the conversion to volume at the selected reference pressure and temperature conditions. The resulting methods and facilities in service today are discussed below.

Table 2
Carbon Dioxide Pipeline Metering Options

	Orifice Meter	Turbine Meter
Advantages	Industry acceptance Proven equipment and operations Least sensitive to density determination accuracy	Accuracy ±0.5% Rangeability ±10:1
Disadvantages	Limited rangeability Accuracy ±1% Costly installation	Performance unknown with no experience in CO_2 No experience with proving in CO_2

Orifice measurement practices

The measurement system that provided basic simplicity, reliability, wide acceptance, and the capability of handling variable mix streams without breaking new frontiers in measurement methods was resolved to be the orifice flow measuring element with online density meter and microprocessor flow computer. This basic assembly of three proven devices provides real time-mass flow output. The orifice is the key to the success of the measurement system.

The orifice plate has a 50-year record of widespread use in the measurement of all manner of fluids. It is a static device generally inert to the measured fluid conditions, and calibration consists of simple dimensional measurement and conformance to specified physical tolerances. This, along with the periodic calibration of the associated differential pressure transmitter, completes the verification of this part of the measurement system. The orifice meter is one of the few measurement devices accepted for custody transfer that does not require in-service proving.

The second element, the density meter, has been selected to be an external unit that is easily isolated from the flowing stream for calibration, inspection, and maintenance. The relative insensitivity of CO_2 density to small changes in pressure in the primary flow to the orifice meters permits locating the density meter upstream of the meter manifold, thereby serving several meters. All installations are insulated to assure that pipeline density is maintained in the density meter.

The microprocessor flow computer, or third element of the system, is essential to achieving the advantages of integrated mass flow. This advantage comes from the ability of the computer to make computations in essentially "real time." How this advantage comes about is seen upon examination of the flow equation.

Orifice flow equation

The *API Manual of Petroleum Measurement Standards*, Chapter 14, Section 3 (also API 2530, AGA Report No. 3,

ANSI/API 2530-1985, and GPA 8185-85) provides the widely accepted formulation for mass flow. Mass flow is considered by this source to be the fundamental orifice meter flow equation.

Equation 38 of this publication, "Equation for Mass Flow Rate when Density of Fluid Flowing is Known", is shown in Figure. 3.

The terms used in Figure. 3 are:

q_m is the mass flow rate.

K and Y, the flow coefficient and the expansion factor, are the empirical elements of the orifice equation, based upon the work reported by Buckingham and Bean in 1934.

The flow coefficient, K, is in part dependent upon the viscosity of the fluid. Pipeline CO_2 viscosity ranges from approximately 0.05 to 0.08 mPa's (0.05–0.08 cP) in the measurement envelope.

The expansion factor, Y, is a function of the ratio of the orifice differential pressure to static pressure and to the specific heat ratio of the CO_2. The specific heat ratio ranges from 2.9 to 4.0 for CO_2 in the measurement envelope.

ρ_{fl} is the density output of the density meter.

ΔP is the differential pressure across the orifice plate.

Two correction factors, F_a and F_{pwl}, from Appendix E of Chapter 14, Section 3 are also utilized. F_a provides for the thermal expansion of the orifice plate, and F_{pwl} provides for local gravitational correction for deadweight testers.

Note that errors in density are taken at half their real value by integrating the density value with the orifice differential before taking the square root.

Figure 4 illustrates the accumulation of individual tolerances to arrive at the mass measurement system uncertainty tolerance. Individual element tolerances are modified by an

14.3.5.4.3 "Equation for Mass Flow Rate when Density of Fluid Flowing is Known"

$$q_m = \frac{\pi}{4}[2g_c]^{0.5}KY_1d^2[\rho_{f1}\Delta P]^{0.5}$$

Correction Factors in Accordance with Appendix E

E.3: Orifice Thermal Expansion Factor

$$F_a = 1 + [0.0000185(t_f - t_{meas})]$$

E.7: Local Gravitational Correction for Deadweight Tester

$$F_{pwl} = \left[\frac{g}{g_o}\right]^{0.5}$$

Figure 3. CO_2-rich mixture orifice measurement equation. From the *API Manual of Petroleum Measurement Standards*, 2nd Edition, September 1985.

VARIABLE	TOLER- ANCE %	EFFECT	SQ OF TOL & EFF	SUM OF SQUARES	SQ. RT. OF SUM %
Basic Orifice, F(B)	0.50	1.00	.2500	.2500	.5000
Diffntl Press, H	0.50	0.50	.0625	.3125	.5590
Density	1.00	0.50	.2500	.5625	.7500
Processor Error	0.01	1.00	.0001	.5626	.7500
Reynolds, F(R)	0.03	1.00	.0009	.5635	.7506
Expansion, Y	0.02	1.00	.0004	.5639	.7509
Orifice, F(A)	0.05	1.00	.0025	.5664	.7523

Individual factors with tolerance as above will provide system measurement with tolerance of (plus or minus) 0.75%

Figure 4. Orifice meter system tolerance (root-sum-square method).

effect factor, then the square root of the sum of the squares of these products is computed. The resulting overall measurement uncertainty for the Cortez mass measurement system is ±0.75%.

Pipeline operation

Full Cortez pipeline operation is accomplished using the mass flow output of the measurement system. This system also forms the basis for all custody transfer measurements in volumetric units.

It is necessary to introduce a mass-to-volume conversion step to satisfy the volumetric contracts. This conversion is accomplished using an analysis of the composite stream with a gas–liquid chromatograph maintained at each measurement area. The CO_2, N_2, and CH_4 composition of the CO_2-rich stream is determined for each measurement ticket period. An off-line equation of state computer program then calculates the appropriate standard volumetric units. A custody transfer measurement ticket (see Figure 5) is produced to document all pertinent data, including mass units, three-component composition, and the standard volumetric units.

The equation of state computer program utilized by Cortez Pipeline for this conversion was developed specifically for CO_2-rich mixtures. It is based upon the International Union of Pure and Applied Chemists (IUPAC) publication, "CO_2 International Thermodynamic Tables of the Fluid State," Volume III, compiled by Angus, Armstrong, and DeReuch. The method of extended corresponding states is utilized for the stated properties of the CO_2-rich mixture.

Turbine metering

The authors do not have firsthand experience or knowledge with turbine metering of carbon dioxide and therefore will not delve into the details of the application. In 1981, turbine metering was considered as one of the two major alternatives and rejected for a variety of reasons. Interestingly, other companies did select turbine metering, and experience is now being accumulated on large-scale carbon dioxide pipelines.

Turbine meters were not selected for Cortez or subsequent Shell systems, because the experience with mechanical displacement provers at such high pressure with compressible fluids was nonexistent. Likewise, the application of liquid or gas turbine meters in supercritical carbon dioxide service was a complete unknown.

The volumetric accuracy of turbine meters was attractive and no doubt led others to select them. However, the direct influence of density in the mass flow calculation is a disadvantage (as compared to orifice metering where density enters the equation as the square root of density and effectively halves the sensitivity of mass flow accuracy to density determination errors) (See Figure 6).

The proving of turbine meters was a concern and has been a significant commissioning problem on systems utilizing turbine meters. High-pressure provers in the 900 to 1500 ANSI flange ratings were scarce and expensive. The seals, design, and materials were considered potential problems. Significant progress has been made by those utilizing the equipment. In the long term, turbine meters and provers will likely be more widely applied as seal systems are

```
FIELD LOC       CO2 DENVER UNIT STATION                 DELIVERY
LEASE/COMP
FOR ACCT OF
CREDIT
ACCTG CODE  74222502 01   TICKET NO. 0004   ACCTG MO/YR   05 85
TICKET FROM DATE 04 01 85      TICKET TO DATE 05 01 85
METER NO.     1
CLOSE(MLBS)          226998
OPEN(MLBS)           186118
SUBTOTAL              40880
TOTAL ALL METERS (MT)              40,880  SAMPLE NO. 4
COMPOSITION,%MOL              CO2 98.350   CH4  .136   N2  1.514
SPEC. VOL(SCF/LB)        8.6568048
VOL = SPEC VOL X MT =          353,890    MSCF AT 14.650 PSIA/60F
OPEN    7:00  A M
CLOSE   7:00  A M                ISSUED BY:                DATE:
COMMENTS:
```

Figure 5. Cortez Pipeline Company. CO_2 meter ticket.

```
FOR ORIFICE METERING:

MASS = (CONSTANT) (DENSITY * DIFFERENTIAL) 0.5

FOR TURBINE OR VORTEX METERING:

MASS = (VOLUME) (DENSITY)
```

Figure 6. Mass flow calculation for orifice metering.

demonstrated to be reliable and as the other operational aspects are demonstrated to be manageable.

Potential measurement methods

Two additional meter types appear attractive and are coming into use in carbon dioxide service for various applications. Cortez Pipeline now utilizes the vortex shedding meter for line integrity/leak detection purposes, and the coriolis meter is attractive due to its simplicity and apparent accuracy.

The vortex shedding meter offers advantages in higher rangeability, requiring fewer meter runs than orifice meters, and thus lower installation costs. We believe the rangeability to be between 10 and 20 to 1 and the off-the-shelf accuracy to be 1% or better. Operating personnel have made online comparisons with adjacent, though several miles distant, orifice metering systems, and observe agreement within less than 1% for extended periods. No detailed studies have been made, however.

Currently, no custody transfer applications of the vortex shedding meter are known, although consideration is becoming more common. Application will first occur on low-volume service between agreeable parties or at the opportunity that assures a proving system will be available. (Proving via master meter and master prover techniques is one possibility that has been tested on other similar products.) Once the proving methods are applied, it is expected that accuracies better than 0.5% mass basis would be achievable.

Application of the coriolis meter lags the vortex shedding meter since it is a newer development. The expected accuracy of better than 1% and rangeability of 10 to 1 are attractive. The coriolis meter also provides direct mass measurement, which removes the need for a separate density determination and should significantly reduce both the installed and operating costs. The meter currently lacks widespread industry acceptance for custody transfer measurement applications, and the high-pressure operation of carbon dioxide systems currently limits selection to the smaller meter sizes. No custody transfer metering applications are known at this time.

Future application of the coriolis meter in carbon dioxide service will be much the same as described earlier for the vortex shedding meter. Industry experience, acceptance, application of proving methods, and availability in larger sizes with higher pressure ratings will open the routes for increased use.

The future

Carbon dioxide measurement technology is advancing rapidly, and the major difficulties in accuracy and operation are now identified. One accuracy enhancement that is progressing is the proprietary development of a flow computer with an accurate equation of state for carbon dioxide density determination from on-line/real time pressure and temperature inputs and keypad, or possibly analog, input of stream composition.

At this time, the Shell equation of state is a proprietary unit and is being used in its mainframe form for conversion of mass to volumetric units on Cortez Pipeline. Industry acceptance and verification in field practice has been excellent. The equation of state is a modification of the IUPAC equation for pure carbon dioxide and incorporates the effects of nitrogen and methane impurities over a range of 80–100% carbon dioxide. The equation of state is considered to be accurate to ±0.2% for density determination in the supercritical regions. The program is being applied to a flow computer with density update calculations every 1–2 min. Pressure and temperature are real-time analog inputs, while composition will be manually input via the keypad on a periodic basis dependent upon compositional stability of the fluid at each location. The planned operation region is shown in Table 3.

It is recognized that the National Bureau of Standards and others are pursuing similar equation-of-state expressions. Early investigation, however, determined that an industry-accepted equation of state was in the distant future and the earliest implementation could be achieved on a proprietary development with industry acceptance to follow. This decision was also supported from the view that numerous inhouse

Table 3
Region of Operation for CO_2 Equation of State

Fluid	Single Phase CO_2		
		m^3/m^3	Vol %
Composition	CO_2	0.8–1.0	80–100
	Methane	0–0.02	0–2
	Nitrogen	0–0.20	0–20
Pressure	3,400–20,500 kPa absolute (500–3000 psia)		
Temperature	273.16–333.16 K (32–140 °F)		

uses could be made of the computers where outside acceptance was not required. These applications included process operations at the carbon dioxide production, separation, and reinjection plants where density meter operation is a considerable burden and optimized accuracy is not justified. Likewise, the computer is attractive for comparison against critical density meters and for line integrity/leak detection flow measurement facilities. We are hoping to implement some of these applications in 1986.

In addition to the equation of state, it is our intent to remain flexible and consider the use of the potential meter types discussed earlier. The meter manifolds on the Cortez system have been designed with flexibility for other meter types and proving operations if these become sufficiently attractive.

The technology of carbon dioxide pipeline measurement is gaining uniformity via conferences such as this and joint industry development projects such as the equation-of-state work at NBS. In the commercial area, it is desirable that similar efforts be directed toward the use of mass measurement for operation and custody transfer purposes and to do away with the processes necessary for conversion to volumetric units. Mass units are obtained with simpler systems, at a lower cost, and more accurately. We should all work toward converting the contracts and agreements to mass units.

Reprinted with permission - *Oil & Gas Journal*, November 3, 1986.

Master meter proving orifice meters in dense phase ethylene

James E. Gallagher, Shell Oil Company, Houston, Texas

Shell Pipeline Corporation, cognizant of the importance of highly accurate orifice measurement for polymer-grade ethylene, embarked on a developmental program to improve the overall uncertainty of orifice metering. The program's goal was to successfully apply master meter proving techniques to dense phase ethylene orifice meter facilities operated on a mass basis.

This paper presents the techniques and results of mass proving concentric, square-edged, flange-tapped orifice meters. All data were obtained on commercially operated metering facilities that are in compliance with ANSI 2530 requirements.

Results indicate an overall orifice meter uncertainty tolerance of approximately $\pm 0.31\%$ is attainable when orifice meters are mass proved in situ using master meter proving techniques.

Goal of development program

During the past decade, increasing commercial activity for ethylene demanded improvement of measurement technology. Significant measurement uncertainties were being experienced on ethylene pipeline systems. In response, a developmental program to improve the overall uncertainty of orifice metering was initiated by Shell Pipeline Corporation.

The program's goal was to develop an economical method for proving ethylene orifice meters under actual operating conditions. Master meter proving on a mass basis was selected as the only viable approach.

Pipeline transportation and measurement systems

An ethylene distribution system consists of a pipeline network and salt dome storage facility linking producers and consumers. Since producers and consumers are not equipped with on-site storage facilities, the pipeline is designed with maximum flexibility to satisfy the continually changing demands on the operations.

Shell's ethylene systems are operated in the dense phase fluid region due to lower transportation costs. The ethylene meter stations operate in two regions depicted in Figure 1—the dense-phase fluid and single-phase gas regions.

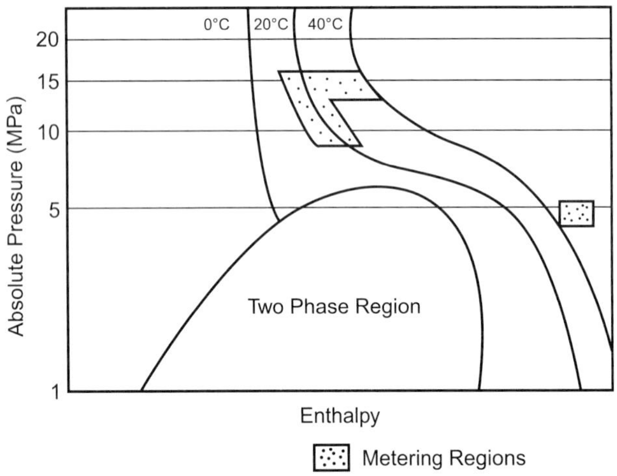

Figure 1. Ethylene pressure-enthalpy diagram.

Velocity Profile
- Obstructions, Deposits, Etc.
- Piping Configuration
- Pre & Post Run Lengths
- Pipe Condition
  - Wall Roughness
  - Roundness
- Plate Condition
  - Concentric
  - Edge Sharpness
  - Nicks or Burrs
  - Flatness
  - Roughness
- D.P. Sensing Taps
  - Burrs

Operating Conditions
- Pulsating Flow
- Widely Varying Flow

Fluid Properties
- Density
  - Equation of State Uncertainty
  - Transmitters' Sensitivity
- Expansion Factor
- Viscosity

Instrumentation
- Ambient Effects
- Static Pressure on "DP" Sensing
- Calibration Accuracy

Figure 2. Influence quantities on orifice meters.

The quantity of ethylene measured by concentric, square-edged, flange-tapped orifice meters, shown in Figure 3, is calculated through the combined use of the mass orifice equation and API's ethylene density equation. In the United States, the applicable standards are API, Chapter 14, Section 3—Orifice Metering of Natural Gas and Other Related Hydrocarbon Fluids, Second Edition, September 1985 (ANSI 2530), and API Chapter 11.3.2.1—Ethylene Density.

In Shell, the concentric square-edged, flange-tapped orifice predominates as the primary element. Most installations are equipped with a dual chamber orifice fitting to allow retrieval of the orifice plate under pressure. For the most part, the orifice plates are sized to maintain a 0.3 to 0.6 beta ratio. Solid state electronic flow computers and transmitters are installed to maximize metering accuracy. Parallel meter runs and stacked "dp" transmitters maximize rangeability while maintaining a high degree of accuracy.

Historically, the industry has presumed that compliance with the reference standard piping conditions infers accuracy. No consideration was given to deviation from the empirical equation's criteria upon initial installation or

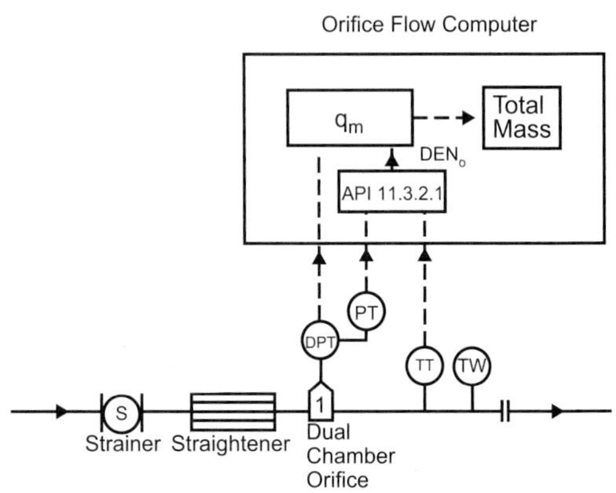

Figure 3. Ethylene orifice meter.

further deviation over the life of the facility. The major weakness of orifice metering has been the inability to prove the meter under actual operating conditions.

In 1982, the Industry Ethylene Measurement Project conducted at Mt. Belvieu, Texas, demonstrated the feasibility of gas turbine meters, conventional piston pipe provers, and API's equation of state for custody transfer. The project also demonstrated the turbine meter and pipe prover approach produced a lower uncertainty than orifice meters. The ability to accurately prove the turbine meter, shown in Figure 4, on a mass basis under actual operating conditions eliminated the reliance on reference conditions.

However, installation costs could not justify this technique for most custody transfer installations. As a result, the majority of Shell's ethylene custody transfers continues to be based on the traditional orifice meter.

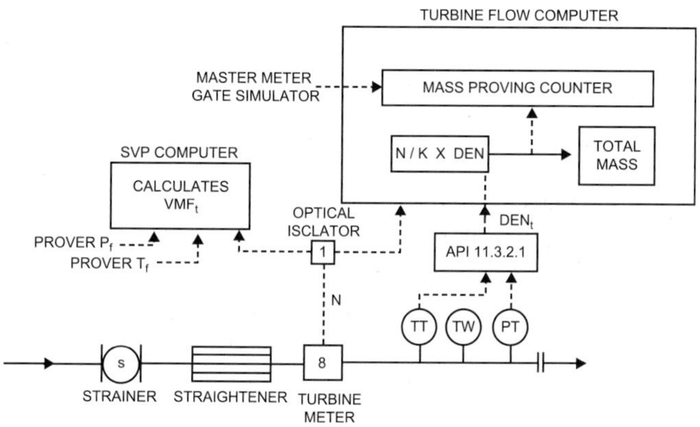

Figure 4. Ethylene turbine meter.

Development program

Development of the master meter proving technique was conducted on commercially operating ethylene pipelines by field personnel. Orifice meters associated with these systems were proven under actual operating conditions.

Fluid properties. The thermodynamic and physical properties of polymer-grade ethylene have been the subject of intensive investigations. Comprehensive treatises on the density of pure ethylene have been published by two highly reputable sources—API and the NBS. The pipeline industry in the United States presently utilizes API.

Chapter 11.3.2–Ethylene Density (API 2565). Ethylene densities are calculated using measured temperatures and pressures. The accuracy of the calculated density depends on the accuracy of the equation of state, the accuracy of the measuring devices, and the sensitivity of the calculated density to small temperature and pressure variations.

Portable proving facility. A truck-mounted 900 ANSI proving system was developed to allow maximum utilization of capital investment. The truck is equipped with a small volume prover, master turbine meter, and associated instrumentation, as shown in Figures 5 and 6.

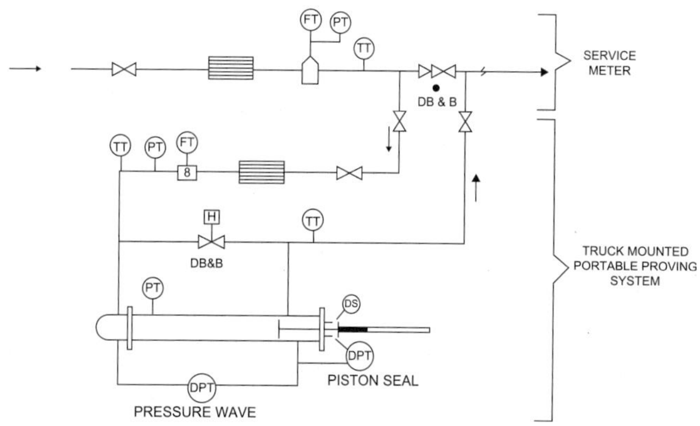

Figure 5. Master meter method—mechanical.

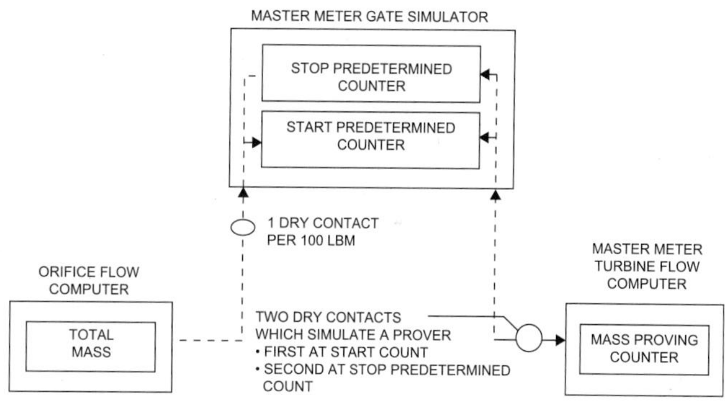

Figure 6. Master meter method—electronic.

The small volume prover, SVP, operates on the principle of repeatable displacement of a known volume of fluid in the calibrated section. The mass of the fluid in the prover is determined through the combination of the prover's base volume, the fluid's pressure and temperature, and the API ethylene density equation.

The SVP was selected based on the following criteria:

- Minimize capital investment by using portable system
- Unidirectional flow design
- Honed and calibrated pipe section
- Dynamic piston seal leak detection
- Single fiber optic piston position detector
- Control of launching pressure wave
- Double chronometry pulse interpolation

Principal parts include an outer housing, a honed inner measuring cylinder, a piston displacer with an Invar extension rod, and an automatic hydraulic system. The hydraulic system returns the piston displacer to the upstream launch position after each proving run and actuates the prover bypass valve to allow the fluid to bypass the prover while the piston is being returned to the launch position.

The pressure wave that occurs when launching the prover's piston is monitored by a high-speed recorder, which was added during initial testing. The differential pressure, or pressure wave, across the prover is recorded to assure the piston does not act as a compressor or brake. The hydraulic system controls the piston to maintain a constant velocity through the calibrated section.

A gas turbine meter specially modified by the manufacturer for ethylene service was selected for the master turbine meter. The meter is equipped with a Lexan rotor, ball bearings, external lubricator, and conventional preamplified electronic pulse output. The master meter has a stated linearity of ±0.25% over a 5:1 turndown. The truck is equipped with two master turbine meters to cover normal operating flow rates—10 mm (4 in) and 50 mm (2 in) turbine meters.

The SVP is controlled by a microprocessor-based computer. The computer monitors repeatability, piston position, and prover bypass valve position; performs dual chronometric pulse interpolation; and calculates/prints a hard copy report for each successful proving.

The computer receives input from the prover's temperature and pressure transmitters, the turbine meter's pulse train, the fiber optic gate detector, and various hydraulic controls.

Two microprocessor-based flow computers are used to calculate the turbine meter's mass flow. The first computer calculates the API ethylene density based on the turbine meter's pressure and temperature transmitters. The second computer calculates the turbine's mass flow based on the turbine meter's pulse input and contains an internal mass proving counter.

Solid state temperature transmitters, equipped with high-accuracy 100 ohm platinum RTD sensors, are installed 4–5 pipe diameters downstream of the turbine meter and at the exit of the prover. The calibrated accuracy is ±0.08 °C (0.15 °F).

Solid state pressure transmitters, equipped with a sealed capacitance sensing element, are installed 4–5 pipe diameters downstream of the turbine meter and at the inlet to the prover. The calibrated accuracy is ±6.89 kPa (1.0 psi).

Solid state differential pressure (dp) transmitters, equipped with a sealed capacitance sensing element, are installed across the prover and the piston's seal. The calibrated accuracy is ±12 Pa (0.05 in H_2O).

A high-speed chart recorder monitors the prover's "dp" transmitters:

- To assure that the prover does not induce a pressure wave greater than 6.89 kPa (1 psi) while the piston is in the calibrated section.
- To assure that the piston's seal does not leak in the dynamic state.

A solid state counter, equipped to receive a contact closure from the orifice's computer and send a dry contact closure output to the turbine meter's flow computer, is installed to start and stop the master meter proving run. This device triggers the running start/stop master meter proving.

Master meter proving principle. The proving is based on the concept that a "master meter" can act as a transfer standard between the primary standard and another meter.

Meter proving by the master meter method requires two distinct steps:

- Mass prove the master turbine meter on site under operating conditions.
- Master meter prove the orifice meter on a mass basis using the master turbine meter.

To assure the devices are experiencing identical conditions, the orifice meter, turbine meter, and SVP are placed in series, as shown in Figure 5, during the entire calibration. Before proving the "master meter," the flow and/or density conditions through the orifice meter, turbine meter, and SVP must approach steady state conditions.

In the "master meter" proving method, the turbine meter is calibrated on a mass basis under actual operating conditions by the SVP. Once the turbine meter has been proved, a comparison of the 2 m mass outputs forms the basis for determining the orifice meter factor as shown in Figure 6.

Proving procedures

In order to assure accuracy, the operator must be certain the system approaches steady state conditions (flow rate, density, pressure, temperature) and that no leakage is occurring between the orifice, turbine, and the prover. If steady state conditions do not exist, accuracy and/or repeatability will be extremely difficult to obtain. The orifice meter's instrumentation is calibrated prior to proving.

Preparation.

1.1 Initiate flow through the SVP. Check for leakage on all critical valves, relief valves, and piston displacer seal.
1.2 Calibrate all instrumentation associated with the prover truck (pressures, temperatures, flow computers).
1.3 Check operation of assembled instrumentation by performing preliminary turbine meter proving runs.

1.4 If necessary, adjust hydraulic system to assure the prover does not act as a compressor or a brake. Adjust the hydraulic system if the piston displacer's differential pressure exceeds ±6.89 kPa (1 psi).

1.5 Allow ethylene to circulate through the prover truck until the outlet temperature is stabilized.

Proving run.

2.1 Initiate the SVP proving sequence by commanding the prover computer/controller to obtain a VMFt. (Acceptable repeatability for the turbine meter is five consecutive runs in which the volumetric meter factor, VMFt, does not exceed a tolerance of 0.05%.) The computer will print a proving report when this criterion has been met.

2.2 Calculate the turbine meter's mass meter factor, MMFt.

2.3 Clear the turbine meter flow computer's internal mass prover counter.

2.4 Reset and program the master meter gate simulator to allow the master meter proving to occur for a minimum of 1 h. Record the turbine meter's mass flow rate. (The simulator initiates a master meter proving after receiving five dry contacts from the orifice computer. The simulator stops the proving after approximately 1 h. Example—for flow at 50,000 lb/h, the gates would be set at 5 and 505.)

2.5 When the master meter run is complete, record the turbine meter's mass counts.

2.6 Calculate the orifice's mass meter factor (MMFo).

2.7 Repeat steps 2.1 thru 2.6 until two consecutive master meter runs satisfy repeatability criteria. (Acceptable repeatability for orifice meter is two consecutive runs in which the MMFo does not exceed a tolerance of 0.10%.)

2.8 Average the results of the master meter runs. Enter the average MMFo into the orifice flow computer for use in continuing operations.

Figure 7 is a typical orifice proving report.

Analysis of data

Analysis of the data is contained in this section, along with the calculation methods for reduction of the data.

Proving calculations. The prover base volume, PBV, is stated at 101.325 kPa (14.696 psia) and 15.6 °C (60.0 °F). Therefore, correction for the effect of operating pressure and temperature shall be applied to determine the prover's volume at operating conditions. These corrections are as follows:

Correction for Effect of Pressure on Steel Prover (Cpsp)

$$\text{Cpsp} = 1.000$$

Correction for Effect of Temperature on Steel Prover (Ctsp)

$$\text{Ctsp} = \left[1 + 0.000024\left(T_f - 15.6\right)\right] \text{ SI Units}$$

$$\text{Ctsp} = \left[1 + 0.0000134\left(T_f - 60\right)\right] \text{ IP Units}$$

LOSS UNIT: ETHYLENE SYSTEM
LOCATION: TYPICAL
METER: 1

DATE: APRIL 10. 1986
REPORT NO.: 1
MCF: 1.0018

TURBINE METER DATA

FLOWRATE: 9100 LBM/HR
FLOWRATE: 11.466 KG/S

ORIFICE METER DATA

MODEL	SENIOR	FA	1.0000		
I.D.	5.761	FB	2479.1	I.D.:	146.33 MM
PLATE	3.375	FR	1.0002	PLATE:	85.73 MM
BETA	0.5858	Y2	1.0000	BETA:	0.5858
		FDP	1.0055		

TURBINE DATA				ORIFICE DATA			
RUN NO.	TOTALIZER (LBM)	MASS FACTOR	NET MASS (LBM)	TOTALIZER (100 LBM)	MASS (LBM)	MMFO	RUN NO.
1	91646.2	1.0006	91701.2	920 5	91500	1.00219	1
2	91594.9	1.0004	91631.5	920 5	91500	1.00143	2
	91620.6	1.0005	91666.4		91500.0	1.00181	
			REPEATABILITY	0.076%			

MMFO = 1.0018

REYNOLDS NUMBER		DISCHARGE COEFFICIENT		
PIPE	BORE	STOLZ	BUCKINGHAM	MASS
4.99E+06	8.52E+06	0.6046	0.6047	0.6057

REMARKS:

Figure 7. Shell Pipeline Corporation orifice master meter proving report version 86.2A.

The calibrated volume of the prover at operating conditions is:

$$PVtp = [PBV * Cpsp * Ctsp]$$

The mass of fluid in the prover's calibrated section at operating conditions is:

$$PMtp = [PVtp * DENp]$$

In the "master meter" proving method, the turbine meter is calibrated on a mass basis under actual operating conditions. The master meter's Mass Meter Factor, MMFt, can be calculated using the following equations:

$$TVtp = [(\text{Proving Run Pulses})/(K\ \text{Factor})]$$
$$TMtp = [TVtp*DENt]$$
$$VMFt = [PVtp/TVtp]$$
$$MMFt = [PMtp/TMtp]$$

The average mass meter factor, MMFt, for the turbine meter is calculated along with the repeatability for five consecutive runs. Acceptable repeatability is based on a maximum tolerance of 0.05% for the turbine's volumetric meter factor, VMFt. The master turbine meter is reproved at the start of each master meter proving run. This mass factor is applied forward.

In the "master meter" proving method, the orifice meter is proved on a mass basis under actual operating conditions. The average mass meter factor for the orifice meter, MMFo, is calculated along with the repeatability for two consecutive runs. Acceptable repeatability is based on a maximum deviation of 0.15%. An MMFo is calculated for each master meter proving run using the following equation:

$$MMFo = [(\text{Turbine Mass}) * (MMFt)]/[(\text{Orifice Mass})]$$

Discharge coefficients. When an orifice proving report was calculated, the empirically derived mass discharge coefficient was compared with two empirical predictions—the Stolz and Buckingham equations. The Mass Discharge Coefficient was calculated from the master meter proving results using the following formula:

$$\text{Mass Cd} = [(MMFo) * (\text{Buckinghams Cd})]$$

Influence quantities. Orifice meter applications are based on steady flow of a homogeneous, single-phase, Newtonian fluid with a velocity profile that follows the reference standard piping conditions. Departures from these reference piping conditions, the effect of the dynamic properties of the fluid, instrumentation effects, and operating conditions are termed influence quantities. Influence quantities are summarized in Figure 2.

Repeatability and reproducibility. Repeatability is the ability to obtain the same result when measuring the same quantity several times in succession. The turbine meter's VMFt has a required repeatability range of 0.05%. The orifice meter's MMFo has a required repeatability range of 0.10%. Repeatability has been expressed as a percentage on each orifice proving report.

Reproducibility is the ability to obtain the same result when measuring the same quantity under different conditions (i.e. time, observer, operating conditions, etc.). With respect to the turbine meter, reproducibility is not significant, since the meter is proved immediately prior to proving an orifice meter. The orifice meter's MMFo reproducibility was demonstrated over a time period of several days at different flow rates and was generally within 0.10%.

Uncertainty of orifice calibrations. The generally accepted statistical method of the quadrature sum (square root of the sum of the squares) has been applied in deriving the overall measurement accuracy.

In calculating the uncertainties, the following assumptions were made:

- The universe is homogeneous (statistically controlled) with respect to MMFo.
- The distribution is approximately Normal or Gaussian.
- The sample is random.
- A 95% confidence limit is required.

The PBV was established to within ±0.031% by water draw calibration using NBS certified standards.

When proved separately as a volume meter, the turbine meter's VMFt demonstrated a standard deviation of ±0.025% for five consecutive proving runs. An assigned uncertainty of ±0.05%, or two standard deviations, was used in calculating the overall uncertainty.

The flow computer has a stated uncertainty of 0.10%. The density of ethylene was determined using API Chapter 11.3.2.1—Ethylene Density, using measured pressures and temperatures. The uncertainty of the equation of state is stated as ±0.20%.

Additional uncertainties when calculating density using an equation of state, no matter how accurate the equation of state might be, are errors resulting from the uncertainties of the measured temperature and pressure. Transmitter uncertainties of ±0.08 °C (0.15 °F) and ±6.89 kPa (1 psi) at metering conditions of 8 MPa (1160 psia) and 20 °C (68 °F) would result in a calculated density uncertainty of ±0.15%.

Following is a summary of the master meter uncertainty equation:

	Tolerance (%)	Effect	Square of tol & eff
Prover calibration	0.03	1.0000	0.0009
Turbine repeatability	0.05	1.0000	0.0025
Turbine computer	0.10	1.0000	0.0100
Density computer	0.10	1.0000	0.0100
Equation of state	0.20	1.0000	0.0400
Density sensing	0.15	1.0000	0.0225
		Sum	0.0861
		Square root of sum	±0.2934%

When the orifice meters were proved on a mass basis, they demonstrated a standard deviation of ±0.05% for two consecutive proving runs. An assigned uncertainty of ±0.10%, or two standard deviations, was used in calculating the overall uncertainty.

The following is a summary of the overall orifice meter uncertainty equation as a result of proving on a mass basis using the master meter method:

	Tolerance (%)	Effect	Square of tol & eff
Master turbine	0.2934	1.0000	0.0861
Master meter proving	0.10	1.0000	0.0100
		Sum	0.0961
		Square root of sum	±0.3100%

The uncertainties calculated are indicative at the time of proving and do not reflect additional uncertainties associated with measurements occurring beyond the proving period.

Many factors contribute an element of uncertainty to the flow rate indicated by the flow meter in relation to the actual or true flow. Orifice metering systems are made of an assembly of parts originating from different manufacturers. To determine the uncertainty with which the total assembly can measure, the tolerances on the different components and the way they affect the metering accuracy have to be considered. Since the orifice meter has been proved in situ, the practical working formula for the uncertainty can be expressed as follows:

$$\% = \left(\left[(\text{orifice calibration})^2\right] \pm 0.25\left[(\text{dp error})^2\right] + 0.25\left[(\text{density sensing error})^2\right]^{0.5}\right)$$

Conclusions

The results of the orifice calibrations were rather interesting. Installations that were not in compliance with ANSI 2530 requirements were discovered, as well as human errors associated with flow computer programming.

The master meter method identified and measured errors associated with the following problems:

- Deformed orifice plates
- Eccentrically mounted orifice plates
- Orifice plates installed backwards
- Improperly programmed computers
- Effect of compressor pulsations.

Additional information was no doubt masked by the influence quantities. From an application standpoint, identifying the source of error is not as important as measuring and correcting for the error.

Based upon the results, the majority of orifice meters have indicated a general understatement of measured quantities prior to proving.

In summary, our original goal was achieved—to develop an economically viable method for proving ethylene orifice meters under actual operating conditions. The master meter proving method has demonstrated to be a reliable measurement tool in identifying the quantifying measurement problems.

References

1. *Orifice Metering of Natural Gas and Other Related Hydrocarbon Fluids*, API MPMS Chapter 14.3, Second Edition, September 1985 (AGA Report No. 3), (GPA 8185-85), (ANSUAPI 2530).
2. *Ethylene Density*, API MPMS Chapter 11.3.2.1, 1974.
3. *Shell Flow Meter Engineering Handbook*, Second Edition, McGraw-Hill Book Company (UK) Limited, 1985.
4. R.W. Miller, *Flow Measurement Engineering Handbook*, McGraw-Hill Book Company, 1983.
5. J. Stolz, *A Universal Equation for the Calculation of Discharge Coefficients of Orifice Plates*, Flow Measurements of Fluids, IMEKO Conference on Flow Measurement of Fluids, Groningen, The Netherlands, 1978.
6. R.H. Kolb, *Ethylene Mass Measurement System with Prover*, The Second Petroleum Measurement Seminar, 1980.
7. R.H. Pfrehm, "Improved Turbine Meter System Measures Ethylene Accurately," *Oil & Gas Journal*, April, 1981.
8. H. Bellinga, A. Kooi, C.P. Hoeks, F.A.L. van Laak, and P.J. Orbons, "Piston Prover Used to Calibrate Gas Meters," *Oil & Gas Journal*, August, 1985.
9. F.S. Boncher, L.M. Shipman, and L.C. Yen, *Hydrocarbon Processing*, 54 (2), 89 (1975).
10. *ASTM Manual on Presentation of Data and Control Chart Analysis*, ASTM Special Technical Publication 15D, 1976.
11. Measurement Uncertainty for Fluid Flow in Closed Conduits, ANSI/ASME MFC-2M-1983.

Notation

β = ratio of orifice plate bore to internal meter diameter
Cd = orifice discharge coefficient
Cpsp = correction for effect of pressure on prover
Ctsp = correction for effect of temperature on prover
d = orifice plate bore, in (mm)
dp = differential pressure across orifice plate, in H_2O @ 60 (Pa)
D = meter internal diameter, in (mm)
DENo = density of fluid at orifice, lbm/ft^3 (kg/m^3)
DENp = density of fluid at prover, lbm/ft^3 (kg/m^3)
DENt = density of fluid at turbine, lbm/ft^3 (kg/m^3)
F_a = orifice thermal expansion factor
Fb = basic orifice factor
Fdp = correction for static pressure on dp transmitter
Fr = Reynolds number factor
IP = Inch-Pound System of Units
MMFt = mass meter factor for turbine
MMFo = mass meter factor for orifice
PBV = prover base volume, ft^3 (m^3)
PMtp = prover mass, lbm (kg)
PVtp = prover volume, ft^3 (m^3)
q_m = mass flow rate, lbm/s (kg/s)
Q_m = mass flow rate, lbm/hr (kg/hr)
Rd = orifice bore Reynolds number
R_D = pipe Reynolds number
SI = International System of Units
SVP = small volume prover
T_f = flowing temperature, °F (°C)
TMtp = turbine meter's mass, lbm (kg)
TVtp = turbine meter's volume, ft^3 (m^3)
VMFt = volumetric meter factor for turbine meter
Y2 = expansion factor based on downstream static pressure

Reprinted with permission - *Oil & Gas Journal*, December 15, 1986.

Gas or vapor flow measurement—orifice plate flange taps

Sequence of calculations

1. Calculate $K_0\beta^2$ value.

$$K_0\beta^2 = \frac{W_m}{D^2\left(358.5 + .007T_f\right)\sqrt{\gamma_f h_m}}$$

2. Iterate to find uncorrected beta ratio.

$$K_0\beta^2 = \beta^2\left[\left(0.5993 + \frac{0.007}{D}\right) + \left(0.364 + \frac{0.0076}{\sqrt{D}}\right)\beta^4\right]$$
$$+[0.4\left(1.6 - \frac{1}{D}\right)^5$$
$$\times\left(0.07 + \frac{0.5}{D} - \beta\right)^{5/2}\left(0.009 + \frac{0.034}{D}\right)$$
$$\times(0.5-\beta)^{3/2} + \left(\frac{65}{D^2} + 3\right)(\beta - 0.7)^{5/2}]$$
$$\div\left[1 + 0.000015\left(830 - 5000\beta + 9000\beta^2\right.\right.$$
$$\left.\left. - 4200\beta^2 + \frac{530}{\sqrt{D}}\right)\right]$$

Any negative factor, such as $(\beta - 0.7)^{5/2}$ when β is less than 0.7, shall be taken as equal to zero, and thus the whole term $(65/D^2 + 3)(\beta - 0.7)^{5/2}$ will drop out of the equation.

Let fx equal the equation in step 2—$K_o\beta^2$, and let:

$$f^1x = 2\left(\left(0.5993 + \frac{0.007}{D}\right)\right.$$
$$+4\left(1.6 - \frac{1}{D}\right)^5\left(0.07 + \frac{5}{D} - \beta\right)^{5/2}$$
$$-\left(0.009 + \frac{0.034}{D}\right)(0.5-\beta)^{3/2}$$
$$\left.+\left(\frac{65}{D^2} + 3\right)(\beta - 0.7)^{5/2}\beta + 6\left(0.364 + \frac{0.076}{\sqrt{D}}\right)\beta^6\right)$$

Start with $\beta = 0.7$ and calculate fx/f^{1}x. If greater than 0.000015, subtract fx/f^{1}x from current value of β and repeat until the error is less than 0.000015.

3. Calculate the expansion factor.

$$Y = 1 - \frac{(0.41 + 0.35\beta^4)h_m}{39.6kP_f}$$

4. Divide $K_0\beta^2$ by Y.
5. Calculate the Reynolds number correction factor.

$$F_r = 1 + \left(830 - 5000\beta + 9000\beta^3 - 4200\beta^3 + \frac{530}{\sqrt{D}}\right) \times \frac{(D\beta)^2 \mu_{cp}}{4.42 W_m}$$

6. Divide $K_0\beta^2$ by F_r.
7. Repeat step 2 to find corrected beta ratio.
8. Repeat steps 3, 4, 5, and 6 using corrected beta ratio.
9. Repeat step 2 for final beta ratio.
10. Calculate orifice bore.

$$d = BD \text{ in}$$

11. Calculate flow rate for an even-sized orifice bore.

$$\beta = \frac{d}{D}$$

Calculate $K_o\beta^2$ using the equation in step 2.

$$\text{new flow} = \frac{\text{new } K_o\beta^2}{\text{original } K_o\beta^2} \times \text{original flow, lb/h}$$

Notation

D = inside diameter of pipe, in
d = bore diameter of primary device, in
F_r = Reynolds number correction factor for steam, vapor, or gas, dimensionless
h_m = maximum differential or range of differential gauge, in of water, dry calibrated
K_o = coefficient of discharge, including velocity of approach at a hypothetical Reynolds number of infinity, dimensionless
k = ratio of specific heats C_p/C_r, dimensionless
P_f = flowing pressure, psia
T_f = flowing temperature, °F
W_m = maximum rate of flow, lb/h
Y = expansion factor, dimensionless
β = diameter ratio d/D, dimensionless
γ_f = specific weight of flowing conditions, lb/ft^3
μ_{cp} = absolute viscosity, centipoises

Source

Instrument Engineering Programs, Gulf Publishing Company, Houston, Texas.

Reference

L.K. Spink, *The Principles and Practice of Flowmeter Engineering*, The Foxboro Company.

Properties of gas and vapors

Table 1
Physical Properties

No.	Compound	Formula	Critical Constants: Pressure, psia (P_c)	Critical Constants: Temperature, °R (T_c)	Molecular Weight (M)	Specific Heat, C_p/C_v (k)	Heat of Vaporization, Btu/lb (ΔHV)
1	Methane	CH_4	667.8	343.4	16.043	1.311	219.22
2	Ethane	C_2H_6	707.8	550.0	30.070	1.193	210.41
3	Propane	C_3H_8	616.3	666.0	44.097	1.136	183.05
4	*n*-Butane	C_4H_{10}	550.7	765.0	58.124	1.094	165.65
5	Isobutane	C_4H_{10}	529.1	735.0	58.124	1.097	157.53
6	*n*-Pentane	C_5H_{12}	488.6	845.7	72.151	1.074	153.59
7	Isopentane	C_5H_{12}	490.4	829.1	72.151	1.075	147.13
8	Neopentane	C_5H_{12}	464.0	781.0	72.151	1.074	135.58
9	*n*-Hexane	C_6H_{14}	436.9	913.7	86.178	1.062	143,95
10	2-Methylpentane	C_6H_{14}	436.6	895.8	86.178	1.063	138.67
11	3-Methylpentane	C_6H_{14}	453.1	908.3	86.178	1.064	140.09
12	Neohexane	C_6H_{14}	446.8	860.1	86.178	1.064	131.24
13	2,3-Dimethylbutane	C_6H_{14}	453.5	900.3	86.178	1.065	136.08
14	*n*-Heptane	C_7H_{16}	396.8	972.8	100.205	1.054	136.01
15	2-Methylhexane	C_7H_{16}	396.5	955.0	100.205	1.054	131.59
16	3-Methylhexane	C_7H_{16}	408.1	963.8	100.205	1.054	132.11
17	3-Ethylpentane	C_7H_{16}	419.3	973.5	100.205	1.054	132.83
18	2,2-Dimethylpentane	C_7H_{16}	402.2	937.2	100.205	1.053	125.13
19	2,4-DimethylperUane	C_7H_{16}	396.9	936.0	100.205	1.053	126.58
20	3,3-Dimethylpentane	C_7H_{16}	427.2	965.9	100.205	1.053	127.21
21	Triptane	C_7H_{16}	428.4	956.4	100.205	1.055	124.21
22	*n*-Octane	C_8H_{18}	360.6	1024.2	114.232	1.046	129.53
23	Diisobutyl	C_8H_{18}	360.6	990.4	114.232	1.049	122.80
24	Isooctane	C_8H_{18}	372.4	979.5	114.232	1.049	116.71
25	*n*-Nonane	C_9H_{20}	332.0	1070.7	128.259	1.042	123.76
26	*n*-Decane	$C_{10}H_{22}$	304.0	1112.1	142.286	1.038	118.68
27	Cyclopentane	C_5H_{10}	653.8	921.5	70.135	1.117	167.34
28	Methylcyclopentane	C_6H_{12}	548.9	959.4	84.162	1.085	147.83
29	Cyclohexane	C_6H_{12}	591.0	996.7	84.162	1.089	153.00
30	Methylcyclohexane	C_7H_{14}	503.5	1030.3	98.189	1.068	136.30
31	Ethylene	C_2H_4	729.8	506.6	28.054	1.243	207.57
32	Propene	C_3H_6	669.0	656.9	42.081	1.154	188.18
33	1-Butene	C_4H_8	583.0	755.6	56.108	1.111	167.94
34	*cis*-2-Butene	C_4H_8	610.0	784.4	56.108	1.121	178.91
35	*trans*-2-Butene	C_4H_8	595.0	771.9	56.108	1.107	174.39
36	Isobutene	C_4H_8	580.0	752.6	56.108	1.106	169.48
37	1-Pentene	C_5H_{10}	590.0	836.9	70.135	1.084	154.46
38	1,2-Butadiene	C_4H_6	653.0	799.0	54.092	1.118	181.00
39	1,5-Butadiene	C_4H_6	628.0	766.0	54.092	1.120	174.00
40	Isoprene	C_5H_8	558.4	872.0	68.119	1.089	153.00

No.	Compound	Formula	Critical Constants Pressure, psia (P_c)	Critical Constants Temperature, °R (T_c)	Molecular Weight (M)	Specific Heat, C_p/C_v (k)	Heat of Vaporization, Btu/lb (Δ HV)
41	Acetylene	C_2H_2	890.4	555.3	26.038	1.238	—
42	Benzene	C_6H_6	710.4	1012.2	78.114	1.117	169.31
43	Toluene	C_7H_8	595.9	1065.6	92.141	1.091	154.84
44	Ethylbenzene	C_8H_{10}	523.5	1111.2	106.168	1.072	144.00
45	*o*-Xylene	C_8H_{10}	541.4	1135.0	106.168	1.069	149.10
46	*m*-Xylene	C_8H_{10}	513.6	1111.0	106.168	1.072	147.20
47	*p*-Xylene	C_8H_{10}	509.2	1109.6	106.168	1.072	144.52
48	Styrene	C8H8	580.0	1166.0	104.152	1.076	151.00
49	Isopropylbenzene	C_9H_{12}	465.4	1136.4	120.195	1.060	134.30
50	Methyl alcohol	CH_4O	1174.2	923.0	32.042	1.237	473.0
51	Ethyl alcohol	C_2H_6O	925.3	929.6	46.069	1.149	367.0
52	Carbon monoxide	CO	507.0	240.0	28.010	1.403	92.7
53	Carbon dioxide	CO_2	1071.0	547.9	44.010	1.293	238.2
54	Hydrogen sulfide	H_2S	1306.0	672.7	34.076	1.323	235.6
55	Sulfur dioxide	SO_2	1145.1	775.5	64.059	1.272	166.7
56	Ammonia	NH_3	1636.1	730.3	17.031	1.304	587.2
57	Air	N_2O_2	547.0	238.7	28.964	1.395	92.0
58	Hydrogen	H_2	188.1	60.2	2.016	1.408	193.9
59	Oxygen	O_2	736.9	278.9	31.999	1.401	91.6
60	Nitrogen	N_2	493.0	227.4	28.013	1.402	87.8
61	Chloride	$C1_2$	1118.4	751.0	70.906	1.330	123.8
62	Water	H_2O	3208.0	1165.6	18.015	1.335	970.3
63	Helium	He	—	—	4.003	1.660	—
64	Hydrogen chloride	HCl	1198.0	584.5	36.461	1.410	185.5

Note: C_p/C_v calculated by the formula $k = C_p/(C_p - 1.1987)/M$ at 1.7 psia and 60°F. Heat of vaporization at 14.696 psia at boiling point, Btu/lb (NBPΔHV).

Sequence of calculations

1. Mixture values are calculated by the mixture law, which assumes that the property of the mixture is the sum of the partial fractions.
2. The compressibility factor is iterated from the Redlich–Kwong equation:

$$\text{fx} = Z^3 - Z^2 + BP_f\left[\left(\left(\frac{A^2}{B}\right) - BP_f - 1\right)Z - \left(\frac{A^2}{B}\right)BP_f\right]^2$$

$$\text{f}^1\text{x} = 3Z^2 - 2Z + BP_f\left(\left(\frac{A^2}{B}\right) - BP_f - 1\right)$$

$$\text{where } A = \frac{0.42748T_c^{2.5}}{P_c T_f^{2.5}}$$

$$BP_f = 0.0866403\frac{T_c P_f}{P_c T_f}$$

$$\frac{A^2}{B} = 4.93396\left(\frac{T_c}{T_f}\right)^{1.5}$$

Start with Z = 1 and calculate fx/f¹x. If greater than 0.00001, subtract fx/f¹x from the current value of Z and repeat until the error is less than 0.00001.

3. Calculate the specific weight:

$$\gamma 3_f = \frac{P_f M}{T_f 1.314Z} \text{lb/ft}^3$$

Notation

A, B = partial calculations for Z
M = molecular weight, dimensionless
P_f = flowing pressure, atmospheres
P_c = critical pressure, atmospheres
T_f = flowing temperature, °K
T_c = critical temperature, °K
Z = compressibility, dimensionless
γ_f = specific weight, lb/ft^3

Source

Instrument Engineering Programs, Gulf Publishing Company, Houston.

Determine required orifice diameter for any required differential when the orifice and differential are known

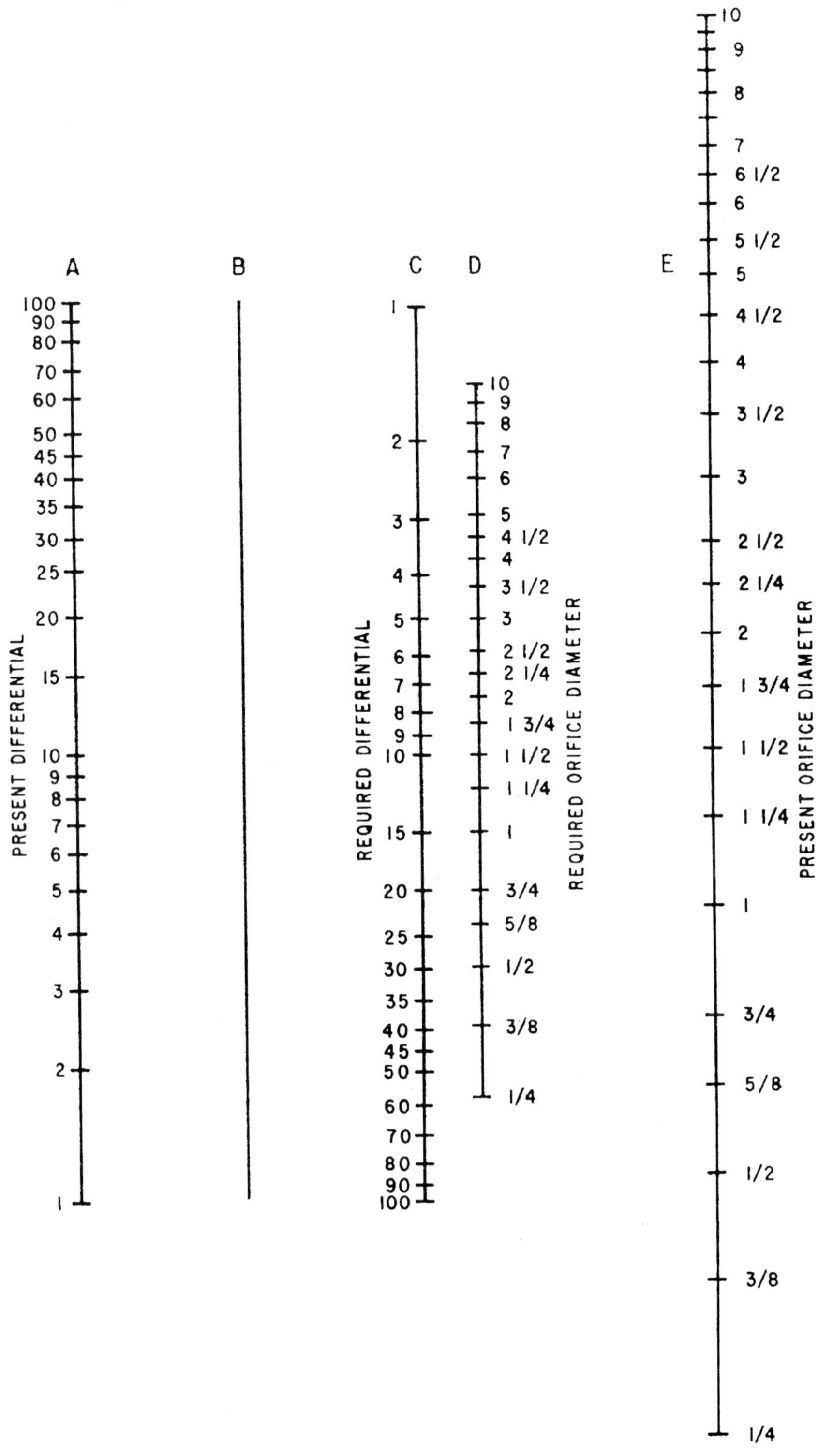

Procedure. Draw a straight line between the values of the present and required differentials on scales "A" and "C." From the intersection of this line on scale "B", draw second line to connect with the size of the present orifice diameter on scale "E." The size of the required orifice will be found at the intersection of his second line with scale "D."

Estimate the temperature drop across a regulator

The temperature drop that will be experienced when gas is passed through a regulator may be estimated as 1° drop for each atmosphere (15 psi) of pressure drop.

Example. Determine the temperature drop that will be experienced when reducing a gas from 750 to 600 psig.

Solution.

$$750 - 600 \text{ psig} = 150 \text{ psig}$$

$$\frac{150}{15} = 10 \text{ atmospheres}$$

$$10 \times 1^\circ/\text{atm} = 10^\circ \text{ of drop}$$

Estimate natural gas flow rates

This chart provides a good estimate of natural gas flow rates through 4-in meter runs equipped with flange taps. No calculations are required to obtain a flow rate. Orifice plate size and differential and static pressures from the orifice meter chart are used to make the estimate.

To use the Gas Flow Rate Estimator, enter the chart at the proper differential value at the indicated lower horizontal scale. Project this value vertically upward to the appropriate static pressure line, and then horizontally to the proper plate size line. Drop vertically from the plate size line to the lowermost horizontal rate scale at the base of the chart to obtain the estimated flow rate.

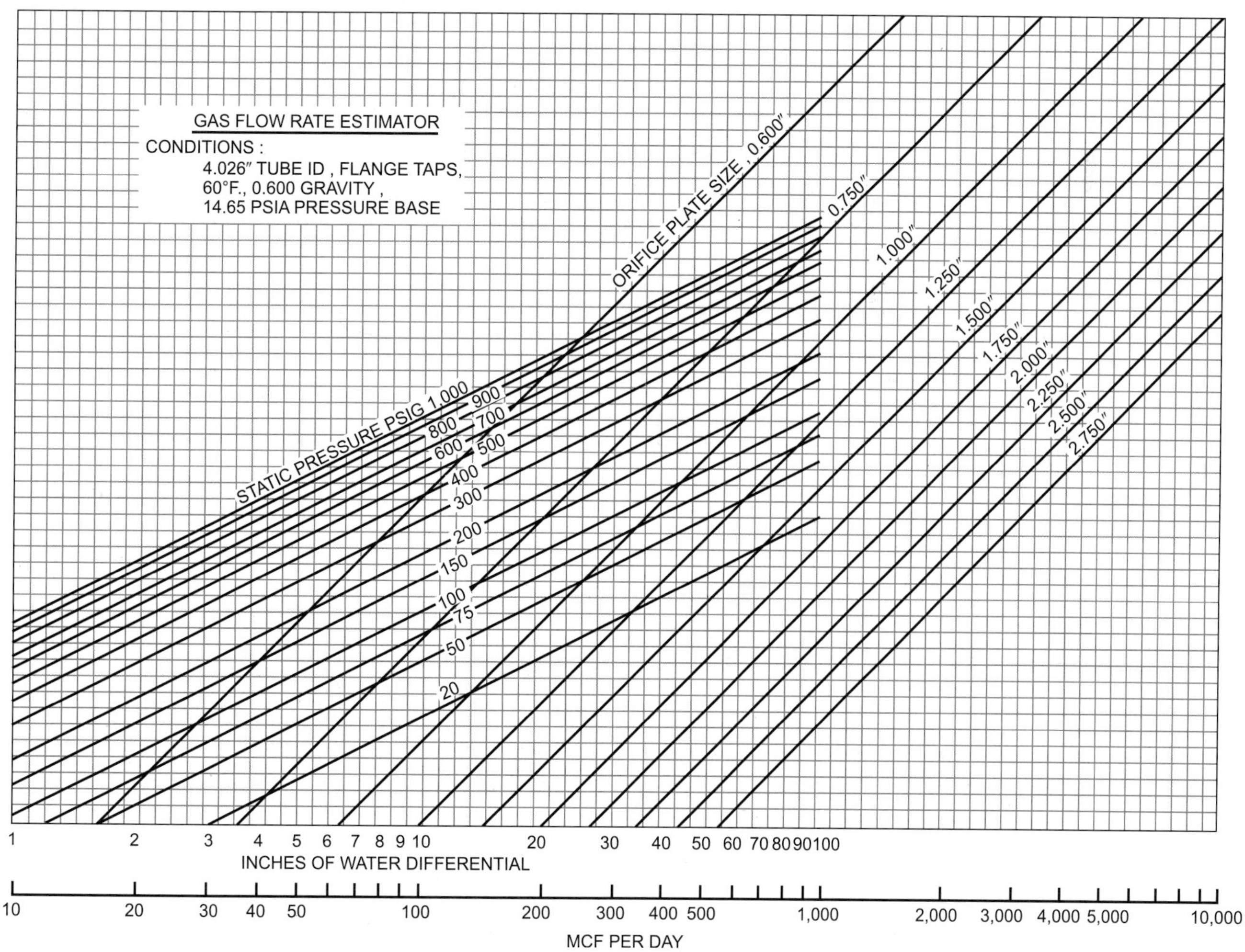

How to estimate the average pressure differential on the remaining meter runs of a parallel system

On gas systems where inferential meters are installed in parallel, at times it may be necessary to shut down one or more of the runs temporarily for repair or maintenance work. When this occurs, it is important to know beforehand what the pressure differential on the remaining runs will be so that the meters will not be over-ranged. The equation is:

Average differential for runs remaining in service

$$(A) = \left[\frac{(\text{No. of runs in service})\sqrt{(\text{Average differential})}}{(\text{No. of runs remaining in service})}\right]^2.$$

Example. A measurement engineer wishes to shut off 1 m run on a 4-m parallel system to inspect the orifice plate. The average differential on the four runs is 49 in and his meters have a 100-in range. He wishes to determine whether shutting down 1 m will over-range the three remaining meters.

$$(A) = \left(\frac{4 \times \sqrt{49}}{3}\right)^3 = (9.3)^2 = 86.5 \text{ in}$$

Since the average differential for the remaining runs is 86.5 in, the operator may safely shut the meter down as planned to inspect its orifice plate.

Sizing a gas metering run

Multiply the absolute pressure by the desired differential pressure (the differential in inches of water should not be greater than the absolute pressure in psia). Take the square root of the product. Divide this figure into the maximum flow rate in cubic ft/hr to obtain the approximate coefficient. Then divide the maximum coefficient by 250 and take the square root to determine the approximate inside diameter of the orifice in inches. The metering run diameter should not be less than one-third larger than the orifice diameter.

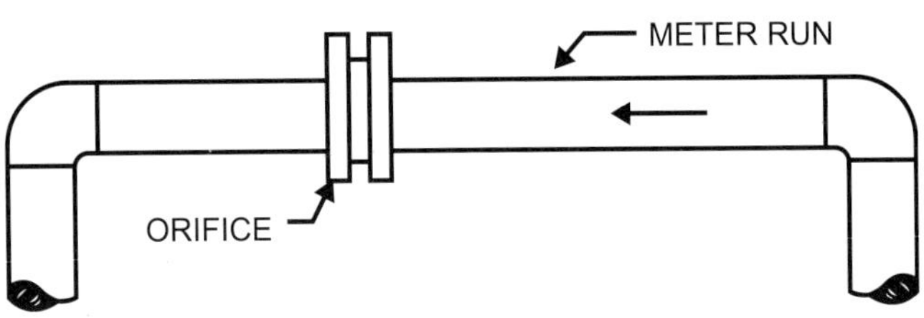

Example. Find the orifice diameter and meter run for a line designed to carry a maximum flow of 2 MMcf/h at a pressure of 600 psia. Use a differential pressure of 50 in of water.

$$\sqrt{600 \times 50} = \sqrt{30,000} = 173.2$$

$$\frac{2,000,000}{173.2} = 11,540$$

$$\frac{11,540}{250} = 46.2$$

$$\sqrt{46.2} = 6.8 \text{ in, approximate orifice diamerter}$$

Meter run should not be less than 10-in in diameter.

Determine the number of purges for sample cylinders

To determine the proper number of times a sample cylinder should be pressured to ensure removal of the original gas, multiply the reciprocal of each filling pressure (in atmospheres) by each succeeding reciprocal. The final filling will determine what fraction of the original gas is still in the cylinder.

Example. The problem is to fill a gas sample for analysis to obtain an accuracy of 0.05%. Source pressure is 450 psia.

Solution. First filling to line pressure and emptying

$$\frac{1}{450/15} = \frac{1}{30} \times 100 = 3.33\%$$

Second filling to line pressure and emptying:

$$\frac{1}{30} \times \frac{1}{30} \times 100 = 0.11\%$$

Third filling to line pressure:

$$\frac{1}{30} \times \frac{1}{30} \times \frac{1}{30} \times 100 = 0.0037\%$$

Find the British thermal units when the specific gravity of a pipeline gas is known

Use the graph below prepared from the Headlee formula:

Btu @ 30 in Hg and 60° saturated = 1545 (specific gravity) + 140 − 16.33 (% N_2)

Example. Assume a gas having a gravity of 0.670, which contains 1.0% nitrogen. From the chart, read up from 0.670 gravity on the baseline to the intercept with the curve marked 1% nitrogen. Reading horizontally across, the total Btu of the gas is 1160 Btu at 30 in, 60° saturated.

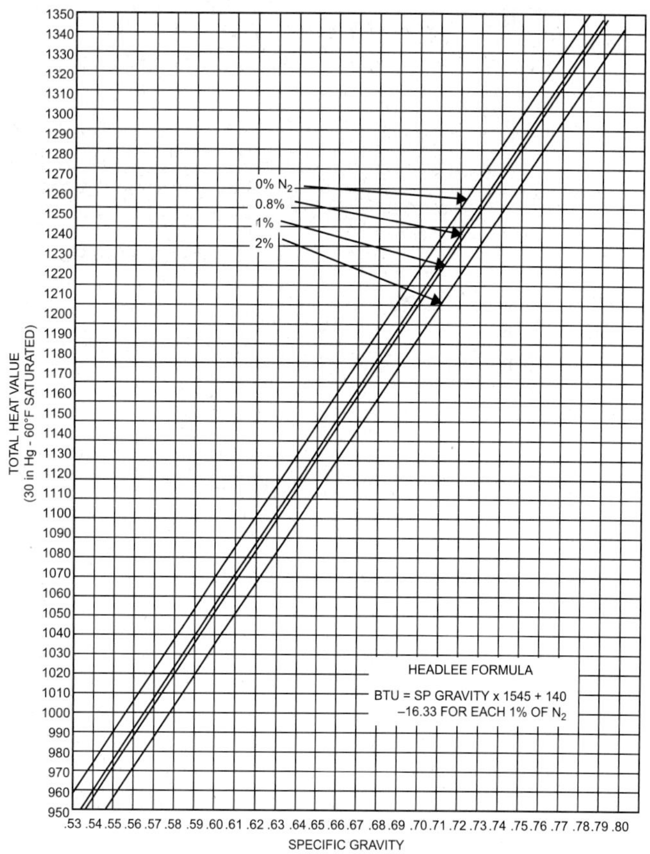

Make quick calculation of the gas flow through orifice meters

Where the static pressure remains constant, to reduce the flow by one half, lower the differential reading to one fourth. For example, if the differential is reading 64 in, to reduce the flow by one half, reduce the differential to 16 in.

For better measurement, keep the beta ratio between 0.2 and 0.6.

Headers on meter stations should be twice the internal diameter of the meter runs they manifold together.

Measurement stations should be installed on the suction side of a compressor station rather than the discharge side where possible, to reduce effect of pulsation.

Use the wine size orifice plate in meter runs that are manifolded together where possible; recordings can then be used to check against the other, since their differentials should be approximately the same.

On meter runs up to and including 10-in, this can be accomplished by multiplying the coefficient for a 1-in orifice times the actual orifice installed squared times the square root of the extension (the product of the differential in inch of water times the static pressure in lb/in^2 absolute). AGA-3 coefficients for a 1-in orifice in 2, 3, 4, 6, 8, and 10 in standard orifice meter runs using flange taps should be committed to memory. Base conditions would be for a 0.60 specific gravity 60 °F, and 14.73 psia pressure base. The answer would be in standard cubic ft/hr or day depending upon the coefficient used.

AGA Committee Report Number 3 coefficients at the conditions shown are:

2 in meter run, 1 in orifice—271.4 cfh, 0.6G, 60 °F, and 14.73 psia
3 in meter run, 1 in orifice—262.8
4 in meter run, 1 in orifice—260.8
6 in meter run, 1 in orifice—259.3
8 in meter run, 1 in orifice—258.7
10 in meter run, 1 in orifice—258.5 (heavy)

Example. Find the hourly gas flow through an 8-in meter run having a 2.00-in orifice when the differential pen is recording 16 in of water and the static pressure is 900 lb/in^2 absolute.

Solution. From memory the coefficient for an 8-in meter run, 1-in orifice is 258.7. Multiplying 258.7 times 4 (or 2 squared) times 120 (the square root of the product of 900 times 16) equals 124,176 cfh, the approximate flow in cubic ft of gas per hour at 0.6 gravity, 60 °F temperature, and 14.73 psia base pressure.

How to measure high pressure gas

T. J. Lewis, Measurement Engineer, Columbia Gulf Transmission Co., Houston, Texas

The rapidly increasing value of natural gas, along with rising operating costs, has made accurate gas measurement vital for the natural gas industry.

Accurate gas measurement is obtained only through an efficient maintenance and operating schedule, adhered to rigidly, after a properly designed measuring station has been installed. Any error in the measurement of gas at high pressures and large volumes magnifies itself, so that there can be either a considerable gain or loss in revenue.

Field measurement at high pressure is defined as the measurement of gas at a specific point in a gas producing area at a pressure higher than that of contract base pressure.

Contracts

The gas purchase contract is of prime importance in determining the requirements for gas measurement at a given location.

Information such as gas analysis, gas/oil ratios, specific gravity determinations, flowing temperatures, pressures, etc., should be gathered before a gas purchase contract is written. This collection of data should be used to eliminate any assumptions regarding the requirements for performing accurate measurement. It also provides information for the engineering and accounting functions, so that they are made aware of their responsibilities and limitations in helping provide good measurement.

The contract should also state as clearly as possible all parameters of the gas measurement operation to prevent possible disputes by having a clear understanding between the buyer and the seller.

Design and planning

In the planning and designing of a natural gas measuring station, many items of equal importance must be considered if accurate measurement is to be obtained.

The design and fabrication of the orifice meter runs must meet the specifications of the AGA Gas Measurement Committee Report No. 3, latest revision. In addition to these requirements, consideration must be given to safety, volume and quality of gas, term of contract, flexibility of station design, and physical location of the station. The most important point in this whole procedure is remembering that the measuring station is similar to a large cash register with money flowing both ways for the company.

A meter tube should be designed to allow easy internal inspection. Shutoff valves should be placed on each end of the meter, so that it may be depressured independently of any other piping. The instrumentation and/or control piping should be located so and supported to furnish maximum protection against accidental breakage.

Meter runs, piping, etc., should be designed as a Class III area facility. If possible, a properly vented walk-in type meter building should be used, with the total installation laid out with the safety of the operating personnel firmly in mind (Figure 1). As periodic orifice plate inspections are necessary, provisions should be made so that these inspections can be made without interruption of the gas flow. This can be done by installing a bypass line or by using a senior type orifice fitting.

The flow and temperature recording instruments should be located in a walk-in type meter building. This arrangement provides a year-round working area for chart changing, meter tests, maintenance work, etc.

Installation

The measuring or metering components in a measuring station are divided into two elements: primary and auxiliary.

The *primary element* of an orifice-type measuring station consists of the meter tube, the orifice plate fitting, the orifice plate, pressure taps, and straightening vanes. While the primary element is being fabricated, its components should be checked against the specifications listed in AGA Gas Measurement Committee Report No. 3. Any discrepancies found can then be corrected before the meter tube is finally assembled and shipped to the field.

Figure 1. Meter runs, piping, etc., should be designed as a Class III area facility. If possible, a properly vented walk-in type meter building should be used, with the total installation laid out with the safety of the operating personnel firmly in mind.

When the meter tube is being installed in the field, the orifice plate is checked to make sure it is centered in the meter tube, and that it meets AGA Report No. 3 specifications. Meter tubes are normally installed in a horizontal or vertical plane. Pressure taps in a vertical plane are preferred, since this allows the gauge lines to be installed from the top of the orifice fitting. The pressure taps located on the bottom of the orifice fittings then can be used for installing drips or drains.

The *auxiliary element* consists of the meters (recording devices) used for measuring the static pressure, differential pressure, and temperature of the gas in the meter tube. Meters are placed as close as possible to the orifice or flange fitting. Gauge lines are connected from the meter to the orifice fitting in a way that will best provide a drain to the meter tube. Liquids trapped in a gauge line can cause significant errors in the recorded pressures.

There are two principal types of orifice meter gauges: the mercury manometer-type and the bellows-type meters. Each of these meters is suitable for field measurement, with the bellows-type meter becoming the favorite for wet gas measurement.

Installation of these elements in a measuring station should be under the supervision of experienced measurement personnel.

Operation and maintenance

After a measuring station has been properly installed, accurate measurement is obtainable only by conscientious care and maintenance by the field measurement technicians.

Any discrepancy noticed on the chart should be explained on the chart, and a correction should be made immediately. Since immediate correction is not always possible, a report should then accompany the chart describing the condition of the meter and when the correction was completed. These reports, along with test and inspection reports, are necessary for the calculation of accurate volumes.

The accuracy of recording instruments depends on the men who test and operate the measuring station equipment. Good test equipment, properly maintained, is absolutely necessary in checking the calibration of temperature and pressure recording instruments and the quality of the gas being measured. The frequency of meter tests, gas quality tests, and meter tube inspections depends on the individual company and its contractual agreements with other companies.

Monthly calibration tests and inspections on the flow and temperature recorders are normally adequate for accurate measurement. In special cases, however, a weekly inspection may be required. Monthly calibration tests and inspections are vital to accurate measurement practices. Figures 2 and 3 show what happens to volume calculations when there are small errors in static pressure, differential pressure, temperature, and specific gravity measurements.

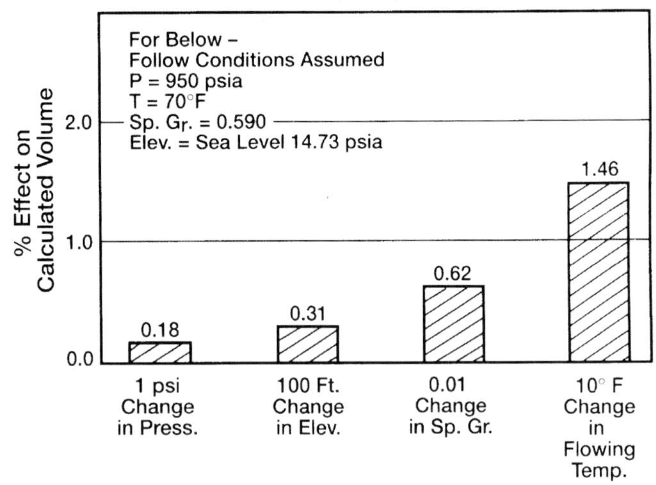

Figure 2. Relative effect of various factors on calculated flow volume (also Figure 3).

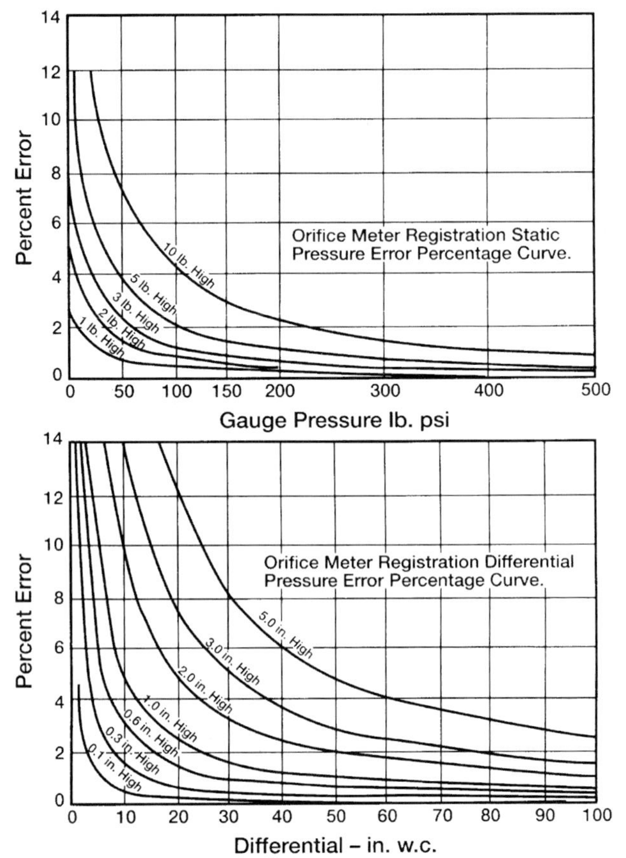

Figure 3. The impact to volume calculations when there are small errors in static pressure, differential pressure, temperature, and specific gravity measurements.

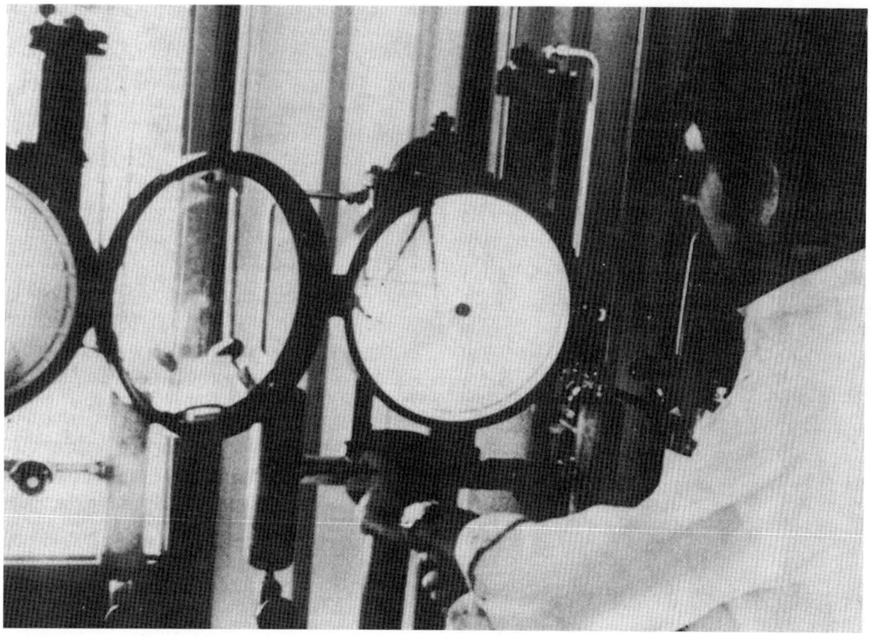

Figure 4. Charts are shown being changed, and clocks wound and adjusted. Each measuring station has its own flow pattern, which is recognizable by experienced chart changers.

Since external tests and inspections do not reveal the internal condition of a meter tube, the meter tube and its components must be inspected internally on an annual basis for possible buildups and internal corrosion. This inspection ensures continuing accuracy.

Chart changing

Charts must be changed on time, and clocks must be kept wound and adjusted (Figure 4).

Each measuring station has its own flow pattern, which is recognizable by experienced chart changers. Any change in this flow pattern indicates a possible error, which should be corrected if possible. If the correction cannot be made by the chart changer, he should notify the field measurement personnel who are qualified to perform required adjustment and/or repairs.

Each chart must be properly dated and identified along with field information necessary for calculating the required volumes. Automatic chart changers have become increasingly popular, especially in remote areas such as offshore Louisiana.

Periodic pulsation tests should be made at each measuring station, particularly where a measuring station is located on the suction or discharge side of a compressor station. In this instance, pulsation dampeners must be installed between the compressor and the measuring station. Errors caused by pulsation can be very large, but they will not necessarily show in calibration tests.

A comprehensive maintenance program, faithfully followed, can detect and prevent costly errors before they occur. Leaking valves, gauge lines, etc., will cause errors in an increase or decrease in company revenue. Such leaks should be corrected as soon as possible.

Wet gas measurement

The problems encountered in measuring wet gas are considerably different from dry gas measurement.

Wet gas contains entrained liquids consisting of water and hydrocarbon condensates that can cause hydrate formations in the orifice plate fitting, flow recorders, and instrument lines. These hydrates are formed with the aid of free water under certain pressure and temperature conditions. Formation of hydrates can be reduced or eliminated in several different ways.

One successful solution to prevent hydrate formation is to use drip or condensate bottles in instrument and gauge lines. Infrared gas heaters can be used in meter buildings for heating gauge lines and flow recorders. This type of heater has become very popular lately, since it does not produce an open flame.

Errors caused by hydrate formations in the meter tube and around the orifice plate are common in wet gas measurement. The formation of these hydrates can be prevented by continuously injecting methanol into the gas stream ahead of the meter run. Experience and experimentation will determine the amount of methanol to be injected.

The use of methanol-containing inhibitors is not recommended as the inhibitor often solidifies in the meter tubes.

Four ways to calculate orifice flow in field

Rogers G. Thompson, Measurement Technician, Texas Eastern Transmission Corp., Shreveport, Louisiana

Several methods are available for orifice flow field calculations.

The particular method used is determined by such factors as company policy, advance warning, time available, individual preference, accuracy required, and the information and equipment available at the time of need. A good rule to follow is to calculate as precise a volume as possible at all times.

Any volume calculated manually will be considered a field calculation.

Most companies now use some type of computer system, and rarely will a manually calculated volume match a computer volume. Field calculations are usually made under conditions that eliminate the use of adding machines and calculators. Accuracies can vary from an extremely accurate spot check to a very basic calculation. External factors usually determine what procedure to use.

Uses for the calculations determine the degree of accuracy required for each particular instance. Since time is usually at a premium, it would be absurd to calculate to the cubic ft when calculating to the Mcf would suffice. A good example is

Table 1
Combined Specific Gravity and Temperature Correction Factors—Orifice Meters Combined Factor $= \left(\sqrt{\frac{1}{G}}\sqrt{\frac{520}{460^{\circ}F}}\right)$

Flow Temp. °F	Specific Gravity								
	.590	.591	.592	.593	.594	.595	.596	.597	.598
35	1.3343	1.3332	1.3321	1.3309	1.3298	1.3287	1.3276	1.3264	1.3254
36	1.3330	1.3319	1.3208	1.3296	1.3285	1.3274	1.3263	1.3251	1.3241
37	1.3317	1.3306	1.3295	1.3283	1.3272	1.3261	1.3250	1.3238	1.3228
38	1.3303	1.3292	1.3280	1.3269	1.3258	1.3247	1.3235	1.3224	1.3214
39	1.3290	1.3279	1.3267	1.3256	1.3245	1.3234	1.3222	1.3211	1.3201
40	1.3277	1.3266	1.3254	1.3243	1.3232	1.3221	1.3209	1.3198	1.3188
41	1.3264	1.3253	1.3241	1.3230	1.3219	1.3208	1.3197	1.3185	1.3175
42	1.3251	1.3240	1.3228	1.3217	1.3206	1.3195	1.3184	1.3172	1.3162
43	1.3238	1.3227	1.3215	1.3204	1.3193	1.3182	1.3171	1.3159	1.3149
44	1.3223	1.3212	1.3201	1.3190	1.3179	1.3168	1.3156	1.3145	1.3135
45	1.3210	1.3199	1.3188	1.3177	1.3166	1.3155	1.3143	1.3132	1.3122
46	1.3197	1.3186	1.3175	1.3164	1.3153	1.3142	1.3130	1.3119	1.3109
47	1.3184	1.3173	1.3162	1.3151	1.3140	1.3129	1.3118	1.3106	1.3096
48	1.3171	1.3160	1.3149	1.3138	1.3127	1.3116	1.3105	1.3093	1.3083
49	1.3158	1.3147	1.3136	1.3125	1.3114	1.3103	1.3092	1.3080	1.3070
50	1.3147	1.3135	1.3124	1.3113	1.3102	1.3091	1.3080	1.3069	1.3059
51	1.3134	1.3122	1.3111	1.3100	1.3089	1.3078	1.3067	1.3056	1.3046
52	1.3121	1.3109	1.3098	1.3087	1.3076	1.3065	1.3054	1.3043	1.3033
53	1.3108	1.3096	1.3085	1.3074	1.3063	1.3052	1.3041	1.3030	1.3020
54	1.3095	1.3083	1.3072	1.3061	1.3050	1.3039	1.3028	1.3017	1.3007
55	1.3081	1.3070	1.3059	1.3048	1.3037	1.3026	1.3015	1.3004	1.2994
56	1.3070	1.3059	1.3048	1.3037	1.3026	1.3015	1.3004	1.2992	1.2982
57	1.3057	1.3046	1.3035	1.3024	1.3013	1.3002	1.2991	1.2980	1.2970
58	1.3044	1.3033	1.3022	1.3011	1.3000	1.2989	1.2978	1.2967	1.2957
59	1.3032	1.3021	1.3010	1.2999	1.2988	1.2977	1.2966	1.2955	1.2945
60	1.3019	1.3008	1.2997	1.2986	1.2975	1.2964	1.2953	1.2942	1.2932
61	1.3006	1.2995	1.2984	1.2973	1.2962	1.2951	1.2940	1.2929	1.2919
62	1.2994	1.2983	1.2972	1.2961	1.2950	1.2939	1.2928	1.2917	1.2907
63	1.2981	1.2970	1.2959	1.2948	1.2937	1.2926	1.2915	1.2904	1.2894
64	1.2970	1.2959	1.2948	1.2937	1.2926	1.2915	1.2904	1.2893	1.2883
65	1.2957	1.2946	1.2935	1.2924	1.2913	1.2902	1.2891	1.2880	1.2870
66	1.2945	1.2934	1.2923	1.2912	1.2901	1.2890	1.2879	1.2868	1.2858
67	1.2932	1.2921	1.2910	1.2899	1.2888	1.2877	1.2866	1.2855	1.2845
68	1.2920	1.2909	1.2898	1.2887	1.2876	1.2865	1.2855	1.2844	1.2834
69	1.2908	1.2897	1.2887	1.2876	1.2865	1.2854	1.2843	1.2832	1.2822
70	1.2895	1.2884	1.2874	1.2863	1.2852	1.2841	1.2830	1.2819	1.2809
71	1.2884	1.2873	1.2862	1.2851	1.2840	1.2829	1.2818	1.2807	1.2798
72	1.2872	1.2861	1.2850	1.2839	1.2828	1.2818	1.2807	1.2796	1.2786
73	1.2859	1.2848	1.2837	1.2826	1.2815	1.2805	1.2794	1.2783	1.2773
74	1.2847	1.2836	1.2825	1.2815	1.2804	1.2793	1.2782	1.2771	1.2761
75	1.2835	1.2825	1.2814	1.2803	1.2792	1.2781	1.2770	1.2760	1.2750
76	1.2824	1.2813	1.2802	1.2791	1.2780	1.2770	1.2759	1.2748	1.2738
77	1.2811	1.2800	1.2789	1.2778	1.2767	1.2757	1.2746	1.2735	1.2725
78	1.2799	1.2788	1.2777	1.2767	1.2756	1.2745	1.2734	1.2723	1.2713
79	1.2787	1.2776	1.2766	1.2755	1.2744	1.2733	1.2722	1.2712	1.2702

automatic control of a compressor station where rates are given in terms of MMcf.

Field calculations are used for such things as determining purge gas volumes, relief valve losses, set point determination, and sizing of plates. In general, volume, rate, differential, and plate size are the form variables that are determined by field calculations.

This discussion will be based on the AGA Committee Report No. 3 as it is the most commonly used. It is also the most complicated due to the large number of corrections applied.

Four categories have been devised to aid in discussing field calculations: the AGA formula, graphs, computers, and the quick method.

AGA formula

In making field calculations, a good rule to follow is—the simpler, the better. Detailed calculations are difficult to perform under usual field conditions.

Another good rule to follow is cool and check. Always allow some time to elapse between calculating and checking. This can be from 5 min to 5 h. To have someone else check your work is even better. Following these rules will eliminate most errors.

The gas flow equation as stated in the AGA Committee Report No. 3 is:

$$Qh = Fb\ Fr\ Y\ Fm\ Fpb\ Ftb\ Fm\ Fa\ F1\ \ Ftf\ \ Fg\ \ Fpv\sqrt{h_w pf}$$

Obviously, the most accurate method of volume calculation would use all of the above factors. Most field calculations will use some modified form of the AGA equation.

In order to present a logical discussion, these correction factors are broken down into three groups—major, minor, and intermediate. This grouping is based upon the effect each factor has on the volume.

The major factors are the basic orifice factor Fb and the specific gravity factor Fg. These two factors have to be included in all computations.

The intermediate factors are the pressure base correction Fpb, flowing temperature correction Ftf, and supercompressibility correction Fpv. The value of these factors is variable. Usually, they are determined by operating conditions at each specific location or are set by contract as in the case of pressure base.

Supercompressibility can become a large correction with transmission companies and producers where gas is measured at elevated pressures. For field calculation purposes, intermediate factors will be considered as either major or minor.

The Fr, Y, Fm, Ftb, Fa, and Fl corrections are minor factors for field calculation because of their minimal effect upon the final volume. Most of these factors are usually ignored.

Condensed formula

As previously stated, the most accurate method of field calculation is the AGA formula using all corrections. However, this is usually not possible because of insufficient time, advance warning, or operating information.

If a little work is done prior to the time of use, the formula can be condensed to three or four calculations. However, some operating information must be available. The static pressure, differential, run and plate size, specific gravity, and gas composition must either be known or assumed.

Assumptions should be based on knowledge of normal operating procedures and characteristics of similar M&R stations. As an example, the inert composition of gas is assumed to be 0% nitrogen and 0% carbon dioxide until an analysis is performed. This procedure permits volume calculations until more definite information is available.

By using known and assumed data, an hourly coefficient composed of Fb, Fr, Y, Fm, Ftb, Fpb, Fa, and Fl can be computed. The only corrections left to be made are Fg, Ftf, and Fpv. By giving some thought, time, and prior work, the formula can be reduced to:

$$Qh = C\ \ Fg\ \ Ftf\ \ Fg\sqrt{h_w pf}$$

Tables are available that further reduce the amount of calculation that has to be done. Combined specific gravity and temperature tables are available from different sources. Also available are extension values for $\sqrt{h_w pf}$ or the individual values of $\sqrt{h_w}$ and $\sqrt{pf}$. These tables eliminate what is probably the hardest part of a field calculation.

When following these guidelines, calculations become fairly simple. One note of caution: always make computations in a legible and logical order. This makes everything fairly simple to check, because many errors are caused by illogical layout of calculations as any other factor.

From this point on, all methods are an adaptation of the AGA Report No. 3 equation.

Flow graphs

Typically, the next most accurate method is the company-developed flow graphs.

Although the graphs necessarily have to be based on average data, they are tailored to each particular company's operating conditions. Thus, you get more accurate results for

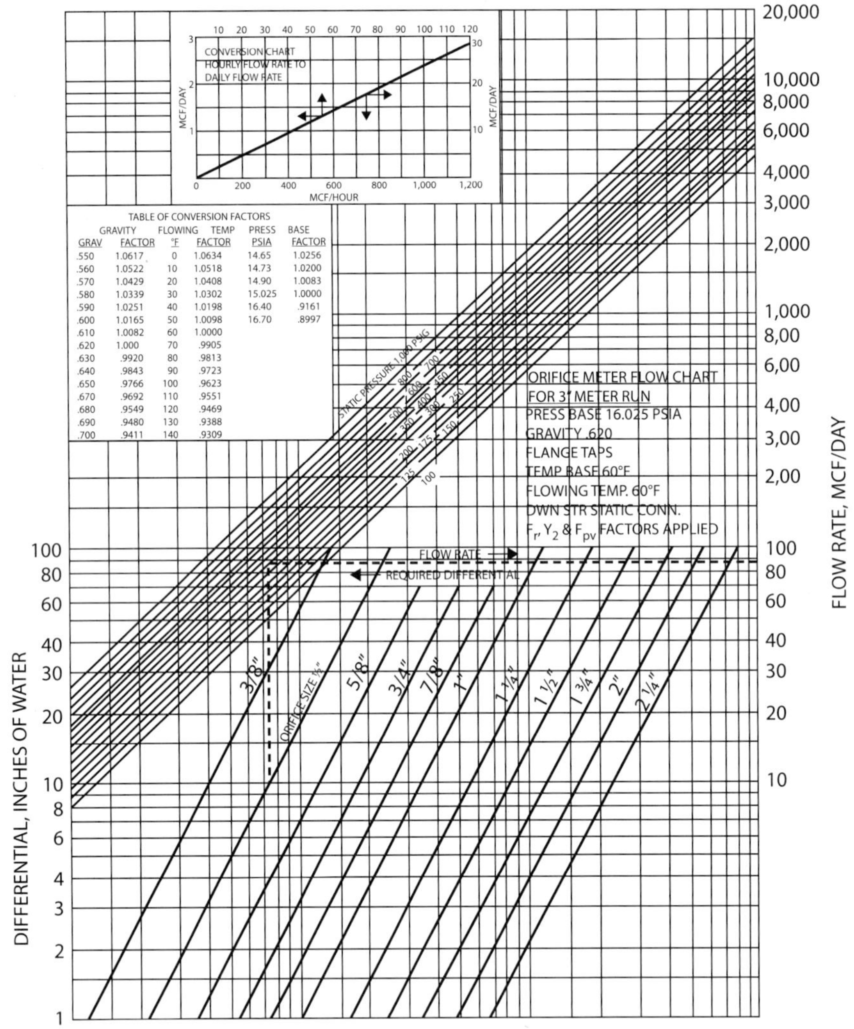

Figure 1. Example of orifice meter flow graph used by Texas Eastern.

your particular requirements. Corrections such as gravity can be made to the final figure if desired, but usually are not.

The graph has the advantage of giving considerable information quickly. All plate size and differential combinations are shown at a glance. Daily and hourly flow rates are also given. This is the fastest and easiest of all methods to use, and accuracy is usually good enough to use under most conditions. No calculations are necessary under most conditions.

Most paper graphs are bulky, requiring a straight edge and space to use. Humidity will destroy them. However, replacements are usually available upon request. Oddly enough, this is probably the least used of all methods.

Most companies use the AGA equation to develop their graphs. The newer factors Fa and Fl are almost certainly not included in current data. Most factors are based on a particular company's average data.

Some factors are given a value of 1.000 and therefore ignored. Supercompressibility corrections are usually by average specific gravity and inert composition.

Computers

Flow computers are probably the most widely used method of field calculation. This is due to their compactness and simplicity of operation. Accuracy is generally in the acceptable range for field calculations.

Most manufacturers of measurement equipment furnish some type of flow computer either free or at a nominal cost. Construction usually is of paper and plastic.

All of these flow computers are based on the AGA Report No. 3 flow equation. Most of the minor corrections are not used, with the omissions listed in the instructions. Since computers are designed for use throughout the industry, they must be kept standard. Consequently, they will not fit a particular set of conditions without making additional corrections.

Most flow computers incorporate Fb, Fg, Ftf, Ftb, and sometimes a generalized form of Fpv. The minor factors are usually ignored. The pressure base is usually set at 14.7 or 14.73, but some computers do incorporate a pressure base correction feature.

The biggest single drawback to computers is the fact that they must be general in scope and not slanted to one particular method of calculating volumes. Corrections are kept too general to fully correct for a specific set of conditions. There must be enough use of a computer for a person to maintain speed and accuracy of operation. If not, the instruction booklet will have to be studied each time the computer is used.

Flow computers are compact, easy to carry and store, and are very quick to use. As long as their limits are recognized, they can do an excellent job.

Quick method

Primarily, the quick method uses the major factors Fb and Fg only.

Inclusion of the intermediate factors is desirable if sufficient information is available. These intermediate corrections are made in a general way only. For example, if the pressure base is 15.025, subtract 2% of the volume. Algebraically, add 1% of the volume for each 10° variation from the temperature base. For each 100 psi of static pressure, add 8% of the volume.

This procedure will suffice for a lot of field calculation requirements. Keep in mind that this method is a rough calculation and is not intended to be used as a precise quantity.

Some mention must be made of the extension that probably causes more trouble than any other part of a volume calculation. In determining average differential and pressure, spot readings are the easiest to get and the most commonly used method. Sometimes predetermined values are used. Quite often, differential and pressure are averaged over several hours. When averaging, care should be exercised for differentials under 25 in, especially for differentials under 10 in. Due to the square root effect, errors are magnified in these ranges.

Extension books are available either in the form $\sqrt{h_w pf}$ or the individual $\sqrt{h_w}$ and $\sqrt{pf}$ tables. Although the combined tables are bulky, the individual tables can be limited to a few small sheets. Use of these tables will improve accuracy and speed.

Practical maintenance tips for positive displacement meters

J. A. Wilson, Meter Shop Manager, Northern District, Skelly Oil Co., Pampa, Texas

Thorough understanding of a meter's function is necessary to properly repair any mechanism within it.

There is a definite reason for each meter part, and each part has a definite relation to the other. The alignment of any one part with another has been precisely established for each type and size of meter.

The functions of a positive displacement meter are similar to a four-cylinder engine. For one revolution of the tangent, each chamber must fill and exhaust once.

One revolution of the tangent is 360°. Therefore, to make a smooth running meter with uniform displacement, one of the chambers must start to exhaust every 90°. Since the valves control the time at which the chamber starts to fill and exhaust, they must be coordinated, one with the other, so that this 90° relationship is maintained.

Similarly, the diaphragms must also be connected to assure this relationship, at least in their extreme positions. Both valves and diaphragms must be connected to coordinate both 90° relationships.

The following are practical tips for PD meter repair and testing.

PDM repairs

The repairman cannot always carry parts for every type meter. Also, testers who only do an occasional repair job are not as efficient as men performing repairs daily. Therefore, most major repairs and all major overhauls are more efficiently handled in the meter shop.

Operating conditions in the field vary and are not always the best. Some of the operating problems in the field are gas saturated with water and hydrocarbons, iron oxide, sour gas, and lubrication.

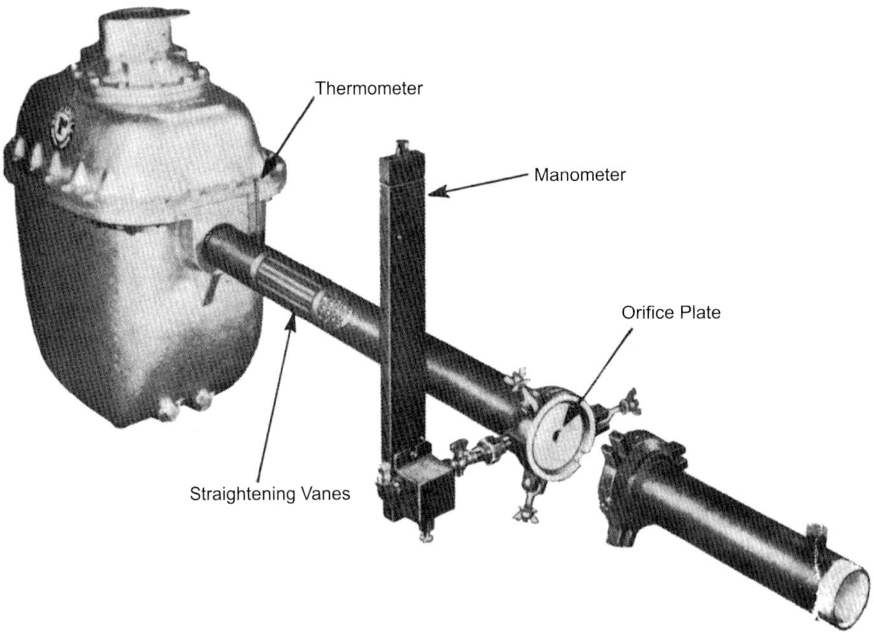

Figure 1. Low pressure flow provers for PD meters can be used in the field or shop. The principle is essentially that of the orifice meter, except the orifice discs are calibrated for different flow rates instead of using various coefficients. Shown above is nomenclature for the Rockwell prover.

Shop repair. Meter shops, no matter how well equipped, cannot perform satisfactory meter repair without qualified and dedicated personnel.

When an experienced man disassembles a meter for repair, he inspects each part as he tears it down, considering each as a possible cause of the malfunction. When the problem is found, if possible, it is corrected before proceeding

Some of the most common causes of malfunction and points to watch on the more widely used meters are:

- Exhaust port gaskets should be inspected to make sure the seal is good. Some meters have metal spacers under the seal and must be in place when reassembled.
- The clearance of the flag rods at the top of the meter should be checked. After a meter has been time checked and balanced, it will sometimes run rough. Look for a bright spot right over the top of the flag rods in the top cover. This is not a common problem, but is difficult to locate, and a few strokes with a file will take care of it.
- The reassembly of the tangent gear between the valve gears of meters, as well as other gear-driven valves, should be carefully centered. There is usually a small amount of side play as well as end play on the tangent shaft bracket. A little extra care here will contribute to a quiet operating meter.
- The body seal O-rings on the later models should be placed carefully in the slot when the top cover is put into place.

Flag arms. Because new materials are being used in meters and the new aluminum-case meters are designed differently, there have been changes in some repair procedures. Some meters use an aluminum, long flag arm with a brass bushing cast in the center, so the flag arm can be soldered to the flag rod.

Most repairmen use a wet sponge to rapidly cool a solder joint on flag rods when flag arms are attached, but because of the different expansion coefficients of brass and aluminum, the repairman must use extra precaution in cooling these solder connections.

Tests have proven that cooling the aluminum part of the flag arms too quickly will cause fractures in the aluminum surrounding the brass bushing. Most of these fractured flag arms will go undetected and later break down in operation. However, if the repairman cools the solder joints by easing a wet sponge slowly up the flag rod, the joint is cooled from the inside out and no damage is done.

Valves and seats. Careful grinding of the valves and valve seats is very important. Number 80 grit may be used to clean up bad valves and seats, finishing with number 120 grit. Most repairmen use the figure eight motion in grinding, letting the weight of the glass block control the pressure on the valve seat. Use a very light touch on the valves. Where it is practical, both valve seats should be ground together. Grinding paper that is self adhesive is certainly recommended over grinding paper that has to be cemented to a glass block.

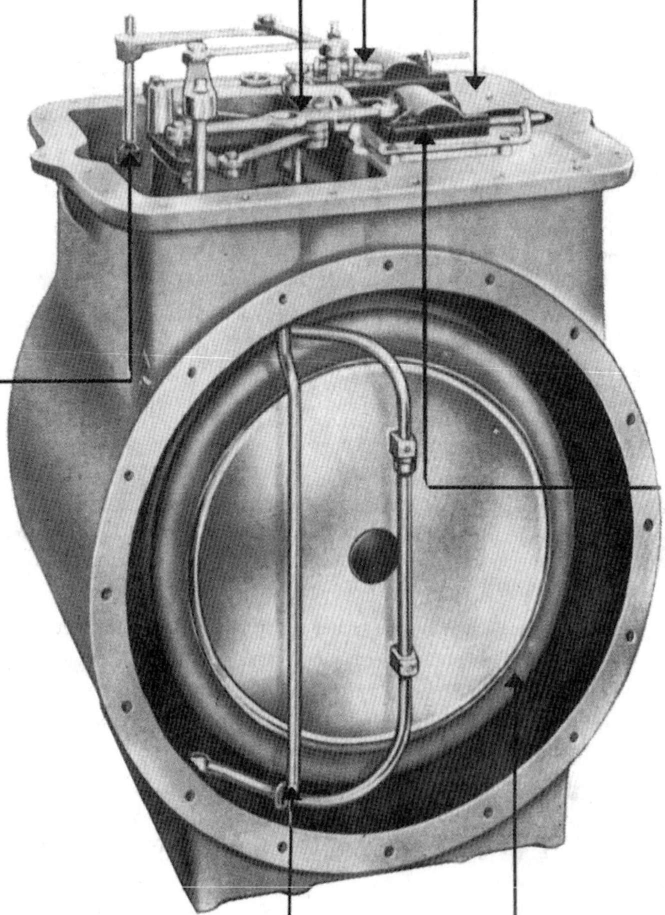

Figure 2. Thousands of ironcase PD meters, such as the American 25B and 35B shown above, have been in service for many years. Valve seats should be carefully checked to make sure they are still bonded to the meter body.

There are thousands of ironcase meters that have been in service for a great number of years. The valve seats on these meters should be carefully checked to make sure they are still bonded to the meter body. Remove and reseal the seats if there is any doubt about the strength of the bond. Use an old grinding block to true up the bottom of the valve seats and buff the meter case clean to ensure a good bond.

Clean all surplus cement from the intersurfaces of the valve seat opening. Be careful not to use screws that are longer than standard in valve guides that screw into the valve seat. An overlength screw can jack the valve seat off the body and break the seal.

Watch the clearance on the new plastic parts when they are installed in old style meters. The raised reinforcing rib on some of these parts may pose clearance problems in many meters. Soldering acid will dissolve some plastic, so if you must use acid core solder, be careful to keep it off the plastic.

Balance and time check. Take plenty of time to balance and time the valves and diaphragms to ensure a smooth operating meter. Use care in setting the tangent in proper relation to the valves. Time spent here will be regained manyfold when the meter is calibrated.

Most shops have a utility-type vacuum sweeper. This is an excellent means of rotating a meter before the top cover is installed. This gives the repairman a good chance to watch the valve action and make a final check on timing and balance, gauge, and depth of the undergear driving arm in relation to the tangent. Make sure the tangent wrist is engaged, but do not touch the short flag arm.

Installing a long pointer on the index driver dog will help proof test the meter, since the circumference of the circle will be much longer than the test-hand on the index. It is a good practice to rotate the meter at least two rotations of the test-hand after the index is installed. This gives a final check to make certain the meter is operating smoothly. A vacuum cleaner can also be used to good advantage here.

Pan-type diaphragms can be rebuilt with a minimum amount of investment in jigs and other equipment. Many parts, such as crank brackets and flag arms that have pressed-in bushings, are easily repaired. Rebuilding parts can save up to 75% of the cost of a new item.

When it comes to repairing meters, "about right" is not good enough. The meter repairman who wants to make his company money will strive to make all checks and adjustments as accurately as possible.

PDM testing

Positive displacement meters may be tested by any of the following methods: (1) bell type meter prover, (2) low pressure flow prover, (3) critical flow prover, and (4) test meter. All of these methods, when correctly used under proper conditions, will give similar results.

Bell prover. Meter shop personnel are more familiar with the bell prover than with the other three types.

When a bell of sufficient size is available, large-capacity meters can be rapidly calibrated in the shop. The use of a slope gauge will certainly reveal if the repairman has assembled a well-balanced, timed meter with a minimum of internal friction.

Bell provers should be in a temperature-controlled area, and all meters should be at room temperature before testing. Normally both a 20% rate and capacity rate are run when proof testing flow rate on the meter. To test small domestic meters, 2 in of water pressure on the bell is used instead of the 1 1/2 in used.

Low pressure flow prover. The low pressure flow prover is used both in the field and in the shop. It is very compact and portable, consisting of two pieces of aluminum pipe with straightening vanes in the upstream sections.

The principle is essentially that of the orifice meter, except the orifice discs are calibrated for different flow rates instead of using various coefficients. Many shops use the low pressure flow prover to test large capacity meters when a bell prover of sufficient size is not available. One or two blowers may be used to supply the necessary volume of air, or low pressure gas may be used where safe.

Critical flow prover. Critical flow is a condition occurring in the gas flow when the downstream pressure is approximately one half the upstream pressure.

The critical flow prover consists of a piece of 2-in, extra heavy pipe approximately 18-in long. One end is threaded to screw into a test connection placed as close to the meter outlet as possible. The other end has a special orifice plate holder. The critical flow prover can be used to test positive displacement meters operating on pressure in excess of 15 psi by passing gas or air through the meter and the prover and discharging it into the air.

The orifice plates of a critical flow prover are stamped on the downstream side with the time (in seconds) required for one cubic ft of air to pass through the orifice at critical flow conditions. Plates used in critical flow provers have rounded edges on the upstream side, whereas orifice plates in a low pressure flow prover are sharp on the upstream edge.

It is vitally important that both the critical flow and low pressure orifice plate be handled with care to prevent damage, which would change the time stamped on the plate.

The stopwatch, which is important in timing prover tests, should be serviced and checked at regular intervals. Such as many pieces of mechanical equipment, the stopwatch will operate for a long time before it completely gives out, and there is a very good possibility that it will be operating inaccurately during this time. Remember that tests taken with a defective timer are inaccurate.

Test meter method. Test meters have become popular in recent years with larger companies and contract testers. Continued development of rotary displacement has made the test meter a practical method for field use.

Although the shop tester normally runs only two rates—check rate and capacity rate—on a bell prover, most testers run three rates on the critical flow, low pressure, and test meter methods. These rates are usually about 20, 50, and 100%.

Testing frequency is a much-discussed subject. Sales through a large-capacity meter represent a great deal of money; thus, it is important to regularly check the meter's accuracy. Some of the larger meters should be tested at 90-day intervals. Others of intermediate size could perhaps operate for two, five, or seven years without being tested and at the end of the scheduled time, simply be changed out.

Many factors affect a testing and change-out schedule. The meter's mileage, its type of service, the type and amount of gas the meter must measure, and the length of time the meter is in service, are important factors in setting up a testing and change-out schedule. Only after all of these factors have been considered can an adequate testing and change-out program be established.

Sizing headers for meter stations

Glenn E. Peck, Senior Engineer, Pipe Line Division, Bechtel Corp., San Francisco, California

Meter and regulator station design for natural gas transmission systems continues to grow more complex and demand more of the design engineer. Basic elements of today's station remain about the same, but higher pressures and ever-increasing capacities have replaced the positive displacement meter with multi-tube orifice meters and automatic chart changers. Buildings are often prefabricated structures specifically designed to house telemetering transmitters and supervisory and communications equipment in addition to basic recording instruments. The multi-run setting may have a computer giving the dispatcher instantaneous rate and totalized flow for the station. Electrical equipment may cost several times that of the entire station of a few years ago.

There is a sparsity of published criteria for sizing of orifice meter headers in the multirun station. A discussion of factors that affect the sizing of the header and arrangement of component parts for satisfactory operating conditions should be in order.

We shall assume that the orifice tube and its component parts are designed in accordance with AGA Gas Measurement Report No. 3, that piping meets all the requirements of USASI B31.8 for pressure, materials, and class of construction, that all valves and fittings meet appropriate codes or standards, and that the differential pressure across the orifice plate will not exceed 100 in of water.

The design goals may be listed as follows:

1. Commercially available equipment will be used that is specified to the closest reproducible economic tolerances.
2. The arrangement of the equipment shall conform to accepted practice for good measurement and allow fullest use of the available meter tubes.
3. Fittings and connections to the header shall minimize the flow turbulence in so far as is practical.

Meeting the above goals allows considerable latitude in the choice of components according to the judgment and experience of the designer without contradicting code, contract, or economic requirements.

In sizing the header, a number of factors must be considered, namely:

- The number of orifice tubes that will eventually be connected to the header, or the ultimate flow capacity.
- The setting layout, or arrangement, which governs the resistance flow paths the gas must take from entering the inlet header and dividing through the meter tubes before recombining in the outlet header and exiting.
- The type of connections which the tubes make with the header.

Station layout

There are as many opinions of what constitutes good meter station layout as there are engineers working with gas measurement. Four of the more common arrangements are illustrated in Figures 1–4 and will be discussed relative to the flow-resistance paths offered.

Header connections

Not many years ago, the header was a common piece of pipe with "orange-peel" ends and notched-miter nipple connections. The next step was a shaped nipple reinforced by a welding saddle or pad. Higher pressures brought forged steel tees into use. On larger-sized tees, the spacing between runs or tubes depended on the measurement for the tee and required a number of girth welds. With the introduction of the extruded nozzle header, the limitation of tees was eliminated for a generally more economical header.

The radius of curvature or flaring of the nozzle as extruded has superior flow characteristics over the reduced branch of a forged tee, and both are vast improvements over the square cut opening for a shaped nipple. Some designers have gone a step further and reduced the 90° header nozzle by the use of a 45° or 60° lateral nipple to the axis of the header. This reduces the turbulence entering or leaving the header and could reduce the diameter of the header. This method can be much more costly for fabrication, is suspect for structural integrity at higher pressures and extreme ambient conditions, and goes beyond the configurations used in establishing the proportions, beta ratios, and coefficients for the orifice meter.

Runs per header

The number of runs, or meter tubes, for a given header is limited by the velocity criterion selected, the operating pressure, the size of fittings obtainable, and the relationship of the peak load to the normal load or capacity for which the meter setting is designed. The operating pressure has an important bearing on the number of orifice tubes used with a given header, because the capacity of the tube is proportional to the square root of the pressure while the capacity

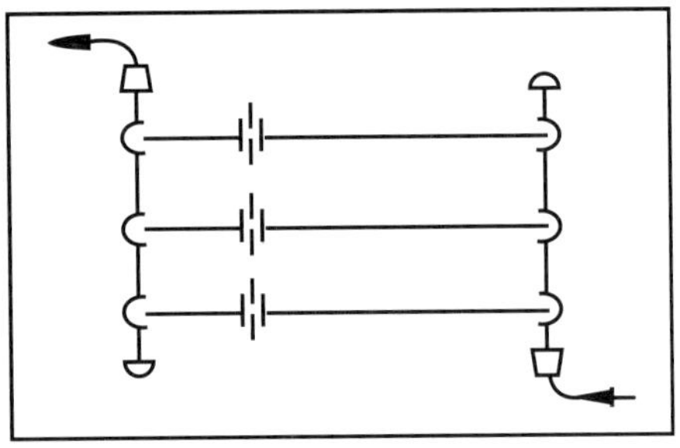

Figure 1. The flow-resistant paths are identical for this layout and each of the two runs will furnish its full capacity with near identical differential readings on the meters. However, when additional capacity is required, an additional run will have a higher or longer resistance-flow path and will furnish a lower differential reading and capacity as compared with the first two runs.

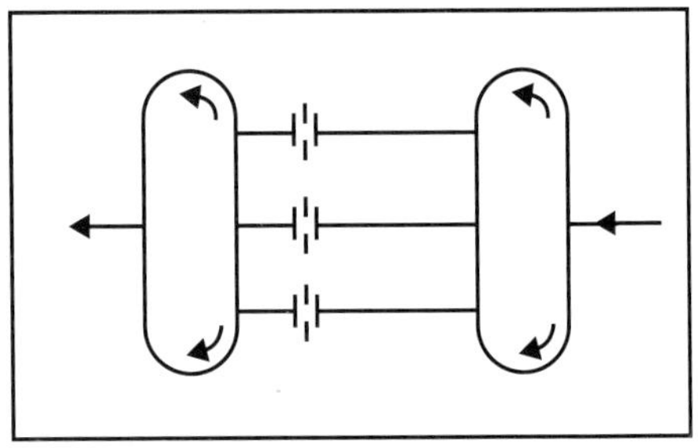

Figure 2. In this arrangement, the middle tube has the lower resistance-flow path compared with the outer tubes and will affect the meter reading accordingly. The difference between meter readings will be more noticeable if the headers are undersized.

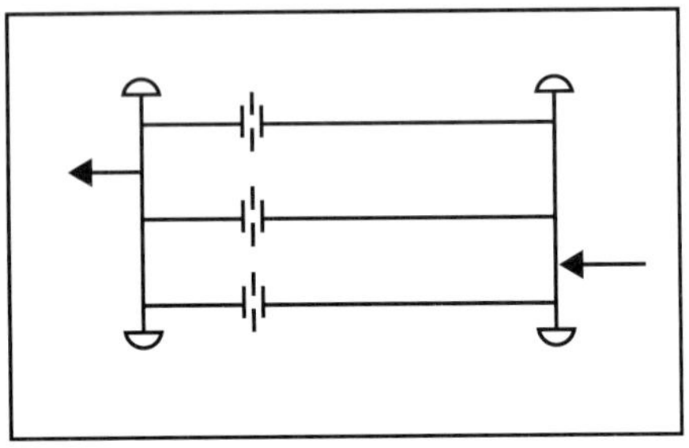

Figure 3. The "donut" header has been frequently used when large headers could be utilized. At higher pressures and capacities, this layout places a high economic burden on the station and requires more space, which can be a problem when a limited site is available.

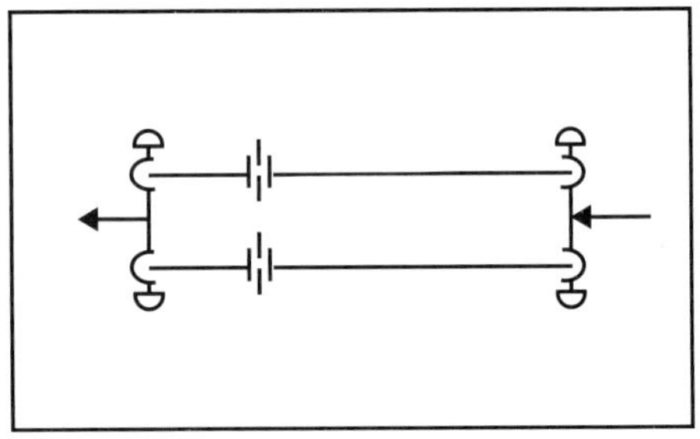

Figure 4. This layout has been a favorite for many since the advent of the extruded or drawn nozzle header and because the resistance-flow paths are as nearly identical as is possible using commercially available equipment. It is limited to three or 4 m tubes before the header begins to be "out-sized."

(velocity) of the header is proportional to the pressure in atmospheres. At lower pressures the number of runs might be limited to three, while at higher pressures the limit would be four runs. Size of fittings is generally less limiting than other factors unless the meter tube is greater than 12 in in diameter. For the very large tube, the extruded nozzle header is required, since reducing branch tees are not generally available. The relationship of the peak load to the sizing of the header is discussed under the velocity sizing criterion.

Velocity sizing

The following velocity criterion is suggested for sizing of headers for the sales orifice meter setting (expressed at ft per minute):

Target	Normal Limit	Maximum Limit
1500	2000	2500

The relationship of the peak hour load to the average hour for the peak day designed for the station determines the range of velocity used in sizing the header. For example, the peak hour which is 10% of the peak day and one that is only 1% higher than the average hour for the peak day obviously demands that different limits be used for economic justification. The 10% peak hour suggests that for higher fluctuation in demand, a higher velocity be allowed in the header for the short time involved.

Keeping the velocity as low as possible within this criterion will provide the basis for good measurement and minimize the difference between adjacent meter readings. The way that gas enters and leaves the header connections to the meter tube is important in considering the velocity criterion that is chosen.

Sizing procedure. The normal procedure for sizing the station follows:

1. Take the peak day forecast or contract load and calculate the appropriate peak hour figure—provided the peak hour figure has not been fixed as a part of the forecast or contract. (Suggested period: 10 years, minimum).

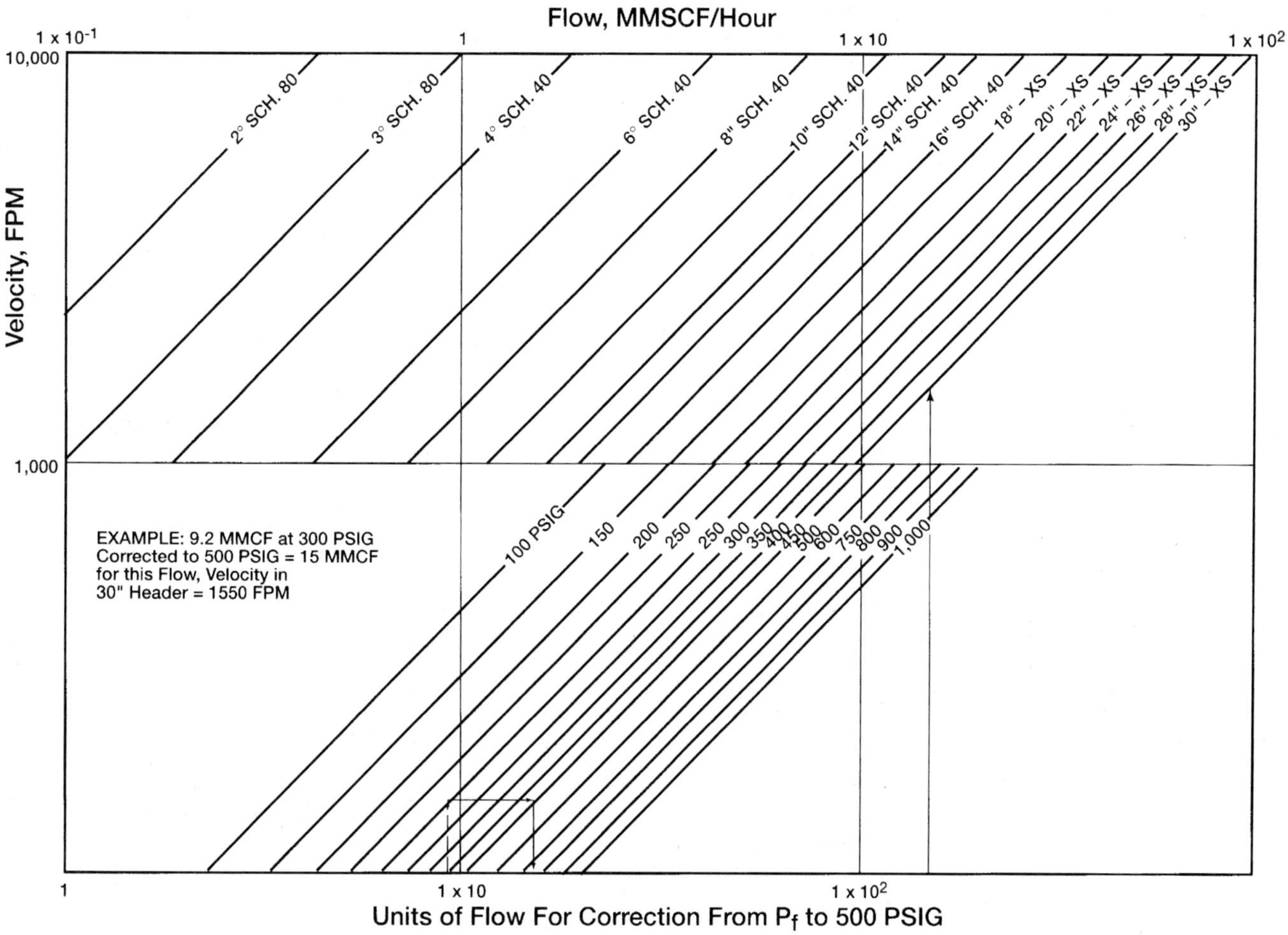

Figure 5. Pipe and header velocity sizing chart.

2. Divide the peak hour by the number of parallel tubes desired.
3. Use a flow rule or chart to size the orifice tube. Check the initial minimum hour against the minimum beta ratio and differential desirable, since this may affect the size and number of tubes.
4. Use a chart, such as Figure 5, to calculate the velocity in the header for the proper criterion. Figure 5 has been found useful since the capacity can be traced up to the desirable velocity and the nearest appropriate size of header chosen.

Figure 5 was prepared for 500 psig or 514.65 psia at a temperature of 60 °F in order that the sizing of pipe or headers be made against velocity and capacity in Mscf/h on the upper part of the chart. The lower part of the chart is used to correct the flow at other pressures to the proportionate flow at 500 psig. Since the sizing procedure is an approximate one, it is felt that corrections for temperatures other than the standard condition and for compressibility are not necessary. Use of Figure 5 or a chart prepared to your own physical conditions will quickly and accurately allow sizing of headers for the orifice meter setting within the present state of the art.

Measuring flow of high-viscosity liquids

Peter B. Jakubenas, Manager of Installed Product Services FMC Technologies Measurement Solutions, Houston, Texas

Measuring custody transfer of high-viscosity fluids has always presented challenges, and as the world increasingly relies on heavy crudes with various amounts of wax, bitumen, and sand, the challenge increases. Crude oils typically flow at rates up to 50,000 barrels per hour (BPH) at export terminals, up to 1500 BPH at ship and truck loading facilities and to 7000 BPH in pipelines.

In addition to heavy crude, other challenging high-viscosity products include heavy fuel oils, In addition to heavy crude, other challenging high-viscosity products include heavy fuel oil, bunker fuels, lubricating oils, grease components, and asphalts. In most cases, flow requires elevated temperatures—up to 400 °F (205 °C)—and flow meters must be reliable in this environment.

Viscosity of water and liquid petroleum products

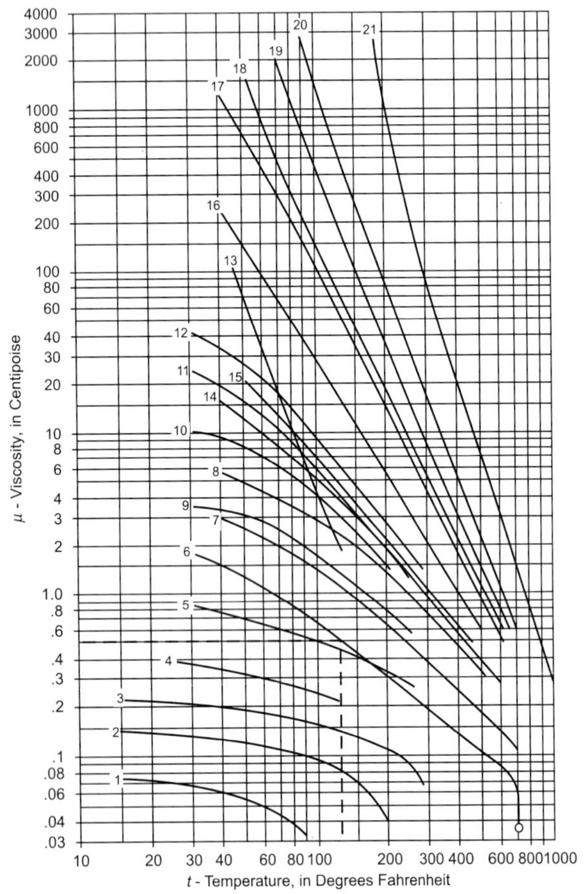

1. Ethane (C_2H_6)
2. Propane (C_3H_8)
3. Butane (C_4H_{10})
4. Natural Gasoline
5. Gasoline
6. Water
7. Kerosene
8. Distillate
9. 48 Deg. API Crude
10. 40 Deg. API Crude
11. 40 Deg. API Crude
12. 32.6 Deg. API Crude
13. Salt Creek Crude
14. Fuel 3 (Max.)
15. Fuel 5 (Min.)
16. SAE 10 Lube (100 V. I.)
17. SAE 30 Lube (100 V. I.)
18. Fuel 5 (Max.) or Fuel 6 (Min.)
19. SAE 70 Lube (100 V. I.)
20. Bunker C Fuel (Max.) and M. C. Residum
21. Asphalt

Viscosity, specific gravity and temperature

A viscous hydrocarbon is considered to be any liquid hydrocarbon that requires special treatment or equipment in order to be handled or stored.

Viscosity (resistance to flow) is related to specific gravity (density), although the relationship between the two is not direct, as the table below shows.

In addition, each hydrocarbon has a characteristic viscosity versus temperature curve making it possible to infer viscosity for known stable hydrocarbon mixtures using a curve developed in the laboratory. In cases where mixtures of hydrocarbons vary over time, viscosity can be measured using one of the many on-line viscometers available.

The figure in this article shows viscosity—temperature curves for the same fluids. In all cases, viscosity decreases with increasing temperatures, but the decline may be more or less dramatic.

Measurement alternatives

Measuring flow of fluids with viscosities above 100 cP requires special consideration by those charged with designing, operating, and maintaining the equipment. The sliding vane positive displacement has been the first choice for measurement since the 1930s; however, in the last 10–15 years, other technologies have emerged that can offer improved cost/benefit ratios, depending on the viscosity and temperature of the fluid and the required accuracy of the meter. Most custody transfer flow meters fall into one of four categories:

- Positive displacement (PD) meters
- Helical turbine meters
- Coriolis meters
- Liquid ultrasonic meters

Typical Viscosity and Specific Gravity of Petroleum Products

Typical Viscosity in cP @ degrees F (degrees C)							S.G. @ degrees F (degrees C)	
Product	30 (−1)	60(15)	100(38)	150 (66)	300(150)	400 (205)	60(15)	50 (66)
Ethane LPG	0.07	0.05	0.03				38	0
Propane LPG	0.14	0.12	0.09	0.07			0.51	0.41
Butane LPG	0.20	0.18	0.15	0.13			0.56	0.49
Gasoline	0.83	0.63	0.49	0.38			0.75	0.70
Water	1.8	1.2	0.7	0.4	0.2	0.1	1.00	0.95
Kerosene	3.5	2.2	1.7	0.9	0.4	0.2	0.82	0.77
Jet fuel	3.5	2.2	1.7	0.9	0.4	0.2	0.82	0.77
48 API crude	3.5	2.7	1.7	1.1			0.79	0.74
40 API crude	10	7	4	2			0.81	0.77
35.6 API crude	25	16	6	3			0.85	0.80
32.6 API crude	42	21	9	5			0.86	0.81
Fuel 3 (max.)		10	5	3	1	0.5	0.90	0.85
Fuel 5 (min.)		16	7	4	1	0.6	0.97	0.92
Fuel 6 (min.)		820	150	43	5	2	0.99	0.94
SAE 10 Lube		68	29	11	2	1	0.88	0.83
SAE 30 Lube		450	105	31	4	2	0.90	0.85
SAE 70 Lube			450	95	8	3	0.92	0.87
Bunker C			1500	290	16	5	1.01	0.96
Asphalt				>3000	80	19	0	0

Data from Crane Co.
Centipoise (cP) = Specific Gravity × Centistokes (cSt).

The positive displacement meter is the only direct measuring alternative, in that every molecule of fluid passes through the meter. The other technologies infer total flow by measuring velocity within a separate flow conduit, where changes in the physical properties of the fluid or of the conduit itself increase the uncertainty of the measurement.

This disadvantage can be overcome by proving the meters on a regular basis and monitoring fluid characteristics to be sure the meter readout has not gone beyond acceptable tolerances. Inferential meters can also be influenced by upstream (and to a lesser extent downstream) flow conditions, such as elbows, strainer debris, or partially open valves.

In defining appropriate flow meters for given conditions, the total turndown range is the range of flow rates over which a meter will perform within a specified linearity, usually ±0.15%.

There are two components of the total turndown range: the viscosity turndown range (which depends largely on the type of meter). Each of the four technologies noted above has a characteristic turndown range, defined as the ratio of the maximum flow divided by the minimum flow rate. The higher the turndown range, the greater the range of flow rates over which the meter will be linear.

The total turndown range is equal to the flow turndown range multiplied by the viscosity turndown range. If meters are operated within a narrow band of flow rate and viscosity and proved at flow rates and viscosities close to these, linearity becomes less of a factor, and meter performance becomes more repeatable. Meter linearity is less of an issue with large parcels such as pipeline and ship loading/unloading and applications, but becomes more important in truck loading, rail-car loading, bunkering, and other small-parcel applications.

Most of the very high-viscosity refined products are handled at temperatures above ambient to facilitate pumping, transportation, and metering, so it is important to specify measuring component materials that can withstand the elevated temperatures provided by steam, hot oil, or electric heat-traced and recirculation lines. In cases where it is uneconomical to heat large quantities of high-viscosity crudes, they are treated with diluents to enable transportation and metering at near-ambient temperatures.

Meter selection and sizing

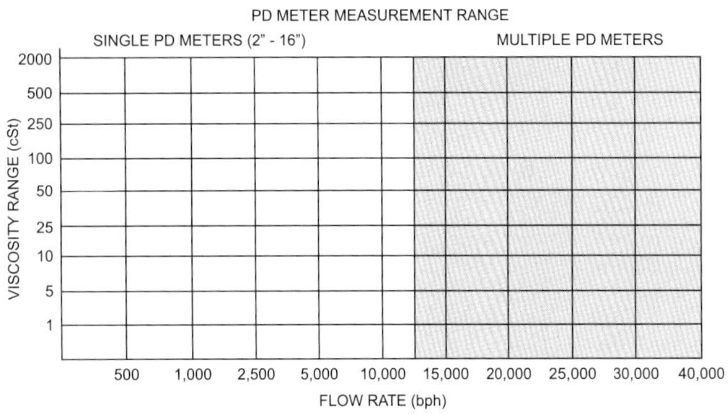

This section lists the advantages and disadvantages of the four meter types used for high-viscosity fluids and shows graphs of suitability depending on viscosity and flow rate.

Positive displacement meters

Advantages

- High accuracy over a wide range of viscosities and flow rates up to 2000 cSt with proper clearances
- Extremely good repeatability on high-viscosity fluids, very low slippage, long life if there is little or no abrasive material in the fluid
- Low pressure drop
- Functions without external power
- Special construction available for high viscosities and temperatures
- Can register near zero flow rate
- Flow conditioning not required
- Measures directly, not an inferential device

Disadvantages

- More moving parts leads to increased maintenance compared to other meters
- May become damaged by flow surges and gas slugs
- Chance of corrosion and erosion from chemicals and abrasive materials
- Derated flow rate capacity for high viscosities and temperatures
- Relatively high cost for large meter sizes, since all fluid must pass through the meter.

Helical turbine meters

Advantages

- Higher turndown range on high-viscosity crudes than conventional turbine meters
- Very good repeatability
- Reduced susceptibility to fouling, abrasives, and deposits
- Less sensitive to viscosity changes
- Lower pressure drop than conventional turbine meters
- Available in large sizes, providing good value for high flow rates

Disadvantages

- Requires flow conditioning
- Back pressure required
- Requires pulse interpolation due to low-resolution pulses
- An inferential device

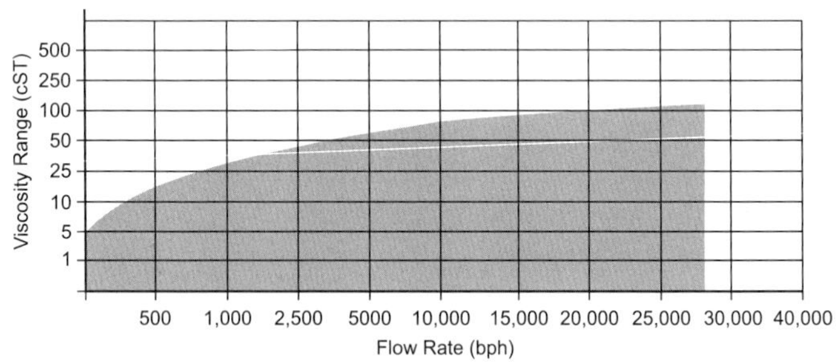

Coriolis meters

Advantages

- Low maintenance, minimally affected *y* abrasives and corrosives

- Not susceptible to damage by gas slugging
- Registers near zero flow rate
- Minimally affected by viscosity changes
- Flow conditioning not normally required

Disadvantages

- Sensitivity to installation conditions, including shock, vibration, pulsations, and effects of adjacent parallel runs
- Deposits can affect accuracy
- Difficult to prove due to time lag of the pulse output
- Requires periodic 43-zeroing
- Needs back pressure control
- High pressure drop that drastically increases with viscosity
- An inferential device

Liquid ultrasonic meters

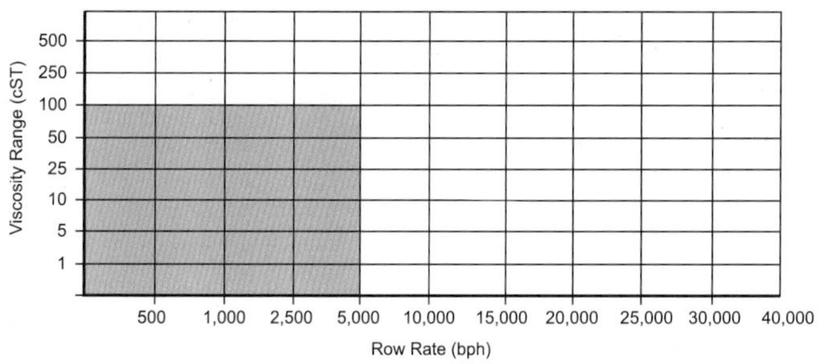

Advantages

- High accuracy
- Wide dynamic flow range
- Negligible pressure drop
- Nonintrusive and no moving parts, making this the least influenced by abrasive materials
- Diagnostic capabilities
- No need for upstream strainer
- Self diagnostic capabilities

Disadvantages

- Flow conditioning recommended
- Susceptible to fouling or deposits
- Sensitive to installed conditions
- Sampling and microprocessor-based output contributes to difficulty in proving
- High cost for small sizes
- Maximum size limited by proving capability
- Back pressure required
- An inferential device

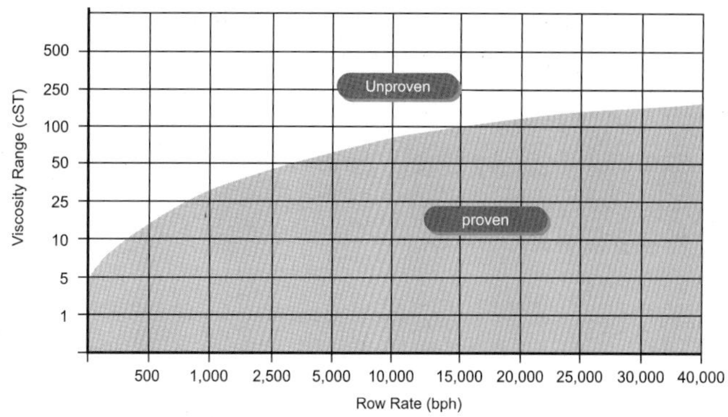

Operational issues

Air removal. All of the meter types discussed require the removal on entrained air and/or vapor which may be difficult with most viscous liquids. As viscosity increases, so does the time required to separate fine bubbles of air or vapor from the liquid. A vapor removal system may be required and a number of such systems are available.

Proving the meters. For custody transfer applications, all of the inferential meter types described above must be periodically proved at conditions close to operating temperature, viscosity, gravity, flow rate, and pressure. For high flow rates, in situ pipe provers are recommended. For lower flow rates, portable proving equipment or master meters may be used.

Although all proving methods may be used with high-viscosity fluids, pipe provers have an inherent advantage over tank provers because the displacer seal against the pipe wall ensures that all fluid is measured. When a tank prover is used, the operator should be alert to high-viscosity fluid clinging to the tank wall because this will affect the measurement. The amount of clingage will vary with the temperature and it is difficult to control temperature within a tank.

Pipe provers used in above ambient temperatures should be insulated and possibly even steam or electrically heat traced. Make sure that the maximum operating temperatures of the elastomer prover displacer sphere and the internal prover coatings are sufficient to handle fluid temperatures.

Readout and control equipment. Since most high viscosity liquids are handled at line temperatures far above the 60 °F reference temperature, accurate electronic temperature compensation of integrated volumes is essential, and we recommend electronic flow computers. For large systems where several meters operate in parallel, we recommend supervisory computers capable of automatic proving plus temperature monitoring and control.

For small stand-alone PD meters deployed where no power is available, mechanical wheel and disk temperature and gravity compensators (ATG and ATC) are still available.

Special operational and maintenance considerations. For viscous hydrocarbons heated above ambient, it is essential carefully to monitor and control temperature because viscosity drastically increases with a drop in temperature. If the temperature of the system decreases enough that the pumps cannot sustain flow, pump damage can result. Removing solidified product from the pipelines and metering equipment can be very difficult and expensive.

If systems are used intermittently, it may be more practical to evacuate or flush them with a low-viscosity solvent between uses. Upon restarting, slowly fill the system and completely evacuate air or vapor before resuming flow, since the meters and other components can be damaged by gaseous slugs.

If the temperature control system fails and temperatures rise too high, serious damage to valves, meters, and provers can occur. In addition, overheating the fluid can cause coking or chemical changes—and some overheated petroleum liquids may ignite on exposure to air.

Hydraulic considerations are also important since the pressure drop in the measuring system will dramatically increase with an increase in fluid viscosity. Pump sizing assumptions become very important. For example, consider a crude oil with 12 API gravity and viscosity of 300 cP @ 180 °F and 6500 cP @ 60 °F. the maximum flow rate desired is 6000 BPH, but will this really be achieved when the product is heated? This needs to be confirmed. If this flow rate can be achieved, what is the maximum flow rate that is hydraulically possible when the viscosity falls to 6500 cP?

Conclusions

The figure below shows a composite of the four technologies. While there are many choices of flow meters for high-viscosity liquids, no single metering technology is best for all applications.

For abrasive-free liquids, positive displacement meters can handle the widest range of flow rates and viscosities and will have a very long life. They have a track record of infield service of over 70 years, and some meters are still in operation after 30 years on heavy, waxy crudes. For applications where future liquid viscosities and flow rates are uncertain, positive displacement meters are the most flexible.

Helical turbine meters offer the lowest cost alternative for high flow rates on high-viscosity, abrasive-free fluids.

Coriolis meters perform well within their limited flow and viscosity range although they have high pressure drops.

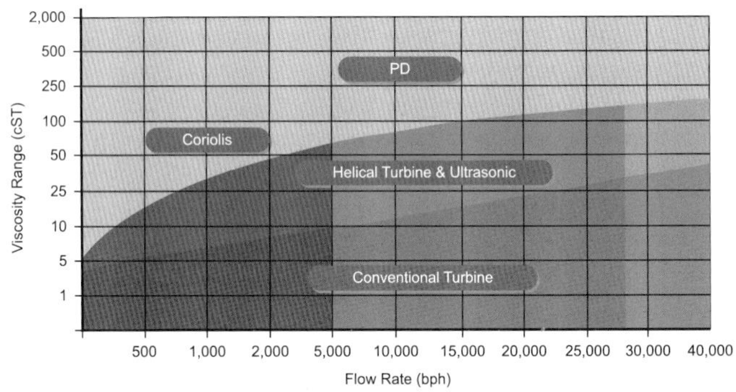

Ultrasonic meters only become economically attractive for high flow rates and are the best at accommodating abrasive fluids.

Selecting the proper ancillary equipment is essential to meter performance and equipment operation and maintenance of high-viscosity flow meters may require special techniques to each application.

References

1. API Manual of Petroleum Measurement Standards Chapter 1 "Vocabulary" provides a comprehensive list of definitions for many of the terms used in this article.
2. API Manual of Petroleum Measurement Standards Chapter 6, "Metering Viscous Hydrocarbons" provides a comprehensive guide for the design, installation, operation, and proving of meters and their auxiliary equipment used in metering viscous hydrocarbons.

Reprinted with permission - *Pipeline & Gas Journal*, July 2007.

Matching the flow meter to the application

Greg Livelli

Two significant specifications for a flow meter are accuracy and turndown (or rangeability). The accuracy of all flowmeters to some degree depends on the circumstances of their application. These effects include pressure, temperature, fluid dynamic influences, and external influences. Pressure and temperature cause dimensional and physical changes. Fluid flow and pipe roughness can shift meter performance. Mass meters and displacement meters are the least affected, while turbine, vortex, and ultrasonic meters are the most affected. The upshot is that a traceable and complete calibration in a lab is no guarantee that the same performance and accuracies will be obtained or maintained in the field.

A flow meter's turndown is the ratio of the maximum to the minimum flow rate capable of being measured at a specified accuracy, providing a measure of rangeability. High values of turndown can be irrelevant because typical processing flow velocities often fall within relatively narrow ranges. For example, most practical liquid pipeline flow rates in the processing industries range from 0.5 to 12 ft/s, a turndown ration of 24-to-1. Lower rates are difficult to accurately measure and higher rates result in high pressure drops, pumping energy costs, and erosion (if solids are present). In the case of pipelines carrying gases in the processing industries, the practical flow velocities range from 15 to 200 ft/s—a turndown ratio of about 13-to-1. Many actual applications have flow rate ranges well within these extremes.

Instrument engineers commonly apply volumetric flowmeters to measure liquids. When the process fluid is a gas or steam, measuring mass flow becomes especially important.

Figure 1. The orifice plate is a popular DP element.

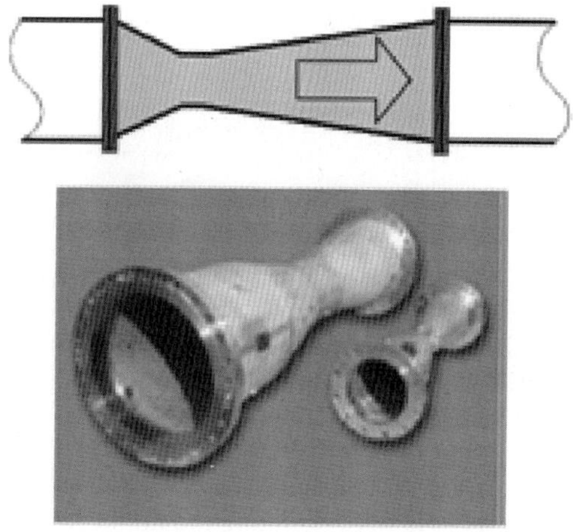

Figure 2. The Venturi DP element minimizes pressure loss.

Because these fluids are compressible, their density changes with pressure and temperature. Volumetric flow measurement is often meaningless in these cases. However, volumetric flow meters work if the application also includes measurement of temperatures and pressures at the meter or these values are known quantities. Simple calculations based on the three variables can infer mass flow rates.

Differential-pressure flow meters

Flow meters based on differential pressure insert an element within the piped flow. The unrecoverable pressure loss across the element—typically an orifice plate, Venturi, nozzle, pitot tube, or wedge—is a measure of the volumetric flow rate.

Flow meters based on differential pressure represent a popular choice in the processing industries, constituting nearly 30% of the installations. They have good application flexibility since they can measure liquid, gas, and steam flows and are suitable for extreme temperatures and pressures with moderate pressure losses. Accuracy ranges from 1 to 5%. Compensation techniques can improve accuracy to 0.5–1.5%.

On the other hand, DP flow meters are expensive to install and have limited rangeability (turndown) compared to other types. They require a separate transmitter to compute the volumetric flow rate and develop a standard signal. Changes in density, pressure, and viscosity can significantly affect accuracy of DP flow meters. And while they have no moving parts, maintenance can be intensive.

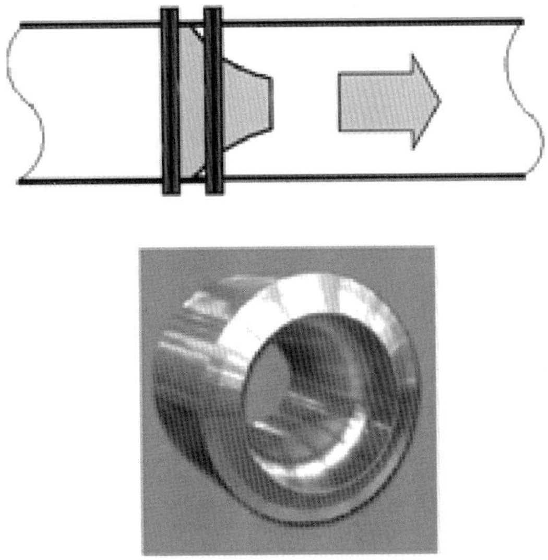

Figure 3. The nozzle has Venturi-like properties.

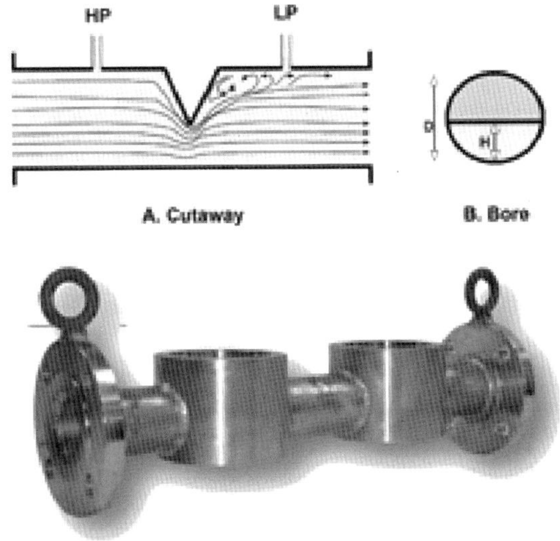

Figure 5. The wedge DP element handles tough, dirty fluids.

Orifice plate. Instrument engineers can employ a variety of elements to create the differential pressure, with the orifice plate being the most common. Orifice plates are inexpensive and available in a variety of materials. The turndown ratio, however, is less than 5-to-1, and accuracy is moderate—2–4% of full scale. Maintenance of good accuracy requires a sharp edge to the upstream side. This edge will wear and degrade over time. Pressure loss for orifice plates is high relative to other DP elements.

Venturi. This element finds use primarily in water and wastewater applications. It handles dirty fluids and does not require upstream flow profiling. Pressure loss is minimal, making it a good choice when little pressure head is available. Rangeability, while better than orifice plates, is less than 6-to-1, with accuracy of ±1–2% of full scale. Viscosity effects on accuracy are high. Flow must be turbulent (Reynolds numbers > 10,000). The Venturi has limited acceptance within the processing industries.

Nozzle. Characteristics of nozzle elements mimic those of the Venturi. Nozzles come in three types: ISA 1932 nozzle (common outside of the United States); the long radius nozzle; and the Venturi nozzle.

Pitot tube. This element is another low-cost DP element used to measure fluid flow, especially air flow in ventilation and HVAC systems. Pitots commonly measure the speed of airplanes. This element works by converting the kinetic energy of the flow velocity into potential energy (pressure). Engineers can easily insert the pitot tube into existing piping,

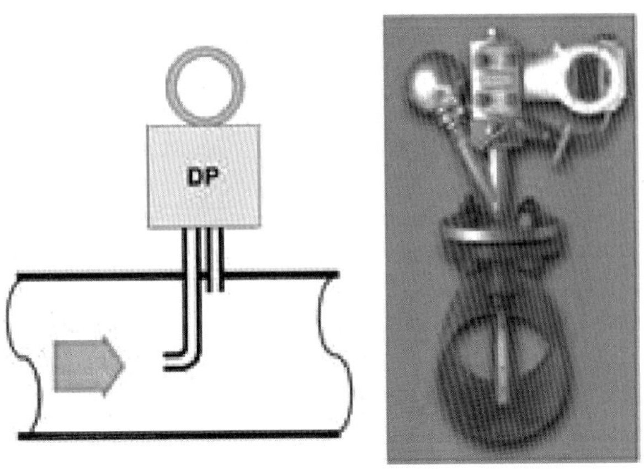

Figure 4. The pitot tube converts fluid kinetic energy into pressure.

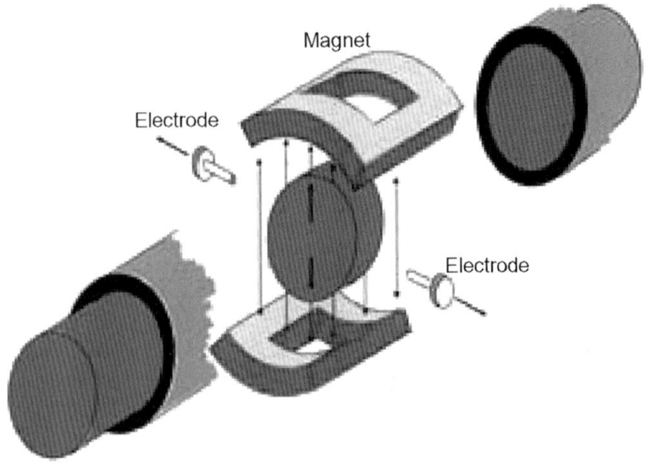

Figure 6. Electromagnetic flow meters require conductive fluids.

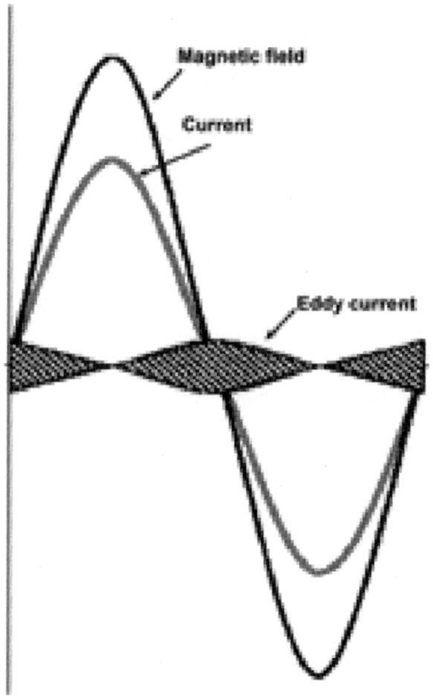

Figure 7. AC magmeters have good signal-to-noise ratios, but require frequent calibrations to eliminate zero drift.

minimizing installation costs. It cannot handle dirty flow and has limited rangeability.

Such as an insertion meter, the pitot tube makes a measurement at a point within the pipeline or ductwork. An exception is the annubar pitot tube, which contains multiple orifices. This element sees the dynamic pressure across the velocity profile, providing an averaging effect.

Wedge. This element consists of a V-shaped restriction molded into the meter body. In profile, the wedge looks similar to a segmental orifice plate to the incoming fluid, but it is more expensive than an orifice plate. This basic meter has been on the market for over 40 years, demonstrating its ability to handle tough, dirty fluids. The slanted faces of the wedge provide self-scouring action and minimize damage from impact with secondary phases. Rangeability of 8-to-1 is relatively high for a DP element and accuracies are possible to ±0.5 percent of full scale.

The wedge flow meter, however, is not approved for flow measurement by the American Gas Association (*www.aga.org*) or the American Petroleum Institute (*www.api.org*). Despite this, it is a popular choice in oil and gas applications, especially in production fields. For difficult fluids, it can be equipped with a pair of remote seals that effectively isolate the metered fluid from the DP transmitter without affecting accuracy.

Electromagnetic flow meters

Electromagnetic flow meters (magmeters) are a popular choice among instrument engineers, making up about 20% of flow meter installations. Faraday's law says that a conductor moving through a magnetic field produces an electric signal. In this case, the fluid is the conductor and electromagnetic coils surrounding the meter body generate the magnetic field.

Magmeters have no moving parts. They offer wide rangeability and an unobstructed flow path. Magmeters are ideal

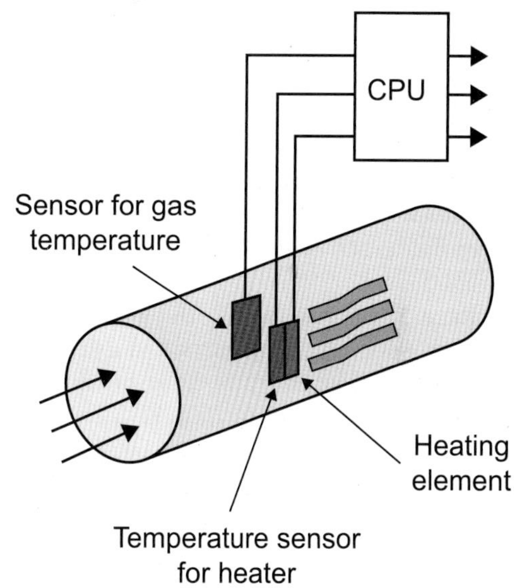

Figure 8. Coriolis mass flow meters are ideal, but relatively costly.

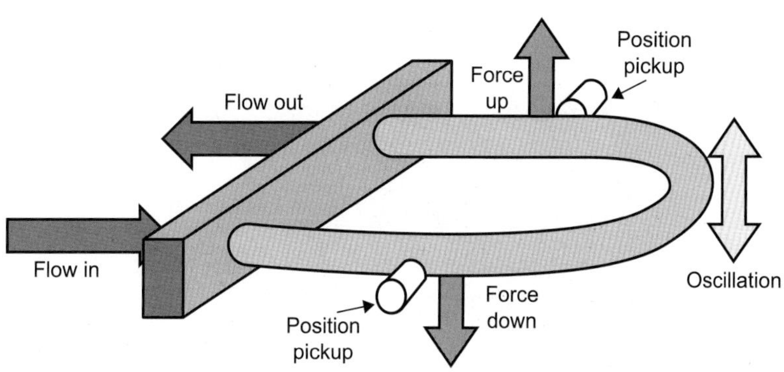

Figure 9. Thermal mass flow meters excel at measuring low flow rates.

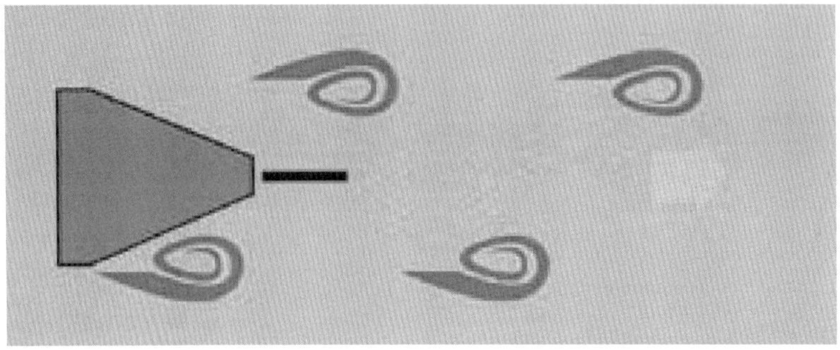

Figure 10. Vortex meters have no moving parts, but may be affected by vibration.

for slurries, and perform well with corrosive and erosive fluids. Engineers can pick a nonconductive liner material for the flow tube—Teflon, Tefzel, rubbers, ceramic, etc.—to suit the fluid measured. Magmeters require minimum lengths of straight pipe to condition the flow profile. On the downside, the fluid must be conductive, which rules out most petroleum-based flows. Magmeters also have physical pressure and temperature limits.

Magmeters have rangeability up to 1000-to-1, depending on the maximum tolerable measurement error. For flow velocities of 0.5–50 ft/s, accuracies are usually stated as a percent of rate, for lower velocities, accuracies are stated as a percent of span. Typical velocities measured range from three to 15 ft/s for water and clean chemicals, three to six ft/s for abrasive fluids, and six to 12 ft/s for coatings and liquids with entrained air.

The voltage developed across the electrodes is in the millivolt range and should be carried via shielded cable to a nearby converter. The converter near the flow tube boosts the signal to a standard (four to 20 mA) or frequency output (zero to 10,000 Hz) for transmission to a display or controller.

The coils producing the magnetic field may be excited by AC, DC, or pulsed DC. DC excitation accounts for most magmeter installations today. With AC excitation, line voltage powers the magnetic coils, and the resulting signal has a sine wave shape of line frequency whose amplitude is a linear function of flow velocity. AC designs produce a high coil current (typically 3.2 amp) and a good signal-to-noise ratio. They are relatively insensitive to media noise. However, they consume much power and can be plagued with zero drift caused by eddy current noise. To compensate, operators stop the flow and set the transmitter output to zero. Frequent operations to eliminate eddy current noise are sometimes necessary to maintain accuracy, which is about ±1 percent of rate.

With DC excitation designs, a lower frequency DC pulse excites the magnetic coils. The converter reads both the flow and noise signals during a pulse. In between pulses, however, it sees only noise, permitting noise cancellation after each cycle. Therefore, zero drift is not a problem for pulsed-DC designs. These magmeters offer 0.15–0.5% accuracy and low power consumption. The low coil current (0.1–0.5 amp) comes with smaller signal and low signal-to-noise ratio. Low frequency designs have longer response times.

A newer DC design develops a stronger flow signal by increasing the DC coil excitation current (0.7 to five amp).

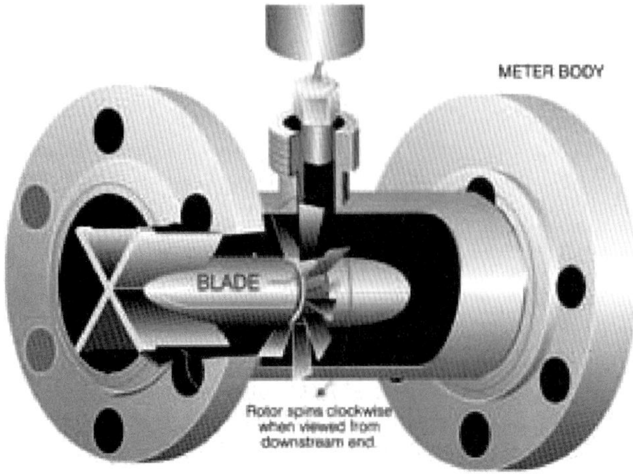

Figure 11. Turbine meters are well-accepted in the processing industries.

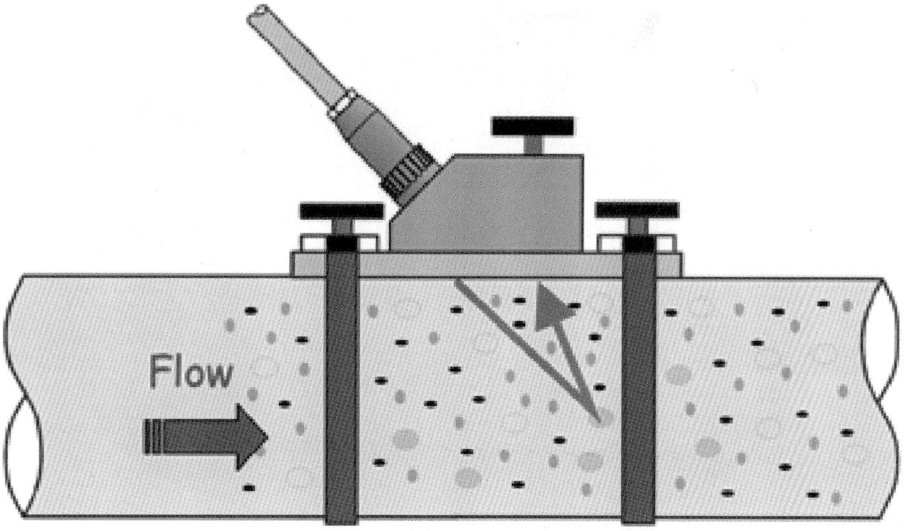

Figure 12. Doppler ultrasonic flow-meters work best on slurries and dirty liquids.

This design improves the signal-to-noise ratio, but increases power consumption and cost. Flow tubes must generally be less than 20 in in diameter. Another new design employs dual excitation, pulsing the coils at about seven Hz for zero stability and also at 70 Hz for a stronger signal and improved response times (0.1 s). But, this design raises costs and applies only to flow tubes smaller than 16 in in diameter. Eddy current instability increases proportionally as sensor size increases.

Mass flow meters

Mass flow meters directly measure mass flow rate, as compared to other flow meter types, which measure flow rate based on volumetric principles. Two primary types of mass flow meters, Coriolis and thermal mass flow meters, account for about 18% of installations.

Coriolis flow meters represent the ideal flow meter. They can be applied to virtually any fluid—liquids, gasses, and slurries. They are immune to changes in viscosity, pressure, temperature, or density of the piped fluid and are unaffected by flow profile disturbances. They offer high accuracy (±0.1%) and rangeability (100-to-1), and provide an additional measurement of fluid density. However, their purchase and installation costs are relatively high, as well as their sensitivity to vibration. In addition, they come in a limited selection of sizes, rarely over two inches.

In these meters, the fluid measured runs through one or two vibrating tubes. Tube designs take on a multitude of shapes, depending on the manufacturer. Coriolis forces resulting from the vibration and flowing fluid act on the tubes, causing them to twist or deform. Sensors pick up the change in tube shapes. Sensor signals may be directly related to mass flow rate.

The thickness of the tubing walls varies considerably from design to design. But, even the sturdiest tubing will be thinner than the process piping. Long, bent tubes twist more easily than do short, straight tubes, and so they will generate stronger signals under the same conditions. Curved-tube designs provide wider rangeability (up to 200-to-1), while straight-tube meters are limited to 30-to-1 to 50-to-1, with lower accuracy. Straight-tube meters are more immune to pipeline stresses and vibration.

Thermal mass flow meters add heat to the flow stream to make measurements. One or more temperature sensors

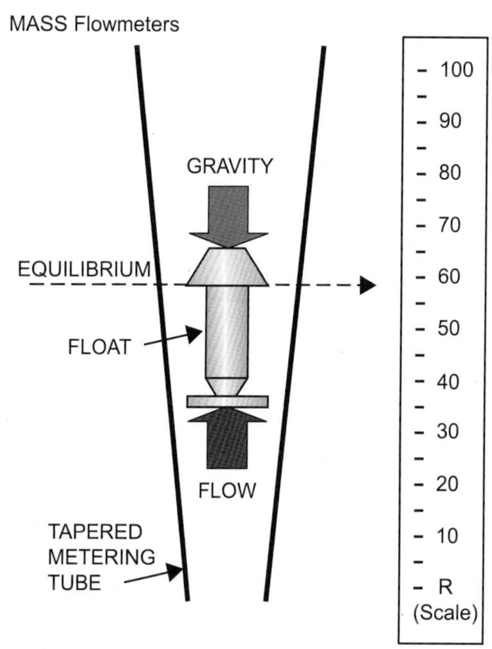

Figure 13. Transit-time ultrasonic flow meters work best on clean fluids.

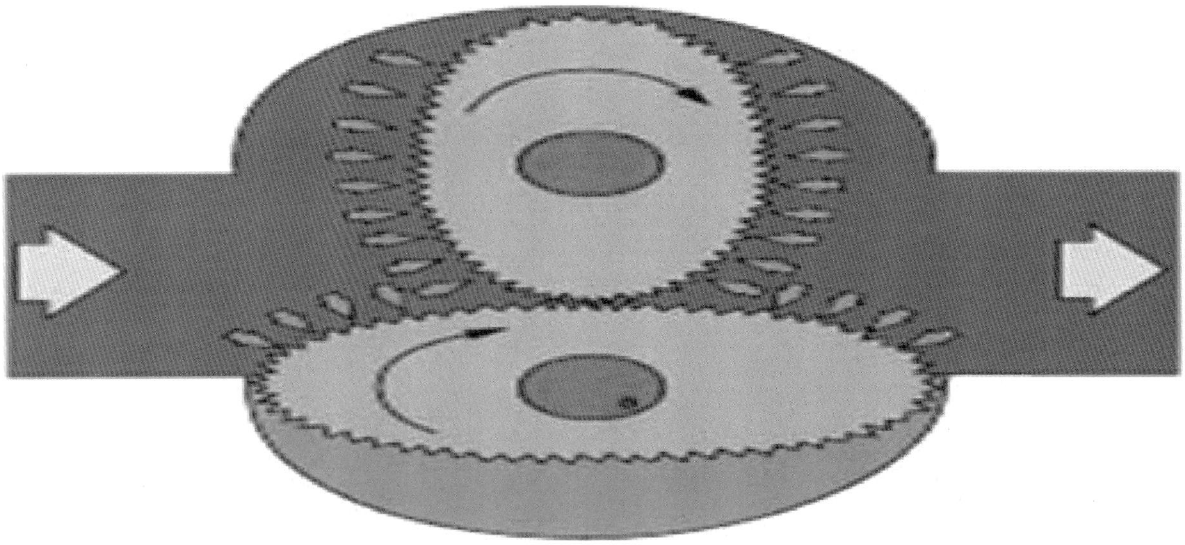

Figure 14. Positive-displacement flow meters are unaffected by poor flow profiles.

measure how quickly this heat dissipates. One technique maintains a heated sensor at a constant temperature, measuring the current required. Another way is to measure the temperature difference between a sensor in the flow stream and a heated sensor outside the stream. Higher flow rates result in increased cooling. The cooling effects are a function of mass flow.

These meters are compact, and offer wide rangeability (50-to-1) and good repeatability. They have a fast response time, especially for gasses, and excel at measuring low mass flow rates. They assume a constant specific heat and thermal conductivity of the fluid. Changes in density cause calibration shift, and coating of the sensor can cause drift. Additionally, they measure a point within the pipe diameter, so the flow profile must be known. Accuracy levels typically range from 1 to 3%. However, manufacturers expect to improve their accuracy, which will widen their application.

Vortex vs. Swirl Meters	
Vortex	Swirl
Lower cost	Minimal inlet/outlet straight piping
All cast construction	Better low-flow capability
Water and flanged meter bodies	Flanged bodies only
Higher maximum flow capability	Better rangeability on more viscous liquids
Measures liquids (viscosity < 8 cp) gases and steam	Measures liquids (viscosity <30 cp) gases and stream
Typical sizes ½ to 12 inches	Sizes ½ to 16 inches flanged
316L SS or Hastelloy	316L SS
Accuracy +/– 0.75% of rate for liquids and +/– 1.0% of rate for gases/stream	Higher accuracy +/–0.50% of rate for liquids, gases and stream

Vortex/swirl flow meters

Place a bluff body within a fluid pipeline and the stream will alternately produce whirling vortices on either side of the body. The spacing between the vortices remains constant, regardless of the fluid velocity. Within a pipeline, the vortice effect attenuates a few pipe diameters downstream from the bluff body. Typically, piezoelectric or capacitance sensors count the pressure oscillations caused by the vortices. The vortex-shedding frequency provides a measure of fluid velocity.

Such as a vortex meter, swirl meters detect pressure pulsations, whose frequency depends on the flow rate. Vanes within the meter cause the incoming fluid to take the shape of a rotating helical coil of relatively low-pressure zones. At low flow rates the low pressure swirls or coils are far apart. As flow rates increase, the coils become compressed, coming closer together. Detectors within the meter measure the frequency of the swirling low-pressure zones as they pass through the meter. A deswirler at the meter's exit eliminates any effects on downstream instruments.

Engineers can apply these meters to liquid, gas, or steam, but should avoid dirty, corrosive, or abrasive fluids. The meters have no moving parts, have wide rangeability, and are immune to pressure, temperature, or density changes. These meters, however, lose effectiveness as viscosity increases. They are also affected by vibration or pulsation.

Swirl meters are less sensitive to flow profiles than vortex meters. While vortex meters may require a minimum straight-pipe run of 15 or more pipe diameters upstream and five pipe diameters downstream, swirl meters require only three diameters upstream and one diameter downstream. Both may be installed horizontally or vertically, but must be kept full of fluid during operation. About half of all vortex meter installations require "necking down" oversized process piping by concentric reducers and expanders.

Turbine flow meters

A piped fluid that acts on turbine vanes will rotate the turbine with a rotational rate proportional to flow rate. Turbine meters make up about eight percent of the installed base. Rangeability can reach 100-to-1 if the meter measures the rate of a single fluid at constant conditions. Accuracy is high and may top $\pm 0.1\%$.

Turbine meters are suitable for extreme temperatures and pressures and can be applied to liquids and gasses. They install easily and are well accepted in processing applications, especially the oil and gas industries. On the other hand, turbine meters suit only low viscosities and are sensitive to flow profile and vibration. Having moving parts, they usually require frequent calibration. Partially open valves upstream from a turbine meter can cause significant errors. Typically, turbine meters require flow upstream straightening vanes. Additional considerations include:

- Possible flashing or cavitations from pressure drops. Flashing causes the meter to read high while cavitations result in rotor damage. The downstream pressure must equal 1.25 times the vapor pressure plus 2 times the pressure drop.
- Small amounts of air entrainment (<100 mg/1) will make the meter read high, while large quantities can destroy the rotor.
- Solids entrained in the fluid can damage the meter. In these cases, a flushing strainer or filter must be installed upstream of the flow meter.

Ultrasonic flow meters

Doppler flow meters, limited to liquids, send an ultrasonic beam into the fluid and look at reflections from discontinuities such as entrained particles or bubbles. Reflections from the moving objects will experience a frequency shift that is a measure of the flow rate. Transit-time flow meters have beam transmitters and receivers separated by a known distance. They look at the differences in time of flight for ultrasonic beams moving with and against the flow. Doppler meters work best on slurries or dirty liquids, while transit-time meters work best on clean fluids.

These flow meters have no moving parts and offer an unobstructed passage for fluid flow. Ultrasonic flow meters have wide rangeability, and some designs can simply be strapped on to the outside of pipeline. Typical accuracy of these clamp-on flow meters is about five percent. Generally, clamp-on units will not work well if the pipe has a lining or if the pipe material is stainless steel, PVC, or fiberglass. In addition, the inside of the pipe must be free of sound-absorbing material, such as dirty grease or scale. Multibeam, transit-time units in direct contact with the fluid, however, now have sufficient accuracy to be used for custody transfer of gases and liquids.

Published data on ultrasonic flow meters is limited. They have a relatively high initial cost and performance is sensitive to multiple installation factors. Deposit buildup on the pipe walls can significantly affect accuracy, as can extreme vibrations.

Variable-area flow meters

Commonly known as rotameters, these devices provide practical solutions for many flowmetering applications. They are simple and inexpensive, performing as well or better than many high-tech flow meters on the market. Every rotameter basically consists of two components—a tapered metering tube made of glass, metal, or plastic and a float that rides within the tube. These components come in a wide variety of shapes, sizes, weights, and materials of construction, making them adaptable to a wide range of application needs. They are easy to install and maintain, but must be mounted vertically and plumb. Some designs can handle high pressures and viscosities. Today, most manufacturers have taken all the factors and data involved in sizing a rotameter and incorporated them in a sophisticated software program.

Accuracy, however, is relatively low ($\pm 2\%$) and depends on precise knowledge of the fluid and process. Rotameters are susceptible to vibration and plugging by solids and are affected by density and temperature changes. They primarily apply to flow rates below 200 gpm and pipe sizes less than three inches.

Target flow meters

This meter inserts a physical target within the fluid flow. The moving fluid deflects a force bar attached to the target. The deflection caused by the force depends on the target area, as well as the fluid density and velocity. These flow meters have no moving parts such as bearings and can measure fluid flow rates for any line size above 0.5 in. Engineers can adapt the flow meter to different fluids or flow rate ranges by changing the target size and material. Rangeability is about 15-to-1. One disadvantage is that the target meter's calibration must be verified in the field.

Positive-displacement flow meters

Precision-fitted rotors measure fluid flow in positive-displacement flow meters. The volumes between the rotors are known and fixed. The rotor rotation varies directly with the volume of the fluid being displaced. Counters monitor the number of rotations and a transmitter converts the counts to volumetric flow rate. Principal types include:

- Single or multiple reciprocating piston meters.
- Oval-gear meters with synchronized, close-fitting teeth.
- Movable nutating discs mounted on a concentric sphere located in spherical side-walled chambers.
- Rotary vanes creating two or more compartments inside the meter's housing.

These flow meters are applicable to a wide range of nonabrasive fluids, including those with high viscosities. Examples include heating oils, lubrication oils, polymer additives, ink, and many more. They are unaffected by poor flow profiles. Accuracy may be up to ±0.1% of full scale with a rangeability of 70-to-1 or better. Moreover, they require no power and can handle high pressures. Positive-displacement flow meters cannot handle solids, entrapped air in liquids, or entrained liquids in gases. They are expensive to install and maintain, having many moving parts. Pressure drop across the meters is high.

Reprinted with permission - *Flow Control*, August 2007.

Use liquid ultrasonic meters for custody transfer

P. Syrnyk and D. Seiler, Daniel Division, Emerson Process Management Houston, Texas

Using liquid ultrasonic meters for liquid petroleum applications such as custody transfer or allocation measurement is gaining global acceptance by the hydrocarbon processing industry. Ultrasonic technology is well established, but using it for custody transfer and allocation measurement is relatively new. End users often try to utilize the same measurement practices that apply to turbine technology.

While the two approaches share some similarities—such as the need for flow conditioning and upstream and downstream piping requirements—there are also differences. Several in situ proving methods can be used to successfully field calibrate a liquid ultrasonic meter.

Transit time ultrasonic flow meters. This device uses the transit times of the signal between two transducers to determine the fluid velocity. The transducer transmits ultrasonic pulses with the flow and against the flow to a corresponding receiver (Figure 1). Each transducer alternates as a transmitter and a receiver.

Consider stationary fluid in a full meter spool. In theory, it will take precisely the same amount of time for a pulse to travel through the fluid in each direction since the speed of sound is constant within the fluid. If fluid is flowing through the pipe, then a pulse traveling with the flow traverses the pipe faster than the pulse traveling against the flow. The resulting time difference is proportional to the fluid velocity passing through the meter spool. Single and multiple acoustic paths can be used to measure fluid velocity. Multipath meters tend to be more accurate since they collect velocity information at several points in the flow profile.

The transit time of the ultrasonic signal is measured and used with other variables to calculate flow. Although the ultrasonic signal is traveling in a straight line, it is traveling at an angle, Θ, to the pipe axis (Figure 2). Equations 1 and 2

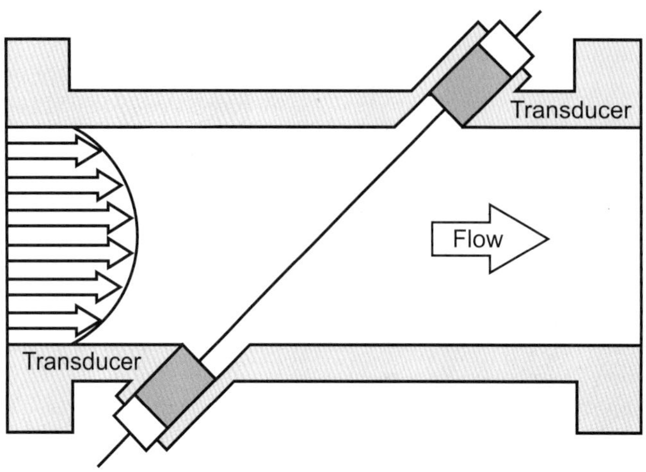

Figure 1. Transit-time principle for ultrasonic flow meters.

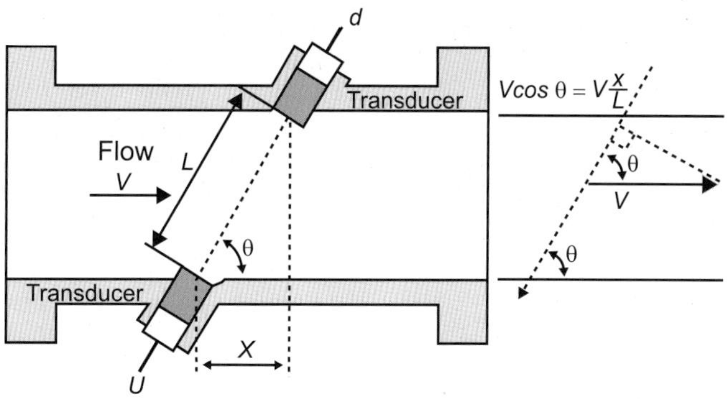

Figure 2. Variables for calculating transit time.

define the flow rate between two transducers located at positions u (upstream) and d (downstream):

$$t_{ud} = \frac{L}{C + V_i \cos \Theta} \tag{1}$$

$$t_{du} = \frac{L}{C + V_i \cos \Theta} \tag{2}$$

Solving Equations 3 and 4 simultaneously yields the following results for V_i and C (notice that V_x/L can be substituted for "Vcos Θ" to simplify Equation 4):

$$V_i = \frac{L}{2 \cos \Theta} - \frac{t_{du} - t_{ud}}{(t_{ud})(t_{du})} = \frac{L^2 t_{du} - t_{ud}}{2X(t_{ud})(t_{du})} \tag{3}$$

$$C = \frac{Lt_{du} + t_{ud}}{2(t_{ud})(t_{du})} \tag{4}$$

where:

t_{ud} = transit time from transducer u to d
t_{du} = transit time from transducer d to u
L = path length between transducer faces u and d
X = axial length between transducer faces
C = velocity of sound in the liquid in still condition
V_i = mean chord velocity of the flowing liquid
Θ = acoustic transmission angle.

Since the equations are valid for fluid flowing in either direction, the method is inherently bidirectional. Also, the speed of sound term through the medium drops out of the velocity equation. Consequently, velocity is determined from the transit times through the predetermined distances, and is independent of factors such as temperature, pressure, and fluid composition.

Field proving. In custody transfer or allocation measurement applications, it is common practice to periodically prove liquid meters in the field. This is done to reduce uncertainty and maintain accuracy for fiscal measurement, establish meter factors at working conditions, and determine meter factors for different fluids. The measurement application or the customer contract generally determines the frequency at which it is proved.

Meters in a pipeline that transports several liquids may be proved at every product change. The American Petroleum Institute provides guidelines and requirements. There are four basic requirements for successful proving:

1. The meter should be proved under the normal conditions in which it is expected to operate. Flow variables, including flow rate, pressure, and temperature must be stable and unchanging.
2. Adequate prover capacity is required to provide proving runs of sufficient duration. Larger volumes and longer run times will improve repeatability performance. Upstream flow conditions can affect the run time and volume required.
3. A sufficient number of runs are required to establish a valid proving. If the run-to-run variation is small, then an adequate meter can be found by averaging a few runs. As the run-to-run variation increases, the average of more runs is required for an adequate meter factor.
4. Results should be traceable to the National Institute of Standards and Technology (NIST). The meter prover calibration must be traceable to NIST-calibrated test measures.

Proving principle. Displacement prover systems operate on the displacement principle. This is the repeatable displacement of a known volume of liquid from a calibrated section of pipe between two detectors (Figure 3).

Liquid displacement is achieved by a sphere or a piston traveling through the pipe. A corresponding volume of liquid is simultaneously measured by a meter installed in series with the prover. A meter that is being proved on a continuous-flow

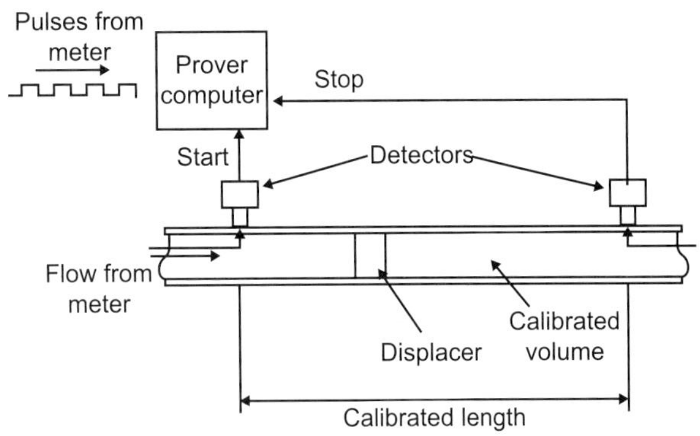

Figure 3. Proving a meter by displacement.

basis must be connected at the time of proof to a proving counter. The counter is started and stopped when the displacing device actuates the two detectors at either end of the calibrated section of the prover.

A displacement-type prover must be installed so that the flow of the liquid stream through a meter being proved will fully pass through the prover. When flow through the meter and prover has stabilized, the proving sequence is initiated. The displacer then travels through the pipe, while maintaining a seal between it and the pipe wall. Each of the detector switches is individually actuated upon displacer passage.

A high-resolution pulse counter connected to the meter output is gated on and off by the detectors. In the event the meter being proven cannot generate a minimum of 10,000 whole meter pulses between detector switches, some form of pulse interpolation must be used to allow discrimination at a rate higher than 1 part per 10,000.

Types of provers. A displacement prover may utilize either a sphere or a piston as a displacer and may be either unidirectional or bidirectional. Liquid always flows in the same direction through a unidirectional prover which has a mechanism that returns the displacer to its initial position in the prover.

A bidirectional prover utilizes a flow reversal mechanism (such as a four-way valve) that causes the displacer to alternate its direction of passage through the prover pipe. Unidirectional proving runs consist of single trips of the displacer past the detectors in the same direction. Bidirectional proving runs consist of "round trips" of the displacer (a complete cycle past the detectors in both directions). A typical bidirectional prover is depicted in Figure 4; a typical unidirectional "small volume" prover is depicted in Figure 5.

Small-volume displacement provers are defined as those with a calibrated volume that permits a minimum of 10,000 pulses to be collected by the meter under test. Its attributes include smaller size and weight, shorter proving cycles, and extremely wide rangeability (in excess of 1000:1).

Meter characteristics. Flow meter characteristics are important in developing a proving methodology. For example, both positive displacement and turbine meters have mechanical elements that rotate as fluid passes through the meter. It is convenient to use electrical sensors to detect the passing of gear teeth or rotor blades. These pulses are totaled to determine the flow volume through the meter.

Since these mechanical elements rotate as a result of the flow passing through them, it tends to average out eddies associated with turbulent flow. The frequency output, representing pulses per unit volume, is quite steady and is considered to be a uniform pulse output. These meters

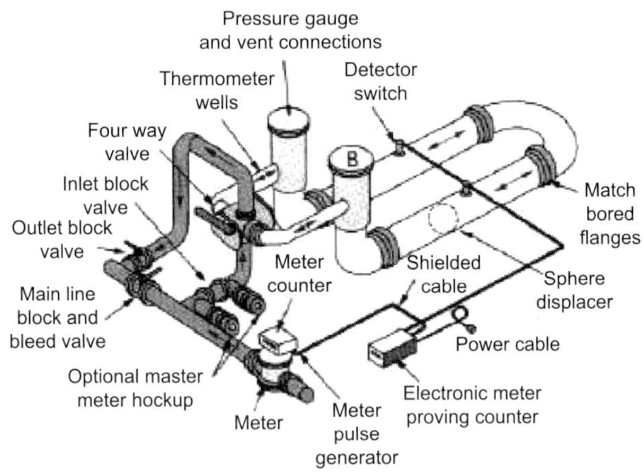

Figure 4. A bidirectional ball prover.

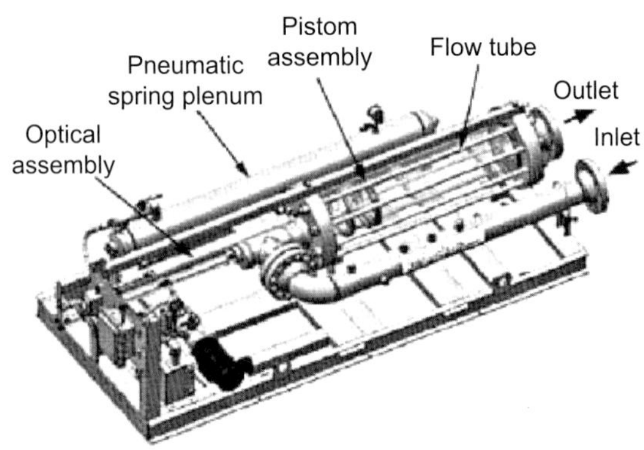

Figure 5. A small volume prover.

exhibit good proving repeatability with fairly short volumes and run times.

An ultrasonic meter has no moving parts. It determines the average velocity over one or more acoustic paths. The actual measurement made is the transit time of an ultrasonic pulse over the path in both the upstream and downstream directions. From this, an average velocity is calculated for each path. The path averages are then combined to form an average velocity for the whole meter, and the volume flow through the meter can be determined based on the cross-sectional area of the meter.

Since an ultrasonic meter has no moving parts, there is nothing to rotate to provide pulse counts. Pulse generation is done by means of a digital-to-frequency converter. Since the frequency is generated as the result of a measurement, it always lags the flow by the update period.

Also, since the meter is sampling the flow at several locations, turbulence in the flow will be measured. The standard deviation of the path velocity measurements is typically 2–5%. This implies a peak-to-peak variance of 12–30%. The turbulence is noise in the flow measurement that averages to zero. As a result, these meters have varying frequency output and are said to have a nonuniform pulse output. In proving ultrasonic meters, volumes and run times must be sufficient to average the turbulence to an acceptably low value.

Regardless of the calibration method, repeatability is an important criterion used to determine the acceptability of proving results. Ultrasonic meter performance verification can be ascertained by conventional means and to a consistent level. Table 1 provides guidance in determining the consecutive number of proving runs that may be required to obtain a meter uncertainty factor of ±0.027% at a 95% confidence level.

Any of the number of runs chosen from this table will produce results that verify a meter's performance to ±0.027% uncertainty. There is no difference between the traditional 5 consecutive runs that repeat within 0.05% and 10 consecutive runs that repeat within 0.12%, they both demonstrate the exact same uncertainty. The operator in the field will attempt to prove the meter in the smallest number of runs possible.

How proving volume has a direct effect on repeatability is shown in Figure 6. The example shows a turbine and ultrasonic meter being proven simultaneously with a small-volume displacement prover. Each of the three proving cycles could be as short as 0.5 s, depending upon flow rate. As a result of the small proving volume, it can be seen that the turbine meter, with its uniform pulse output, displays excellent repeatability (0.011%) over the 3 runs, while the ultrasonic meter does not—it shows a repeatability of 0.23%.

This does not mean that the turbine meter is more accurate or repeatable than the ultrasonic meter since both meters would provide excellent results over a longer time period. An ultrasonic meter proven using a prover with a larger volume, which effectively allows more time for the nonuniform pulse variations to average out, is depicted in Figure 7. The important point is that increasing proving volumes improves repeatability.

Table 1
Number of Runs and Repeatability Needed to Achieve ±0.027% Uncertainty in Proving an Ultrasonic Flow Meter

Runs	Repeatability	Uncertainty
3	0.02%	±0.027%
4	0.03%	±0.027%
5	0.05%	±0.027%
6	0.06%	±0.027%
7	0.08%	±0.027%
8	0.09%	±0.027%
9	0.10%	±0.027%
10	0.12%	±0.027%
11	0.13%	±0.027%
12	0.14%	±0.027%
13	0.15%	±0.027%
14	0.16%	±0.027%
15	0.17%	±0.027%
16	0.18%	±0.027%
17	0.19%	±0.027%
18	0.20%	±0.027%
19	0.21%	±0.027%
20	0.22%	±0.027%

$$\text{Repeatability} = \frac{\text{Highcounts} - \text{lowcounts}}{\text{lowcounts}} \times 100$$

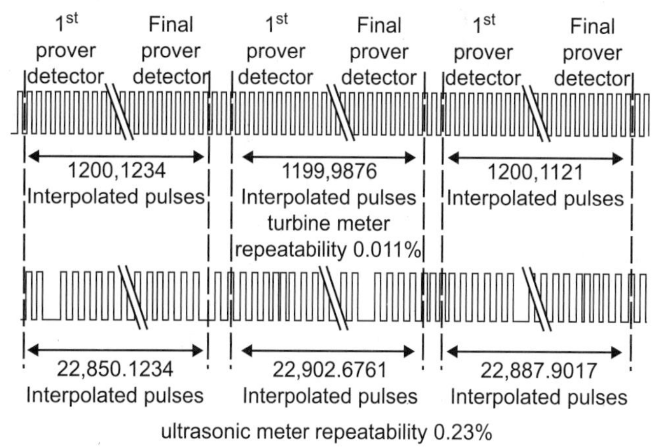

Figure 6. Typical results for turbine and ultrasonic meters using small volume provers.

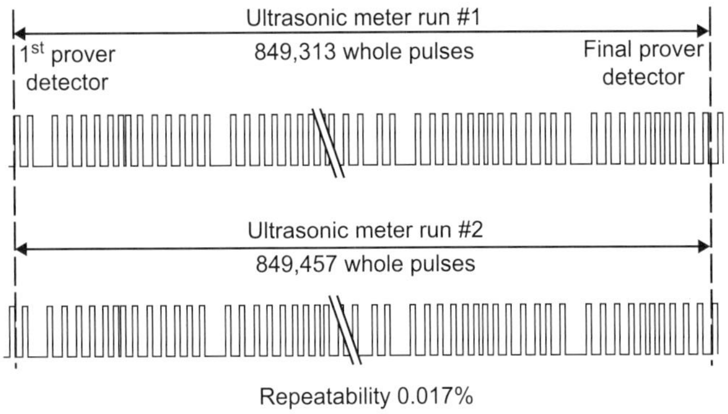

Figure 7. Typical proving results for ultrasonic meters using a larger proving volume.

Increasing proving volume. It can be seen that repeatability has improved significantly from 0.23% in Figure 6 over three consecutive runs to 0.017% in Figure 7 over two consecutive runs. This is a direct result of increasing the proving volume, thus allowing additional time for the meter to average the signal output. Increasing the proving volume can be accomplished by using a larger-volume prover or by using another method that is being adapted by many end users—the prover-master method (Figure 8).

Master meter proving is an accepted and recognized method of proving flow meters. However, master meters must also be calibrated on a regular basis to minimize errors. Calibrating a master meter in any situation other than the actual operating conditions can introduce uncertainties that must be compensated for. Changes in fluid composition, viscosity, density, and pressure can affect a meter's performance as can installation.

Using a prover-master meter combination installed in series with the meter under test eliminates all of these uncertainties. The master meter is first calibrated by the meter prover in situ under the actual operating conditions and using the actual fluid being measured. Once accomplished, then the master meter is used to calibrate the ultrasonic meter using the master meter calibration methodology. The advantage of this method is that it allows much larger volumes since the volume can be as large as necessary, thus allowing an acceptable level of repeatability.

Existing provers, whether ball-type or small-volume can be fitted with master meters in the field if it is determined that the existing prover does not have sufficient volume to provide acceptable repeatability. An actual calibration curve on a 6-in ultrasonic meter that was calibrated using a bidirectional ball prover with a volume in accordance with API recommendations, is shown in Figure 9. The same 6-in meter calibrated using the prover-master meter method, is depicted in Figure 10.

A small-volume prover was used to calibrate an 8-in turbine meter, which was then used as the master meter to calibrate the 6-in ultrasonic meter. The results are virtually identical to those above, illustrating the viability of both methods and the reproducibility of the meter itself.

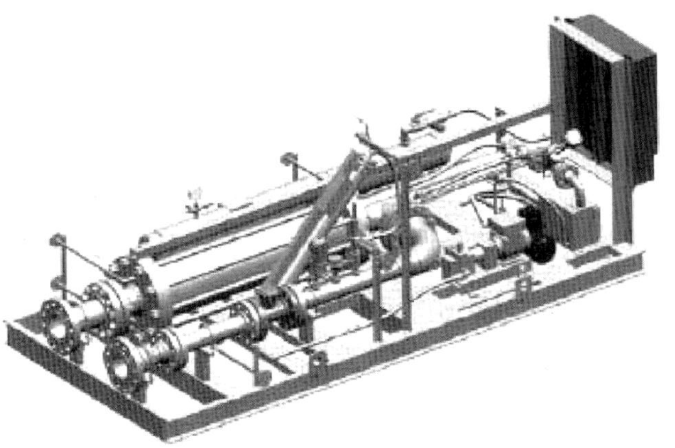

Figure 8. Small volume prover with turbine master meter prover.

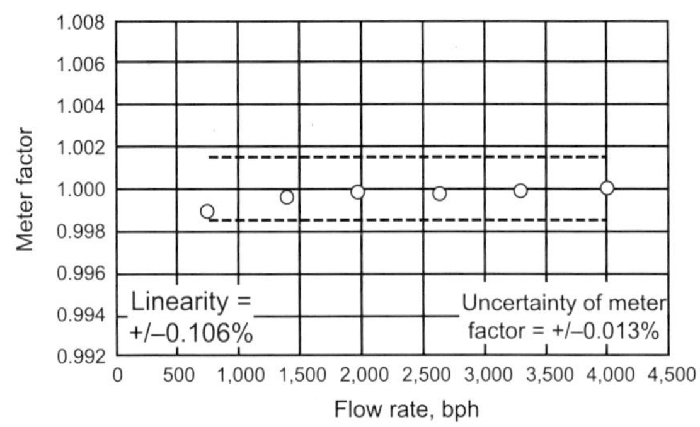

Figure 9. Liquid ultrasonic meter 6-in calibration chart—using Ball prover with ample volume.

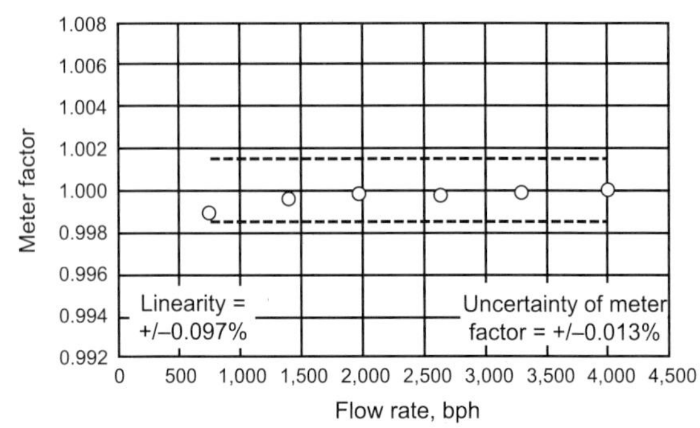

Figure 10. Liquid ultrasonic meter 6-in calibration chart—using in situ master-meter small volume.

Summary. Liquid ultrasonic meters are a relatively new technology and they continue to gain acceptance in a variety of applications. Proving a meter per API guidelines requires a ±0.027% meter uncertainty factor with a 95% confidence level. A key to meeting this requirement is having sufficient proving volume, which has a direct impact on repeatability. Generally, the longer the proving cycle, the larger the proving volume and the more repeatable the results.

Despite the relatively recent use of ultrasonic technology in this application, oil and gas companies with a good understanding of the issues are successfully proving liquid ultrasonic meters in the field on a regular basis.

Bibliography

1. API Manual of Petroleum Measurement Standards (MPMS), Chapter 5—Metering, Section 8- Measurement of Liquid Hydrocarbons by Ultrasonic Flow meters Using Transit Time Technology.
2. API Manual of Petroleum Measurement Standards (MPMS), Chapter 4—Proving Systems, Section 2—Displacement Provers.
3. API Manual of Petroleum Measurement Standards (MPMS), Chapter 4—Proving Systems, Section 5—Master-Meter Prover.

Reprinted with permission - *Hydracarbon Processing*, August 2007.

Handling entrained gas

Tim Patton, Director of Measurement Technology, Micro Motion, Emerson Process Management, Houston, Texas

Gas and liquid coexist in many processes. This mixing of dissimilar states can manifest itself in many ways and with many names—transient, two-phase, flashing, and slugs, to name a few. In a primarily liquid process, it is perhaps best known as entrained gas. By any name, entrained gas is often a troublesome challenge to accurate and reliable flow measurement.

Entrained gas is present in many liquid flow measurement applications, such as the manufacture of film emulsions, beer (wort), glucose, liquid fertilizers, soaps and lotions, hair-care products, phenols, dairy products, polyethylene, and ethylene oxide. Flow measurement performance in such environments can be unpredictable, depending upon the application, as measurement accuracy can be affected by many factors, such as the positioning of the measuring devices or the characteristics of the fluid being measured.

Characterizing entrained gas

To understand how recent technologies have largely overcome problems related to entrained gas, it is important to understand the different categories of entrainment: slug flow, bubble flow, and Empty-Full-Empty (EFE).

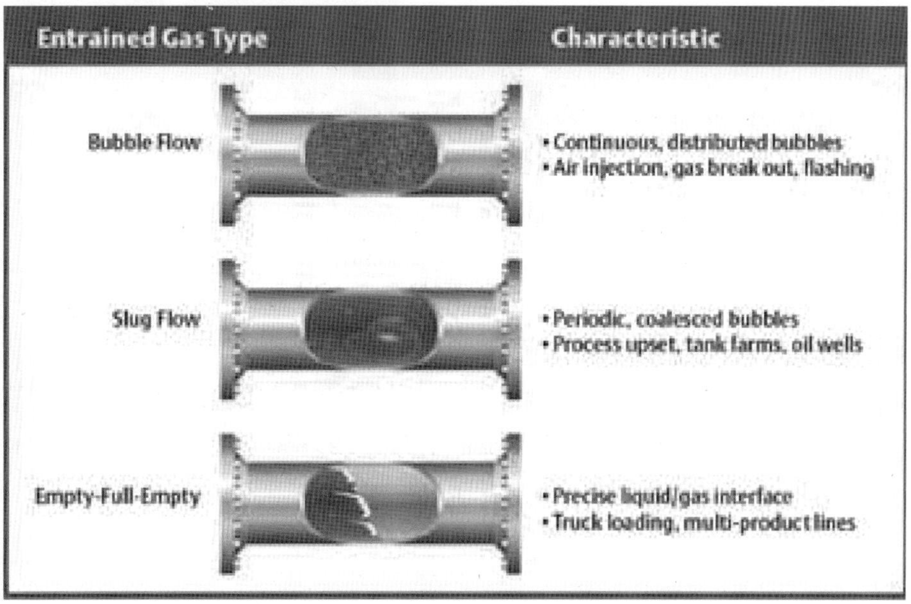

Figure 1. Types of entrained gas.

Slug flow is the occurrence of periodic large, coalesced gas bubbles in a liquid process. Slugs are usually not intentional and can be disruptive to flow and density measurement. Slug flow is commonly found in long-distance piping that does not fill properly. It may occur as a result of leaks in pump suction or tank agitation, allowing air to be pulled into a line.

Bubble flow is continuous distribution of gas bubbles (usually air) in a liquid process. The gas could be intentionally injected or could be caused by faulty equipment somewhere in the piping. Bubble flow is commonly found in high-viscosity liquids, such as toothpaste or peanut butter. It can also be present in products that are agitated or pumped at high speeds, as well as in environments where cavitating pumps or broken pump seals are a factor in the process. Most processes with bubble flow have less than five percent entrained gas, though in rare cases the value can be higher.

Empty-full-empty (EFE): In the case of EFE, piping starts empty, fills with liquid, remains full for the duration of the batch, and then empties upon completion of the batch. This batching is commonly found with railcar and tanker truck loading. Because the piping and the meter must be emptied between each fill, measurement can be a challenge when using a flow meter that is configured for only liquid processes.

A subset of empty-full-empty batching is a condition referred to as multiphase flow—processes with multiple density phases with some degree of separation and a mixing layer between them. The width of the mixing layer, dispersion of bubbles, and pipe orientation are all large contributors to the flow structure. Multiphase flow is commonly found in gas injections and well-head testing applications where multiple fluids coexist in the pipe—namely oil, water, and gas.

The design challenge

The biggest design challenge for Coriolis flow measurement is that gas entrainment in a pipeline is actually a continuum offering an infinite variety of conditions. To determine how these conditions affect the performance of the meter, development engineers at Emerson Process Management (*www.emersonprocess.com*) used flow meters with clear tubing to observe what actually occurred inside the meter. These observations, along with simplification of gas entrainment into bubble, slug, and EFE, enabled the engineers to develop a flow meter sensor and electronics that significantly expand the meter's capabilities under entrained-gas conditions.

These investigations concluded that four key elements played a role in the entrained-gas performance for Coriolis meters—the signal processing speed, processing algorithms, sensor design, and meter stability independent of environmental changes.

Improving slug flow & EFE performance

Emerson's Micro Motion Coriolis flow meters feature multivariable digital technology. The accuracy and stability of the Coriolis signals are considerably improved by digitally processing the signal from the flow sensor. Digital signal processing also delivers online monitoring and diagnostics data, providing predictive process and flow meter health information.

The characteristics of gas entrainment in slug flow and EFE batches cause frequent and large disturbances within the Coriolis meter that require the signal processing to be able to control and interpret a chaotic vibration. By increasing the processing speed in the Coriolis electronics, the meter is able to measure all these variables in synchronization with the disturbance. The improved signal processing and drive control optimize performance for transient events.

These developments have helped Micro Motion significantly reduce the flow rate error during the entrained-gas state from 20% with traditional technologies to just 1% using the company's latest-generation ELITE flow meter with enhanced core processor.

Overcoming bubble-flow conditions

Under bubble-flow conditions, the Coriolis sensor design and sensor stability dictate measurement performance. Sensor design is important because the critical bit of information is the relative difference in motion of the bubbles and fluid. Sensor stability is important because sensor vibration during bubble flow can be noisy and can cause the sensor to couple to the environment. With added stability, Coriolis flow meters can ensure that noise introduced by bubbles causes

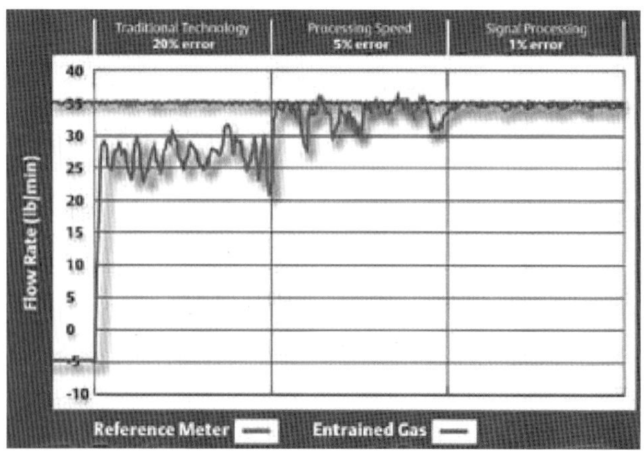

Figure 2. Accuracy graph showing performance of new Coriolis flowmetering technology under entrained gas conditions.

minimal flow measurement errors. Measurement is further compromised when bubbles change the compressibility of the fluid. As the level of gas entrainment increases, so does the measurement error. Tests show that for a given amount of entrained gas, the measurement error becomes more pronounced as the meter drive frequency is increased. By operating at lower frequencies, Coriolis flow meters can minimize these errors.

Final results

By addressing the four key design elements identified during its development research for entrained-gas applications—processing speed, signal processing, sensor stability, and sensor design—Micro Motion believes that many measurement applications once deemed impossible are now well within today's measurement capabilities. Regarding gas/liquid mixture applications specifically—whether the combination of gas and liquid flow is by design, by flaw, or as a consequence of day-to-day operations—Micro Motion is confident that it has developed methods that are capable of providing measurement accuracy and minimizing waste under such conditions.

How do Coriolis flow meters work?

Jesse Yoder, President, Flow Research, Inc., Wakefield, Massachusetts

The Coriolis flow meter market has been one of the fastest growing flow meter markets over the past five years, and despite their higher initial purchase price, many users find them a good investment when cost of ownership is considered. As such, process plants are increasingly selecting Coriolis meters to replace DP devices, and their use is also growing in the oil and gas industry.

One of the most important features of Coriolis flow meters is that they measure mass flow. While in many cases volumetric flow is sufficient, it is also desirable at times to measure mass flow. Many products are sold by weight rather than by volume, and in these cases it is often desirable to measure mass flow. Chemical reactions are often based on mass rather than volume, so mass flow measurement is often required in the chemical industry.

While both volumetric and mass flow apply to liquids and ases, mass flow is especially appropriate for measuring gases. This is because gases are much more affected by temperature and pressure than are liquids. Pressure has minimal effect on liquids in terms of compressibility, and is often ignored in making volumetric measurements of flow. The effects of temperature on liquids are also often disregarded, except where high accuracy is desired.

Coriolis flow meters are the most accurate flow meters made. While many magnetic flow meters have accuracies in the range of 0.5%, many Coriolis flow meters achieve accuracy of 0.1%. It is the high accuracy of Coriolis flow meters that is one of the major reasons for their extremely rapid growth over the past five years. Companies that need flow meters for custody transfer, or want highly accurate measurements in terms of mass, have a good reason to select Coriolis flow meters.

What is the Coriolis effect?

Coriolis flow meters are named after Gaspard Gustave de Coriolis, a French mathematician and engineer. In 1835, he wrote a paper in which he described the behavior of objects in a rotating frame of reference. While this is sometimes called the Coriolis force, it is more accurately called the Coriolis effect, since it is not the result of a force acting directly on the object, but rather the perceived motion of a body moving in a straight line over a rotating body or frame of reference.

One common example given to illustrate the Coriolis effect is that of a ball propelled through the air a long distance in a straight line from the North Pole toward a target on the equator. By the time the ball arrives at the equator, it will not land at its apparent target, because the earth will have sufficiently rotated underneath the moving ball so that it will land some distance away from the perceived target on the equator. From the perspective of the person standing where the ball is thrown," the ball will appear to have curved.

A similar example is someone throwing a ball from the center of a rotating merry-go-round. If she throws the ball at a horse on the edge of the merry-go-round, assuming the

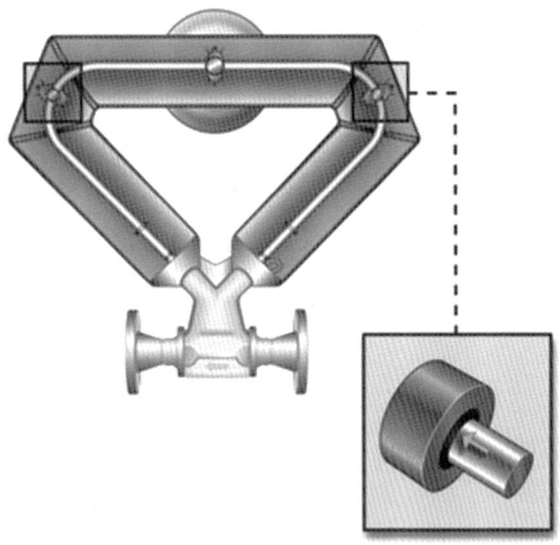

Figure 1. A diagram showing the location of the inlet and outlet sensors in a Coriolis flow tube. *Image courtesy of Micro Motion.*

merry-go-round is rotating at a rapid pace, she will not hit the horse she is aiming at, but perhaps the one after it, since the horse she aimed at will have moved by the time the ball reaches the edge of the merry-go-round.

In both these cases, there is no force acting to push the ball in a curved direction. Instead, its motion will appear to be curved from the perspective of the ball-thrower because the frame of reference is moving underneath the ball. This is why the Coriolis "force" is more appropriately called the Coriolis effect. It refers to the apparent effect on the motion of an object passing over a rotating frame of reference when viewed from the perspective of the point of origin of that moving object.

Do Coriolis flow meters embody the Coriolis effect?

If Coriolis flow meters embody the Coriolis effect, it is logical to assume that they must have rotating tubes. Yet, it is not clear why fluid passing through rotating tubes would have a Coriolis effect on the motion of the tubes, since the effect of the fluid is simply the result of inertial motion on the tubes. However, in the Coriolis effect, there is no force exerted on the ball thrown on the earth or the merry-go-round apart from the force exerted in throwing the ball. The ball simply appears to curve when viewed from the perspective of the rotating body.

Proponents of the Coriolis theory also cite a similar example when trying to explain the Coriolis effect. They point to a hose full of water that is held in someone's hand. When no water is running through the hose, it remains still. But as water runs through the hose, the hose starts to twist and turn, due to the inertial effect of the running water on the hose. This is very much such as the action of fluid on Coriolis tubes, but it is not clear why this is an example of the Coriolis effect. Of course rapidly flowing fluid through a narrow flexible object like a hose will cause it to twist and turn, due to the inertia of the fluid. If the pipes that carry many of the liquids in a plant were flexible instead of rigid, and the flow velocity was high enough, they would twist and turn as well. But, what does this have to do with the Coriolis effect?

How Coriolis flow meters work

Instead of having a rotating frame of reference, Coriolis flow meters work on the principle that the inertia created by fluid flowing through an oscillating tube causes the tube to twist in proportion to mass flow rate. Many Coriolis meters have two tubes. They are made to vibrate in opposition to each other by means of a magnetic coil. Sensors in the form of magnet and coil assemblies are mounted on the inlet and the outlet of both flow tubes. As the coils move through the magnetic field created by the magnet, they create a voltage in the form of a sine wave. These sine waves are the key to measuring mass flow.

Under no-flow conditions, the inlet and outlet sine waves are in phase with each other. When fluid is moving through the tubes, the tubes twist in proportion to mass flow rate. The amount of this twist is detected by the inlet and outlet sensors, based on a phase shift (time difference) that occurs in the sine waves formed by the two sensors. The mass flow rate is derived from the difference in phase shift in the sine waves formed by the inlet and outlet sensors.

Coriolis flow meters or inertial mass flow meters?

It is difficult to see the relation between the Coriolis effect and the inertial principle that Coriolis flow meters actually operate on. However, this does not cast doubt on their effectiveness or undermine their theory of operation. It simply calls into question whether they are appropriately named, and what connection the inventors of this meter saw between the Coriolis effect and the inertial operation of Coriolis meters. Perhaps they saw the Coriolis effect as a type of metaphor or analogue to the inertial force exerted by fluid flowing through a vibrating tube.

Coriolis meters do have a kind of "mystique" about them that no other flow meter has. It may be due in part to the difficulty people have in understanding how they operate. Also, manufacturers typically enclose the sensors inside a container, concealing their inner workings. No doubt, part of this mystique is that they are the most accurate flow meter and they do the measurement with no moving parts, apart from the vibration of the tubes.

Proving liquid ultrasonic flow meters for custody transfer measurement

Raymond J. Kalivoda, FMC Technologies, Erie, Pennsylvania

Introduction

Liquid Ultrasonic Flow Meters (LUFMs) continue to gain popularity in petroleum measurement with the promise of high accuracy and low maintenance. These are favorable features, but because of the high volume and value of petroleum products buyers and sellers must have a high level of confidence in the measurement. This assurance in Custody Transfer measurement is gained by adherence to the standards, procedures, and methods that define the measurement process.

There are two fundamental ways that petroleum products are measured—statically or dynamically. The measured volume determined by either method must be validated at operating conditions and traced to a fundamental standard. In a static system, the product is transferred to a tank or similar container. After the transaction is completed, the volume can be measured and validated by a suitable method. In a dynamic system PD, turbine or ultrasonic meters provide instantaneous information on the rate and volume of the transfer. As with static methods, the measurement results must be validated by a suitable method.

A key difference between a static and dynamic measurement is time. A static measurement is similar to a bank transaction, where the cash draw can be pulled and verified at any time. When a static transaction is complete, there is time to review the measured volume, and if an error is suspected, the volume can be re-measured. With a dynamic system the measurement must be right the first time; there is no means to re-measure the volume.

The proving system and process are essential in either static or dynamic measurement systems to validate the accuracy of volume measured. There are various worldwide standards and regulations that define the measurement system requirements, but they are fundamentally the same. Each is traceable to a national standard which in turn is traceable to the international standard. In dynamic measurement systems, this link is the prover volume. There are gravimetric and volumetric methods where the prover volume is traceable the International Bureau of Legal Metrology (BIML). The technical arm of the bureau is OIML (International Organization for Legal Metrology). This is an intergovernmental organization that deals with all aspects of legal metrology. It has 59 member states and 50 corresponding members. In this way, all the countries are tied to the same weights and measurement standards, which facilitates world trade.

The various measurement standards either directly or indirectly define the accuracy requirements over a range of operating conditions which include: flow rate, viscosity, temperature, pressure, piping configurations, etc. It is therefore important in the proving process to define both the operating conditions and the accuracy requirements because they are interdependent. A change in operating conditions can affect the meter's accuracy. As a general rule, the meter must be proven at operating conditions to validate its accuracy with the highest degree of confidence. The greater the change in operating conditions, the more uncertainty is introduced into the measurement.

The object of all custody transfer meters is the same, highly accurate measurement, but the procedures to achieve the results may vary with the different meter technologies. This paper will focus on the standards, procedures, and methods used to prove ultrasonic meters.

Custody transfer requirements

Custody Transfer refers to the fiscal measurement used to determine the quantity and financial value of a petroleum product transaction (delivery). The custody transfer requirements can be of two types:

Legal, which is defined by Weights & Measures (W&M) in the country or jurisdiction in which the sale is conducted. The various W&M codes and regulations control the wholesale and retail trade requirements to facilitate fair trade. The regulations and accuracy requirements widely vary between countries and commodities, but they all have one common characteristic—traceability. There is always a procedure that defines the validation process where the duty meter is compared to a standard that is traceable to the legal metrology agency of the respective region. The meters for wholesale and retail trade are normally smaller (4 in and under) to handle volume flow rates of under 2800 l/m (750 gpm). These applications are outside the range of current liquid ultrasonic meters and therefore will not be addressed in this paper.

Contract, is a written agreement between buyers and sellers that defines the measurement requirements. These are large-volume sales between operating companies where refined products and crude oils are transported by marine, pipeline, or rail. Since even a small error in measurement can amount to a large financial difference, custody transfer measurement must be at the highest level of accuracy possible. Because of the critical nature of these measurements, petroleum companies around the world have developed and adopted standards to meet the industry's needs.

A typical contract may define a specific measurement standard such as the American Petroleum Industry (API) Petroleum Measurement Standards or an OIML standard that is tied to the International Standards Organization (ISO). These standards include all the required equipment and details on the process to achieve an acceptable level of measurement.

Specifying a standard in a contract eliminates the need to enumerate many details concerning the equipment and measurement process that are common industry practice. For example, API standards are based on "best practice." They define the proper application parameters of a specific flow meter from field experience. OIML standards are "performance based." They set quantitative conditions that any custody transfer meter must meet. To gain approval, a specific size and type of meter must be tested over the operating range by a notified body (qualified testing agency) that issues a "pattern test report." The "pattern test report" is then submitted to the respective weights and measures authority as part of the application for approval.

Even though the standards may appear to widely vary, they have two fundamental points in common: (1) They all strive to maintain a minimum measurement error for a specific application and (2) They require traceability to national/international metrology standards by proving the meter. In terms of accuracy, which is expressed as uncertainty, a good API system would typically have a ±0.1% extended uncertainty. OIML requires a minimum system uncertainty of ±0.3% over a wide operating range. For large volume transfers, a better accuracy is required, so the systems are proven over a narrower range of operating conditions and typically achieve a ±0.1% extended uncertainty, the same as a system using API standards.

Measurement definitions

The terminology associated with measurement can be confusing, but is important in understanding the measurement process. Two good sources for measurement terminology are:

- **API** Manual of Petroleum Measurement Standards Chapter 1—"Vocabulary" and Chapter 13—"Statistical Aspects of Measuring and Sampling"
- **VIM** "International Vocabulary of Basic and General Terms in Metrology," commonly known as VIM, which is accepted by the international community of legal metrology

Included below is a summary of the key terminology used in evaluating a meter's performance. Definitions were selected from both sources. As expected the terminology is similar because accurate measurement is essentially the same worldwide.

Accuracy [API 1.0]: the ability of the measuring instrument to indicate values closely approximating the true value of the quantity measured.

Accuracy of measurement [VIM 3.5]: closeness of the agreement between the result of a measurement and the value of the measurand.

Measurand [API 1.0] is a physical quantity that has been or is to be measured.

Repeatability (of results of measurements) [VIM 3.6]: closeness of the agreement between the results of successive measurements of the same measurand carried out under the same conditions of measurement which include the same: measurement procedure, observer, measuring instrument, conditions and location over a short period of time.

API [1.0] adds "...the ability of a meter and prover system to repeat its registered volume during a series of consecutive proving runs under constant operating conditions."

Reproducibility (of results of measurements) [VIM 3.7]: closeness of the agreement between the results of measurements of the same measurand carried out under changed conditions of measurement. A valid statement of reproducibility requires specification of the conditions changed. The changed conditions may include: principle of measurement, method of measurement, observer, measuring instrument, reference standard, location, conditions of use and time. Results here are usually understood to be corrected results.

API [1.0] adds "...the ability of a meter and prover system to reproduce results over a long period of time in service where the range of pressure temperature, flow rate, and physical properties of the metered liquid is negligibly small."

Linearity of a meter [API 1.0]: The ideal accuracy curve of volume meters is a straight line denoting a constant meter factor. Meter linearity is expressed as the total range of deviation of the accuracy curve from a straight line between the minimum and maximum recommended flow rates.

Bias [API 13] is any influence on a result that produces an incorrect approximation of the true value of the variable being measured. Bias is the result of systemic error.

Error (of measurement) [VIM 3.10] is the result of a measurement minus the value of the measurand. Since the value of the measurand cannot be determined, in practice a conventional value is used.

Types of Error [API 13 edited]: The difference between the measured quantity and the true value of the quantity includes all the errors associated with the measurement process—the person taking the measurement, the process, the instruments and changes in conditions during the period of measurement. In evaluating a system there are three (3) types of errors:

Spurious error [API] is a gross mistake or blunder that must be identified and eliminated. There are statistical methods for testing for outliners (e.g. Dixon Outline Test) but data should not be discarded unless there are sound reasons for its elimination.

Random error [API] is a variation at constant conditions, normally evenly distributed about a mean, which can be statistically analyzed and the uncertainty defined within a given confidence limit.

Random error [VIM 3.13]: result of a measurement minus the mean that would result from an infinite number of measurements of the same measurand carried out under repeatability conditions. Because only a finite number of measurements can be made, it is possible to determine only an estimate of random error.

Systemic error [API] is one that remains constant as a positive or negative bias resulting in over or under estimating the true value of the measurement. In many liquid measurement applications, systemic error may make a larger contribution to the overall uncertainty than random error. There are two general classes of systemic error:

Constant systemic error is a bias that is particular to an installation and operating conditions. These errors include hydraulic and zero calibration effects.

Variable systemic error is a bias that varies with time and includes bearing wear or changes in meter tolerances.

Systematic error [VIM 3.14]: mean that would result from an infinite number of measurements of the same measurand carried out under repeatability conditions minus the value of the measurand.

Uncertainty of a Measurement [API 13]: the true value of a measurement cannot be determined, but a valid estimator can be obtained by the statistical analysis. The range or interval within which the true value can be expected to lie is the uncertainty range.

Confidence Limit [API 13 edited] It is normally impossible to absolutely set a range of uncertainty. It is more practical to indicate a degree of confidence on this range. This degree of confidence indicates the probability that the uncertainty range contains the "true value." The most common practice is to use a 95% confidence level. This level implies that there is a 95% probability (19 chances in 20) that the "true value" lies within the stated range of uncertainty. The 95% confidence level is recommended for all commercial applications in petroleum measurement. It should also be noted that this is the same confidence level used by legal metrology worldwide.

Uncertainty Analysis: encompasses the statistical analysis of the data derived from a measurement process to determine uncertainty range and confidence limits. It is the highly-mathematical and rigorous process defined in The Guide to the Expression of Uncertainty in Measurement or GUM, as it is commonly called. GUM is the result of ISO/TAG/WG 3 and printed in 1995 by ISO; the terminology is defined by VIM. It was fully supported by OIML and BIPM all worldwide legal metrology. In general the analytical method for calculating uncertainty and confidence limit of 95% is fully agreed upon throughout the world.

It is the uncertainty limits that are set by W&M that are used in the respective countries for the purpose of legal trade. For custody transfer transactions these limits are set by contract. Most times the uncertainty is not precisely stated, but based on precedence. An expected level of accuracy has been established over many transactions based on a specific measurement practice. In the US and many other parts of the world, the best measurement practice for pipeline or marine transfers of larger petroleum volumes is based on proving the custody transfer meter in situ with a displacement-type prover. Typically, a custody meter is proven at the start of a batch once the maximum transfer flow rate is established. Since the product is not changing and the variation in flow rate is limited (e.g. ±10%), the measurement is at optimum accuracy if the measurement process is sound.

For the purpose of this paper;

$$\text{Total Uncertainty} = \text{Random Uncertainty} + \text{System Uncertainty}$$

By viewing uncertainty in this form, it is easier to understand the measurement process and identify the potential factors that can affect uncertainty. There are a number of factors that can affect the measurement, but we need only be concerned with the key factors which include installation conditions, flow rate, viscosity, and temperature for most petroleum products, and also pressure for the lighter products like LPGs. It also should be noted from a review of the definitions that random errors are evenly distributed and will cancel out with a large enough sample. Systemic errors, on the other hand, "remain[s] constant as a positive or negative bias resulting in over or under estimating the true value of the measurement." Since systemic error or measurement bias is a business risk, identifying and reducing these errors is essential. Proving has always been, and remains, fundamental to custody transfer measurement because it is the only sure method to determine and correct systemic errors.

Unique problems in proving liquid ultrasonic meters

Proving, as described, is highly important to the custody transfer measurement and in detail is defined in the respective measurement standards. These standards, especially API "Best Practice" standards, were developed around the available technology. PD meters were the only custody transfer meters until the early 1970s when API 5.3 "Measurement of Liquid Hydrocarbons by Turbine Meters" was issued. Coriolis meters were introduced for petroleum service in the early 1990s, but API 5.6 "Measurement of Liquid Hydrocarbons by Coriolis Meters" was not issued until 2002. API 5.8 "Measurement of Liquid Hydrocarbons by Ultrasonic Flow meters Using Transit Time Technology" was issued in 2005.

Each of these meters uses fundamentally different technologies to determine the custody transfer measurement. The measurement technology can dictate the application and proving technology. Since the PD and turbine meters have been used for custody transfer measurement over a long period, the proving technology and procedures have been developed around these technologies.

There is a fundamental difference between PD and turbine meter technology and Coriolis and Ultrasonic meter technology that affects proving. The PD and turbine meters produce a discreet volume output for every volume unit measured. For example, a PD meter with electronic output or a turbine meter can produce 1000 pulses/barrel. Therefore, each time a barrel passes through the measuring chamber a pulse will be transmitted. Coriolis and Ultrasonic meters use a sampling technique. They sample the flow for a period of time, typically 50–500 ms, and then the microprocessor transmits the correct number of pulses for the volume that passed through the meter. The measurement by this method is accurate, but the microprocessor-based flow meters have two unique problems in regard to proving: (1) **sample time** required to achieve an accurate measurement and (2) **time delay** in the pulse per unit volume output. There is a further difference between Coriolis and Ultrasonic meter in the sampling process. This paper only focuses on the issues affecting ultrasonic meters.

Sample time

Proving LUFMs can be difficult because of their unique measurement method. A fluid flow stream is a complex flow field with numerous eddies, nonaxial components, and changing flow profiles. Unlike PD and turbine flow meters which integrate the flow field by mechanical convergence, ultrasonic meters measure all the numerous eddies and nonaxial velocity components in a turbulent flow field. To provide an accurate measurement, the ultrasonic meter must take a number of readings (or snapshots) of the fluid velocity at multiple locations along the product's flow path. The time it takes to acquire a sufficient number of samples for an accurate measurement depends on the flow regime and meter processing time, but, in general, the longer the sample time, the better the measurement. These sample times are in the milliseconds which don't effect the batch measurement, but become a problem with proving. Proving technology, as indicated, is based on discreet pulses for PD or turbine meters. Ultrasonic meters normally require larger prove volumes for the same flow rates than traditional custody transfer meters. However, with proper provers and sound proving practice, highly accurate measurement can be achieved.

Time delay

Because of the sampling method used in ultrasonic meters, the output at any instant of time will represent a volume that has already passed thought the meter. This lag poses no problem under steady flow conditions. But, if the flow rate is changed, which is typical with displacement provers, this lag can cause poor run repeatability and even bias error in the meter factor. This problem was identified during the development of the API ultrasonic meter standard 5.8. In 2003, API formed a task group and funded testing to determine the affect of this pulse delay on meter proving. The group was headed by Kenneth Elliott, who reported the preliminary finding at the 2004 North Sea Measurement Workshop in paper 5.4, "API's Microprocessor Based Flow meter Testing Program." There were a number of interesting findings, but the most important results are that: (1) the potential meter factor bias is proportional to the time delay and (2) no accurate factor, based on available data, could be used to correct this systemic error. The practical application of this data is to reduce the meter sample time, and when possible, minimize flow variations during proving and use larger-volume displacement provers.

Proving ultrasonic meters

The problems with proving ultrasonic meters can be resolved by either or both of the two methods:

1. Increase the number of runs. The data scatter for an ultrasonic meter is greater than a conventional meter, therefore it may not repeat in 0.05% in five consecutive runs as required for custody transfer measurement. A more general statement would state the repeatability in terms of uncertainty, which is ±0.027% at 95% confidence level. This more general statement allows a wider repeatability test tolerance. For example, a repeatability range of 0.12% in 10 consecutive runs or a repeatability of ±0.17% in 15 consecutive runs both meet ±0.027% at a 95% confidence level. Table 1 is an abridged version of the number of runs at different repeatability ranges to achieve ±0.027% uncertainty.
2. Increase the "prove volume" to allow a sufficient number of samples. This will also increase the time it takes to complete each prover run, which reduces the time lag to a relatively insignificant uncertainty. As a rule, an SVP will not have sufficient "prove volume" or "prove cycle time" for proving an ultrasonic meter unless it is oversized and the change in flow rate during the prove is minimal. In the ultrasonic meter standard, API proves suggested "prover volume" for different size meters. Table 2 shows these suggested prover volumes compared to the typical "prove volume" for turbine meters.

Proving systems

Proving systems can be divided into:

Direct Proving, where the custody transfer meter is proven against:

- Displacement Provers—Conventional or Small Volume.
- Tank Provers—Volumetric or Gravimetric. This method is only suitable for PD meters and Coriolis mass meters and will not be included in this paper.

Indirect Proving, where the custody transfer meter is proven against:

Table 1
Runs vs. Repeatability

Runs	Repeatability	Uncertainty*
3	0.02%	0.027%
5	0.05%	0.027%
8	0.09%	0.027%
10	0.12%	0.027%
15	0.17%	0.027%
20	0.22%	0.027%

*Uncertainty of Meter Factor at 95% confidence level.

Table 2
Prover Volume

Meter Size	5 runs 0.05%		8 runs 0.09%	10 runs 0.12%
	Prover Size (barrels)			
	TM	Ultrasonic Meters		
4″	5	33	15	10
6″	12	73	34	22
8″	20	130	60	40
10″	24	203	94	62
12″	48	293	135	89
16″	100	521	241	158

• Master Meter Provers—Transfer Standard Proven, on-site or laboratory-proven.

Conventional displacement provers are well-established in this sector of the petroleum measurement industry. SVPs, because of their size and flow range, have recently gained the industry's interest. Master meters, which were rarely used on large volume custody transfer systems, are being evaluated because of technical and economic advantages.

Proving systems and proving is costly. On large pipelines and marine loading/unloading facilities, the flowline is normally divided into a number of meter runs, each handling a percentage of the flow. Metering systems are designed in this manner to facilitate proving and increase reliability. There is a technical and economic limit on the size of a prover for a given system. For example, it may be more economical on a 12,000 m^3/h (75,000 bph) measurement system to use 4–12″ flow meters and a 6.5 m^3 (40 bbl) prover than 3–16″ meters with a 15 m^3 (95 bbl) prover. Also, by adding an additional meter run, the system's reliability can be increased because the full flow could be handled by any of the 3 m.

In economic terms, design of custody transfer measurement systems becomes a cost versus risk relationship. Table 3 provides the potential cost of systemic measurement errors at different levels of uncertainty. The table clearly illustrates that the cost of a systemic error on a large system can be a substantial risk. A more subtle implication of this table is that using a number of smaller meters is less risky than using a larger meter.

Proving has always been, and remains, fundamental to custody transfer measurement. Choosing the proper prover and method is critical. Below is a review of each method prover with its advantages and disadvantages.

Displacement provers

Displacement provers are the fundamental device used to verify the performance of large-volume custody transfer meters (≥4″ PD) and all types of turbine meters. There are various methods to calibrate displacement provers, but all are traceable to a primary metrology standard recognized by BIML for international trade. Displacement provers can be built into a measurement system or a portable prover can be connected into the system. This allows the meter to be validated at the operating conditions.

Conventional Displacement Provers. These are the most common provers in the petroleum transportation market because of their accuracy and reliability. There are two types of conventional displacement provers—uni-directional (Figure 1) and bi-directional (Figure 2). The displacer, which is either a sphere or piston, is installed in a coated pipe of a specific length. The difference between an uni-

Table 3
Systemic Uncertainty Risk*

Systemic Uncertainty (±U_s)	Daily Value (US $)			Annual Value (US $ millions)		
	Barrels per hour					
	10,000	50,000	100,000	10,000	50,000	100,000
0.1%	$11,520	$57,600	$115,200	$4	$21	$42
0.2%	$23,040	$115,200	$230,400	$8	$42	$84
0.3%	$34,560	$172,800	$345,600	$13	$63	$126
0.4%	$46,080	$230,400	$460,800	$17	$84	$168
0.5%	$57,600	$288,000	$576,000	$21	$105	$210

*Based on crude oil @ US $60/bbl and 80% duty cycle.

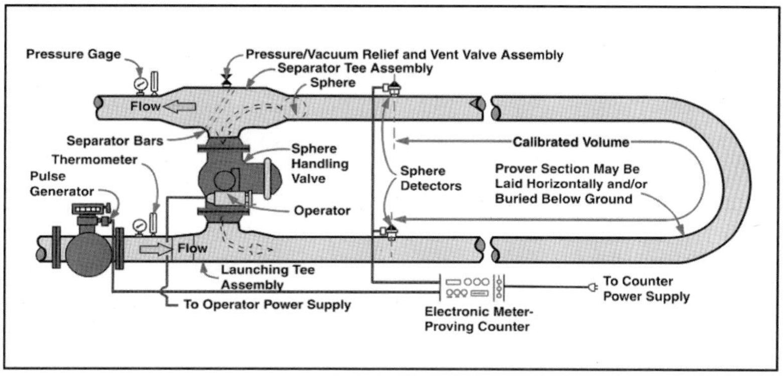

Figure 1. Uni-directional prover.

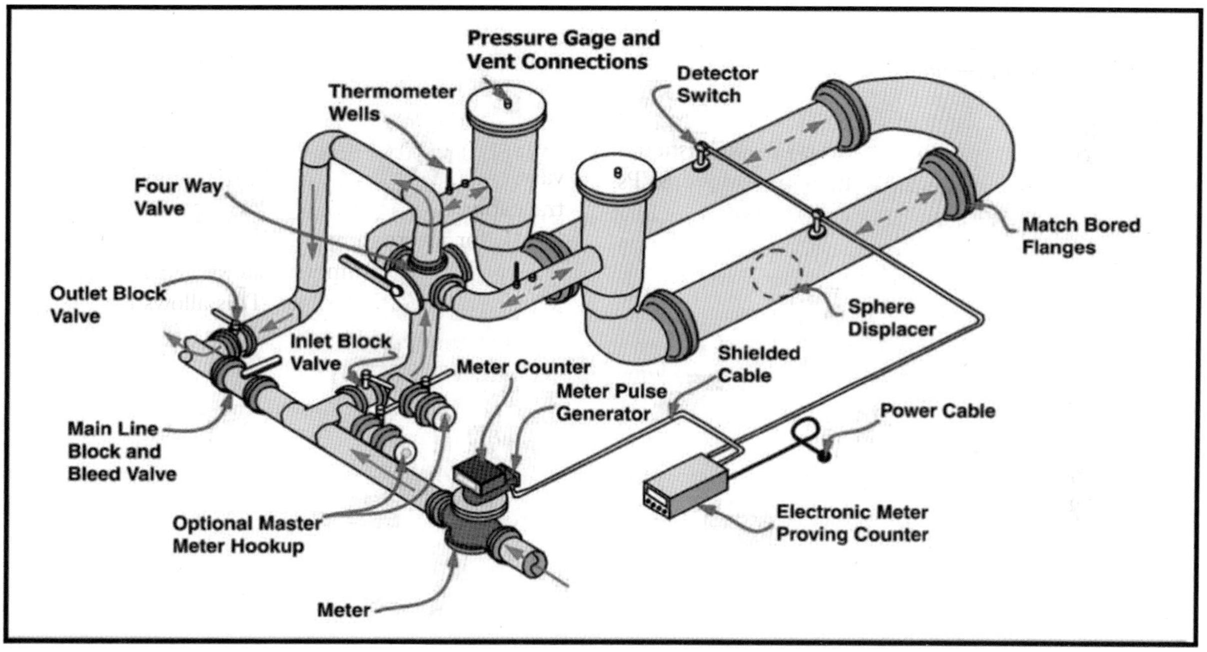

Figure 2. Bi-directional prover.

directional and bi-directional prover is how the displacer is launched and retrieved. The pipe is also fitted with detector switches at two or more points in the pipe. When activated by the displacer, the detector switches define a precise volume which is known as the "prove volume." There are API and other standards that define methods to calibrate this "prove volume" to a high degree of accuracy which is traceable to a national/international metrology standard. When a conventional displacement prover is connected in series with a meter, the displacer is launched and travels at the same speed as the liquid flowing in the pipe.

The switches control the high-speed pulse stream transmitted by the meter. The first detector switch gates an electronic prover counter "on" to collect meter pulses and the second detector switch gates the electronic prover counter "off". The total number of pulses on the electronic prover counter represents the volume recorded by the meter. The indicated meter volume is then compared to the precise volume between the prover detector switches. It is possible to calibrate the "prover volume" to an expanded uncertainty of ±0.02%.

To achieve the required level of accuracy for a custody transfer measurement, the meter must generate at least 10,000 m pulses between the detector switches. The 10,000 pulse minimum is required because there can be ±1 pulse uncertainty at either detector switch, so the resolution is ±2 pulses/10,000 pulses or ±0.02%. An uni-directional-type prover normally collects a minimum of 10,000 pulses per proving run. The proving run for a bi-directional prover is both the forward and reverse directions, so a minimum of 20,000 pulses are collected.

Advantages of Conventional Displacement Provers. Ultrasonic meters can be successfully proven with the properly-sized ball-type displacement prover because of the large "prove volume."

The "time delay" of the ultrasonic meter has little effect on ball provers because of the large "prove volume."

It is relatively inexpensive to increase the "prover volume" of ball provers.

If a ball prover is on site with a smaller than required "prover volume," the meter can still be successfully proven by increasing the number of prover runs. Increasing the number of runs compensates for the smaller "prove volume" provided that the required number is not excessive, typically over 20. Normally, doubling the number of runs (e.g. 5 runs to 10 runs) with a ball prover is sufficient to achieve a custody transfer uncertainty level of ±0.027%.

Ball provers are accepted technology that many users know how to operate and maintain. Ball provers cost 20–50% less than SVP-type displacement provers.

Disadvantages of Conventional Displacement Provers. Size and limited operating range are the key disadvantages. Bi-directional provers have an operating range which is limited to 10:1 (10–100% of max flow) for well-lubricated products and 5:1 (20–100% of max flow) for dry products. The operating range of uni-directional provers is twice that of bi-directional provers.

Small Volume Type Displacement Provers. SVPs (Figure 3) are displacement provers that have a volume between detector switches which does not permit collecting 10,000 m pulses. To achieve the required level of proving uncertainty, pulse interpolation techniques are used. These techniques divide the initial and final pulse by the use of chronometry. This allows the volume between detector switches to be substantially less than a conventional prover.

Advantages of Small Volume Provers. Size and flow range are the key advantages of SVP's. They have a much smaller footprint than comparable ball provers and have a 100:1 (1–100%) flow range.

The small "prover volume" in the SVP makes it easier to clean and reduces the contamination between products when used for proving meters on multi-products.

SVPs are preferred for offshore platforms because of their size and flow range. The limitation for platforms is flow rate at 17,500 bph for a 12″ turbine and 12,500 bph for a 16″ PD or 10″ ultrasonic meter (2).

They are ideal as portable provers because of size and flow range. The limitation is the weight of the prover. The largest

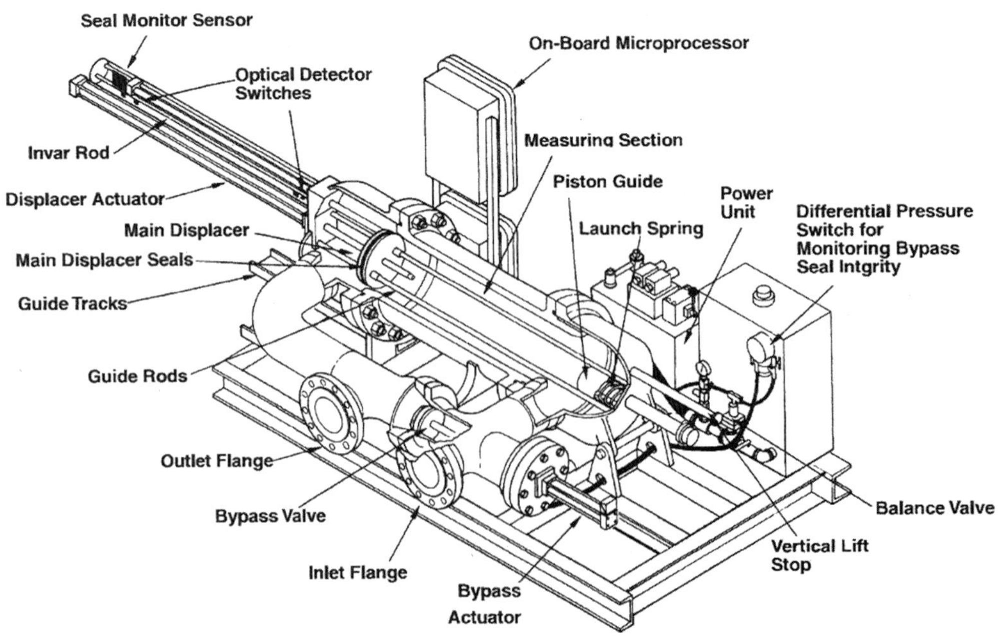

Figure 3. Small volume prover.

practical size has a 40 gal "prove volume," which is limited to proving an 8″ turbine meter (7200 bph) or a 10″ PD (4800 bph) or 6″ ultrasonic meter (2) (4000 bph).

SVPs are also becoming popular at pipeline terminals that are being automated, which is the goal of operators. Though in many cases, this excludes ultrasonic meters until the proving problems are resolved.

Disadvantages of Small Volume Provers. The small "prove volume" is a disadvantage for proving ultrasonic meters because the "time delay" and "sample time," as previously explained, are the major disadvantages of SVP's. In most, if not all cases, issues are demonstrated in the API Microprocessor Based Flow meter.

Testing programs severely limit the use of SVPs for proving ultrasonic meters. SVPs are also much more complex than pipe provers, and they are more susceptible to damage from fluids contaminated by debris, particularly abrasives. One method that is developed in the "Master Meter Proving" section is Transfer Meter Proving.

Master meter proving systems

A master meter is termed an indirect meter proving method as compared to a displacement prover, which is a direct proving method. API 4.5 "Master Meters Provers" only recognizes the use of PD meters or turbine meters as master meters. Currently, there is an API working group that is adding Coriolis Flow Meters and Liquid Ultrasonic Flow Meters to this standard. A revision to the standard is expected for balloting by this spring. Until the ballot results are reviewed it is difficult to predict when the revised standard will be completed, but there is not any apparent technical reason why these meters would not be added to the standard.

Since the master meter proving is a secondary proving method, it has the highest uncertainty of all meter-proving methods. As previously defined, the total or extended uncertainty of a proving system is the combination of random and systemic uncertainty. The expanded uncertainty for proving a master meter with an SVP can be expressed as:

$$U_T = U_R + \sqrt{U_{MM}^2 + U_{SVP}^2}$$

where:

U_T—Total uncertainty of a typical prove
U_R—Random uncertainty which is the repeatability of the meter runs
U_S—Systemic uncertainty which is the combined uncertainty of the master meter and SVP
U_{MM}—Master meter uncertainty (systematic)
U_{SVP}—SVP volume uncertainty (systematic)

Systemic error, as defined, is one that remains constant as a positive or negative bias in the measurement. Since it can make a large contribution to the overall uncertainty (U_T), it is imperative to determine and correct for this bias by proving the meter. The wider the variation in installation and operating conditions, the greater the potential systemic error and associated uncertainty.

Two separate sequences of events are necessary in master meter proving. First, the master meter must be calibrated using a prover that is traceable to a national/international standard. The master meter can then be used to calibrate the field-operating meter.

Factors That Contribute to Uncertainty in Master Meter Proving. To best understand the uncertainty in master meter proving it is helpful to distinguish between the Constant Systemic Errors introduced by installation and operating conditions and Variable Systemic Errors that vary with time, which include bearing wear, changes in meter tolerances, strainers, and piping conditions. In general, ultrasonic meters are more sensitive to installation and operating conditions than PD and turbine meters, but less sensitive to the Variable System Errors because there are no moving parts.

Constant Systemic Error Contributions. **Installation Effects**—refers to anything in the measurement system set-up that can cause a systemic change in the meter factor. Certain metering technologies like PD and Coriolis meters are only slightly affected by installation conditions whereas velocity meters such as turbine and ultrasonic meters can be highly affected. Multipath ultrasonic meters may have algorithms that detect and compensate for these effects, but it is problematic on how robust the specific meter can handle these anomalies.

Currently with transfer proving, the total uncertainty (U_T) of a liquid custody transfer measurement can be made in the range of $U_T = \pm 0.1\%$ at a 95% confidence level. This is industry norm for most large volume liquid transactions. If a large systemic installation error (e.g.$\pm 0.5\%$) goes undetected for even a small period of time, it increases the business risk for over or under delivery.

Operating Effects—The most important operating conditions that affect the accuracy of liquid measurement are: flow range, viscosity range, and temperature.

Flow Range is the minimum/maximum flow rate at which a meter/measurement system can operate and maintain the stated accuracy. Custody transfer meters are normally specified to operate over a 10:1 (10–100% of maximum flow) flow range with a linearity of ±0.15%. This means that if all the operating conditions are held constant and only the flow rate is varied, then the maximum meter factor variation will be ±0.15% over the flow range. This variation in meter factor does not include any systemic error due to installation conditions or changes in the other operating conditions. The accuracy of the measurement can be improved by reducing the flow range or establishing a separate meter factor for each flow rate. With today's flow computers, a number of meter factors versus flow rates can be programmed and the meter factor will automatically be linearized. At a given flow rate, the measurement uncertainty (U_R) becomes the repeatability of the meter which must be $\leq \pm 0.027\%$ for custody transfer measurement.

Viscosity Range of petroleum products can vary from less than 0.1 cP for LPG's to over 1000 cP for heavy crude oils (see Table 4). All meters are sensitive to viscosity, but each metering technology is affected differently. Because of the slippage principle, PD meters are affected on low viscosity products (<2 cP) whereas turbine and ultrasonic meters are sensitive to high viscosity products. Multi-path ultrasonic meters may have algorithms that compensate for viscosity, but on high viscosity crude oils, the other factors such as particles and entrained gaws may also affect the measurement accuracy.

Temperature affects the measurement in several ways: (1) the viscosity can change as illustrated in Table 4; (2) changes in liquid volume as defined by the volumetric coefficient of expansion; (3) changes in meter tolerances that affect the MF; and (4) a change in the prover volume. There are correction factors for the volumetric expansion of liquids and prover volumes, but master meter factors must be established for specific temperatures within the operating range. Since the repeatability of a meter during proving is based on a gross volume (nontemperature compensated volume), it is highly important to maintain a minimum temperature difference between the meter being proved and the prover. For example, a 2 °F change in temperature between the meter under test and the prover will impact meter factor and custody volume by 0.1% across the entire product range.

Table 4
Viscosity of Selected Products

Product	API Gravity	Viscosity in cP		
		60 °F (15 °C)	100 °F (38 °C)	150 °F (66 °C)
Propane—LPG	145	0.12	0.09	0.07
Gasoline	56	0.63	0.49	0.38
Kerosene	40	2.2	1.7	0.9
Light crude	48	2.7	1.7	1.1
Light crude	32	21	9	5
Medium crude	25	1442	243	93
Heavy crude	16	3440*	574	230*
Heavy crude	10	5100*	1294	520*

*Estimated.

Variable Systemic Error Contributions. The variable systemic errors are associated with the stability of the master meter over time. Wear in a mechanical meter can cause a shift in the factor due to bearing wear or changes in critical tolerances. There can also be waxing or other coatings that similarly cause shifts in the calibration factor. Debris caught in the strainer basket and/or on flow conditioning elements can have a significant effect on the master meter's calibration.

The only way to confidently detect and correct for these systematic bias is by periodically reproving the meter and comparing the proving results. This raises the question, how often does the master meter need reproving? Because of the variation in master meters, service conditions and acceptable level of uncertainty (risk), it is impractical to establish a fixed time. The answer to this question can only be gained from careful analysis of the data. API 13.2 provides statistical methods to evaluate and track proving data, but a practical approach would be:

- Establish the baseline meter factor for the master meter on each test fluid under consideration. This would consist of ideally six runs at different times and conditions over which the master meter will operate.
- Develop a control chart based on the uncertainty tolerance that is acceptable. Typically for custody transfer measurement of $U_T = \pm 0.1\%$ the current MMF should be within 0.02% of the last meter factor and the baseline meter factor. The maximum allowance from the baseline should be 0.04%.
- The time between provings can be increased based on the master meter factor's stability.

Typically, a good master meter should remain within the stated tolerances for a least one year or 200 h of service. The precise criteria, though, is based on the system and the agreement of the parties affected by the measurement.

Transfer (Master Meter) Proving. With "Transfer Proving," the master meter is proven on site, which eliminates systemic errors due to installation effects. It also provides the ability to prove over the complete operating range, and if the conditions change or if measurement is challenged, the master meter can be easily reproven.

There is an added uncertainty of the master meter to the total uncertainty budget, but it can be precisely calculated. In practice, the uncertainty added by a master meter should be far less than the bias caused by a change in conditions.

Currently, this method is used to prove terminal load rack meters at large terminals. Normally, a PD master meter is installed in series with one of the line meters and validated using a volumetric can prover. The other delivery meters on the same product are then calibrated against the master meter with the product being pumped directly into the tank truck compartment. This is far more efficient than proving each meter with a volumetric tank and pumping the product back to storage.

The same method, if applied to proving ultrasonic meters, would eliminate both the "time delay" and "sample time" issues. The master meter, whether it is a PD or turbine meter, can be proven with a conventional or SVP and then used to calibrate the ultrasonic meter. Since the master meter is not limited by "prove volume," an adequate volume can be run through the master meter to eliminate both the "time delay" and "sample time" issues.

Laboratory Calibrated Master Meter. Laboratory calibrated master meters have the highest level of uncertainty because of the bias introduced by installation and operating conditions. All meters have some sensitivity to installation and operating conditions and the resulting bias must be quantified for custody transfer measurement. The difficulty of determining this bias depends on the total uncertainty (U_T) required (level of acceptable risk) and the degree of variation in the installation and operating conditions. Obviously, a lower level of uncertainty (e.g. $U_T = \pm 0.1\%$) with a wide variation in operating conditions is significantly more difficult to determine than a higher level of uncertainty (e.g. $U_T = \pm 0.5\%$) over a limited flow range, with one product at a stable operating temperature. In either case, the analytical methods used to determine uncertainty analysis are dependent on test data for the specific type of meter and operating conditions. Specific data is required for each type of meter, brand, meter model, and size to develop an analytical model. The calibration factors developed by the model should then be validated in the field by proving the master meter in situ over the operating conditions. As more data is gathered, the confidence in the measurement will increase and the factors affecting the measurement can be identified and improved to reduce the total uncertainty (U_T).

Laboratory-proven meters are not normally used for large volume custody transfers because of the limited knowledge of the installation and operating effects. However, even today the economical advantages may far outweigh the uncertainty risk. In the long run, improving the diagnostic capabilities of master meters and defining the installation conditions will lower the uncertainty to an acceptable level.

Choosing a Master Meter for an Ultrasonic Application. The use of master meters for large custody transfer applications (6″ and larger meter, over 4000 bph) is not common practice because of the increased risk with large volume transfers. Normally, the meters are directly proven with a displacement prover, either a Conventional Displacement Prover or more recently an SVP. The latter provers are becoming more popular because of their smaller space requirement and recent improvements in reliability. The use of SVP to prove ultrasonic meters is difficult because of the time delay and sampling requirements. On-site or portable conventional provers may be too small for the respective ultrasonic meter. Master meters for these applications may offer a technically sound alternative.

For the purpose of this paper, it will be assumed that Coriolis Flow Meters (CFMs) and Liquid Ultrasonic Flow Meters will be added to API Standard 4.5—Master Meter Provers. Since CFMs are applied at low flow rates, they have limited application as a master meter for ultrasonic applications. PD meters, conventional turbine meters, helical turbine meters, and ultrasonic meters are possible selections.

Traditionally, PD meters were applied for high viscosity products and turbine meters for low viscosity products. This rule-of-thumb has a sound technical foundation. The PD direct measurement principal is based on slippage theory. That is, the volume through-put is segmented and directly measured in the measurement chamber, except for a small amount that slips though the clearances. This slippage is a constant for a given product, flow rate, and temperature. By proving the meter, the bias due to slippage can be identified and compensated with a calibration factor. As the product viscosity increases, the slippage decreases. At a given viscosity (about 10 cP for a 10″ PD meter) the slip is essentially "zero." Furthermore, the meter factor remains constant for all higher viscosity products and over a wide flow range.

PD meters directly measure volumetric flow and as such are not highly affected by installation conditions. Coriolis meters are likewise not affected by installation conditions as long as a stable "zero" can be established and maintained. Turbine and ultrasonic meters measure velocity and are therefore affected by the dynamic flow conditions established by the installation conditions.

Turbine meters are far more sensitive to viscosity. A conventional multibladed turbine meter is normally applied to liquid viscosity of two times the meter's size in inches and a helical turbine meter to 10 times the meter's size in inches. In either case, as the viscosity increases, the minimum flow rate must be increased, limiting the meter's overall flow range.

Ultrasonic meters, such as turbine meters, are sensitive to viscosity effects. They are not independent of these effects as some suppliers claim, but they can be designed to compensate for viscosity profile effects.

Conclusion

When properly proven, liquid ultrasonic flow meters can provide custody transfer accuracy. Conventional uni-directional and bi-directional provers provide the best direct proving method for custody transfer meters. A direct method has the least uncertainty because it eliminates the added uncertainty of a master meter.

It is difficult to directly prove a liquid ultrasonic meter with an SVP because of its small "prove volume". The time lag, with changes in flow rate, becomes a significant percentage of the run time per pass and can cause a bias error. The small "prove volume" also limits the number of samples which affects repeatability. An SVP can be used in a transfer (Master Meter) proving system. In these applications, a PD or turbine would normally be used for the master meter.

Master meter proving can be used to calibrate a liquid ultrasonic meter, but it does add uncertainty to the measurement. The transfer master meter proving method is preferred because it eliminates the bias caused by installation effects and allows the master meter to be calibrated at operating conditions. The master meter can also be easily reproven if there is any doubt about the measurement.

The greatest uncertainty is introduced when the laboratory proven master meter method is used. In this method, the systemic bias caused by the installation effects is difficult to estimate. Similar operating conditions are often difficult to simulate in a laboratory and any differences in operating conditions add to the expanded uncertainty budget. When the operating conditions change, the master meter will need to be recalibrated at these new conditions.

Liquid ultrasonic meters can accurately be proven by a number of different methods. The best proving method depends on properly assessing the "risk cost" of a measurement error compared to the cost of the measurement system. In most high-volume custody transfer applications, the proving system with the least uncertainty and corresponding lowest risk is the best.

References

1. Barry N. Taylor, Chris E. Kuyatt, NIST Technical Note 1297, 1994 Edition, Guidelines for Evaluating and Expressing the Uncertainty of NIST Measurement Results, Appendix D, Terminology from International Vocabulary of Basic and General Terms in Metrology (abbreviated VIM), second edition. ISO (International Organization for Standardization, Geneva, Switzerland, 1993.
2. American Petroleum Institute (API), Manual of Petroleum Measurement Chapter 1 – Vocabulary, second edition (July 1994).
3. OIML R 117 (Organisation Internationale De Metrologie Legale, French), International Recommendation, Measuring systems for liquids other than water, edition 1995 (E).
4. API Manual of Petroleum Measurement Chapter 4 - Proving Systems, Section 2 – Displacement Provers, third edition (September 2003).
5. API Manual of Petroleum Measurement Chapter 4 - Proving Systems, Section 5 – Master- Meter Provers, second edition (May 2000).
6. API Manual of Petroleum Measurement Chapter 5 - Metering, Section 8 – Measurement of Liquid Hydrocarbons by Ultrasonic Flow Meters Using Transit Time Technology, first edition (February 2005).
7. API Manual of Petroleum Measurement Chapter 13 – Statistical Aspects of Measuring and Sampling, Section 1 – Statistical Concepts and Procedures in Measurement, first edition (June 1985).
8. Kenneth Elliott.APIs Microprocessor based Flowmeter Testing Program, paper 5.4, North Sea Measurement Workshop in 2004.

This paper was originally written and presented at the 7th South Asia Hydrocarbon Flow Measurement Workshop, 2008 by Raymond Kalivoda, Petroleum Measurement Marketing Manager, FMC Technologies, Erie, Pennsylvania USA.

V-Cone® flow meter

The McCrometer V-Cone® flow meter is a patented technology that accurately measures flow over a wide range of Reynolds numbers, under all kinds of conditions and for a variety of fluids. It operates on the same physical principle as other differential pressure-type flow meters, using the theorem of conservation of energy in fluid flow through a pipe. The V-Cone's remarkable performance characteristics, however, are the result of its unique design. It features a centrally-located cone inside the tube. The cone interacts with the fluid flow, reshaping the fluids velocity profile and creating a region of lower pressure immediately downstream of itself. The pressure difference, exhibited between the static line pressure and the low pressure created downstream of the cone, can be measured via two pressure sensing taps. One tap is placed slightly upstream of the cone, the other is located in the downstream face of the cone itself. The pressure difference can then be incorporated into a derivation of the Bernoulli equation to determine the fluid flow rate. The

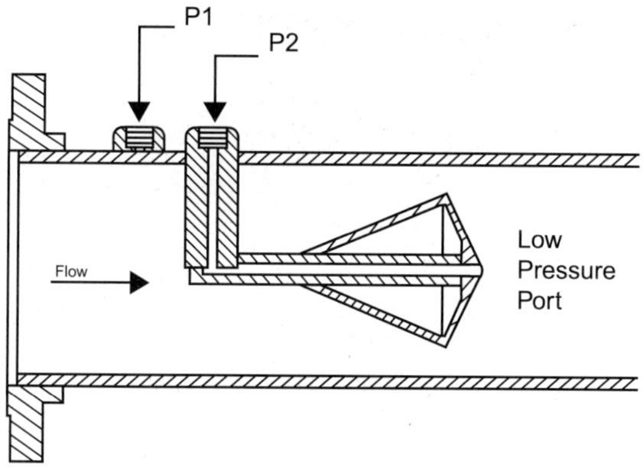

Figure 1. V-Cone flow meter.

cone's central position in the line optimizes the velocity profile of the flow at the point of measurement, assuring highly accurate, reliable flow measurement regardless of the condition of the flow upstream of the meter.

The V-Cone flow meter is a differential pressure type flow meter. Basic theories behind differential pressure type flow meters have existed for over a century. The principal theory among these is Bernoulli's theorem for the conservation of energy in a closed pipe. This states that for a constant flow, the pressure in a pipe is inversely proportional to the square of the velocity in the pipe. Simply, the pressure decreases as the velocity increases. For instance, as the fluid approaches the V-Cone flow meter, it will have a pressure of P_1. As the fluid velocity increases at the constricted area of the V-Cone, the pressure drops to P_2, as shown in Figure 1.

Both P_1 and P_2 are measured at the V-Cone flow meter's taps using a variety of differential pressure transducers. The Δp created by a V-Cone flow meter will increase and decrease exponentially with the flow velocity. As the constriction takes up more of the pipe cross-sectional area, more differential pressure will be created at the same flow rate. The beta ratio equals the flow area at the largest cross section of the cone (converted to an equivalent diameter) divided by the meter's inside diameter.

The V-Cone flow meter is similar to other differential pressure (Δp) meters in the equations of flow that it uses. V-Cone flow meter geometry, however, is quite different from traditional Δp meters. The V-Cone constricts the flow by positioning a cone in the center of the pipe.

This forces the flow in the center of the pipe to flow around the cone. This geometry presents many advantages over the traditional concentric Δp meter. The actual shape of the cone has been continuously evaluated and tested for over 10 years to provide the best performance under differing circumstances.

One must understand the idea of a flow profile in a pipe to understand the performance of the V-Cone flow meter. If the flow in a long pipe is not subject to any obstructions or

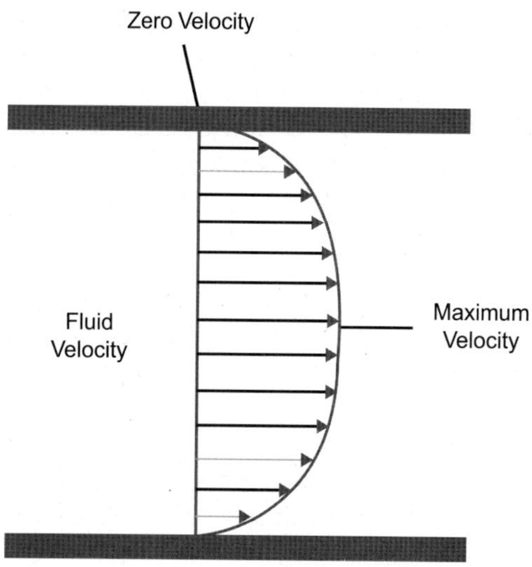

Figure 2. Velocity profile.

disturbances, it is well-developed flow. If a line passes across the diameter of this well-developed flow, the velocity at each point on that line would be different. The velocity would be zero at the wall of the pipe, maximum at the center of the pipe, and zero again at the opposite wall. This is due to friction at the pipe walls that slows the fluid as it passes. Since the cone is suspended in the center of the pipe, it directly interacts with the "high velocity core" of the flow. The cone forces the high velocity core to mix with the lower velocity flows closer to the pipe walls. Other Δp meters have centrally located openings and do not interact with this high velocity core. This is an important advantage to the V-Cone at lower flow rates. As the flow rate decreases, the V-Cone continues to interact with the highest velocity in the pipe. Other Δp meters may lose their useful Δp signal at flows where the V-Cone can still produce one.

The pipe flow profile in actual installations is rarely ideal. There are many installations where a flow meter exists in flow that is not well developed. Practically, any changes to the piping, such as elbows, valves, reductions, expansions, pumps, and tees can disturb well-developed flow. Trying to measure disturbed flow can create substantial errors for other flow meter technologies. The V-Cone flow meter overcomes this by reshaping the velocity profile upstream of the cone. This is a benefit derived from the cone's contoured shape and position in the line. As the flow approaches the cone, the flow profile "flattens" toward the shape of a well-developed profile.

High accuracy

The V-Cone flow meter primary element can be accurate to ±0.5% of reading and the Wafer-Cone® can be accurate to ±1.0%. The level of accuracy is dependent to a degree on application parameters and secondary instrumentation.

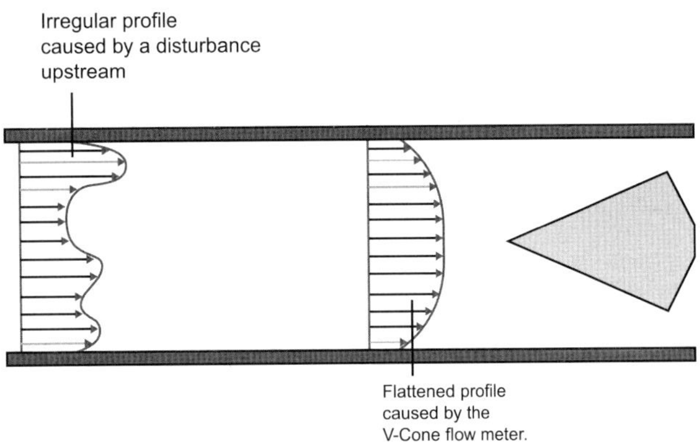

Figure 3. Flattened velocity profile.

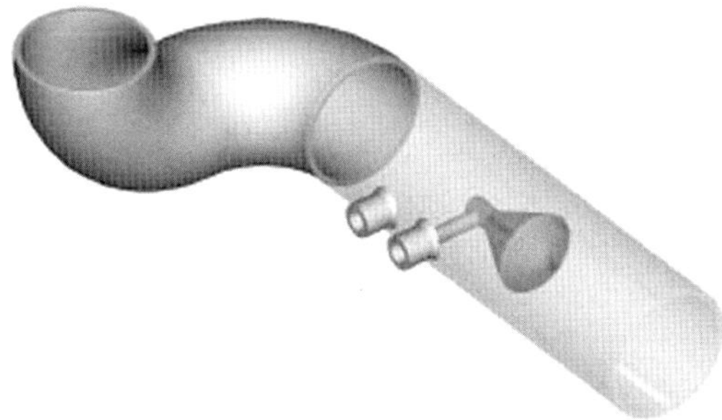

Figure 5. Double elbow and V-Cone flow meter.

Repeatability

The V-Cone flow meter and the Wafer-Cone flow meter primary element exhibits excellent repeatability of ±0.1% or better.

Turndown

The turndown of the V-Cone flow meter can reach far beyond traditional Δp meters. A typical turndown for a V-Cone flow meter is 10 to 1. Greater turndowns are attainable. Flows with Reynolds numbers as low as 8000 will produce a linear signal. Lower Reynolds number ranges are measurable and are repeatable by applying a curve fit to the measured Δp, derived from calibration over a specific Reynolds number range.

Since the V-Cone flow meter can flatten the velocity profile, it can function much closer to upstream disturbances than other Δp meters. The recommended installation for the V-Cone flow meter is zero to three diameters of straight run upstream and zero to one diameters downstream. This can be a major benefit to users with larger, more expensive line sizes or users which have small run lengths. McCrometer conducted performance tests of the V-Cone downstream of a single 90° elbow and two close coupled 90° elbows out of plane. These tests show that the V-Cone flow meter can be installed adjacent to either single elbows or two elbows out of plane without sacrificing accuracy.

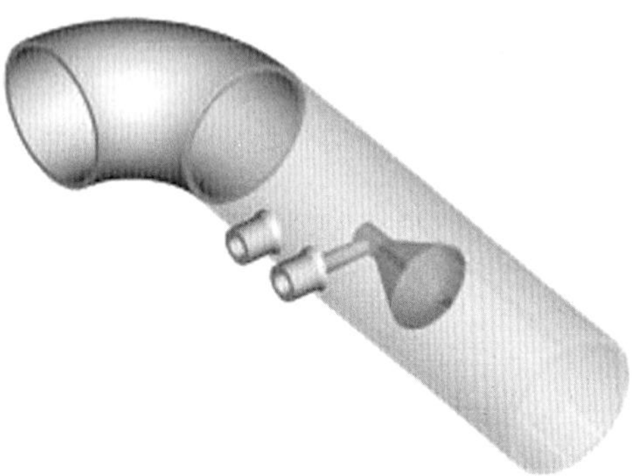

Figure 4. Single elbow and V-Cone flow meter.

Minimum Straight Pipe Run Requirements For Gas Metering at Reynolds Numbers > 200,000

Obstruction	Upstream	Downstream
1 Elbow	1D	1D
2 Elbows	1D	1D
Tees	1D	1D
Butterfly valve (control valve)	Not the preferred position	Valve downstream 1D
Butterfly valve (Shutoff valve)	2D	1D
Full port ball valve (Shutoff)	1D	1D
Heat exchanger (depends on type)	1D	OD
Expander (0.67D to D) over a length of 2.5D	2D	1D
Reducer (3D to D) over a length of 3.5D	OD	OD

The meter and the adjoining pipe should have equal IDs.
For $\beta \geq 0.7$, add 1D

Long term performance

The contoured shape of the cone constricts the flow without impacting the flow against an abrupt surface. A boundary layer forms along the cone and directs the fluid away from the beta edge. This means the beta edge will not be as subject to the usual wear by unclean fluids, as is the case with an orifice plate. The beta ratio will then remain unchanged and the calibration of the meter will be accurate for a much longer time.

Signal stability

Every Δp meter has a "signal bounce." This means that even in steady flow, the signal generated by the primary element will fluctuate a certain amount. On a typical orifice plate, the vortices that form just after the plate are long. These long vortices create a high amplitude, low frequency signal from the orifice plate. This could disturb the Δp readings from the meter. The V-Cone forms very short vortices as the flow passes the cone. These short vortices create a low amplitude, high frequency signal. This translates into a signal with high stability from the V-Cone. Representative signals from a V-Cone flow meter and from a typical orifice plate are shown in Figure 6.

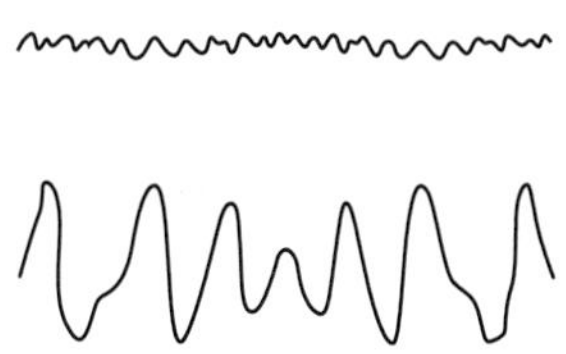

Figure 6. Signal stability.

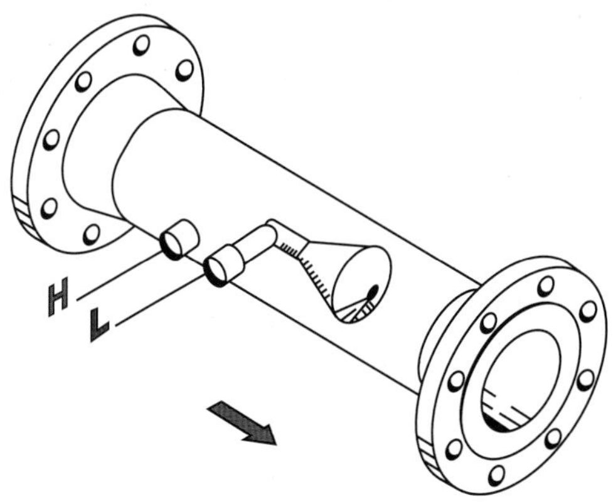

Figure 7. Precision tube V-Cone flow meter.

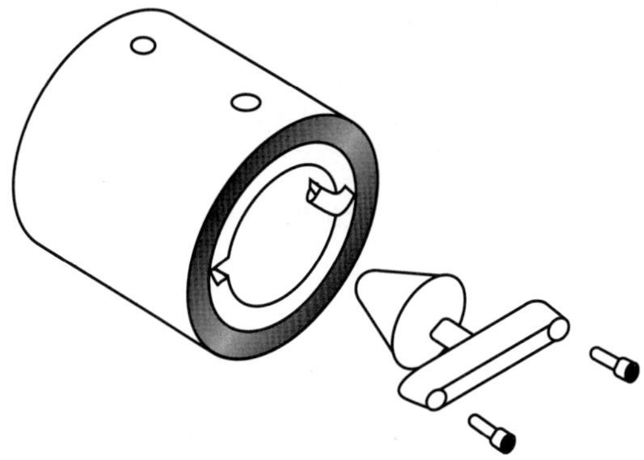

Figure 8. Wafer-Cone flow meter.

Low permanent pressure loss

Without the impact of an abrupt surface, the permanent pressure loss is lower than a typical orifice plate meter. Also, the signal stability of the V-Cone flow meter allows the recommended full scale Δp signal to be lower for the V-Cone flow meter than other Δp meters. This will lower the permanent pressure loss.

Sizing

The unique geometry of the V-Cone flow meter allows for a wide range of beta ratios. Standard beta ratios range from 0.45, 0.55, 0.65, 0.75, and 0.80.

No areas of stagnation

The "swept through" design of the cone does not allow for areas of stagnation where debris, condensation, or particles from the fluid could accumulate.

Mixing

The short vortices described earlier mix the fluid thoroughly just downstream of the cone. The V-Cone flow meter is currently used in many applications as a static mixer where instant and complete mixing is necessary.

Application data

The user must provide application parameters so that the appropriate V-Cone flow meter may be selected. McCrometer has an extensive meter performance database of fluid properties which can be utilized for sizing purposes.

Flow Calculations Nomenclature (All references to "V-Cone" are to the V-Cone flow meter.)

Symbol	Description	English Units	Metric Units
α	Material thermal expansion α or $\alpha_{cone,}$ α_{pipe} (alpha)	$^{\circ}R^{-1}$	$^{\circ}R^{-1}$
β	Beta ratio	—	—
C_p	Flow meter coefficient	—	—
d	Cone outside diameter	In	mm
D	Pipe inside diameter	In	mm
ΔP	Differential pressure (dp)	In WC	Mbar
ΔP_{max}	Maximum differential pressure on sizing	Refer to note 4	Refer to note 4
F_a	Material thermal expansion factor	—	—
K	Gas isentropic exponent	—	—
K_1	Flow constant	$\sqrt{\frac{lbm \times ft^3}{s^2 \times inWC}}$	$\sqrt{\frac{kg \times m^3}{s^2 \times mbar}}$
K_2	Simplified liquid flow constant	See note 4	See note 4
μ	Viscosity	cP	cP
P	Operating pressure	psiA	barA
P_b	Base pressure	psiA	barA
Q	Actual volume flow	ACFS	m^3/s
Q_{max}	Maximum flow rate on sizing	See note 4	See note 4
Q_{STD}	Standard gas volume flow	SCFS	Nm^3/s
Re	Reynolds number	—	—
ρ	Flowing density (rho)	lbm/ft^3	kg/m^3
ρ_{water}	Water density	62.3663 lbm/ft^3	999.012 kg/m^3
S_g	Specific gravity of the gas	—	—
T	Operating temperature	°R	K
T_b	Base temperature	°R	K
T_d	Deviation from standard temperature (°R)	$T_d = T - 527.67$	$T_d = \frac{9}{5}T - 527.65$
U_1	Unit conversion	0.0360912 psiA/inWC	0.001 barA/mBar
U_2	Unit conversion	144 in^2/ft^2	1,000,000 mm^2/m^2
U_3	Unit conversion	167.213 lbm/s^2 ft in WC	100 kg/ms@ mbar
U_4		124.0137 cP ft s/lbm	1
U_5	Unit conversion	2.6988 °R lbm/ft^3 psiA	348.338 K kg/m^3 barA
V	Velocity	ft/s	m/s
Y	Gas expansion factor	—	—
Z	Gas compressibility	—	—
Z_b	Base gas compressibility	—	—

General Flow Calculations

3.2.1	V-Cone Beta Ratio	$\beta = \sqrt{1 - \frac{a^2}{D^2}}$	β from sizing report
3.2.2	Flow constant	$k_1 = \frac{\Pi\sqrt{2 \times U_3}}{4 \times U_2} \times \frac{D^2 \times \beta^2}{\sqrt{1-\beta^4}}$	
3.2.3	Material thermal expansion factor	$F_a = 1 + 2 \times \alpha \times T_d$	See note 1
3.2.4	Material thermal expansion factor if cone and main pipe are made of different materials	$F_a = \frac{D^2 - d^2}{[(1-\alpha_{pipe} \times T_d) \times D]^2 - [(1-\alpha_{cone} \times T_d) \times d^2]^2}$	See note 1
3.2.5	Pipeline velocity	$v = \frac{4 \times U_2 \times Q}{\Pi \times D^2}$	
3.2.6	Reynolds number	$\mathrm{Re} = U_4 \times \frac{v \times D \times \rho}{\mu}$	A dimensionless number that can be used to correlate meter calibration in different fluids
3.2.7	V-cone gas expansion factor	$Y = 1 - (0.649 + 0.696 \times \beta^4)\frac{U_1 \times \Delta P}{k \times P}$	For liquids, Y= 1
3.2.8	Wafer gas expansion factor	$Y = 1 - (0.755 + 0.678 \times \beta^8)\frac{U_1 \times \Delta P}{k \times P}$	For liquids, Y= 1
3.2.9	Liquid density	$\rho = \rho_{water} \times S_L$	
3.2.10	Gas density	$\rho = U_5 \times \frac{S_g \times P}{Z \times T}$	
3.2.11	Actual volume flowrate	$Q = F_a \times C_D \times Y \times k_1 \times \sqrt{\frac{\Delta P}{\rho}}$	See notes 2, 3 & 5
3.2.12	Standard gas volume flowrate	$Q_{STD} = Q \times \left(\frac{P \times T_b \times Z_b}{P_b \times T \times Z}\right)$	Converts actual flow to standard flow at base conditions

Simplified liquid calculation

3.3.1	Simplified Liquid Flow Constant	$k_2 = \frac{Q_{max}}{\sqrt{\Delta P_{max}}}$	See note 4
3.3.2	Simplified liquid flow rate	$Q = k_2\sqrt{\Delta P}$	See note 4

Notes:

1. Material Thermal Expansion—The thermal expansion equations correct for dimensional changes which occur as the operating temperature deviates from the base temperature of 68 °F (see 3.2.3 and 3.2.4) The F_a factor can be excluded from the flow equation if the operating temperature is <100 °F, <559.67 °R, <37.78 °C, <310.93 K. If the F_a factor is significant and the operating temperature is stable, then a constant F_a value can be used. If the F_a factor is significant and the temperature varies, then an F_a factor should be calculated prior to every flow calculation.
2. Discharge Coefficient—Discharge coefficients can be implemented in the flow equations via several different methods. Typical methods are average C_D, C_D look up table, or C_D fitted data. If a C_D look up table or fitted data is utilized, additional calculations must be made based on the Reynolds number [see example process 3(d) and 5(b)].
3. Liquid—Typical Calculation Process
   a. Given: D, β, ρ, C_D, and input of ΔP Calculate: 3.2.2, 3.2.11
   b. Given: D, β, ρ, C_D, and input of ΔP, T Calculate: 3.2.2, 3.2.3 or 3.2.4 if req., 3.2.11
   c. Given: D, β, S_l, C_D, and input of ΔP, T Calculate: 3.2.2, 3.2.3 or 3.2.4 if req., 3.2.9, 3.2.11
   d. Given: D, β, μ, ρ, C_D look up, and input of ΔP Calculate: initially set $C_D = 0.8$, 3.2.2, 3.2.3 or 3.2.4 if required 3.2.11

Iterate until flowrate is <0.01% different from last calculation

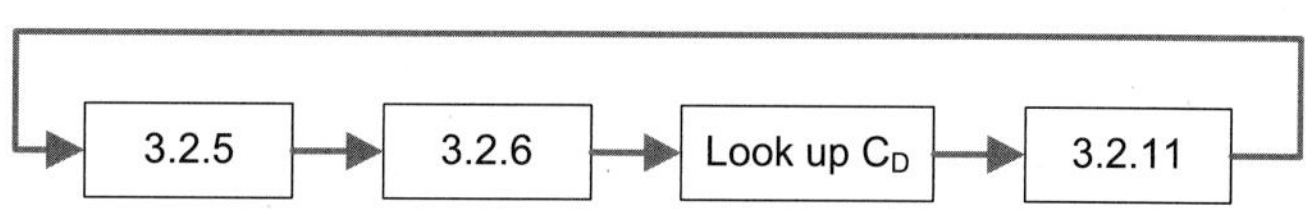

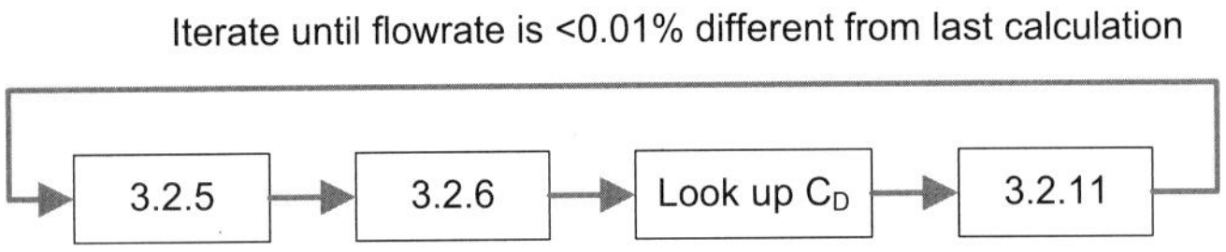

4. Simplified Liquid Calculation—The simplified liquid calculation can be used if the operating temperature is stable and the C_D is constant. The simplified flow constant (k_2) can be calculated from Equation 3.3.1 using the V-Cone Flow Meter Application Sizing sheet. The flow rate can then be calculated using Equation 3.3.2. Units of measure will be the same as those listed on the V-Cone Flow Meter Application Sizing sheet.
5. Gases and steam—Typical Calculation Process: Given: D, β, μ, S_g, Z, k, C_D, and inputs of ΔP, P, T Calculate: 3.2.2, 3.2.3 or 3.2.4 if req., 3.2.7 or 3.2.8, 3.2.10, 3.2.11 b. Given: D, β, μ, S_g, Z, k, C_D look up, and inputs of ΔP, P, T Calculate: initially set $C_D = 0.8$, 3.2.2, 3.2.3 or 3.2.4 if req., 3.2.7 or 3.2.8, 3.2.10, 3.2.11.

Fluid Properties—Fluid properties such as velocity, compressibility, and isentropic exponent vary with temperature and to some extent pressure. The viscosity in the calculations above could affect the selected C_D value, the compressibility directly affects the density and the isentropic exponent effects the Y factor, although to a small degree. The instrumentation industry uses many different approaches to calculate flow. McCrometer application engineering and the customer must determine which fluid properties are calculated at each set of flow conditions and which properties are constant.

Sampling and testing program for unrefined petroleum-based products

In new regulations adopted in the Hazardous Materials Regulations (49 CFR Parts 100-180) on May 8, 2015 (80 FR 26643), PHMSA established requirements for shippers in all modes of transportation to develop and carry out a sampling and testing program for all unrefined petroleum-based products.

Considering the challenges posed by materials with variable composition and potentially variable properties—such as crude oil—providing criteria for sampling and testing of unrefined petroleum-based products is a critical first step in the safe transportation of these materials. Proper classification and the assignment of a packing group for a hazardous material (hazmat) plays a key role in the selection of an appropriate packaging for hazmat.

Requirements

The physical and chemical properties of unrefined petroleum-based products are complex and can vary by region, time of year, and method of extraction. In accordance with §173.41, shippers in all modes of transportation must develop and carry out a sampling and testing program for all unrefined petroleum-based products by:

Sufficiently testing and documenting the physical and chemical characteristics of hazmat to ensure proper classification;

Certifying that the sampling and testing program is in place, and ensuring this information is available to DOT personnel upon request; and

Properly classing and describing unrefined petroleum-based products offered for transportation as prescribed in §173.22 (shippers' responsibility).

A person may offer hazmat for transportation in a packaging or container required by §173.22 if that person:

Classes and describes the hazmat in accordance with parts 172 and 173 of the subchapter, and

Determines that the packaging or container is an authorized packaging including part 173 requirements.

Components of the sampling and testing plan

The shipper must properly classify unrefined petroleum-based products based on the following minimum requirements:

- Frequency
  - History,
  - Temperature,
  - Method of extraction (including chemical use),
  - Location of extraction,
  - Time of year, and
  - Length of time between shipments.

- Sampling prior to the initial offering for transportation and when changes that may affect the properties of the material occur (i.e., mixing of the material from multiple sources);
- Sampling methods that ensure a representative sample of the entire mixture;
- Testing methods that enable classification of the material;
- Quality control of measure sample frequencies;
- Duplicate sampling methods for quality assurance;
- Criteria for modifying the sampling; and
- Testing or other methods used to identify properties of the mixture relevant to packaging requirements.

Certification

When certifying a shipment, the shipper is stating that the requirements in §173.41 have been met. The shipper must enter the certification on the shipping paper and sign either manually, by typewriter, or by other mechanical means. An acceptable certification (see §172.204) is as follows:

This is to certify that the above-named materials are properly classified, described, packaged, marked, and labeled, and are in proper condition for transportation according to the applicable regulations of the U.S. Department of Transportation.

Documentation

Persons required to develop and implement a sampling and testing program must:

Maintain a copy of the sampling and testing program documentation in writing;
Retain documentation as long as the program is in effect (minimum of one year);
Revise or update as necessary;
Review annually;
Make the most recent version (changes) available to employees responsible for implementing the plan; and
Make documentation available, upon request, to authorized official(s).

Responsibility

As stated in §173.1(b), shipments of hazmat not prepared for shipment in accordance with this subchapter may not be transported by air, rail, highway, or water.

It is the responsibility of the hazmat employer to ensure that each hazmat employee is trained in accordance with the requirements of this subchapter.

It is the responsibility of the offeror to instruct each of his officers, agents, and employees—having any responsibility for preparing hazmat for shipment—on the applicable HMR requirements.

Failure to comply with these regulations may result in civil penalty actions. Federal law has set $75,000 as the maximum civil penalty that may be assessed for a violation.

Source

The Hazardous Materials Info Center, http://hazmat.dot.gov.

16: Instrumentation

Types of control systems 710
Developments in pipeline instrumentation 712
Automatic tank gauging systems and proof testing 726
Instrumentation and sensors for automatic tank gauging 737

Types of control systems

Early pipeline systems depended upon manual control and the use of tanks to maintain a pipeline system in balance. On long pipelines that required multiple pump stations, a surge tank was used to catch excess flow or make up the deficiency. Engine speed was varied to maintain the level in the surge tank at an acceptable level. This type of operation was fine as long as one grade of product was being pumped but was not satisfactory for batching different products unless there was a tank available for each grade of product handled.

It was obvious that to build a tight line system, one that would not require surge tankage, some kind of automatic control system was needed to keep the booster stations in balance. Before the widespread use of centrifugal pumps, engine speed (for reciprocating pumps) was controlled to maintain the suction and discharge pressure within limits by sensing discharge and suction pressure and directing a control signal to either the engine governor or a control valve. When control valves are used with reciprocating pumps, they are connected to recycle fluid from the discharge of the pump to the pump section.

When centrifugal pumps came into use, the same type of control system was used except that the final control element was a control valve installed in the pump discharge line as shown in Figure 1. The control system at origin pump stations normally controls discharge pressure or system flow rate. Intermediate booster stations require a control system that will control suction or discharge pressure and in some cases, motor load. An examination of the system line friction and pump head curve will show that simultaneous control of suction and discharge pressure and high flow rate is not required since the pipeline system characteristics will not permit the coexistence of these events except in case of extreme system upset. As will be seen later, the control system is configured so that only one controller will be in control at a given time.

There are several modes of control available to the pipeline engineer. Mode of control is the manner in which control systems make corrections relative to the deviation of the controlled variable. The modes of control commonly used by pipeliners include:

- Proportional position
- Proportional plus reset
- Proportional-speed floating

Other modes such as proportional plus reset plus derivative are more commonly used in the process industries. This mode of control may, however, be used in pipeline measurement stations where it is necessary to control the temperature of a heater used to heat the product being pumped or measured.

In the proportional-position mode, there is a constant linear relation between the controlled value and the position of the final control element. With this type of control system,

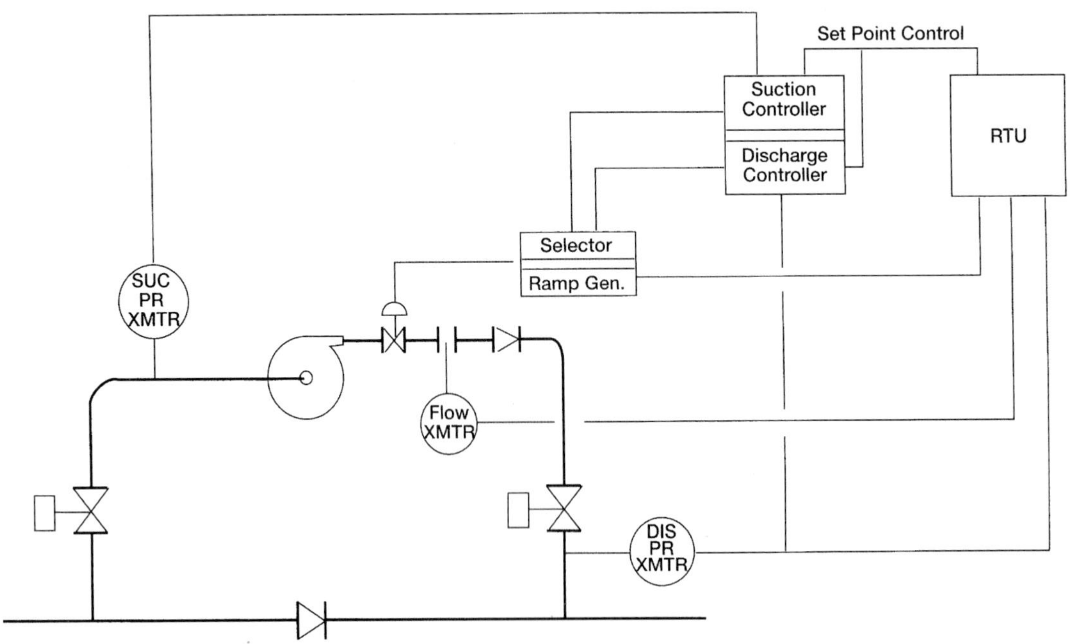

Figure 1. Typical control scheme for a pipeline booster station.

the amount of correction will be the same regardless of the amount of deviation of the controlled variable. A proportional-position control system will not necessarily return the controlled variable to the exact value of the set point when an upset occurs. This may be satisfactory for some types of control systems, but not for pipeline pump station control.

The offset deviation noted in the proportional-position system can be overcome by adding reset to the controller. The reset action will continue to correct valve position until no deviation occurs between the controlled variable and controller setpoint.

Proportional-speed floating causes the final control element to change at a rate proportional to the amount of deviation between the setpoint and the measured variable. Correction to the final control element will continue until there is no deviation between the setpoint and the measured variable. The amount of the correction signal is proportional to the amount of deviation between the setpoint and the measured variable. This mode of control will produce an exact correction for any system load change.

Proportional-position and proportional-speed floating are the control modes most commonly used for pipeline control.

Figure 1 shows a typical control scheme for a single unit booster station. A minimum control system will include a suction pressure and discharge pressure controller. Other controllers such as motor current controllers and a ramp generator may be connected in the control loop.

The suction pressure controller is set to maintain the suction pressure above the low suction pressure shut down setting while the discharge pressure controller limits the station discharge pressure to a value that will not exceed the maximum allowable operating pressure. The output from each of the controllers is fed to a signal selector that will select either the high or low signal, depending upon the control action selected, and direct this signal to the control valve. Some control systems are designed so that a selector module may not be required, but they will include a means for selecting the proper signal. In a properly designed system, the transfer of control from one controller to another will be smooth and bumpless.

A ramp generator is used for ramping the control valve to a new position. The control valve may be closed when the station is at rest, and upon station startup the ramp generator will cause the valve to open slowly until one of the controllers is in control of the system or until the pump is "floating"—that is, operating at a condition where neither suction nor discharge pressure control is required. It may also be used to ramp the valve partially closed prior to unit shut down to minimize system upset. This feature is desirable in a multi-unit booster station since when one unit is shut down, the remaining operating units will be operating momentarily at a no-flow condition. This condition will continue to exist until the system pressure decays to a value equal to the discharge pressure of the remaining units. Ramp speed should be adjustable so that it can be fine-tuned to the system requirements.

The control valve action, i.e., valve opens on increasing signal or valve closes on decreasing signal, will determine how the controller action is set. The suction pressure controller will be set opposite of the discharge controller.

Although not shown, the control system is usually equipped with a manual/automatic switch so that the control valve may be controlled in case one of the controllers fails or during system maintenance.

The controllers may also be equipped with remote setpoint control so that controller setpoint may be varied from a remote location to accommodate a changed condition in the system.

Figure 2 shows a control system for a multi-unit booster station. The control systems for single- and multi-unit stations are identical except that in a multi-unit station the station pressure transmitters are located so that one control system will suffice regardless of the number of units installed. It should also be noted that one control valve serves both units.

The control valve may be operated electrically, pneumatically, or hydraulically. Hydraulic systems offer the advantage of being able to operate the valve at a high rate of speed at virtually any system load. Most hydraulic operators are equipped with dual cylinders and may be equipped with quad cylinders where the system loads are high. The selection of control valves is covered in Section 7.

Variable speed drives or controls are becoming an economically attractive choice for the operation of pumps. The use of these devices will virtually eliminate throttling excess head and the attendant energy costs. Few systems are operated at a constant rate, and therefore some throttling is inevitable. Also, systems transporting multiple grades will require throttling to keep the system in balance as the batches move through the system.

Most control systems used for pipeline control are made up of standard control system elements. Some applications may, however, have unique characteristics that will require a system that is just a little bit different from a standard control system. Most of these types of systems can be put together using standard modules, or a programmable logic controller may be used.

In Section 13—Liquid Hydraulics, the discussion on hydraulic gradients illustrated a case where a hill required maintaining back pressure at the end of the line. It was shown that the amount of back pressure required varied with the system flow rate. A conventional back pressure controller and control valve installed at the end of the line and set for the worst case (maximum back pressure) would work; however, it would not be economical since more pressure than required

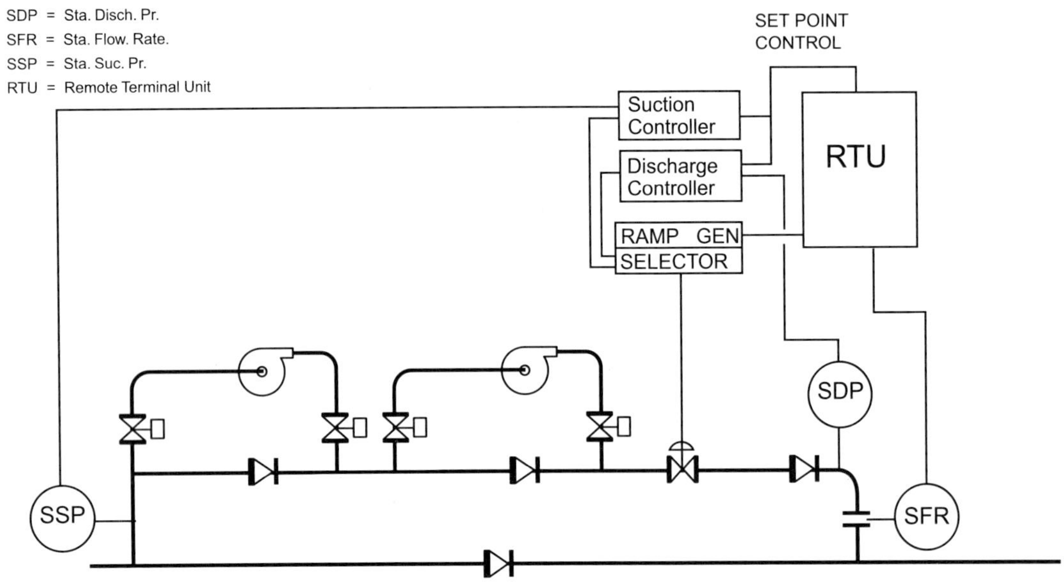

Figure 2. Multi-unit booster station.

would be maintained at the higher flow rates. The pressure at the hill could be telemetered to the end of the line, where it could be used as the measured variable for the controller. The setpoint would be set to the amount of pressure required at the hill which in this case is 200 psig. This type of control system would not penalize system economics since it would always maintain the required 200 psig at the hill. However, it would require the installation and maintenance of a telemetry link.

Another option is to use a relationship controller that would not require telemetry of the hill pressure. The relationship controller would use the system flow rate to drive the setpoint of a pressure controller. The controller would need to be equipped with limit stops so that the setpoint could only be changed between the upper and lower limits. High system flow rates would drive the controller setpoint to the minimum allowable value of back pressure and low system flow rates would drive the controller to the allowable maximum limit. Intermediate system flow rates would drive the setpoint to an intermediate pressure setting.

With today's technology and hardware, the control engineer is limited only by his imagination in solving simple as well as complex control problems. Programmable logic controllers (PLC) and self-tuning controllers are being used more and more in the control of pipelines and pipeline stations. Control systems are being incorporated as a part of the PLC. With the addition of a modem and a suitable communication protocol, the PLC may also be used as the remote terminal unit (RTU) for the supervisory control and data acquisition system (SCADA).

Developments in pipeline instrumentation

R. A. Furness, Department of Fluid Engineering and Instrumentation, Cranfield Institute of Technology, Bedford, UK

Introduction

Our need to transport fluids from the point of production to the area of end use has led in the past 25 years to a large increase in the number of pipelines being designed and constructed. Many of these carry hazardous or toxic products, often through areas of high environmental sensitivity or close to centers of population. The cost of oil production rose sharply to more than $35 per barrel before the recent sudden fall, and the cost of manufactured chemicals and industrial gases has also risen in the same period. With the need to safeguard the cost of the transported fluid and also the population and the environment close to the pipeline, on-line monitoring and in some cases 24-hour surveillance is now becoming routine in many systems throughout the world.

When one sees maps of the pipeline network in some parts of the world, the number and length of pipelines is staggering. At the recent Aberdeen conference, one speaker gave just two of the many statistics that can be produced to show how vast the network is. One line runs from central Canada eastwards to the Toronto/Montreal area. This line alone has 54 million square meters of pipe surface to survey. Also in 1984 some 6 billion barrels of oil and 4.5 billion barrels of manufactured liquids were transported in pipelines in the USA. The actual world volume of pipeline fluids is many times this amount and the value immense.

Pipelines are not new, indeed almost 50% of the world's total network is 30 or more years old. These have been subject to corrosion and wear, and some have carried fluids for which they were not designed. Others have been operated outside design limits. Monitoring and control systems therefore are not just required for measurement purposes but also for integrity surveillance and on-line fault diagnostics. Some of the many developments that can be applied in modern day line operation are briefly reviewed, and some design pointers for their integration into complete systems are given.

Flow measurements

Flow measurement is the most important process variable in the operation and control of pipelines. Simple line flow balances are used to check for discrepancies on hourly, daily, or weekly bases. This may identify that a problem exists, but flowmeters at the end of a line will not identify the location of the problem. For this, pressure transducers along the pipeline will be required together with the flowmeters to establish where the pressure gradient has deviated from that predicted from the basic equations of fluid flow. These equations use pressure/flow values at inlet and outlet to confirm the line is in balance. Such a simple system is described in more detail later in this paper.

Most flowmeters sold are of the differential pressure type, and of these the predominant meter is the sharp-edged orifice plate. Many hundreds if not thousands of these are installed on pipelines around the world to measure gas, liquid, and mixed fluid streams. With the arrival of the modern flow computer these types of flowmeter have almost received a new lease on life. One fact not appreciated, though, is the total measurement uncertainty of such devices. Frequently we read performance claims of ±0.5% and better for these pipeline metering systems. This is very difficult to believe when it must be remembered there is an uncertainty of ±0.6% on the data used to derive the flow coefficient in the basic calculation. When all the other variables that can affect the uncertainty of measurement are included, such as fluid composition changes, ambient and process temperature and pressure variations, errors in temperature and pressure measurement, A/D conversion and computational errors, to name but a few, then it is unreasonable to expect uncertainties much better than ±1% FSD. If wide flow rates are also experienced, then the uncertainty of the low flow determination could be ±5% FSD or worse. This makes their use as accurate pipeline flow devices somewhat questionable, although they are very common in the metering of natural gas in this country.

Turbine flowmeters also find widespread use in pipeline applications, particularly hydrocarbons. They all suffer from bearing limitations, however, and this means they require regular calibration to ensure good performance. Recent developments have included twin rotor designs, where measurements of the speed of both rotors simultaneously has led to the meter becoming self diagnostic. These meters can now detect their own bearing wear and also deviations from initial calibration due to blade damage. The same on-board microprocessors can also check the integrity of the data generated by the turbine meter, as well as providing alarm output for loss of signal and other problems.

Two different designs have been applied in pipeline systems with good success. The first design, from Quantum Dynamics and shown in Figure 1, has two rotors with two independent sets of bearings. The shaft and slave rotor ride on one set of bearings with the measurement rotor upstream of the slave motor riding on a second set of bearings on the rotating shaft.

Relative motion between the indicating turbine and the shaft therefore remains close to zero. This plus a drag-free RF pick-off results in an almost frictionless flowmeter. Should the shaft or measuring motor bearings begin to wear, this will result in a change in the ratio of the rotor speeds for a fixed flow rate. Thus, bearing wear can be detected on-line and action taken. This design of meter is also different from all other turbines because of the slender center body and axially short blades. Because friction is very low, the fluid does not have to be accelerated past the rotor to obtain sufficient torque to overcome retarding forces. Thus, pressure drop is also low and pressure and temperature measurements can be carried out at the point of volume flow rate determination. They have been shown to be outstanding meters for leak detection purposes on gas lines.

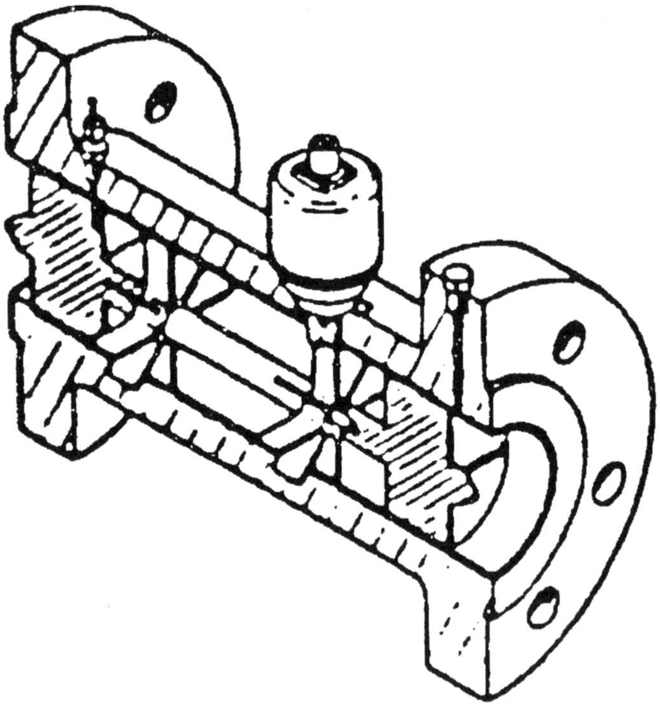

Figure 1. Twin rotor turbine meter. (Courtesy Quantum Dynamics Inc).

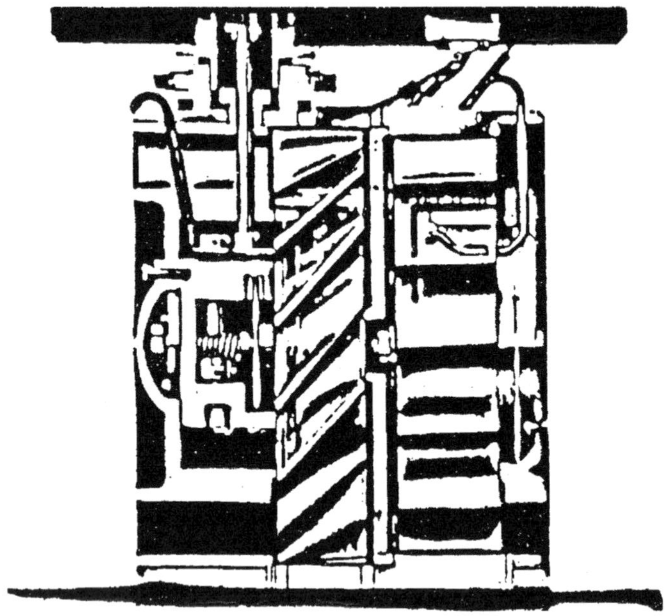

Figure 2. Auto adjust gas pipeline meter. (Courtesy Rockwell International, Inc).

The Auto Adjust turbo-meter from Rockwell International is also self-checking but is in addition self-adjusting. The design shown in Figure 2 is quite different from the Quantum meter. The rotors are next to each other with the measuring rotor downstream of the main rotor. Constant performance is obtained because the sensor rotor detects deviations in blade angle of fluid leaving the main rotor. With a very small angle, the sense rotor runs at a much lower speed than the upstream rotor. Performance is then given by the difference in the speed of the two rotors rather than from a single rotor. When bearings begin to wear, the main rotor will spin slightly slower and so the exit angle will also change. This is sensed by the second rotor, which adjusts speed by an equal amount. Bearing wear is therefore automatically adjusted out. These meters have been installed in many gas transmission lines in North America with generally very good performance. The Quantum meter, on the other hand, is a newcomer to pipeline applications, but personal firsthand experiences in a gas pipeline were favorable.

Pipelines are generally operated on material balances (mass), but most flowmeters are volumetric devices. This means that for pure fluids measurements of temperature and pressure plus an appropriate equation of state are required to enable density to be determined. For mixtures of gas or liquids or indeed as an alternative to the above method, on-line densitometers are used. The product of volume flow and density gives mass flow. Until recently pipeline mass flowmeters were not available, but recently a number of Coriolis type direct mass meters have appeared. These have found slow but steady acceptance in industrial applications and are beginning to find applications in pipeline monitoring. One such design is shown in Figure 3, and recently four other designs were launched at the INTERKAMA trade show in West Germany. This type of meter is applicable to both liquids and gases, and some have been used on liquid/liquid and gas/liquid mixtures but with mixed success. They are currently undergoing rigorous evaluation and analysis in many chemical, process, and small pipeline applications, and indications are that they are standing up well in some applications but not others. Accuracy is around $\pm 0.5\%$ of reading or better, and they do not need additional measurements to determine flowing density. Indeed, by monitoring the output signal, they can be used to indicate density as well as mass flow. The output is in digital and/or analog form making them compatible with many types of flow computer.

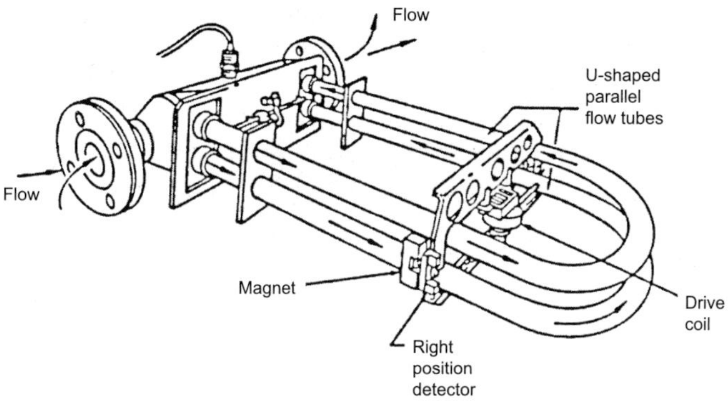

Figure 3. Micro motion direct mass flowmeter. (Courtesy Brooks Instruments).

Direct mass measurement has one main advantage over volume measurement, since an integration of the signal over a period of time will indicate the pipeline fluid inventory directly. The disadvantage for pipeline operators is the current size range that is available, the largest being 150 mm bore. As most major pipelines are in the range of 500–2,000 mm bore, these Coriolis meters must be installed in parallel to meet the larger applications. Mass metering is relatively new, and the next decade should see a steady increase in sales and acceptance of these meters.

The ideal flowmeter is one that is installed on the outside of the line but gives the performance of the best flowmeters installed inside the line. Ultrasonic time-of-flight meters are slowly developing toward these apparently conflicting but demanding criteria. Multiple beam systems have been installed on many pipelines, the most notable of which is the Alaskan oil pipeline. Some 23 meters, each 4 ft in diameter, look for gross imbalances along sections of this important installation. They permit the early detection of product spillage and also indicate the section of line giving the problem. This system has been operating since the late 1970s with millions of barrels of fluid being shipped through the line in that time. Experience of these meters has been very favorable in this application.

Ultrasonic meters, like most modern types of meter, were initially oversold, but the main problem has been to convince users that the two types available, the Doppler and transit time meters, are completely different animals that both use ultrasonics to infer flow. When ultrasonic meters are mentioned, people tend to think of Doppler types. Most of these are relatively crude flow indicating devices. The modern clamp-on transit time meter can indicate flow to around ±2% of reading or better depending on design, compared to the average Doppler meter, which is a ±5% FSD device. The advantage of ultrasonic methods is the negligible headloss and the ability to install other portable or dedicated systems without line shutdown. It is surprising that more have not been installed in pipeline systems, but as discussed later in this paper, this is an example of the conservatism often shown in the application of new technology.

Flow measurement has really been transformed by the application of digital techniques to on-line signal analysis. Much more data than just flow rate is now available, and with the latest systems, operators are taking advantage of the new generation of micro and flow computers to control and run their pipelines. It is unusual these days not to install a local flow computer with each flowmeter, with temperature, pressure, and differential pressure signals being fed into the machine for local "number crunching."

Proving devices

The problem with all flowmeters is that they are affected to some degree by the fluid they are metering. For the least uncertainty of measurement, the flowmeter, whatever type is chosen, should be calibrated on-line under actual operating conditions. In the past this was performed with large and expensive pieces of hardware called ball provers. These had to be large to provide adequate volume within the calibrated section of pipe to ensure sufficient resolution and accuracy for the flowmeter they were intended to calibrate.

The latest generation of proving devices, called compact provers, are essentially time/volume calibrators that eliminate the ±1 count error present on their "big brothers." It is physically much smaller but with the same or better performance. There are currently 10 different versions on the market, of which five are available in the UK. The prover, one design of which is shown in Figure 4, uses pneumatics to launch a piston, which is then driven by the fluid through a calibrated chamber, displaying a known volume. The key to their excellent performance is the data acquisition system illustrated in the lower part of Figure 4. As the piston passes through the calibrated section, pulses from a high frequency internal clock are counted to determine the time, A, to displace the calibrated volume D. Once the counting has started, a second internal timer is gated to respond to the first input pulse from the flowmeter. This timer then records the time interval B to collect C whole flowmeter pulses.

During the proving run there were then C × A/B meter pulses. The meter factor is found by dividing this count by the volume displaced:

$$K = \frac{A}{B} \times \frac{C}{D}$$

Some designs are uni-directional and some bi-directional. In the uni-directional types pneumatics or hydraulics launch and retrieve the piston; in the bi-directional designs the

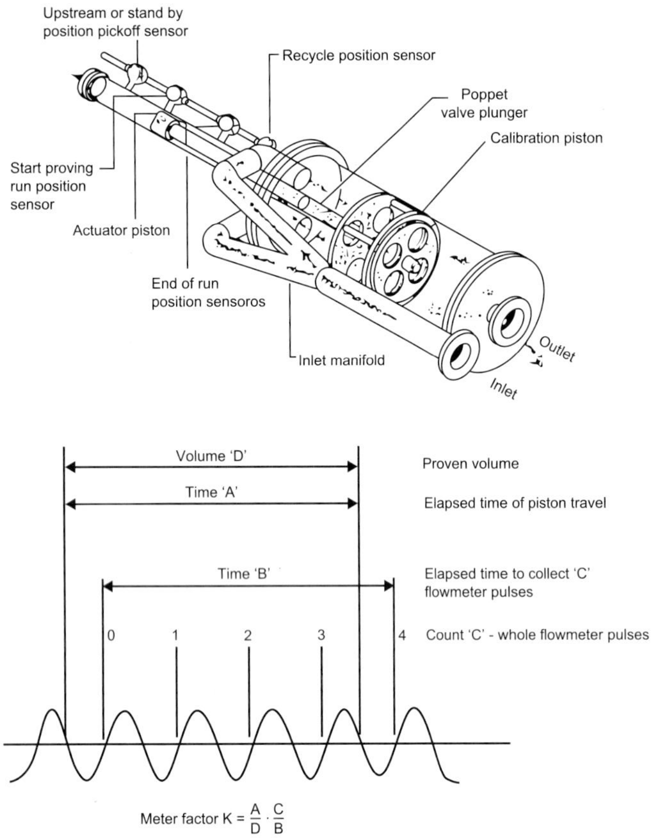

Figure 4. Modern compact meter prover. (Courtesy Brooks Instruments).

process fluid is used to drive the piston. Other versions incorporate double-barrelled liners to remove the need for pressure compensation and also reduce the effect of temperature variations.

In one proving run there may be as few as 100 pulses collected, so timing error must be minimized. If the pulse output is frequency modulated, from a positive displacement bulk meter, for example, then these few pulses may not represent a true statistical sample of the pulse train and multiple passes may be required. Alternatively, pulse interpolation techniques may be required, several of which are applicable to give the accuracy required. These are currently the subject of intense activity in standards committees who are producing a new general code of practice on meter provers.

The compact prover is better suited to offshore platform applications and to field use on operating pipelines. Being portable, it can be easily coupled in series with the meter to be proved. A suitable choice of materials allows proving on a vast range of fluids to be carried out. Operational experience has generally been favorable, and several direct comparison tests with ball provers have shown the new generation of compact provers to be superior in most respects. However, their acceptance in custody transfer applications has been slower than had been expected, even though several countries have certified some designs for Weights and Measures approval. Some international calibration centers are now using this technology for routine traceable calibrations, and confidence is growing as operational experience is accumulated.

By far the most economic method of measuring flows in pipelines is to install pairs of flowmeters in custom-designed metering stations. By monitoring the ratio of the outputs, a measure of wear or drift is obtained. Provided the outputs of the meters remain within preset tight limits, uncertainty of flow rate measurement is low. The disadvantage of this on pipeline operation is that there are two meters to be maintained and checked, and should faults develop, the pipeline may have to be shut down more frequently, unless by-pass loops are provided when the metering station is designed. Turbine meters and PD meters with pulse outputs give the best performance, particularly if they are regularly proved with either a dedicated or portable prover of the type described.

For large bore lines, particularly gas lines, the proving of flow meters is very expensive if it can be performed at all. The majority of installations rely on orifice plates built to the standards. Recently a new high-pressure gas proving station was completed by British Gas at Bishop Auckland. Here turbine and other meters can be calibrated in actual operating conditions on natural gas. Such an installation has previously only been available in Holland, where gas calibration standards have been developed to a high level of confidence over many years of patient data collection. With the new British Gas facility we should see the ability to cross check our own standards against international standards.

Valves

Valves are an important part of any pipeline system. They come in many shapes and sizes and perform numerous functions. Some act as primary flow regulators, some as check valves for flow direction control, and some as pressure reducers for safety purposes. Developments have centered in two main areas, materials and actuators (both simple and intelligent). One of the development sources is again the North Sea, where the demands for better reliability, better in-field performance, and reduced cost have forced many manufacturers to look critically at designs. As one of the most expensive components in the valve is the body, effort has been concentrated in this area, both in terms of new materials to meet more demanding specifications and of quality and price. These last two items are critical to the survival of many companies, and in a market sector that is growing slowly or even declining, as suggested by one leading instrument journal, valve body cost savings, however small, may mean the difference between life and death.

Slimline butterfly valves have recently been introduced by a number of manufacturers to meet the demand for lower cost reliable devices. These can be installed between existing pipeline flanges. They can be manufactured in a wide range of materials with special seal materials chosen for the most demanding of duties. With suitable actuators, even valves as large as 60 in. in diameter do not take up much space.

Other designs of control valve provide straight-through flow patterns to minimize flow turbulence, vibration, and noise. Should a pipeline be fitted with acoustic line break detectors (see section following), line noise will be important in setting the sensitivity of the detectors, and hence the type of valve fitted will be important to system operation. Other operational problems such as "hunting" in globe valves have prompted the designers to work more closely with the operators to define full specifications. Pipeline operators often run their systems with rapid transients present. Demand changes of both a positive and negative value can subject the control valves to very severe step changes in flow and pressure, and initially many conventional valves could not withstand such treatment.

This is where intelligent sensors on pressure relief valves have come into their own. They are able to give better speed of response and meet some stringent operating criteria. Pressure may also rise gradually due to a steady rise in line temperature, as well as the rapid transient caused by block valve closure or pump failure/surge. These transients travel at sonic velocity and even tenths of a second's advantage in operating the valves can affect safety on the line. Microprocessor-based detectors fitted to the valves can be used to sense line conditions and provide suitable inputs to the valve. In some advanced designs manufactured in the United States, it is possible to program intelligence into the unit to begin closure/opening based on measurements of an anticipated condition such as pump surge further down the line. If the signals from one part of the line can be transmitted to other sectors, corrective action can be taken before safety blowoff valves are activated, releasing product into the environment.

A pipeline pressure relief system based on this principle is shown in Figure 5. A strain gauge sensor fitted on the suction side of the compressor is monitored by a microprocessor that linearizes and conditions the signal. The microprocessor operates a pair of solenoid valves acting as the pilot mode with the line pressure acting as the actuating medium. By switching solenoids, the relief valve can be opened, closed, or programmed to act as a regulator at the pressure trapped in the valve body. Such a system can provide adequate response to step changes on the order of 70 bar in 3 milliseconds. In this system the status of the pipeline pumps is provided as an additional input to the microprocessor to trigger action before the pressure surge occurs. Tests on large lines have shown such intelligence to be vital in the safe operation of systems when 5,000 hp pumps were suddenly lost.

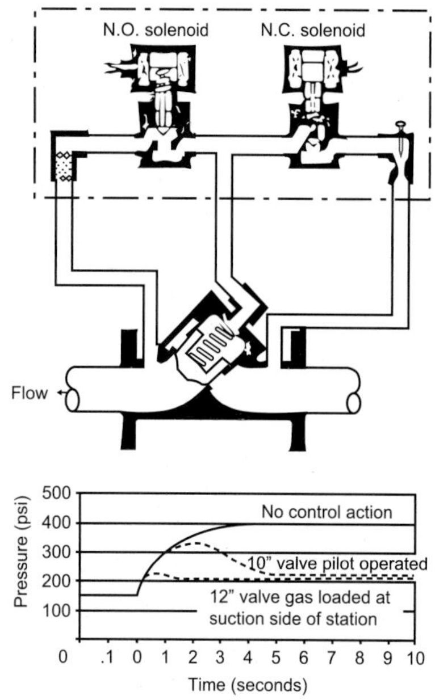

Figure 5. Intelligent pressure relief valve.

Line block valves can also be actuated by output signals from line break detectors. This is important in isolating problem sections of line carrying toxic products. The wave detector will sense problems in seconds, and the line can be closed to minimize the amount of toxic product released. That section of the line may well depressure, but the release effects will be reduced. Such systems are fitted as anti-terrorist devices on lines in problem areas of the world such as the Middle East and Central Africa, where all too often pipelines are political targets.

The choice of materials for valve bodies and the styles of valve available is now very wide. A survey in *Processing* (January 1986) listed 15 different types supplied by 110 companies. These range from the cryogenic applications up to line temperatures in excess of 700°C. Selecting the correct valve for the duty is now the biggest problem facing the user, and where it is necessary to take outputs into computers for safety and control purposes, the choice becomes more difficult. In such a competitive market, developments are bound to come thick and fast, and it will be interesting to watch the development of the next generation of intelligent pipeline valves.

Acoustic line break detectors

Problems with the pipeline are frequently accompanied by the generation of noise, either from problem valves (see above) or line breaks. If the flow becomes too high from a surge or a leak occurs through corrosion or third-party intervention, then there are instances where vibrations in excess of 20 kHz are produced. These frequencies are in the ultrasonic range and can be detected with suitable transducers. Such devices have been made portable so that pipeline crews can clamp them at any point along the line to check for noise. By noting the signal strength, the source of the problem can be pinpointed. British Gas has used this principle to develop one of two types of pipeline monitoring tools that are described later.

A similar technique, though based on a different principle, is the acoustic "wavealert" monitor, though more correctly called a negative pressure wave detector. The central element is a piezo-electric sensor that gives an output when stressed. When a line rupture or leak occurs, there is a sudden drop in line pressure at the source of the problem, followed by line repressurization a few milliseconds later. The rarefaction wave generated by these events moves away from the leak in both directions at the speed of sound. The pipewalls act as wave guides enabling this wave to travel for great distances, attentuating in amplitude as it does so. Sensors placed at known points along the line can detect the passage of this pressure transient. From the triggering of internal timing circuits and a knowledge of the prevailing acoustic velocity for the line, the leak location can be calculated.

Such devices are particularly useful for detecting large breaks in lines very rapidly. The transient wave moves at 1 mile per second in liquids and 1 mile in 5 seconds in gases. If the detectors are positioned at intervals of 2 to 5 miles in critical areas, then the break is detected in a few seconds. If the output signal is fed into a computer, then this can be used to trigger block valves to isolate the problem section of the line very easily. Figure 6 shows one further development where the instrument can be made to cancel line noise and use the full measuring capacity of the sensor for signal detection. There is, however, the practical problem in setting the background threshold correctly. Pipeline noise is generated by bends, valves, pumps, regulators, etc., and it is crucial

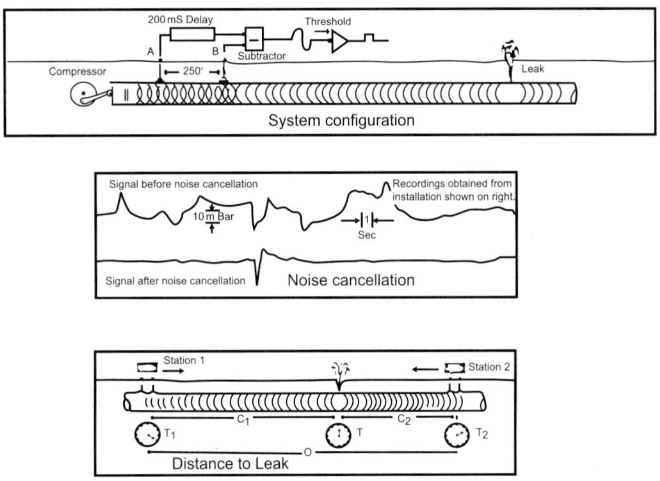

Figure 6. Negative pressure wave "wavealert" monitor. (Courtesy Spectra-Tek).

to be able to remove this background noise to avoid unnecessary line shutdowns when the line noise exceeds the detector setting. The Reynolds number of the flow has been shown to be the important parameter in installations both in the United States and the United Kingdom. Experience has also shown that the installation of these devices is critical to reliable performance. It is important to have the sensor as close as possible to the line to avoid attenuation effects and improve sensitivity. Both this and the background line noise will influence the number of "false alarms" from the instrument.

ICI has been experimenting with strain gauge-type sensors on their cross country ethylene line, and these are showing more promise. These experiments will be closely monitored in the next few years to see whether we can improve the reliability and sensitivity of line break detectors.

"Smart" pressure sensors

One area where the impact of microprocessors has been felt is in the development of smart sensors, particularly pressure. Many physical effects respond directly or indirectly to pressure, and as a result, the first generation of "smart" process transducers appeared on the market some years ago. Surprisingly, however, market penetration has been slow, mainly, I suspect, because hand-picked conventional pressure and differential pressure cells perform almost as well over the limited range often required. There is virtually no limit to the amount of intelligence and diagnostics that can be incorporated into pressure transducers, but so far only those that are reasonably linear or that have well defined characteristics have been commercially exploited. The need for linearity is now no longer there because all signal processing the digital and algorithms can be written to cope with any signal/pressure curve, provided it is repeatable. Even hysteresis can be taken account of.

It is in the communications area that the full potential of intelligent transducers is likely to be felt. Hand held field communicators enable the user to rerange, configure, and display process output values without the need to be close to the transmitters. On-line reranging is particularly useful when the pressure transducer is located in a very inaccessible place. It is possible to use two-way communication links to examine the outputs from a variety of intelligent transducers located at remote outstations, but the cost of such a feature should be carefully considered as this is likely to be expensive. Intelligent transducers are more expensive than the commercial instrumentation, but in certain applications cost is not a consideration.

The general features of the new generation of smart sensors, in addition to those mentioned above, include:

- On-line temperature compensation
- Built in-diagnostics
- Claimed accuracies of $\pm 0.1\%$ of span

Leading companies in the field include Honeywell, Bailey, and SattControl, but other established companies are working on similar devices. Most of these sensors use the properties of silicon plus the addition of micro electronics to give the performance. Silicon is an elastic drift-free material, and pressure sensing elements are diffused onto it. A cavity is usually etched on the reverse side to form the diaphragm. Where the line fluid is very aggressive, a second insulating diaphragm is required to protect the silicon from corrosion. Generally speaking, the cost of these sensors is coming down, and it appears we can look forward to many more instruments appearing in the more difficult applications. Provided we can overcome the general conservatism about applying new technology, this is one section of the instrumentation industry where microelectronics will have a great future influence, particularly when used with the computer and real-time models. This will give the pipeline operator more confidence and more information on both the condition of the instrument and the accuracy of the measurement. Interfacing problems should also reduce.

Another type of pressure transducer that looks promising for future applications in pipelines is the vibrating wire sensor. One such instrument from Foxboro has shown considerable reduction in size and manufacturing costs from other sensors. A tungsten wire encapsulated inside a cell is vibrated at its natural frequency. As pressure is applied to the silicon barrier diaphragms, this changes the tension in the wire and so alters the resonant frequency. This is sensed and amplified and gives the transducer excellent resolution characteristics. Pressure and temperature compensation is accomplished within the cell, and although the instrument is not strictly a "smart" transducer, it does possess basic characteristics that should be attractive to pipeline applications.

Densitometers

Continuous high accuracy on-line density measurement has become increasingly important in pipeline operations. The composition of natural gas and hydrocarbon streams are different from each production well, and assumptions about constant composition cannot be made. In fiscal measurement and where the energy content of the stream is required, a knowledge of density is essential, particularly for gases.

Recently great strides in the development of rugged on-line densitometers have been made with the introduction of the vibrating element transducer. This type of instrument is fast becoming the industry standard for pipeline applications. One such instrument is shown in Figure 7. The operation is very simple, being based on the spring mass principle. A flat plate, cylinder, or fork is maintained at resonance by an electronic circuit. A fluid flowing over or through the vibrating element adds to the vibrating mass and causes a decrease in natural frequency. The higher the fluid density,

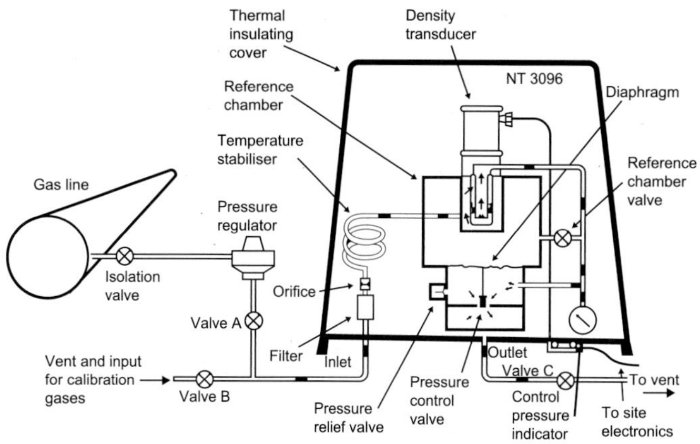

Figure 7. Pipeline densitometer. (Courtesy Solartron Inst and Control and Instrumentation).

the lower the frequency and vice versa. By calibration using fluids of precisely known density, the relationship between frequency and density for each element can be determined very accurately. The characteristic equation for most vibrating element densitometers is of the form:

$$\rho = At^2 + Bt + C$$

where A, B, and C are calibration constants, t is the period of oscillation, and ρ the fluid density. Once the constants have been determined by the laboratory tests, the characteristic equation can be programmed into the system computer for on-line calculation of density. In some cases it may be possible to check the performance *in situ* by passing a fluid of known density through the transducer. This will establish if there has been any drift from the original manufacturer's calibration and will help to minimize the uncertainty of the operating system. The latest generation of instruments can approach accuracies of ±0.1% of reading under ideal conditions and are typically ±0.2% of reading under actual operating conditions.

When used in conjunction with orifice meters, for example, the flow through the densitometer is maintained by taking fluid from a tapping downstream of the orifice and feeding this back at the downstream tapping of the plate. This method, known as pressure recovery, is very popular in pipeline applications, and a typical installation is shown in Figure 8. Other methods include pumping a small percentage of the stream flow through a flowmeter and densitometer before feeding back into the main stream.

There are other types of fluid densitometer such as nuclear or force balance, but these do not find the same widespread use on pipelines as the vibrating element devices. The choice of transducer will depend on the application, since fluid properties, flow rate, sensor element design, and materials of

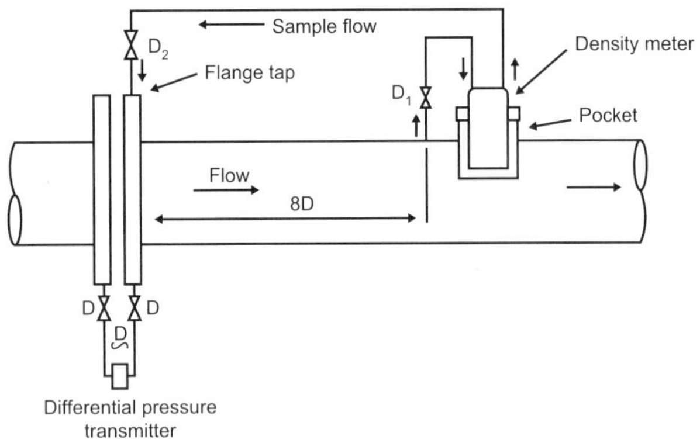

Figure 8. Standard installation for pipeline density measurement. (Courtesy IP manual section 6.2 Nov 1983).

construction all influence the type of densitometer used. For example, if the fluid contains solid particles or is of a "dirty" nature, then straight pipe vibrating cylinder elements are more likely to give better long-term performance. The actual elements are manufactured in a wide range of materials. Ni-Span-C is used for high accuracy metering, and this material also has a low temperature coefficient. Stainless steel, on the other hand, has better corrosion characteristics and is cheaper to use. Materials such as hastelloy C-276 offer excellent resistance to the most corrosive of fluids.

It is probable that the use of these instruments will increase as general experience on pipelines has been favorable. They must, however, be operated in single-phase fluids, as the presence of either droplets in gas lines or vapor bubbles in liquid lines will produce erroneous outputs of unknown magnitudes. The straight tube cylinder-type designs could be used in low second-phase concentration applications if the densitometer is installed vertically with upward flow. This will ensure closer approximation to homogeneous flow. Experience has shown, however, that two-phase flows are significantly more troublesome to handle, and the output of the transducer should be used with some caution.

Pipeline samplers

Oil lines, particularly crude oil lines, often contain significant amounts of water. This has to be monitored and allowed for when the oil is being metered for custody transfer or taxation purposes. It is a growing and important area of pipeline technology to install on-line samplers for the measurement of water content, and work is also under way to use statistical methods to estimate how many samples are representative of the composition of the flowing stream. There are currently two different basic methods being used on pipelines to monitor the water content. These are the capacitance method and grab samplers.

The first method is based on the fact that the difference in the relative permittivities of water and oil makes a capacitive transducer sensitive to the water content flowing between two electrodes. In the past 3 years, commercial instruments using this method have appeared and have attracted a great deal of interest, particularly in North Sea applications where water content could be a problem. They are basically also non-intrusive sensors with the potential for in-line installation. There are two basic designs, by-pass and main line types. By placing electrodes in a by-pass line from the main flow, a measure of water content in the main stream can be made. The accuracy of these meters, however, is highly dependent on the extent to which the flow in the by-pass loop is representative of the main stream. With the main line type, this source of error is reduced, but the problem of flow regime effects still remains. Both types are also affected by the excitation frequency, the salinity of the water content, the water droplet size, the thickness of the isolation material between electrode and fluid, and the homogeneity of the electric field.

Field experience has shown that meaningful measurements can only be made if the flow regime is constant and the best results are obtained when the water is evenly dispersed and in droplet form in the liquid. Such criteria often do not exist, with slugs of water passing down the pipeline at irregular intervals. One such commercial instrument is shown in Figure 9. The important feature of the design is the short distance between the electrodes, which results in high sensitivity. The electrodes are intrusive and therefore liable to either coating or blockage effects if the oil is particularly dirty.

Grab samplers are devices that take small volumes of fluid from the line at regular intervals to build up a representative sample of the mixture over a period of time. An international standard is currently being reassessed in the light of operating experiences by many oil producers. The current practice states that 10,000 "grabs" of around 1 cc each should be taken, and there are guidelines laid down for the sampling frequency and other criteria to be considered. Work at Cranfield and elsewhere has been evaluating this standard, and some of the limitations have been identified. As with the capacitance devices discussed previously, these may be installed either in the main stream or in a by-pass loop. Often signals from flowmeters in the line are used to set the sampling frequency, and the use of jet mixers, bends, pumps, and other devices that produce a more homogeneous stream is desirable to improve performance and accuracy.

Another variation, the fast loop sampler, is used for continuous monitoring of water content, and such a system is shown in Figure 10. The main advantage of the fast loop sampler is that flow is continuously extracted, and the sampler itself seems to have a smaller perturbing effect on the stream characteristics. A great deal of work is currently under way to examine the effect that the presence of the sampling probe has on the fluid dynamics of the stream. Early work has indicated that the flow streamlines around the probe head do have an influence on the droplet distribution and

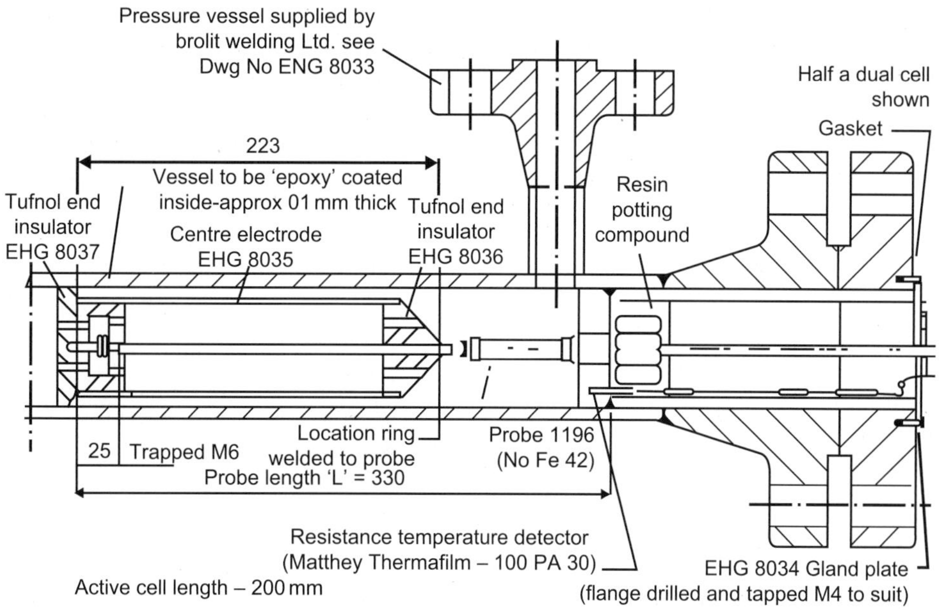

Figure 9. Capacitance-type water content transducer. (Courtesy Endress and Hauser).

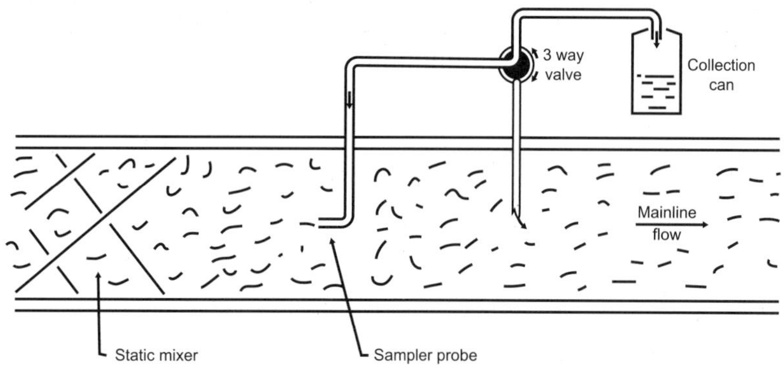

Figure 10. Fast loop pipeline sampling system.

hence the oil/water ratio. This is developing technology, and we should see increased activity in this area as some of the older North Sea fields begin to produce greater concentrations of water.

Pipeline monitoring systems

On-line inspection techniques have recently been developed to a very high level of sophistication. British Gas has spent some 12 years and more than £100 M producing intelligent monitoring tools and the data analysis software to inspect line sizes from 8-in. to 42-in. bore. There is a demand worldwide for their expertise, and with a potential 1 million miles plus of line to survey, they should be busy for years to come! These inspection tools are commonly called "pigs" but should be distinguished from the pipeline pigs used, for cleaning, dewaxing, batching, filling, and dewatering of pipelines. I feel the term *monitoring tool* should be used, because these devices are very advanced indeed.

Two types of monitoring tools have been developed, one based on magnetic sensors to detect metal loss and the second based on ultrasonics to sense defects and cracks in pipe walls. The two used together give a complete picture of the mechanical state of the pipeline. They are not continuous monitors like pressure, temperature, and flow transducers but are useful in supplementing the on-line measurements. If the lines are inspected at regular intervals, then changes can be identified and possible trouble spots pinpointed. The data can provide a very useful historical record to be used when maintenance, cleaning, or line repairs are needed.

The magnetic pipeline inspection vehicle operates by magnetizing a short section of line as it travels down the pipe, driven by the process fluid. If a section of pipe has suffered metal loss, it will cause a local distortion of the magnetic field. The data are recorded on 14 channel tape systems, and onboard software reduces the amount of data saved to only that where significant signals are noted. The information from the clean or undamaged pipe is not saved. Once the inspection run is complete, the data are analyzed to identify the problem areas. The first offshore run was completed in early 1986 when the Rough and Morecambe gas field lines were surveyed. This run took around 5 hours. The method can be used to survey sections of line around 300 km long, although a more typical single pass would be around 80 km (50 miles) (*Monitor*, No. 2, 1986).

The ultrasonic system is virtually identical except for the transducers and detectors. The sensors are ultrasonic, operating in the megahertz region. They are housed in wheels that press against the pipe walls. They are pulsed from an on-board electronic control system, and the reflected waves are stored in the same way as in the magnetic system. Both inspection vehicles travel at around 3 meters per second. If a line problem is detected by an imbalance between flowmeters, for example, the location could be confirmed by the acoustic detectors on board the inspection vehicle. This could be arranged to alarm when the detector output reaches a maximum, and the precise location can be given by radio transmitters mounted on board.

Other problems that can be identified include corrosion, fatigue, ground movement (causing pipe buckling), and third-party damage. The inspection center at Cramlington has already undertaken a number of overseas surveys and has experience in gas, crude oil, kerosene, and gasoline lines. The impressive part of these pieces of equipment is the engineering. They have to negotiate bends, valves, crossovers, etc. and still protect the relatively fragile electronic recording equipment from damage. They have been used both in on-shore and off-shore lines with equal success. This is one area of pipeline instrumentation development that I will be watching over the next 2 years as new on-line vehicles are developed for pipe coat-less monitoring.

Computer systems

It is in computer systems that the greatest amount of data can be gathered, analyzed, and acted upon in the shortest amount of time. Most operators now have some form of computer-based monitoring system employing either commercially available or custom-developed software to run the system. Programs can calculate the inventory of the line at any point in time and compare this with line measurements to provide independent cross checking of pipeline measurement data. These data can be transmitted by telemetry, radio, or telephone links to a central computer that continuously monitors the "health" of the line. The central operations room is now more likely to contain VDUs rather than banks of process indicators and enables the engineers to have rapid and accurate updates of conditions along the entire line or network. By changing programs and subroutines, a large number of tasks and functions can be performed easily, simultaneously, and cost effectively.

The many functions that can be performed by computer-based systems include:

- Leak detection and location
- Batch tracking of fluids
- Pig tracking
- On-line flow compensation
- Real-time pipeline modeling
- Instrument data and malfunction checking
- Inventory and material balance accounting

Such systems have very rapid responses but are dependent on the accuracy of the input data to be effective. Most computer-based systems consist of two major elements, namely:

1. A supervisory computer plus associated software
2. A number of independent pipeline monitoring stations plus data transmission equipment for communication with the computer

The complexity of the software, the number of input variables, and the pipeline instrumentation selected all influence the size and the choice of computer. Some systems are capable of running on the compact powerful micros such as the IBM AT or XT, while the more complex pipeline networks will require a mainframe machine, possibly communicating with satellite micros installed close to the measurement stations. The overall design time varies with each system but should not be underestimated, especially if SCADA systems are being developed in house (see section following).

At the most basic level the computer system can take flowmeter inputs to perform gross balances along the pipeline. Such a method is very useful in identifying small but consistent loss of product from corrosion pits in the pipe wall. This will arise if the integrated flowmeter outputs diverge with time when the flow in the line is maintained constant. If the line flow rate varies with time, then imbalances are more difficult to detect, since the meters at each end may have different characteristics or their outputs may vary nonlinearly with flow rate. A loss of product will be identified simply as the difference between the steady state inventory of the system and the instantaneous inlet and outlet flows. Mathematically, this is:

$$\Delta V = V_{in} - V_{out} - V_1$$

where ΔV is the leakage volume, V_{in} and V_{out} are the metered inlet and outlet flows, and V_1 is the inventory of the

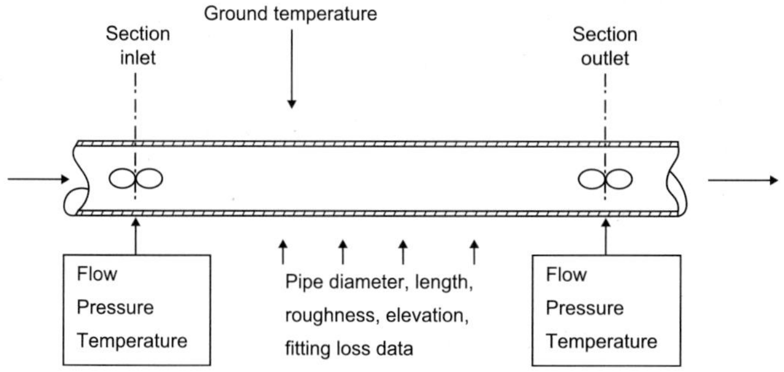

Figure 11. Input data for computer inventory calculation.

pipeline. This last term can be calculated as the average of the integrated flowmeter signals but is dependent on variations in temperature, pressure, and many other variables. It is the use of computers that enables this last term to be calculated on line and as a function of pipeline elevation, pipe material characteristics, process variations, pipeline fitting loss data among the many variables. Temperature and pressure variations change both the density of the fluid and the actual dimensions of the line. Both will give rise to sources of errors in pipeline inventory if compensation is not included in the calculations. Figure 11 shows typical input data requirements for simple on-line inventory calculations.

At the other extreme, computer-based systems can control entire networks with almost no need for the operators to intervene unless serious problems arise. Figure 12, taken from *Oil and Gas Journal*, March 3, 1986, shows two examples from the many hundreds that exist of computer-based monitoring. The benefits from the addition of real-time pipeline models cannot be overemphasized. Even when operating transients occur, some commercially available programs enable problems to be detected. An example is shown in Figure 13. Here a pipeline leak was induced following pump failure and restart, and although the line was operating under highly transient conditions, a material imbalance was detected. Real-time modeling also allows density profiles to be calculated along the line. It is fairly common practice to have amounts of different fluids in the same pipeline separated by a pig or sphere. From the pressure drop/flow characteristics, the real time can track the progress of these batches of fluids along the pipe.

SCADA systems

Real time computer modeling of pipelines is a technique that uses the full data gathering capabilities of modern digital systems and the computational power of computers to give accurate "snapshots" of the line. These enable the process operator to act quickly when problems develop. The whole system is under the control of the SCADA (Supervisory Control And Data Acquisition) system. This suite of programs polls the data stations on the line in turn, processes the data, controls the running of any pipeline model, and activates alarm and other monitoring routines. Most modern SCADA systems now include checking routines to determine the validity of the data before any calculations are made. It has been found that the use of continuous quality and cross checking of data improves the sensitivity of the system. In addition to these basic functions, some of the more advanced SCADA systems might include predictive models to analyze "what if" operating scenarios, include automatic startup and shutdown procedures, or provide optimization routines for least-cost operating strategies.

There are a number of companies supplying standard SCADA packages, but the majority of systems use custom written master routines to enable the commercially available programs to be used. Each pipeline is different, in line characteristics, instrumentation installed, fluid properties, and need of the user to name but four requirements. Experience has shown that the adaptation of standard software modules to specific needs takes much longer than is first estimated and also gives rise to the greatest number of problems when pipeline monitoring and control systems are being commissioned.

It is very important for the software specialist to communicate directly with the instrument engineer who specified the hardware, to ensure proper integration of the various parts of the system. All too frequently these two key members of the design team do not meet until the project is well under way.

Cathodic protection

Protecting a pipeline from corrosion is an important and expensive part of building and running a pipeline system safely. Every year, operators spend millions of pounds on pipe coatings, gas detectors, and sacrificial anodes to lengthen the effective life of pipelines by reducing the effects of corrosion. Whenever a steel structure of any type comes into contact with the soil and/or water there is almost certainly going to be corrosion. The structure, in our case the pipeline, divides itself into anodes and cathodes with corrosion taking place at the anode. This chemical process is accompanied by the flow of electric current from one point on the line to another via the ground or water. There are two basic preventative methods, sacrificial anodes and impressed currents (more commonly called cathodic protection). This second method is the more common and preferred method for long pipelines to identify the problem spots.

If a voltage is applied to a section of line, a profile of the current flow can be drawn up by measuring the current flow at intermediate points. This gradient will show areas where current is escaping, indicating a pipe defect. Recently, techniques using AC current have been perfected where the

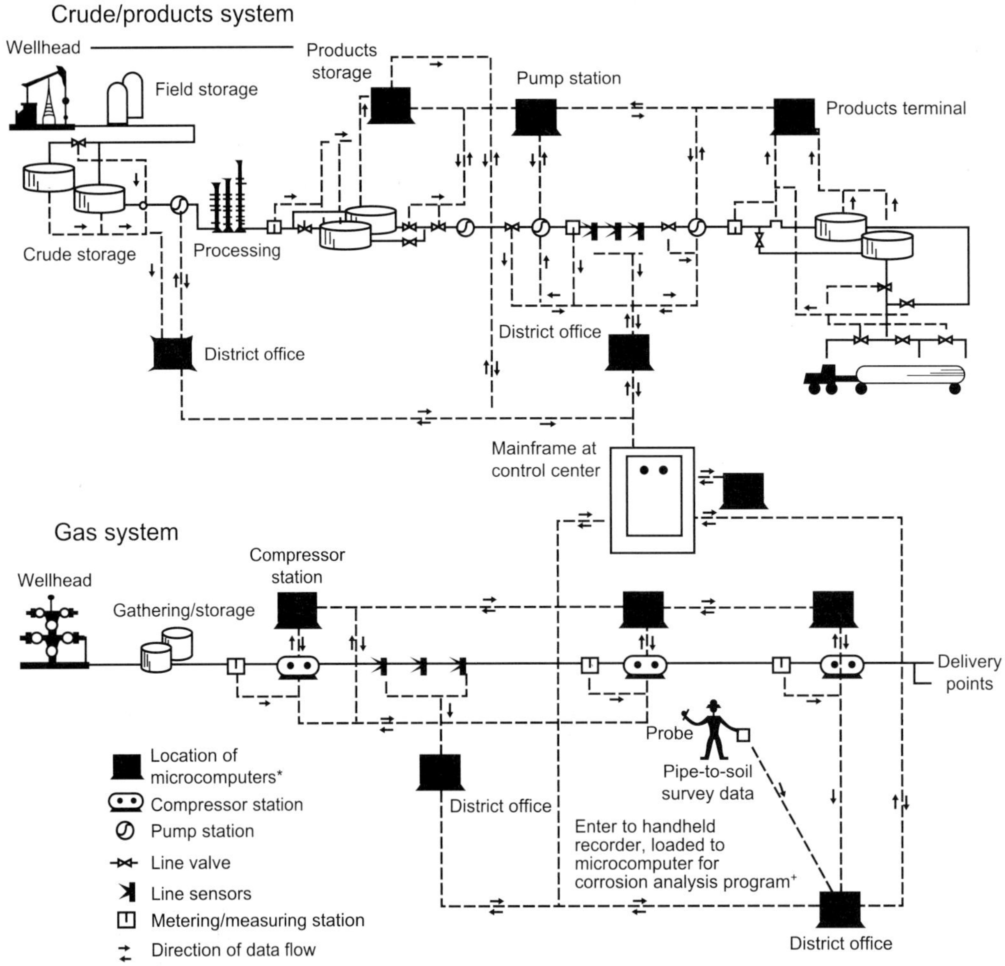

Figure 12. Examples of modern computer-based pipeline. (Courtesy Oil and Gas Journal, 1986).

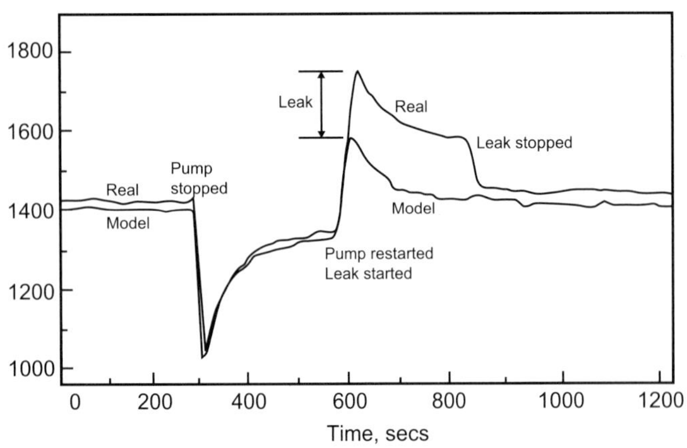

Figure 13. Example of transient real time modelling of pipelines. Data from Ref 9.

magnitude of the current can be sensed inductively from above ground.

This method can be used by pipeline crews walking the line, and compact lightweight handheld units store data on the mechanical condition of the line in lengths as short as 5 meters. On long cross-country lines the profile can be plotted at intervals of a few kilometers, and from historical cross-comparisons, line changes can be identified.

The object of cathodic protection is to prevent current flow from the pipe. The operating principle is to apply a voltage in the opposite direction to the natural potential of the pipe. The pipeline voltage is made negative with respect to earth so that in theory the voltage and current flowing from the cathodic protection system balance the corrosion current that would otherwise flow. Recent developments include the use of solar panels and converters so that energy from the sun is used to run the system. One of the largest examples of this new technique is the Sarir to Tobruk in Libya, where a 32-in. diameter, 319-mile pipeline is protected by such an installation.

Remote station monitoring is easily accomplished with the data being telemetered back to central control rooms many miles from the location of the actual installation. Continuous automatic control is often performed by means of thyristors. Such continuous monitoring is now routine on new lines that are being constructed, and many older lines are being fitted with some form of cathodic protection system. In most cases it is probably more important for the older lines to be surveyed than the newer ones, especially where different fluids have been transported over the lifetime of the line.

System design guidelines

The aims and requirements of a project should be specified as fully as possible as early as possible when designing a pipeline monitoring system. Based on personal experience, the following points should be borne in mind. They are not in any special order but do form a useful checklist when starting system development.

1. Make the system as simple as is practical but do not underspecify the needs for the sake of simplicity.
2. Select instrumentation and pipeline equipment based on performance and not economic grounds. It is better to install fewer high-quality pieces of equipment than numerous poor ones.
3. Equipment compatibility is important. Use transducers, interface modules, and other pieces or hardware that have standard communications protocol. Custom-designed systems invariably put up development costs and do lead to endless sources of trouble and frustration.
4. When calibration and maintenance are due on the system, do not put them off. Wherever possible use computer software to cross check data to aid in the identification of problems. This will help to reduce operating and maintenance costs.
5. Where system availability is important, install devices that are self checking, self diagnosing, or dual systems where appropriate. The initial cost may be higher but the benefits may be more economical in the longer term.
6. Seek independent references, user experience, or validation of the equipment chosen. Most pieces of hardware perform differently in real applications to the specification published under ideal conditions.

Future trends

Future trends are in one respect easy to assess, but it is very difficult to estimate where we will be in 10 years' time. There is no doubt that the application of computer-based technology is going to proceed at an ever-increasing speed to both the instrumentation and pipeline monitoring systems. There will be reliable and rugged intelligent flow, pressure, valve, and other hardware components appearing, and once the initial testing over the application has occurred, the second-generation systems will give the desired performance, if previous history is anything to go on.

By far the biggest problem facing the whole spectrum of control and instrumentation is to develop standard protocols to facilitate communication between instruments. The situation at present means that most, if not all, monitoring computers must have a multitude of interface boxes scattered through the system to enable some elements of the system to "talk" to others. It is sincerely hoped that manufacturers will develop microcomputers that are able to communicate directly with each other, with full-scale industrial process monitoring and control systems and with the individual process sensors and associated controllers. The more elements there are in the system, the greater the potential of line downtime. Some moves are being made in this area, but it could be some time before we see commonality of process protocol.

The most rapid area of advancement has been in the microchip. Oil pipeline operators are among the many people who have been watching the development of 32-bit machines. The Intel 80386 chip for example, now contains more than 275,000 transistors in one piece of silicon. This device has an operating speed of 4 million instructions per second and has the capability to address over 4 gigabytes of main memory. One other manufacturer is working on a computer that will perform 10 billion calculations per second, more than 10 times faster than any other computer on the current market. Such machines will permit easier real-time modeling of pipelines and will present the line operator with many extra pieces of data on the condition of the line. Of course, complete new suites of software will be required for these fifth generation machines.

Conclusion

In such a short space, the paper cannot do justice or even mention the vast number of recent instrumentation developments that have had an impact on pipeline operation and control. The technology of measurement and control is proceeding at an ever-increasing pace, and it is now quite difficult to keep up with developments in one area. Pipeline instrumentation covers a very wide spectrum from pressure and surge control to intelligent instrumentation and advanced software. Certainly our "friend" the microchip has totally transformed our thinking in the past decade, and the poor pipeline operator now needs specialized help in specifying a system to meet his needs. There are many options open: control using PLCs at sources, control using software in the process computer, and manual override if required. With the development of expert systems, systems can now be programmed to almost preempt problems before they arise and take appropriate corrective action. The flexibility in the industry is vast, but one thing is sure, you cannot control unless you measure accurately, and all too frequently the primary elements of a system are under-specified compared to the rest of the system. It is of very little use, for example, to pressure and temperature compensate transducer outputs when the basic measurement uncertainty is poor.

References

Adams R.B., "Buckeye modernizes computer operations and SCADA system," *Oil and Gas Journal* 11 (August 1986) 55–59.

Anderson C.C., "The Ekofisk pipeline control system," *Pipes and Pipelines International* 30 (No. 3) (1985) 21–25.

Baker R.C., Hayes E.R., "Multiphase measurement problems and techniques for crude oil production systems," *Petroleum Review* (November 1985) 18–22.

Chester J.R., Phillips R.D., "The best pressure relief valves: simple or intelligent," *In Tech* (January 1985) 51–55.

Covington M.T., "Transient models permit quick leak identification," *Int Pipeline Industry* (August 1979).

Davenport S.T., "Forecast of major international pipeline projects for the 1986–2000 period," *Pipes and Pipelines International* 31 (No. 1) (1986) 9–14.

Furness R.A., "Modern pipeline monitoring techniques—Parts 1 and 2," *Pipes and Pipelines International* 30 (No. 3) (May and September 1985) 7–11 Vol. 30, No. 5, 14–18.

Kershaw C., "Pig tracking and location—Parts 1 and 2," *Pipes and Pipelines International* 31 (No. 2) (1986) 9–15 Vol. 31, No. 3, 13–16.

Stewart T.L., "Operating experience using a computer model for pipeline leak detector," *Journal of Pipelines* 3 (1983) 233–237.

Taffe P., "Intelligent pigs crack pipe corrosion problems," *Processing* (September 1986) 19–20.

Tinham B., "Micros and sensors make smart transmitters," *Control and Instrumentation* (September 1986) 55–57.

Williams R.I., "Fundamentals of pipeline instrumentation, automation and supervisory control," ISA *Transactions* 21 (No. 1) (1982) 45–54.

Reprinted with permission - *Measurement + Control*, Vol. 20, February 1987.

Automatic tank gauging systems and proof testing

Excerpts from Chapter 6 of the Endress+Hauser Guidebook for Overfill Prevention & Tank Gauging.

Tank gauging systems are employed in large storage tanks, pipelines, depots, refineries, airports, power plants, chemical plants, mines and elsewhere (Figures 1, 2). Making a liquid level measurement in a tank is called tank gauging. When the measurement is made by permanently fixed tank instrumentation, it is called automatic tank gauging. Automatic tank gauging (ATG) can provide continuous level measurement and record the level changes with time. ATGs can transmit the level data to remote locations but this feature is not always used. The measurement can usually be read at the side of the tank at grade level but is also commonly transmitted to a central control room where all tank measurements are readily available on computer display screens.

API standards for automatic tank gauging

The most widely used tank gauging standards in the petroleum industry are API Ch.3.1B[1]. Other relevant automatic tank gauging API documents are:

- Chapter 3.3. Standard Practice for Level Measurement of Liquid Hydrocarbons in Stationary Pressurized Storage Tanks by Automatic Tank Gauging
- Chapter 3.6. Measurement of Liquid Hydrocarbons by Hybrid Tank Measurement Systems
- Chapter 16.2. Mass Measurement of Liquid Hydrocarbons in Vertical Cylindrical Storage Tanks by Hydrostatic Tank Gauging

An automatic tank gauge or ATG is a method of determining the level of liquid within a tank without human intervention within a tank. Automatic tank gauging tends to be more reliable than manual tank gauging because of the reduction/elimination of human error. The reliability of an ATG is dependent on the instrumentation and system, their calibration, installation and maintenance of that instrumentation and system.

Where an automatic tank gauge is used, an automatic tank thermometer (ATT) system should be installed. As in manual tank gauging, the temperature must be known to convert the experimental volume to standard temperature volume. Remote monitoring systems use data from the ATG and ATT to provide inventory information in real-time. This system and information is valuable and critical for inventory control and tax purposes. An ATG system generally includes:

Figure 1. Storage tanks at a petrochemical facility.

Figure 2. Spherical storage tanks.

- A contact or non-contact level sensor which measures either the height of the liquid in the tank (innage) or distance between the liquid level and the reference point (ullage)
- A local display for operators to see the level
- A transmitter that sends data to monitoring computer
- A remote readout device that transmits information to a central data system

Factors influencing the accuracy of tank measurements:

- Accuracy of tank capacity/strapping tables
- Bottom deformations
- Thermal expansion of tank diameter
- Errors in density, temperature, or level measurements
- Incrustation
- Movement of the reference point
- Temperature affects the density of the liquid and therefore affects the volume of liquid but not the accuracy of the level measurement (the thermal expansion is accounted for in custody transfers)
- If an ATG is used, then it must utilize appropriate measurement techniques for the liquid product it is

measuring (a float and tape gauge should not be used to measure viscous liquids)
- Operators should allow for settling time prior to taking a gauge reading to prevent inconsistencies, especially when monitoring a custody transfer

It is essential to accurately calibrate and verify the performance of an ATG. Verification if performed by operators who will adjust the liquid level in the tank to three distinct levels and record the results of the ATG and manual gauge. Note, the manual gauge should be calibrated and verified based upon MPMS Chapter 3.1A [2] prior to the ATG verification process.

At each level, the ATG measurement shall be recorded and a manual tank gauge measurement shall be recorded. This process is repeated for the three liquid levels selected by the operator. After this process is concluded, the difference between the manual gauge readings and ATG readings shall be evaluated. An ATG used for internal inventory purposes should be verified on a quarterly basis, and an ATG used for custody transfers should be verified on a monthly base.

Verification of an ATG is defined by API MPMS Ch3.1B as the process of manual gauging the liquid level in a tank at three positions and recording the measurements obtained for the ATG and manual gauge. The ATG must be within 1 inch (25 mm). The difference between the manual gauge and ATG should be less one inch (25 mm). If the ATG does not match the manual gauge reading, consider a different type of ATG, installation parameters, or accuracy of the manual gauge.

Overfill prevention systems

An Overfill Prevention System detects, indicates, and prevents hazardous overflow levels in storage tanks. After the Buncefield and other incidents, systematic overhauls to overfill prevention systems took place and industry best practice for managing storage tanks now combines the existing API 2350 prescriptive standards with the IEC 61511 functional safety standards.

The IEC 61511 functional safety standards describe Safety Instrumented Functions (SIF) that are designed to prevent or mitigate a hazardous event by taking a process to a tolerable risk level. This is the core function of every Overfill Prevention System. A SIF is composed of a combination of sensor(s), logic solver(s), and final control element(s). A SIF has an assigned Safety Integrity Level (SIL) depending on the amount of risk that needs to be reduced.

One or more SIFs comprise a Safety Instrumented System (SIS). To ensure safe functionality and meet the highest safety standards, every SIS must be independent of all other control systems controlling the equipment. All control elements including level devices, controls and alarming devices must be dedicated exclusively to the SIS. Therefore, an Overfill Prevention System must be completely independent of all other process control systems and devices.

Safety instrumented systems

Mostly because of the Buncefield incident and the UK HSE demand for better tank overfill performance through regulatory channels of API 2350 included the use of Safety Instrumented Systems (SISs) which are called AOPS (automated overfill protection systems) in API 2350. These systems are like what has been done in the past where tank receipts have been terminated by a valve closure. However, SISs compliant with industry standards such as IEC 61511 safety standards is a major challenge for many owners and operators. Using AOPS warrants a careful and thorough review by management and engineers to ensure that the implementation is realistic and will work.

When the risks are sufficiently high given a conventional overfill protection system and risk mitigation beyond the ordinary and traditional tank overfill protection methods is not sufficient to reduce risks to acceptable levels, then AOPS is one way to further reduce risks. The detailed discussions of AOPS can be found in industry standards such as IEC 61508 and IEC 61511. These standards are 'safety standards' and follow the ideas of ISA S84 which address the lifecycle management of 'Safety Instrumented Systems'. These systems for tank overfill protection are simple in concept but a monumental challenge to accomplish in fact. This is because of the complexity and comprehensive implementation requirements required by these standards. Without going into all the details, it can be said that it takes from 1 to 2 years for large companies to develop an appropriate tank overfill protection standard that incorporates the safety component of API 2350.

A major consideration for these safety systems is that for a pipeline company the risks of hydraulic transients resulting from closure against flow increases the risks of more incidents. So, a balance between the various risks must be obtained by considering all risks and optimizing the risk reduction solutions. For example, in some cases the risk may be sufficiently low that AOPS is not needed and the risks can be controlled by operations. In other cases, the scenario that is preferred is to have an overfill occur in the terminal as opposed to rupturing the pipeline at some unknown upstream location. Still, in other cases, the best solution to prevent a spill over water may be to properly design the AOPS for all conditions so that the receipt may be safely terminated in an emergency without overfills and without blowing piping components. This may require extra and unused tank ullage, longer response times and so forth. There is no single optimal answer and each specific situation for the tank and its overfill scenarios must be addressed individually if optimal benefits are to be derived.

One should be able to see that there is much analysis and optimization required to accomplish these goals. This takes much discussion with knowledgeable stakeholders and appropriate analyses to be able to have robust systems that have been well thought out and universally agreed to by stakeholders and which can stand up to the questions of regulators. Because there are tradeoffs in costs and benefits, the solution usually requires these basic optimization elements of discussion, analysis, more discussion, consensus, and a plan for implementation.

This is achieved by consulting all relevant parties including independent subject matter surveyors as well as suppliers. Some activities that are required to accomplish a uniform standard for tank overfill protection that would stand up to the scrutiny of internal corporate managers, regulators and the public or peer level organizations the following should be considered:

- Define the management system in detail for tank overfills. This management system could be incorporated into the corporate management systems or a separate system could be implemented under the umbrella of the larger corporate management system.
- Select and develop the type of risk assessment systems that will be applied to the tanks. This involves probabilistic estimates of overfills as well as consequence analysis. Because the fluids in many cases are Class 1 liquids, vapor cloud formation and the size of the vapor clouds must be estimated as a function of time. Many other fire related conditions must also be considered. In addition, if there is no fire then the impact of crude spills on the environment and impacts on other receptors must be developed.
- Set up a meeting to discuss exactly how tank overfills are controlled today and what the changes will be for the new systems.
- Implement a new standard that addresses all the current issues that arise because of using the current edition of API 2350 including a rationale for why AOPS was used or not used, under what conditions, and how risks are mitigated in other ways.

Addressing the entire safety loop

The API 2350 standard prescribes methods by which owners and operating personnel can prevent tank overfills. It requires the use of a risk assessment system. Functional safety itself can only be applied to complete functional loops. Therefore, the complete Overfill Prevention System elements (including level devices, controls, and alarming devices) must be certified according to functional safety standards. Such system usually fulfills the requirements of levels SIL2 and SIL3.

An Automated Overfill Prevention System (AOPS) per definition is directly connected to the valve or pump. This means, if the system detects a High-High level alarm, the valve will be closed and/or the pump will be stopped immediately. Usually this happens quickly (typically less than 1 minute).

An Manual Overfill Prevention System (MOPS) per definition is a system which requires human intervention to terminate the flow of liquid into a tank. For a MOPS the worst-case Response Time must be calculated by considering the following additional factors:

- the maximum communication time needed for the notification of the personnel who can respond (close the inlet flow)
- the time the personnel need to analyze the situation
- time needed to initiate the shutdown (close the valve)
- time required to complete the response action (termination of receipt)
- time to verify that all elements are reacting correctly and the required time needed if they don't
- a safety factor.

According to API 2350 minimum response times are defined depending on the tank category. For a "fully attended" Category 1 tank the minimum response time is defined as 45 minutes. A Category 1 tank is a so called "fully attended" which means:

- Attend at least 1h prior to start of filling
- Attend at least 1h after the receipt
- + periodically visual check the tank gauge on site, but do not manual hand dip until at least 30 min after a product receipt!

This means there are always people present in the field. Typically, there is only one transfer at a time in a tank farm with Category 1 tanks. For a Category 3 tank "Automated Tank Gauging with an independent Overfill Prevention System" the minimum response time is defined at 15 minutes, if it can be proved that the personnel are able to terminate the inlet flow in all circumstances within this time!

Proof testing

Even though rigorous testing of instrumentation control loops is often considered in the context of safety instrumented systems, the principles apply to any control system and these kinds of tests should be used for any overfill protection control system regardless of vintage.

The purpose of proof testing is to ensure that a control system works under realistic conditions and to find faults and problems so that they can be corrected. The act of proof testing also provides information that enables the owner/operator to re-assess the system reliability over time. Common sensor variables are pressure (or differential pressure), temperature, level, pH, density, speed (RPMs), etc. Proof testing is commonly referred to as a "wet probe test" because the probe or sensor experiences a process change that is sufficient to drive the loop into its alarm or other designated function. Ideally, a test of the complete control system should be done by changing the process variable (i.e. level) sufficiently to trip the control system sensor. However, this is ill advised for reasons discussed below.

The very first problem encountered with proof testing is to test a control loop from sensor to final element under realistic conditions. With instrumentation loops that use pressure, the loop test is relatively easy to conduct since the pressure sensor can be subjected to the appropriate test pressure which will then activate the control system and valves to see if they operate correctly. However, this should not be entertained as a practical methodology if the process is driven into a demand state or a state where an actual process hazard exists without a risk assessment to ensure that the risks are acceptable.

But most instrument loops for tanks are using liquid level height in the tank as the primary means of performing the alarm or final element function. One way to implement the proof test is to increase the physical liquid level to at least the alarm or final element set point. But this can be hazardous and is usually not easy to perform and can be also time consuming.

First, consider changing the liquid level in a tank to trigger an HH alarm. Running the liquid level into a restricted region (above the maximum normal working level) itself can be hazards and can result in an overfill. Strict control over this process would be tightly controlled by procedures and is certainly one way to conduct a proof test. It should be noted that most major oil company owners/operators consider this too hazardous to allow, especially where flammable liquids are involved.

Another option is to set the liquid level sensor trip at a lower liquid level to test the alarm or final element operation. This procedure is often the only way to proof test a continuous level sensor. But this has a few problems as well. Changing the set point of the sensor introduces the distinct possibility that the correct original set point is not properly reset. In addition, it may be possible that the alarm function works in the lower contrived set point but does not actually work at the original set point depending on the technology.

In general, a proof test of a high-level alarm function performed with a continuous level sensor can be complex and involves the risk of leaving incorrect settings after the test. Most issues with a measuring loop are connected to wrong installations and wrong configurations. Therefore, the simpler a sensor is to install and to configure the smaller is the chance that something is done wrong.

Yet another option is to test parts of the system at different points in time (something allowed by API 2350). For example, suppose that we have an alarm system which uses a displacer or float which hangs down on a wire from a sensor head which is mounted on top of the tank. If we did not want to increase the level process variable to the actual set point, we could test just the functionality of the displacer by lifting the wire connecting the displacer to the control head, which should activate the switch and the alarm. Indeed, manufacturers of these devices have often built into the device a physical lever that lifts the displacer so that a switch is actuated. But note that this approach can have failure modes. Since the lever lifts the displacement wire and displace physically upward, the test will be successful. However, if the displacer is made of a hollow metal object weighted to float on the surface or an interface, it could be that the weight is incorrect, and it will not actually rise with the liquid level. A common example of this is a corrosion hole that penetrates the displacer and causes it to fill with liquid. In this case, the displacer will never float and even though the test lever simulated most of the functionality of the alarm loop it missed the problem with the sensor. This type of failure is called an "dangerous undetected" meaning that it is hidden until a full test from the sensor to the output is conducted. It must be emphasized that each technology has its own set of hidden failure modes and subject matter surveyors and manufacturers should be brought to the table to discuss these.

To perform a "bucket test" the sensor will be removed (typically at the top of the tank) and put in a bucket with a fluid of similar density (Figure 3). The problem with all "bucket tests" are that the sensor can be damaged or lose its calibration and this type of failure will not be detected because of the removal/reinstallation process. Furthermore, the operator must climb tanks and is exposed to dangerous environments as the tank is "open" during the test.

State of the art point level sensors like a tuning fork level switch are equipped with continuous checking online diagnostic functions. Such sensors do not only detect and alert cable breaks and open circuits but also the correct operation of the electronics and of the vibrating fork itself.

Build up and corrosion can be detected by analyzing the frequency. Therefore, such sensors are very reliable and are available with the requested SIL certifications. Proof testing

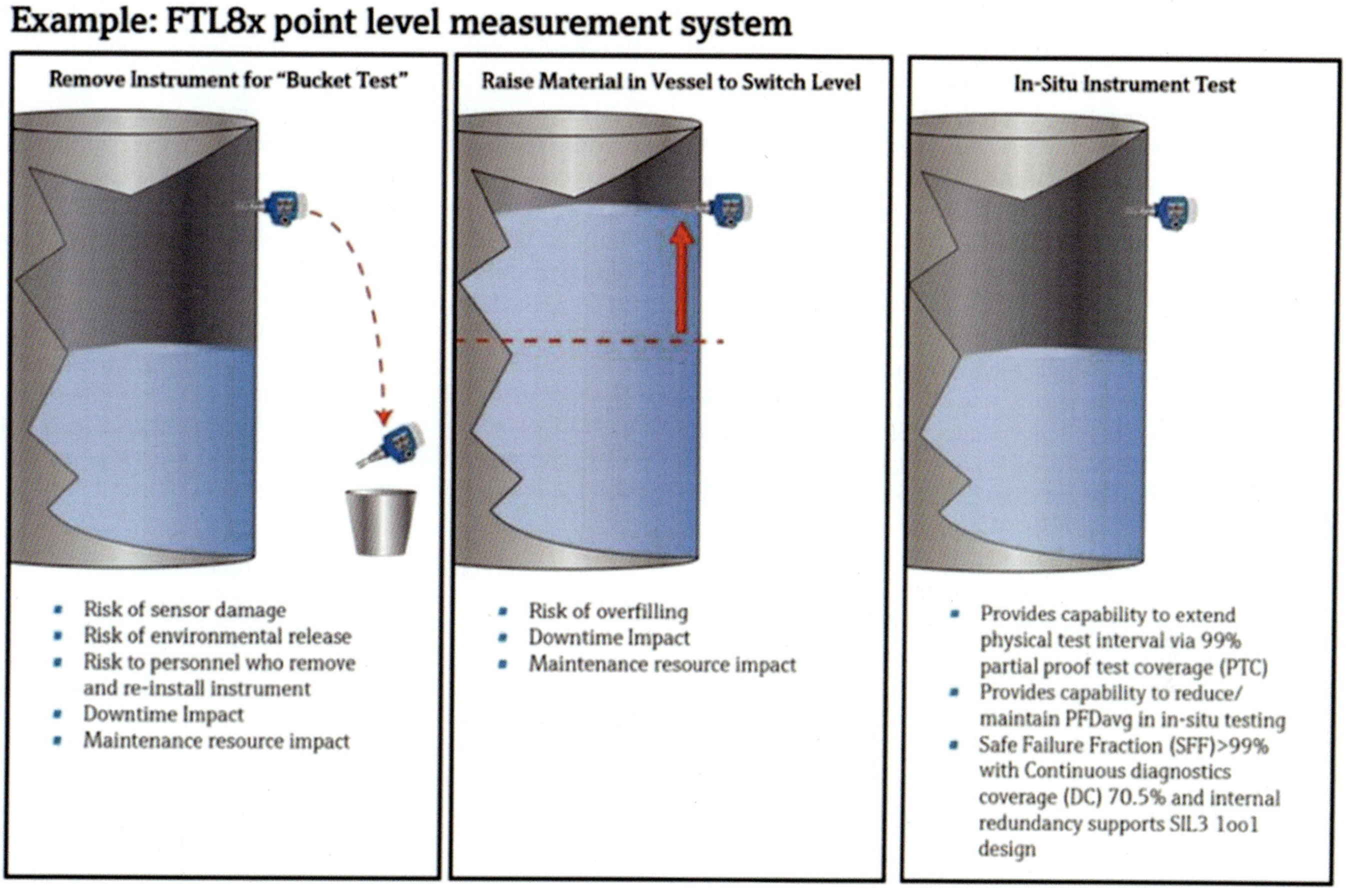

Figure 3. Methods of proof testing sensors.

is an important part of tank overfill protection and it can be technical. Subject matter experts and the manufacturers should be consulted so that the proof testing procedures produce the desired results.

Another part is the documentation of these proof tests. Any proof test needs to be formally documented in a report. In these reports the proof test results (pass or fail) as well as the date and time when the proof test was performed must be noted. These reports should be easily accessible and protected against tampering at the same time. Special attention must be given to the backup and restore concept of such report as they are needed most after any kind of event.

Independence and layers of protection

The idea of independence arose as far back as the first edition of NFPA 30-1984 and API 2350-1987 where it was required that tanks must be "equipped with a high-level detection device which is independent of any tank gaging equipment". API 2350 further defined that "An independent level alarm is an alarm function actuated by a primary level-sensing device that is separate and independent from any automatic tank gauging equipment on the tank." These were required on unattended facilities because it was recognized that overall system reliability was enhanced by the independence concept.

In the early days of API 2350 the level sensors or "detectors" which operated the ATGs were typically mechanical devices that had a high rate of failure. By employing an independent alarm, the failure of the ATG would not impact the alarm function. Moreover, the alarm function could be independently proof tested to ensure that it worked. However, there are still numerous systems that do not have an independent alarm and when there is a failure of the level sensor everything associated with overfill protection, except for the operator has failed.

The independence concept is still in use in the current edition of API 2350 as it is fundamental to system reliability.

Another good example of independence is the use of an Automated Overfill Prevention System (AOPS). Because an AOPS requires no human intervention to fulfill its duty it is independent of operations. For this reason, in various risk assessment methodologies, risk reduction credit is given to these systems as a "layer of protection", independent of other layers such as the operating layer or the emergency response layer. In all cases, this allows one to take full probability credit for reduced likelihood of system failure.

These ideas apply when the failure rates of the equipment (such as the Automated Tank Gauging System (ATG) are high. But with today's highly reliable equipment and sensors, the idea of independence is not so straight forward. So, suppose that you had perfectly reliable ATG that never failed, then there would be no need for the independent alarm. It could operate from the ATG and its sensor would never fail because the ATG never fails. While this is hypothetical, the idea shows that independence has variable value depending on how good or bad the instrumentation is.

A good example of this idea in practice is the removal of the requirement starting with the 4th edition of API 2350 to have both H and HH alarms; only a single HH alarm is required. This was justified by the concept that the use of redundant alarms was to enhance overall alarm reliability, but with the highly reliable and self-diagnostic alarm sensors and alarms today, there is no need to require redundancy in the physical alarm system function. The concept should obviously be supported by the tank owner/operator and the type and reliability of level sensing equipment judged to be sufficiently reliable for this configuration. To judge whether it is better to use a single alarm or the pre-4th edition really requires detailed calculations and management philosophy and decision making.

Another often debated topic involves one of degree. If the power that feeds alarm sensors is common meaning that a failure of the power system will fail any and all sensors connected to the power, then this is a "common cause" failure which is amenable to simplistic methods of analysis. However, if there are independent power sources supplying the sensors then are they truly independent if they run in the same conduit? If the conduit is destroyed in a fire or severed by a mistaken operation such as by a back hoe then the failure will cut both power supplies and they would not be considered "independent". These types of arguments can only be resolved by computation and common sense. For example, if the conduit is well protected and unlikely to be involved in a fire, then one could consider the electrical or signal circuits independent enough so that the two circuits could be considered "independent'.

One example of independence that arises is whether putting two radar level sensors in the same manway can be considered "independent" since they are in the same manway. One can argue that they are indeed independent because the reliability of a manway is so much more reliable than an electrical device. However, these debates come up it is important to understand the relative failure rates to make common sense judgements and decisions about what is most appropriate for the overfill protection systems.

Compliance

That an Overfill Prevention System complies to the current regulations and Standards and therefore fulfills the RAGAGEP (Recognized and Accepted Good Engineering Practices) at least the following points should be considered:

- The system hast to be completely (physically and electrically) independent of any other system.
- The system should be designed according to IEC 61511: "Functional safety" standards
- The system should offer an easy proof test function that can be operated remotely from the "outside of the storage tank" without the need of filling or emptying the tank and that covers the complete system from the sensor to the final element
- The system should be equipped with self-monitoring and failsafe sensors that offer highest on-line diagnostics.
- The system (including the sensors and actors) should be supplied by an uninterruptible power supply (UPS) to ensure highest availability
- The system should use diverse measurement devices (e.g., continuous level sensor for tank gauging and point level for overfill prevention) to reduces common cause failures

Typical applications for Overfill Prevention Systems are shown in Figures 4-8. When installing an emergency shutdown system like an overfill prevention system, it is essential to consider the complete process. For example, if a vessel is equipped with a pump to deliver hazardous fluids into a tank that may be located at a remote site, it is important to switch off the pump before closing the inlet valve to prevent overpressure in the pipeline. Sometimes the inlet flow will be diverted into another tank. Any use case must be examined with a safety expert during a risk assessment. As an outcome of such a risk assessment, clear working procedures and interface descriptions should be created to have a mutual understanding of the safety measures.

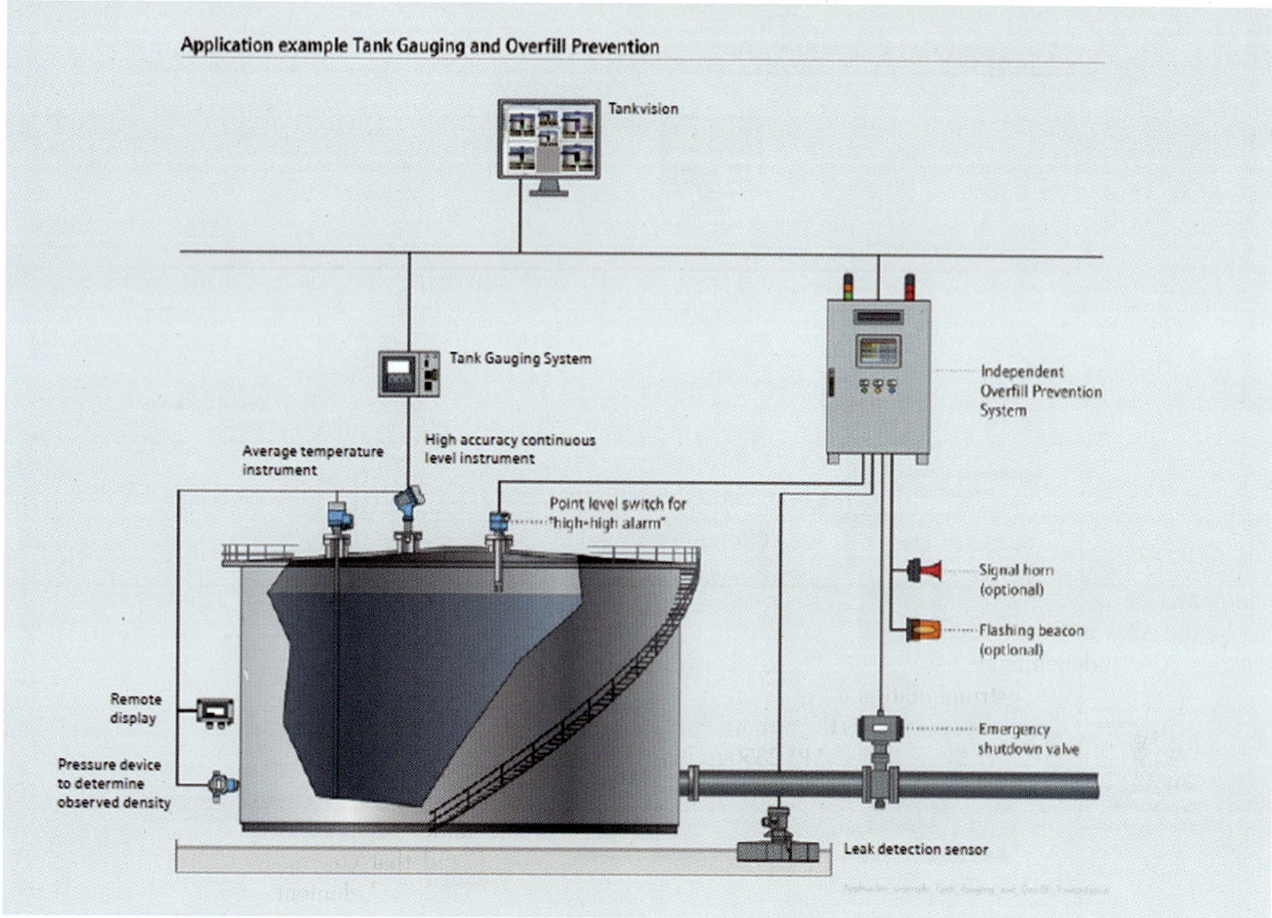

Figure 4. Typical architecture of a tank management system using a tank gauging system and independent overfill prevention system.

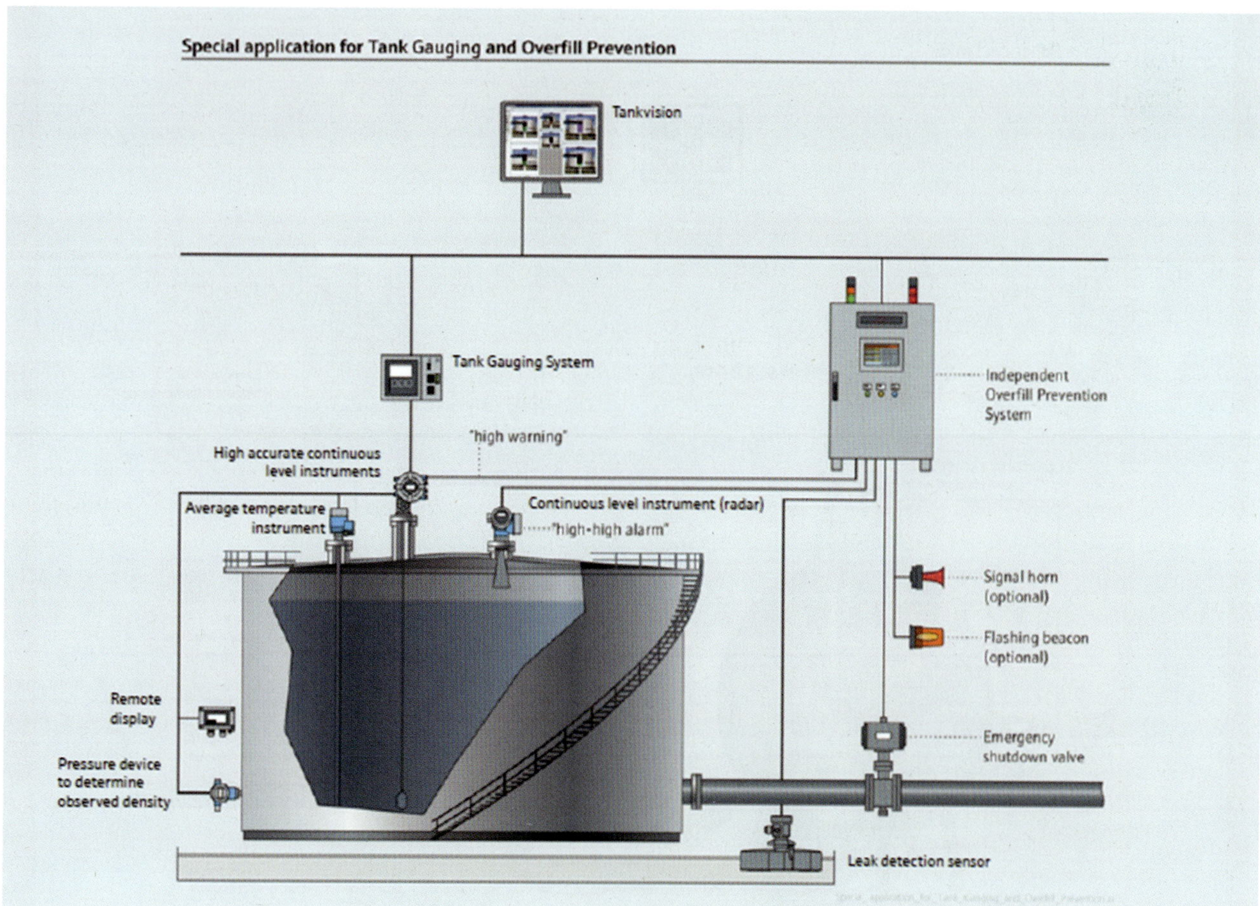

Figure 5. Setup for continuous level measurement with integrated tank gauging.

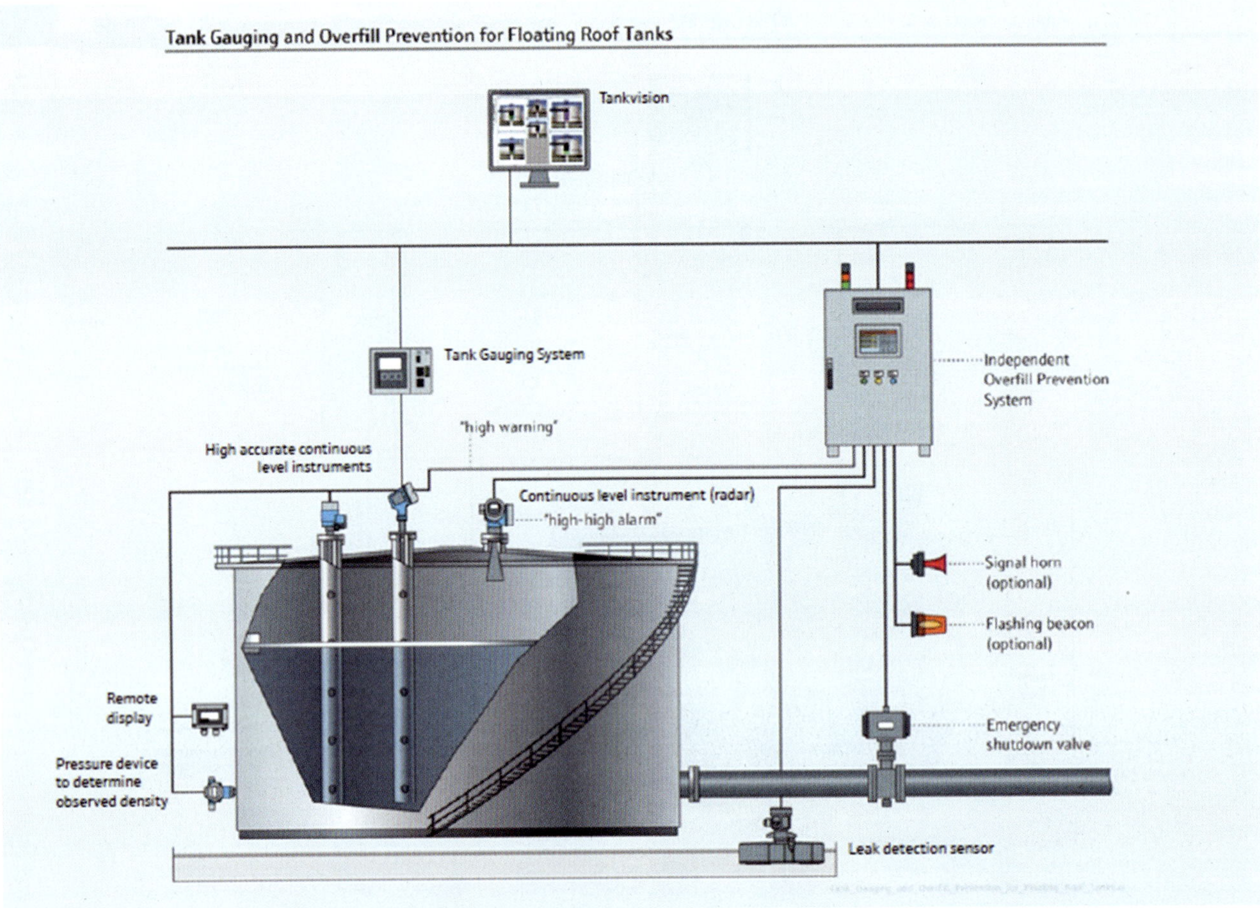

Figure 6. Setup for a floating roof tank.

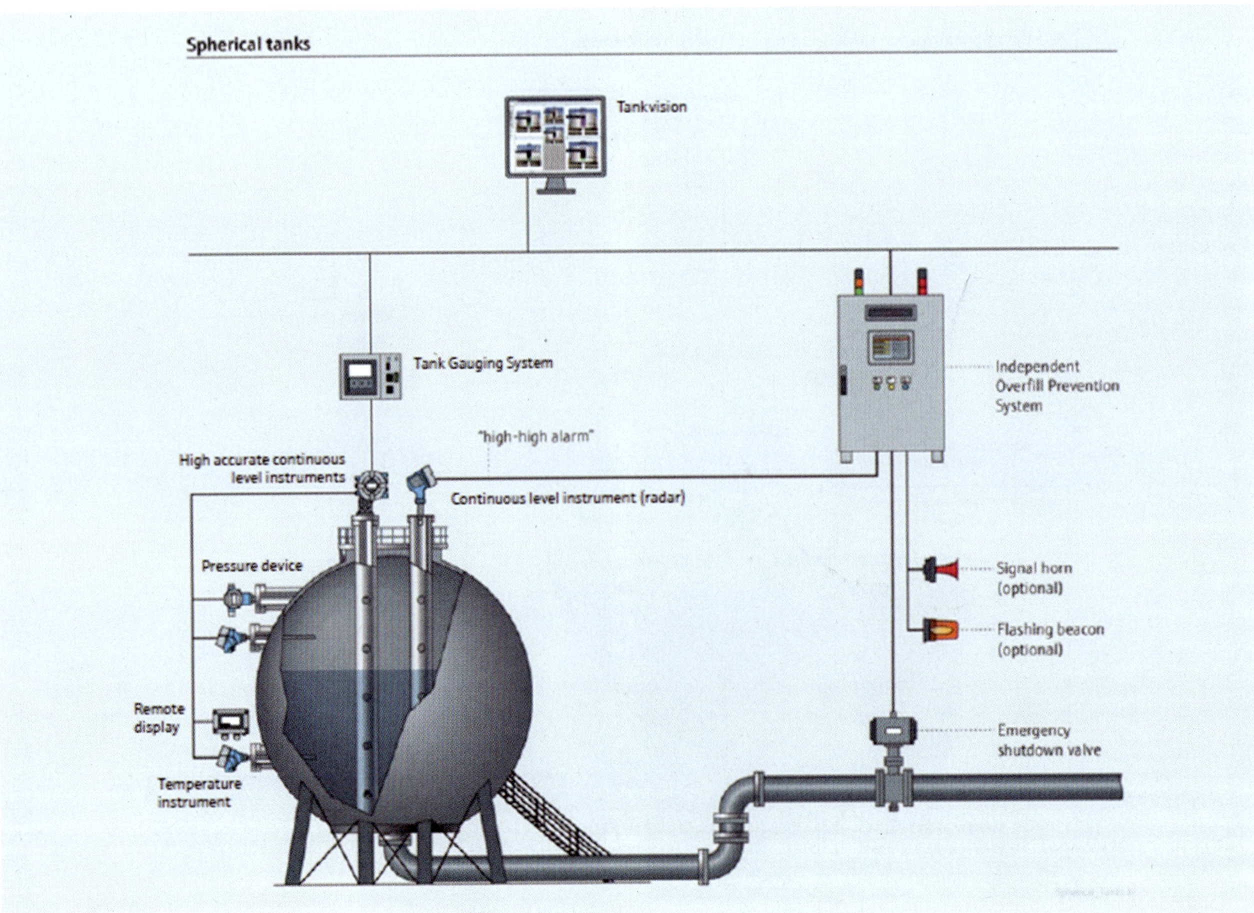

Figure 7. Setup for a spherical tank.

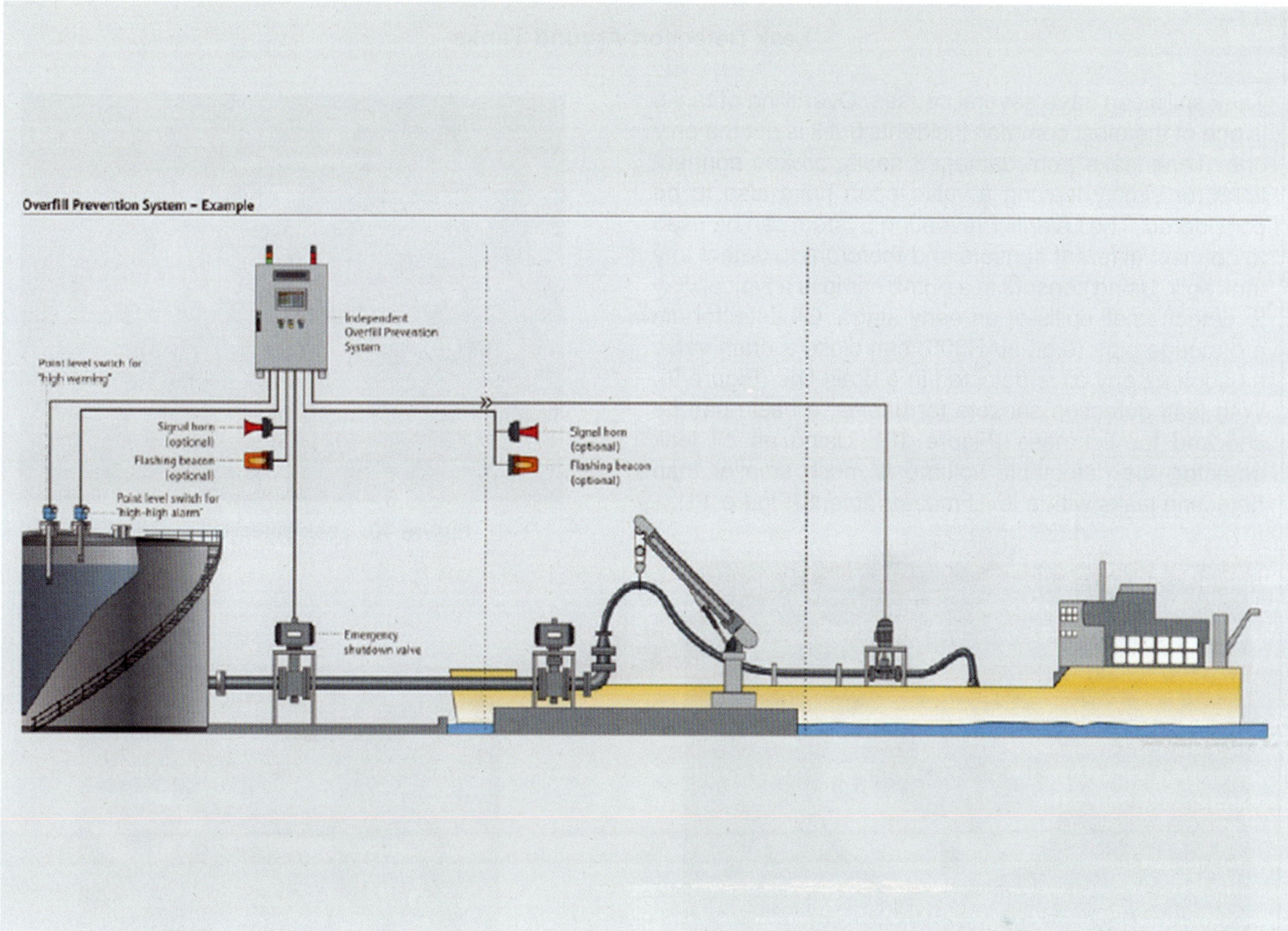

Figure 8. Complete setup for a tank farm.

Leak Detection Around Tanks

Tank spills can have several causes. Overfilling of tanks is one of the most common incidents but it is not the only one. Tank leaks from damaged seals, broken connections, or simply leaving a valve open have also to be considered. The Overfill Prevention System can be used to connect different sensors and therefore to detect any tank spill. Using sensors in a pump sump is a typical way to detect small spills at an early stage. Oil detectors in a drainage pipe (e.g., NAR300) can close a drain valve as soon as any oil is detected in a drain line (Figure 9). With leak detection sensors for basins, a basin can be checked for tightness (Figure 10). Using an oil leak detector, the detectable volume is much smaller than detecting leaks with a level measurement (Figure 11).

Figure 10. Leak detection sensors.

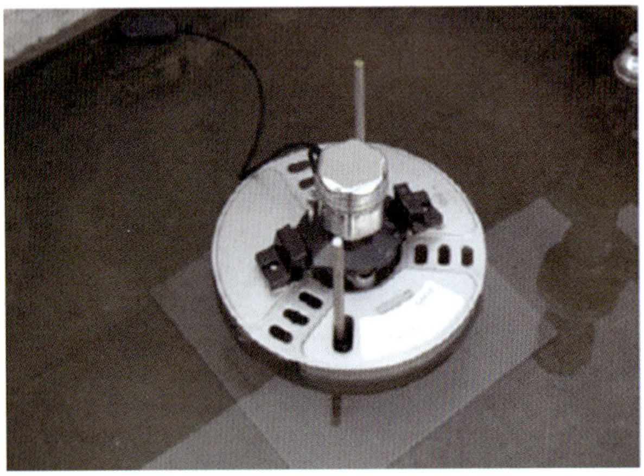

Figure 9. Oil detectors

Figure 11. Oil leak detector.

References

[1]. Manual of Petroleum Measurement Standards Chapter 3.1B, Standard Practice for Level Measurement of Liquid Hydrocarbons in Stationary Tanks by Automatic Tank Gauging, 3rd Edition (April 2018).

[2]. Manual of Petroleum Measurement Standards Chapter 3.1A, Standard Practice for the Manual Gauging of Petroleum and Petroleum Products, 3rd Edition (2013).

Source

Guidebook for Overfill Prevention & Tank Gauging, PEMY Consulting and Endress+Hauser: Philip Myers, Brock Trotter, Dean Mallon, Hai-Thuy Ngoand Markus Wohlgemuth. For the complete report, go online at: www.addresses.endress.com or www.pemyconsulting.com.

Instrumentation and sensors for automatic tank gauging

Excerpts from Chapter 6 of the Endress+Hauser Guidebook for Overfill Prevention & Tank Gauging.

Contact vs. non-contact level sensors

Automatic tank gauging relies solely on sensors. Contact measurement devices use a probe that must "touch" the liquid to make a measurement. Contact measurement technologies include capacitance, mechanical, electromechanical, hydrostatic, tuning fork, float, and servo measurement methods. Contact methods may be point-level or continuous.

Non-contacting measurements typically measure the time-of-flight (TOF) taken by energy waves to emit from the transmitter, bounce off the liquid surface and return to the instrument. These methods include ultrasonic, radar and radiometric. These measurement methods can be point-level or continuous. Some manufacturers are working with sonar to allow mounting the transmitter at grade and sending the signal to the liquid service where the TOF is measured to determine the liquid level.

Categorization of level sensors

Several level measurement techniques are shown in Table 1. The earliest measurement systems were Float and Tape, or else hand gauging tapes. Even today many tanks are manually measured by a tape measure inserted into a manway and measuring how much of the tape is wetted by product to obtain the liquid level.

Appropriate level sensor for common applications

The properties of the liquid being measured play a significant role in selection of the appropriate level control sensor. For example, highly viscous fluids such as asphalts and condensates are sticky. The sticking of the product fluid caused mechanical measurement methods to fail rapidly. Because of the sticking failure, a noncontact level measurement method, such as radar, is more appropriate sensor.

The cost associated for the available level sensors and technologies depends on the tank system and corporate desires. The cost of the sensor generally increases with technological complexity. The cheapest option tends to be a mechanical probe such as a float and tape system. The higher cost options tend to be electric noncontacting technologies such as radar sensors.

Each type of level gauging technology has a capital expenditure and operating expenditure. The capital expenditure is the cost of purchase, and the operating expenditure

Table 1
Categorization of Level Sensors

Sensor	Contacting	Non-Contacting	Mass	Point Level	Continuous Level	Interface	Density
Mechanical							
Float Switch	✓			✓		✓	
Float and Tape	✓				✓		
Electromechanical							
Magnetic Float	✓				✓		
Tuning Fork	✓			✓			
Servo	✓				✓	✓	✓
Electronic							
Capacitance	✓			✓	✓	✓	
Differential Pressure	✓		✓		✓		
Guided Wave Radar	✓				✓	✓	
Free Space Radar		✓			✓		
Ultrasonic		✓			✓		
Radiometric		✓		✓	✓	✓	✓

Table 2
Appropriate Level Sensor for Various Applications

Sensor	Viscous Asphalt	Crudes	Black Oils	White Oil	LPG/LNG	High Temp
Mechanical						
Float and Tape		Poor	Poor	✓		✓
Electromechanical						
Magnetic Float		Poor	Poor	✓		✓
Servo		✓	✓	✓	✓	✓
Electronic						
Differential Pressure		Poor	Poor	Poor		
Guided Wave Radar		✓	✓	✓	✓	✓
Free Space Radar	✓	✓	✓	✓	✓	✓

is the cost to operate and maintain the system, such as the electric costs to operate a guided wave radar system. Systems which use continuous electricity must purchase electricity for operation, but continuous level monitoring improves the reliability of the system and reduces the potential for an overfill event.

Consequently, more complex technologies tend to have higher reliability because contact with product fluid and mechanical contact is minimized. Product fluid can corrode and mechanical contact points can become lodged. Because of this, the trend between the reliability of the level measurement and sensor technology is like the trend in cost (Figure 1).

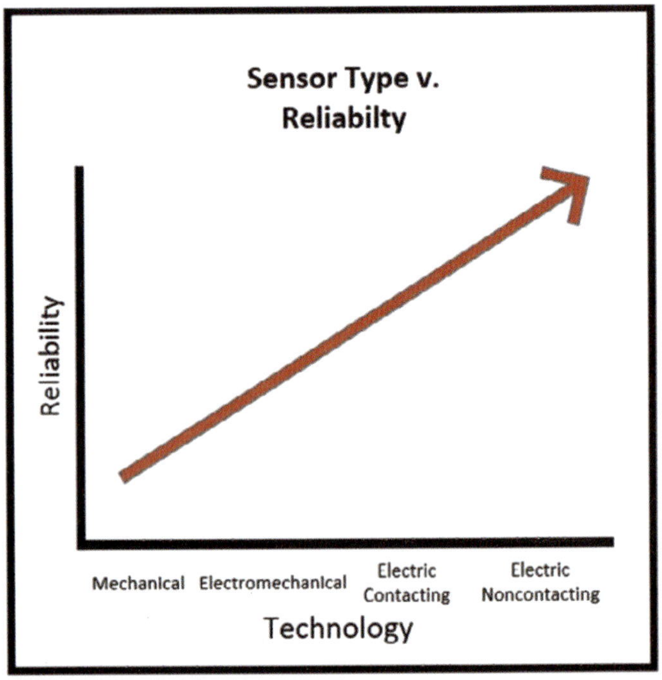

Figure 1. Generalized trend of technology and reliability.

Point level sensors

These are devices that display if the liquid level reaches a certain height in the tank or vessel. Point level measurement is binary information (i.e., the level is either below the sensor, or the level is at or above the sensor). So, point level sensors deliver binary information (ON/OFF), displaying whether the level has passed a specified point. Point level indicators have the benefit of being lower cost and simple, without the need for complex signal transmission protocols.

Point levels can be used as alarm switches that activate when setpoint conditions are achieved (liquid levels become too high or too low). Point levels are used in addition to continuous level transmitters to notify operators of critical process conditions. In all cases, the sensor must be immersed by the liquid and will trigger a contact at that elevation point. These devices are useful for level alarms.

Mechanical float switch

With a mechanically actuated float switch, switching occurs because of the movement of a float against a miniature (micro) switch (Figure 2). For such level sensors, chemical compatibility, temperature, specific gravity (density), buoyancy, and viscosity affect the functionality. The choice of float material is also influenced by temperature-induced changes in specific gravity and viscosity – changes that directly affect buoyancy. Some mechanical float switches offer a test function where the float is manually moved into the alarm state. These manual test functions are generally error-prone and unreliable. Usually such basic sensors have no diagnostic functionality but float switches are popular for simplicity and low cost.

Figure 2. Mechanical float switch.

Tuning fork level sensor

Tuning fork (vibronic) point level sensors are electronic and contacting (Figure 3). A piezoelectric motor vibrates two metal protrusions, referred to as a "tuning fork". The tuning fork is excited to its resonant frequency by piezoelectric material. When the tuning fork enters the product the frequency of vibrations is damped by the liquid immersion of the tuning fork. The change in frequency is detected and used to determine if the liquid level is above the tuning fork position. The tuning fork method is advantageous when compared to a mechanical displacement because the output of the tuning fork sensor is diagnostic and continuous. The tuning fork is a reliable sensing method because if the liquid is in contact with it, or the sensor requires maintenance, then the setpoint conditions would be achieved and an alarm would activate. Operators would receive a signal and act accordingly.

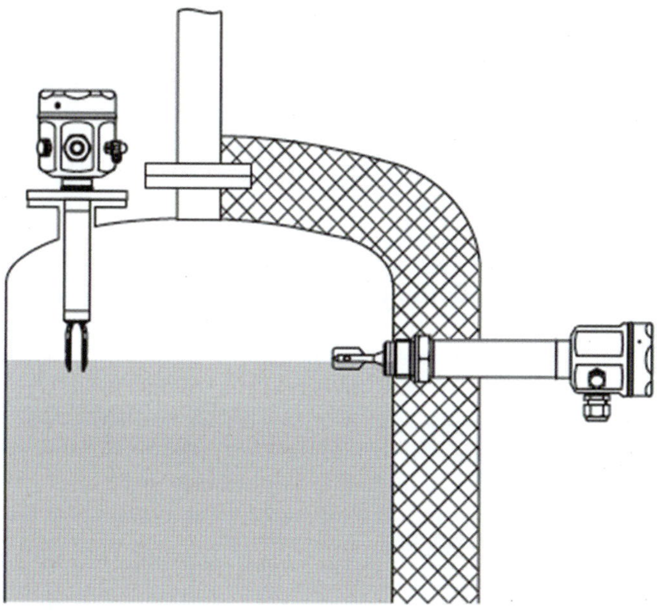

Figure 3. Endress+Hauser Liquiphant FTL50 point level sensor.

Tuning fork sensors are typically used for maximum or minimum detection for liquids in tanks or pipes (leak monitoring, dry running protection/pump protection or overfill protection), particularly for the chemical, energy, and oil & gas industry.

Compared to other physical principles, e.g., floats, the vibronic point level switches can offer a decisive advantage with the frequency evaluation facilitating automatic self-monitoring of the fork (Figure 4). Such sensors do not only detect and alert cable breaks and open circuits but also the correct operation of the electronics and of the vibrating fork itself. Build up and corrosion can be detected by analyzing the frequency (Figure 5). Cable breaks and open circuits as well as the correct operation of the electronics are continuously checked by using current pulses (PFM signals = pulse-frequency modulation or LIVE signal) between the sensor and the logic solver.

Capacitive level sensors used as level switch

The capacitive measuring principle is based on the method of the operation of a capacitor (Figure 6). A capacitor is formed by two differently charged electrodes isolated from each other. Applying an alternating current between the electrodes will create an electric field. This electrical field depends on the distance between the electrodes, the size of electrodes surface, and the isolating medium between the electrodes.

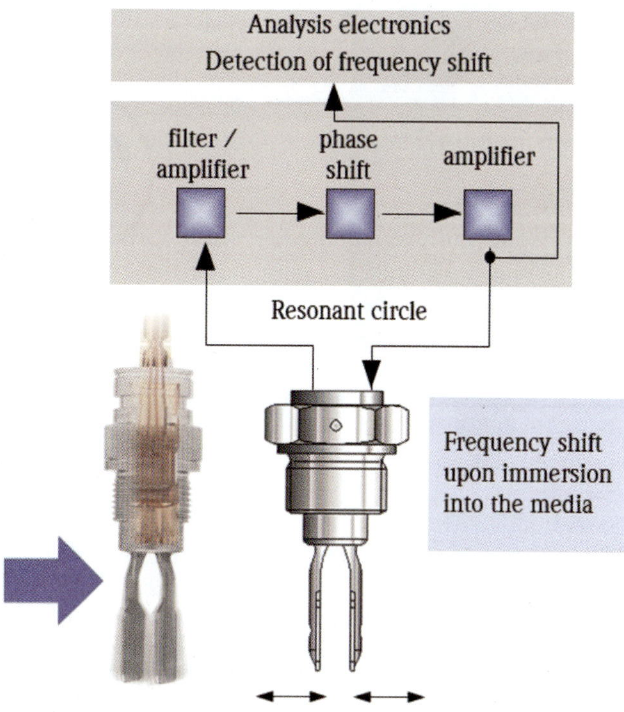

Figure 4. Detection of frequency shift.

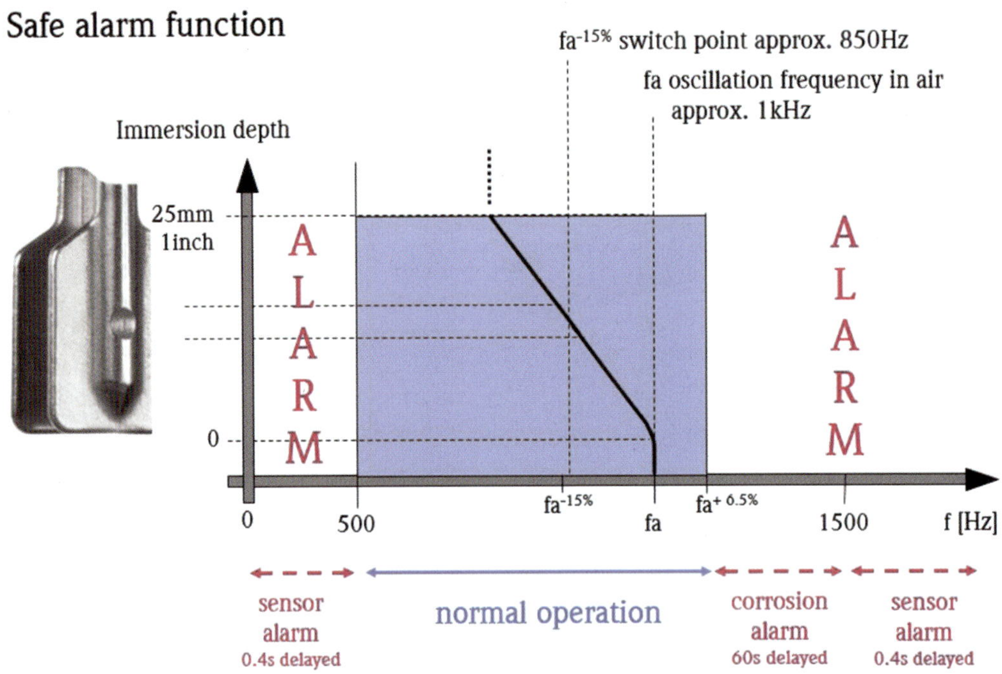

Figure 5. Safe alarm function. fa = failure alarm.

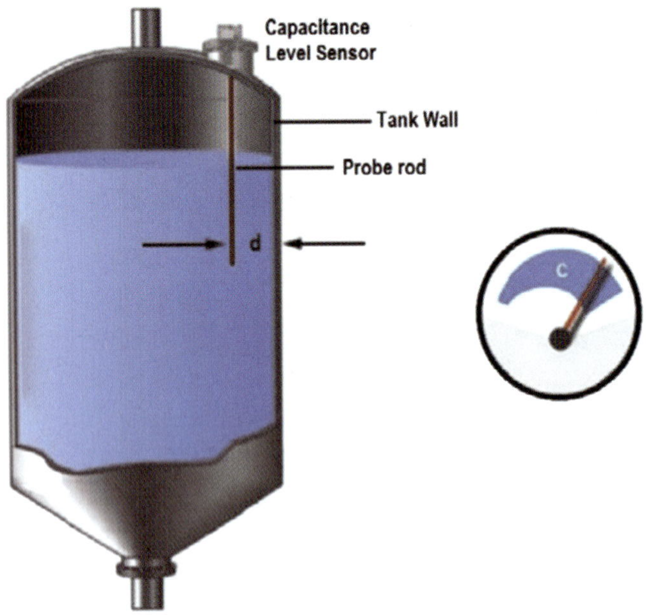

Figure 6. Capacitance measurement principle.

If the distance between electrodes and size of surface of the electrodes are kept constant, only the medium would influence the electrical capacitance. When the medium changes the electrical field changes also consequently the capacitance evolves as follows:

$$\begin{aligned}\text{Capacitance (C)} &= \text{Dielectric constant } (\epsilon 0)\\ &\times \text{Relative Dielectric constant (DC)}\\ &\times \text{Electrode Surface Area}\end{aligned}$$

The capacitive measuring principle may be used as level switch for floating roofs (Figure 7). The capacitive sensor is installed on a suspension arm at the tank wall. The sensor is selected with a long rod weight and a short flexible robe and a long inactive length, which makes the measurement unsusceptible of wind. On the floating roof a detector plate is mounted which can be designed in a hemispheric shape to avoid deposits (Figure 8). If the floating roof rises to the defined high level, the sensor touches the detector plate. This leads to a change of the capacitance and triggers the sensor to switch.

Continuous level sensors

Continuous level sensors perpetually monitor the height of product within a tank. Whereas the point level sensor provides only a yes/no answer to whether the liquid is at or above the sensor, the continuous level sensor outputs real time level. More information is obviously available to the control systems and the operator with continuous level measurement. While continuous level measurement is important for both inventory and custody transfer operations, it provides the ability for operations to project or estimate when a tank will be filled or emptied because of the extra

Figure 7. Floating roof detection.

Figure 8. Floating roof detector plate

information provided beyond just point level readings. In custody transfer applications, level transmitters need to be accurate within 4 mm.

Older technologies, such as mechanical point level indicators, may be less accurate and reliable than newer electronic technologies. Continuous level transmitters range in complexity, but best practices such as redundant alarm sensors and self-diagnostic systems can significantly improve process reliability and safety.

Float level sensors

A float transmitter is a mechanical or magnetic device which moves vertically with the liquid surface. The float device moves up only when the level increases because the buoyancy of the float is just enough to suspend it in the liquid, much like a fishing bob. This method of level monitoring has the benefit of being simple and usually low cost but poses some risks such as sticking, buoyancy failure, and lack of redundancy. While there are many applications which are ideally suited to this technology those applications using it should avoid services which can degrade service such as sticky fluids, dirty liquids, or highly variable densities which change the relative float location with respect to level.

Float and tape sensor

Float and Tape sensors are mechanical and have moving parts in addition to requiring liquid contact to function. A tape measure attaches to a float, and the float rises with the liquid level. The float rising creates slack at the counter weight. This slack is pulled by the external housing which is at grade and which winds up the tape, displaying the height of the liquid level. This method of level measurement is simple and inexpensive, however, unless the installation is properly made and the device used in the appropriate service, there can be risks of sticking and failure to operate correctly. Devices with moving components require periodic maintenance and may have lower reliability when compared to recently developed electronic measurement systems.

Magnetorestrictive float level sensor

Magnetorestrictive float level sensor (magnetic float) measurement is an electromechanical contacting method because a float moves up and down with the liquid level, but the level is determined by an electric signal (Figure 9). The float contains magnets which produce a magnetic field. The voltage of the resistor chain changes with position of the magnetic field created by the magnetic float. The voltage of the resistor chain is calibrated at known product levels. The change in voltage across the resistor chain is used to continuously calculate to the product level.

Magnetic float indicators are superior to the tape float transmitters because they have less moving parts. Less moving parts reduces the risk of sticking, however magnetic float indicators are not considered highly reliable. Again, unless the installation is properly made and the device used in the appropriate service, there can be risks of sticking and failure to operate correctly. Devices with moving components require periodic maintenance and may have lower reliability when compared to recently developed electronic measurement systems.

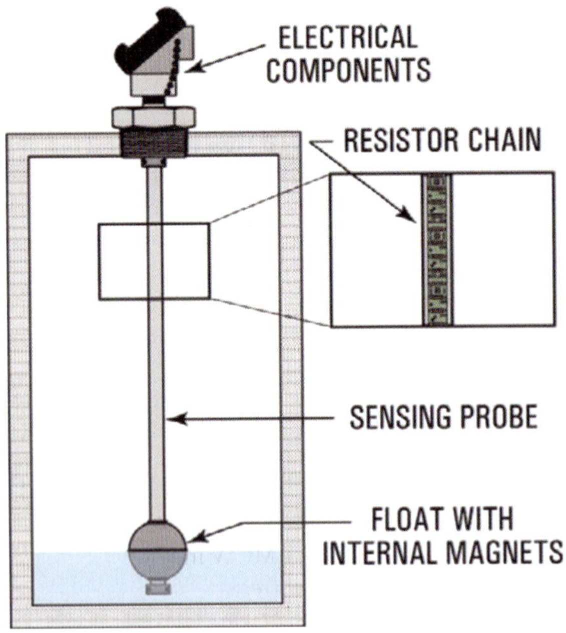

Figure 9. Diagram of magnetic float level sensor.

Electromechanical servo transmitter

Servo measurement technology is based on the partially immersed in a stationary fluid, is buoyed up by a force equal to the weight of the fluid displaced by the object." – Archimedes, 250 BC. Servo level transmitters are electronic and contacting.

Servo measurement works through a displacer (which is not a float) that is attached to a servo motor. The displacer is suspended on a fine wire that is lowered into the liquid product. When the displacer contacts the liquid surface, buoyant forces reduce the weight of the displacer. Because the servo motor is suspending the displacer, the force on the servo motor is reduced by the buoyant forces acting on the displacer. That change in force changes the torque within the magnetic coupling of the servo motor. Electromagnetic sensors measure the change in torque of the servo motor and the position of the displacer. The position of the displacer when the torque change occurs is converted to a product height.

The change in torque is based upon the density of the fluid. This allows servo measurement to determine a liquid level, interface level, and product density. Servo measurement is reliable and often used in custody transfers. The benefits of servo measurement include high reliability, unaffected by dielectric constants and the vapor phase, capable of determining an interface, capable of detecting water bottom, and capable of determining product density (both points and profiles). The ability to determine mass within a tank makes servo measurement one of the most robust and valuable form of level measurement. Further, servo level sensors are versatile enough for application in a crude oil vessel or liquified gas (LNG/LPG) tank.

Capacitance level sensors

Capacitance level sensors use an electronic contacting measurement method. Capacitance transmitters work by lowering an electric probe into the tank. The tank wall and the probe create a current, and that flow of current is dependent on the liquid level within the tank. The probe detects little to no current when the tank is primarily full of air because air has a high resistance to electric current. As the liquid level rises, more current flows between the probe and the tank walls. This is because water and oil are less resistive than air.

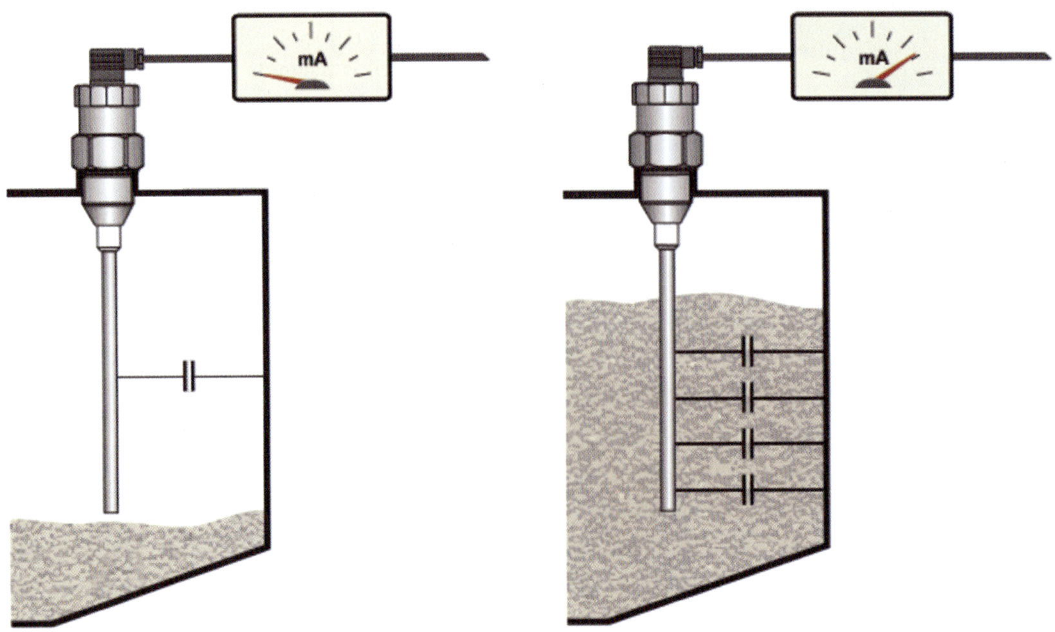

Figure 10. Diagram of capacitance level.

Likewise, hydrocarbons are more resistive to an electric current than water. This is because the dielectric constants are different for the species. The dielectric constant is the measured ability of a substance to store electric energy created by an electric field. The dielectric phenomenon is utilized in capacitance measurement devices to gauge both the liquid level and interfacial level. The interfacial level is crucial to many separation and storage processes. In industrial application, continuous capacitance level sensors are not used to measure the hydrocarbon product level in large tanks but are effective at monitoring the level of water bottoms.

Ultrasonic level transmitters (TOF sensors)

Ultrasonic level transmitters are electronic and can contact the liquid surface or not depending on the sensor type and application. This method works by measuring the free space within a tank by emitting ultrasonic pulses towards the liquid level. Ultrasonic pulse frequencies are above 20KHz, which is inaudible by the human ear. Ultrasonic waves bounce off the liquid surface and the reflections return to the sensor.

The time required for the ultrasonic signal to emit and return to the sensor, is the "Time-of-Flight", or TOF. The TOF is used to determine the distance the signal travels using the speed of sound in air as the velocity of the ultrasonic pulse. The time for the pulse to emit from the sensor and return to it is inversely proportional to the distance between the sensor and the liquid level. Therefore, as the liquid level drops the signal response time increases. The speed of sound in air at 20°C is 343 m/s. Therefore, the distance travelled is:

$$Distance_{ultrasonic\ pulse}[m] = \frac{\left(343\left[\frac{m}{s}\right]\right) \times \left(Time\ of\ Flight[s]\right)}{2}$$

The height of the liquid is then obtained by the difference between the height of the tank and distance travelled by the ultrasonic pulse.

$$h_{liquid} = Distance_{sensor\ to\ base} - Distance_{ultrasonic\ pulse}$$

Ultrasonic level sensors tend to be easy to install, reliable, and lower cost than alternative electric noncontacting sensors. Ultrasonic vibrations are vulnerable to changes in the density of the medium it propagates through. Therefore, it is difficult to have high accuracy level measurement of volatile liquids using the ultrasonic method.

Radar level transmitters (TOF/FMCW sensors)

Radar level transmitters are electronic and noncontacting, using the principle of TOF. The primary difference between the radar method and ultrasonic method is the type of wave used to measure the TOF. Radar level transmitters emit radio waves rather than ultrasonic pulses. Radio waves are generated electromagnetically by an antenna and directed towards the product surface in the tank. The radio waves contact the product surface and bounce back towards the top of the tank. The sensor on the top of the tank then registers the incoming radio waves.

Radio waves travel at the speed of light, so the time required to return can be converted into a distance. Like the ultrasonic method, the time required by the radio wave to return to the sensor is inversely proportional to the liquid level. Therefore, as the time between emission and return increases, the liquid level decreases. Because the speed of light is extremely fast, the frequency of the radar must be large to obtain high resolution from the sensor.

Frequency-Modulated Continuous Wave (FMCW) radar is different from normal continuous wave (CW) radar methods. FMCW can change the frequency of the pulses from the radar emitter. The FMCW method accounts for the delay caused by the wave frequency and creates higher accuracy measurements than CW. The radio waves can be free to propagate through a tank or restricted to a guide tube (Figure 11). Emitting and receiving radar waves through a guide tube can further reduce the noise in the measurements.

The success of a radar level sensors is dependent on good installation. The radar needs an unrestricted path to the liquid surface, as all other continuous level measurements require. Any obstruction, such as piping, can cause unwanted echoes in radar measurement. Although, modern radar technology has software algorithms to mask and compensate

Figure 11. Endress+Hauser Micropilot FMR540 TOF radar level sensor.

for these echoes. Radar level sensors have both high reliability and require little maintenance. Radar waves are unaffected by vapors in the tank, unlike sonar waves.

Free space radar sensors that are used for custody transfer and inventory control applications usually need a NMi- and PTB-approval and must meet the requirements according to OIML R85 and API 3.1B. Modern devices offer drip-off lens antennas a sharply focused beam angle of to avoid obstacles even close to tank wall by using a higher transmitting frequency.

Radiometric (nuclear) level transmitters

Radiometric level transmitters are electronic and non-contacting, but operate differently from the radar and ultrasonic methods. The largest difference is that radiometric measurement can be used for continuous measurement, point level measurement, and product density measurement. Because of the ability to determine product density, radiometric measurement devices can determine the mass of material within a tank. The abilities of the radiometric measurement, including the ability to measure liquid density, makes radiometric measurement a robust forms of level measurement.

Radiometric transmitters emit gamma radiation from radioactive isotopes of Cesium (137) or Cobalt (60). These isotopes only emit β (particle) and γ (wave) radiation. The isotopes are secured in a double walled stainless-steel vessel. This vessel prevents the emission of particle radiation, specifically β radiation. Therefore, only γ radiation can penetrate the stainless-steel vessel. The radioactive source is surrounded by an external container such that the γ waves are directed in one direction, towards the storage tank. A contact transmitter is positioned on the opposite side of the tank to receive the gamma radiation (Figure 12).

The radiation adsorbed is dependent on the density of the medium which the radiation passes through. High density mediums absorb more radiation. Therefore, more radiation is absorbed by product fluid than air. The contact transmitter measures the strength of radiation across the vertical length of the tank, determining the level. Water and hydrocarbons are immiscible and have different densities. Therefore, radiometric methods can also determine the level of an interface.

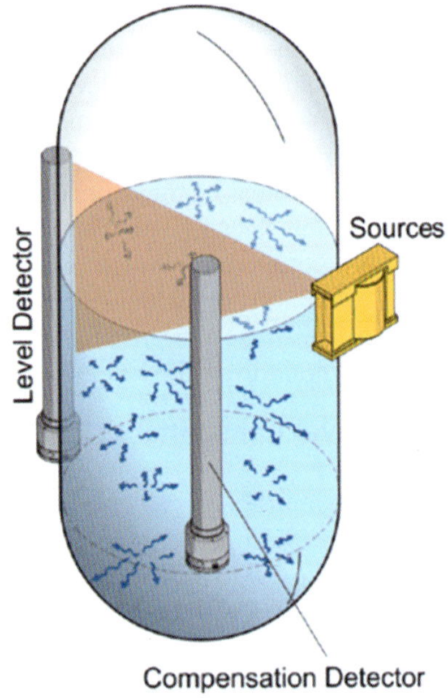

Figure 12. Diagram of radiometric level transmitters.

Oil leak sensors

Because overfilling and spilling events are dangerous and punishable by law, it is important to detect product liquid early during a spill event. Innovative technology combines different sensors to distinguish water and oil in a dike, sump, or floating roof. The different sensors within the NAR300 can distinguish hazardous liquids such as gasoline from water, preventing a false alarm. The sensor can be placed within a drainage dike, runoff containment, underneath a tank bottom or wherever the application is required. The benefit of an oil leak sensor is early detection. Early detection is becoming aware of a risk event very quickly after it initiates. Early detection can be the difference between a small leak and a massive overfill event.

References

[1]. Manual of Petroleum Measurement Standards Chapter 3.1 B, Standard Practice for Level Measurement of Liquid Hydrocarbons in Stationary Tanks by Automatic Tank Gauging, 3rd Edition (April 2018).

[2]. Manual of Petroleum Measurement Standards Chapter 3.1A, Standard Practice for the Manual Gauging of Petroleum and Petroleum Products, 3rd Edition (2013).

Source

Guidebook for Overfill Prevention & Tank Gauging, PEMY Consulting and Endress+Hauser: Philip Myers, Brock Trotter, Dean Mallon, Hai-Thuy Ngoand Markus Wohlgemuth. For the complete report, go online at: www.addresses.endress.com or www.pemyconsulting.com.

17: Integrity Management

Overview of leak detection technologies 746
Inspection tools and case studies 750
Factors to consider when evaluating damage on pipelines 760
Understanding mechanical damage in pipelines 766
Understanding stress corrosion cracking in pipelines 774
When does a pipeline need revalidation? 783
Modeling pipeline risk assessment 792
A risk management tool for establishing budget priorities 797
Floating roof failure modes and risk 802
Floating roof tanks rainfall risk 812

Overview of leak detection technologies

Excerpts from the report "Hazardous Liquid Leak Detection Techniques and Processes," by General Physics Corporation for the U.S Department of Transportation, Office of Pipeline Safety, June 2002.

The methodologies used for leak detection cover a wide spectrum of technologies and processes and are based on a number of different detection principles. They vary from intermittent aerial inspections to hydrocarbon sensors to sophisticated real-time monitoring. Each approach has its strengths and weaknesses. These strengths and weaknesses are dependent on the application and the complexity of the pipeline system to which the leak detection is applied. In combination with processes and procedures, applying the appropriate technology (or technologies) is the key to an effective leak detection system.

Regulatory background

The Office of Pipeline Safety published the Final Rule on Pipeline Integrity Management in High Consequence Areas for Hazardous Liquid Operators on November 2, 2000 (65 FR 75378). The rule applies to operators who own or operate 500 or more miles of hazardous liquid or carbon dioxide pipelines (large liquid operators) and went in to effect on May 29, 2001. The rule establishes requirements for pipelines that can affect High Consequence Areas (HCA), which include populated areas defined by the Census Bureau as urbanized areas or places, unusually sensitive environmental areas, and commercially navigable waterways. Included in this rule is a requirement for operators to have a means to detect leaks on their pipeline systems.

The Office of Pipeline Safety has extended the regulations requiring hazardous liquid pipeline operators with 500 or more miles to establish an integrity management program for pipelines that could affect a high consequence area, to operators with less than 500 miles. This rule "Pipeline Integrity Management in High Consequence Areas for Hazardous Liquid Operators with Less Than 500 Miles of Pipeline" was published on January 16, 2002, and went in to effect on February 15, 2002 (67 FR 2136).

Classification of leak detection technologies

Leak detection technologies can be classified according to the physical principles involved in the leak detection. Using this type of classification, leak detection systems can be divided into the following four groups:

Physical Inspection - This type of leak detection involves either direct or remote visual inspection to detect a leak.

Manual Tabulation - This type of leak detection includes direct monitoring of pipeline flow and/or pressure for evidence of a leak. This may also involve manual calculations to identify lost product.

Discrete Sensor-Based Technologies - Sensor-based technologies rely on the use of an external sensor to detect the escaping hydrocarbon liquid. These systems include, but are not limited to:

- Liquid Sensing
- Vapor Sensing
- Acoustic emissions

Computational Pipeline Monitoring - Computational Pipeline Monitoring (CPM) systems are distinguished from other leak detection systems by the use of an algorithm that uses input from field sensors that monitor the internal pipeline parameters (e.g. pressure, flow, temperature, frictional pressure drop, density, batch interfaces) to determine when a leak has occurred. These systems include, but are not limited to:

- Over and short comparison
- Mass balance with line pack correction:
  - line pack correction based on pressure and temperature sensors
  - line pack correction based on transient flow modeling
- Pattern of discrepancy in pressure/flow between model and measurement
- Rate of pressure/flow change
- Statistical methods that are not model-based
- System identification methods based on digital signal analysis

Evaluation of leak detection systems

Each leak detection system is unique based on the pipeline on which it is used. As such, the capabilities of the system and the degree to which it mitigates risk to HCAs must be evaluated for each pipeline system. More sophisticated systems will have more unique capabilities. The criteria used to evaluate the capability of an installed leak detection technology may include, but are not limited to, the following:

Leak Size or Leak Flow Rate - What is the minimum leak size that the system is capable of detecting? A leak is detectable only when its' effect rises above uncertainties in the variables being monitored (see Response Time below). The size of a leak is usually expressed as a percentage of the throughput of the pipeline. Leak size is a function of the size and shape of the opening (leak area) and the pipeline pressure. A leak can be either constant in size, such as a pre-existing small leak, or variable over time, such as a sizable leak that diminishes as the pipeline is depressurized.

Response Time - What is the time needed to detect a leak of a given size? Depending on the leak detection methodology used, the response time can vary over a wide range.

For algorithms based on volumetric balance, the response time is related to the leak size. This is because of the uncertainties in the variables involved. Uncertainties, or noise in the variables used for leak detection, are always present. A leak can be detected only when its effect, herein called leak signal, is discernable amongst noise. Since noise is random in nature while a leak signal is not, over time, the accumulated noise remains at a noise level while the accumulated leak signal grows in size. Eventually, the accumulated leak signal rises above the noise and becomes detectable in a probabilistic sense (see False Alarms and Misses below). A minimum time period exists for each minimum detectable leak. A curve that relates minimum detectable leak size to response time is a leak threshold curve for this leak detection methodology. Two such leak threshold curves are shown in Figure 1 to illustrate the general trend. Given an uncertainty level, larger minimum detectable leaks have a shorter response time. A smaller uncertainty in the variable results in a tighter threshold. It takes less time to detect for a given size leak if the uncertainty is reduced. Small leaks with size approaching the combined non repeatability of instrumentation have a very long response time. Such leaks can only be determined by physical observations.

For leak detection methods based on discrepancy patterns generated from a real-time transient flow model, the response time is not a function of leak size. Instead, it is a function of the propagation speed (about 3000 to 4000 ft/s) of a pressure disturbance and the distance between the leak and the nearest pressure or flow sensors.

Leak Location Estimation - Can the system locate a leak and what is the accuracy of the location estimate? The relevance of this criterion is to aid pipeline operator response to a leak in leak mitigation. Location can be estimated based on the time of arrival of a leak disturbance at a pair of sensors. Figure 2 indicates a leak occurring at time t0. This leak generates a local pressure drop, which then propagates both upstream and downstream. If this signal is picked up by pressure transducer A at time t1 and by pressure transducer B at time t2, then the leak can be located. This approach requires either a fast data scan rate or the time of arrival at the transducers is registered by data collectors and later transmitted to the control center.

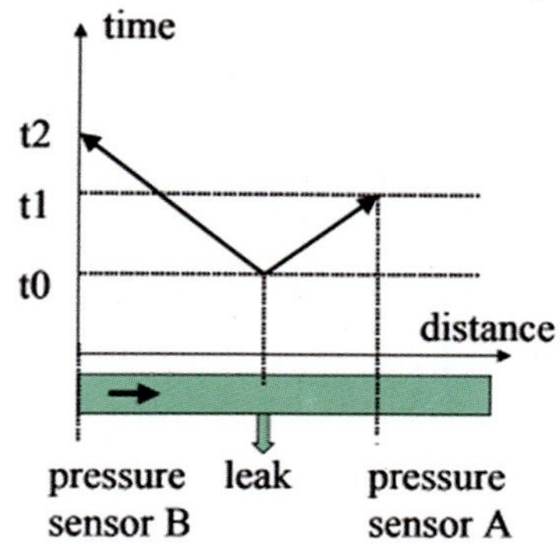

Figure 2. Locating a leak by the time of arrival of a leak signal.

Alternatively, a leak can be located by the profile of the piezometric head, also known as the hydraulic grade line. Figure 3 shows a pipeline with its inlet and outlet pressures held constant. The dotted profile is associated with the steady state flow prior to a leak. The solid profile is the hydraulic grade line after the transients caused by the leak have damped out and new steady state is established. The leak steepens the upstream hydraulic grade and flattens the downstream hydraulic grade. The effectiveness of this approach relies on multiple pressure sensors along the pipeline so that segments of the hydraulic grade line can be defined after a leak has occurred.

Release Volume Estimation - Does the system have the ability to determine the volume of liquid released? Reasonably accurate release volume estimation is possible for CPM methods where a mathematical model for transient flows is used. By using the measured pressure and flow from each

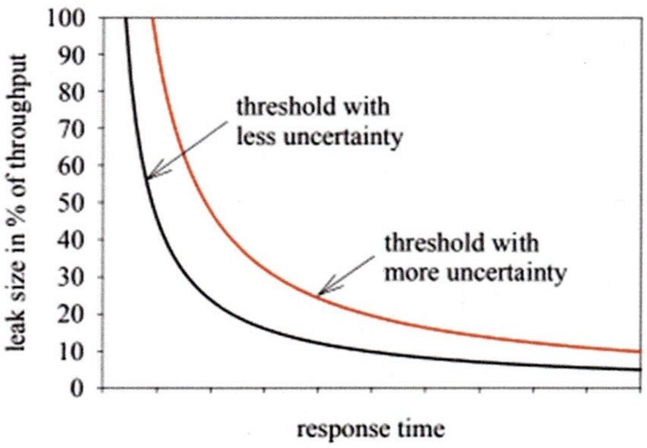

Figure 1. Detectable leak size versus response time.

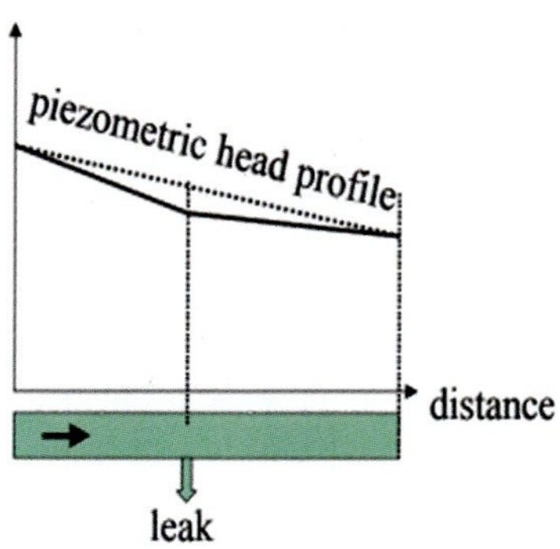

Figure 3. Locating a leak by the piezometric head profile.

end of a pipeline segment, the leak flow rate as a function of time can be calculated.

Less accurate release volume can be estimated if a CPM method tracks the mean volume or mass imbalance (linefill change minus the difference between inflow and outflow). When a leak is detected, the volume or mass imbalance prior to and after the leak can be used to estimate the release volume over time.

Detecting Pre-existing Leaks - Does the system have the ability to detect between pre-existing leaks, as well as, the onset of a new leak? Some CPM approaches depend on a change in one or several parameters to detect the onset of a leak. Such approaches will not be able to detect a leak (usually small) that is in existence before the CPM is activated.

Detecting a Leak in Shut-in Pipeline Segments - Does the system have the ability to detect the onset of a leak in a shut-in pipeline segment? The detection of a leak under such a situation is a matter of monitoring linefill change and discerning variations due to environmental temperature variations and/or due to a leak. CPM methods based only on metered inflow outflow comparison will not be able to detect a leak in a shut-in pipeline segment.

Detecting a Leak in Pipelines in a Slack Condition During Transients - Does the system have the ability to detect a leak in pipelines under a slack condition during transients? Liquid vaporizes when its pressure is sufficiently low. A pipeline is slack if vaporization occurs. A pipeline can be slack under both steady state and transient flow conditions. Leak detection on a slack line under transient conditions is difficult because the uncertainty in line pack change due to vaporization is large.

Rate of False Alarms and Misses - What is the false alarm rate for the system? There are many sources of uncertainty in the data that drive the CPM algorithm. These sources include hydraulic noise, non-repeatability of field sensors, uncertainties introduced by the data collection and communication system (analog-to-digital conversions, data timing), uncertainties in batch positions for product lines, and the state of flow (steady, drifting, or transient). As a result, the output from the algorithm is also uncertain. This uncertainty can be a significant issue facing the CPM technologies.

To illustrate this issue, consider the volume imbalance as the algorithm output. In terms of standardized volumes, subtract the change of linefill over a time period from the difference between inflow volume and outflow volume over the same period. The result is the volume imbalance. A positive imbalance means a leak. Refer to Figure 4 where the estimated imbalance is plotted against the true imbalance. Had the estimations been perfect, all points should fall on the 45-degree (diagonal) line. However, because of uncertainties, the points will be scattered around the diagonal line. Points above the diagonal represent under-estimation of the imbalances, while points below the diagonal represent over-estimated imbalances.

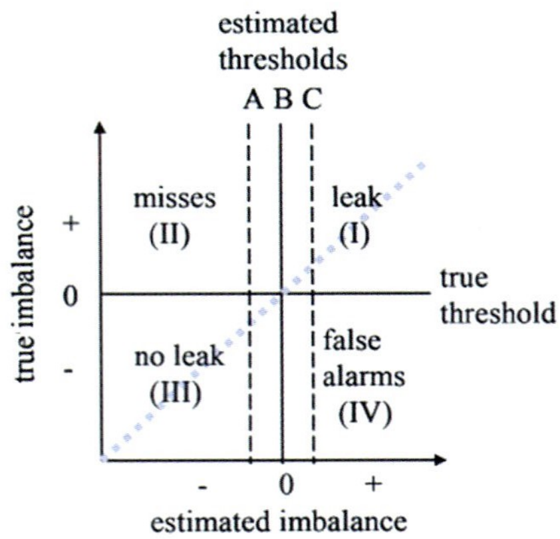

Figure 4. False alarms, misses, and leak thresholds.

The estimated imbalance versus true imbalance plot in Figure 4 is divided into four quadrants by the horizontal line labeled "true threshold" and the vertical line B which is the "perfectly estimated threshold." In reality, the true threshold is unknowable and the estimated threshold is determined empirically (by tuning, for example). Scatter of the points near the center of the plot gives rise to false alarms (for those points falling into quadrant IV) and misses (for those points falling into quadrant II). Notice that false alarms and misses occur even when the estimated threshold is perfect. For this reason, and given the fact that variable uncertainties are unavoidable, CPM is not the appropriate technology for detecting very small leaks. However, given the practical limitations of various other technologies, CPMs may be applied as long as their performance limitations are understood and acceptable to the pipeline operator.

Given the scatter in the estimates, the frequency of false alarms can be reduced by raising the estimated threshold (vertical line C). In so doing, the chances of misses (leaks not detected) increases. Lowering the threshold (vertical line A) reduces the chances of misses at the expense of increasing the frequency of false alarms.

Periods of greater linefill uncertainty occur when the pipeline is undergoing transients due to planned pipeline operations, such as pump startup and valve swings. To reduce the occurrence of false alarms, the leak threshold may be raised temporarily during such periods. Having the flexibility to raise the leak threshold can be an advantage, provided the operator understands that this is done at the expense of increased chances for misses (see Availability below).

Sensitivity to Flow Conditions - Will operational transients (such as those caused by pump startups or valve swings) degrade the ability to detect a leak? A pipeline seldom operates at a true steady state. This is especially true for long lines with numerous booster pump stations and delivery terminals. The linefill changes as a result of transients. Volume balance methods that do not compensate for linefill change accurately will be excessively sensitive to the flow conditions. The uncertainty in linefill induced by even mild transients can routinely exceed the combined non-repeatability of flow measurements in short time intervals.

Transients generated by pump startups, shutdowns, and valve swings also put extra demands on the data collection system since data polling frequency and timing skew can become issues of concern. Shorter data sampling periods help to discern leaks in such a situation.

Robustness - Will degradation or malfunction of a system component cause catastrophic loss of leak detection ability? This criterion measures how gracefully the leak detection capability degrades when system components malfunction. It also measures a system's ability to function in complex pipeline configurations when not all the needed information is available. Pipeline operators should be alerted at the first sign of degradation so that restoration efforts can be initiated, and catastrophic loss of leak detection ability can be avoided.

System Self Check - Will the leak detection system have the capability to automatically check and possibly rectify parameters that affect leak detection performance? Will it have the capability to detect and locate non-functional or degrading field sensors and alert pipeline operators?

Ability to Handle Complex Pipeline Configurations - What is the ability of the system to handle complex pipeline configurations as well as complex operations? Complex systems may include multiple injection and delivery points, or multiple modes of operation. These may complicate CPM due to needed model (i.e. algorithm) refinements, increased data requirements, and increased uncertainty when the needed data is not available.

Availability - Is the leak detection algorithm active around the clock? To avoid false alarms, some CPM systems that can not handle transient flow conditions usually increase the detection threshold until the operational transients have passed. Since a leak is equally likely (or even more likely) to occur when a pipeline is experiencing transients, the leak detection function is considered unavailable during periods of raised leak threshold. The percentage of time during which operational transients exist is an important factor in selecting the appropriate CPM method.

Retrofit Feasibility - What is required to install a new leak detection system and/or methodology on an existing pipeline? An upgrade requiring modification to, or addition of, field sensors may be less feasible than one that only requires software modifications. Algorithms that require a prolonged period of on-line parameter tuning are more difficult to retrofit.

Ease of Testing - API 1130 "Computational Pipeline Monitoring" recommends that a leak detection system be tested during commissioning and every 5 years thereafter. As a result, ease of testing to affirm leak detection capability is a relevant criterion. Can the system be tested with pre-existing leak test data, as well as, by actual withdrawal?

Cost - What is the cost of the system including capital and operational expenses, as well as, data and equipment requirements?

Ease of Personnel Training - How are personnel trained on the operation and maintenance of the system? Is the system easy to operate? Complex systems requiring a high level of training may not afford the same level of leak detection capability when the human interface is considered.

Ease of Maintenance - What are the maintenance requirements for the system? Will the system degrade with improper or missed maintenance tasks?

Standards for leak detection systems

Several industry consensus standards have been developed to address the selection, design, operation, maintenance of leak detection systems (Table 1).

Of the consensus standards, API 1130 has been incorporated by reference into 49 CFR 195. Specifically, section 195.134, "CPM leak detection" under Subpart C - Design Requirements, states:

> *"This section applies to each hazardous liquid pipeline transporting liquid in single phase (without gas in the liquid). On such systems, each new computational pipeline monitoring (CPM) leak detection system and each replaced component of an existing CPM system must comply with section 4.2 of API 1130 in its design and with any other design criteria addressed in API 1130 for components of the CPM leak detection system."*

Table 1
Industry Leak Detection Standards

Standard	Revision	Title
API 1130	2nd Edition Nov. 2002	Computational Pipeline Monitoring
API 1149	First Edition Nov. 1993	Pipeline Variable Uncertainties and Their Effects on Leak Detectability
API 1155	First Edition Feb. 1995	Evaluation Methodology for Software Based Leak Detection Systems
API 1160	First Edition Nov. 2001	Managing System Integrity for Hazardous Liquid Pipelines

Section 195.444, "CPM leak detection" under Subpart F - Operation and Maintenance Requirements, states:

"Each computational pipeline monitoring (CPM) leak detection system installed on a hazardous liquid pipeline transporting liquid in single phase (without gas in the liquid) must comply with API 1130 in operating, maintaining, testing, record keeping, and dispatcher training of the system."

For the purpose of the regulations 49 CFR 195.2, "Definitions" defines CPM as follows:

"Computation Pipeline Monitoring (CPM) means a software-based monitoring tool that alerts the pipeline dispatcher of a possible pipeline operating anomaly that may be indicative of a commodity release."

Source

Dr. Jim C. P. Liou, Robert J. Hall, Mona C. McMahon, General Physics Corporation, Report on Hazardous Liquid Leak Detection Techniques and Processes, April 2003. U.S. Department of Transportation, Research and Special Programs Administration, Office of Pipeline Safety. Report No. DTRS56-02-D-70037-01. For the complete report, go online at: www.phmsa.dot.gov.

Inspection tools and case studies

Excerpts from the report "On the Use of In-line Inspection Tools for the Assessment of Pipeline Integrity" by General Physics Corporation for the U.S Department of Transportation, Office of Pipeline Safety, June 2002.

The Office of Pipeline Safety (OPS) has allowed the use of newly developed in-line inspection tools, the integration of data from different inspection tool runs, and the reinterpretation of raw data from previous inspections to assess the integrity of pipelines. In this article, three case studies document how in-line inspection tools have been used to detect a number of common pipeline defects including: Longitudinal Seam Cracks, Stress Corrosion Cracking, Internal Corrosion, External Corrosion, Top-Side Anomalies, Dents, and Gouges.

Magnetic flux leakage inspection tools

Magnetic Flux Leakage (MFL) is a well-established in-process method of non-destructive examination (NDE). In standard MFL, a powerful, axially oriented, magnetic field is created in the pipeline walls by magnets mounted on the MFL in-line inspection tool. Loss of wall metal due to corrosion or to circumferential gouging can be readily detected, as can many problem features in the circumferential (girth) welds. All of these features have substantial components that are perpendicular to the magnetic field lines. By forcing the magnetic field in the pipe wall to pass through an area of reduced wall metal, they cause "leakage" of the field into the interior of the pipeline, where the increase in magnetic flux can be detected by sensors mounted on the in-line inspection tool. Data from the sensors is stored and analyzed off-line after the in-line inspection tool is recovered. Combined with the orientation and axial position data also logged by the in-line inspection tool, a comprehensive survey of the pipeline can be obtained with minimal disruption of operations.

Three different types of MFL tools have been developed for the in-line inspection of pipelines. These included the Standard MFL Tool, the High Resolution MFL Tool, and the Transverse Field Inspection or Circumferential MFL Tool.

Standard MFL Tool

The Standard MFL Tool was the first MFL tool developed. Generally, the corrosion anomalies detected are reported as light (10% to 30% wall loss), moderate (30% to 50% wall loss), or severe (greater than 50% wall loss). There is significant industry experience with the use of the Standard MFL Tool for detecting corrosion.

Standard MFL Tools have limited ability to detect and characterize narrow, axially oriented, features, such as longitudinal gouges and cracks. This makes many of the features associated with ERW pipe problems, such as hook cracks and preferential corrosion, nearly "invisible" to standard MFL. The magnetic field in the pipe wall "sees" only a slight restriction in the cross sectional area of metal available to carry field lines. As a result, very little field leakage occurs and the signals generated by these features do not identify them as significant anomalies.

High resolution MFL tool

The High Resolution MFL Tool uses the same technology as the Standard MFL Tool. The major difference between the High Resolution MFL and Standard MFL Tools is that the High Resolution MFL Tool uses a larger number of closely spaced sensors. With the higher number of closely spaced sensors, the High Resolution MFL Tool can more precisely measure the length, width and depth of corrosion anomalies than the Standard MFL Tool. These measurements are sufficiently accurate for performing remaining strength calculations. High Resolution MFL Tools have been used to measure wall loss as low as 3% and have also been successful in detecting axial defects (selective seam corrosion of ERW pipe).

TFI Tool

In TFI, the direction of the magnetic field is rotated 90° with respect to standard MFL so that the field lines are circumferentially oriented, forming a magnetic "hoop" around the pipeline. In this orientation, axially oriented features such as hook cracks cause a significant restriction in the metal area available to the magnetic field lines, causing significant flux leakage and easy detection by the sensors on the in-line inspection tool. This makes TFI much more effective in locating ERW pipe seam problems than standard MFL.

TFI's field orientation makes it insensitive to circumferentially oriented features, such as girth weld anomalies and circumferential gouging. The two methods are thus complimentary in many respects.

A modified form of MFL, known as Transverse Field Inspection or Circumferential MFL, addresses many of these limitations. Pipeline Integrity International (PII) introduced the first commercial TFI system in 1996 specifically to detect Narrow Axial External Corrosion (NAEC) in a 34 inch pipeline in Canada. PPI developed a 20-inch TFI in-line inspection tool in early 1998 for the inspection of the Platte Pipeline. To establish the ability of the TFI in-line inspection tool to detect the specific types of anomalies, PPI conducted laboratory tests and pull tests in their facilities in April, 1998. The tests confirmed the ability of the tool to detect longitudinal cracks.

Standard and transverse technologies both have comparable abilities to detect metal loss due to general or pitting corrosion. Standard MFL is easier to implement in practice than TFI, as more space is available in the tool for the necessary equipment.

Comparison of MFL Technologies

Both standard and transverse MFL have limited abilities to characterize the dimensions of pipe wall anomalies, although the greater operating experience with standard MFL has allowed refinement of the software used to process the data generated. All forms of MFL are sensitive to background noise in the pipe metal, and can suffer from degraded performance at high operating speeds due to eddy current formation in the pipe.

The MFL methods are also effective in non-liquid service, making them suitable for gas pipeline examination. TFI appears to have particular value in the examination of ERW pipe, due to its ability to detect features such as hook cracks that have the potential to grow through fatigue mechanisms, at sizes below those detectable under normal hydrostatic test conditions.

Ultrasonic inspection tools

The basic function of a ultrasonic inspection instrument is of the pulse-echo principle, similar to sonar. A short ultrasonic pulse is transmitted into the object by a probe (transducer). The pulse travels through the material until it encounters an interface. An interface is a material with substantially different physical characteristics, such as crack or the back surface of the object being inspected. At the interface, the pulse is reflected back to the probe where it is detected. The time needed for the pulse to make this round trip is divided by two and multiplied by the velocity of sound in the material by the ultrasonic inspection instrument. The result is the thickness of the material between the surface and the interference. Figure 3 shows a graphic representation of an ultrasonic instrument.

Ultrasonic Testing (UT) is most sensitive to defects and material boundaries that are 90° to the to the incident sound

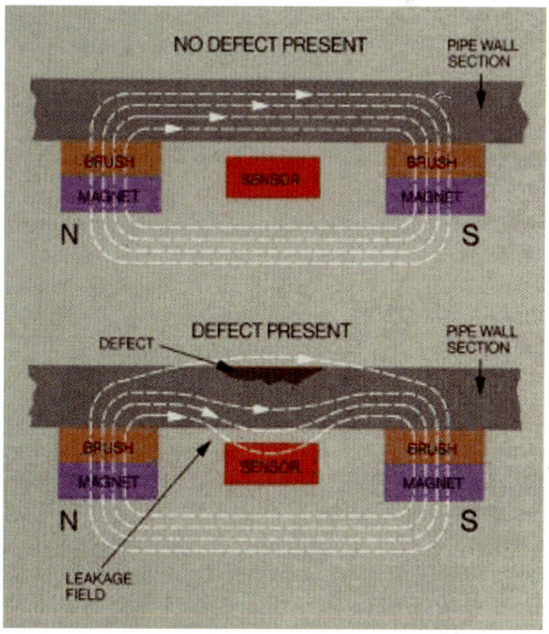

Figure 1. Magnetic flux defect detection.

Figure 2. TranScan TFI tool.

Figure 4. Standard UT tool.

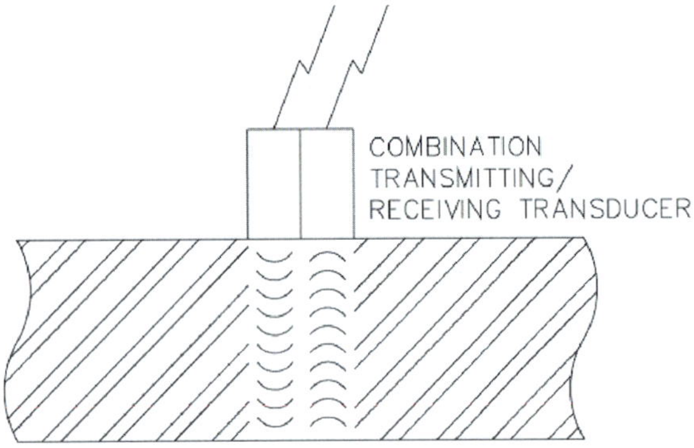

Figure 3. Ultrasonic inspection.

wave. For this reason two basic applications of UT have been used in in-line inspection tools; Standard UT and Shear Wave UT. In Standard UT the ultrasonic sound wave is transmitted through the pipe wall at 90° to the surface. Standard UT is used to detect corrosion (metal loss) and lamination in the pipe wall. In Shear Wave UT, the ultrasonic sound wave is transmitted through the pipe wall at an angle less than 90° to the surface. Shear Wave UT is used to detect cracks that are 90° to the surface of the pipe.

Standard UT Tools

Standard UT tools are principally used to measure wall loss. The tools consist of a large number of sensors mounted at 90° to the surface. The applications of standard UT in an in-line inspection tool requires a clean internal surface. Standard UT also must be performed in a liquid medium. The liquid is necessary to couple the transducers to the surface permitting the ultrasonic sound waves to travel from the transducer through the liquid and into the pipe wall.

Standard UT can be performed in natural gas pipelines if the in-line inspection tool is run in a liquid batch. The inspection data obtained from a Standard UT is sufficiently detailed and accurate to assess the remaining pipe strength. Figure 4 is a photograph of a typical Standard UT in-line inspection tool.

Shear Wave UT Tools

Shear Wave UT tools are principally used to detect cracks. The tools consist of a large number of sensors mounted at an angle to the surface (typically 45°). Like Standard UT the application of Shear Wave UT in an in-line inspection tool requires a clean internal surface and must be performed in a liquid media. Shear wave UT is generally used in pipelines to inspect for seam weld defects that could comprise integrity (e.g. hook cracks in ERW pipe) and for stress corrosion cracking.

To overcome the need to run a Shear Wave UT tool in a liquid medium, the Elastic Wave Tool was developed. In the elastic wave tool the UT sensors are mounted in wheels that are maintained in tight contact with the pipe wall. The Elastic Wave Tool is not as accurate as the Standard Shear Wave Tool but can be used without the need for a liquid batch. Elastic Wave Tools have also been shown to be effective in detecting disbonding of certain types of coatings. Figure 5 is a photograph of an elastic wave in-line inspection tool.

Geometry tools

Geometry Inspection Tools are used to detect and measure mechanical deformation of the pipe. Sources of mechanical deformation can include third part damage, construction damage, and movement of the pipeline from natural causes (frost heave, floods, subsidence and land slides).

Figure 5. Elastic wave UT tool.

Figure 6. Geometry inspection tool.

Geometry Inspection Tools use a series of lever arms to follow the pipe wall and measure changes in the pipe wall geometry as the tools moves through the pipeline. The lever arms are connected to a potentiometers to create a position signal that is recorded. The tool is also equipped with an odometer to measure the distance traveled by the tool. Geometry tools can detect and accurately locate circumferential welds, valves and tees.

Pipeline anomalies that are detected and characterized using Geometry Inspection Tools generally include dents, wrinkles, buckles, and pipe ovality. Some Geometry Inspection Tools have been equipped to measure changes in slope that occur with movement of the pipe and/or measure bend radius. A photograph of a typical geometry inspection tools is included as Figure 6.

Platte Pipe Line

The Platte Pipe Line operated by AEC Pipelines (USA) Inc. conducted an inspection program using Transverse Field Inspection (TFI) to identify seam weld defects in their 932 mile system. This inspection method was determined to have less impact on pipeline operations and offered greater reliability assurance than hydrostatic testing.

Background

The Platte Pipe Line System experienced a failure in July 1997 due to a longitudinal seam split initiated by the fatigue growth of hook crack, resulting in a spill of more than 7,000 barrels of crude oil with an estimated incident cost of $325,000. The failure occurred in a segment manufactured by A. O. Smith that was flash welded.

In response to this accident, the Office of Pipeline Safety issued a Corrective Action Order (CAO) that required Marathon Pipe Line (operator at the time of the accident) to verify the integrity of the pipeline through hydrostatic testing or in-line inspection.

Description of Integrity Issue

The pipe segment that failed in 1997 was manufactured by A. O. Smith Corporation from plate and used a flash welding process to produce the longitudinal seam (Table 1). The flash welding process is an application of electric resistance welding (ERW) where the joint is heated above its melting point by its resistance to the flow of the electric current. A force is then applied immediately following the heating that causes an expulsion of metal and the formation of a flash. The weld is made simultaneously over the entire area of abutting surfaces by the application of pressure after the heating. The strength of the joint is affected by the tolerances on the edges of the joint and the cleanliness of the surfaces to be joined.

In the late 1960's and early 1970's, the pipeline industry experienced a number of failures of longitudinal seam welds in ERW pipe. Metallurgical testing of these failures showed that the failures resulted from cracks that grew from defects in the ERW seams.

Inspection Tool Selection

Marathon Pipe Line was required to perform either a hydrostatic test or an in-line inspection, as part of a Corrective Action Order. There were a number of disadvantages to performing a hydrostatic test, including cost, removal of the pipeline from service, and disposal of the test water. Ultrasonic inspection could have been used but a tool appropriate for the 20-inch line was not available. Recent advances in inspection technology indicated that Transverse

Table 1
Platte Pipe Line Parameters

Platte Pipeline Parameters	
Commodity	Crude Oil
Total Length	932 miles
Size	20 inches OD
Wall Thickness	0.312 and 0.344 inches
Manufacturer	A. O. Smith
	Miles – 550 miles
	Seam – ERW (flash welded)
	Youngstown
	Miles – 255 miles
	Seam – ERW
	Kaiser
	Miles – 127 miles
	Seam – SAW

Field Inspection (TFI) might be a viable method for detecting defects in ERW seams. Marathon Pipe Line contacted Pipeline Integrity International (PII) to work together to develop a 20-inch Transverse Field Inspection (TFI) in-line inspection tool.

For Platte Pipe Line to establish pipeline integrity based on TFI technology, three items needed to be established:

- The ability of the tool to detect the suspected anomalies on the line.
- Data analysis techniques to sort and discriminate the data collected by the tool.
- The ability to evaluate the suspected anomalies against the established criteria to determine if they required excavation.

Field Verification

To validate the TFI in-line inspection tool in the field, suspected anomalies needed to be confirmed through non-destructive examination (NDE). It was determined that sheer wave would be inadequate due to sensitivity issues, so the Flaw Assessment and Sizing Technique (FAST) was selected based on the technique's ability to detect crack-like anomalies. Use of FAST was then proposed to OPS. The FAST technique is a multimodal ultrasonic sensor probe. To validate the FAST technique, a field trial run was conducted. The TFI in-line inspection tool was directed through a section of pipe and suspected anomalies identified. Anomalies detected by the TFI tool and characterized by the FAST method were validated based on destructive examination. Platte initially ran the TFI in-line inspection tool through a 50-mile section of pipe. From the data, 50 TFI signals were selected within 10 excavation sites that represented a variety of signal characteristics. At each site, a section of the pipe was removed to perform destructive examination. The destructive examination confirmed the NDE findings and validated the ability of the TFI in-line inspection tool to identify these longitudinal-type anomalies.

TFI Tool Validation Program

The Validation Program was conducted on a 73-mile section of pipe and included the following steps:

- Perform a TFI in-line inspection run
- Analyze the data to select suspected seam weld anomalies
- Excavate all selected sites
- Perform NDE at excavated sites
- Perform pipe replacement based on NDE findings for seam-weld cracks greater than 10% of wall thickness

A total of 262 excavations were performed as part of the Validation Program. From these excavations, 63 anomalies were removed by pipe replacement. Several of the pipe sections were then hydrostatically tested to failure. Of the defects that were hydrostatically tested to failure, only one defect failed at a pressure less than 100% of specified minimum yield strength (SMYS). This defect failed at 1550 psi or 95.9% SMYS. If the pipeline integrity had been validated solely on hydrostatic testing, this defect could have gone undetected due to its size, and would probably have failed at a future date while in-service. The ability to detect smaller defects than can be detected by hydrotest is a major advantage of in-process inspection.

After all the identified defects had been repaired through pipe replacement, the 73-mile section of pipe was hydrotested to within 90% - 95% SMYS with no failures. Because the section passed the hydrotest, there was a high degree of confidence that no critical defects remained in the section following the TFI inspection. This, combined with the detection of numerous defects that failed at greater than SMYS, led the TFI in-line inspection tool methodology to be considered an acceptable alternative to hydrotesting for the entire line.

Implementation

Inspection on the entire line began in late summer of 1998 and continued through March 1999. Inspecting the entire line required 12 cleaning tool runs, 12 geometry tool runs, and 22 TFI tool runs. Each section, approximately 100 miles in length, required 25-30 days of preparation work and was removed from service for approximately 10 days for repairs. Once the data from the TFI in-line tool was received, the data was analyzed to identify excavation sites. After the pipeline was excavated, the external surface was cleaned to perform FAST NDE. The NDE was performed to confirm the presence of a rejectable defect. These defects were then repaired through pipe replacement. A total of 1,262 excavations were made and 381 crack-like defects were removed. The ratio of excavations to defects was 3.3:1. This is high, but the number should improve as the TFI data analysis techniques are refined through additional inspections.

Summary

The use of the TFI in-line inspection tool on the Platte Pipe Line to detect longitudinal cracks was proven to be at least as effective as hydrostatic testing. It offered the additional benefit of detecting many defects that may "just survive" a hydrostatic test and possibly fail in the future. The TFI in-line inspection tool can be used in-service and reduces pipeline outages. Due to its ability to perform the inspection in-service, with modest cleaning requirements, the costs associated with the TFI in-line inspection tool are less than a hydrostatic test. As TFL in-line inspection runs increase, the accuracy of data assessments should improve to reduce the number of excavations required.

Trans Alaska Pipeline System

The Trans Alaska Pipeline System (TAPS) operated by Alyeska Pipeline Service Company (Alyeska) began their inspection program to identify differential settlement due to melting permafrost that had resulted in two leaks in 1979. Alyeska also identified a problem with external corrosion at the time. Alyeska conducted 58 separate inspection runs using eight different tool designs between 1978 and 2001. As part of their inspection program, Alyeska has integrated inspection data from multiple inspections to develop an Integrity Management System.

Background

The TAPS consists of an 800-mile (1287 km), 48-inch (122 cm) diameter pipeline extending from Prudhoe Bay to Valdez, Alaska. A majority of the pipeline route is underlain by permafrost, which when thawed may subject pipeline structures to subsidence. Therefore 420 miles (676 km) of the pipeline was constructed aboveground to avoid the adverse effects of thaw-settlement. Conventional buried pipeline construction is used for 376 miles (605 km) in thawed or thaw-stable ground, with the remaining 4 miles (6.4 km) refrigerated, insulated and buried in thaw-unstable ground.

The TAPS began operation in June of 1977, following three years of construction. Table 2 details the design parameters for this system. Table 3 details the MOP based on the pipe grade and wall thickness.

Table 2
TAPS Line Parameters

TAPS Line Parameters	
Commodity	Crude Oil
Length Analyzed	800 miles
Size	48 inches OD
Material	API 5L Grade X60, X65, and X70
Wall Thickness	0.562 and 0.462 inches
Manufacturer	NKK, Nippon, Sumitomo, Ilva
Seam Type	DSAW, includes 100 miles of spiral
Coating	Fusion Bonded Epoxy, tape overcoat

Table 3
TAPS Maximum Operating Pressure

Wall Thickness (in)	Pipe Grade	MOP (psig)
0.462	X60	832
0.462	X65	901
0.462	X70	970
0.562	X60	1012
0.562	X65	1096
0.562	X70	970*
0.562	X70	1180

*Atigun reroute.

Description of Integrity Issue

In 1979, TAPS experienced two leaks that were related to differential settlement due to melting permafrost. Following these failures, Alyeska routinely ran internal inspection tools of various types to monitor locations of known problems and to identify new occurrences. In 1987, external corrosion was also identified as an integrity issue based on a run of an improved MFL corrosion tool.

Inspection Tool Selection

Since Alyeska began their Integrity Management Program in 1978, they have conducted 58 separate in-line inspection runs using eight different tool designs. Each tool was selected to provide the best possible results based on the available tools at the time of the inspection for the suspected problem. Table 4 lists the eight tools and the year(s) they were used on TAPS.

Table 4
In-Line Inspection Tools Used on TAPS

Tool	Year(s) Used
TDW Kaliper	1979, 1981
Vetco MFL Corrosion	1978, 1987
Vetco Corrosion/Deformation	1981, 1987
Vetco Deformation	1987, 1998
IPEL (Pipetronix) MFL Corrosion	1987, 1988
IPEL (Pipetronix) HRMFL Corrosion	1992
NKK Ultrasonic Corrosion	1989, 2001
BJ Inertial Navigation Curvature	1992, 2000

Inspection and Assessment

Alyeska has performed over 800 excavations and physical examinations of the pipeline in response to ILI results and as part of other maintenance activities. The in-line inspections and excavations have provided significant knowledge about the types of defects that can occur on the TAPS. Alyeska has performed a critical review of this information to identify potential integrity issues and evaluate the risk associated with each.

Table 5 provides a listing of integrity issues observed over the operating life of the pipeline, along with an estimate of the numerical magnitude of the occurrence of these defects.

In addition to excavations and in-line inspection tool indications, Alyeska has conducted screening risk assessments or informational analysis to systematically assess integrity-related threats. Corrosion, mechanical defects, washout settlement, and excavation damage are the primary integrity-related issues mitigated by integrity testing and account for approximately 44 percent of the total perceived risk on the pipeline. The use of ILI identifies these potentially threatening defects, so they can be assessed, inspected, and repaired if necessary.

Data Analysis and Integration

Alyeska has developed and implemented a comprehensive Integrity Management Program (IMP). The IMP is based on the requirements of 49 CFR 195.452 and API Standard 1160. Key parts of this program are the data management and risk management systems. These systems are designed to support the gathering, review, and integration of pipeline integrity data and perform a line-wide risk assessment. The overall objectives of the IMP is to:

- Integrate critical integrity-related data
- Support integrity maintenance decisions in a timely, accurate, secure, and readily accessible manner
- Assess operational and environmental risks on the TAPS
- Prioritize maintenance resources in high consequence and other areas commensurate with the risks perceived

The data management and risk management systems are based on, and contain "all available information" associated with Alyeska's operating history, the regulations, and industry practice. This body of knowledge includes but is not limited to the following:

- Pipeline configuration data contained in the as-built engineering database
- Screening risk assessments completed on 5-year intervals (1991, 1995, and 2001)

Table 5
Assessment of Integrity Issues for the Entire Length of the TAPS

Observation	Number of Observations
External corrosion	25,000 joints with indications > 50 mils
Internal corrosion	250–300 bottom of pipe indications
Microbiologically-influenced corrosion	5–10 external indications
Girth-weld corrosion	<10 significant locations
Seam-weld corrosion	2 observed in heat affected zone at animal crossings
Narrow axial external corrosion	2
Dents	100 (>2%) constrained rock dents on bottom of pipe
Mechanical damage	10–100 repaired, includes dents with metal loss
Fatigue	None known
Corrosion fatigue	None known
Stress corrosion cracking	1 (sleeved), unique circumstances
H_2S (hydrogen-induced stress cracking)	None known
Curvature	24 (>85% critical curvature)
Insulated animal crossing	2 of 23 non-refrigerated crossings had significant base metal corrosion; 4 had undercutting of double submerged arc weld
Monitoring rods	~ 2,200 monitoring rods, potential for corrosion at cadweld
River weights	~ 7,000 locations, potential for longitudinal corrosion
Casings	32 locations, potential for shorted casing corrosion
Sleeves	100 locations, potential for under sleeve corrosion
Structural overfill	150 locations, potential for elevated curvature
Vandalism or bullet strikes	10–50 locations observed during line-walk

Note: Data as of March 2002.

- ILI experience with 58 ILI runs as of March 31, 2002
- Direct examination of the pipeline in excavations at over 800 locations as of March 31, 2002
- Leak experience, based on eight pipeline leaks over 50 bbls as of March 31, 2002
- RCM analysis of the TAPS mainline
- Defects defined by 49 CFR 195.452 (h)
- Defects defined by industry practice, ASME B31.4, and API Standard 1160

Alyeska's risk management system encompasses all aspects of the company's day-to-day business, such as pipeline and facility integrity, operations and maintenance, safety, and environment. The system:

- Identifies hazards and risks
- Evaluates and assesses the risks
- Recommends remedial or mitigating actions to reduce the risks to an acceptable level
- Complies with appropriate regulations and standards

Data Management System Description

Alyeska's data management system is designed to support all required integrity-related maintenance decisions in a timely, accurate, secure, and readily accessible manner. Operating experience is supplemented by risk assessments in order to define pipeline failure modes and required maintenance decisions. The data management system is capable of maintaining records of decisions. The following methods, processes, and procedures describe Alyeska's data management system.

Alyeska has developed an Oracle^{tm}-based, central database repository to store the large volume of data gathered over its operating history and provide a secure environment to maintain historical data. In addition to the central repository, two customized portals (entry points into the data system) have been developed to provide data entry, query, and analysis capabilities. These portals, referred to as Engineering Data Management/Corrosion Data Management (EDM/CDM), provide secure data entry and allow Alyeska engineers easy access to information needed to make decisions from their desktop or portable computers. The following is the key information contained in the data management system:

- Location of High Consequence Areas and could-affect pipe segments
- As-built survey information for significant features found along the pipe
- Pipe design and construction information by joint, including the location of all welds, seam type, coating type and wall thickness, yield strength, maximum operating pressure, and manufacturer
- Monitoring history of the above-ground system, including the vertical support members (VSMs) and heat pipes
- Monitoring history of the below-ground pipe, including thermistor readings and monitoring rod data
- Cathodic protection (CP) system and monitoring history, including test station and CP coupon readings, rectifier readings, and close interval survey data
- Corrosion pig data
- Dent and curvature pig data
- Pipe investigation history and results, including wall thickness found in the field, sleeve locations, and derates in operating pressure
- Corrosion inhibitor program and associated internal coupon monitoring results
- Significant features that may increase risk such as river weights, floodplains, etc.
- Graphic representation of integrated data

The integrated monitoring overlays (Figure 7) is one example of Alyeska's ability to stack (integrate) and graphically display data for improved analysis and planning. The overlays provide graphical representation of corrosion pigging data, CP system and monitoring data, and significant features overlaid to the same scale for every mile of below-ground pipe. Other data sets are routinely stacked on an ad hoc basis for analysis purposes without the graphic component. As part of continual improvement further data stacking may be implemented.

The engineering portal (EDM) provides access to two major classifications of data, reference data and pipe monitoring data. Reference data includes configuration data gathered during pipeline construction and subsequent maintenance activities. Monitoring data includes data gathered on the above-ground support assemblies, elevation surveys, and thermistor readings.

The corrosion portal (CDM) provides access for tracking and analyzing corrosion-related monitoring, including test stations and CP coupon monitoring, bi-monthly rectifier performance, corrosion pig calls, and the corrosion investigation program. Additional monitoring data includes tank inspection results, the internal coupon monitoring program, and inhibitor injection.

Data Analysis

Alyeska's data management system is designed to support all integrity-related maintenance decisions required by Alyeska's IMP. The results of the analysis are used to reach one of three decisions described below:

- **Remediation required -** Two conditions must occur to determine that repair/mitigation is required: (1) the anomaly does not meet the minimum standard as set

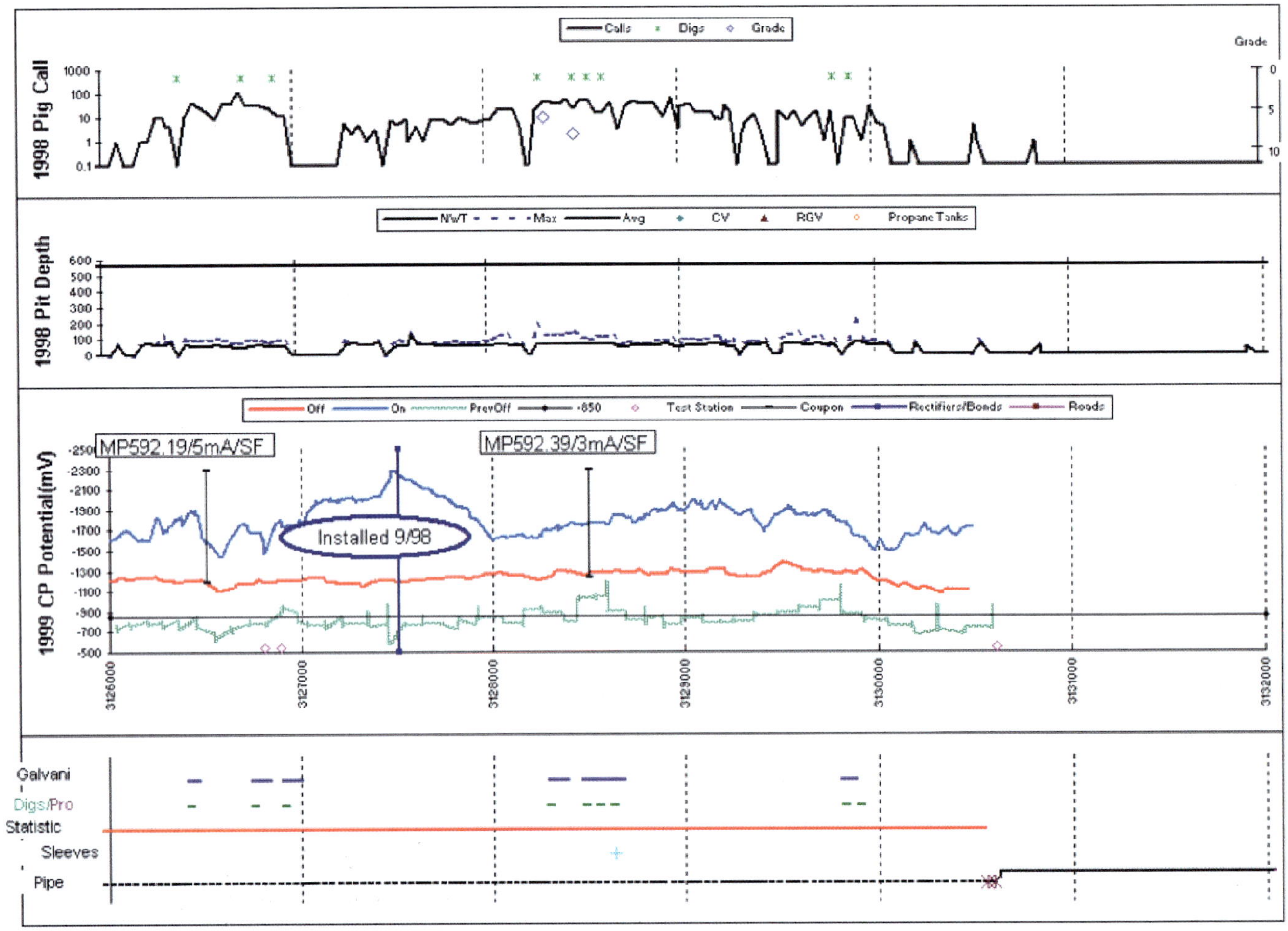

Figure 7. Example of integrated monitoring overlay.

forth by Alyeska procedures and (2) adequate information is available to make a decision.

- **Engineering evaluation required -** If (1) the anomaly has not yet reached the level requiring immediate action but has reached a level requiring further analysis, or (2) adequate information is not yet available to determine that remedial action is required, then further engineering analysis or data acquisition is required.
- **No remediation required** – If an anomaly does not meet the criteria cited above, repair/mitigation or further engineering analysis is not needed

The locations of anomalies requiring either remediation or engineering evaluation are maintained in EDM/CDM. Locations requiring remedial action are additionally tracked in Alyeska's maintenance management system. At the end of each year, Alyeska reviews the status of recommended maintenance and overall success of the program and updates the procedures to reflect changes, if any.

Summary

The Integrity Management Program developed by Alyeska is unique in its integration of in-line inspection data from multiple inspections, engineering data, cathodic protection data, and other inspection data. The data management system is a key feature of Alyeska's comprehensive risk management approach to maintaining line integrity.

Columbia Gas

Columbia Gas, operator of a natural gas pipeline, conducted an inspection of a 39-mile segment of the line in Virginia using an in-line inspection tool to identify corrosion. The in-line inspection was accomplished using a high resolution Magnetic Flux Leakage (MFL) tool. The high resolution MFL tool was selected for its ability to detect both internal and external wall loss in natural gas pipelines without the use of a liquid couplant.

Background

The Richmond, VA to Petersburg, VA natural gas line was constructed in 1957 of electric resistance welded pipe. The manufacturer of the pipe is not known. Both Polychem Tape and Semastic Asphalt were used to coat the pipe. Cathodic protection is used to protect against external corrosion, however the date on installation of the cathodic protection is unknown. The MAOP for this line segment is 639 psi. Table 6 details the design parameters for this system.

Description of Integrity Issue

During bell-hole inspections, Columbia Gas found evidence of disbonded coating. This combined with low close interval survey (CIS) readings and anodic to anodic DCVG, led pipeline personnel to believe that the line may be susceptible to external corrosion. Since similar ell-hole inspection findings, cathodic protection readings, and soil composition were observed along this line segment, the potential for external corrosion was identified as a possible systemic problem that could affect the entire segment.

Inspection Tool Selection

Columbia Gas selected a high resolution MFL tool because of its ability to accurately locate and detail both internal and external wall loss anomalies. The MFL tool was selected over the ultrasonic tool because it does not require the use of a liquid couplant.

Tool Performance Specifications

Pipeline Integrity International (PII), the company contracted to supply the high resolution MFL tool and evaluate the inspection results, made the performance specifications for the tool described in Table 7.

Table 6
Columbia Gas Pipeline Parameters

Columbia Gas Pipeline Parameters	
Commodity	Natural Gas
Total Length	39 miles
Size	18 inches OD
Material	X52
Wall Thickness	0.250 inches
Manufacturer	Unknown
Seam Type	ERW
Coating	Polychem tape and semastic asphalt

Table 7
Tool Performance Specifications

Parameter	Specification
Smallest bend radius for passage	1.5 D
Minimum metal loss size	0% wall loss for 2t × 2t anomaly
Accuracy of metal loss measurement (for 3t x 3t anomaly)	±10% of nominal wall thickness
Location accuracy of detected anomalies (linear measurement)	±0.5 m (1.6 ft)
Tool speed	0.1 to 2.25 m/sec (1.3 to 7.4 ft/sec)
Maximum wall thickness	1 inch
Maximum pressure	1750 psi

Note: D = pipe diameter, t = nominal wall thickness.

Inspection and Assessment

Columbia Gas conducted two inspections of the Richmond, VA to Petersburg, VA natural gas line. The first inspection, conducted in June 2001, was made using the Enduro DDL digital tool. This inspection was conducted to obtain geometry data for the line. A second inspection using the PII High Resolution MFL tool was conducted in November 2001. The data from each of the inspections was analyzed by the tool vendor. Columbia Gas performed an independent analysis of a majority of the anomalies.

Using the MFL tool, 2151 corrosion anomalies were identified based on the measured wall loss of greater than 10%. Of the 2151 anomalies, 1989 were observed to have a measured wall loss between 10 and 20%, and 162 anomalies had a wall loss greater than 20%.

Based on current planning, Columbia Gas anticipates that the follow-up investigations will include an excavation and direct local inspection of the pipe. The direct local inspection of the pipe will include measurement of the defect size and shape and the use of ultrasonic thickness measurement to determine the pipe wall thickness. Additionally, Columbia Gas will use Magnetic Particle inspection techniques if dents or cracks are evident or suspected.

Columbia Gas has adopted the following criteria to establish when an anomaly is considered significant and follow-up action is required:

- Anomalies with a measured wall loss greater than 80%
- Anomalies that could potentially lower operating pressures
- Large groups of anomalies with a 10% or greater wall loss that may be indicative of a problem (coating or cathodic protection failure)

Summary

The use of the MFL inspection tool on the Richmond, VA to Petersburg, VA natural gas line has identified a significant number of anomalies for follow-up analysis. When the follow-up investigations are conducted, this inspection program should provide valuable insight to the capabilities of the High Resolution MFL tool and its usefulness as a part of an integrity management program.

Source

Robert J. Hall, Mona C. McMahon, General Physics Corporation, Report on the Use of In-line Inspection Tools for the Assessment of Pipeline Integrity, June 2002. U.S. Department of Transportation, Research and Special Programs Administration, Office of Pipeline Safety. Report No. DTRS56-96-C-0002-008. For the complete report, go online at: www.phmsa.dot.gov.

Factors to consider when evaluating damage on pipelines

Michael J. Rosenfeld, P.E., Kiefner & Associates, Inc.

Mechanical damage has been widely recognized as a leading cause of significant incidents and failures in pipelines. It's the number one cause of pipeline incidents reportable to the US DOT, accounting for at least 1 of every 3 pipe-related incidents in gas and liquid products pipelines.

The characteristics of mechanical damage have been a subject of study since the mid-1950's. While a simple damage assessment criterion has so far eluded researcher's best efforts, the results of tests and service experience have revealed many factors that are important for the pipeline operator to consider when responding to damage on a pipeline.

Two forms of damage

Mechanical damage commonly manifests itself in one of two forms. The first and most serious is encroachment damage (or "third party" damage) that occurs when the pipe is struck by earthmoving equipment. It usually consists of a shallow residual dent plus a gouge, sometimes called a "combined defect" because it consists of both a geometry distortion and a stress- concentrator or notch.

The other important form of damage arises principally during construction of the pipeline. Most often the damage is seen as prominent dents on the bottom half of the pipeline, often called "plain dents". The traits of these two forms of damage, and the appropriate treatment for them when the damage is detected, will be discussed in more detail subsequently.

Encroachment damage

Examples of encroachment damage are shown in Figure 1. Common causes include contact from excavators, graders, ditchers, plows, directional drillers, and just about any other sort of equipment used to penetrate or move soil. General excavation is the most common source of damage, but everything from roadwork to farming has caused incidents, including the pipeline operator or his contractors hitting his own line. Offshore, encroachment damage arises from contact by ship keels and anchors.

Two aspects of encroachment damage interact to render the combined defects particularly serious for pipelines:

- metallurgical damage in the scrape or gouge, and

Figure 1. Examples of mechanical damage as found in the field.

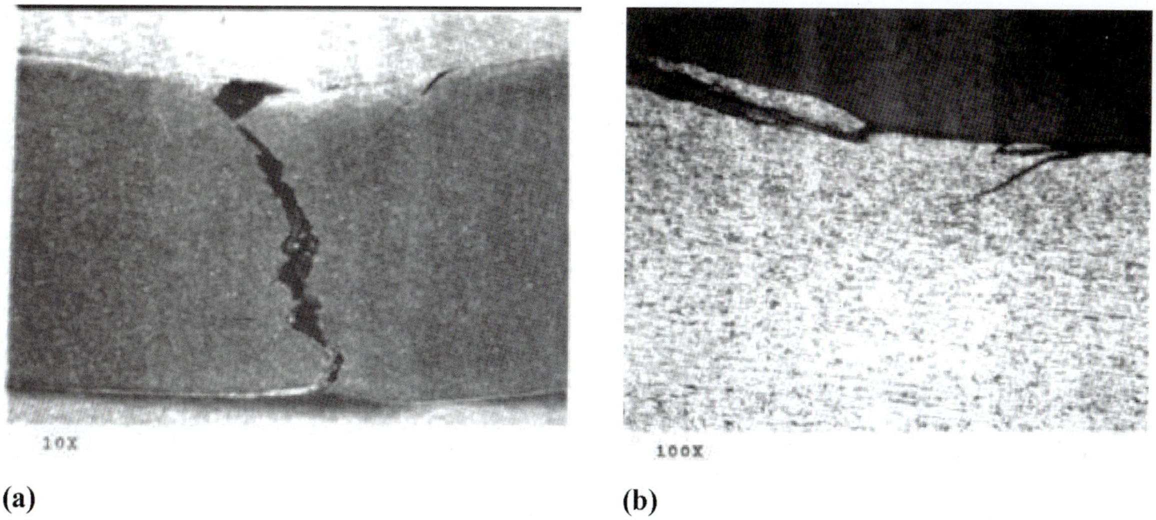

Figure 2. Cross section of cracking in mechanical damage.

- rerounding of the dent under the influence of internal pressure.

Metallurgical damage occurs where the striking object contacts and slides along the pipeline surface. Visually, the damage appears as a scrape or gouge, and might not look very severe. But within the scrape or gouge, plastic flow, metal transfer, and even remelting due to the heat of friction would have occurred at the contact point. For some depth below that, the microstructure is crushed and severely cold-worked, locally degrading the ductility and toughness. This type of damage is evident in a cross sectio n at the root of a gouge. Figure 2(a) shows a cross section of a fracture at a gouge. Figure 2(b) shows the severely crushed microstructure in the gouge just below the pipe surface, as well as cracks which often form due to "rerounding".

As the contacting object indents and scrapes along the pipe surface, the pressure in the pipe immediately pushes the dent back out (in other words, the dent "rerounds") behind it. As the dent pushes out, very high tensile strains develop in the root of the scraped or gouged area which may cause cracks to form in the cold-worked, reduced-ductility metal. If the pipe doesn't fail right away (which actually occurs in about 80% of encroachment damage failures), a crack is left which can grow in service. The crack may not be visible to the naked eye.

Although the residual dent depth is usually very shallow, tests have shown that the dent never fully rerounds, which means that it continues to fluctuate in and out with changes in operating pressure. It can take many cycles of pressure for the dent to "shake down" to elastic action, which means that plastic strain can continue to accumulate in the damaged area. If the damage survives shakedown, the dent will subsequently cycle elastically over a smaller range of motion,

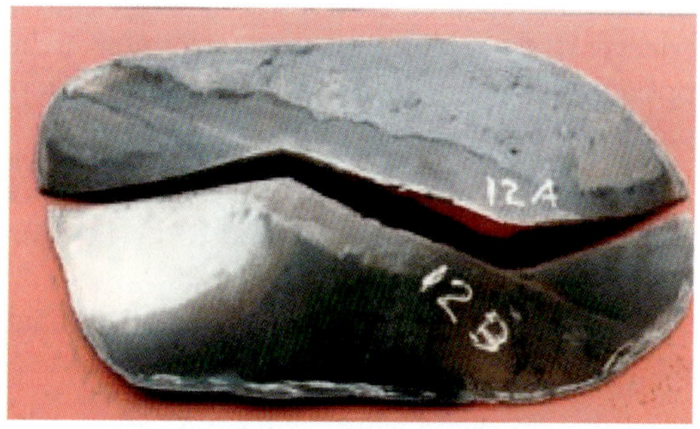

Figure 3. Failure in a gouge crease.

providing a mechanism for continued crack growth by fatigue in service.

The interaction of metallurgical damage with rerounding behavior in the manner described above seems to explain the delayed nature of many mechanical damage failures. Encroachment damage failures typically occur in the crease left by the striking equipment, as in the example shown in Figure 3. Often multiple creases are present where such damage occurred, such as that shown from the inside of a pipe in Figure 4.

Evaluation and repair of encroachment damage

There remains no reliable method for evaluating the safe operating pressure or the safe time to failure of pipe affected by encroachment damage. When such damage is found, the operator should be prepared to make a repair as soon as possible. Moreover, when entering a ditch to investigate possible encroachment damage, the operator should reduce the line pressure by at least 20 percent to provide a margin of

Figure 4. Multiple creases from backhoe.

safety for the worker, since the damage could literally fail at any time.

A number of repair options are available to the operator. Steel sleeves designed for pressure containment are always an appropriate repair method. Composite wrap repairs are acceptable, provided the following steps are taken:

- the scrape or gouge is ground down to a smooth contour,
- the damaged area is inspected to verify that any cracks have been removed by grinding, and
- the residual indentation is filled with a hardenable filler under the sleeve.

Where the damage is light, it may be sufficient to simply grind the damage out and repair the coating. The safe length and depth of grinding may be determined from the following equation:

$$L \leq 1.12\left[(Dt)\left[\left(\frac{a/t}{1.1(a/t) - 0.11}\right)^2 - 1\right]\right]^{1/2}$$

where L is the length of grinding, a is the depth of grinding, D is the pipe diameter, and t is the pipe wall thickness, in any consistent system of units. The maximum safe depth of grinding is 40% of the wall thickness. The pipe surface should be inspected to verify removal of any cracking, and to check the remaining wall thickness. The effectiveness of this repair technique was demonstrated by testing where

- the burst strength of pipe containing combined defects severe enough to cause failure at low stress was increased to well above 100% of SMYS by grinding them out, Figure 5(a), and
- the pressure-cycle fatigue life of pipe containing combined defects was increased by a factor of 2 to 10 by grinding them out, Figure 5(b).

Figure 6 shows a pair of 36- inch OD gas flare lines that were damaged when they fell off their pipe rack during a plant upset. Also shown is the repair being made by careful grinding to a smooth contour, accompanied by surface NDE.

Construction damage

Examples of damage resulting from poor pipe installation practices are shown in Figure 9. In every case, the pipe was inadequately protected from settlement onto rocks in the ditch bottom. In general, urethane or styrofoam pillows are inadequate to support the pipe under the combined weights of the pipe, overburden, and hydrotest water or product inside the pipe.

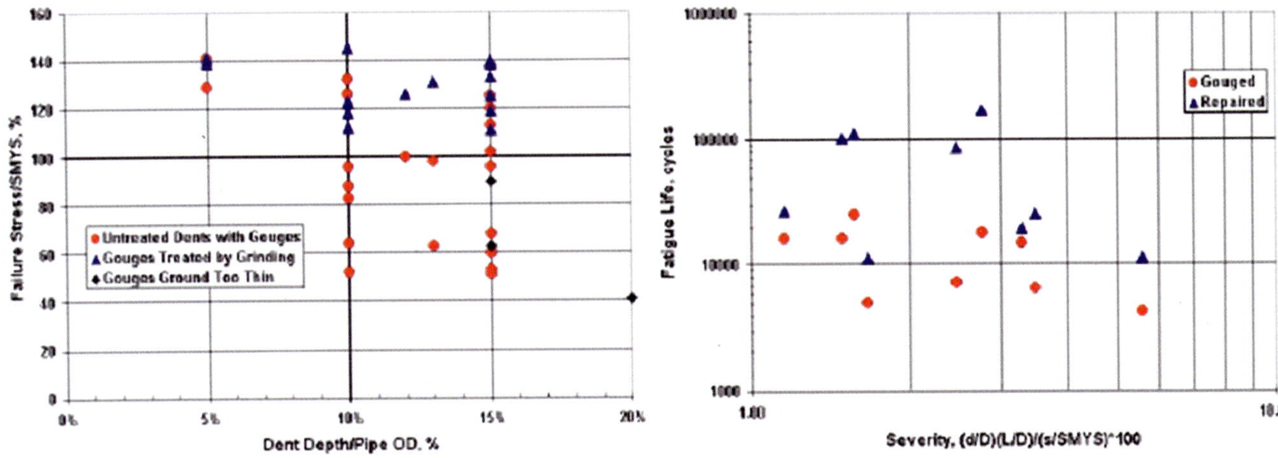

(a) (b)

Figure 5. Beneficial effects of grinding damage on burst strength (a) and fatigue life (b).

Figure 6. Mechanical damage on rack piping and repair by grinding.

Figure 7. Typical rock indentation.

Rock shield wraps and thin-setting epoxy-based concrete coatings provide effective protection for the corrosion-prevention coatings in rocky soils, but cannot prevent indentation of the pipe due to settlement. Heavy concrete coating (1 inch or thicker) provides more effective protection from rock, though it can be penetrated by sharp rocks. Graded fill materials (sand, for example) can provide adequate padding when properly installed. Setting pipe on regularly spaced sandbags in the ditch and pouring the padding over the pipe will leave an unfilled gap below the pipe, resulting in spanning over the supports and settlement of the pipe into the gap. The most effective way to prevent damage in the ditch is to diligently remove rocks from the ditch bottom or else pour in a generous bed of padding sand or fines before laying the pipe in.

Other forms of construction damage besides indentation can occur, such as scrapes, dents, or buckles. The scrapes and dents are essentially similar to the encroachment damage suffered in service. They should be cut out if found before the pipeline enters service, or repaired the same as encroachment damage if found later on.

Rocks in the ditch or backfill tend not to produce metallurgical damage of the extent observed in backhoe hits, though some localized metal damage is possible if the rock is hard and sharp. Also, the rock- induced indentations are restrained from rerounding by the surrounding earth. So unlike encroachment damage, rock damage does not pose an immediate threat to pipeline integrity. Rather, conditions are established for potential long-term problems such as:

- coating damage
- shielding from cathodic protection
- corrosion
- stress-corrosion cracking (SCC) or hydrogen cracking
- punctures due to continued settlement

Examples of these problems are shown in Figure 8. Failures caused by rock punctures occur as leaks; failures caused by corrosion or environmental cracking that develops as a result of rock damage may occur as ruptures. Rocks have been documented in at least 5% of reportable incidents caused by corrosion. The actual contribution of rocks to corrosion problems is likely to be higher than that.

A couple of other surprising problems associated with rock dents have been observed on operating pipelines. One is the occurrence of fatigue cracks or SCC in the saddle-shaped area between closely-spaced dents. Tests showed that the saddle area has a large radius of curvature and is unrestrained from flexure, creating a mechanism for cyclical crack growth.

A second problem has been the occurrence of fatigue cracks in a rock dent that was excavated, the rock removed, and the coating repaired. Removal of the rock left the dent unrestrained and free to fluctuate in response to changes in operating pressure, leading to fatigue. This has only been observed in liquids pipelines, due to their more intensely cyclic operating spectrum compared to gas pipelines. In fact, tests have shown that the pressure-cycle fatigue life of dents restrained by rocks is at least an order of magnitude greater than that of unrestrained rocks. This suggests that operators of liquids pipelines might be better off leaving rock dents in place and focusing on corrosion inspection and control.

Figure 8. Forms of damage from rocks: puncture, coating damage and corrosion, local mechanical damage, and stress corrosion cracking (clockwise from upper left).

Evaluation and repair of construction damage

Rock- induced damage can be repaired by a number of methods, including:

- welded steel sleeves, either designed for internal pressure, or with the ends left unwelded (a hardenable filler in the dent space is recommended in either case)
- composite wrap repair with hardenable filler in the dent space
- repair to the coating only (appropriate for gas pipelines).

Present regulations and industry codes indicate that dents deeper than 6% of the pipe OD should be repaired. Historically, the main consideration behind the 6% criterion was to accommodate pigging. While the 6% criterion probably leads to repair of many of the most damaging rock dents, the severity of the dent is not uniquely a function of depth alone. Punctures and coating damage have occurred in dents shallower than 6% of the OD; conversely, deformations spread out over large areas of the pipe caused by the pipe sitting on a flat rock ledge may represent negligible damage to the pipe metal and coating even where the reduction in diameter exceeds 6%.

An alternative to dent depth that is gaining some usage is to make use of the actual dent geometry (as measured in the field or by high-resolution deformation ILI tools) to estimate the strains associated with the dent, since strain is a measure of material damage. As Figure 9 shows, dent depth does not correlate to strain, but thresholds for some kinds of failure do correlate to strain. The question is, how much strain is too much? We know that strain up to 3% is acceptable because that is what is incurred in allowable cold field bends. Evidence is given by the corroded dents that coal tar may disbond at strains of 4% to 5%. Test specifications for fusion-bonded epoxy require surviving strains between 2% and 3%, though test straps have been observed showing crack resistance to strains of 7%. Finally, cracks have been observed in pipeline buckles at strains that are a large fraction of the uniaxial elongation. Hence a maximum strain of 6% would provide a reasonable factor of safety against metal failure in a restrained dent, with the onus that continued monitoring for corrosion may be necessary in the long-term, depending on coating type. The ASME B31.8 Code is adopting new allowances for strains in deformations up to 6% as an optional alternative to depth-based limits.

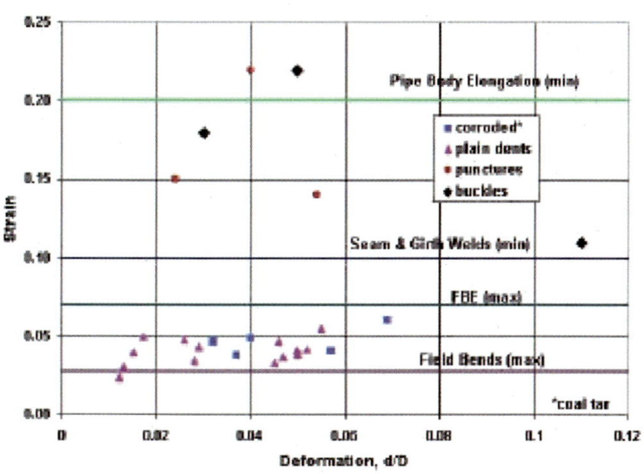

Figure 9. Strain levels associated with damage thresholds.

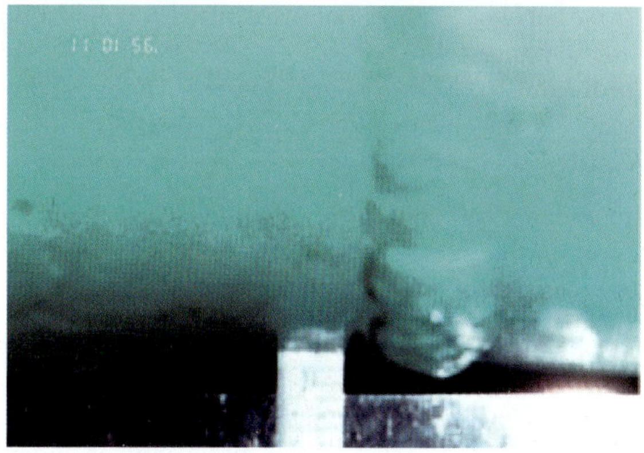

Figure 10. Rock dent on a girthweld.

Occasionally, rock dents are discovered on pipeline girth or seam welds, suc h as what is shown in Figure 10. Present ASME Code requirements are that such a dent must be repaired. However, tests and analysis demonstrate that if a weld is ductile and free of gross defects, and the operating pressures are not severely cyclical, the weld can tolerate a dent. The ASME B31.8 Code is adopting new allowances to permit shallow dents in welds of sound quality.

In-line inspection

With encroachment damage, the residual dent may have a depth of as little as 0.5% of the pipe OD, posing a serious challenge to reliable and consistent detection using geometry-type in- line inspection (ILI) tools. Sometimes the damage produces a signal that, with experience, can be interpreted as such using magnetic flux leakage or ultrasonic ILI tools.

As an aid to interpretation of possible encroachment damage indications, the operator should consider:

- shape – long and narrow aspect ratio, or several parallel or adjacent similar features, crease- like shape;
- clock position – most encroachment hits come from above;
- the physical lay of the pipeline – an indicated "sag bend" where the line crosses flat terrain may represent creases

from backhoe scrapes being confused for mild ripples in a field bend;
- proximity to above-ground features where earthmoving may occur — nearby road crossings or recent construction, for example.

Rock dents can usually be recognized by size, shape, and position on the pipe. Typical traits include:

- depth – 2% of the OD or greater;
- clock position — most rock- induced damage occurs on the lower half of the pipe;
- shape — rounded profile rather than long and narrow;
- extent — significant amounts of ovality or indentation extending over several feet of pipe. The data from a high-resolution deformation tool can be used to estimate strain as a way of evaluating the severity of the dent.

Conclusion

The foregoing discussion highlighted the fact that mechanical damage in pipelines is a complex issue. The severity of damage, potential failure modes, and appropriate repair choices all depend greatly on how the damage was introduced. Encroachment damage presents an immediate hazard potential, whereas construction damage presents a long-term maintenance concern.

Reprinted with permission - Applus+, Energy & Industry Division, 3 Sugar Creek Center Blvd. Suite 600, Sugar Land, TX, and *Oil & Gas Journal*, September 2002. Online at www.applus.com.

Understanding mechanical damage in pipelines

Excerpts from the Mechanical Damage Study performed by Michael Baker Jr., Inc. for the U.S Department of Transportation, Office of Pipeline Safety, April 2009.

Defining mechanical damage

Mechanical damage is defined as follows:

> *Localized damage to the pipe resulting from contact with an object.*

"Localized" means that the damage is confined to some portion of the pipe cross section or that it has a limited extent along the pipe's length (for example, less than five pipe diameters).

Contact with an object results in a load or stress experienced by the pipe that has not been considered during the design of the pipeline. Such loads and stresses can be of short or long duration. The contact is usually unintentional, but may be deliberate (e.g., vandalism). Contact may occur during manufacture (excluding pipe-forming activities), transport, construction, operations, or maintenance. (Note that modifications to the pipe due to properly conducted maintenance activities are excluded, but that damage due to incorrectly conducted maintenance activities is included.) Most mechanical damage that could lead to a failure occurs as excavation damage, either by the operator or its contractor, or, more prevalently, from third-party activity over or near the pipeline.

Mechanical damage cannot be typified as a singular condition. The damage introduced by various kinds of equipment or particular circumstances may result in a broad range of impacts to the pipe. The physical attributes of the damage itself — length, depth, width, orientation, and surface appearance — can vary greatly. As with any defect, the severity of the condition is dependent on the defect geometry, the properties of the pipe steel, and the stress level at which the pipeline operates. Ultimately, the significance of the damage is "path dependent," meaning it is strongly affected by all of the circumstances of the damage event and the affected pipe. Many of the inciting causes of a particular occurrence of mechanical damage may be unknown because the damage event was not witnessed or reported. In addition, the effects of particular types of damage on pipeline integrity are not fully understood from a scientific standpoint.

Mechanical damage is classified in three broad categories: dents, gouges, and combined dent/gouge defects. Figure 1 shows examples. Each mechanical damage defect reflects a unique combination of conditions, such as the indenter size and shape, as well as force, angle, orientation, and duration of impact. The different forms of damage can be generalized to facilitate assessment and prioritization of repairs, as explained below[1].

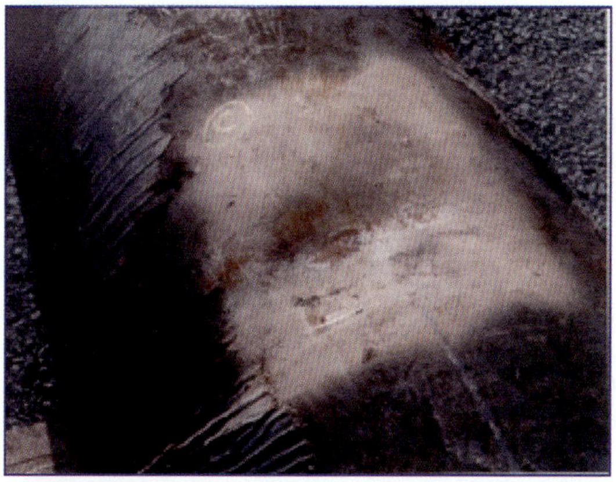

A. Dent

B. Gouge

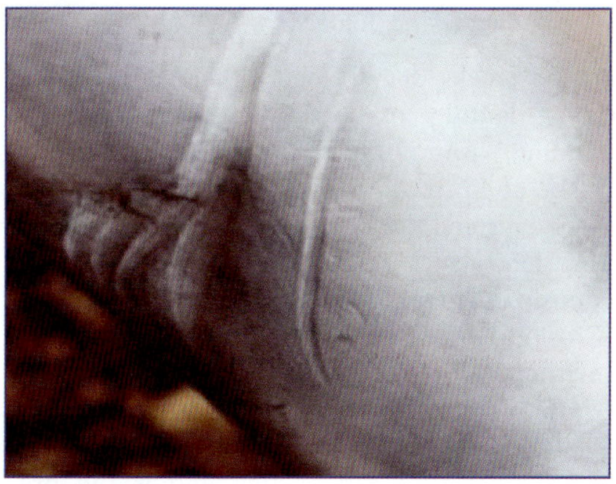

C. Dent with Gouge

Figure 1. Examples of mechanical damage features.

Dents

A dent is defined as a depression which produces a gross disturbance in the curvature of the pipe wall, caused by contact with a foreign body, and resulting in plastic deformation of the pipe wall.

A dent can be further categorized as:

Smooth or Plain Dent: A dent which causes a smooth change in the curvature of the pipe wall, does not contain a stress concentrator or wall thickness reduction, and does not change the curvature of an adjacent girth weld or seam weld.

Unconstrained Dent: A dent that is free to rebound elastically (spring back) when the indenter is removed and is free to reround as the internal pressure varies.

Constrained Dent: A dent that is not free to rebound or reround, because the indenter is not removed. A rock dent is an example of a constrained dent.

Complex Dent: Any dent with a gouge, groove, scratch, stress riser, or other secondary defect or a dent that affects the curvature at a nearby weld.

The depth of a dent depends on the force of the indenting object, the diameter and wall thickness of the pipe, internal pressure, material properties, and any constraint of the pipe during deformation. Pipes of larger diameter and lower wall thickness have a higher compliance ratio, and so are able to deform more elastically under an applied force. However, if the pipeline is constrained by stiff soil, then the damage may be greater due to the thinner wall. Clearly, a pipe with lower yield strength will plastically deform sooner under applied force. If a pipeline is internally pressurized during the damage process, then the effective stiffening will act against the applied force and possibly reduce dent depth. These parameters all affect the resulting depth of indent and, therefore, the pipeline's structural integrity[2].

Gouges

A gouge is defined as surface damage to a pipeline caused by contact with a foreign object that has moved or removed material from the pipe, resulting in metal loss, metal movement, microstructure damage, crack, and/or other detrimental feature created during the damage process.

Gouges are typically formed when an object such as an excavator tooth is drawn across the surface of the pipe. The gouge acts as a stress concentrator which affects structural integrity. In addition, the gouge typically results in metallurgical damage in the form of strain hardening of the material and residual stress. These features differentiate gouges from many other metal loss defects, such as corrosion, which may result in reduced wall thickness of the pipe but do not involve metallurgical damage. The length and orientation of a gouge, as well as other characteristics, can vary significantly at a mechanical damage location, as shown in Figure 2.

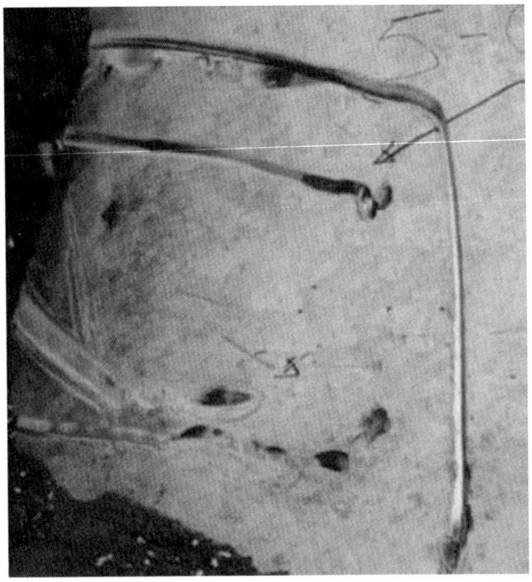

Figure 2. Gouge characteristics.

Combined Defects

When dents and gouges are found together, containing cracks or damaged metal, or when one or both are near a seam or girth weld, the damage should be treated with caution. Complex mechanical damage is not easy to evaluate, nor are simple "rules of thumb" applicable.

Mechanical damage features and characteristics

Mechanical damage can degrade or reduce the serviceability (ability to function as designed) of a pipeline. In order to understand and evaluate the severity and structural implications of mechanical damage, it is first necessary to identify the features associated with damage, to categorize them and, where appropriate, to measure them. Many features or characteristics can affect the severity of damage. These features may include one, several, or all of the following:

Changes in pipe shape.

- A localized dent, change in curvature in the circumferential direction, and/or change in straightness along the pipe axis.
- Ovalization of the cross section near or at positions removed from the point of contact or area of most significant indentation.

Changes in wall thickness.

- Removed metal due to cutting, gouging, scraping, plowing or similar mechanisms.
- Thinning due to stretching, usually in a sharp dent or near a scrape, gouge, or similar anomaly.
- Increases in wall thickness, as sometimes occurs where an indenter ushes material to the side as it moves along the pipe.
- Re-deposited metal, as sometimes occurs when an indenter rides up and over metal that has been cut or scraped off during damage.

Changes in localized stresses and strains.

- Residual stresses and strains. Stress and strain fields have axial, circumferential, and radial components. The local stresses and strains are "path dependent," which means they are influenced by the order and magnitude of loads and deformations.
- Stress concentrations at, for example, the corners of a gouge.
- Other increases or decreases in stresses due to changes in shape.
- Re-rounding, created when an indenting object is removed from a pressurized pipe.

Changes of material properties.

- Modification of mechanical (and magnetic) properties of the pipe material due to plastic strain. Strain hardening may increase yield strength, reduce ductility (i.e., tensile elongation), and/or affect toughness. These changes can range from modest, in areas where the local strains are the same order of magnitude as yield strains, to extreme, immediately under a sliding indenter.
- Changes due to phase transformations resulting from rapid heating and cooling.

Other anomalies or defects.

- Surface cracks that formed either at the inside surface during denting, or at the outside surface during re-rounding. Re-rounding cracks are often angled on a plane at 45° to the radial direction and/or "Z-shaped."
- Microvoids and buried cracks, especially in the deformed subsurface layer.
- Extension of cracks or microcracks by subsequent fatigue, corrosion fatigue, or stable crack growth.
- Environmental cracking, such as superficial stress corrosion cracking (SCC)
- Local or generalized corrosion, sometimes related to shielding by damaged coating and/or the presence of rocks immediately adjacent to the pipe wall.
- Other defects or anomalies, such as latent defects in an adjacent strain-sensitive seam or girth weld.

Elastic Spring-Back and Re-Rounding

The process of introducing a dent into a pipeline involves both elastic and plastic deformation. When the indenter is removed, the dent will "spring back" to some degree due to the stored elastic energy being released. The spring-back phenomenon is also referred to as "elastic rebound." Elastic distortions or flexing may occur when the pressure is increased or decreased at a dent.

Because the pipeline is pressurized during the deformation, or if the pipeline is subsequently re-pressurized during a return to service or pressurized above the level present when the damage was formed, the internal pressure will also reduce the dent depth via a process referred to as "re-rounding." Re-rounding refers to plastic or permanent distortion in and around a dent due to internal pressure. Often, the effects of re-rounding due to internal pressure are greater than those due to the elastic response.

Spring-back and re-rounding behavior depend on pipe geometry, material properties, operating pressure, the pipeline's external support stiffness, and dent shape. Typically, a long dent will spring back and re-round more readily than a short dent, and recovery occurs more towards the middle of the dent than at the ends. Spring back and re-rounding is greater in thin-walled pipes than in thick-walled pipes, due to higher geometric compliance. Pipes operating at higher pressures will exhibit more re-rounding due to the higher driving force to re-round.

Stress/Strain Fields

The initial indentation of a pipe leads to tension in the pipe's internal surface and compression of its external surface. The material yields locally, causing a plastic strain on both surfaces. When the indenter is removed and the pipe re-rounds, the external surface experiences residual tension. If the residual tension is severe enough, cracking may initiate. In many cases, cracking may lead to immediate failure of the pipe during the re-rounding process[8].

In some circumstances, the dent is prevented from re-rounding due to constraint of the pipe material. If damage is caused when the pipeline is laid against a rock, or ground movement leads to the pipeline being pressed against a fixed object, the dent is "constrained" against elastic rebound and re-rounding. If damage is caused by mechanized equipment and the indenting object is removed during the damage event, then the dent is typically "unconstrained" against elastic rebound and re-rounding. The re-rounding process leads to significant changes in localized stress/strain fields, and may lead to cracking.

Denting and re-rounding processes are associated with plastic deformation of the pipe material. The plastic strain incurred in the material during the damaging process may reduce the toughness or remaining ductility of the material. This strain is superimposed on any mechanical strain introduced during pipe manufacture and pipeline installation. The final stress and strain state near a dent and gouge depends on the strain history. It is triaxial in nature, though this depends on the specific details of the denting process[3].

Reduction of Toughness and Ductility Due to Strain

The deformation associated with mechanical damage can overload (greater than Ultimate Tensile Strength) and significantly affect the properties of the pipe material. When a gouge is formed, the area immediately beneath the indenter is highly deformed and its crack-initiation resistance significantly reduced. This type of damage extends into the pipe material to a shallow depth. The deformed layer is usually harder than the base metal and may contain transformed microstructures resulting from rapid heating and cooling during its formation.

In addition, the denting process may strain-harden the material away from the location that was contacted by the indenter. Strain-hardening or cold work during plastic deformation can increase hardness and decrease toughness, but to a lesser extent than under the indenter. The effects of prior strain on mechanical properties, particularly toughness, have been reviewed by Leis and Hopkins (2003).

Few experimental studies have addressed reductions in the material properties issue; the results appear to indicate that crack initiation resistance reduces rapidly, falling to about half the initial value on reaching 10 percent prior strain, but that the tearing resistance remains unchanged. Other researchers[4,5] have shown that tensile ductility is also reduced by prior strain, but not to the same extent as for fracture initiation. However, there appear to be significant variations among different pipe materials, linked to the original toughness and ductility of the unstrained material.

Mechanical damage can also affect the magnetic properties of the material. Changes in magnetic properties affect the magnetic in-line inspection results.

Geometry and characterization of damage

Dent depth is usually measured by reference to the original outside contour of the undamaged pipe, and often expressed as a percentage of the original pipe diameter. Dent shape can be defined in terms of the local radius of curvature in the circumferential and/or axial direction. (Figure 3). Distinction is made between whether the measurement is obtained when the pipe is pressurized or depressurized. Such measurements include a degree of ovality or flattening of the pipe surface as an unavoidable component of depth.

Dent length, width, or size may be defined by the envelope within which the radius of the damaged pipe is different than that of the undamaged pipe, but this is sometimes difficult to

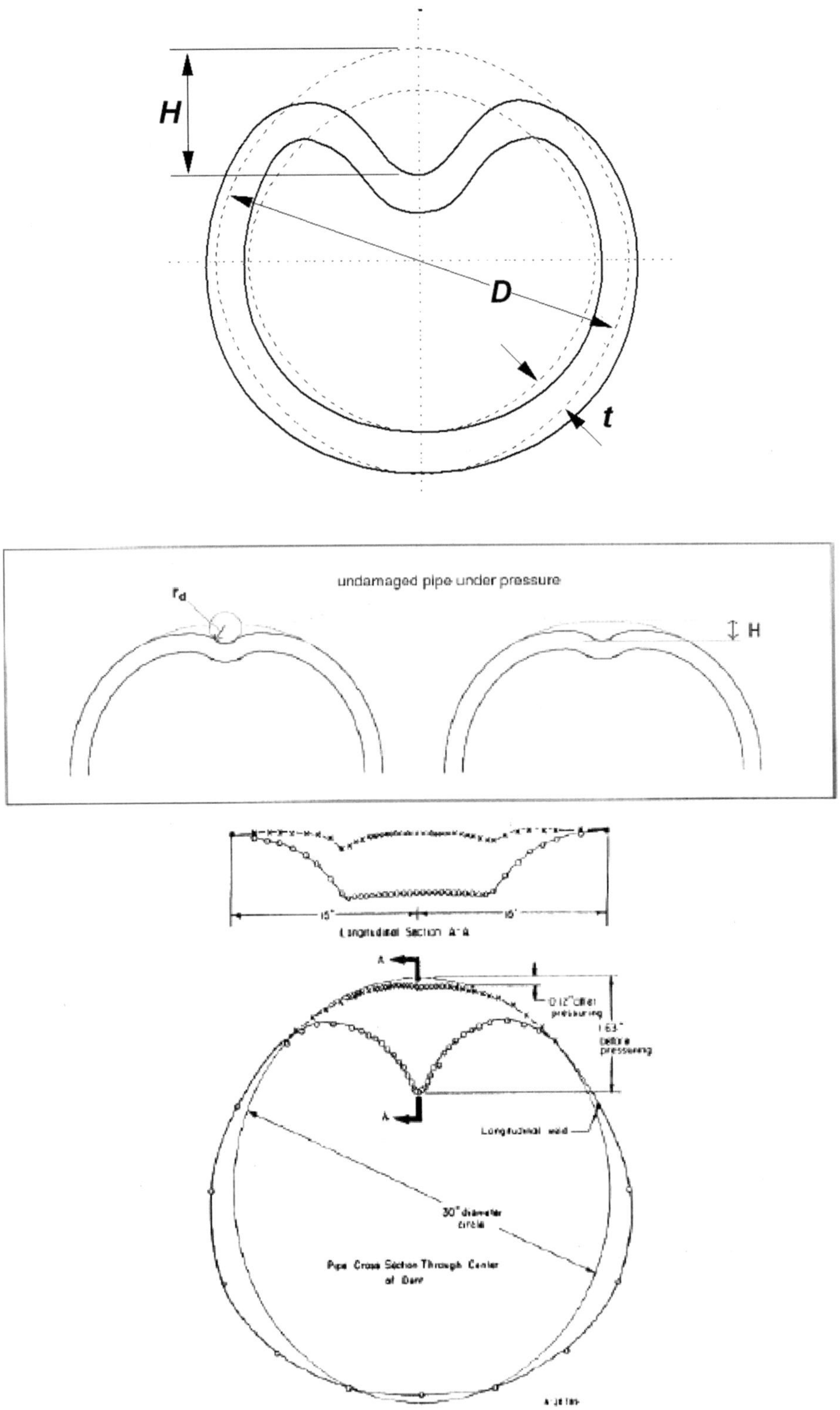

Figure 3. Dent depth measurement.

determine. Dent size has also been defined in terms of the locus of points of half-depth[2,6–10,11].

Dents can be described as either non-reentrant or reentrant (Figure 4). A positive radius value defines a non-reentrant dent, in which the outer surface of the pipe is flattened but retains its convexity. A negative radius value defines a reentrant dent, in which the outer surface of the pipe is dented deeply enough that it becomes concave. Dent shape has also been described in terms of the maximum angle subtended by the flanks of the dent[11].

The geometric parameters used to characterize a gouge are the depth, length, width, and angle to axis, etc. These are usually measured from the undeformed outer radius of the pipe (Figure 5).

For combined dents and gouges, the gouge length and depth are usually measured from the metal surface at the shoulder of the gouge – that is, not including the depth or length of the dent (Figure 6). Similarly, the depths of cracks are measured from the metal surface, excluding the depth or length of the associated dent and/or gouge. For assessment purposes, it sometimes may be appropriate to add the gouge and crack depths.

The locations of other secondary features associated with mechanical damage (including corrosion, SCC, and welds) are usually defined by measuring the axial and/or circumferential separations of the deepest point of the dent/gouge and the deepest/midpoint of the secondary feature. The location of damage is usually referenced to the circumferential orientation and to the girth weld immediately upstream.

Repair conditions and failure modes

The operating conditions of the pipeline and the ductility of the pipe material have profound effects on the impact of damage to a pipeline. Mechanical damage poses a range of possible threat to the integrity of an affected pipeline, ranging from dire and urgent, to tolerable today but only for some limited period of time, to benign.

All forms of mechanical damage that do not cause immediate failure of the pipe are potential delayed failures. Mechanical damage which does not fail right away may someday fail, for example, due to the effects of several possible mechanisms:

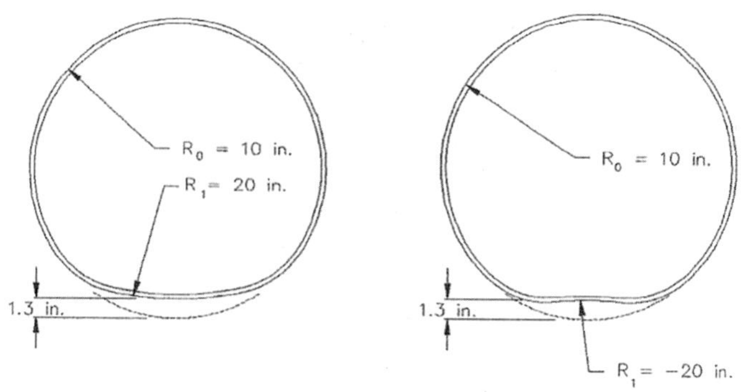

Figure 4. Measurement of re-entrant and non-re-entrant dent depth.

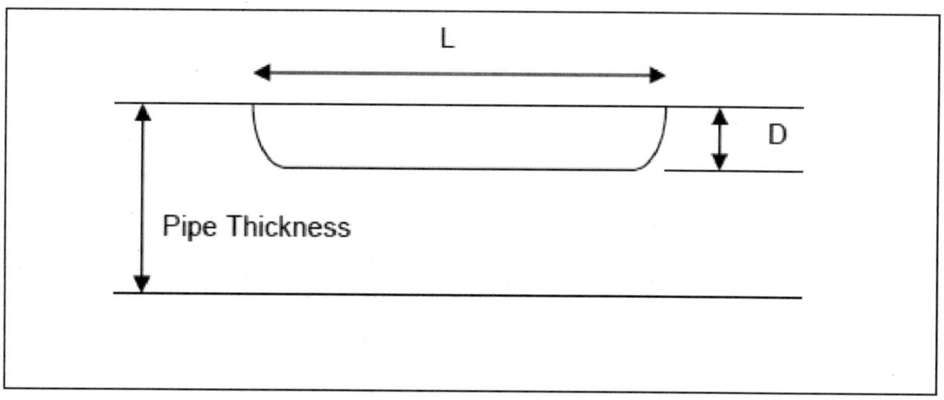

Figure 5. Gouge depth measurement.

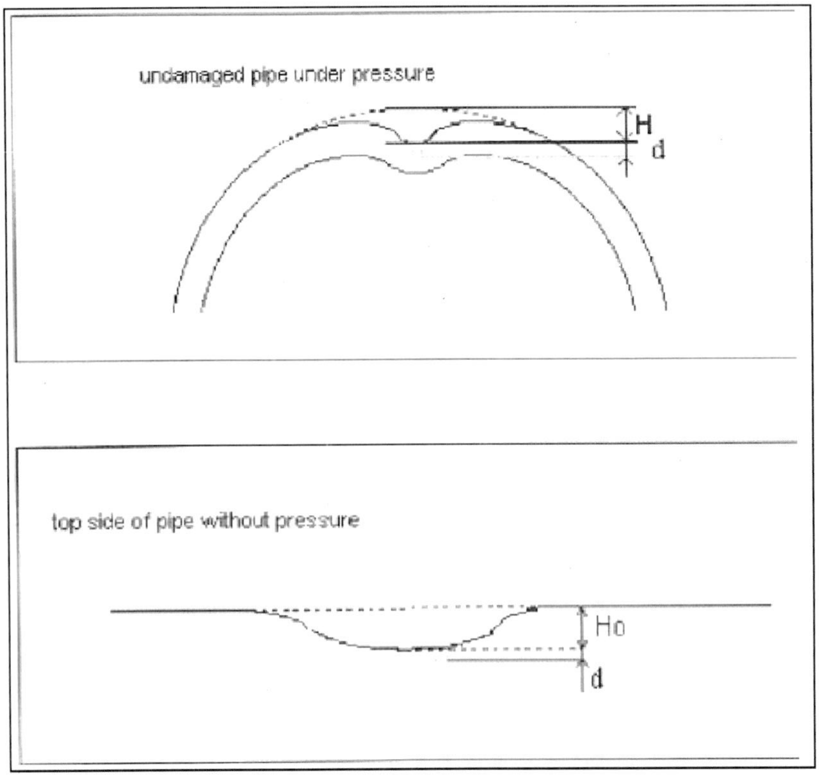

Figure 6. Depth measurement for combined dents and gouges.

- The damage may fail at a higher pipeline pressure than the pressure within the pipeline at the time of the damage event, or than what is normally present during a normal pressure rise or surge condition;
- The damage may fail as a result of accumulation of local plastic strain and exhaustion of ductility in the damage zone, due to shakedown[1], creep, and low-cycle fatigue;
- The damage may fail from high-cycle fatigue crack growth caused by the effects of many pressure cycles acting over time on the residual deformation and stress concentrator.
- The damage may serve as a site for the initiation of corrosion or SCC.

To overcome these threats, pipeline assessment and the mitigation or repair must be carried out. The need for repair can be categorized as "immediate" or "delayed." The critical distinction between the immediate and delayed pipeline repair categories is in the urgency of the response as dictated by the current factor of safety above the pipeline's operating pressure and the anticipated time to failure. Times to failure may range from a few weeks or months (warranting an immediate response), to years, decades, or possibly never.

Immediate Repair Conditions

The Immediate Repair category includes damage sufficiently severe that the pipe is thought to be unsafe to operate at the current pressure level. So even though the pipeline is currently holding pressure, the presumption is that the factor of safety is inadequate and a delayed failure could occur at any time. This category may also include damage that is insufficiently characterized to determine whether the pipe can operate safely at the current pressure level.

Such urgent conditions require that the operator respond immediately to investigate and act to restore the margin of safety of the pipeline. Response options could include shutting down the pipeline, reducing the pressure of the pipeline by some significant amount so as to provide a known minimum factor of safety, or performing a repair capable of effectively restoring the strength of the pipe[3].

[1] Although the residual dent depth of a "rerounded" dent is usually very shallow, tests have shown that the dent never fully rerounds, which means that it continues to fluctuate in and out with changes in operating pressure. It can take many cycles of pressure for the dent to "shake down" to elastic action, which means that plastic strain can continue to accumulate in the damaged area. If the damage survives shakedown, the dent will subsequently cycle elastically over a smaller range of motion, providing a mechanism for continued crack growth by fatigue in service[12].

Current integrity management planning standards such as ASME B31.8S, 49 CFR Part 192 Subpart O, and 49 CFR Part 195.452 consider that any combination of indentation with a gouge poses an immediate threat. A principal reason for this is that such damage has been known to fail at operating stress levels. In addition, there is no simple, straightforward method an operator can use to accurately calculate the safe operating pressure (in other words, there is no "B31G" for dents) or the time to failure. Involved analytic methods derived from research are used by experts to obtain estimated times to failure.

Delayed (or Scheduled) Repair Conditions

The Delayed Repair category includes all forms of damage not deemed to be severe enough to fall into the immediate threat category and not otherwise shown to be tolerable. The length of the predicted time to failure is sufficiently long to afford opportunities to monitor the pipeline condition and/or schedule the investigation and make the decision to repair the damage[3].

A dent with a minor scrape or gouge need not impair the safety of the pipeline, and the pipeline's remaining life can be many years. This is particularly true where the pipeline operates at a moderate or low stress level and experiences only infrequent pressure changes. It may be true for slight damage in a pipeline that operates at higher stresses. However, the possibility of minor or non-threatening mechanical damage is not yet recognized by integrity management planning standards or regulations. Current integrity management planning standards generally consider only plain deformations containing no scrapes, gouges, or cracks to qualify as delayed repair conditions if they do not exceed the allowable strain limit.

Growth of Cracking from Mechanical Damage

Ref. 2 describe the growth of cracking from mechanical damage as driven by two forces:

- The over load plastic strain tension at the inside surface of the pipe during contact with the indenter. If sufficient, it can lead to cracking or plastic collapse of the remaining pipe wall and possibly to immediate failure.
- The tension at the outside surface of the pipe as the indenter is removed and the pipe re-rounds. Again, if this plastic strain magnitude is sufficient, it can lead to immediate failure.

Cracking at the outside surface of the pipe may also be promoted by the presence of microcracks generated during sliding contact with the indenter. As mentioned earlier, high plastic strains incurred during deformation increase the hardness and reduce the toughness, sometimes to the extent that cracks or microcracks develop along the internal or external surface as the pipe deforms. These cracks are often oriented perpendicularly to the motion of the indenter and penetrate at a 45 degree angle to the pipe surface.

Cracking can develop further as a result of stable tearing, fatigue, environmental stresses, cyclic-stress activated low temperature creep, and/or as a result of other delta strain mechanisms. Thus, dents and gouges can show both time-dependent and pressure-cycle-dependent crack development. These phenomena may diminish as the local stresses redistribute and/or the steel strain hardens sufficiently to blunt the damage, but may nevertheless continue until delayed failure occurs. In-service failures have occurred at any time from a few days to tens of years after the damage event.

Features associated with high-cycle fatigue are more often seen in oil or hazardous liquid product pipelines than in natural gas pipelines. The more severe fatigue experienced by liquid pipelines is attributed to higher frequency load fluctuations. This condition is also exacerbated by the larger stress cycles and higher rates of stress change than seen in natural gas pipelines.

Leak versus Rupture

A mechanical damage defect can fail as a leak or a rupture. In a leak, the full-penetration crack length is similar to that of the initial damage feature, and the release is controlled or limited by the opening size. In a rupture, the initial opening extends or grows rapidly along the pipe axis due to crack propagation at one or both ends. The critical flaw size, which is the length at which the failure mode changes from leak to rupture, depends upon the geometry of the pipe, the length of the fully penetrating defect, the pipe properties, the operating stress level, and other factors.

Determination of the boundaries that discriminate between leak and rupture behavior is similar to that for other axial defects, except that there may be additional stresses associated with the geometry of the damage, and the material in the immediate vicinity may exhibit reduced toughness. For both these reasons, rupture may be more likely at the point of mechanical damage than at the site of equivalent length corrosion defects or axial cracks. Such considerations influence the consequences of mechanical damage and the priority for repairs.

Source

Mechanical Damage Study, U.S Department of Transportation, Office of Pipeline Safety, performed by Michael Baker Jr., Inc., April 2009. For the complete report and citations to the literature, go online at: www.phmsa.dot.gov.

References

1. American Society of Mechanical Engineers, ANSI/ASME B31.8S-2004, Managing system integrity of gas pipelines. New York (2004).
2. API 1156 (1997, November). Effects of smooth and rock dents on liquid petroleum pipelines.
3. Common Ground Alliance (CGA), CGA DIRT Analysis and Recommendations, Volume IV (2008, December).
4. A. Cosham, P. Hopkins, The pipeline defect assessment manual. Project IPC02-27067, International Pipeline Conference, Alberta, Canada (2002).
5. A. Cosham, P. Hopkins, A. Palmer, May). An experimental and numerical study of the effect of prestrain on the fracture toughness of line pipe steel, in: Proceedings of the International Pipeline Technology Conference, Ostend (2004), pp. 1833–1856.
6. A. Dinovitzer, R. Lazor, L.B. Carroll, J. Zhou, F. McCarver, S.D. Ironside, D. Raghu, K. Keith, Geometric dent characterization, in: Paper 27076 presented at 4th International Conference on Pipeline Technology, Calgary (2002, October).
7. EGIG. (2008). Available from: http://www.egig.nl/downloads/7th_report_EGIG.pdf.
8. EPRG, Methods for assessing the tolerance and resistance of pipelines to external damage, Parts I & II, 3R International 739-744 (1999) 806–811.
9. P. Hopkins, B.N. Leis, Mechanical damage gaps analysis. Final Report to Pipeline Research Council International, Battelle, 2003, March.
10. S.D. Ironside, L.B. Carroll, Pipeline dent management program, in: Paper 27260 presented at 4th International Conference on Pipeline Technology, Calgary (2002, October).
11. Michael Baker Jr., Inc. (2004, November). Dent Study, OPS TTO10, Office of Pipeline Safety.
12. National Energy Board, Focus on safety: A comparative analysis of pipeline safety performance 2000-2006 (2008, July).
13. M.J. Rosenfeld, Factors to consider when evaluating damage on pipelines, Oil and Gas Journal. September 9 (2002) 2002.
14. M.J. Rosenfeld, J.W. Pepper, K. Leewis, Basis of the new criteria in ASME B31.8 for prioritization and repair of mechanical damage, in: Paper 27122 presented at 4th International Conference on Pipeline Technology, Calgary (2002, October).
15. T. Swankie, M. Martin, R.M. Andrews, Mechanisms and kinetics of crack growth in areas of mechanical damage (2005, March) GRI Report 8545.

Understanding stress corrosion cracking in pipelines

Excerpts from the report "Stress Corrosion Cracking" performed by Michael Baker Jr., Inc. for the U.S Department of Transportation, Office of Pipeline Safety, January 2005.

General characterization

Stress Corrosion Cracking (SCC) in pipelines is a type of environmentally-assisted cracking (EAC). EAC is a generic term that describes the formation of cracks caused by various factors combined with the environment surrounding the pipeline. Together these factors reduce the pressure carrying capacity of the pipeline. When water (electrolyte) comes into contact with steel, the minerals, ions, and gases in the water can attack or corrode the steel. These chemical or electrochemical reactions may result in general wall thinning, corrosion pits, and/or cracks.

EAC includes two mechanisms that should be distinguished: Corrosion fatigue and SCC. "Corrosion fatigue" occurs when chemically reactive agents penetrate fatigue cracks. These agents can accelerate crack progression. The chemical condition within the crack can be more aggressive than on the free surface. Even if the metal surface at the crack tip passivates (forms an inert barrier) the next fatigue loading cycle can crack the brittle deposit and reactivate the process. Thus, corrosion fatigue is the joint action of a cyclic stress and a corrosive environment that decreases the number of cycles to failure. Compared to the life of the pipeline when no corrosion is present, the corrosive environment generally decreases the life of the component. Similarly, SCC involves corrosive mechanisms and depends on both an aggressive environment and tensile stress. The tensile stress opens up cracks in the material and can be either directly applied or residual in form. Therefore, SCC occurs under tension loads, while corrosion fatigue occurs under cyclic loading.

SCC in pipelines is further characterized as "high pH SCC" or "near-neutral pH SCC," with the "pH" referring to the environment on the pipe surface at the crack location and not the soil pH. (pH is the measure of the relative acidity or alkalinity of water. It is defined as the negative log (base 10) of the hydrogen ion concentration. Water with a pH of 7 is neutral; lower pH levels indicate an increasing acidity, while pH levels above 7 indicate increasingly basic solutions.)

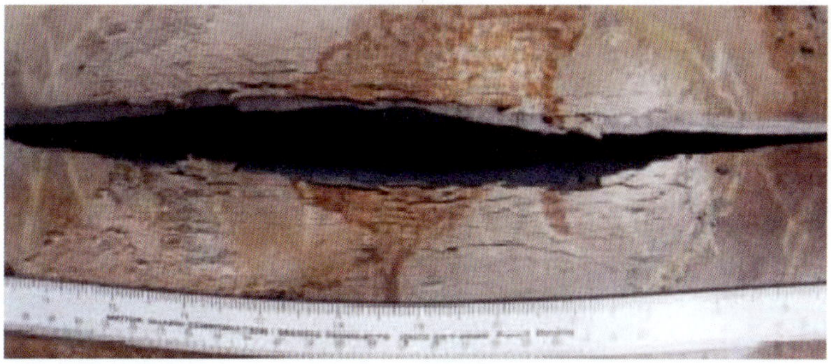

Figure 1. SCC colony on a large-diameter, high-pressure transmission gas pipeline.

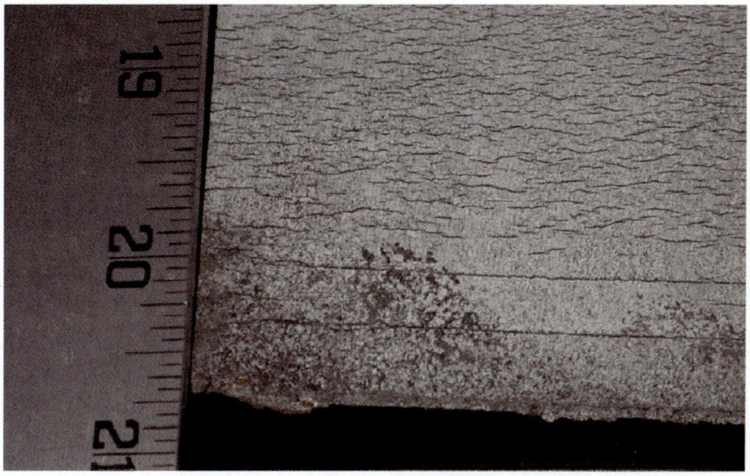

Figure 2. SCC colony – example one.

The most obvious identifying characteristics of SCC in pipelines, regardless of pH, is the appearance of patches or colonies of parallel cracks on the external surface of the pipe. There may be several of these colonies on a single joint of pipe and many joints of pipe may be involved. The cracks are closely spaced and of varying length and depth. These cracks may coalesce to form larger and longer cracks, which in some cases can lead to rupture. If the cracks are sparsely spaced, they might grow through the wall and leak, before they reach a length that is sufficient to cause a rupture.

In order for SCC to occur, three conditions must be satisfied simultaneously:

1. A tensile stress higher than the threshold stress, frequently including some dynamic or cyclic component to the stress;
2. A material that is susceptible to SCC; and
3. A potent cracking environment.

Figure 3. SCC colony – example two.

Historically, SCC has been found on onshore buried pipelines. SCC cracking is usually oriented longitudinally, normal to the hoop stress of the pipeline, which is usually the dominant stress component resulting from the internal pressure (Figure 1). Additional examples of SCC colonies on steel pipe are presented as Figures 2-5. However, SCC may

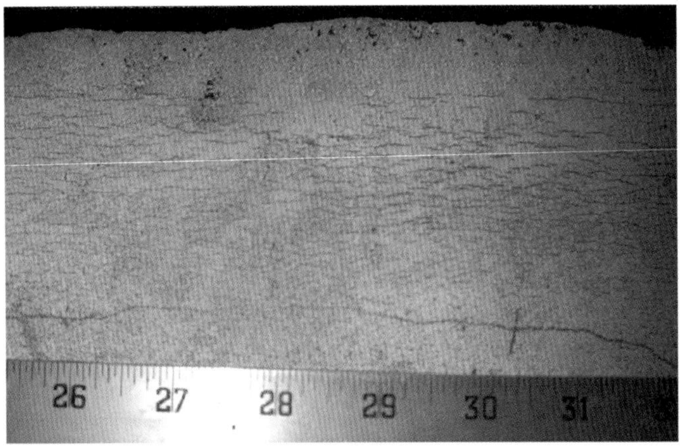

Figure 4. SCC colony – example three.

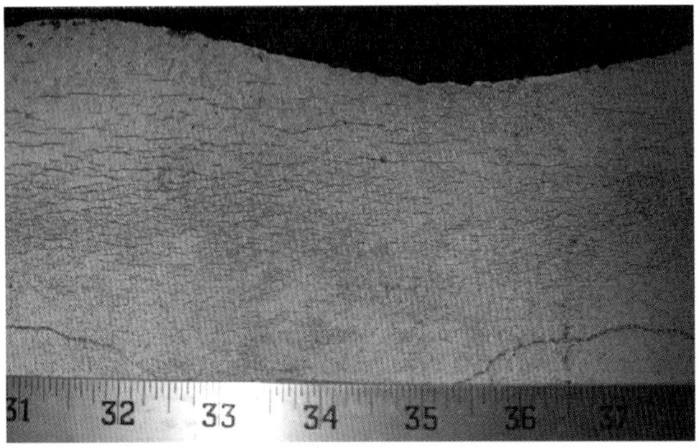

Figure 5. SCC colony – example four.

also occur in the circumferential direction (C-SCC) when the predominant stress is an axial stress. Incidents resulting from C-SCC have been reported due to stresses induced by soil creep and localized bending from slope movements, and rock dents. Residual stresses at girth welds may also produce a resultant axial load within a pipeline.

High pH SCC

When pipeline steel is exposed to the surrounding environment due to some form of coating failure, it is vulnerable to corrosion. Because soil corrosion is an electrochemical reaction, CP is used to mitigate corrosion by passing an electrical current through the soil thus giving the pipeline a cathodic potential. A concentrated carbonate-bicarbonate (CO_3-HCO_3) solution has been identified as the most probable environment responsible for high pH SCC. This environment may develop as a result of the interaction between hydroxyl ions produced by the cathode reaction and carbon dioxide (CO_2) in the soil generated by the decay of organic matter. CP current causes the pH of the electrolyte beneath disbonded coatings to increase, and the carbon dioxide readily dissolves in the elevated pH electrolyte, resulting in the generation of the concentrated carbonate-bicarbonate electrolyte. The pH of this electrolyte depends on the relative concentration of carbonate and bicarbonate, and the cracking range is between pH 8 and 11.

The fractured surface of the cracks normally exhibits a dark, discolored coating of oxidized material (primarily magnetite) at the mouth of the crack. The last portion of the pipe wall to fracture, i.e., the rapid fracture region, remains a shiny silver color. The presence of multiple black thumbnail-like flaws on the fracture surface normally is an indicator that SCC caused the failure (Figure 6).

Analysis of the liquid trapped in the disbonded area or in the crack itself indicates a carbonate-bicarbonate solution with a pH of 8 to 9, or higher. Metallographic examination of a section across the crack shows the fracture path to be intergranular, often with small branches, as shown in Figure 4-8. Laboratory simulation with small test specimens indicates that this form of SCC is temperature sensitive and occurs more frequently at higher temperature locations above 100°F. This supports field reports that demonstrate a greater likelihood of SCC immediately downstream of the compressor stations where the operating temperature might reach 150°F.

Figure 6. Matching fracture faces showing several co-linear high pH stress corrosion cracks broken open to reveal the black oxide on the crack.

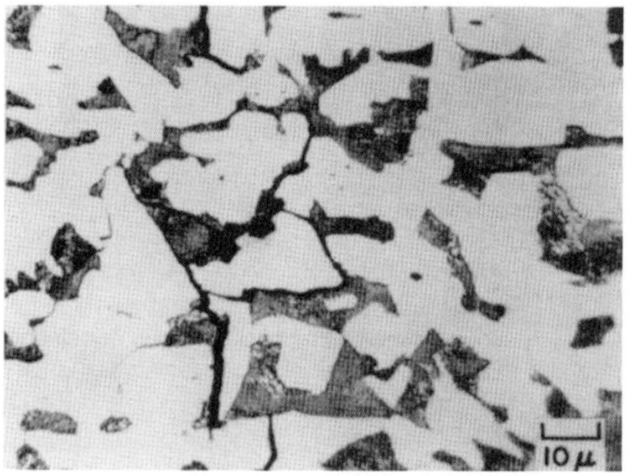

Figure 7. An example of intergranular cracking of pipeline steel.

Near-Neutral pH SCC

This form of SCC was not documented until the mid 1980s and was first identified on buried pipelines in Canada coincident with water having a pH between 5.5 and 7.5 trapped in wrinkles in the tape wrapped pipe coating. In the case of near-neutral pH SCC, the cracking environment appears to be a diluted groundwater containing dissolved carbon dioxide. The source of the carbon dioxide is typically the decay of organic matter and geochemical reactions in the soil. This form of cracking occurs under conditions where there is little, if any, CP current reaching the pipe surface over a prolonged period, either because of the presence of a shielding coating, a highly resistive soil, or ineffective CP. Typically, the SCC colonies initiate at locations on the outside surface where there is already pitting or general corrosion, which is sometimes obvious to the unaided eye and other times very difficult to observe.

Metallographic examination of near-neutral pH SCC reveals the cracks are predominately transgranular (Figure 8) and are wider (more open) than the high pH form, i.e. the crack sides have experienced metal loss from corrosion. This morphology implies that the fracture mechanism is different; however, the direct visual appearance of a pipe fracture surface is similar to that of high pH SCC.

Crack Characteristics

There are many similarities between the two forms of SCC. Both occur as colonies of multiple parallel cracks that are generally perpendicular to the direction of the highest stress on the external pipe surface. These cracks can vary in depth and length and grow in two directions. They increase in depth and length and tend to coalesce, or link together, to form longer cracks. At some point these cracks may reach a critical depth and length combination that can result in a rupture. A leak will occur if a crack grows through the pipe wall before it reaches a critical length for rupture. Note that critical size stress corrosion cracks do not need to fully penetrate the pipe wall for a rupture to occur, i.e., a shallow crack may reach a length that becomes critical. The strength and ductility of the remaining wall determines the critical size at which the crack behavior changes from a slowly growing stress corrosion mechanism to an extremely rapid brittle or ductile stress overload.

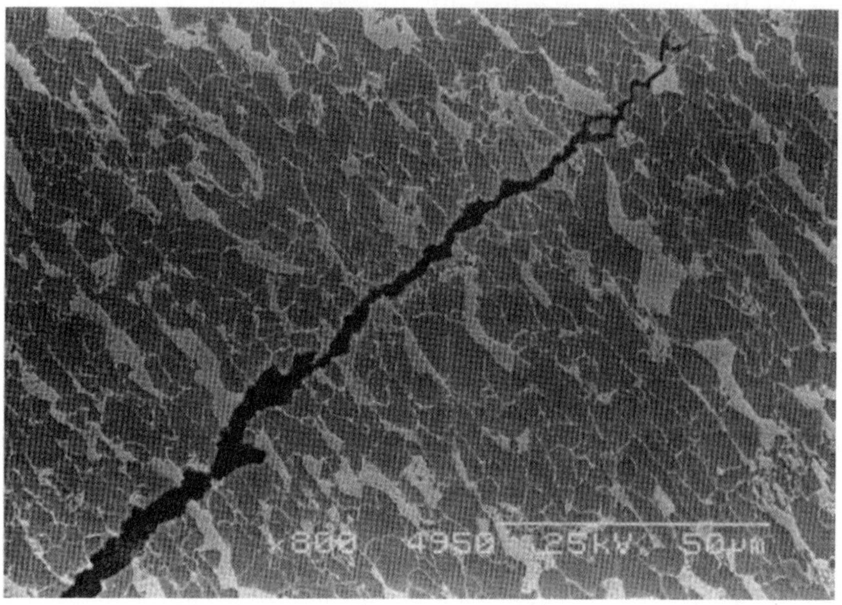

Figure 8. Transgranular cracking in pipeline steel.

Table 1
Characteristics of High pH and Near-Neutral SCC in Pipelines

Factor	Near-neutral pH SCC	High pH SCC (Classical)
Location	65 percent occurred between the compressor station and the 1st downstream block valve (distances between valves are typically 16 to 30 km) 12 percent occurred between the 1st and 2nd valves 5 percent occurred between the 2nd and 3rd values 18 percent occurred downstream of the 3rd valve SCC associated with specific terrain conditions, often alternate wet-dry soils, and soils that tend to disbond or damage coatings	Typically within 20 km downstream of pump or compressor station Number of failures falls markedly with increased distance from compressor/pump and lower pipe temperature SCC associated with specific terrain conditions, often alternate wet-dry soils, and soils that tend to disbond or damage coatings
Temperature	No apparent correlation with temperature of pipe Appear to occur more frequently in the colder climates where CO2 concentration in groundwater is higher	Growth rate increases exponentially with temperature increase
Associated Electrolyte	Dilute bicarbonate solution with a neutral pH in the range of 5.5 to 7.5	Concentrated carbonate-bicarbonate solution with an alkaline pH greater than 9.3
Electrochemical Potential	At free corrosion potential: −760 to −790 mV (Cu/CuSO4) Cathodic protection does not reach pipe surface at SCC sites	−600 to −750 mV (Cu/CuSO4) Cathodic protection is effective to achieve these potentials
Crack Path and Morphology	Primarily transgranular (across the steel grains) Wide cracks with evidence of substantial corrosion of crack side wall	Primarily intergranular (between the steel grains) Narrow, tight cracks with almost no evidence of secondary corrosion of crack wall

The most obvious differences between the two forms of SCC are the temperature sensitivity of high pH SCC, the fracture morphology, and the pH of the environment in contact with the pipe surface. The characteristics of high pH and near-neutral pH SCC are summarized in Table 1.

Crack Growth

The cycle of SCC crack growth is normally modeled as a four-stage process as shown in Figure 9. The first stage is the development of conditions conducive to SCC and is followed by the crack "initiation" stage. These cracks then continue to grow and coalesce, while additional crack initiation also occurs during stage 3. Finally, in stage 4, large cracks coalesce and failure occurs.

While a single crack might grow large enough to cause a leak, coalescence typically is necessary for the defect to grow long enough to cause a rupture. If cracks form close to one another, crack growth may be dominated by coalescence into collinear cracks and can occur throughout the SCC life cycle. A combination of environmental and mechanical forces can cause cracks to grow. In the final stage of growth, after cracks have coalesced sufficiently for tearing to begin, the environment no longer plays a role. In some cases, tearing is preceded by a stage of crack growth in which fatigue is the dominant crack propagation mechanism.

The geometry of the crack colonies resulting from near-neutral pH SCC is important in determining whether the cracks coalesce and grow to failure (NEB 1996). Colonies of cracks that are long in the longitudinal direction yet narrow in the circumferential direction are a greater risk to pipeline integrity than colonies of cracks that are shorter in the longitudinal direction and wide in the circumferential direction. The individual cracks in long, narrow colonies are oriented head to tail and tend to link together, leading to rupture. However, for colonies that are about as long as they are wide, growth occurs mainly near the edges. Cracks located deeper within these colonies with circumferential spacing less than 20 percent of the wall thickness are generally shielded from stress and become dormant (NACE 2003).

Contributing factors to SCC in pipelines

Metallurgy

Metallurgy can affect SCC through chemical composition and microstructure. However, pipeline steels, and certainly the conventional steels that have historically been used in the last 50 years, do not typically contain elements not found in

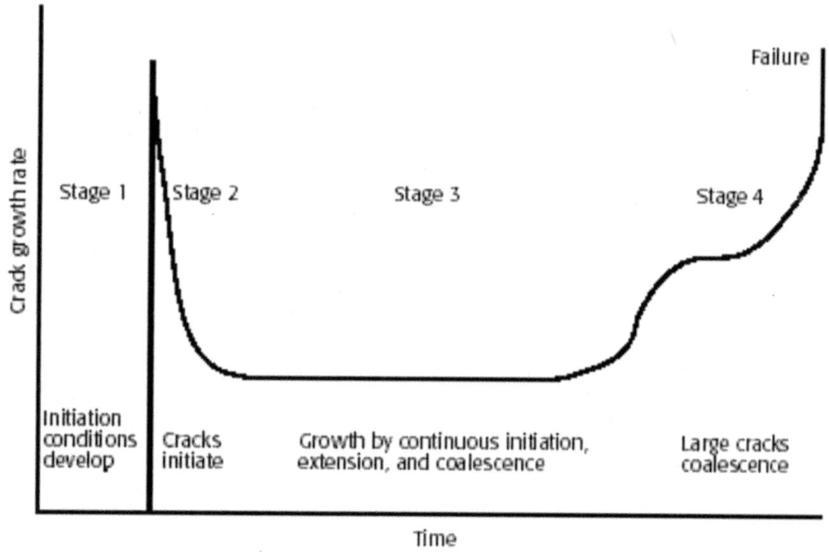

Figure 9. Four stage process of SCC growth.

similar carbon-manganese steels used in literally hundreds of construction applications without reports of SCC.

More recently, the yield strength of line pipe has gradually increased by the addition of micro alloying elements such as vanadium, columbium, and/or titanium. The addition of these elements tends to produce a finer grain in the microstructure, increasing both strength and toughness. Controlled rolling and cooling of the steel plate used to manufacture line pipe has resulted in finer grained bainitic microstructures.

A number of research investigations involving small-scale, laboratory-reproduced SCC and using both high and near-neutral pH environments have been conducted without achieving a meaningful correlation between steel composition or microstructure and susceptibility to SCC. Danielson and Jones (2001) discuss the high pH SCC testing of six different heats of API 5L X52, as well as three heats (X65, X70, and X80) of modern steels. Their paper concludes: "In general, the microstructure/microchemistry had a small effect on the SCC behavior."

Nevertheless, certain batches of pipeline steel have been found to be much more susceptible to SCC than other batches with similar compositions and microstructures (Beavers and Harper 2004). A full understanding of this remains to be developed, but current research suggests that other characteristics of the steel, such as creep response to cyclic loading, may be important.

Manufacturing

Carbon steel line pipe is manufactured using multiple forming and joining processes and may be seamless or have a welded seam. Seamless line pipe is hot formed into cylinders by piercing solid bars of steel. The obsolete manufacturing process of forming and lap welding plates into cylinders was also performed at an elevated temperature to permit pressure welding of the overlapped edges of a steel plate. Modern seam-welded line pipe is formed at ambient temperature from flat-rolled plates or strip from coils. Longitudinal seams may be electric-resistance welded (ERW) or double submerged-arc welded (DSAW) in line pipe formed from strip and plate, respectively. An obsolete process, electric-flash welded, was also used to produce longitudinal seams in line pipe formed from plate. DSAW helical seams are also used to produce line pipe rolled from both strip and plate. The vast majority of line pipe for large diameter pipelines is produced by one of the seam-welded processes, while smaller-diameter pipeline are typically constructed from either seamless or ERW line pipe.

The strength of hot-formed seamless and lap-welded line pipe is primarily due to alloy additions during steelmaking, but seamless line pipe with SMYS 60 ksi and higher is typically quenched and tempered (QT) after forming to achieve the required strength. The strength of vintage line pipe with welded seams was also primarily due to alloy additions, but advances in both steelmaking and rolling practices over several decades permitted strip and plate to achieve some of its strength from controlled rolling procedures rather than alloy additions. Production of modern strip and plate for manufacture of line pipe may be described as a Thermo-mechanical Control Process (TMCP), in which both rolling temperature and degree of thickness reduction are controlled. Although TMCP was developed to produce line pipe grades with higher strength and toughness, some producers may employ TMCP on lower strength grades.

Because of the evolution of both steelmaking and forming practices for carbon steels converted to line pipe, the microstructure of existing line pipe may contain varying amounts of ferrite, pearlite, and bainite with significant variation in the grain size. In spite of both significant and subtle differences in the chemical composition and microstructure of existing line pipe, there is no strong evidence that any of the differences either promote or inhibit SCC.

The first reported cases of SCC exhibited intergranular cracking (high pH SCC) in steels with a microstructure that consisted of grains of low-carbon ferrite and higher carbon colonies of pearlite. Typically, the more recently detected near-neutral pH SCC has occurred on slightly higher yield strength steels with a much finer grain size and a higher toughness. However, there are a number of cases of near-neutral pH SCC in older, large grained, lower-strength steels and cases of high pH SCC in newer, fine-grained steels. Thus, regarding reported cases of SCC, no generalizations regarding manufacturing process can be made.

Mechanical Properties

The mechanical properties of greatest interest for onshore gas transmission pipelines are the specified minimum yield strength (SMYS) and the toughness. Generally, the most cost-effective design of large-diameter onshore pipelines results from selecting the highest strength line pipe grade that still provides a wall thickness and diameter-to-thickness ratio that are constructible. As improved manufacturing procedures are being developed, higher-strength grades of line pipe are being purchased when the combination of diameter and maximum allowable operating pressure (MAOP) or maximum operating pressure (MOP) is suitable for the specific application. There is no strong evidence that increasing SMYS through 70 ksi increases susceptibility to SCC initiation or growth.

Increases in toughness, which have generally occurred in parallel with increasing SMYS, have significantly increased the critical size of cracks that can result in ruptures. There is a growing body of evidence to suggest that tensile residual stresses in the line pipe play a significant role in SCC and that cracking can be minimized or prevented by reducing these stresses during manufacturing, as well as during installation and operation.

Pipeline Operating Conditions

SCC requires three conditions to be satisfied simultaneously: 1) a tensile stress above the threshold stress, 2) an appropriate environment at the steel surface, and 3) a susceptible material.

Below some value of tensile stress, referred to as the threshold stress, crack initiation does not occur. The threshold stress is difficult to accurately define but, depending on the range of stress fluctuation, is on the order of 60 to 100 percent of the yield strength for high pH SCC based on laboratory results of tapered tensile tests. A threshold stress for near-neutral pH SCC has not been established (Beavers 1993).

Relating threshold stress determined by laboratory testing to stresses in service is challenging. Stresses reported for laboratory testing are generally uniaxial tension applied by the test fixture. Actual services stresses that can promote SCC include residual tension stresses and stresses from external forces, in addition to the stress applied by internal pressure. Consequently, comparison of the stress applied by internal pressure, as a percent of SMYS, with a threshold stress determined by laboratory testing is inappropriate, and may be misleading.

The operator has some control of the applied tension stress that is the result of internal pressure. Unfortunately, residual tension stresses from manufacture and field bending, bending stresses from pipeline movement, overburden loads from soil, dents or gouges, or from heavy equipment can cause as much or more tension stress as that caused by internal pressure, all of which is beyond the control of the operator.

The longitudinal stress caused by pressure can be up to half of the hoop stress. However, pipeline flexure can result in additional longitudinal stress, so that the toal longitudinal stress can exceed the hoop stress, with the maximum/minimum values at the extreme fibers of bending. The C-SCC cases reported by CEPA are associated with undulating terrain where pipeline loading resulted from soil creep or localized bending. Localized bending may also occur at dents resulting in higher longitudinal stresses in the local region.

A pipeline that is exposed to cyclic pressure fluctuations may experience cyclic softening. Cyclic softening is a phenomenon in which the application of stress cycles at maximum stress levels below the yield stress causes the steel to exhibit local micro-plastic deformation after a period of load cycles. This phenomenon manifests itself as a loss of yield strength and can significantly reduce the threshold stress. The operator has little control over the metallurgical susceptibility of a line pipe steel to cyclic softening but can, in some instances, monitor the magnitude and frequency of pressure cycles on a pipeline.

In addition, the operator has little control over the pH of the groundwater and is unable to control the aggressiveness of the environment, except for new construction by installing a premium coating system applied under carefully controlled and monitored conditions in a coating plant. Unfortunately, these mill applied coating systems require preparation and application methods that may not be considered applicable for recoating in the ditch. Note also that the pH of the groundwater will be modified by the electrochemical reaction at the pipeline surface.

The operator does have some control over the operating temperature. For example, some operators have installed cooling towers to help control high pH SCC.

Coating

Coating type and condition have a profound effect on SCC. SCC will not occur beneath an intact coating that prevents contact of groundwater with the pipe surface. Coatings can fail by disbonding from the pipe surface (i.e. the coating comes away from the pipe but does not break), the formation of holidays (i.e., breaks or gaps in the coating), or general degradation. Disbondment is the most severe form of degradation with respect to SCC susceptibility because the cathodic protection current can be shielded beneath the disbonded coatings.

Construction and operating practices can have a significant influence on coating performance. For example, coatings typically will not bond well to poorly prepared surfaces, leading to increased susceptibility to disbondment. Damage during construction can create coating holidays. Operating a pipeline at elevated temperatures can accelerate general coating degradation. Cathodic disbondment of a susceptible coating can occur in the presence of excessive CP current.

Tape coatings, such as the polyethylene-backed tapes, which were used predominantly in the early 1960s to 1980s, have particularly poor resistance to disbondment. These tapes are spirally wrapped around the pipe with an overlap at the helix line. "Tenting" occurs between the pipe surface and the tape along the ridge created by longitudinal, spiral, and girth welds. Tenting also occurs at the overlap between the helix of the wrap as illustrated in Figure 10.

When the tape disbonds from the pipeline, moisture can accumulate beneath the tape surface. The tape itself has fairly high electrical insulation properties, thus preventing CP current from reaching entrapped moisture between the tape and the pipe surface. At the time of the NEB SCC inquiry, it was reported that about three-quarters of near-neutral pH SCC-related service incidents occurred under these tape coatings. The cracks tend to occur at or near the toe of the seam weld where stress is concentrated and water has access, as well as where the coating has been damaged or disbonded (NEB 1996).

Asphalt and coal tar coatings may also disbond, especially due to poor surface preparation. However, since these coatings tend to become saturated with moisture, or if brittle may break into pieces, CP current is often able to reach the pipe surface in the disbonded area. SCC might still occur when the soil is so resistive that the CP current cannot reach the pipeline. For these coating types, there is no preferential

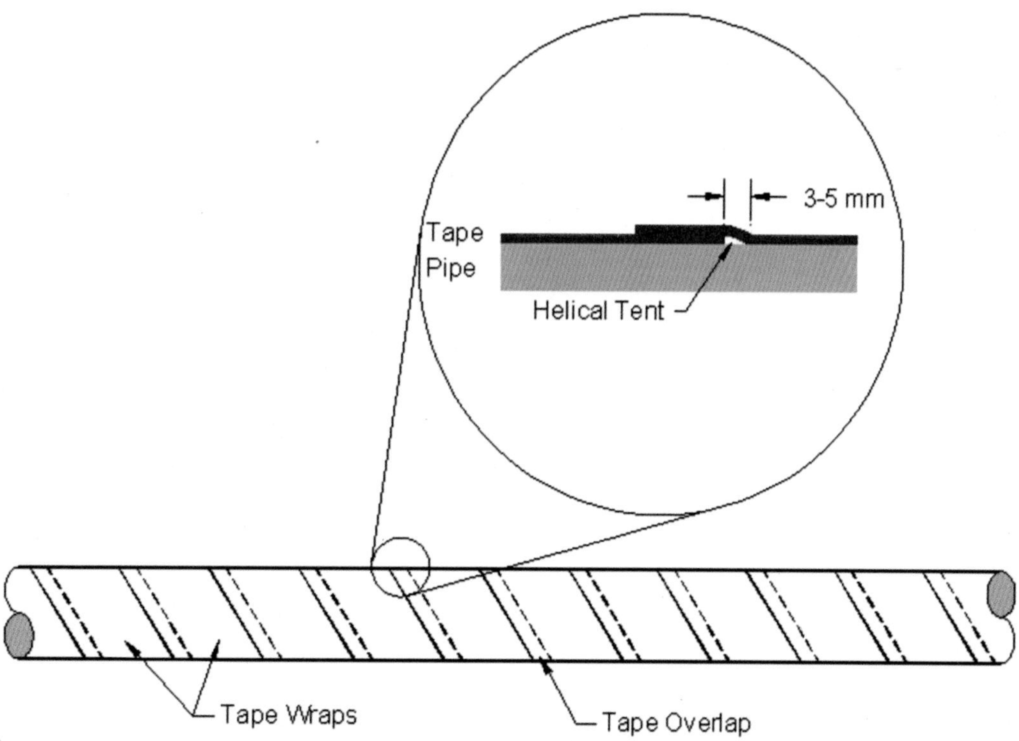

Figure 10. Polyethylene tape helical tent.

location, but SCC might occur wherever disbondment occur (NEB 1996).

It is generally agreed that fusion-bonded epoxy (FBE) coatings, which are often the coating of choice for newly installed pipelines in the United States, are an effective protection against SCC. Extruded polyethylene, because the coating system is monolithic, also appears to be effective, except possibly at tape-wrapped girth welds.

Soil Conditions

In 1973, Wenk described results of analyses of soil and water extracts (from the soil) taken from high pH SCC locations (Wenk 1974). While supporting data were not provided, it was stated that SCC had occurred in a wide variety of soils, covering a range in color, texture, and pH. No single characteristic was found to be common to all of the soil samples. Similarly, the compositions of the water extracts did not show any more consistency than did the physical descriptions of the soils, according to Wenk. On several occasions, small quantities of electrolytes were found beneath disbonded coatings near locations at which stress corrosion cracks were detected. The principal components of the electrolytes were sodium carbonate and bicarbonate. Sodium bicarbonate crystals were also found on pipe surfaces near some SCC colonies (Fessler, et al. 1973). Based on the presence of the sodium-based carbonates and bicarbonates, it is likely that these were high pH SCC sites. Therefore, it is not surprising that these results are not consistent with the results of the TCPL studies performed in the 1980s and 1990s, when near-neutral pH SCC was found.

Mercer described the results of a field study conducted by British Gas Corporation in 1979 (Mercer 1979). Soil data from both the UK and U.S. were collected and analyzed. As in the study by Wenk, detailed information on the soil analyses was not provided, but it was concluded that soil chemistry had no obvious direct influence on high pH SCC. The moisture content of the soil, the ability of the soil to cause coating damage, and localized variation in the level of CP were the primary soil-related factors identified.

Delanty and O'Beirne (1991, 1992) reported on the results of more than 450 investigative excavations performed on TCPL's system in the mid- to late-1980s. In the tape-coated portions of the system, near-neutral pH SCC was found in all of the various types of terrains and soils (e.g., muskeg, clay, silt, sand, and bedrock) present on the system. There was no apparent difference in the soil chemistry for the SCC and non-SCC sites. However, the SCC was predominantly located in imperfectly to poorly drained soils in which anaerobic and seasonally reducing environmental conditions were present.

In the same system, near-neutral pH SCC was found in the asphalt-coated portions of the system, predominantly (83 percent) in extremely dry terrains consisting of either sandy soils or a mixture of sand and bedrock. There was inadequate CP in these locations, based on pipe-to-soil potential measurements or pH measurements of electrolytes found beneath disbonded coatings. The remainder of the SCC sites on the asphalt-coated portions of the system had localized areas of inadequate CP, based on pH measurements of electrolytes.

Delanty and Marr developed an SCC severity rating model for near-neutral pH SCC for the tape coated portions of TCPL's system in eastern Canada (Delanty and Marr 1992; Marr 1990). The predictors in that model were soil type, drainage, and topography. The soil classifications were based on method of deposition. The most aggressive soil types were lacustrine (formed by deposits in lakes), followed by organics over glaciofluvial (formed by deposits in streams fed by melting glaciers), and organics over lacustrine. The prevalence of SCC in glaciofluvial soils was about 13 percent of that in lacustrine soils, and about 17 percent of that in soils with organics over glaciofluvial or lacustrine. Very poorly or poorly drained soils were found to be the most aggressive, while level-depressed soil was found to be the most aggressive topography. The SCC model did not contain parameters associated with soil chemistry because the results of previous geochemical projects were inconclusive.

As described above, neither the early field studies conducted on high pH SCC, nor the later field studies conducted on near-neutral pH SCC, detected a correlation between the occurrence of SCC and soil chemistry. On the other hand, high pH SCC was not reported where the extensive field study of near-neutral pH SCC was performed in Northern Ontario (Delanty and O'Beirne 1991, 1992), suggesting that the soil conditions were not conducive to this form of cracking. Furthermore, no near-neutral or high pH SCC was found in Northern Ontario where elevated pH electrolytes were detected, possibly because the soil conditions could not support the development of concentrated carbonate-bicarbonate solutions, even when the CP conditions were conducive to such development. These observations suggest that a further analysis of field soil data might provide insight into the role of soil/groundwater chemistry on the occurrence of SCC (Beavers and Garrity 2001).

Near-neutral pH SCC may be associated with local topographical depressions, e.g., at the base of hills or streams, where the groundwater either channels along the pipeline or across it. Flowing water may help to maintain the near-neutral pH environment by supplying carbon dioxide to the electrolytic solution in a disbonded area. The majority of laboratory investigation has been performed in an NS4 electrolyte solution containing 5 percent carbon dioxide. NS4 is a simulated trap water that is typical of liquids found beneath disbonded polyethylene tape coatings at locations where near-neutral pH SCC was found. Research shows that the crack growth rate increases with increasing carbon dioxide concentrations, and that the cracking becomes dormant in carbon dioxide free environments (Beavers, et al. 2001).

References

J.A. Beavers, On the Mechanism of Stress Corrosion Cracking of Natural Gas Pipelines, in: Eighth Line Pipe Symposium. PRCI. L51680. Paper 17 (1993).

J.A. Beavers, K.C. Garrity, 100 mV Polarization Criterion and External SCC of Underground Pipelines. NACE International CORROSION/2001, 2001, Paper No. 01592.

J.A. Beavers, C.L. Durr, B.S. Delanty, D.M. Owen, R.L. Sutherby, Near-Neutral pH SCC: Crack Propagation in Susceptible Soil Environments. NACE International CORROSION/2001, 2001, Paper No. 01217.

J.A. Beavers, W.V. Harper, Stress Corrosion Cracking Prediction Model. NACE International CORROSION/2004, 2004, Paper 04189.

M.J. Danielson, R.H. Jones, Effect of Microstructure and Microchemistry on the SCC Behavior of Archival and Modern Pipeline Steels in a High pH Environment. NACE International CORROSION/2001, 2001, Paper No. 01211.

B.S. Delanty, J.E. Marr, Stress Corrosion Cracking Severity Rating Model, in: Proceedings from the International Conference on Pipeline Reliability. CANMET, 1992 June.

B.S. Delanty, J. O'Beirne, Low-pH Stress Corrosion Cracking, in: Proceedings from the 6th Symposium on Line Pipe Research. PRCI, 1991, p. L30175.

B.S. Delanty, J. O'Beirne, Major Field Study Compares Pipeline SCC with Coatings, Oil and Gas Journal 90 (1992) 24.

R.R. Fessler, Stress-Corrosion-Cracking, in: Proceedings from the Fourth Symposium on Line Pipe Research. PRCI. L30075, 1969 p. F-1 to F-18.

R.R. Fessler, T. Groeneveld, A. Elsea, Stress-Corrosion and Hydrogen-Stress Cracking in Buried Pipelines, in: International Conference on Stress Corrosion Cracking and Hydrogen Embrittlement of Iron Base Alloys. June. Published in Stress Corrosion Cracking and Hydrogen Embrittlement of Iron Base Alloys. NACE. 1977 (1973).

J.E. Marr, The Relationship Between External Pipeline Stress Corrosion Cracking and the Environment Within TCPL's Facilities Across Canada. Unpublished TCPL Research Report, 1990, May.

W.L. Mercer, Stress Corrosion Cracking — Control Through Understanding, in: Proceedings from the 6th Symposium on Line Pipe Research. PRCI (1979), p. L30175.

NACE, External Stress Corrosion Cracking of Underground Pipelines. NACE International. Publication 35103, 2003, October.

NEB, Stress Corrosion Cracking on Canadian Oil and Gas Pipelines. Report of the Inquiry. National Energy Board. MH-2-95, 1996, December.

R. Winston Revie (Ed.), Uhlig's Corrosion Handbook, 2nd Edition. Sponsored by the Electrochemical Society, Inc. John Wiley & Sons, Inc, 2000.

R.L. Wenk, Field Investigation of Stress Corrosion Cracking, in: Proceedings of the Fifth Symposium on Line Pipe Research. PRCI, 1974, p. L30174.

Source

"Stress Corrosion Cracking" performed by Michael Baker Jr., Inc. for the U.S Department of Transportation, Office of Pipeline Safety, January 2005. For the complete report, go online at: www.phmsa.dot.gov.

When does a pipeline need revalidation?

John F. Kiefner and **Patrick H. Vieth**, Kiefner and Associates, Inc. Worthington, Ohio

Introduction

Buried, high-pressure pipelines are subjected to loads and environmental effects that may cause them to become degraded with the passage of time. All pipeline operators are well aware of this, and the prudent operators have active programs to mitigate deterioration and to remediate defective pipe. The efficiency of an operator's revalidation program can be optimized if the necessary responses can be made when needed, but not until then. The authors have assisted several pipeline operators in establishing timely intervention programs to assure continuing pipeline serviceability. A variety of techniques are used depending on the nature of the pipeline and the perceived problems. Some of the basic techniques are described in this document.

Background

Pipelines may suffer degradation due to a variety of causes, including corrosion, mechanical damage, fatigue, and stress-corrosion cracking. The appropriate remedies for these problems are well known, but other than the routine patrolling of the rights-of-way and monitoring of cathodic-protection potentials and rectifier currents, such remedies are too expensive to be applied on a regular or routine basis. By that we mean that most operators cannot afford to routinely and periodically utilize in-line inspection and/or hydrostatic testing to revalidate their pipelines. Usually, these techniques are invoked when some special circumstances exist. The special circumstances may be the existence of excessive amounts of low pipe-to-soil potential readings, the occurrence of leaks or ruptures, or just an intuitive feeling that it is time to check the condition of a pipeline. Alternatively, as we are finding, more and more operators are coming to depend on more sophisticated models to determine when intervention is needed. The types of models we have used are described.

Failure pressure versus defect size

Pipeline integrity is usually defined in terms of pressure-carrying capacity. For a given diameter, wall thickness, and grade of material, one can expect that a sound piece of line pipe will be able to sustain an internal pressure level of at least 100% of its specified minimum yield strength (SMYS). And, that pressure level can readily be calculated by means of the Barlow formula. For example, if one has a 16-in O.D. by 0.250-in wall thickness X52 pipeline, the pressure level corresponding to 100% of SMYS is 1625 psig. For comparison, the 72% of SMYS maximum operating pressure level is 1170 psig. Thus, one has the expectation that a new piece of this particular pipe has a safety margin against failure of 1625/1170, or 1.39. In fact, if it is free of defects, it will have a failure pressure of at least 1938 psig. The latter is based upon research conducted by the pipeline research Committee of the American Gas Association.[1]

Realistically, not all pipe is defect-free; otherwise there would never be preservice hydrostatic test breaks. But, after a preservice test to a pressure level of 90–100% of SMYS, one can expect a pipeline to perform satisfactorily at an operating stress level up to 72% of SMYS, that is, until or unless it becomes degraded in service by corrosion, mechanical damage, fatigue, or stress-corrosion cracking.

Thanks to extensive research by the previously mentioned Pipeline Research Committee of the American Gas Association (AGA),[1] an experimentally validated model exists for calculating the effects of a longitudinally oriented part-through-the-wall defect on the pressure-carrying capacity of the pipe. For any given piece of pipe, the model can be used to generate the relationships between failure pressure and flaw size. For our example, namely the 16-in O.D. by 0.250-in wall thickness X52 material, the relationships for selected depths of flaws are shown in Figure 1. Each of the nine parallel curves represents the failure-pressure-versus-flaw-length relationship for a particular depth (d) to wall thickness (t) ratio. The curve that cuts across the others is the dividing line between leaks and ruptures. Defects with failure at pressures above and to the right of this curve will fail as ruptures, whereas those that have failure pressures below and to the left of this curve will fail as leaks.

A horizontal line is drawn in Figure 1 to represent the maximum operating pressure (MOP) of this pipe material, 1170 psig, corresponding to a hoop stress level of 72% of SMYS. All flaws with length and d/t combinations that lie below this line will fail at pressure levels below the MOP. All flaws that lie above this line could exist in the pipeline in service. For example, the curve representing a d/t of 0.5 crosses the line at a flaw length of 5.8 in. Flaws of that length that are more than halfway through the wall will fail at the MOP; flaws of that length that are less than halfway through will not.

The family of curves shown in Figure 1 is quite useful in the kinds of models we use to determine the time-dependent degradation of a pipeline. Two features of the curves should be kept in mind as the discussion proceeds. First, these curves represent failure pressures that are achievable with steadily increasing pressure over a relatively short period of time (minutes). Because of the phenomenon of time-dependent growth, it is possible to observe failures at pressure levels of 5–10% lower than these curves predict if pressure levels 5–10% below the predicted levels are held long enough.

Secondly, these curves are affected by the ductile-fracture toughness of the material. The curves shown in Figure 1 are characteristic of a garden-variety X52 line-pipe material having a Charpy V-notch upper-shelf energy (full-size specimens) of 25 ft-lb. The set of curves shown in Figure 2 was generated for a hypothetical material of the same pipe geometry and grade but with a Charpy energy of 250 ft-lb (a material of optimum toughness). Note that for a given defect size, the material of optimum toughness has a higher failure pressure. To illustrate this, it is recalled from Figure 1 that

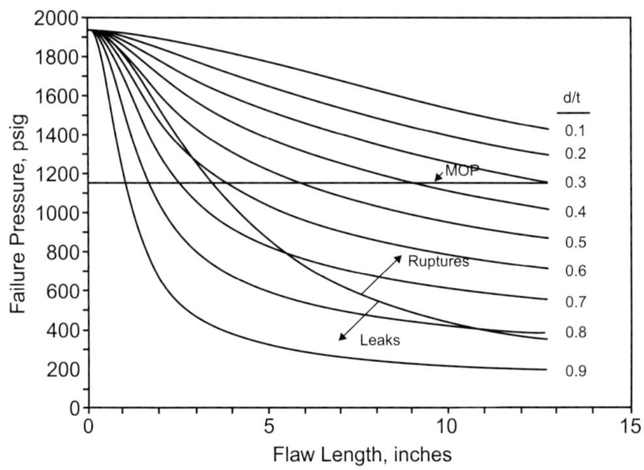

Figure 1. Failure pressure versus flaw size for a material of normal toughness (16 in by 0.250 in, X52, normal toughness).

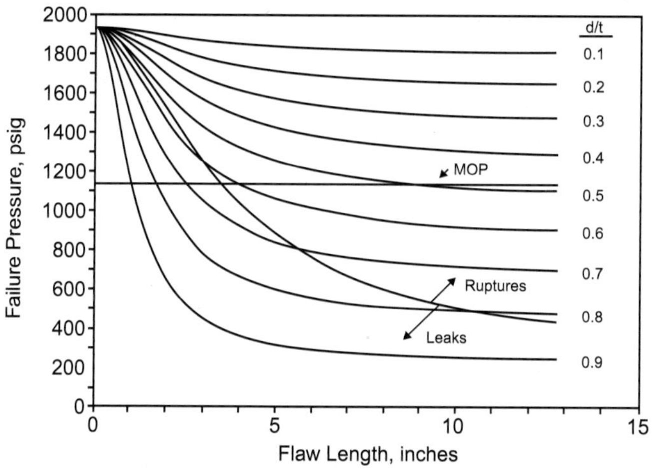

Figure 2. Failure pressure versus flaw size for a material of optimum toughness or for blunt defects (16 in by 0.250 in, X52, optimum toughness or blunt defects).

a 5.8-in-long flaw halfway through the wall fails at 1170 psig. In contrast, the same 5.8-in-long flaw with a *d/t* of 0.5 in the very tough material has a failure pressure of 1210 psig. It is noted that blunt flaws such as corrosion-caused metal loss tend to behave as illustrated in Figure 2 rather than Figure 1 regardless of the toughness of the material. This is because the material in a corrosion pit is strained over a large "gauge length", unlike the situation of a crack-like flaw. For the latter, toughness becomes a very important parameter.

Representing the other extreme, Figure 3 presents the failure-pressure-versus-flaw-size relationship for a very low toughness material such as the electric resistance welded (ERW) bond line in an older, low-frequency welded material. Even though the diameter, thickness, and grade are the same as that represented in Figures 1 and 2, the critical flaw sizes are much smaller.

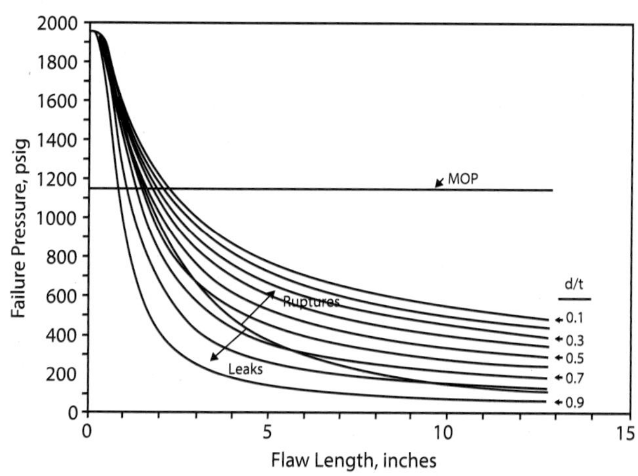

Figure 3. Failure pressure versus flaw size for a material of low toughness (16 in by 0.250 in, X52, low toughness).

Benefits of hydrostatic testing and in-line inspection

Another useful concept in evaluating pipelines for revalidation arises from the unique failure-pressure-versus-flaw-size relationships of the type described in Figures 1–3, that is, the manner in which those relationships illustrate the benefits of hydrostatic testing and in-line inspection. For illustrative purposes, let us consider the 25 ft-lb material for which Figure 1 gives the failure-pressure-versus-flaw-size relationship. This relationship is replotted in Figure 4, but with additional information. It contains horizontal lines at 1170 psig (maximum operating pressure), 1463 psig (90% of SMYS), and 1625 psig (100% of SMYS). As illustrated previously, no flaws having failure pressures below 1170 psig can exist if the pipeline is operating at 72% of SMYS; otherwise, they would fail (i.e. their particular lengths and *d/t* values would lie below the horizontal line at 1170 psig). Similarly, if the pressure in the pipeline is raised to 1463 psig (90% of SMYS), any flaws with length and *d/t* combinations lying above the line at 1170 psig but below that at 1463 psig will fail and be eliminated. If no failures occur, the absence of flaws in this size range is proven. And, if the pressure in the pipeline is raised to 1625 psig (100% of SMYS), the absence of additional even smaller flaws will be proven. Hence, the benefit of a hydrostatic test can be expressed as the ratio of test pressure to operating pressure. The higher the ratio, the more beneficial the test, because the flaws that can remain will be smaller.

With respect to in-line inspection, most pipeliners are aware that the conventional magnetic-flux leakage tools are reasonably capable of sorting metal-loss anomalies into the following categories: depth of light, moderate, and severe. Lights generally are those with depths of less than 30% of the

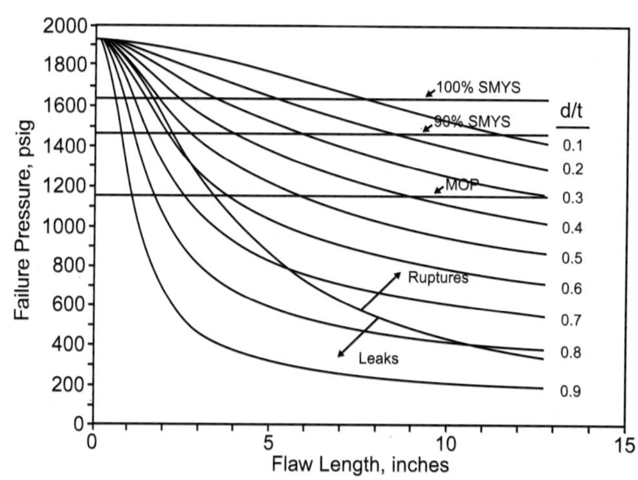

Figure 4. Sizes of flaws located by hydrostatic testing (all types except gouge and dent combinations) (16 in by 0.250 in, X52, normal toughness).

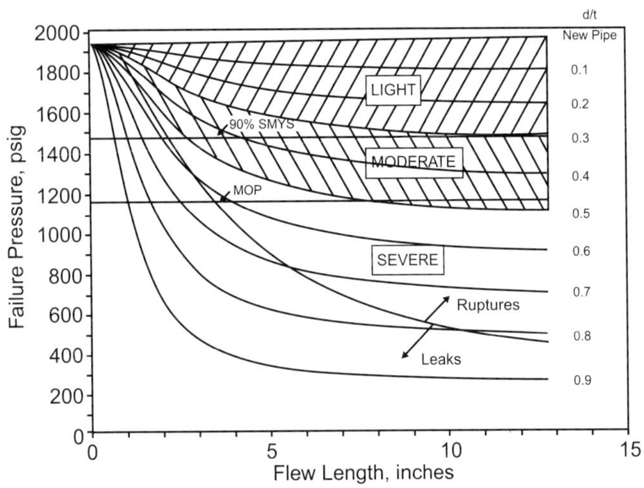

Figure 5. Sizes of flaws located by in-line inspection (corrosion) (16 in by 0.250 in, X52, blunt defects).

wall thickness ($d/t < 0.3$). Moderates are those with depths of 30–50% through the wall thickness ($d/t = 0.3–0.5$). Severes have depths greater than 50% of the wall thickness ($d/t > 0.5$). Recalling that the behavior of blunt metal-loss flaws is best represented by the optimum toughness relationships of Figure 2, one can see in Figure 5 (based on Figure 2) how in-line inspection data can be interpreted in terms of the effects on pressure-carrying capacity. The top cross-hatched area corresponds to light, the lower cross-hatched area corresponds to moderate, and the unshaded area corresponds to severe. Note that for long flaws, the failure pressures corresponding to moderates become low enough to intersect the MOP level. This illustrates the need to excavate and examine moderates in a program of conventional in-line inspection. With the use of the more expensive, high-resolution in-line inspection tools, it is possible to define flaw length and, hence, to avoid many excavations.

Figure 5 provides an important conceptual comparison between in-line inspection and hydrostatic testing. With the aid of this figure, one can see where each has its advantages and disadvantages. The hydrostatic test represented by the 90% SMYS line provides a demonstration of the immediate pressure-carrying capacity up to a pressure level of 1463 psig. The use of conventional magnetic-flux tools cannot provide this kind of assurance unless all severe and moderate anomalies are excavated, leaving only the lights. However, the hydrostatic test does not locate any of the anomalies that do not fail. Hence, very deep pits can survive the test and develop leaks shortly thereafter (as illustrated by the amount of unshaded area—severe anomalies—lying above the horizontal line at 1463 psig). At least with in-line inspection, one can locate and remove or repair the anomalies (especially the deep ones). After having removed or repaired all moderates and severes, one can have a high degree of confidence that the pipeline will not fail or leak for a long time as the result of corrosion-caused metal loss. The same cannot be said after a hydrostatic test. Very short but deep pits could have survived the test and may become leaks (not ruptures) within a short time after the test.

On the other hand, because none of the existing in-line inspection tools is capable of finding cracklike flaws, hydrostatic testing is the only means to assure the integrity of a pipeline that may have sustained cracking in the form of fatigue-crack growth or stress-corrosion cracking.

Determining when to revalidate

Having established the concepts illustrated previously, we can now proceed to the determination of when to revalidate a pipeline. This is done by different means depending on the type of anomaly involved. First, let us consider corrosion-caused metal loss.

A key parameter for scheduling the revalidation of an externally or internally corroded pipeline is the worst-case corrosion rate. A pipeline operator's corrosion control personnel may be able to estimate such a rate. Alternatively, one can obtain a reasonable estimate if prior corrosion leaks have occurred. If leaks have occurred, the decision to revalidate may be obvious without the use of a model. Nevertheless, having some idea of the corrosion rate is useful for scheduling the "next" revalidation. The rate can be estimated by dividing the nominal wall thickness of the pipeline by the number of years between the time of the original installation and the time of the first leak. This involves assuming a constant rate of corrosion over the life of the pipeline. For any pit that is not a leak, the corrosion rate will be less; it will be proportional to the d/t ratio of the pit. For our example pipeline (0.250-in wall thickness), if we assume that it first developed a leak after 25 years of service, its worst-case corrosion rate is 0.25/25 or 0.010 in per year (10 mils per year). If we postulate that there is a pit on this pipeline that is 80% through the wall, the corrosion rate for that pit is $0.8 \times 0.01 = 0.008$ in per year (8 mils per year). With this kind of a corrosion-rate rationale, we can utilize a figure like Figure 2 to plan the revalidation of a corroded pipeline.

The family of curves represented in Figure 2 is reproduced in Figure 6 for the purpose of examining the effects of time-dependent metal loss. In Figure 6, the pipeline when it was installed is represented by the horizontal line at 1938 psig ($d/t = 0$). If this pipeline is losing metal to corrosion, the effect of the metal loss is to lower the failure pressure.[a] In terms of Figure 6, one can see that as metal loss proceeds, the horizontal line originally at 1938 psig would move downward as

[a] The ASME B31G method for evaluating corroded pipe is based on the same empirical equation that is used to derive the curves of Figure 6. The only difference is that the flaws represented in Figure 6 have been idealized as being rectangular in profile.

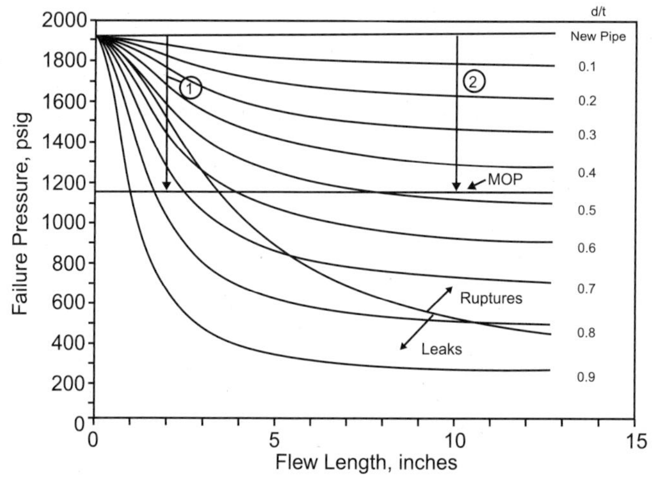

Figure 6. Determination of need for revalidation of a pipeline affected by corrosion (16 in by 0.250 in, X52, blunt flaws).

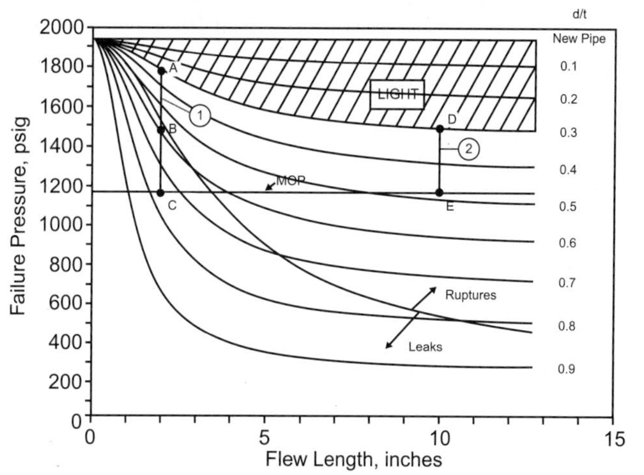

Figure 7. Determination of time to next revalidation after in-line inspection or hydrostatic test (corrosion) (16 in by 0.250 in, X52, blunt flaws).

indicated by the arrows until it would eventually reach the 1170 psig operating level, and a failure would occur. It is important to note that this occurs at a *d/t* of 0.75 for a 2-in-long flaw (Arrow 1) and at a *d/t* of 0.47 for a 10-in-long flaw (Arrow 2). If the company's corrosion specialist has determined that the worst-case corrosion rate is 8 mils per year, then 75% of the wall thickness (250 mils) will be penetrated in less than 23 years, and 47% of the wall thickness will be penetrated in less than 14 years. If the type of metal loss that is occurring is long (10 in or more along the axis of the pipe), this pipeline will require revalidation in less than 14 years. On the other hand, if experience shows that the corrosion is occurring as isolated pits (less than 2 in in axial extent), the time for revalidation would be less than 23 years.

Perhaps a more practical case is that corresponding to a pipeline which has either been hydrostatically tested or has been repaired following in-line inspection. Now the question becomes: When will revalidation be needed again? The question can be answered in terms of Figure 7. Let us consider two alternatives simultaneously. In one case, let us assume that our 16-in O.D. by 0.250-in wall thickness pipeline has been inspected by a conventional magnetic-flux leakage tool and that all anomalies except the lights have been repaired (see the cross-hatched area in Figure 7). In the other case, let us assume that the pipeline has been hydrostatically tested to 1463 psig (90% of SMYS).

With respect to a 2-in-long isolated pit (Arrow 1), the deepest pit remaining after the in-line inspection is represented by Point A (*d/t* = 0.3, pit depth = 75 mils). From the standpoint of the hydrostatic test, the deepest pit that can exist with a 2-in length is represented by Point B (*d/t* = 0.6, pit depth = 150 mils). To cause a leak in service (and it will be a leak rather than a rupture if the length remains less than about 3 in), the pit must grow to Point C (*d/t* = 0.76, pit depth = 190 mils). If the corrosion rate remains at 8 mils per year, the time to grow from A to C is (190−75)/8 = 14.4 years, and the time to grow from B to C is (190−150)/8 = 5 years. This shows again the advantage of in-line inspection over hydrostatic testing with respect to dealing with pitting corrosion.

With respect to a long corrosion anomaly (Arrow 2), the deepest possible remaining anomaly is the same whether the present revalidation was by in-line inspection or by hydrostatic test. The anomaly must grow from Point D to Point E to fail in service (and it will fail as a rupture rather than a leak, because it is well above the leak/rupture dividing line). To do so, *d/t* must change from 0.3 to 0.47. This is a change in wall thickness of 42.5 mils and will take place in 5.3 years at a corrosion rate of 8 mils per year. From this example, it appears that if the corrosion is occurring over long regions as opposed to isolated pits, the advantage of in-line inspection over hydrostatic testing is eroded. This is not quite correct for two reasons. First, the high-resolution tools give reasonably accurate information on the length of the anomaly as well as the depth. Hence, one can focus the repairs on the basis of the remaining pressure-carrying capacity. Second, even with the use of conventional tools, the advantage of locating the corrosion and having the option to examine and repair it provides benefits that do not accrue with hydrostatic testing.

Returning to our analysis, we see that for our example, a revalidation interval of less than 5 years is suggested. An appropriate factor of safety should be applied to this interval, so if this were a real case, the pipeline operator might arbitrarily decide to take action after 2 or 3 years. This probably is an excessively conservative example. Hopefully, with adequate corrosion control, a corrosion rate lower than 8 mils per year can be assured. If so, the revalidation interval would be longer than shown in this example. However, the principles are the same regardless of the corrosion rate.

Fatigue crack growth model

As we have seen in terms of Figure 4, if a pipeline is hydrostatically tested to a given pressure, defects of a certain size that are present at the time of the test will fail. Smaller defects would be expected to survive the test but would not be expected to grow further at a static pressure level much below the test pressure level. However, defects of size a_T limited by pressure level PT in Figure 8 could grow during cyclic-pressure service until they reach the size a_S. At that point, failure will occur at the maximum operating pressure (MOP). If a defect as large as a_T remains after a hydrostatic test and if the pipeline is subjected to sufficient pressure fluctuations, the defect will fail after time t_1.

It should be obvious from Figure 8 that flaw enlargement by fatigue-crack growth will erode the effective safety margin provided by a hydrostatic test. This situation may be addressed by analyzing the effects of the service-pressure cycles in terms of the size of the margin between test pressure and operating pressure and assessing when it may be necessary to conduct a subsequent hydrostatic test in the event that flaws that survive the first test become enlarged.

For this purpose, one may use a crack-growth model developed explicitly for pipelines.[2] The model is based upon linear-elastic fracture mechanics. It assumes that crack growth takes place under fluctuating stress levels according to the relationship

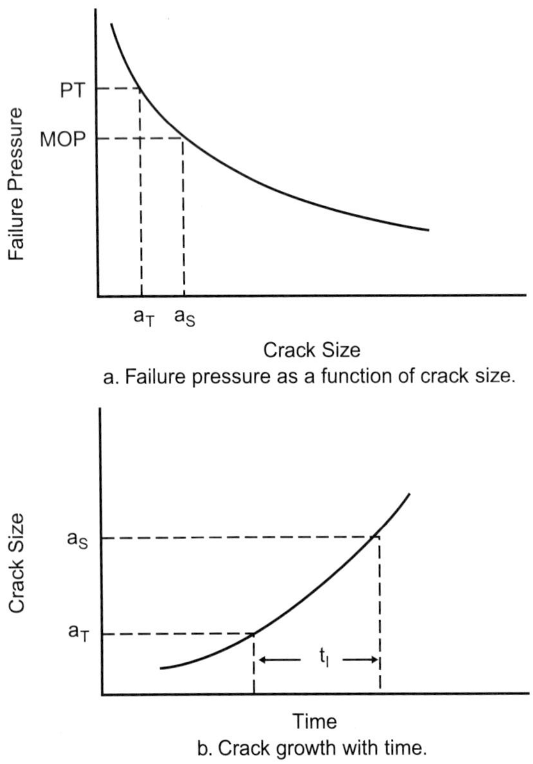

a. Failure pressure as a function of crack size.

b. Crack growth with time.

Figure 8. Effect of crack growth on pipeline integrity.

$$da / dN = C\Delta K^n \qquad (1)$$

where

da/dN is the increment of crack advance per cycle of loading, and

ΔK is the range of the stress intensity factor, which is a function of the change in stress and the initial crack size.

C and n are material properties determined through crack-growth-rate tests.

In the use of the model, the initial crack size is fixed by the hydrostatic test. A defect that just barely survives the test is assumed to be the initial crack. The stress range is determined by the maximum and minimum pressure levels. The stress-intensity factor that provides the cracking-driving force is calculated via a Newman-Raju algorithm.[3] The model is capable of handling irregular-size pressure variations. For example, we have analyzed pressure spectra as complex as that shown in Figure 9. Equation 1 is numerically summed over a number of cycles until the crack becomes large enough to fail in service. The failure criterion for flaw size at failure is that used to generate Figures 1–3.

To demonstrate the importance of a high test-pressure-to-operating-pressure ratio in a situation in which service pressure cycles can cause flaw growth, one can consider an example based on an actual cyclic pressure test. The results of a full-scale pipe experiment involving crack growth are described in Ref. 4 (Figure 47 of Ref. 4) and are shown in Figure 10. In this experiment, a 2-in-long part-through (0.095-in-deep) flaw was placed in the fusion line of the ERW seam of a 12/-3/4-in O.D. by 0.250-in wall thickness X52 material. It was then subjected to constant pressure cycles from 0 to 1020 psig until it grew through the wall and failed (after about 8000 cycles). The actual failure pressure levels are not important, so for purposes of illustration, we can consider the implications of Figure 10 for a material having an ERW seam toughness equivalent to 2 ft-lb of Charpy V-notch impact energy. If this were a 2-ft-lb Charpy-energy material, the flaw would have had a failure pressure level of 1300 psig after growing to a depth of 0.1 in as indicated by the 1300 psi "test-pressure" line. Similarly, it would have had failure pressure levels of 1250 and 1200 psig after growing to depths of 0.12 and 0.13 in, respectively. These values were calculated on the basis of the failure-pressure flaw-size relationship of the types shown in Figures 1–3 but for this particular pipe geometry.

Consider the following scenarios based on Figure 10. First, as service cycles are applied and the flaw has grown from 0.095 to 0.1 in ($N = 2300$ cycles corresponding to Point A), the remaining flaw is exactly the size that would just survive a test to 1300 psig. Since the number of cycles to failure is 8000, the flaw has a remaining life of 5700 cycles. Second, after the flaw has grown to 0.12 in ($N = 4800$ cycles corresponding to Point B), the remaining flaw is exactly the size that would just survive a test to 1250 psig. It has a remaining

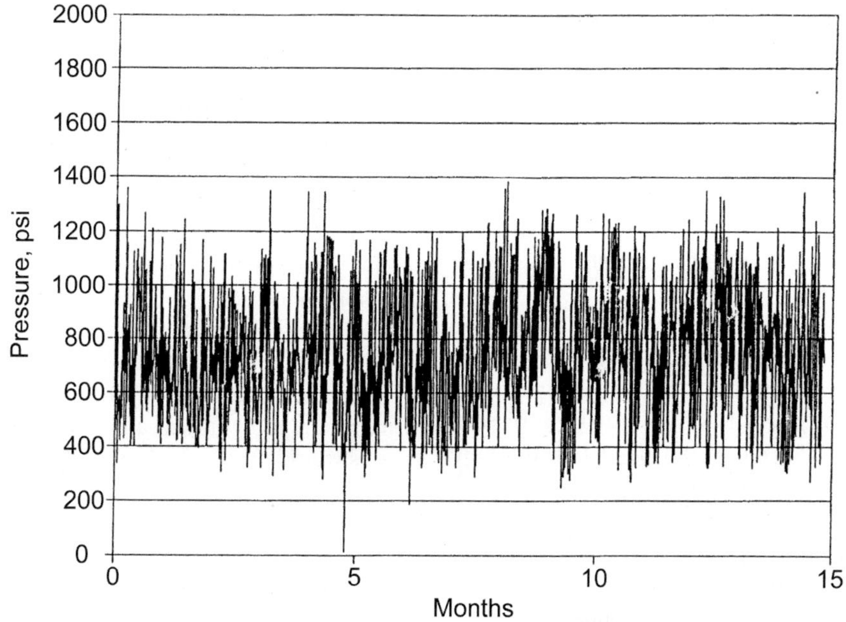

Figure 9. Pressure versus time for a typical petroleum products pipeline.

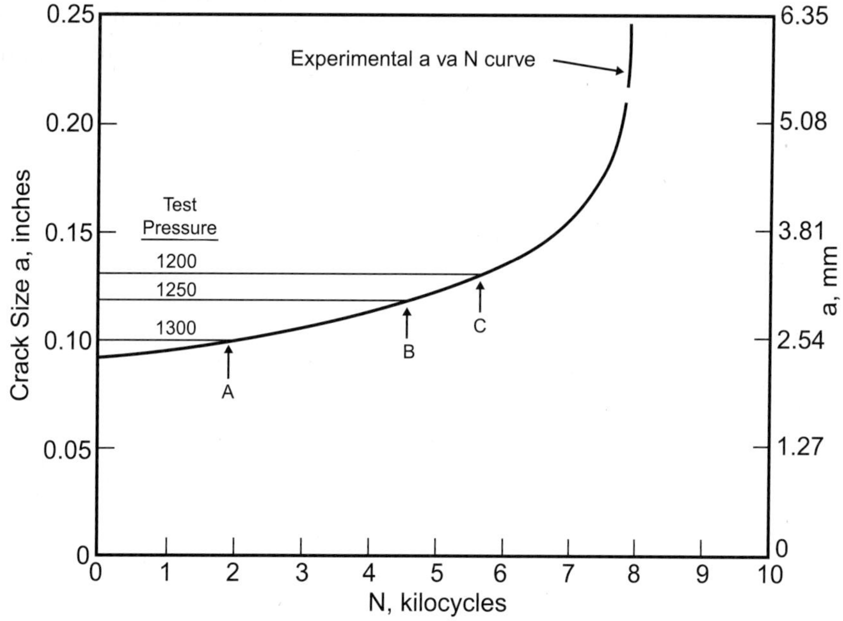

Figure 10. Example of an actual a versus N relationship used to compare the effects of test pressure.

life of 3200 cycles. Third, after the flaw has grown to 0.13 in ($N = 6000$ cycles corresponding to Point C), the remaining flaw is exactly the size that would just survive a test to 1200 psig. It has a remaining life of 2000 cycles. What this shows is the effectiveness of maximizing the test-pressure-to-operating pressure ratio. In terms of test pressure versus cycles to failure, the results would be as follows:

This exercise illustrates the valuable gain that can be made by small increases in test pressure. For a 4% increase over the base case (1200 psig), the gain in time to failure is 60%. For an 8% increase, the gain is nearly 300%. The above-demonstrated effect of test-pressure-to-operating-pressure ratio on service life provides one of the strongest incentives for high-pressure testing.

Test Pressure (psig)	Cycles to Failure after the Test	Relative Time to Failure
1200	2000	1.0
1250	3200	1.6
1300	5700	2.9

Calculating a retest interval

In the example shown above, let us assume that the pipeline experiences 200 cycles of pressure from 0 to 1020 psig per year. Let us also assume that it was tested to 1200 psig. Such a test probably does not give an adequate margin of safety and, in fact, it does not meet the D.O.T.'s minimum requirement of 1.25 × MOP for a liquid pipeline (Part 195). But it serves as a useful example. As the example shows, the worst-case flaw that might survive the test could be expected to fail in 2000 cycles or 10 years at 200 cycles per year. Prudence dictates that one use a factor of safety as was suggested in our previous discussion on corrosion. The recommended factor of safety is 2, and it is based upon an examination of the "*a*" versus "*N*" relationship of Figure 10.

Shown in Figure 11 is a replot of the "*a*" versus "*N*" relationship of Figure 10. Early in the life of a crack, the growth is slow. It accelerates, however, as the crack grows deeper, because the stress intensity factor driving the crack is a function of both applied stress and crack size. The implications of this behavior are illustrated as follows. Consider the crack represented by the curve in Figure 11. This crack is initially 0.095 in in depth and will grow to failure in 8000 cycles of the service pressure.

The hydrostatic test and retest level of 1200 psig is represented by the horizontal line at "*a*" = 0.13. This line is arbitrarily divided into four increments represented by Points 1, 2, 3, 4, and 5. The horizontal line intersects the *a* versus *N* relationship at about 5500 cycles. Any time the pipeline is tested after 5500 cycles, the flaw will fail in the test. Suppose one chooses a retest interval equal to the distance from Point 1 to Point 3 (5000 cycles). If the pipeline is tested at Point 3, the flaw will not fail. But it will also not survive until the next test at Point 5 (10,000 cycles). Such a retest interval is too long even though no test failure occurred in the first test.

Now consider what happens if the test interval is cut in half. The retests are then conducted at Points 2, 3, and 4, and the flaw fails during the test at Point 4. Obviously, this is a somewhat contrived example. However, the point is that if the model is reasonably accurate, the times to failure it predicts should be adjusted by a substantial factor of safety to establish an appropriate retest interval. Because of this, we normally recommend a factor of safety of 2. In our example, this would correspond to a retest interval of only 5 years. For a higher test pressure, say 1300 psig, the interval could be 4700/(2 × 200) = 14 years where the 5700 cycles come from our previous analysis based on Figure 10.

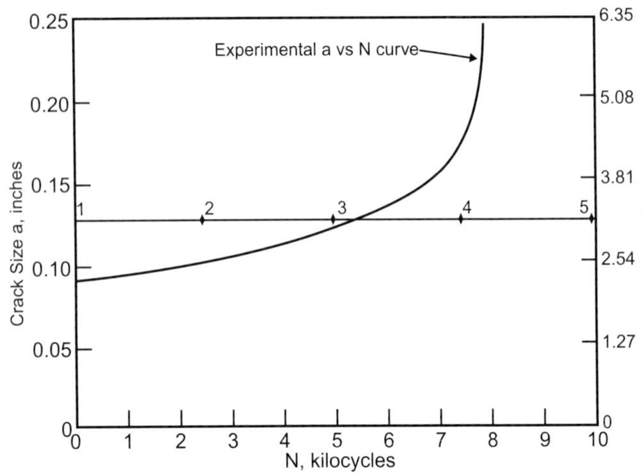

Figure 11. Example of a versus N relationship used to compare the effects of retest interval.

Mechanical damage

Although mechanical damage is frequently inflicted on pipelines and can cause severe degradation of their pressure-carrying capacities, we have not applied any revalidation model to this type of defect for three reasons. First, the state of analysis of mechanical-damage flaws is not as advanced as that for other types of flaws, although both Maxey[5] and British Gas[6] have made attempts to derive algorithms for dents and gouges. Second, the time-dependent nature of mechanical damage is nearly impossible to model. Since its occurrence is often concealed, the "zero" time is seldom known. The rate at which it degrades may depend on creeplike behavior and/or environmental effects. Third, we feel that the best action with respect to mechanical damage is preventative action, patrolling the rights-of-way, participating in one-call systems, conducting public education programs, and providing on-site locating of pipelines at excavation sites. In fact, our primary activity in this respect has been the development of ranking algorithms to calculate relative risk numbers for damage-prevention activities.

It should be mentioned that both in-line inspection and hydrostatic testing are viable remedies for mechanical damage. The problem is that since the existence of concealed damage is difficult to detect short of doing either of these, it is difficult to know when or even if such remedies are needed.

Stress-corrosion cracking

Perhaps the hardest type of defect to deal with is stress-corrosion cracking. With this type of cracking, the main problem is that no reliable model exists to define the crack growth rate. Furthermore, it is suspected that the rate is highly variable with changes in environmental conditions. However, unlike the case for mechanical damage, it is possible to establish a "zero" time based on the date of installation. And if the cracks exist, it is almost certain that they will grow.

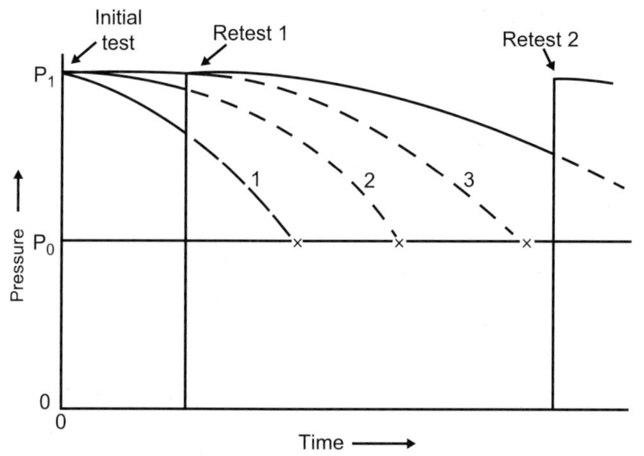

Figure 12. Concept of retesting when defect growth rate is unknown.

A scheme used by several pipeline operators to control stress-corrosion cracking is periodic hydrostatic retesting. In the absence of a crack-growth-rate model, the retest interval has to be set arbitrarily at first and corrected as required by the results of the test. A generalized scheme for setting retest intervals to deal with defect growth at unknown rates is illustrated in Figure 12. First, an initial test is performed in which defects may or may not be discovered. But, in either case, the maximum size flaw remaining is fixed by the test pressure, P_f in Figure 12, and the margin of effectiveness is proportional to P_f/P_0, where P_0 is the MOP. It is assumed that the defects will grow with the passage of time. Retest 1 is done after some arbitrary period of time short enough to assure that none of the remaining flaws can grow to a size that will fail in service. If this is done correctly, the growing flaws, 1, 2, and 3 in Figure 12, will fail in the test instead of in service at the times suggested by the dashed curves. If no service failures have occurred before Retest 1, then the first interval was adequate (based on the assumption that the crack-growth rate is fairly constant on average). Depending on the outcome of Retest 1, the pipeline operator may decide to increase the interval between retests. If no test failures occur or if they occur only at levels at or near the test pressure, then it makes sense to consider a longer interval. If Retest 2 is then conducted before any flaws grow to failure in service, the new interval is apparently valid and can be lengthened depending again on the failure pressure levels of flaws in Retest 2. In this manner, the pipeline operator is able to control the phenomenon by preventing service failures, and in effect determine the average flaw-growth rates.

It should be emphasized that the scheme portrayed in Figure 12 is valid only if the crack-growth rate is reasonably constant. For example, such a scheme would not be appropriate for fatigue cracks because of the shape of the "*a*" versus "*N*" curve as shown in Figures 10 and 11. Where the growth of the flaw accelerates rapidly, it is necessary to determine the actual growth rate. The setting of retests based upon arbitrarily assumed intervals could easily be defeated by the crack being in a rapid stage of growth, as was illustrated in the discussion of Figure 11.

Summary

We have presented concepts that are frequently used by us and others to attempt to determine when it is time to be concerned about the integrity of a pipeline. The models we use rely on well-established principles and require the use of data that are usually readily available. The models we have discussed can be used to assess revalidation needs for pipelines affected by corrosion, fatigue-crack growth, and stress-corrosion cracking. By means we alluded to but did not discuss, the risks associated with mechanical-damage defects can also be evaluated. While there is never an absolute certainty about the correctness of these types of models, we recommend them to our clients to provide guidance for establishing revalidation policies. We also recommend that they be used in conjunction with sound engineering judgment and generous factors of safety.

References

1. W.A. Maxey, J.E. Kiefner, R.J. Eiber, A.R. Duffy, Ductile Fracture Initiation, Propagation, and Arrest in Cylindrical Vessels, Fracture Toughness Proceedings of 1971 National Symposium on Fracture Mechanics, Part II. ASTM STP 514, American Society for Testing and Materials, 1972 70–81.
2. Kiefner, J. F., and Forte, T P., "Model Predicts Hydrostatic Retest Intervals," Oil and Gas Journal (January 7 and January 14, 1985).
3. J.C. Newman, I.S. Raju, An Empirical Stress Intensity Factor Equation for Surface Crack, Engineering Fracture Mechanics 15 (1–2) (1981).
4. Mayfield, M. E., and Maxey, W. A., ERW Weld Zone Characteristics, NG-18 Report No. 130 to American Gas Association, Catalog No. L51427 (June 18, 1982).
5. W.A. Maxey, "Outside Force Defect Behavior," NG-18 Report No. 162, Line Pipe Research Supervisory Committee, Pipeline Research Committee. American Gas Association, August 15, 1986.
6. P. Hopkins, D.G. Jones, A.J. Clyne, "The Significance of Dents and Defects in Transmission Pipelines," Conference on Pipework Engineering and Operations. I Mech E., London, February 1989, Paper C376/049.

Reprinted with permission of Systems Integrity and Specialty Pipeline Inspection & Engineering, Inc., organizers of the International Pipeline Rehabilitation Seminar Paper presented at the 1992 International Pipeline Rehabilitation Seminar, Houston, Jan. 28–31, 1992

Modeling pipeline risk assessment

W. Kent Muhlbauer, P.E., WKM Consultancy, LLC, Austin, Texas

Pipeline risk assessment

The safety of pipelines relative to other modes of transportation is generally not in question: By most measures, moving products by pipeline is safer than by alternate means. Yet even as pipelines claim a superior safety record, catastrophic failures are a real possibility. With an aging infrastructure, ever-encroaching population areas, and increasing economic pressures, the perceived, if not actual, risk of pipeline failure is increasing. Pipeline safety probably can and should be improved.

The ability to predict pipeline failures—where and when the next is to occur—would obviously be a great advantage in reducing risk. Unfortunately, this is not possible except in extreme cases. So, modern risk assessment methodologies provide a surrogate for such predictions. Our risk efforts are *not* attempts to predict how many failures will occur or where the next failure will occur. Rather, the efforts are designed so that we can systematically and objectively capture everything we know and use this information to make better decisions.

Formal versus informal risk management

Although formal pipeline risk management is growing in popularity among pipeline operators, it is important to note that risk management has always been practiced by these pipeline operators. Every time a decision is made to spend resources in a certain way, a risk management decision has been made. This informal approach to risk management has served us well, as evidenced by the very good safety record of pipelines versus other modes of transportation. An informal approach to risk management can have the additional advantages of being simple, being easy to comprehend and to communicate, and being the product of expert engineering consensus built upon solid experience.

However, an informal approach to risk management does not hold up well to close scrutiny, because the process is often poorly documented and not structured to ensure objectivity and consistency of decision making. Expanding public concerns over human safety and environmental protection have contributed significantly to raising the visibility of risk management. Although the pipeline safety record is good, the violent intensity and dramatic consequences of some accidents, an aging pipeline infrastructure, and the continued urbanization of formerly rural areas have increased perceived if not actual risks.

Historical (informal) risk management, therefore, has these pluses and minuses:

Advantages:

- Simple/intuitive
- Consensus is often sought
- Utilizes experience and engineering judgment
- Somewhat successful, based upon pipeline safety record

Reasons to change:

- More at stake from mistakes
- Inefficiencies/subjectivities inherent in informal systems
- Lack of consistency and continuity in a changing workforce unless decision-support systems are in place
- Need to better consider complicated risk factors and their interactions

The beginning of the twenty-first century has witnessed the introduction of more formalized risk programs and studies, as well as an increasing number of private pipeline risk management ventures. Although the general conclusion is typically a recommendation to move the industry toward such formal risk management programs, there is considerably less consensus as to the type and extent of assessment programs to adopt.

Beginning the risk modeling process

Successful risk assessment modeling involves a balance among various issues:

- Identifying an exhaustive list of contributing factors versus choosing the critical few to incorporate in a model (complex versus simple)
- "Hard" data and engineering judgment (how to incorporate widely held beliefs that do not have supporting statistical data)
- Uncertainty versus statistics (how much reliance to place on the predictive power of limited data)
- Flexibility versus situation-specific model (ability to use the same model for a variety of products, geographical locations, facility types, etc.)

It is important that all risk variables be considered, even if only to conclude that certain variables will not be included in the model. In fact, many variables will not be included when such variables do not add significant value but reduce the usability of the model. These decisions are done carefully and with full understanding of the role of the variables in the risk picture.

Note that many simplifying assumptions are often made, especially in complex phenomena such as dispersion modeling, fire and explosion potentials, etc., in order to make the risk model easy to use and still relatively robust.

Both probability variables and consequence variables are examined in most formal risk models. This is consistent with the most widely accepted definition of risk:

$$(\text{event risk}) = (\text{event probability}) \times (\text{event consequence})$$

Several questions to the pipeline operator may better focus a risk management effort and direct the choice of a formal risk assessment technique:

- What data do you have?
- What is your confidence in the predictive value of those data?
- What are the resource demands (and availability) in terms of costs, man hours, and time to set up and maintain a risk model?
- What benefits do you expect to accrue, in terms of cost savings, reduced regulatory burdens, improved public support, and operational efficiency?

Choices in risk-assessment techniques

The goal of any risk-assessment technique is to quantify the risks in either a relative or an absolute sense. The risk-assessment phase is the critical first step in practicing risk management. It is also the most difficult phase. Although we understand engineering concepts about corrosion and fluids flow, predicting failures beyond the laboratory in a complex "real" environment can prove impossible. No one can definitively state where or when an accidental pipeline failure will occur. However, the more likely failure mechanisms, locations, and frequencies can be estimated in order to focus risk efforts.

There is no universally accepted way to assess risks from a pipeline. It is apparent, however, that simply using historical accident counts is not enough. Because of the rare occurrence of pipeline failures, there is a high level of uncertainty associated with pipeline failure prediction and risk assessment. Although this is not always apparent, all risk models, from the simplest to the most complex, make use of probability theory and statistics. In a very simple application, these tools manifest themselves in hard-to-quantify experience factors and engineering judgments, that is, they are the underlying basis of sound judgments. In the more mathematically rigorous models, historical data, statistical theory, and/or mechanistic models (hard data) primarily drive the calculations.

Three general categories of more formal pipeline risk-assessment models can be found in use today:

Simple decision support: Matrix models. One of the simplest risk-assessment approaches is the matrix decision-analysis system. It ranks pipeline risks according to the likelihood and potential consequences of an event by a simple scale, such as high, medium, and low. Each threat to the pipeline is assigned to a cell of the matrix, i.e. the intersection of the perceived likelihood and perceived consequence. Events with both a high likelihood and a high consequence appear higher on the resulting prioritized list. This approach may be as simple as using an expert's opinion or as complicated as using quantitative data to rank risks.

The rigorous approach: Probabilistic/mechanistic models. The more rigorous and complex risk assessment is often called a Probabilistic Risk Assessment (PRA). It usually refers to a technique employed in the nuclear and aerospace industries that generally uses event trees and fault trees to model every aspect of a system. Initiating events are flow-charted forward to all possible concluding events, with probabilities being assigned to each branch along the way. Failures are backward flowcharted to all possible initiating events, again with probabilities assigned to all branches. All possible paths can then be quantified based upon the branch probabilities along the way. This technique usually relies heavily on historical failure rate data. It yields absolute risk assessments for all possible failure events. These more elaborate models are more costly than simpler risk assessments. They are technologically more demanding to develop and integrate via computer applications, they require trained operators, and they benefit from extensive databases.

The hybrid approach: Indexing models. Perhaps the most popular pipeline risk-assessment technique in current use is the "index model". In this approach, a relative weight is assigned to every important condition and activity on the pipeline. This includes both risk-reducing and risk-increasing items. This relative weight reflects the importance of the item in the risk assessment and is based upon statistics where available and upon engineering judgment where data are not available. The risk component scores are subsequently summed for each pipeline section to obtain a relative risk ranking of all pipe sections. The various pipe segments may then be ranked according to their relative risk scores in order to prioritize repairs and inspections. Among pipeline operators today, this technique is widely used and ranges from a very simple 5–20 factor model (where only factors such as leak history and population density are considered) to models with hundreds of factors considering virtually every item that impacts risk. When an indexing model is created from a probabilistic approach—using scenarios, event/fault trees, and all available historical data—this approach is often a solution with the best cost–benefit ratio.

Other issues in risk modeling

In comparing risk-assessment approaches, some issues arise that often lead to confusion. The following section discusses some of those issues:

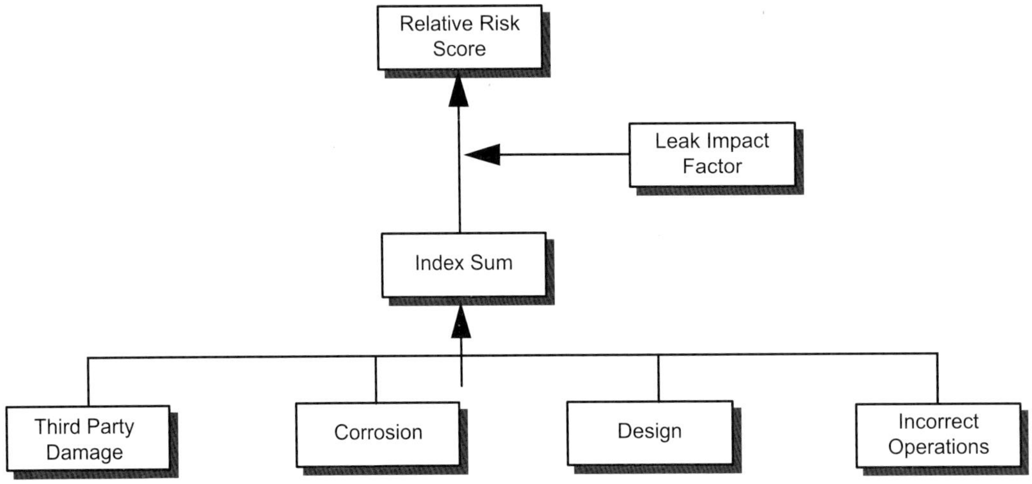

Figure 1. Typical pipeline risk model structure.

Absolute versus relative risk. Absolute risk implies a universal scale, usually in terms of a measurable consequence such as fatalities or costs per time period by which all risks can be compared. An example would be a risk assessment that produces a value such as 1.25 E-08 fatalities per thousand mile-years of pipeline. Relative risk is designed to prioritize risks by saying that pipeline section B is riskier than section A by an amount equal to X. The former is a predictive measure yielding the probability and consequence of failure, whereas the latter is a comparative tool yielding a ranking of current risks. A criticism of the relative scale is its inability to compare risks of several dissimilar systems (e.g. pipelines versus trucks) and its inability to directly provide failure predictions. The absolute scale fails in that it relies heavily on point estimates, particularly for rare events that are extremely difficult to quantify, and in the unwieldy numbers that often generate a negative reaction from the public.

For purposes of communications and in practical applications, this is not really an important issue. Either scale can be readily converted to the other scale if the circumstances so warrant. A relative risk scale can be converted to an absolute scale by correlating the relative risk scores with known failure rates and ensuring that the relative scales appropriately reflect real-world processes. The absolute scale is converted to more manageable and understandable (nontechnical) relative scales by simple mathematical relationships.

A danger underlying this issue is the possible assumption that a complex number expressed in scientific notation is more accurate than a simple number. In reality, either method should use the same available data pool and be forced to make the same number of assumptions when data are not available.

Quantitative versus qualitative models. These terms are often used to distinguish the amount of historical failure-related data analyzed in the model and the amount of mathematical calculations employed in arriving at a risk answer. A model that exclusively uses hard data is sometimes referred to as quantitative, whereas a model employing relative scales, even if numbers are assigned later, is referred to as qualitative or semiquantitative. The danger in such labels is that they imply a level of accuracy. In reality, the labels often tell more about the level of modeling effort, cost, and data sources. All complete risk models must use both quantitative and qualitative information.

Subjectivity versus objectivity. In theory, a purely objective model will strictly adhere to statistical information and will have no opinion data. A purely subjective model implies complete reliance upon expert opinion. In practice, no pipeline risk model fully adheres to either. Objectivity cannot be purely maintained when dealing with the real-world situation of missing data and variables that are highly confounded. On the other hand, subjective models certainly use "hard" data to form or support judgments.

The use of unquantifiable evidence. In a difficult-to-quantify item, some would argue that nonstatistical analyses are potentially damaging. Although there is always this danger of misunderstanding the role of a factor in a risk assessment, there is similarly the more immediate danger of an incomplete analysis by omission of a factor. For example, public education is seen by most pipeline professionals to be a very important aspect of reducing the number of third-party damages, leak reporting, and emergency response. Quantifying this level of importance and correlating with the many varied approaches to public education is quite difficult. A concerted effort to study these data is needed to best assess the effect on risk, but in the absence of such a study, most would agree that a company that practices public education will be more likely to achieve some level of risk reduction than a company that does not.

Uncertainty. False precision is often assigned to analysis yielding a specific number, especially when such a number appears to be the result of extensive calculations. In reality, uncertainty always exists. Uncertainty can be thought of as a property of the situation. We add to that uncertainty when we measure the situation. The measurement process itself adds uncertainty as measurement errors and inaccuracies accumulate. Separating the noise (the random variations caused in part by measurement uncertainties) from the signal (the true differences in risk) is an arduous task. At best, we hope for a high signal-to-noise ratio. The higher the ratio, the higher our confidence in identifying true signals. All risk-assessment techniques must address uncertainty, and no method can reliably eliminate it or even reduce it significantly.

Choosing a risk-assessment technique

Any or all of the previously described techniques might have a place in risk assessment/management. The choice may well depend on the situation. Understanding the strengths and weaknesses of the different risk assessment methodologies gives the user the basis for choosing one. A case can be made for using each in certain situations. For example, a simple matrix approach crystallizes thinking and is a step above informal risk assessment. However, if the need is to evaluate specific events at any point in time, a narrow focus assessment with well-known event probabilities, then probabilistic risk analysis might be the tool of choice. If the need is to weigh immediate risk tradeoffs or perform inexpensive overall assessments, indexing models might be the best choice. The following shows some tradeoffs associated with each choice:

Pros and cons I (PRA)

\+ Rigorous, scientific approach
\+ Accepted in other industries
\+ Uses all available information
– Costly
– Difficult to do resource allocation
– Difficult to do overall risk management
– Might create the “illusion of knowledge”
– Intimidating to nontechnical audience

Pros and cons II (indexing)

\+ Uses all available information
\+ Intuitive
\+ Flexible
– Possibly more subjective
– Must be well documented
– Often requires a subsequent linkage to absolute values

Pros and cons III (matrix)

\+ Improvement over informal techniques
\+ Forces more logical examination of situation
\+ Inexpensive approach
– Limited number of risk factors (not comprehensive)
– Subjective

In addition to the three general categories of risk models, supporting tools such as fault trees, event trees, hazard and operability analysis (HAZOPS), what-if scenarios, and others play a role in performing risk assessment. These tools are almost always the foundation and building blocks of the complete risk assessment model.

The following are some ways a particular technique can support a specific need:

Application choices I (PRA, scenarios, event/fault trees)

- Study specific events and/or create risk models
- Perform postincident investigations
- Compare risks of specific failures
- Calculate specific event probabilities

Application choices II (indexing model built from scenarios, event/fault trees, PRA)

- Obtain an inexpensive overall risk model
- Create a resource allocation model
- Model the interaction of many potential failure mechanisms
- Study or create an operating discipline

Application choices III (matrix)

- Better quantify a belief
- Create a simple decision support tool
- Combine several beliefs into a single solution
- Document choices in resource allocation

The ideal risk-assessment model

Regardless of the technique used, some properties of the ideal risk-assessment tool can be identified. A partial list of the characteristics of the “ideal model” includes the following:

Costs. The value or benefits derived from the risk-assessment process should clearly outweigh the costs of setting up, implementing, and maintaining the program.

Learning ability. Because risk is not constant over the length of a pipeline or over a period of time, the model must be able to “learn” as information changes. This means that new data should be easy to incorporate into the model.

Signal-to-noise ratio. Because the model is in effect a measurement tool, it must have a suitable signal-to-noise ratio. This means that the “noise”, i.e. the amount of uncertainty in the measurement (due to numerous causes), must be low enough that the "signal", i.e. the risk value of interest, can be read. This is similar to the "accuracy" of the model but involves additional considerations that surround the high level of uncertainty associated with risk management.

In examining a proposed risk-assessment effort, it may be wise to evaluate the risk-assessment model to ensure the following:

- All failure modes are considered
- The model appropriately treats both time-dependent and -independent failure mechanisms
- All risk elements are considered and the most critical are included
- Failure modes are considered, independently as well as in aggregate
- All available information is being appropriately utilized
- Real-world processes are appropriately represented
- Relative risk values are anchored by real failure rates and consequence potential
- Provisions exist for regular updates of information, including new types of data
- Consequence factors are separable from probability factors
- Weightings, or other methods to recognize the relative importance of factors, are established
- The rationale behind weightings is well documented and consistent
- A sensitivity analysis has been performed
- The model reacts appropriately to failures of any type
- Risk elements are combined appropriately ("and" versus "or" combinations)
- Steps are taken to ensure consistency of evaluation
- Risk-assessment results form a reasonable statistical distribution (outliers?)
- There is adequate discrimination in the measured results (signal-to-noise ratio)
- Comparisons can be made against fixed or floating "standards" or benchmarks

Finally, a view to the next step, risk management, should be taken. A good risk-assessment technique will allow a smooth transition into the management of the observed risks. This means that provisions for resource allocation modeling and the evolution of the overall risk model must be made.

Managing risks

In some sense, we have near-complete control of the risk. We can spend nothing on preventing accidents, or we can spend enormous sums of money overdesigning facilities, employing an army of inspectors and routinely shutting down lines for preventive maintenance and replacement. Pragmatically, operators spending too little on preventing accidents will be out of business from regulatory intervention or from the cost of accidents. On the other hand, if an operator spends too much on accident prevention, he can be driven out of business by the competition, even if, perhaps, that competition has more accidents. Risk management, to a large extent, revolves around the central process of making choices in the design and day-to-day operations of a pipeline system. Many of these choices are mandated by regulations, whereas others are economically (budget) constrained. By assigning a cost to pipeline accidents (a sometimes difficult and controversial thing to do) and including this in the cost of operations, the optimum balance point is the lowest cost of operations.

No operator will ever have all of the relevant information needed to guarantee safe operations. There will always be an element of the unknown. Managers must control the "right" risks with limited resources, as there will always be limits on the amount of time, manpower, or money to apply to a risk situation. Managers must weigh their decisions carefully in light of what is known and unknown. It is, perhaps, this assimilation of complex data and the subsequent integration of competing risk reduction-profit goals that lies at the heart of the debate about how best to manage pipeline risks.

Conclusion

The move to more formal risk techniques is intended to increase operational consistency and credibility, especially when such techniques are offered for public viewing. In a concerted effort to demonstrate its dedication to public and environmental safety while simultaneously seeking to reduce potential liabilities and regulatory requirements, the federal government and the pipeline industry (both gas and liquid) are embracing risk management techniques through integrity management programs.

Even without such broad-based support, risk management has long been recognized as a valuable effort in pipeline operations. Risk-assessment approaches are available to serve many needs, although no single approach is the best at addressing all needs. It is important that those practicing (or planning to practice) risk management understand the strengths and limitations of various approaches and set their expectations accordingly.

Reference

W. Kent Muhlbauer, Pipeline Risk Management Manual, 3rd ed. Gulf Publishing Co., Houston, 2004.

A risk management tool for establishing budget priorities

John F. Kiefner, President, Kiefner and Associates, Inc., Worthington, Ohio

Pipeline operators have obligations to their owners or stockholders to protect their assets and to their employees and the public to minimize the potential for injury-producing or property-loss accidents. Pipeline risk management programs are generally instituted to meet these obligations. One technique that has proven its value in pipeline risk management is "prioritizing" based on a ranking model. By means of prioritizing, a pipeline operator can expect to enhance or optimize the effectiveness of a maintenance program in the sense that the areas of potential risk will be identified and ranked in the order, which achieves the most safety for the least cost. Such models typically contain a "feedback" feature, which allows reranking in response to risk reduction activities such as removal of anomalies, replacement of questionable components, or implementation of improved procedures. "What if" scenarios can be examined for planning and budgeting purposes using this feature. Ranking models can be linked to the operator's GIS system and/or system parameters database. The generic concepts of ranking models are discussed herein.

Background

A fair number of ranking models for prioritizing pipeline maintenance exist. Some have been created by individual operators[1-7] while others have been formulated through industry groups[8,9] or by contractors serving the industry[10,11]. The concepts embodied in many of these will be addressed. First, I would like to point out that ranking for maintenance purposes is a tool for managing risk. It does not provide a true risk assessment, but it can be the basis for reducing risk and/or efficiently allocating maintenance expenditures.

For an operating system risk arises from the probability that a loss-producing accident can occur. The risk associated with such an accident can be calculated as the product of two quantities: the probability that a failure of the system can occur and the consequences that can arise as a result. Well-known techniques such as fault-tree analyses may be used to calculate probabilities of failure provided that failure rates are known. Consequences can be assessed by means of complex computer programs, which are capable of assessing various failure scenarios.

Probabilistic risk assessments of this type are made for chemical and refining processes, for nuclear power plants, and for aerospace systems. Such assessments are not done routinely for pipelines for two reasons. First, the accident-producing factors for pipelines vary so widely that reliable failure rate data are virtually nonexistent. Second, to a large extent pipeline safety improvements can be effected without knowing the true risk associated with a given pipeline operation. Therefore, the industry has tended to rely on ranking models to augment their risk management activities and to prioritize maintenance. The concepts involved in these relatively simple techniques are explained below.

Basic ranking concepts

Most ranking models emulate probabilistic risk assessment by calculating risk as a "relative" number. The latter is the product of "pseudo-probabilities" of failure and "relative" consequence scenarios. As in probabilistic risk assessment, the perceived causes of failure are postulated. In the absence of reliable failure rates, however, the analyst creates equations based upon system and operating parameters, which are known to influence the likelihood of failure for a given section of a pipeline.

Instead of calculating realistic outcomes of various accident scenarios to evaluate true consequences, the creators of ranking models usually construct simple equations, which embody relative outcomes of accidents based on factors such as population density, product release size, and loss of service for a given section of a pipeline. The section of pipeline is then given a "relative risk" or ranking number, which is the product of its pseudo-probability of failure and its relative consequence of failure. Sections of a pipeline or pipelines within a system may then be ranked according to relative risk. Those having the highest relative risk can be allotted the most attention with respect to maintenance or remedial measures.

Failure probabilities

Causes of Failure

Ranking models for pipeline maintenance can involve several causes of failure. Almost always they address outside force (including damage by excavating equipment), external corrosion, and manufacturing or construction defects. Other causes that may be considered include soil instability, internal corrosion, equipment failure, and operator error.

Construction of Equations by Cause

Equations for the probabilities of failure by cause are constructed from factors perceived to influence the occurrence of a failure. For example, a typical equation for the probability of failure from outside force might consider the following:

Pipe geometry (the thinner the pipe or the smaller its diameter, the more easily it is damaged)

Material (strength and toughness contribute to damage resistance)

Depth of Cover (the deeper the pipe, the less likely it will be encroached upon) Exposure (the higher the level of construction activity the greater the chance a pipeline will be damaged)

Damage prevention (the probability of damage is reduced by responses to one-call systems, active patrolling, public education, communications with land owners, robust right-of-way maintenance and marking programs).

Hence the typical equation for the probability of failure from outside force, P_{of}, might look something like this. Bear in mind that these are only examples. Also, please refer to Figure 1 to see the details of each equation.

$$P_{of} = \left(\begin{array}{c}\text{Pipe}\\ \text{Geometry}\end{array}\right)\left(\begin{array}{c}\text{Material}\\ \text{Factor}\end{array}\right)\left(\begin{array}{c}\text{Depth of}\\ \text{Cover}\end{array}\right)(\text{Exposure})\left(\begin{array}{c}\text{Damage}\\ \text{Prevention}\end{array}\right) \tag{1}$$

where

Pipe Geometry is a function of $1/t^2 + 12/D$, t is wall thickness; D is diameter both in inches.
Material Factor is $100{,}000/\text{SMYS} + (10/\sqrt{\text{CVN}}) + C_1$, SMYS is specified minimum yield strength in psi; CVN is the Charpy V-notch upper shelf energy in ft-lb, and C_1 is zero if the pipe is never exposed to a temperature below its ductile-to-brittle transition temperature, one if the pipe is sometimes exposed to temperatures up to 60°F below its transition temperature, and ten if the pipe is frequently exposed to temperatures of 60°F or more below its transition temperature or if its transition temperature is unknown.
Depth of Cover is $(3/d_c)^{1.5}$, d_c is the depth of cover in feet.
Exposure is $(1+0.1C_2+C_3+100C_4)$, C_2 is the number of foreign line crossings, road crossings, and river crossings in the segment; C_3 is a number ranging from 1 to 5 depending on the perceived construction activity; and C_4 is the number of outside force incidents per mile per year on the segment.
Damage Prevention is $(1+5C_5+5C_6+C_7+C_8)$, C_5 ranges from 0 to 1 and represents the one- call response, i.e., 1 for locating and marking the pipeline, 0.5 for locating, marking and providing an on-site representative, and 0.1 if the operator exposes the line for the excavating contractor; C_6 ranges from 0 to 1 and represents patrolling, i.e., 1.0 for bi-weekly overflights, 0.5 for 2 overflights per week, 0.1 for 3 or more overflights per week; C_7 ranges from 0 to 1 and represents communication with land-owners, i.e., 1 for none, 0.8 for periodic contacts by mail, 0.5 for periodic face-to-face meetings; C_8 represents right-of-way maintenance and marking and ranges from 0.5 to 1, i.e., 0.5 for robust program, 1.0 for normal program.

Factors which one would expect to affect the probability of failure from corrosion would be the thickness of the pipe (the thicker it is, the longer it takes for corrosion to penetrate the wall thickness) the presence and effectiveness of a cathodic protection system, the presence and effectiveness of an anti-corrosion coating, the effect of the soil on the coating and the ability to cathodically protect the pipe, and the operating experience associated with a particular segment of pipe including its leak history.

Accordingly, the equation for the probability of failure from corrosion P_{cor}, might look like this:

$$P_{cot} = \left(\frac{1}{t}\right)\left(\begin{array}{c}\text{Cathodic}\\ \text{Protection}\end{array}\right)(\text{Coating})\left(\begin{array}{c}\text{Soil}\\ \text{Factor}\end{array}\right)\left(\begin{array}{c}\text{Experience}\\ \text{Factor}\end{array}\right) \tag{2}$$

where

t is the wall thickness of the pipe, inches.
Cathodic Protection is $8T_1 + \frac{E_{cp}}{3} T_2$, T_1 represents the time in years that the pipeline was buried without cathodic protection being applied and T_2 represents the time in years that the pipeline has been subjected to cathodic protection. The coefficient, E_{cp}, is based upon the perceived effectiveness of the cathodic protection. E_{cp} could be 1.0 for an adequately protected, coated pipeline, but it could range as high as 4 for a poorly protected, bare pipeline.
Coating (CT) is usually a number ranging between 1 and 20 where 1 represents the best coating available and 20 represents bare pipe. Companies typically represent fusion-bonded epoxy coating as 1 or 1.5, coal tar enamel as 1, 1.5 or 2, and single-layer polyethylene tapes as 8.
Soil Factor (SF) might range from 5 to 10 where 5 represents a benign soil which neither damages coating nor shields the pipe from cathodic protection and where 10 represents a coating-damaging soil and/or one which tends to shield the pipe from cathodic protection.
Experience Factor should include a constant of 1.0 to prevent it from being zero and the number of corrosion leaks per mile per year, C_9 (multiplied by 50). It may also include observations of the pipe or coating condition based on bell-hole inspections, cathodic protection surveys, and/or cathodic current consumption.

The probability of failure from manufacturing or construction defects is strongly influenced by the operating hoop stress level and its relationship to the preservice hydrostatic test level. In addition, the age of the pipe material and known material deficiencies may be taken into account. For example, the probability of failure from manufacturing and construction defects, P_{mc}, might be

$$P_{mc} = \left(\frac{\text{MAOP}}{\text{HTP} - \text{MAOP}}\right)\left(\begin{array}{c}\text{Age}\\ \text{Factor}\end{array}\right)\left(\begin{array}{c}\text{Seam}\\ \text{Factor}\end{array}\right)\left(\begin{array}{c}\text{Girth Weld}\\ \text{Factor}\end{array}\right) \tag{3}$$

where

MAOP is the maximum allowable operating pressure.
HTP is the hydrostatic test pressure (mill test pressure may be used if the actual preservice test pressure is unknown).

Age Factor could be as simple as the number of years in service but if the pipeline operator feels that this is too extreme, the effect of years in service could be tempered by adding a constant (e.g., 100).
Seam Factor (C_{10}) is generally taken to be the reciprocal of the longitudinal joint design factor (e.g., 1/0.6 for butt-welded pipe, 1/0.8 for lap-welded pipe).
Girth Weld Factor (C_{11}) can be set at 1 for high-quality shield-metal-arc girth welds; 1.2 for adequate-quality girth welds; and 1.5 for poor-quality girth welds, acetylene welds, or mechanical joints.

Consequences

The relative consequences of an incident are usually based on the perceived chances that people will be killed or injured, that property losses will occur, and that loss of the product and loss of service will be costly to the pipeline operator. These outcomes logically would seem to be a function of the population and property exposed, the violence of the incidence (including sudden pressure release as well as possible explosion and fire), the amount of product lost, the effect on other company operations, and the customers or shippers affected. The exposure of population and property value is often simply assumed to be proportional to the "class location" as defined for natural gas pipelines by the ASME B31.8 code and Part 192 of the federal regulations. The violence of an incident can conveniently be expressed in terms of the hoop stress level in the pipe or the pressure and the size of the release (i.e., pinhole leak, small hole, or significant rupture). The probability of ignition can be considered as well as the type of thermal event (i.e., torch fire, deflagration, or detonation). However, the latter two are usually relevant to products such as ethylene or propane and not natural gas. Losses to the company, the customers or the shippers can be considered in terms as simple as the amount of throughput lost, and/or the number of customers disrupted. Lastly, in the case of many liquid petroleum products, damage to the environment may be a significant factor. An example is shown below in which "Q" represents the consequences.

$$Q = \begin{pmatrix}\text{Class}\\ \text{Location}\end{pmatrix}(\text{MAOP})(\text{Diameter})^n\begin{pmatrix}\text{Release}\\ \text{Size}\end{pmatrix}\begin{pmatrix}\text{Throughput}\\ \text{Loss}\end{pmatrix}\begin{pmatrix}\text{Customers}\\ \text{Interrupted}\end{pmatrix}\begin{pmatrix}\text{Environmental}\\ \text{Damage}\end{pmatrix} \quad (4)$$

The "class location" factor can be chosen as 1, 2, 3, or 4 corresponding to the defined class location. Alternative formats include using the number of houses in the definitions of class location (10, 46, 100, and an arbitrary number, say 200, for Class 4) or the reciprocal of the design factor for each class location (1/0.72, 1/0.60, 1/0.50, and 1/0.40). The appropriate choice can be ascertained in the type of sensitivity analysis described later herein. MAOP or maximum allowable operating pressure is usually used as a factor, and diameter is used raised to a power. The power "n" may be fractional or one or a number greater than one. Release size is often not used at all for natural gas pipelines because small leaks usually do not pose significant risks. For motor fuels, crude oil, or highly volatile liquids, leaks can be damaging, however, and release size takes on more importance. Throughput loss (C_{12}) and customers interrupted (C_{13}) are highly operator-specific. These terms are usually handled by integers ranging from 1 to 10 based largely upon the operators' judgment. The environmental damage term can also be handled in terms of a range of integers depending on the potential for water pollution or destruction of wildlife.

Relative risk calculations

The calculation of relative risk, R, for ranking pipeline segments can be expressed as follows

$$R = \left(m_1 P_{of} + m_2 P_{cor} + m_3 P_{mc}\right) Q/1000 \quad (5)$$

where

$\mathbf{P_{of}}$, $\mathbf{P_{cor}}$, $\mathbf{P_{mc}}$ are the probabilities of failure for the causes considered
Q is the consequence number
$\mathbf{m_1}$, $\mathbf{m_2}$, $\mathbf{m_3}$ are coefficients, if needed, to balance the relative effects of the probabilities.

Other options that can be considered are as follows. If an operator wishes to consider the effects of release size, an R_{Leak}, for "leaks" may be calculated and an $R_{Rupture}$ for "ruptures" may be calculated. In these cases the relevant Q values would be based upon the different values of the release-size term, and the "m" coefficients of Equation (5) might by adjusted for release size effects. For example, since a higher percentage of outside force incidents than corrosion incidents are known to occur as ruptures, the release size versions of Equation (5) might look like this:

$$R_{Leak} = \left(0.75P_{of} + 0.95P_{cor} + 0.80P_{mc}\right) Q_{leak}/1000 \quad (5a)$$

$$R_{Rupture} = \left(0.25P_{of} + 0.05P_{cor} + 0.20P_{mc}\right) Q_{rupture}/1000 \quad (5b)$$

This level of sophistication is probably not necessary for natural gas pipelines because small leaks usually have no significant consequences.

Mitigation of risk

A pipeline operator may use the foregoing equations to obtain "R" values for any number of pipeline segments. The list of segments ranked by R values provides a rational basis for carrying out a rehabilitation or maintenance program. The model also can provide a means for assessing mitigative measures and for reranking the segments as the measures are applied. For example, for a pipeline exposed to a high risk of failure from outside force, deeper burial (by lowering the line at critical locations), more frequent patrolling, better right-of-way maintenance and marking, more active response to one-calls, and better communication are all options for reducing risk which have values represented in Equation (1). Changes in relative risk in response to any or all of these mitigative measures can be readily calculated.

Another kind of assessment of mitigative action arises in conjunction with in-line inspection and hydrostatic testing. In these cases a risk reduction factor can be applied to the Pcor term. These factors are usually of the following form.

$$F_{ILI} = \frac{T_{NOW} - T_{ILI} + 1}{10} \quad (6)$$

$$F_{HRT} = \frac{T_{NOW} - T_{HRT} + 1}{5} \quad (7)$$

where

F_{ILI} is the factor applied to P_{cor} if in-line inspection is carried out followed up by appropriate repairs

F_{HRT} is the factor applied to P_{cor} if a hydrostatic test to at least 1.25 × MAOP is applied to the pipeline.

T_{ILI} is the year of the repair response following an in-line inspection

T_{HRT} is the year of the hydrostatic retest

T_{NOW} is the current year.

In effect the credit for these mitigative steps is time-dependent. The credit is substantial immediately. P_{cor} is reduced by 90 percent following the in-line inspection repairs or by 80 percent following a hydrostatic test. In ten years the credit for in-line inspection is gone and in five years the credit for hydrostatic retesting is gone.

Examples

The equations of the model are summarized in Figure 1, and two pipelines are compared using the model in Table 1. The use of the model is demonstrated by these examples.

Discussion

The ranking model embodied in Equations 1-7 is simple and generic. It could be useful if customized for a given operator. It needs two essential preliminary evaluations, however. First, it needs to be subjected to a rigorous sensitivity analysis. The effect on each of the major terms: P_{of}, P_{cor}, P_{mc}, Q, and R of changes in the input parameters over realistic ranges should be made to assure that all terms and coefficients lead to

(1) $$P_{of} = \left(\frac{1}{t^2} + \frac{12}{D}\right)\left(\frac{100{,}000}{SMYS} + \frac{10}{\sqrt{CVN}} + C_1\right)\left(\frac{3}{d_c}\right)^{1.5}(1 + 0.1C_2 + C_3 + 100C_4)\ (1+5C_5 + 5C_6 + C_7 + C_8)$$

(2) $$P_{cot} + \left(\frac{1}{t}\right)\left(8T_1 + \frac{E_{cp}}{3}T_2\right)\left(\frac{CT}{6}\right)(SF)(1+50C_9)$$

(3) $$P_{mc} = \left(\frac{MAOP}{HTP - MAOP}\right)(100 + T_1 + T_2)(C_{10})(C_{11})$$

(4) $$Q = (CL)(MAOP)(D)^{0.5}(C_{12})(C_{13})$$

(5) $$R = (P_{of} + P_{cor} + P_{mc})Q/1000$$

(6) $$F_{ILI} = \frac{T_{NOW} - T_{ILI} + 1}{10}$$

(7) $$F_{HRT} = \frac{T_{NOW} - T_{HRT} + 1}{5}$$

Figure 1. Summary of equations for a simple, generic ranking model.

Table 1
Two Pipelines Compared on the Basis of the Generic Ranking Model

Parameter	Symbol	Value for Line 1	Value for Line 2
Diameter, inches	D	12.75	30
Wall Thickness, inches	t	0.250	0.375
Yield Strength, psi	SMYS	42,000	52,000
Charpy V-notch upper-shelf energy, ft-lb	CVN	25	25
Transition Temperature Factor	C_1	1	1
Depth of Cover, feet	d_c	3	3
Number of Crossings	C_2	2	2
Construction Activity Factor	C_3	4	2
Outside Force Incidents/Mi/Yr	C_4	0.02	0
One-Call Response Factor	C_5	0.5	0.5
Patrolling Frequency Factor	C_6	1	1
Communications Factor	C_7	0.8	0.8
R.O.W. Maintenance/Marking Factor	C_8	1	1
Years in Service without CP	T_1	0	10
Years in Service with CP	T_2	40	30
CP Effectiveness Factor	E_{cp}	1	2
Coating Type Factor	CT	2	2
Soil Factor	SF	5	10
Corrosion Incidents/Mi/Yr	C_9	0	0.02
Maximum Allowable Operating Pressure, psig	MAOP	824	936
Hydrostatic Test Pressure, psig	HTP	1482	1170
Seam Factor	C_{10}	1	1
Girth Weld Factor	C_{11}	1.25	1
Class Location	CL	3	1
Throughout Loss Factor	C_{12}	1	2
Customer Interrupt Factor	C_{13}	2	1
Probability of Failure, Outside Force	P_{of}	6760	1560
Probability of Failure, Corrosion	P_{cor}	89	1778
Probability of Failure, Manufacturing/Construction	P_{mc}	224	560
Consequences	Q	17654	10253
Rank	R	124864	39966

reasonable answers. Second, a reality check is needed in which the model's output is compared to historical experience. In this manner a pipeline operator establishes confidence in the model and is more likely to consider it a useful decision-making tool. Finally, the following advice should be kept in mind. Accidents can occur in pipeline systems in spite of the availability of a ranking model because no model can be expected to perfectly describe the performance of a system and because the values used in model, although they may be the best available, may not adequately describe the physical reality. Other reasons why a risk assessment might not prevent an accident could be that the user does not or cannot respond fast enough to the list of needed maintenance, and that a ranking model has no means for assessing when all of the essential responses have been carried out.

Nevertheless, with a ranking model the chances are very good that a pipeline operator can prioritize correctly and thus spend the available maintenance dollars wisely.

References

1. R.M. Jamieson, J.S. McDonald, Pipeline Monitoring, in: Pipeline Engineering Symposium-1986. ASME, February 23-27, 1986.
2. M.J. Davis, Tenneco's Efforts for Verifying Pipeline Integrity, in: American Gas Association Operating Section, Distribution and Transmission Conference (May 16, 1988).
3. S. Nunez, H.B. Martch, A Proposal for Risk Assessment of EPNG Pipelines, in: A report for the Renovation Task Force and the Pipeline Safety Committee (October 7, 1988).
4. British Gas, Anon, "Allocation of Points for Pipeline Inspection Scheduling Scheme".
5. G.G. Mouser, Evaluation Criteria for Replacement of Transmission Pipelines, in: American Gas Association Operating Section, Distribution and Transmission Conference (May 16, 1988).
6. H. Ahmad, Economic and Risk Analysis for Gas Line Replacement, Pipe Line Industry (February, 1988) 17−19.
7. W.K. Muhlbauer, "RIPS Identifies Pipeline Risks", Pipeline Risk Assessment, Rehabilitation, and Repair, Houston, Texas (May 20-23, 1991).
8. J.F. Kiefner, P.H. Vieth, Methods for Prioritizing Pipeline Maintenance and Repair, in: Pipeline Research Committee (A.G.A.) Catalog No. L61631 (1991).
9. P.M. Scott, J.E. Orban, W. Goldwaite, J.F. Kiefner, F.L. Leverenz, "Development of Data Base Parameters to Aid in Making Repair/Replace Decisions", a report to the American Gas Association (March 5, 1986).
10. G.R. Mayer, H.J. Van Dyke, C.E. Myrick, Risk Analysis Determines Priorities Among Pipeline-Replacement Projects, Oil & Gas Journal (September 21, 1987) 100−106.
11. J.B. Cornwell, W.E. Martinsen, and D.W. Johnson, "Risk Analysis Methodology for Gas Pipelines", Quest Consultants, Inc. Norman, Oklahoma.

Reprinted with permission - Applus+, Energy & Industry Division, 3 Sugar Creek Center Blvd. Suite 600, Sugar Land, TX. Online at www.applus.com. Presented at the Risk Assessment/Management of Regulated Pipelines. A NACE TechEdge Series Program, Houston, Texas, February 10-12, 1997.

Floating roof failure modes and risk

Philip Myers, P.E., Founder and Director, PEMY Consulting, Pleasant Hill, California

There are many ways that a floating roof can fail during its life cycle stages. Figure 1 shows the concept of failure as it applies to the various life cycle stages of a floating roof. The grey boxes are not explored in detail here, but these failure modes occur, and they should be considered in any tank safety management program. As a direct result of improper design, maintenance, and construction, there have been many serious incidents involving floating roofs throughout the history of the oil and gas industry. The focus of operational tank safety, especially floating roofs, is on many different and complex failure mechanisms.

Failure causation

Floating roofs are the most complex component of a petroleum storage tank and they can fail in a variety of ways. Proper tank safety management requires asking questions like: "How can my floating roof fail? Where is it most vulnerable? What can I do to systematically prevent failure of a floating roof? Am I willing to take the chance of a serious incident caused by a floating roof fire?".

There are many points of potential mechanical failure of floating roofs, so many, in fact, that determining causation is a significant problem. One method to conceptualize floating roof failure is using diagrammatic fault trees. Fault trees can be used in either a qualitative, or quantitative manner, and are particularly useful qualitatively to determine floating roof failure modes.

Fault trees

Fault trees use a top-down logic diagram approach to analyze system failures. The methodology of the fault tree is deductive, and maps causation in the tree. The basic ideas of the

Figure 1. Failure domains of floating roofs.

fault tree are most easily understood by examining the simple fault tree diagram in Figure 2. Because of its diagrammatic nature, as well as the computational methodology used, fault trees can serve as both a qualitative and quantitative model in risk assessment.

In Figure 2, events are classified as the top event, intermediate events, or basic events. We start at the top of the diagram with the top event, which is the hazard, or negative outcome that we are concerned about. For examining floating roofs with fault trees, the top event might be Partial Sinking of Floating Roof. Rectangles represent top events or intermediate events. Basic events are circles and represent the lowest level of resolution we are working at. They represent the components or system elements which can initiate the sequence of events leading to the top event.

In order for the top event, Partial Sinking of Floating Roof, to occur, both the rain must occur AND drainage from the floating roof must be inadequate. The AND gate represents

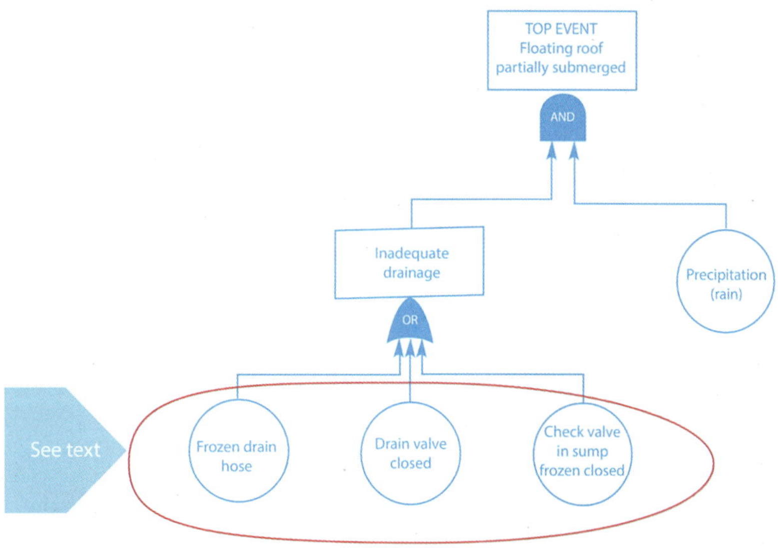

Figure 2. Partial sinking of floating roof.

the logic that requires both events to be true before the path upward can continue. The OR gate shows that the open valve condition can occur if either the valve was left open or if the valve fails. Although AND/OR gates are the workhorses of the logic used in fault trees, they are not the only type of gates. Note, at the base level of the tree, basic events are shown as circles (but other symbols are used). The events which are either a top event or events to be further analyzed are shown as rectangles. The 'inadequate drainage' box shows an OR gate below, indicating that any one of the initiating causes could lead to the inadequate drainage event.

Notice the red ellipse. These events are all able to be continuously monitored by tilt monitoring systems. Safety management of floating roofs using roof tilt monitoring system dramatically reduces the chance of an event. Fault trees can be built for many floating roof failure scenarios.

Loss of buoyancy events

Most of the failures that lead to serious top events, such as fires, business interruption, or injuries and fatalities, begin with a loss of buoyancy (LOB) event. Although the LOB event is not the only category leading to serious incidents, it is by far the most common failure mode for EFRs and IFRs.

APFR LOB Event from Corrosion

One of the most common and serious failure modes for a floating roof is the loss of buoyancy caused by corrosion due to corrosion holes or weld cracks as in Figure 3. These damage mechanisms lead to loss of buoyancy of either a pontoon compartment or the center deck in an annular pontoon floating roof (APFR).

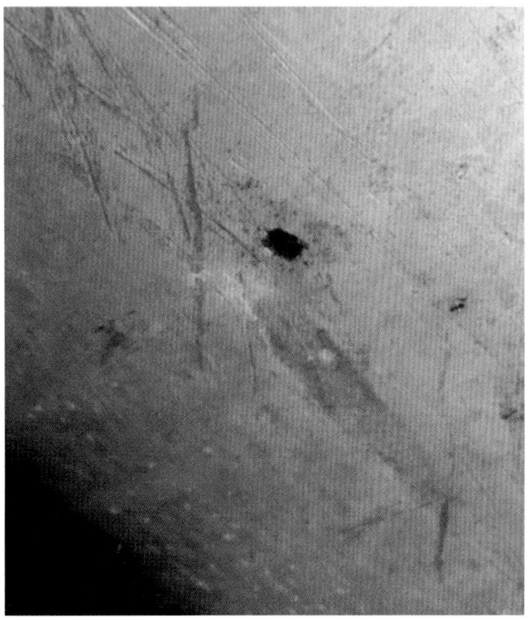

Figure 3. Corrosion hole in floating roof pontoon.

The buoyancy forces of the APFR support the deck and the compartmented pontoon ring. API 650 Annex C allows for any APFR to have a punctured center deck and two adjacent punctured pontoon compartments without the roof sinking or ceasing function as intended. However, it is unlikely that the effects of corrosion will occur within these parameters. When corrosion failures occur, they tend to occur in numerous locations almost simultaneously. For example, if there is a fatigue crack near a floating roof leg, due to landing the roof many times or otherwise, it is likely that other similar locations will imminently fatigue crack within just a few weeks or months later. Moreover, the design criteria of API 650 are often not met, because many tank manufacturers neglect to design floating roofs to meet these criteria. Checking that API 650 criteria are fulfilled requires advanced structural engineering, and that is problematic for even major integrated oil companies.

The major effect of the center deck filling with product fluid is that the circumferential stresses become extremely high. Even if the design criteria are met, the potential for sudden buckling and sinking of the floating roof exists. The net forces causing the potential for buckling are shown in Figure 4. The photo in Figure 5 shows the inner rim of the pontoon ring section completely failed by buckling. When buckling occurs it is unlikely the floating roof will maintain buoyancy, and a LOB event will occur.

Loss of Buoyancy in Pontoon Section

A punctured pontoon compartment schematic is depicted in Figure 6. A punctured compartment causes varying effects, but generally results in tilting the deck surface. This subjects the floating roof to an asymmetrical load, and the roof begins to lisp, as shown schematically in Figure 7.

In addition to lisping, or tilting, the pontoon weight that is unsupported by the buoyancy force will sag, creating uneven stress in the pontoon ring section, further weakening the roof structure. The combined lisp and tilt is often sufficient to cause the edge of the floating roof to dip into the liquid, often causing a hazardous and costly cascading failure of the roof, as in Figure 8. Should the floating roof not fail in this event, a common emergency response action is to foam the roof. The additional weight of the foam and water, on the already compromised deck, can trigger a buckling and subsequent sinking of the floating roof.

Loss of Buoyancy DDFR

DDFRs are the most expensive roof to construct. They are used either on very small diameter tanks about 30 feet and

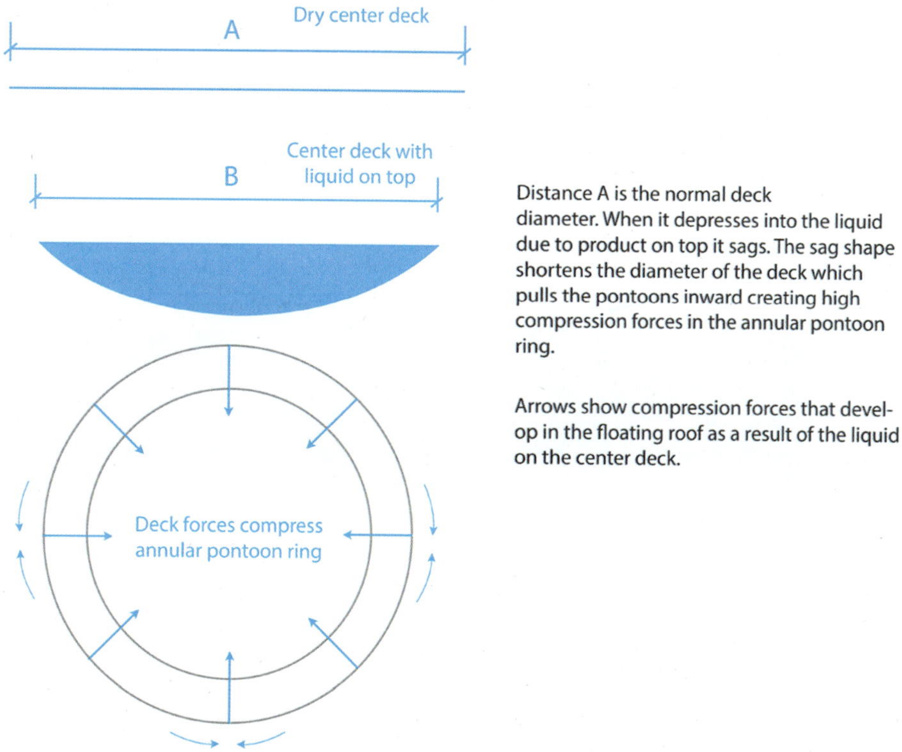

Figure 4. Forces experience by an APFR due to product or water on center deck.

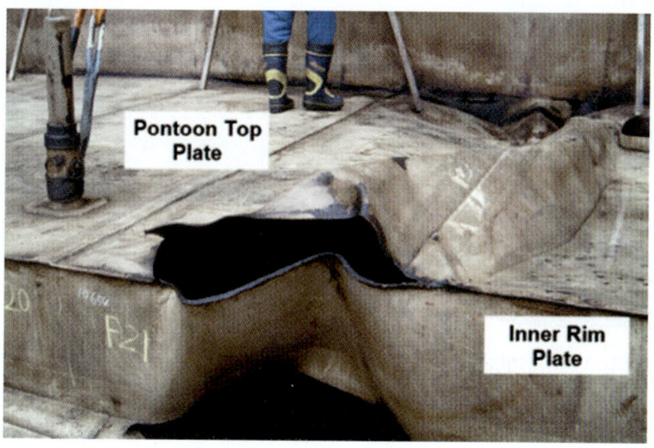

Figure 5. Buckled inner rim section of an APFR.

under or on tanks that are typically over 280 to 300 feet in diameter. For the small roof the cost of construction of a double deck is actually cheaper than that of an annular pontoon. However, this cost advantage quickly evaporates at diameters above 30 feet. When tanks are over about 280 to 300 feet in diameter APFRs are just too flexible and flimsy to remain robust against wind flutter, liquid turbulence and heavy precipitation. When the deck can vibrate from wind, fatigue becomes a common mode of failure. Repeated stress fatigue can cause deck cracks, which lead to flooding of the deck with product. Most companies with experience recognize the need for more robust DDFR designs on tanks over about 280 feet in diameter. Experience shows that DDFRs have survived earthquakes, hurricanes and other loads that would typically result in sunken floating roofs.

Because double deck floating roofs can have numerous compartments and the entire floating roof is partitioned into bulkheads, it is very stable and hard to sink. Moreover, large volumes of water cannot accumulate on the decks of floating roofs because there is not a ring acting as containment for rain water as there is with an APFR. As a result, it is very rare to have double deck floating roofs sink as a result of water or product overloading. However, poor maintenance has led to some serious incidents. There are cases where allowing the pontoons of double deck floating roofs to become corroded and filled with product and vapor have resulted in explosions caused by lightning. In several cases, as a result of poor maintenance and failure to repair leaking pontoons, lightning ignited the vents or pontoon covers and exploded the DDFR top skin. Portions of the pontoons were blown out of the top of the open top tank. The bottom portions of the roof were pushed down, settling to the bottom and resulting in a major incident and fire.

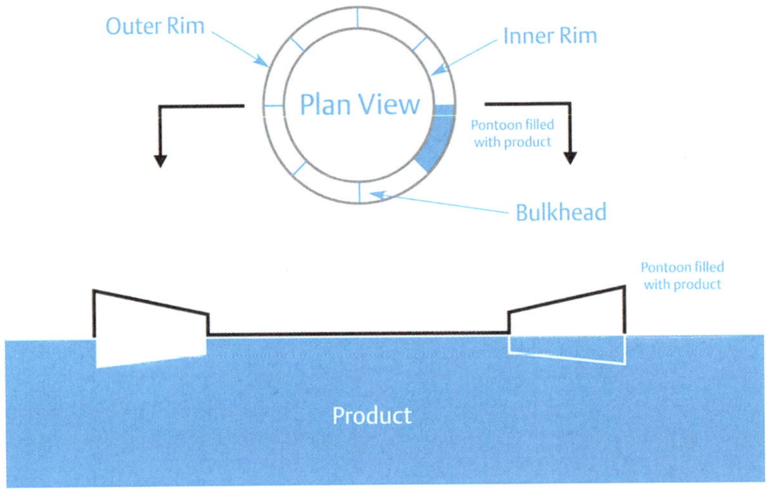

Figure 6. Schematic of a punctured pontoon compartment.

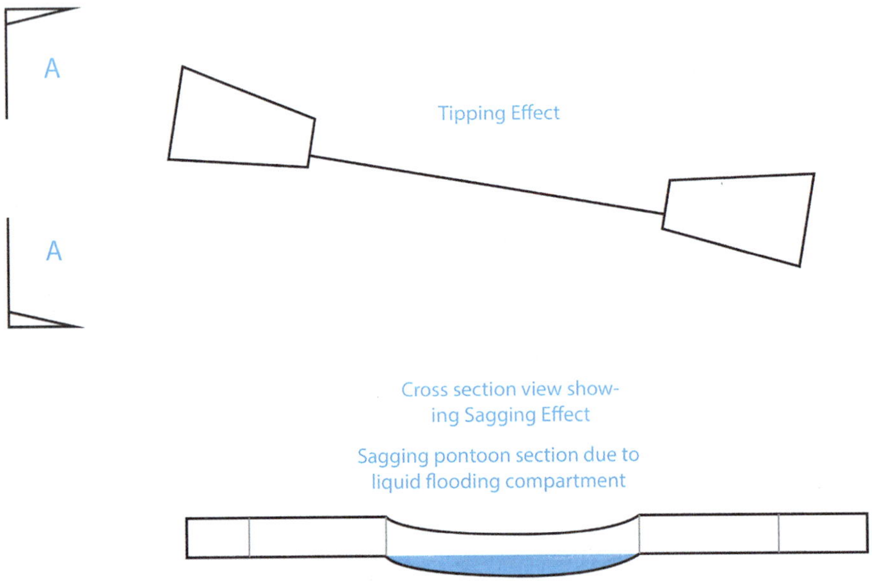

Figure 7. Lisping due to an asymmetrical load.

Figure 8. Effects of a loss of buoyancy event.

Floating roof failure modes

Excessive Flexure of Floating Roofs

Double deck floating roofs (DDFRs) are generally too stiff to have excessive flexural failure modes. In contrast, APFRs are highly flexible structures. Because of the APFRs flexibility, especially in the larger sizes, it is critical that the designer considers the freeboard height when the roof is distorted by liquid in pontoons or rainwater on the deck under various loading conditions. For example, a center deck with a hole in it deflects downward into the product. Some large APFRs can contain depths of rain water of up to 4 or 5 feet. If the floating roof support legs are not designed with sleeves that are above the liquid level, then product can well up through the leg

Figure 9. Bad - pontoon cover is loose.

Figure 10. Good - pontoon access hatch has locked down lid.

sleeves and continue the flooding process eventually sinking the floating roof. This freeboard concept also applies to APFR fittings such as vacuum breakers and vents.

Another common issue can occur at the pontoon compartments lid or cover. The covers go over the pontoon compartment access hatches. The purpose of access hatches is for internal inspection of the pontoon compartments. Many existing tanks have loose covers, as shown in Figure 9. While hatches may appear suitable from inspection, problems arise when an initiating event occurs such as excessive wind or precipitation. It has also been observed that local wind eddy currents inside the shell of an EFRT can be of sufficient velocity to dislodge these covers.

A floating roof that is partially submerged is better than a totally sunk floating roof, from a safety and incident cost perspective. There have been cases where the loose covers have blown off pontoons because of firefighting during a rim seal fire. While the rim fire could have been a relatively minor incident, it escalates to a major full surface fire because the lids are missing. The product then gushes into the pontoons, eliminating its buoyancy and causing it to sink.

The proper design for floating roof access hatch is a sealed and gasketed cover, which is bolted or affixed to the neck of the entry hatch. Proper use and maintenance of a pontoon access hatch is shown in Figure 10. A gooseneck type small diameter pipe is used to ventilate the pontoon since it must equalize any changes in pressure due to the atmosphere or leaks. The turned down vent maintains buoyancy, even if that part of the floating roof submerges below the product surface, maintaining as much buoyancy as possible.

Mechanical Impact

If a rolling roof ladder, the primary access to a floating roof, becomes jammed or stuck, it can force the floating roof to tip sufficiently to sink or collapse shown in Figure 11. This can

Figure 11. Failure by ladder impact or sticking.

result from a floating suction, or internal piping sticking in a position that when the roof descends, it cannot move and tips over sideways. If the tank becomes too out of round, or the design of the seal system is inadequate, the floating roof can bind on the shell, fixed roof columns, or floating roof guide poles. A binding floating roof can result in:

- Prematurely wearing seals
- Hanging floating roof and possible submergence or sinking
- Damage to tank internals

Seal Failure

Seals confine the annular area between the floating roof and the shell, by applying a fabric to cover the open space. The rim seal is the part of the roof responsible for preventing the evaporation of product, so it is an important consideration for floating roofs. A schematic for a rim seal is shown in Figure 12.

Rim seals are a moving part, and can wear quickly and they can bind, causing the floating roof to tip. There have been extreme cases where the seal has frozen in cold weather to the shell, and refused to move allowing the roof to stick and suddenly drop, causing failure.

If the floating roof is not level, then the rim seal will not be sufficient to trap the vapors below. The rim seal is designed to fit perfectly when the roof is operating on a level plane. If the plane of the roof begins to tilt, then vapors can rise through open spaces between the seal and shell. Poorly maintained rim seals lead to evaporation of product, a significant waste and hazard and possible violations from the local regulatory air quality management district.

Gas Blowing

Some terminal operations require that pipelines connected to the terminal be cleared for maintenance. One way is to use nitrogen and pigging operations (or gas blowing) to force liquid back into tanks (Figure 13). Pigs, which are foam or ball like plugs used to clear lines, are pushed by an inert gas (usually nitrogen). This allows the lines to be cleared of most liquid and allows maintenance to make repairs and modifications to a process.

For one reason or another gas can get past the pig or the pig can be stuck. If gas enters the tank it accumulates under a floating roof and has caused many floating roofs to sink, as shown schematically in Figure 14. The gas tends to accumulate under the floating roof and is trapped by the floating roof seals. The roof lifts and can tip enough to sink. Gas entry can also cause gurgling of liquid and vapors through the seal area. This tends to create an eruption of product onto the floating roof.

High Flow Operations

Because flow rates in, or out of tanks can be extremely high, vibration of the floating roof can cause fatigue failures. These

Figure 13. Pipeline and pipeline pigs.

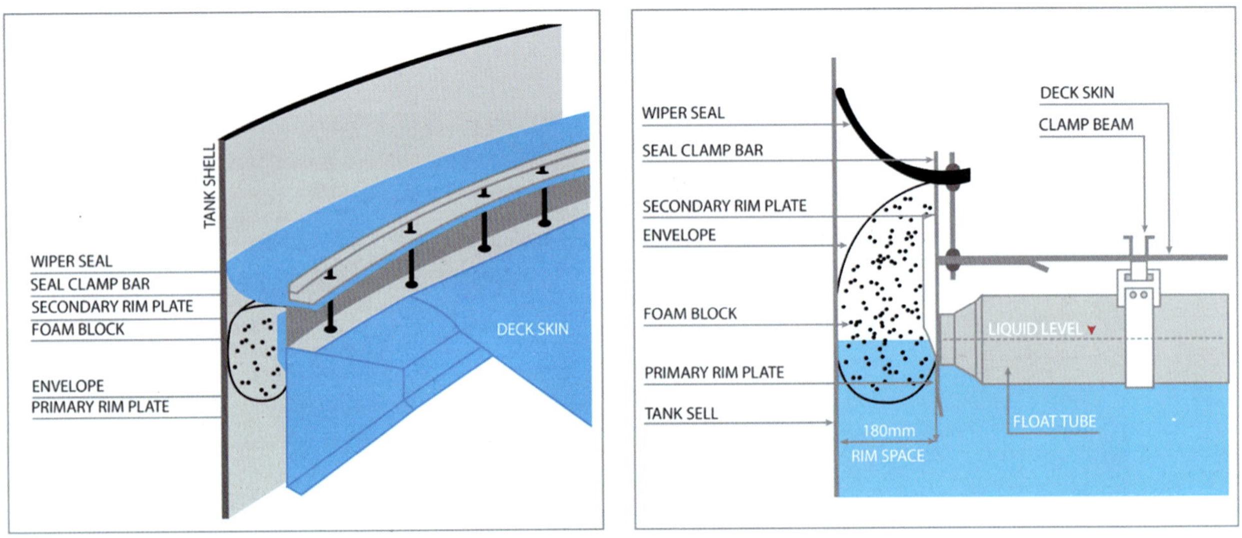

Figure 12. Rim seal diagram.

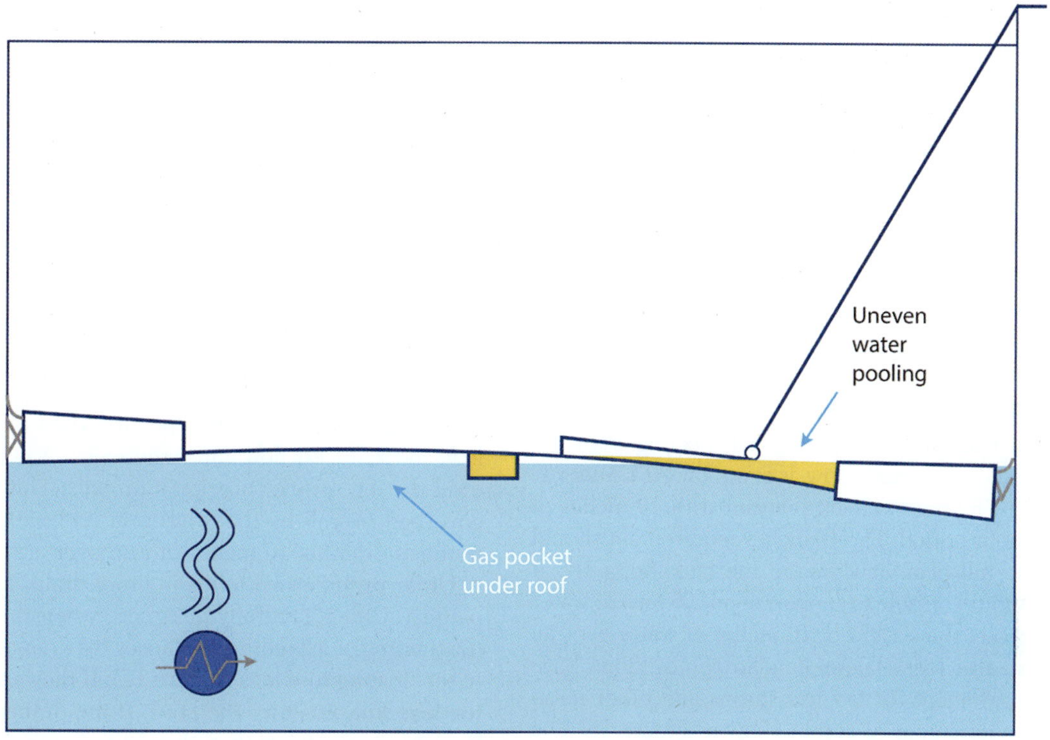

Figure 14. Failure by gas blowing.

are cracks that occur in deck plates welds and around fittings such as legs. The cracking leads to product entry into the compartments or the deck of floating roofs.

Exceeding Travel Limits

Floating roofs are designed to remain below the maximum safe filling height of a tank. If the fill limit is exceeded the seals can catch on the top edge of the tank, or on internal trappings, such as firefighting foam chambers. When the liquid goes down the floating roof will hang up and tip or buckle.

Stress Corrosion Cracking

Although uncommon, the introduction of ethanol for motor fuel oxygenates has caused some floating roofs to crack at welds due to stress corrosion. API 939 has details covering this failure mode.

Design Flaws

The most serious and overlooked failure mode for floating roofs is bad design. Because the stability of a floating roof is a complex structural engineering problem, many tank manufacturers, as well as owners, do not know how to evaluate them. All issues related to loss of buoyancy are directly dependent on the efficacy of the design.

A largely overlooked design problem is the sizing of the floating roof drainage system. Because a floating roof operates from levels as low as 5 to 6-ft up to more than 40-ft, the available head or driving pressure for drainage is limited to the gravitational head. A floating roof may expel water at the API intended flow rate when full, but it will have a significantly reduced flow capacity when the roof is low. Due to the dynamic nature of rain intensity, and the flow rate at the drain sumps, computations of required flow versus actual flow capacity are often either overlooked, or oversimplified utilizing rules of thumb. These methods may work due to chance, but they are not a systematic solution to reducing risk and protecting personnel.

Non-Operational State Failures

There are many maintenance failures that result from failure to conduct an adequate job hazard analysis, including fires and explosions from residual fuel in hidden locations, during hot work. Floating roof supports are another source of common failures. When landing floating roofs, the legs are pinned in a 'high' position so workers can enter below the floating roof. If operations fail to pin every leg at the same position, then some legs will be taking excessive loads and collapse is possible. There have been many other failures resulting in collapse of floating roofs during maintenance, due to instability. API 2016 provides guidance on how to stabilize landed floating roofs during maintenance.

Maintenance fires are very common because flammable liquids have many places to 'hide' in the floating roof components such as support legs, pontoons, the seal, and any space that is not visible within the floating roof construction.

Insufficient Draining of Precipitation

Although overloading the floating roof with product is uncommon, overloading of an external floating roof by precipitation, such as rain and snow, is common. API Annex C requires that APFRs must remain buoyant with 10 inches of rainfall in a 24-hour period. The 10 inches requirement should be based on the full plan view area of the tank. So, a 100 ft diameter tank should be able to support about 25 tons of water without sinking (i.e. the weight of 10 inches of water over an area of 7854 square feet. These standard precautions have often proven insufficient for tropical storms and other naturally occurring phenomena as results show in Figure 15. The effects of precipitation are like those of product entry on the floating roof- the additional weight can overload the available floatation buoyancy forces. When sufficient overloading occurs, a bucking failure or a sinking event may occur.

In Figure 16 an EFRT is shown which drains precipitation from the roof through an articulated piping system or a flexible hose which is typically 3 to 6-inches in diameter. There may be multiple hoses for multiple drains. Failure of the hose could lead to a major oil release. A manually operated valve is usually located on the outside of the tank so that it can be kept closed except when draining water off the floating roof.

Double deck floating roofs are not subject to the API rain requirement because DDFRs cannot accumulate as much volume of water. The double deck drains precipitation off by a system of roof drains, and they also have overflow emergency drains which dump rainwater directly into the product if the water level rises too high. The APFR, on the other hand, must carry the full amount of rain that falls on the footprint of the tank, specifically on the center deck. API 650 has the 24-hour time limit in the requirement because the floating roof drainage system has a limited capacity, and it may take a significant time to drain the water off the roof.

A key principle that is often unrecognized by the tank engineer is that the driving force for precipitation drainage from a floating roof is strictly limited to the height of the floating roof above the drainage nozzle. A typical tank would have less than 40 feet of head. But if the floating roof is not a maximum level, then the head for driving flow may be less than 10 feet. If a heavy rain occurs when the floating roof is low, the reduced flow rate may cause water to accumulate until the intensity of precipitation falls. This balancing act is governed overall by the criteria in API 650 Annex C which requires draining 10 inches of rain over a 24-hour period.

Other problems with floating roof drains are the 'laydown' pattern. Unless carefully designed, when the floating roof is descending to a 'landing' the hoses for drainage must be clear of the floating roof legs (Figure 17). If they are not clear, then the legs will puncture the hose. If the drain valve exterior to the tank is not closed then the entire contents of the tank can drain out.

Each type of floating roof drain hose has its own failure modes. For example, a rubber hose, if heated beyond its specified temperature range in a crude oil service can actually continue to vulcanize so that the rubber is 'cooked' which results in reduced life or even premature cracking. An articulated design can fail by excessive movement if one of the chains used to restrain flexibility breaks. Yet other hoses constructed with compression fittings or ferrules have in the past slipped off of the hose allowing a complete break of the hose system. It is up to the owners and users to review and study the failure modes and effects and make a decision as to what type of drain hose is optimal in the intended service for the intended operating interval.

Figure 15. Inadequate precipitation drainage.

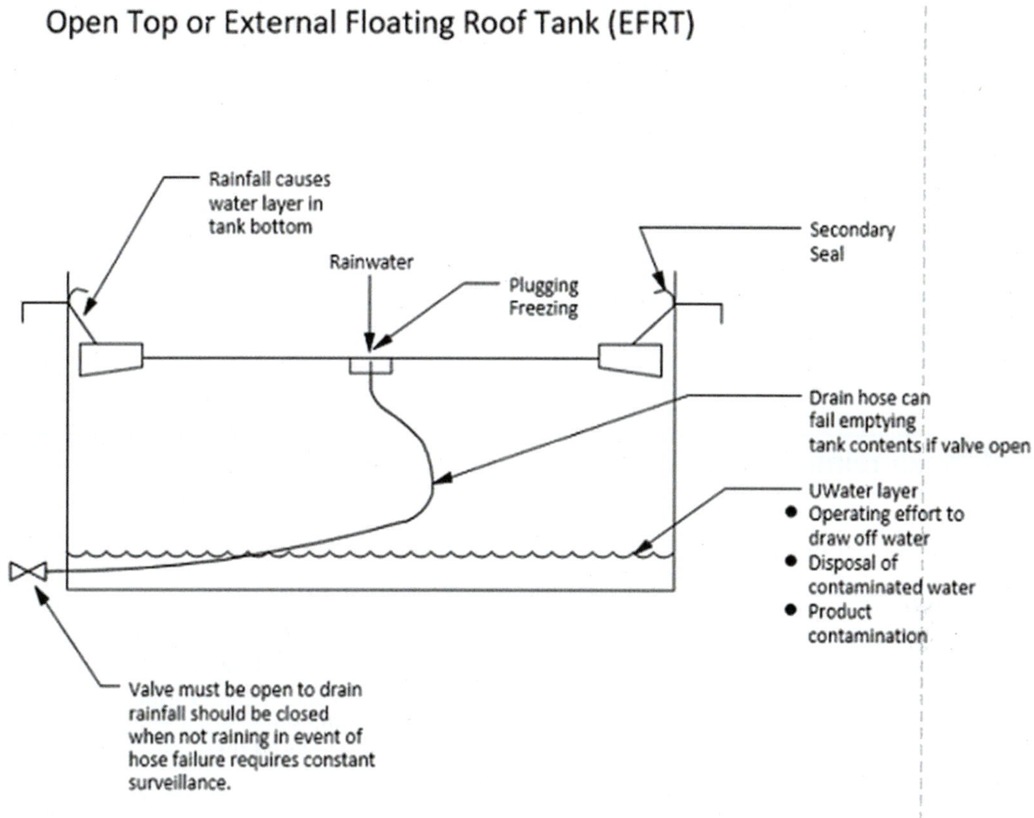

Figure 16. Schematic diagram of APFR drainage systems.

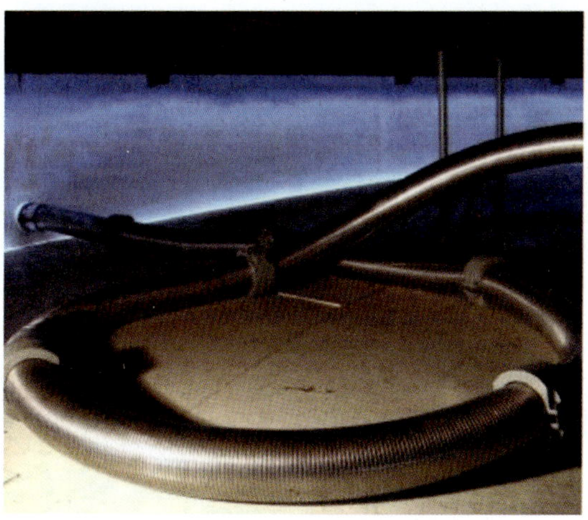

Figure 17. Floating roof's drain hose.

Loss of Buoyancy Due to Corrosion

APFRs and DDFRs are subject to various environments depending on location. For example, the top of the floating roof is exposed to the elements for an EFR. Because the deck will often be subjected to rain water for significant amounts of time, the coating system should generally be suitable for 'immersion service'. It is common to see corrosion on these exposed surfaces in EFR's.

For internal or external service, when sour products are involved, the corrosion may be accelerated. This can be on the outer rim of the floating roofs under the seals where the closed space can maintain high humidity for long periods of time. Corrosion of the outer rim of floating roofs is a common problem but aggravated by services which tend to be corrosive.

Typically, the underside of a floating roof is not welded as the seams are lap welded above. This means that crevice corrosion can occur at the crevices between the joints near the topside fillet welds. Some owner/operators have seal welded the lapped plates on the underside to reduce crevice corrosion.

Anywhere product comes into contact with metal the surfaces of the tank components, corrosion may be accelerated, especially when gas accumulates under the deck of the floating roof. Many failure modes are associated with penetration of the steel membrane in floating roofs by corrosion. Common corrosion failures are:

- Moisture and debris corrode the deck surfaces from the topside, and penetrates the deck, allowing oil to flood the deck. Corrosion from the underside is common where the product is corrosive.

- Corrosion in the pontoons leads to flooded pontoons. Because most companies are not rigorous about pontoon inspections, the pontoons can not only be flooded, but also fill with flammable vapors. Even though API allows for two pontoons to be flooded, it is common to have more than two flooded, and depending on the specific roof details, the roof can tilt, buckle, or collapse.

Source

Philip E. Myers, Floating Roof Tanks in Petroleum Storage: An Overview of Roof Types, Fault Modes, Failure Causes and Technology for Incident Prevention, White Paper. PEMY Consulting. Go online at: www.pemyconsulting.com.

Floating roof tanks rainfall risk

George G Woodworth, Data Analyst, and **Philip Myers**, P.E., Director, PEMY Consulting, Pleasant Hill, California

In this paper we quantify how a compromised roof drain raises the probability of high levels of standing water. Since standing water adds extra stresses to a roof, it will raise the probability of failure; however, engineering analysis is needed to quantify how standing water–induced stress influences the probability of roof failure by each of the possible failure modes. We present a way to approximate roof flooding risk from widely available hydrological intensity–duration–frequency tables and demonstrate the accuracy of our approximation using data from the Harris County, TX, Flood Warning System rain gauge network.

Standing Water Events

The space above a floating roof is a broad, shallow, open-topped cylinder with a drain at bottom-center. Standing water will build up if rainfall intensity (inches per hour, iph) exceeds roof-drain drawdown rate (ddr) expressed in inches per hour.

Roof-drain flow rate is generally expressed as volume per time unit; for example, gravity flow through 4″ pipe is about 15,000 gallons per hour. Our forecasting system requires reexpressing drain rate (volume per hour) as drawdown rate (DDR) expressed as inches of water depth per hour. Thus, for example, a 124-foot diameter tank contains 7528 gallons per vertical inch $\left(7.48 \cdot \pi \cdot (62)^2 / 12\right)$, so the drawdown rate would be about 2 inches per hour (iph) (15,000/7528).

Floating roofs are required by API Standard 650 to withstand 10 inches of rainfall over 24 hours *with primary drains inoperative*. The standard is hard to interpret; 10 inches of water accumulated due to *partly inoperative* drains exposes a roof to the same stress, as does 10 inches accumulated over a longer or shorter time interval. For that reason, we have developed statistical tables to predict the depth and frequency of standing water events for fully operational as well as impaired roof drains.

For our hypothetical 124-foot tank with 2 iph nominal drawdown, rain would have to fall at an average rate of 12 inches per hour (iph) for at least one hour to exceed the 10-inch limit. As of August 28, 2017, the US one-hour rainfall record was 13.5 inches per hour (Burnsville, WV, 1943-08-04), so 10 inches of standing water is possible but very unlikely for properly sized, fully operative roof drains. See also International Plumbing Code[1].

In fact, Hurricane Harvey's 60-minute maximum near the Exxon storage terminal at Baytown, TX, was about 4.12 inches per hour, Figure 1.

At that rate, our hypothetical tank would have accumulated a little over 4 inches of standing water, as shown in Figure 2.

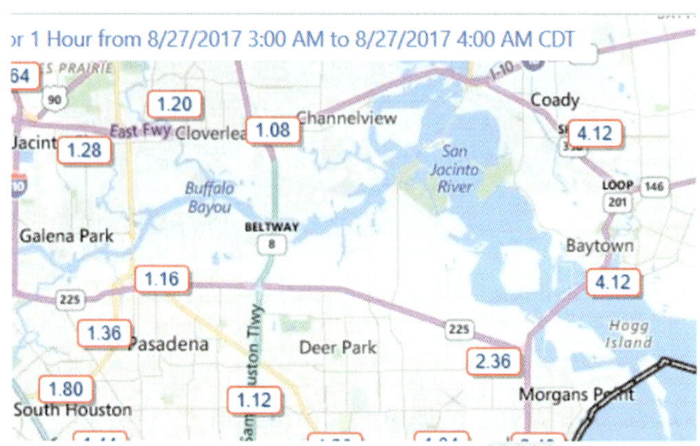

Figure 1. One hour of rainfall data during Hurricane Harvey, Harris County, TX.

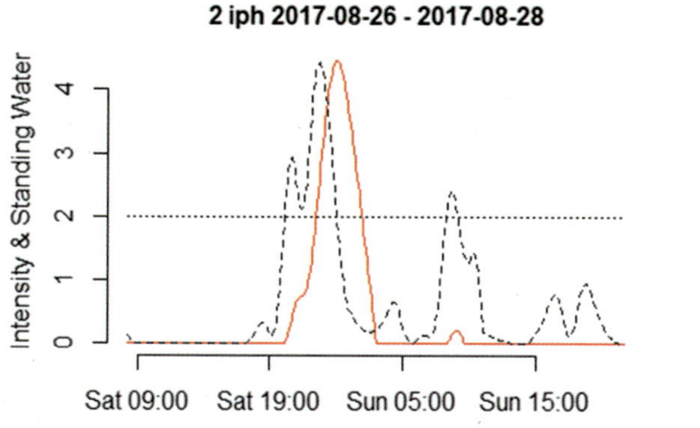

Figure 2. Gage 1540 Harvey Rain Event, Harris Co. TX, August 26–28, 2017. *Dotted*: drawdown rate (in/hr), *dashed*: rain intensity (in/hr), *solid*: standing water level (in).

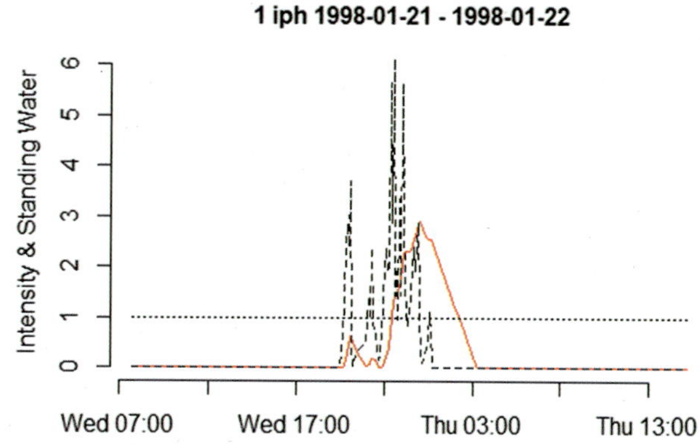

Figure 3. Rainfall hyetograph and resulting standing water assuming 1 iph drawdown. *Dotted*: drawdown rate (in/hr), *dashed*: rain intensity (in/hr), *solid*: standing water level (in).

Why Do Floating Roofs Sink?

It appears that a well-maintained floating roof with drain operating at full capacity is very unlikely to experience 10 inches of standing water. So why did 15 floating roofs fail in Harris County, TX, during Harvey? Based on the few published analyses of roof failures we have found, we think the answer is, either the drain was partially clogged, improperly sized, or one or more pontoons were sufficiently corroded to take on water. Other failure modes are referenced in an endnote[2].

Rain Gauge Data

The rain gauge in Figure 2 is part of a grid of 133 rain gauges maintained by the Harris County, TX, Flood Warning System (HCFWS). Continuous 5-minute data for years 1986 to the present are available for 61 of the gauges.

Identifying Standing Water Events

Standing water accumulates when rainfall intensity (in/hr) exceeds the (possibly compromised) drawdown rate (in/hr). We define a Standing Water Event as a continuous period of positive standing water beginning and ending with a dry roof (zero inches of standing water).

Figure 3 shows the history of a standing water event in a hypothetical tank located near HCFWS gauge 1540. The event began at 22:00 on Wednesday, Jan 21, 1998 and ended just after 03:00 on Jan 22. We assumed that the roof drain was compromised to the extent that the drawdown rate was only 1 in/hr. The *dashed line* is a hyetograph: a graph of rainfall intensity in inches per hour (iph). Note that standing water increases only when intensity exceeds draw down rate.

Depth-Drawdown-Frequency Table

We summarize the statistical risk of various depths of standing water in a Depth-Drawdown-Frequency (DDF) table, analogous to Intensity-Duration-Frequency (IDF) tables used in infrastructure planning[3].

Table 1 shows estimated maximum standing water at various drawdown rates and return periods based on pre-Harvey data. For example, a partly clogged drain with 2 in/hr drawdown rate (DDR) would experience 1.21 or more inches of standing water once in 10 years. Tabled values are percentiles of Generalized Extreme Value distributions fitted to annual standing water maxima at each draw down rate.

Figure 4 is a plot of the fitted standing water values in Table 1.

Table 1
Gauge 1540 Maximum Standing Water by Return Period and Drawdown Rate

DDR (in)	0.25	0.5	1	2	3
RP (yrs)	Maximum standing water (in)				
2	2.34	1.71	1.14	0.62	0.32
5	3.12	2.35	1.63	0.91	0.51
10	3.89	2.99	2.14	1.21	0.73
25	5.20	4.10	3.07	1.77	1.16
50	6.48	5.21	4.03	2.36	1.65
100	8.08	6.62	5.29	3.14	2.35
1000	16.78	14.69	13.05	8.11	7.56

Note: DDR = drawdown rate; RP = return period.

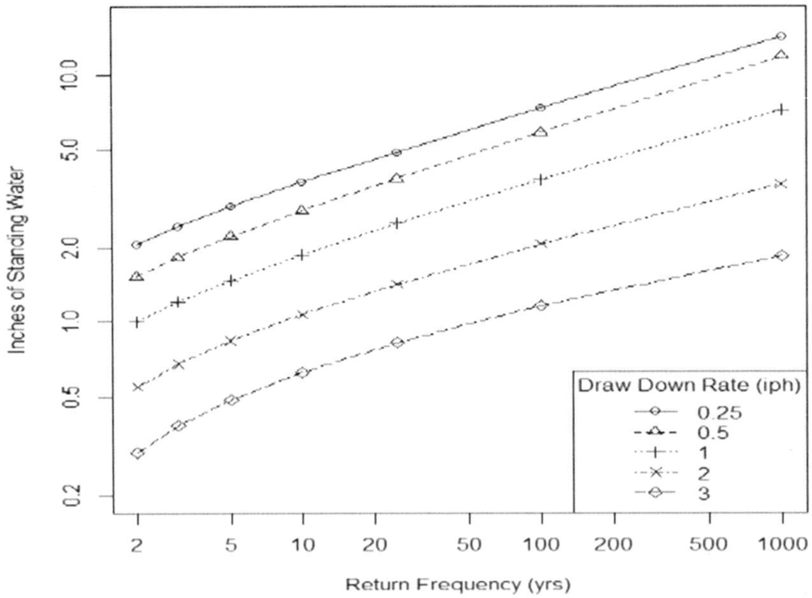

Figure 4. DDF: max standing water depth by draw down rate and return frequency.

IDF Approximation of DDF

It is difficult to find the three decades of high quality 15-minute rainfall data needed to compute a DDF table for a given location. For that reason, we investigated using rainfall Intensity-Duration-Frequency (IDF) tables from NOAA's Precipitation Frequency Data Server[4] to approximate standing water DDF tables.

NOAA Precipitation Frequency Server

NOAA has IDF tables for 43 of the lower 48 states[4] computed from rainfall data through April 2017 over a grid of rain gauges in each state. Table 2 is an excerpt of the IDF table for NOAA's Goose Creek gauging station near Baytown terminal in Houston, TX. Rain gauge data underlying this table included Hurricane Harvey.

Table 2
IDF Table for Goose Creek Station, Texas

Duration	Return period (years)						
	2	5	10	25	50	100	1000
	Inches of rain						
5-m	0.61	0.76	0.89	1.08	1.22	1.37	1.96
10-m	0.96	1.21	1.42	1.72	1.95	2.19	3.05
15-m	1.22	1.53	1.79	2.15	2.43	2.73	3.88
30-m	1.75	2.18	2.54	3.05	3.43	3.83	5.56
1-hr	2.33	2.93	3.44	4.14	4.68	5.27	7.90
2-hr	2.92	3.77	4.53	5.66	6.58	7.61	12.0
3-hr	3.27	4.30	5.26	6.71	7.95	9.35	15.2
6-hr	3.90	5.24	6.55	8.56	10.3	12.4	20.9
12-hr	4.59	6.23	7.85	10.3	12.6	15.1	26.7
24-hr	5.32	7.29	9.24	12.3	14.9	18.1	32.5

Duration	Hours	25 year Rainfall	Drawdown ddr x hours	Net standing water
5-m	0.08	1.08	-0.13	0.96
10-m	0.17	1.72	-0.25	1.47
15-m	0.25	2.15	-0.38	1.78
30-m	0.50	3.05	-0.75	2.30
1-h	1.00	4.14	-1.50	2.64
2-h	2.00	5.66	-3.00	**2.66**
3-h	3.00	6.71	-4.50	2.21
6-h	6.00	8.56	-9.00	0.00
12-h	12.00	10.3	-18.00	0.00
24-h	14.00	12.3	-21.00	0.00

Figure 5. Approximate 25 year standing water is 2.66 inches when ddr = 1.5 iph.

Given the wide availability of up-to-date IDF tables, it would be useful to be able to approximate tank roof flooding risk from IDF data.

How the Approximation Works

To illustrate how our approximation works, consider a tank that can drain 1.5 inches of water per hour (ddr = 1.5). We can approximate the 25-year standing water event for that tank from the 25-year rainfall maxima in Table 2. For example, the 3-hour 25-year maximum is 6.71 inches of rain. In 3 hours, 4.5 inches of that accumulation would drain off leaving 2.21 inches of standing water.

That is fine if the 25-year standing water event happened to last 3 hours from dry to maximum. In general, we need to do the same calculation for other durations and take the maximum of those. Figure 5 repeats this calculation at each duration. Maximum standing water is 2.66 inches at two hours. So, we know that about once in 25 years, there will be a two-hour interval in which 2.66 inches of standing water are added to any existing standing water; consequently the 25-year standing water maximum must be at least 2.66 inches.

Figure 6 is the approximate DDF table computed by this method; the nominal drawdown rate is 1.5 inches per hour. The table shows standing water risk at full capacity as well as 75, 50, 25, and 10% capacity. If the drain is operating at 0.15 inches per hour (10% capacity), 10 inches of standing water is a 25-to-50-year event.

Draw down rate	Return period (years)						
	2	5	10	25	50	100	1000
	Inches of Standing Water						
1.500	1.00	1.43	1.94	**2.66**	3.58	4.85	11.9
1.125	1.21	1.81	2.32	3.41	4.58	5.98	14.2
0.750	1.58	2.27	3.03	4.46	5.80	7.90	17.7
0.375	2.17	3.18	4.30	6.31	8.10	10.6	23.5
0.150	3.00	4.43	6.05	8.70	11.3	14.5	30.7

Figure 6. Approximate maximum standing water by drawdown rate and return period, Goose Creek.

Potential Accuracy of the Approximation

Figure 7 shows a standing water event at HCFWS gauge 1540 assuming drawdown rate (DDR) 1.0 inches per hour. Maximum standing water was 2.92 inches at 00:05 on 1/22/1998 starting from no standing water at 21:55 on the previous day, an interval of two hours.

The *dashed line* in Figure 8 is the three-hour interval with maximum average rainfall intensity (ARI), 1.9 inches per hour. The diagonal line shows theoretical standing water assuming constant intensity 1.9 iph. Total accumulation is:

$$(ARI - DDR) \times duration = \left(1.9\frac{in}{hr} - 1\frac{in}{hr}\right) \times 3\,hr = 2.7\,in$$

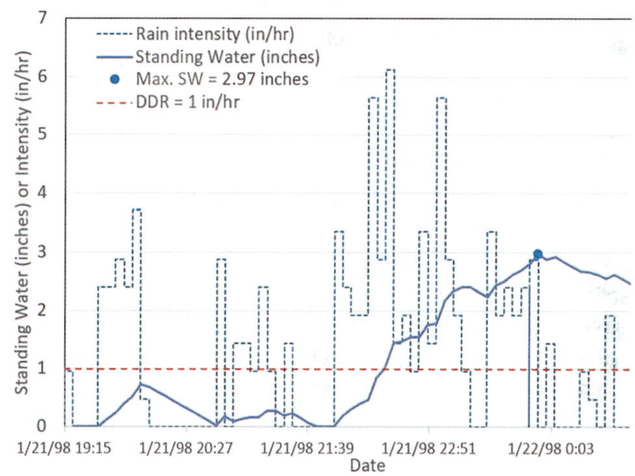

Figure 7. Rain event of 21-Jan-1998.

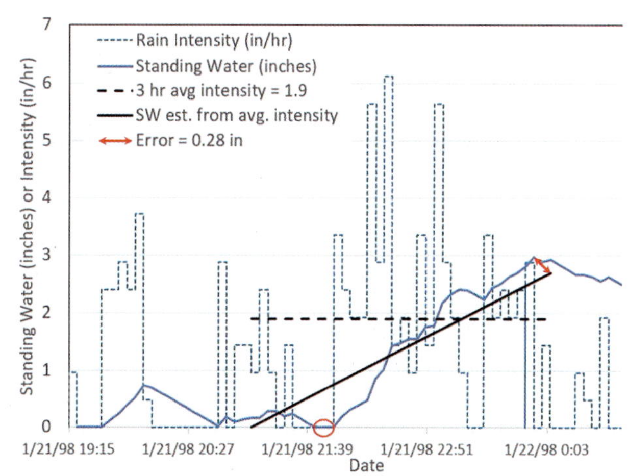

Figure 8. Maximum standing water approximated by three-hour average intensity.

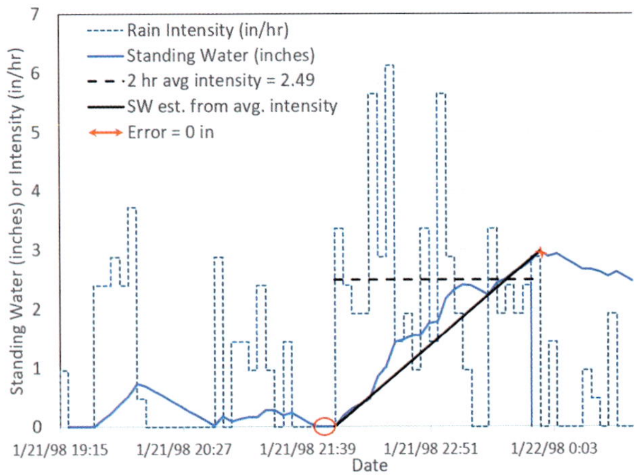

Figure 9. Maximum standing water approximated by two-hour average intensity.

Actual maximum standing water is 2.97 inches, so the approximation is about 9% low.

There are two reasons for the error. First, during the circled interval on the time axis, there was no standing water and therefore no drawdown; however, the approximation assumes constant rainfall intensity and therefore continuous drawdown. Second, there happened to be positive standing water at the beginning of the three-hour window which is not included in the estimate.

Figure 9 shows that it is possible for the estimate to be perfect when the interval of maximum constant intensity starts with zero standing water and includes no intermediate stretches of zero standing water. Unfortunately, without access to raw data, there is no way to know if or when this is true.

Error Analysis of the IDF Approximation

While it is true that for any standing water maximum there is a corresponding maximum average rain intensity interval that produces a perfect approximation, another source of error emerges when estimating standing water percentiles from rain intensity percentiles.

This component of error can be computed at a given site from raw five-minute data or it could be estimated using simulated five-minute data, generated with a Poisson cluster stochastic rainfall generator[5]. We illustrate the raw data method with five-minute data from the average of Harris County, TX.

Our error analysis used the average five-minute rain accumulation of HCFWS gauges 1540 and 1520, north and south of Baytown in Figure 1.

We computed annual maximum rain accumulation for years 1986–2013 at durations five minutes through four days and fit Gumbel distributions to each duration to produce the IDF table graphed in Figure 10.

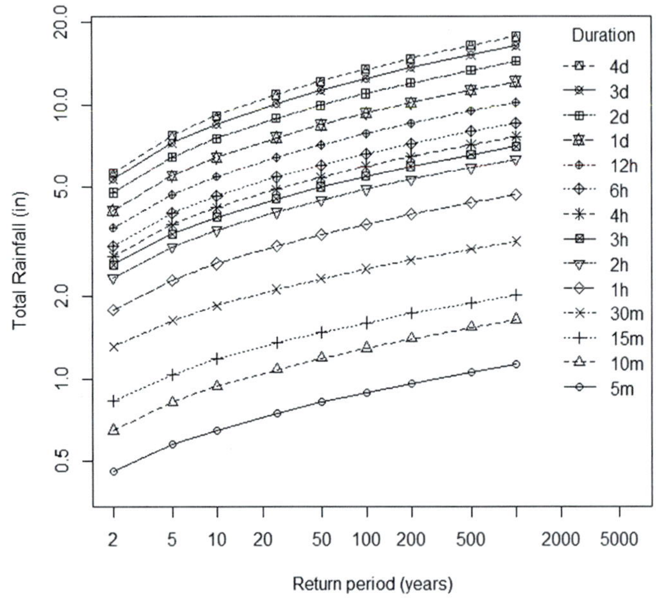

Figure 10. Baytown IDF, Gumbel-smoothed.

Next, we computed, directly from raw rainfall data, annual maximum standing water at drawdown rates from 1 to 4 inches per hour and fit Gumbel distributions to each drawdown to produce the true DDF table. Approximate (symbol) and exact (curve) DDF values are in Figure 11

Approximation errors for Figure 11 are listed in Table 3. Errors for extreme standing water events (5 or more inches of standing water) are highlighted; for extreme events, absolute errors are less than ± 0.55 inches and percent errors are less than ± 7%.

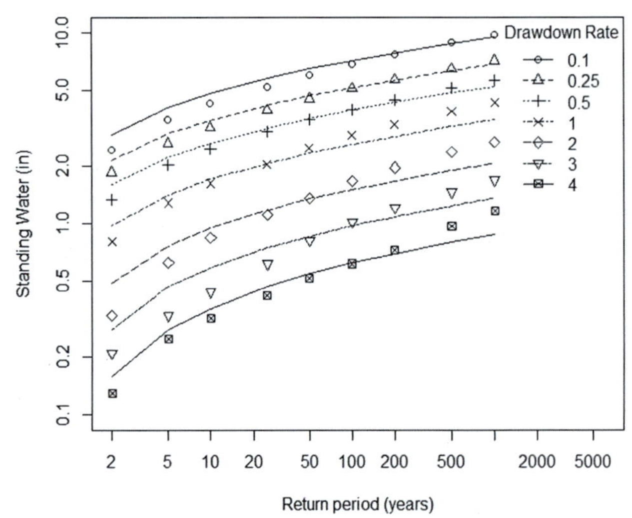

Figure 11. Approximate and Gumbel-smoothed exact DDF at Baytown.

Table 3
Drawdown Rate Approximation Errors

Drawdown (in/hr)		Return period (years)								
		2	5	10	25	50	100	200	500	1000
0.1	%	.19	.16	.13	.11	.7	.4	.2	1	2
	in	.47	.56	.55	**.55**	.44	.28	.13	.07	.22
0.25	%	.16	.12	.9	.6	.3	.1	1	2	5
	in	.29	.31	.28	.23	.15	.06	.04	.15	.33
0.5	%	.19	.10	.7	.4	.2	1	3	6	**7**
	in	.26	.20	.17	.12	.06	.02	.12	.29	.42
1	%	.21	.10	.5	0	6	11	14	17	19
	in	.17	.13	.08	.01	.14	.31	.46	.66	.82
2	%	.48	.21	.12	.4	1	10	15	20	23
	in	.16	.13	.10	.05	.01	.16	.29	.47	.62
3	in	.33	.42	.34	.21	.5	4	10	16	20
	in	.07	.14	.15	.13	.04	.04	.12	.23	.33
4	%	.23	.12	.13	.12	.6	.2	4	18	25
	in	.03	.03	.04	.05	.03	.01	.03	.17	.29

Note: Highlighted entries are standing water events of 5 or more inches.

Conclusion

Approximate standing water depths can be computed from NOAA IDF tables. These estimates, accurate to about a half inch, are used in sizing floating roof tank drains, but drainage calculations are based on ideal conditions that are not representative of major weather events.

The failure of floating roofs during major weather events is a fairly common failure mechanism for floating roof tanks. The ability to drain water from the roof is dependent on the hydraulic potential of the water on the roof.

The hydraulic potential is dependent upon the drain hose size, floating roof height, height of water in the berm area, valve position, and obstructions present in the floating roof drain hose.

These factor in the ability to flow water off the roof point to the need to clear drain intakes regularly to maintain adequate drawdown rates. Dust and debris, birds' nests, and more can get inside of floating roof hoses and prevent or reduce the flow of water from floating roofs. It is a good practice to oversize floating roof drain hoses during the design stage. Floating roof drain hoses that are already in use can be tested to determine the current flow rate, and cleaning can improve those drain hoses that are underperforming.

References

1. International Plumbing Code, Roof Drains.
2. Maintenance & Reliability of Floating Roofs. https://onestopndt.com/blog/storage-tanks-maintenance-reliability-of-floating-roofs/.
3. API 650 C.3.4.1. Floating roofs shall have sufficient buoyancy to remain afloat on liquid with a specific gravity of the lower of the product specific gravity or 0.7 and with primary drains inoperative for the following conditions: 250 mm (10 in.) of rainfall in a 24-hour period over the full horizontal tank area with the roofs intact.
4. NOAA Atlas 14, point precipitation server. https://hdsc.nws.noaa.gov/hdsc/pfds/pfds_map_cont.html.
5. A Poisson Cluster Stochastic Rainfall Generator That Accounts for the Interannual Variability of Rainfall Statistics: Validation at Various Geographic Locations across the United States. https://www.hindawi.com/journals/jam/2014/560390/.

18: Tanks

Classification of storage tanks 820
What is the difference between fixed and floating tank types? 826
Tank styles and appurtenances 830
Floating roof applications and components 834
Summary of federal requirements for aboveground storage tanks 844
Design and containment for atmospheric storage tanks 845
Overview of petroleum storage tank standards 851
Tank weight estimation tool using diameter as input variable 870
How to estimate roof and framing weight for fixed cone tank 871
Budgetary pricing cost curve of cone roof tanks 872
How to estimate flow rate from a leaking tank hole 873
Charts give vapor loss from internal floating-roof tanks 875
Estimating the contents of horizontal cylindrical tanks 878
How to gauge a horizontal cylindrical tank 879
Correct the volume of light fuels from actual temperature to a base of 60°F 880
Volume of liquid in vertical cylindrical tanks 880
Chart gives tank's vapor formation rate 881
Tank farm line sizing 882
Tank farm management 884
Managing oil movement and storage in tank farms 892
Standard construction practices for LNG storage tanks 902
Choosing the best containment for LNG storage 908
A review of API 2350 overfill protection, 4th and 5th edition 911
Manual tank gauging procedures 916
How to apply vapor cloud analysis methodology to tank overfills 923
Simplified procedure to calculate emissions for liquid organic storage tanks 932
Calculating tank emissions with BREEZE TankESP 936

Classification of storage tanks

Sunil Pullarcot, Independent Consultant, India

A storage tank is a container, usually for holding liquids, sometimes for compressed gases. The liquids may be flammable and corrosive, and the tank may be constructed of ferrous or non-ferrous metals, alloys, reinforced concrete, etc., depending on the use for which it is to be built.

API specification for storage tanks

Storage tanks are used extensively in production, refining, transportation, and marketing of petroleum products. Storage tanks operate under no (or very little) internal pressure, distinguishing them from pressure vessels. Storage tanks are often cylindrical in shape, perpendicular to the ground with flat bottoms, and a fixed or floating roof. There are several environmental regulations applied to the design and operation of storage tanks, often depending on nature of fluid contained within. Though there are many codes prevailing across the world for design, manufacture, inspection and testing of storage tanks, the most widely used is that by American Petroleum Institute (API).

Storage tanks can be divided into two basic types: above ground (AST) and underground (UST). While AST's are used in production, refining, marketing, and pipeline operations, UST's are typically used in marketing of gasoline at retail service stations. AST differ from UST with regard to design considerations and applicable regulations.

Classification of storage tanks

One of the fundamental classifications employed in tankages is based on location of the tank, whether it is above ground or below, whose design and manufacturing features vary drastically (Figure 1). Above ground tanks have their bottom directly resting on earthen or concrete foundation, and in

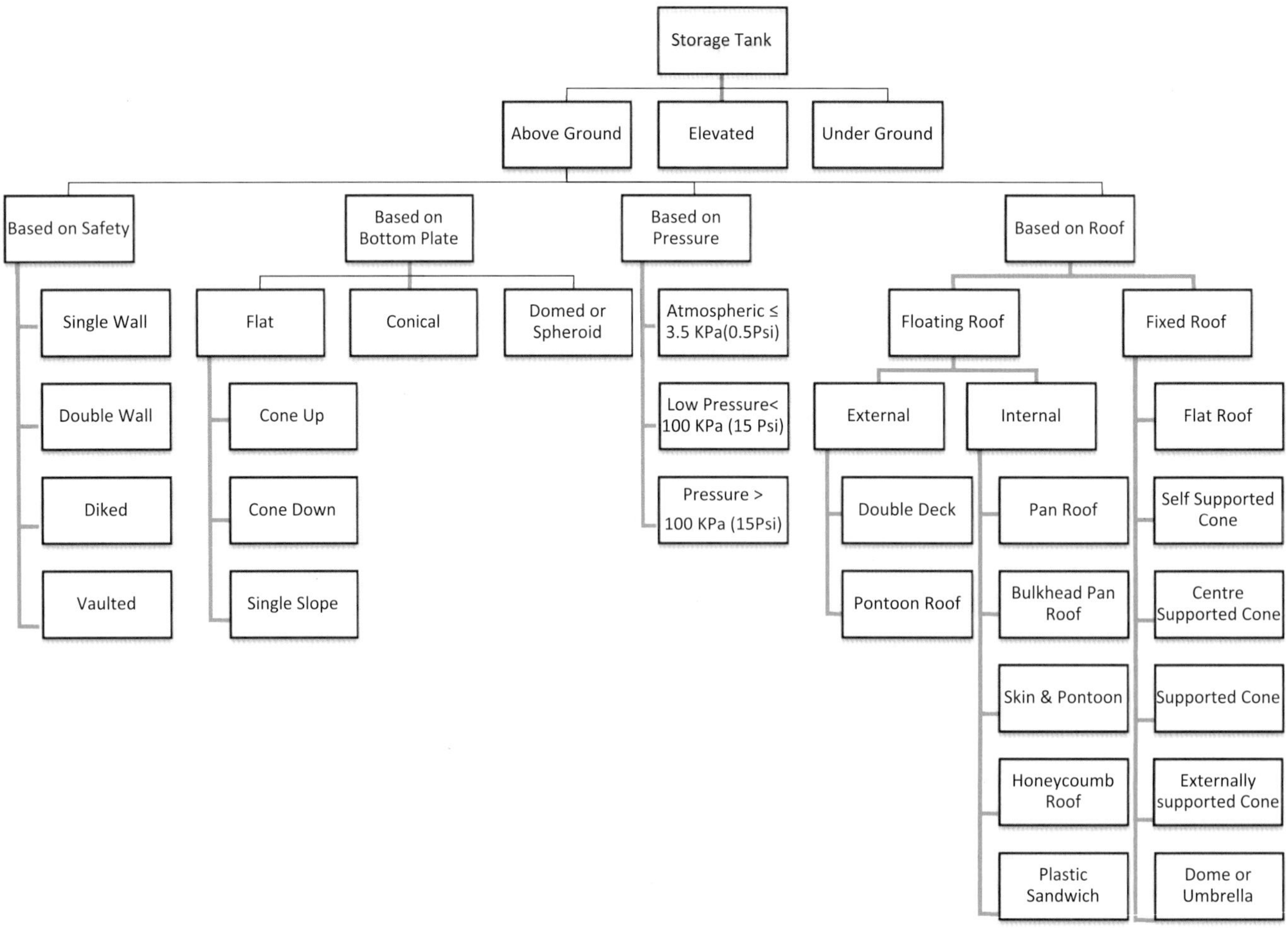

Figure 1. Storage tank classification.

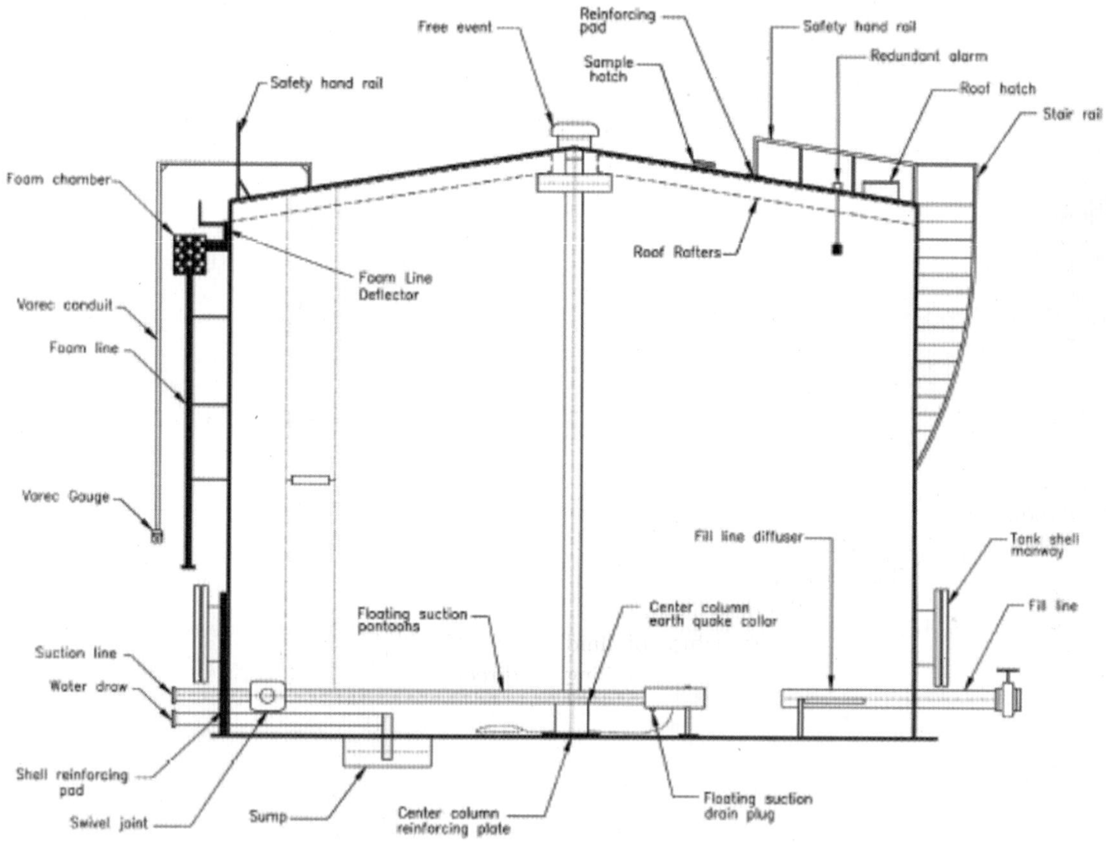

Figure 2. Cone roof tank.

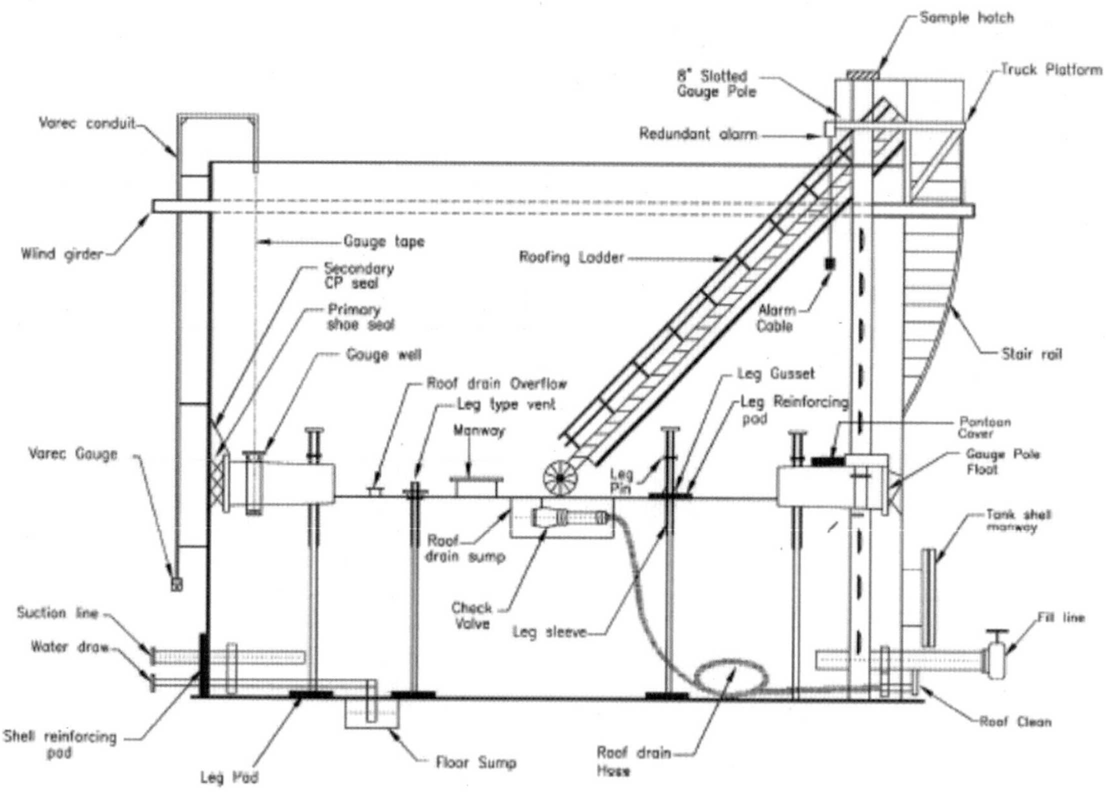

Figure 3. External floating roof tank.

some specific cases, tanks may be placed on grillages which facilitates inspection of bottom plates. Another classification is elevated tanks placed on structural supports. Water supply tanks placed on steel structures belong to this category.

Underground storage tanks are usually designed for capacities in the range of 23 to 91 m^3 (5000-20,000 gallons), with majority falling under 55 m^3 (12,000 gallons). Underground tanks are mainly used to store fuels and a variety of chemicals.

Classification of above ground storage tanks

Prevailing codes, standards and regulations classify storage tanks according to their internal pressure. Tanks operating at internal pressure slightly above atmospheric, say up to 3.5 kPa (0.5 psig), are known as atmospheric pressure tanks, and tanks operating at a higher pressure than atmospheric storage tanks, up to a pressure of 100 kPa (15 psig), are known as low pressure tanks. Tanks operating at pressure above 100 kPa (15psig) are normally called pressure vessels. The term high pressure tank is not used by those working with tanks as its falls under the specialized category of pressure vessels.

Another way of classification is based on type of construction of the storage tank (Table 1).

Two other classifications are based on the bottom plate provided for the tank (Table 2) and the safety feature of the tank (Table 3).

Selection of type

The decision to select a particular type of tank is principally based on process requirements, but cost also plays a role. For example, for storage of volatile fluids floating roof tanks are preferred.

In the design and manufacture of storage tanks, the following considerations are needed:

1. Earth load is to be considered while designing the tank and its foundation.
2. Buoyancy forces need to be considered, especially in the case of buried tanks, as well as anchoring.
3. As the tanks are prone to external corrosion,
   a. Back fill material and its properties
   b. Necessity of cathodic protection system to be employed
   c. Necessity for coating / lining of exterior surface
4. Necessity of leakage monitors as per regulatory requirements

API maintains several documents that address the design, fabrication, operation, inspection, and maintenance of aboveground (AST) and underground (UST) storage tanks. API storage tank standards are developed by committees of experienced tank designers, fabricators, owners, and operators that bring a wealth of accumulated knowledge from their respective companies. Although API standards cover many

Table 1
Storage Tank Classification Based on Type of Construction

Classification	Sub-Classification	Further Classification	Description
Fixed Roof Tanks	Flat Roof		For small diameter storage tanks without any supports
	Self-supported cone roof		Roof is conical in shape but is self-supporting due to rigidity offered by shape of the roof
	Centre supported cone roof		Cone roof supported at center usually used for medium sized tanks
	Supported cone roof		Supported at many points to take care of the load of roof as well as other mobile and static loads expected
	Externally supported cone roof		The roof of this type is externally supported even by means of external columns outside tank shell.
	Dome or umbrella roof		The peculiar shape makes the roof a bit more rigid but more expensive compared to other types of roofs
Floating Roof Tanks	External FR	Pontoon Roof	Common for floating roof for sizes in the range of 10 to 30m (30-100ft) in diameter. The roof is simply a steel deck with an annular compartment that provides buoyancy
		Double Deck Roof	Double deck roof are built for wide range of diameters including those above 30m (100ft) in diameter. They are very strong and durable because of the double deck, and hence suitable for large diameter tanks.

Classification	Sub-Classification	Further Classification	Description
	Internal FR	Pan Roof	Pan roof are simple sheet of steel disks with edge turned up for buoyancy. These roofs are prone to capsizing and sinking because small leak can cause them to sink.
		Bulk Head Pan Roof	Bulkhead pan roof has open annular compartment at the periphery to prevent the roof from sinking, in the event of a leak.
		Skin & Pontoon Roof	Skin & pontoon roof usually constructed of an aluminium skin supported on a series of tubular aluminium pontoons. These tanks have a vapour space between the deck and the liquid surface.
		Honey Comb Roof	Honeycomb roof is made from hexagonal cell pattern similar to a beehive in appearance. The honeycomb is glued to a top and bottom aluminium skin that seals it. This roof directly rests on the liquid.
		Plastic Sandwich Roof	Plastic sandwich roof is made from rigid polyurethane foam panels sandwiched inside a plastic coating.

Table 2
Storage Tank Classification Based on Type of Bottom Plate

Classification	Sub-Classification	Description
Flat Bottom	Flat	For tanks up to 6 to 10m (20 to 30ft) diameter, inclusion of small slope does not provide benefit and hence flat bottom is permitted.
	Cone Up	Cone up with a high point at the center and with a slope of about 25 to 50mm per 3m (1 to 2” per 10ft) diameter.
	Cone Down	Cone down slopes towards center. A collection sump is usually provided at the center with piping under the tank to a well for draining purpose.
	Single Slope	Single slope bottom is tilted to only one side. Drainage is taken from the low point. As the diameter increases, the difference in elevation of the bottom plate increases and hence the use of the same is limited to diameter up to 30m (100 ft)
Conical Bottom		Conical bottom tanks provide complete drainage of even residues. Since these types are costly, they are limited to small sizes and are often found in chemical industry or in processing plants.
Domed or Spheroid		Domed or spheroid bottom also provides complete & effective drainage. Being costly and difficult to construct, use is restricted to smaller sizes.

Table 3
Storage Tank Classification Based on Safety Features

Classification	Description
Single Wall Tank	Usually cylindrical with either vertical or horizontal.
Double Wall Tank	Common for both above ground and underground applications. As outer tank can contain leaks from inner tank, it serves as leak detection as well.
Diked or Unitized Secondary Containment Tank	The primary containment aboveground storage tank is housed within a rectangular steel dike which contains the product spill in the event of a leak or rupture of primary tank. Dike can be open or closed type
Vaulted Tank	Refers to tanks installed inside concrete vaults

aspects of AST and UST design and operation, they are not all-inclusive. Apart from API, several other organizations also publish standards on tank design, fabrication, installation, inspection, and repair. Wherever API specifications are silent, related specifications are often referred for more clarity on requirements.

- American Society of Mechanical Engineers (ASME) – www.asme.org
- American Society for Testing and Materials (ASTM) – www.astm.org
- American Water Works Association (AWWA) – www.awwa.org
- Building Officials and Code Administrators International (BOCA) – www.bocai.org
- NACE International (Corrosion Engineers) – www.nace.org
- National Fire Protection Association (NFPA) – www.nfpa.org
- Petroleum Equipment Institute (PEI) – www.pei.org
- Steel Tank Institute (STI) – www.steeltank.com
- Underwriters Laboratories (UL) – www.ulonet.ul.com
- International Fire Code Institute (Uniform Fire Code) – www.ifci.com

For AST's, the following API standards have been developed.

1. API Spec. 12B, Bolted Tanks for Storage of Production Liquids;
2. API Spec. 12D, Field Welded Tanks for Storage of Production Liquids;
3. API Spec. 12F, Shop Welded Tanks for Storage of Production Liquids;
4. API Spec. 12P, Fiberglass Reinforced Plastic Tanks;
5. API 620, Design and Construction of Large, Welded, Low-Pressure Storage Tanks;
6. API 650, Welded Steel Tanks for Oil Storage.

Tanks manufactured according to API 12 are much smaller than API 650 tanks and are often subject to different operating conditions. Among the above mentioned API standards, one of the predominantly used standards among them is the API 650, the standard for Welded Steel Tanks for Oil Storage.

Standards for storages

API 650 - Welded Steel Tanks for Oil Storage

This standard provides minimum requirements for material, design, fabrication, erection, and testing for vertical, cylindrical, aboveground, closed and open-top, welded storage tanks in various sizes and capacities for internal pressures approximating atmospheric pressure (internal pressures not exceeding the weight of the roof plates). Its Annex F specifies requirements for tanks with mild positive internal pressures.

Since the purpose of this standard is to construct storage tanks of adequate safety and reasonable economy, for petroleum product storage, this standard does not provide any specific sizes for the tanks, which falls in the domain of end users, depending on their specific use.

API 620 - Design and Construction of Large, Welded, Low-Pressure Storage Tanks

This standard covers large, field-assembled storage tanks intended to contain petroleum /petrochemical intermediates or products, such as LNG, LPG, Ammonia etc, in liquefied stage. Since the internal pressure has limitations on account of large diameters, these gases (at normal atmospheric temperatures) need to be refrigerated to sub-zero temperatures to store as liquid under low positive pressures. This poses challenges with regard to suitable materials for such low temperatures and suitable welding and construction methodologies. Though such low temperature storage tanks usually provided with a refrigeration system, (in addition to insulation), the scope of API 620 covers only tank proper. Here also, the purpose of the standard is to provide a safe and cost effective storage and hence the standard does not specify the sizes. The mandatory sections of the standard specifies requirements for storage at temperatures as low as −46°C (−50°F). Whereas, its Annex S covers stainless steel low-pressure storage tanks in ambient temperature service in all areas, without limit on low temperatures. Similarly, Annex R covers low-pressure storage tanks for refrigerated products at temperatures in the range of 4 to −51°C (40 to −60°F) and Annex Q covers low-pressure storage tanks for liquefied hydrocarbon gases at temperatures not lower than −168°C (−270°F).

ASME Section VIII Div(1) Rules for Construction of Pressure Vessels

ASME Section VII Div(1) covers all types of pressure vessels, (containers for containment of pressure), either internal or external, applied by an external source, or by application of heat from a direct or indirect source, or by any combination thereof. The code lists mandatory requirements, specific prohibitions, and non-mandatory guidance for pressure vessel materials, design, fabrication, examination, inspection, testing, certification, and pressure relief devices. In this case as well, the code does not address all aspects of above activities, and hence it shall not be construed that those aspects which are not specifically addressed are prohibited. In such instances, engineering judgment applied shall be consistent with the under laying philosophy of the code and such judgments shall never be used to overrule mandatory requirements or specific prohibitions.

The scope of ASME Section Viii Div(1) defines the coverage in detail, by excluding vessels of certain types and pressures. However, on a broader perspective, vessels for containing internal or external pressures above 15 psi (100 kPa) and with dimension (inside diameter, width, height, or cross section diagonal) exceeding 6 in (152 mm) are generally covered by this code.

Classification of storage liquids

Flammable and Combustible Liquids

Flammability of a liquid is the fundamental hazardous property of flammable and combustible liquid, which is its ability to produce ignitable vapors. Properties that have an influence on flammability are:

Vapor pressure & boiling point
Flash Point
Fire Point
Auto-ignition temperature
Vapor-air density
Liquid density
Water miscibility

NFPA 30 Classification

National Fire Protection Association (NFPA) in their code for Flammable & Combustible Liquids, (NFPA 30), classifies liquids within its ambit generally as either a flammable liquid or a combustible liquid and are defined and classified as follows.

Broadly, liquids that have a closed-cup flash point at or above 37.8°C (100°F) are classified as combustible liquids, whereas, liquids that have a closed-cup flash point below 37.8°C (100°F) are classified as flammable liquids. These broad classifications are categorized in Table 4 with a few commercially known liquids.

Institute of Petroleum Classification

Institute of Petroleum uses a slightly different methodology in classifying crude oil and its derivatives, which are potentially hazardous materials. The degree of the hazard is determined essentially by volatility and flash point.

The purpose of above classifications is to arrive at specific requirements for storage facilities with regard to health and safety of operating personnel as well as inhabitants within the vicinity, based on associated hazards of each of the products stored. Hazard characteristics vary with physical properties of stored fluid. NFPA and Institute for Petroleum have developed criteria for storage tanks based on nature of fluid stored and classifications. Usually, a group of fluids are categorized under each class considering their physical and chemical characteristics.

Source

Pullarcot, S. Aboveground Storage Tanks - Practical Guide to Construction Inspection & Testing, CRC Press, 2015.

Myers, P.E. Aboveground Storage Tanks, McGraw Hill Companies, 1997.

Table 4
NFPA Classification of Storage Liquids

Classification		Flash point	Boiling Point	Examples of Fluids
Flammable Liquids	IA	<22.8°C (<73°F)	<37.8°C (<100° F)	Diethyl ether, ethylene oxide, some light crude oils
	IB	<22.8°C (<73°F)	≥37.8°C (≥100°F)	Motor and aviation gasolines, toluene, lacquers, lacquer thinner
	IC	≥22.8°C, ≤37.8°C (≥73°F, ≤ 100°F)		Xylene, some paints, some solvent-based cements
Combustible Liquids	II	≥37.8°C, <60°C (≥100°F, <140°F)		Diesel fuel, paint thinner
	III	≥60°C (≥140°F)		
	IIIA	≥60°C, <93.0°C (≥140°F, <200°F)		Home heating oils
	IIIB	≥93°C (≥200°F)		Cooking oils, lubricating oils, motor oil

Source: NFPA

Table 5
Institute of Petroleum Classification Storage Liquids

Class		Fluid
Class 0		Liquefied Petroleum Gases (LPG)
Class I		Liquids which have flash points below 21°C (69.8°F)
Class II	(1)	Liquids which have flash points from 21°C (69.8°F) up to and including 55°C (131°F) handled, below flash point
Class II	(2)	Liquids which have flash points from 21°C (69.8°F) up to and including 55°C (131°F) handled, at or above flash point
Class III	(1)	Liquids which have flash points above 55°C (131°F) up to and including 100°C (212°F) handled, below flash point
Class III	(2)	Liquids which have flash points above 55°C (131°F) up to and including 100°C (212°F) handled, above flash point
Unclassified		Liquids with flash points above 100°C (212°F)

Source: IP

What is the difference between fixed and floating tank types?

Flat bottom, vertical, cylindrical tanks used to store oil can be classified into three basic types:

1. Fixed roof tank (FRT)
2. External floating roof tank (EFRT)
3. Internal floating roof tank (IFRT)

Fixed roof tank

A fixed roof tank schematic is shown in Figure 1. Note that they have a lid or a cover which is a shallow cone. Thus, these tanks are often called a 'cone roof tank'. Another type of fixed roof is a geodesic dome (Figure 2).

PV valves (short for pressure-vacuum) are used on fixed roof tanks. They are simple devices attached to fixed roof tanks to minimize air flow into and out of the tank (Figure 3). In this way evaporation is minimized. While not as efficient, in general, as a floating roof to reduce emissions, they are far more efficient than simply leaving an opening in a fixed roof tank.

Figure 4 shows the simple mechanism which relies on gravity to hold the stoppers or plates called 'pallets' atop the

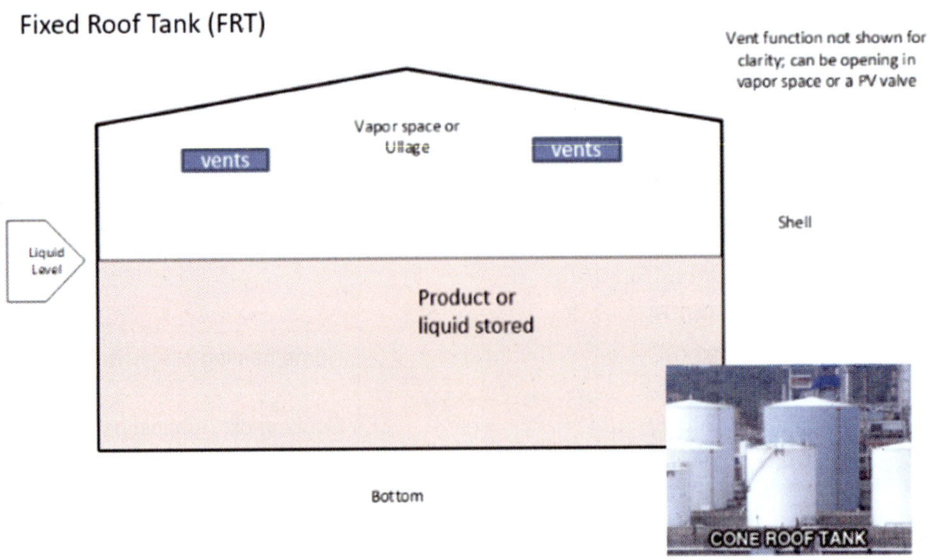

Figure 1. Fixed roof tank.

Figure 2. Tanks with aluminum geodesic domes.

Figure 3. A pressure vacuum vent valve.

open hole. When the pressure is sufficient to lift the weight of the pallet, the air or gas can pass through the opening. Notice the clever design where two such pallets can be arranged so that when one lifts up it reduces pressure inside the tank and the other allows air to flow into the tank when a vacuum is developing. API 2000 covers the detailed requirements for venting of fixed roof tanks.

Floating roof tanks

There are two types of floating roof tanks: external and internal (EFRT and IFRT, respectively). Figure 5 shows a schematic of EFRT.

A floating roof tank (either EFRT or IFRT) uses a floating disk that rests on the liquid surface, which rises and lowers with the liquid level (see Figure 5 and Figure 6). The main function of the floating roof in most cases is to prevent vaporization of the stored liquid. In fact, the US EPA requires the use of a floating roof for controlling evaporative emissions when the vapor pressure of volatile organic liquids exceeds 11 psia. Table 1 shows some examples of common liquids that are stored with and without a floating roof. Note that any of the floating roof tank types (EFRT or IFRT) may store any of the fuels, but not all fuels or organic liquids should be stored in FRTs, noting the exceptions in the table notes.

External Floating Roof Tank

Tanks with open tops are not used without a floating roof (Figure 6). That is, when product is stored in an open top tank, then there is always a floating roof, and therefore it is an EFRT. The 'E' means external and that the floating roof is exposed to the atmosphere. Even though the tank itself is an open top tank, the term is not used because an open top tank always has a floating roof in it. When a tank has a fixed roof constructed on top of the cylindrical shell, then it is an IFRT.

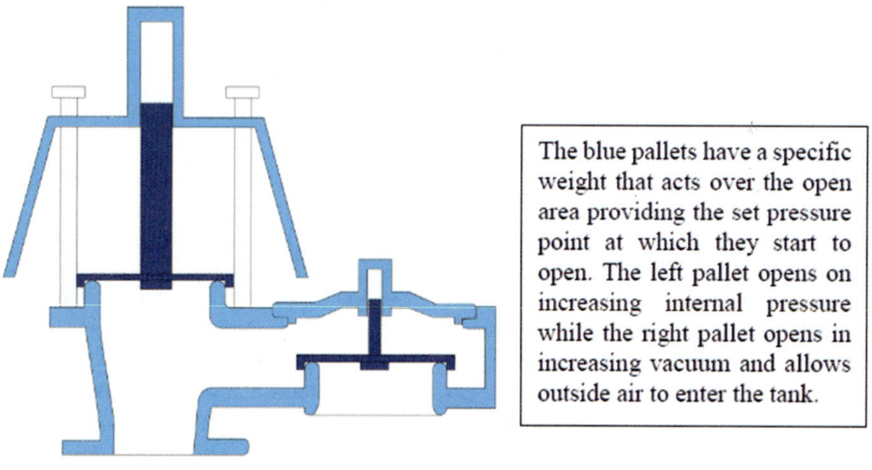

Figure 4. Cut away diagram of pressure vacuum vent valve.

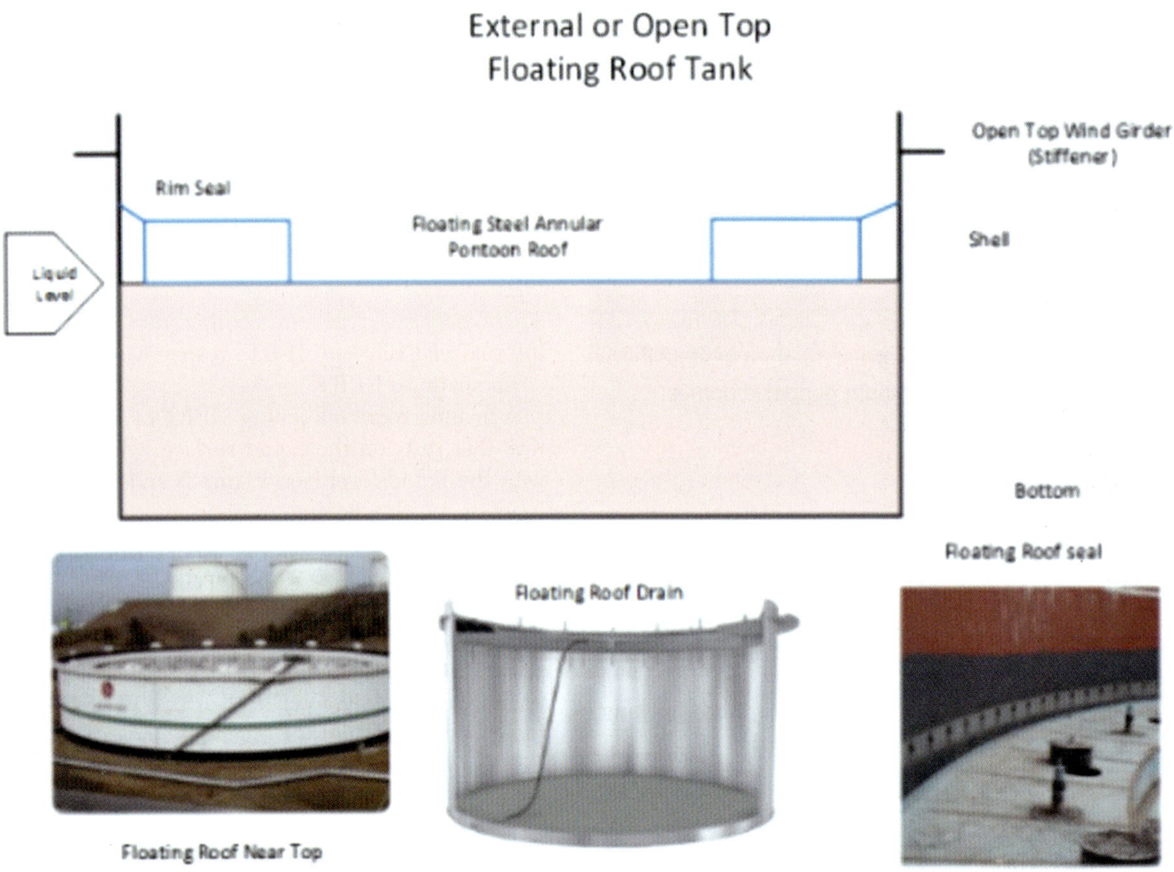

Figure 5. Open top or external floating roof tank.

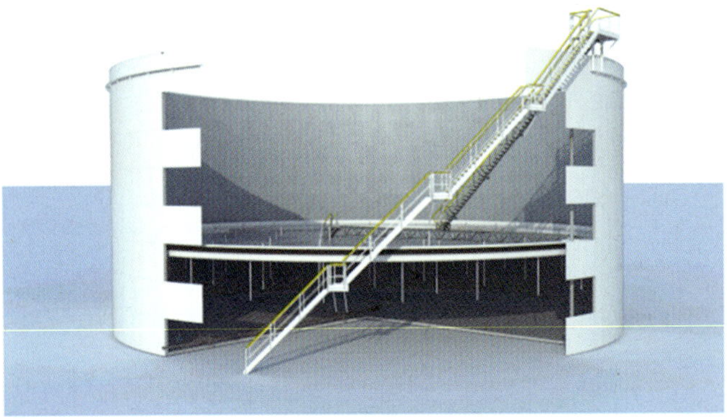

Figure 6. External floating roof tank cut away.

Table 1
Storage Options for Bulk Liquids

Product	FRT	EFRT	IFRT
Crude Oil (low vapor pressure)	Yes	Yes	Yes
Crude Oil (high vapor pressure)	No*	Yes	Yes
Gasoline	No*	Yes	Yes
Diesel Fuel	Yes	Yes	Yes
Ethanol	No	Yes	Yes

*In upstream tanks, volatile crude oil is often stored in small tanks (< 30 ft diameter). When regulations require or when the owner must control vapors, then vapor recovery systems can be applied to these fixed roof tanks. If storing volatile organic liquids in a fixed roof tank, the pressure vacuum or PV vent valve should be the vapor recovery system used.

A tank which has no floating roof is an FRT. An FRT always has a fixed cover to protect the liquid from the weather.

Perhaps the most common tank type is the EFRT. Additional information on the rules for the design of these floating roofs is provided in API 650 Annex C. Appendix C governs the EFR design requirements. For example, EFRs must be constructed of carbon steel. The relatively high-level design requirements appear in this Annex because the tank can be constructed without a fixed roof (reducing capital costs for the tank itself), and then the floating roof constructed to float on the liquid surface of the open top tank.

The dominant types of floating roofs are the annular pontoon floating roof (APFR) and double deck floating roof (DDFR). These must be able to withstand the weather conditions from its exposure through the open top. These floating roofs must also have inherent buoyancy, in the form of closed compartments referred to as pontoons. Compartments are separated from each other by bulk heads. Annex C requires that the floating roof be able to operate with no more than two adjacent pontoons compartments combined with a punctured center deck. The most common designs of external floating roof tanks are the APFR and the DDFR. These roofs are fitted with drainage hoses that pass through the rain water to the base of the tank where it can exit through one or more of the tank shell nozzles.

Internal Floating Roof Tank

The internal floating roof tank (Figure 7) provides the best protection against tank floating roof incidents. This is because the IFRTs prevents wind and weather from contacting the deck of a floating roof. Preventing weather reduces the risk of failure from overload precipitation, corrosion, and blocking wind reduces the evaporation of liquid. As shown in Figure 7, the internal floating roof is protected from the weather by the fixed roof above it. Fires in IFR tanks are rare compared to EFRTs, because the cover prevents lighting and static charges from being an ignition source.

A tank with a low vapor pressure liquid, such as diesel fuel, does not need a floating roof since there is little evaporation. If the tank service is changed to something like gasoline, then a floating roof may be required by regulations, or the tank owner may wish to control evaporation with a floating roof. In the case of a fixed roof tank, an internal floating roof can be installed, and the tank is then classified as an internal floating roof tank. Many times, an EFRT is converted into an IFRT by retrofitting an aluminum geodesic dome.

Source

Philip E. Myers, Floating Roof Tanks in Petroleum Storage: An Overview of Roof Types, Fault Modes, Failure Causes and Technology for Incident Prevention, White Paper. PEMY Consulting. Go online at: www.pemyconsulting.com.

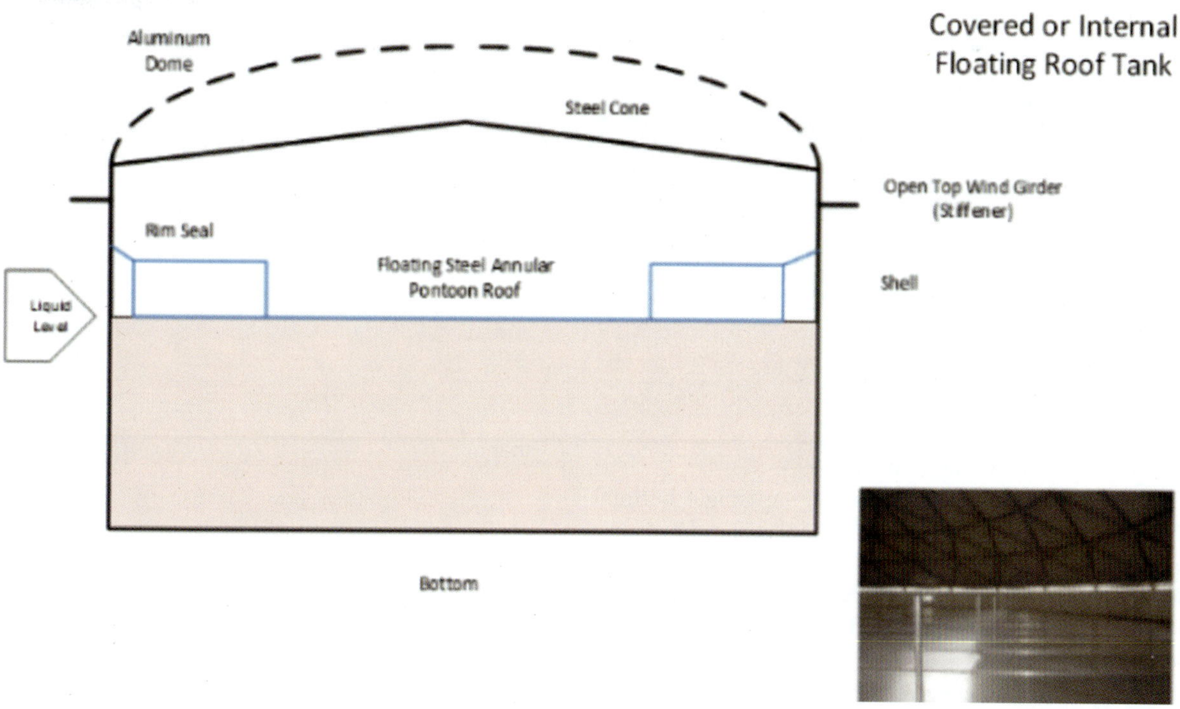

Figure 7. Internal floating roof tank.

Tank styles and appurtenances

Open Top Tank with Back Slope Floating Roof

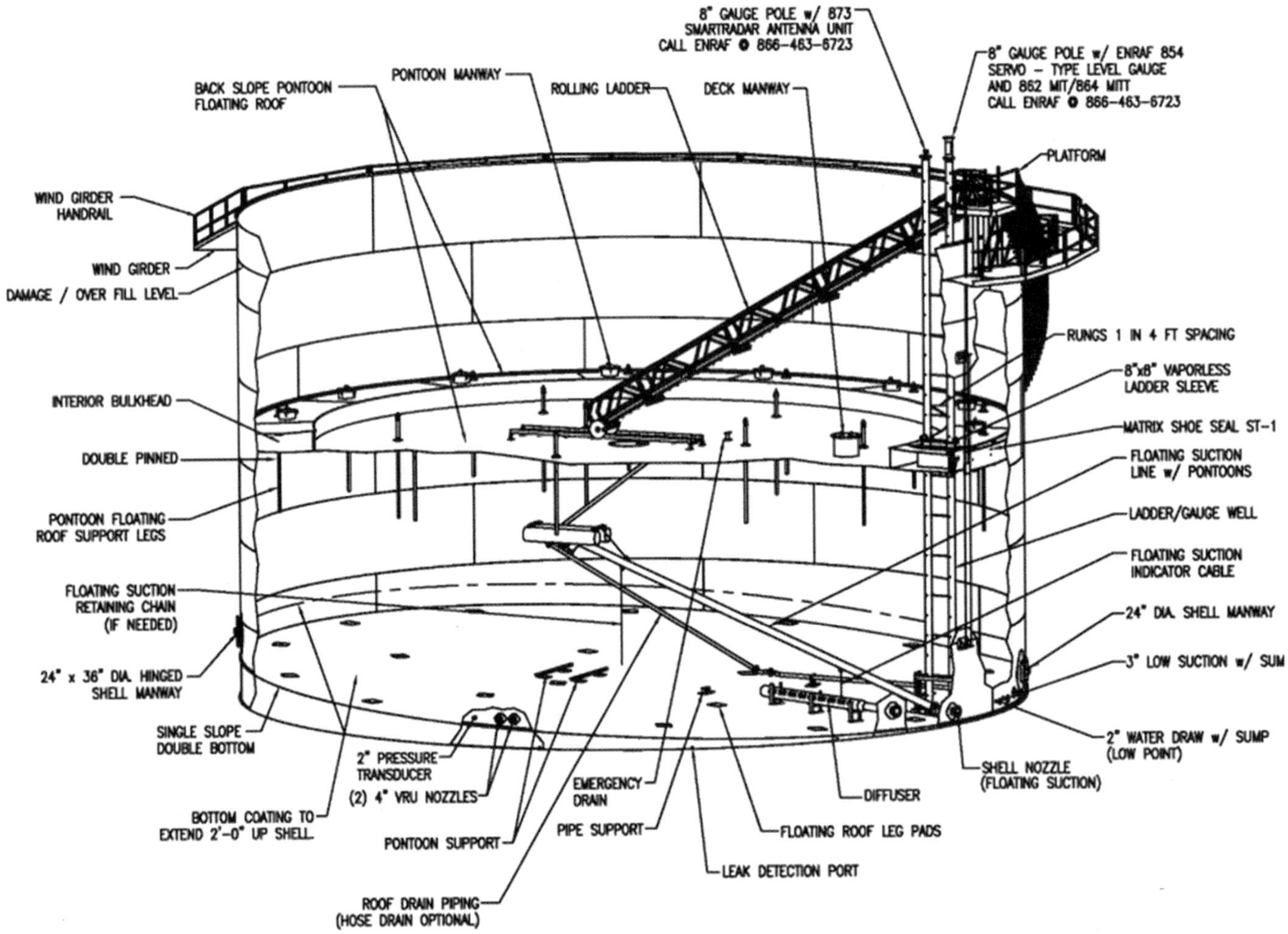

- Generally most economical choice.
- Shovel bottom shown. Typical is cone-up single bottom (allows for thinnest sand pad). Flat or cone-down are also possible.
- Cone down usually only seen with center sump for drain-dry operation. Not common with OTT / EFR because of roof landings.
- Diffusers limit incoming velocity to reduce static discharge. Same nozzle can still be used as outlet depending on drain velocity

Reprinted with permission - Matrix PDM Engineering

Open Top Tank with Double Deck Floating Roof

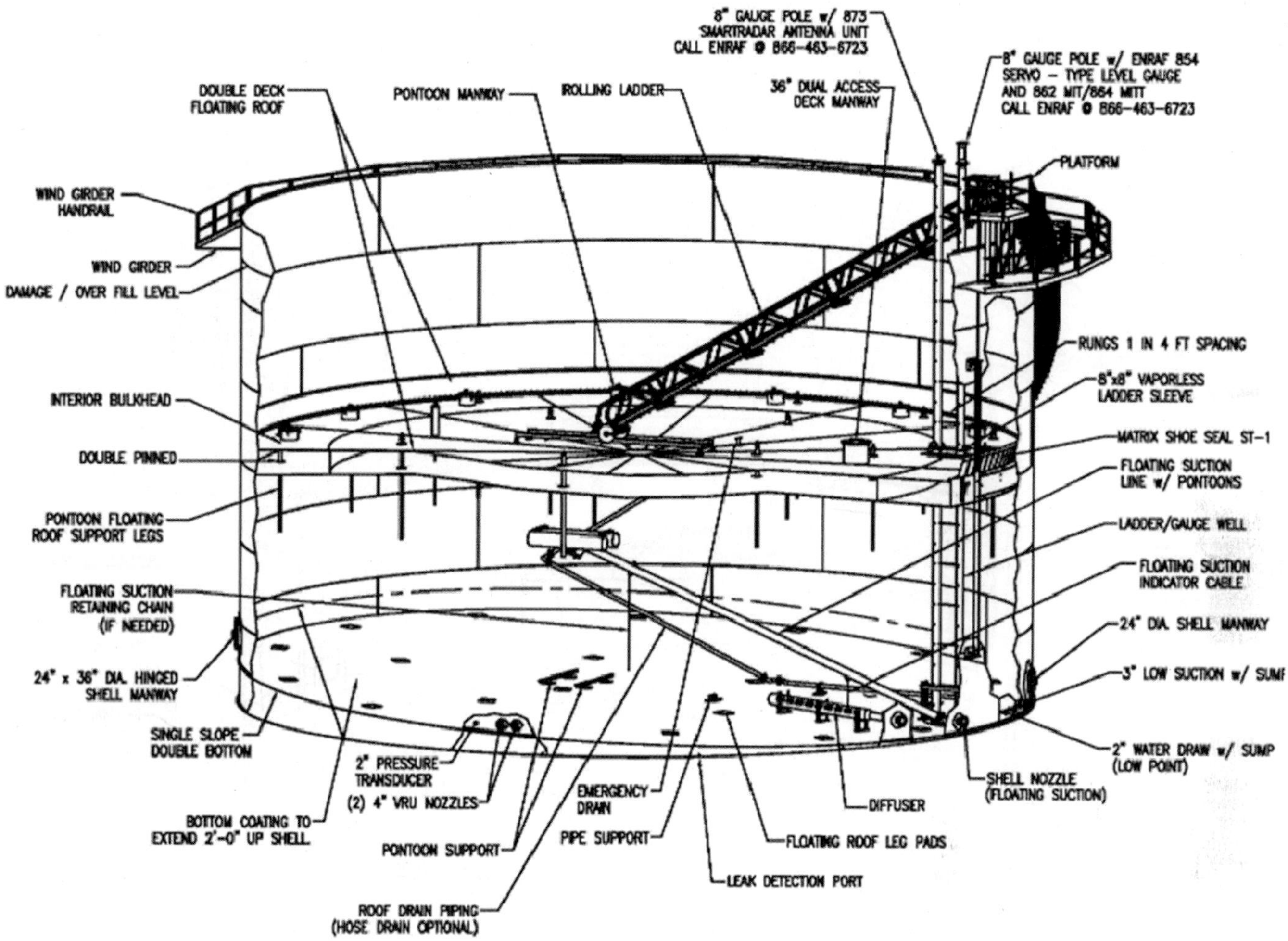

- Generally most economical choice.
- Shovel bottom shown. Typical is cone-up single bottom (allows for thinnest sand pad. Flat or cone-down are also possible.
- Cone down usually only seen with center sump for drain-dry operation. Not common with OTT / EFR because of roof landings.
- Diffusers limit incoming velocity to reduce static discharge. Same nozzle can still be used as outlet depending on drain velocity.

Reprinted with permission - Matrix PDM Engineering

Open Top Tank with Pontoon Floating Roof, Geodesic Dome

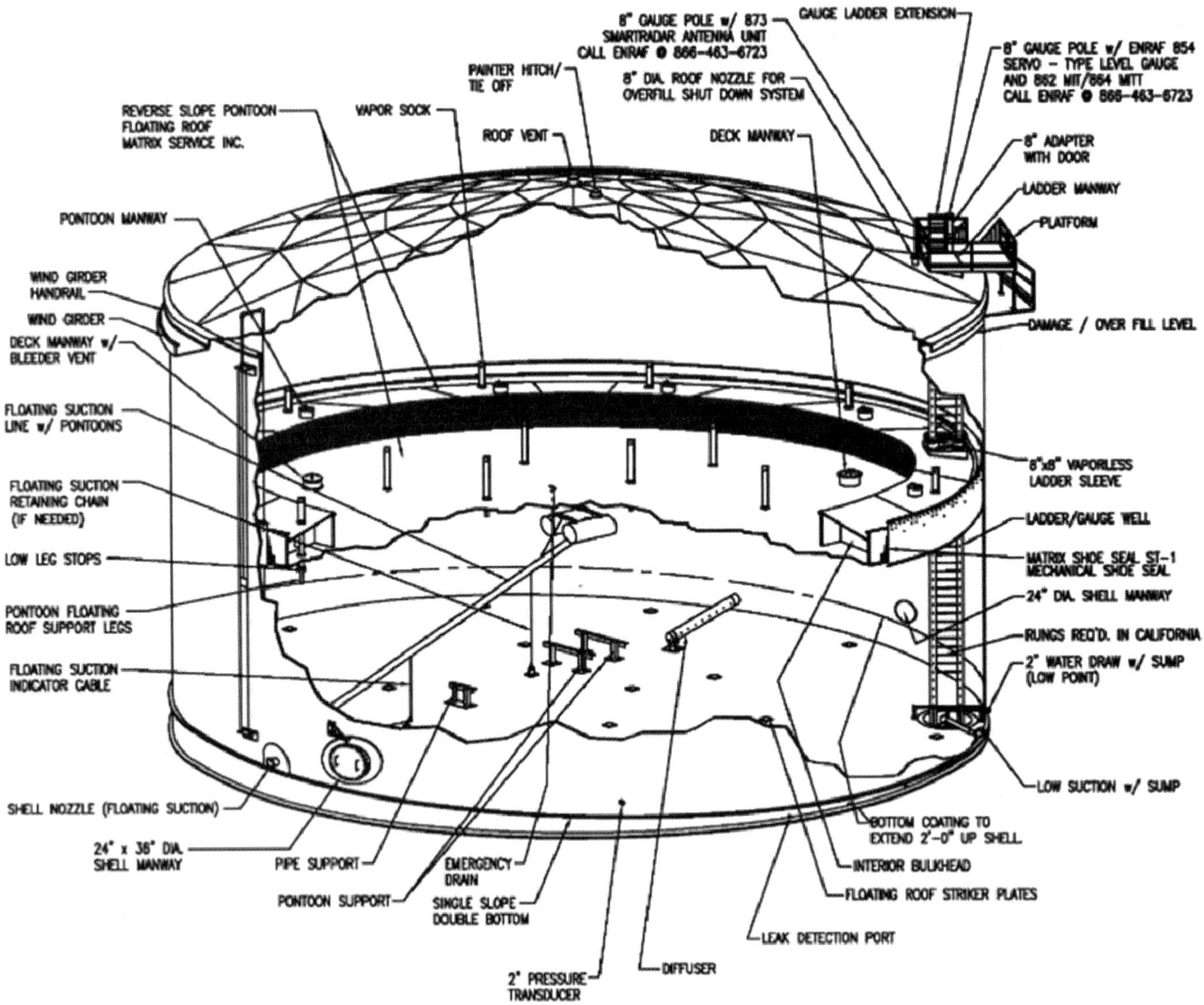

- Most expensive option.
- Pontoons across entire deck.
- Reserved for very small diameter tanks (consider fixed roof) or low specific gravity products.

Reprinted with permission - Matrix PDM Engineering

Cone Roof Tank with Internal Floating Roof

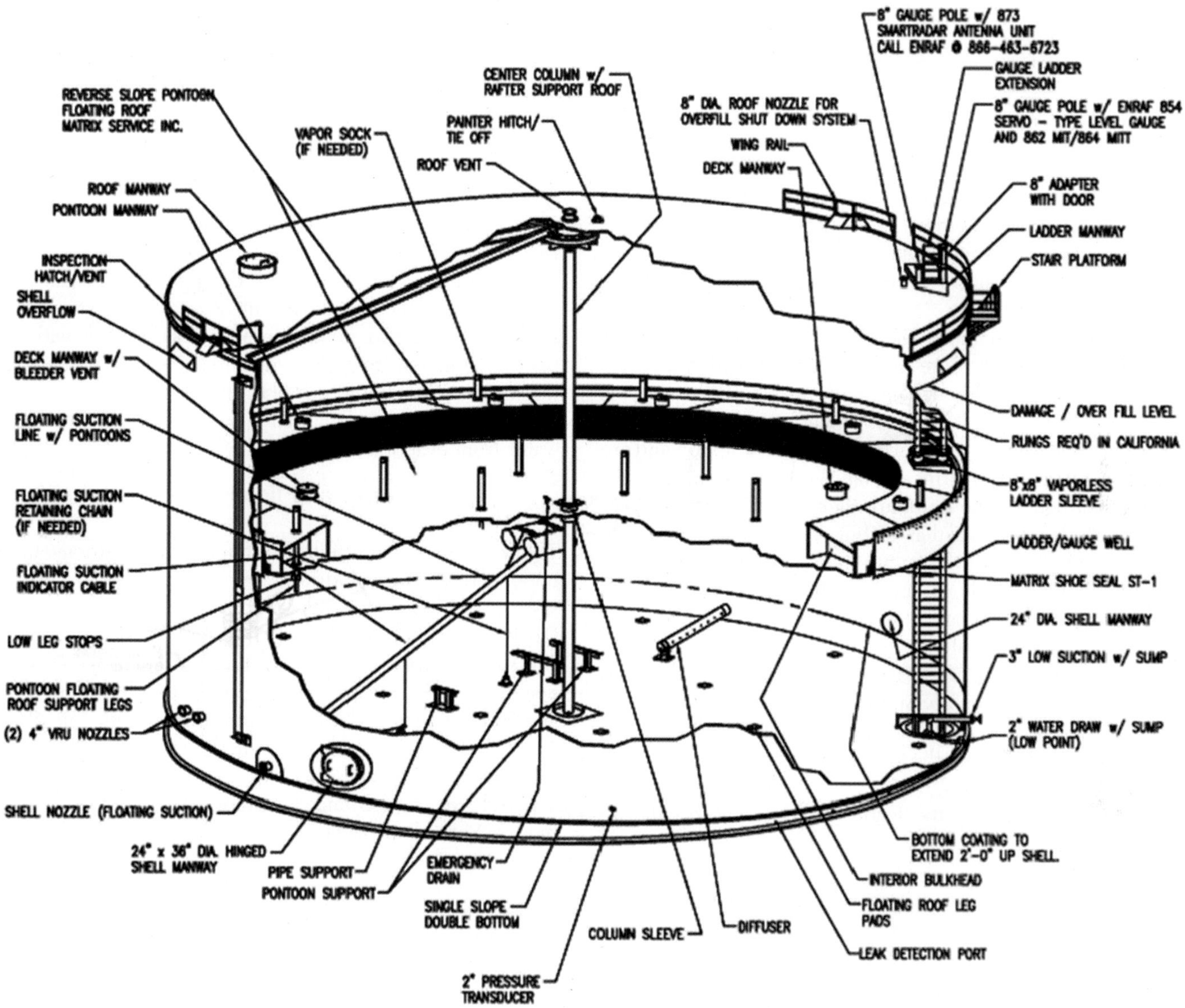

- Reduced emissions by lowering prevailing wind. Circulation vents reduce fume build-up between floating and fixed roofs.
- Small diameter tanks can use a steeper roof pitch to eliminate the center column. Above that a dome or umbrella may be used to eliminate columns (up to a point).
- Steel pontoon floating roofs are similar in design to external but do not require wastewater drains or standing rainwater check.
- Secondary seals often not required (depends on permit).

Reprinted with permission - Matrix PDM Engineering

Floating roof applications and components

Philip Myers, P.E., Founder and Director, PEMY Consulting, Pleasant Hill, California

Background

Owning and managing a petroleum storage facility requires numerous activities, skills, operations, and equipment, many of which have been captured in documents such as API 2610[1]. In spite of all the experience and accumulated industry knowledge, serious incidents continue to happen. Rim fires occur at a rate of about 1 – 10 per 1000 tank years. Of these, 1 in about 50 escalate to tanks fully involved in fire[2].

Floating roofs are one of the most common sources of tank incidents because they are the 'moving part' within a petroleum storage tank. Although incidents involving floating roofs may not always result in what could be called a 'serious incident', they always result in a business interruption and are usually quite costly in terms of financial and operational goals.

Interestingly, in the literature there is little about how to cope with the problems that floating roofs create. We believe that because many of the failure modes of floating roofs are 'latent', that is, they do not manifest themselves in an obvious way, little attention is paid to reducing the risk of floating roof incidents, which are almost always treated as an 'accident'. But far from being an accident, the risk assessment and management sciences are sufficiently advanced today such that these incidents can be entirely avoided. Moreover, new technology makes the prevention much easier than it used to be.

Even though most of the floating roof incidents do not rise to the level of a serious fire, injury or fatalities, any floating roof incident involving failure can wipe out months or years of effort. In causal analysis there may be no difference in the accident sequence leading to a minor or serious incident. However, serious incidents almost always result in litigation due to property and environmental damage or injuries and fatalities. Therefore, the tank owner/operator should consider becoming proactive in preventing these failures before they occur.

In the early days of crude oil storage, tanks were constructed of wood, many of which burned down and resulted in fires, losses, and business interruptions. Steel tanks reduced the problem, and would soon become the dominant choice of construction material for oil storage tanks. Still, many of these steel tanks had no roof and were subject to lightning strikes and other ignition sources, which resulted in tank fires. These events continued to cause headaches and losses for oil men. Not long after the advent of steel storage tanks, designers came up with the idea of a floating roof. Floating roofs prevent evaporation, which is beneficial for several reasons:

1. To reduce the probability of fires and explosions.
2. To reduce evaporative oil loss.
3. Keep precipitation and debris out of the product.

A diagram of a tank with a floating roof is shown in Figure 1.

The roof rises and lowers with the liquid level, to maintain a barrier between the environment and the volatile liquid surface. In short, a floating roof is a large disk with this primary function: to reduce or eliminate evaporation from the surface of the liquid stored within the tank. One can think of a floating roof like a swimming pool cover, preventing pool water from evaporating too quickly.

Tanks and floating roofs

There are three types of flat bottom petroleum storage tanks, both vertical and cylindrical, which are commonly used to store oil: the fixed roof tank (FRT), external floating roof tank (EFRT), and internal floating roof tank (IFRT). The latter two storage tank types utilize a floating roof. FRTs have no floating roof. However, a floating roof can be retrofitted into an FRT so that the FRT is transformed into an IFRT. Additional tank type acronyms are described in Table 1.

Floating roofs are the most complex component of a petroleum storage tank and they have numerous failure modes. For volatile organic petroleum liquids, floating roofs

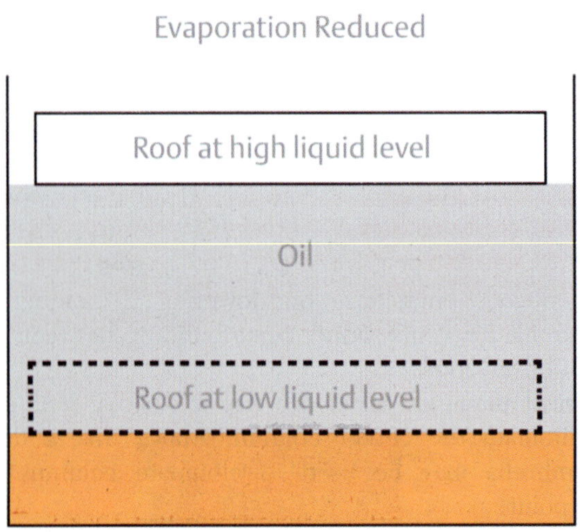

Figure 1. Conceptual diagram of floating roof tank.

Table 1
Tank Type Acronyms

EFRT	External Floating Roof Tank
IFRT	Internal Floating Roof Tank
FRT	Fixed Roof Tank (no floating roof inside)
EFR	External Floating Roof
IFR	Internal Floating Roof
FR	Fixed Roof (no floating roof inside)
APFR	Annular Pontoon Floating Roof
DDFR	Double Deck Floating Roof
EFRM	Electronic Floating Roof Monitoring System
LOB	Loss Of Buoyancy

play a vital role to reduce the reduction of evaporation losses. This has three major benefits:

1. Reducing emissions, resulting in financial, environmental, and health savings.
2. Reducing the probability of tank fires and explosions, protecting personnel and capital.
3. Keeping debris and precipitation out of the product.

Floating roof failures must be actively avoided, as high potential exists for severe tank related incidents such as fires, injuries, and fatalities when failure of a floating roof is involved.

Having a tool that allows the tank owner/operator to preemptively act to prevent a floating roof incident from occurring would be highly valuable, and in the last few years, new technology has evolved to make such tools available. For example, an Electronic Floating Roof Monitoring (EFRM) system can provide early detection of floating roof failure, preventing a variety of serious floating roof failure modes. The implementation of an EFRM system allows oil and chemical storage terminals assurance that a floating roof is operating as specified (Quick Facts 1).

Our journey into the domain of petroleum tank floating roofs begins with a discussion on the types of floating roofs.

The floating roof design criteria include (but are not limited to) the construction materials of the roof and required material thicknesses. EFRs are used for external applications (i.e., open top tanks with the EFRs floating on the liquid surface), where they are exposed to the weather. For IFR applications, the floating roof is protected from the weather by a fixed roof that is mounted on top of the tank. Precipitation factors heavily into the design requirements for floating roofs. As one might expect, the EFR must be heavier-duty construction than the IFR because of exposure to weather. Thus, all tank standards such as API 6503[3] (or EN 1993-4-2: Tanks) have more stringent design requirements for EFRs than for IFRs.

Why are there so many different types of floating roofs? A glimpse of the answer is in Figure 3. Floating roofs first are divided by whether their service is internal (IFRT) or external (EFRT). Within the external branch there are only the annular pontoon floating roof (APFR) and the double deck floating roof (DDFR) designs. However, historically, pan roofs were used despite their disadvantages. In addition, there were floating roof designs where a center compartment was in the middle of what would be an APFR, but these have fallen into disuse. Nonetheless, in the population of existing tanks there are many such old floating roof designs still in existence.

QUICK FACTS 1: Electronic Floating Roof Monitoring (EFRM) Systems

Floating roofs serve a vital role in the storage of highly volatile and dangerous chemicals, and roofs are an inherent source of risk at every chemical storage terminal. Failure of floating roofs can lead to a wide variety of devastating events, but new technology can significantly aid in the prevention of roof failure. A multi-point roof tilt monitoring system is a safety tool that monitors the difference in height between multiple points on a floating roof surface (Figure 2). These points are converted into a distance, and a user specifies the distance parameters that will trigger an alarm or alert on the system computer. This early alert system will prevent the tilt of a floating roof from reaching critical levels before action can be taken to avert an event.

For example, if an EFRM system uses three radars that read 55, 54.5, and 55.5 inches, this measurement would be designated as nonhazardous. In this scenario, operations and maintenance can be confident in the performance and stability of the floating roof of that storage tank. Given another scenario where the radars measure a significant difference in the distance to the roof, operations can be certain that the roof is tilting. Operations can then inform maintenance to inspect and fix the cause of tilt.

Preventive maintenance is the best barrier between a company and an incident. In short, the use of multi-point roof tilt monitoring systems is a progressive step towards personnel, asset, and corporate image security. A multi-point tilt monitoring system applied to all floating roofs at a terminal ensures much safer operations, prevention of emissions, better risk management, and peace of mind.

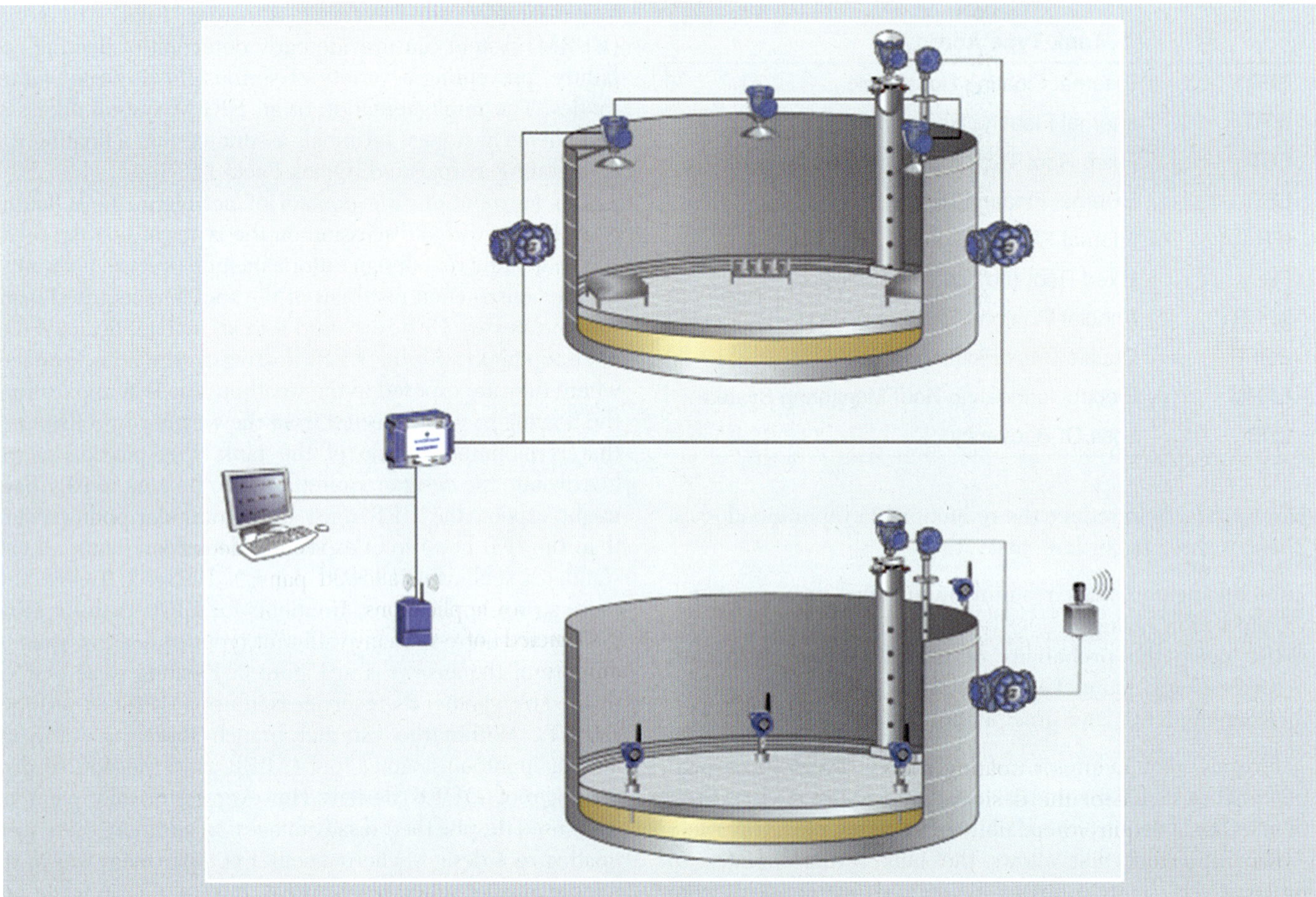

Figure 2. Electronic floating roof monitoring system.

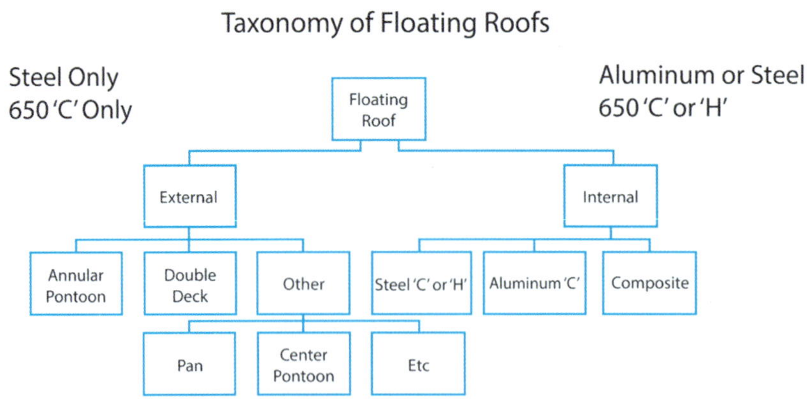

Figure 3. Floating roof taxonomy.

On the *internal* branch of Figure 3 there is great variety. The first point to be understood is that an EFR may be used in an IFRT, however, an IFR may not be used in an EFRT. It is mostly economics that limit the use of EFRs in IFRT applications. This is due to the usually much higher costs of the EFR as compared with the IFR. IFRs allow for different materials of construction, and for different means to achieve buoyancy such as light weight panels or tubular pontoons. IFRs can use steel, aluminum, non-metallic and even composite materials in their designs.

By far, the most robust and most expensive of all floating roofs is the DDFR. Juxtaposed against the DDFR, the pan roof is at the opposite side of the cost and reliability spectrum. Unsurprisingly, the pan roof was historically the most economic choice for floating roof applications. However, industry members and standards development organizations recognized that the pan roof brought grief to owners and operators due to its high failure rates. The industry was willing to pay more for the better roofs. In order of robustness against failure (for EFRs), the DDFR is the best, followed by the APFR. Today, the primary EFRT applications are APFR and DDFR types which are the gold standard for EFRs. The rules for constructing EFRs are given in API 650 Annex C.

Because IFRTs reduce the weather exposure, and provide a benign environment for floating roofs, API 650 provides separate Annex H for the design and construction requirements. These requirements are less stringent and allow for construction materials other than steel- the costs of which are far less than those of EFRT floating roofs. IFRs may be constructed of different materials and there are advantages and disadvantages for each design and manufacturer.

Ranking these roofs for suitability for service is a challenging task that is company and service specific. There is no universally suitable ranking method for the IFRs as being more robust or best for a particular application. While IFRs are typically designed and manufactured according to API 650 Annex H, there is a small percentage of IFRTs that use floating roof designs of Annex C.

Naturally, there are tradeoffs in costs and performance that govern what type of floating roof should be selected or used for a given application, and that is the job of the design engineer and business unit to determine.

Applications and details

As discussed earlier there are two floating roof applications:

- External Floating Roof (API 650 Annex C)
- Internal Floating Roof (API 650 Annex H)

Floating roofs are separated into internal and external categories (Figure 3). IFRTs have a fixed roof that prevents weather and wind from contacting the deck of the floating roof, whereas EFRTs are environmentally exposed from the top and they are often called 'open top tanks'. EFRs have several different designs: double deck, annular pontoon, and older designs such as the high center deck floating roof. Older editions of API 650 allowed pans to be used for external floating roof applications. Unlike the EFRs, which must be constructed of steel, the IFRs may be constructed of aluminum, steel, or composite materials.

One major difference between the Annexes in API 650 is that Annex C for EFRs requires steel construction and has more stringent design loading criteria to account for weather loads, such as rain, snow, and ice. Annex H covers the design of IFRs for which the design criteria are less stringent since these floating roofs are always protected from the weather by either a fixed steel roof- sometimes called 'cone roof'- or by an aluminum geodesic dome (Quick Facts 2).

Bear in mind that standards such as API 650 apply only to the design of new tanks and floating roofs. So, there are many EFRs built to API 650 Annex C standards, even though they are enclosed by a fixed roof or geodesic dome- this is due to a change in service, in which an FRT was converted to an IFRT. A legislative example of this is the Los Angeles region of California mandating that all oil-refining tanks shall be 'covered'. This meant that the many existing EFRTs had to be

QUICK FACTS 2: Cone versus Dome Roofs

FRTs and IFRTs most often have fixed roofs of two types:

1. Cone
2. Dome

The cone is a shallow cone with a small slope usually 3/4-inch rise over a 12-inch run up to 2 inches in 12. The cone is 3/16-inch thick steel plate resting on steel beams (rafters). For smaller tanks up to 50 or 60 feet the rafters can be supported with a single central column which transfers the roof weight to the foundation.

The dome is typically an aluminum geodesic structure. These do not have to be painted as do steel cones. However, their first cost is usually higher than steel. One major difference is that a dome can span the tank diameter without the need for central support columns or systems of columns that are needed for cone roofs, especially in the larger tank sizes (Figure 4). This is an advantage because there is no need to have openings in the floating roof to allow the roof to move up and down with roof support columns.

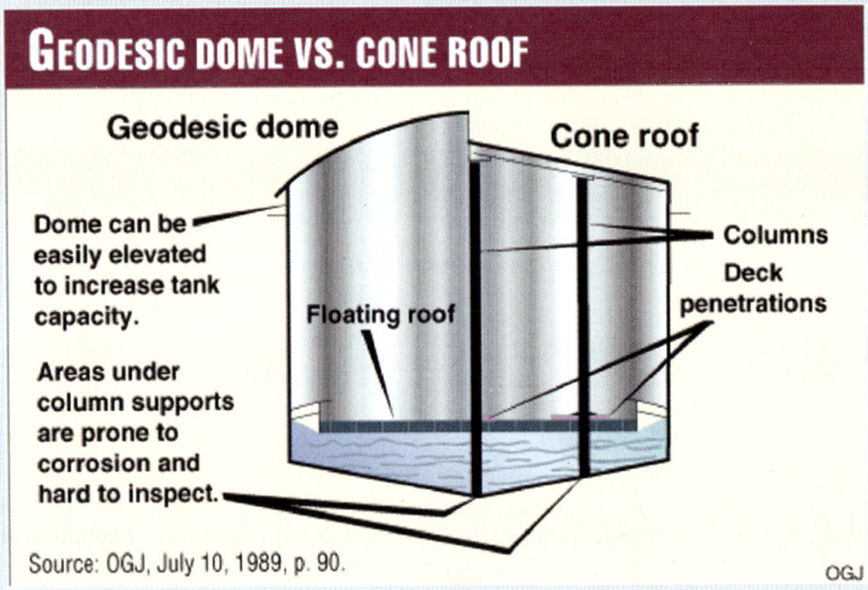

Figure 4. Contrast between geodesic and cone roofs.

converted to IFRTs by the addition of either a steel cone, or an aluminum geodesic dome. In most cases, owners/operators chose to retrofit their tanks with geodesic domes. The floating roofs were then under cover and therefore installed in IFRTs. This is an example of how an API 650 Annex C floating roof ends up in an IFRT, which only requires an API 650 Annex H design. It is rare for a new IFRT to use an EFR such as an APFR or DDFR.

In our journey through floating roofs we will limit the discussion primarily to annular pontoon floating roofs (APFR) and double deck floating roofs (DDFR).

The workhorses of floating roofs

Annular Pontoon Floating Roof

The APFR is by far the most common floating roof for EFRTs. A schematic view for APFRs is given in Figure 5. The APFR floats because of the buoyant forces provided by the outer pontoon ring section as well as the deck. The ring section is made up of closed compartments, which have 11nothing in them but air. The tops of the pontoon ring sections close off the pontoon sections and are generally made of sheet steel. Figure 6 shows an image of an APFR in application. Note the vertical floating roof leg sleeves. The pipes in these sleeves can be pinned into a low or high position so that when the roof comes to rest on the tank bottom (called landing), the roof position can be set at two different levels (usually about 3 ft or 7 ft). The deck is constructed of 3/16 inch steel and the rectangular plates can be

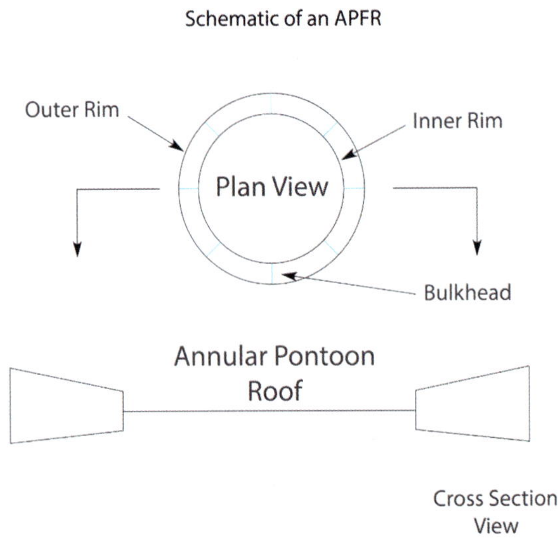

Figure 5. Schematic of an annular pontoon floating roof.

seen. The deck is composed of these plates lap welded together. In the distance, the step up represents the inner rim of the annular pontoon ring section which is covered on the top. Also, note the rolling roof ladder that allows access to the floating roof from the gauger's platform.

Figure 7 shows a close up of the annular pontoon ring section. Note the pontoon access hatches that allow entry into the pontoons for inspection as well as the leg sleeves with legs inside the sleeves. A view of the underside of this roof is shown in Figure 8 along with the legs which hold the roof up in the high position.

Figure 6. Annular pontoon floating roof in application.

Figure 7. Close up of annular pontoon ring section.

Figure 8. Underside of an annular pontoon floating roof in 'high' position.

Double Deck Floating Roof

The schematic diagram for the DDFR is shown in Figure 9. It is the heaviest construction and most robust design against failure because of its strength. Similar to the APFR, it is constructed of steel. The pontoon compartments are covered with two skins or decks that form the top and bottom surfaces of the floating roof.

Other floating roof components

Floating Roof Legs

On occasion, all the product from a tank must be removed, for example, during an internal inspection or a change of service. At this time, the floating roof must be landed, either in high or low position. The leg position is changed by setting steel pins about 3/4 to an inch in diameter into the leg sleeve and leg as can be seen in Figure 7. The high position is needed so that a worker can move around under the floating roof with sufficient space to comfortably walk around. During procedures that do not require personnel to be under the roof, such as a change of service, the floating roof is typically set to the low position for landing. An additional consideration is useable volume. In order to maximize the useable volume of a floating roof tank it is important to run the floating roof to a position as high and as low as possible during normal operations. Normally, the position of a landed floating roof is set to the low position which maximizes the useable tank volume.

Figure 10 shows the typical leg, leg sleeve, and pin used to assign these positions to the landed floating roof. In reviewing Figure 8 it is apparent that legs seem skinny and flimsy. Even though the legs are 3 to 6-inch diameter schedule 40 or 80 pipe, they can fail for many reasons:

1. Corrosion thinning of the wall thickness.
2. Legs not vertical which allows the eccentric load to cause the legs to slip sideways.
3. Uneven tank bottom caused by settlement puts indeterminate loads on some legs (some legs carry much more weight than others).

In addition to the basic loading problems listed above operational events can cause failure:

1. Failure to pin all leg sleeves (from above the deck) can excessively load adjacent legs.
2. Failure to ensure all water and product is removed prior to landing the roof can overload the legs.
3. Allowing the landed floating roof to collect precipitation without ensuring drainage.

There have been fatalities as a result of floating roof collapses while landed. API 2015[4] and API 2016[5] provide guidance on how to stabilize and ensure that a floating roof does not collapse on entrants.

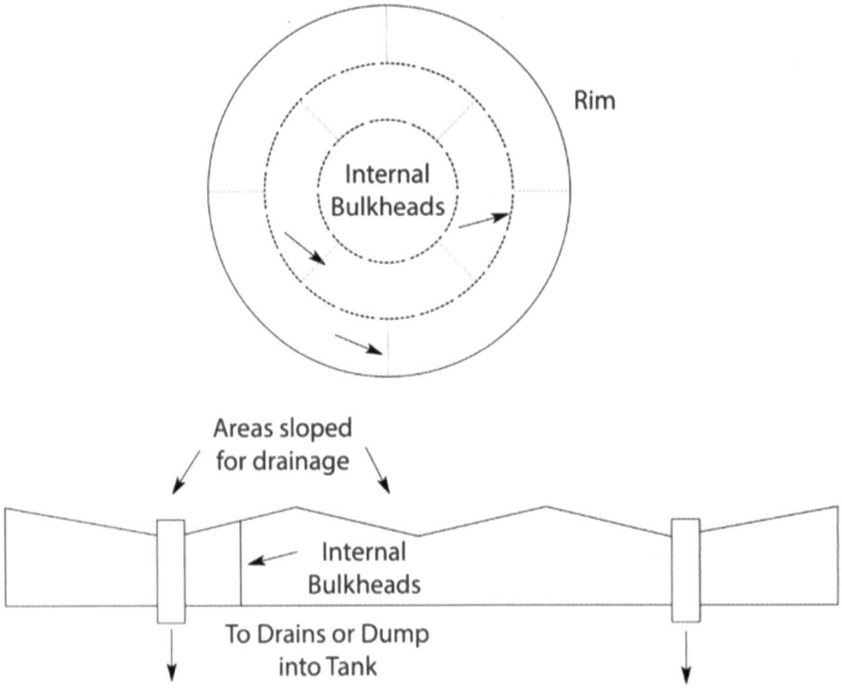

Figure 9. Schematic of a double deck floating roof.

QUICK FACTS 3: Summary of Tank & Roof Types

- There are three basic tank types (FRT, IFRT, EFRT)
- There are two basic floating roofs that can be used
  - Steel APFR or DDFR constructed to API 650 Appendix C in EFRT or 'open top tanks'
  - Steel, Aluminum, or Composite material floating roofs constructed to API 650 Appendix H in IFRTs. An FRT has no floating roof, but if one is put inside it then it becomes an IFRT.
- Be warned that pan roofs have a high probability of failure and should be phased out of use as quickly as possible.

Pontoon and Compartment Access Hatches

Pontoon compartments provide the inherent buoyancy needed to keep the roof afloat. An access hatch is needed for each compartment in APFR. This allows inspection of the pontoons for leakage. The hatch necks and covers are visible in Figure 7 and give a sense for the number of compartments needed for a robust floating roof. The integrity of floating roofs is highly dependent on the properties of these compartments and on the covers. Many roofs have sunk as a result of inadequate compartment design, inspection, covers, and more.

Vacuum Breakers

When liquid is rapidly pumped into the tank or out of the tank, the fitting shown in Figure 11 allows air movement to pass through the floating roof with little resistance. This is so that excessive internal or external pressure is not applied to the floating roof due to the displacement of air caused by the liquid's transfer. This fitting is nothing more than a pipe which is long enough to open the fitting cover when the roof is landed. Once floating, this fitting is closed.

The floating roof could rotate if the guidepole that passes through it were not present. The guidepole (Figure 12) is a pipe, usually 8 inches in diameter, that comes up vertically through the gauger's platform. It doubles as a sampling device so that liquid samples from any level in the tank can be pulled up in a small container called a 'sample bomb' and taken from the guidepole by the operator for lab analysis.

Seals

Floating roof seals are varied but the basic idea is a mechanism that performs as follows. It reduces emission by providing a barrier between the liquid in the annular space between the floating roof and the tank shell. Tanks are never perfectly round, but an annular gap of 4 – 8 inches is typical so that the floating roof does not bind on the shell as it moves up and down. The seal typically has a plastic sheet that prevents the vapors from freely moving up out of the exposed liquid annular area between the seal and the

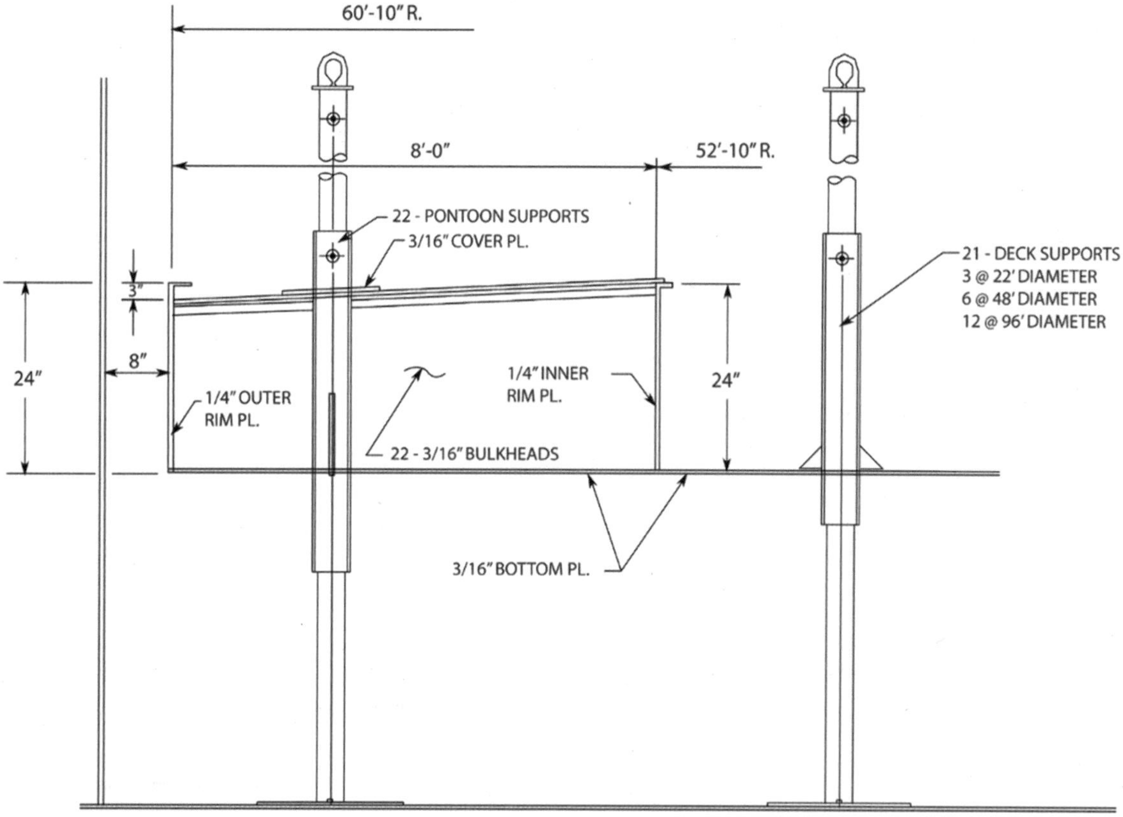

Figure 10. Diagram of leg, leg sleeve, and pin.

QUICK FACTS 4: API Pontoon Buoyancy Requirements Overview

The American Petroleum Institute (API) sets standards for the buoyancy requirements for floating roof in API 650 Annex C and Annex H, for external and internal floating roofs, respectively. See API 650 Annex C.3.4 and Annex H.4.2.1 for more details. Both floating roofs must be designed to be buoyant in liquids of specific gravities (SG) at least as low as the SG of the product stored or 0.7, whichever is lowest.

API 650 Annex C – External Floating Roofs

EFRs must be operative and intact while subject to the following scenarios:

- under 10-inches of rainfall, over the entire area of the roof, over a period of 24 hours without drainage; OR
- with two adjacent pontoon/buoyancy compartments are punctured and flooded, while the roof is under no water or live load; OR
- under unbalanced snow loading

An APFR must be designed with enough strength to prevent permanent distortion resulting from a rainwater load in the center of the roof. Penetrations or fittings such as leg sleeves of the EFR must not allow product to flow onto the roof.

API 650 Annex H – Internal Floating Roofs

IFRs must by designed to be buoyant enough to support at least twice its own weight, and to offset friction exerted by the seals on the edges of the roof.

IFRs with multiple flotation compartments must be floating even after any two compartments are punctured and flooded.

IFRs must be designed to be meet the above puncture requirement and support at least two men while it is floating, without allowing product on the roof.

Figure 11. Vacuum breaker.

Figure 12. Guide pole.

floating roof. Good, tight fitting seals are important to reduce the likelihood of tank seal fires due to lightning or maintenance activities.

Pan roofs

Figure 13 shows the concept of a pan roof. Note the simplicity. The difference between a pan roof and an APFR, or DDFR is what we call 'inherent buoyancy'. Although pan roofs are obsolete, and no reputable vendor would supply them today, there are many in the existing tank populations. A single small hole, or a moderate amount of liquid on the surface can sink a pan roof. It has no inherent buoyancy meaning that it has an 'Achilles heel'. Although some API 650 members have advocated outright banning of the pan roof, some committee members wanted to ensure maximum choice and prevented the removal of pans from the inventory of acceptable external floating roof designs but did include some warnings.

Figure 14 shows a photo of a pan roof in use. Pan roofs are so large that it would be far too flexible and flimsy to construct and operate without the system of guy wires and beams depicted. Nonetheless, a single pin hole from corrosion or a crack can cause flooding of the entire surface of this floating roof and sink it.

Pan roofs cannot generally be made strong enough to resist the liquid loads from even just a few inches of product. The deck becomes stretched into an arc shape, and as it does, exerts compressive forces on the rim. The rim is very weak structure, and it can buckle when subjected to these compressive forces, leading to a sudden and catastrophic failure.

A better, but still highly vulnerable design, is the bulkhead pan roof (Figure 15). The bulkheads provide additional floatation should a hole or cracks develop in the roof, and

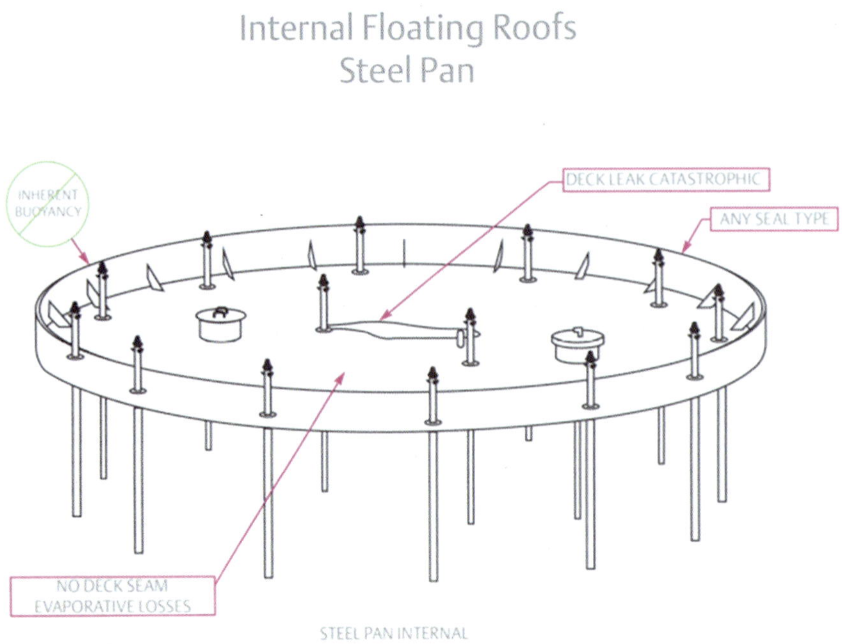

Figure 13. Pan roof diagram.

Figure 14. Truss supported pan roof.

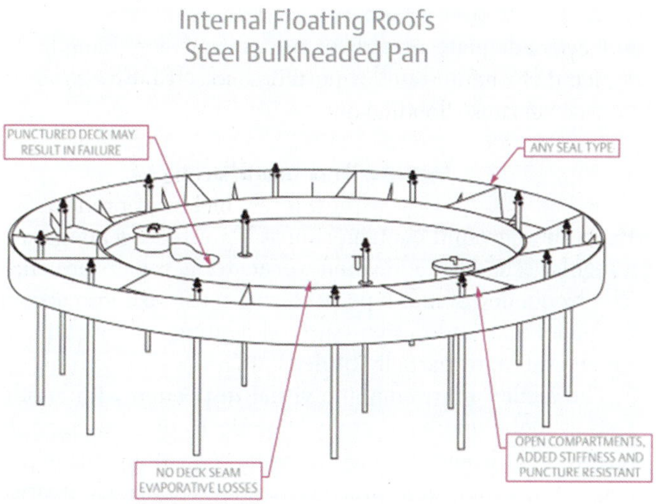

Figure 15. Schematic of steel bulkhead floating pan roof.

therefore, they do have more robust buoyancy. Unlike the annular pontoon roof, the top is open, and there is very little strength for this type of roof to handle oil, water, or firefighting foam that could enter the bulkhead areas.

The incident illustrated in Figure 16 shows the outer rim buckled as a result of gas blowing into a pipeline breakout tank. The bubbling of nitrogen gas used to push liquid out of a pipeline and into a tank causes overflow of diesel over the outer rim. The excessive forces from the liquid weight in the open pontoons caused the rim to buckle and the roof to sink.

Figure 16. Bulkhead pan roof failed by liquid in pontoons buckling the outer rim.

References

1. Design, Construction, Operation, Maintenance and Inspection of Terminal Tank Facilities, 2nd Edition (2005).
2. Marsh Risk Engineering Position Paper-01 Atmospheric Storage Tanks, Marsh and McLennan Companies
3. Welded Tanks for Oil Storage, 12th Edition (2013).
4. Requirements for Safe Entry and Cleaning of Petroleum Storage Tanks, 7th Edition (2014).
5. Guidelines and Procedures for Entering and Cleaning Petroleum Storage Tanks, 1st Edition (2001).

Source

Philip E. Myers, Floating Roof Tanks in Petroleum Storage: An Overview of Roof Types, Fault Modes, Failure Causes and Technology for Incident Prevention, White Paper. PEMY Consulting. Go online at: www.pemyconsulting.com.

Summary of federal requirements for aboveground storage tanks

Aboveground storage tanks (ASTs) used for the store of petroleum products is regulated primarily under 40 CFR 112. The regulation does not actually use the terms "aboveground storage tank." Instead the term "bulk storage container" is used and is defined as "any container used to store oil. These containers are used for purposes including, but not limited to, the storage of oil prior to use, while being used, or prior to further distribution in commerce. Oil-filled electrical, operating, or manufacturing equipment is not a bulk storage container."

A bulk storage container is 55 gal. or greater and may be aboveground, partially buried, bunkered, or completely buried. "Bunkered tanks" are defined as "a container constructed or placed in the ground by cutting the earth and recovering the container in a manner that breaks the surrounding natural grade, or that lies above grade, and is covered with earth, sand, gravel, asphalt, or other material. A bunkered tank is considered an aboveground storage container for purposes of 40 CFR 112."

Design

A container (i.e. storage tank) cannot be used for the storage of oil unless its material and construction are compatible with the material stored and conditions of storage such as pressure and temperature.

All bulk storage container installations must be constructed so a secondary means of containment is provided for the entire capacity of the largest single container and sufficient freeboard to contain precipitation. Diked areas must be sufficiently impervious to contain discharged oil. While dikes, containment curbs, and pits are commonly employed for this purpose, an alternative system consisting of a drainage trench enclosure that must be arranged so that any discharge will terminate and be safely confined in a facility catchment basin or holding pond may also be used.

Each bulk storage container installation must be engineered or updated in accordance with good engineering practice to avoid discharges, including at least one of the following devices:

- high liquid level alarms with an audible or visual signal at a constantly attended operation or surveillance station (NOTE: In smaller facilities an audible air vent may suffice.)
- high liquid level pump cutoff devices set to stop flow at a predetermined container content level
- direct audible or code signal communication between the container gauge and the pumping station
- a fast response system for determining the liquid level of each bulk storage container such as digital computers, telepulse, or direct vision gauges (NOTE: In this method, a person must be present to monitor gauges and the overall filling of bulk storage containers.)

Management

Control leakage through defective internal heating coils is by monitoring the steam return and exhaust lines for contamination from internal heating coils that discharge into an open watercourse, or pass the steam return or exhaust lines through a settling tank, skimmer, or other separation or retention system.

The drainage of uncontaminated rainwater is not allowed from the diked area into a storm drain or discharge of an effluent into an open watercourse, lake, or pond, bypassing the facility treatment system unless the facility:

- normally keeps the bypass valve sealed closed
- inspects the retained rainwater to ensure that its presence will not cause a discharge
- opens the bypass valve and reseals it following drainage under responsible supervision
- keeps adequate records of such events, for example, any records required under permits (i.e., NPDES).

Testing and monitoring

Each aboveground container must be tested for integrity on a regular schedule, and whenever material repairs are made. The frequency of and type of testing must take into account container size and design (such as floating roof, skid-mounted, elevated, or partially buried).

The facility must combine visual inspection with another testing technique such as hydrostatic testing, radiographic testing, ultrasonic testing, acoustic emissions testing, or another system of non-destructive shell testing. Comparison records must be kept and the container's supports and foundations also inspected. Personnel must frequently inspect the outside of the container for signs of deterioration, discharges, or accumulation of oil inside diked areas.

Liquid level sensing devices must be regularly tested to ensure proper operation.

Effluent treatment facilities should be observed frequently enough to detect possible system upsets that could cause a discharge.

If field-constructed aboveground containers undergo a repair, alteration, reconstruction, or a change in service that might affect the risk of a discharge or failure due to brittle fracture or other catastrophe, or has discharged oil or failed due to brittle fracture failure or other catastrophe, the container is evaluated for risk of discharge or failure due to brittle fracture or other catastrophe and appropriate actions taken.

Regulatory applicability

40 CFR 112 applies to petroleum oils and non-petroleum oils, animal fats and oils and greases, and fish and marine mammal oils; and vegetable oils (including oils from seeds, nuts, fruits, and kernels).

40 CFR 112 also applies to any owner or operator of a non-transportation-related onshore or offshore facility engaged in drilling, producing, gathering, storing, processing, refining, transferring, distributing, using, or consuming oil and oil products, which due to its location, could reasonably be expected to discharge oil in quantities that may be harmful, into or upon the navigable waters of the United States or adjoining shorelines, or into or upon the waters of the contiguous zone, or in connection with activities under the *Outer Continental Shelf Lands Act or the Deepwater Port Act of 1974*, or that may affect natural resources belonging to, appertaining to, or under the exclusive management authority of the United States (including resources under the *Magnuson Fishery Conservation and Management Act*) that has oil in:

- any aboveground container
- any completely buried tank (see definitions)
- any container that is used for standby storage, for seasonal storage, or for temporary storage, or not otherwise "permanently closed" (see definitions)
- any "bunkered tank" or "partially buried tank" (see definitions), or any container in a vault, each of which is considered an aboveground storage container for purposes of 40 CFR 112.

Hazardous materials storage

Depending on the type and amount of hazardous materials (i.e., gasoline, diesel fuel, other liquids) stored and/or used at the facility, EPA reporting requirements may apply.

Summary of state requirements

ASTs are primarily regulated by states. The requirements can usually found in environmental regulations and/or the fire code. There may be a requirement to paint the tank, or a fill pipe, a certain color to identify its contents, provide alarms in addition to those Federally required, and registration of the tank.

Source

Fed Center. Federal Facilities Environmental Stewardship & Compliance Assistance Center. Facility Regulatory Tour. Online at fedcenter.gov.

Design and containment for atmospheric storage tanks

Excerpts from the Marsh Risk Engineering Position Paper 01 - Atmospheric Storage Tanks, 2015.

Selection of atmospheric storage tanks

The layout and general design of a storage facility should be based upon considerations of safety, operational efficiency, and environmental protection. A primary consideration is ensuring that the design of the storage tanks themselves is suitable for the classification of the hydrocarbon being stored. Table 1 summarizes the design of tank suitable for each class of hydrocarbon.

Layout and spacing

Ideally, tank layout should be optimized to ensure that there is sufficient access to tanks for firefighting and to minimize the risk of escalation in the event of a tank fire. Minimum spacing for tanks is specified in Table 2, although Marsh would advocate a minimum separation distance of 1× the diameter of the largest tank with an absolute minimum of 15 meters, as such a distance provides sufficient access for fire fighters.

One notable exception is the separation between crude tanks, where the destructive effects of a boil over can extend from 5 to 10 diameters. Therefore, consideration should be given to locating crude tanks at the edge of tank farm installations and with the largest practical separation from adjacent tanks.

These distances should only be used in conjunction with appropriate levels of fire protection. In the event that tanks are existing and do not conform to the above spacing, then additional fire protection should be considered.

Floating roof tanks, with external metal domed roofs extending over the entire roof area (i.e., internal floating roof tanks or tanks fitted with geodesic domes), may be considered as fixed roof tanks for the purpose of tank location and spacing.

Table 1
Recommended Storage Tank Design

Crude Oil Derivative Class	Definition	Recommended Storage Tank	
		Floating Roof	Fixed Roof
Class 0	Liquefied petroleum gases (LPG)	×	×
Class I	Liquids which have flash points below 21°C	✓	×
Class II (1)	Liquids which have flash points from 21°C up to and including 55°C, handled below flash point	✓	✓
Class II (2)	Liquids which have flash points from 21°C up to and including 55°C, handled at or above flash point	✓	×
Class III (1)	Liquids which have flash points above 55°C up to and including 100°C, handled below flash point	×	✓
Class III (2)	Liquid which have flash points above 55°C up to and including 100°C, handled at or above flash point	×	✓
Unclassified	Liquids which have flash points above 100°C	×	✓

Figure 1. Tank layout is optimized to ensure sufficient access for firefighting and to minimize risk of escalation in event of a tank fire.

Figure 2. A minimum separation distance of 1 × the diameter of the largest tank with an absolute minimum of 15 meters is a good rule of thumb for tank spacing.

Bunds (Dykes)

Above-ground tanks should be surrounded by bund. These should be designed to offer protection to fire fighters. Therefore, the bund wall should be located so that a reasonably close approach can be made to a tank fire. Firefighting screens and steps should be located at various points around the bund wall to assist the positioning and protection of firefighting personnel and equipment.

Bunds would normally be constructed from earth or preferably concrete and should be largely impervious to liquid and capable of withstanding hydrostatic and hydrodynamic pressures to which they could be subjected. Earthen bunds should be fitted with a water-resistant liner and whilst grass growth is acceptable, it should be regularly maintained and kept short.

Tanks should ideally be located alone within their own bunds, although whenever tanks share a common bund, intermediate walls up to half the height of the main bund walls and no more than 0.5 meters high should be provided to control small spillages from one tank affecting another.

Bund floors should drain to a single location complete with sump for the regular removal of water from rainfall or firewater testing. Drains should normally be kept closed, with the drain isolation valve situated outside the bund.

Table 2
Minimum Separation Distance Between Tanks

Factor	Minimum Separation from Any Part of the Tank
Between adjacent fixed-roof tanks	Equal to the smaller of the following: (a) The diameter of the smaller tank. (b) Half the diameter of the larger tank. (c) Not less than 10m (Marsh advocate 15m).
Between adjacent floating-roof tanks	10m for tanks up to and including 45m diameter. 15m for tanks over 45m diameter. (The spacing is determined by the size of the larger tank).
Between a floating-roof tank and a fixed-roof tank	Equal to the smaller of the following: (a) The diameter of the smaller tank. (b) Half the diameter of the larger tank. (c) Not less than 10m (Marsh advocate 15m).
Between a group of small tanks and any tank outside the group	15m
Between a tank and the site boundary, any designated non-hazardous area, process area or any fixed source of ignition	15m

Source: HSG 176. The storage of flammable liquids in tanks.

Figure 3. Aboveground tanks should be surrounded by bund and bund floors should drain to a single location complete with sump for regular water removal.

Primary containment

It is essential that the risk of loss of containment is properly managed. This can be achieved by the proper design, operation, maintenance, and inspection of tanks. Ideal practice is outlined below.

The most effective way to prevent a major accident at any site is the continued provision of suitable primary containment of the flammable materials. This is achieved through the suitable design, construction, and maintenance of the storage systems in accordance with standards.

For guidance on the most appropriate standards to use for the design and construction of atmospheric storage tanks, refer to the UK Health and Safety Executive's review of standards, Mechanical integrity management of bulk storage tanks, which can be found at http://www.hse.gov.uk.

Features considered to be the most appropriate for a modern facility include:

- Double seals on floating roof tanks (NB foam dam heights should be above the height of the upper rim seal to provide suitable protection against a rim seal fire).
- High level alarm and independent back-up high level instrument (with executive action only if appropriate).
- Continuous automatic monitoring of tank contents, including "rate of change" during filling/emptying.
- Anti-rotation devices on floating roof tanks.
- Closed water drains.
- Single skin floors.
- Tank floor leak detection on double-floor tanks. This is normally comprised of an instrument to detect loss of vacuum in the inter-floor space.

Tank design and operation

Tanks should be designed to a relevant standard, such as API 650 or BS EN 14015. In some instances, single skin bottom tanks are a better option than double-bottomed tanks as these provide the optimum conditions for ensuring integrity of the tank floor by the inspection of tank floor plates. Disadvantages of double-bottom designs include settlement, product entrapment, and modification to nozzle compensating plates. Where double-bottomed tanks are provided, additional inspection measures should be provided in accordance with a relevant standard such as EEMUA 183 or BS EN 14015. Leak detection should also be provided on double-bottomed tanks.

Tank Roofs

Double deck roofs on floating roof tanks are preferred to single deck roofs with external pontoons as they are more stable and less likely to lose buoyancy.

For floating roof tanks, the water drains to remove rainwater from the roof should be normally closed. To ensure roofs are emptied of rainwater, procedures should be in place and followed for the opening and re-closure of valves.

Whenever valves are left normally open, then the following measures should be put in place:

- Hydrocarbon detection in drained water with automatic isolation.
- Testing of operability of isolation valves and detection systems.

Vents

Tanks should be provided with a weak seam/frangible roof construction, or with an emergency vent suitably sized for the worst-case relieving scenario, to prevent overpressure under all relief conditions. Emergency vents should comply with relevant standards, such as API 2000.

Wherever a fixed roof tank is used for the storage of materials with a flash point below 21°C (generally on day tanks, not recommended for large bulk storage tanks), then flame arresters should be provided on the vents to prevent ignition of the flammable vapors burning back into the tank. In warmer climates, where product surface temperatures could exceed 21°C, flame arresters should be used wherever maximum surface temperatures could be within 10°C of the flash point.

Flame arresters should be included in preventative maintenance routines to ensure they do not become blocked by scale, paint, ice or other materials. Flame arresters are not recommended for use when the material being stored is liable to polymerize or foul the arrester. Due to the potential for blockage, conservation vents (vacuum and pressure relief) should not be fitted with flame arresters.

Fixed roof tanks can be fitted with a gas blanket (normally nitrogen) to maintain an inert atmosphere in the vapor space. Nitrogen supply pressure should be just above atmospheric, but sufficient to displace any liquid pumped out. As a backup, a vacuum breaker should be provided if nitrogen supply is lost.

Vents or vapor recovery systems (often venting back to the source vessel) are required. These should be designed to relieve pressure slightly above that of the nitrogen and at a suitable margin below the design pressure of the storage tank.

Rim Seals and Foam Dams

Double rim seals (of fire-resistant construction) are preferable to single seals. Due to irregularities during the construction phase of large floating roof tanks, it is likely that a single seal will not maintain a 100% seal between the tank roof and tank wall along the entire height of the tank wall. Therefore, a second seal will improve the likelihood of achieving a tight seal, will reduce emissions, and minimize the risk of rim seal fire.

When installing double rim seals, the height of the second, upper rim seal should be below the height of the foam dam to ensure coverage whenever rim seal pourers are activated.

Foam dams should be provided with intermittently spaced gaps to allow the drainage of rainwater. However, if the gaps are too large or too numerous, they may affect the ability to form a single continuous foam barrier in the event of a fire.

Overflow

Consideration should be given to providing suitable overflow systems to ensure that in the event of a tank overfilling, the tank contents are safely routed into suitable secondary containment. Additionally, the overflow route should be designed to minimize turbulent flow, reduce the surface area of flammable hydrocarbon, and reduce the generation of flammable vapors, such as through splashing. Particular care needs to be taken with strengthening rings and firewater dispersion rails around tanks which are specifically designed to maximize spread of fluids over a tank's surface.

For existing tanks, consideration should be given to modifications of tank top design and to the safe rerouting of overflowing liquids.

Figure 4. All important valves on atmospheric tanks should be labeled and their function indicated.

Valves

All important valves on atmospheric tanks should be labelled and their function indicated.

Fire-safe shut off valves should be provided, preferably automatic and remotely operated shut off/isolation valves (ROSOV/ROIV). They must be fitted close to the tank on both the inlet and outlet connections. Valves must either conform to an appropriate standard, such as BS EN ISO 10497 or equivalent international design, or should be of intrinsically fire-safe design. Such features include:

- Metal-to-metal seats (secondary metal seats on soft-seated valves are acceptable).
- Must not be of cast-iron construction.
- Should not be wafer-bolted (sometimes referred to as long bolt flanges).

In regions subject to a high risk of earthquake and/or rapid ground-acceleration, the isolation valves and actuators should also be secured directly to the tank wall and not to the ground. Anchoring the valve to surrounding structures or the ground could result in the valve separating from the tank during excessive vibrations.

Features of the ROIV should be:

- Fail safe (or if not fail safe have a back-up power supply, especially if the emergency plan requires the tank to be drawn down in an emergency).
- It should not be possible to auto reset the ROIV.
- Adequate margin of safety for shutting off the valve, with at least 150% torque available from fully open to fully closed.
- No manual operation or override (e.g., hand wheels) which may inhibit the operation of the ROIV.
- Suitable integrity and performance to satisfy the safety integrity level (SIL) requirements.
- Designed to minimize pressure surges on system pipework and couplings, particularly ship to shore flexible pipes.

Tank drainage valves should be blanked off when not in use. Whenever operations to remove accumulations of water from underneath the product are to be conducted, isolation of the drain should be achieved using two valves in series. The second valve can be a temporary installation.

Piping

In areas where earthquake is a significant exposure, all associated piping on storage tanks should have step geometry to allow flexing in the piping and prevent puncturing of the tank during an earthquake.

Suitably designed seismic hangers should be provided in earthquake zones, and pipe supports should have fire protection within fire hazard zones, e.g., within the bunded area.

Testing

Before filling tanks with flammable liquids, leak testing of the installed tank and associated pipework is required. Hydraulic testing should only be used (i.e. not pneumatic) as stored energy in hydrotesting is substantially lower, and inherently safer. However, air may be used as a means of applying pressure to waterfilled tanks and piping. Salt water should not be used to hydraulically test systems containing stainless steel.

Housekeeping

Debris within bunds should be kept at an absolute minimum. In addition to the checks within formalized inspection and maintenance routines, operators should also conduct frequent checks of tanks and their components as part of their routines to ensure that the tanks are kept in a reasonable condition for the complete report.

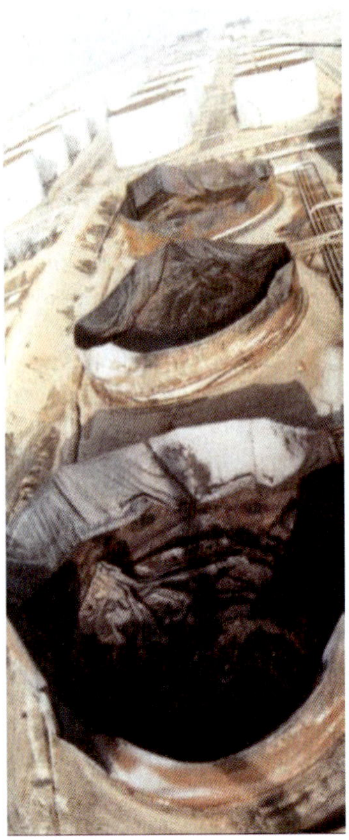

Figure 5. Double rim seals of fire-resistant construction are preferable to single seals on large floating roof tanks.

Secondary containment

Whilst priority should be given to preventing loss of primary containment, adequate secondary (and sometimes tertiary) containment is necessary for the protection of the environment and to contain any spillages. In atmospheric tank applications, secondary containment will be provided by bund walls.

Bund capacity should be sufficient to contain the largest predictable spillage. A bund capacity of 110% of the capacity of the largest storage vessel within the bund will normally be sufficient. When estimating bund capacity, the space occupied by other tanks should be considered.

Bund Integrity (Leak Tightness)

Bund wall and floor construction should be leak tight. This includes the provision of leak tight expansion joints between different casts of concrete in bund walls and wherever there are penetrations in the wall for pipes. Surfaces should be maintained crack free, and without any discontinuities, and without any failed joints that may allow liquid migration.

As observed in major incidents, such as the Buncefield fire of 2005, the joints in concrete bund walls are particularly susceptible to the effects of fire and/or subsequent cooling. Therefore, to maintain integrity, joints should be capable of resisting fire. Ideally, steel plates (waterstops) should be fitted across the inner surface of bund joints and fire sealants should be used to replace or augment non-fire-resistant materials. Similarly, joints to wall and floor penetrations need to be protected against the effects of fire.

Whenever designing protection plate, consideration should be given to avoiding weakening the wall structure in relation to resistance to fire, hydrostatic, and hydrodynamic forces.

Bund penetrations should be avoided unless alternative over-wall routings are not practical. On existing bunds, fitting steel collars or bellows to improve fire resistance at pipework penetrations may introduce local corrosion initiation sites in the pipework and is not recommended unless corrosion prevention can be assured. Where corrosion can not be prevented, joints should be improved by replacing existing sealants with fire-resistant sealants.

Bund floor penetration joints are inherently weak as failure of the integrity is difficult to predict and detect and may continue for some time unnoticed. Consequently, floor penetrations should not be incorporated into new bund designs. Existing bund floor penetrations should be eliminated wherever practicable. Where flexible sealants are used in floor penetration joints, these should be replaced with fire-resistant sealants.

There have also been several significant accidents resulting from leaks of hydrocarbons from tanks through the base of storage tanks. Therefore, the provision of a suitable impervious base (ideally concrete with a membrane liner) or floor leak detection on double floor configurations should be provided.

Tertiary Containment

Tertiary containment need only be concerned if there is a significant risk of secondary containment being insufficient to prevent an escalation to a major accident affecting personnel, assets, or the environment. Firewater containment is one such example. Implementation of tertiary containment should therefore be a risk-based decision. However, tertiary containment should be:

- Independent of secondary containment.
- Capable of fully containing foreseeable firewater and liquid pollutant volumes on the failure of secondary containment.
- Impermeable to water and foreseeable entrained pollutants/hydrocarbons.
- Of cellular configuration to allow segregation and limit the extent of spread of pollutants and/or fire.
- Robust under emergency conditions, e.g. loss of electrical supply.
- Capable of allowing the controlled movement of contained liquids under normal and emergency conditions.
- Capable of aiding the separation of water from pollutants (e.g., oil/water separator).
- Able to manage rainwater and surface waters.

More guidance on tertiary containment can be found in documentation including CIRIA 164 and PPG18.

Overview of petroleum storage tank standards

Philip Myers, P.E., Founder and Director, PEMY Consulting, Pleasant Hill, California, and **Brock Trotter**, E.I.T., PEMY Consulting, Oklahoma City, Oklahoma

Overview

To give you an idea of how broadly a term like "tank" can be used consider that an electrical engineer could be talking about a tank intending the meaning to be a capacitor for storing electrical charge. Another engineer may be talking about fuel tanks in an aircraft, while yet another may be talking about petroleum frac tanks. In essence, a tank means any form of storage and is therefore too general to communicate ideas for a particular industry. Therefore, we discuss tanks in the context of the oil, gas and chemicals businesses. Many of the ideas discussed here also apply to tanks applied in the potable water tank industry as well. Figure 1 shows the basic industry standards for "tanks" and what is meant by the word "tank" when one uses the term.

When liquid storage tank capacities become larger than a few thousand barrels (1 oil bbl = 42 US liquid gallons), stresses in the containment walls becomes an important factor governing the thickness of the tank containment walls. To optimize the use of steel which is the most common structural material for storage tanks, one of the most efficient ways to store liquids for this case is the so called "flat bottom, vertical cylindrical" storage tank. The flat bottom is fully supported by a ¼ inch thick steel membrane uniformly over the foundation. The cylinder provides an efficient structure to contain the hydrostatic pressure which increases as the liquid depth increases. Since most petroleum storage tanks operate near atmospheric pressure the roof only needs to handle minimal pressure loads. It turns out that the supported cone roof tank is one of the most cost-efficient ways to construct tanks. Typical steel cone roof tanks are 3/16 in thick steel. When higher pressures are needed then steel domes can be used. Figure 2 shows the basic components of vertical cylindrical tanks such as those constructed to API Standard 650.

In the "good old days" tanks were constructed with anything available and wooden tanks were used. However,

Tank standard	API 650	API 653	API 12F	API 510	API 620
intended meaning of "tank"	new welded shell atmospheric tank	used welded shell atmospheric tank	shop fabricated upstream tank, usually for crude oil	pressure vessel, tank storing high vapor pressure fluid	low-pressure storage tanks, usually greater than 300 ft in diameter

Figure 1. Oil and gas industry tank standards.

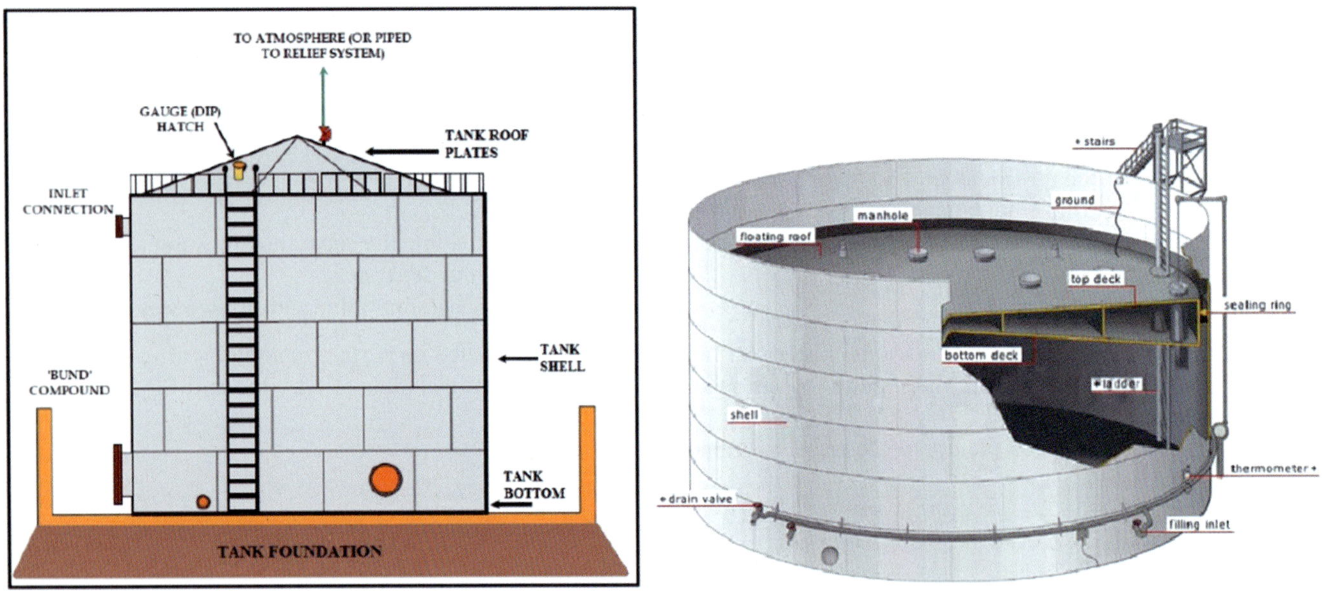

Figure 2. Basic components of an API 650 tank (atmospheric).

numerous and frequent tank fires pushed the industry to move to a better construction material – steel. Oil companies created the American Petroleum Institute (API) to work out the biggest problems the industry had with commerce such as defining common units such as the oil barrel, agreed upon methods for buying and selling oil, standardizing on oil drilling equipment. The natural result of the API was to produce standards that met the common and agreed upon minimum requirements for equipment such as tanks, pumps, piping and so on. Today, the API tank standards are the premier standards for constructing and maintaining petroleum and chemical storage tanks throughout the world.

Most petroleum storage tanks operate with low pressures of up to a few inches of water column. But the most widely used API Standard 650 covers tanks with internal design pressures ranging from a small partial vacuum to pressures as high as 2.5 psig. When higher pressures are needed then API Standard 620 covers pressures from atmospheric to 15 psig. Most cryogenic tanks used to store liquified gases as well as LNG are constructed to API 620.

One of the key factors in storage tank design is governed by the vapor pressure of the stored liquid. Lube oil has a very low vapor pressure and emits hardly any vapor. A simple container called a *fixed roof tank* or *FRT* can be used. A FRT is pictured in Figure 3 with more details than shown in Figure 1. The ullage or vapor space can be vented to the atmosphere to ensure that there is no differential pressure between it and the atmosphere since such differential over the large surface area can easily destroy a tank.

However, if the vapor pressure is higher than atmospheric pressure then evaporation becomes a pollution problem and fire safety hazard. Gasoline is a good example. While in the old days' gasoline was stored in fixed roof tanks, too many lightning induced fires and external ignition sources caused numerous fires. As a result, the floating roof tank was invented (Figure 4). This will be covered in more detail later. For now, a floating roof can be considered equivalent to a swimming pool cover. It is a barrier for evaporation and prevents the volatile organic compounds from escaping into the atmosphere.

Because petroleum storage tanks have three major but negative effects, they are heavily regulated. First, they are a fire hazard. Second, they result in groundwater contamination if they leak. It is worth noting that almost every terminal and refinery built before the 80s and 90s in North America is contaminated by leaking petroleum. Third, they evaporate and eject organic vapors into the atmosphere. The primary governing regulations are twofold. The SPCC regulation governs financial responsibility and oil pollution prevention related to navigable waters of the United States. On the air side, the Clean Air Act governs the hydrocarbon emissions that result from the required venting of fixed roof tanks and the leakage from floating roof tanks.

The oil industry has made great strides forward in the safe handling of oil and gas and improved the safety of these facilities and operations in recent decades. But natural hazards such as earthquakes, hurricanes, tsunamis and other disasters have shown that storage facilities are one of the major culprits in what are called NATECH events (Natural Hazard Triggered Technological Disasters). API is currently developing a publication that will attempt to address petroleum and chemical Natech events.

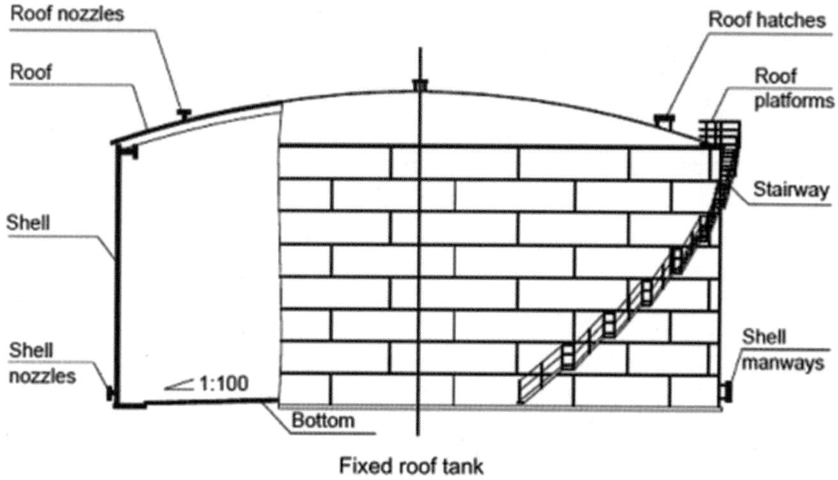

Figure 3. Basic components of a fixed roof tank.

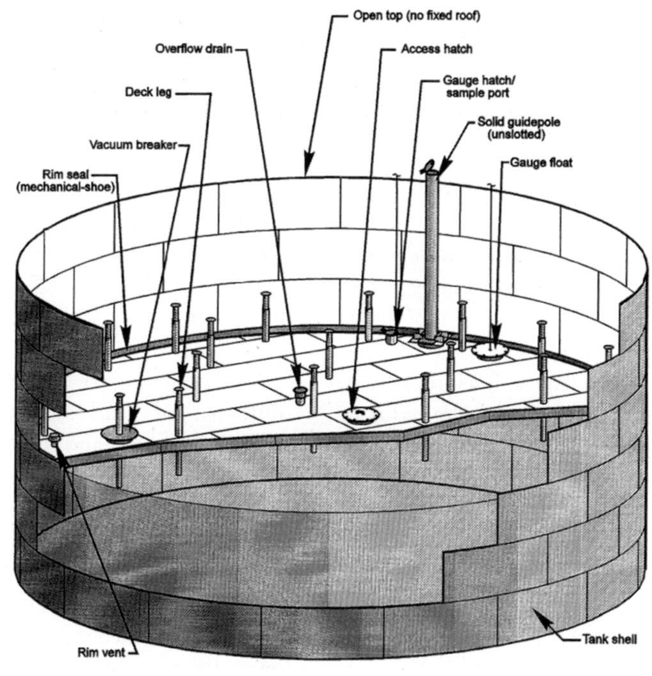

Figure 4. Basic components of a floating roof tank.

Petroleum storage tank facility standards

Figure 5 shows the most important industry standards related to tanks and pressure vessel and how they relate to one another. Naturally, the equipment is categorized according to equipment type as in the columns of the figure. But interestingly, they have been created in such a way that they self-organize themselves into publications on new equipment, existing equipment and supporting publications. The tank standards are shown in the middle of the diagram.

There are numerous standards that touch on storage tanks or their issues. Perhaps the most relevant is API 2610 below followed by summaries of some additional important tank standards.

API 2610

API 2610-Design, construction, operation, maintenance, and inspection of terminal and tank facilities traces its origins back to the US Senate Bill S.1761 of the 102nd Congress

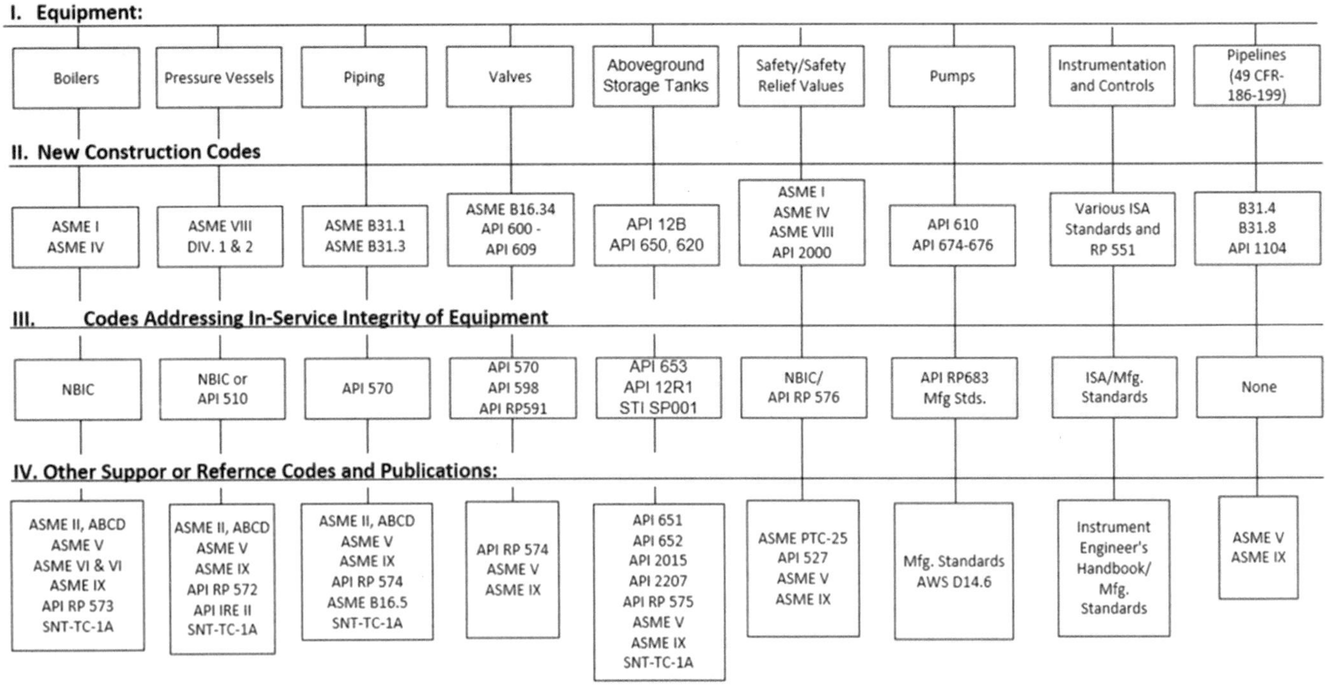

Figure 5. Self organizing nature of industry standards.

(1991-1992) promoted by Senator Daschle and a few others. The senate bill failed in becoming law, but it was aimed to amend the Solid Waste Disposal Act to regulate aboveground storage tanks used to store regulated substances. The amendment would have required the typical regulatory burden of financial responsibility, documentation of tank specific data, release detection, corrective actions, reporting, closure, upgrades, enforcement, etc. Of course, the industry was largely disagreeable to this bill and in response, API and the industry discussed what could be done to head off such universal regulations and requirements.

At that time there were several pivotal incidents that were in the public's recent memory:

1. The Ashland Oil Spill (Floreffe, Pennsylvania) occurred on January 2, 1988, spilling one million gallons of diesel oil into the Monongahela River in Pennsylvania, contaminating the potable water supplies of over a million people throughout Ohio, West Virginia and Pennsylvania. Senators Arlen Specter and John Heinz, both Republicans of Pennsylvania, and Senator Thomas A. Daschle, Democrat of South Dakota, also proposed new legislation to deal with oil storage tanks.
2. Shell Oil Martinez Refinery released 400,000 gallons of crude oil from a 12.5-million-gallon storage tank. The oil flowed into the Carquinez Strait which connects San Pablo and Suison Bays of the San Francisco Bay and Sacramento Delta. Large areas of wetlands were destroyed, and the largest fines ever (up to this time) were levied.

These two incidents were specially related to petroleum storage tanks and probably had more to do with the development of the SPCC regulations than anything else. But there were other related events on the minds of industry and regulators such as the 11-million-gallon Exxon Valdez oil tanker spill of March 23, 1989, the 1990 Mega Borg spilling 5.1 million gallons of oil in the Gulf of Mexico, 57 miles south-southeast of Galveston, Texas, following a pump-room explosion and fire and the 1993 Bouchard B155, Balsa 37, and Ocean 255 colliding in Tampa Bay, causing the Bouchard to spill 336,000 gallons of No. 6 fuel oil.

As a result of these perceived threats to the industry API set about to create API 2610 by conducting a review of approximately 60 standards, identifying at least 18 needing revision and development of several new standards. It was beyond the capability of most terminal owners/operators to collect, assess, and apply the hundreds of recommended practices, specifications, publications, and procedures into a single, comprehensive management guide to ensure safe, reliable and cost-effective measures to avert potential tank incidents and accidents. But API Standard 2610 did this through an enormous, focused effort which truly represents the accumulated knowledge of the related segments of industry and everything that is known about accepted best practices for storage tank facilities. This makes API 2610 a valuable component to the available API tank and facility standards. It also can serve as a training component for anyone who will operate or maintain petroleum storage tanks.

API 650

As mentioned, the most widely used standard in the world for petroleum storage and chemical tanks is standard API 650-Welded tanks for oil storage (the 12 edition is most current). It governs the construction materials, fabrication, welding, and the design of the field erected tanks. API 650 covers the envelope of the tank up to the first flange, threaded or welded connection but not beyond. It does cover the internal piping, however. API 650 is applicable as a procurement standard, but as soon as the tank is hydrostatically tested and commissioned for service, it no longer applies. The ongoing service integrity becomes the domain of API 650.

One of the important aspects of tanks built to API 650 is the credibility of the entire tank manufacturing and erection process. This is in large part driven by API 650 Section 10 – Marking. This section basically provides a certification process that the tank not only meets API 650 but the quality control measures in API 650 have been followed, all tests met, and who is responsible for certification. In addition, the Manufacturer's Nameplate specifies basic design and operation criteria that are engraved onto the nameplate and which serves the indicate that all the requirements of API 650 have been met. An API 650 nameplate must not be used if the tank does not meet all the API 650 requirements. It is surprising how useful the nameplate is when engineering analyses such as fitness for service must be done but there are no written documents (a common occurrence). From the nameplate critical summary information about the tank is available.

The Annexes in API 650 play an important role (Figure 6). When an Annex is used it overlays the requirements on top of the basic main body requirements. Here are a few key features of the API 650 Annexes:

- Some Annexes are optional but must be followed under certain conditions. A good example if Annex E. Although it is designated as normative, other methods may be used and there is no requirement that it specifically must be used. Another example is Annex G for structurally supported aluminum dome roof tanks. This annex would not apply to an open top tank or steel cone roof tank but would apply to a tank with an aluminum dome cover. Other normative annexes such as Annex F for tanks with a small internal pressure are normative only when the pressure in the tank is sufficiently high to trigger the requirement. This type of normative annex is automatically triggered.
- Annexes are either "normative" meaning mandatory or "informative" meaning that there are no compliance requirements to meet API 650. For example, if a shop-built tank is needed then Annex J can be specified. While the main body of API 650 covers field erected

Annex	Addresses
A	Optional Design Basis for Small Tanks
AL	Aluminum Storage Tanks
B	Recommendations for Design and Construction of Foundations for Aboveground Storage Tanks
C	External Floating Roofs
D	Inquiries and Suggestions for Change
E	Seismic Design of Storage Tanks
EC	Commentary on Annex E
F	Design of Tanks for Small Internal Pressures
G	Structurally-Supported Aluminum Dome Roofs
H	Internal Floating Roofs
I	Undertank Leak Detection and Subgrade Protection
J	Shop-Assembled Storage Tanks
K	Sample Applications of the Variable-Design-Point Method to Determine Shell-Plate
L	API Standard 650 Storage Tank Data Sheet
M	Requirements for Tanks Operating at Elevated
N	Use of New Materials That Are Not Identified
O	Underbottom Connections
P	Allowable External Loads on Tank Shell Openings
R	References for Tanks in Non-petroleum Product Service
S	Austenitic Stainless Steel Storage Tanks
SC	Stainless and Carbon Steel Mixed Materials Storage Tanks
T	NDE Requirements Summary
U	Ultrasonic Examination in Lieu of Radiography
V	Design of Storage Tanks for External Pressure
W	Commercial and Documentation Recommendations
X	Duplex Stainless Steel Storage Tanks
Y	API Monogram

Figure 6. API 650 Annexes.

tanks Annex J provides all the differences in fabrication, testing, and materials needed to ensure a quality shop-built tank. It is therefore normative when invoked. However, Annex B is for tank foundations. It has highly valuable information about tank foundations and many of these principles must be applied to prevent foundation induced problems. But since the standard only covers the tank envelope there are no mandatory requirements for foundations. Therefore, it is up to the Purchaser to ensure that they carefully spell out the requirements desired in the procurement documents.

- Annexes only state what is different, not applicable or an addition the requirements of the main body. For example, if using Annex S to construct a stainless-steel tank, the material that can be used as well as their allowable stresses are different and thus these are called out, specifically in the Annex. But all the rules applicable to constructing the cones of a stainless-steel cone roof tank are not given because they are already covered by the rules given in the main body (except for materials).

See Figure 6 for as summary of the contents of API 650 and its Annexes.

API 650 uses a system of bullets to indicate when a specific user input is required. Annex L called the *data sheet* captures the Purchaser's specifications and requirements for the tanks and in conjunction with the bulleting ensures that the specification for the tank is completely defined. All too often Purchasers fail to put in the effort to specify the tank completely or ask the manufacturer to do it for them and have unexpected and disappointing results.

Tank Design

API 650 specifies what criteria must be used to establish the design basis of the tank. Some of the variable associated with design are:

- Design metal temperature
- Dead load
- Internal design pressure
- External design pressure (vacuum)
- Design liquid level
- Specific gravity of product
- Roof loads
- Load combinations

Materials

The main body of API 650 deals with the construction of steel tanks which in the structural analysis domain are called *plate structures*. Much specificity is therefore applied to the selection and use of steel plate and other components used in tanks. Section 4 for materials, API 650 limits the materials to a specific list of carbon steels and these are further subdivided into Group I, IIA, III, IIIA, IV, IVA, V, VI and VIA which establishes the minimum permissible design metal temperature for tank shells that do not required impact testing.

Although there are different steels such as carbon steels, alloy steels, stainless steels and tool steels API tanks are mostly constructed from carbon steel plate. Carbon is the principal element added to iron to create "carbon steel". In general, as the carbon content increases so does the hardness and tensile strength. A mild carbon steel has a range of 0.04% to 0.3% by weight. Increasing the carbon range is 0.31% to 0.6% and manganese content of 0.06% to 1.65% increases strength. The addition of both carbon and manganese provide for greater ductility and improved toughness as compared to steels hardened by carbon only.

The control over chemistry and the metallurgical properties of steel is a complex process which has led to greatly improved steels of the last 50 years. Fortunately, many of the important steel chemistry related issues related to the use of the various steels in tanks has been addressed by the list of allowed materials given in API 650 Section 4 – Materials. The first limit on materials is list of ASTM, CSA, ISO and EN or other recognized national standard specifications for acceptable materials. Further restrictions are placed on the material related to heat treatment of plates. Impact testing requirements further restrict the use of steel plate and similar components such as reinforcing pads using Figure 4.1b, see Figure 7.

An important parameter to be established for the tank is the Design Metal Temperature (DMT). This is the lowest temperature considered in the design. This is defined as the lowest one-day mean temperature (LODMT) given in Figure 4.2 (Figure 8) plus 15F. The 15F assumes that the liquid temperature lags the LODMT because of the thermal mass of the liquid which cannot cool as fast as the ambient air temperature does. The procedure is to find the location of the tank in Figure 4.2, interpolating between isotherms or contours to establish the LODMT and then adding 15F.

One the DMT is established then Figure 4.1 can be applied to determine requirements for impact toughness testing. Based on the point located by thickness on the abscissa and design metal temperature (DMT) on the ordinate, impact toughness requirements will apply if the point is below the line representing the specific material and group.

Tank Shell

In determining thicknesses for the shell courses the maximum allowable stress for the product which is based on the specific gravity of the product and the hydrostatic pressure of the tank which depends on maximum liquid level

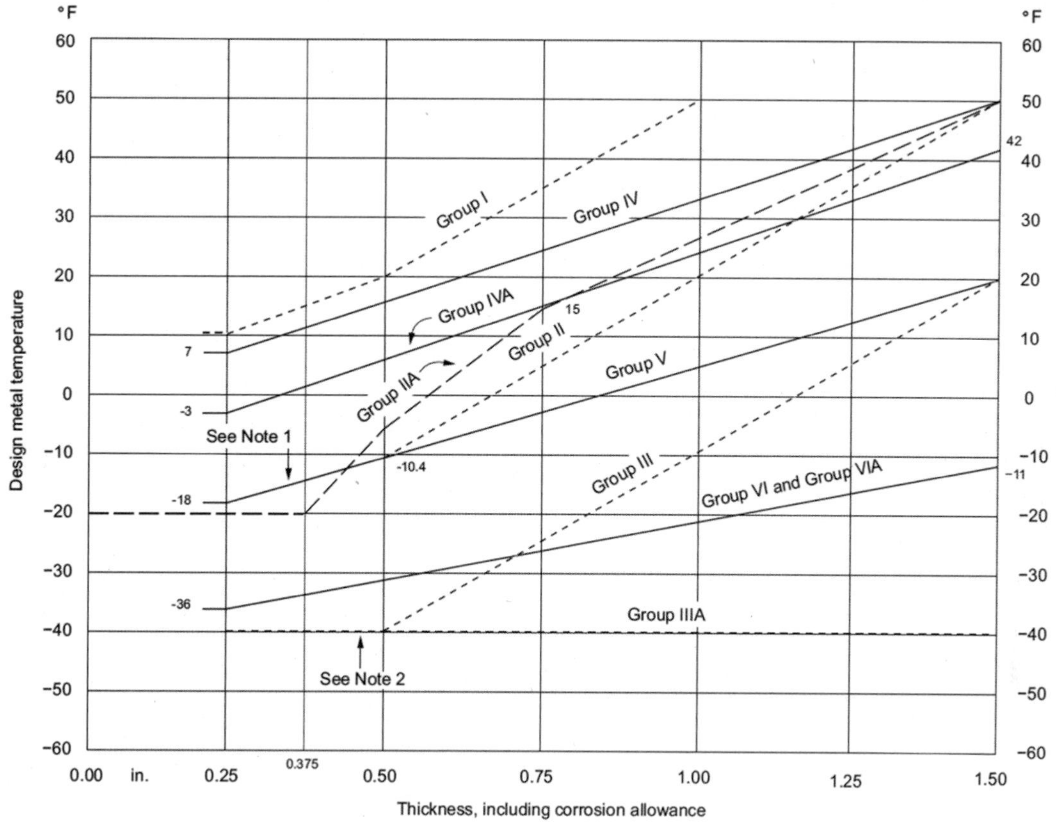

NOTE 1 The Group II and Group V lines coincide at thicknesses less than 1/2 in.

NOTE 2 The Group III and Group IIIA lines coincide at thicknesses less than 1/2 in.

NOTE 3 The materials in each group are listed in Table 4.4a and Table 4.4b.

NOTE 4 **Deleted.**

NOTE 5 Use the Group IIA and Group VIA curves for pipe and flanges (see 4.5.4.2 and 4.5.4.3).

NOTE 6 Linear equations provided in Table 4.3b can be used to calculate Design Metal Temperature (DMT) for each API material group and the thickness range.

Figure 7. API 650 Figure 4.1b.

is taken as 2/3 yield strength or 2/5 of the tensile strength whichever is less. In addition, because water for the hydrostatic test will be used to test the tank, the hydrostatic test allowable stresses are somewhat higher at 3/4 of the yield strength or 3/7 of the tensile strength whichever is less. API 650 provides tables of design and hydrostatic allowable stresses for the acceptable materials.

The shell plate thicknesses are computed according to one of two methods.

The first method is called the *one-foot method*. The thickness required for shell courses is the greater of the two thicknesses computed by:

$$t_c = \frac{2.6D(H-1)G}{S_d} + CA$$

$$t_t = \frac{2.6D(H-1)}{S_t}$$

where t_d is the design shell thickness, t_t is the hydrostatic test shell thickness, D is the tank diameter, H the design liquid level, G the product specific gravity, CA corrosion allowance, S_d the allowable stress for design and S_t the allowable stress for hydrostatic testing.

The second method is the *variable design point* (VDP) method. Details are given in API 650. It is more accurate than the one-foot method and often results in slightly lower thicknesses. The theory is to find the actual point in the shell course where the hoop stress is the highest and based the thickness on this location whereas the one-foot method assumes that the highest hoop stress is located one foot above the bottom of the course being considered.

The VDP method may only be used when the Purchaser has not specified the one-foot method and:

- when $L/H \leq 2$,

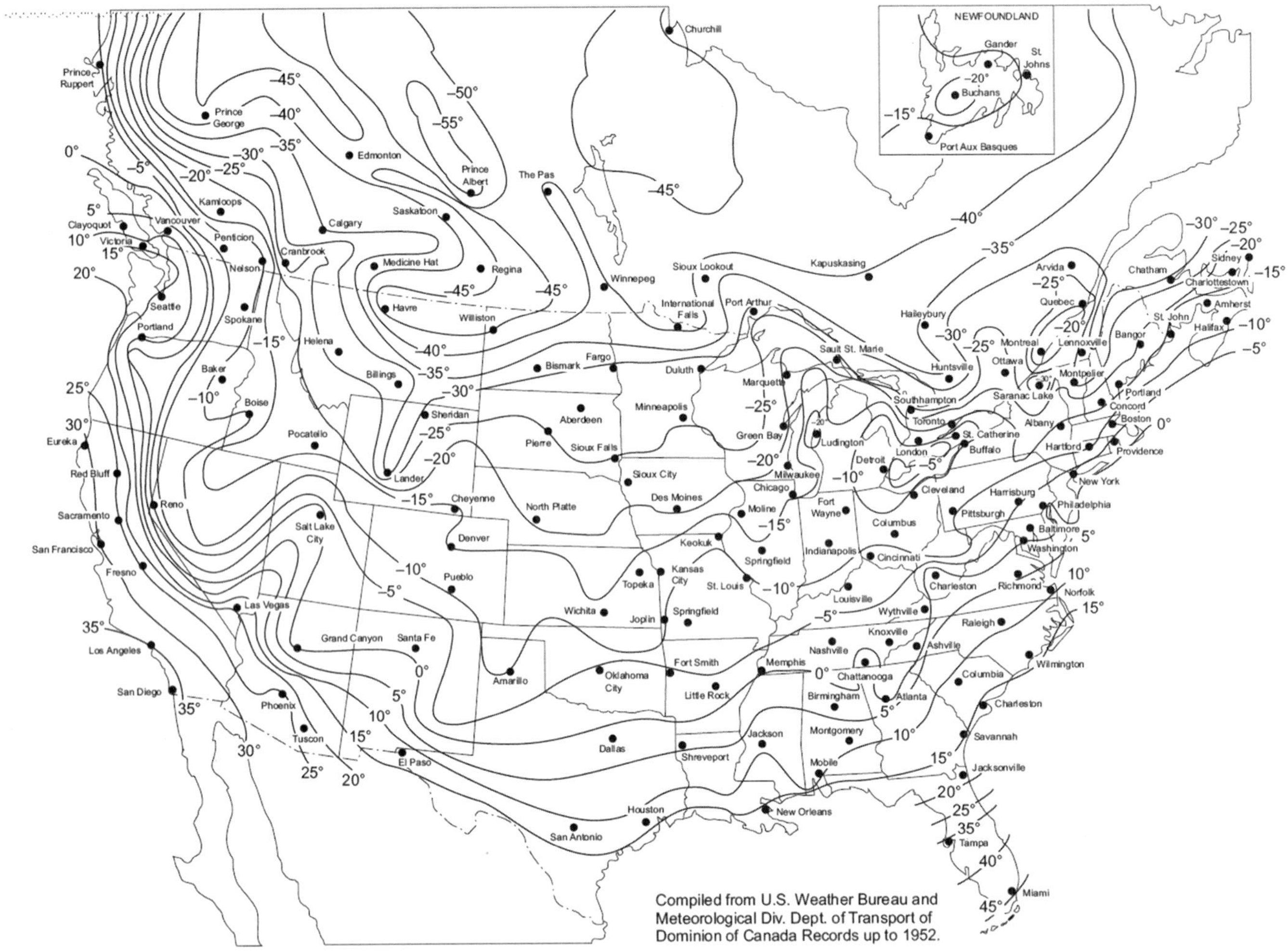

NOTE °C = (°F – 32)/1.8

Figure 8. The lowest one-day mean temperature isotherms (API 650 Figure 4.2).

where $L = (6Dt)^{0.5}$, D diameter in feet, t is the bottom course corroded shell thickness in inches and H is the maximum design liquid level in feet.

Shell

Most tanks are built with 8 foot wide by 40-foot-long plates and use six courses resulting in a height of 48 feet. But other widths and heights are used if they are greater than 6 ft width. Because the hydrostatic pressure is decreasing as the height above ground increases the courses can be progressively thinner. The minimum thickness of plates is governed by the greater of a constructability minimum thickness or the required thickness to meet allowable stresses coved by design (see above). The minimum constructability thicknesses are based on diameter as shown in Figure 9. Tanks shell plates must also be shaped meaning rolled to the curvature of the tank prior to installation based on diameter and thickness.

Nominal Tank Diameter (ft)	Nominal Plate Thickness (in)
< 50	3/16
50 to < 120	1/4
120 to 200	5/16
> 200	3/8

Figure 9. API 650 tank diameter to shell plate thickness.

Tank Bottom

Although the tank bottom has the maximum hydrostatic pressure acting on it, interestingly, it experiences very little stress with some exceptions discussed in a moment. Because the bottom is flat on the foundation, the hydrostatic pressure is almost always less than 20 psi, and this is a negligeable stress on the steel plates which have allowable stresses of at least 20 ksi. Indeed, some tanks built to other standards such as AWWA have no bottom plate in the sense that the tank has a shell which is directly bolted to a concrete foundation slab. Alternatively, some tanks have used plastic liners for the tank bottom.

But API 650 tanks require at least a 6mm or ¼ inch thick bottom where the thickness is primarily for the purpose of providing time before corrosion can produce a hole and cause a leak. Bottom plates are typically rectangular lap welded plates configured as in Figures 10-12. Note that almost all tanks use 6mm or ¼ in thick bottoms. But there are exceptions. People use thicker plates occasionally to increase the lift and protect against corrosion holes. In some commercial

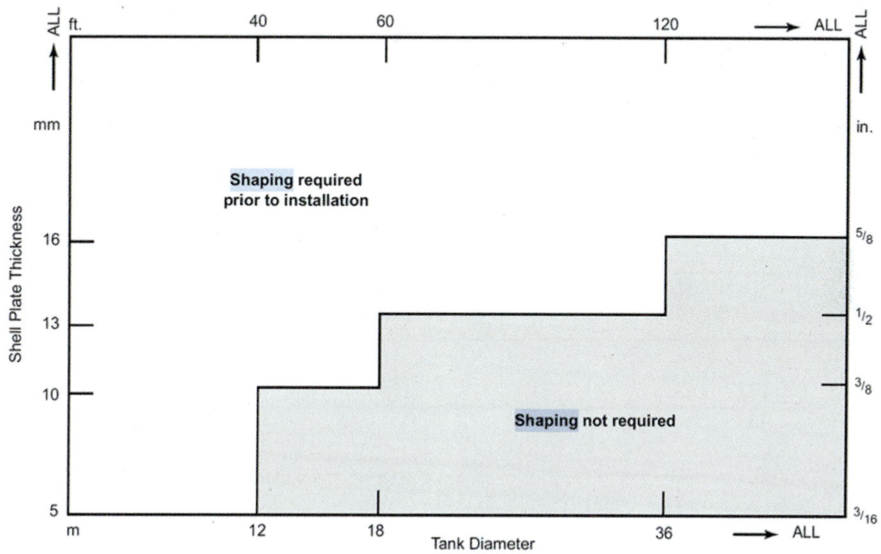

NOTE Any combination of diameter and thickness falling on or above the solid line requires shaping prior to installation.

Figure 10. API 650 Figure 6.1 - Shaping of Plates.

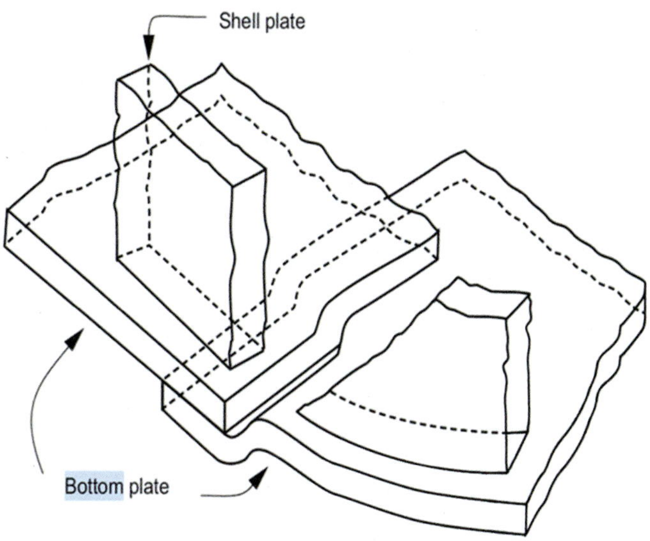

Figure 11. API 650 Figure 5.3b.

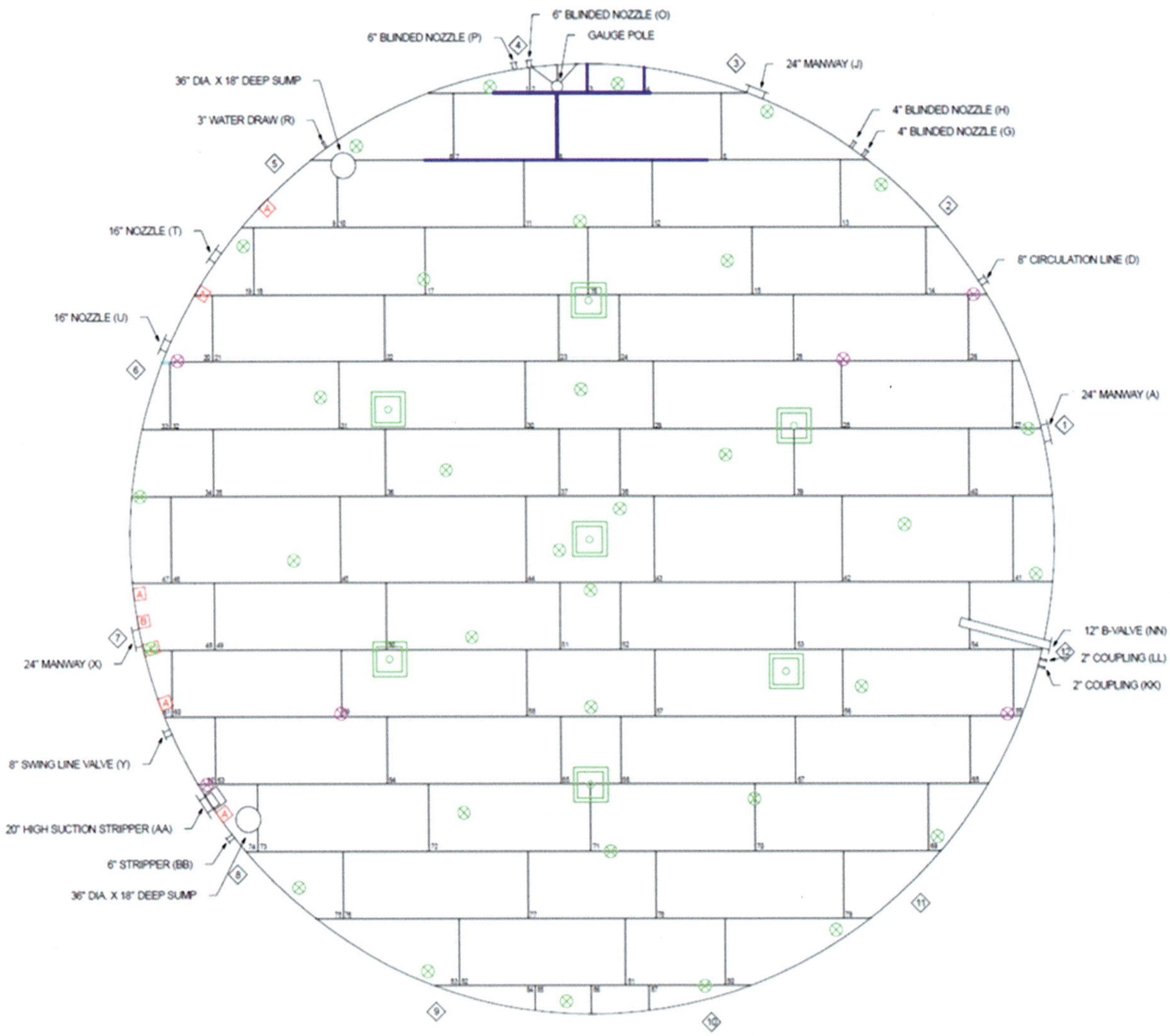

Figure 12. Tank bottom plate layout.

sulfuric acid storage tanks, it is common to use bottoms are ½ in thick. In Japan, the regulations requires that all tanks have ½ in thick bottoms.

For a tank using all lap welded sketch plates it can easily see that there are potential problems. Where the bottom plates must be bent to fit under the shell both in the radial and circumferential direction it is difficult to get good contact and obtaining high quality welds is difficult let alone the impossibility of getting complete penetration joints. In addition to many welding defects that are inevitable in this region the stresses are very high due to the bending forces acting from the shell into the bottom and chime area. "Chime" or "chine" is slang for the bottom extension beyond the shell which is typically 1 − 2 inches.

API 650 provide requirements for distances between three plate laps (where three plates meet), bottom to annular plate joints and the critical zone which is the bottom or annular plate for a distance of 3 inches from the inside of the shell to the center. This is one of the most highly stressed regions of the tank bottom and therefore welding and patching must be controlled so that failure does not occur.

A better but more costly alternative to sketch plates is annular plates. Annular plates are butt welded and required complete penetration and fusion welds. This is accomplished by using a backing strip as shown in Figure 13. The projection of the annular plate beyond the exterior of the shell must be a minimum of 2 inches.

Annular plates are required when the bottom course is designed using allowable stress for materials in groups IV and higher and the maximum first course shell stress is higher than 23,200 psi or the hydrostatic test stress is greater than 24,900 psi. Annular plates for this case must be at least 24 inches wide but may need to be larger based on formulas in API 650.

An annular plate is also required when wind overturning requires a thickness under the shell which is thicker than the

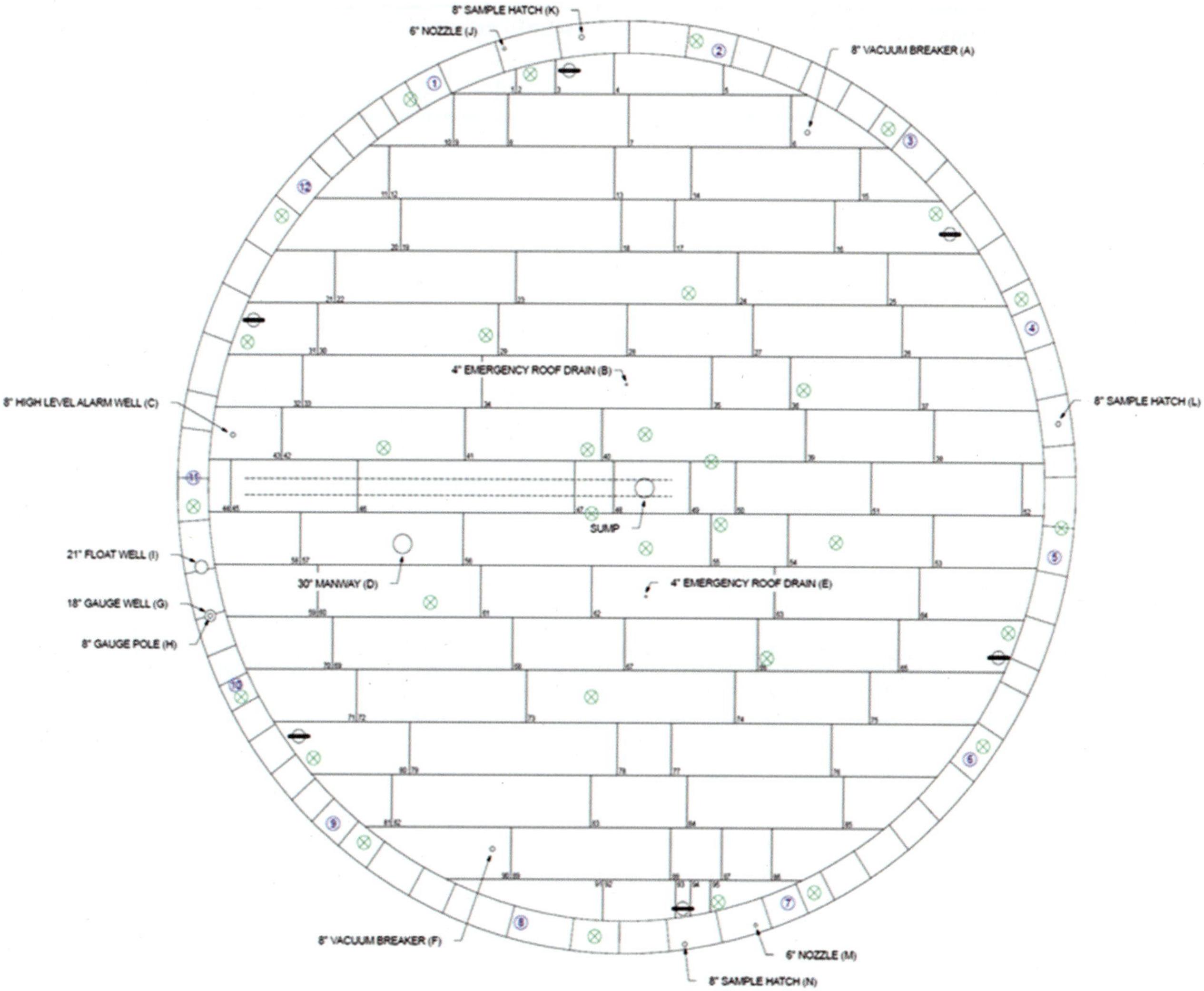

Figure 13. Tank bottom with annular plate.

nominal thickness of the bottom. An annular ring must be at least 18 inches wide. Annular plates may be required for resistance to seismic overturning and formulas are given in API 650 Annex E. See Figure 14 (API 650 – Table 5.1a – Annular Bottom-Plate Thickness) for additional details.

Tank Roof

In the early days of the petroleum industry, crude oil was sometimes stored in drained or manmade ponds, exposed to the air and environment. Fire were frequent as flammable petroleum vapors naturally travelled outside the pond and reached an ignition source. People then began storing crude oil and other petroleum fluids in enclosed storage tank, but different fluids require different roofs to effective transport and manage them.

Fixed Roof Tank

Fixed roof storage (FRT) tanks are used to store low vapor pressure fluids such as diesel or water. Low vapor pressure fluids will not expand or contract considerably within normal atmospheric conditions, and flammable vapors will not reach the lower explosive limit (LEL) within normal atmospheric conditions. Because the pressure in fixed roof tanks is low and fairly constant the volume of the tank can remain constant. See Figure 3 for an example of an FRT.

External Floating Roof Tank

External floating roof tanks (EFRTs) are used to store fluids with a moderate vapor pressure under normal atmospheric conditions, see Figure 15. For example, the vapor pressure of

Plate Thickness[a] of First Shell Course (mm)	Stress[b] in First Shell Course (MPa)			
	≤ 190	≤ 210	≤ 220	≤ 250
$t \leq 19$	6	6	7	9
$19 < t \leq 25$	6	7	10	11
$25 < t \leq 32$	6	9	12	14
$32 < t \leq 40$	8	11	14	17
$40 < t \leq 45$	9	13	16	19

[a] Plate thickness refers to the corroded shell plate thickness for product design and nominal thickness for hydrostatic test design.

[b] The stress to be used is the maximum stress in the first shell course (greater of product or hydrostatic test stress). The stress may be determined using the required thickness divided by the thickness from "a" then multiplied by the applicable allowable stress:

Product Stress = $((t_d - CA)/\text{ corroded } t)\ (S_d)$

Hydrostatic Test Stress = $(t_t / \text{ nominal } t)\ (S_t)$

NOTE The thicknesses specified in the table, as well as the width specified in 5.5.2, are based on the foundation providing uniform support under the full width of the annular plate. Unless the foundation is properly compacted, particularly at the inside of a concrete ringwall, settlement will produce additional stresses in the annular plate.

Figure 14. API 650 – Table 5.1a – Annular Bottom-Plate Thickness.

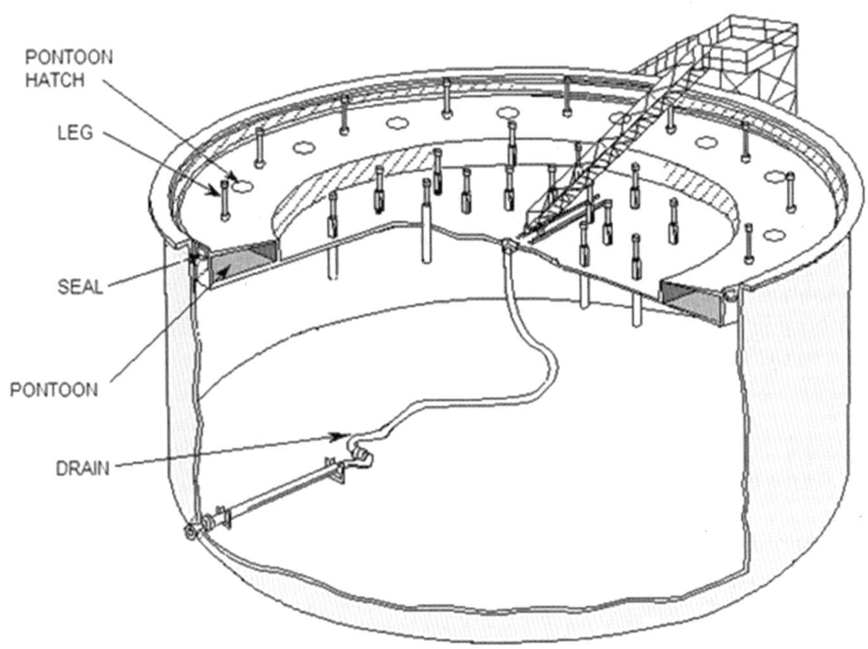

External Floating Roof Tank, single skin roof with pontoons

Figure 15. External floating roof tank diagram.

gasoline exceeds atmospheric pressure on hot days, and gasoline would rapidly vaporize in a fixed roof storage tank losing money and increasing the risk of an explosion. Instead, external floating roofs will adjust position with the liquid level in the tank and the vapor pressure of the fluid stored. Because the roof moves with the liquid surface, the pressure within the vapor space of the tank is maintained, preventing the vaporization or loss of the stored fluid.

External floating roof tanks are less efficient than internal floating roof tanks because wind and other environmental factors. External floating roof tanks must relieve the water on the roof and have drainage sufficient to prevent 10 inches of standing water for the worst rain event to occur in a 100-year period (1-hr rainfall, MRI=100).

Internal Floating Roof Tank

Internal floating roof tanks (IFRTs) are also used to store fluids with a moderate vapor pressure under normal atmospheric conditions. Internal floating roofs adjust position with the liquid level in the tank, just like the EFRT, but the IFRT has an additional fixed roof above the floating roof to prevent wind and rain from affecting the internal floating roof, see Figure 16.

Because the roof moves with the liquid surface, the pressure within the vapor space of the tank is maintained, preventing the vaporization or loss of the stored fluid. Internal floating roof tanks are more efficient than external floating roof tanks because wind and other environmental factors.

Release Prevention Barriers

One of the most important environmental initiatives in the API tanks standards is the use of an RPB or release prevention barrier. This can take different forms as illustrated by Annex I of API 650. Basically, these systems function as two bottom membranes where the space between the membranes acts as monitoring component. If a leak develops it will be detected. Functionally, there are many ways to constructed RPBs. For example, a double steel bottom with a spacer of some sort can be used. This would typically not be done for a new tank since installation of a simple plastic membrane (usually some elastomeric material such as HDPE) under the new bottom and integrated into the tank foundation could be used at far lower costs than installing two steel bottoms.

However, for existing tanks, when the bottom has reached the end of its useful life, it is often the lowest cost option to use a double steel bottom tank. Why? The old bottom can be left in place and there will typically not be hazardous soil waste removal and associated costs. The old bottom serves as a form to contain the space material which can be sand, concrete or other materials.

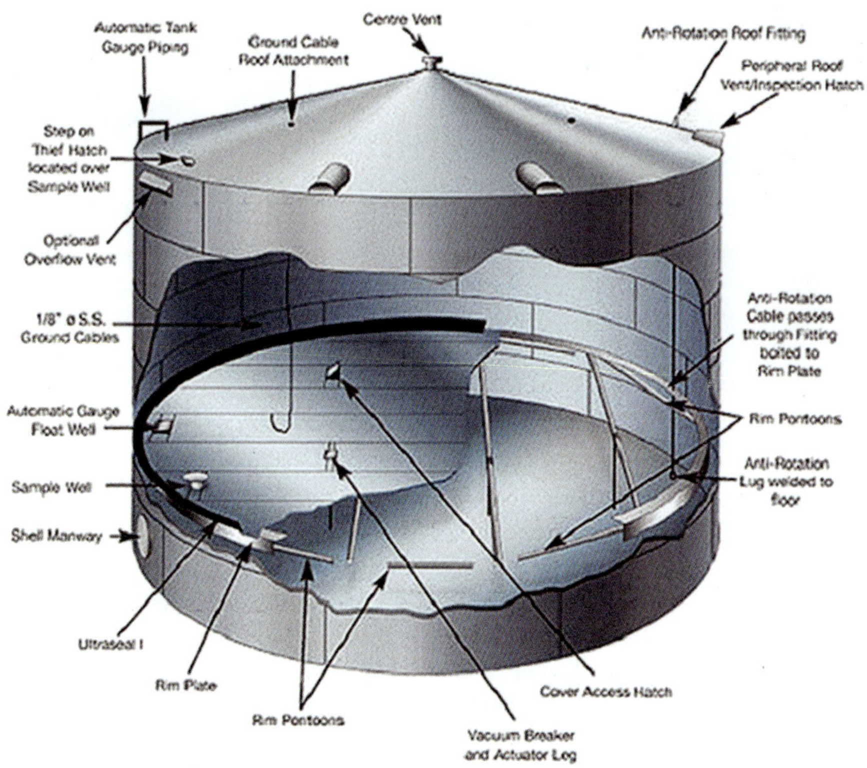

Figure 16. Internal floating roof tank diagram.

Plain Hemispheroid, This is a low pressure type tank with internal pressures greater than 2.5 psig.

Noded Spheroidal Tank, a spherical roof with a knuckle radius or smooth transition at the intersection of the shell and top head.

Figure 17. Hemispheroid and noded spheroidal tank.

Figure 18. Noded spheroidal tank.

API 620

API 620—Design and construction of large welded low pressure storage tanks covers tanks that have internal pressures that range from approximately atmospheric to 15 psig. Many petroleum and organic liquids have higher vapor pressures that cannot be stored in API 650 tanks and API 620 provides the storage capabilities for them. A 620 tank has a shape is generated by a single vertical axis of revolution and so it covers tanks shaped like vertical cylindrical tanks such as API 650 but also covers other design such as tanks with dished bottom heads or spherical tanks or nodded spheroid tanks.

The shell design for API 620 tanks is based on membrane theory where force may act in two directions. The basic equation of statics is

$$\frac{\sigma_1}{r_1}+\frac{\sigma_2}{r_2}=\frac{p}{t}$$

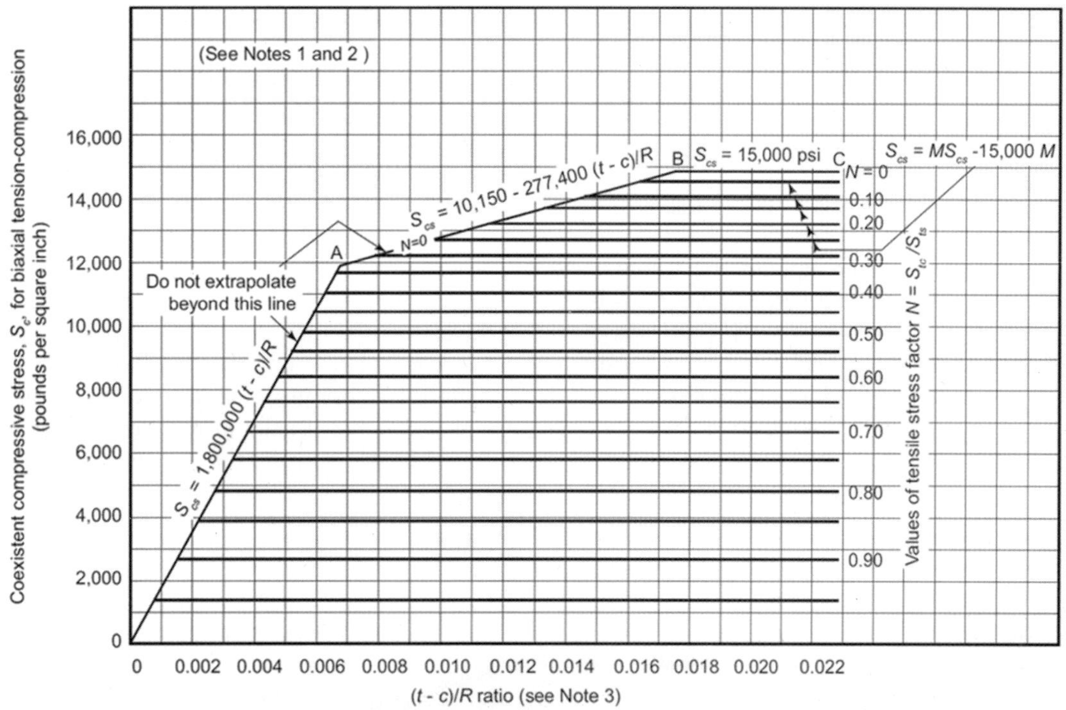

NOTE 1 At no time can a compressive stress for a particular value of $(t - c)/R$ exceed S_{cs} represented by curve 0ABC, no values of compressive stress or N are permitted to fall to the left or above this curve.

NOTE 2 See Figure F-1 for relationship between factors M and N.

NOTE 3 If compressive stress is latitudinal, use $R = R_1$; if compressive stress is meridional, use $R = R_2$.

Figure 19. Coexistent compressive stress for biaxial tension-compression.

Tank Safeguard	Add to Initial Interval
i. Fiberglass-reinforced lining of the product-side of the tank bottom installed per API RP 652.	5 yrs
i Installation of an internal thin-film coating as installed per API RP 652.	2 yrs
iii. Cathodic protection of the soil-side of the tank bottom installed, maintained, and inspected per API RP 651.	5 yrs
iv. Release prevention barrier installed per API Std 650, Annex I.	10 yrs
v. Bottom corrosion allowance greater than 0.150 in.	(Actual corrosion allowance -150 mils)/corrosion rate*
vi. Bottom constructed from stainless steel material that meets requirements of API 650, Annex SC, and either Annex S or Annex X; and internal and external environments have been determined by a qualified corrosion specialist to present very low risk of cracking or corrosion failure.	10 yrs
* Corrosion rate to be 15 mpy, or as determined from Appendix H, Similar Service	

Figure 20. API 650 Table 6.1 – Tank Safeguard.

where:

σ_1 – tensile stress in meridian direction
σ_2 – tensile stress in hoop or axis of rotation direction
t – thickness of shell
r_1– radius of curvature in direction 1
r_2– radius of curvature in direction 2
p – pressure

A vertical cylindrical tank is the simplest tank where one of the radii is infinite (the longitudinal axis) and thus the formula becomes the hoop stress formula used for API 650 tanks (with slight modifications). Another simple shape is the sphere were r1 and r2 are equal and the formula for stress becomes

$$\sigma = \frac{pr}{2t} \text{ since } r_1 = r_2.$$

API 620 allowable stress formulas are more complex that for API 650 tanks because of the biaxial stress conditions that are possible. The meridianal force T1 and the latitudinal force T2 may be both tensile or tensile and compressive. If both tensile then the allowable stresses listed in Table 5-1 apply. If there are coexistent compressive and tensile stresses, then the biaxial stress chart in API 620 Fig 5-1 is given. Compressive stresses have additional limitations given in the standard.

The main allowable stress table for membranes in tension (API 620 Table 5-1) is based on an allowable stress which is the least of 60% of the specified minimum yield point or 30% of the specified minimum ultimate tensile strength and when other limitations do not apply. Contrast this to API 650 where the design stress for shell plates is the least of 2/3 yield (67%) or 2/5 (40%) tensile strength. This is one reason why most vertical cylindrical storage tanks are constructed to API 650, even though it is entirely acceptable to use API 620 for this design.

API 653

Perhaps the most widely used tank standard in the world today is API 653 – Inspection of Welded Storage Tanks (currently in the 5th edition). API 653 addresses the mechanical integrity of a steel storage tanks primarily through inspection, but it also covers ensuring that the materials are sufficiently tough so that brittle failure does not occur. Note that the Ashland Oil Incident described earlier was a brittle failure in the steel shall of the tank. In addition, it covers hydrostatic testing for in-service tanks and conditions including metal thinning from corrosion, settlement, distortions of the shell or roof, and many others. However, it does not address the risks arising from operations such as tank cleaning and gas freeing, overfilling or other operational scenarios which we discuss later.

API 653 uses three types of inspections:

1. Routine informal monthly inspection
2. Formal external inspection (FEI)
3. Formal internal inspection (FII)

Routine Informal Monthly Inspection

The inspection can be done by anyone, but it is best to be done by the operator who sees the tank frequently because in this type of inspection a change is the best indicator of a potential problem. For example, new leaks from mixer shafts, a new patch of weeds growing near the perimeter, severe deterioration of tank components by corrosion, or an observation that the floating roof is partially submerged in oil are all important indicators of potentially serious threats to the tank or its components. The inspection is informal and is not required to be documented although many operators and some regulatory jurisdictions require proof that the inspection was done, usually in the form of a checklist which is signed by whoever did the inspection.

Formal External Inspection

For the remaining two formal inspections, they must be done by a certified API 653 inspector. The FEI is done without removing the tank from service and must be done at least every 5 years. API 653 does not prescribe how the inspection is to be done but most companies use several tools such as ultrasonic thickness testing, magnetic flux exclusion and other non-destructive examination (NDE) methods to provide as much data as possible about the tank. Most inspectors and inspection companies also typically provide a photographic log of what the tank looks like and take photos of all components of interest. It should be clear that if the tank is operating, then the primary component which cannot be fully examined is the tank bottom, and hence the need for the Formal Internal Inspection.

Formal Internal Inspection

The FII requires that the tank be cleaned and gas free so that it is safe for inspection personnel to safely enter the tank which is a confined space. API 653 states that there are 3 reasons for the FII:

1. Ensure the bottom is not severely corroded and leaking.
2. Gather the data necessary for the minimum bottom and shell thickness assessments.
3. Identify and evaluate any tank bottom settlement.

Because an FII usually requires a costly cleaning effort, a business interruption, and generates large amounts of air pollution and hazardous wastes, an important goal is to maximize the internal inspection interval or in-service run period between FIIs.

In many, if not most cases, the length of the operating service interval is governed by the action of the corrosion rate and the requirement to have a minimum remaining thickness (MRT) on the tank bottom. An FII may be triggered by other problems such as failure of the floating roof seals, belief that there is a potential threat to tank integrity due to shell thinning, or other causes.

Since the greatest number of FIIs are triggered by the tank bottom corrosion it is important to understand the basis of how the FII interval is set. Almost all the nationally recognized recommended practices, guidelines and standards set the maximum interval allowed between the time the tank is inspected to the time that it must again be taken out of service for the next inspection based on tank bottoms as follows:

1. Maximum FII intervals specified by API 653.
2. Adjustments to the maximum interval based on APIs tank corrosion rate algorithm.
3. Minimum remaining thickness (MRT).

The default FII interval for a new tank is 10 years meaning that it cannot operate longer than 10 years before having its first FII. However, API has applied a deterministic risk management approach to this short interval by giving time credits for the following safeguards.

From this table the maximum FII interval is 30 years but only if the tank has a release prevention barrier (RPB). If not, the limit is capped at 20 years which is when most tanks are scheduled in an API 653 tank program to be taken out of service. In general API allows 2 approaches for determining the FII interval:

Determining the FII Interval

By far the most common method of establishing the FII is based on the MRT. The MRT is the limit of corrosion thinning at which the tank must be internally inspected. Since this limit is projected at the beginning of tank operation or the preceding FII, the future MRT cannot be known with certainty and is a random variable.

Figure 21 graphically depicts the relationship between corrosion, MRT and the FII interval. Considering corrosion from the topside, the maximum corrosion rate must be established to set the FII interval. Note that for a constant corrosion rate the depth of corrosion is equal to the thickness divided by corrosion rate. For example, if the tank bottom remaining thickness at the thinnest location is 200 mils thick, the MRT is 100 mils, the corrosion can proceed until it reaches 100 mils. Assuming a linear corrosion rate, the interval would be (200-100) divided by the linear corrosion rate. So, if the corrosion rate is 10 mils per year then the tank would be allowed to operate for 100/10=10 years. While linear corrosion rates are simple and straightforward to use, the actual corrosion rate may be less than linear, following a square root law. In this case the linear corrosion rate assumption would result in conservative estimates of the interval.

API 653 does not mandate a method of computing the corrosion rate, but states that an acceptable method is the one given in Section 4.4.5 Minimum Thickness of Tank Bottom Plates. We call this method an algorithm because it computes the interval based on minimum remaining thickness after repairs. This means that API is not using a corrosion rate at all since the thickness after repairs is an arbitrary value based on a decision made by the tank owner/operator and the inspection agency. Unfortunately, while this algorithm is identified as "an acceptable method" almost all companies (owners and inspections) use this algorithm to determine the FII interval. Please visit www.pemyconsulting.com for more information on determining the FII interval.

It turns out for the most part that tank owners/operators can achieve the 20-year FII interval by adjusting the repair-to thickness and because of large conservatism built into the API 653 assumptions in this method that the deepest topside pit lines up with the deepest underside pit which is clearly not the case. Unfortunately, the algorithm is not transparent as to how it performs under different actual corrosion rates and repair-to thicknesses.

Risk Based Inspection

Risk based inspection (RBI) considers the likelihood and consequences of a failure and adjusts the inspection interval based on the risk associated with failure of the tank. The risk is often viewed as the product of likelihood and consequence but there are other ways of determining risk from these two factors. A common difficulty with RBI is transparency, repeatability and understandability of the applied methods and results obtained which has led to a lack of acceptance by some regulatory authorities and a distrust of the methodology even by people in the oil industry. In fact, some states as well as PHMSA (define) have prohibited the use of RBI in the tank inspection programs they regulate. However, if done properly, without bias or intent to mislead, using risk to establish FII intervals is entirely scientifically sound and should not be rejected out-of-hand as many regulatory agencies have done. The RBI procedure also has a maximum cap at 30 years.

API 2350

Most tank filling and receiving operations are manual in nature. Although operator awareness of the tank filling operation and communication with the transporter who is delivering the petroleum batch to the operator is the most fundamental aspect of overfill protection, an alert/alarm system that grabs the operator's attention is usually needed as well. Many overfills have their causes in faulty alarm systems that did not work when they were supposed to. But it has long been understood that using redundant alarms which ensure that secondary alarm is triggered, even if there is an alarm failure is important. To ensure high reliability these systems must be independent and redundant and be proof tested periodically.

Although this has long been understood, the 4th edition of API 2350 – Tank Overfill Protection published in May of

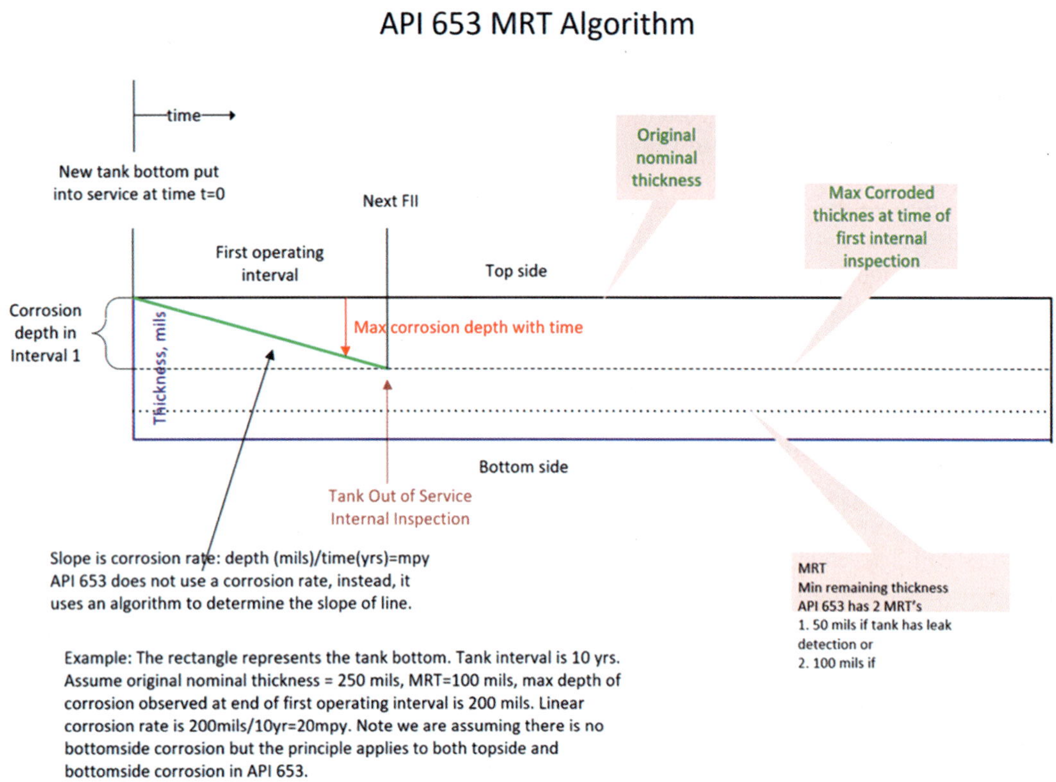

Figure 21. Setting the formal internal inspection interval.

2012 not only clarified how redundancy and independence of alarm systems should be set up for high reliability in the context of modern instrumentation and control systems but represents a generational change by incorporating management system principles into the standard to make the operations of overfill prevention as robust as possible. Recall that a management system incorporates a systematic set of elements required to achieve business goals efficiently and safely. Management system elements include personnel competency and training, procedures for normal and abnormal operations, functional equipment, maintenance and testing, management of change procedures, a system to identify near misses and incidents, risk assessment and risk management and most important, leadership messaging and accountability regarding the importance of ensuring the management system remains robust.

The risk assessment and management system components are new in the 4th edition of API 2350. There are significant challenges with its implementation as an industry standard since risk is unique and site specific but is also different for each company. Although there are many overfills annually, it is the high-profile incidents that bring calls for actions as well as regulations. In 2005 and 2009, respectively, the Buncefield and Puerto Rico tank overfills occurred. This put significant pressure on the petroleum industry standards to address risk in API 2350 and to adopt state-of-the-art-standards that could be applied to prevent overfills. At the same time, the pressure in the UK (as a result of the Buncefield incident) evolved into the requirement that tank operators must use a risk assessment process such as LOPA[1] and must adopt the international safety standard IEC61511[2].

IEC 61511 is the recognized instrumentation and control systems standard that applies to safety instrumented systems applicable to the process safety industry where automation is the primary control over safety requirements. IEC61511 is intended to apply only to new systems. Retroactivity by application of this standard to existing systems is always viable. The 4th edition of API 2350 used the best ideas and practices to address existing tank systems. For new systems API 2350 simply says to apply IEC 61511 Safety Instrumented Systems (SISs) as warranted and as needed.

The use of SISs is new to the tank and terminal business. Because of the complexity and engineering expertise and infrastructure needed to implement these systems, most of the industry has not adopted the use of SISs for tank overfill

[1] Center for Chemical Process Safety coined the term 'Layers of Protection Analysis' as LOPA
[2] IEC 61511 International Electrotechnical Commission "Functional safety - Safety instrumented systems for the process industry sector".

protection in its fully prescribed form as delineated by IEC 61511. Moreover, this implementation is not possible when existing systems are under consideration. Because of this problem, the API 2350 4th edition includes Annex A which provides best practices for ensuring that the existing systems are indeed fit for purpose.

Some owner issues related to inclusion/adoption of SIS standards are:

- Complexity and workability of implementation.
- Definition and setting of risk methodologies and criteria.
- Reconciliation risk tolerance criteria with outside stakeholders, especially regulatory.
- Determining if and when SISs are required in a simple and practical manner.
- The potential increases in risk when they are used due to implementation faults in management systems.
- The potential increases in risk due to hydraulic transients or other physical causes.
- Contractual problems between transporters and receivers.
- Pressure to use wireless technology which is in an unresolved state as to applicability for safety systems.
- Ensuring that the standard is not just US 'centric' but applicable worldwide.

It is possible that better industry safety results may be achieved by more focus on the basic reliability issues rather than turning to higher complexity systems that while in theory work but fail in their implementation in practice.

API 579

API and ASME Codes provide construction standards for new equipment. However, societal tolerance for spills, fires, health and safety issues decreased sharply in the 80s and 90s and standards aimed to ensure the integrity of mechanical equipment emerged. One of the first important standards for tanks was API Standard 653; for piping it was API 570 and API 2611.

In order to make a finer line between equipment that has been damaged by corrosion or other damage mechanisms (for a complete list see API RP 571 - Damage Mechanisms Affecting Fixed Equipment in the Refining Industry) which is safe to operate and to optimize the criteria between equipment which is degraded from its original state and the current state, API 579 – Fitness-for-Service was developed. Since it applies to essentially all fixed pressurized refinery equipment which cannot simply be just shut down it has aided in the ability to safety delay repairs until such time as the repairs can be done in an orderly manner.

The RP starts out with an administrative section that provides a way to implement the assessment for fitness for service using the current infrastructure of non-destructive testing, inspection, inspector certifications as well as breaking the assessment into three fundamental levels.

Level 1 – the most conservative assessment but the easiest to use. Special qualifications not required beyond plant or engineering personnel.

Level 2 – less conservative than Level 1. Typically requires personnel knowledgeable in performing fitness for service assessments.

Level 3 – least conservative but requires the most data, analysis and implemented by specialists due to the highly technical nature of the analyses.

API 579 relies heavily on the concept of remaining strength factor (RSF) which is defined as

$$RSF = \frac{\text{limit or plastic collapse load of the damaged component (component with flaws)}}{\text{limit or plastic collapse load of the undamaged component}}$$

For example, for a tank the RSF might be used this way.

$$MFH_r = H_f + \left(MFH - H_f\right)\left(\frac{RSF}{RSF_a}\right) \quad for\ RSF < RSF_a$$

where:

MFH maximum fill height of the undamaged component, see Annex 2C, paragraph 2C.1.2.

MFH_r reduced maximum fill height of the damaged tank course.

RSF remaining strength factor computed based on the flaw and damage mechanism in the component.

RSF_a allowable remaing strength factor.

Common uses of API 579 in tank work are for:

- settlement analysis that does not meet API 653
- hydrostatic test exemptions
- dents, buckles, tilting, out-of-roundness
- brittle failure analysis
- cracks, general or local metal loss
- fabrication defects that do not meet the original construction codes
- fire damage, environmental damage or other damage mechanisms
- problems with cyclic service

The use of API 579 relies heavily on the materials sciences, linear and nonlinear fracture mechanics analyses as well as finite element structural analysis and requires engineering specialists who can correctly apply the modeling techniques with sufficient judgement to properly interpret and apply the results from these types of analyses.

Conclusion

The operation of storage tank facilities is much more than just about standards. The industry standards provide the ground floor level of understanding required to be in the business. The challenges of operating efficiently, safely and reliably with a good reputation are the milestones of business acumen and longer-range thinking which is often not what drives companies as metrics for excellence. Probably the single most important element of doing business excellently with petroleum storage facilities is the proper application of management systems. A good way to find out more about management systems is by a Google search for "Operational Excellence" or "Operations Integrity Management System" which are the management systems applied by Chevron and ExxonMobil, respectively.

Tanks are unique in the sense that the carry much more flammable, toxic or dangerous fluids than other equipment used in the oil and chemical businesses. Rigorous application of safety management systems and pressure by the public and regulatory stakeholders has created a much safer industry. Because the potential to leak product over long periods of time results from the contact of the bottom with soil where leaks are not visible, piping and tanks have resulted in the contamination of almost all petroleum storage facilities older than a few decades.

Tank weight estimation tool using diameter as input variable

Philip Myers, P.E., Founder and Director, PEMY Consulting, Pleasant Hill, California

Estimating the weight of a tank can be important for seismic work, consideration of tank overturning during high winds, flooding and storms, or for cost estimating purposes. In this note, a tank weight estimation tool is presented that eliminates the need for detailed specifications, designs, and use of tank design software. The tool only uses diameter as input variable. The output yields an approximation of the tank weight.

How it works

When developing the tank weight calculator, four main components were considered:

1. weight of the bottom
2. shell
3. roof/framing
4. product

The weights of each of these components were estimated independently, and the weight of the tank was approximated as the sum total of these different weights. Other weights can be determined by using a subset of the four components (e.g., the weight of the bare metal tank can be determined by excluding the weight of the product). To minimize the amount of data or inputs needed to calculate the tank weight, several assumptions were made as to the design of the tank:

- steel is A36 (with density of 0.284 lb/in^3);
- product of the tank was assumed to have a specific gravity of 0.7
- shell course heights of 8 ft
- bottom thickness 0.25" (250 mils) with no annular ring
- roof thickness 3/16" and slope of 3/4" per foot

These assumptions are typical for storage tanks and if different values are used the model requires reparameterization. Differences between the model tank assumptions and actual tank parameters will lead to differences in the accuracy of the estimate.

Methodology

For the tank shell, the thickness, hydrostatic thickness, and the minimum thicknesses were calculated as per API 650 Section 5.6, based on the tank diameter, height, material (A36) and specific gravity (0.7). Each of the thicknesses were rounded up to the nearest 1/16th of an inch. Tank thicknesses were calculated for each course, where the height of each course was taken to be 8 ft for each shell course.

The most difficult part of the tank weight estimation comes from determining the weight of the tank roof/framing. At this stage, the type of roof that the tank will have is determined, which then determine which weights to calculate (Figure 1).

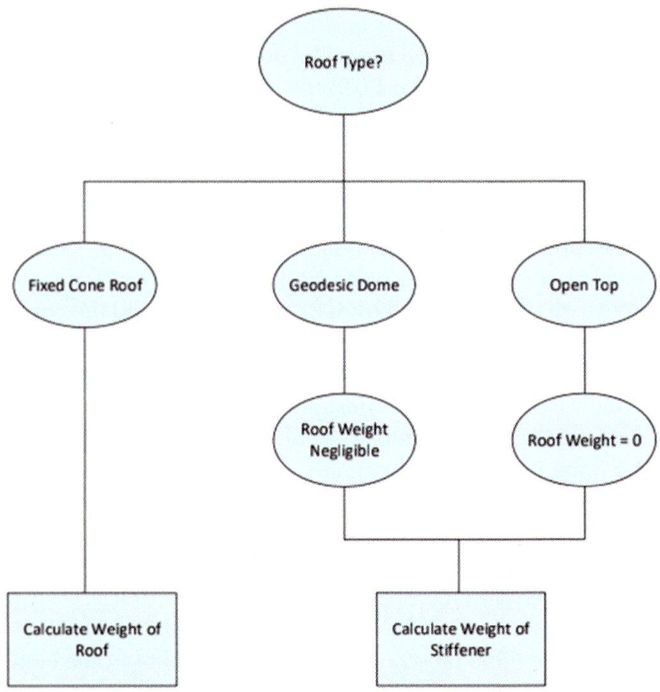

Figure 1. Flow chart of roof weight calculation.

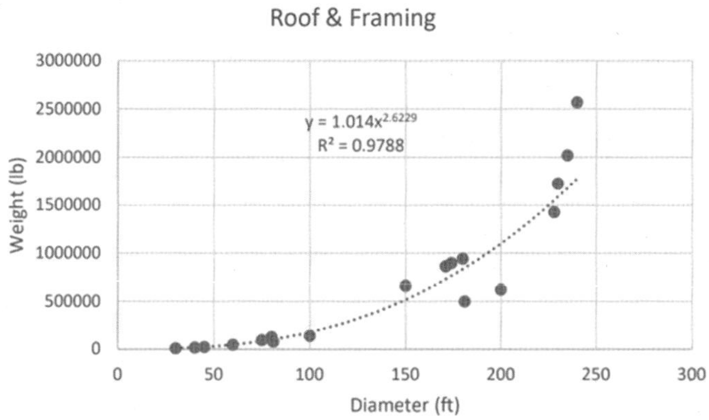

Figure 2. Roof and framing weight estimation for fixed cone roofs.

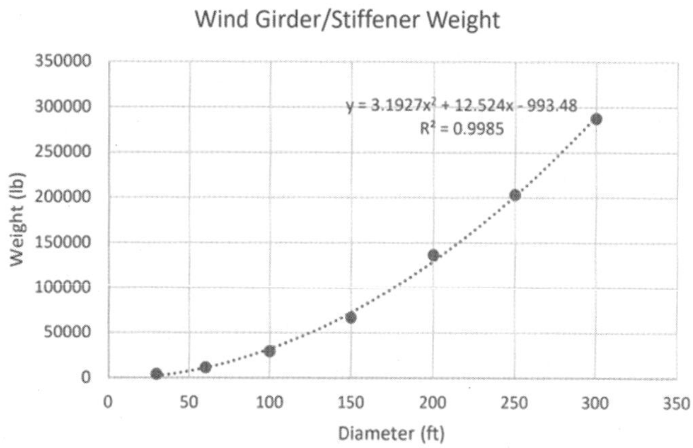

Figure 3. Wind girder and stiffener weight estimation for geodesic domes and open top roofs.

For fixed cone roof, the weight of the roof itself is straightforward, but complications arise when the addition of columns, girders and rafters need to be considered. Additional calculations and significant design considerations would need to be taken into account to determine the appropriate size, number, and spacing for these different components.

For the roof and framing, detailed design calculations were performed for several different cases for the appropriate designs for columns, girders and rafters; each case varied in diameter, and the corresponding weights of the columns, girders and rafters were then computed at a given diameter. These data were plotted and a trendline determined (Figure 2).

The result: for a given diameter, the weight of the roof and framing in pounds is approximated as $1.014D^{2.6229}$.

Geodesic domes and open top tanks have negligible roof weights. However, when considering these tanks, the weight of the roof stiffener must be considered. Using a similar approach to the calculation of fixed coned roofs, several tanks were designed with varying diameters and appropriate stiffener sizes. These values were then plotted to determine the weight of the roof stiffener as a function of tank diameter (Figure 3).

The result: for a given diameter, the weight of the wind girder and roof stiffener in pounds is approximated as $3.193D^2 + 12.524D - 933.48$.

Conclusion

Combining the tank weights, the total weight of the tank is estimated by summing the different components. A user of the equation need only to provide the tank diameter and height in order to get a quick and approximate value for the tank weight, which can then be used for seismic design considerations (as well as other considerations that require the tank weight, like the tank foundation). A more sophisticated application was also developed that allowed more control over different aspects of the tank—being able to choose the specific gravity, bottom thickness, roof thickness, roof slope, and so on. This provides greater accuracy in the tank weight estimation tool, but also increases the amount of inputs and data that is needed to run the tool. Regardless of which tool users would want to use, the benefits and advantages are clear—quick and efficient means for determining the weight of the tank.

How to estimate roof and framing weight for fixed cone tank

Estimate the roof and framing weight of a fixed cone roof of diameter 150 ft. Steel is A36, product of the tank has specific gravity of 0.7, shell course heights are 8 ft, bottom thickness 0.25 in, with no annular ring, and roof thickness 3/16" with slope of 3/4" per foot. Estimate girder and stiffener weight for a geodesic dome roof under the same conditions.

Solution.

Weight FCR $= 1.014\ (150)^{2.6229} = 517{,}300$ lbs

Weight GDR $= 3.193\ (150)^2 + 12.524\ (150) - 933.48 =$ 72,800 lbs

Budgetary pricing cost curve of cone roof tanks

API 650 Cone Roof Tanks
Budgetary Pricing (+/- 30%)

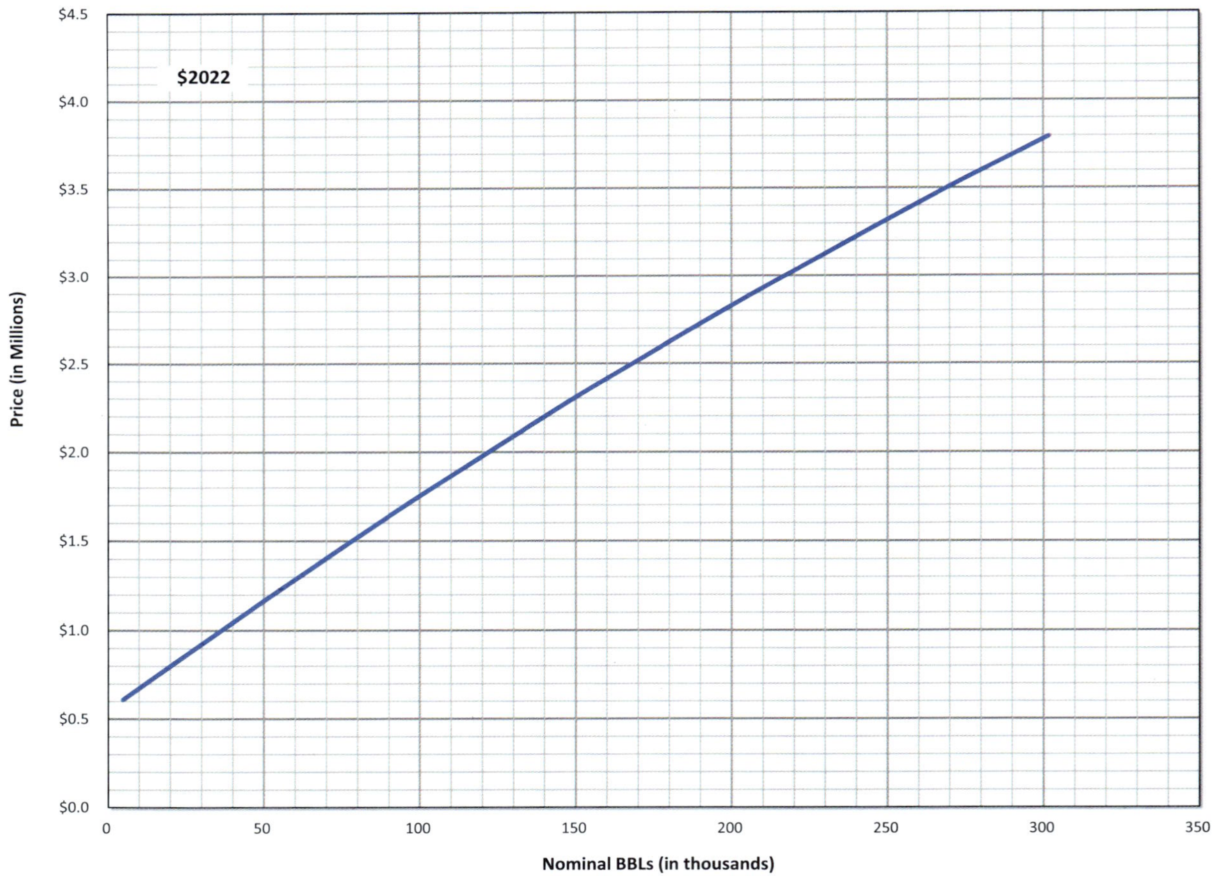

Workscope Basis:
-Engineering & Drawings
-Materials & Shop Fabrication
-Field Installation
-Ringwall Foundation, Liner, CP System
-Foam Fire Supression System
-Standard NDE & Hydrotest
-Exterior Coatings
-Interior Coatings (Bottom and 3' Up Shell)

Nominal BBLs	Tank Dimensions
5,000	30' Dia. x 40' Ht.
10,000	43' Dia. x 40' Ht.
15,000	52' Dia. x 40' Ht.
20,000	60' Dia. x 40' Ht.
25,000	67' Dia. x 40' Ht.
30,000	67' Dia. x 48' Ht.
40,000	78' Dia. x 48' Ht.
50,000	87' Dia. x 48' Ht.
60,000	95' Dia. x 48' Ht.
70,000	103' Dia. x 48' Ht.
80,000	110' Dia. x 48' Ht.
90,000	116' Dia. x 48' Ht.
100,000	123' Dia. x 48' Ht.
120,000	134' Dia. x 48' Ht.
140,000	145' Dia. x 48' Ht.
160,000	155' Dia. x 48' Ht.
180,000	164' Dia. x 48' Ht.
200,000	173' Dia. x 48' Ht.
220,000	182' Dia. x 48' Ht.
240,000	190' Dia. x 48' Ht.
260,000	197' Dia. x 48' Ht.
280,000	205' Dia. x 48' Ht.
300,000	212' Dia. x 48' Ht.

How to estimate flow rate from a leaking tank hole

Philip Myers, P.E., Founder and Director, PEMY Consulting, Pleasant Hill, California

When a corrosion hole penetrates a pipe, tank, or pressure vessel where the leak is in contact with the ground and not visible or evident, the natural question arises as to how much fluid has or can escape into the environment.

In this article, a simple model is presented assuming a hemispherical pit grows from the process side surface to the external surface and as soon as it breeches the steel wall liquid flow occurs. We assume that the hemispherical pit is growing at a constant rate which provides a function that yields the diameter of the hole with time. Application of Bernoulli flow is used to model the flow rate. Because of the uncertainties in the actual geometric shape as a function of time we consider that the assumption of an inviscid fluid is adequately conservative for most petroleum liquids.

The application of this kind of model has significant utility in cases where a spill has already occurred, and where an estimate is needed to determine the expected loss or in a predictive manner, when a risk assessment is being conducted to determine the scale of contamination that might occur because of a breeched tank wall in contact with the soil.

Hole diameter

Figure 1 shows the model schematic. Simple geometry gives the hole diameter as

$$(t+h)^2 = t^2 + (d_s/2)^2$$

Since $h = r\theta$ we find that

$$d_s{}^2 = 8tr\theta + r^2\theta^2$$

Any consistent units can be used but the following are convenient:

- t - mils, the original nominal thickness, usually 250 mils
- r - mil/yr, corrosion rate
- θ - yrs, time from hole penetration through wall thickness

Fluid flow

For risk assessment where the accident scenario details are not known including the permeabilities of the underlying layers of soil it should be sufficient to use Bernoulli flow as an upper bound. The actual flow could range from nothing if the hole is plugged by debris, to a low flow where the hole is forced through a porous medium such as soil, limited to a maximum value, which is Bernoulli flow, because there is no resistance to flow beyond that of fluid getting through the hole.

The Bernoulli flow assumptions are:

- The fluid is incompressible
- There is no friction loss
- The fluid is inviscid (no viscosity)
- The discharge coefficient is taken to be 0.6

Bernoulli's flow from an orifice in a tank bottom is

$$q_b = C_d A\sqrt{2gh}$$

where:

- C_d is a constant usually equal to 0.6
- A is the open area of the hole
- g is the gravitation constant 9.8 m/s^2
- h is the head in the tank relative to the location of the hole

Examples

Example 1. The maximum typical pitting rate is 5 mpy. The tank bottom is 1/4 inch thick (250 mils) and it has been one year since the pit penetrated the tank bottom. What is the diameter of the hole?

From the equation for the soil side pit diameter

$$d_s = \sqrt{8th\theta + r^2\theta^2} = \sqrt{8(250)(5)(1) + (5^2)(1^2)} = 100.5 \text{ mils},$$

slightly over 0.1 inch. Note that the function $d_s(\theta)$ is dominated by the first term for the first few years and its shape is therefore similar to a square root function (Figure 2), which initially rises rapidly but continues to slow with time. This is also intuitively correct from a geometric perspective.

Example 2. What is the flow rate of fuel in Example 1 exiting the tank?

Using the Bernoulli flow equation determine the flow in gallons per day from a hole that is 100.5 mils in diameter. Assume that the liquid head above the hole is 15 meters.

While the actual flow will be slightly less due to the assumptions, the Bernoulli flow provides a reasonable bound on flow rates.

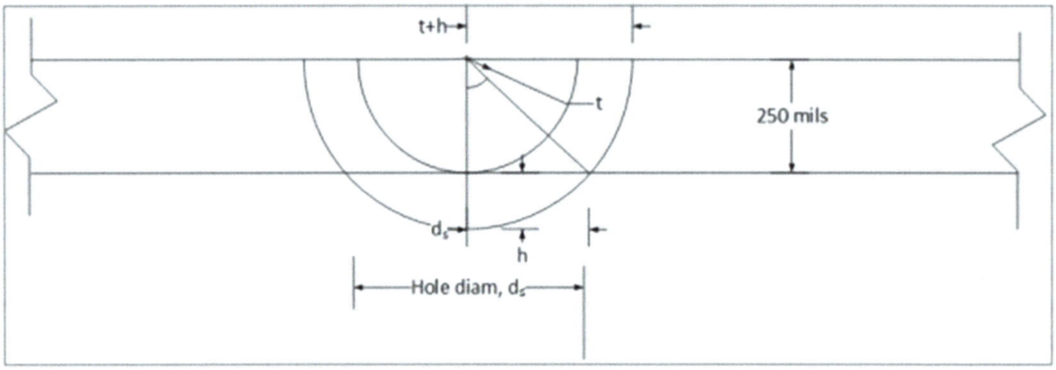

Figure 1. Tank bottom corroded profile.

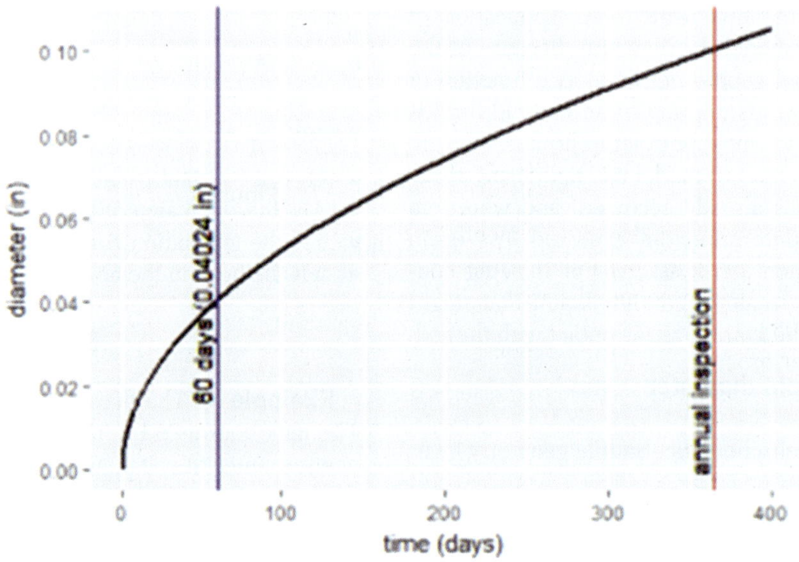

Figure 2. Tank hole growth, t = 250 mils, r = 5 mils/yr, θ = 1 yr.

We must use consistent units with the equation $q_b = C_d A \sqrt{2gh}$, and we apply the basic dimensions of meters and meters per sec squared. We then convert to gallons per day, yielding 1202 gallons/day.

The flow is continuously increasing, so the flow per day is approximate, but well within engineering accuracy (Figure 3).

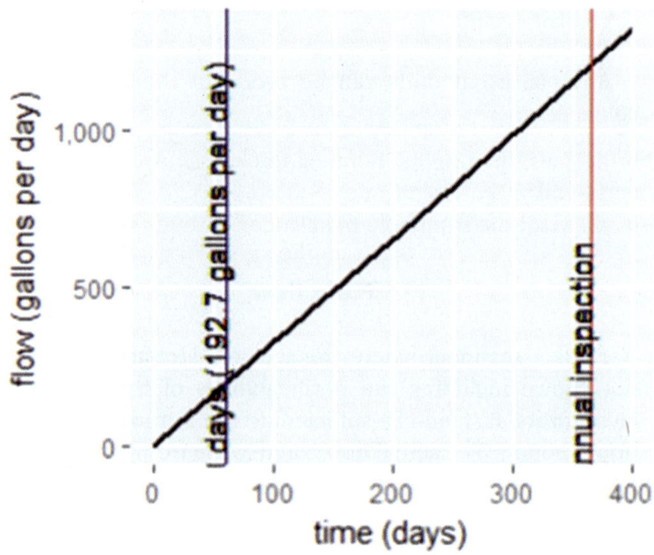

Figure 3. Daily flow in gallons per day, t = 250 mils, r = 5 mils/yr, θ = 1 yr.

Example 3. Ultimately, the quantity of real interest is the cumulative leakage that could occur over time due to the penetration of the tank wall by the pit. Taking the same conditions as in the previous example, what will be the total leaked volume at day 365? Moreover, what will be the cumulative leakage 1, 2 and 3 months after the hole penetrates the tank wall?

For a given corrosion rate both the hole size model and the Bernoulli flow model can be combined into a single equation that is easily integrated so that the cumulative flow for any given time θ can be estimated.

The form of the combined equation is

$$q_b = a + \theta + b + \theta^2$$

where a and b are constants.

```
a <- (0.6*pi*sqrt(2*9.8*15)*10000/4/39370.1^2)*365*1440
*60*264.172;a
## [1] 434285.3
b <- (0.6*pi*sqrt(2*9.8*15)*25/4/39370.1^2)*365*1440*
60*264.172;b
## [1] 1085.713
```

The constants a and b allow for writing the cumulative flow as an integral

$$\int_0^1 f(\theta) = \int_0^1 a\theta + b\theta^2 \tag{1}$$

$$= a\theta^2/2 + b\theta^3/3 \tag{2}$$

$$= 217142.6\theta^2 + 361.9043\theta^3 \tag{3}$$

For example, the cumulative flow in gallons on day 365 is:
[1] 217504.5

The plot of the cumulative flow as a function of time is shown in Figure 4.

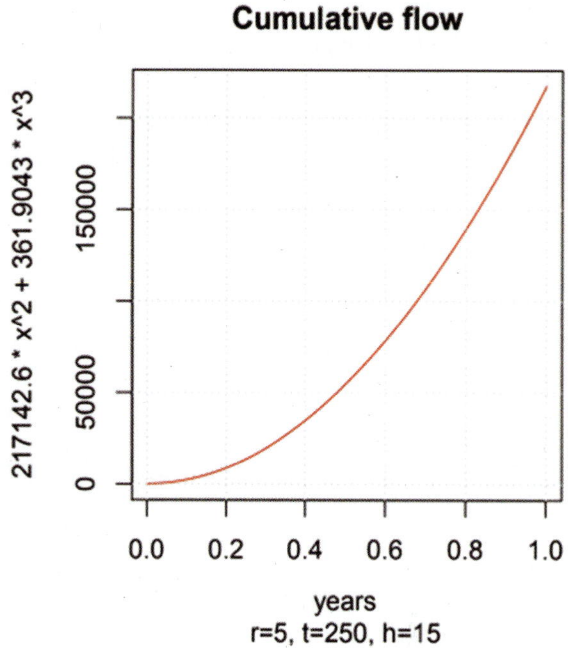

Figure 4. Cumulative flow as a function of time.

We can easily compute the cumulative flow at month 1, 2 and 3 respectively. Assuming 30 days per month, the results in gallons at the end of the first three months are:
[1] 1467 5869 13208

A spreadsheet or other program can be used to determine the cumulative flow with time under different scenarios and multiple cases should be reviewed.

Conclusions

The model is a crude one but much better than nothing. There have been cases where a hole has been found by regulatory authorities and fines based on Bernoulli flow based on the current hole size and for the estimated time to hole though have been levied. In cases like this, the use of this model would have direct financial value since the fines based on cumulative leaked petroleum liquids as described by the regulators might be excessively high.

One way to improve the model is to further reduce the flow by modeling the resistance to flow offered by the soil into which the leak is occurring. For flat bottom tanks some of these methods have been applied to leachate ponds.

Finally, for new facilities or where leaks could occur from tanks, piping or pressure vessels, this model can be used to provide inputs to a risk assessment process.

Charts give vapor loss from internal floating-roof tanks

S. Sivaraman, Exxon Research & Engineering Co., Florham Park, New Jersey

Nomographs, based on the guidelines presented in American Petroleum Institute (API) Publication No. 2519, have been constructed to estimate the average evaporation loss from internal floating-roof tanks.[1] Loss determined from the charts can be used to evaluate the economics of seal conversion and to reconcile refinery, petrochemical plant, and storage terminal losses.

The losses represent average standing losses only. They do not cover losses associated with the movement of product into or out of the tank.

The average standing evaporation loss from an internal floating-roof tank depends on:

- Vapor pressure of the product
- Type and condition of roof seal

- Tank diameter
- Type of fixed roof support

The nomographs (Figures 1-4) can estimate evaporation loss for product true vapor pressures (TVP) ranging from 1.5 to 14 psia, the most commonly used seals for average and tight fit conditions, tank diameters ranging from 50 to 250 ft, welded and bolted designs, and both self-supporting and column-supported fixed roof designs. The charts are purposely limited to tank diameters 250 ft and less, because internal floating-roof tanks are generally below this diameter.

Typical values of the deck fitting loss factors presented as a function of tank diameters in the API Publication 2519 have been used in the preparation of these nomographs. In addition, for the calculations of the evaporation loss for the bolted deck design, a typical deck seam loss factor value of 0.2 has been assumed.

Table 1 gives the proper axis to use for various seal designs and fits.

Use of these nomographs is illustrated by the following example.

Example. Determine the evaporation loss for an internal floating roof tank given the following:

- Tank diameter 200 ft
- Liquid-mounted primary seal only and an average seal fit
- Product true vapor pressure of 10 psia
- Welded deck with self-supporting fixed roof

Table 1
Selection of Seal Axis

Seal type	Seal axis	
	Average fit	**Tight fit**
Vapor-mounted primary seal only	H	G
Liquid-mounted primary seal only	F	E
Vapor-mounted primary seal plus secondary seal	D	C
Liquid-mounted primary seal plus secondary seal	B	A

Solution

1. Use Figure 1 for the welded deck and self-supporting fixed roof.
2. From Table 1 select the seal axis. The seal axis for the example problem is F.
3. Locate the point of intersection F1 between the seal axis F and the tank diameter contour for the 200-ft diameter tank.
4. From the point F1 traverse horizontally to intersect the reference axis R at R1.
5. Locate the true vapor pressure point P1 corresponding to 10 psia on the pressure axis P.

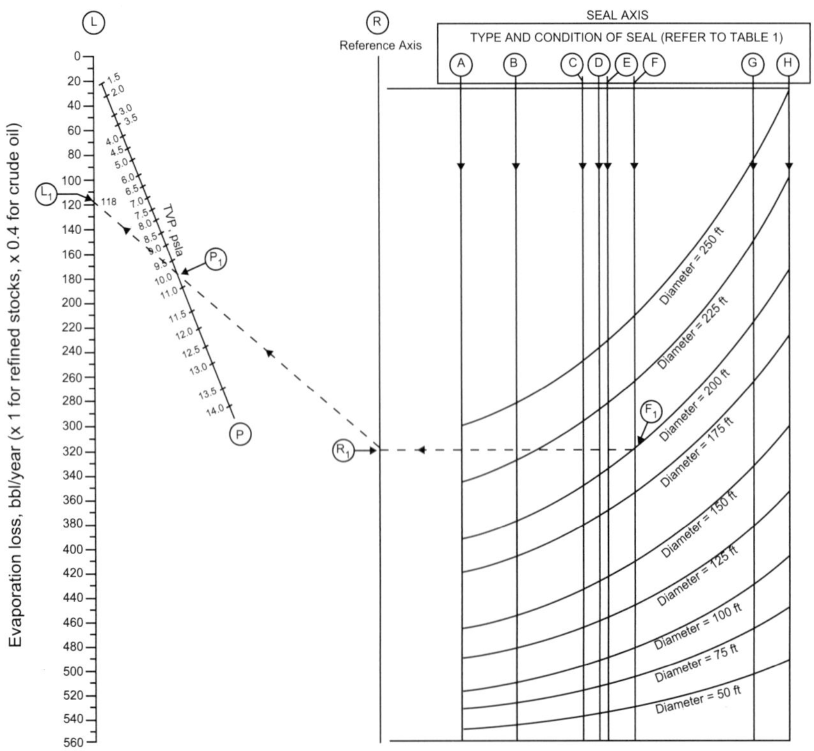

Figure 1. Loss from welded deck, self-supporting fixed roofs.

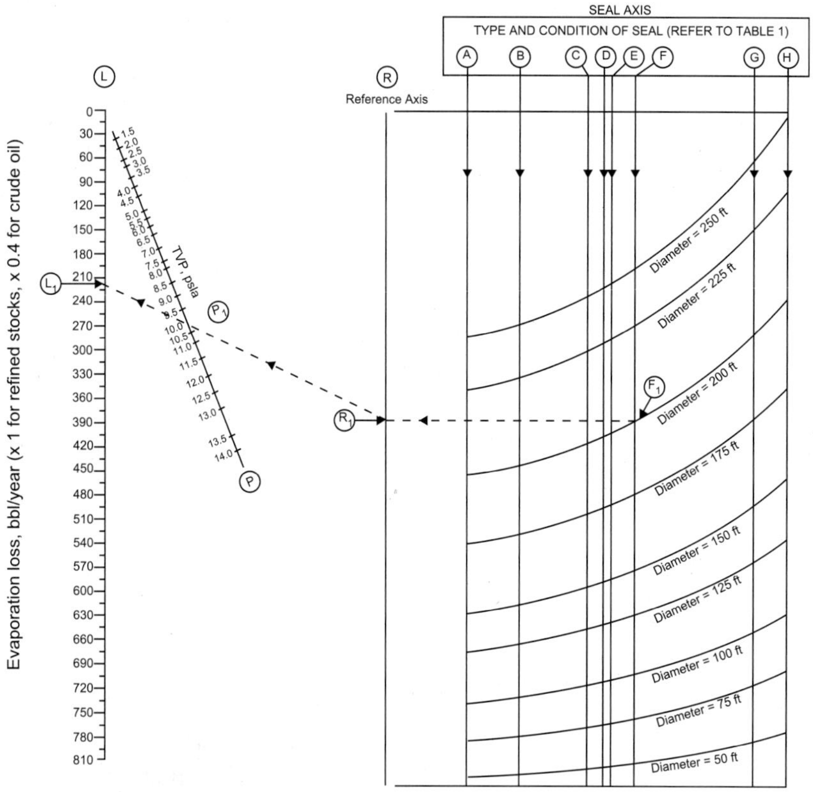

Figure 2. Welded deck, column-supported fixed roofs.

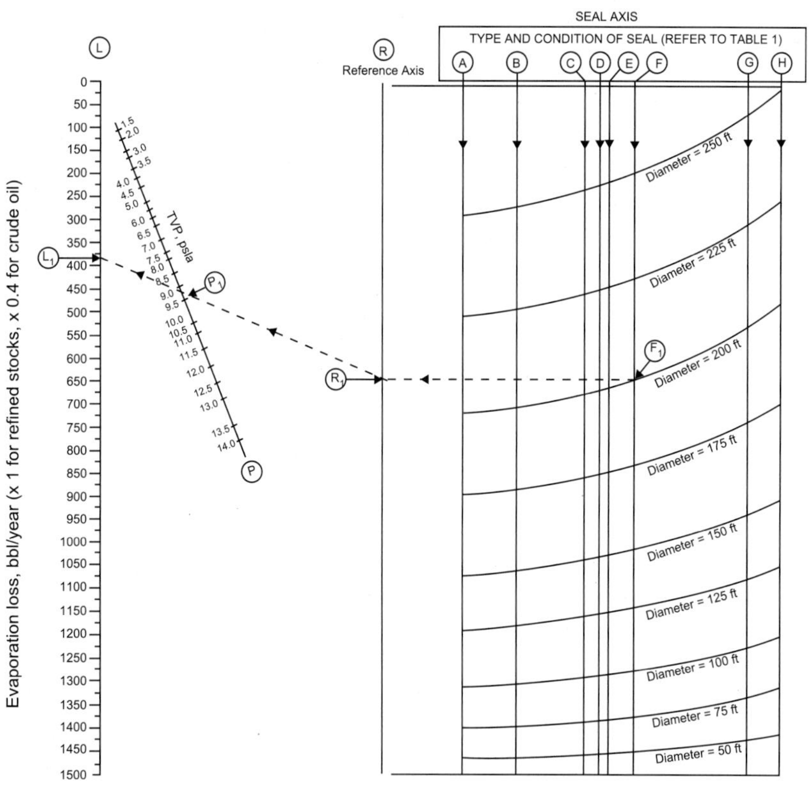

Figure 3. Bolted deck, self-supporting fixed roofs.

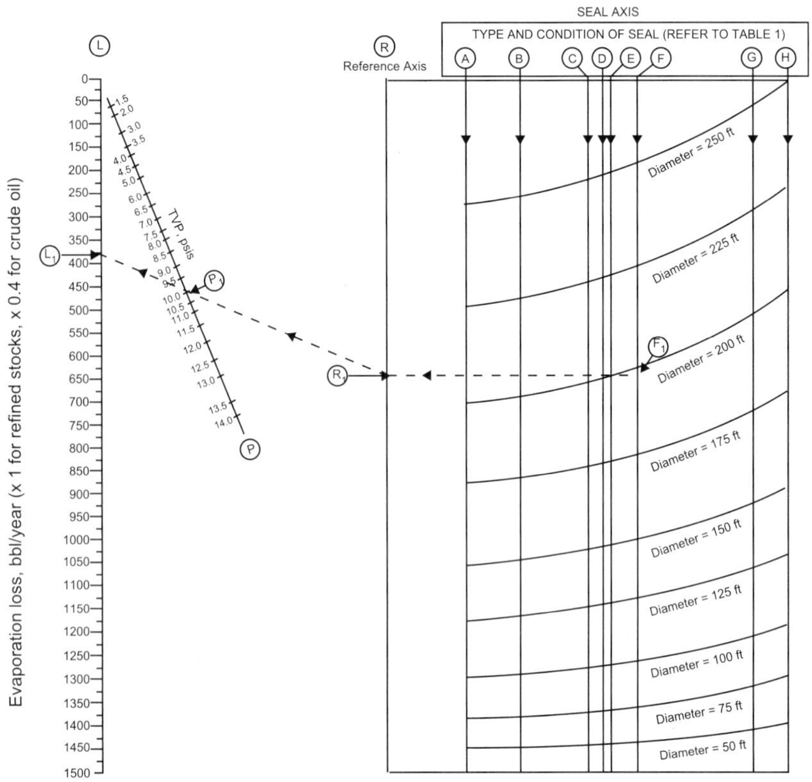

Figure 4. Bolted deck, column-supported fixed roofs.

6. Connect the point R1 on the reference axis R and the point P1 on the pressure axis P and extend in to intersect the evaporation loss axis L at L1.

Read the evaporation loss in bbl/year at L1. The average evaporation loss is 188 bbl/year for this example. The same example is shown in Figures 2, 3, and 4 for other deck designs and roof supports.

Source

Oil & Gas Journal, March 9, 1987.

Reference

Evaporation Loss from Internal Floating-Roof Tanks, American Petroleum Institute Publication No. 2519.

Estimating the contents of horizontal cylindrical tanks

Horizontal cylindrical tanks are frequently used for water and fuel storage, and in many cases it is important to be able to gauge these vessels to determine the volume of liquid contained in them. However, it is normally much more difficult to establish a volume-per-inch scale for a horizontal tank than for one in a vertical position. The accompanying nomograph simplifies this problem.

To use the nomograph, it is necessary to gauge the tank and determine the ratio of the depth of liquid in the tank to the tank diameter. After this is found, draw a straight line from the point on the "ratio" scale through the known point on the "diameter of tank" scale and read the intercept on the "gallons per ft of length" scale. From this point, draw a second line through the known point on the "length of tank" scale and read the intercept on the "gallons (or barrels) in total length" scale.

Example. Find the volume of liquid contained in a horizontal cylindrical tank 7 ft in diameter and 20 ft long when the depth of the liquid is 4 ft, 10.8 in.

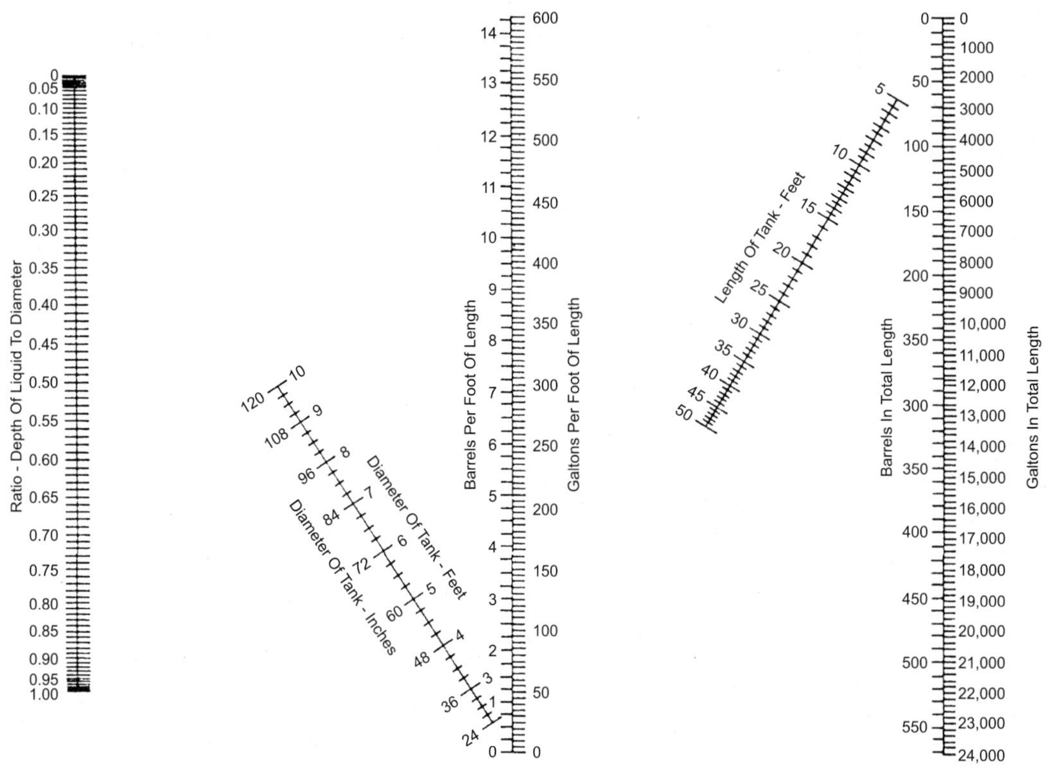

The ratio of depth of liquid to tank diameter is:

$58.8/84 = 0.70$

Connect 0.70 on the ratio scale with 7 ft on the diameter scale and continue the straight line to obtain the intercept 215 on the gallons per ft of length scale. Draw a second line from the point 215 through the 20 on the length of tank scale and continue the line to obtain the intercept 4,300 on the gallons in total length scale. Total liquid content of the tank is 4,300 on the gallons.

How to gauge a horizontal cylindrical tank

Express the depth in % of the diameter; then the result will be given in % of total capacity of the tank.

Rule 1. For depth up to 30; multiply the square root of the depth by the depth, and then by 0.155.

Example. Liquid depth is 16% of tank diameter

$16 \times 16 \times 0.155 = 4 \times 16 \times 0.155 = 9.9\%$

The correct answer is 10.3%; error is about 0.4%.

Rule 2. For depth between 30 and 50; subtract 10 from the depth, multiply by 1.25.

Example. Liquid depth is 44% of tank diameter

$(44 - 10) \times 1.25 = 34 \times 1.25 = 42.5\%$

The correct answer is 42.4%.

The maximum error for depths less than 5% may be as great as 10%; gauging in this range is always difficult, and a very small slope can introduce a much larger error than this. When the depth is greater than 50%, apply the same rule to get the volume of the empty space above the fluid, and subtract.

Correct the volume of light fuels from actual temperature to a base of 60°F

To approximate quickly the volume of gasoline or other light liquid fuel at 60°F from a known volume at any temperature in the atmospheric range, use the formula:

$$V_a - V_{60} = 0.0006(T - 60)V_{60}$$

where:

V_a = Volume at actual temperature
V_{60} = Volume corrected to 60°F
T_a = Actual temperature of fuel

Example. A tank contains 5,500 gallons of gasoline at 46°F. Correct the volume to a base of 60°F.

$(5{,}500 - V_{60}) = 0.0006(46 - 60)V_{60}$
$(5{,}500 - V_{60}) = 0.0006(-14)V_{60}$
$5{,}500 = V_{60} - 0.0084V_{60}$
$5{,}500 = 0.9916V_{60}$
Volume at 60°F = 5,546.6 gallons

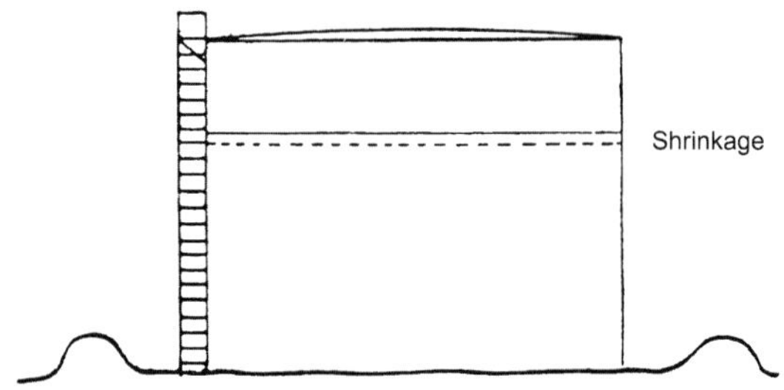

To approximate the shrinkage or expansion, obtain the difference between the actual volume measured and the corrected volume. In this case:

Shrinkage = 5,546.6 − 5,500 = 46.6 gallons

Volume of liquid in vertical cylindrical tanks

Measure the depth of the liquid and either the diameter or circumference of the tank, then the volume in:

Gallons = $0.0034\ d^2h$ or $0.00034\ c^2h$
Barrels = $0.000081\ d^2h$ or $0.00082\ c^2h$
Gallons = $5.88\ D^2H$ or $0.595\ C^2H$
Barrels = $0.140\ D^2H$ or $0.0142\ C^2H$

where:

d = Diameter, inches
c = Circumference, inches
h = Depth, inches
D = Diameter, feet
C = Circumference, feet
H = Depth, feet

If the circumference is measured on the outside, then three times the thickness of the tank wall should be subtracted before using the formula. Naturally, these rules cannot supplant the results of accurate tank strapping, which take many other factors into account.

Example. How many gallons will a tank 12 ft in diameter and 16 ft high hold when full?

$$\begin{aligned}\text{Gallons} &= 5.88D^2H \\ &= (5.88)(144)(16) \\ &= 13{,}548 \text{ gallons}\end{aligned}$$

Example. How many barrels will a tank 8 ft in diameter and 16 ft high hold when full?

$$\begin{aligned}\text{Barrels} &= 0.140D^2H \\ &= (0.140)(64)(16) \\ &= 143 \text{ barrels}\end{aligned}$$

Chart gives tank's vapor formation rate

When sizing the vapor piping for a manifolded expansion-roof tank system, the rate of vapor formation must be known. While the rate of vapor formation can be computed using longhand methods, the calculation is tedious. A nomograph quickly gives an approximate answer.

Example. Determine the rate of formation of vapor in a 140,000 barrel capacity tank when it is filled at the rate of 8,000 barrels per hour.

Solution. Enter the chart on the left at a capacity of 140,000 barrels and draw a straight line through the filling rate of 8,000 barrels per hour on the right. At the intersection with the central scale read the vapor formation rate as 55,000 cu. ft per hour. The vapor piping for this tank would have to be designed for this formation rate if the maximum filling rate anticipated were 8,000 barrels per hour. If a higher filling rate were expected, the vapor formed would have to be computed for the higher rate. The chart could, of course, also be used for this computation.

This chart is based on the following equation:

$$V = \frac{\text{tank capacity, bbl}}{14.3} + \frac{\text{filling rate, bbl per hr}}{0.178}$$

where V = vapor formed, cubic feet per hour

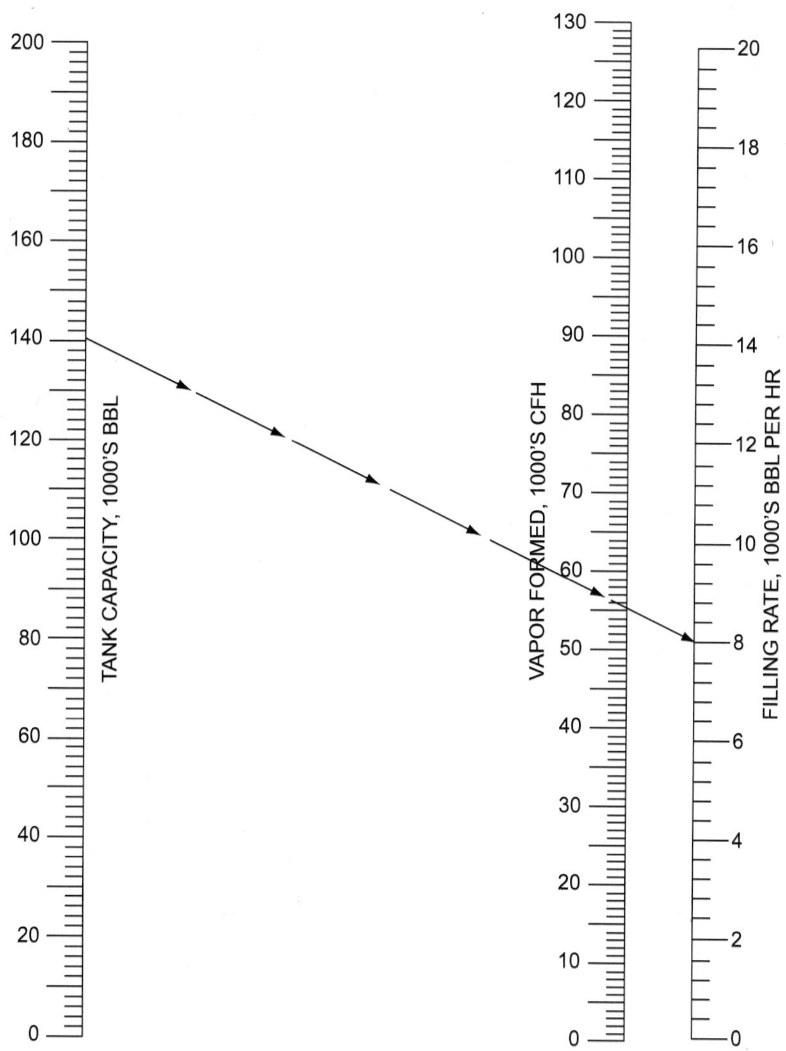

Tank farm line sizing

J. C. Stivers

The designer must locate the tanks, size the tank lines and select pumps. The designer must know the receiving flow rates, discharge flow rates, liquids to be handled along with the properties of the liquid such as viscosity, specific gravity and vapor pressure. The minimum and maximum tank levels must be determined which will set the available volume that can be pumped. Other factors include determining the number of tanks which will depend upon the type of material being stored, tank turnover frequency, system load factor (which takes into consideration system outages) and environmental considerations.

If possible, the tanks should have combination suction and fill lines. To size the lines, it is necessary to consider the fill and discharge flow rates. If a tank farm can receive from a number of small systems and has one out-going pipeline, the size of the tank lines will probably be determined by the outgoing flow rate. If a tank farm is fed by one stream and delivers to a number of smaller systems, the line sizes will probably be determined by the incoming stream rate. It is necessary to check both to confirm this generalization.

The criteria for sizing a tank line based on filling rates is somewhat arbitrary. When there are a number of sources, the operator may specify a maximum pressure of 25 to 50 psig required at the connection. The piping manifold, meters and other equipment must then be sized so the pressure will not exceed 25 to 50 psig at the maximum flow rate. If a metering system is to be used on the incoming stream and a back pressure valve is required for the meters, the tank line size may be sized to produce the minimum back pressure at the maximum flow rate. When the tank farm owner develops the pressure required to fill tanks, an economic analysis should be made to determine line sizes. Large diameter pipelines require less pressure and therefore less power cost but require greater investments than smaller diameter lines.

Determine pipeline size

The equation for determining the line size from the tanks and to fill a pump based on discharge flow rates is as follows:

NPSH available must be ≧ NPSH required by the pump

$$\frac{P_a \times 1.135}{sg} - \frac{P_v \times 2.31}{sg} \pm \text{static}$$

where:

P_a = Atmospheric pressure, in Hg
P_v = Vapor pressure, psia
sg = specific gravity

Pressure (ft of head) – Pipe friction loss (psi/mile) × (equiv. feet of suction line × 2.31)/(5280 × sg)

Atmospheric Pressure—The lowest atmospheric pressure for the area should be considered. This is particularly important for tank farms where product is transferred from one tank to another using the pump. This usually occurs when the atmospheric pressure is low

Vapor Pressure—Vapor pressure increases with temperature and therefore reduces the NPSH available. Conversely liquid viscosity decreases with temperature increases and therefore reduces pipe friction losses. Therefore, it becomes necessary to determine the liquid temperature that produces the lowest NPSH available.

Static Pressure—This is the elevation difference between the minimum liquid level in the tank and the centerline of the pump suction. If the ground elevation in the tank farm

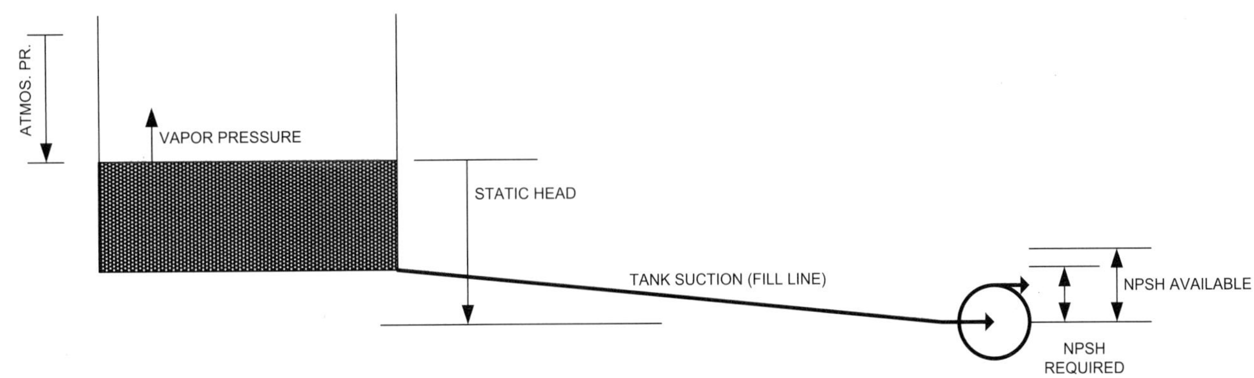

Figure 1. Tank farm line sizing.

changes, the designer can take advantage of the terrain and locate the tanks accordingly.

Pipe Friction Loss—There are many equations used to determine pipe friction loss. See Section 13.

Most tank lines are not designed to be pigged. Some crudes, particularly sour crudes, may cause a build up in the tank line, in turn reducing the pipe i. d. Any buildup will cause the friction factor to be higher and if build up is expected, the designer may wish to increase the pipe size—particularly if NPSH available to the pump is marginal.

NPSH required by pump

The pump manufacturer supplies the NPSH requirement for each pumps. Some positive displacement pump manufacturers may not publish these requirements due to a wide range of operating speeds. They will however, provide requirements for specific conditions. If the NPSH requirements of the pump are not met, the pump will cavitate and a loss of pressure and flow will occur. Mechanical damage may also be possible.

Types of booster pumps

The pumps most typically used include:

Horizontal split case
Vertical can
Vertical inline

A horizontal split case pump may be used in some instances and will be preferred by most maintenance and operations personnel because it is easy to maintain. The NPSH requirement for a horizontal pump may be 7 to 10 feet and the suction nozzle may be 2 to 4 feet above grade. The vertical can pump can have a 0 NPSH requirement with the suction nozzle three or more feet below grade. The NPSH advantage for the can pump can range from 12 to 17 feet which may permit a size reduction of the tank lines. NPSH requirements for the vertical inline pump are similar to the horizontal pump. The advantage of the inline pump is pump and installation costs.

Design considerations

The designer should not automatically assume that a booster pump is required. Most positive displacement pumps have low NPSH requirements and a therefore a booster pump is seldom required unless metering is installed between the tank and mainline pump(s). If the terrain permits the tanks to be located considerably higher than the main pumps, a booster pump may not be required for centrifugal pumps.

If a series of main line pumps are used and the pumping rate is determined by the number of pumps operating, it may be possible to obtain a pump with a lower NPSH requirement for the first pump and thus eliminate the need for a booster pump. This is shown by the following example.

Assume three main line pumps are connected in a series. Pump combinations produce the following flow rates:

No. of Pumps Operating	Flow Rate
3	100%
2	75%
1	50%

NPSH available:

Flow Rate	PSH Available
100%	15 feet
75%	20 feet
50%	27 feet

At a flow rate of 100%, it may be necessary to pay a premium for the number one pump to obtain a 15 feet NPSH requirement. However, a minimum cost second unit sized for 100% flow rate may be able to operate at 75% flow rate and 20 feet of NPSH.

Low NPSH requirement booster pumps are normally less efficient than mainline pumps. When this is the case, a booster pump should be selected to develop just enough head to feed the main line or delivery pumps.

If a metering system is required, it would normally be installed between the tanks and mainline pumps. In this case, a booster pump would almost certainly be required to overcome the pressure losses in the metering system. The booster pump then must develop enough head to produce the minimum back pressure required by the meters or the minimum NPSH required by the mainline pumps.

If the booster pump is used to transfer liquid from one tank to another, it will probably develop more head than required. This allows the designer to reduce the size of the transfer header and valves.

Once the minimum atmospheric pressure and vapor pressure of the liquid to be pumped is determined, there are three variables on the left side (NPSH available) and one on the right (NPSH required by the pump). The solution is trial and error and the following procedure is suggested:

- Select a tentative location for the tanks and determine the minimum liquid level elevation.
- Determine the equivalent pipe length of the suction line.
- Locate the booster pump and assume it can operate at a specific NPSH.

This allows the solution of the equation for maximum pipe friction loss. Application of pipe pressure loss formula will give the minimum size pipe that will result in a specific NPSH. Select the next larger standard pipe size and rework the equation to determine the actual minimum NPSH available at the pump. Generally, the cost of the booster pump will decrease as the NPSH requirement increases. Higher NPSH will also allow the designer to select a different style pump. For these reasons, the equation should be solved using the next two larger pipe sizes. Pumps can now be selected for three different pipe sizes. The costs of the three systems should be estimated and an economic analysis made to select the system.

The designer must be aware that a low NPSH pump can pull a vacuum which must be considered in determining pump differential. (Pipe collapse?) Frequently, the required pump discharge pressure is used to specify pump differential head which assumes zero pressure at the pump suction. The following example assumes the booster pump is designed to operate at 7 feet NPSH and 7 feet NPSH is available. The liquid is fuel oil with a vapor pressure of 0.2 psig and 0.85 s.g. and atmospheric pressure is 14.7 psia.

(2.31/s.g.) × (press. avail. + atmos. press − vapor press) = NPSH

Press. avail. = [(s.g. × NPSH)/2.31] − atmos. press. + vapor press.

Press. avail. = [(0.84 × 7)/2.31] − 14.7 + 0.2 = −11.95 psig

If a pump discharge pressure of 25 psig is required, the pump differential is 25 − (−11.95) = 36.95 psig or 102 feet.

Tank farm management

Excerpted from Tank Farm Management by Suresh S. Agrawal, ASTM Manual MNL58 Petroleum Refining and Natural Gas Processing, 2013.

Any liquid-based manufacturing facility, such as a refinery, petrochemical complex, beverage plant, etc., uses various types and sizes of tanks to receive, store, and dispatch products. In a refinery with 100,000 to 600,000 bbl/day of crude oil processing capacity, a tank farm normally consists of 50 to 300 or more tanks. The efficiency and timely available information about the quality and contents of tanks can affect the operations and profitability of the plant. This article will discuss various aspects of tank farm management from initial design methodologies to information management technology for a crude oil refinery.

Tank farm design

A typical refinery or liquid-based plant uses tanks for feed receipts, intermediate products, and dispatch of final products (Table 1). Although plants are designed and built with adequate storage capacity for design and future production, there are situations with serious economic effects on the refinery bottom-line:

- Crude oil is stocked up in excessive inventory.
- Shipment is delayed because of insufficient products, product storage capacity, or both; and
- Process units are operating at less than optimal capacity.

There are methods to estimate the storage requirement either at the front-end engineering design (FEED) level or for expansion purposes after the plant is already operating.

Required tank information and methods

The management of a tank farm requires that the following information be available to various departments/personnel to manage the plant efficiently, safely, and economically:

- Tank inventory,
- Tank qualities, and
- Estimate of fugitive emission.

Tank inventory

It is important that the inventory of each tank in the tank farm be available in near real time for inventory control/management, to avoid spills during loading, and to safely operate the plant. This requires the measurement of the following five tank process parameters:

1. Level (mandatory),
2. Pressure (mandatory for high-pressured tanks),
3. Temperature (mandatory for custody transfer and for heated tanks),
4. Density (mandatory for mass-based custody transfer), and
5. Water level (optional but recommended for crude and fuel tanks).

These parameters are measured manually or by an automatic tank gauging (ATG) system.

Table 1
Tank Farm Storage for a Typical Refinery

Name	Total Number of Tanks	Total Tank Capacities (bbl)	Average Tank Capacity (kbbl)	Production Factor	Daily Production (kbbl/day)	Days of Storage
Aromatics	13	170,000	13	3.5%	22	8
Ballasts and slop	7	240,000	34	4.8%	2	105
Butane + propane	7	230,000	34	4.8%	9	26
Crude oil	9	5,000,000	546	100.0%	225	22
Fuel oils	16	3,200,000	198	64.4%	9	352
Gasoil	19	1,400,000	73	28.1%	95	15
Gasoline	30	2,300,000	78	47.4%	47	49
Kerosine	5	460,000	92	9.4%	23	20
Naphtha	2	80,000	41	1.7%	5	18
Process units	6	5,000	1	0.1%	7	1
Propylene	3	35,000	12	0.7%	7	5
Truck terminal	24	130,000	6	2.7%	173	1
Total	**141**	**13,200,000**	**1,128**		**624**	**622**

Manual measurements

Although manual measurements of tank parameters are almost nonexistent for more than 50 kbbl/day capacity refineries, they are still in practice for smaller refineries. In this method, two people climb up a tank at least two times a day and measure the level by using a calibrated gage tape and visual observation. They would then enter this information in the tank information system (TIS). This method of level measurement not only poses safety concerns but also illustrates that an inaccurate level measurement of ±6 mm multiplied by all tanks will considerably deviate the actual inventory estimate from such measurement.

Automatic measurements

Hybrid Gauging System - A hybrid tank gauging system uses different instruments and technology for the measurement of tank parameters.

Tank Level Measurement - There are two categories of methods for the measurement of tank level: the direct method and the inferential method. The direct method directly measures the liquid height compared with a datum reference line. The indirect method measures effects caused by the changing liquid position. We have listed some of the most common techniques used in the industry for direct and inferential (indirect) methods.

Direct method techniques:

- Direct visual observation on a calibrated scale (e.g., gage stick, hook gage, or gage glass).
- Direct position of a float on the liquid surface.
- Contact of electrode probe with liquid surface.
- Interruption of light beam to photoelectric cell; and
- Reflection of radio, radar frequency, or some waves from a liquid surface.

Indirect or inferential method techniques:

- Measurement of hydrostatic pressure,
- Measurement of buoyant forces exerted by partially immersed object,
- Thermal determination between liquid and vapor phases,
- Based on physical or electrical properties of liquid to infer surface position, and
- Attenuation of radiation through liquid and vapor phases.

Two of the most common direct techniques are the float type gauge and the reflection of waves from the liquid surface. Two of the most common indirect measurements are hydrostatic pressure and buoyancy forces by a partially immersed object. Figure 1 shows the schematic diagrams of these two methods by direct and indirect methods, and Table 2 compares the pros and cons of the same.

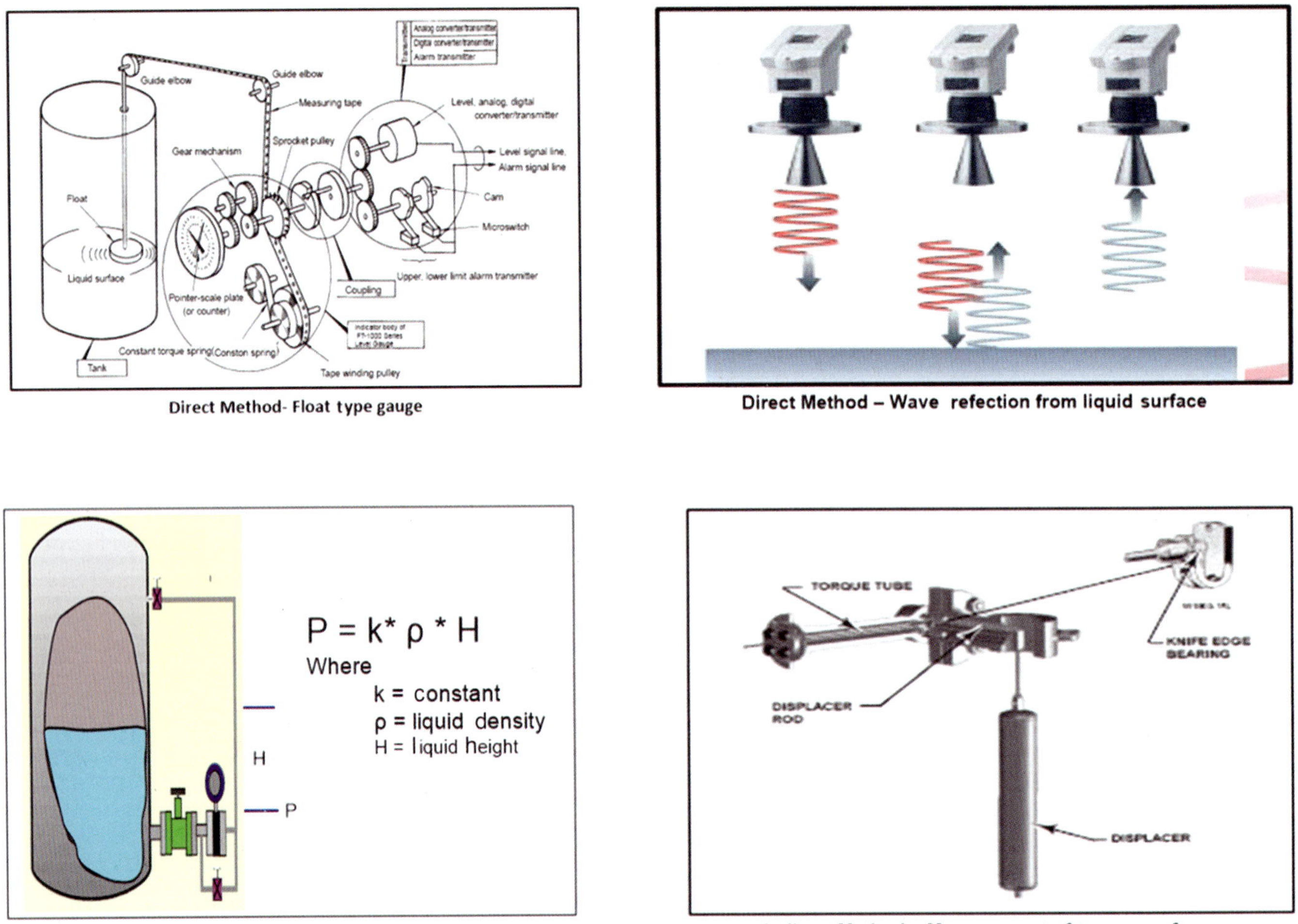

Direct Method- Float type gauge

Direct Method – Wave refection from liquid surface

Indirect Method – Measurement of hydrostatic pressure

Indirect Method – Measurement of buoyancy force

Figure 1. Direct and indirect methods of level measurements.

Table 2
Summary of Pros and Cons of Direct and Indirect Methods of Level Measurement

Method	Measurement	Advantages	Disadvantages
Direct method	Float type gauge	• Can be used for unlimited tank height. • Excellent accuracy of 4–5 mm • Comparatively low cost • Can be used for low- to medium-pressured tanks	• Intrusive sensors • Mechanical devices are subject to wear, corrosion, failure, and "getting stuck." • Subject to material buildup causing loss of accuracy
	Wave reflection from liquid surface	• High accuracy of 0.5–1 mm • No moving parts, nonintrusive components • Low maintenance, operational, and ownership cost; high MTBF of >60 years • Modular design to install in operation	• Can be applied for still liquid and not turbulent liquids. • Higher cost compared with other technologies. • Cannot measure interface. • Effective for liquid and slurries but requires caution for granular material
Indirect or inferential method	Measurement of hydrostatic pressure	• Simple and familiar technology • Reliable method to measure level, density, mass, and volume. • Can be calibrated from pressure standard. • Reasonable cost • Typical accuracy is ±0.03%	• Must be corrected for density. • Susceptible to fugitive emission • Requires additional valves and pipes. • Useful only for clear liquid applications • Not practical for slurries
	Measurement of buoyant forces by partially immersed object	• Limited motion and few moving parts • Proven for difficult hydrocarbon applications. • Can be used for high temperature and pressure. • Can measure interfaces	• Calibration is density-dependent. • High cost with long spans • Awkward to install and requires careful and constant maintenance. • Buildup on displacer can affect accuracy

Table 3
Methods and Purpose of Tank Process Variables

Process Variable	Method(s) of Measurement	Purpose of Measurement
Level	Manual, ATG	• Issue alarms to avoid spills. • Inventory calculations
Temperature	Manual, ATG	• Fugitive emission control and calculations • Temperature compensation in inventory calculations
Density	Laboratory, ATG, mass flow meters	• Mass-based inventory calculations
Pressure	ATG or none	• Fugitive emission control and calculations • Safety for pressurized tanks
Water level	ATG or none	• Limit water content and purge excessive water because it is detrimental to process units

Temperature Measurement - The temperature in a tank can be measured either at one point for a well-mixed tank or at multiple points for a stratified tank. In the case of the latter, tank temperature is typically measured at a dozen or more points using multi-resistor thermometers and then averaged over all of those points to calculate an average tank temperature.

Density Measurement - The measurement of density is measured by the hydrostatic (HTG) method to measure pressures at two points and then infers the average liquid density from their values using equation 1:

$$P_2 - P_1 = k \times \rho \times H \tag{1}$$

where:

P = Pressure at point 1 and 2,
k = Constant,
ρ = Density, and
H = Distance between points 1 and 2.

For this equation to be valid, the liquid density needs to be the same at points 1 and 2, which may not be valid in a stratified tank.

Pressure Measurement - The tank pressure is measured traditionally by a hydrostatic method using pressure transmitters at a lower point above the water and sediment level.

Multifunction gauging system

A multifunction gage uses a single instrument to measure all required parameters such as level, multi-profile temperatures and density, pressures, water levels, and tank empty space vapor pressure without the use of multiple instruments for each measurement. It consists of 4-12 measurement modules installed on a single steel pipe installed through one tank opening.

A multifunction gage is a single instrument to provide all necessary data to calculate volume by either the hybrid method (level, temperature, density, and water) or the hydrostatic method (mass, density, and water) in real time directly from the gage. Its transmitter can support over 1000 strapping points and can provide diagnostic measurements, raw measured data, and calculated data. It is the perfect gage for use with distributed control system (DCS), supervisory control and data acquisition (SCADA), and HMI software [1].

A multifunction tank gauge has the following features:

- It provides the most accurate volume measurement for custody transfer, inventory, or loss control on the market by a tank gage.
- It provides real-time volume by hybrid or hydrostatic methods (all product variables are measured from one instrument, one tank opening, one electrical connection, etc.).
- It provides real-time profiling of product within a liquid storage tank (water, density, and temperature stratification). It does not require additional instruments, tank entries, electrical connections, etc.
- It determines the product quality (i.e., specific gravity, free water, and entrained water content in product).
- It provides vapor pressure and vapor temperature for hydrocarbon emissions reporting, gas blanketing regulation, or alarm indication of a stuck pressure relief vent.
- It is certified for leak detection (mass sensitivity).
- It provides volume calculations using measured variables for multistate density (not reference/average density), free and entrained water (not just free water), actual vapor pressure and temperature, actual atmospheric pressure and temperature, actual mass, etc.
- It has no moving parts.
- It is self-diagnostic and self-calibrating.
- It is bottom referenced (standard). No reference point error as with roof-mounted level technologies.
- It installs in service with or without a gage well.
- When installed within the gage well, it provides a direct comparison of the ATG with the metrology reference point (i.e., the manual hand line and samples at the same physical location on the tank for level, temperature, density, and water. Thus, reducing measurement errors caused by multiple data locations from other hybrid-level gages.
- It has the lowest cost of ownership among inventory tank gauging technologies and the most benefits.

Inventory calculations

Methodology

In this section, we will discuss the steps required to calculate tank inventory, our ultimate goal, in a step-by-step procedure re. However, to understand the various terms that will be used in this section, let us examine a tank and its physical characteristics as shown in Figure 2.

The tank inventory calculations require tank parameters (height, diameter, and unusable volume), characteristics data (strapping table), and dynamic process parameters (level, temperature, density, pressure, water level). This process is shown in Figure 3.

The following steps are required, either manually or via software, to calculate tank inventory and other information needed for the real-time management of tank farm.

Step I—Calculation of raw volume from liquid level measurement: There are three methods available to convert liquid level to tank raw volume (i.e., without any temperature adjustment):

- Strapping table: This is a table of tank volume versus tank level provided by the tank manufacturer in the form of

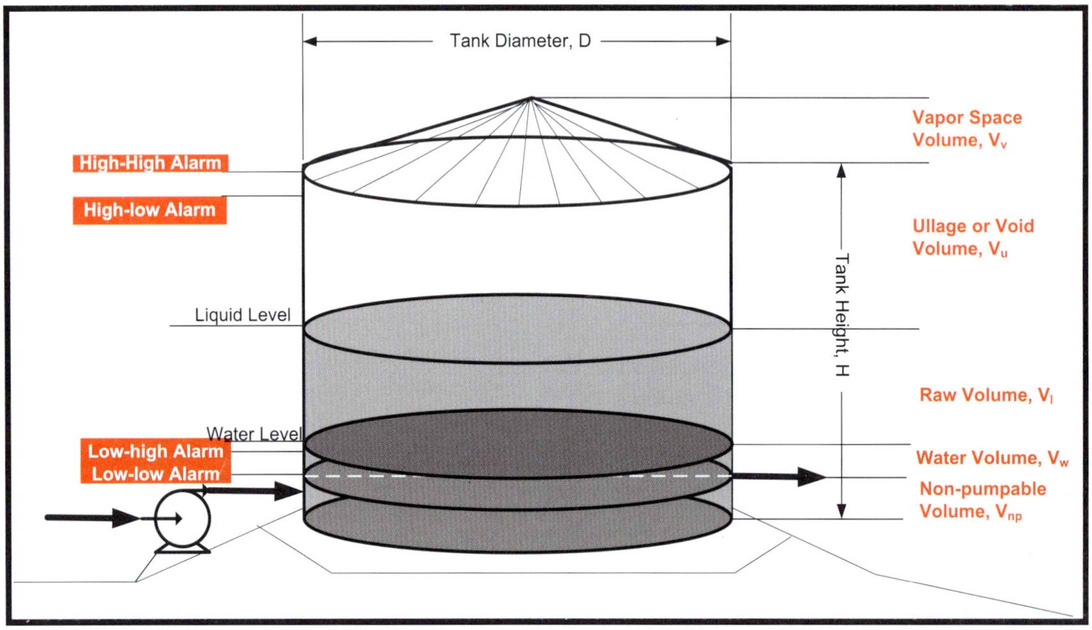

Figure 2. Tank parameters.

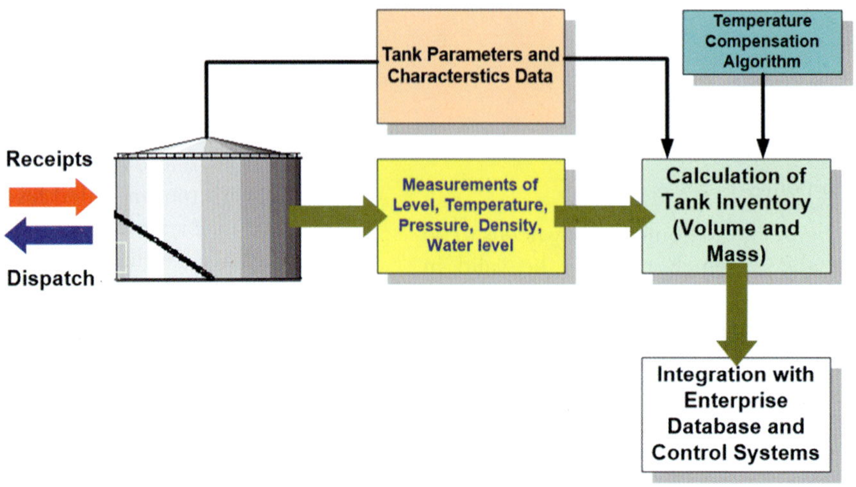

Figure 3. Flow diagram for the measurement and calculation of tank inventory calculation.

a printed tabular format. It is useful for a deformed and aged tank and is also necessary for noncylindrical tanks because tank geometry is not uniform with respect to height. Because this table is very long, it is resource-consuming for programming in a tank information automation system. This is especially true if the plant has old tanks and the tank tables were only provided on papers and not in electronic format.

- Strapping factor: This can be calculated very easily to use as a rough estimate of volume at a certain level using a simple calculator. It is accurate for cylindrical tanks and is easier to program. It is useful for calculating flow rates in and out of the tank on the basis of level changes and to estimate time for emptying and filling.
- Mathematical formulas: This method uses the mathematical equations to calculate for a given tank geometry, but it cannot be used easily with a calculator. However, it is accurate for all tank geometries and may have errors for deformed and aged tanks. It is not difficult to program and is more accurate than the strapping factor. Figure 4 illustrates the mathematical formulas used for common tank shapes in a plant.

Step 2—Temperature compensation of raw volume: Because the tank volume is dependent on the temperature, the raw volume cannot be used for custody transfer purposes and hence must be standardized to either 15 or 20°C using ASTM Table 54 and ASTM D125-80, respectively. This procedure ca n be used by using the FORTRAN program available from ASTM and is not a manual calculation. The temperature used for the calculation of compensated volume must be the average temperature to account for the temperature stratification in the tank.

Step 3—Calculations of tank mass: This step calculates the tank mass using the temperature-compensated volume from step 2 and density measurement at 15 or 20°C in the laboratory or using the automatic measurement from the tank gauging system or mass flow meter.

Step 4—Calculation of net volume, mass, and other information: This step uses the tank geometry using the non-pumpable volume to calculate the net mass and net volume of the tank for custody transfer, loading, and unloading purposes. Historical information is also used at this step to calculate the status of tank movements.

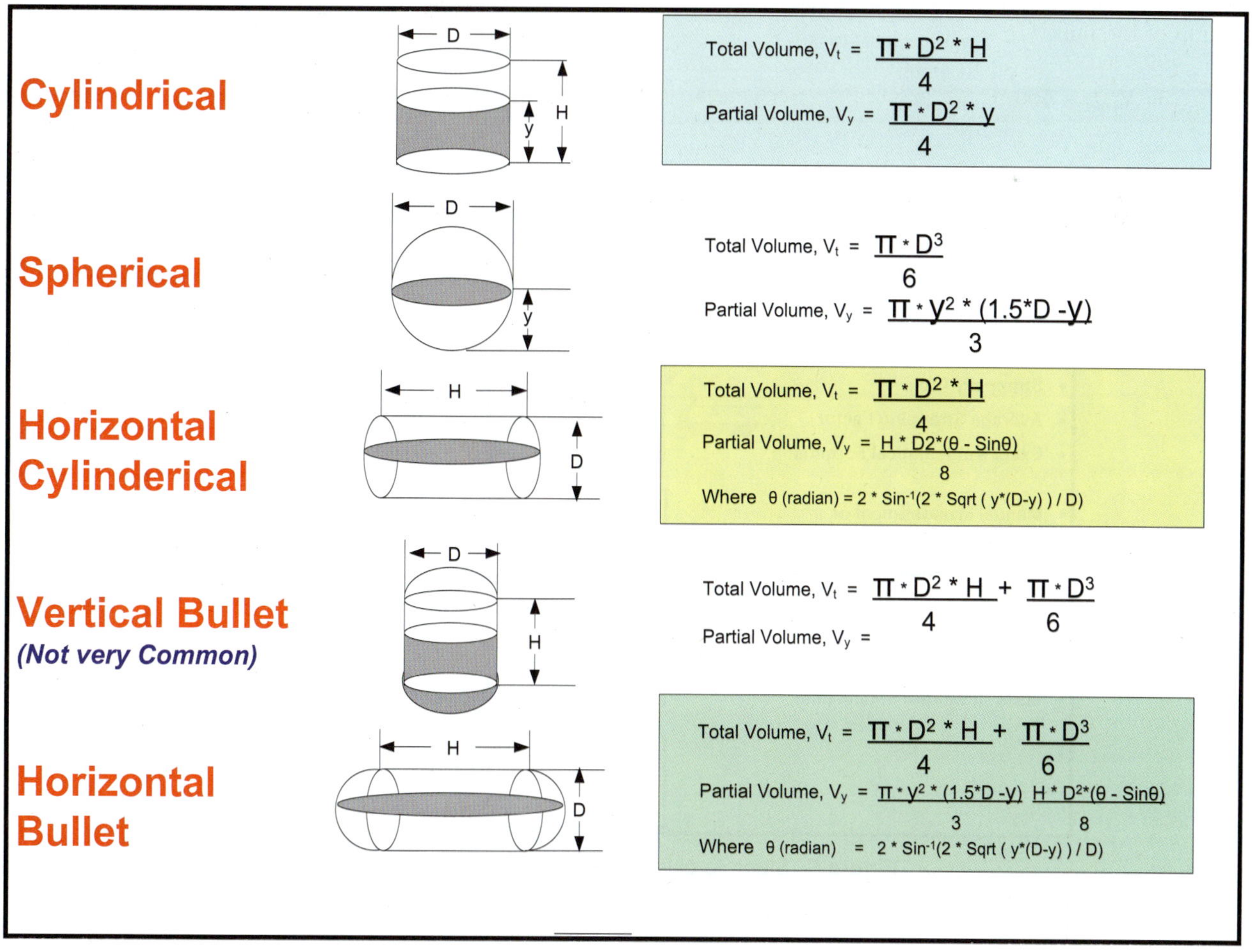

Figure 4. Calculation of tank volumes by mathematical formulas.

Equations

The following equations are used in tank inventory calculation and management systems:

- Gross or raw volume,

$$V_{gross} = \text{Level} \times \text{strapping factor} \quad (2)$$

or look up from strapping table or calculate using tank geometry and calculation equation.

- Compensated volume,

$$V_{comp} = V_{gross} \times \text{VCFACT} \quad (3)$$

where VCFACT is the temperature compensation factor calculated using ASTM methods at the reference temperature of 15 or 20°C.

- Void volume,

$$V_{void} = V_{maxcapacity} \times V_{gross} \quad (4)$$

- Percent void volume, which measures the void space at the top of the tank,

$$\%V_{void} = V_{void} \times 100/V_{maxcapacity} \quad (5)$$

- Non-pumpable volume (V_{npv}) is the fixed and known characteristic of the tank and it represents the volume at the bottom of the tank that cannot be pumped out because it is below the suction line of the pump.
- Maximum available volume,

$$V_{max} = V_{maxcapacity} \times V_{npv} \quad (6)$$

- Gross weight,

$$W_{gross} = V_{gross} \times \rho \text{ at reference temperature} \quad (7)$$

- Net available volume,

$$V_{net} = V_{gross} - V_{npv} \quad (8)$$

- Net weight,

$$W_{net} = V_{net} \times \rho \text{ at reference temperature} \quad (9)$$

Tank status is estimated by comparing the old level versus the current level within a measurement error tolerance. Rate of change of level and estimation of tank fill-up or emptying can be estimated as shown in Figure 5.

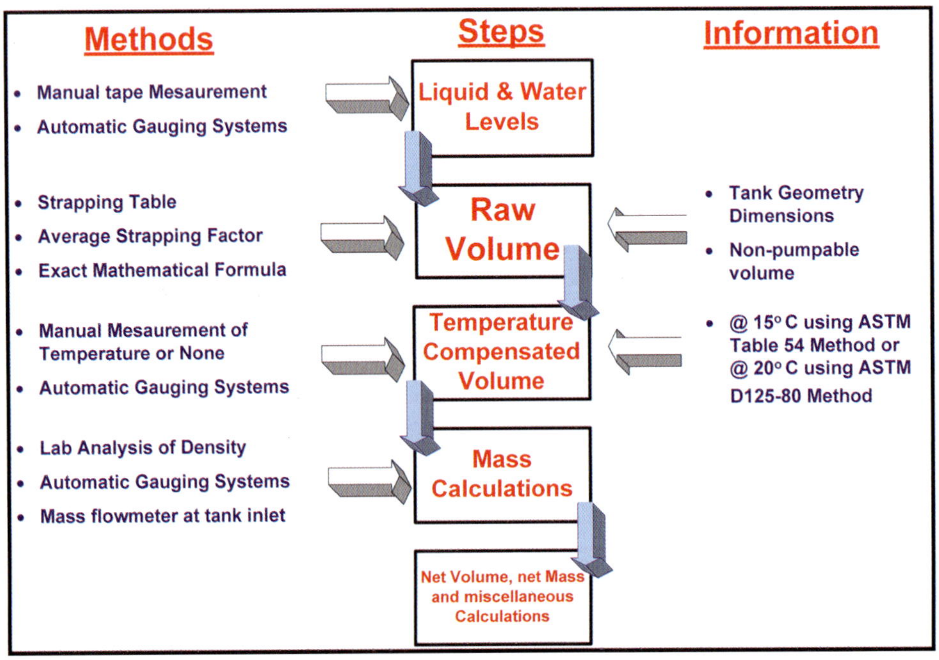

Figure 5. Steps to calculate tank inventory parameters.

$$\text{Time left} = \text{Void Vol} / \text{rate of Change of Volume}$$

$$\text{Rate of Change of Volume} = \text{Rate of Change of Level} * \text{Cross Sectional area of tank}$$

$$\text{Cross Sectional area of tank} = \pi * (\text{Diameter})^2 / 4$$

$$\text{Rate of change of level} = \frac{(\text{Level @ Time } T_2 - \text{Level @ Time } T_1)}{(T_2 - T_1)}$$

Figure 6. Method to estimate time for cylindrical tank emptying and filling.

Advanced inventory management

- Tank leak detection: If there are flow meters installed at tank inlet and outlets, then any leak can be detected by making a balance of total input/output flow and volume changes from the tank gauging system. However, variation in tank levels due to surface wave ripples should be considered before calculating actual volume changes.
- Flow meter calibration offsets: Just like leak detection, calibration of flow meters installed at upstream and downstream points of the tank can also be checked if the flow balance around the tank does not close within 2 % for a steady tank.
- Mass reconciliation and oil loss estimation: By balancing input and output flow and changes in tank inventories around a boundary of the tank farm, an estimate of oil loss can be arrived at in real time using a well-featured tank inventory management software system.

Tank qualities

In addition to tank inventory, it is important to know the qualities of tank for the following purposes:

- Tank certification before final sale to customer,
- Qualities of feed material purchased from external sources,
- Input feed qualities to process units, and
- Qualities of the blending component tanks to control the blend recipe.

These tank qualities are obtained in two modes: offline mode by laboratory samples and analysis and online mode by using the online analyzers.

Storage tanks emissions

Fugitive emissions are the unintentional release of material from equipment and they can occur from any leaking equipment such as pumps and compressors, storage and processing vessels, flow control and pressure relief valves, pipelines, tanks, etc. Although the fugitive emissions are small in quantity, their origin exposes workers and they cumulatively cause concern for environmental pollution.

Fixed Roof Tanks

Although they are the least expensive, fixed roof tanks are mainly used lor storing low-pressure liquids. The emissions from this tank are caused by the changes in temperature, pressure, and liquid levels while filling and emptying tanks. When the tank is filling, the displaced vapor vents directly to the atmosphere and fresh air fills in when the tank is emptying. This tank has minimal control over the fugitive emissions, and refineries are burdened with estimating the total vapor discharged per year on the basis of tank movements. This type of tank is also fitted with a pressure vacuum vent to bring in the fresh air on cooler evenings to balance the pressure below the vent pressure, thus minimizing the vapor discharge to the atmosphere.

Floating Roof

There are three types of floating roof tanks used to overcome the shortcomings of the fixed roof tanks.

- External floating roof tanks: They have an open-ended cylindrical steel shell that floats on the liquid surface to minimize the vapor space above the liquid level. The emissions from this tank are less than the fixed roof tanks and limit losses from the rim seal system, fittings, exposed liquid on the tank walls after the liquid is withdrawn (emptying tank), and the roof lowers.
- Internal floating roof tanks: These have an improvement over the external floating tank by installing a fixed roof to protect from sun and rain. They have limited emissions from docks fittings, docks seams, and the annular space between the floating internal roof deck and the tank wall. Top and side vents on the fixed roof allow for the discharge of vapor. The fixed roof may be dome shaped to block winds and minimize evaporative losses.
- Closed floating roof lank: This type of tank has closed vents on the fixed roof and a pressure-vacuum vent uses injection of an inert gas (e.g., nitrogen) to balance the pressure variation and provide essentially a zero-emission tank.

Pressure Tanks

Pressure tanks are spherical or bullet-shaped used for storing gases with high pressure such as butane and pentane. These tanks are equipped with pressure-vacuum vents to minimize losses from boiling and breathing loss from temperature and pressure changes.

Detection methods

Some common methods to detect the fugitive emissions from field equipment such as valves, relief valves, compressors, pumps, drains, etc., are

- Portable gas detectors,
- Catalytic beads,
- Nondispersive infrared detectors,
- Photoionization detectors,
- Combustion analyzers, and
- Standard gas chromatography with flame ionization detectors.

References

1. Offsite Management Systems LLC (OMSLLC), *Training Manual*, OMSLLC, Sugar Land, TX (2009).

Excerpted from Ch 20 - Tank Farm Management by Suresh S. Agrawal, ASTM Manual MNL58 Petroleum Refining and Natural Gas Processing, 2013. For full text and citations, see original paper. Reprinted with permission of ASTM.

Managing oil movement and storage in tank farms

Excerpted from Tank Farm Management by Suresh S. Agrawal, ASTM Manual MNL58 Petroleum Refining and Natural Gas Processing, 2013.

One of the most manpower-intensive activities in a typical refinery or petrochemical facility is managing the movement of liquids in and out of tanks in and across the plant boundaries. Equipment management and efficient operation of material movement activities is termed oil movement and storage (OM&S), and in this article we will discuss the following aspects:

- Problems and challenges,
- Operational data analysis,
- Estimation of automation incentives,
- Automation technology and strategy, and
- Terminal operations.

Problems and challenges

Problems typically associated with OM&S activities can be classified into the following four areas:

1. Decrease in plant profitability,
2. Operating losses,
3. Product losses, and
4. Operating costs.

Decrease in Plant Profitability

Plant profitability may decline because of various factors:

- Product quality, quantity giveaway, or both: This is encountered in activities such as blending and custody transfer because of a lack of inline blending and accurate measurements.
- Product degradation due to displacement of previous line material: Transfer of products of different qualities in activities such as tank-to-tank transfer, tank-to-ship transfer, etc., in the same pipeline segment degrades the products. The product degradation usually requires selling the product at a lower price or quality correction by reblending or addition of corrective components.

Operating Problems

Operating problems vary from areas related to man/machine interfaces to errors by field operators and they may result in economic losses. Some of most common operating problems are:

- Multiple operator interfaces: One of the main concerns of OM&S operations in a refinery is the operator having to work with multiple types of displays and systems such as tank gauging system, DCS workstations, blend control, MIS computer, laboratory information system which at times reduces his or her effectiveness and may result in operational errors which at times may be very expensive for the refinery's profitability.
- Line-up errors: The refinery has a very complex network of pipelines and field equipment. To line up various valves, pumps, etc., for a transfer activity requires careful execution and may result in product contamination if any errors are made. Line-up activity is also time-consuming and requires the availability of field operators in a timely fashion.

- Product contamination: This may be the result of line-up errors or unplanned or unauthorized movements.
- Pump cavitation: This situation could occur if the valve before the pump is stuck or does not open fully. Failure of the valve downstream of the pump could also damage the pump.
- Floating roof sinking: This problem may occur if the liquid in the tank falls below a certain level because of a lack of timely action by the operator due to failure of a low-level alarm.

Product Losses

The direct effect of product losses is the decrease in plant profitability, and the following factors contribute to this problem area:

- Inadequate or erroneous measurements or both: OM&S activities involving movement of products across the refinery boundaries such as custody transfer and marine loading/ unloading may incur product losses because of inadequate or erroneous measurements.
- Unplanned and unauthorized movements: These are related to human error by the operator, field operator, or both. The product loss may occur in terms of the need to reprocess the contaminated products or sell at a lower price because of its downgraded quality.
- Leaks: Product leaks from valves and tanks result not only in products losses but also present environmental hazards. An unattended tank during its filling or failure of a high-level alarm may cause spills and product losses, environmental hazards, and clean-up cost associated with it.

Operating Costs

In addition to economic loss to a plant, a plant may also incur an increase in operating cost due to the following factors:

- Large plant inventory: In the absence of timely information on the availability of raw material and finished products, plants resort to holding a large plant inventory. Storage cost, unnecessary movement of products, not being able to respond to all market demands and fluctuation are some of the factors attributing to increased operating cost.
- Underutilized tank capacities: Operators may fill tanks to levels lower their maximum due to fear of tank spills and failure of alarm. This results in underutilization of tank capacities and increase in operating cost.
- Wasted energy: Electric utility cost increase if transfers are repeated because of human error or from execution of avoidable transfer activities.
- Demurrage charges: Demurrage charges payable to shipping companies because of delays in marine loading and unloading can be a substantial loss for the refinery.
- Large manpower: Manual operations of opening and closing of valves and pump start/stop sequences require many field operators for the transfer activities. In addition, more maintenance personnel are needed in repair.
- Improper and untimely reports: Lack of any automation and implementation of information system or both may not provide inventory reports in a timely and accurate fashion for the day-to-day planning cycle of operations.

Most of these problems can be eliminated or reduced by implementing computer automation of OM&S. However, before implementing an OM&S automation project, a first step would be analyzing the OM&S operational data.

Operational data analysis

A refinery typically performs between 80-120 movement tasks on a daily basis. These tasks have various combinations of sources and destinations and vary in execution frequency and duration. Agrawal (1996) pointed out that a benefit study must establish the basis to recognize the estimated benefits from automation of OM&S and this is done by analyzing the operational data for a specific process unit or an entire area's operations.

The period for such operational data must be between at least four months and preferably up to one year. In the case of offsite operations in a refinery, it is necessary to collect the following information for each OM&S task:

- Identification number
- Start date, end date, and time
- Source/destination identification
- Source/product name
- Initial/final levels of source
- Initial/final levels of destination
- Initial/final and total quantity transferred and
- Planned volume or weight for transfer.

Automation technology and strategy

Requirements

The automation of an OM&S operation requires that the computer system respond to the needs of the operator to perform the major functions. These include the following:

- Use remote readings from tank gages to
  - Calculate inventories, conditional events, and total volumes shipped.
  - Monitor tanks for improper movements and alarm on high or low levels.

- Respond to field element status:
  - Monitor remotely operated valves (ROVs), pumps, and mixers for alarm conditions and proper response operator requests.
  - Keep a list of out-of-service equipment.
  - Provide status information to operator.
- Monitor and control movements for receipts, rundowns, transfers, and shipments.
- Select equipment (tanks, valves, pumps, etc.) for movements required in the schedule.
- Automatically sequence equipment to implement and terminate movements.
- Terminate movements, swing tanks, start other movements, etc., when specified conditions are met.
- Provide interfaces between DCS and supervisory-level controls of inline blenders and equipment sequencing, movement monitoring, and customers blend control strategy programs.
- Provide summary reports and movement logs and profiles that are defined by the customer.
- Use networked computers to receive operating data and to transmit historical data.

The characteristics of a well designed and developed OM&S computer system provide interfaces and interaction among all activities of the OM&S operations in a refinery

OM&S Control Functions

The functionality of an automated OM&S system can be translated into various control levels, as shown in Figure 1:

- Level 0: This is the lowest level at which all field equipment such as pumps, ROVs, mixers, and control valves are controlled by the electronic system. Tank levels and temperature are also monitored at this level.
- Level 1: This level is called the supervisor control in which computer systems monitor the status of process variables or field equipment or send the command for remote control operation. For example, instead of starting or stopping a pump from the field or remote panel the operator could issue a command from the computer to operate a valve or pump. Tank gaging systems such as Wesson, Varec, Enraf, and Saab provide computer interfaces to monitor the tank levels and temperatures.
- Level 2: This level involves extensive application programming and offers the maximal benefits of automation. Various steps of executing an OM&S activity are automated at these levels and are categorized into separate modules such as task definition, sequence generator and control, task monitor, etc. Blending is also implemented at this level.
- Level 3: This is the topmost level and is usually termed as the information system level. Information about various refinery activities such as planning, scheduling, shipping, and dispatching is collected at this level and is interfaced with OM&S control modules for overall integration of the system.

Categories Of OM&S Control Functions

OM&S control functions can be divided into three basic categories:

1. What must be done?
2. What is going on now?
3. What has happened?

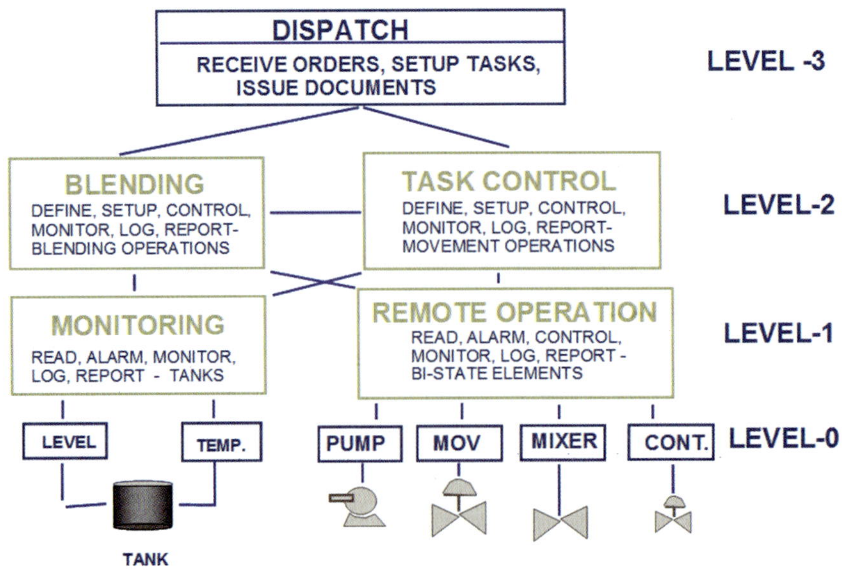

Figure 1. Control levels of oil movement system.

Almost all functionality of an OM&S computer belongs to a category answering one of these three questions.

Steps to Execute an OM&S Activity

The following steps to execute an OM&S activity arc valid regardless of their automation; that is, the operator goes through these steps even in manual operations.

- Selection of activity type
- Task definition
- Path selection
- Sequence generation, and
- Sequence control

We shall walk through an example of an OM&S task with all these steps, along with representative displays, to illustrate how an OM&S operator would use an OM&S computer system to perform his daily transfer activities. Figure 2 shows the functional modules of a typical OM&S system.

Selection of Activity Type. This is the first step for the operator to select the type of transfer activity he wants to execute by choosing a task type from the menu. The operator can select the task type (ID) if he knows it already.

Task Definition. After selecting live activity type, tank-to-tank transfer in Figure 2, the next step is the task definition step. This step requires three inputs: source, destination, and stop condition or size of transfer. Because the task database has other information such as material, pump ID and task ID. this information is shown only upon entry of source and destination. The operator can also enter the task ID if this information is known. The operator can always override the inferred data shown by the system.

Path Selection. The next step of task implementation is the selection of a path through the refinery network. The path selected is governed by two specifications: the data defined for the task and the path/network database. A path for a specified task can be defined in one of two ways: the predefined or "canned" path or the dynamic and optimal path. The approach followed for a given refinery depends on the preferences of the organization. The canned path algorithm is not very flexible compared with the dynamic path algorithm, but it takes levs resources and is more efficient. The dynamic path algorithm is recommended for a complex and interactive network of field equipment. In Figure 3, the path selected for the specified task is displayed in tabular and graphical formats.

Sequence Generation. After the path is selected and approved by the operator, a sequence of operating the field equipment needs to be generated. This sequence consists of field equipment for the isolation of path and for flow path and startup (i.e., pump) of the task. Each piece of field equipment in this sequence of operations has a relevant action (of valve close or open) and its own individual modes of operation (i.e., manual, automatic, or required operator initiation) as shown in Figure 4.

Task Status Monitoring. After the OM&S task is implemented, the operator can monitor the status of tasks in several ways, each displaying information from different perspectives as follows:

- Status of single task.
- Status of all tasks (active, defined, scheduled, committed, unassigned, etc.).

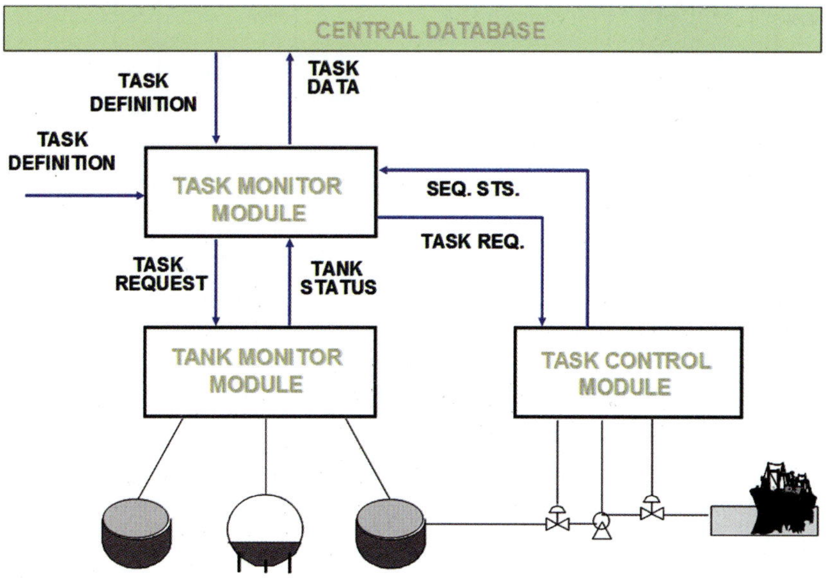

Figure 2. Functional modules of an oil movement control system.

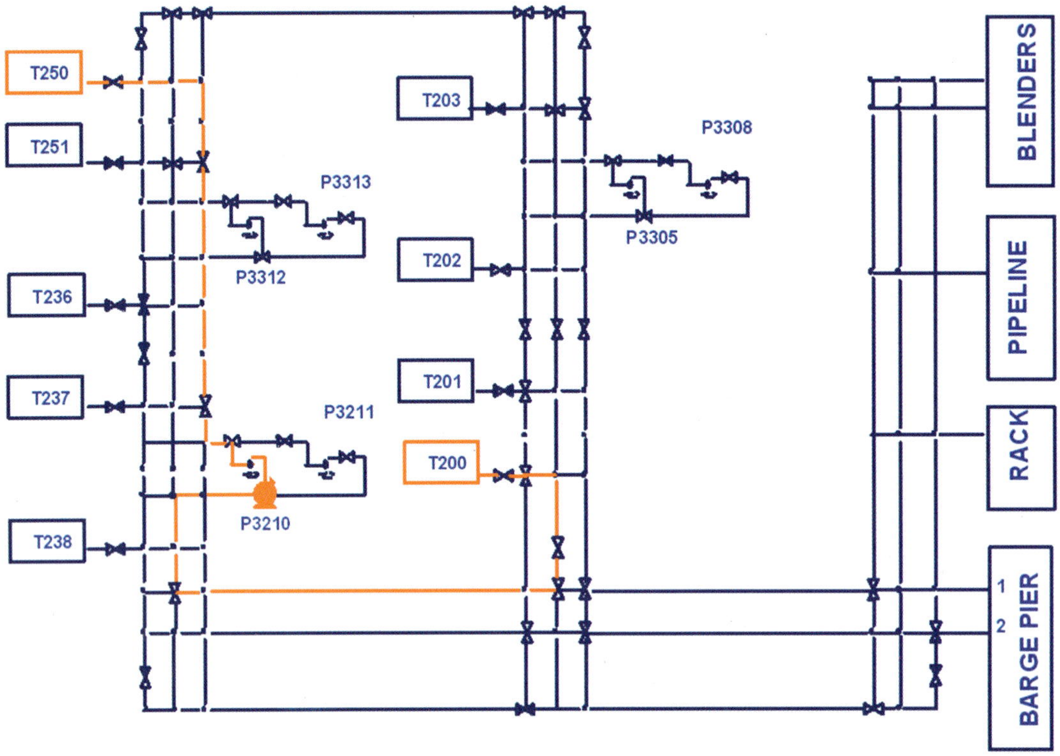

Figure 3. Path selection of an oil movement control system.

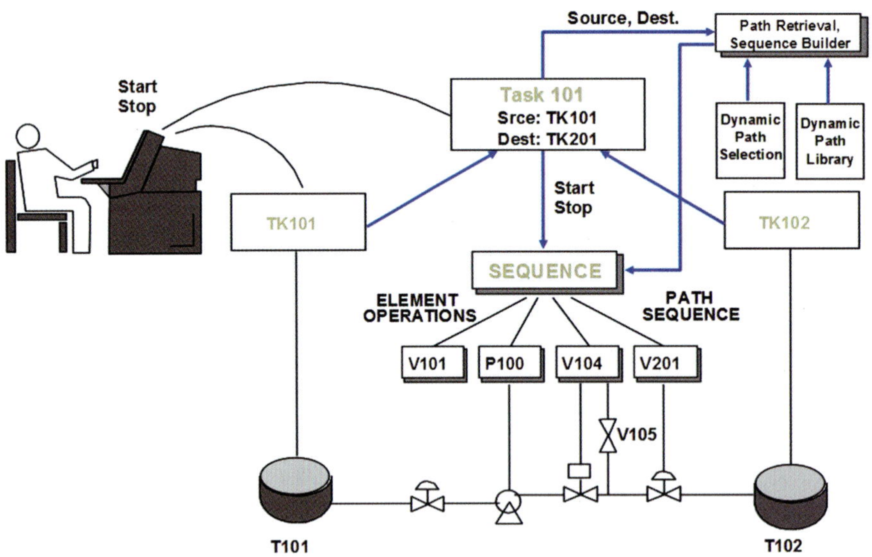

Figure 4. Sequence generation of an oil movement control system.

- Status of a group of tasks such as all those related to gasoline activities; and
- Screen report of tank inventory, which can again be grouped by products or area or tanks.

Integrated OM&S System. The components of an integrated OM&S computer system depend on the status of automation, the refinery information system in a particular refinery, and the computing platforms of the OM&S system. For example, some commercially available OM&S systems include blending as one of the modules. The OM&S algorithm as such does not include any linear programming (LP) algorithm, whereas the inline blending uses either LP or the nonlinear equation solver and optimization using nonlinear blend models depending on the type of product blend.

A truly integrated OM&S system must interface with the refinery's information management system for history, databases, reports, reconciliation, etc. This results in the customized integration of an OM&S system for each refinery and may involve developing many interfaces for various modules and computers (Figure 5).

Terminal operations

Terminal operations handle crude/product receipts and product dispatch at marine, truck, railcar or wagon, and pipeline terminals, with each having their own characteristics from an operational point of view as well as their own problems and challenges. Custody transfer is the reconciliation of accounting of the amount of refinery feed or product exchanged between the refinery and an external party (buyer or seller). Delayed reconciliation and documentation results in transfer disputes between supplier and refinery and a loss of revenues. Generally speaking, marine and pipeline terminals pose the most challenges compared with trucks and railcar terminal operations (Figure 6).

Truck Terminals

Truck terminal operations face the following challenges:

- They require efficient planning and scheduling.
- Tank compartments are manually gaged, leaving room for custody transfer discrepancies.
- Human errors can cause product contamination in truck compartments.
- Requires planning product inventory to schedule incoming trucks.
- The trucks queues must be coordinated efficiently.

Figure 7 shows an example of truck terminal operation.

Marine Terminals

The challenges of marine terminal operations include:

- Planning and scheduling,
- Demurrage charges,
- Accurate flow measurement,
- Flow meter proving,
- Water content analysis,
- Crude oil sampling point location,

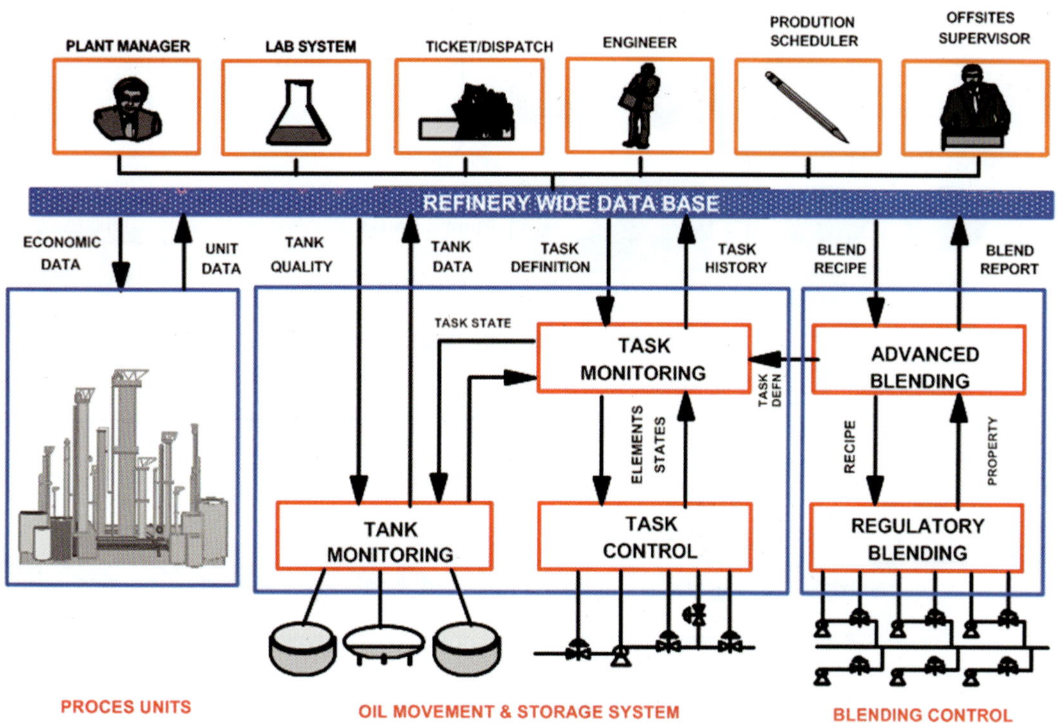

Figure 5. Integration of OM&S system with refinery-wide planning and scheduling system.

Challenge Severity Matrix

	Marine	Trucks / RailCar	Pipeline
• Demurrage Charges	High	None	None
• Opportunity loss due to product unavailability	High	Small	High
• Quantity giveaway due to faulty meters	High	Small	High
• Product degradation	Small	Negligible	High
• Contamination	Small	Negligible	Medium
• Lack of proper scheduling of resources and Products shipment or feed receipts	High	Medium	High
• In-time billings	High	Small	Small
• Improper feed / Products preparation due to lack of tanks	High	Negligible	High
• Custody Transfer reconciliation Errors	High	Small	High

Figure 6. Challenge severity matrix of terminal operations.

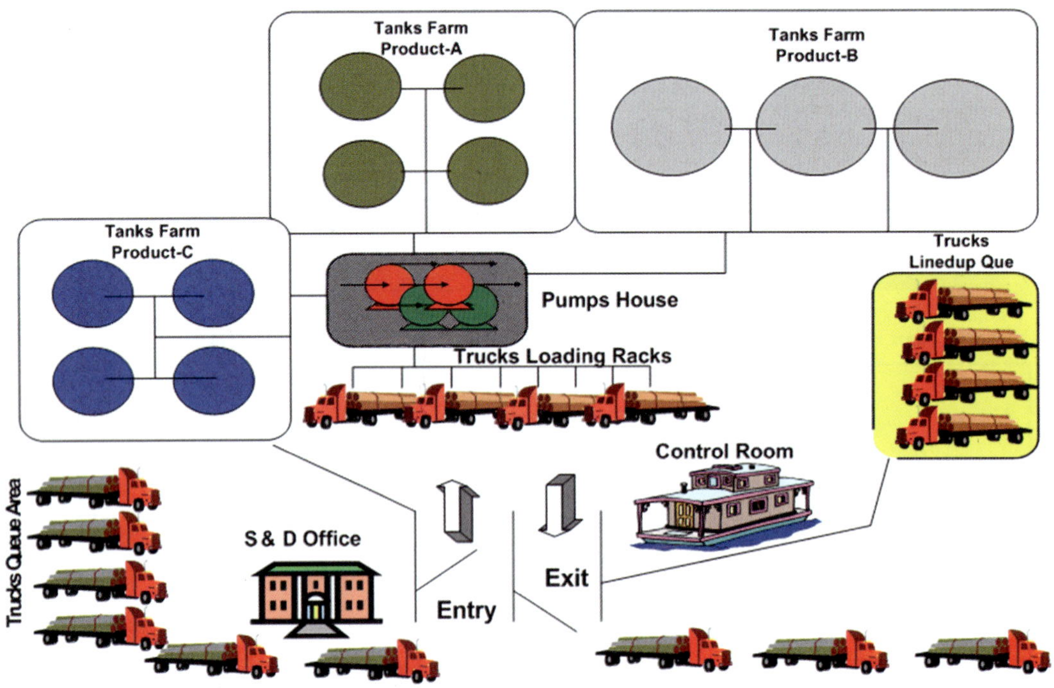

Figure 7. Example of truck terminal operation in a typical refinery.

- Representative sampling of crude oil, and
- Sampling system proving.

Railcar Terminals

Railcar terminal operations are similar in operations to the truck terminals. However, their activities are for longer periods of time and the scheduled wagons are less frequent than truck terminals on a daily basis.

Pipeline Terminals

Pipeline terminal challenges include:

- They are the most difficult to manage because of distance.
- They need elaborate equipment (pumping stations) to send material over varying heights.
- Leaks may disrupt the service and cause safety hazards.
- Product contamination.
- They involve multiple enterprises and vendors.
- They require robust control systems and sophisticated remote monitoring instruments.

Blending

Blending is the process of mixing two or more ingredients to produce a product with certain desirable qualities or attributes. The final product qualities must not violate the specified qualities or attributes and must be produced with minimal cost or maximal profit. End-users require stringent product qualities, efficient and low-emission fuels.

Crude Blending

The main objectives of a crude unit in a refinery are to optimize the quality and yields of cuts and operate the unit in a stable and safe manner. It is important that crude feed has uniform quality and availability. Crudes are blended from different sources with varying qualities to feed to the crude unit to maintain a consistent quality. The objectives of crude blending are

- Minimization of dependency on crude feed quality for stable and optimal operation,
- Maximization of product yields and quality,
- Improve crude oil inventory for "just-in-time" production, and
- Reduction in demurrage by better planning.

These objectives can be achieved by the following feed and product quality-control systems:

- Feed quality control
  - Marine planning and scheduling
  - Movement tasks monitor and control
  - Crude tank composition, monitoring, and optimization
- Product quality control
  - Crude unit model
  - Crude unit LP optimizer
  - Product blending optimizer

Figure 8 shows a schematic diagram that integrates all systems required to control the quality of crude feed. Crude can be blended in any of the following modes:

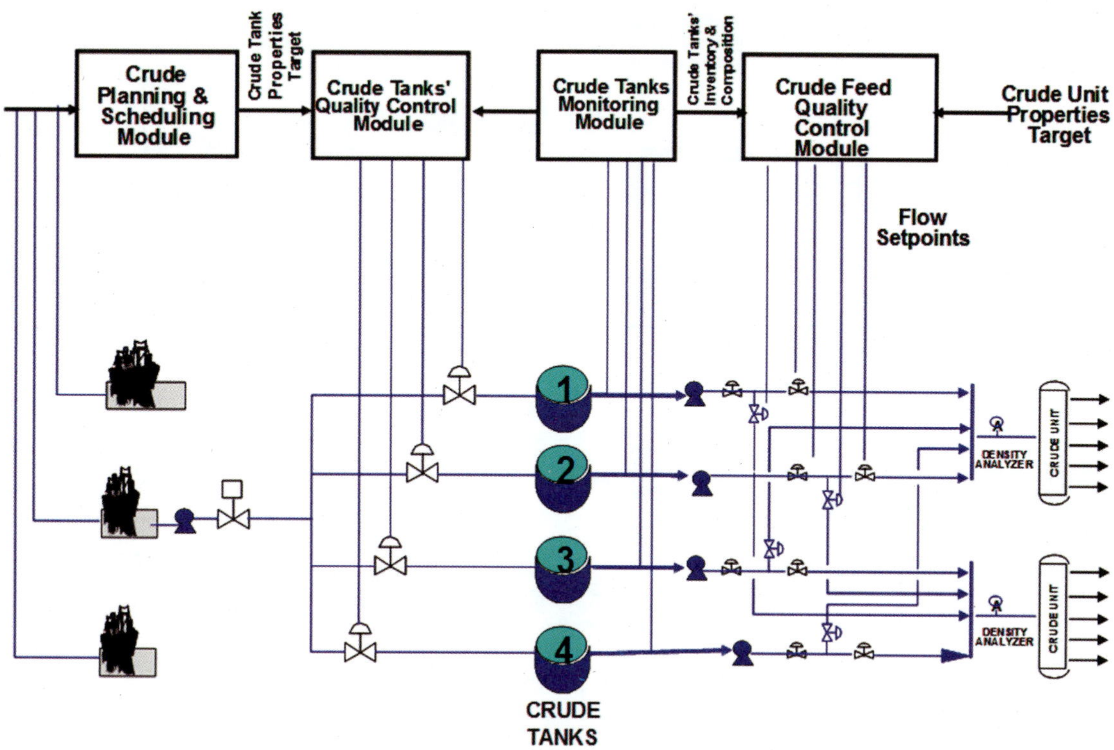

Figure 8. Integrated crude feed quality-control systems.

- "Pipelines-to-tanks" blending,
- "Ships-to-tanks" blending,
- "Tanks-to-tanks" blending, and
- "Tanks-to-units" blending.

Crude blending can be complex if the crude supply is via pipelines, and their source, types, and qualities vary extensively. Crudes from pipelines can be stored in tanks keeping their individuality, or they can be blended upon receipt in the tanks to achieve uniform qualities for a particular type of crude (e.g., light, heavy, low- or high-sulfur, paraffinic, naphthenic, etc.). Figure 9 shows an example of a crude blending configuration receiving crude via pipelines and two-stage blending to prepare specific crude unit mixes, and in Figure 10 multi-header crude blending is depicted.

Lube Oil Blending

Lube blending is different than crude or fuel blending. Lube blending is characterized by 300-400 types of finished products with smaller batches of 15-25 bbl. On the other hand, fuel blending produces 10-15 types of grades with 50,000 to 100,000 bbl batch size. Lube blending has a higher risk of contamination leading to sending the batch to burners and loss of revenue. Inline blending for the lube plant is economical for plants producing more than 150,000 bbl/year. There are two methods of inline lube blending:

1. Stationery batch tanks: In this mode, the batch holding tanks remain stationery and can meet product demands rather quickly. On the downside, the mode requires a great deal of pipe and instrumentation to control the batch quality.
2. Moving batch tanks: This mode keeps the batch tanks moving and as they pass the filling stations, recipe components are blended. The mode requires less manpower, piping, and valves and it has better quality controls.

Naphtha Blending

Naphtha is a byproduct of crude distillation, and its boiling point is between 30 and 200°C. It typically constitutes 15-30 % of crude oil by weight. Light naphtha is the fraction boiling between 30 and 90°C and consists of molecules with five to six carbon atoms. Heavy naphtha boils between 90 and 200°C and consists of molecules with 6-12 carbons. Naphtha is used primarily as feedstock for producing high-octane gasoline (via the catalytic reforming process). It is also used in the petrochemical industry for producing ethylene.

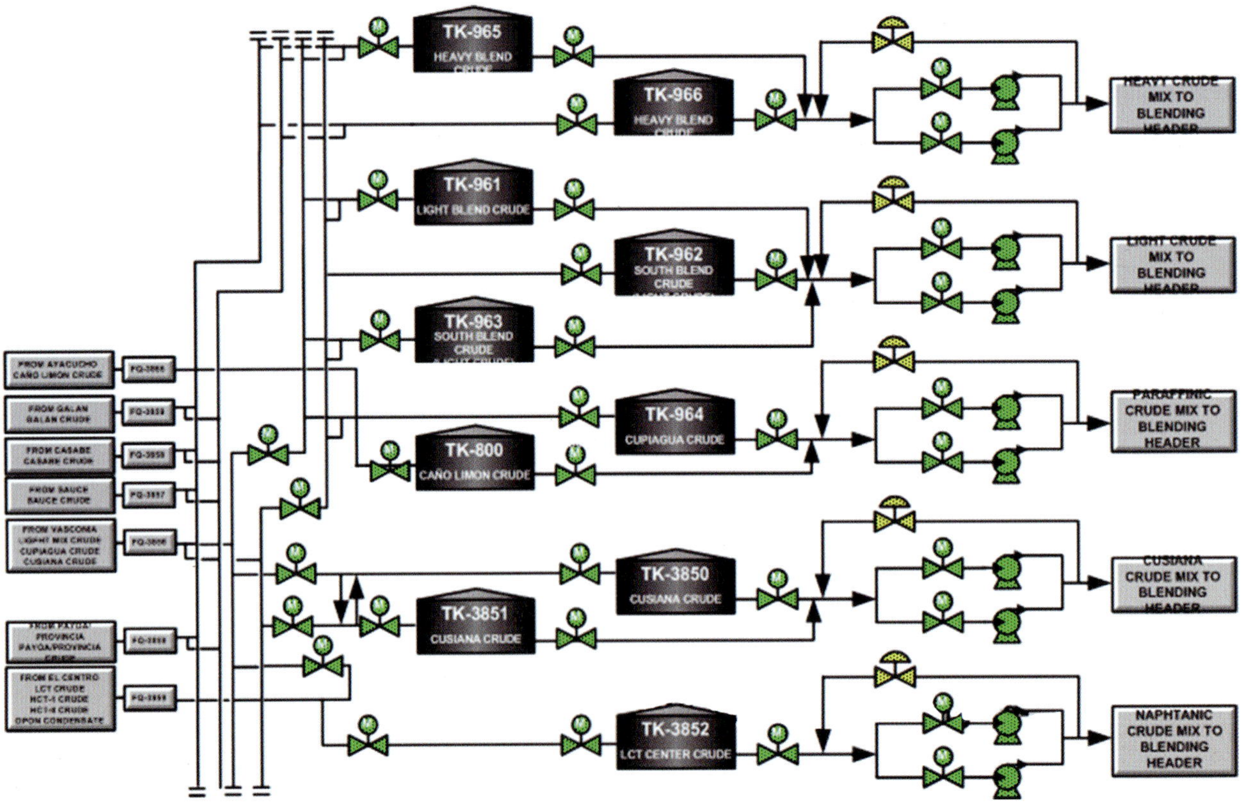

Figure 9. Typical example of pipelines-to-tanks crude blending.

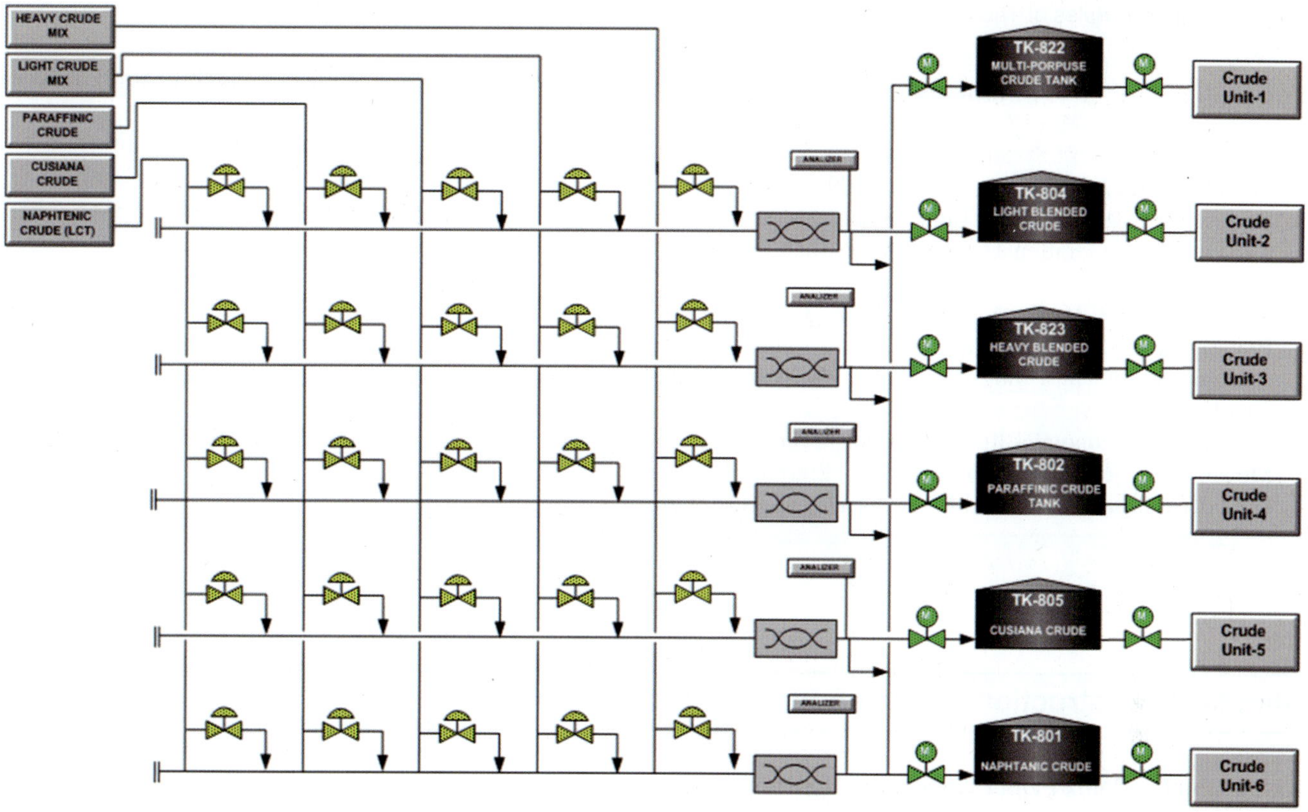

Figure 10. Multi-header crude blending.

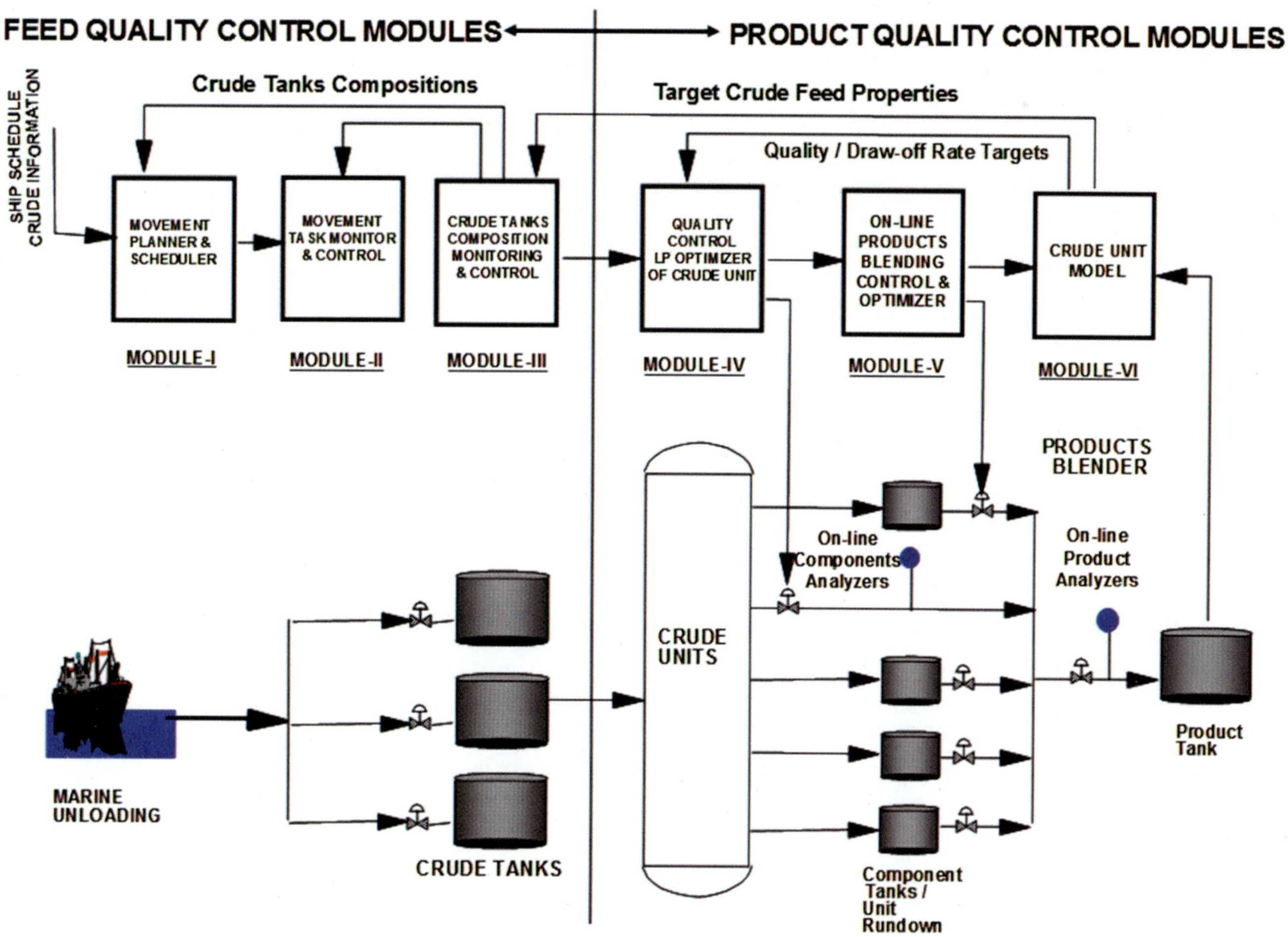

Figure 11. Integrated crude and product blending and optimization strategy.

References

1. Offsite Management Systems, LLC (OMSLLC), Training Manual, OMSLLC, Sugar Land, TX, 2009.
2. S.S. Agrawal, Advances in Tank Quality Measurements Can Help Cut Operational Costs, Hydrocarbon Process 86 (2007) 67.
3. S.S. Agrawal, K.M. Leong, L.H. Wee, C.T.J. James, Implementation and Benefits of Online Tanks Quality Tracking System in a Singapore Refinery, Hydrocarbon Asia 15 (2005) 36.
4. S.S. Agrawal, Economics Justifications of an Integrated Oil Movement & Storage Control System, in: Presented at the ISA'96 Conference and Exhibit, Chicago, IL (October 9–12, 1996).
5. "Oil Movement and Storage," (sales brochure), Honeywell, Inc., Morristown, NJ.
6. S.S. Agrawal, Scope of an Integrated Oil Movement and Storage (OM&S) Control System, in: Proceedings of ISA Conference, New Orleans, LA (October 1995).
7. S.S. Agrawal, Scope and Feasibility of Integrated Crude Blending Control and Optimization System, in: Presented at the ISA'94 Conference and Exhibit, Philadelphia, PA (May 9–12, 1994).

Excerpted from Tank Farm Management by Suresh S. Agrawal, ASTM Manual MNL58 Petroleum Refining and Natural Gas Processing, 2013. For full text and citations, see original paper. Reprinted with permission of ASTM.

Standard construction practices for LNG storage tanks

Rama Challa, Matrix PDM Engineering, Houston, Texas

Standard practice

At a time when small and large scale LNG facilities are in great demand, it is especially important to recognize their complexity. Owners and contractors have a responsibility to understand the equipment and nuances involved in design, construction, commissioning and operation. Umbrella standards such as API 625 ensure that systems provide a uniform level of performance, safety, reliability and quality, and provide an overall framework. API 625 addresses all aspects of a tank system and differs from component standards, such as API 620 for metallic materials.

This article introduces components of LNG storage tank systems, and covers storage types, construction materials, accessories and appurtenances. Certain aspects, such as performance criteria, are also included. In addition to this, a few key construction and post-construction activities are highlighted. Appropriate sections of API Standard 625, 'Tank Systems for Refrigerated Liquefied Gas Storage' are referenced.

Codes and standards

Storage, though critical, forms only a part of an LNG facility (Figure 1). Regulatory standards within the US, such as 49 CFR 193 and NFPA 59A, invoke API 625, which, in turn, defers to standards such as API 620 for metallic materials and ACI 376 for concrete parts (Figure 2). Figure 3 indicates API 625's scope of coverage.

API 625 is handled by the Refrigerated Tank Task Group (RTTG) of the Subcommittee on Aboveground Storage Tanks (SCAST) within the Committee on Refining Equipment (CRE). API standards are consensus standards developed by industry experts drawn from owners and contractors. The RTTG was created in spring 2006 due to a perceived need for an umbrella standard. The first edition was published in August 2010. Some RTTG members participate in NFPA 59A, ACI 376, as well as other standards.

A typical LNG storage tank (Figure 4) can be considered as an inner tank with refrigerated liquid insulated from ambient conditions by a concrete or steel outer container. The 11 sections and five appendices of API 625 define the various components (Figure 5). The appendices address gas properties, foundation settlement, risk assessment and selection of storage, as well as inquiries.

Containment types

The three main storage concepts are covered in API 625 (Chapter 5 and Appendix C).

A single containment system consisting of an inner and outer container, designed and constructed so that only the inner container is required to contain liquid and be

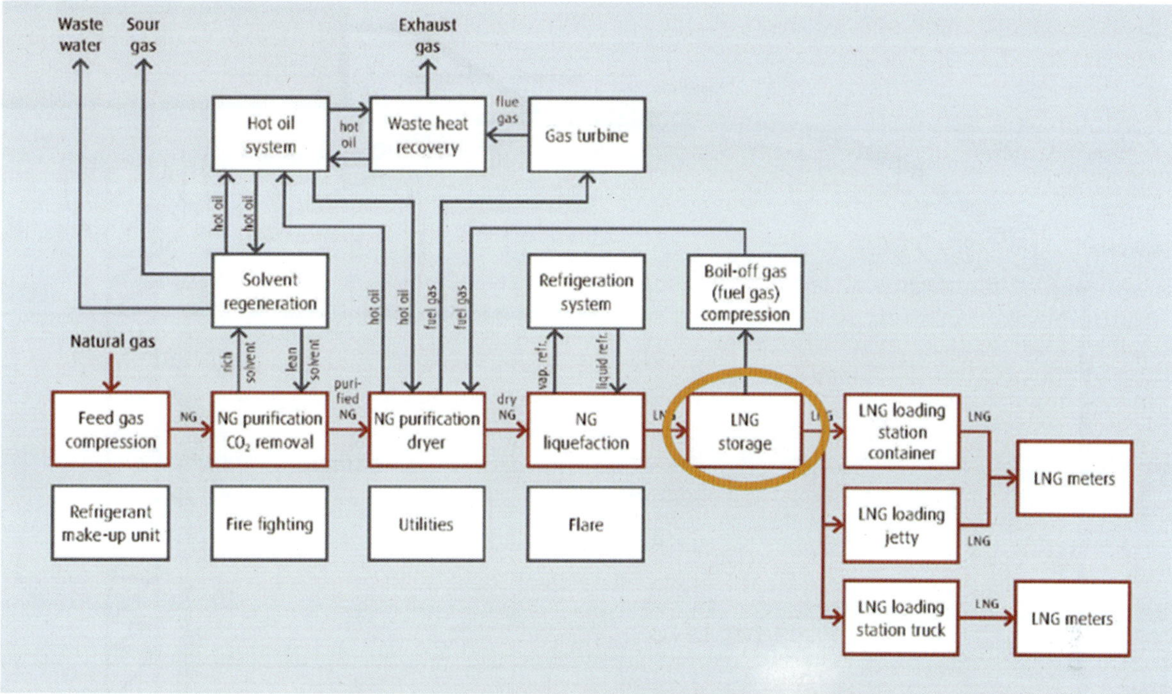

Figure 1. Storage: one component in the system.

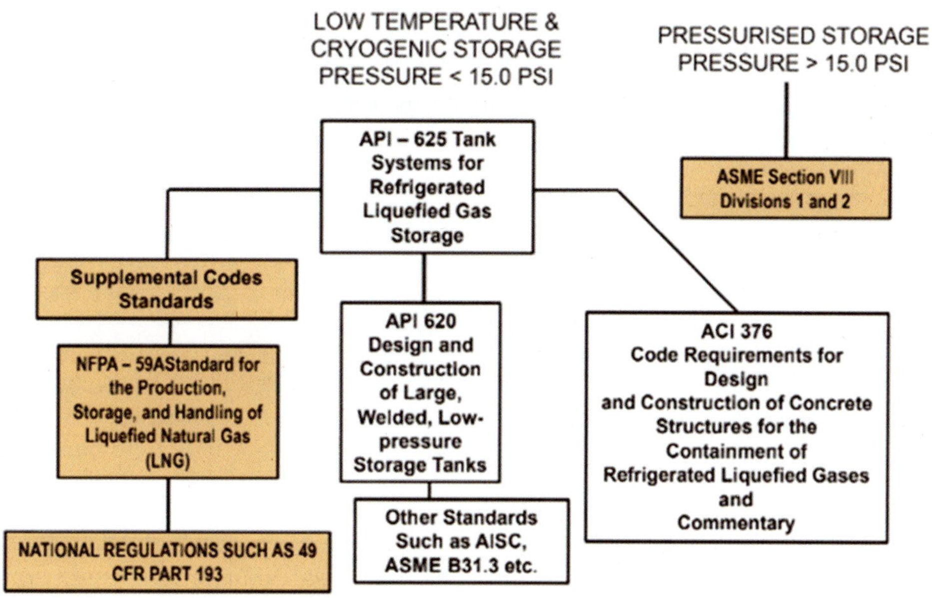

Figure 2. LNG tank U.S. codes and standards structure.

liquid-tight. The outer container, designed for product vapor pressure, retains and protects the insulation, but does not contain liquid if the inner container leaks. These systems require dikes that impact facility layout.

The double containment tank system consisting of a liquid and vapor-tight primary tank system built inside of a liquid-tight secondary liquid container. The secondary liquid container is liquid-tight, but not vapor-tight.

The full containment tank system consisting of a liquid-tight primary container and a liquid and vapor-tight secondary container. Both can independently contain the product. The secondary container controls vapor release in

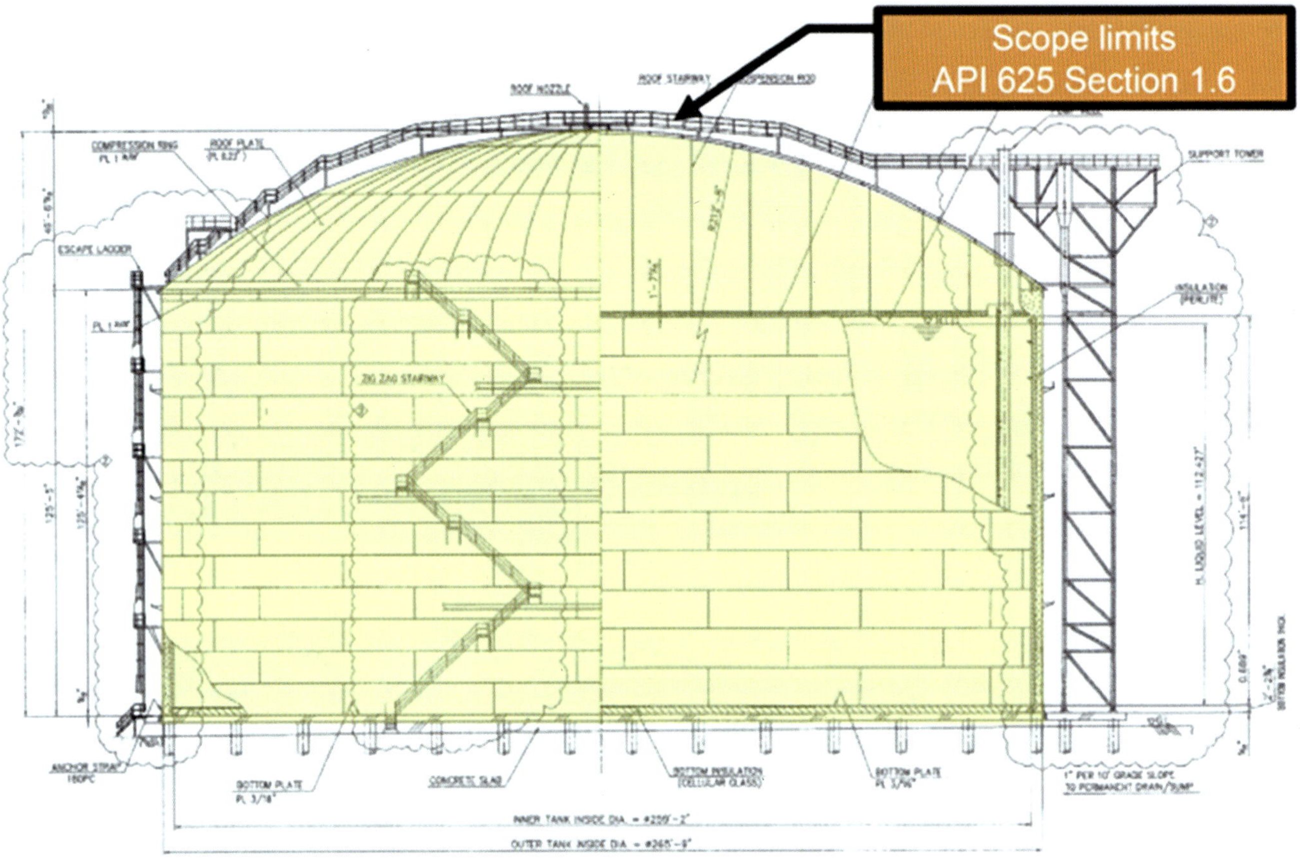

Figure 3. API 625 boundaries.

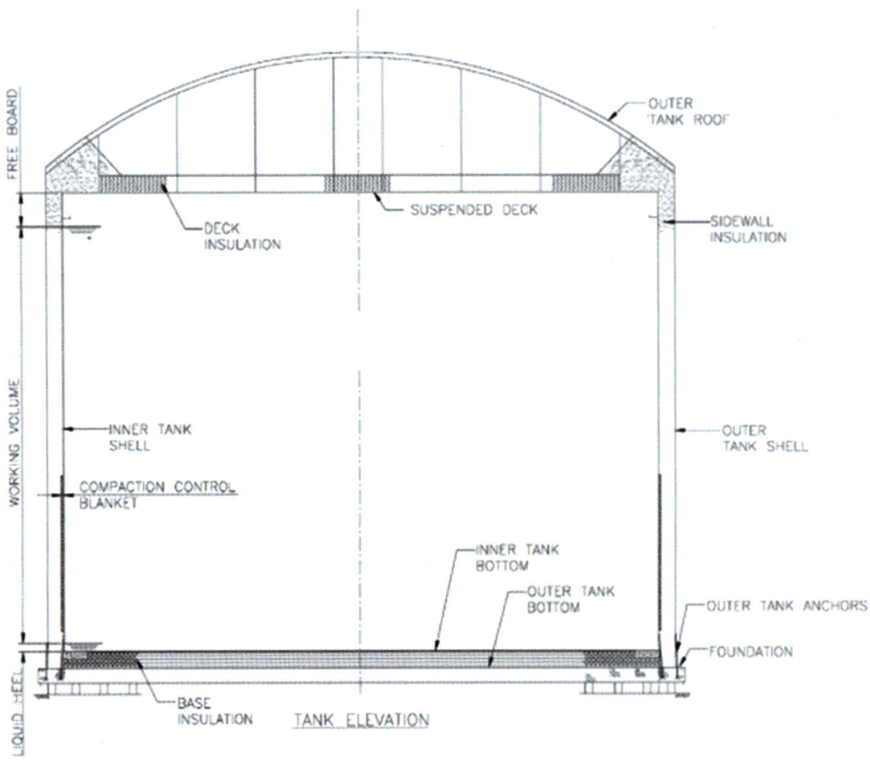

Figure 4. Nomenclature – basic LNG storage tank.

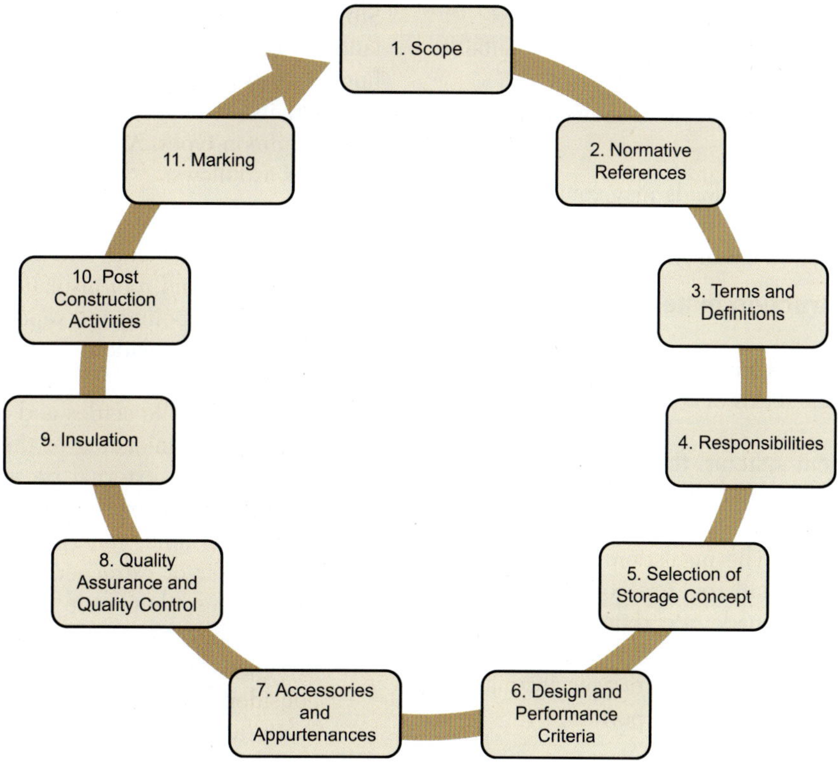

Figure 5. API 625 structure.

the event of a primary product leak. A key design feature is the thermal corner protection (TCP) at the outer tank base-to-wall joint. The TCP provides liquid tightness and thermally isolates this joint from the cold liquid in the event of a primary container leak.

Performance criteria

Once an owner has selected the tank and containment configuration, the design organization establishes performance requirements. These requirements, in Section 6.4 of API 625, define the following:

- Conditions to calculate boil-off rate (625-6.4.4).
- Product rollover provisions by active temperature and density monitoring with in-service product mixing (625-6.4.5).
- Component design temperatures for material selection and allowable stresses (625-6.4.6).
- Details for the tank insulation system and attached piping for differential movement control during cooldown (625-6.4.7).
- Foundation settlement and effect on piping systems and insulation (625-6.4.8).
- Soil freezing mitigation by providing air gap or foundation heating (625-6.4.9).
- Seismic loading (625-6.4.10).

Of these items, boil-off and the seismic hazard study are typically owner activities.

Structural design

Structural design loads and load combinations follow API 625 Section 6.5 requirements. Design requires consideration of both normal and abnormal loads. Normal loads include dead load, liquid, seismic operating base earthquake (OBE), decommissioning and differential settlement effects. Abnormal loads include liquid spill, risk assessment loads (fire; external projectile; blast wave) and seismic safe shutdown earthquake (SSE) and aftershock level earthquake (ALE).

A seismic hazard study, per Section 6.5.2 of API 625, is required to define the ground motions associated with three earthquake levels. Performance requirements are as follows:

- OBE:
  - System operational during the event.
  - System to be operational with minor damage after multiple events.
- SSE:
  - Primary containment capability.
  - System isolation and maintainability during the event.
  - Survival of primary containment with minor leaks after singular event (with an expectation of extensive damage).

- ALE:
  - Secondary containment of the primary container volume at maximum normal operating level.
  - Secondary containment with minor damage and leaks after multiple events.

For certain locations, seismic loads may control the tank and foundation design. Seismic design, including sloshing effects, is covered in Section 6.6 of API 625.

Construction materials

The materials used are influenced by the design Materials must comply with the standards of construction, such as API 620, Appendix Q.

For single containment systems, the inner tank is usually constructed of 9% or 7% nickel carbon steel. Stainless steels or aluminum that are rated for the -260°F design temperature are an option as well. The insulation deck is typically aluminum. The outer tank and roof are not rated for cryogenic temperatures and are made of carbon steel.

Full containment tank systems use the same inner tank materials. The outer tank and roof are typically made of carbon steel with the TCP at the lower portion.

Insulation

The tank systems maintain the internal liquid below its boiling point to limit boil-off gas (BOG). The key to ensure this function is proper insulation (Section 9, API 625). Storage tank system design build contractors design insulation by coupling data from insulation manufacturers and in-house tests. Parameters considered include thermal conductivity, strength, density, etc. Figure 6 demonstrates some common types. A broad classification of the insulation system is as follows:

- Load bearing bottom and TCP insulation: cellular glass with open surface cells and interleaving material.
- External wall and roof insulation: rigid insulation covered by a weatherproofing and vapor barrier.
- Internal wall insulation: loose fill perlite, an expanded volcanic ore.
  - Loose fill perlite settles and compacts during repeated loading and unloading of the inner tank. As it settles, it will impose external pressure on the inner tank. A compaction control system in the form of a resilient blanket on inner tank is used to minimize the effect. A reservoir of extra perlite is maintained to cater for long-term settlement in the interstitial space.
  - If the resilient blanket is not installed, the inner tank is designed for the uncontrolled perlite pressure considering the designated fill/empty cycles of the inner tank.
- Suspended deck insulation: the system typically consists of normal fiberglass blankets.
- Penetration and internal piping insulation: cold vapor or liquid process piping is isolated from the roof by utilizing thermal distance pieces with insulation.

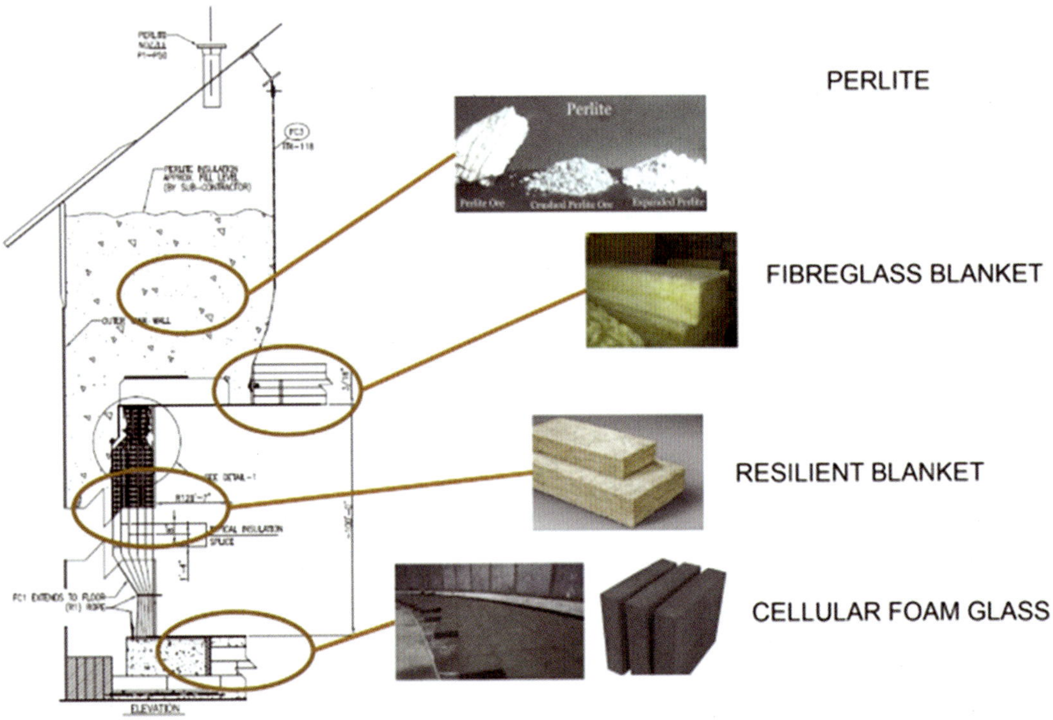

Figure 6. Insulation materials.

Accessories and appurtenances

The tank system accessories and appurtenances provide interfaces with the rest of the facility. These subsystems include the following (Section 7.0, API 625):

- External piping for tank purge, cooldown, fill and LNG send out (625-7.3).
- Tank pumping systems (625-7.3.3).
- Internal shut-off valves for single containment tanks with bottom or shell penetrations (625-7.3.1.4.2).
- External pressure relief and vacuum valves (625-7.4).
- Pressure, temperature and product density monitoring (625-7.5).
- External platforms and personnel access (625-7.2.3).
- Foundation heating, settlement monitoring and seismic motion monitoring (625-7.6).
- Hazard detection and mitigation (625-7.7).
- Lightning protection, tank grounding and aviation warning lights (625-7.8).

Additional requirements for these subsystems are addressed in API 620, ASME B

Foundations

The type of foundation selected is affected by site soil properties, seismicity, climate, and owner-selected tank containment type. Good soil properties may lead to usage of a ring wall or slab on grade. For poor soils, piles and a pile cap are used. An elevated pile cap eliminates the foundation heating system due to the presence of an air gap preventing heat transfer to and from the soil.

Quality control and quality assurance

These systems require that design, procurement, shop fabrication, site construction and commissioning are performed by the tank system contractor under a quality management plan (QP) (Section 8, API 625) for compliance with design documents, codes and standards. Contractors employ a corporate level quality management system (QMS) that includes the concepts of quality planning (identification of standards), quality assurance (evaluation of overall project performance) and quality control (specific project result monitoring). A project specific quality plan (QP), derived from the QMS, includes project specific requirements. The QP has two components: quality assurance (QA) and quality control (QC).

QA is an offline activity that ensures that QC is completed per the plan. For example, it generally involves regular audit of the fabrication and construction QC for correctness and completeness. It also involves surveillance to ensure compliance with the project quality plan, drawings, specifications and procedures.

QC is an online activity that provides an operational means to fulfil quality requirements and produces results that comply with the QP. For example, the API 620 requirement that welds in the primary containers are examined using radiography (RT) or an ultrasonic method (UT) is a QC activity. Similarly, the dye penetrant examination (DPT) or vacuum box (VB) are QC activities as well. QC ensures that these types of activity are carried out.

Tank construction

Field construction of the LNG tank system depends upon the containment type selected. A general sequence for a single containment tank consists of the following:

- Steel fabrication.
- Foundation pour.
- Outer tank floor installation.
- Outer tank wall erection.
- Outer tank roof assembly and air raise.
- Inner tank floor insulation.
- Inner tank floor installation.
- Inner tank wall erection.
- Insulation deck installation.
- Hydrostatic and pneumatic testing.
- Sidewall and roof insulation installation.
- Installation of appurtenances and accessories.

Construction of double and full containment tanks are more complex and may include outer tank wall post-tensioning.

Post-construction activities

Activities after welding completion include hydrostatic and pneumatic testing, drying and tank purging (Section 10, API 625). Hydrostatic testing is carried out to ensure foundation adequacy and leak tightness of the completed inner tank. Pneumatic testing of the outer tank verifies leak tightness and anchor integrity.

Drying and purging prepares the completed tank system for cooldown and introduction of cryogenic LNG. This consists of the following:

- Removal of standing residual water.
- Purging with warm air followed by warm nitrogen.
- Purging with nitrogen to reduce oxygen followed by purging with methane to 80-90% concentration.

After the warm methane purge, the tank is gradually cooled down with LNG. During this process, the cooldown rate is carefully controlled to minimize the inner tank wall thermal gradients. Once cooled down, the tank system is ready to receive liquid and is placed into service.

Conclusions

LNG storage tank systems are complex entities with highly engineered components that, when combined together, constitute the storage system. The technology for these systems has evolved over the last 100 years with the first natural gas liquefaction plant in the US built in West Virginia in 1917, and the first LNG safety standards in the US coming into being in the late 1960s and early 1970s. Understanding is enhanced by advances in materials and operating history. Umbrella standards, such as API 625, reflect the collective experience of the contributors and ensure that the systems are built with safety and quality considerations.

References

1. API 625, Tank Systems for Refrigerated Liquefied Gas Storage, 1st Edition Addendum 2. American Petroleum Institute, November 2014.
2. API 620, Design and Construction of Large, Welded, Low-pressure Storage Tanks, 12th Edition Addendum 1. American Petroleum Institute, November 2014.
3. 49 CFR Part 193, Liquefied Natural Gas Facilities: Federal Safety Standards. US Code of Federal Regulations, 2016.
4. NFPA 59A, 'Standard for the Production, Storage, and Handling of Liquefied Natural Gas (LNG)', 2013 Edition, National Fire Protection Association.
5. ACI 376-11, Code Requirements for Design and Construction of Concrete Structures for the Containment of Refrigerated Liquefied Gases and Commentary. American Concrete Institute, May 2013.

Reprinted with permission – *LNG Industry*, November 2016.

Choosing the best containment for LNG storage

Bob Watson, Matrix PDM Engineering, Houston, Texas

In the small and medium LNG facility market, various approaches are available for storing natural gas after liquefaction but before delivery to its intended use point. Depending upon the capacity required, this could be as small as a number of high pressure shop-built bullet tanks linked by a piping system. For larger needs, use of a single low pressure flat bottom double-wall tank is the most cost effective. These tanks, with an integrated insulation system, can be configured as single, double, or full containment in order to conform to the regulatory spill containment requirements at a given site and within the available space.

This article reviews flat-bottom tank containment types and highlight two single containment tank. The Eagle LNG is located at Maxville, Florida, just west of the Jacksonville harbor area. JAX LNG LLC is also in the Jacksonville, Florida area, but this tank is at Dames Point directly on the Saint Johns River.

Storage tank types

A single containment system consists of an inner primary liquid container and outer insulation and vapor container, designed and constructed so that only the inner container is required to contain liquid and be liquid-tight. The outer container, designed to contain product vapor pressure of the inner tank contents, retains and protects the insulation but does not contain liquid if the inner container leaks. These systems necessitate spill impoundment dikes that require onsite space that is remote from adjacent landowners and occupied structures, and from other tanks. Typical construction materials for the major components of a single containment tank configuration are cryogenic rated steels (such as 9% nickel or stainless steel alloys) for the inner tank and product piping, and general carbon steel for the outer tank which meet the ambient conditions of the tank site.

The double containment tank system consists of a liquid and vapor tight primary tank system built inside a liquid tight secondary liquid container. The secondary liquid container is liquid tight but not vapor tight.

The full containment tank system consists of a liquid tight primary container and a liquid and vapor tight secondary container. Both can independently contain the product. The secondary container also controls vapor release in case of a primary product leak. A key design feature of a full containment tank system is the Thermal Corner Protection (TCP) at the outer tank base-to-wall joint. The TCP provides liquid tightness and thermally isolates this joint from the cold liquid in case of a primary container leak. Regulatory requirements such as 49 CFR 193 may also require the secondary container to be concrete.

Double and full containment tanks are normally not required to have significant spill impoundments and, as a result, require less land. However, FERC-regulated facilities may be required to have full capacity spill impoundments even for full and double containment tanks.

Both are more costly to build however, due to their added complexities. As both the primary and secondary containers of a double or full containment tank are designed to be liquid tight, they are both typically constructed from cryogenic rated steels. Full containment tanks are often known to include an integrated concrete outer wall to help mitigate external objects from impacting or damaging the internal liquid container.

LNG tank size ranges

Field-erected flat-bottom LNG tanks have been installed in the US in varying sizes, up to 52.8 million gallons (1.25 million bbls). Single containment tanks are generally economical down to about 1 million gallon capacity (25,000 bbl).

Foundation options

Choice of the foundation type is affected by site soil properties, seismicity and climate, and owner selected tank containment type.

Good soil properties may lead to use of a ring wall or slab on grade. A ground heating system is typically utilized for storage tanks that utilize a ring wall and/or slab on grade. The ground heating system ensures creep between the tank and the soil does not occur that could result in foundation issues caused by freeze-thaw effects to the soil under the tank. In areas where poor soils exist, elevated tank foundations are used which incorporate piles and a concrete pile cap. Additionally, an elevated pile cap eliminates the need for a foundation heating system due to the air gap which prevents heat transfer to and from the soil. The final selection will depend upon an evaluation of the required installation costs and the operation and maintenance costs over the life of the tank.

Selection considerations

For sites with ample space for spill impoundments and sufficient distance to adjacent landowners and structures to accommodate calculated thermal and vapor dispersion effects due to a tank leak, the single containment tank type is the most economical. On smaller sites with less buffer distance, double or full containment tanks are normally required. Accordingly, calculations of thermal and vapor dispersion effects need to be addressed early in the design of the facility so that the most economical solution can be determined.

Maxville LNG facility

Location

The LNG tank for Eagle LNG is located at Maxville, Florida, just west of the Jacksonville harbor area.

Function

The Maxville LNG Facility and tank will be used to liquefy and store LNG to be used for ship bunkering. The tank will be filled using natural gas liquefaction systems on site and then LNG will be transferred by truck to the harbor for transfer to ships.

Capacity

This tank is small-to-medium size and will have a 1 million gallons (24,000 bbls) working capacity.

Configuration

As is typical, this tank has an outer cylinder that is at ambient temperature and an inner cylinder that contains the LNG at −270°F. Between the two cylinders is an expanded perlite insulation system designed to limit boil-off to 0.15% of the tank contents per day.

The outer tank has a domed roof and is designed for an internal vapor pressure of 2.5 psig. The inner tank has a suspended insulation deck at the top and the deck is not pressure retaining. The insulation on the deck is fiberglass.

Pumps for transferring the LNG to the trucks are mounted inside the inner tank. Discharge from the in-tank pumps is piped upward through the insulation deck and dome roof and back down to grade through insulated lines mounted outside of the outer tank.

The tank foundation is a concrete ring wall on grade, with stone column subsurface soil improvement. Since the foundation is on grade, it is electrically heated to prevent possible freezing of the soil and frost heave.

Regulatory jurisdiction

Since this facility is not associated with LNG import or export, the authority with jurisdiction is the US Pipeline and Hazardous Material Safety Administration (PHMSA). The facility is designed in accordance with 49 CFR 193 and NFPA-59A.

Construction

The common field erection sequence for tanks of this type starts with welded assembly of the outer tank cylinder, followed by air-raising of the dome roof and insulation deck. Then, the inner tank is assembled using the outer tank as weather protection. In November 2017, the Maxville LNG facility tank was erected, hydro-tested, and insulated, and then subsequently commissioned by purging the inner and outer tank with gaseous nitrogen prior to the first introduction of LNG and cooldown.

JAX LNG facility

Location

The LNG tank for JAX LNG LLC (a joint venture of Northstar Midstream and Southern Company Gas) is also in Jacksonville, Florida area, at Dames Point directly on the Saint Johns River.

Function

The JAX LNG facility and tank will be used for ship bunkering, as well as trucking and small-scale power generation via ISO containers. The tank will be filled using natural gas liquefaction systems on site and then LNG will be transferred to bunkering barges, which will refuel ships.

Capacity

This tank is 2 million gallons (48,000 bbls) working capacity.

Configuration

The layout of this tank is very similar to the Maxville LNG facility tank, except for the larger diameters of the inner and out tanks and the larger overall height. The design vapor pressure of the outer tank is 2.0 psig. Once again, the LNG transfer pumps are mounted inside the inner tank and discharge through piping to grade without penetrating the sidewalls of the tank. Unlike the Maxville tank, this tank is supported on driven pre-stressed piles and an elevated concrete pile cap. With this arrangement, heating of the pile cap is not necessary since it is not in contact with the soil below.

Regulatory jurisdiction

The JAX LNG facility is also not associated with LNG import or export or interstate transfers. As a result, the authority having jurisdiction is the PHMSA. The facility is again designed in accordance with 49 CFR 193 and NFPA-59A.

Construction

This tank was constructed in a similar manner to the LNG tank for the Maxville facility. The design and construction duration for small and midscale LNG liquefaction facilities has in the past been limited by the time to install the LNG storage tank. In both of these facilities, the LNG single containment storage tank was completed prior to the commissioning of the facility.

Reprinted with permission – *LNG Industry*, March 2018.

Figure 1. The domed roof on the single containment tank at Eagle's LNG's Maxville LNG facility is air-raised into place.

Figure 2. Eagle LNG's Maxville LNG facility near Jacksonville, Florida, will be used to liquefy and store LNG used for ship bunkering.

A review of API 2350 overfill protection, 4th and 5th edition

Philip Myers, P.E., Founder and Director, PEMY Consulting, Pleasant Hill, California, and **Brock Trotter**, E.I.T., PEMY Consulting, Oklahoma City, Oklahoma

The overfilling of a petroleum storage tank is potentially among the most devastating events that can occur at a terminal or refining facility. The loss of product is only a minor consequence of spilling petroleum liquids from a tank compared to potential outcomes such as lawsuits, fines, damage to reputation, fires, injury to personnel, a vapor cloud explosion (VCE) and possible facility closure. Figure 1 shows the destruction of the 2005 Buncefield terminal VCE caused by overfilling a petrol tank.

Filling a storage tank seems uncomplicated but repeating the process thousands of time, flawlessly, for hundreds of tanks over decades requires robust procedures, training, equipment, a good management of change process as well as the right corporate culture. API/ANSI 2350 4th and 5th edition guides users on these prerequisites for best practices that can ward off overfills. The new 4th and 5th editions of API 2350 radically change that way the tank receipts are supposed to be handled from a best practices point of view and from lessons of the past.

Evolution of overfill standards

API/ANSI Standard 2350 provides petroleum facilities with the minimum standards for overfill protection of storage tanks. The 4th edition of the standard was published in May of 2012, and the 5th edition was published in 2019.

The differences between the 4th and 5th editions of API 2350 are slight and are expected to clarify some problems of interpretation as well as to make the document more user-friendly. In terms of procedural or technical issues, not much has changed between the 4th and 5th editions.

Originally, the scope of API 2350 applied only to flammable liquids storage, that is, NFPA Class I liquids which have a flash point below 100°F. This was because significant losses were occurring too often, caused by fires. Later, environmental protection was applied to overfill prevention practices in the 3rd edition of API 2350 published in January 2005. API 2350-3 applies to all above ground tanks which store 1,320 gallons or more of Class I or Class II petroleum liquids. The Class II liquids have a flash point between 100°F-140°F which includes jet fuels, kerosene and most diesel fuels.

API/ANSI 2350 does not apply to underground tanks, solid storage tanks, small tanks (less than 1,320 gallons), or the storage of Class III liquids (e.g., lubricating oils), or pressure vessels. The principles can be applied to other tanks not covered by the standard, however, and this is encouraged where appropriate.

Figure 1. Result of the Buncefield vapor cloud explosion.

The original concepts of API 2350 were informal best practices and therefore labeled 'recommended practices', however a seminal change to the standard occurred with the 4th edition. The 4th edition of the document is a 'standard', and no longer a 'recommended practice'. The most important changes from previous editions, by far, were the introduction and formalization of the 'management system' concepts that must be applied to tank filling operations. API 2350-4 calls this 'overfill prevention process' (OPP).

Overfill prevention process approaches

OPP, like any safety management system, requires companies to use a risk assessment program to prioritize and understand the risks of overfill. The application of a maintenance program must include training and written procedures, specific rules for communication between parties involved with the tank filling operations, measures that address normal and abnormal conditions, and initiation and maintenance of tank parameters (i.e., management of change) as well as other components. Because there are so many types of companies and managements systems, integration of OPP into the corporate DNA is an important task to overcome.

Categories

OPP is the systematic approach to shutting down a pump, diverting the incoming flow, closing a receipt valve or alternative means of terminating the flow of liquid into a tank. There are three basic equipment configurations or categories of OPP addressed by API/ANSI 2350 which should be considered in developing an OPP (Figure 2). Categories serve as a starting point for understanding system risk and for categorizing the numerous tanks that a company has into manageable groups.

A Category I system does not have transmittable level or alarms and is entirely dependent on the operator and manual gauging to prevent an overfill. While this type of operation has been around a long time and can be highly reliable, it is limited to a low frequency and low rates of transfer that an operator is able to cope with.

The Category II configuration uses level and alarm data, which can be transmitted to a control room. Thus, the operator can deal with many more receipts at a high frequency when tank level and alarm data are transmitted to a control room. The Category II configuration dominated the large petroleum facilities up to and including the 1990s. The weakness of the Category II system was that it had no redundancy in terms of reliability. If the level alarm failed, then most likely there would be an overfill.

Category III systems are the same as Category II, but they are much more reliable because of the redundancy in the alarm system. Category II requires an independent second alarm usually called the 'high-high' to operate should the first alarm be failed or because an operator was not able to react to terminate the receipt at the first or 'high' alarm. The failure of the high-high cannot be caused by any failure of the automatic level gauge or by the high alarm.

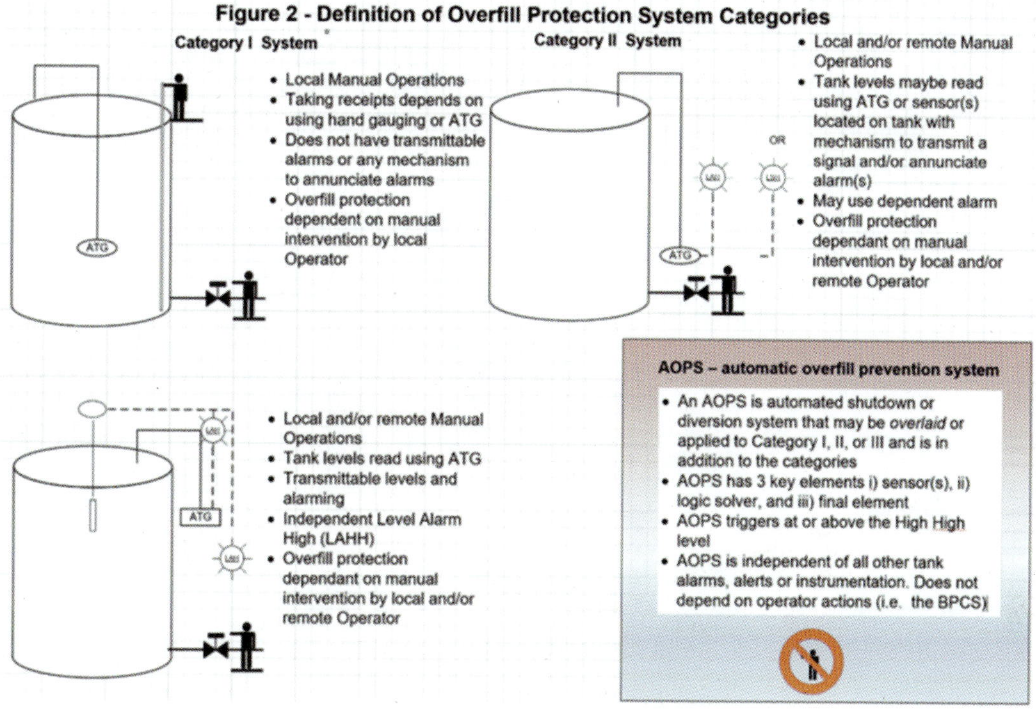

Figure 2. Overfill equipment categories.

AOPS

The 4th edition of API 2350 introduced the idea of Automatic Overfill Prevention Systems (AOPS). Most overfill prevention systems use the operator to control the receipt and terminate it before an overfill occurs. In the jargon of IEC 61511 for safety instrumented systems, this control is referred to as the Basic Process Control System (BPCS) and forms the most basic control over the process, whether it is an operator or some combination of operator and instrumentation. A safety system can be applied to a hazardous process such as filling a tank, but it should be independent of the BPCS and no initiating event in the BPCS should affect the safety system.

This system, called AOPS, is independent of the BPCS and requires no human intervention in its ability to terminate a receipt. The design and operating framework for AOPS arise from industry trends in controlling highly hazardous processes and the use of recent safety standards such as ISA S80 and IEC 61511. AOPS, being an independent system, may be used in addition to the three categories of OPP. AOPS is characterized by the automatic termination of product flow into a tank upon activation of a Level Switch High (LSH). AOPS provides redundancy to the filling process, reducing the risk of an overfill event.

AOPS can provide additional redundancy, but it can present many practical issues with implementation. For example, an incorrectly designed AOPS applied to a pipeline could cause a pipeline rupture due to hydraulic shock. The use of safety instrumented systems and function in typical oil distribution terminals is often beyond the typical capabilities of the people and companies associated with designing and implementing these systems. A glaring example of AOPS implementation failure was, in fact, the Buncefield incident.

Also, AOPS costs are high. The expense is not because of equipment or installation costs, but because a whole new process that applies to the entire lifecycle of the equipment including documentation and other management system elements such as testing and auditing functions. Fortunately, AOPS is only one way to reduce risks. It may be the best solution in many cases, but certainly not all or even most. There are many considerations when deciding to implement an AOPS and alternatives must be considered.

LOC

A key step in OPP is to establish LOCs (levels of concern) as shown in Figure 3. OD is the liquid level we never want to reach. It is defined as overfill, damage, or any other problem associated with a too high liquid level as specified by the operator/owner. In API 2350 there are two important LOCs:

1. LAH
2. LAHH

The LAH is mandatory. Regular practice tends for most owner/operators to use an LAH and LAHH. API 2350 now requires only one alarm level. It has been demonstrated that adding more alarms does not necessarily improve

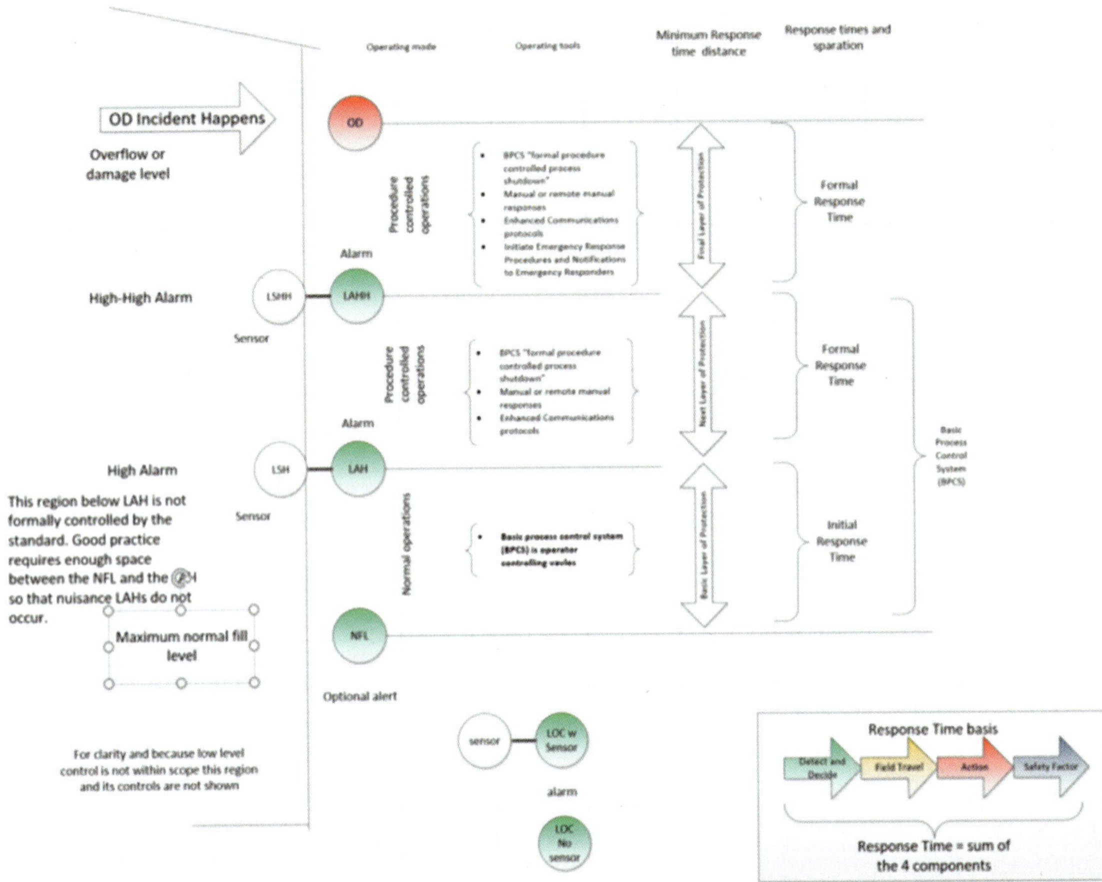

Figure 3. Levels of concern and response times.

operator performance. Historically, two alarms were required because the alarms were so unreliable.

Today however, alarm systems can be incredibly reliable. For these reasons, API 2350 allows for a single alarm. The owner operator must ensure that their alarm system is engineered to be highly reliable. So, the use of a second alarm (LAHH) is optional. However, when used, the LAHH or second alarm must then meet all the rules specified in API 2350.

Obviously, alarms must be initiated using sensors which are shown in Figure 3. The other levels such as OD and NFL must still be determined but they are just elevations; no physical sensor or equipment is needed at these levels.

Take note that API 2350 does not address the requirements for alarms at the normal maximum fill level (NFL). API 2350 enforces the principle that an alarm is actionable, must be procedure driven and is always to be considered an emergency. Therefore, if an operator wants to be notified when the level exceeds the NFL it can use an 'alert', which is not within the scope of API 2350. Differentiating between alerts and alarms will be an important cultural shift in some organizations.

The high alarm should never be used as an alert or notification that the level is too high. That opportunity has already passed when the NFL is exceeded. Companies which do not use this process must rethink their approach and understand the benefit of the API 2350 methodology, which is aimed to reduce issues related to faults in human factors.

Response time

Response time is the time between LOCs. It is, of course, based on the flow rate into the tank which is a maximum flow value. The inset in Figure 3 shows that response time (RT) is the reasonable and necessary time for an operator to detect the alarm, physically go to the field and turn the appropriate valves or shut off pumps, with the addition of a safety factor.

Measuring the response time is a critical step to preventing an overfill event. There is usually pressure to maximize tank storage capacity and response time works against this. Do not reduce the LOCs until the remaining tank space is not viable for overfill prevention.

Other management system components

Like any management system, written procedures are required. These procedures should address normal and abnormal conditions. For example, loss of power can affect the level indication and alarm systems. There must be procedures to address issues such as power failure. Facility personnel must be competent in performing their duties. This comes from experience and training.

Managing change for both equipment and personnel is important. The management of change process is mature in the industry so will not be covered here. Proof testing is a critical element of reliability and thus must be incorporated into written procedures. API 2350 discusses the requirements and principles for proof testing. As a point to reiterate the importance of proof testing, a significant causation for the Buncefield event was due to improper proof testing of the alarm system.

Risk assessment

One of the key elements of API 2350 is risk assessment of a tank facility. Since all tanks cannot be simultaneous, it is important to have a method for determining which tanks pose the most overfill risk to deal with those risks first. Many factors contribute to risk and the entire topic of risk can be complex. Because there are so many ways of dealing with risk and each owner has a unique perspective and perhaps tolerance for risk. API 2350 only requires that a risk assessment be done, it does not address how to do it.

Risk assessment, as a requirement for overfill protection, requires a robust approach and strategy. If a tank facility owner has only one or two tanks the problem is relatively uncomplicated. However, if the tank facility owner has 1,800 tanks then a more strategic approach is needed to deal with the resources and time required to appropriately address the risk assessment process. Moreover, if the facility owner wants to accomplish a specific management objective, such as complying with the 5th edition of API 2350 within a specific period, then a strategy that ranks and prioritizes risks for accomplishing this goal is essential.

In a complex environment, such as those portrayed in API 2350 for overfill protection, everything can't be assessed or measured at the same time; data collection, assessments and risk assessment must prioritize resources. The most effective approach is to address the most critical risk assessments first – those addressing the most undesirable threats first. Prioritizing risk assessment or risk screening studies are the quickest way to prioritize actions that reduce larger, more imminent risk concerns.

The many tools for doing risk assessments range from simple what-if analyses, to check list, hazard and operability studies (HAZOPS), even trees, fault trees, failure modes and effects, etc. Because of the labor involved, it is usually infeasible to apply complex methods to every single tank in the tank population. Rather, a risk screening process is usually appropriate in the early stages, and it gives a moving window that focuses on those high priority tanks that need to be addressed now. A simulated risk study on 1,000 tanks is shown in Figure 4.

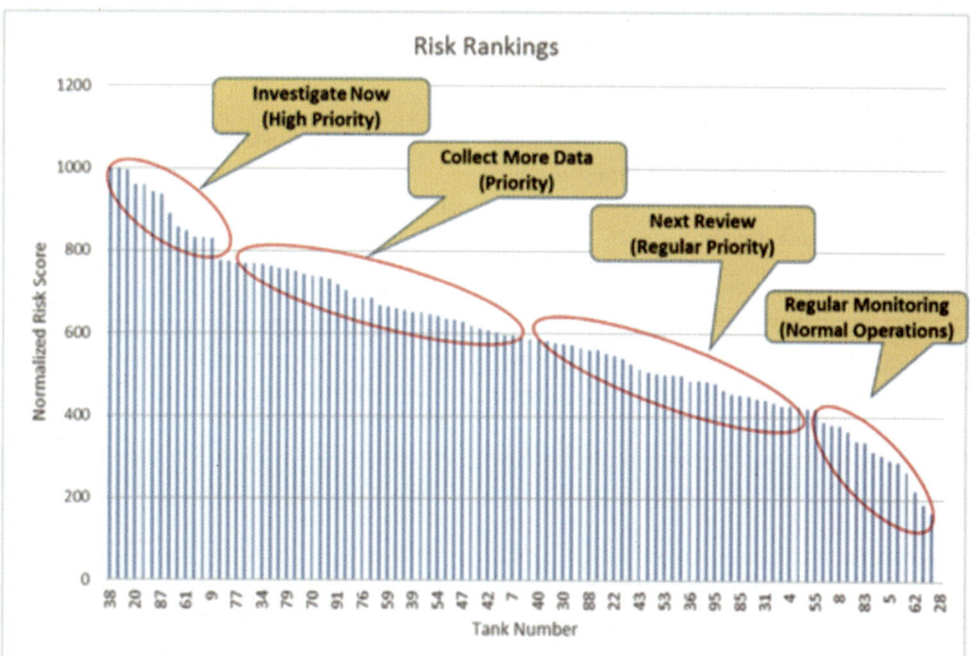

Example ranking of tanks by relative risk (0-best to 1000-worst)

Figure 4. Risk screening results.

A screening exercise can be done by using metrics for likelihood and consequence and then combining them in an appropriate manner. Because risks can arise from different types of threats, such as those of human health and safety or damage to the environment, a logical system that incorporates the owner/operator values must be applied to the risk assessment metrics.

Conclusion

A standard, law, regulation or policy is only that – paperwork and penalties for noncompliance. By themselves or with absentminded compliance, standards cannot create safe and effective practices to reduce risks. Corporate leadership and messaging supersede standards because the belief and value systems, as well as resources allocated for improvement, are all determined by senior level management. Senior level management sets the tone within the minds and hearts of employees who make the business run as it does, creating change or stagnation. Senior level management determines the corporate safety culture regardless of what it claims publicly.

No regulation, policy or law can create the kind of environment where safety or environmental protection are a top corporate priority without the continued messaging and accountable of leadership within the organization. Unfortunately, this is not something that API 2350 can influence and bolster.

Blind compliance with API 2350 (or for that matter any important safety or environmental standard) will not be as effective in reducing risk without the right corporate values and leadership philosophy, accountability and messaging to back it up. This is the hard and difficult journey for getting API 2350 to work for the industry.

Reprinted with permission – *Tank Storage*, February/March 2018, 14(1); April/May, 14(2).

Manual tank gauging procedures

Excerpts from Chapter 6 of the Endress+Hauser Guidebook for Overfill Prevention & Tank Gauging, 2018.

A simple and important operation for companies that use liquid storage tanks is the measurement of the liquid level within a tank. Making a liquid level measurement in a tank is called tank gauging. When an operator makes this measurement manually by use of a measuring tape or electronic hand line, it is called manual gauging. Making a manual measurement requires the operator to climb to the top of the tank, open a port or hatch that has line of site to the liquid, and drop a measuring device such as a weighted tape measure directly into the liquid.

When the measurement is made by permanently fixed tank instrumentation, it is called automatic tank gauging. The requirements for tank gauging are specified in the API standards. API MPMS Chapter 3.1A[1] covers manual tank gauging and API MPMS Chapter 3.1B[2] overs automatic tank gauging. These are the most widely used tank gauging standards in the petroleum industry.

The product level is the distance between the reference point at the bottom of a tank to the product level. The level within a tank determines the volume of product within the tank, and this information is required to safely fill tanks, execute customer orders, and inform production of the demand. The level measurement also fulfills an important alternative role of controlling the risk of overfilling. Overfill events can range from a simple clean up to a vapor cloud explosion. Overfill events endanger a company, their personnel, the environment, and the petroleum industry. The technology available to monitor the level within a tank has improved significantly since the dawn of the oil industry with ever improving instrumentation reliability.

Tank gauging systems

A tank gauging system measures and displays the level within a tank. Modern systems automatically calculate the volume of product from the liquid level. Crude oil and other petroleum products are traded by volume, so the measurement of level is integral to business transaction accuracy. A tank gauging system is necessary at facilities or plants for inventory management and custody transfers. Inventory management is the internal corporate audit of the volume or mass within a tank or tank system. Custody transfer is the purchase and transfer of liquid between parties. Both custody transfers and inventory management have standard accuracy requirements discussed in API MPMS 3.1A&B, the European MID Directive, or comparable national standards documents.

Tank gauging systems are employed in large storage tanks, pipelines, depots, refineries, airports, power plants, chemical

Figure 1. Example of a tank gauging system.

plants, mines and elsewhere. A tank gauging system is typically capable of measuring the liquid level (Figure 1), temperature, pressure, and water bottoms (thickness of the water layer at the bottom of the tank).

Peripheral equipment to tank gauging systems includes communications technology, computer hardware and software and wiring/signaling. The water and oil levels, temperature and pressure data are necessary to calculate the corrected or standardized volume of fluid within the tank.

Gauging systems support functions related to inventory control, custody transfer and overfill prevention. Custody transfer tank gauge systems are used to calculate the net volume of oil which enters or exits a tank. Customers are billed based upon the standardized net measured volume of oil transferred, therefore the sensing technology must be very accurate. Failure to accurately gauge and bill for a transfer can result in lawsuits, loss of revenue and loss of consumer confidence. Government authorities or intercompany contractual arrangements hold jurisdiction over typical custody transfer standards such as API MPMS 3.1A&B and the European MID Directive and they are usually based upon the country of operation.

Innage vs. ullage measurements (liquid length vs. empty length measurements)

An innage measurement extends a sensor or tape through the entire vertical distance of the product at the tank bottom to surface of the liquid level. A dipstick in an engine is a simple analogy for an innage measurement. Innage gauging is a direct liquid level measurement but it does have some drawbacks (Figure 2). It may be impractical if the liquid is highly viscous. Another inconvenience of innage gauging is the tape can be coated with sticky oily products that must be cleaned when reeled in. This can also pose a potential hazard to tank operators.

An ullage measurement is the measurement of empty air space as a vertical measure from a fixed point usually located on the tank roof to the liquid surface. The reference point is the uppermost measurement point in a manual tank level measurement. The reference point is usually a metal slit or extension inside the gauge hatch at the top of a tank.

The innage measurement is the difference between the datum plate (strike plate) and the top of liquid surface. A datum plate is a metal extension that is attached above the tank bottom and underneath the reference point. The liquid volume in the tank is inferred by measuring the difference in ullage to obtain the change in liquid volume.

Brief History of Tank Gauging

At the beginning of the oil industry in the 1860s, the accuracy of level measurement was very poor. Customers were growing concerned that they were being overcharged for the volume of oil purchased. In order maintain consumer confidence, oil companies agreed to create the standard 42-gal barrel. Although the volume is 42-gal, oil companies agreed to charge customers at the rate of 40-gal per barrel to compensate for spillage, evaporation, and measurement error.

In the 1960s the accuracy of custody transfer was $\pm 0.5\%$. This error was attributed to error in temperature measurement, sampling, and other factors. The accuracy was built into the API standards of the time. This meant that the purchaser of the crude oil was losing a barrel for every 200 barrels purchased. This became an extreme concern in the 1970s due to the dramatic increase in the price of crude oil. The increased price per barrel pushed the petroleum industry to adopt new methods, the API Manual of Petroleum Measurement Standards (MPMS). This method of transfer practice has an accuracy of 0.25%. MPMS protocol has been revised several times since its conception.

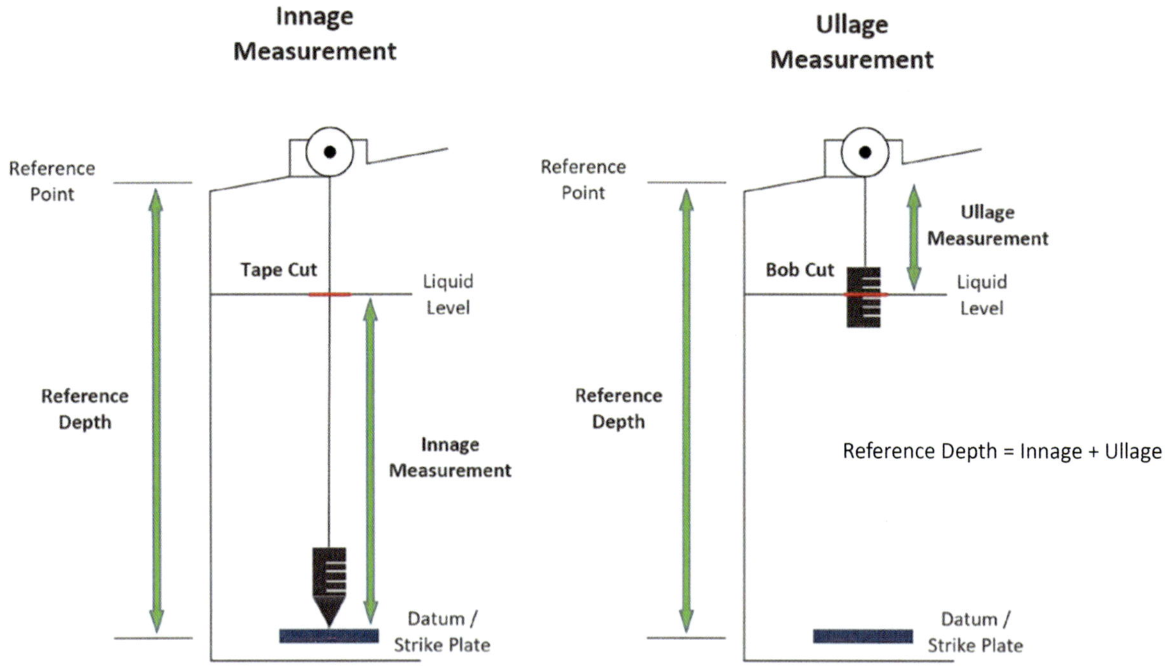

Figure 2. Innage and outage (ullage) diagram.

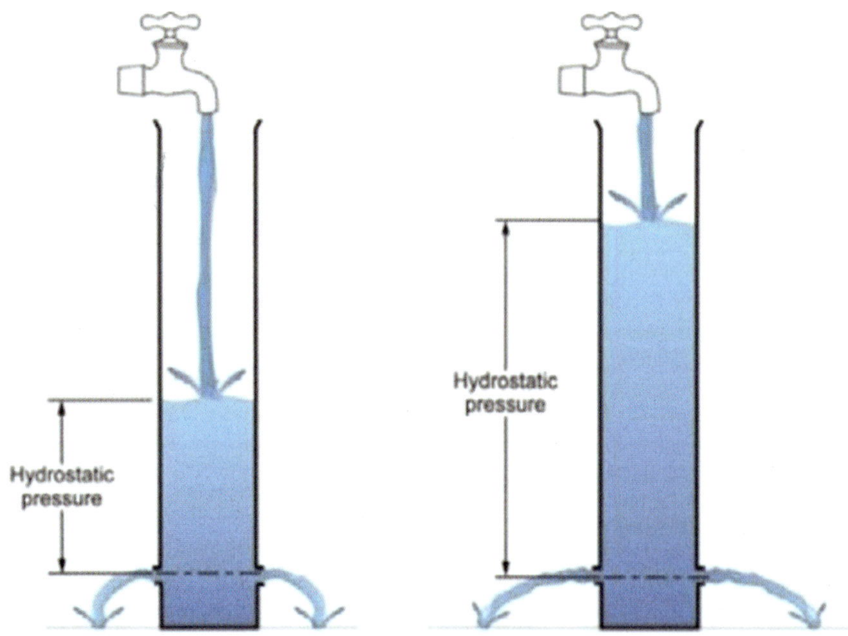

Figure 3. Mechanism for hydrostatic pressure due to liquid column.

Level and mass measurement technology concepts

Both innage and ullage measurements use the simple distance measurement to relate it to the concept of volume in a cylinder and its relationship to the height of the liquid column. Most of the information contained in API Chapters 3.1A and 3.1B are based on this principle.

A different technique for determining the amount of product within a tank is based upon the product mass and its effect on pressure at the base of the liquid column (Figure 3). API Standards governing this type of measurement are covered by a section in API Chapter 3.6 and 16.2. Mass measurement techniques use the hydrostatic pressure created by the vertical column of product liquid above the sensor or pressure gauge. The hydrostatic pressure is then related to the stored liquid volume through the density.

Level measurements directly determine the amount of product within a tank, whereas mass measurements are a measurement of pressure. If density is known, then pressure can be used to determine level which allows for determination of the volume within a tank.

Manual tank gauging

Manual tank gauging is accomplished when an operator measures the product level and the water bottoms from the top of a tank. The measurement can be done either with non-electronic measuring tapes or by portable electronic liquid measuring devices. There are some hazards associated with manually tank gauging. The measurement must occur from a fixed reference point from the top of the tank, account for temperature of the petroleum and water phases and account for water bottoms depth. The process of manual tank gauging is specified in API MPMS, Chapter 3.1A, "Standard Practice for the Manual Gauging of Petroleum and Petroleum Products"[1]. The scope of API MPMS, Chapter 3.1A is limited to the manual gauging of aboveground storage tanks which apply to API 2350-4. This includes the use of portable electronic gauging technology (PEG).

Chapter 3.1A describes:

1. Procedures for manual gauging of atmospheric flat-bottom vertical cylindrical tanks.
2. Methods applied when there is a need to measure water bottoms or sediment and emulsions that may exist at the bottom of the tank.
3. Methods to verify the accuracy of the measuring devices being used.
4. How the accuracy is affected by the selection of the reference and datum points.

Variables include:

- ***Innage level***
- ***Water/sediment bottoms level***
- ***Ullage level***
- ***Temperature of both phases***
- ***Tank dimensions (provided by the manufacturer)***

Open gauge is the liquid level prior to a custody transfer. Closing gauge is the liquid level after a custody transfer. The difference in liquid level before and after a transfer is used to verify or quantify the amount of petroleum transferred or sold. In Figure 4, an example of open and closing gauge are depicted, the liquid levels would be reversed if the tank was receiving liquid.

Manual level measurement process

Discover the Reference Point (Mark)

The reference point is a position near the top of the tank which is used for accurate level gauging. The reference point is usually a fixed plate inside the gauging hatch, a groove cut horizontally inside a gauging hatch, or the edge of a fixed metal arm that is not in contact with the gauging hatch. See Figure 5 for an example of reference point.

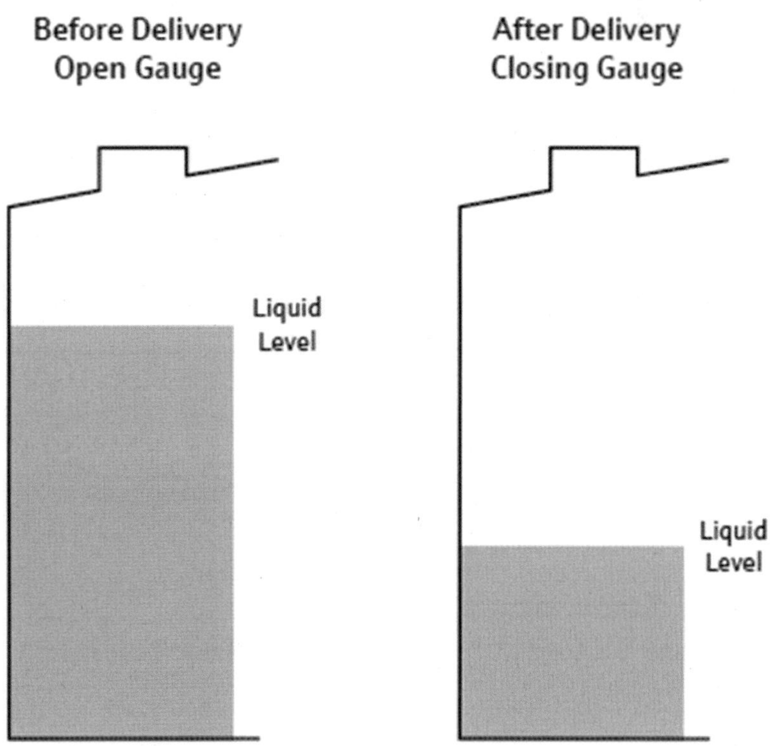

Figure 4. Open and close gauging diagram.

Determine the Reference Height (Depth)

The reference height is the length from the reference point at the top of the tank to the datum plate at the bottom of the plate (Figure 6). Measuring the reference height requires an operator to lower a tape gauge from the reference point to the datum plate and record the length.

Datum/Strike Plate

A datum plate is a metal plate attached to the tank shell or bottom that is located below the reference point (Figure 7). The reference point and datum plate create a straight, unobstructed line that is useful for manual tank gauging. Although the bottom is sometimes used, the bottom can build up sludge and cause error measurement. In addition, the bottom tends to flex and settle over time and so would lead to inaccurate measurements.

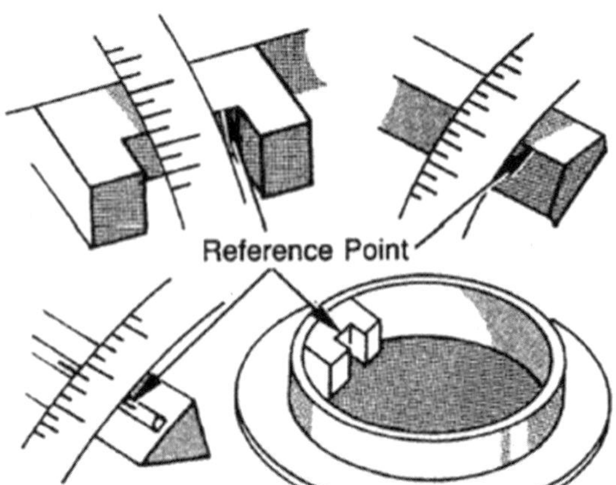

Figure 5. Reference point diagram.

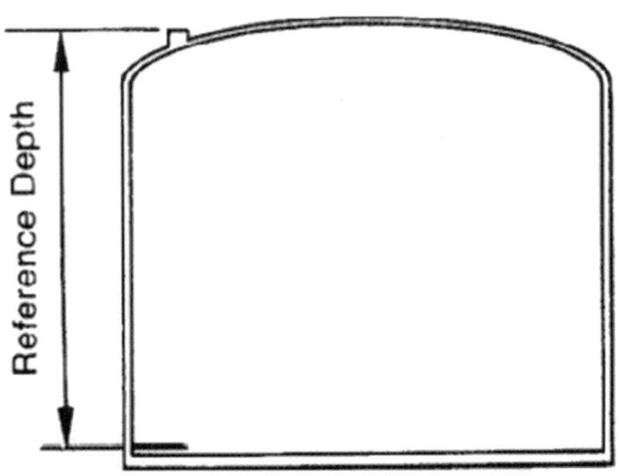

Figure 6. Reference depth diagram.

Measuring the Cut (Total Liquid Level)

When using a measuring tape, a first step in the process is to apply an oil paste, or paste to the tape which dissolves or changes color in petroleum, at the expected product position on the manual tape (Figure 8).

The reference height is the length of the tape and bob from the reference point after striking the datum plate. The height between the datum plate and product cut is the ***Product level.*** The ***Ullage level*** is the height between the reference point and the product cut.

An operator will then slowly unreel an innage tape and bob or Portable Electronic Gauge (PEG, Figure 9) into the product through the gauge hatch while the tape lies along the reference point. A PEG is an electronic tool sometimes used to manually gauge the level in a tank. The PEG bob is lowered into the product and displays the length on a digital display. When the bob strikes the datum plate the operator should slowly reel up the tape. The height of the product liquid should create a clear line on the oil paste, called a product cut (Figure 10).

Gauging standards are set by Manual of Petroleum Measurement Standards Chapter 3.1A[1]. The standard requires two identical consecutive measurements of the cut or three consecutive measurements within 3 mm of each reading (Table 1). A minimum of two measurements are required and if they are the same, then the reading is recorded. If they differ, then three trials are conducted (within 3 mm) and the readings are averaged. The equation for averaging level measurements is:

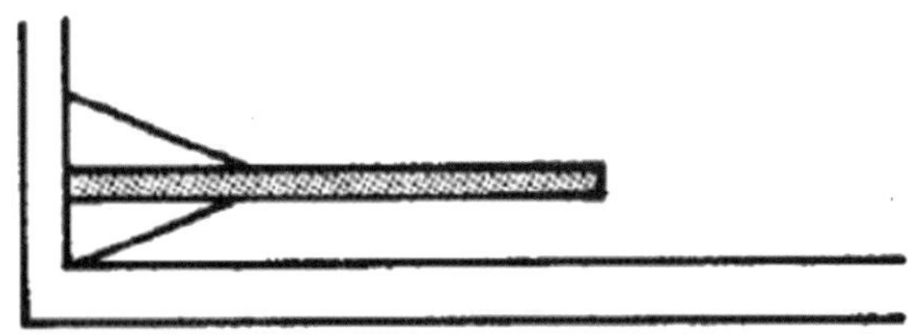

Figure 7. Strike plate diagram.

Figure 8. Oil paste (determine the product cut).

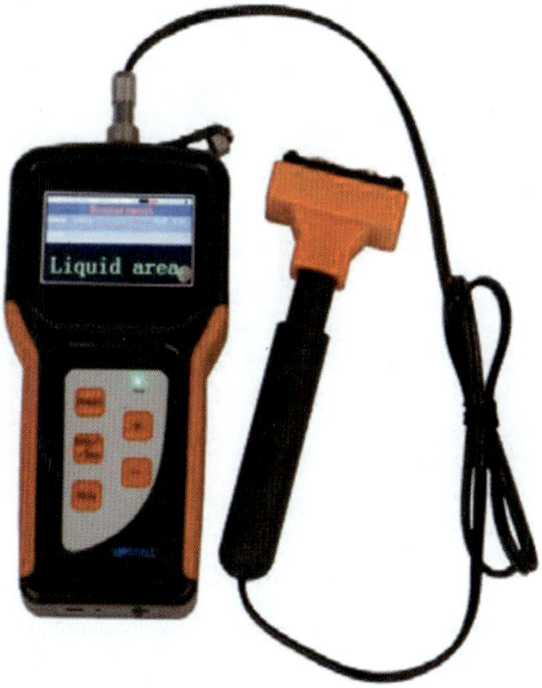

Figure 9. Example of portable electronic gauge (PEG).

Figure 10. Product/water cut schematic.

Table 1
Manual Gauging Requirements by API Chapter 3.1A

Number of Level Measurements	Accuracy Requirements
2	Both trials are the same
3	Each trial within 3 mm

$$\text{Level}_{\text{Avg}} = \frac{\sum Level_i}{i_{total}}$$

Measuring Water and Sediment Bottoms

A first step, if water bottoms exist and will be measured, is to apply a water finding paste, or paste which dissolves or changes color in water (Figure 11), at the expected water position on the bob of the manual tape. The paste should not be soluble in the product. The ***Water/sediment bottoms level*** is the height between the datum plate and the water cut.

Figure 11. Water finding paste (determine the water cut).

Table 2
Minimum Number of Temperature Measurements Based on Height of the Liquid by API Chapter 3.1A

Depth of Liquid	Min. Number Measurements	Measurement Level
> 10 feet (>3.05 meters)	3	Center of upper, middle, and lower thirds
≤ 10 feet (≤3.05 meters)	1	Center of liquid

Temperature Gauging

Determination of temperature, density, API gravity, suspended sediment and water are not addressed by Chapter 3.1A or 3.1B. These requirements are established in API Chapter 7 and the appropriate requirements from this standard are covered here as relevant to tank gauging.

The minimum number of temperature measurements that are required per liquid level is shown in Table 2. The temperature of each layer, usually water and product layers, must be measured separately and recorded. For example, if the ***Water/sediment bottoms level*** is 2 ft in height then the water layer would require one temperature measurement. If the ***Product level*** is 25 ft in height then the product layer would require three temperature measurements. This concept is displayed in Figure 12, and Figure 13 illustrates a typical temperature sensor.

Once the operator has determined the necessary number of temperature measurements per phase and knows the levels of those phases, they can begin to take temperature measurements using a Cup-Case Thermometer, Armored Case Assembly, Angle-Stem Thermometer, or Portable Electronic Thermometer (PET).

Portable electronic gauges (PEGs) should be calibrated or tested prior to use for inventory management or custody transfer purposes under the requirements of MPMS Chapter 3.1A[1]. They should be manufactured in accordance with ASTM E-1.

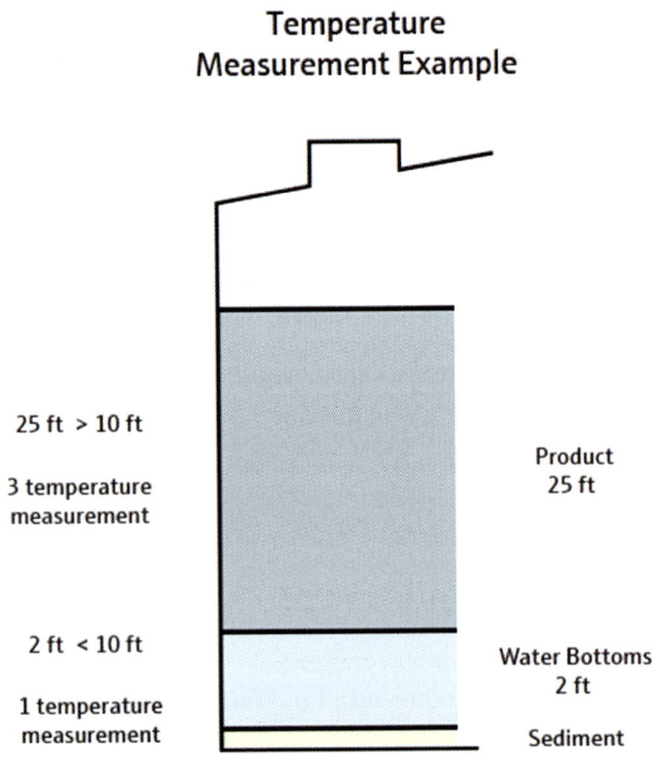

Figure 12. Temperature measurement diagram.

Figure 13. Multipoint Prothermo NMT539 for average temperature measurement.

The general specifications for each type of PET tank thermometer is shown in Table 3. The immersion time is dependent on the thermometer type and the API gravity of the liquid temperature being measured. Tables for the suggested immersion times for various thermometers and API gravities are given in Tables 4 and 5.

One temperature measurement trial requires lowering, immersing, and receiving the temperature reading of the thermometer. Repeat these steps and record the temperature data for each recommended trial. The average liquid temperature is computed from:

$$T_{Avg} = \frac{\sum T_i}{i_{total}}$$

Upon completing these steps, you will have obtained the ***Temperature of both phases***. The temperatures and level measurements are then used to determine the volume of petroleum at standard temperature conditions. The density of a liquid is dependent on the temperature, so it is necessary to standardize the temperature used for custody transfers.

$$\rho = \frac{m}{V}$$

Because the mass inside the tank is constant:

$$\rho_0(T) \times V_0 = \rho(T) \times V$$

Table 3
General Specifications by Thermometer Type

Name	ASTM Thermometer	Range	Length (inches)	Graduation	Accuracy
ASTM Tank	50°F–80°F	−30°F to +120°F	12	1°F	±0.5°F
ASTM Tank	50°F–80°F	0°F to +120°F	12	1°F	±0.5°F
ASTM Tank	50°F–80°F	0°F to +180°F	12	1°F	±0.5°F
ASTM Tank	50°F–80°F	60°F to +180°F	12	1°F	±0.5°F
ASTM Tank	50°F–80°F	170°F to +500°F	12	1°F	±0.5°F
Angle-stem		Suitable Range	12	1°F	±0.5°F
Tank thermometer		20°F to +220°F	12	1°F	±0.5°F

Table 4
Recommended Immersion Times for Woodback Cup-Case Thermometer

API Gravity at 60°F	Immersion Time in Minutes	
	In Motion	Stationary
> 50	5	10
40 to 49	5	15
30 to 39	12	25
20 to 29	20	45
< 20	45	80

Table 5
Recommended Immersion Times for Portable Electronic Thermometer

API Gravity at 60°F	Immersion Time	
	In Motion	Stationary
> 40	30 Seconds	25 Minutes
20 to 40	45 Seconds	45 Minutes
< 20	75 Seconds	80 Minutes

The density within the tank is dependent on the temperature, therefore the volume will fluctuate with temperature.

$$V_0 = \frac{\rho(T) \times V}{\rho_0(T)}$$

References

[1]. Manual of Petroleum Measurement Standards Chapter 3.1A, Standard Practice for the Manual Gauging of Petroleum and Petroleum Products, 3rd Edition (2013).

[2]. Manual of Petroleum Measurement Standards Chapter 3.1B, Standard Practice for Level Measurement of Liquid Hydrocarbons in Stationary Tanks by Automatic Tank Gauging, 3rd Edition (April 2018).

Source

Guidebook for Overfill Prevention & Tank Gauging, PEMY Consulting and Endress+Hauser: Philip Myers, Brock Trotter, Dean Mallon, Hai-Thuy Ngo and Markus Wohlgemuth. For the complete report, go online at: www.addresses.endress.com or www.pemyconsulting.com.

How to apply vapor cloud analysis methodology to tank overfills

Philip Myers, P.E., Founder and Director, PEMY Consulting, Pleasant Hill, California, and **Brock Trotter**, E.I.T., PEMY Consulting, Oklahoma City, Oklahoma

The Buncefield explosion was one of the worst post World War 2 fire/explosion incidents. The fire, which shut down Heathrow and Gatwick airports in the UK, resulted in a regulatory tidal wave for aboveground petroleum storage tanks. Because of HSE vapor cloud explosion studies on petroleum storage tanks by the Fire and Blast Information Group (FABIG), new models have been proposed to understand and quantify whether vapor cloud explosions (VCEs) can occur.

The following article builds on the vapor cloud analysis (VCA) proposed by the UK Health and Safety Executive (UK HSE) as documented in its Research Report 908 (RR908) and FABIG Technical Note 12 (TN12). Any company that transfers a flammable liquid into a storage tank is vulnerable to the vapor cloud that is generated by a tank overfill. Because the liquid typically pours out from the top of the tank and falls into the secondary containment, the liquid may be contained but the vapor can easily traverse the secondary containment wall and reach an ignition source where either a vapor cloud explosion (VCE) or a flash fire (deflagration) may result. In either case, it is important to understand and prevent this type of incident.

Although recent gasoline tank overfill VCEs have made the news, much larger volumes of crude oil are shipped throughout the world. Therefore, this article investigates a novel approach to adapt and apply the VCA methodology to crude oil tank overfills.

Typical tank and terminal operations resulting in ground spills and leaks do not generate VCEs. Several prerequisites are required to create a VCE:

- A substantial flow of a volatile and flammable organic liquid, such as gasoline in the form of a spray or mist, that can quickly evaporate; in tank overfills, the energy for creating the spray is caused from fuel spilling from the top of the tank where it can be ejected into a cascade that is a few feet away from the shell—e.g., the flow at the Buncefield fuel depot was 115 kg/s (about 2,400 gpm)

when a series of explosions occurred on December 11, 2005
- A substantial time for the overflow to occur (usually more than 10 min)
- Calm, still air or very low wind conditions.

Flow rate proxy

Whether it is an owner, an operator, a manager or a regulator, understanding tank overfills begins with an estimate of the potential flow rate. Obtaining this information can be surprisingly difficult. If many tanks must be assessed or managed simultaneously, obtaining potential tank overfill flow rates quickly is a must for personnel.

Although tank filling flow rates vary considerably, a reasonable estimate for flow is the incoming pipeline size (Table 1). Several flow velocities are given that represent the limits of economic line sizes. For general purposes, the authors suggest that estimating flow should be based on the velocity column of 10 ft/sec.

For example, at the Buncefield fuel depot, the fuels arrived at the site in batches through a system of three pipelines: (1) one 10-in. diameter pipeline from the Lindsey Oil Refinery in Humberside, terminating in the HOSL West site; (2) one UKOP 10-in.diameter pipeline from the Stanlow refinery in Merseyside, terminating in the BPA North site; and (3) one UKOP 14-in. diameter pipeline (Thames-Buncefield) from Shell's Haven and Coryton Refinery, terminating in the BPA main site. Based on the limiting velocities and economic line sizing considerations, the use of a 10-in. line would be reasonable to flow fuel at 150 kg/s—240 kg/s through an 8-in. or 10-in. pipeline. The actual overfill at the Buncefield site started at 115 kg/hr, but ramped upward in the last minutes before the VCE.

Vapor cloud analysis methodology (FABIG TN12)

As background, the following provides a synopsis on the content and material presented in the RR908 and FABIG TN12 documents. After the 2005 Buncefield explosion and fire, a UK HSE-funded research team was assembled to understand how a VCE could initiate due to tank overfill. The basic mechanism of the vapor cloud generation is shown in Figure 1.

The fuel overfill spills from the open top tank or overflow openings. When it hits an obstruction, such as a wind girder or a stiffening ring, the fuel is thrown outward a few feet where a cascade is formed (Figure 2). The cascade entrains air, which becomes nearly saturated with fuel vapor near the ground. The splashing of liquid on the ground further increases the saturation levels to a final level of at least 90%, given an adiabatic assumption, where the only heat input is from the entrained air.

The adiabatic assumption is fair, given no wind conditions. The latest technical work leading to the VCA method is given in the RR908 document, where a remarkable amount of

Table 1
Flow Rate Proxy (bbl/hr)

Incoming pipe size, NPS	Velocity, ft/s				
	5	8	10	15	20
3	219.4	329.1	438.7	658.1	877.5
4	377.9	566.8	755.8	1133.7	1511.5
6	857.9	1286.8	1715.8	2573.7	3431.5
8	1484.2	2226.3	2968.5	4452.7	5936.9
10	2342.1	3513.2	4684.2	7026.3	9368.5
12	3357.3	5036	6714.7	10,072	13,429.3
14	4016.3	6024.5	8032.7	12,049	16,065.3
16	5245.3	7867.9	10,490.5	15,735.8	20,981.1
18	6640.4	9960.7	13,280.9	19,921.3	26,561.8
20	8252.3	12,378.5	16,504.6	24,757	33,009.3
22	10,538	15,807.1	21,076.1	31,614.1	42,152.1
24	11,933.2	17,899.8	23,866.4	35,799.6	47,732.8
36	28,972.2	43,458.3	57,944.3	86,916.5	115,888.7
42	39,658.6	59,487.9	79,317.3	118,975.9	158,634.5

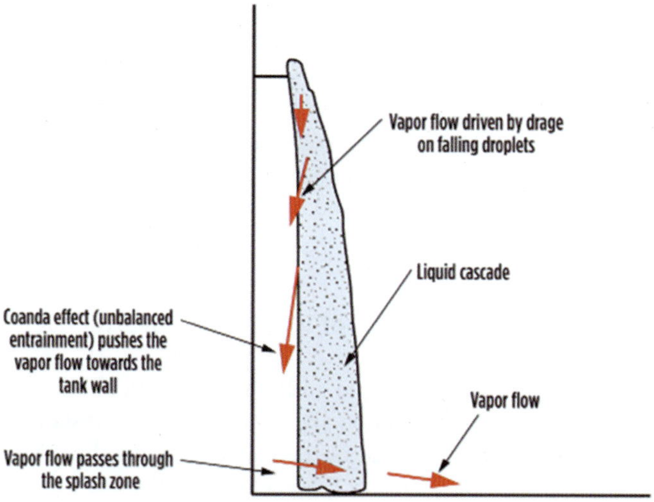

Figure 1. Basic test setup and results.

Figure 2. Fixed-roof tank showing stiffening rings and spill trajectory into a spray cascade.

testing was done. This testing included careful measurement of the temperature of the falling fuel, the entrained air and the vapor streams near and leaving the cascade.

Tests were developed to simulate the cascading fuel. The details of the experimental setup are given in RR908. The inference on evaporation was largely supported by a system of thermocouples that measured the temperature changes of the liquid and vapor in and around the cascade. From these measurements, reasonably accurate estimates of vaporization were constructed.

Computation fluid dynamics (CFD) analysis methods were used to support this work and describe how large vapor clouds can be formed. The bunds, or secondary containment walls, were found to redirect vapor flow back to the cascade and to inhibit large amounts of fresh air from being drawn into the cascade, allowing high-concentration vapors to build within the bunded areas.

If the duration of the overfill cascade is sufficiently long (Buncefield was 23.3 min), then it serves as a vapor generation source, and the heavy vapor cloud can slump by gravity flowing outward. In the case of a flat terrain, as was the case at all three overfills mentioned in the introduction, the "pancake cloud" flows outward, hunting for an ignition source.

VCA key input parameters

The key input parameters for the VCA method include the following:

- **Air entrainment into the cascade of falling liquid.** The air entrainment is primarily sensitive to the cross sectional area of the cascade. However, this depends on how the flow of fuel overtops the tank and over what distance. Since tanks are never exactly flat at the top, the cross-sectional area of the overfill cascade is uncertain, but its general effect can be examined by simulation. The VCA method assumes that 30% of the circumference forms the cascade. Other variables affecting the vaporization are liquid surface tension, liquid mass flow and height of free fall for the cascade.
- **Fuel vapor concentration at the foot of the cascade.** The VCA method assumes that the vaporization is enough to bring the air fuel mixture at the foot of the tank to about 70% of thermal equilibrium. This was demonstrated by measuring the concentration of hexane vapor in a replicated cascade.
- **Additional vaporization from fine splash products.** Additional vaporization occurs when the cascade hits the ground and splashes, creating a further push toward vapor equilibrium. The air-vapor mixture is essentially at thermal equilibrium and is saturated with fuel. The main variables controlling the equilibrium conditions are the fuel-air ratio created by the cascade, the fuel temperature and the ambient air temperature. The VCA method suggests that hydrocarbon mixtures, such as crude oil, the only fraction evaporating in the splash zone, are C8 (octane) and lighter. The mass evaporated is experimentally determined, but the VCA method suggests using $0.02 \times F \times$ (fraction less than or equal to C8), where F is the mass fuel flow rate.
- **Near-field (within bund) dilution.** The movement of the vapors within the bunded areas causes dilution. The VCA method chooses a value of 2 for the ratio of the concentration in the cascade to that in the outflowing vapor cloud. This is reasonable, since most bunds are 5 ft or 6 ft in height. The report results are listed in Table 2.
- **Volume flow rate and concentration of the cloud leaving the bund.** The mass of the vapor cloud is calculated as: Mcloud = 2(Mair + Mvaporized + Msplash). The volume of the cloud is assumed to be that of ambient air, even though warming from the ground and heat transfer from the vapor cloud will change the volume slightly.
- **Idealized hazard ranges for clouds spreading in "zero" wind speed conditions.** Despite the many assumptions made for a well-mixed cloud at the lower flammable limit (LFL), the model still fits the data. The

Table 2
Dilution Factors for Constrained Vapor Flows

Barrier height, m	Cloud depth, m	Distance to barrier, m	Dilution factor*
4	5	30	1.5
2.5	3.6	30	2.1
2	–	5	1.8
2	–	10	2
2	–	15	2

*Ratio of concentration in cascade to concentration in cloud

method assumes that the terrain is flat and that the vapor flows out over the secondary containment as a disk that is 2-m (6.5-ft) thick. This assumption, while not very accurate, was, in part, validated by the video recordings made at Buncefield and elsewhere.

- **Extended VCA methodology for crude oil tank overfilling.** The VCA method was originally published in RR908. However, in the RR908 document, the method was only applicable to a specific grade of gasoline and was unsuitable for any other determination. With the publication of FABIG TN12, the VCA method was extended to those compounds shown in Table 3.

Note: The only crude oil shown is for Brent. However, most of the world's tanks are filled with other types of crude oil containing other properties. FABIG TN12 is not transparent with respect to the parameters required for the VCA. As shown in Table 3, the correction factor contains four fitted parameters (α, β, γ and δ) that are not functions of thermodynamic properties that are readily computed.

Calculating the radius of ignition for overfill events

The industry lacks a simple and effective method for calculating the radius of ignition (R_{ign}) and minimum distance to evacuate personnel (R_{esc}) in the event of an overfill. Therefore, the authors' goal was to develop a method to calculate the radius of ignition for a wide variety of overfill events and parameters. These parameters include the ambient temperature (T_{amb}), the fuel temperature (T_{fuel}), the height of the tank (H), the diameter of the tank (D), the mass flow rate of the overfill event (M), the time length of the overfill event (t) and, most uniquely, the type of fuel being overfilled. The following calculations make some critical assumptions, including: no wind, a disk-shaped vapor cloud, and a lower explosive limit based upon hexane. Crude oil data is provided by Canada's National Energy Board (NEB) statistics.

Table 3
Concentration at the Foot of the Tank and Greek Constants for Various FABIG TN12 Fuels

	C^{θ}_{fuel}	α	β	γ	δ
Hexane	14.24	0.946	−0.225	0.0133	0.0212
Acetone	12.21	0.941	−0.262	0.0128	0.0192
Ethyl acetone	9.89	0.957	−0.181	0.0177	0.0242
Benzene	9.29	0.959	−0.176	0.0182	0.0222
Methyl ethyl ketone	8.71	0.955	−0.182	0.0163	0.0255
Toluene	4.7	0.981	−0.061	0.025	0.03
Methanol	4.53	0.95	−0.215	0.0167	0.0287
Ethanol	3.63	0.967	−0.133	0.0212	0.0345
Naphtha	20.3	0.928	−0.312	0.0093	0.0142
Winter-grade gasoline	17.25	0.888	−0.454	0.0074	0.0131
Raw gasoline	11.43	0.936	−0.264	0.0137	0.0179
F3 condensate	11.43	0.936	−0.264	0.0137	0.0179
Brent	6.88	0.876	−0.511	0.0088	0.0136
Reformate	5.29	0.967	−0.114	0.021	0.0295
Heavy reformate	5.29	0.967	−0.114	0.021	0.0295

This source was selected because it supplies the volume fraction of many of the vapor-contributing components within crude oil.

Establish overflow event parameters

The following example is a crude data set supplied by Canadian NEB statistics:

- Crude name: Mixed Sweet Blend (MSW)
- Date: August 6, 2017 (Batch #MSW-814)
- API gravity: 42.3
- Location: Edmonton, Alberta.

The parameters include:

$D = 75$ m
$H = 25$ m
$M_{fuel} = 800$ kg/s
$T_{amb} = 20°C$
$T_{fuel} = 14°C$
$t = 1400$ s

Convert from volume or mass fraction to molar fraction

Data from the Canadian NEB statistics is given in a volume fraction. This is the most common method of displaying crude component fractions (Table 4).

The lighter components will be the majority of what evaporates into the air when crude is being overfilled. This action occurs because the lighter components have a higher vapor pressure than heavier components. We assume the ideal gas law on the vapor phase and the liquid phase behaves as an ideal mixture. We then convert the volume fraction data in Table 4 into a molar fraction by using known molar density data.

The following molar densities (mol/L) are assumed at standard temperature and pressure conditions:

- C_3: 0.04228
- Butane: 0.04294
- Pentane: 8.6731
- Hexane: 7.6514
- Heptane: 6.8244
- Octane: 6.1481
- Nonane: 5.5985
- Decane: 5.1336
- Benzene: 11.252
- Toluene: 9.4086
- Ethylbenzene: 8.46774
- Xylenes: 7.620966
- Unaccounted components: 5

Regarding "unaccounted components" for crude oil, the entire molar composition is usually unknown. Therefore, it is necessary to group unknown components together into an "unaccounted component" group. A value is selected with properties like the heavy alkanes. The molar density for unaccounted components will vary for every crude oil, but the value of 5 optimizes the results for a wide variety of crude oils for the vapor pressure trend, as shown in Figure 3.

Hexane example of 1 L of sample:

$$\text{mol}_{\text{hexane}} = (100\ \text{L})(7.05\ \text{vol\%})(7.7514\ \text{mol/L}) = 53.942\ \text{mol}$$

$$\text{mol}_{\text{total}} = \sum \text{mol}_i$$

$$\text{mol}_{\text{total}} = 546.17\ \text{mol}$$

$$\text{mol}_i\% = \frac{\text{mol}_i}{\text{mol}_{\text{total}}} = \frac{53.942}{546.17} = 9.9\ \text{mol}_{\text{hexane}}\%$$

In application, we must apply the method above to each component. This process yields the values shown in Table 5 for molar composition of the crude sample.

Pure component vapor pressure calculation

To find the total vapor pressure of the crude oil, one must calculate the pure vapor pressure at ambient temperature ($T_{amb} = 20°C$). The P_{vap} of pure substances is achieved by the following equation and information from literature (Green and Perry, 2017):

Table 4
Crude Component Fractions

Component	Vol %
C_3	0.34
Butane	4.37
Pentane	5.08
Hexane	7.05
Heptane	7.3
Octane	5.67
Nonane	4.06
Decane	2.64
Benzene	0.26
Toluene	0.92
Ethylbenzene	0.18
Xylenes	1.24
Unaccounted	60.9
Total	**100**

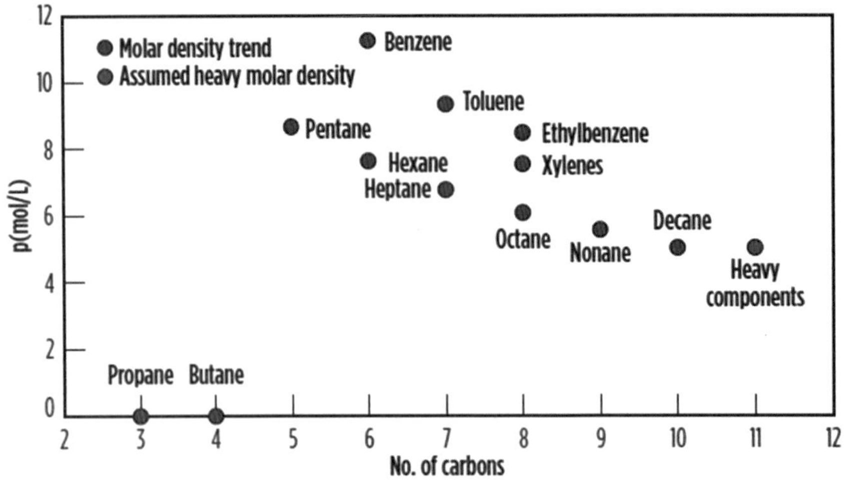

Figure 3. Carbon number vs. molar density of hydrocarbons.

Table 5
Molar Composition of Crude Sample MSW-814

Paraffins	
C_3	0.002632
C_4	0.034357
C_5	8.066969
C_6	9.876483
C_7	9.121361
C_8	6.38258
C_9	4.161693
C_{10}	2.481408
Aromatics	
C_6	0.535643
C_7	1.584839
C_8	0.279069
Xylenes	1.730231
Heavy components	55.74273533

$$\ln(P_{vap\ pure}) = C_1 + C_2/T + C_3 \times \ln(T) + C_4 \times T^{C5},$$

where the pure vapor pressure is described in Pascals.

The results at $T_{amb} = 20°C$ are shown in Table 6.

In application, one can ignore the vapor pressure contributions of components with more carbons than C8 (octane). These heavier components can be ignored, since their vapor pressures are very low relative to the light components of crude oil, such as propane and butane. Unknown heavy components contribute negligible vapor pressure, thus simplifying the model without losing much accuracy.

With pure vapor pressure data and molar composition of a crude oil, we can calculate an idealized total vapor pressure through Raoult's law:

$$P_{vap\ total} = \sum P_{ivap}$$

$$P_{ivap} = P_i^* \times x_i.$$

The MSW-814 crude oil sample yields a $P_{vap\ total} = 6816$ Pa. The total vapor pressure of the components is crucial for calculating the R_{ign} and R_{esc}.

Vapor pressure and Cθ correlation

To calculate the R_{ign} and R_{esc}, one needs a concentration of vapor at the base of the storage tank. The concentration at the foot of the tank can be determined by using a linear regression for the $\ln(P_{vap\ total})$ vs Cθ from the FABIG TN12 documents (Tables 7 and 8). The trend relates the natural logarithm of the total vapor pressure to the concentration at the foot of the tank as follows:

$$C\theta = 3.3073 \times \ln(P_{vap\ total}) - 21.696.$$

This trend is determined by a least squares regression fit of the data supplied by the FABIG TN12 document. The trend yields a strong fit with an $R^2 = 0.9097$ (Figure 4). The difference in the two values for Brent crude is explained by the different molar fractions of each sample, as shown in Table 9, when using the total vapor pressure from components:

$$C\theta = 3.3073 \times \ln(6816) - 21.696 = 7.4976\ kg/m^3.$$

Table 6
Pure Vapor Pressure Components at T_{amb} = 20°C

T(K)		C_1	C_2	C_3	C_4	C_5	P_{vap} (Pa)
Paraffins	C_3	59.078	−3492.6	−6.0669	1.09E−05	2	834184.2
	C_4	66.343	−4363.2	−7.046	9.45E−06	2	206981.9
	C_5	78.741	−5420.3	−8.8253	9.62E−06	2	56278.83
	C_6	104.65	−6995.5	−12.702	1.24E−05	2	16121.63
	C_7	87.829	−6996.4	−9.8802	7.21E−06	2	4666.773
	C_8	96.084	−7900.2	−11.003	7.18E−06	2	1391.773
	C_9	109.35	−9030.4	−12.882	7.85E−06	2	416.4769
	C_{10}	112.73	−9749.6	−13.245	7.13E−06	2	125.5671
Aromatics	C_6	83.107	−6486.2	−9.2194	6.98E−06	2	9913.166
	C_7	76.945	−6729.8	−8.179	5.3E−06	2	2902.748
	C_8	89.063	−7733.7	−9.917	5.99E−06	2	945.5245
	C_9	91.379	−8672.8	−10.176	5.62E−06	2	334.2583
Naphthenes	C_5	66.341	−5198.5	−6.8103	6.19E−06	2	34429.28
	C_6	51.087	−5226.4	−4.2278	9.76E−18	6	10311.3

Table 7
Molar Compositions from FABIG TN12 Samples

	Paraffins							**Aromatics**				**Naphthenes**	
	C_3	C_4	C_5	C_6	C_7	C_8	C_{10}	C_6	C_7	C_8	C_9	C_5	C_6
Winter-grade gasoline	9.6	17.2	16				57.2						
Naphtha		2	56	21	7			4				5	5
Raw gasoline		1	9	21				35	13	7	14		
Reformate				4	20	19		5	24	23	5		
Heavy reformate				4	5	3		1	31	34	22		
F3 condensate		4.7	6.5	4.1	6.5		69.3	4.7	1.4			2.8	
Brent	0.81	1.75	2.65	2.27	2.84		84.4	3.78				1.5	

Correction factor

The value for Cθ must then be multiplied by a correction factor (F). The value of this correction factor has an average of 0.07834679 and a standard deviation of 0.336811. These values are only appropriate under the following conditions:

- T_{amb}: −20°C to 30°C
- T_{fuel}: −20°C to 30°C
- M_{air}/M_{fuel}: 0.3 to 3.

The FABIG TN states, "over these ranges, the corrections factor (F) has an average accuracy of ±2% with large deviations of up to 15%."

$$F = \alpha\left(\frac{M_{air}}{M_{fuel}}\right)^{\beta} \exp(\gamma(T_{air} - 10))\exp\left(\delta\left(T_{fuel} - 10\right)\right)$$

$$C_{foot} = C\theta \times F$$

For the example scenario: T_{amb} = 20°C, T_{fuel} = 14°C, Greek variables = $\bar{x}$ F = 1.3237.

Table 8
C(θ) and Correction Factors for FABIG TN12 Samples

	C_{fuel}^{θ}	α	β	γ	δ
Hexane	14.24	0.946	−0.225	0.0133	0.0212
Acetone	12.21	0.941	−0.262	0.0128	0.0192
Ethyl acetate	9.89	0.957	−0.181	0.0177	0.0242
Benzene	9.29	0.959	−0.176	0.0182	0.0222
Methyl ethyl ketone	8.71	0.955	−0.182	0.0163	0.0255
Toluene	4.7	0.981	−0.061	0.025	0.03
Methanol	4.53	0.95	−0.215	0.0167	0.0287
Ethanol	3.63	0.967	−0.133	0.0212	0.0345
Naphtha	20.3	0.928	−0.312	0.0093	0.0142
Winter-grade gasoline	17.25	0.888	−0.454	0.0074	0.0131
Raw gasoline	11.43	0.936	−0.264	0.0137	0.0179
F3 condensate	11.10	0.881	−0.476	0.0085	0.0132
Brent	6.88	0.876	−0.511	0.0088	0.0136
Reformate	5.39	0.967	−0.114	0.021	0.0295
Heavy reformate	4.02	0.97	−0.095	0.0234	0.0302

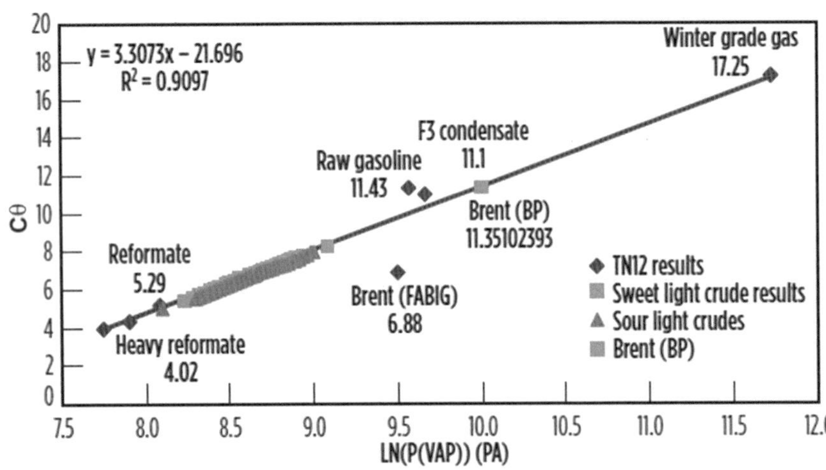

Figure 4. FABIG TN12 linear regression and crude results.

Table 9
The Difference in the Two Values for Brent Crude is Explained by the Different Molar Fractions of Each Sample

	C_3	C_4	C_5	C_6	C_7	C_8	C_9	C_{10}
Brent (FABIG TN12)	0.81	1.75	2.65	2.27	2.84	0	0	84.4
Brent (BP)	1.433409	3.41828	3.692294	3.26147027	0	0	0	0

Composition chart FABIG TN12 and BP

Remaining steps in calculating the radius of ignition and escape

For the equation:

$$C_{foot} = C\theta \times F = 7.4976 \times 1.323759 = 9.9251 \text{ kg/m}^3$$

This is the concentration value needed to continue the process of calculating the R_{ign} and R_{esc}.

$$M_{air} = 90\left(\frac{D}{25 \text{ m}}\right)^{0.75}\left(\frac{H}{10 \text{ m}}\right)^{0.45}\left(\frac{M_{fuel}}{115 \text{ kg/s}}\right)^{0.25}$$

$$M_{air} = 503.2 \text{ kg/s}$$

It is now necessary to consider the mass of fuel vaporized and splashed. Vaporized fuel is based on the concentration on the foot of the tank and M_{air}:

$$M_{vap} = M_{air} \times C_{foot}/\left(100 - C_{foot}\right)$$

$$M_{vap} = 55.45 \text{ kg/s}$$

Only components that are equal to, or lighter than, octane will contribute to M_{splash}. To calculate this component fraction, add the component fractions that are below C_8:

$$M_{splash} = 0.02 \times M_{fuel} \times x_{\leq C_8}$$

$$M_{splash} = 574.14 \text{ kg/s}$$

Now that we have calculated the mass contributions by air, vaporized fuel and fine droplet splash, one can calculate the total mass of the vapor cloud:

$$M_{cloud} = 2\left(M_{splash} + M_{vap} + M_{air}\right)$$

$$M_{cloud} = 2265.6 \text{ kg/s.}$$

The density of air at the ambient conditions, $\rho_{air} = 1.3015$ kg/m^3, is then used to determine the volume of the vapor cloud:

$$V_{cloud} = M_{cloud}/\rho_{air} = 1740.8 \text{ m}^3$$

Using the mass of explosive components and volume of the cloud, one can calculate the concentration of explosive vapor in the cloud:

$$C_{cloud} = \frac{M_{splash} + M_{vap}}{V_{cloud}} = 0.3617 \text{ kg/m}^3$$

All values used to calculate R_{ign} and R_{esc} are calculated:

$$R_{ign} = \left(\frac{1}{\pi} \times V_{cloud} \times t\right)^{1/2}$$

$$R_{esc} = \left(\frac{1}{2\pi} \times V_{cloud} \times t\right)^{1/2}$$

$$R_{ign} = 880.8 \text{ m}$$

$$R_{esc} = 622.8 \text{ m}$$

The overall process requires first defining the components within the fuel system and the other parameters, and then determining the vapor pressure due to the contributing components. This vapor pressure is used with the linear regression from the FABIG TN12 data to obtain the $C\theta$ value. $C\theta$ is then corrected by the correction factor F to obtain C_{foot}. Calculate the M_{air}, M_{splash} and M_{vap} to get M_{cloud}. Use the M_{cloud} to obtain the V_{cloud}. Then determine the C_{cloud}, R_{ign} and R_{esc} from V_{cloud}.

Eliminating the potential for a VCE

It is surprising that the authors found no references as to how to positively eliminate the large vapor cloud from forming. In the past, many tanks were constructed with an "overflow pipe" that directed an overflow to the ground. While this has typically been done for reasons unrelated to vapor cloud formation, it was applied to smaller tanks. This idea has not been used on large tanks above 50 m. The problem with applying this idea to tanks constructed to a standard is that the standards do not anticipate a design aimed at mitigating potential overfill vapor cloud events.

UK Health and Safety Executive research has resulted in two documents that provide a method for deter mining the size of a potential vapor cloud that might develop from a petroleum tank overfill. Research Report 908 was the first document to be published, followed by the FAB IG TN12 document. In RR908, the VCA method was applied through examples of overfills involving gasoline and ethanol.

FABIG TN12 extended the method to a list of 15 fluids, but only one of the fluids was crude oil. That fluid was Brent crude oil, and, in an example given in the document, a vapor cloud with an ignition radius of 612 m was computed.

Reference

D.W. Green, R.H. Perry, Perry's Chemical Engineers' Handbook. McGraw Hill, 2017.

Reprinted with permission – *Hydrocarbon Processing*, December 2018, January 2019.

Simplified procedure to calculate emissions for liquid organic storage tanks

Title V of the 1990 Clean Air Act required the accurate estimation of emissions from all U.S. manufacturing processes, and placed the burden of proof for that estimate on the process owner. The Emission Factor and Inventory Group in the U.S. Environmental Protection Agency (EPA) Office of Air Quality Planning and Standards develops and maintains a database of emission factors and a series of calculation methods for estimating air emissions from manufacturing processes. These emission factors are published in a series known as AP-42.

Here, a simplified procedure to calculate tank emissions associated with floating and fixed roof storage tanks is provided based on the assumptions and conditions applicable to the South Coast Air Basin, which includes Orange County, California, the non-desert portions of Los Angeles and San Bernardino counties, and the Riverside county areas west of the Palo Verde Valley.

For underground liquid fuel and aboveground small tanks (less than 10,000 gallons capacity), and for special cases such as horizontal, square, or rectangular tanks, additional formula are given in the South Coast Air Quality Management District Supplemental Instructions for Liquid Organic Storage Tanks, October 2019. For other areas and basins of the U.S. and for other tank designs, a different set of assumptions and conditions should be applied. Detailed descriptions and emission calculation procedures are provided in Chapter 7 of AP-42.

Tank emissions

Emissions from organic liquids storage tanks occur because of evaporative loss of the liquid during its storage and as a result of changes in the liquid level.

Emissions from fixed roof tanks are a result of evaporative losses during storage (known as breathing losses or standing storage losses) and evaporative losses during filling and emptying operations (known as working losses). Storage loss is the expulsion of vapor from a tank through vapor expansion and contraction, which are the results of changes in temperature and barometric pressure. This loss occurs without any liquid level change in the tank.

The combined loss from filling and emptying is called working loss. Evaporation during filling operations is a result of an increase in the liquid level in the tank. As the liquid level increases, the pressure inside the tank exceeds the relief pressure and vapors are expelled from the tank. Evaporative loss during emptying occurs when air drawn into the tank during liquid removal becomes saturated with organic vapor and expands, thus exceeding the capacity of the vapor space.

For floating roof tanks, standing storage losses are a result of evaporative losses through rim seals, deck fittings, including deck seams losses, in case of internal floating roof tanks, for construction other than welded decks. Withdrawal losses occur as the liquid levels, and thus the floating roof is lowered. Some liquid remains on the inner tank wall surface and evaporates. For an internal floating roof tank that has a column supported fixed roof, some liquid also clings to the columns and evaporates. Evaporative loss occurs until the tank is filled and the exposed surfaces are again covered.

Assumptions

The calculation methodologies are the simplified versions of those outlined in the AP-42 and based on the following assumptions:

• average ambient temperature:	520 degree R
• daily ambient temperature range:	17.9 degree R
• atmospheric pressure:	14.7 psia
• tank color:	white
• shell condition:	light rust
• other default parameters assumed in the EPA AP-42 document	

Other parameters needed for calculating emissions from storage tanks, including the properties of the stored material such as vapor or liquid molecular weights and true vapor pressure, can be found in AP-42, and are shown in abridged form in Figures 1-3.

Emissions from floating roof tanks

Total emissions or losses from floating roof tanks may be calculated using the following equation:

$$L_T = L_W + L_R + L_F + L_D + L_X \qquad (1)$$

where,

L_T = Total loss, lbs/yr
L_W = Working loss, lbs/yr
L_R = Rim seal loss, lbs/yr
L_F = Deck fitting loss, lbs/yr
L_D = Deck seam loss, lbs/yr (applicable to internal floating roof tanks only)
L_X = Loss from process upset, lbs/yr

Product		Chemical Name	CAS	Vapor		Liquid Density	P_{VA}	F_P	K_E	S_A	S_B	Small Tank Loss Factors		
Category	Code			M_V	W_V	W_L						f	a	b
Crude Oils	001	Crude oil (RVP 5)		50	0.028	7.10	3.181	0.061	0.094	0.0843	0.0115	2.863	0.378	0.084
Petroleum Dist	002	Distillate fuel oil #2		130	0.00018	7.10	0.0077	0.00013	0.034	0.0002	0.00003	0.0241	0.0009	0.0002
Petroleum Dist	003	Residual oil #6		190	0.000002	7.90	0.00005	0.000001	0.034	0.000001	0.0000002	0.000241	0.00001	0.000001
Petroleum Dist	004	Jet naphtha (JP-4)		80	0.02	6.40	1.419	0.025	0.056	0.0376	0.0052	2.725	0.162	0.038
Petroleum Dist	005	Jet kerosene		130	0.0002	7.00	0.0098	0.00017	0.034	0.0003	0.00004	0.0306	0.0011	0.0003
Petroleum Dist	006	Gasoline (RVP 6)		69	0.04	5.60	3.275	0.063	0.091	0.0868	0.0119	5.423	0.522	0.087
Petroleum Dist	007	Gasoline (RVP 7)		68	0.047	5.60	3.880	0.076	0.106	0.1028	0.0141	6.332	0.714	0.103
Petroleum Dist	008	Gasoline (RVP 8)		68	0.054	5.60	4.494	0.091	0.118	0.1191	0.0163	7.334	0.918	0.119
Petroleum Dist	009	Gasoline (RVP 9)		67	0.061	5.60	5.116	0.107	0.134	0.1356	0.0186	8.226	1.171	0.136
Petroleum Dist	010	Gasoline (RVP 10)		66	0.067	5.60	5.744	0.123	0.143	0.1522	0.0209	9.099	1.379	0.152
Petroleum Dist	011	Gasoline (RVP 11)		65	0.074	5.60	6.379	0.141	0.174	0.1691	0.0232	9.952	1.835	0.169
Petroleum Dist	012	Gasoline (RVP 12)		64	0.08	5.60	7.020	0.161	0.199	0.1860	0.0255	10.783	2.272	0.186
Petroleum Dist	013	Gasoline (RVP 13)		62	0.084	5.60	7.667	0.182	0.229	0.2032	0.0278	11.408	2.764	0.203
Organic Liquids	014	Acetaldehyde	75070	44.0	0.106	6.58	13.568	0.566	2.542	0.3596	0.0493	14.328	38.608	0.360
Organic Liquids	015	Acetic acid	64197	60.1	0.0021	8.79	0.200	0.0034	0.039	0.0053	0.0007	0.289	0.012	0.005
Organic Liquids	016	Acetic anhydride	108247	102.1	0.0012	9.01	0.0650	0.0011	0.036	0.0017	0.0002	0.159	0.0060	0.0017
Organic Liquids	017	Acetone	67641	58.1	0.034	6.63	3.327	0.064	0.108	0.0882	0.0121	4.637	0.529	0.088
Organic Liquids	018	Acetonitrile	75058	41.1	0.0094	6.56	1.296	0.023	0.058	0.0343	0.0047	1.277	0.079	0.034
Organic Liquids	019	Acrylamide	79061	71.1	0.000001	9.36	0.00012	0.000002	0.034	0.000003	0.0000004	0.0002	0.000007	0.000003
Organic Liquids	020	Acrylic acid	79107	72.1	0.0007	8.86	0.0538	0.00092	0.035	0.0014	0.0002	0.0932	0.0035	0.0014
Organic Liquids	021	Acrylonitrile	107131	53.1	0.015	6.76	1.607	0.029	0.066	0.0426	0.0058	2.047	0.143	0.043
Organic Liquids	022	Allyl alcohol	107186	58.1	0.0033	7.13	0.324	0.0056	0.043	0.0086	0.0012	0.452	0.020	0.0086
Organic Liquids	023	Allyl chloride	107051	76.5	0.073	7.86	5.406	0.114	0.163	0.1433	0.0196	9.929	1.711	0.143
Organic Liquids	024	Aniline	62533	93.1	0.00013	8.53	0.0078	0.00013	0.034	0.0002	0.00003	0.0173	0.0006	0.0002
Organic Liquids	025	Benzene	71432	78.1	0.019	7.37	1.340	0.024	0.060	0.0355	0.0049	2.512	0.159	0.036
Organic Liquids	026	Butanol-(1)	71363	74.1	0.00093	6.76	0.0710	0.0012	0.036	0.0019	0.0003	0.126	0.0048	0.0019

Figure 1. Properties and parameters for selected materials (for complete table, see SCAQMD 2019).

TANK DIAMETER (ft)	FLOATING											FIXED ROOF
	External Roof					Internal Roof						
			F_F					F_F		K_D	S_D	
	N_C	K_R	Pontoon	Double	K_D	N_C	K_R	Bolted	Welded	Bolted	Bolded	V_F
10	0.0	20.1	1628.6	1634.5	0.0	0.100	6.7	245.5	244.5	0.14	0.2	54
12	0.0	20.1	1630.2	1634.8	0.0	0.083	6.7	245.3	243.9	0.14	0.2	93
14	0.0	20.1	1631.8	1635.1	0.0	0.071	6.7	245.6	243.7	0.14	0.2	148
16	0.0	20.1	1633.4	1635.5	0.0	0.063	6.7	246.4	243.9	0.14	0.2	220
18	0.0	20.1	1635.1	1635.8	0.0	0.056	6.7	247.6	244.5	0.14	0.2	314
20	0.0	20.1	1636.8	1636.2	0.0	0.050	6.7	249.3	245.5	0.14	0.2	430
22	0.0	20.1	1638.5	1636.6	0.0	0.045	6.7	251.5	246.8	0.14	0.2	573
24	0.0	20.1	1640.2	1637.1	0.0	0.042	6.7	254.1	248.6	0.14	0.2	744
26	0.0	20.1	1641.9	1637.6	0.0	0.038	6.7	257.2	250.8	0.14	0.2	946
28	0.0	20.1	1643.7	1638.1	0.0	0.036	6.7	260.8	253.3	0.14	0.2	1181
30	0.0	20.1	1645.5	1638.7	0.0	0.033	6.7	264.9	256.2	0.14	0.2	1453
32	0.0	20.1	1647.3	1639.2	0.0	0.031	6.7	269.4	259.6	0.14	0.2	1763
34	0.0	20.1	1649.1	1639.8	0.0	0.029	6.7	274.4	263.3	0.14	0.2	2115
36	0.0	20.1	1650.9	1640.5	0.0	0.028	6.7	279.9	267.4	0.14	0.2	2510
38	0.0	20.1	1652.8	1641.2	0.0	0.026	6.7	285.8	271.9	0.14	0.2	2952
40	0.0	20.1	1654.7	1641.9	0.0	0.025	6.7	292.2	276.9	0.14	0.2	3443
42	0.0	20.1	1656.6	1642.6	0.0	0.024	6.7	299.1	282.2	0.14	0.2	3986
44	0.0	20.1	1658.5	1643.4	0.0	0.023	6.7	306.4	287.9	0.14	0.2	4583
46	0.0	20.1	1660.4	1644.2	0.0	0.022	6.7	314.3	293.9	0.14	0.2	5237
48	0.0	20.1	1662.4	1645.0	0.0	0.021	6.7	322.5	300.4	0.14	0.2	5950
50	0.0	20.1	1664.4	1645.8	0.0	0.020	6.7	331.3	307.3	0.14	0.2	6725

Figure 2. Storage tank factors by diameter (for complete table, see SCAQMD 2019).

Crude Oil			
Chemical Name	**CAS Number**	**Liquid Weight (%)**	**Molecular Weight**
Hexane (-n)	110543	0.40	86.17
Benzene	71432	0.60	78.11
Isooctane	26635643	0.10	114.22
Toluene	108883	1.00	92.13
Ethylbenzene	100414	0.40	106.17
Xylene (-m)	1330207	1.40	106.17
Isopropyl benzene	98828	0.10	120.20
1,2,4-Trimethylbenzene	95636	0.33	120.19
Cyclohexane	110827	0.70	84.16

Distillate Fuel Oil #2 (Diesel)			
Chemical Name	**CAS Number**	**Liquid Weight (%)**	**Molecular Weight**
Hexane (-n)	110543	0.00	86.17
Benzene	71432	0.00	78.11
Toluene	108883	0.03	92.13
Ethylbenzene	100414	0.01	106.17
Xylene (-m)	1330207	0.29	106.17
1,2,4-Trimethylbenzene	95636	1.00	120.19

Figure 3. Default TAC profile for select petroleum products (for complete table, see SCAQMD 2019).

Working Loss. The working loss from floating roof tanks can be estimated using Equation 2.

$$L_W = (1 - N_C)^*\left(\frac{22.45^*Q^*S_C^*W_L}{D}\right) \quad (2)$$

where,

L_W = Working loss, lbs/yr
N_C = Roof support factor, dimensionless; for external floating roof tank, $N_C = 0$; for internal floating roof tanks, depending on tank diameter
Q = Annual throughput in units of 1,000 gallons, M gal/yr
S_C = Shell clinage factor, $S_C = 0.006$ for crude oil; $S_C = 0.0015$ for other materials
W_L = Average organic liquid density, lbs/gal
D = Tank diameter, ft

Rim Seal Loss. Rim seal loss from floating roof tanks are estimated using the following equation:

$$L_R = K_R{}^*D^*F_P{}^*M_V{}^*K_C \quad (3)$$

where,

L_R = Rim seal loss, lbs/yr
K_R = Rim seal loss factor, lb-mole/ft-yr; $K_R = 20.1$ for external floating roof tank; $K_R = 6.7$ for internal floating roof tanks
D = Tank diameter, ft
F_P = Vapor pressure function, dimensionless
M_V = Average vapor molecular weight, lb/lb-mole
K_C = Product factor; $K_C = 0.4$ for crude oil; $K_C = 1$ for all other materials

Deck Fitting Loss. Deck fitting loss from floating roof tanks can be estimated using the following equation:

$$L_F = F_F{}^*F_P{}^*M_V{}^*K_C \quad (4)$$

where,

L_F = Deck fitting loss, lbs/yr
K_R = Total deck fitting loss factor, lb-mole/ft-yr, depending on tank diameter and type of rook deck
F_P = Vapor pressure function, dimensionless
M_V = Average vapor molecular weight, lb/lb-mole
K_C = Product factor; $K_C = 0.4$ for crude oil; $K_C = 1$ for all other materials

Deck Seam Loss. Deck seam loss, L_D, is applicable only to internal floating roof tanks with bolted decks and can be estimated by the following equation:

$$L_D = S_D{}^*K_D{}^*D^{2*}F_P{}^*M_V{}^*K_C \quad (5)$$

where,

L_D = Deck seam loss, lbs/yr
S_D = Deck seam length factor, ft/ft^2 ; for bolted deck, $S_D = 0.2$; for others, $S_D = 0.0$
K_D = Deck seam loss per unit seam length factor, lb-mole/ft-yr; for bolted deck, $K_D = 0.14$; for others, $K_D = 0.0$
D = Tank diameter, ft
F_P = Vapor pressure function, dimensionless
M_V = Average vapor molecular weight, lb/lb-mole
K_C = Product factor; $K_C = 0.4$ for crude oil; $K_C = 1$ for all other materials

Emissions from fixed roof tanks

Total emissions or losses from fixed roof tanks are equal to the sum of the working loss and standing loss:

$$L_T = L_W + L_S + L_X \quad (6)$$

where,

L_T = Total loss, lbs/yr
L_W = Working loss, lbs/yr
L_S = Standing loss, lbs/yr
L_X = Loss from process upset, lbs/yr

Working Loss. The working loss from fixed roof tanks can be estimated from

$$L_W = 0.024^*M_V{}^*P_{VA}{}^*Q^*K_N{}^*K_P \quad (7)$$

where,

L_W = Working loss, lbs/yr
M_V = Average vapor molecular weight, lb/lb-mole
P_{VA} = True vapor pressure of stored liquid at average liquid surface temperature, psia
Q = Annual throughput, M gal/yr
K_N = Turn over factor, dimensionless, dependent of annual throughput, Q (M gallon/yr), and tank capacity, C (M gallons). K_N is calculated as follows: If $Q/C \leq 36$ then $K_N = 1.0$; if $Q/C > 36$ then

$$K_N = \frac{180^*C + Q}{6^*Q} \quad (8)$$

K_P = Working loss product factor, dimensionless, $K_P = 0.75$ for crude oil; $K_P = 1.0$ for all other materials

Standing Loss. Fixed roof tank standing loss can be estimated from:

$$L_S = U^*V_V{}^*W_V{}^*K_E{}^*K_S \quad (9)$$

where,

L_S = Standing storage loss, lbs/yr
U = Number of days of the year that the tank is used to store liquid material. This number must not be more than 365 days. If not known, assume 365 days.
V_V = Vapor space volume can be calculated as:

$$V_V = 66.84*C + V_F \tag{10}$$

where:

C = Tank capacity, M gallons
V_F = Vapor space function, depending on tank diameter
W_v = Vapor density, lb/ft^3
K_E = Vapor space expansion factor, dimensionless
K_S = Vented vapor saturation factor, dimensionless, can be calculated as follows:

$$K_S = \frac{1}{1 + (S_A*H) + (S_D*D)} \tag{11}$$

where:

S_A, S_B = Vapor saturation functions
D = Tank diameter, ft
H = Tank height, ft

Toxic air contaminant calculation

Toxic air contaminant (TAC) emissions associated with storage tanks must be calculated and reported. In general, the emission rate for each component can be estimated by:

$$L_{TAC} = Z_{TAC}*L_T \tag{12}$$

where,

L_{TAC} = Emission rate of TAC component, lbs/yr
Z_{TAC} = Weight fraction of TAC component
L_T = Total losses, lbs/yr

Source

South Coast Air Quality Management District (SCAQMD). Supplemental Instructions For Liquid Organic Storage Tanks Annual Emissions Reporting Program, October 2019. Online at: www.aqmd.gov.

Calculating tank emissions with BREEZE TankESP

Christine Heath, Rob Ferry, Tiffany Stefanescu, and **Qiguo Jing**, Trinity Consultants, Philadelphia, Pennsylvania

On November 20, 2019, the US EPA released the final revisions to AP-42 Chapter 7, proposed on July 25, 2018, which provides guidance for calculating air emissions from fixed and floating roof storage tanks. With the release of the final revisions to AP-42 Chapter 7, TANKS 4.09d is no longer a valid tool[1] for tank emission calculations.

As a result, many states are no longer accepting TANKS 4.09d for permitting and emission inventories. The Louisiana Department of Environmental Quality announced it will no longer accept permit applications containing TANKS 4.09d emission estimates at the end of February 2020. The Alabama Department of Environmental Management has indicated that TANKS 4.09d may not be acceptable in some cases. Other states have indicated they plan to continue accepting TANKS 4.09d depending upon circumstance.

[1] The US EPA website for TANKS 4.09d has a warning in bold red font which states that the program is "outdated [and] not reliably functional," that EPA no longer assists with its use, and that it is to be used "at your own risk."

AP-42 Chapter 7 Revisions

The primary driver for the July 2018 revisions was the need to address emissions from tank cleaning events. Other revisions included proposed changes to the temperature equations, which impact estimated emissions for all tanks regardless of tank type, and guidance for estimating emissions from tanks storing hot stocks, taking into account different configurations of insulation. The November 2019 revisions include new calculation methodology for tank cleaning and degassing losses, updated meteorological data, and the upward revision to the default vapor pressure for No. 6 Oil. The impact of these final AP-42 revisions on estimated emissions may vary greatly, depending upon the tank scenario. For those who had not been accounting for tank cleaning emissions previously, then beginning to account for them will have a significant impact. The changes to the temperature equations will typically result in only single-digit percent differences, unless the tank stores hot stock and the elevated temperature had not been previously considered.

BREEZE TankESP software

In response to the November 2019 final revisions, BREEZE Software from Trinity Consultants released a new version (5.1) of the tank emissions calculation software product suite, BREEZE TankESP, that uses the revised AP-42 Chapter 7 emission estimation procedures. This program was designed by Rob Ferry, one of the authors of AP-42 Chapter 7.1 and American Petroleum Institute (API) documents 19.1, 19.2 and 19.4.

BREEZE TankESP also includes several additional enhancements, including the inclusion of three short-term emission calculation methods based on technical guidance from the API (technical report 2576) and the Texas Commission on Environmental Quality (guidance documents APGD 6250 and APGD 6419), and resolves the technical deficiencies associated with the US EPA's TANKS 4.09d software, which is no longer supported by the US EPA.

Challenges in calculating tank emissions

One key challenge and potential source of error in previous tank emission calculations is outdated or incorrect tank design and liquid product characterization data. To resolve these issues, environmental staff with need to adjust as follows:

- Update product chemical/physical properties with reasonable HAP speciation;
- Record and report tank turnaround inspection and maintenance activities; and
- Update tank seal and design parameters.

Additionally, it is important to check regulatory requirements and operating permit limits (e.g., NSPS Subpart Ka, Kb, and BWON requirements). Finally, and perhaps most significantly, due to the technical issues with TANKS 4.09d, owner/operators may need to revisit historical emission limits or permitting analyses when updating emission calculation methodologies.

Deficiencies and resolution

Table 1 illustrates how some of the technical deficiencies of TANKS 4.09d are resolved in TankESP.

Table 1
TANKS 4.09d Issues and Resolution in BREEZE TankESP

TANKS 4.09d Issue	Resolution in BREEZE TankESP
Guidepole. TANKS allows defaulting the guidepole configuration (with the default assumption being an uncontrolled, unslotted guidepole). The potential emissions from a guidepole in an EFRT, however, are such that any default for this fitting may result in a large error in the overall estimate of emissions.	Both TANKS and TankESP require the user to enter the rim seal configuration, rather than offering a default entry, due to the potential magnitude of the resulting emissions. TankESP similarly requires the user to make entries for the guidepole quantity and status (while allowing defaults for all other deck fittings).
Recalculate Deck Fittings. When a change has been made to the tank diameter, TANKS only recalculates deck fitting quantities if the deck fittings are shown as "Typical." Furthermore, if a change is made to the type of deck, TANKS only adjusts the deck fitting selections if "Typical" is indicated. If a change has been made to any deck fitting (such as changing the control status from Ungasketed to Gasketed), then none of the deck fitting quantities will recalculate with a change in tank diameter, and none of the deck fitting selections will adjust with a change in the deck type.	When a change is made to the tank diameter, TankESP recalculates the quantity of each deck fitting that has not had an override entered for the quantity of that particular deck fitting. When a change is made to the type of deck, TankESP adjusts each deck fitting selection and quantity that does not have an override entry.
IFRT Deck Support Legs. TANKS allows changing the IFR deck support legs from "Roof Leg or Hanger Well" to "Roof Leg (3-in. Diameter)" without explaining that the latter is only appropriate for EFR-type deck legs {the 3-inch diameter is not the critical parameter; it is the 30-inch long or longer leg housing of an EFR-type deck (versus the 12-inch long housing typical of an IFR-type deck leg) that matters}.	TankESP does not allow combining EFR-type deck legs with an IFR-type deck. In order to model the deck legs as the taller EFR-type, the user must select an EFR-type deck (i.e., a steel pontoon or double-deck type, which are available options for a tank with a fixed roof, resulting in the configuration that TANKS labels "Domed External Floating-Roof Tank").

Table 1
TANKS 4.09d Issues and Resolution in BREEZE TankESP (continued)

TANKS 4.09d Issue	Resolution in BREEZE TankESP
EFRT Deck Support Legs. TANKS does not explain that deck legs for double-deck EFRs are similar to the center-area legs of pontoon EFRs (i.e., an assumed 48-inch housing length), and thus the factors for "center area, sock" may be used when a double-deck EFR is equipped with leg socks (cf. API MPMS 19.2 Table 6).	TankESP allows the selection of leg socks as a control level for a double-deck EFR, and appropriately applies the same emission factor as for a center-area leg of a pontoon-type EFR.
Change of Tank Type. If an alteration to a tank changes the tank type (e.g., installing an IFR in a Fixed-Roof Tank, or installing a dome on an EFRT), TANKS requires the creation of a new tank record.	TankESP allows changing the type of fixed roof or floating roof within a tank record, and automatically recognizes the change in all calculations of future emissions (while retaining records for prior emissions based on the prior tank details).
Monthly emissions. TANKS does not account for the actual number of days in each month, but rather simply divides the annual period by 12. This causes emissions for February to be overestimated by about 10%, with smaller errors in the other months.	TankESP bases monthly emission estimates on the actual number of days in the month.
Vent settings. TANKS does not account for fixed-roof tank vent settings in the calculation of working loss. API, in cooperation with EPA, revised the working loss tion to account for the vent settings. This revision is included in the Third Edition of API 19.1 and in the 2006 revisions to AP-42 Section 7.1, but it has not been incorporated into EPA's TANKS program.	TankESP includes the vent setting term in the calculation of fixed-roof tank working loss.
Riveted tanks. TANKS does not have Tank Construction as a variable for IFRTs, thereby not accommodating a selection of "Riveted" with a Mechanical-shoe seal.	TankESP allows the selection of the Riveted factor for Mechanical-shoe seals for all types of floating-roof tanks.
Sample Pipe or Well. This Fitting Type in TANKS is not found in AP-42, and it poses unnecessary confusion. Two of the options refer to "Slotted Pipe," and are duplications of "Slotted Guide-Pole" options. The other option, "Slit Fabric Seal," should be an option under "Gauge-Hatch/Sample Well."	TankESP lists deck fitting options that clearly correspond to the selections provided in AP-42.
Horizontal Tanks. TANKS incorrectly calculates the height of the vapor space, Hvo, in a horizontal tank as (D/2), rather than as (He/2) per AP-42 Equation 1-15.	TankESP correctly calculates Hvo as (He/2) for horizontal tanks.

References

1. Public Comment on EPA's National Enforcement Initiatives for Fiscal Years 2017-2019, published in the Federal Register on September 15, 2015, 80 FR 55352.
2. AP 42, Chapter 7: Liquid Storage Tanks, Fifth Edition, Volume I, (November 2006). http://www3.epa.gov/ttn/chief/ap42/ch07/index.html.
3. Storage Tank Emissions Determination Challenges, in: Russ Nettles TCEQ Air Quality Division, presented to the 2015 Emissions Inventory Workshop (January 14, 2015).

19: Economics

Time Value of Money **940**
Simple interest versus compound interest 940
Nominal interest rate versus effective annual interest rate 941
Present value of a single cash flow to be received in the future 941
Future value of a single investment 942
The importance of cash flow diagrams 942
Analyzing and valuing investments/projects with multiple or irregular cash flows 942
Perpetuities 943
Future value of a periodic series of investments 944
Annuities, loans, and leases 944
Gradients (payouts/payments with constant growth rates) 945
Analyzing complex investments and cash flow problems 946

Decision and Evaluation Criteria **947**
Payback method 947
Accounting rate of return (ROR) method 948
Internal rate of return (IRR) method 949
Net present value (NPV) method 950

Sensitivity Analysis **951**

Decision Tree Analysis **952**

Accounting Fundamentals **956**

TIME VALUE OF MONEY

The value of money is not constant but changes with time. A dollar received today is worth more than a dollar received a month from now, for two reasons. First, the dollar received today can be invested immediately and earn interest. Second, the purchasing power of each dollar will decrease during times of inflation, and consequently a dollar received today will likely purchase more than a dollar next month. Therefore, in the capital budgeting and decision-making process for any engineering or financial project, it is important to understand the concepts of *present value*, *future value*, and *interest*. Present value is the "current worth" of a dollar amount to be received or paid out in the future. Future value is the "future worth" of an amount invested today at some future time. Interest is the *cost* of borrowing money or the *return* from lending money. Interest rates vary with time and are dependent upon both risk and inflation. The rate of interest charged, as well as the rate of return expected, will be higher for any project with considerable risk than for a "safe" investment or project. Similarly, banks and investors will also demand a higher rate of return during periods of monetary inflation, and interest rates will be adjusted to reflect the effects of inflation.

Simple interest versus compound interest

Simple interest is interest that is only earned on the original principal borrowed or loaned. Simple interest is calculated as follows:

$$I = (P)(i)(n),$$

where

I = Simple interest,
P = Principal (money) borrowed or loaned,
i = Interest rate per time period (usually years), and
n = Number of time periods (usually years).

Compound interest is interest that is earned on both the principal and interest. When interest is compounded, interest is earned each time period on the original principal *and* on interest accumulated from preceding time periods. Table 1 and Figure 1 illustrates the difference between simple and compound interest and show what a dramatic effect interest compounding can have on an investment after a few years.

Table 1
Comparison between \$5000 Invested for 3 Years at 12% Simple and 12% Compound Interest

Simple Interest			
End Year	Interest Earned	Cumulative Interest	Investment Balance
0	\$0.00	\$0.00	\$5000.00
1	\$5000.00 × 0.12 = \$600.00	\$600.00	\$5600.00
2	\$5000.00 × 0.12 = \$600.00	\$1200.00	\$6200.00
3	\$5000.00 × 0.12 = \$600.00	\$1800.00	\$6800.00
Compound Interest (Annual Compounding)			
End Year	Interest Earned	Cumulative Interest	Investment Balance
0	\$0.00	\$0.00	\$5000.00
1	\$5000.00 × 0.12 = \$600.00	\$600.00	\$5600.00
2	\$5600.00 × 0.12 = \$672.00	\$1272.00	\$6272.00
3	\$6272.00 × 0.12 = \$752.74	\$2024.64	\$7024.64

Figure 1
Comparison of Simple versus Compound Interest

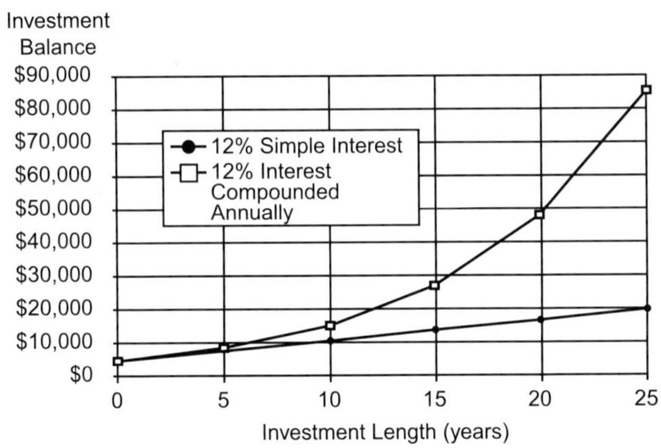

Nominal interest rate versus effective annual interest rate

Almost all interest rates in the financial world now involve compound rates of interest, rather than simple interest. Compound interest rates can be quoted either as a *nominal rate* of interest or as an *effective annual rate* of interest. The nominal rate is the stated annual interest rate compounded periodically (at the stated compounding time interval). The effective annual rate is the rate that produces the same final value as the nominal rate, when compounded only once per year. When the compounding period for a stated nominal rate is one year, then the nominal rate *is the same* as the effective annual rate of interest. The following formula can be used to convert nominal interest rates to effective annual rates:

$$i_{\text{eff}} = \left(1 - \frac{i_{\text{nom}}}{c}\right)^{c} - 1,$$

where

i_{eff} = Effective annual rate,
i_{nom} = Nominal rate, and
c = Number of compounding periods per year.

Note: Interest rates are expressed as fractions (0.12) rather than percentage (12%).

For *continuous* compounding, natural logarithms may be used to convert to an effective annual rate:

$$i_{\text{eff}} = e^{i_{\text{cont}}} - 1,$$

where

i_{cont} = Continuous compounding rate.

Note: As in the discrete case, in continuous compounding interest rates are expressed as fractions (0.12) rather than percentage (12%).

Present value of a single cash flow to be received in the future

How much is an amount of money to be received in the future worth today? Stated differently, how much money would you invest today to receive this cash flow in the future? To answer these questions, you need to know the *present value* of the amount to be received in the future. The present value (PV) can easily be calculated by *discounting* the future amount by an appropriate compound interest rate. The interest rate used to determine PV is often referred to as the *discount rate, hurdle rate*, or *opportunity cost of capital*. It represents the opportunity cost (rate of return) foregone by making this particular investment rather than other alternatives of comparable risk.

The formula for calculating present value is:

$$\text{PV} = \frac{\text{FV}}{(1+r)^{n}}$$

where

PV = Present value,
FV = Future value,
r = Discount rate (per compounding period), and
n = Number of compounding periods.

Example. What is the present value of an investment that guarantees to pay you $100,000 three years from now? The first important step is to understand what risks are involved in this investment, and then choose a discount rate equal to the rate of return on investments of comparable risk. Let's say you decide that 9% effective annual rate of interest (9% compounded annually) is an appropriate discount rate.

$$\text{PV} = \frac{\$100,000}{(1+0.09)^{3}} = \$77.218.$$

Thus, you would be willing to invest $77,218 today in this investment to receive $100,000 in 3 years.

Future value of a single investment

Rearranging the relationship between PV and FV, we have a simple formula to allow us to calculate the future value of an amount of money invested today:

$$FV = PV(1 + r)^n$$

where

FV = Future value,
PV = Present value,
R = Rate of return (per compounding period), and
n = Number of compounding periods.

Example. What is the future value of a single $10,000 lump sum invested at a rate of return of 10% compounded monthly (10.47 effective annual rate) for 5 years? Note that monthly compounding is used in this example.

$$FV = \$10,000(1 + 0.10/12)^{60} = \$16,543$$

The $10,000 investment grows to $16,453 in 5 years.

The importance of cash flow diagrams

Many financial problems are not nearly as simple as the two previous examples, which involved only one *cash outflow* and one *cash inflow*. To help analyze financial problems, it is extremely useful to diagram cash flows, since most problems are considerably more complex than these previous examples and may have multiple, repetitive, or irregular cash flows. The *cash flow diagram* is an important tool that will help you understand an investment project and the timing of its cash flows and help with calculating its PV or FV. The cash flow diagram shown represents the earlier example (in which PV was calculated for an investment that grew to $100,000 after annual compounding at 9% for 3 years).

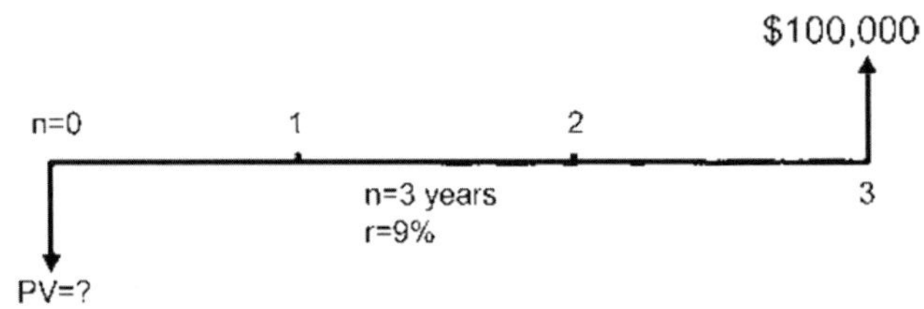

The cash flow diagram is divided into equal time periods, which correspond to the compounding (or payment) periods of the cash flows. Note that this diagram is also presented from the investor's (lender's) point of view, since the initial investment is a cash outflow (negative) and the final payoff is a cash inflow (positive) to the investor. The diagram could also have been drawn from the borrower's point of view, in which case the PV would be a cash inflow and therefore positive. Before drawing a cash flow diagram, you must first choose which viewpoint (lender or borrower) the diagram will represent. It is then extremely important to give each cash flow the correct sign (positive or negative), based upon whichever point of view is chosen.

Analyzing and valuing investments/projects with multiple or irregular cash flows

Many investments and financial projects have multiple, periodic, and irregular cash flows. A cash flow diagram will help in understanding the cash flows and with their analysis. The value of the investment or project (in current dollars) may then be found by discounting the cash flows to determine their present value. Each cash flow must be discounted individually back to time zero and then summed to obtain the overall PV of the investment or project. This *discounted cash flow formula* may be written as

$$\mathrm{PV} = \sum_{i=1}^{n} \frac{C_i}{(1+r)^i}$$

where

PV = Present value,
C_i = Cash flow at time i,
n = Number of periods, and
r = Discount rate of return (opportunity cost of capital).

Example. How much would you be willing to invest today in a project that will have generated positive annual cash flows of \$1,000, \$1,500, \$2,400, and \$1,600, respectively, at the end of each of the first 4 years? (Assume a 10% annual rate of return can be earned on other projects of similar risk.)

Step 1: Cash flow diagram:

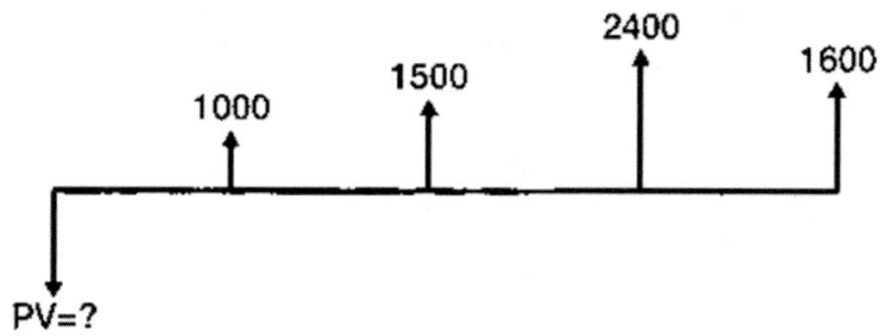

Step 2: Discount cash flows into current dollars (present value):

$$\mathrm{PV} = \frac{1.000}{(1+0.1)^1} + \frac{1,500}{(1+0.1)^2} + \frac{2,400}{(1+0.1)^3} + \frac{1,600}{(1+0.1)^4}$$
$$= \$5044.74$$

Thus, you should be willing to invest a maximum of \$5044.74 in this project.

Perpetuities

A *perpetuity* is an asset that provides a periodic fixed amount of cash flow in perpetuity (forever). Endowments and trust funds are often set up as perpetuities so that they pay out a fixed monetary amount each year to the beneficiary, without depleting the original principal. For valuation purposes, some physical assets such as an oil well, natural gas field, or other long-lived asset can also be treated as a perpetuity, if the asset can provide a steady cash flow stream for many years to come. The cash flow diagram for a perpetuity is represented here:

The value of a perpetuity can be found by this simple formula

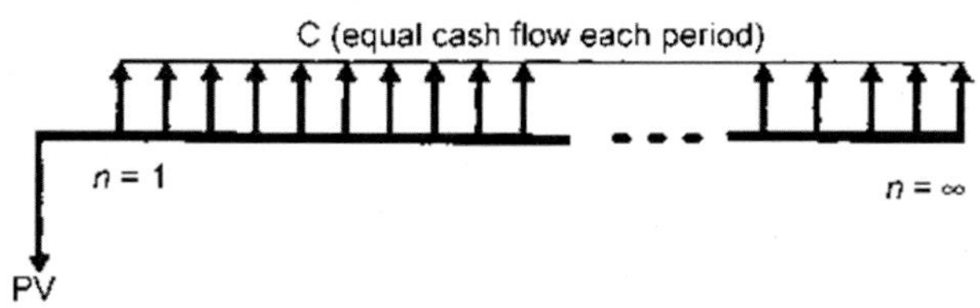

$$\mathrm{PV}_{\text{perpetuity}} = \frac{C}{r}$$

where

C = Cash flow per time period, and
r = Discount rate of return (per time period).

A *growing perpetuity* is a cash flow stream that grows at a constant rate. The formula to value a growing perpetuity expands the basic perpetuity formula to include a growth rate for the cash flow stream:

$$\mathrm{PV} = \frac{C_1}{r-g},$$

where

C_1 = Initial cash flow,
r = Discount rate of return (per time period), and
g = Growth rate of cash flows (per time period).

Example. An existing oil well has been producing an average of 10,000 barrels per year for several years. Recent estimates by geologists show that this well's oil field is not large enough to justify additional drilling but can maintain the existing well's production rate for hundreds of years to come. Assume that the recent price of oil has averaged $20 per barrel, prices are historically estimated to grow 3% per year forever, and your discount rate is 10%. Also assume that maintenance costs are insignificant, and the land has no value other than for the oil. What is the current monetary value of this oil field and well, if the owner were to sell it?

Solution:

$$PV = \left(\frac{10,000 \times 20}{0.10 - 0.03}\right) = \$2,857,143.$$

The value of the oil field and well should be $2,857,143 based upon these assumptions.

Future value of a periodic series of investments

Both individual investors and companies often set aside a fixed quantity of money every month, quarter, or year, for some specific purpose in the future. This quantity of money then grows with compound interest until it is withdrawn and used at some future date. A cash flow diagram for a periodic investment is shown.

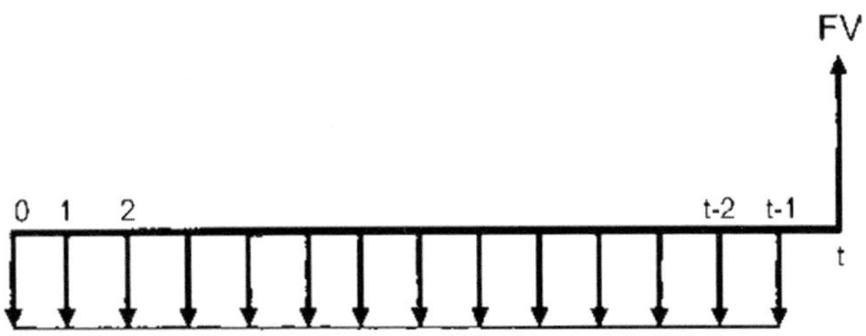

It can easily be calculated how much these periodic, fixed investments each time period will accumulate with compounding interest to be in the future. In other words, we want to determine the future value (FV), which can be calculated using the following formula:

$$FV = \frac{C}{r}\left[(1+r)^t - 1\right],$$

where

C = Cash flow invested each time period,
r = Fractional compound interest rate (each time period), and
t = Number of time periods (length of investment).

Example. A self-employed contract engineer has no company-paid retirement plan, so he decides to invest $2000 per year in his own individual retirement account (IRA), which will allow the money to accumulate tax-free until it is withdrawn in the future. If he invests this amount every year for 25 years, and the investment earns 14% a year, how much will the investment be worth in 25 years?

Solution:

$$FV = \frac{\$2,000}{0.14}\left[(1+0.14)^{25} - 1\right] = \$363,742.$$

The periodic investment of $2000 each year over 25 years will grow to $363,742 if a 14% rate of return is achieved. This example illustrates the "power of compounding." A total of only $50,000 was invested over the 25-year period, yet the future value grew to more than seven times this amount.

Annuities, loans, and leases

An *annuity* is an asset that pays a constant periodic sum of money for a fixed length of time. Bank loans, mortgages, and many leasing contracts are similar to an annuity in that they usually require the borrower to pay back the lender a constant, periodic sum of money (principal and interest) for a fixed length of time. Loans, mortgages, and leases can therefore be considered "reverse" annuities, as shown by the cash flow diagrams. The relationship between the value (PV) of the annuity/loan/lease, its periodic cash flows, and the interest or discount rate is

Cash Flow Diagrams: Annuity

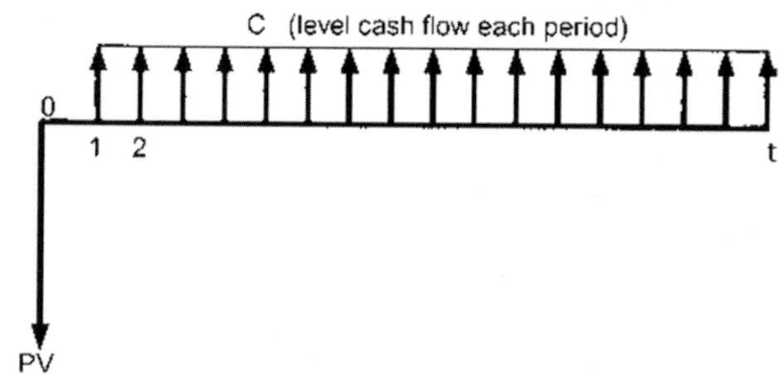

Cash Flow Diagrams: Reverse Annuity

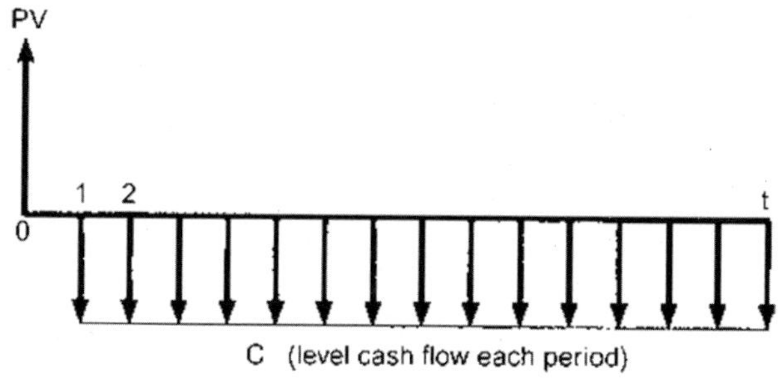

$$\mathrm{PV} = \frac{C}{r}\left[1 - \frac{1}{(1+r)^t}\right],$$

where

C = Cash flow received/paid each time period,
r = Interest or discount rate (per time period), and
t = Number of time periods (time length of the annuity).

This formula may be used to determine the PV of the annuity, or conversely, to determine the periodic payouts or payments when the PV is known. It should be noted that the future value (FV) at time *t* will be zero. In other words, all the principal of the annuity will have been expended, or all the principal of the loan or lease will have been paid off, at the final time period. Financial calculators are also pre-programmed with this formula and can simplify and expedite annuity, loan, and other cash flow problem calculations.

Example. Oil pipeline equipment is to be purchased and installed on a working pipeline in a remote location. This new equipment will have an operational life of 5 years and will require routine maintenance once per month at a fixed cost of $100. How much would the oil company have to deposit now with a bank at the remote location to set up an annuity to pay the $100 a month maintenance expense for 5 years to a local contractor? Assume annual interest rates are 9%.

Solution:

$$\mathrm{PV} = \frac{100}{0.09/12}\left[1 - \frac{1}{(1+0.09/12)^{60}}\right] = \$4,817.34.$$

Example. Apex Corp. borrows $200,000 to finance the purchase of tools for a new line of gears it's producing. If the terms of the loan are annual payments over 4 years at 14% annual interest, *how much are the payments*?

$$\$200,000 = \frac{C}{0.14}\left[1 - \frac{1}{(1+0.14)^4}\right]$$

Solution:

$C = \$68,640.96$ per year.

Gradients (payouts/payments with constant growth rates)

A *gradient* is similar to an annuity, but its periodic payouts increase at a constant rate of growth, as illustrated by its cash flow diagram. A gradient is also similar to a growing perpetuity, except that it has a fixed time length. The cash flow diagram for a gradient loan or lease is simply the inverse of this cash flow diagram, since the loan/lease payments will be negative from the borrower's point of view. The PV of a gradient may be found by

$$\mathrm{PV} = \frac{C_1(1-x^n)}{(1+r)(1-x)} \quad \text{when } g \neq r$$

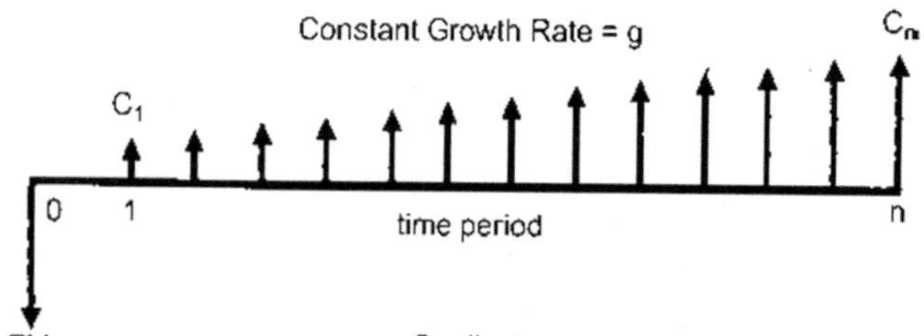

Gradient

or

$$PV = \frac{C_1 n}{(1+r)} \quad \text{when } g = r$$

where

C_1 = Initial cash flow at the end of first time period,
R = Rate of return (discount rate),
n = Number of time periods,
$x = (1 + g)/(1 + r)$, and
g = Fractional growth rate of cash flow period.

Analyzing complex investments and cash flow problems

No matter how complex a financial project or investment is, the cash flows can always be diagrammed and split apart into separate cash flows if necessary. Using the discounted cash flow (DCF) formula, each cash flow can be discounted individually back to time zero to obtain its present value (PV), and then summed to determine the overall valuation of the project or investment. It may also be possible to simplify things if you can identify a series of cash flows in the complex investment. Cash flow series can be discounted using either annuity or perpetuity formulas, and then added together with the PV of any singular cash flow.

When analyzing any complex cash flow problem, it is important to remember the basic time value of money (TVM) concept (i.e. *the value of money changes with time*). Therefore, if you wish to add or compare any two cash flows, they both must be from the same time period. Cash flows in the future must be discounted to their present value (PV) before they can be added or compared with current cash flows. Alternatively, current cash flows can be compounded into the future (FV) and then added or compared with future cash flows.

Example. An automotive components company has been asked by a large automobile manufacturing company to increase its production rate of components for one particular car model. Sales of this car model have been much higher than expected. In order to increase its production rate, the components company will have to purchase additional specialized machinery, but this machinery will easily fit into its existing facility. The automobile manufacturer has offered to sign a contract to purchase all the additional output that can be produced for 4 years, which will increase annual net revenues by $300,000 without any additional administrative overhead or personnel expense. The added machinery will cost $450,000 and will require maintenance every 6 months at an estimated cost of $20,000. The machinery will be sold at the end of the 4-year period, with an estimated salvage value of $80,000. *If the company requires a minimum rate of return of 20% on all investments, should it sign the contract, purchase the machinery, and increase its production rate?* (Depreciation and tax effects are not considered in this example.)

Step 1: From the point of view of the automotive components company, the cash flow diagram will be:

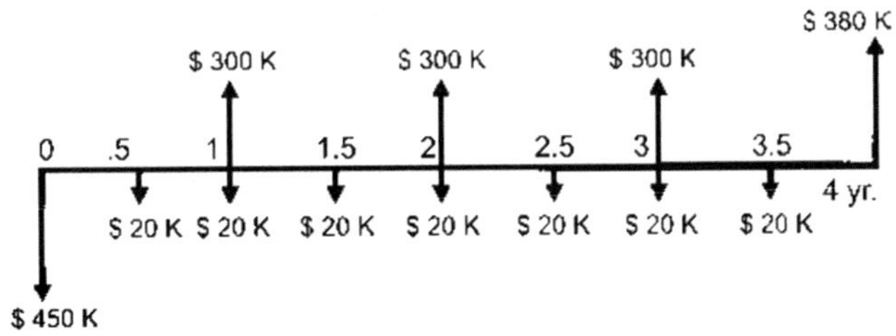

Step 2: Separate the overall cash flow diagram into individual components and series, and then determine the PV of each future cash flow or series of flows by discounting at the required annual rate of return of 20%:

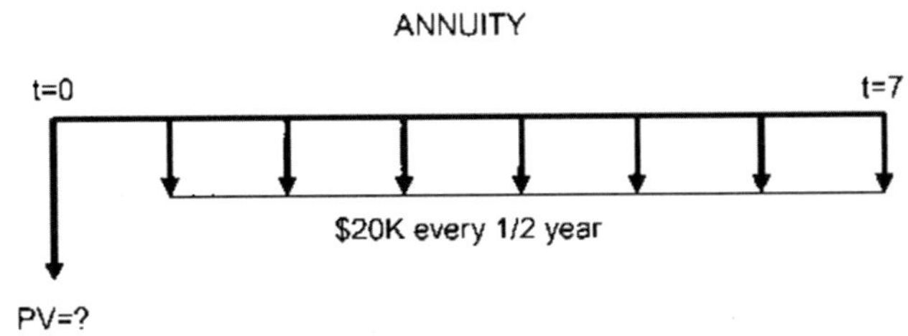

$$PV = \frac{\$20,000}{0.20/2}\left[1 - \frac{1}{(1+0.20/2)^7}\right]$$

$$= \$97,368 \quad \text{(negative)}$$

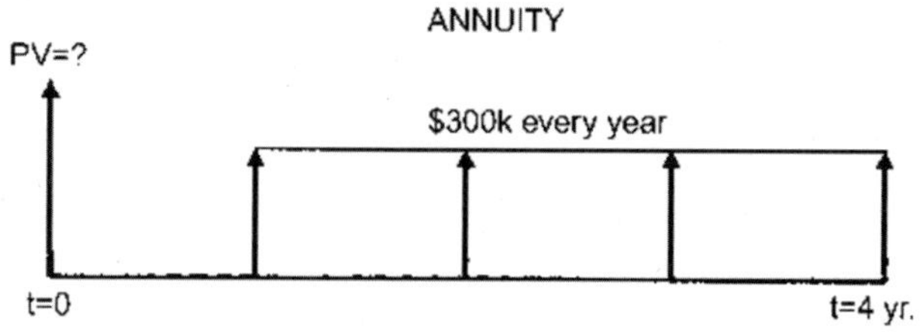

$$PV = \frac{\$30,000}{0.2}\left[1 - \frac{1}{(1+0.2)^4}\right] = \$776,620 \quad \text{(positive)}$$

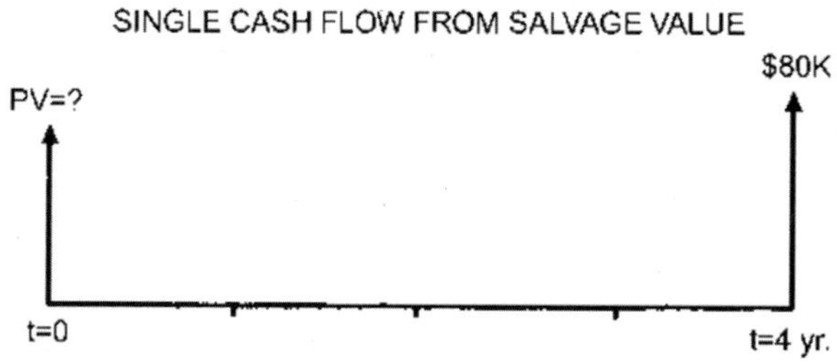

$$PV = \frac{\$80,000}{(1+0.2)^4} = \$38,580 \quad \text{(positive)}$$

Step 3: Sum the PV of all cash flows to obtain the project's overall PV from all future cash flows:

$$\begin{aligned} PV &= \$776,620 + \$38,580 - \$97,368 \\ &= \$717,832 \quad \text{(positive)} \end{aligned}$$

Step 4: Compare the PV of the future cash flows with the project's initial cash flow (initial investment):

The $717,832 PV of the project's future cash inflows (positive) is clearly greater than the $450,000 (negative) initial investment. Thus, this project has a positive *net present value* (NPV), because the PV of the future cash flows are greater than the initial investment. A positive NPV means that the rate of return on the investment is greater than the rate (20%) used to discount the cash flows.

Conclusion: The company should buy the additional machinery and increase its production rate.

DECISION AND EVALUATION CRITERIA

How should you decide whether or not to make an investment or go forward with a financial project? If you also have a number of alternative investment options, how do you decide which is the best alternative? How do you evaluate an investment's performance some time period after the investment has been made and the project is up and running? There are a number of analysis techniques that can help with these questions. Four separate methods are presented here, although each method varies in its effectiveness and complexity. There are also additional methods and techniques that can be found in other engineering economics, finance, and accounting textbooks. However, most modern books and courses in these fields of study recommend the *net present value* (NPV) method as the most effective and accurate technique of evaluating potential investments and financial projects. Accounting measures such as *rate of return* (ROR) on an investment or *return on equity* (ROE) are useful to evaluate a project's financial results over a specific time period or to compare results against competitors.

Payback method

The payback method is a simple technique that determines the number of years before the cash flow from an investment or project "pays back" the initial investment. Obviously, the shorter the length of time it takes a project to pay off the initial investment, the more lucrative the project.

Example. A company wants to add a new product line and is evaluating two types of manufacturing equipment to produce this new line. The first equipment type costs $700,000 and can produce enough product to result in an estimated $225,000 in additional after-tax cash flow per year.

The second type of equipment is more expensive at $900,000, but has a higher output that can bring an estimated $420,000 per year in after-tax cash flows to the company. Which type of equipment should the company buy?

Solution:

Equipment Type 1

$$\text{Payback period} = \$700,000/\$225,000 \text{ per year} = 3.11 \text{ years}$$

Equipment Type 2

$$\text{Payback period} = \$900,000/\$420,000 \text{ per year} = 2.14 \text{ years}$$

Based upon this payback analysis, equipment type 2 should be purchased.

The advantage of the payback method is its simplicity, as shown in the example. However, there are several disadvantages that need to be mentioned. First, the payback method does not evaluate any cash flows that occur after the payback date, and therefore ignores long-term results and salvage values. Secondly, the payback method does not consider the time value of money (TVM) but assumes that each dollar of cash flow received in the second year and beyond is equal to that received the first year. Thirdly, the payback method has no way to take into account and evaluate any risk factors involved with the project.

Accounting rate of return (ROR) method

The *accounting rate of return* (*ROR*) for an existing company or project can be easily calculated directly from the company's or project's quarterly or annual financial accounting statements. If a potential (rather than existing) project is to be evaluated, financial results will have to be projected. Since ROR is an accounting measure, it is calculated for the same time period as the company's or project's income statement. The ROR on an investment is simply the net income generated during this time period, divided by the book value of the investment at the start of the time period.

$$\text{ROR} = \frac{\text{Net Income}}{\text{Net Investment (book value)}}$$

Net income is an accounting term defined as "revenues minus expenses plus gains minus losses" (i.e. the "bottom line" profit or loss on the income statement). Net investment is the "book value" of the investment, which is the original investment amount (or purchase price) less any depreciation in value of the investment.

Example. Company XYZ invested $500,000 in plant and equipment 4 years ago. At the end of year 3, the company's financial statements show that the plant and equipment's value had depreciated $100,000 and had a listed book value of $400,000. If company XYZ has a net income of $60,000 during its fourth year of operation, what is its ROR for this time period?

Solution:

$$\text{ROR} = \frac{60,000}{400,000} = 0.15 = 15\%$$

Like the payback method, the advantage of the accounting ROR is that it is a simple calculation. It is especially simple for an existing company or project when the accounting statements are already completed and don't have to be projected. One disadvantage of ROR is that it doesn't consider TVM when it is used for more than one time period. It also uses accounting incomes rather than cash flows, which are not the same (see Accounting Fundamentals in this chapter). Another disadvantage is that ROR is a very subjective method, because the ROR is likely to vary considerably each time period due to revenue and expense fluctuations, accumulating depreciation, and declining book values. (Some textbooks recommend averaging ROR over a number of time periods to compensate for fluctuations and declining book values, but an "averaged" ROR still does not consider TVM.)

In general, accounting measurements such as ROR are best used for evaluating the performance of an *existing* company over a specific annual or quarterly time period or for comparing one company's performance against another. Net present value (NPV) and internal rate of return (IRR) are superior methods for evaluating *potential* investments and financial projects and in making the final decision whether to go ahead with them.

Internal rate of return (IRR) method

Internal rate of return (IRR) is an investment profitability measure that is closely related to net present value (NPV). The IRR of an investment is that rate of return which, when used to discount an investment's future cash flows, makes the NPV of an investment equal zero. In other words, when the future cash flows of an investment or project are discounted using the IRR, their PV will exactly equal the initial investment amount. Therefore, the IRR is an extremely useful quantity to know when you are evaluating a potential investment project. The IRR tells you the exact rate of return that will be earned on the original investment (or overall project with any additional investments) if the projected future cash flows occur. Similarly, when analyzing past investments, the IRR tells you the exact rate of return that was earned on the overall investment. The IRR decision rule for whether or not to go ahead with any potential investment or project being considered is simple: *If the IRR exceeds your opportunity cost of capital (rate of return that can be earned elsewhere), you should accept the project.*

The following example will make it clear how the IRR of an investment is calculated. It is highly recommended that a cash flow diagram be drawn first before trying to calculate the IRR. Particular attention should be paid to make sure cash inflows are drawn as positive (from the investor's point of view) and cash outflows are drawn as negative. It is also recommended that a financial calculator (or computer) be used to save time; otherwise, an iteration procedure will have to be used to solve for the IRR.

Example. Acme Tool and Die Company is considering a project that will require the purchase of special machinery to produce precision molds for a major customer, Gigantic Motors. The project will last only 4 years, after which Gigantic Motors plans to manufacture its own precision molds. The initial investment for the special machinery is \$450,000, and the machinery will also require a partial overhaul in 3 years at an estimated cost of \$70,000. Projected after-tax cash flows resulting from the project at the end of each of the 4 years are \$140,000, \$165,000, \$185,000, and \$150,000, respectively. These projections are based upon a purchase agreement that Gigantic is prepared to sign, which outlines the quantities of molds it will buy. Expected salvage value from the machinery at the end of the 4-year period is \$80,000. Should Acme proceed with this project, if its opportunity cost of capital is 23%?

Step 1: Cash flow diagram:

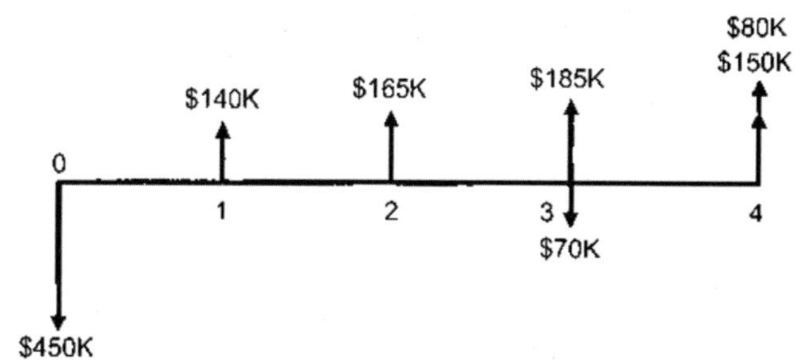

Step 2: Write an equation that sets the original investment equal to the PV of future cash flows, using the discounted cash flow formula. Solve for r, which is the IRR. (This is a simple process with a financial calculator or computer but otherwise requires iteration with any other calculator.)

$$450,000 = \frac{140,00}{(1+r)} + \frac{165,000}{(1+r)^2} + \frac{(185,000 - 70,000)}{(1+r)^3} + \frac{(150,000 + 80,000)}{(1+r)^4}$$

Solving for r (IRR):

$$\text{IRR} = 0.1537 = 15.37\%$$

Step 3: Apply decision rule to results: Since the IRR for this project is 15.37%, and Acme's required rate of return (opportunity cost of capital) is 23%, Acme should not proceed with this project.

IRR has a number of distinct advantages (as does NPV) over both the payback and ROR methods as a decision-making tool for evaluating potential investments. The first advantage is that the IRR calculation takes into account TVM. Secondly, IRR considers all cash flows throughout the life of the project (including salvage values), rather than looking at only one time period's results like accounting ROR does, or the years until the initial investment is paid off like the payback method. Therefore, IRR is an objective criterion, rather than a subjective criterion, for making decisions about investment projects. Thirdly, IRR uses actual cash flows rather than accounting incomes like the ROR method. Finally, since IRR is calculated for the entire life of the project, it provides a basis for evaluating whether the overall risk of the project is justified by the IRR, since the IRR may be compared with the IRR of projects of similar risk and the opportunity cost of capital.

The IRR method has several disadvantages compared to the NPV method, though only one disadvantage is mentioned here for purposes of brevity. Further information about potential problems with the IRR method (compared to NPV) may be obtained from most finance textbooks. One major problem with IRR is the possibility of obtaining multiple rates of return (multiple "roots") when solving for the IRR of an investment. This can occur in the unusual case where cash flows change erratically from positive to negative (in large quantities or for sustained time periods) *more than once* during the life of the investment. The graph in the following example illustrates how multiple IRR roots can occur for investments with these types of cash flows. The second graph is for comparison purposes and typifies the IRR calculation and graph for most investments. (Normally NPV declines with increasing discount rate, thus giving only one IRR "root".) When multiple roots are obtained using the IRR method, it is best to switch to the NPV method and calculate NPV by discounting the cash flows with the opportunity cost of capital.

Example. A project under consideration requires an initial investment/cash flow of $3000 (negative) and then has cash flows of $26,000 (positive) and $30,000 (negative) at the end of the following 2 years. A graph of NPV versus discount rate shows that two different IRR "roots" exist. (IRR is the point on the curve that intersects the horizontal axis at NPV = 0.) The second graph is from the previous example, showing that most "normal" investments have only one IRR root.

The IRR for this example is both 37.1% and 629.6%, as NPV equals zero at these two discount rates.

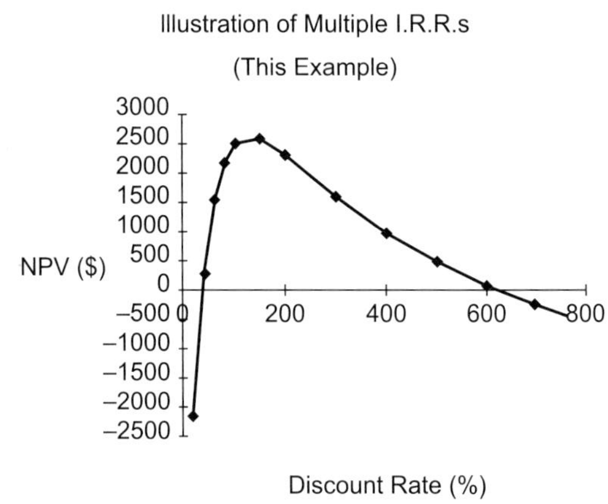

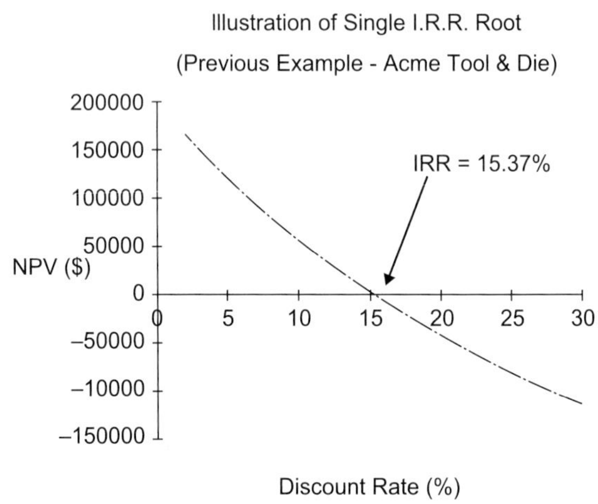

Net present value (NPV) method

Net present value (NPV), as its name suggests, calculates the net amount that the discounted cash flows of an investment exceed the initial investment. Using the discounted cash flow (DCF) formula, the future cash flows are discounted by the rate of return offered by comparable investment alternatives (i.e. the opportunity cost of capital) and then summed and added to the initial investment amount (which is usually negative).

$$\text{NPV} = C_0 + \sum_{i=1}^{n} \frac{C_i}{(1+r)^i}$$

where

C_0 = Initial cash flow (negative for a cash "outflow"),
C_i = Cash flow in time period i,
n = Number of time periods, and
r = Opportunity cost of capital/discount rate.

Though similar to the IRR method, NPV does not calculate an investment's exact rate of return but instead calculates the exact dollar amount that an investment exceeds, or fails to meet, the *expected* rate of return. In other words, if an investment provides a rate of return exactly equal to the opportunity cost of capital, then the NPV of this investment will be zero, because the discounted future cash flows equal the initial investment. Thus, NPV provides an excellent decision criterion for investments. An investment with a positive NPV should be accepted, since it provides a rate of return above the opportunity cost of capital. By the same reasoning, an investment with a negative NPV should be rejected. NPV also does *not* suffer from any of the drawbacks of the payback, accounting ROR, or IRR methods. Because of this, NPV is the method most recommended by financial experts for making investment decisions. IRR is still useful to determine the exact rate of return for an investment, but NPV has none of the problems that IRR may have with unusual investments (such as the "multiple root" problem illustrated previously).

The following example shows how NPV is used to decide between investment alternatives.

Project	C_0	C_1	C_2	C_3	C_4
A	−$125,000	$31,000	$43,000	$48,000	$51,000
B	−$210,000	$71,000	$74,000	$76,000	$79,000

Example. A company must choose between two alternative manufacturing projects, which require different amounts of capital investment and have different cash flow patterns. The company's opportunity cost of capital for projects of similar risk is 15%. Which project should it choose?

Solving with the NPV formula:

$$\text{NPV(A)} = -\$4809$$
$$\text{NPV(B)} = \$2833$$

Conclusion: The company should choose project B. Despite a higher initial cost than A, Project B earns $2833 more than its required rate of return of 15%. Incidentally, IRR also could have been used, since future cash flows are all positive, and therefore no multiple roots exist. Project As IRR is 13.23%; Project B's IRR is 15.65%.

SENSITIVITY ANALYSIS

The previous sections of this chapter show how to analyze investments and financial projects. These analysis methods can help quantify whether or not an investment is worthwhile and also help choose between alternative investments. However, no matter what method is used to analyze a financial project, that method and the decision about the project will ultimately rely upon the *projections* of the project's future cash flows. It is, therefore, extremely important that the future cash flows of any financial project or investment under consideration be forecast as accurately and realistically as possible. Unfortunately, no matter how precise and realistic you try to be with your projections of cash flows, the future is never certain, and actual results may vary from what is projected. Thus, it is often useful to perform a *sensitivity analysis* on the investment or financial project.

Sensitivity analysis is a procedure used to describe analytically the effects of uncertainty on one or more of the parameters involved in the analysis of a financial project. The objective of a sensitivity analysis is to provide the decision maker with quantitative information about what financial effects will be caused by variations from what was projected. A sensitivity analysis, once performed, may influence the decision about the financial project, or at least show the decision maker which parameter is the most critical to the financial success of the project. The NPV method lends itself nicely to sensitivity analysis, since the discount rate is already fixed at the opportunity cost of capital. One parameter at a time can be varied in steps (while the other parameters are held constant) to see how much effect this variance has on NPV. A financial calculator or computer spreadsheet program will greatly expedite the multiple, repetitive calculations involved in this analysis. Plotting the results graphically will help show the sensitivity of NPV to changes in each variable, as illustrated in the following example.

Example. A-1 Engineering Co. is contemplating a new project to manufacture sheet metal parts for the aircraft industry. Analysis of this project shows that a $100,000 investment will be required up front to purchase additional machinery. The project is expected to run for 4 years, resulting in estimated after-tax cash flows of $30,000 per year. Salvage value of the machinery at the end of the 4 years is estimated at $32,000. A-1's opportunity cost of capital is 15%. Perform a sensitivity analysis to determine how variations in these estimates will affect the NPV of the project.

Step 1: Cash flow diagram:

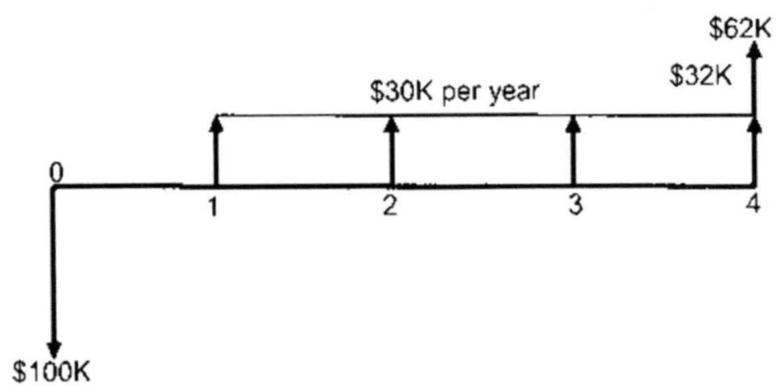

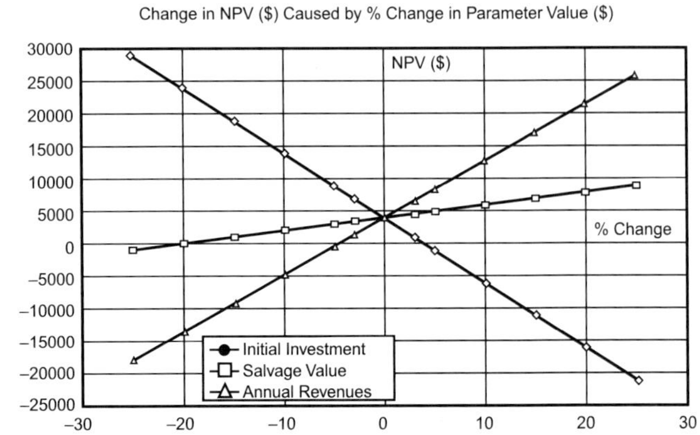

Step 2: Calculate baseline NPV:

$$
\begin{aligned}
\text{NPV} &= -100,000 + \frac{30,000}{(1+0.15)^1} + \frac{30,000}{(1+0.15)^2} \\
&\quad + \frac{30,000}{(1+0.15)^3} + \frac{62,000}{(1+0.15)^4} \\
&= \$3.945.45
\end{aligned}
$$

Step 3: Change salvage value, annual cash flow revenues, and initial investment amount in steps. Calculate NPV for each step change.

Step 4: Plot values to illustrate the NPV sensitivity to changes in each variable.

Conclusion: The graph shows that this project's NPV is equally sensitive to changes in the required initial investment amount (inversely related) and the annual revenues (after-tax cash flow) from the project. This is easily seen from the slope of the lines in the graph. The graph also shows that NPV is not affected nearly as much by changes in salvage value, primarily because the revenues from salvage value occur at the end of the fourth year and are substantially discounted at the 15% cost of capital. In conclusion, this sensitivity analysis shows A-1 Machinery that it cannot accept anything more than a 5% increase in the initial cost of the machinery or a 5% decrease in the annual cash flows. If these parameters change more than this amount, NPV will be less than zero and the project will not be profitable. Salvage value, however, is not nearly as critical, as it will take more than a 20% decrease from its estimated value before the project is unprofitable.

DECISION TREE ANALYSIS

Financial managers often use *decision trees* to help with the analysis of large projects involving sequential decisions and variable outcomes over time. Decision trees are useful to managers because they graphically portray a large, complicated problem in terms of a series of smaller problems and decision branches. Decision trees reduce abstract thinking about a project by producing a logical diagram that shows the decision options available and the corresponding results of choosing each option. Once all the possible outcomes of a project are diagrammed in a decision tree, objective analysis and decisions about the project can be made.

Decision trees also allow *probability theory* to be used, in conjunction with NPV, to analyze financial projects and investments. It is often possible to estimate the probability of success or failure for a new venture or project, based upon either historical data or business experience. If the new venture is a success, then the projected cash flows will be considerably higher than what they will be if the project is not successful. The projected cash flows for both of these possible outcomes (success or failure) can then be multiplied by their probability estimates and added to determine an *expected* outcome for the project. This expected outcome is the *average* payoff that can be expected (for multiple projects of the same type) based upon these probability estimates. Of course, the *actual* payoff for a single financial project will either be one or the other (success or failure), and not this average payoff. However, this average expected payoff is a useful number for decision making, since it incorporates the probability of both outcomes.

If there is a substantial time lapse between the time the decision is made and the time of the payoff, then the time value of money must also be considered. The NPV method can be used to discount the expected payoff to its present value and to see if the payoff exceeds the initial investment amount. The standard NPV decision rule applies: *Accept only projects that have positive NPVs.*

The following example illustrates the use of probability estimates (and NPV) in a simple decision tree:

Example. A ticket scalper is considering an investment of \$10,000 to purchase tickets to the finals of a major outdoor tennis tournament. He must order the tickets 1 year in advance to get choice seats at list price. He plans to resell the tickets at the gate on the day of the finals, which is on a Sunday. Past experience tells him that he can sell the tickets and double his money, but only if the weather on the day of the event is good. If the weather is bad, he knows from past experience that he will only be able to sell the tickets at an average of 70% of his purchase price, even if the event is canceled and rescheduled for the following day (Monday). Historical data from the *Farmer's Almanac* shows that there is a 20% probability of rain at this time of year. His opportunity cost of capital is 15%. Should the ticket scalper accept this financial project and purchase the tickets?

Step 1: Construct the decision tree (very simple in this example).

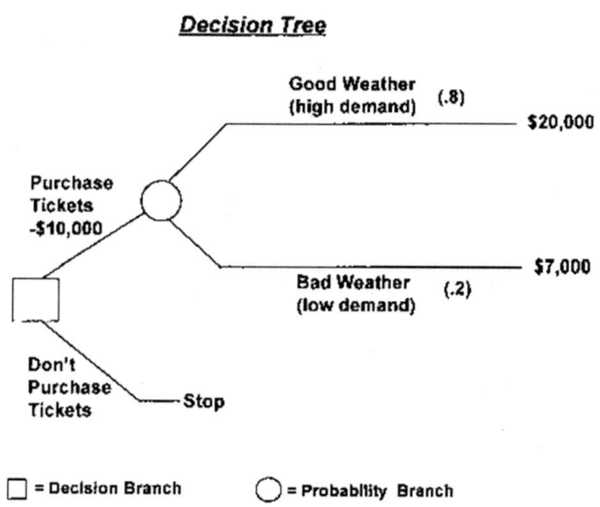

Step 2: Calculate the expected payoff from purchasing the tickets:

$$\begin{aligned}\text{Expected payoff} &= (\text{probability of high demand}) \\ &\quad \times (\text{payoff with high demand}) \\ &\quad + (\text{probability of low demand}) \\ &\quad \times (\text{payoff with low demand}) \\ &= (0.8 \times 20,000) + (0.2 \times 7000) \\ &= \$17,400\end{aligned}$$

Step 3: Calculate NPV of expected payoff:

$$\text{NPV} = -10,000 + \frac{\$17,400}{(1+0.15)^1} = \$5130$$

Remember that the cash flow from selling the tickets occurs 1 year after the purchase. NPV is used to discount the cash flow to its present value and see if it exceeds the initial investment.

Conclusion: The ticket scalper should buy the tickets, because this project has a positive NPV.

The preceding example was relatively simple. Most financial projects in engineering and manufacturing applications are considerably more complex than this example; therefore, it is useful to have a procedure for setting up a decision tree and analyzing a complex project.

Here is a procedure that can be used:

1. Identify the decisions to be made and alternatives available throughout the expected life of the project.
2. Draw the decision tree, with branches drawn first for each alternative at every decision point in the project. Secondly, draw "probability branches" for each possible outcome from each decision alternative.
3. Estimate the probability of each possible outcome. (Probabilities at each "probability branch" should add up to 100%.)
4. Estimate the cash flow (payoff) for each possible outcome.
5. Analyze the alternatives, starting with the most distant decision point and working backwards. Remember the time value of money and, if needed, use the NPV method to discount the expected cash flows to their present values. Determine if the PV of each alternative's expected payoff exceeds the initial investment (i.e. positive NPV). Choose the alternative with the highest NPV.

The following example illustrates the use of this procedure.

Example. Space-Age Products believes there is considerable demand in the marketplace for an electric version of its barbecue grill, which is the company's main product. Currently, all major brands of grills produced by Space-Age and its competitors are either propane gas grills or charcoal grills. Recent innovations by Space-Age have allowed the development of an electric grill that can cook and sear meat as well as any gas or charcoal grill, but without the problems of an open flame or the hassle of purchasing propane gas bottles or bags of charcoal.

Based upon test marketing, Space-Age believes the probability of the demand for its new grill being high in its first year of production will be 60%, with a 40% probability of low demand. If demand is high the first year, Space-Age estimates the probability of high demand in subsequent years to be 80%. If the demand the first year is low, it estimates the probability of high demand in subsequent years at only 30%.

Space-Age cannot decide whether to set up a large production line to produce 50,000 grills a month or to take a more conservative approach with a small production line that can produce 25,000 grills a month. If demand is high, Space-Age estimates it should be able to sell all the grills the large production line can make. However, the large production line puts considerably more investment capital at risk, since it costs $350,000 versus only $200,000 for the small line.

The smaller production line can be expanded with a second production line after the first year (if demand is high) to double the production rate, at a cost of an additional $200,000.

Should Space-Age initially invest in a large or small production line or none at all? The opportunity cost of capital for Space-Age is 15%, and cash flows it projects for each outcome are shown in the decision tree. Cash flows shown at the end of year 2 are the PV of the cash flows of that and all subsequent years.

Steps 1–4 of procedure: (Construct decision tree and label with probabilities and cash flow projections).

Step 5: Analysis of alternatives:

(a) Start the analysis by working backwards from the R/H side of the decision tree. The only decision that Space-Age needs to make at the start of year 2 is deciding

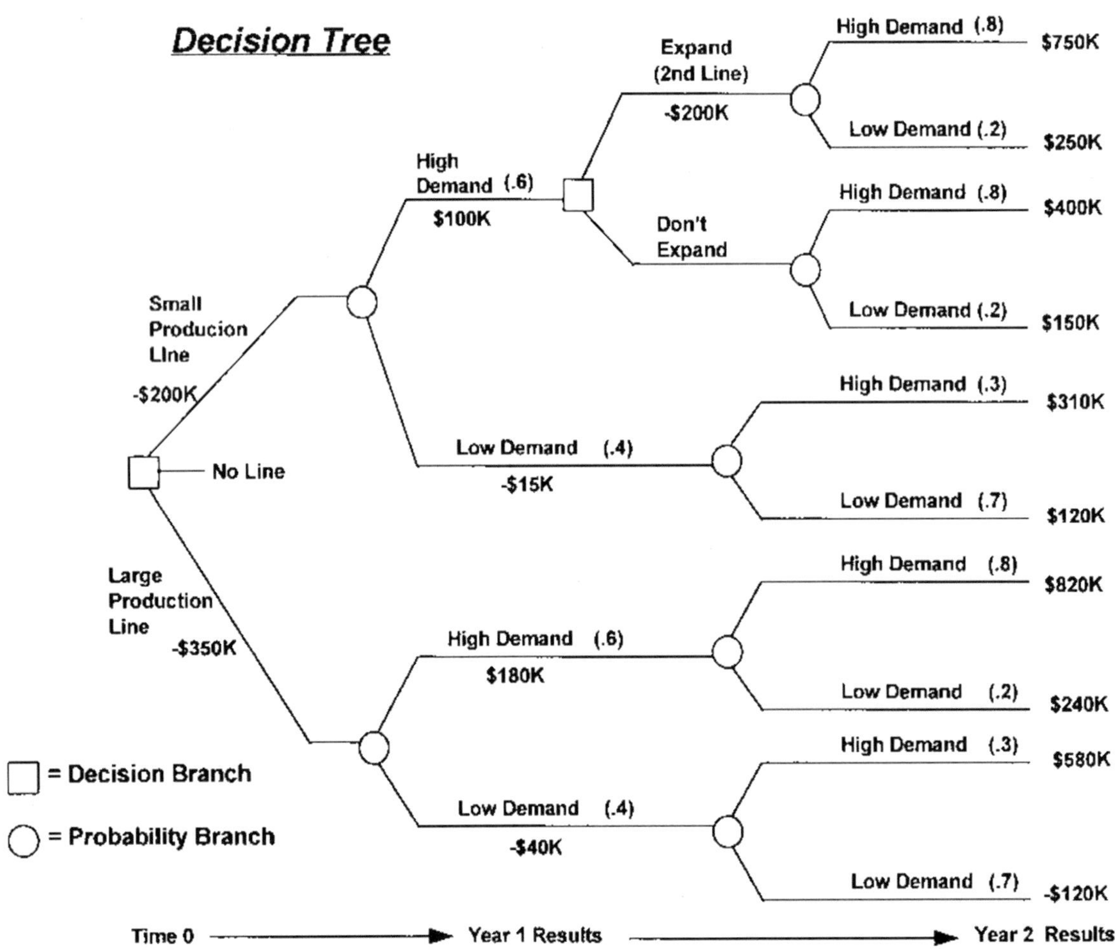

whether or not to expand and install a second production line if demand is high the first year (if the initial decision was for a small production line).

$$= (0.8 \times \$750\text{K}) + (0.2 \times \$250\text{K}) = \$650,000$$

Expected payoff without the expansion (at the end of year 2):

$$= (0.8 \times \$400\text{K}) + (0.2 \times \$150\text{K}) = \$350,000$$

To decide if expansion is worthwhile, the NPVs of each alternative (at end of year 1) must be compared:

$$\begin{aligned}\text{NPV(with 2nd line)} &= -\$200\text{K} + \frac{\$650\text{K}}{(1+0.15)^1}\\ &= \$365,217\end{aligned}$$

$$\text{NPV(without 2nd line)} = \$0 + \frac{\$350\text{K}}{(1+0.15)^1} = \$304,348$$

These NPV amounts show that Space-Age *should clearly expand with a second small production line* (if its initial decision at time 0 was to open a small production line and the demand was high the first year). Now that this decision is made, the \$365,217 NPV of this expansion decision will be used to work backwards in time and calculate the overall NPV of choosing the small production line.

(b) The expected payoff (at the end of year 2) if the demand is low the first year with the small production line:

$$= (0.3 \times \$310\text{K}) + (0.7 \times \$120\text{K}) = \$177,000$$

This expected payoff must be discounted so that it can be added to the projected cash flow of year 1:

$$\text{PV} = \frac{\$177,000}{(1+0.15)} = \$153,193$$

(c) The expected payoff (at the end of year 1) for the small production line can now be determined:

$$\begin{aligned}&= [0.6 \times (\$100,000 + \$365,217)]\\ &\quad +[0.4 \times (-\$15,000 + \$153,193)]\\ &= \$334,695\end{aligned}$$

Notice that the cash flow for year 1 is the sum of the first year's projected cash flow plus the discounted expected payoff from year 2.

(d) Calculate the NPV (at time 0) of the small production line option:

$$\text{NPV} = \$200,000 + \frac{\$334,695}{(1+0.15)} = \$91,039$$

(e) Calculate the NPV (at time 0) of the large production line option. Since there are no decisions at the end of year 1, there is no need to determine intermediate NPVs, and the probabilities from each branch of each year can be multiplied through in the NPV calculation as follows:

$$\begin{aligned}\text{NPV} &= -\$350\text{K} + \frac{(0.6)(\$180\text{K}) + (0.4)(\$40\text{K})}{(1+0.15)^1}\\ &\quad + \frac{(0.6)[(0.8)(\$820\text{K}) + (0.2)(\$240\text{K})] \quad (0.4)[(0.3)(\$580\text{K}) + (0.7)(-\$120\text{K})]}{(1+0.15)^2}\end{aligned}$$

Solving: NPV = \$76,616

(f) *Conclusion*: Both the NPVs of the large and small production lines are positive, and thus both production lines would be profitable based upon the expected payoffs. However, the NPV of the small production line is higher than the NPV of the large production line (\$91,039 versus \$76,616). Therefore, Space-Age's decision should be to set up a small production line, and then expand with a second line if the demand for electric grills is high in the first year of production.

ACCOUNTING FUNDAMENTALS

Regardless of whether a person is managing the finances of a small engineering project or running a large corporation, it is important to have a basic understanding of accounting. Likewise, it is also impossible to study engineering economics without overlapping into the field of accounting. This section, therefore, provides a brief overview of accounting fundamentals and terminology.

Accounting is the process of measuring, recording, and communicating information about the economic activities of accounting entities (financial projects, partnerships, companies, corporations, etc.). Public corporations, and most other types of companies, issue financial statements from their accounting data on a periodic basis (usually quarterly and yearly). These financial statements are necessary because they provide information to outside investors and creditors and assist a company's own management with decision making. Four primary types of financial statements are normally issued:

- Balance sheet
- Income statement
- Statement of changes in retained earnings
- Statement of cash flows

The *balance sheet* provides information (at one particular instant in time) about financial resources of a company and claims against those resources by creditors and owners. The balance sheet's format reflects the *fundamental accounting equation:*

$$\text{Assets} = \text{Liabilities} + \text{Owners Equity}$$

Assets are items of monetary value that a company possesses, such as plant, land, equipment, cash, accounts receivable, and inventory. *Liabilities* are what the company owes its creditors, or conversely, the claims the creditors have against the company's assets if the company were to go bankrupt. Typical liabilities include outstanding loan balances, accounts payable, and income tax payable. *Owner's equity* is the monetary amount of the owner's claim against the company's assets. (Owner's equity is also referred to as stockholder's equity for corporations that issue and sell stock to raise capital.) Owner's equity comes from two sources: (1) the original cash investment the owner used to start the business and (2) the amount of profits made by the company that are not distributed (via *dividends*) back to the owners. These profits left in the company are called *retained earnings*.

There are numerous asset, liability, and owner's equity *accounts* that are used by accountants to keep track of a company's economic resources. These accounts are shown in the following "balance sheet" illustration, which is a typical balance sheet for a small manufacturing corporation.

Notice in this illustration that the balance sheet's total assets "balance" with the sum of liabilities and owner's equity, as required by the fundamental accounting equation. There are also a number of additional *revenue* and *expense* accounts that are maintained but not shown in the balance sheet. The balances of these revenue and expense accounts are closed out (zeroed) when the financial statements are prepared, and their amounts are carried over into the income statement. The *income statement* is the financial statement that reports the profit (or loss) of a company over a specific period of time (quarterly or annually). The income statement reflects another basic and fundamental accounting relationship:

$$\text{Revenues} - \text{Expenses} = \text{Profit(or Loss)}$$

A sample "income statement" for our typical small corporation is illustrated.

The income statement simply subtracts the company's expenses from its revenues to arrive at its *net income* for the period. Large companies and corporations usually separate their income statements into several sections, such as income from operations, income or loss from financial activities and discontinuing operations, and income or loss from extraordinary items and changes in accounting principles.

Companies frequently show an expense for depreciation on their income statements. *Depreciation* is an accounting method used to allocate the cost of asset "consumption" over the life of the asset. The value of all assets (building, equipment, patents, etc.) depreciate with time and use, and depreciation provides a method to recover investment capital by expensing the portion of each asset "consumed" during each income statement's time period. Depreciation is a *noncash expense*, since no cash flow actually occurs to pay for this expense. The *book value* of an asset is shown in the balance sheet and is simply the purchase price of the asset minus its accumulated depreciation. There are several methods used to calculate depreciation, but the easiest and most frequently used method is *straight-line depreciation*, which is calculated with the following formula:

$$\text{Depreciation Expense} = \frac{\text{Cost} - \text{Salvage Value}}{\text{Expected Life of Asset}}$$

Straight-line depreciation allocates the same amount of depreciation expense each year throughout the life of the asset. *Salvage value* (residual value) is the amount that a company can sell (or trade in) an asset for at the end of its useful life. "Accelerated" depreciation methods (such as sum-of-the-years'-digits and declining balance methods) are available and may sometimes be used to expense the cost of an asset faster than straight-line depreciation. However, the

XYZ Company
Balance Sheet
As of March 31, 2020

ASSETS			
Current Assets:			
Cash		$25,000.00	
Accounts Receivable		$268,000.00	
Inventory		$163,000.00	
Total Current Assets			$456,000.00
Property, Plant, & Equipment:			
Building	$600,000.00		
less Accumulated Depreciation	($250,000.00)	$350,000.00	
Equipment	$300,000.00		
less Accumulated Depreciation	($125,000.00)	$175,000.00	
Land		$100,000.00	
Total Property, Plant, & Equipment			$625,000.00
Total Assets			***$1,081,000.00***
Liabilities			
Current Liabilties:			
Accounts Payable	$275,000.00		
Salaries Payable	$195,000.00		
Interest Payable	$13,000.00		
Income Tax Payable	$36,000.00		
Total Current Liabilities		$519,000.00	
Long-Term Liabilities:			
Notes Payable		$260,000.00	
Total Liabilities			$779,000.00
Owner's (Stockholder's) Equity			
Capital Stock	$200,000.00		
Retained Earnings	$102,000.00		
Total Owner's Equity			$302,000.00
Total Liabilities & Owner's Equity			***$1,081,000.00***

use of accelerated depreciation methods is governed by the tax laws applicable to various types of assets. A recent accounting textbook and tax laws can be consulted for further information about the use of accelerated depreciation methods.

The profit (or loss) for the time period of the income statement is carried over as retained earnings and added to (or subtracted from) owner's equity, in the "statement of changes in retained earnings" illustration.

If the company paid out any dividends to its stockholders during this time period, that amount would be shown as a deduction to retained earnings in this financial statement. This new retained earnings amount is also reflected as the retained earnings account balance in the current balance sheet.

The last financial statement is the *statement of cash flows*, which summarizes the cash inflows and outflows of the company (or other accounting entity) for the same time period as the income statement. This statement's purpose is to show where the company has acquired cash (inflows) and to what activities its cash has been utilized (outflows) during this time period. The statement divides and classifies cash flows into three categories: (1) cash provided by or used by operating activities; (2) cash provided by or used by investing activities; and (3) cash provided by or used by financing activities. Cash flow from operations is the cash generated (or lost) from the normal day-to-day operations of a company, such as producing and selling goods or services. Cash flow from investing activities includes selling or purchasing long-term assets and sales or purchases of investments. Cash flow

XYZ Company
Income Statement
For the Quarter Ended March 31, 2020

Sales Revenues		$2,450,000.00
Less: Cost of Goods sold		($1,375,000.00)
Gross Margin		$1,075,000.00
Less: Operating Expenses		
Selling expenses	$295,000.00	
Salaries expense	$526,000.00	
Insurance expense	$32,000.00	
Property Taxes	$27,000.00	
Depreciation, Equipment	$35,000.00	
Depreciation, Building	$70,000.00	($985,000.00)
Income from Operations		$90,000.00
Other Income		
Sale of Assets	$16,000.00	
Interest Revenue	$2,000.00	$18,000.00
		$108,000.00
Less: Interest Expense		($22,000.00)
Income before Taxes		$86,000.00
Less: Income Tax Expense		($34,400.00)
Net Income		*$51,600.00*

XYZ Company
Statement of Changes in Retained Earnings
For the Quarter Ended March 31, 2020

Retained Earnings, Dec. 31, 2019	$50,400.00
Net Income, Quarter 1, 2020	$51,600.00
Retained Earnings, March 31, 2020	$102,000.00

from financing activities covers the issuance of long-term debt or stock to acquire capital, payment of dividends to stockholders, and repayment of principal on long-term debt.

The statement of cash flows complements the income statement, because although income and cash flow are related, their amounts are seldom equal. Cash flow is also a better measure of a company's performance than net income, because net income relies upon arbitrary expense allocations (such as depreciation expense). Net income can also be manipulated by a company (e.g. by reducing R&D spending, reducing inventory levels, or changing accounting methods to make its "bottom line" look better in the short term. Therefore, analysis of a company's cash flows and cash flow trends is frequently the best method to evaluate a company's performance. Additionally, cash flow rather than net income should be used to determine the rate of return earned on the initial investment or to determine the value of a company or financial project. (Value is determined by discounting cash flows at the required rate of return.)

The statement of cash flows can be prepared directly by reporting a company's gross cash inflows and outflows over a time period, or indirectly by adjusting a company's net income to net cash flow for the same time period. Adjustments to net income are required for (1) changes in assets and liability account balances, and (2) the effects of noncash revenues and expenses in the income statement.

For example, in comparing XYZ Company's current balance sheet with the previous quarter, the following information is obtained about current assets and liabilities:

- Cash increased by $10,000
- Accounts receivable decreased by $5000
- Inventory increased by $15,000
- Accounts payable increased by $25,000
- Salaries payable increased by $5000

Additionally, the balance sheets show that the company has paid off $166,600 of long-term debt this quarter. This information from the balance sheets, along with the noncash expenses (depreciation) from the current income statement, is used to prepare the "statement of cash flows." The example demonstrates the value of the statement of cash flows, because it shows precisely the sources and uses of XYZ Company's cash. Additionally, it reconciles the cash account balance from the previous quarter to the amount on the current quarter's balance sheet.

XYZ Company

Statement of Cash Flows

For the Quarter Ended March 31, 2020

Net Income		$51,600.00	
Adjustments to reconcile net income to cash flow from op. activities:			
Decrease in accounts receivable	$5,000.00		
Increase in inventory	($15,000.00)		
Increase in accounts payable	$25,000.00		
Decrease in salaries payable	$5,000.00		
Depreciation expense, equipment	$35,000.00		
Depreciation expense, building	$70,000.00		
Total Adjustments:		$125,000.00	
Net cash flow from operating activities:			$176,600.00
Cash flows from financing activities:			
Retirement of long-term debt (notes payable)	($166,600.00)		
Net cash flow from financing activities:			($166,600.00)
Net Increase in cash:			$10,000.00
Cash account balance, Dec. 31, 2019			$15,000.00
Cash account balance, March 31, 2020			$25,000.00

20: Cost

Cost Estimation 962
Cost estimation methods, normalization and data 962
How to use time and location cost indices 966
How to adjust pipeline construction cost across time 967

Onshore Construction 968
Onshore pipeline construction process 968
Onshore pipeline construction cost factors 971
Onshore FERC gas pipeline cost statistics in the United States, 2014–2019 974
How to estimate onshore pipeline construction cost using unit costs? 977
How to estimate cost bounds in onshore pipeline construction in the United States? 978
Onshore gas pipeline construction cost in the United States, 2017–2019 978
How to develop construction cost models using regression 980

Offshore Construction 984
Offshore pipeline construction process 984
Offshore pipeline construction cost factors 986
U.S. Gulf of Mexico pipeline cost statistics, 1995–2019 991
How to estimate shallow water pipeline construction cost bounds in the U.S. Gulf of Mexico using empirical data? 995
Offshore pipeline construction cost estimation using work decomposition 996

Offshore Decommissioning 1002
Offshore pipeline decommissioning process in the U.S. Gulf of Mexico 1002
Offshore pipeline decommissioning cost factors 1004
Offshore pipeline decommissioning cost statistics in the U.S. Gulf of Mexico, 1980–2019 1005
How to estimate the cost to decommission shallow water pipeline in the U.S. Gulf of Mexico? 1007
How to estimate the cost to decommission hurricane-damaged pipeline in the U.S. Gulf of Mexico? 1007
How to estimate deepwater pipeline decommissioning cost in the U.S. Gulf of Mexico? 1008
How to estimate deepwater umbilical, flowline, and riser decommissioning cost in the U.S. Gulf of Mexico? 1008
Typical workflows for offshore pipeline decommissioning 1009

Gas Compression Stations 1010
Onshore FERC gas transmission compressor station cost statistics in the United States, 2014–2019 1010
How to estimate gas transmission compressor station cost in the United States? 1013
How to compare gas transmission compressor station cost in the United States? 1013

Scaling Rules 1013
Equipment and process scaling rules 1013
How to estimate equipment cost curves 1014
How to estimate equipment cost using two-factor models 1015
How to estimate equipment cost curves using two data points 1017

Tariff Rates 1017
What are reservation and commodity charges? 1017
How to calculate firm service with fixed maximum daily quantity and reservation charge? 1019
How is the cost of service formula applied? 1019
How to compute price adjustments in pipeline system quality banks 1023

COST ESTIMATION

Cost estimation methods, normalization and data

All costs are time, location, and project-specific and must be normalized prior to application. Capital costs that date from earlier years need to be adjusted for material and labor changes. If the capital costs are associated with equipment of different capacity than the equipment you are considering, an adjustment for size will have to be made. Regional and technology differences also need to be considered, and the impact of learning may be important in operations that involve a large number of similar activities such as constructing storage tanks. There are usually significant limitations on our ability to collect cost data, and unlike statistical analysis that deals with large data sets, engineers and project managers have to understand the pitfalls of small data sets. The project scope and expected time to complete activity are typically the most important factors in determining cost, while the experience of the evaluator(s) and the time available to perform the estimate, as well as the time between when the estimate was performed and the project executed, will also impact the reliability of the estimation.

Top-down versus bottom-up

The two basic methods of cost estimation are often referred to as the "top-down" and the "bottom-up" approaches.

In the bottom-up approach, also referred to as a work breakdown structure or work decomposition, project tasks are broken down into discrete and identifiable units and the costs of each unit are estimated and added together, often with contingencies, to obtain the cost estimate for the project. A task is an assigned piece of work identified to be finished within a certain time. Each descending level represents an increasingly detailed definition of a project component. Work decomposition is the most common estimation method used in engineering, and the standard method used in pipeline construction and decommissioning cost estimation.

The top-down approach uses historical data and normalizes for project size, complexity, time, location, and other factors (Figure 1). Statistical methods are applied or regression models may be adopted using project attributes to develop unit cost and correlations. Top-down approaches are rarely used in engineering studies except as an order of magnitude estimate, or when required by legislation or regulation, such as in the insurance industry where use of historical data are prevalent. In engineering studies, historical data are frequently used for due diligence, forecasting, and market studies.

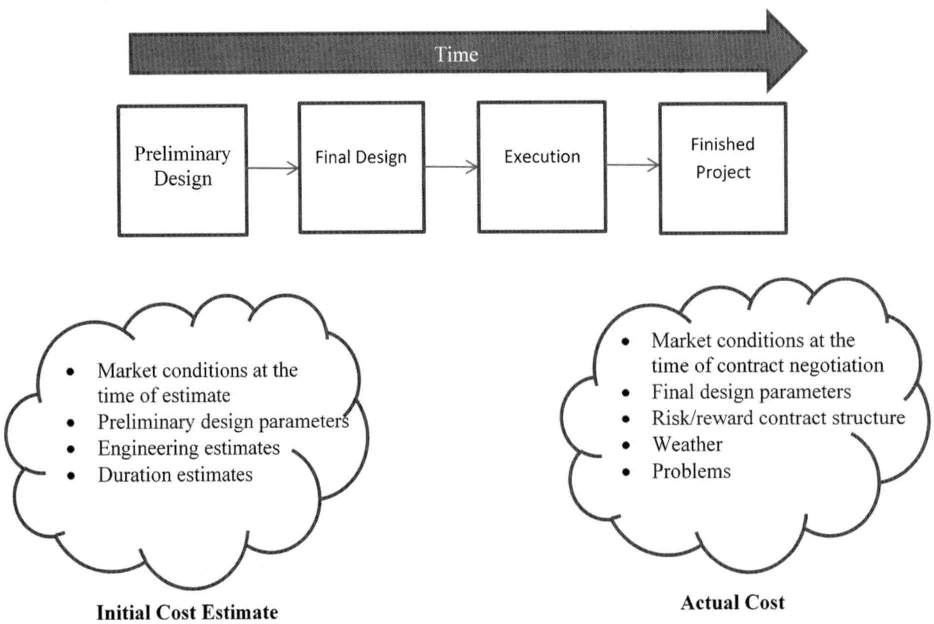

Figure 1. Some of the main factors that impact cost estimation accuracy and reliability.

Uncertainty range

The American Association of Cost Engineers recommends three basic categories of cost estimates according to detail and accuracy required by their intended use:

- Order of magnitude estimates are considered accurate within −30% to +50% and are usually based on cost-capacity curves, ratios, and historical data.
- Preliminary estimates are considered accurate within −15% to 30% and based on work breakdown schedules, flow sheets, and historical data.
- Definitive estimates are considered accurate within −5% to 15% and based on detailed design and engineering data using work breakdown schedules and market data.

Each estimate is performed at its appropriate time during the execution of a project and has its own relative accuracy. Order of magnitude estimates are used for investment screening purposes, preliminary estimates are used for budgeting and expenditure approval, and definitive estimates are used in procurement and contract negotiation. Each estimate, whether factored or detailed, has a scope and usually an execution plan established. To account for technology, location, and market risks, a contingency is added. Contingencies are for risks associated with events which may or may not occur but where impact cannot be identified at the time the estimate is performed. Examples of contingency include labor rate changes, estimating inaccuracies, unexpected construction problems including weather, and schedule impacts.

Normalization

Costs are normalized based on project characteristics and the preferences of the user. Available data often define the level of granularity of analysis. When project attributes are available, user preferences often differentiate normalization variables.

Onshore construction. In onshore pipeline construction, costs are normalized by route mileage and pipeline diameter since these are primary construction attributes. Aggregate average unit mileage costs are computed by dividing the sum of the costs of all projects during the period by the total mileage installed and a composite diameter-mileage factor as follows:

$$\text{Aggregate}_{\text{mile}} = \frac{\Sigma\text{Project cost}}{\Sigma\text{Project mileage}}$$

$$\text{Aggregate}_{\text{inch-mile}} = \frac{\Sigma\text{Project cost}}{\Sigma\text{Project diameter} \times \text{mileage}}.$$

For these statistics, since all project cost and mileage are added together, the composite represents an average aggregate unit cost metric. Average aggregate cost statistics provide no information on cost deviation since all projects are combined in a cumulative fashion.

Using projects as the basis for averaging, an average project cost per mile and per diameter-mile is defined as follows:

$$\text{Project}_{\text{mile}} = \frac{1}{N}\sum\left(\frac{\text{Project cost}}{\text{Project mileage}}\right).$$

$$\text{Project}_{\text{inch-mile}} = \frac{1}{N}\sum\left(\frac{\text{Project cost}}{\text{Project diameter} \times \text{mileage}}\right).$$

where N represents the number of projects.

Project statistics are distinct from aggregate statistics and represents an upper bound on the aggregate average due to the method of computation. The following relation applies:

$$\text{Aggregate}_{\text{mile}} < \text{Project}_{\text{mile}}$$

$$\text{Aggregate}_{\text{inch-mile}} < \text{Project}_{\text{inch-mile}}$$

Offshore construction. In offshore pipeline construction, projects are grouped according to export systems, infield flowline systems, and export/infield systems and normalized by route length. The purpose of the classification, as with most categorizations in statistical assessment, is to group projects of similar type for evaluation. Export systems include only export pipeline, and infield flowlines include only flowlines and umbilicals. Export/infield systems is a composite category that includes both export and infield pipe. An aggregate all-in pipeline cost consolidates total pipeline mileage and cost regardless of project type. Unit cost normalizes by the total mileage of the system.

Water depth and contract type plays no role in differentiating projects, and no distinction is made between line diameter and wall thickness, although these factors may be important in some cases. Infield contracts typically include umbilicals, which may or may not be broken out separately from flowline cost, but, if included with the system costs, are included in the mileage total. Ideally, flowline and umbilical materials and construction cost should be broken out separately, but this is usually not possible with public data sources. Pipeline end terminals and manifolds (PLETs and PLEMs), subsea sleds and jumpers, tie-in assemblies, flying leads, and other subsea components are typically included in pipeline construction cost but cannot be evaluated separately; fortunately, they are usually a small percentage of total cost. Riser work and additional system components may be included in the work scope but are also not considered.

Offshore decommissioning. Pipeline decommissioning costs are normalized by mileage, mile-cuts, and flush volume in cubic feet. A mile-cut is defined as the length of the pipeline segment(s) decommissioned multiplied by the number of cuts involved to perform the operation, and flush volume is computed in the obvious way through route length and pipeline inner diameter. All offshore projects require at least two cuts (one at each endpoint) and some may require more. Special conditions are identified, such as leaking pipelines and hurricane-damaged facilities, and projects with onshore work. Project complexity is proxied by the number of interconnects, which translate roughly to the number of cuts required, and enters the normalization as mile-cuts. Water depth and contract type are not used to differentiate projects.

Regulatory agencies

Onshore, the Federal Energy Regulatory Commission (FERC) and Department of Transportation (DOT) are the primary pipeline regulatory authorities in the United States. In U.S. federal waters, also referred to as the outer continental shelf (OCS), the Bureau of Safety and Environmental Enforcement (BSEE), the Bureau of Ocean Energy Management (BOEM), the DOT, and the FERC are the main regulatory authorities for pipelines. The BSEE and BOEM are branches within the U.S. Department of Interior.

Data sources

Federal Energy Regulatory Commission. The FERC approves the construction, operation, and decommissioning of interstate natural gas transmission pipelines in the United States under Section 7 of the Natural Gas Act (1978). FERC determines whether gas pipelines are gathering lines or transportation lines on a case-by-case basis using the "modified primary function test." If a line is determined to be gathering, it is nonjurisdictional and is not regulated by FERC.

When a FERC-regulated natural gas pipeline company wants to modify its system, to add new pipeline or to expand capacity, or to remove capacity, it must apply for a "Certificate of Public Convenience and Necessity." This filing documents and justifies the construction and specifies what the company estimates construction cost will be. If a company receives its certificate and builds its facilities, it must report back to FERC at the completion of the project the actual cost of construction and how the original estimates compared with what it spent. The filing occurs within six months after successful hydrostatic testing or compression operation.

When the Federal Power Commission (later renamed FERC) was first organized to regulate the sector in the 1960s, it released annual reports summarizing onshore and offshore pipeline construction. FERC eventually stopped releasing their annual surveys, but the *Oil and Gas Journal* (OGJ) had by that time begun reporting FERC summaries which continues to the present. FERC is not the only source of construction data on pipeline projects, but it is one of the best sources, since it uses a standardized set of cost categories and reports final, actual cost of operations. Private data sources and consultancies are another potential source of data.

FERC data for pipeline construction is reported for final (completed) costs using standardized categories: ROW, damages, surveys, materials, labor, engineering, line pack, taxes, freight, pipe coating, cathodic protection, telecom, overheads, AFUDC, contingencies, legal fees, other services, and regulatory fees. There are differences in how companies allocate cost across categories, and FERC dockets must frequently be reviewed to understand these variations. The OGJ summarizes pipeline construction cost data from FERC filings using material, labor, miscellaneous, and right-of-way (ROW)/damage cost categories. All FERC filings between July 1, 201X and June 30, 201(X + 1) are summarized annually, either in the early fall (September or October) or late winter (February) OGJ issue each year.

Companies with FERC jurisdictional pipelines are required to request permission to abandon transportation facilities. These requests are reviewed by FERC in public dockets and applications include abbreviated and full descriptions. Applications include a description of the system and services to be abandoned, a discussion of Public Convenience and Necessity, a system map, ownership interest, pro forma accounting entries, and a list of affected customers. Abbreviated applications contain less detailed information. Each company reports the accounting treatment of abandonment, along with pro forma estimated cost of activity. Companies do not report the final actual cost of decommissioning.

Industry publications. Industry conferences and publications, such as the Offshore Technology Conference (OTC) and Society of Petroleum Engineers (SPE), *Offshore Engineer*, and *Offshore Magazine*, frequently report on the engineering, construction, hardware systems, installation techniques, and contracting strategies used in offshore field development, but only occasionally report costs. Oil and gas companies, contractors, and service providers also occasionally announce development costs and describe awards won via press releases. Press release data is usually not as detailed or descriptive as OTC and SPE publications, but project scopes are frequently described, along with contract award and methods used or expected to be used, and locations where the engineering and fabrication occurred or are expected to occur.

System descriptions in OTC and SPE publications are most often as-built at the time of installation. When pipeline construction is described, contracting philosophy, vendor selection, problems, and installation techniques are usually discussed. In the 1980s and early 1990s, operators occasionally reported development costs for offshore projects in the Gulf of Mexico, but post-2000 cost descriptions have been much less frequent. Most project descriptions do not include cost data, and therefore, case studies assembled from industry publications do not represent a random selection of projects, but rather projects where cost data are available.

There is no reason to believe projects that report cost data are unusual in any significant way (i.e., top performers, worst case, etc.) and therefore the costs reported are considered representative. For some projects, operators report cost across specific categories according to export pipeline, infield flowlines, subsea risers, umbilicals, and flowlines, while for other projects, operators report only total development cost or use different combinations of cost categories, making comparisons at a granular level impossible to perform. Detailed cost comparisons are not possible using industry and trade publications, but reported data are still useful as point estimates for comparative purposes.

Press releases. Oil and gas companies, contractors, and service providers occasionally announce development costs and describe awards won via press releases. Press release data are usually not as detailed or descriptive as OTC and SPE publications, but project scopes are frequently described, along with contract award, vessels, and methods used or expected to be used. In some announcements, contract values are reported, but there is frequent ambiguity on what is or is not included in the project scope and most descriptions do not provide cost data. The trade press and other organizations repeat or paraphrase the announcements, the latter sometimes introducing errors in the account. Contract terms and conditions are never discussed in detail, but useful information can often be gleaned.

Engineering procurement construction and installation (EPCI) contracts are typically fixed price, lump-sum awards, and for these contracts, useful data can be inferred from press releases if the work scope is not too broad. Besides pipeline, contracts typically include PLETs, PLEMs, in-line assemblies, and related appurtenances, and may or may not include risers and umbilicals. It is usually not possible to separate individual component costs. Press release data are only useful for aggregate cost and as a summary check on other data results.

Data quality

Comparison. FERC pipelines represent high-quality cost data because they represent actual final construction costs reported using a detailed common categorization (Figure 2). FERC pipelines most often serve single phase processed gas export flow and are generally large diameter systems that use rigid steel pipeline and connect fixed platforms via risers to subsea tie-in assemblies, shelf platforms, or to shore gas plants. Gas pipelines are not significantly different from oil pipelines, but the configurations and installation requirements are simpler, and gas line export projects are usually more homogeneous than infield flowlines and deepwater projects.

Industry publications and press releases report deepwater development projects encompassing oil and gas flowlines, risers, export pipelines, and related equipment. These projects are more complex and varied than FERC pipelines, and may occur over multiple work seasons in congested areas

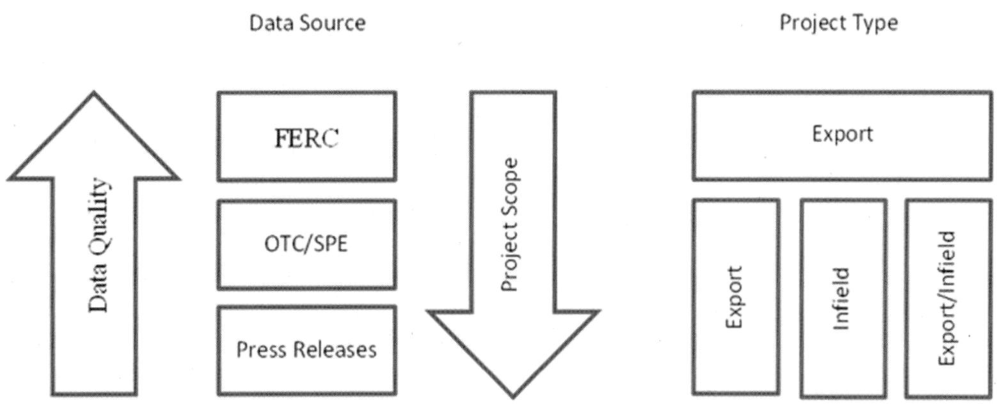

Figure 2. Data quality and project scopes for Federal Energy Regulatory Commission, Offshore Technology Conference/Society of Petroleum Engineers, and press release projects.

with complicated subsea arrangements using a variety of pipe sizes and types, vendors, and installation methods. Project scopes and contractual agreements are more complicated, and the cost data are less detailed and presented at higher and less uniform levels of aggregation. System components are usually described according to the contract requirements set forth by the operator to meet schedule and ambiguity in system descriptions are not uncommon. Cost comparisons are more complicated because of differences in project scope.

Representative statistics. Whenever the total number of projects available for analysis is small and the number of influencing factors large, granular analysis (i.e., categorized cost statistics) will be problematic due to sample size bias. Average or median cost is usually the most meaningful statistical measure of data sets when performed in aggregate, but without an adequate sample size, it is impossible to understand if project data are "representative" or an "outlier" (i.e., nonrepresentative) project.

Representative data are useful for predictive purposes and understanding industry trends, but outliers are also useful as indicators of cost range and project complexity. Representative cost statistics normally require at least four or five data points per category to maintain a minimum level of confidence in summary statistics. It is possible that categories with only one or two projects are representative, but failure to have a minimum number of data points means that the reliability of the statistics should (always) be suspect. Average aggregate cost is usually the most stable and representative sample statistic in evaluation followed by the normalized median project cost.

All other things equal. When multiple factors influence a system, it is common to discuss the impact of one factor at a time holding all other factors constant. In economic parlance, this is referred to as "for all other things equal," since all other things are held constant in the assessment regardless of whether they are constant or not. In the real world, of course, rarely are all other things equal, and factors may interact in ways that are complicated or impossible to model. Because of such conditions, cost statistics computed from sample data are inherently variable. If cost factors are correlated or interact in meaningful ways, holding all other things equal in evaluation may lead to erroneous results and interpretations.

How to use time and location cost indices

Time adjustment

Cost may increase over time because of inflation, regulatory changes, fewer qualified contractors, etc., but may also decrease with technology improvements, better materials, controls and instrumentation, market competition, and other factors. Which factors dominate and determine the direction of cost trends over time depends upon the service or technology under consideration, general economic conditions, region of interest, and many other factors that are often hard to discern. Many factors enter into the impact of long-term improvements in technology and construction efficiency, as well as short-term impacts such as the cost of steel and labor requirements. Time impacts vary across regions and technology application, and frequently enter evaluations in a complicated manner that is handled (simplistically) using indices.

Cost indices permit the estimator a means to adjust cost from one period to another by applying the cost of a similar design and construction from the past to the present or future period without detailed costing. It is important to recognize, however, that cost indices are limited in their application to construction activities and are better suited to equipment such as compressors and pumps and pipeline. If historic data are not carefully processed or categorized, for example, no adjustment will improve its reliability or accuracy. Cost indices are generally not suitable for determining costs for equipment beyond more than a few years because of technological impacts and market conditions. For construction activity, cost indices may simply not be applicable.

The cost at time t_2 to construct a pipeline project, $C(t_2)$, is obtained by multiplying the estimated or actual cost of a similar unit in the same region at time t_1, $C(t_1)$, by a ratio of cost indices $I(t)$ valid for the project location:

$$C(t_2) = C(t_1)\left(\frac{I(t_2)}{I(t_1)}\right)$$

For pipeline activity in the United States, the Bureau of Labor Statistics (BLS) Producer Price Index for steel pipes and tubes (WPU101706), finished goods (WPSFD49207), and support activities for oil and gas operations (WPU60110401) are used to inflation-adjust cost to a common base year (Figure 3). The steel pipe and tube index were used to inflate material cost, and the oil and gas services index was used for all other inflation adjustments. The year of contract award is used as the baseline year, and if not available, the year of installation or first production serves as a proxy.

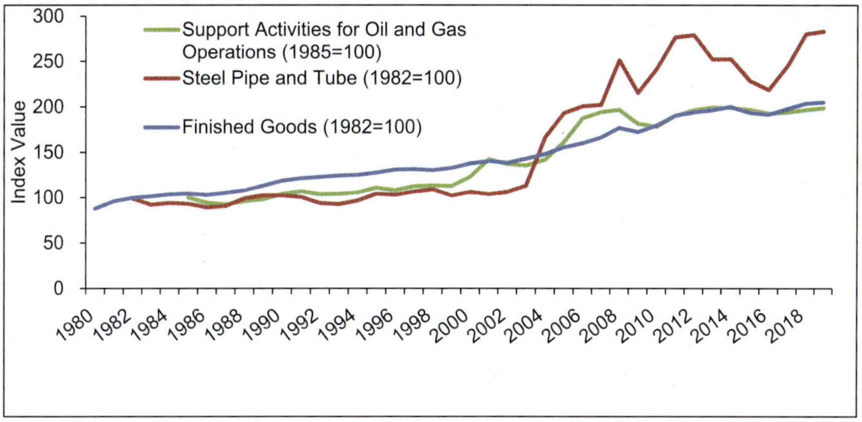

PPI	2014	2015	2016	2017	2018	2019
Support Activities for Oil and Gas Operations	199.2	196.8	192.8	194.1	196.7	198.8
Steel Pipe and Tube	253.0	228.7	218.8	245.1	280.6	283.5
Finished Goods	200.4	193.8	191.9	198.0	204.1	205.4

Figure 3. U.S. Bureau of Labor Statistics Producer Price Indices for steel pipes and tubes, finished goods, and support activities for oil and gas operations, 1980–2019.

From 1980 to 1995, the BLS indices increased at rates reflective of general inflationary trends in the U.S. economy, and both indices were relatively stable over the 1995–2003 period before increasing dramatically over the 2004–08 period, due in large part to increased demand for pipes and services due to intense hurricane destruction and cleanup activity in the U.S. Gulf of Mexico. The selection of cost indices is not unique, and a variety of indices can be used.

Location adjustment

Location is an important consideration in pipeline construction because the cost of development is influenced and impacted by the terrain, environmental conditions, quality and type of infrastructure, labor availability and productivity, level of competition, and other regional factors. In extreme or isolated environments, location also impacts the nature and efficiency of work and influences design requirements.

If the same project is constructed in different regions of a country, in different countries or on different continents, location differences become increasingly important. The global nature of equipment procurement and the competitive forces involved limits the impact of location differences for equipment, but differences in labor practices, availability and experience, tax benefits, the cost of steel and other basic materials, utilities, and transportation are location-specific and impact construction cost. Location factors are normally based on a composite index that includes labor rate, exchange rate, and import material.

An estimate of the construction cost at location l_2, $C(l_2)$, given the cost of the same or similar unit at l_1, $C(l_1)$, is obtained by multiplying the cost of the unit by a ratio of location cost indices $L(l)$ for the time period of interest holding all other factors equal:

$$C(l_2) = C(l_1)\left(\frac{L(l_2)}{L(l_1)}\right)$$

In the United States, location factors are available from various sources such as CDOL Richardson Cost Factors and Global Cost and Construction Yearbook, and if regional projects are well represented, project cost data may be used to infer location factors. Outside the United States, government or private sources must be employed if available.

How to adjust pipeline construction cost across time

Example

The U.S. onshore construction cost of a pipeline built in 2014–15 was \$3.63 million per mile. What is the inflation-adjusted cost of a 40-mile pipeline built in the United States in 2017–18 using the BLS cost-index adjustment? Compare with the actual average unit cost of \$4.35 million/mi in 2017–18. Would the procedure change if the pipeline was constructed in New York?

Solution

The construction cost of a 40-mi pipeline built in 2017–18 is estimated by applying the ratio of the BLS cost indices for oil and gas operations between the two periods:

$$\text{Cost}(2017-2018) = \text{Cost}(2014-2015) \cdot \frac{\text{Index}(2017-2018)}{\text{Index}(2014-2015)}.$$

According to the BLS adjustment WPU60110401, the estimated unit cost would be the following:

$$\$145 \text{ million} \cdot \frac{195.4}{198} = \$143 \text{ million}.$$

This is a simple, indeed trivial calculation, and one would suspect that such calculations across long time periods (say, greater than five years or so) will not provide an accurate assessment.

The average cost of pipeline activities in the United States in 2017–18 was $4.35 million/mi or $174 million. A more detailed evaluation of the individual activities and materials of pipeline construction cost would provide a better assessment, and if reliable region-specific indices were available, these could be applied. Region-specific indices may not be easily available but can usually be constructed with relevant proxy variables.

ONSHORE CONSTRUCTION

Onshore pipeline construction process

In the United States, gathering, transmission and distribution pipelines are governed by regulations and laws that are administered by the U.S. DOT. The regulations are contained in the Code of Federal Regulations (CFR), Title 49, Part 190 Enforcement Procedures, Parts 191 and 192 Natural Gas Pipelines, Parts 193 Liquefied Natural Gas Pipelines, Part 194 Oil Pipeline Response Plans, Part 195 Hazardous Liquid Pipelines, Part 198 State Grants, and Part 199 Drug Testing. The regulations incorporate the industry codes, guidelines, and standards including the American National Standards Institute and the American Society of Mechanical Engineers. Internationally, many countries have adopted the U.S. regulations, codes, and guidelines, and even for countries with different requirements and practices, they are still quite similar to the U.S. regulations.

Construction follows a well-defined sequence of events, where each portion of the operation depends upon the preceding portion like in an automobile assembly plant (Figure 1). Site preparation is an integral part of pipeline construction, varying with the soil conditions, terrain, crossings, and other factors. Bulldozers are frequently used to clear and grade the ROW, remove trees and brush, strip and stockpile topsoil, and build temporary access roads (Bauer, 2011a). The ditch is excavated using ditching machines, excavators, backhoes, and related equipment. In open terrain with long lengths and limited obstructions, trench machines are used, while close to obstructions backhoes are common.

The geology, topography, and physical characteristics along the pipeline route are important features (Ozanne, 2011). The physical characteristics of soil determine trenching efficiency, corrosion protection requirements, and other attributes that impact construction cost. Trench width and depth is designed for specific project conditions (Figure 2). For example, contract conditions may specify that the ditch be excavated to a width not less than six inches greater than the nominal outside pipe diameter and cut to a uniform grade for a minimum cover of 36 inches measured from the top of the pipe to the ground unless rock is encountered (Wininger, 2011b). In consolidated rock, the minimum cover usually varies between 18 and 24 in. For Class 1 locations, the cover is 18 in; for Classes 2 to 4 locations, the cover is 24 in.

Larger pipe requires larger and deeper trenches contributing to increased material and construction cost compared to smaller diameter pipe, for all other things equal. Using small diameter, high strength, and thinner wall pipe can provide cost savings because joints use less filler metal (welding rod) to weld the thinner wall pipe and less welding time is needed. Shipping costs are also lower because of the lower tonnage. The greater the wall thickness of pipe, the more welding passes required which increases the time and cost of welding. Higher grades of steel have some drawbacks, however, such as more stringent welding procedures, more costly welding tools, and greater welder skill requirements. High strength steel such as X-70 and X-80 also have greater corrosion problems such as hydrogen cracking.

The number and length of crossings across roads, railroad tracks, rivers, streams, utilities, and other pipelines, as well as obstructions such as above-ground structures, residential communities, and businesses, add to the complexity of the

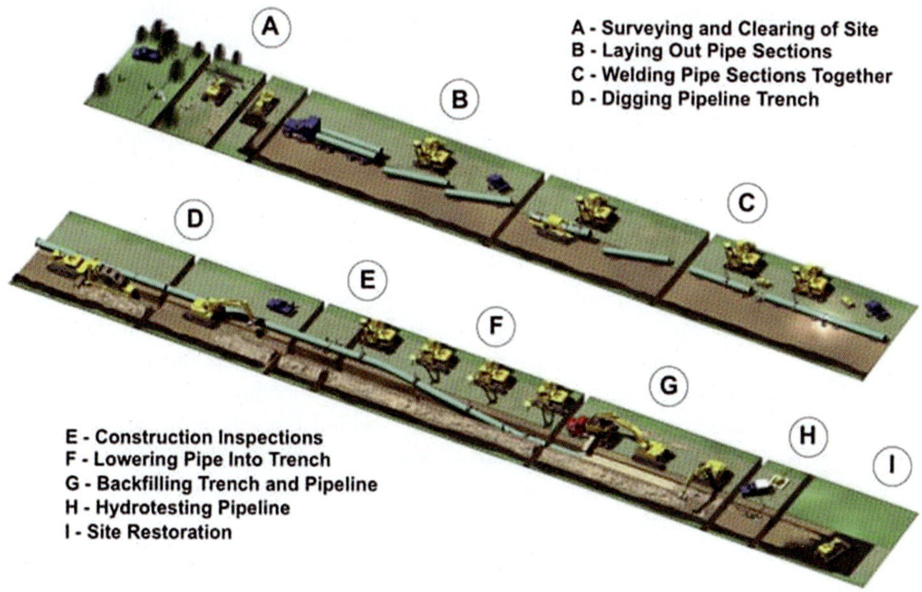

Figure 1. Typical pipeline construction sequence. Source: AOPL.

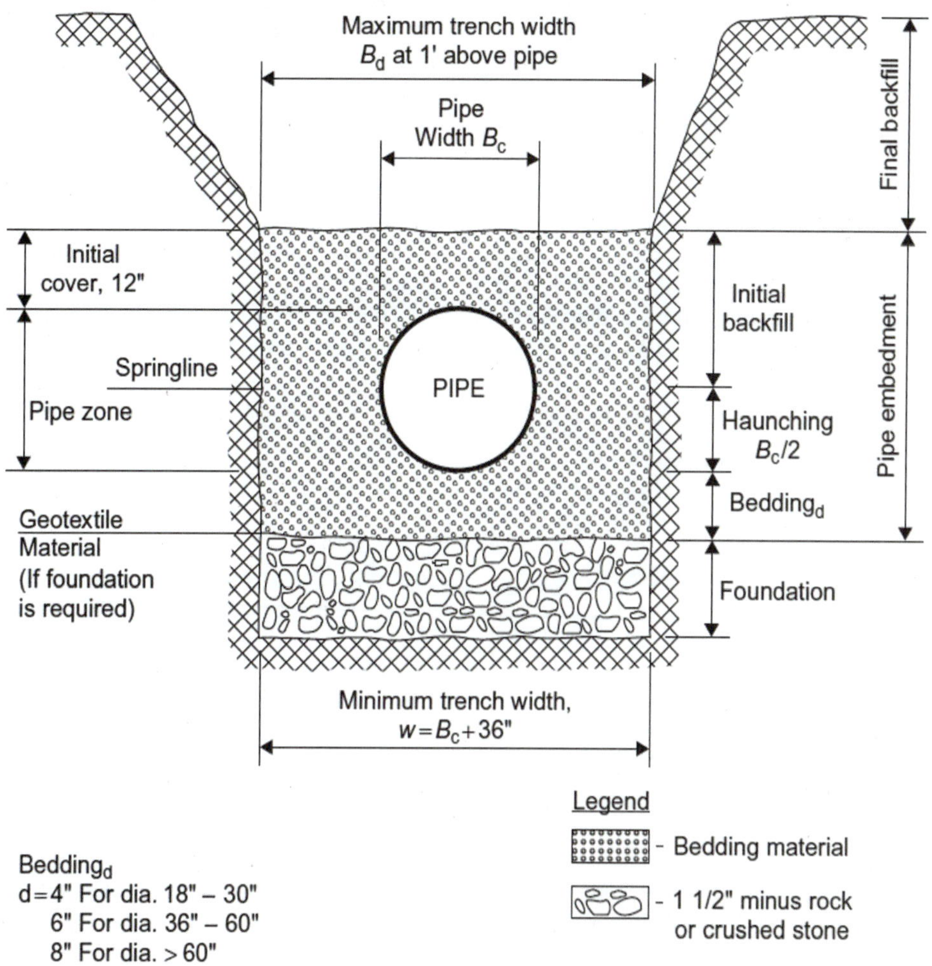

Figure 2. Trench width specifications are site and project-specific. Source: Menon, E.S., 2011. Pipeline Planning and Construction Field Manual. Gulf Professional Publishing, Waltham, MA.

design, permitting, and construction process. Construction methods vary with terrain conditions and are major cost factors. For example, laying pipeline across a rocky crossing will require boring machines and river crossings typically apply horizontal directional drilling which increase equipment, labor, and material costs. Design practices in bored crossings normally specify thicker wall pipe and concrete coating to protect the cathodic systems, and equipment and labor cost increase because of the specialized spreads and crew requirements. Differences in the type and extent of crossings contribute to differences in project cost and complexity.

ROWs or easements are purchased for the right to use the subsurface. The ROW allows the contractor access to the land to lay the pipeline and sets the width on each side of the pipeline route for the construction equipment to operate efficiently and safely (Figure 3). Normally, the working side of the construction equipment should remain on one side of the ROW to maintain productivity, and there should be adequate space within the working area to allow for vehicles to move in both directions safely (Wininger, 2011a). Pipeline sited within populated areas, federal lands, protected areas, and rough terrain require a more rigorous and time-consuming permitting process and may constrain operations (Miesner and Leffler, 2006).

The greater the diameter of the pipeline, the wider the ROW that will be required for construction and maintenance. Larger-diameter pipe requires wider ROWs because the larger pipe requires bigger pipe-handling equipment (side-booms), wider ditches, and wider spoil piles (Figure 3). During construction, temporary use of a larger ROW typically 50–100 ft on each side of pipeline centerline is normally acquired, which reverts to widths of 25–30 ft after construction is complete for maintenance activity. About 80- to 100-ft wide construction working ROWs are typical for 4- to 12-in pipe, and 200-ft plus construction ROWs are common for pipe up to 36 in. In the event the surface cannot be restored to its original condition before construction, companies may pay damages for the loss of use or loss of income. ROWs are more expensive to obtain in congested and high-population areas.

Welding is one of the most critical operations to construction efficiency (Kennedy, 1993). Weld types and procedures are set by federal regulation, and pipe welders must be qualified in accordance with industry standards. Welding procedure specifications are developed for the pipeline and welders must be qualified on these procedures. Welding procedure specifications describe the type of pipe to be welded (specification, grade, etc.), the type and specification of the pipe joint (bevel(s), angle, shoulder, spacing, etc.), the material thickness or range of thickness, the type and size of welding rods, the position and direction of the weld, etc. The most popular techniques to inspect welds after completion are radiography and ultrasonic. When weld defects are found, they are ground or cut out, repaired and retested. Welding defects add to material and labor cost and may delay project completion.

After passing inspection, weld joints receive the same anticorrosion protective coating as the pipe body received at the pipe mill. Field joint coatings are applied after the joints are welded to avoid heat damage. Shrink sleeves are wrapped around the joint and after heating with a propane torch, provides a waterproof seal. Tape is another type of joint coating. Following completion of the field joint application and coating inspection, the pipe is lowered into the trench by sidebooms or other lowering equipment and covered with backfill.

Hydrostatic testing is the final quality check before the pipeline is placed in service, and any leak indications require

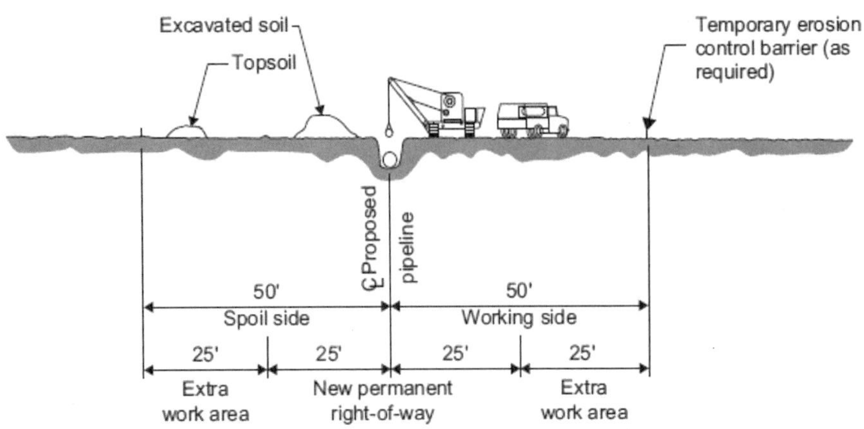

Figure 3. Construction right of way 50 ft either side of pipeline with topsoil segregation on spoil side. Source: Menon, E.S., 2011. Pipeline Planning and Construction Field Manual. Gulf Professional Publishing, Waltham, MA.

finding and repairing the leak, which can be time consuming. To ensure the pipeline can be operated safely at its minimum acceptable operating pressure (MAOP), the pipeline is tested at a higher internal pressure prior to operation. Each test section of the pipeline is subjected to the required hydrostatic test pressure and maintained for a period of 4 h (above ground piping) or 8 h (buried pipeline) or as required by the pipeline design code or by city, county, state or federal regulations. Generally, this is done using water and a minimum test pressure of 125% of the MAOP. Long pipelines are normally tested in sections of 10–20 mi depending on the configuration and elevation change. Significant elevation changes often require short test segments. Testing short sections increases sensitivity in detecting small leaks and reduces the amount of water required and disposal cost but increases time on site.

Onshore pipeline construction cost factors

Pipeline construction costs are typically categorized according to material factors, construction services, engineering, and inspection.

Material factors

Material. Material costs include all the facilities and equipment to construct and operate the pipeline, pipeline coatings, anodes, valve assemblies, fittings, pumps, compressors, pig launchers and receivers, instrumentation, and meters. Inspection of the material, transportation cost, and related miscellaneous cost may be included in the material cost category.

Design specifications. Engineers perform hydraulic calculations for the pipeline to determine the pipe diameter, wall thickness, and operating pressure based on the length of the pipeline and volume of product to be transported subject to federal regulations (Menon, 2011, Stevens and May, 2007). The operating pressure (maximum and normal), operating temperature and other design factors, and pipe material will determine the wall thickness. An economic evaluation is made to determine the most economic grade of steel compared with the pipe diameter and wall thickness. It is usually more economic to use a higher grade of steel and/or wall thickness and operate at a higher pressure than to use a large diameter pipe and a lower operating pressure.

Dimensions and tolerances. Pipelines are designed for each application according to regulatory requirements and are described by their diameter, wall thickness, material of construction, and other parameters. Outside diameter and wall thickness for API 5L pipe need to be within the tolerance specifications. During manufacturing, any pipe found to be out of tolerance is cause for individual diameter measurement of all pipe back to the last, and up to the next two sequential pipes measured and found to be within tolerance. Pipe is normally furnished in nominal lengths and within length tolerances specified on the purchase order. All pipes are randomly checked for straightness and should not exceed 0.2% of the length.

Grade. Grade designations are dimensionless. Grades A and B do not include reference to the specified minimum yield strength, and other grades are composed of the letter A or X, followed by the first two digits of the specified minimum yield strength in U.S. customary units. API 5L X42, for example, has a specified minimum yield strength equal to 42,000 psi. Standard grades include A25, A, B, X42, X46, X52, X56, X60, X65, X70, and X80. PSL 1 pipe is supplied in Grades A25 through X70; PSL 2 pipe can be supplied in Grades B through X80. Intermediate grades higher than X42 between two sequential standard grades may be agreed by the purchaser and manufacturer. For high-pressure pipelines, higher-grade pipe is normally selected because thinner-wall pipe can be used.

Manufacturing process. Oil and gas industry pipe is made using seamless manufacturing methods, electrical resistance welding, helical or spiral welding, and submerged arc welding (SAW). Welded pipes are classified according to the method of welding. Submerged arc welds use filler metal in welding, whereas electric resistance welded and electric fusion welding are no-filler metal processes. SAW is further classified into longitudinal (or straight) welding (L-SAW) and spiral (or helical) welding (S-SAW). Normally, medium diameter straight L-SAW has a single seam, and large diameter L-SAW uses a double seam. The manufacturing process has a direct impact on costs, with seamless pipe typically being the most expensive, and helically welded pipe being the cheapest. ERW pipe is more expensive than hot rolled coiled pipe. Double SAW pipe is the most common for large diameter lines.

Contract structure. Pipeline construction requires procuring all materials required in development and constructing the pipeline. Bidding is used to secure the best pricing for both construction and materials. In cases where materials and construction bids are combined, the contractor is responsible for procuring materials. In other cases, the owner(s) of the pipeline may prefer to separate the materials and construction bids and manage the interface to reduce costs. There are advantages and disadvantages to each approach.

Procurement contracts. Pipelines are purchased on terms that vary with the grade, diameter and wall thickness specification, lot size, composition, alloy content, delivery location, mechanical or plastic lining, and various other factors. Supply and demand conditions at the time contracts are an important factor in procurement cost. Steel prices are frequently reported on a per ton basis, but pipeline prices are typically specified using footage. Pipeline mills normally charge a premium for small batch sizes, so projects that have significant pipe diameter or wall thickness variation are expected to be greater than projects with only one or two specifications.

Coating. Industry standards and federal regulations require that pipelines, appurtenances, and facilities be protected from corrosion. Pipe coating is normally based on soil conditions and moisture content of the soil, and fusion bonded epoxy (FBE) or dual FBE, normally 12–14 mils thick, is a common choice. Other coatings that are used in approximate order of cost are asphalt, coal tar enamel, FBE, polyethylene (PE), extruded thermoplastic PE and polypropylene (PP), and elastomeric coatings (EPDM).

FBE is a powder-based thermoset polymer coating where bonding occurs after the hardener and epoxy react and cross-link. Although FBE coating is common, where soil moisture content is high, FBE coating may not be the most effective. If the pipe is to be pushed or pulled through a bore hole or a directional bore, and in areas where boulders or rock are encountered, an additional abrasive-resistant overcoat is typically placed over the fusion-bonded coating. PP systems and three-layer PP allow robust handling.

Cathodic protection. Coatings are the primary barrier to corrosion, and cathodic protection prevents corrosion in areas of permeable, missing, or damaged coating. The purpose of cathodic protection is to transfer electrons into the pipe by draining an electric current from it using impressed current or sacrificial anodes.

Ancillaries. Ancillaries include all equipment required for construction such as valves, fittings, assemblies, pig launchers and receivers, cathodic protection, and anodes. The number and type of valves, fittings, and ancillary equipment will reflect site-specific conditions. Collectively, they usually do not account for more than 10% of total material costs. Instrumentation and pipeline monitoring systems are additional cost factors.

Construction factors

Construction costs include all the equipment and labor to perform the installation, including field joint coating, crossings, hazards survey, hydrotesting and dewatering, and commissioning.

Route length and corridor. Route length is a primary factor in material and construction cost, and the amount and type of terrain and route complexity will contribute to differences in project cost. Topographic maps, aerial photographs, satellite imagery, and property ownership maps are used in the routing process. Structures along the route corridor such as power or communication lines and trees will add to the cost of construction. Environmental features such as wetlands, marshes, swamps, nesting areas, cultural resources, and endangered species will add to permitting time, construction cost, and potential delays. In some instances, pipelines may be rerouted to avoid sensitive environments and ecologies. The U.S. Army Corps of Engineers have the primary jurisdictional authority, but other federal agencies such as the U.S. Fish and Wildlife Services are also involved. If mitigation requirements are imposed by the regulating agency, cost will increase.

Market conditions. Pipeline projects are typically contracted on a lump-sum basis to the low-cost bidder, and thus contract requirements, market conditions, and competitions are key elements in determining price. Contractors estimate the time and cost to complete the job based on the complexity of the project and prepares a bid to win the work. Contract structure and the degree of competition in the market at the time the contract is let will impact how the contractor bids, and ultimately, the construction cost of the project.

Soil type. Soil type and soil water content, saturation level, void ratio, and porosity are used to determine load analysis in construction. Soil types will determine if a trench can be cut with standard equipment without blasting, or if special equipment will be required. Soil characteristics determine trenching efficiency, corrosion protection requirements, and other attributes that impact cost. If the trench has a high concentration of rock, the ditch will be padded with sand or other fine soil to protect the pipe and coating from being damaged.

Class locations. Federal regulations for natural gas pipelines mandate classifications based on population density along the route using class locations and will determine the yield strength and wall thickness needed to operate at a given pressure. Broadly speaking, Class 1 locations usually include deserts, mountains, grazing land, farmland, and sparsely populated areas. Class 2 locations include areas around cities and industrial areas, Class 3 locations include suburban housing developments, shopping centers, and residential areas, and where multistory buildings are prevalent, traffic is heavy, or where there are other utilities, Class 4 locations arise.

In CFR, Title 49, Part 192, a class location unit is defined as an onshore unit that extends 220 yards on either side of the centerline of a continuous 1-mile length of pipeline:

Class 1: Class location unit that has 10 or less buildings for human occupancy. Each building unit in a multiple-dwelling unit building is counted as a separate building intended for occupancy.

Class 2: Class location unit that has more than 10 but less than 46 buildings intended for human occupancy.

Class 3: Class location unit that has 46 or more buildings intended for human occupancy or an area where the pipeline is within 100 yards of a building or a playground, recreation area, outdoor theater, or other places of public assembly that is occupied by 20 or more people on at least five days a week for 10 weeks in any 12-month period. The days and weeks need not be consecutive.

Class 4: Class location unit where buildings with four or more stories exist above the ground.

The design factor F used in the calculation of design pressure depends on the class location and range from 0.72 for Class 1 to 0.40 for Class 4 pipe (see Chapter 2). If development occurs within a region after a pipeline is installed and class location changes, the pipeline operator may be required to replace the pipe in the impacted area with higher yield strength or wall thickness or reduce the operating pressure to achieve the higher safety factor.

Design codes also specify the spacing of valve stations for gas pipelines based on class location. For Class 1 locations, each point on the pipeline must be within 10 miles of a valve, or in other words, the code allows a maximum of 20 mi value spacing. For high-impact areas, the valve spacing requirement is reduced for safety reasons. For Class 2 locations, the valve spacing is 15 miles, 10 miles for Class 3 locations, and 5 miles for Class 4 locations.

Crossings. Permits are required to install pipelines across public highways and roads, and special easements or permits are needed from railroads and other pipelines. Higher construction costs are associated with the number and extent of crossings. Horizontal directional drill crossings for waterways and bored crossings under roads and railways add significantly to construction cost.

Constructability. Constructability refers to issues that impact the ease in which a pipeline is constructed. Typical factors that play a role in constructability include accessibility, existing infrastructure, construction techniques, soil type, terrain, and other physical features along the route. Constructability ratings are expected to broadly correlate with construction cost.

Rural versus suburban. Urban and suburban areas usually take a longer time in construction due to the need to minimize work space, limited ditching operations, and requiring the contractor to build the pipeline in the ditch one joint at a time (called "stovepiping"). All these issues contribute to a higher cost of construction. Mitigation and damages are also usually higher since most communities require the landscape be returned to as good or better condition than before the installation.

Established corridors. Regulatory agencies may encourage or mandate a pipeline route that follows or parallels an existing pipeline or other infrastructure (e.g., electric transmission) corridor (Bauer, 2011b). This may reduce some costs such as planning and clearing, but there may be disadvantages with the added restriction. Construction costs generally increase with additional constraints and restrictions.

Engineering and inspection factors

Engineering and inspection include the portion of the pay and expenses of engineers, surveyors, draftsmen, inspectors, superintendents, and their assistants applicable to construction work.

References

W.E. Bauer, Right-of-way. in: E.S. Menon (Ed.), Pipeline Planning and Construction Field Manual. Gulf Professional Publishing, Waltham, MA, 2011a, pp. 67–79.

W.E. Bauer, Pipeline regulatory and environmental permits. in: E.S. Menon (Ed.), Pipeline Planning and Construction Field Manual. Gulf Professional Publishing, Waltham, MA, 2011b, pp. 57–65.

J.L. Kennedy, Oil and Gas Pipeline Fundamentals. PennWell Books, Tulsa, OK, 1993.

E.S. Menon, Pipeline Planning and Construction Field Manual. Gulf Professional Publishing, Waltham, MA, 2011.

T.O. Miesner, W.L. Leffler, Oil and Gas Pipelines. PennWell, Tulsa, 2006.

H.S. Ozanne, Route selection. in: E.S. Menon (Ed.), Pipeline Planning and Construction Field Manual. Gulf Professional Publishing, Waltham, MA, 2011, pp. 43–56.

R.S. Stevens, D. May, Piping and pipelines. in: K.E. Arnold (Ed.), Facilities and Construction Engineering, Vol III. Petroleum Engineering Handbook. Society of Petroleum Engineers, Richardson, TX, 2007, pp. 318–394.

G.A. Wininger, Pipeline construction. in: E.S. Menon (Ed.), Pipeline Planning and Construction Field Manual, 1st ed. Gulf Professional Publishing, Waltham, MA, 2011a, pp. 341–355.

G.A. Wininger, Specification writing, data sheet production, requisition development, and bid analysis. in: E.S. Menon (Ed.), Pipeline Planning and Construction Field Manual, 1st ed. Gulf Professional Publishing, Waltham, MA, 2011b, pp. 421–463.

Onshore FERC gas pipeline cost statistics in the United States, 2014–2019

Federal Energy Regulatory Commission pipelines

The FERC approves the construction, operation, and decommissioning of interstate natural gas transmission pipelines under Section 7 of the Natural Gas Act (15 U.S. Code 717). FERC determines whether gas pipelines are gathering lines or transportation lines on a case-by-case basis using the "modified primary function test." If a line is determined to be gathering, it is classified as nonjurisdictional and is not regulated by FERC.

When a FERC-regulated natural gas pipeline company wants to modify its system, to add new pipeline or to expand capacity, or to remove capacity, it must apply for a "Certificate of Public Convenience and Necessity." This filing documents and justifies the construction and specifies what the company estimates construction cost will be. If a company receives its certificate and builds its facilities, it must report back to FERC at the completion of the project the actual cost of construction and how the original estimates compared with what it spent. The filing occurs within six months after successful hydrostatic testing.

Summary statistics

From 2014 to 2019, 99 projects totaling 2063 miles of FERC pipeline were installed in the United States at a cost of about \$11.8 billion (Table 1). In total, 18 states hosted pipeline projects during the five-year period: Alabama, Connecticut, Florida, Illinois, Kentucky, Louisiana, Maryland, Massachusetts, Minnesota, Nevada, New Jersey, New York, Ohio, Pennsylvania, South Carolina, Texas, West Virginia, and Virginia. Less than two dozen projects were completed per year and more than half of the total cost and almost half of total mileage installed occurred in 2018–19.

Aggregate average cost

Annual aggregate construction cost ranged from \$3.63 to \$7.26 million/mi from 2014 to 2019 and averaged \$5.72 million/mi over the period. Annual aggregate per inch-mile construction cost ranged from \$132,000 to \$372,000/in-mi per year and averaged \$207,000/in-mi over the period.

Materials, labor, miscellaneous, and ROW/damages averaged 13%, 49%, 30%, and 9%, respectively. Labor cost was about half of the total construction cost, and by far the largest cost category, indicating the labor-intensive nature of onshore pipeline construction. The miscellaneous category is the second largest category representing about a third of total cost. Materials and ROW/damages represent similar and much smaller contributions to total cost at about \$743,000 and \$492,000 per mile, respectively.

Large diameter pipes have significantly greater construction cost per mile and variability from year-to-year, but when normalized by diameter show less variation (Table 2). High variability in the large diameter class is due in part to small samples; in the two small diameter classes, the statistics are all based on five or more projects.

In 2017–18, for example, construction cost was \$2.26 million per mile for <24 in diameter pipe, \$4.83 million/mi for 24–36 in diameter pipe, and \$7.41 million/mi for >36 in diameter pipe (Figure 4). When pipeline diameter is used for normalization, some of this variability is attenuated, with construction cost averaging about \$155,000/in-mi in the <24 in diameter and 24-36 in diameter categories, increasing to \$170,000/in-mi for >36 in diameter pipelines (Figure 5).

Table 1
Federal Energy Regulatory Commission Onshore Pipeline Aggregate Average Cost, 2014–19

Year	Number of projects	Total length (mi)	Total cost (\$million)	Aggregate cost (\$million/mi)	Aggregate cost (\$1000/in-mi)
2018–19	23	957	6271	6.55	226
2017–18	22	364	1581	4.35	148
2016–17	19	198	1441	7.26	372
2015–16	15	331	1617	4.88	198
2014–15	20	243	881	3.63	132
2014–19	99	2063	11,791	5.72	207

Note: Aggregate cost based on all pipeline miles and weighted average pipeline diameter.

Table 2
Federal Energy Regulatory Commission Onshore Pipeline Aggregate Average Cost by Diameter, 2014–19

Year	Aggregate (\$million/mi)			Aggregate (\$1000/in-mi)		
	<24 in	24–36 in	>36 in	<24 in	24–36 in	>36 in
2018–19	3.37 (12)	6.18 (9)	12.7 (2)	198 (12)	204 (9)	303 (2)
2017–18	2.26 (6)	4.83 (12)	7.41 (4)	155 (6)	156 (12)	170 (4)
2016–17	3.71 (10)	8.25 (6)	27.9 (3)	237 (10)	282 (6)	664 (3)
2015–16	2.14 (5)	4.74 (6)	13.2 (4)	131 (5)	177 (6)	315 (4)
2014–15	2.75 (7)	3.75 (12)	42.0 (1)	168 (7)	118 (12)	999 (1)
2014–19	3.03 (40)	5.41 (45)	13.8 (14)	211 (40)	180 (45)	328 (14)

Note: Number of projects shown in parenthesis. Categories with less than five projects are subject to greater variability and interpretation uncertainty.

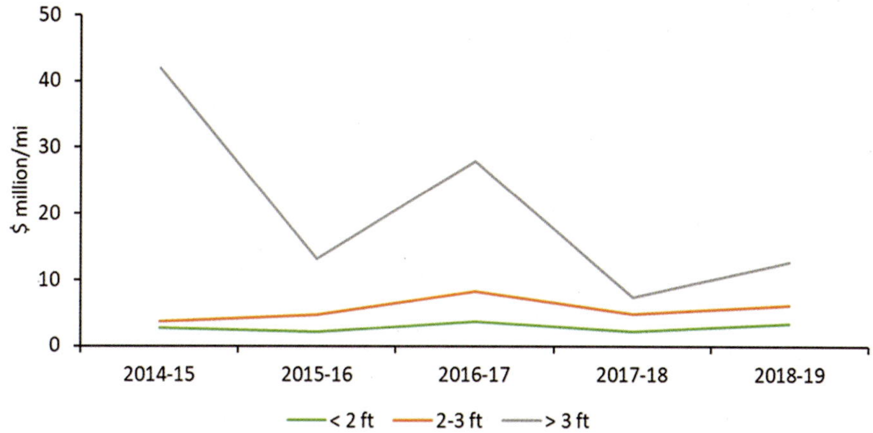

Figure 4. Federal Energy Regulatory Commission onshore gas pipeline aggregate average cost by diameter class, $ million per mile.

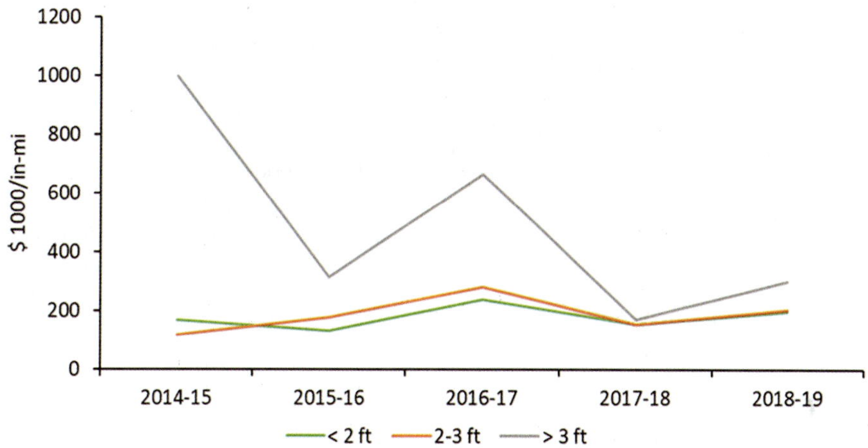

Figure 5. Federal Energy Regulatory Commission onshore gas pipeline aggregate average cost by diameter class, $1000 per inch-mile.

In aggregate, pipeline construction cost from 2014 to 2019 was $3.03 million/mi for <24 in diameter projects, $5.41 million/mi for 24–36 in diameter, and $13.8 million/mi for >36 in diameter, approximately doubling across each category. In inch-mile units, pipeline construction cost was $211,000/in-mi for <24 in diameter, $180,000/in-mi for 24–36 in diameter, and $328,000/in-mi for >36 in diameter.

Holding all other factors fixed, as diameters increase unit construction cost increase. All <24 in and 24–36 in diameter class statistics are based upon five or more projects, and are considered representative, whereas the >36 in diameter category has far fewer projects and is subject to greater variability and interpretation uncertainty.

Project average cost

Average project unit cost is about one-and-a-half times the aggregate unit cost and standard deviations are on the order of the mean for both the per mile and per inch-mile categories, providing no useful data to bound the averages (Table 3, Figure 6). Categories with fewer data will exhibit greater variability and interpretation uncertainty.

Table 3
Federal Energy Regulatory Commission Onshore Pipeline Project Average Cost, 2014–19

Year	Number of projects	Project ($million/mi)	Project ($1000/in-mi)
2018–19	23	6.81 (4.4)	228 (127)
2017–18	22	7.99 (8.2)	286 (325)
2016–17	19	11.4 (9.9)	489 (424)
2015–16	15	12.2 (16.2)	441 (605)
2014–15	20	9.5 (12.3)	352 (383)
2014–19	99	9.3 (10.4)	360 (386)

Note: Standard deviation shown in parenthesis.

In 2017–18, for example, 22 projects were completed at an average cost of about $8 million/mi compared to $4.4 million/mi on a composite basis. In diameter-miles, project average

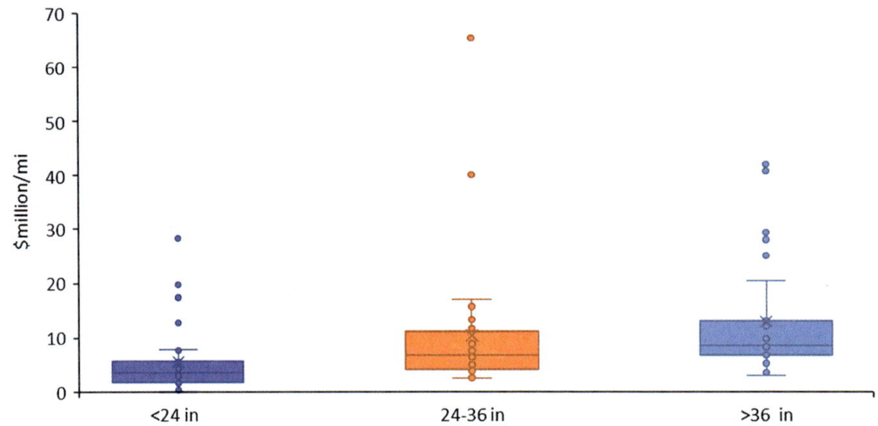

Figure 6. Federal Energy Regulatory Commission onshore gas pipeline project cost by diameter class, 2014–19.

Table 4
Federal Energy Regulatory Commission Onshore Pipeline Project Average Cost by Diameter, 2014–19

Year	Project ($million/mi)			Project ($1000/in-mi)		
	<24 in	24–36 in	>36 in	<24 in	24–36 in	>36 in
2018–19	4.10 (12)	9.51 (9)	11.4 (2)	252 (12)	314 (9)	270 (2)
2017–18	3.18 (6)	10.6 (12)	7.39 (4)	218 (6)	359 (12)	170 (4)
2016–17	8.34 (10)	8.46 (6)	27.5 (3)	550 (10)	305 (6)	656 (3)
2015–16	5.68 (5)	16.5 (6)	13.9 (4)	306 (5)	626 (6)	332 (4)
2014–15	6.79 (7)	8.34 (10)	42.0 (1)	430 (7)	254 (10)	999 (1)
2014–19	5.69 (40)	10.3 (45)	16.6 (14)	360 (40)	350 (45)	394 (14)

Note: Number of projects shown in parenthesis. Categories with less than five projects are subject to greater variability and interpretation uncertainty.

cost was $286,000/in-mi compared to $148,000/in-mi in aggregate. Average project cost is always greater than the aggregate average cost.

From 2014 to 2019, project average cost was $9.3 million/mi compared to $5.72 million/mi on an aggregate basis. On a diameter-mile basis, project average cost was $360,000/in-mi compared to $207,000/in-mi in aggregate.

By diameter class, project average cost ranged from $5.60 million/mi (<24 in) to $16.6 million/mi (>36 in), and from about $355,000/in-mi (<24 in, 24–36 in) to $394,000/in-mi (>36 in). Project average costs are between 20% and 80% greater than aggregate cost (Table 4).

Regression models

Regression model results based on pipeline length and diameter-length for 99 projects performed between 2014 and 2019 are shown in Table 5. Length and diameter-length

Table 5
Cost Models for Federal Energy Regulatory Commission Onshore Pipeline Construction, 2014–19

	Cost ($million) = *a* + *b* Length (mi) + *c* Diameter (in) + *d* Diameter × Length (in-mi)				
Model	*a*	Length *b*	Diameter *c*	Dia × Len *d*	R^2_{adj}
I	10.2	5.15			0.83
	(0.8)	(22,1)			
I_o		5.21			0.84
		(24.0)			
II	18.5			0.17	0.90
	(1.8)			(30)	
II_o				0.17	0.90
				(30)	
III	−96.9	5.11	4.16		0.85
	(−3.1)	(23.3)	(3.7)		
III_o		5.04	1.02		0.85
		(22.2)	(2.1)		
IV	−22.6		1.61	0.17	0.90
	(−0.9)		(1.7)	(29.4)	
IV_o			0.87	0.17	0.90
			(2.3)	(29.4)	
V	61.1	−5.75	−1.09	0.36	0.92
	(2.1)	(−4.8)	(−1.9)	(9.1)	
V_o		−4.23	0.91	0.31	0.92
		(−4.3)	(2.6)	(3.6)	

Note: t-statistics shown in parenthesis.

variables are not the only attributes of pipeline projects, of course, but they are probably the most useful characteristics and are publicly available. Other variables such as class location, wall thickness, terrain profile, etc., could be included in the specification, but they are usually tedious to collect and/or unavailable for all projects.

Total project cost as a function of mileage follow robust linear relationships:

Cost (\$million) = 5.15 Length (mi) + 10.2, $R^2 = 0.83$.
Cost (\$million) = 5.21 Length (mi), $R^2 = 0.83$.

Route length is a primary factor in construction cost and is a consistently good variable to apply over most time periods. For short pipelines, the cost model without fixed-term is preferred since short distance projects will only incur a small portion of the \$10.2 million fixed cost.

If diameter and diameter-length variables are added to the one-factor specification, model fits increase modestly, and since the variables are statistically significant, the two- and three-factor models are also significant (Table 5).

Limitations

In recent years, significantly fewer pipeline projects have been built in the United States, and average statistics may be less representative than in years past because of sample bias. Whenever sample sizes are small and the number of influencing factors is large, granular analysis (i.e., categorized cost statistics) needs to be approached cautiously or will lead to interpretation problems and poor reliability. Statistical results based on small samples without discussion of their limitations can lead to significant interpretation problems. Small samples are prone to high levels of uncertainty in granular evaluations because project outliers, when they exist, cannot be detected in a meaningful way nor averaged with other project data to reduce their influence. Hence, users should be cautious if sample data are not clearly and precisely presented and should be aware that granular analysis of small data sets may not be representative of industry activity.

Pipeline costs are site- and project-dependent based on the length and diameter of the line, population density and terrain along the route, and the number of rivers, rail, and road crossings, and many other factors. Only two factors were applied in evaluation, namely route length and pipeline diameter because these are accessible and available publicly, but these factors may not be the main factors responsible for cost in every project. On average and in aggregate, however, these two factors are probably the most important in determining project cost, so the unit statistics are meaningful over specific time periods. Information on class location, terrain, and other factors may be useful descriptors.

Very short pipelines less than a mile or two in length tend to have high unit cost and are not typical. These projects can be considered separately or not at all since they will distort the average if sample size is small. The unit cost statistics and regression models for pipeline construction apply to U.S. gas transmission projects. They do not apply to gathering pipeline construction (typically small diameter), which may be more than half as expensive as transmission, or on international projects or domestic projects expected to be performed greater than five years from the end of the data evaluation period.

How to estimate onshore pipeline construction cost using unit costs?

Example

Compare the expected cost of a 40 mile, 16-in oil transmission pipeline installed in Tennessee in 2018 using two different unit cost statistics and a two-factor regression model.

Solution

Although FERC projects are based on gas transmission pipelines, the unit cost statistics should be applicable to oil transmission pipelines. Using all the information provided with the aggregate average U.S. unit cost statistic for 2018 yields:

Cost = \$226,000/in-mi · 16 in · 40 mi = \$145 million.

If the diameter of the line is not applied, the cost estimate is given by:

Cost = \$6.55 million/mi · 40 mi = \$262 million.

Using the two-factor cost model without fixed term yields:

$\text{Cost}_{\text{IIIo}} = 5.04\ (40) + 1.02\ (16) = \$218$ million.

The range varies from \$145 to \$262 million and bounds the expected construction cost. Applying more information (i.e., project characteristics) usually ensures more reliable estimates.

How to estimate cost bounds in onshore pipeline construction in the United States?

Example

Estimate the pipeline construction costs for 5-mile and 60-mile gas transmission projects expected to be performed in Florida in 2022. The evaluation year in 2020. Provide cost bounds of the estimate. Should cost be inflation adjusted? Do the results apply to the Middle East?

Solution

Since only route length is given in the problem statement, the aggregate U.S. cost statistic from 2014 to 2019, $5.72 million/mi, or a one-factor regression model should be applied. If four or more projects were constructed in Florida over the past five years, regional unit cost could be applied.

For the 5-mi line, the regression model without a fixed-term yields:

$\text{Cost}_{\text{Io}} = 5.21\ (5) = \$26.1$ million.

If the fixed-term model is applied, which is not advisable since the route length is relatively small, the cost estimate would be about 40% larger at $36 million with the fixed-term contributing about one-quarter of the estimated cost (10.2/36 = 27%).

For the 60-mi line, regression models with or without a fixed-term yield nearly identical results:

$\text{Cost}_{\text{I}} = 5.15\ (60) + 10.2 = \$319$ million.
$\text{Cost}_{\text{Io}} = 5.21\ (60) = \$313$ million.

In this case, since the pipeline is long, the fixed cost contribution represents a smaller percentage of the total cost, and either model can be used.

Estimation is required three years from the end of the 2014–19 period where data are available and inflation-adjustment is probably not necessary. The uncertainty range of the estimated cost is expected to be ±40% to 50% from the actual cost of construction, and the cost estimates do not apply to international regions.

Onshore gas pipeline construction cost in the United States, 2017–2019

In Tables 1–3, U.S. onshore pipeline project costs reported for FERC-regulated gas pipelines constructed between July 2016 to June 2020 are presented by diameter (in), location (state), project type (loop, lateral, replacement), and length (mi). Total cost and three cost subgroups for materials, labor and miscellaneous are reported in million dollars. The ROW/damage cost category is not depicted in the tables but can be calculated by subtracting the sum of the three subcategories from total cost.

Table 1
U.S. FERC Onshore Gas Pipeline Project Costs, 2017–18

Size (in)	Location (Project Type)	Length (mi)	Materials ($MM)	Labor ($MM)	Misc. ($MM)	Total ($MM)
8	Minnesota (L)	2	0.241	2.07	1.14	3.75
12	Louisiana-Mississippi	55	9.49	37.1	19.0	68.6
12	Minnesota (L)	2.8	0.653	5.94	1.47	8.26
16	Pennsylvania (L)	7.2	2.12	12.4	25.0	42.0
20	Pennsylvania (R)	21	8.09	35.0	24.9	75.3
20	New York	1.2	1.15	2.04	1.15	4.34
24	New York (R)	7	3.57	18.5	3.35	25.7
24	Virginia (lat.)	4	3.05	9.66	5.54	19.9
24	New Jersey (lat.)	0.44	1.74	8.13	6.52	17.6
24, 36	NY-Mass-Conn. (L)	13.26	9.20	47.2	36.33	100
30	West Virginia (R)	3.9	2.53	16.6	10.93	31.9
30, 36	Florida	126	98.9	240.8	93.75	495

Size (in)	Location (Project Type)	Length (mi)	Materials ($MM)	Labor ($MM)	Misc. ($MM)	Total ($MM)
36	Louisiana	66	50.6	2.30	154.71	239
36	Pennsylvania (L)	13	7.46	51.0	27.5	92.9
36	Pennsylvania (L)	8.1	8.42	33.4	14.6	58.4
36	Pennsylvania (L)	7	6.84	30.7	12.7	52.4
36	Louisiana	3.58	3.96	20.9	46.9	73.4
36	Ohio (L)	2.1	3.45	12.1	8.03	25.8
42	Alabama (L)	7.6	6.42	36.8	10.1	55.0
42	Alabama (L)	5.3	4.67	23.9	6.29	35.8
42	Alabama (L)	2.55	2.22	11.3	3.49	17.7
48	Alabama (L)	4.6	5.91	24.9	7.73	39.4
Total		363.63	240	683	521	1582

Note: L = loop, lat. = lateral, R = replacement.
Source: U.S. FERC for completed projects filed between July 1, 2016 and June 30, 2017.

Table 2
U.S. FERC Onshore Gas Pipeline Project Costs, 2018–19

Size (in)	Location (Project Type)	Length (mi)	Materials ($MM)	Labor ($MM)	Misc. ($MM)	Total ($MM)
4	South Carolina	5	0.219	3.42	2.09	6.22
8, 12, 20	Nevada (L, lat. R)	8.45	3.30	15.1	3.70	22.3
12	South Carolina	55	7.59	48.4	23.2	87.6
12, 26	Florida (L, lat.)	7.7	5.50	14.3	7.00	26.9
16	Minnesota-ND	37.3	9.41	20.2	5.05	36.9
16	Minnesota (L)	7.83	5.01	24.9	8.81	44.3
16	New York (lat.)	7.8	2.62	42.8	13.5	60.9
16	Louisiana	0.17	0.302	0	0	0.302
20	Ohio (R)	21.6	9.85	104	41.4	166
20	Pennsylvania (R)	14.2	4.81	0.782	45.9	55.3
20	Pennsylvania (R)	11.7	2.87	2.20	41.1	53.3
20	Pennsylvania (R)	8.1	4.51	1.55	51.5	63.6
24	Illinois (lat. R)	0.54	1.33	1.86	0.935	4.12
24, 36	Alabama-Florida	509	403.8	1087	778	2587
26	Virginia-WV	28.9	19.1	212	106	348
30	Pennsylvania	57.3	35.6	259	129	487
30	Louisiana (lat.)	3.8	3.63	6.76	4.85	16.3
30	Texas (lat.)	0.76	1.05	1.96	1.06	4.28
30, 36	New York (L)	7.8	12.1	106	13.6	133
36	Louisiana	27	69.0	153	91.1	327
36	Pennsylvania (L)	2.9	3.91	17.7	15.6	38.4
42	Pennsylvania	125.2	147	1007	362	1619
42	Pennsylvania (L)	8.6	7.86	49.0	24.2	83.1
Total		956.66	760.7	3177.7	1769.6	6270.6

Note: L = loop, lat. = lateral, R = replacement.
Source: U.S. FERC for completed projects filed between July 1, 2018 and June 30, 2019.

Table 3
U.S. FERC Onshore Gas Pipeline Project Costs, 2019–20

Size (in)	Location (Project Type)	Length (mi)	Materials ($MM)	Labor ($MM)	Misc. ($MM)	Total ($MM)
6	Pennsylvania	5.8	1.61	2.68	1.80	6.10
6	Minnesota (L)	1.6	0.208	1.26	1.22	2.84
8	Minnesota (L)	4.2	0.806	5.07	2.07	8.37
12	Pennsylvania	14.4	2.66	20.2	5.92	30.0
16	Minnesota (lat.)	12	4.11	26.0	6.87	39.3
16	Louisiana	0.3	2.00	0.226	4.86	7.09
20	Washington (lat./R)	5.85	4.99	10.6	12.4	65.5
20	Louisiana (lat.)	0.72	0.558	1.53	2.89	5.08
20, 30	Pennsylvania	7.25	1.90	17.4	12.4	35.7
24	Minnesota (lat.)	10.2	7.68	27.8	8.54	47.1
24	Minnesota (L)	2.7	2.44	5.15	1.08	10.4
24, 30, 36	West Virginia	170.9	155.6	2203	634	3134
30	Texas (lat.)	0.5	1.30	1.99	1.40	4.88
30	Washington (R)	0.32	0.385	16.1	3.58	20.3
36	Ohio (L)	4.4	2.88	32.9	15.3	58.1
Total		241.14	191.1	2401.6	715.7	3475.1

Note: L = loop, lat. = lateral, R = replacement.
Source: U.S. FERC for completed projects filed between July 1, 2019 and June 30, 2020.

How to develop construction cost models using regression

Regression Modeling

Regression is a useful tool for pipeline cost estimation as long as the limitations of the model and interpretation are appreciated. Standard statistical analysis should be performed prior to regression modeling to understand the nature and spread in the data and the relationship between cost and system factors. In small data sets special precautions need to be taken. All assumptions should be stated in clear and concise language, so results are understood by decision makers and reproducible by third parties.

The user needs to select a functional form for the cost model (e.g., linear, nonlinear) and decide how to group and process project data according to spatial, technical, and temporal categories. Usually, the choices are obvious, but proper selection depends upon experience and familiarity with the underlying engineering processes. Choices must be made on categorizing the projects in terms of time and location (e.g., time frame for evaluation, regional subgroups), if adjustments should be performed (say for inflation or location differences), if certain projects should be excluded from analysis (e.g., high-cost projects, low-mileage installations, exceptional events), and what factors should be considered in the model. The factors available for analysis are frequently limited to project data reported, but additional factors may also be collected from other sources. Actual project cost data are strongly preferred to estimated cost, and if common accounting schemes are not employed, the categories used may require adjustment.

Expectations

Linear cost models are often preferred, but nonlinear (log and product) models may also be applied. Nonlinear cost functions often do not present significantly better model fits and suffer from a less intuitive interpretation compared to linear models. Linear models may be constructed with or without a constant term, but variables with negative coefficients, including the fixed term, are generally to be avoided on physical grounds.

Total cost of pipeline projects are almost always dependent on the length of pipe installed, but should also be influenced by the size of the pipe because of ditching, equipment requirements, material cost, and other factors. Nonlinear combinations of length and size variables may be explored such as length × diameter and length × diameter squared terms, the later frequently employed when the model coefficient for the length × diameter term is negative. Subcategory costs usually do not yield useful results and are normally of secondary interest. Not all the variables available are expected to be relevant, and when projects are subdivided according to diameter classes or by region, small sample issues normally arise and play an important role in the validity of the results.

Data

In Tables 1–3 in the previous section, U.S. onshore FERC-regulated gas pipeline construction cost for the years 2017–19 are presented by diameter (in), location (state), project type (loop, lateral, replacement), and length (mi). Total cost and three cost subgroups for materials, labor and miscellaneous are reported in million dollars for materials, labor, and miscellaneous categories.

Assumptions

- A three-year time frame from 2017–19 is selected for evaluation and no inflation adjustment is performed on the projects.
- All project types are grouped together, and no distinction is made between looped, lateral, or replacement projects.
- All projects in all states are aggregated because terrain and other project features are not described to differentiate projects.
- Projects with costs exceeding $500 million are excluded from evaluation. This includes two projects constructed in 2018-19, a 509-mi Alabama-Florida line that cost $2.6 billion and a 125-mi Pennsylvania line that cost $1.6 billion. In 2019-20, a 171-mi West Virginia line that cost $3.1 billion is also excluded.
- Terrain, urban/rural routes, river crossings, mountain routes, and other factors that might be important in project cost differences are not included in the evaluation.
- Operator/contractor characteristics and contract type are not distinguished.
- When multiple pipeline sizes are reported, the average diameter is used in assessment.

One-Variable Models

For one-variable linear models, the total cost data is graphed against pipeline length using Excel's curve fitting routine with trendline. In Figure 1, all 63 projects from Tables 1–3 are included and the cost model is computed with and without the fixed term. Note that forcing the linear model through the origin (i.e., setting the fixed term to zero in the parameterization), the model coefficient increases (as expected) from $6.39 to $6.50 million/mile. In Figure 2, three projects over $500 million are excluded from the data set. In this case, the unit cost coefficients are reduced to $3.9 million/mi with the fixed term and $4.29 million/mi without the fixed term. Excluding high cost projects reduce the unit cost as expected.

In summary, using the complete data set (63 projects) we obtain:

$$\text{Total Cost} = 22.3 + 6.40\ \text{Length},\ R^2 = 0.68$$
$$\text{Total Cost} = 6.50\ \text{Length},\ R^2 = 0.72$$

Using the truncated data set (60 projects) we obtain:

$$\text{Total Cost} = 18.2 + 3.9\ \text{Length},\ R^2 = 0.62$$
$$\text{Total Cost} = 4.29\ \text{Length},\ R^2 = 0.72$$

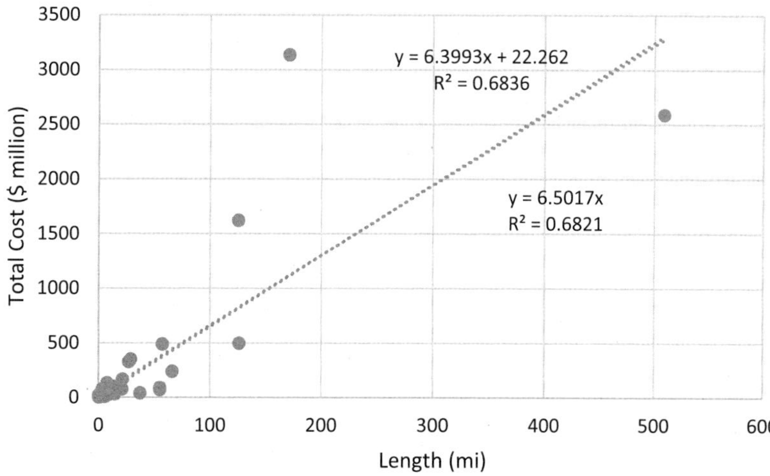

Figure 1. Linear regression cost model for all U.S. FERC onshore gas pipeline projects with and without fixed term, 2017–19.

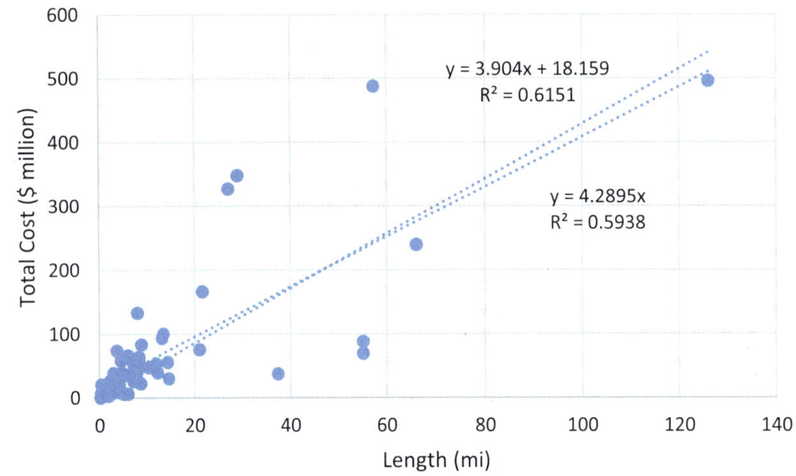

Figure 2. Linear regression cost model for U.S. FERC onshore gas pipeline projects with and without fixed term, excluding projects over $500 million, 2017–19.

Two-Variable Models

For two or more variable cost models, Excel's built-in Data Analysis tool is used for multiple regression evaluation. First, transfer the project data to Excel (Figure 3), and invoke the Data Analysis tab (Figure 4). Select the Regression tool listed and populate the input cells from the spreadsheet representing the cost column (Input Y Range) and project descriptors (Input X Range) as shown in Figure 5. The project descriptor variable columns must be adjacent to one another because of the input range format. Locate where the regression output is to be presented on the sheet in the Output Range cell (here I3) and press "OK".

Results are presented using diameter and length terms with and without fixed term (Figure 6), and length and length × diameter squared terms without fixed term (Figure 7). Three regression models are depicted along with adjusted model fits:

	A	B	C	D	E
1					
2		Size	Length	Length x Diameter^2	Total
3		in.	mi.	mi-in-in	$MM
4		8	2	128	3.75
5		12	55	7920	68.6
6		12	2.8	403.2	8.26
7		16	7.2	1843.2	42
8		20	21	8400	75.3

Figure 3. Project data transferred to Excel spreadsheet. Only five rows of project data are shown.

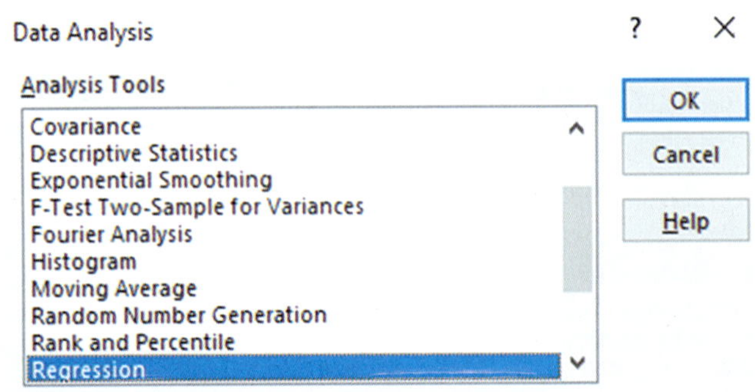

Figure 4. Data Analysis tab in Excel has a Regression tool to perform multiple regression analysis.

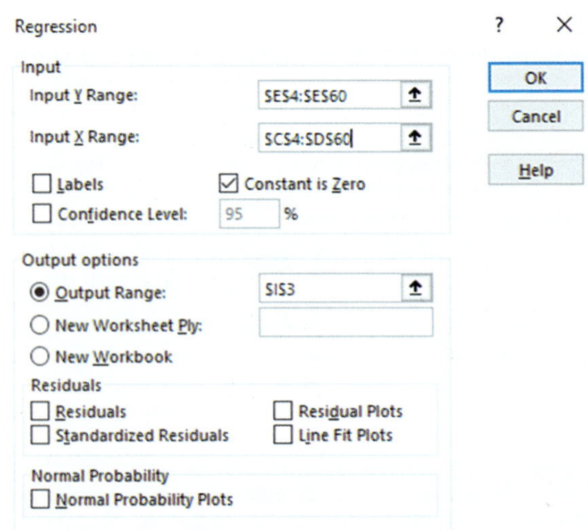

Figure 5. Apply regression tool to the cost data column (Input Y Range) and project descriptors (Input X Range). In this set-up, length and length × diameter squared terms are the input factors, and the constant term of the model is set equal to zero.

Size in.	Length mi.	Length x Diameter^2 mi-in-in	Total $MM
8	2	128	3.75
12	55	7920	68.6
12	2.8	403.2	8.26
16	7.2	1843.2	42
20	21	8400	75.3
20	1.2	480	4.34
24	7	4032	25.7
24	4	2304	19.9
24	0.44	253.44	17.6
30	13.26	11934	100
30	3.9	3510	31.9
33	126	137214	495
36	66	85536	239
36	13	16848	92.9
36	8.1	10497.6	58.4
36	7	9072	52.4
36	3.58	4639.68	73.4
36	2.1	2721.6	25.8
42	7.6	13406.4	55
42	5.3	9349.2	35.8

SUMMARY OUTPUT

Regression Statistics	
Multiple F	0.869882
R Square	0.756695
Adjusted I	0.73409
Standard E	63.55759
Observati	57

ANOVA

	df	*SS*	*MS*	*F*	*gnificance F*
Regressio	2	690984.3	345492.1	85.52702	1.83E-17
Residual	55	222176.2	4039.567		
Total	57	913160.4			

	Coefficient	*andard Err*	*t Stat*	*P-value*	*Lower 95%*	*Upper 95%*	*ower 95.0%*	*pper 95.0%*
Intercept	0	#N/A	#N/A	#N/A	#N/A	#N/A	#N/A	#N/A
X Variable	1.078654	0.364204	2.961675	0.004511	0.348773	1.808535	0.348773	1.808535
X Variable	3.708124	0.389762	9.513826	3.23E-13	2.927024	4.489224	2.927024	4.489224

Figure 6. U.S. FERC gas pipeline cost data and output using diameter and length terms in a cost model without fixed term. Only 20 project data are depicted but model was run with 60 projects.

Size in.	Length mi.	Length x Diameter^2 mi-in-in	Total $MM
8	2	128	3.75
12	55	7920	68.6
12	2.8	403.2	8.26
16	7.2	1843.2	42
20	21	8400	75.3
20	1.2	480	4.34
24	7	4032	25.7
24	4	2304	19.9
24	0.44	253.44	17.6
30	13.26	11934	100
30	3.9	3510	31.9
33	126	137214	495
36	66	85536	239
36	13	16848	92.9
36	8.1	10497.6	58.4
36	7	9072	52.4
36	3.58	4639.68	73.4
36	2.1	2721.6	25.8
42	7.6	13406.4	55
42	5.3	9349.2	35.8

SUMMARY OUTPUT

Regression Statistics	
Multiple F	0.872028
R Square	0.760433
Adjusted I	0.737896
Standard E	63.06745
Observati	57

ANOVA

	df	*SS*	*MS*	*F*	*gnificance F*
Regressio	2	694397.7	347198.9	87.29065	1.2E-17
Residual	55	218762.7	3977.503		
Total	57	913160.4			

	Coefficient	*andard Err*	*t Stat*	*P-value*	*Lower 95%*	*Upper 95%*	*ower 95.0%*	*pper 95.0%*
Intercept	0	#N/A	#N/A	#N/A	#N/A	#N/A	#N/A	#N/A
X Variable	2.165087	0.757454	2.858376	0.006003	0.647116	3.683058	0.647116	3.683058
X Variable	0.002483	0.000794	3.125154	0.002836	0.000891	0.004075	0.000891	0.004075

Figure 7. U.S. FERC gas pipeline cost data and output using length and length × diameter squared cost model without fixed term. Only 20 project data are depicted but model was run with 60 projects.

$$\text{Total Cost} = -38.8 + 2.35\ \text{Diameter} + 3.85\ \text{Length},\ R^2 = 0.66$$

$$\text{Total Cost} = 1.08\ \text{Diameter} + 3.71\ \text{Length},\ R^2 = 0.73$$

$$\text{Total Cost} = 2.17\ \text{Length} + 0.0025\ \text{Length} \times \text{Diameter}^2,\ R^2 = 0.74$$

Only 20 project data are depicted in Figures 6 and 7 but each model was run with 60 projects. The t-statistics for each model are depicted in the output table. For example, in Figure 6, the t-statistic for the Diameter variable is 3.0 and for Length variable is 9.5. In Figure 7, the t-statistic for the Length variable is 2.9 and for Length × Diameter squared variable is 3.1. Normally, t-statistics greater than 2.0 denote statistically significant parameters, and the larger the value the more descriptive the statistic.

OFFSHORE CONSTRUCTION

Offshore pipeline construction process

Route selection

Offshore pipeline route selection is based on consideration of bathymetry, seafloor character, seabed terrain, geohazards, bioenvironmental issues, and existing infrastructure.

The first step in pipeline construction is to perform a survey of the proposed route and contingency routes collecting bathymetry data and core samples to determine soil characteristics (Tootill et al., 2004; Haneberg et al., 2013). The soil information is used to predict potential span and perform stability checks. Oil and gas companies typically perform or oversee the route selection process and are involved to varying extents with the engineering, procurement, and construction processes depending on the nature of the project.

An economic evaluation is performed on various routes considering the installation and material cost of the pipe, burial footage, water depth, pipeline crossings, and subsea terrain to determine the preliminary route. After preliminary routing studies are completed, a hazard survey is performed, which includes core samples of the bottom conditions and a more detailed overview of the bathymetry, escarpments, outcrops, and related features. If leaseholders along the route have drilling or development plans that would conflict with the route or installation schedule, contingency plans and a contingency around the proposed activity need to be in place. Congested areas and areas with known or expected geohazards require greater planning and review.

A good strategy in routing is to avoid hazards, but this is not always possible, and with every new restriction or avoidance area added, pipeline length increases, which increases the cost. Pipeline routes are selected to avoid archeological sites or biological resources, and areas of possible landsliding and faulting, mud seeps, undulations, and rocky outcrops. Routes may be selected to avoid difficult terrain, but the added length needs to be compared with the cost to rectify the spans and in many cases avoidance is not practical or cost-effective.

Laying pipeline

Lay barge construction is by far the most frequently used technique for marine pipeline construction (Palmer and King, 2008, Guo et al., 2014). In S-lay and J-lay systems, sections of pipe are welded together on the deck of the vessel, which is then lowered to the seabed as a continuous string of pipe as the vessel moves forward. In a reel-lay system, the pipeline is assembled onshore and spooled onto the installation vessel. Once on location, the spool is paid out and the pipeline is guided into the water via a tower or ramp equipped with a straightener and a tensioner. Towing fully assembled pipeline from shore to site has also been employed but is not common in the Gulf of Mexico or elsewhere.

S-lay system. In S-lay systems, tensioners apply a tension force to the pipe near the stern, and the pipe is supported on rollers by a stinger structure where it enters the water. As a pipe is laid, it is first bent in one direction at the stinger (overbend) and then moves through the suspended section and the underbend (sagbend), where it is bent in the reverse direction to the touchdown point (Figure 1). The shape of the overbend is controlled primarily by the stinger, a steel framework that provides support for the overbend, while

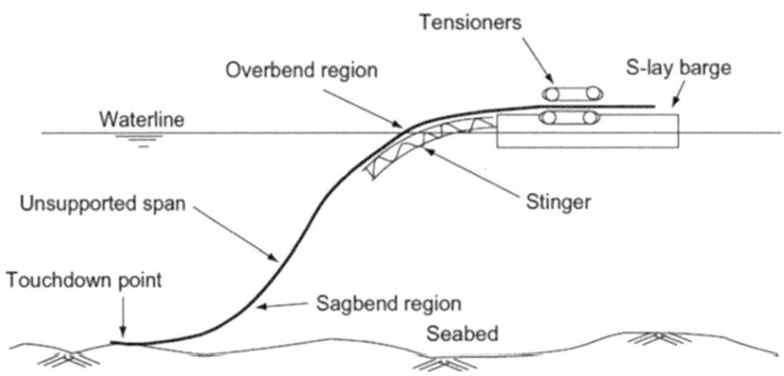

Figure 1. S-lay pipeline installation method. Source: Reproduced from Guo, B., Song, S., Ghalambor, A., Ranlin, T., 2014. Offshore Pipelines: Design, Installation, and Maintenance, second ed. Gulf Professional Publishing, Houston, TX.

tension devices (tensioners) serve to control the pipe in the sagbend.

The shape of the sagbend depends on the interaction between the applied tension and submerged weight of the pipe. If the applied tension is increased, the curvature in the sagbend decreases, and the sagbend becomes longer and flatter and the lift point move up the stinger. Conversely, if the applied tension is reduced, the sagbend curvature increases, and if the bending becomes excessive, the pipe may buckle. Normal installation practice limits the allowable stresses in the overbend and sagbend to about 0.20% strain in the pipe, and from this, the required stinger configuration and pipe tension are calculated (Palmer and King, 2008).

In deepwater, S-lay installations become increasingly constrained by the greater weights and tension requirements. Laying deeper implies getting stronger, since increasing water depth results in an increase in the pipeline spanning the water column, requiring increased holding tensions. Stinger length could be increased, but long stingers are vulnerable to wave and current forces. High tension is undesirable because of damage to the pipe coating and greater power requirements to maintain position. Steep angles are required with greater water depths.

J-lay system. In J-lay, the pipe is assembled on a vertical tower and the pipeline enters the water at a vertical or near vertical angle, hanging like a cable and gently curving toward the horizontal as it approaches the seabed (Figure 2). Tensions are significantly reduced, spans are shorter, and touchdown points are not as far behind the barge, which are easier to monitor and position (Springmann and Hebert, 1994). However, because the ramp is high above the sea level, the line-up, welding, field joint application, and other activities are constrained, reducing the weld and lay rate (Faldini et al., 2014; McDonald et al., 1998). Typical lay rates are 2–3 km/day (1–2 mi/day).

J-lay systems are best suited for ultra-deep (>5000 ft) pipeline installation, and, because it has the smallest bottom tension of all methods, it has high flexibility for route layout in congested areas. J-lay systems can typically handle in-line appurtenances with relative ease but may not be feasible in shallow water.

Reel-lay method. The reel method was first used to lay pipeline across the English Channel during World War II, and applied in the offshore oil and gas industry in the 1960s in the Gulf of Mexico when Gurtler Hubert converted a landing-ship hull into a reel barge and laid 6-inch pipe from a vertical axis, horizontal reel (Timmermans, 2000).

Pipeline is reeled off by passing over an aligner and then through the straightener system located on the vessel ramp or tower (Figure 3). The pipe is reverse bent in the straightener system to remove the residual curvature from the reel. It then passes through tensioners which grip the pipe and apply an upwards force to maintain the catenary of the pipeline as it passes down from the vessel to the seabed.

Reel-lays are well suited for small diameter lines (<18 inches) and small diameter-to-thickness ratios, and, if the entire pipeline can be stored on-board, a very fast installation campaign can be achieved. If recharging trips to the spooling base are needed to reload, lay rates will be reduced, and so the capacity of the reel is critical. Concrete coated pipes cannot be reel-laid but pipe-in-pipe (PIP) systems can be reel-laid.

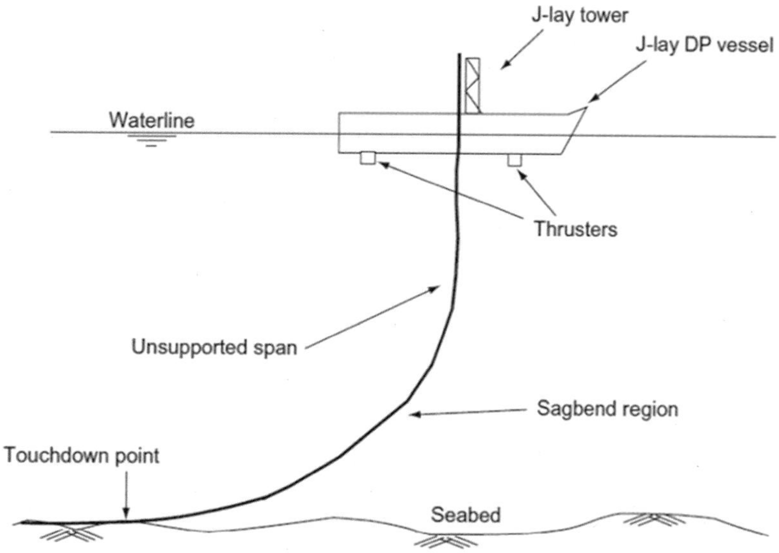

Figure 2. J-lay installation process. Source: Reproduced from Guo, B., Song, S., Ghalambor, A., Ranlin, T., 2014. Offshore Pipelines: Design, Installation, and Maintenance, second ed. Gulf Professional Publishing, Houston, TX.

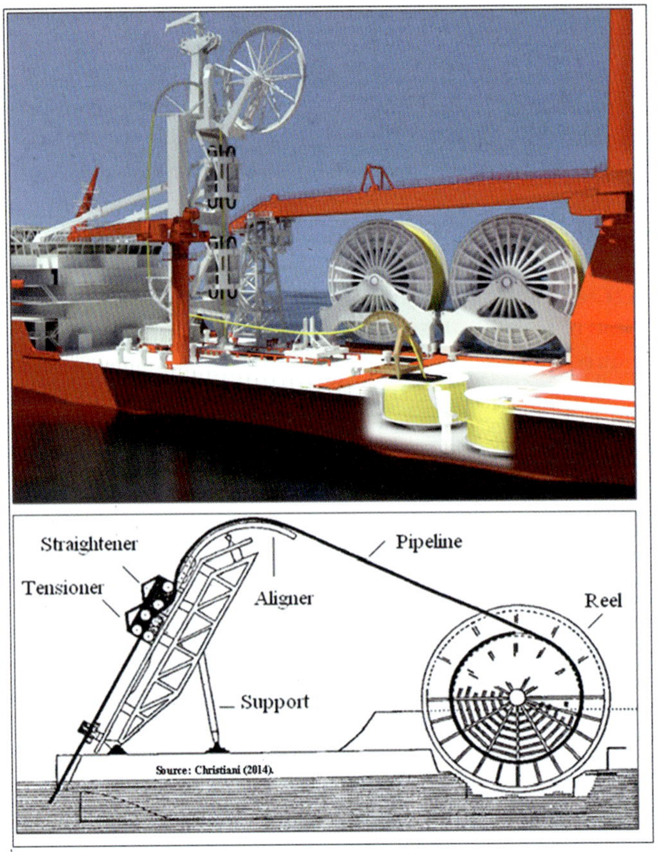

Figure 3. Schematic of a reel-lay system. Source: Adapted from Christiani, E., 2014. Ultra deepwater reel and flexible pipelay vessel for pre-salt field developments. In: OTC 25288. Offshore Technology Conference, Houston, Texas, May 5–8.

Offshore pipeline construction cost factors

Material factors

Material. Offshore pipelines have many similarities with onshore pipelines, but in water pipelines are exposed to hydrostatic pressures and in deepwater reservoir pressures are typically greater than onshore reservoirs. The nature of the installation process also requires pipelines to be manufactured to higher specifications than onshore line pipe.

Length. Short pipelines are expected to be more expensive to install on a per-mile basis, all else being equal, due in part to the costs associated with vessel spread mobilization because costs are allocated over fewer miles. All other factors are rarely equal, however, and there may be project-specific learning effects that increase the rate of installation as a project proceeds. As the length of a pipeline increases, technical complexity may also increase, due to increasing numbers of subsea crossings, the need to connect with or provide future connection points with other pipelines, geological hazards, and greater weather exposure.

Wall thickness. Design engineers determine wall thickness based on the outer diameter of the pipe, the specified minimum yield strength of the steel, the pipeline design operating pressure, external hydrostatic pressure from the water depth, and the need for enough pipeline weight for stability on the seabed (Nogueira and McKeehan, 2005; Palmer and King, 2008). Wall thickness must be adequate to prevent internal pressure containment (burst) during operation and hydrotest, collapse due to external pressure, local buckling due to bending and external pressure, and buckle propagation and its arrest. Typically, the wall thickness will range from 4% to 6% of the outside diameter of the pipe; for an 18-in (457 mm) line, for example, wall thickness usually ranges from 18 to 27 mm.

Grade. Pipelines for offshore service are normally specified as PSL 2, and operators may specify additional requirements negotiated with the manufacturer. PSL 2 pipe has a maximum carbon, phosphorus, and sulfur content of 0.22%, 0.025%, and 0.015% for all welded pipe, and specifies

Figure 4. Simple steel pipe, flexible pipe, and pipe-in-pipe.

a maximum value for carbon equivalent, yield strength, and ultimate yield strength for each grade (Chapter 22). Fracture toughness is also required for all grades. Repair by welding of pipe body, plate, and skelp is prohibited, and repair of welding of weld seams without metal fill is also prohibited.

Type. Offshore pipelines may be rigid steel, flexible line, or PIP systems (Figure 4). All types are used in offshore applications, but rigid steel and flexible lines are by far the most common in terms of miles laid.

Traditional pipe is composed of a simple steel tube wrapped in an anticorrosion coating and possibly covered with a polymer concrete. Traditional pipe is used almost exclusively onshore and is also common offshore.

Flexible pipe is often used for smaller diameter, short distance flowlines, as jumpers from marine wellheads and well manifolds to rigid flowlines, and as risers. The main characteristic of flexible pipe is its ability to bend which is achieved by using several layers of different materials in fabrication. Steel armor layers with high stiffness provide strength and polymer layers with low stiffness provide fluid integrity. These layers can slip past each other when under the influence of external and internal loads and cause pipe failure.

PIP consists of an insulated and sometimes electrically heated pipe placed inside of an outer pipe. Prices for flexible pipes and PIP are significantly greater than prices for conventional pipes.

Weight coating. Pipelines are frequently externally coated with a layer of concrete up to 150 mm thick to add weight to the pipeline to ensure stability on the seabed and protect the line from mechanical damage (Hillenbrand et al., 2003). The concrete weight coating may be applied at the pipe mill or a regional facility. Flexible pipes and PIP are not concrete coated.

The cost of concrete weight coating may add between 15% and 20% to the pipeline cost (e.g., $18/ft for a $60/ft steel pipeline). Weight coatings in the U.S. Gulf of Mexico have historically ranged from $15 to $25/ft (Kaiser, 2017a). Weight coatings are not required on thick-walled or small diameter pipes as these pipes are heavy enough to remain stable on the bottom without additional weight. For thin walled and large diameter pipes, concrete weight coatings are used because they are less expensive than adding to the pipe wall thickness. In deepwater, heavier wall pipe is commonly utilized in lieu of concrete coating so that tie-ins or repairs not accessible by divers do not have to contend with concrete removal. Burial will reduce the hydrodynamic forces acting on the pipelines and reduce concrete coating required for stability, but the tradeoff is the added cost of trenching.

Corrosion coating. Almost all offshore pipelines use some type of factory-applied coating as the primary means of protecting the pipeline from corrosion. Steel pipelines are coated on their outside surfaces to prevent corrosion and protect the pipe from abrasion damage. Pipeline interiors are also sometimes coated for similar purposes and to provide a smooth surface to flowing fluids. Coatings are selected based on price, the expected temperature and composition of the fluid, seawater composition and temperature, and whether a concrete weight coating is required (Roche and Total, 2006a). There are several different coating systems used, but for offshore applications, three-layer polypropylene (3LPP) is one of the most common. 3LPP system consists of an epoxy primer, and a grafted copolymer PP adhesive to bond the epoxy primer with a topcoat.

The principal coatings in approximate order of cost are asphalt, coal tar enamel, FBE, cigarette wrap PE, extruded thermoplastic PE and PP, and elastomeric coatings (EPDM). The cost ratio between coal tar, FBE, PE/PP, and EPDM has been reported as 1:1.5:2:4 (Palmer and King, 2008) but will vary on a regional basis and with changing market conditions. FBE is the most commonly used coating for reeled pipelines and for bundled pipes, and EPDM coatings are used for high temperature pipelines and risers. EPDM requires autoclaving to vulcanize the elastomeric coatings and account for its high cost. Epoxy coating costs in the Gulf of Mexico historically have ranged from $4 to $12/ft.

Cathodic protection. Offshore pipelines are almost always protected by sacrificial anodes attached to the pipe at regular spacing and electrically connected with welded or brazed cables. As the anode corrodes, it provides its electrons to the pipe. Water temperature is generally the most important factor in design since it is correlated to the concentration of dissolved oxygen. Other factors utilized in the design of cathodic protection systems include the water flow rate, fluid temperature, coatings, and design standards. Anodes are typically zinc and manganese component attachments and range in cost from $100 to $500 per anode depending on metal market conditions. Anode costs depend on the pipeline diameter and water depth; smaller diameter pipelines in shallow water require smaller anode attachments. Each anode costs at least as much as the metal weight plus the cost to manufacture and deliver to an onshore base.

Other equipment. PLETs and PLEMs, risers, hot tap assemblies, buckle arrestors, swivels, balls, valves, pig launchers, fittings, anodes, and meters are minor contributors to cost, and collectively do not usually account for more than 10% of material costs. Each piece of equipment costs at least as much as steel weight plus the cost to manufacture and deliver to an onshore base.

Construction factors

Contract terms. The cost to lay offshore pipeline is approximately proportional to the total time on the job, including the time required for laying pipe, installation of risers, subsea assemblies, manifolds, and related components. The time required for these operations depend upon the overall complexity of the project, route length, weather conditions, and the success of the operations. Pipeline projects are typically contracted on a lump-sum basis to the low-cost bidder, and thus market conditions and competition are key elements of determining price. Weather risk is normally assumed by the contractor. If contractors do not add an allowance to win the work and bad weather or unforeseen problems occur, the job may not be profitable. The risk/reward contract structure and the degree of competition in the market at the time the contract is let will impact how the contractor bids (Kaiser, 2018).

Vessel day rates. Vessel day rates are the most important component of construction cost. Day rates vary depending on market conditions and vessel specifications, and because spot markets for pipelay vessels do not exist, it is necessary for the engineer performing the estimate to prognosticate expected values based on prior experience and knowledge of market conditions. In general, small conventional barges have the lowest day rates and the slowest production rate, while high-specification vessels have the largest day rates and the fastest installation rates. Additional vessels that are required for pipelay operations include dive support vessels, supply barges, utility vessels, and tugboats. The costs for these services are considerably less than the pipelay vessel and are often provided by the contractor or subcontracted out depending on the contractor's vessel fleet. Day rates are expected to be higher in periods of high demand and in regions with low levels of competition, and day rates for high-specification vessels are greater than day rates for lower specification vessels. High-specification vessels are required for deeper water or large diameter or particularly heavy pipes.

Mobilization and demobilization. Mobilization cost for the pipelay vessel is a function of the distance the vessel must travel to reach the worksite and drop anchor, the speed of the transport system and mob dayrate, all of which are unknown at the time of the estimate. In the Gulf of Mexico, it is common practice to assume one day to mobilize and one day to demobilize to site for vessels located in the region. Low-spec barges do not typically leave the regions in which they reside, but high-spec vessels are highly mobile and operate in all the world's offshore basins and may be scheduled years in advance. For deepwater projects, mob/demob cost may be a significant portion of the construction cost.

Transition zones. Pipelines that transition from deepwater to shallow water or continue to shore may require two installation vessels which will increase costs due to the extra mobilization and demobilization, interface management, and increased operational complexity. Landfalls increase the complexity of a project due to different environmental impacts and regulations and the use of different installation methods (Howitt et al., 2012).

Geohazards. Examples of geohazards that may be encountered in the deepwater Gulf of Mexico include steep slopes, unstable slopes, hard seabed sediments, bedforms, and scarps. Often, these issues are superimposed upon and interrelated to each other (Hill et al., 2013; Hill and Wood, 2015; Kaiser, 2017b). Integrated geohazards analysis evaluates the possibilities to allow direct avoidance through choice of well or field layout, or by an appropriate mitigation of the desired hazard through engineering design.

Infrastructure located near or on peaks and canyons is at risk if the slope fails during the lifetime of the development. The mechanisms capable of triggering a slope failure are related to changes in the shear stress applied to the soil (Jeanjean et al., 2003). For slope stability issues, the annual probability of occurrence of a failure is commonly estimated, along with the consequences in terms of the debris flows and damage to infrastructure.

Furrows are longitudinal bedforms formed in cohesive sediments that tend to be aligned parallel to bottom currents. Extensive regions of the seafloor in and around the base of the Sigsbee Escarpment are furrowed, which implies high bottom currents in the area (Clukey et al., 2007). Bottom currents impact pipeline routing and the orientation of risers.

Risers are typically oriented in the same direction as the current to increase fatigue life. Some of the furrows can reach 200 ft peak to peak and be up to 30 ft deep, but typical depths vary between 1 and 10 m and widths from 5 to 50 m. Flowlines across furrow fields are often straked to reduce interaction with the bottom current. Potential pipeline hazards associated with deep-sea furrow fields include unsupported pipeline spans, pipeline shoulder scour, vortex induced vibrations, and large-scale effects which may cause lateral instability and pipeline movement (Mannaerts et al., 2005).

Seafloor faults and scarps create risks for pipeline infrastructure from potential offset at fault crossings. To understand the causative processes and quantitatively evaluate the locations, magnitude, and recurrence potential of displacement events, geologic and geotechnical survey data are used in integrated models (Angell et al., 2003).

Lay method. Factors that impact the rate of lay and burial include the type and size of the pipeline, water depth, and the method of installation. Reeled pipes can be laid faster than welded line, and small diameter pipe can be welded faster than larger diameter pipe, leading to lower costs for all other things equal. Differences in the productivity of S-lay and J-lay methods lead to differences in cost. Generally, the capacity of the barge determines the inventory of pipe that can be held and restocking frequency. The number of welding stations determines lay rates.

Burial. Pipeline burial adds cost to the installation process because of the need for route selection and extra equipment to perform burial operations. In deepwater, operators may bury pipe to increase heat retention for flow assurance, and in shallow water may be required by regulation to bury pipe. In the U.S. Gulf of Mexico, pipelines greater than 8 and 5/8 inches and installed in water depths less than 200 feet are required to be buried at least three feet below the mudline. Burial cost may account for a significant portion of the total installation cost. For example, in the Mars export line in the Gulf of Mexico, burial costs were \$16/ft relative to a lay cost of \$56/ft. In the Garden Banks line, burial costs were about half of the lay cost, and in the Nautilus pipeline, burial costs were \$10/ft relative to lay costs of \$35/ft (Kaiser, 2017).

Pipeline crossings. In developed basins, it is often not possible to design a route that avoids all existing pipelines. Installing a new pipeline over an existing pipeline adds costs because the two lines must not touch and installation rates are reduced in sensitive areas. Existing pipeline will either be lowered into the seabed, or a bridge-like structure must be constructed over the existing line. In the Gulf of Mexico, federal regulations require 18-inch separation between the pipelines and in water depths less than 500 ft, the crossed pipelines must be covered by mats from touchdown to touchdown. Typically, two 9-inch thick articulated concrete mats are used for protecting crossed pipeline. Pipeline crossings may add an additional \$100,000 to \$200,000 per crossing.

Field joint coating. Field joint coating refers to the coating placed on the joints formed offshore when two sections of pipe are welded together. Typically, the weld area is wrapped with a corrosion-resistant polymer sleeve and a mold is placed over the weld area. The mold is then filled with a polyurethane and allowed to set. In reporting, field joint coating costs are included in labor costs, but the materials are often procured and allocated separately. Field joint coating is frequently the rate limiting step in the construction sequence and costs are determined by the type of anticorrosion coating employed and pipe diameter.

Riser and tie-in work. Risers add expense to projects, especially in deepwater where they can be technically complex. The type of riser and its associated installation technique are the primary factors influencing costs. Tie-in work is technically demanding and expensive because it involves the use of divers and/or ROVs. A variety of pipeline tie-in structures are utilized in construction including PLETs, PLEMs, and inline tees, and the offshore work and cost depends on the system architecture.

Survey. Survey costs include both preinstallation surveys and postinstallation surveys. Survey costs may be specified on a footage or lump-sum basis and are performed by utility or specialized vessels at day rates much smaller than pipelay vessels.

Commissioning and hydrotesting. Hydrostatic testing is the final quality check before the pipeline is placed in service, and any leak indications require finding and repairing the leak, which can be time consuming and costly. To ensure the pipeline can be operated safely at its MAOP, the pipeline is tested at a higher internal pressure prior to operation.

Design and hydrotest requirements are specified in 30 CFR 250.1003 and require testing with water at a stabilized pressure of at least 1.25 times the MAOP for at least 8 h when installed, relocated, uprated, or reactivated after being out of service for more than one year. A utility boat and dive boat spread may be needed to perform the operations, or pump equipment may be available on the platform. In shallow water pipeline designs, parameters such as product gradient, water depth, hydrostatic pressure, and hydrotest location are well understood and accepted (Mappus and Torstrick, 2007). In deepwater, pipeline commissioning and hydrotesting are significantly more complex and challenging than shallow water.

Engineering and inspection factors

Project management and engineering. Project management and engineering costs scale with the size and complexity of the project. Factors that add to the complexity and engineering work of a project are many and include subsea pipeline crossings, the installation of multiple PLETs or PLEMs, tie-ins to existing infrastructure, deepwater, and difficult terrain.

Inspection. Final inspections are performed to confirm that operator construction standards and specifications according to the contract were followed. ROV video inspection is often performed simultaneously with the pipelay operation and may be specified in the contract, but because of mechanical difficulties, may not be conducted during pipelay. If the contractor performs the final inspection after the lines are on the seafloor costs will be greater if problems are found.

Repair. Diving services are typically required to repair damage, and the time and cost may be significant. The current state of remote intervention-based pipeline repair technology is still in its infancy but continues to increase beyond diving limits.

References

M.M. Angell, K. Hanson, F.H. Swan, R. Young, H. Abramson, Probabilistic fault displacement hazard assessment for flowlines and export pipelines, Mad Dog and Atlantis field developments, deepwater Gulf of Mexico, in: OTC 15402. Offshore Technology Conference, Houston, TX, May 5–8 (2003).

E. Christiani, Ultra deepwater reel and flexible pipelay vessel for pre-salt field developments, in: OTC 25288. Offshore Technology Conference, Houston, Texas, May 5–8 (2014).

E. Clukey, K. Israel, C. Jones, C.K. Ziegler, Pipeline risk assessment in deep-sea furrow regions, in: OTC 18939. Offshore Technology Conference, Houston, TX, April 30–May 3 (2007).

R. Faldini, S. Marchini, A. Oldani, R. Pellegrini, CastorOne and FDS2: getting stronger for laying deeper, in: OTC 25110. Offshore Technology Conference, Houston, TX, May 5–8 (2014).

B. Guo, S. Song, A. Ghalambor, T. Ranlin, Offshore Pipelines: Design, Installation, and Maintenance, second ed. Gulf Professional Publishing, Houston, TX, 2014.

W.C. Haneberg, B. Bruce, M.C. Drazba, Using quantitative slope hazard maps and quantitative probabilistic slope stability models to constrain least-cost pipeline route optimization, in: OTC 23980. Offshore Technology Conference, Houston, TX, May 6–9 (2013).

A.W. Hill, A.H. Walls, L.A. Romo, J.B. Turnbull, L. Rogers, Past, present and future marine geohazard issues: developing processes to address developing industry concerns, in: OTC 24248. Offshore Technology Conference, Houston, TX, May 6–9 (2013).

A.W. Hill, G.A. Wood, Managing marine geohazard risks over the full business cycle, in: SPE/IADC 173139. SPE/IADC Drilling Conference and Exhibition, London, UK, March 17–19 (2015).

H. Hillenbrand, C. Norris, F. Hafner, The Gulfstream project – meeting the pipe supply challenge, in: 14th Joint Technical Meeting on Pipeline Research, Berlin, Germany, May 19–23 (2003).

L. Howitt, M. Lincoln, S.C. Blount, P. Branson, Coastal shoreline crossing using horizontal directional drilling for pipeline installation achieves excellent environmental and social outcomes, in: SPE 156643. SPE/APPEA International Conference on Health, Safety, and Environment in Oil and Gas Exploration and Production. Perth, Australia, September 11–13 (2012).

P. Jeanjean, A. Hill, S. Taylor, The challenges of siting facilities along the Sigsbee Escarpment in the southern Green Canyon area of the Gulf of Mexico: framework for integrated studies, in: OTC 15156. Offshore Technology Conference, Houston, TX, May 5–8 (2003).

M.J. Kaiser, U.S. Gulf of Mexico deepwater pipeline construction – a review of lessons learned, Marine Policy 86 (12) (2017) 214–233.

M.J. Kaiser, Global offshore pipeline construction service market 2017 – part I, Ships and Offshore Structures 13 (1) (2018) 65–95.

H. Mannaerts, J. Oldroyd, M. Williams, R. Fisher, E. Radjef, R. Day, et al., Atlantis, the quest for its stray shallow water flow sands, in: OTC 17658. Offshore Technology Conference, Houston, TX, May 2–5 (2005).

B.E. Mappus, A.G. Torstrick, Pipeline hydrotesting, dewatering, and commissioning, in: OTC 19062. Offshore Technology Conference, Houston, TX, April 30–May 3 (2007).

D.N. McDonald, D.J. Sullivan, H.S. Choi, The next generation J-lay system, in: SPE 39848. International Petroleum Conference and Exhibition of Mexico, Villahermosa, Mexico, March 3–5 (1998).

A.C. Nogueira, D.S. Mckeehan, Design and construction of offshore pipeline. in: S. Chakrabrarti (Ed.), Handbook of Offshore Engineering. Elsevier, Kidlington, Oxford, 2005, pp. 891–937.

A.C. Palmer, R.A. King, Subsea Pipeline Engineering. PennWell, Tulsa, OK, 2008.

M. Roche, S.A. Total, External corrosion of pipeline: what risk?, in: SPE 93600. SPE Middle East Oil and Gas Show and Conference, Bahrain, March 12–15 (2006a).

S.P. Springmann, C.L. Hebert, Deepwater pipelaying operations and techniques utilizing J-lay methods, in: OTC 7559. Offshore Technology Conference, Houston, TX, May 2–5 (1994).

W.J. Timmermans, The past and future of offshore pipelines. Offshore Pipeline Technology Conference, Oslo, Norway, 2000, May 4–6.

N.P. Tootill, M.P. Vandenbossche, M.L. Morrison, Advances in deepwater pipeline route selection – a Gulf of Mexico case study, in: OTC 16633. Offshore Technology Conference, Houston, TX, May 3–6 (2004).

Source

Kaiser, M.J., 2020. The Offshore Pipeline Construction Industry – Activity Modeling and Cost Estimation in the U.S. Gulf of Mexico. Gulf Professional Publishing, Cambridge, MA.

U.S. Gulf of Mexico pipeline cost statistics, 1995–2019

Federal Energy Regulatory Commission data

About 20 FERC pipeline projects with detailed costs over the 1995–2019 period and 173 FERC pipelines with summary costs from 1980 to 1995 were examined (Kaiser, 2017). The 20 pipeline projects from 1995 to 2019 cover 1139 miles and include most offshore projects considered or approved by FERC during this time in the U.S. Gulf of Mexico. About 11 of the 20 projects from 1995 to 2019 were built, and for these projects, final costs are reported.

Aggregate cost. The normalized, inflation-adjusted costs to engineer, procure, and install offshore gas pipelines in the Gulf of Mexico from 1995 to 2019 was approximately $3.31 million/mi with a standard deviation of $2.70 million/mi (Table 1). Normalizing by pipeline diameter yields an average of $136,000/in-mi and a standard deviation of $115,000/in-mi.

Table 1
Average Inflation-Adjusted Costs for Federal Energy Regulatory Commission Offshore Pipelines in the U.S. Gulf of Mexico, 1995–2019 (2014$)

	Excluding zero values ($1000/mi)	Including zero values ($1000/mi)
Right of way	18.2	8.2
Damages	8.9	0.4
Surveys	36.8	29.4
Materials	814.0	814.0
Labor	1866.9	1866.9
Engineering	182.4	182.4
Line pack	13.3	4.0
Taxes	32.3	1.6
Freight	65.3	9.8
Pipe coating	288.3	43.3
Cathodic protection	16.6	1.7
Telecom	9.5	1.4
Overheads	127.9	95.9
AFUDC	154.9	139.4
Contingencies	103.1	46.4
Legal fees	55.4	8.3
Other services	154.6	54.1
Regulatory fees	0.0	0.0
Total	3948.5	3307.3

From 1980 to 1994, the inflation-adjusted cost was $1.40 million/mi with a standard deviation of $1.46 million/mi.

Cost distributions from both periods are skewed right with large standard deviations indicating wide differences between projects (Figure 1). The 1995–2019 data period is more representative of future construction cost than the legacy projects because of the bias associated with the cost adjustment over long periods of time and changes in technology that have occurred as well as the deeper water location of more recent projects.

Labor and contract services. From 1995 to 2019, labor accounted for about half of total costs with materials accounting for between a quarter to a third of total costs (Table 1). Historically, labor and material costs exhibited roughly equal contributions to total cost, but in recent years, labor has played a larger role. Not all firms report data for all cost categories, either because there was no cost in the category or because cost was allocated elsewhere. Some categories are recorded as having zero costs and including these data points as zeros artificially lowers the average cost for the category. Excluding the zero values, the average inflation-adjusted cost for FERC pipelines installed from 1995 to 2019 is $3.95 million/mi.

Materials. Material costs average $814,000 per mile and range from $368,000 to $1.6 million per mile from 1995 to 2019. Material costs are all-inclusive and are closely related to pipeline diameter. Pipe coating ranged from $232,000 to $382,000 per mile with an average of $288,000/mi. Cathodic protection systems cost between $12,000 and $20,000 per mile. Anodes cost between $500 and $1000 each depending on pipeline diameter.

Engineering. Engineering costs ranged broadly from about $19,000 to $709,000 per mile with an average of $182,000/mi and a median of $117,000/mi. Larger more complex projects require greater preparation and planning and exhibit larger staff and engineering expense. Though engineering costs account on average for only 5% of total costs, normalized engineering costs explain a surprisingly large proportion of the variation (about two-thirds), suggesting that other cost factors are linearly related with engineering requirements.

Time trends. The late 1970s and early 1980s was a time of numerous discoveries and intense development activity on the U.S. Gulf of Mexico shelf. During this time, 127 FERC pipelines were installed at an average inflation-adjusted cost of $1.4 million/mi (Table 2). Installation cost declined with

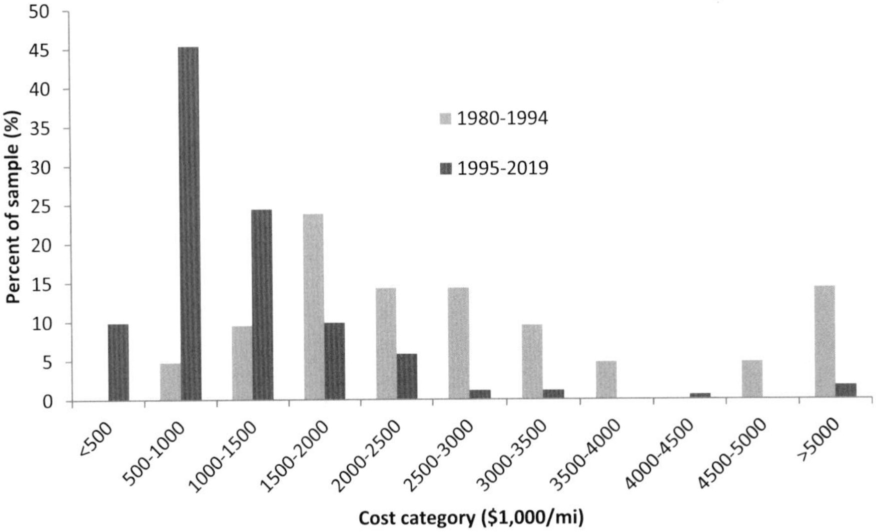

Figure 1. Distribution of normalized inflation-adjusted Federal Energy Regulatory Commission pipeline costs in the U.S. Gulf of Mexico, 1980–1994 and 1995–2019.

Table 2
Federal Energy Regulatory Commission Pipeline Costs in the U.S. Gulf of Mexico by Period ($1000/mi, 2014$)

Period	Sample	Average	SD	Max	Min
1980–84	127	1286	1180	9376	118
1985–89	40	899	546	2973	265
1990–94	6	777	296	1255	454
1995–99	11	1959	786	3914	1086
2000–04	4	3224	1135	4776	2130
2005–09	4	7094	4823	12,185	2595
2010–19	1	3321	NA	3320	3320
All	193	1399	1466	12,185	118

Note: SD = Standard deviation

the decline in oil prices over the 1985–89 and 1990–94 periods, and the number of FERC pipelines has declined significantly with the maturation and production decline in the shallow water region (Figure 2). In recent years, the number of new FERC pipelines has averaged less than one per year.

Larger diameter lines are expected to be more expensive than smaller diameter lines due to the increased material costs and the longer time required for welding, and when category counts are statistically meaningful, these trends are generally observed (Table 3). Small diameter lines were common in the 1980–84 period, but in most other periods, medium diameter lines predominated. Short (<10 mi) pipelines appear slightly more expensive per mile than medium length (10–50 mi) pipelines, and pipelines greater than 50 mi are the most expensive on a unit basis (Table 4).

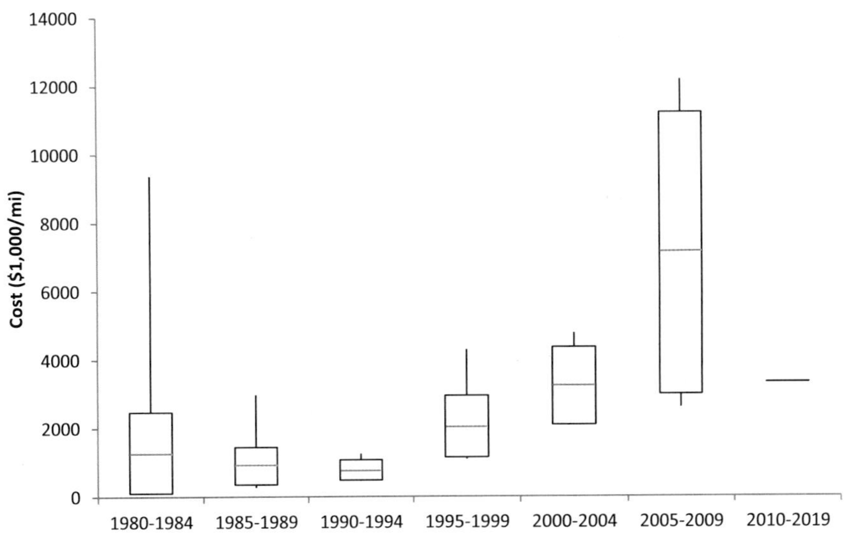

Figure 2. Inflation-adjusted Federal Energy Regulatory Commission pipeline costs in the U.S. Gulf of Mexico. Source: FERC (1995–2019), Oil and Gas Journal (1980–1994).

Table 3
Federal Energy Regulatory Commission Pipeline Costs in the U.S. Gulf of Mexico by Diameter ($1000/mi, 2014$)

Period	<14 in		14–24 in		>24 in	
	Sample	Average	Sample	Average	Sample	Average
1980–84	70	1293	51	1272	6	1318
1985–89	12	531	23	940	3	1679
1990–94	1	631	4	768	1	961
1995–99	0		7	1746	4	2333
2000–04	0		3	3392	1	2719
2005–09	0		3	8594	1	2595
2010–19	0		0		1	3321
All	83	1175	91	1514	17	1875

Table 4
Federal Energy Regulatory Commission Pipeline Costs in the U.S. Gulf of Mexico by Length ($1000/mi, 2014$)

Period	<10 mi		10–50 mi		>50 mi	
	Sample	Average	Sample	Average	Sample	Average
1980–84	86	1480	40	866	1	1364
1985–89	17	995	21	777	2	785
1990–94	2	943	4	694	0	
1995–99	0		5	1615	6	2246
2000–04	1	2130	2	4024	1	2719
2005–09	1	8293	3	6694	0	
2010–19	0		1	3321	0	
All	107	1467	76	1227	10	1913

The standard deviation of unit costs as a percentage of the average has generally decreased over time indicating greater commonality in project execution, and perhaps, greater levels of competition.

Offshore Technology Conference and Society of Petroleum Engineers data

OTC/SPE publications provide cost data for 20 projects in the Gulf of Mexico from 1979 to 2019 (Kaiser, 2017). These projects include 10 oil and gas export pipelines, 13 infield flowlines, and two infield/export systems. All the projects except Mobile Bay are in deepwater. Most installations were constructed from 1990 to 2005 from 1000 to 3000 ft water depth, with three projects reported before 1990 and five projects after 2005. Independence Hub represents the deepest pipeline project in the sample at about 8000 ft, Mardi Gras is the longest line at 489 miles, and Morpeth is the shortest line at about two miles. These data represent most pipeline cost data reported in OTC/SPE publications.

The average aggregate all-in cost for OTC/SPE pipeline installation was $2.8 million/mi and ranged from $520,000/mi to $12.9 million/mi. High cost construction tends to be short segments, deepwater, and specialized pipeline. Excluding the two cost endpoints, most costs ranged from $1 to $6 million/mi. Export and export/infield projects cost about $2 million/mi and infield flowline projects cost about $3.5 million/mi (Table 5). Standard deviations are on the order of the mean, indicating significant variation within each category. Infield flowlines cost almost twice as much as export pipelines, with cost distribution approximately lognormal (Figure 3).

Table 5
Average EPCI Contract Costs by Project Type in the U.S. Gulf of Mexico ($million/mi, 2014$)

	OTC/SPE	Press release	All
Project type			
Export	1.96 (1.17)	3.79 (2.66)	2.67 (2.02)
Infield	3.50 (3.17)	3.84 (3.53)	3.61 (3.19)
Export/Infield	1.98 (NA)	2.80 (1.82)	2.53 (1.37)
All	2.76 (2.49)	3.68 (2.80)	3.11 (2.61)
Sample			
1979–89	3	0	3
1990–2005	17	6	23
2005–2019	5	10	15

Note: Standard deviation denoted in parenthesis.

Press release data

Press release data for 16 Gulf of Mexico projects from 1998 to 2019 were also assembled (Kaiser, 2017). There was no significant overlap with OTC/SPE project data. Cameron Highway was the longest pipeline in the sample, at 390 miles, and several projects exceeded 100 miles. The shortest project was less than 10 miles, and, along with systems in greater water depths, tended to characterize high unit cost construction. Both export and infield networks were represented, and there were two combined export/infield systems. All projects are deepwater and used a breadth of marine contractors, including Allseas, EMAS, McDermott, Subsea 7, and Technip, which were the sources of the data.

The average aggregate cost for press release data was $3.7 million/mi with a range of $0.8 to $10.0 million/mi (Table 5). Export pipeline and infield flowline system cost were reported at about $3.8 million/mi with a standard deviation of the same order of magnitude. Export project costs were greater in part due to a greater number of more recent projects reviewed. Like OTC/SPE projects, the distribution of the press release cost data was approximately lognormal (Figure 3).

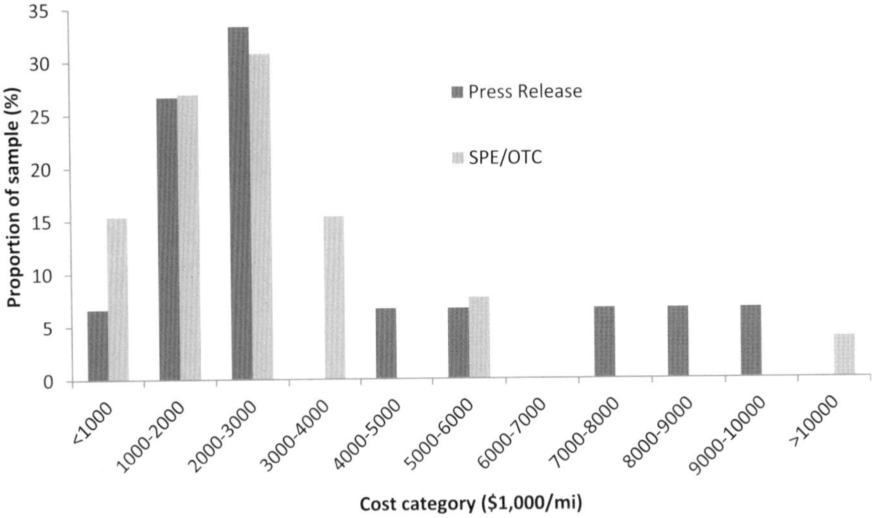

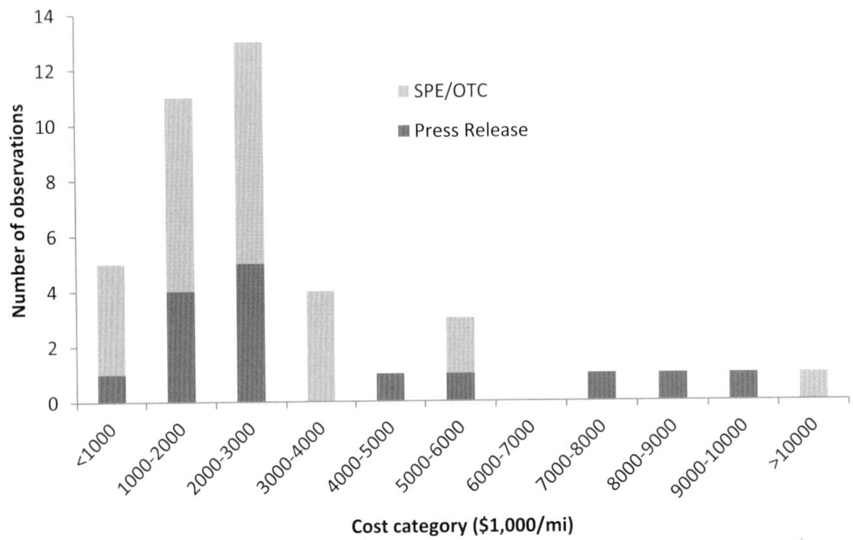

Figure 3. Inflation-adjusted Offshore Technology Conference/Society of Petroleum Engineers and press release pipeline cost distribution in the U.S. Gulf of Mexico (2014$).

Comparison

The OTC/SPE and press release data were combined by project type and in aggregate to form a composite data sample. The average export pipeline installation cost in the composite analysis was $2.7 million/mi, the infield system cost $3.6 million/mi, and the aggregate all-in pipeline cost was $3.1 million/mi. FERC pipeline cost is exclusively export pipeline and so the relevant comparison with OTC/SPE and press release data is with export systems. FERC pipeline inflation-adjusted cost was $3.3 million/mi versus $2.7 million/mi for industry data. Both cost distributions are approximately lognormal (Figure 4).

Limitations

Project data occur over a 20-year time period of technological improvement and market changes. A long period is needed to provide adequate sample sizes, but over long evaluation periods, inflation adjustment is required which introduces uncertainty in the assessment because the BLS indices are not specifically designed for the offshore Gulf of Mexico, nor do other offshore indicators necessarily provide a more reliable gauge of inflation levels. The most specific BLS index available was used to adjust costs, but an argument could also be made for using hybrid indices.

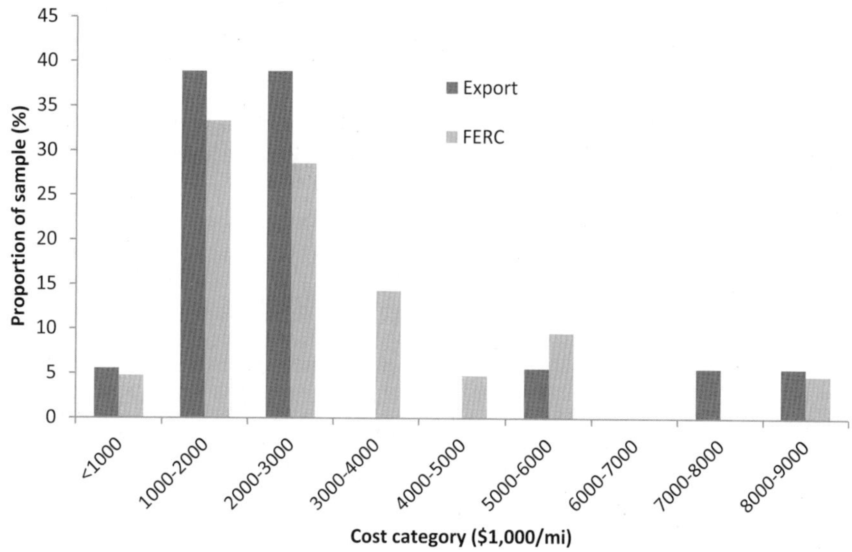

Figure 4. Federal Energy Regulatory Commission pipeline cost versus the composite Offshore Technology Conference/Society of Petroleum Engineers and press release export cost data in the U.S. Gulf of Mexico.

There are several unobserved variables of varying importance that were not included in evaluation. Contract structure, and the manner risk is apportioned between the contractor and oil and gas company, is one of the most important factors not considered. Risk allocation can have important impacts on project costs, but these effects are not observed in the data or captured in the statistics. Firms that adopt lump-sum contracts price risk within the contract, while day rate contracts may yield more favorable economics for resource owners if interfaces are properly managed and projects do not encounter unexpected problems or weather.

Reference

M.J. Kaiser, Offshore pipeline construction cost in the U.S. Gulf of Mexico, Marine Policy 82 (10) (2017) 147–166.

How to estimate shallow water pipeline construction cost bounds in the U.S. Gulf of Mexico using empirical data?

Example

Estimate the expected cost of a 70 mile, 24-inch export pipeline and a 55 mile, 12-inch diameter infield flowline installed in the U.S. Gulf of Mexico in 2022. Describe cost bounds of the estimates.

Solution

Historic inflation-adjusted cost for export pipeline range from \$2.7 million/mi to \$3.3 million/mi from FERC and industry data sources in 2014 dollars. Using these unit cost statistics, the cost of a 70 mi line is estimated to range between \$189 to \$231 million in 2014 dollars. Use indices to inflation-adjust results to the current year.

Historic inflation-adjusted cost for infield pipeline range from \$3.5 million/mi to \$3.8 million/mi for OTC/SPE and press release data sources in 2014 dollars. Using these unit cost statistics, the cost of 55 miles of flowline is estimated to range between \$193 to \$209 million. Use indices to inflation-adjust results to the current year.

Offshore pipeline construction cost estimation using work decomposition

Two offshore pipeline construction cost examples in the U.S. Gulf of Mexico illustrate work breakdown structures and estimation procedures and are compared with actual cost. The data for the Nautilus pipeline example are from FERC Docket CP96-790, and the data for the Discovery pipeline example are from FERC Docket CP96-712.

Work breakdown structure

Work breakdown structures represent the most common bottom-up approaches used in cost estimation. Work decomposition unite a project's scope, cost and schedule management processes, and ensures that all major facets of a project are considered. Engineers and project managers normally describe the project scope by system, and in terms of components, activities, and subactivities:

- Project (Level 0)
  - System (Level 1)
    - Component (Level 2)
      - Activity (Level 3)
        - Subactivity (Level 4)

Level 1 identifies the main stages, and in Level 2, the components are decomposed into subcategories. In Level 3, the functional requirements and operations, durations, and costs associated with each activity are identified. Additional subcategories and levels are specified if necessary.

Activity decomposition captures the primary activities of a project in an organized way by breaking the activities into progressively smaller pieces and estimating the work duration and costs at the level selected. Task identification and estimates are based on knowledge of the project requirements and require experienced engineers and project managers to perform.

Cost components

Material

Material costs include all the facilities and equipment to construct the pipeline and ancillaries, and include the pipeline, pipeline coatings (e.g., thin-film epoxy, concrete, internal epoxy), anodes, subsea valve assemblies, buckle arrestors, PLEMs, PLETs, pig launchers, platform piping, meters, and risers. Inspection of the material, transportation cost, and related miscellaneous cost may also be included in the material cost category.

Material costs are typically estimated on a per foot or unit basis. The number of feet of pipe of each wall thickness is specified and multiplied by an estimated cost per foot based on historical data or industry quotes. The costs of epoxy coating and concrete weight coatings are similarly estimated. The spacing of anodes is estimated based on the environmental conditions and service life of the pipeline. Anodes, buckle arrestors, PLEMs, PLETs, pig launchers, fittings, tees, risers, assemblies, and subsea valves are estimated on a unit basis using historic costs or industry quotes.

Construction

Construction costs include all the vessels, equipment, and labor to perform the installation, including mob/demob cost of the lay vessel and spread, field joint coating, riser work, subsea crossings, hazards survey, hydrotesting and dewatering, installing subsea assemblies, tie-in work, and commissioning. Engineering and construction may be broken out separately or portions may be included within the construction category.

Construction costs are estimated on a footage, time, or job basis. The costs to mobilize and demobilize the vessel spread are estimated as one-time costs unless the work is expected to be performed over multiple work campaigns. The costs to lay and, if necessary, bury the pipe are typically estimated on a per foot basis. The number of pipeline crossings are estimated, and costs are specified per pipeline crossing. Historic or current dayrates for pipelay and dive boats, utility boats, and other vessel spreads are estimated along with the expected duration of operation. Hazards survey and permitting and inspection are usually estimated by foot.

Cost estimates are performed by multiplying the estimated production (lay) rate by the estimated dayrates. Production rate is estimated by historic data or is empirically determined using seasonal curves based on a theoretical rate, weather efficiency, and crew efficiency. In general, small conventional barges have the lowest dayrates and the slowest production rate. They are more limited in their capability compared to high-specification vessels which have the most flexibility in operations and the fastest installation rates but at the highest dayrates. Production rates are uncertain, and at the time of the evaluation, dayrates are based on market conditions; the accuracy of cost estimation depends on the experience of the estimator and their knowledge of the job requirements and market conditions.

Engineering and inspection

Engineering and inspection include the portion of the pay and expenses of engineers, surveyors, draftsmen, inspectors, superintendents, and their assistants applicable to construction work. Engineering and project management costs are estimated on an hourly or monthly basis. The time required for engineering and management tasks is estimated and multiplied by the hourly or monthly wage of engineers and managers.

Nautilus

The Nautilus pipeline system is a 56-mile long, 30-inch diameter, 0.577-inch thick pipeline from Ship Shoal block 207 in approximately 300 ft water depth in the U.S. Gulf of Mexico to an onshore facility in Burns, Louisiana (Figure 1).

The 30-inch OD × 0.577-inch WT X-65 pipe cost $672/ton in 1997 dollars ($59/ft), and the thicker wall 0.875-inch pipe cost $2192/ton or $96/ft and was used sparingly (Figure 2). Epoxy corrosion coating was applied at $5/ft for the 0.577-inch line, and concrete weight coating for the thin pipe cost $19/ft. Anodes costing $441 each were spaced every 615 ft,

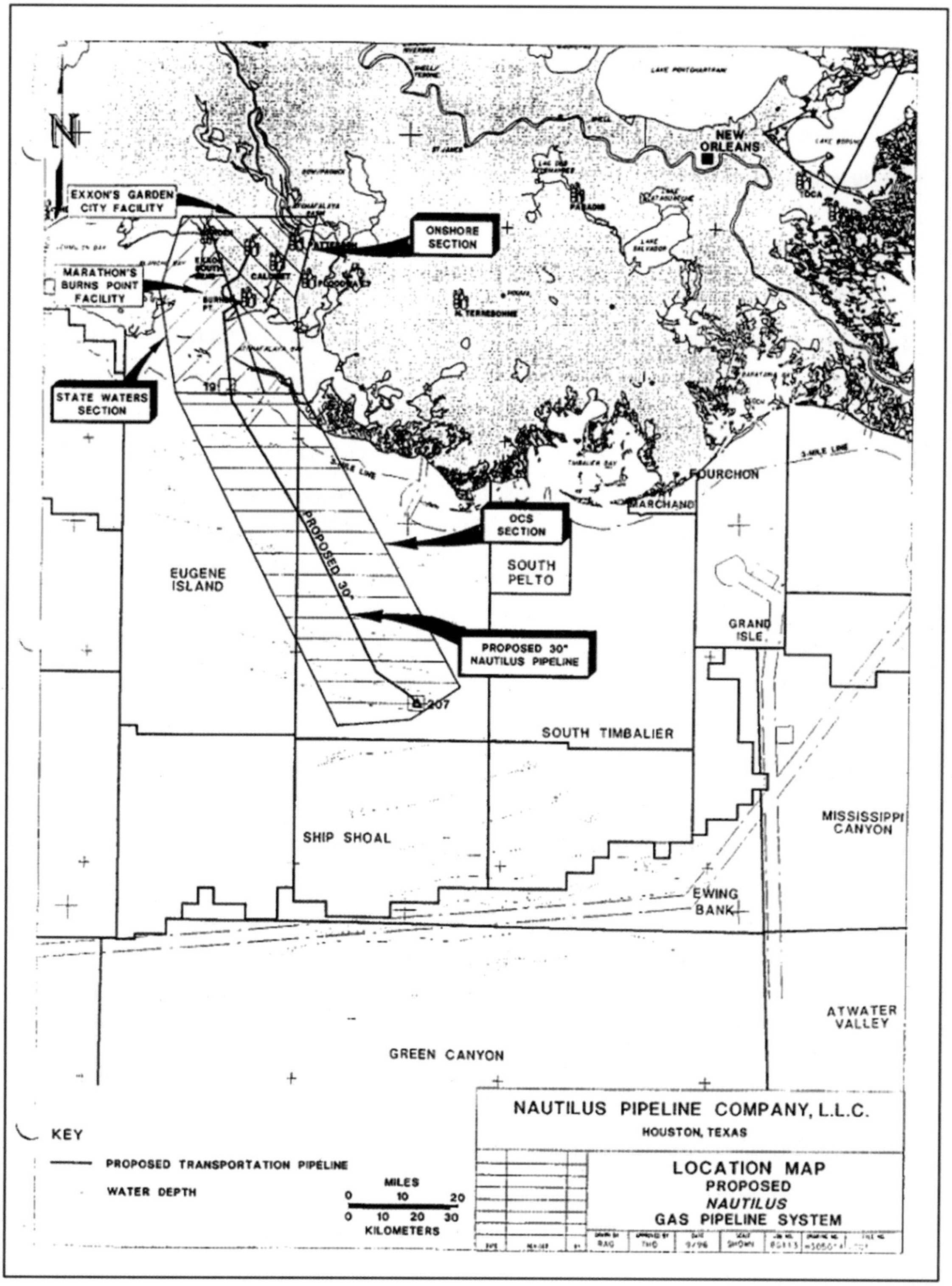

Figure 1. Nautilus pipeline route in the U.S. Gulf of Mexico. Source: FERC Docket CP96-790.

Component	Revised Units Required	Revised $/Unit	Revised Exhibit K	Actual & Estimated Expenditures
Materials				
1 Pipe: 30" OD x 0.577" WT. X-65	460,000	59.20	27,232,000	28,024,449
2 Pipe: 30" OD x 0.875" WT. X-65	500	98.00	49,000	149,187
3 Thin Film Epoxy Coating (0.577 wall)	460,000	5.20	2,392,000	2,441,563
4 Thin Film Epoxy Coating (0.875 wall)	500	6.32	3,160	0
5 Concrete Weight Coating	460,500	17.87	8,229,135	9,279,933
6 Anodes	750	700	525,000	330,856
7 Subsea Valves	3	70,000	210,000	0
8 Pig Launcher, Platform Piping	1	550,000	550,000	1,413,181
9 Meter @ SS 207	1	500,000	500,000	0
10 Internal Epoxy Coating	460,500	4.50	2,072,250	0
			41,762,545	41,639,169
Construction				
11 Contractor Management Cost	2	100,000	200,000	513,586
12 Mobe/Demobe Deep Water	1	5,500,000	5,500,000	3,945,878
13 Mobe/Demobe Shallow Water	1	1,000,000	1,000,000	1,555,866
14 Pipe Lay Deep Water	300,000	35.00	10,500,000	6,320,563
15 Pipe Lay Shallow Water, Burial included	160,000	70	11,200,000	10,861,693
16 Field Joint Coating	1	1,000,000	1,000,000	1,795,333
17 Pipeline Crossings	40	100,000	4,000,000	1,424,218
18 Riser @ 207	1	550,000	550,000	156,619
19 Shore Approach - Directional Drill	1	1,500,000	1,500,000	0
20 NDT/Diving Support	460,000	2.00	920,000	1,049,083
21 Postlay Survey	460,000	1.00	460,000	0
21 Burial	300,000	10	3,000,000	3,145,760
22 Hydrotest	2	500,000	1,000,000	967,972
23 Commissioning	1	500,000	500,000	122,006
24 Tie-in work @ SS207	1	250,000	250,000	640,186
25 Future Tap Installations	3	300,000	900,000	0
			42,480,000	32,498,763

Figure 2. Nautilus pipeline estimated and actual cost. Source: FERC Docket CP96-790.

and a total of 750 anodes were used. Three subsea valves, one pig launcher, and one meter were installed, and all component cost were consolidated in the pig launcher, platform piping category ($1.4 million vs. $1.3 million estimated). Internal epoxy coating was removed from the final design. Estimated materials cost was $41.8 million, and actual materials cost was $41.6 million.

One vessel was used to lay the line in shallow waters, and another was used in deepwater. Mob/demob for the deepwater vessel was much more expensive than for the shallow water vessel and was obtained cheaper than the expected cost, but the shallow water mobilization was more expensive than anticipated. Together, mob/demob represented about 15% of the $32 million construction costs. Approximately 56 miles of pipe were laid by the deepwater vessel at a cost of $21/ft compared to $68/ft for the shallow water pipeline, which partially reflects the need for burial in shallow water, which can add $10 to 15/ft additional cost and the shore transition.

Forty pipeline crossings were anticipated to cost $100,000 per crossing, but final cost was about $35,600 per crossing. Field joint coating was estimated to cost $1 million but final

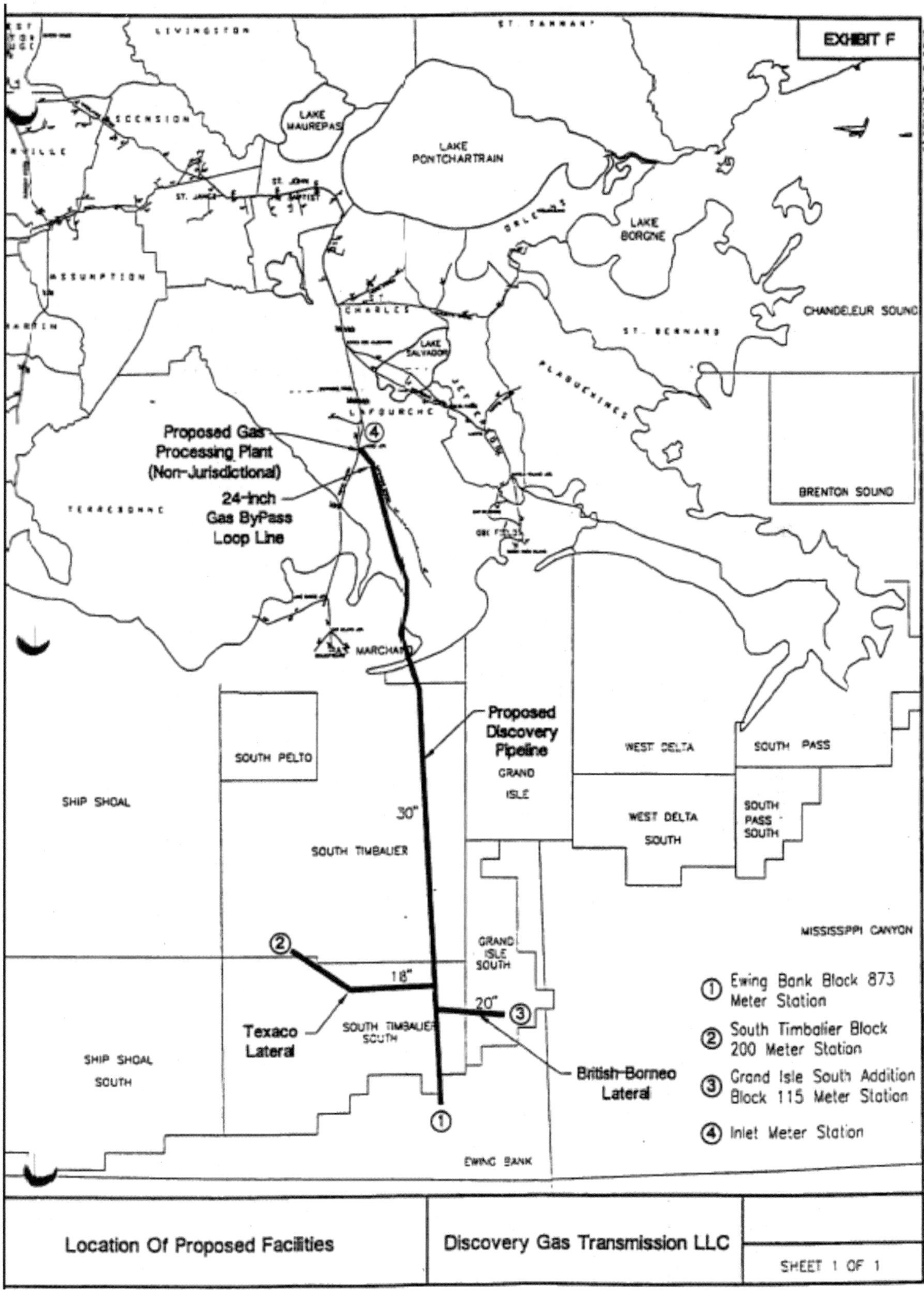

Figure 3. The Discovery Gas Transmission system in the U.S. Gulf of Mexico. Source: FERC Docket CP96-712.

No.	Particular	Exhibit K Estimate	Actuals & Estimated Expenditures
1.0	Land and Land Rights		
	1.1 Right of way (Onshore Mainline)	$ 2,502,500	$ 2,135,393
	1.2 Mitigation/Damages (Onshore Mainline)	$ 2,000,000	$ 2,001,296
1.0	Subtotal	$ 4,502,500	$ 4,136,689
2.0	Pipeline (Mainline and Laterals)		
	2.1 Offshore Mainline		
	a. Pipe and Coatings	$ 38,917,200	$ 35,164,967
	b. Valves, Flanges, Misc. Materials	$ 1,033,200	$ 1,621,230
	c. Inspection	$ 2,657,400	$ 1,628,244
	d. Traceability	$ 100,000	$ 236,873
	e. Construction	$ 30,068,100	$ 31,199,477
	f. Survey, Mapping	$ 270,000	$ 207,981
	Subtotal	$ 73,045,900	$ 70,058,772
	2.2 Onshore Mainline		
	a. Pipe and Coatings	$ 21,573,100	$ 19,836,559
	b. Valves, Flanges, Misc. Materials	$ 2,047,200	$ 1,995,436
	c. Inspection	$ 997,900	$ 1,922,187
	d. Traceability	$ 100,000	$ 196,846
	e. Construction	$ 23,444,400	$ 28,985,413
	f. Survey/Mapping/Permitting	$ 1,125,000	$ 1,700,248
	Subtotal	$ 49,287,600	$ 54,636,689
	2.3 Offshore Lateral to Grand Isle 115 Platform		
	a. Pipe and Coatings	$ 2,523,200	$ 2,682,617
	b. Valves, Flanges, Misc. Materials	$ 442,000	$ 645,221
	c. Inspection	$ 574,300	$ 253,297
	d. Traceability	$ 20,100	$ 37,031
	e. Construction	$ 5,127,800	$ 3,524,172
	f. Survey, Mapping	$ 70,200	$ 57,737
	Subtotal	$ 8,757,600	$ 7,200,075
	2.4 Offshore Lateral to South Timbalier 200 Platform		
	a. Pipe and Coatings	$ 5,616,300	$ 4,741,491
	b. Valves, Flanges, Misc. Materials	$ 498,000	$ 675,608
	c. Inspection	$ 1,594,800	$ 357,649
	d. Traceability	$ 32,100	$ 51,226
	e. Construction	$ 9,863,300	$ 7,429,920
	f. Survey, Mapping	$ 185,000	$ 75,201
	Subtotal	$ 17,789,500	$ 13,331,095
2.0	Subtotal	$ 148,880,600	$ 145,226,631
3.0	Compressor Stations	$ -	$ -
4.0	Meter Stations/SCADA		
	4.1 Offshore Mainline	$ 348,400	$ 572,094
	4.2 Onshore Mainline	$ 972,800	$ 2,941,321
	4.3 Offshore Lateral to GI115	$ 323,200	$ 452,286
	4.4 Offshore Lateral to ST200	$ 292,500	$ 780,769
4.0	Subtotal	$ 1,936,900	$ 4,746,470
5.0	O&M Capital		
	5.1 Spare Parts	$ 200,000	$ 200,000
	5.2 Other	$ 300,000	$ 300,000
5.0	Subtotal	$ 500,000	$ 500,000

Figure 4. The Discovery Gas Transmission system pipeline estimated and actual cost. Source: FERC Docket CP96-712.

No.	Particular	Exhibit K Estimate	Actuals & Estimated Expenditures
6.0	Project Management & Engineering		
	6.1 Project Management/Company Cost	$ 2,013,600	$ 4,714,105
	6.2 Engineering		
	a. Offshore Lateral	$ 350,000	$ 934,406
	b. Onshore Lateral	$ 600,000	$ 634,510
	c. Offshore Lateral to GI115	$ 250,000	$ 160,201
	d. Offshore Lateral to ST200	$ 400,000	$ 213,020
6.0	Subtotal	$ 3,613,600	$ 6,656,242
7.0	State and Local Taxes on Materials	$ 1,810,700	$ 1,884,045
8.0	Pre-Permit	$ 267,600	$ 440,121
9.0	Contingency	$ 16,878,200	
10.0	AFUDC	$ 8,941,000	$ 12,807,939
11.0	Line Pack	$ 549,000	$ 570,483
12.0	TOTAL COST	$ 187,880,100	$ 176,968,620

Figure 4. The Discovery Gas Transmission system pipeline estimated and actual cost. Source: FERC Docket CP96-712. (continued)

cost was $1.8 million. The riser and installation at Ship Shoal 207 was anticipated to cost $550,000 but actually cost $157,000. The original estimated labor and/or contractor cost was $42.5 million, but the actual realized cost was $32.5 million, mostly due to favorable developments in mob/demob and cheaper deepwater pipelay cost.

Total project cost was $74.1 million, and estimated cost was $84.3 million. About 60% of the total cost was due to materials.

Discovery

The Discovery Gas Transmission system was built in 1997 and is a network of 147 miles of pipelines originating at Ewing Bank block 873 platform (Lobster) in the U.S. Gulf of Mexico, and including a 30-inch diameter 104-mile main line, a 20-inch diameter 11-mile lateral, connecting the mainline to a platform in Grand Isle block 115, a 26 mile lateral that connects the mainline to a platform in South Timbalier block 200, and a six-mile bypass line (Figure 3). The main line included approximately 37 miles of onshore lines, which was budgeted separately.

The offshore section of the mainline was expected to cost $73 million and actually cost $70 million, or about $1.0 million/mile (Figure 4). About half of costs were associated with the pipe and coatings ($515,000/mi), and 44% of costs were associated with operations ($459,000/mi). Material cost estimates exceeded actual costs and construction estimates understated actual costs.

The onshore mainline was expected to cost about $49 million, but actual cost was about $55 million, with about half of the cost due to construction and 40% due to materials. Onshore mainline cost was about $1.6 million/mi ($1 million/km), greater than the offshore unit cost.

The 11-mile (18 km) lateral to the GI 115 platform cost $7.2 million, or $650,000/mi. The lower costs were partly associated with lower pipe costs ($245,000/mi), which were likely due to the smaller diameter line used, and the construction costs ($318,000/mi) saved by avoiding burial requirements.

The 26-mile lateral to the ST 200 platform cost $13.3 million ($512,000/mi), 25% less than expected due to material and construction costs savings. Pipe and coatings costs were $180,000/mi and accounted for 35% of total costs. Construction costs were $284,000/mi and accounted for 56% of total costs.

OFFSHORE DECOMMISSIONING

Offshore pipeline decommissioning process in the U.S. Gulf of Mexico

Regulations

Regulations require the lessee to remove all structures and related underwater obstructions within one year after a lease terminates (Federal Register, 2002). Pipeline decommissioning and removal procedures in U.S. federal waters are detailed in 30 CFR 250.

The type and scope of decommissioning work to be performed on a pipeline is dependent on the time the pipeline is scheduled to be out of service. If a pipeline is out of service for one year or less, the pipeline is isolated with a blind flange or a close block valve at each end of the pipeline. For pipelines out of service for more than one year but less than five years, the pipeline is flushed and filled with inhibited seawater. For a pipeline out of service for more than five years, the pipeline is decommissioned according to requirements, found in 30 CFR 250.1750–250.1754.

Pipeline decommissioning regulations found in 30 CFR 250.1754 require a pipeline to be removed, rather than abandoned in place, when the Regional Supervisor of the BSEE determines that the pipeline is an obstruction. If it is determined that a decommissioned pipeline must be removed, the requirements of 30 CFR 250.1752 must be met, which state in part:

"Before removing a pipeline, you (the operator) must:

A. Submit a pipeline removal application in triplicate to the Regional Supervisor for approval that includes the following information:
   1. Proposed removal procedures
   2. If the Regional Supervisor requires it, provide a description, including anchor pattern(s), of the vessel(s) you will use to remove the pipeline
   3. Length (feet) to be removed
   4. Length (feet) of the segment that will remain in place
   5. Plans for transportation of the removed pipe for disposal or salvage
   6. Plans to protect archaeological and sensitive biological features during removal operations, including a brief assessment of the environmental impacts of the removal operations and procedures and mitigation measures that you will take to minimize such impacts
   7. Projected removal schedule and duration
B. Pig the pipeline, unless the Regional Supervisor determines that pigging is not practical
C. Flush the pipeline."

Regulations require that the ends of the pipeline are buried three feet below the seabed or covered by a concrete mat. Most pipelines in the Gulf of Mexico have been abandoned in place, and therefore, the decommissioned pipeline segments are not removed from the seabed. Only small portions of the pipeline, typically the risers and tube turns, are removed, or if a pipeline is damaged or leaking, the damaged portion will need to be repaired or replaced prior to decommissioning. Decommissioning procedures in deepwater are similar to shallow water but more involved (Burke and Stokes, 2015; Phillip et al., 2014).

In OCS areas that are designated OCS Significant Sediment Resources, BOEM requires that pipelines are decommissioned by removal, which is a longer and more complex operation. Companies are required to obtain BSEE authorization before operations to relinquish the ROW, and because no blasting is performed during decommissioning operations, usually no additional environmental clearances are required. Each state within the Gulf of Mexico varies on their Coastal Zone Management Act enforceable policies and management of pipelines.

Pipeline endpoints that lie entirely within federal waters qualify for categorical exclusion from the requirement to prepare an environmental assessment or environmental impact statement, but pipelines that traverse state and federal waters require additional preparation and review, typically involving an environmental impact statement. Pipelines that are not under FERC regulation only require BSEE authorization and permit to abandon.

Workflows

The workflows associated with pipeline decommissioning depend on the requirements, the status and nature of the system at the time of operation, and the procedures established by the company or contractor (Ayoade, 2002). No generalized procedure applies because system configurations are unique, although on a functional level, there is a high degree of commonality of work activities.

Work duration and market rates for the vessels and equipment required vary and lead to different workflows and costs. One of the most important factors that distinguish differences in project cost is the status of the pipeline. If the pipeline is damaged or leaking, decommissioning costs may be two or three (or more) times greater than normal operations.

The basic procedures used in pipeline decommissioning for abandonment-in-place are as follows:

1. Isolate line
2. Pig pipeline and collect contaminants
3. Cut and bury tube turns near platform base; install fitting and caps
4. Remove appurtenant facilities, including valves, pig launchers, meters, and risers
5. Dispose of flush water and pipeline materials to shore

For pipeline decommissioned by removal, step three is replaced by the complete removal of all pipeline in the area specified. Selection of the equipment and procedures in pipeline decommissioning depend on the availability and cost of equipment, and the nature and sequence of operations.

Operations

Cleaning. Pigging is a common operation in the oil and gas industry, done with many objectives, including cleaning deposits such as wax layers and scale, removal of liquids, and flowline inspection (Jayawardena et al., 2002; Tiratsoo, 2013). A pig refers to a solid or semisolid object propelled through the pipeline, and described by its function (purpose), type, and physical characteristics such as diameter, density, and composition. To make pigs more aggressive, brushers and scrapers can be added to clean pipe wall more thoroughly (Pigging Products and Services, 2017). In decommissioning, pigging is required to clean the line unless the Regional Supervisor determines that it is not practical.

For pipelines that have not been pigged for some time, a progressive cleaning program is often implemented, starting with the least aggressive pig to ensure that the pigs do not get stuck in the pipeline. The process involves sending a series of PE foam pigs and cleaning fluids through the pipeline with chemical agents and flush water to remove all hydrocarbons (Culwell and McCarthy, 1998; MacLeod and Pierson, 2000). The progressive nature of pigging refers to the sequence, type, and number of pigs applied.

Cutting. A platform pipeline is cut near the base of the platform by divers using an arc oxygen torch and a cap is installed on the end. The pipeline end is buried below the mudline, typically by diver-operated jetting. The pipeline end may alternatively be covered by a concrete mat, which provides a cover for the pipeline. Concrete mattresses are commonly used in deepwater where it is not practical to bury the ends using divers and are also more common in the North Sea due to differences in practice and soil type (Oil & Gas UK, 2013).

The riser that extends through the water column may be removed or left in place along with platform meters and associated equipment depending on agreement between the pipeline and platform owners, if different (Figure 1). The riser may be partially or wholly removed. A partially removed riser will be cut below the waterline and near the base of the platform. Tube turns are cut, the end on the seafloor plugged and buried three feet or covered, and the remaining section removed.

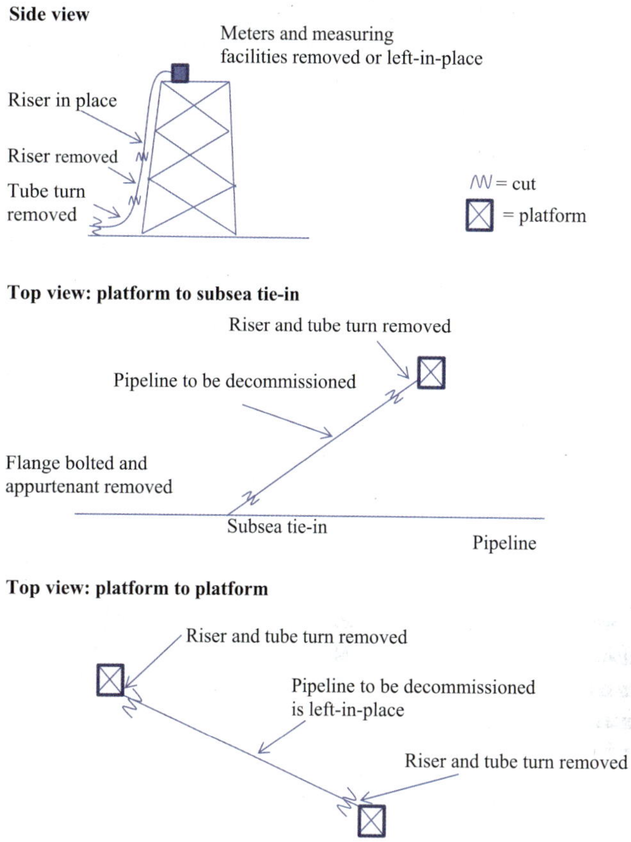

Figure 1. Typical work requirements in pipeline decommissioning operations.

Removal and recovery. Recovering removed pipeline sections is done by rigging a winch wire to the pipeline and lifting it to a barge. A crane may be used in conjunction with the winch to hoist the pipeline onto the recovery vessel. Other methods can be used, but this is the primary and most cost-effective procedure. Excavation may be required to remove the pipeline, or it may be recovered without excavation if enough lifting force can be applied.

Risers, appurtenances, and meter facilities are usually owned by the pipeline company and removed at opportune times, or transferred, sold, or left in place during pipeline decommissioning. If pipelines are damaged by hurricanes or corrosion, they are either first repaired to permit pigging or the damaged portions are removed. Either operation will add significant cost and uncertainty to project operations. Pipelines in an OCS Significant Sediment Resource area in the Gulf of Mexico are removed in decommissioning.

Pipelines that come onshore may be removed through the surf zone. The onshore pipeline may be removed completely, or some sections may be abandoned in place due to their transition through a sensitive environment. The pipeline end seaward of the surf zone is capped and jetted down three feet below mudline by divers. Pipeline crossings may be an obstacle to decommissioning, particularly if the pipeline to be decommissioned crosses under a live (in-service) pipeline.

Disposal. Pipeline materials are transported by truck or barge to an approved dump site. The scrap value of the steel is usually exceeded by the cost of cutting and removing the pipeline coatings, and therefore, the resale value is often negligible. Pipeline materials must be reduced in length according to the dump site requirements and may be as short as six feet. A hydraulic shear is usually an effective way to section the pipeline to meet these requirements.

References

M.A. Ayoade, Disused Offshore Installations and Pipelines: Towards Sustainable Decommissioning. Kluwer Law International, The Hague, 2002.

J. Burke, A. Stokes, Preparation for cost effective decommissioning and abandonment of subsea pipelines, in: SPE 175426. SPE Offshore Conference & Exhibition, Aberdeen, Scotland, UK, September 8–11 (2015).

A.S. Culwell, J. McCarthy, Pipeline and power cable decommissioning. in: F. Manago, B. Williamson (Eds.), Proceedings: Public Workshop, Decommissioning and Removal of Oil and Gas Facilities Offshore California: Recent Experiences and Future Deepwater Challenges, September 1997. OCS Study MMS 98- 0023. Coastal Research Center, Marine Science Institute, University of California, Santa Barbara (CA), 1998, pp. 66–73.

Federal Register 67(96), 35398-35412, Department of Interior, Minerals Management Service, 30 CFR Parts 250, 256, Oil and Gas and Sulphur Operations in the Outer Continental Shelf Decommissioning Activities. Final Rule, 2002.

S. Jayawardena, L. Dykhno, J. Hudson, Challenges in pigging of subsea gas flowlines, in: SPE 77576. SPE Annual Technical Conference and Exhibition, San Antonio, TX, September 29–October 2 (2002).

R.B. MacLeod, J.J. Pierson, Garden Banks 388 abandonment – technical aspects covering pigging and flushing, in: OTC 12124. Offshore Technology Conference, Houston, TX, May 1–4 (2000).

Oil & Gas UK, Decommissioning of Pipelines in the North Sea Region. London (2013).

N.S. Phillip, S. Wilde, R. Arshad, I. Washash, Decommissioning process for subsea pipelines, in: SPE 171898. Abu Dhabi International Petroleum Exhibition and Conference, Abu Dhabi, UAE, November 10–13 (2014).

Pigging Products & Services Association, An Introduction to Pipeline Pigging, 12th ed. Suffolk, UK, 2017.

J.N.H. Tiratsoo, Pipeline Pigging and Integrity Technology, fourth ed. Clarion Technical Conferences, Accrington, UK, 2013.

Offshore pipeline decommissioning cost factors

Project

Pipeline systems are described in terms of their segments and associated features such as length, diameter, endpoint type, and location. The work activities required in decommissioning, and time spent offshore are directly related to the pipeline characteristics and determine the cost of activity.

The length, diameter, and service of the line determine the amount of fluids that are required to flush and clean the line. Oil lines carrying high sulfur crude will likely require more pig runs and fluid volume than gas or condensate lines, time on site will be longer, and there will be greater disposal volumes and cost than sweet gas service. A line that has been kept clean during its service life will require fewer pig runs and less flush water at the end of service than a line that has not been serviced regularly and disposal cost will be higher, all other things being equal.

The scope of work activity, the number of interconnects and appurtenances needing to be cut and removed, and the endpoint and interconnect water depths are important factors in determining the time and complexity of the operation. Complex projects require more planning and labor, will be more expensive, and more uncertain to execute than less complex projects. Lines that come ashore require additional environmental review and scrutiny, which increase costs. Large retrieval and remove requirements will add costs relative to abandon-in-place projects.

Pipelines that are inactive for a long period of time may not maintain their cathodic protection, and have greater exposure to natural disturbances (hurricanes, slope failures, etc.), stress induced motions, and third body impacts. Pigging operations may reveal leaks and repairs will need to be made before pigs can be run to perform decommissioning. Repairs typically involve the use of a clamp placed over the damaged area(s) or the damaged area is removed, and a pup piece is welded in place between two points. Repairs are expensive and add significantly to the cost of pipeline decommissioning.

Hurricane-damaged platforms and pipeline increase project complexity and are associated with greater project costs and risk. Hurricane-damaged platforms and pipelines require significantly greater time to plan, access, and safely perform. If a pipeline is leaking, leaks need to be repaired before abandonment operations can proceed. Stress-induced (bent) pipes are a safety issue. Hurricane-damaged pipelines will require a larger number of cutting operations and may require retrieval, which will increase cost due to the additional support requirements and time spent offshore. The conditions of the work and contract specifications, including the degree of competition at the time the contract was let, are important determinants of project cost. For example, Wild Well Control was awarded a fixed sum contract in 2008 worth $750 million to decommission seven downed platforms in the shallow water Gulf of Mexico for BP, Chevron, and Apache.

As the number of side valves that need to be cut and flanged increases, diver time and set-up time increases, which increases project costs. The diameter, thickness, pipe coating, and water depth will determine access by divers or ROVs, time, and cost to cut.

Time

Time enters project costs according to the time required to perform the work activity, the year and season work is performed, market conditions at the time the service contract is let, and the uncertainty that arises when using a price index to adjust cost over long periods.

The time to perform work relates directly to cost because vessel spreads are based on market day rates and are usually the greatest contributor to project costs. If a fixed price lump-sum contract is used instead of a day rate contract, contractors will use their estimate of the expected time of operations to value the contract and bid on work; in this case, time enters contracts indirectly. Market conditions may change dramatically due to changes in vessel supply and demand, and identical offshore work performed in different years may have dramatically different cost. When comparing project costs performed over long periods (e.g., decade or longer), a reliable means to adjust cost is necessary but the quality of index adjustments is uncertain.

Waiting on weather is usually a function of the season work is performed. In the Gulf of Mexico, winter demand for decommissioning vessels is usually less than in the summer, leading to a reduced day rate, but winter work has greater exposure to adverse sea states that can delay work and increase time offshore. Summer work is exposed to delays due to tropical depressions and hurricanes. Waiting on weather increases time on site and leads to higher costs.

The difference in time when cost estimates are made and when operations are performed is an often-overlooked factor in the reliability of the estimation. As one would expect, when the difference in these times increases, the reliability of estimates decreases because market conditions and project scopes are more likely to change in the interim. Projects that are performed within a few months of the preparation of the estimate are expected to be more reliable, for all other things equal, than when delays exceed a year. In some cases, project delays may work to the operator's advantage, if for example, market conditions weaken and vessel day rates decline from when the cost estimates were prepared, but in other cases, if project scopes expand and market conditions tighten project costs will likely increase relative to original estimates, assuming all else equal.

Location

The water depth of the pipeline endpoints and side valve interconnects impact project complexity and selection of diver and/or ROV options, and, if there is an onshore endpoint, access may require barge or service boat entry, which will add to costs. Pipeline endpoints that lie entirely within federal waters qualify for categorical exclusion from the requirement to prepare an environmental assessment, but pipelines that traverse state and federal waters require additional preparation and review, typically involving an environmental impact statement.

Offshore pipeline decommissioning cost statistics in the U.S. Gulf of Mexico, 1980–2019

Statistics

Offshore pipeline decommissioning in the shallow water U.S. Gulf of Mexico is generally an inexpensive, low-tech operation involving procedures that are relatively easy and quick to perform under normal conditions. Pipeline decommissioning projects are usually less expensive than well plugging and structure removal, but are more involved than site clearance and verification operations.

The average inflation-adjusted cost for 28 pipeline decommissioning projects reviewed between 1995 and 2019 was $3.06 million and ranged over three orders-of-magnitude, from $30,000 to $21.8 million (Kaiser, 2017). All costs were adjusted to 2014 U.S. dollars using the BLS offshore oil field services index. Most projects occurred after 2008. Standard deviations exceeded the mean values in decommissioning due to differences in project scope and complexity.

The average inflation-adjusted unit cost was $301,000/mile or $57/ft, and ranged from $7000 to $2.1 million per mile, or $1.3 to $398/ft. The average cost per mile-cut was $190,000/mile-cut and ranged from $2000 to $2.1 million per mile-cut. The average cost per pipeline volume was $47 per cubic foot and ranged from $0.4 to $124/cf. The cost distribution by pipeline length and volume is approximately exponential with a large tail (Figure 1).

Pipeline length is expected to be positively correlated with decommissioning cost due to the greater time and additional fluids required to clean the pipeline. System complexity is described by the number and type of interconnects, the water depth of the cuts, and if pipelines are hurricane-damaged or leaking. As system complexity increases, decommissioning cost is expected to increase.

The inflation-adjusted decommissioning cost for hurricane-damaged and leaking pipelines was $6.9 million compared to $1.8 million for nondamaged and nonleaking lines. On a normalized basis, hurricane-damaged and leaking lines cost $663,000/mi, $400,000/mile-cut, and $126/cf versus $180,000/mi, $90,000/mile-cut, and $34/cf for nondamaged and nonleaking lines. Hurricane-damaged and leaking lines were about three to four times more expensive than undamaged/nonleaking lines, and relative to the aggregate averages, damaged lines were about two to three times more expensive than the average pipeline decommissioned.

Limitations

Pipeline decommissioning operations are performed using methods and contracting strategies specific to the project characteristics and market conditions in the region at the time of operation. If the project scope or work duration is different than expected, if the equipment and spread requirements are different, or if the vessel day rates are different, cost estimates will differ from the values expected.

FERC public dockets do not report final costs of pipeline decommissioning, and therefore no comparison between estimated and actual cost was possible.

Projects were distinguished by pipeline length, complexity (i.e., damaged, number of cuts required, volume), type, and year. Complexity is a multidimensional attribute which would ideally be incorporated as an additional factor or through subcategory analysis, but the sample set was too small for granular decomposition.

BOEM pipelines are more diverse than FERC pipelines, but the pipeline decommissioning procedures will be quite similar even if oil pipelines are considered, and so these estimates should apply broadly to BOEM pipelines of similar scope, water depth, complexity, and pipeline type. The decommissioning cost of flowlines and deepwater pipelines were not part of the sample data, and extrapolation of the results to these pipeline classes do not apply.

Decommissioning costs estimates are uncertain under the best circumstances, but for deepwater pipelines, because requirements for resources, vessels and work duration are expected to be more variable, higher levels of uncertainty in cost estimation are expected. When estimating deepwater

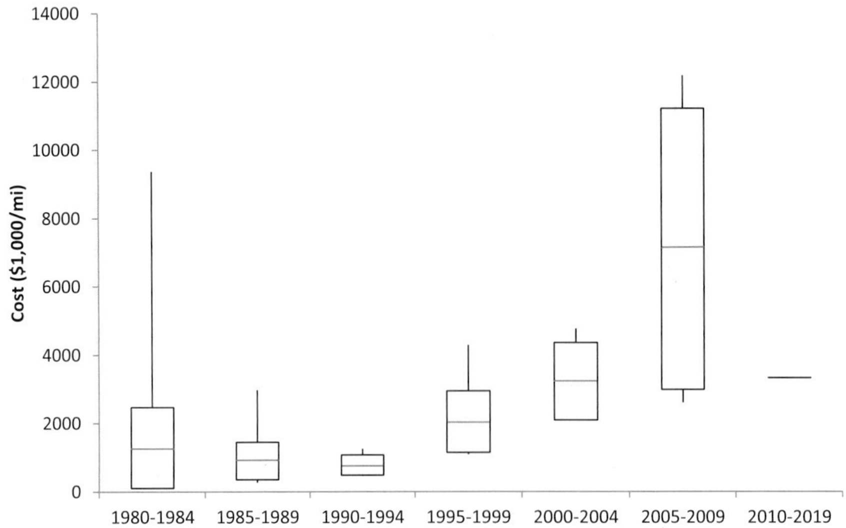

Figure 1. Inflation adjusted normalized pipeline decommissioning cost distribution (2014$).

pipeline decommissioning cost, there is likely to be more than one option, technology, and vessel spread available, and if the activity is expected to occur beyond the near-term, new technologies and new regulations will impact cost.

Different assumptions are employed in decommissioning cost estimation across companies, and unlike the standardized categories, FERC requires in construction, a common set of categories are not used in FERC decommissioning applications. Engineers apply different categories and procedures, including, for example, different weather contingency allowances, work provisions, and contingencies. If estimates are made following good engineering judgment, these differences are not expected to be significant, but other factors will also play a role in reliability, such as limited resources and time constraints, and, of course, good engineering practices may not always be followed, and experienced cost estimators may not have performed the evaluations.

Reference

M.J. Kaiser, FERC pipeline decommissioning cost in the U.S. Gulf of Mexico, 1995–2015, Marine Policy 82 (8) (2017) 167–180.

How to estimate the cost to decommission shallow water pipeline in the U.S. Gulf of Mexico?

Example

Estimate the expected cost to decommission a shallow water pipeline in 250 ft water depth in the U.S. Gulf of Mexico in 2025. The length of the line is 5 miles and three cuts will be required in the operation. How does the estimate change if it is known that the line diameter is 24 in? What is the best estimate?

Solution

Using only information on pipeline length, the average (expected) decommissioning cost is:

$301,000/mi · 5 mi = $1.5 million.

Using the length and number of cuts as data inputs, the average cost increases to:

$190,000/mi-cut · 5 mi · 3 cuts = $2.9 million.

Using the diameter and length of the line, pipeline volume is computed as 82,938 cf and yields the highest and probably best estimate of cost:

$47/cf · 82,900 cf = $3.9 million.

Cost bounds are based on the range for each statistic and represent the maximum and minimum of the project data, and in this instance are not that useful because of the large spreads. For the volume measure, for example, the cost range is $33,000 to $10.3 million.

The best estimate (probably) applies all the available (known) data and the relevant unit statistic, however, this is not necessarily a truism. Note that in this case it is not known if the operation will be based on a damaged or undamaged pipeline, so statistics that have aggregated all these projects is the relevant metric. Alternatively, the user may estimate cost based on an undamaged and damaged line to obtain cost bounds.

How to estimate the cost to decommission hurricane-damaged pipeline in the U.S. Gulf of Mexico?

Example

Estimate the expected cost to decommission a 3-mi hurricane-damaged shallow water pipeline in 250 ft water depth in the U.S. Gulf of Mexico in 2025. Compare with the cost with an undamaged pipeline.

Solution

Using the length input and the damaged pipeline unit cost statistic, the average (expected) cost is:

$663,000/mi · 3 mi = $2.0 million.

Using the length input and the undamaged pipeline unit cost statistic, the expected cost is:

$180,000/mi · 3 mi = $540,000.

How to estimate deepwater pipeline decommissioning cost in the U.S. Gulf of Mexico?

Deepwater (>400 ft) fixed platform decommissioning operations in the U.S. Gulf of Mexico are like shallow water projects in the requirements that must be achieved, but operations are functionally different, involve different system components, and require different vessels cost statistics from shallow water fixed platforms are not suitable analogs and are not expected to scale to deepwater floater structures, and thus a work decomposition approach is required in the absence of historical statistics.

Pipelines that start and end at a platform riser (R-R) require similar abandonment tasks as riser-to-subsea tie-ins (R-SSTI) but will vary in flushing duration and the vessel spread utilized. In deepwater, four-point derrick barges (4PDBs), dynamically positioned derrick barges (DPDBs), and dynamically positioned deepwater subsea intervention (DSI) vessels with ROV capability are typically used in decommissioning operations. Greater water depths and weights limit the number of marine vessels that can perform operations, and the level of competition is also more limited.

Twenty pipeline decommissioning scenarios were assumed using various dive vessels (4PDB, DPDB, DSI), pipeline type (R-R, R-SSTI), water depth (400–1300 ft), pipeline diameter (4–16 in), pipeline length (1–23 nm), mobilization distance (75–193 nm), and flush volume (1000–94,000 cf) ranges. For each scenario, a cost estimate was performed using a work decomposition method and set of suitable assumptions. All pipelines are assumed abandoned in place with the pipeline flushed and the ends severed, plugged and buried with sandbags or concrete mats.

Using ProServ's project cost scenarios across all vessel spreads and pipeline types based on water depth and diameter, length and flushing volume requirements two cost functions were derived through aggregation (Kaiser and Liu, 2014):

$$PLA_1 = 42{,}968 + 1550WD + 43{,}305DIAM + 3036L$$
$$PLA_2 = 228{,}804 + 1703WD + 9511VOL$$

where PLA is the pipeline abandonment cost ($), WD is the water depth (ft), DIAM is the pipeline diameter (in), L is the pipeline length (1000 ft), and VOL the flush volume (1000 cubic feet, Mcf). Flush volume is computed from pipeline diameter and length. The cost algorithms are not supported by empirical data of actual operations, but they will improve as actual project data is collected and analyzed.

Example

Structure 23800 is in 483 ft water depth and its oil and gas production is routed to a transmission line about 20 nm away. Oil is transported 20.8 nm via a 16-inch diameter oil pipeline, and gas is transported 19.5 nm via a 12-inch gas pipeline. Estimate the oil and gas pipeline decommissioning cost.

Solution

Flush volume is 177 Mcf for the oil pipeline and 93 Mcf for the gas pipeline. Pipeline decommissioning cost estimates using the water depth, diameter, and length relation yields $1.87 million for the oil pipeline:

$$PLA_{1,\,Oil} = 42{,}938 + 1550(483) + 43{,}305(16) + 3036\,(126.7)$$
$$= \$1.87 \text{ million}$$

The gas pipeline decommissioning cost estimate is $1.67 million. Using the water depth and flush volume relation, the oil pipeline decommissioning estimate is $2.73 million, and the gas pipeline decommissioning estimate is $1.93 million. Application of either relation provide similar order of magnitude estimates and applying one relation or the other does not imply a better estimate.

Reference

M.J. Kaiser, M. Liu, Decommissioning cost estimation in the deepwater U.S. Gulf of Mexico – Fixed platforms and compliant towers, Marine Structures 37 (7) (2014) 1–32.

How to estimate deepwater umbilical, flowline, and riser decommissioning cost in the U.S. Gulf of Mexico?

Deepwater (>1000 ft) floating platform decommissioning operations in the U.S. Gulf of Mexico are like shallow water projects in the requirements that must be achieved, but operations are functionally different, involve different system components, and require different vessels. Cost statistics from shallow water fixed platforms are not suitable analogs and are not expected to scale to deepwater floater structures, and thus a work decomposition approach is required in the absence of historical statistics.

Umbilicals, flowlines, and risers are associated with subsea wells tied back to the platform and are cut and detached from the structure and removed from the seabed. Umbilicals and flowlines removal will require marine vessels and ROVs. Pull tubes and I-tubes for risers are used extensively to simplify installation and hookup, requiring only a single platform mounted winch to pull up the riser, and during decommissioning, this same winch can be used for removal operations.

All umbilicals and flowlines are assumed to be flushed, cut, and removed from the seabed using anchor handling vessel spreads with ROV capabilities using vessel dayrates at the time of the evaluation. Risers are flushed, cut, and detached from the platform and removed using a platform winch or crane vessel. Mobilization cost for the vessel spread is not included in the cost estimation. Riser length is assumed equal to the water depth of the structure, and removal cost is assumed to be a fixed cost per foot of riser length. Umbilical and flowline removal cost increases with water depth and decreases with segment length because of assumed operational economies in the tasks performed. Diameter and weight are not included in the cost algorithms, and so for similar length segments, umbilical and flowline removal will have similar cost. Riser removal cost is estimated at \$78 per foot of riser length.

A two-factor power relation for umbilical and flowline severance and removal cost was derived from ProServ estimated cost data (Kaiser and Liu, 2015):

$$\mathrm{UFSR} = 1.97\,\mathrm{L}^{-0.85}\mathrm{WD}^{0.31}$$

where UFSR is the umbilical and flowline severance and removal cost per length (\$/ft), L is the segment length (nm), and WD is the water depth (ft). The cost relation is nonlinear, and so each umbilical and flowline segment needs to be considered separately in evaluation. The cost algorithms are not supported by empirical data of actual operations, but they will improve as actual project data is collected and analyzed.

Example

Snapper is in 863 ft water depth, and one subsea well in 1110 ft water depth is tied back using a 4-inch diameter, 15.5 nm flowline controlled via a 2-inch diameter, and 15.8 nm umbilical line. Estimate total pipeline decommissioning cost.

Solution

The unit cost to remove the umbilical is estimated as:

$$\mathrm{UFSR} = 1.97(15.8)^{-0.85}(1110)^{0.31} = \$1.66/\mathrm{ft}.$$

Unit cost for flowline removal is computed at \$1.69/ft. Umbilical removal cost is estimated as (15.8 nm) (6076 ft/nm) (\$1.66/ft) = \$159,361; flowline removal cost is estimated as (15.5 nm) (6076 ft/nm) (\$1.69/ft) = \$159,161. The cost to cut and remove the production riser and the two export risers is estimated as (3) (\$78/ft) (863 ft) = \$201,942. Total umbilical, flowline, and riser removal cost is estimated as \$520,464.

Example

Front Runner is in 3330 ft water depth and in 2019 had nine dry tree wells and one subsea well tieback located 1.2 nm away in 3278 ft water depth and controlled via a 3-inch diameter, 1.3 nm umbilical line. Estimate total pipeline decommissioning cost.

Solution

The unit cost to cut and remove the umbilical is estimated as:

$$\mathrm{UFSR} = 1.97(1.3)^{-0.85}(3278)^{0.31} = \$19.38/\mathrm{ft}.$$

Unit cost for flowline severance and removal is nearly identical at \$20.75/ft. Umbilical removal cost is estimated as (1.3 nm) (6076 ft/nm) (\$19.38/ft) = \$153,113; flowline removal cost is estimated as (1.2 nm) (6076 ft/nm) (\$20.75/ft) = \$151,286. The cost to cut and remove the wet tree riser and the two export pipelines is estimated as (3) (1.6) (\$78/ft) (3330 ft) = \$1,246,752. The cost to cut and remove the dry tree riser is estimated as (9) (\$78/ft) (3330 ft) = \$2,337,660. Total umbilical, flowline, and riser removal cost is estimated as \$3.9 million.

Reference

M.J. Kaiser, M. Liu, Decommissioning cost estimation for floating structures in the deepwater U.S. Gulf of Mexico, Ships Offshore Structure 10 (4) (2015) 436–455.

Typical workflows for offshore pipeline decommissioning

Onshore segment

- Isolate pipeline valves
- Launch cleaning pigs
- Collect and dispose of all contaminants, liquids, etc., after pig(s) receipt confirmed
- Isolate and close valves to relieve residual pressure and confirm gas free
- Stopple fittings on either cut of the pipeline
- Tap and drain portion of pipe to be abandoned/removed
- Once pipe section removed, insert pipe caps

Offshore segment

- Locate, uncover, and operate subsea valves on one or more side taps to pig line
- Load pig at platform and launch
- If no pigs can be launched, flush with seawater from a utility boat or platform
- Launch and run pig with high pressure inhibited sea water
- Collect pig and fluids
- Close subsea valves on side tap
- Remove all appendages (valves, subsea assemblies)
- Cut pipeline at base of riser and remove section
- Open side taps
- Remove risers, if connected to platform
- Remove related piping and appurtenant equipment/meters on platform, if applicable

GAS COMPRESSION STATIONS

Onshore FERC gas transmission compressor station cost statistics in the United States, 2014–2019

Compressor stations are installed on gas pipelines to provide the pressure needed to transport gas from one location to another.

FERC compressor stations

The FERC approves the construction, operation, and decommissioning of interstate natural gas transmission pipelines under Section 7 of the Natural Gas Act (15 U.S. Code 717). FERC determines whether gas pipelines are gathering lines or transportation lines on a case-by-case basis using the "modified primary function test." If a line is determined to be gathering, it is classified as nonjurisdictional and is not regulated by FERC.

When a FERC-regulated natural gas pipeline company wants to modify its system, to add new pipeline or to expand capacity, or to remove capacity, it must apply for a "Certificate of Public Convenience and Necessity." This filing documents and justifies the construction and specifies what the company estimates construction cost will be. If a company receives its certificate and builds its facilities, it must report back to FERC at the completion of the project the actual cost of construction and how the original estimates compared with what it spent. The filing occurs within six months after a compressor station is put in service.

Cost components

The OGJ summarizes compressor station cost data from FERC filings using material, labor, miscellaneous, and ROW/damage cost categories.

Material cost for compressor stations include all equipment and ancillary material, including bulks (e.g., electrical, structural steel, concrete, piping, etc.) and prime movers. Material costs depend on the technical requirements of the project and market conditions during procurement.

Compressor types are either positive displacement or centrifugal. Positive displacement compressors trap a volume of gas within the compressor and by reducing its volume its pressure increases. Piston-operated reciprocating compressors are an example of positive displacement compressors. Centrifugal compressors develop the pressure required by the centrifugal force of rotation, where a compressor wheel translates kinetic energy into gas pressure. Centrifugal compressors are more commonly used in gas transmission systems due to their lower capital cost, lower maintenance expense, and operational flexibility.

Compressor selection (centrifugal or reciprocating) usually dictates the choice of prime mover (electric motors, steam turbines, or gas turbines). Natural gas-powered reciprocating engines are usually limited to driving reciprocating compressors. Natural gas-powered turbines generally are limited to driving centrifugal compressors. Electric motors may be used with either compression technology. The industry operates over 6000 natural gas-fired reciprocating engines, 1000 natural gas-powered turbines, and 200 electric motors, although the number of compressor stations are far fewer than the number of engines and motors because multiple engines and motors are grouped at a single station. Low speed compressor units powered by reciprocating engines are the most expensive option in installation cost ($/hp), while gas-fired combustion turbines and electric motors have approximately the same installed cost.

At the suction side of the compressor, gas flows through a complex piping system within the compressor station, and when leaving the compressor station at the discharge end, gas traverses the discharge piping system consisting of valves and fittings. Heat exchangers are used between stages of compression so that the heated gas is cooled to the original suction temperature before being compressed in the next stage.

Ancillary station equipment may include electrical power and transformers, instruments and control equipment, air compressors for purging and valve actuators, hydraulic pumps and motors to supply hydraulic power valve actuators. Generally, all underground metallic structures are coated, except ground rods which are typically magnesium or zinc anodes. Bare ground wires and rods also require cathodic protection.

Labor comprises the crews required to bring everything together and the specialist involved in equipment hook-up and testing. Foundations and supports are constructed and major pieces of equipment are delivered to the site and installed. Equipment is tied together, and piping, fittings, valves, and related correspondents are tied-in to the flow path. A fuel line from the transmission pipeline is run to a meter and compressor building.

Compressor stations purchase or lease the surface site on a long-term basis, but due to their small physical size and limited site preparation requirements compared to pipelines, ROWs are not expected to be a large component of station cost.

Aggregate average cost

From 2014 to 2019, 95 compressor stations for gas transmission lines consisting of 1.9 million hp capacity was installed in the United States at a total cost of $5.7 billion (Table 1). About a third of total horsepower installed over the past five years was installed in 2018–19, and represented about 40% of the total capital expenditures during the time. Compressor stations were constructed in 18 states during the period, usually (but not always) in conjunction with pipeline construction, including: Alabama, Connecticut, Florida, Illinois, Kentucky, Louisiana, Maryland, Massachusetts, Minnesota, Nevada, New Jersey, New York, Ohio, Pennsylvania, South Carolina, Texas, West Virginia, and Virginia.

Compressor station costs are usually associated with new pipeline construction, but in some cases, expansion (add-on) projects may be included, or existing compressors may be adequate for the project. Once built, pipeline companies monitor system components to determine whether to repair, modify, or replace a compression unit or other system component, but because compressors are long-lived assets and can work cost-effectively for many years, economic comparison for efficiency upgrades or retrofit decisions can be quite complicated.

Table 1
Federal Energy Regulatory Commission Compressor Station Projects, 2014–19

Year	Number projects	Total hp (1000 hp)	Total cost ($million)	Aggregate cost ($/hp)
2018–19	23	682	2285	3352
2017–18	23	425	1154	2744
2016–17	19	320	949	2965
2015–16	14	228	673	2957
2014–15	16	223	629	2816
2014–19	95	1878	5690	3030

Source: OGJ, FERC.

Note: Total cost is comprised of materials, labor, miscellaneous, and land for projects completed between July 1, 201X and June 30, 201(X+1).

Compression cost is computed as an aggregate average or on an average project basis as:

$$\text{Aggregate}_{\text{hp}} = \frac{\Sigma\text{Total cost}}{\Sigma\text{Total horsepower}}$$

$$\text{Project}_{\text{hp}} = \frac{1}{\text{N}}\left(\sum\frac{\text{Project cost}}{\text{Horsepower}}\right)$$

where N represents the number of projects.

Project average statistics are distinct from aggregate average statistics and represents an upper bound on the aggregate average due to the method of computation:

$$\text{Aggregate}_{\text{hp}} < \text{Project}_{\text{hp}}$$

All state data are combined in a composite category. Reciprocating compressors and centrifugal compressors are not differentiated nor are location differences distinguished, and expansion (add-on) projects are excluded. No adjustment for inflation was applied.

From 2014 to 2019, annual aggregate unit horsepower cost ranged between $3352 and $2965/hp (Table 1), a relatively narrow interval. Installation requirements of equipment are subject to fewer variables than laying pipeline and exhibit much less variability in annual average cost. Installation cost also does not depend significantly on multiple factors contributing to well-defined project scopes and uniform execution.

Total cost for materials, labor, miscellaneous, and land for projects completed between July 1, 201X and June 30, 201(X+1) averaged 40%, 32%, 26%, and 1%, respectively, from 2014 to 2019 (Table 2). Miscellaneous and labor represent the largest categories but are reversed in order of percent contribution compared to pipeline construction, which is not unexpected. ROW costs are very small, usually less than 1% of total cost, again not surprising.

Regression models

The total installed cost of compressor stations exhibits a reasonably good fit with half of the project variability explained by total horsepower alone:

$$\text{Cost}_{>5000\text{hp}}\ (\$\text{million}) = 0.0028\ \text{Horsepower} + 3.9,\ R^2 = 0.83.$$

For small projects (e.g., <5000hp), the fixed-cost contribution posited by this model is not realistic, and alternative estimation methods are suggested such as applying a model without the fixed-term component:

$$\text{Cost}_{<5000\text{hp}}\ (\$\text{million}) = 0.0029\ \text{Horsepower},\ R^2 = 0.82.$$

Table 2
Federal Energy Regulatory Commission Compressor Station Category Cost, 2014–19

Year	Materials	Labor	Misc.	ROW	Total
2018–19	849	807	598	32.1	2285
2017–18	459	335	344	16.3	1155
2016–17	414	264	254	4.3	946
2015–16	282	196	159	15.3	663
2014–15	270	217	129	4.4	629
2014–19	2274	1819	1484	72.4	5678

Note: Completed project costs filed between July 1, 201X and June 30, 201(X+1).

Average statistics for small projects could also be utilized in lieu of the general cost model. Note that as the fixed term is eliminated in the parameterization, the model coefficient increases one part in 10000 and the model fit decreases one part in a 100, so this later model can be applied to both small and large compression projects.

Unit horsepower cost exhibit economies of scale with declining unit cost for larger capacity compressors (Figs. 1 and 2):

$$\text{Unit cost (\$/hp)} = 60{,}033\ \text{Horsepower}^{-0.302},\ R^2 = 0.35.$$

Limitations

The cost for compression on gathering systems may be 50% or more than transmission compression on a unit basis.

Historic cost statistics and regression models can only provide order of magnitude (±40%) estimates of pipeline projects because construction costs are sensitive to multiple site, location, and time factors. The best way to estimate cost is through a work decomposition procedure on a well-defined project scope by trained and experienced engineers near the time the project is scheduled to be performed.

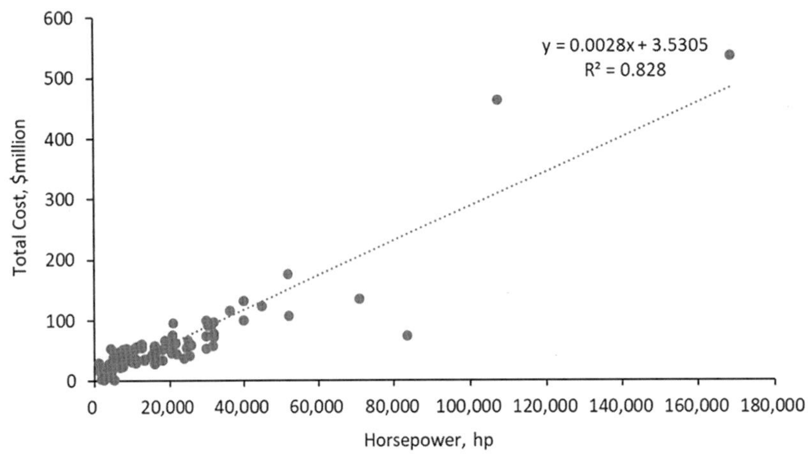

Figure 1. Federal Energy Regulatory Commission compression station project cost by horsepower, 2014–19.

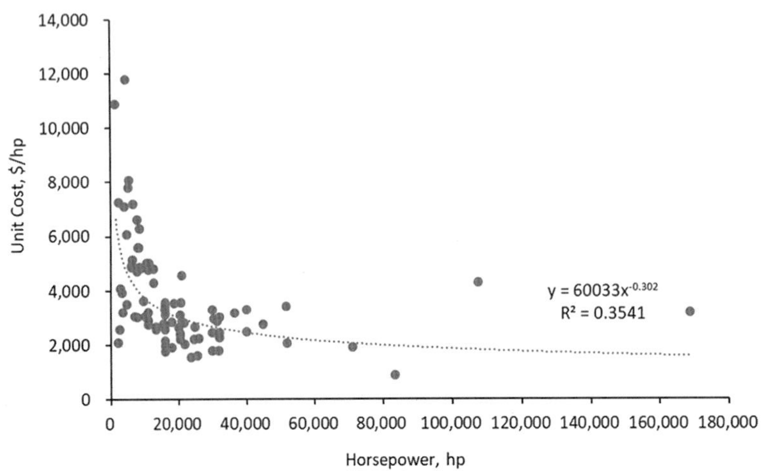

Figure 2. Federal Energy Regulatory Commission compression station unit cost by horsepower, 2014–19.

How to estimate gas transmission compressor station cost in the United States?

Example

Estimate the cost to construct a 30,000 and 2000 hp transmission compression station in 2022 in the United States. Should an inflation adjustment be performed? Will your estimates change for a gathering system?

Solution

Assuming the historic cost relations are a reasonable predictor of future cost, which for equipment and near future time periods are usually a reasonably good assumption, apply the general cost model relationships with the fixed-term coefficient without inflation-adjustment:

$\text{Cost}_{>5000\text{hp}} = 0.0029\ (30{,}000) = \$87$ million.

For the 2000 hp transmission compression-station, apply the historic data without the fixed-term coefficient.

$\text{Cost}_{<5000\text{hp}} = 0.0029\ (2000) = \$5.8$ million.

For near (<5 yr) term time periods, inflation adjustment may be performed if desired and deemed useful, but for longer time periods, inflation adjustment is almost always required. The cost models were derived from FERC compressor stations and do not apply to projects that are not similarly classified, although in some instances useful results may be inferred.

How to compare gas transmission compressor station cost in the United States?

Example

Compression station cost can be compared on a normalized basis using unit cost data. Compare the unit cost for 2000 and 40,000 hp transmission compressor stations built in the United States from 2014 to 2019. What is the rationale for economies of scale?

Solution

Using 2014 to 2019 project data, the average unit cost per horsepower is computed as follows:

Unit cost (2000 hp) $= 60{,}033\ (2000)^{-0.302} = \$6046/\text{hp}$.
Unit cost (40,000 hp) $= 60{,}033\ (40{,}000)^{-0.302} = \$2447/\text{hp}$.

Economies of scale often arise from engineering and material savings and construction practices that are inherent on large projects.

SCALING RULES

Equipment and process scaling rules

It is often necessary to estimate the cost of a piece of equipment or process when cost data are not available for the particular size or capacity involved. Compressors, pumps, and tanks are primary examples where scaling rules apply. A one-dimensional power relationship is often hypothesized to relate cost and capacity, and in special circumstances, a two-dimensional relationship may be applied.

One-dimensional scaling rule

A one-dimensional power relationship that relates construction cost and capacity is described by:

$$C(Q) = AQ^{\alpha}.$$

where $C = C(Q)$ represents the cost, Q represents the capacity, and A and α denote the power coefficients which need to be estimated.

In general, the more data available, the better the cost specification. If only two data points are available, a closed-form expression for A and α is easily derived using a log transformation. If more than two data points are available, the accuracy of the estimation will improve because a wider range of possibilities are considered, and the averaging process will minimize the impact of outlier data. If additional data are employed to enlarge the sample but the data are not properly screened for technology, location, or related conditions, more data will not necessarily increase the robustness of the analysis since other forms of uncertainty will enter the analysis and reduce the reliability of the assessment.

To estimate the unknowns A and α, take the logarithms of the one-dimensional cost equation:

$$\log C = \log A + \alpha \log Q$$

and apply ordinary least squares to the input data $\log C$ and $\log Q$ to obtain the best-fit estimate for $\log A$ and α. Curve fitting routines in Excel may also be applied directly.

Two-dimensional scaling rule

Application of the one-dimensional power law often simplifies the true cost relation since it only considers one factor in the determination of cost when more factors may be involved, such as operating specifications (pressure, temperature), equipment weight, flow rate, fabrication complexity, material selection, etc. The two-dimensional scaling rule is applied when two factors impact the cost function and adequate sample data exist to estimate the function parameters. The functional specification is written in product form using the factor X to denote another attribute that impacts cost:

$$C(Q, X) = AQ^{\alpha}X^{\beta}$$

There are no *a priori* expectations on the values of α and β. In many cases, sufficient data cannot be collected to specify the function even when two factors are known to be relevant. Generalizing the two-dimensional relation to three or more factors is possible but rarely performed because of limitations on sample data.

To estimate the unknowns A, α, and β, take the logarithms of the two-dimensional cost equation:

$$\log C = \log A + \alpha \log Q + \beta \log X$$

and apply ordinary least squares to the input data $\log C$, $\log Q$, and $\log X$ to obtain the best-fit estimate for $\log A$, α, and β.

How to estimate equipment cost curves

Example

Four heater treaters were purchased and installed in a field in Montana between 2005 and 2010. Estimate the cost to purchase and install a 150 bpd heater treater in 2011. Adjust cost data using the Marshall and Swift Equipment Cost Index and compare your result with the nominal (unadjusted) cost.

Year	Cost ($1000)	Capacity (bpd)	Marshall and Swift Equipment cost index
2005	200	250	1245
2007	150	120	1373
2008	80	50	1449
2010	110	75	1510
2011		150	1560

Solution

All cost data are time, location, and site-specific and must be adjusted (normalized) to a common basis to account for differences in these factors before evaluation. Since all the heater treaters were purchased and installed in fields in Montana, regional variations in cost are considered small and data only need to be adjusted for time and capacity. Inflationary effects are adjusted using the Marshall and Swift Equipment Cost Index (Figure 1), and capacity effects are modeled assuming a power function specification.

Input the cost and capacity data into Excel and then apply the built-in power function to plot the data and estimate the model coefficients (Figure 2). The estimated equation is Cost = 6.33 Capacity$^{0.67}$. The inflation-adjusted estimated cost of the 150 bpd heater treater is $184,110. The nominal best-fit equation is Cost = 9.34 Capacity$^{0.56}$, and without inflation adjustment, the cost is estimated at $157,553, a difference of about 15%.

Year	Cost ($1000)	Adjusted Cost ($1000)	Capacity (bpd)	Marshall & Swift Equipment Cost Index
2005	200	251	250	1,245
2007	150	170	120	1,373
2008	80	86	50	1,449
2010	110	114	75	1,510
2011	158	184	150	1,560

Figure 1. Nominal and adjusted cost for heater treaters.

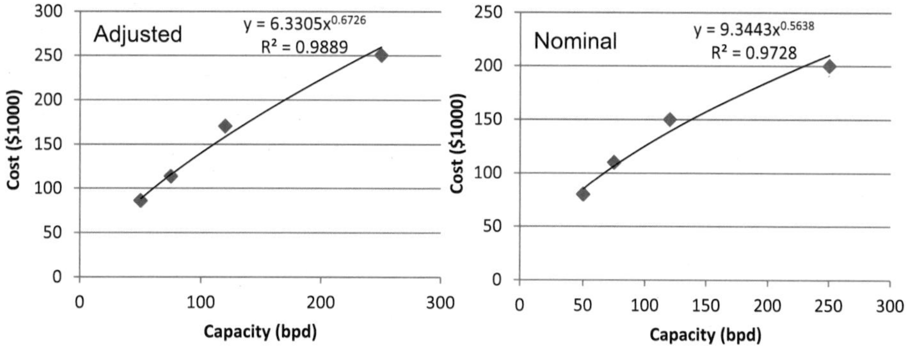

Figure 2. Heater treater cost curves for adjusted and nominal data.

How to estimate equipment cost using two-factor models

Example

The most common field equipment used to separate water out of produced oil is the vertical heater treater. Price quotes were obtained from a Texas vendor in 2011 based on the emulsion characteristics and flow rate of the well stream. Prices are reported in \$1000.

Emulsion (%)	Flow rate		
	875 bpd	1750 bpd	3000 bpd
3	140	250	450
6	160	300	550
8	200	400	740

(a) Estimate the cost function
$\text{Cost} = A(\text{Flow Rate, bpd})^{\alpha}(\text{Emulsion, \%})^{\beta}$
(b) Estimate the cost of a heater treater designed to handle a 10% emulsion and 2500 bpd flow rate.

Solution

Time and location are fixed and only capacity variation needs to be considered in cost estimation. Apply logarithms to the cost function to yield:

$$\log(\text{Cost}) = \log(A) + \alpha\log(\text{Flow Rate}) + \beta\log(\text{Emulsion})$$

and use regression to estimate the coefficients A, α, and β based on the data provided.

There are three unknowns and nine cost data so the functional specification can be reasonably estimated. Note that as the emulsion percentage and the flow rate increases, the cost of the unit increases, because of the greater equipment complexity and size.

Crude oil streams with high emulsions require greater capacity and time to process, and are more energy intensive to separate inside the equipment. A heater treater designed to handle a 10% emulsion falls outside the range of the data provided, and application of the cost function should be applied cautiously. In this case, since the 10% emulsion specification is reasonably close to the available data, extrapolation outside the data range is considered acceptable. The 2500 bpd flow rate falls within the source data and does not lead to additional uncertainty.

Logarithmic transformations of the input data are performed (Figure 1), and the regression coefficients are computed (Figure 2). The intercept refers to log(A), X Variable 1 refers to coefficient α, and X Variable 2 refers to coefficient β. The cost function for the vertical heater treater is $\text{Cost}(\$1000) = 0.094(\text{Flow Rate})^{1.00}(\text{Emulsion})^{0.41}$. The estimated cost of a heater treater capacity of handling a flow rate of 2500 bpd and 10% emulsion rate is \$603,801. Observe that cost scales linearly with flow rate but not emulsion percentage.

Example

Bolted steel tanks were once common for oil storage but have mostly been replaced with welded steel tanks. Average USGC construction costs for bolted steel tanks with roof deck and supports, sand and gravel foundation with steel retaining ring, painting, and basic fitting are reported in 2010 dollars. Estimate USGC 2014 steel tank cost curves as a function of capacity. Apply CEPCI for the time adjustment.

Capacity (bbl)	Diam × height (ft × ft)	Bolted tank (\$1000)	Welded steel (\$1000)
100	9 × 8	11.75	
200	9 × 16	19.25	
750	16 × 24	54.25	
1000	22 × 16	67.50	
2000	30 × 16	112.50	144.5
3000	30 × 24	127.50	163.0
5000	39 × 24	147.50	204.3
7500	39 × 36	181.00	236.8
10,000	55 × 24	219.50	291.8
15,000	55 × 36	285.00	366.8
30,000	80 × 34		570.8
50,000	90 × 44		808.8
75,000	120 × 36		1131
150,000	180 × 33		2073

Inputs

Cost ($1,000)	Flow Rate (bpd)	Emulsion (%)
140	875	3
250	1,750	3
450	3,000	3
160	875	6
300	1,750	6
550	3,000	6
200	875	8
400	1,750	8
740	3,000	8

Logarithms

Log(Cost)	Log(Flow Rate)	Log(Emulsion)
2.15	2.94	0.48
2.40	3.24	0.48
2.65	3.48	0.48
2.20	2.94	0.78
2.48	3.24	0.78
2.74	3.48	0.78
2.30	2.94	0.90
2.60	3.24	0.90
2.87	3.48	0.90

Coefficient	Value
A	0.094
B	1.000
C	0.413

Output

Capacity	Unit	Value
Flow Rate	bpd	2500
Emulsion	%	10
Cost	$1000	604

Figure 1. Estimating two-dimensional cost functions.

SUMMARY OUTPUT

Regression Statistics	
Multiple R	0.990296439
R Square	0.980687038
Adjusted R Square	0.974249384
Standard Error	0.039713508
Observations	9

ANOVA

	df	*SS*	*MS*	*F*	*Significance F*
Regression	2	0.480517597	0.240259	152.3361	7.20355E-06
Residual	6	0.009462976	0.001577		
Total	8	0.489980573			

	Coefficients	*Standard Error*	*t Stat*	*P-value*	*Lower 95%*	*Upper 95%*	*Lower 95.0%*	*Upper 95.0%*
Intercept	-1.028809706	0.202248345	-5.08686	0.00225	-1.523693578	-0.533925834	-1.523693578	-0.533925834
X Variable 1	0.999682979	0.060438959	16.54037	3.11E-06	0.851794175	1.147571783	0.851794175	1.147571783
X Variable 2	0.412840879	0.074043142	5.57568	0.001412	0.231663837	0.594017922	0.231663837	0.594017922

Figure 2. Regression summary output for heater treater cost function.

Solution

The best-fit power law cost models are readily computed using Excel's power curve relation as:

$$\text{Cost(Bolted Steel Tank)} = 0.84\ \text{Capacity}^{0.62}$$

where cost is described in \$1000 and capacity is described in barrels (Figure 3). The CEPCI inflation factor, 580/551 = 1.05, adjusts the multiplicative coefficient to 0.88 but does not impact the power coefficient of the cost curve.

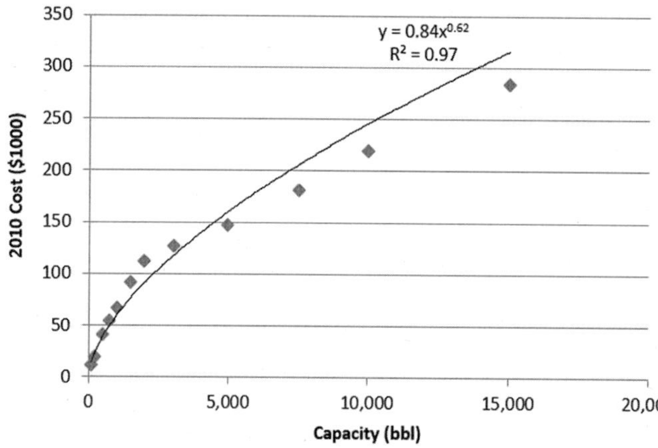

Figure 3. Bolted steel tank construction cost curve circa 2010.

How to estimate equipment cost curves using two data points

For two pieces of equipment such as pumps or compressors with given cost and capacity, the coefficients K and a of the cost curve:

$$\text{Cost} = K\ \text{Capacity}^a$$

can be determined directly using a closed-form expression. Write the cost functions for the two units as:

$$\text{Cost}_1 = K\ \text{Capacity}_1^a,\ \text{Cost}_2 = K\ \text{Capacity}_2^a,$$

take the ratio and apply logarithms, canceling out the K-term:

$$\frac{\text{Cost}_1}{\text{Cost}_2} = \frac{\text{Capacity}_1^a}{\text{Capacity}_2^a} = \left(\frac{\text{Capacity}_1}{\text{Capacity}_2}\right)^a$$

$$\log \text{Cost}_1 - \log \text{Cost}_2 = a\ [\log(\text{Capacity}_1) - \log(\text{Capacity}_2)].$$

Solving for a yields:

$$a = \frac{\log\text{Cost}_1 - \log\text{Cost}_2}{\log(\text{Capacity}_1) - \log(\text{Capacity}_2)}$$

The value of K is found using the cost data from either unit and the value of a:

$$K = \frac{\text{Cost}_1}{\text{Capacity}_1^a} = \frac{\text{Cost}_2}{\text{Capacity}_2^a}$$

Example

Sour water is generated when steam or water are used to process various streams within the refinery. Hydrogen sulphide and ammonia are transferred to the water, and these contaminants must be removed prior to the sour water being processed in the waste water treatment plant. Sour water strippers (SWSs) are used with steam to remove these contaminants. In 2008, two SWS were installed in the Mid-continent U.S., a 150 gpm unit for \$13.9 million, and a 95 gpm unit for \$9 million. Estimate the cost of a 135 gpm SWS unit in 2013.

Solution

From the data given, $K = \$119{,}000$ and $a = 0.95$. The cost of a 135 gpm unit in 2008 is estimated to be \$12.6 million, and adjusted to 2013 dollars using the CEPCI is \$12.4 million.

TARIFF RATES

What are reservation and commodity charges?

Transportation and gathering fees depend on the location of the property, volumes and distance transported, pipeline ownership and regulation, product transported, age of the pipeline(s), capital investment, and operational requirements. Age-related factors can act in favor of or against the shipper. For example, after the pipeline is fully depreciated, and if the pipeline is well maintained and has adequate flow, tariffs may decline after the investment cost are fully depreciated. On the other hand, when volumes in the line decline, corrosion usually increases and maintenance becomes more difficult, and the pipeline owner may raise rates on transporters to recover their cost of service.

Pipelines are built in segments. Each segment has a different cost associated with construction, operation, and maintenance, and different rates apply to transport fluid from one area to another depending on the segment of pipeline where the fluid is injected. The tariff transmission rate schedule indicates where product is received from and where it will be delivered to.

The monthly bill for deliveries usually includes a reservation and commodity charge, and other charges as applicable. The reservation rate is paid to reserve the firm capacity regardless of usage. The commodity rate is paid based on volumes committed and shipped. Interruptible service is based on transportation being on an interruptible basis.

- Reservation charge: Equal to the product of the reservation rate multiplied by the maximum daily quantity (MDQ) specified in the service agreement, multiplied by the shipper specific heating value and the number of days in the month.
- Commodity charge: Equal to the commodity rate multiplied by the quantity of gas allocated to the delivery point in the month.
- Other charges: Any applicable surcharges, new facilities charges, repair charges, and any incidental expenses.

The sum of the charges is equal to the transportation fee for a segment of line. If product is delivered using two or more segments, a separate tariff may apply to each segment.

Historically, ownership of export pipelines in the United States, especially gas trunklines, has seen significant activity from gas transmission companies, and if these lines are considered jurisdictional (gas) or common carriers (oil), FERC oversight will impact tariff rates. If pipelines are regulated by FERC, the transportation agreements are in the public domain, otherwise, the agreements are generally confidential.

In North America, the heat value or energy content of a fuel is often expressed in Btus, a British thermal unit, the amount of energy needed to heat or cool one pound of water by 1° Fahrenheit. When dealing with larger quantities of energy, larger standards of measure are often used. In many natural gas pipelines, energy content is measured in decatherms (Dth), 10 therms, or 1,000,000 Btu (1 MMBtu).

Example. The Mars A platform (Figure 1) in Mississippi Canyon block 801 was one of the first regional hubs in the United States. Gulf of Mexico and circa 2020 serves as a central processing facility for several fields. The Mars A pipeline is a 163-mi line originating approximately 130 miles offshore and delivers crude to salt dome caverns in Clovelly, Louisiana.

Figure 1. Mars A platform. Source: Shell Offshore.

Crude production from Mars A is transported to West Delta 143 at the rate of \$2.61/bbl and then onward to Bay Marchand block 4 at \$1.16/bbl if < 30,000 bbl/mo was delivered and \$0.7/bbl if > 30,000 bbl/mo was transported (Table 1). A discounted rate applies because greater volumes for the pipeline means greater revenue for the transporter to cover fixed cost. From Bay Marchand product is delivered to Fourchon and into storage at Clovelly/Caverns.

Crude production from Mars B (Olympus) to WD 143 is delivered in a newly constructed pipeline and because it is newer and its capital cost of construction greater than the adjacent Mars A pipeline, its tariff rate for the first pipeline segment will be higher.

Table 1
Mars Pipeline Tariff Firm Transportation

From	To	Rate (cents/bbl)
MC 807 (Mars A)	WD 143	¢121.39
WD 143	Bay Marchard 4	¢116.61 if < 30,000 bbl ¢70.26 if > 30,000 bbl
Bay Marchand 4	Clovelly	¢60.24
Bay Marchand 4	Fourchon Terminal	¢1.19
Fourchon Terminal	Clovelly/Caverns	¢43.26

Source: Shell Midstream Partners.

How to calculate firm service with fixed maximum daily quantity and reservation charge?

Example. A traditional firm service with fixed maximum daily quantity and reservation charge (FT-1 service) is depicted in Table 1 for the Destin Pipeline.

Table 1
Destin Pipeline, Rate Schedule FT-1 Firm Transportation Rate

	Monthly reservation (\$/Dth/mo)	Daily reservation (\$/Dth)	Transportation rate (¢/Dth)
Maximum rate	\$7.19	\$0.237	0.3¢
Minimum rate	0.00	0.00	0.3¢
Fuel retention percentage	0.3%		

Source: Destin Pipeline.

Solution. For a shipper reserving 50 MMcf/d capacity with a heat content of 1000 Btu/cf, the monthly reservation rate is computed as:

$$\frac{\$7.19}{\text{Dth/mo}}\cdot\frac{50\ \text{MMcf}}{\text{d}}\cdot\frac{1000\ \text{Btu}}{\text{cf}}\cdot\frac{30\ \text{d}}{\text{mo}}=\$10.785\ \text{million.}$$

If the actual average amount of gas delivered to the pipeline was 48 MMcf/d for the month, the commodity rate is computed as:

$$\frac{\$0.237}{\text{Dth}}\cdot\frac{48\ \text{MMcf}}{\text{d}}\cdot\frac{1000\ \text{Btu}}{\text{cf}}\cdot 30\ \text{d}=\$341,280.$$

The transportation rate 0.3¢/Dth = \$0.03/Dth is computed based on volumes delivered and being an order of magnitude smaller than the commodity rate is negligible in this case. The fuel retention percentage 0.3% represents a percentage of the quantity of gas delivered for transportation and used by the pipeline company for compressor fuel and gas otherwise used, lost, or unaccounted for and is also negligible.

The total cost for service for the month is \$11.126 million, 3% of which is due to the variable cost and 97% for the reservation rate.

How is the cost of service formula applied?

Cost of service is the primary methodology for setting rates on FERC-regulated pipelines, although rates may also be market-based or set by negotiation. The underlying principal of cost recovery is the universal theme in ratemaking.

When a pipeline is installed, the tariff is used to recover the investment, operating, and maintenance expense and various other secondary costs. The investment and owner of the line and the way it is regulated are the most important factors that determine tariffs, but the age of the line, number of customers, volume throughput, and special circumstances (e.g., hurricane damage) are also important in determining rates at different times in the lifecycle of the pipeline.

Producer and producer affiliate owners may charge different rates than third-party owners and, if the line is FERC-regulated, the process is open to public review. If the pipeline is old and the capital investment is mostly recovered (i.e., pipeline construction cost is fully depreciated), fees will need to cover only operating and maintenance cost. If the number of customers and amount of product transported through the pipeline at this time is low, however, then the operator may have to increase rates on remaining subscribers to generate adequate revenue to cover operations and maintenance. The conditions governing each system are unique and therefore many factors potentially impact tariffs. Generalized statements are difficult to make. Any empirical analysis, for example, must consider all relevant factors in comparison to be defensible, which often precludes meaningful evaluations because of the difficulty of analysis.

The cost-of-service (COS) formula is the most common ratemaking basis and is the easiest to understand. COS is defined as the amount of revenue a regulated gas pipeline company must collect from rates charged consumers to recover the cost of doing business. These costs include operating and maintenance expenses, depreciation expense, taxes, and a reasonable return on the pipeline's investment.

The total COS includes the product of the pipeline's Rate Base (investment, or capital expenditures) and the Overall Rate of Return (ROR), which is negotiated with FERC, plus its Operation and Maintenance Expenses (O&M), General and Administrative Expenses (G&A), Depreciation Expenses (DEP), Nonincome Taxes (NTAX) and Income Taxes (TAX), less Revenue Credit (RCD). The COS formula collects all these terms:

$$\begin{aligned} \text{COS} = &\ \text{Rate Base} \times \text{Overall ROR} + \text{OPEX} + \text{GA} + \text{DEP} \\ &+ \text{NTAX} + \text{TAX} - \text{RCD}. \end{aligned}$$

The Rate Base represents the total investment of the pipeline and is used to compute certain components of the COS, including the return on its investment, and the depreciation expense that permits the pipeline to recover its investment.

$$\begin{aligned} \text{Rate Base} = &\ \text{Gross Plant} - \text{Accumulated Depreciation} \\ &- \text{Accumulated Deferred Income Taxes} \\ &+ \text{Working Capital}. \end{aligned}$$

Gross plant is the original cost of the plant or facilities owned by the pipeline and includes the cost of land and land rights, right of way, surveys, line pack, construction costs, overheads, and accumulated funds used during construction (AFUDC). The cost of the investment in gross plant is recovered through the cost of service as depreciation expense.

For a detailed treatment of FERC cost of service methodology, see (FERC, 1999). The procedures are easy-to-follow, but in many cases, calculations become tedious because of special situations. An example illustrates the basic steps.

Example

The Garden Banks Gathering System (GBGS) was constructed in the mid-1990s to serve deepwater and shelf natural gas production developed in the Garden Banks area offshore Louisiana, including the Auger, Baldpate, and Enchilada projects (Figure 1). GBGS is a 30-inch diameter, 50-mile natural gas pipeline extending from the Garden Banks 128 A platform to the P platform in South Marsh Island 76. The line was determined to be a FERC jurisdictional facility, and, under protest, the pipeline owner Garden Banks Gas Pipeline LLC (GBGP)—a limited liability company made up of Shell Enchilada Gas Pipeline Company and Hess Garden Banks Gas Gathering, Inc.—filed an application to section 7(c) of NGA requesting authorization to construct the 30-inch line and appurtenant facilities.

GBGP proposed a maximum reservation rate under two firm transportation services FT-1 and FT-2 and a maximum interruptible service IT-1 commodity rate of \$0.1214 per MMBtu. The proposed initial rates were based on:

- Estimated facility cost of \$108 million, which include AFUDC.
- Straight fixed variable cost classification.
- A proposed return on equity of 13.75%.
- Btu factor of 1100 Btu/cf based on expected average Btu content of Auger, Baldpate, and Enchilada production.
- Initial billing determinant of 241 MMBtu based on a 60% average annual load factor for the 1000 MMcf/d capacity of the 30-inch line.
- Debt/equity structure of 21.5/78.5%, which is the weighted debt/equity structure of Shell Oil Company and Amerada Hess Corporation at the time of application.

The cost of service is calculated using GBGP's estimated cost of facilities (\$108 million), engineering estimates for operation and maintenance expenses, a pretax return of 13.75%, and a depreciation rate of 6.67% based on a 15-year depreciation rate. The pretax return is negotiated with FERC and based on approved settlement rates, and the final cost is based on the actual cost of construction. For calculation purposes, we assume no reserve for depreciation, \$1 million working capital, operation and maintenance expense of \$3 million per year, and taxes other than income tax at \$2 million.

Based on the contract quantity of 600 MMcf/d, the annual heat content delivered is computed as:

$$\frac{600\ \text{MMcf}}{\text{d}} \cdot \frac{1100\ \text{Btu}}{\text{cf}} \cdot \frac{\text{Dth}}{\text{MMBtu}} \cdot \frac{365\ \text{d}}{\text{y}} = 241\ \text{MMDth}.$$

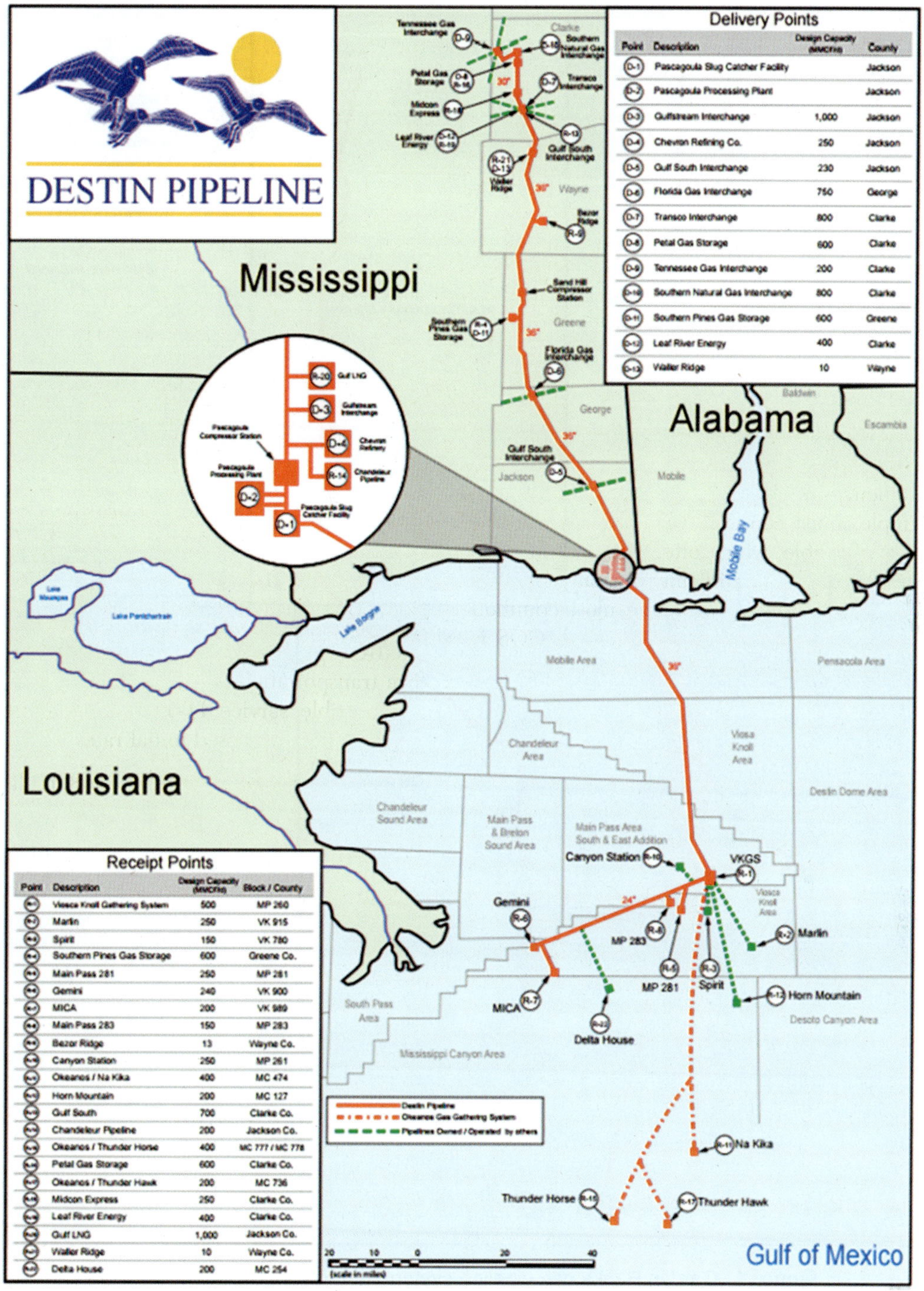

Figure 1. Destin Pipeline system map circa 2020. Source: FERC.

The total cost of service is computed to be $26.9 million and is used to deliver 241 MMDth per year, yielding a reservation rate of $0.112/Dth. The calculation steps are shown in Table 1.

Statement of Rates varies based on business practices and the drafter of the instrument. Despite the wording used, all costs usually indicate the maximum rate unless specifically designated as minimum rates. In addition, certain additional charges or surcharges may be negotiated for and will be indicated in the rate schedule.

Interruptible transportation service and firm transportation services are usually available, and the terms will vary depending on the type of service provided.

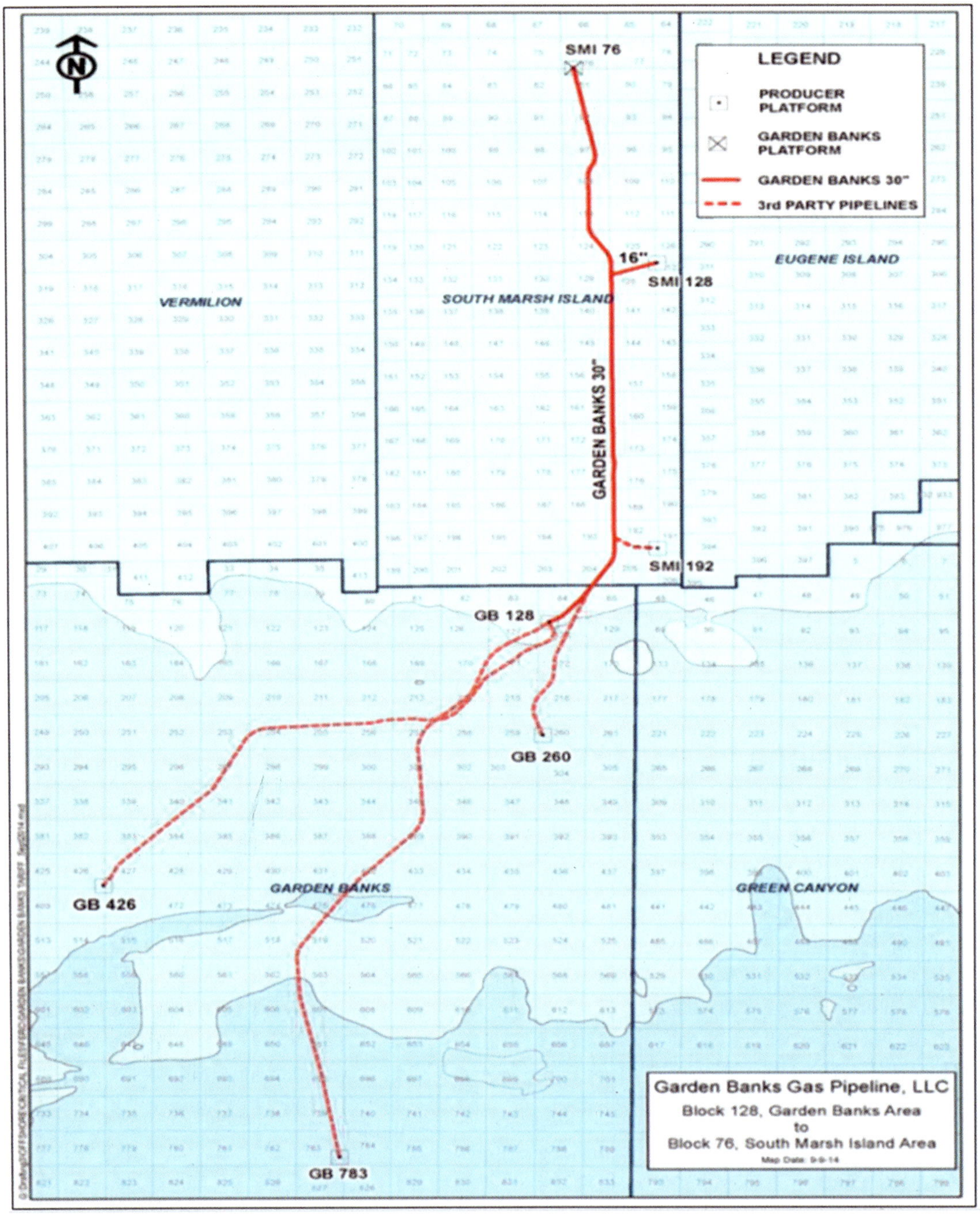

Figure 1. Garden Banks gas pipeline system map circa 2020. Source: FERC.

IT-1 service

Interruptible transportation service for GBGS is available under Rate Schedule IT-1 to anyone who requests transportation of natural gas on an interruptible basis through the 30-inch line and who has executed an IT-1 agreement with GBGP. The rate schedule applies to the maximum daily quantity (MDQ) and offered at the maximum commodity rate of $0.1168 per MMBtu.

Firm transportation

FT-1 service describes a traditional firm service with fixed MDQ and reservation charge while FT-2 service describes flexible firm service with variable MDQ and rates determined, in part, based on volumes committed and shipped.

FT-1 service uses a fixed MDQ and a reservation charge regardless of actual throughput and is offered at the maximum reservation rate of $0.1168 per MMBtu of demand. There is no commodity rate.

Table 1
Garden Banks Gathering System Rate Base and Cost of Service

Description	Units	Costs
Rate base and return		
Gas plant in service	$	108,000,000
Working capital	$	1,000,000
Rate base	$	107,000,000
Cost of service		
Pretax return (@13.75% of rate base)	$	14,712,500
Operation and maintenance expenses	$	3,000,000
Depreciation expense (@ 6.67% gas plant)	$	7,200,000
Taxes other than income taxes	$	2,000,000
Total cost of service	$	26,912,000
Annual contract quantity	Dth	241,000,000
Daily reservation rate	$/Dth	0.112

Source: FERC.

FT-2 service provides flexible firm service with variable MDQs and rates determined on the basis of volumes committed and shipped. To be eligible for FT-2 service, a shipper must execute a Reserve Commitment Agreement making a life of reserves commitment of its share of production from specified OCS blocks. The shipper is not required to warrant or guarantee that any quantity of gas will be produced.

Reference

Federal Energy Regulatory Commission, Cost-of-service rates manual. Federal Energy Regulatory Commission, Washington, DC, 1999.

How to compute price adjustments in pipeline system quality banks

To account for the difference in quality and value of crude oil streams entering and exiting different pipeline systems, quality banks are commonly used to make monetary adjustments among shippers to account for the difference in quality between the crude oil tendered by the shipper and the crude oil received by that shipper at the destination point. The terms and conditions of quality banks are specific to the streams transported, and for illustration, GEL Offshore Pipeline LLC's Eugene Island Pipeline System Quality Bank Policy in the U.S. Gulf of Mexico issued January 11, 2012 and effective February 11, 2012 is described.

Three large tables in the Eugene Island Quality Bank document, referred to as Exhibits A, B, and C, are used to adjust for differences in API gravity (Exhibit A), sulfur content (Exhibit B), and a ratio normalizing to a reference 35.5°API common stream (Exhibit C). The tables are not shown here, and instead the tabled data are transformed into formula for easier presentation and application:

$$\text{Gravity differential per bbl} = 1.25 + 0.15\ (°\text{API}-10)$$
$$\text{Sulfur differential per bbl} = 1 + \%\text{S}$$
$$\text{Ratio factor} = 0.00003\ (°\text{API})^2 - 0.0085\ (°\text{API}) + 1.2596.$$

For illustration assume three shippers transport 650 total barrels (Table 1). Shipper A transports 250 bbl using two carriers; Shipper B transports 300 bbl using one carrier, and Shipper C transports 100 barrels using one carrier. The barrels received from each carrier and the crude quality for each are depicted in the first five columns, and the mostly

Table 1
Data for Quality Bank Calculations

Shipper	Carrier	Barrels Received	Recent Sulfur	API Gravity	From Exhibit C Ratio to 35.5°WT	From Sulfur Times Ratio	From Exhibit B Sulfur Differential	From Exhibit A Gravity Differential	Barrels Received × Sulfur Differential	Barrels Received × Gravity Differential
A	1	100.00	0.92	29.8	1.03544	0.95	1.950	4.550	195.00	422.00
A	2	150.00	0.36	38.6	0.98172	0.35	1.350	5.060	202.50	759.00
B	1	100.00	0.42	36.4	0.99461	0.42	1.420	5.020	142.00	502.00
B	1	200.00	0.78	46.2	0.93976	0.73	1.730	4.920	346.00	984.00
C	2	50.00	0.66	32.8	1.01644	0.67	1.670	4.670	83.50	233.50
C	2	50.00	0.81	30.1	1.03343	0.84	1.840	4.265	92.00	213.25
Total		650.00							1061.00	3113.758

Note: (1) Common stream weighted average gravity value is equal to 3113.75/650.00 = 4.79038.
(2) Common stream weighted average sulfur value is equal to 1061.00/650.00 = 1.63231.

self-explanatory calculations are depicted in the rest of the table.

A weighted aggregate common gravity and common sulfur term is first computed from the total products transported. Thus, from the last two columns of Table 1, the common stream weighted average gravity value is equal to 3113.75/650 = 4.79038, and the common stream weighted average sulfur value is equal to 1061/650 = 1.63231, and then each shipper's payment into the bank is computed.

Shipper A needs to pay the bank $6.02, shipper B receives from the bank $50.58, and shipper C pays the bank $44.56. The net transaction is zero dollars. Shipper A calculations for each carrier are shown in Table 2. Shipper B and C calculations are performed similarly.

Table 2
Quality Bank Calculations for Shipper A, Carriers 1 and 2

Shipper A; Carrier 1		
Weighted average GRAVITY value: 422.00/100 = 4.22000		
Calculation (4.79038–4.22) × 100 =	$57.04	
Weighted average SULFUR value: 195.00/100 = 1.95000		
Calculation (1.95–1.63231) × 100 =	$31.77	
Shipper A; Carrier 1 total:		$88.41
Shipper A; Carrier 2		
Weighted average GRAVITY value: 759.00/150 = 5.06000		
Calculation (4.79038–5.06) × 150 =	($40.44)	
Weighted average SULFUR value: 202.50/150 = 1.35000		
Calculation (1.35–1.63231) × 150 =	($42.35)	
Shipper A; Carrier 2 total:		($882.79)
Shipper A pays the bank:		$6.02

21: Health, Safety and Environment

Health 1026
Grading the nation: state disclosure policies for lead pipes 1026

Safety 1031
Trenching and excavation safety 1031
Crane and hoist safety 1033
Electrical welding safety 1035
Tank fire types and causes 1038
Firefighting protection systems for atmospheric storage tanks 1040
Carbon dioxide pipeline safety and integrity 1042

Environment 1044
U.S. natural gas standards and regulations development 1044
What are the key performance indicators in the U.S. pipeline industry? 1045
Oil spills in the United States, 2002–2016 1047
Estimating oil spill thickness on water visually 1049
Fast discharge volume calculator for offshore leaking pipeline 1051
How to estimate discharge volume for offshore leaking pipeline (fast method) 1053
Advanced discharge volume calculator for offshore leaking pipeline 1053
How to estimate discharge volume for offshore leaking pipeline (advanced method) 1056
Guidelines on oil characterization to inform spill planning and decision making 1058
Oil spills inland response good practice guidelines 1066
Calculating volatile organic compounds flash emissions from crude oil and condensate tanks at oil and gas production sites 1083

HEALTH

Grading the nation: state disclosure policies for lead pipes

Lindsay McCormick, Chemicals and Health Project Manager, **Samantha Lovell**, Health Project Specialist, **Thomas Neltner**, Chemicals Policy Director, Environmental Defense Fund, Washington, DC

When purchasing a home, buyers expect to be informed about deficiencies, defects, or environmental hazards on the property. Since 1996, they have been told about lead in paint. However, the likelihood that a buyer will be told their prospective home has lead pipes, including a lead service line, depends on where they live.

Lead service lines (LSLs)—the lead pipes connecting water mains under the street to homes and other buildings—are the main source of lead in contact with drinking water. Homebuyers deserve to know about this liability when they choose a home and negotiate a price. When done properly, removing the full LSL significantly reduces the risk of lead exposure.

In 2017, Environmental Defense Fund (EDF) analyzed and graded the housing disclosure policies of all US states and the District of Columbia (D.C.) according to their ability to help homebuyers make informed decisions about LSLs before they sign a sales contract. We did not address the extent to which LSLs are actively being disclosed under each policy. Four states—Connecticut, Delaware, New York, and Pennsylvania—and the District of Columbia scored an A-. Twenty-one states scored D or F. The remaining 25 states scored a B or C because they help buyers but are silent or ambiguous on lead pipes or disclosure of lead pipes is voluntary.

EDF sees significant opportunities for states to help protect homeowners from lead by improving their disclosure requirements. Water utilities can help this effort by informing all property owners if they are likely to have an LSL. Home inspectors can also help by checking the service line as it comes into the home and letting the buyer know whether or not it is lead and recommending replacement.

Lead in drinking water

LSLs are the largest potential source of lead in drinking water. Other sources include lead solder and brass fixtures. An estimated 6–10 million homes across the United States have an LSL. Water utilities typically do not know how many LSLs are in the system or where they are located. Many utilities address the problem of LSLs by treating the water to build a protective coating on the inside of the pipe to prevent the leaching of lead, a process known as corrosion control. However, corrosion control can fail when the line is disturbed, resulting in the release of unpredictable levels of lead into drinking water. The most effective way to deal with LSLs in the long term is to locate and fully replace them using methods shown to protect residents. Housing disclosure policies can help create market incentives for removal.

Property disclosures

In the 1980s, many states began requiring sellers to proactively disclose to buyers information about known property defects. Requirements differ by state, and some states do not have disclosure requirements at all. Disclosure laws are intended to protect buyers from purchasing a property without full knowledge of potential defects. They also help protect sellers from legal liability.

Why is reducing exposure to lead important?

A portion of any lead in the water we drink is absorbed into our blood. The science now makes clear there is no safe level of lead in the blood of children. The lead is likely to impair children's normal brain development, contribute to learning and behavioral problems, and lower IQs. Despite decades of progress in reducing children's exposure to lead, America continues to have a toxic legacy of lead. Approximately 500,000 children have elevated blood lead levels and poor and minority children remain at the greatest risk.

While the only federal housing disclosure requirement for environmental hazards is for lead-based paint, many states have requirements or policies that would trigger disclosure for LSLs. As noted earlier, this report does not analyze the extent to which policies are actively being used to provide buyers with such information. The Lead Service Line Replacement Collaborative, a diverse group of 28 organizations that aims to accelerate full LSL replacement, identified expanding federal, state, and voluntary disclosure policies to include LSLs as an opportunity to help consumers make informed decisions. EDF believes that buyers deserve to know about the presence of environmental hazards, including the presence of an LSL, on property they are considering buying. An informed buyer can decide how to value the property and take appropriate precautions. If the property has an LSL, the buyer can decide to add the cost of replacement to the mortgage, deduct the estimated cost from the sale price, demand replacement prior to purchase, or plan to replace it later. Over time, increased transparency can increase market incentives to replace LSLs.

1. Does the state law require any disclosures of deficiencies, defects, or environmental hazards at sale?
2. Is there a required or voluntary form for disclosure?
3. Does a seller need to disclose knowledge of lead pipes or pipe material?
4. Does a seller need to disclose knowledge of environmental hazards generally?

Using these questions, we defined the conditions for each letter grade. The highest grade we gave was an A- to reflect the fact that even the top performing states have room for improvement. States with mandatory disclosure specifically asking about lead pipes received an A-. All other states were compared to these top performers. We did not assess enforcement or confirm that disclosure of lead pipes is actually occurring in each state. A summary of state ratings are shown in Fig. 1 (last updated March 2019).

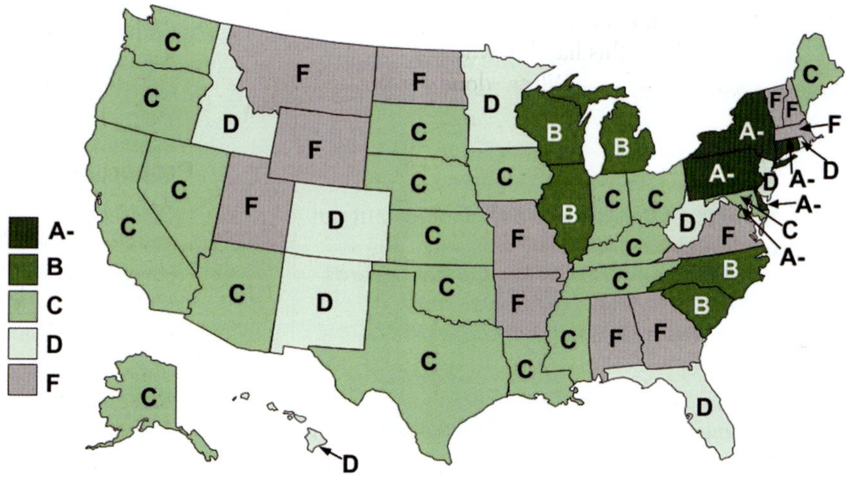

Figure 1. Summary of state ratings for lead service line disclosure circa March 2019.

How did we grade the nation?

EDF graded the housing disclosure policies of all states and D.C. according to their ability to help homebuyers make informed decisions about LSLs before they sign a sales contract. We started our search using the database of legal articles at www.Nolo.com to identify state disclosure laws, required disclosure forms, and voluntary disclosure forms. Where we could not identify a publicly available form on a state government website, we looked at the state realtor association website to give credit for voluntary programs not supported by the state. If we could not find a form, we followed-up with state realtor associations to access additional voluntary disclosure forms. Ultimately, if the form was only accessible to members, we did not give them credit (noted with c in the table). We assessed the following aspects of the seller disclosure policies:

What did we find?

Four states (Connecticut, Delaware, New York, Pennsylvania) and D.C. scored an A- because they have a state-required disclosure form that specifically asks if the home has lead plumbing.

New York: A *(Mandatory disclosure of lead pipes)*
Mandatory form asks, "Is lead plumbing present? If yes, state location or locations below."

North Carolina: B *(Mandatory disclosure of material)*
Mandatory form asks, "Water pipes are: Copper/Galvanized/Copper/Plastic/Polybutylene/Other."

Grade	Description	No. of States	States	
A-	**Mandatory disclosure of lead pipes** (*Example*: *State-required form asks, "Is lead plumbing present? If yes, state location or locations."*)	5	Connecticut D.C. Delaware	Pennsylvania New York
B	**Mandatory disclosure (M) of pipe material (lead not specifically addressed) or lead pipes if seller determines conditions unsafe** (*Example*: *State-required form asks, "Type of plumbing system: Copper/Galvanized/Plastic/Polybutylene/Unknown/Other"*) **Voluntary disclosure (V) of lead pipes** (*Example*: *Voluntary form asks, "Type of water supply pipes: Lead/Galvanized Copper/Polybutylene/Other/Don't know"*)	5	Illinois (M)[a] Michigan (M) North Carolina (M) South Carolina (M) Wisconsin (M)[a]	
C	**Mandatory disclosure (M) of general environmental hazards** (*Example*: *State required form asks, "Have there been or are there any hazardous conditions on the property, such as methane gas, lead paint, radon …"*) **Voluntary disclosure (V) of pipe material (lead not specifically addressed)** (*Example*: *Voluntary form asks, "Are you aware of the type of water pipes, such as galvanized, copper, PVC, CPVC, or polybutylene?"*)	20	Alaska (M) Arizona (V) California (M) Indiana (M) Iowa (M) Kansas (M) Kentucky (M) Louisiana (M) Maine (M) Maryland (M) Mississippi (M)[b]	Nebraska (M) Nevada (M) Ohio (M) Oklahoma (M) Oregon (M) South Dakota (M) Tennessee (M) Texas (M) Washington (M)
D	**Mandatory disclosure (M) of defects and deficiencies but not specifically environmental hazards** **Voluntary disclosure (V) of general environmental hazards** (*Example*: *Voluntary form asks, "Have there ever been substances, materials, or products which may be an environmental hazard …"*)	9	Colorado (V) Florida (M)[c] Hawaii (M)[c] Idaho (M)[c] Minnesota (V) (M)	New Jersey (V) New Mexico (M)[c] Rhode Island (M)[c] West Virginia (V)
F	**Limited or no disclosure requirements** (*Example*: *Disclosure only required if seller "knows the home may pose a health or safety risk to the buyer …"*)	12	Alabama Arkansas[c] Georgia[c] New Hampshire[c] North Dakota Massachusetts[c]	Missouri[c] Montana[c] Utah[c] Vermont[c] Virginia Wyoming[c]

[a]Disclosure if unsafe concentrations or unsafe conditions related to lead in water pipes.
[b]State also has less detailed mandatory requirements.
[c]Voluntary state realtor association disclosure form does address lead pipes or environmental hazards generally, but is not made public by the association.

Five states received a B. Of these, three states require disclosure of pipe material (lead is not explicitly addressed) and two states require disclosure of unsafe conditions or unsafe concentrations related to lead in water pipes.

Twenty states scored a C because they require disclosure of environmental hazards generally but do not address lead pipes specifically. One state, Arizona, scored a C because the voluntary disclosure used by realtors requires identification of pipe material.

Nine states scored a D because they have mandatory disclosure requirements that do not broadly address environmental hazards and/or they have a voluntary disclosure form that asks about environmental hazards generally.

Twelve states failed because they lack or have extremely limited disclosure requirements.

Fourteen states have voluntary disclosure forms developed by the real estate association that are available only to member licensed agents or available for purchase (noted with c). If these associations were to make their forms publicly available, it would have the potential to benefit more people. For example, Missouri, Rhode Island, New Mexico, and Vermont appear to have voluntary real estate association forms not made public by the associations that specifically address lead pipes or lead in water.

Variation and limitations of state requirements

We found a remarkable amount of variance in state disclosure laws regarding LSLs—from states that require sellers to fill out detailed, 10+ page disclosure forms to "buyer beware" states, where the responsibility is on the buyer, not the owner, to investigate hazards.

California: C *(Mandatory disclosure of env. hazards)*

Mandatory form asks, "Are you ... aware ... of any of the following substances, materials, or products which may be an environmental hazards such as, but not limited to, asbestos, formaldehyde, radon gas, lead-based paint, fuel or chemical storage tanks, and contaminated soil or water on the subject property?"

Colorado: D *(Voluntary disclosure of env. hazards)*

Voluntary form asks if there are/is, "Hazardous materials on the property, such as radioactive, toxic, or biohazardous materials, asbestos, pesticides, herbicides, wastewater sludge, radon, methane, mill tailings, solvents, or petroleum products?"

North Dakota: F *(Limited disclosure requirements)*

Buyer beware: The responsibility is on the buyer to investigate hazards.

A buyer's likelihood of being informed of whether a home has an LSL before signing a sales contract depends entirely on the state. However, it is important to note that there are limitations to the effectiveness of disclosure even in those states where it is required by law. A seller only needs to disclose information they know, and sellers are usually given the option to select "unknown" on forms. States generally do not require the seller to examine the property for hazards, and a real estate agent may discourage a seller from performing testing to avoid disclosing negative information to buyers. Presumably, if a utility notifies a property owner that there is an LSL or makes maps of LSL locations available online, it should result in a more effective disclosure.

None of the state laws specifically address LSLs, but rather "lead pipes" or "lead plumbing." A seller may interpret this language as referring only to a house's internal plumbing, as opposed to the service line connecting the water main in the street to the house. While EDF believes the language in these laws is broad enough to create a framework for LSL disclosure, actual practice may differ.

The timing of providing disclosures to buyers is critical. Unfortunately, state laws often do not prescribe timing beyond requiring delivery of a disclosure statement *before* the purchase agreement is fully executed. Ideally, the presence of an LSL would be disclosed as early in the process as possible, including on any online listing of the property and on-site during open houses.

States should require disclosure of lead pipes and LSLs

States have significant opportunities to more effectively inform and protect buyers by integrating lead pipe disclosure into state disclosure requirements and policies. Connecticut, Delaware, New York, and Pennsylvania are useful examples, which each earned an A- by requiring sellers to specifically inform a buyer if the property has lead pipes. However, all states can do better, including those earning an A-. Those five states that earned a B by requiring identification of pipe material could improve by specifically listing lead as a potential material. Over 20 states scored a C because they only call for identification of environmental hazards broadly; they could improve by calling out one specific hazard: lead pipes, including LSLs. State laws and policies could generally be improved by specifically requiring disclosure of LSLs, in addition to lead pipes or plumbing.

In states that do not already require housing disclosure, a change of statute, rule, or common law would be necessary to ensure improved transparency. Short of changing state law, however, the state or real estate association could update the forms that realtors use voluntarily.

Buyers and sellers can seek information through inspections

Buyers and sellers can be proactive by hiring an inspector certified by the American Society of Home Inspectors (ASHI) to identify the pipe material. The ASHI Standard of Practice for Home Inspectors requires inspection and description of the pipe material of the interior water supply and distribution system. It is our understanding from ASHI that this includes any lead pipe visible from within the building. Buyers can also check for information from previous property inspections. Water utilities can help increase transparency.

Water utilities can proactively inform property owners if they are likely to have an LSL and create a responsibility for the property owner to disclose what is known about the service line when they sell their home. The utility could directly notify the property owner, through a letter, or indirectly through online mapping tools that disclose what is known and unknown about LSLs in the service area. Several cities, including D.C., Cincinnati, Boston, Tacoma, and others, have taken steps to increase transparency by creating online maps and search tools for the public. Utilities could go a step further by recommending that property owners disclose if their property has an LSL to tenants or at sale.

State disclosure requirements

The following legislative statutes indicate state disclosure requirements.

Alaska: Alaska Statutes 34.70 et seq.
California: California Civil Code Section 1102
Connecticut: Connecticut General Statute § 20-327b
Delaware: Delaware Code Chapter 25, Title 6
District of Columbia: Lead Pipe Replacement and Disclosure Amendment Act of 2017
Florida: Johnson v. Davis, 480 So.2d 625 (Florida 1985)
Hawaii: Hawaii Statute 508D
Idaho: Idaho Statutes 55-2501, et seq.
Illinois: 765 Illinois Compiled Statutes § 77/35
Indiana: Indiana Code §32-21-5-2
Iowa: Iowa Code § 558A
Kansas: Kansas Statute Annotated 58-30.106
Kentucky: Kentucky Revised Statutes §324.360
Louisiana: Louisiana Revised Statutes §9:3198
Maine: Maine Revised Statutes Title 33, §173
Maryland: Maryland Code Annotated, Real Property § 10-702
Michigan: Michigan Compiled Laws Annotated § 565.951
Minnesota: Minnesota Statutes § 513.55 et seq.
Mississippi: Mississippi State Code § 89-1-501 et seq.
Nebraska: Nebraska Revised Statutes § 76-2,120
Nevada: Nevada Revised Statutes 113.130
New Mexico: New Mexico Statutes § 47-13 et seq.
New York: New York Real Property Law §§ 460-467
North Carolina: North Carolina G.S. 47E
Ohio: Ohio Revised Code 5302.30
Oklahoma: 60 Oklahoma Statutes § 833
Oregon: Oregon Revised Statutes 105.465(2)
Pennsylvania: Pennsylvania Code § 35.284a and 35.335a
Rhode Island: Rhode Island Code § 5-20.8-2
South Carolina: South Carolina Code § 27-50-10 et seq.
South Dakota: South Dakota Statutes § 43-4-37 et seq.
Tennessee: Tennessee Code Annotated § 66-5-201 et seq.
Texas: Texas Property Code Section § 5.008
Washington: Washington Code § 64.06 et seq.
Wisconsin: Wisconsin Statutes Chapter 709

Article appeared online on March 14, 2017, and updated on March 2019 at edf.org. For the full report, see https://www.edf.org/grading-nation-lead-pipe-disclosure-policies. Reproduced with permission by Environmental Defense Fund, 1875 Connecticut Ave, NW Washington, DC 20009.

SAFETY

Trenching and excavation safety

Excerpted from OSHA 2226-10R 2015, Trenching and Excavation Safety, U.S. Department of Labor Occupational Safety and Health Administration.

Excavation and trenching are among the most hazardous construction operations. The Occupational Safety and Health Administration's (OSHA) Excavation standards, 29 Code of Federal Regulations (CFR) Part 1926, Subpart P, contain requirements for excavation and trenching operations.

What is the difference between an excavation and a trench?

OSHA defines an excavation as any man-made cut, cavity, trench, or depression in the Earth's surface formed by earth removal. A trench is defined as a narrow excavation (in relation to its length) made below the surface of the ground. In general, the depth of a trench is greater than its width, but the width of a trench (measured at the bottom) is not greater than 15 feet (4.6 m).

What are the dangers of trenching and excavation operations?

Trenching and excavation work presents serious hazards to all workers involved. Cave-ins pose the greatest risk and are more likely than some other excavation-related incidents to result in worker fatalities. One cubic yard of soil can weigh as much as a car. An unprotected trench can be an early grave. Employers must ensure that workers enter trenches only after adequate protections are in place to address cave-in hazards. Other potential hazards associated with trenching work include falling loads, hazardous atmospheres, and hazards from mobile equipment.

OSHA Excavation standards

OSHA's Excavations standards at 29 CFR Part 1926, Subpart P. The *Occupational Safety and Health Act* requires employers to comply with safety and health standards promulgated by OSHA or by a state with an OSHA-approved state plan. The standards apply to all open excavations made in the Earth's surface, including trenches. Following the requirements of the standards will prevent or greatly reduce the risk of cave-ins and other excavation-related incidents.

What are the soil classification categories?

Some of the compliance methods permitted under the Excavation standards require a competent person to classify soil and rock deposits as:

- Stable rock;
- Type A soil;
- Type B soil; or
- Type C soil.

Stable rock: Natural solid mineral matter that can be excavated with vertical sides and remain intact while exposed.

Type A: Cohesive soils with an unconfined compressive strength of 1.5 tons per square foot (tsf) (144 kPa) or greater.

Examples include clay, silty clay, sandy clay, and clay loam. Certain conditions preclude soil from being classified as Type A. For example, no soil is Type A if it is fissured or has been previously disturbed. See Appendix A to Subpart P of Part 1926, paragraph (b)—Definitions (Type A), for a detailed definition of Type A soil.

Type B: Type B soil includes cohesive soil with an unconfined compressive strength greater than 0.5 tsf (48 kPa) but less than 1.5 tsf (144 kPa) and granular cohesionless soils (such as angular gravel, similar to crushed rock, silt, silt loam, sandy loam, and, in some cases, silty clay loam and sandy clay loam). See Appendix A to Subpart P of Part 1926, paragraph (b)—Definitions (Type B), for a detailed definition of Type B soil.

Type C: Cohesive soil with an unconfined compressive strength of 0.5 tsf (48 kPa) or less, granular soils (including gravel, sand, and loamy sand), submerged soil or soil from which water is freely seeping, submerged rock that is not stable, or material in a sloped, layered system where the layers dip into the excavation or with a slope of four horizontal to one vertical (4H:1V) or steeper. See Appendix A to Subpart P of Part 1926, paragraph (b)—Definitions (Type C), for a detailed definition of Type C soil.

Protective systems

OSHA generally requires that employers protect workers from cave-ins by

- Sloping and benching the sides of the excavation;
- Supporting the sides of the excavation; or
- Placing a shield between the side of the excavation and the work area.

What types of protective systems can employers use to protect workers from cave-ins?

In many cases the type of protective system needed is well known and simple to use. At other times employers will undertake the more complex process of designing a protective system. Designing a protective system requires consideration of many factors, including soil classification, depth of cut, water content of soil, weather and climate, and other operations in the vicinity. Employers are free to choose the most practical design that will provide the necessary protections. Any system used must meet the required performance criteria. The Excavation standards describe various methods and approaches that can be used to design protective systems.

Examples of protective systems that can be used to comply with the Excavation standards include:

- Sloping the sides of the excavation to an angle not steeper than 1½:1 (for every foot of depth, the trench must be excavated back 1½ feet). A slope of this gradation is safe for any type of soil (Fig. 1).
- Designing a sloping and benching system in accordance with tabulated data, such as tables and charts, approved by a registered professional engineer. These data must be in writing and must include enough explanatory information (including the criteria for selecting a system and the limits on the use of the data) for the user to be able to select an appropriate protective system. At least one copy of the data, which identifies the registered professional engineer who approved it, must be kept at the worksite during construction of the protective system. After the system is completed, the data may be stored away from the jobsite, but a copy must be provided upon request to an OSHA compliance officer.
- Using a trench box or shield approved by a registered professional engineer or designed in accordance with tabulated data approved by a registered professional engineer (Fig. 2).

Trenching and Excavation safety

The Excavation standards do not require a protective system when an excavation is made entirely in stable rock or when an excavation is less than 5 feet (1.52 m) deep and a competent person has examined the ground and found no indication of a potential cave-in.

What other precautions do employers need to take to protect workers from cave-ins?

The Excavation standards require employers to provide support systems, such as shoring, bracing, or underpinning,

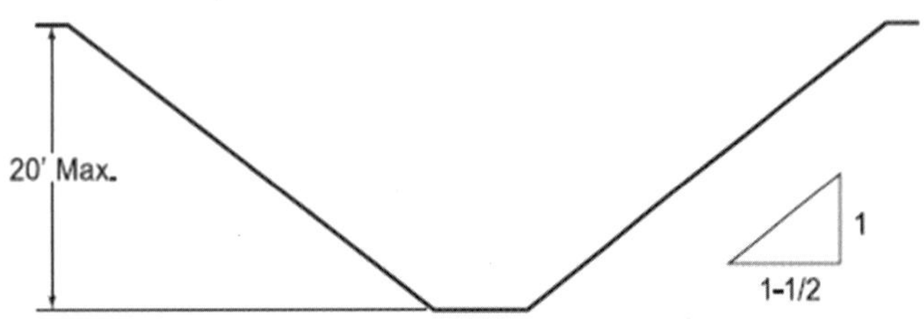

Figure 1. Slope of 1½:1.

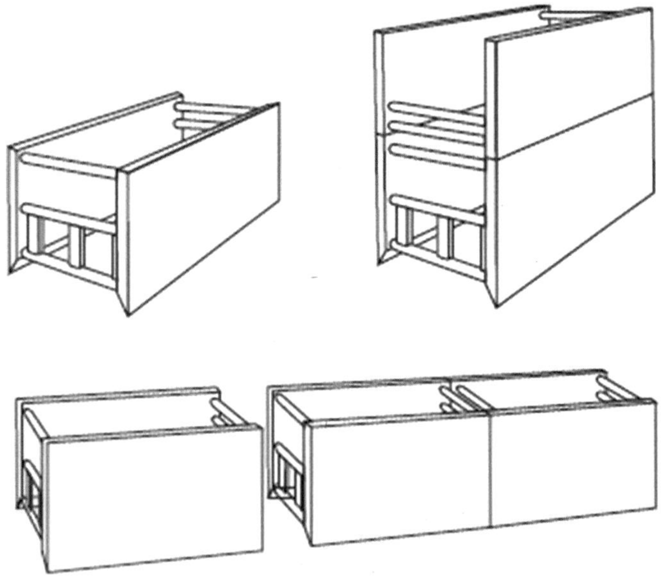

Figure 2. Trench shields.

when necessary to ensure that adjacent structures (including adjoining buildings, walls, sidewalks, and pavements) remain stable for the protection of workers. The standards also prohibit excavation below the base or footing of any foundation or retaining wall that could be reasonably expected to pose a hazard to workers unless

- The employer provides a support system, such as underpinning;
- The excavation is in stable rock; or
- A registered professional engineer determines that the structure is far enough away from the excavation that it would not be affected by the excavation activity or that the excavation work will not pose a hazard to workers.

Excavations that would undermine sidewalks, pavement, and appurtenant structures are prohibited unless the employer provides an appropriately designed support system or another effective method of protecting workers from the possible collapse of those structures.

What are the requirements for safely installing and removing protective systems?

The Excavation standards require employers to take certain steps to protect workers when installing and removing support systems. For example:

- Members of support systems must be securely connected to prevent sliding, falling, kickouts, or predictable failure.
- Support systems must be installed and removed in a manner that protects workers from cave-ins and structural collapses and from being struck by members of the support system.

- Members of support systems must not be overloaded.
- Before temporary removal of individual members, additional precautions are required, such as installing other structural members to carry loads imposed on the support system.
- Removal must begin at, and progress from, the bottom of the excavation.
- Backfilling must progress together with the removal of support systems from excavations.

In addition, the standards permit excavation of 2 feet (0.61 m) or less below the bottom of the members of a support system, but only if the system is designed to resist the forces calculated for the full depth of the trench and there are no indications, while the trench is open, of a possible loss of soil from behind or below the bottom of the support system. Employers must coordinate the installation of support systems with the excavation work.

Trenching and Excavation safety

Employers are responsible for maintaining materials and equipment used for protective systems. Defective and damaged materials and equipment can cause protective systems to fail and lead to other excavation hazards.

Employers must ensure that:

- Materials and equipment are free from damage or defects that might impair their proper function;
- Manufactured materials and equipment are used and maintained consistent with the manufacturer's recommendations and are used so as to prevent worker exposure to hazards;
- A competent person examines any damaged materials or equipment to evaluate its suitability for continued use; and
- If a competent person cannot assure that damaged material or equipment can support the intended loads or is otherwise suitable for use, the materials and equipment are removed from service until evaluated and approved by a registered professional engineer.

Crane and hoist safety

PART 1926—SAFETY AND HEALTH REGULATIONS FOR CONSTRUCTION.

Subpart S—Underground Construction, Caissons, Cofferdams, and Compressed Air.

Sec. 1926.800 Underground construction.

(1) General requirements for cranes and hoists.

(i) Materials, tools, and supplies being raised or lowered, whether within a cage or otherwise, shall be secured or stacked in a manner to prevent the load from shifting, snagging, or falling into the shaft.

(ii) A warning light suitably located to warn employees at the shaft bottom and subsurface shaft entrances shall flash whenever a load is above the shaft bottom or subsurface entrances, or the load is being moved in the shaft. This paragraph does not apply to fully enclosed hoistways.

(iii) Whenever a hoistway is not fully enclosed and employees are at the shaft bottom, conveyances or equipment shall be stopped at least 15 feet (4.57 m) above the bottom of the shaft and held there until the signalman at the bottom of the shaft directs the operator to continue lowering the load, except that the load may be lowered without stopping if the load or conveyance is within full view of a bottom signalman who is in constant voice communication with the operator.

(iv) (A) Before maintenance, repairs, or other work is commenced in the shaft served by a cage, skip, or bucket, the operator and other employees in the area shall be informed and given suitable instructions. (B) A sign warning that work is being done in the shaft shall be installed at the shaft collar, at the operator's station, and at each underground landing.

(v) Any connection between the hoisting rope and the cage or skip shall be compatible with the type of wire rope used for hoisting.

(vi) Spin-type connections, where used, shall be maintained in a clean condition and protected from foreign matter that could affect their operation.

(vii) Cage, skip, and load connections to the hoist rope shall be made so that the force of the hoist pull, vibration, misalignment, release of lift force, or impact will not disengage the connection. Moused or latched open-throat hooks do not meet this requirement.

(viii) When using wire rope wedge sockets, means shall be provided to prevent wedge escapement and to ensure that the wedge is properly seated.

(2) Additional requirements for cranes: Cranes shall be equipped with a limit switch to prevent overtravel at the boom tip. Limit switches are to be used only to limit travel of loads when operational controls malfunction and shall not be used as a substitute for other operational controls.

(3) Additional requirements for hoists.

(i) Hoists shall be designed so that the load hoist drum is powered in both directions of rotation and so that brakes are automatically applied upon power release or failure.

(ii) Control levers shall be of the "deadman type" which return automatically to their center (neutral) position upon release.

(iii) When a hoist is used for both personnel hoisting and material hoisting, load and speed ratings for personnel and for materials shall be assigned to the equipment.

(iv) Material hoisting may be performed at speeds higher than the rated speed for personnel hoisting if the hoist and components have been designed for such higher speeds and if shaft conditions permit.

(v) Employees shall not ride on top of any cage, skip, or bucket except when necessary to perform inspection or maintenance of the hoisting system, in which case they shall be protected by a body belt/harness system to prevent falling.

(vi) Personnel and materials (other than small tools and supplies secured in a manner that will not create a hazard to employees) shall not be hoisted together in the same conveyance. However, if the operator is protected from the shifting of materials, then the operator may ride with materials in cages or skips which are designed to be controlled by an operator within the cage or skip.

(vii) Line speed shall not exceed the design limitations of the systems.

(viii) Hoists shall be equipped with landing level indicators at the operator's station. Marking the hoist rope does not satisfy this requirement.

(ix) Whenever glazing is used in the hoist house, it shall be safety glass, or its equivalent, and be free of distortions and obstructions.

(x) A fire extinguisher that is rated at least 2A:10B:C (multipurpose, dry chemical) shall be mounted in each hoist house.

(xi) Hoist controls shall be arranged so that the operator can perform all operating cycle functions and reach the emergency power cutoff without having to reach beyond the operator's normal operating position.

(xii) Hoists shall be equipped with limit switches to prevent overtravel at the top and bottom of the hoistway.

(xiii) Limit switches are to be used only to limit travel of loads when operational controls malfunction and shall not be used as a substitute for other operational controls.

(xiv) Hoist operators shall be provided with a closed-circuit voice communication system to each landing station, with speaker microphones so located that the operator can communicate with individual landing stations during hoist use.

(xv) When sinking shafts 75 feet (22.86 m) or less in depth, cages, skips, and buckets that may swing, bump, or snag against shaft sides or other structural protrusions shall be guided by fenders, rails, ropes, or a combination of those means.

(xvi) When sinking shafts more than 75 feet (22.86 m) in depth, all cages, skips, and buckets shall be rope or rail-guided to within a rail length from the sinking operation.

(xvii) Cages, skips, and buckets in all completed shafts, or in all shafts being used as completed shafts, shall be rope or rail-guided for the full length of their travel.

(xviii) Wire rope used in load lines of material hoists shall be capable of supporting, without failure, at least five times the maximum intended load or the factor recommended by the rope manufacturer, whichever is greater. Refer to Sec. 1926.552(c)(14)(iii) of this part for design factors for wire rope used in personnel hoists. The design factor shall be calculated by dividing the breaking strength of wire rope, as reported in the manufacturer's rating tables, by the total static load, including the weight of the wire rope in the shaft when fully extended.

(xix) A competent person shall visually check all hoisting machinery, equipment, anchorages, and hoisting rope at the beginning of each shift and during hoist use, as necessary.

(xx) Each safety device shall be checked by a competent person at least weekly during hoist use to ensure suitable operation and safe condition.

(xxi) In order to ensure suitable operation and safe condition of all functions and safety devices, each hoist assembly shall be inspected and load-tested to 100% of its rated capacity: at the time of installation; after any repairs or alterations affecting its structural integrity; after the operation of any safety device; and annually when in use. The employer shall prepare a certification record which includes the date each inspection and load test was performed; the signature of the person who performed the inspection and test; and a serial number or other identifier for the hoist that was inspected and tested. The most recent certification record shall be maintained on file until completion of the project.

(xxii) Before hoisting personnel or material, the operator shall perform a test run of any cage or skip whenever it has been out of service for one complete shift, and whenever the assembly or components have been repaired or adjusted.

(xxiii) Unsafe conditions shall be corrected before using the equipment.

(4) Additional requirements for personnel hoists.

(i) Hoist drum systems shall be equipped with at least two means of stopping the load, each of which shall be capable of stopping and holding 150% of the hoist's rated line pull. A broken-rope safety, safety catch, or arrestment device is not a permissible means of stopping under this paragraph (t).

(ii) The operator shall remain within sight and sound of the signals at the operator's station.

(iii) All sides of personnel cages shall be enclosed by one-half inch (12.70 mm) wire mesh (not less than No. 14 gauge or equivalent) to a height of not less than 6 feet (1.83 m). However, when the cage or skip is being used as a work platform, its sides may be reduced in height to 42 inches (1.07 m) when the conveyance is not in motion.

(iv) All personnel cages shall be provided with a positive locking door that does not open outward.

(v) All personnel cages shall be provided with a protective canopy. The canopy shall be made of steel plate, at least 3/16-inch (4.763 mm) in thickness, or material of equivalent strength and impact resistance. The canopy shall be sloped to the outside, and so designed that a section may be readily pushed upward to afford emergency egress. The canopy shall cover the top in such a manner as to protect those inside from objects falling in the shaft.

(vi) Personnel platforms operating on guide rails or guide ropes shall be equipped with broken-rope safety devices, safety catches, or arrestment devices that will stop and hold 150% of the weight of the personnel platform and its maximum rated load.

(vii) During sinking operations in shafts where guides and safeties are not yet used, the travel speed of the personnel platform shall not exceed 200 feet (60.96 m) per minute. Governor controls set for 200 feet (60.96 m) per minute shall be installed in the control system and shall be used during personnel hoisting.

(viii) The personnel platform may travel over the controlled length of the hoistway at rated speeds up to 600 feet (182.88 m) per minute during sinking operations in shafts where guides and safeties are used.

(ix) The personnel platform may travel at rated speeds greater than 600 feet (182.88 m) per minute in completed shafts.

Electrical welding safety

Article 630 Electric Welders

Part I. General

630.1 Scope

This article covers apparatus for electric arc welding, resistance welding, plasma cutting, and other similar welding and cutting process equipment that is connected to an electrical supply system.

630.6 Listing

All welding and cutting power equipment under the scope of this article shall be listed.

Part II. Arc welders

630.11 Ampacity of supply conductors

The ampacity of conductors for arc welders shall be in accordance with 630.11(A) and (B).

630.11(A) Individual welders. The ampacity of the supply conductors shall be not less than the I_{1eff} value on the rating plate. Alternatively, if the I_{1eff} is not given, the ampacity of the supply conductors shall not be less than the current value determined by multiplying the rated primary current in amperes given on the welder rating plate by the factor shown in Table 630.11A based on the duty cycle of the welder.

630.11(B) Group of welders. Minimum conductor ampacity shall be based on the individual currents determined in 630.11(A) as the sum of 100% of the two largest welders, plus 85% of the third largest welder, plus 70% of the fourth largest welder, plus 60% of all remaining welders.

Exception: Percentage values lower than those given in 630.11(B) shall be permitted in cases where the work is such that a high-operating duty cycle for individual welders is impossible.

Informational Note: Duty cycle considers welder loading based on the use to be made of each welder and the number of welders supplied by the conductors that will be in use at the same time. The load value used for each welder considers

Table 630.11(A)
Duty Cycle Multiplication Factors for Arc Welders

Duty cycle	Multiplier for arc welders	
	Nonmotor Generator	Motor Generator
100	1.00	1.00
90	0.95	0.96
80	0.89	0.91
70	0.84	0.86
60	0.78	0.81
50	0.71	0.75
40	0.63	0.69
30	0.55	0.62
20 or less	0.45	0.55

both the magnitude and the duration of the load while the welder is in use.

630.12 Overcurrent protection

Overcurrent protection for arc welders shall be as provided in 630.12(A) and (B). Where the values as determined by this section do not correspond to the standard ampere ratings provided in 240.6 or where the rating or setting specified results in unnecessary opening of the overcurrent device, the next higher standard rating or setting shall be permitted.

630.12(A) For welders. Each welder shall have overcurrent protection rated or set at not more than 200% of I_{1max}. Alternatively, if the I_{1max} is not given, the overcurrent protection shall be rated or set at not more than 200% of the rated primary current of the welder.

An overcurrent device shall not be required for a welder that has supply conductors protected by an overcurrent device rated or set at not more than 200% of I_{1max} or at the rated primary current of the welder.

If the supply conductors for a welder are protected by an overcurrent device rated or set at not more than 200% of I_{1max} or at the rated primary current of the welder, a separate overcurrent device shall not be required.

630.12(B) For conductors. Conductors that supply one or more welders shall be protected by an overcurrent device rated or set at not more than 200% of the conductor ampacity.

Informational Note: I_{1max} is the maximum value of the rated supply current at maximum rated output. I_{1eff} is the maximum value of the effective supply current, calculated from the rated supply current (I_1), the corresponding duty cycle (duty factor) (X), and the supply current at no-load (I_0) by the following equation:

$$I_{1eff} = \sqrt{I_1^2 X + I_0^2(1 - X)}$$

630.13 Disconnecting means

A disconnecting means shall be provided in the supply circuit for each arc welder that is not equipped with a disconnect mounted as an integral part of the welder. The disconnecting means identity shall be marked in accordance with 110.22(A).

The disconnecting means shall be a switch or circuit breaker, and its rating shall be not less than that necessary to accommodate overcurrent protection as specified under 630.12.

630.14 Marking

A rating plate shall be provided for arc welders giving the following information:

(1) Name of manufacturer
(2) Frequency
(3) Number of phases
(4) Primary voltage
(5) I_{1max} and I_{1eff}, or rated primary current
(6) Maximum open-circuit voltage
(7) Rated secondary current
(8) Basis of rating, such as the duty cycle

630.15 Grounding of welder secondary circuit

The secondary circuit conductors of an arc welder, consisting of the electrode conductor and the work conductor, shall not be considered as premises wiring for the purpose of applying Article 250.

Informational Note: Connecting welder secondary circuits to grounded objects can create parallel paths and can cause objectionable current over equipment grounding conductors.

Part III. Resistance welders

630.31 Ampacity of supply conductors

The ampacity of the supply conductors for resistance welders shall be in accordance with 630.31(A) and (B).

Informational Note: The ampacity of the supply conductors for resistance welders necessary to limit the voltage drop to a value permissible for the satisfactory performance of the welder is usually greater than that required to prevent overheating.

630.31(A) Individual welders. The ampacity of conductors for individual welders shall comply with the following:

(1) The ampacity of the supply conductors for a welder that can be operated at different times at different values of primary current or duty cycle shall not be less than 70% of the rated primary current for seam and automatically fed welders and 50% of the rated primary current for manually operated nonautomatic welders.
(2) The ampacity of the supply conductors for a welder wired for a specific operation for which the actual primary current and duty cycle are known and remain unchanged shall not be less than the product of the actual primary current and the multiplier specified in Table 630.31A for the duty cycle at which the welder will be operated.

630.31(B) Groups of welders. The ampacity of conductors that supply two or more welders shall not be less than the sum of the value obtained in accordance with 630.31(A) for the largest welder supplied and 60% of the values obtained for all the other welders supplied.

Informational Note: Explanation of terms:

(1) The *rated primary current* is the rated kilovolt-amperes (kVA) multiplied by 1000 and divided by the rated primary voltage, using values given on the nameplate.
(2) The *actual primary current* is the current drawn from the supply circuit during each welder operation at the particular heat tap and control setting used.
(3) The *duty cycle* is the percentage of the time during which the welder is loaded. For instance, a spot welder supplied by a 60-Hz system (216,000 cycles per hour) and making 400 15-cycle welds per hour would have a duty cycle of 2.8% (400 multiplied by 15, divided by 216,000, multiplied by 100). A seam welder operating two cycles "on" and two cycles "off" would have a duty cycle of 50%.

Table 630.31(A)
Duty Cycle Multiplication Factors for Resistance Welders

Duty cycle (%)	Multiplier
50	0.71
40	0.63
30	0.55
25	0.50
20	0.45
15	0.39
10	0.32
7.5	0.27
5 or less	0.22

630.32 Overcurrent protection

Overcurrent protection for resistance welders shall be as provided in 630.32(A) and (B). Where the values as determined by this section do not correspond with the standard ampere ratings provided in 240.6 or where the rating or setting specified results in unnecessary opening of the overcurrent device, a higher rating or setting that does not exceed the next higher standard ampere rating shall be permitted.

630.32(A) For welders. Each welder shall have an overcurrent device rated or set at not more than 300% of the rated primary current of the welder. If the supply conductors for a welder are protected by an overcurrent device rated or set at not more than 200% of the rated primary current of the welder, a separate overcurrent device shall not be required.

630.32(B) For conductors. Conductors that supply one or more welders shall be protected by an overcurrent device rated or set at not more than 300% of the conductor ampacity.

630.33 Disconnecting means

A switch or circuit breaker shall be provided by which each resistance welder and its control equipment can be disconnected from the supply circuit. The ampere rating of this disconnecting means shall not be less than the supply conductor ampacity determined in accordance with 630.31. The supply circuit switch shall be permitted as the welder disconnecting means where the circuit supplies only one welder.

630.34 Marking

A nameplate shall be provided for each resistance welder, giving the following information:

(1) Name of manufacturer
(2) Frequency
(3) Primary voltage
(4) Rated kilovolt-amperes (kVA) at 50% duty cycle
(5) Maximum and minimum open-circuit secondary voltage
(6) Short-circuit secondary current at maximum secondary voltage
(7) Specified throat and gap setting

Tank fire types and causes

There are many types of fires associated with tanks. A fire can occur anywhere on the floating roof where vapors can emit from it. They can even result from failure to clean oil spills on the top of the roof resulting from oil which has seeped, spilled, or bubbled up through the seals onto the deck or by gurgling through leg sleeves or fittings. When not cleaned this house keeping issue is a fire hazard. There are two commonly recognized major fire categories for storage tanks: rim seal fire and full surface or fully involved fire.

Rim seal fire

If a fire is initiated by maintenance or by lightning on an external floating roof, then the space between the floating roof and the tank shell can burn, usually burning away the seal fabric. There is often an explosion under the seal, which bends the weather shields away from the floating roof, as shown in Figs. 1 and 2.

Although the floating roof rim fire can appear to be catastrophic, it is typically easy to extinguish. If foam is properly applied over the rim area, the fire can be extinguished quickly. One strategy is to spray foam over the edge of the interior shell, and let it run down into the seal space where it extinguishes the fire. Another strategy is to lob foam from the gauger's platform using foam handlines. However, there are many variables that govern the outcome of tank rim seal firefighting. Success of the firefight often depends on the fire department's experience, skill, knowledge, and practice drills.

Figure 1. Rim seal fire diagram.

Figure 2. Rim seal fire.

Full surface or fully involved fire

A fully involved fire occurs if there is a fire in a fixed roof tank (without a floating roof), or if either an external or internal floating roof sinks partially or completely. A common scenario is a rim seal fire that escalates to a fully involved fire, which can happen if some of the roof pontoons explode, or the roof is sunken by improper firefighting techniques. The fully involved fire is orders of magnitude more difficult to extinguish for several reasons.

Most of the foam use and fire preplanning for tanks with floating roofs assumes that the roof is intact and that the exposed maximum liquid surface area that can burn is only that of the annular strip near the tank shell. If the liquid area increases based on partial or total sinking of the floating roof, then the design of the foam application systems as well as the foam available will typically be insufficient to fight the fire. The devastating results of a full surface fire are shown in Fig. 3.

Other serious fire events

Crude oil with boilover properties stored in fixed roof tanks are subject to the "boilover event." When crude oil has a wide boiling range, it will have boilover properties. This means that if there is a fully involved fire, the crude will burn off the light ends initially. A thick layer, several feet thick, begins to form at the top which is devoid of the light ends but which has denser heavy ends. Because the top several feet of the burning tank oil is of a higher density due to the burning off of

Figure 3. Full surface tank fire.

light ends, the heavy layer begins moving downward through the crude oil at a rate of several feet per hour.

After several hours, the hot layer may be over 500°F (260°C) and because crude oil tanks always have water or brine at the bottom of the tank, the hot layer contacting the water layer will cause the water layer to vaporize. As it does, the burning froth consisting of oil and steam starts to eject from the tank perhaps hundreds of feet into the air. Because the crude cover on top of the water layer is reduced by the burning ejected crude, the steam at the bottom at the water layers has less weight to contain it. The process is a positive feedback system where the process of the boilover makes the burning ejecta worse as it continues. The end result is that the entire crude contents of the tank can be ejected as a burning two-phase oil/steam mixture that is almost explosion like (Fig. 4).

The burning crude volume can be much greater than the tank volume because it is a steam—oil mix where the superheated steam can expand to about 1700 times the original water volume. Secondary containment may not contain the burning flames. The most notable boilover was the event on December 19, 1982 at a power plant in Tacoa, Venezuela, where over 150 firefighters, emergency responders, law enforcement agents, news media personnel, and neighbors

Figure 4. Simulated boilover event.

Figure 5. Tacoa, Venezuela, power plant boilover event on December 19, 1982.

were killed (Fig. 5). The boilover is the most dangerous type of terminal or refinery fire, and it greatly endangers personnel and firefighters.

Note: You cannot get a boilover unless you are storing crude oil in fixed roof tanks.

Source

Philip E. Myers, Floating Roof Tanks in Petroleum Storage: An Overview of Roof Types, Fault Modes, Failure Causes and Technology for Incident Prevention, White Paper. PEMY Consulting. Go online at: www.pemyconsulting.com.

Firefighting protection systems for atmospheric storage tanks

Excerpts from the Marsh Risk Engineering Position Paper 01—Atmospheric Storage Tanks, 2015.

Protection systems and fixed firefighting installations for atmospheric storage tanks should comprise fixed/semifixed foam systems and cooling water sprays. Additionally, access to the tanks for mobile firefighting appliances, equipment, and materials should be provided giving multiple points of attack with no dead ends.

In many cases it is safer and more practical to allow a tank fire to burn itself out; for instance, when there is no chance of escalation to other tanks, little environmental exposure or where the risks of fighting the tank fire are greater than not fighting it. Two notable exceptions are

- Rim seal fires—these will take a long time to burn out.
- Crude tank fires—these are susceptible to boilovers due to the presence of water.

Broad principles of water and foam application are described here, with more details given in the Energy Institute's Model Code of Safe Practice Part 19: Fire precautions at petroleum refineries and bulk storage installations. Alternatively, NFPA 11 Standard for Low-, Medium-, and High-Expansion Foam can be used as a reference.

As for process plants, dedicated fire preplans should also be compiled for storage installations. These should specify type of attack, firewater and foam application rates, location of hydrants, semifixed installations and mobile apparatus, firewater containment, and worst case scenarios. API 2021, Management of Atmospheric Storage Tank Fires provides additional guidance in this matter.

Emergency response personnel should receive training on the fire preplans, and regular exercises should be conducted to assess the suitability of the preplan.

Water systems

Provision should be made for the application of cooling water to fixed roof tanks containing products adjacent to those containing Class I or II (petroleum) or wherever there is less than 1× diameter of the largest tank between adjacent tanks. In particular, there should be cooling wherever there is less than 15 m access between tanks.

According to IP 19 Model Code of Safe Practice Part 19, where water is to be applied to tanks, water application rates should be

- 2 lpm/m^2 for exposed tanks.
- 10 lpm/m^2 for engulfed tanks (to vertical and inclined surfaces).

Water application should be provided by either fixed or semifixed systems. It is unlikely that mobile tenders would have sufficient capacity for water storage to mount a prolonged cooling water attack on a storage tank.

Water should not be applied to the roof area of floating roof tanks as this may result in the destabilization and sinking of the roof.

Foam firefighting systems

While being a good cooling medium for tanks, water is not a suitable fire extinguishing medium for hydrocarbon fires. Therefore, foam is used for such applications. The design of foam firefighting systems, be they fixed or mobile, should be suitable for the types of fires they would be required to extinguish (or suitable for the cooling of adjacent equipment/tanks). Therefore, each facility needs to have its own tailored firefighting system. Important issues to consider when designing and installing a firefighting system include the following:

- Foam and the foam injection/proportioning systems must be compatible (also important if the supplier of foam is changed).
- Hydraulic design of foam systems must be specified.
- Foam selection must be appropriate for the type of fire likely, e.g., aqueous film forming foam should not be selected for polar solvents.
- Correct proportioning of foam that may otherwise result in foam that does not flow (too little water) or that quickly goes back to solution (too much water).
- Provision of facilities to aid the maintenance or testing of firefighting systems.
- Location of equipment, e.g., monitors not too far from tanks for foam application to tank surface.
- Attention to any faults in the fabrication and installation of hardware, e.g., check valves installed the wrong way round, transport packing still present around air inductors.

Once a system has been designed and installed, it is critical that it is tested. While this may be expensive, it is critical to ensure that the system completes the job it is intended to do. Ideally, systems should be tested at the manufacturers as it is then cheaper, quicker, and easier to rectify any mistakes.

Floating roof tank fires

Rim seal fires are the most common type of fire on floating roof tanks. There is little chance of these escalating to other tanks or turning to full surface tank fires if the design, maintenance, and layout of the tanks are suitable. This should not be considered a rule, however, as a poorly maintained tank could lose its integrity or be vulnerable to subsequent explosions, perhaps as a result of vapor ingress into one of the pontoons.

Rim seal protection should only be used on certain types of roofs (i.e., double deck, steel pontoon, etc.). Pan roofs, for example, should have full surface protection. The application of foam from a remote monitor (i.e., one outside the bund) would not represent best practice as it can create roof instability and potentially escalate the incident. Additionally, remote monitors are wasteful of foam and cannot direct the jet so accurately as the objective is to apply foam to the rim seal as quickly and gently as possible (irrespective of the roof level) and to retain a complete foam seal for as long as possible.

The most common and (arguably) most effective method to control a rim seal fire is the application of foam via fixed rim seal pourers on the top of the shell wall. Even with simple, fixed purpose-built systems, such as rim seal pourers, there is the danger that basic errors can be made so that the system does not operate as intended.

Such examples that should be avoided are:

- Top of foam dam below the top of the rim seal: this is especially an issue where secondary rim seals have been added to the tank to reduce tank emissions. The top of the foam dam should be at least 50 mm above the top of the rim seal, but ideally about 150 mm.
- Too many or too large gaps at the bottom of the foam dam. The purpose of the dam is to hold the foam in position around the circumference of the tank to reduce the amount of foam needed. This ensures efficient application to the rim seal and prevents unnecessary loads being placed on the tank roof. With large gaps at the bottom of the dam, a significant quantity of foam could leak to the roof. However, small, intermittently spaced drains should be provided to allow rainwater to drain away.
- Incorrect spacing of the foam pourers. Modern foams can flow in excess of 30 m if proportioned correctly, although if there are changes to the foam type, foam proportion, or even water supply, then the flow characteristics could be affected so that a foam seal is not possible.
- Forced/irregular addition of foam. The intention is to apply foam gently down the side of the tank shell so that it forms an even blanket within the rim seal area. If the pourer has been incorrectly positioned, it may not apply the foam in such a way, e.g., catching on the top edge of the seal, resulting in splashing and loss of foam to the roof.
- Foam pourers located too high on the tank, exposing them to wind and impairing their passage to the rim seal itself.

To overcome these issues, tests should be conducted on the system to ensure it operates correctly. Depending upon the number of tanks being protected, annual testing frequency would be appropriate.

Inspections should be conducted regularly to ensure air inlets to foam generators are clear and that foam pourers are not blocked by, for example, birds' nests. Other good features to aid inspection or firefighting in the event of a system failure are the following:

- Walkway around the top of the floating roof tank: this can be the wind girder with a handrail around the circumference. This allows for the pourers to be accessed for maintenance or for firefighters to access the roof with manual hoses.
- Provide a foam hydrant at the top of the tank (preferably at the top of the stairs) for firefighters in the event of rim seal pourers becoming blocked. This also requires less

equipment to be carried up the tank in the event of a rim seal fire. Careful consideration needs to be given as to the safety of personnel when attempting such a tactic.

- It is important that foam can be applied equally well when the floating roof is at high level and when it is at low level.

Other foam delivery systems that are more complex to varying degrees are the following:

Catenary systems: foam applied to the rim seal by local generators located directly within the foam dam through a supply line attached to the roof via the rolling ladder.

Coflexip system: foam solution is applied in the same way as a catenary, although the supply pipe runs up from low level through the tank itself via a "Coflexip" flexible pipe.

"One-shot" systems: an extinguishing agent (such as obsolete Halon) is applied at the rim seal in a single burst extinguishing the flames. While potentially effective as an initial extinguisher, these do not prevent re-ignition and require all the parts, including cylinders, being placed on the roof. For that reason, Marsh does not recommend such an approach.

Fixed roof tank

Similar systems can be applied to fixed roof tanks, with foam application to the surface of the product through a fixed or semifixed system, or even subsurface allowing for the foam to float to the surface of the product. NFPA 11 Standard for Low-, Medium-, and High-Expansion Foam provides a guide as to which products and materials can be protected by such systems.

Application rates

A minimum of 40 min attack should be envisaged to have a reasonable chance of success in extinguishing a fire, and foam stocks should be sufficient for this duration; a duration of 60 min is preferred for crude oil tank fires (Fig. 1).

Typical application rates for foam solution are shown in Table 1, although foam manufacturer's advice on the type of foam and application should always be sought. NFPA 11 Standard for Low-, Medium-, and High-Expansion Foam also provides similar application rates, although there are some minor differences.

Figure 1. Example of subsurface foam application apparatus.

Table 1
Application Rates for Foam Solution

Type	Rate ($L/min/m^2$)
Applied with Mobile Equipment	
Pool fires	5–10[a]
Fixed roof tanks, Class I product	5–10[a]
Fixed roof tanks, Class II/III product	5
Applied with Fixed or Semifixed Equipment	
Fixed roof tank with pourer outlet	4
Base injection (oils only)	4
Floating roof rim seal fires	12
Floating roof tanks (roof area), projected foam	6.5
Floating roof tanks (roof area), fixed pourers	4

[a]The higher rate may be applied to Class I products, including gasoline containing up to 10% of oxygenates.

Energy Institute's Model Code of Safe Practice Part 19: Fire precautions at petroleum refineries and bulk storage installations.

Carbon dioxide pipeline safety and integrity

Excerpts from World Resources Institute (WRI). CCS Guidelines: Guidelines for Carbon Dioxide Capture, Transport, and Storage. Washington, DC: WRI, 2008.

Pipelines are the dominant mode of transporting CO_2. In the United States, there is an estimated 3900 miles of CO_2 pipelines transporting CO_2 for enhanced oil recovery operations circa 2008. Tanker and ship CO_2 transportation is mainly found in the food and beverage industries. About 100,000 tons of CO_2 are transported annually for these industries—far less than the amounts expected to be associated with a commercial-scale power plant, or even ethanol, cement, or natural gas refining output. The advantage of pipeline transportation of CO_2 is that it can deliver a constant and steady supply of CO_2 without the need for intermediate storage along a distribution route. Ship transportation of large quantities of CO_2 may be feasible when it needs to be transported over long distances or overseas; however, many anthropogenic CO_2 sources are located far from navigable waterways, so such a scheme will still most likely require pipeline construction between CO_2 sources and port terminals.

CO_2 pipeline safety regulations

The Office of Pipeline Safety (OPS) administers a national regulatory program to ensure the safety of pipelines transporting natural gas and other gases, liquefied natural gas, hazardous liquids, and CO_2. OPS safety jurisdiction over pipelines covers approximately 1.6 million miles of gas pipelines and an estimated 155,000 miles of hazardous liquid pipelines. CO_2 pipelines are regulated under the same rules as hazardous liquid pipelines codified in 49 CFR 195.

Although CO_2 pipelines are classified as hazardous, CO_2 is not defined as a hazardous substance. It is a Class L, highly volatile, nonflammable/nontoxic material. CO_2 pipelines are treated as hazardous and are reviewed as high-risk hazardous pipelines when they have a diameter greater than 457 mm (18 in) or when they pass through high-consequence areas.

While OPS is primarily responsible for developing, issuing, and enforcing pipeline safety regulations for interstate pipelines, it may authorize a state to act as its agent to inspect interstate pipelines. For hazardous liquid pipelines, currently six states maintain oversight of the interstate pipelines that cross their state. Even where the oversight is delegated to a state, OPS retains responsibility for enforcement of the regulations for interstate pipelines.

Pipeline safety statutes do allow for individual states to assume the intrastate regulatory, inspection, and enforcement responsibilities under an annual certification. To qualify for certification, a state must adopt the minimum federal regulations and may adopt additional or more stringent regulations, so long as they are not incompatible. A state must also provide for injunctive and monetary sanctions that are substantially the same as those authorized by the pipeline safety statutes. Circa 2008, 14 states are certified to regulate intrastate CO_2 pipelines. A state that does not satisfy the criteria for certification may enter into an agreement to undertake certain aspects of the pipeline safety program for intrastate facilities on behalf of OPS. While the state agency under such an agreement will inspect pipeline operators to ascertain their compliance with federal safety regulations, any probable violations are reported to OPS for enforcement action. Kentucky and South Carolina operate their intrastate CO_2 pipelines under such an agreement.

For states that are not certified or that have not entered into an agreement with OPS to regulate CO_2 pipelines at the state level, the safety regulations in 49 CFR 195 apply.

Comparative safety of CO_2 pipelines

The risks posed by increasing CO_2 pipelines should be manageable based on the extensive CO_2 pipeline operating experience of industry. The DOT data suggest that the impacts from CO_2 pipeline incidents are typically less than those from natural gas and hazardous liquid pipelines. As measured by the lack of fatalities and injuries, and significantly lower property damage, impacts from CO_2 pipeline incidents are typically less than those from natural gas and hazardous liquid pipelines.

The main cause for CO_2 pipeline incidents appears to be material failure (i.e., relief valve failure, valve/gasket/weld, or packing failure), followed by corrosion and outside force. While CO_2 is more benign than many other fluids transported through pipelines, it is important to note that the CO_2 pipeline incident statistics are also probably related to the fact that there are many fewer miles of CO_2 pipelines than pipelines transporting other fluids, and they tend to be located in less populated areas.

Environmental health and safety

Surface risks of CO_2 leakage from pipelines are covered by state environmental, health, and safety regulations, which are established by OSHA and the U.S. Department of Labor and adopted and enforced mainly by the states. The National Institute for Occupational Safety and Health (NIOSH) also carries out research and publishes its own recommendations for workplace safety. However, only those promulgated by OSHA have the force of law. The permissible exposure and air contamination limit for CO_2 as specified by OSHA is 5000 ppm.

Reprinted with permission. For complete citations and full text, see the report World Resources Institute (WRI). CCS Guidelines: Guidelines for Carbon Dioxide Capture, Transport, and Storage. 2008. Washington, DC: WRI. Available at www.wri.org.

ENVIRONMENT

U.S. natural gas standards and regulations development

Michael J Rosenfeld, Kiefner & Associates Inc./Applus-RTD, Worthington, Ohio and **Rick W Gailing**, Southern California Gas Co, Los Angeles, CA

Evolutionary steps

The evolution of modern gas pipeline standards can be traced to the B31 Code for Pressure Piping, Standard B31.1, first published as a tentative standard by the American Standards Association (ASA), a predecessor to the American National Standards Institute (ANSI), with sponsorship of the American Society of Mechanical Engineers (ASME). This standard covered the materials, design, and fabrication of piping systems with industry-specific sections for power piping, gas and air piping, oil piping, and district heating piping. The scope of Section 2 covering gas and air piping systems included city gas distribution systems and cross-country gas pipelines and compressor stations. ASA B31.1 was updated and republished in 1942, 1947, and 1951.

The gas pipeline industry desired to further develop the standard to better address the technical requirements for buried natural gas pipelines, which differ substantially from the technical issues associated with piping systems within power and process facilities that tended to dominate technical development of the standard. This desire was further stimulated by a widely publicized gas distribution system incident in Rochester, New York, in 1950 and concern for a consequent regulatory response[1]. In response, Section 8 of the 1951 B31.1 addressing only natural gas pipelines was approved and published as a stand-alone document in 1952. Although it drew largely on the technical requirements for gas and air piping in Section 2 and selected fabrication details from Section 6 of the 1951 B31.1 standard, the publication separately from B31.1 provided the platform for further development of a more comprehensive pipeline-specific technical standard.

The 1955 edition of Section 8, designated B31.1.8, represented a significant technical advancement in requirements for natural gas transmission and distribution piping systems. It incorporated a risk-informed design basis in the form of a location class scheme based on the density of development near the pipeline, significantly more guidance relevant to the design and installation of cross-country transmission pipelines and gas distribution systems, and rigorous new pressure testing requirements. It was thought that a well-conceived technical standard for pipelines could be useful to state pipeline safety regulations[2]. Elements of the 1955 standard are still evident in the current edition. The standard was revised and republished as B31.8 in 1958, 1963, and 1968 prior to the issuance of pipeline safety regulations by the Department of Transportation (DOT) in 1970. Addenda were issued in some years between editions. B31.8 continued to be revised and periodically republished from 1974 to the present time.

In response to a significant gas pipeline incident in Natchitoches, LA, in 1965, the Natural Gas Pipeline Safety Act (NGPSA) of 1968 authorized DOT to create the Office of Pipeline Safety (OPS, predecessor to the present Pipeline and Hazardous Materials Safety Administration, or PHMSA), enact interim safety standards within 3 months consisting of existing state safety standards, and issue federal pipeline safety regulations within 24 months. Interim regulations comprised of existing standards were imposed until complete regulations were adopted as Part 192, effective July 1, 1970. A review of the technical content of Part 192 shows a clear influence of B31.8, with revisions in language and additional content for clarity and enforcement. Part 192 does not make specific reference to B31.8 on most technical matters because it was the belief of the then-director of OPS that a regulation may be potentially compromised by referring to industry-developed standards[3].

The NGPSA also required the establishment of the Technical Pipeline Safety Standards Committee (TPSSC). The purpose of the TPSSC was to review all proposed pipeline regulations for "technical feasibility, reasonableness, and practicality"[4].

In 1970, by agreement with OPS, ASME began publishing language from Part 192 supplemented with practices from B31.8 and other sources to guide operators in meeting the regulatory requirements. The publication was prepared by the Gas Piping Standards Committee (GPSC) and known as the "GPSC Guide." In 1982, the administrative support was transferred to the American Gas Association (although it continued to be published by ASME), the committee name changed to the present Gas Piping Technology Committee (GPTC), and it acquired recognition as ANSI Z380[5].

Standards are not regulations

The foregoing discussion explains the origin of present-day regulations in contemporaneous industry-developed standards. Standards exist to provide technical guidance and promote uniformity in practices. In particular, ASME B31.8 was intended to be a statement of what is generally accepted to be good practice[6], written by engineers for an audience of other engineers, designers, managers, and regulators. Hence the standard cannot include practices that are not generally accepted even if they are superior, nor should it include practices that are considered unnecessary. The requirements set forth in B31.8 are considered adequate under normally encountered conditions, while unusual conditions are not specifically provided for. Also, the standard is not law. The standard was intended to improve public safety through compliance by pipeline operators voluntarily and in good faith[7].

A regulation is a legally enforceable requirement as a government response to a problem. Regulations are written by regulators for an audience of inspectors and the regulated entities for the purpose of enforcement. The regulation embodied in Title 49 of the Code of Federal Regulations Part 192 (49 CFR 192) was intended to prescribe the level of performance that must be met, while leaving industry free to develop the specific means of meeting the prescribed level of performance. In other words, regulations prescribe "what" while industry standards describe "how."

Even though technical provisions in the regulations (GO 112 and 49 CFR 192) have their origins in technical provisions in the standard (B31.8), there are many areas in which the regulations and the standard do not agree, both historically and today. These include matters of pressure design, material properties, hydrostatic pressure test requirements, valve spacing, recordkeeping, and various elements of operation and maintenance.

References

1. L.L. Elder, The History of the Gas Piping Standards/ Technology Committee. GPTC/GPSRC 25th Anniversary Meeting, 1995.
2. F.A. Hough, The Gas Industry has Approved its New Safety Code. Gas Magazine, November 1954.
3. W.C. Jennings, The Regulator's Handbook (June 1971).
4. Federal Register, Vol. 35, No. 161, Wednesday, August 19, 1970, pg. 13256.
5. Elder.
6. F.A. Hough, The New Gas Transmission and Distribution Piping Code (ASA B31 Section 8)", Gas Magazine, Series in 8 Parts, January through September (1955).
7. Hough, 1954.

Excerpted with permission - Applus+, Energy & Industry Division, 3 Sugar Creek Center Blvd. Suite 600, Sugar Land, TX. For full paper and citations, go online at www.applus.com. Paper "Pressure testing and recordkeeping: reconciling historic pipeline practices with new requirements" presented at Pipeline Pigging and Integrity, Management Conference, Houston, Texas, 13–14 February, 2013.

What are the key performance indicators in the U.S. pipeline industry?

Excerpts from the Pipeline Safety Excellence Performanced—2020 Annual Liquids Report by AOPL and API, 2021.

Key performance indicators

Measuring the performance of pipelines is a key way to determine how safe they are and whether their safety is improving. Particularly useful measures of pipeline safety include incident size, location, commodity, and cause.

As a sign of overall pipeline safety performance, the liquids pipeline industry tracks a core set of key performance indicators (KPIs). These KPIs are based primarily on incidents impacting people or the environment. They were created through a recommendation of the U.S. National Transportation Safety Board in a collaborative effort between PHMSA, pipeline operators, and public pipeline safety advocates represented by the Pipeline Safety Trust.

The four industry-wide KPIs are

(1) Total incidents impacting people or the environment
(2) Integrity management incidents impacting people or the environment
(3) Operations and maintenance incidents impacting people or the environment
(4) Participation in Pipeline Safety Management System (PSMS) programs

Integrity management incidents are those of the pipeline itself, such as corrosion, cracking, or weld failure. Operations and maintenance causes include equipment failure or incorrect operations.

2015–2019 highlights

Pipeline incidents impacting people or the environment decreased 36% from 2015 through 2019 (Fig. 1). Total pipeline incidents were down as well, dropping 17% over five years with 77 fewer incidents in 2019 compared to 2015.

Liquids pipeline incidents occurring in high-consequence areas (HCAs) declined 15% from 2015 through 2019 (Fig. 2). Through federal regulation, PHMSA defines HCAs as areas of population concentration, commercially navigable waterways, or sensitive environmental locations. Fewer than half (43%) of pipeline incidents occurred in HCAs in 2019. HCA data differ from incidents impacting people or the environment, because under PHMSA regulation an incident can

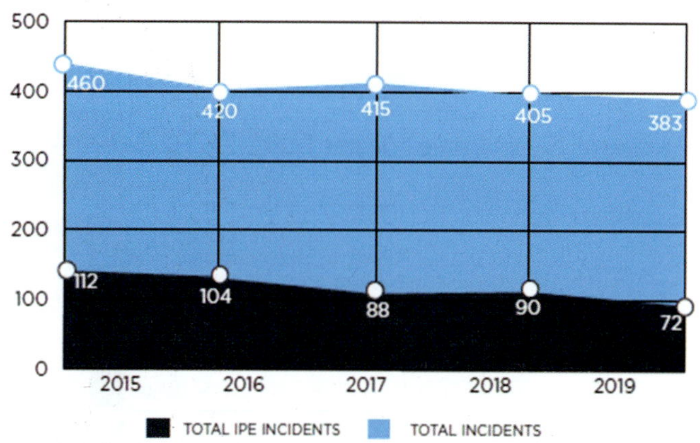

Figure 1. Total incidents and incidents impacting people and the environment, 2015–19. Source: AOPI, API.

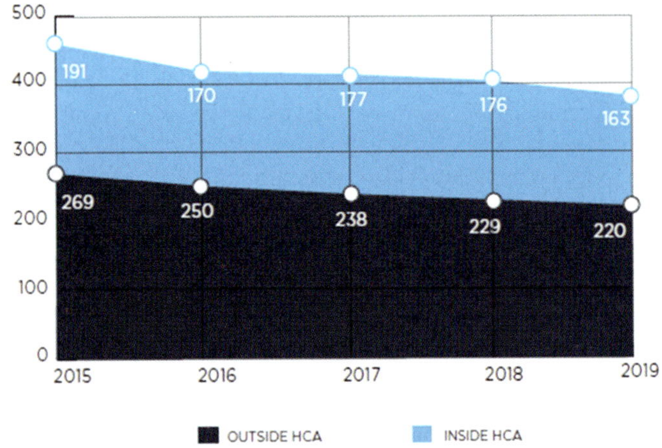

Figure 2. Total incidents inside and outside high-consequence areas, 2015–19. Source: AOPI, API.

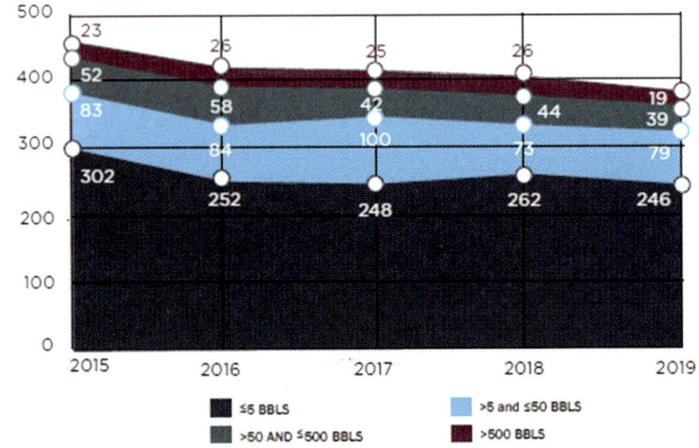

Figure 3. Total incidents by size, 2015–19. Source: AOPI, API.

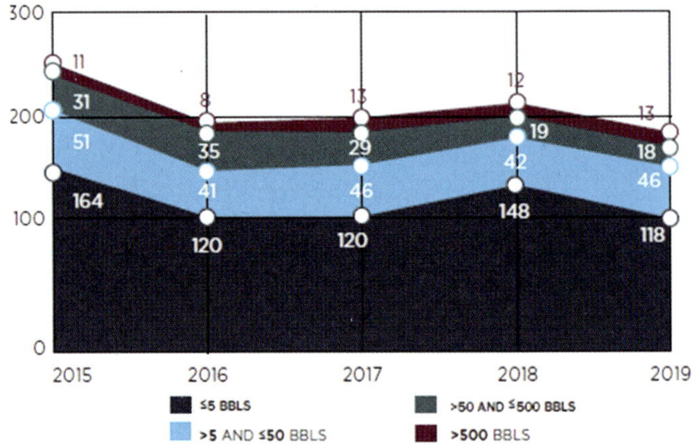

Figure 4. Crude oil incidents by size, 2015–19. Source: AOPI, API.

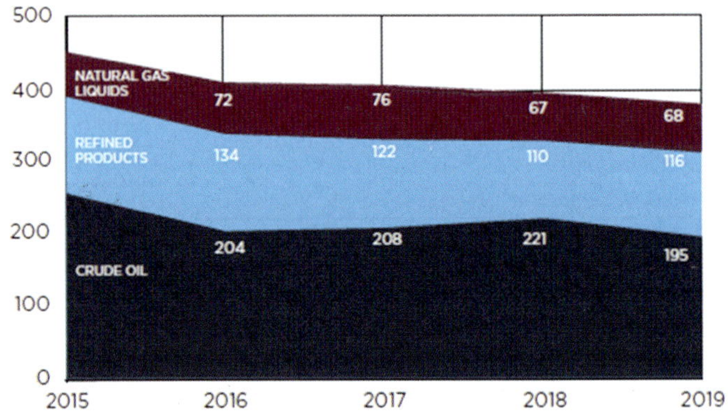

Figure 5. All incidents by commodity, 2015–19. Source: AOPI, API.

have no impact on people or the environment, remain wholly within an operator's facility, and still count as an HCA if that facility is surrounded by an HCA.

Most pipeline incidents are small in size. In 2019, 64% of incidents were less than five barrels and 85% were less than 50 barrels (Fig. 3). Large pipeline incidents are also the rarest. In 2019, only 5% of incidents were 500 barrels or larger.

Similar to total incident trends, the majority of crude oil pipeline incidents are small in size. In 2019, 61% of crude oil incidents were 5 barrels or smaller and 84% of crude oil incidents were smaller than 50 barrels (Fig. 4). From 2015–19, only 7% of crude oil incidents were over 500 barrels. Crude oil incidents greater than 50 barrels decreased 26%, from 42 to 31 incidents.

In 2019, crude oil incidents represented 50% of total incidents, with refined products at 30% and natural gas liquids at 18% of total incidents (Fig. 5). The number of annual crude oil incidents is down 24% from 2015.

Equipment failure is the most frequent cause of liquids pipeline incidents (Fig. 6). Equipment failure represented

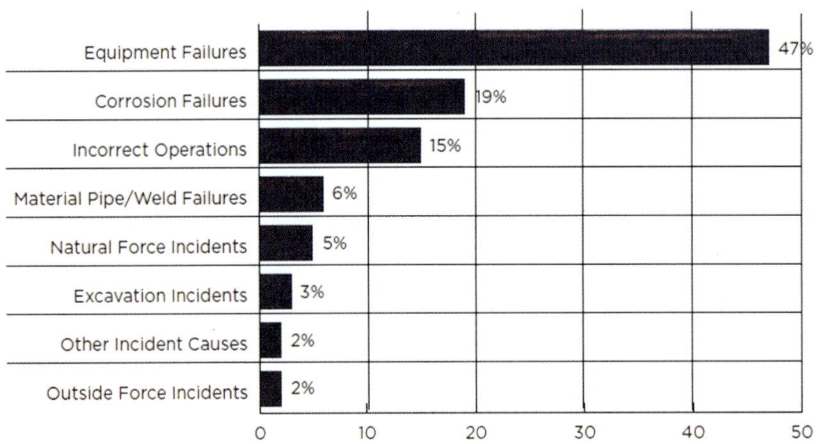

Figure 6. Liquid pipeline incidents by cause in 2019. Source: AOPI, API.

47% of incidents, corrosion failure 19% of incidents, and incorrect operation 15% of incidents. Material pipe/weld failures, which include cracking, a primary source of large volume releases, represented only 6% of incidents from 2015–19.

Reprinted with permission. A full description of the specific types of incidents impacting people or the environment can be found in the Pipeline Safety Excellence Performance—2020 Annual Liquids Report by AOPL and API. See www.api.org and www.aopl.org.

Oil spills in the United States, 2002–2016

Excerpted and edited from J Ramseur's "Oil Spills: Background and Governance", Congressional Research Service. Washington, D.C., September 2017.

Background

Oil spills occur from a wide variety of sources. Some sources release relatively minor amounts per individual release but, in aggregate, contribute a significant annual volume (e.g., recreational vessels). Other sources, such oil tankers or offshore oil wells, release oil on a less frequent basis but have the potential to release a significant volume in one incident. These variances in frequency and volume of oil releases create different environmental impacts as well as different challenges for responders and policymakers.

Major oil well blowouts are relatively uncommon but have accounted for the largest unintentional oil spills in world history. In 1979, the *IXTOC I* oil well blowout released an estimated 140 million gallons in Mexican Gulf Coast waters. By comparison, the largest oil tanker spill in world history—the *Atlantic Empress* off the coast of Tobago in 1979—was estimated at approximately 84 million gallons.

Over the past few decades, two major US oil spills have had lasting repercussions that transcended local environmental and economic effects:

1. 2010 *Deepwater Horizon* oil spill: On April 20, 2010, an explosion occurred at the *Deepwater Horizon* drilling platform in the Gulf of Mexico, resulting in 11 fatalities. The platform had been attached to the Macondo oil well approximately 5000 feet below sea level. Two days later the platform sank into the Gulf, and responders discovered that the well was releasing oil at a significant rate. According to estimates, the well released more than 100 million gallons of oil before it was contained 86 days later.
2. 1989 *Exxon Valdez* oil spill: On March 24, 1989, the *Exxon Valdez* oil tanker ran aground on Bligh Reef in Prince William Sound, Alaska, releasing approximately 11 million gallons of crude oil. Cleanup efforts lasted for six months in 1989 until the US Coast Guard suspended operations due to weather and climatic conditions. Cleanup efforts resumed during the warmer months of 1990 and 1991. The *Exxon Valdez* spill produced extensive consequences beyond Alaska. According to the National Academies, the *Exxon Valdez* disaster caused "fundamental changes in the way the U.S. public thought about oil, the oil industry, and the transport of petroleum products by tankers. ... 'Big oil' was suddenly seen as a necessary evil, something to be feared and mistrusted."

Oil spill response authority

Oil spill response authority is determined by the location of the spill: the US Coast Guard has response authority in the US coastal zone and the Environmental Protection Agency (EPA) cover the inland zone. The Clean Water Act, as amended by the Oil Pollution Act (OPA) in 1990, provides the federal authority to perform cleanup immediately using federal resources, monitor the response efforts of the spiller, or direct the spiller's cleanup activities. The lead federal responder (either from Coast Guard or EPA) determines the level of cleanup required. Federal responders have immediate access to funds in the Oil Spill Liability Trust Fund to support cleanup activities. The trust fund is primarily financed by a per-barrel tax on domestic crude oil and imported petroleum products. The fund's balance was estimated at $5.4 billion at the end of FY2017.

Parties responsible for an oil spill may be liable for cleanup costs, natural resource damages, and specific economic damages, including personal property damage and lost profits or earning capacity. OPA provided (1) limited defenses from liability (act of God, act of war, and act or omission of certain third parties) and (2) conditional liability limits (or caps) for cleanup costs and other damages.

Jurisdiction over oil spill prevention and preparedness duties is determined by the potential sources (e.g., vessels, facilities, pipelines) of oil spills. A series of executive orders, coupled with memoranda of understanding, have established the various agency responsibilities. For example, EPA oversees onshore facilities, the Coast Guard oversees vessels, the Department of Transportation oversees pipelines and rail transportation, and the Department of the Interior's Bureau of Ocean Energy Management oversees offshore facilities (e.g., oil platforms).

Coast Guard Data

The Coast Guard has maintained an Oil Spill Compendium with spill data for various sources within its jurisdiction for responding to oil spills. Pursuant to the National Contingency

Plan, the Coast Guard's oil spill response jurisdiction is the "coastal zone," defined in regulations to include all United States waters subject to the tide, United States waters of the Great Lakes, specified ports and harbors on inland rivers, waters of the contiguous zone, other waters of the high seas subject to the NCP, and the land surface or land substrata, ground waters, and ambient air proximal to those waters.

Fig. 1 illustrates the number of oil spill incidents and spill volume between 2002 and 2016 from data provided by the Coast Guard. The spill data include incidents from vessels, facilities, and pipelines. The 2010 *Deepwater Horizon* oil spill is not included in the figure, because the magnitude of its spill volume—more than 100 million gallons—would make it difficult to compare the other year's annual spill volumes. The figure does include an estimate of oil released on the surface (approximately 400,000 gallons) from the *Deepwater Horizon* mobile offshore drilling unit.

The figure indicates that the number of incidents has decreased over time. Except for several large incidents in 2005 and 2006 (and the *Deepwater Horizon* incident in 2010), the volume of spilled oil has remained relatively consistent. In 2005, approximately 8 million gallons of oil were released from Louisiana facilities damaged during Hurricane Katrina; in 2006, approximately 2 million gallons spilled from a refinery in Louisiana.

Fig. 2 compares the volume of spills over a longer time period from the same Coast Guard data identified in Fig. 1. As Fig. 2 illustrates, the annual oil spill volumes from all sources—particularly tankers and barges—declined dramatically in the 1990s compared to previous decades. This historical decline is likely related, at least in part, to the Oil Pollution Act of 1990 (OPA), which was enacted after the 1989 *Exxon Valdez* oil spill. The 1990 act made comprehensive changes to US oil pollution law by expanding federal response authority and increasing spill liability. The high costs associated with the *Exxon Valdez* spill (approximately $2 billion in cleanup cost and $1 billion in natural resource damage in 1990 dollars, and $500 million in punitive damages and $500 million in interest awarded in 2008), and the threat of broad liability imposed by OPA (in some scenarios, unlimited liability), were likely significant drivers for the spill volume decline seen in the 1990s. The "other" category shown in Fig. 2 includes a wide spectrum of sources, such as deepwater ports, onshore and offshore platforms, facilities, aircraft, land vehicles, and railroad equipment.

PHMSA data

The Department of Transportation's Pipeline and Hazardous Materials Safety Administration (PHMSA) collects oil spill data for pipelines and rail transportation. Fig. 3 illustrates the number of oil incidents and spill volume by mode of transportation between 2002 and 2016. The pipeline and rail data illustrated include incidents that involve spills as small as one gallon, and the pipeline spill volumes do not account for oil that may have been recovered.

Pipeline incidents increased between 2011 and 2014 but have decreased since then and in 2019 totaled 400. Spill volumes fluctuate over time depending on the number and size of major spills. For example, in 2016, three pipeline incidents resulted in spills of over 300,000 gallons. In 2015, two spills resulted in releases of over 100,000 gallons.

Spills from rail transportation increased dramatically between 2009 and 2014 largely due to the increased transportation of crude oil by rail. In 2015 and 2016, the number of incidents by rail decreased as crude by rail transportation decreased. As with pipeline spill volume, the annual spill volumes from rail transportation are a function of the number and size of major spills. For example, the increased rail volume in 2013 is the result of two major incidents of over 450,000 gallons. The increased rail volume in 2015 is the result of three major incidents of over 100,000 gallons.

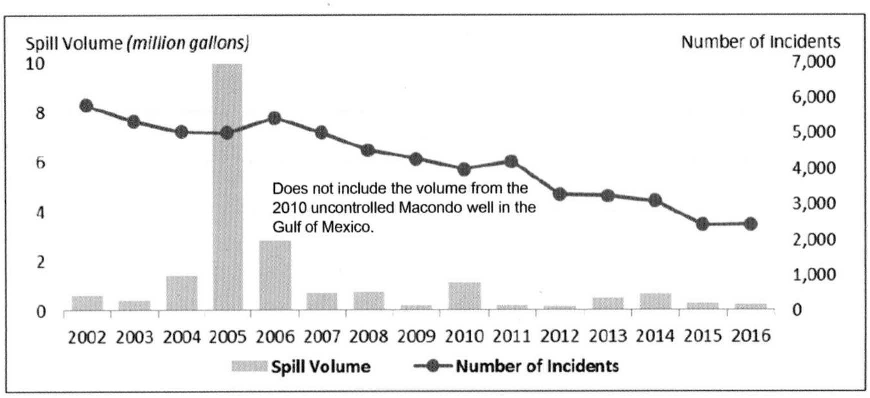

Figure 1. Oil spills from all sources in U.S. Coast Guard's jurisdiction: 2002–2016. Source: Prepared by CRS; 2002–2011 data from the USCG Oil Spill Compendium; 2012–2015 data from the Bureau of Transportation Statistics, "Petroleum Oil Spills Impacting Navigable U.S. Waterways," which cites the Coast Guard as its data sources; 2016 data provided to CRS in personal correspondence with Coast Guard (June 2017).

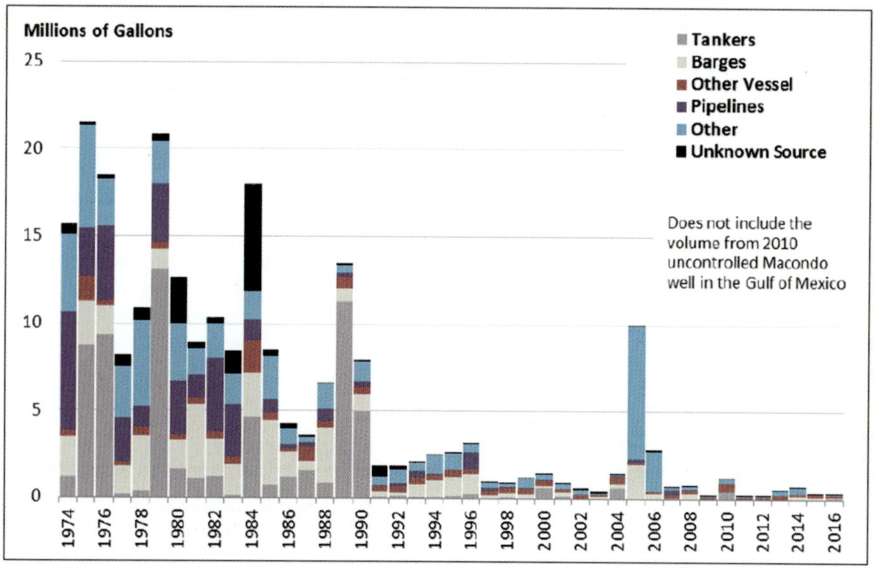

Figure 2. Volume of oil spills by source in the Coast Guard's jurisdiction: 1974–2016. Source: Prepared by CRS; 2002–2011 data from the USCG Oil Spill Compendium; 2012–2015 data from the Bureau of Transportation Statistics, "Petroleum Oil Spills Impacting Navigable U.S. Waterways," which cites the Coast Guard as its data sources; 2016 data provided to CRS in personal correspondence with Coast Guard (June 2017).

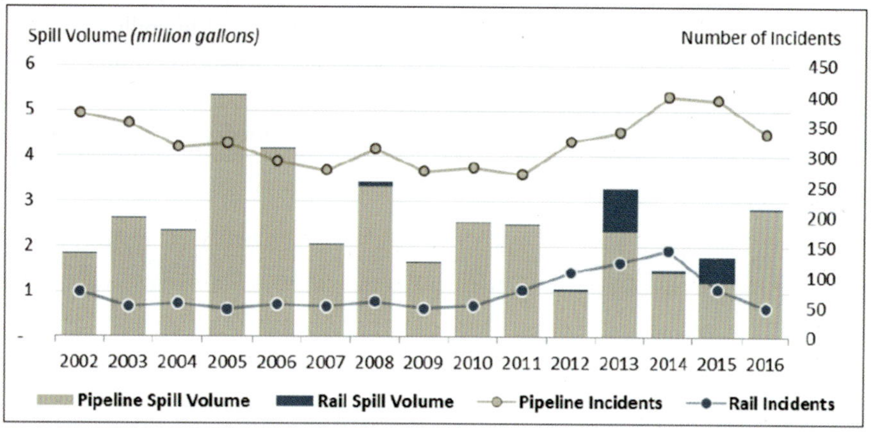

Figure 3. Oil spill volume and number of Incidents from pipeline and rail transportation: 2002–2016. Source: Prepared by CRS; data from PHMSA, "Incidents Statistics," September 14, 2017, http://www.phmsa.dot.gov/hazmat/library/data-stats/incidents.

Estimating oil spill thickness on water visually

The change in visual appearance of an oil slick on water provides a means to estimate oil slick thickness. Only the appearance of rainbow colors at 0.5 to 3 μm is a strong indication of slick thickness and only in the range noted. Other appearances change with numerous variables (e.g., weather, visibility conditions, viewing angle, oil type, water conditions and color, presence of waves, and the presence of other material on the water surface) and thus should be used with caution. Visual methods are estimates and where possible ranges of thicknesses should be provided. If volume is to be calculated, it should also be given as a range of values.

It was recognized early on that slicks on water had somewhat consistent appearances. A series of experiments were conducted in the 1930s to establish the relationship between appearance and thickness and resulted in charts that are still used. Most thickness-appearance charts date from the 1930s and little research has been done on the topic in recent times.

Estimating oil slick thickness is useful for estimating the volume of oil in an area, positioning oil spill countermeasures in optimal locations, evaluating a spill situation, estimating volume for legal purposes, and developing spill control strategies.

The only physical change in appearance that is reliable is the onset of rainbow colors, at 0.5 to 3 μm thickness (Table 1). ASTM F2534-17 is only applicable to thin sheens (sheen and rainbow sheen up to about 3 μm). Thick oil and water-in-oil emulsions do not show visual differences with respect to thickness. Oil thickness estimates using visual clues are highly uncertain because the appearance of oil on water may be affected by a number of factors including oil type, sea state, visibility conditions, view angle, and weather.

In ASTM F2534-17, criteria for estimating the thickness of oil on water using only visual clues are summarized from a variety of literature sources (Table 2), and the physical/theoretical basis for observations related to wavelengths is described.

Source

ASTM F2534-17. Standard Guide for Visually Estimating Oil Spill Thickness on Water.

Table 1
Visibility Characteristics (Appearance)

Minimum Observable Thickness		Minimum Onset Thickness (μm)			
		Silvery	Rainbow	Dark Rainbow	Dark[a]
Average	0.08	0.1	0.5	3	>3
Typical range	0.05–0.2	0.1–0.3	0.2–3	0.2–3	>3

[a]This color is sometimes called "oil-like," "dark colored," "brown," "black," or "metallic."

Table 2
Relationships Between Appearance and Slick Thickness

Year	Oil	Type	#	Height (m)	Viewing Angle	Visibility Thresholds (μm)					
						Min	Silvery	Rainbow	Darkening Colors	Dull Colors	Dark[a]
1930	Various including bunker, fuel oil	e	>15	Ship	Oblique	0.1					
1969	Crude—Santa Barbara, CA	e	Multiple	ns	ns		0.05–0.18	0.23–0.75	1–2.5	2.5–5.5	
1969	General	l		ns	ns	0.04	0.08	0.15–0.3	1	2	
1972	Arabian and Louisiana crudes	e	>20	1–2	Various	<0.15	≤0.15	0.15–0.9	0.9–1.5	1.5–3	
1973	Various	e, l		Ship, aerial	Various	0.038	0.076	0.15–0.31	1	2	
1979	North Sea and Arabian crudes	e	2	Ship, aerial	Various	0.1					
1981	General	l		Aerial	ns	0.1	0.1	0.3	0.1		
1987	General	e, l		Aerial	Various	0.05	0.1	0.15	0.3	1	2
1987	General	e, l		Aerial	Various		0.1	0.3	1	5	15
1993	Various crudes	e	Several	ns	ns	0.1	0.1	0.1–1			
1995	Crude—Norman, OK	e	32	30 m	Nadir	0.094					
	Diesel	e	25	30 m	Nadir	0.165					
	Lubricating oil	e	16	30 m	Nadir	0.077					
	Hydraulic oil	e	13	30 m	Nadir	0.159					
1996	Tar bodies	l				0.04	0.075	0.15	0.3	1	3
2003	Bonn agreement	l				0.04	0.04–0.3	0.3		5	5–50[b]
					Average	0.09	0.1	0.6	0.9	2.7	8.5

e = experiment; l = literature; ns = not specified. For references to table rows, see ASTM F2534-17.

[a]Dark is sometimes stated as true oil color, black, brown, darker colors, or metallic.

[b]The Bonn agreement document has two thicknesses in addition, based on oil distribution: 50 to 200 for patchy, discontinuous distribution, and >200 μm for continuous slicks.

Adapted from ASTM F2534-17.

Fast discharge volume calculator for offshore leaking pipeline

An offshore pipeline oil spill volume estimator known as the "Pocket Guide" was prepared by SINTEF Applied Chemistry for the Minerals Management Service to estimate the amount of oil that will escape from an offshore leaking pipeline due to a rupture (OCS Study MMS 2002-033).

The "Initial" or "Fast" volume calculation method is intended as a first best guess on the amount of oil that has been released from a leaking pipeline and is to be used when data on the event are limited and quick decisions on response strategy are required to minimize spill impacts.

The "Initial" ("Fast") method assumes a single horizontal pipeline segment and a full pipeline break or rupture. This method is not applicable to pinhole leaks or other small pipeline fractures.

Required data

This calculation procedure requires the following data:

- Pipeline internal diameter, ID_{pipe} [in]
- Pipeline length, L_{pipe} [ft]
- Pipeline pressure, P_{pipe} [psi]
- Gas-oil ratio, GOR [scf/stb]
- Water depth at rupture location, d [ft]
- Pipeline flow rate, Q [stb/d]
- Time before shut-in, t [min]

Calculation procedure

Simple equation

The total released volume of oil, Vrel, is found from Eq. (1):

$$V_{rel} = 0.1781 \cdot V_{pipe} \cdot f_{rel} \cdot f_{GOR} + V_{pre\text{-}shut} \quad (1)$$

where

V_{rel} is the total volume released [bbls]
V_{pipe} is the volume of pipeline [ft^3]
f_{rel} is the maximum release volume fraction
f_{GOR} is the GOR reduction factor
$V_{pre\text{-}shut}$ is the volume of oil released prior to shut-in [bbls].

Volume of oil released prior to pipe shut-in, $V_{pre\text{-}shut}$

The oil volume released before shut-in is calculated by using Eq. (2):

$$V_{pre\text{-}shut} = \frac{Q \cdot t}{1440} [\text{bbls}] \quad (2)$$

where

Q is the pipeline flow rate [stb/d]
t is the time before shut-in [min]

Pipeline volume

The pipeline volume is calculated by using Eq. (3):

$$V_{pipe} = \left(\frac{ID_{pipe}}{24}\right)^2 \cdot L_{pipe} \cdot \pi \quad (3)$$

where

V_{pipe} is the volume of pipeline [ft^3]
ID_{pipe} is the internal diameter of pipeline [in]
L_{pipe} is the pipeline length [ft]

Example. A 10,000 ft 12″ ID pipeline has ruptured. Find the volume of the pipeline.

Solution. From Eq. (3):

$$V_{pipe} = \left(\frac{12}{24}\right)^2 \cdot 10,000 \cdot 3.14 = 7850 \text{ ft}^3$$

Release volume fraction, f_{rel}

The maximum release fraction, f_{rel}, represents the maximum ratio of oil released relative to the volume of the pipeline.

To find the release volume fraction, the relative pressure difference over the leak point has to be known. The relative pressure ratio, P_{rel}, is found from Eq. (4):

$$\Delta P_{rel} = \frac{P_{pipe}}{P_{amb}} \quad (4)$$

The ambient pressure P_{amb} is determined from Eq. (5):

$$P_{amb} = 0.446533 \cdot d \quad (5)$$

where

P_{amb} is the pressure outside leakage [psi]
d is the water depth at leak point [ft].

Use Table 1 to find the maximum release fraction, f_{rel}, and the G_{max}. G_{max} is used to find the GOR reduction factor (f_{GOR}).

Table 1
Maximum Released Volume Fraction, f_{rel}

Relative Pressure Ratio, ΔP_{rel}	Maximum Release Fraction, f_{rel}	Maximum Release Occurs for a GOR of G_{max} [scf/stb]
1	0.0	Not applicable (no leak)
1.1–1.2	0.08	140
1.2–1.5	0.17	225
1.5–2	0.30	337
2–3	0.40	449
3–4	0.47	505
4–5	0.50	560
5–10	0.55	505
10–20	0.64	337
20–30	0.71	168
30–50	0.74	140
50–200	0.76	112
>200	0.77	112

Example. The pipeline pressure is 950 psi. The GOR of the fluid in the pipeline is 450 scf/stb. A rupture occurs at a water depth of 100 ft. Find the maximum release volume fraction, f_{rel}.

Solution.

Step 1. Find ambient pressure at leak point, 100 ft water depth. From Eq. (5):

$$P_{amb} = 0.446533 \cdot 100 = 44.65 \text{ psi.}$$

Step 2. Find the relative pressure ratio. Use Eq. (4):

$$\Delta P_{rel} = \frac{P_{pipe}}{P_{amb}} = \frac{950}{44.65} = 21.$$

Step 3. Find the maximum release volume fraction and the G_{max}. Use Table 1. The value of ΔP_{rel} was determined to be 21, and so from Table 1 pick the f_{rel} and G_{max} corresponding to this value:

$$f_{rel} = 0.71 \text{ and } G_{max} = 168 \text{ scf/stb.}$$

GOR reduction factor, f_{GOR}

The total released volume of oil is strongly connected to the gas-oil ratio of the oil. Table 2 gives the volume reduction factor to be used in Eq. (1) as function of the GOR.

In Table 2, columns two and three give reduction factors. Column two is used if GOR of the pipeline fluid is less than the G_{max}. The maximum G_{max} that can be found from Table 1 is 561 scf/stb. If the GOR of the fluid is higher than this value, the GOR will always be higher than G_{max}. Column three is used for GOR higher than G_{max}. G_{max} is found in Table 1.

Example *Fluid 1*: The GOR of the fluid in the pipeline is 450 scf/stb. Find the GOR reduction factor, f_{GOR}. All the other data are the same as in the examples above.

Solution. The release volume fraction f_{rel} and G_{max} has been found to be $f_{rel} = 0.71$ and $G_{max} = 168$ scf/stb. Since GOR > G_{max}, use Table 2 to obtain $f_{GOR} = 0.9$.

Example *Fluid 2*: The GOR of the fluid in the pipeline is 150 scf/stb. Find the GOR reduction factor, f_{GOR}. All the other data are the same as in the examples above.

Solution. The release volume fraction f_{rel} and G_{max} has been found to be $f_{rel} = 0.71$ and $G_{max} = 168$ scf/stb. Since GOR < G_{max}, use Table 2 to compute

Table 2
GOR Reduction Factors, f_{GOR}

GOR [scf/stb]	GOR Reduction Factors, f_{GOR}	
	GOR < G_{max}	GOR > G_{max}
0–225	$f_{GOR} = \frac{GOR}{G_{max}}$	1
225–280		0.98
280–340		0.97
340–420		0.95
420–560		0.9
560–1100	Will not occur	0.85
1100–1700		0.82
1700–2800		0.63
2800–5600		0.43
5,600–11,300		0.26

$$f_{GOR} = \frac{GOR}{G_{max}} = \frac{150}{168} = 0.89.$$

Example *Fluid 3*: The GOR of the fluid in the pipeline is 700 scf/stb. Find the GOR reduction factor, f_{GOR}. All the other data are the same as in the examples above.

Solution. The release volume fraction f_{rel} and G_{max} has been found to be $f_{rel} = 0.71$ and $G_{max} = 168$ scf/stb. Since GOR > G_{max}, use Table 2 to obtain $f_{GOR} = 0.85$.

General trends regarding the reduction factors

- An increase in the pressure difference between the pipe pressure, P_{pipe}, and the ambient pressure at the leak point, P_{amb}, will result in an increased oil release.
- An increase in the fluid GOR will result in reduced oil release as long as GOR > G_{max}. If GOR < G_{max}, the oil release will increase.
- Higher pipe volume V_{pipe} will result in a higher amount of oil released.

Source

Edited from Pocket Guide: Pipeline Oil Spill Volume Estimator. OCS Study MMS 2002-033, New Orleans, LA. 2002.

Note: A computer model can be downloaded from the BOEM website (https://www.boem.gov) that allows the user to input pipeline leak hole size, fluid properties, and variable water depths. Users can also create a pipeline network containing many pipeline segments. Model output includes reports that show pipeline leakage rates versus time and cumulative leak rates with reports being in both tabular and graphic formats.

How to estimate discharge volume for offshore leaking pipeline (fast method)

Example

A 10,000 ft 12″ ID pipeline is shut-in with a pressure of 950 psi. A rupture occurs at a water depth of 100 ft and the GOR of the fluid in the pipeline is 450 scf/stb. Calculate the total oil release. The pipeline is shut-in 2 min after the leakage occurred. The oil flow rate before the spill was 18,000 stb/d.

Solution

1. Volume released prior to shut-in

$$V_{pre\text{-}shut} = \frac{Q \cdot t}{1440} = \frac{18000 \cdot 2}{1440} = 25 \text{ bbls}$$

2. Total pipeline volume is 7850 ft^3 (from Eq. 3)
3. Ambient pressure at leak point is 44.65 psi (from Eq. 5).
4. Relative pressure difference is

$$950 \text{ psi}/44.65 \text{ psi} = 21$$

5. From Table 1, f_{rel} is 0.71 and G_{max} is 168 scf/stb.
6. From Table 2, f_{GOR} is 0.9 since GOR > G_{max}.

Total release of oil from Eq. (1):

$$V_{rel} = 0.1781 \cdot 7850 \cdot 0.71 \cdot 0.9 + 25 = 918 \text{ bbls.}$$

Advanced discharge volume calculator for offshore leaking pipeline

An offshore pipeline oil spill volume estimator known as the "Pocket Guide" was prepared by SINTEF Applied Chemistry for the Minerals Management Service to provide a tool to estimate the amount of oil that will escape from an offshore leaking pipeline (OCS Study MMS 2002-033).

The "Advanced" method refines the "Initial/Fast" spill volume estimate as the spill response proceeds, and more variables are required. The "Advanced" spill volume method allows a more realistic volume estimate than the "Initial/Fast" estimate for assisting in developing response strategies and revising incident action plans. The approach is based on the principle of conservation of mass and the equilibrium of the oil fluid, and the fluid equations are based on experience and observations from a limited quantity of data. As with all spill estimation techniques, appropriate caution is warranted in interpretation.

The "Advanced" method assumes a single horizontal pipeline segment and a full pipeline break or rupture as illustrated in Fig. 1. The calculation approach is based on the principle of conservation of mass and the equilibrium of the oil fluid. The method does not apply to very light crude oil or gas condensates and is not applicable to pinhole leaks or other small pipeline fractures.

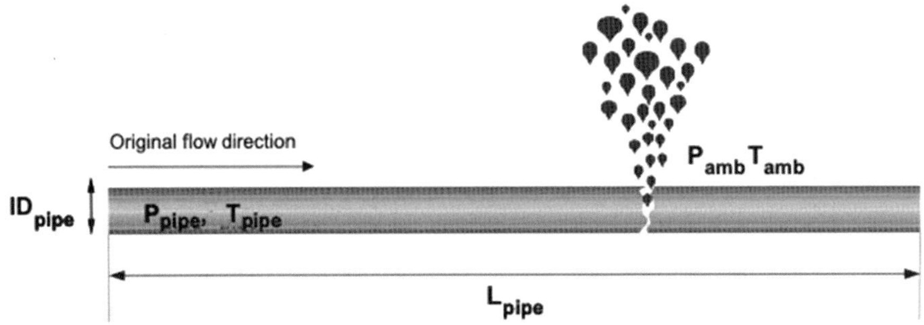

Figure 1. Problem schematic for pipeline leak estimation assuming a single horizontal pipe segment and a full break or rupture.

Required data

The following data are required to perform the calculations:

- Pipeline internal diameter, ID_{pipe} [in]
- Pipeline length, L_{pipe} [ft]
- Pipeline pressure, P_{pipe} [psi]
- Pipeline temperature, T_{pipe} [°F]
- Ambient pressure, P_{amb} [psi]
- Ambient temperature, T_{amb} [°F]
- Gas-oil ratio at standard conditions, GOR [scf/stb]
- Oil density at standard conditions, γ_{API} (API gravity), ρ_l^{stc} [lb/ft^3], and γ_o (specific gravity)
- Gas density at standard conditions, ρ_g^{stc} [lb/ft^3], γ_g (specific gravity)
- Pipeline flow rate, Q [stb/d]
- Time before shut-in, t [min]

Step-by-step procedure

1. Calculate total volume of pipe, V_{pipe}:

$$V_{pipe} = \left(\frac{ID_{pipe}}{24}\right)^2 \pi \cdot L_{pipe} \left[ft^3\right]$$

2. Calculate the oil and gas densities at initial operational conditions, $\rho_o^{initial}$ and $\rho_g^{initial}$. Units: [lb/ft^3].
3. Calculate oil volume fraction at initial operational conditions (P_{pipe}, T_{pipe}), $X_{ovf}^{initial}$.
4. Calculate the initial mass in the pipeline:

$$m_{tot}^{initial} = \left(\rho_o^{initial} \cdot V_{pipe} \cdot X_{ovf}^{initial}\right) + \left(\rho_g^{initial} \cdot V_{pipe} \cdot \left(1 - X_{ovf}^{initial}\right)\right)[lb]$$

5. Calculate the oil and gas densities at ambient pressure and temperature, ρ_o^{amb} and ρ_g^{amb}. Units: [lb/ft^3].
6. Calculate oil volume fraction at ambient pressure and temperature (P_{amb}, T_{amb}), X_{ovf}^{amb}.
7. Calculate the total mass in the pipeline when pressure in pipeline has reached ambient pressure:

$$m_{tot}^{amb} = \left(\rho_o^{amb} \cdot V_{pipe} \cdot X_{ovf}^{amb}\right) + \left(\rho_g^{amb} \cdot V_{pipe} \cdot \left(1 - X_{ovf}^{amb}\right)\right)[lb]$$

8. Calculate the total mass released.

$$m_{tot\_rel} = m_{tot}^{initial} - m_{tot}^{amb}[lb]$$

9. Calculate the gas mass fraction at standard conditions.

$$X_{gmf}^{stc} = \frac{1}{1 + \frac{\rho_L^{stc} \cdot 5.614583}{\left(GOR \cdot \rho_g^{stc}\right)}}$$

10. Calculate the volume of oil released prior to shut-in.

$$V_{pre\text{-}shut} = \frac{Q \cdot t}{1440}[bbls]$$

11. Calculate the volume of oil released at standard conditions.

$$V_{rel}^{stc} = 0.1781 \frac{m_{rel}\left(1 - X_{gmf}^{stc}\right)}{\rho_o^{stc}} + V_{pre\text{-}shut}[bbls]$$

Fluid calculations

This section gives a procedure to calculate fluid properties at a given pressure and temperature (P, T). The gas and oil densities at standard conditions (60°F, 14.7psi) and the GOR of the fluid have to be known to perform the calculations.

Find constants needed to calculate gas solubility

Coefficient	$\gamma_{API} \leq 30$	$\gamma_{API} > 30$
C_1	0.0362	0.0178
C_2	1.09	1.19
C_3	25.7	23.9

Calculate gas gravity at 100 psi

$$\gamma_{g100} = \gamma_g\left(1.0 - 3.1126 \cdot 10^{-3}\gamma_{API}\right)$$

Calculate gas solubility

$$R_s = C_1 \gamma_{g100} P^{C_2} \exp\left[C_3 \frac{\gamma_{API}}{T+460}\right]$$

Find coefficients needed to calculate bubble point pressure

Coefficient	$\gamma_{API} \le 30$	$\gamma_{API} > 30$
K_1	27.62	56.18
K_2	0.9143	0.8424
K_3	11.13	10.39

Calculate coefficient needed to calculate bubble point pressure

$$a = \frac{-K_3 \gamma_{API}}{T+460}$$

Calculate bubble point pressure [psi]

$$P_b = \left[\left(\frac{K_1 GOR}{\gamma_{g100}}\right) 10^a\right]^{K_2}$$

Find coefficients needed to calculate formation volume factor

Coefficient	$\gamma_{API} \le 30$	$\gamma_{API} > 30$
X_1	4.677×10^{-4}	4.670×10^{-4}
X_2	1.751×10^{-5}	1.100×10^{-5}
X_3	-1.811×10^{-8}	1.337×10^{-9}

Calculate the formation volume factor, B_o

If pressure is below bubble point pressure, P_b, then

$$B_o = 1.0 + X_1 R_s + (T-60)\left(\frac{\gamma_{API}}{\gamma_{g100}}\right)(X_2 + X_3 R_s)$$

If pressure is above bubble point pressure, P_b, then

Calculate the isothermal compressibility of oil

$$c_o = \frac{6.83 \times 10^{-6}}{P} \cdot R_s^{0.5} \cdot \gamma_{API}^{0.36} \cdot T^{0.77} \cdot \gamma_{g100}^{-0.0355}$$

Calculate the B_o factor

$$B_o' = 1.0 + X_1 GOR + (T-60)\left(\frac{\gamma_{API}}{\gamma_{g100}}\right)(X_2 + X_3 GOR)$$

$$B_o = B_o' \exp[-c_o(P - P_b)]$$

If $B_o < 1$ then $B_o = 1$.

Calculate specific gravity of dissolved gas in the oil phase

$$\gamma_{gd} = \left(-3.57 \cdot 10^{-6} \cdot \gamma_{API} - 2.86 \cdot 10^{-9}\right) \cdot R_s + 0.02\gamma_{API} + 0.25$$

If $\gamma_{gd} < 0.56$ then $\gamma_{gd} = 0.56$

If $\gamma_{gd} < \gamma_g$ then $\gamma_{gd} = \gamma_g$

Calculate specific gravity of free gas

$$\gamma_{gf} = \frac{GOR \cdot \gamma_g - R_s \gamma_{gd}}{GOR - R_s}$$

If $\gamma_{gf} < 0.56$ then $\gamma_{gf} = 0.56$

If $\gamma_{gf} > \gamma_g$ then $\gamma_{gf} = \gamma_g$

Calculate the oil density $\left[lb/ft^3\right]$

If pressure is below bubble point pressure, then

$$\rho_o = \frac{62.4\,\gamma_o + 0.0136 R_s \gamma_{gd}}{B_o}$$

If pressure is above bubble point pressure, then

$$\rho_o = \left[\frac{62.4\,\gamma_o + 0.0136\,GOR\gamma_{gd}}{B_o}\right] \cdot e^{c_o(P-P_o)}$$

Calculate pseudocritical temperature [°R]

$$T_{pc} = 187 + 330\gamma_g - 71.5\gamma_g^2$$

Calculate reduced temperature, T_{pr}

$$T_{pr} = (T + 460)/T_{pc}$$

Calculate pseudocritical pressure [psi]

$$P_{pc} = 706 - 51.7\gamma_g - 11.1\gamma_g^2$$

Calculate reduced pressure, P_{pr}

$$P_{pr} = P/P_{pc}$$

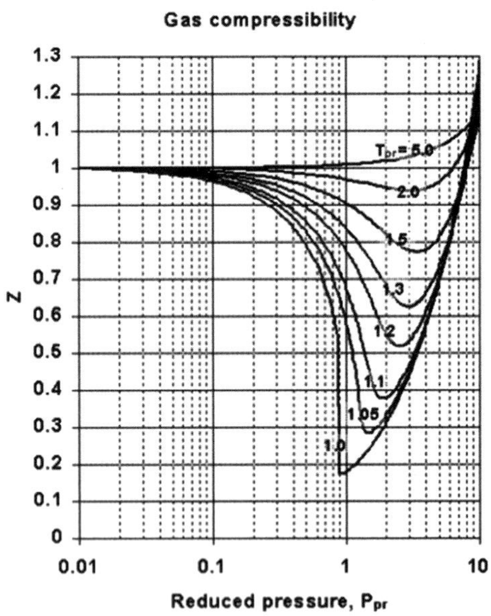

Calculate the gas density $\left[\text{lb}/\text{ft}^3\right]$

$$\rho_g = \frac{2.7 \cdot \gamma_{gf} \cdot P}{Z(T + 460)}$$

Pick the gas compressibility factor Z from the chart below.

Calculate the gas mass fraction of free gas

$$R_{sfree} = GOR - R_s$$

Calculate the gas mass fraction

$$X_{gmf} = \frac{1}{1 + \frac{\rho_L^{stc} \cdot 5.615}{R_{sfree} \cdot \rho_g^{stc}}}$$

If $X_{gmf} < 0$ then $X_{gmf} = 0$

Calculate the volume fraction of oil

$$X_{ovf} = \frac{\left(1 - X_{gmf}\right) \cdot \rho_g}{X_{gmf} \cdot \rho_o + \left(1 - X_{gmf}\right) \cdot \rho_g}$$

Source

Edited from Pocket Guide: Pipeline Oil Spill Volume Estimator. OCS Study MMS 2002-033, New Orleans, LA. 2002.

Note: A computer model can be downloaded from the BOEM website (https://www.boem.gov) that allows the user to input pipeline leak hole size, fluid properties, and variable water depths. Users can also create a pipeline network containing many pipeline segments. Model output includes reports that show pipeline leakage rates versus time and cumulative leak rates with reports being in both tabular and graphic formats.

How to estimate discharge volume for offshore leaking pipeline (advanced method)

Example

A 10,000 ft 12″ ID pipeline transporting oil ruptures and is shut-in 2 min after the leakage occurs. The pipeline was flowing at a rate of 18,000 stb/d and was shut-in with a pressure of 3000 psi and an operational temperature of 80°F. The rupture occurs at a water depth of 100 ft. The GOR of the fluid in the pipeline is 450 scf/stb. The oil and gas densities at standard conditions are 50 lb/ft^3 and 0.05 lb/ft^3, respectively. The seabed temperature is 65°F. Calculate the total oil release.

Solution

Gas density is described as 0.05 lb/ft^3. In the fluid equations the gas density needs to be expressed in specific gravity. To convert natural gas from lb/ft^3 to specific gravity, multiply by 12.98408, to obtain: $\gamma_g = 0.649$ (gas gravity).

Oil density is described as 50 lb/ft^3. In the fluid equations the oil density is expressed in specific gravity and API gravity. To convert crude from lb/ft^3 to specific gravity, multiply by

0.016018 to obtain: $\gamma_o = 0.8009$ (specific gravity). From specific gravity to API gravity, the standard formula applies:

$$API = \frac{141.5}{\text{specific gravity}} - 131.5 = \frac{141.5}{0.8009} - 131.5 = 45.2$$

1. Calculate total volume of pipe, V_{tot}.

$$V_{pipe} = \left(\frac{ID_{pipe}}{24}\right)^2 \cdot \pi \cdot L_{pipe} = \left(\frac{12}{24}\right)^2 \cdot 3.14 \cdot 10000$$
$$= 7850\ ft^3$$

2. Calculate the oil and gas densities at initial operational conditions.

γ_{g100}	0.5579
R_s	251.7
a	−0.87
P_b	1548
B_o	1.14
γ_{gd}	1.11
γ_{gf}	0.56
ρ_o	**47.4**
T_{pc}	371
T_{pr}	1.45
P_{pc}	668
P_{pr}	1.42
Z	0.81
ρ_g	**3.28**

Oil density, $\rho_o^{initial} = 47.4\ lb/ft^3$
Gas density, $\rho_g^{initial} = 3.28\ lb/ft^3$

3. Calculate oil volume fraction at initial operational conditions.

R_{sfree}	198.3
X_{gmf}	0.03411
X_{ovf}	**0.6628**

Volume fraction of oil, $X_{ovf}^{initial} = 0.729$.

4. Calculate the initial mass in the pipeline.

$$m_{tot}^{initial} = \left(\rho_o^{initial} \cdot V_{pipe} \cdot X_{ovf}^{initial}\right) + \left(\rho_g^{initial} \cdot V_{pipe} \cdot \left(1 - X_{ovf}^{initial}\right)\right)$$

$$m_{tot}^{initial} = \left(47.4\ lb/ft^3 \cdot 7850\ ft^3 \cdot 0.6628\right) + \left(3.28\ lb/ft^3 \cdot 7850\ ft^3 \cdot (1 - 0.6628)\right)$$

$$m_{tot}^{initial} = 2.55 \cdot 10^5\ lb$$

5. Calculate the oil and gas densities at ambient pressure and temperature.

γ_{g100}	0.5579
R_s	7.1
a	−0.90
P_b	1475
B_o	1.008
γ_{gd}	1.15
γ_{gf}	0.64
ρ_o	**49.7**
T_{pc}	371.1
T_{pr}	1.14
P_{pc}	668
P_{pr}	0.067
Z	0.99
ρ_g	**0.149**

Oil density, $\rho_o^{end} = 49.7\ lb/ft^3$
Gas density, $\rho_g^{end} = 0.149\ lb/ft^3$

6. Calculate oil volume fraction at ambient pressure and temperature.

R_{sfree}	442.9
X_{gmf}	0.073
X_{ovf}	**0.037**

Volume fraction of oil, $X_{ovf}^{end} = 0.037$.

7. Calculate the total mass in the pipeline when pressure in pipeline has reached ambient pressure.

$$m_{tot}^{end} = \left(\rho_o^{end} \cdot V_{pipe} \cdot X_{ovf}^{end}\right) + \left(\rho_g^{end} \cdot V_{pipe} \cdot \left(1 - X_{ovf}^{end}\right)\right)$$

$$m_{tot}^{end} = \left(49.7\ lb/ft^3 \cdot 7850\ ft^3 \cdot 0.037\right) + \left(0.149\ lb/ft^3 \cdot 7850\ ft^3 \cdot (1 - 0.037)\right)$$

$$m_{tot}^{end} = 15,561\ lb$$

8. Calculate the total mass released.

$$m_{tot\_rel}^{released} = m_{tot}^{initial} - m_{tot}^{end}$$

$$m_{tot\text{-}rel}^{released} = 2.55 \cdot 10^5 lb - 15,561 lb = 239,439\ lb$$

9. Calculate the gas mass fraction at standard conditions.

$$X_{gmf}^{stc} = \frac{1}{1 + \frac{\rho_L^{stc} \cdot 5.614583}{GOR \cdot \rho_g^{stc}}} = \frac{1}{1 + \frac{50 \cdot 5.614583}{450 \cdot 0.05}} = 0.0742$$

10. Calculate the volume of oil released prior to shut-in.

$$V_{pre\text{-}shut} = \frac{Q \cdot t}{1440} = \frac{18000 \cdot 2}{1440} = 25\ bbls$$

11. Calculate the total volume released at standard conditions.

$$V_{rel}^{stc} = 0.7181 \cdot \frac{m_{tot\_rel}^{released}\left(1 - X_{gmf}^{stc}\right)}{\rho_o^{stc}} + V_{pre\text{-}shut}$$

$$V_{rel}^{stc} = 0.7181 \cdot \frac{239,439\ lb\ (1 - 0.0742)}{50\ lb/ft^3} + 25$$

$$V_{rel}^{stc} = 815\ bbls$$

Guidelines on oil characterization to inform spill planning and decision making

Excerpts from the IPIECA/IOGP report "Guidelines on oil characterization to inform spill planning and decision making."

What is oil characterization, and why is it important?

Oils can act very differently once released/spilled in the marine environment; however, an understanding of an oil's physical and chemical characteristics will indicate its likely behavior. Characterization is, in basic terms, a description of an oil's distinctiveness in relation to other oils. It commonly involves the analysis of the source oil's physical and chemical properties, which can be followed by laboratory and meso-scale studies (or, much more rarely, field studies) focusing on the changes in the oil's properties once it is released to the marine environment and subjected to weathering processes. Characterization is a key tool in operational practice and contingency planning, for example, in the case of an accidental release to the environment.

Oil characterization is important in the case of a release to the environment because

- a properly characterized oil may be utilized to prove or disprove a suspected spill source;
- the determination of an oil's properties can help to prevent delays in decision making for an oil spill response;
- information on the properties of oil can help to ensure the effectiveness of decisions made with regard to response measures;
- knowledge of the physical and chemical properties of an oil can be used to predict the severity of environmental damage and the oil's persistence in the environment; and
- characterization data is a requirement for spill trajectory and fate models.

How often should oils be characterized?

Oil characterization is an important aspect in the development of contingency plans for oil spill response. When developing a response plan, it is vital to understand the range of response options available and which of those options are most suited to the properties of the oil in question. The data obtained from characterization assessments are therefore extremely useful for guiding planning decisions.

Crude oils or condensates

For crude oils or condensates, the samples tested should be representative of the oil being produced, handled, stored, or transported at that time. During production, the properties of the oil have the potential to change over a period of years. The parameters could therefore be analyzed every few years, depending on the rate of physical and chemical changes over time. However, the weathering tests carried out in the laboratory would not need to be repeated as frequently provided that the physical and chemical characterization assessments do not illustrate a significant change in key parameters, such as viscosity and density.

Refined products

Samples of refined products (such as bunker fuel oil or distillation products transported in bulk) can be characterized using the same methods as those used for crude oils, although some of the characterization methods for lighter/heavier products have been modified to cope with the differences in viscosity. Characterization of refined products should not be a frequent requirement as long as the properties of the final product remain unchanged.

Environmental processes at work on oil released to the marine environment

Whenever an oil spill occurs, questions are raised relating to the potential fate and behavior of the crude oil, condensate or refined product involved. Spilled oil undergoes a number of physical and chemical changes, some of which lead to its removal from the marine environment while others cause it to persist. These changes depend on factors such as the quantity of oil spilled, the oil's initial physical and chemical characteristics, the prevailing climatic and sea conditions, and whether the oil remains at sea or is washed ashore.

Spilled oil is subjected to weathering processes such as spreading, evaporation, emulsification, dispersion, dissolution, or photo-oxidation (Fig. 1). These processes occur under natural conditions due to agitation at the sea surface by wind, waves, and currents and the exposure of the oil to the sun. As weathering progresses, the physical state of the oil is continuously changing; the light fractions evaporate gradually, its density increases, and part of the oil disperses naturally into the water column while the remaining oil on the water surface may become emulsified with seawater (which tends to increase the volume of the oil and its persistence) and is oxidized by ultraviolet radiation. Some components of the oil partially dissolve and may evaporate. Finally, residues (or high-density oils to begin with) can be subjected to the process of sedimentation.

The behavior of weathered oil is often different from the behavior of the source oil. Oils generally become increasingly viscous over time and can therefore become a more persistent pollutant. Weathering can also have an impact on the physical and chemical characteristics of an oil to the point where the oil is no longer easily identifiable when compared to the source oil.

Understanding these transformations is a key element in evaluating the potential impacts of the spilled oil; optimizing the response options according to the effects of weathering on the oil; and implementing the emergency response plan.

Characterization strategies

In relation to spill response, an oil is commonly characterized by its properties as a source oil, in addition to its behavior once spilled and its subsequent amenability to response techniques.

Some of the more basic characteristics of an oil can provide a reliable view of its behavior when spilled without considering interactions with the environment. Examples of standardized methods for characterizing the basic physical and chemical parameters of oil are shown in Table 1. The parameters which can provide useful insight into the fate of spilled oil include the following:

- **Distillation characteristics:** The basic distillation characteristics (boiling point distribution) are expressed as the proportions of the parent oil that distil within given temperature ranges. Some oils contain bituminous, asphaltenic, or waxy residues which do not readily distil, even at high temperatures, and which are likely to persist in the marine environment for extended periods of time. The basic distillation characteristics can also give an indication of the maximum evaporation expected (which is generally considered to correspond with the fraction that distils below 250°C) (Fig. 2).
- **Wax content and pour point:** The pour point, which is directly related to the wax and asphaltene content, will mark the difference between the solid and liquid state of the oil. If the ambient temperature is below the pour point, the oil will essentially behave as a solid and will not spread over the sea surface as a film but fragment into discrete patches (Fig. 3). This has the potential to limit the effectiveness of dispersants, and care will be needed when choosing appropriate recovery equipment.

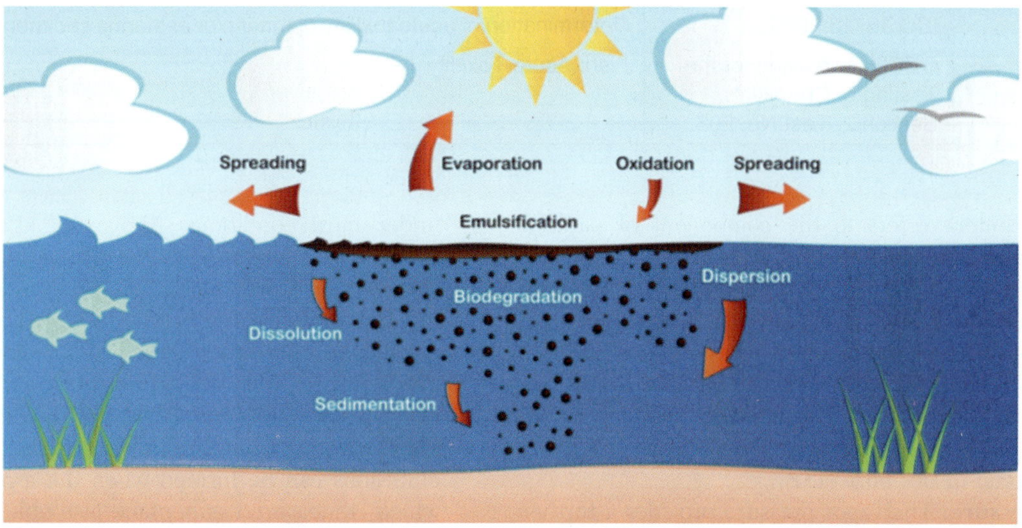

Figure 1. Overview of the weathering process.

Table 1
Characterization Methods

Characterization	Examples of Appropriate Standardized Methods	
Physical and chemical parameter tests	ASTM D6560	Standard Test Method for Determination of Asphaltenes (Heptane Insolubles) in Crude Petroleum and Petroleum Products
	ASTM D1298—12b	Standard Test Method for Density, Relative Density, or API Gravity of Crude Petroleum and Liquid Petroleum Products by Hydrometer Method
	ASTM D5002—99(2010)	Standard Test Method for Density and Relative Density of Crude Oils by Digital Density Analyser
	ASTM D287—12b	Standard Test Method for API Gravity of Crude Petroleum and Petroleum Products (Hydrometer Method)
	ASTM D97; D5853	Standard Test Methods for Pour Point of Petroleum Products/Crude Oils
	ASTM D323; D5191; D6377	Standard Test Methods for Determination of Vapour Pressure of Petroleum Products/Crude Oils
	ASTM D445	Standard Test Method for Kinematic Viscosity of Transparent and Opaque Liquids (and Calculation of Dynamic Viscosity)
	ASTM 95—05	Standard Test Method for Water content in Petroleum Products and Bituminous Materials by Distillation
	ASTM D3278—96(2011) ASTM D7215—08	Standard Test Method for Flash Point of Liquids by Small Scale Closed-Cup Apparatus /Standard Test Method for Calculated Flash Point from Simulated Distillation Analysis of Distillate Fuels
	ASTM D7169	High Temperature Simulated Distillation
	ASTM D5739	Standard Practice for Oil Spill Source Identification by Gas Chromatography and Positive Ion Electron Impact Low Resolution Mass Spectrometry
	CEN/TR 15522–2:2012	Oil spill identification—Waterborne petroleum and petroleum products—Part 2: Analytical methodology and interpretation of results based on GC-FID and GC-MS low resolution analyses
Dispersant efficacy tests	ASTM F2059	Swirling Flask Test
	EPA Appendix C to part 300, 40 CFR: Section 2.0	EPA Swirling Flask Dispersant Effectiveness Test
ISB tests	ISO 9038:2013	Determination of sustained combustibility of liquids
Toxicity tests	ISO 10253:2006	Marine algal growth inhibition test with Skeletonema costatum and Phaeodactylum tricornutum
	ISO 14669:1999	Determination of acute lethal toxicity to marine copepods
	ISO 16712:2005	Determination of acute toxicity of marine or estuarine sediment to amphipods
	OECD Guidelines for the Testing of Chemicals, Section 2, Test No. 203	Fish, Acute Toxicity Test

The wax content refers to the component of paraffin hydrocarbon within the oil. When the temperature of a waxy crude oil is lowered, the heavier fractions may begin to crystallize.

- **Flash point:** The flash point represents a reliable parameter when assessing the risks of ignition. This has obvious implications for the health and safety of response personnel and may be particularly useful if considering an in-situ burn (ISB).
- **Vapor pressure:** The vapor pressure provides a further indication of the volatility of the oil. A vapor pressure greater than 3 kPa is the criteria for evaporation to occur under most conditions. Above 100 kPa, the substance behaves like a gas. For example, gasoline has a vapor pressure of 40–80 kPa; Cossack crude of 44 kPa; and the much less volatile Boscan crude oil of 1.7 kPa.
- **Asphaltene content:** These heavy molecules have active properties and play a significant role in the formation of water-in-oil emulsions. They are qualified as "hard" or "soft" according to testing. It is generally considered that emulsions can be formed when the asphaltene content, which is measured as heptane insolubles, is greater than 0.5%, with the resultant emulsions containing up to 80% water.

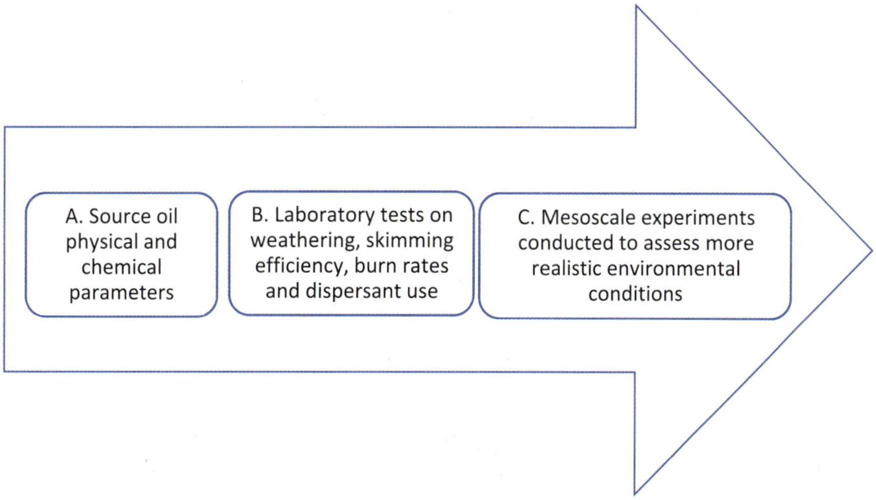

Figure 2. Common approaches to oil characterization.

Figure 3. A heavy fuel oil spill fragments into patches.

- **Viscosity:** The viscosity of an oil is its resistance to flow and will provide information in terms of spreading and whether the application of dispersants would be appropriate. High-viscosity oils flow less easily than low-viscosity oils and are less susceptible to the effects of dispersant chemicals. All oils become more viscous as the temperature falls and as the volatile component evaporates over time.
- **Specific gravity:** The specific gravity, or relative density, is the density in relation to water. Most oils are less dense (lighter than) seawater which has a specific gravity of approximately 1.025 at 20°C. Heavy or very dense oils may sink, partially or fully. The American Petroleum Institute gravity (°API) is commonly used to describe the specific gravity of crude oils and petroleum products. In addition to determining whether or not the oil will float, the specific gravity can also give a general indication of other properties of the oil. For example, oils with a low specific gravity (high °API) tend to contain a high proportion of volatile components and be of low viscosity.

Fingerprinting

The term "fingerprinting" refers to the examination of a sample of spilled oil, to determine whether or not it originates from the suspected source, rather than a parameter to be used in response decisions. It involves the quantification of the organic constituents of an oil sample, most commonly by the use of capillary column gas chromatography coupled with flame ionization detection. For a more accurate and detailed analysis, mass spectrometry may also be used. The term "biomarker" is used in fingerprinting and refers to the calculation of peak area ratios for a range of the more stable hydrocarbon compounds including hopanes, isoprenoids, and steranes. When conducting fingerprint characterization of weathered oils, the distribution patterns and ratios of biomarkers can be extremely useful in comparing samples with suspected source oils despite the weathering effects (differences in the oil's appearance and properties).

Laboratory tests on weathering and dispersant effectiveness

Weathering studies

Laboratory simulation of the main weathering processes at sea can be carried out for crude oils, condensates, or refined products by exposing them to a systematic, stepwise procedure involving distillation, emulsification processes, and, occasionally, photo-oxidation.

- **Topping:** The evaporation process is simulated by distilling the oil at various temperatures (typically 150°C, 200°C, and 250°C) for periods of 6, 12, and 36 h. This provides information on the true boiling point as well as the asphaltene and wax content. The residues can be analyzed to determine the density, viscosity, flash point, and pour point.

- **Emulsification:** After evaporation, the residues can be emulsified with seawater at various water contents (50%, 75%, and the maximum) and characterized by recording a number of physical measurements (e.g., viscosity, water content, emulsion stability, etc.). For example, a modified version of the Mackay–Zagorsky method, using rotary funnels, can be used to simulate the emulsification process. The rotation simulates the mixing energy of wave activity at sea over the first 6, 12, or 72 h, starting with a 50% water content.
- **Photo-oxidation:** Finally, samples may be exposed to ultraviolet light to recreate the process of photo-oxidation if there is a particular emphasis on longer-term weathering processes. As this method provides data for just one test temperature, modeling software can then be used to extrapolate the data for various temperatures and wind speeds, which can then be utilized in fate and trajectory models.

Dispersant effectiveness

Approximately 50 different laboratory test methods are currently available for determining the effectiveness of dispersants on oil, although not all are currently in use. They can be categorized into the following groups:

- Laboratory tests include both those that use standard laboratory equipment, such as the warren springs rotating flask test; the Labofina test; the swirling flask test (or baffled flask test); the Exxon wrist action shaker method test; and those involving specifically constructed laboratory equipment, such as the Mackay–Nadeau–Steelman test; the Revised Standard Dispersant Effectiveness test; the Institut Français du Pétrole test; the oscillating hoop test; and the flowing cylinder test.
- Mesoscale tests, such as flume tests, the cascading weir test, and the delft flume test.
- More recent tests designed to understand specific aspects of dispersant use, such as subsea injection of dispersants using the SINTEF/CEDRE MiniTower (Fig. 4), are currently being developed. These do not comprise part of a standard approach to oil characterization but can be useful to understand specific aspects of oil spill response options.

The test methods commonly focus on agitation/mixing, and record the success or failure to disperse the oil, thereby providing an indication of the conditions in which it would be effective during an actual spill response. To confirm the accuracy of laboratory testing, the results need to be correlated with dispersant effectiveness in field tests and at real incidents.

The oil samples tested for dispersibility using these test methods should be "weathered" in the laboratory to provide a range of viscosities that correspond to different weathering times; this will help to define the "window of opportunity" for dispersant use in the field. Previous studies have illustrated the limits of dispersibility with increasing viscosity over time for a range of crudes; the effectiveness of the applied dispersant does not decrease progressively but drops quickly to a value close to zero when the oil reaches its viscosity limit for dispersant use.

Figure 4. The SINTEF/CEDRE MiniTower bench scale test to investigate subsea injection of dispersants.

Mesoscale experiments and calibration with field tests

Mesoscale experiments simulate different weathering conditions simultaneously and can be conducted in a purpose-built facility. To simulate the various weathering processes as realistically as possible, hydraulic canals (wave tanks) or flumes are used to recreate different wind, current, wave, and solar conditions. Evaporation and emulsification are recreated as the oil weathers as a slick at the water surface with the addition of realistic mixing energy.

There are a number of wave tanks, including the Texas A&M SERF tank at Corpus Christi (previously called COSS or Coastal Oil Studies System); the EPA/BIO tank at the Bedford Institute of Oceanography in Dartmouth, Nova Scotia; the Ohmsett (national oil spill response test facility) large wave tank in New Jersey; and the smaller S. L. Ross (SLR) wave tank (Fig. 5).

A flume, such as the facilities at CEDRE and SINTEF, is a looped tank in which water is continuously circulated and waves can be generated (Fig. 6). Some flume tanks have facilities for specialized testing, for example, the analysis of long-term behavior of oil in cold environments. All environmental parameters can be controlled to ensure reliable comparisons between tests. Test runs can last for up to a week using around 10 L of oil. Oil samples taken throughout the test runs can be used to determine the physical and chemical properties and to evaluate the likely effectiveness of available response options. For example, a dispersability test might be carried out in the flume to provide a realistic assessment of the likely efficacy.

Mesoscale tests can also be carried out in the open environment, by containing an area in which the experiments are to be performed. A major disadvantage of such testing is that it does not provide any control of the environmental conditions.

Field tests are less commonly carried out due to the cost and difficulty obtaining the necessary authorizations and hence would not be part of a standard oil characterization. However, calibration of laboratory and mesoscale experiments against field experiments provides valuable information on the scalability of the lab and mesoscale tests. Most recent field testing has focused on oil in ice environments, such as those conducted by SINTEF. Some field-based dispersant effectiveness tests that have been undertaken include the US EPA field dispersant effectiveness test, API field dispersant effectiveness test, the Fina spill test kit, the Mackay simple field test, and the Pelletier screen test.

Response options

The results of laboratory scale or mesoscale experiments performed on fresh or weathered samples of oil can help to guide decisions on the most appropriate choice of spill response options. For example, emulsions and heavily weathered oils cannot normally be chemically dispersed or ignited; a well-characterized oil will therefore enable confident decisions to be made with regard to the effectiveness of the response options in such cases.

Importantly, the oil characterization may indicate that the particular oil spilt is not persistent under the prevailing environmental conditions at the time of the spill. Hence, an oil characterized as nonpersistent and of low viscosity may not

Figure 5. The Texas A&M SERF tank (left) and the Ohmsett test tank (right).

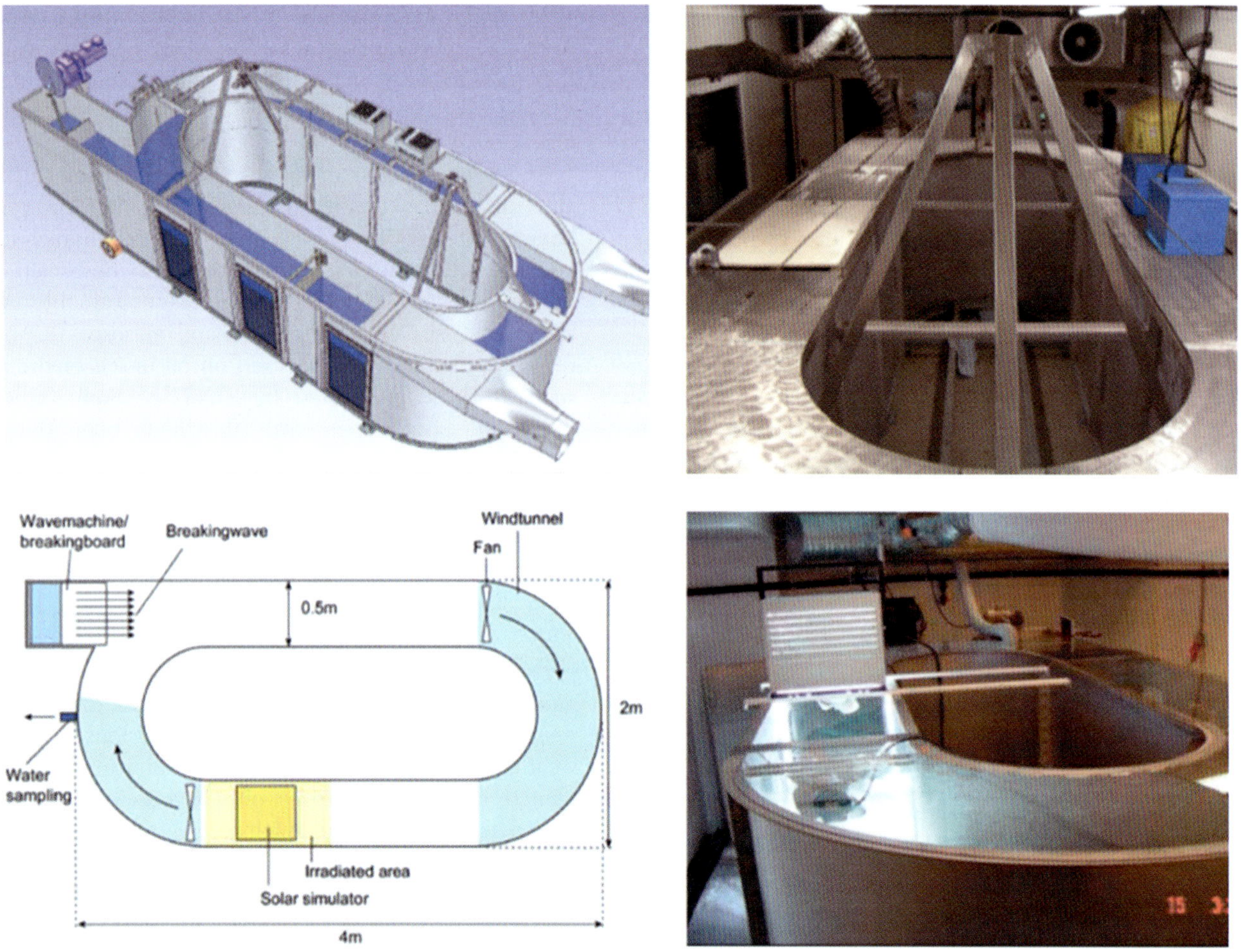

Figure 6. Examples of flumes used for mesostudies.

require an active response if, through monitoring, it is predicted that there is little risk of the spilt oil having an impact on sensitive resources before it dissipates at sea. In such a case, the primary response option would be to closely monitor the spill. It should be borne in mind that, as environmental conditions vary, the behavior of the oil may alter, for example, in extreme cold a usually nonpersistent oil may exhibit persistent traits.

Dispersants

It is important to differentiate between natural and chemical dispersion. Natural dispersion occurs when wave energy breaks up a slick (as occurred during the Braer Oil Spill in 1993). Chemical dispersants greatly increase the rate of natural dispersion of oil into the water column by breaking up the oil slick on the surface into small droplets that disperse more easily through the water column.

The effectiveness of chemical dispersants depends on the oil's initial chemical composition, its weathering stage, and the intrinsic efficiency of the dispersant when used with that particular source oil. The prevailing environmental conditions will also have significant influence, not only in terms of the extent of weathering but also with regard to providing sufficient mixing energy required to achieve active dispersion; calm seas are not ideal in this respect.

Laboratory tests for efficacy should be representative of the real conditions at sea. The dispersibility of the source oil in question should be assessed at the laboratory scale when evaluating a contingency plan (and when planning for the choice of dispersant) prior to an incident. However, smaller-scale tests on subsequently weathered samples and a test spray run on site are more practical and can serve as a useful planning tool before full-scale use of dispersants at sea. Samples tested in the laboratory should reflect the various stages of weathering, so that the most effective "window of opportunity" for dispersant application can be determined. Chemical dispersants are generally more effective when applied to oils that have low viscosity and are unweathered/unemulsified.

In-situ controlled burning

In-situ burning of spilled oil can be an effective response option under certain conditions. However, the window of opportunity for a burn depends on the oil's "ignitability" which, in turn, is limited by weathering and emulsification of the oil. While more research is still needed, tests have been developed to determine the ignitability of oil under different conditions. By performing these tests on a fresh crude oil and on weathered samples of the same oil, it is possible to specify a window of opportunity as a function of weathering. In-situ burning has been studied extensively for arctic conditions, where the limited water uptake due to ice coverage, combined with the slow evaporation of light fractions due to low temperatures, provides a longer window of opportunity for burning. Conversely, if the water content of the oil increases, for example, due to emulsification, its ability to ignite will decrease and the window of opportunity may be lost.

Containment and recovery

The efficacy rates of skimmers and pumps are generally published by the respective manufacturers and are normally dependent on viscosity. As the viscosity of heavy oil increases, it will become more difficult to pump. This is because an increase in viscosity is generally accompanied by an increase in adhesion. Waxy oils, for example, can clog pipes and valves and have a significant impact on pump flow rates; a heavily emulsified oil will be similarly difficult to pump, although the entrainment of water by emulsification may reduce adhesion. Some types of skimmers rely on oleophilic attraction to collect the spilled oil, and the effectiveness of this specific type of skimmer can be reduced significantly for the same reasons.

Tests which measure the degree of oil adhesion to an oleophilic plate are a reliable means of assessing whether the use of oleophilic skimmers would be an effective response option. These tests are generally simple and can easily be performed in combination with weathering experiments. They are particularly relevant in mesoscale experiments as they can be performed directly on the slick at various times and degrees of weathering.

Characterization parameters

Various modeling software have been developed by different organizations (for example, ASA, BMT, NOAA, and SINTEF) and the characterization data required is fairly standard. The most basic trajectory models require information regarding specific gravity (°API), pour point, viscosity, and density. Other properties such as distillation information and asphaltene and trace metal concentrations may also be useful to the operator. Models that can accommodate characterization data for weathered samples provide increased reliability for longer-term predictions. Some models can also combine data on the weathering of oil slicks at the sea surface with information on the solubilization of certain compounds in the water column.

Conclusion

The characterization of oils is considered to be a necessary step in both preparing for and responding to an oil spill. Adequate knowledge regarding the likely behavior of the oil, and the most effective response strategies according to changes in the oil over time, is invaluable for both the responsible party and the oil spill response personnel involved in the clean-up operations.

Excerpted from the IPIECA/IOGP report "Guidelines on oil characterization to inform spill planning and decision making". 14th Floor, City Tower, 40 Basinghall Street, London EC2V 5DE, United Kingdom. For the full report and citations, see www.ipieca.org and www.iogp.org. Reprinted with permission.

Oil spills inland response good practice guidelines

David Fritz, HDR Inc., **Alexis Steen**, ExxonMobil, and **Peter Taylor**, Petronia

Excerpts from the IPIECA/IOGP report "Oil spills: inland response—Good practice guidelines for incident management and emergency response personnel."

Most large, well-known oil spills have occurred in the marine environment; however, inland spills outnumber marine spills. Many of the classic oil spill response techniques were originally developed for use in offshore and coastal spill settings. While some basic principles of oil spill response are the same no matter where oil is spilled, techniques for inland oil spill response operations require some degree of adaptation. The objective of this paper is to present an overview of inland oil spill response, identifying similarities to marine response and highlighting unique issues pertinent to inland oil spills.

This guide addresses the response phase of inland oil spill incidents, where actions are undertaken to ensure safety, minimize the immediate spread and threat of a spill, and deploy techniques to clean-up spilled oil. Its principal focus is on spills in the aquatic environment, although some additional information on adjacent terrestrial environments is also included. It does not address the possible remediation actions which may be considered where oil has contaminated the ground.

Inland aquatic environments in this guide encompass freshwater rivers and streams, lakes and ponds, wetlands and estuarine water bodies, and their associated shores and banksides. Excluding the infrequent large oil spill events, more oil is spilled on inland habitats than on marine coastal or offshore habitats. Inland spills most frequently involve refined products, though almost half of the largest inland spills involve crude oil. In general, more inland spills originate from fixed facilities than from transportation, though pipeline breaks have caused many of the largest inland spills.

Oil fate and weathering

Crude oils are composed of a large number of individual chemical compounds. Almost all of these are hydrocarbons, composed of only hydrogen and carbon. Hydrocarbons can be classified by molecular weight or carbon chain length, and the majority of hydrocarbons in crude oil contain from 5 to 35 carbon atoms.

Hydrocarbon categories include:

- paraffins (alkanes)—consisting of saturated carbon chains with no double bonds;
- olefins (alkenes)—consisting of carbon chains with at least one double bond;
- naphthenes (cycloalkanes)—consisting of carbon rings, with up to four rings in one compound; and
- aromatics—consisting of one or more unsaturated carbon rings with alternating double bonds.

The relative proportions of these chemical compounds differ between crude oils, and are responsible for the range of physical properties that crude oils exhibit. The majority of hydrocarbons in most crude oils are alkanes and cycloalkanes, and these can range from volatile liquids to nonvolatile liquids or solids (waxes) depending on their size (number of carbon atoms) and the prevailing temperature. Oil products represent different oil fractions derived from refining crude oil.

The light, single-ring aromatic compounds are the most toxic compounds in oil and consist of benzene, toluene, ethylbenzene, and xylenes (BTEX). Aromatics with two or more rings are called polycyclic aromatic hydrocarbons (PAHs), and many are also toxic. Examples include naphthalene (two rings), anthracene (three rings), and benzo(a) pyrene (five rings). The extremely large aromatics are called asphaltenes. These compounds can have sulfur, oxygen, and nitrogen derivatives in the mixture and may also have low levels of metals.

Weathering processes

Spilled oil is affected by natural processes that transform the oil (weather) and modify its fate and behavior.

Movement

Liquid oil spreads out over the water to form very thin slicks with an average thickness of about 0.1 mm, although these can range from as high as 1–2 mm for thick oil to as low as 0.1 μm for sheens in open waters. Low-density oils spread faster than heavier oils, which means lighter, refined products will spread faster than most crude oils. Depending on the location and spill volume, spreading may be constrained by available surface area. Wind and water flows may fragment the oil as it spreads, and fast flowing rivers may move oil downstream relatively quickly. Smaller inland water bodies result in rapid oil stranding, and their size makes it easier to locate and track oil.

Evaporation

The smaller, lighter hydrocarbons (typically with less than 12 carbon atoms) are likely to vaporize under ambient conditions. The process is temperature-dependent, and large percentages of gasoline, light crude oils, and light fuel oils can evaporate in the first hours and days of a spill. The heavy hydrocarbon components in crude oils and heavy fuel oils have a low potential for evaporation, so the amounts lost to evaporation during a spill can be small. Initially, in a spill with light to medium crude oils or refined products, the vapor concentrations can be so high that health and fire hazards may result. Any remaining oil after evaporation is more viscous, which can complicate response efforts.

Dissolution

Some hydrocarbons dissolve in water. However, less than 1% of crude oil will dissolve and many of the hydrocarbons that do dissolve are the light aromatics and polar compounds containing oxygen, sulfur, or nitrogen. These compounds are also highly volatile and therefore tend to evaporate rather than remain in the water. The solubility of oil is greater in freshwater compared to the sea, and increases with higher concentrations of dissolved organic matter.

Natural dispersion

Wave and current energy causes the oil to form into small droplets in a water body. Most of these droplets are large enough to quickly resurface. It is therefore typical for only a small percentage of oil to remain naturally dispersed, although there have been a few cases of marine spills during major storms with very high winds and waves where virtually 100% of the spilled oil dispersed naturally. Inland lakes rarely experience these conditions.

Emulsification

Naturally dispersed oil droplets that return to the surface can trap water in the surface oil slick to form a water-in-oil emulsion. Emulsification greatly increases a spilled oil's viscosity and thereby complicates recovery efforts. Emulsified oils are also less affected by other weathering processes such as evaporation, dispersion, and biodegradation. An emulsion may consist of up to 80% water, and the apparent volume of the oil pollution can increase by 4–5 times. The formation of stable (persistent) emulsions requires heavy components such as asphaltenes and resins. Many lighter refined products (gasoline and light fuel oils), therefore, will either not emulsify or will form unstable emulsions (not persistent). Emulsification is a major weathering process affecting marine spills of crude and heavy fuel oils because the mixing energy in the ocean is extremely effective in incorporating water into oil. Emulsification is less likely in freshwater, even for spills of heavier oils, because significant mixing energy often does not persist long enough to generate the stable emulsions seen at sea.

Photo-oxidation

Sunlight oxidizes oil. This process occurs with less than 1% of spilled oil and is what makes heavily weathered black oil turn greyish over time. The process makes the oxidized compounds more soluble.

Sedimentation

Oil can adhere to solids in the water. In waters with high silt content, spilled oil can bind to the suspended sediments, becoming neutrally buoyant in a water body and/or sinking to the bottom. Very fine particles, especially clays, can bind with oil and provide a platform for microbes to biodegrade the oil. Sedimentation can be significant for oil spills during flooding events when waters contain high levels of suspended solids; some spills have seen dramatic decreases in surface oil due to this process.

Biodegradation

Microbes use spilled oil as food and degrade the compounds to simpler hydrocarbons and, ultimately, to carbon dioxide and water. This process requires nutrients (primarily nitrogen and phosphorus) and oxygen. The very large hydrocarbons (such as asphaltenes) found in heavier oils degrade very slowly or not at all. Biodegradation in calm freshwater bodies with low oxygen replacement rates may cause oxygen depletion, slowing subsequent oil breakdown.

Sinking

Fresh oils are typically less dense than water, and will therefore float on the surface. However, the weathering processes tend to increase the density of the oil and lead to situations where the oil may sink. This is most likely to occur with very heavy oils, whose initial density is relatively high; sinking is also more likely in freshwater, which is less dense than seawater.

Terrestrial spills

Inland oil spills can come into contact with soils and groundwater as well as surface water bodies. On the ground surface, some weathering processes, such as evaporation, are active while others, such as emulsification and dispersion, are nonexistent. Spreading does occur, but the degree is highly dependent on the topography and surface roughness. The amount of biodegradation is dependent on moisture and nutrient levels. Underground spills contaminating soil and groundwater undergo little weathering. Oxygen levels are usually limiting and biodegradation proceeds slowly. The oil does adhere to the soil, but other weathering processes are virtually nonexistent.

Inland aquatic environments

Inland aquatic environments or habitats encompass:

- freshwater rivers and streams, lakes and ponds, and wetlands;
- estuarine water bodies and shorelines; and
- shore or river bank habitats including bedrock, man-made structures, sand, mixed sand and gravel, vegetated shores and mud.

Large inland water bodies, such as the North American Great Lakes and Lake Victoria in Africa, have waves and currents. However, compared to oceans, the currents are usually weak, and waves fluctuate with local wind conditions with no oceanic swell. Tides are low or nonexistent, and water level fluctuations may be driven by water inputs and evaporation rates. Shores are usually narrower than marine beaches. Smaller lakes and ponds may have limited flushing rates and also be subject to variations in water levels due to seasonal effects and/or flooding. If water exchange rates are low, high nutrient levels and low oxygen levels can occur which can affect the biodegradation rates of the spilled oil. Lakes in higher latitudes can be subject to winter icing. Even large lakes can freeze over substantially, and ice cover along the shore is common (Fig. 1).

Large rivers can have variable flow rates; these may be seasonally high which may lead to substantial flood events. River flows in estuarine environments are usually subject to tides and have varying salinity. Some rivers have multiple channels with midchannel islands, and some have controlled flow due to the construction of locks and dams. Sediment loads can be high. There will be floodplains of varying widths with associated wetlands and backwaters.

Small rivers and streams can be shallow, with flow characteristics ranging from fast currents with falls and rapids to slow winding channels. All can freeze over during winter in colder climates. The channels can contain varying amounts of debris and sediment load.

Ecological effects

Exposure of the environment to oil can lead to several types of effects. There are two key concepts to consider: vulnerability (able to be exposed) and sensitivity (reactivity to exposure). It is the sensitivity of exposed habitats and organisms that dominates considerations of the environmental effects of a spill. Habitat characteristics which influence sensitivity include:

- species diversity and abundance;
- climate extremes and types of precipitation; and
- the amount of water circulation and flushing.

Exposure to oil can have two types of effects on animals and plants:

1. Many compounds in oil are acutely (short-term and lethal) or chronically (long-term and sublethal) toxic, and exposure to sufficient concentrations for a long enough

Figure 1. Heavy rains caused a spill that led to the oiling of vegetation in the resulting flood.

duration can cause sickness or death in plants and animals. Toxicity from ingestion and inhalation mostly results from the BTEX and PAH compounds because they are more soluble in water, especially the BTEX compounds. Light crude oils and petroleum products (gasoline, kerosene, diesel, and No. 2 fuel oil) contain more of these compounds and are therefore more toxic than heavier oils that contain fewer of these compounds. These aromatic compounds are also the most easily weathered compounds in oil, and as spilled oil weathers, it becomes less toxic as a result. The PAHs persist in the environment longer than the BTEX compounds, and much longer-term toxicity results from their presence.

2. Physical smothering can occur if plants and animals become coated in thick layers of oil. Physical smothering is more likely with heavier crude and fuel oils, especially if they become emulsified, because of their greater viscosity and adhesion potential. These heavier oils can persist for long periods of time and cause longer-term environmental disruption, but their toxicity is minimal.

There can be indirect effects from the above through alterations in species diversity and abundance after a spill. For example, removal of grazing animals within a habitat due to toxicity or smothering by oil may allow a subsequent increase in vegetation until ecological balance is restored. Aggressive or invasive clean-up actions can also alter habitats. Care needs to be taken when choosing response techniques to ensure that the overall result is beneficial compared to taking no action.

Impacts on inland and marine habitats are judged by the relative severity of the impact and speed of recovery.

Habitat recovery is dependent on the:

- degree and persistence of oiling;
- characteristics and weathering of an oil;
- circumstances of a spill (especially the weather); and
- response techniques and tactics.

The recovery of species is dependent on:

- rates of reproduction; and
- the rate of recolonization from adjacent habitats.

The effects of oil on various classes of organisms are summarized below. Generally, organisms can accumulate hydrocarbons from surrounding water and air, and by the ingestion of food and contaminated sediment. The concentrations of hydrocarbons which may be found within organisms are a reflection of the physiochemical properties of the hydrocarbon, environmental conditions, characteristics of the exposed organisms, and the food chain.

- **Bacteria**: While some oils can be initially toxic to bacteria, many bacteria use oil as a food source and, consequently, remove spills from the environment. In agricultural areas, and urban or suburban areas with high runoff, nutrient levels from fertilizers are likely to be high, which increases the potential for eutrophication and plummeting levels of dissolved oxygen even without a spill.
- **Algae**: Freshwater and estuarine algae are affected by exposure to the lighter weight, acutely toxic volatile compounds in many oils, but algal populations usually recover very quickly. The acutely toxic effects decline quickly as the toxic components weather. An increase in cyanobacteria (which fix nitrogen) provides nutrients to spur new algal growth. Microorganisms (zooplankton) which feed on algae are usually suppressed by the toxicity of spilled oil, and the subsequent lower predation rate helps algae to repopulate.
- **Invertebrates**: These animals can be subject to acute and chronic toxic effects from exposure to spilled oil. Sublethal (chronic) effects include impaired growth and reproduction. After a spill, the invertebrate population density and diversity can be significantly altered and population changes can indirectly affect the survival of organisms higher up the food chain.
- **Fish**: Unlike marine species in the open ocean, which can avoid a slick, freshwater fish may be unable to avoid spilled oil in a smaller water body. Fish kills have occurred after spills in constrained freshwaters and can be a direct result of hydrocarbon toxicity or may occur indirectly due to low dissolved oxygen. Fish can accumulate hydrocarbons in their tissues but can also metabolize them; short-term exposures may therefore result in few long-term effects.
- **Amphibians**: The permeable skin of amphibians makes them more at risk than reptiles to oiling. The greatest risk to these animals is from smothering by physical coating, and toxicity due to ingestion of the oil.
- **Mammals and birds**: These animals are similarly susceptible to physical coating by oil, and to ingestion of oil through preening and consumption of oiled prey and plants. Physical coating of oil can severely affect the ability of fur and feathers to insulate an animal, resulting in hypothermia and possibly death. These injuries are often highly visible during a spill response and can lead to much public concern. Some species may smell the oil and avoid it.
- **Vegetation**: Oiling can affect more than just the plants themselves. Animals that live among the plants can become exposed to the oil by interacting with the plants. The oil on plants can remobilize during changes in water level or rain events, or can be transferred by animals and cause additional exposure effects.

Sensitivity

Habitat sensitivity to oil takes into account several factors, including: the potential for natural removal processes;

biological productivity and the ability to recover following oiling; socioeconomic use of the habitat; and the ease of oil removal. These factors can be used to rank the overall sensitivity of habitats as part of an Environmental Sensitivity Index (ESI). Use of an ESI is common for marine habitats, and has been used for freshwater shores.

An ESI typically uses a scale from 1 to 10 to indicate increasing sensitivity. An example is provided in Table 1.

Ecologists differentiate between types of freshwater wetlands, which include marshes, swamps, bogs, and fens. To response personnel, these differences often have low relevance. Inland wetlands (ESI 10A and 10B) consist of marshes, bogs, fens, and swamps. In any of these habitats, the water can be stagnant, oxygen levels can be low, and nutrient levels can be high. Low oxygen levels, uncommon in marine wetlands because of tidal water exchange, can make it much more difficult for spilled oil to degrade aerobically. The soft sediment that comprises these wetlands brings specific challenges to response techniques, as there is heightened risk of clean-up activities exacerbating damage.

Table 1
Environmental Sensitivity Index Scale I Shore Type

Environmental sensitivity index	Shore type
A	Exposed rocky cliffs
1B	Exposed solid walls
2	Shelving bedrock shores
3	Eroding scarps in unconsolidated sediments
4	Sand beaches
5	Mixed sand and gravel beaches
6A	Gravel beaches
6B	Riprap structures
7	Exposed tidal flats
8A	Sheltered rocky shores
8B	Sheltered, solid, man-made structures
9A	Sheltered vegetated low banks/ bluffs
9B	Sheltered sand/mud flats
10A	Freshwater marshes (herbaceous vegetation)
10B	Freshwater swamps (woody vegetation)

Response management

The basic principles of incident management are the same for all spills:

- current and future actions are proposed based on an incident assessment within the framework of oil spill contingency plans;
- response objectives and actions are set and approved by those in command;
- field crews are directed by operations personnel; and
- logistical and financial activities are performed to support the operations.

These response activities take place regardless of incident complexity, whether a large formal organization is created or a small group of responders handles a spill.

In many countries, government agencies lead and coordinate the response to an oil spill, and the government expects the spiller to compensate them for their efforts and to provide support. Management of an incident is often carried out by a government official from a regulatory, law enforcement, or military organization. In some countries, the spiller is required to respond at their own expense and under the direction of government officials. Ideally, everyone involved works together in a coordinated and unified effort to limit the ecological and socioeconomic effects.

Since the majority of inland oil spills are relatively small, the scale of most inland response operations is also proportionately small, and such operations are usually handled by small teams on site. One notable difference between managing inland spills versus marine spills is that the initial responders are often from local municipalities, i.e., from fire and law enforcement agencies. For oil spills resulting from vehicle rollovers, these local agencies often conduct the whole response without any assistance from external resources (such as regional, provincial, state, or national response organizations). However, a larger incident will often quickly overwhelm local capabilities and expertise, and local officials may be supplemented or even replaced by external responders. Regional and national agencies responsible for inland oil spill response may differ from those responsible for marine spills.

Inland spills can occur at varied locations—along a highway, railroad, river or pipeline—virtually everywhere oil is produced, used or transported. Response organizations need to be prepared to cover large geographic areas with a greater variety of habitats than offshore.

- Many inland spills occur at fixed facilities, so oil spill contingency planning will be specific to a location. Scenarios can be developed, detailed contingency plans written, suitable response capability can be procured or identified, communities can be engaged, and exercises can be performed that specifically address potential incidents for that facility.

- Transportation-related spills are more difficult to predict and plan for. Pipelines, railways, and roadways represent a route over which spills can occur. Oil spill contingency planning will need to encompass greater geographic areas in a manner suitable to provide response resources.
- Response resources tend to be concentrated at or near to major facilities or ports and responders may be a considerable distance from a potential spill site. The probability of inland oil spills occurring near populated areas is relatively high, and impacts on the public can occur immediately. Public demands for a quick, robust response can be intense and can outstrip local governments' resources.
- For smaller oil spills, simple containment (keeping the oil from spreading or from reaching flowing water) is the overriding initial goal, and simple construction equipment (bulldozers, front-end loaders, etc.) are often more than adequate to contain the spill. Vacuum trucks can be used to suck up the oil, and these are usually available because they are used for other purposes.
- Specialized oil spill response equipment, such as containment booms and skimmers, may be located hours away from the spill site, resulting in a potential for considerable delays in initiating a comprehensive response.

Care should be taken when considering the use of oil spill trajectory models as part of an inland response. Most oil spill trajectory models were developed for forecasting the fate and transport of spills on open waters (ocean or inland seas) and are not appropriate for spills on rivers, streams, and lakes, or on land. Simple calculations tracking the progress of a slick downriver as a function of the water current do not account for loss by sinking or stranding as oil encounters the river banks, especially at bends. Models can be applied to rivers to predict the timing and extent of the oiling downstream, but their resolution may not yield reliable results.

The process of setting response priorities using shoreline clean-up and assessment techniques applies to both marine and inland habitats. Assessment methodologies may have to be expanded because of the greater number of shore types encountered during an inland spill, and as tidal influences are replaced by possible fluctuations due to varied water flows (Fig. 2).

Safety and health issues

The safety of response teams and the community affected is the top priority during an oil spill response. Hazards to responders include physical ones (such as slips, trips, and falls) and chemical hazards (such as spilled oil). Four dominant factors that influence the degree of hazard to responders are:

- the properties and composition of the spilled oil;
- environmental circumstances at the time of an oil spill and during the response;
- the location and types of tasks (including duration); and
- measures to minimize exposures.

Evaporation can produce high levels of flammable gases, and the risk of igniting a fire is a serious consideration, particularly for spills of light crude oils and light refined

Figure 2. An oil spill response operation is conducted around a private residence.

products (especially gasoline). An early task during an oil spill response is often to monitor for explosive limits of volatile hydrocarbons, as crude oils and gasoline can often contain high levels of light aromatics.

The public (and responders) near oil spills can be exposed to volatile compounds evaporating from the oil. Chemicals of concern are the light aromatics, plus hydrogen sulphide and other mercaptans (organic sulfides). The light aromatics evaporate readily, are acutely toxic, and can cause cancer in the case of prolonged exposure at high doses. The light aromatics are so volatile that hazardous levels usually only last for a few hours, except at very low temperatures and in still air conditions. Heavier fuel oils, however, contain much lower concentrations of light aromatics, therefore spills of these oil types rarely cause hazardous levels in the atmosphere.

Hydrogen sulphide and other sulfur compounds are not carcinogenic, but they are highly acutely toxic. The human nose can detect these compounds at very low levels. Sulfur compounds cause irritation well before toxic levels are reached. However, since our ability to detect odors is quickly deadened, spilled oils with sulfur compounds should be approached carefully. Sulphide levels in crude oils vary widely. Some crudes pose few problems, but the ones that do can present serious health issues even with concentrations at nontoxic levels, due to their highly irritating effects (e.g., headaches, nausea etc.).

Further information and details on managing safety and on how responders are protected from potential hazards is provided in the IPIECA-IOGP publication Oil Spill Responder Health and Safety.

Net environmental benefit analysis

Net environmental benefit analysis (NEBA) is a process used by the response community for making the best choice of response options to minimize the impacts of oil spills on people and the environment—see the IPIECA-IOGP Good Practice Guide on NEBA. It involves consideration and judgment to compare the likely outcomes of using different oil spill response techniques alongside recommendations as to the preferred tactics from experienced response/NEBA practitioners. NEBA typically involves the steps shown in Table 2, which should be carried out prior to a spill incident as an integral part of oil spill contingency planning.

Net environmental benefit analysis in practice

The simple NEBA example below raises the question of whether or not to use controlled in-situ burning in an oiled wetland:

- The oil may be pooled in substantial amounts, so it would be easily ignitable. Yet, burning would create unaesthetic black smoke, and combustion would damage and/or destroy the oiled vegetation above the surface water level.
- Traditional manual/mechanical clean-up operations could destroy the same vegetation (e.g., after oiled sediment and oiled vegetation removal). Equipment operations and responders could unintentionally push or trample oil down toward root systems where the effects of the oil exposure could create long-term damage.

Table 2
Typical Steps Involved in the Net Environmental Benefit Analysis Process

Net environmental benefit analysis step	Description
Evaluate data	The first stage is to consider where the spilled oil is and to where it will drift under the influence of currents and wind—various oil spill trajectory models exist to support this. It is also useful to know how an oil will "weather" as it drifts. This is part of evaluating the available data.
Predict outcomes	The second stage is to assess what is likely to be affected by the spilled oil if no response is undertaken. This may include ecological resources offshore, nearshore, and on shorelines, alongside socioeconomic resources. The efficiency and feasibility of the response toolkit should also be reviewed. This covers the response techniques, the practicalities of their utilization, and how much oil they can recover or treat. If areas under threat include oil-sensitive coastal habitats, the role of oil spill response at sea is to either prevent or limit the spilled oil from reaching these habitats. Previous experience can help assess which oil spill response techniques are likely to be effective. Pragmatic, operational considerations should form a very important part of the net environmental benefit analysis process applied to all feasible response techniques.
Balance trade-offs	The advantages and disadvantages of the potential response options are considered and weighed against the ecological and socioeconomic impacts of each to understand and balance the trade-offs.
Select best options	The process concludes with the adoption of response technique(s) within oil spill contingency plans that minimize potential spills' impacts on the environment and promote the most rapid recovery and restoration of the affected area.

- What is the degree of oiling and the forecast recovery potential for the oiled wetland?
- What is the predicted oil removal efficiency of the response options?
- How unique is this habitat and what is the seasonal sensitivity of these plants?
- Both the controlled in-situ burning and mechanical recovery options need to be evaluated against each other, and also against the additional option of no clean-up with monitoring of the natural recovery of the wetland.

These NEBA considerations can be assessed beforehand and incorporated into oil spill contingency planning. This planning then facilitates effective decision making, with equipment and personnel identified, staging areas selected and priority protection needs in place. NEBA considerations are also applicable during a specific incident to guide the process of assessing the spill circumstances and response options, and estimating the potential outcomes.

Response techniques

The suite of response techniques appropriate to inland aquatic spills is not the same as for marine spills. The larger containment and recovery equipment (e.g., ocean booms and skimmers) used on marine spills are generally inappropriate for use on the smaller water bodies encountered inland. However, smaller-scale containment and recovery equipment, as typically used nearshore in coastal spills, will be suitable for many inland water bodies. If the use of booms is considered, it should be noted that there are specific deployment tactics for their use in fast flowing waters. For oil stranded on freshwater shores, manual recovery of the oil is likely to be the dominant approach in many countries. The use of dispersants is typically limited to marine waters deeper than a specific minimum depth and beyond a specified distance from the coastline, where dilution can quickly mitigate the impact of dispersed oil and facilitate biodegradation; furthermore, dispersants are not usually formulated for use in freshwater and, as such, are much less effective in a freshwater context. This technique is therefore not normally considered for an inland response. Controlled in-situ burning, however, may be suitable for use in a wide variety of inland habitats.

Oiled wildlife response procedures and processes are almost identical for both inland and marine oil spills. Oiled animals in inland spills may include domesticated livestock or pets.

Oiled waste must be collected, stored, and disposed of in accordance with applicable government regulations. The principles of waste minimization and segregation apply. In contrast to offshore spills:

- liquids recovered from spills at fixed facilities may be recycled back into storage tanks at the facilities; and
- contaminated soils may be removed or remediated in place.

Oil from smaller spills is often recovered using sorbents, and relatively large amounts of oily solid waste may be generated for disposal. Most countries have regulations that treat these wastes as hazardous, requiring greater levels of analysis and care in handling.

Isolating sites for safety reasons and keeping the public away from the clean-up crews and the oil may be difficult if the oil is on, or adjacent to, their land or property. Participation in the decision making process by the public and their elected local representatives can be intense and can easily divert response leadership from the operational aspects of cleaning up the oil. Challenges facing the incident management team can include relocating families, supplying drinking water, reopening roads, and handling claims for damages and lost income.

Containment and recovery

Containing spilled oil and then manually or mechanically recovering it is by far the most common inland response technique. The concepts are simple and many operations do not require special equipment—an important factor when an oil spill might occur at a considerable distance from a stockpile of specialized response equipment. Containment is used to prevent the spilled oil from spreading and to concentrate it in quantities that can be more easily recovered. The types of equipment used range from local construction materials and earthmoving equipment to smaller versions of the booms and skimmers used in offshore and nearshore marine spills (Fig. 3).

Most aquatic inland oil spills originate as spills on land. Responders initially want to keep the oil from flowing into surface waters and from penetrating into the ground. If spills are on impenetrable surfaces, responders will construct berms or other barriers to keep the oil contained. Sorbent booms have been used for the same purpose on small spills. Drains for storm water runoff are often protected by blocking them off with specially designed covers or by placing sheets of plastic over them weighted down with sand piles or sandbags.

Small streams and ditches are often dammed with local construction materials such as clay and gravel. This is especially useful when containment booms are not readily available or cannot be deployed because the stream is too narrow, or the current is too fast or the depth too shallow. Underflow dams use pipes installed through the dam to allow water to pass while retaining surface oil for collection. In deep streams, pipes just need to be placed at the bottom of the stream bed to allow the water to pass. Care must be taken to provide enough capacity for the flow of water, otherwise the dams may be washed away. For shallower streams where the constructed dam raises the water level, the pipes, or at least some of them, may be angled with the high ends on the downstream side of the dam providing a means to regulate the water level. Increased runoff due to rain or snow melt can overwhelm the flow-through capacity of these temporary

Figure 3. Left: piles of soil used to contain spilled gasoline. Right: sorbents being used to contain oil for recovery.

Figure 4. Left: underflow dams with angled pipes to raise the water level and allow surface oil to be retained for collection. Right: underflow weirs using plywood sheets.

dams; hence, it is advisable to monitor the weather to ensure dams are appropriately constructed, maintained, and able to retain their integrity. Locations for these dams should be chosen so that water does not spill out of the stream onto adjacent terrain and result in unintended exposures upstream of the dam (Fig. 4).

Sheets of plywood can be used effectively to block culverts, ditches, and very small streams. These need to be cut into the stream bank and secured. If there is flow in the stream or ditch, they can be raised off the bottom to create a simple underflow dam. Some wooden dams can be preconstructed and inserted where needed or they can be built on-site. They will often have a center section which can be raised and lowered to control the water level and trap oil. As with the other damming techniques, the function of these structures needs to be monitored for flow changes and maintained for structural integrity.

Containment booms are designed to control the oil floating on water. Booms of different sizes are readily available for a variety of conditions. Small booms are usually used on rivers and streams because they possess relatively low freeboard and are easier to handle (Fig. 5).

The physics of water flowing under a containment boom is such that very low water velocities (even below one knot or 0.5 m per second) can cause oil to become entrained in the water flowing underneath a boom. As a result, containment booms are usually deployed at an angle to the water current direction so that the effective water velocity is reduced and no oil is lost. However, even sharply angled booms will fail to hold back oil in currents greater than three knots (1.5 m per second). When current speed and direction varies due to estuarine tidal cycles, flood events, and other normal water variations, it can be challenging to deploy containment booms that will continue to hold back flowing oil. The forces exerted on containment booms by flowing water are high, and a skilled crew is required for successful deployment of a boom (or booms) in the proper configuration, and for regular boom maintenance. In some instances, such as floods, it may not be practical to deploy booms due to worker safety considerations. Deployed booms should be monitored periodically to ensure proper function and adjusted as required (Fig. 6).

Booms can be deployed to exclude oil from sensitive resources or to divert the oil to a collection point. For example, deflection booming may be used to deflect oil away from relatively small but particularly sensitive areas such as a water intakes. Multiple booms can be cascaded down a watercourse to achieve these aims. Booms need to be anchored to hold them in place. Deploying long sections of boom in a proper configuration with anchors and ropes can be difficult and time consuming. Boom deflectors have been developed that use the force of the flowing water to hold a boom in place.

Figure 5. Above booms used to exclude oil from side channels.

Figure 6. Left: booms used to divert oil to a collection point. Right: a deflector is used to position the boom at the desired angle.

For some incidents on large rivers, where it has proven difficult to deploy a boom effectively, barges have been driven onto the river bank to act as barriers. Oil can still entrain beneath the barges, and they may need to be angled into the current, as with a containment boom (Fig. 7)

Recovery of spilled oil can be as simple as using vacuum trucks. Vacuum trucks are best used when connected to an oil skimmer or other recovery device, though they can be used alone when oil is accessible and pooled to larger thicknesses. However, an open suction hose on a vacuum truck can recover large volumes of water; this recovered water can significantly increase waste handling requirements and associated disposal costs.

The most common skimmers are the oleophilic types in which oil is adsorbed onto the skimmer surface with reduced water recovery. Of these, drum and disc skimmers are very

Figure 7. A vacuum truck used in conjunction with a drum skimmer to recover the spilled oil.

Figure 8. Left: oleophilic (oil-loving) drum skimmer. Right: oleophilic rope mop in heavy debris.

popular, and rope mops are desirable in locations with heavy debris. For simple suction operations, suction head skimmers that lay flat can be used to limit the amount of water recovered. Response stockpiles typically contain the types of skimmers most likely to be effective on the oils and in the conditions found in the local area (Fig. 8).

Oil trapped on quiescent water can be physically herded to a skimming location by spraying water. The intent is to push oil toward containment booms and skimmers. Care must be taken not to mix the oil into the water column where it may come into contact with, and attach to, sediments and thereby become more difficult to remove.

Spills on snow and ice

Locating oil beneath snow and ice can be challenging, and boring holes through the ice may be the only viable technique to find it. Remote sensing technologies to find the oil beneath the ice have been researched, and a number of new promising techniques have been identified but none have yet made it into the response community.

Figure 9. A slot cut into ice allowing oil to rise and flow to a collection point.

Once located, oil in moving water under ice can be recovered by cutting slots in the ice or by inserting plywood sheets into the ice at appropriate angles to divert the oil (analogous to containment booms) to where it can be recovered. Chainsaws can be used to cut into the ice, and the safety risks when using these saws in adverse conditions while wearing bulky cold-weather protective clothing should be addressed (Fig. 9).

Sorbents

Sorbents (materials that can absorb or adsorb oil) are frequently used for collecting oil on hard surfaces. For small spills, they may be the only recovery method needed. Oil spilled on roads can be very slippery and must be removed before the road can be reopened. Often, mineral particulates (such as sand, vermiculite, and clay mixtures) are used because they are inexpensive and can be easily applied and then collected.

The use of straw bales and other sorbent barriers is a viable recovery technique, especially if only natural materials are available. Wire mesh fencing can be installed across narrow, slow moving waterways to support and anchor sorbent materials placed on the upstream side to contain and absorb floating oil while the water continues to flow through the barrier. These barriers will not stop the flow of high levels of oil but can adsorb light quantities of oil. While straw is most commonly used, other sorbent materials can also be substituted. A major operational challenge is the recovery, handling, and disposal of the oiled sorbent. Controlled burning at the site is sometimes an option.

Traditional synthetic sorbents are often used to help contain and recover oil. They may be placed outside drains and along containment booms to adsorb small amounts of oil. They can be overused and create a recovery and waste disposal problem when it might have been better to use a skimmer or controlled in-situ burning. Used, oily sorbents create waste and usually have to be disposed of as hazardous waste. The logistics of moving them to the spill site, using them, recovering them, and then removing them from the site need to be considered in advance (Figs. 10 and 11).

Figure 10. Workers recover sorbent material that has been used to clean an oiled road.

Solidifiers

Solidifiers are products which mix with, and immobilize, oil. They consist of dry, granular polymers that partially melt when contacting petroleum oils, and which act to bind or encapsulate oil. Heavy, viscous oils and emulsified oils do not react well due to poor penetration and mixing. Manufacturers often recommend that they be broadcast over spilled oil as powders. These powders can be difficult to recover on water but can be easier to deploy and recover for small spills on land or hard surfaces.

Solidifiers can also be placed in disposable booms, pillows, sausages, etc., and deployed and recovered like sorbents. They create stronger bonds with oil than sorbents and will not leak oil as sorbents may do. They can remove oil sheens from water. There may be regulations restricting the use of solidifiers, particularly those in loose powder form.

Controlled in-situ burning

There have been numerous cases of spilled oil accidently catching fire and damaging facilities and vessels. However, deliberate burns have succeeded in limiting the spread of the spilled oil, rapidly removing oil from water, snow, or ice surface and thereby mitigating its consequences from a NEBA perspective. Inland burning has been successfully and routinely used in North America.

Controlled in-situ burning is usually considered when access is limited or when it is necessary to remove the oil quickly. Mechanical equipment may not be able to access a spill site if, for example, the terrain is too steep, too forested, or water-logged. While mechanical clean-up can take considerable time, controlled burns can remove large amounts of oil in a few hours. This saved time may be critical if heavy rains are forecast that could flush oil to wider or more sensitive areas. Another consideration is waste disposal. If large amounts of waste will be generated and/or there are no convenient disposal sites available, controlled in-situ burning may be the best option to remove the oil (Fig. 12).

Wetlands and other sensitive environments are highly susceptible to damage by the intrusive actions of clean-up workers. Wet conditions may prevent equipment from accessing the site and can make responders' actions more damaging, but the moisture will protect plant roots and soil from the heat of a burn.

When considering the use of a controlled burn, the following should be borne in mind:

- Safety: responders and the public should be protected from any risk of spreading fire. Often local fire departments are present to wet surrounding vegetation to protect against the fire spreading beyond the spilled oil.
- Ignition: simple and safe methods are preferred. Flares, flare guns, torches, and propane burners have been used successfully.
- Moisture: many ecosystems are tolerant of the heat from fires because fire is part of their life cycle (e.g., wetlands). However, heat from burning oil may be more intense than even these plants can tolerate. High moisture levels are therefore desirable, in particular to protect roots from heat stress and to limit the risk of the fire spreading beyond the oiled area.

Figure 11. Left: straw used to hold back and absorb low quantities of oil. Right: oiled straw is disposed of by burning on site.

Figure 12. Controlled in-situ burning in a peat bog; the white powder is fire retardant.

- Season: in higher latitudes, winter burning often creates little environmental damage because many plants are dormant and could be covered in snow. Burns in late summer conditions can cause stress to plants as they are still building food reserves.
- Weather conditions: unpredictable weather can produce safety-related hazards at a controlled burn. Low, steady winds with no threatening storms or weather fronts are preferred because such conditions reduce the risk of the fire spreading. Atmospheric temperature inversions can trap the smoke plume and are not desirable because they restrict the dispersion of the smoke.
- Residue collection: some unburned oil or oily residue is almost always left after burning. If quantities are sufficient, the unburned oil could be collected and burned. Residues from crude oil burns can have varying consistencies from tarry to brittle; this residue needs to be collected. On water, oils subject to intense burns may create residues that sink after they cool, and it may be unacceptable to leave these behind (Fig. 13).

Shoreline clean-up/treatment

Shoreline assessment. Oiled shoreline assessment surveys—also known as Shoreline Clean-up Assessment Technique (SCAT) surveys—are a critical component of a response operation. The information gathered by the survey teams is used by the response managers to set objectives, priorities, constraints, and end points, all of which are essential in supporting the planning, decision making, and implementation of an effective shoreline response program.

Oiled shoreline assessment surveys of inland aquatic environments are carried out to:

- define and document the scale and character of shoreline oiling;
- identify and document the shoreline type and riverine or lacustrine character within the affected area;
- develop recommendations for treatment end points and treatment techniques which provide a net environmental benefit based on sound science;
- provide support throughout the treatment program so that shore clean-up operations personnel understand the expectations and concerns of the response managers;
- provide a process for closure once treatment has been completed; and
- involve appropriate representatives to ensure consensus throughout the shoreline response program.

Further details on how SCAT surveys are developed and implemented are provided in the IPIECA-IOGP Good Practice Guide on oiled shoreline assessment (SCAT) surveys. The principles of SCAT can be applied equally to coastal or freshwater incidents. To facilitate standardization or a systematic approach to assessment and capturing field data, a series of forms for recording data on inland habitats have been produced.

For successful implementation of SCAT, a team of dedicated personnel is needed who are familiar with its aims and

Figure 13. Snow and ice protects this wetland from oil exposure and aids fire control. A large smoke plume rises from a wetland burn.

terminology. This team should be fully integrated within an incident management team to ensure that their data are utilized to support the decision making process.

Response options. Techniques for cleaning oiled river banks and inland shorelines share similarities with marine shorelines for more information on the latter, see the IPIECA-IOGP Good Practice Guide on Oiled Shoreline Clean-up Techniques. The term "cleaning," in this context, is applied broadly and includes various water-flushing options, manual and mechanical oil removal, and the removal of oiled vegetation and oiled debris; it may or may not remove all of the oil from the shoreline. The scale of response operations may be different for smaller inland spills, with the techniques requiring adjustment according to the specific circumstances of a spill.

While tides and waves typically dominate marine shoreline-oil interaction, it is waves in lakes, and water flow in streams and rivers that dominate freshwater shoreline-oil interactions. The tendency for freshwater systems to be more confined than marine environments leads to less dilution of oil in water, and less spreading. As a result, oil concentrations can be greater in freshwater habitats (Fig. 14).

Many inland habitats have been transformed by human activity, including urban and suburban construction, the development of commercial and industrial facilities, and agricultural terraforming. As a result, the original landscape may have disappeared or been substantially altered. These factors can result in less concern being shown for the ecological sensitivity of such habitats to oil exposure, and more aggressive clean-up techniques may be considered acceptable.

While marine shorelines are classified based on tidal zone, inland habitats are classified as follows:

- Lake shores are classified based on swash zone from wave action. The swash zones are analogous to marine tidal zones, and include the supra-wash, upper swash, lower swash, and submerged zone (down to where rooted plants stop growing).
- River bank zones are classified based on water levels, and include the over-bank (flood plain), upper bank, lower bank, and mid-stream (exposed bars or shoals in the channel).

Marine tidal effects typically create wider shorelines than the swash zones do on freshwater shores. In a freshwater setting, a 1-m wide shoreline would be considered wide, whereas in a marine setting, it would be considered narrow. Freshwater shores include various mud, clay, and other sediment banks and vegetated shorelines.

Water level variations in the marine environment are often predictable (e.g., those caused by tides) while variations in freshwater levels can be unpredictable (e.g., those caused by storm events and runoff from precipitation). These unpredictable changes in water levels can dramatically affect clean-up operations as well as the natural oil removal. Small streams and shallow water may become oiled across the whole channel, resulting in the need to clean the entire bottom of the water course from bank to bank. The ecological effects from this spreading of oil must be balanced against potential damage from manual and mechanical removal and the option of leaving some oil behind to degrade naturally.

All cleaning techniques can be intrusive and can damage habitats, even in the absence of oil. It is best to identify

Figure 14. Removing oil from a wetland contaminated with heavy fuel oil—planking is used to reduce trampling.

preferred techniques in advance when preparing oil spill contingency plans, and guidance is available to help make these judgments for freshwater habitats and associated shorelines. During a response, an informal and rapid NEBA can be used to confirm that appropriate decisions are made regarding the choice of response techniques, thereby ensuring that such actions are operationally feasible, effective, and least intrusive to the environment. A summary of the relative effects of physical response techniques for use in freshwater habitats and shorelines in the absence of spilled oil is provided in Table 3.

Surface washing or shore cleaning agents. Shore cleaning agents are products which increase the ease or efficiency of oil removal when flushed with water. The oil is washed, or mobilized, off a hard surface, corralled by booms, and collected by skimmers or sorbents. These products are designed to be used on very heavy oils on hard substrates that cannot be cleaned by water flushing alone, and their use is not an initial response technique. They do not work well on porous habitats like sand shores and marshes.

Many cleaning products are sold for the purpose of surface washing, some of which are also suitable for household use. However, only a few have low aquatic toxicity and avoid causing the removed oil to disperse. Because these agents are good cleaners, some are used to clean oiled response equipment at decontamination facilities (Fig. 15).

There is a long history of their use by local responders, such as fire departments, for removing oil quickly from roads and other hard surfaces. Their priority is to quickly reopen paved areas for safe public use. However, care should also be taken to constrain the further spread of mobilized or dispersed oil.

The best products are those that have low aquatic toxicity while remaining effective. They often require soak time to interact well with the oil before being flushed with water. As a result, they cannot be successfully used if waves or rain are present because the water will wash the product off the oiled surface before it has time to interact with, and mobilize, the oil.

Biodegradation

Although not a technique for spill response to limit the spread of oil and recover it, subsequent longer-term clean-up activities may include bioremediation tasks. Microbes that degrade oil exist everywhere and need adequate oxygen and nutrients, primarily nitrogen and phosphorus, to promote their growth. In quiescent inland waters with limited flow, such as some wetlands, swamps, ponds, and lakes, a major influx of hydrocarbons can be overwhelming to microbes, and oxygen and nutrients may become limiting. Microbial degradation of spilled hydrocarbons is easier and faster if oil components are soluble in water. However, this is atypical since most oil components are not very soluble in water, or may be insoluble. The more soluble components are the lighter-weight compounds found in refined products, very light crude oils, and ethanol fuels.

If soluble components (e.g., refined products) are available to microbes in large quantities, their rapid growth could deplete the dissolved oxygen supply in smaller water bodies and cause hypoxia, slowing down their growth and affecting other aquatic species. Extreme reductions in dissolved oxygen can lead to fish deaths. Studies have shown that the addition of fertilizers accelerates the rate of degradation if nutrients are lacking; however, this can also lead to eutrophication, thereby increasing the potential for hypoxia.

Biodegradation is not a new response technology. It occurs naturally, regardless of whether it is enhanced by the addition of nutrients. The rate of biodegradation depends on four key factors: nutrients, oxygen, ambient temperatures, and the local degree of oiling. Light oiling can be degraded within weeks or a growing season. Because it may take months to degrade thick layers of stranded oil, this is not regarded as a response technique, but rather a process that can be relied upon after the majority of the oil has been removed by other clean-up techniques.

It can be a challenge to reach agreement on the end points of clean-up activities, as stakeholders may have the initial perception that removing all oil through clean-up activities is required and appropriate. However, the ability for the natural process of biodegradation to break down the majority of oil compounds is an important consideration in the framework of NEBA and should be factored into the analysis alongside the potential harm of other clean-up operations.

Sunken oil

There are specific challenges to be faced where very dense oils sink in freshwaters. The identification of sunken oil may be difficult as it is likely to be hidden from view. Initial predictions of depositional areas can be made by studying the water body (bathymetry) and its surface water hydrology to determine where natural areas of deposition may occur (for example, the pools of a meander in a river or around the edges of a lake, or depressions in the stream bed). This would be followed by probing or surveying for oil. In some cases, changes in temperature or seasonal factors may lead to sheens appearing, which could provide indicators of sunken oil. The use of sorbent snares (pom poms) or sorbent pads fastened to a weight and dragged along the bottom can also be useful in locating the presence of sunken oil.

Clean-up actions may include agitating or aerating the river or lake bed to encourage the release of oil, with subsequent recovery of released oil as it reaches the surface. Dredging of oiled sediment may also be considered. In all cases, the treatment of sunken oil is likely to be a long-term project extending beyond the emergency phase of an incident (Fig. 16).

Table 3
Relative Effects of Physical Response Techniques for Use in Freshwater Habitats and Shorelines in the Absence of Spilled Oil

Physical response method	Water environment				Shoreline habitat							
	Open water	Small lakes/ ponds	Large river	Small rivers/ streams	Bedrock	Man-made	Sand	Vegetated shores	Sand and gravel	Gravel	Mud	Wetlands
Natural recovery	–	–	–	–	–	–	–	–	–	–	–	–
Booming	L	L	L	L	–	–	–		–	–	–	–
Skimming	L	L	L	L	–	–	–		–	–	–	–
Barriers/berms	–	–	–	H	–	–	–		–	–	–	–
Physical herding	L	L	L	L	–	–	–		–	–	–	–
Manual oil removal/ cleaning	L	H	L	M	L	L	L	H	M	M	H	H
Mechanical removal	L	H	H	H	–	M	M	H	M	M	H	H
Sorbents	L	L	L	L	L	L	L	L	L	L	M	M
Vacuum	L	L	L	L	L	L	L	M	L	L	H	M
Debris removal	–	L	L	L	L	L	L	L	L	L	M	M
Sediment reworking	–	H	–	H	–	–	M	H	M	M	H	H
Vegetation removal	L	H	M	H	–	–	–	H	–	–	–	H
Controlled in-situ burning	L	M	L	M	L	L	M	M	M	M	H	M
Flooding	–	–	–	–	L	L	L	L	M	L	L	L
Low-pressure, cold-water flushing	–	–	–	–	L	L	M	L	L	M	H	L
High-pressure, cold-water flushing	–	–	–	–	L	L	H	H	H	H	H	H
Low-pressure, hot-water flushing	–	–	–	–	M	L	H	H	M	M	H	H
High-pressure, hot-water flushing	–	–	–	–	M	L	H	H	H	H	H	H
Steam cleaning	–	–	–	–	M	L	H	H	M	M	H	H
Sand blasting	–	–	–	–	H	M	–	–	–	–	–	–

H, High; *L*, Low; *M*, Moderate.

Figure 15. (a) Oiled riprap with sorbent boom at water's edge. (b) Spraying shoreline cleaner. (c) Flushing after soak time. (d) Cleaned riprap.

Figure 16. Pumping water into river sediment to release sunken oil and flush it to the surface.

Summary

Spills to inland environments are typically much smaller in volume but more frequent than spills to marine waters and their shorelines. Inland spills do not garner the degree of broad public and media attention compared to large offshore spills.

The response to oil spills in inland aquatic environments shares similar principles to a marine spill response, but there are significant differences. Inland spills, even small ones, often directly affect the public in a more intimate manner than similar-sized marine spills. They are also more likely to involve oil products, and raise immediate safety concerns due not only to their potential proximity to communities but also because of specific fears about toxic vapors and fire risks. Spills may occur near homes and businesses, along highways, and in cities and towns. This proximity can, and often should, drive response priorities, including:

- preventing the exposure to oil of drinking water supplies;
- monitoring hydrocarbon vapors and smoke plumes for responder and public health and safety, and securing operational exclusion zones;
- preventing exposure to oil of livestock and pets; and
- limiting the disruption of waterborne traffic, and arranging the temporary relocation of residents and closure of businesses, as necessary.

Response activities, especially for smaller spills, may be conducted and/or supervised solely by the local authorities. A response to small inland spills can be achieved by relatively few responders and can often be completed more quickly than marine spills, as smaller geographic areas may be affected. Where potential inland oil spills originate from fixed facilities or other assets, it is feasible to engage in detailed oil spill contingency planning based on realistic spill scenarios, leading to improved communication and coordination during a response.

The industry has evaluated inland spills and associated response operations, and has developed guidance for response techniques which account for oil type and habitat sensitivity. This guidance aims to inform the selection of response techniques which are both effective and provide a net environmental benefit.

Excerpted from the IPIECA/IOGP report "Oil spills: inland response—Good practice guidelines for incident management and emergency response personnel." 14th Floor, City Tower, 40 Basinghall Street, London EC2V 5DE, United Kingdom. For the full report and citations, see: www.ipieca.org and www.iogp.org. Reprinted with permission.

Calculating volatile organic compounds flash emissions from crude oil and condensate tanks at oil and gas production sites

One of the largest sources of volatile organic compound (VOC) emissions from an oil and gas production site is oil or condensate storage tanks. This guidance from the Air Permits Division, Texas Commission on Environmental Quality (Air Permit Reference Guide, APDG 5942), can be used to help evaluate flash emissions.

Storage tank emissions at production sites

There are three types of emissions generated by a storage tank:

(1) Breathing losses (also called standing losses) which is the normal evaporation of liquid in a tank. Breathing losses will increase if the temperature increases.
(2) Working losses are an increase in evaporation due to agitation of liquid from activities such as filling the tank.
(3) Flash losses occur when the pressure of a liquid is decreased or the temperature is increased.

Flash emissions occur when produced liquid (crude oil or condensate) is exposed to temperature increases or pressure decreases during the transfer from the production separators (or similar sources) into atmospheric storage tanks.

Generic oil and gas site

A typical oil and gas production site is shown in Fig. 1. For a tank that receives condensate or oil from a separator, emissions will include working, breathing, and flash losses (Fig. 2). If a site has multiple tanks, the types of emissions from each tank will depend on whether the tanks are in series or parallel (Fig. 3).

Estimation methods

There are several methods to calculate or measure emissions from storage tanks, and some are more accurate than others (Table 1). Even though working, breathing, and flash losses

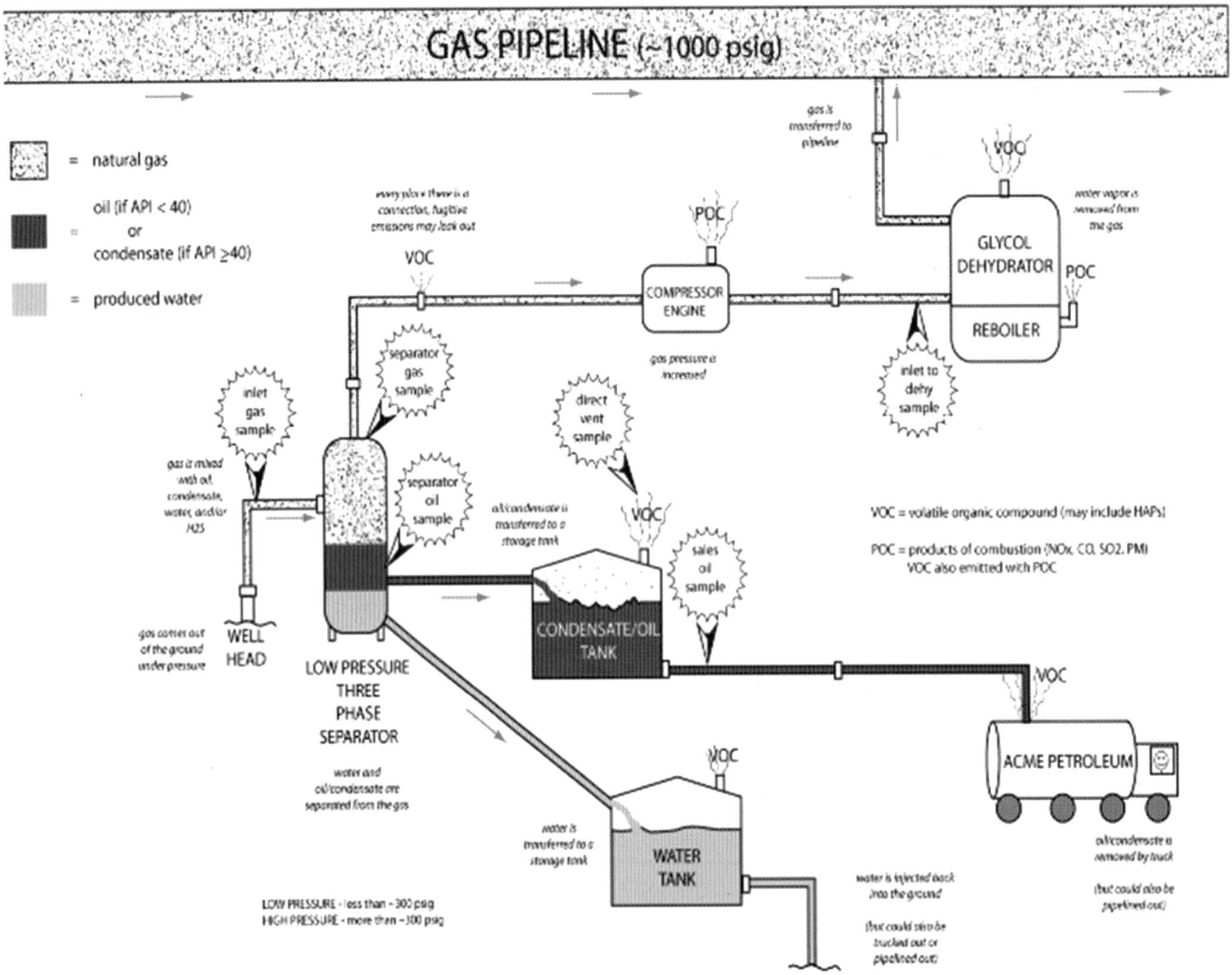

Figure 1. Schematic of a simple oil and gas production site.

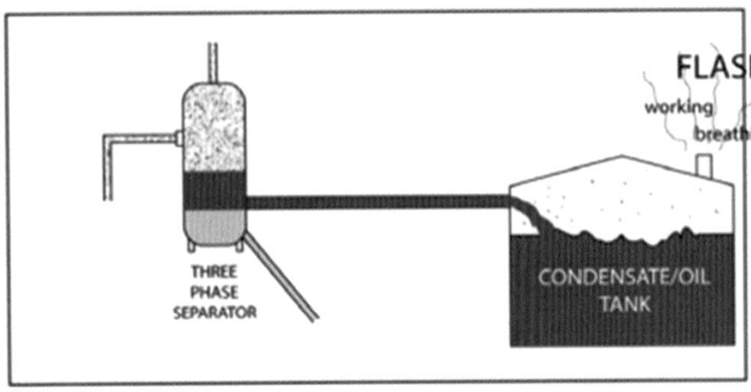

Figure 2. Working, breathing, and flash losses occur when tanks receive crude oil or condensate from production separators.

are almost always mixed together and exit the tank vent at the same time, some methods only calculate working and breathing losses, while some methods only calculate flash losses. There are also several methods to calculate all three types of emissions simultaneously.

Each method for estimating emissions has specific constraints. Regardless of which method is used, however, all supporting data used to calculate the emissions, including identification of the calculation method, description of sampling methods, and copies of lab sampling analysis, must be provided with the emissions estimate.

Most state agencies prefer that the most accurate emission estimates be submitted, based on site-specific, representative worst-case data when possible. The first four methods

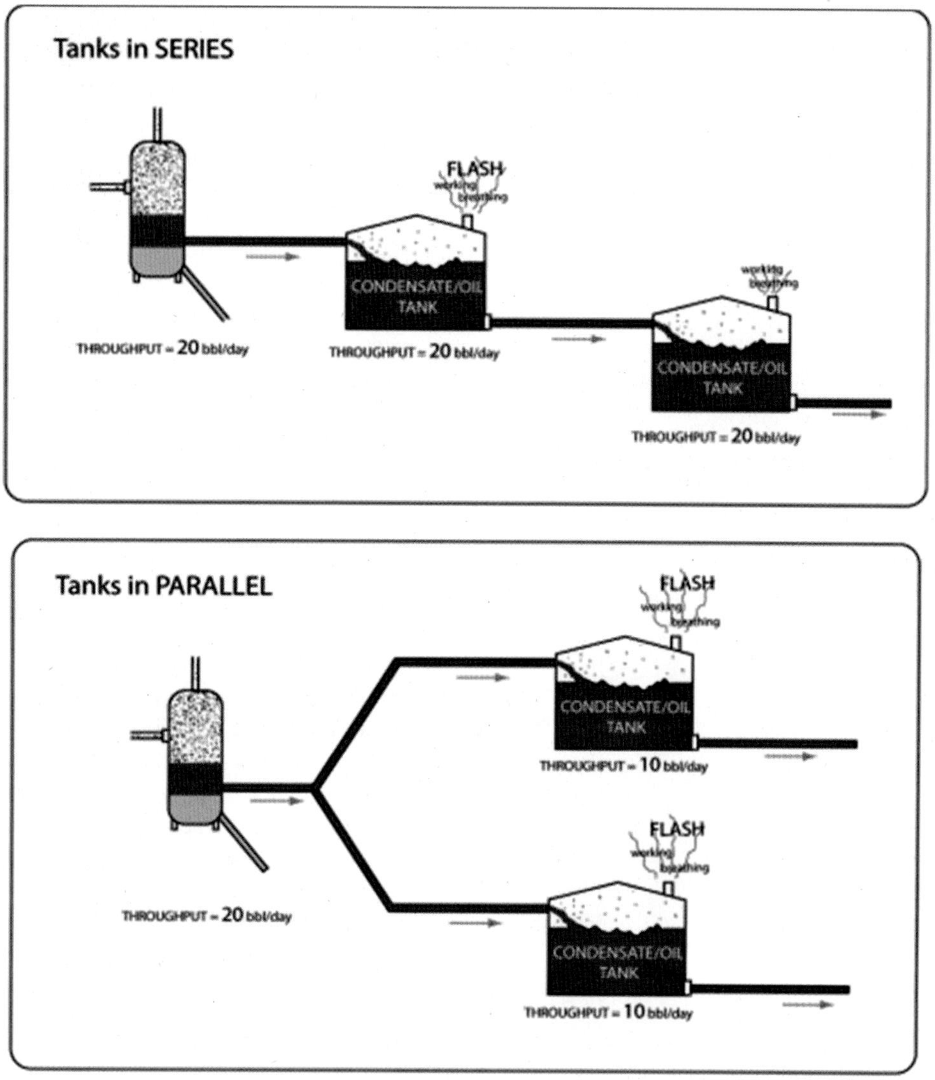

Figure 3. Tanks in series and parallel.

described in Table 1 – direct measurement, process simulation, E&P Tanks Software V2.0, and laboratory measurement of GOR from a pressurized liquid sample – are considered the most accurate. Applicants using a less accurate method should be aware of the risk of potentially underestimating emissions at a site.

No matter which method is used to calculate flash emissions, verification of the inputs and calculation methods is required. State the calculation method used and any critical parameters in the project description so they are available to program personnel. If at an existing production site, the emission calculations should be determined from site-specific sampling or analysis. If a site is not yet in operation, information from sister sites, nearby sites on the same field, or other empirical data may be used with a justification as to why that information is appropriate (Table 2).

Griswold and Ambler gas-oil ratio chart method

A graph developed by Griswold and Ambler (1978) can be used to approximate total potential vapor emissions from a barrel of oil based on pressure differentials. The curves were constructed based on empirical flash data from lab studies and field measurements. It requires three inputs:

- API gravity;
- Separator pressure;
- Oil/condensate throughput.

Knowing the oil throughput, the total volume of flash gas emitted can be determined. To calculate VOC emissions, more information about the composition of the flash gas would need to be known (the molecular weight and VOC content). This method will not give these data.

Table 1
Flash Loss Estimation Methods

Method	Emissions calculated	Comments
Direct measurement of tank emissions	Working, Breathing, Flash	Sampling and analysis are expensive, but the results are relatively accurate.
Process simulator computer programs	Flash only	There are several different process simulators (e.g., WinSim, Designer II, HYSIM, HYSIS, VMG, and PROMAX, etc.). The software is expensive, but the results are accurate when based on site-specific sample and analysis.
E&P Tanks Software, V2.0, using an option that requires site-specific sampling	Working, Breathing, Flash	A pressurized liquid and/or gas sample analysis from a separator is needed. This choice does not include the geographical database option.
Laboratory measurement of the gas-oil ratio (GOR) from a pressurized liquid sample	Flash only	This is a direct laboratory analysis of flash gas emitted from a pressurized oil/condensate sample.
Vasquez-Beggs equation (VBE)	Flash only	A calculation method based on empirical data. The VBE variables must be supported with a lab sampling analysis that verifies the API gravity, separator gas gravity, stock tank gas molecular weight, and VOC fraction. If an operating variable used in the VBE calculations falls outside of the parameter limits, the applicant must use another method to calculate flash emissions.
E&P Tanks Software, V2.0, geographical database option	Working, Breathing, Flash	Emissions are based on choosing an example base that closely matches operating parameters at the site in question. The geographical database is based on 103 sampled sites and is a very poor estimate of emissions from any particular storage tank.
Griswold and Ambler GOR Chart Method	Flash only	A graph developed by Griswold and Ambler (1978) can be used to approximate total potential vapor emissions from a barrel of oil based on pressure differential. The curves were constructed using empirical flash data from laboratory studies and field measurements.

Table 2
Volatile Organic Compound (VOC) Emissions Throughout the United States at Conditions Shown

E&P National case No.	E&P Regional case No.[a]	API gravity (°API)	RVP (psia)	Separator pressure (psig)	Separator temperature (°F)	VOC emission if throughput = 10 bbl/day (tpy)
86	SW case 27	57	13.1	54	60	15.006
87	SW case 28	57	13.1	870	78	48.132
88	NW case 33	57	7.5	600	70	37.853
89	SW case 29	58	8.0	780	70	28.792
90	NW case 34	58	8.0	60	56	6.643
91	NW case 35	58	9.1	500	84	29.895
92	NE case 4	58	10.6	300	80	41.859
93	SW case 30	59	10.0	110	72	7.896
94	NW case 36	60	9.4	750	90	57.537
95	SW case 31	61	7.0	85	85	4.475
96	NW case 37	61	9.8	730	84	73.168
97	SW case 32	62	10.4	57	82	14.693
98	SW case 33	63	7.0	72	80	3.167

Table 2
Volatile Organic Compound (VOC) Emissions Throughout the United States at Conditions Shown (continued)

E&P National case No.	E&P Regional case No.[a]	API gravity (°API)	RVP (psia)	Separator pressure (psig)	Separator temperature (°F)	VOC emission if throughput = 10 bbl/day (tpy)
99	NW case 38	63	11.9	730	80	54.645
100	NW case 39	64	6.4	580	77	21.334
101	NW case 40	64	11.0	730	80	113.571
102	NW case 41	66	11.8	807	96	85.523
103	NE case 5	68	12.5	170	75	12.461

[a]The gray cases are in the northwestern and northeastern United States.

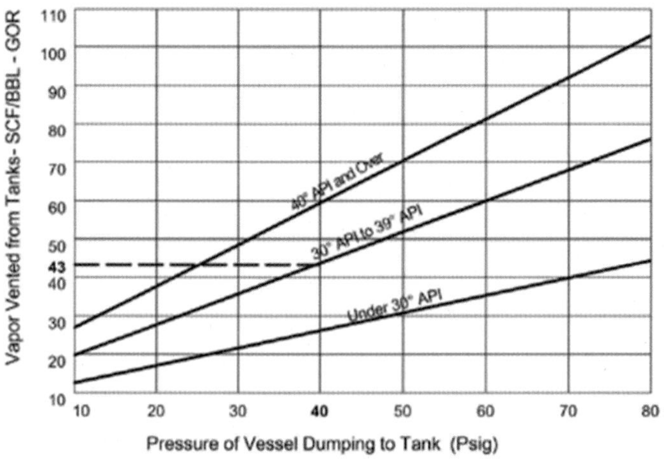

Figure 4. Vapor vented from tanks in cubic feet per barrel throughput as function of separator pressure and API gravity.

To calculate flash using this method:

1. Determine the range that the actual API gravity falls into (under 30, 30–39, or 40+);
2. Consult the graph (Fig. 4) and find the line that corresponds to the correct API gravity range;
3. Find the point on the line that corresponds to the separator pressure (on X axis);
4. Determine where on the Y axis this point is; this is the gas-oil ratio (GOR) described in field units scf/bbl.

Example. Given API gravity = 38°API,
Separator pressure = 40 psig, and
Throughput = 1000 bbl/day.
Then from Fig. 4, GOR = 43 scf/bbl.
Therefore, emission volume is computed as (43 scf/bbl) * (1000 bbl/day) = 43,000 scf/day of flash gas.

Source

Calculating Volatile Organic Compounds (VOC) Flash Emissions from Crude Oil and Condensate Tanks at Oil and Gas Production Sites, Air Permits Division, Texas Commission on Environmental Quality, Air Permit Reference Guide, APDG 5942, Revised May 2012.

Reference

J.A. Griswold, T.C. Ambler, A Practical Approach To Crude Oil Stock Tank Vapor Recovery. SPE-7175-MS. Paper presented at the SPE Rocky Mountain Regional Meeting. Wyoming, Cody, May 1978.

22: Specifications

Pipeline Steel Specifications 1090
API 5L pipeline specifications 1090
Main differences between API 5L PSL 1 and PSL 2 pipeline 1091
Chemical composition for API 5L PSL 1 welded pipe 1092
Chemical composition for API 5L PSL 2 welded pipe 1093
Tensile property requirements for pipe body of seamless and welded pipe for PSL 1 1094
Tensile property requirements for pipe body of seamless and welded pipe for PSL 2 1094

Pipeline Material Coating Specifications 1095
External application procedures for plant applied fusion bonded epoxy coatings and abrasion resistant overlay coatings to steel pipe 1095
External application procedures for coal tar epoxy protective coatings to steel pipe 1099
External application procedures for polyolefin pipe coating applied by the cross head extrusion method or the side extrusion method to steel pipe 1102
Exterior application procedures for plant applied tape coatings to steel pipe 1107
Suggested procedures for coating field joints, fittings, connections, and prefabricated sections using tape coatings 1110
Suggested procedures for coating field joints using heat-shrinkable materials 1111
External application procedures for hot applied coal tar coatings to steel pipe 1113

Fuel Specifications 1117
Types of fuel specifications 1117
Transportation fuel specifications 1119
Fuel quality 1119
Gasoline fuel specifications 1121
Aviation gasoline fuel specifications 1124
Jet fuels specifications 1126
Diesel fuel specifications 1131

Pipeline Product Specifications 1136
NGL products, propane 1136
NGL products, butane, normal butane 1137
NGL products, butane, refinery grade butane 1138
NGL products, iso-butane 1139
Gasoline, conventional, 85.5 octane (no ethanol) 1140
Gasoline, conventional, 87 octane (no ethanol) 1141
Gasoline, subgrade, 87 octane after 10% ETOH addition 1142
Distillate, #1 diesel fuel, ultra-low sulfur (15 ppm max) 1143
Distillate, #1 diesel fuel, ultra-low sulfur (15 ppm max) 1144
Distillate, #2 diesel fuel, ultra-low sulfur (15 ppm max) 1145

PIPELINE STEEL SPECIFICATIONS

API 5L pipeline specifications

API 5L provides requirements for the manufacture of seamless and welded pipes and consists of historical, scientifically validated criteria. It is not a "how to" instructional manual or recipe on manufacturing line pipe. In general, API standards are reviewed and revised, reaffirmed or withdrawn at least every five years, and the 46th edition of API 5L, effective May 1, 2019, was over 150 pages long with more than 40 tables and 12 appendices. For complete reference to the exact pipeline specifications, see API 5L, 46th edition.

Product specification levels

API Specification 5L covers line pipe grades from A25 to X120, diameter from ½ inch to 84 inches, and wall thickness up to 2 inches. There are two product specification levels, PSL 1 and PSL 2, for the manufacture of seamless and welded steel pipe for use in the petroleum and natural gas industries. PSL 1 covers grades from A25 through X70 and a size range from 0.405 through 80. PSL 2 covers grades from B through X80. A summary of the main differences between PSL 1 and PSL 2 pipe is shown in Table 1.

PSL 1 is a standard quality level, and PSL 2 applies tighter controls on chemical composition, mechanical properties, and quality. For example, PSL 2 has mandatory requirements for carbon equivalent, notch toughness, maximum yield strength, and maximum tensile strength. PSL 2 also limits maximum carbon, phosphorus, and sulfur content at lower levels than PSL 1, provides for a maximum carbon equivalent, yield strength, and tensile strength for each grade, and prohibits repair by welding of pipe body, plate, or skelp. Onshore, both PSL 1 and PSL 2 are common, but offshore PSL 2 is usually specified.

Grade

Steel is an iron-based alloy that contains up to 2% carbon, and in the oil and gas industry, steel pipeline is a low carbon-manganese alloy specified by several characteristics that include chemical composition, mechanical properties, surface condition, grain size, hardenability, tolerances, and production process (API, 2009). Yield strength, defined as the maximum stress the material can experience without failure, is measured in units of force per area as 1000 pounds per square inch (ksi) or Newtons per square meter (N/m^2). The two-digit number following the "X" indicates the minimum yield strength (in ksi) of pipe produced to this grade.

The addition of alloying elements like carbon, manganese, silicon, niobium, vanadium, and titanium strengthens the steel; however, the addition of alloying elements may reduce the weldability and other parameters. The carbon and manganese contents are the primary elements that determine weldability. Corrosion resistance is another important specification and elements such as chromium and nickel are used to improve corrosive properties. The precise mixture of alloys in the steel will depend on the pipe manufacturer and the specifications of the project including wall thickness, strength required, and the corrosive characteristics and temperature of the fluid.

Chemical properties

The composition of steel used for the manufacture of pipe furnished to API 5L specifications conform to the chemical requirements given in Table 2A for PSL 1 and Table 2B for PSL 2. The composition of intermediate grades higher than X42 need to conform to the chemical requirements of the next higher grade.

At a minimum, the following elements are required to be evaluated: carbon, manganese, sulfur, chromium, columbium (niobium), copper, molybdenum, nickel, silicon, titanium, and vanadium. Boron also needs to be evaluated unless heat analysis indicates content is less than 0.001%.

For PSL 2 pipe, carbon equivalent calculations are calculated and reported using formula based on carbon content. For carbon content less than or equal to 0.12%, the *CE(Pem)* equation, also known as the Ito-Bessyo formula, is applied:

$$CE(Pem) = C + \frac{Si}{30} + \frac{Mn}{20} + \frac{Cu}{20} + \frac{Ni}{60} + \frac{Cr}{20} + \frac{Mo}{15} + \frac{V}{10} + 5B$$

When the carbon content is greater than 0.12%, the carbon equivalent is calculated using the *CE(DW)* equation, also called the (International Institute of Welding) DW formula:

$$CE(DW) = C + \frac{Mn}{6} + \frac{(Cr + Mo + V)}{5} + \frac{(Ni + Cu)}{15}$$

Mechanical properties

PSL 1 Grades A25 to X70 tensile requirements are specified in Table 3A, and PSL 2 Grades B to X80 tensile requirements are specified in Table 3B. Grades intermediate to the listed grades have tensile requirements agreed upon between the purchaser and manufacturer and consistent with the values shown in the tables.

In Table 3, the minimum elongation in 2 in (50.8 mm) is determined by the following equation:

$$e = 625{,}000 \frac{\mathbf{A}^{0.2}}{\mathbf{U}^{0.9}} \text{ (US customary units)}$$

or

$$e = 1944 \frac{\mathbf{A}^{0.2}}{\mathbf{U}^{0.9}} \text{ (SI units)},$$

where:

e = minimum elongation in 2 in (50.8 mm) in percent rounded to the nearest percent.

A = applicable tensile test specimen area for round bar specimens 0.20 in^2 (130 mm^2) for the 0.500 in and 0.350 in specimens, and 0.10 in^2 (63 mm^2) for the 0.250 in specimens.
U = specified minimum ultimate tensile strength in psi (MPa).

For full section and strip specifications, see API 5L, 46th edition.

Other specifications

In addition to the chemical and mechanical material requirements, specifications refer to dimensions, weights, lengths, defects and end finishes, couplings, inspection and testing, marking, coating and protection, documentation, and pipe loading. There are flattening tests, fracture toughness tests, impact tests, hydrostatic tests, and dimensional tests (tolerances) that must be satisfied. For a full list and reference to the exact specifications, see API 5L, 46th edition.

References

American Petroleum Institute, Design, construction, operation and maintenance of offshore hydrocarbon pipelines (Limit State Design). API-RP 1111. American Petroleum Institute, Washington, DC, 2009.

American Petroleum Institute, Specification for line pipe, 46th ed. American Petroleum Institute, Washington, DC, 2019.

Main differences between API 5L PSL 1 and PSL 2 pipeline

Table 1
Differences Between API 5L PSL 1 and PSL 2 Pipeline

Parameter	PSL 1	PSL 2
Electric welds: welder frequency	No minimum	100 kHz minimum
Heat treatment of electric welds	Required for grades > X42	Required for all grades
Chemistry: max C for seamless pipe	0.28% for grades ≥ B	0.24%
Chemistry: max C for welded pipe	0.26% for grades ≥ B	0.22%
Chemistry: max P	0.030% for grades ≥ A	0.025%
Chemistry: max S	0.030%	0.015%
Carbon equivalent	Only when purchaser specifies SR18	Maximum required for each grade
Yield strength, maximum	None	Maximum for each grade
UTS, maximum	None	Maximum for each grade
Fracture toughness	None required	Required for all grades
Repair by welding of pipe body, plate, and skelp	Permitted	Prohibited
Repair by welding of weld seams without filler metal	Permitted by agreement	Prohibited

Source: Adapted from API 5L.

Chemical composition for API 5L PSL 1 welded pipe

Table 2A
Chemical Composition for API 5L PSL 1 Welded Pipe, Wall Thickness <25.0 mm (0.984 in)

Steel grade	Mass fraction, % based on heat and product analyses[a]						
	C max[b]	Mn max[b]	P max	S max	V max	Nb max	Ti max
A	0.22	0.90	0.30	0.30	—	—	—
B	0.28	1.2	0.30	0.30	c,d	c,d	d
X42	0.28	1.3	0.30	0.30	d	d	d
X46	0.28	1.4	0.30	0.30	d	d	d
X52	0.28	1.4	0.30	0.30	d	d	d
X56	0.28	1.4	0.30	0.30	d	d	d
X60	0.28[e]	1.40[e]	0.30	0.30	f	f	f
X65	0.28[e]	1.40[e]	0.30	0.30	f	f	f
X70	0.28[e]	1.40[e]	0.30	0.30	f	f	f
A	0.22	0.90	0.30	0.30	—	—	—
B	0.26	1.2	0.30	0.30	c,d	c,d	d
X42	0.26	1.3	0.30	0.30	d	d	d
X46	0.26	1.4	0.30	0.30	d	d	d
X52	0.26	1.4	0.30	0.30	d	d	d
X56	0.26	1.4	0.30	0.30	d	d	d
X60	0.26[e]	1.40[e]	0.30	0.30	f	f	f
X65	0.26[e]	1.45[e]	0.30	0.30	f	f	f
X70	0.26[e]	1.65[e]	0.30	0.30	f	f	f

[a]Cu ≤ 0.50%; Ni ≤ 0.30%; Cr ≤ 0.50%; and Mo ≤ 0.15%.
[b]For each reduction of 0.01% below the specified max concentration for carbon, an increase of 0.05% above the specified max concentration for Mn is permissible, up to a max of 1.65% for grades ≥ B, but ≤ X52; up to a max of 1.75% for grades > X52, but < X70; and up to a maximum of 2.00% for X70.
[c]Unless otherwise agreed Nb + V ≤ 0.06%.
[d]Nb + V + Ti ≤ 0.15%.
[e]Unless otherwise agreed.
[f]Unless otherwise agreed, Nb + V + Ti ≤ 0.15%.

Adapted from API 5L.

Chemical composition for API 5L PSL 2 welded pipe

Table 2B
Chemical Composition for API 5L PSL 2 Welded Pipe, Wall Thickness <25.0 mm (0.984 in)

	Mass fraction, % based on heat and product analyses									Carbon equivalent[a]	
Steel grade	C max	Si max	Mn max[b]	P max	S max	V max	Nb max	Ti max	Other	CE aw max	CE Pcm max
					Seamless and welded pipe						
BR	0.24	0.40	1.20	0.025	0.015	c	c	0.04	e,l	0.043	0.25
X42R	0.24	0.40	1.20	0.025	0.015	0.06	0.05	0.04	e,l	0.043	0.25
BN	0.24	0.40	1.20	0.025	0.015	c	c	0.04	e,l	0.043	0.25
X42N	0.24	0.40	1.20	0.025	0.015	0.06	0.05	0.04	e,l	0.043	0.25
X46N	0.24	0.40	1.40	0.025	0.015	0.07	0.05	0.04	d,e,l	0.043	0.25
X52N	0.24	0.45	1.40	0.025	0.015	0.10	0.05	0.04	d,e,l	0.043	0.25
X56N	0.24	0.45	1.40	0.025	0.015	0.10[f]	0.05	0.04	d,e,l	0.043	0.25
X60N	0.24[f]	0.45[f]	1.40[f]	0.025	0.015	0.10[f]	0.05[f]	0.04[f]	g,h,l	As agreed	
BQ	0.18	0.45	1.40	0.025	0.015	0.05	0.05	0.04	e,l	0.043	0.25
X42Q	0.18	0.45	1.40	0.025	0.015	0.05	0.05	0.04	e,l	0.043	0.25
X46Q	0.18	0.45	1.40	0.025	0.015	0.05	0.05	0.04	e,l	0.043	0.25
X52Q	0.18	0.45	1.50	0.025	0.015	0.05	0.05	0.04	e,l	0.043	0.25
X56Q	0.18	0.45[f]	1.50	0.025	0.015	0.07	0.05	0.04	e,l	0.043	0.25
X60Q	0.18[f]	0.45[f]	1.70[f]	0.025	0.015	g	g	g	h,l	0.043	0.25
X65Q	0.18[f]	0.45[f]	1.70[f]	0.025	0.015	g	g	g	h,l	0.043	0.25
X70Q	0.18[f]	0.45[f]	1.80[f]	0.025	0.015	g	g	g	h,l	0.043	0.25
X80Q	0.18[f]	0.45[f]	1.90[f]	0.025	0.015	g	g	g	i,j	As agreed	
X90Q	0.16[f]	0.45[f]	1.90	0.020	0.010	g	g	g	j,k	As agreed	
X100Q	0.16[f]	0.45[f]	1.90	0.020	0.010	g	g	g	j,k	As agreed	
					Welded pipe						
BM	0.22	0.45	1.20	0.025	0.015	0.05	0.05	0.04	e,l	0.043	0.25
X42M	0.22	0.45	1.30	0.025	0.015	0.05	0.05	0.04	e,l	0.043	0.25
X46M	0.22	0.45	1.30	0.025	0.015	0.05	0.05	0.04	e,l	0.043	0.25
X52M	0.22	0.45	1.40	0.025	0.015	d	d	d	e,l	0.043	0.25
X56M	0.22	0.45[f]	1.40	0.025	0.015	d	d	d	e,l	0.043	0.25
X60M	0.12[f]	0.45[f]	1.60[f]	0.025	0.015	g	g	g	h,l	0.043	0.25
X65M	0.12[f]	0.45[f]	1.60[f]	0.025	0.015	g	g	g	h,l	0.043	0.25
X70M	0.12[f]	0.45[f]	1.70[f]	0.025	0.015	g	g	g	h,l	0.043	0.25
X80M	0.12[f]	0.45[f]	1.85[f]	0.025	0.015	g	g	g	i,j	0.043[f]	0.25
X90M	0.10	0.55[f]	2.10[f]	0.020	0.010	g	g	g	j,k	–	0.25
X100M	0.10	0.55[f]	2.10[f]	0.020	0.010	g	g	g	j,k	–	0.25

[a]SMYS > 0.737; CE limits as agreed. The CE aw limits applied if $C > 0.12\%$ and the CE Pcm limits apply if $C \leq 0.12\%$.
[b]For each reduction of 0.01% below the specified max concentration for carbon, an increase of 0.05% above the specified max concentration for Mn is permissible, up to a max of 1.65% for grades $\geq$ B, but $\leq$ X52; up to a max of 1.75% for grades $>$ X52, but $<$ X70; and up to a maximum of 2.00% for X70.
[c]Unless otherwise agreed $Nb + V \leq 0.06\%$.
[d]$Nb + V + Ti \leq 0.15\%$.
[e]Unless otherwise agreed, $Cu \leq 0.50\%$; $Ni \leq 0.30\%$; $Cr \leq 0.50\%$; and $Mo \leq 0.15\%$.
[f]Unless otherwise agreed.
[g]Unless otherwise agreed, $Nb + V + Ti \leq 0.15\%$.
[h]Unless otherwise agreed, $Cu \leq 0.50\%$; $Ni \leq 0.50\%$; $Cr \leq 0.50\%$; and $Mo \leq 0.50\%$.
[i]Unless otherwise agreed, $Cu \leq 0.50\%$; $Ni \leq 0.50\%$; $Cr \leq 0.50\%$; and $Mo \leq 0.50\%$.
[j]$B \leq 0.004\%$.
[k]Unless otherwise agreed, $Cu \leq 0.50\%$; $Ni \leq 1.00\%$; $Cr \leq 0.55\%$; and $Mo \leq 0.80\%$.
[l]Unless otherwise agreed, no intentional addition of B is permitted and residual $B \leq 0.001\%$.

Source: Adapted from API 5L.

Tensile property requirements for pipe body of seamless and welded pipe for PSL 1

Table 3A
Tensile Property Requirements for Pipe Body of Seamless and Welded Pipe for PSL 1

	Yield strength, Minimum		Tensile strength, Minimum		Elongation in 2 in
Grade	**Psi**	**MPa**	**Psi**	**MPa**	**Minimum, %**
A25	25,000	172	45,000	310	a
A	30,500	207	48,600	331	a
B	35,500	241	60,200	414	a
X42	42,100	290	60,200	414	a
X46	46,400	317	63,100	434	a
X52	52,200	359	66,700	455	a
X56	56,600	386	71,100	490	a
X60	60,200	414	75,400	517	a
X65	65,300	448	77,500	531	a
X70	70,300	483	82,700	565	a

[a]The minimum elongation in 2 in is determined using $e = 625{,}000\,\frac{A^{0.2}}{U^{0.9}}$ (US units) or $e = 1944\,\frac{A^{0.2}}{U^{0.9}}$ (SI units).
Adapted from API 5L.

Tensile property requirements for pipe body of seamless and welded pipe for PSL 2

Table 3B
Tensile Property Requirements for Pipe Body of Seamless and Welded Pipe for PSL 2

Grade	Yield strength, minimum		Yield strength, maximum[b]		Tensile strength, minimum		Tensile strength, maximum[c]		Elongation in 2 in
	Psi	**MPa**	**Psi**	**MPa**	**Psi**	**MPa**	**Psi**	**MPa**	**Minimum, %**
B	35,500	241	65,300	448	60,200	414	95,000	758	a
X42	42,100	290	71,800	496	60,200	414	95,000	758	a
X46	46,400	317	76,100	524	63,100	434	95,000	758	a
X52	52,200	359	76,500	531	66,700	455	110,200	758	a
X56	56,600	386	79,000	544	71,100	490	110,200	758	a
X60	60,200	414	81,900	565	75,400	517	110,200	758	a
X65	65,300	448	87,200	600	77,500	531	110,200	758	a
X70	70,300	483	92,100	621	82,700	565	110,200	758	a
X80	80,500	552	102,300	690	90,600	621	119,700	827	a

[a]The minimum elongation in 2 in is determined using $e = 625{,}000\,\frac{A^{0.2}}{U^{0.9}}$ (US units) or $e = 1944\,\frac{A^{0.2}}{U^{0.9}}$ (SI units).
[b]Maximum yield strength for an intermediate grade shall be the maximum for the next higher listed grade.
[c]All intermediate grades have a maximum ultimate tensile strength of 110,000 psi (758 MPa).
Adapted from API 5L.

PIPELINE MATERIAL COATING SPECIFICATIONS

External application procedures for plant applied fusion bonded epoxy coatings and abrasion resistant overlay coatings to steel pipe

NAPCA Bulletin 12-78-04

1. **General**
   a. These specifications may be used in whole or in part by anyone without prejudice, if recognition of the source is included. The National Association of Pipe Coating Applicators (NAPCA) assumes no responsibility for the interpretation or use of these specifications.
   b. The intended use of these coatings is to provide corrosion protection for buried pipelines. Above ground storage of coated pipe in excess of six months without additional ultraviolet protection is not recommended.
   c. The following definitions apply:
      i. Applicator—The contractor who applies the coating to the pipe.
      ii. Company—The purchaser of the coated pipe or the entity for whom the Applicator coats the pipe.
      iii. SSPC—The Steel Structures Painting Council.
      iv. NACE—NACE International.
      v. Manufacturer—The company that makes the coating materials which are applied to the pipe.
      vi. FBE—Fusion bonded epoxy pipe coating
      vii. ARO—Abrasion resistant overlay coating
2. **Scope**
   a. The Applicator shall furnish all labor, equipment, and material required, shall prepare all surfaces to be coated, and shall apply the coating to all surfaces to be coated.
   b. Corrosion protection, as provided under this specification, is furnished by the application of fusion bonded epoxy to the exterior of pipe to be placed underground.
3. **Pipe Conditions**
   a. Pipe delivered to the Applicator for coating shall be free of protective oils, lacquers, mill primer, dirt, or any other deleterious surface contamination which may affect the application of the coating. The pipe surface shall be as free as possible from scabs, slivers, and laminations. Removal of such contaminants shall be as agreed between the Applicator and the Company.
   b. Any paint markings or stenciling of the pipe surface shall be of the type and thickness that can be removed easily during normal surface preparation.
4. **Handling of Bare Pipe**
   a. Proper equipment for unloading, handling, and temporary storage of bare pipe shall be used to avoid any damage to the pipe or pipe ends.
   b. If internally coated pipe is received at the Applicator's plant, care shall be taken to avoid damage to the internal coating or the obliteration of the internal pipe markings during any phases of work covered by this specification. Internal coatings must be capable of withstanding the processing conditions necessary for the application of the external coating.
   c. The Applicator shall visibly inspect the pipe upon receipt for damage such as dents, flat ends, and bevel damage. Any damage observed at this point shall be noted on the inbound tally, and the Company shall be informed within 24 h of receipt of the pipe. Any nonvisible defects such as slivers, scabs, laminations, burrs, dents, etc., will be observed after the pipe is blast cleaned and at the Company's request, removed as an extra work item.
   d. When the roughness of the pipe surface is such that the normal coating rate will be impaired or will result in coated pipe that will be in nonconformance with these specifications, at the Company's option, the Applicator shall take whatever steps necessary to rework the pipe surface and/or add additional coating thickness to the pipe so as to provide coated pipe which can be processed at a normal rate and which will comply with these specifications. The cost of reworking the pipe surface and/or applying additional coating thickness will be agreed upon in advance by the Applicator and Company and paid for by the Company.
5. **Materials and Workmanship**
   All material furnished by the Applicator shall be of the specified quality. All work shall be done in a thorough workmanlike manner. The entire operation of pipe receiving, stockpiling, surface preparation, coating application, storage, and loadout shall be performed under the supervision of and by experienced personnel skilled in the application of protective coating.
6. **Equipment**
   The Applicator's equipment shall be in such condition as to permit the Applicator to follow the procedure and obtain results prescribed in these specifications.
7. **Coating Material**
   a. All coating materials, including repair or patch materials, purchased or used under these

specifications, shall be packaged in suitable and approved containers. The containers shall be plainly marked with the name of the Manufacturer, type of material, and batch or lot number where applicable. Bulk shipments shall be allowed provided the above information is included in the bill of lading.

b. The coating material shall be packaged in containers suitable to keep the contents clean and dry during handling, shipping, and storage. Storage and handling conditions shall be in accordance with the Manufacturer's recommendations.
c. Precautions shall be taken during the handling, shipping, and storage of all materials to prevent damage to the containers that would result in contamination of the coating materials. All contaminated, or otherwise damaged materials shall be discarded.
d. Time/temperature limitations for FBE powders shall be in accordance with the Manufacturer's recommendations.

8. **Surface Preparation**
   a. Before blasting, all oil, grease, mill lacquer, and other deleterious material on the surfaces of the metal to be coated shall be removed by suitable means.
   b. In cold weather or any time when moisture tends to collect on the steel, the pipe shall be uniformly warmed for sufficient time to dry the pipe prior to cleaning. The pipe temperatures shall be maintained at least 5°F above the dew point during the cleaning and coating operations. Pipe temperature shall not exceed 160°F as a result of preheat.
   c. Pipe surfaces shall be blast cleaned to a near-white metal finish in accordance with SSPC-SP-10 or NACE #2 requirements.
   d. NACE, Swedish Pictorial, SSPC, or other mutually agreed upon standards shall be used to judge the degree of cleaning.
   e. A consistent abrasive working mix shall be maintained by frequent additions of small quantities of new abrasive commensurate with consumption. Infrequent large quantity additions of abrasive shall be avoided.
   f. Following cleaning and prior to coating the pipe, abrasive remaining on the outside of the pipe shall be removed by air blast, vacuum, or other suitable methods. If air is used, the air should be dry and free of contaminants, and all particles removed from the surface shall be collected in such a manner as not to contaminate clean pipe.
   g. Following cleaning and prior to coating, the pipe surface shall be inspected for adequate cleaning and surface condition. Pipe not properly cleaned shall be rejected and recleaned.
   h. Blast cleaned pipe surfaces shall be protected from conditions that would allow the pipe to flash rust before coating. If flash rusting occurs, affected pipe shall be recleaned.
   i. Surface imperfections such as slivers, scabs, laminations, burrs, and weld spatter shall be removed by hand filing or light grinding as long as such removal does not:
      i. Adversely affect the quality of pipe cleaning.
      ii. Adversely affect the normal production rate of the plant.
9. **Coating Application (FBE)**
   a. The pipe shall be heated to a temperature within the tolerances recommended by the Manufacturer of the coating material to be applied. The pipe shall not be heated to a temperature in excess of 600°F at any time during the process. Blue oxide formation shall not be used as an indicator that maximum temperature has been exceeded or that damage to steel properties has occurred.
   b. The pipe shall be monitored for proper temperature prior to coating by use of a temperature sensitive crayon.
   c. Coating material shall be applied uniformly utilizing the best commercial practices and in such a manner so that the minimum cured film thickness is that specified by the Company.
   d. The coating shall be applied to the full length of each pipe except for a "cutback" of not less than 0.5 inches (1.27 cm).
   e. Use of recycled coating material is permitted if adequate recovery and screening equipment is used and maintained. An adequate recycle system must properly blend recycled and virgin coating material into the delivery system.
   f. The Applicator shall supply samples of the coating material to the Company, at any time requested, for such test as the Company may wish to run, to assure that the quality of the coating material is being maintained. Company inspection and acceptance of coating materials must be completed prior to application of said materials. Subsequent testing may be conducted to confirm that the coating material has not suffered from improper handling or storage.
   g. During coating and curing periods, the coated pipe shall be handled so as to avoid any damage to the coating.
   h. After the coating is cured, the pipe may be force cooled to facilitate coating inspection and repair,

provided care is taken to avoid any damage from thermal shock to the newly applied coating. The maximum temperature for inspection and repair is 250°F.

10. **Coating Application (ARO)**
    a. The corrosion barrier and the abrasion resistant overlay shall utilize a separate fluidized bed.
    b. Two separate spray booths are preferred. Reclaimed FBE and ARO powder may be used if they are applied in two separate spray booths, but may not be used if they are applied in the same booth. Powder from the topcoat shall not be recycled into the base coat powder except as directed by the ARO powder manufacturer.
    c. The ARO topcoat shall be applied in accordance with the coating manufacturer's recommended procedure and prior to gelling of the FBE corrosion base coat.
11. **Inspection and Testing**
    a. The entire procedure of applying the protective coating material as herein specified will be rigidly inspected from the time the bare pipe is received until the coated pipe is loaded on the carrier for shipment.
    b. If the Company designates an Inspector, the Inspector shall be provided free access to the Applicator's plant at any time during any operation involving the pipe, with the right to inspect and to accept or reject work performed.
    c. The Applicator's Quality Control Inspector shall be responsible for stopping operations when conditions develop which could adversely affect the quality of the completed work.
    d. Although the principal purpose of the coating inspection by the Company and Applicator is to insure compliance of the coating with these specifications, such inspection shall also include examination for previously undetected defects in the pipe, pipe surface, or on the pipe ends. Pipes having such defects shall be set aside for subsequent repair or replacement by the pipe supplier and for any necessary coating repair. Recoating or coating repair that may be necessary by reason of these defects in the pipe which do not involve fault on the part of the Applicator shall be done at the Company's expense.
    e. When Company's Representative exercises Company's right of approval at the Applicator's plant, the Company's Representative shall conduct final Inspection on the Applicator's out-bound rack. Accepted pipe shall be presumed to be produced as specified unless test results indicate a discrepancy.
    f. Coating Thickness Measurements
        i. An appropriate film thickness gauge, calibrated to the National Bureau of Standards' Certified Coating Thickness Calibration Standards shall be used to perform coating thickness measurements on FBE and ARO coatings. A Tooke gauge or grinding/filing down the ARO to the base FBE shall be used to perform measurements of base coat and the top coat or a piece of masking tape can be applied to the FBE before the ARO is applied and the measurements can be taken with an appropriate dry film thickness gauge.
        ii. The agreed upon coating specification shall state the absolute minimum thickness to be applied. All thickness readings measured at or above the absolute minimum shall be accepted for both FBE and ARO coatings.
        iii. The range of coating thickness measurements shall be recorded for each joint of pipe.
        iv. NAPCA recommends that no milage be specified below 10 mils minimum.
    g. Electrical Inspection
        i. Holiday inspection of the entire coated surface shall be performed with an approved high voltage Holiday Detector to indicate any flaws, holes, breaks, or conductive particles in the protective coating.
        ii. The Holiday Detector shall have sufficient DC voltage and be equipped with a positive signaling device. The search electrode shall be made of conductive rubber, or other applicable material. The Holiday Detector shall be operated in such a way as to audibly and/or visually detect the presence of all holidays.
        iii. The voltage to be used shall be per the guidelines of NACE RPO490 or the ARO powder manufacturer.
    h. Adhesion of the cured coating shall be checked by pushing a sharp knife blade through the cured coating to the pipe surface. The coating will not strip or peel when the knife is moved in a "whittling" motion against the steel surface, if proper adhesion of the cured coating to the pipe surface has been attained.
    i. A minimum of one bend test shall be performed for each day of production. The acceptance criteria shall be 1.5° per pipe diameter permanent strain with no tears, cracks, or disbonding of the coating as outlined in the following procedure. Either the mandrel or four point bend methods are acceptable for producing the permanent strain. Bend testing is not applicable for ARO coatings.
        i. A ring sample at least 12 inches long shall be cut from production coated pipe and strips measuring 1 inch by 8 inches shall be prepared. The 8 inch dimension shall be parallel to the longitudinal axis of the pipe.

ii. Condition the straps to 32°F in a freezer or with dry ice.

iii. If the mandrel method is to be used, the required mandrel radius shall be calculated by the following equation.

$$R = 57.3\ t/s - t/2$$

where t = effective strap thickness
R = bend radius of outer curve of strap

When a mandrel of the calculated radius is not available, the mandrel of the next smaller size shall be used.

iv. For the four point method, estimate the required bend using the equation explained in the Mandrel method.

v. The straps shall be bent so that the uncoated side is in contact with the mandrel/support pins. The bend shall take at least 10 s and not more than 30 s to complete.

vi. Remove the strap from the bend apparatus and measure the permanent radius by matching the strap against a chart of known radius curves.

vii. Calculate the permanent strain in degrees per pipe diameter by the following equation:

$$deg/pd = 57.3t/R - t/2$$

where t = effective strap thickness
R = bend radius of outer curve of strap

viii. Allow the straps to warm to room temperature and visually inspect them for cracks, tears in the coating, and disbonding of the coating. Disregard any defects within 0.1 inches of the strap edge or within 0.5 inches of the support pins. The presence of strain marks alone does not constitute a failure.

j. A minimum of one 24-hour cathodic disbondment test shall be performed for each day of production. The acceptance criteria shall be a maximum of 12 mm radius from the edge of the intentional holiday as outlined in the following procedure. Disbondment greater than 8 mm radius generally indicates that problems exist in either the process or powder and shall be investigated.

i. The sample shall be a four inch square segment cut from the test ring.

ii. Drill a 1/8 inch diameter holiday in the coating at the center of the sample.

iii. Glue a plastic cylinder onto the specimen with the holiday at the center of the cylinder.

iv. Pour approximately 350 ml of electrolyte into the cylinder. The electrolyte shall be composed of 3% by weight sodium chloride (NaCl) in distilled water.

v. Place the test cell on a hot plate controlled heat transfer medium (steel shot or grid in a metal pan), insert a thermometer so that it is immersed and resting on the sample, and adjust the temperature to 150 ± 5°F.

vi. Connect the negative lead from a variable voltage DC power supply to the specimen. Attach the positive lead to a platinum or platinum-coated wire immersed at the center of the cell.

vii. Adjust the voltage to 3.5 V DC using a calomel reference electrode.

viii. After 24 h, remove the test cell and immediately drain the electrolyte. Rinse with tap water, dismantle the cell, and air cool to room temperature.

ix. Within one hour of removal from the hot plate, make radial cuts from the edge of the holiday outward using a utility knife. The cuts shall be at least 0.8 inches in length and through the coating to the substrate.

x. Immediately insert the blade of the utility knife under the coating and, using a prying action, chip off the coating. Continue until the coating demonstrates a definite resistance to the prying action.

xi. Measure the radius of the disbonded area from the holiday edge along each radial cut and average the measured results.

12. **Repair Procedures**

a. All defects disclosed by the Holiday Detector and other obvious defects shall be repaired by the Applicator.

b. Holidays which are the result of slivers, scabs, laminations, or other steel conditions beyond the control of the Applicator shall be repaired at the Company's expense.

c. Areas of repair to the coating shall be holiday inspected by the Applicator on a 100% basis.

d. Pinhole type holidays may be patched using the hot melt patch stick method. Abrade the adjacent coating surface with either a hand file or coarse sand paper. The abrasion should be on the surface only and should not remove a significant thickness of the coating. The surface to be touched up will be heated with a small torch until the stick starts to melt, then the stick should be rubbed over the heated surface, building up a small puddle of patching compound over the entire area being patched. Such holidays may also be patched by use of a two part, 100% solids, liquid epoxy compound specified by the Manufacturer using the method set forth in the following paragraph. Liquid epoxy repairs of pinhole type holidays shall be considered an extra work item to be performed at a price agreed upon between the Company and Applicator.

e. Where larger areas of damaged coating are to be repaired and the use of patch sticks is not practical, a two part, 100% solids, liquid epoxy compound specified by the Manufacturer shall be used. The damaged area shall be abraded by hand filing or use of carborundum cloth. Application shall be made to a minimum thickness of 25 mils (0.64 mm) and shall overlap the undamaged area a minimum of 0.5 inches (1.27 cm).

f. The liquid patch compounds shall not be applied when the pipe temperature is below 50 F unless provisions are made for heat curing the patch material using methods and temperatures in accordance with procedures recommended by the coating Manufacturer.

g. All repairs performed on ARO coatings shall be made with an accepted 100% solids liquid epoxy compound.

13. **Coated Pipe Handling, Storage, and Loading Requirements**

a. Pipe shall be stored, handled, and transported in a manner to prevent damage to the pipe walls, beveled ends, and the coating.

b. Storage racks shall be so designed as to protect the coated pipe from standing water, direct soil contact, and sharp or hard objects that might damage the coating.

c. All individual coated pipes, that are handled, stored, or shipped, shall be protected by padding, separators, or dividers. These separators shall be affixed to the exterior surface of the pipe. Suggested separators consist of (1) polypropylene rope rings for larger and smaller diameters of pipe joints, with rope thickness varying from 3/8 to 5/8 inches in diameter depending on pipe wall thickness; (2) spiral cardboard rings with waxed outside surface for 4 inches nominal and smaller diameter pipe joints, with ring being 2 inches wide and 0.25 inch thickness; (3) 0.25 inch minimum thickness rubber; or (4) other material approved as separators and acceptable to the Company.

i. Each 40 foot joint of coated pipe shall have affixed to its exterior surface a minimum of three (3) separators, randomly spaced, but spaced relatively close to the ends of each joint and near the center of the pipe. The minimum number of separators for nominal 60 foot pipe shall be four (4); and for nominal 80 foot pipe, the minimum number of separators shall be five (5).

ii. The type and thickness of separators shall be chosen to prevent joints of pipe from coming in contact with each other.

d. The coated pipe shall be shipped using sufficient and proper dunnage to adequately protect the pipe and coating.

e. All pipe shipped by rail shall be loaded in accordance with API Specifications RP 5L1, Latest Edition.

14. **Supplementary Details Supplied by the Company**

When possible, the Company shall supply the following supplemental information:

a. Length and diameter of pipe.
b. Grade, wall thickness, and/or weight per foot of pipe.
c. Source and approximate shipping date from the pipe mill.
d. Method of shipment from the mill.
e. Approximate shipping date to the destination.
f. If pipe is to be stored, the approximate length of time it is to be stored.
g. Length, style, and postpreparation of cutback.
h. Minimum weight per car or truck required to protect lowest outbound rate.
i. Name and type of carrier.
j. Stacking and/or loading instructions.
k. Pipe manifest, preferably electronic.

External application procedures for coal tar epoxy protective coatings to steel pipe

NAPCA Bulletin 13-79-94

1. **General**

a. These specifications may be used in whole or in part by anyone without prejudice, if recognition of the source is included. The National Association of Pipe Coating Applicators (NAPCA) assumes no responsibility for the interpretation or use of these specifications.

b. The intended use of these coatings is to provide corrosion protection for buried pipelines. Above ground storage of coated pipe in excess of six months without additional ultraviolet protection is not recommended.

c. The following definitions apply:

i. Applicator—The contractor who applies the coating to the pipe.

   ii. Company—The purchaser of the coated pipe or the entity for whom the Applicator coats the pipe.
   iii. SSPC—The Steel Structures Painting Council.
   iv. NACE—NACE International.
   v. Manufacturer—The company that makes the coating materials which are applied to the pipe.

2. **Scope**
   a. The Applicator shall furnish all labor, equipment, and material required, shall prepare all surfaces to be coated, and shall apply the coating to all surfaces to be coated.
   b. Corrosion protection, as provided under this specification, is furnished by the application of coal tar epoxy to the exterior of pipe to be placed underground.
3. **Pipe Conditions**
   a. Pipe delivered to the Applicator for coating shall be free of protective oils, lacquers, mill primer, dirt, or any other deleterious surface contamination which may affect the application of the coating. The pipe surface shall be as free as possible from scabs, slivers, and laminations.
   b. Any paint markings or stenciling of the pipe surface shall be of the type and thickness that can be removed easily during normal surface preparation.
4. **Handling of Bare Pipe**
   a. Proper equipment for unloading, handling, and temporary storage of bare pipe shall be used to avoid any damage to the pipe or pipe ends.
   b. If internally coated pipe is received at the Applicator's plant, care shall be taken to avoid damage to the internal coating or the obliteration of the internal pipe markings during any phases of work covered by this specification. Internal coatings must be capable of withstanding the processing conditions necessary for the application of the external coating.
   c. The Applicator shall visibly inspect the pipe upon receipt for damage such as dents, flat ends, and bevel damage. Any damage observed at this point shall be noted on the inbound tally, and the Company shall be informed within 24 h of receipt of the pipe. Any nonvisible defects such as slivers, scabs, laminations, burrs, dents, etc., will be observed after the pipe is blast cleaned and at the Company's request, removed as an extra work item.
5. **Material and Workmanship**
   All material furnished by the Applicator shall be of the specified quality. All work shall be done in a thorough workmanlike manner. The entire operation of pipe receiving, stockpiling, surface preparation, coating application, storage, and loadout shall be performed under the supervision of and by experienced personnel skilled in the application of protective coating.
6. **Equipment**
   The Applicator's equipment shall be in such condition as to permit the Applicator to follow the procedure and obtain results prescribed in these specifications.
7. **Coating Material**
   a. All coating materials, including repair or patch materials, purchased or used under these specifications, shall be packaged in suitable and approved containers. The containers shall be plainly marked with the name of the Manufacturer, type of material, and batch or lot number where applicable. Bulk shipments shall be allowed provided the above information is included in the bill of lading.
   b. The coating material shall be packaged in containers suitable to keep the contents clean and dry during handling, shipping, and storage. Storage and handling conditions shall be in accordance with the Manufacturer's recommendations.
   c. Precautions shall be taken during the handling, shipping, and storage of all materials to prevent damage to the containers that would result in contamination of the coating materials. All contaminated, or otherwise damaged materials shall be discarded.
8. **Surface Preparation**
   a. Before blasting, all oil, grease, mill lacquer, and other deleterious material on the surfaces of the metal to be coated shall be removed by suitable means.
   b. In cold weather or any time when moisture tends to collect on the steel, the pipe shall be uniformly warmed for sufficient time to dry the pipe prior to cleaning. The pipe temperatures shall be maintained at least 5°F above the dew point during the cleaning and coating operations. Pipe temperature shall not exceed 160°F as a result of preheat.
   c. Pipe surfaces shall be blast cleaned to a commercial blast metal finish in accordance with SSPC-SP-6 or NACE #3 requirements. Certain coating systems require a greater degree of cleanliness; in such cases, the degree of cleaning shall be as required by the coating system.
   d. NACE, Swedish Pictorial, SSPC, or other mutually agreed upon standards shall be used to judge the degree of cleaning.
   e. A consistent abrasive working mix shall be maintained by frequent additions of small quantities of new abrasive commensurate with consumption. Infrequent large quantity additions of abrasive shall be avoided.

f. Following cleaning and prior to coating the pipe, abrasive remaining on the outside and loose contamination on the inside of the pipe shall be removed by air blast, vacuum, or other suitable methods. If air is used, the air should be dry and free of contaminants, and all particles removed from the surface shall be collected in such a manner as not to contaminate clean pipe.
g. Following cleaning and prior to coating, the pipe surface shall be inspected for adequate cleaning and surface condition. Pipe not properly cleaned shall be rejected and recleaned.
h. Blast cleaned pipe surfaces shall be protected from conditions that would allow the pipe to flash rust before coating. If flash rusting occurs, affected pipe shall be recleaned.

9. **Coating Application**
   a. After surface preparation and prior to coating, each pipe end shall be masked to allow for a cutback in accordance with the Company's specifications.
   b. If system selected requires a primer, it shall be applied as soon after cleaning as possible to avoid surface cleanliness deterioration. All cleaned surfaces shall be primed the same work day.
   c. Mixing shall be done with a powered mixing device to insure proper blending of the components.
   d. Mixed material that has exceeded the Manufacturer's published pot life shall be discarded.
   e. Application of the coating shall be by standard industrial spray equipment, airless or conventional, as per Manufacturer's recommendations.
   f. Coal tar epoxies shall be applied in one or more coats to obtain the Manufacturer's recommended thickness. Manufacturer's recommendations for coating application shall be strictly followed.
   g. Particular attention must be paid to the allowable time interval between coats so as to prevent coating delamination. If an Applicator has exceeded the allowable time interval between coats, then the first coat shall be "sweep blasted" or chemically treated prior to recoating so that a mechanical bond will develop between coats.
   h. Cure conditions may vary widely depending on application parameters. Actual cure conditions shall be established after consultation with the Manufacturer.

10. **Inspection and Testing**
   a. The entire procedure of applying the protective coating material as herein specified will be rigidly inspected from the time the bare pipe is received until the coated pipe is loaded on the carrier for shipment.
   b. If the Company designates an Inspector, the Inspector shall be provided free access to the Applicator's plant at any time during any operation involving the pipe, with the right to inspect and to accept or reject work performed.
   c. The Applicator's Quality Control Inspector shall be responsible for stopping operations when conditions develop which could adversely affect the quality of the completed work.
   d. Although the principal purpose of the coating inspection by the Company and Applicator is to insure compliance of the coating with these specifications, such inspection shall also include examination for previously undetected defects in the pipe, pipe surface, or on the pipe ends. Pipes having such defects shall be set aside for subsequent repair or replacement by the pipe supplier and for any necessary coating repair. Recoating or coating repair that may be necessary by reason of these defects in the pipe which do not involve fault on the part of the Applicator shall be done at the Company's expense.
   e. When Company's Representative exercises Company's right of approval at the Applicator's plant, the Company's Representative shall conduct final inspection on the Applicator's out-bound rack. Accepted pipe shall be presumed to be produced as specified unless test results indicate a discrepancy.
   f. Coating Thickness Measurements
      i. An appropriate film thickness gauge, calibrated to the National Bureau of Standards' Certified Coating Thickness Calibration Standards shall be used to perform coating thickness measurement.
      ii. The coating thickness shall meet or exceed the agreed upon minimum coating thickness. All joints which fail to meet the minimum coating thickness test shall be recoated or repaired.
   g. Electrical Inspection (when applicable)
      i. Holiday inspection of the entire coated surface shall be performed with an approved high voltage Holiday Detector to indicate any flaws, holes, breaks, or conductive particles in the protective coating.
      ii. The Holiday Detector shall have sufficient DC voltage and be equipped with a positive signaling device. The search electrode shall be made of conductive rubber or other applicable material. The Holiday Detector shall be operated in such a way as to audibly and/or visually detect the presence of all holidays.
      iii. The voltage to be used shall not exceed 125 V per mil of nominal coating thickness. If a low voltage, wet sponge, type detector is used, the test voltage shall be 67.5 V.

11. **Repair Procedures**
    a. All defects disclosed by the Holiday Detector and other obvious defects shall be repaired by the Applicator.
    b. Holidays which are the result of slivers, scabs, laminations, or other steel conditions beyond the control of the Applicator shall be repaired at the Company's expense.
    c. Areas of repair to the coating shall be holiday inspected by the Applicator on a 100% basis. Holiday inspection of repairs is only required when also required for the coating.
    d. The damaged coating shall be removed from the steel surface. The edge of the cured coating adjacent to the damaged area shall be "feathered" and coated with an appropriate surface conditioning solvent recommended by the coating Manufacturer. This solvent shall be applied in accordance with the Manufacturer's recommendations.
    e. The damaged area shall then be recoated.
12. **Coated Pipe Handling, Storage, and Loading Requirements**
    a. Pipe shall be stored, handled, and transported in a manner to prevent damage to the pipe walls, beveled ends, and the coating.
    b. Storage racks shall be so designed as to protect the coated pipe from standing water, direct soil contact, and sharp or hard objects that might damage the coating.
    c. The coated pipe shall be shipped using sufficient and proper dunnage to adequately protect the pipe and coating.
    d. All pipe shipped by rail shall be loaded in accordance with API Specifications RP 5L1, Latest Edition.
13. **Supplementary Details Supplied by the Company**
    When possible, the Company shall supply the following supplemental information:
    a. Length and diameter of pipe.
    b. Grade, wall thickness, and/or weight per foot of pipe.
    c. Source and approximate shipping date from the pipe mill.
    d. Method of shipment from the mill.
    e. Approximate shipping date to the destination.
    f. If pipe is to be stored, the approximate length of time it is to be stored.
    g. Length, style, and postpreparation of cutback.
    h. Minimum weight per car or truck required to protect lowest outbound rate.
    i. Name and type of carrier.
    j. Stacking and/or loading instructions.

External application procedures for polyolefin pipe coating applied by the cross head extrusion method or the side extrusion method to steel pipe

NAPCA Bulletin 14-83-94

1. **General**
    a. These specifications may be used in whole or in part by anyone without prejudice, if recognition of the source is included. The National Association of Pipe Coating Applicators (NAPCA) assumes no responsibility for the interpretation or use of these specifications.
    b. The intended use of these coatings is to provide corrosion protection for buried pipelines. Above ground storage of coated pipe in excess of six months without additional ultraviolet protection is not recommended.
    c. The following definitions apply:
        i. Applicator—The contractor who applies the coating to the pipe.
        ii. Company—The purchaser of the coated pipe or the entity for whom the Applicator coats the pipe.
        iii. SSPC—The Steel Structures Painting Council.
        iv. NACE—NACE International.
        v. Manufacturer—The company that makes the coating materials which are applied to the pipe.
        vi. AWWA—American Water Works Association.
2. **Scope**
    a. The Applicator shall furnish all labor, equipment, and material required, shall prepare all surfaces to be coated, and shall apply the coating to all surfaces to be coated.
    b. Corrosion protection, as provided under this specification, is furnished by the application of extruded polyolefins to the exterior of pipe to be placed underground.
    c. As a variation of the Side Extrusion method described in this specification, the procedure for Tape and Side Extrusion Polyolefin Sheath coatings described in AWWA C215 is available.
3. **Pipe Conditions**
    a. Pipe delivered to the Applicator for coating shall be free of protective oils, lacquers, mill primer, dirt, or any other deleterious surface contamination which

may affect the application of the coating. The pipe surface shall be as free as possible from scabs, slivers, and laminations.

b. Any paint markings or stenciling of the pipe surface shall be of the type and thickness that can be removed easily during normal surface preparation.

4. **Handling of Bare Pipe**

a. Proper equipment for unloading, handling, and temporary storage of bare pipe shall be used to avoid any damage to the pipe or pipe ends.

b. If internally coated pipe is received at the Applicator's plant, care shall be taken to avoid damage to the internal coating or the obliteration of the internal pipe markings during any phases of work covered by this specification. Internal coatings must be capable of withstanding the processing conditions necessary for the application of the external coating.

c The Applicator shall visibly inspect the pipe upon receipt for damage such as dents, flat ends, and bevel damage. Any damage observed at this point shall be noted on the inbound tally, and the Company shall be informed within 24 h of receipt of the pipe. Any nonvisible defects such as slivers, scabs, laminations, burrs, dents, etc., will be observed after the pipe is blast cleaned and at the Company's request, removed as an extra work item.

5. **Material and Workmanship**

All material furnished by the Applicator shall be of the specified quality. All work shall be done in a thorough workmanlike manner. The entire operation of pipe receiving, stockpiling, surface preparation, coating application, storage, and loadout shall be performed under the supervision of and by experienced personnel skilled in the application of protective coating.

6. **Equipment**

The Applicator's equipment shall be in such condition as to permit the Applicator to follow the procedure and obtain results prescribed in these specifications.

7. **Coating Material**

a. All coating materials, including repair or patch materials, purchased or used under these specifications, shall be packaged in suitable and approved containers. The containers shall be plainly marked with the name of the Manufacturer, type of material, and batch or lot number where applicable. Bulk shipments shall be allowed provided the above information is included in the bill of lading.

b. The coating material shall be packaged in containers suitable to keep the contents clean and dry during handling, shipping, and storage. Storage and handling conditions shall be in accordance with the Manufacturer's recommendations.

c. Precautions shall be taken during the handling, shipping, and storage of all materials to prevent damage to the containers that would result in contamination of the coating materials. All contaminated, or otherwise damaged materials shall be discarded.

8. **Surface Preparation**

a. Before blasting, all oil, grease, mill lacquer, and other deleterious material on the surfaces of the metal to be coated shall be removed by suitable means.

b. In cold weather or any time when moisture tends to collect on the steel, the pipe shall be uniformly warmed for sufficient time to dry the pipe prior to cleaning. The pipe temperatures shall be maintained at least 5°F above the dew point during the cleaning and coating operations. Pipe temperature shall not exceed 160°F as a result of preheat.

c. Pipe surfaces shall be blast cleaned to a commercial metal finish in accordance with SSPC-SP-6 or NACE #3 requirements.

d. NACE, Swedish Pictorial, SSPC, or other mutually agreed upon standards shall be used to judge the degree of cleaning.

e. A consistent abrasive working mix shall be maintained by frequent additions of small quantities of new abrasive commensurate with consumption. Infrequent large quantity additions of abrasive shall be avoided.

f. Following cleaning and prior to coating the pipe, abrasive remaining on the outside and loose contamination on the inside of the pipe shall be removed by air blast, vacuum, or other suitable methods. If air is used, the air should be dry and free of contaminants, and all particles removed from the surface shall be collected in such a manner as not to contaminate clean pipe.

g. Following cleaning and prior to coating, the pipe surface shall be inspected for adequate cleaning and surface condition. Pipe not properly cleaned shall be rejected and recleaned.

h. Blast cleaned pipe surfaces shall be protected from conditions that would allow the pipe to flash rust before coating. If flash rusting occurs, affected pipe shall be recleaned.

9. **Coating Application**

a. Cross Head Extrusion Method

i. The adhesive undercoating shall be applied to the clean pipe surface. It shall consist of a blend of rubber, asphalt, and high molecular weight resins and shall conform to the physical properties specified in Table I, under the heading "Cross Head Extrusion Data." The adhesive shall be applied to the pipe by flood-coating, at a temperature of 265 to 325°F, to a minimum thickness of 8 mils.

ii. Immediately following the application of the adhesive, the polyolefin sheath shall be extruded over the undercoating. The seamless sheath must be smooth, free of pinholes, bubbles, blisters, wrinkles, cracks, or mechanical voids, and must conform to the physical properties specified in Tables II–IV, under the heading "Cross Head Extrusion Data."

   iii. Immediately following the extrusion, the polyolefin sheath shall be water-quenched so that it will shrink tightly around the pipe and adhesive undercoating.
   iv. Unless otherwise specified by the Company, the polyolefin sheath shall extend to approximately 4 to 6 inches from each end of plain end pipe and approximately 2 to 3 inches from the last thread on threaded pipe.
   v. Unless otherwise specified by the Company, the polyolefin sheath shall be applied to the thickness specified in Table V, under the heading "Cross Head Extrusion Data."

b. Side Extrusion Method
   i. The adhesive material shall be a butyl rubber compound and shall be made from isobutylene and 1.5% to 2.0% by weight of isoprene. The butyl rubber compound shall contain no asphaltic, coal tar, or other property-degrading extenders. The butyl rubber compound shall conform to the physical properties specified in Table I, under the heading "Side Extrusion Data."
   ii. The adhesive shall be applied by side extrusion at approximately 300°F in single or multiple layers to specified thickness.
   iii. Immediately following the extrusion of the adhesive, a high molecular weight polyolefin shall be applied by side extrusion at approximately 500°F in single or multiple layers to specified thickness to produce a tightly bonded, seamless coating between 30 and 250 mils thick, as specified by the Company (minimum coating thickness allows 90% of specified thickness). The polyolefin material shall conform to the physical properties specified in Tables II under the heading "Side Extrusion Data."

10. **Inspection and Testing**
   a. The entire procedure of applying the protective coating material as herein specified will be rigidly inspected from the time the bare pipe is received until the coated pipe is loaded on the carrier for shipment.
   b. If the Company designates an Inspector, the Inspector shall be provided free access to the Applicator's plant at any time during any operation involving the pipe, with the right to inspect and to accept or reject the work performed.
   c. The Applicator's Quality Control Inspector shall be responsible for stopping operations when conditions develop which could adversely affect the quality of the completed work.
   d. Although the principal purpose of the coating inspection by the Company and Applicator is to insure compliance of the coating with these specifications, such inspection shall also include examination for previously undetected defects in the pipe, pipe surface, or on the pipe ends. Pipes having such defects shall be set aside for subsequent repair or replacement by the pipe supplier and for any necessary coating repair. Recoating or coating repair that may be necessary by reason of these defects in the pipe which do not involve fault on the part of the Applicator shall be done at the Company's expense.
   e. When Company's Representative exercises Company's right of approval at the Applicator's plant, the Company's Representative shall conduct final inspection on the Applicator's out-bound rack. Accepted pipe shall be presumed to be produced as specified unless test results indicate a discrepancy.
   f. Coating Thickness Measurements
      i. An appropriate film thickness gauge, calibrated to the National Bureau of Standards' Certified Coating Thickness Calibration Standards shall be used to perform coating thickness measurement.
      ii. The coating thickness shall meet or exceed the agreed upon minimum coating thickness. All joints which fail to meet the minimum coating thickness test shall be recoated or repaired.
   g. Electrical Inspection
      i. Holiday inspection of the entire coated surface shall be performed with an approved high voltage Holiday Detector to indicate any flaws, holes, breaks, or conductive particles in the protective coating.
      ii. The Holiday Detector shall have sufficient AC voltage and be equipped with a positive signaling device. The search electrode shall be made of conductive rubber or other applicable material. The Holiday Detector shall be operated in such a way as to audibly and/or visually detect the presence of all holidays.
      iii. The minimum testing voltage for a particular coating thickness shall be within 20% of the value determined from the following formula:

      1250 × (square root of coating thickness (mils))

11. **Repair Procedures**
   a. All defects disclosed by the Holiday Detector and other obvious defects shall be repaired by the Applicator.
   b. Holidays which are the result of slivers, scabs, laminations, or other steel conditions beyond the control of the Applicator shall be repaired at the Company's expense.
   c. Areas of repair to the coating shall be holiday inspected by the Applicator on a 100% basis.
   d. Repairs shall be made by using heat shrinkable sleeves, pressure sensitive polyolefin/butyl rubber tapes, and other materials approved by the Company for this purpose.

12. **Coated Pipe Handling, Storage, and Loading Requirements**
    a. Pipe shall be stored, handled, and transported in a manner to prevent damage to the pipe walls, beveled ends, and the coating.
    b. Storage racks shall be so designed as to protect the coated pipe from standing water, direct soil contact, and sharp or hard objects that might damage the coating.
    c. The coated pipe shall be shipped using sufficient and proper dunnage to adequately protect the pipe and coating.
    d. All pipe shipped by rail shall be loaded in accordance with API Specifications RP 5L1, Latest Edition.
13. **Supplementary Details Supplied by the Company**
    When possible, the Company shall supply the following supplemental information:
    a. Length and diameter of pipe.
    b. Grade, wall thickness, and/or weight per foot of pipe.
    c. Source and approximate shipping date from the pipe mill.
    d. Method of shipment from the mill.
    e. Approximate shipping date to the destination.
    f. If pipe is to be stored, the approximate length of time it is to be stored.
    g. Length, style, and postpreparation of cutback.
    h. Minimum weight per car or truck required to protect lowest outbound rate.
    i. Name and type of carrier.
    j. Stacking and/or loading instructions.

Cross head extrusion data

Table I
Physical Properties of Adhesive Undercoating

Properties	Test method	Requirement
Softening point, F	ASTM E 28	140 to 190
Penetration, mm, 77F	ASTM D 5	3 to 12
Stormer viscosity, sec for 100 revs	ASTM D 562	
at 270F		120 to 180
at 300F		60 to 80
Specific gravity, 77F	ASTM D 71	1.0 to 1.2

Table II
Properties of Polyethylene

Properties	Test method	Requirement
Density, g/cm^3	ASTM D 1505	Min. 0.941
Flow rate, g/10 min	ASTM D 1238, condition E	Max. 0.75
Pigment content, % by wt		Min 2.0
Environmental stress crack resistance, F(50) hrs	Condition B, 100% Igepal	Min. 100

Table III
Properties of Polypropylene Resin

Properties	Test method	Requirement
Density, g/cm^3	ASTM D 1505	Min. 0.897
Flow rate, g/10 min	ASTM D 1238, condition E	Max. 1.0
Pigment content, % by wt		Min. 0.95
Environment stress crack resistance, F(10) hrs	ASTM D 1693	Not applicable

Table IV
Thickness of Plastic Sheath of Pipe Coating * Sheath Thickness

Pipe size	Standard (mils)		60 mil sheath	
	Nom.	*Min.*	*Nom.*	*Min.*
1/2″ through 1-1/2″ Nom.	25	22.5	60	54
2-3/8″ through 2-7/8″ OD	30	27.0	60	54
3-1/2″ through 4-1/2″ OD	35	31.5	60	54
5-9/16″ through 24″ OD	40	36.0	60	54

Side extrusion data

Table I
Physical Properties of Butyl Rubber

Property	Test method	Requirement
Density, g/cm^3	ASTM D 1505	1.0 to 1.1
Flow rate, g/10 min	ASTM D 1283	1.8 to 8.0

Table II
Physical Properties of White or Black Polyethylene

Property	Test method	Requirement
Tensile elongation	ASTM D 638	Min. 500%
Tensile strength	ASTM D 638	Min. 2700 lpsi
Dielectric strength	ASTM D 149	700–800 volts/mil
Water penetration	ASTM G 9	0.0046 in dissipation factor
Environmental stress crack resistance, F(50) hrs	ASTM D 1693; condition B 100% Igepal	Min. 1000 h
Density (pigmented), g/cm^3	ASTM D 1505	Min. 0.941
Flow rate, g/10 min	ASTM S 1238; conditon E	0.2 to 1.0

Table III
Recommended Nominal Coating Thickness (mils)

Pipe size	Adhesive	Polyolefin	Total
Up to 16″	10	40	50
Up to 42″	10	50	60
Up to 120″	10	60	70

NOTE: The above are the nominal thicknesses to be specified. Tolerance would be less 10%. Additional thickness of adhesive and/or polyolefin may be specified for specific requirements.

Exterior application procedures for plant applied tape coatings to steel pipe

NAPCA Bulletin 15-83-94

1. **General**
   a. These specifications may be used in whole or in part by anyone without prejudice, if recognition of the source is included. The National Association of Pipe Coating Applicators (NAPCA) assumes no responsibility for the interpretation or use of these specifications.
   b. The intended use of these coatings is to provide corrosion protection for buried pipelines. Above ground storage of coated pipe in excess of six months without additional ultraviolet protection is not recommended.
   c. The following definitions apply:
      i. Applicator—The contractor who applies the coating to the pipe.
      ii. Company—The purchaser of the coated pipe or the entity for whom the Applicator coats the pipe.
      iii. SSPC—The Steel Structures Painting Council.
      iv. NACE—NACE International.
      v. Manufacturer—The company that makes the coating materials which are applied to the pipe.
2. **Scope**
   a. The Applicator shall furnish all labor, equipment, and material required, shall prepare all surfaces to be coated, and shall apply the coating to all surfaces to be coated.
   b. This specification covers the material and application for in-plant application of cold applied tapes to the exterior of all diameters of steel pipe through mechanical methods.
   c. The total thickness of the system is variable and is achieved through multilayer build-up of the innerwrap and outerwrap. The ultimate thickness of the coating is dictated by and dependant upon pipe diameter, pipe weight, operating conditions, and conditions along the right of way. In some cases, more than one application of the outerwrap may be desired.
3. **Pipe Conditions**
   a. Pipe delivered to the Applicator for coating shall be free of protective oils, lacquers, mill primer, dirt, or any other deleterious surface contamination which may affect the application of the coating. The pipe surface shall be as free as possible from scabs, slivers, and laminations.
   b. Any paint markings or stenciling of the pipe surface shall be of the type and thickness that can be removed easily during normal surface preparation.
4. **Handling of Bare Pipe**
   a. Proper equipment for unloading, handling, and temporary storage of bare pipe shall be used to avoid any damage to the pipe or pipe ends.
   b. If internally coated pipe is received at the Applicator's plant, care shall be taken to avoid damage to the internal coating or the obliteration of the internal pipe markings during any phases of work covered by this specification. Internal coatings must be capable of withstanding the processing conditions necessary for the application of the external coating.
   c. The Applicator shall visibly inspect the pipe upon receipt for damage such as dents, flat ends, and bevel damage. Any damage observed at this point shall be noted on the inbound tally, and the Company shall be informed within 24 h of receipt of the pipe. Any nonvisible defects such as slivers, scabs, laminations, burrs, dents, etc., will be observed after the pipe is blast cleaned and at the Company's request, removed as an extra work item.
5. **Material and Workmanship**
   a. All material furnished by the Applicator shall be of the specified quality. All work shall be done in a thorough workmanlike manner. The entire operation of pipe receiving, stockpiling, surface preparation, coating application, storage, and loadout shall be performed under the supervision of and by experienced personnel skilled in the application of protective coating.
6. **Equipment**
   The Applicator's equipment shall be in such condition as to permit the Applicator to follow the procedure and obtain results prescribed in these specifications.
7. **Coating Material**
   a. All coating materials, including repair or patch materials, purchased or used under these specifications, shall be packaged in suitable and approved containers. The containers shall be plainly marked with the name of the Manufacturer, type of material, and batch or lot number where applicable. Bulk shipments shall be allowed provided the above information is included in the bill of lading.
   b. The coating material shall be packaged in containers suitable to keep the contents clean and dry during handling, shipping, and storage. Storage and handling conditions shall be in accordance with the Manufacturer's recommendations.
   c. Precautions shall be taken during the handling, shipping, and storage of all materials to prevent damage to the containers that would result in

contamination of the coating materials. All contaminated, or otherwise damaged materials shall be discarded.

d. The tape system shall be composed of the following components.
   i. Primer supplied in liquid form consisting of solid ingredients carried in a solvent. The primer shall be compatible with the inner- and outerwraps. The function of the primer is to provide a highly effective bonding medium between the pipe surface and the corrosion protection innerwrap.
   ii. Corrosion protection innerwrap tape consisting of a polyethylene backing and a butyl rubber adhesive supplied in suitable thicknesses in roll form. The innerwrap is applied after the primer and prior to the outerwrap. The main function of the innerwrap is to serve as the corrosion protective coating.
   iii. Mechanical protection outerwrap tape consisting of a polyethylene backing and an adhesive mass composed of the same types of materials used in the innerwrap adhesive and is supplied in suitable thicknesses in roll form. The function of the outerwrap is to provide mechanical protection to the innerwrap and to protect the system from environmental hazards.

8. **Surface Preparation**
   a. Before blasting, all oil, grease, mill lacquer, and other deleterious material on the surfaces of the metal to be coated shall be removed by suitable means.
   b. In cold weather or any time when moisture tends to collect on the steel, the pipe shall be uniformly warmed for sufficient time to dry the pipe prior to cleaning. The pipe temperatures shall be maintained at least 5°F above the dew point during the cleaning and coating operations. Pipe temperature shall not exceed 160°F as a result of preheat.
   c. Pipe surfaces shall be blast cleaned to a commercial blast metal finish in accordance with SSPC-SP-6 or NACE #3 requirements.
   d. NACE, Swedish Pictorial, SSPC, or other mutually agreed upon standards shall be used to judge the degree of cleaning.
   e. A consistent abrasive working mix shall be maintained by frequent additions of small quantities of new abrasive commensurate with consumption. Infrequent large quantity additions of abrasive shall be avoided.
   f. Following cleaning and prior to coating the pipe, abrasive remaining on the outside and loose contamination on the inside of the pipe shall be removed by air blast, vacuum, or other suitable methods. If air is used, the air should be dry and free of contaminants, and all particles removed from the surface shall be collected in such a manner as not to contaminate clean pipe.
   g. Following cleaning and prior to coating, the pipe surface shall be inspected for adequate cleaning and surface condition. Pipe not properly cleaned shall be rejected and recleaned.
   h. Blast cleaned pipe surfaces shall be protected from conditions that would allow the pipe to flash rust before coating. If flash rusting occurs, affected pipe shall be recleaned.

9. **Coating Application**
   a. If specified, prior to application of the coating materials, each longitudinal weld seam may be covered with a 4 inch (minimum) wide strip of innerwrap tape, or suitable alternative.
   b. The coating application should be a continuous three step operation starting with a properly prepared pipe surface. The three steps following one after the other consist of: (1) priming; (2) application of innerwrap tape directly onto the primed pipe surface; and (3) application of outerwrap directly over the innerwrap.
   c. The outerwrap overlap should not coincide with the innerwrap overlap.
   d. The above materials should be applied within the temperature range recommended by the Manufacturer.
   e. Application of Primer
      i. The primer shall be applied in a uniform thickness.
      ii. The primer shall be thoroughly mixed and agitated as needed during application to prevent settling.
      iii. The primer must coat the entire exterior pipe surface.
      iv. The primer coat on the pipe shall be uniform and free of floods, runs, sags, and drips. The primed surface shall be free of any foreign substances.
   f. Application of Tape Innerwrap
      i. The innerwrap shall be applied directly onto the primed pipe surface with appropriate dispensing and tensioning equipment.
      ii. The innerwrap shall be spirally applied with overlap widths and application tensions required and/or recommended by the Manufacturer. The minimum overlap shall not be less than 1/2 inch. The minimum overlap at the end of a roll of tape shall be 6 inches or more.
      iii. The applied innerwrap shall conform tightly to the pipe surface, shall be free of voids and wrinkles, and shall be smooth.

iv. When a new roll of tape is started, the end lap of the splice shall overlap the previous roll. This splice shall be smooth and placed so as to maintain the continuity of the innerwrap coating.

g. Application of the Tape Outerwrap
   i. The outerwrap is applied directly over the innerwrap using the same type of mechanical tape tensioning equipment.
   ii. The overlap of the outerwrap shall not be applied directly over the lap of the innerwrap under any circumstances.
   iii. The outer layer may be applied at the same time as the inner layer.

h. Cutbacks shall be as specified by the Company.

10. **Inspection and Testing**
   a. The entire procedure of applying the protective coating material as herein specified will be rigidly inspected from the time the bare pipe is received until the coated pipe is loaded on the carrier for shipment.
   b. If the Company designates an Inspector, the Inspector shall be provided free access to the Applicator's plant at any time during any operation involving the pipe, with the right to inspect and to accept or reject work performed.
   c. The Applicator's Quality Control Inspector shall be responsible for stopping operations when conditions develop which could adversely affect the quality of the completed work.
   d. Although the principal purpose of the coating inspection by the Company and Applicator is to insure compliance of the coating with these specifications, such inspection shall also include examination for previously undetected defects in the pipe, pipe surface, or on the pipe ends. Pipes having such defects shall be set aside for subsequent repair or replacement by the pipe supplier and for any necessary coating repair. Recoating or coating repair that may be necessary by reason of these defects in the pipe which do not involve fault on the part of the Applicator shall be done at the Company's expense.
   e. When Company's Representative exercises Company's right of approval at the Applicator's plant, the Company's Representative shall conduct final inspection on the Applicator's out-bound rack. Accepted pipe shall be presumed to be produced as specified unless test results indicate a discrepancy.
   f. Coating Thickness Measurements
      i. An appropriate film thickness gauge, calibrated to the National Bureau of Standards' Certified Coating Thickness Calibration Standards shall be used to perform coating thickness measurement.
      ii. The coating thickness shall meet or exceed the agreed upon minimum coating thickness. All joints which fail to meet the minimum coating thickness test shall be recoated or repaired.
   g. Electrical Inspection
      i. Holiday inspection of the entire coating surface shall be performed with an approved high voltage Holiday Detector to indicate any flaws, holes, breaks, or conductive particles in the protective coating.
      ii. The Holiday Detector shall have sufficient DC voltage and be equipped with a positive signaling device. The search electrode shall be made of conductive rubber or other applicable material. The Holiday Detector shall be operated in such a way as to audibly and/or visually detect the presence of all holidays.
      iii. The minimum testing voltage for a particular coating thickness shall be within 20% of the value determined from the following formula:

      1250 × (square root of coating thickness (mils))

11. **Repair Procedures**
   a. All defects disclosed by the Holiday Detector and other obvious defects shall be repaired by the Applicator.
   b. Holidays which are the result of slivers, scabs, laminations, or other steel conditions beyond the control of the Applicator shall be repaired at the Company's expense.
   c. Areas of repair to the coating shall be holiday inspected by the Applicator on a 100% basis.
   d. Production Line Repairs
      i. All holidays detected, such as flaws or mislaps, shall be repaired prior to application of the outerwrap tape.
      ii. The flawed or mislapped area shall be cleared of all affected material and reprimed, if necessary.
      iii. The area should be recoated with inner wrap tape, being careful to cover the entire area and overlapping onto the surrounding inner wrap coating by at least 2 inches.
      iv. The affected area should then be rechecked with a Holiday Detector.
   e. Field Repairs
      i. All holidays detected such as damaged or flawed areas shall be repaired by peeling back and removal of the outer and inner layers from the damaged area.
      ii. The holiday area shall then be brushed with primer and a patching tape with a polyethylene backing, and butyl adhesive should be applied directly to the defective area in the manner specified by the Manufacturer.

iii. The minimum lap at the damaged area shall be 4 inches all around.
iv. The repaired area shall be tested with a Holiday Detector after the repair is completed.
v. The repaired area may be covered in cigarette wrap fashion with the outerwrap to a minimum lap of 4 inches beyond the patching tape.

12. **Coated Pipe Handling, Storage, and Loading Requirements**
a. Pipe shall be stored, handled, and transported in a manner to prevent damage to the pipe walls, beveled ends, and the coating.
b. Storage racks shall be so designed as to protect the coated pipe from standing water, direct soil contact, and sharp or hard objects that might damage the coating.
c. The coated pipe shall be shipped using sufficient and proper dunnage to adequately protect the pipe and coating.
d. All pipe shipped by rail shall be loaded in accordance with API Specifications RP5L1, Latest Edition.

13. **Supplementary Details Supplied by the Company**
When possible, the Company shall supply the following supplemental information:
a. Length and diameter of pipe.
b. Grade, wall thickness, and/or weight per foot of pipe.
c. Source and approximate shipping date from the pipe mill.
d. Method of shipment from the mill.
e. Approximate shipping date to the destination.
f. If pipe is to be stored, the approximate length of time it is to be stored.
g. Length, style, and postpreparation of cutback.
h. Minimum weight per car or truck required to protect lowest outbound rate.
i. Name and type of carrier.
j. Stacking and/or loading instructions.

Suggested procedures for coating field joints, fittings, connections, and prefabricated sections using tape coatings

NAPCA Bulletin 6-69-94-3

1. **General**
a. These specifications may be used in whole or in part by anyone without prejudice, if recognition of the source is included. The National Association of Pipe Coating Applicators (NAPCA) assumes no responsibility for the interpretation or use of these specifications.
b. The intended use of these coatings is to provide corrosion protection for buried pipelines. Above ground storage of coated pipe in excess of six months without additional ultraviolet protection is not recommended.
c. The following definitions apply:
i. Applicator—The contractor who applies the coating to the pipe.
ii. Company—The purchaser of the coated pipe or the entity for whom the Applicator coats the pipe.
iii. SSPC—The Steel Structures Painting Council.
iv. NACE—NACE International.
v. Manufacturer—The company that makes the coating materials which are applied to the pipe.

2. **Scope**
a. The Applicator shall furnish all labor, equipment, and material required, shall prepare all surfaces to be coated, and shall apply the coating to all surfaces to be coated.
b. Tape coating products may generally be classified into two categories as follows:
i. Hot-applied tapes which are manufactured with either coal tar or asphalt bitumen material saturated onto a reinforcement fabric with either a removable separator liner or a separator liner that becomes an exterior protective film in the finished application.
ii. Cold-applied tapes which are manufactured with a plastic film and an adhesive material, either with or without a separator liner in the roll.
c. Since installation conditions vary, specific recomm-?tjl] endations from the Manufacturer may be required; however, the application procedures set forth hereafter are generally recommended.

3. **Coating Material**
a. All coating materials, including repair or patch materials, purchased or used under these specifications, shall be packaged in suitable and

approved containers. The containers shall be plainly marked with the name of the Manufacturer, type of material, and batch or lot number where applicable. Bulk shipments shall be allowed provided the above information is included in the bill of lading.

b. The coating material shall be packaged in containers suitable to keep the contents clean and dry during handling, shipping, and storage. Storage and handling conditions shall be in accordance with the Manufacturer's recommendations.

c. Precautions shall be taken during the handling, shipping, and storage of all materials to prevent damage to the containers that would result in contamination of the coating materials. All contaminated, or otherwise damaged materials shall be discarded.

4. **Surface Preparation**
   a. The surface to be coated must be cleaned of all rust, mud, oil, grease, moisture, mill lacquer, or other deleterious substances. Wire brushing and/or solvent washing is sufficient in most instances. Weld splatter should be removed by filing.
   b. When required, kraft paper shall be removed from the mill wrapped pipe in an area equivalent to the width of tape being used.
5. **Primer Application**
   a. Welded joints shall be allowed to cool prior to application of primer.
   b. A uniform and continuous coat of primer shall be applied in accordance with the Manufacturer's recommendation for the specific tape and primer system being used.
   c. The primer coverage and curing or drying time shall be sufficient to insure an effective bond between the substrate and the coating.
6. **Application Procedures**
   a. Hot-Applied Tapes
      i. Hot-applied tapes are applied by hand or machine, spirally or in a cigarette wrap, after heating to obtain a softening of the coating material.
      ii. Only enough tape should be heated to insure that it will remain in a liquefied state during application.
      iii. When being applied by hand, a propane fueled torch with a wide mouth tip is recommended.
      iv. Application proceeds by alternately heating and wrapping the tape to obtain the tension and the overlap recommended by the Manufacturer.
      v. Hot-applied tapes with removable separators require a light bleeding over the exterior surface to insure lap seal. Exterior surface heating is not required for hot-applied tapes designed for machine application or for hot-applied tapes that incorporate the separator liner as a part of the finished application.
   b. Cold-Applied Tapes
      Cold-applied tapes are applied by hand or machine spirally or in a cigarette method with tension and overlap as recommended by the Manufacturer.
7. **Inspection and Testing**
   Test with Holiday Detector, voltage not to exceed that used by plant coating application.
8. **Repair Procedures**
   When plant applied coatings require repair, wire brush and reprime the exposed metal surfaces. Next, the perimeter of the damaged area should be feathered. Apply either hot or cold-applied tape to the exposed area pursuant to the Manufacturer's specific recommendations.

Suggested procedures for coating field joints using heat-shrinkable materials

NAPCA Bulletin 6-69-94-5

1. **General**
   a. These specifications may be used in whole or in part by anyone without prejudice, if recognition of the source is included. The National Association of Pipe Coating Applicators (NAPCA) assumes no responsibility for the interpretation or use of these specifications.
   b. The intended use of these coatings is to provide corrosion protection for buried pipelines. Above ground storage of coated pipe in excess of six months without additional ultraviolet protection is not recommended.
   c. The following definitions apply:
      i. Applicator—The contractor who applies the coating to the pipe.

ii. Company—The purchaser of the coated pipe or the entity for whom the Applicator coats the pipe.
iii. SSPC—The Steel Structures Painting Council.
iv. NACE—NACE International.
v. Manufacturer—The company that makes the coating materials which are applied to the pipe.

2. **Scope**
   a. The Applicator shall furnish all labor, equipment, and material required, shall prepare all surfaces to be coated, and shall apply the coating to all surfaces to be coated.
   b. Heat shrinkable sleeves are designed to provide moisture penetration resistance and corrosion prevention for pipeline field joints.
   c. The product is a heat-shrinkable material fabricated from cross-linked polyolefin precoated with a mastic or adhesive.
   d. These adhesives display properties not possible with pressure-sensitive tapes.
   e. The sleeves shrink when heated and force the mastic (or adhesive) to flow into the irregularities of the weld.
   f. The shrinking action provides a simple, reproducible delivery system for the mastic or adhesive.
   g. When shrunk, the polyolefin sleeve provides penetration and abrasion resistance during handling, backfill, and operation of the pipe. It also provides resistance against soil stresses, alkalis, acids, microbes, and fungi.
   h. Wrap around sleeves can be installed after welding and inspection.
   i. Heat-shrinkable sleeves are also self-priming in nature and bond to virtually all substrates including heavy wall pipe, polyethylene, epoxy powder, and bituminous coatings.
3. **Coating Material**
   a. All coating materials, including repair or patch materials, purchased or used under these specifications, shall be packaged in suitable and approved containers. The containers shall be plainly marked with the name of the Manufacturer, type of material, and batch or lot number where applicable. Bulk shipments shall be allowed provided the above information is included in the bill of lading.
   b. The coating material shall be packaged in containers suitable to keep the contents clean and dry during handling, shipping, and storage. Storage and handling conditions shall be in accordance with the Manufacturer's recommendations.
   c. Precautions shall be taken during the handling, shipping, and storage of all materials to prevent damage to the containers that would result in contamination of the coating materials. All contaminated, or otherwise damaged materials shall be discarded.
4. **Surface Preparation**
   a. For tube type sleeves, before welding the pipe, slide the sleeves over one end and move it at least 3 feet (1 m) away from the joint to prevent damage from weld spatter.
   b. The surface to be coated must be cleaned of all rust, mud, oil, grease, moisture, mill lacquer, or other deleterious substances. Wire brushing and/or solvent washing is sufficient in most instances. Weld splatter should be removed by filing.
   c. Coal tar and asphalt pipe coatings require removal of kraft paper prior to installation of the sleeve.
   d. Preheat the exposed metal area to a hand warm temperature—approximately 120°F—taking care to remove all existing moisture from the pipe surface.
   e. Preheating is accomplished with the use of a butane or propane torch capable of providing a high output yellow or blue flame. Other heating sources are available for special applications. Consult the Manufacturer for specific instructions.
5. **Tube Type Sleeve Application**
   a. Center the sleeve over the joint and begin heating it with a propane or butane torch, having a high output yellow or blue flame.
   b. Start at the top center of the sleeve and shrink it all around the diameter of the pipe working toward one end.
   c. Keep the torch moving to avoid burning the sleeve.
   d. When completed, work circumferentially from the center to the opposite end.
   e. The sleeve is fully recovered when conformed to the pipe diameter and mastic flow is evident at the sleeve edges.
6. **Installation—Wrap Around Sleeves**
   a. If the sleeve is packaged in a poly bag, remove and discard the packing. If the closure seal is included, put it in an area where it will not be contaminated by dirt or moisture.
   b. Remove the protective release paper. Hold the sleeve vertically and center it on the weld joint.
   c. Press the sleeve to the far side of the top of the pipe (approximately 11 o'clock). When applied correctly, the leading edge of the wrap will face down toward the bottom of the ditch.
   d. The release paper may be removed as the sleeve is wrapped around the joint to insure against mastic contamination.
   e. Wrap the sleeve around the joint. Some sleeves require heating of overlap prior to contact with the pipe; some do not.

f. If a closure seal is required, center the closure seal on the overlap and press down. Some sleeves require heating of the closure prior to sleeve recovery.
g. After proper closure application, begin sleeve recovery. With the torch, start at the center of the sleeve and shrink it all around the diameter of the pipe.
h. Keep the torch moving using broad circumferential strokes to avoid burning. Continue shrinking the sleeve toward one end.
i. Repeat this operation on the other end of the sleeve. The sleeve is fully recovered when conformed to the pipe diameter and mastic is evident at the sleeve's edges.
j. Consult the Manufacturer for instructions on specific procedures.

7. **Inspection and Testing**
   Test with Holiday Detector, voltage not to exceed that used by plant coating application.
8. **Lowering and Backfilling**
   a. Allow the joint and sleeve to cool at least 30 min before lowering in.
   b. Selected backfill, containing no large or sharp stones, should be used to avoid damaging the sleeve or coating during the backfilling operation.
   c. A water quench may be used to reduce backfill time, if required.

External application procedures for hot applied coal tar coatings to steel pipe

NAPCA Bulletin 3-67-94

1. **General**
   a. These specifications may be used in whole or in part by anyone without prejudice, if recognition of the source is included. The National Association of Pipe Coating Applicators (NAPCA) assumes no responsibility for the interpretation or use of these specifications.
   b. The intended use of these coatings is to provide corrosion protection for buried pipelines. Above ground storage of coated pipe in excess of six months without additional ultraviolet protection is not recommended.
   c. The following definitions apply:
      i. Applicator—The contractor who applies the coating to the pipe.
      ii. Company—The purchaser of the coated pipe or the entity for whom the Applicator coats the pipe.
      iii. SSPC—The Steel Structures Painting Council.
      iv. NACE—NACE International.
      v. Manufacturer—The company that makes the coating materials which are applied to the pipe.
   d. The Company, when placing orders, shall specify his requirements by referring to the applicable specification symbol in NAPCA Specification Designation for Enamel Coating 1-65-94.
2. **Scope**
   a. The Applicator shall furnish all labor, equipment, and material required, shall prepare all surfaces to be coated, and shall apply the coating to all surfaces to be coated.
   b. Corrosion protection, as provided under this specification, is furnished by the application of hot-applied enamel to the exterior of pipe to be placed underground. The Company shall specify the reinforcement or wrapper material by referring to the applicable NAPCA specification symbol in Bulletin 1-65-94 and, where desired, the Company may define the exact wrapper to be used.
3. **Pipe Conditions**
   a. Pipe delivered to the Applicator for coating shall be free of protective oils, lacquers, mill primer, dirt, or any other deleterious surface contamination which may affect the application of the coating. The pipe surface shall be as free as possible from scabs, slivers, and laminations.
   b. Any paint markings or stenciling of the pipe surface shall be of the type and thickness that can be removed easily during normal surface preparation.
4. **Handling of Bare Pipe**
   a. Proper equipment for unloading, handling, and temporary storage of bare pipe shall be used to avoid any damage to the pipe or pipe ends.

b. If internally coated pipe is received at the Applicator's plant, care shall be taken to avoid damage to the internal coating or the obliteration of the internal pipe markings during any phases of work covered by this specification. Internal coatings must be capable of withstanding the processing conditions necessary for the application of the external coating.

c. The Applicator shall visibly inspect the pipe upon receipt for damage such as dents, flat ends, and bevel damage. Any damage observed at this point shall be noted on the inbound tally, and the Company shall be informed within 24 h of receipt of the pipe. Any nonvisible defects such as slivers, scabs, laminations, burrs, dents, etc., will be observed after the pipe is blast cleaned and at the Company's request, removed as an extra work item.

5. **Material and Workmanship**

a. All material furnished by the Applicator shall be of the specified quality. All work shall be done in a thorough workmanlike manner. The entire operation of pipe receiving, stockpiling, surface preparation, coating application, storage, and loadout shall be performed under the supervision of and by experienced personnel skilled in the application of protective coating.

6. **Equipment**

a. The Applicator's equipment shall be in such condition as to permit the Applicator to follow the procedure and obtain results prescribed in these specifications.

7. **Coating Material**

a. All coating materials, including repair or patch materials, purchased or used under these specifications, shall be packaged in suitable and approved containers. The containers shall be plainly marked with the name of the Manufacturer, type of material, and batch or lot number where applicable. Bulk shipments shall be allowed provided the above information is included in the bill of lading.

b. The coating material shall be packaged in containers suitable to keep the contents clean and dry during handling, shipping, and storage. Storage and handling conditions shall be in accordance with the Manufacturer's recommendations.

c. Precautions shall be taken during the handling, shipping, and storage of all materials to prevent damage to the containers that would result in contamination of the coating materials. All contaminated, or otherwise damaged materials shall be discarded.

8. **Surface Preparation**

a. Before blasting, all oil, grease, mill lacquer, and other deleterious material on the surfaces of the metal to be coated shall be removed by suitable means.

b. In cold weather or any time when moisture tends to collect on the steel, the pipe shall be uniformly warmed for sufficient time to dry the pipe prior to cleaning. The pipe temperatures shall be maintained at least 5°F above the dew point during the cleaning and coating operations. Pipe temperature shall not exceed 160°F as a result of preheat.

c. Pipe surfaces shall be blast cleaned to a commercial blast metal finish in accordance with SSPC-SP-6 or NACE #3 requirements.

d. NACE, Swedish Pictorial, SSPC, or other mutually agreed upon standards shall be used to judge the degree of cleaning.

e. A consistent abrasive working mix shall be maintained by frequent additions of small quantities of new abrasive commensurate with consumption. Infrequent large quantity additions of abrasive shall be avoided.

f. Following cleaning and prior to coating the pipe, abrasive remaining on the outside and loose contamination on the inside of the pipe shall be removed by air blast, vacuum, or other suitable methods. If air is used, the air should be dry and free of contaminants, and all particles removed from the surface shall be collected in such a manner as not to contaminate clean pipe.

g. Following cleaning and prior to coating, the pipe surface shall be inspected for adequate cleaning and surface condition. Pipe not properly cleaned shall be rejected and recleaned.

h. Blast cleaned pipe surfaces shall be protected from conditions that would allow the pipe to flash rust before coating. If flash rusting occurs, affected pipe shall be recleaned.

9. **Priming**

a. Immediately after cleaning, all blasted surfaces shall be free of dust, grit, and other foreign material and shall be primed by brushing, spraying, or other suitable means to produce an effective bond between the metal and subsequent coating. No primer shall be applied to a wet surface.

b. The use of primer that becomes contaminated with foreign substances or has thickened through evaporation of the solvents shall not be permitted.

c. The minimum and maximum allowable drying time of the primer between application of primer and application of the coating shall be in accordance with instructions issued by the Manufacturer of the primer. If the coating is not applied within the maximum allowable time after priming as required, the pipe shall be recleaned and reprimed.

d. After application, the primer coat shall be uniform and free from floods, runs, sags, drips, or bare spots. Any bare spots shall be reprimed. All runs, sags, floods, or drips shall be removed by brushing or scraping and the bare areas, if any, shall be retouched. Suitable measures shall be taken to protect wet primer from contact with moisture, dust, or other foreign matter.

10. **Coating Application**
    a. Coating materials may be delivered in hot bulk or solid forms. The coating shall be heated in approved heating kettles equipped with accurate and easily read thermometers, mechanically agitated, and covered with proper lids.
    b. The coating shall be maintained moisture and dirt free at all times prior to and at the time of heating and application.
    c. The coating shall be loaded into the kettles in pieces of suitable size for the melting equipment used. When the Applicator receives hot material in bulk, the holding kettles shall be operated in accordance with the Manufacturer's instructions.
    d. In heating the coating material, the charge will be melted and brought to application temperature without injury to the coating material. The temperature of application shall not exceed that which is recommended by the Manufacturer.
    e. The primed pipe surface shall be dry and clean at the time the coating is applied.
    f. External coating shall be applied by spraying, flooding, or pouring in such a manner as to maintain the specified thickness. The coating shall overlap the preceding spiral, producing a continuous coat free from defects. The thickness of the coating shall be as specified in NAPCA Specification Designation for Enamel Coating 1-65-94.
    g. If specified, the fiberglass innerwrap shall be of uniform width for spiral application. It shall be applied simultaneously with coating at a nominal lap of 1/2-inch and a minimum lap of 1/4-inch. All fibers shall be completely covered. Sufficient tension shall be applied to the glass to embed it in the outer ½ of the coating. Fiberglass shall meet the specifications of NAPCA Bulletin 5-69-94, Table 1.
    h. If specified, the pipeline felt shall be applied mechanically to the pipe so as to be positively bonded to the coating. It shall be of uniform width for spiral application. It shall be applied neatly and smoothly with a nominal lap of 1/2-inch and a minimum lap of 1/4-inch and shall be free of wrinkles and buckles. Pipeline felt shall meet the specifications of NAPCA Bulletin 5-69-94, Table 2.
    i. Unless otherwise specified, a spiral wrapping of heavy kraft paper shall be applied over the coating surface prior to final inspection.
    j. Other outerwraps specified shall be applied in accordance with the Manufacturer's specifications. Refer to the specifications in NAPCA Bulletin 5-69-94, Table 3 for some common outerwraps.
11. **Inspection and Testing**
    a. The entire procedure of applying the protective coating material as herein specified will be rigidly inspected from the time the bare pipe is received until the coated pipe is loaded on the carrier for shipment.
    b. If the Company designates an Inspector, the Inspector shall be provided free access to the Applicator's plant at any time during any operation involving the pipe with the right to inspect and to accept or reject work performed.
    c. The Applicator's Quality Control Inspector shall be responsible for stopping operations when conditions develop which could adversely affect the quality of the completed work.
    d. Although the principal purpose of the coating inspection by the Company and the Applicator is to insure compliance of the coating with these specifications, such inspection shall also include examination for previously undetected defects in the pipe, pipe surface, or on the pipe ends. Pipe having such defects shall be set aside for subsequent repair or replacement by the pipe supplier and for any necessary coating repair. Recoating or coating repair that may be necessary by reason of these defects in the pipe which do not involve fault on the part of the Applicator shall be done at the Company's expense.
    e. When Company's Representative exercises Company's right of approval at the Applicator's plant, the Company's Representative shall conduct final inspection on the Applicator's out-bound rack. Accepted pipe shall be presumed to be produced as specified unless test results indicate a discrepancy.
    f. Coating Thickness Measurements
        i. An appropriate film thickness gauge, calibrated to the National Bureau of Standards' Certified Coating Thickness Calibration Standards shall be used to perform coating thickness measurement.
        ii. The coating thickness shall meet or exceed the agreed upon minimum coating thickness. All joints which fail to meet the minimum coating thickness test shall be recoated or repaired.
    g. Electrical Inspection
        i. Holiday inspection of the entire coated surface shall be performed with an approved high voltage Holiday Detector to indicate any flaws, holes, breaks, or conductive particles in the protective coating.

ii. The Holiday Detector shall have sufficient DC voltage and be equipped with a positive signaling device. The search electrode shall be made of conductive rubber or other applicable material. The Holiday Detector shall be operated in such a way as to audibly and/or visually detect the presence of all holidays.

iii. The voltage to be used shall be as set forth in the table below.

Minimum test voltages		
32nd inch	**Mils**	**Test volts**
2	62	9,800
3	94	12,100
4	125	14,000
5	156	15,000
6	188	17,100

12. **Repair Procedures**
   a. All defects disclosed by the Holiday Detector and other obvious defects shall be repaired by the Applicator.
   b. Holidays which are the result of slivers, scabs, laminations, or other steel conditions beyond the control of the Applicator shall be repaired at the Company's expense.
   c. Areas of repair to the coating shall be holiday inspected by the Applicator on a 100% basis.
13. **Coated Pipe Handling, Storage, and Loading Requirements**
   a. Pipe shall be stored, handled, and transported in a manner to prevent damage to the pipe walls, beveled ends, and the coating.
   b. Storage racks shall be so designed as to protect the coated pipe from standing water, direct soil contact, and sharp or hard objects that might damage the coating.
   c. The coated pipe shall be shipped using sufficient and proper dunnage to adequately protect the pipe and coating.
   d. All pipe shipped by rail shall be loaded in accordance with API Specifications RP 5L1, Latest Edition.
14. **Supplementary Details Supplied by the Company**
   When possible, the Company shall supply the following supplemental information:
   a. Length and diameter of pipe.
   b. Grade, wall thickness, and/or weight per foot of pipe.
   c. Source and approximate shipping date from the pipe mill.
   d. Method of shipment from the mill.
   e. Approximate shipping date to the destination.
   f. If pipe is to be stored, the approximate length of time it is to be stored.
   g. Length, style, and postpreparation of cutback.
   h. Minimum weight per car or truck required to protect lowest outbound rate.
   i. Name and type of carrier.
   j. Stacking and/or loading instructions.

FUEL SPECIFICATIONS

Fuels vary widely in their physical and chemical properties to match their applications and environmental exposure. The properties of fuels must be balanced to provide satisfactory performance over a wide range of operating conditions. Specifications are based on physical, mechanical, or chemical properties, and often include safety, quality, and performance criteria. Standards are differentiated by fuel type (e.g., gasoline, diesel, and heating oil), transportation mode (e.g., automobile, aircraft, and marine), and usage (e.g., highway and nonroad application), and frequently vary by region and season (climate). The standards are periodically updated to reflect changes in specifications.

Types of fuel specifications

Fuel properties are specified to give satisfactory performance of engines over a wide range of operating conditions. Different types and grades of fuels are manufactured because there are many different types of engines and operating conditions. Requirements either prescribe a balance of properties to ensure satisfactory engine performance, or limit the concentrations of components that could have an adverse effect on engine performance. Specifications differ between fuel types and across grades of fuels. Other requirements are common to all grades.

There are a variety of detailed sources of information on fuel specifications. The latest fuel specification for motor vehicles and commercial and military aviation fuels can be found in ASTM Specifications, UK Ministry of Defense DEF STAN Specifications, US Military Specifications, and IATA Guidance Material. The Coordinating Research Council report is useful for designers of aircraft and engine fuel systems (CRC, 2014). The Annual Book of ASTM Standards and technical reviews from Chevron (2007a,b, 2009), Shell Aviation (2012), and ExxonMobil Aviation (2005) provide detailed information and are a good place to start. Reference texts such as (Owen and Coley, 1995) and (Blackmore, 2001) and petroleum refining texts (Kaiser et al., 2020) may also be consulted.

Consensus specifications

Consensus specifications are written and published by standards development organizations such as ASTM International, the Committee European de Normalization (CEN, also known as the European Committee for Standardization), and the International Organization for Standards. As a general rule, these specifications contain the minimum requirements for a fuel to provide acceptable performance under the scope of the specification. Specifications are not legal documents, but are often cited in contracts or by regulatory bodies in which case they obtain legal status.

Pipeline companies and terminal operators also have specifications for the products that flow through their systems to ensure the quality of the fuel and to protect the systems from corrosion. Engine and car/truck/aircraft manufacturers have specifications which are typically based on consensus standards. Regulatory agencies at all government levels issue fuel regulations, either in limited aspects (e.g., sulfur content) or simply by reference to an existing consensus standard such as ASTM D4814 (gasoline).

The Regulations Government ASTM Technical Committees define specification, standard, guide, practice, and test methods as follows:

Specification	An explicit set of requirements to be satisfied by a material, product, system, or service.
Standard	A document that has been developed and established within the consensus principles of the ASTM and meets the requirements of ASTM procedures and regulations.
Guide	An organized collection of information or series of options that does not recommend a specific course of action.
Practice	A set of instructions for performing one or more specific operations that does not produce a test result.
Test Method	A definite procedure that produces a test result.

Test methods

ASTM will not adopt a requirement for a property until a standard test method is developed to measure that property. The process starts with a technical review of the method, and then a test protocol is conducted on a common set of samples sent to a group of labs, the results of which are compiled and statistically reviewed. A precision statement is developed that contains the maximum difference to be expected between consecutive tests in one laboratory (repeatability), and the maximum difference between results obtained by different laboratories on the same sample (reproducibility). Every ASTM test method is reviewed every five years for reapproval, revision, or cancellation. The year of last review appears after the hyphen in each test method.

Mandatory and suggested specifications

Each ASTM standard contains one or more tables, with one or more annexes and appendices. The tables outline the specification properties and limits. Along with each property, there is an assigned ASTM test method (or methods) detailing how the property is to be measured. Test methods are essential so that everyone measuring the property uses the same procedure and can obtain the same answer within the defined precision. Limits (maximum or minimum or range) are associated with each property. When more than one method is allowed, one of the methods is designated a referee method in cases of dispute. Usually, detailed footnotes are associated with the entries.

Annexes are mandatory information whereas appendices provide additional information but are not mandatory. The wording of specifications is precise and used to qualify requirements and action. "Shall" is used when the provision is mandatory; "should" denotes a provision is recommended but not mandatory; and "may" is used when the provision is optional.

Enforcement

The two parties to a fuel purchase will specify procedures/agreements regarding compliance to specifications on the chemical, physical, and performance characteristics of the fuel. The enforcement of a specification depends on the type of specification and the nature of the agreement between the buyer and seller. Often times, samples of the fuel are tested to confirm specification compliance. When the transaction is between a seller and the general public, the enforcement is usually performed by governmental or regulatory agencies.

References

Annual book of ASTM standards, Sec. 5. Petroleum products, lubricants, and fossil fuels, Vol 05.03. Petroleum products and lubricants (III). ASTM International, West Conshohocken, PA, 2015.

ASTM international regulations governing ASTM technical committees. 2015. ASTM International: West Conshohocken, PA.

C.J.S. Bartlett, Standards, specifications, test methods and codes of practice, in: A.G. Lucas (Ed.), Modern petroleum technology, Vol. 2. John Wiley & Sons, Chichester, England, 2001, pp. 457–465.

D.R. Blackmore, Gasoline and related fuels, in: A.G. Lucas (Ed.), Modern Petroleum Technology, Vol. 2. John Wiley & Sons, Chichester, England, 2001, pp. 217–254.

Chevron Products Company, Aviation fuels technical review. Chevron International, San Ramon, CA, 2007a.

Chevron Products Company, Diesel fuels technical review. Chevron International, San Ramon, CA, 2007b.

Chevron Products Company, Motor gasolines technical review. Chevron International, San Ramon, CA, 2009.

Coordinating Research Council, Inc, Handbook of Aviation Fuel Properties, 4th Ed. CRC, Alpharetta, GA, 2014, CRC Report No. 663.

ExxonMobil Aviation, World jet fuel specifications with avgas supplement. Leatherhead, UK, 2005, Surry KT22 8UX.

R.F. Jennings, N.D. Smith, R.G. Hayes, S.D. Benjamin, Specifications – what they are, why we have them, and how they are used, in: ASTM International 2016, Chapter 2. ASTM International, West Conshohocken, PA, 2016.

M.J. Kaiser, A. de Klerk, J.H. Gary, G.E. Handwerk, Petroleum Refining: Technology, Economics, and Markets. CRC Press, Boca Raton, FL, 2020.

K. Owen, T. Coley, Automotive Fuels Reference Book, 2nd Ed. Society of Automotive Engineers, Warrendale, PA, 1995.

Shell Aviation, The aeroshell book, 19th Ed. Shell International Petroleum Co. Ltd. Shell Centre: York Road. SE1 7NA, London, 2012.

Transportation fuel specifications

Most states adopt ASTM standard specifications to ensure quality products are consumed within their borders, but the scope of products regulated and tests performed varies from state to state. The role of the ASTM differs from most standardization bodies in that the ASTM develops and publishes both test methods and specifications.

Gasoline—ASTM D4814

The ASTM specification for gasoline, ASTM D4814, *Standard Specification for Automotive Spark-Ignition Engine Fuel*, is written into US gasoline regulations, both federal and state. D4814 specifies 10 gasoline properties and references more than 30 ASTM standards that describe how to run specific test methods. One of the most important property categories of gasoline is fuel volatility, which is controlled by vapor pressure, distillation, and a vapor-liquid ratio. D4814 does not include octane number or antiknock index as a specification which are set by refiners and marketers based on their perception of the technical and competitive needs in the market. In the United States, the Federal Trade Commission requires gasoline to be certified throughout the distribution system for octane rating. The US EPA and some states set additional standards for gasoline that are not included in D4814 unless they involve properties traditionally covered by the method such as vapor pressure.

Jet fuel—ASTM D1653

Jet fuel used by commercial airlines is defined by ASTM D1655, *Standard Specification for Aviation Turbine Fuels*. Early aircraft engines were similar to those used in automobiles and burned the same fuel. The need for increased power led to the development of specialized engines and aviation gasolines tailored to their requirements. In the 1940s, the turbine engine emerged and kerosene was the first fuel used in aircraft turbine engines, which was eventually replaced by specialized aviation turbine fuels known today as jet fuels. The most important specifications for jet fuel are related to performance and include flash point, freeze point, aromatics content, and smoke point. Less important specifications such as density, final boiling point (FBP), and net heat of combustion are frequently used as control over quality check variables.

Diesel—ASTM D975

Diesel fuel is defined by ASTM D975, *Standard Specification for Diesel Fuel*. There are seven grades of diesel fuel oils that are classified according to viscosity, sulfur content, and application. Distillation is a routine test performed that provides the boiling range of the compounds, and the Cetane Index is computed and verified because it only requires a calculation utilizing the distillation curve. Flash point and conductivity tests are monitored for safety, and water and sediment are monitored to ensure adequate workmanship and handling practices.

Fuel quality

Transportation fuels are complex mixtures composed predominately of hydrocarbons which varies depending on crude source and manufacturing process. It is impossible to define the exact composition of fuels and, therefore, specifications have evolved primarily as a performance measure.

Fuel properties are grouped according to the following categories:

- Composition
- Performance
- Volatility
- Fluidity
- Combustion
- Corrosion
- Stability
- Contaminants
- Other
- Additives

The significance of the properties within each group vary with the fuel and operating conditions (Table 1). There are spark ignition and diesel automotive engines, piston and turbine aircraft engines, commercial and military aircraft, and numerous marine vessels such as recreational boats, tugs, and ocean going vessels. Operating conditions and requirements vary with each engine type.

Table 1
Fuel Property and Significance

Composition	Appearance	Assessed in a qualitative pass/fail test of fuel cleanliness to preclude free water sediment and suspended matter
	Total acidity	Indicates the corrosive potential of fuel to metals. Trace organic acids can affect water separation properties
	Aromatics content	Relates directly to flame radiation, carbon deposition, and smoke. Also affects swelling of elastomers in the fuel system
	Hydrogen content	Contributes to combustion cleanliness and is broadly related to aromatics consents
	Olefins	Unsaturated hydrocarbons which are potential contributors to instability in storage
	Total sulfur	Sulfur oxides formed during combustion can cause corrosion of turbine metal parts
	Mercaptan sulfur	Have a very unpleasant odor and attack certain elastomer materials
	Doctor test	Detects the presence of reactive sulfur compounds and is an alternative measure of mercaptans sulfur
Volatility	Distillation curve	Defines the kerosene boiling range which needs to be appropriate for balanced vaporization of the whole fuel volume
	Flash point	Related to volatility and therefore affects combustibility. It is a leading factor in determining fire safety and fuel handling.
	Density	Must be known for aircraft weight loading calculations since fuel is typically metered by volume. Also relates to specific energy
	Vapor pressure	Significant for wide-cut fuels. Indicates venting loss of light ends at altitude and in hot climates. Also relates to cold starting
Fluidity	Freezing point	Limits higher molecular weight hydrocarbons that crystalize at low temperature and influences low temperature pumpability during flight
	Viscosity	Affects fuel pumpability over the operating temperature range and relates to droplet size in sprays produced by burner nozzles
Combustion	Specific energy	Net heat of combustion denotes the amount of heat energy obtainable from a fuel to provide power (value is calculated)
	Smoke point	Indicates the tendency of a fuel to form soot, which is related to the type of hydrocarbons making up its composition
	Naphthalenes	Polycyclic aromatics high in carbon content exacerbating the problems of carbon formation, flame radiation, and smoke
Corrosion	Copper strip corrosion	Ensures that organic sulfur compounds will not corrode copper components in the fuel system
Stability	Thermal stability	JFTOT measures relate to the amount of deposits formed at high temperature in the engine fuel system
Contaminants	Existent gum	A nonvolatile residue left on evaporation of a fuel. Also serves as a check for fuel contamination within product distribution systems
	Particulates	Dirt and rust are undesirable and detected by filtration through a membrane filter
	Filtration time	Measured by the same test procedure as particulates
	Water reaction	Determines the presence of materials that react with water and affect the stability of the fuel-water interface
	Water separation	MSEP index rates the ability of fuel to release entrained or emulsified water when passed through a fiberglass filter coalesce
Other	Electrical conductivity	Needs to be high enough to dissipate any electrostatic charges generated during fuel handling operations, so as to prevent fire or exploration hazards
	Lubricity (BOCLE)	Refers to the effectiveness of lubricating moving parts in engine fuel system components such as pumps and control units

Source: ExxonMobil Aviation. 2005. *World jet fuel specifications with avgas supplement.* Leatherhead: UK, Surry KT22 8UX.

Fuel quality as it relates to emissions is regulated by one or more government agencies, and as it relates to performance by industry standards developed by industry boards. In the United States, environmental standards are issued by the US EPA, and industry standards are developed by the ASTM. In Europe, environmental directives are issued by the European Union (European Commission, Parliament, Council and Member States), and industry standards are developed by the European Standards Organization (CEN).

Specifications vary depending on the fuel type and the manner in which the fuel is combusted and the characteristics of its use. Premium fuels should have better (lower/higher) specifications within the fuel class relative to specific characteristics. The terminology employed varies by region and time, of course, and can cause confusion when the same word can mean different fuels, and so proper definitions are essential. Usually, applications, usage, and specifications are the best way to compare fuel type.

Gasoline fuel specifications

Fuel properties are specified to give satisfactory performance of engines over a wide range of operating conditions. Different types and grades of fuels are manufactured because there are many different types of engines and operating conditions. Requirements either prescribe a balance of properties to ensure satisfactory engine performance or limit the concentrations of components that could have an adverse effect on engine performance. Specifications differ between fuel types and across grades of fuels. Other requirements are common to all grades.

Motor gasoline is a mixture of primarily naphtha-range hydrocarbon molecules from C5 to C12 blended together to meet a tight range of physical specifications on volatility, distillation temperature profile, sulfur, aromatics, and olefins content. Restrictions also arise on the content of specific octane rating enhancers, such as lead and MTBE, and for US automotive gasoline grades, requirements on oxygenates such as ethanol. ASTM D4814 does not include an octane number as a specification which are set by refiners and marketers based on the technical and competitive needs in the market. Volatility specifications are more complicated and difficult to comprehend than other properties, but are a key specification for gasoline and sets it apart from other transportation fuels in the cost to manufacture.

Volatility

Volatility refers to a fuel's tendency to vaporize, that is, move from a liquid to a gas state as the temperature is raised. The volatility characteristics of transportation fuels are very important, especially for gasoline and jet fuel, and to a lesser extent for diesel. Motor and aviation gasolines are manufactured as liquids but are consumed/combusted in the vapor phase. Consequently, gasoline volatility must be high enough to assure acceptable engine performance under normal conditions, but must be restricted to avoid vaporization losses, air pollution, and unsafe storage and handling.

In the United States, three properties are used to measure gasoline volatility: vapor pressure, distillation profile, and vapor-liquid ratio. A fourth property, drivability index, is calculated from the distillation profile to proxy a fuel's cold-start and warm-up performance. Outside the United States, a vapor lock index is often used instead of a vapor-liquid ratio to control hot fuel handling problems.

Vapor pressure. Vapor pressure is the most important property for cold-start and warm-up drivability. Vapor pressure requirements in the United States vary with the season and geographic location, and normally ranges from 48.2 to 103 kPa (7 to 15 psi) measured at 37.8°C (100°F). Higher values of vapor pressure generally result in better cold-start performance, whereas lower values are better to prevent vapor lock and other hot fuel handling problems. Gasoline is designed to balance these and other trade-offs.

Reid vapor pressure (RVP) is defined as the absolute vapor pressure exerted by a liquid at 100°F. In ASTM D4814, RVP is no longer used because the test method involves the fuel making contact with water, which does not work for gasoline-ethanol blends which are now prevalent. Today, there are several vapor pressure test methods (D4953, D5191, D5482, and D6378) that report the dry vapor pressure equivalent, a value calculated by a correlation equation that is comparable to the vapor pressure value obtained by the test method D4953.

High levels of vaporization are desirable for winter starting and operation, and lower levels are desirable in avoiding vapor lock during summer heat. Fuel cannot be pumped when there is vapor in the fuel line (summer), and winter starting will be more difficult when liquid gasoline in the combustion chambers has not vaporized. Hence, it is common in countries with large seasonal variations in temperature to specify summer and winter gasoline blends. The EPA currently specifies RVP seasonally from 7.8 to 9 psi (54 to 62 kPa) in the summer and 13.5 to 15 psi (93 to 103 kPa) in the winter, and such "tight" specifications are the primary reason why manufacturing gasoline is more complicated and more expensive than other transportation fuels.

Table 1
ASTM Seasonal Vapor Pressure Classes

Summer	Spring/autumn	Winter
AA 7.8 psi	B 10.0 psi	D 13.5 psi
A 9.0 psi	C 11.5 psi	E 15.0 psi

Source: ASTM D4814-14b.

In the United States, there are six ASTM seasonal vapor pressure classes that vary by month and by geography (Table 1). The EPA regulates RVP for gasolines and gasoline-ethanol blends during the summer season, June 1—September 15.

Distillation profile. The distillation curve of gasoline is comprised of three segments, the front-end volatility (up to 25% evaporated), the midrange volatility (up to about 50% evaporated), and the back-end volatility (more than 90% evaporated). Each segment comprises distinct properties that affects the performance of the engine and are balanced to achieve suitable trade-offs.

A distillation profile represents the set of increasing temperatures at which gasoline evaporates for a series of increasing volume percentages under specific test conditions. Alternatively, the profile may be the set of increasing evaporation volume percentages for a fixed series of increasing temperatures. Two equivalent systems of volatility parameters are used, Exxx and Txxx numbers. For example, midrange volatility may be quoted as E100, that is, percent fuel evaporated at 100°C, or as T50, the temperature at which 50% of the fuel is evaporated. Front-end volatility is usually quoted as E180 or T90.

ASTM D4814 specification distillation limits include the maximum temperature for distilled gasoline at 10% recovered (T10), as well as both minimum and maximum temperature limits at 50% recovered (T50), the maximum at 90% recovered (T90), and the maximum FBP for the six vapor pressure classes (Table 2).

Gasoline needs the correct volatility properties to work in engines across the entire distillation curve at the front end, midrange, and tail-end distillation, which entails specifying RVP, distillation cutoffs, vapor-liquid ratios, and drivability. Matching the distillation curve is usually sufficient to satisfy the other volatility properties required.

Distillation limits are tightly controlled to balance engine performance over a wide range of conditions. Front-end volatility (< 20% vol evaporated) is adjusted to provide easy cold starting and freedom from vapor lock, while midrange volatility (20%—90% vol evaporated) is adjusted to provide rapid warm-up, good power and acceleration, and good short-trip fuel economy. Tail-end volatility (>90% vol evaporated) determines fuel economy after warm-up and is adjusted to minimize engine deposits and volatile organic compound exhaust emission.

Table 2
ASTM Vapor Pressure and Distillation Class Requirements

Property	Vapor pressure/distillation class					
	AA	A	B	C	D	E
Vapor pressure, at 37.8°C (100°F) max, kPa (psi)	54 (7.8)	62 (9.0)	69 (10.0)	79 (11.5)	93 (13.5)	103 (15)
Distillation temperature, °C (°F), at % evaporated						
10 vol%, max	70 (158)	70 (158)	65 (149)	60 (140)	55 (131)	50 (122)
50 vol%						
Min	77 (170)	77 (170)	77 (170)	77 (170)	77 (170)	77 (170)
Max	121 (250)	121 (250)	118 (245)	116 (240)	113 (235)	110 (230)
90 vol%, max	190 (374)	190 (374)	190 (374)	185 (365)	185 (365)	185 (365)
End point, max	225 (437)	225 (437)	225 (437)	225 (437)	225 (437)	225 (437)
Distillation residue, vol%, max	2	2	2	2	2	2
Drivability index, max°C (°F)	597 (1250)	597 (1250)	591 (1240)	586 (1230)	580 (1220)	569 (1200)

Source: ASTM D4814-14b.

Volatility specifications and schedules. Gasoline volatility in the United States is specified using six vapor pressure/distillation classes (Table 2) and six vapor-liquid ratio classes. The specification, except during transition months, assigns one vapor pressure/distillation class and one vapor-liquid ratio class each month to each geographic area (state or portion of state) based on altitude and the expected ambient temperature range.

Composition

Several components of gasoline are limited by specifications to deliver certain performance and environmental objectives (Table 3). Sulfur is limited by the EPA, the European Union, and many other countries for environmental reasons. When sulfur is burned in an engine, it creates sulfur dioxide gas which absorbs on the metal catalyst in a vehicle's exhaust system and makes the catalyst less effective, increasing emissions of pollutants.

In April 2014, EPA finalized new gasoline sulfur standards (79 Federal Register 23618). In 2017, the average sulfur content in most gasoline was reduced from 30 to 10 ppm as part of a larger strategy to reduce emissions from both new and existing passenger vehicles. The sulfur standards were issued along the new "Tier 3" emissions standards for new vehicles.

Benzene, toluene, and xylenes have very high octane rating, but their content is usually limited by legislation because they are highly reactive and can cause health issues. Benzene is a known carcinogen and toxic chemical and is often required to be limited to under 1%–2% by weight. In the United States, aromatics are limited to 25 vol%, with benzene limited to 1.3 vol% (0.95 vol% on average).

Olefin molecules are prized for their high octane rating but can often leave deposits on engine surfaces which build up and prevent efficient burning. Many parts of the world limit olefin content for this reason. In Europe, for example, olefins are limited to 18 vol%.

Metals in gasoline are sometimes limited by regulation. Lead alkyls (tetramethyl lead, tetraethyl lead, and their mixtures) were discovered in the early 1910s to be an excellent octane number enhancing additive, but once the harmful impact of lead on humans was discovered, regulations prohibiting the addition of lead to gasoline were promulgated. The intentional addition of lead or phosphorus compounds to unleaded fuel for street/highway vehicles is not permitted by the EPA, and regulations limit their maximum concentrations to 0.05 g lead/gal. The regulations do not prohibit the use of lead additives in fuel for aircraft, racing cars, and nonroad engines such as farm equipment and marine engines. In some parts of the world, lead is still used as a gasoline additive.

Corrosion

Gasoline and gasoline-ethanol blends are shipped, stored, and delivered in metallic pipes and tanks, and the fuel handling system in vehicles and engine cylinders are mostly made of metals. Thus, the ability of fuel to be noncorrosive to metals is an important property. Sulfur compounds in fuel can corrode and tarnish silver and copper alloys, and because some fuel tank level sensing components are made from silver alloys, there are ASTM specifications for the corrosion propensity of gasoline to silver (ASTM D7667, ASTM D7671) and copper (ASTM D130). Corrosion tests expose metallic strips to fuel for a specific time under controlled conditions and are visually inspected and compared to scales.

Table 3
ASTM Gasoline Requirements for All Volatility Classes

Property	Limit	Test method
Lead content, max g/L (mg/US gal)		
Unleaded	0.013 (0.05)	D3237 or D5059
Leaded	1.1 (4.2)	D3341 or D5059
Sulfur, max, mass %		D1266, D2622, D3120, D5453, D6920
Unleaded	0.0080	
Leaded	0.15	
Copper strip corrosion, max	No. 1	D130
Silver strip corrosion, max	No. 1	D7667 or D7671
Solvent-washed gum, mg/100 mL, max	5	D381
Oxidation stability, minimum, minutes	240	D525

Source: ASTM D4814-14b.

In the copper strip corrosion test, a standard polished copper strip is dipped into the liquid and kept at constant temperature of 30°C. After 3 h, the copper strip is removed and compared to the brightness of standard copper strips to indicate its degree of corrosion as ASTM 1, 1a, 1b, 2, 2a, etc. The desired specification is that the surface brightness of the strip should not be worse than ASTM 1.

Storage and stability

It is common to leave gasoline in fuel tanks of recreational and lawn equipment over the winter, but because gasoline can oxidize over time by reacting with oxygen in the air, the reacting materials can clog engine fuel distribution equipment and render an engine hard or impossible to start. Higher temperatures and metals such as copper, zinc, or iron can accelerate these reactions and form higher boiling materials called gums.

To test the oxidation resistance of gasoline, there is a specification limit described by ASTM D525, *Oxidation Stability of Gasoline (Induction Period Method)*. In this test method, 50 mL of gasoline is placed in a glass container and oxygen is introduced to a pressure of 700 kPa (100 psi). The charged pressure vessel is placed in a 100°C (212°F) bath and the pressure continuously monitored. The break point of the fuel is determined to be the point just prior to a specified drop of pressure within a 15-minute interval. The number of minutes to the break point is reported, and according to ASTM D4814, 240 min minimum is the limit.

Once gasoline has begun to oxidize, there is a test to detect the resulting gum content, ASTM D381, *Gum Content in Fuels by Jet Evaporation*. Gums are tested by evaporating a sample of gasoline, leaving residue. The residue is washed with a nonpolar solvent such as n-heptane, and the washed material that is left is called solvent-washed gums. The residue is weighed before and after extraction with n-heptane. ASTM D4814 has a maximum solvent-washed gum limit of 5 mg/100 mL.

Energy content

The heating value of gasoline is measured by its heat of combustion, that is, the heat released when a known quantity of fuel is burned under specific conditions. In the United States, heating value is expressed as British thermal units (Btus) per pound or per gallon at 60°F. The SI metric units are joules per kilogram or per cubic meter at 15°C.

For gross (or higher) heating value, the water produced by the combustion is assumed to be recondensed to liquid. For the lower (or net) heating value, the water is assumed to remain as gas. Since all engines exhaust water as a gas, net heating value is used for comparing fuels, while assays typically report gross heating value.

Additives and blending components

Additives improve or enhance a specific property and are typically introduced at a very low level as parts per million (ppm). Oxidization stability improvers, engine-cleaning detergents, and corrosion inhibitors are examples of fuel additives. Blending components such as MTBE and alcohols are added at the percentage level.

All motor gasoline sold in the United States must contain an additive that provides a minimum level of deposit control performance. Engine deposits may form at various locations (e.g., intake valves, injector, and combustion chamber) for various reasons (e.g., gasoline stability) and will have different impacts on performance. Additive manufacturers are required to obtain EPA certification for their additives on the effectiveness in specified fuel injector and intake valve keep-clean tests.

In the United States, the European Union, and other countries, there are requirements for blending renewable components into transportation fuels. In the United States, the EPA created the Renewable Fuel Standard to provide the framework for this mandate and has required increasing amounts of renewables to be included in US fuel. The programs goal was to encourage the use of corn-based ethanol and biodiesel and to provide incentives for cellulosic biofuels. Almost all US gasoline contains 10% ethanol to meet the federal specification for use of renewables.

In the 1990s, the EPA required a special type of gasoline in areas that did not meet ambient ozone standards. The gasoline is known as reformulated gasoline, and one of the requirements was that it contains an oxygenate, such as MTBE or ethanol. The use of MTBE in US gasoline has ceased due to many state laws and bans, but is still used in Europe and elsewhere.

Aviation gasoline fuel specifications

The first engines used to power flight were automotive engines fueled with gasoline. As aviation engines and fuels improved with the commercial jet industry in the 1950 and 60s, wide-cut and kerosene-type jet fuels were developed. Today, aviation gasoline (often abbreviated avgas) is used mainly by small airplanes and light helicopters and by aircraft powered by piston engines. The two primary physical properties of avgas volatility are vapor pressure and distillation, and two specifications cover avgas, ASTM D910 the UK Ministry of Defense DEF STAN 90–91.

The ASTM D910 avgas specification is very tight and specific, and some special components such as lead and dye are required to be manufactured outside the refinery and blended to make the finished product.

ASTM D910 requirements are shown in Table 4. The boiling range runs from about 30 to 150°C, which confines the hydrocarbons to the C5 to C10 range. The constraints imposed by the distillation range, freezing point, vapor pressure, and antiknock properties significantly limit the hydrocarbon molecules that make up avgas.

Today, there is only one grade available, Grade 100LL, which refers to a 100-octane avgas with a low lead content, and allows a maximum of 0.53 mL of tetraethyl lead per liter. The fuel is designated "low lead" because the previous fuel for 100/130 octane contained twice as much lead. The lead in aviation gasoline is considered essential to flight safety.

The primary function of aviation gasoline is to power an aircraft, and so energy content is a key performance property. Knock resistance, volatility, fluidity, stability, and non-corrosivity are also important properties. Transportation fuels begin to freeze when cooled to a low enough temperature, but must retain its fluidity at altitude and on the ground at locations subject to cold weather.

The development of high performance aircraft engines in the 1930 and 40s created the need for higher octane avgas. The alkylation process was developed specifically to produce an avgas component with high antiknock ratings. Grades 100 and 100LL avgas are based on alkylate and toluene is often added to meet the rich-mixture antiknock requirement. Butane or isopentane are typically added to meet the minimum vapor pressure requirement. Tetraethyl lead, ethylene dibromide, and dye additives are also employed.

Table 4
ASTM Grade 100LL Avgas Requirements

Property	ASTM D910	Test method
Knock rating, lean mixture		D2700
Motor method octane number, min	99.5	
Knock rating, rich mixture		D909
Supercharge rating		
Performance number, min	130.0	
Tetraethyl lead concentration		D3341 or D5059
mL TEL/L, max	0.53	
g Pb/L, max	0.56	
Color	Blue	D2392
Dye content		
Blue dye, mg/L max	2.7	
Yellow dye, mg/L max	None	
Red dye, mg/L max	None	
Density at 15°C, kg/m^3	Report	D1298 or D4052
Distillation, °C		D86
Initial boiling point	Report	
Volume percent evaporated (VPE)		
10%, max	75	
40%, min	75	
50%, max	105	
90%, max	135	
Final boiling point	170	
Sum of temperatures at 10 and 50 VPE, min	135	D86
Distillation yields, % vol		

Table 4
ASTM Grade 100LL Avgas Requirements (continued)

Property	ASTM D910	Test method
Recovery, min	97	
Residue, max	1.5	
Loss, max	1.5	
Vapor pressure at 38°C, kPa		D323, D5190 or D5191
Min	38.0	
Max	49.0	
Freezing point, °C, max	−58	D2386
Sulfur content, % mass, max	0.05	D1266 or D2622
Net heat of combustion, MJ/kg, min	43.5	D4529 or D3338
Corrosion, copper strip, 2 h at 100°C, max	No. 1	D130
Oxidation stability, after 5 h aging, mg/100 mL		D873
Potential gum, max	6	
Lead precipitate, max	3	
Water reaction, volume change, mL, max	±2	D1094

Source: ASTM D910.

Jet fuels specifications

Jet fuel has about the same boiling range as ordinary kerosene but is more highly refined to satisfy performance requirements. The main issues of concern in jet fuel performance is low temperature behavior, smoke formation, thermal stability, and lubricity. In a jet engine, combustion is continuous and therefore combustion timing issues (i.e., antiknock properties) are not relevant. Volatility specifications of jet fuel are important, but because other specifications are more stringent (e.g., freeze point, flash point), the distillation requirements are usually readily satisfied and are used mostly as a quality control parameter. The operation of an aircraft turbine engine dictates some but not all of the performance requirements of jet fuel.

From the beginning of powered flight in 1903 until the end of World War II, there have been tremendous advances in aircraft engines. The first planes modified automotive internal combustion engines and used the same motor gasolines as fuel. The development of the turbine engine began in the 1930s in Germany and in Britain, and jet powered aircraft flew late in World War II. Illuminating kerosene produced for wick lamps was used to power the first turbine engines.

When the commercial jet industry was developing in the 1950s, kerosene-type fuel (Jet A and Jet A-1) was chosen as having the best combination of properties, although wide-cut fuel (Jet B) is still used in some parts of Canada and Alaska because it is well suited to cold climates. Jet A-1 has a slightly lower freezing point specification (−47°C) compared to Jet A (−40°C), which make it more suitable for long international flights, especially on polar routes during the winter, but is otherwise almost identical to Jet A (Table 5).

Specifications

ASTM D1655, *Standard Specification for Aviation Turbine Fuels*, include specifications for two kerosene-type jet fuels, Jet A and Jet A-1, and the UK Technical Authority for Defense Standard 91-91 maintains a specification for Jet A-1 that is similar to the ASTM specification ASTM D6615 that covers wide-cut jet fuel (Jet B). The standard specification for jet fuel has evolved primarily as a performance specification rather than as a compositional specification.

Table 5 summarizes the main requirements of commercial jet fuels. There are many footnotes to the requirements that are not listed and readers should consult the current version of the ASTM D1655 standards for more detail. ASTM standards include a suffix indicating the year of the last revision, e.g., D1655-15 was last revised in 2015. Military jet fuel specifications are maintained separate from civilian use because of operational and logistical differences, and the additional demands of high performance jet fighter engines. The US military use kerosene-type fuels JP-5 (Navy) and JP-8 (Air Force).

Table 5
ASTM Requirements of Aviation Turbine Fuels

Property		Jet A or Jet A-1
Composition		
Acidity total mg KOH/g	Max	0.10
1. Aromatics, percent by volume	Max	25
2. Aromatics, percent by volume	Max	26.5
Sulfur, mercaptan percent by mass	Max	0.003
Sulfur, total percent by mass	Max	0.30
Volatility		
Distillation temperature, °C		
10% recovered, temperature	Max	205
50% recovered, temperature		Report
90% recovered, temperature		Report
Final boiling point, temperature	Max	300
Distillation residue, %	Max	1.5
Distillation loss, %	Max	1.5
Flash point, °C	Min	38
Density at 15°C, kg/m^3		775 to 840
Fluidity		
Freezing point, °C	Max	−40 Jet A −47 Jet A-1
Viscosity, −20°C, cSt	Max	8.0
Combustion		
Net heat of combustion, MJ/kg	Min	42.8
One of the following requirements shall be met:		
1. Smoke point, mm, or	Min	25.0
2. Smoke point, mm, and	Min	18.0
Naphthalenes, vol, %	Max	3.0
Corrosion		
Copper strip 2 h at 100°C	Max	No. 1
Thermal stability		
(2.5 h at control temperature of 260°C min)		
Filter pressure drop, mm Hg	Max	23
Tube rating—one of the following requirements shall be met:		
1. Annex A1 VTR, VTR color code	Less than	3, No peacock or abnormal color deposits
2. Annex A2 ITR or annex 3 ETR, nm avg. over area of 2.5 mm^2	Max	85
Contaminants		
Existing gum, mg/100 mL	Max	7
Microseparometer, rating		
Without electrical conductivity additive	Min	85
With electrical conductivity additive	Min	70

Source: ASTM D1655-15.

Fluidity

Jet fuel is exposed to very low temperatures, both at altitude and on the ground at locations subject to cold weather extremes, and the ability of jet fuel to flow is one of its most important specifications. If jet fuel cannot freely flow through its distribution system, the results can be devastating. Viscosity and freezing points are the physical properties used to characterize the fluidity of jet fuel.

Jet fuel is injected into the combustion chamber of the turbine engine through nozzles. The system is designed to produce a fine spray of fuel droplets that evaporate quickly as they mix with air before combustion. Fuel viscosity influences the spray pattern and droplet size, and if it is too high, the fluid flow will be reduced or even stop. For this reason, jet fuel standards place an upper limit on viscosity, specified as 8 cSt at −20°C and measured according to test method D445/IP 71. Viscosity also affects fuel pumpability over the operating temperature range.

When jet fuel is cooled, the hydrocarbon components with the highest freezing points solidify first, forming wax crystals, and then a slush of fuel and hydrocarbon crystals form. The freezing point of jet fuel is not the temperature hydrocarbon crystals first form, but rather, the temperature at which the crystals completely disappear after forming. The maximum freezing point temperature is specified at −47°C for Jet A-1 and −40°C for Jet A which is the primary distinguishing characteristic between the fuels. Freezing point limits higher molecular weight hydrocarbons that crystalize at low temperatures.

Volatility

Volatility is a fuel's tendency to vaporize and for jet fuel is measured by its distillation profile, vapor pressure, and flash point. A fuel must vaporize before it can burn, but high volatility can promote loss of light ends at altitude and in hot climate, which is undesirable. A more volatile fuel has a higher vapor pressure and lower distillation temperatures.

Kerosene-type jet fuel volatility is normally measured by flash point and the 10% recovery temperature position of the distillation curve. Wide-cut jet fuel has a sufficiently high volatility that the flash point is usually not even measured, and instead, volatility is evaluated using vapor pressure.

Volatility is one of the major differences between kerosene-type (Jet A, A-1) and wide-cut (Jet B) fuel. Kerosene-type jet fuel is relatively nonvolatile with an RVP of about 1 kPa (0.14 psi) compared to wide-cut jet fuel with an RVP as high as 21 kPa (3 psi). ASTM D1655 jet fuel specifications for Jet A and Jet A-1 does not have a requirement for vapor pressure, but D6615 specifies a maximum vapor pressure of 21 kPa for Jet B.

Flash point is a leading factor determining safety in fuel handling. The flash point is the lowest temperature at which the vapors above a flammable liquid will ignite on the application of an ignition source. The minimum flash point of Jet A and Jet A-1 jet fuels is 38°C (100°F). The US military requires a minimum flash point of 60°C (140°F) for JP-5.

The ease of evaporation at different temperatures are determined by distillation, but only the 10% distilled temperature and FBP are specified for Jet A and Jet A-1:

10% recovered, 205°C maximum
50% recovered, report
90% recovered, report
FBP, 300°C maximum.

The 10% distilled temperatures are limited to ensure easy starting and the FBP limit excludes heavier fractions that would be difficult to vaporize. Normally, if the freezing point specification is satisfied, the FBP will also be satisfied. For jet fuels, the distillation thresholds are normally of secondary importance and frequently used as a quality control test, to show that there has not been any contamination with heavier boiling materials such as diesel fuel or lighter boiling materials such as gasoline.

Stability

A stable fuel is one whose properties remain unchanged over time (storage stability) or when exposed to high temperature (thermal stability). Storage stability is usually not a problem because most jet fuel is used within weeks or months of its manufacture. Thermal stability on the other hand is an important property because jet fuel serves as both the source of power and as a heat exchange medium in the engine and airframe. Jet fuel is used to remove heat from engine oil, hydraulic fluid, and air conditioning equipment, and because of its exposure to high temperatures, oxidation and polymerization reactions often lead to fuel deposits.

Commercial jet fuels are designed to be thermally stable at temperatures as high as 163°C (325°F). Jet fuel instability involves chemical reactions and products that remain dissolved in the jet fuel and attack and shorten the life of fuel system elastomers, and result in the formation of gums and particulates which may clog fuel filters and deposit on fuel systems restricting flow.

Heat content

The minimum heat content of jet fuel is set at a net heat of combustion of 42.8 MJ/kg. It may come as a surprise that the energy content limit of jet fuel is not more dominant or a significant concern in jet fuel specifications, but this is because the molecular makeup of the jet fuel is such that the hydrogen to carbon ratio is high enough that this specification is almost always met. It would be hard for jet fuel to fail its density specification.

The net heat of combustion of a fuel is measured in the laboratory using ASTM D4809 or IP 12, but the test is time consuming and requires precise thermometry and calorimetry, and is rarely used anymore. Instead, heat of combustion is usually estimated based on equations using density, distillation temperatures, aromatics content, and a sulfur content correction (Box 18.3).

The energy content and density of jet fuel is related. Density specifications for Jet A and Jet A-1 are allowed to range from 775 to 840 kg/m^3 at 15°C, JP-5 range from 38 to 47°API, and JP-8 range from 38 to 52°API. Generally, less dense fuels have a higher gravimetric energy content, and more dense fuel have a higher volumetric energy content. Which is preferred for aviation fuel, a higher density fuel with a higher volumetric energy content, or a lower density fuel with a higher gravimetric energy content? Weight and volume are primary constraints in aircraft design, but because of the fixed volume restriction of fuel storage tanks, high volumetric energy content is generally preferred for both military and civilian airplanes.

Combustion characteristics

The combustion characteristics of jet fuels are described by the net heat of combustion, smoke point, and naphthalenes content. Jet fuels are burned by injecting liquid fuel into a rapidly flowing stream of hot air. The fuel is vaporized and burned in a primary zone and the product gases are diluted with excess air to lower their temperature to a safe operating level for the turbine.

Fuel combustion characteristics relating to soot formation are emphasized in specifications. Small carbonaceous particles are formed early in the combustion process, and if these particles are not completely consumed by the flame, they can be harmful if they impinge on turbine blades and stators, causing erosion, or they may create hot spots that can lead to cracks and engine failures. Fuels with high aromatics content, and especially fuels with high naphthalene content, form more of these carbonaceous particles and thus specifications on their content are controlled. Jet A and Jet A-1 specifications set maximum aromatics content at 25 vol% using D1319 or IP156 test methods.

Smoke point provides an indication of the relative smoke-producing properties of jet fuels and is related to hydrocarbon composition. Generally, the more aromatic the jet fuel, the smokier the flame. A high smoke point indicates a fuel with low smoke-producing tendency. Smoke point is correlated with flame radiation, hot parts life, engine soot, and smoke formation. As the smoke point goes up, part life increases and sometimes significantly. ASTM D1322, *Smoke Point of Kerosine and Aviation Turbine Fuel*, is based on burning a sample in a wick-fed lamp. The smoke point is the maximum height of flame that can be achieved without smoking. Military jet fuels require a smoke point of 19 mm minimum for JP-5 and JP-8.

The smoke point specification for Jet A is 25 mm minimum, although a lower smoke point of 18 mm minimum is permitted provided naphthalene content is less than 3 vol %. In ASTM D1840, *Naphthalene Hydrocarbons in Aviation Turbine Fuels by Ultraviolet Spectrophotometry*, a sample is dissolved in isooctane at a known concentration, and the absorbance of the solution at 285 nm is measured and used to calculate the naphthalene content.

Composition

The composition requirements of jet fuel are mostly related to performance specifications. In Jet A and Jet A-1, total aromatics content and naphthalenes content is limited to 25 vol% and 3 vol%, respectively, to constrain smoke and carbon/soot deposition, while sulfur content is limited to 0.30% mass to minimize the sulfur oxides formed during combustion which may be corrosive to turbine metal parts. Mercaptans are limited to 0.003% mass to reduce chemical reactions with certain elastomers.

There are several test methods to determine the sulfur content and mercaptans sulfur in petroleum products: D1266 (lamp method), D2622 (wavelength-dispersive X-ray spectrometry), D4294 (energy-dispersive X-ray spectrometry), and D5453 (ultraviolet fluorescence) for sulfur, and D3227 (potentiometric method) for mercaptans sulfur.

Some petroleum products are treated with mineral acid or caustic, or both, as part of refining, and some crude oils are especially acidic. Any residual mineral acid or caustic is undesirable and a determination of total acidity confirms this when inspecting fuel. The acidity of Jet A and Jet B is limited to 0.10 mg KOH/g (ASTM D1655, D6615), while the acidity of Jet A-1 is limited to 0.015 mg KOH/g (DEF STAN 91-91). In ASTM D1655, acidity is measured by test method D3342/IP 353, *Acidity in Aviation Turbine Fuel*, where a weighed amount of sample is dissolved in titration solvent and titrated with potassium hydroxide to a calorimetric endpoint.

Lubricity

The fuel system components and control units in aircraft rely on the fuel to lubricate their moving parts. The effectiveness of a jet fuel as a lubricant in such equipment is referred to as its lubricity.

Fuel viscosity alone may not prevent metal-to-metal contact, and a layer of adsorbed polar materials several microns thick may be required to prevent excessive wear. Straight run jet fuel generally has sufficient polar compounds to provide acceptable lubricity, but severely hydrocracked fuel from a single refinery may not, in which case an additive may be added to improve lubricity. Corrosion inhibitor/

lubricity improver additives are routinely used to improve the lubricity of military fuels.

The boundary lubricity of a jet fuel cannot be predicted from bulk physical or chemical properties but can be measured in a test apparatus referred to as the Ball-on-Cylinder Lubricity Evaluator, BOCLE (ASTM D5001). In this test, a nonrotating steel ball is held against a cylindrical ring which is rotated while partially immersed in the fuel. At the end of the test, the ball is removed and examined for wear. The size of the wear scar is related to the lubricity of the fuel. Lower lubricity fuels give large BOCLE wear scar diameters. Obviously, BOCLE measures a very specific type of wear, and may not accurately model all types of wear that cause in-service problems, so other methods may also be utilized.

Corrosion

Jet fuel contacts a variety of materials during distribution and use, and it is essential that fuel not corrode any of these materials, especially those in aircraft fuel systems. A requirement that jet fuel must pass the copper strip corrosion test helps ensure that organic sulfur compounds will not corrode any copper or copper-based alloys in the fuel systems.

In ASTM D130/IP 154, *Detection of Copper Corrosion from Petroleum Product*, a polished copper strip is immersed in a sample for two hours at 100°C (212°F) and then removed and washed. The condition of the copper surface is visually evaluated by comparing it to standards. Corrosive compounds potentially present in jet fuel include organic acids and mercaptans and by-products of microbial growth.

Contaminants

A contaminate is any material that was originally not part of the fuel, and the general philosophy in all fuel specifications is that they should not be present, or present at very low levels. The general mantra is "keep it clean and keep it dry." Examples of contaminants include free water, solid materials (rust, dust, and sand), microbial debris, other fuels, unapproved additives, solvents, and surfactant materials.

The cleanliness of aviation turbine fuel is an essential performance requirement and is protected in part by allowing time for dirt and water to settle during fuel distribution and by the routine and frequent use of filtration systems. Particulates can plug fuel filters and increase pump wear. Water will freeze at low temperatures and may plug filters and impede fuel flow, and may facilitate corrosion of some metals and growth of microorganisms.

Gum is limited to 7 mg/100 mL as determined by ASTM D381. In ASTM D381, *Gum Content in Fuels by Jet Evaporation*, a measured quantity of fuel is transferred to a weighted beaker, placed in a heated bath, and evaporated under a flow of steam. The resulting residue is weighted and reported as existent gum.

The presence of surfactants can compromise the ability of fuel handling systems to remove dirt and water. For example, surfactants can reduce the particle size of suspended solids and water droplets, which slows removal by settling. Surfactants can disperse dirt and water, so they pass through filters, can adsorb on the surfaces of filters and interfere with water removal, and can also lift rust from surfaces. The presence of surfactants at the point of manufacture should meet a minimum microseparometer rating of 85, when electrical conductivity additives are not used, or a minimum of 70, if an electrical conductivity is used. The difference in the ratings with and without the electricity conductivity additive is because the additive contains a surfactant.

Additives

Additives are fuel-soluble chemicals added in small quantities to enhance or maintain properties important to fuel performance or fuel handling. Their function and chemistry is highly specialized, and they produce the desired effect in the ppm concentration range. Some additives are required to be added, some are optional, and others are approved for use by agreement between buyer and seller. Protocols exist for testing additives, but there may be unintended consequences when not used appropriately, sometimes referred to as the "dark side" of additives.

ASTM D1655 requirements specify that all additives used in aircraft fuel systems must be approved by the airframe and engine manufacturers, and may only be added within the concentration limit. Additives that are allowed in civil jet fuel include antioxidants, metal deactivator, fuel system icing inhibitor, electric conductivity improver, leak detection, biocides, and corrosion inhibitors/lubricity improvers.

Use of additives is the principal difference between commercial and military jet fuels. Jet A usually contains no additives except an antioxidant. US military jet fuels contain three or more additives. Jet A-1 is very similar to JP-8 but without the fuel system icing inhibitor, corrosion control additives, and static dissipater additive required in military contracts.

If an electrical conductivity additive is used, then it is necessary to measure the electrical conductivity, which is limited to 50 to 600 pS/m (ASTM D2624). Antioxidants and metal deactivator additives are used to prevent the formation of oxidation deposits in aircraft engine fuel systems, to counteract the catalytic effects of active metals, and to improve the oxidation stability of fuels in storage.

Diesel fuel specifications

Diesel fuel powers diesel engines and is used throughout the world for on- and off-road uses, farming, rail transportation, and marine shipping. Diesel fuel is a mixture of thousands of hydrocarbon compounds, most with carbon numbers between 10 and 22, and the relative proportion of the paraffinic, naphthenic, and aromatic classes is one of the factors that distinguish diesel fuels from one another. Crude oils provide a wide diesel cut, and diesel fuels are generally easier and cheaper to manufacture than gasoline. Although diesel fuel properties may be similar to gasoline, its effect on performance is likely to be different since diesel fuel is combusted in a manner specific to diesel engines, which differs markedly from gasoline engines.

Diesel engines are the workhorse of industry, favored for their high torque output, durability, exceptional fuel economy, and ability to provide power under a wide range of conditions. Diesel engines are dominant in trucking, construction, farming, and mining. They are used extensively for marine propulsion and in passenger vehicles in many regions of the world.

In the United States, there are seven grades of diesel fuel oils classified according to viscosity, sulfur content, and application (Table 6). A three-number designation (1-D, 2-D, and 4-D) refers to viscosity and engine applications. With increasing number corresponding to increasing viscosity. The Sxxx designation was first adopted in the ASTM D975-04 edition to distinguish grades by sulfur content in parts per million. S15 grades are commonly referred to as ultra-low sulfur diesel or ULSD with 15 ppm sulfur content. S500 grades correspond to low sulfur diesel, and S5000 grades correspond to regular diesel.

No. 1-D diesel is used in high-speed engines where there is a wide variation in speed and load. No. 2-D diesel fuel is used in high-speed engines where there is less variation in speed and load. In the United States, No. 2-D diesel fuel is the most common. No 4-D diesel is used in low or medium-speed diesel engines with predominately constant speed and load. There are four classes of marine distillate fuels, some of which contain small amounts of resid, and 15 marine residual fuels in which resid is the majority constituent. The shipping industry uses high viscosity residual fuel because they are less expensive than distillate fuels and safer to handle.

Specification

ASTM D975 requirements set the limits for the values of properties for seven grades of diesel fuel (Table 7). The specifications for each product are developed to ensure that it is suitable for its intended use to guarantee an acceptable performance for the majority of users. The standard test methods used to measure the values of the properties are also listed.

Diesel fuel specifications in the European Union are similar to those in the United States, but are not identical. For example, the minimum cetane number in Europe is 51 compared to 40 in the United States, and in Europe, a density range for diesel fuels is specified while in the United States, density limits are not constrained.

Cetane number

Cetane number is a measure of the ignition quality of the fuel and influences combustion roughness and emissions. A fuel with a high cetane number starts to burn shortly after it is injected into the cylinder and has a short ignition delay period, both of which are desirable characteristics in a diesel engine. High cetane number corresponds to fuels that auto-ignite readily.

ASTM D975 specifies a minimum cetane number requirement of 40 for the six grades of No. 1-D and No. 2-D, and 30 for No. 4-D. The minimum cetane number in Europe

Table 6
ASTM D975 Diesel Fuel Grades

Grade	Description
1-D S15 1-D S500 1-D S5000	Special purpose, light middle distillate fuel with 15 ppm, 500 ppm, and 5000 ppm sulfur (max) with higher volatility than 2D grades. Flash point requirement is 38°C (min).
2-D S15 2-D S500 2-D S5000	General purpose, middle distillate fuel with 15 ppm, 500 ppm, and 5000 ppm sulfur (max), suitable for use in applications with varying speed and load. Flash point requirement is 52°C (min).
4-D	High distillate fuel or a blend of distillate and residual oil for use in low or medium-speed diesel engines involving predominately constant speed and load.

Table 7
ASTM D975 Diesel Fuel Requirements

Property	ASTM Test method	Grade No. 1-D S500	No. 2-D S500	No. 4-D
Flash point, °C, min	D93	38	52	55
Water and sediment, % vol, max	D2709	0.05	0.05	
	D1796			0.50
Distillation				
1. Physical distillation	D86			
Distillation temperature, °C 90%, % vol				
min			282	
max		288	338	
2. Simulated distillation	D2887			
Distillation temperature, °C 90%, % vol				
min			300	
max		304	356	
Kinematic viscosity, cSt at 40°C	D445			
min		1.3	1.9	5.5
max		2.4	4.1	24.0
Ash % mass, max	D482	0.01	0.01	0.10
Sulfur, ppm (μg/g) max	D5453			
% mass, max	D2622	0.05	0.05	
% mass, max	D129			2.00
Copper strip corrosion rating max 2 h at 50°C	D130	No. 3	No. 3	
Cetane number, min	D613	40	40	30
a. Cetane index, min	D976-80	40	40	
b. Aromaticity, % vol, max	D1319	35	35	
Operability requirements				
Cloud point, °C, max	D2500			
or				
LTFT/CFPP, °C, max	D4539/S6371			
RCR on 10% residue % mass, max	D524	0.15	0.35	
Lubricity, HFRR at 60°C micron max	D6079	520	520	

Note: For distillation, cetane number, and operability one of the two requirements shall be met.
Source: ASTM D975.

is 51. Cetane number requirements depend on engine design, size, nature of speed and load variations, and on starting and atmospheric conditions. An alternative to minimum cetane number requirement was specified by EPA at 35 vol% maximum aromatics as a means to limit aromatics content of the fuel.

The cetane number of a diesel fuel oil is determined by comparing its combustion characteristics in a test engine with those for blends of reference fuels of known cetane number under standard operating conditions (ASTM D613), analogous to the manner octane ratings of gasoline are performed but with different reference fuels and test methods (ASTM D6890). A derived cetane number (DCN) is computed by injecting the sample into a heated, temperature-controlled constant volume chamber which has been previously charged with compressed air. Each injection produces a single-shot, compression ignition combustion cycle, whereby the ignition delay is measured using sensors. This test is repeated several times, and the average ignition delay is correlated to a DCN.

Distillation

Fuel volatility requirements depend on engine design, size, nature of speed and load variations, and on starting and atmospheric conditions. The more volatile the fuel, the better the performance when loads and speeds have wide fluctuations, but the best fuel economy is generally obtained from heavier fuels because of greater energy densities.

The only distillation temperature requirement for 1-D grades specifies 90 %vol maximum recovery at 288°C. For No. 2-D grades, maximum and minimum thresholds at 90% vol recovery are specified as 282°C (min) and 338°C (max). Compared to gasoline, which has five distillation recovery temperatures specified, diesel fuels are relatively easy to satisfy and significantly less constrained.

The T50 of a typical automotive diesel fuel is between 230 and 280°C. A fuel with a T50 above 300°C will result in smoke formation in high-speed engines, while one below 200°C will not have adequate viscosity and energy content for acceptable engine performance. In general, there is a high correlation between viscosity and T50.

Fuel volatility has a major effect on engine performance. Similar to gasoline, too high a volatility can result in vapor lock in the fuel system and poor penetration of the fuel spray into the combustion chamber. Conversely, if volatility is too low the fuel spray will be poorly atomized and burned. In general, volatility should be as high as possible without adversely affecting other characteristics.

Flash point

Flash point indicates how easily a chemical will burn. Materials with higher flash points are less flammable or hazardous than chemicals with lower flash points. By definition, the flash point is the lowest temperature at which the vapor formed above a pool of liquid can be ignited in air at atmospheric pressure. At the flash point, the flame will stop as soon as the ignition source is removed. Small impurities of low-boiling flammable liquids or gases will lower the flash point, and interestingly, the flash point of a mixture may be lower than that of the individual components.

The fire point is the lowest temperature at which the sample will support combustion for at least 5 s. Above the fire point, the temperature is high enough to keep the flame alive with freshly formed vapors. Hence, the fire point is always a few degrees above the flash point. The fire point of a mixture depends on the vapor pressure of the substances. The spread between flash point and fire point depends on the volatility of the fuel. Higher flash points will usually give rise to a higher delta to the fire point.

Lubricity

Diesel fuel functions as a lubricant in most components of fuel injection equipment such as pumps and injectors. Lubricity is the ability to reduce friction between solid surfaces in relative motion. Two fuel characteristics affect equipment wear, low viscosity, and lack of sufficient quantities of trace components that have an affinity for metal surfaces.

The lubrication mechanism is a combination of hydrodynamic lubrication and boundary lubrication. In hydrodynamic lubrication, a layer of liquid prevents contact between surfaces, and viscosity is the key property. Boundary lubrication is important when high loads and/or low speed have squeezed out much of the liquid that provides hydrodynamic lubrication leaving small areas of the opposing surfaces in contact. Boundary lubricants are compounds that form a protective antiwear layer by adhering to the surface.

If fuel viscosity meets the requirements of a particular engine, a fuel film is maintained between the moving surfaces and prevents excessive metal-to-metal contact. ASTM D975 viscosity specifications for 1-D grades range between 1.3 cSt (min) and 2.4 cSt (max) at 40°C; 2-D grade viscosity limits are 1.9 cSt (min) and 4.1 cSt (max); 4-D grade viscosity ranges from 5.5 cSt (min) to 24 cSt (max). All tests are performed at 40°C using ASTM D445 test method.

Fuels that have been processed in a manner that results in a severe reduction of the trace levels of sulfur, nitrogen, and oxygen contaminants can reduce the beneficial impact of the surface active molecules that produce a protective film. Two test methods which evaluate lubricity are the BOCLE method (ASTM D6078), and the high frequency reciprocating rig (HFRR) method (ASTM D6079). In order to protect diesel fuel injection equipment, an HFRR wear scar diameter of 520 microns maximum is specified for all 1-D and 2-D grades.

Ash content

ASTM D975 limits the ash content of most diesel fuels to a maximum of 100 ppm (0.01% mass), but does not specifically address particle size. Inorganic particles contribute to ash content and cause wear of the fuel system and piston rings because they are manufactured to close tolerances to maintain the high pressures employed. In ASTM D482, *Ash from Petroleum Products*, the sample is placed in a crucible, ignited, and allowed to burn. The residue is heated further in a furnace, converts all the carbon to carbon dioxide, and all the mineral salts to oxides (ash). The ash is then cooled and weighted.

Carbon residue

No. 1-D fuel oils limit Ramsbottom carbon residue (RCR) to a maximum 0.15% mass; No. 2-D fuel oils limit RCR to maximum 0.35% mass. In ASTM D524, *Ramsbottom Carbon Residue of Petroleum Products*, the sample is distilled until 90% has been recovered, and then the residue is placed in a glass bulb and heated in a furnace to 550°C (1022°F). Most of the sample evaporates or decomposes under these conditions. The difference in the sample weights before and after heating yields the carbon residue mass percent.

Low temperature operability

All middle distillate fuels contain straight and branched chain hydrocarbons that become solid at ambient winter temperatures in colder geographic areas. Cloud point defines the temperature at which a cloud or haze of wax crystals appears (precipitate) in the fuel under prescribed test conditions.

The climate across the United States varies widely depending on location and time of year, and it is not possible or practical to recommend a single cloud point for all diesel fuels year-round. Therefore, suggested levels are based on seasonal and regional differences in the form of guidelines using the tenth-percentile minimum ambient air temperatures, defined as the lowest ambient temperature that will not go lower on average more than 10% of the time. The daily minimum ambient air temperature would on average not be expected to go below the monthly 10th percentile minimum ambient air temperature more than three days in a 30-day month.

The tests used to characterize a fuel's low temperature operability include cloud point (ASTM D2500), pour point (ASTM D97), cold filter plugging point (ASTM D6371), and low temperature flow test (ASTM D4539).

Stability

Diesel fuel stability is not a specification property within ASTM D975 but is presented as suggested practices in an accompanying appendix on storage and thermal stability. Unstable diesel fuels can form soluble gums or insoluble organic particulates. Both gums and particulates can contribute to injector deposits and clog fuel filters. Thermal (high temperature) stability is usually the primary issue, since diesel fuel must function effectively as a heat transfer fluid. Low sulfur diesel tends to be more stable than high sulfur diesel because hydrotreating tends to destroy the precursors of insoluble organic particulates. The formation of gums and particulates may occur gradually during long-term storage or quickly during fuel system recirculation.

Other middle distillate products

There are several other mid-distillate products similar to No. 1-D and No. 2-D diesel fuel. These include fuel oils, gas turbine fuel oils, aviation turbine fuels and kerosene. The specifications are nearly identical to No. 1-D and No. 2-D diesel fuels, except that a density pour point and limit are commonly specified, the threshold on kinematic viscosity is different, and the 10% volume recovery distillation is specified (Table 8).

Table 8
ASTM D396 Fuel Oils Requirements

Property	ASTM	Grade	
	Test method	No. 1-D	No. 2-D
		S500	S500
Flash point, °C (°F), min	D93—proc. A	38 (100)	38 (100)
Water and sediment, % vol, max	D2709	0.05	0.05
Distillation			
1. Physical distillation, °C (°F)	D86		
10% volume recovered, max		215 (420)	

(table continued on next page)

Table 8
ASTM D396 Fuel Oils Requirements (continued)

Property	ASTM Test method	Grade No. 1-D	No. 2-D
		S500	S500
90% volume recovered, min			282 (540)
90% volume recovered, max		288 (550)	338 (60)
2. Simulated distillation, °C (°F)	D2887		
10% volume recovered, max		195 (380)	
90% volume recovered, min			300 (570)
90% volume recovered, max		304 (580)	356 (670)
Kinematic viscosity, cSt at 40°C (104°F)	D445		
Min		1.3	1.9
Max		2.1	3.4
Ramsbottom carbon residue, max (% mass on 10% distillation residue)	D524	0.15	0.35
Sulfur, % mass, max	D2622 D129	0.05 0.50	0.05 0.50
Copper strip corrosion rating, max, after 3 h at 50°C	D130	No. 3	No. 3
Density at 15°C, kg/m^3	D1298		
Min			
Max		850	876
Pour point, °C (°F), max	D97	−18 (0)	−6 (21)

Note: For distillation one of the two requirements shall be met.
Source: ASTM D396.

PIPELINE PRODUCT SPECIFICATIONS

Pipeline property specifications vary with product, location, season, operator, year, and other factors. Here, pipeline specifications for two operators and two destinations are used to illustrate typical property specifications for products. For complete specifications and notes, see company website.

NGL products, propane

Operator: Phillips 66 Pipeline LLC.
Destinations: Blue Pipeline.
Effective Date: January 1, 2019.
Product: NGL Products, Propane.

Property	Test method	Units	Min	Max
Composition—butane and heavier	D2163	lv%		2.50
Composition—propane	D2163	lv%	90.00	
Composition—propylene	D2163	lv%		5.00
Copper strip corrosion	D1838	Rating		1
Fluorides	ASTM D7359	ppmw		1.0
Hydrogen sulfide	D2420			Neg
Moisture content—freeze valve	D2713		Pass	
Most current version	See note			
Other deleterious substances	Inspection		See note	
Residual matter—GC	D7756	ppmw		350
Residual matter—oil stain	D2158		Pass	
Residual matter—residue	D2158	ml		0.05
Specific gravity @ 60/60F	D1657		Report	
Specifications	See note			
Sulfide—carbonyl	D5623	ppmw		20
Sulfur	D2784 or D6667	ppmw		123
Vapor pressure @ 100°F	D1267, D2598, or D6897	psig		208
Volatile res—95% evap	D1837	Deg F		−37
Test methods	See note			

Source: Phillips 66 Pipeline LLC. For complete specification and notes, see company website.

NGL products, butane, normal butane

Operator: Phillips 66 Pipeline LLC.
Destinations: Blue Pipeline.
Effective Date: January 1, 2019.
Product: NGL Products, Butane, Normal Butane.

Property	Test method	Units	Min	Max
Composition—butadiene	D2163	lv%		0.01
Composition—hexanes and heavier	D2163	lv%		0.05
Composition—iso-butane	D2163	lv%		5.00
Composition—normal butane	D2163	lv%	95.00	
Composition—olefins	D2163	lv%		0.30
Composition—pentanes and heavier	D2163	lv%		1.50
Composition—propylene, propane, and lighter	D2163	lv%		0.50
Copper strip corrosion	D1838			1
Fluorides	UOP 619, ASTM D7359	ppmw		1.0
Hydrogen sulfide	D2420		Pass	
Moisture content—free water	Visual			None
Most current version	See note			
Other deleterious substances	Inspection		See note	
Oxygenates—IPA and heavier alcohols	ASTM D7423, UOP-845	ppmw		5.00
Oxygenates—methanol	ASTM D7423, UOP-845	ppmw		50.00
Oxygenates—MTBE, and other ethers	ASTM D7423, UOP-845	ppmw		2.00
Oxygenates—other oxygenates	ASTM D7423, UOP-845	ppmw		5.00
Oxygenates—total	ASTM D7423, UOP-845	ppmw		50.00
Specifications	See note			
Sulfur	D2784 or D6667	ppmw		93
Vapor pressure @ 100°F	D1267, D2598, or D6897	psig		50
Test methods	See note			

Source: Phillips 66 Pipeline LLC. For complete specification and notes, see company website.

NGL products, butane, refinery grade butane

Operator: Phillips 66 Pipeline LLC.
Destinations: Blue Pipeline.
Effective Date: January 1, 2019.
Product: NGL Products, Butane, Refinery Grade Butane (RGB).

Property	Test method	Units	Min	Max
Color	D2158			See note
Composition—normal butane	D2163	lv%	65.0	
Composition—propane	D2163	lv%		1.0
Composition—total olefins	D2163	ppm		4000
Copper strip corrosion	D1838			1
Fluorides	ASTM D7359	ppmw		Report
Hydrogen sulfide	D2420		Pass	
Moisture content—free water	Visual			None
Most current version	See note			
Other deleterious substances	Inspection		See note	
Specifications	See note			
Sulfur	D2784 or D6667	ppmw		160
Vapor pressure @ 100°F	D1267, D2598, or D6897	psig		43.0
Volatile res—95% evap	D1837 or D2163	Deg F		+36.0
Test methods	See note			

Source: Phillips 66 Pipeline LLC. For complete specification and notes, see company website.

NGL products, iso-butane

Operator: Phillips 66 Pipeline LLC.
Destinations: Blue Pipeline.
Effective Date: January 1, 2019.
Product: NGL Products, Iso-Butane.

Property	Test method	Units	Min	Max
Composition—iso-butane	D2163	lv%	95.00	
Composition—normal butane and heavier	D2163	lv%		5.00
Composition—propane and lighter	D2163	lv%		3.00
Copper strip corrosion	D1838			1
Fluorides	ASTM D7359	ppmw		Report
Hydrogen sulfide	D2420		Pass	
Moisture content—free water	Visual			None
Most current version	See note			
Other deleterious substances	Inspection		See note	
Specifications	See note			
Sulfur	D2784 or D5623	ppmw		93
Vapor pressure @ 100°F	D1267, D2598, or D6897	psig	56.00	60.00
Volatile res—95% evap	D1837	Deg F		+16
Test methods	See note			

Source: Phillips 66 Pipeline LLC. For complete specification and notes, see company website.

Gasoline, conventional, 85.5 octane (no ethanol)

Operator: Yellowstone Pipeline Company.
Destinations: Montana.
Effective Date: 11/16/2020.
Product: Gasoline, Conventional, 85.5 octane (no ethanol).

Property	Test method	Units	Min	Max
Additives	General note			
API gravity (60°F)	D287, D1298, D4052	API	Report	
Appearance	Visual		Clear and Bright	
Basicity	D1093, modified, see note		Report	
Benzene	D3606	Vol%		4.0
Color	Visual		Undyed	
Copper strip corrosion	D130 3 Hr @ 122°F	Rating		1
Ethanol blends	General note			
Lead (Pb)	D3237, D5059	gPb/gal		0.01
Mercaptan sulfur	D3227	Wt%		0.002
Mercaptan sulfur	D4952	Rating	Sweet	
NACE rust test	NACE rust TM0172	Rating	B+	
Octane, (R + M)/2	D2699 and 2700 or D2885		85.5	
Octane, motor	D2700 or D2885		80.5	
Octane, research	D2699 or D2885		Report	
Oxidation stability	D525	minutes	240	
Oxygenates prohibited	See note			
Phosphorus	D3231	g/gal		0.003
Product description	See note			
Referee methods	See note			
Silver strip corrosion	D7667, D7671	Rating		1
Solvent washed gum	D381	mg/100 mL		5
Sulfur	D2622, D5453, D7039	ppm		80
Volatility and distillation	See D4814		See table	

Source: Yellowstone Pipeline Company. For complete specification and notes, see company website.

Gasoline, conventional, 87 octane (no ethanol)

Operator: Yellowstone Pipeline Company.
Destinations: Montana and Eastern Washington.
Effective Date: 11/16/2020.
Product: Gasoline, Conventional, 87 octane (no ethanol).

Property	Test method	Units	Min	Max
Additives	General note			
API gravity (60°F)	D287, D1298, D4052	API	Report	
Appearance	Visual		Clear and Bright	
Basicity	D1093, modified, see note		Report	
Benzene	D3606	Vol%		4.0
Color	Visual		Undyed	
Copper strip corrosion	D130 3 Hr @ 122°F	Rating		1
Ethanol blends	General note			
Lead (Pb)	D3237, D5059	gPb/gal		0.01
Mercaptan sulfur	D3227	Wt%		0.002
Mercaptan sulfur	D4952	Rating	Sweet	
NACE rust test	NACE rust TM0172	Rating	B+	
Octane, (R + M)/2	D2699 and 2700 or D2885		87.0	
Octane, motor	D2700 or D2885		82.0	
Octane, research	D2699 or D2885		Report	
Oxidation stability	D525	minutes	240	
Oxygenates prohibited	See note			
Phosphorus	D3231	g/gal		0.003
Product description	See note			
Referee methods	See note			
Silver strip corrosion	D7667, D7671	Rating		1
Solvent washed gum	D381	mg/100 mL		5
Sulfur	D2622, D5453, D7039	ppm		80
Volatility and distillation	See D4814		See table	

Source: Yellowstone Pipeline Company. For complete specification and notes, see company website.

Gasoline, subgrade, 87 octane after 10% ETOH addition

Operator: Yellowstone Pipeline Company.
Destinations: Montana and Eastern Washington.
Effective Date: 11/16/2020.
Product: Gasoline, Subgrade, 87 octane after 10% ETOH addition.

Property	Test method	Units	Min	Max
Additives	General note			
API gravity (60°F)	D287, D1298, D4052	API	Report	
Appearance	Visual		Clear and Bright	
Basicity	D1093, modified, see note		Report	
Benzene	D3606	Vol%		4.0
Color	Visual		Undyed	
Copper strip corrosion	D130 3 Hr @ 122°F	Rating		1
Ethanol blends	General note			
Lead (Pb)	D3237, D5059	gPb/gal		0.01
Mercaptan sulfur	D3227	Wt%		0.002
Mercaptan sulfur	D4952	Rating	Sweet	
NACE rust test	NACE rust TM0172	Rating	B+	
Octane, E10, (R + M)/2	D2699 and 2700		87.0	
Octane, E10, motor	D2700 or D2885		82.0	
Octane, NEAT, (R + M)/2	D2699 and 2700		83.0	
Octane, NEAT, motor	D2700 or D2885		Report	
Octane, research	D2699 or D2885		Report	
Oxidation stability	D525	minutes	240	
Oxygenates prohibited	See note			
Phosphorus	D3231	g/gal		0.003
Product description	See note			
Referee methods	See note			
Silver strip corrosion	D7667, D7671	Rating		1
Solvent washed gum	D381	mg/100 mL		5
Sulfur	D2622, D5453, D7039	ppm		80
Volatility and distillation	See D4814		See table	

Source: Yellowstone Pipeline Company. For complete specification and notes, see company website.

Distillate, #1 diesel fuel, ultra-low sulfur (15 ppm max)

Operator: Yellowstone Pipeline Company.
Destinations: For CHS Missoula and XOM Helena Destinations Only.
Effective Date: 11/16/2020.
Product: Distillate, #1 Diesel Fuel, Ultra-Low Sulfur (15 ppm max), (does NOT meet Jet A specifications).

	Test method	Units	Min	Max
Additives	General note			
API gravity (60°F)	D287, D1298, D4052	API	Report	
Appearance	Visual		Clear and Bright	
Ash	D482	Wt%		0.01
Carbon res 10% Btms	D524	Wt%		0.15
Cetane by DSC	D6890, D7170, D7668		40	
Cetane Ind-2 or arom	D1319	Vol%		35.0
Cetane Ind-2 or arom	D976		40	
Cetane number or Ind-4	D4737, D613		40.0	
Color	Visual		Undyed	
Color, ASTM	D1500			2.5
Copper strip corrosion	D130 3 Hr @ 122°F	Rating		1
Dist 10 vol% rec	D2887	Deg F	Report	
Dist 10 vol% rec, corr	D86	Deg F	Report	
Dist 50 vol% rec	D2887	Deg F	Report	
Dist 50 vol% rec, corr	D86	Deg F	Report	
Dist 90 vol% rec	D2887	Deg F		579
Dist 90 vol% rec, corr	D86	Deg F		550
Dist end Pt	D2887	Deg F	Report	
Dist end Pt	D86	Deg F	Report	
Dist IBP	D2887	Deg F	Report	
Dist IBP, corr	D86	Deg F	Report	
Dist loss, corr	D86	Vol%	Report	
Dist residue	D86	Vol%	Report	
Flash Pt	D93	Deg F	110	
Haze @ 77°F	D4176 Procedure 2	Rating		2
NACE rust test	NACE rust TM0172	Rating	B+	
Product description	See note			
Referee methods	See note			
Sulfur @ origin	D2622, D5453, D7039	ppm		11
Viscosity @ 104°F	D445	cSt	1.3	2.4
Water and sed, total	D2709	Vol%		0.05

Source: Yellowstone Pipeline Company. For complete specification and notes, see company website.

Distillate, #1 diesel fuel, ultra-low sulfur (15 ppm max)

Operator: Yellowstone Pipeline Company.
Destinations: Montana and Eastern Washington.
Effective Date: 11/16/2020.
Product: Distillate, #1 Diesel Fuel, Ultra-Low Sulfur (15 ppm max), (also meet Jet A specifications).

Property	Test method	Units	Min	Max
Acid number	D3242	mg KOH/g		0.10
Additives	General note			
API gravity (60°F)	D287, D1298, D4052	API	37.5	50.5
Appearance	Visual		Clear and Bright	
Aromatics	D1319	Vol%		24.9
Ash	D482	Wt%		0.01
Carbon res 10% Btms	D524	Wt%		0.15
Cetane by DSC	D6890, D7170, D7668		40	
Cetane number or Ind-4	D4737, D613		40.0	
Color	Visual		Undyed	
Color, saybolt	D156, D6045		+18	
Copper strip corrosion	D130 2 Hr @ 212°F	Rating		1
Dist 10 vol% rec	D2887	Deg F		361
Dist 10 vol% rec, corr	D86	Deg F		397
Dist 50 vol% rec	D2887	Deg F	Report	
Dist 50 vol% rec, corr	D86	Deg F	Report	
Dist 90 vol% rec	D2887	Deg F		579
Dist 90 vol% rec, corr	D86	Deg F		550
Dist end Pt	D86	Deg F		566
Dist end Pt, corr	D2887	Deg F		638
Dist IBP	D2887	Deg F	Report	
Dist IBP, corr	D86	Deg F	Report	
Dist loss, corr	D86	Vol%		1.5
Dist residue	D86	Vol%		1.5
Existent gum	D381	mg/100 mL		6
Flash Pt	D56	Deg F	110	
Freeze Pt	D2386, D5972	Deg C		−41
Haze @ 77°F	D4176 Procedure 2	Rating		2
JFTOT press drop	D3241 @ 260°C	mm Hg		25
JFTOT tube rating	D3241 @ 260°C	Rating		<3
Mercaptan sulfur	D3227	Wt%		0.003
Mercaptan sulfur	D4952	Rating	Sweet	
MSEP	D3948	Rating	85	

(table continued on next page)

Property	Test method	Units	Min	Max
Naphthalenes or smoke Pt	D1322	mm	19	
Naphthalenes or smoke Pt	D1840	Vol%		2.9
Net heat of combustion	D3338	BTU/lb	18,410	
Net heat of combustion	D3338	MJ/kg	42.9	
Net heat of combustion	D4809	BTU/lb	18,410	
Net heat of combustion	D4809	MJ/kg	42.9	
Particulates	D2276, D5452	mg/gal	Report	
Product description	See note			
Referee methods	See note			
Specification severity	See note			
Sulfur @ origin	D2622, D5453, D7039	ppm		11–13[a]
Test tolerances	See note			
Viscosity @ −4°F	D445	cSt		8.0
Viscosity @ 104°F	D445	cSt	1.3	2.4
Water and sed, total	D2709	Vol%		0.05

Note: a. Depending on location.
Source: Yellowstone Pipeline Company. For complete specification and notes, see company website.

Distillate, #2 diesel fuel, ultra-low sulfur (15 ppm max)

Operator: Yellowstone Pipeline Company.
Destinations: Montana and Eastern Washington.
Effective Date: 11/16/2020.
Product: Distillate, #2 Diesel Fuel, Ultra-Low Sulfur (15 ppm max), (also meet #2 Fuel Oil specifications).

Property	Test method	Units	Min	Max
Additives	General note			
API gravity (60°F)	D287, D1298, D4052	API	30.0	
Appearance	Visual		Clear and Bright	
Ash	D482	Wt%		0.01
Carbon res 10% Btms	D524	Wt%		0.35
Cetane by DSC	D6890, D7170, D7668		40	
Cetane Ind-2 or arom	D1319	Vol%		35.0
Cetane Ind-2 or arom	D976		40	
Cetane number or Ind-4	D4737, D613		40.0	
Cloud Pt	D2500, D5773	Deg F		6–16[a]
Color	Visual		Undyed	
Color, ASTM	D1500			2.5
Copper strip corrosion	D130 3 Hr @ 122 °F	Rating		1

(table continued on next page)

Property	Test method	Units	Min	Max
Dist 10 vol% rec	D2887	Deg F	Report	
Dist 10 vol% rec, corr	D86	Deg F	Report	
Dist 50 vol% rec	D2887	Deg F	Report	
Dist 50 vol% rec, corr	D86	Deg F	Report	
Dist 90 vol% rec	D2887	Deg F	572	672
Dist 90 vol% rec, corr	D86	Deg F	540	640
Dist end Pt	D2887	Deg F	Report	
Dist end Pt, corr	D86	Deg F	Report	
Dist IBP	D2887	Deg F	Report	
Dist IBP, corr	D86	Deg F	Report	
Flash Pt	D93	Deg F	130–135[b]	
Haze @ 77°F	D4176 Procedure 2	Rating		2
NACE rust test	NACE rust TM0172	Rating	B+	
Pour Pt	D5949	Deg F		−5–20[a]
Product description	See note			
Referee methods	See note			
Sulfur @ origin	D2622, D5453, D7039	ppm		11–13[b]
Thermal stability	D6468-90 min	% Refl	75–82[c]	
Thermal stability	DuPont 90 min @ 300°F	Pad rating		4
Viscosity @ 104°F	D445	cSt	1.9	3.4
Water and sed, total	D2709	Vol%		0.05

Note: a. Depending on season, b. Depending on location, c. Depending on test unit.

Source: Yellowstone Pipeline Company. For complete specification and notes, see company website.

23: Conversion Factors

Metric conversions—metric to English, English to metric 1148
Density conversions 1149
General liquid density nomograph 1149
Chart gives specific gravity/temperature relationship for petroleum oils 1151
Weight density and specific gravity of various liquids 1151
True vapor pressure of crude oil stocks with a Reid vapor pressure of 2–15 psi 1152
Low temperature vapor pressures for light hydrocarbons 1153
High temperature vapor pressures for light hydrocarbons 1154
Hydrocarbon gas viscosity 1155
Viscosity—equivalents of kinematic and Saybolt Universal Viscosity 1156
Viscosity—equivalents of kinematic and Saybolt Furol Viscosity at 122 °F and 210 °F 1156
Viscosity—general conversions 1157
ASTM standard viscosity temperature chart for liquid petroleum products 1159
Pressure conversion chart 1160
Energy conversion chart 1161
Flow conversion chart 1162
Petroleum fuels volume, weight and calorific value factors 1162
Crude oil approximate conversion factors 1163
Refined products approximate conversion factors 1164
Natural gas and LNG approximate conversion factors 1164
Units and calorific equivalents 1165
Weight of water per cubic foot at various temperatures 1165

Metric conversions—metric to English, English to metric

To convert from	To	Multiply by	To convert from	To	Multiply by
Length			**Length**		
mm	inches	.03937	inches	mm	25.40
cm	inches	.3937	inches	cm	2.540
meters	inches	39.37	inches	meters	.0254
meters	feet	3.281	feet	meters	.3048
meters	yards	1.0936	feet	km	.0003048
km	feet	3280.8	yards	meters	.9144
km	yards	1093.6	yards	km	.0009144
km	miles	.6214	miles	km	1.609
Area			**Area**		
sq mm	sq inches	.00155	sq inches	sq mm	645.2
sq cm	sq inches	.155	sq inches	sq cm	6.452
sq meters	sq feet	10.764	sq feet	sq meters	.09290
sq meters	sq yards	1.196	sq yards	sq meters	.8361
sq km	sq miles	.3861	sq miles	sq km	2.590
hectares	acres	2.471	acres	hectares	.4047
Volume			**Volume**		
cu cm	cu inches	.06102	cu inches	cu cm	16.387
cu cm	fl ounces	.03381	cu inches	liters	.01639
cu meters	cu feet	35.314	cu feet	cu meters	.02832
cu meters	cu yards	1.308	cu feet	liters	28.317
cu meters	US gal	264.2	cu yards	cu meters	.7646
liters	cu inches	61.023	fl ounces	cu cm	29.57
liters	cu feet	.03531	US gal	cu meters	.003785
liters	US gal	.2642	US gal	liters	3.785
Weight			**Weight**		
grams	grains	15.432	grains	grams	.0648
grams	ounces†	.0353	ounces†	grams	28.350
kg	ounces†	35.27	ounces†	kg	.02835
kg	pounds†	2.2046	pounds†	kg	.4536
kg	US tons	.001102	pounds†	tonnes	.000454
kg	long tons	.000984	US tons	kg	907.2
tonnes	pounds†	2204.6	US tons	tonnes	.9072
tonnes	US tons	1.1023	long tons	kg	1016
tonnes	long tons	.9842	long tons	tonnes	1.0160

†avoirdupois pounds and ounces.

Courtesy Cameron Hydraulic Data Book, *Ingersoll-Rand Company.*

To convert from	To	Multiply by	To convert from	To	Multiply by
Unit weight			**Unit weight**		
gr/sq cm	lb/sq in	.01422	lb/ft	kg/m	1.4881
gr/cu cm	lb/cu in	.0361	lb/sq in	gr/sq cm	70.31
kg/sq cm	lb/sq in	14.22	lb/sq in	kg/sq cm	.07031
kg/cu m	lb/cu ft	.0624	lb/cu in	gr/cu cm	27.68
kg/m	lb/ft	.6720	lb/cu ft	kg/cu m	16.018
Unit volume			**Unit volume**		
liters/min	US gpm	.2642	US gpm	liters/min	3.785
liters/min	cfm	.03531	US gpm	liters/hr	227.1
liters/hr	US gpm	.0044	US gpm	cu m/hr	.2271
cu m/min	cfm	35.314	cfm	liters/min	28.317
cu m/hr	cfm	.5886	cfm	cu m/min	.02832
cu m/hr	US gpm	4.4028	cfm	cu m/hr	1.6992
Power			**Power**		
watts	ft-lb/sec	.7376	ft-lb/sec	watts	1.356
watts	hp	.00134	hp	watts	745.7
kw	hp	1.3410	hp	kw	.7457
cheval-vap	hp	.9863	hp	cheval-vap	1.0139
Heat			**Heat**		
gr-cal	Btu	.003969	Btu	gr-cal	252
kg-cal	Btu	3.9693	Btu	kg-cal	.252
kg-cal/kg	Btu/lb	1.800	Btu/lb	kg-cal/kg	.5556
gr-cal/sq cm	Btu/sq ft	3.687	Btu/sq ft	gr-cal/sq cm	.2713
kg-cal/cu m	Btu/cu ft	.1124	Btu/cu ft	kg-cal/cu m	8.899
Work or energy			**Work or energy**		
joule	ft-lb	.7376	ft-lb	joule	1.356
meter-kg	ft-lb	7.2330	ft-lb	meter-kg	.1383
gr-cal	ft-lb	3.087	ft-lb	gr-cal	.3239
kg-cal	ft-lb	3087	ft-lb	kg-cal	.000324
hp-hr	ft-lb	1,980,000	ft-lb	hp-hr	5.05×10^{-7}
kwhr	ft-lb	2,655,000	ft-lb	kwhr	3.766×10^{-7}
Btu	ft-lb	778.0	ft-lb	Btu	.001285

Density conversions

Density Conversions

To convert the numerical value of a property expressed in one of the units in the left-hand column of the table to the numerical value expressed in one of the units in the top row of the table, multiply the former value by the factor in the block common to both units.

Units→ ↓	g/cu cm	g/ml	lb/cu in.	lb/cu ft	lb/gal (U.S.)
g/cu cm	1	1.000028	0.036128	62.428	8.3455
g/ml	0.99997	1	0.036126	62.427	8.3452
lb/cu in.	27.680	27.681	1	1,728	231
lb/cu ft	0.016018	0.016019	5.7870×10^{-4}	1	0.13368
lb/gal (U.S.)	0.11983	0.11983	4.3290×10^{-3}	7.4805	1

General liquid density nomograph

***X* and *Y* Values for Density Nomograph**

Compound	*X*	*Y*	Compound	*X*	*Y*	Compound	*X*	*Y*
Acetic acid	40.6	93.5	Ethyl chloride	42.7	62.4	Methyl sulfide	31.9	57.4
Acetone	26.1	47.8	Ethylene	17.0	3.5	*n*-Nonane	16.2	36.5
Acetonitrile	21.8	44.9	Ethyl ether	22.6	35.8	*n*-Octadecane	16.2	46.5
Acetylene	20.8	10.1	Ethyl formate	37.6	68.4	*n*-Octane	12.7	32.5
Ammonia	22.4	24.6	Ethyl propionate	32.1	63.9	*n*-Pentadecane	15.8	44.2
Isoamyl alcohol	20.5	52.0	Ethyl propyl ether	20.0	37.0	*n*-Pentane	12.6	22.6
Ammine	33.5	92.5	Ethyl auindo	25.7	55.3	*n*-Nonadecane	14.9	47.0
Benzene	32.7	63.0	Fluorobenzene	41.9	87.6	Isopentane	13.5	22.5
n-Butyric acid	31.3	78.7	*n*-Heptadecane	15.6	45.7	Phenol	36.7	103.8
Isobutane	13.7	16.5	*n*-Heptane	12.6	29.8	Phosphine	28.0	22.1
Isobutyric acid	31.5	75.9	*n*-Hexadecane	15.8	45.0	Propane	14.2	12.2
Carbon dioxide	78.6	45.4	*n*-Hexane	13.5	27.0	Propionic acid	35.0	83.5
Chlorobenzene	41.7	105.0	Methanethiol	37.3	59.5	Piperidene	27.5	60.0
Cyclohexane	19.6	44.0	Methyl acetate	40.1	70.3	Propionitrile	20.1	44.6
n-Decane	16.0	38.2	Methyl alcohol	25.8	49.1	Propyl acetate	33.0	65.5
n-Dedecane	14.3	41.4	Methyl n-butyrate	31.5	65.5	Propyl alcohol	23.8	50.8
Diethylamine	17.8	33.5	Methyl isobutyrate	33.0	64.1	Propyl formate	33.8	66.7
n-Elconane	14.8	47.5	Methyl chloride	52.3	62.9	*n*-Tetradecane	15.8	43.3
Ethane	10.8	4.4	Methyl ether	27.2	30.1	*n*-Tridecane	15.3	42.4
Ethanethiol	32.0	55.5	Methyl ethyl ether	25.0	34.4	Triethylamine	17.9	37.0
Ethyl acetate	35.0	95.0	Methyl formate	46.4	74.6	*n*-Undecane	14.4	39.2
Ethyl alcohol	24.2	48.6	Methyl propionate	36.5	68.3			

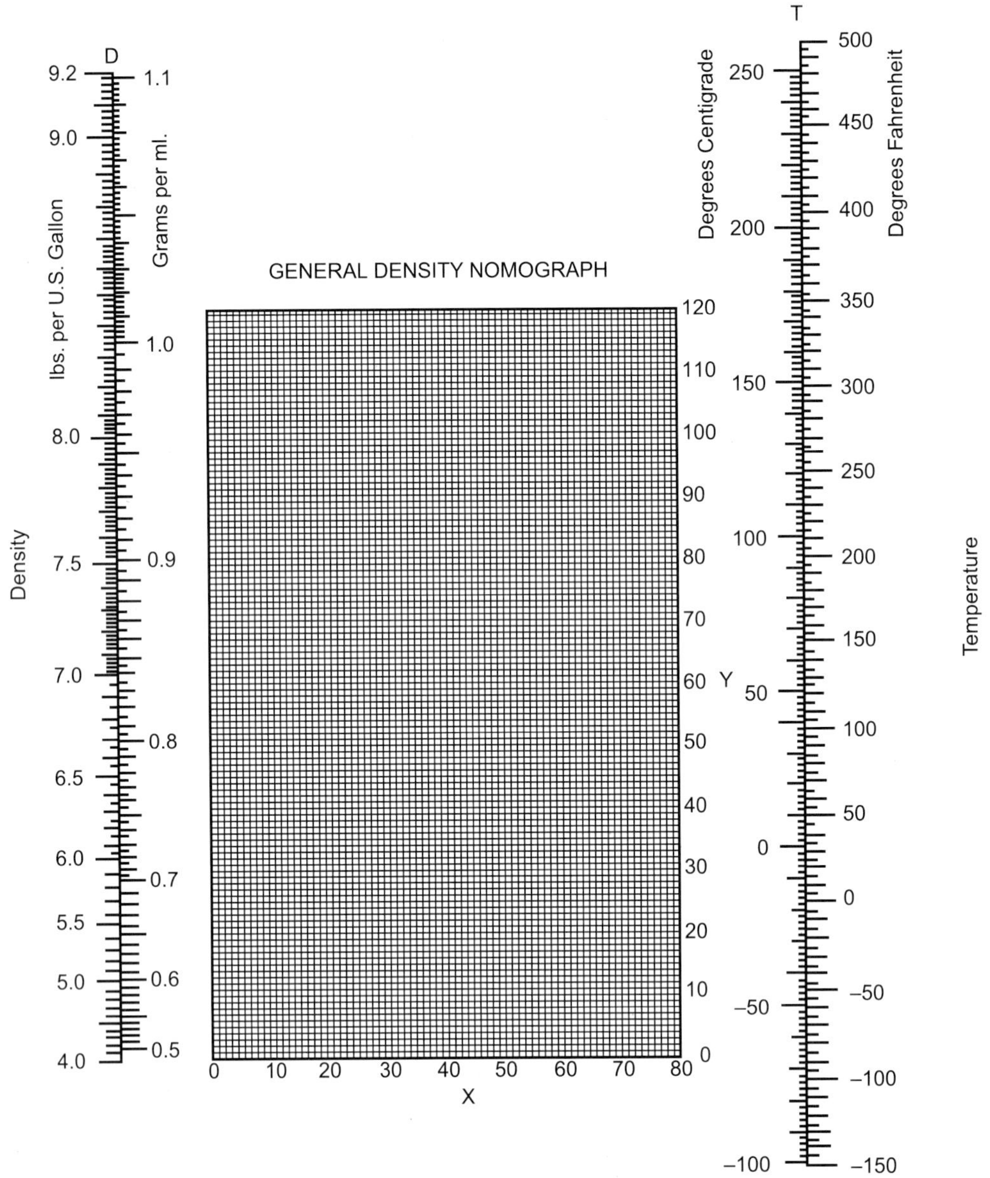
GENERAL DENSITY NOMOGRAPH
D
Density
lbs. per U.S. Gallon
Grams per ml.
9.2
9.0
8.0
7.5
7.0
6.5
6.0
5.5
5.0
4.0
1.1
1.0
0.9
0.8
0.7
0.6
0.5
X
0 10 20 30 40 50 60 70 80
Y
0 10 20 30 40 50 60 70 80 90 100 110 120
T
Temperature
Degrees Centigrade
250 200 150 100 50 0 −50 −100
Degrees Fahrenheit
500 450 400 350 300 250 200 150 100 50 0 −50 −100 −150

Chart gives specific gravity/temperature relationship for petroleum oils

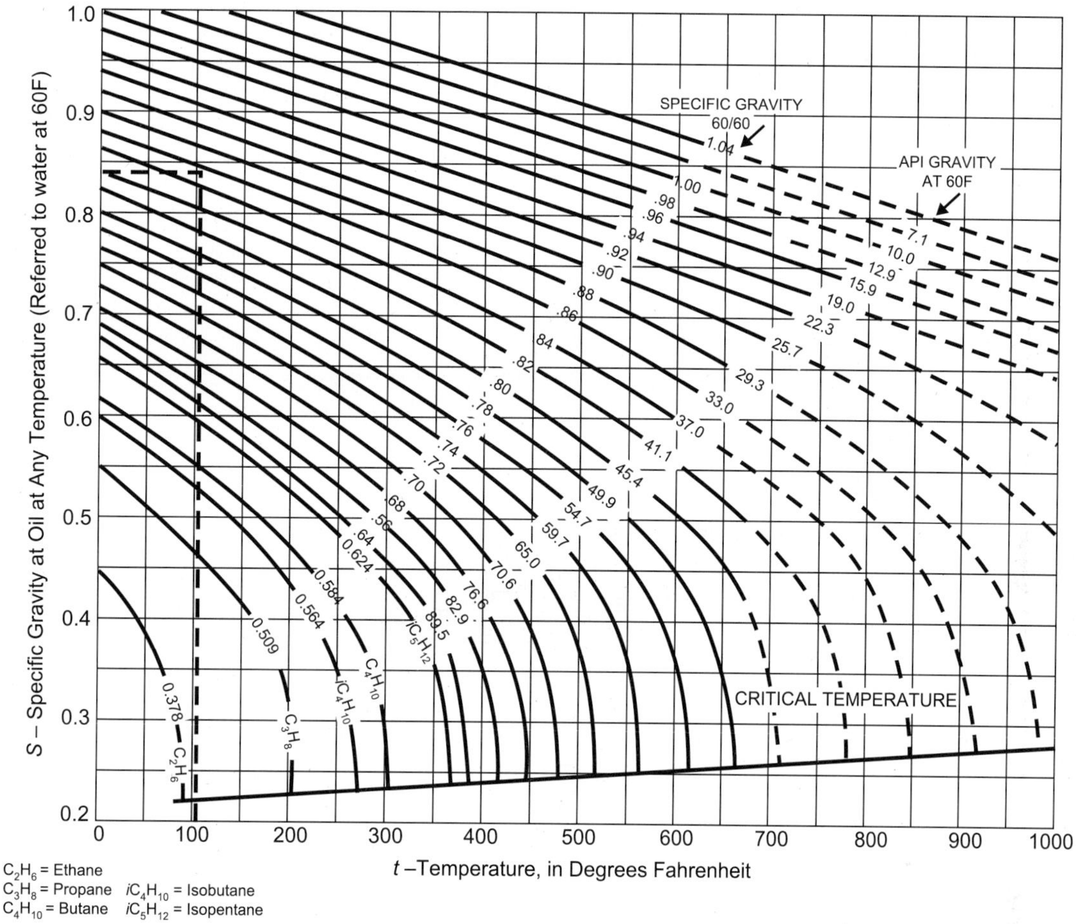

Example. The specific gravity of an oil at 60 °F is 0.85. The specific gravity at 100 °F = 0.83.

To find the weight density of a petroleum oil at its flowing temperature when the specific gravity at 60 °F/60 °F is known, multiply the specific gravity of the oil at flowing temperature by 62.4, the density of water at 60 °F.

Courtesy Crane Co. (Technical Paper #410).

Weight density and specific gravity of various liquids

Liquid	Temp. t Deg. Fahr.	Weight Density ρ Lbs. per Cu. Ft.	Specific Gravity S	Liquid	Temp. t Deg. Fahr.	Weight Density ρ Lbs. per Cu. Ft.	Specific Gravity S
Acetone	60	49.4	0.792	Mercury	20	849.74	. . .
Ammonia, Saturated	10	40.9	. . .	Mercury	40	848.03	. . .
Benzene	32	56.1	. . .	Mercury	60	846.32	13.570
Brine, 10% Ca Cl	32	68.05	. . .	Mercury	80	844.62	. . .
Brine, 10% Na Cl	32	67.24	. . .	Mercury	100	842.93	. . .
Bunkers C Fuel Max.	60	63.25	1.014	Milk	. . .	†	. . .
Carbon Disulphide	32	80.6	. . .	Olive Oil	59	57.3	0.919
Distillate	60	52.99	0.850	Pentane	59	38.9	0.624
Fuel 3 Max.	60	56.02	0.898	SAE 10 Lube‡	60	54.64	0.876
Fuel 5 Min.	60	60.23	0.966	SAE 30 Lube‡	60	56.02	0.898
Fuel 5 Max.	60	61.92	0.993	SAE 70 Lube‡	60	57.12	0.916
Fuel 6 Min.	60	61.92	0.993	Salt Creek Crude	60	52.56	0.843
Gasoline	60	46.81	0.751	32.6° API Crude	60	53.77	0.862
Gasoline, Natural	60	42.42	0.680	35.6° API Crude	60	52.81	0.847
Kerosene	60	50.85	0.815	40° API Crude	60	51.45	0.825
M. C. Residuum	60	58.32	0.935	48° API Crude	60	49.16	0.788

*Liquid at 60 F referred to water at 60 F.

†Milk has a weight density of 64.2 to 64.6.

‡100 Viscosity Index.

Courtesy Crane Co. (Technical Paper #410).

True vapor pressure of crude oil stocks with a Reid vapor pressure of 2–15 psi

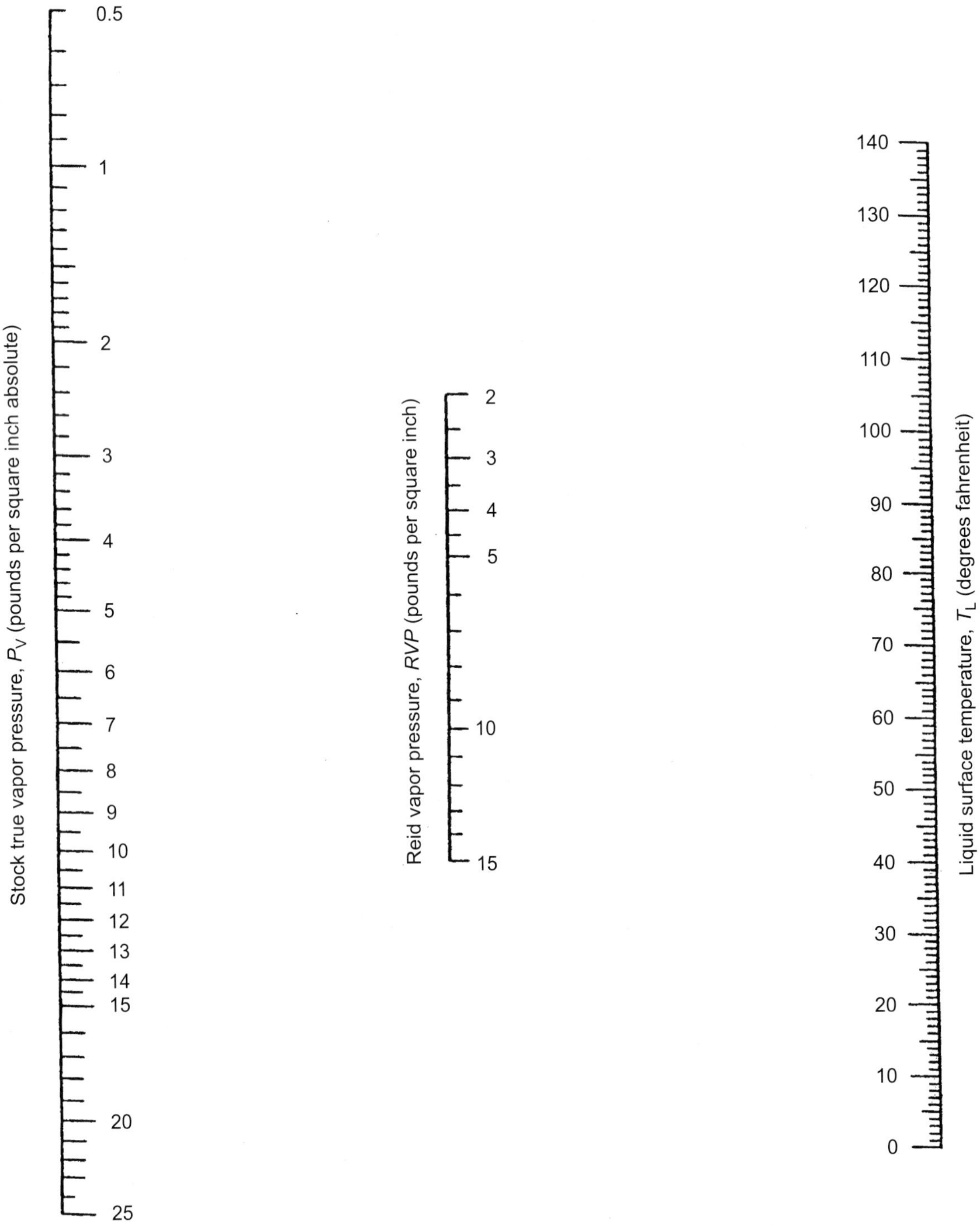

True vapor pressure (P_V) of crude oil stocks with a Reid vapor pressure of 2–15 psi. Reprinted courtesy of the American Petroleum Institute.

Low temperature vapor pressures for light hydrocarbons

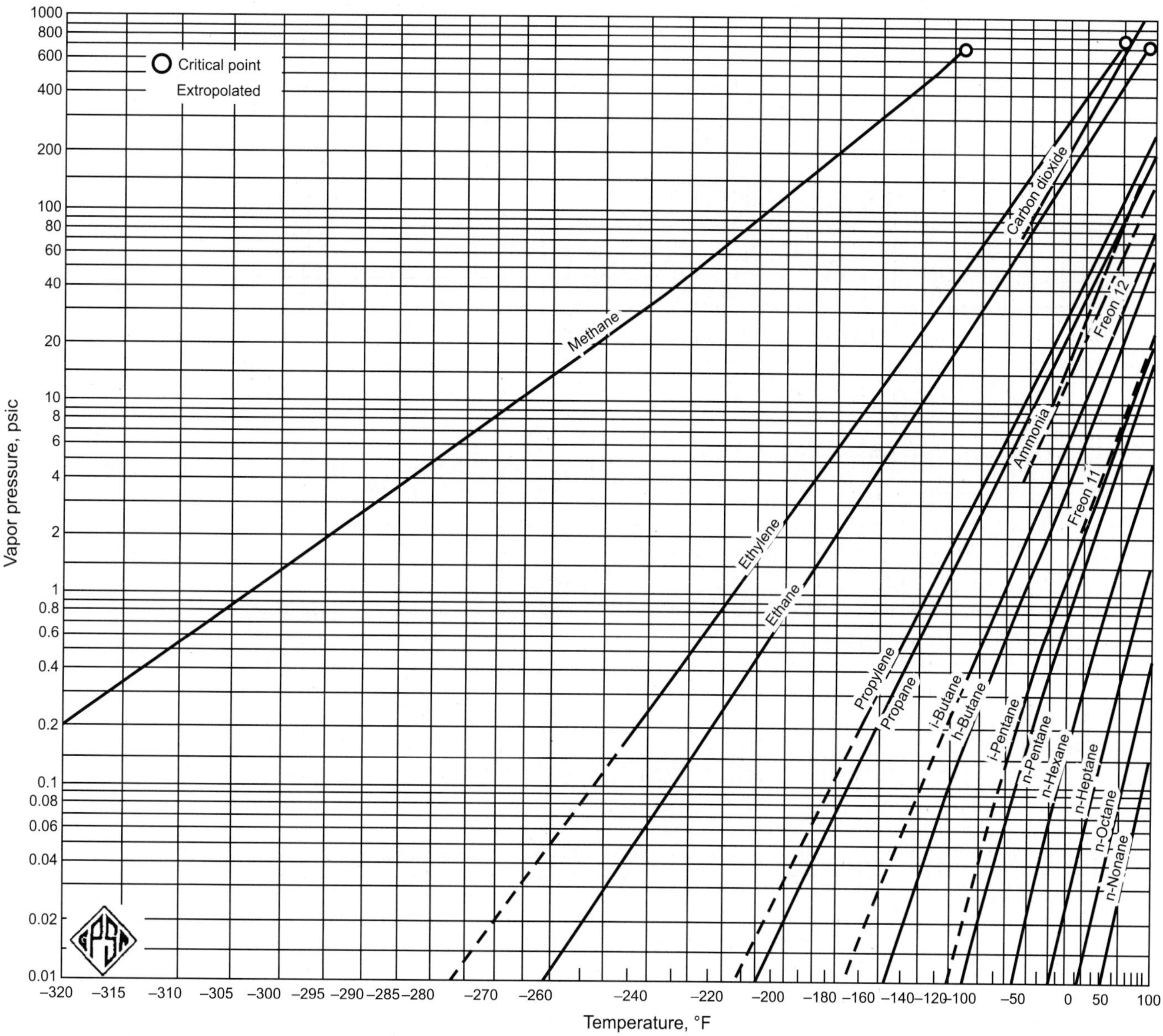

Low-temperature vapor pressures for light hydrocarbons. Reprinted with permission—Gas Processors Suppliers Association.

High temperature vapor pressures for light hydrocarbons

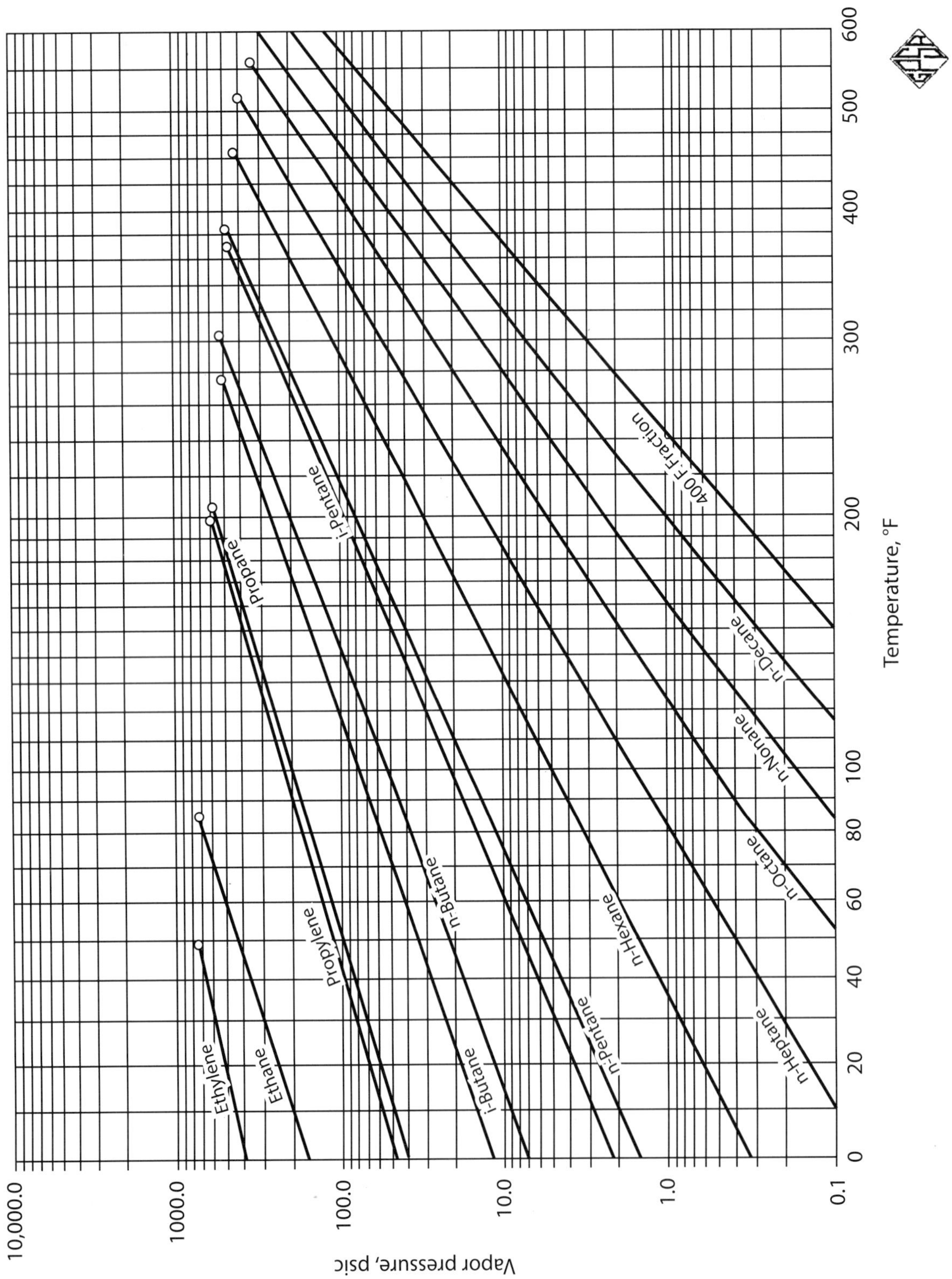

High-temperature vapor pressures for light hydrocarbons. Reprinted with permission—Gas Processors Suppliers Association.

Hydrocarbon gas viscosity

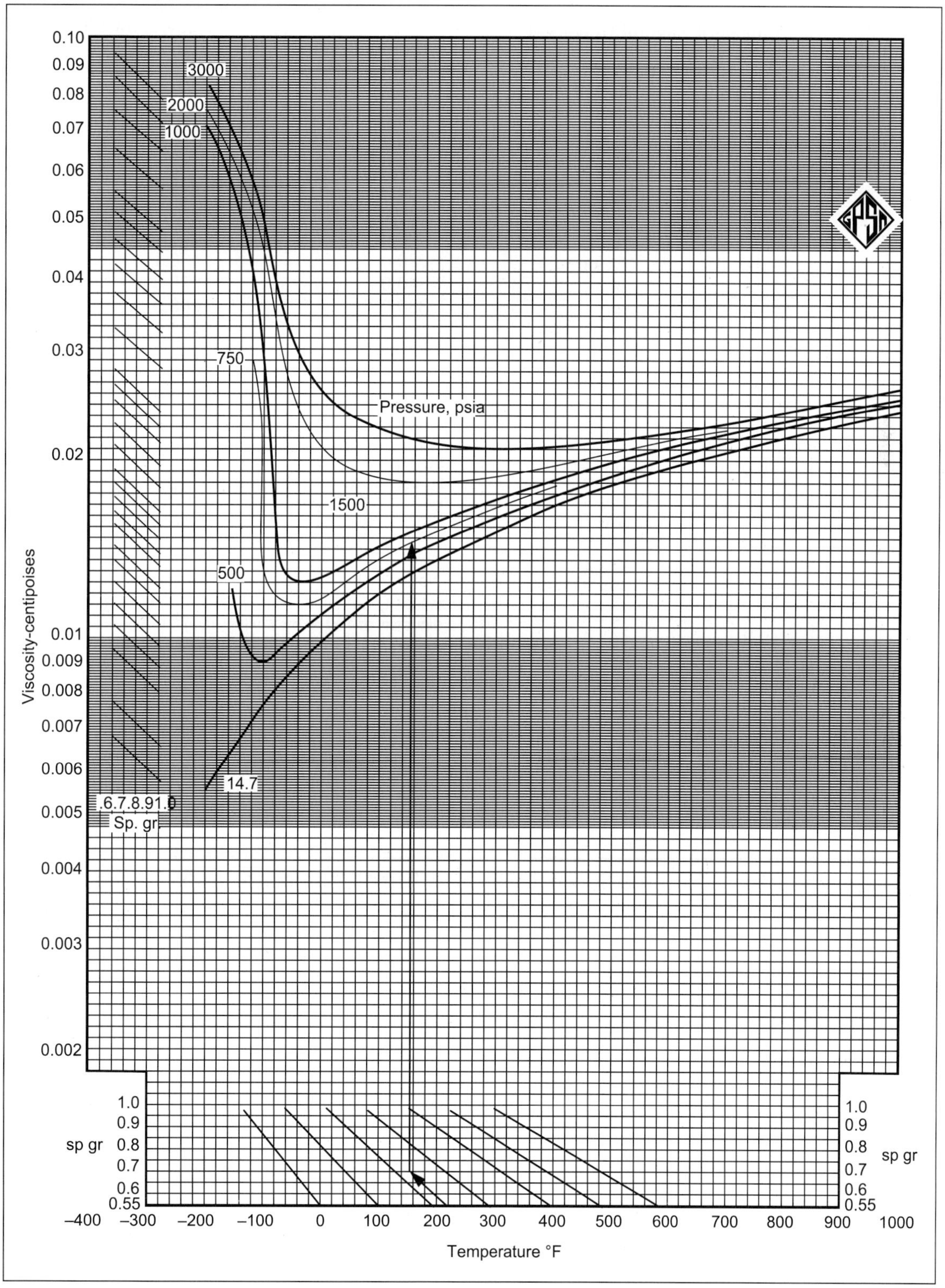

Hydrocarbon gas viscosity. Courtesy of Western Supply Co., Tulsa.

Viscosity—equivalents of kinematic and Saybolt Universal Viscosity

Kinematic Viscosity, Centistokes v	Equivalent Saybolt Universal Viscosity, s	
	At 100 °F	At 210 °F
1.83	32.01	32.23
2.0	32.62	32.85
4.0	39.14	39.41
6.0	45.56	45.88
8.0	52.09	52.45
10.0	58.91	59.32
15.0	77.39	77.93
20.0	97.77	98.45
25.0	119.3	120.1
30.0	141.3	142.3
35.0	163.7	164.9
40.0	186.3	187.6
45.0	209.1	210.5
50.0	232.1	233.8
55.0	255.2	257.0
60.0	278.3	280.2
65.0	301.4	303.5
70.0	324.4	326.7
75.0	347.6	350.0
80.0	370.8	373.4
85.0	393.9	396.7
90.0	417.1	420.0
95.0	440.3	443.4
100.0	463.5	466.7
120.0	556.2	560.1
140.0	648.9	653.4
160.0	741.6	
180.0	834.2	

Kinematic Viscosity, Centistokes v	Equivalent Saybolt Universal Viscosity, s	
	At 100 °F	At 210 °F
200.0	926.9	
220.0	1019.6	
240.0	1112.3	
260.0	1205.0	
280.0	1297.7	
300.0	1390.4	
320.0	1483.1	
340.0	1575.8	Saybolt seconds equal centistokes times 4.6347
360.0	1668.5	
380.0	1761.2	
400.0	1853.9	
420.0	1946.6	
440.0	2039.3	
460.0	2132.0	
480.0	2224.7	
500.0	2317.4	
Over 500	Saybolt seconds equal centistokes times 4.6347	

Note: To obtain the Saybolt Universal viscosity equivalent to a kinematic viscosity determined at *t*, multiply the equivalent Saybolt Universal viscosity at 100 °F by 1 + (t − 100) 0.000064. For example, 10*v* at 210 °F are equivalent to 58.91 multiplied by 1.0070 or 59.32 see Saybolt Universal at 210 °F.

Courtesy Crane Co. (Technical Paper #410).

Viscosity–equivalents of kinematic and Saybolt Furol Viscosity at 122 °F and 210 °F

Kinematic Viscosity, Centistokes v	Equivalent Saybolt Furol Viscosity, s	
	At 122 °F	At 210 °F
48	25.3	
50	26.1	25.2
60	30.6	29.8
70	35.1	34.4
80	39.6	39.0
90	44.1	43.7
100	48.6	48.3
125	60.1	60.1
150	71.7	71.8
175	83.8	83.7
200	95.0	95.6

Kinematic Viscosity, Centistokes v	Equivalent Saybolt Furol Viscosity, s	
	At 122 °F	At 210 °F
225	106.7	107.5
250	118.4	119.4
275	130.1	131.4
300	141.8	143.5
325	153.6	155.5
350	165.3	167.6
375	177.0	179.7
400	188.8	191.8
425	200.6	204.0
450	212.4	216.1
475	224.1	228.3

Kinematic Viscosity, Centistokes *v*	Equivalent Saybolt Furol Viscosity, s	
	At 122 °F	At 210 °F
500	235.9	240.5
525	247.7	252.8
550	259.5	265.0
575	271.3	277.2
600	283.1	289.5
625	294.9	301.8
650	306.7	314.1
675	318.4	326.4
700	330.2	338.7
725	342.0	351.0
750	353.8	363.4
775	365.5	375.7
800	377.4	388.1
825	389.2	400.5
850	400.9	412.9
875	412.7	425.3
900	424.5	437.7
925	436.3	450.1
950	448.1	462.5
975	459.9	474.9

Kinematic Viscosity, Centistokes *v*	Equivalent Saybolt Furol Viscosity, s	
	At 122 °F	At 210 °F
1000	471.7	487.4
1025	483.5	499.8
1050	495.2	512.3
1075	507.0	524.8
1100	518.8	537.2
1125	530.6	549.7
1150	542.4	562.2
1175	554.2	574.7
1200	566.0	587.2
1225	577.8	599.7
1250	589.5	612.2
1275	601.3	624.8
1300	613.1	637.3
Over 1300	*	†

*Over 1300 cSt at 122 °F: Saybolt Furol Sec = Centistokes × 0.4717.

†Over 1300 cSt at 210 °F: Log (Saybolt Furol Sec − 2.87) = 1.0276 [Log (Centistokes)] − 0.3975.

Courtesy Crane Co. and American Society for Testing and Materials.

Viscosity—general conversions

Approximate Conversion Table

Seconds Saybolt Universal	Seconds Saybolt Furol	Engler Degrees	Engler Time	Barbey	Redwood Standard	Redwood Admiralty	Kinematic Viscosity Stokes
35		1.18	60	2800	32		0.026
50		1.60	82	880	44		0.074
75		2.30	102	460	65		0.141
100	15	3.00	153	320	88		0.202
150	19	4.40	230	205	128		0.318
200	23	5.90	305	148	170	18	0.431
250	28	7.60	375	118	212	23	0.543
300	33	8.90	450	98	254	27	0.651
400	42	11.80	550	72	338	36	0.876
500	52	14.50	750	59	423	45	1.10
600	61	17.50	900	48	518	53	1.32
700	71	20.6	1050	41	592	62	1.54
800	81	23	1200	36.5	677	71	1.76
900	91	27	1300	32.0	762	78	1.98
1000	100	29	1500	29.5	846	87	2.20
1500	150	42	2300	19.5	1270	135	3.30

Seconds Saybolt Universal	Seconds Saybolt Furol	Engler Degrees	Engler Time	Barbey	Redwood Standard	Redwood Admiralty	Kinematic Viscosity Stokes
2000	200	59	3000	14.5	1695	175	4.40
2500	250	73	3750	11.5	2120	230	5.50
3000	300	87	4500	9.6	2540	260	6.60
4000	400	117	6000	7.4	3380	350	8.80
5000	500	145	7500	6.0	4230	435	11.00
6000	600	175	9000	5.2	5080	530	13.20
7000	700	205	10,500	4.1	5925	610	15.40
8000	800	230	12,000	3.7	6770	700	17.60
9000	900	260	13,500	3.2	7620	780	19.8
10,000	1000	290	15,000	2.9	8460	870	22
20,000	2000	590	30,000	1.4	16,920	1760	44
40,000	4000	1170	60,000		33,850	3600	88
60,000	6000	1750	90,000		50,800	5300	132
80,000	8000	2300	120,000		67,700	7000	176
100,000	10,000	2900	150,000		84,600	8700	220
200,000	20,000	5900	300,000		169,200	17,600	440
400,000	40,000	11,700	600,000		338,500	36,000	880
600,000	60,000	17,500	900,000		508,000	53,000	1320
800,000	80,000	23,000	1,200,000		677,000	70,000	1760
1,000,000	100,000	29,000	1,500,000		846,000	87,000	2200

Absolute Viscosity is the force required to move a unit plane surface over another plane surface at a unit velocity when surfaces are separated by a layer of fluid of unit thickness.

Kinematic Viscosity is the absolute viscosity divided by the density of the fluid.

SAE Viscosity Numbers are arbitrary numbers that have been assigned to classify lubricating oils and do not have any functional relationship with other systems.

Viscosity Index is an arbitrary system relating the effect of temperature change to viscosity. It is the slope of a plot of viscosity versus temperature for a given fluid.

1 St = 1 P/Density in Grams

$1\ \text{cSt} = 1076 \times 10^{-5}\text{ft}^2/\text{s}$

1 P = 1 g/(cm)(s)

1 cP = 1/100 P

1 cSt = 1/100 St

1 cP = 0.000672 lbs/(ft)(s)

$1\ \text{cP} = 0.000672(\text{poundals})(\text{s})/\text{ft}^2$

$1\ \text{cP} = 0.0000209(\text{lbs})(\text{s})/\text{ft}^2$

1 cP = 0.0000209 slugs/(ft)(s)

1 cP = 2.42 lbs/(ft)(h)

1 slug/(ft)(s) = 1/0.002089 P

$1\text{ft}^2/\text{s} = 92,903\text{cSt}$

ASTM standard viscosity temperature chart for liquid petroleum products

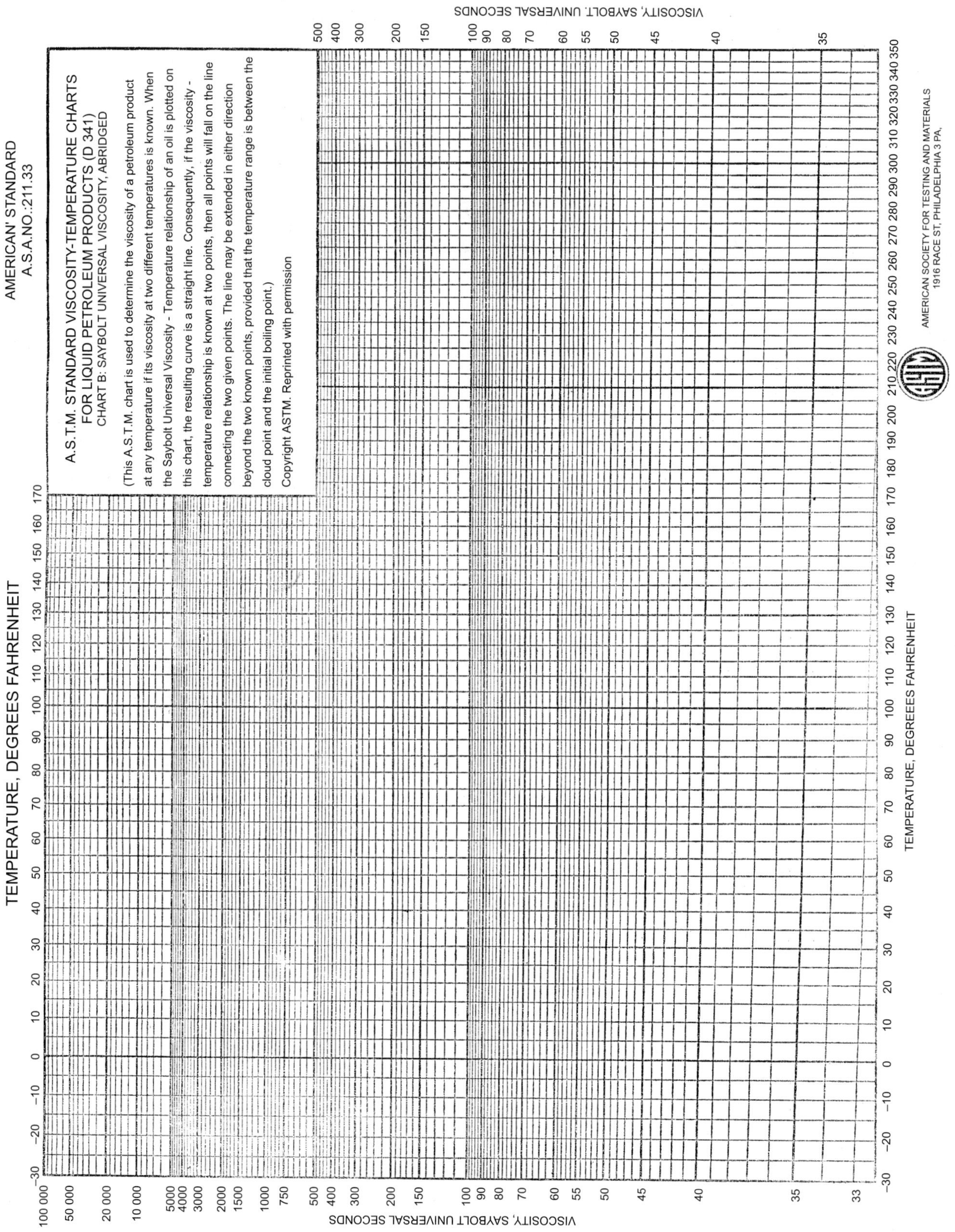

Pressure conversion chart

This line chart provides an easy method for converting units of pressure. Draw a line perpendicular to the scale lines through a known value of flow and read the equivalent value on any of the other scales. For values off the scale, known values can be read in multiples of 10 or 100.

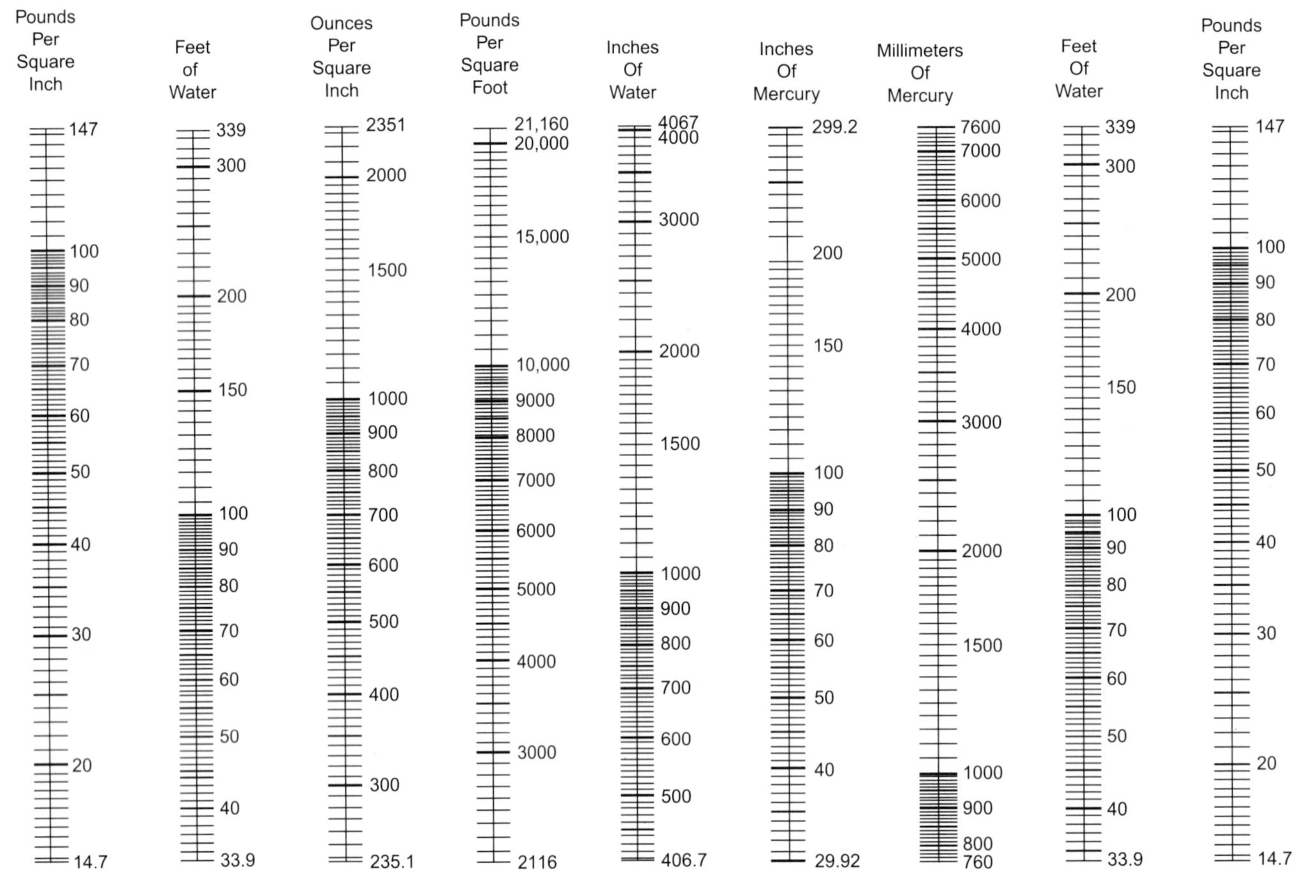

Energy conversion chart

This nomograph provides an easy method for determining the fuel requirements for various types of fuel. Since the Btu content per unit volume of natural gas and crude oil vary over a limited but significant range of values, nomographs for the energy value of several gases and oils are included rather than using an average value.

Example. 30 cu ft of a gas with 1100 Btu per cu ft yields 33,000 Btu. Conversion of 33,000 Btu to the other units of energy is obtained by drawing a straight line connecting the values 33,000 on two end Btu scales. Using the right-hand nomograph, an oil with 11,000 Btu per gallon (about 4.6 million Btu per barrel) will produce 33,000 Btu from 0.3 gallon (about 0.0072 barrel).

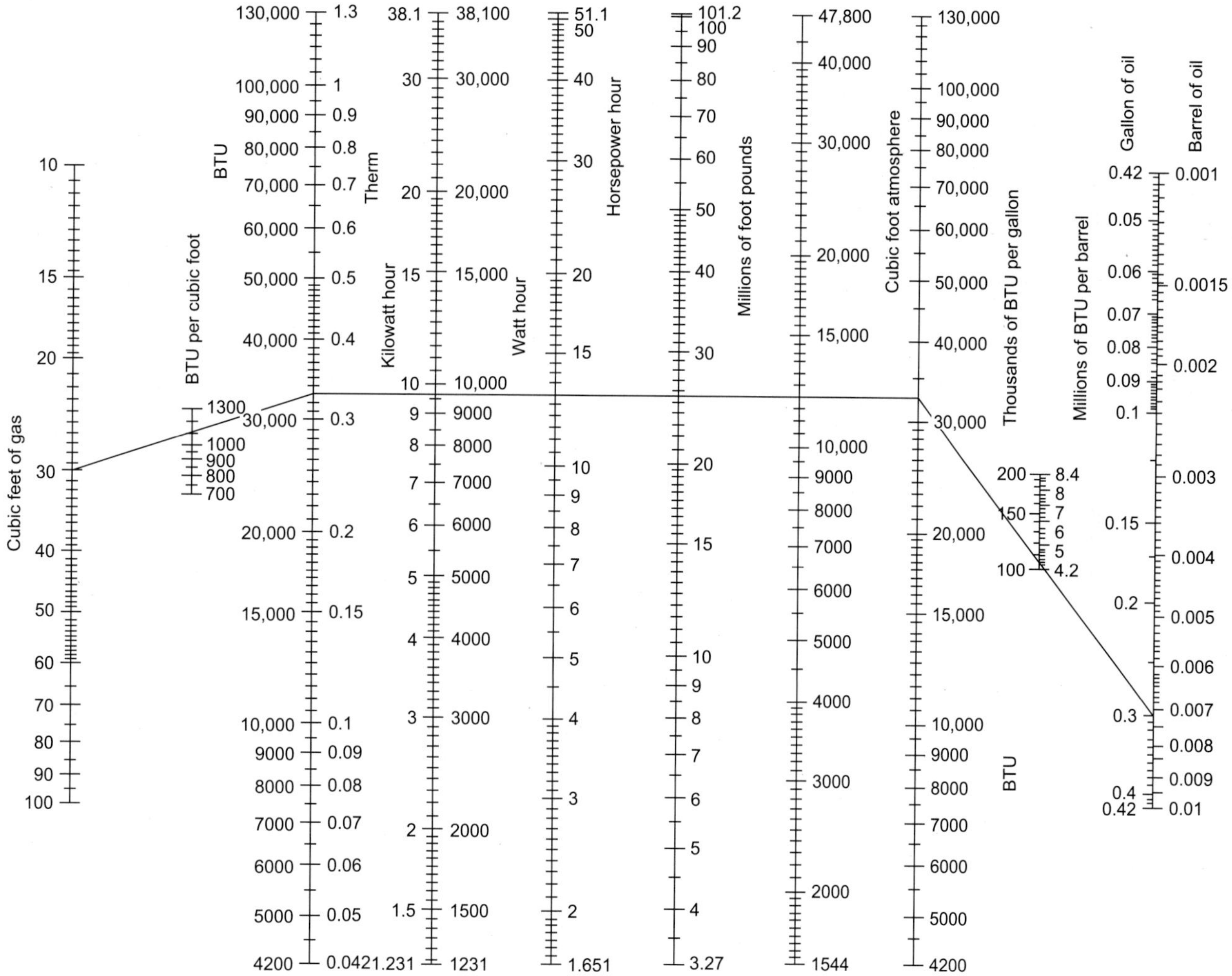

Flow conversion chart

This line chart provides an easy method for converting units of volume flow. Simply draw a line perpendicular to the scale lines through a known value of flow and read the equivalent value on any of the other scales.

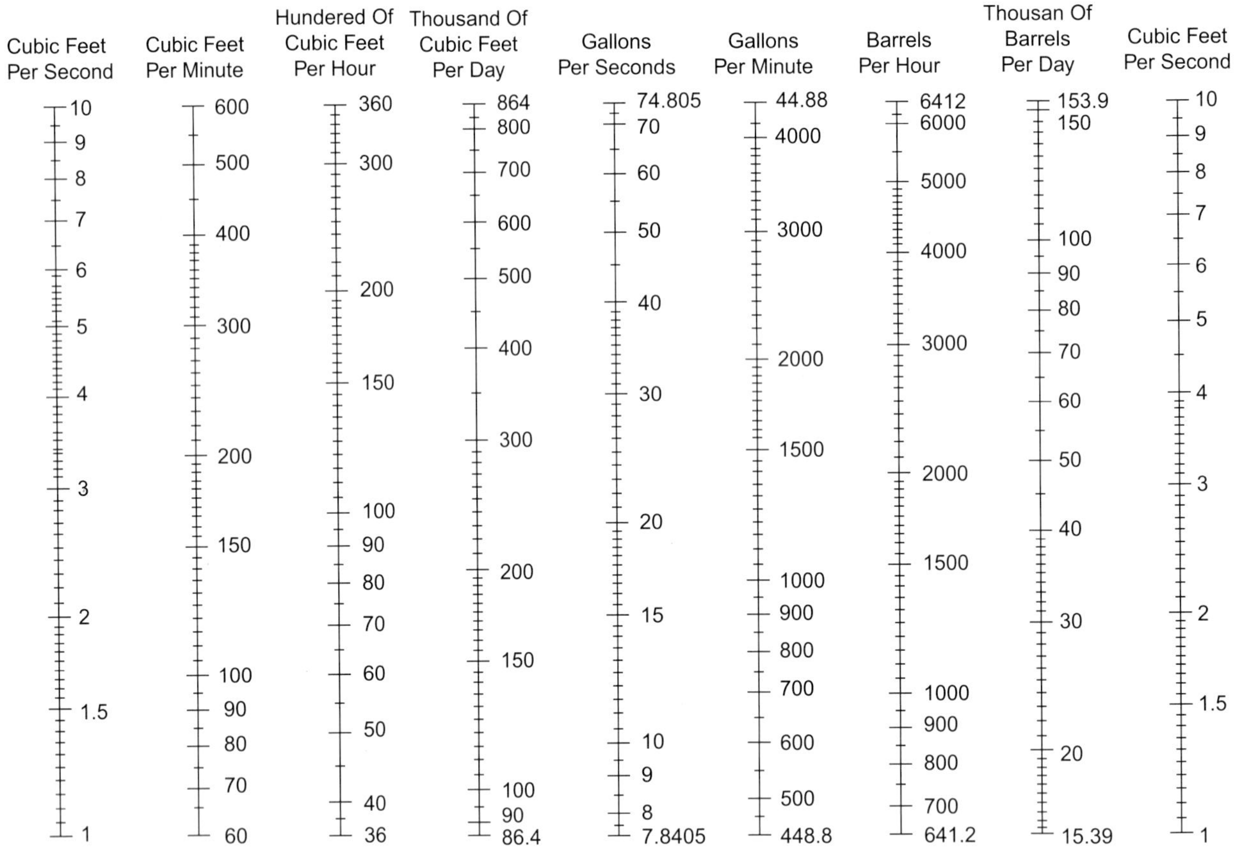

Petroleum fuels volume, weight and calorific value factors

The calorific value of natural gas varies according to its composition, and no two fields produce gas of identical quality. The composition of gas from any one field can also vary over its producing life. Thus, there is no such thing as a "standard" natural gas, nor a corresponding "standard" calorific value. However, when referring to natural gas it is quite common to indicate an "average" gross calorific value of 1000 Btu/ft^3, the equivalent of which is taken to be 9400 kcal/Nm^3 gross. In United States most natural gas is at or above this heat value, with coal bed methane fields lower and shale gas production higher.

In the case of liquid fuels, the actual calorific value of the particular fuel under consideration should be used where possible for conversions to other units.

Petroleum Fuels Volume/Weight/Calorific Value Factors

Hydrocarbons	Specific Gravity	Gross Calorific Value					
		bbl/ton	Btu/Imp.gal	Btu/lb	kcal/kg	10^6 Btu/bbl	10^6 kcal/bbl
Propane	0.501	12.790	...	21,500	11,943	3.795	0.956
Butane	0.579	11.060	...	21,140	11,742	4.315	1.088
Motor gasoline	0.733	8.740	148,000	20,160	11,200	5.176	1.304
Kerosene	0.780	8.213	156,000	20,000	11,110	5.456	1.377
Vaporizing oil	0.823	7.783	163,000	19,800	11,000	5.701	1.442
Gas oil	0.840	7.625	164,000	19,500	10,835	5.736	1.445
Fuel oils	0.850	7.536	165,660	19,490	10,830	5.794	1.460
	0.860	7.448	166,840	19,400	10,780	5.835	1.470
	0.870	7.362	168,080	19,320	10,730	5.878	1.481
	0.880	7.278	169,310	19,240	10,690	5.921	1.492
	0.890	7.196	170,440	19,150	10,640	5.961	1.502
	0.900	7.116	171,640	19,070	10,590	6.002	1.513
	0.910	7.038	172,730	18,980	10,540	6.041	1.522
	0.920	6.961	173,800	18,890	10,490	6.078	1.532
	0.930	6.887	174,860	18,800	10,440	6.115	1.541
	0.940	6.813	175,890	18,710	10,390	6.151	1.550
	0.950	6.741	176,910	18,620	10,340	6.187	1.559
	0.960	6.671	177,910	18,530	10,290	6.222	1.568
	0.970	6.602	178,890	18,440	10,240	6.256	1.577
	0.980	6.535	179,860	18,350	10,190	6.290	1.585
	0.990	6.469	180,800	18,260	10,140	6.323	1.593
	1.000	6.404	181,730	18,170	10,090	6.356	1.602

Sources: W.P.C. Publication "Technical Data on Fuel" (Editor H. M. Spiers) 6th Edition 1961; all products except fuel oil. Latter from H. Hyams, "Petroleum Cargoes, Measurement and Sampling, 1966."

Notes:

1. Btu/gal × 34.9726 = Btu/bbl.
2. Btu/bbl × 0.252 = kcal/bbl.
3. $\text{boe/bbl} = \dfrac{10^6\,\text{Btu/bbl}}{5.8}$.
4. kcal/kg × 1.8 = Btu/lb.
5. To convert 10^6 Btu/bbl to ft^3 natural gas (1000 Btu), multiply by 10^3.
6. To convert 10^6 kcal/bbl to m^3 natural gas (9400 kcal), multiply by 106.4.

Crude oil approximate conversion factors

	To convert				
	Tonnes (metric)	Kilolitres	Barrels	US gallons	Tonnes/year
From	Multiply by				
Tonnes (metric)	1	1.165	7.33	307.86	–
Kilolitres	0.8581	1	6.2898	264.17	–
Barrels	0.1364	0.159	1	42	–
US gallons	0.00325	0.0038	0.0238	1	–
Barrels/day	–	–	–	–	49.8

Note: Based on the worldwide average gravity

Source: BP Statistical Annual

Refined products approximate conversion factors

	To convert					
	Barrels to tonnes	**Tonnes to barrels**	**Kilolitres to tonnes**	**Tonnes to kilolitres**	**Tonnes to gigajoules**	**Tonnes to barrels oil equivalent**
From	**Multiply by**					
Ethane	0.059	16.850	0.373	2.679	49.400	8.073
LPG	0.086	11.600	0.541	1.849	46.150	7.542
Gasoline	0.120	8.350	0.753	1.328	44.750	7.313
Kerosene	0.127	7.880	0.798	1.253	43.920	7.177
Gas oil/diesel	0.134	7.460	0.843	1.186	43.380	7.089
Residual fuel oil	0.157	6.350	0.991	1.010	41.570	6.793
Product basket	0.124	8.058	0.781	1.281	43.076	7.039

Source: BP Statistical Annual

Natural gas and LNG approximate conversion factors

	To convert						
	Billion cubic metres NG	**Billion cubic feet NG**	**Petajoules NG**	**Million tonnes oil equivalent**	**Million tonnes LNG**	**Trillion British thermal units**	**Million barrels oil equivalent**
From	**Multiply by**						
1 billion cubic metres NG	1.000	35.315	36.000	0.860	0.735	34.121	5.883
1 billion cubic feet NG	0.028	1.000	1.019	0.024	0.021	0.966	0.167
1 petajoule NG	0.028	0.981	1.000	0.024	0.021	0.952	0.164
1 million tonnes oil equivalent	1.163	41.071	41.868	1.000	0.855	39.683	6.842
1 million tonnes LNG	1.360	48.028	48.747	1.169	1.000	46.405	8.001
1 trillion British thermal units	0.029	1.035	1.050	0.025	0.022	1.000	0.172
1 million barrels oil equivalent	0.170	6.003	6.093	0.146	0.125	5.800	1.000

Source: BP Statistical Annual

Units and calorific equivalents

Units

1 metric tonne	= 2204.62 lb = 1.1023 short tons
1 kilolitre	= 6.2898 barrels = 1 cubic metre
1 kilocalorie (kcal)	= 4.1868 kJ = 3.968 Btu
1 kilojoule (kJ)	= 1,000 joules = 0.239 kcal = 0.948 Btu
1 petajoule (PJ)	= 1 quadrillion joules (1 x 10^{15})
1 exajoule (EJ)	= 1 quintillion joules (1 x 10^{18})
1 British thermal unit (Btu)	= 0.252 kcal = 1.055 kJ
1 tonne of oil equivalent (toe)	= 39.683 million Btu = 41.868 million kJ
1 barrel of oil equivalent (boe)	= 5.8 million Btu = 6.119 million kJ
1 kilowatt-hour (kWh)	= 860 kcal = 3600 kJ = 3412 Btu

Calorific equivalents

One exajoule equals approximately:

Heat units	239 trillion kilocalories 948 trillion Btu
Solid fuels	40 million tonnes of hard coal 95 million tonnes of lignite and subbituminous coal
Gaseous fuels	See Natural gas and LNG table
Electricity	278 terawatt-hours

All fuel energy content is net or lower heating value (i.e., net of heat of vaporization of water generated from combustion)

1 barrel of ethanol = 0.58 barrels of oil equivalent
1 barrel of biodiesel = 0.86 barrels of oil equivalent
1 tonne of ethanol = 0.68 tonne of oil equivalent
1 tonne of biodiesel = 0.88 tonne of oil equivalent

Weight of water per cubic foot at various temperatures

Temperature °F	lb/ft^3	Temperature °F	lb/ft^3	Temperature °F	lb/ft^3	Temperature °F	lb/ft^3	Temperature °F	lb/ft^3
32	62.42	150	61.18	260	58.55	380	54.36	500	48.7
40	62.42	160	60.98	270	58.26	390	53.94	510	48.1
50	62.41	170	60.77	280	57.96	400	53.5	520	47.6
60	62.37	180	60.55	290	57.65	410	53.0	530	47.0
70	62.31	190	60.32	300	57.33	420	52.6	540	46.3
80	62.23	200	60.12	310	57.00	430	52.2	550	45.6
90	62.13	210	59.88	320	56.66	440	51.7	560	44.9
100	62.02	212	59.83	330	56.30	450	51.2	570	44.1
110	61.89	220	59.63	340	55.94	460	50.7	580	43.3
120	61.74	230	59.37	350	55.57	470	50.2	590	42.6
130	61.56	240	59.11	360	55.18	480	49.7	600	41.8
140	61.37	250	58.83	370	54.78	490	49.2	...	...

From "National Pipe Standards," The National Tube Co.

Index

Note: Page numbers followed by "*f*" indicate figures and "*t*" indicate tables.

A

Above ground storage tanks (AST), 820, 822, 822t–823t
  federal requirements for, 844–845
    design, 844–845
    hazardous materials storage, 845
    regulatory applicability, 845
    state requirements, 845
    testing/monitoring, 844
Abrasion resistant overlay coating (ARO coating), 1095
Absolute pressure, 380
Absolute viscosity, 1158
Absolute *vs*. relative risk, 794
Absorb-and-return theory, 522
AC voltages, induced, hazards of, 342
Acceleration, 567
  head, 589–591
Accounting fundamentals, 388f, 956–959, 957f–958f
Accounting rate of return method, 948
Accumulated funds used during construction (AFUDC), 1020
Accuracy *vs*. cost, 628
Accurate field data, 421
Accurate meter, 627–628
Acid-producing bacteria (APB), 289
Acoustic line break detectors, 717–718, 718f
Actual capacity, 400
Additives, 1124
Additives of jet fuels, 1130
Adiabatic horsepower, 400
Adiabatic process, 409
Aerodynamic losses, 424
Aerodynamic Noise Prediction, 269
Age factor, 799
Aggressivity, 231
Air
  compressors, 201
  locks, 201, 232–233
  moisture content of, 236–237, 236f
  removal, 671
Air Products & Chemicals Inc (APCI), 382
Allowable operating region (AOR), 615–616, 618, 622
Allowable stress values, 128t–129t, 137
Alloying elements, 1090
Alternating current (AC) motors, 160
Alternating current attenuation (ACCA), 347
Alternating current voltage gradient (ACVG), 347
American Conference of Governmental Industrial Hygienists (ACGIH), 74
American Gas Association (AGA), 784
American Petroleum Institute (API), 156–157, 690, 820
  API 2350 4th and 5th edition, 911–916
    automatic overfill prevention systems, 913
    other management system components, 915

American Petroleum Institute (API) *(continued)*
overfill prevention process approaches, 912
overfill standards, 911–912
response time, 914
risk assessment, 915–916, 915f
API 5L pipeline specifications, 1090–1091
chemical composition for API 5L PSL 1 welded pipe, 1092, 1092t
chemical composition for API 5L PSL 2 welded pipe, 1093, 1093t
chemical properties, 1090, 1092t
differences between API 5L PSL 1 and PSL 2 pipeline, 1091, 1091t
grade, 1090, 1091t, 1093t
mechanical properties, 1090–1091
other specifications, 1091
product specification levels, 1090, 1091t
tensile property requirements for pipe body of seamless and welded pipe for PSL 1, 1094, 1094t
tensile property requirements for pipe body of seamless and welded pipe for PSL 2, 1094, 1094t
standards for automatic tank gauging, 726–728
American Society for Testing and Materials (ASTM), 824
ASTM D1653, 1119
ASTM D1655, 1126, 1127t
ASTM D396 fuel oils requirements, 1134, 1134t–1135t
ASTM D4814, 1119, 1125t–1126t
ASTM D910, 1125, 1125t
ASTM D975, 1119
diesel fuel requirements, 1131–1133, 1132t
standard reinforcing bars, 27, 27t
standard viscosity temperature chart for liquid petroleum products, 1158, 1159f
American Society of Mechanical Engineers (ASME), 824
American Water Works Association (AWWA), 824
Amine system, 385
Amortization, 173
Analyzer interface unit (AIU), 375
Annuities, 944–945, 945f
Annular pontoon floating roof (APFR), 804, 829, 835, 838, 838f–839f
LOB event from corrosion, 804, 804f–805f
Annular sections, pressure loss calculation, 505, 505f
Anodes, graphite, 325
Anodes, magnesium, 325
Anodes, sacrificial, group installation of, 326
Anodic reaction, 288
ANSI forged steel flanges, 39, 39f
Anticavitation trim, 268
Antifreeze, 28, 28f
Area formula, 24f
Aromatics, 1030
Ash content of diesel fuel, 1134
Asphalt, 287
Atmospheric pressure, 882
Atmospheric storage tanks
bund integrity, 850
bunds, 847
design and containment for, 845–851
layout and spacing, 845–847, 847t
overflow, 849, 849f
piping, 850
primary containment, 848
rim seals and foam dams, 849
secondary containment, 850–851
selection, 845, 846t
tank design and operation, 848–850
tank roofs, 848
tertiary containment, 851
testing, 850
valves, 849
vents, 848–849
Aude, T. R., 498
Auto Adjust turbo-meter, 714
Autoignition temperature (AIT), 239
Automated overfill protection systems (AOPS), 728
Automatic Overfill Prevention Systems (AOPS), 913
Automatic tank gauging (ATG), 726, 732, 884
addressing entire safety loop, 729
API standards for automatic tank gauging, 726–728
compliance, 732, 733f–736f
independence and layers of protection, 731–732
overfill prevention systems, 728
proof testing, 729–731, 731f
SIS, 728–729
systems and proof testing, 726–732, 727f
Automatic tank thermometer system (ATT system), 726–727
Availability, leak detection technologies, 749
Aviation gasoline fuel specifications, 1124–1125, 1125t–1126t
ASTM grade 100LL avgas requirements, 1125t–1126t
Avogadro's law, 355

B

Bare casing, number of magnesium anodes needed for protection to, 325–326
Backpressure regulators and accumulation, 283
Balance sheet, 956
Ball provers, 715
Bare steel pipe
determine buoyancy of, 145
in water and mud, 145
Barrels per hour (BPH), 668
Barrels per mile, 477
Baseline assessment plan (BAP), 211
Basic Process Control System (BPCS), 913
Basic ultrasonic flow equations, derivation, 475–476
Basic valve flow-capacity coefficient (C_v), 268–269
choked flow, 269
compressibility factor (Z), 268–269
low-noise trim, 269, 271t
valve outlet velocity, 269
Batch injection rate as per enclosure, determination of, 508
Batching pigs, 223

Battelle staff recommended hydrostatic testing, 189
Beam-lumen design method, 178–179
Bell provers, 664
Belts, 22
calculate stress in shaft key, 23
determining length of V-belt, 22–23
estimate horsepower that transmitted by shaft, 23
Benching, soil type, 72–73, 72f–73f
Bend, 78
Bending radius, minimum, 170, 171t
Benedict–Webb–Rubin equation, 378
Benjamin Miller formula for calculating pressure drop and flow rates. *See* Miller, Benjamin, pressure drop and flow rates formula
Benzene, toluene, ethylbenzene, and xylenes (BTEX), 1045
Bernoulli's theorem, 701
Best efficiency point (BEP), 618, 621–622
'Bi-di' pigs, 227
Billion barrels delivered by liquids pipeline in U.S., 14f
Biocides, 298
Biodegradation, 1031, 1042–1043
Bitumen, 287
Bladder type pulsation dampeners, advantages of, 566
Blending, 899–900
components, 1124
index, viscosity of mixture of liquids, 465
Blends, 466
Blow-down system, determining throat pressure in, 445–446, 445f
Boil-off gas (BOG), 906
Boiling, trench excavations, 67
Bolted deck, 876
Bolting dimensions for flanges, 33–36, 33t–36t
Bomb calorimetry, 371
Bonding grounding system to building and structure foundations, 181–182
"Bottom-up" approaches, 962
Boyle's law, 354
Brake horsepower (bhp), 599, 622–623
required to compress gas, 419
Breakdown Insulation Level (BIL), 342
Breakdown torque, 160
BREEZE TankESP, tank emissions with, 936–937
AP-42 Chapter 7 Revisions, 936–937
challenges in calculating tank emissions, 937
deficiencies and resolution, 937, 937t–938t
British thermal units (Btu), 653, 1124
Brush pig run
with gas, 233
with liquid, 233
Btu, calculation of, 375–376
Bubble flow, 686
Bucket test, 730
Building Officials and Code Administrators International (BOCA), 824
Bulging, trench excavations, 67
Bulk modulus, calculation of, 462
Bunds, 847
Buoyancy
of bare and concrete-coated steel pipe in water and mud, 145
of bare steel pipe, 145
Bureau of Labor Statistics (BLS), 966
Bureau of Ocean Energy Management (BOEM), 964
Bureau of Safety and Environmental Enforcement (BSEE), 964
Burn-through, 97
Butane, 1137t
Butane, refinery grade (RGB), 1138t

C

Cables, 16
capacity of drums, 22, 22t
with different outside diameters, 170
drums and sheaves be for wire rope, 16, 16t
estimating strength of cable, 16
find advantages of block and tackle with pull out friction, 17
find working strength of manila rope, 16
safe loads for wire rope, 17, 17f
strength and weight of popular wire rope, 20, 20t
stress in guy wires, 18–19, 18f
wire rope, 21–22, 21t–22t
Calculations
allowable working pressure, 147, 147t
bends, length of, 134, 134t
brake horsepower, 622–623
Btu, 375–376
bulk modulus, 469
buoyancy, 145
centrifugal compressor performance, 409–412, 410t, 411f
contraction or expansion, 143
floodlighting, 177
gas flow, 371
gas pipeline hydraulics, 434–435
gas velocity, 626
line loss using cross-sectional areas table when testing mains with air or gas, 443, 443t
microwave hops on level ground, 174–175
orifice flow in field, 658–661
performance, for reciprocating compressors, 401–403
pressure and temperature loss for viscous crudes, 505–507, 505f, 506t, 507f, 507t
pressure drop, 439
pressure loss in annular sections, 505, 505f
Reynolds number, 515–517, 515f–516f
rubber hose, pressure drop for, 497, 497f

Calculations *(continued)*
  specific gravity of blend, 466
  speed torque, 558, 558f
  square root, 665–666
  stress, 115
  velocity, 141–142
  viscosity, 155
  weight, 114
Caliper pigging, 221
CALM system, 492
Calorific values of gases, conversion factors for, 1162
Capacitance level sensors, 742–743
Capacities of gas pipelines, comparative, 439
Capacitive coupling, 341
Capacitive level sensors as level switch, 739–740, 740f–741f
Capacity curve, head curve *vs*., 620
Capacity increases, crude oil and products lines for, 527
Carbon capture and sequestration processes, 478
  capture, 478
  CO_2 pipeline infrastructure in U.S., 479–486, 480t, 481f
  sequestration in geological formations, 479
  transportation, 478
  U.S. CO_2 pipelines, 479–486, 480t, 481f
Carbon dioxide (CO_2), 296, 633, 776
  CO_2-EOR operations and infrastructure in U.S. circa 2020, 8f
  composition, 154
  measurement of supercritical, 633–639, 633f
  pipeline CO_2 composition, 154–155
  pipeline design, 154–157
  pipeline infrastructure in U.S., 479–486, 480t, 481f
    Gulf Coast, 480, 483f, 483t
    Mid-Continent, 480–483, 485f, 485t
    Permian Basin, 480, 481f, 482t
    Rocky Mountains, 480, 484f, 484t
    U. S. CO_2 pipeline networks, 485, 486t
  pipeline operating experience, 154
  pipeline operating temperature and pressure, 155–157
  pipeline operation, 486
    capture facilities—power plant load variations, 487, 488t
    effects of corrosion, 489
    effects of impurities, 487–489
    effect of flow rate variability, 487
    offshore *vs*. onshore pipelines, 486
    pipeline first fill, start-up, and shutdown procedures, 489–490
  primary types of CO_2 pipelines, 157
    acceptable CO_2 and co-constituent concentrations, 156t
  systems in United States, 12f
Carbon residue of diesel fuel, 1134
Carbon steel
  allowable working pressure for carbon steel pipe, 147, 147t
  color and approximate temperature, 32
  line pipe, 779
  pressure rating for carbon steel flanges, 27
Carbonate-bicarbonate (CO_3-HCO_3), 776
Carbonic acid, 489
Cash flow
  importance of cash flow diagrams, 942
  problems, 946–947
  statement, 957–958
Cathodic protection (CP), 211, 287, 287f–288f, 293, 723–725, 757
  efficiency of cathodic protection rectifier, 322
  methods, 341
  for pipelines, 309–317, 309f, 309t, 310f
    galvanic anodes, 313–317
    source, 317
  risk management tool, 798
Cathodic reactions, 288–289
Cavitation, 249, 563
  cavitation in butterfly valves, avoiding, 278–281
Cavitation index (K_c), 268
Centrifugal compressors, 407
  performance calculations, 409–412, 410t, 411f
    sample performance calculations, 409–412
  performance testing guidelines for, 421–431, 422f, 423t, 424f–428f, 429t
Centrifugal pumps, 546–557, 546f–549f, 549t, 550f–552f, 552t, 553f–557f, 710 (*see also* Reciprocating pumps)
  estimating discharge of centrifugal pump at speeds, 584
  estimating head for average, 585
  parameters and applications, 574–579
    control valve losses, 578, 578f
    friction component, 577
    horsepower, 578–579
    minor losses, 578
    pipe losses, 577–578
    for pressure component, 577
    pump curve, 574–575, 574f–576f
    for static head component, 577
    system head curves, 576–577, 576f
    for velocity head components, 577
  reading centrifugal pump curve, 617–620
    other pump curve elements, 618
    pump curve, BEP, and system curves, 618
    variable speed pumps, 618–620, 619f
  systems and examples, 579–584
Cetane number, 1131–1133
Charles' law, 354
Chemical analysis, resistivity of freshwater from, 324
Chemical cleaning, 233–234
Chemicals, physical properties of selected (chart), 360–367, 360t–367t
Choked flow, 267–269, 282–283
Chromatograph controller, 375

Circumferential direction-stress corrosion cracking (C-SCC), 775–776
Cleaning, 221
pigs, 297
pipelines, 233
Clearance drills for machine screws, 31, 31t
Close interval potential survey, 335, 336, 337–338
Close Interval Survey (CIS), 346–347, 759
Cloud point, 1134
Coal tar, 287
epoxy protective coatings, 1099–1102
Coal-tar enamel, estimating, 305
Coated lines, determining current requirements for, 327
Coating (CT), 291, 297, 798
defects, 335
field joints using heat-shrinkable materials, 1111–1113
materials, 1095–1096
SCC, 781–782, 781f
soil conditions, 782
selection, 224–225
Coating thickness, minimum test voltages for, 225
Code for pressure piping, 137
Code of Federal Regulations (CFR), 968
Coefficient of beam utilization (CBU), 178
Coefficient of discharge (K_D), 263
Coefficient of friction for pipe coating materials, 306–308
results, 307–308, 307f
Color codes for locating underground utilities, 49f
Columbia Gas, 758
background, 759, 759t
Columbia gas pipeline parameters, 759t
description of integrity issue, 759
inspection and assessment, 759
inspection tool selection, 759
tool performance specifications, 759, 759t
Combating pulsation, 560–561
Combination Capacity Factor, 263
Commissioning, 238
Committee European de Normalization (CEN), 1117
Committee on Refining Equipment (CRE), 902
Commodity charges, 1017–1019, 1018f, 1019t
Compact provers, 715–716
Compacted soil, shrinkage and expansion of, 66
Comparative capacities of gas pipelines, 439
Complex investments, analyzing, 946–947
Compliance matrix of pressure testing, 187, 188f
Compound interest, 940
Compressibility factor (Z), 268–269
Compressible fluids in pipes, 426
Compression efficiency, 401
Compression horsepower determination, 405–407, 405f–406f
centrifugal compressors, 407
Compressors, 400–401, 400f, 403–404
calculate brake horsepower required to compress gas, 419
centrifugal compressor performance calculations, 409–412, 410t, 411f
compression horsepower determination, 405–407, 405f–406f
cylinders, 408
estimate compressor horsepower, 418
estimate engine cooling water requirements, 420
estimate horsepower required to compress natural gas, 418
fuel gas line for compressor station, 419–420
generalized compressibility factor, 407
horsepower, 418
nomograph aids in diagnosing compressor cylinder ills, 408
nomographs for estimating compressor performance, 413–417, 413f–417f
performance testing guidelines for centrifugal compressors, 421–431, 422f, 423t, 424f–428f, 429t
source, 404
Computation fluid dynamics (CFD), 925
Computational pipeline monitoring (CPM), 157, 746, 749
Concrete-coated steel pipe in water and mud, 145
Condensate drop-out, 223
Condensate stablization, 384
Conductor size conversion chart—metric to AWG, 179
Conduit size for combinations of cables with different outside diameters, 170
maximum allowable diameter, 170t
Confirmatory Direct Assessment (CDA), 211
Consensus specifications, 1117
Constrained dent, 767
Construction
cross-country pipeline, 109, 109t
directional crossing, 110
ditching, 66
equipment, 970
pipe laying
allowable pipe span between supports, 84
degrees of bend in pipe, 78
engineers make pipe fit ditch, 85–87, 86f
length of pipe bend, 82
length of pipe in arc subtended by any angle, 82
maximum bend on cold pipe, 80
pipe bending computations, 79
pipe lowering, 78, 88
polyethylene piping, 118
project scoping data, 62
right-of-way, 64
trenching, 66, 66t
determination of soil type, 68
test equipment and methods for evaluating soil type, 68

Construction *(continued)*
trench excavations, 67–68
welding, 91
Construction cost, 963–964
Construction damage, pipelines, 762–764, 765f
evaluation and repair of, 765
Contact measurement devices, 737
Contact *vs.* non-contact level sensors, 737
Contingency plan, 1030
Continuous level sensors, 740–741
Continuous wave (CW), 743
Control systems for pipeline control, 711
Control valves, 257
action, 711
avoid cavitation in butterfly valves, 278–281
basic valve flow-capacity coefficient, 268–269
losses, 578, 578f
regulator flow curve, 281–284
relief valve sizing, selection, installation, and testing, 259–263
discharge piping, 262
inlet piping, 261
installation procedures, 261
rapid cycling, 261–262, 261f
reactive force, 262–263, 262f
resonant chatter, 262
selection, 260, 260f
sizing, 259
sonic flow, 259
subsonic flow, 260
testing, 263
rupture disk sizing methodologies, 263–264
rupture disk sizing using resistance to flow method, 264–266
selection, 257–258, 257f, 257t, 258f
sizing control valves for throughput, 252–255
examples, 255, 255f–256f
K factors for valves and fittings, 252–255, 255t
sizing formulas, 248–251, 248f–249f, 251f
gases and vapors, 249–251
liquid formulas, 248–251
liquids, 248–249
special conditions, 248
visualize pump and control valve interaction, 272–276
Controlled in-situ burning, 1040–1041
Conventional displacement provers, 694–695
advantages of, 696
disadvantages of, 696
Conversion factors
conversion factors for calorific values of gases, 1162, 1163t
conversions involving different types of fuel, 1162, 1163t
energy conversion chart, 1161f
flow conversion chart, 1162f
general liquid density nomograph, 1148, 1149f
high temperature vapor pressures for light hydrocarbons, 1153
hydrocarbon gas viscosity, 1155
LNG conversion table, 1164t
low temperature vapor pressures for light hydrocarbons, 1152
metric conversions, 1148, 1148f
pressure conversion chart, 1160, 1160f
specific gravity/temperature relationship for petroleum oils, 1151f
true vapor pressure of crude oil stocks with Reid vapor pressure, 1152
viscosity
equivalents of absolute viscosity, 1158t
equivalents of kinematic and Saybolt Furol Viscosity at 122°F, 1156t–1157t
equivalents of kinematic and Saybolt Universal viscosity, 1156t–1157t
equivalents of kinematic viscosity, 1158t
general conversions, 1157t–1158t
weight density and specific gravity of liquids, 1151f
weight of water per cubic foot at various temperatures, 1165t
Copper conductors in steel conduit, voltage drop, 172
Copper grounding electrode cables, 182
Copper sulfate electrode (CSE), 295
Copper wires, resistance and weight of, 174
Copper-based grounding system, 181
Coriolis effect, 687–688
Coriolis flow meters (CFMs), 677, 686–689, 699
coriolis effect, 687–688, 688f
coriolis flow meters or inertial mass flow meters, 688–689
coriolis flow meters work, 688
embody coriolis effect, 688
Coriolis meters, 670–671
Coriolis meters, 715
Corrective Action Order (CAO), 753
Corrosion, 1123–1124
advances in pipeline protection, 332
cathodic protection for pipelines, 309–317, 309f, 309t, 310f
cell, 181
chart aids in calculating ground bed resistance and rectifier power cost, 320f, 321
coal-tar enamel for mile of pipe, 305
coefficient of friction for pipe coating materials, 306–308
comparison of other reference electrode potentials of copper-copper sulfate, 321
control corrosion at compressor stations, 332–333
coatings, 332
grounding, 332
insulation, 332–333

materials selection, 333
retirements, 332
shielding, 332
water systems, 333
control methods, 286–288
cathodic protection, 287, 287f–288f
corrosion inhibitors, 288
materials selection and design, 287–288
protective coatings, 286–287
corrosion of low-velocity, high water cut oil emulsion pipelines, 344–346
current output of magnesium rod used for cathodic protection of heat exchanger shells, 325
current requirements for coated lines when pipe-to-soil potential values, 326f, 327
determining current requirements for coated lines, 327
efficiency of cathodic protection rectifier, 322
estimate amount of primer required for tape, 340
estimate number of squares of tape for pipe coating, 339
estimate pounds of sacrificial anode material required for offshore pipelines, 317–318
typical anode specifications, 318, 318t–319t
estimating coating and wrapping materials per mile of pipe, 305
fatigue, 300, 774
graphite anodes in terms of resistance of single anode, 325
group installation of sacrificial anodes, 326
HVDC effects on pipelines, 327–331
induced AC voltages on pipelines present serious hazard, 341–343
inhibitors, 288
of jet fuels, 1130
life of magnesium anodes, 314
loss of buoyancy to, 811–812
of low-velocity, high water cut oil emulsion pipelines, 344–346
development of strategy, 344–345
effect of hydrogen sulfide and water chemistry, 345
limitations of strategy, 344
potential applications of this strategy, 345–346
significance of fluid velocity, 344
validation of corrosion mitigation strategy, 345
magnesium anodes, 322
for supplementary protection to short-circuited bare casing, 325–326
required and spacing for bare line or corrosion "hot spot", 323–324
minimizing shock hazards on pipelines near HVAC lines, 332
NAPCA
specification designations for coal tar enamel coatings, 302
specifications for pipeline wrappers, 304
standard applied pipe coating weights, 302–303
pipeline external corrosion direct assessment process, 346–351
in pipelines, 288–301
environmentally assisted cracking, 299–301
external corrosion, 289–295
internal corrosion, 295–299
monitoring internal corrosion, 298–299
process, 288–289, 288f
uniform. *vs*. localized corrosion, 289
primer for mile of pipe, 304
project leak growth, 333–334
rectifier-ground bed, 331
resistance, 1090
resistivity of freshwater from chemical analysis, 324
single graphite anode, 324
tape requirements for fittings, 340
test leads to measure current on pipeline, 323
troubleshooting magnesium anode cathodic protection systems, 308
types of, 286–288
dealloying, 286
environmental cracking, 286
flow-assisted corrosion, 286
fretting corrosion, 286
galvanic corrosion, 286
general attack corrosion, 286
high-temperature corrosion, 286
intergranular corrosion, 286
localized corrosion, 286
types of corrosion and control methods, 286–288
voltage drop in ground bed cable, 322–323
voltage rating of rectifier, 327
wrapping for mile of pipe, 305
Corrosion "hot spot" and magnesium anodes, 323–324
Corrosion-caused metal loss, 784–786
Cost(s)
estimation
data quality, 965–966
data sources, 964–965
methods and data, 962–966, 962f, 965f
normalization, 963–964
regulatory agencies, 964
"top-down" and "bottom-up" approaches, 962
uncertainty range, 963
gas compression, 1010
ideal risk-assessment model, 795
indices, 966–967, 967f
location adjustment, 967
time adjustment, 966–967
offshore pipeline
construction, 963
decommissioning, 1002
onshore pipeline construction, 968

Cost(s) *(continued)*
pipeline construction cost across time, 967–968
price adjustments in pipeline system quality banks, 1023–1024, 1024t
scaling rules, 1013
of service, 1019–1023, 1021f–1022f
tariff rates, 1017
Crack growth, model for, 788
Cracking from mechanical damage, growth of, 773
Crane TP-410 methods, 264–265
Crevice corrosion, 286
Critical flow prover, 664
Cross head extrusion data, 1105
Cross head extrusion method, 1103–1104
"Crowding" effect, 325
Crude
blending, 899–900, 899f–901f
gathering line throughput, estimating, 526
viscosity of, 462, 462f
Crude oils, 1044
line, 233
true vapor pressure of crude oil stocks, 1152
Cryogenic process, 385
Cubic spline regression, 610, 610f
Current measurement, and spacing for test, 323
Custody transfer requirements, 690
Cylinder inlet capacity, 401
Cylinders, sample, purges for, 653

D

Damage prevention, risk management tool, 798
Dampeners, pulsation control with, 564–565
Darcy formula, 577
Deadweight pressure and temperature readings, 201
Dealloying, 286
Debris removal
pigging during construction, 220–221
pigging during operation, 222
Decision and evaluation criteria for investments and financial projects, 947
accounting rate of return method, 948
internal rate of return method, 949–950, 949f–950f
net present value method, 950–951
payback method, 947–948
Decision tree analysis of investments and financial projects, 952–955, 953f–954f
Defect growth rates, influence of, 783–791
Defect size, failure pressure *vs.*, 784
Dehydration, 297
Dense phase ethylene, master meter proving orifice meters in, 639–646
Densitometers, 719–720, 719f–720f
Density
conversion chart, 463, 463f
and specific volume of gases and vapors, calculation of, 370, 370f
Dents, 750, 767
depth, 769
length, 769–771
shape, 769
Depreciation, 956
Depth of cover, risk management tool, 798
Derived cetane number (DCN), 1133
Design Metal Temperature (DMT), 856
Desired protection, determine glycol requirements for, 28
Desulfurization, 384
Detectors, acoustic line break, 717–718, 718f
Dew-point readings, 234
Dewatering, pipeline, 232–233
Diesel fuel specifications, 1131–1134, 1131t–1132t, 1134t–1135t
ash content, 1134
carbon residue, 1134
cetane number, 1131–1133
distillation, 1133
flash point, 1133
low temperature operability, 1134
lubricity, 1133
other middle distillate products, 1134
specification, 1131
stability, 1134
Differential-pressure flow meters, 673–675, 674f
Diffusivity, 379
Direct current voltage gradient (DCVG), 347
Direct mass measurement, 715
Directional crossing, 110
guidelines for successful directional crossing bid package, 110–118
advantages, 110
allocation of risk of loss, 117
bid package, 117
conditions of contract, 117
cost mitigation for owner, 117
development and uses, 110
differing ground conditions and walkaway provision, 117
dispute resolution, 118
drilling slurry, 116–117
environmental concerns, 117
geotechnical investigation, 114–115
layout and design, 112
pipe material selection, 115–116
pipeline coating, 116
profile design parameters, 112–114
profile survey, 112
technique, 110, 111f
technology limits, 110
work space, 112, 112f

Directional Crossing Contractors Association (DCCA), 110
Discharge piping, 262
Discharge pressure (Pd), 592
Discharge rate, 599
Discharge systems, pulsations in, 563, 564f
Discharge temperature, 401
Discharge volume bottle sizes for pulsation control for reciprocating compressors, 403–404
Discounted cash flow (DCF), 946, 950
Discounted cash flow formula, 942–943
Discrete sensor-based technologies, 746
Dissolution, 1048
Distillate
  #1 diesel fuel, ultra-low sulfur, 1143, 1143t
  #1 diesel fuel, ultra-low sulfur, 1144, 1144t–1145t
  #2 diesel fuel, ultra-low sulfur, 1145, 1145t–1146t
Distributed control system (DCS), 887
Ditching, 66
  cubic yards of excavation in a mile of ditch, 66
  ditching and trenching, 66, 66t
  shrinkage and expansion of excavated and compacted soil, 66
Doppler flow meters, 679
Doppler meter, 715
Double deck floating roofs (DDFRs), 806–807, 810, 829, 835, 838–839, 840f
  loss of buoyancy, 804–805
Double Mixed Refrigerant Process (Double MR Process), 390
Double submerged-arc welded (DSAW), 779
Drag-reducing agents (DRAs), 521–523
  activity, 522–523
  early applications, 521
Drill sizes for pipe taps, 32
Drip legs, 298
Drip logs. *See* Drip legs
Droop, 282–283
Dry air, 368
Dry strength test, 69
Drying, 222
  with natural gas, 235
  with nitrogen, 235
  pipeline, 234–235
Dryness, of pipeline, 242–243, 243f, 243t
Dual-diameter pigs, 230, 230f
Dye penetrant examination (DPT), 907

E

Economics, 238
  accounting fundamentals, 388f, 956–959, 957f–958f
  decision and evaluation criteria for investments and financial projects, 947
  decision tree analysis of investments and financial projects, 952–955, 953f–954f
  sensitivity analysis, 951–952, 952f
  time value of money, 940
Effective annual interest rate, 941
Efficiencies (n), 593
Elastic re-rounding process, 768–769
Elastic spring-back process, 768–769
Electric motor
  horsepower for pumping units, 175
  selection, 160–161
Electric power conductors, most economical size determination for, 173
Electric resistance welding (ERW), 55, 104, 289, 753, 785
Electrical design, 160–161
  beam-lumen design method, 178–179
  bonding grounding system to building and structure foundations, 181–182
  chart of electric motor horsepower for pumping units, 175
  conductor size conversion chart—metric to AWG, 179
  conduit size for combinations of cables with different outside diameters, 170
  determine economical size for electric power conductors, 173
  electric motor selection, 160–161
  electrical formulas, 169, 174
  floodlighting calculations, 177
  full-load currents—single-phase transformers, 169, 169t
  full-load currents—three-phase transformers, 172
  hazardous locations, 161–162
  knockout dimensions, 164
  microwave hops on level ground, 174–175
  minimum bending radius, 170, 171t
  motor controller sizes, 172
  national electrical code tables, 165, 165t–167t, 169t
  NEMA enclosure types, 162–163
  point-by-point method, 177–178, 177f–178f
  pumping stations, 176
  quick determination of hp per ampere for induction motors at different voltages, 175
  resistance and weight of copper wires, 174
  size portable electric generators, 163
  switchgear device numbers, 180–181
  voltage drop on circuits using copper conductors in steel conduit, 172
  wattages for tools and appliances, 164
Electrical formulas, 169, 174
Electrochemical process, 181
Electromagnetic coupling, 341
Electromagnetic flow meters, 675–677, 675f
Electromechanical servo transmitter, 742
Electronic Floating Roof Monitoring (EFRM), 835
Electrostatic coupling, 341

Elevated pressure, 387
Ells with tangents, dimensions of, 153
Emergency rescue equipment, 74
Emerson's Micro Motion Coriolis flow meters, 686
Empty-Full-Empty (EFE), 685–686
Emulsification, 1031–1033, 1032f
Encroachment damage, pipelines, 760–761, 760f–762f
  evaluation and repair of, 761–762, 762f–763f
Energy conversion chart, 1161, 1161f
Engine cooling water requirements, estimation of, 420
Engine starter, air receiver for, 47
English to metric conversions, 1148f
Enhanced oil recovery (EOR), 157, 478
Entrained gas, 685–687
  characterizing entrained gas, 685–686
  design challenge, 686
  overcoming bubble-flow conditions, 686–687
  slug flow & EFE performance, 686
Environmental cracking, 286
Environmentally assisted cracking (EAC), 299–301, 774
  corrosion fatigue, 300
  hydrogen embrittlement, 300–301
  SCC, 299–300, 299f
  sulfide-stress cracking, 301
Epoxies, 287
Equivalent lengths, 435–436, 502, 502f
Erosional velocity, 344
"Error free" polarized potential, 333
Ethylene, 234, 238
Ethylene, dense phase, orifice meters in, 639–646
Ethylene decomposition, 239
Ethylene distribution system, 640
Ethylene pipelines, commissioning, 238, 239f–240f
Ethylene propylene (EPR), 568
European Committee for Standardization. *See* Committee European de Normalization (CEN)
Evaporation, 241, 1047–1048
Event trees tool, 795
Excavated soil, shrinkage and expansion of, 66
Excavations, angle of repose for sloping sides of, 49, 49f
Experience factor, risk management tool, 798
Exposure, risk management tool, 798
External corrosion, 289–295, 750 (*see also* Internal corrosion)
  factors affect, 289–291
  monitoring techniques for, 295, 296f
External Corrosion Direct Assessment (ECDA), 346
External floating roof tank (EFRT), 826–829, 828f, 828t, 834–835, 861–863, 862f
External SCC, 299

F

Failure pressure *vs*. defect size, 784
Failure risk curve, 345
False alarms and misses, rate of, 748, 748f
Fanning equation for gas flow, 448–450, 449t–452t
Fatigue crack growth, 788–789, 788f–789f
Fault trees tool, 795, 802–804, 803f
Federal Energy Regulatory Commission (FERC), 964
  compressor stations, 1010
Federal Power Commission (FPC), 187
Ferrous debris, 230–231
Field-gathering pumps, selecting motors for, 588
Financial projects, decision and evaluation criteria for, 947
Firm service with fixed maximum daily quantity and reservation charge, 1019
Fixed roof tank (FRT), 826–827, 826f–827f, 834, 861
Flanged fittings, series number identification, 153
Flanges
  ANSI forged steel, 39, 39f
  bolt tightening sequence, 37, 37f
  bolting dimensions for, 33–36, 33t–36t
Flash point, 1133
Flash welded seam defects (FW seam defects), 104
Flashing service, 248
Flaw Assessment and Sizing Technique (FAST), 754
Float and tape sensor, 741
Float level sensors, 741
Float transmitter, 741
Floating LNG (FLNG), 382
Floating Production, Storage and Offloading (FPSO), 394
Floating roof, 834–843
  applications and details, 837–838
  components, 839–842
  failure modes and risk, 802–812
    APFR LOB event from corrosion, 804, 804f–805f
    common failure mode, 804–805
    design flaws, 809
    exceeding travel limits, 809
    excessive flexure of floating roofs, 806–807, 807f
    failure causation, 802
    fault trees, 802–804, 803f
    gas blowing, 808, 808f–809f
    high flow operations, 808–809
    insufficient draining of precipitation, 810, 810f–811f
    loss of buoyancy DDFR, 804–805
    loss of buoyancy in pontoon section, 804, 806f
    loss of buoyancy to corrosion, 811–812
    mechanical impact, 807–808, 807f
    non-operational state failures, 809–810
    seal failure, 808, 808f
    stress corrosion cracking, 809
  legs, 839, 839f
  pan roofs, 842–843, 842f–843f

pontoon and compartment access hatches, 839f, 840
seals, 840–842
tanks, 827–829, 828f
tanks and floating roofs, 834–837, 835t
vacuum breakers, 840, 842f
workhorses, 838–839, 838f–839f
Floating storage and offtake (FSO), 394
Floodlighting calculations, 177
Flow
coefficient, 266
conversion chart, 1162, 1162f
curve, 281
flow-assisted corrosion, 286
flow-capacity coefficient, 269
of fuel gases in pipelines, 444, 444f
graphs, 659–660, 660f
measurements, 713–715, 714f–715f
meter to application, 673–680
additional considerations include, 679
differential-pressure flow meters, 673–675, 674f
electromagnetic flow meters, 675–677, 675f
mass flow meters, 677–678, 677f
positive-displacement flow meters, 679–680
target flow meters, 679
turbine flow meters, 679
ultrasonic flow meters, 679
variable-area flow meters, 679
vortex/swirl flow meters, 678–679
pattern, 539
rate estimation from leaking tank hole, 872, 872f
examples, 873–875
fluid flow, 873
hole diameter, 873, 874f
rate
effect of flow rate variability, 487
proxy, 924, 924t
regimes, 453–454
velocity, 495
Flow-induced localized corrosion (FILC), 344
Flow-through dampeners, 564–565, 565f
Fluid
flow stream, 693
fluid–pipe–equipment interaction, 529, 530f, 530t–532t
power formulas, 51
pressure pulsations, 561, 561f–562f
properties, 528
significance of fluid velocity, 344
Fluidity of jet fuels, 1128
Foam pigs
cleaning pigs, 227
sealing pigs, 229
Footcandle light (fc light), 177
Form rollers techniques, 55
Formal external inspection (FEI), 866
Formal internal inspection (FII), 866–867
determining FII interval, 867, 868f
Formulas
Aude, T. R, 498
Benjamin Miller, 498
fluid power, 51
Hazen-Williams, 498
Shell/MIT, 498
tubular goods, 142
valve sizing, 248–251, 248f–249f, 251f
Weymouth's gas-flow formula, 449t
Foxboro equation, 250
Frequency-Modulated Continuous Wave (FMCW), 743
Fretting corrosion, 286
Friction component, 577
Friction horsepower, 602
Friction loss, 500
Friction-dominated systems, 605
Front-end engineering design (FEED), 884
Fuel
conversions involving different types of, 1162, 1163t
gas line for compressor station, 419–420
quality, 1119–1121, 1120t
fuel property and significance, 1120t
specifications, 1117
types, 1117–1118
consensus specifications, 1117
enforcement, 1118
mandatory and suggested specifications, 1118
test methods, 1117–1118
Full-load currents
single-phase transformers, 169, 169t
three-phase transformers, 172
Full-load-torque, 160
Fusion bonded epoxy (FBE), 782, 1095–1099
pipe coating, 1095
resistant overlay coatings to steel pipe, 1095–1099
Fusion bonding, 292
Future value (FV), 945
of periodic series of investments, 944
of single investment, 942

G

Galvanic anodes, 313–317
Galvanic corrosion, 286
Garden Banks Gathering System (GBGS), 1020

Gas
  analysis, 356–357
  blowing, floating roof failure modes, 808, 808f–809f
  chromatograph oven, 374, 374f
  composition, 422
  compression, 1010
    gas transmission compressor station cost in United States, 1013
    gas transmission compressor station unit cost in United States, 1013
    U.S. Federal Energy Regulatory Commission transmission compressor station cost, 2014–19, 1010–1012, 1011t, 1012f, 1012t
  conversion table for pure methane, 381t
  distribution systems in U.S., 14, 14f
  energy measurement equipment
    Btu, 375–376
    chromatograph controller, 375
    chromatographs work, 373
    gas chromatograph oven, 374, 374f
    general methods, 371–373, 372f–373f
    input/output chassis, 375, 375f
    sample conditioning system, 374, 374f
  flow measurement—orifice plate flange taps, 646–647
    properties of, 648–649, 648t–649t
    sequence of calculations, 646–647, 649
  flow through orifice meters, make quick calculation of, 654
  gas-gathering systems, 228
  glossary of common gas industry terms, 396–398
  heating value of natural gas, 369, 369t
  LNG, 382–395
  methane and behavior, 377
  nomograph for calculating density and specific volume of gases and vapors, 370f
  properties, 356–357
  properties from gas analysis, 356–357
  slugs, 156
  specific gravity of natural gas, 368
  transmission pipelines
    circa, 4, 4f
    incident heat map, 6, 6f
  types of heating value, 368
  and vapors, 249–251
Gas chromatographs (GCs), 372
Gas pipelines, 296
  ANSI/ASME, 126, 126t–127t
    longitudinal joint factor, 127t
    specified minimum yield strength, 126t–127t
    temperature derating factor for steel pipe, 127t
    values of design factor, 127t
  hydraulics calculations, 434–435
    equations commonly used for, 434
    nomenclature for Panhandle equations, 434
    nomenclature for Weymouth equation, 434
    sample calculations, 434–435
    source, 435
  pressure test requirements, 185f, 204
    49 CFR 192, 187
    advent of hydrostatic testing, 184
    ASA B31.1 prior to 1955, 184–186
    B31.8 Post-1955, 186
    compliance matrix of pressure testing, 187, 188f
    grandfathered pipelines, 187
    standard, predecessors, and sequels, 184
Gas-saturated crude oil, viscosity of, 462
Gasoline
  ASTM specification for, 1119
  conventional, 85.5 octane, 1140t
  conventional, 87 octane, 1141t
  fuel specifications, 1121–1124, 1122t–1123t
    additives and blending components, 1124
    composition, 1123
    corrosion, 1123–1124
    energy content, 1124
    storage and stability, 1124
    volatility, 1121–1123
  subgrade, 87 octane after 10% ETOH addition, 1142t
Gathering lines, gas, size determination, 440
Gathering systems in U.S., 13, 13f
Gauging pigs, 221, 230
Gauging plate, 221, 230
Gay-Lussac's Law. *See* Charles' law
Gel pigging, 224
Gel pigs, 224
General Attack Corrosion, 286
General liquid density nomograph, 1148, 1149f, 1149t
Generalized compressibility factor, 407
GENNET-M program, 505–506, 538, 539t
Geometry Inspection Tools, 752–753, 753f
Geometry pigging, 221
Girth Weld Factor (C_{11}), 799
Glossary of common gas industry terms, 396–398
Glycol, 222, 297
Gouges, 750, 767, 768f
Governmental effects on flow measurement, 629
Grab samplers, 720
Gradients, 945–946, 945f
"Grandfather" rule, 212
"Granny ragged" hot tar coating, 292
Graphite anodes, 325
Gross heating value, 368
Ground bed resistance calculation (chart), 320f, 321
Ground temperatures, 524
Grounding system components, 182
Growing perpetuity, 943
Guided Wave Ultrasonic Testing (GWUT), 347

Gulf Coast, 480, 483f, 483t
Gulf coast CO_2 pipeline infrastructure circa, 10f
Guy wires, stress in, 18–19, 18f

H

Hard-spot failures, 300, 300f
Hatschek equation, 469
Hazard and operability analysis tool (HAZOPS tool), 795, 915
Hazardous Air Pollutants (HAPs), 382
Hazardous liquid pipelines
  accident heat map, 5, 5f
  circa, 4, 4f
  schematic of hazardous liquid pipeline system networks, 2, 2f
Hazardous liquid systems in United States, 12, 12f
Hazardous locations, 161–162
  definition of class locations, 161–162
Hazen-Williams formula for calculating pressure drop andflow rates, 498, 577
  examples, 498
Head loss
  factor, 252
  due to friction, determination of, 526
Head *vs.* capacity curve, 620
Header connections, 665
Heat capacities, 379–380
Heat-transfer coefficients, 490
Heating, ventilation and air conditioning system (HVAC system), 618–620
Heating value
  of natural gas, 369, 369t
  types of, 368
Helical turbine meters, 670
Hex jam nuts, dimensions of, 48t
Hex nuts, dimensions of, 48t
High consequence areas (HCAs), 211, 346, 746
High frequency reciprocating rig method (HFRR method), 1133
High frequency welding (HFW), 55
High pressure gas, 654–657
  chart changing, 657f
  contracts, 654–655
  design and planning, 655, 655f
  installation, 655–656
  operation and maintenance, 656–657, 656f
  wet gas measurement, 657
High resolution MFL tool, 751
High temperature vapor pressures for light hydrocarbons, 1153
High Voltage Direct Current (HVDC), 328
High-Density Polyethylene (HDPE), 345
High-Temperature Corrosion, 286
High-viscosity liquids
  coriolis meters, 670–671
  helical turbine meters, 670
  liquid ultrasonic meters, 671, 671f
  measurement alternatives, 668–670
  measuring flow of, 668–672
  meter selection and sizing, 670
  operational issues, 671–672
  positive displacement meters, 670
  viscosity, specific gravity and temperature, 668
Horizontal cylindrical tanks
  contents, 878–879, 879f
  gauge, 879
Horizontal three-phase flow, 456
Hose string, pressure loss in, 495–496, 496t
HVDC effects on pipelines, 328f, 330–331
  electric field, 327f, 328–330
  protective system limits, 329–331, 329f–330f
Hydraulic grade line, 747
Hydraulic gradients, 499–502, 499f–501f
Hydraulic shoring, 69
Hydraulics
  gas
    calculate line loss using cross-sectional areas table, 443, 443t
    calculate velocity of gas in pipeline, 445
    calculate weight of gas in pipeline, 447
    chart for determining viscosity of natural gas, 448, 448f
    determination of leakage from gas line using pressure drop method, 439
    determine pressure loss for low-pressure gas system, 436
    determining throat pressure in blow-down system, 445–446, 445f
    energy conversion data for estimating, 440
    equivalent lengths for multiple lines based on Panhandle A, 435–436
    estimate amount of gas blown off through line puncture, 446
    estimate average pressure in gas pipeline, 447
    estimate equivalent length factors for gas lines, 438
    estimate pipe size requirements for increasing throughput volumes of natural gas, 442, 442f
    estimate time required to get shut-in test on gas transmission lines, 441
    estimating comparative capacities of gas pipelines, 439
    fanning equation for gas flow, 448–450, 449t–452t
    flow of fuel gases in pipelines, 444, 444f
    gas contained in line section, 438
    gas pipeline hydraulics calculations, 434–435
    maximum acceptable pressure loss for new lines, 441

Hydraulics (*continued*)
multiphase flow, 452–456
nomograph for calculating Reynolds number, 457f–458f
nomograph for determining pipe-equivalent factors, 437f
practical way to calculate gas flow for pipelines, 446
quick way to determine size of gas gathering lines, 440
liquids
approximate throughput of line estimating from pipe size, 525
calculating pressure and temperature loss for viscous crudes, 505–507, 505f, 506t, 507f, 507t
calculating pressure loss in annular sections, 505, 505f
determining batch injection rate as per enclosure, 508
determining maximum surge pressure in liquid-filled pipeline, 528
DRAs, 521–523
equivalent lengths, 502, 502f
estimating crude gathering line throughput for given pipe diameter, 526
estimating rate of liquid discharge from pipe, 524, 524f
flow pattern, 539
head loss due to friction in ordinary iron pipeline carrying clear water, 526
hydraulic gradients, 499–502, 499f–501f
hydraulics calculations for multiphase systems, 538–539, 538f, 539f, 540f
hydrostatic pressure due to column of liquid H feet in height, 527
looped systems, 504, 504f
marine hose pressure loss, 492–493
nomograph for calculating pressure drop of liquids in lines for turbulent flow, 517, 518f–520f
nomograph for calculating Reynolds number, 515–517, 515f–516f
predicting subsurface temperature ranges, 524–525, 524f
pressure drop calculations for rubber hose, 497
pressure loss in hose string, 495–496, 496t
pressure loss through valves and fittings, 509–514, 509f–514f
series systems, 503, 503f
sizing crude oil and products lines for capacity increases, 527
sizing lines, estimate pressure drop, and estimate optimum station spacing for crude systems, 527, 527t
sizing pipelines for water flow, 525
transient pressure analysis, 528–529
typical formulas used for calculating pressure drop and flow rates for pipelines, 497–499
Hydraulics calculations for multiphase systems, including networks, 538–539, 538f–540f
Hydrocarbon(s), 355, 1044
classification of gases and products, 3, 3f
gas viscosity, 1155
hydrocarbon-carrying pipeline, 226
liquids, 384
viscosities of, 467, 467f–468f
products, 297
of viscosity, 465
volumetric efficiency for, 595–596, 596t, 597f
Hydrogen (H_2), 300–301
embrittlement, 300–301
Hydrogen sulfide (H_2S), 157, 296, 345
Hydrogen-induced cracking (HIC), 299
Hydropneumatic bladder pulsation dampeners, 565
Hydropneumatic pulsation dampeners, 564
Hydrostatic design stress (HDS), 57
Hydrostatic pressure, 527
testing, 184
Hydrostatic test pressure (HTP), 798
Hydrostatic testing
benefits and limitations of
background, 189–190
practical considerations, 197–198
pressure reversals, 195, 195t, 196f–197f
pressure–volume plots, 192, 193f
test-pressure-to-operating-pressure ratio, 190–191, 190f
testing existing pipelines, 192–195
testing to levels above of SMYS, 191, 191f
compressibility factor for water, 209
estimating amount of pressure change for change in test water temperature, 206, 208f
history of gas pipeline pressure test requirements, 185f, 204
hydrostatic test records, 209
pipeline, 785–786
for pipelines, 199–203, 199f–204f
pressure testing and recordkeeping requirements, 209
volume of water required at test pressure, 205–206
volume of water required to fill test section, 185f, 204
Hyperbaric spheres, 224
Hysteresis, 283

I

Ideal flowmeter, 715
Ideal gas law, 355
Ideal risk-assessment model, 786–787
IEC 61511, 868
Impeller trim, 559–560, 559f–560f
Impressed current method, 287
In situ coatings, 224–225
In-line inspection (ILI), 223, 765–766
pipeline, 785–786
tools, 750

Incident management, 1070
Inclined two-phase flow, 456
Income statement, 956
Incorrect flowmeter, 422
Indexing models, pipeline risk assessment, 793
Indicated horsepower, 400
Indirect Inspection tools (IDI tools), 346–347
Individual retirement account (IRA), 944
Induced AC voltages on pipelines, 333
  electromagnetic coupling, 341
  induced AC mechanisms, 341
  mitigation techniques, 342–343
  personnel safety, 343
  pipeline hazards, 341
  predicting AC voltages on pipeline, 342
  resistive coupling, 342
Industrial pigging systems, 230
Industry Ethylene Measurement Project, 641
Inertial mass flow meters, 688–689
Inhibitors, 297
Injection rate, batch, as per enclosure, determination of, 508
Inland aquatic environments, 1031–1032
  ecological effects, 1032
  sensitivity, 1032
Inlet cu ft per minute (ICFM), 409
Inlet Gas Reception, 384
Inlet Gas Treating, 384
Inlet piping, 261
Inlet valve velocity, 402–403
Inner diameter (ID), 435
Input/output chassis, 375, 375f
Inspection pigging, 223
Inspection tools and case studies, 750–760
  Columbia Gas, 758
  comparison of MFL technologies, 751
  description of integrity issue, 753, 753t
    field verification, 754
    implementation, 754
    inspection tool selection, 753–754
    Platte pipe line parameters, 753t
    TFI tool validation program, 754
  geometry tools, 752–753, 753f
  high resolution MFL tool, 751
  MFL inspection tools, 750–751
  Platte pipe line, 753–755
  shear wave UT tools, 752, 752f
  standard MFL tool, 750
  standard UT tools, 752, 752f
  TAPS, 755–760
  TFI tool, 751
  ultrasonic inspection tools, 751–753, 752f
"Instant off potential". *See* "Off potential"
Institute of Petroleum Classification, 825, 826t
Instrumentation
  automatic tank gauging systems and proof testing, 726–732, 727f
  developments in pipeline instrumentation, 712–726
  instrumentation and sensors for automatic tank gauging, 737–744
  types of control systems, 710–712, 710f, 712f
Insulated cables for permanent training during installation, 170, 171t
Integrity management
  factors to consider evaluating damage on pipelines, 760–766
  floating roof failure modes and risk, 802–812
  inspection tools and case studies, 750–760
  leak detection technologies, 746–750
  modeling for pipeline risk assessment, 792–796
  pipeline revalidation, 783–791
  pipeline risk assessment, 792
  planning, 211
  risk management, 792
    tool for establishing budget priorities, 797–802
  understanding mechanical damage in pipelines, 766–773
  understanding stress corrosion cracking in pipelines, 774–782
Integrity Management Program (IMP), 756
Intelligent pigs', 218
Interest, 940
Interface mixing, 223
Intergranular corrosion, 286
Internal corrosion, 295–299 (*see also* External corrosion)
  gas pipelines, 296
  liquid Pipelines, 296, 296f
  monitoring, 298–299
  preventive/mitigative measures for, 297–298
Internal corrosion, 750
Internal corrosion direct assessment (ICDA), 297
Internal floating roof tank (IFRT), 826, 829, 829f, 834–835, 863, 863f
Internal rate of return method (IRR method), 948–950, 949f–950f
Internal sand blasting, 233, 235
Internal SCC, 299–300
International Bureau of Legal Metrology (BIML), 689
International Organization for Legal Metrology (OIML), 689
International Standards Organization (ISO), 618, 690
International Union of Pure and Applied Chemists (IUPAC), 637
Investments, decision and evaluation criteria for, 947

Investments/projects with multiple or irregular cash flows, analyzing and valuing, 942–943, 943f
Irregular cash flows, 942–943, 943f
Iso-butane, 1139, 1139t
Isothermal horsepower, 400

J

J-C-O forming techniques, 54
J-lay system, 985
JAX LNG facility, 910
Jet fuels specifications, 1126–1130, 1127t
  additives, 1130
  ASTM requirements of aviation turbine fuels, 1127t
  combustion characteristics, 1129
  composition, 1129
  contaminants, 1130
  corrosion, 1130
  fluidity, 1128
  heat content, 1128–1129
  lubricity, 1129–1130
  specifications, 1126
  stability, 1128
  volatility, 1128
Joint efficiency factor, 186

K

K factors for valves and fittings, 252–255, 255t
Key performance indicators, 1045–1047
  total incidents, 2015–19, 1045f
  total incidents, inside and outside HCA, 1046f
  total incidents, by size, 1046f
  total incidents, by commodity, 1046f
Kinematic viscosity, 1158
  at 122°F, 1156
Knockout dimensions, 164

L

Laboratory calibrated master meter, 699
Laboratory-proven meters, 699
Land descriptions, 45–47, 45t
  descriptions show directions in terms of degrees, 47, 47f
  meandered water and government lots, 45
  metes and bounds descriptions, 46
  reading land descriptions, 45
Lead pipe service lines state ratings, 1045f
Leak
  classification of, 746
  detecting leaks around tanks, 736f
  detecting leak in pipelines in slack condition during transients, 748
  detecting leak in shut-in pipeline segments, 748
  detection technologies, 746–750
  evaluation of, 746–749
  flow rate, 746
  location estimation, 747, 747f
  regulatory background, 746
  signal, 747, 747f
  size system, 746
  standards for, 749–750, 749t
    industry leak detection standards, 749t
Learning ability, ideal risk-assessment model, 795
Lee-Kesler equations, 423
Lengths, equivalent, 435–436, 502, 502f
Lens shade selector, 93
Light hydrocarbons
  high temperature vapor pressures for, 1153
  low temperature vapor pressures for, 1152
Light Loss Factor (LLF), 177–178
Line block valves, 717
Line break detectors, acoustic, 717–718, 718f
Line loss using cross-sectional areas table when testing mains with air or gas, calculation of, 443, 443t
Line puncture, estimation of gas blown off through a, 446
Linefill in barrels per mile, 477
Liquefaction process, 392–393
Liquefaction Technology and Cost, 387–390, 387f–392f
Liquefied natural gas (LNG), 377, 382–395
  background, 383–384, 383f–384f
  conversion table, 1164
  floating LNG facility, 394–395, 394f–395f
  liquefaction technology and cost, 387–390, 387f–392f
  plant components, 384–386
  plants connected to natural gas pipeline systems, 7, 7f
  storage tanks, 902–908
Liquefied natural gas facilities in U.S., 15f
Liquid Ultrasonic Flow Meters (LUFMs), 689
  for custody transfer measurement, 689–700
  custody transfer requirements, 690
  measurement definitions, 690–692
  unique problems in proving liquid ultrasonic meters, 692
Liquid ultrasonic meters, 671, 671f, 700
  advantages, 671
  for custody transfer, 680–685
    transit time ultrasonic flow meters, 680–685, 680f–682f, 683t, 684f
  disadvantages, 671
Liquids, 248–249
  calculate bulk modulus, 469
  calculate specific gravity of blend, 466
  calculate viscosity of blend, 465
  carbon capture and sequestration processes, 478
  convert specific gravity to API gravity, 466
  convert viscosity units, 466
  derivation of basic ultrasonic flow equations, 475–476
  determining viscosity of crude, 462, 462f
  estimate leakage amount small holes in pipeline, 478

estimate volume of pipeline per linear foot inside diameter, 477
formulas, 248–251
gravity and density conversion chart, 463, 463f
hydraulics, 711–712
linefill of given pipe in barrels per mile, 477
measurement orifice plate flange taps, 630–631
sequence of calculations, 630–631, 631f
nomograph for calculating viscosity
of compressible fluids in pipes, 471
liquid hydrocarbons high pressure, 464, 464f
of liquids in pipes, 471, 472, 472f–473f
of slurries, 469, 470f
oil move in pipeline, 477
pipelines, 189, 296, 296f
ANSI/ASME, 128, 128t–130t
tabulation of examples of allowable stresses, 128t–129t
weld joint factor, 130t
simplified procedure to calculate emissions for liquid organic storage tanks, 932–936
assumptions, 932
emissions from fixed roof tanks, 935–936
emissions from floating roof tanks, 932–935
tank emissions, 932, 933f–934f
toxic air contaminant calculation, 936
effect of variability on CO_2 pipeline operation, 486
velocity heads for, 474, 474f
velocity of liquids in pipes, 472, 472, 472f–473f
viscosities of hydrocarbon liquids, 467, 467f–468f
viscosity of mixture of liquids blending index, 465
weight density and specific gravity of, 1151, 1151f
Localized corrosion, 286
Location adjustment, 967
Lock-up, 283
Locked-rotor current, 161
Locked-rotor torque, 160
Longitudinal Seam Cracks, 750
Longitudinal-submerged arc welding (L-SAW), 971
Loop lines, 502
Looped systems, 504, 504f
Loss of buoyancy (LOB), 804
to corrosion, 811–812
DDFR, 804–805
in pontoon section, 804, 806f
Loss of ductility, 301
Losses, 424
Low pressure flow prover, 664
Low temperature vapor pressures for light hydrocarbons, 1152
Low-noise trim, 269, 271t
valve reference data, 271t
Low-pressure gas system, determine pressure loss for, 436
Lower Explosive Limit (LEL), 861
Lowest one-day mean temperature (LODMT), 856
LPG Fractionation, 385
LPG Recovery, 385
Lube oil blending, 900
Lubricity
of diesel fuel, 1133
of jet fuels, 1129–1130
Lumen light, 177
Lux light (lx light), 177

M

Machine screws, tap drills and clearance drills for, 31, 31t
Magnesium anodes, 322
Magnesium rod, current output of, used for cathodic protection, 325
Magnetic float indicators, 741
Magnetic Flux Leakage (MFL), 750, 758
comparison of MFL technologies, 751
inspection tools, 750–751
Magnetic-cleaning pig, 230–231
Magnetite, 224
Magnetorestrictive float level sensor, 741, 742f
Maintenance tips for positive displacement meters, 661–664
Mandrel pigs
cleaning pigs, 227, 227f
sealing pigs, 228–229, 228f
Manila rope, find working strength of, 16
Manual Overfill Prevention System (MOPS), 729
Manual tabulation, leak detection technologies, 746
Manual tank gauging procedures, 916–923, 919f
innage *vs.* ullage measurements, 917, 918f
level and mass measurement technology concepts, 918–919, 918f
manual level measurement process, 919–923
datum/strike plate, 920, 920f
measuring cut, 920–921, 920f–921f
measuring water and sediment bottoms, 921, 921f
reference height, 920, 920f
reference point, 919
temperature gauging, 921–923
tank gauging systems, 916–917, 917f
Marathon Pipe Line, 753–754
Marine hose pressure loss, 492–493
CALM system, 492
multi-point mooring system, 493, 493f
SALM system, 492
Tandem system, 493, 493f
Marine terminals, 897–899
Mass diffusivity (D), 379
Mass flow meters, 677–678, 677f
Mass measurement light hydrocarbons, 632, 632t

Mass measurement light hydrocarbons *(continued)*
composite sample analysis, 632
sample calculation, 632
Master meter for ultrasonic applipcation, 699
Master meter proving, 684, 700
principle, 642
systems, 697–699
Master meter proving orifice meters in dense phase ethylene, 639–646
analysis of data, 643–645
development program, 641–642
goal of development program, 639
pipeline transportation and measurement systems, 640–641, 641f
proving procedures, 642–645
Master meter proving orifice meters in dense phase ethylene, 639–646
Material factor, risk management tool, 798
Matrix models, pipeline risk assessment, 793
Maximum allowable operating pressure (MAOP), 199, 780, 798
Maximum operating pressure (MOP), 190, 780, 784
Maximum surge pressure determining in liquid-filled pipeline, 528
Maxville LNG facility, 909
Meandered water, 45
Measurement
Btu, 653
coriolis flow meter, 687–689
determine number of purges for sample cylinders, 653
determine required orifice diameter, 650, 650f
estimate natural gas flow rates, 651, 651f
estimate temperature drop across regulator, 651
four ways to calculate orifice flow in field, 658–661
gas or vapor flow measurement—orifice plate flange taps, 646–647
handling entrained gas, 685–687
high pressure gas, 654–657
liquid measurement orifice plate flange taps, 630–631
liquid ultrasonic meters for custody transfer, 680–685
make quick calculation of gas flow through orifice meters, 654
mass measurement light hydrocarbons, 632, 632t
master meter proving orifice meters in dense phase ethylene, 639–646
matching flow meter to application, 673–680
measuring flow of high-viscosity liquids, 668–672
multiphase flow meter, 626–627, 626f
pipeline flow measurement factors, 627–630
pipeline measurement of supercritical carbon dioxide, 633–639, 633f
practical maintenance tips for positive displacement meters, 661–664
pressure differential on remaining meter runs of parallel system, 652
properties of gas and vapors, 648–649, 648t–649t
proving liquid ultrasonic flow meters for custody transfer measurement, 689–700
sampling and testing program for unrefined petroleum-based products, 706–707
sizing gas metering run, 652, 652f
sizing headers for meter stations, 665–667
V-Cone® flow meter, 700–706, 701f
Mechanical advantage, 17
Mechanical damage in pipelines, 766–773
combined defects, 768
defining mechanical damage, 766–768, 767f
delayed repair conditions, 773
dents, 767
elastic spring-back and re-rounding, 768–769
geometry and characterization of damage, 769–771, 770f–772f
gouges, 767, 768f
growth of cracking from mechanical damage, 773
immediate repair conditions, 772–773
leak *vs.* rupture, 773
mechanical damage features and characteristics, 768–769
reduction of toughness and ductility strain, 769
repair conditions and failure modes, 771–773
stress/strain fields, 769
Mechanical efficiency, 401, 603
Mechanical float switch, 738, 739f
Mechanical losses, 424–425
Mercury, 43
Metal loss, corrosion-caused, 194
Meter selection and sizing, 670
Meter stations, sizing headers for, 665–667
Meters, maintenance tips for positive displacement, 661–664
Metes and bounds descriptions, 46
Methane (CH_4), 154–155, 377, 380
critical constants, 377
diffusivity, 379
equations of state, 377–378
flammability characteristics, 380–381
heat capacities, 379–380
heat capacity ratio, 380
physical, 377
thermal conductivity (liquid), 378
thermal conductivity (vapor), 378t–379t
thermodynamic, 377
properties, 379
transport, 377
properties, 378
vapor pressure[2], 378
Methanol, 222
drying, 234

Metric conversions, 1148f
Metric tons per kilometer, estimate weight of pipe in, 143t
Microbial corrosion, 289, 290f
Microwave hops on level ground, 174–175
Mid-continent CO_2 pipeline infrastructure, 11f
Miller, Benjamin, pressure drop and flow rates formula, 498
Minimum acceptable operating pressure (MAOP), 970–971
Minimum bending radius, 170, 171t
  for conductor cable, 171t
  power and control cables, 171t
  twisted pair instrumentation cable, 171t
Minimum Net Flow Area (MNFA), 263
Minimum remaining thickness (MRT), 867
Mirror hygrometer, 242
Mitigation techniques, 342–343
Mixed Fluid Cascade technology, 388
Mixed phase flow, 250
Modified Beggs and Brill-Moody correlations, 539
Modulus, bulk, calculation of, 469
Moisture content of air, 236–237, 236f
"Mol-volume", 355
Money, time value of, 940
Monitoring tool, 721
Mooring system, multi-point, 493, 493f
Motor controller sizes, 172
Motor gasoline, 1121
Motor type, selection of, 160
Motors for field gathering pumps, selection of, 588
Multi-diameter pigs. *See* Dual-diameter pigs
Multi-point mooring system, 493, 493f
Multiphase flow, 452–456
  flow regimes, 453–454
  horizontal three-phase flow, 456
  inclined two-phase flow, 456
  two-phase horizontal flow, 453, 453f–455f
Multiphase flow meter, 626–627, 626f
  principle of operation, 626–627, 626f
Multiphase systems, hydraulics calculations for, 538–539, 538f, 539f, 540f, 541f, 542f

N

Nails, common, 32
Naphtha blending, 900
Naphthenes, 1040
Narrow Axial External Corrosion (NAEC), 751
NATASHA PLUS, 529
National Association of Pipe Coating Applicators (NAPCA), 1095, 1099, 1107, 1110
  NAPCA bulletin 12–78–04, 1095–1099
    external application procedures for plant applied FBE coatings and abrasion, 1095–1099
  NAPCA bulletin 13–79–94, 1099–1102
    external application procedures for coal tar epoxy protective coatings, 1099–1102
  NAPCA bulletin 14–83–94, 1102–1106, 1105t–1106t
    cross head extrusion data, 1105
    external application procedures for polyolefin pipe coating, 1102–1105
    side extrusion data, 1106
  NAPCA bulletin 15–83–94, 1107–1110
    exterior application procedures for plant applied tape coatings to steel pipe, 1107–1110
  NAPCA bulletin 3–67–94, 1113–1116
    external application procedures for hot applied coal tar coatings, 1113–1116
  NAPCA bulletin 6–69–94–3, 1110–1111
  NAPCA bulletin 6–69–94–5, 1111–1113
    procedures for coating field joints using heat-shrinkable materials, 1111–1113
  procedures for coating field joints, fittings, connections, and preffabricated sections, 1110–1111
  specifications for pipeline wrappers, 304
  standard applied pipe coating weights, 302–303
National Electrical Code cycle (NEC cycle), 181
National Electrical Code tables, 165, 165t–167t, 169t
  allowable ampacities of insulated conductors rated, 166t
  article 344–rigid metal conduit, 169t
  maximum number of compact conductors, 167t
  maximum number of conductors or fixture wires, 167t–168t
National Electrical Manufacturers Association (NEMA), 160
  enclosure types, 162–163
  hazardous, 163
  nonhazardous locations, 162–163
National Fire Protection Association (NFPA), 162, 824–825, 825t
National Institute of Standards and Technology (NIST), 681
Natural dispersion, 1067
Natural gas, 371, 1164
  liquefaction, 382, 387
  schematic of natural gas pipeline system networks, 2, 2f
  specific gravity of, 368
  transmission in U.S., 13f
Natural Gas Pipeline Safety Act, 187
Nautilus pipeline system, 997–1001, 997f–998f
Navigating pump curves, 620–623
  BEP, 621–622
  brake horsepower, 622–623
  head *vs.* capacity curve, 620
  implied or title block information, 622
  minimum flow, 621, 621f
  POR *vs.* AOR, 622

Navigating pump curves *(continued)*
pump curve from manufacturer's perspective, 623
runout or end of curve, 622
shut-off head, 620–621
understanding specific speed and suction specific speed, 623
Near-neutral pH SCC, 777, 777f, 778t
Negative pressure wave detector, 718
NEMA enclosure types, 162–163
Net environmental benefit analysis (NEBA), 1033–1035
in practice, 1038–1040, 1038f–1040f
Net heating value, 368
Net positive suction head (NPSH), 547, 560, 614
conversion factors and formulas, 549t
formulas for recalculating pump performance with impeller diameter or speed change, 552t
and pump operating regions, 614–617
industry standards, 614–615
NPSH margin & operating region, 616–617, 616f–617f
operating regions, 615–616
pressure & NPSH terminology, 615, 615f
Net Positive Suction Head available (NPSHa), 563, 567
Net positive suction head curve, 618
Net Positive Suction Head required (NPSHr), 563, 567, 592, 608
Net present value method (NPV method), 947–951, 951t
NGL products
butane, normal butane, 1137, 1137t
iso-butane, 1139, 1139t
propane, 1136, 1136t
RGB, 1138, 1138t
Nominal interest rate, 941
Nomograph
aids in diagnosing compressor cylinder ills, 408
for calculating density and specific volume of gases and vapors, 370f
for calculating pressure drop of liquids in lines for turbulent flow, 517, 518f–520f
for calculating Reynolds number, 515–517, 515f–516f
for calculating viscosities of liquid hydrocarbons high pressure, 464f
for determining horsepower requirement of pumps, 588f
for determining pipe-equivalent factors, 437f
for determining reciprocating pump capacity, 586f
for determining specific speed of pumps, 587f
Non-contacting measurements devices, 737
Non-destructive examination (NDE), 750, 754
Normalization, 963–964
North East Jackson Dome (NEJD), 480
Nozzle, 674

O

Occupational Safety and Health Act, 40
Odd-angle long radius elbows, 44, 44f
"Off potential", 295, 337
Office of Pipeline Safety (OPS), 154, 750
Offshore pipeline construction, 984
cost estimation using work decomposition
cost components, 996
Discovery Gas Transmission system, 999f, 1001
Nautilus pipeline system, 997–1001, 997f–998f
work breakdown structure, 996
cost in U.S. Gulf of Mexico, 984–985, 984f–986f
construction process, 988–990
cost factors, 986–990, 987f
engineering and inspection, 989–990
limitations, 994–995
Offshore Technology Conference/Society of Petroleum Engineers data, 993
statistics, 991–995, 991t, 992f, 992t–993t, 994f–995f
shallow water pipeline construction cost bounds in US Gulf of Mexico using empirical data, 995
Offshore pipeline decommissioning, 1002
cost to decommission hurricane-damaged pipeline in U.S. Gulf of Mexico, 1007
cost to decommission shallow water pipeline in US Gulf of Mexico, 1007
deepwater pipeline decommissioning cost in U.S. Gulf of Mexico, 1008
deepwater umbilical, flowline, and riser decommissioning cost in U.S. Gulf of Mexico, 1008–1009
in U.S. Gulf of Mexico, 1002–1004, 1003f
cost factors, 1004–1005
limitations, 1006–1007
operations, 1003–1004
statistics, 1005–1007, 1006f
workflows, 1009–1010
Oil fate and weathering, 1047–1048, 1048f–1049f
biodegradation, 1031
dissolution, 1048
emulsification, 1031–1033, 1032f
evaporation, 1047–1048
movement, 1047
natural dispersion, 1067
photo-oxidation, 1031
sedimentation, 1031
sinking, 1031
terrestrial spills, 1031
weathering processes, 1047
Oil leak sensors, 744
Oil movement control systems, 892–902
control level, 894f
crude blending, 901f
functional modules, 895f

path selection, 896f
sequence generation, 896f
Oil movement and storage in tank farms, 892–902
automation technology and strategy, 893–897
blending, 899–900
OM&S control functions, 894, 894f
operating costs, 893
operating problems, 892–893
product losses, 893
terminal operations, 897–899, 898f
Oil spills, 1058–1083, 1068f, 1070t, 1071f, 1072t, 1074f–1079f, 1081t, 1082f
characterization methods, 1060t
net env. benefit analysis, 1072t
physical response techniques, 1081t
thickness, estimating visually, 1050t
U.S.C.G. juris.,
all sources, 1045f
volume, 1046f
volume and number of incidents, 1046f
Oiled shoreline assessment surveys, 1040
Olefins, 1040
"On potential", 295, 332
On-line inspection techniques, 721
One-dimensional scaling rule, 1013–1014
One-foot method, 857
Onshore buried pipelines, 289–290
Onshore pipeline construction, 968
cost, 968–971, 969f–970f
aggregate average cost, 974–975, 974t, 975f
construction process, 968–971, 969f–970f
cost factors, 971–973
Federal Energy Regulatory Commission pipelines, 964
limitations, 977
project average cost, 975–976, 975t, 976f, 976t
regression models, 976–977, 976t
in United States using unit costs and regression models, 977
cost bounds in onshore pipeline construction in United States, 978
Open top tank
with back slope floating roof, 830
with double deck floating roof, 831, 831f
with pontoon floating roof, geodesic dome, 832, 832f
Operating costs, 609
Operator qualification, 211
Optimum working pressures estimating in crude oil transmission lines, 527
Orifice
diameter, 650, 650f
flow equation, 635–637, 636f
measurement practices, 635
plate, 674
Orimulsion, 570–571
Outer continental shelf (OCS), 964
Outside influences on flow measurement, 629
Overfill prevention process approaches (OPP approaches), 912
Overfill prevention systems, 728
Oxyfuel combustion power plant, 478

P

Pacific Gas and Electric (PG&E), 213
Pacific Northwest (PNW), 330
Pacific Southwest (PSW), 330
Pan roofs, 842–843, 842f–843f
Panhandle equations, nomenclature for, 434
Paraffins, 1040
Parallel system, remaining meter runs of, 652
Payback method, 947–948
PE piping material designation code, 57–58
Pearson survey, 335
Penetrometers, 68–69
Peng-Robinson equations, 423
Periodic series of investments, future value of, 944, 944f
Permanent spoil, 73
Permanent training during installation, insulated cables for, 170, 171t
Permian Basin, 480, 481f, 482t
CO_2 pipeline infrastructure circa, 9f
Perpetuities, 943–944, 943f
Personnel requirements, 628
Personnel safety, 343
Petroleum Equipment Institute (PEI), 824
Petroleum oils
specific gravity/temperature relationship for, 1151, 1151f
Petroleum storage tank standards, 851–870
API 579, 869
API 620, 864–866, 865f
API 650, 855–863, 855f
API 653, 866–867
API 2350, 867–869
API 2610, 853–854
determining FII interval, 867, 868f
external floating roof tank, 861–863, 862f
facility standards, 853, 854f
fixed roof tank, 861
formal external inspection, 866
formal internal inspection, 866–867
internal floating roof tank, 863, 863f
materials, 856, 857f–858f
release prevention barriers, 863, 864f
Risk Based Inspection, 867
routine informal monthly inspection, 866
shell, 858, 858f
tank bottom, 859–861, 859f–862f
tank design, 856

Petroleum storage tank standards *(continued)*
tank roof, 861
tank shell, 856–858
pH SCC, 776, 778t
Phillips Optimized Cascade process, 388
Physical inspection, leak detection technologies, 746
Pig pipeline, 219–226
cleaning and construction utility pigs, 226–232, 226f
decommissioning, 226
government regulations, 220
inspection pigging, 223
pigging during construction, 220–222
cleaning, 221
commissioning, 222
debris removal, 220–221
dewatering, 221–222
drying, 222
geometry or caliper pigging, 221
hydrostatic testing, 221
initial gauging, 221
pigging during operation, 222–223
batch inhibition, 223
condensate removal, 223
debris removal, 222
product separation, 223
pigging during renovation/rehabilitation, 224–225
pigging for general maintenance and repair, 223–224
isolation, 224
pre-inspection cleaning, 223–224
practical requirements, 220
safeguarding assets and optimizing efficiency, 219–220
Pig train, 225
Pig trap design, 244–246, 244f–246f
Pilot operated relief valves, 260
Pin-wheel pig, 225, 231, 231f
Pipe design
allowable working pressure for carbon steel pipe, 147, 147t
American standard taper pipe threads, 139f
British standard taper pipe threads, 140f
calculate contraction or expansion of pipeline, 143
calculate pipe weight, contents, velocity, 141–142
formulas, 141–142
calculation of pipe bends, 135–136, 135t–136t
carbon dioxide pipeline design, 154–157
determine buoyancy of bare and concrete-coated steel pipe in water and mud, 145
determine buoyancy of bare steel pipe, 145
dimensions of three-diameter ells with tangents, 153
estimate weight of pipe in metric tons per kilometer, 143, 143t
formulas and constants for tubular goods problems, 142
gas pipelines–ANSI/ASME, 126, 126t–127t
identify schedule number of pipe by direct measurement, 144
identify series number of flanged fittings, 153
length of pipe in bends, 134, 134t
liquid pipelines–ANSI/ASME, 128, 128t–130t
maximum allowable length of unsupported line pipe, 144
normal engagement between male and female threads to make tight joints, 141
pipe weight from outside diameter and wall thickness, 144
properties of pipe, 131–133, 131t–133t
spacing of pipe supports, 137–138, 138f
suggested maximum spacing between pipe supports, 137
spectacle blind thicknesses, 153
steel pipe design, 126–128
stress in above ground/below ground transitions, 148–151, 149f–150f, 150t, 151f
stress in pipe wall internal pressure, 148
weights of piping materials, 146
Pipe diameters and different rates of discharge, 474, 474f
Pipe friction loss, 883
factor, 567
Pipe geometry, 798
Pipe laying
allowable pipe span between supports, 84
degrees of bend in pipe, 78
engineers make pipe fit ditch, 85–87, 86f
length of pipe bend, 82
length of pipe in arc subtended by any angle, 82
maximum bend on cold pipe, 80
pipe bending computations, 79
Pipe losses, 577–578
Pipe lowering, 78, 88
lower existing pipeline, 88–91
Pipe provers, 671
Pipe roughness, 522
Pipe size requirements for increasing throughput volumes of natural gas, estimating, 442, 442f
Pipe taps, drill sizes for, 32
Pipe threads
American standard taper pipe threads (NPT), 139f
British standard taper pipe threads, 140f
Pipe weight from outside diameter and wall thickness, 144
"Pipe-in-pipe" method, 204–205
Pipe-to-soil (P/S), 346–347
potential, 295
Pipeline and Hazardous Materials Safety Administration (PHMSA), 298–299
Pipeline Integrity International (PII), 751, 753–754, 759
Pipeline(s), 51–55, 52f, 191
advances in pipeline protection, 334
case histories, 338–339
close interval potential survey, 335, 336f, 337–338
Pearson survey, 335
signal attenuation coating survey, 335, 336f, 337
books, 58
calculate contraction or expansion of, 143

construction cost across time, 967–968
corrosion in, 288–301
environmentally assisted cracking, 299–301
external corrosion, 289–295
internal corrosion, 295–299
monitoring internal corrosion, 298–299
process, 288–289, 288f
uniform *vs*. localized corrosion, 289
dehydrator, 239
depressurization, 490
developments in pipeline instrumentation, 712–726
acoustic line break detectors, 717–718, 718f
cathodic protection, 723–725
computer systems, 722–723, 723f–724f
densitometers, 719–720, 719f–720f
flow measurements, 713–715, 714f–715f
future trends, 725
pipeline monitoring systems, 721–722
pipeline samplers, 720–721, 721f
proving devices, 715–716, 716f
SCADA systems, 723
smart pressure sensors, 718–719
system design guidelines, 725
valves, 716–717, 717f
dewatering, cleaning, and drying
brush pig run with gas, 233
brush pig run with liquid, 233
chemical cleaning, 233–234
cleaning pipelines, 233
dewatering, 232–233
internal sand blasting, 233
pipeline drying, 234–235
distillate
#1 diesel fuel, ultra-low sulfur, 1143t
#1 diesel fuel, ultra-low sulfur, 15 ppm, 1144t–1145t
#2 diesel fuel, ultra-low sulfur, 1145t–1146t
drying, 234–235
estimate leakage amount small holes in, 478
external corrosion direct assessment process, 346–351
direct examination, 347, 348f
indirect inspection, 347
post assessment, 347
potentials in paved areas, 350–351
preassessment, 346–347
results, 348, 350–351
severity ranking, 348–350, 349t–350t
facilities, 160
factors to consider evaluating damage on, 760–766
construction damage, 762–764, 765f
encroachment damage, 760–761, 760f–762f
evaluation and repair of construction damage, 765
evaluation and repair of encroachment damage, 761–762, 762f–763f
ILI, 765–766
mechanical damage in, 766–773
stress corrosion cracking in, 774–782
two forms of damage, 760
first fill, start-up, and shutdown procedures, 489–490
initial start-up, 489
planned shutdown, 489–490
transient conditions, 490
unplanned shutdown, 490
flow measurement factors, 627–630
accuracy, 627–628
economics of flow measurement, 628–629
fluid characteristics, 629
standards and governmental effects on flow measurement, 629
testing or proving flow meters, 629–630
gasoline
conventional, 85.5 octane, 1140t
conventional, 87 octane, 1141t
subgrade, 87 octane after 10% ETOH addition, 1142t
hazards, 334
hydrostatic testing for, 199–203, 199f–204f
material coating specifications, 1095
NAPCA bulletin 12–78–04, 1095–1099
NAPCA bulletin 6–69–94–5, 1111–1113
measurement of supercritical carbon dioxide, 633–639, 633f
definitions, 633–634
future, 638–639, 639t
measurement methods in today, 635, 635t
orifice flow equation, 635–637, 636f
orifice measurement practices, 635
physical properties, 634, 634f
pipeline operation, 637, 637f
potential measurement methods, 638
turbine metering, 637–638, 638f
monitoring systems, 721–722
NGL products
butane, normal butane, 1137t
iso-butane, 1139t
propane, 1136t
RGB, 1138t
oil move in, 477
per linear foot inside diameter, 477
pipeline manufacturing, 52–55
process systems, 219
product specifications, 1136, 1136t
revalidation, 783–791
background, 783–784
benefits of hydrostatic testing and in-line inspection, 785–786
determining to revalidate, 786–787

Pipeline(s) *(continued)*
fatigue crack growth, 788–789, 788f–789f
general concepts, 784–785, 784f–785f
mechanical damage, 790
stress-corrosion cracking, 790–791
risk assessment, 792
beginning risk modeling process, 792–793
choices in risk-assessment techniques, 793
choosing risk-assessment technique, 795
formal *vs*. informal risk management, 792
ideal risk-assessment model, 786–787
managing risks, 796
modeling for, 792–796
other issues in risk modeling, 793–795
samplers, 720–721, 721f
seamless pipe, 52
services
maintenance, 573
pump heating and/or cooling, 572
pump protection, 572–573
pumping stations, 569–570, 569f–570f
rotary pumps on, 569–573
specified operating requirements, 571–572
system considerations, 573, 573f
types, 569
types of pumps applied, 570–571, 570f, 570t, 571f
shaping techniques, 54–55, 54f
in slack condition during transients, detecting, 748
steel production, 51
terminals, 899
test temperature, 201
welded pipe, 54
welding techniques, 55
Pipes, 186, 529
environment, 529
estimating coating and wrapping materials per mile of, 305
primer and coal-tar enamel, 305
wrappings, 305
velocity of compressible fluids in, 467
velocity of liquids in, 469
Piping geometry factor (F_P), 267
Piping materials, weights of, 146
Piston displacement, 400–401
Piston rod load, 593–594, 594t
Piston speed, 401, 592
Pitot tube, 674–675
Pitting, 289, 289f
Plain dents, 760
Plant Pigging Systems, 230
Plasticity test, 69
Platte Pipe Line, 753–755
Plunger, 592
load, 593
Pneumatic shoring, 69–71, 71f
Pocket penetrometer, 68–69
Point level sensors, 738
Point-by-point method, 177–178, 177f–178f
Poisson's ratio, 191
Polly Pig, 218
Poly-tropic head equation, 411
Polyamide-cured epoxy, 224–225
Polycyclic aromatic hydrocarbons (PAHs), 1045
Polyethylene piping, 118
simplified installation guidelines for PE pipe, 118–120
trench for PE piping, 120–123
Polyolefin pipe coating, external application procedures for, 1102–1105
Polyphase motors, 160
Polytropic process, 409
Polyurethane foam, 218
Pontoon and compartment access hatches, 839f, 840
Portable electronic gauging technology (PEG), 919
Portable proving facility, 641, 641f
Positive displacement meters, 669–670
advantages, 670
disadvantages, 670
practical maintenance tips for, 661–664
positive displacement meters repairs, 661–664
positive displacement meters testing, 664
repairs, 661–664
testing, 664
Positive-displacement flow meters, 679–680
Post-installation testing, 185
Potential measurement methods, 638
Power (P), 401–402, 402f, 592
curve, 618
and efficiency curves, 610–613, 611f–614f
supply motor, 160
Pre-existing leaks, detecting, 748
Precision-fitted rotors measure fluid flow, 679–680
Predominantly friction, 604–605
Preferred operating region (POR), 615–616, 618, 622
Present value (PV), 946
of single cash flow to be received in future, 941
Pressure, 592
by-pass pig, 231
change for change in test water temperature, 206, 208f
example, 206
conversion chart, 1160, 1160f
hydrostatic pressure testing, 184
pressure-drop ratio, 269
pressure–volume plots, 192, 193f
recovery, 719
reversals, 195, 195t, 196f–197f
test pressure failure of flaw as percentage of calculated failure pressure, 195t

Pressure drop
calculations for rubber hose, 497
example of, 497, 497f
estimating for crude systems, 527, 527t
formulas used for calculating pressure drop and flow rates for pipelines, 497–499
gas line using, 439
nomograph for calculating pressure drop of liquids in lines for turbulent flow, 517, 518f–520f
Pressure loss
calculation for viscous crudes, 505–507, 505f, 506t, 507f, 507t
calculation in annular sections, 505, 505f
in hose string, 495–496, 496t
through valves and fittings, 509–514, 509f–514f
equivalent lengths L and L/D and resistance coefficient *K*, 513
equivalents of resistance coefficient *K* and flow coefficient C_v, 514
Preventive/Mitigative Measures for Internal Corrosion, 297–298
Price adjustments in pipeline system quality banks, 1023–1024, 1024t
PRICO® process, 388–389
Primer and coal-tar enamel, 305
Primer for mile of pipe, 304
Probabilistic Risk Assessment (PRA), 793, 797
Probabilistic/mechanistic models, pipeline risk assessment, 793
Product leak detection, 572
Programmable logic controllers (PLC), 712
Progressive cavity pumps (PC pumps), 606–607
difficulties, 606–607
Project scoping data worksheet for major facilities, 62–63
Properties of gases and vapors, 648t–649t
Proportional-position mode, 710–711
Proportional-speed floating mode, 711
Propylene, 234
Protective coatings, 286–287
Protective systems to protect workers from cave-ins, 41
Prover-master meter, 684
Provers, types of, 682
Proving devices, 715–716, 716f
Proving flow meters, 629–630
Proving head meters, 629
Proving liquid ultrasonic meters, 692
displacement provers, 694–697
master meter proving systems, 697–699
proving systems, 693–694, 693t–694t, 695f–696f
proving ultrasonic meters, 693
sample time, 693
time delay, 693
Proving meters, 671
Pulley speed, 26
Pulsation control for reciprocating pumps, 560–568, 561f
abilities of pulsation dampeners, 565–566
advantages of bladder type pulsation dampeners, 566
combating pulsation, 560–561
control of pulsation with dampeners, 564–565
fluid pressure pulsations, 561, 561f–562f
major problems caused by pressure pulsations, 562
pressure waves summary, 562, 562f
pulsations in discharge systems, 563, 564f
pulsations in suction systems, 563
sizing considerations, 566
system data, 566–568
material and design specification guidelines, 568
piping, 566–568
sizing of discharge pulsation dampeners, 568
Pulsation control tools (PCT), 589, 590f
Pumping stations, 176
Pumps, 200–201
capacity, 592
centrifugal pumps, 546–557, 546f–549f, 549t, 550f–552f, 552t, 553f–557f
parameters and applications, 574–579
systems and examples, 579–584
curve, 574–575, 574f–576f, 618
curve accuracy, 607–613
cubic spline regression, 610, 610f
polynomial regression, 609, 609f
power and efficiency curves, 610–613, 611f–614f
pump selection programs, 607–609, 608f
displacement, 592
estimating discharge of centrifugal pump at various speeds, 584
estimating head for average centrifugal pump, 585
estimating hp required to pump at given rate at desired discharge pressure, 585
evaluating VFD speed on hydraulics pumps, 604–605, 604f–605f, 605t
heating and/or cooling, 572
navigating pump curves, 620–623
nomograph for determining
horsepower requirement of pumps, 588f
reciprocating pump capacity, 586f
specific speed of pumps, 587f
NPSH and pump operating regions, 614–617
progressive cavity pumps, 606–607
pulsation control for reciprocating pumps, 560–568, 561f
reading centrifugal pump curve, 617–620
reciprocating pump capacity, 585
reciprocating pumps, 589–596
rotary pumps on pipeline services, 569–573
selecting motors for field-gathering pumps, 588

Pumps *(continued)*
  speed torque calculation, 558, 558f
  variable speed or impeller trim, 559–560, 559f–560f
Purging
  with air, 233
  with gas, 233
Purification process, 240
Pyramid rolling techniques, 54

Q

Quality management plan (QP), 907
Quality management system (QMS), 907
Quantitative *vs.* qualitative models, 794
Quantity of floodlights (N), 179
Quantum meter, 714
Quenched and tempered (QT), 779–780

R

Radar level transmitters, 743–744, 743f
Radio telemetry, 722
Radiometric level transmitters, 744, 744f
Railcar terminals, 899
Ramp generator, 711
Ramsbottom carbon residue (RCR), 1134
Rapid cycling, 261–262, 261f
Rapid valve cycling, 261
Rate of return method (ROR method), 947
Ratio of compression, 400
Re-rounding process, 769
Reactive force, 262–263, 262f
Reading land descriptions, 45
Readout and control equipment, 671
Reciprocating compressors
  cylinder inlet capacity, 401
  discharge temperature, 401
  estimate suction for, 403–404
  inlet valve velocity, 402–403
  performance calculations for, 401–403
  piston displacement, 401
  piston speed, 401
  power, 401–402, 402f
  volumetric efficiency, 401
Reciprocating pumps, 589–596 (*see also* Centrifugal pumps)
  acceleration head, 589–591
  calculations of volumetric efficiency, 589
    for hydrocarbons, 595–596, 596t, 597f
    for water, 594
  capacity, 585
    nomograph for determining, 586f
  piston rod load, 593–594, 594t
  plunger load, 593
  plunger or piston rod load, 589
  plunger or piston speed, 592
  power, 592
  pressures, 592
  pulsation control for, 560–568, 561f
  pump capacity, 592
  pump displacement, 592
  slip, 592
  stroke, 592
  tools for liquid pulsation control, inlet and discharge, 589, 590f
  works, 589–596
Recommissioning, 238
Recordkeeping requirements, hydrostatic, 209
  recordkeeping requirements 1955 to 1961, 210
  recordkeeping requirements 1961 to 1970, 210
  recordkeeping requirements post-1970, 210–211
  recordkeeping requirements prior to 1955, 209–210
  "traceable, verifiable, and complete" requirements on recordkeeping, 213–214
  unbroken chain of documentation not rule, 212
Redlich-Kwong equations, 423
Reduced pressure (P_R), 268–269, 407
Reduced temperature (T_R), 268–269, 407
Reel-lay method, 985
Reference electrode potentials, comparison to copper–copper sulfate, 321
Refrigerated Tank Task Group (RTTG), 902
Regulator flow curve, 281–284
  backpressure regulators and accumulation, 283
  basics, 281–282
  droop, choked flow, and other complexities, 282–283
  hysteresis, 283
  terminology, 284
Reid vapor pressure (RVP), 384, 1121, 1152
Relative humidity (R. H.), 241
Release prevention barriers (RPB), 863, 864f
Release volume estimation, 747–748
Remaining strength factor (RSF), 869
Remote station monitoring, 725
Remote terminal unit (RTU), 712
Remotely operated shut off/isolation valves (ROSOV/ROIV), 849
Renovation/rehabilitation, pigging during, 224–225
  applying in situ coatings, 224–225
  chemical cleaning, 225
  cleaning for product conversion, 225
  gel pigging, 224
  scale removal, 225
Reservation, 1017–1019, 1018f, 1019t
Resistance to Flow Factor (K_R), 263
Resistant overlay coatings to steel pipe, 1095–1099

Resistive coupling, 342
Resonant chatter, 262
Response management, 1032–1033
  NEBA, 1033–1035
    in practice, 1038–1040, 1038f–1040f
  safety and health issues, 1033
Response techniques, 1038
  containment and recovery, 1038
  controlled in-situ burning, 1040–1041
  shoreline clean-up/treatment, 1041
  solidifiers, 1040–1042, 1042f, 1042t
  sorbents, 1040
  spills on snow and ice, 1038–1040
  Sunken oil, 1043
Response time, 747
Retained earnings, 956
Retest interval, calculating, 790
Retrofit feasibility, leak detection technologies, 749
Return on equity (ROE), 947
Revalidation, 783–791
Reynold's number (REv), 267
  nomograph for calculating, 457f–458f
  nomograph for calculating, 515f–516f
Reynold's number factor (F_R), 267
  valve reference data, 270t
RGB. *See* Butane, refinery grade (RGB)
Right-of-way
  clearing and grading right-of-way, 64
  estimating man hours for removing tree stumps, 65, 65t
  estimating man hours for removing trees, 64, 64t
Risk Based Inspection (RBI), 867
Risk management, 792–796
  basic ranking concepts, 797
  examples, 800, 800f, 801t
  failure probabilities, 797–799
    causes of failure, 797
    construction of equations by cause, 797–799
  mitigation of risk, 799–800
  relative risk calculations, 799
  tool for establishing budget priorities, 797–802
Risk modeling process, 792–793
Risk-assessment techniques
  choices in, 793
  choosing, 795
Robustness, leak detection technologies, 749
Rock shield wraps, 763
Rocky Mountains, 480, 484f, 484t
  CO_2 pipeline infrastructure circa, 11f
Roof estimation and framing weight for fixed cone tank, 871
Ropes, 16
  capacity of drums, 22, 22t
  drums and sheaves be for wire rope, 16, 16t
  estimating strength of cable, 16
  find advantages of block and tackle with pull out friction, 17
  find working strength of manila rope, 16
  safe loads for wire rope, 17, 17f
  strength and weight of popular wire rope, 20, 20t
  stress in guy wires, 18–19, 18f
  wire rope, 21–22, 21t–22t
Rotameters, 679
Rotary pumps on pipeline services, 569–573
Rotary screw pumps, 598–604, 598f–603f
Round steel shafting, weight in pounds of, 29, 29t
Rubber hose
  pressure drop calculations for, 497
    example of, 497, 497f
Runout (RO), 622
Rupture disk devices, 264
Rupture disk sizing
  methodologies, 263–264
    sizing methodologies, 263–264
    terminology, 263
  using resistance to flow method (K_R), 264–266
    anticavitation trim, 268
    cavitation index (Kc), 268
    choked flow, 267–268
    piping geometry factor (F_P), 267
    Reynold's number factor (F_R), 267
    sample calculation using K_R method, 264–266
    variable orifice rotary control valves, 266–268
Rupture Pressure Ratio (RPR), 348–350

S

S-lay system, 984–985
Sacrificial anodes, 287
  systems, 293–294
Safe loads for wire rope, 17, 17f
Safety and environment
  inland aquatic environments, 1031–1032
  oil fate and weathering, 1047–1048, 1048f–1049f
  oil spills, 1066–1083, 1068f, 1070t, 1071f, 1072t, 1074f–1079f, 1081t, 1082f
  response management, 1032–1033
  response techniques, 1038
Safety Instrumented Functions (SIF), 726
Safety Instrumented System (SIS), 726–729, 868
Safety Integrity Level (SIL), 726, 849
Safety Recommendation (SR), 213
SALM system, 492
Sample calculation using K_R method, 264–266
Sample conditioning system, 374, 374f
Samplers, 720–721, 721f
Sampling program for unrefined petroleum-based products, 706–707
  certification, 707
  components of sampling and testing plan, 706–707
  documentation, 707

Sampling program for unrefined petroleum-based products *(continued)*
requirements, 706
responsibility, 707
Sand blasting, internal, 233
Saybolt Universal viscosity, 1156
at 122°F, 1156
SCADA systems, 723
Scaling rules, 1013
equipment and process, 1013–1014
equipment cost curves, 1014, 1014f, 1014t, 1015f
equipment cost curves using two data points, 1017
equipment cost using two-factor models, 1015–1017, 1015t, 1016f–1017f
Schedule number, identification by direct measurement, 144
Scrapers, 226–227
Screw jack systems, 69
Screw pumps, 599
Scurry Area Canyon Reef Operators Committee (SACROC), 480
Seal failure, floating roof failure modes, 808, 808f
Seal losses, 425
Sealing pigs, 227
Seals, 840–842
Seam Factor (C_{10}), 799
Seamless line pipe, 779
Seamless pipe, 52
Seat load drop, 282–283
Selective Seam Corrosion, 289
Self-tapping screws, recommended drill sizes for, 26, 26t
Sensitivity analysis, 951–952, 952f
Sensitivity of habitat, 1032
environmental sensitivity index scale I shore type, 1042t
Sensors for automatic tank gauging, instrumentation and, 737–744
appropriate level sensor for common applications, 737–738
capacitance level sensors, 742–743
capacitive level sensors as level switch, 739–740, 740f–741f
categorization of level sensors, 737, 737t–738t
contact *vs.* non-contact level sensors, 737
continuous level sensors, 740–741
direct and indirect, 886f
electromechanical servo transmitter, 719f, 742
float and tape sensor, 718f, 741
float level sensors, 718f, 741
magnetorestrictive float level sensor, 741, 742f
mechanical float switch, 738, 739f
oil leak sensors, 721f, 744
point level sensors, 738
process variability, 887t
pros and cons, 886t
proof testing, 731f
radar level transmitters, 743–744, 743f
radiometric level transmitters, 744, 744f
tank inventory parameters, 890f
tuning fork level sensor, 739, 739f–740f
ultrasonic level transmitters, 743
Series number, flanged fittings, 153
Series systems, 503, 503f
Service factors, 160
Severity ranking, 348–350, 349t–350t
Shaft horsepower, 423
Shaft key, calculate stress in, 23
Shafting, properties of, 30–31, 30t
Shafts, 22
calculate stress in shaft key, 23
determine length of V-belt, 22–23
estimate horsepower that transmitted by shaft, 23
Shaping techniques, 54–55, 54f
Shear wave UT tools, 752, 752f
Shearvane, soil, 69
Shell/MIT formula, 497–499, 502
examples, 498
Shock hazards, minimization of, 343–344
Shore cleaning agents, 1042
Shoreline Clean-up Assessment Technique surveys (SCAT surveys). *See* Oiled shoreline assessment surveys
Shoreline clean-up/treatment, 1041
biodegradation, 1042–1043
response options, 1042
Shoreline assessment, 1041–1042
Sunken oil, 1043
surface washing or shore cleaning agents, 1042
Shoring types, 69
Shunting pig, 231–232
Shut-off head (SO head), 620–621
Side extrusion data, 1106
Sieve analysis of granular materials, 115
Signal attenuation coating survey, 334
Signal attenuation coating survey, 335, 336f, 337
Signal-to-noise ratio, ideal risk-assessment model, 795–796
Simple interest, 940, 940t
Single cash flow
future value, 942
present value, 941
Single graphite anode, 324
Single-cylinder hydraulic shores, 69
Single-phase alternating-current motors, 160
Sizing, 259
Sizing gas metering run, 652, 652f
Sizing headers for meter stations, 665–667
header connections, 665
station layout, 665, 666f

velocity sizing, 666–667, 667f
Sizing pipelines for water flow, 525
Sliding, trench excavations, 67
Slimline butterfly valves, 717
Slip (S), 592
Sloping, soil type, 72–73, 72f–73f
Slow crack growth (SCG), 57
Slug flow, 686
Slurries, viscosity of, calculating, 469, 470f
Small volume type displacement provers, 696
advantages of, 696
disadvantages of, 697
Small-volume prover, 684
Smaller-gauge cable, 181
"Smart" pressure sensors, 718–719
Smooth dent, 767
Society of Petroleum Engineers (SPE), 395, 964–965
Soil classification categories, 40
Soil conditions, 782
Soil factor (SF), 798
Soil mechanics, 67–68, 67f–68f
Soil type
determination of, 68
evaluating, test equipment and methods for, 68
hydraulic shoring, 69
permanent spoil, 73
plasticity or wet thread test, 69
pneumatic shoring, 69–71, 71f
pocket penetrometer, 68–69
shielding types, 71, 71f
shoring types, 69
sloping and benching, 72–73, 72f–73f
special health and safety considerations, 73–75
competent person, 73
emergency rescue equipment, 74
exposure to falling loads, 74
exposure to vehicles, 74
hazardous atmospheres and confined spaces, 74
ingress and egress, 74
inspections, 75
standing water and water accumulation, 74–75
surface crossing of trenches, 73
warning systems for mobile equipment, 74
temporary spoil, 73, 73f
visual test, 69
Solid-cast pigs
cleaning pigs, 227–228
sealing pigs, 229
Solidifiers, 1040–1042, 1042f, 1042t
Solve flow measurement problems, 627–628
Sonic flow, 259
Sonic velocity, 269
Sorbents, 1040
Span-C, 719–720
Specialist utility pigs, 230–232
magnetic-cleaning pig, 230–231
pin-wheel pig, 231, 231f
pressure by-pass pig, 231
shunting pig, 231–232
Specific gravity, 495, 496t
of blend, 466
of liquids, 1151, 1151f
specific gravity/temperature relationship for petroleum oils, 1151, 1151f
Specific speed, 623
Specific volume of gases and vapors, 370, 370f
Specifications
API 5L pipeline specifications, 1090–1091
aviation gasoline fuel specifications, 1124–1125, 1125t–1126t
chemical composition for API 5L PSL 1 welded pipe, 1092t
chemical composition for API 5L PSL 2 welded pipe, 1093t
diesel fuel specifications, 1131–1134, 1131t–1132t, 1134t–1135t
differences between API 5L PSL 1 and PSL 2 pipeline, 1091t
fuel quality, 1119–1121, 1120t
fuel specifications, 1117
gasoline fuel specifications, 1121–1124, 1122t–1123t
jet fuels specifications, 1126–1130, 1127t
NAPCA bulletin 13–79–94, 1099–1102
NAPCA bulletin 14–83–94, 1102–1106, 1105t–1106t
pipeline material coating specifications, 1095
pipeline product specifications, 1136
tensile property requirements
for pipe body of seamless and welded pipe for PSL 1, 1094t
for pipe body of seamless and welded pipe for PSL 2, 1094t
transportation fuel specifications, 1119
Specified minimum yield strength (SMYS), 186, 189, 199, 501, 754, 780, 784
testing to levels above of, 191, 191f
Spectacle blind thicknesses, 153
Speed motors, 160
Speed of pumps, determination of, 587, 587f
Speed torque calculation, 558, 558f
Spheres
cleaning pigs, 228
sealing pigs, 229–230
Spiral rolling techniques, 55
Spiral-submerged arc welding (S-SAW), 971
Spot samples, 372
Spring-back process, 769

Spring-loaded relief valves, 260
Stability
  of diesel fuel, 1134
  of jet fuels, 1128
Stabilizers. *See* Pulsation control tools (PCT)
Stable rock soil, 68
Standard MFL tool, 750
Standard Penetration Test (SPT), 114
Standard pigs, 227
Standard UT tools, 752, 752f
Standards effects on flow measurement, 629
Static friction tests, 306
Static pressure, 882–883
Static-dominated systems, 605
Station layout, 665, 666f
Steady-state conditions, 528
Steel, 1090
Steel conduit, voltage drop on circuits using copper conductors in, 172
Steel fitting dimensions, 38, 38f
Steel pipe, 186
  allowable stresses, 128t–129t
  contracts, 143
  design, 126–128
  design factor, 127t
  exterior application procedures
    for hot applied coal tar coatings to, 1113–1116
    for plant applied tape coatings to, 1107–1110
  longitudinal joint factor, 127t
  properties, 131t–133t
  resistant overlay coatings to, 1095–1099
  specified minimum yield strength, 1091t
  temperature derate factor, 127t
  weld joint factor, 130t
Steel production, 51
Steel Structures Painting Council (SSPC), 1095
Steel Tank Institute (STI), 824
Storage tanks, 820–825
  API specification for, 820
  classification, 820–822, 820f–821f
    safety features, 823t
    type of bottom plate, 823t
    type of construction, 822t–823t
  classification of storages, 825
    flammable and combustible liquids, 825
    Institute of Petroleum Classification, 825
    NFPA 30 classification, 825, 825t
  above ground storage tanks, 822, 822t–823t
  selection of type, 822–824
  standards for storages, 824–825
    API 620-design and construction of large, welded, low-pressure storage tanks, 824
    API 650-welded steel tanks for oil storage, 824
    ASME Section VIII Div(1) rules for construction of pressure vessels, 824–825
Straight-line depreciation, 956
Strength of popular wire rope, 20, 20t
Stress
  in above ground/below ground transitions, 148–151, 149f–150f, 150t, 151f
  in guy wires, 18–19, 18f
  in pipe wall internal pressure, 148
  results for high pressure gas transmission lines under design in Iran, 150t
Stress corrosion cracking (SCC), 299–300, 299f, 750, 764, 774, 790–791
  coating, 781–782, 781f
  contributing factors to SCC in pipelines, 778–782
    manufacturing, 779–780
    mechanical properties, 780
    metallurgy, 778–779
    pipeline operating conditions, 780–781
  crack characteristics, 777–778, 778t
  crack growth, 778
  general characterization, 774–778, 775f–777f
  high pH SCC, 776
  near-neutral pH SCC, 777, 777f
  in pipelines, 774–782
Studded foam pig, 225
Subcommittee on Aboveground Storage Tanks (SCAST), 902
Subjectivity *vs*. objectivity model, 794
Submerged arc welding (SAW), 55, 103–104, 971
Subsidence, trench excavations, 67
Subsonic flow, 260
  gas constant based on ratio of heats, 260t
Subsurface temperature ranges, prediction of, 524–525, 524f
Suction control, 563
Suction pressure (Ps), 592
  controller, 711
Suction pulsation dampeners, 566
Suction specific speed, 623
Suction stabilizers, 566
  material and design specification guidelines for pulsation dampeners and, 568
Suction systems, pulsations in, 563
Sulfate-reducing bacteria (SRB), 289, 301
Sulfide-stress cracking (SSC), 299, 301
Sulfur recovery unit (SRU), 384
Sulphate-reducing bacteria (SRBs), 222
Supercritical carbon dioxide, pipeline measurement of, 633–639, 633f
Supervisory Control And Data Acquisition systems (SCADA systems), 157, 712, 723, 887

Surface formula, 25, 25f
Surface washing, 1042
Surge pressure, determination of, 528
Surges, perturbations that cause, 529
Swab pigs, 222
Swirl flow meters, 678–679
Switchgear device numbers, 180–181
System, 2
  hydrocarbon classification of gases and products, 3, 3f
  pipeline construction sequence, 3, 3f
  schematic of hazardous liquid pipeline system networks, 2, 2f
  schematic of natural gas pipeline system networks, 2, 2f
System curves, 618
System design guidelines, 725
System head curves, 576–577, 576f
System self check, leak detection technologies, 749

T

Tandem system, 493, 493f
Tanks
  API 2350 4th and 5th edition, 911–916
  best containment for LNG storage, 908–910
  design and containment for atmospheric storage tanks, 845–851
  farm line sizing, 882–884
    booster pumps, 883
    design considerations, 883–884
    NPSH required by pump, 883
    pipeline size, 882–883
  farm management
    automatic measurements, 885–887, 886f, 886t–887t
    design, 884, 885t
    inventory calculations, 888–890
    manual measurements, 885
    multifunction gauging system, 887
    required tank information and methods, 884
    tank inventory, 884
  federal requirements for above ground storage tanks, 844–845
  fixed and floating tank types, 826–829
  floating roof applications and components, 834–843
  flow rate estimation from leaking tank hole, 872, 872f
  gauging systems, 726
  horizontal cylindrical tanks
    contents, 878–879, 879f
    gauge, 879
  manual tank gauging procedures, 916–923
  oil movement and storage in tank farms, 892–902
  petroleum storage tank standards, 851–870
  roof estimation and framing weight for fixed cone tank, 871
  simplified procedure to calculate emissions for liquid organic storage tanks, 932–936
  standard construction practices for LNG storage tanks, 902–908
  styles and appurtenances, 830–833, 830f
  tank emissions with BREEZE TankESP, 936–937
  vapor cloud analysis methodology to tank overfills, 923–931
  vapor formation rate, 881, 881f
  vapor loss from internal floating-roof tanks, 875–878, 876f, 876t, 877f–878f
  volume of light fuels from actual temperature to a base of 60°F, 880, 880f
  volume of liquid in vertical cylindrical tanks, 880
  weight estimation tool using diameter as input variable, 870–871
Tank management systems
  floating roof, 716f
  integrated tank gauging, 715f
  spherical tank, 717f
  tank farm, 714f
Tank types
  classification, 820f, 836f
  cone roof, 821f, 833f
  external floating roof, 821f, 828f, 853f
  fixed roof, 825t, 853f
  fixed roof, geodesic domes, 855f
  hemispheroid and noded spheroid, 864f
  internal floating roof, 829f, 864f
  open top, backsloping floating roof, 830f
  open top, doubledeck floating, 831f
  open top, pontoon floating roof, 832f
Tap drills for machine screws, 31, 31t
Tape for pipe coating and fittings, 339
Target flow meters, 679
Tariff rates, 1017
  cost of service, 1019–1023, 1021f–1022f
  firm service with fixed maximum daily quantity and reservation charge, 1019
  firm transportation, 1022–1023
  IT-1 service, 1022
  reservation and commodity charges, 1017–1019, 1018f, 1019t
"Taylor" bubbles, 156
Temperature loss calculation for viscous crudes, 505–507, 505f, 506t, 507f, 507t
Temporary spoil, 73, 73f
Tensile property
  requirements for pipe body of seamless and welded pipe for PSL 1, 1094, 1094t

Tensile property *(continued)*
requirements for pipe body of seamless and welded pipe for PSL 2, 1094, 1094t
Tensile test, 191
Tension crack, trench excavations, 67
Terminology
regulator flow curve, 284
rupture disk sizing methodologies, 263
Terrestrial spills, 1031
Test leads, spacing to measure current on, 323
Test meter method, 664
Test-pressure-to-operating-pressure ratio, 190–191, 190f
Testing existing pipelines, 192–195
Testing flow meters, 629–630
Testing program for unrefined petroleum-based products, 706–707
Texas Eastern Transmission Corporation (TETCO), 184
Thermal conductivity (liquid), 378
Thermal conductivity (vapor), 378t–379t
Thermal decomposition, 238–239
Thermo mechanical Control Process (TMCP), 779–780
Thermometers, reuniting separated mercury in, 43
Thin-film epoxy, 292, 307
to soils, 308
Thin-setting epoxy-based concrete coatings, 763
Thin-walled tube sampling, 114
Threshold stress, 301
Throat pressure in a blow-down system, determination of, 445–446, 445f
Throughput, 477
Time of-flight (TOF), 737, 743
Time value of money, 940
analyzing and valuing investments/projects with multiple or irregular cash flows, 942–943, 943f
analyzing complex investments and cash flow problems, 946–947, 946f–947f
annuities, loans, and leases, 944–945, 945f
future value of periodic series of investments, 944, 944f
future value of single investment, 942
gradients, 945–946, 945f
importance of cash flow diagrams, 942, 942f
nominal interest rate *vs*. effective annual interest rate, 941
perpetuities, 943–944, 943f
present value of single cash flow to be received in future, 941
simple interest *vs*. compound interest, 940, 940t
"Top-down" approaches, 962
Top-Side Anomalies, 750
TopFlow meter, 626
Toppling, trench excavations, 67
Torque definitions, 160–161
"Traceable, verifiable, and complete" requirements on recordkeeping, 213–214
Traceable records, 214
Trans Alaska Pipeline System (TAPS), 755–760
background, 755, 755t
TAPS line parameters, 755t
TAPS maximum operating pressure, 755t
data analysis, 757–758
and integration, 756–757
data management system description, 757, 758f
description of integrity issue, 755
inspection and assessment, 756, 756t
inspection tool selection, 755, 755t
Transient conditions, 528
Transient pressure analysis, 528–529
equipment, 529
fluid properties, 528
simulation, 529, 536t–537t
data required by program, 530t–532t
maximum pressure summary, 536t–537t
pressure relief valve summary, 537t
typical time page report, 534t–535t
steady-state *vs*. transient conditions, 528
Transmission lines, crude oil, the optimum working pressures estimation, 527
Transportation fuel specifications, 1119
ASTM D975, 1119
ASTM D1653, 1119
ASTM D4814, 1119
Transportation-related spills, 1030
Transverse Field Inspection (TFI), 753–754
tool, 751
validation program, 754
Trench boxes, 71
Trench excavations, 67–68
soil mechanics, 67–68, 67f–68f
Trench shoring, 120
Troubleshooting magnesium anode cathodic protection systems, 308
Truck terminals, 897, 898f
True vapor pressure of crude oil stocks with Reid vapor pressure, 1152
True vapor pressures (TVP), 876
Tubular goods problems, formulas and constants for, 142
Tuning fork level sensor, 739, 739f–740f
Turbine flow meters, 679, 713
Turbine metering, 637–638, 638f
Turbine meters, 699
Turbulent flow
nomograph for calculating pressure drop of liquids in lines for, 517, 518f–520f
in pipelines, 522
"Two level" protection system, 322
Two-dimensional scaling rule, 1014
Two-phase horizontal flow, 453, 453f–455f

U

U-O-E forming techniques, 54
U.S. CO_2 pipeline networks, 485, 486t
U.S. Department of Agriculture (USDA), 348–350
U.S. Department of transportation (DOT), 154, 964
U.S. Federal Energy Regulatory Commission transmission compressor station cost, 2014–19, 1010–1012, 1011t, 1012f, 1012t
  aggregate average cost, 1011
  cost components, 1010–1011
  Federal Energy Regulatory Commission compressor stations, 1010
  limitations, 1012
  regression models, 1011–1012
U.S. Federal Safety Regulations for Pipelines, 199
U.S. pipeline maps, 4f
  CO_2-EOR operations and infrastructure in U.S., 8f
  gas transmission and hazardous liquid pipelines, 4f
  gas transmission pipeline incident heat map, 6f
  gulf coast CO_2 pipeline infrastructure, 10f
  hazardous liquid pipeline accident heat map, 5f
  LNG plants connected to natural gas pipeline systems, 7f
  mid-continent CO_2 pipeline infrastructure, 11f
  Permian basin CO_2 pipeline infrastructure, 9f
  Rocky mountain CO_2 pipeline infrastructure, 11f
U.S. statistics, 12f
  annual report mileage
    for gas distribution systems in U.S., 14f
    for hazardous liquid or carbon dioxide systems in U.S., 12f
    for natural gas transmission and gathering systems in US, 13f
  billion barrels delivered by liquids pipeline in U.S., 14f
  liquefied natural gas facilities in U.S., 15f
Ultrasonic flow equations, 475–476
Ultrasonic flow meters, 679
Ultrasonic inspection tools, 751–753, 752f
Ultrasonic level transmitters, 743
Ultrasonic meters, 683, 699, 715
Ultrasonic system, 722
Ultrasonic testing (UT), 751–752
Uncertainty
  pipeline risk assessment, 795
  range, 963
Unconstrained dent, 767
Underground storage tanks (UST), 820
Underpinning process, 71
Underwriters Laboratories (UL), 824
Uniform *vs.* localized corrosion, 289
  microbial Corrosion, 289, 290f
  pitting, 289, 289f
  selective Seam Corrosion, 289
US Department of Transportation, Pipeline and Hazardous Materials Administration, 220
Utility pigs, 226–232, 226f
  cleaning pigs, 227–228
    foam pigs, 227
    mandrel pigs, 227, 227f
    solid-cast pigs, 227–228
    spheres, 228
  dual-diameter pigs, 230, 230f
  gauging pigs, 230
  industrial pigging systems, 230
  sealing pigs, 228–230
    foam pigs, 229
    mandrel pigs, 228–229, 228f
    solid-cast pigs, 229
    spheres, 229–230
  specialist utility pigs, 230–232

V

V-belt, determine length of, 22–23
V-Cone® flow meter, 700–706, 701f
Vacuum box (VB), 907
Vacuum breakers, 840, 842f
Vacuum drying, 235, 241, 241f
  dryness, 242–243, 243f, 243t
  evacuation, 241
  evaporation, 241
  final drying, 242, 242f
  purging and commissioning, 243
Validation of corrosion mitigation strategy, 345
Valves, 662, 716–717, 717f
  outlet velocity, 269
Vapor cloud analysis methodology to tank overfills, 923–931
  correction factor, 929
  eliminating potential for VCE, 931
  flow rate proxy, 924, 924t
  key input parameters, 925–926
  overflow event parameters, 927
  pure component vapor pressure calculation, 927–928
  radius of ignition and escape, 931
  radius of ignition for overfill events, 926–927
  vapor pressure and $C\theta$ correlation, 928, 929t–930t
  volume or mass fraction to molar fraction, 927, 927t
Vapor flow measurement—orifice plate flange taps, 646–647
  properties of, 648–649, 648t–649t
Vapor loss from internal floating-roof tanks, 875–878, 876f, 876t, 877f–878f
Vapor pressure, 378, 501, 882, 1121
  low temperature vapor pressures for light hydrocarbons, 1152
  true vapor pressure of crude oil stocks with Reid vapor pressure, 1152
Vapors, nomograph for calculating density and specific volume of, 370, 370f

Variable design point (VDP), 857
Variable frequency drives (VFDs), 604
  evaluating VFD speed on hydraulics pumps, 604–605, 604f–605f, 605t
    observations, 605
Variable orifice rotary control valves, 266–268
  maximum Flow Coefficients (Cv), 267t
Variable speed, 559–560, 559f–560f
  pumps, 618–620, 619f
Variable-area flow meters, 679
Velocity
  heads for various pipe diameters and different rates of discharge, 474, 474f
  of liquids in pipes, 472, 471, 472f–473f
  sizing, 666–667, 667f
Venturi meter, 626, 674
Vertical support members (VSMs), 757
Viscosity
  of blend, 465
  correction, 249
  equivalents of absolute viscosity, 1158
  equivalents of kinematic viscosity, 1158
  of mixture of liquids blending index, 465
  of slurries, 469, 470f
Viscous crudes, 505–507, 505f, 506t, 507f, 507t
VOC flash emission estimation Volatility, 1086t, 1121–1123
  of jet fuels, 1128
Voltage drop on circuits using copper conductors in steel conduit, 172
Volume formula, 25f
Volume of light fuels from actual temperature to a base of 60°F, 880f
Volume of liquid in vertical cylindrical tanks, 880
Volumetric efficiency, 400–401, 603
  calculations, 589
  for hydrocarbons, 595–596, 596t, 597f
  for water, 594
Vortex flow meters, 678–679

W

Water, 296
  compressibility, 594t
  compressibility factor for, 209
  to fill test section, volume of, 185f, 204
  horsepower, 603
  at test pressure, volume of, 205–206
    sample calculation, 205–206
  treatment, 345
  volumetric efficiency for, 594
  weight per cubic foot at various temperatures, 1165, 1165t
Weathering processes, 1047
Wedge, 675
Weight
  density of liquids, 1151, 1151f
  of popular wire rope, 20, 20t
Weld joint factor, 128t–129t
Welded joints, 185
Welded pipe, 54
Welding
  average welder make per hour, 102
  and brazing temperatures, 92
  criteria permit safe and effective pipeline repair, 103–107
    burn-through, 106–107
    choice of repair method, 103
    types of defects, 103–106, 105t
    underbead cracking, 107
  lens shade selector, 93
  mechanical properties of pipe welding rods, 92, 92t
  pounds of electrodes, 102
  reasons for rejecting pipeweld and suggestions to reduce reject rates, 93–101
    change in attitude, 93
    filler, stripper, and cap, 100
    hollow bead, 100
    importance of joint preparation, 93
    internal chamfer, 97, 97f
    internal undercut, 94–96, 95f–96f
    keyhole size *vs.* current and undercut, 97
    recommended procedures for properly cleaning pipe, 93–94
    significance of arc voltage, 97–98
    undercut, 98–99, 99f
    undercut *vs.* welding current, 97
    varying current/varying gap, 98
    wagon tracks, 99–100, 99f
    welding cracks, 100–101, 101f
  steel preheated before, 91
  techniques, 55
  welding rod, 102
Wet gas measurement, 657
Wet thread test, 69
Weymouth equation, nomenclature for, 434
Weymouth's gas-flow formula, 449t
Wind chill chart, 50, 50t
Winterizing cooling systems, convenient antifreeze chart for, 28, 28f
Wire line drums, capacity of, 22, 22t
Wire rope, 21t–22t
  drums and sheaves be for, 16t
  strength and weight of, 20t
Wire-brush pig, 221
Wrappings, 305
  for mile of pipe, 305